2014 IEEE 6th India International Conference on Power Electronics

(IICPE 2014)

Kurukshetra, India
8-10 December 2014

IEEE Catalog Number:	CFP1494B-POD
ISBN:	978-1-4799-6047-7

Copyright © 2014 by the Institute of Electrical and Electronic Engineers, Inc
All Rights Reserved

Copyright and Reprint Permissions: Abstracting is permitted with credit to the source. Libraries are permitted to photocopy beyond the limit of U.S. copyright law for private use of patrons those articles in this volume that carry a code at the bottom of the first page, provided the per-copy fee indicated in the code is paid through Copyright Clearance Center, 222 Rosewood Drive, Danvers, MA 01923.

For other copying, reprint or republication permission, write to IEEE Copyrights Manager, IEEE Service Center, 445 Hoes Lane, Piscataway, NJ 08854. All rights reserved.

***This publication is a representation of what appears in the IEEE Digital Libraries. Some format issues inherent in the e-media version may also appear in this print version.**

IEEE Catalog Number: CFP1494B-POD
ISBN 13:   978-1-4799-6047-7

Additional Copies of This Publication Are Available From:

Curran Associates, Inc
57 Morehouse Lane
Red Hook, NY 12571 USA
Phone: (845) 758-0400
Fax: (845) 758-2633
E-mail: curran@proceedings.com
Web: www.proceedings.com

IEEE Sixth India International Conference on Power Electronics
IICPE 2014
08-10, December 2014
NIT Kurukshetra, India

Table of contents

Track 1–Application of Power Electronics in Power Systems

Residential Demand Response from PV Panel and Energy Storage device....1
By Monika Arora; Saurabh Chanana

PI Controller Tuning & Stability Analysis of The Flyback SMPS....7
By Tapas Halder

Optimization of PMU placement by performing Observability Analysis....13
By Mini Thomas; ShrutiRanjan

A Study of Static Synchronous Compensator in two Power Flow Models....18
By Shagufta Khan; Suman Bhowmick

Optimal Capacitor Placement in a Radial Distribution System using combined Fuzzy & Novel power loss sensitivity method....N/A
By Vijay BabuPamshetti; RatnaDahiya

Performance Evaluation of Conventional and Computationally Intelligent Controller based Shunt Active Power Filter....N/A
By Archana Sharma; Bharat Singh Rajpurohit; Anshul Mishra

Comprehensive Stability Analysis of Radial Distribution System with Load Growth....25
By Pawan Kumar; Surjit Singh

Current Control Methodology for PV in both Standalone & Grid Connected Mode....31
By Lovely Goyal; Sarthak Jain

Increase Efficiency of Solar Photovoltaic System by Data Acquisition Process....37
By Sunil Parashar

Design, Implementation and Performance Analysis of a Single Phase PWM Inverter....42
By Narendra Kumar; SachinSinghal; Dheeraj Joshi

Smart Home Energy Management by Demand Response Controller Design....48
By Praveen Bansal; Mini Thomas

Mitigation of Sub Synchronous Resonance using STATCOM controlled by PID and FL Controllers....54
By Eswar Chinnari; Shelly Vadhera

Design and Analysis of Stationary Frame PR Current Controller for Performance Improvement of Grid Tied PV Inverters....59
By Aditi Chatterjee

An efficient method for D-STATCOM placement in radial distribution system....65
By Abhinav Jain; Atma Ram Gupta, Ashwani Kumar

Voltage Stability Assessment using Phasor Measurement Technology....71
By Priti Prabhakar; Ashwani Sharma

Application of SSSC for Compensation Assessment of Interconnected Power System....77

By Shashi Gandhar; Jyoti Ohri; Mukhtiar Singh

Simultaneous Coordination of Power System Stabilizer and UPFC for improving Dynamic stability of multimachine system....82
By ChintuMakkar; Lillie Dewan

Mitigation of Reactive Power Requirements of Large Scale Wind Power Generation in Western Rajasthan Using Static VAR Compensator....86
By Ashok Kumar Pathak; Mahaveer Sharma

Transient Stability Improvement in Transmission Line using SVC with Fuzzy Logic based TID Controller....92
By Tarang Sharma, Anil Kumar Dahiya

Optimal Placement of Distributed Generator in Transmission System Using an Algorithmic Approach....97
By Satish Kansal; Navpreet Kaur; Kanwardeep Singh

A Novel Method to Focus Intensity by Constructing a Mechanically Controlled Bluetooth based Light Director....103
By Abhishek Gulia; Ravi Prakash; Sandeep Banerjee; Rahul Sharma

Track 2–Power Quality Issues

Neural Network Based Enhancement of Power Quality in Distribution System....107
By Priya Nayar; Bhim Singh; Sukumar Mishra

Improved Power Quality SMPS for Computers Using Bridgeless PFC Converter at Front End....112
By Shikha Singh; Bhim Singh; G. Bhuvaneswari

A Survey on Active Power Filters Control Strategies....118
By Khoisnam Steela; Bharat Singh Rajpurohit

Analysis of Power Quality Disturbances Using Wavelet Packet Transform....124
By PrasantaKundu; Chirag Naik

Modified Series Compensation for Voltage Sag / Swell Mitigation....128
By Kirti Mathuria; ArunVerma; Bhim Singh; G Bhuvaneswari

A Three-Leg Inverter Based DSTATCOM Topology for Compensating Unbalanced and Nonlinear Loads....133
By Manoj Kumar M. V; Mahesh Kumar Mishra

Control Strategy for Voltage Sag- Swell/Harmonic/Flicker Compensation With Conventional And Fuzzy Controller(UPQC)....139
By Mukul Chourasia; S. P. Srivastava; Aurobinda Panda

Performance Evaluation of Shunt Active Power filter Control Strategies under Different Supply Conditions....145
By KhoisnamSteela; Bharat Singh Rajpurohit

Compartive Analysis of Conventional and IRP theory based Shunt APF for 3P3W system....151
By Anil Gambhir; Sharmili Das; Aurobinda Panda

A Modified Switching Scheme for a New Multi-level Inverter Topology for Fuel-Cell Microgrid....156
By G K Naveen Kumar; Yash Pal

Time Domain Electric Arc Furnace Model for Power Quality Study....162
By Deepak Bhonsle

Mathematical Modeling of Composite Filter for Power Quality improvement of Electric Arc Furnace Distribution Network....168
By Deepak Bhonsle

Optimal Capacitor Placement in Radial Distribution System....173
By Atul Kumar; Ravinder Bhatia

An Improved Three Level Hysteresis Current Controller for Single Phase Shunt Active Power Filter....179
By Venkata Subrahmanya Raghavendra Varaprasad Oruganti; Siva Sarma Dvss
Effectiveness of UPQC in a Distribution Network with DTC Drive as Load....184
By Parag Nijhawan; Ravinder Bhatia; Dinesh Jain
Improved performance of active power filter using type-2 fuzzy logic controller....190
By Suresh Dhanavath; Sajjan Pal Singh
Enhancing performance of APF by Fuzzy Controller....196
By Khoisnam Steela; Bharat Singh Rajpurohit
Improved Performance of IPDPWM CHBMLI Based DSTATCOM in a Distribution Network with Induction Furnace Load....202
By Parag Nijhawan; Ravinder Bhatia; Dinesh Jain

Track 3–Renewable energy systems

A Hybrid Optimization Technique of Wind Power Exploration & Utilization....207
By Tapas Halder
BLDC Motor Driven Solar PV Array Fed Water Pumping System Employing Zeta Converter....213
By Rajan Kumar; Bhim Singh
A Practice of Power Paste with the Isolated Flyback Converters....219
By Tapas Halder
Experimental Implementation of Doubly Fed Induction Generator Based Standalone Wind Energy Conversion System....225
By N Krishna Swami Naidu; Bhim Singh
Cascaded DC-DC Converter for a Reliable Standalone PV fed DC load....231
By Malay Bhunia; Rajesh Gupta; BidyadharSubudhi
A SOGI-Q Based Control Algorithm for Multifunctional Grid Connected SECS....237
By Chinmay Jain; Bhim Singh
A Novel control strategy for power extraction from Photo Voltaic panels based on One Cycle Control....243
By Anoop K
Consequences of Dust on Solar Photovoltaic Module and Its Generation....249
By AnirudhDube; Pathik Chamaria; Ruchika Mittal; Alok Mittal
Power Quality Control of SEIG based Isolated Pico Hydro Power Plant Feeding Non-Linear Load....254
By Sanjeev Singh; UmeshRathore
Isolated Micro Grid Employing PMBLDCG for Wind Power Generation and Synchronous Reluctance Generator for DG System....259
By Geeta Pathak; Bhim Singh; BijayaPanigrahi
Performance of Grid Interfaced Solar PV System under Variable Solar Intensity....265
By Sanjay Kumar; ArunVerma; Ikhlaq Hussain; Bhim Singh
Real Time Implementation and Comparison of PI and Modified Inc Cond Control Algorithms for Solar Applications....271
By Venkata Kolluru; Kamalakanta Mahapatra; Bidyadhar Subudhi; Tejavathu Ramesh

Real-Time Simulation of Photovoltaic Panel Using Miniature Full Spectrum Simulator....277
By Ritika S; G Narayanan; Arjun Yadav
Design and Development of Real-Time Small-Scale Wind Turbine Simulator....283
By Jakeer Hussain; Mahesh Kumar Mishra
Comparative study of stochastic wind speed prediction models....288

By Alok Agrawal; Kanwarjit Singh Sandhu

Grid Connected Photovoltaic System with Data-based MPPT and Fuzzy Controlled DVR....294
By Akhil Gupta; SaurabhChanana; Tilak Thakur

Effect of non-uniform irradiance on electrical characteristics of an assembly of PV panels....300
By Md Hasan; Sanjay Parida

Performance Monitoring of 43 kW Thin-film Grid-Connected Roof-top Solar PV System....303
By Suresh Singh; Rakesh Kumar; Vivek Vijay

Grid Integration of Single Stage Solar PV Power Generating System using 12-Pulse VSC....308
By Bhim Singh; Ikhlaq Hussain

Analysis and Control of an Isolated SPV-DG-BESS Hybrid System....314
By Jincy Philip; Bhim Singh; Sukumar Mishra

Maximum Power of PV Plant for SP and TCT Topologies under Different Shading Conditions....320
By Jitendra Kumar; Shelly Vadhera

Modeling of PV Module to Study the Performance of MPPT Controller Under Partial Shading Condition....324
By Malik Sameeullah; A Swarup

Multi Diode Modelling of PV Cell....330
By Pawan Pandey; Kanwarjit Singh Sandhu

LVCMOS Based Energy Efficient Solar Charge Sensor Design on FPGA....334
By AnuSingla; Amanpreet Kaur; Bishwajeet Pandey

Survey on Hybrid (Wind/solar) Renewable Energy System and Associated Control Issues....339
By Rahul Sharma; SathansSuhag

A Novel Method of Generating Electricity by setting up Turbines over Rail Locomotives....345
By Pawan Kumar Sharma; Sahil Budhiraja; Narayan Hari; Sandeep Banerjee; Rahul Sharma

Simulation analysis and THD Measurements of Integrated PV and Wind as Hybrid System Connected to Grid....350
By Manish Khanagwal; Ashwani Sharma; Kanwarjit Singh Sandhu

Maximum Power Point Tracking Scheme for Variable Speed Wind Generator....356
By Sumit Chauhan; RatnaDahiya; Malik Sameeullah

Wind turbine economics: A Study....361
By Sahil Bajaj; Kanwarjit Singh Sandhu

Grid Voltage Monitoring Techniques for Single Phase Grid Connected Solar PV system....366
By LakshmananArunagiri; Amit Jain; Bharat Singh Rajpurohit

Control Strategy for Frequency Regulation using Battery Energy Storage with Optimal Utilization....372
By Manish Singh

Direct Duty Ratio Controlled MPPT Algorithm for Boost Converter in Continuous and Discontinuous Modes of Operation....376
By Pallavi Bharadwaj; Vinod John

Effect of Reliability of Wind Power Converters In Productivity of Wind Turbine....382
By Sanjay Jaiswal; GlPahuja

Hybrid Differential Evolution with BBO for Genco's multi-hourly strategic bidding....388
By Prerna Jain; Rohit Bhakar; S N Singh

On the Control and Design Issues of Single Phase Transformer less Inverters for Photovoltaic Applications....394
By Amit Gupta; Vivek Agarwal; Madhuwanti Joshi

Behavior of Wind Turbine under Different Operating Modes....400

By Navjot Sandhu; SaurabhChanana
Operating temperature of PV module modified with surface cooling unit in real time condition....405
By Madhu Sharma

Track 4–Power Electronics, Machine Control and Its Industrial Applications

Direct Torque Control of Open-End-Winding Induction Motor Using Matrix Converter....410
By Kalyan Govindarajan; Divakhar Anbazhagan; Senthil kumaran Mahadevan

Bridgeless Single-Ended Primary Inductance Converter with Improved Power Quality for Welding Power Supplies....416
By Swati Narula; Bhim Singh; G. Bhuvaneswari

Reliability Analysis with Parametric and Non Parametric Ranking of Buck Converter Components....422
By Ramakoteswara Rao Alla; G L Pahuja; Jagdeep Singh Lather

Direct Torque Control of Matrix Converter fed BLDC Motor....427
By Geeth Prajwal Reddy P.; Ranganath Muthu; S.Mahadevan; Abishek Rajaraman Lakshminarasimhan; Ganesh Palaniappan

Leverrier Algorithm based Reduced Order Modeling of dc-dc Converters....433
By Man Mohan Garg; YogeshHote

Automated Precise Liquid Transferring System....439
By MeeraChitra Sunil; Pinisetti Swami Sairam; Roushan

Design and Implementation of hydraulic motor based elevator system....445
By Roushan Kumar; Prashant Dwivedi; Praveen Reddy; Amiya Das

Hardware Design and Implementation of Unity Power Factor Rectifiers using Microcontroller....451
By Bidyut Mahato; ParashuramThakura; Kartick Jana

Performance Assessment of different Control Strategies for five level DCMLIs supplying Static loads and Dynamic loads....456
By Sandeep Banerjee; Dheeraj Joshi; Madhusudan Singh; Rahul Sharma

An Improved Dead-Time Compensation Scheme for Voltage Source Inverters Considering the Device Switching Transition Times....462
By AnirudhGuha; G Narayanan

Robust Nonlinear Observer Design for Twin Rotor Control System....468
By Abhinav Singh; Bhanu Pratap

Investigations on Optimal Pulse-Width Modulation to Minimize Total Harmonic Distortion in the Line Current....474
By Avanish Tripathi; G Narayanan

Induction Machine Efficiency Estimation Using Population Based Algorithm....480
By Gurinderbir Grewal; Bharat Singh Rajpurohit

Voltage Regulation Enhancement in a Buck Type DC-DC Converter Using Queen Bee Evolution Based Genetic Algorithm....486
By Tousif Khan. N

Class-D/E Resonant Inverter for Multiple Load Domestic Induction Cooking Appliances....492
By P. Sharath Kumar

Linearised Modelling of Switched Reluctance Motor for Closed Loop Current Control....498
By Shahjahan Ahmad Syed; G Narayanan

Design and Development of Improved Power Quality Conditioner Fed PMSM Drive....N/A
By Shailendra Sharma; Menka Dubey; Bhim Singh

A Novel Eighteen-Level Inverter for an Open-end Winding Induction Motor....504
By Sanjiv Kumar; Pramod Agarwal

A Comparative Study of Scalar Controlled IM Using Two and Three Level Bus Clamped SVM....N/A
By VenkataramanaNaik N

Performance Analysis of Surface Permanent Magnet Synchronous Motor....N/A
By Shivarajappa Soukar

Experimental Comparison of Conventional and Bus-Clamping PWM Methods Based on Electrical and Acoustic Noise Spectra....510
By A C Binojkumar; Saritha Balathandayuthapani; G Narayanan

Power Factor Correction in Sensorless BLDC Motor Drive....516
By Vashist Bist; Bhim Singh

Analysis of a Robotic System with two DOF using Haar Wavelet....522
By Atul Kumar Pandey; Monika Mittal

Single Stage Single Phase Solar Inverter with Improved Fault Ride Through Capability....527
By Jalaj Arya; Lalit Mohan Saini

Power Quality Improvement of a Position Sensorless Controlled PMBLDCM Drive using Boost Converter....532
By Sachin Singh; Sanjeev Singh

Type-2 Fuzzy Logic based Controllers for Indirect Vector controlled SVPWM based Two-Level Inverter fed Induction Motor Drive....538
By Abhiram Tikkani; PVN Prasad

On-Line Monitoring of Winding Parameters for Single-Phase Transformers....544
By Abhinay Reddy; Bharat Singh Rajpurohit

Hardware Development and Implementation of Single Phase Matrix Converter as a Cycloconverter and as an Inverter....548
By Anand Kumar

A ZVT-PWM Multiphase Synchronous Buck Converter with an Active Auxiliary Circuit for Portable Applications....554
By Shiva Sarode, A. K. Panda, Tejavathu Ramesh

Flux Weakening Control Algorithm with MTPA Control of PMSM Drive....560
By Sukanta Halder; Satya P Srivastava; Pramod Agarwal

A DC Moter Driver Consisting of a Single MOSFET with Capability of Speed and Direction Control....565
By Gururaj Mulay, Akshay Yembarwar, Surabhi Raje

GA Tuned LQR and PID Controller for Aircraft Pitch Control....568
By Vishal Chugh; JyotiOhri

Fault Identification of Power Transformers Using Proximal Support Vector Machine (PSVM)....574
Hasmat Malik

Implementation of Three -Leg VSC based DVR using IRPT Control Algorithm....579
By Bangarraju Jampana; V Rajagopal; Jaya LaxmiAskani

Power Quality Improvement Using Solar PV H-bridge Based Hybrid MLI....585
By Bangarraju Jampana; V Rajagopal; Jaya LaxmiAskani, N Bhoopal, Priyanka M

Increase Efficinecy of Solar PV System by Data Aquisistion System....N/A
By Sunil Parashar Sunil Dhankar

Track 5–Application of Power Electronics to Transportation

An Elimination Technique of Cross Regulations in The Flyback Converters....590
By Tapas Halder

Design and Simulation of H-Bridge Fed Direct Torque Controlled Electric Traction Drive....596
By Nikita Gupta; Rachana Garg; Priya Mahajan; Parmod Kumar

Optimized Rotor Slot Shape for Squirrel Cage Induction Motor in Electric Propulsion Application....602
By Mohammad Akhtar; RanjanBehera; Sanjay Parida

Secured charging of electric vehicles at unattended public charging stations using Verilog HDL....607
By Ridhi Saini

A Hybrid Electric Vehicle with Incorporation of VaReB Technology....611
By Tanmay Parashar; Aseem Mathur; Dhruv Kohli; Anirudh Dube

Energy Efficiency Performance of Reconfigured Radial Networks with Reactive Power Injection....617
By Pawan Kumar; Surjit Singh

Load Frequency Control of Three Area Hydro Thermal Power System with HFTID Controller....623
By Ajay Kumar, Ratna Dahiya

Track 6– Electronics Devices and Circuits

Advanced Pulse Width Modulation Technique for Z-Source Inverter....629
By Sangeeta Deb Barman; Tapas Roy

Design, Development and Relaibility Analysis of a Four-Quadrant Electromagnet Power Supply....635
By Siddharth Varshney; Manohar Koli; Mangesh Borage; Sunil Tiwari; Amalesh Thakurta

Design of Frequency Selective Surfaces Embedded Broadband Multilayered Microwave Absorbing Structures....N/A
By Ravi Panwar; Dharmendra Singh; Smitha Puthucheri; Vijaya Agarwala

Modified Sine Wave Phase Disposition PWM Technique for Harmonic Reduction in Multilevel Inverter fed Drives....641
By Punit Acharya; Vishal Rangras

4H-Silicon Carbide (SiC) Based Dopant Segregated Schottky Barrier Cylindrical Gate All Around (CGAA) MOSFET for Harsh Environment, High Speed and High Power Microwave Applications....646
By Manoj Kumar; Subhasis Haldar; Mridula Gupta; R S Gupta

A New Technique to Implement Conventional as well as Advanced Pulse Width Techniques for Multi-level Inverter....650
By Debanjan Roy; Tapas Roy

Variation of IGBT Switching Energy Loss with Device Current: An Experimental Investigation....656
By Subash Chandra Das; G Naryanan; Arvind Tiwari

Silicon Carbide based DSG MOSFET for High Power, High Speed and High Frequency Applications....661
By SonamRewari; Vandana Nath; Satvir Singh Deswal; R S Gupta

Comparative Study of Enhancement-Mode Gallium Nitride FETs and Silicon MOSFETs for Power Electronic Applications....665
By Anirban Pal; G Narayanan

Current Regulated Induction Motor Drive with IFOC....671
By Rohit Kumar; Saurabh Kumar; Loveleen Kaur

Design and Analysis of Charge-pump Based Buck Converter....676
By Veerachary M

Design and Analysis of Zero-voltage switching Fifth-order Boost Converter....682
By Veerachary M

Selective Harmonic Elimination: Comparative Analysis by Different Optimization Methods....688
By Sreedhar Madichetty; Dasgupta Abhijit; Rambabu M

A NF Based Direct Torque Control of Induction Motor Drive Using BCSVM....694
By Venkataramana Naik N

Energy Efficient Flip Flop Design Using Voltage Scaling On FPGA....699
By Amanpreet Kaur; Bishwajeet Pandey; Sunny Singh

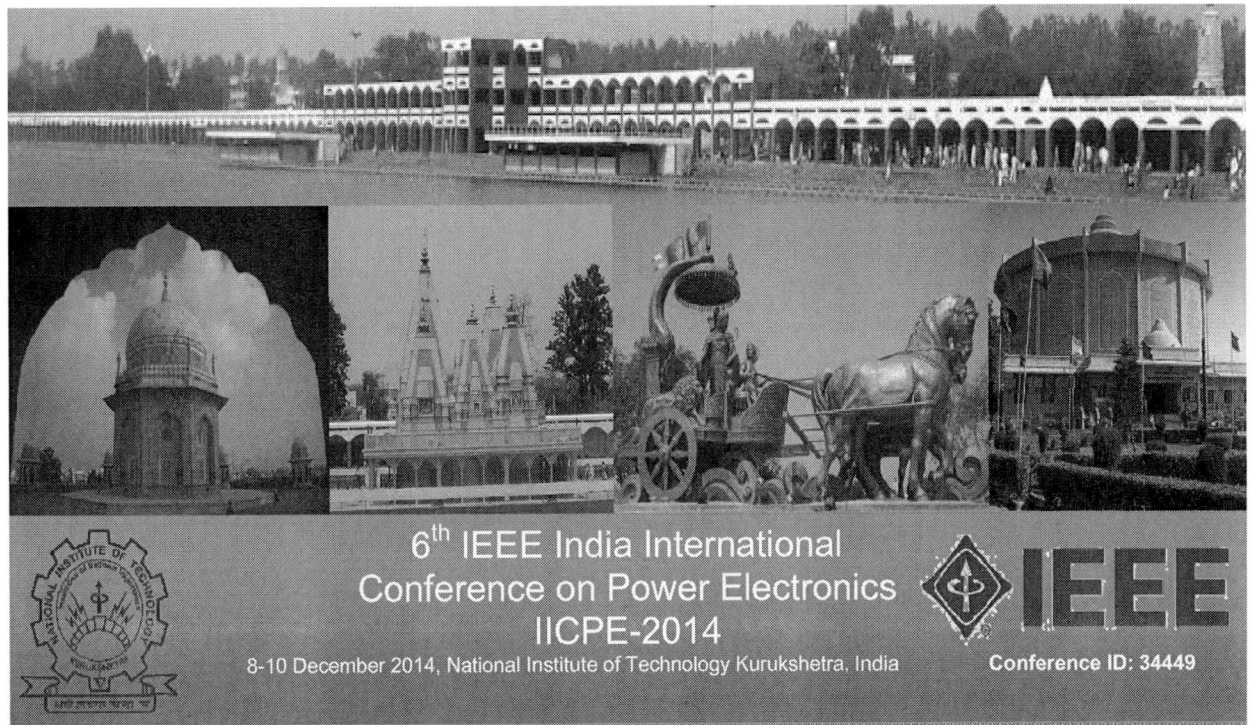

6th IEEE India International Conference on Power Electronics
IICPE-2014

8-10 December 2014, National Institute of Technology Kurukshetra, India

Conference ID: 34449

Residential Demand Response from PV Panel and Energy Storage Device

Monika Arora, Saurabh Chanana
Electrical Engineering Department
National Institute of Technology
Kurukshetra, India
monika.gaba@gmail.com, s_chanana@rediffmail.com

Abstract— This study proposes an optimal scheduling mechanism for residential appliances in a smart home equipped with photovoltaic panel and energy storage device. The optimal mechanism aims to manage the operation of centralized air conditioner, solar energy input and charging/discharging of energy storage device such that energy consumption cost is minimized. Mixed integer programming technique is applied to obtain optimal solution. For evaluating the effectiveness of proposed strategy, the results obtained from this technique are compared with previously established optimization technique using linear sequential algorithm in combination with price-aware storage algorithm. Results indicate that the proposed strategy have the capability of maximizing the savings in electricity cost. Contribution of this work is the introduction of a real time price based scheduling mechanism that utilizes smart grid capability of feeding power back to the grid.

Keywords— *Air conditioner; demand response; energy storage device; home energy management; pv panel; real time price*

I. INTRODUCTION

As a result of economic and industrial growth, world energy demand has increased enormously in recent years. Till now, all the countries have been relying primarily upon on fossil fuels for generation of electricity. Due to global concerns such as cleaner environment and depletion of fossil fuel resources, focus has now shifted towards sustainable sources of energy. Wind and solar resources have the largest share in the energy coming from renewable energy resources. With the increasing competition and growth in photovoltaic (PV) installations, a significant reduction in the module cost has been observed over the past few years. It is predicted that by 2020, power from PV panel will become cheaper than power from the grid [1]. Thus PV panels have become an attractive option for the residential consumers.

Adoption of distributed energy resources (DER) calls for a need to transform the power sector. Centralized generation model of existing power system is not capable of fulfilling the changing needs of 21st century. These problems can be overcome with the implementation of a new power grid called smart grid. In smart grid, advanced technologies of communication, monitoring and control are integrated with existing power system. Demand response (DR) technique is an important tool of smart grid that has the potential of addressing many power grid issues. DR is capable of handling increasing energy demand, deferring the need to expand the power system and integration of DERs etc. DR has been effective in managing the energy consumption of consumers in response to different pricing schemes. Many DR programs have been implemented for industrial and commercial sector in past few years. Now the focus is towards developing efficient DR programs for residential sector. With the integration of DERs at residential level, scheduling of home appliances needs to be managed so that the benefits drawn from DERs are maximized. A suitable scheduling strategy needs to be prepared for the same. DERs viz. PV panel and energy storage device (ESD) are used in this study.

There are various incentive based and price based schemes available in the market. Success of DR program essentially depends upon the selection of a suitable pricing scheme. According to [2], real time price (RTP) scheme is considered to be most suitable among other pricing schemes.

A variety of work done has been done in the area of residential DR in recent years. In [3], a control scheme for an air conditioner (AC) to reduce peak demand by load shifting is proposed. In [4], load model of an AC based on stochastic difference equation is used to study the effects of the load parameters and the direct load control actions. Various set point control strategies have been explained in [5] by Nu and Katipamula for thermostatically controlled appliances (water heater taken as an example). Reference [6] demonstrates the inclusion of smart thermostat into Home Energy Management (HEM) system for residential DR implementation. Some advantages of temperature set point control over direct load control have also been discussed. Various HEM algorithms for peak shaving or achieving cost benefits have been proposed in [7], [8], [9] and [10]. A number of researches are going on for examining the benefits of integrating PV system and ESD at residential level and their contribution in DR programs. A methodology for managing domestic energy consumption in presence of storage device has been presented in [11] under three types of whole sale prices. In [12], a battery storage mechanism is devised to achieve residential cost saving from undertaking energy arbitrage. An energy management algorithm with renewable resources and a battery has been proposed in [13] using time of use (TOU) program. Reference [14] further proposed a storage control algorithm for PV

978-1-4799-6047-7/14 $31.00 © 2014 IEEE

modules with electricity pricing function as a combination of TOU and peak power component. A very few RTP programs have been proposed utilizing both PV modules and energy storage. Recent work [15] implements RTP programs for renewable energy resources and energy storage. However, this program consider PV panel and ESD for storing and supplying energy but their ability to sell power back to the grid is not investigated.

In this paper, problem of optimally scheduling the residential appliances in the presence of PV panel and ESD is considered. The proposed technique aims to reduce electricity consumption cost by appropriately deciding the schedule of AC load, ESD storage level, charging/discharging schedule of ESD and power drawn from power grid. Mixed integer programming (MIP) is used to solve this optimization problem. DR program uses RTP scheme in this case. Utilizing the capability of smart grid, user is allowed to sell surplus power back to the grid. In this problem, residential load comprises of two parts: heating ventilating and air-conditioning (HVAC) load and non-HVAC load. During summer, AC constitutes the major portion of the peak load. AC is considered as the HVAC appliance present in the residential house. Typical power consumption pattern of non-HVAC appliances e.g. clothes washer, clothes dryer, dishwasher and lighting etc. obtained in [16] is used. ESD with basic characteristics of charging/discharging capacity and maximum/minimum storage level is considered. Energy produced from PV panel is used to charge ESD.

To show the effectiveness of proposed scheme, the results of this scheme are compared with previously established techniques: Linear Sequential algorithm (LSA) and Price-aware Storage algorithm (PSA). LSA [17] has been implemented for AC load. PSA [15] strategy aims at managing the energy of ESD and PV panel. A combination of these technologies is carried out to obtain optimal schedule of AC while effectively utilizing the energy from PV panel and ESD.

The paper is organized in V sections. Section I represents the overview of DR and DERs. Section II describes the modeling of AC and ESD. Section III presents implementation PSA for PV panel and ESD. This section also presents the proposed scheduling mechanism. Section IV displays a comparison of the results of these techniques. Section V contains the conclusion followed by references.

II. SYSTEM MODEL

With the implementation of smart grid, advanced metering system, two-way communication and control system are assumed to be in place. RTP signals sent by utility are received by smart meter. Smart meter passes these signals to HEM system so that required action can be taken based on the price of electricity at that time. Smart meter enables consumers to monitor total energy consumed by home appliances. Smart grid should possess the capability of absorbing the energy fed back to the grid by smart home.

Smart home considered in this study, houses AC load as HVAC candidate, non-HVAC load, PV panel and ESD (Battery in this case). Load draws power from PV panel or

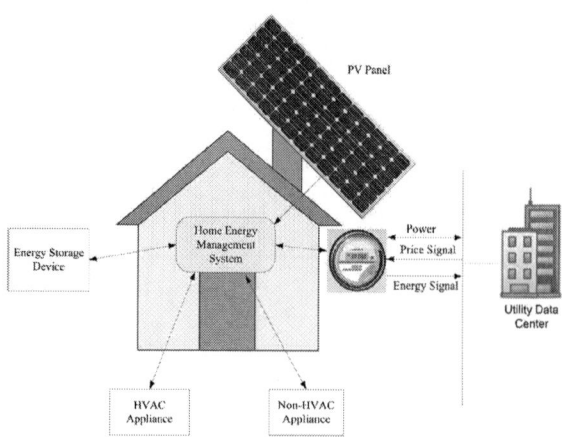

Fig. 1. Interaction of Smart Grid with Smart Home equipped with PV panel and energy storage device

battery on priority. If not available, power grid provides the required load power. ESD is charged from PV panel or from power grid during low cost period. When local energy generated is in surplus, it is fed back to the grid. Interaction of smart home with smart grid has been shown in Fig 1.

Suitable load models of AC and ESD should be used for better scheduling results. Load models of AC and ESD used in this study have been presented below.

A. Air conditioner Model

In this work, discrete time model of AC explained in [18] is used. This model explains the variation of inside temperature of the house with time. Inside temperature at any instant t depends on inside temperature at previous time slot t-1, the outside temperature and ON/OFF state of AC thermostat at time slot t. The model is as follows:

$$T_{in}(t) = T_{in}(t-1) + \tau(E(T_{out}(t) - T_{in}(t)) - \phi S_{AC}(t)) \qquad (1)$$

where

$T_{in}(t)$ – Inside temperature of a house during time slot t

$T_{out}(t)$ – Forecasted outdoor temperature during time slot t

$S_{AC}(t)$ – Status of AC thermostat during time slot t

E – Warming effect of difference in outside and inside temperature on inside temperature

ϕ – Cooling effect of ON state of AC on inside temperature

τ – Time interval duration

AC control strategy should be such that comfortable temperature is maintained inside the house. Therefore temperature limits need to be specified in the scheduling problem. Thermostat set point control strategies (Normal and

Programmable) explained in [19] are used along with model of AC load.

B. Energy storage device

A number of storage devices are available these days but for residential purpose, batteries are preferred as they can be easily placed anywhere inside a house. The dynamics of ESD can be described by following equation:

$$es(t) = es(t-1) + \tau(P_S(t) + P_{ESD}S_{ESD\_CH}(t) - P_{ESD}S_{ESD\_DSCH}(t)) \quad (2)$$

where

$es(t)$ – Energy storage level of ESD during time slot t

$P_S(t)$ – Solar power input to ESD during time slot t

P_{ESD} – Rated charging/discharging power of ESD

$S_{ESD\_CH}(t)$ –Status representing charging of ESD from power grid during time slot t; binary (0/1)

$S_{ESD\_DSCH}(t)$ –Status representing discharging of ESD during time slot t; binary (0/1)

τ – Time interval duration

Assumptions made in this regard are charging/discharging power of battery to be equal and battery is either charging or discharging. ESDs are important components in DR programs as they can help in reducing peak demand and improving the utilization of renewable energy.

III. Scheduling Approach

In this section, both the scheduling strategies used for solving the scheduling problem have been explained.

A. Price-aware Storage Algorithm for Energy Storage Device

The algorithm aims at determining the charging and discharging decisions of the storage such that solar energy is utilized effectively and electricity cost is minimized.

For this, concept of threshold price is adopted from [15]. Threshold price provides a boundary for categorizing low and high price. If price during a time slot increases above threshold price, battery is discharged otherwise it is charged either from solar or from grid. In case solar energy is not present, battery is charged from grid. Threshold price is obtained from (3).

$$\lambda_{threshold} = k\lambda_{MAX} \quad (3)$$

where

λ_{MAX} –Maximum price during the day

$\lambda_{threshold}$ –Threshold price of electricity

k –Threshold factor ($0 < k < 1$)

Value of k is varied between 0 and 1 to visualize the behavior of cost with threshold factor. Due to limited capacity, all solar energy can't be stored in the battery. Surplus power can be sold back to the grid for financial benefits.

B. Optimization Technique

Second technique uses MIP for solving the scheduling optimization problem. Scheduling problem aims to efficiently utilize power sources grid and solar, to supply residential load with the help of apt charging and discharging decisions of ESD. Energy consumption is managed in such a manner that the electricity cost is minimized. Hence the objective function is specified as follows:

$$min \ C = \sum_t P_{Grid}(t)\lambda(t) \quad (4)$$

for all $t = [1,2,........N]$;

where

t -   Time slot

N-   Total no of time slots.

C-   Cost of electricity over the entire time period

$\lambda(t)$ -   RTP of electricity during time slot t

$P_{Grid}(t)$ - Power drawn from grid during time slot t

subject to

1) Equality Constraint:
    a) Thermal dynamics of centralized AC

Thermal dynamics of AC has been explained by (1).

    b) AC Thermostat Setting

Thermostat setting model using Normal thermostat and Programmable thermostat [19] for AC scheduling has been mentioned below:

Normal thermostat control

$$T_{st}(t) = const. \quad (5)$$

where

$T_{st}(t)$ – Thermostat set point during time slot t

Programmable thermostat control

$$T_{st}(t) = a\lambda(t) + b \quad (6)$$

where

a, b are constants.

    c) Comfort Settings

Eq. (7) and (8) present the comfort settings specified by user in terms of minimum and maximum limit of temperature.

Normal thermostat control

$$T_{min}(t) = T_{st}(t) - \Delta T \tag{7}$$
$$T_{max}(t) = T_{st}(t) + \Delta T \tag{8}$$

where

$T_{min}(t)$ –Minimum limit for inside temperature during time slot t

$T_{max}(t)$ –Maximum limit for inside temperature during time slot t

ΔT –Temperature band around thermostat set point in which inside temperature is allowed to vary

Programmable thermostat control

Minimum limit is kept fixed in this case as shown in (9).

$$T_{min}(t) = Const. \tag{9}$$

However, representation of maximum limit is same as shown in (8).

d) Storage dynamics of ESD

Energy storage in ESD is governed by (2). The equation relates the energy stored in ESD at any time slot t to that at time slot t-1, solar input P_S, grid input and discharging status at time slot t.

e) ESD Status constraint

ESD is assumed to be either charging or discharging during any time slot.

$$S_{ESD\_CH}(t) + S_{ESD\_DSCH}(t) = 1 \tag{10}$$

f) Power balance equation

Eq. 11 specifies the net power drawn from grid.

$$P_{Grid}(t) = P_{AC}S_{AC}(t) + P_{non\_HVAC}(t) + P_{ESD}S_{ESD\_CH}(t) - P_{ESD}S_{ESD\_DSCH}(t) \tag{11}$$

where

P_{AC} – Rated power of AC

$P_{non\_HVAC}(t)$ – Power demand of non-HVAC appliances

Other symbols have the same meaning as explained earlier in section 2.

2) Inequality Constraint:

a) State of AC thermostat

AC thermostat can be either in ON state or in OFF state. Binary variable $S_{AC}(t)$ represents the state of AC thermostat in the problem. $S_{AC}(t)$ = '1' represents thermostat is in ON state, '0' represents thermostat is in OFF state.

$$0 \le S_{AC}(t) \le 1; \tag{12}$$

b) Inside Temperature constraints

Inside temperature should vary within minimum and maximum limit specified by user.

$$T_{min}(t) \le T_{in}(t) \tag{13}$$
$$T_{in}(t) \le T_{max}(t) \tag{14}$$

c) State of ESD

ESD is charged from grid at low cost time intervals. Status S_{ESD_CH} = 1 represents charging from grid during time slot t. Status S_{ESD_DSCH} = 1 represents the discharging state of ESD. Charging and discharging status both are binary variables.

$$0 \le S_{ESD\_CH}(t) \le 1; \tag{15}$$
$$0 \le S_{ESD\_DSCH}(t) \le 1; \tag{16}$$

d) Storage Capacity constraints

Energy storage level should remain within a minimum and a maximum limit.

$$E_{min} \le es(t) \tag{17}$$
$$es(t) \le E_{max} \tag{18}$$

where

E_{min} – Minimum energy storage level of ESD

E_{max} – Maximum energy storage level of ESD

IV. RESULTS

This section presents the solution of home appliances scheduling problem as a result of application of both the techniques presented in section 3. Firstly, a combination of LSA and PSA is used to obtain the scheduling results. In this approach, threshold values of RTP determine the operation of AC and operation of ESD. Secondly, MIP optimization technique is applied to obtain the results. For both the cases, we assume the surplus generation from PV panel will be injected to the grid. Price at which surplus energy is sold to the grid is same as RTP of electricity at that time. The results show that MIP scheme gives higher reduction in electricity cost. GAMS and MATLAB software are used for implementing the approach.

Smart home is installed with centralized AC of 11.5 kW, PV Panel of 3 kW and an ESD of 3 kW. Maximum and minimum energy storage level is considered as 30 kWh and 6 kWh respectively [20]. Input data required for executing the program is ambient temperature, RTP of electricity, power generated from PV panel and non-HVAC appliances power demand shown in Fig. 2 to Fig. 5 respectively. RTP data is

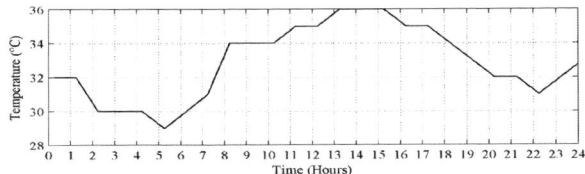

Fig. 2. Variation of ambient temperature with time.

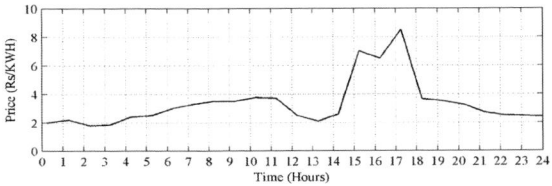

Fig. 3. Variation of RTP with time

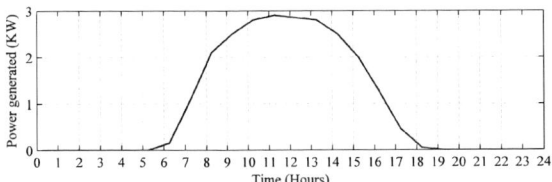

Fig. 4. Power generated from PV panel

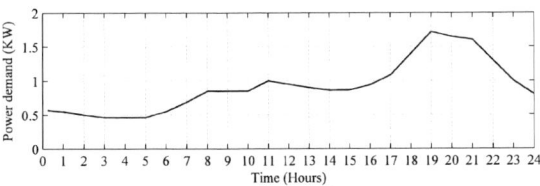

Fig. 5. Power demand of non-HVAC appliances in the smart home

based on day-ahead price variation at Indian Energy exchange and is represented in terms of Rs[1] per kWh [21].

The program is run for a day i.e 24 hours. Each day is divided into 96 time slots of 15 minute each.

Results of scheduling problem are obtained for different thermostat settings of centralized AC.

A. Normal Thermostat

Thermostat set point of AC in this case is taken as 22 °C. Minimum and maximum temperature limit has been taken as 20 °C and 24 °C respectively. As a result of application of LSA, RTP threshold value comes out to be 2.049 Rs per kWh. AC thermostat is turned OFF if RTP of electricity exceeds this threshold and vice-versa. In case of violation of constraints,

status should be changed such that temperature remains within a comfortable range. In case of PSA, threshold price determines the operation of ESD charging or discharging. Threshold price is obtained corresponding to different threshold factors varying from 0 to 1. For each threshold factor, ESD operation results are obtained. A combination of LSA and PSA then gives electricity cost of home appliances. A plot presenting the variation of electricity cost with threshold factors has been shown in Fig 6.

Fig. 6 shows that least cost is observed at k = 0.3. As a result of application of this technique, optimal electricity cost of home appliances comes out to be Rs 431.73 at a threshold factor of 0.3.

Solution of scheduling problem by MIP technique, results in electricity payment of Rs 246.54. A comparison of stored energy level of battery obtained from both the techniques shows that in case of MIP technique, battery is discharged to its minimum level in the evening. As a result, less power is drawn from the grid and cost is reduced. Energy storage level of battery in case of both the techniques is shown in Fig 7.

B. Programmable Thermostat

Maximum temperature limit is varied according to (8). Values of a and b in (6) are taken as 0.89 and 22 respectively. Minimum temperature limit is taken as 20 °C. ΔT is taken as 2 °C. As a result of application of LSA and PSA technique, electricity cost of home appliance obtained with different threshold factors has been displayed in Fig 8.

It is again observed that minimum electricity cost is achieved at threshold factor of 0.3. Electricity cost comes out to be Rs 274.93. Solution of scheduling problem by MIP technique, results in electricity payment of Rs 140.06.

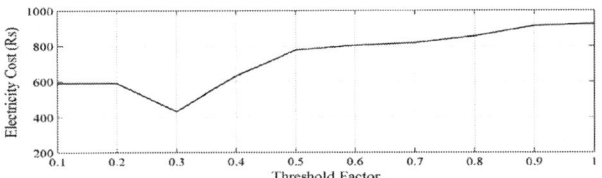

Fig. 6. Variation of electricity cost with threshold factor

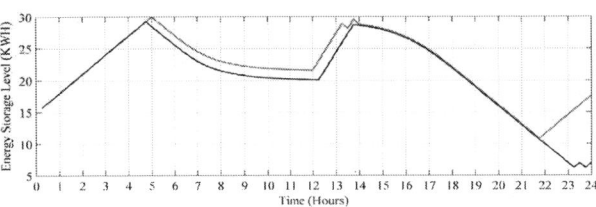

Fig. 7. Comparison of energy storage level in case of PSA and MIP technique (Normal Thermostat)

---

[1] Rs stands for Indian Rupees. Conversion rate: 1 Indian Rupee = 0.016 US Dollar.

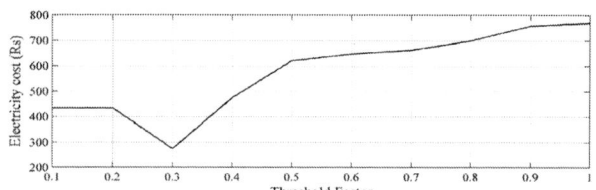

Fig. 8. Variation of electricity cost with threshold factor

Table I presents the comparison of electricity cost obtained via both the techniques.

TABLE I.   COMPARISON OF ELECTRICITY COST OF SMART HOME IN BOTH THE SCHEDULING TECHNIQUES

Cost	ESD, PV, AC-Normal Thermostat	ESD, PV, AC-Programmable Thermostat
LSA & PSA	431.73	274.93
MIP Technique	246.54	140.06

Results show that MIP optimization strategy is more effective in utilizing solar energy and energy from ESD. The proposed scheduling strategy gives lesser cost compared to LSA and PSA techniques.

V. CONCLUSION

This paper presents RTP based optimal scheduling technique for managing the energy consumption of residential appliance in smart home in the presence of an ESD and a local generation source i.e. PV panel. Firstly, simplified model of AC and ESD are explained. Two AC thermostat control strategies corresponding to normal thermostat and programmable thermostat are considered. Next, MIP based scheduling strategy is proposed and results are presented. A comparative study is performed with a previously established technique: LSA in combination with PSA. Both the techniques take advantage of smart grid capability of feeding power back to the grid. Results show that MIP technique presents greater savings in electricity bills as compared to LSA and PSA.

REFERENCES

[1] Ernst & Young Global Ltd., http://www.ey.com/US/en/Services/Advisory/EY-Distributed-energy-the-challenge-for-utilities.

[2] M. H. Albadi and E. F. El-Saadany, "A summary of demand response in electricity market," Electric Power System Research, Vol. 78, No. 11, pp. 1989-96, November 2008.

[3] N. Zhu, X. Bai, and J. Meng, "Benefits analysis of all parties participating in demand response," Power and Energy Engineering Conference (APPEEC), pp. 1-4, March 2011.

[4] C. Ucak and R. Caglar, "The effects of load parameter dispersion and direct load control action on aggregated load", Proc. of International Conference on Power System Technology (POWERCON), vol. 1, pp. 280-284, August 1998.

[5] N. Lu and S. Katipamula "Control strategies of thermostatically controlled appliances in a competitive electricity market", Proc. IEEE Power Eng. Soc. Gen. Meet., vol. 1, pp. 202-207, June 2005.

[6] A. Saha, M. Kuzlu, and M. Pipattanasomporn, "Demonstration of a home energy management system with Smart Thermostat control", IEEE PES Innovative Smart Grid Technologies (ISGT), pp. 1-8, February 2013.

[7] M. Pipattanasomporn, M. Kuzlu, and S. Rahman, "An algorithm for intelligent home energy management and demand response analysis", IEEE Trans. on Smart Grid, vol. 3, no. 4, pp. 2166-73, December 2012.

[8] Z. Zhu, J. Tang, S. Lambotharan, W. H. Chin, and Z. Fan, "An Integer Linear Programming based optimization for home demand-side management in smart grid", in Proc. IEEE PES innovative Smart Grid Technologies Conf., pp. 1-5, January 2012.

[9] S. Nistor, J. Wu, M. Sooriyabandara, and J. Ekanayake, "Cost optimization of smart appliances", Proc. of the 2nd IEEE PES International Conf. and Exhibition on Innovative Smart Grid Technologies (ISGT Europe), pp. 1- 5, December 2011.

[10] G. T. Costanzo, J. Kheir, and G. Zhu, "Peak load shaving in smart homes via Online Scheduling", IEEE International Symposium on Industrial Electronics (ISIE), pp. 1347-1352, June 2011.

[11] Z. Wang, F. Li, and Z. Li, "Active household energy storage management in distribution network to facilitate demand side response", IEEE Power Eng. Soc. Gen. Meet., pp. 1-6, July 2012.

[12] C. Byrne and G. Verbic, "Feasibility of residential battery storage for energy arbitrage", Australasian Universities Power Engineering Conference (AUPEC), pp. 1-7, September 2013.

[13] A. R. Boynuegri, B. Yagcitekin, M. Baysal, A. Karakas, and M. Uzunoglu, "Energy management algorithm for smart home with renewable energy resources", 4th International conference on Power Engineering, Energy and Electrical drives, pp. 1753-1758, May 2013.

[14] Y. Wang, X. Lin, and M. Pedram, "Adaptive control of energy storage systems in household with photovoltaic module", IEEE Trans. on Smart Grid, vol. 5, no. 2, pp. 992-1001, March 2014.

[15] M. Liu, W. J. Lee, and L. K. Lee, "Financial opportunities by implementing renewable resources and storage devices for household under ERCOT demand response program design", IEEE Industry Application Society Annual Meeting, pp. 1-7, October 2013.

[16] IEA energy conservation in buildings and community systems, http://www.ecbcs.org/docs/ Annex_42_Domestic_Energy_Profiles.pdf.

[17] P. Du and N. Lu, "Appliance commitment for household load scheduling", IEEE Trans. on Smart Grid, vol. 2, no. 2, pp. 411-419, June 2011.

[18] M. C. Bozchalui, S. A. Hashmi, H. Hassen, C. A. Canizares, and K. Bhattacharya, "Optimal operation of residential energy hubs in smart grids ," IEEE Trans. on Smart Grid, vol. 3, no. 4, pp. 1755-1766, December 2012.

[19] S. Chanana and M. Arora, "Demand Response from Residential Air Conditioning Load using a Programmable Communication Thermostat", International Journal of Electrical, Electronic Science and Engineering, vol. 7, no. 12, December 2013.

[20] S. A. Hashmi, "Evaluation and improvement of the residential energy hub management system", M.S. thesis, University of Waterloo, Waterloo, Canada, 2010.

[21] Indian Energy Exchange http://www.iexindia.com/Reports/AreaPrice.aspx

PI Controller Tuning & Stability Analysis of the Flyback SMPS

T. Halder
Assistant Professor
Electrical Engineering Department, Member IEEE
Kalyani Government Engineering College
Kalyani, District-Nadia, West Bengal,
PIN-741235, INDIA

Abstract—**The paper presents that the adept tuning process of a proportional and integral (PI) controller to make conforms the load voltage of the flyback converters at which the proficient computerized tuning enlargement places of interest the stability and performance analysis. Software tuning processes are used to maintain the tight load regulations and to restrain the anomalous transient parts of the load voltage to study stability analysis making initial design criteria. The shape of the feedback loop negotiates the predictions of the stabilized zone to formulate the industrial products and generic tuning process**

Keywords—*PI controller; Tuning process; Operational mode; The flyback SMPS; Control loop ; Stability Analysis; and Results*

I. INTRODUCTION

A grouping of proportional controller and integral control (PI) is now a common method to regulate the load voltage of the converter where feedback loop for the most part is used in engineering control techniques. A PI controller enables to carry out as error rate and the variation with respect to the set point. The controller may adjust the error signal in the feedback control loop. The computational algorithm involves two breaks up invariable parameters, and is consequently from time to time called two terms control, the proportional (P) and the integral (I) controller [1]

The error ethics may be incorporated in the loop to stabilize the converter focusing the time domain analysis. The weighted summations of two controller actions are used to normalize the inspection via a control section such as the setting of a control unit. In the absence of relationship of the basic SMPS, a PI controller has unadventurously been used to be the best option of controller [2]-[3].

Having tuned, the PI controller gains and the controller is competent to administer control actions for tight regulations for instant response at which the controller overshoots the setpoint and the degree of scheme oscillations are to be adjusted. The implementation the PI algorithms are for optimal prediction and elucidation of the converter's stability in wider modes.

A numeral of substantial techniques invite to suppress the abnormal oscillations using the trial and error process which is really toughs in practical situations and it also takes more time to implement but erroneous process. It necessitates only one or two control actions to modify again and again to control the load by virtue of which the setting parameters are not to be zero at steady state [4]-[5].

A PI controller may be accredited for a PI, P or I controller in the nonattendance of the individual and transactions of the control at which PI controller is the proverbial and simple as cost effective implementations.

Since, the derivative action does not use alone due to noise formations, whereas, I controller works in absence of P term. It keeps away from the system from accomplishment its mark significance due to the P control actions. A plain picture of a control loop is taken adjusting loop gain and to keep up the load voltage of the SMPS at undesired output of the converters due to sustained oscillations [6]-[7].

This occupation of two control actions based on feedback, the controller performs a control action to regulate the SMPS to stabilize. The load control irregularities or process of value (PV). The reference voltage is called the setpoint (SP). The input to the reference is called the influence of variable (MV) not excellent in common features. The imbalance between the voltage and the reference is the error (e) and enumerates whether the load voltage is too high or low error and by how to a large extent. After a cautious assessment, the voltage (PV), and then plot of the errors, the controller straightens out on when to make corrections the tap position (MV) and by how greatly. When the controller initial turns on the main switch, it may switched on the high or to some extent if far above the ground is preferred, or it may unchain the main switch every one of the practice if very high error is most wanted. This is a paradigm of a straightforward proportional control. The controller occurrence that high value does not pull in hurriedly may face up to go faster for the tight regulations of SMPS. The switching actions are either on or off the main switch again and again as time goes by [8]-[9].

This is a paradigm for an integral control format. Edifice of an adjustment that is too outsized when the error is microscopic is consequent to a far above the ground gain controller and will guide to just about the overshoots. If the controller is to repeatedly commence versions that are too big and chronically overshoot the purposes. The load voltage would be retreated and surge roughly the setpoint in a steady value, mounting, or

This work is solely completed by the author: tapas_haldar@yahoo.com

978-1-4799-6047-7/14 $31.00 © 2014 IEEE

putrefying loop signals. If the oscillations add to with time, then the harmonization is tremulous, whereas the controller actions reduce the system is marginal stable. If the oscillations stay at the back at a reputable magnitude the system is slightly stable. In the significance of achievement an ongoing union at the preferred voltage (SP), the controller may have to to clammy liable stance oscillations. So, in order to compensate the loop for this upshot, the controller may straighten out on to inflict its regulations. Variables that brunt on the SMPS other than the MV are acknowledged as turbulences. The controllers are smashed usually to turn down disorders and setpoint changes which changes in load voltage include a disturbance to the converter control procedure.

In this document, the controller enables to systematize any SMPS at which PV documented agreeable significance for the setting output (SP) and input to the process (MV) that affects the significant response of the PV.

The controllers are utilized in the flyback converter to normalize voltage, current, power flow, and apiece additional capricious for which the online capacity is compulsory for the computer copying platforms.

II. MODEL OF PI CONTROLLER WITH THE FLYBACK SMPS

In this sector, the wherewithal of PI controller principally are coupled for the regulations purposes, the evaluations of the controller and the power circuit interpretation and explanation with a flyback converter making action of a PI controller is shown in the rectangular portion of the Fig. 1.

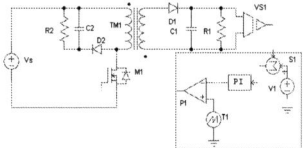

Fig. 1. The basic flyback converter

The Fig. 2 also demonstrates current fed flyback converter operating either discontinuous conduction mode (DCM) or continuous conduction mode of operations for the tight load or line regulations with software tuning process of PI controller unit.

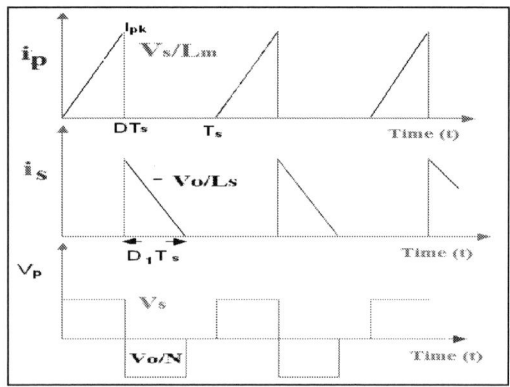

Fig. 2. Waveforms of the flyback converter under DCM operations

The relevant waveforms of the current fed flyback converter in DCM operations are shown in Fig. 2, where the main switch, M1, is turned on by applying the gate pulse for the pulse train of the switching cycle, and is also particular in this certificate for the fresh software simulation processes.

III. OPTIMAL PERFORMANCE OF THE SYSTEM

The optimal performance on a practice revolutionizes and reference adjustment fluctuates relying on the relevancies of undulations of the indications. Two simple requirements are regulations and intermission denunciations carry on with the particular reference, directive pathway, and implementations reference and online modifications.

These highly refer to how strong the prohibited variable tracks the preferred significances. Distinctive criteria enforces upon domination of the roadway which join together with rise time settling time and peak time. A number of approaches of the converter operations envisage the peak overshoot of the converters. The reference for a paradigm, it may be treated as perilous. Other practice must make in light of the power exhausted as per accomplishment and automatic adjustment of reference to accomplish excellent regulations.

IV. THE TUNING METHOD OF PI CONTROLLER

There are considerable customs to tune a PI round. The majority and flourishing techniques across the methods holds good huge concentrations and various improvement silhouette of converter to opt for the P and I based on the dynamic model of the SMPS. Corporal adaptation and modus operandi enables somewhat ineffectual. The loops fundamentally have response time's order of minutes.

The favorite route relies more often than not on whether or not the loop can be taken offline for auto tuning, and on the response time of the format. If the technique uses offline, the most outstanding tuning technique will frequently fit into place imposing to a step change at input assessing the output as a function of time. The response comes across the control parameters. The error of the PI controller is given as:

$$E(s) = K_P + \frac{K_I}{s} \qquad (1)$$

Where K_p and K_I is the integral gain of the controller and the steady state errors are found from its step response for the different values K_p and K_I. At last, the gains of the controller are too adjusted for achieving the stout stability and to regulate the load voltage and the tight line regulations of the flyback converters.

V. DEAD BAND OF THE PI CONTROLLER

A PI loop and maim switch, the electronics perpetuations may be a most important for the cost and costume to control appalling situation in the sketched out of either satiation or a deadband in the dynamic reaction process to input suggestion. The rate of reaction is normally an effectiveness function of how bigger than a main switch. It adeptly makes active to generate error signals at which power losses and thermal limits are very urgent and apprehensions.

The PI ring may perhaps a harvest deadband to lacerate the frequency of investiture of the load. This is accomplished by adaption of PI controller to carry out the load balanced if the change be pocket-sized within the lucid deadband assortments. The on line soft computing data output must make tracks the deadband before the specific loads will change at once.

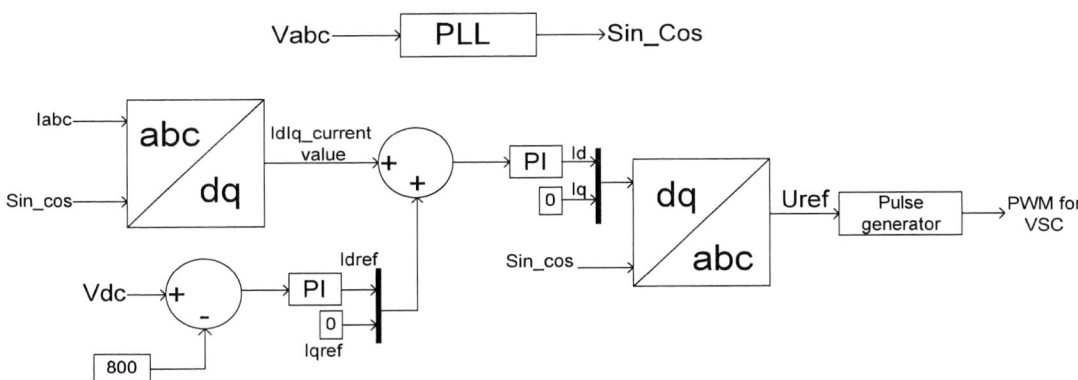

Fig. 3. Control Strategy of Grid Connected System

The error in the DC voltage if keep on falling further, indicates the overloading conditions and signifies the incapability of the PV source to continue supporting the maintenance of DC bus at desired level.

The moment the error voltage hits the tolerance limit (switching on limit) of the negative band, the switch gets on, and start supporting the DC bus by releasing the battery energy till it reaches again to the switching off limit of the negative band where it switches off, presuming that PV source would take over and support the DC bus further and the transient/sag is over. If overloading/under voltage/sag is sustaining further and the error voltage keeps drooping again up to the switching on limit of the negative band the battery is switched in again, and the process will construct a PWM sequence to operate the bidirectional switch for release of energy/isolation of battery till the PV source become capable of handling the loading conditions. Similarly for under loading/overvoltage/swell conditions, the positive band is operated the battery to absorb the energy from DC bus/isolate itself as per the PWM sequence so generated. If the error DC voltage remains exclusive of both positive and negative bands, signifying normal operating conditions the battery stays in deactivation mode waiting for getting charged on occurrence of under loading/overvoltage/swell or otherwise ready to support the DC bus under overloading/under voltage.

IV. RESULTS

A. Results for grid connected system

The grid connected system was simulated for a time period of 1 sec. It had two cases, case 1 doesn't constitutes Non-Linear loading whereas in Case 2 grid connected mode was simulated with Non-Linear load to explore the system with or without harmonics.

It can be seen that DC bus voltage has been maintained constant at required DC voltage level of 800 V in fig. 5.1(a). Also fig. 5.1(b) exhibits the AC voltage after conversion from DC voltage via VSC. The V_{abc} is maintained at 1100 V rms(ph-ph).

Fig. 5.2 contains the results for output at load and Point of common coupling (PCC). Fig. 5.2 (a) shows the active power of load, fig 5.2(b) shows the reactive power of the load. Fig. 5.2 (c) and (d) are 3 phase current (Iabc) and voltage (Vabc) at the PCC, respectively. Fig 5.2(e) represents the zoomed in version of Vabc to confirm the sinusoidal nature of the result.

Fig. 5.3 shows the results related to PV panel. Here, fig 5.3 (a) shows the power of PV after MPPT, fig 5.3(b) shows the Active power by PV panel to grid, and fig 5.3(c) shows the reactive power by PV panel to grid. As it can be seen, the reactive power is 0. Also the developed model and control scheme lets the system settle within 0.1 seconds.

Fig. 5.4 shows the results of case 2. i.e. System with the linear load. In this case, fig 5.4(a) shows the active power by grid, fig 5.4(b) shows the reactive power by grid, fig. 5.4(c) shows Vabc and fig 5.4(d) is zoomed in version of Vabc to confirm the sinusoidal nature of the result.

Vdc > Vdcmax(700V) => Battery Charge => Battery Controller   Charging Mode

Vdc < Vdcmax(700V) => Battery Charge => Battery Controller   discharging Mode

Vmin(690) <Vdc < Vdcmax(700V) => Battery Charge => Battery Controller   idle Mode

Fig. 4 Configuration of Standalone PV System

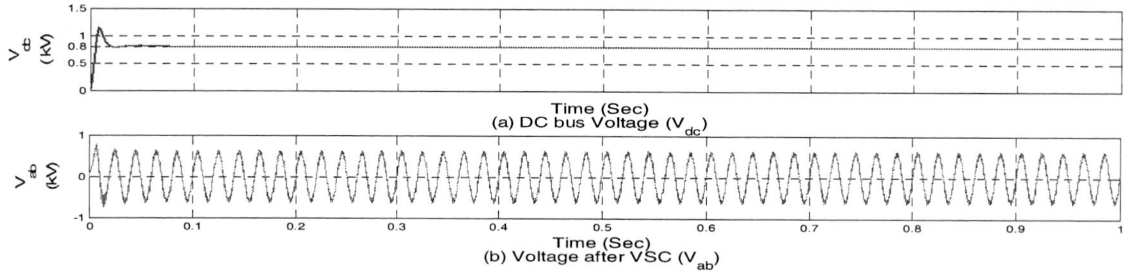

Fig.5.1. DC bus voltage and Voltage after VSC

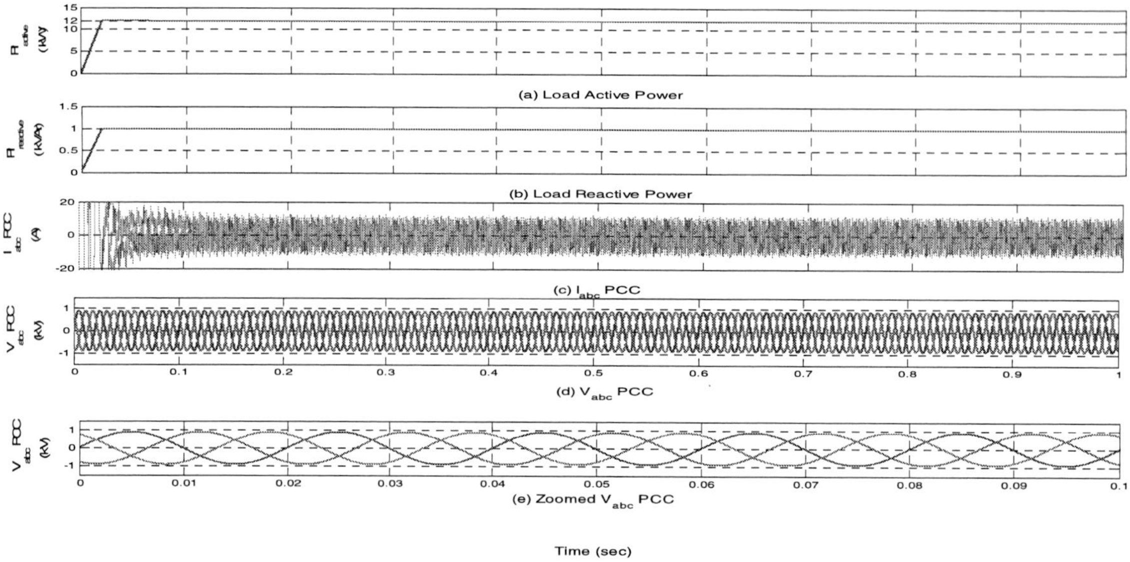

Fig.5.2 . Output At Load and PCC

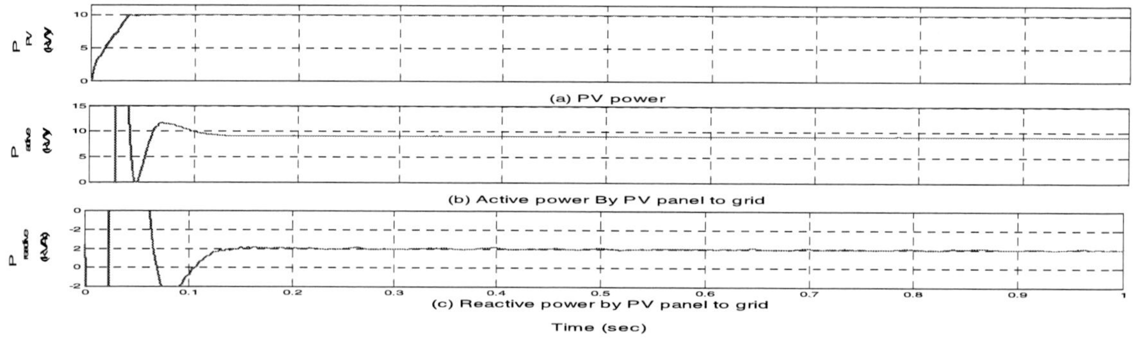

Fig.5.3 . Power of PV Panel (active and Reactive)

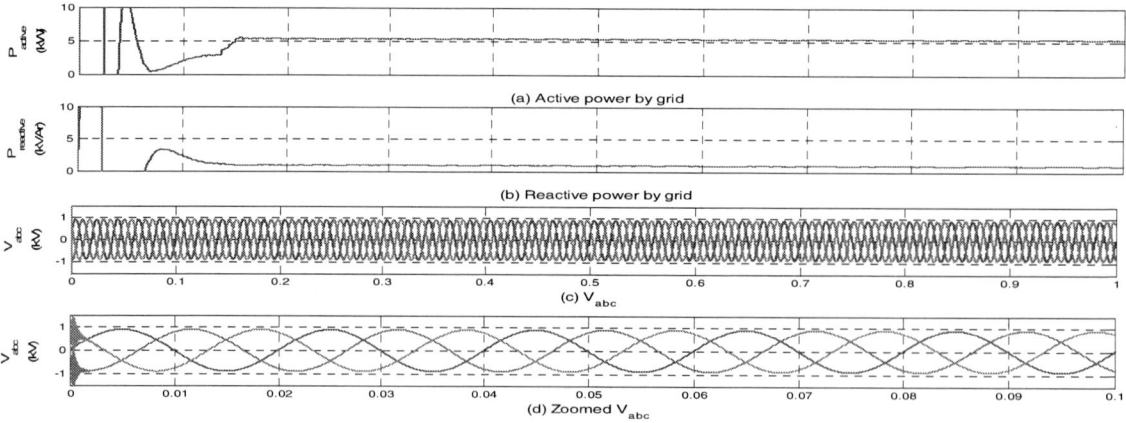

Fig.5.4 . Output At Load and PCC

B. Result of Standalone System

The standalone system model was simulated for 1.2 seconds and fig 6.1, fig 6.2 and fig 6.3 show the results of the standalone system.

Fig 6.1(a) shows DC bus voltage with battery controller. The load has been increased at 0.5 sec and decreased at 1 sec as shown in fig. 6.1(b), which leads to sag and swell of DC bus voltage without battery controller respectively. The battery provides extra power when load is increased and it

V. CONCLUSION

The results presented validates the control algorithm for both grid connected and standalone system with different loading conditions. The grid connected system confirms that the PV works at it's MPP point at all the time and push only real power through VSC into the grid with both linear and non-linear loading conditions along with R-L load of 12kW and 1kVAR. The extra required power is always supported by the grid. Also the triplen harmonics by the VSC has been removed

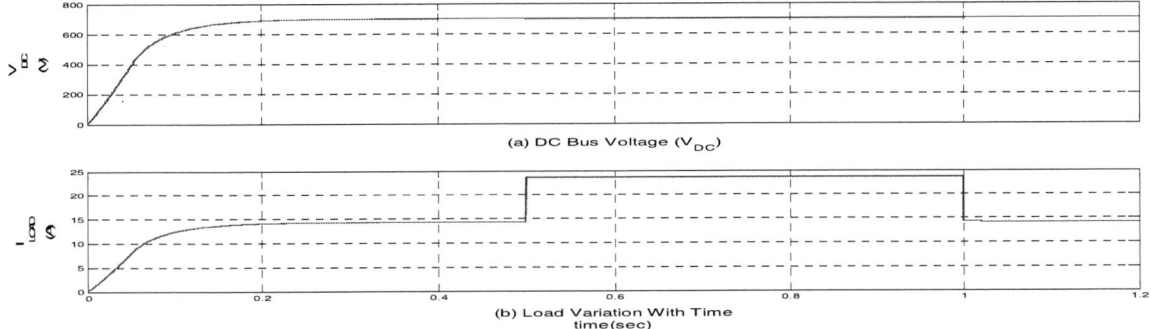

Fig.6.1 Results of stand-alone system with battery controller

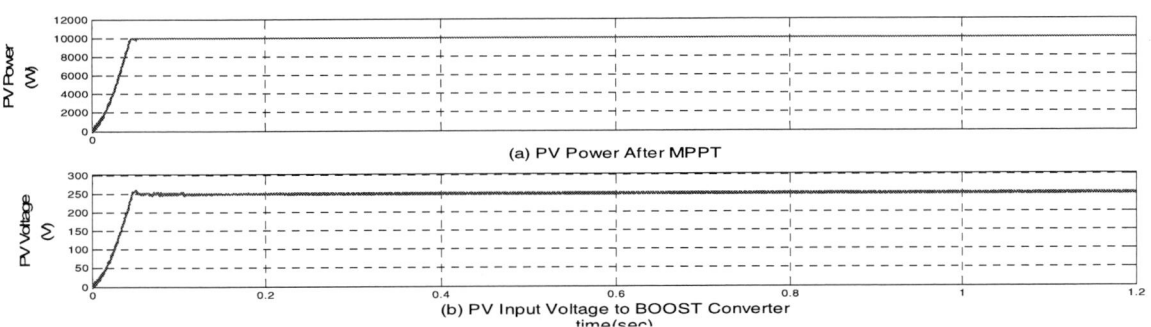

Fig.6.2. Power and Voltage outputs of PV panel

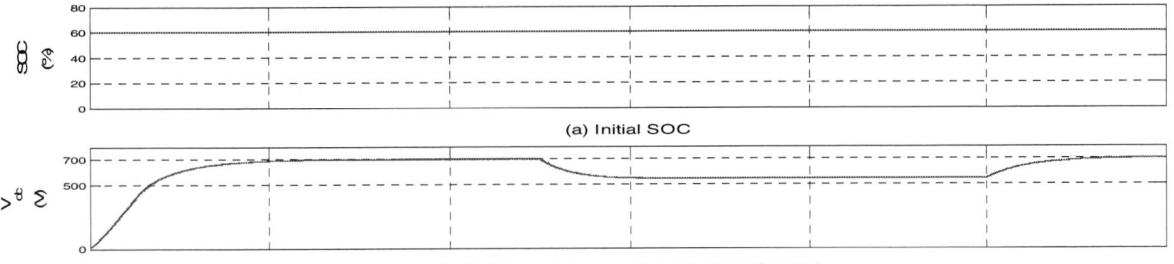

Fig6.3. SOC and DC bus Voltage Without battery controller

absorbs power when load is decreased, thus maintaining the constant DC bus voltage at 700 V.

Fig. 6.2(a) shows the power supplied by PV at STC after MPPT controller and fig. 6.2(b) shows the PV voltage.

Fig. 6.3(a) shows the initial SOC of battery and fig. 6.3(b) shows the DC bus voltage without battery controller.

by the delta-star transformer. In standalone mode, results are shown with the battery controller and without the battery control which validates the control scheme. It shows that in absence of battery charge controller voltage bus fluctuates due to energy imbalance which is eliminated in the presence of battery charge controller which works on the energy balance principle.

978-1-4799-6047-7/14 $31.00 © 2014 IEEE

REFERENCES

[1] The European Union climate and energy package, (The "20 – 20 – 20" package), http://ec.europa.eu/clima/policies/eu/package_en.htm

[2] "Trends in photovoltaic applications. Survey report of selected IEA countries between 1992 and 2009", International Energy Agency, Report IEA-PVPS Task 1 T1-19:2010, 2010. [Online].

[3] S. Jain, V. Agarwal, "Comparison of the performance of maximum power point tracking schemes applied to single-stage grid-connected photovoltaic systems," *Electric Power Applications*, IET, vol. 1, no. 5, pp. 753-762, Sept. 2007.

[4] Marcelo Gradella Villalva, Jonas Rafael Gazoli, Ernesto Ruppert Filho, " MODELING AND CIRCUIT-BASED SIMULATION OF PHOTOVOLTAIC ARRAYS"-IEEE-conf.

[5] P. A. Lynn, *Electricity from Sunlight: An Introduction to Photovoltaics*, John Wiley & Sons, 2010, p. 238.

[6] *"Sunny Family 2010/2011 – The Future of Solar Technology"*, SMA product catalogue, 2010. [Online]. Available: http://download.sma.de/smaprosa/dateien/2485/SOLARKATKUS103936 W. pdf

[7] L. Piegari, R. Rizzo, *"Adaptive perturb and observe algorithm for photovoltaic maximum power point tracking,"* Renewable Power Generation, IET, vol. 4, no. 4, pp. 317-328, July 2010.

[8] T. Esram, P.L. Chapman, *"Comparison of Photovoltaic Array Maximum Power Point Tracking Techniques,"* IEEE Transactions on Energy Conversion, vol. 22, no. 2, pp. 439-449, June 2007.

[9] *Overall efficiency of grid connected photovoltaic inverters*, European Standard EN 50530,2010.

[10] T. Esram, P.L. Chapman, "Comparison of Photovoltaic Array Maximum Power PointTracking Techniques," *IEEE Transactions on Energy Conversion*, vol. 22, no. 2, pp. 439-449, June 2007

[11] X. Fan, S. Misra, G. Xue, D. Yang,"Smart Grid — The New and Improved Power Grid: A Survey", *IEEE Communications Surveys & Tutorials* Vol. 14, pp. 944-980, 2012.

[12] Solar direct,"PV Battery Photovoltaic System Component",http://www.solardirect.com/pv/batteries/batteries.htm,2013.

[13] Miss. Sangita R Nandurkar and Mrs. Mini Rajeev, " Design and Simulation of three phase Inverter for grid connected Photovoltaic systems", Proceedings of Third Biennial National Conference, NCNTE-2012, 2012.

[14] Chandra Bajracharya , Marta Molinas , Jon Are Suul and Tore M Undeland , "Understanding of tuning techniques of converter controllers for VSC-HVDC", Nordic Workshop on Power and Industrial Electronics, NORPIE/2008, 2008

[15] Soeren Baekhoej, John K Pedersen and Frede Blaabjerg, "A Review of single phase grid connected inverter for photovoltaic modules", IEEE transaction on Industry Application , Vol. 41,pp. 55 – 68, Sept 2005

[16] A. Rosenthal, S. Durand, M. Thomas and H. Post, "Economic analysis of PV hybrid power system: Pinnacles National Monument", Conference Record of the Twenty-Sixth IEEE Photovoltaic Specialists Conference, 1997, Publication Year: 1997 , Page(s): 1269- 1272, Anaheim, CA

978-1-4799-6047-7/14 $31.00 © 2014 IEEE

Increase Efficiency of Solar Photovoltaic System by Data Acquisition Process

Sunil Parashar, Sunil Dhankhar

Department of Computer Science & Engineering
SKIT, Jaipur, India
Snlprshr5@gmail.com

Abstract- **In current scenario, renewable energy sources helps to make the environment greener and better. Currently, solar energy is the most available resource of renewable energy. This paper mainly represents the working principle of solar photovoltaic system and way to increase efficiency of the solar system using a method known as the data acquisition system. Data acquisition is a process to monitor changes in system, collect corresponding data and analyze data to make a decision. The data acquisition system has two parts, data acquisition board that contains Microcontroller Kit, ADC, and RS-232 and Zigbee communication modules. Another part is running on a host computer that has a set of software tools as Keil, Proteus, and Visual Basic 6.0 help to analyze and store data. Zigbee communication helps in wireless data transfer without any communication cost. Complete experiment and the result show the data acquisition designed is simple and stable.**

Keywords: **solar energy, green energy, photovoltaic cell, ADC, data acquisition, microcontroller, communication, serial communication, Zigbee communication.**

I. INTRODUCTION

Electricity is the main factor for social, economic and industrial development. In the generation of electric energy, a generation can be possible in two ways renewable and non-renewable energy types that depend upon generation resources. Renewable energy sources are replenished automatically time to time means they may be natural and cannot be replaced as fast as they are being consumed. Example of renewable type is sun, wind, etc. Development and use of large scale solar energy is not only great style of utilization of energy resource in the future, but also efficient measures accommodating to energy resource frame and enhancing energy resource disaster. The data acquisition system is a process used to collect information that can be stored and processed by a computer to analyze some special phenomenon. To increase accuracy of solar energy resource data acquisition is very important, different instruments to obtain data on the solar energy resource can produce different results and have a significant impact on the large-scale solar energy development and utilization. Zigbee is IEEE international standard 802.15.4 based communication protocol for personal area network using small, low power, digital radios.

II. STRUCTURE OF DATA ACQUISITION SYSTEM

Renewable energy sources provide a new area to generate energy by keeping the environment green. The solar energy system is most available and useful renewable resource today. Sun provides enough energy, but a mechanism needs to convert this solar energy into electric energy. The photovoltaic effect provides a way to use this solar energy in the best way by converting it into appropriate electrical form. Solar energy system mainly contains a set of solar cells or solar panel that takes sunlight as input and convert into electric type. Energy generated through photovoltaic system is not continuing because this system is affected by different environmental factors as dust, cloud, rain, etc. Charge controller used to regulate this discontinuous voltage; it can be said as a voltage regulator. Electricity generated by the photovoltaic panel is a DC type so it can be used directly for DC application, for AC application an inverter is used to convert generated DC power into AC power. The Ssorage system can be used to store energy for future use. Batteries are conventional storage of energy.

The data acquisition system is mainly divided into two parts; the first part is a data acquisition board that contains acquisition hardware modules as Analog-to-Digital converter (ADC), microcontroller and communication modules as RS-232 serial communication system and Zigbee wireless communication system. Another part of data acquisition is a host computer that mainly contains application software and database system. All decisions should be made in the host system.

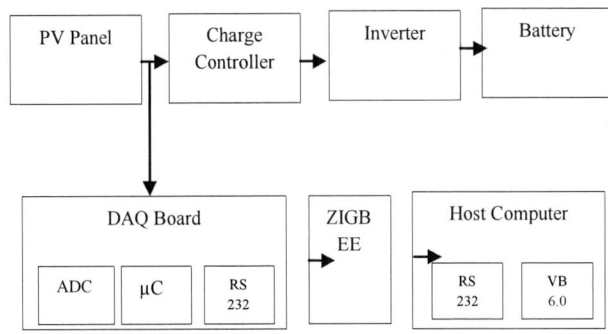

Figure 1. Photovoltaic Cell Power Generation Structure

978-1-4799-6047-7/14 $31.00 © 2014 IEEE

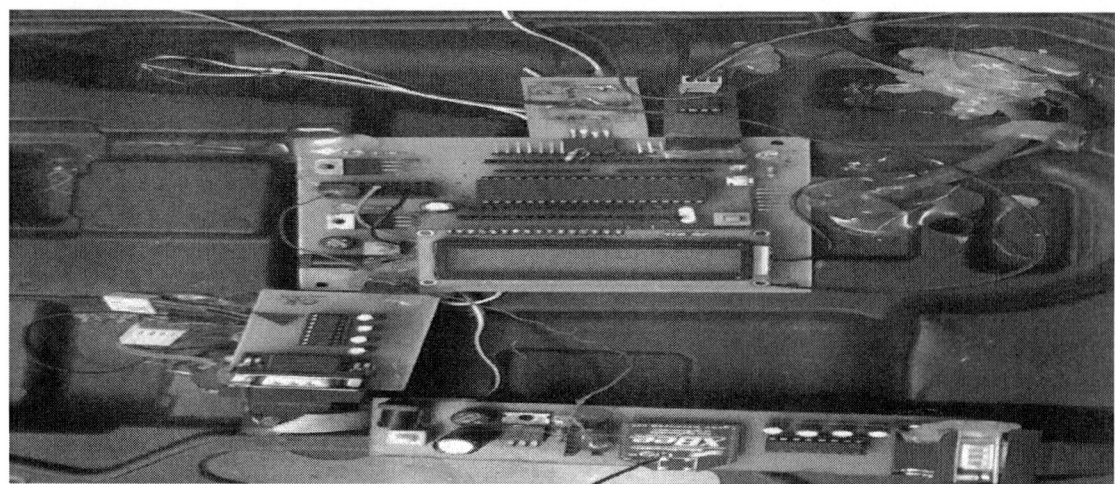

Figure 2. Data Acquisition Board

III. HARDWARES AND SOFTWARES IN DATA ACQUISITION SYSTEM

Data acquisition board comprises a set of electronic components used to control data acquisition process. Data acquisition board mainly contains Microcontroller, Analog-to-Digital Converter, and liquid crystal display (LCD), Zigbee and RS-232 serial modules. The figure 2 shows all components and connections of the data acquisition board.

A. Microcontroller AT89C52:

AT89C52, Atmel designed 8-bit microcontroller with 4K bytes of flash programmable and erasable read only memory (PEROM). The operating frequency of this controller varies from 0 to 24MHz and 11.0592MHz crystal frequency is set to currently. To make compatible with other peripheral devices as ADC, LCD or serial communication module it has 32 programmable I/O lines with 4 ports and 16-bit timer/counter with 6 interrupt sources. Figure 2 shows the peripheral device (LCD) connection with microcontroller.

B. Analog to Digital Converter ADC0831:

ADC0831 is 8 pins, 8-bit serial data output IC used to convert the continuous analog signal into a digital data.
It accepts incoming analog data from panel and converts it into an appropriate digital form and sends this data to microcontroller port.

Figure 3. Microcontroller Kit in DAQ Board

C. Zigbee and RS232:

Zigbee is an IEEE 802.15 standard based communication protocol used to create a wireless personal area network between solar panel and host computer to transmit data from the data acquisition board to the host system. Zigbee works better in a small geographical area without any communication cost.

RS-232 is a serial communication standard used to make the communication link between data terminal equipment as host computer and data circuit terminating or communication equipment Zigbee. Figure 4 shows Zigbee and RS-232 communication boards. MAX232 IC used to make communication more convenient.

Figure 4. Zigbee and RS-232 in DAQ Board

D. KEIL:

The Keil µvision integrated development environment (IDE) from Keil provides support for embedded software development. It provides a way to instruct microcontroller AT89C51. By using this tool, it can be easy to edit source code, debug programs and allow complete simulation with more options. To start with, on startup select appropriate controller type, then set all parameters of it as crystal frequency. .Hex type file is created through this tool and uploaded to a microcontroller that instructs microcontroller.

978-1-4799-6047-7/14 $31.00 © 2014 IEEE

E. Proteus:

Proteus virtual system modeling (VSM) is a simulation tool that includes all electronics and electrical components. It allows a developer to design a virtual prototype and simulate it as real. It also provides facility to interact with code written in Keil and VB. It has a serial communication option that enhances its simulation and communication capability.

F. Visual Basic 6.0:

Visual basic has an integrated development environment that provides some special functionality as rapid application development, remote object access and database access. It is an event driven language means on any event there is an option defined in it. It helps to monitor any changes in the solar system by gathering data. To make data comparison more understandable different graph and charts can be created. Microsoft Excel tables used to store incoming data.

To test data acquisition system Eltima software is used to make the communication link between Proteus VSM tool and Visual Basic program.

IV. DESIGN OF DATA ACQUISITION SYSTEM

A. Zigbee Communication Module:

Zigbee is IEEE international standard 802.15.4 based wireless communication protocol works over 2.4 GHz operating frequency. Networks in Zigbee are as personal area networks so each network has a unique identification as PAN identifier. 64-bit and 16-bit type IDs used in Zigbee network, where 64-bit ID used to join and remove any conflicts and 16-bit ID for data transmission. So 16-bit address is required for packet transmission in Zigbee network and 64-bit addresses for module identification. Figure 5 shows X-CTU application to connect Zigbee devices.

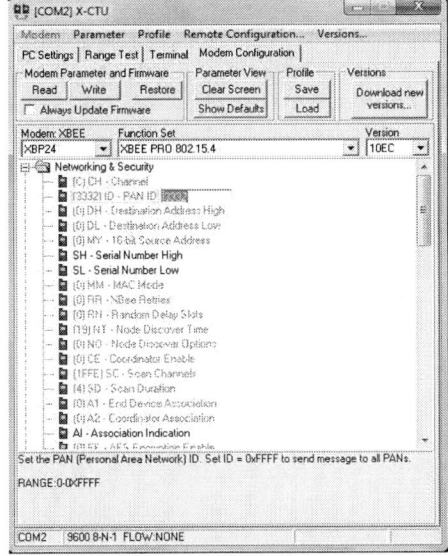

Figure 5. X-CTU Application for Zigbee Communication

Before starting communication needs to configure Zigbee devices in a proper way. Figure 5 shows different properties of modem configuration.

B. Data Acquisition System Model:

Figure 6 shows a flow chart of the data acquisition process. Data acquisition starts by initializing all components at begin level. Analog data are received at ADC to convert it into digital form. If conversion is failed need to convert again else send data to the controller and then move data to a host system to store. Zigbee and RS-232 help to move data from microcontroller to the host system.

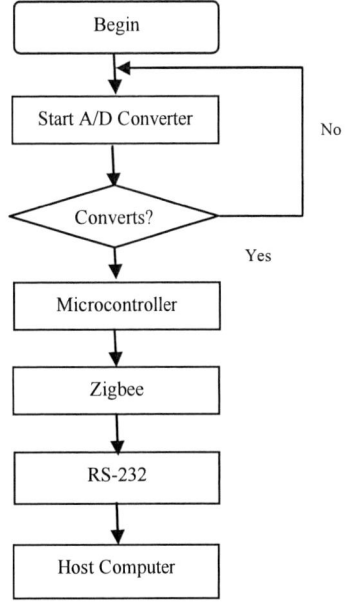

Figure 6. Block Diagram of Data Acquisition Process

V. SIMULATION AND EXPERIMENT RESULT

The completed solar energy system is working with some predefined values. To understand complete working, some basic calculation needed as

The number of solar panels used = 10 (connected parallel)

Solar cells in each panel = 36 cells (connected serially)

The potential of each cell = 1V

So potential of a solar panel = 36V (shown in figure 4.15)

Current of each panel = 4A

Potential of complete solar systems = 36V

Current of completed solar system = 4×10 = 40A

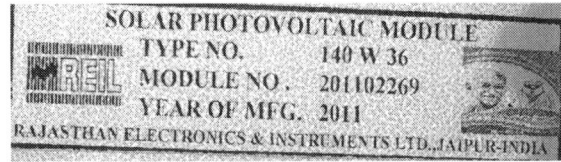

Figure 7. Solar Photovoltaic Module

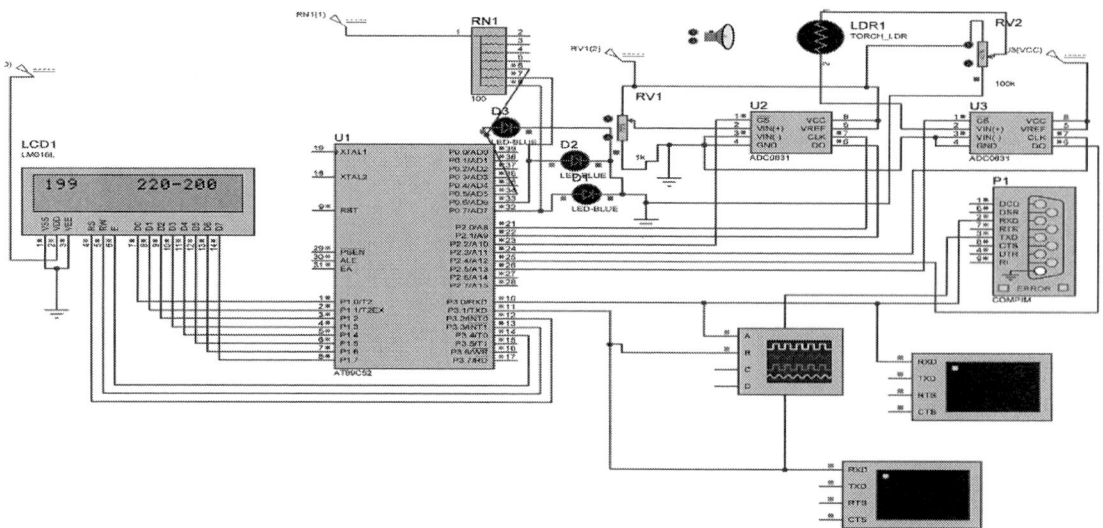

Figure 8. Simulation of DAQ in Proteus VSM

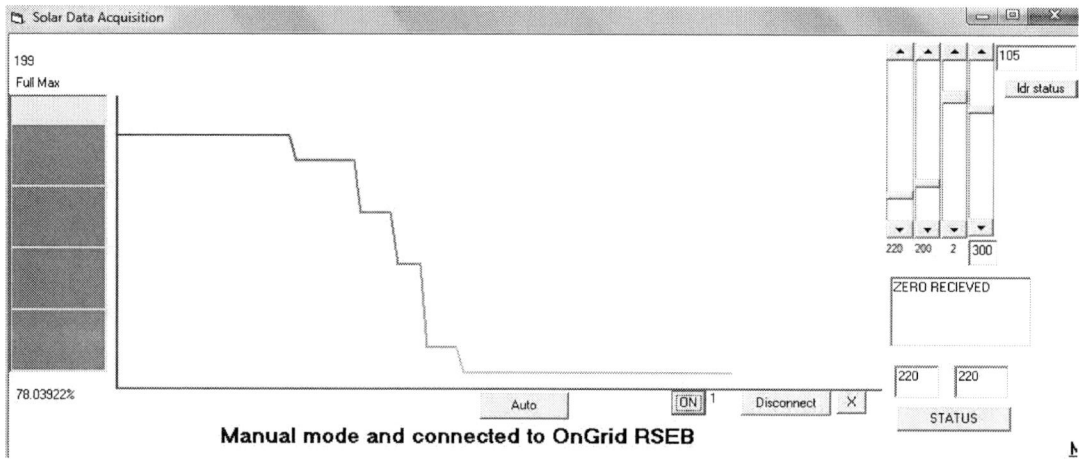

Figure 9. Data Acquisition Data Chart in Visual Basic at Host System

Simulation of complete data acquisition process is designed in Proteus virtual system modeling tool as shown in figure 8. This virtual system comprises a microcontroller AT89C52, ADC0831 and LCD and other circuit elements. An LDR (light dependent resistor) is also added to the model to check status of the environment. Input coming from RV1 is applied at VIN (+) pin of ADC0831 (U2) and output send to microcontroller's   1 pin of P2 port. To display the current status of solar system, an LCD is added at P1 port of the microcontroller. RS-232 communication terminals are also in circuit (P1) to transmit data to another terminal point as visual basic to draw a graph or chart. By getting data, system trigger between two modes, auto mode and manual mode with options for off-grid solar system or on-grid RSEB connected. Figure 9 shows the manual mode with on-grid RSEB connected and status of complete solar systems with chart of incoming data. Complete experiment is simulated at TISHITU research center, Jaipur (26.8533, 75.8113 positions).

To check the reliability of data acquisition system a comparison table between the actual reading (taken using a multimeter) and DAQ system reading provided in table 1.

TABLE 1
EXPERIMENTED VALUE VERSUS ACTUAL VALUE

Time (PM)	12:15	1:15	2:15	4:15	6:35
Voltage (Measured)	31.2	32.0	31.6	29.0	19.0
Voltage (Actual)	31.5	32.2	31.5	29.1	19.3
Current (Measured)	15.1	15.5	15.3	11.4	4.3
Current (Actual)	15.2	15.7	15.3	11.8	4.5

Note: All voltage values in Volts and current values in Ampere.

After collecting data from data acquisition system, it can be shown that the data collected through a data acquisition system meets the requirement of accuracy as: voltage error < 1V and current error < 1A. So it can be proved that the designed DAQ system meets with all requirements.

VI. CONCLUSION

This designed data acquisition system makes use of the microcontroller based data acquisition board hardware, software modules and communication modules as Zigbee. So it can be easier to reduce complexity of the entire system in terms of hardware, software and communication. Maintenance cost and upgradation cost of the system are also reducible because of the use of software.

Zigbee provides a new way of wireless communication by reducing communication cost in a small geographical area. Compared to other implementations, this data acquisition model is more reliable and uses less energy and meets the goal of data acquisition system as high speed, real time calculation.

REFERENCES

[1] Yonghui Xing, Wenzhuo Chen, Tao Xing, "Design of the Solar Photovoltaic System Data Acquisition Board," *Computer Science & Service System (CSSS)*, Aug. 2012, pp. 54-57.

[2] Mukaro, R, Carelse, Xavier Francis, "A microcontroller-based data acquisition system for solar radiation and environmental monitoring," *Instrumentation and Measurement,* Volume: 48, Dec 1999, pp. 1232-1238.

[3] Mingzhi Zhao, Zhizhang Liu ; Mingjun Yu, "Data acquisition and analyzing of solar energy resource," *Information and Automation (ICIA)*, 20-23 June 2010, pp. 2205-2208.

[4] Yatendra Yadav, Rajiv Roshan, Umashankar S, D.Vijaykumar, Kothari D P, "Real time simulation of solar photovoltaic module using labview data acquisition card," *Energy Efficient Technologies for Sustainability*, 10-12 April 2013, pp. 512 – 523.

Design, Implementation and Performance Analysis of a Single Phase PWM Inverter

Prof. Narendra Kumar
Department of Electrical Engineering
Delhi Technological University
Delhi, India
narendrakumar@dce.edu

Dr. Dheeraj Joshi
Department of Electrical Engineering
Delhi Technological University
Delhi, India
joshidheeraj@dce.ac.in

Sachin Singhal
Department of Electrical Engineering
Krishna Engineering College
Ghaziabad, India
sachinsinghal9@gmail.com

Abstract— **Power Electronic equipment that converts a DC power into AC power at required voltage and frequency level is known as Inverter. Voltage Source inverters produce an output voltage or current with levels either zero to positive voltage or zero to negative voltage that means two levels. These bi-level inverters are powered by different type of gating signals like square, quasi – square and sinusoidal Pulse width modulated signals. Different gating signals result in different harmonic levels in the output signal. In this paper, a hardware prototype of a single phase inverter using MOSFETs as the power switches has been developed. The MOSFETs are driven by square and quasi-square gating signals. These gating signals have been generated by designing the driver circuit using a MOSFET driver IC. In order to maintain the voltage level, the conduction time interval of MOSFETs has been maintained by controlling the pulse width of the gating pulses. Performance of the inverter has been studied for RL load. Simulation has been done in SIMULINK, results of simulation and experimental model has been compared on the basis of various parameters like output voltage, load current and THD levels.**

Keywords—Single Phase Inverter; Hardware; Simulation; Experimental; THD.

I. INTRODUCTION

Ac loads require constant or adjustable voltages at their input terminals. When such loads are fed by inverters, it is essential that the output voltage of inverters is so controlled as to fulfil the requirements of AC loads. This involves coping with the variation of the DC input voltage, for voltage regulation of the inverters and for the constant voltage/frequency requirement. If the input to the inverter is a DC Voltage source, it is referred to as voltage source inverter (VSI). If the input is a DC current source, it is referred as Current Source Inverter (CSI). The current source inverter is commonly used for high power applications. The voltage source inverters are further classified into single phase and multi-phase inverters. Chang and Wang have made a highly compact AC-AC converter which is also known as matrix inverter [1]. Due to recent growth of Digital signal processing and microcontrollers, real time control of these converters have become easy and economical [2]. Yaosuo, et al. have reviewed the topologies of the single phase inverter working in a distributed generation system, they have analysed the single

and multistage single phase inverter and given an overview of four switch and six switch inverter topologies [3]. M. A. Al-Nema et al. in [4] have proposed a new topology of inverter in which with single power stage the output voltage of higher magnitude than the input can be achieved. They have connected the load differentially across two dc-dc converters, the output voltage of these converters was modulated sinusoidally, and each converter produces only unipolar voltage. Authors found that the output waveform was nearly sinusoidal with harmonic content less than five percent but at the expense of using large value of inductance and capacitance in the implementation of the circuit. In all the work that has been done, we see that design of a single phase inverter requires working on different functional blocks such as mosfet gate driving circuits, the modulation techniques to regulate the harmonic content of the resultant output. While driving the MOSFET in full bridge topology, driving the upper MOSFET of both legs is quite critically as the source of the MOSFET switch floats between the positive rail voltage and supply ground, various methods of driving these drives are available in industry [5]. Recent trend in inverter design has been increase in modulating frequency to increase the power density and reduce the size of the passive components such as inductors and capacitors, however increase in frequency results in more switching losses and gate capacitance losses. Wilson et al. in [6] have proposed a resonant gate drive circuit which achieves quick turn on and turn off transient times to reduce switching losses and conduction losses, in addition to this it also helps in recovering any heat loss that occurs due to MOSFET drive. V.V Graczkowski et al. in [7] have discussed a gate drive technique using bootstrap capacitor; this technique eradicates the need for isolated power supplies in multilevel inverters. Ian D. de Varies in [8] has devised a low loss capacitance driver circuit topology. Full bridge converter are very sensitive to input switching waveforms, it contains two legs of two switches each, if one of the switches in one legs turn on before the other has switched off, then the dc rail connected across the converter may get short and may burn the device and hence the equipment itself, hence a dead time is introduced between the turn on time of one of the switches in bridge leg and turn on time of another switch of the same leg. Chen in [9] has implemented a new gate driver circuit using gate bias for inductive loads. This methods prevents the use of hardware circuit for dead time generation, hence it reduces the hardware complexity and increase the system efficiency.

978-1-4799-6047-7/14 $31.00 © 2014 IEEE

Gate signals to the switches for the bridge determine the harmonic content of the output. Various types of gating signals have been used in the literature; most primitive of them has been Pulse Width modulation schemes. Joachim Holtz has carried out a comprehensive analysis of the various types of PWM techniques such as carrier based PWM and non-carrier based PWM, various parameters namely harmonic spectrum, torque harmonics and dynamic performance for analysing the performance for different PWM techniques have been discussed [10] [11]. Apart from these modulation techniques, Vorperian in [12] and Yousefazadeh et al. in [13] have reviewed new techniques namely zero voltage switching and zero current switching, ZVS (zero voltage switching) produces minimum voltage stress on a switch under transient conditions whereas ZCS (zero current switching) produces minimum current. With the introduction of switch mode devices in the power system comes the problem of harmonics, all power electronic converter generate non-sinusoidal or distorted waveforms which contain signals of multiples of the fundamental frequency. The performance of the load degrades at such high frequencies. Various techniques have been purposed in the literature to reduce the harmonic content, V. d. Broeck et al. in [14] have analysed the effects of using unipolar PWM on voltage and current within an inverter. In their work, they have evaluated the peak and RMS value of current ripple and voltage ripple using both time domain and frequency domain methods, in addition to this EMI problems related with different PWM schemes have been compared. In reducing the harmonics, the lower order harmonics possess greater difficulty as the size of the filter required to filter out these harmonics is large and the system tends to be bulky and expensive, in view of this F.G Turnbull in [15] has introduced a control technique which eliminates the third and fifth harmonic voltage present in a single leg centre tap (half bridge) single phase inverter. He has stated that with addition of two more switches or one leg(full bridge), the fundamental frequency component of the output ac voltage can be controlled from maximum to zero without reintroducing the third and fifth harmonics. It Bau Huang et al. in [16] have developed a new technique for harmonic reduction in inverters using Sinusoidal Pulse Width Modulation, in this technique a the duty cycle of a high frequency square signal is varied in accordance with the a sinusoidal modulating signal, the frequency of this signal is same as that the fundamental frequency desired. Authors have carried out a detailed analysis of the resultant signal. It is further purposed that the use of LC filter at the output of such an inverter results in nearly perfect sinusoidal output [17]. Z. H. Pankaj et al. in [18] have implemented the SPWM technology in their work on a single phase full bridge circuit. With the advent of microprocessor in early 70's the generation of SPWM for the gating of control switches has become easy, tweaking the pieces of code here and there results in generation different type of gate pulses. B. Ismail et al. in [19] have described the method of generating SPWM signals using an 8 bit Atmel microcontroller. A Mamun et al. in [20] also used an 8 bit microcontroller to generate the SPWM gating pulses. In microcontrollers the sampling frequency is limited by interrupt latency, also the constraints imposed by the finite word length limits the microprocessor's computation capability. Hence authors have sought the use of Digital Signal processor to implement the modulation needed for the inverters [21] [22] [23]. M. Tumay et al. in [24] have created an experimental set up for controlling a single phase inverter using Texas instrument DSP Kit. Though the present work does not incorporate the closed loop control of the inverter but when these inverters need to be connected to the grid in case of distributed power field, the inverter need to be controlled using feedback control. The voltage and frequency of the output voltage need to be controlled in order to synchronize with the grid. A. I. Maswood et al. in [25] have analysed the performance of a PI current control scheme on a single phase inverter under varying conditions in the grid viz. normal condition , unbalanced, load outage and load short circuit conditions. P. A. Michael et al. in [26] have used sliding mode control for simulating the behaviour of single phase inverter under different conditions. N. Kapadia et al. in [27] have simulated a push pull inverter technology for a developing a low cost single phase inverter for solar applications. Apart from designing the inverter, present work also focusses the effects of using different type of gating signals on resistive load. B. V. Guynes et al. in [28] have investigated the relative merits and demerits of a quasi-wave fed single phase induction motor and a sinusoidal fed motor. Authors found that when the motor was fed by quasi square wave, the motor incurred more losses due to harmonics, there was a reduction in the rms voltage of the input, and the waveform shape had deteriorated due to effects of harmonics stress on a switch.

II. SINGLE PHASE INVERTER ARCHITECTURE AND HARDWARE REALIZATION

Block Diagram of the developed inverter system is shown in Figure 1. The block diagram consists of following parts:

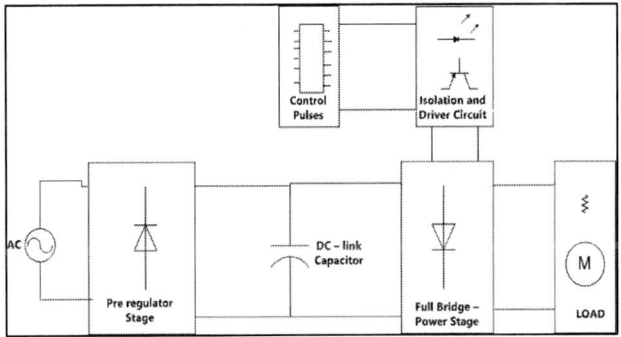

Figure 1 Block Diagram of a typical Single phase inverter system

- Pre-regulator Stage
- Power stage
- Control Stage
- Driver Stage

A. Pre-regulator Stage

Pre regulator stage is the intermediate conversion stage which

Converts the input AC input to DC input, this conversion process is more commonly called as Rectification. Here, we have developed a uncontrolled rectifier using power diodes. DC-link capacitance of 3000 μF, 400 V is used to keep the voltage ripples in the limits. Figure 2 shows the pre-regulator circuit.

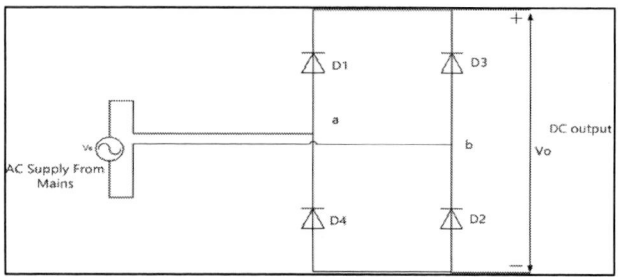

Figure 2 Uncontrolled Rectifier

B. Power Stage

Power Circuit is used for the conversion stage. H bridge inverter is used to convert DC voltage to AC voltage, and as shown in Figure 3, it consist from four mosfet transistors and we use IRF840 mosfet as the power switch.

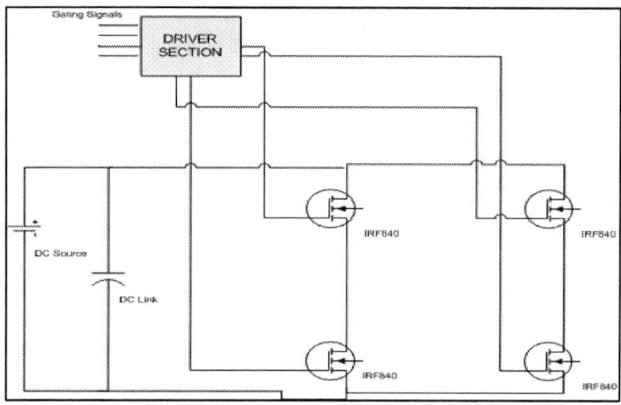

Figure 3 H-Bridge Power Circuit

C. Control Stage

The control signals are generated by the Arduino development board [29]. These signals are further isolated from the power stage by a 6N137 optocoupler and transistor gain circuit.

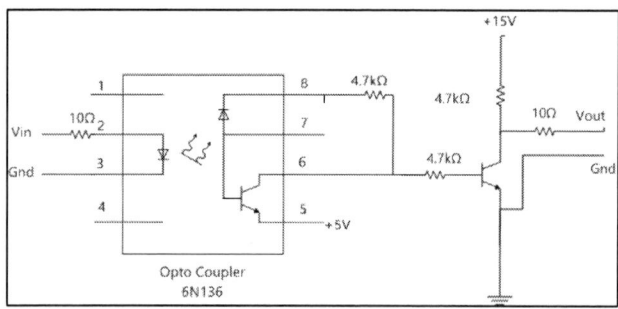

Figure 4 Isolation and Gain Circuit

D. Driver Stage

Gate drive is required to supply the switches such as IGBTs and MOSFETs with required voltages and currents since the microcontroller couldn't supply the required value [7] [30]. In present work IR2110 is used as the driver IC for gate drive. Circuit diagram for IR2110 high side and low side driver is shown in Figure 5.

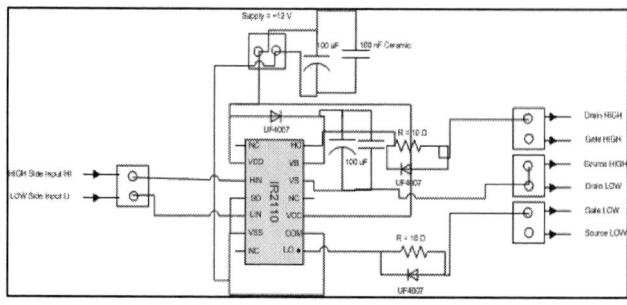

Figure 5 IR2110 Driver Circuit

Based on the architecture discussed in the previous section, hardware was developed for experimental analysis. Figure 6 shows the hardware developed.

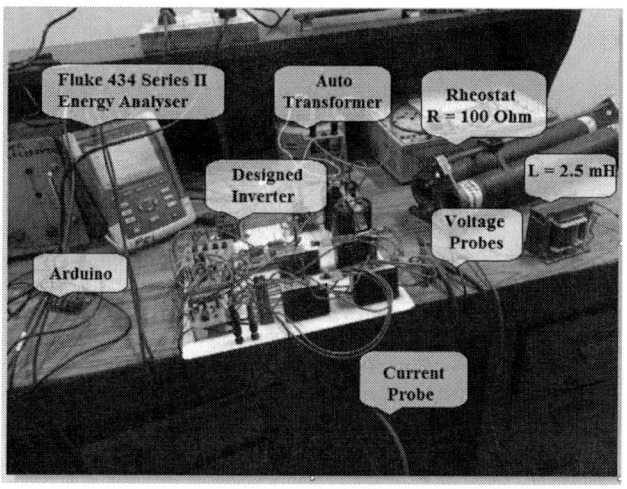

Figure 6 Developed Hardware

III. SIMULATION RESULTS

Simulation was carried out in Simulink, a square wave and quasi square wave inverter with conduction angle α = 30, 45, 60 and results were reported for RL load.

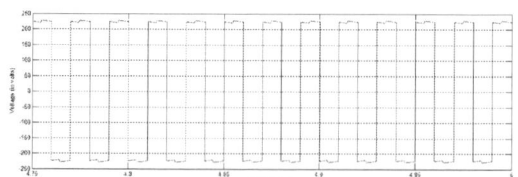

Figure 7 Voltage output at R = 10 Ω and L = 2.5 mH

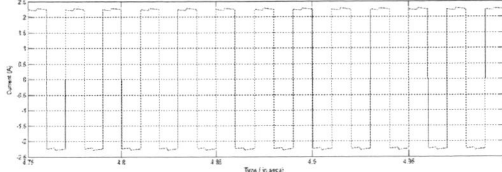

Figure 8 Load Current at R = 10 Ω and L = 2.5 mH

As shown in Figure 7 and 8, the load voltage and current for a square wave inverter are square waveforms; THD for this inverter comes out to be 48.09 percent for voltage and for current it comes out to be 47.45. By changing the timing of the gating signals for the MOSFETs, quasi square wave inverter was realized with different conduction angles.

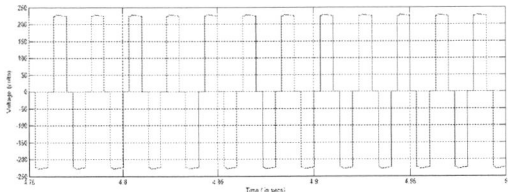

Figure 9 Quasi Square wave voltage output with α = 30

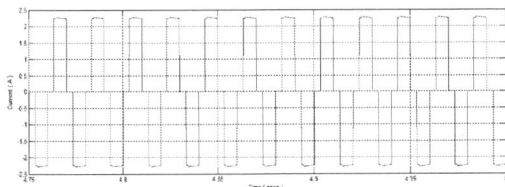

Figure 10 Quasi Square wave output current with α = 30

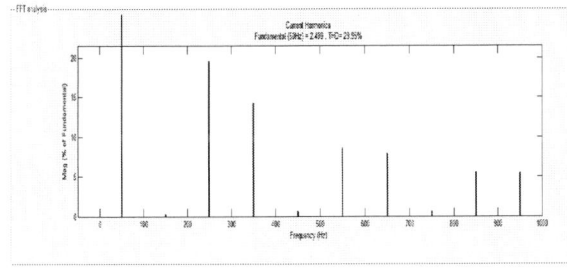

Figure 11 Current Harmonics for RL load with α = 30

As shown by Figure 9 and 10, in quasi square output there is a delay in the output waveform by 15 degrees. Figure 11 shows that current has become less distorted with the addition of inductance and current THD comes out to be 29.95 percent.

IV. EXPERIMENTAL RESULTS

After simulating the inverter model in the Simulink software, developed hardware was tested for a resistive load of 100 Ω and L = 2.5 mH results were recorded using Fluke 430 series II energy analyzer.

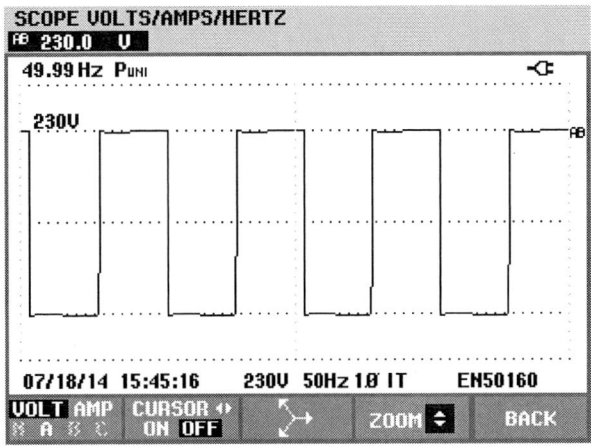

Figure 12 Square Wave output at R = 100 Ω and L = 2.5 mH

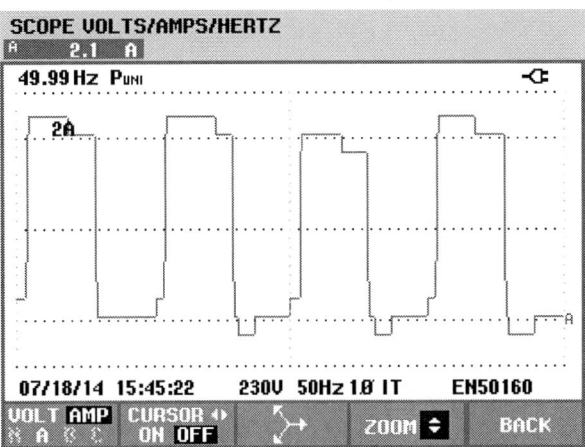

Figure 13 Load Current at R = 100 Ω and L = 2.5 mH

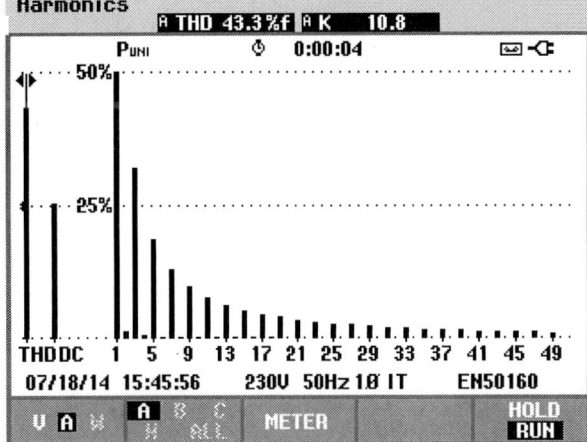

Figure 14 Current Harmonics at R = 100 Ω and L = 2.5 mH

In experimental results too as shown in Figure 12 to 14, the voltage and current THD for the developed square wave inverter came out to be nearly 46 percent. Inverter was also tested for quasi wave output and results were reported.

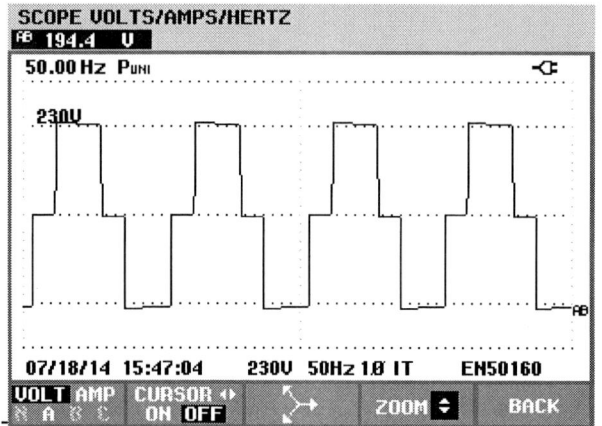

Figure 15 Quasi Square Voltage output at R = 100 Ω and L = 2.5 mH

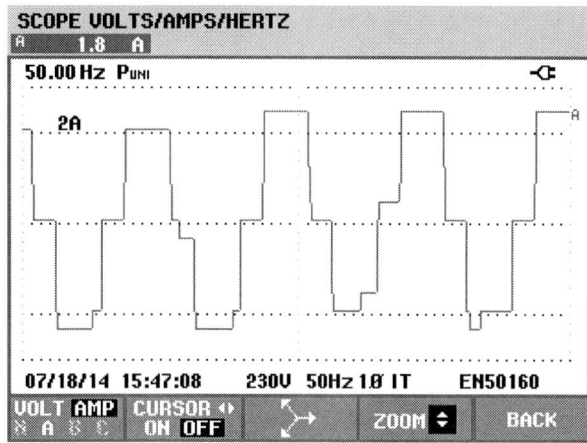

Figure 16 Current Waveform at R = 100 Ω and L = 2.5 mH

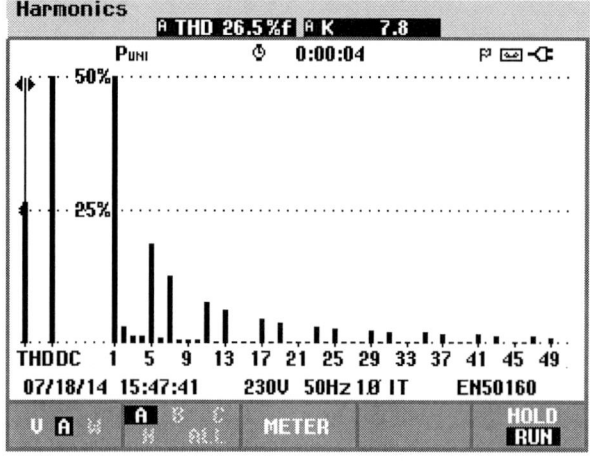

Figure 17 Voltage Harmonics at R = 100 Ω and L = 2.5 mH

Figure 17 shows that with a conduction angle of 30 degrees the voltage and current THD reduces to 26.5 percent.

V. COMPARATIVE ANALYSIS OF HARDWARE AND SIMULATION RESULTS

Simulation and Hardware results were taken as discussed in the previous sections. It is seen that square wave inverter has more harmonic content as compared to a quasi-square wave inverter with a conduction angle of thirty degrees.

Square Wave Inverter				
Type of Load	Simulation Results		Experimental Results	
	THDv	THDi	THDv	THDi
R = 100 Ω and L = 2.5mH	48.09	47.46	47.2	43.3
Quasi-Square Wave Inverter with α = 30				
R = 100 Ω and L = 2.5mH	30.59	29.95	30	26.5
Quasi-Square Wave Inverter with α = 45				
R = 100 Ω and L = 2.5mH	48.34	47.71	47.7	44.3
Quasi-Square Wave Inverter with α = 60				
R = 100 Ω and L = 2.5mH	79.88	79.13	80.1	75.1

Table 1 Performance Analysis of Inverter

As shown in Table 1 and bar graph comparison in Figure 17 and 18. Square wave inverter has a THDv of 48.09 percent in simulation; experimentally it comes out to be 47.46 percent.

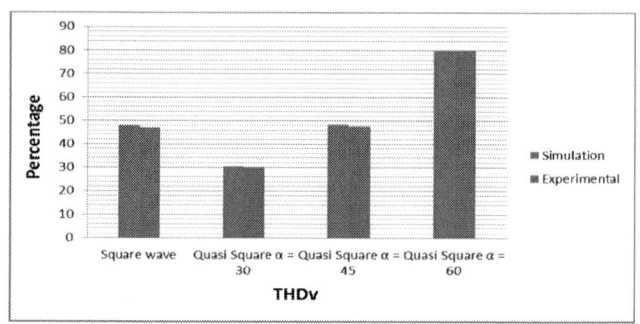

Figure 18 THDv with RL-load for various inverter types

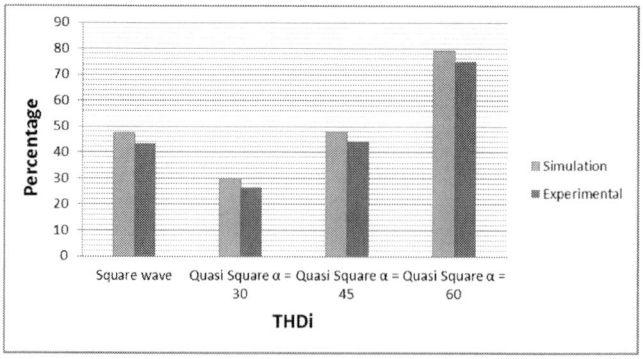

Figure 19 THDi with RL-load for various Inverter types

978-1-4799-6047-7/14 $31.00 © 2014 IEEE

Quasi-Square wave inverter with a conduction angle of thirty degrees, THDv is 30.59 percent in simulation; experimentally it comes out to be 30 percent. Quasi-Square wave inverter with a conduction angle of forty five degrees, THDv in simulation is 48.83 percent, whereas in experiment it is 47.7 percent. THDi for RL load in simulation comes out to be 47.71 percent whereas experimentally it is 44.3 percent. Quasi-Square wave inverter with a conduction angle of sixty degrees gives worst performance in terms of THDv and THDi. THDv in simulation is 79.88 percent; experimentally it is 80.1 percent.

VI. CONCLUSION

This paper has evaluated two types of PWM control schemes for single phase DC-AC converter applications i.e. Square wave and modified Square (Quasi- Square) Wave. The evaluation is based on comparative studies using several methods. 1) Theoretical discussion on Fourier analysis of square wave and quasi square wave, 2) Computational simulation in SIMULINK; 3) Microcontroller based Hardware experimental implementation and verification. Based on the results shown in the previous sections, first, we can conclude that using quasi square output inverter; we have harmonic and amplitude control over the output waveform as compared to the square wave inverter which does not have harmonic or amplitude control. As demonstrated and analyzed in this paper, the modified square wave inverter with a conduction angle of 30 degrees has an advantage in the reduction of THDv and THDi because of the removal of the third harmonic component from the output waveform.

REFERENCES

[1] J. Chang and S. T. Anhua Wang, "Highly Compact AC–AC Converter Achieving a High Voltage Transfer Ratio," *IEEE transactions on Industrial Electronics,* pp. 345-352, 2002.

[2] A. LaxmiKanth and M. M. Morcos, "A Power Quality Monitoring System: A Case Study in DSP-Based Solutions for Power Electronics," *IEEE transactions on Instrumentation and Measurement,* pp. 724-731, 2001.

[3] Y. Xue, L. Chang, J. Bordonau, T. Shimizu and S. Kjaer, "Topologies of Single Phase inverters for small Distributed Power Generators: An overview," *IEEE transactions on Power Electronics,* pp. Vol 19, No 5, 2004.

[4] M. A. Al-Nema and S. M. Al-Layla, "Analysis,Design and implementation of modified single phase inverter," in *IEEE Proceedings,* 2007.

[5] I. Rectifier, "AN-978," International Rectifier.

[6] W. Eberle, Y. F. Liu and P. C. Sen, "A Resonant Gate Drive Circuit with Reduced MOSFET Switching and Gate Losses," in *IEEE Industrial Electronics, IECON 2006 - 32nd Annual Conference on ,* Paris, 2006.

[7] J. J. Graczkowski, K. L. Neff and X. Kou, "A Low-cost Gate Driver Design using Bootstrap Capacitors for Multi-level Mosfet Inverters," in *IPEMC,IEEE,* 2006.

[8] I. D. Vries De, "A Resonant Power MOSFET / IGBT Gate Driver," in *IEEE Conference,* 2002.

[9] H. C. Chen, "An H-Bridge Driver Using Gate Bias for DC Motor Control," in *Consumer Electronics (ISCE), 2013 IEEE 17th International Symposium on ,* Hsinchu, 2013.

[10] J. Holtz, "Pulsewidth Modulation - A survey," *IEEE transactions on Industrial Electronics,* pp. 410-420, 1992.

[11] P. Bhimbra, Power Electronics, New Delhi: Khanna Publishers, 2004.

[12] V. Vorperian, "Quasi-Square-Wave Converters: Topologies and Analysis," *IEEE Transactions on Power Electronics,* pp. 183-191, 1988.

[13] V. Yousefazadeh, D. Maksimovic and Q. Li, "A Zero Voltage switching single phase inverter using Hybrid Pulse Width Modulation Techniques," in *IEEE Power Electronics Specialists,* 2004.

[14] V. D. Broeck and M. Miller, "Harmonics in DC to AC converters of single phase uniinterriuptible power supplies," in *Telecommunications Energy Conference, INTELEC'95,* 1995.

[15] F. G. Turnbull, "Selected Harmonic Reduction in Static D-C—A-C Inverters," IEEE , New York, 1964.

[16] I. B. Huang and W. S. Lin, "Harmonic Reduction in Inverters by use of Sinusoidal Pulse Width Modulation," *IEEE transactions on Industrial Electronics and Control Instrumentation,* 1980.

[17] H. Kim and S. K. Sul, "Analysis on output LC filters for PWM inverters," in *IPEMEC , IEEE,* 2009.

[18] Z. H. Pankaj, B. G. Pravin, P. Sonare and S. R. Suralkar, "Design and Implementation of carried based Sinusoidal PWM Inverter," *International Journal of Advanced research in Electrical, Electronics and Instumentation Engineering,* pp. Volume 1, Issue 4, 2012.

[19] B. Ismail, S. Taib, A. M. Sahd, M. Isa and I. Daut, "Development of Control Circuit for Single Phase Inverter using Atmel Microcontroller," *IEEE.*

[20] A. A. Mamun, M. F. Elahi, M. Quamaruzzaman and M. U. Tomal, "Design and implementation of single phase inverter," *International Journal of Science and Research,* pp. Vol 2, Issue 2, 2013.

[21] R. SenthilKumar, "Design of Single Phase inverter using dsPIC30F4013," *International Journal of Engineering Science and Technology,* pp. 6500-6506, 2010.

[22] N. AphiratSakun, S. R. Baghanagarapu and K. Techakittiroz, "Implementation of Single phase unipolar inverter using DSP TMS320F241," *AU. JT,* pp. 191-195, 2005.

[23] H. Zhou, C. T. M. M. and C. G. , "Development of Single Phase Photovoltaic Grid - connected Inverter based on DSP control," in *IEEE Symposium on Power Electronics for Distributed Generation Systems,* Beijing, 2010.

[24] M. Tumay, K. C. Bayindir, M. U. Cuma and A. Teke, "Experimental Set up for a DSP based Single phase PWM inverter".

[25] A. I. Maswood and E. A. Emmar, "Analaysis of PWM voltage source inverter with PI controller under Non-ideal Conditions," in *IEEE Conference,* 2010.

[26] J. M. V and P. A. Michael, "Design and Analysis of a Single Phase Unipolar Inverter using Sliding mode Control," *International Journal of Engineering and advanced technology,* pp. Vol 2, Issue 2, 2012.

[27] N. Kapadia, A. Patel and D. Kapadia, "Simulation and Design of low cost single phase solar inverter," *International Journal of Emerging Technology and Advanced Engineering ,* pp. 158-163, 2012.

[28] B. V. Guynes, R. L. Haggard and J. R. Lanier, "Evaluation of Quasi Square Inverter as a Power Source to Single Phase Induction Motor," *NASA technical Note,* 1977.

[29] "Arduino Home page," Arduino, [Online]. Available: http://www.arduino.cc [Accessed 09 July 2014].

[30] T. Shimizu and K. Wada, "A Gate Drive Circuit of Power MOSFETs and IGBTs for Low Switching Losses," in *IEEE Conference on Power Electronics ,* Daegu, Korea, 2007.

978-1-4799-6047-7/14 $31.00 © 2014 IEEE

Smart Home Energy Management by Demand Response Controller Design

Mini S. Thomas
Professor, Jamia Millia Islamia
New Delhi, India

Praveen Bansal
GET, Mitsubishi Electric
New Delhi , India

Prateek Taneja
GET, PGCIL
New Delhi , India

Abstract— **This paper includes an HEM algorithm for managing household power-intensive appliances. The highlight of the proposed HEM algorithm is its ability to control selected appliances and keep the total household power consumption below a certain limit, while considering customer preferences and allowing the customer more flexibility to operate their appliances. Advanced Metering Infrastructure (AMI) coupled with Demand Response enables to relieve the stress on the Power System at the times of peak load while simultaneously shifting it to non-peak periods which enables us to achieve a higher load factor and consequently, cheaper cost of production of electricity.**

This paper presents a prototype of a Smart Controller working on the principle of Demand Response for managing high power consumption household appliances. The controller manages household loads according to their preset priority and guarantees the total house-hold power consumption below certain levels. A hardware model has been developed on an Arduino board to showcase the applicability of the proposed algorithm in performing DR at a lumped load level.

Index Terms- **Customer choice, demand Response (DR), home energy management (HEM), load priority, smart controller (SC)**

I. INTRODUCTION

Recently, large quantity of fossil fuel is being consumed due to an increase in the standard of living causing an increase in consumption of energy. Fossil fuels like petroleum are being continuously exhausted and see the ever high oil price era. In addition, the burning of the fossil fuel produces greenhouse gases like the carbon dioxide causing global warming. Thus the concept of "Sustainable Development" is introduced for the preservation of fossil fuel and reduction in the emission of greenhouse gases [20][21]. The sustainable development concept refers that we have to minimize consumption of our natural resources and plan positively for the development of mankind. For this, the amount of energy consumption should be reduced through efficient use of the available energy[8]. The development of new and renewable energy technologies like the solar and wind force should be promoted and consequently the greenhouse gases and waste exhaustion can be reduced. In order to achieve these goals, the energy plan to raise efficiency for the energy-efficient use and reduction of greenhouse gases is proposed and amongst these, the smart grid has attained much appreciation and attention[9][10].

The smart grid is the intelligent power network which fuses the IT technology into the existing power network and optimizes the energy efficiency[6]. It is the energy network which traces the development, power transmission, and information like the power consumption technique through the communication network, the sensor system, and software to efficiently control demand and supply[7]. If the demand and supply of energy are generally controlled in homes, it will not only enhance energy efficiency of the system but also control load drop in peak time and reduction in the emission of greenhouse gases. Then, in order to develop an energy efficient operating system, a smart grid-base AMI system, which is a check meter and controls the electric energy used with the demand reaction technique, is necessary for remotely managing the load of the energy and power consumed[19].

Use of Smart Controller will make our system smarter as unlike before, the utility has direct control over each consumers load. Utility operators are relying on the convergence of power transmission and information technologies to support the bidirectional energy flow, while increasing availability, predictability and coefficients. This paper suggests a DRSC (Demand Response Smart Controller) which can be used as a central hub which receives an external input from the utility in the form of a signal containing two components – the amount to be reduced and the time for which it's to be reduced and provides demand reaction function[17][18]. The smart controller connects with the electro-mechanical gear outlet for measuring electricity consumption and delivers this amount to AMI through the network. The electric charge imposed from the smart meter is also received and the demand reaction function limits the use of unnecessary power when a price is high [3]. The efficient use energy for lighting and other appliances installed in home can be planned through this function. Thus Having a general idea about DR and HEM System enables us to understand the motivations, goals and advantages of employing such a technique , which is explained in the next sections. The role of smart Controller, its hardware and software design , its algorithm will be define in subsequent section.

II. DEMAND RESPONSE

According to Federal Energy Regulatory Commission, Demand Response (DR) is defined as: "Changes in electric usage by end-use customers from their normal consumption patterns in response to changes in the price of electricity over time, or to incentive payments designed to induce lower electricity use at times of high wholesale market prices or when system reliability is jeopardized." DR includes all intentional modifications to consumption patterns of electricity of end use

978-1-4799-6047-7/14 $31.00 © 2014 IEEE

customers that are intended to alter the timing, level of instantaneous demand, or the total electricity consumption DR enables consumers to manage their consumption according to the available generation which means electric supply is generation oriented rather than demand oriented [13][14].

In electricity grids, demand response (DR) is similar to dynamic demand mechanisms to manage customer consumption of electricity in response to supply conditions, for example, electricity customers reduce their consumption at critical times or in response to market prices. The difference is that demand response mechanisms respond to explicit requests to shut off, whereas dynamic demand devices passively shut off when stress in the grid is sensed. Demand response can involve actually curtailing power used or by starting on-site generation which may or may not be connected in parallel with the grid [11][12]. This is a quite different concept from energy efficiency, which means using less power to perform the same tasks, on a continuous basis or whenever that task is performed. At the same time, demand response is a component of smart energy demand, which also includes energy efficiency, home and building energy management, distributed renewable resources, and electric vehicle charging.

The main objective of the work done in this paper is peak load shifting for reducing the maximum demand and other is a increased load factor for increasing the uniformity of load curve.

III. HOME ENERGY MANAGEMENT SYSTEM

The bi-directional, decentralized electricity grid and its associated industrial control plane are at the heart of the smart-grid, but will however only be able to maximize its benefits if communication to the home and among appliances within the home can be ensured[16][18]. The promises of the smart-grid can therefore come to fruition only if greater response, greater engagement and active participation from end consumers within homes can be ensured. However the above mentioned smart energy services and applications, enabled by the smart electricity grid will not be hosted on the smart-meter, the grid's endpoint into residential homes, but need to be hosted on complementary devices like the home energy management box. This modular device will help provide comprehensive management of energy within customer premises. The smart meter makes it possible for consumers to save on their energy bill, but doesn't do that by itself; smart meters mainly help utility companies get better readings on electricity use and help utility companies save energy and money[12].

A. The Home Energy Management box:

The Home Energy Management box is responsible for following operations

- It receives price events or demand-response events through the Advanced Metering Infrastructure (AMI) network and its smart-meter interface to the home (or potentially through the broadband interface)
- It monitors and controls a set of demand-response enabled appliances (e.g. thermostat, water boiler, heater, …)

- It monitors and controls a set of home-automation enabled appliances (e.g. washing machine, dimming lights, …)
- It reads power figures out of the various meters and loads on a periodic basis
- It serves rich analytics (e.g. load disaggregation results) to various online, mobile or local displays. and is ready to
- run certified 3rd party value-add widgets (e.g. energy saving widgets)
- monitor micro-generation unit production
- monitor electric vehicle charging

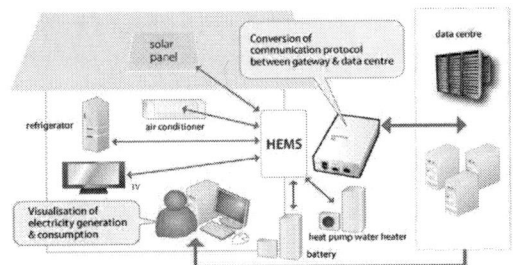

Fig. 1. Home Energy Management Development Platform

The Home Energy Management platform is powerful enough to collect real-time data delivered by a network of intelligent meters and sensors, but also to run a framework processing and delivering analytical visualizations in an intuitive and powerful visual way through a number of access methods like touch screen, mobile phones and web browsers[5]. The platform can serve as the basis for devices targeting consumers willing to manage and control, in real time, consumption of electricity and other energy loads in a building or a house.

B. HEM Algorithm

The following is the proposed algorithm for a Home Energy System employing Demand Response by considering four critical loads[1] [2]

Mostly the loads falling under the category of non-Critical Loads.

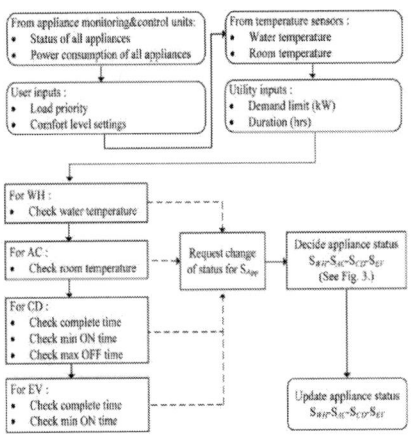

Fig. 2. HEM algorithm flow chart – overall algorithm. SAPP is the status of appliance APP that has the comfort level violation.

After a specified period of time (refresh interval), the appliance monitoring and control units gather data on the status of all the appliances. This status would include their current state of operation and their power consumption in kW. From this data, the current usage may be obtained. The HEM unit also receives the utility input, which includes the desired reduction in demand and the duration for which the reduction needs to be done. Once the entire data has been accumulated, the process of Demand Response begins keeping in mind the user's load priority and his comfort level settings. The load having the least priority is considered first and by using the sensors and/or timing units attached to it, information is obtained which is then compared with the comfort level settings of the user. If they are in conflict, the appliance continues to operate, in which case the control moves to the appliance upwards in the priority list to that of the current appliance. If the comfort level settings of the user are not violated, then the appliance is turned off or may be allowed to run at a reduced capacity. In any case, the HEM unit continues to keep a track of the amount to be reduced and the amount that has already been reduced by turning off the appliances. If considering all possible conditions, the reduction in demand cannot possibly match or exceed the desired reduction, the Utility is sent back a message indicating inability to achieve the desired results. The change in operating condition of a load, if any, is then changed in the HEM unit. This process goes on indefinitely [1] [6].

IV. SMART CONTROLLER

A. The Role of Smart Controller

In order to apply the smart grid system in the house, three components are necessary; AMI, EMS and smart controller. By using the bi-direction communication, the AMI communicates with the utility company. The information like the cost and electric consumption status is exchanged. The EMS reports the electricity amount of energy consumption between AMI and SC; or controls the smart controller. The smart controller (SC) calculates the amount of electricity consumed by customer on real time basis from the utility supplier like electric supply company. The AMI base is required to apply demand response techniques in home along with SC for connecting electric appliance and HAN (Home Area Network) connection. The SC has the network function to transfer information as the power state or the power consumption by controlling the electrical appliance.

The SC sets up the peripheral of the electric appliance. The SC informs the power state or the amount of power used by the electric appliance to the EMS. It also receives commands like the power management from the AMI. However this product is still not widely supplied to the market although efficient use of the energy is possible. The main reason is that the changing cost is much high and causes more environmental contamination in case of discarding the existing electronics along with the loss of investment.

B. AMI and Demand Response

Fig.3 shows the configuration of the smart grid system. As the energy producers produce and supply the gas, electricity, water etc, the AMI system checks a meter and reports the present consumption state to the utility company. By using the network system, it also regularly keeps the price of energy manufacture from the utility system on a real time basis. As for the energy consumer, the DR controller is directly connected to PC, TV, Light and all other electric appliances. The DR controller is connected to the AMI system through the usage of the energy server with ZigBee network.

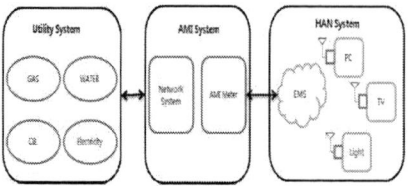

Fig. 3 The configuration of the smart grid

The smart controller is the central hub for the HEM system. In order for it to perform intelligently, it must act along the lines of an algorithm. It would include the conditions under which the controller would operate and the corresponding action it would take.

V. DESIGN OF SMART CONTROLLER

A. Hardware Design

The Hardware Design of the controller involves using a microcontroller and interfacing it with loads through relays. The commands to the microcontroller are given through serial communication and the current status of the loads is also relayed through serial transmission.

1) *Hardware Components Description*: The Smart Controller uses the following components

   i. Arduino Board: Arduino is a single-board microcontroller intended to make the application of interactive objects or environments more accessible. The hardware consists of an open-source hardware board designed around an 8-bit Atmel AVR microcontroller. The model feature a USB interface for serial communication, 6 analog input pins, as well as 14 digital I/O pins. It uses a FT232 chip for the USB interface.

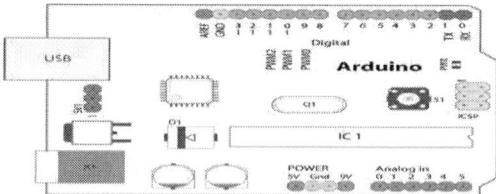

Fig. 4. Pin Diagram of the Arduino Board

   ii. SPDT Relay (6V DC Coil Voltage, 10A, 227V AC

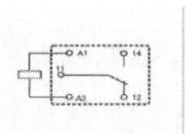

Fig. 5. SPDT Relay Pin Diagram

978-1-4799-6047-7/14 $31.00 © 2014 IEEE

The above relay is used to perform switching of the load connected to the controller. It is an electromechanical switch that deviates from its normally closed position to its normally open position when the coil of the relay is excited by a voltage.

Here, terminals A1-A2 are used for coil excitation. Pin 11 is the common terminal. 12 is the normally closed terminal and 14 is normally open one. On giving a voltage between A1 and A2, the switch moves from 12 to 14, breaking one circuit and making the other.

B. Circuit Diagram

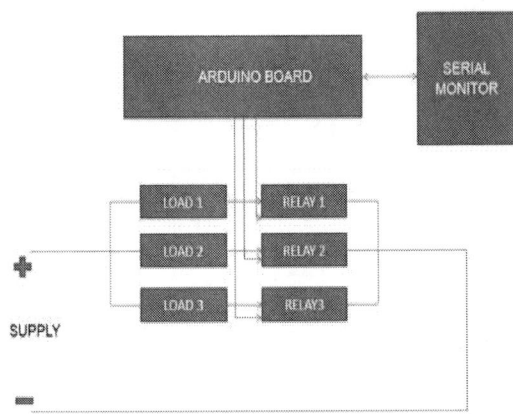

Fig. 6. Block Diagram of Smart Controller

The smart controller is manufactured in the form of the exterior module. It comprises of the power supply unit, controller, sensor unit, and driving part. The power supply unit is provided power from the external power source so that the smart controller can operate independently. The direct currents have to be supplied in order to operate the digital elements such as a microprocessor and sensor. The sensor unit is connected to the power plug of the electric appliance used in the homes such as the air conditioner and TV. A sensor calculates the total current and the voltage consumed by the electric appliance. The driving part is also connected inside the electric load of the home through the relay circuit. If a large amount of electricity is consumed and the electricity charge rises, the electrical energy supplied to the electric appliance is cut off then the energy consumption is reduced. The controller includes the network module and the microprocessor. The microcontroller manages the sensor unit and driving part, and delivers the required information to the AMI through the network module.The smart controller decides the use of an electric appliance according to the change of energy rental fee on real time basis. In case the energy consumption of the electric appliance grows and the rental fee is high, the AMI broadcasts this information to the smart controller. The rate of communication between the devices using this smart controller and the AMI is not so high; therefore we can use the broadcast method. The maintenance and repair can be convenient and exchange period of data can be also consistently maintained.

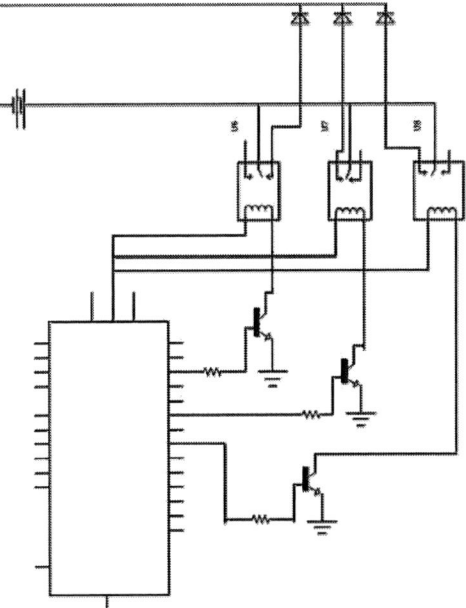

Fig. 7. Model atmega-8 microcontroller

In this model, Atmega-8 microcontroller is the processor for the operation of the SC device which works on the coding of the hardware model. The Serial Monitor of the Arduino Software is used to control the load priorities and create Demand Response events. Relays play an important role of switching the load according to the priority settings and balance condition. Here we have connected three relays which are connected to three loads individually. These relays will switch off their individual load only when customer is exceeding his predefined load limit. And the control of switching off the individual load is depends on the priority decided by the customer. This command signal of switching of the relays is sent by the micro-controller to the relays. And this signal generation is depends on the status of the system. Here the normal condition is taken as when all the loads are operating. Now when this system is under operation then it monitors the serial port to check if there is any signal available on it. As soon as it's available, it begins to switch off the loads in the decreasing order of priority. And after the turning off of each load, it again checks if the new usage is less than the recommended usage. If not, it again switches off the next load having the highest priority and the process continues. As soon as the consumption is equal to or less than the recommended level, it stops switching off of the loads.

Here, 833H-1C-C SPDT relays were used which require a minimum voltage of 6V DC to operate. However, the Arduino board can output a maximum voltage of 5V DC, which would make the controlling of the relays impossible on a direct connection with the board. To counter this, a simple transistor circuit was used that steps up the voltage from 5V to a level on which the relay can operate satisfactorily. Diodes were connected which prevent the backward flow of current. Thus, the following schematic was used for each pin.

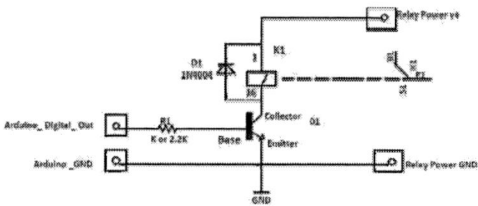

Fig. 8. Relay connections with Arduino

C. Software Design

The code for the Arduino board has been designed using the Arduino Integrated Development Environment which makes it easy to code and upload it to the I/O board. The environment is written in Java and based on Processing, AVR-GCC, and other open source software. The code for any Arduino program comprises of two functions – setup() and loop(). As the name suggests, the setup() is used to define the pins as input or output pins. It may include code that is only designed to run once. It also includes defining the serial rate of transmission which is 9600 bauds in our case.

The 'core' of the DR program is kept in the loop() function, which includes monitoring of the current usage and

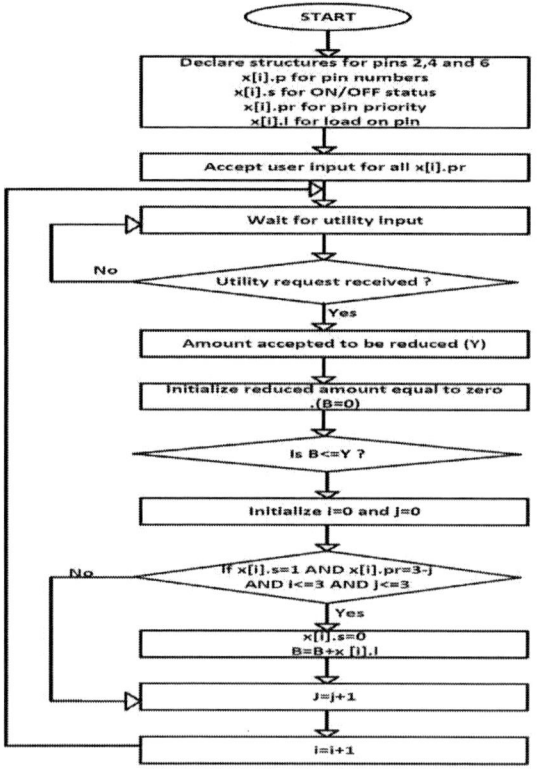

Fig. 9. Algorithm for implemented Smart Controller

the conditions for the switching of loads. Using the in-built Serial Monitor of the Arduino IDE, data can be sent to and from the board to the board or computer respectively. Use of the Serial Monitor helps in avoiding the use of a separate LCD display and a keypad and offers more flexibility and ease in display and input. Figure 10 shows the algorithm for working

of SC. This flow chart shows continuous automatic checking of the power consumption and controlling the load by utility.

VI. RESULT AND DISCUSSIONS

The proposed HEM algorithm can effectively keep the total household power consumption below the demand limit requirement. By following the procedure for hardware and software design, the controller was designed. On starting the controller, the user has to enter the priority of pins 2, 4 and 6 in the form of a three digit number. E.g. if the user wishes to set priority 1 for pin 2, priority 3 for pin 4 and priority 2 for pin 6, the user enters the figure "132."

On receiving the above information, the controller informs the user about the current state of the system.

The user then enters the amount of load that he he wants reduced. Here, in this case it will be "4 KW" Since the least priority is of pin 2, it is switched off the first. The controller thus displays the message "Reduced amount 2 kW" But since the load reduction to be done is 4kW, it again checks for the least priority. This time, it is of pin 3. Thus, the 3kW load at pin 3 is switched off. If the total reduced amount is now 5kW which is displayed. Since now the reduced amount is more than 4kW, the cycle now stops. Thus the final result of the DR cycle is pin 4 and 6 have been switched off. The observations table shown below defines the above observation

Fig 10. Observation of the above Implimentation

Table 1. Result and observations

Total load= 6 KW (Pin 2= 1KW , Pin 4= 2KW , Pin 6=3KW	
Priority (Pin 2 has 1st , Pin 4 has 3rd and Pin 6 has 2nd priority	
Condition	Shut down load According to priority
Normal	None (All loads are in on condition)
Reduced load "2 KW"	Pin 4 load
Reduced load " 3 KW"	Pin 6 load
Reduced load " 4 KW"	Pin 4 and pin 6

As a result, equipment manufacturers are employing this technology that offer equipment developers new capabilities for simplifying software consolidation, managing equipment remotely and increasing system security. In addition, distributed computing architecture based on smart controller technology can help drive distributed intelligence throughout the grid and its end points, which enables optimized levels of automation and decision making at each link in the chain. These solutions can be applied across the energy network, from energy generation facilities to consumer dwellings and business site.

VII. CONCLUSIONS

In light of the results obtained, it can be said that the Demand Response Smart Controller has been successfully implemented. It can receive an external signal from the utility which is in the form of a serial communication from the computer. It then talks back to the utility which is again in the form of a serial communication to the meter installed in the home. Also, the concept of prioritization of the load has been successfully implemented. The loads which are switched off can be switched on later, resulting in the same energy consumption but an overall decrease in the peak load on the system. Using the proposed system with the described algorithm would result in savings at the consumer side, lesser stress on the power system and lesser cost of generation.

It would also mean a lower installed capacity as the maximum demand on the system would be less, resulting in savings. The demonstration proved the efficient use of smart controller and showed that it can be used in the smart grid system. This system supports all home appliances as there is no need to change it for new products. It can be easily attached outside of the existing equipment. It is not only a cheap solution but also prevent the wastage of the resources and saves the investment of customer.We have developed the hardware model and software coding for implementation of smart controller with demand response for Advanced Metering Infrastructure.

VIII. FUTURE WORK

The Smart Controller (SC) can be modified to control loads with larger ratings. Though the circuit and the relays used remain the same, since they have sufficient rating, but the board would need to be modified to handle the possibility of a large current flowing back into it, damaging it. Also, instead of communicating with a computer, the controller would now require communication with an actual metering equipment. The Energy meters will be smart meters which can communicate on real time basis with SC and utility. Through this utility can monitor the loads directly and control instantaneously. The power distribution system will be well controlled, managed and secure.

REFERENCES

[1] Pipattanasomporn, M.; Kuzlu, M.; Rahman, S. "An Algorithm for Intelligent Home Energy Management and Demand Response Analysis,"Smart Grid, IEEE Transactions on, vol.3, no.4, pp.2166-2173, Dec. 2012

[2] A. H. Mohsenian-Rad, V. W. S. Wong, J. Jatskevich, and R. chober,"Optimal and autonomous incentive-based energy consumption scheduling algorithm for smart grid," in Proc. IEEE Innov. Smart Grid Technol., Jan. 2010, pp. 1–6.

[3] Yee Wei Law; Alpcan, T.; Lee, V.C.S.; Lo, A.; Marusic, S.;

Palaniswami, M., "Demand Response Architectures and Load Management Algorithms for Energy-Efficient Power Grids: A Survey," Knowledge, Information and Creativity Support Systems (KICSS), 2012 Seventh International Conference on , vol., no., pp.134-141, 8-10 Nov. 2012.

[4] P. Du and N. Lu, "Appliance commitment for household load scheduling,"IEEE Trans. Smart Grid, vol. 2, pp. 411–419, Jun. 2011.

[5] In-Ho Choi, Joung-Han Lee, "Development Of Smart Controller With Demand Response For AMI System", International Conference on Control, Automation and Systems Vol. 1, Pp. 120–133, Sep. 2010.

[6] S. Shao, M. Pipattanasomporn, and S. Rahman, "Demand response as an load shaping tool in an intelligent grid with electric vehicles," IEEE Trans. Smart Grid, vol. 2, no. 4, pp. 624–631, Dec. 2011.

[7] M. Erol-Kantarci and H. T. Mouftah, "Wireless sensor networks for cost efficient residential energy management in the smart grid," IEEE Trans. Smart Grid, vol. 2, no. 2, pp. 314–325, Jun. 2011.

[8] A. Gomes, C. H. Antunes, and A. G. Martins, "Physically-based load demand models for assessing electric load control actions," in Proc. IEEE Bucharest PowerTech, Jul. 2009.

[9] Department of Energy—Energy Savers Tips. [Online]. Available: http://www1.eere.energy.gov/consumer/tips/water_heating.html

[10] V. Hamidi, F. Li, and F. Robinson, "Demand response in the UK's domestic sector," Elect. Power Syst. Res., vol. 79, no. 12, pp. 1722–1726, Dec. 2009

[11] J. Han, C. S. Choi, W. K. Park, and I. Lee, "Green home energy management system through comparisonof energy usage between the same kinds of home appliances," in Proc. 15th IEEE Int. Symp. Consum. Electron. (ISCE), Jun. 2011, pp. 1–4.

[12] F. C. Schweppe, B. Daryanian, and R. D. Tabors, "Algorithms for a spot price responding residential load controller," IEEE Trans. Power Syst., vol. 4, pp. 507–516, May 1989.

[13] B. Daryanian, R. E. Bohn, and R. D. Tabors, "Optimal demand-side response to electricity spot prices for storage-type customers," IEEE Trans. Power Syst., vol. 4, pp. 897–903, 1989.

[14] A.-H. Mohsenian-Rad and A. Leon-Garcia, "Optimal residential load control with price prediction in real-time electricity pricing environments, IEEE Trans. Smart Grid, vol. 1, no. 2, pp. 120–133, Sep. 2010.

[15] M. A. A. Pedrasa, T. D. Spooner, and I. F. MacGill, "Coordinated scheduling of residential distributed energy resources to optimize smart home energy services," IEEE Trans. Smart Grid, vol. 1, no. 2, pp. 134–143, Sep. 2010.

[16] J. C. van Tonder and I. E. Lane, "A load model to support demand management decisions on domestic storage water heater control strategy," IEEE Trans. Power Syst., vol. 11, no. 4, pp. 1844–1849, 1996.

[17] P. Constantopoulos, F. Schweppe, and R. Larson, "ESTIA: A real-time consumer control scheme for space conditioning usage under spot elec tricity pricing,"Comput. Oper. Res., vol. 19, no. 8, pp. 751–765, 1991.

[18] N. Chowdhury and R. Billinton, "Interruptible Load Considerations in Spinning Reserve Assessment of Isolated and Interconnected Generating Systems," IEE Proceedings, vol. 137, pp. 159-167, March 1990.

[19] H. Jorge, C. H. Antunes, and A. Martins, "A multiple objective decision support model for the selection of remote load control strategies," IEEE Trans. Power Systems, vol. 5, pp. 865-872, May 2000.

[20] A. Molina, A. Gabaldon, C. Alvarez, J. A. Fuentes, and E. Gómez; "A Physically Based Load Model for Residential Electric Thermal Storage: Applications to LM Programs," International Journal of Power and Energy Systems, vol. 24, 1, 2004.

[21] C. Alvarez, R.P. Malhamé, and A. Gabaldón; "A Class of Models for Load management application and evaluation revisited," IEEE Trans. Power Systems, vol. 7, pp. 1435-1443, Nov. 1992.

[22] H. Jorge, C. H. Antunes, and A. Martins, "A multiple objective decision support model for the selection of remote load control strategies," IEEE Trans. Power Systems, vol. 5, pp. 865-872, May 2000.

Mitigation of Sub Synchronous Resonance using STATCOM Controlled by PID and FL Controllers

Chinnari Eswar Prasad, Shelly Vadhera
Department of Electrical Engineering
NIT Kurukshetra
Haryana, India
eswarchinnari@gmail.com, shelly_vadhera@rediffmail.com

Abstract — **This paper deals with the use of fuzzy logic controller (FLC) and proportional integrator and derivative (PID) controller to control static synchronous compensator (STATCOM) which is used for damping oscillations caused by sub synchronous resonance (SSR). The results obtained by both the controllers are compared and it is found that FLC is better controller as compared to PID for controlling STATCOM to mitigate SSR. The study in this paper is particularly based on the torque amplification produced on the turbine generator shaft assembly after a fault takes place on a series-compensated power system. An IEEE benchmark model is used to study the mitigation of sub synchronous resonance (SSR) characteristics using both FL and PID controllers along with STATCOM in Matlab/Simulink.**

Keywords — Fuzzy logic control; Proportional integral derivative; Static synchronous compensator; Sub-synchronous resonance.

I. INTRODUCTION

The rise in threat to power system stability has made power engineers think about the requirement of series compensation and its emphasis on power system. In this situation, the long transmission lines are series compensated to improve transient stability and maximum power transfer capability [1]. It was believed that 70% series compensation can be used for long transmission lines without much concern till 1970 [2]. This series capacitive compensated network causes sub synchronous resonance (SSR) problem whose outcome is the interactions between both mechanical shaft of turbo-generator set and electrical mode of series compensated network. After extensive research this phenomenon was discovered and henceforth the utilities started taking care to avoid the SSR problem.

The interactions between the turbine section of generating units and series compensated transmission lines causes SSR problem. It is a resonant condition which operates at sub synchronous frequency (less than system frequency). The exchange of energy takes place between both electrical and mechanical system during SSR. The SSR phenomenon can be further divided into two sub categories [3]: Transient torque which is also known as torque amplification (TA) and steady state SSR. The torsional interaction (TI) and induction generator effect (IEG) are included under steady state SSR. This paper incorporates the TA and TI effects which lead to shaft failure. These effects are believed to be harmful and thus they are required to be mitigated from the system.

Numerous flexible a.c. transmission system (FACTS) controller devices are used for reactive power compensation, power quality improvement, regulating power flow, SSR mitigation and hence improving overall power system stability. Some of the FACTS devices which can be used for SSR mitigation are static series synchronous compensator (SSSC), static synchronous compensator (STATCOM) and thyristor controlled series capacitor (TCSC). In this paper STATCOM has been used to mitigate the SSR problem because of its several advantages over other FACTS controllers. A proportional integral derivative (PID) controller and fuzzy logic controller (FLC) are used to control STATCOM to achieve robust and dynamic control for mitigation of SSR problem. The FLC is selected as a controller because of its distinct advantages over other controllers like robustness, requirement of less computational space and time. Moreover there is no requirement of mathematical model to explain the system under study.

In this paper an IEEE benchmark power system model for computer simulation is used to study the sub synchronous resonance characteristics. The paper further deals with the mitigation of the SSR problem using STATCOM which is controlled by fuzzy logic and PID controllers. Here the study is particularly based on the series-compensated power system's torque amplification and torsional interaction after a three phase fault is applied. The comparison of the controllers is based on analysis of different parameters of the system such as generator speed deviation, turbine speed deviation and torque deviation. The results obtained in this paper are better as compared to [4] in terms of reduction in magnitude of torque deviation in turbine and speed deviations in generator.

II. SUB SYNCHRONOUS RESONANCE PHENOMENON

The sub synchronous phenomenon is described as the sustained oscillations caused below the fundamental system frequency. The use of series capacitor compensation to increase line loading in long lines brings about the danger of sub synchronous resonance in which electromagnetic forces in the generator can produce torque which corresponds to torsional frequencies in the shaft and results in mechanical damage [5].

In series compensation the capacitor connected in series with the line along with the inductance of system forms a resonating circuit with natural frequency given by:

$$f_e = f_s \sqrt{\frac{x_c}{x}} \qquad (1)$$

Where,

f_e= Electrical resonant frequency

f_s= Fundamental system frequency

x_c= Series capacitor reactance

x = Total reactance of the line (including transformer leakage reactance and generator reactance)

$$f_r = f_s - f_e \qquad (2)$$

Where

f_r= Sub synchronous frequency component

This resonant frequency f_r is also known as sub synchronous harmonic frequency. The value of electrical resonant frequency f_e is less as compared to fundamental power system frequency f_s. The electric circuit oscillates when a network or system disturbance occurs. The presence of sub harmonic line current generates a sub harmonic field in the machine. The sub harmonic field makes itself rotate in backward direction relative to main field (since f_r is positive) producing a torque on rotor at a speed of f_r. If any one of the torsional speed of turbine generator set matches with differential speed of sub synchronous harmonic frequency f_r, then mechanical torsional oscillations are excited which further excites the electrical resonance.

III. TEST MODEL

A fault is applied on a series compensated power system model for the analysis of SSR. The test model used for analysis is 'IEEE benchmark model of power system for SSR' and is used for the study of torque amplification phenomenon. The power system model consists of a single generator connected to an infinite bus via two transmission lines, out of which one line is 55% series-compensated as shown in Fig. 1 [6]. The corresponding simulink model is shown in Appendix (Fig. 14). A three phase fault is introduced and cleared from the system for the analysis of sub synchronous resonance in a series compensated line (with capacitor). It excites the torsional oscillations in the turbine generator set (multi-mass shaft system), where the torque amplification phenomenon can be observed. The mechanical system model constitutes of four sub-systems such as generator (G), low pressure (LP) turbine, high pressure (HP) turbine and excitation system (EX).

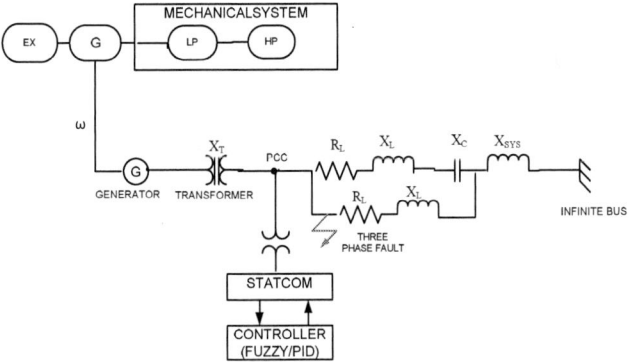

Fig. 1. Test model

The need for linearizing the power system model is eliminated as the fuzzy logic system doesn't depend on the system model. So the model uses all the power system parameters with their non linear equations.

IV. STATIC SYNCHRONOUS COMPENSATOR

STATCOM is a static var generator, categorized under flexible a.c. transmission system (FACTS) devices and is used as a shunt compensator in power system as shown in Fig. 2. It is operated without any external energy source. A static var generator can be controllable reactive impedance or a synchronous voltage source or a combination of both [7]. It uses power electronics devices for improving transient stability and power quality in power system network. It controls the reactive power of the system depending upon the voltage profile. It injects reactive power if the system voltage is low and absorbs reactive power when the system voltage is high. The STATCOM injects current into the system and the injected current opposes the sub harmonic line current which is the root cause for SSR problem. The STATCOM used in this paper uses voltage source converter (VSC) for controlling the voltage. Forced commutated power electronics devices (GTO, IGBT) are used in VSC. The VSC used can be of two level (six pulse), three level (twelve pulse) and so on. Here a 48 pulse converter is used for superior performance over other pulse converters. The output of the eight level converter resembles a sine wave.

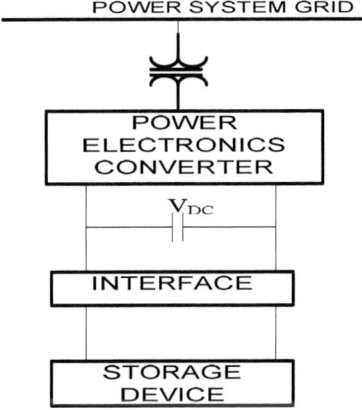

Fig. 2. Block diagram representation of STATCOM

V. FUZZY LOGIC CONTROLLER

Fuzzy logic controller is a mathematical tool used in power system models with uncertain and ambiguous data and is specially used for non linear system since it is independent of system model. It is one of the best controllers for converters and FACTS devices to improve power system stability. It is a heuristic control technique and is based on human experience. It also enhances the operation of closed loop control system. It can work taking in as many inputs at a time and the output obtained is reliable. The only issue with FLC is that the solution is entirely dependent on the membership function which has to be defined by the user.

There are four main blocks describing the operational function of FLC as shown in Fig. 3 [8]:

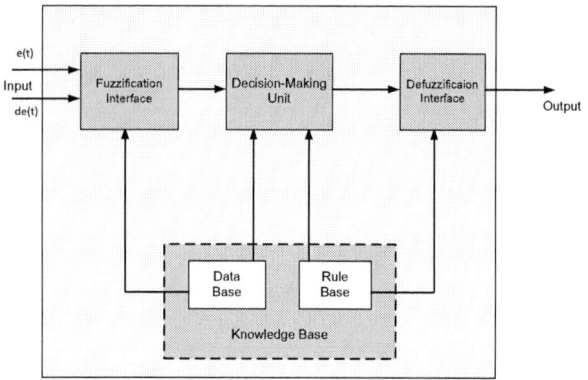

Fig. 3. Block diagram of FLC

The block diagram of fuzzy logic based control structure used in this paper is shown in Fig. 4.

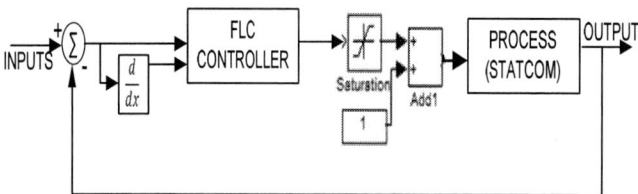

Fig. 4. Control structure using FLC.

Fuzzification block performs fuzzification of the input signals given to FLC. The fuzzy inference engine block constitutes of decision making logic used in FLC. Fuzzy knowledge module consists of set of basic rules which are used by the fuzzy interference engine for decision making. Defuzzification block converts the fuzzified signal into the required output format [9].

In this paper, inputs given to FLC are speed deviation of low pressure turbine ($\Delta\omega l$) and its derivative ($\Delta\omega h$). The input and output membership functions of FLC are tri (triangular shaped memberships function). The inputs to the FLC are $\Delta\omega l$ and $\Delta\omega h$ and output is ΔU. The corresponding membership functions are shown in Fig. 5 [4]. The fuzzy set used is defined in Table 1 whereas the surface plot of fuzzy logic is shown in Fig. 6.

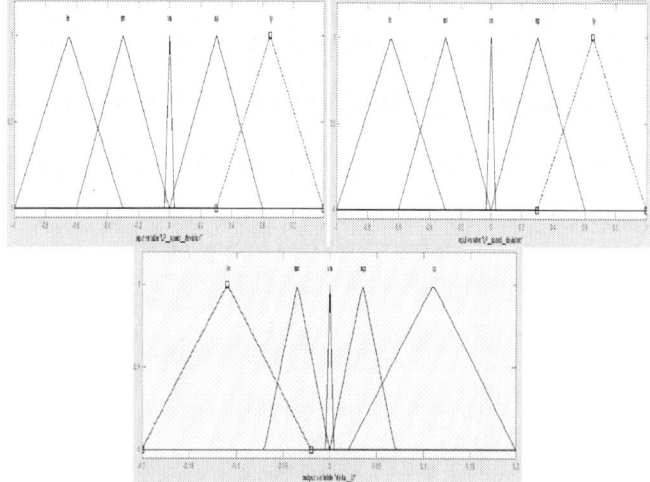

Fig. 5. Input and output membership function

TABLE I.   FUZZY SET

$\Delta\omega_l$	$\Delta\omega_h$				
	LN	SN	VS	SP	LP
LN	SN	SN	LN	LN	LN
SN	SN	SN	SN	LN	SN
VS	SP	SP	VS	SN	SN
SP	LP	SP	SP	SP	SP
LP	LP	LP	LP	SP	SP

a. LN = Large negative, SN = Small negative, VS = Very small, SP = Small positive, LP = Large positive

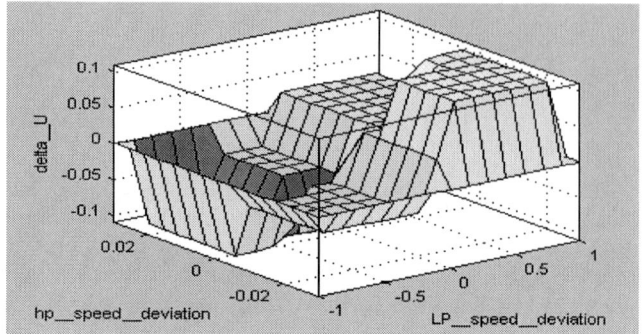

Fig. 6. Surface plot of FLC

VI. PID CONTROLLER

A PID controller provides design flexibility since it involves three adjustable gains. In many cases, the nominal plant transfer function is unknown, and the PID design is based on a step response analysis of the process, with the proportional gain K_p, the integral gain K_i, and the derivative gain K_d, manually adjusted or tuned on-line to obtain the best performance. The block diagram of PID is shown in Fig. 7 and transfer function [10] is written as:

$$H_{PID}(s) = K_p + \frac{K_i}{s} + K_d s = \text{P+I+D} = \frac{Y(s)}{E(s)} \qquad (3)$$

The significance of proportional part in PID controller is to reduce the error responses to disturbances. The integral part of PID eliminates the steady-state error and the derivative part of PID dampens the dynamic response and thereby improves the stability of system. The process of selecting the controller parameters (that is the three gains of the PID controller) to meet the desired performance is achieved through tuning of the controller. The desired step response for a closed loop control system should have minimum settling time with a very small or nearly zero overshoot.

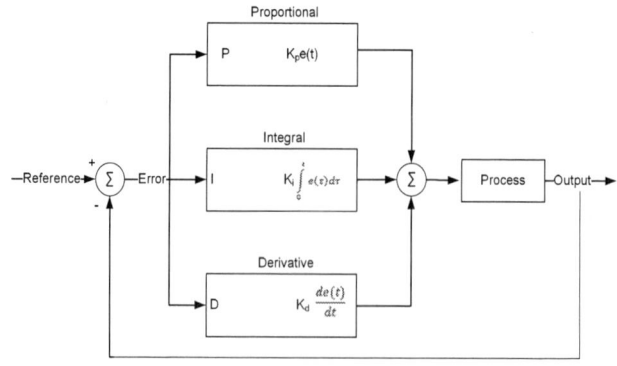

Fig. 7. Block diagram of PID controller

978-1-4799-6047-7/14 $31.00 © 2014 IEEE

In this paper a PID controller is used for minimizing speed deviation and torque deviation of the synchronous machine. The best results are obtained by using the following values for the PID gain constants as K_p=0.02, K_i=1.6, K_d=0.06.

VII. SIMULATION RESULTS

The results obtained from the simulation shows the impact of STATCOM controlled by fuzzy logic on the IEEE benchmark model to mitigate SSR problem. Firstly, the plots of generator speed deviation, low pressure (LP) turbine speed deviation and high pressure (HP) turbine speed deviation shows an increased damping and decreased steady state error with the use of FLC and PID as shown in Fig. 8 to Fig. 10. In case of rotor speed, with the use of FLC and PID the time taken by it to approach its rated value (1p.u.) is reduced as shown in Fig. 11. The torque deviation in LP-HP turbines and generator are minimized by the use of FLC and PID as shown in Fig. 12 and Fig. 13 respectively.

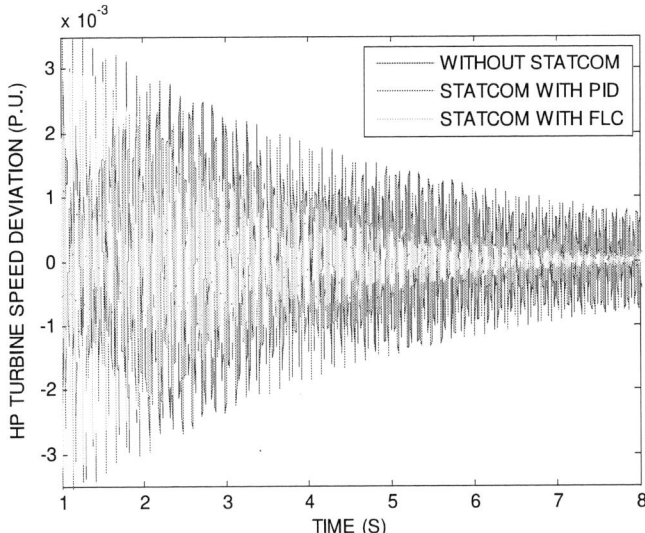

Fig. 10. HP turbine speed deviation

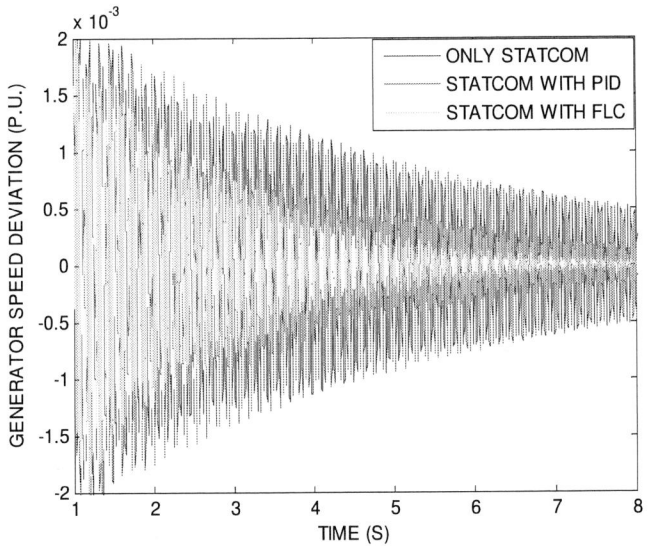

Fig. 8. Generator speed deviation

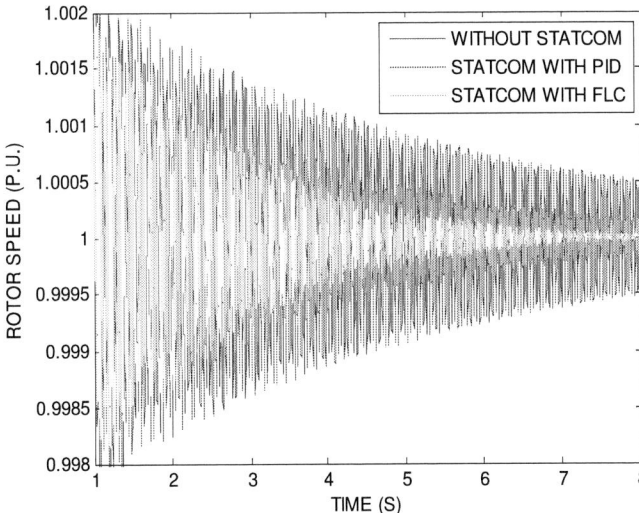

Fig. 11. Rotor speed

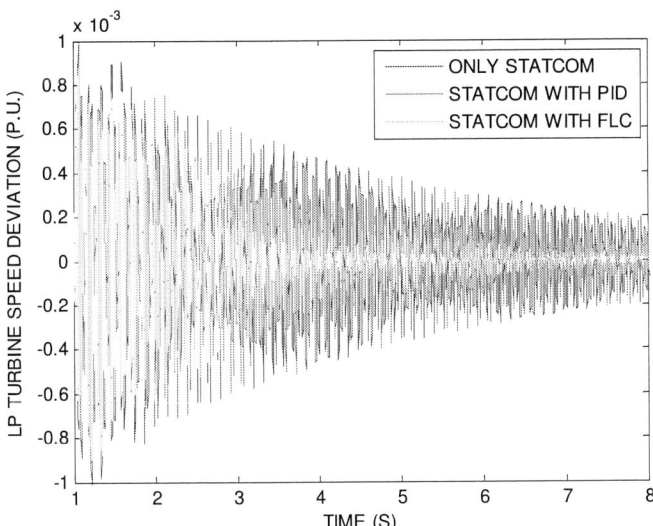

Fig. 9. LP turbine speed deviation

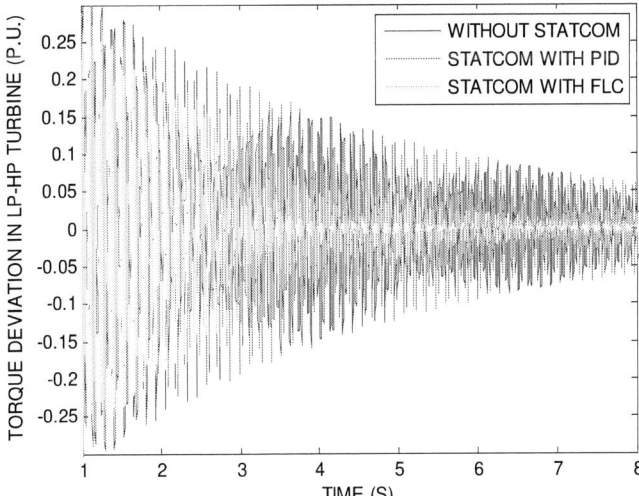

Fig. 12. Torque deviation in LP-HP turbine

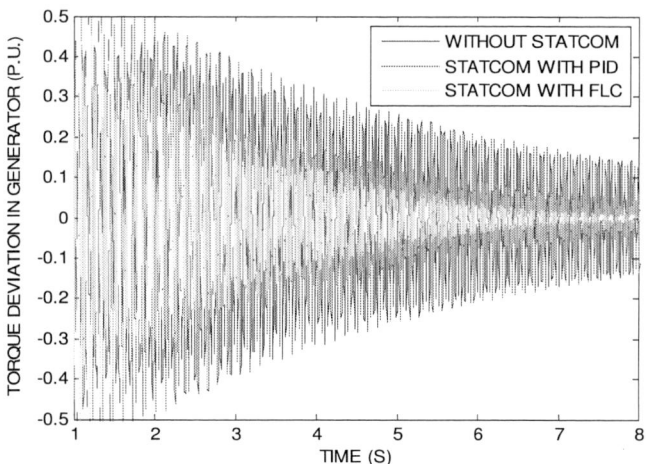

Fig. 13. Torque deviation in generator

VIII. CONCLUSION

In this paper, STATCOM controlled by FLC and PID are successfully used to mitigate the SSR problem. The use of STATCOM, along with FLC and PID, provides better results in minimizing speed deviation(generator) and torque deviation (LP-HP turbines) by increasing the damping and decreasing the steady state error. In case of generator speed it is observed that it reaches the rated speed value (1p.u.) in less time when FLC is used as compared to the PID controller. Furthermore, by analyzing the overall simulation results it can be inferred that the use of STATCOM controlled by FLC brings a better improvement in the stability of the power system by minimizing the SSR problem caused by series compensated transmission line as compared to STATCOM controlled by PID.

APPENDIX

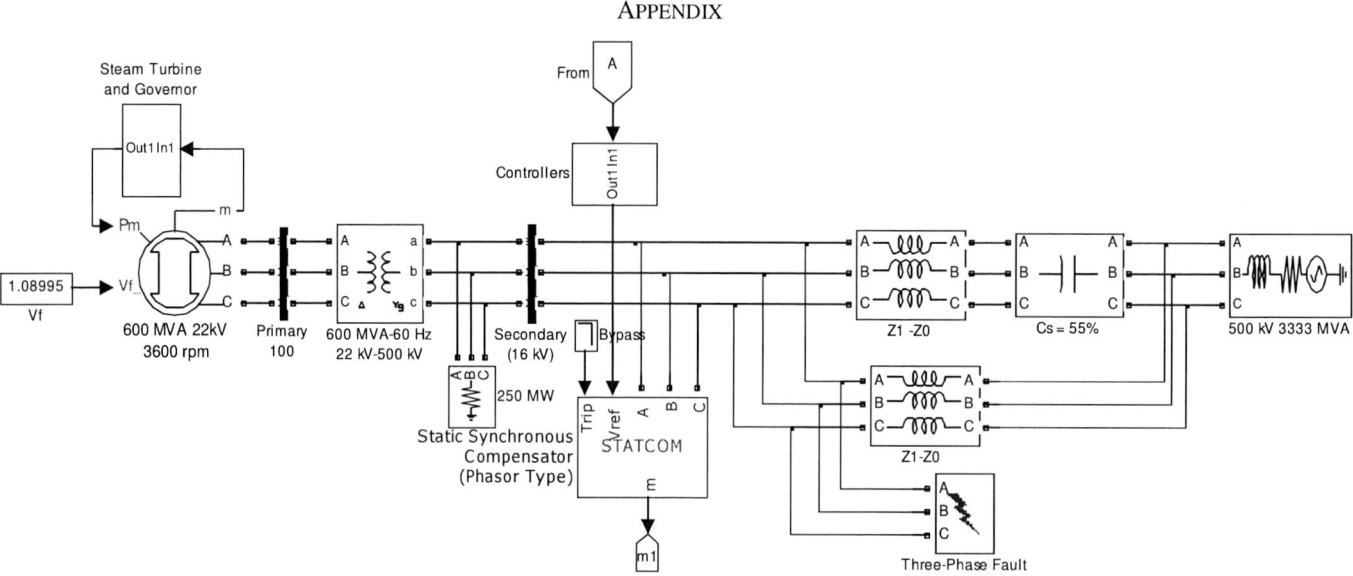

Fig. 14. Matlab simulation model.

REFERENCES

[1] Hossein Ali Mohammadpour, Seyyed Mohammad Hosein Mirhoseini, and Abbas Shoulaie, "Comparative study of proportional and TS fuzzy controlled GCSC for SSR mitigation," POWERENG International Conference on Power Engineering, Energy and Electric Drives, Lisbon, Portugal, pp. 564-569, March 18-20, 2009.

[2] Massimo Bongiorno, Jan Svensson, and Lennart Ängquist, "On control of static synchronous series compensator for SSR mitigation," IEEE Tranc. Power Electronics, vol. 23, no. 2, pp. 735-743, March 2008.

[3] Massimo Bongiorno, Lennart Ängquist, and Jan Svensson, "A novel control strategy for subsynchronous resonance mitigation using SSSC," IEEE Tranc. Power Delivery, vol. 23, no. 2, pp. 1033-1041, April 2008.

[4] S. T. Nagarajan and Narendra Kumar, "Fuzzy logic based control of STATCOM for mitigation of SSR," 5th Indian International Conference on Power Electronics (IICPE), Delhi, pp. 1-6, 6-8 Dec., 2012.

[5] Rakesh Das Begamudre, Extra high voltage AC transmission engineering, second edition, New Age International (P) Limited, Publishers, October 1997.

[6] IEEE Subsynchronous resonance working group, "Second benchmark model for computer simulation of subsynchronous

resonance," IEEE Tranc. Power Apparatus and Systems, vol. 104, no. 5, pp. 1057-1066, 1985.

[7] Narain G. Hingorani and Laszlo Gyugyi, Understanding FACTS concepts and technology of flexible AC transmission systems, IEEE Press, Standard Publishers Distributors, New Delhi, 2001.

[8] K. A. Ellithy and K. A. El-Metwally, "Design of decentralized fuzzy logic load frequency controller," I. J. Intelligent Systems and Applications, vol. 2, pp.66-75, 2012.

[9] Akbar Lak, Daryoush Nazarpour, and Hasan Ghahramani, "Novel methods with fuzzy logic and ANFIS controller based SVC for damping sub-synchronous resonance and low-frequency power oscillation," 20th Iranian Conference on Electrical Engineering, (ICEE2012), Tehran, Iran, pp. 450-455, May 15-17, 2012.

[10] I. J. Nagarath, M. Gopal, Control system engineering, 3rd edition, New Age International Publications, New Delhi, India, 1999.

Design and Analysis of Stationary Frame PR Current Controller for Performance Improvement of Grid Tied PV Inverters

A.Chatterjee
Department of Electrical Engineering
National Institute of Technology
Rourkela, India
contactaditi247@gmail.com

K.B. Mohanty
Department of Electrical Engineering
National Institute of Technology
Rourkela, India
kbmohanty@nitrkl.ac.in

Abstract— In this paper a single phase grid connected photovoltaic (PV) system has been modeled and simulated using Matlab/Simulink. The PV generator is interfaced with the utility grid by a single phase voltage source inverter (VSI). A maximum power point tracking (MPPT) control algorithm is implemented to extract maximum power from the PV generator irrespective of operating conditions. The DC-DC boost converter placed in between PV generator and VSI performs MPPT and amplifies the output voltage from PV array to desired level. A control strategy to regulate the quality of power injected by the grid connected VSI is proposed. Two controllers are designed for the grid side DC-AC inverter one is the proportional integral (PI) voltage controller which regulates the DC link voltage, the other is the Proportional + Resonant (PR) current controller which maintains the current injected by the inverter into grid in phase with the grid voltage so that unity power factor can be achieved. A harmonic compensator (HC) is cascaded with PR controller to mitigate low order odd harmonic components present in the output current of VSI and minimize the total harmonic distortion (THD). Simulation results validate the effectiveness of the proposed control strategy.

Keywords— DC-DC converter, Harmonic compensator Maximum power point tracking, Proportional Integral controller, Proportional Resonant controller, Total harmonic distortion.

I. INTRODUCTION

Diminution of fossil fuel reserves and increased concern about environmental pollution has amplified the demand of renewable energy sources (RES) for power generation. Moreover to meet the escalating electricity demand more distributed generation (DG) plants based on wind and solar energy are integrated into power distribution system. The form in which power is generated by the DG plants may not be compatible with the conventional distribution system such as DC output from fuel cells, photovoltaic modules and batteries, variable frequency AC output from wind turbines etc. To alter the generated power into the required format power electronics converters are required. Single phase distribution system is often used to serve the rural and residential areas and the PV plants are mostly installed in these areas. To interface the PV units with the grid single phase inverters are required.

Different converter topologies have been proposed for single phase grid connection of PV modules [1]. One is the single stage topology where the PV module is connected to the grid through a single DC-AC inverter. In that case the inverter performs MPPT to extract out the maximum power from the module and also voltage amplification to step up the output voltage to desired value. The other one is the dual stage topology where there is a DC-DC converter after the PV unit. In this case DC-DC converter handles the task of maximum power extraction and also boosts the voltage as per requirement. In this paper two power processing stage topology has been implemented. The grid side DC-AC converter which injects power into the grid has to ensure that the quality of power is good. The basic tasks which the grid side converter handles includes control of active and reactive power exchange between the grid and PV system and synchronization of grid current with grid voltage.

Various control strategies have been proposed to enhance current control of the grid tied VSIs. [2]. Basically these controllers are categorized into three major classes: 1)synchronous frame controller 2)stationary-frame controller 3)natural-frame controller. In synchronous or d-q frame control the control variables are transformed from natural frame to a synchronous rotating frame (frame rotates synchronously with grid voltage). Hence the control variables appear as dc quantities and the control is better. The PI controllers are associated with this control structure. The major drawback associated with this structure is the necessity to extract the phase angle of grid voltage and incorporating the voltage feed forward and cross coupling terms in the control loop. In stationary or α-β frame control structure the control variables are time varying. The Proportional Resonant (PR) controllers falls under the category of stationary frame controllers are simple to design and has excellent reference signal tracking capabilities. The PR controllers can achieve very high gain at resonant frequency thus reducing the steady state error to zero [3-5]. More over harmonic compensators can be used to mitigate low order harmonic without influencing behavior of the current controller. Hence they are superior than PI controllers in terms of eliminating steady state errors and harmonic current rejection. The natural or abc frame control structure is usually implemented for non linear controllers such as hysteresis and dead beat control. The drawback associated with these controllers is variable switching frequency.

II. SINGLE PHASE GRID TIED PHOTOVOLTAIC SYSTEM

Based on the number of power processing stages the possible inverter topologies are given below [1]: (i) Single stage (ii) Dual stage

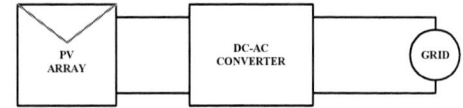

Fig.1 (a) Single stage grid tied PV array

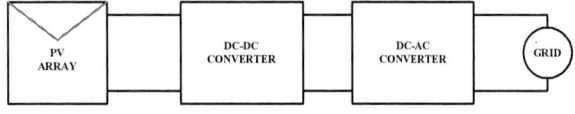

Fig.1 (b) Dual stage grid tied PV array

In single power processing stage as shown in Fig.1(a) the DC-AC converter is responsible for voltage amplification, MPPT and grid current control and in dual power processing stage shown in Fig.1(b) the DC-DC converter handles MPPT, voltage amplification and the DC-AC converter handles the grid current control. The former topology is used for past centralized PV inverter technology. In this case the inverter has to be designed to handle peak power of twice the nominal power of the system. The later topology is mostly used in present day multi-string technology. Here the PV inverter is designed to handle only the nominal system power. In this research work the dual stage topology is implemented.

A. Dual Power Processing Structure for Single Phase Grid Tied PV System

Figure.2. shows the schematic diagram of a single phase grid tied PV system where the PV generator is interfaced with the grid via two converters [2]. Two controllers are designed.

a)Input side controller: a MPPT controller is designed to estimate the output voltage and current from the PV array and extract maximum power from the source. It generates reference voltage for the front end DC-DC converter.

b)Grid side controller: the task of the grid side controller is to control power flow between the DG source and grid, ensure good quality of power injected by the inverter into the grid and grid synchronization.

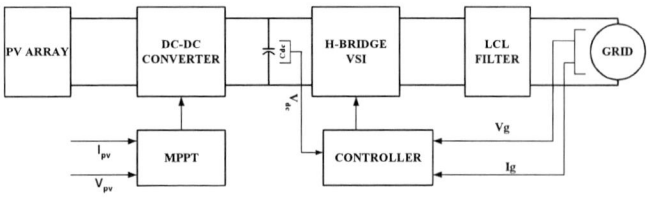

Fig. 2. Schematic diagram of grid tied photovoltaic system

B. Modeling of PV Array

The selection of PV array depends on the power rating of the system. Solar cell is basically a p-n junction fabricated on a layer of semiconductor, when sun rays strikes the surface of cell, the solar energy gets converted to electrical energy. The output of a single PV cell usually vary between 0.8 volts to 1.2 volts which is not suitable for practical applications. The output of a single cell can be scaled up as desired to meet the system power requisite. Cells are connected in series to raise the voltage level and in parallel to increase the current level. Individual PV cells allied in series and parallel form a PV module and appropriate association of modules make a PV array. The solar panel used in this paper is KC200GT and the parameters of the panel at 25^0C, AM 1.5, and irradiance 1000 W/m^2 is tabulate in Table I. 10 number of such panels are connected in series to obtain a power output of 2000 Watts. Fig.3 and Fig.4 demonstrate the PV curve and IV curve obtained by simulating the PV array model in MATLAB [6].

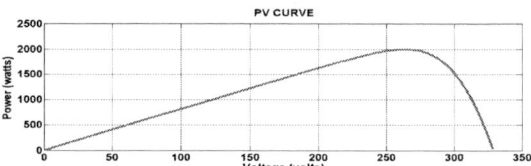

Fig. 3. PV curve at STC

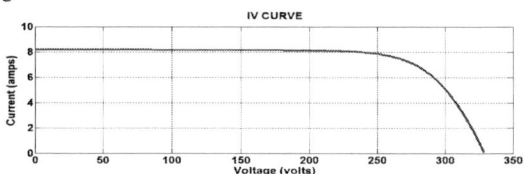

Fig.4. IV curve at STC

TABLE I. Parameters of KC200GT ,PV array at STC

Parameters	Values
Maximum Power, P$_{max}$ (W)	200.143
Current at P$_{max}$, I$_{mp}$ (A)	7.61
Voltage at P$_{max}$, V$_{mp}$(V)	26.3
Short circuit current I$_{sc}$ (A)	8.21
Open circuit voltage V$_{oc}$ (V)	32.9
Number of cells in series	54

C. Maximum Power Point Tracking

There is a unique point on the PV curve where the photovoltaic panel delivers maximum power which is known as the maximum power point (MPP). The power output of the PV panel may vary with the atmospheric conditions like temperature, irradiance etc. So in order to draw maximum power from the solar array a MPP tracking algorithm is integrated with the PV system. Various MPPT algorithms have been reported in literature [7], among which the incremental conductance (IC) algorithm is applied in this work. This algorithm is based on the observation that at MPP, the condition that occurs is given by (1a):

$$\frac{dp}{dv} = 0 \tag{1a}$$

$$\frac{dp}{dv} = I + v\frac{dI}{dv} \tag{1b}$$

$$I + v\frac{dI}{dv} > 0 \tag{1c}$$

$$I + v\frac{dI}{dv} < 0 \tag{1d}$$

When (1b) is satisfied the operating point is at MPP. When the condition given by (1c) and (1d) is met the operating point is at left and right of MPP respectively. From (1c) and (1d) it can be determined in which direction perturbation must occur to budge the operating point towards the MPP and the perturbation is repeated until (1a) is satisfied. Once the MPP is reached, the algorithm continues to operate at this point until an alteration in current is detected.

The MPPT controller estimates the output voltage and current of the PV array and generates the reference voltage for the front end DC-DC boost converter. The front end converter maintains a constant DC link voltage irrespective of the input voltage level from the PV array. The value of the boost inductor can be calculated from (2) and the DC link capacitance can be calculated from (3) [8].

$$L_b = \frac{D(1-D)^2 T_s V_0}{2*I_0} \tag{2}$$

$$C_{dc} = \frac{P_{dc}}{2*\omega*V_{dc}*\Delta V_{dc}} \tag{3}$$

Where D is duty ratio, T_s is switching time and V_0 and I_0 are the output voltage and current respectively. ω is grid frequency P_{dc} and V_{dc} are the DC link power and voltage respectively. ΔV_{dc} is amplitude of voltage ripple which varies between 1% to 5%.

III. CONTROLLER DESIGN FOR SINGLE PHASE GRID TIED VSI

Two controllers are developed for the single phase grid tied inverter: The inner current controller and the outer voltage controller. The current controller takes care of the quality of current injected into the grid and the power exchange between the system and grid. The voltage controller regulates the DC link voltage and generates reference current signal for current loop. The VSI is connected to grid via an LCL filter whose parameters are designed from [9] as shown in Fig. 5.The task of the passive filter is to reduce the output current distortion.

A. Orthogonal Signal Generation

To design controller for single phase inverter a signal orthogonal to the original singe phase signal has to be created. Researchers have proposed various orthogonal signal generation methods. Among all, the all pass filter (APF) method [10] is selected for this study because it is not complex and does not attenuate the input signal. The transfer function of the APF is given by (4).

$$\frac{V_{g\beta}}{V_{g\alpha}} = \frac{\omega_f - s}{\omega_f + s} \tag{4}$$

The output of the APF, $V_{g\beta}$ is the voltage signal perpendicular to the original grid voltage and the input $V_{g\alpha}$ is voltage signal aligned with grid voltage, ω_f is the fundamental frequency. The measured grid voltage signal V_g is the input to the filter.

B. DC Link Voltage Controller

The reference dc link voltage is compared with measured voltage of the DC link capacitor and the error is fed to a PI controller, which controls the DC link voltage and generates reference current signal for the inner current control loop as shown in Fig.5. The voltage controller is of the form given by (5).

$$G_v = K_p + \frac{K_i}{s} \tag{5}$$

Where K_p and K_i are the proportional and integral gain of the controller.

C. Synchronized Reference Current Generation

The single phase grid voltage signal is split into two signals orthogonal to each other by APF method as mentioned earlier. By considering the output of the voltage control loop which is the reference current $I_{g\alpha}^{ref}$, parallel to the original voltage signal and the current reference $I_{g\beta}^{ref}$, perpendicular to the grid voltage which is a user input command a reference current synchronized with the grid voltage can be calculated by (6).

$$I_g^{ref} = \frac{I_{g\alpha}^{ref} V_{g\alpha} + I_{g\beta}^{ref} V_{g\beta}}{V_{gmag}} \tag{6}$$

Where V_{gmag} is the magnitude of the grid voltage and given by (7).

$$V_{gmag} = \sqrt{V_{g\alpha}^2 + V_{g\beta}^2} \tag{7}$$

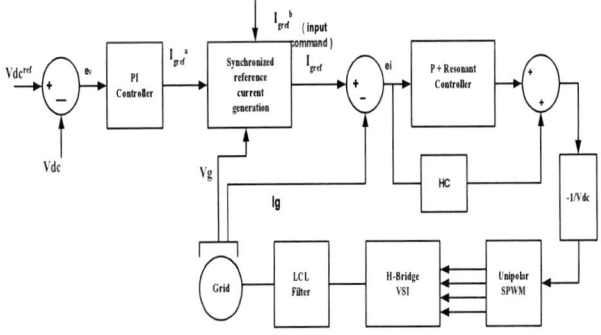

Fig.5. Block diagram for closed loop DC link voltage and grid current control

$I_{g\alpha}^{ref}$ which is in phase with the grid voltage controls the real power of the system and the orthogonal component $I_{g\beta}^{ref}$ controls the reactive power exchange of the system with the grid. Hence a decoupled control of real and reactive power can be achieved [11-12].

D. Proportional Resonant Current Controller

The input to the PR controller is the current error which is obtained by comparing the reference current and the grid current and the output is the voltage signal which is passed to the PWM modulator as shown in Fig.5. Unipolar pulse width modulation technique is implemented to control the switching of the IGBT switches of the single phase VSI. The transfer function of an ideal PR compensator is given by (8) [4-5].

$$G_{pr}(s) = K_p + \frac{2K_i s}{s^2 + \omega^2} \quad (8)$$

Where k_p and k_i are the controller gains and ω is the grid frequency. From frequency response analysis of an ideal PR compensator in Fig.6(a) it was observed that the gain is infinitely high at resonant frequency (grid frequency) and much low at other frequencies. Hence a high gain low pass filter is introduced to reduce the gain and broaden the bandwidth of the controller. The transfer function of non-ideal PR controller is mentioned by (9). From the bode diagram in Fig.6(b) it can be observed that the gain of the controller is much reduced at resonant frequency.

$$G_{npr}(s) = K_p + \frac{2K_i \omega_c s}{s^2 + 2\omega_c s + \omega^2} \quad (9)$$

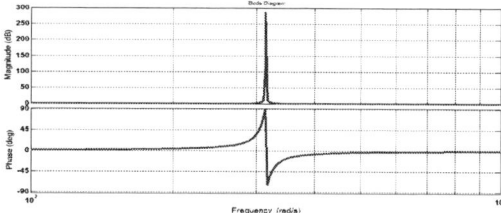

Fig. 6(a) Bode plot for ideal PR compensator

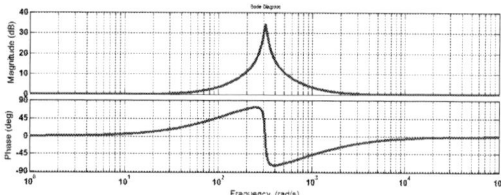

Fig.6(b) Bode plot for non-ideal PR compensator

The parameters of the PR controller are adjusted by performing frequency response analysis. A harmonic compensator is cascaded with the controller to mitigate the odd harmonics of 3rd, 5th and 7th order and to reduce the THD of grid current below 5% as per IEEE 1547 standard for interconnecting distributed generation. The transfer function of the HC is given by (10).

$$G_{hc}(s) = \sum_{n=3,5,7} \frac{2K_{in} s}{s^2 + (n\omega)^2} \quad (10)$$

Where n is the order of harmonics.

E. Stability Analysis of PR Current Controller

The block diagram of current control loop is shown in Fig. 7. The output of the current loop is given by (11) [5].

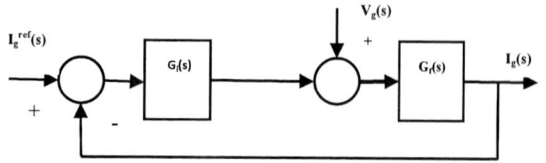

Fig.7. Block diagram of current controller

$$I_g(s) = H_i(s) I_g^{ref}(s) + H_v V_g(s) \quad (11)$$

$$H_i(s) = \frac{G_i(s)G_f(s)}{G_i(s)G_f(s) + 1} \quad (12)$$

$$H_v(s) = \frac{G_f(s)}{1 + G_i(s)G_f(s)} \quad (13)$$

Where $G_i(s)$ and $G_f(s)$ are the transfer functions of the current controller and LCL filter respectively. Where the former is given by (9) and the later is given by (14).

$$G_f = \frac{C_f R_d s + 1}{s^3 L_i L_g C_f + s^2 C_f R_d (L_i + L_g) + s(L_i + L_g)} \quad (14)$$

Where the values of L_g, L_i and C_f are given in Table II.

To obtain zero steady state error and track the reference current signal I_g^{ref} effectively the PR controller introduces very high gain hence $G_i(j\omega)$ value is infinite for ideal PR controller at fundamental frequency. The magnitude of $H_i(j\omega)$ and $H_v(j\omega)$ approaches unity and zero at the same frequency. So, the grid voltage feed forward in the current loop can be detached. The transfer function of the current controller can be given by (15).

$$I_g(s) = H_i(s) I_g^{ref}(s) \quad (15)$$

The closed loop gain of the current control loop with the PR controller is given by (16).

$$G(s) = G_i(s) G_f(s) \quad (16)$$

Where $G_i(s)$ and $G_f(s)$ are given by (9) and (14) respectively. Fig.8(a) and Fig.8(b) shows the bode plot of the current loop gain with the PR compensator and with the harmonic compensator respectively.

From the frequency response plot it can be observed that the phase margin and gain margin have large positive value, which demonstrate the closed loop stability of the system. On adding the harmonic compensator the only change that can be

978-1-4799-6047-7/14 $31.00 © 2014 IEEE

observed is appearance of gain peaks at harmonic frequencies but the dynamics of the controller, in terms of bandwidth and stability margin remains unaltered.

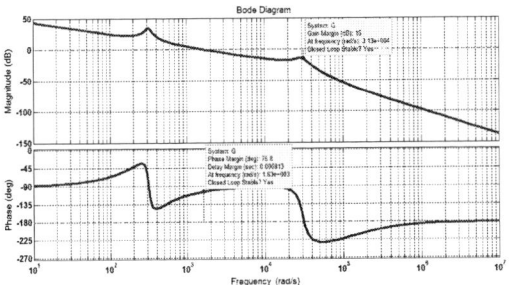

Fig. 8(a). Bode plot of compensated current loop gain

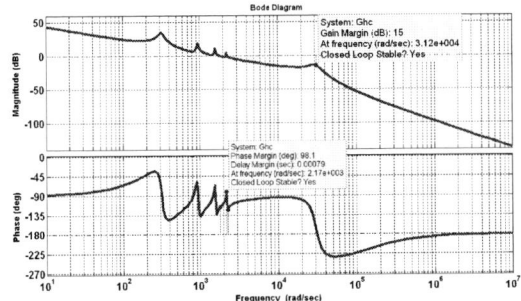

Fig. 8(b). Bode plot of compensated current controller loop gain with HC

TABLE II. Simulation parameters

Grid voltage (V_g)	230 V (rms)
DC link voltage (V_{dc})	400 V
DC link capacitor (C_{dc})	2 mF
Inverter side inductance (L_i)	12.5 mF
Grid side inductance (L_g)	0.312 mF
Filter capacitance (C_f)	3.612 µF
Filter damping resistance (R_d)	2.38 Ohms
Boost inductor (L_b)	0.5 mH
Switching frequency (f_{sw})	10 Hz

IV. SIMULATION RESULTS AND DISCUSSION

Simulation study is performed on a 2 kW single phase grid tied PV system. The simulation parameters are tabulated in Table II. The simulation results confirm the efficacy of the proposed control strategy.

A. Steady State Operation

The reactive current command is set to zero and the system is simulated for 0.5 sec the objective is to extract 2 kW of power from the PV system and inject it to grid. Fig. 9(a) gives the grid voltage waveform and current waveform superimposed on it. From which it can be observed that the voltage and current are in same phase. Fig. 9(b) shows the power waveforms which reveal that the system only delivers a constant active power and the reactive power is almost zero.

From FFT analysis waveform given in Fig.9(c) it can be observed that the THD of the grid current is 2.66% which is much below the prerequisite standard.

B. Dynamic Operation

To study the dynamic response of the system the reference reactive current command is changed from +6.14 A to -6.14 A at 0.45 s. Fig. 10(a) shows the grid voltage waveform with current waveform superimposed on it. From the figure it can be observed that when reactive power command is changed, the current starts lagging the voltage waveform. Hence the operating power factor also changes. The power waveforms given in Fig. 10(b) reveal that when there is a step change in reactive power command at 0.45 s the reactive power output changes from +1000 W to -1000 W at 0.45 s but the real power output remains constant. This demonstrates the decouple control of power.

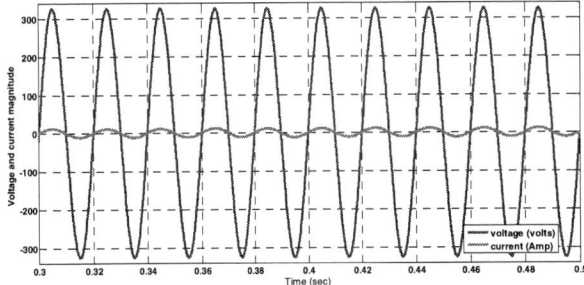

Fig.9(a). Grid voltage and current waveform

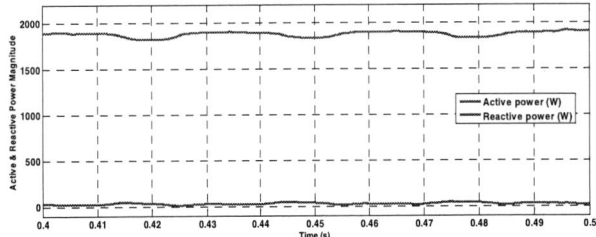

Fig. 9(b). Real and reactive power waveforms.

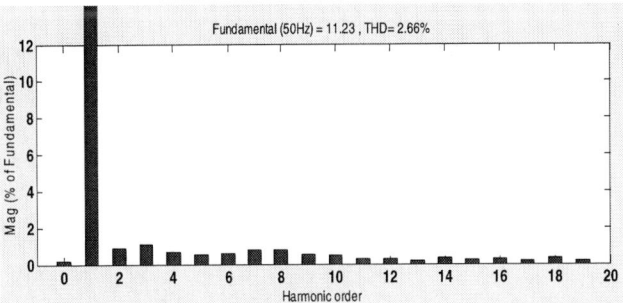

Fig. 9(c). FFT analysis waveform of grid current

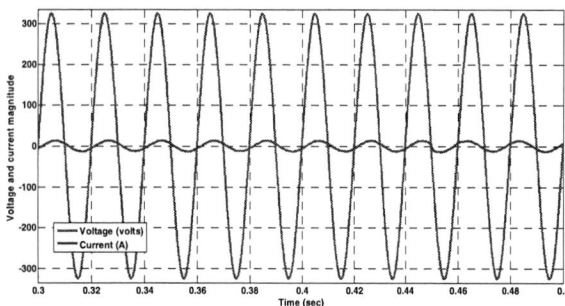

Fig. 10(a). Grid current waveform

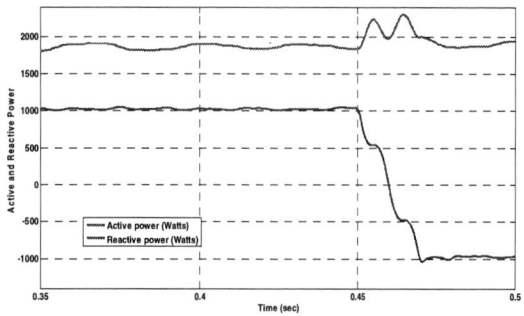

Fig. 10(b). Real and reactive power waveforms

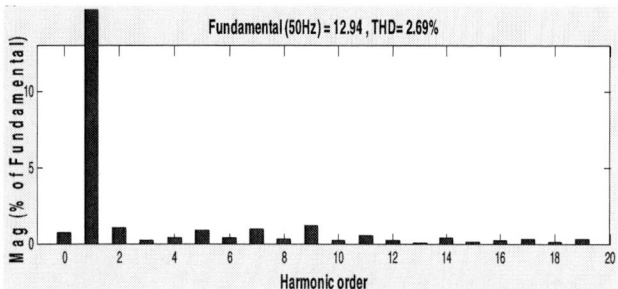

Fig. 10(c). FFT analysis waveform of grid current

Fig. 10(c) shows the FFT analysis waveform of the grid current during dynamic operation. From the waveform it can be observed that the THD is 2.69% which is below 5%.

V. CONCLUSION

This paper has proposed a stationary frame PR controller to regulate the power exchange between the PV inverter and grid. The current injected by inverter into the grid is aligned with the grid voltage. A harmonic compensator cascaded with the PR current controller reduces the THD of grid current. Stability analysis of the current controller is performed which reveals that the current controller is stable.

REFERENCES

[1] S. B. Kjaer, J. K. Pedersen, and F. Blaabjerg, "A review of single-phase grid-connected inverters for photovoltaic modules," IEEE Trans. Ind. Appl., vol. 41, no. 5, pp. 1292–1306, Sep./Oct. 2005.

[2] F. Blaabjerg, R. Teodorescu, M. Liserre, and A. V. Timbus, "Overview of control and grid synchronization for distributed power generation systems," IEEE Trans. Ind. Electron., vol. 53, no. 5, pp. 1398–1409, Oct. 2006.

[3] D. N. Zmood and D. G. Holmes, "Stationary frame current regulation of PWM inverters with zero steady-state error," IEEE Trans. Power Electron., vol. 18, no. 3, pp. 814–822, May 2003.

[4] M. Castilla, J. Miret, J. Matas," Linear Current Control Scheme With Series Resonant Harmonic Compensator for Single-Phase Grid-Connected Photovoltaic Inverters" IEEE Trans. on Ind. Electronics, vol. 55, no. 7, pp. 2724-2733, July 2008.

[5] H. Cha, T.-K. Vu, and l.-E. Kim, "Design and control of proportional resonant controller based photovoltaic power conditioning system," in Energy Conversion Congress and Exposition, ECCE, IEEE, pp. 2198 - 2205, Sept. 2009.

[6] Marcelo Gradella Villalva, Jonas Rafael Gazoli, and Ernesto Ruppert Filho, "Comprehensive Approach to Modeling and Simulation of Photovoltaic Arrays," IEEE Trans. Power Electron., Vol. 24, no. 5, pp.1198-1208, May 2009.

[7] Mohamed A. Eltawil, Zhengming Zhao, "MPPT techniques for Photovoltaic applications", Renewable and Sustainable Energy Reviews, Volume 25, Pages 793-813, September 2013.

[8] N. Mohan, T. M. Undeland, and W. P. Robbins, Power Electronics - Converters, Applications, and Design, 2nd ed: John Wiley & Sons, Inc.,1995.

[9] Hyosung Kim, Kyoung-Hwan Kim, "Filter design for grid connected PV inverters", IEEE International Conference on Sustainable Energy Technologies (ICSET2008), pp.1070-1075, 2008.

[10] R. Y. Kim, S. Y. Choi, and I. Y. Suh, "Instantaneous control of average power for grid tie inverter using single phase DQ rotating frame with all pass filter," in Proc. IEEE Annu. Conf. Ind. Electron. Soc., pp. 274–279, Nov 2004.

[11] B. Bahrani, A. Rufer, S. Kenzelmann, and L. A. C. Lopes, "Vector Control of Single-Phase Voltage-Source Converters Based on Fictive-Axis Emulation", IEEE Trans. on Ind. Electronics, vol. 47, no. 2, pp. 831-840, March/April 2011.

[12] S.Samerchur, S.Premrudeepreechacharm, Y.Kumsuwun, K.Higuchi, "Power control of single phase Voltage control Inverter for Grid Connected Photovoltaic Systems", IEEE Proceeding, 978-1-61284- 788-7/11, 2011.

An Efficient Method for D-STATCOM Placement in Radial Distribution System

Abhinav Jain ,A.R. Gupta, Ashwani Kumar
Department of Electrical Engineering
National Institute of Technology, Kurukshetra
Kurukshetra, India
jain.abhinav.999@gmail.com, arguptanitd@gmail.com, ashwa_ks@yahoo.co.in

Abstract—**This paper presents an effective method for the identification of candidate bus for DSTATCOM placement for the minimization of power losses and improvement of voltage profile in radial distribution system with load modeling. The D-STATCOM is modeled for determination of its size by assuming the voltage magnitude as 1 p.u. at the candidate bus. The validity of the method is tested on the standard IEEE 33-bus radial distribution system by performing load flow analysis after compensating the candidate bus using MATLAB software. The results obtained are compared without and with the D-STATCOM for all load models. The voltage profile and the losses reduction is obtained for IEEE 33 bus test system with the optimal placement of D-STATCOM based on the sensitivity index.**

Keywords—*D-STATCOM; Load flow analysis; Radial distribution system; Optimum location ; Optimum size.*

I. INTRODUCTION

Among the three sections of the power system which consists of Generation, transmission and distribution, distribution system has most portion of power loss and the lowest reliability due to various reasons such as low voltage levels, high flowing current, multiplicity of faults and radial structure of distribution network. Unlike transmission systems distribution systems have high R/X ratio which results in high power loss which leads to voltage instability. Series voltage regulator and shunt capacitors are the two conventional ways of maintaining voltages of the distribution system at an acceptable range [1-3]. But these devices have some disadvantages that are conventional series voltage regulators cannot generate reactive power and have quite slow response because of their step by step operations [4]. And the disadvantage with the shunt capacitors is that they cannot generate continuously variable reactive power. Another problem faced on the application of distribution capacitors is their natural oscillatory behavior when these are used in the same circuit with inductive components.

To overcome the problems faced by series voltage regulators and shunt capacitors concept of FACTS devices were introduced. FACTS devices are playing a leading role in efficiently reducing the line power losses, correcting power factor and improving voltage profile of power system. Similar ideas have been applied to the distribution system. D-

STATCOM (Distribution STATCOM) which is a shunt connected voltage source converter has been utilized to increase the reliability and efficiency of distribution systems. These devices play an important role in improving voltage profile, reducing power losses etc of distribution systems under both steady state and dynamic conditions. Now a days lots of research work is carried out for finding the optimal location of FACTS devices [5-7]. Authors in [5] proposed immune algorithm for optimal location and sizing of D-STATCOM. Authors in [6] determine the location and size of D-STATCOM by using particle swarm optimization algorithm.

Now-a-days, lot of attention is also given to load flow analysis. Teng in [8] introduced two developed matrices BIBC and BCBV and a simple matrix multiplication is applied to obtain load flow solution. Haque [9,10] proposed the analysis for both radial and mesh networks. Unlike other methods it also includes shunt admittances. Ghosh and Das [12] presented a simple method for solving distribution networks.

In this paper an effective method for improving the voltage profile and reduction of power losses is presented utilizing the presence of D-STATCOM based on the sensitivity index. Initially the voltage stability of all the buses is examined by calculating stability index values. The bus which is found to be most unstable is selected as candidate bus. The D-STATCOM is modeled to find the optimal size by assuming voltage at candidate bus as 1 p.u. and the required reactive power and voltage phase angle are calculated for compensating the candidate bus. Finally the load flow is carried out by implementing the compensating values for constant power (CP), constant current (CI), constant impedance (CZ) and finally for composite load of 30 % constant impedance, 20% constant current , 50% of constant power (ZIP) load [17] . The results have been obtained for IEEE 33 bus test system using MATLAB 7.8 version [18].

Section 2 describes the load flow method used in this paper. Section 3 explains the method to find the most unstable bus. Section 4 consists of mathematical modeling of D-STATCOM for determination of optimal size of D-STATCOM. In Section 5 proposed methodology is presented. The proposed method is tested on standard IEEE 33-bus radial distribution system and the results comparisons are discussed in section 6. Finally the conclusion is made.

II. LOAD FLOW ANALYSIS

There are various load flow methods in literatue [8-12]. The load flow method used in this paper consist of calculation of load current, formation of BIBC matrix and forward sweep across the line. The circuit of simple distribution system is shown in figure.1. The load flow analysis which is carried out in three parts is explained as follows:

A. Calculation of load current:

The load current at any bus n is given as:

$$ILn = \left(\frac{P_n + jQ_n}{V_n} \right)^* \qquad (1)$$

$$= \frac{P_n - jQ_n}{V_n^*} \qquad where \ n = 1,2,.........N$$

Where

N = total no. of buses
IL = load currents
P_n = real power load demand
Q_n = reactive power load demand
V_n = bus voltage

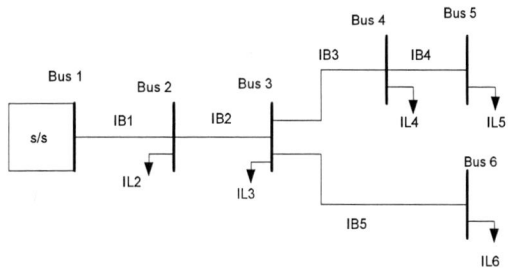

Fig. 1. Simple 6-bus distribution system

B. BIBC matrix formation:

The relationship between load currents (*IL*) and branch currents(*IB*) can be found by using simple KCL equations as follows:

$$IB1 = IL2 + IL3 + IL4 + IL5 + IL6 \qquad (2)$$
$$IB2 = IL3 + IL4 + IL5 + IL6 \qquad (3)$$
$$IB3 = IL4 + IL5 \qquad (4)$$
$$IB4 = IL5 \qquad (5)$$
$$IB5 = IL6 \qquad (6)$$

Thus the relationship between load currents and branch currents can be expressed in matrix form as shown below:

$$\begin{bmatrix} IB1 \\ IB2 \\ IB3 \\ IB4 \\ IB5 \end{bmatrix} = \begin{bmatrix} 1 & 1 & 1 & 1 & 1 \\ 0 & 1 & 1 & 1 & 1 \\ 0 & 0 & 1 & 1 & 0 \\ 0 & 0 & 0 & 1 & 0 \\ 0 & 0 & 0 & 0 & 1 \end{bmatrix} \begin{bmatrix} IL2 \\ IL3 \\ IL4 \\ IL5 \\ IL6 \end{bmatrix} \qquad (7)$$

The matrix can also be expressed as:

$$[IB] = [BIBC][IL] \qquad (8)$$

Here BIBC is load current to branch current matrix and contains only 0 and 1 values.

C. Forward sweep:

The receiving end voltages can be calculated by forward sweeping across the line by subtracting the line section drop from the sending end voltages of the line section.

$$V_n(K) = V_m(K) - IB(K) * Z_m(K) \qquad (9)$$

Where

$K = 1,2,.......Nb$
Nb = total no. of branches = $N - 1$
$V_n(K)$ = receiving end voltage of branch K
$V_m(K)$ = sending end voltage of branch K
$IB(K)$ = branch current flowing in branch K
$Z_m(K)$ = impedance of branch k

D. Power losses:

The active and reactive power losses of branch K are given as follows:

$$Active \ power \ losses = |IB(K)|^2 * real(Z_m(K)) \qquad (10)$$

$$Reactive \ power \ losses = |IB(K)|^2 * imag(Z_m(K)) \qquad (11)$$

The total active power loss for a radial distribution system is calculated by taking summation of individual active power losses for K=1,2,.....,Nb using equation (10). Similarly total reactive power loss is calculated using equation (11).

III. OPTIMAL LOCATION OF D-STATCOM

Optimal location of D-STATCOM is found by calculating the stability index of all the buses [13-15]. The bus with maximum value of stability index is selected as candidate bus. Fig.2 shows single line diagram of a two bus distribution system where V_m & V_n are sending and receiving end voltages respectively, I_m is the branch current, R_m & X_m are branch resistance and reactance respectively.

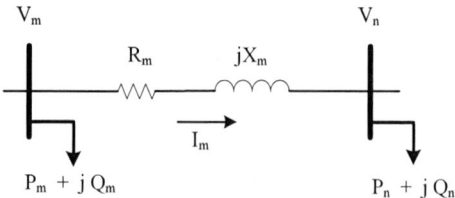

Fig. 2. single line diagram of 2-bus distribution system

The expression for stability index is derived as follows. From fig.2 the current in branch is given by:

$$I_m = \frac{V_m\angle\theta_m - V_n\angle\theta_n}{R_m + jX_m} \tag{12}$$

Here θ_m and θ_n are the phase angles of V_m and V_n respectively.

The complex power is expressed as:

$$S = VI^* \tag{13}$$

Which gives

$$I_m = \frac{P_n - jQ_n}{V_n^*} = \frac{P_n - jQ_n}{V_n\angle-\theta_n} \tag{14}$$

On equating equations (12) and (14) and cross multiplying, we get

$$V_m V_n \angle(\theta_m - \theta_n) - V_n^2 = P_n R_m + Q_n X_m - j(R_m Q_n - P_n X_m) \tag{15}$$

On equating real and imaginary parts by assuming, $(\theta_m - \theta_n) \cong 0$ as voltage phase angles are negligible in radial distribution systems, we get

$$V_m V_n - V_n^2 = P_n R_m + Q_n X_m \tag{16}$$

$$P_n X_m = R_m Q_n \Rightarrow X_m = \frac{R_m Q_n}{P_n} \tag{17}$$

On putting eqn (17) in eqn (16) and arranging, we get

$$V_n^2 - V_m V_n + \frac{R_m(P_n^2 + Q_n^2)}{P_n} = 0 \tag{18}$$

For roots of eqn (18) to be real,

$$V_m^2 - \frac{4R_m(P_n^2 + Q_n^2)}{P_n} \geq 0 \tag{19}$$

$$\Rightarrow \frac{4R_m(P_n^2 + Q_n^2)}{V_m^2 P_n} \leq 1 \tag{20}$$

So Stability Index(SI) is defined as

$$SI = \frac{4R_m(P_n^2 + Q_n^2)}{V_m^2 P_n} \tag{21}$$

Stability index is calculated for all the buses. The value of SI should be ≤ 1 for stability. The bus with highest value of SI is most unstable and is selected as candidate bus.

IV. Mathematical Modeling of D-statcom

For steady state modeling of D-STATCOM [5,16], it is installed at the bus as shown in Fig.3 . By installing D-STATCOM the voltage values at the bus where it is installed and at the neighboring buses changes. The new voltages are V_n' at the candidate bus and V_m' at the previous bus. Current changes to I_m' which is summation of I_m and I_{DS}. Here I_{DS} is the current injected by D-STATCOM and is in quadrature with the voltage. Therefore the expression for new voltage after installing D-STATCOM is given as

$$V_n'\angle\theta_n' = V_m'\angle\theta_m' - (R_m + jX_m)(I_m\angle\delta)$$
$$-(R_m + jX_m)(I_{DS}\angle(\frac{\pi}{2} + \theta_n')) \tag{22}$$

Here $\theta_n', \theta_m' \& \delta$ are the phase angles of $V_n', V_m' \& I_m$ respectively.

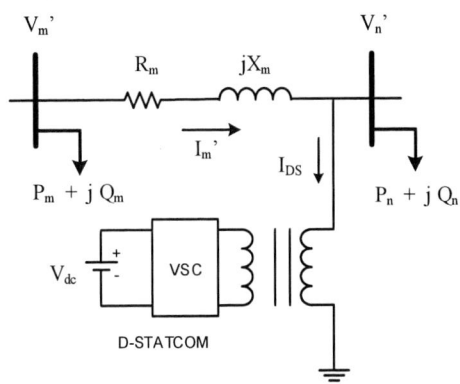

Fig. 3. Single line diagram of 2-bus distribution system with D-STATCOM

By separating real and imaginary parts of eqn (22) and manipulating the equations we get:

$$t = \frac{-B\pm\sqrt{D}}{2A} \tag{23}$$

where

$$t = \sin\theta_n' \tag{24}$$

$$A = (h_1 h_3 - h_2 h_4)^2 + (h_1 h_4 + h_2 h_3)^2 \tag{25}$$

$$B = 2(h_1 h_3 - h_2 h_4).(V_n').(h_4) \tag{26}$$

$$C = (V_n'.R_m)^2 - (h_1 h_4 + h_2 h_3)^2 \tag{27}$$

$$D = B^2 - 4AC \tag{28}$$

Where

$$h_1 = real(V_m'\angle\theta_m') - real(Z_m.I_m\angle\delta) \tag{29}$$

$$h_2 = imag(V_m'\angle\theta_m') - imag(Z_m.I_m\angle\delta) \tag{30}$$

$$h_3 = -X_m \tag{31}$$

$$h_4 = -R_m \tag{32}$$

Now there are two roots of t. For determining the correct value of root, the boundary conditions are examined as:

$$V_n' = V_n \Rightarrow I_{DS} = 0 \& \theta_n' = \theta_n$$

Results show that $t = \dfrac{-B+\sqrt{D}}{2A}$ is the desired root of the equation (23). Therefore, the phase angle and magnitude of D-STATCOM current & reactive power injected to the system by the D-STATCOM are given by the expressions:

$$\angle I_{DS} = \frac{\pi}{2} + \theta_n' = \frac{\pi}{2} + \sin^{-1} t \tag{33}$$

$$|I_{DS}| = \frac{V_n'\cos\theta_n' - h_1}{-h_4\sin\theta_n' - h_3\cos\theta_n'} \tag{34}$$

$$jQ_{DS} = (V'_n \angle \theta'_n).(I_{DS} \angle (\frac{\pi}{2} + \theta'_n))^* \qquad (35)$$

Where * denotes the complex conjugate.

V. PROPOSED METHODOLOGY

Algorithm for radial distribution system load flow:

Step 1: Read the radial distribution system line data and bus data.

Step 2: Perform the load flow as shown in section 2 to calculate voltages for all the buses and power losses for all the branches.

Step 3: Calculate stability index (S.I.) for all the buses using equation (21).

Step 4: Select the candidate bus with highest value of stability index.

Step 5: Calculate I_{DS}, voltage phase angle and injected reactive power of D-STATCOM by assuming voltage at the candidate bus where D-STATCOM is located as 1 p.u. using equations (23), (33), (34) & (35).

Step 6: Run the load flow by compensating the reactive power and voltage phase angle at the candidate bus to calculate the voltages and power losses.

Step 7: End

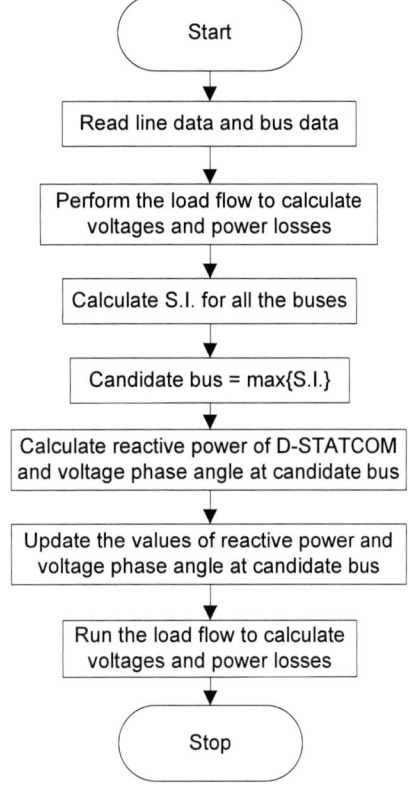

Fig. 4. Flow chart of proposed method

VI. RESULTS AND COMPARISON

The performance of this method is tested on standard IEEE 33-bus radial distribution system using MATLAB software.

The single line diagram of 12.66 KV, 10 MVA 33-bus radial distribution system is shown in the figure 5 below.

TABLE 1. STABILITY INDEX VALUES FOR BUSES

Bus number	Stability index for CP	Stability index for CI	Stability index for CZ	Stability index for ZIP
2	0.0003	0.0003	0.0003	0.0003
3	0.0013	0.0013	0.0013	0.0013
4	0.0016	0.0016	0.0016	0.0016
5	0.0007	0.0007	0.0007	0.0007
6	0.0015	0.0014	0.0014	0.0014
7	0.0013	0.0013	0.0013	0.0013
8	0.0119	0.0118	0.0118	0.0119
9	0.002	0.002	0.0019	0.002
10	0.002	0.002	0.002	0.002
11	0.0004	0.0004	0.0004	0.0004
12	0.0009	0.0009	0.0009	0.0009
13	0.0035	0.0035	0.0034	0.0035
14	0.0028	0.0028	0.0027	0.0028
15	0.0011	0.0011	0.0011	0.0011
16	0.0015	0.0015	0.0015	0.0015
17	0.0026	0.0026	0.0025	0.0026
18	0.0024	0.0024	0.0023	0.0024
19	0.0005	0.0005	0.0005	0.0005
20	0.0041	0.0041	0.0041	0.0041
21	0.0011	0.0011	0.0011	0.0011
22	0.0019	0.0019	0.0019	0.0019
23	0.0013	0.0013	0.0013	0.0013
24	0.012	0.012	0.012	0.012
25	0.0122	0.0121	0.0121	0.0122
26	0.0004	0.0004	0.0004	0.0004
27	0.0006	0.0006	0.0005	0.0006
28	0.002	0.002	0.0019	0.002
29	0.0037	0.0037	0.0036	0.0037
30	**0.0296**	**0.0292**	**0.029**	**0.0293**
31	0.0052	0.0052	0.0051	0.0052
32	0.0024	0.0023	0.0023	0.0023
33	0.0009	0.0009	0.0009	0.0009

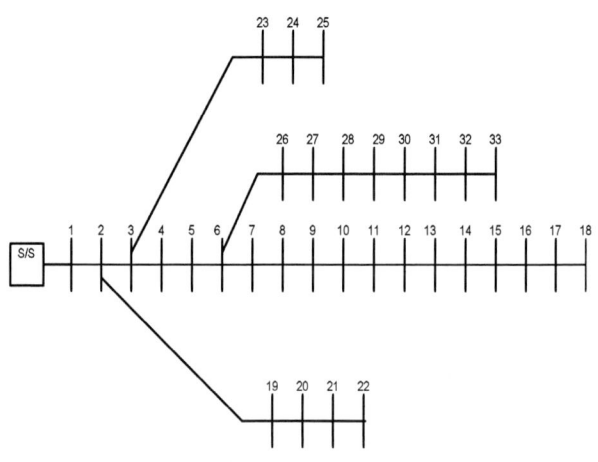

Fig. 5. IEEE 33-bus radial distribution system

Load flow analysis is performed on the 33-bus radial distribution system as described in section 2 to determine the voltages and power losses. Line and load data is taken from reference [5].

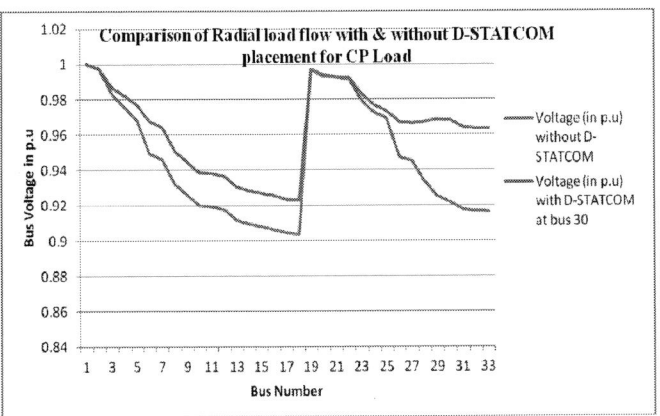

Fig. 6. Voltages with and without D-STATCOM for CP Load

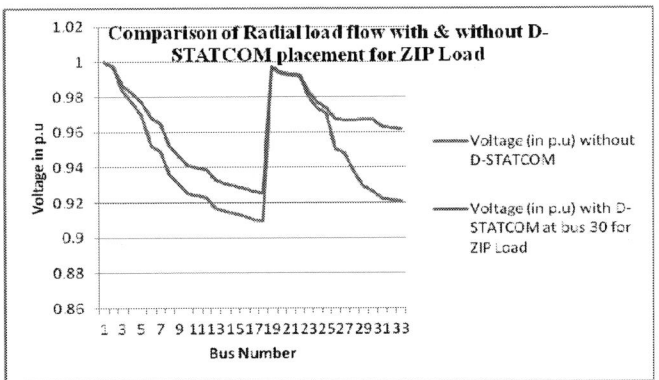

Fig. 9. Voltages with and without D-STATCOM for ZIP Load

Stability index values for all the buses are calculated as described in section 3. The corresponding results of SI are shown in table 1. The bus number 30 is having the highest value of SI , which is selected as candidate bus. D-STATCOM is modeled as shown in section 4 to calculate the voltage phase angle and injected reactive power by assuming voltage at 30th bus as 1 p.u. The load flow analysis is carried out by compensating the reactive power and voltage phase angle at the candidate bus which is 30th bus. The results are shown in fig.6-9. From these figures improvement in voltage profile can be clearly seen.

Table 2-5 shows the comparison of results with and without D-STATCOM for CP, CI, CZ and ZIP loads respectively. The size of D-STATCOM required at 30th bus is 1.993 MVAR. The upper and lower limits of voltages are taken as 5% that is $0.95 \leq V_{pu} \leq 1.05$ and the results are derived. Table 2 shows that the no. of buses having under voltage problem reduced to 10 after D-STATCOM installation at the 30th bus as compared to 21 without D-STATCOM for CP load. Total active power losses reduces by 19.52 % that is from 210.997 KW to 169.795 KW. Total reactive power losses reduced by 17.39 % that is from 143.032 KVAR to 118.148 KVAR. The minimum bus voltage increases to 0.923 as compared to 0.903 previously. Similar improvements can be seen for CI, CZ & ZIP loads from tables 3, 4 & 5 respectively.

TABLE 2. COMPARISON OF RESULTS WITH AND WITHOUT D-STATCOM PLACEMENT IN RADIAL LOAD FLOW WITH CONSTANT POWER LOAD

Description	Without D-STATCOM	With D-STATCOM
Total Active Power Load in KW	3715	3715
Total Reactive Power Load in KVAr	2300	2300
No. of buses having under/over voltage problem	21	10
Minimum voltage (p.u.)	0.9038(at 18th bus)	0.923(at 18th bus)
Optimal location of D-STATCOM	-	30th bus
Optimal Size of DSTATCOM	-	1.993 MVAR
Total Active Power Loss in KW	210.9970	169.795
Total Reactive Power Loss in KVAr	143.0320	118.1488
No. of Iterations	5	4
CPU Time in sec	0.028040	0.058832

TABLE 3. COMPARISON OF RESULTS WITH AND WITHOUT D-STATCOM PLACEMENT IN RADIAL LOAD FLOW WITH CONSTANT CURRENT LOAD

Description	Without D-STATCOM	With D-STATCOM
Total Active Power Load in KW	3536	3536
Total Reactive Power Load in KVAr	2178	2178
No. of buses having under/over voltage problem	19	10
Minimum voltage (p.u.)	0.9113(at 18th bus)	0.9264(at 18th bus)
Optimal location of D-STATCOM	-	30th bus
Optimal Size of DSTATCOM	-	1.8302MVAR
Total Active Power Loss in KW	182.4910	148.3276

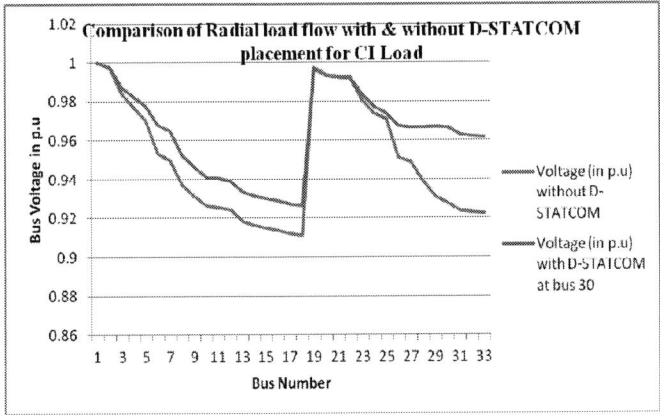

Fig. 7. Voltages with and without D-STATCOM for CI Load

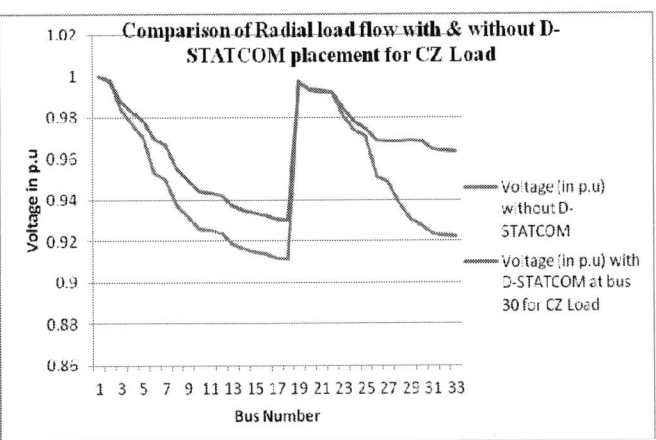

Fig. 8. Voltages with and without D-STATCOM for CZ Load

Total Reactive Power Loss in KVAr	123.3644	102.6844
No. of Iterations	4	4
CPU Time in sec	0.030858	0.064114

TABLE 4. COMPARISON OF RESULTS WITH AND WITHOUT D-STATCOM PLACEMENT IN RADIAL LOAD FLOW WITH CONSTANT IMPEDANCE LOAD

Description	Without D-STATCOM	With D-STATCOM
Total Active Power Load in KW	3389	3389
Total Reactive Power Load in KVAr	2077	2077
No. of buses having under/over voltage problem	17	10
Minimum voltage (p.u.)	0.9173 (at 18th bus)	0.9295 (at 18th bus)
Optimal location of D-STATCOM	-	30th bus
Optimal Size of DSTATCOM	-	1.7010 MVAR
Total Active Power Loss in KW	161.1981	131.9037
Total Reactive Power Loss in KVAr	108.7140	90.8923
No. of Iterations	4	4
CPU Time in sec	0.028052	0.058836

TABLE 5. COMPARISON OF RESULTS WITH AND WITHOUT D-STATCOM PLACEMENT IN RADIAL LOAD FLOW WITH ZIP LOAD

Description	Without D-STATCOM	With D-STATCOM
Total Active Power Load in KW	3573	3573
Total Reactive Power Load in KVAr	2203	2203
No. of buses having under/over voltage problem	19	10
Minimum voltage (p.u.)	0.9098 (at 18th bus)	0.9257 (at 18th bus)
Optimal location of D-STATCOM	-	30th bus
Optimal Size of DSTATCOM	-	1.8641 MVAR
Total Active Power Loss in KW	188.2386	152.5979
Total Reactive Power Loss in KVAr	127.3298	105.7613
No. of Iterations	4	4
CPU Time in sec	0.026812	0.055348

VII. CONCLUSIONS

This paper presents an effective method for the placement of D-STATCOM in radial distribution system. Voltage stability index values for all the buses are calculated to find the most suitable bus which is selected as candidate bus for placement of D-STATCOM. The D-STATCOM is modeled to calculate the voltage phase angle and injected reactive power by assuming voltage at candidate bus as 1 p.u. The load flow analysis is carried out by compensating the reactive power and voltage phase angle at the candidate bus. The result shows the improvement in voltage profile and reduction in active power losses. The D-STATCOM in distribution system can play an important role for better voltage profile and can provide effective solution of loss savings that in turn can earn profit for utilities.

REFERENCES

[1] M.T. Bishop, J.D. Foster and D.A. Down, "The application of single-phase voltage regulators on three-phase distribution systems", Rural Electric Power Conference, the 38th Annual Conference, pp. C2/1-C2/7, April 1994.

[2] Z. Gu and D.T. Rizy, "Neural networks for combined control of capacitor banks and voltage regulators in distribution systems", IEEE Transactions on Power Delivery, Vol. 11, pp. 1921-1928, October 1996.

[3] L.A. Kojovic, "Coordination of distributed generation and step voltage regulator operations for improved distribution system voltage regulation", IEEE Power Engineering Society General Meeting, pp. 232-237, June 2006.

[4] S.M. Ramsay, P.E. Cronin, R.J. Nelson, J. Bian and F.E. Menendez, "Using distribution static compensators(D-STATCOMs) to extend the capability of voltage-limited distribution feeders", Rural Electric Power Conference, the 39th Annual Conference, pp. A4/1-A4/7,April 1996.

[5] S. A. Taher, S. A. Afsari, "Optimal location and sizing of DSTATCOM in distribution systems by immune algorithm", Electrical Power and Energy Systems, vol. 60, pp. 34–44, February 2014.

[6] S. Devi, M. Geethanjali, "Optimal location and sizing determination of Distributed Generation and DSTATCOM using Particle Swarm Optimization algorithm", Electrical Power and Energy Systems, vol. 62 ,pp. 562–570, May 2014.

[7] S. M. S. Hussain, M. Subbaramiah, "An analytical approach for optimal location of D-STATCOM in radial distribution system", IEEE, 2013.

[8] J. H. Teng, "A direct approach for distribution system load flow solutions", IEEE Transactions on Power Delivery, Vol. 18, NO. 3, pp. 882-887, JULY 2003.

[9] M.H. Haque, "Efficient load flow method for distribution systems with radial or meshed configuration", IEE Proc.-Gener. Transm. Distrb. , Vol. 143, No. 1, pp. 33-38, January 1996.

[10] M.H. Haque, "A general load flow method for distribution systems", Electric Power Systems Research, vol. 54, pp. 47–54, 2000.

[11] K. Prakash, M. Sydulu, "An effective topological and primitive impedance based distribution load flow method for radial distribution systems", IEEE, DRPT, Nanjing China, pp. 1044-1049, April 2008.

[12] S. Ghosh and D. Das, "Method for load-flow solution of radial distribution networks", IEE Proc.-Gener. Transm. Distrb. , Vol. 146, No. 6, pp. 641-648, November 1999.

[13] S. Chanda, B. Das, "Identification of weak buses in a power network using novel voltage stability indicator in radial distribution System", International Conference on Power Electronics, IICPE.2011.5728121 pp.1-4, 28-30 Jan. 2011.

[14] S. Banerjee, C. K. Chanda, S. C. Konar, "Determination of the weakest branch in a radial distribution network using local Voltage Stability Indicator at the proximity of the Voltage Collapse Point", Third International Conference on Power Systems, Kharagpur, INDIA December 27-29, 2009.

[15] S. M. S. Hussain, N. Visali, "Identification of weak buses using Voltage Stability Indicator and its voltage profile improvement by using DSTATCOM in radial distribution systems", IOSR Journal of Electrical And Electronics Engineering (IOSRJEEE), Volume 2, Issue 4, pp. 17-23, Sep.-Oct. 2012.

[16] M. Hosseini, H. A. Shayanfar, M. F. Firuzabad, "Modeling of series and shunt distribution FACTS devices in distribution systems load flow", Regular paper, J. Electrical Systems 4-4:1-12, 2008.

[17] S. Shivanagaraju, N. Vishali, V. Sankar and T. Ramana, "Enhancing Voltage stability of Radial Distribution systems by Network Reconfiguration", Electrical Power Component and Systems, vol. 33, pp 539-550,August2006.

[18] The MATLAB by Mathworks corporation, SIMULINK toolbox of MATLAB version 7.8, 2009

Voltage Stability Assessment using Phasor Measurement Technology

Priti Prabhakar
Printing Technology Department
GJU S&T, Hisar

Dr. Ashwani Kumar
Electrical Engineering Department
NIT, Kuruskhetra

Abstract—The Phasor Measurement Units (PMU) installations worldwide along with the advancements in computing and communications enabled the time tagged phasor measurements of voltages and currents. These real time phasor measurements have become the backbone of wide-area monitoring, protection, and control for various applications in power systems. This paper highlights the monitoring of voltages and phase angles across different nodes that provides very useful information indicating the system state and its proximity to stability limit. The most crucial task is placement of PMUs at minimum number of buses for complete observability. Recursive spanning tree algorithm of PSAT is applied to find out the minimal placement locations for observability of all buses. The Thevenin's equivalent parameters have been obtained from the measured and estimated voltages at the load buses and impedance matrix Zbus. The parameters obtained are used to find the voltage stability boundary. Results on IEEE-14 bus system and IEEE-30 bus system are presented to illustrate the proposed approach.

Keywords— voltage collapse; phasor measurement unit; optimal placement; wide area monitoring

I. INTRODUCTION

Voltage stability is the most important aspect of today's power systems operations as several incidents of blackouts have been reported due to voltage instability. Large group of customers are left without supply for hours because the system requires longer restoration time. Maintaining adequate network voltage with increased congestion levels under restructured environment has been the major source of vulnerability.

Prior to sharp changes in voltages symptoms observed are high reactive power flows and low voltage profiles leading to uncontrollable decrease or voltage collapse (VC) [1] - [3]. The schemes that mitigate against VC need to include these symptoms to diagnose/ estimate the system state and initiate proper preventive and corrective control in time. This can only be possible by the advanced technologies like fibre base telecommunications and phasor measurement technology facilitated through LANs and WANs. The system can be monitored on line with high accuracy by using data transfer at a time resolution of 100 milli seconds. The major component of phasor technology is the Phasor Measurement Unit (PMU) that converts the analogous voltage and current signals into phasors. PMUs deliver the dymamic time tagged voltage and current phasors to the phasor data concentrators through fast wired or wireless communication links with minimum latency.

The data is converted into useful information at the system protection centre. The cost of PMU is quite high and hence optimal placement for complete observability is the most crucial task. Therefore, the primary goal is optimal and phase wise deployment of PMUs to have real time visualization of the overall system.

This paper presents the basic concepts of phasor technology and a brief overview of various approaches to find the optimal placement of PMUs. A discussion on voltage stability assessment using PMUs is also presented. The methods based on Thevenin's equivalent have some limitations even then a continuously updated equivalent provides practical and fairly accurate system model for predicting system voltage stability. Therefore, a novel concept of P-Q boundary [4] to assess voltage stability margin of load buses using Thevenin's equivalent is implemented. The recursive tree algorithm of PSAT [26] is applied to find optimal locations of PMUs. Thevenin's parameters are calculated with measured and estimated voltage phasors and line parameters. Voltage stability boundaries are drawn for the load buses to find the weakest bus close to the stability limit. The results are also compared with the PV curves of the load buses to identify the weak buses. Two test systems IEEE 14 bus and IEEE 30 bus systems are taken for the proposed study.

II. PHASOR TECHNOLOGY

The steady state supervision and security of power system is done through open loop system consisting of Remote Sensing Units (RTUs) and the conventional SCADA systems. In the present system only quasi steady states can be monitored and the transient phenomenon remains uncaptured. PMUs along with the advanced communication technology and global positioning system (GPS) enabled the recording of the time tagged transients at high data sampling rate. This is faster than real time and has the ability to create revolution in power system control in diverting the system away from transient and voltage instability.

A. Phasor Measurement Unit

In electric power networks phasors are indispensable in analysing the steady state and transient behaviour [5], [6]. The a.c. input signal is converted into its phasor representation by phasor measurement unit and provides data at rates upto 60 hertz for transmission to remote locations. A sinusoidal voltage

978-1-4799-6047-7/14 $31.00 © 2014 IEEE

or current wave can be represented by a vector called as a phasor.

A sinusoidal signal is represented by (1).

$$x(t) = X_m \sin(\omega t + \phi) \qquad (1)$$

Its phasor equivalent is then given by (2) and (3).

$$X = \frac{X_m}{\sqrt{2}}(\cos\phi + j\sin\phi) \qquad (2)$$

$$= \frac{X_m}{\sqrt{2}} \angle \phi \qquad (3)$$

Phasors are indispensable in analysing the steady stae and transient behaviour. The discrete Fourier transform (DFT) to most common technique for determining the phasor representation of an input signal. If N samples of the input X_k {k = 0, 1, ..., N} are the taken over one period, then its phasor presentation is given by

$$X = \frac{\sqrt{2}}{N}\sum_{k=0}^{N-1} X_k e^{-\frac{jk\lambda}{N}} \qquad (4)$$

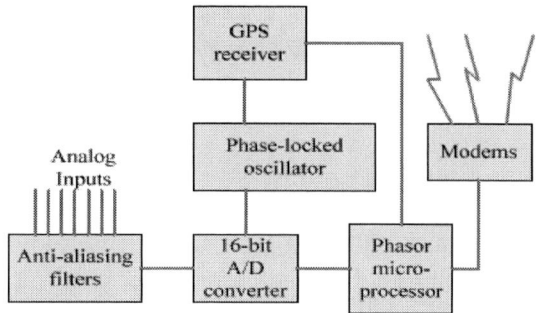

Fig. 1 Block Diagram of Phasor Measurement Unit

Fig. 1 represents the block diagram of PMU [7]. The input signals are applied to aliasing filter through the secondaries of voltage and current transformers. The function of anti aliasing filter is to limit the bandwidth of the pass band and pass these signals to 16-bit analog to digital converter.

The GPS receiver provides the one pulse per second (pps) signal and a time tag. The pps signal from GPS is divided into number of pulses per second for sampling of the analog signals by a phase oscillator. The microprocessor outputs are the positive sequence phasors with the time tags that are transmitted to a local or remote receiver at rates up to 60 samples per second.

B. Advantages

The main advantages are:

- Enhanced post disturbance evaluation capability

- Precise monitoring with improvements in state estimation
- Low latency resulting in advanced anticipation of abnormal conditions and timely control action
- Coordinated optimized control

C. Implementations

Due to low availability and high cost of communication channels post event monitoring was their only application initially but now many utilities worldwide have implemented phasor technology owing to above mentioned advantages.

- The first pilot project on a PMU-based WAMS in India was executed in the northern region in 2010-11 to install 26 PMUs and 2 phasor data concentrators in 3 stages. The National Load Despatch Centre (NRLDC) with 9 PMUs and 1 phasor data concentrator (PDC) has completed in second stage, and in the third phase another 18 PMUs, 1 PDC would be installed. Thus, there would be 53 PMUs and 6 PDCs in India after the completion of all the pilot projects.
- North American Synchro-Phasor Initiative (NASPI) at Tennessee Valley Authority, Bonneville Power Administration and more than twenty utilities
- China State Grid Dynamic Monitoring Systems
- Washington State University
- The French system EdF ("Electricite de France")
- Spain Sevillana de Electricidad (CSE)
- Brazil established the MedFassee project
- Scandinavia, Italy, Japan, Korea, Switzerland, Croatia, and Greece

The primary goal for synchronised phasor data use has been to monitor system dynamics and build a data base for the applications in post-disturbance analysis with a perspective for many control and protection schemes.

III. VOLTAGE STABILITY ASSESSMENT USING PMUs

The two approaches used for voltage stability (VS) monitoring and control are [9]:

1. Local Measurements
2. Wide Area Monitoring

Local Measurements:

Local measurements methods uses voltage and current phasors of the monitoring bus only and are based on Thevenin's impedance matching condition for stability limit. Critical buses are replaced by the network equivalent with only one line and checked independently without the need of synchronization and information exchange with other buses (or PMU locations) [10]-[14].

The Thevenin's Impedance matching approach has some limitations:

- Singularity of J that depends on maximum power limit over a set of loads can not be judged by observing a single load.

978-1-4799-6047-7/14 $31.00 © 2014 IEEE

- There is contradiction between the size of time windows for Thevenin's impedance estimation and time window to satisfy constant Thevenin's assumptions. This causes inaccuracy and difficulties in actual implementations

Wide Area Monitoring:

The dynamic monitoring of the power system is quite evident, but to convert the phasors into indices predicting the state is more complex [15], [16]. A systematic approach for monitoring, control and protection is shown in Fig. 2. The time tagged measurements from PMUs are sent to phasor data concentrators and converted into meaningful information. These devices are designed for network operation in combination with other devices and tool sets for alarm generation, displays and analysis.

IV. OPTIMAL PMU PLACEMENT(OPP)

Various placement methodologies have been proposed in literature using two approaches:

1. Coherent regions
2. Observability

Coherenency based Approach:

PQ buses exhibiting the same trends (sign) in voltage phasor variation following changes in the power injections at load or generator buses, or changes in the topology outside a given region form a coherent region. The most important buses considered are either the bus with highest short circuit capacity or the Electrical centre or the largest load bus. Therefore, the PMUs are placed at these pilot points of the coherent regions of the power system [17]. Though this approach gives satisfactory results but to form meaningful clusters and dynamic behaviour of coherent regions poses difficulties in implementation.

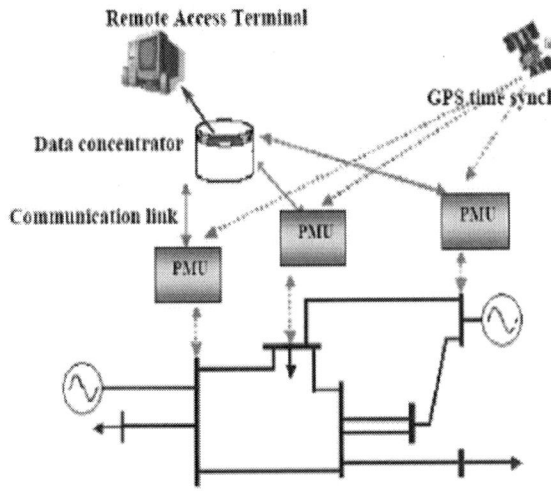

Fig. 2. Typical Power System with Phasor Measurement Units

Observability based Approach:

In this approach PMUs are placed at the minimum number of locations based on the observability theory.

Two methods used for observability are:
1. Numerical observability
2. Topological observability

Numerical observability based methods involving huge matrix manipulations may become ill conditioned in case of large power systems. Topological observability based methods utilize graph theory (Depth first search, Spanning tree based).

The Approaches reported in literature can be categorized in two groups:

1. Meta-heuristic Optimization Methods
   (Genetic Algorithm, Simulated Annealing, Tabu search, Tabu search combined with tree search, Particle Swarm Optimization, Binary search, Discrete Particle Swarm Optimization [18]-[22]
2. Deterministic Approaches
   (Integer Linear Programming)

The drawbacks of applying meta-heuristic methods provide local minima and have large execution time where as the methods based on integer linear programming are more flexible and can take into account following constraints [22] - [24]:

- Depth of unobservability (incomplete observability)
- Effects of zero injection buses
- Measurement redundancy
- Multiphase framework
- Ranking of contingencies

V. VOLTAGE STABILITY ASSESSMENT USING VOLTAGE STABILITY BOUNDARY(VSB)

The VSB in P-Q plane shows the relationship between the active power and reactive power at the point of voltage collapse [19]. The VSB can be determined very easily from a two bus equivalent of the original system from the solution of a simple polynomial. The active, reactive and apparent power margins of the load buses can then be directly determined from the VSB without successive load flow solutions.

Two bus equivalent of original power system:

The parameters of the Thevenin's equivalent are obtained by the method well documented in [19] using load flow solution considering all the loads and bus impedance matrix (Z_{bus}). The k_{th} diagonal element (Z_{kk}) of Z_{bus} represents the Thevenin's impedance of bus k. In this paper the voltages at that bus are obtained from the PMU measurements and Zbus from the line parameters of network. The Thevenin's parameters for the kth bus are then obtained from the following relations:

$$Z_{th} = (Z_{kk}Z_L)/ (Z_L - Z_{kk}) \qquad (5)$$

$$E = ((Z_L + Z_{th}\ )/\ Z_L)V_L \qquad (6)$$

Where; Z_{th} : Thevenin's equivalent impedance

E: Thevenin's equivalent voltage

V_L: k^{th} bus voltage

Z_{kk}: k^{th} diagonal element of Z_{bus}

Here Z_L and V_L are load impedance and load voltage of the kth load bus. These two parameters can be very easily obtained from the PMUs installed at the load bus.

VI. SIMULATIONS AND RESULTS

The proposed method of determining the VSB is applied to IEEE-14 bus IEEE-30 bus system. The bus impedance matrix is obtained using Zbus building algorithm and MATLAB program [27]. The PMUs are placed on the three buses (bus numbers: 2, 6 and 9) using PMU placement tool of PSAT (Power System Analysis Tool)[26], with recursive N security constrained spanning tree algorithm[25] as shown in table 1.

TABLE 1

OPTIMAL LOCATIONS OF PMUS

Test systems	Number of PMUs	PMU locations
IEEE 14 bus	3	2, 6, 9
IEEE 30 bus	7	1, 5, 10, 12, 19, 24, 30

Fig. 3 and 4 shows the spanning tree presentation with PMU placements for IEEE 14 and IEEE 30 bus systems.

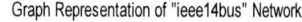

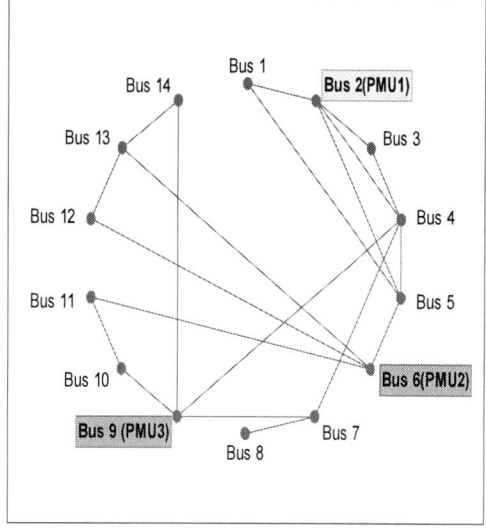

Fig.3. Graph presentation of IEEE_14 bus system

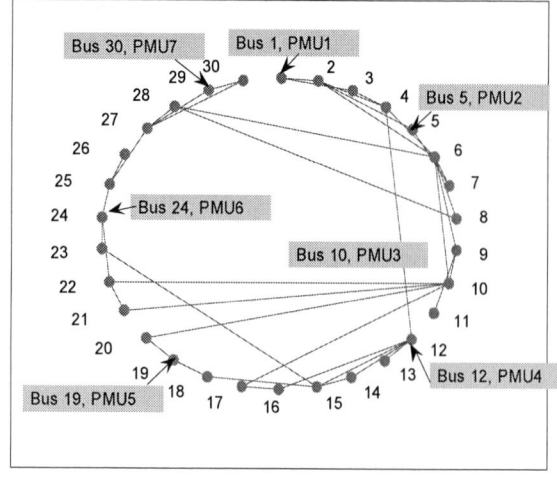

Fig.4. Graph presentation of IEEE_30 bus system

The linear state estimations of voltages from PMU measurement are obtained and Thevenin's parameters are calculated. The VSB from these parameters are drawn as shown in Fig 5 and 6. From the VSBs of load buses, it is observed that the bus number 14 and bus 30 of IEEE 14 bus and IEEE 30 bus respectively are the weakest buses. These buses have minimum distances from its normal operating point to their VSBs.

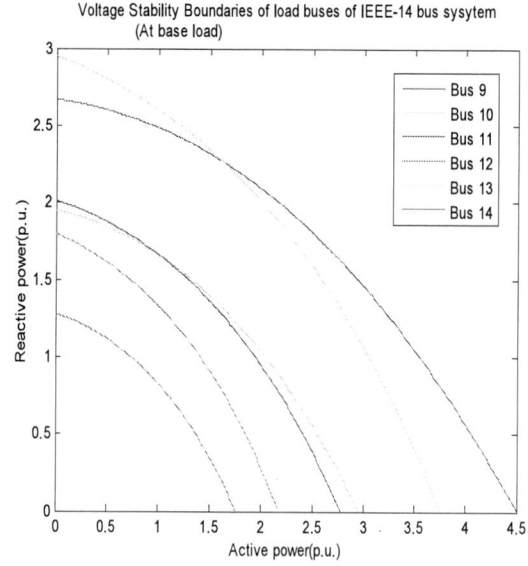

Fig. 5. Voltage stability boundaries at load buses of IEEE_14 bus system

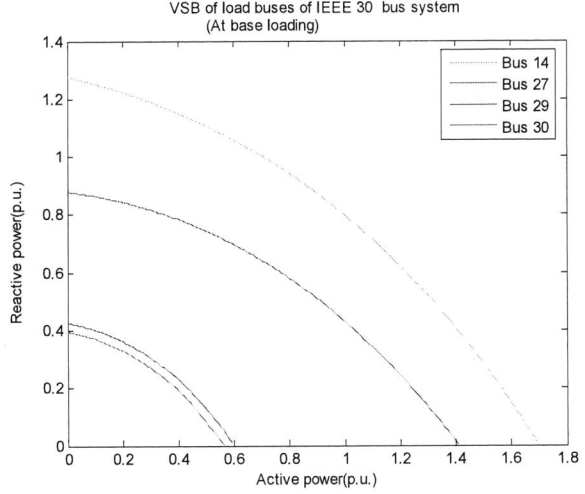

Fig.6. Voltage stability boundaries at week buses of IEEE_30 bus system

PV curves are also drawn at weak load buses using continuation power flow (Fig. 7 and 8) for the comparison. It is observed that bus 14 of IEEE 14 bus system has the most drooping voltage profile as compared to other buses before the nose point. Similarly bus 30 of IEEE 30 bus system is found to be the weakest bus.

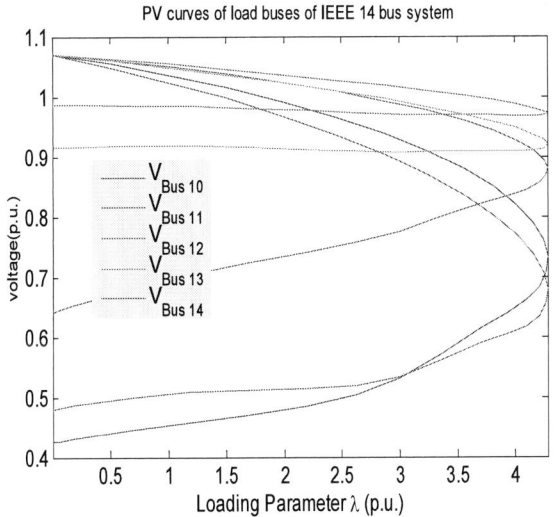

Fig.7. PV_Curves of Weak Buses of IEEE_14 Bus Systems

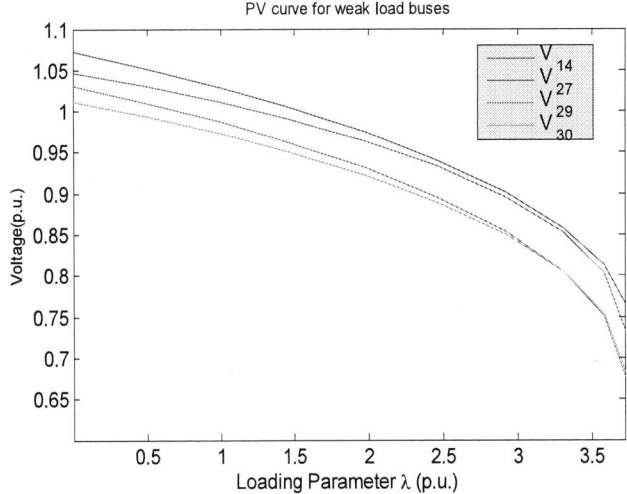

Fig.8. PV_Curves of Weak Buses Of IEEE_30 Bus System

VII. CONCLUSIONS

In this paper, a phasor measurement driven voltage stability monitoring is proposed. With the high data sampling rate transient events can be captured and more accurate dynamic state can be monitored. It eliminates the need of detailed power system modeling and hence does not suffer from inaccuracy in modeling parameters. Therefore, phasor measurements along with line parameters have been used for finding the voltage stability limit of load buses. Optimal PMU locations are found using the recursive N security constrained tree spanning algorithm for two systems. Voltage stability boundaries are drawn for various load buses of IEEE-14 bus and IEEE 30 bus systems using the measured and estimated values of voltages at the buses. It is observed that bus number 14 and 30 of the respective systems are the most critical as these are nearest to their stability limits. Also PV curves are drawn at weak load buses using continuation power flow for comparison. It is observed that buses 14 and 30 of IEEE-14 bus and IEEE 30 bus systems have the weakest voltage profiles.

REFERENCES

[1] P. Kundur, Power System Stability and Control , New York : McGraw-Hill, 1994.

[2] T. V. Cutsem, and C. Vournas, Voltage Stability of Electric Power Systems, Norwell, MA : Kluwer, 1998

[3] A.K Panchal , "Voltage Stability of a Large Grid", Proc. of 7th International Power Engg Conference , pp 1 - 550, Nov. 29 -Dec. 2 2005.

[4] M. H. Haque, "Novel method of assessing voltage stability of a power system using Stability boundary in P-Q plane", Electric PowerSystems Research, vol. 64, pp. 35-40, 2003.

978-1-4799-6047-7/14 $31.00 © 2014 IEEE

[5] A. G. Phadke, "Synchronized phasor measurements in power systems", Proc. IEEE Computer Applications Power Conf., vol. 6, pp.10-15, 1993.

[6] J. D. l. Ree, V. Centeno, J. S. Thorp, and A. G. Phadke, "Synchronized phasor measurement applications in power systems", IEEE Trans. Power Grid, vol. 1, no. 1,pp. 20-27, Jun. 2010.

[7] N. H. El-Amary, Y. G. Mostafa, M. M. Mansour, S. F. Mekhamer, and M. A. L. Badr, "Phasor measurement units allocation using discrete particle swrm for voltage stability monitoring", Proc. IEEE PowerTech. Conf., Bucharest, pp. 1-6, June 28-July 2, 2009.

[8] S. R. Samantaray, "Letter to the Editor: Smart Grid Initiatives in India", Electric Power Components and Systems, vol. 42(3–4), pp. 262–266, Feb 2014

[9] A. M. Chebbo, M. R. Irving, and M. J. H. Sterling, "Voltage collapse proximity indicator: Behavior and implications", IEE Proc. Generation Transmission Distribution, vol. 139, pp. 241-252, 1992.

[10] F. Gubina, and B. Strmcnik, "A simple approach to Voltage stability assessment in radial networks", IEEE Trans. Power System, vol. 12, no. 2, pp. 1121-1128, May 1997.

[11] K. Vu, and M. M. Begovic, "Use of local measurements to estimate voltage stability margin", IEEE Trans. Power System, vol. 14, no.3, pp. 1029-1035, August 1999.

[12] D. E. Julian, R. P. Schultz, K. T. Vu, W. H. Quaintance, N. B. Bhatt, and D. Novosal,"Quantifying proximity to voltage collapse using the voltage instability predictor (VIP)", IEEE Power Eng. Soc., vol. 2, pp. 931-936, July 2000..

[13] Y.Wan , W. Li and J. Lu, "A new node voltage stability index based on local phasors", Electric Power Systems Research ,Volume 79, Issue 1, pp. 265 – 271, January 2009.

[14] B. Genet, J. C. Maun, "Voltage-stability monitoring using wide area measurement systems", Power Tech., Lausanne, pp. 1712-1717, July 1-5, 2007.

[15] S. Corsi, "Wide area voltage regulation & protection", Proc. IEEE Power Tech. Conf., Bucharest, pp. 1-7, June 28-July 2, 2009.

[16] M. Glavic, and T. V. Cutsem, "Wide – area detection of voltage instability from synchronized phasor measurements. Part11: Simulation results", IEEE Trans. Power System, vol. 24, no.3, pp. 1408-1416, August 2009.

[17] R. Sodhi, S. C. Srivastava, and S. N. Singh, "A Simple scheme for wide area detection of impending voltage instability", IEEE Transactions on Smart Grid, Vol. 3, no. 2, pp. 818-827, June 2012.

[18] L. Mili, T. Baldwin, and R. Adapa, "Phasor measurement placement for voltage stability analysis of power systems", Proc. IEEE 29th Conf. on Decision and Control, Honolulu, pp. 3033-3038, 1990.

[19] A. B. Antonio, and R. A. TorreBo, "Meter placement for power system state estimation using simulated annealing," Proc. of the IEEE Porto Power Tech Conference, , pp. 10-13, Sep. 2001.

[20] H. Mori, and Y. Sone, "Tabu search based meter placement for topological observability in power system state estimation," Proc. of the IEEE *Transmission and Distribution Conf.*, vol. 1, , pp. 172-177, 1999.

[21] B. Milosevic, and M. Begovic, "Nondominated sorting genetic algorithm for optimal phasor measurement placement," IEEE Transactions on. Power Systems, vol. 18, no. 1, pp. 69–75, Feb. 2003.

[22] R. F. Nuqui, and A. G. Phadke, "Phasor measurement unit placement techniques for complete and incomplete observability," IEEE Trans on Power Delivery, vol. 20, no. 4, pp. 2381- 2388, Oct. 2005.

[23] S. Chakrabarti, and E. Kyriakides, "Optimal placement of phasor measurement units for power system observability," IEEE Transaction. on Power Systems, vol. 23, no. 3, pp. 1433-1440, Aug. 2008.

[24] R. Sodhi, S C Srivastava and S N Singh, "Optimal PMU placement to ensure system observability under contingencies", Proc. of IEEE PES General Meeting, July 2009.

[25] G. B. Denegri et al, "A security oriented approach to PMU positioning for advanced monitoring of a transmission grid," Proceedings of Power System Technology, PowerCon, 2002, vol. 2, pp. 798-803.

[26] F. Milano, Power System Analysis Toolbox (PSAT), 2005. [Online]. Available: http://www.power.uwaterloo.ca/˜fmilano/.

[27] H. Saadat, Power System Analysis, New York : Tata McGraw-Hill,2002.

Application of SSSC for Compensation Assessment of Interconnected Power System

Shashi Gandhar
Research scholar
NIT, Kurushetra
India
shashi.abhi@gmail.com

Jyoti Ohri
Professor
NIT, Kurushetra
India
ohrijyoti@rediffmail.com

Mukhtiar Singh
Associate Professor
DTU, New Delhi
India
smukhtiar_79@yahoo.co.in

Abstract—Present paper describes the application and study of SSSC (Static Series Synchronous Compensator), a flexible Alternating Current Transmission System (FACTS) Controller for providing dynamic compensation to an interconnected power system. A combination of conventional sources 'diesel generating system' and a non-conventional energy source 'wind turbine generator system' has been used to design the power system. A SSSC is inserted in the middle of this combination and thus acting as a shunt connected static compensator therefore not affecting active power significantly but improving reactive power profiles directly. In this paper time Simulation studies investigate the effect of SSSC on the reactive power profiles of different generators and it also consider for stabilizing voltages of different loads. Simulations are carried out in Sim Power System toolbox of MATLAB software.

Keywords—FACTS; SSSC; MATLAB

I. INTRODUCTION

In India Power systems are probably the largest machines designed by power engineers. Existing power systems consists of more than hundreds generators, thousands lines and substations, and millions of consumers. Still there are many locations which are deprived of electric power because of economic and other factors .Electric power can be provided to these areas either by commissioning of new power plants at those locations or by extension of existing power system and transmission networks but both requires a huge amount for design and establishment . A very considerable amount is also required for the running expenses like fuel cost. Best solution to these increasing demands and making electricity cost effective is to design distributed generation system using nonconventional resources like wind, hydro etc. Presently, the wind power emerges as prominent renewable energy resource to play an important role in the modern power systems. Wind power is providing a solution to these power crisis problems to a large extent, but the wind power system has associated with some technical problems, one of the problems is that the fluctuations in the wind energy resources are very frequent and it can't be neglected. This affects the quality of supply. Voltage is the main factor which determines the quality of supply. The variations in voltage clearly produce the mismatches in the reactive power demand of system. This situation has resulted in an increased possibility of voltage instability which becomes a worry concern of many utilities especially in planning and operation [1-3].

Voltage variations may be the cause of system voltage collapse which can paralyze the system from which they are unable to recover. In result of voltage collapse a partial or full power interruption may occur in the system. The only way to save the system from voltage collapse is to reduce the reactive power load or add additional reactive power prior to reaching the point of voltage collapse. Introducing the sources of reactive power, i.e., shunt capacitors and/or Flexible AC Transmission System (FACTS) controllers at the appropriate location is the most effective way for utilities to improve voltage stability of the system [4-5]. Further these power electronics devices can be used for increasing the system efficiency and stability of the system.

The recent development and use of FACTS controllers in power transmission system lead to many applications in power systems. These are not only capable of improving the reactive power profiles of the existing power network resources but also to provide operating flexibility to the power system. Many FACTS devices, such as STATCOM (static compensator), SVC (static VAR compensator), SSSC (static synchronous series compensator) and UPFC (Unified Power Flow Controller) can be inserted in series or shunt, or a combination of the two, to control the different parameters of power system FACTS devices have been defined by the IEEE as "alternating current transmission system incorporating power electronic-based and other static controllers to enhance controllability and increase power transfer capability" [6-8].

This paper presents an application and study for compensation assessment of a hybrid /interconnected system with SSSC is presented with the use of MATLAB software package. Here test system comprises of a balanced 4-Bus system connected with a combination of six wind turbine generators and a diesel generator system. A SSSC is inserted in between these sources and a brief introduction of the SSSC is presented in section II. In section III test system is designed and analyzed in Sim Power System toolbox of MATLAB. Simulation Test results are clearly presented in section IV. This paper also discusses the voltage stabilization capability of SSSC.

978-1-4799-6047-7/14 $31.00 © 2014 IEEE

II. BASIC CONFIGURATION OF SSSC

The SSSC is a static series synchronous generator and very similar to a series compensator. On the contrary, this Controller has the same structure as that of a STATCOM. The only difference between them is the coupling transformer of an SSSC is connected in series with the transmission line while in case of STATCOM; it is connected in parallel with line. It can be termed as a synchronous voltage source (SVS) as it can inject a variable and controllable sinusoidal voltage in series with a transmission line [9-10]. Its output voltage is in quadrature with the line current, which is independent of line current. It can be varied according to requirement of the system. Thus the objective of the controller is to increase or decrease the total reactive voltage drop across the line and thereby can vary the transmitted power. A small part of the injected voltage that is in phase with the line current provides the losses in the inverter. The basic configuration of a SSSC is shown in Fig. 1 [10, 11, and 12].

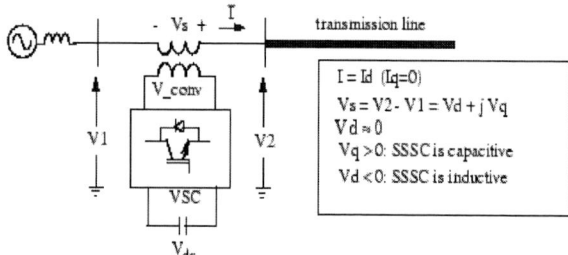

Fig. 1. The Single line diagram Of SSSC

Presented control system of SSSC has a phase-locked loop (PLL) which synchronizes on the positive-sequence component of the current I. The output of the PLL (angle =ωt) is utilized to get the d-axis and q-axis components of the 3-Θ voltages and currents (labeled as V_d, V_q or I_d, I_q). Measurement systems measuring the q components of AC positive-sequence of voltages V_1 and V_2 (V_{1q} and V_{2q}) as well as the DC voltage V_{dc}. AC and DC voltage regulators which compute the two components of the converter voltage V_{d_conv} and V_{q_conv}) required to obtain the desired DC voltage (V_{dcref}) and the injected voltage (V_{qref}). The V_q voltage regulator is assisted by a feed forward type regulator which predicts the V_{conv} voltage from the Id current measurement.

$$P = \frac{V_s V_r}{x_L} \sin(\delta_s - \delta_r) = \frac{v^2}{x_L} \sin \delta \quad (1)$$

$$Q = \frac{V_s V_r}{x_L}[1 - \cos(\delta_s - \delta_r)] = \frac{v^2}{x_L}(1 - \cos \delta) \quad (2)$$

$$\delta = (\delta_s - \delta_r) \quad (3)$$

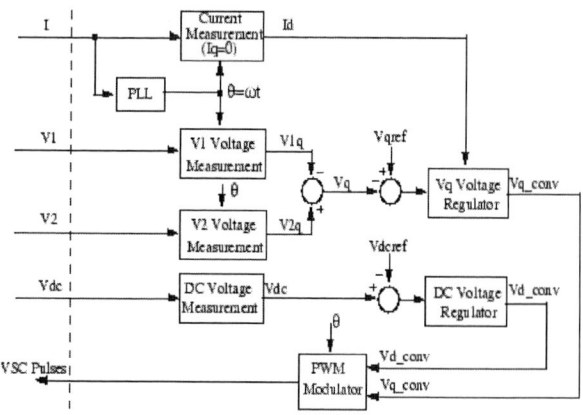

Fig. 2. control system block diagram of SSSC [13]

$$|V_s| = |V_r| = |V| \quad (4)$$

$$P_q = \frac{v^2}{x_{eff}} \sin \delta = \frac{v^2}{x_L \left[1 - \frac{x_q}{x_L}\right]} \quad (5)$$

$$Q_q = \frac{v^2}{x_{eff}}[1 - \cos \delta] = \frac{v^2}{x_L \left[1 - \frac{x_q}{x_L}\right]}[1 - \cos \delta] \quad (6)$$

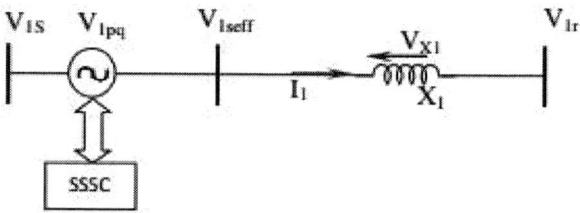

Fig. 3. Effect of injected voltage

III. MODEL OF TEST SYSTEM

A 4-bus system is considered for the application and study of dynamic compensation capabilities of SSSC [13] is shown in figure 3. Detailed test system is redesigned in the sim power system toolbox of MATLAB software for the analysis and time simulations are carried out in this very effective power system software .It is connected with a six Wind turbine generators and with a synchronous generator. This combination is representing a wind power system and a diesel power system respectively. The data for the system is given in Appendix. In this paper, the investigations are executed on the assumption that the wind turbine is operating at rated speed (9 m/s). The six Induction Generators of wind power systems of rating 1.5 MVA nominal power and connected at bus 'WT'.

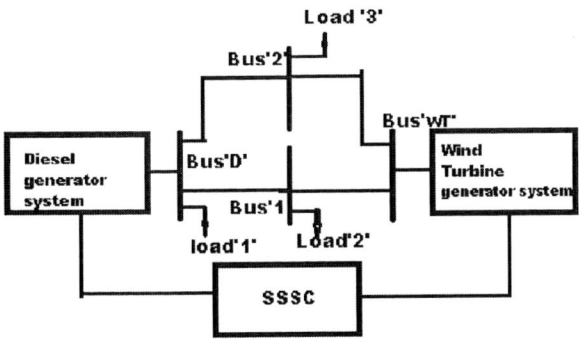

Fig. 4. Block Diagram of Test System [14]

The diesel generator is having nominal power of 3.0 MVA and connected at bus 'D'. Three different loads L-1(Resistive only), L-2 (R-L in series), L-3(R-L in series) are connected in between these two buses Therefore whole test system is divided in two parts/sections i.e. Wind induction generator section and Diesel Synchronous Generator section. The FACT controller SSSC of rating 1 MVA has been implemented in between. Thus the SSSC Controller is relieving both types of systems in a very effective manner. Waveforms presented in next section clearly indicating the reactive power control capability of SSSC and also it is stabilizing the voltage profiles of all three connected loads.

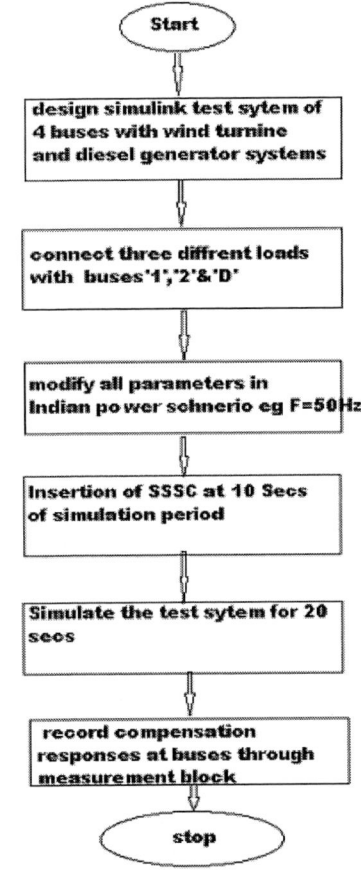

Fig. 5. The Flow line diagram Of Methodology

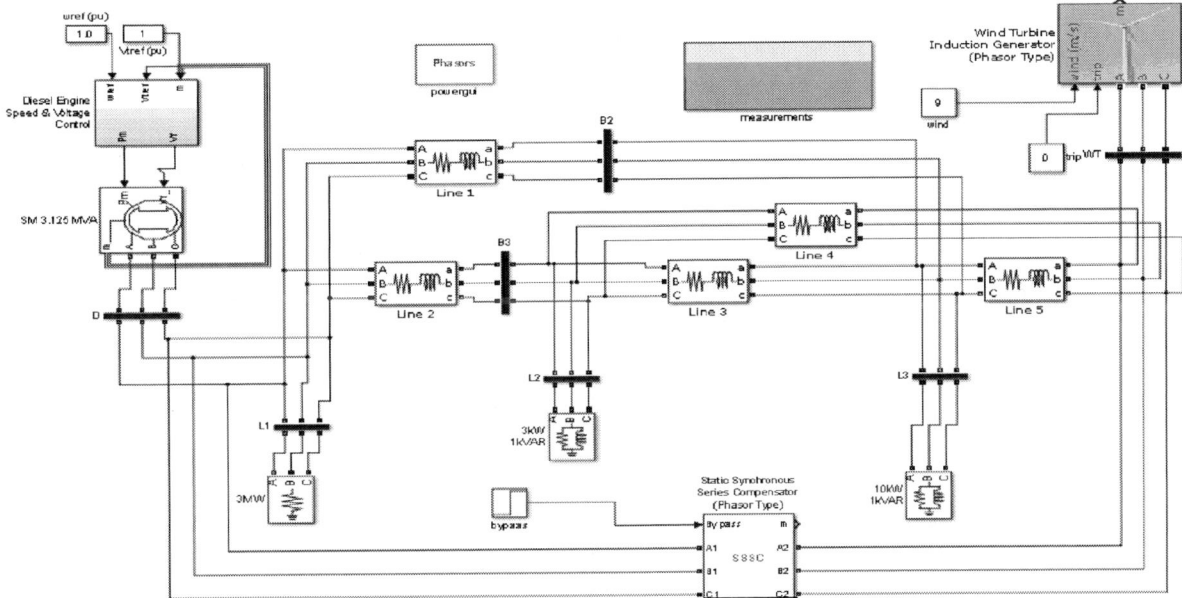

Fig. 6. Simulated Test system with SSSC

978-1-4799-6047-7/14 $31.00 © 2014 IEEE   79

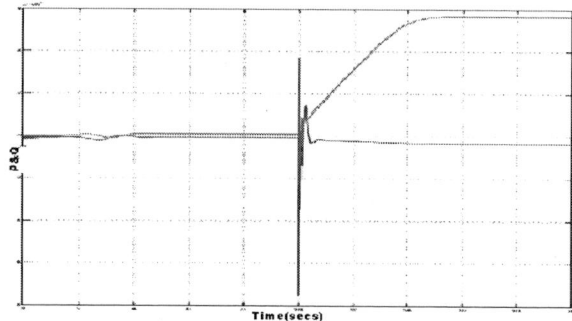

Fig. 7. Reactive Power (green) and Active Power (blue) profiles at bus 'W'

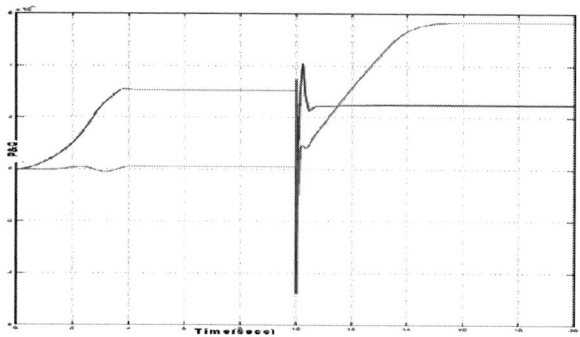

Fig. 8. Reactive Power (green) and Active Power (blue) profiles at bus 'd'

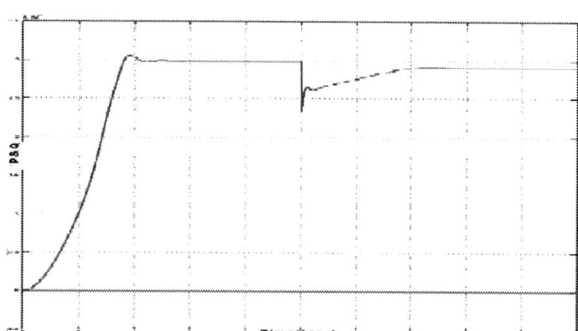

Fig. 9. Reactive Power (green) and Active Power (blue) profiles of load '1'

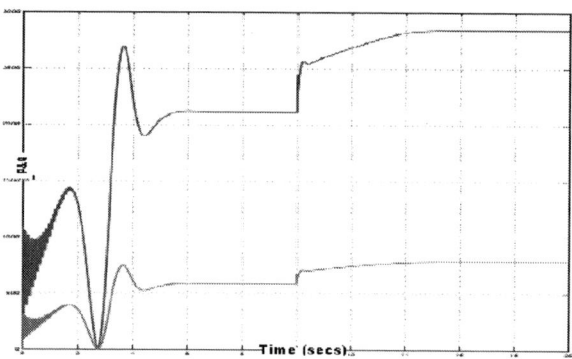

Fig. 10. Reactive Power (green) and Active Power(blue) profiles of load '2'

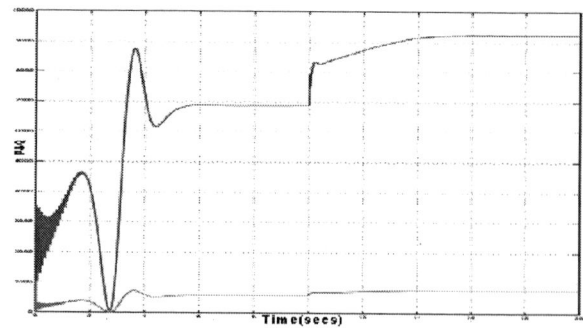

Fig. 11. Reactive Power (green) and Active Power(blue) profiles of load '3'

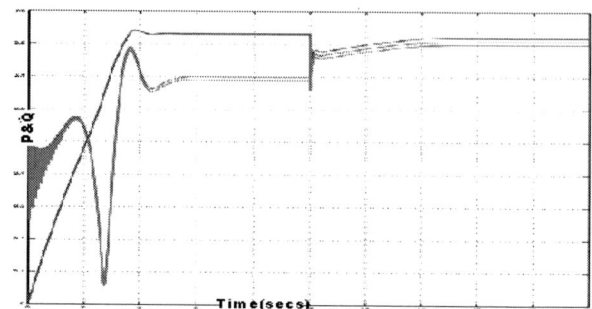

Fig. 12. Voltage profiles of three loads

IV. SIMULATION RESULTS

The Active power (blue curves) and reactive power (green curves) profiles at different buses are presented in this section where wind power system is connected with bus 'wt' and diesel system is connected with bus'd'. The active and reactive profiles of three loads L1, L2 and L3 are shown in figures 7-9 .Here in the test system the SSSC has been inserted into system at 10 seconds where total simulation time is 20 seconds and all result variations with the SSSC controller are clearly presented in this section. A significant amount of reactive power supplied by the SSSC at buses 'd'and'wt' is shown in figure 4 & 5. Reactive power demand of loads L2 and L3 is also fulfilled by SSSC as the increment in the reactive power profile after the insertion of SSSC at time 10 sec shown in figures 7-9 .Here L1 does not require any type of compensation as this is purely resistive load of 10 KW It is fact that the reactive power demand varies continuously with the variation in wind speed. So connecting fixed capacitors for reactive power compensation will not yield better performance. Then, the dynamic VAR requirement at these terminals that has to be supplied by SSSC is selected as the difference between full-load and no-load compensation. Thus a SSSC has been employed in between two types of power system. The voltage profiles of three loads '1, 2&3' are also presented. The SSSC is inserted at '10' sec and with in no time. It stabilizes the voltage profiles of these loads shown in figure '9' (load'1'with blue curve, load'2'with red curve,

978-1-4799-6047-7/14 $31.00 © 2014 IEEE

load '3' with green curve) .Thus SSSC is playing important role for controlling powers of given test system and to be proved as one of most important and powerful FACT controller of the family.

V. CONCLUSION

In this paper, the improvement in the Voltage profiles of three connected loads is presented. It is also concluded that dynamic compensation with '1 MVA' SSSC connected with designed 4-bus system with hybrid combination of wind and diesel power systems. It is clearly presented that SSSC helps in regulation of the active and reactive power for the considered system. The result reveals that in such cases, dynamic VAR compensation is required and it can be achieved by using the FACTS controllers like SSSC.

Appendix
Load '1'=10KW, 400 V, 50 Hz
Load '2'=3KW, 1KVAR, 400 V, 50 Hz
Load '3'=3MW, 1KVAR, 400 V, 50 Hz

Wind Turbine generator data:
Six generators of 1.5 MVA, 400 V, 50 Hz
Stator resistance=0.004843 p.u,
Rotor resistance= 0.004377 p.u
Stator leakage reactance=0.01248 p.u,
Rotor leakage reactance= 0.0179 p.u
Magnetizing reactance=6.77 p.u
Inertia constant= 5.04s, friction Factor=0.01 p.u
Pole pairs=03
Base wind speed=9m/s
Nominal mechanical o/p power= 9KW,
Pitch angle regulator Gains,K_p = 5,K_i =25,
Max Pitch angle=45°

Synchronous generator data :(Salient Pole Type):
P $_{nom}$=3MVA, V=400V,50Hz
stator resistance=0.0036Pu
X_{d_i}' =0.296 , X_{d_i}'' =0.177, X_q =1.04, X_q'' =0.177 ,
T_d' =3.7 Sec, T_d'' =0.05 Sec, T_{qo} =0.05 Sec
Inertia constant= 1.07s, friction Factor=0
Pole pairs=02
Nominal mechanical o/p power= 9KW,
Pitch angle regulator Gains,K_p = 5,K_i =25,
Max Pitch angle=45°

SSSC data:
Series controller data: 1MVA,400V,50Hz,
Max Injected Voltage = 0.1 p.u
Rse: 0.16 p.u, Xse: 0.016 p.u
Injected voltage regulator gains
K_{pse}: 0.03, K_{ise}: 1.5

REFERENCES

[1] T. Aekcrmann, "Distributed generation: a definition," Electrical Power Systems Research, vol. 57, 2001, pp. 195-204.

[2] Al-Majed, S.I. and T. Fujigaki,"Wind power generation: An overview," Proceedings of the International Symposium: Modern Electric Power Systems (MEPS), 2010 Sept. 20-22, IEEE Xplore Press, Wroclaw, Poland, pp: 1-6.

[3] R,C.Bansal,T.S.Bhatti,and D.P.Kothari,"Automatic Reactive Power Control Of Wind-Diesel-Micro Hydro Autonomous Hybrid Power Systems Using ANN Tuned Static Var Compensator," Proc.Int.Conf.on Large engineering System conference on Power Engineering(LESCOPE),Montreal, Canada,May 7-9,2003, pp182-188.

[4] Sandeep Gupta, R. K. Tripathi, and Rishabh Dev Shukla,"Voltage Stability Improvement in Power Systems using Facts Controllers: State-of-the-Art Review," IEEE Trans. On Power Systems, 2010

[5] A. Sode-Yome, N. Mithulananthan, Kwang Y. Lee,"A Comprehensive Comparison of FACTS Devices for Enhancing Static Voltage Stability," 1-4244-1298-6/07, 2007, IEEE.

[6] John J. Paserba, Fellow, IEEE, "How FACTS Controllers Benefit AC Transmission Systems," IEEE Transactions on Power Delivery,Vol.1 , No 1 pages 949-957,2003.

[7] N.G. Hingorani, L. Gyugyi,"Understanding FACTS, Concepts and Technology of Flexible AC Transmission systems," IEEE Press.

[8] W. Breuer, D. Povh, D. Retzmann, E. Teltsch,"Role of HVDC and FACTS in Future Power Systems,"Shanghai Power Conference, Shanghai CEPSI- 2004.

[9] Eskandar Gholipour, Shahrokh Saadate,"Improving of Transient Stability of Power Systems Using SSSC," IEEE Transactions On Power Delivery, Vol. 20, No. 2, April 2005.

[10] R. Natesan, G. Radman,"Effects of STATCOM, SSSC and SSSC on Voltage Stability," IEEE Spectrum 2004.

[11] Abhishek Gandhar, Balwinder Singh , Rintu Khanna,"Impacts of FACTS technology-A state of Art Review," International Journal of Innovative Technology and Exploring Engineering (IJITEE),ISSN: 2278-3075, Vol.1, No.4, pp. 28-31,Sept. 2012.

[12] L. Gyugyi, C. D. Schauder and K. K. Sen,"Static synchronous series compensator: a solid-state approach to the series compensation of transmission lines," IEEE Trans. Power Del., Vol. 12, No. 1, pp. 406-417, 1997

[13] Deepak Divan,Harjeet Johal, "Distributed FACTS-A New Concept For Realizing Grid Power Flow Control," IEEE Trans. on Power Electronics , Vol. 22, No.6, Nov 2007.

[14] Sidhartha Panda, N. P. Padhy , "A PSO-based SSSC Controller for Improvement of Transient Stability Performance",International Journal of Electrical, Robotics, Electronics and Communications Engineering Vol.1 No.9, 2007.

Simultaneous Coordination of Power System Stabilizer and UPFC for Improving Dynamic Stability of Multimachine System

Chintu Rza Makkar
Department of Electrical Engineering
D.A.V.Institute of Engg & Technology
Jalandhar, India
chinturza78@yahoo.co.in

Dr. Lillie Dewan
Department of Electrical Engineering
National Institute of Technology
Kurukshetra, India
l_dewan@nitkkr.ac.in

Abstract—**The coordination of FACTS devices and stabilizers can be utilized for better dynamic stability of the power system. In this paper, the effectiveness of UPFC and Power System Stabilizer in simultaneous coordination is investigated to manage the low frequency oscillations in a single machine and multimachine power system. The proposed controller of PSS and POD of UPFC are tuned to overcome disturbances caused due to faults in the power system. The proposed system is studied for single machine and multi-machine system and found effective for enhancing the dynamic stability of system. The simulations have been carried out in time domain frame.**

Keywords— Unified Power Flow Controllers (UPFC; Stability; Damping Control; Power System Stabilizers (PSS).

I. INTRODUCTION

Transient stability is a main restrictive factor for loading a transmission line as with the enlarged loading of large transmission line after disturbances, the difficulty of power transmission increases [1]. The introduction of "Flexible AC Transmission System" (FACTS) controllers/devices into the power systems network facilitated to direct the transmission line power flow, damping of low frequency oscillations and transient stability improvement [2]. The effectiveness of various FACTS devices/controllers such as static var compensator (SVC), Unified Power Flow Controller (UPFC), fixed-capacitor thyristor-controlled reactor (FC-TCR) type etc was investigated by many researchers as an efficient and cost-effective means to damp the power system oscillations. Depending upon the applications, appropriate FACTS devices were used for improving the steady state stability and dynamic stability [3-5]. So the most capable power system network controllers used for bulk power transmission are FACTS devices. Work related to applications of UPFC to improve the small signal stability of single machine and multi-machine system has been reported by many authors [6 &7]. Further the various control modes such as constant voltage injection mode, power flow mode of the series Voltage Source Controller of UPFC and their effect to lower the initial transient swings was reported by Huang et al [8]. It was observed that the constant injected voltage control mode was found much effective when compared with the power flow control mode with a contingency [9]. Low frequency and small magnitude oscillations often remained for a extensive time in power systems. These types of oscillations were controlled by controlling the excitation signal of automatic voltage regulator of the generator. Generally, Low frequency oscillations lead to the various problems in interconnected power systems [10]. The most economical damping controller is the power system stabilizer (PSS) and is extensively applied to damp the electromechanical oscillations [11]. The tuning and location of PSS were found to be important parameters to suppress low frequency oscillations [12]. Due to the use of regulators in Voltage Source Converter, impact of the different signals to UPFC POD control may alter. Hence certain methodologies had also developed by authors to select appropriate input signals of damping controller [13]. The basic principle of operation of UPFC was discussed by Gyugi and Hingorani et.al [14]. The choice of various parameters of the controllers was discussed and the improvement of power system stability with PSS and UPFC based on Lyapunov methods was presented by Robak et.al [15]. The Power System Stabilizer was represented through block diagram by Rogers [16].

In this paper, simultaneous coordination of UPFC and PSS for enhancement of dynamic stability of a single machine and multi-machine system is investigated. An attempt has made to damp low frequency system oscillations by tuning the parameters in a coordinated set of UPFC damping controller and PSS.

This paper is organized in five sections; Section-I gives the Introduction, Section–II presents the basic operating principle of UPFC. In Section–III PSS and UPFC Power Oscillation Damping Controller and their tuning have been discussed. The simulation results and Conclusion are given in Section IV & V respectively.

II. BASIC OPERATING PRINCIPLE OF UPFC

978-1-4799-6047-7/14 $31.00 © 2014 IEEE

A simplified Schematic diagram of Unified Power Flow Controller installed in the power system network is shown in Fig. 1. The Unified Power Flow Controller consists of a shunt and series part and each part has its own shunt and series transformer, power-electronic based converter with turn-off capable semiconductor devices which are connected through the dc link. The function of inverter 1 is to provide the active power required by inverter 2 via the common dc link capacitor as shown. Inverter 2 is linked in series through series transformer with the transmission line. The Inverter 2 controls the reactive and real line flows in the line by the variation of the phase angle and magnitude of the added voltage generated by inverter 2. The voltage source inverter 1 can also absorb or generate reactive power [13 &14].

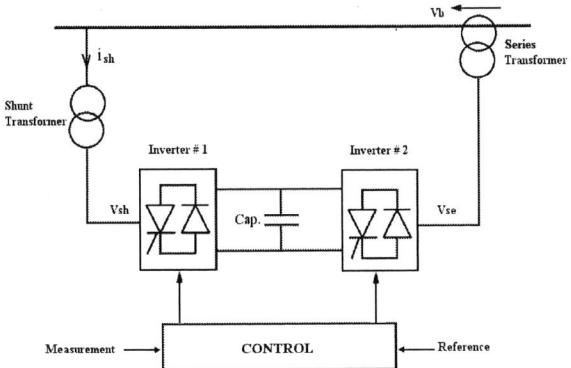

Fig. 1. UPFC and its control in a transmission line.

III. DAMPING CONTROLLERS

A. Power System Stabilizer

In order to regulate the damping caused by disturbances in power system, Power system stabilizer (PSS) adds a modulated signal to a reference input of automatic voltage regulator of generator. Figure 2 presents the structure of Power System Stabilizer and AVR (Automatic Voltage Regulator) [14]. The various inputs required for PSS are accelerating power, generator shaft speed, or terminal bus frequency.

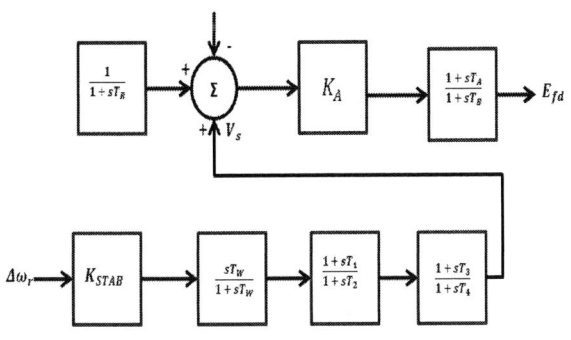

Fig 2. Block diagram of the Power System Stabilizer and Automatic Voltage Regulator.

To provide required damping of oscillations, the PSS must generate a part of electrical torque which is in proportion to the rotor speed variation. It comprises of a wash out block, an amplification block and lead-lag compensator blocks as shown in block diagram in Fig. 3[15&16].

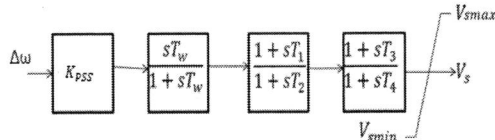

Fig. 3. Block Diagram representation of Power System Stabilizer

B. UPFC Power Oscillation Damping (POD) Controller

A POD controller is developed using swing equations to damp the oscillations caused by the disturbances in the power system. The output of this controller is Reference Real Power which is compared and fed as input to the UPFC. The configuration of Power Oscillation Damping Controller is shown in Fig 4 which is much similar to PSS controller.

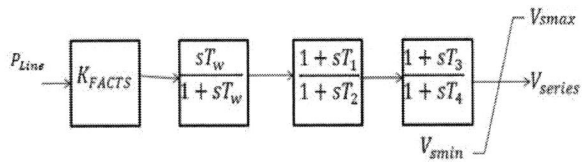

Fig. 4. UPFC POD Controller

C. Effect of simultaneous coordination of Power System Stability and Unified Power Flow Controller on Power System Stability

The Effects of simultaneous coordination of Power System Stabilizer and UPFC on the improvement of the Power System Stability is discussed in section IV. With the addition of input into the excitation system, PSS control loop helps to recover the damping of transients. The optimal designed PSS and UPFC POD controllers are found effecting/ improving damping in the extensive series of operating conditions. To see the effects on stability, a three phase short circuit fault is simulated for several cycles at the receiving bus to initiate the power system oscillations. To attain the effectiveness of the coordinated tuning, the gains of the controller require to be optimized with various operating conditions [16].

IV. NON-LINEAR SIMULATION RESULTS

A. Power Oscillation Damping in SMIB Systems

A test system of single machine with infinite bus system (SMIB) is presented in Fig. 5. The system has been designed in MATLAB/ SIMULINK to analyze the effect of the control algorithm of the UPFC and PSS on the transient stability. The SMIB consists of a synchronous generator of 1200 MVA

978-1-4799-6047-7/14 $31.00 © 2014 IEEE   83

rating connected to a transmission system through 500 kV transformers of rating 1000 MVA and 800MVA. The transmission system consists of 230 kV double circuit transmission line of length 65 km and 500kV two single transmission lines of 50 km each. Synchronous generator comprises an excitation system, speed regulator, a power system stabilizer (PSS) etc.

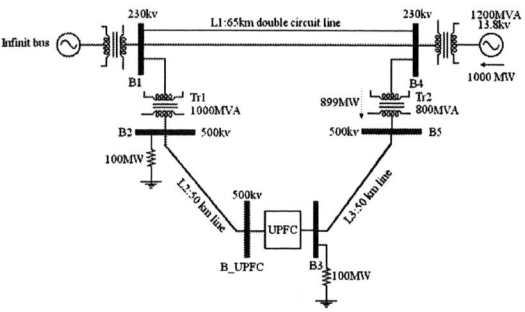

Fig. 5 Single machine Infinite bus system installed with PSS and UPFC

The system is simulated with a three-phase short circuit fault at the generator end of one transmission line. The fault instant for all simulations is considered between 5.0 s to 5.2s i.e. the fault duration is 0.2 s. The SMIB System shown in fig. 5 is investigated for variation in rotor angle difference and real power flow. The system is simulated with and without coordinated tuning of PSS and UPFC.

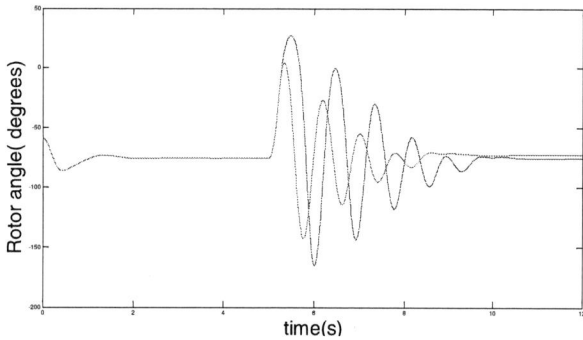

Fig. 6. Variations in Rotor Angle Difference
------ System behavior without coordinated tuning
—— System behavior with coordinated tuning

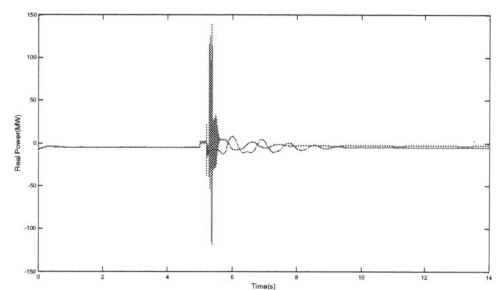

Fig7. Variations in Real Power flow (MW)
------ System behavior without coordinated tuning

------ System behavior with coordinated tuning

With the coordinated tuning, rotor angle difference variations are found to be reduced effectively as shown in fig.6. The operation of system with coordinated tuning of controllers shows that real power flow is increased as well as the oscillations are damped more quickly as shown in Fig.7. The parameters of PSS tuning are optimized and results in damping of oscillations. The parameters used for simulation are:

$T_1 = 0.8$ s, $T_2 = 0.08$ s, $T_3 = 0.2$ s, $T_4 = 0.5$s, $K_{PSS} = 2$

B. Power Oscillation damping in Multimachine Systems

The schematic diagram of multi-machine system which consists of a weak tie connecting two similar areas is shown in Fig 8 [10]. The rating of coupled units of two areas has 20 kV, 900 MVA each. The parameters of the line (pu) are as follows:
R = 0.0001 pu / km
B_C = 0.00175 pu/km
X_L = 0.001 pu/km

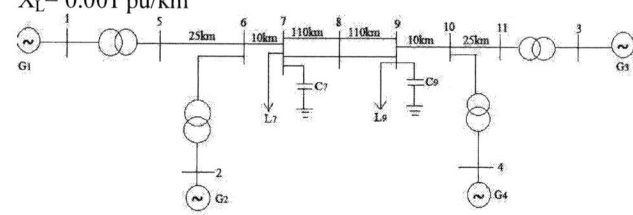

Fig. 8. Two area system model

Transformers (step-up) used for simulation are of 900MVA rating having voltage ratio 20kV/230 kV as shown in Fig.8. The impedance of each transformer is 0+j0.15 pu and has an off-nominal ratio of 1.0. The transmission line lengths are shown in Fig.8. Rated voltage of transmission line is 230 kV. The time duration of three-phase short-circuit fault is 0.2 sec and taken from the instant 5s and 5.2s. The non linear simulation results are plotted in Fig.9

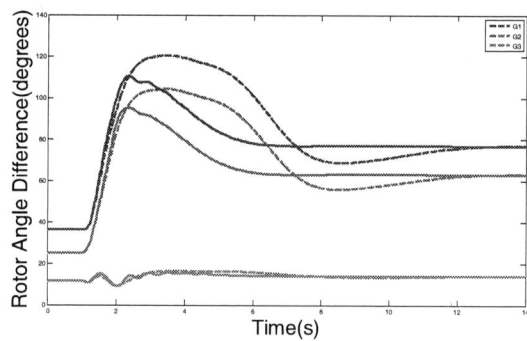

Fig. 9. Non-linear simulation results
--- System behavior without coordinated tuning
—— System behavior with coordinated tuning

The results are plotted for variation in rotor angle difference in case of multi-machine system i.e. for three generators w. r. t. to fourth generator which is considered as reference generator. Dotted lines show the variations in rotor angle difference without coordinated tuning of PSS and UPFC controllers for three generators i.e. G_1, G_2 and G_3 where as Solid lines show the variations with coordinated tuning of PSS and UPFC. It is clearly observed from Fig.9 that oscillations are damped out quickly with coordinated tuning in case of G_1 and G_2 where as no change is observed in G_3.

V. CONCLUSION

In this paper the effect of coordination of UPFC and Power System Stabilizer has been studied using MATLAB simulation realizing multi-machine system. The simulations are performed in MATLAB/ SIMULINK on a Single Machine Infinite bus System and Multi-machine System taking into account the effect of UPFC with three phase fault in the network. The duration of the fault is kept 200 ms. It is found that with the simultaneous use of POD controller of UPFC and PSS in a network with the tuned values of kp & ki, the best results are obtained in terms of rotor angle difference with a three phase fault in system. The designed controller is able to control oscillations effectively in the single and multimachine system.

REFERENCES

[1] R. Mihalic and P. Zunko, D. Povh, "Improvement Of Transient Stability Using Unified Power Flow Controller", IEEE Transactions on Power Delivery, vol. 11, No. 1, January 1996

[2] Guo, M. L. Crow, and Jagannathan Sarangapani, "An Improved UPFC Control for Oscillation Damping", IEEE Transactions on Power Systems, vol. 24, No. 1, 2009.

[3] Hingorani N G, , "Flexible ac transmission", IEEE Spectrum, April, 40-45,1993.

[4] M. Noroozian, L. Angquist, M. Ghandhari and G. Andersson, "Use of UPFC for optimal power control", IEEE Trans. on Power Delivery vol. 12, No. 4, pp. 1629-1634, 1997.

[5] M. Ghandhahi, G.Adersson, I.A.Hiskens, "Control Lyapunov functions for series devices", IEEE Transactions on Power Delivery, vol. 16, No.4, pp 689-694, 2005.

[6] E. Gholipour, S.Saasate, "Improving of Transient Stability of Power Systems Using UPFC", IEEE Trans. On Power Delivery, vol. 20, No.2, pp. 1677-1682, 2005.

[7] P.Kumkratug, M.H.Haque, "Versatile model of a unified power flow controller in simple system", IEE Proceedings.-Generation, Transmission,Distribution, vol.150, No.2, pp. 155-161, 2003.

[8] Z.Y.Huang, Y.Ni, C.M.Shen, F.F.Wu, S. Chen, B.Zhang, "Application of Unified Power Flow Controller in Interconnected Power Systems-Modeling, Interface, Control Strategy, and Case Study", IEEE Transactions on Power Systems, vol.15, pp.817-824, May 2000.

[9] Xia Jiang, Joe H. Chow, Abdel-Aty Edris, Bruce Fardanesh, Edvina Uzunovic, "Dynamic Control Modes of Unified Power Flow Controllers for Transmission Reinforcement", IEEE 2008

[10] Kundur P., Balu N.J.,Lauby M.G., Power System Stability and Control, 1994.

[11] Larsen E. Swann D.,Applying Power System Stabilizers, Part I.

[12] Nelson Martins, Leonardo T.G.L, "Determination of Suitable locations for Power Systems Stabilizers and Static Var Compensators for damping electromechanical oscillations in large Scale Power Systems", IEEE Transactions on Power Systems, Vol. 5, No.4, November 1990.

[13] L.Gyugi, "Unified power flow control concept for flexible AC transmission systems", Proc. Inst. Electrical Engg C, vol.139, No.4, pp 323-331, July 1992.

[14] N.G.Hingorani, L.Gyugi, Understanding FACTS. IEEE Press, 2000.

[15] S.Robak, M.Januszwski, D.D.Rasolomampionona, "Power system stability enhancement using PSS and UPFC Lyapunov based controllers: A comparative study", Power Tech Conference, Bologna, IEEE, 2003.

[16] G.Rogers, Power System Oscillations, Kluver Academic Publishers,1999.

APPENDIX

Ka -AVR gain

Tw –time constant of wash out block

Ta -time constant of AVR

T_1, T_2, T_3, T_4-time constants of phase compensator

K_{STAB}, K_{PSS}, K_{FACTS} –Stabilizer, PSS and Facts POD gains

Cap- dc link capacitance

978-1-4799-6047-7/14 $31.00 © 2014 IEEE

Mitigation of Reactive Power Requirements of Large Scale Wind Power Generation in Western Rajasthan Using Static Var Compensator

A. K. Pathak
Ph.D Scholar Poornima University
Jaipur India
pathak287@gmail.com

Dr. M. P. Sharma
Assistant Engineer (RRVPNL)
Jaipur, India
mahavir_sh@rediffmail

Abstract-Shunt capacitor banks are installed by developers pooling stations for management of reactive power for a wind farm. Operation of shunt capacitor banks is manually and accordingly there is a wide variation in wind power plant power factor of 0.9 lagging to 0.9 leading. Due to large variations in wind generation power factor, reactive power flows on transmission lines are also variable and accordingly there is a wide variation in power transmission system voltage from minimum 0.8 p.u. to a maximum of 1.20 p.u. Due to low & high power system voltages, transmission lines are tripped resulting constrained in a wind power evacuation. Considering the high penetration of wind power in the system, in this paper, Static Var Compensator (SVC) is proposed at 400kV GSS Jaisalmerfor large scale wind power penetrated power system, for mitigation of wind power generators reactive power. Simulation studies have been carried out to validate the effectiveness of the SVC for voltage control with the variation in wind power generation power factor. Case studies are carried out in 19-bus system in Western Rajasthan area where wind power penetration is even more than 2000MW to demonstrate the performance of the SVC taking the consolidated effect of voltage behavior with and without SVC at a different power factor. Wind power, penetrated part of Rajasthan power system has been modeled using Mi.-Power system analysis software. Results of tests conducted on the model system in various possible field conditions are presented and discussed.

Keywords—static var compensator; fixed speed induction generator; fixed capacitor; thyristor controlled reactor; thyristor switched capacitor; wind power

I. INTRODUCTION

Conventional generation is reliable nowadays, but there are lots of disadvantages and negative environmental impact is predominant amongst all which limit future development. Several renewable energy resources are introduced to replace conventional generation to a certain extent. Wind energy is one of the feasible choices, presently which has led to the great expansion of wind power generation during the past decade. The worldwide installed capacity reached from 59GW in 2005, to 318GW in end 2013. Wind power is important distributed renewable energy resources, however, voltage stability issues with wind farms may require to be augmented with reactive power compensation devices the increasing penetration of wind energy, into the traditional power grid, with its characteristics challenge to the operation and dispatching of power in the grid. The conventional power system provides of high intermittent and uncertainty brought a

huge necessary reactive power and voltage support to the Grid as compared to wind power generation system.

The continuous increase of the wind power penetration level is likely to influence the operation of the existing utility networks, especially the power system voltage stability. In general, there is Fixed-Capacitors (FC) mounted on the terminal of Fixed Speed Induction Generator (FSIG) to provide reactive power support [1], [2]. FC can support the needed reactive power under normal circumstance, but the terminal voltage will fluctuate when there are gust wind turbulences. For reactive power compensation Static Var Compensator is a dynamicreactivepowercompensator whose reactive power output depends upon the system voltage. With the help of Static Var Compensator (SVC) devices [3], [5] it is possible to regulate power system voltage with the variability of wind power generation power factor. This paper studied the impact of voltage stability when SVC was connected to wind farm[4].

In Rajasthan, most of wind power plants are concentrated in the Western part, which is far away from load centers (Fig.1). Presently, more than total 2000 MW capacity wind power plants are installed in Jaisalmer area. For evacuation of wind power generation from Jaisalmer to load centers, 2 nos. 400 kV lines, 3 nos. 220 kV lines & 2 nos. 132 kV lines have been constructed. Shunt capacitor banks are installed at developers pooling sub-stations for mitigation of wind power generators reactive power. Operation of shunt capacitor banks is manually, therefore, there is a wide variation in wind power plant power factor from minimum 0.9 lagging to 0.9 leading. Most of windpower plants are installed far away from load centers. Wind Generators do not provide the required VAR support. Due to large variations in wind generation power factor, reactive power flows on transmission lines are also variable and accordingly, there is a wide variation in system voltage from 0.8p.u.to 1.2p.u. Due to low & high power system voltages, transmission lines are tripped resulting constrained in a wind power evacuation.

Over-voltage causes over-fluxing in transformers resulting in tripping of transformers. Due to over voltages, transmission lines also tripped. Static Var compensators (SVCs) have played important roles in voltage support and stability improvement of power systems[6], [7].

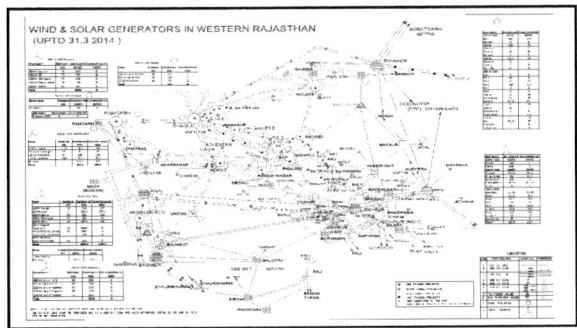

Fig.1 Renewable & Nonrenewable Power Generation & Transmission in Western Rajasthan

SVC characteristic is discussed in section II, section III deals with simulation study. Detailed results are analyzed in section IV by discussing the bus voltages in steady state & fault condition and transmission losses with and without SVC and conclusion is given in section V.

II. SVC AND POWER SYSTEM V-I CHARACTERISTICS

SVC composed of a Thyrister Controlled Reactor and three of Thyrister Switched Capacitor which can control reactive capacitive compensation [8]. The schematic diagram of SVC is given in Fig.2

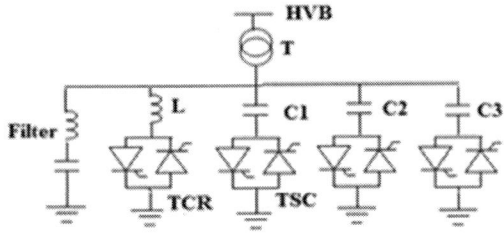

Fig.2 SVC Schematic

The susceptance of SVC with conduction angle β is given by
$$B_{SVC} = (2\beta - \text{Sin}2\beta - \pi\omega^2 LC)/\pi\omega L \quad (1)$$
Where $\beta = (\pi - \alpha)$ indicating firing angle. The reactive power compensation for the system is expressed $Q_{SVC} = B_{SVC} * V^2$
The composite characteristic of SVC is derived by adding the individual characteristics of the components. SVC is defined by the slope reactance when the controlled voltage is within the control range. SVC, V-I characteristic is drawn in Fig-3.

The system characteristic may be expressed as
$$V = Eth - Xth\ Is \quad (2)$$
Where
V = Power system bus voltage, Is = Bus load current
Eth = Source voltage, Xth = System Thevenin reactance.

For inductive load current I_s is positive and for capacitive load current I_s is negative.

The SVC characteristic may be expressed as
$$V = V0 + XSL\ Is \quad (3)$$
Where
V = Power system bus voltage, Is = SVC current

Vo = SVC reference voltage where net SVC current is zero
XSL = SVC slope reactance

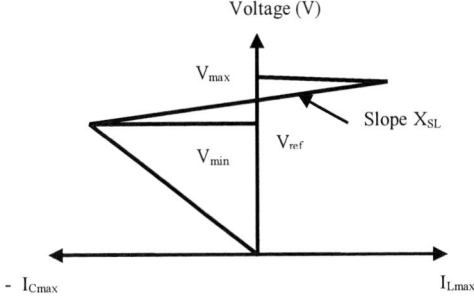

Fig 3: SVC V-I characteristics

For voltage outside the control range, the ratio V/I_s is determined by the ratings of the inductor and capacitor. The solution, of SVC and power system characteristic equations are graphically illustrated in Fig.4. Three system characteristics are detailed corresponding to three values of the source voltage. The middle characteristic represents the nominal system conditions and is assumed to intersect the SVC characteristic at Point A where $V = \mathbf{V_0}$ and $I = \mathbf{I_s}$. If the system voltage increases by ΔE_{th} due to decrease of system load level, bus voltage V will increase to V_1 without an SVC. With the SVC, the operating point moves to B, by absorbing inductive current I_3. Therefore, SVC holds the voltage V_3 instead of V_1 without the SVC. Similarly, if the system voltage decreases by ΔE_{th} due to increase of the system load level, bus voltage V will decrease to V_2 without an SVC. With the SVC, the operating point moves to C, by injecting capacitive current I_4. Therefore, SVC holds the voltage V_4 instead of V_2 without the SVC.

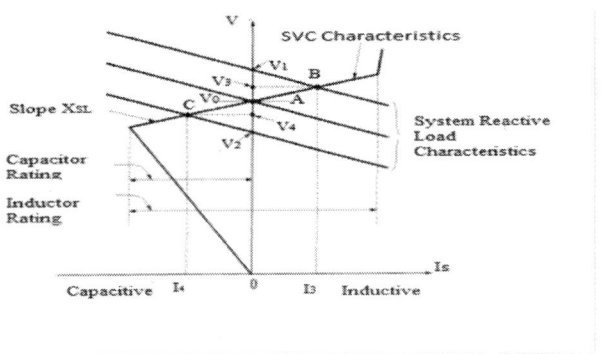

Fig.4: Graphical solution of SVC operating point for given system conditions

III. SIMULATION STUDY

In Western Rajasthan the wind power plants of different capacity are installed, and comprising of various types of wind generators (DFIG, PMIG, and SCIG). For study purposes, these wind turbines, complete capacity in consolidation are assumed to be connected with main bus of receiving sub-station as a single wind generator. In modeling and simulation these are assumed as negative load at various strategic buses. On actual scenario, the wind power generation with the unpredictable nature of wind causes fluctuating wind power which gave rise of instability problem to already existing network, along with other associated problem. In case of

978-1-4799-6047-7/14 $31.00 © 2014 IEEE

negative load presentation of wind farm wind generation/degeneration can be considered with incremental uniform steps up to voltage stability margin limits. In this paper, the study of dynamic reactive power requirement of the system at a different power factor of wind power generation condition, with and without SVC and its effect on system losses and voltage variation of different buses under steady state and fault condition have been carried out.

SVC of (+) 150/ (-) 150 MVAR capacity is connected at 400 KV GSS Jaisalmer through 400/33 kV transformer (Bus-19). The purpose of connecting SVC at Bus number 19 is to mitigate wind power generators reactive power [10], [11]. To demonstrate the effect of SVC on reactive power flow, voltage and system losses, load flow studies have been carried out without and with (+)150/(-)150 MVAR SVC at Bus-19 for various wind power(WP) generation power factor while wind power plants MW generation remains same. Five cases have been simulated in the Load flow studies as per details given in Table I.Wind power plants penetrated part of Rajasthan power system has been modeled using Mi.-Power system analysis software designed by the M/s PRDC Bangalore.

TABLE I – CASES SIMULATED FOR LOAD FLOW STUDIES

GSS	WP Gen. (MW)	Reactive Power of Wind Power Generation (MVAR)				
		I .95lag	II .98lag.	III 1.0	IV .98lead.	V .95lead
400kV Jaisalmer at 220kV level	600	+200	+120	0	-120	-220
220 kV Amarsagar	200	+65	+40	0	-40	-65
132kV Amarsagar	200	+65	+40	0	-40	-65
Total	1000	+ 330	+200	0	-200	-330

Case-I: 0.95 lagging power factor of wind power generation

Power plots of load flow study without SVC at Bus-19 is placed at Fig-5A and Power plots of LFS with (+) 150/(-)150 MVAR SVC at Bus-19 while other conditions remain unchanged is placed at the Fig-5B.

Case-II: 0.98 lagging power factor of wind power generation

Power plots of load flow study without SVC at Bus-19 is placed at Fig.6A and Power plots of LFS with (+)150/(-) 150 MVAR SVC at Bus-19 while other conditions remain unchanged is placed at Fig.6B

Case-III: Unity power factor of wind power generation

Power plots of load flow study without SVC at Bus-19 is placed at Fig.7A and Power plots of LFS with (+)150/(-)150 MVAR SVC at Bus-19while other conditions remain unchanged is placed at Fig.7B.

Case-IV: 0.98 leading power factor of wind power generation

Power plots of load flow study without SVC at 400 kV GSS Jaisalmeris placed at Fig.8A and Power plots of LFS with

(+)150/(-) 150 MVAR SVC at Bus-19 while other conditions remain unchanged is placed at Fig.8B.

Case-V: 0.95 leading power factor of wind power generation

Power plots of load flow study without SVC at 400 kV GSS Jaisalmeris placed at Fig.9A and Power plots of LFS with (+)150/(-) 150 MVAR SVC at Bus-19 while other conditions remain unchanged is placed at Fig.9B.

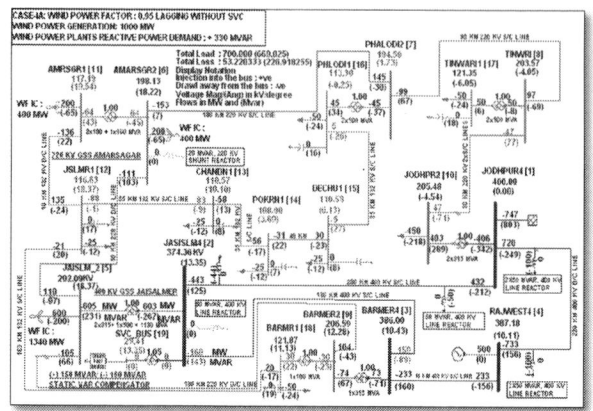

Fig.5 (A) Wind power generators power factor 0.95 lagging (Without SVC at Bus-19)(Case-IA)

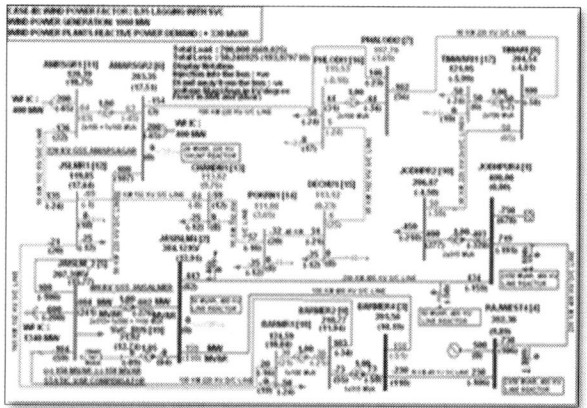

Fig.5 (B) Wind power generator power factor 0.95 lagging (With SVC at Bus-19)(Case-IB)

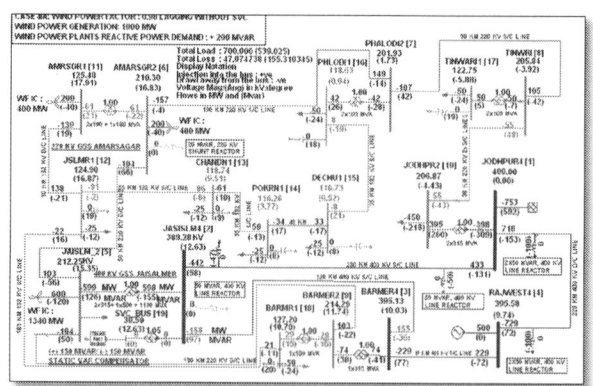

Fig.6A: Wind power generators power factor 0.98 lagging (Without SVC at Bus-19) (Case-IIA)

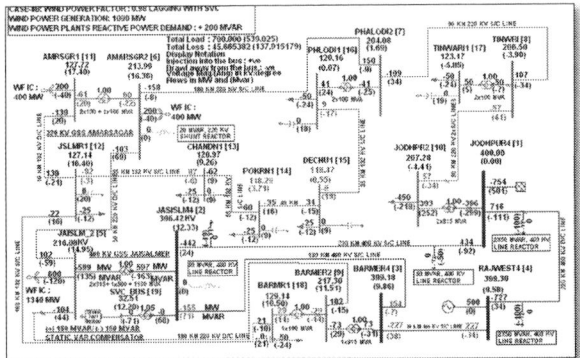

Fig.6B: Wind power generators power factor 0.98 lagging
(With SVC at Bus-19) (Case-IIB)

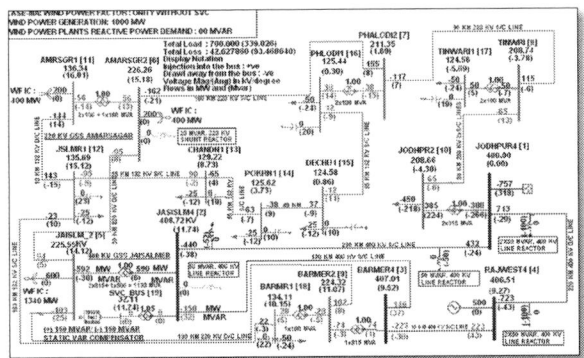

Fig.7A: Wind power generators power factor Unity
(Without SVC at Bus-19) (Case-IIIA)

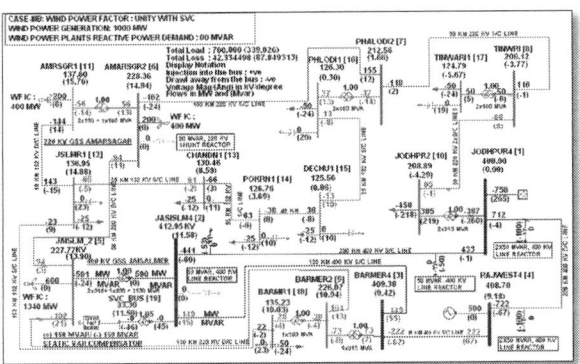

Fig.7B: Wind power generators power factor Unity
(With SVC at Bus-19) (Case-IIIB)

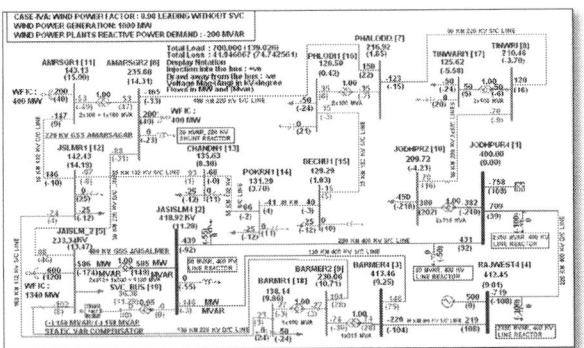

Fig.8A: Wind power generators power factor 0.98 leading
(Without SVC at Bus-19) (Case-IVA)

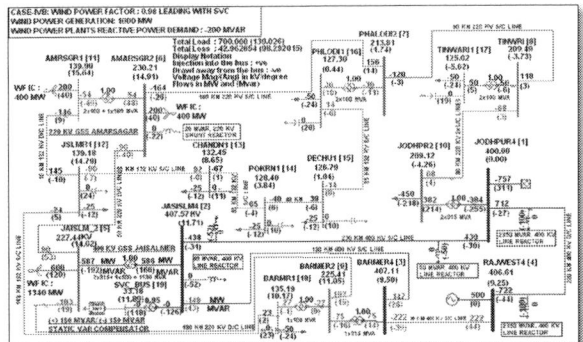

Fig.8B: Wind power generators power factor 0.98 leading
(With SVC at Bus-19) (Case-IVB)

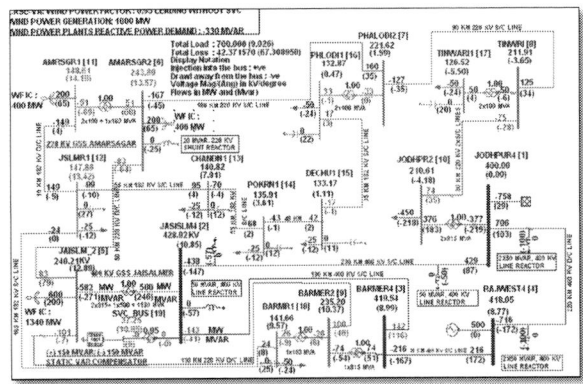

Fig.9A: Wind power generators power factor 0.95 leading
(Without SVC at Bus-19) (Case-VA)

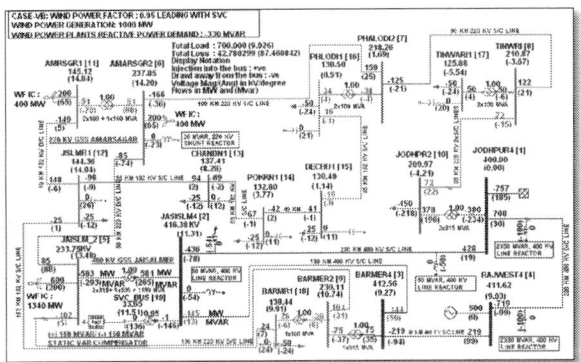

Fig.9B: Wind power generators power factor 0.95 leading
(With SVC at Bus-19) (Case-VB)

Power system voltage without and with SVC for lagging and leading power factor of wind power generation obtained at the different bus voltage level at various represented GSS and detail graphical representation and its variation profile is given in Fig.10 at 400kV bus voltage level, Fig.11 at 220 kV bus voltage level and Fig.12 at 132 kV bus voltage level.

To analyze the bus voltage under fault condition, with and without SVC, three phase to ground fault is created in transmission system at 400kV GSS Barmer for the duration of 100ms and bus voltage analysis is carried out in transient conditions at wind buses, 220kV GSSJaisalmer Fig.14, 220kV GSS Amarsagar Fig.15 and 132 kV GSS Amarsagar Fig. 16.

978-1-4799-6047-7/14 $31.00 © 2014 IEEE

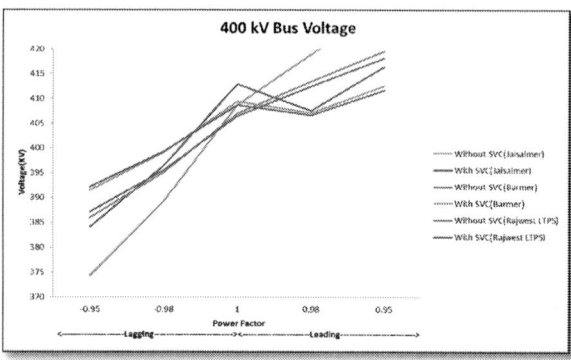

Fig.10 Busvoltages underdifferentpowerfactor conditions at 400kV Level

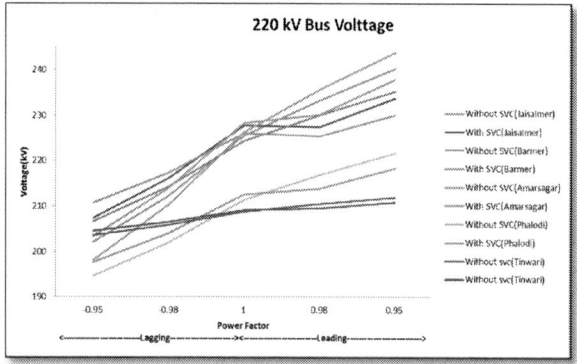

Fig.11 Bus voltages under differentpower factor conditionsat 220kV Level

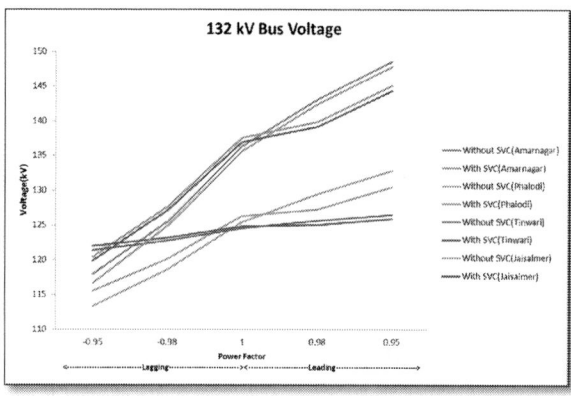

Fig.12 Bus voltages underdifferent power factorconditions at 132 KV Level

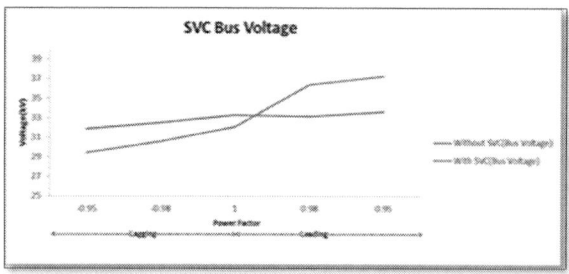

Fig.13 SVC Bus Voltage

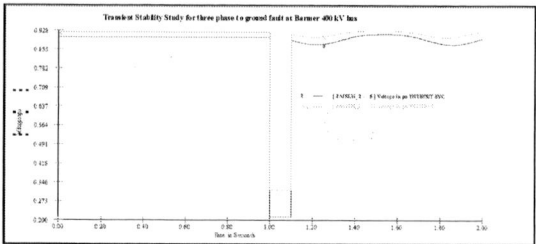

Fig.14: Jaisalmer 220 kV bus voltage for three phase fault at Barmer 400 kV bus

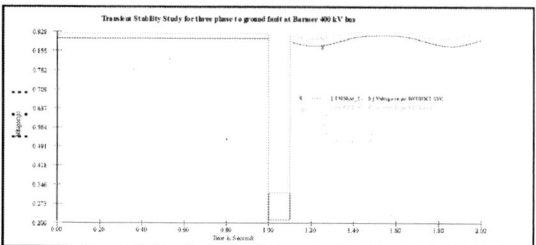

Fig.15 Amarasager 220 kV bus voltage for three phase fault at Barmer 400 kV bus

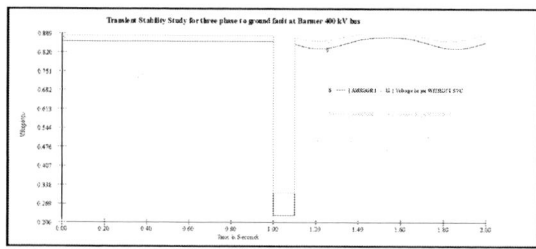

Fig.16 Amarasager 132 kV bus voltage for three phase fault at Barmer 400 kV bus

IV. RESULT ANALYSIS

A Effect of SVC on Power System Voltage under steady state condition

The bus voltages shown from fig.10 to Fig. 12 are the various bus voltages under steady state condition.The analysis of result of load flow, with and without SVC carried out at different buses and it is observed that without SVC, power system voltages are low in lagging power factor condition of wind power generation and power system voltages are high in leading power factor condition of wind power generation. The SVC bus voltage variation is given in Fig. 13. It is observed that system voltages are within limits with SVC even under variation in reactive power requirement of wind power generation. Voltage stability margin is also increased at wind farm node with SVC.

B Effect of SVC on Transmission Losses

Power system losses without and with SVC for different reactive requirement of 1000MW wind power generation are tabulated in Table-II. The simulation results show that with the use of SVC transmission losses are reduced in lagging power factor of wind power generation due to increase of system voltage and decrease of reactive power flow on transmission lines[12].

978-1-4799-6047-7/14 $31.00 © 2014 IEEE

TABLE II - POWER SYSTEM LOSSES WITHOUT AND WITH SVC

Particulars	Case I	Case II	Case III	Case IV	Case V
Reactive power requirement (MVAR)	328	200	0.00	- 200	- 325
Losses Without SVC (MW)	53.22	47.07	42.62	41.94	42.37
Losses With SVC (MW)	50.24	45.66	42.33	42.96	42.78

However, in leading power factor of wind power generation, transmission losses are slightly increased with SVC but for control of high voltage and to mitigate the reactive power requirement in the system, SVC is essential.

C Effect of SVC under fault condition

The wind buses behavior with and without SVC under three phase to ground fault at 400 KV GSS Barmer are tabulated in Table III.

TABLE III - WIND BUS VOLTAGE DURING FAULT

GSS	Voltage dip level without SVC	Voltage dip level with SVC
220 kVJaisalmer	0.212 p.u.	0.316 p.u.
220 kVAmarsagar	0.234 p.u.	0.326 p.u.
132kVAmarsagar	0.227 p.u.	0.311 p.u.

The results of effect of fault on the transmission system at wind buses indicate that with SVC there is reduction of voltage dip under transient condition. SVC improves the bus voltage considerably in fault conditions.

V. CONCLUSION

In this paper SVC is proposed in wind power generation penetrated power system for mitigation of wind power generators reactive power. Simulation studies have been carried out to validate the effectiveness of the SVC for voltage control with the variation in reactive power requirement of wind power generation. Case studies are carried out on 19-bus power system of WesternRajasthan, to demonstrate the performance of the SVC with the variability of power factor. Studies indicate that without SVC, power system voltages are low in lagging power factor of wind power generation and high in leading power factor.   With SVC, power system voltages are within limits under variation in reactive power requirement of wind power generation.

Simulation studies indicate that by SVC transmission losses are reduced in lagging power factor of wind power generation due to increase of system voltage and decrease of reactive power flow on transmission lines. However, in leading power factor of wind power generation, transmission losses are slightly increased with SVC but for high voltage control SVC is essential. Under fault conditions in transmission system SVC supports the bus voltage of wind buses. Along with the improvement of bus voltages under steady state and transient conditions, it also improves voltage stability margins; reduce transmission losses.

However the further studies are needed with detailed modeling of different type of wind generators at different system loading conditions for large power system in steady state and transient conditions.

REFERENCES

[1] P. Vuorenpää, P. Järventausta, *"Enhancing the Grid Compliance of Wind Farms by means of Hybrid SVC"*, Proc. of 2011 IEEE Trondheim PowerTech, Trondheim (Norway), 19-23 June 2011, pp. 1-8, ISBN 978-1-4244-8417-1.

[2] L. G. Meegahapola, D. Flynn, T. Littler, *"Decoupled-DFIG fault ride-through strategy for enhanced stability performance during grid faults"*, IEEE Transactions of Sustainable Energy, vol. 1, (3) pp. 152-162, 2010.

[3] A. Daneshi, N. S. Momtazi, H. Daneshi, J. Javan, *"Impact of SVC and STATCOM on Power System Including a Wind Farm"*, Proc. of 2010 9th International Conference on Environment and Electrical Engineering (EEEIC), Prague (Czech Rep.u.blic), 11-19 May 2010, pp. 117-120, ISBN 978-1-4244-5370-2.

[4] A. Öztürk, K. Dö_o_lu, *"Investigation of the control voltage and reactive power in wind farm load bus by STATCOM and SVC"*, Proc. of ELECO 2009 5-8 Nov. 2009, pp. I-60 – I-64.

[5] Energinet Technical Regulation 3.2.5, *Wind Turbines Connected to Grids with Voltages above 100 kV – Technical Regulation for the properties and the regulation of wind turbines*, published by Elkraft Systems and Eltra, 3rd December 2004

[6] D. Jovcic, Pahalawaththa, N., Zavahir, M. & Hassan, H.A. (2003) *"SVC Dynamic analytical Model"_* IEEE Trans. On Power Delivery, Vol. 18, No. 4, (October), pp. 1455 - 1461.

[7] K. M. Son, K. S. Moon, S. K. Lee, and J. K. Park, *"Coordination of an SVC with a ULTC reserving compensation margin for emergency control,"* IEEE Trans. Power Del., vol. 15, no. 4, pp. 1193–1198, Oct. 2000.

[8] N.G. Hingorani and L. Gyugy, *Understanding FACTS, Concepts and Technology of Flexible AC Transmission System*. New York: Inst. Elect. Electron. Eng., Inc., 2000.

[9] A. D. Hansen, *Generators and Power Electronics for Wind Turbines*, in T. Ackermann, *Wind Power in Power Systems*, Wiley, 2005, pp.53-78.

[10] J. G. Slootweg, H. Polinder, W. L. Kling, *"Reduced-order Modelling of Wind Turbines"*, in T. Ackerman, "Wind Power in Power Systems", ed. Wiley, 2005, ISBN 0-470-85508-8.

[11] S. K. M. Kodsi and C. A. Cañizares, *"Modeling and simulation of IEEE 14 bus system with Facts controllers,"* Tech. Rep., Univ. Waterloo, ON, Canada,.

[12] F. Milano *"Power System Analysis Toolbox Documentation for PSAT"* version 1.3.0, 2004.

Prof. A.K. Pathak has carried out his B.E (Electrical Engineering) Degree in 1972 and MSc .Engineering (PS) in 1982. He has joined R.S.E.B, in 1973 as Assistant Engineer and retired as Chief Engineer (PPM, Fuel &H&GP) in 2010.He also remains as Executive Director in Rajasthan Renewable Energy Corporation Jaipur. He joined as Professor in Rajasthan Institute of Engineering & Technology, Jaipur in 2010 & presently working as Dean Academic in Rajasthan College of Engineering for Women Jaipur. He involved in 400 KV System & Line Design, Renewable Energy. He worked in different capacities in Thermal, Gas, Combined Cycle Power Plants, Hydro, Solar and Wind Power plants. He also worked in Beas Construction Board & Bhakhra Beas Management Board. He is carrying out research work in renewable energy areas. He is Ph.D Scholar in Poornima University, JaipurElectrical Engineering field.

Dr. M. P. Sharma received the B.E. Degree in Electrical Engineering in 1996 Govt.Engineering College, Kota, Rajasthan and M.E. Degree in Power Systems in 2001 and Ph.D. degree in 2009 from the Malaviya Regional Engineering College, Jaipur (Now name as MNIT). He is presently working as Assistant Engineer, Rajasthan RajyaVidhyutPrasaran Nigam Ltd., Jaipur. He is involved in the system studies of Rajasthan power system for development of power transmission system in Rajasthan and planning of the power evacuation system for new power plants.

Transient Stability Improvement in Transmission Line using SVC with Fuzzy Logic based TID Controller

Tarang Sharma
Department of Electrical Engineering
National Institute of Technology, Kurukshetra
Kurukshetra, Haryana, India
Sharmam.tarang90@gmail.com

Anil Dahiya
Department of Electrical Engineering
National Institute of Technology, Kurukshetra
Kurukshetra, Haryana, India
anildau@yahoo.co.in

Abstract—**In this paper a new approach is taken to solve the stability problem in the power system with the help of Flexible AC Transmission System (FACTS) devices and Tilt Integral Derivative (TID) controller. The FACTS devices can enhance the power system capability of transmission. A number of FACTS devices are working in the power system. Among these FACTS devices, Static VAR Compensator (SVC) is a shunt connected FACTS device which can be used effectively in Voltage Control mode. In the MATLAB power system model, SVC is operating with a PI controller, which has simple functioning and is cheaper in operation but has poor performance with non-linear characteristics. So to overcome this problem the controller is designed as a combination of both TID controller and Fuzzy Logic controller. TID controller has a tilted component, which replaces the proportional component in the PID controller. In this paper the suggested controller has been tested on a 2 machine 3 bus power system using MATLAB/SIMULINK software. The simulation consumes lesser run time and there are fewer oscillations in the system with the hybrid TID and Fuzzy Logic controller.**

Keywords— *FACTS; SVC; Fuzzy Logic Controller; TID controller; Voltage Control*

I. INTRODUCTION

As the power system network is growing in size and covering a vast geographical area, it has become essential to maintain stability between various parts of interconnected systems. The study of transient stability involves the effect of severe disturbances (short circuit, sudden loss of load etc.) on the power system. When stability is lost, there will be a wild fluctuation in voltage, current, and power. During this time parallel operation may not work properly and the units may start tripping.

The stability problem begins whenever there is an imbalance between the mechanical input and electrical power output or vice versa. Due to this disturbance, the tendency of the rotor is to accelerate or decelerate. When the rotor is accelerating or decelerating the speed is above or below the synchronous speed. Due to this variation in speed, the parallel operation may not happen satisfactorily. The objective of transient stability is to retain the load angle to its steady state value after the clearance of the disturbance.

In this paper we have introduced a new model to improve the control mechanism of SVC with the help of hybrid TID and Fuzzy logic controller. The designed model is tested on a 2 machine 3 bus power system model using MATLAB Simulation. We have analysed it by using the new controller subjecting it to a three phase fault.

It has been identified that the transmission power capability can be increased and the voltage profile can be improved by using reactive var compensation [1],[2],[3]. Most of the reactive var compensation incorporate power electronics based static controllers and FACTS devices [4],[5],[6]. FACTS devices are characterized as series, shunt and combination of these two controllers [7],[8]. Reactive var compensation is generally used at the mid point of the systems to prevent voltage instability as well as to damp power oscillations [9].The PI based controllers are not suitable for non-linearity. Unlike PI controller Fuzzy logic based controller has an advantage over PI controller under non-linear characteristics [10],[11]. The Tilt Integral Derivative controller is a newly designed controller which has good performance and gives reliable results over PI or PID based controller [12].

The content is as follows: section II talks about the operation of SVC and its operating waveforms. Section III introduces the basics of Fuzzy logic controller; its input output membership functions, Rule base and Fuzzy Inference system. Section IV talks about the newly designed Tilt Integral Derivative controller (TID controller). Section V shows the results obtained through the TID and Fuzzy logic based controller using SVC. Section VI shows the conclusion obtained from the above results and future work to be done in the area of neural network based fuzzy logic controller with TID controller.

II. SVC OPERATION AND WAVEFORMS

Static Var Compensator (SVC) is a shunt connected FACTS device, which can be used in Voltage control mode or Var control mode. Here the term static is used to indicate that SVC has no rotating part unlike synchronous machine. SVC is used in voltage control mode by controlling the reactive var in the system where it is connected. SVC can draw leading or lagging var to control the voltage fluctuation or voltage regulation in the system. If there is a dip in the voltage then it supplies reactive power and if there is a rise in the voltage then it absorbs reactive power. So the SVC can be used as a source or sink of reactive Var in accordance to the need of system.

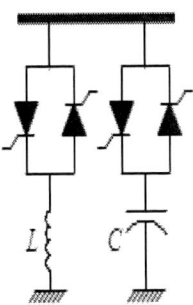

Fig. 1.   Structure of SVC

As shown in Fig. 1, the SVC comprises of fixed or switched capacitor bank in parallel with switched reactor bank. The reactive power can be controlled by switching the capacitor bank with the help of Thyristor switched capacitor (TSC) and by controlling the reactor bank with the help of Thyristor controlled reactor (TCR). The operating waveforms are given below:

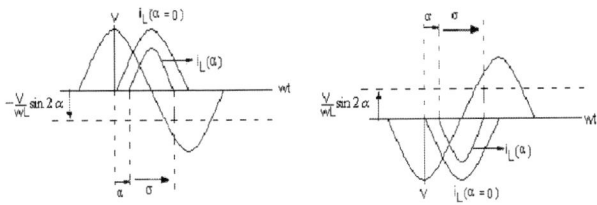

Fig. 2.   Firing angle control

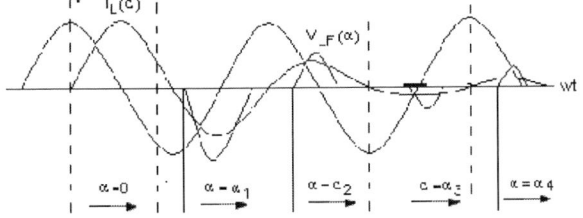

Fig. 3.Operating waveforms

Fig.3 shows the reactor current varied from its peak value (at α = 0) to zero value (at α = $\pi/2$). Here the terms $i_L(\alpha)$ and $i_{LF}(\alpha)$ represent the reactor current and its fundamental component. The reactive current injected into the system is associated with SVC susceptance B_L. The susceptance variation for SVC operating region is given in Fig. 4

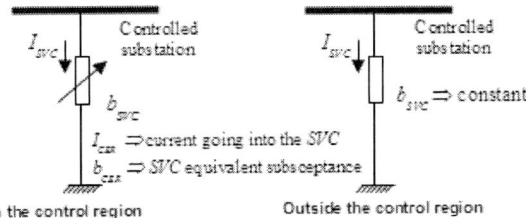

Fig. 4.Operating region of svc

Fig. 5. Shows the basic configuration of SVC control system which is used to analyse the impact the SVC on transient condition of power system.

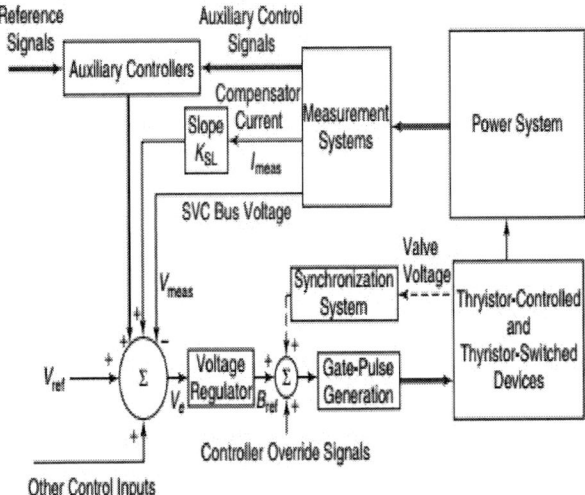

Fig. 5.Basic configuration for SVC control system

III. FUZZY LOGIC CONTROLLER

The Fuzzy Logic controller works as a rule based controller, where a set of rules shows a control mechanism to solve the effect of certain problems coming from power system. It is a fuzzy code designed to control something, usually mechanical. They can be in software or hardware and can be used in anything from small circuits to large mainframes.

Typically a fuzzy controller has at least 2 inputs and one output. In this paper, the inputs to the fuzzy logic controller are the error in the voltage and change of error, while its output acts as the control signal.

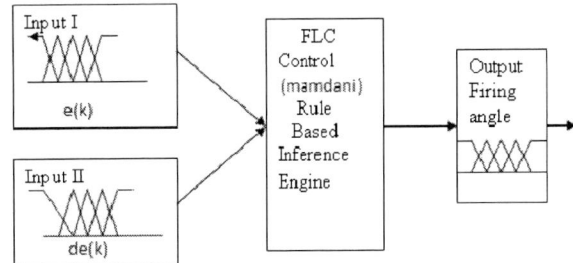

Fig. 6.Structure of Fuzzy Logic Controller

978-1-4799-6047-7/14 $31.00 © 2014 IEEE

A. Block Diagram of Fuzzy Logic Controller

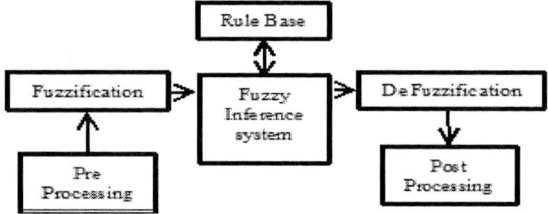

Fig. 7. Block diagram of Fuzzy logic controller

FIS (Fuzzy Inference System) consists of the following components:

1. Input variables

2. Fuzzification through membership functions

3. Rule-base (if/then rules)

4. Defuzzification through membership functions

5. Output variables

A. Input Variable

Input variables are used as controlled variables, which have a certain value called set point to make the system oscillation free and stable.

B. Fuzzification

Fuzzification is defined as a conversion of real input to fuzzy set values. For eg.

Medium(x) = {0 if X >= 1.90 or X < 1.70;

(1.90 – X)/0.1 if X >= 1.80 and X < 1.90;

(X – 1.70)/0.1 if X >= 1.70 and X < 1.80}

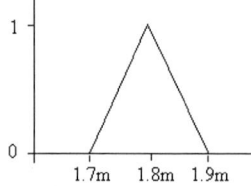

Fig. 8. Fuzzification of input

Different types of parameterized membership functions commonly used are:

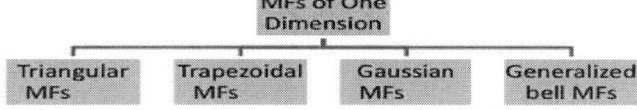

Fig. 9. Types of membership functions

C. Rule Base

The Rule Base is described as a "If Then" format wherein the 'If' side represents the condition and the 'Then' side represents conclusions.

Rule 1: If error in voltage, e(k) is LN and error change de(k) is OZ, then the output is LN.

Rule 2: If error in voltage, e(k) is OZ and error change de(k) is OZ, then the output is OZ.

TABLE I. RULE BASE OF FUZZY CONTROLLER

e[k]	ė[k]Δu[k]		
	LN	OZ	HN
LN	LN	LN	OZ
OZ	LN	OZ	HN
HN	OZ	HN	HN

a. LN -= LOW, OZ = OK, HN = HIGH

D. Defuzzification

The task of Defuzzification is to find one single crisp value which summarizes the fuzzy set which is obtained from the inference block. There are several mathematical techniques that are used to perform this.

(i) Centroid Method, (ii) Centre of Sums, (iii) Weighted Average Method

The Centre of Gravity (COG) law is employed here, equation is given by

$$COG(FAST) = \frac{\sum_{i=1}^{N} A_i \int B_i}{\int B_i} \qquad (1)$$

Where A_i symbolizes the membership function and B_i symbolizes the membership of member i of output fuzzy set.

IV. TILT INTEGRAL DERIVATIVE CONTROLLER

A Tilt Integral Derivative controller is characterized as a feedback control compensator of PID type wherein proportional component is replaced by tilted component having transfer function 1/s of power n. The equivalent transfer function of the suggested compensator provides the improved control mechanism for feedback control system. Moreover, TID compensator has a good transient rejection ratio, simpler tuning, and a few effect on the closed loop system response[12].

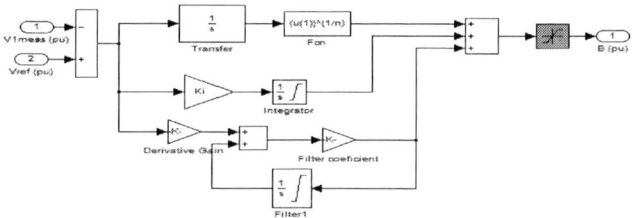

Fig. 10. Tid controller

V. RESULTS

The designed controller is tested on a 2 machine 3 bus power system model in MATLAB. The power rating of machine 1 is 1000 MVA which is connected to load center with the help of a 700km transmission line, operating at 500 KV. Load is of resistive type with a rating of 5000MW, supplied by machine 2 of rating 5000MVA. The single line diagram of a 2 machine 3 bus system is given below.

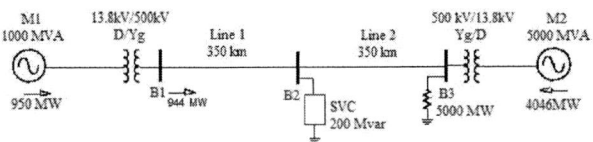

Fig. 11. single line diagram of 2 machine 3 bus system

A three phase fault occurs at bus 1 for a period of 0.1s. The designed controller has a precise control to make the system stable after the fault is cleared. During the fault when there is a dip in the bus voltage (high current), svc injects the reactive var into the transmission line where it is connected to regulate the voltage. The MATLAB Simulink model and the designed controller is given below:

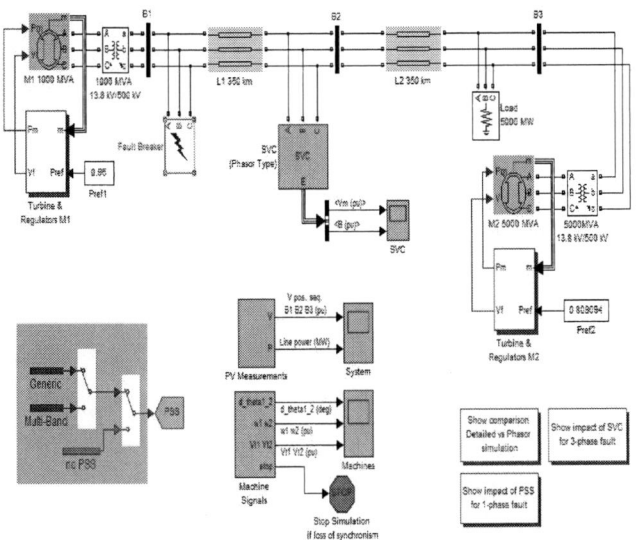

Fig. 12. Simulink Model

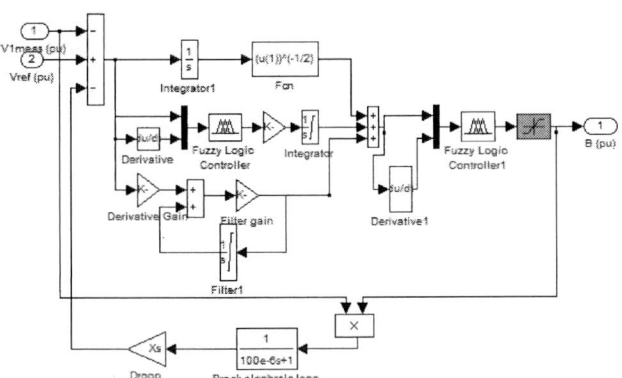

Fig. 13. Fuzzy logic based TID controller

When a three phase fault occurs at 0.1s, there is a loss of synchronism between machines 1 and 2. In Fig. 14, when the fault occurs, the first overshoot in the rotor angle is reduced

with fuzzy and TID based SVC with respect to conventional SVC. In conventional SVC, rotor angle magnitude is 65 p.u and goes to its steady state value of 53 p.u in 5.3s , but in the designed controller it decreases from 61 p.u to 50 p.u in 5.2s.

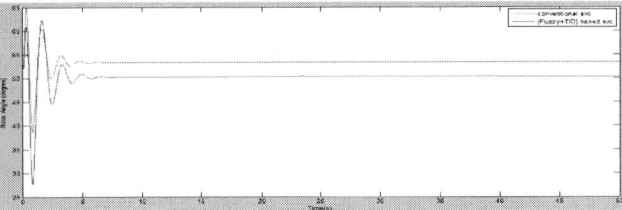

Fig. 14. Rotor Angle Difference

Fig. 15 shows that the generator G1 synchronizes faster at 5.2s with designed controller as against 5.3s in conventional SVC. The designed controller has lesser speed deviations as compared to conventional SVC.

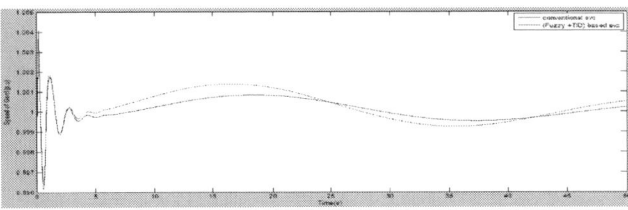

Fig. 15. Speed of Gen 1

In Fig. 16, during the fault there are lesser oscillations in the bus voltage and it stabilizes quickly with the designed controller.

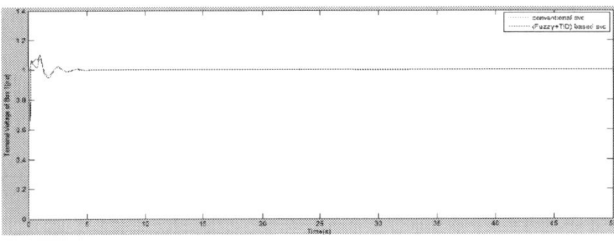

Fig. 16. Bus1terminal voltage

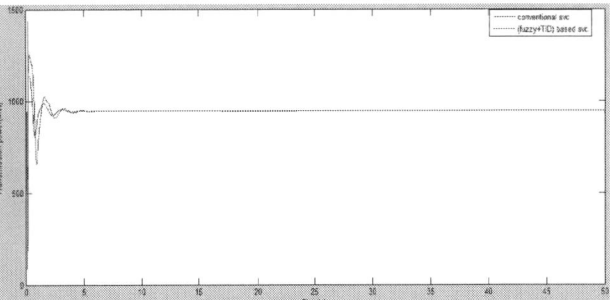

Fig. 17. Transmission line power

VI. CONCLUSION

In this paper, the SVC mechanism is controlled with fuzzy and TID based controller. This controller combines the advantages of both TID controller and Fuzzy logic controller. The designed controller is tested on a 2 machine 3 bus Simulink model in MATLAB. The Simulation tests are performed on various parameters like rotor angle difference, speed of generator, bus terminal voltage and transmission line power. The performance of designed controller is compared with the conventional SVC and it is seen that the fuzzy and TID based controller has enhanced the transmission line power stability during the disturbances. The performance of the designed controller is reliable and is quite stable. Future work will be done on improving the transient stability of SVC using neural network based TID controller.

REFERENCES

[1] P. Kundur 1994, "Power system stability and control", McGraw Hill.

[2] IEEE Committee Report: 'Bibliography of static VAr compensators—Working Group 79.2 on static VAr compensators in A.C. substations committee', IEEE Trans., 1983, PAS-102, (12), pp. 3744-3752

[3] IEEE Committee Report: 'Bibliography of static VAr compensators—Working Group 79.2 on static VAr compensators in A.C. substations committee', IEEE Trans., 1983, PAS-102, (12), pp. 3744-3752

[4] G. Hingorani and L. Gyugyi, "Understanding FACTS, Concepts, andTechnology of Flexible AC Transmission Systems", Piscataway, NJ:IEEE Press, 2000.

[5] CIGRE working group 01 of study Committee 31: 'Modelling of static shunt VAR systems (SVS) for system analysis', Electra, 1977, 51

[6] HAUTH, R.L., HUMANN, T., and NEWELL, R.J.: 'Application of a static VAR system to regulate system voltage in Western Neb- raska', IEEE Trans., 1978, PAS-97, (5), pp. 1955-1964

[7] A.A Edris, R Aapa, M H Baker, L Bohman, K Clark, "Proposed terms and definitions for flexible ac transmission system (FACTS)", IEEE Trans. on Power Delivery, Vol. 12, No.4, 1997,pp.1848-1853

[8] P.L. So, T. Yu, "Coordination of TCSC and SVC for Interarea stability Enhancement" IEEE Trans. on Power Delivery , Vol. 9, No. 1,2000.

[9] B T Ooi, M Kazerrani, R Marcean, Z Wolanski, F D Galiana, D.Megillis, G. Joos, "Mid point sitting of FACTS devices in transmission lines", IEEE Tran. On Power Delivery, Vol.1 No.4, 1997, pp.1717-1722.

[10] H. Ying, "TIT0 Mamdami Fuzzy PI/PD Controllers as Nonlinear, Variable Gain PI/PD controllers, International Journal of Fuzzy System, Vol. 2, No. 3, September 2000.

[11] A.Kazemi and M.V.Sohrforouzani, "Power system damping using Fuzzy controlled FACTS Devices", International conference on Power system Technology, POWERCON 2004, Singapore.

[12] The United States of America as represented by the Administrator of Jones; Guy Miller the National Aeronautics and Soace Administration, Washington, D.C.

Optimal Placement of Distributed Generator in Transmission System Using an Algorithmic Approach

Navpreet Kaur, Kanwardeep Singh
Department of Electrical Engineering
Guru Nanak Dev Engineering College, Ludhiana
Ludhiana, India
matharoo.navpreet90@gmail.com, kds97dee@gmail.com

Satish Kansal
Department of Electrical Engineering
Baba Hira Singh Bhattal IET Lehragaga
Sangrur, India
kansal.bhsb@gmail.com

Abstract—This paper presents the optimal placement of distributed generation (DG) in locational marginal price (LMP) based electricity market. The problem formulation for optimal placement of DG is based on maximization of social welfare subject to operational and security constraints. In this paper, an algorithmic approach has been used based on LMPs and CPs for optimal placement of DG. LMP at a particular location provides an economic signal to the consumers and suppliers for their power consumption and generation, respectively. CP at load bus is defined as the product of LMP and real power demand at that load bus. The system buses have been ranked for optimal placement of DG based on decreasing order of LMPs and CPs. Various cost characteristics are assumed in test system to examine the proposed methodology. The proposed methodology has been tested on IEEE 14-bus system. The simulation results clearly present the effectiveness of the proposed methodology.

Keywords—electricity market; social welfare; distributed generation; locational marginal price; consumer payment; optimal power flow

I. INTRODUCTION

With the liberalization of electricity market, stronger competitive technologies are developing in the field of generation. Due to which, importance of power supply security is increasing, now-a-days. So, the supplier has developed a great interest in the distribution generation (DG) for the production of energy in the future [1]. DGs are the small scale generation that provides additional capacity to generation system. Now-a-days, these small scale generators are getting wide spread acceptance as there are constrains in building new transmission lines, raised customer demand for highly reliable electricity, climate changes and energy security [2].

DG provides generation of power near to customer end using different technologies. It is considered as a new concept on an old one that plays an important role to reduce pressure on an already overloaded electric power system.

DGs can reduce:

- The transmission and distributed (T&D) system and investment cost for installing them at optimal location.

- Losses in the system.

- Congestion on transmission line and wholesale power prizes.

- Burden on the heavy load feeder, increases equipment's life and minimizes the un-served consumer's power (load curtailment).

- Reduces the greenhouse effect and air contaminations.

Various researches have been carried out for the placement of DG in the power system. LMP has also being used to decline line losses and provide the right price signals to locate DG in the distribution network. The optimal placement of DG for minimizing real power losses can be done with the Genetic algorithm (GA), Tabu search and so on [3]. GA approach is intensely explained in the paper [4], [5]. In reference [6], optimal DG size is calculated using an analytical approach by which optimal location for the DG placement is also determined. In reference [7], optimal DG size is calculated using a PSO approach by which optimal location for different types of DG placement is also determined in the radial distribution system. Optimal DG location can be done with the fuzzy method and is explained in reference [8]. In reference [9], an analytical approach for the optimal DG placement in the distribution system is developed, whereas economical and graphical factors are not considered. Hence, DG size is not optimized. In reference [10] determined the optimum size and location of DG by using simple approach; the authors take the weights of energy loss and cost of DG. The weights have been changed to achieve the objective. Reference [11] presented the optimal placement of wind DGs in the distribution system to minimize the annual energy losses using generation-load model. The optimization problem has been solved by using mixed integer non-linear programming in GAMS software considering various system constraints.

In this paper, an algorithmic approach is developed which is more accurate to determine the optimal size and location of DG in the transmission system. The main aim of this paper is to get maximum social welfare which is given by reducing the peak LMP and CP. Hence two rankings LMP based and CP based are being used to attain the objective of social welfare maximization with the optimal placement and size of DG.

This paper is divided into five sections. Section II gives the complete understanding about the OPF. Section III presents the methodology employed to evaluate the DG placement in which two rankings are defined to identify the candidate nodes for the placement of DG. The results obtained are discussed in the

Section IV. The conclusion of the analysis is presented in the Section V.

II. PROBLEM FORMULATION

Social welfare maximization/total generation cost minimization is considered to be the main objective while formulating the problem. The optimal power flow (OPF) is widely accepted for the minimization of cost and is modified to establishing the demand bids, in addition to the generation bids.

Locational marginal price is introduced as Lagrangian multiplier used in power balanced equation in case of optimal power flow (OPF). The input has been taken by the OPF are generation bids and demand (customer) bids. Firstly, the base case OPF algorithm is calculated on the basis of social welfare maximization which evaluates the generation dispatch, price and demand at each node. So, the various nodes for the DG placement are identified from the nodal prices. Here, the placement means to obtain the lowest price demand by changing the dispatch scenario.

The objective function obtained after having OPF is formed as quadratic benefit curve submitted by the buyer (DISCOS) minus quadratic bid curve supplied by the seller (GENCO) minus the quadratic cost function supplied by the DG owner.

$$max \sum_{i=1}^{N}(B_i(P_{Di}) - C_i(P_{Gi})) - C(P_{DGi}) \tag{1}$$

Minimization problem formulation is also shown in (2)

$$min \sum_{i=1}^{N}(C_i(P_{Gi}) - B_i(P_{Di})) + C(P_{DGi}) \tag{2}$$

Subject to

A. Equality Constraints

Transmission network is established via power balance equation at each node in network. The active power plus reactive power injected into the node minus the power flow extracted from the node should be zero as shown in (3) and (4) respectively.

$$P_i = P_{Gi} + P_{DGi} - P_{Di} = v_i \sum_{j=1}^{N}[v_j\{G_{ij}\cos(\delta_i - \delta_j) + B_{ij}\sin(\delta_i - \delta_j)\}] \tag{3}$$

$$Q_i = Q_{Gi} - Q_{Di} = v_i \sum_{j=1}^{N}[v_j\{G_{ij}\sin(\delta_i - \delta_j) - B_{ij}\cos(\delta_i - \delta_j)\}] \tag{4}$$

B. Inequality Constraints

1) Real and reactive generation limits: There are maximum and minimum generating capacities available for every generating plant. Beyond and before the maximum and minimum capacities respectively, it is not feasible to generate electricity because of economic fact. Upper and lower limits for real and reactive power outputs given for real and reactive power separately as follows:

$$P_{Gi}^{min} \leq P_{Gi} \leq P_{Gi}^{max} \tag{5}$$

$$Q_{Gi}^{min} \leq Q_{Gi} \leq Q_{Gi}^{max} \tag{6}$$

2) Line flow limits: Line flow limits refers the maximum power which a transmission line is able to deliver power under the thermal and various stability considerations:

$$S_{ij} \leq S_{ij}^{max} \tag{7}$$

$$s_{ji} \leq S_{ji}^{max} \tag{8}$$

3) Bus voltage limits: Voltage limits refer to the maximum and minimum values of voltage that are required for the system:

$$v_i^{min} \leq v_i \leq v_i^{max} \tag{9}$$

Where,

N = the total number of buses.

P_{Gi} = power generation at i bus.

P_{Di} = real power demand at i bus.

P_{DGi} = real power supplied by DG at i bus.

For base case OPF,

$$P_{DGi} = 0 \tag{10}$$

For load bus,

$$P_{Gi} = 0 \tag{11}$$

For generator bus,

$$P_{Di} = 0 \tag{12}$$

$$B_i(P_{Di}) = a_{Di} + b_{Di}P_{Di} - c_{Di}(P_{Di})^2 \tag{13}$$

$B_i(P_{Di})$ is purchaser benefit function at bus i

$$C_i(P_{Gi}) = a_{Gi} + b_{Gi}P_{Gi} + c_{Gi}(P_{Gi})^2 \tag{14}$$

$C_i(P_{Gi})$ is the producer offer bid price at bus i

$$C(P_{DGi}) = a_{DGi} + b_{DGi}P_{DGi} + c_{DGi}(P_{DGi})^2 \tag{15}$$

$C(P_{DGi})$ is the cost characteristics of DG at bus i

V_i = the voltage at bus i

δ_i = the power angle at bus i

B_{ij} = the susceptance of line ij

G_{ij} = the conductance of line ij

Q_{Gi} = reactive power generated at bus i.

P_{Gi}^{max} and P_{Gi}^{min} are upper and lower real power generation limits of generator at bus i.

Q_{Gi}^{max} and Q_{Gi}^{min} are upper and lower reactive power generation limits of generator at bus i

v_i^{max} and v_i^{min} are upper and lower limits of voltage at bus i

S_{ij} are complex power transferred from bus i to bus j

S_{ji} are complex power transferred from bus j to bus i

S_{ij}^{max} and S_{ij}^{min} are the complex power flow limits for line ij and line ji.

III. METHODOLOGY

Electricity prices at different nodes of power system are calculated by the base case OPF. These prices are defined as Lagrangian multiplier of non-linear equality constraints. Due to active line constraints and losses in transmission system, difference in prices will result. An algorithmic approach has been used in this paper based on LMPs and CPs as given below for optimal placement of DG.

A. Algorithm

This algorithm depicts the procedure for the optimal placement of DG in transmission system with the following steps:

1) Data used to carry out OPF are taken from existing literature.

2) Base case OPF is conducted to calculate LMP at each node.

3) Ranking of buses based on LMP is made initiating with the bus having highest LMP and further considering in decreasing order of LMPs.

4) Calculating CP on the same buses considered in the case of LMP ranking and arranging in descending order, is called ranking based on LMP.

5) Assumptions are made for the DG cost characteristics.

6) Generation cost of central generation is calculated by proposed placement of DG (10-50MW) at first selected bus.

7) Computing DG cost bus from the cost characteristics with respect to the size of the DG.

8) Now, total generation cost is formed by summing up the generation cost of central generators and generation cost of DG.

9) Optimal size of DG is selected for that bus by considering lowest total generation cost and applying same procedure for the rest of buses.

10) Forming a table of optimal size and location of DG at each bus.

11) Optimal placement of DG is done on the bus which has the lowest total generation cost among the all buses.

12) Same procedure is followed for the placement of other DGs.

13) Results are achieved in the form of LMP after optimal placement of one DG at a time.

Two rankings are adopted namely, LMP based and CP based to identify the candidate nodes for the DG placement.

B. The LMP based Ranking

Lagrangian multiplier associated with real power flow equations for each bus in the system is considered as LMP of real power at each system bus. Actually, LMP is sum of three components namely, a marginal energy component, a marginal loss component and a congestion component; the marginal energy component being same for all buses. For the real power spot price at bus i, LMP is

$$LMP_i = \lambda + \lambda \frac{\partial P_L}{\partial P_i} + \sum_{ij=1}^{N_L} \mu L_{ij} \frac{\partial P_{ij}}{\partial P_i} \quad (16)$$

$$LMP_i = \lambda + \lambda_{L,i} + \lambda_{C,i} \quad (17)$$

The spot price at each bus differs by the loss component and congestion component. As the LMP is higher, means the active power flow equation of the node has greater impact on the social welfare of the system. In another words, if LMP goes higher at a particular node, then the more electric power

subjected to the demand on the same node has to be generated. Thus, this gives clue that by integrating active power at that bus will satisfy the objective of social welfare maximization. Hence, the net social welfare gets improved. The node which has highest LMP would formerly prefer for the DG placement because DGs are meant to inject real power only. Similarly, load buses are arranged in descending order of LMP with the initial node in the order as given below.

$$LMP = \begin{bmatrix} LMP_1 \\ LMP_2 \\ LMP_3 \\ . \\ . \\ . \\ LMP_n \end{bmatrix} \quad (18)$$

Where n = no. of load buses.
Best location = index (max LMP)   (19)

C. CP based Ranking

CP determined as the product of LMP and the real power demand (load) is adopted as another criterion to identify the nodes for the placement of DG. Therefore, the CP evaluated at each node by applying the formula given below.

$$CP = LMP_i * Load_i = \begin{bmatrix} CP_1 \\ CP_2 \\ CP_3 \\ . \\ . \\ . \\ CP_n \end{bmatrix} \quad (20)$$

Best location = index {max(CP)}   (21)

Thus, CP at particular node determines the amount of payment that consumer at that node have to pay for the power (electricity) they purchased. CP based ranking follows the pattern which satisfies the consumer in terms of economy wherein the total nodal payment is given the priority than the higher price. With this, LMP get decreases and the dominant consumer would be wealthier. And of course, the amount they will have to pay would be less in comparison to no DG placement. As the main purpose is to lower down the LMP, DG with higher operating cost than LMP will have no incentive for placement.

The DG having lower operating cost than the supplier bid is considered to have higher penetration whereas the DG with higher operating having smaller penetration.

IV. RESULTS AND DISCUSSIONS

Simulation results of proposed placement techniques are being discussed in this section. The proposed algorithmic approach is tested on modified IEEE 14 bus test system [11]. DG cost characteristics are taken from [2]. Modified IEEE 14 bus test system is shown in Figure1.

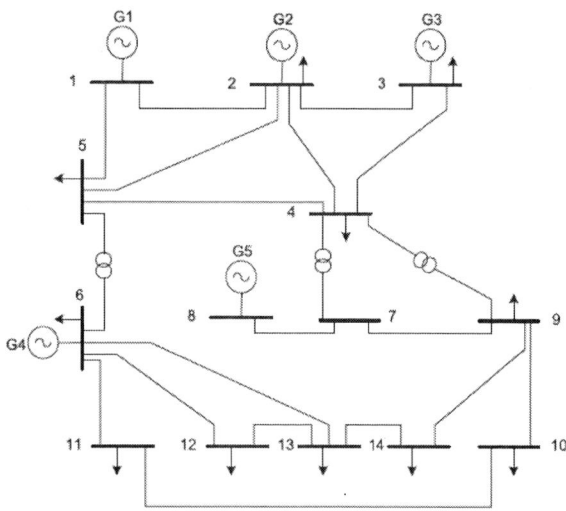

Fig. 1. Single line diagram of IEEE 14 bus test system

A. Base case OPF

The social welfare maximization problem includes the welfare of consumers as well as producers. The IEEE 14 bus test system used in this work has 9 load busses and 5 generation buses. After the base case OPF, the total generating cost is found to be 8081.53 $/hr. The total real power loss in the system is 9.282 MW. For the base case, the generation, load and LMP are determined at each node. According to results, generation buses have lower values of LMP as compared to the load buses.

B. Nodes for DG placement

The maximum load is 47.80 at node 4, whereas the maximum LMP of 41.198 $/MWh is recorded at bus 14 as shown in table 3. This is seen from the Table 1 that the bus with high load not necessarily have high LMP. Sometimes, load at other nodes and overall network arrangement play a load in determining LMP. LMP and CP based ranking of buses are shown in Table 1 and 2 respectively. Optimal DG size for each bus is determined from the stand point of social welfare maximization/total generation cost minimization. There is always an optimal size of DG by which net social welfare is being maximized. The net social welfare obtained from optimal DG sizes is different of one bus from another one. Placement and penetration of DG is found to change with the cost characteristics with the DG. There are different optimal sizes are obtained when different cost characteristics are used even for the same load.

Fig. 2 and Fig. 3 show the result after placing DG at optimal location (node 3). By varying DG size, the respective LMP has also varied. As the DG size increases, the peak LMP is reduced that gives economic benefit to the consumer and also it is the achievement of objective of this paper. A study for the DG has been carried out to identify the optimal placement and penetration, when DG is cheaper or expensive than the existing central generation. Actually, central generations are more preferred than the DG technologies as they contend in terms of economy of scale. Result in details shown only with DG5 and

final results of other DGs are only being discussed in this section.

TABLE I. LMP BASED RANKING

Rank	Bus	Power Demand (MW)	LMP ($/MWh)
1	14	14.9	41.198
2	13	13.5	40.575
3	3	94.2	40.575
4	12	6.1	40.379
5	10	9	40.318
6	4	47.8	40.190
7	7	0	40.172
8	8	0	40.170
9	9	29.5	40.166

TABLE II. CP BASED RANKING

Rank	Bus	Power Demand (MW)	LMP ($/MWh)
1	3	94.2	3822.17
2	4	47.8	1921.09
3	9	29.5	1184.90
4	14	14.9	613.85
5	13	13.5	547.76

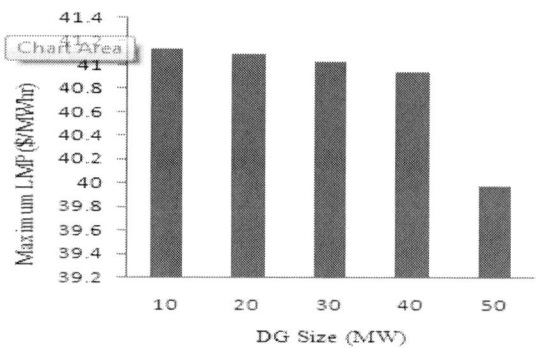

Fig. 2. Variation of maximum LMP with size of DG at bus 3

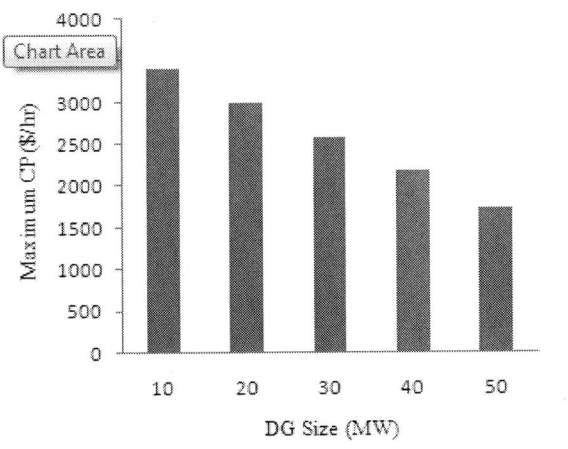

Fig. 3. Variation of maximum CP with size of DG at bus 3

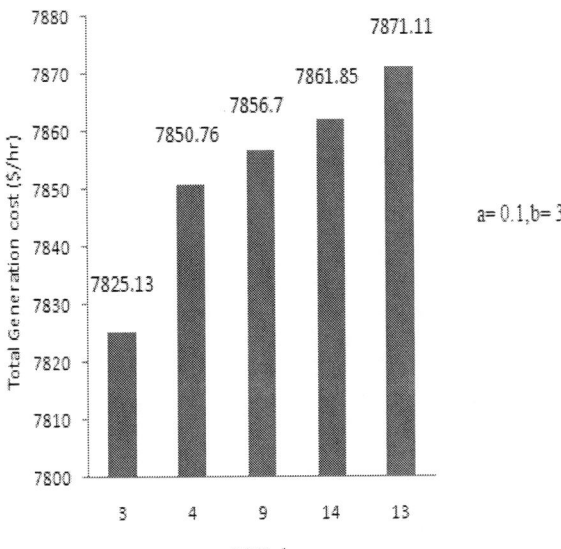

Fig. 4. Variation of total generation cost with placement of DG5 at different nodes

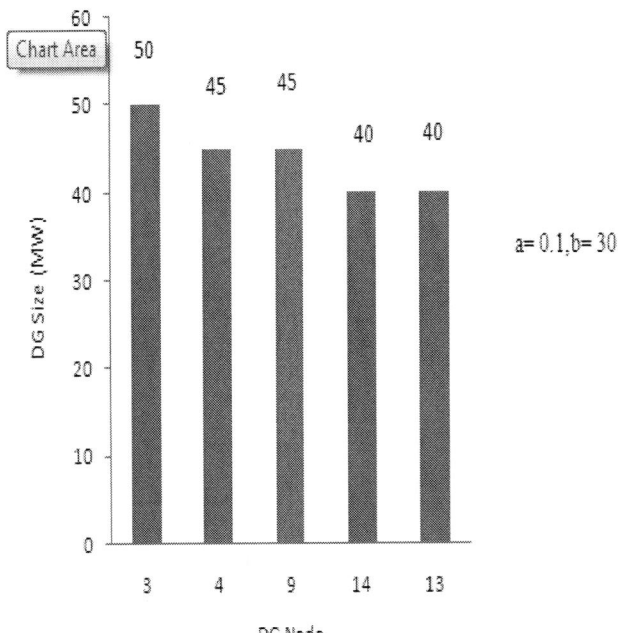

Fig. 5. Optimal size of DG 5 for placement at different nodes

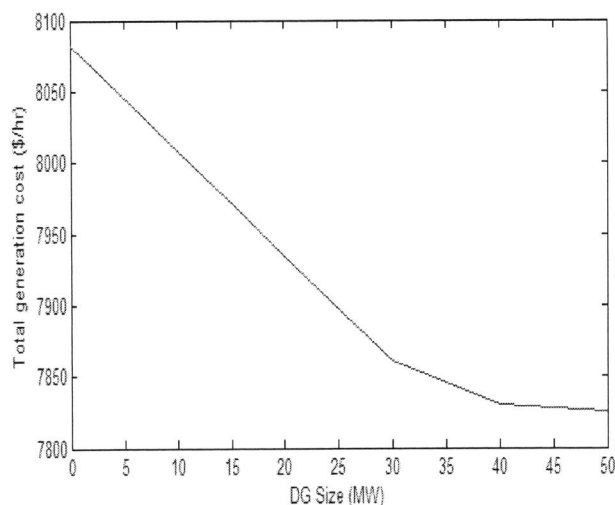

Fig. 6. Variation of total generating cost with DG size for the placement of DG5 at node 3

C. Placement of DG5

Results of placement of DG5 corresponding to total cost minimization carried out at different nodes is shown in Fig. 4. The optimal DG size correspondence to total generation cost minimization at each node is shown in Fig. 5. For example, if the placement is done at node 3, optimal size of DG5 is 50 MW for the total generation cost of 7825.13 $/hr. Similarly, if placement is carried out at node 4, optimal size of DG5 is 45 MW for the total generation cost of 7850.76 $/hr. Therefore, the node selected for the optimal placement of DG5 is 3 with the size of 42.84 MW. Total generation cost minimization is found to capture the first candidate node of CP ranking shown in Table II. Total generation cost variations w.r.t. DG size is shown in Fig. 6.

Moreover, it is seen from the Table III, that all DGs except DG7 are placed according to the CP based ranking. But it is not feasible to place DG7 because of its high incremental cost than the LMP.

V. CONCLUSION

In this paper, an algorithmic approach has been presented based on LMPs and CPs rankings for optimal placement of DG in OPF based formulation. The following points can be concluded from the results obtained.

- DGs are placed in transmission system by which the consumer and supplier both get economic benefit.

978-1-4799-6047-7/14 $31.00 © 2014 IEEE

- Optimal size and location for DG is identified in such a way that the net social welfare will increase.

- The optimal placement and DG size varies with varying cost characteristics of DG. Placement and penetration is found to change with the cost characteristics of DG and is considered as the function of cost characteristics of central generators.

- Placement of DG can only become feasible in electricity market if the incremental cost of DG is less than the maximum LMP of the system.

- CP based ranking gives more optimal results than the LMP based ranking.

TABLE III. COMPARISON OF RESULTS FOR PLACEMENT OF DIFFERENT COST DGS

DG	Best location	Optimal DG size (MW)	Total generation cost ($/hr)	Remarks
DG1	Bus 3	50	6825.13	CP based ranking
DG2	Bus 3	50	7035.13	CP based ranking
DG3	Bus3	50	7182.71	CP based ranking
DG4	Bus3	40	7669.66	CP based ranking
DG5	Bus3	50	7825.13	CP based ranking
DG6	Bus3	17	8076.65	CP based ranking
DG7	—	—	—	—

REFERENCES

[1] A. Kumar and W. Gao, "Optimal distributed generation location using mixed integer non-linear programming in hybrid electricity markets," IET Gener. Transm. Distrib., vol. 4, no. 2, pp. 281–98, 2010.

[2] D. Gautam and N. Mithulananthan, "Influence of Distributed Generation on Congestion and LMP in Competitive Electricity Market", 2010.

[3] D. Gautam and N. Mithulananthan, "Locating Distributed Generator in the LMP-based Electricity Market for Social Welfare Maximization" Elect. Power Comp. Syst., vol. 35, pp.489-503, 2007.

[4] V. Kumar, Rohit Kumar, I. Gupta, and H.O. Gupta, "DG integrated approach for service restoration under cold load pickup," IEEE Trans. Power Del., vol. 25, no.1, pp. 398-406, January 2010.

[5] G. Celli, G., and F. Pilo, "Penetration level assessment of distributed generation by means of genetic algorithms," IEEE Proc. of Power System Conference, Clemson, SC, 2002.

[6] N. Acharya, P. Mahat, and N. Mithulananthan, "An analytical approach for DG allocation in primary distribution network", Elect. Power and Energy Syst., vol. 28, no. 10, pp. 669-678, December 2006.

[7] S. Kansal, V. Kumar, B. Taygi, "Optimal placement of different type of DG sources in distribution networks," Elect Power & Energy Syst, vol.53. pp. 752-760, 2013.

[8] M.R. Haghifam, H. Falaghi, and O.P. Malik, "Risk-based distributed generation placement," IET Gener. Transm. & Distrib., vol. 2, no. 2, pp. 252-260, 2008.

[9] S. Biswas and S.K. Goswami, "Genetic Algorithm based Optimal Placement of Distributed Generation Reducing Loss and Improving Voltage Sag Performance," ACEEE Int. J. on Electrical and Power Engineering, Vol. 02, No. 01, February 2011.

[10] S. Ghosh, S.P. Ghoshal, and S. Ghosh, "Optimal sizing and placement of distributed generation in a network system," Elect. Power and Energy Syst., vol. 32, pp. 849-856, January 2010.

[11] Y.M. Atwa and E.F. El-Saadany, "Probabilistic approach for optimal allocation of wind-based distributed generation in distribution systems," IET Ren. Power Gener., vol. 5, no. 1, pp. 79–88, November 2010.

[12] N. Jain, S. N. Singh, and S. C. Srivastava, "A Generalized Approach for DG Planning and Viability Analysis under Market Scenario," IEEE Trans. Industrial Electr., vol. 60, issue 11, November 2013.

A Novel Method to Focus Intensity by Constructing a Mechanically Controlled Bluetooth based Light Director

A. Gulia
Student, EEED
BVCOE
New Delhi, India

R. Prakash
Student, EEED
BVCOE
New Delhi, India

S. Banerjee
Assistant Professor, EEED
BVCOE
New Delhi, India

R. Sharma
Assistant Professor, EED
NIT
Kurukshetra, India

Abstract—This paper concerns with the control of intensity and direction of light for the purpose of saving electrical energy. This work presents a strategy by virtue of which a user can vary position of light and amount of light with the help of a cell phone. The main issue with current lighting schemes is that they do not have a provision for saving energy. Moreover, these schemes do not focus light on the target. Instead, light gets spread all over. The idea is to control the intensity and direction of light in a room. It can be implemented by using an array of ultra bright Luxeon LEDs, microcontroller, power converter, mechanical slider type control, Bluetooth module and Android Mobile App.

Keywords—android; bluetooth; direction; intensity; LEDs; micro- controller; slider; strategy

I. INTRODUCTION

Due to the increasing demand on electrical energy, huge effort is being made to generate it using renewable sources. Efforts are also being made to minimize the wastage of power. Light is one of the essential necessities of life. Artificial light is important in homes as well as in offices. A large amount of energy can be saved by optimizing the usage of artificial lighting. The existing lighting schemes spread light all over a room, instead of focusing it on a target. To overcome this limitation, an innovative solution has been described in this work. Currently, this paper describes the setup for a room which can later be extended to other spheres too.

These days, LED lights are being increasingly used because they offer the advantage of power saving and more lumens per watt. However, there is no control over light intensity or direction, once the product is designed. Requirement of light is generally at the place where the person is sitting in the room. Rest of the light energy gets wasted. This work presents a method which will help overcome energy wastage and result in a vast power saving.

II. COMPONENTS AND THEIR WORKING

A. Arduino Microcontroller

a) Hardware: Arduino [1] is a platform intended to make the application of interactive objects or environments more accessible. The hardware consists of a board designed around an 8-bit microcontroller, or a 32-bit Atmel. Pre-programmed into the on-board chip is a provision that allows uploading programs into the microcontroller memory without needing a chip (device) programmer (refer Fig. 1).

b) Software: The Arduino (IDE) is an application program written in C. It is designed to introduce programming to unskilled users, unfamiliar with software development and functioning. It includes code editor with features such as

Fig. 1. Arduino Micro-controller Board

automatic indentation. It compiles and uploads programs to the board easily. A program written for Arduino is called a sketch (refer Fig. 2). Arduino programs are written in C.

B. Luxeon LEDs

An LED [2] (Light Emitting Diode) is a type of solid state device. It has properties of a diode which allows unidirectional flow of current. As the electrons cross from negative charge dominating region to positive carrier dominated region, recombination occurs which results in release of energy. This energy is emitted in the form light as photons of a particular wavelength. The colour of emitted light depends on the semiconductor used for manufacturing the LED.

Luxeon LEDs have very high Lumen output which is used for large applications [3].

There are 5 types of Luxeon LEDs:

a) Luxeon I is a 1-watt LED capable of producing 30-60 lumens at 350 mA. It is used in Fenix L1S.

b) Luxeon III a 3-watt LED capable of producing 60-90 lumens at currents up to 1400 mA. It is used in Fenix L1T.

c) Luxeon V is a 5 watt LED.

d) Luxeon K2 is smaller in size but provides greater output of upto 140 Lumens at 1500 mA.

978-1-4799-6047-7/14 $31.00 © 2014 IEEE

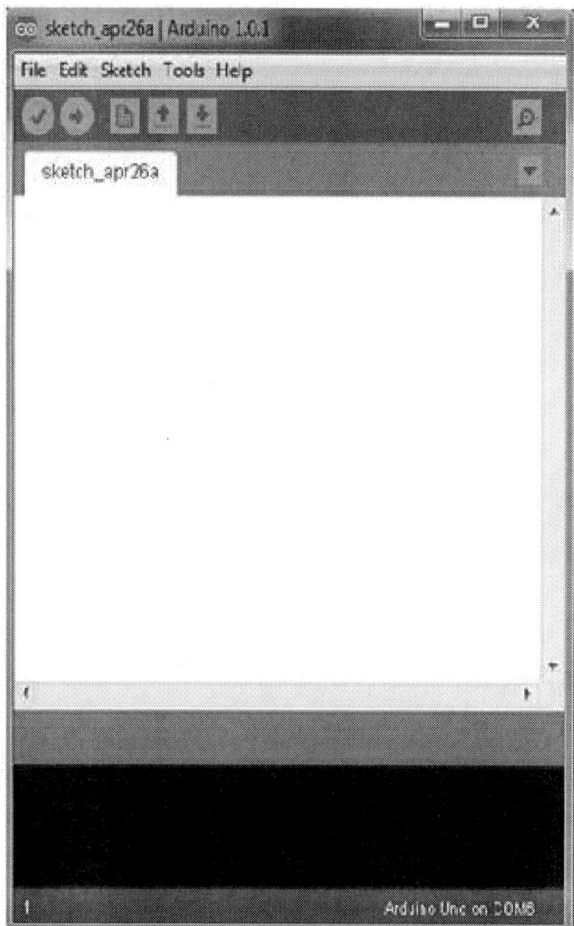

Fig. 2.   Arduino Software IDE

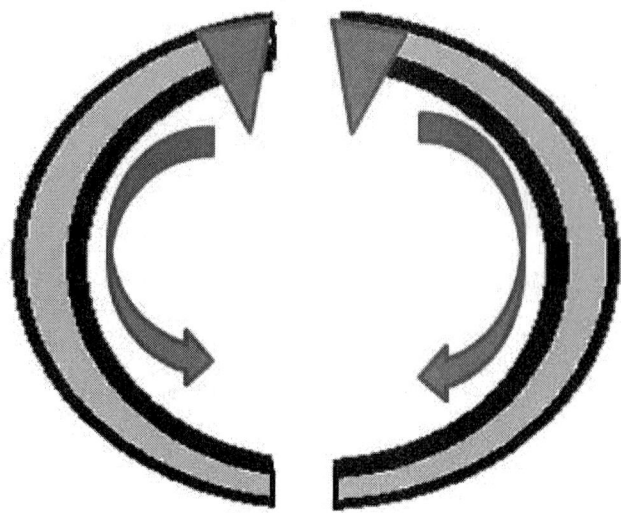

Fig. 3.   Rotatory Switch

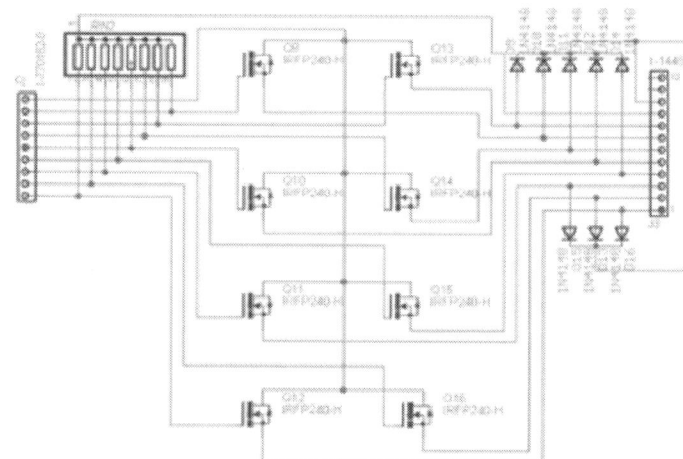

Fig. 4.   MOSFET assembly

e) Luxeon K2 TFFC is also one of the above types. TFFC stands for thin film flip chip.

In this project, Luxeon 1 type LEDs have been used.

C. Mechanically rotating Switch

It is a two input two output slider based C shaped switch. The switch gives the position of the slider in terms of voltage. It is based on the principle of potentiometer. This arrangement is used to set the position of the LEDs the user wants to light up, as shown in fig. 3.

A potentiometer placed in the centre of the arrangement is used for controlling of intensity of the light from the device.

D. MOSFETs

The metal oxide semiconductor field-effect transistor [4] (MOSFET, MOS-FET, or MOS FET) is used for amplification or switching purposes. It is a voltage controlled three-terminal device with source (S), gate (G) and drain (D) terminals. MOSFET is one of the most common transistors, finding application in both analog and digital circuits.

In the arrangement worked upon in this paper, eight MOSFETs [5] are used for controlling eight individual LED strips which face different directions (illustrated in figure 4).

E. Bluetooth [6]

Bluetooth is a wireless technology available for short-range voice and data communication. It is a low-cost and low-power technology. It acts as a platform for easy communication with smart phones. As shown in figure 5, Model HC- 05 is used for this project.

Fig. 5.   HC-05 Bluetooth Module

978-1-4799-6047-7/14 $31.00 © 2014 IEEE   104

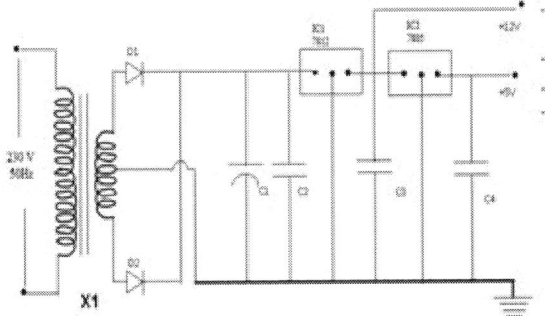

Fig. 6.   Power Circuit

F. DC supply

The circuit of this project needs two different voltages, +5V and +12V, to work. These dual voltages are supplied by a specially designed power supply [7] as illustrated in detail in figure 6.

G. Bluetooth Application

ARDUDROID is a simple android app to wirelessly control the pins of Arduino UNO microcontroller from an android based Smartphone. It is both an android app (Fig. 7) and an Arduino program. Bluetooth module is connected to Arduino through proper channel. ArduDroid employs a simple user interface to

1) Easily control Digital and PWM pins of Arduino Uno.

2) Send numeric and text commands.

3) Receive data over Bluetooth using HC -05 Bluetooth serial module from Arduino.

III.   LIGHTING CONTROL SYSTEMS

The proposed lighting control system delivers light as and when required, according to demand. Lights can turn on, off or dim or brighten under set conditions automatically.

Users can exercise a control over their lighting levels to provide desirable working conditions. Lighting control helps in cost reduction and thus results in energy conservation. The block diagram and working of the system is explained next.

IV.   WORKING

The block diagram of the proposed system is exhibited in figure 8. The switching operation of the MOSFETs is governed by the microcontroller. The microcontroller can be controlled mechanically using a potentiometer. Alternately, a Bluetooth module can be tasked to operate the same. A step by step algorithm detailing the process is mentioned henceforth.

1. Luxeon led array is connected to the power supply via a MOSFET.

2. MOSFET is used to control the on/off of Luxeon array.

3. Gating pulse is controlled by Arduino which is programmed accordingly.

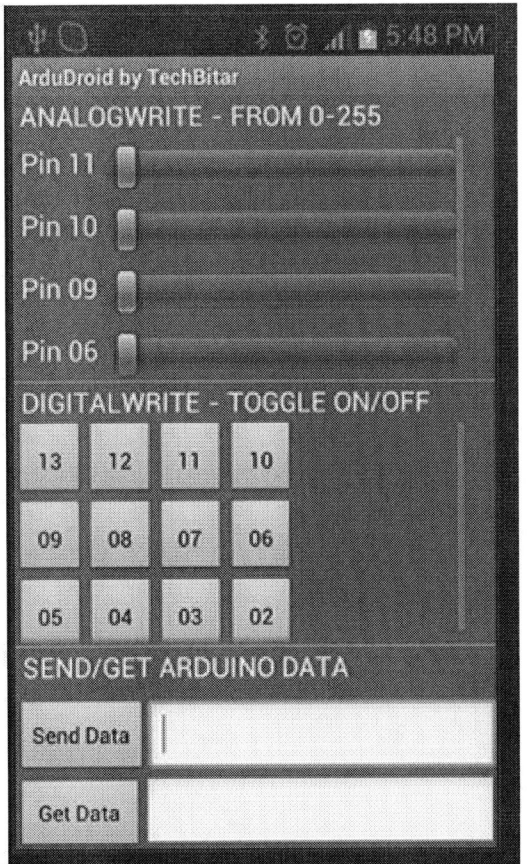

Fig. 7.   Ardudriod Smartphone App

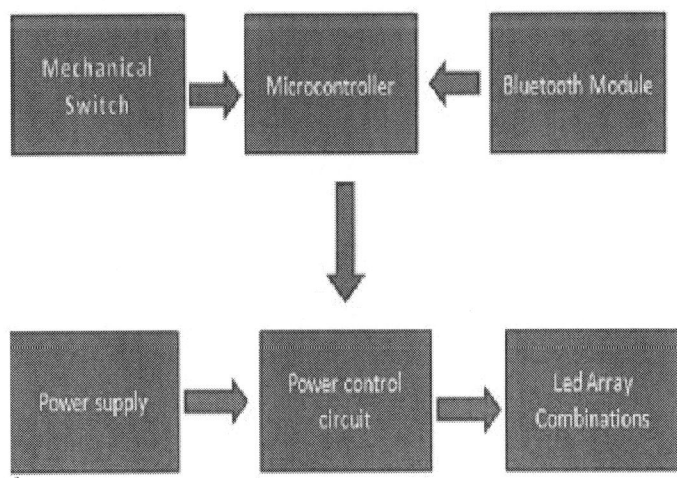

Fig. 8.   Block Diagram

978-1-4799-6047-7/14 $31.00 © 2014 IEEE

The microcontroller works in two modes:

I. Use based on Potentiometer Based Mechanical Switch

1) The Mechanical Switch is adjusted by the user as per the requirement.

2) This information is acquired by the Arduino microcontroller by its analog to digital converter. A particular value of resistance corresponds to a definite switching signal pattern. This helps control the lights as required, by giving gating pulses to MOSFET corresponding to certain LED Arrays.

II. Use based on Bluetooth based Android App

1) The user decides the lighting requirements and sets the priorities to be conveyed to the Arduino via Bluetooth.

2) The information is received by the Arduino via the HC-05 module and processed. The control signals are generated for the MOSFET and accordingly the light arrays are controlled.

Fig. 9 Illustrates the fabricated hardware

V. RESULT

Any strip can be lighted or put off by applying the control either by using the potentiometer or through Bluetooth. Fig. 10 shows a snapshot of a glowing LED strip which has been turned on for verifying the proposed control strategy. All other strips can be individually or collectively lighted by the aforementioned approach. So intensity of light can be focused on a specific area according to the user need.

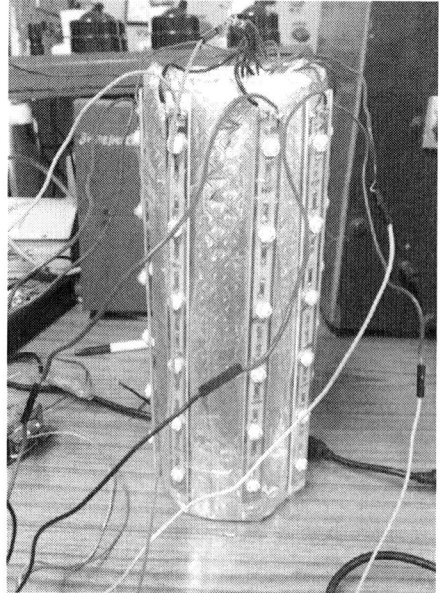

Fig.10   Snapshot of glowing LED strip

VI.   CONCLUSION

A device has been developed which provides saving in electrical energy along with complete control of lighting parameters such as intensity and direction. This is achieved by either mechanical control or a Bluetooth based control via phone.

Thus the project discussed presents a new way to inculcate the habit of energy saving by providing efficient and effortless method for the same. The product described here is service- free, robust and power saving and has a huge potential in forthcoming years.

REFERENCES

[1]   Massimo Banzi. Getting Started with Arduino. Maker Media Inc, October,2011.

[2]   Patrick Mottier.   LED for Lighting Application.   ISTE Ltd, 27-37 St.George's Road, LondonSW 19 4EU UK, 2009.

[3] Steve Winder. Power Supplies for LED Driving. Elsevier Inc, 2008.

[4]   B.Jayant Baliga. Advanced Power MOSFET Concepts. Springer Science,2010.

[5]   Laszlo Balogh. Design and application guide for high speed mosfet gate drive circuits.

[6]   Nathan J Muller. Bluetooth Demystified. Tata McGraw-Hill Education, 2001.

[7] Li Zhang William Shepherd. Power Converter Circuits. Marcel Dekker

Neural Network Based Enhancement of Power Quality in Distribution System

Priya Nayar
Department of Electrical Engineering,
Indian Institute of Technology,
New Delhi-110016, India.
E-mail: priya1.p6@gmail.com

Bhim Singh
Department of Electrical Engineering,
Indian Institute of Technology,
New Delhi-110016, India.
E-mail: bsingh@ee.iitd.ac.in

Sukumar Mishra
Department of Electrical Engineering,
Indian Institute of Technology,
New Delhi-110016, India.
E-mail: Sukumar@ee.iitd.ac.in

Abstract—**An application of artificial intelligence is presented in solving power quality problems using a distribution static compensator (DSTATCOM) in the distribution system involving lagging pf loads. A set of three nonlinear neurons is used to obtain the three phase compensating currents. Another set of three layered feed-forward neural network controls the compensating currents. The developed model works accurately under varying load conditions and provides good dynamic response to the step changes in the load currents. A real time performance is achieved using SIMULINKR/Sim-powersystem (SPS) toolboxes and simulated results adhere to the IEEE–519standard for improvement of power quality.**

Keywords—Power quality, DSTATCOM, ANN, PCC, power factor regulation.

I. INTRODUCTION

The main concerns of an electric utility are to supply clean and reliable power while maintaining a sinusoidal voltage [1]. There is a sharp rise of nonlinear loads in modern supply system like rectifier devices employed in telecommunications, household appliances like adjustable speed drivers, TVs and ovens. Non-sinusoidal currents are drawn from the utility resulting in injection of harmonics content at the point of common coupling (PCC). This results in overheating of the transformer, cables, motors and power factor correction capacitors. This may also lead to noise and severe vibrations in the induction motors [2-3] connected in the system that may decrease machine capacity, efficiency and a reduced life span [4]. Hence, DSTATCOM (Distribution Static Compensator) has been reported for power quality enhancement in the past [5-9].These approaches have been possible to be used due to the presence of upgraded micro-controllers and Hall-effect sensors.

The neural networks based approaches are reported of improving the control in power electronic based systems [10-11]. This paper presents an application of artificial neural network (ANN) based extended delta control algorithm for power quality enhancement [12-13]. It is an indirect control technique involving extraction of positive sequence active power component of load currents with no shift in its phase. It involves the weights being trained for a given load and system parameters. It is self adapting and has extremely fast computing characteristics making it apt to handle system uncertainties, nonlinearities and parameter variations. It has

two stages viz., the three adaptive nonlinear neurons (Extended Delta) to separate the active power component from nonsinusoidal load currents and the second to control by a three layered feed-forward network the gate pulse generation for the VSC based DSTATCOM. As compared to the classical techniques which involve reasonable transformation and computation, it is simple, thus imposes less computation burden and is robust. In addition, it is flexible and adapts itself to the parameter changes and fluctuations in the load. The analysis of working of the DSTATCOM has been validated subsequently for mitigation of reactive power demand and harmonics alongwith load leveling as an integral part for the proposed system under linear/nonlinear consumer loads. The guidelines set by an IEEE-519 standard for power quality enhancement [14] are strictly adhered to in this system.

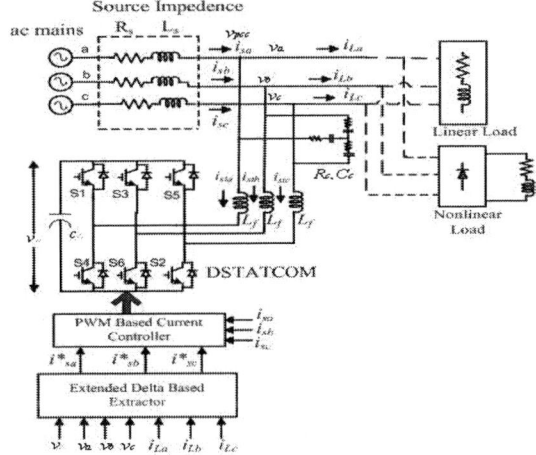

Fig.1 Schematic of ac supply with DSTATCOM

II. SYSTEM CONFIGURATION

The developed DSTATCOM based distribution system under linear (resistive-inductive) and nonlinear consumer loads has been illustrated in Fig.1. A high pass ripple filter (R_c, C_c) at the PCC eliminates high frequency switching noise in PCC voltage. The DSTATCOM is connected in shunt through interfacing inductors that smoothens out the compensating DSTATCOM currents. The negative feedback PI controller regulates the ac voltage which decides the reactive power control provided by the DSTATCOM. The

978-1-4799-6047-7/14 $31.00 © 2014 IEEE

PWM based current controller employed in the algorithm enables the supply currents to track the extracted reference currents. The DSTATCOM injects compensating currents to neutralize the load current harmonics in order to render clean and undistorted supply currents.

The design specifications of the proposed auxiliary constituents of DSTATCOM (25kVA) like interfacing inductors (L_f) and capacitor value (C_{dc}) are given as,

A. DC Link Capacitor

The governing equation of dc bus capacitor design is given by the expression and depends on the fall in the dc link voltage on addition of loads and the corresponding increase in voltage on load removal. According to the law of energy conservation, the expression for C_{dc} is given as,

$$k1\{3V_{ph}(a_1i_1)t\} = 0.5\{(v_{dc})^2 - (v_{dc1})^2\}$$

where, v_{dc} depicts the nominal value of dc link voltage and v_{dc1} depicts the lowest level of dc link, "a_1" depicts the factor of overloading, V_{ph} depicts phase voltage, i_1 depicts phase current of the DSTATCOM and t depicts time taken for recovery of dc link voltage. Considering the parameters as, v_{dc} = 700V, v_{dc1}= 680V, V_{ph}= 220V, I= 40 A, t= 0.5s, a1= 1.3, the value of k1 factor is between 0.5 and 0.9. The selection of the capacitor at the dc link is done on the basis of the permissible variation in the dc link voltage and also on the supply current variation. A change of energy Δe_{dc} occurs when dc bus voltage changes from v_{dc0} to v_{dc1}. This energy is exchanged with the energy absorbed or dissipated by the filter inductance.

$$C_{dc} = \frac{3L(i_1^2 - i_0^2)}{(2\varepsilon v_{dc}^2)} \tag{1}$$

where, ε denotes the maximum permissible ripple in the dc link voltage. On the basis of 2% ripple in the voltage, v_{dc} is taken as 700V and a current variation of 5A to 75A, giving a capacitance value of 6410 μF which is taken as 12,000μF.

B. AC Interfacing Inductors

The AC interfacing inductors design depends on the peak - current ripple (i_{peak}), DC link voltage and the frequency of modulation (f_{mod}) of the VSC are as,

$$L = \frac{\sqrt{3}\,m\,V_{dc}}{12\,af_{mod}i_{peak}} \tag{2}$$

where, m is factor of modulation, V_{dc} is the dc link voltage, a depicts overloading factor, f_{mod} depicts switching frequency of the VSC and i_{peak} depicts measured peak-peak VSC ripple current.

For example, in an extreme case, m is considered as 1, dc link voltage as 700V, a=1.2, f_{mod}=10 kHz. Generally i_{peak} is 10% of the VSC current thus giving the value of interfacing inductor L= 3.5 mH which is rounded to 3mH which is closest to the designed value in the system.

III. CONTROL ALGORITHM

The control technique is of utmost importance so that the system can sustain sudden disturbances. An indirect control based ANN is employed to generate the reference supply currents taking the three phase voltages (v_a, v_b, v_c) and load currents (i_{la}, i_{lb}, i_{lc}) as the feedback signals as illustrated in Fig.2.

As the supply voltages are sinusoidal, their magnitude [11] is derived as,

$$V_m = [\frac{2}{3}\{(v_a^2) + (v_b^2) + (v_c^2)\}]^{1/2} \tag{3}$$

where, v_a, v_b and v_c depict phase voltages estimated as,

$$v_a = V_m\sin(\omega t),\ v_b = V_m\sin(\omega t-120°),\ v_c= V_m\sin(\omega t-240°) \tag{4}$$

A. In-phase Component of Reference Supply Currents Estimation

The derivation of unit templates is as,

$$u_{Ap} = v_a / V_m,\ u_{Bp} = v_b / V_m,\ u_{Cp} = v_c / V_m \tag{5}$$

The weights computed for the load current active fundamental component extracted and trained through Extended Delta rule as,

$$W_{ap(n+1)} = W_{ap(n)} + \mu(n)\{i_{La(n)} - W_{ap(n)}u_{Ap(n)}\}u_{Ap(n)} * \frac{\partial(W_{ap(n)}u_{Ap(n)})}{\partial(W_{ap(n)})}$$

$$W_{bp(n+1)} = W_{bp(n)} + \mu(n)\{i_{Lb(n)} - W_{bp(n)}u_{Bp(n)}\}u_{Bp(n)} * \frac{\partial(W_{bp(n)}u_{Bp(n)})}{\partial(W_{bp(n)})}$$

$$W_{cp(n+1)} = W_{cp(n)} + \mu(n)\{i_{Lc(n)} - W_{cp(n)}u_{Cp(n)}\}u_{Cp(n)} * \frac{\partial(W_{cp(n)}u_{Cp(n)})}{\partial(W_{cp(n)})} \tag{6}$$

The value of μ (the coefficient of convergence) conversely affects the convergence rate and precision in accuracy and is between 0.01 and 1.0. Hence a tradeoff between precision and speed of convergence is observed for a value of μ equal to 0.2. Averaging of weights is given to eliminate current components unbalance as,

$$W_{p(n)} = \{ W_{ap(n)} + W_{bp(n)} + W_{cp(n)} \}/3 \tag{7}$$

A low pass filter is used for smoothening the active power components of load current giving an output W_{plpf}.

The reference real power component is computed following the comparison of the reference and the sensed dc link voltage of DSTATCOM. This gives an error ($v_{dcer(n)}$) which at the nth sampling interval given as,

$$v_{dcer(n)} = v^*_{dc(n)} - v_{dc(n)} \tag{8}$$

This error signal ($v_{dcer(n)}$) is fed to a PI (Proportional-Integral) regulator which regulates the dc link voltage of the DSTATCOM.

At the nth sampling time interval, the dc voltage PI controller gives an output as,

$$W_{loss(n)} = W_{loss(n-1)} + K_{pdc}\{v_{dcer(n)} - v_{dcer(n-1)}\} + K_{idc}v_{dcer(n)} \tag{9}$$

where, $W_{loss(n)}$ depicts loss component of the active power component of supply current which adds with the weight estimated above. K_{pdc} and K_{idc} depict the corresponding proportional and integral gains for dc-link voltage PI regulator.

$$W_{p1(n)} = [W_{loss} + W_{plpf}] \tag{10}$$

Three phase fundamental reference active power components of supply current components are thus derived as,

$$i^*_{spa} = W_p u_{Ap}\quad i^*_{spb} = W_p u_{Bp}\quad i^*_{spc} = W_p u_{Cp} \tag{11}$$

Similarly, the quadrature component of the reference supply currents is estimated as below.

B. Quadrature Component of Reference Supply Currents Estimation

There is a voltage drop at the line impedance even under unity power factor loads; hence to get zero voltage regulation, the reactive power is fed from some other source like a DSTATCOM to the system [8]. In order to control the terminal voltage, a PI regulator is employed to which are fed the magnitude of the sensed terminal voltage and its reference value. The ac voltage error, $V_{er(n)}$ is derived as,

$$V_{er(n)} = V^*_{m(n)} - V_{m(n)} \tag{12}$$

where $V_{m(n)}$ denotes the nth instant sensed terminal voltage while $V^*_{m(n)}$ is its reference magnitude.

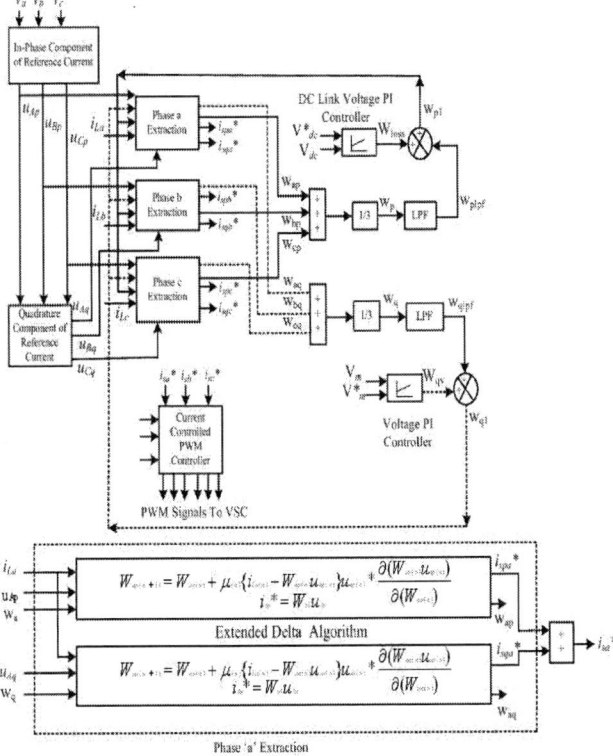

Fig.2 Block diagram of Extended Delta based reference current estimation

The terminal voltage PI regulator for the terminal voltage gives an output $W_{qv(n)}$ at the nth interval to bring the voltage amplitude to a constant value.

$$W_{q1V(n)} = W_{q1V(n-1)} + K_{pa}\{V_{er(n)} - V_{er(n-1)}\} + K_{ia}V_{er(n)} \tag{13}$$

where K_{pa} and K_{ia} depict proportional gain and the integral gain respectively of the voltage PI regulator while $v_{er(n)}$ and $v_{er(n-1)}$ are the corresponding errors incurred in the voltage at the nth and (n-1)th sampling time intervals and $W_{q1}^*{}_{vr(n-1)}$ is the quadrature component magnitude of reference fundamental current at the (n-1)th sampling time interval.

The quadrature unit templates v_a, v_b and v_c are derived as,

$$u_{Aq} = (u_{AP} + u_{CP})/\sqrt{3},$$

$$u_{Bq} = (3u_{AP} + u_{BP} - u_{CP})/2\sqrt{3},$$

$$u_{Cq} = (-3u_{AP} + u_{BP} - u_{CP})/2\sqrt{3} \tag{14}$$

The weights of the reactive load current component are computed as,

$$W_{aq(n+1)} = W_{aq(n)} + \mu(n)\{i_{La(n)} - W_{aq(n)}u_{Aq(n)}\}u_{Aq(n)} * \frac{\partial(W_{aq(n)}u_{Aq(n)})}{\partial(W_{aq(n)})}$$

$$W_{bq(n+1)} = W_{bq(n)} + \mu(n)\{i_{Lb(n)} - W_{bq(n)}u_{Bq(n)}\}u_{Bq(n)} * \frac{\partial(W_{bq(n)}u_{Bq(n)})}{\partial(W_{bq(n)})}$$

$$W_{cq(n+1)} = W_{cq(n)} + \mu(n)\{i_{Lc(n)} - W_{cq(n)}u_{Cq(n)}\}u_{Cq(n)} * \frac{\partial(W_{cq(n)}u_{Cq(n)})}{\partial(W_{cq(n)})} \tag{15}$$

Hence the mean weight of the fundamental reactive component of reference currents is as,

$$W_{q(n)} = \{W_{aq(n)} + W_{bq(n)} + W_{cq(n)}\}/3 \tag{16}$$

A low pass filter (LPF) to eliminate the low frequency components giving a ripple free output W_{qlpf}.

$$W_{q1(n)} = [W_{qv(n)} - W_{qlpf}] \tag{17}$$

Henceforth, the three phase fundamental reference reactive power component of supply currents is given as,

$$i^*{}_{saq} = W_{q1}u_{Aq}, \; i^*{}_{sbq} = W_{q1}u_{Bq} \text{ and } i^*{}_{scq} = W_{q1}u_{Cq} \tag{18}$$

C. Reference Supply Currents Estimation

The summation over in-phase and quadrature components of the reference supply currents gives the total reference supply currents as,

$$i^*{}_{sa} = i^*{}_{sqa} + i^*{}_{spa}, \; i^*{}_{sb} = i^*{}_{sqb} + i^*{}_{spb} \text{ and } i^*{}_{sc} = i^*{}_{sqc} + i^*{}_{spc} \tag{19}$$

These supply reference currents ($i^*{}_{sa}$, $i^*{}_{sb}$, $i^*{}_{sc}$) and the supply currents sensed (i_{sa}, i_{sb}, i_{sc}) are compared for individual phases and the resulting phase current error is amplified by PI current regulator. Thereafter, outputs of these current controllers on comparison with a 10 kHz carrier signal (triangular wave) generate unipolar gating signals for VSC based DSTATCOM.

Thus the ANN based current control scheme (an indirect control) results in slow–varying supply currents (unlike compensating currents) and hence requires reduced computations. In addition, this scheme compensates the supply currents instantaneously (with a negligible computational delay) [9].

IV. MATLAB BASED MODELING OF THE SYSTEM

Fig.3 illustrates the developed MATLAB model of proposed extended delta control scheme employing the selective compensation done by the DSTATCOM is simulated in MATLAB environment employing sim-power system (SPS) toolbox and a discrete step based solver. The DSTATCOM is modeled along with the non-stiff supply, a ripple filter connected in shunt and balanced/unbalanced linear/nonlinear consumer loads. The three-phase voltage supply along with impedance is considered as a non-stiff supply. A set of load consisting of diode rectifier and star-connected resistive-inductive load are used. The controller

978-1-4799-6047-7/14 $31.00 © 2014 IEEE

employs a DSTATCOM and a capacitor on the dc link that makes it a self supporting dc bus.

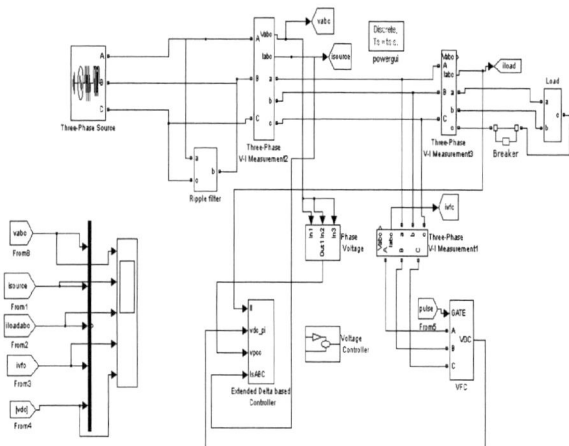

Fig.3 MATLAB Simulation of the Model

V. RESULTS AND DISCUSSION

Figs. 4-6 demonstrate the working of the DSTATCOM compensated utility system under linear (RL) and nonlinear loads for voltage regulation, reactive power mitigation, and harmonics compensation. The variation of the electrical quantities are shown viz. three phase PCC voltages (v_{pcc}), supply currents as i_s, ac load line currents as i_L, The DSTATCOM currents as i_c and dc link voltage of the DSTATCOM as v_{dc}.

A. Performance of DSTATCOM under Linear Load

Fig.4 shows the transient response when the load is removed from one phase at t=1s and load balancing is demonstrated. The DSTATCOM compensates for the load unbalance thus keeping the supply currents balanced with less harmonic content in them and these current are in phase with the PCC voltages thus validating the DSTATCOM performance using the proposed algorithm.

B. Performance of DSTATCOM under Nonlinear Load

Fig. 5 demonstrates the working of the concerned system under nonlinear load. A balanced bridge rectifier with 28kW dc link resistive load is employed. At t=1s, the load is shed from one phase and the subsequent load perturbations are studied. It has been observed that in spite of the load currents being distorted; the PCC voltage and supply currents are always balanced and sinusoidal thus proving the efficacy of the control algorithm for the DSTATCOM.

C. Power Quality Improvement in the System

Fig.6 illustrates the THD and harmonic spectra of the PCC voltage, supply current and load current. The supply current has a %THD equal to 2.65% as compared to the %THD of the load current of 23.45%. Thus it is shown that by employing the proposed control algorithm, effective mitigation of

harmonics and reactive power demand is satisfactorily done by the DSTATCOM along with load balancing.

VI. CONCLUSIONS

A new current decomposition neural network based technique with an indirect control of current has been proposed for a prioritized selective compensation of the components of power quality. Performance of DSTATCOM with proposed control scheme has been demonstrated for effective compensation of current harmonics, priority based mitigation of reactive power demand along with load balancing. The scheme has been found effective in maintaining practically stable PC voltages and supply currents even under the presence of load harmonics and single-phasing. The algorithm is simple and a total harmonic distortion of as low as 2.65% is achieved. The proposed indirect current control extracts reference current with zero phase shift as the technique is flexible and adapts extended delta to track in a closed loop.

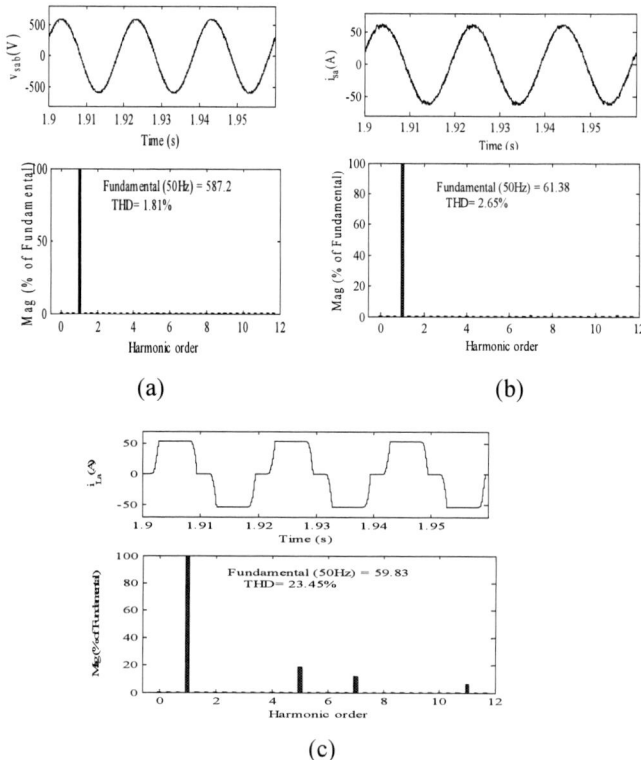

Fig.6 Harmonic spectra and THD for (a) supply voltage, (b) supply current and (c) load current

REFERENCES

[1] J. Mazumdar, R.G.Harley, F.C.Lambert and G.K.Venayagamoorthy, "Neural network based method for predicting nonlinear load harmonics", *IEEE Trans. Pow. Electron.*, vol. 22, no. 3 pp.1036-1045, 2007.

[2] B. Singh and J.Solanki, "A comparative study of control algorithms for DSTATCOM for load compensation", in Proc. *IEEE Conf. Ind. Tech. ICIT*, 2006, pp.1492-1497.

[3] A. Dell' Aquila, M. Marinelli, V.G. Monopoli and P Zanchetta, "New power -quality assessment criteria for supply systems under unbalanced and nonsinusoidal conditions", *IEEE Trans. Pow. Del.*, vol. 19, no. 3, pp.1284-1290, July 2004.

[4] M.M.A. Aziz, A.F. Zobaa and A.A. Hosni, "Neural network controlled shunt active filter for nonlinear loads", in *Proc. IEEE Conf. on Pow. Sys., MEPCON*, pp.180-188, 2006.

[5] B.Singh, J.Solanki and V.Verma, "Neural network based control of reduced rating DSTATCOM", in *Proc. of IEEE Conf., INDICON*, pp.516-520, 2005.

[6] S. Kang, H. Zhang and Y. Kang, "Simulation analysis of time-frequency based on waveform detection technique for power quality application", in *Proc. IEEE Int. Conf. Control and Decision*, pp.2529-2532, 2010.

[7] Y.A.-R.I Mohamed, M.A.-Rahman and R. Seethapathy, "Robust line-voltage sensorless control and synchronization of LCL-filtered distributed generation inverters for high power quality grid connection", *IEEE Trans. Pow. Electronics.*, vol. 27, no. 1, pp.87-98, Jan. 2012.

[8] Y.A.-R.I Mohamed, M.A.-Rahman and R. Seethapathy, "Robust line-voltage sensorless control and synchronization of LCL-filtered distributed generation inverters for high power quality grid connection", *IEEE Trans. Pow. Electronics.*, vol. 27, no. 1, pp.87-98, Jan. 2012.

[9] K. Ilango, A. Bhargav, A. Trivikram and P.S. Kavya, "Power quality improvement using STATCOM with renewable energy sources", in *Proc. IEEE Int. Conf. Pow. Electronics. (IICPE)*, pp.1-6, Dec. 2012.

[10] X. Jing and L. Cheng, "An optimal-PID control algorithm for training feed-forward neural networks," *IEEE Trans. on Ind. Electron.*, vol. 60, no. 6, pp. 2273–2283, Jun. 2013.

[11] B. Singh, J.Solanki and V. Verma, "Neural network based control of DSTATCOM with rating reduction for three phase four wire system", in *Proc. of IEEE Int. Conf. Pow. Electron. and Drives Sys.*, pp.920-925, 2005.

[12] Laurene Fausett, *Fundamentals of Neural Networks*, Pearson Publication, 2006.

[13] S.O. Haykin, *Neural Networks and Learning Machines*, 3rd ed., Pearson Prentice Hall, Canada, 2008.

[14] *IEEE Recommended Practices and requirement for Harmonic control on electric power system*, IEEE Std.519, 1992 standard.

Improved Power Quality SMPS for Computers Using Bridgeless PFC Converter at Front End

Shikha Singh
Department of Electrical Engineering
Indian Institute of Technology, Delhi
New Delhi, India, 110016
Ishikha.singh@gmail.com

Bhim Singh
Department of Electrical Engineering
Indian Institute of Technology, Delhi
New Delhi, India, 110016
bsingh@ee.iitd.ac.in

G. Bhuvaneswari
Department of Electrical Engineering
Indian Institute of Technology, Delhi
New Delhi, India, 110016
bhuvan@ee.iitd.ac.in

Abstract— **This paper envisages to design, analyze and assess the performance of a new Power Factor Corrected (PFC) bridgeless converter based power supply for Personal Computers. The proposed Switched Mode Power Supply is a combination of a bridgeless converter operating as a PFC at the front end and an isolated converter at the back end for obtaining multiple dc output voltages. The proposed configuration eliminates the diode bridge at the front end thereby offering low conduction losses and high power quality in line with IEC-61000 standard norms. The front end converter offers fast dynamic response compounded with good output voltage regulation where the output is characterized by a low 100 Hz ripple. The performance of the proposed power supply is analyzed to validate the design and its improved performance is demonstrated by means of simulations.**

Keywords— *Bridgeless PFC converter, Switched mode power supply (SMPS), Discontinuous inductor current mode (DICM), Power quality*

I. INTRODUCTION

In the present era, Personal Computers (PCs) have become ubiquitous for improving work productivity with reasonable accuracy. A switched mode power supply (SMPS) with multiple dc outputs is used for supplying power to different parts of a PC [1-2]. Fig.1 shows the system configuration of one such SMPS. It employs ac-dc conversion at the front end using a single-phase diode-bridge rectifier (DBR) followed by an isolated converter with multi-winding high frequency transformer (HFT) and filters. The existing power supply has power quality (PQ) problems like periodically dense harmonic rich input current, high crest factor, low power factor, high voltage and current stresses and poor output voltage regulation. Further, it uses high values of capacitors to reduce 100Hz ripple that affects the distribution system adversely. The neutral current in the distribution system increases which overloads the neutral conductor, transformer and distorts the voltage [3-4]. This violates the limits set by various PQ standards [5]. To attain improved efficiency with enhanced output voltage regulation in an SMPS, employing high frequency switching and power factor correction (PFC) technique has become essential [2].

Various PFC converters are used in modern SMPS circuits in single stage or two stage configurations to fix these PQ problems [6]. These modern power supplies are able to

This work was supported by the Department of Science and Technology (DST), Govt. of India. Grant Number RP02506.

maintain low harmonic content and almost unity power factor at the point of common coupling (PCC) and also capable of regulating the output voltages stiffly even under varying supply voltage and load conditions. In single stage power supplies, various isolated PFC ac-dc converters are used for achieving multiple outputs. The single stage conversion offers light weight, high efficiency and low component count. However, these power supplies are bound to use bulky output capacitors; they also suffer from high device stresses, complex control, slow dynamic response and poor output voltage regulation [7]. To overcome these problems, two stage SMPSs are preferred in PCs. These offer regulated output voltages over a wide variation in input ac mains voltage, fast dynamic response and reduction in the component stress; also, the output capacitors are not as bulky as their single-stage counterparts. However, the efficiency of these power supplies is slightly lower than the single stage power supplies.

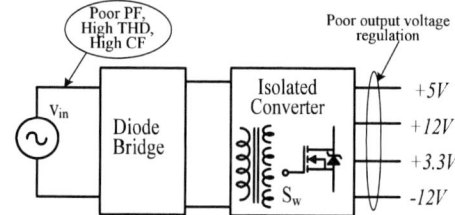

Fig. 1 Circuit configuration of single-stage SMPS system for computers

To eliminate the DBR, PFC bridgeless ac-dc converter configurations are now in demand because of their multiple advantages such as improved PF, stiffly regulated output dc voltages and low harmonic content in the input current. However, the component count is increased in any PFC bridgeless converter; although, the conduction losses are reduced. Various bridgeless converter configurations are reported in the literature which eliminate the diode drop in the current path, thereby improving the efficiency [8-12]. However, in these buck and boost configurations, the output voltage range is limited which makes it unsuitable for a computer power supply. Therefore, buck-boost converter configurations are widely used in the computer power supplies. Wei et al [13] have proposed a bridgeless buck-boost converter that uses three switches in the conduction path which enhances the functionality of the circuit, but increases the overall cost of the power supply. Various configurations of other buck-boost converters such as Cuk converter and SEPIC are widely used for PQ improvement [14-15]. However, the bridgeless zeta converter based SMPS is still un-explored. Therefore, an attempt is made here to design a computer power supply which incorporate features like low cost, high efficiency and satisfactory performance with improved power quality at the single ac mains for a wide range of input voltage and load change.

II. SYSTEM CONFIGURATION AND OPERATING PRINCIPLE

The circuit configuration of the proposed SMPS system is shown in Fig. 2. The first stage consists of a non-isolated PFC bridgeless converter while the second stage employs an isolated converter. The PFC converter is made up of two bridgeless zeta converters; the upper one operates during the positive half cycle and the lower one operates during the negative half cycle of the single-phase ac supply to eliminate the diode bridge at the front end. The inductors L_{p2} and L_{n2} are connected together at the output side and are chosen such that they operate in Discontinuous inductor conduction mode (DICM) to eliminate the need for an input current sensor. The output of this PFC converter is connected to the isolated dc-dc converter to achieve multiple dc voltage outputs. Two

capacitors are used at the primary side of the HFT to produce a constant mid-point voltage. So, the switching elements have to withstand only half of the input voltage. It alternately switches the voltage across the primary winding such that primary winding experiences a positive and negative voltage swing, thereby, fully utilizing the core flux and the secondary winding. Additionally, due to its full-wave nature, the secondary side of the isolated converter operates at twice the switching frequency. So, the size of the inductors and capacitors on the secondary windings are small leading to savings in cost and space. The operation of a bridgeless zeta converter in the positive and negative cycles of the ac voltage is shown in Fig. 3. In the first mode of the switching cycle called as inductor charging mode, switch S_p turns on and inductor L_{p2} starts storing energy making its current change from zero to maximum value. Diode D_{p2} remains off while diode D_{p1} is on to complete the current path as shown in Fig. 4a. The second mode called as inductor discharging mode starts when the inductor current reaches its peak value; S_p turns off at the beginning of the second mode. The inductor stored energy is discharged and the inductor current i_{Lp2} decreases to zero as shown in Fig. 4b. Now, Diode D_{p2} turns on to complete the current path. In the last mode of the PWM cycle, called as inductor zero current mode, the inductor

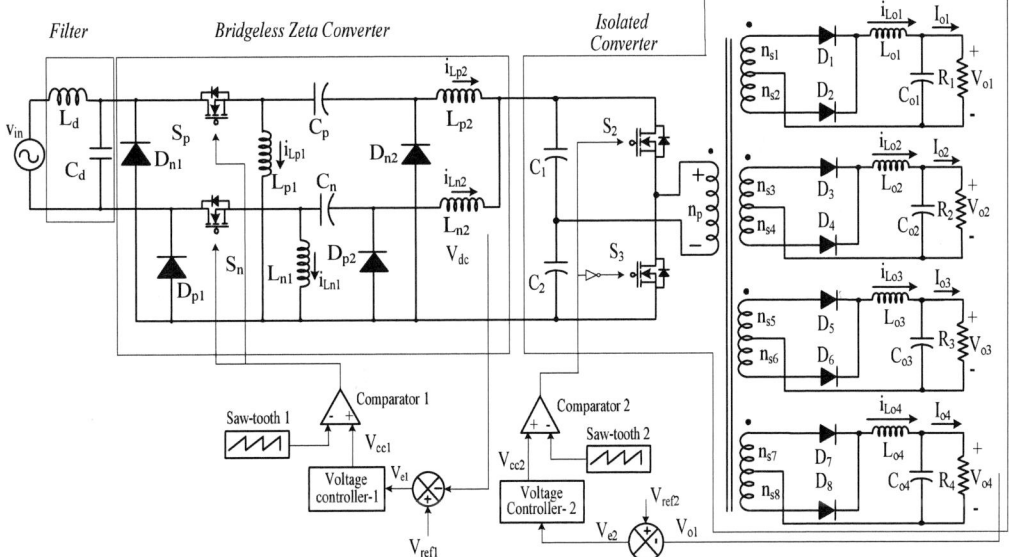

Fig. 2 Circuit configuration of the SMPS system using PFC bridgeless zeta converter

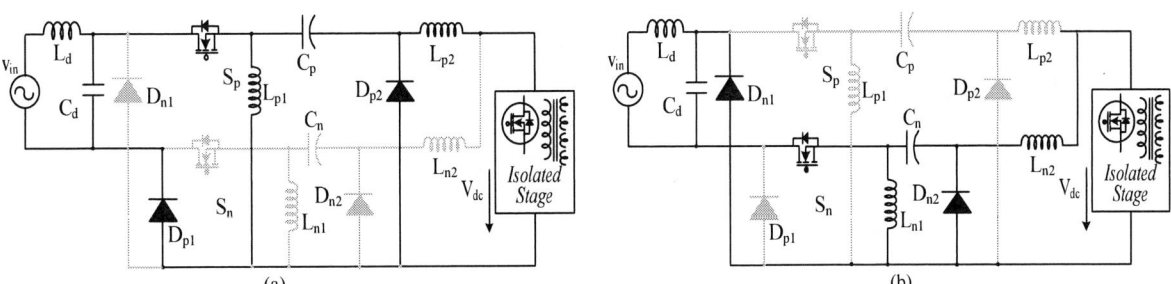

Fig. 3 Positive half cycle operated and negative half cycle operated PFC zeta converter

current i_{Lp2} remains zero for the rest of the switching cycle. Neither the diode nor the switch is on during this period as shown in Fig. 4c. The same operation repeats in each switching cycle until the ac voltage remains positive. When the ac voltage becomes negative, the lower zeta converter operates in the same way as it was described so far for the upper one.

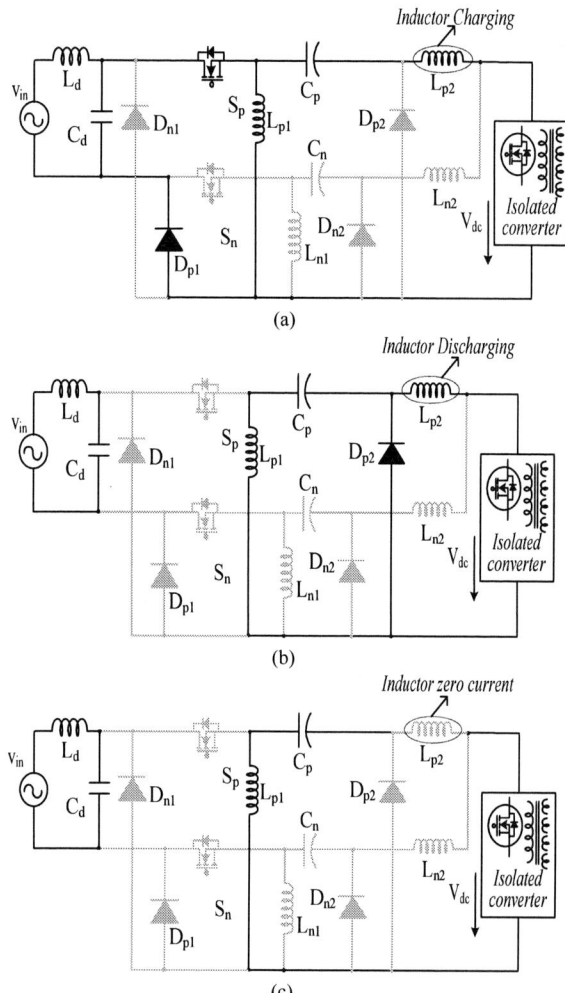

Fig. 4 Operating modes of positive half cycle operated PFC bridgeless zeta converter

The isolated dc-dc converter has two high frequency switches and four secondary windings with filters. Both the switches are turned on and off twice in one PWM period with sufficient dead time to avoid shoot-through. Only one of the output voltages is sensed; it is compared with the set reference value. The error thus generated is passed through a voltage controller which, in turn, generates the PWM pulses through a comparator which are given to the switches of the isolated converter. The voltage delivered at any output is affected by the changes in the demand of power in the other outputs and therefore the duty ratio has to be changed to regulate the affected output.

III. DESIGN AND ANALYSIS OF SMPS SYSTEM USING PFC BRIDGELESS ZETA CONVERTER

The component values of proposed power supply are estimated by calculating the changes in the inductor current during switch on and off conditions. The input inductor of the PFC bridgeless converter is designed in CCM while the output inductor is designed in DCM. Table-I shows the design equations and selected values while designing the proposed power supply with following specifications.

Design specifications: input voltage v_{in}=220V_{rms}, average voltage V_d=198V, front end converter switching frequency=20kHz, capacitor voltage ripple=10%, V_{dc}=300V (same as isolated converter input voltage), output power ≈ 350W, isolated converter switching frequency 60 kHz, output voltages and currents=12V 12.5 A, 5V 23A, 3.3V 16A, -12V 0.8A.

TABLE I
DESIGN EQUATIONS AND SELECTED DESIGN VALUES

Design Equation	Parameters	Calculated values	Selected Value
A. Front end converter			
$L_{p1min}=L_{n1min}=\dfrac{DV_d}{2f_zI_{Lp1}}$	$\dfrac{0.25*198V}{2*20kHz*1.59A}$	0.77mH	1.5mH
$C_1=\dfrac{DV_{dc}}{Rf_z\Delta V_{C1}}$	$\dfrac{0.25*300V}{257\Omega*20kHz*31.1V}$	0.47mF	0.46mF
$L_{p2min}=L_{n2min}=\dfrac{D_i V_{dc}}{2I_{dc}*f_z}$ $L_{p2}==L_{n2}=\dfrac{L_{p2min}}{10}$	$\dfrac{0.25*300V}{2*1.16A*20kHz}$	0.16mH	100µH
B. Isolated converter			
$\dfrac{C}{2}=\dfrac{I_{dc}}{2\omega\Delta V_{dc}}$	$\dfrac{1.16A}{314*6V}$	0.615mF	630 µF
$n=\dfrac{V_{o1}}{D_d V_{dc}}$	$\dfrac{12V}{0.45*300V}$	0.088	0.09
$L_1=\dfrac{V_{o1}(0.5-D_h)T}{\Delta i_{L1}}$	$\dfrac{12V(0.5-0.45)*16.66\mu S}{0.25A}$	0.049mH	0.05mH
$L_2=\dfrac{V_{o2}(0.5-D_h)T}{\Delta i_{L2}}$	$\dfrac{5V(0.5-0.45)*16.66\mu S}{0.46A}$	0.009mH	0.01mH
$L_2=\dfrac{V_{o2}(0.5-D_h)T}{\Delta i_{L2}}$	$\dfrac{3.3V(0.5-0.45)*16.66\mu S}{0.32A}$	0.0086mH	9µH
$L_2=\dfrac{V_{o2}(0.5-D_h)T}{\Delta i_{L2}}$	$\dfrac{12V(0.5-0.45)*16.66\mu S}{0.017A}$	0.588mH	0.6mH

IV. PERFORMANCE OF THE PROPOSED SMPS

The proposed SMPS has been modeled and simulated in MATLAB environment and its simulated model is shown in Fig. 5. The dynamic performance of the SMPS system under different input voltages and load perturbations is presented along with the input current waveform and its harmonic spectra at full load and light load.

Fig. 6a shows the performance of the proposed SMPS at full load condition. The dc voltage of the PFC converter is kept constant at 300V using closed loop control. The input voltage and input current of the power supply are sinusoidal and in phase, confirming unity PF. All the dc output voltages are maintained constant. The input current THD at full load condition is shown in Fig. 6b which is observed to be 3.37% at 220V condition, confirming the PQ improvement. The stress on different components of the front end PFC converter is shown in Fig. 7. The input inductor currents are continuous whereas the output inductor currents are maintained in DCM even under low voltage conditions validating the design carried out. It is seen that the upper converter operates in the positive half cycle while the lower converter operates in the negative half cycle of the input voltage. The switch voltage at full load condition is observed in the range of 550V which is within the satisfactory limits. The input voltage is varied from 110 V to 270V (Figs. 8 & 9) and the PQ indices of the SMPS system under different input voltages are tabulated in Table-II. From Table-II, it is clear that the PQ is maintained exceedingly well under wide range of input voltages.

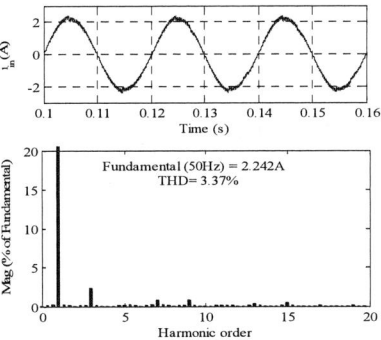

Fig. 6b Waveform of input current and its harmonic spectrum at 220V

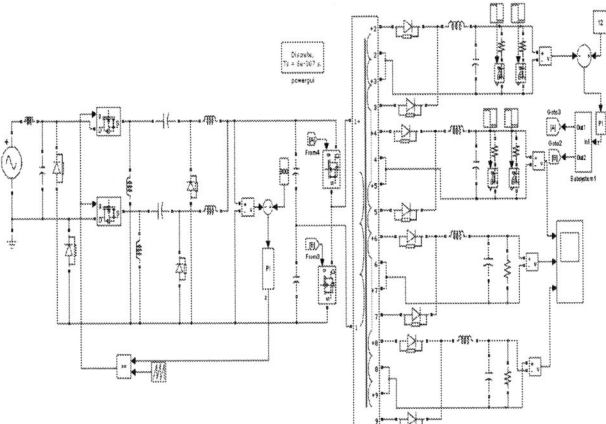

Fig. 5. MATLAB model of the SMPS system using PFC bridgeless zeta converter

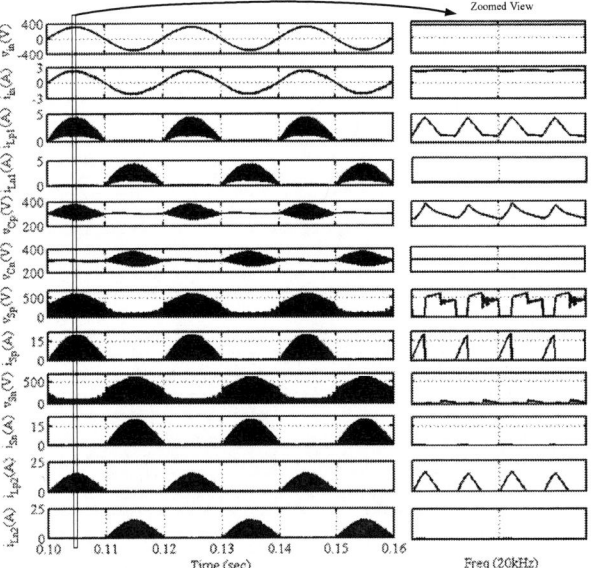

Fig. 7 Input voltage, current, input inductor currents, switch voltages, currents, capacitor voltages and output inductor currents of the bridgeless zeta converter at 220V

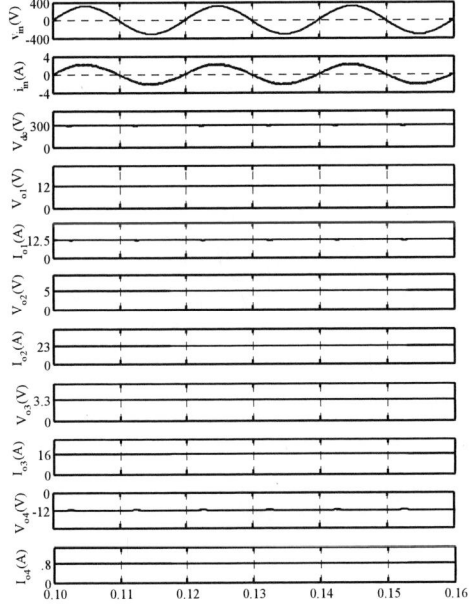

Fig. 6a Performance of bridgeless zeta converter based SMPS system at full load

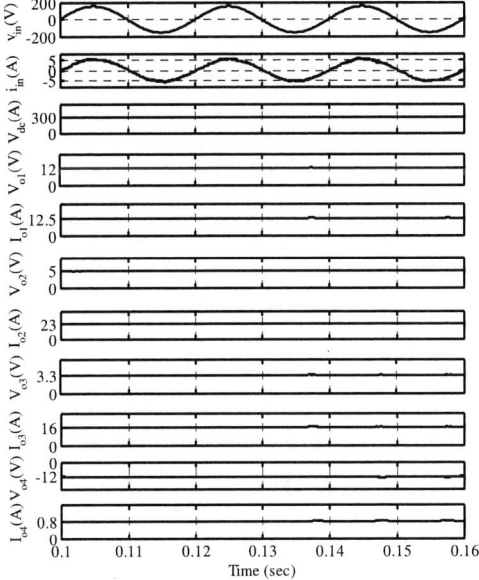

Fig. 8a Performance of bridgeless zeta converter based SMPS system at 110V

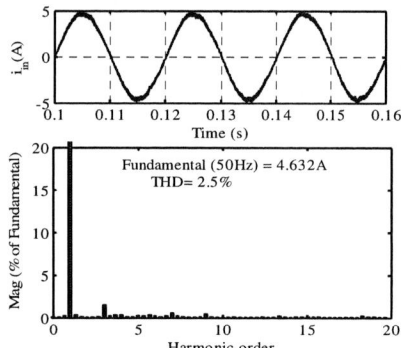

Fig. 8b Waveform of input current and its harmonic spectrum at 110V

TABLE II
POWER QUALITY INDICES OF PROPOSED POWER SUPPLY

Input voltage (v_{in})	Input current (A)	THD (%)	DPF	DF	PF
110	3.27	2.5	1	0.9997	0.9997
130	2.81	2.6	1	0.9996	0.9996
150	2.45	2.9	1	0.9995	0.9995
170	2.13	3	1	0.9995	0.9995
190	1.91	3.1	1	0.9994	0.9994
210	1.75	3.2	1	0.9994	0.9994
230	1.62	3.4	1	0.9994	0.9994
250	1.45	3.7	1	0.9993	0.9993
270	1.32	4.2	1	0.9991	0.9991

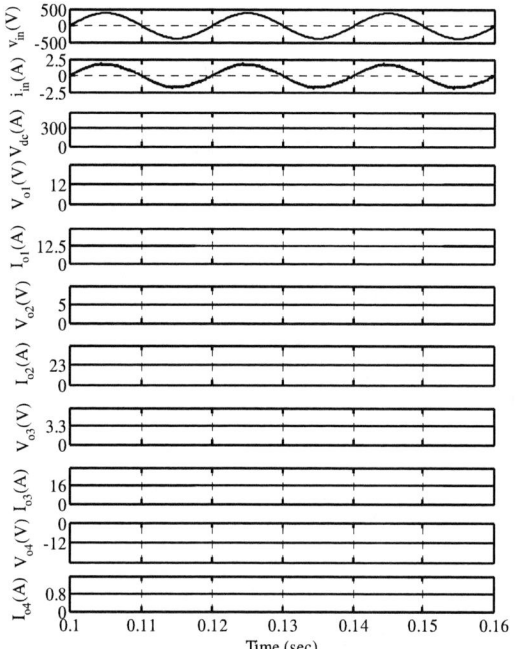

Fig. 9a Performance of bridgeless zeta converter based SMPS system at 270V

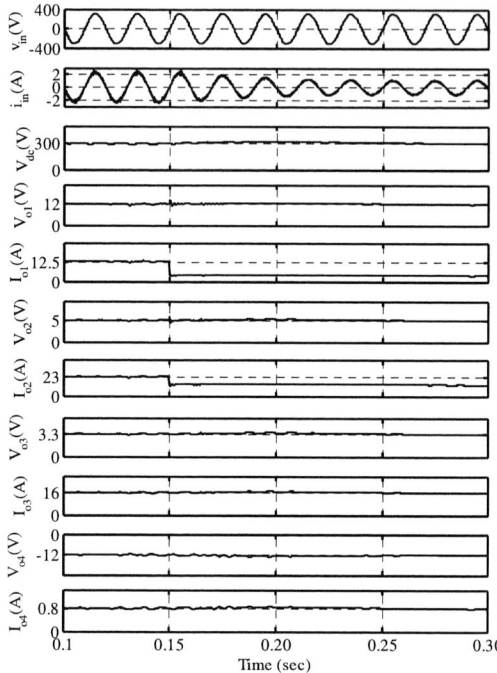

Fig. 10a Performance of bridgeless zeta converter based SMPS system at load variation in +12V and +5V outputs

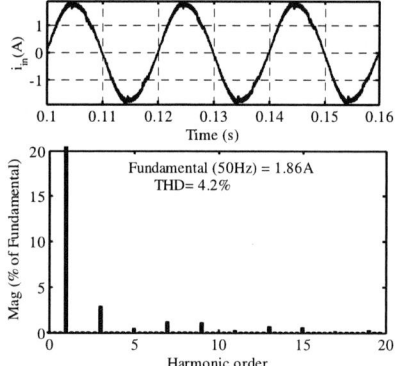

Fig. 9b Waveform of input current and its harmonic spectrum at 270V

The dynamic performance of the SMPS system has been simulated by switching the load suddenly after the power supply has reached steady state condition. Fig. 10a shows the dynamic performance of the SMPS system under step load variations from 100% to 30% in +12V and 100% to 60% in

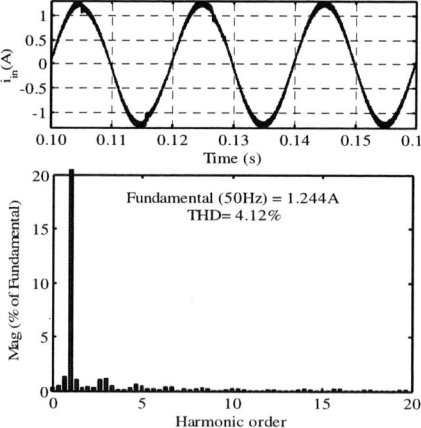

Fig. 10b Waveform of input current and its harmonic spectrum at light load

+5V. From Fig. 10a, it can be seen that the load currents are suddenly decreased from 12.5A to 3.75A and 23A to 13.8A at

0.15s in +12V and +5V outputs, respectively. The waveforms of input current and load current respond in less than half-a-cycle. The dc output voltages are also regulated very well by the PI voltage controller-2. Fig. 10b shows the input ac mains current (i_{in}) for the first few cycles along with its frequency spectrum for the SMPS system under light load condition. The THD of ac mains current is 4.12% which is within the limit specified in an IEEE-519 standard. The variation of input current THD and PF under varying input voltages is shown in Fig. 11a and 11b. At overvoltage (270V) condition, the PF is slightly reduced but it is still close to unity. Thus, the feasibility of the PFC bridgeless converter based SMPS system is demonstrated and validated under dynamic operating conditions like varying input voltages and loads. The THD of ac mains current is found to be lying between 2.5%-4.2% for the wide ac mains voltage variations from 110V-270V. It is found to be capable of mitigating the PQ problems at the front-end effectively and can be recommended as a solution for an SMPS for a PC.

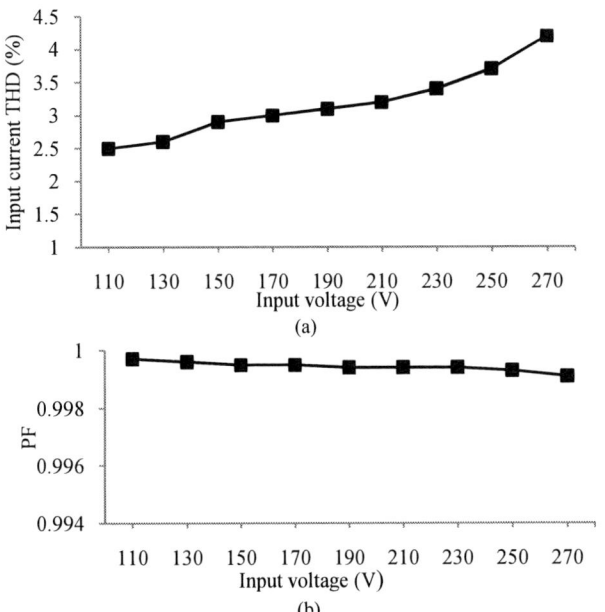

Fig. 11(a)Variation of input current THD under varying input voltages, (b) Variation of PF under varying input voltages

V. CONCLUSIONS

A PFC bridgeless zeta converter based SMPS system for a PC has been designed, simulated and its performance has been assessed in this paper. The proposed power supply has exceptionally good power quality indices under both supply voltage variations and load variations. The supply current

THD is always less than 5% and the power factor close to unity and hence it can be stated without hesitation that the proposed power supply conforms to IEEE and IEC power quality norms. Based on the observed performance of the PFC bridgeless converter fed SMPS system for PCs, it is concluded that the system is capable of mitigating the power quality problems faced by the conventional SMPS systems and hence can be a promising candidate for future use in the SMPS market for PCs.

REFERENCES

[1] Abraham I. Pressman, Keith Billings and Taylor Morey, "*Switching Power Supply Design*," 3rd ed., McGraw Hill, New York, 2009.

[2] Slobodan Cuk and R. D. Middlebrook, "Advances in switched-mode power conversion Part I," *IEEE Trans. on Industrial Electronics,* vol. IE-30, no. 1, pp. 10-19, Feb. 1983.

[3] P. J. Moore and I. E. Portugues, "The influence of personal computer processing modes on line current harmonics," *IEEE Trans. on Power Delivery*, vol. 18, no. 4, pp. 1363-1368, Oct 2003.

[4] T. M. Gruzs, "A survey of neutral currents in three-phase computer power systems," *IEEE Trans. on Industry Applications*, vol. 26, no. 4, pp. 114-122, July/August 1990.

[5] Limits for Harmonic Current Emissions, International Electro technical Commission Standard, 61000-3-2, 2004.

[6] B. Singh, S. Singh, A. Chandra and K. Al-Haddad, "Comprehensive study of single-phase ac-dc power factor corrected converters with high-frequency isolation," *IEEE Trans. on Industrial Informatics*, vol. 7, no. 4, pp. 540-556, Nov. 2011.

[7] J. Reddy, G. Bhuvaneswari and B. Singh, "A single dc-dc converter based multiple output SMPS with fully regulated and isolated outputs," in *Annual IEEE Proc. of INDICON*, 2005, pp.585-589

[8] Wenfei Wang, D.D.-C. Lu and G.M. Chu, "Digital control of bridgeless buck PFC converter in discontinuous-input-voltage-mode," in *37th IEEE Annual Conf. on Industrial Electronics Society, IECON*, 2011, pp. 1312-1317.

[9] L. Huber, Y. Jang and M. M Jovanovic, "Performance evaluation of bridgeless PFC boost rectifiers," *IEEE Trans. on Power Electronics*, vol.23, no.3, pp.1381-1390, May 2008.

[10] K. Mino, H. Matsumoto, S. Fujita, Y. Nemoto, D. Kawasaki, R. Yamada and N. Tawada, "Novel bridgeless PFC converters with low inrush current stress and high efficiency," in *International Conf. on Power Electronics Conference (IPEC)*, 2010 pp. 1733-1739.

[11] Suja C. Rajappan and John Neetha "An efficient bridgeless power factor correction boost converter," in *7th International Conf. on Intelligent Systems and Control (ISCO)*, 2013, pp. 55-59.

[12] Y. Jang and M. M. Jovanović, "Bridgeless high-power-factor buck converter," *IEEE Trans. on Power Electronics*, vol. 26, no. 2, pp.602-611, Feb. 2011.

[13] Wang Wei, Liu Hongpeng, Jiang Shigong and Xu Dianguo, "A novel bridgeless buck-boost PFC converter," in *Conf. IEEE Power Electron. Spec. PESC'08*, 15-19 June 2008, pp. 1304-1308.

[14] A. A. Fardoun, E. H. Ismail, A. J. Sabzali and M. A. Al-Saffar, "New efficient bridgeless Cuk rectifiers for PFC applications," *IEEE Trans. on Power Electronics*, vol. 27, no. 7, pp. 3292-3301, July 2012.

[15] M. Mahdavi and H. Farzaneh-Fard, "Bridgeless Cuk power factor correction rectifier with reduced conduction losses," *IET Power Electronics*, vol. 5, no. 9, pp. 1733-1740, Sept. 2012.

A Survey on Active power Filters Control Strategies

Khoisnam Steela and Bharat Singh Rajpurohit
School of Computing and Electrical Engineering
Indian Institute of Technology Mandi
Mandi, H.P., India
email id: khoisnam_steela@students.iitmandi.ac.in, bsr@iitmandi.ac.in

Abstract— **Due to increased level of harmonic pollution in present day power system, power quality conditioning has become indispensable, and therefore, various power conditioning techniques were introduced. Among them, active power filters appear to be the most viable device used for mitigating power quality issues. Various strategies of controlling active power filter are proposed and implemented by researchers and engineers in aim of achieving near perfect compensation. This paper presents a comprehensive review of state-of-art control strategies of active filters. Various strategies are discussed and compared in terms of implementation and performance.**

Index Terms— *Active power filter, harmonic mitigation, control strategy, PI controller, Fuzzy Controller*

I. INTRODUCTION

In recent years, there has been intensive increase in the use of power converters and switching devices, various non linear loads and equipments like high-power diode/thyristor rectifiers, arc furnaces, cyclo-converters, variable speed drives, computers, faxes, printers, etc., which draw highly non-linear current from the source or ac mains. They inject harmonics in the system. These harmonics greatly pollute the power system and cause serious damage. They deteriorate power quality by increasing total harmonic distortion (THD) and reactive power and by causing poor power factor, voltage flicker, bad voltage regulation, voltage sags and swells, unbalanced load etc.[1], [4], [5]. Harmonics also causes disturbances in communication network present in the vicinity of the system. Problem associated with power quality gradually became more and more severe; therefore, mitigation of the same has become an inevitable task for maintaining a good power system.

To provide clean-harmonic free power at consumer ends power filters are used as solution. Conventionally, passive L-C filters were employed for harmonics mitigation and capacitors for power factor improvement. But, the need for replacing them with better alternative soon arose because of their inherent disadvantages of bulky size, mistuning, instability, resonance with load impedance or utility impedance and lack of flexibility (i.e. they are best suited only for fixed compensation). Active filters (AFs) became a valuable alternative which give dynamic and versatile solution to the problem. AFs are also know by alternate names active power line conditioners (APLCs), instantaneous reactive power compensators (IRPCs), active power filters (APFs), and active power quality conditioners (APQCs). Soon, research on APF technology boomed up. A large number of control techniques are being proposed and implemented. At present there is an appreciably large number of control methods used for power compensation by APFs.

This paper aims at presenting an exhaustive review on the various state-of-the-art control strategies employed on APFs. The first section of this paper covers overall introduction. Next section briefly explains the working principle of APF followed subsequently by description of various control methodologies and then the concluding remarks.

II. COMPENSATION SCHEME

APFs are basically pulse width modulated (PWM) inverter(s) connected in series or shunt or series-parallel (active series and active shunt or hybrid of active series and passive shunt) configuration with respect to load or supply. The basic system configuration of shunt APF, series APF and unified power quality conditioner (UPQC) are shown in Fig.1, Fig.2 and Fig.3 respectively. These PWM inverters can either be voltage source inverter (VSI) type which has a dc-link capacitor or a current source inverter (CSI) type which has a dc-link inductor. Among the two types VSI type has been used more extensively in APF technology because of their better efficiency and dynamics [1], [2]. Shunt connected APFs generate a compensating signal which is actually an exact opposite of the distortion produced by the load to the line current. The effect of distortion produced by load is nullified by injecting the compensator power. They are mostly used at load end.

Series APFs are connected in series to the mains before the load by using appropriate transformer. They act as harmonic isolator by providing high impedance to harmonics and zero impedance to fundamental. Series APFs can also regulate the voltage at point of common coupling (PCC) by controlling inverter output [3].

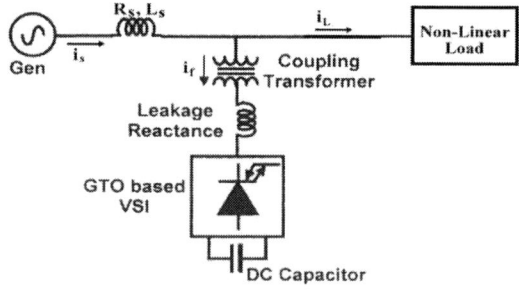

Fig. 1. Single-phase shunt connected compensator [64].

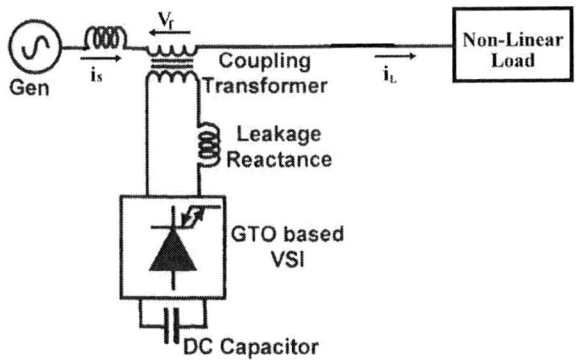

Fig. 2. Single-phase series compensator [64].

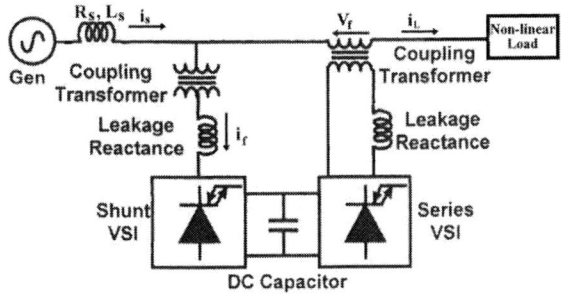

Fig. 3. Single-phase UPQC [64].

III. CONTROL STRATEGIES

Control strategy employed to an APF plays the crucial role of maintaining a proper compensation to the system to be compensated. Control of an APF includes overall system control, extraction of reference signals, controlling dc-link capacitor voltage or inductor current and gating signal generation. It is realized in three stages, first, the required voltage and current signals are sensed with the help of current transformers (CTs), potential transformers (PTs), hall-effect sensors or isolation amplifier [4]. Second, deriving compensating signals in terms of voltage and current level and regulating dc-link. Third step is generation of gating signal for solid-state devices of the APF.

Very accurate and fast sensing of signals, fast processing of reference signals and high dynamic response of controller are the key requirement of a good compensating device [5]. Keeping this in mind, researchers have proposed and implemented large number of control techniques for reference signal generation, dc-link voltage balance and switching pattern. These methods are discussed under three heads below.

A. Reference Signal Extraction Schemes

Using a proper method for current/voltage reference generation is the key to successful implementation of APF compensation [6]. Reference signal generation affects the rating and transient of APF as well as steady-state performance. Control strategy for generating compensating command can be extracted either by frequency based techniques or time domain techniques.

1) Frequency-domain compensation schemes: Frequency domain compensation is based on principle of Fourier analysis, wavelet analysis and infinite impulse response and the periodicity of distorted signal to be compensated. Among frequency domain schemes, conventional Fourier based method and modified Fourier method have gained popularity. They are explained in sub-sections (a) and (b) given below. Apart from this two methods, another approach which takes the advantage of decomposing three-phase signals into synchronously rotating direct and quadrature components has also been reported [7].

a) Conventional Fourier based techniques: In conventional Fourier based techniques conventional FT or FFT is applied to capture voltage or current signals. Compensating signal in time domain is obtained by subtracting the fundamental component from the FT and then applying inverse Fourier transform to the resultant, which is the compensating harmonic signal. Sampling time delay is a major demerit associated with this method [8]-[10].

b) Modified Fourier Transform technique: Modified Fourier transform was introduced to overcome the issue of sampling time delay in conventional method. It is based on calculating only fundamental component of current which is used to separate total harmonic signal from sampled load current waveform [11]. For practical implementation it requires modifying the main Fourier series equation to generate recurring formula with a sliding window. The components of sine and cosine coefficient computed at each sub-cycle are stored in two different circular arrays. Coefficient values are continuously updated. It is applicable to both single and three-phase. Computation is fast.

2) Time-domain compensation schemes: Time-domain approaches are mainly based on instantaneous derivation of reference signals in terms of voltage or current signals from distorted/harmonic polluted signals. There are a large number of time domain approaches. Some of them are briefly discussed here.

a) Instantaneous reactive power (p-q) theory: Instantaneous reactive power theory or *p-q* theory relies on Clarke's Transformation (transformation of *abc* to *αβ0* in stationary reference frame). Real and reactive power of load consisting of dc-component and oscillating component are calculated from the transformed quantities. These components are separated by passing through filter and the compensating components are selected for reference current generation. This method is only applicable to three-phase system. It give good performance for balanced sinusoidal supply condition. For contaminated supply, performance is poor [12].

The original '*p-q*' theory was proposed by Akagi *et al* [13] and then revised by Marshal *et al* [14]. Thereafter it was further modified/extended by Nabae *et al* [15] to make it applicable to eliminate neutral current of 3-phase, 4-wire systems where source voltages are unbalanced and loads are non-linear and unbalanced. These modified forms are known as 'modified *p-q* theory' [16] 'extended pq theory [17] and '*p-q-r* theory' [18].

b) Synchronous reference frame (d-q) theory: SRFT [19] or d-q theory relies on Park's Transformation where three-phase signals are transformed to synchronously rotating reference frame. The active and reactive components of the signal are represented by the direct and quadrature component respectively. The fundamental quantities become dc-quantities. The system is very stable since the controller deals mainly with dc-quantities. To generate the transformation angle 'θ' required for determining the synchronous reference frame a PLL is used. This method is ideally designed for three-phase system but recently it has been implemented in single-phase systems [60],[61].

A modified version of SRFT which calculate the transformation angle directly from voltage waveform without using PLL is proposed in [20]. This method is named Modified-SRFT or i_d-i_q method. It has the advantage of eliminating the PLL and phase error [37]-[39] and therefore has gained more popularity than the conventional SRFT.

c) Instantaneous power balance theory: Instantaneous power balance method [21] is based on principle that power supplied from mains equals power demanded by load at steady state. So, mains current magnitude is determined from power balance of mains, APF and the load. The compensating signal is calculated by processing the error signal generated at time of unbalance. The component of peak of supply currents is computed using the difference of average dc-bus voltage and the reference value. The main component of peak of supply currents to feed load currents is computed using sensed load currents and supply voltages. These two components add up to give the total peak value of supply current. Three-phase instantaneous reference supply currents are obtained by multiplying this peak magnitude with unit current templates derived in phase with supply voltages. This technique is simple and easy to implement. It gives fast dynamic response if proportional and integral constants are tuned properly [5].

d) Synchronous detection method (SDM): In the synchronous detection method [22],[23], the active current in the power line tracks the same wave-shape and maintains in-phase with the voltage waveform. The average power is obtained by monitoring the line currents and voltage. Average power is distributed proportionately among the three phases according to their instantaneous phase voltage. Instantaneous current for the three phases are calculated using these powers. Compensating current is obtained on subtracting this current from measured current. These signals are then synchronized relative to the mains voltage. Major drawback of this method is that it is highly dependent on voltage signal harmonics.

e) Direct detection method: Direct methods are those techniques that avoid transformation to α-β co-ordinates or d-q frame of reference for the generation of reference signal. Instead, in these methods the reference is calculated by using the abc phase voltage and line current directly. References [24],[25] uses direct method of reference generation.

f) Notch filter method: In notch filter based method the fundamental component of distorted current or voltage signals are removed by passing through a notch filter. The resultant signal is used as compensating command for reference signal [5].

g) Flux based control: The flux-based controller enables direct implementation of a current regulator without explicit generation of reference voltage [26],[27]. In this approach, harmonic components of load current are extracted using synchronous d-q transformation technique and the inductor reference flux (φ^*) is derived from the value of this current component. The AF terminal flux (φ_F) is estimated by integrating measured AF terminal voltage and integrating it through a special integrator. This flux is also used as feed forward quantity in generation of inductor reference flux. The inverter flux (φ_{inv}) is calculated by pure integration of measured inverter three-phase output voltage. The actual inductor flux (φ) is given by the difference of inverter flux (φ_{inv}) and AF terminal flux (φ_F). The inverter switching logics is obtained by using the value of the deviation between the actual inductor flux (φ) and the reference inductor flux (φ^*) in the synchronously rotating frame. This technique is associated with time delay due to presence of separate current and voltage loop.

h) Fictitious power detection method: This techniques is founded upon the basis that the apparent power (S) comprises only two real orthogonal subdivided components termed as active power (P) and fictitious power (F) of which fictitious power as can be sub-devided into orthogonal components reactive power (Q) and deactive power (D), where in active power hold the usual meaning. Reactive power (Q) represents the reciprocating component that contributes no net transfer between source and load. Deactive components accounts for non-similarity and un-correlation between voltage and current waveforms [28].

Here the compensation system forms an integral part of the load connected to network. The reactive power is compensated by means of cost effective low dynamic response compensation system like capacitor bank, TCR, synchronous condenser etc. The deactive power portion is compensated by high dynamic response APFs which are either PWM based VSI or CSI. The instantaneous representation of the deactive current is derived directly from the cross-correlation between the measured system voltage and load current. The system controller generates a reference current signal, which minimizes the undesired components of the power (fictitious power). This method finds application in single-phase systems. It has a drawback of large computation requirements that need microcontroller or DSP for implementation

i) Adaptive detection method: This is a method based on adaptive interference canceling theory. In this technique the total harmonics components and reactive components present in current is extracted by taking undistorted voltage signal as the reference and distorted current signal as the primary input. It has the ability to continuously self-study and self-adjust from start to end in order to maintain suitable operating state [29], [30]. This system is highly robust. This method shows advantage in situations where measurement error is a plausibility. But it has a drawback of implementation complexity.

j) Other techniques: There are various other control techniques, apart from those mentioned in earlier sections, in the field of APF technology. Some of the techniques not covered earlier are unity power factor method[31] employed to achieve UPF condition. This method is suitable for combined system of current harmonic and VAR compensation. Another method proposed in[32] multiplies load current and sine wave fundamental frequency. The resultant is integrated. The real fundamental component of load is calculated from this value so obtained. APF command current is obtained by subtracting the instantaneous load current from the fundamental. This technique is known as Sine-multiplication technique. Another approach is the Delta Modulation Method [33]-[34] which is a variation of traditional hysteresis current regulator. Use of soft computing based techniques like Neuro Controller and Fuzzy Logic Controller are also reported[35].

B. DC- Link Voltage Regulation Schemes

The dc-link voltage regulation refers to regulating the voltage across the dc-link capacitor of the VSI, which is being used as compensator. The dc-link capacitor plays two important roles in APF control, first it maintains a dc voltage with small ripple in steady state and second it serves as energy balancing medium that supplies or absorbs real power at times of transient period. So the load current variation is reflected upon dc-link voltage in terms of a rise or a fall. For balanced loading this voltage should remain constant. Therefore, a preset reference value is fixed and the rise or fall is regulated in such a way that it tracks the reference.

The rise or fall in value of dc-link voltage occuring due to transients or load change can be translated in terms of requirement for change i.e. requirement for increasing or decreasing the amplitude of supply current. Hence, the peak value of reference current of the system under compensation is related to value of dc-link voltage. The regulator that regulates this voltage to a fixed value, in doing so, also gives the command for calculation of peak value of reference current. PI regulator and soft computing technique based controller like Fuzzy Logic Controller are used vastly for this purpose.

In regulation scheme involving PI controller [36]-[39], the actual capacitor voltage is compared with reference value of link voltage. The error signal is then processed through a PI controller, which contributes to zero steady error in tracking the reference current signal. The output of the PI controller is considered as peak value of the supply current. This peak current has two components fundamental active power component of load current and the component that accounts for losses in the APF to maintain the average capacitor voltage to a constant value.

Regulation scheme employing fuzzy controller [36]-[40] also uses the error signal between actual capacitor voltage and reference voltage to compute the peak of reference signal. In a fuzzy controller, linguistic control strategies are converted to automatic control strategy. The fuzzy rules are constructed by expert knowledge. Input error and change in input error are fed as numerical input to the controller. They are then converted to linguistic variable by using predefined number of fuzzy sets or levels (a set of seven levels is commonly used). Fuzzy logic is characterized by (1) some fuzzy sets for each input and output variables (2) a particular membership function (3) Implication using Mamdani-type min-operator and (4) Defuzzification.

C. Switching Control Schemes

Generation of the gating signals for the solid state devices of the APF based on the derived compensating command is the final step of APF control. Gating signal can be generated either in open loop or closed loop. Popular scheme of open loop type are PWM and SPWM [41]. For closed loop schemes hysteresis control is the most widely used scheme for lower order system. For second and higher order systems sliding mode [42], linear quadratic regulator (LQR) [43], pole shift control[63], dead bead control[44], Kalman Filter[45] are used. Further, advancement in microprocessors, microcontrollers and DSP has rendered possible the implementation of complex algorithms like fuzzy logic, neural network and genetic algorithm also for improving the dynamic and steady state performance [4]. Here three most extensively used methods are discussed.

1) Carrier based PWM technique: Use of carrier base PWM switching technique has been reported in many papers including [46]-[52]. In this method the reference signal (i_{ref}) is compared with the actual filter current (i_f). The difference is recorded as an error (e), i.e. ($e = i_{ref} - i_f$) This error is then amplified and compared with triangular carrier wave. When e > triangular carrier wave signal the upper switch of a particular phase leg (S_a or S_b or S_c) of the VSI is turned OFF and lower switch (S_a' or S_b' or S_c') is turned ON. So current in that phase decays. When e < carrier signal lower switch is turned OFF and Upper Switch is turned ON so current increases. Carrier based PWM technique has been extensively used for gating signal generation owing to its fast response, simplicity and ease in implementation.

2) Hysteresis band current control technique: Hysteresis band current control (HBCC), is the most widely used [37]-[39], [53], [54] switching technique in APF technology. In this method the actual current is forced to follow the reference current by trapping the actual current within a certain preset limit or tolerance band, around the reference current, referred to as the Hysteresis Band (HB). When the actual current crosses or hits the upper limit of the band, the upper switch of a particular phase leg (S_a or S_b or S_c) of the VSI is turned OFF and lower switch (S_a' or S_b' or S_c') is turned ON. So current in that phase decays until it reach the lower limit of the HB. As it strikes lower limit of HB, lower switch is turned OFF and Upper Switch is turned ON. Further detail and generation of switching logics are described in [54]. The major drawback of this method is that the switching frequency is not constant and its working is rough and it produces large current ripple in steady state.

3) *Adaptive hysteresis band current-control technique:* Adaptive Hysteresis Band Current Control Technique (Adaptive-HBCC) is an enhanced version of HBCC. Switching logic generation principle remains the same as in HBCC. The improvement here is that the band-width of the hysteresis band is made variable by modulating it as a function of dc-link voltage and slope of reference current so that switching frequency can be held nearly constant. Formulation of the adaptive-HBCC is explained in [55]-[59]

IV. CONCLUSION

A comprehensive review of state-of-the-art control strategies of APFs has been presented in this paper to give an insight to the current stand-point of various control techniques employed on APFs. This paper also addressed the issue of growing harmonics pollution attributing to the substantial booming in the use of solid-state power electronics devices. The performances of a large number of control methods are discussed.

ACKNOWLEDGMENT

The authors would like to acknowledge the support provided by the Department of Science & Technology, Government of India, Under Science and Engineering Research Board Fast Track Scheme for Young Scientists (SERC/ET-0123/2012).

REFERENCES

[1] B. Singh, K. AL. Haddad and A. Chandra, "A review of active filters for power quality improvement," *IEEE Trans. Ind Elect.*, vol.46, pp. 960-971, 1999.

[2] Z. Peng, "Harmonic Sources and Filtering Approaches", *IEEE Industry Applications Magazine.* pp- 18-25, July/August 2001

[3] G. M. Lee. D.C. Lee and J. K. Seok, "Control of series active power filter compensating for source voltage unbalance and current harmonics", *IEEE Transactions on Industrial Electronics*, vol. 51, No.1, pp..132-139,Feb. 2004.

[4] Y. Pal, A. Swarup and B. Singh, "A review of compensating type custom power devices for power quality improvement, " *in proc of Joint International Conference on Power System Technology and IEEE Power India Conference, POWER CON*, 2008, pp. 1 -8

[5] N. Maruin, A. Alam and S. Mahmod, and H. Hizam, "Review of control strategies for power quality conditioners," in *proc. of IEEE PECon 2004*, pp.109-115, Nov. 2004

[6] L. Asiminoaei, F. Blaabjerg, and S. Hansen, "Detection is key—Harmonic detection methods for active power filter applications," *IEEE Ind. Appl. Mag.*, vol. 13, no. 4, pp. 22–33, Jul.Aug. 2007.

[7] Blajszczak, G "Non-active power compensation using time-window method," *Eur. Trans. Electr. Power Eng*, Vol. 2. No. 5, pp.285-290, Sept/Oct. 1992

[8] M.El-Habrouk, M. K Darwish and P. Mehta, "Active power filters: a review," *IEE Proc- Electric Power Applicrrtions*. vol: 147, Issue: 5 , pp. 403-413, Sept. 2000.

[9] M.El-Habrouk, M. K Darwish and P. Mehta ''A survey of active filters and reactive power compensation techniques," in *proc. 8th Int Conf on Power Electronics and Variabie Speed Drives, 2000, (IEE Conf Publ. No. 475)* , pp. 7-12, Sept. 2000.

[10] W. M. Grady, M. J. Samotyj, and A. H. Noyola, "Survey of active power line conditioning methodologies" *IEEE Trans on Power Delivery*, vol. *5*, Issue: 3. pp. 1536 - 1542, July 1990.

[11] M.El-Habrouk and M. K. Darwish "Design and implementation of a modified Fourier analysis harmonic current computation technique for power active filters using DSPs," *IEE Proceedings- Electric Powcr Applications*, Vol. 148, Issue: 1 ,Jan. 2001 Pages:21 - 28

[12] M. K.Mishra, A. Ghosh, and A. Joshi, "Operation of a DSTATCOM in voltage control mode," *IEEE Tran. Pow .Del.*,vol. 18, pp.258-264, Jan.2003

[13] H. Akagi, Y. Kanazawa and A. Nabae, "Instantaneous reactive power compensators comprising of switching devices without energy storage components," *IEEE Trans. Ind Appl.* Vol. *20.* no. *3*, pp.625-630, May/Junc 1984.

[14] D. A. Marshal and J. D. van Wyk, "An evaluation of the real-time compensation of fictitious power in electric energy networks", *IEEE Transaction on Power Delivery*, vol. 6. no. 4. pp. 1774-1780. October 1991.

[15] A. Nabae, H. Nakano and S. Togasawa, "An instantaneous distortion current compensator without any coordinate transformation," in *Proc. IEEJ Int. Power Electro. Conf. (IPEC-Yokohama)*, 1995, pp. 1651 1655.

[16] H. Akagi, S. Ogasawara, and H. Kim "The theory of instantaneous power in three-phase four-wire systems: A comprehensive approach," in *proc. of Industry Application Conf. 34th IAS Annual Meeting*, 1999, vol. 1, pp. 431-439.

[17] M. T. Haque, S. H. Hosseini and T. Ise, "A control strategy for parallel active filters using extended p-q theory and quasi instantaneous positive sequence extraction method," in *proc. ISIE 2001. Puson, Korea.* pp. 348-353.

[18] H. Kim and H. Akagi, "The instantaneous power theory on the Rotating p-q-r Reference Frame,," *IEEE 1999 Int. Conf. On Power Electronics and Drive Systems, PEDS, 1999*, Hong Kong, pp. 422- 427

[19] S. Bhattacharya. D. M. Divan and B. Banerjee. "Synchronous Framc harmonics Isolator using active series filter," *European Power Electronics Conf. EPE 1991* ,vol. 3, pp. 30-35, Fircnzi, Italy

[20] V. Soares. P. Verdelho, C. D. Marques, "Aclive power filter control circuit based on the instantaneous active and reactive current i_d-i_q method," *Power Elecfronrcs Specialists Conference. PESC '91.* St Louis. Missouri, 1997, pp. 1096-1101.

[21] *G. W. Chang;* "A new method for determining reference compensating currents of the three-phase shunt active power filter," *IEEE Power Engineering Review*, pp, *63-65,* March 2001.

[22] C. L. Chen, C. E. Lin, C. L. Huang, "The reference active sourcc current for active power filter in an unbalanced three-phase power system via the synchronous detection method," *in Conf Proc. IMTU 94.* 1994, pp. 502-505 vol.2.

[23] C. E. Lin, C. L. Chen, and C. L. Huang, "Calculating approach, implementation for active filters in unbalanced three-phase system using synchronous detection method," in *Proc. IEEE IECON'92*, 1992, pp. 374-380

[24] L. F. C. Monteiro, M. Aredes, J. A. Moor Neto, "A control strategy lor unified power quality conditioner," *in IEEE int. Symp. on Industrial Elecrronics. 2003*, 2003, vol. 1 . pp..391-396.

[25] P. Li, Y. H. Yang; Z. Ma, H. Li; "The study and simulation verifying of the controlling signal detecting method about UPQC," in Proc. 6th Int. Conf. on Electrical Machines and systems, 2003, vol. 2, pp 657-660.

[26] S. Bhattacharya, A. Veltman, D. M. Divan, and R. D. Lorenz, "Flux based active filter controller," in *Conf. Rec. IEEE-IAS Annu. Meeting*, 1995, pp. 2483-2491.

[27] S Bhattachaya, A. Vellman. D. M. Divan, R. D. Lorenz, "Flux-bascd active filter controller," *IEEE Transaction on Industrial Applications*, Vol. 32, Issue: 3 pp. 491-502, 1996.

[28] J. H. R Enslin, J. D. Van Wyk, "A new control philosophy for power electronic converters as fictitious power compensators," *IEEE Transaction on Power Electronics.* Vol. 5, issue. 1, pp 88-97, Jan. I990

[29] S. Luo, Z. Hou "An adaptive detecting method for harmonic and reactive currents" *IEEE Trans. on Industrial Electronics*, Vol. 42, Issue. 1, pp. 85 –89, Feb. 1995.

[30] H. Karimi, M. Karimi-Ghartemani. M. R. Iravani, A. R Bakhshai. "An adaptive filter for synchronous extraction of harmonics and distortions," *IEEE Transactions an Power Delivery*, Vol. 18, Issue.4, pp. 1350-1356, Oct. 2003.

[31] A. Abellan. G. Garccra, M. Pascual, E. Figucrcs, "A new current controller applied to Four-brach inverter shunt activc filters with UPF control method," *IEEE 32nd Annual Power Elccrronics Specialist Conference, 2001,* vol. 3. pp. 1402-1407.

[32] C. Y. Hsu, H. Y. Wu, " A new single-phase active filter with reduced energy storage capacity," *IEE proc. Electric Power Applications*, vol. 143, issue 1, pp. 25-30, Jan 1996.

[33] R. Kazemzadeh "Sigma-Delta Modulation Applied to a 3-Phase Shunt Active Power Filter Using Compensation with Instantaneous Power Theory," in *proc of IEEE international conference on Computer and Automation Engineering*, 2010, vol. 5, pp. 88-92.

[34] E. Wiebe-Quintana, J. L. Duran-Gomez, P. R. Acosta-Cano, "Delta-Sigma Integral Sliding-Mode Control Strategy of a Three-Phase Active Power Filter using d-q Frame Theory," in proc. *of IEEE conf on Electronics, Robotics and Automotive Mechanics*, 2006, pp. 291-298

[35] R. Zahira and A. P. Fathima, "A technical survey on control strategies of active filter for harmonic suppression," in *proc of ICCTSD 2011, procedia Engineering*, vol. 30, pp. 686-693, 2012

[36] M. Fatiha, M. Mohamed and A Nadia, "New hysteresis control band of an unified power quality conditioner," in *Electric Power System Research*, vol. 81, pp. 1743-1753, 2001

[37] M. Suresh, A. K. Panda, S. S. Patnaik and Y. Suresh, "Comparison of two compensation control strategies for SHAF in 3ph 4wire system By using PI controller," in *proc. of the IICPE 2010*, pp. 1-7, 2012

[38] M. Suresh, A. K. Panda , "Instantaneous Active and Reactive Power and Current Strategies for Current Harmonics Cancellation in 3-ph 4-Wire SHAF with Both PI and Fuzzy Controllers," *Journal of energy and power engineering* vol. 3, pp.285-298, 2011

[39] S. Mikkili, A. K. Panda "PI and Fuzzy Logic Controller based 3-phase 4-wire Shunt active filter for mitigation of Current harmonics with Id-Iq Control Strategy," *Journal of power Electronics (JPE)*, vol. 11, No. 6, Nov. 2011

[40] S. K. Jain, P. Agarwal and H. O. Gupta, "Fuzzy Logic Controlled Shunt Active Power Filter for Power Quality Improvement," *IEEE Proceedings of Electric Power Applications*, Vol. 149, No. 5, pp. 317-328, 2002.

[41] Z. Changjiang *et al.* "Dynamic voltage restorer based on voltage-space-vector PWM control," IEEE Trans. on Ind. Application vol.37, pp.1855-1863, Dec. 2001.

[42] R. R. Errabelli, Y. Y. Kolhatkar and S. P. Das, "Experimental investigation of sliding mode control of inverter for custom power applications," in *Proc.IEEE PES General Meeting*, pp. 1-8, June2006.

[43] G. Ledwich and A. Ghosh, "A flexible DSTATCOM operating in voltage or current control mode," Proc. *IEE Generation, Transmission and Distribution*, vol. 149, pp.215-224, March 2002.

[44] M. K. Mishra, A. Ghosh, and A. Joshi, "Operation of a DSTATCOM in voltage control mode," *IEEE Tran. Pow. Del.*, vol. 18, pp. 258-264, Jan. 2003.

[45] S. Bhattacharya and D. M. Divan, "Hybrid series active/parallel passive power line conditioner with controlled harmonic injection," U.S. Patent 5 465 203, Nov. 1995.

[46] C. K. Sao, P. W. Lehn, M. R. Iravani and J. A. Martinez, "A benchmark system for digital time-domain simulation of a pulse-width-modulated D-STATCOM," *IEEE Trans. Pow. Del.*, vol. 17, pp. 1113-1120,Oct. 2002.

[47] B. Singh, S. S. Murthy and S. Gupta, "Statcom-based voltage regulator for self-excited induction generator feeding nonlinear loads*," IEEE Transaction on Industrial Electronics*, vol. 53, issue 5, pp. 1437-1452, 2006.

[48] B. Singh, K. Al Haddad and A. Chandra, "Harmonic elimination, reactive power compensation and load balancing in three-phase, four-wire electric distribution systems supplying non-linear loads," Electric Power Systems Research , vol. 44, pp. 93–100, 1998.

[49] A. Banerji, S. K. Biswas and B. Singh, "DSTATCOM control algorithms: A review," *International Journal of Power Electronics and Drive System (IJPEDS)* Vol.2, No.3, pp. 285-296, September 2012.

[50] B. Singh, S.S. Murthy and S. Gupta, "STATCOM-Based Voltage Regulator for Self-Excited Induction Generator Feeding Nonlinear Loads", *IEEE Transactions on Industrial Electronics*, Vol. 53, No. 5, October 2006.

[51] B. Singh, S. S. Murthy and S. Gupta, "Modelling of STATCOM based voltage regulator for self-excited induction generator with dynamic loads," in *proc. of IEEE int. conf. on PEDES '06.*, 2006, vol., no., pp.1-6.

[52] A. Chandra, B. Singh, B. N. Singh and K. Al-Haddad, "An improved control algorithm of shunt active filter for voltage regulation, harmonic elimination, power-factor correction, and balancing of nonlinear loads", *IEEE Transactions On Power Electronics*, vol. 15, No. 3, pp. 495-507, May 2000.

[53] B. Singh, K. Al Haddad and A. chandra, "A new control approach to three-phase activt: filter for harmonics and reactive power compensation", *IEEE Transactions on Power Systems*, Vol. 13, No. 1, pp. 133-138, February 1998.

[54] Y. K. Chauhan, S. K. Jain, and B. Singh, "A prospective on voltage regulation of self-excitedninduction generators for industry applications," *IEEE Transactions On Industry Applications*, vol. 46, no. 2, pp. 720-730, march/april 2010.

[55] B. K. Bose "An adaptive hysteresis-band current control technique of voltage-fed PWM inverter for machine drive system," *IEEE Transaction on Industrial Electronics*, vol.337, no.5, pp.402-406, October 1990.

[56] M. Kale and E. Ozdemir, "A novel adaptive hysteresis band current controller for shunt active power filter" *Electric Power Systems Research*, vol. 73, Issue 2, pp 113-119, February 2005.

[57] B.S. Rajpurohit and S.N. Singh "An efficient adaptive hysteresis band current controller for active power filters", *IEEE-Bangalore Secton-15th Annual Symposiium on Technological Advances and IT Applications for Indian Power Sector*, Nov. 2006.

[58] B.S. Rajpurohit and S.N. Singh "Performance Evaluation of Current Control Algorithms Used for Active Power Filters", in *proc of EUROCON 2007, The International Conference on "Computer as a Tool"*, pp. 2570-2575, 2007

[59] S. R. Prusty, S. K. Ram, B. D. Subudhi and K. K. Mahapatra, "Performance analysis of adaptive band hysteresis current controller for shunt active power filter" in *proc. of IEEE int conf on Emerging Trends in Electrical Computer Technology (ICETECT) 2011*, pp. 425-429.

[60] V. Khadkikar, M. Singh, Chandra A and B. Singh, "Implementation of signal-phase synchronous d-q reference frame controller for shunt active filter under distorted voltage conditioner, " In *Proc. of Int. Conf. on Power Electron. Drives and Energy syst.*, pp. 1-6, 2010.

[61] Anzari M, R. Chandran and A. Kumar R, "Single-Phase Shunt Active Power Filter Using Indirect Control Method," *Advance in Electronic and Electric Engineering*, vol. 3, No. 1, pp. 81-90, 2013.

[62] A. Ghosh, A. K. Jindal and A. Joshi, "Inverter Control Using Output Feedback for Power Compensating Devices," in *Proc. TENCON 2003*, vol. 1, pp.48-52, Oct.2003.

[63] A. Ghosh and G. Ledwich, "Characterization of electric power," in *Power quality enhancement using custom power devices.* Kluwer Academic Publishers, 2002.

Analysis of Power Quality Disturbances Using Wavelet Packet Transform

Chirag A. Naik and Prasanta Kundu

Department of Electrical Engineering
S. V. National Institute of Technology
Surat, India.
e-mail: naik.chirag@gmail.com

Abstract— **The term Power Quality (PQ) has gained lots of interest in the changed power distribution system scenario. The increased use of power electronic switches based equipment, nonlinear loads and the changed power system regulations has made PQ an important issue. The non-stationary PQ disturbances such as; voltage sag, swell, momentary interruptions, transients, along with the stationary harmonics have become more frequent in today's power system. The conventionally used Fourier transform is found unsuitable for the analysis of non-stationary PQ disturbances. Different time –frequency analysis like, short time Fourier transform (STFT), discrete wavelet transform (DWT) and S- transform are in use for the analysis of non-stationary PQ disturbances. In this paper, the abilities of wavelet packet transform (WPT) are utilized to analyze different PQ disturbances. A new statistical variable is defined and computed for different PQ disturbances in order to analyze them. The newly defined variable is based on the average deviation of the WPT decompositions. The proposed variable is computed and analyzed for different variations of PQ disturbances to establish the usefulness of the WPT.**

Index Terms— *Time-Frequency Distribution, Power Quality, Wavelet Packet Transform, Feature extraction.*

I. INTRODUCTION

The concern for the high quality of power has increased rapidly in today's distribution system. The deregulation and liberalization in the power distribution market has allowed competition between the utilities to supply power in the same area. Consequently, the requirement for the good power quality (PQ) has become essential [1]. Poor PQ affects both utility and their consumer's equally. Because of the recent developments in power electronic switches based equipment and increased use of non-linear loads, non-stationary PQ disturbances such as; voltage sag, swell, momentary interruptions and transients have become more frequent along with the stationary harmonics [2]. Apart from that, the regular operations of the utilities such as; load switching, fault clearing, also adds to the further PQ deterioration [3]. Poor PQ may lead to partial or complete failure of equipment, loss of important data etc. The mitigation strategy can also be planned only if these PQ disturbances are correctly monitored [4].

Many digital signal processing techniques are used for the identification and monitoring of the PQ disturbances in recent years. Fourier Transform (FT) has been very effectively used for the analysis of stationary PQ disturbances like; harmonics. The established PQ indices total harmonic distortion (THD) and total demand distortion (TDD) are based on FT. The FT

converts the time domain signal to frequency domain signal where the time information gets void. The non-stationary disturbances are defined in terms of their frequency, magnitude and duration in IEEE 1159-2009 standards [5]. Thus, to monitor these disturbances, the time information is also essential to be preserved. Hence, the FT cannot be used for the analysis of the non-stationary disturbances [6]. The windowed version of FT; short time Fourier transform (STFT) is capable of fetching time information; but suffers with the compromised time resolution caused by the fixed window width [7]. Wavelet transform (WT) can simultaneously give time and frequency information of the PQ signals and has been utilized in number of research [7]-[11]. WT also exhibit disadvantages like; dependence on the basis wavelet, less immunity to noise etc [12]. A modified version of STFT and WT, the S-transform is used in most recent articles to overcome these disadvantages in [13]-[16].

In this paper, Wavelet Packet Transform (WPT) is employed to compute a unique variable; $AVG_{DEV}(m,n)$ based on the average deviation of WPT decompositions. The proposed variable is computed for the analysis of various PQ disturbance signals. PQ signals are simulated by using MATLAB.

After this introduction, brief theory of WPT is explained in section II. Section III summarizes the computation of the proposed statistical variable. The results and discussions are covered in section IV and finally the conclusions are made in section V.

II. THE WAVELET PACKET TRANSFORM

The time domain signals give amplitude-time representation giving zero frequency resolution. The FT gives magnitude-frequency representation giving zero time resolution. The STFT suffers with the issue of being a compromise between time and frequency resolution caused by the fixed window width. WT solves the issue of time and frequency resolution to a great extent, but suffers with the feasibility aspect. Complex computation of WT makes it unsuitable for the practical applications. The discrete wavelet transform (DWT) reduces the computational complexity of WT and is used in this was been utilized in many PQ analysis applications.

WPT used in this work is an extension of the idea of DWT. In DWT transform, the sampled signals are analyzed by half frequency high pass and low pass filter at the first level called details cD_1 and approximations cA_1, then further at every level

978-1-4799-6047-7/14 $31.00 © 2014 IEEE

only approximations (cA$_j$) are further analyzed in the same way. In WPT both details and approximations are further analyzed at every level of decomposition. Fig. 1 shows the WPT algorithm. At level (0,0) the sampled signal is passed through first level of half band high pass and half band low pass filters to get detail cD(1,0) and approximation cA(1,1) and the process is continued further till the desired level is decomposed. In WPT, similar to DWT, at every level of decomposition the signal is downsampled by the factor of 2. WPT decomposition tree up to three levels is shown in Fig. 1. The original signal is at level (0,0). (1,0) and (1,1) are respectively the details and approximations of the first level of decomposition and so on.

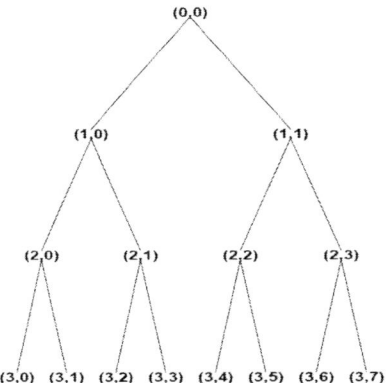

Fig. 1: Three level WPT decomposition

This makes WPT a strong contender for PQ analysis as it can simultaneously analyze both high and low frequency content of the time series. Voltage sag, swell and interruptions are related with low frequencies while oscillatory transients and harmonics might contain higher frequencies.

III. COMPUTATION OF WPT BASED VARIABLE

Different PQ disturbances such as; voltage sag, voltage swell, momentary interruptions, harmonics and oscillatory transients are simulated using MATLAB / SIMULINK block and analyzed with WPT.

Figs. 2(a) to 2(c) shows respectively the voltage sag, voltage swell and momentary interruption signals acquired with 20 kHz sampling frequency for the analysis. Voltage sag and voltage swell are simulated for 2 cycles to 5 cycles durations and 20% to 50% of their magnitudes. Momentary interruptions are simulated for 2 cycles to 5 cycles durations. Oscillatory transient signals are simulated to have transient frequencies of 3.5 kHz to 8 kHz in four steps.

Stationary harmonics are amongst very common PQ disturbances. The harmonic signals having THDs of 4.5 % to 23.97 % in four steps are also simulated. The simulated harmonic signal and oscillatory transient signal are shown in Figs. 3(a) and 3(b) respectively. All the signals are analyzed using WPT based statistical features.

A statistical variable $AVG_{DEV}(m,n)$ is defined using WPT in this work to analyze different PQ disturbances in order to get their specific nature.

The parameter $AVG_{DEV}(m,n)$; based on the average deviation of the WPT decomposition is computed for each disturbances as equation (1),

$$AVG_{DEV}(m,n) = \frac{1}{N}\sum_{i=1}^{N}\left(x(i) - \bar{x}\right) \qquad (1)$$

Where, N is the length of the (m,n) level WPT decomposition

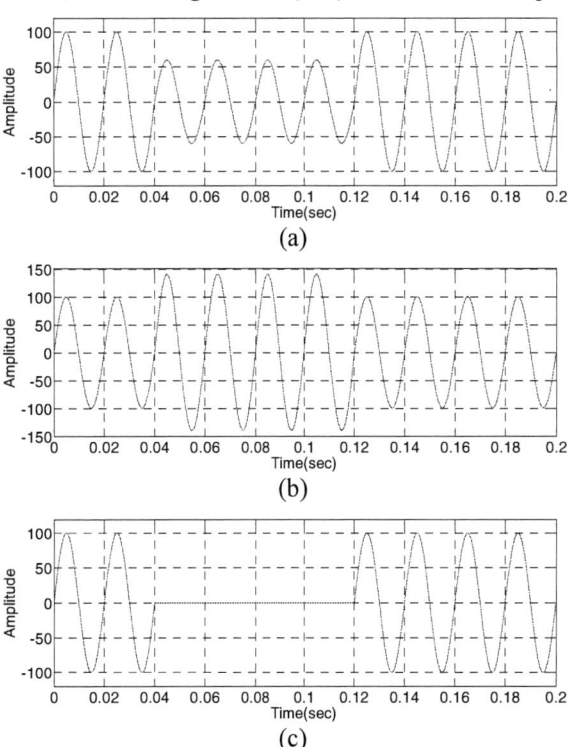

Fig. 2: (a) voltage sag signal (b) voltage swell signal, (c) momentary interruption

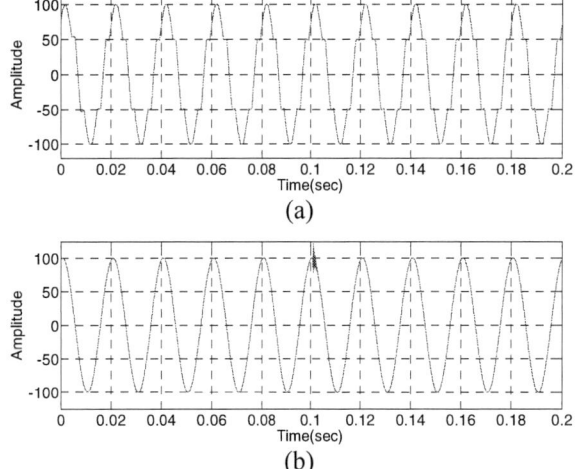

Fig. 3: (a) voltage harmonics signal, (b) oscillatory transient signal

$AVG_{DEV}(m,n)$ is computed for all the WPT decomposition up to level three as shown in Fig. 1. Further its values for each

978-1-4799-6047-7/14 $31.00 © 2014 IEEE

specific decompositions and for different types of disturbances is analyzed.

IV. RESULTS AND DISCUSSIONS

The commonly occurring PQ disturbances such as; voltage sag, voltage swell, momentary interruptions, harmonics and transients are simulated using MATLAB environment and the proposed parameter is computed for the same.

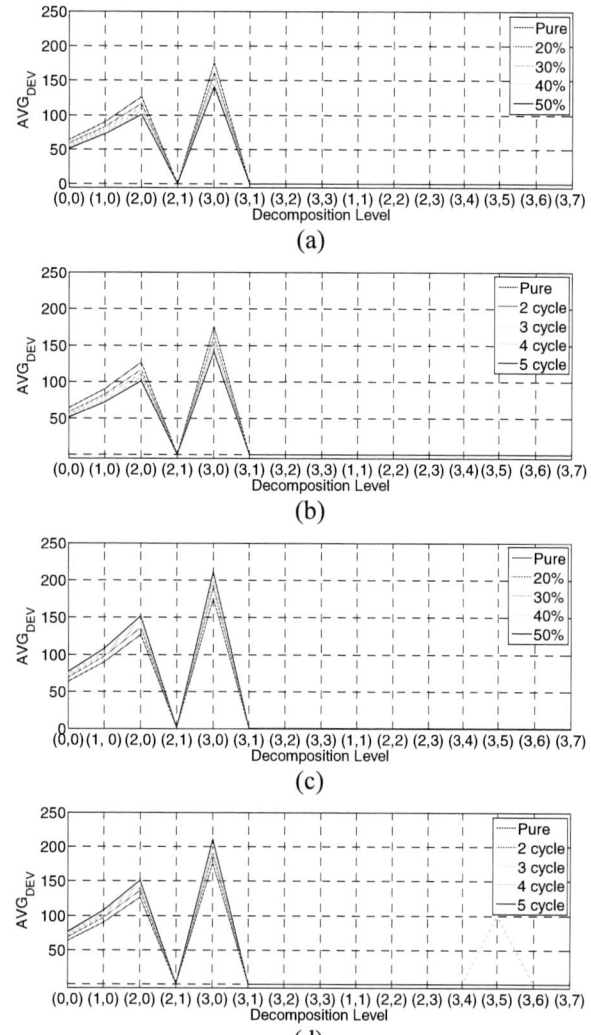

(a)

(b)

(c)

(d)

Fig. 4. AVG_{DEV} of (a) sag of 20% to 50% magnitudes and 4 cycles duration, (b) sag of 2 cycles to 5 cycles durations and 40 % magnitude, (c) swell of 20% to 50% magnitudes and 4 cycles duration, (d) swell of 2 cycles to 5 cycles durations and 40 % magnitude.

Fig.4 (a) shows the variations of $AVG_{DEV}(m,n)$ for the voltage sag of 20%, 30%, 40% and 50 % magnitudes respectively along with the pure high quality signal. It can be observed here that, the parameter shows the corresponding reduction in their magnitudes for specific decompositions associated with the low frequencies. Similarly, Fig. 4(b) shows the variation of the proposed parameter for different cycle durations of the voltage sag; i.e. 2 cycles to 5 cycles durations respectively. Here also it can be seen that as the duration of the sag increases the proposed parameter also show corresponding variation in their magnitudes but only for low frequency side decompositions.

For the decompositions on the high frequency side; i.e. decompositions after level (1, 1) the proposed parameter does not show any variations. Thus, by using the value of the computed parameter at specific decomposition only, it is possible to comment about the sag event. The $AVG_{DEV}(2,0)$ and $AVG_{DEV}(3,0)$ are showing the corresponding variation to identify voltage sag as shown in Table 1.

The results for voltage swell are shown in Fig. 4(c) and 4(d). as the voltage swell are associated with increase in magnitude the proposed parameter also show increased magnitudes for 20%, 30%, 40%, 50% swell magnitudes and also for 2 cycles to 5 cycles durations of the voltage swell disturbances. Again similar to voltage sag, voltage swell are also associated with low frequency disturbances and are picked up only in low frequency side WPT decompositions only (as per Fig. 1).

Table I:
Parameters extracted from $AVG_{DEV}(m,n)$ plots of voltage sag

Sr. No.	Sag magnitude (%)	Duration in cycles	Maximum of $AVG_{DEV}(2,0)$	Maximum of $AVG_{DEV}(3,0)$
1.	20	4	115.8816	161.6143
2.	30	4	110.8532	154.6192
3.	40	4	105.8247	147.6241
4.	50	4	100.7963	140.6290
5.	40	2	115.8816	161.6143
6.	40	3	110.8532	154.6192
7.	40	4	105.8247	147.6241
8.	40	5	100.7963	140.6290

Table II:
Parameters extracted from $AVG_{DEV}(m,n)$ plots of voltage swell

Sr. No.	Sag magnitude (%)	Duration in cycles	Maximum of $AVG_{DEV}(2,0)$	Maximum of $AVG_{DEV}(3,0)$
1.	20	4	135.9952	189.5947
2.	30	4	141.0236	196.5898
3.	40	4	146.0520	203.5849
4.	50	4	151.0804	210.5800
5.	40	2	135.9952	189.5947
6.	40	3	141.0236	196.5898
7.	40	4	146.0520	203.5849
8.	40	5	151.0804	210.5800

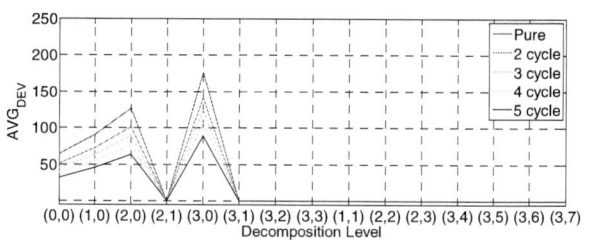

Fig. 5. AVG_{DEV} of momentary interruptions of 2 cycles to 5 cycles durations.

Table III:
Parameters extracted from $AVG_{DEV}(m,n)$ plots of interruptions

| Sr. No. | Momentary Interruption | | |
	Duration in cycles	$AVG_{DEV}(2,0)$	$AVG_{DEV}(3,0)$
1.	2	100.8415	140.7716
2.	3	88.2926	123.3547
3.	4	75.7438	105.9377
4.	5	63.1949	88.5207

The results for momentary interruptions of 2 cycle to 5 cycle duration also shown in Fig. 5 and the values of the proposed variable for levels (2,0) and (3,0) are shown in Table III. These results also confirm the effectiveness of the proposed variable.

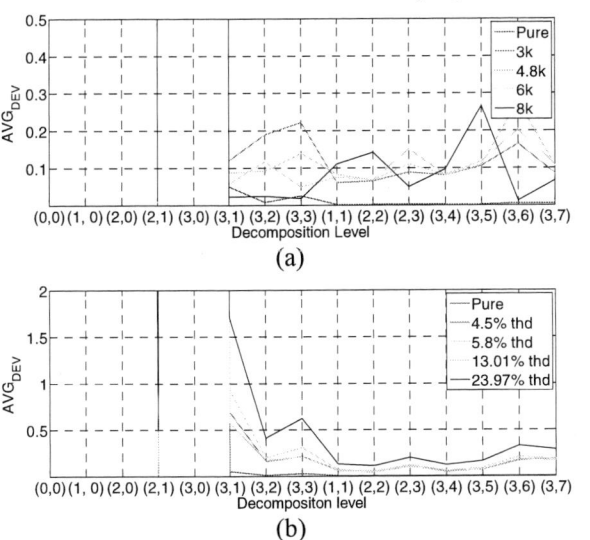

(a)

(b)

Fig. 6. AVG_{DEV} of (a) Oscillatory transient signals of 3kHz to 8kHz natural frequencies, (b) Harmonics of 4.5% to 23.97% THDs.

The PQ disturbances oscillatory transients and harmonics are associated with high or medium frequency contents. The oscillatory transients are simulated with 3 kHz to 8 kHz natural frequencies and the harmonics are simulated with 4.5% to 23.97% THD in four steps. The proposed variable show constant values for low frequency decomposition and variation only in the high frequency decomposition levels as shown in the Figs. 6(a) and 6(b) for the oscillatory transients and harmonics respectively.

The values of the decomposition level (3,5) is tabulated in Table IV. These variations are in accordance to the intensity of the disturbance that the PQ signals are carrying.

Table IV:
Parameters extracted from $AVG_{DEV}(m,n)$ plots of Oscillatory transients and harmonics

Sr. No.	Oscillatory Transients		Harmonics	
	Natural Frequency	$AVG_{DEV}(3,5)$	% THD	$AVG_{DEV}(3,5)$
1.	3 kHz	0.1042	4.5%	0.0683
2.	4.8 kHz	0.1075	5.8%	0.0723
3.	6 kHz	0.1164	13.01%	0.0851
4.	8 kHz	0.2655	23.97%	0.1584

V. CONCLUSIONS

The WPT based statistical variable $AVG_{DEV}(m,n)$ is defined and verified with the simulated stationary and non-stationary power quality disturbances such as; voltage, sag, swell, interruptions, harmonics and transient disturbances. The plot of the proposed variable at different levels of decompositions shows its effectiveness for PQ monitoring applications like identification and classification. The value of proposed variable at specific decomposition level gives information about the type of PQ disturbance present in the signal.

This parameter and some more parameters are further required to be defined and computed in order to effectively use them for classification of PQ disturbances. This work can be further extended for the automatic identification and classification of the power quality disturbances using intelligent classifiers such as; artificial neural networks, fuzzy logic controllers etc.

ACKNOWLEDGMENT

The authors acknowledge the support provided by S.V. National Institute of Technology, Surat, India.

REFERENCES

[1] R. Billinton and P. Zhaoming, "Incorporating reliability index probability distributions in performance based regulation," in *Electrical and Computer Engineering, 2002. IEEE CCECE 2002. Canadian Conference on*, 2002, pp. 12-17 vol.1.

[2] S. Yong-June, E. J. Powers, M. Grady, and A. Arapostathis, "Power quality indices for transient disturbances," *Power Delivery, IEEE Transactions on*, vol. 21, pp. 253-261, 2006.

[3] A. Rodríguez, J. A. Aguado, F. Martín, J. J. López, F. Muñoz, and J. E. Ruiz, "Rule-based classification of power quality disturbances using S-transform," *Electric Power Systems Research*, vol. 86, pp. 113-121, 2012.

[4] M. Uyar, S. Yildirim, and M. T. Gencoglu, "An expert system based on S-transform and neural network for automatic classification of power quality disturbances," *Expert Systems with Applications*, vol. 36, pp. 5962-5975, 2009.

[5] IEEE Recommended Practice for Monitoring Electric Power Quality, IEEE Standard 1159-2009, June 2009.

[6] G. T. Heydt and W. T. Jewell, "Pitfalls of electric power quality indices," *Power Delivery, IEEE Transactions on*, vol. 13, pp. 570-578, 1998.

[7] S. H. Jaramillo, G. T. Heydt, and E. O'Neill-Carrillo, "Power quality indices for aperiodic voltages and currents," *Power Delivery, IEEE Transactions on*, vol. 15, pp. 784-790, 2000.

[8] G. Zwe-Lee, "Wavelet-based neural network for power disturbance recognition and classification," *Power Delivery, IEEE Transactions on*, vol. 19, pp. 1560-1568, 2004.

[9] H. Erişti and Y. Demir, "A new algorithm for automatic classification of power quality events based on wavelet transform and SVM," *Expert Systems with Applications*, vol. 37, pp. 4094-4102, 2010.

[10] M. S. Kandil, S. A. Farghal, and A. Elmitwally, "Refined power quality indices," *Generation, Transmission and Distribution, IEE Proceedings-*, vol. 148, pp. 590-596, 2001.

[11] A. R. Abdullah, A. Z. Sha'ameri, A. R. M. Sidek, and M. R. Shaari, "Detection and Classification of Power Quality Disturbances Using Time-Frequency Analysis Technique," in *Research and Development, 2007. SCOReD 2007. 5th Student Conference on*, 2007, pp. 1-6.

[12] W. G. Morsi and M. E. El-Hawary, "Novel power quality indices based on wavelet packet transform for non-stationary sinusoidal and non-sinusoidal disturbances," *Electric Power Systems Research*, vol. 80, pp. 753-759, 2010.

[13] M. V. Chilukuri and P. K. Dash, "Multiresolution S-transform-based fuzzy recognition system for power quality events," *Power Delivery, IEEE Transactions on*, vol. 19, pp. 323-330, 2004.

[14] P. K. Dash, K. B. Panigrahi, and G. Panda, "Power quality analysis using S-transform," *Power Delivery, IEEE Transactions on*, vol. 18, pp. 406-411, 2003.

[15] Naik, C. A.; Kundu, P., "Identification of short duration power quality disturbances employing S-transform", *In Proc. Of International Conference on Power and Energy Systems*, pp. 1-5, 2011.

[16] Z. Fengzhan and Y. Rengang, "Power-Quality Disturbance Recognition Using S-Transform," *Power Delivery, IEEE Transactions on*, vol. 22, pp. 944-950, 2007.

Modified Series Compensation for Voltage Sag / Swell Mitigation

Kirti Mathuria, Arun Kumar Verma, *Member, IEEE,* Bhim Singh, *Fellow, IEEE* and G. Bhuvaneswari, *Senior Member, IEEE,*
Department of Electrical Engineering, Indian Institute of Technology Delhi,

Abstarct: A voltage source inverter (VSI) is employed to implement a variable frequency drive (VFD) for obtaining desired speed-torque characteristics of an AC motor which is normally fed by a diode rectifier at the front end. These uncontrolled converters inject excessive harmonic currents in to AC mains resulting in detoriated power quality in terms of AC voltage waveform distortionat the point of common coupling (PCC) and increase losses in the distribution system. Because of voltage sag and swells, the DC link voltage is reduced and the drive needs to be derated. An active series compensator is proposed here which is capable of reducing harmonic currents at the PCC and regulating the DC link voltage at the input of the VSI, thus offering dual functionality.

Keywords— Three-phase active series compensator, harmonics compensation, voltage control.

I. INTRODUCTION

The proliferation of Variable-Frequency Drives (VFDs) in a number of applications such as cement plants, furnace fans, mills, oil and gas extraction, mining, waste water plants, high voltage alternating current (HVAC) has accentuated many power-quality related issues in a distribution system and due to the VFDs are themselves prone to malfunctions [1-3]. The distortion in the input voltage and current waveforms is a major issue especially, in case of weak grid system [4]. The voltage harmonic distortion affects other loads connected at the PCC (Point of Common Coupling). Voltage distortions also affect the performance of these VFDs.[5,6]. These VFDs generally employ a three phase diode bridge rectifier with a filter capacitor at the front-end. Due to harmonics, it inherits the problem of poor power factor, which deteriorates the power quality for the other loads connected to the same line. Further, there is a voltage drop due to source impedance producing voltage distortions [7-10].

This paper proposes a series active compensator to eliminate voltage quality problems with a common DC link configuration of VFD. A feed forward based control is used to regulate the voltage of the common DC link configuration of VFD. Under varying AC mains voltages, the DC link voltage is regulated thus it avoids the derating of the AC motor drive. The proposed series active compensator provides dual functions of improving the power quality at the AC mains and regulating the DC link voltage.

II. SYSTEM CONFIGURATION

The basic configuration of the system is shown in Fig. 1. The active series compensator is connected in series with the power line. The DC link of the active series compensator and

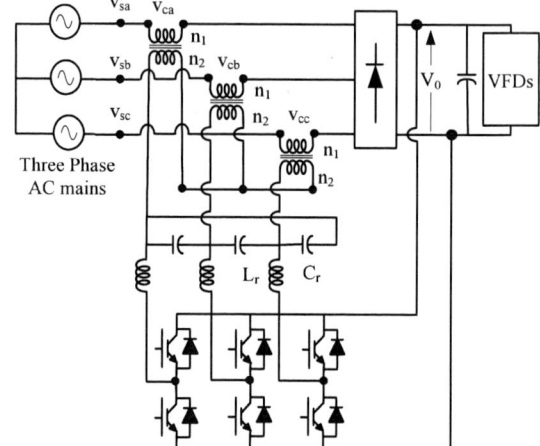

Fig. 1 Proposed configuration of the series active compensator

that of the diode rectifier are connected together, thus forming a common DC link. In the series active compensator, injection transformers (T_r) inject the voltage in series with the supply. The series active compensator acts as a harmonic isolator between the supply and VFD. It has a ripple filter to absorb voltage ripples generated by the VSC. The voltage fed type harmonic producing load is realized by using a three-phase full bridge diode rectifier, which has a capacitor filter at its DC link along with an equivalent resistive load of appropriate power rating.

III. DESIGN OF PROPOSED SYSTEM

The design of various components of the proposed system such as DC link capacitor, interfacing inductor, and ripple filter is presented in this section. The rating of the VFD is taken as 22kW. As the DC link is common for the both VSI and the active compensator, a nominal DC link voltage of 560V is selected here.

A. Design of VFD

The design of VFD is an important aspect. A VFD of 22 kW is considered here which is represented by an equivalent resistive load at 560V of DC link voltage with 5% ripple factor (RF) [9, 10].

978-1-4799-6047-7/14 $31.00 © 2014 IEEE

An equivalent load resistance, R is the given as,

$$R = \left(\frac{V_o^2}{P_o} \right) = \left(\frac{560^2}{22000} \right) = 14.25\Omega \tag{1}$$

Where V_o is the output voltage of the diode rectifier which is taken as 560V. P_o is the kVA rating of the VFD, The value of, R is estimated 14.25 Ω from eq. (1).

The value of DC bus filter capacitance is given as [10],

$$C = \left[\frac{1}{12*f*R} \left\{ 1 + \left(\frac{1}{\sqrt{2}*RF} \right) \right\} \right]$$
$$= \left[\frac{1}{12*50*14.25} \left\{ 1 + \left(\frac{1}{\sqrt{2}*0.05} \right) \right\} \right] = 1771\mu F \tag{2}$$

Where f is the supply frequency (50 Hz), RF is the ripple factor which is considered as 5%.

From eq. (2) the value of C is calculated as $1771\mu F$ and it is selected as 2000 μF for implementing the series compensator. The average load voltage V_{dc} is given as [10],

$$V_{dc} = \left[V_o - \left(\frac{V_o}{12*f*R*C} \right) \right] = 559.9V \tag{3}$$

V_{dc} from Eq. (3) is calculated as 559.9V and it is selected as 560V. Using this average load voltage, the value of R is calculated as 14.25 Ω and it is used to calculate C and V_{dc}. The value of C is selected as 2000 μF and equivalent load resistance R is selected as 14.25 Ω for 22 kW VFD.

B. Estimation of the Transformer Injected Voltage

The injected voltage (v_{in}) of the transformer of the series active compensator is obtained from the difference between the PCC voltage (v_{sa}) and load voltage (v_{La}) as follows. The AC voltage at the input of the diode bridge rectifier (v_{La}) is a stepped waveform having multiples/sub-multiples of DC link voltage. Hence the primary winding voltage or injected voltage (v_{in}) is estimated as,

$$v_{in}^2 = \frac{1}{\pi} \int_0^{\frac{\pi}{3}} \left(\sqrt{2}\, v_s \sin\theta - \frac{V_{dc}}{3} \right)^2 d\theta + \frac{1}{\pi} \int_{\frac{\pi}{3}}^{\frac{2\pi}{3}} \left(\sqrt{2}\, v_s \sin\theta - \frac{2\,V_{dc}}{3} \right)^2 d\theta$$
$$+ \frac{1}{\pi} \int_{\frac{2\pi}{3}}^{\pi} \left(\sqrt{2}\, v_s \sin\theta - \frac{V_{dc}}{3} \right)^2 d\theta \tag{4}$$

Where v_s is the per phase PCC voltage and V_{dc} is the DC link voltage. The estimation of the injected voltage by using (4) at different value of the DC link voltage is listed in Table I

C. Design of Ripple Filter

A ripple filter is designed for eliminating the switching frequency ripples in the injected voltage of active series compensator. The filter elements, capacitor and inductor, offer a low and high impedance path respectively to the switching ripples and fundamental frequency. The reactance at half of the switching frequency is given as [12],

$$C_{(r)} = \frac{1}{2*\pi*f_r*X_{crsw}}$$
$$C_{(r)} = \frac{1}{(2*3.14*10000*3)} = 5.53\mu F \tag{5}$$

$$L_{(r)} = \frac{X_{Lrsw}}{2*\pi*f_r}$$
$$L_{(r)} = \frac{200}{(2*3.14*10000)} = 3.18mH \tag{6}$$

In (5) and (6), $f_r = f_s/2$ Where f_s(20 kHz) is the switching frequency, X_{crsw}=3Ω and X_{Lrsw} =200Ω respectively at fundamental frequency. The calculated values of $C_{(r)}$ and $L_{(r)}$ from (5) and (6) are 5 μF and 3 mH. These ripple filter elements are found suitable for maintaining minimum ripple at the output of the series active compensator.

IV. CONTROL ALGORITHM

The control algorithm for the proposed configuration of series active compensator is given in Fig.2. The DC link voltage (V_{dc}) is sensed and filtered using a low pass filter (LPF). A Proportional and Integral (PI) DC link voltage controller generates a control signal (I^*_{sm}) to eliminate the voltage error $V_e(k)$ by processing the error estimated by comparing the reference voltage V^*_{dc} (k) and a sensed voltage $V_{dc}(k)$ at k^{th} instant [11, 12].

$$V_e(k) = V^*_{dc} - V_{dc}(k) \tag{7}$$

The output of the PI controller, $I_{sm}^*(k)$ at the kth instant is

$$I^*_{sm}(k) = I^*_m(k-1) + K_{pv}\left\{ V_e(k) - V_e(k-1) \right\} + K_{iv}\left\{ V_e(k) \right\} \tag{8}$$

The voltage amplitude is estimated as [13],

$$V_t = \sqrt{\frac{2}{3}\left(v_{sa}^2 + v_{sb}^2 + v_{sc}^2 \right)} \tag{9}$$

The DC power is given as,

$$P_{dc} = V_{dc} * I_{dc} \tag{10}$$

$$I_{sdc}(k) = \frac{(2*P_{dc})}{(3*V_t)} \tag{11}$$

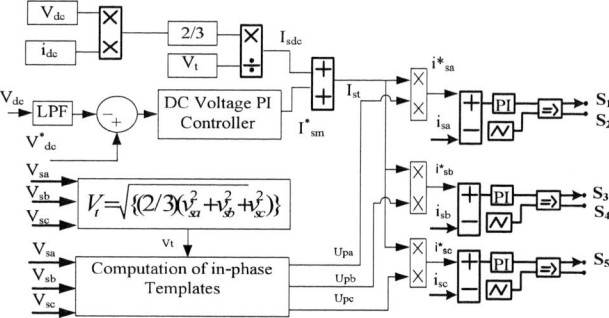

Fig. 2 Control Algorithm for Proposed Configuration of the Series active compensator

Total active power component of the supply current is given as,

$$I_{st}(k) = I_{sm}^*(k) + I_{sdc}(k) \qquad (12)$$

Now, in phase template of PCC voltage are given by

$$u_{pa} = \frac{v_{sa}}{V_t}, \ u_{pb} = \frac{v_{sb}}{V_t}, \ u_{pc} = \frac{v_{sc}}{V_t} \qquad (13)$$

The reference supply currents are calculated as,

$$i_{sa}^* = I_{st}^* * u_{pa}, \ i_{sb}^* = I_{st}^* * u_{pb}, \ i_{sc}^* = I_{st}^* * u_{pc} \qquad (14)$$

These reference supply currents (i*$_{sa}$, i*$_{sb}$, i*$_{sc}$) are compared with the sensed supply currents (i$_{sa}$, i$_{sb}$, i$_{sc}$), and corresponding currents errors are processed in current controllers. The outputs of current controllers are compared with a fixed frequency (10 kHz) triangular wave to generate PWM switching signals for the series active compensator devices.

V. MATLAB BASED MODELLING AND SIMULATION

The series active compensator for the three phase diode bridge rectifier with filter capacitor feeding an equivalent load is modelled and its performance is simulated for power quality enhancement with feed forward based control method. In this system, the considered load is a three-phase diode rectifier with the filter capacitor and an equivalent load resistance of 22 kW capacity. A 3-leg VSC with DC link capacitor employing IGBTs as switches is used as an active series compensator.

VI. RESULTS AND DISCUSSION

The performances of the proposed common DC link based active series power compensator are discussed here in detail. The simulated results of the proposed system are depicted in Figs.3-5. The performances of the proposed system are simulated in Matlab with Simulink SPS toolboxes and discrete step solver of 1e-6. The results are describe in terms of PCC voltage (v$_{abc}$), supply current (i$_{abc}$), injected voltage (v$_n$), voltage across load (v$_{La}$), and DC link voltage V$_{dc.}$

A. Performance of an Uncompensated Distribution System Employing VFDs

The performance of the uncompensated distribution system employing VFD is demonstrated in Fig.3(a) and corresponding harmonic spectrum is shown in Fig.3(b). The analysis of an uncompensated distribution system, shows that without compensation the AC mains current THD reaching 62.07% and the value of PF beingquite low [10].

B. Performance of the Proposed System

Figs.4-5 show the performance of the common DC link based proposed configuration under voltage sag/swell. Fig.4

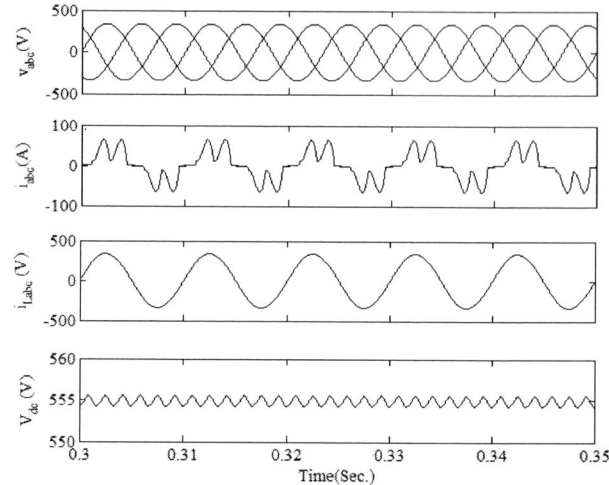

Fig. 3(a). Performance of uncompensated distribution system employing VFDs at AC main voltage of 415V

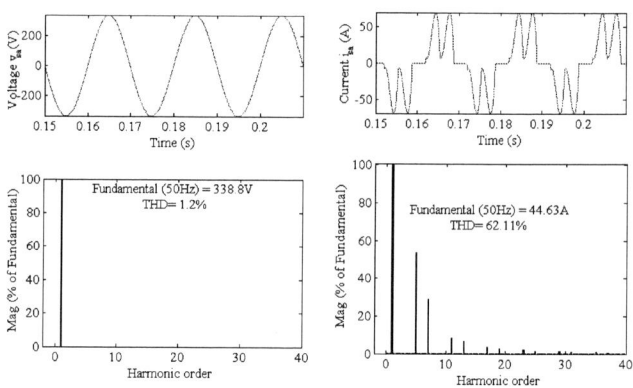

Fig.3 (b) Harmonic spectra uncompensated system employing VFDs

shows that the 20% voltage sag is appeared at 0.2s to 0.25s.During voltage sag, the supply currents are balanced and sinusoidal. The DC link voltage has some disturbances but remains constant. As the sag is rectified at 0.25s, the system regains it steady state performance rapidly.

Fig. 5(a) shows the performances of the system under voltage swell of 20% at 0.2s and it ends at 0.25s.

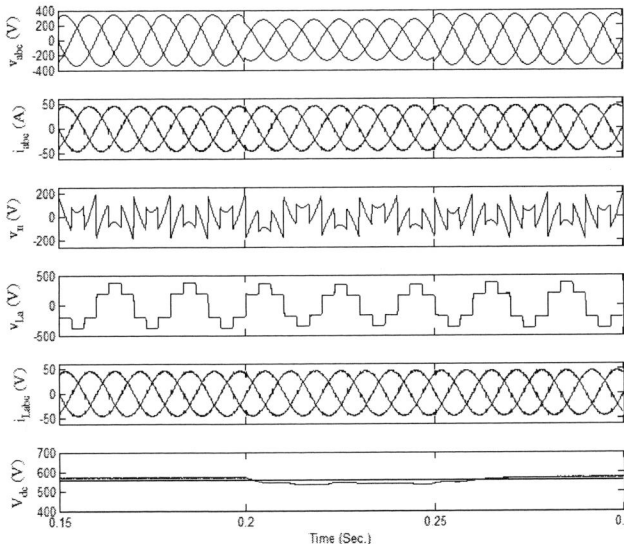

Fig 4. Performance of proposed series active compensator employing system employing VFDs at AC main voltage of 415V and DC link Voltage of 560V

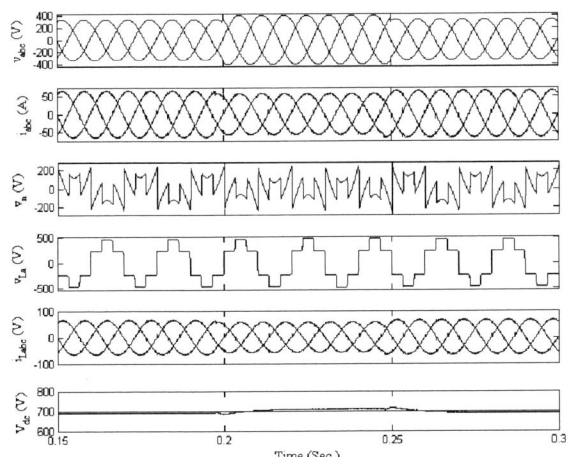

Fig 5(a). Performance of proposed series active compensator employing VFDs at AC main voltage of 415V and DC link Voltage of 700V

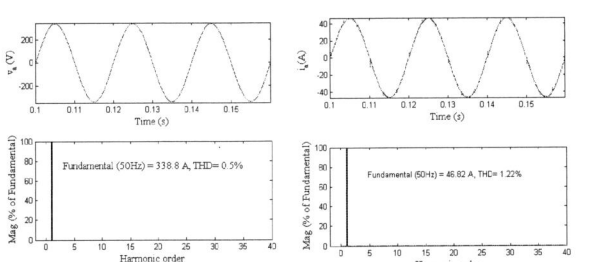

Fig.5 (b) Harmonic spectra for proposed series active compensator employing system employing VFDs

During voltage sag, the supply current is balanced and sinusoidal injected voltage reduces slightly and the DC link voltage remains constant with some small voltage disturbance. Fig. 5(b) shows the harmonics spectra and it shows that supply current THD is 1.22% and PCC voltage THD is of 0.5% which are well within the range of IEEE 519 Standard.

VII. CONCLUSION

This investigation has demonstrated that the proposed active series compensator has regulated the DC link voltage of a diode bridge rectifier feeding a VSI controlling an AC motor drive. The proposed active series compensator has immuned the syste from voltage sag/voltage swell. It has resulted in reduced supply current for the same output power of the load which further reduces the losses in the distribution system. The component count is less as compared to the conventional active compensator thus economical. The THD of supply current and PCC voltage are well within the IEEE 519Standard limits.

TABLE-I PERFORMANCE OF SERIES ACTIVE COMPENSATOR UNDER VARIABLE DC LINK VOLTAGES

Dc link voltage (V)	Calculated injected voltage (V)	Simulated Injected voltage (V)	Estimated kVA Rating (kVA)	Turn ratio	KVA (% of load)
310	112.67	113.2	10.25	0.61	48.01
325	103.78	101.9	9.35	0.71	42.5
360	96.54	97.88	8.71	0.85	38.59
375	88.11	87.1	7.99	0.98	36.31
400	81.71	79.80	7.32	1.05	33.27
420	75.59	74.6	6.6	1.16	32.54
450	73.04	71.5	6.70	1.44	30.45
480	71.28	71.5	6.47	1.49	29.40
500	71.45	69.41	6.39	1.58	27.95
525	73.53	72.5	6.65	1.55	30.22
550	77.37	76.5	7.1	1.53	30.90
575	82.72	81.5	7.48	1.47	33.00
600	89.32	87.19	8.00	1.41	36.36
620	96.91	96.4	8.75	1.35	38.77
650	105.27	103.9	9.53	1.34	42.31
675	114.23	113.2	10.4	1.35	47.22
700	124.68	151.6	11.16	1.34	50.72

APPENDIX

A. Data of Active Series Compensator

Line impedance: L_s =1.1mH, R_s=0.12Ω, Ripple filter: L_r=2mH, C_r=5µF, DC link voltage: V_{dc}=560V, DC bus capacitance: C=2000µF, AC voltage: V_{sa}= 220V. Frequency: 50Hz, PWM switching frequency: 20 kHz. DC PI controller gains: K_{pvd} = 1.1, k_{ivd} = 0.29, Current PI controller: K_{pvi} = 1.5, k_{ivi} = 0.32.

REFERENCES

[1] M. H. J. Bollen, Understanding Power Quality Problems: Voltage Sags and Interruptions, IEEE Press Series on Power Engineering, New York, 2000.

[2] B. Singh, K. Al-Haddad and A. Chandra, "A Universal Active Power Filter for Single-Phase Reactive Power and Harmonic

Compensation," *Power Quality'98*, Hyderabad, India, 1998, pp. 81-87.

[3] Parag Kanjiya and Bhim Singh, "A Robust Control Algorithm for Self Supported Dynamic Voltage Restorer (DVR)," *in proc. IEEE International Conference on Power Electronics (IICPE), 2010 India*, pp. 28-30 Jan. 2010.

[4] Hideaki Fujita, and Hirofumi Akagi, "An Approach to Harmonic Current-Free AC/DC Power Conversion for Large Industrial Loads: The Integration of a Series Active Filter with a Double-Series Diode Rectifier," *IEEE Trans. on Industry Application*,vol.33,no.5,pp. 1233-1240, Sept./Oct.1997.

[5] Sunt Srianthumrong, Hideaki Fujita, and Hirofumi Akagi, "Stability Analysis of a Series Active Filter Integrated with a Double-Series Diode Rectifier," *IEEE Trans. on Industry Application*, vol.17, no.1, pp. 117-124, Jan2002.

[6] E. Babaei, M. Farhadi Kangarlu and M. Sabahi, "Mitigation of voltage disturbances using dynamic voltage restorer based on direct converters," *IEEE Trans. on Power Delivery*., vol. 25, no. 4, pp. 2676–2683, Oct. 2010.

[7] Chnis Fitzer, Mike Barns and Peter Green, "Voltage Sag Detection Technique for a Dynamic Voltage Restorer," *IEEE Trans. Power Delivery, vol. 40, no.1 pp. 203-212, Jan/Feb 2004.*

[8] Brian R. Pelly, "Active filter for reduction of common mode current," US Patent 6,636,107 B2, Oct. 21, 2003.

[9] I.R Filho, J.L. Silva Neto, L.G. Rolim and M. Aredes, "Design and implementation of a low cost series compensator for voltage sags," in proc. *IEEE International Symposium on Industrial Electronics2006*, pp. 1353-1357.

[10] P. Jayaprakash, Bhim Singh and D.P.Kothari, "Control strategies of series active filter for harmonic current compensation in voltage fed nonlinear loads," in Proc. NPEC 2007 Dec. 17-19, 2007, IISC, Bangalore.

[11] Bhim Singh and Vishal Verma, "Selective Compensation of Power Quality problems through active power filter by Current decomposition," *IEEE Tran. Power Del.*, vol.23, no.2, pp.792-799, April 2008.

[12] Bhim Singh, P. Jayaprakash and D. P. Kothari, "Adaline Based Control of Capacitor Supported DVR for Distribution Systems," *Journal of Power Electronics,* vol. 9, no. 3, pp.386-395, May 2009.

[13] B. N. Singh, P. Rastgoufard, B. Singh, A. Chandra, and K. Al. Haddad, "Design, simulation and implementation of three pole/four pole topologies for active filters," *Proc. Inst. Electr. Eng. Electr. Power Appl.*, vol. 151, no. 4, pp. 467–476, Jul. 2004.

A Three-Leg Inverter Based DSTATCOM Topology for Compensating Unbalanced and Nonlinear Loads

M.V Manoj Kumar, *Student Member, IEEE* and Mahesh K. Mishra, *Senior Member, IEEE*
Department of Electrical Engineering,
Indian Institute of Technology Madras, Chennai, India.
*Corresponding author: Tel.:+91 44 22575459; Fax:+91 44 22574402; E-mail: yemvee1975@gmail.com

Abstract—This paper proposes a new three-leg voltage source inverter (VSI) based distribution static compensator (DSTAT-COM) topology to compensate unbalanced and nonlinear loads in low voltage three phase four wire distribution systems. The proposed topology uses a three-leg VSI with a single DC storage capacitor and an additional capacitor which is connected between negative DC bus to the system neutral. As the proposed topology uses a single DC link capacitor, the DC voltage balancing issues associated with the popular split capacitor neutral clamped VSI topology is avoided. Moreover, the topology is able to compensate neutral current without using four-leg inverter topologies. Analysis and modelling of the proposed topology is explained in detail. Simulation studies are carried out to verify the performance of the proposed schemes and results are presented.

Index Terms—Distribution static compensator, harmonics, topology, instantaneous symmetrical component theory.

I. INTRODUCTION

Distribution static compensators (DSTATCOMs) are being used for reactive, harmonic and unbalanced load compensation in distribution systems. Most of the loads in low voltage three phase four wire distribution systems include single phase loads like lighting ballasts, variable speed drives for small motors and office automation units etc.. The use of these single phase unbalanced and nonlinear loads create excessive neutral current in addition to their other power quality effects. This will overload the neutral conductor and has many other serious impacts on the distribution system such as, overloading of feeders and transformers, malfunctioning of sensitive equipments, and distortions in terminal voltage. These power quality issues and their remedial measures using DSTATCOMs are well analyzed and reported in many literature. The mitigation methods mainly differs on the topology of the voltage source inverter (VSI) used and their control algorithms.

The transformer based topologies for the mitigation of neutral current along with power quality compensation use three-leg VSI combined with zig-zag transformer [1] or T-connected transformer [2] or star-delta transformer etc.. In these schemes, the zero sequence current is compensated by the use of transformer and the positive and negative current compensation is achieved through the use of three-leg VSIs. A comprehensive review of various DSTATCOM topologies have been reported in [3], [4]. Inverter based topologies include four-leg VSI [5], [6], three single phase VSIs [3], three-leg

VSI with split capacitors [7] and three-leg VSI with neutral terminal at the positive or negative of dc bus [8]. All these methods are having their own advantages and disadvantages. In the above methods, four-leg topology outweighed the others in terms of DC link voltage utilization and compensation performance. However, it has some disadvantages such as the usage of more number of VSI switches and lack of independent control of different inverter legs. On the other hand, the split capacitor active filter topology has gained popularity because of its less number of semiconductor switches and independent control of inverter legs. One of the serious issue with this topology is the voltage unbalance between two DC link storage capacitors [9], [10]. The voltage unbalance is due to the flow of neutral current through one of the capacitor which causes one capacitor voltage to rise and the other to fall. This will results in unequal voltage stress across semiconductor switches and also degrade the performance of DSTATCOM. There are many inverter control schemes and topologies to balance the voltage across the capacitors [11], [12]. All these methods require additional control circuitry and hence increases the overall cost.

This paper proposes a new topology of DSTATCOM for a three-phase four-wire distribution system which can compensate unbalanced and nonlinear load. The proposed scheme uses a three-leg VSI with a single DC capacitor. In this scheme, the negative terminal of the DC link is connected to the system neutral through a small capacitor. The rating of this additional capacitor depends on the neutral current to be compensated. The average voltage across the neutral capacitor will be always clamped to half of the DC link voltage. In the proposed scheme, the DC link voltage unbalance issues are fully avoided and at the same time, the advantages of four-leg and split capacitor neutral clamped topologies are simultaneously achieved. The reference filter currents are generated using instantaneous symmetrical component theory (ISCT) [13] and hysteresis band current control method [14] are used to control VSI switches to track these currents. The steady state and transient performance of the proposed DSTATCOM topology for a three phase four wire distribution system has been verified through simulation studies.

The remaining part of this paper is organized as follows. The proposed topology has been described in Section II. The dynamic modelling of DSTATCOM is explained in Section III. The design of various VSI parameters is described in Section IV. Control algorithm using ISCT is given in Section

This work is supported by Department of Science and Technology, India under the project grant DST/TM/SERI/2k10/47(G).

978-1-4799-6047-7/14 $31.00 © 2014 IEEE

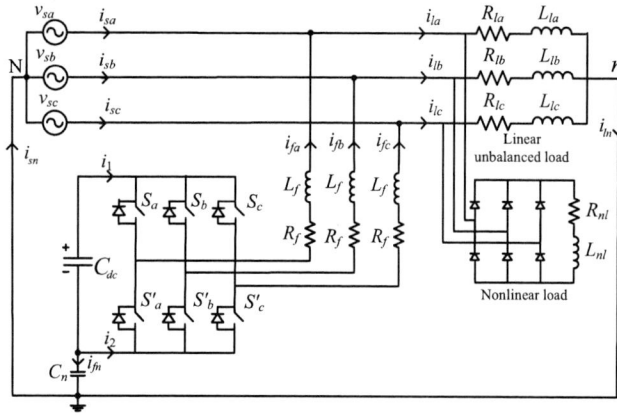

Fig. 1. Proposed DSTATCOM topology.

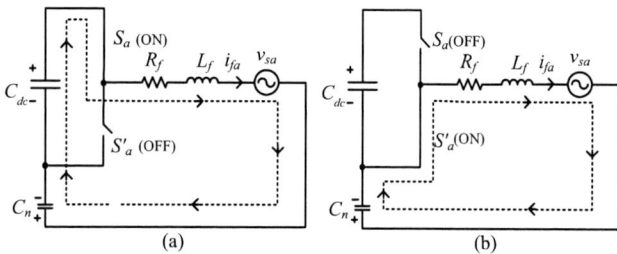

Fig. 2. Equivalent circuit: conditions (a) S_a is ON and S'_a is OFF (b) S_a is OFF and S'_a is ON.

V. Section VI describes the simulation studies and conclusions are given in Section VII.

II. PROPOSED TOPOLOGY

Fig. 1 shows the power circuit of the proposed DSTATCOM topology. Here, v_{sa}, v_{sb} and v_{sc} are source voltages of a, b and c phases respectively. The three phase source currents are represented by i_{sa}, i_{sb} and i_{sc}, load currents by i_{la}, i_{lb} and i_{lc} respectively. The load consists of both linear unbalanced and nonlinear currents. The terms, i_{ln}, i_{sn} and i_{fn} represent the load neutral current, compensated source neutral current and filter neutral current respectively. In this analysis, the distribution feeder is considered to be stiff and hence feeder impedance is neglected. The inductance and resistance of interfacing filter are represented by L_f and R_f respectively. The DC link capacitor and its voltage are represented by C_{dc} and $2V_{dc}$ respectively. In this topology, the voltage V_{dc} is selected as 1.6 times the peak value of the source voltage as given in [15]. A capacitor C_n is connected between the lower leg of VSI and neutral. The voltage across the capacitor C_n is taken as v_{cn} with polarities as in Fig. 2. It will be proved in the next section that, the average voltage V_{cn} across C_n is half the DC link voltage of VSI.

III. MODELLING AND ANALYSIS

As the proposed system is a three phase four wire, each phase of VSI can be controlled independently [16]. Therefore,

to analyze the circuit, phase-a is considered and the corresponding equivalent circuit is shown in Fig. 2. A DC voltage of $2V_{dc}$ is maintained across C_{dc} using a DC link voltage proportional and integral (PI) controller.

Let the average voltage across the neutral capacitor (C_n) be V_{cn}. It can be proved that the voltage, V_{cn} depends on the DC link voltage ($2V_{dc}$) and their relation is derived before modelling the proposed DSTATCOM topology.

Applying KVL for the equivalent circuit of Fig. 2(a), we get

$$V_{cn} = 2V_{dc} - v_{sa} - R_f i_{fa} - L_f \frac{di_{fa}}{dt} \quad (1)$$

where, V_{cn} is the average voltage across C_n with polarities as shown in figure.

Applying KVL for the equivalent circuit of Fig. 2(b), we get

$$V_{cn} = v_{sa} + R_f i_{fa} + L_f \frac{di_{fa}}{dt}. \quad (2)$$

Adding (1) and (2),

$$V_{cn} = V_{dc} \quad (3)$$

Now, the model of the proposed DSTATCOM topology is derived. Consider the equivalent circuit as shown in Fig. 2. The KVL equation when S_a is closed and S'_a is open, can be written as

$$\frac{di_{fa}}{dt} = -\frac{R_f}{L_f} i_{fa} - \frac{v_{sa}}{L_f} + \frac{V_{dc}}{L_f}. \quad (4)$$

Similarly, the KVL equation when S_a is open and S'_a is closed, can be written as

$$\frac{di_{fa}}{dt} = -\frac{R_f}{L_f} i_{fa} - \frac{v_{sa}}{L_f} - \frac{V_{dc}}{L_f}. \quad (5)$$

These two equations can be combined by using a switching function u (when $u_a = 1$, $\overline{u}_a = 0$ and when $u_a = 0$, $\overline{u}_a = 1$) and its complementary $\overline{u}$ as follows.

$$\frac{di_{fa}}{dt} = -\frac{R_f}{L_f} i_{fa} - \frac{v_{sa}}{L_f} + (u_a - \overline{u}_a)\frac{V_{dc}}{L_f}. \quad (6)$$

Similar equations can be written for b and c phases as follows

$$\frac{di_{fb}}{dt} = -\frac{R_f}{L_f} i_{fb} - \frac{v_{sb}}{L_f} + (u_b - \overline{u}_b)\frac{V_{dc}}{L_f}$$
$$\frac{di_{fc}}{dt} = -\frac{R_f}{L_f} i_{fc} - \frac{v_{sc}}{L_f} + (u_c - \overline{u}_c)\frac{V_{dc}}{L_f}. \quad (7)$$

The VSI currents i_1 and i_2 can be represented as

$$i_1 = u_a i_{fa} + u_b i_{fb} + u_c i_{fc}$$
$$i_2 = \overline{u}_a i_{fa} + \overline{u}_b i_{fb} + \overline{u}_c i_{fc}. \quad (8)$$

These VSI currents can be expressed in terms of capacitor voltages as follows

$$(2C_{dc} - C_n)\frac{dV_{dc}}{dt} = -i_1$$
$$C_n \frac{dV_{dc}}{dt} = i_2 \quad (9)$$

978-1-4799-6047-7/14 $31.00 © 2014 IEEE

where, C_{dc} is the DC link capacitor and C_n is the capacitor connected between the negative of DC link to neutral. Using (8) and (9), we get

$$\frac{dV_{dc}}{dt} = -\frac{(u_a - \overline{u}_a)i_{fa}}{2C_{dc}} - \frac{(u_b - \overline{u}_b)i_{fb}}{2C_{dc}} - \frac{(u_c - \overline{u}_c)i_{fc}}{2C_{dc}}. \tag{10}$$

Equations (6), (7) and (10) are used to model the proposed inverter topology.

IV. DESIGN OF VSI PARAMETERS

The different parameters of the VSI need to be designed suitably for better performance of the DSTATCOM. The important parameters those need to be considered are DC link voltage (V_{dc}), DC link capacitor (C_{dc}), interfacing inductance (L_f), hysteresis band ($\pm h$) and the neutral capacitor (C_n). The procedure to select neutral capacitor (C_n) is explained in detail and the other parameters are selected based on the design parameters for conventional DSTATCOM which is given in [15].

A. Design of Neutral Capacitor (C_n)

The neutral capacitor C_n is designed based on the neutral current which is to be compensated by the DSTATCOM. In a survey in the United States, reported in [17] shows that, in some loads the neutral currents varies from 0 to 1.73 times the phase current and in many systems have neutral currents exceeding the full load phase currents. Subsequently therefore, the capacitor C_n is designed based on the assumption that the neutral current is equal to phase current. So, the reactive part of phase-a load current is equated to the filter phase-a current. The load current, $\overline{I}_{la}$ and fundamental filter current $\overline{I}_{fa}$ are obtained as

$$\overline{I}_{la} = \frac{\overline{V}_{sa}}{(R_{la} + jX_{la})} \tag{11}$$

$$\overline{I}_{fa} = \frac{\overline{V}_1 - \overline{V}_{sa}}{R_f + j(X_f - X_{cn})} \tag{12}$$

where, V_{sa} is the phase-a source voltage. R_{la} and X_{la} represent the phase-a load resistance and reactance respectively. The phase-a resistance and reactance of interfacing filter are indicated by R_f and X_f respectively. V_1 is the fundamental inverter output voltage and X_{cn} is the reactance of neutral capacitor. In this case,

$$V_1 = \frac{0.612V_{dc}}{2\sqrt{3}}$$

By equating the imaginary part of (11) and (12), we get

$$\frac{V_{sa}X_{la}}{(R_{la}^2 + X_{la}^2)} = \frac{V_1 - V_{sa}}{R_f^2 + (X_{lf} - X_{cn})^2}(X_{lf} - X_{cn})$$

Substituting the values, $V_{sa} = 230$ V, $V_{dc} = 520$ V, $R_{la} = 17.5$ Ω, $X_{la} = 7.8$ Ω, $R_f = 0.5$ Ω and $X_{lf} = 3.14$ Ω, the value of C_n is calculated to be 100 μF.

B. Design of Other VSI Parameters

The other important parameters of VSI like, DC link voltage (V_{dc}), DC storage capacitor (C_{dc}), interfacing inductance (L_f), and hysteresis band ($\pm h$) are selected based on the design method of conventional DSTATCOM topology [15]. The DC link voltage (V_{dc}) is taken as 1.6 times the peak of phase voltage. In this topology, the DC link voltage reference (V_{dcref}) is taken as ($2V_{dc}$) and is found to be 1040 V. The DC link capacitor value is given by

$$C_{dc} = \frac{(2X - X/2)nT}{(1.8V_m)^2 - (1.4V_m)^2}. \tag{13}$$

where, V_m is the peak value of the source voltage, X is the kVA rating of the system, n is number of cycles during which the voltage across C_{dc} has to be resumed to the original and T is the time period. Here, $V_m = 325.27$ V, $X = 15$ kVA, $n = 1$ and $T = 0.02$ s. On substitution, the value of C_{dc} is found to be 3322 μF.

A proper hysteresis band ($\pm h$) is to be selected for better tracking performance. In this topology, it is taken as 5% of load current and is approximately obtained as ±0.5 A.

The interfacing inductance is given by

$$L_f = \frac{1.6V_m}{4hf_{swmax}}. \tag{14}$$

Assuming a maximum switching frequency (f_{swmax}) of 12 kHz, the interfacing inductance (L_f) is found to be 20 mH.

V. CONTROL ALGORITHM

The reference currents for the active filter are generated using ISCT [13], [18] and are given as,

$$i_{fa}^* = i_{la} - \frac{v_{sa} + \beta(v_{sb} - v_{sc})}{\Delta}(P_l + P_{loss})$$

$$i_{fb}^* = i_{lb} - \frac{v_{sb} + \beta(v_{sc} - v_{sa})}{\Delta}(P_l + P_{loss}) \tag{15}$$

$$i_{fc}^* = i_{lc} - \frac{v_{sc} + \beta(v_{sa} - v_{sb})}{\Delta}(P_l + P_{loss})$$

where,

$$\Delta = \sum_{j=a,b,c} v_{sj}^2, \beta = tan\varphi/\sqrt{3}.$$

In the above equations, P_l is the average load power and is obtained using a moving average filter of one cycle window of time T in seconds as in (16)

$$P_l = \frac{1}{T}\int_{t_1-T}^{t_1}(v_{sa}i_{la} + v_{sb}i_{lb} + v_{sc}i_{lc})\,dt. \tag{16}$$

In (15), P_{loss} denotes the switching and ohmic losses in actual compensator. An expression for P_{loss} is derived based on the condition that, average DC capacitor current is zero to maintain a constant capacitor voltage [13]. The deviation of average capacitor current from zero will reflect as a change in capacitor voltage from a steady state value. PI controller is used to generate P_{loss} term as given by

$$P_{loss} = K_p e_{vdc} + K_i \int e_{vdc}dt. \tag{17}$$

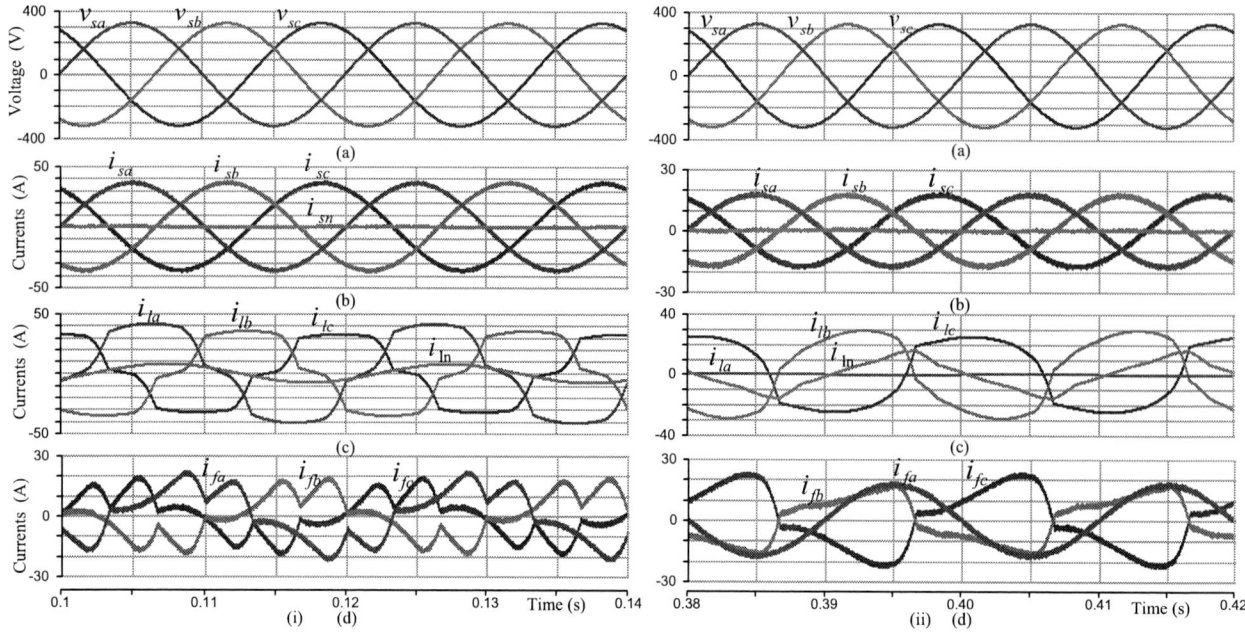

Fig. 3. Steady state performance: i(a) & ii(a) filter currents i(b) & ii(b) source currents i(c) & ii(c) load currents.

Where $e_{vdc} = V_{dcref} - v_{dc}$, v_{dc} represents the actual voltage sensed and updated once in a cycle.

The term φ is the desired phase angle between the source voltage and current. In this analysis, it considered as zero. The switching signals for the VSI switches are generated using hysteresis band current control method [14] by comparing the actual filter currents (i_{fabc}) with the reference currents generated by using ISCT (i^*_{fabc}). These controllers issue the switching commands whenever the error signal exceeds a specified tolerance band '$\pm h$'. The switching control law is obtained as follows.

If $i_{fa} \geq i^*_{fa} + h$ then bottom switch is turned ON whereas top switch is turned OFF ($S_a = 0$, $S'_a = 1$)

If $i_{fa} \leq i^*_{fa} - h$ then top switch is turned ON whereas bottom switch is turned OFF ($S_a = 1$, $S'_a = 0$).

A hysteresis controller is a high gain proportional controller. This controller adds certain phase lag in the operation based on the hysteresis band. As the proposed DSTATCOM uses a first order filter, the controller will not make the closed loop system unstable [19]. Thus, the stability analysis is not being described for the proposed topology.

In the case of distribution feeders with considerable feeder impedance and are loaded by unbalanced nonlinear load, the terminal voltages become distorted. In this case, the fundamental positive sequence of terminal voltage has to be extracted and to be used for generating the reference filter currents.

VI. SIMULATION STUDIES

In order to investigate the compensation capability of proposed DSTATCOM as in Fig. 1, it is simulated using PSCAD 4.2.1 with the circuit parameters given in Table. I.

The steady state performance of the proposed topology for two types of load conditions are verified and are given in

Fig.3. In the simulation, it is considered that, a nonlinear unbalanced load given in Table. I is applied at $t = 0$ s. At $t = 0.3$ s, phase-a is open circuited and again reconnected at $t = 0.5$ s. Also a load change to half the given load occurs at $t = 0.7$ s. The source voltage (v_{sa}, v_{sb} and v_{sc}), source currents after compensation (i_{sa}, i_{sb} and i_{sc}), load currents (i_{la}, i_{lb} and i_{lc}) and filter currents (i_{fa}, i_{fb} and i_{fc}) are shown in Fig. 3(i)(a-d). Fig. 3(ii)(a-d) represent the corresponding waveforms when phase-a is open circuited. It is observed that the proposed DSTATCOM compensate the reactive, nonlinear and unbalanced components of load currents and thus makes the source current balanced sinusoidal and in phase with source voltage. The load neutral current (i_{ln}), filter neutral current (i_{fn}) and compensated source neutral currents (i_{sn}) are shown in Fig. 4. It indicates that, load neutral currents are fully compensated, hence the source neutral current becomes zero.

The transient performances of the proposed scheme during

TABLE I
SYSTEM PARAMETERS FOR SIMULATION STUDY

Parameter	Value
Grid voltage	400 V L-L (rms), 50 Hz
VSI	DC capacitor, C_{dc}= 3300 μF DC link voltage, $2V_{dc}$ = 1040 V Neutral capacitor, C_n = 100 μF Interfacing inductor, L_f = 20 mH Inductor resistance, R_f = 0.5 Ω Hysteresis band ($\pm h$) = 0.5 A
Voltage controller PI gains	K_p=1, K_i=100
Linear load	Z_{la} = 17.5 + j7.8 Ω Z_{lb} = 27.5 + j12.5 Ω Z_{lc} = 37.5 + j31.4 Ω
Nonlinear load	3ϕ diode bridge rectifier with a DC load of 20 Ω

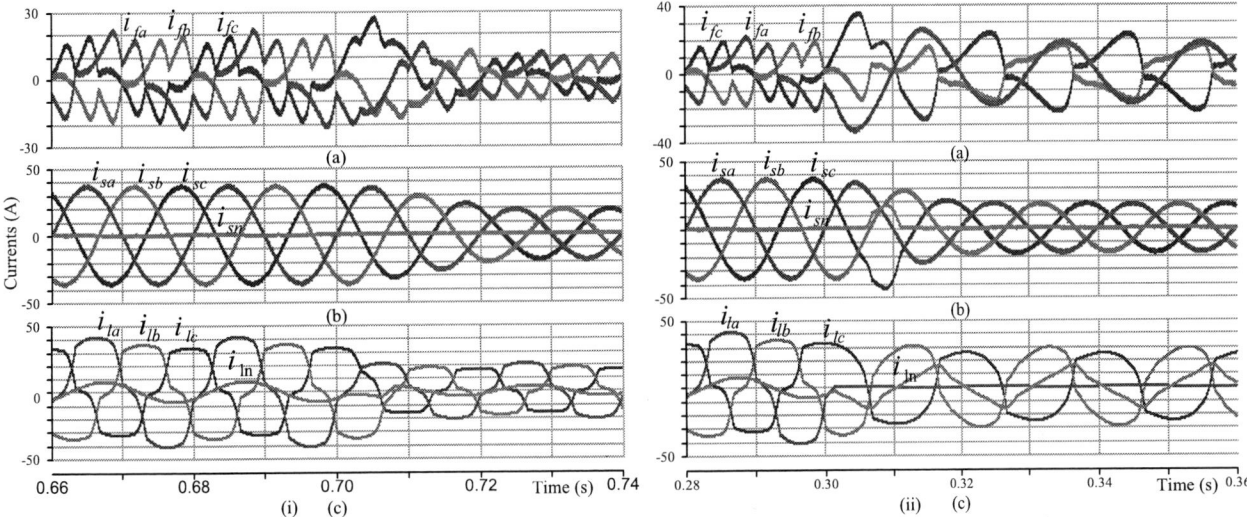

Fig. 5. Dynamic performance under load change: i(a) & ii(a) filter currents i(b) & ii(b) source currents i(c) & ii(c) load currents.

a dynamic load change is shown in Fig. 5. The characteristics when the load changes from full load to half the full load at $t = 0.7$ s is shown in Fig. 5(i). Fig. 5(ii) indicates the same when phase-a opens at $t = 0.3$ s. These results show that all currents are instantaneously changed to the new steady state value. The DC link voltage variation during a load change is shown in Fig. 6. It is observed that, during a decrement in the load, even though there is a slight increase in DC link voltage, it brings back to the reference value within a few cycles. The voltage across the neutral capacitor is shown in Fig. 7 and it indicates that, the capacitor voltage contains an average voltage which is equal to V_{dc} and sinusoidal component which is proportional to neutral current.

Through the different simulation results, it is thus verified the efficacy of the proposed topology for power factor correction, load balancing, harmonic compensation and neutral current compensation.

VII. CONCLUSION

A new three-leg voltage source inverter (VSI) based distribution static compensator (DSTATCOM) topology to compensate unbalanced and nonlinear loads in three phase four wire distribution systems is proposed. It uses a single DC storage capacitor and a low rating neutral capacitor connected between the negative DC bus to the system neutral. The detailed modelling and analysis is provided. The performance of the proposed topology for power factor correction, load balancing,

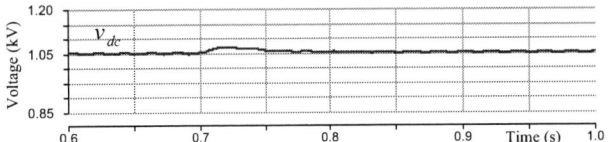

Fig. 6. DC link voltage.

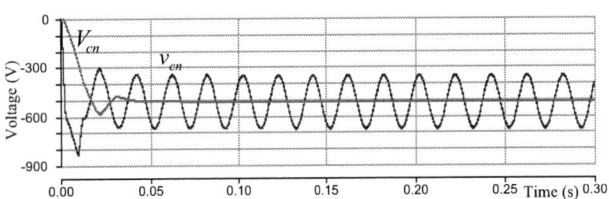

Fig. 7. Voltage across the neutral capacitor.

harmonic compensation and neutral current compensation are demonstrated through various simulation results. The proposed scheme is able to achieve the comparable performance of existing DSTATCOMs while avoiding the problem of DC capacitor voltage unbalancing of popular split capacitor three-leg DSTATCOM topology.

REFERENCES

[1] B. Singh, P. Jayaprakash, T. R. Somayajulu, and D. Kothari, "Reduced rating VSC with a zig-zag transformer for current compensation in a three-phase four-wire distribution system," *IEEE Trans. on Power Del.*, vol. 24, no. 1, pp. 249–259, Jan. 2009.

[2] B. Singh, P. Jayaprakash, and D. Kothari, "A T-connected transformer and three-leg VSC based DSTATCOM for power quality improvement," *IEEE Trans. on Power Electron.*, vol. 23, no. 6, pp. 2710–2718, Nov. 2008.

[3] B. Singh, P. Jayaprakash, D. Kothari, A. Chandra, and K. Al Haddad, "Comprehensive study of DSTATCOM configurations," *IEEE Trans. on Ind. Informatics*, vol. 10, no. 2, pp. 854–870, May 2014.

[4] D. Sreenivasarao, P. Agarwal, and B. Das, "Neutral current compensation in three-phase, four-wire systems: A review," *Electric Power Systems Research*, vol. 86, no. 0, pp. 170 – 180, 2012.

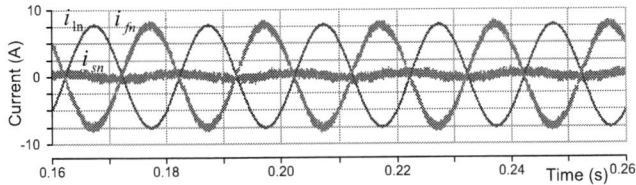

Fig. 4. Neutral current compensation.

978-1-4799-6047-7/14 $31.00 © 2014 IEEE

[5] B. Singh, S. Arya, C. Jain, and S. Goel, "Implementation of four-leg distribution static compensator," *IET Generation Transmission Distribution*,, vol. 8, no. 6, pp. 1127–1139, Jun. 2014.

[6] Mahesh K. Mishra, A. Ghosh, A. Joshi, and H. Suryawanshi, "A novel method of load compensation under unbalanced and distorted voltages," *IEEE Trans. on Power Del.*, vol. 22, no. 1, pp. 288–295, Jan. 2007.

[7] S. Karanki, N. Geddada, Mahesh K. Mishra, and B. Kumar, "A DSTATCOM topology with reduced dc-link voltage rating for load compensation with nonstiff source," *IEEE Trans. on Power Electron.*, vol. 27, no. 3, pp. 1201–1211, Mar. 2012.

[8] A. Bhattacharya, C. Chakraborty, and S. Bhattacharya, "Parallel-connected shunt hybrid active power filters operating at different switching frequencies for improved performance," *IEEE Trans. on Ind. Electron.*, vol. 59, no. 11, pp. 4007–4019, Nov. 2012.

[9] Mahesh K. Mishra., A. Joshi, and A. Ghosh, "Control schemes for equalization of capacitor voltages in neutral clamped shunt compensator," *IEEE Trans. on Power Del.*, vol. 18, no. 2, pp. 538–544, Apr. 2003.

[10] A. Shukla, A. Ghosh, and A. Joshi, "Control schemes for DC capacitor voltages equalization in diode-clamped multilevel inverter-based DSTATCOM," *IEEE Trans. on Power Del.*, vol. 23, no. 2, pp. 1139–1149, Apr. 2008.

[11] S. Srikanthan and Mahesh K. Mishra, "DC capacitor voltage equalization in neutral clamped inverters for DSTATCOM application," *IEEE Trans. on Ind. Electron.*, vol. 57, no. 8, pp. 2768–2775, Aug. 2010.

[12] Y. Chen, B. Mwinyiwiwa, Z. Wolanski, and B.-T. Ooi, "Regulating and equalizing DC capacitance voltages in multilevel STATCOM," *IEEE Tran. on Power Del.*, vol. 12, no. 2, pp. 901–907, Apr. 1997.

[13] A. Ghosh and A. Joshi, "A new approach to load balancing and power factor correction in power distribution system," *IEEE Trans. on Power Del.*, vol. 15, no. 1, pp. 417–422, Jan. 2000.

[14] D. M. Brod and D. Novotny, "Current control of VSI-PWM inverters," *IEEE Trans. on Ind. Appl.*, vol. IA-21, no. 3, pp. 562–570, May 1985.

[15] U. Rao, Mahesh K. Mishra, and A. Ghosh, "Control strategies for load compensation using instantaneous symmetrical component theory under different supply voltages," *IEEE Trans. on Power Del.*, vol. 23, no. 4, pp. 2310–2317, 2008.

[16] Mahesh K. Mishra., A. Ghosh, and A. Joshi, "A new STATCOM topology to compensate loads containing AC and DC components," in *Power Engineering Society Winter Meeting, 2000. IEEE*, vol. 4, 2000, pp. 2636–2641.

[17] T. Gruzs, "A survey of neutral currents in three-phase computer power systems," *IEEE Trans. on Industry Appl.*, vol. 26, no. 4, pp. 719–725, Jul. 1990.

[18] A. Ghosh and G. Ledwich, *Power Quality Enhancement Using Custom Power Devices*, ser. Kluwer international series in engineering and computer science. Springer US, 2002.

[19] Ghosh, A and Ledwich, G., "Load compensating dstatcom in weak ac systems," *IEEE Trans. on Power Del.*, vol. 18, no. 4, pp. 1302–1309, Oct. 2003.

Control strategy for voltage sag/swell/harmonic/flicker compensation with conventional and fuzzy controller(UPQC)

Mukul Chourasia[*], S.P.Srivastava, Aurobinda Panda

Department of Electrical Engineering,
IIT Roorkee, Uttrakhand, India-247667
Email id[*] mukul.chourasia@gmail.com

Abstract— **This post provides a uniform system of power quality control strategy. This control strategy is in 3-phase and 3-wire system. The device combines a UPQC, shunt active filter with an active series filters in the back to back configuration. UPQC designed for simultaneous compensation of current and voltage harmonics. This paper focus on simple control strategy with voltage sag-swell/harmonic/flicker compensation, fuzzy controller benefits over PI controller. Simulation results based on SIMPOWERSYSTEM by MATLAB / Simulink are presented, and verified the effectiveness of the proposed control technology.**

Keywords— Active Power Filter(APF), Series power filter(SPF), Point Of Common Coupling(PCC), Unified Power Quality Conditioner(UPQC), Simpowersystem(SPS), Harmonic, Power Quality, THD.

I. INTRODUCTION

In past few years, the uses of power electronic have increased rapidly in industries as well as in domestic application, right from few watts to mega watts. These electronic devices for power, suffering harmonic injection and reactive power flow problem, they have highly nonlinear characteristics. In addition to these induction motor, transformer, lamps, ballast, CFL, welding machine, arc furnace have characteristics non-linear and also responsible for harmonic and reactive power problems. These problems are reflected in different transient disturbances as well as in stable condition. These interruptions are resulting in a voltage sag / swell / flicker / fluctuation / transitional imbalance current and voltage, mains frequency variation, under voltage and overvoltage problems, phase angle jumps etc among all these void disturbance voltage sag and swell, unbalanced voltage and waveform distortion, harmonics are cause for concern. Any significant deviation from the value the voltage, current and frequency or the purity of the waveform can lead to power quality problems. The generation of harmonics and flow of reactive power in power system result in "ELECTRIC POWER QUALITY" problems. These power quality problems reflected in the reduced effectiveness system, interference, neutral combustion, evil of relay operation etc, uses of AC-DC drives, motors with current commutation, SMPS are highly responsible for the injection of harmonic and reactive power flow.

To overcome the problem, two possible solutions are there (i) make the equipment less sensitive towards the power variation. (ii) Use of filters. The first one is impossible, second can implemented. The use of filters starts with the first name passive filters, which are not popular due to some drawbacks as resonance problem, static in nature, bulky size and many more. These drawbacks are overcome by the active filters so as popular nowadays. With some variation then comes the hybrid filter (combination of passive and active filter) they do popular. But the matter of concern is they all deal with the current variation that is they concern for current only. Now what if the voltage has to be take care, then came the word UPQC (unified power quality conditioner). It take care of voltage as well as current.

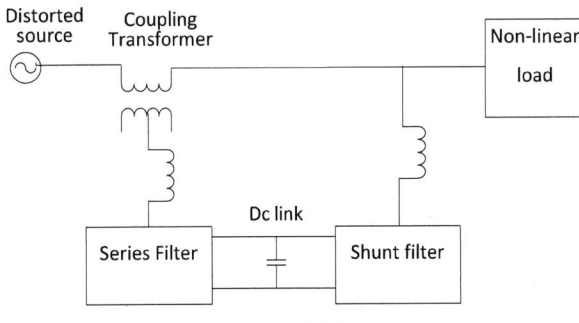

Figure 1: UPQC

It is the combination of shunt and series active filter, shunt APF filter act as a current source and inject compensating harmonic current in order to have sinusoidal as well as in phase current. The series APF filter acts as a voltage source and inject the voltage in series with the harmonic source in order to have sinusoidal voltage at PCC [1]. In the past few years solution based on combination of series and shut active filter have appeared. Its main motto is to compensate for supply voltage and load current imperfections, as sags, swell etc. in most of the articles control technique are complex requiring different kinds of transformations. The control technique pressed here is simple and easy to implement.

978-1-4799-6047-7/14 $31.00 © 2014 IEEE

II. CONTROL TECHNIQUE

Figure 1 show the basic block diagram of UPQC. The control strategy proposed here is a conventional or the unit template as called sine template control strategy. In this we generate the sine template from the source and use as reference for the both current and the voltage compensation.

The sine template generated in matlab is with the help of PLL block (phase locked loop). It provides the sine and cosine component of fundamental frequency through the distorted waveform. Distorted waveform in the sense contains the harmonics. PLL block has the capability to extract the fundamental component from distorted source. The template extracted from PLL block are unity in nature. This template then redistributed as three phase system that resemble with the actual source. Now this three phase system template act as a reference.

For series filter the template get multiplied by the rated voltage, become the reference voltage waveform then compared with the actual one and the error get process by hysteresis controller and the switching pulse get generated and supplied to the series inverter. The voltage generated by the inverter is injected to the PCC by the coupling transformer. The sum of source voltage and the coupling transformer voltage results in sinusoidal voltage at the PCC. Similarly for the shunt inverter the rated current is multiplied to the three phase unit template signal. Again compare the reference and the actual signal, error is process by hysteresis controller and operates the shunt inverter. The benefit for the extraction of unit template from voltage source is to get current in the phase with the voltage.

Figure 2 represents the block diagram of unit template method. Now how to get these rated values. For voltage it define by the system but for the current it is variable. So to get the current rated value, we use pi controller. This pi controller is to maintain the constant voltage of the dc link. The reference value of the dc link is set with the help of calculation [1]. Then the error is process by the PI controller and the output considers as rated current that to be drawn from the source. With the help of sine template system it is easy to control the functioning of the UPQC in comparison with the instantaneous pq theory. In pq theory lots of calculation are there, first transform the abc to αβØ quantities after that calculate the power which is virtual in nature with the help of virtual current. Then again calculate the αβØ quantities and after all again transform to abc, a huge process it is. This all avoided by the proposed control strategy. Control strategy help to avoid sag/swell problems along with flicker compensation.

III. FUZZY CONTROLLER

If on realizing the results, notice that the PI controller works properly but the settling time and the rise/dip of the capacitor voltage are good, but can be make more appropiate. This is done by fuzzy controller, instead of PI apply fuzzy. Here by analyzing nine rules, develop the fuzzy controller.

Above one are the nine rules. The first two linguistic values are associated with the input variable error and change in error, whereas third linguistic value is for the output. In conventional controller, the gain parameters are the combination of numerical values. The equivalent term in FLC is 'rule' and they are linguistic in nature. The FLC performance greatly depends on the designing of the rules and their collection. The control rules relate pairs of fuzzy inputs error and change in error to fuzzy outputs. Basically rules are derived from the general knowledge of the system behavior. Here consider three membership functions for each inputs to FLC, error and change in error.

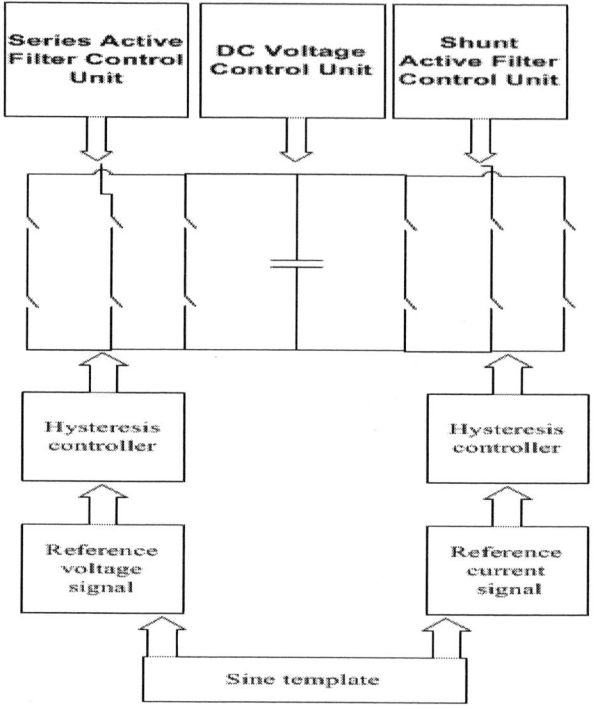

Figure 2: Block diagram

The flicker compensation is also done by the same nine rules. Here 10% of flicker are introduce in the simulation.

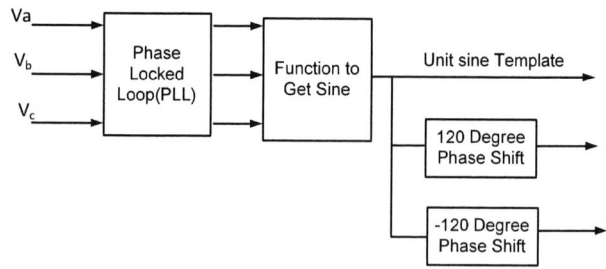

Figure 3: PLL Block

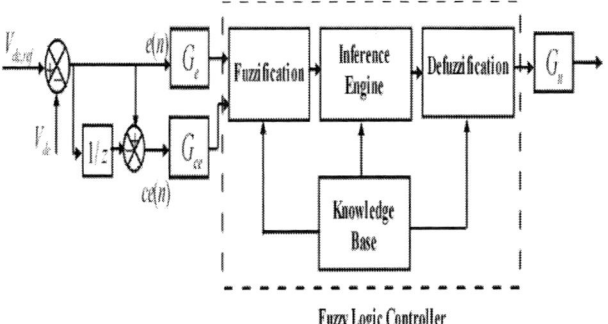

Figure 4: Fuzzy block

The fuzzy controller is developed based on the linguistic description and does not require mathematical model of the closed loop system. Exhaustive simulations are carried out with the help of the developed models in matlab/simulink environment for three phase and the flexibility of control algorithm under variety of load is established.

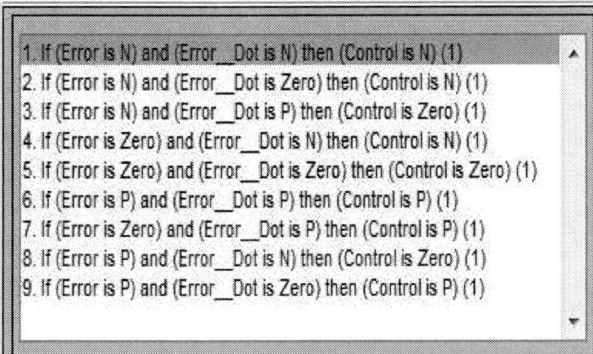

Figure 5: Rules

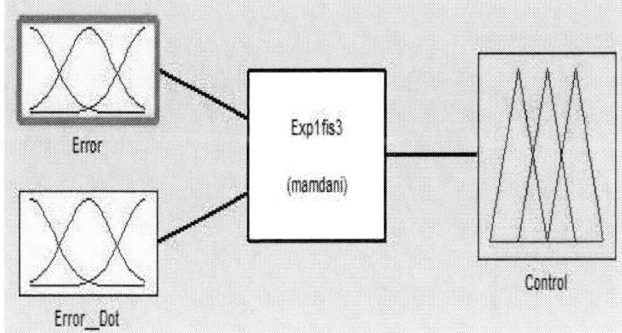

Figure 6: Fuzzy matlab

Hence the fuzzy controller used is characterized as follow:

➢ Three fuzzy sets for each input and output
➢ Triangular membership function for simplicity

➢ Fuzzification implication using Mamdani's "min" operator and defuzzification

IV. SIMULATION RESULTS

During the simulation following parameter are assumed:

- Source voltage: 100V peak + 20% 5th harmonic+10% 7th harmonic.
- Filter inductance for shunt APF: 3 mH
- Capacitance for DC link capacitor: 4700 µF
- PI parameter: kp=0.5 + ki= 9
- Coupling transformer magnetizing parameter: .8 ohm + 90 mH.
- Both the VSI are realize by mosfet switches, each contain six mosfet switches.
- All voltage are in Volts, current are in Amps, time are in Seconds.

Case 1: Harmonics (15% of 5th , 10% of 7th)

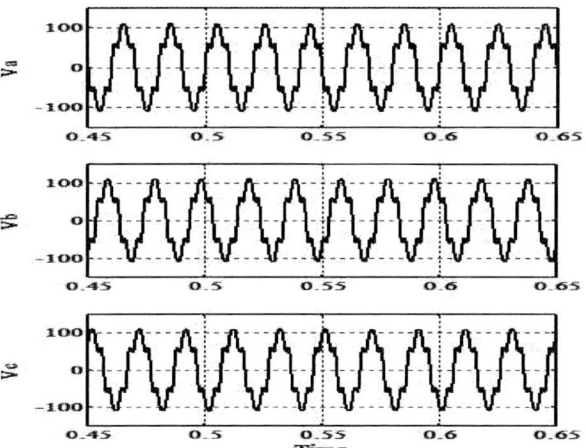

Figure 7: Source Voltage

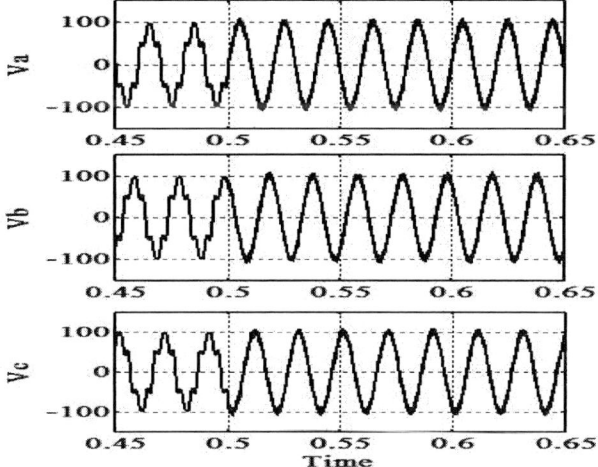

Figure 8: Voltage UPQC ON

978-1-4799-6047-7/14 $31.00 © 2014 IEEE

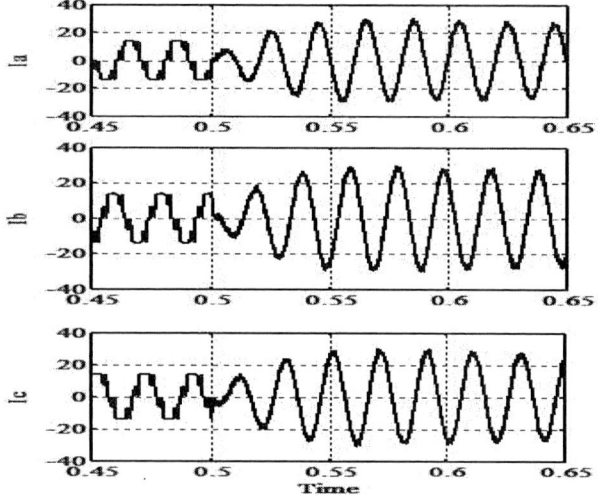

Figure 9: Current

Case 2: Sag Compensation (20% w.r.t. Source Voltage)

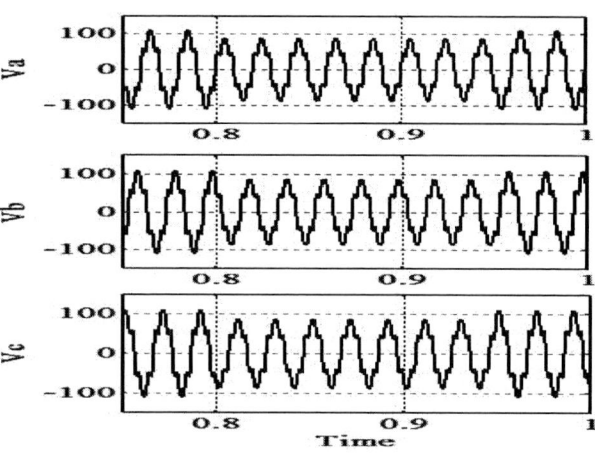

Figure 10: Source Voltage

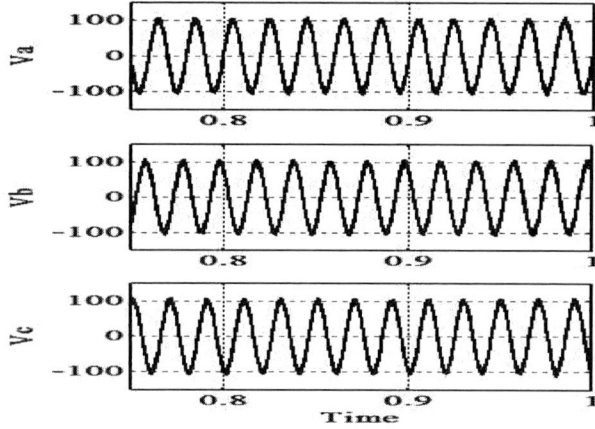

Figure 11: Voltage UPQC ON

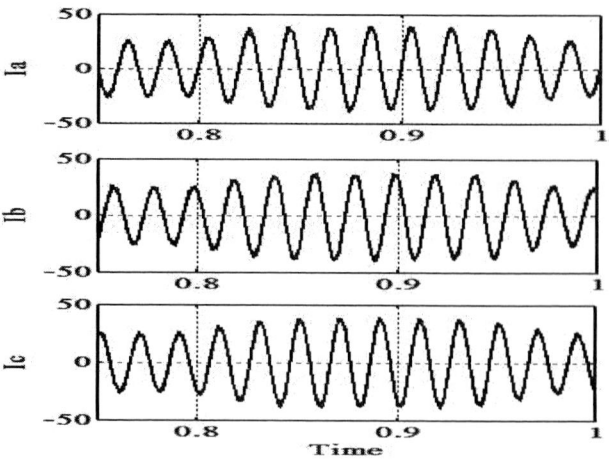

Figure 12: Current

Case 3: Swell Compensation (20% w.r.t. Source Voltage)

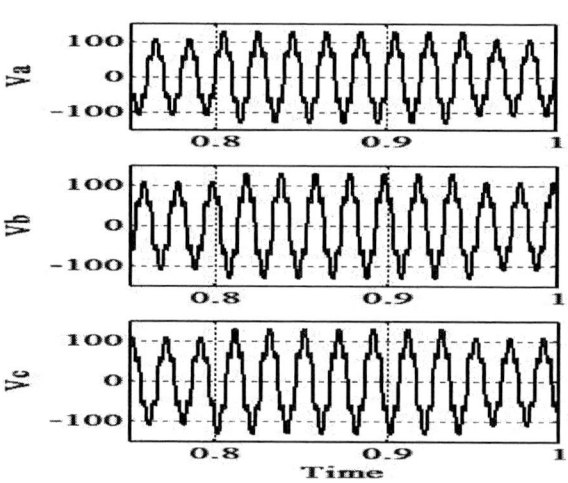

Figure 13: Source Voltage

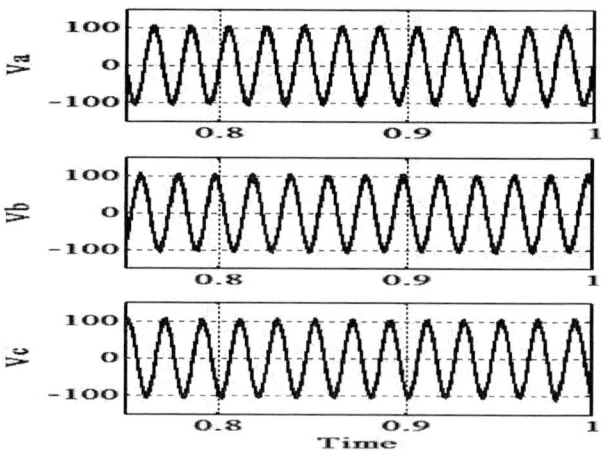

Figure 14: Voltage UPQC ON

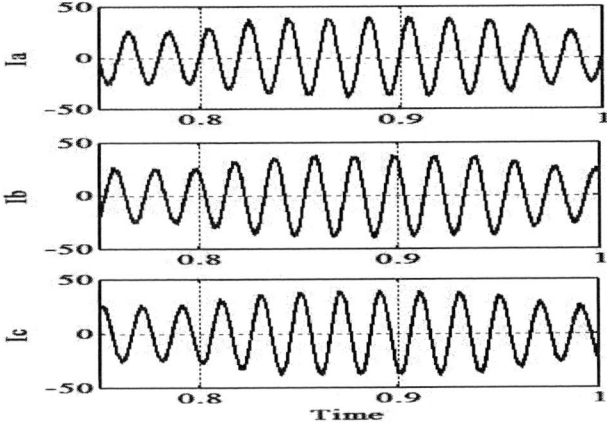

Figure 15: Current

Now on realizing the result, notice that voltage at PCC remain constant by UPQC but the current during sag/swell period slightly swell up, is due to the voltage difference between source side and the PCC. This increase in current can be handle by inserting additional source impedance between source and PCC. Further this problem is reduce by using fuzzy controller shown later.

Lets talk about the THD. Intially the distorted voltage source has 24.38% THD, and that of source current is 27.25% THD. After the UPQC made ON source voltage THD at the PCC remains only 3.29%, and that of current is 2.93%.

Case 4: Fuzzy Compensation

Now fuzzy controller is implemented, all results are same except sag and swell compensation.If you see the difference between fuzzy and PI results, it's a dramatic change in the current value.

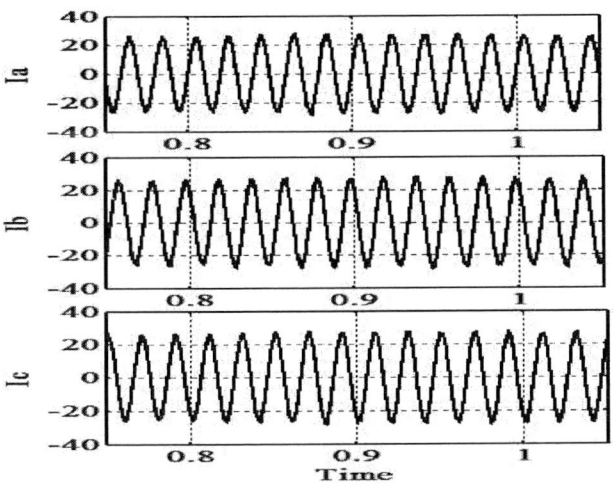

Figure 16: Swell Current

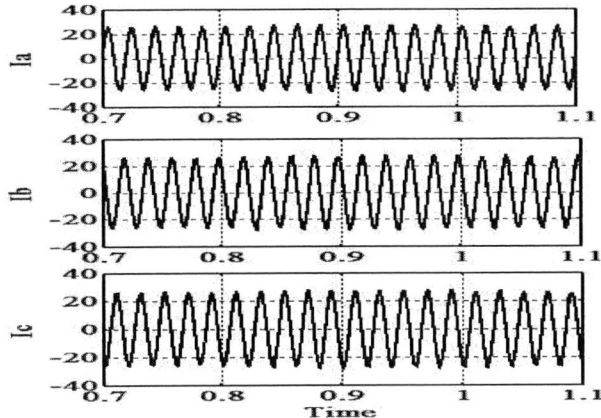

Figure 17: Sag current

Case 5: Flicker Compensation (20% of. Source voltage)

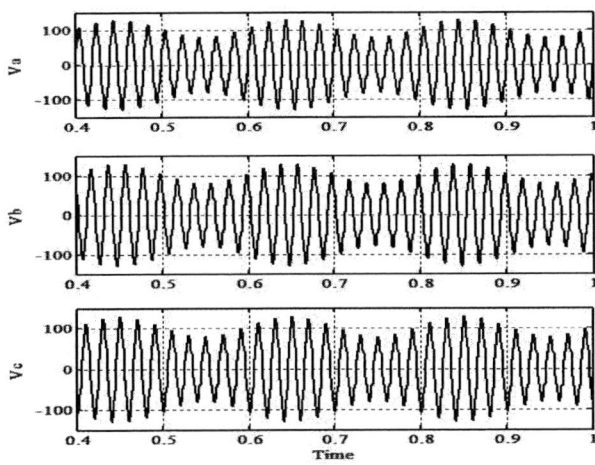

Figure 18: Source Voltage

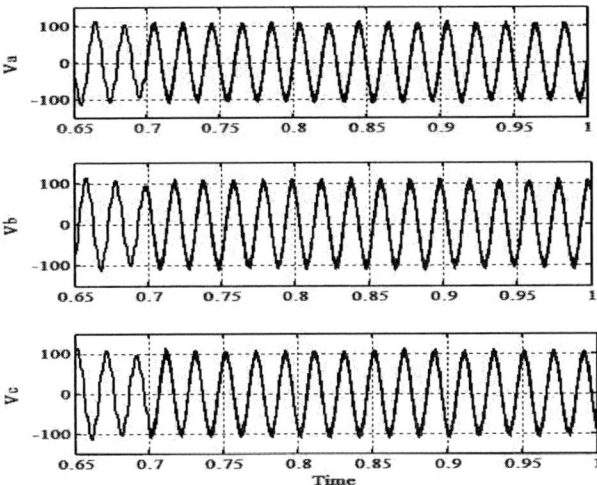

Figure 19: Voltage UPQC ON

978-1-4799-6047-7/14 $31.00 © 2014 IEEE   143

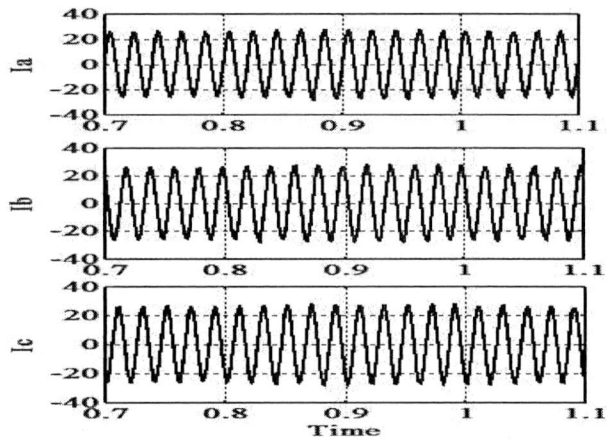

Figure 20: Current

This flicker compensation is done by fuzzy controller only, because results associate with PI are not good.

Now the purpose for using fuzzy instead of PI.

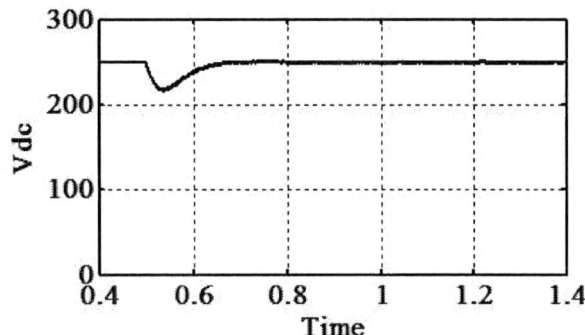

Figure 21: PI Controller

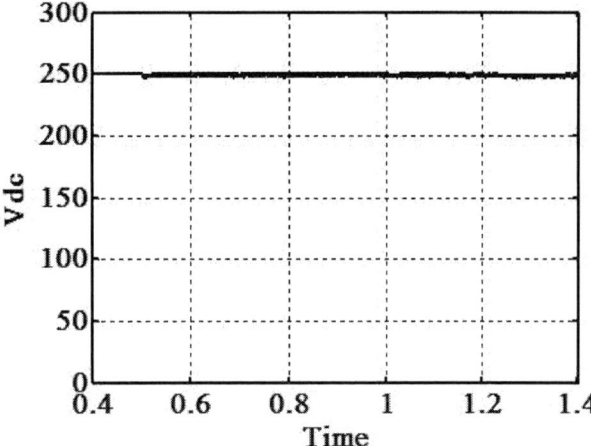

Figure 22: Fuzzy Controller

It is clearly identified the difference between PI an Fuzzy controller. That is settling time and the percentage rise /dip in capacitor voltage status, that's why fuzzy is more reliable than PI. Not only these benefits but also the fuzzy controller is robust in nature, handle the parameter variation in comparison with PI.

V. CONCLUSION

A simple control strategy for power conditioning by UPQC is suggested. Simulation results have confirmed the proposed control strategy for use in three phase three wire system, but for the three phase four wire system we need to take care for neutral current, for that three phase four wire converter is required.

The computational effort get reduced in compare with instantaneous PQ theory based strategy. The power of the suggested approach is comparable with instantaneous PQ theory without loss of stability.

VI. REFERENCES

1: V. Khadkikar, P. Aganval, A. Chandra, A.O. Bany and T.D. Nguyen, "A Simple New Control Technique For Unified Power Quality Conditioner (UPQC)", 2004 11th International Conference on Harmonics and Quality of Power.
2: H. Akagi, "Now trends in active filters for improving power quality" Procceding of the 1996 International Conference, Vol.1, Jan 1996, pp.417-425.
3: A.Chandra, B. Singh, B.N. Singh, K. Al-Haddad, "An Improved Control Algorithm of Shunt Active Filter for Voltage Regulation, Harmonic Elimination, Power Factor Correction, and Balancing of Non-linear Loads", IEEE Trans on Power Electronics, Vol.15, No.3, May 2000, pp.495-507.
4: C. Pahmer, G.A. Capolino, H. Henao, "Computer-aided design for control of shunt active filter", IECON 94, Vol.1, Sept 1994, pp.669-674.
5: C N Srivatsav, Dr. Pramod Agarwal, Dr. Sharmili Das, —Comparative Analysis Hybrid Power Filter Topologies with distorted sorce voltage□, IEEE Fifth India International Conference on Power Electronics (IICPE 2012), Delhi, December 2012.
6: M. Aredes, J. Hafner. K. Heuniann. " A Combined Series and Shunt Active Power Filter." /€€E/KTH Stuckllolrii Power. Twh Cor!$. vol. Power Electronics. pp. 237-242. Stockholm, Sweden. June 1995.
7: L. Malesani. L. Rosseto. P. Tenti. "Active Filter for Reactive Power and Harmonics Compensation", IEEE - PESC 1986. pp. 321-330.
8: Akagi H., Fujita H.; "A new power line conditioner for harmonic conpensation in power systems", Power Delivery, IEEE Transactions on, Volume: IOIssue:3, July 1995.p~1. 570-1575.
9: Fujita H., Akagi H., "The unified power quality conditioner: the integration of series and shunt-active filters." Powa Electronics, IEEE Transactions on, Volume: 13 Issue: 2, March 1998, pp.315 - 322.
10: Vilathgamuwa, M.; Zhang, Y.H.; Choi, S.S.; " Modelling, analysis and control of unified power quality conditions", Proceedings, Harmonics And Quality of Power, Vol. 2, Oct. 1998, pp.1035 -1040.

Performance Evaluation of Shunt Active Power filter Control Strategies under Different Supply Conditions

Khoisnam Steela and Bharat Singh Rajpurohit

School of Computing and Electrical Engineering
Indian Institute of Technology Mandi
Mandi, H.P., India
email id: khoisnam_steela@students.iitmandi.ac.in, bsr@iitmandi.ac.in

Abstract—**This paper presents a comparative dynamic performance evaluation of three most extensively used control algorithms for shunt active power filter (APF), namely, *p-q* theory, *d-q* method and i_d-i_q method under various conditions supply voltage. The effectiveness of these control methods under different operating conditions are validated and compared in terms of total harmonic distortion (THD) content in supply line current and dc-link voltage regulation.**

Index Terms— **power quality, shunt APF, *p-q* theory, *d-q* method, i_d-i_q method.**

I. INTRODUCTION

Since the past few decades, there has been intensive increase in the use of power converters and switching devices, various non linear loads and equipments like high-power diode/thyristor rectifiers, arc furnaces, cyclo-converters, variable speed drives, computers, etc., which draw highly non-linear current from the source and thereby inject harmonics in the system. These harmonics are perilous and greatly pollute the power system. They degrade power quality by increasing total harmonic distortion (THD) and reactive power consumption and also by causing poor power factor, voltage flicker, bad voltage regulation, voltage sags and swells, unbalanced load etc[1]-[3]. Harmonics also cause electromagnetic interference (EMI) in communication network present within certain proximity. Issues pertaining to power quality are increasing steeply, therefore, tackling these problems become an indispensable task for maintaining a good power system.

Power filters came up as a viable solution for providing clean-harmonic free power at consumer ends. Conventionally, passive L-C filters were employed for harmonics mitigation and capacitors for power factor improvement. But soon, the need to replace them with better alternative arose because they are associated with certain inherent disadvantages like bulky size, mistuning, instability, resonance with load impedance or utility impedance and lack of flexibility (i.e. they are best suited only for fixed compensation). Active power filters (APFs) became a viable alternative which give dynamic and versatile solution to the problem of power quality. Since then, research on APF technology boomed up. A large number of control techniques are being proposed and implemented. At present there is an appreciably large number of control methods used for power compensation by APFs.

This paper presents a comparative dynamic performance evaluation of shunt APFs operating with three of the most extensively used control methods for APF technology, viz. *p-q* theory, *d-q* method and i_d-i_q method. Detailed implementation scheme for these three algorithms is also presented.

II. SYSTEM CONFIGURATION

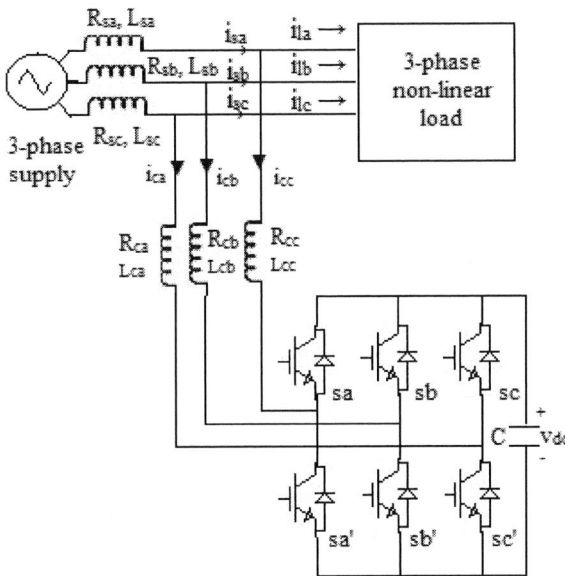

Fig. 1. System configuration of 3-phase compensated system

Fig.1. Shows the system configuration of a power system compensated by shunt connected APF. The APF is basically pulse width modulated (PWM) inverter connected in shunt with respect to load or supply and the non-linear load is composed of a three phase diode rectifier with R-L load. The PWM inverter used in this work is an IGBT based voltage source inverter (VSI) which has a dc-link capacitor. The values of system parameters are given in Appendix. The shunt connected APF generates a compensating signal which is an exact opposite of the distortion produced by the load to the line current. The effect of distortion produced by load is nullified by injecting the compensator power. The amount of compensating power injected by the filter has to be controlled in order to get perfect compensation. There are a large number of control schemes available. Control of an APF includes

978-1-4799-6047-7/14 $31.00 © 2014 IEEE   145

extraction of reference signals, controlling dc-link capacitor voltage and gating signal generation. In this work three schemes, *p-q* theory, *d-q* method and i_d-i_q method, are used for reference current extraction. The dc-link voltage is regulated with PI controller and hysteresis band current controller(HBCC) is employed for gating signal generation.

III. CONTROL ALGORITHMS

A. p-q theory

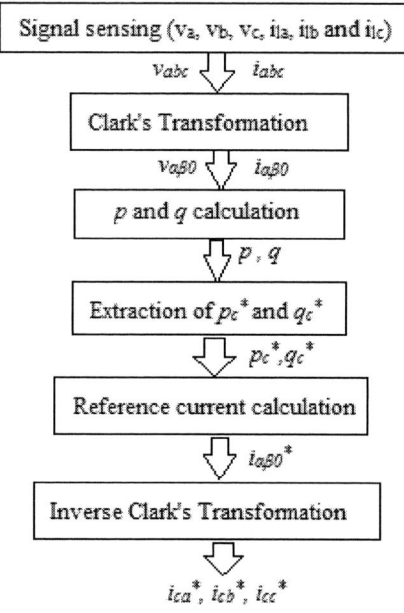

Fig. 2. Reference current generation with *p-q* theory

Fig. 2. shows the reference current extraction algorithm for *p-q* theory based compensation [4]. The three phase voltages (v_a, v_b and v_c) and load currents (i_{la}, i_{lb} and i_{lc}) are transformed from *abc* co-ordinates to *αβ0* co-ordinates using Clark's transformation as per Eq. 1 and 2. The instantaneous active power (p) and reactive power (q) are then computed in the *α-β* co-ordinates as per Eq. 3.

$$\begin{bmatrix} v_\alpha \\ v_\beta \end{bmatrix} = \sqrt{\frac{2}{3}} \begin{bmatrix} 1 & -\frac{1}{2} & -\frac{1}{2} \\ 0 & \frac{\sqrt{3}}{2} & -\frac{\sqrt{3}}{2} \end{bmatrix} \begin{bmatrix} v_a \\ v_b \\ v_c \end{bmatrix} \quad (1)$$

$$\begin{bmatrix} i_\alpha \\ i_\beta \end{bmatrix} = \sqrt{\frac{2}{3}} \begin{bmatrix} 1 & -\frac{1}{2} & -\frac{1}{2} \\ 0 & \frac{\sqrt{3}}{2} & -\frac{\sqrt{3}}{2} \end{bmatrix} \begin{bmatrix} i_{la} \\ i_{lb} \\ i_{lc} \end{bmatrix} \quad (2)$$

$$\begin{bmatrix} p \\ q \end{bmatrix} = \begin{bmatrix} v_\alpha & v_\beta \\ v_\beta & -v_\alpha \end{bmatrix} \begin{bmatrix} i_\alpha \\ i_\beta \end{bmatrix} \quad (3)$$

Both these instantaneous active and reactive powers consist of a dc or average component and an oscillating component as shown in Eq. 4.

$$p = \bar{p} + \tilde{p} \ \text{ and } \ q = \bar{q} + \tilde{q} \quad (4)$$

The instantaneous active power (p) and reactive power (q) are then decomposed into the average and oscillating components by passing through a low pass filter (LPF). Once these components are separated the reference compensating powers (p_c^* and q_c^*) are decided based on compensation requirements and then selected. The reference current in *α-β* co-ordinates are then computed using Eq. 5. The minus sign in the compensating power indicates that the compensator draws a current that produces the exact inverse of the undesirable power drawn by the by the load.

$$\begin{bmatrix} i_\alpha^* \\ i_\beta^* \end{bmatrix} = \frac{1}{v_\alpha^2 + v_\beta^2} \begin{bmatrix} v_\alpha & v_\beta \\ v_\beta & -v_\alpha \end{bmatrix} \begin{bmatrix} -p_c^* \\ -q_c^* \end{bmatrix} \quad (5)$$

Finally, the three phase reference current in *abc* co-ordinates is obtained by employing Inverse Clark's transformation of reference current in *α-β* co-ordinates using transformation Eq. 6.

$$\begin{bmatrix} i_{ca}^* \\ i_{cb}^* \\ i_{cc}^* \end{bmatrix} = \sqrt{\frac{2}{3}} \begin{bmatrix} 1 & 0 \\ -\frac{1}{2} & \frac{\sqrt{3}}{2} \\ -\frac{1}{2} & -\frac{\sqrt{3}}{2} \end{bmatrix} \begin{bmatrix} i_{c\alpha}^* \\ i_{c\beta}^* \end{bmatrix} \quad (6)$$

B. d-q method

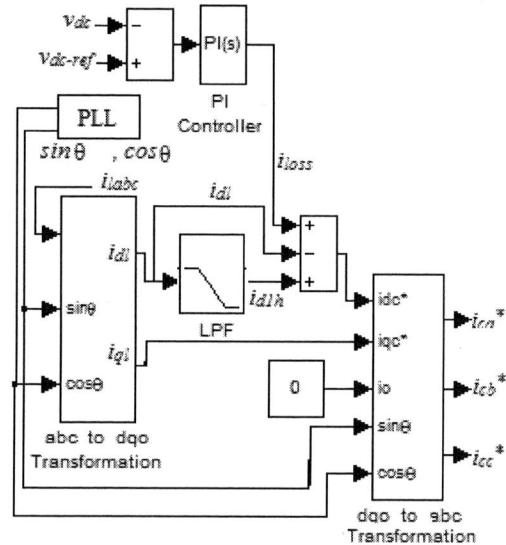

Fig. 3. Reference current generation with *d-q* method

The *d-q* method of control was first proposed in [5]. Fig. 3. shows the reference current extraction scheme for *d-q* method based compensation. In this approach reference signal extraction begins with first measuring load current at PCC. These load currents i_{la}, i_{lb} and i_{lc} are transformed to *α-β* co-ordinates using Clark's transformation given by Eq. 7. These currents in stationary *α-β* co-ordinates are then transformed to *d-q* co-ordinates, using transformation equation as given in Eq. 8. Here the reference frame *d-q* is determined by an angle "*θ*" w.r.t. *α-β* frame. The angle (*θ*) is determined through a phase lock loop (PLL).

978-1-4799-6047-7/14 $31.00 © 2014 IEEE   146

$$\begin{bmatrix} i_\alpha \\ i_\beta \end{bmatrix} = \sqrt{\frac{2}{3}} \begin{bmatrix} 1 & -\dfrac{1}{2} & -\dfrac{1}{2} \\ 0 & \dfrac{\sqrt{3}}{2} & -\dfrac{\sqrt{3}}{2} \end{bmatrix} \begin{bmatrix} i_{la} \\ i_{lb} \\ i_{lc} \end{bmatrix} \quad (7)$$

$$\begin{bmatrix} i_{ld} \\ i_{lq} \end{bmatrix} = \begin{bmatrix} \cos\theta & \sin\theta \\ -\sin\theta & \cos\theta \end{bmatrix} \begin{bmatrix} i_\alpha \\ i_\beta \end{bmatrix} \quad (8)$$

The currents i_{ld} and i_{lq} consists of an average or dc component and an oscillating or ac component as represented in Eq. 9. The dc-component corresponds to the fundamental component and the ac component corresponds to harmonics present in the total current.

$$\begin{bmatrix} i_{ld} \\ i_{lq} \end{bmatrix} = \begin{bmatrix} i_{ld1h} + i_{ldnh} \\ i_{lq1h} + i_{lqnh} \end{bmatrix} \quad (9)$$

The currents i_{ld} and i_{lq} is decomposed into these two above mentioned components by passing through a low pass filter. Once the components are separated, compensating signals are selected based on type of compensation requirements. The *d-axis* and *q-axis* reference currents are represented by i_{cd}^* and i_{cq}^* respectively. To obtain reference current in *abc* co-ordinates i_{cd}^* and i_{cq}^* are first transformed back to *α-β* co-ordinates using transform equation as given in Eq. 10 and then further transformed to *abc* using Clark's inverse transformation as given by Eq. 11.

$$\begin{bmatrix} i_{c\alpha}^* \\ i_{c\beta}^* \end{bmatrix} = \begin{bmatrix} \cos\theta & -\sin\theta \\ \sin\theta & \cos\theta \end{bmatrix} \begin{bmatrix} i_{cd}^* \\ i_{cq}^* \end{bmatrix} \quad (10)$$

$$\begin{bmatrix} i_{ca}^* \\ i_{cb}^* \\ i_{cc}^* \end{bmatrix} = \sqrt{\frac{2}{3}} \begin{bmatrix} -\dfrac{1}{2} & \dfrac{\sqrt{3}}{2} \\ \dfrac{1}{2} & \dfrac{\sqrt{3}}{2} \\ -\dfrac{1}{2} & \dfrac{\sqrt{3}}{2} \end{bmatrix} \begin{bmatrix} i_{c\alpha}^* \\ i_{c\beta}^* \end{bmatrix} \quad (11)$$

C. id-iq method

Instantaneous active and reactive current (i_d-i_q) method [6] is an SRF based compensation scheme which follows same control steps as the *d-q* method. Here, the *α-β* to *d-q* and vice-versa transformation matrices are modified based on the geometry of voltage and current space vectors in *α-β* and *d-q* co-ordinates, shown in Fig.4. This figure shows that the angle 'θ' can be represented by $\theta = tan^{-1}\left(\dfrac{v_\beta}{v_\alpha}\right)$ [7].

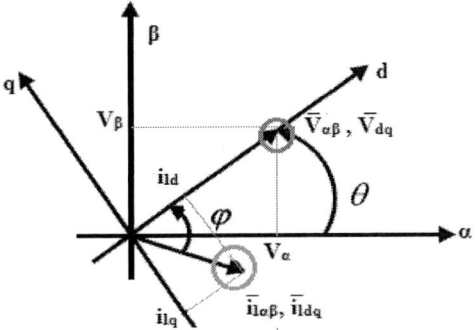

Fig. 4. Instantaneous voltage and current space vectors in *α-β* and *d-q* co-ordinates.

Under balanced-sinusoidal mains voltage conditions, the transformation angle 'θ' is a function of time and it increases uniformly [7]. But, it is sensitive to voltage harmonics and unbalanced voltage sources, therefore, $\frac{d\theta}{dt}$ does not remain constant over a period of supply voltage [6]. If *d-axis* is in the direction of the voltage space vector then the direct and quadrature components of voltage can be written as Eq. 12 and $sin\theta$ and $cos\theta$ can be expressed as Eq. 13.

$$v_d = |\bar{v}_{dq}| = |\bar{v}_{\alpha\beta}| = \sqrt{v_\alpha^2 + v_\beta^2} \ and \ v_q = 0 \quad (12)$$

$$\cos\theta = \frac{v_\alpha}{\sqrt{v_\alpha^2 + v_\beta^2}} \ and \ \sin\theta = \frac{v_\beta}{\sqrt{v_\alpha^2 + v_\beta^2}} \quad (13)$$

Using Eq. 12 and 13, now Eq. 8 and Eq. 10 can be expressed as Eq.14 and 15 respectively

$$\begin{bmatrix} i_{ld} \\ i_{lq} \end{bmatrix} = \frac{1}{\sqrt{v_\alpha^2 + v_\beta^2}} \begin{bmatrix} v_\alpha & v_\beta \\ -v_\beta & v_\alpha \end{bmatrix} \begin{bmatrix} i_\alpha \\ i_\beta \end{bmatrix} \quad (14)$$

$$\begin{bmatrix} i_{c\alpha}^* \\ i_{c\beta}^* \end{bmatrix} = \frac{1}{\sqrt{v_\alpha^2 + v_\beta^2}} \begin{bmatrix} v_\alpha & -v_\beta \\ v_\beta & v_\alpha \end{bmatrix} \begin{bmatrix} i_{cd}^* \\ i_{cq}^* \end{bmatrix} \quad (15)$$

For the complete algorithm, same sequences of steps as in *d-q* method are followed with these transformation equations. The advantage here is that the PLL used for obtaining transformation angle (θ) can be omitted since "θ" is obtained directly from supply voltage, thereby reducing chances of phase lag error.

IV. SWITCHING CONTROL SCHEME

In this paper HBCC is used for controlling the switching patterns in the VSI. In this method the actual current is maintained within a certain preset limit or tolerance band, around the reference current, referred to as the Hysteresis Band (HB). When the actual current crosses or hits the upper limit of the band, the upper switch of a particular phase leg (S_a or S_b or S_c) of the VSI is turned OFF and lower switch (S_a' or S_b' or S_c') is turned ON. So current in that phase decays until it reach the lower limit of the HB. As it strikes lower limit of HB lower switch is turned OFF and Upper Switch is turned ON. The upper device and the lower device in one phase leg of VSI are switched in complementary manner. The switching logic for "phase-a" is formulated as follows:

If $i_a < (i_a^* - H^+)$, Sa=0 and Sa'=1 in phase-a leg (i.e. upper switch is OFF and lower switch is ON)

If $i_a > (i_a^* + H^-)$, Sa=1 and Sa'=0 in phase-a leg (i.e. upper switch is ON and lower switch is OFF)

where, H^+ and H^- are the upper and lower Half-Bandwidth.

V. SIMULATION RESULTS AND PERFORMANCE EVALUATION

The compensation system described in section II is configured in MATLAB/SIMULINK environment with the

978-1-4799-6047-7/14 $31.00 © 2014 IEEE

three control algorithms given in section III. The system is analyzed under two conditions which are more prominently present in practical supply system, ***Case 1: Unbalanced-sinusoidal supply and Case 2: Non-sinusoidal supply.*** The algorithms are analyzed and compared under these operating conditions in order to evaluate the effectiveness for these conditions. The simulation results under these cases are shown in fig. 5-11. In this work, the compensator is switched ON after 1 cycle of supply (20ms) to exhibit the quick change in shape of waveform after start of compensation. The THD content in supply current and the regulation of dc-link voltage are taken as index for evaluating the performance of the system. The THD of supply current is calculated through FFT analysis. FFT analysis is performed at sampling rate of 512samples/cycle. The THD in supply current without compensation are (a) for ***Unbalanced supply***: i_{sa} = 30.28%, i_{sb} = 29.48% and i_{sc} = 29.89% (b) for ***Non-sinusoidal supply***: i_{sa} = 29.60%, i_{sb} = 30.38% and i_{sc} = 30.39%. The system is analyzed under ***balanced supply*** also for all the three control algorithms and THD is calculated. The THD content of supply current post compensation is presented in Table I. The simulation results for all three control techniques are found to be quite satisfactory in terms of eliminating the harmonic components and compensating for reactive components of load current. These control techniques are effective for keeping supply line sinusoidal and in UPF condition and also found to meet the IEEE-519 standard on harmonics level.

Case 1: Unbalanced sinusoidal supply

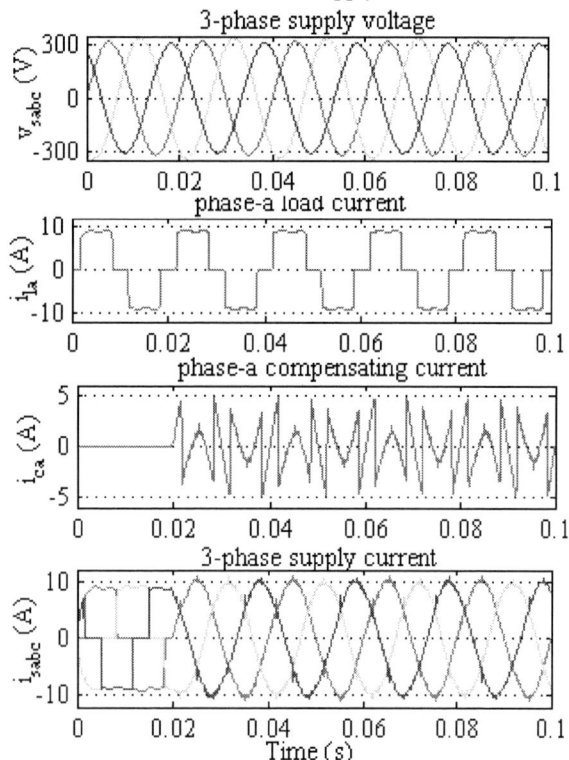

Fig. 5. Performance of shunt APF with *p-q* theory

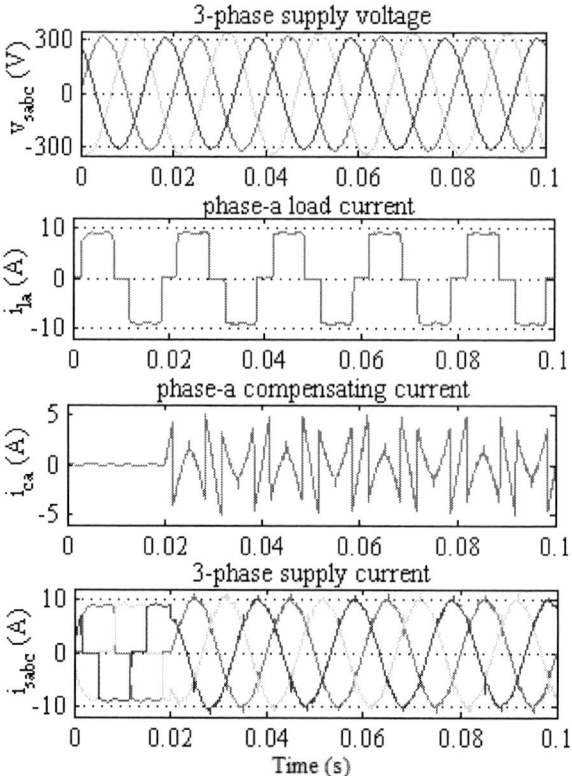

Fig. 6. Performance of shunt APF with *d-q* method

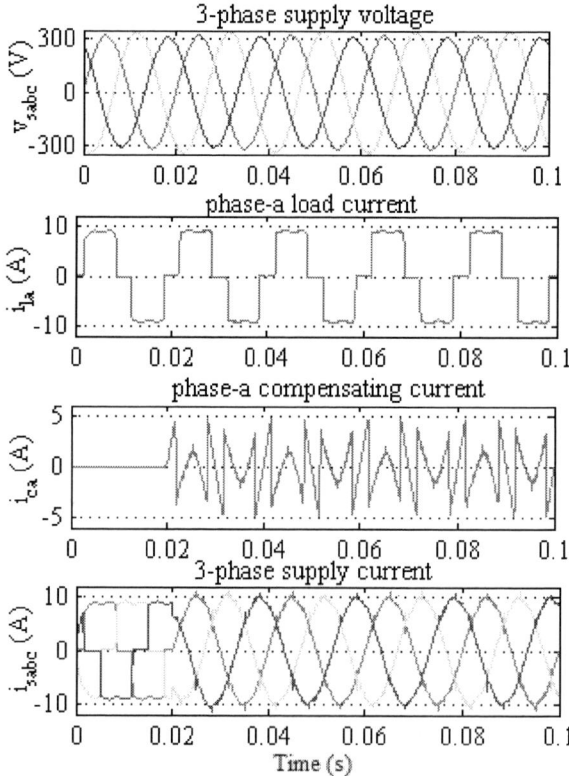

Fig. 7. Performance of shunt APF with i_d-i_q method

Case 2: Non-sinusoidal supply

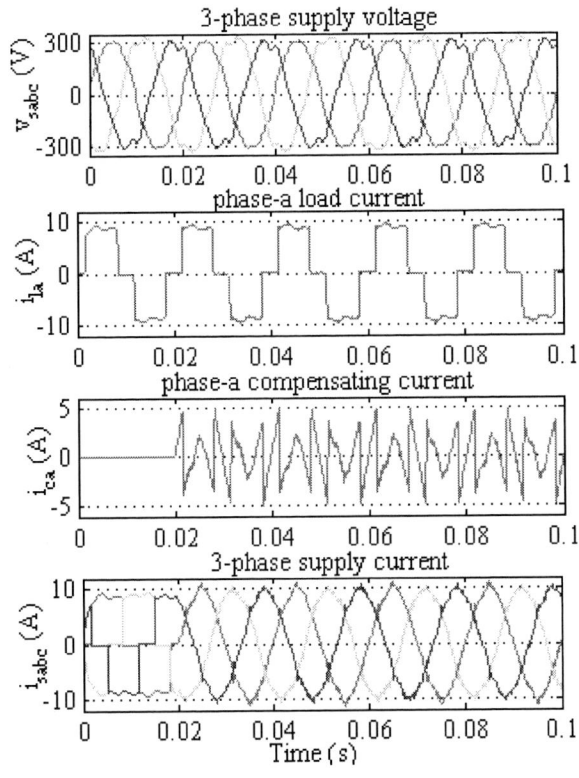

Fig. 8. Performance of shunt APF with *p-q* theory

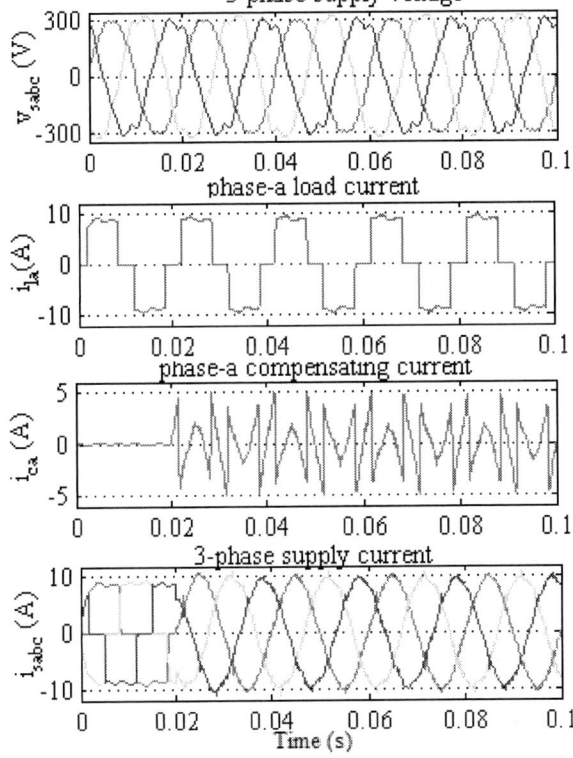

Fig. 9. Performance of shunt APF with *d-q* method

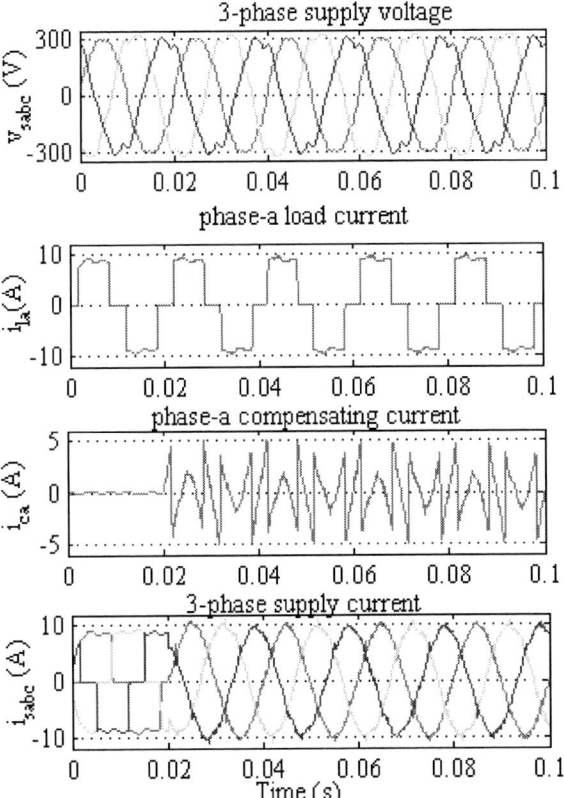

Fig. 10. Performance of shunt APF with i_d-i_q method

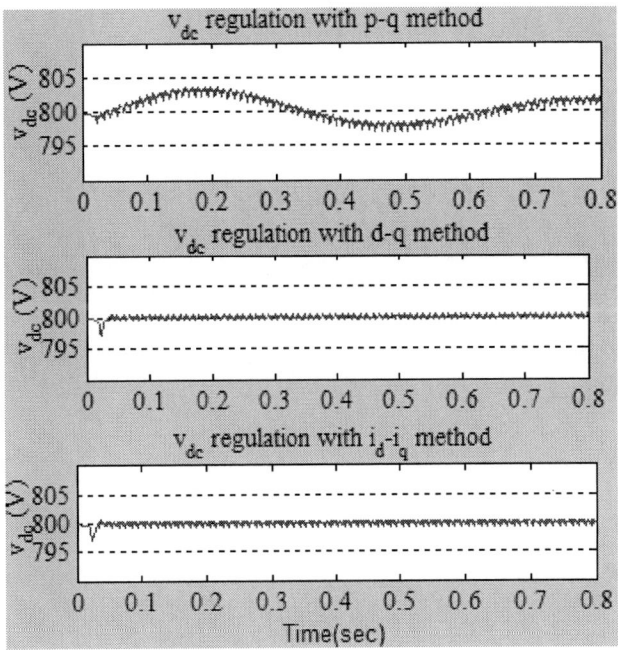

Fig. 11. dc-link voltage regulation

978-1-4799-6047-7/14 $31.00 © 2014 IEEE   149

TABLE I.   SUMMARY OF PERFORMANCE ANALYSIS IN TERMS OF THD CONTENT IN SUPPLY CURRENT

case→	Balanced supply			Unbalanced supply			Non-sinusoidal supply		
Method↓	i_{sa}	i_{sb}	i_{sc}	i_{sa}	i_{sb}	i_{sc}	i_{sa}	i_{sb}	i_{sc}
p-q theory	2.16	2.30	2.02	3.27	3.42	3.45	4.92	4.82	3.95
d-q method	2.61	2.70	2.51	2.89	3.18	3.09	3.49	4.27	4.08
i_d-i_q method	2.62	2.73	2.48	2.87	3.02	3.07	3.25	4.03	3.85

Under balanced supply, THD present in the system after compensation for d-q and i_d-i_q methods are slightly higher than that obtained from p-q theory, but, for cases of unbalanced and non-sinusoidal supply which are more common in practical system (utility supply), these two methods give great improvement over p-q theory. So, from practical implementation point of view, d-q and i_d-i_q methods are more promising. Furthermore, the regulation of dc-link voltage shows great improvement over the p-q theory. Variation of dc-link voltage for different control methods under the three cases of supply condition is detailed in Table II.

TABLE II.   VARIATION OF DC-LINK VOLTAGE

Control method	Balanced Supply	Unbalanced supply	Non-sinusoidal supply
p-q theory	(797.5 -803.4) V	(797.5-804.4)V	(797-803.5)V
d-q and i_d-i_q methods	(799.2-800.2)V	(799.6-800.4)V	(799.5-800.4)V

VI. CONCLUSION

In this paper the dynamic performance of three most extensively used control algorithms for shunt active power filter (APF), viz. p-q theory, d-q method and i_d-i_q methods are evaluated. It is observed from this analysis that p-q method shows excellent performance at ideal supply conditions. However, under unbalanced and non-sinusoidal supply conditions its performance drops down more steeply while that of d-q and i_d-i_q methods are relatively less affected. Since practical supplies are usually non-ideal, d-q and i_d-i_q methods are better suited for practical implementation. The d-q and i_d-i_q methods give better performance in terms of dc-link voltage regulation also. All the three control methods are found to be satisfactory in maintain supply line THD below the limit of 5% as specified by IEEE-519 standard. However, i_d-i_q method shows superiority in overall performance among the three control methods.

APPENDIX

Supply Voltage:- (a) *Balanced (rms-line)*: 400V; (b) *Unbalanced (peak)*: V_{sa}=320V, V_{sb}=340V, V_{sc}=310V; (c) *Non-sinusoidal (peak, %THD)*: V_{sa}=320V,7.21%THD, V_{sb}=340V,8.51%THD; V_{sc}=310V,8.75%THD, f=50Hz, R_s=0.1Ω, L_s=0.1mH; **APF**:- C=1100μF, V_{dc}=800V, R_{on}=1mΩ, R_{snub}=100kΩ; R_c=0.1Ω, L_c=3.75mH; **Load**:- R_L=60Ω, L_L=60mH.

ACKNOWLEDGMENT

The authors would like to acknowledge the support provided by the Department of Science & Technology, Government of India, Under Science and Engineering Research Board Fast Track Scheme for Young Scientists (SERC/ET-0123/2012).

REFERENCES

[1] B. Singh, K. AL. Haddad and A. Chandra, "A review of active filters for power quality improvement," *IEEE Trans. Ind Elect.*, vol.46, pp. 960-971, 1999.

[2] Y. Pal, A. Swarup and B. Singh, "A review of compensating type custom power devices for power quality improvement," in proc. *POWER CON*, 2008, pp. 1-8

[3] N. Maruin, A. Alam and S. Mahmod, and H. Hizam, "Review of control strategies for power quality conditioners," in proc. *of IEEE PECon 2004*, pp.109-115.

[4] H. Akagi, Y. Kanazawa and A. Nabae, "Instantaneous reactive power compensators comprising of switching devices without energy storage components," *IEEE Trans. Ind Appl.* Vol. 20. no. 3, pp.625-630, May/June 1984.

[5] S. Bhattacharya. D. M. Divan and B. Banerjee. "Synchronous Frame harmonics Isolator using active series filter," proc. of *EPE 1991* ,vol. 3, pp. 30-35, Firenzi, Italy

[6] V. Soares. P. Verdelho, C. D. Marques, "Active power filter control circuit based on the instantaneous active and reactive current i_d-i_q method", in proc. *PESC '91*. St Louis. Missouri, 1997, pp. 1096-1101.

[7] N. Zaveri, A. Chudasama, "Control strategies for harmonic mirigation and power factor correction using shunt active filter under various source voltage conditions," in *Electrical Power and Energy System*, vol. 42, pp. 661-671, 2012.

Comparative Analysis of Conventional and IRP theory based Shunt APF for 3P3W system

Anil Gambhir*, Sharmili Das, Aurobinda Panda
Department of Electrical Engineering,
IIT Roorkee, Uttrakhand, India-247667
Email id*. anilgiitr@gmail.com

Abstract— **Harmonics and Reactive current in transmission line causes additional transmission losses in the line. To mitigate this problem, Shunt Active Power Filter (SAPF) is presented for the compensation of harmonics and reactive power with balanced and unbalanced non linear load. In this paper two control topology such as conventional control scheme and instantaneous reactive power theory are presented. Finally, a comparative analysis is made based on the simulation results.**

Keywords— *Shunt active power filter, instantaneous reactive power theory, harmonics, reactive power, voltage source inverter.*

I. INTRODUCTION

Ideally, all the power utilities should provide a good quality of supply to their customer. Good quality means constant magnitude and frequency of sinusoidal voltage. Due to nonlinear load it is difficult to maintain quality of supply of constant magnitude and frequency The common nonlinear loads draw current from the sources which do not follow the voltage shape and hence introduce harmonic in the power line. Three phase network with a non-linear load causes unbalance and excessive neutral currents in the system. The injected harmonics, reactive power burden, unbalance, and excessive neutral currents cause low system efficiency and poor power factor, thereby affecting the power quality. Different methods has already been discussed in the literature to mitigate these harmonics. Instantaneous reactive power theory (p-q theory) is one of the best solution to suppress the current harmonics[1].

II. CONTROL TECHNIQUE

A. Conventional Theory

In this method a sinusoidal reference waveform is generated using Phase Locked Loop (PLL) and the fundamental component is calculated by multiplying the load current signal by a sine wave of the fundamental frequency and the resulting signal is integrated. This give the active component of the signal which is assumed to be come from source side. Then it is compare with actual signal and switching pulses are generated. Analysis of shunt active filter for the generation of reference current is given in Fig.2. Let source voltage is sinusoidal that is free from harmonics [2]. This is given by:

$$V_S(t) = V_p \sin wt$$

$$i_L(t) = \sum_{n=1}^{\infty} I_n \sin(nwt + \theta_n)$$

$$i_L(t) = I_1 \sin(wt + \theta_1) + \sum_{n=2}^{\infty} I_n \sin(nwt + \theta_n) \quad (1)$$

For both harmonics and reactive power compensation reference current is given as: $i_r(t) = 1.\sin(wt)$ (Unit amplitude current template)

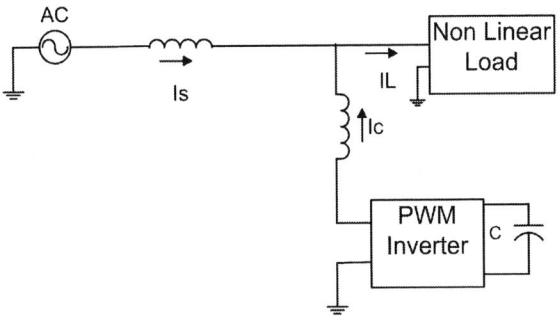

Fig. 1. Basic Block diagram of SAPF

Let the active component in load current is I_x

$$
\begin{aligned}
I_x &= \frac{2}{T} \int_0^T i_L(t).i_r(t)dt \\
&= \frac{2}{T} \int_0^T \begin{matrix} [I_1(\sin wt \cos\theta_1 + \cos wt \sin\theta_1).\sin wt \\ + \sum_2^{\infty} I_n(\sin nwt + \theta_n).\sin wt]dt \end{matrix} \quad (2) \\
&= \frac{2}{T} \int_0^T I_1 \sin^2 wt.\cos\theta_1 dt \\
&= I_1 \cos\theta_1
\end{aligned}
$$

This Ix is the active component and is suppose to come from the source side. The Instantaneous value of desired component can be given as:

$$i_{sd}(t) = I_x * i_r(t) = I_1 \cos(\theta_1)\sin(\omega t) \quad (3)$$

Compensator reference current can be written as (Fig.2),

$$i_{cr}(t) = i_1(t) - i_{sd}(t) \quad (4)$$

This compensation reference current is only when there are no losses in filter. But practically there will be some losses in filter due to this capacitor will discharge. So active filter work properly if it should be charge with a constant voltage. For that a compensating component is added in active current as shown in Fig.3.

978-1-4799-6047-7/14 $31.00 © 2014 IEEE

$$S = (e_0 i_0) + (e_\alpha i_\alpha + e_\beta i_\beta) + j(e_\beta i_\alpha - e_\alpha i_\beta) \quad (7)$$

1) Instantaneous Zero-sequence power(Po)

$$P_0 = E_0 I_0 = \overline{p}_0 + \tilde{p}_0 \quad (8)$$

$\overline{p}_0$: Mean value of the instantaneous Zero-Sequence Power.

$\tilde{p}_0$: Alternating value of instantaneous Zero-Sequence Power. (it only exchanges energy with the load, and does not transfer any energy to the load)

The zero-sequence power exists only in three-phase systems with neutral wire

2) Instantaneous Real power(p)

$$p = e_\alpha . i_\alpha + e_\beta . i_\beta = \overline{p} + \tilde{p} \quad (9)$$

$\overline{p}$: Mean value of the instantaneous real power. It corresponds to the Power that transferred from the power source to the load, in a balanced way, through the a-b-c coordinates (This is only active power component that is to be supplied by the power source).

$\tilde{p}$: Alternating value of the instantaneous real power. Since $\tilde{p}$ does not involve any energy transfer from the power source to load, it must be compensated.

3) Instantaneous Imaginary Power(q)

$$q = e_\alpha . i_\beta - e_\beta . i_\alpha = \overline{q} + \tilde{q} \quad (10)$$

$\overline{q}$: Mean value of instantaneous imaginary power.

$\tilde{q}$: Alternating value of instantaneous imaginary power. It must be compensated.

The instantaneous imaginary power q does not involve power transfer from source to load it exchanged between the system phases. It must be compensated for Harmonic and Reactive power.

The oscillating powers are responsible for the harmonics in the currents which are demanded by the loads. To separate out these components of powers from the p, q and p_0. These p, q, p_0 are passed through the low pass filter by which we can get average powers $\overline{p}_0, \overline{q}, \overline{p}$ and then it is subtracted through the p_0, q, p will gives $\tilde{p}_0, \tilde{q}, \tilde{p}$ or we can use high pass filter to separate out this Power [3].

Desired power component for harmonics ($\overline{p}_0, \tilde{p}, \tilde{q}$) and for reactive power ($\overline{q}$) is to be selected. From these powers the reference values of the compensating currents in the αβ0 domain is find out by the expression given below-

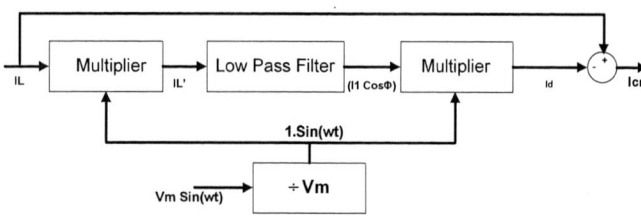

Fig. 2. Block diagram representation for calculating compensating refrence current

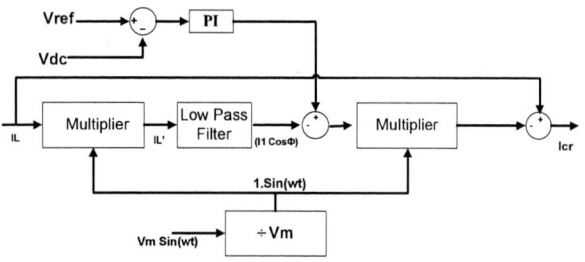

Fig. 3. Block diagram representation for calculating reference current with adding loss component

B. Instantaneous Reactive Power Theory(IRP theory)

The p-q theory based on a set of instantaneous powers defined in the time domain without any restriction on the current and voltage waveform. This theory can be applied to three-phase systems with or without neutral and this theory valid not only for steady state but also valid for transient state. Calculation of this theory is easy as it required only algebraic operation. The p-q theory first transforms voltages and currents from ABC to αβo coordinates and then defines the instantaneous power on these coordinates. It corresponds to an algebraic transformation, known as *Clarke* transformation, which also produces a stationary reference system, where coordinates α,β are orthogonal to each other, and coordinate 0 corresponds to the zero-sequence component. The zero-sequence component calculated here differs from the one obtained by the symmetrical components transformation. The voltages and currents in αβ0 coordinates are calculated as follows(excluding Zero-phase sequence components)

$$\begin{bmatrix} e_\alpha \\ e_\beta \end{bmatrix} = \sqrt{\frac{2}{3}} \begin{bmatrix} 1 & \frac{-1}{2} & \frac{-1}{2} \\ 0 & \sqrt{3}/2 & -\sqrt{3}/2 \end{bmatrix} \begin{bmatrix} e_a \\ e_b \\ e_c \end{bmatrix} \quad (5)$$

$$\begin{bmatrix} i_\alpha \\ i_\beta \end{bmatrix} = \sqrt{\frac{2}{3}} \begin{bmatrix} 1 & \frac{-1}{2} & \frac{-1}{2} \\ 0 & \sqrt{3}/2 & -\sqrt{3}/2 \end{bmatrix} \begin{bmatrix} i_a \\ i_b \\ i_c \end{bmatrix} \quad (6)$$

$$\begin{bmatrix} i_{c\alpha} \\ i_{c\beta} \end{bmatrix} = \begin{bmatrix} e_{\alpha} & e_{\beta} \\ -e_{\beta} & e_{\alpha} \end{bmatrix}^{-1} \begin{bmatrix} -\bar{p} + \tilde{p}_{loss} \\ -q \end{bmatrix} \quad (11)$$

Compensating current in $\alpha\beta0$ is to be calculated from equation (11) & by using the inverse Clark's transformation, we can find out the reference value of compensating current in ABC domain [4].

Inverse transform i.e. from $\alpha\beta o$ to ABC coordinates for currents are-

$$\begin{bmatrix} i_a^* \\ i_b^* \\ i_c^* \end{bmatrix} = \sqrt{\frac{2}{3}} \begin{bmatrix} 1 & 0 \\ -1/2 & \sqrt{3}/2 \\ -1/2 & -\sqrt{3}/2 \end{bmatrix} \begin{bmatrix} i_{c\alpha} \\ i_{c\beta} \end{bmatrix} \quad (12)$$

Now these values of reference compensating currents and actual compensator currents of filter are given to the pulse generation circuit which generate the pulses for the compensator. The main advantage of using the instantaneous reactive power theory is to control the power component instantaneously and dynamically. The basic block diagram for control of 3P3W SAPF is shown in Fig.4 which is compensating the harmonics present in the supply, as well as the reactive power in Transmission Lines.

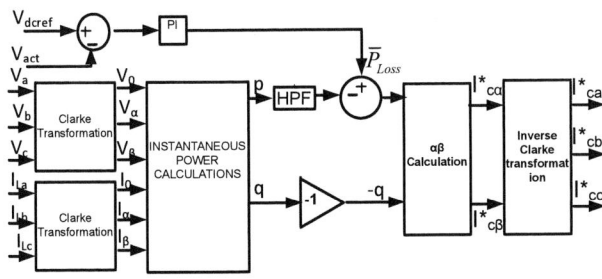

Fig. 4.   Schematic Repersantation SAPF with instantaneous reactive power theory control.

III. RESULTS AND DISCUSSION

With the help of Matlab Simulink, simulation for the three-phase three wire system is carried out. For the non-linear load a single-phase fully controlled rectifier with R-L element on its dc side is used with firing angle of 20 deg. For the simulation result, parameters used are shown in Table-I

Fig.5(a) Shows harmonic spectrum of Phase A without compensation, Fig.5(b) Shows Harmonic spectrum with compensation using IRP theory based scheme. In this case before compensation, load is drawing current from the source having 27.39% THD, but after compensation THD become only 1.68% .

TABLE I.   PARAMETER USED IN SHUNT ACTIVE FILTER

Supply voltage (phase)	230Volts (Line to Line)
Supply frequency	50Hz
Source impedance	$R_s = 0.1\Omega, L_s = 0.01mH$
Coupling inductor	$R_c = 0.01\Omega, L_c = 4mH$
Smoothing inductor	$R_{smoothing} = 0.1\Omega,$ $L_{smoothing} = 0.8mH$
Sampling time	50 µs
Reference DC link voltage	700 Volts
DC link Capacitor values	$C_1 = 2200\mu F$
Hysteresis band limits	Lower = -0.01, Upper = 0.01

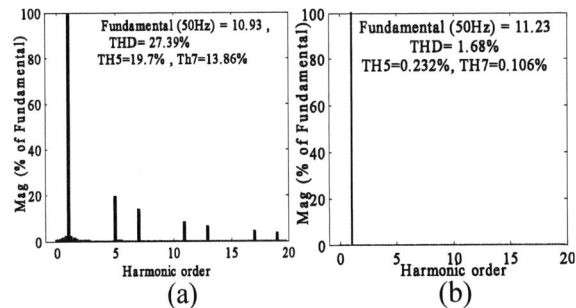

Fig. 5. Harmonic spectrum of PHASE A (a) Before compensation (b) After compensation with IRP theory

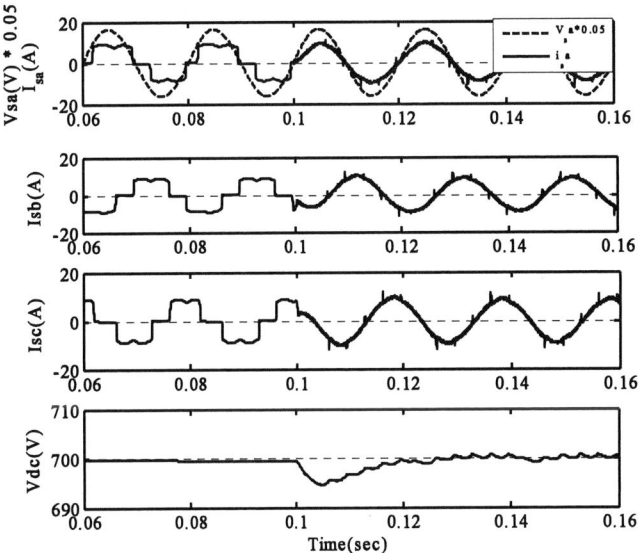

Fig. 6.   During switching on of Shunt APF with Conventional control Trace1 :Phase voltage and source current of phase A,Trace2: Source current of phase B,Trace3: Source current of phase C and Trace4: Capacitor Voltage.

The source currents and Capacitor voltage under disable and enable of APF with conventional control and IRP theory based control technique respectively are shows in Fig.6 & Fig.7. It can be observed from this simulated result that at that instant APF is switched on, capacitor changes its reference value to accommodate the load current harmonic. This drop in voltage

978-1-4799-6047-7/14 $31.00 © 2014 IEEE   153

is restored in 2-3 cycle in conventional control where as it is only half cycle in case of Instantaneous Reactive Power theory based control. This shows the dynamic behaviour of shunt active filter with Instantaneous Reactive Power Theory based is fast.

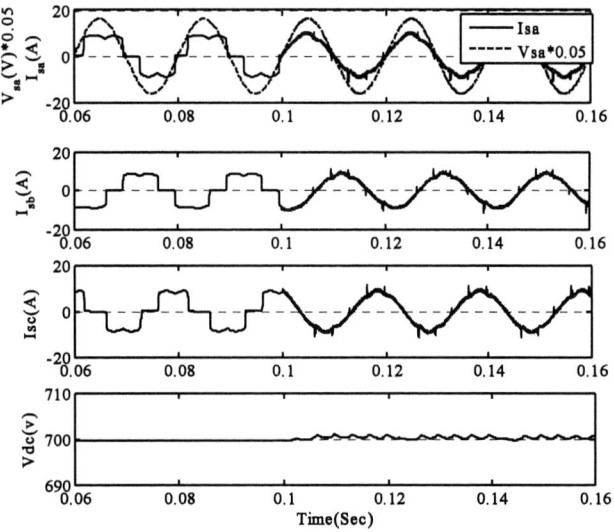

Fig. 7. During switching on of Shunt APF with IRP theory based control Trace1:Phase voltage and source current of phase A,Trace2: Source current of phase B,Trace3: Source current of phase C and Trace4: Capacitor Voltage.

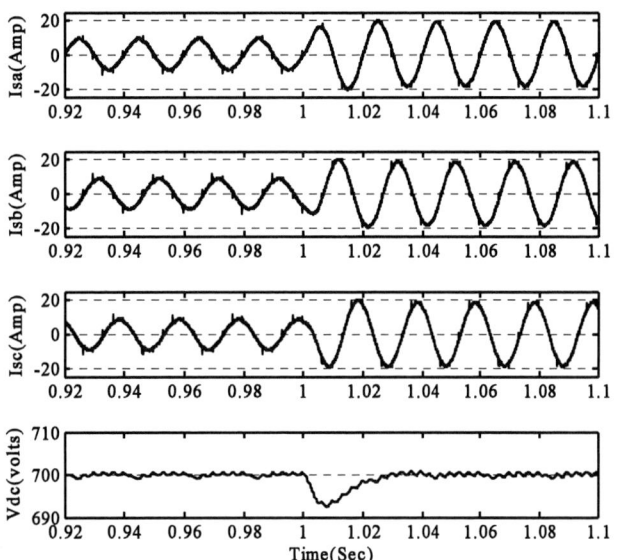

Fig. 8. During suuden load change with Conventional theory based control Trace1:Phase voltage and source current of phase A,Trace2: Source current of phase B,Trace3: Source current of phase C and Trace4: Capacitor Voltage.

Fig.8 & Fig.9 shows the simulation results of SAPF under varying load condition with conventional and IRP theory based control scheme respectively. It is observed that with IRP theory based technique the system is found to reach the steady state much early compare to conventional technique.

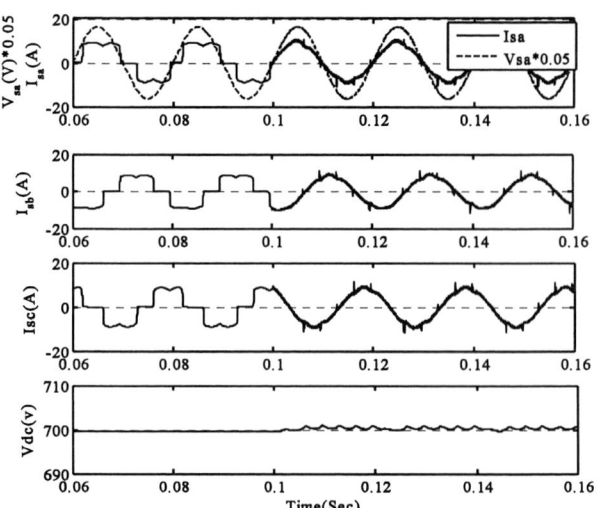

Fig. 9. Trace1:Phase voltage and source current of phase A,Trace2: Source current of phase B,Trace3: Source current of phase C and Trace4: Capacitor Voltage. During suuden load change with IRP theory based control

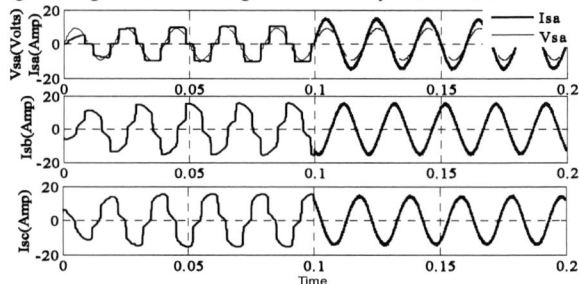

Fig. 10. During unbalace loading with IRP theory based control. Trace1:source current of phase A,Trace2: Source current of phase B,Trace3: Source current of phase C.

This Fig.10 shows the waveform of a,b,c phase current during unbalance load condition with IRP control technique. It can be observed that before switching on the filter at 0.1 sec, all the three phase are unbalance however after switching on all the phase share the load equally.

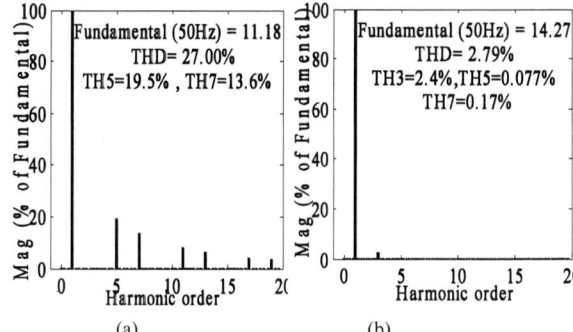

Fig. 11. Harmonic spectrum of PHASE A(a) Before (b) After compensation

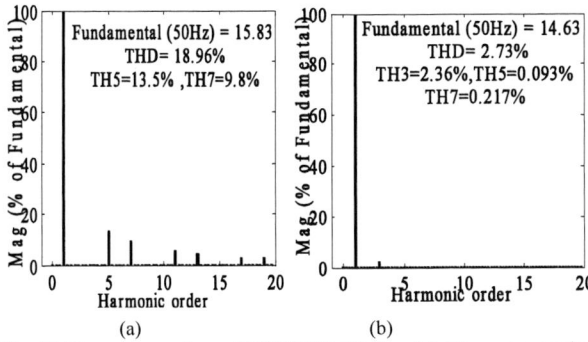

Fig. 12. Harmonic spectrum of PHASE B(a) Before (b) After compensation

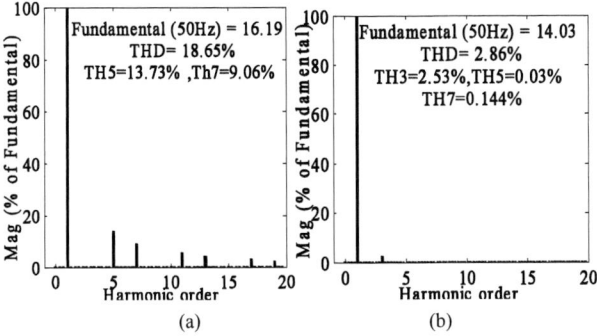

Fig. 13. Harmonic spectrum of PHASE C(a) Before (b) After compensation

Harmonic spectrum of a,b & c phase current is shown in Fig.11, Fig.12 and Fig.13 respectively during unbalance load condition. After compensation the fundamental current of each phase is approximately equal. Which show that SAPF reduces Harmonics as well as Balances the load current as seen from source side[5]

TABLE II.   HARMONICS ANALYSIS ON BALANCE AND UNBALANCED LOAD.

PHASE	Balance Load		Unbalanced Load	
	Before compensation	After compensation	Before compensation	After compensation
A	27.06%	1.66%	27%	2.79%
B	27.39%	1.68%	18.96%	2.73%
C	26.88%	1.65%	18.65%	2.86%

IV. CONCLUSION

A three-phase three wire shunt active power filter with instantaneous control is discussed for the elimination of harmonics and compensation of reactive power. we can also see that in unbalance system, three phases can be balance if we use Active filter with this control technique. Its transient behavior shows that with PQ technique, dynamic response is much better than conventional control. With the above discussion it is clear that dynamic response of SAPF using IRP based technique is better with improved harmonic spectrum. As load on grid changes continuously in a whole day for that instantaneous reactive power theory is best suitable solution for Shunt Active Power Filter.

References

[1] S. K. Jain, P. Agrawal, and H. O. Gupta, "Fuzzy logic controlled shunt active power filter for power quality improvement," *IEE Proceedings-Electric Power Applications,* vol. 149, pp. 317-328, 2002.

[2] P. Karuppanan and K. Mahapatra, "PLL with PI, PID and Fuzzy Logic Controllers based Shunt Active Power Line Conditioners," in *Drives and Energy Systems (PEDES) & 2010 Power India, 2010 Joint International Conference on Power Electronics*, pp. 1-6.

[3] B. Singh and J. Solanki, "A comparison of control algorithms for DSTATCOM," *IEEE transactions on Industrial electronics,,* vol. 56, pp. 2738-2745, 2009.

[4] R. S. Herrera, P. SalmerÃ³n, and H. Kim, "Instantaneous reactive power theory applied to active power filter compensation: Different approaches, assessment, and experimental results," *IEEE transactions on Industrial electronics,,* vol. 55, pp. 184-196, 2008.

[5] M. I. M. s. Montero, E. R. Cadaval, and F. B. GonzÃ¡lez, "Comparison of control strategies for shunt active power filters in three-phase four-wire systems," *IEEE Transactions on Power Electronics, ,* vol. 22, pp. 229-236, 2007.

[6] B. Singh, K. Al-Haddad, and A. Chandra, "A review of active filters for power quality improvement," *IEEE transactions on Industrial electronics,,* vol. 46, pp. 960-971, 1999.

A Modified Switching Scheme for a New Multi level Inverter Topology for Fuel-Cell Microgrid

G K Naveen Kumar, Yash Pal
Department of Electrical Engineering
National Institute of Technology, Kurukshetra
Kurukshetra, India
gknavinkmr712@gmail.com, yash_pal1971@yahoo.com

Abstract —Recently the use of multilevel inverters (MLIs) in modern drives and for interfacing the renewable generation systems to grid have given wide scope for designing new topologies of MLIs. Fuel cell generation systems are expected to see practical usage due to several advantages over conventional generation systems. This paper introduces a new switching scheme for a new topology of MLI with reduced number of switches for interfacing Fuel-cell with the grid. The unipolar pulse width modulation (PWM) technique is used for the switching of H-bridge switches, while for the outer switches sinusoidal modulating wave is compared with DC component. With this modified switching scheme the Total Harmonic Distortion (THD) in the output voltage of MLI decreased. It is also observed from the simulation results that the THD of output voltage of 21-level MLI is within limits as specified by IEEE 519 standard. A comparative analysis of proposed switching scheme with the reported switching scheme is carried out for the seven and thirteen levels of new topology in terms of the THD in the output voltages. Moreover, a comparison of new topology with conventional reported MLI topologies is also made in terms of the number of switches and required DC sources for different levels. It is also observed from simulation results that during the integration of the Fuel-cell with the grid, the grid requirements such as phase angle, frequency and amplitude of grid voltage are also satisfied.

Keywords – Fuel cell, Multi level inverter, Micro grid, Total Harmonic Distortion.

I. INTRODUCTION

In view of public concern about global warming and climatic conditions many efforts are afforded on environmentally friendly distributed energy resources like photovoltaic, wind power, micro turbines, and fuel cells into micro grid system. For employing transformer less grid connection inverter is employed between energy resource and the micro grid. Ideally, the desired output of an inverter is a sinusoidal waveform which is a continuous function of time. Multi level inverters are an alternative choice to improve output by synthesizing a staircase waveform imitating a sinusoidal waveform to obtain low distortion with less dv/dt stress [1]. In order to obtain sinusoidal output voltage conventional MLIs use Pulse-width modulation (PWM) technique to eliminate lower order harmonics. But with high switching frequency in PWM not only leads to switching losses but also limits its use due to unavailability of high power switching devices [2].

Conventional MLI topologies such as: neutral point clamped (NPC) converters [3-4]. Flying Capacitor converters [5] and cascade H-bridge (CHB) converters [6],

when used for increased number of levels leads to higher device count, cost and complexity in its implementation. To overcome these limitations various new topologies [7-13] have been proposed in literature with more output levels, but at the same time having less numbers of active and passive devices. A novel topology of high level MLI, and cascaded proposed in [7], [9] with less number of passive devices and insulated gate driver circuit. For implementation this proposed topology requires four high rating switches for output side H-bridge and bi-directional switches. One more topology of MLI is proposed in literature [8] with minimum number of switches, DC voltage sources and standing voltage on the switches along with two new algorithms are proposed to find the amplitude of DC voltage sources, but it requires bi-directional switches.

Another class of multilevel inverters based on a multilevel DC link (MLDCL) and a bridge inverter has been proposed in [10]. The realization of MLDCL can be possible using cascaded half-bridge cells or a flying-capacitor phase leg or diode-clamped phase leg having their own DC sources. Despite a higher total VA rating of the switches, the output side H-bridge needs four high rating switches. A twenty seven levels MLI topology is proposed in [11], to maximize the number of output voltage levels with 4.4% THD of its output voltage. But the drawback of this topology is that it requires four bidirectional switches and three voltage sources in the ratio of 1: 3: 9, which increase the voltage stress. Nineteen levels MLI topology proposed in literature [12] based on the series connection of sub multilevel inverters comprising the basic unit and full bridge converter. Recently, seven levels and thirteen levels of a new topology of MLI is proposed in literature [13] having less number of power switches and DC sources, as compared to topology reported in [12]. In this new topology [13] of MLI, the switching is carried out using the simple pulse generator.

In this paper, the new topology [13] of MLI is extended for 21 levels with a modified switching technique to obtain more sinusoidal output voltage of the inverter to inter face the Alkaline Fuel Cell (AFC) with the Grid by satisfying all conditions such as phase angle, frequency and amplitude of Grid voltage. Moreover, the effectiveness of modified switching scheme is also evaluated in terms of THD of the output voltage for seven-levels and thirteen-levels. A comparative analysis of new topology of MLI with conventional MLIs topologies is also carried out in terms of the hardware requirement for their realization.

978-1-4799-6047-7/14 $31.00 © 2014 IEEE

II. THE NEW TOPOLOGY [13]

The general structure of new multilevel inverter topology is shown in Fig.1. It consists of one H-bridge inverter on right part of its structure and 'N' number of cascaded cells with integral multiple of V_{dc}. The number of levels that can be obtained by this topology depends on number of cascaded cells in the circuit and is given by:

$$\text{Number of Levels} = [N (N+1) +1] \quad (1)$$

where N=Number of cells cascaded excluding outer H-bridge. In order to obtain integral multiple of $+V_{dc}$ level in the output voltage switches S_1 and S_2 should be turned 'ON' and for $-V_{dc}$ level switches S_3 and S_4 should be turned 'ON'. For zero level either S_1 or S_2 and either S_3 or S_4 should be turned 'ON'.

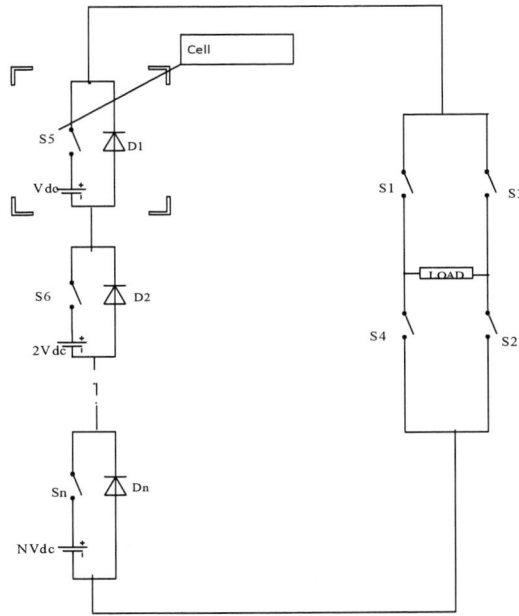

Fig.1 General Structure of a new MLI Topology

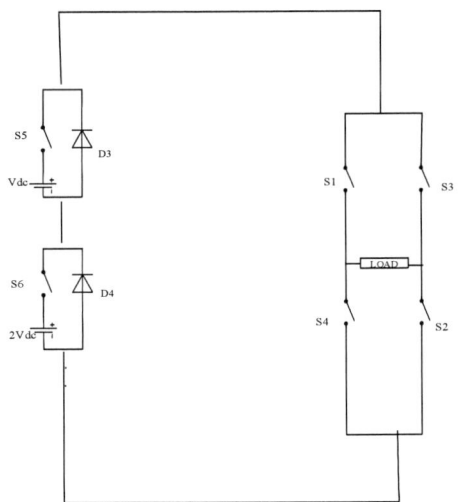

Fig.2 Circuit diagram for a seven levels MLI

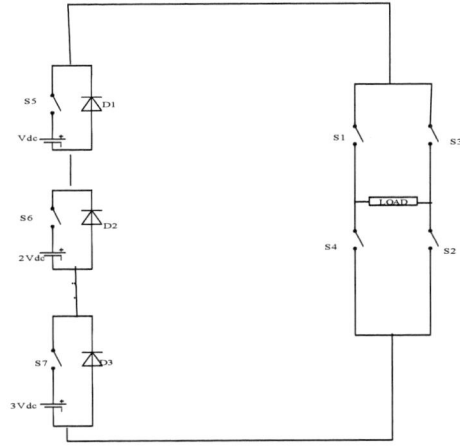

Fig.3 Circuit diagram for thirteen levels MLI

Based on the concept of the new topology [13] the circuit diagrams of the seven and thirteen levels MLI topologies are shown in Fig. 2 and 3 respectively. It is observed from Fig. 2 that the seven levels MLI topology requires six switches to obtain seven level output voltage, whereas the conventional H-bridge MLI requires 12 switches and a separate DC sources for each quad for the same output levels, which leads to more switching losses. Similarly, it is observed from the Fig. 3 that the thirteen levels MLI topology requires only seven switches, while the conventional H-bridge MLI topology requires twenty four switches and six separate DC sources to obtain an output of thirteen level. The ideal output voltages of the seven and thirteen levels are shown in Fig.4 and 5, respectively. Fig. 4 shows that the output voltage contains seven levels, such as: $+/-3V_{dc}$, $+/-2V_{dc}$, $+/-V_{dc}$ and 0, while Fig. 5 shows the output of the thirteen level inverter contains thirteen level such as: $+/-6V_{dc}$, $+/-5V_{dc}$, $+/-4V_{dc}$, $+/-3V_{dc}$, $+/-2V_{dc}$, $+/-V_{dc}$ and zero.

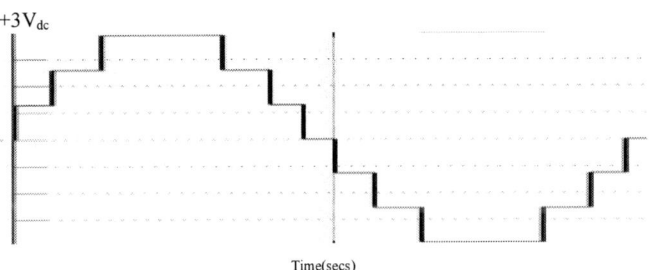

Fig.4 Ideal seven level output voltage waveform

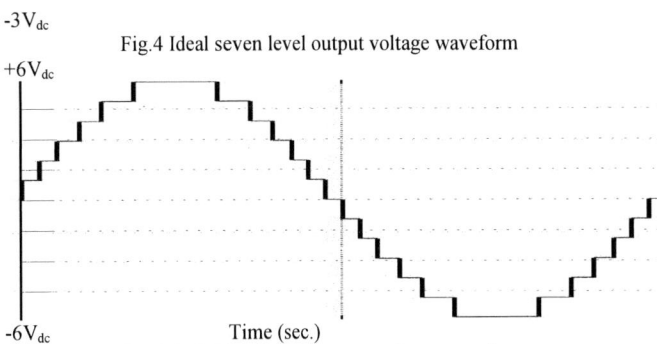

Fig.5 Ideal thirteen level output voltage waveform

The output voltage obtained in thirteen levels MLI has low harmonics content, hence; reduced distortion as compared to seven levels MLI due to increase in number of voltage levels in the output. The number of DC sources (V_{dc}, $2V_{dc}$, $3V_{dc}$) and switches used for thirteen level is increased by one compared to seven level. By using proper switching sequence the seven and thirteen levels output voltage can be achieved. Tables I and II, show the switching sequence for obtaining seven levels and thirteen levels respectively.

TABLE I: SWITCHING SEQUENCE FOR SEVEN LEVELS

S_1	S_2	S_3	S_4	S_5	S_6	Load Voltage
On	on	off	off	on	on	$+3V_{dc}$
On	on	off	off	off	on	$+2V_{dc}$
On	on	off	off	on	off	$+V_{dc}$
Off	on	off	on	off	off	0
Off	off	on	on	on	off	$-V_{dc}$
Off	off	on	on	off	on	$-2V_{dc}$
Off	off	on	on	on	on	$-3V_{dc}$

TABLE II: SWITCHING SEQUENCE FOR THIRTEEN LEVEL INVERTER

S_1	S_2	S_3	S_4	S_5	S_6	S_7	Load Voltage
on	on	off	off	on	on	on	$+6V_{dc}$
on	on	off	off	off	on	on	$+5V_{dc}$
on	on	off	off	on	off	on	$+4V_{dc}$
on	on	off	off	off	off	on	$+3V_{dc}$
on	on	off	off	off	on	off	$+2V_{dc}$
on	on	off	off	on	off	off	$+V_{dc}$
off	on	off	on	off	off	off	0
off	off	on	on	on	off	off	$-V_{dc}$
off	off	on	on	off	on	off	$-2V_{dc}$
off	off	on	on	off	off	on	$-3V_{dc}$
off	off	on	on	on	off	on	$-4V_{dc}$
off	off	on	on	off	on	on	$-5V_{dc}$
off	off	on	on	on	on	on	$-6V_{dc}$

III. PROPOSED SWITCHING SCHEME

In the reported MLI topology [13], a simple pulse generator is used for the switching of power devices. This topology of MLI inverter consists of a cascaded H-bridge comprising four switches at the output and a series of switches at the input of the inverter, as shown in Fig.1. In this paper, a new switching pattern is proposed this MLI topology. For the switches forming cascaded H-bridge are switched by Unipolar PWM technique and for the outer switches from S_5 to S_n are switched by comparing a sinusoidal modulating wave with DC component. For the Unipolar PWM technique the carrier frequency is equal to the line frequency as shown in Fig.6.

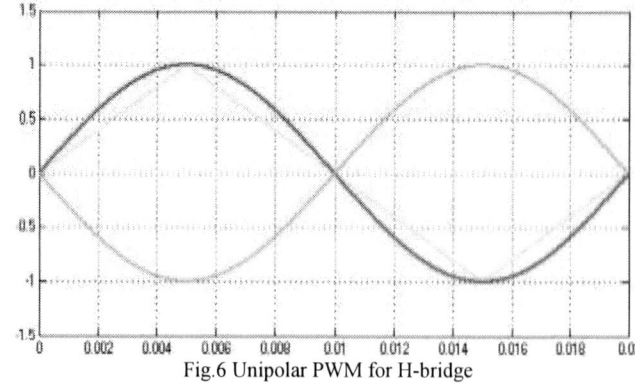

Fig.6 Unipolar PWM for H-bridge

IV. THE PROPOSED TOPOLOGY INTERFACED TO THE GRID

A twenty one levels MLI topology is proposed for the integration of the Fuel cell with the grid. This topology requires only eight switches and four separate DC sources, the conventional H-bridge MLI requires forty switches and ten separate DC sources for twenty one levels. Table III shows the switching sequence for the twenty one levels output voltage. The circuit diagram for twenty one levels inverter is shown in Fig. 7. In twenty one levels MLI, the number of switches and DC sources are increased by one as compared to thirteen levels MLI topology. The ideal output voltage waveform for this twenty one levels inverter is shown in Fig.8.

TABLE III: SWITCHING SEQUENCE FOR NEW TWENTY ONE LEVEL INVERTER

S_1	S_2	S_3	S_4	S_5	S_6	S_7	S_8	Load Voltage
On	on	off	off	on	on	on	on	$+10V_{dc}$
On	on	off	off	off	on	on	on	$+9V_{dc}$
On	on	off	off	on	off	on	on	$+8V_{dc}$
On	on	off	off	off	off	on	on	$+7V_{dc}$
On	on	off	off	on	on	off	on	$+6V_{dc}$
On	on	off	off	on	off	off	on	$+5V_{dc}$
On	on	off	off	off	off	off	on	$+4V_{dc}$
On	on	off	off	off	off	on	off	$+3V_{dc}$
On	on	off	off	off	on	off	off	$+2V_{dc}$
On	on	off	off	on	off	off	off	$+V_{dc}$
Off	on	off	on	off	off	off	off	0
Off	off	on	on	on	off	off	off	$-V_{dc}$
Off	off	on	on	off	on	off	off	$-2V_{dc}$
Off	off	on	on	off	off	on	off	$-3V_{dc}$
Off	off	on	on	off	off	off	on	$-4V_{dc}$
Off	off	on	on	on	off	off	on	$-5V_{dc}$
Off	off	on	on	off	on	off	on	$-6V_{dc}$
Off	off	on	on	off	off	on	on	$-7V_{dc}$
Off	off	on	on	on	off	on	on	$-8V_{dc}$
Off	off	on	on	off	on	on	on	$-9V_{dc}$
Off	off	on	on	on	on	on	on	$-10V_{dc}$

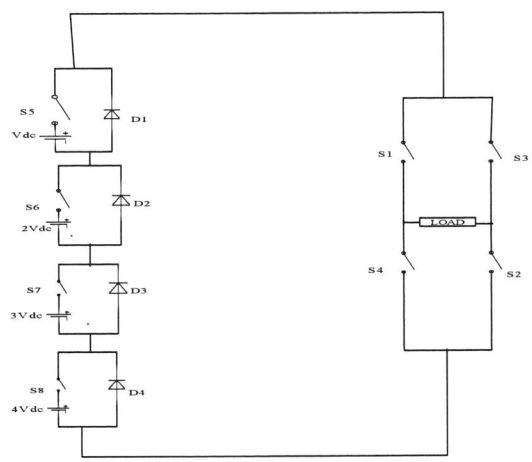

Fig. 7 Circuit diagram for twenty one level MLI

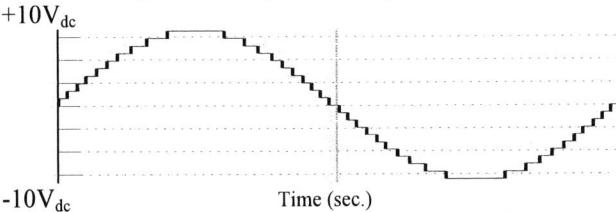

Fig.8 Ideal twenty one level output voltage waveform

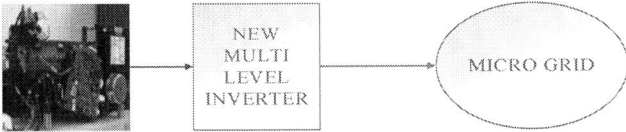

Fig.9 Fuel cell module and new MLI interfaced to Micro grid

The block diagram shown in Fig. 9 depicts the fuel cell module interconnected to grid through a new MLI topology. The MLI receives DC from the Feel cell and convert to AC for feeding to the micro grid satisfying grid conditions such as: amplitude of grid voltage, phase angle and frequency [14]. In this paper, Alkaline Fuel cell (AFC) is considered as an energy resource to the MLI for feeding energy to micro grid, whose parameter details as given in Table V.

TABLE V: ALKALINE FUEL CELL PARAMETER DETAILS

Voltage at 0A and 1A	[64.6 64]
Nominal operating point[Inom(A),Vnom(A)]	[50, 48]
Maximum operating point[Iend(A),Vend(A)]	[62, 46]
Number of cells	68
Nominal Stack Efficiency	56
Operating temperature(Celsius)	65
Nominal Air flow rate	300
Nominal Composition (%) [H2 O2 H2O(Air)]	[99.95 21 1]

V. SIMULATION RESULTS

The Matlab/Simulink platform is used for the modeling of different level MLIs and simulation results are considered for the analysis and comparative analysis of different MLIs topologies.

A. Seven level multilevel inverter

The output voltage and its THD spectrum of seven levels inverter using the pulse generator switching scheme is given in Fig.10 and 11 respectively, while the output voltage and its THD spectrum with modified switching technique is given in Fig. 12 and 13, respectively. From Fig. 11, it is observed that the THD of the seven levels inverter output voltage with pulse generator switching scheme is 17.34 %, while with the modified switching it is 14.56 % as shown in Fig.13.

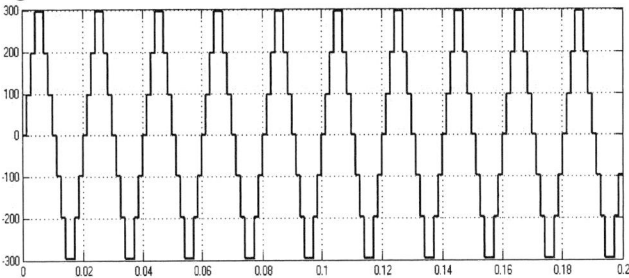

Fig.10 Simulated output voltage of seven levels inverter with pulse generator switching

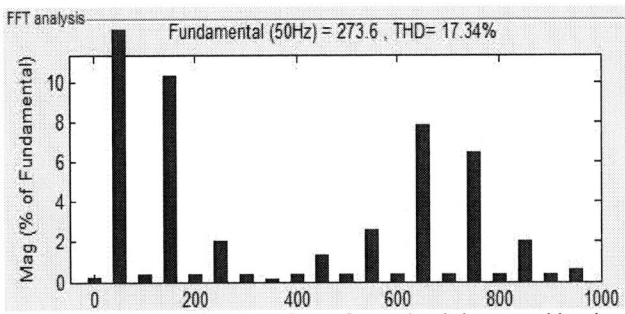

Fig.11 THD spectrum of output voltage of seven levels inverter with pulse generator switching

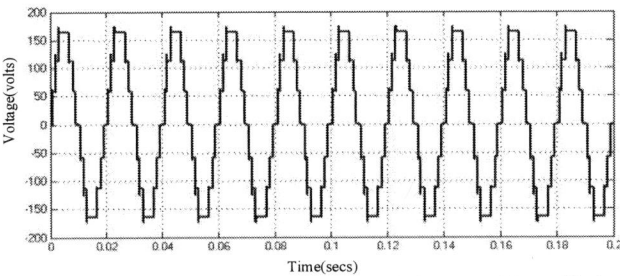

Fig.12 Simulated output voltage of seven levels inverter with modified switching

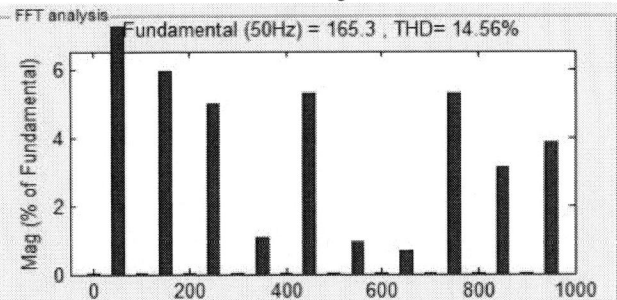

Fig.13 THD spectrum of output voltage of seven levels inverter with modified switching

B. Thirteen level multilevel inverter

The output voltage and its THD spectrum of thirteen levels inverter using the pulse generator switching scheme is given in Fig.14 and 15 respectively, while the output voltage and its THD spectrum with modified switching technique is given in Fig. 16 and 17, respectively. It is observed from Fig. 15 that the THD of the thirteen levels inverter output voltage with pulse generator switching scheme is 14.08 %, while with the modified switching it is 7.40 % as shown in Fig.17.

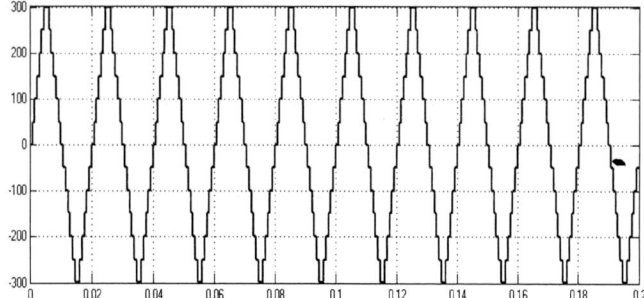

Fig.14 Simulated output voltage of thirteen levels inverter with pulse generator switching

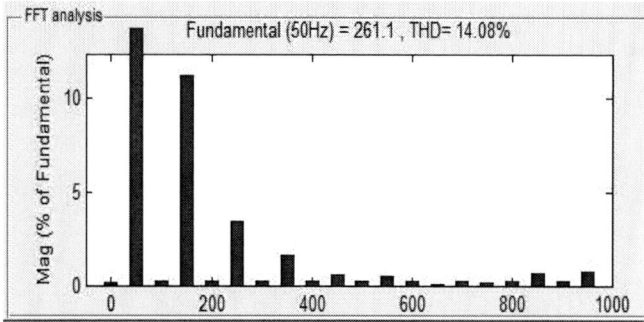

Fig.15 THD spectrum of output voltage of thirteen levels inverter with pulse generator switching

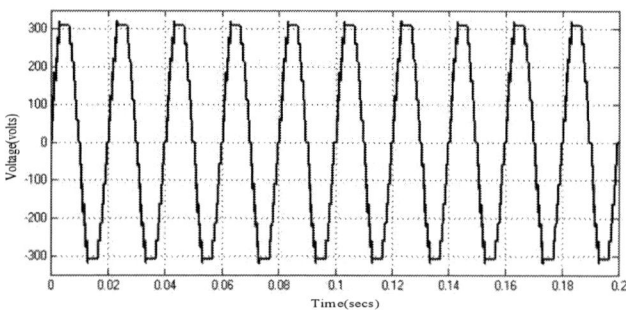

Fig.16 Simulated output voltage of thirteen levels inverter with modified switching

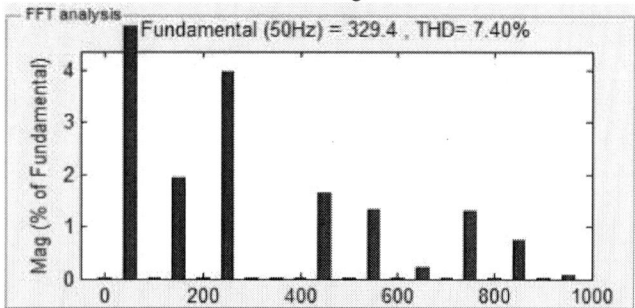

Fig.17 THD spectrum of output voltage of thirteen levels inverter with modified switching

C. Twenty one level multilevel inverter

The twenty one levels topology of the MLI is utilized for the interfacing of the AFC with the Grid by satisfying all conditions such as phase angle, frequency and amplitude of Grid voltage. The output voltage and its THD spectrum of twenty one levels MLI with modified switching are shown in Fig. 18 and 19, respectively. It is observed from Fig. 19 that the THD in the output voltage of the twenty levels inverter is 4.64%, which is within limits as per the standard IEEE 519, hence this topology is used for interfacing its output voltage to grid through a filter. The Inductive filter of 1mH is used while interfacing inverter to grid. The corresponding grid voltage and grid current are shown in Fig. 20. Ideally the grid voltage and the current should be in phase to satisfy grid conditions, but in this case the voltage and current are having a phase angle difference of 13.75°. The power factor obtained is 0.9713, which is approximately tending towards UPF. It is observed form the THD spectrum of the filtered output voltage that there is some amount of third harmonics exists in its output, which may be neutralized when implemented for the three phase.

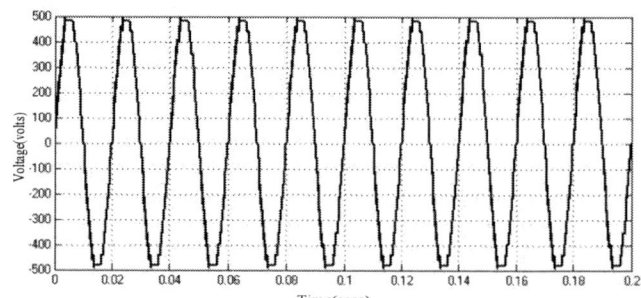

Fig.18 Simulated output voltage of twenty one levels inverter with modifie

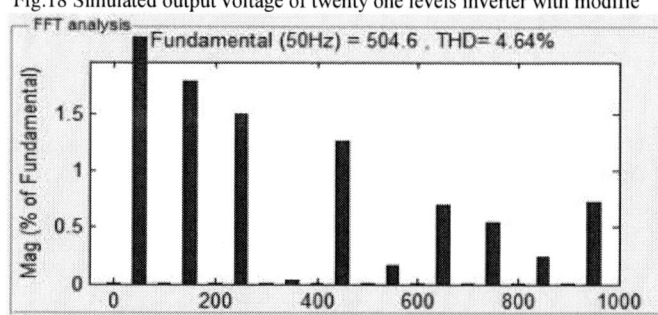

Fig.19 THD spectrum of output voltage of twenty one levels inverter with modified switching

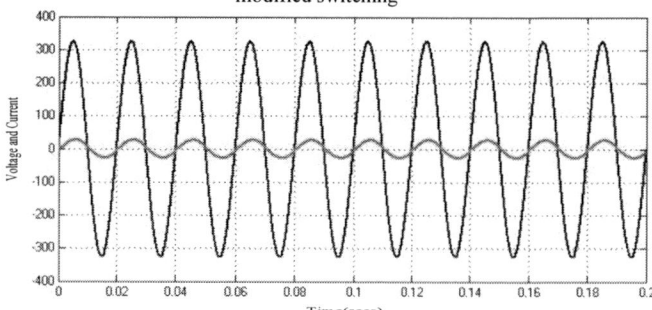

Fig.20 Grid connected Voltage and Current of twenty one levels inverter

TABLE VI: COMPARATIVE ANALYSIS OF THD OF SEVEN AND THIRTEEN LEVEL INVERTERS

MLI	THD in Output voltage with pulse generator	THD in Output voltage with modified switching
Seven level MLI	17.34 %	14.56 %
Thirteen Level MLI	14.08 %	7.40 %

TABLEVII: COMPARISON FOR HARDWARE FOR DIFFERENT TOPOLOGIES

Topology	7-level MLI		13-level MLI		21-level MLI	
	Number of switches	Number of DC Sources	Number of switches	Number of DC Sources	Number of switches	Number of DC Sources
Diode Clamped MLI	12	6	24	12	40	20
Flying Capacitor MLI	12	6	24	12	40	20
Cascaded H-bridge MLI	12	6	24	12	40	20
NEW MLI	6	2	7	3	8	4

VI. COMPARATIVE ANALYSIS OF DIFFERENT TOPOLOGIES

In this section, a comparative analysis of different topologies is carried out in terms of the % THD of the output voltage and the hardware required for their realization. It is observer from the simulation results that by modifying the switching technique the THD output of the seven and thirteen levels inverter topologies have been reduced as compared to the reported switching technique in [13]. Table VI gives a comparative analysis of THD in the output voltage of the seven and thirteen levels inverter topologies with modified switching technique and conventional switching technique. The comparative analysis of seven, thirteen and twenty one levels different MLIs topologies is also carried out in terms of the required power switches and DC sources for their realization as given in Table VII.

VI. CONCLUSIONS

This paper proposes a twenty one level new topology with modified switching technique for the interfacing of Fuel cell with micro grid. The proposed topology of the MLI requires very less numbers of power switches and DC sources as compared to conventional, resulting in less switching losses. Because of reduced hardware requirement the cost of the proposed MLI topology is very less as compared to conventional MLI topologies. Moreover, the % THD of the output voltage of the proposed topology is less than 5%, which is within limits as per the standard IEEE 519. It is also observed from the frequency spectrum of the output voltage that lower order harmonics are very less, whose filtering is not so easy. It is also observed from the simulation results that after interfacing a Fuel cell to the micro grid, the grid voltage and grid connected current are almost in phase with each other.

REFERENCES

[1] Power Electronics Handbook. 2nd ed. Academic Press; 2007. p 451.

[2] Franquelo LG, Rodriguez JL, Leon J, Kouro S, Portillo R, Prats MA. The age of multilevel converters arrives. IEEE Ind Electron Mag 2008;2(2):28–39.

[3] Nabae A, Takahashi I, Akagim H. A new neutral point clamped PWM inverter.IEEE Trans Ind Appl 1981;IA-17(5):518–23.

[4] Zhiguo Pan, Peng FZ, Stefanovic V, Leuthen M. A diode-clamped multilevel converter with reduced number of clamping diodes. in: Applied power electronics conference and exposition, 2004. APEC '04. 19th annual IEEE. vol.2, 2004. p. 820–24.

[5] Meynard T, Foch H. Multi-level conversion: high voltage choppers and voltagesource inverters. In: Power electronics specialists conference. PESC '92 Record.,23rd Annual; 29 June–3 July 1992. p. 397–403.

[6] L.M. Tolbert, F.Z. Peng, et al., Multilevel converters for Large Electric Drives,IEEE Transactions on Industry Applications on Industrial Applications, Vol. 35, No. 1, pp. 36-44,1999.

[7] E. Babaei, A cascade multilevel converter topology with reduced number of switches, IEEE Trans. Power Electron. 23 (6) (2008) 2657–2664.

[8] E. Babaei, Optimal topologies for cascaded sub-multilevel converters, J. PowerElectron. 10 (3) (2010) 251–261.

[9] J. Dixon, L. Morán, High-level multistep inverter optimization using a minimum number of power transistors, IEEE Trans. Power Electron. 21 (2) (2006)330–337.

[10] S. Gui-Jia, Multilevel dc-link inverter, IEEE Trans. Ind. Appl. 41 (3) (2005),848–845.

[11] Krishna Kumar Gupta, Shailendra Jain A multilevel Voltage Source Inverter (VSI) to maximize the number of levels in output waveform, Electrical Power and Energy Systems 44 (2013) 25–36.

[12] Ebrahim Babaei, Ali Dehqan, Mehran Sabahi A new topology for multilevel inverter considering its optimal structures, Electric Power Systems Research 103 (2013) 145– 156.

[13] PrasadaraoK V.S, Sudha Rani, P, Tabita, G, A new multi level inverter topology for grid interconnection of PV systems, Power and Energy Systems Conference 2014.

[14] Yi-Hung Liao, Ching-Ming Lai, Newly-Constructed Single-Phase Multistring Multilevel Inverter for Fuel-Cell Microgrid, International Conference on Power Electronics, May-June 2011.

New Time Domain Electric Arc Furnace Model for Power Quality Study

Deepak C. Bhonsle
Electrical Engineering Department
Maharaja Sayajirao University of Baroda
Vadodara, INDIA
dcbhonsle@gmail.com

Ramesh B. Kelkar
Electrical Engineering Department
Maharaja Sayajirao University of Baroda
Vadodara, INDIA

Abstract— **Power quality is becoming a more concern of today's power system engineer due to rapid growth of non-linear loads in distribution network. Electric arc furnace (EAF) is one of the typical industrial non-linear loads responsible for deteriorating the power quality in the distribution network by-introducing harmonics, propagating voltage flicker and causing unbalance-in voltages and currents. Hence electric arc furnace model is needed to study and to analyze the power quality in the distribution network. This paper presents a new time domain model of electric arc furnace to study power quality problems. The proposed model is a combination of two previous EAF models called-Exponential and hyperbolic model-using transition functions. The functioning of the proposed model has been validated by comparing its performance characteristics with the existing Cassie-Mayr EAF model. Simulation carried out in SIMULINK/MATLAB environment.**

Index Terms—**Power quality, harmonic distortion, harmonic analysis**

I. NOMENCLATURE

i =Arc current

v =Arc voltage

g =Arc conductance

E_0 =Momentarily constant steady state arc voltage

θ =Arc time constant

θ_0 =Constant

θ_1 = Constant

α = Constant

P_0 =Momentarily power loss

I_0 =Transition current

g_{min} =Minimum conductance

THD_I =Total Current Harmonic distortion

THD_V = Total Voltage Harmonic distortion

II. INTRODUCTION

Electric arc furnace (EAF) is an inherently non-linear, time-variant load and it can cause power quality problems such as harmonics and voltage flicker. Odd and even harmonic currents are generated by EAF operation. These harmonic currents, when circulated in the electric network can generate harmonic voltages which in turn can affect other users connected in the distribution network. Flicker is the sensation that is experienced by human eye when subjected to changes in the illumination intensity. The maximum sensitivity to change in illumination is in the frequency range of 5 to 15 Hz [4, 5]. As an EAF is a large source of flicker, causes voltage fluctuation in the connected electric network which is a major power quality issue which affects operation of other connected load in the distribution network. Hence, modeling of EAF has attracted attention of power system engineers to solve these power quality issues pertaining to EAF.

The important issue in the modeling of the EAF is the simulation of arc. There are several methods used to describe the electric arc [1-4, 7-8]. On the basis of actual measured samples of an electric arc in several functioning cycles of EAF, different operating points are generated in the form of statistical probability, corresponding to hidden Markov theory in [1]. This requires actual measurement of an electric arc. The time domain methods based on the differential equations are also presented [2]. Variation of power transmitted to the load by the arc furnace during the cycle of operation is considered in [3]. Comparison of EAF modeling in time domain and frequency domain shows that he time domain is more useful in studying the EAF [4, 8]. The balanced steady state equations are used in [7]. Other methods such as frequency response, V-I characteristic are employed to analyze the behavior of the EAF [8]. The above methods suffer from limitations such as knowledge of initial conditions for the differential equations, balanced situation of thee phase currents, actual arc measurement and use of

978-1-4799-6047-7/14 $31.00 © 2014 IEEE

complicated mathematical equation for the modeling of EAF. This paper presents simulation of the EAF model in the time domain using MATLAB. The main feature of the proposed model is good approximation without need of initial conditions of the EAF. Also, the proposed method can be used to describe different operating situations of the EAF and its effect of the connected electric network.

III. EAF Modeling as Non-linear Load

A. Model1:Cassie-Mayr EAF model

Mathematical model of Cassie-Mayr EAF model expressed as in [1, 5]:

$$g = g_{min} + \left[1 - \exp\left(-\frac{i^2}{I_0}\right)\right] \cdot \frac{v \cdot i}{E_0^2} + \exp\left(-\frac{i^2}{I_0}\right) \cdot \frac{i^2}{P_0} - \theta \cdot \frac{dg}{dt} \quad (1)$$

$$\theta = \theta_0 + \theta_1 \cdot \exp(-\alpha \cdot |i|) \quad (2)$$

$$v = \frac{i}{g} \quad (3)$$

Typical values of and E_0, θ_0, θ_1, α, P_0, I_0, and g_{min} are tabulated in Table 1[4-6].

TABLE I.   CASSIE-MAYR EAF MODEL PARAMETERS

Parameter Description	Parameter	Value
Mimimum arc conductance	g_{min}	0.008
Tansition current	I_0	10 A
Momentarily constant steady state arc voltage	E_0	250 V
Momenttarily power loss	P_0	110 W
Time Consatnat	θ_0	110 μs
Time Consatnat	θ_1	100 μs
Constant	α	0.0005

B. Model 2: Proposed EAF model

Proposed EAF model is a combination of Hyperbolic and Exponential EAF models. The $v - i$ characteristic of hyperbolic EAF model is considered to be in the form of $v = v(i)$ and it can be described as [5, 7]:

$$v_{hyp}(i) = V_{at} + \left(\frac{C}{D+i}\right) \quad (4)$$

In (4) variable v and i are arc voltage and arc current per phase respectively. V_{at} is the magnitude of the voltage threshold to which the voltage approaches as current increases. This voltage is dependent on the arc length which is defined by constants C and D taking care of arc power and arc current respectively. Typical values of these constants are tabulated in Table I.

The V-I characteristic of exponential EAF model is approximated by exponential function as [4, 6]:

$$v_{exp}(i) = V_{at}\left(1 - e^{\left(i/I_o\right)}\right) \quad (5)$$

In (5) current constant I_o is employed to model the steepness of positive and negative currents. A typical value of I_o is tabulated in Table 2.

TABLE II.   PROPOSED EAF MODEL PARAMETERS

Parameter Description	Parameter	Value
Voltage threshold	V_{at}	200 V
Arc power condtant	C	19 kW
Arc current constant	D	5 kA
Current steepness constant	I_o	20 kA

Exponential and hyperbolic models can be combined into single model by defining a transition function $O(i)$, which is a function of arc current and is given by:

$$v_{com}(i) = \underbrace{[1 - O(i)] \cdot v_{exp}}_{Higher\ Current} + \underbrace{O(i) \cdot v_{hyp}}_{Lower\ Current} \quad (6)$$

In (6), v_{hyp} and v_{exp} are the arc voltages given by (4) and (5) respectively. A satisfactory form of $O(i)$ used in this combination is given in [2, 6]:

$$O(i) = e^{\left(-\frac{i^2}{I_o^2}\right)} \quad (7)$$

In (7) I_o is the transition current. When arc current (i) is small, value of $O(i)$ is approaching unity which yields arc voltage value v_{com} is dominated by v_{hyp} and when arc current value is large, $O(i)$ is approaching zero yields arc voltage value v_{com} is dominated by v_{exp}. The combined model voltage follows exponential model characteristic during high arc currents and follows hyperbolic model characteristic during low arc currents.

Thus the V-I characteristic of the proposed model is described by following equation:

$$v_{com}(i) = \begin{cases} V_{at}\left[1 - e^{\left(i/I_o\right)}\right] & for\ higher\ arc\ current \\ V_{at} + \left(\dfrac{C}{D+i}\right) & for\ lower\ arc\ current \end{cases} \quad (8)$$

The combined model has the capability of describing the EAF behavior in time domain. Also the combined model can explains various operating conditions of the EAF such as scrap meltdown stage, refining stage from power quality point of view. The refining stage contributes harmonics in

978-1-4799-6047-7/14 $31.00 © 2014 IEEE

current and voltage at point of common coupling (PCC) while scrap meltdown stage yields voltage flicker majorly.

IV. EAF DYNAMIC BEHAVIOR

Dynamic EAF model is required for real time analysis of the effect of the arc. The dynamic arc characteristic is simulated by varying arc voltage. In general the variation is of random nature. However two types of variation are considered for the study-sinusoidal variation and random variation. In order to study the effect of voltage flicker on the power system of EAF, V_{at} is varied sinusoidally and randomly. In this regard V_{at} is modulated as follows:

A. The sinusoidal variation

The sinusoidal variation is assumed as [6-7],

$$v_{at}(t) = V_{at0}[1 + m \cdot sin(\omega_f \cdot t)] \qquad (9)$$

In (9) m is modulation index and ω_f is a flicker frequency.

B. The random variation

The random variation is assumed as [6-7],

$$v_{at}(t) = V_{at0}[1 + m \cdot N(t)] \qquad (10)$$

In (10) $N(t)$ is a band limited white noise with zero mean and variance of one. The parameters used for sinusoidal variation and random variation are tabulated in Table III.

TABLE III.   FLCIKER GENERATION PARAMETERS

Parameter Description	Parameter	Value
Sinusoidal variation		
Arc voltage threshold	V_{at0}	250 V
Modulation index	m	0.8
Flciker frequency	ω_f	4 Hz
Random variation		
Time Consatnat	V_{at0}	240 V
Modulation index	m	0.7
Band limited white noise	$N(t)$	4-14 Hz

V. VOLTAGE FLICKER ASSESSMENT

Voltage flicker assessment is also one of the important aspects of power quality study. The assessment of voltage flicker involves the derivation of system RMS voltage variation and the frequency at which the variation occurs. The voltage flicker usually expressed as the RMS value of the modulating waveform divided by the RMS value of the fundamental value, as follows [17-19]:

$$\% \, Voltage \, Flickr = \frac{V_{2P} + V_{1P}}{V_{2P} - V_{1P}} \qquad (11)$$

Equation (4) is useful for estimating voltage flicker. A variety of perceptible/limit curves are available in published literature which can be used as general guidelines to verify whether the amount of flicker is a problem [17].

VI. EAF MODELING WITH POWER SYSTEM

Fig. 1 shows a simple single phase equivalent electric network of a source which supplies an EAF. It consists of voltage source, source impedance, furnace transformer impedance and EAF.

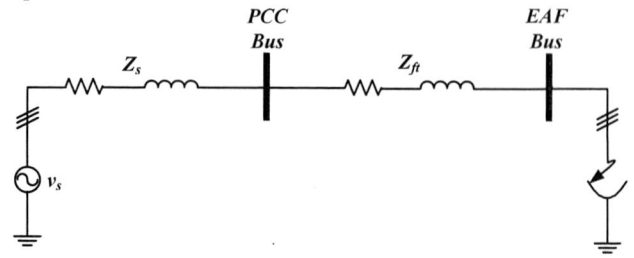

Figure 1. EAF connected in power system

In Fig. 1, the system impedance is represented as Z_s and the furnace transformer whose impedance is given by Z_{ft}. The system parameters along with proposed EAF Model are tabulated in Table IV [7].

TABLE IV.   POWER SYSTEM PARAMETERS

Parameter Description	Parameter	Value
Source voltage	V	415 V
Supply frequency	f	50 Hz
Source impedance	Z_s	(0.0528+j0.468) mΩ
Furnace transformer impedance	Z_{ft}	(0.3366+j3.22) mΩ

VII. SIMUALTAION RESULTS

Simulated results are presented as a comparison of performance of Model 2 (Proposed-Exponential-Hyperbolic) with that of Model 1 (Cassie-Mayr). The performance of EAF includes various performance characteristics such as arc current, arc voltage, harmonic spectrum, arc conductance variation, arc voltage-current characteristic (VIC), variation in active & reactive power, etc. For better comparison, each performance characteristic of EAF model 1 and model 2 is presented together.

A. Steady State Characteristics

1) Arc Current and Arc Voltage

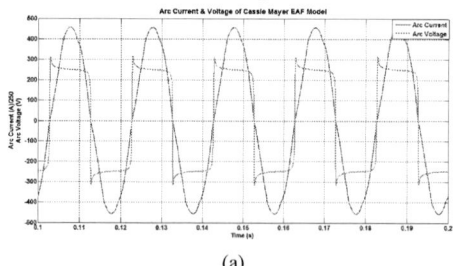

(a)

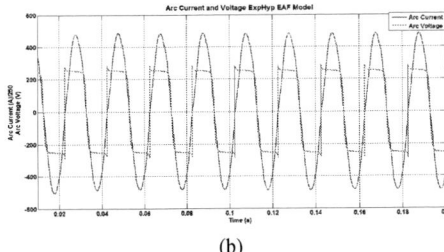

(b)

Fig. 2 Arc Current and Arc Voltage of (a) Model 1 (b) Model 2

Fig. 2 represents steady state characteristics of two models of EAF i.e. arc length is kept constant, which demonstrates refining condition of an EAF. In this condition, the level of molted material is constant and melting is uniform in the furnace. Hence behavior of V-I characteristic is also uniform. This condition does not produce any flicker at PCC but produces harmonics in voltage and current.

2) Voltage-Current Characteristic (VIC)

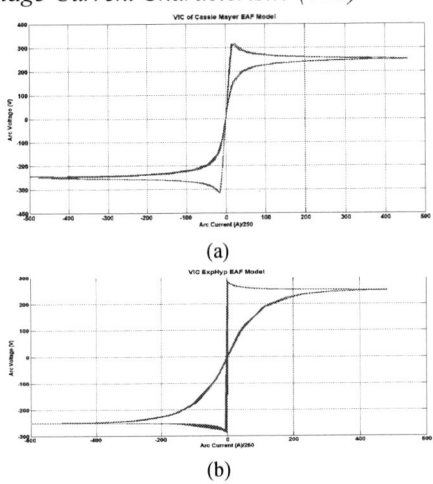

(a)

(b)

Fig. 3 VIC of (a) Model 1 (b) Model 2

3) Current Harmonics at PCC

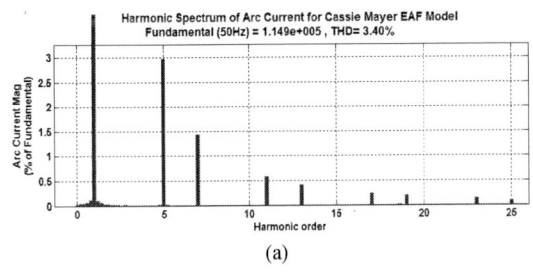

(a)

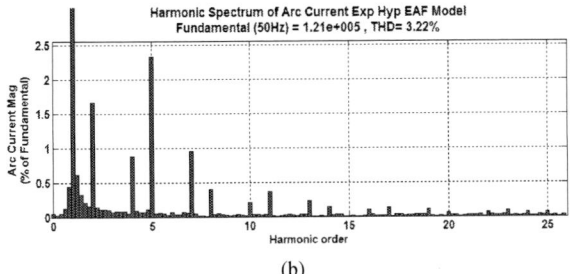

(b)

Fig. 4 Harmonic Spectrum of Current at PCC of (a) Model 1 (b) Model 2

It can be seen from Fig. 4 that total harmonic distortion (THD_I) observed in both the models is quite same (3.40 % for model 1 and 3.22 % for model 2). This shows validity of model 2 for refining cycle.

4) Voltage Harmonics at PCC

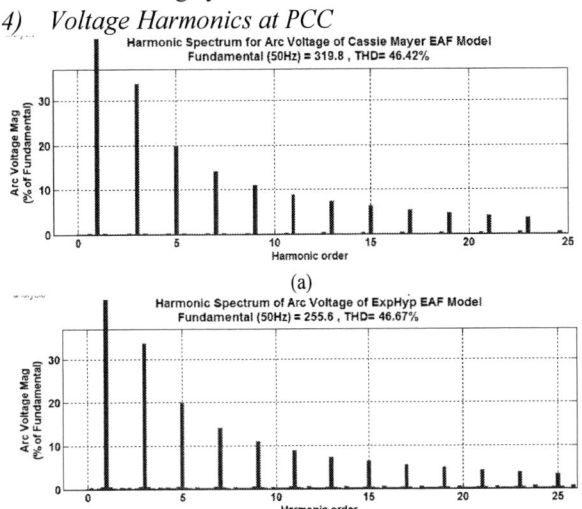

(a)

(b)

Fig. 5 Harmonic Spectrum of Current at PCC of (a) Model 1 (b) Model 2

Similarly, it can be seen from Fig. 5 that total harmonic distortion (THD_V) observed in both the models is quite same (46.42 % for model 1 and 46.67 % for model 2). This shows validity of model 2 for refining cycle

5) Active and Reactive Power (P-Q)

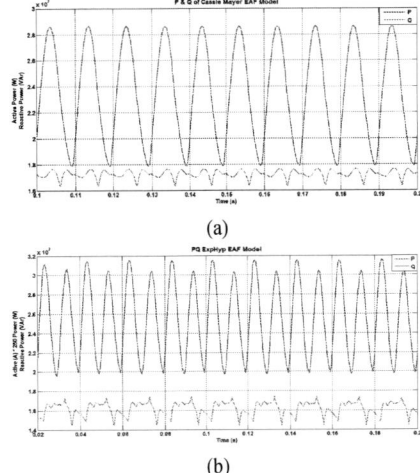

(a)

(b)

Fig. 6 Active (P) and reactive (Q) of (a) Model 1 (b) Model 2

Fig. 6 shows active and reactive power consumption by EAF.

B. Dynamic Characteristics

Dynamic characteristic represents melting cycle of EAF. In this operation the furnace is charged with scrap, after that the electrodes could be lowered, each of which has its own regulator and mechanical drive. This operation exhibits severe voltage flickers. The effect of voltage flicker on the system with EAF can be studied using voltage variation with reference to time. As described in previous section, the effects of two types of flicker on the dynamic characteristic of the EAF are studied. Results of the simulation are obtained using (9) and (10) with values given in Table III.

C. Sinusoidal Flicker

Results for sinusoidal flickers are presented in Fig. 12, which show the variation of arc voltage and arc current. It can be seen that if the furnace load generates sinusoidal flicker, the arc voltage and arc current, are varied sinusoidally with the flicker frequency.

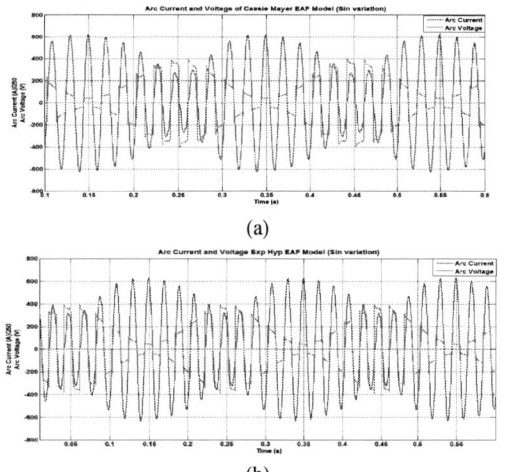

(a)

(b)

Fig. 7 Arc voltage and current of (a) Model 1 (b) Model 2 for Sinusoidal Flicker

D. Random Flicker

The simulation results for VIC for model 1 and model 2 under random flicker condition are presented in Fig. 8. The proposed EAF model 2 provides identical VIC as model 1.

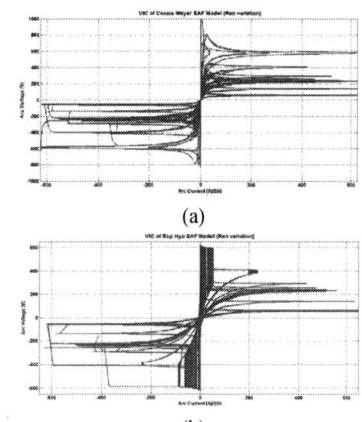

(a)

(b)

Fig. 8 VIC of (a) Model 1 (b) Model 2 Random Flicker

VIII. COMPARISION AND ANALYSIS

Table V shows comparison of voltage harmonic analysis between EAF Model 1 and Model 2. Total harmonic distortion observed in Model 1 and Model 2 is 46.42 % and 46.67 % respectively, which is violating IEEE 419-1992 Limits of 5%. % Error observed in THD of Model 2 (Proposed) with respect to Model 1 is -0.54 %, which is less than 10 %. It makes Model 2 acceptable.

TABLE V. VOLTAGE HARMONIC ANALYSY @ PCC

Harmonics (% of Fund.)	Model 1 (Cassie-Mayer)	Model 2 (Proposed)	% Error	% Error (Average)
V_{peak} (V)	305	288	5.57	
THD_V	46.42	46.67	-0.54	
3rd	33.69	33.64	0.15	
5th	19.98	20.07	-0.45	
7th	14.12	14.19	-0.49	
9th	10.99	11.05	-0.54	
11th	8.78	8.92	-1.57	1.85
13th	7.3	7.41	-1.48	
15th	6.27	6.39	-1.88	
17th	5.36	5.54	-3.25	
19th	4.67	4.82	-3.11	
21st	4.16	4.29	-3.03	
23rd	3.65	3.82	-4.45	
25th	3.26	3.38	-3.55	

Harmonic distortion of each harmonic order is expressed as % of fundamental. Harmonic distortion observed in almost all harmonic orders (3rd to 25th) is more than IEEE 419-1992 Limit of 3 % for individual harmonic order. % Error for each harmonic order is calculated by taking Model 1 (Cassie-Mayer) to be the reference. Maximum error observed in model 2 with respect to model 4 is +0.15 % (3rd order) and -4.45 % (23rd order) on positive and negative side respectively. Average error observed is -1.86 % which is less than 10 %, which makes model 2 (Proposed) acceptable. % Error observed in the voltage magnitude at PCC of Model 2 with respect to Model 1 is 5.57 %, which again confirms validity of Model 2.

Table VI shows comparison of current harmonic analysis between EAF Model 1 and Model 2. % Error observed in THD of Model 2 (Proposed) with respect to Model 1 is 5.29 %, which is less than 10 %. It makes Model 2 acceptable.

TABLE VI. CURRENT HARMONIC ANALYSY @ PCC

Harmonics (% of Fund.)	Model 1 (Cassie-Mayr)	Model 2 (Proposed)	% Error	% Error (Average)
I_{peak} (kA)	117.5	120	-2.13	
THD_I	3.4	3.22	5.29	
5th	2.98	2.79	6.38	
7th	1.43	1.24	13.29	
11th	0.58	0.56	3.45	7.85
13th	0.42	0.39	7.14	
17th	0.24	0.22	8.33	
19th	0.19	0.17	10.53	
23rd	0.14	0.13	7.14	
25th	0.09	0.08	11.11	

Harmonic distortion of each harmonic order is expressed as % of fundamental. % Error for each harmonic order is calculated by taking Model 1 (Cassie-Mayr) to be the reference. Average error observed is 8.07 % which is less than 10 %, which makes model 2 (Proposed) acceptable. % Error observed in the current magnitude at PCC of Model 3 with respect to Model 4 is -2.13 %, which again confirms validity of Model 3.

Table VII shows comparison of active power, reactive power and power factor between Model 1 and Model 2. % errors calculated are less than 10%.

TABLE VII.   POWER ANALYSY @ PCC

Parameter	Model 1 (Cassie-Mayr)	Model 2 (Proposed)	% Error (w. r. t. model 1)
Active Power (P)	23280 kW	24900 kW	-6.96
Reactive Power (Q)	17250 kVAR	16130 kVAR	+6.49
Power Factor (PF)	0.574	0.606	-5.57

Table VIII shows comparison of sinusoidal flicker generated by Model 1 and Model 2. % Errors is +1.36 %, which is again less than 10%.

TABLE VIII.   VOLTAGE FLICKER ANALYSY @ PCC

Parameter	Model 1 (Cassie-Mayr)	Model 2 (Proposed)	% Error (w. r. t. model 1)
Voltage measurement			
V_{1P} (Volts)	65	64	1.54
V_{2P} (Volts)	390	400	-2.56
Flicker Calculations			
% Voltage Flicker	1.40	1.38	+1.36

Comparison of various performance characteristics of EAF Model 2 (Proposed) with existing EAF Model 1 (Cassie-Mayer), as shown in Fig. 7 to Fig. 24, validates Model 2. Comparison of voltage harmonic analysis, current harmonic analysis, power analysis and voltage flicker analysis at PCC of EAF Model 2 (Proposed) with respect to EAF Model 1 (Cassie-Mayer) shows that the % error observed is less than 10 %. This again confirms the validity of EAF Model 2 (Proposed).

IX.   CONCLUSIONS

This paper describes performance evaluation of composite filter for power quality improvement of electrical electric arc furnace distribution network. First of all, distribution network is simulated using Cassie-Mayr EAF model. The simulated EAF distribution network is used for power quality analysis including voltage-current harmonics, voltage flicker and voltage unbalance. Next, a control strategy for a composite filter, which is connected with the existing passive filter, is proposed for taking care of the unbalance, non-sinusoidal and randomly varying EAF. The control strategy is based on the dual vectorial theory of power. Finally, detail performance of composite filter is evaluated by comparing its performance with passive filter for various operation cycles of EAFs connected distribution network. Performance comparison shows that, the proposed composite filter performs better than the passive filter alone for harmonic compensation, voltage flicker mitigation, and for clearing voltage unbalance on EAF load side.

X.   REFERENCES

[1] Esfahani, M.T. and Vahidi, B., "A New Stochastic Model of Electric Arc Furnace Based on Hidden Markov Model: A Study of Its Effects on the Power System", IEEE Transactions on Power Delivery, Vol. 27, Issue-4, pp. 1893-1901, October 2012

[2] Tavakkoli, M. Ehsan, S. M. T. Batahiee and M. Marzband, "A SIMULINK Study of Electric Arc Furnace Power Quality Improvement by Using STATCOM", IEEE International Conference on Industrial Technology 2008, ICIT 2008, 21-24 April 2008, pp. 1-6.

[3] Golkar, M.A. , Meschi, S., "MATLAB modeling of arc furnace for flicker study", IEEE International Conference on Industrial Technology, 2008. ICIT 2008. , 21-24 April 2008, pp. 1-6

[4] K. Anuradha, B. P. Muni and A. D. Raj Kumar, "Modeling of Electric Arc Furnace & Control Algorithms for voltage flicker mitigation using DSTATCOM", IPEMC, 1123-1129, 2009.

[5] Mahdi Banejad, Rahmat-Allah Hooshmand and Mahdi Torabian Esfahani, "Exponential-Hyperbolic Model for Actual Operating conditions of Three Phase Arc Furnaces", American Journal of Applied Scinces 6, pp.1539-1547, 2009.

[6] Mokhtari H. and Heiri M., "A New Three Phase Time-Domain Model for Electric Arc Furnace Using MATLAB", Transmission and Distribution Conference and Exhibition 6-10 October 2002: Asia Pacific, IEEE/PES, Vol. 3, pp. 20787-283

[7] Rahmatallah Hooshmand, Mahdi Banejad and Mahdi Torabian Esfahani, "A New Time Domain Model for Electric Arc Furnace", Journal of Electrical Engineering, Vol. 59, No. 4, 195-202, 2008.

[8] Zheng T., Makram E. B. and Girgis A. A., "Effect of different arc furnace models on voltage distortion", IEEE Transactions , International Conference on Harmonics and Quality of Power, 14-18 October 1998, Volume 2, pp. 1079-1085

[9] Haruni A. M. O., Muttaqi K. M. and Negnevitsky M., "Analysis of harmonics and voltage fluctuation using different models of Arc furnace", IEEE Transactions, Power Engineering Conference, 9-12 December 2007, AUPEC 2007, Australasian Universities, pp. 1-6.

[10] E. A. Cano Plata and H. E. Tacca, "Arc Furnace Modeling in ATP-EMPT", International Conference on Power Systems Transients (IPST'05), Montreal, Canada, 19-23 June 2005, Paper No. IPST05-067.

[11] S. R. Mendis, M. T. Bishop and J. F. Witte, "Investigations of Voltage Flicker in Electric Arc Furnace Power Systems", IEEE Industry Applications Magazine, January/February 1996, pp. 28-34.

[12] Z. Zhang, N. R. Fahmi and W. T. Norris, "Flicker Analysis Methods for Electric Arc Furnace Flicker (EAF) Mitigation (A Survey)", IEEE Porto Power Tech Conference (PPT 2001), 10th -13th September 2001, Porto, Portugal.

[13] M. Walker, "Electric Utility Flicker Limitations", IEEE Transactions on Industry Applications, Vol. 1A-15, No. 6, November/December 1979.

Mathematical Modeling of Composite Filter for Power Quality improvement of Electric Arc Furnace Distribution Network

Deepak C. Bhonsle
Electrical Engineering Department
Maharaja Sayajirao University of Baroda
Vadodara, INDIA
dcbhonsle@gmail.com

Dr. Ramesh B. Kelkar
Electrical Engineering Department
Maharaja Sayajirao University of Baroda
Vadodara, INDIA

Abstract— **Electrical Arc Furnace (EAF) is one of the responsible cause for deteriorating power quality in the distribution network by, introducing harmonics, propagating voltage flicker and causing unbalance in voltages and currents. This paper presents mathematical modeling of composite filter (CF) for power quality improvement of EAF distribution network. The composite filter is consisting of a shunt LC passive filter connected with a lower rated voltage source PWM converter based series active power filter (SAPF). The control strategy adopted for composite filter operation is based on simultaneous detection of source current and load voltage harmonic based on the vectorial theory dual formulation of instantaneous reactive power. A state-space averaging model of a composite filter constructed to analyze its system stability by traditional control strategy taking into account the effect of the time delay. Simulation for a typical EAF distribution network with a composite power filter has been carried out to validate the performance. Simulation results are shown in an attempt to verify the mathematical model of the filter. The simulations have been carried out in MATLAB environment using SIMULINK and power system block set toolboxes.**

Index Terms-- **Active filter, harmonics, harmonic distortion, modeling**

I. INTRODUCTION

The increasing popularity of EAF in metallurgical industries to melt scrap causes significant impacts on power system and on electrical power quality. EAF is one of responsible source for deteriorating the power quality in the connected network. An EAF is chosen as an industrial non-linear load to demonstrate power quality problems.

Harmonic distortion in power distribution systems can be suppressed using two approaches namely, passive and active powering. The passive filtering is the simplest conventional solution to mitigate the harmonic distortion [1]. Although simple, the use passive elements do not always respond correctly to the dynamics of the power distribution systems. Although simple and least expensive, the passive filter inherits several shortcomings [2]. Combined operation of series APF with traditional passive filter i.e. composite filter (CF) is found one of the solutions to overcome the disadvantages of the existing passive filters. It controlled to act as a harmonic isolator between the source and nonlinear load by injection of a controlled harmonic voltage source [3-5]. A control algorithm for a three-phase series APF is proposed. The control strategy is based on the vectorial theory dual formulation of instantaneous reactive power [5], so that the voltage waveform injected by the active filter is able to compensate the reactive power and the load current harmonics and to balance asymmetrical loads. Here an attempt is made to apply the same theory for unbalanced and non-sinusoidal voltage conditions for randomly varying load as an electric arc furnace (EAF).

System stability during CF operation is also one of the important factors from control aspects. Choice of a suitable gain to ensure a good compensation is a tedious and difficult task while keeping the system stable for different load and source condition [6-8]. The analysis and experimental results have pointed out that a time delay can cause a stability problem [9]. Hence, a state-space averaging model of composite filter is constructed to analyze the stability problem by traditional control strategy taking into account the delay time. Bode plot of open-loop transfer function concludes that larger time delay could results into phase angle loss seriously leading to system unstable and poor filter performance. A new leading compensating control strategy discussed in [10] is applied for enhancing the system stability, for improving the filter performance and for reducing the power rating of the active part. The simulations have been carried out in MATLAB environment using SIMULINK and power system block set toolboxes to validate the same.

978-1-4799-6047-7/14 $31.00 © 2014 IEEE

II. STATE-SPACE MODELING OF COMPOSITE FILTER

A. Main circuit configuration

The system configuration of a composite filter (CF) consists of a Series Active Filter with a Shunt Passive Filter is shown in Fig. 1.

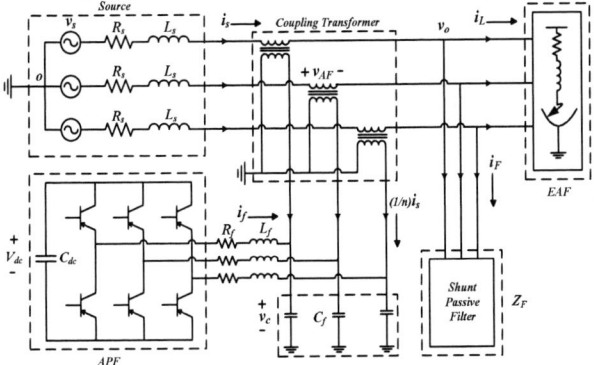

Figure 1. Main system configuration

Main circuit configuration can be divided into four parts: (a) an ac main (b) a series active filter (SAPF) (c) a shunt passive filter (d) an electric arc furnace EAF) as non-linear load. SAPF can also be further sub-divided into three parts: (a) a three phase bridge Inverter (generating harmonic voltage for compensation) (b) a Second order Carrier-wave filter (applied to filter the switching frequency component) (c) a Coupling transformer (connected between the electric network and the load with transformation ratio= n).

B. System Modeling

The output of voltage of the inverter v_c changes between $\pm V_{dc}$ when the IGBTs are switching. The inverter output voltage is represented by:

$$v_c = V_{dc} \cdot (2 \cdot V - 1) \tag{1}$$

Where, $V = [0,1]$, represents the input discrete values of IGBTs. The controlled output voltage v_c so generated will constrain the harmonics produced by the load. The state-Space averaged model in the stationary frame of reference is described as follows:

Applying KCL at node we get:

$$C_f \cdot \frac{dv_c}{dt} = \frac{1}{n} \cdot i_s + i_f \tag{2}$$

Applying KVL at node we get:

$$V_{dc} \cdot (2 \cdot V - 1) - i_f \cdot R_f - v_c - L_f \cdot \frac{di_f}{dt} = 0 \tag{3}$$

$$\therefore L_f \cdot \frac{di_f}{dt} = [v_c + i_f \cdot R_f - V_{dc} \cdot (2 \cdot V - 1)]$$

$$v_s - \frac{1}{n} \cdot v_c - v_o - L_s \cdot \frac{di_s}{dt} = 0$$

$$\therefore L_s \cdot \frac{di_s}{dt} = v_s - \frac{1}{n} \cdot v_c - v_o \tag{4}$$

$$i_s = i_F + i_L = \frac{v_o}{Z_F} + i_L$$

$$\therefore v_o = (i_s - i_L) \cdot Z_F \tag{5}$$

$$V = \frac{1}{2} \cdot v_c^*(t) + \frac{1}{2} \tag{6}$$

Where, i_s =source current, i_L =load current, i_f =inductor current through L_f, v_s =source voltage, V_{dc} =DC bus voltage, v_c =voltage in the secondary of the coupling transformer, $v_c^*(t)$ =input reference voltage, v_o =output voltage seen by the load, R_f =carrier filter resistance, L_f =carrier filter inductance, C_f =carrier filter capacitance. Let the sate variables are,

$$x_1 = v_c \quad \Rightarrow \quad \overset{\bullet}{x_1} = \frac{dv_c}{dt} \tag{7}$$

$$x_2 = i_r \quad \Rightarrow \quad \overset{\bullet}{x_2} = \frac{di_f}{dt} \tag{8}$$

$$x_3 = i_s \quad \Rightarrow \quad \overset{\bullet}{x_3} = \frac{di_s}{dt} \tag{9}$$

Re-writing the equations (2-4) using equations (7-9):

$$\overset{\bullet}{x_1} = 0 \cdot x_1 + \frac{1}{C_f} \cdot x_2 + \frac{1}{n \cdot C_f} \cdot x_3 \tag{10}$$

$$\overset{\bullet}{x_2} = \left(-\frac{1}{L_f} \right) \cdot x_1 + \left(-\frac{R_f}{L_f} \right) \cdot x_2 + 0 \cdot x_3 + \frac{V_{dc}}{L_f} \cdot (2 \cdot V - 1) \tag{11}$$

$$\overset{\bullet}{x_3} = \frac{1}{n \cdot L_s} \cdot x_1 + 0 \cdot x_2 + \left(-\frac{Z_F}{L_s} \right) \cdot x_3 + \left(\frac{v_s}{L_s} + \frac{i_L \cdot Z_F}{L_s} \right) \tag{12}$$

The equivalent state-space model can be presented as:

$$\begin{bmatrix} \overset{\bullet}{x_1} \\ \overset{\bullet}{x_2} \\ \overset{\bullet}{x_3} \end{bmatrix} = \begin{bmatrix} 0 & \frac{1}{C_f} & \frac{1}{n \cdot C_f} \\ -\frac{1}{L_f} & -\frac{R_f}{L_f} & 0 \\ \frac{1}{n \cdot L_s} & 0 & \frac{Z_F}{L_s} \end{bmatrix} \begin{bmatrix} x_1 \\ x_2 \\ x_3 \end{bmatrix} + \begin{bmatrix} 0 & 0 & 0 \\ \frac{1}{L_f} & 0 & 0 \\ 0 & \frac{1}{L_f} & \frac{Z_F}{L_s} \end{bmatrix} \begin{bmatrix} V_{dc} \\ v_s \\ i_L \end{bmatrix} \tag{13}$$

$$y = x_1 = v_c = \begin{bmatrix} 1 & 0 & 0 \end{bmatrix} \begin{bmatrix} x_1 \\ x_2 \\ x_3 \end{bmatrix} + \begin{bmatrix} 0 & 0 & 0 \end{bmatrix} \begin{bmatrix} V_{dc} \\ v_s \\ i_L \end{bmatrix} \tag{14}$$

Equation (13) is a state equation and equation (14) is an output equation. Comparing them with the standard form,

$$\overset{\bullet}{\overline{x}} = \overline{A}\overline{x} + \overline{B}\overline{u} \tag{15}$$

$$\overline{y} = C \cdot \overline{x} + D \cdot \overline{u} \tag{16}$$

These yields:

$$A = \begin{bmatrix} 0 & \frac{1}{C_f} & \frac{1}{n \cdot C_f} \\ -\frac{1}{L_f} & -\frac{R_f}{L_f} & 0 \\ \frac{1}{n \cdot L_s} & 0 & \frac{Z_F}{L_s} \end{bmatrix} \tag{17}$$

978-1-4799-6047-7/14 $31.00 © 2014 IEEE

$$B = \begin{bmatrix} 0 & 0 & 0 \\ \dfrac{1}{L_f} & 0 & 0 \\ 0 & \dfrac{1}{L_f} & \dfrac{Z_F}{L_s} \end{bmatrix} \tag{18}$$

$$C = \begin{bmatrix} 1 & 0 & 0 \end{bmatrix} \tag{19}$$

$$D = \begin{bmatrix} 0 & 0 & 0 \end{bmatrix} \tag{20}$$

Now Transfer Function of the system is given by:

$$G(s) = C[sI - A]^{-1}B + D \tag{21}$$

III. STABILITY ANALYSIS

Fig. 2 shows an equivalent single-phase circuit of the system considering current and voltage harmonics produced by EAF. Here, v_{sf} and v_{sh} are fundamental and harmonics components of source voltage respectively. v_{lf} and v_{lf} are the fundamental and harmonics components of load voltage respectively.

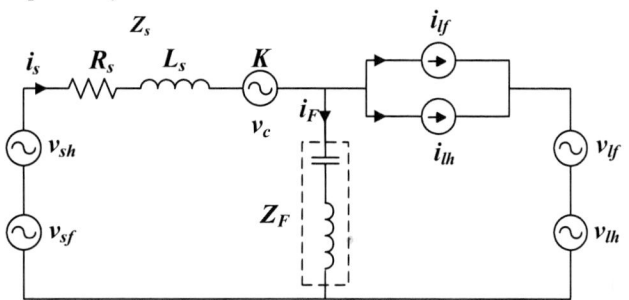

Figure 2. Single phase equivalent circuit of main system

Similarly i_{lf} and i_{lh} are the fundamental and harmonics components of current injected by load respectively. Z_S and Z_F are the source and the passive filter impedance respectively. Fig. 3 shows single-phase equivalence harmonic circuit of system under consideration.

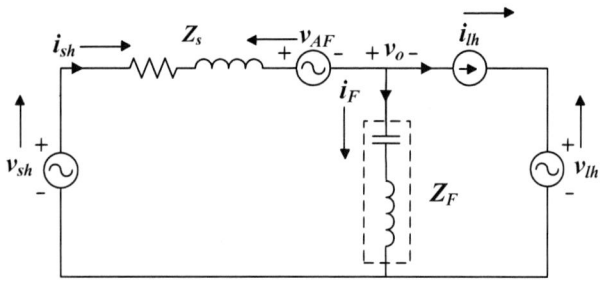

Figure 3. Single phase equivalent harmonic circuit of main system

The EAF is represented by harmonic current and harmonic voltage source. From the equivalent circuit shown in Fig. 3, it is obtained:

$$v_o = (i_{sh} - i_{lh}) \cdot Z_F \tag{22}$$

$$v_{AF} = K \cdot i_{sh} \tag{23}$$

Applying KVL:

$$v_{sh} - i_{sh} \cdot Z_s - v_{AF} - v_o - v_{lh} = 0$$

$$\therefore v_{sh} - i_{sh} \cdot Z_s - (K \cdot i_{sh}) - (i_{sh} - i_{lh}) \cdot Z_{Fo} - v_{lh} = 0$$

$$\therefore v_{sh} - i_{sh} \cdot Z_s - K \cdot i_{sh} - i_{sh} \cdot Z_F + i_{lh} \cdot Z_F - v_{lh} = 0$$

$$\therefore (v_{sh} - v_{lh}) - i_{sh}(Z_s + K + Z_F) + i_{lh} \cdot Z_F = 0$$

$$\therefore i_{sh} = \frac{(v_{sh} - v_{lh})}{(Z_s + K + Z_F)} + \frac{Z_F}{(Z_s + K + Z_F)} \cdot i_{lh} \tag{24}$$

With respect to the performance of the compensation, the system behaves like a closed-loop control system. Equation (24) can be represented by traditional closed-loop model of composite filter as shown in Fig. 4. Therefore, the analysis in the 's' domain could be developed with the help of block diagram shown in Fig. 5.

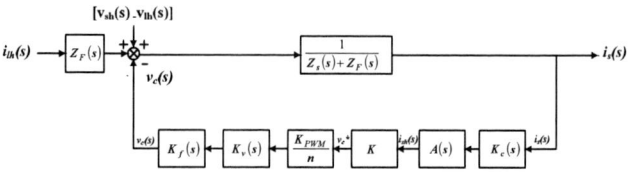

Figure 4. Block diagram of closed loop model of CF

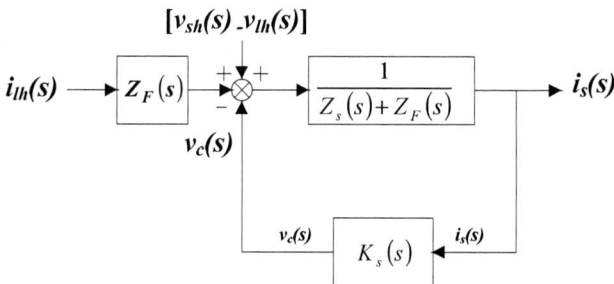

Figure 5. Reduced block diagram of closed loop model of CF

The open-loop transfer function $k(s)$ from input source current to output voltage can be obtained as:

$$k(s) = \frac{v_c(s)}{I_s(s)} = k_c(s) \cdot A(s) \cdot \frac{K_{PWM}}{n} \cdot k_v(s) \cdot k_f(s) \tag{25}$$

Here, $k_c(s)$=Transfer function of the sensor modulating circuit=$\dfrac{k_c}{T_c \cdot s + 1}$, where k_c and T_c are the gain and the time constant respectively. $A(s)$=Transfer function of the harmonics calculating circuit=$\dfrac{s^2 + \omega^2}{s^2 + m \cdot s + \omega^2}$, where m is the internal gain, K_{PWM}=Inverter gain, $k_v(s)$=Transfer function of the inverter=$\dfrac{k_v}{T_v \cdot s + 1}$, where k_v and T_v are the gain and the time constant determined by the speed of the processor/related software respectively (Generally, $T_v = \dfrac{1}{5}$ of carrier wave period (neglecting the sidebands of the switching frequency and the dead time), $k_f(s)$= Transfer function of the output carrier filter circuit=$\dfrac{R_f \cdot C_f \cdot s}{L_f \cdot C_f \cdot s^2 + R_f \cdot C_f \cdot s + 1}$, it can be alternatively

978-1-4799-6047-7/14 $31.00 © 2014 IEEE

expressed as $= \dfrac{1}{T^2 \cdot s^2 + 2 \cdot \xi \cdot T \cdot s + 1}$, where $T = \sqrt{L_f \cdot C_f}$ and ξ is the damping factor (that controls the shape of the transfer function in the vicinity of the $f_0 = \dfrac{1}{2 \cdot \pi \cdot T}$ respectively.

$$L_f = \frac{n \cdot k}{2 \cdot \pi \cdot f_v} \tag{26}$$

Where, $k = \left(\dfrac{3 \cdot V^2}{S} \right)$, S =apparent output of the load and V =phase voltage.

$$f_{cf} = \frac{1}{2 \cdot \pi \cdot \sqrt{L_f \cdot C_f}} \Rightarrow C_f = \frac{1}{(2 \cdot \pi)^2 \cdot L_f \cdot (f_{cf})^2} \tag{27}$$

f_{cf} =cross-over frequency of the filter $= \dfrac{f_v}{10}$, f_v =carrier frequency.

$$\xi = \frac{R_f}{2} \cdot \sqrt{\frac{C_f}{L_f}} \Rightarrow R_f = 2 \cdot \xi \cdot \sqrt{\frac{L_f}{C_f}} \tag{28}$$

Where, ξ =damping factor (chosen between 0.4 to 0.6).

$$\therefore 1.6 \sqrt{\frac{L_f}{C_f}} \leq R_f \leq 0.8 \sqrt{\frac{L_f}{C_f}} \tag{29}$$

IV. CALCULATIONS

A three-phase 550/50 Hz, 100 MVA power rating an electric arc furnace is considered as a non-linear load. The detailed specifications are as follows:

TABLE I.   EAF DTRIBUTION NETWORK PARAMETERS

Parameter	Value	Parameter	Value
V	550 V	S	1 MVA
f	50 Hz	C_{dc}	200 μF
S	10 MVA	V_{dc}	400 V
R_s	0.1 Ω	f_v	25 kHz
L_s	110 μs	n	25
R_c	100 μs	ξ	0.6
L_c	0.23 mH	k_c	0.2
τ_c	100 μs	m	100
k_v	40	K_{PWM}	200/16
τ_v	8 μs		

Neglecting Kelvin effect, eddy current effect and attenuation of filter's gain, the maximum gain will be calculated using (26) to (29) [10]:

$$k = \left(\frac{3 \cdot V^2}{S} \right) = 3 \cdot \left(\frac{\left(\frac{550}{\sqrt{3}} \right)^2}{10^7} \right) = 0.0303$$

$$k_c(s) = \frac{k_c}{\tau_c \cdot s + 1} = \frac{0.2}{100 \times 10^{-6} \cdot s + 1} = \frac{0.2}{1 \times 10^{-4} \cdot s + 1}$$

$$A(s) = \frac{s^2 + \omega^2}{s^2 + m \cdot s + \omega^2} = \frac{s^2 + (2 \cdot \pi \cdot 50)^2}{s^2 + 100 \cdot s + (2 \cdot \pi \cdot 50)^2}$$

$$\therefore A(s) = \frac{s^2 + 314^2}{s^2 + 100 \cdot s + 314^2}$$

$$k_v(s) = \frac{k_v}{T_v \cdot s + 1} = \frac{40}{8 \times 10^{-6} \cdot s + 1}$$

$$L_f = \frac{n \cdot k}{2 \cdot \pi \cdot f_v} = \frac{25 \times 0.03025}{2 \cdot \pi \cdot 25 \times 10^3} = 4.8144 \times 10^{-6} \ H$$

$$C_f = \frac{1}{(2 \cdot \pi)^2 \cdot L_f \cdot (f_{cf})^2} = \frac{1}{(2 \cdot \pi)^2 \cdot (4.816 \times 10^{-6}) \cdot \left(\frac{25 \times 10^3}{10} \right)^2}$$

$$\therefore C_f = 8.4181 \times 10^{-4} \ F$$

$$R_f = 2 \cdot \xi \cdot \sqrt{\frac{L_f}{C_f}} = 2 \cdot (0.6) \cdot \sqrt{\frac{4.816 \times 10^{-6}}{8.424 \times 10^{-4}}} = 0.0908 \ \Omega$$

$$k_f(s) = \frac{R_f \cdot C_f \cdot s}{L_f \cdot C_f \cdot s^2 + R_f \cdot C_f \cdot s + 1}$$

$$= \frac{(0.09073) \cdot (8.424 \times 10^{-4}) \cdot s}{(4.816 \times 10^{-6}) \cdot (8.424 \times 10^{-4}) \cdot s^2 + (0.09073) \cdot (8.424 \times 10^{-4}) \cdot s + 1}$$

$$= \frac{(7.639 \times 10^{-5}) \cdot s}{(4.053 \times 10^{-9}) \cdot s^2 + (7.639 \times 10^{-5}) \cdot s + 1}$$

Finally the feedback path transfer function given by (25):

$$k_s(s) = \frac{0.2}{1 \times 10^{-4} \cdot s + 1} \cdot \frac{s^2 + 314^2}{s^2 + 100 \cdot s + 314^2} \cdot \frac{200}{16} \cdot \frac{40}{8 \times 10^{-6} \cdot s + 1}$$

$$\cdot \frac{(7.639 \times 10^{-5}) \cdot s}{(4.053 \times 10^{-9}) \cdot s^2 + (7.639 \times 10^{-5}) \cdot s + 1} \tag{30}$$

As per [24], the time constant of leading compensator can be calculated as:

$$K_P = \frac{(4.053 \times 10^{-9})}{(7.639 \times 10^{-5})} = 0.530 \times 10^{-4}$$

Transfer function of phase-lead compensator can be given by:

$$K_P(s) = 0.530 \times 10^{-4} \cdot s + 1 \tag{31}$$

Now the transfer function of feedback path with phase-lead compensator can be given by:

$$K'_s(s) = (0.530 \times 10^{-4} \cdot s + 1) \cdot K_s(s) \tag{32}$$

V. SIMALUATION RESULTS

Simulation results of Bode plot for functions $K_s(s)$ and $K'_s(s)$ are obtained by (30) and (32) respectively. Obtained simulation results are shown in Fig. 6 and Fig. 7 respectively.

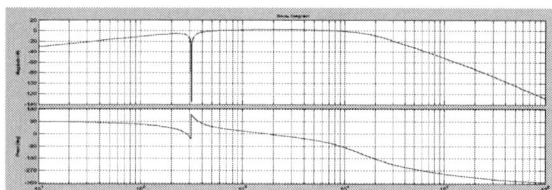

Figure 6. Magnitude and Phase Bode Plot for function k(s)

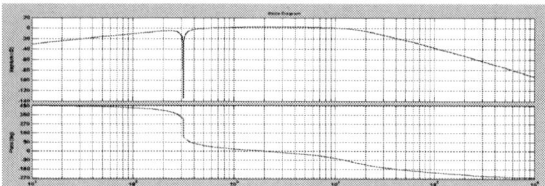

Figure 7. Magnitude and Phase Bode Plot for function $k^\square(s)$

Comparison of simulated results shown in Fig. 7 with Fig. 6 shows that the gain margin and the phase margin of the open loop transfer function can be improved a lot. Also, the cut-off frequency boosted when, the lead compensation is adopted. Quantified results of gain margin and phase margin are tabulated in Table II.

TABLE II.   BODE PLOT OUTPUT

Parameter	Gain margin	Phase margin
$K_S(s)$	0.506 dB	1.45 deg
$K'_S(s)$	50.1 dB	40.9 deg

It can be seen that the gain margin and the phase margin has increased considerably by lead compensation strategy.

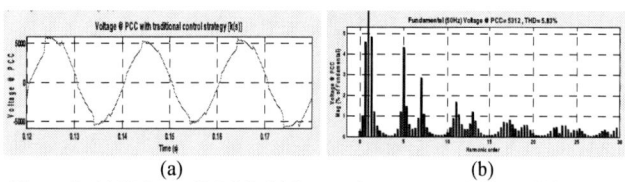

(a)   (b)

Figure 8. (a) Voltage @ PCC (b) harmonic spectrum (with tradition control strategy k(s))

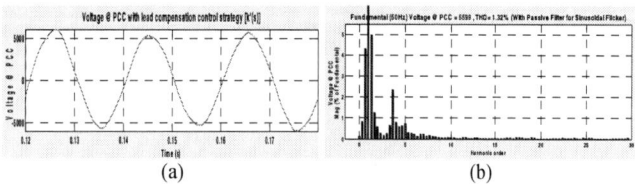

(a)   (b)

Figure 9. (a) Voltage @ PCC (b) harmonic spectrum (with lead compensation control strategy k'(s))

TABLE III.   THD IMPROVEMENT

Parameter	THD	% THD improvement
$K_S(s)$	5.83	77.35
$K'_S(s)$	1.32	

Comparison of Fig. 8 and Fig. 9 shows improved voltage wave at PCC as well as total harmonic distortion (THD) after lead compensation application.   Quantified results are tabulated in Table III. THD has improved by 77.35% after lead compensation application.

VI. CONCLUSIONS

In this paper, a state-space averaging model has been constructed to analyze the stability of CF used for PQ improvement of EAF distribution network. It is difficult to obtain a better filter performance confined to the system stability by the traditional control strategy, taking into account the time delay in the control circuit. The analysis has indicated that the time delay can cause a stability problem. Therefore, a phase-lead compensation strategy has been proposed to eliminate the effect of time delay. It suggests method of lead compensation for $K_s(s)$ to increase the value of the cut-off frequency. The increase of gain margin and phase margin can clearly clarify that the proposed control strategy improves the system stability compared with the traditional control strategy. Simulation verifies the validity of the developed theory.

VII. REFERENCES

[1] Douglas Andrews, Martin T. Bishop and John F. Witte, "Harmonic Measurements, Analysis, and Power Factor Correction in a Modern Steel Manufacturing Facility", IEEE TRANSACTIONS ON INDUSTRY APPLICATIONS, VOL. 32, NO. 3, MAY-JUNE 1996, pp. 617-624.

[2] Janusz Mindykowski, Tomasz Tarasiuk and Piotr Rupnik, "PROBLEMS OF PASSIVE FILTERS APPLICATION IN SYSTEM WITH VARYING FREQUENCY ", 9th International Conference, Electrical Power Quality and Utilization, 9-11 October 2007, Barcelona.

[3] Juan W. Dixon, Gustavo Venegas and Luis A. Mor´an, "A Series Active Power Filter Based on a Sinusoidal Current-Controlled Voltage-Source Inverter",   IEEE TRANSACTIONS ON INDUSTRIAL ELECTRONICS, VOL. 44, NO. 5, OCTOBER 1997 pp. 612-620.

[4] G. Carpinelli, Member and A. Russo, " Comparison of some Active Devices for the Compensation of DC Arc Furnaces", IEEE Bologna Power Tech Conference, June 23-26, Bologna, Italy.

[5] P. Salmerón and S. P. Litrán, "Improvement of the Electric Power Quality Using Series Active and Shunt Passive Filters", IEEE TRANSACTIONS ON POWER DELIVERY, VOL. 25, NO. 2, APRIL 2010, pp. 1058-1067.

[6] F. Z. Peng, H. Akagi and A. Nabae, "A New approach to Harmonic Compensation in Power System-a combined system of shunt passive and series active filters", IEEE Transactions on Industrial Applications, Vol. 26, Issue 1, pp. 983-990, Nov/Dec'90.

[7] F. Z. Peng, H. Akagi and A. Nabae, "Compensation characteristics of the combined system of Shunt Passive and Series active Filters ", IEEE Transactions on Industrial Applications, Vol. 29, Issue 6, pp. 144-152, Jan/Feb'93.

[8] L. Xu, E. Acha and V. G. Agelidis, "A New Synchronous Frame-base control strategy for a Series Voltage and Harmonic Compensator", Proceedings of Applied Power Electronics Conf., 2001 IEEE. Vol. 2. pp. 1274-1280, 2001.

[9] S. rianthumrong, H. Fujita and H. Akagi, "Stability Analysis of a Series Active Filter integrated with a double-series diode rectifier", IEEE 31st Annual Power Electronics Specialists Conf., Vol. 3, pp. 1305-1311, 2000.

[10] WeiMinWu, LiQuing Tong, MingYue Li, Z. M. Quian, Zheng Yu Lu and F. Z Peng, "A New Control Strategy for Series Type Active Power Filter", IEEE 35st Annual Power Electronics Specialists Conf., Aachen, Germany, pp. 3054-3059, 2004.

Optimal Capacitor Placement in Radial Distribution System

Atul Kumar, R. S. Bhatia
Department of Electrical Engineering
NIT Kurukshetra
Haryana, India
atulkrj132@gmail.com, rsibhatia@yahoo.co.in

Abstract—**This paper presents the optimal placement of capacitor banks in radial distribution system. For the purpose, capacitor placement optimized based upon the cost, size and location. In the past research work it has been noticed that the power factor correction improve the stability and efficiency of electrical power delivery. Also, in most of the research work the objective function is the reduction of power losses and the improvement in voltage profile. The reduction in power loss and the system voltage profile further depends upon the size and location of the capacitor placement. In this paper the capacitor placement is optimized based upon the methods for cost, size and location of capacitor bank to be installed in order to compensate for the reactive power demand by the load. For the purpose of capacitor placement in radial distribution system ETAP tool is to be used, which will optimize the capacitor placement problem for maximum cost saving and is computationally efficient. The proposed method is to be demonstrated on a standard IEEE-10 bus radial distribution system.**

Keywords-Optimal capacitor placement, power loss, voltage profile, optimization techniques.

I. INTRODUCTION

Substantial research has been carried out on the solution of optimal capacitor placement planning in the distribution systems for the purpose of power factor correction, voltage profile improvement and loss reduction. Especially, industrial plant with variable load conditions has large inductive loads and its power factor is very poor. These industries benefit most from capacitor banks. These capacitor banks provide improved power factor, improved voltage profile and minimize the electric utility bills. In most of the cases, the main reason for installation of capacitor banks by consumer is to avoid penalty in the electricity bill. The installations of the capacitor in distribution system need to pay a very sincere effort to optimize the size and location using different methods. Therefore, in past the researcher has given the different methods to optimize the capacitor placement problem by which the customer and the power utilities both are get benefitted simultaneously. Reactive power compensation devices are discrete in nature in the real scale system, so optimal capacitor placement is a nonlinear programming problem with integer (discrete) variables.

Srinivas Rao [1] presented an approach for OCP in RDS, which consists of two parts: in the first part loss sensitivity factors are calculated to determine the optimal candidate locations for capacitor placement and in the second part a new algorithm which employs Plant Growth Simulation Algorithm (PGSA) is used to determine the optimal size of capacitors to be placed at candidate buses. It handles the constraints and the objective function separately, and avoids the difficulty to determine the barrier factors. It doesn't require any external parameter. The proposed method has a guiding search direction which changes continuously as the objective function changes.

Murthy [2] presented a real and reactive combined power loss sensitivity (PLS) index-based approach to determine the optimal locations for capacitor placement in the radial distribution system (RDS). This approach provides better result as compared with the existing methods of index vector (IV) and power loss index (PLI). Load growth factor is essential for planning and expansion of existing system which is considered in this approach. PLS provides best results in terms of power loss and overall cost benefit.

Kannan [3] presented an enhanced approach for capacitor placement in radial distribution feeders to reduce the power loss and to improve the voltage profile. The optimal capacitor placement method involves two steps, one is to identify the optimal candidate bus location for capacitor placement and the other is to determine the optimal size of the capacitor to be installed at the optimal candidate location. The optimal location for capacitor placement is decided by a set of rules, after that the sizing of the capacitor banks is modelled as an optimization problem.

Tzong Su [4] proposed a relatively new metaheuristic approach for difficult combinational optimization problems. This approach is based on population that uses exploration of positive feedback and greedy search. By simultaneously operating the population of agents, the process stagnation problem can be effectively prevented. This approach gives an optimal solution like as the exhaustive search but with little computational burden.

Yan Xu [5] proposed a method for deployment of shunt capacitor banks in a RDS to reduce the power loss and to provide additional benefits. The net present value (NPV) criterion is used to determine the cost saving of the capacitor placement. Mixed-integer-programming (MIP) model is used to maximize the NPV of the optimal capacitor placement based on certain constraints. It uses commercial MIP package GUROBI to solve the optimization model. The proposed method has been implemented in Macau distribution system. This method is computationally efficient and provides a considerable positive NPV for optimal capacitor placement.

M.A.S. Masoum [6] proposed a fuzzy approach for optimal capacitor placement in RDS in the presence of harmonic.

978-1-4799-6047-7/14 $31.00 © 2014 IEEE

Fuzzy set theory optimizes the combination of objective function and constraints for capacitor placement problem in the presence of harmonics.

Diana P. Montoya [7] presents a heuristic approach for OCP and a deterministic technique for network reconfiguration. A Genetic Algorithm (GA) is used for OCP and a Minimum Spanning Tree Algorithm (MST) is used for reconfiguration to achieve the minimum losses and maximum cost saving. These two techniques provide complimentarily quite efficient results. It shows that after the reconfiguration, capacitor placement is not necessary under some conditions for greater savings. MST algorithm is deterministic and is easy to apply.

Sudha Rani [8] presents a recently developed Harmony Search Algorithm (HSA) for OCP. It achieves a perfect state of harmony using musical search. This approach employs self-adaptive harmony search algorithm and loss sensitivity factors for OCP in RDS. It provides better results as compared with Plant Growth Simulation Algorithm (PGSA) and Particle Swarm Optimization (PSO) algorithm.

In this paper optimal capacitor placement (OCP) in a 10 bus radial distribution system is done using ETAP. It place capacitors on optimal nodes and provides minimal losses and maximum cost saving. The result obtained from the proposed method is compared with the result obtained with fuzzy reasoning approach [9]. Proposed method gives more loss reduction and also gives more cost benefit as compare to the fuzzy reasoning approach.

II. PROBLEM FORMULATION

The problem is to determine the best capacitor size and location in a radial distribution system by minimizing the costs incurred by power loss and capacitor installation. The basic formulae [10] can be derived as follow;

A. Imposing the correction limits

To improve our power factor from a certain low value to desired value the correction limits is to be decided and the corresponding size of the capacitor can be calculated as follow:

$$VAr = P * (\tan \phi_1 - \tan \phi_2) \tag{1}$$

Here, P is the active power at the respective node, Φ_1 is initial power factor angle and Φ_2 is improved power factor angle at the same node.

B. Imposing the power flow limit

In power distribution system the power flow in the line depends upon the active and reactive power characteristics of the load. Therefore, to limit the resultant power flow in the line the following relation has been derived:

Amount of kVA reduced is given by

$$kVA_{nominal} = kVA_{original} \left(1 - \frac{PF_{new}}{PF_{org}}\right) \tag{2}$$

C. Imposing the gestation period

In particular load demand the where the load operates in defined period for very short span of time the capacitor placement becomes important in terms of the return or gestation period. Therefore, the relationship for installation cost of capacitor versus saving, defined as gestation period (GP), can be expressed as:

$$Gestation\ Period (GP) = \frac{Cost\ of\ Installation}{Cost\ of\ Savings} \tag{3}$$

Higher the gestation period more will pay the customer or utilities in terms of the interest on the installation cost.

D. Imposing the cost of installation and demand charge

This criterion basically depends on the cost of power as charged to consumers. Considering tariff to the consumer is Rs A per kVA + Rs. X per kWh and the load is P kWh at corresponding power factor of $\cos\Phi_1$. The power factor is to be improved to $\cos\Phi_2$ to effect reduction of maximum kVA demand. Reduction in maximum demand when the power factor is changed from $\cos\Phi_1$ to $\cos\Phi_2$

$$= \frac{P}{\cos\Phi 1} - \frac{P}{\cos\Phi 2}\ kVA \tag{4}$$

The size of the power factor improvement apparatus will be corresponding to the reduction in reactive power of the load is:

$$Q_{reduction} = P * (\tan \phi_1 - \tan \phi_2) \tag{5}$$

Further, if the apparatus for the purpose incurs an annual cost of Rs. B per kVA, the net savings by the apparatus can be obtained as,

$$A \left(\frac{P}{\cos\Phi 1} - \frac{P}{\cos\Phi 2}\right) - BP(\tan\phi 1 - \tan\phi 2) \tag{6}$$

The savings will be maximum when the differentiation of "(6)," with respect to Φ_2 is zero and it is given by:

$$-A * P \sec \phi_2 \tan \phi_2 + B * P \sec \phi_2 = 0 \tag{7}$$

Therefore, the limit of the improvement in power factor of the load is given can be obtained as from "(7),"

$$\sin \phi_2 = \frac{B}{A} \tag{8}$$

Thus from "(8)," the maximum economic limit for power factor improvement depends upon the costs of installation of the capacitor and the demand charge in the region for particular time of operation i.e. if the demand charge varies the optimal size of the capacitor can vary at the same location. Therefore, the improvement in power factor to unity is not the only choice to improve the system power factor always.

E. Imposing the constraint by developing Indices

In power distribution the method of capacitor placement by developing the indices depends upon the overall improvement in the system parameters. These parameters may be considered as voltage profile or resultant power loss of the system under consideration. Further, the implementation of these index based techniques do requires the imposition of constraint to find the size and location of the capacitor to be installed. The constraint could be voltage limit at particular node or in a group of node or the partial system configuration e.g. Considering the voltage profile,

Improvement in voltage profile of the nodes
OR
Maximizing the power loss reduction,

Subject to:

$$V_{i,min} < V_i < V_{i,max} \text{ and}$$

$$I_i < I_{limit}$$

Imposing the constraints considered above the optimal size or location of the capacitor placement may vary from one node to another node in the system considered.

III. METHODS OF CAPACITOR PLACEMENT

In power distribution system the optimal capacitor placement based upon the problem formulation can be obtained considering the size, location and cost of the installation to benefit the most, subject to the operating constraints. Therefore, in order to compensate the reactive power demand of load there are four methods of capacitor installation [11] as follows:

A. Capacitor at load end

Install a single capacitor at each motor and energize it whenever the motor is in operation. This method usually offers the greatest advantage for all, and the capacitors can be placed either before or after the starter of the loading arrangement. Normally placement after the starter is used for most of the motor applications and placement before is preferred only when motors are plugged, jogged or reversed; for multi-speed motors, as reduced voltage start motors.

B. Fixed capacitor banks

Install a fixed capacity of kVAR electrically connected at one or more locations in the plant's electrical distribution network, and energize them at all times. This method is often used when the system has few motors of variable size horsepower to which capacitors can economically be added. But it has certain disadvantages that when the system is lightly loaded, and the amount of kVAR supplied by the capacitor banks is too large, the voltage can be so high that the motors, lamps, and other controls equipment can burn out.

C. Combination of methods

Since no two electrical distribution systems are identical, each must be carefully analysed to arrive at the most cost effective solution, using one or more combination of the method

D. Automatic capacitor banks

It is installed at the motor control centre at the service entrance. This bank will closely maintain a preselected value of power factor. This is accomplished by timing a controller switch step of kVAR on, or off, as needed. Automatic switching ensures exact amount of power factor correction, eliminates over capacitance and resulting over voltages.

IV. SELECTION CRITERION FOR OPTIMIZATION TECHNIQUES AND RESULT DISCUSSIONS

The sample system under study is a 23kV, IEEE10-node radial distribution network as shown in Fig. 1. The data of the system are obtained from [9] and given in **Appendix-A.** The substation voltage is maintained at |1.0| p.u. The network is to be analyzing using different optimization techniques for size, location and cost of installation. The loading is considered fixed, and the following assumptions are made: a) At distribution level of voltage the effect of shunt capacitance is considered negligible. (b) The balanced radial distribution system can be represented by a single line diagram. (c) The substation feeder is capable of supplying the load demand for the network considered. (d) All the calculations are obtained considering that the losses in the line are negligibly small as compare to the load.

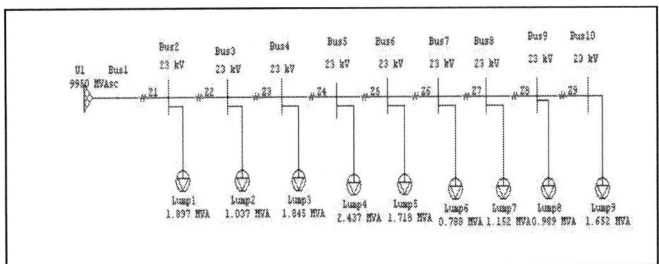

Fig. 1.   IEEE-10 bus radial distribution system

A. On the bases of location

In the 10 bus RDS as shown in the Fig. 1 we have to determine the optimal locations for capacitor bank installation. In the proposed method there is an option to select the candidate node for capacitor placement, based on which it determines the most optimal solution for capacitor placement in RDS.

TABLE I.   CAPACITOR BANKS PLACED AT CANDIDATE NODES

Candidate Buses	No. of Banks Placed	Total kVAR
Bus 2	5	335
Bus 4	2	134
Bus 5	28	1876
Bus 6	8	536
Bus 7	1	67
Bus 9	1	67
Bus 10	4	268
Total	49	3283

From the data in **Appendix-A** for the said system it is given that initial power factor is 0.92 at swing bus. Further, this is considered that the improvement in power factor above 0.95 is required at particular location. The corresponding reactive power demand is calculated at each candidate node and tabulated in Table-I. From the observation it is noticed that the reactive power demand at each node differ from each other.

On the observations of data at each node it is very clear that at node '5' there is maximum demand of reactive power. As these demands of reactive powers are to be fulfilled by using capacitor bank it can be noticed that there is maximum reduction of reactive power flow from the supply in case of node '5', so it is better to install a capacitor bank at node '5' for the maximum loss reduction in the line of supply side. Here we can see that overall 49 capacitor banks are placed at six different nodes which will provide maximum cost benefit and minimum losses.

TABLE II.   OPTIMAL CAPACITOR PLACEMENT RESULTS FOR 67 KVAR

Capacitor size (kVAR)	Total losses		No. of Capacitor Bank	Total kVAR
	kW	kVAR		
00	783.8	1036.7	00	00
67	702.7	913.9	49	3283

Now as far as location of capacitor bank is concerned, the best location should be as near as possible to the load for the maximum power factor improvement of the load. Sometimes, capacitor banks are installed to act as group correction i.e. it is connected with the main bus bar and the main bus has many feeders, since it not always economical to install the capacitor for individual feeders. For example, if the motors of different rating are installed, many capacitors of different ratings will be needed resulting in increase in cost. The another important point, which should be taken care of, is that the improvement takes place only from point of application towards the source of power and not in the opposite direction. Therefore, it is always advantageous to place the capacitor bank near the load for the load power factor improvement that reduces the losses in the circuit between the load and the supply feeder and thus improves the voltage profile.

B. On the bases of cost

The upper limit of power factor improvement depends upon the installation cost & gestation period. Gestation period is defined as the period after which the investor completely recovers his investment and obtains the first return. From section II and part D it is observed that the optimal capacitor placement depends upon the cost of installation and the demand charge, therefore, considering the candidate locations the size of capacitor can be calculated using "(8),".

One example is taken here considering the tariff of the consumer is Rs.5/- per kVA and the cost of installation is Rs.20/- at an annual rate of interest of 10%.

$$\sin \phi_1 = \frac{B}{A} = \frac{2}{5} = 0.40$$

$$\Phi_2 = 23.6^0,$$

$$\cos \Phi_2 = 0.9165$$

TABLE III.   ACCUMULATIVE PROFIT IN 5 YEAR PLANNING PERIOD

Capacitor size (kVAr)	Loss reduction (Rs)	1st year profit (Rs)	Accumulative profit (Rs)
67	1841546	806266.5	6996449

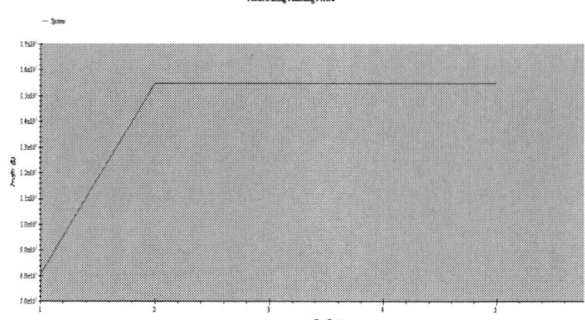

Fig. 2.   Profit during planning period

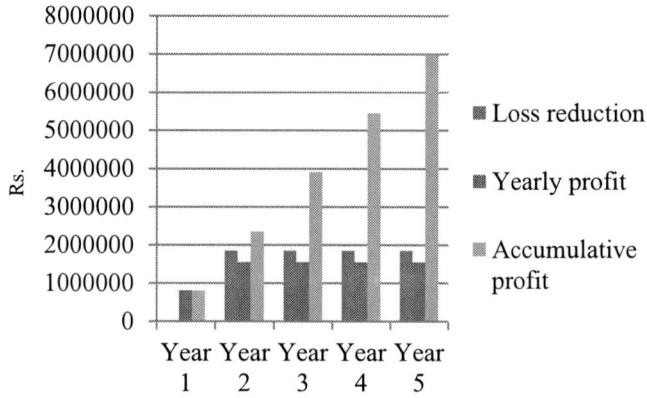

Fig. 3.   Bar diagram analysis for OCP cost report

Table-III shows that the capacitor placed at different location provides a significant loss reduction and 1st year cost benefit and subsequently the accumulative profit of Rs.6996449 over five year planning period. Though, the cost of installation at each node is same but the reactive power demand at candidate node causes the gestation period varied for different sizes of capacitor placements.

C. On the bases of Size

In distribution system the loads at each node may vary and the corresponding reactive power demands. In case the reactive power demand by the load at particular location is very high then capacitor placement based upon the size will becomes the ultimate choice.

TABLE IV.   : POWER FACTOR OF SWING BUS (BUS 1) ON OCP

Capacitor size(kVAR)	MW	MVAr	MVA	PF(Lagging)
00	13.152	5.223	14.151	0.9294
67	13.071	2.120	13.242	0.9871

Table-IV shows the power factor of swing bus on OCP in radial distribution system. Here it is shown that for a particular size of capacitor bank unit we will get optimal power factor with minimal losses and maximum cost saving. Here on placing capacitor banks of 67 kVAR size improves the power factor of swing bus to 0.9871 from 0.9294.

V. RESULT AND ANALYSIS

For the test feeder average energy cost is selected to be Rs.2.7/ (kW-hr), capacitor purchase cost Rs.213/ kVAR [11], installation cost Rs.6000 at each node and capacitor bank operation cost Rs.6000/ (Bank-year).

OCP results in the improvement of power factor of each and individual node, as a result of this the overall power factor of the swing bus get improved. The power factor before capacitor placement and after capacitor placement is shown in Table-V.

TABLE V. % POWER FACTOR OF EACH INDIVIDUAL NODE

Buses	Before capacitor placement pf	After capacitor placement pf
Bus1	0.929	0.987
Bus2	0.933	0.998
Bus3	0.930	0.945
Bus4	0.936	0.985
Bus5	0.929	0.998
Bus6	0.981	0.997
Bus7	0.992	0.998
Bus8	0.993	0.999
Bus9	0.991	0.997
Bus10	0.993	0.100

SUMMARY OF TOTAL GENERATION, LOADING & DEMAND

	MW	Mvar	MVA	% PF
Source (Swing Buses):	13.152	5.223	14.151	92.94 Lagging
Total Demand:	13.152	5.223	14.151	92.94 Lagging
Total Motor Load:	12.368	4.186	13.057	94.72 Lagging
Total Static Load:	0.000	0.000	0.000	
Apparent Losses:	0.784	1.037		

Fig. 4. Summary for total generation, loading and demand without capacitor placement

SUMMARY OF TOTAL GENERATION, LOADING & DEMAND (Max. Loadi

	MW	Mvar	MVA	% PF
Source (Swing Buses):	13.071	2.120	13.242	98.71 Lagging
Total Demand:	13.071	2.120	13.242	98.71 Lagging
Total Motor Load:	12.368	4.186	13.057	94.72 Lagging
Total Static Load:	0.000	-2.980	2.980	0.00 Lagging
Apparent Losses:	0.703	0.914		

Fig. 5. Summary for total generation, loading and demand with capacitor placement

As a result of optimal capacitor placement in RDS the voltage drop in the line impedance will reduce, voltage profile will improve (see Fig. 6) and the power factor will also get improve.

Comparison of power loss and percentage voltage drop is shown in the Table-VI. The minimum and maximum voltages before capacitor installation are 0.8375 p.u (bus 10) and 0.9929 p.u (bus 2), and after capacitor installation are 0.8623 p.u (bus 10) and 0.9953 p.u respectively.

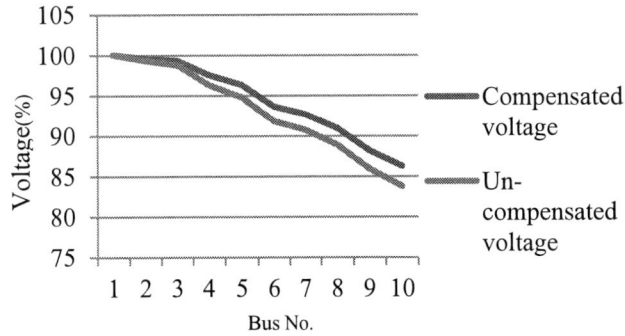

Fig. 6. Voltage profile of 10 bus system

TABLE VI. BRANCH LOSSES AND % VOLTAGE DROP

ID	Before capacitor placement		After capacitor placement	
	Loss (kW)	% Voltage drop	Loss (kW)	% Voltage drop
Z1	46.7	0.71	40.9	0.47
Z2	4.0	0.55	3.4	0.23
Z3	177.2	2.40	151.7	1.74
Z4	114.4	1.54	95.6	1.21
Z5	190.2	3.08	175.9	2.75
Z6	47.8	1.00	44.9	0.93
Z7	75.7	1.82	71.2	1.71
Z8	88.5	3.03	82.4	2.80
Z9	39.3	2.12	36.5	1.93
Total	783.8		702.7	

TABLE VII. COMPARISON OF RESULT OBTAINED

	Without compensation	Proposed	Fuzzy Reasoning [1]
Active power loss (kW)	783.77	702.7	704.883
% Active power loss reduction	-	10.34	10.065
Reactive power loss (kVAR)	1037	914	-
% Reactive power loss reduction	-	11.86	-
Total kVAR placed	-	3283	4950
Annual Profit (Rs.)	-	806266	269514
Accumulative Profit (Rs.)	-	6996449	5660971

978-1-4799-6047-7/14 $31.00 © 2014 IEEE

Thus, the overall OCP programme analysis of the system implies that, the optimal capacitor placement result is obtained for the capacitor size of 67 kVAR, which places 49 capacitor banks in 10 bus RDS and thus improves the power factor to 0.9871 from 0.9294. There is a real power loss reduction to 702.7kW from 783.8 kW and Reactive power loss reduction to 914kVAR from 1037kVAR (see Fig. 4 and Fig. 5). The result is compared with the existing fuzzy reasoning approach in the Table-VII.

It also reduces the loss by an amount Rs.1841546 per year, provides an annual profit of Rs.806266.5 and an accumulative profit of Rs.6996449 for a planning period of 5 years. Thus this method is observed to provide an enhanced cost saving over its planning period.

VI. Conclusions

From the test results it can be observed that the improvement in power factor up to unity is not always economical. Also, it can be noticed that the power factor correction is the size, location and cost of installation dependent. The optimum value of the capacitor depending upon the any of the criterion chosen does not guarantee that relative parameter in the system configuration under consideration will improve simultaneously.

Therefore, it can be concluded from the test results that the main advantages of power factor improvement is reduction in kVA demand, which results in reduced per unit power cost. In addition, the losses in the system get reduced and a saving in energy is realized. But as compared to saving which is obtained in KVA charges, it may be small and installation of capacitor bank just for saving due to reduction in losses is not justified. It should always be looked as additional benefit e.g. improvement in system loadability so that the maximum consumers can be served from the same feeder and this will reduces the addition cost of installation of the power structure.

VII. References

[1] R. Srinivasas, "Optimal capacitor placement in radial distribution system using plant growth simulation algorithm," Elsevier Journal of Electrical Power and Energy Systems, vol. 33, pp. 1133-1139, 2011.

[2] Murty, "Comparison of optimal capacitor placement methods in radial distribution system with load growth and ZIP load model," Research Article in Springer- Verlag Berlin Heidelberg, vol. 7(2), pp. 197-213, 2013.

[3] S.M. Kannan, "Optimal capacitor placement and sizing using combined fuzzy-HPSO method," International Journal of Engineering,Science and Technology, vol. 2, no. 6, pp. 75-84, 2010.

[4] Tzong Su, "Optimal capacitor placement in distribution systems employing ant colony search algorithm," Taylor and Francis Journal of Electrical Power Components and Systems, vol. 33, pp. 931-946, 2005.

[5] Yan Xu,, Zhao Yang Dong, Kit Po Wong, Evan Liu and Benjamin Yue, "Optimal capacitor placement to distribution transformers for power loss reduction in radial distribution systems," IEEE Transactions on Power Systems, vol. 28, no. 4, pp. 4072-4079 November 2013.

[6] M.A.S. Masoum, "Fuzzy approach for optimal placement and sizing of capacitor banks in the presence of harmonics", IEEE Transactions in Power Delivery, vol.19, no. 2, pp. 822-829, 2004.

[7] Diana P. Montoya,Juan M. Ramirez,"Reconfiguration and optimal capacitor placement for losses reduction", Transmission and Distribution: Latin America Conference and Exposition (T&D-LA), sixth IEEE/PES, pp1-6, 2012

[8] Sudha Rani, N. Subrahmanyamand M. Sydulu, "Self adaptive harmony search algorithm for optimal capacitor placement on radial distribution systems," International Conference onEnergy Efficient Technologies for Sustainability (ICEETS), pp.1330-1335, 2013.

[9] Tzong Su , "A new fuzzy-reasoning approach to optimum capacitor allocation for primary distribution systems," IEEE International Conference on Industrial Technology, pp. 237-241, 1996.

[10] Atul, "Economics of power factor correction in radial distribution systems," International Journal of Applied Engineering Research (IJAER), vol.8, no.7, ISSN:0973-4562, pp. 52-56, 2013.

[11] ABB Group, Technical papers on power factor correction, Issue 3, 2008.

Appendix-A

TABLE VIII.   Line Data of IEEE-10 Bus Radial Distribution System

Se	Re	R, in ohm	X, in ohm	PL in kW	QL in kVAR
1	2	0.1233	0.4127	1840	460
2	3	0.0140	0.6057	980	340
3	4	0.7463	1.2050	1790	446
4	5	0.6984	0.6084	1598	1840
5	6	1.9831	1.7276	1610	600
6	7	0.9053	0.7886	780	110
7	8	2.0552	1.1640	1150	60
8	9	4.7943	2.7160	980	130
9	10	5.3434	3.0264	1640	200

Power factor = 0.9294, Base kV= 23kV, Se-sending end node, Re-receiving end node, R – Line resistance, X - Line reactance, P_L- Active load power, Q_L – Reactive load power

An Improved Three Level Hysteresis Current Controller for Single Phase Shunt Active Power Filter

O V S R Varaprasad, *Student Member, IEEE,* D V S S Siva Sarma, *Senior Member, IEEE*

Electrical Engineering Department
National Institute of Technology, Warangal, INDIA
varaprasad.oruganti@gmail.com, sivasarma@ieee.org

Abstract—Hysteresis Current Control (HCC) technique is one of the popular current control techniques used for active power filtering application. However the conventional two level HCC have limitations of variable switching frequency and harmonic behavior that varies with interfacing inductance. To resolve these issues an improved three level HCC for single phase shunt active power filter (APF) has been proposed in this paper. The three level HCC is designed by considering two hysteresis band comparators. The hysteresis bands of the three level HCC are varied by using zero crossing time error of the actual source current error to attain least variation in switching frequency, In order to reduce the switching stress for all the switches of the APF. A single phase welding machine load is considered as a non linear load and its equivalent model is considered in the simulation. The efficacy of the proposed three level HCC based single phase shunt APF to mitigate the current harmonics and reactive power with reduced switching frequency up to 40% to that of conventional two level HCC are substantiated by MATLAB/Simulink© simulation studies.

Keywords—Harmonics, Reactive power, Single phase shunt APF, Single phase welding machine, Switching frequency, Two level HCC, Three level HCC

I. INTRODUCTION

The widespread use of nonlinear loads such as air conditioners, compact fluorescent lamps, computers, micro wave ovens, printers, televisions, uninterrupted power supplies, welding machines, Xerox machines, electronic fluorescent lamp ballasts, LED lighting and high rating traction systems induce extensive harmonics into the single phase supply system. This results in malfunctioning of electronic equipment, failure of power factor correction capacitors, overheating of motors, transformers and cables. Therefore, it is essential to install suitable mitigating devices to compensate the current harmonics injected by the nonlinear loads. Normally, passive filters have been widely employed to compensate the current harmonics and reactive power by increasing the power factor. However these passive filters are bulky and have the limitations of resonance and fixed frequency compensation, so this solution is not valid for all the applications. Active power filters (APFs) have been confirmed successful in overcoming the above limitations of passive filters [1].

APFs are feasible solution for compensating these current harmonics. From past two decades there has been increased awareness in the design and development of APFs because of the major concern over power quality, at both distribution and consumer levels. Major research is concentrated on appropriate control algorithm design for APF. Significant progress has been made in the area of APFs, especially with the introduction of advanced control techniques and topologies. The control algorithm decides the rating, steady state and dynamic performance of the APF [1]-[3].

In general, shunt APFs are current controlled voltage source inverters (VSI) which are proposed to inject a compensating current into the supply system to mitigate the harmonics and reactive power and improve power quality indices [2]. The APF is designed in such a way that it is capable of compensating the complex and randomly varying load current harmonics with high control accuracy. There are two key segments in APF. The first segment is current estimation circuit that estimates the compensating current to be injected at point of common coupling (PCC) and the essential active component of the current to be absorbed, which is necessary to retain the DC bus voltage. Different compensating current estimation techniques are discussed in [4]-[5]. In this paper, a simple technique based on the extraction of Unit Vector Template (UVT) from the input supply has been used to generate the reference current to the modulation technique [6].

The second key segment of an APF is a current controlled voltage source inverter (VSI). A range of current control methods are reported in related literature [4]-[5]. Among the existing current control techniques, HCC technique is considered by its simplicity of implementation with high precision and dynamic response. Utilization of conventional HCC is restricted due to large variations in switching frequency and its related effects. To make sure safety and efficiency of APF operation, the switching frequency of current controlled VSI and the DC link voltage should be minimum. In this paper, a three level HCC based single phase shunt APF has been developed which make sure slight variation in switching frequency by considering an additional offset hysteresis band (HB) over the existing two level HB. This method uses the inverter zero state switching condition, which reduces the switching stress on the inverter switches. But the conventional two level HCC does not consider the inverter zero state switching condition. The three level hysteresis modulation achieves a considerable reduction in the magnitude and deviation of the switching frequency, by maintaining the benefits identified with two level HCC.

978-1-4799-6047-7/14 $31.00 © 2014 IEEE

II. SINGLE PHASE SHUNT ACTIVE POWER FILTER CONFIGURATION

The fundamental structure of the single phase shunt APF connected to a single phase AC system feeding non linear loads are shown in Fig.1.

Where,
V_s = Supply voltage

I_s = Source current

L_s = Source impedance

I_L = Load current

The single phase shunt APF is built by an H-Bridge VSI, a DC link capacitor (C_{dc}) and an interfacing filter inductance (L_f). Its current controller must determine the compensation reference current, which is composed by the load current component and an instantaneous compensation current component corresponds to the reference current [6].

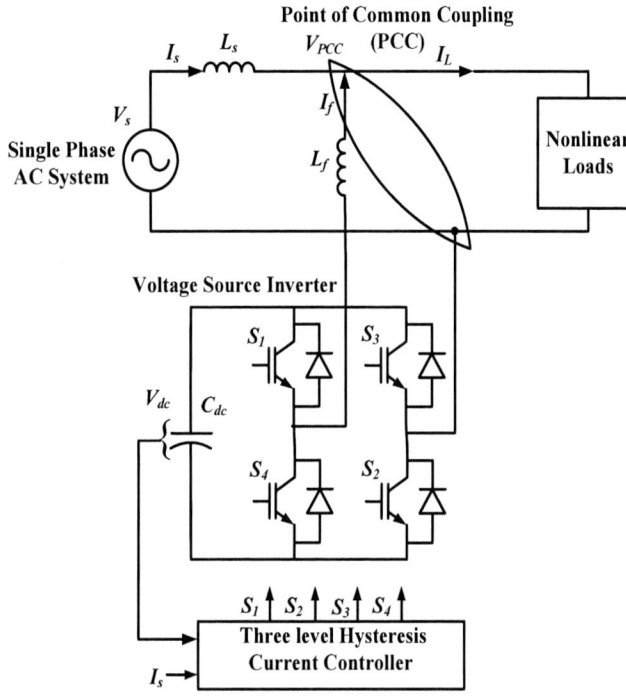

Voltage Source Inverter

Fig. 1.   Three level HCC based single phase shunt APF configuration

The APF behaves as a variable current source. It is used to compensate the current harmonics and to exchange the essential reactive power required by the nonlinear load. A single-phase welding machine load is considered as a nonlinear load, the current harmonics wave form of the welding machine load is measured using *FULKE 435 II* power quality analyzer, In the simulation studies the current harmonic waveform of the welding machine is obtained by considering a diode rectifier along with RL elements.

III. PROPOSED THREE LEVEL HYSTERESIS CURRENT CONTROL

HCC is a widely used current control technique in majority of the current controlled VSI applications, because of its simple configuration and fast dynamic response with peak current-limiting capability [4]-[5]. The conventional two-level HCC is also a better solution for most of the current control applications. However, the two level HCC technique has a limitation of wide deviation in switching frequency, which can be prohibitively high and it leads to higher switching losses of the inverter because of increased number of switching per cycle. A suitable selection of the HB width is necessary to limit the maximum switching frequency. A narrow HB reduces the current tracking error which leads to increased switching frequency whereas a wider HB increases the current tracking error with a reduced switching frequency.

A. Two level hysteresis current control

Traditionally, two level HCC technique is used for single phase system because of its simplicity. The two level HCC is operated by comparing a current error (i.e. the difference between the reference and the actual current aligned with fixed HBs.) Fig.2. illustrates the conventional two level hysteresis modulation. When the current tracking error crosses the upper HB, the VSI output is switched low, and when the current error crosses the lower HB, the inverter output switches high. Therefore each inverter phase leg output is the replica of the other. This approach does not employ the inverter zero state condition.

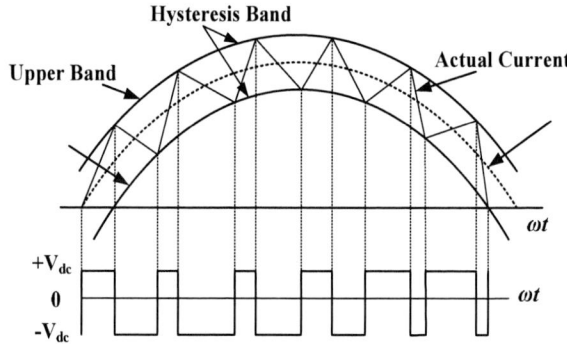

Fig. 2.   Conventional two level hysteresis modulation

B. Improved three level hysteresis current control

The proposed control technique for single phase shunt APF is shown in Fig.3.

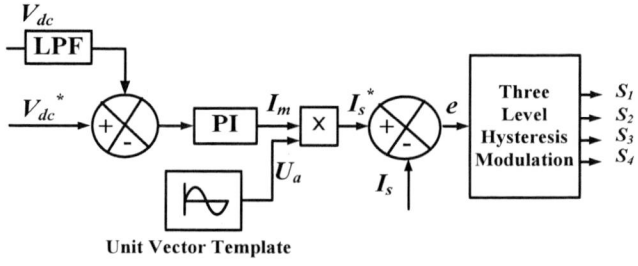

Fig. 3.   Proposed three level hysteresis current control

The DC bus voltage (V_{dc}) is fed through a low pass filter (LPF) in order to reduce the switching ripples in the measured voltage and compared with its reference voltage value $(V_{dc}{}^*)$. The error voltage is passed through the PI controller, which

978-1-4799-6047-7/14 $31.00 © 2014 IEEE

restrict the output to an allowable maximum current value. This PI controller output is considered as a source current peak value (I_m). In this control technique unit vector template of the supply voltage has been generated [7]. The source current peak magnitude is multiplied by U_a to generate sinusoidal reference source current which is in phase with source voltage. This reference source current (I_s) is compared with instantaneous source current (I_s^*) to extract the current tracking error (e). The current tracking error is fed to the HB comparator to generate the switching signals to the inverter. The instantaneous source current (I_s^*) is written as

$$I_s^* = I_m \times U_a \qquad (1)$$

The current tracking error is written as

$$e = I_s^* - I_s \qquad (2)$$

The three level hysteresis modulation is obtained by using two hysteresis band comparators, with small offset between upper and lower hysteresis band [8]-[9] as shown in Fig.4.

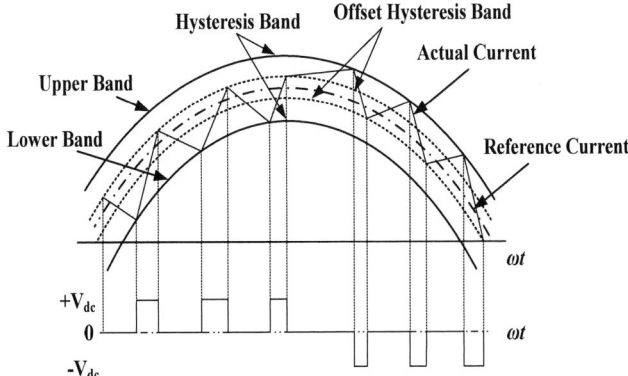

Fig. 4.   Improved three level hysteresis modulation

The rate of change of filter current (I_f) for positive, negative and zero states are written as follows:

$$\frac{dI_f}{dt} = \begin{cases} \dfrac{V_{PCC} + V_{dc}}{L_f} \\[2mm] \dfrac{V_{PCC} - V_{dc}}{L_f} \\[2mm] \dfrac{V_{PCC}}{L_f} \end{cases} \qquad (3)$$

When the source current tracking error crosses an outer HB, the inverter output is set to an active positive or negative state. Whenever the source current tracking error crosses an inner HB, the inverter output is set to zero state condition. The switching procedure is shown in Fig.4, where the current tracking error is bounded between the upper inner and lower outer HB for a positive inverter output, and between the lower inner and upper outer HB for a negative inverter output [7].

In [8] the basic three level HCC for a single phase VSI is presented. Here the maximum switching frequency has been reduced to $1/4^{th}$ that of conventional two level HCC for a motor load application. In this paper an enhanced three level HCC for active filtering application is developed.

The Inverter switching logic of the proposed three level HCC technique for active filtering application is given below: Here the values of actual source current (I_s) and reference current (I_s^*) decides the inverter switching. The proposed improved three level HCC switching logic is given below.

i.   if $(I_s < (I_s^* + (0.01 \times I_s^*)))$,
      $S_1 = 0$ and $S_4 = 0$,
      end.
      In this state the inverter will be on zero state.

ii.   if $(I_s < (I_s^* - (0.1 \times I_s^*)))$,
      $S_1 = 0$ and $S_4 = 1$,
      end.
      In this state the inverter lower switch is turned ON.

iii. if $(I_s > (I_s^* - (0.01 \times I_s^*)))$,
      $S_1 = 0$ and $S_4 = 0$,
      end.
      In this state again the inverter will lies on zero state.

iv.   if $(I_s > (I_s^* + (0.1 \times I_s^*)))$,
      $S_1 = 1$ and $S_4 = 0$,
      end.
      In this state the inverter upper switch is turned ON

The proposed three level HCC effectively uses the inverter zero state condition to reduce the switching stress on the inverter which in turn reduces the switching frequency.

IV. HARMONIC MEASUREMENT CASE STUDY

Harmonic measurements were conducted on a single phase welding machine using *FLUKE 435 II* power quality analyzer to determine the current harmonics, % THD (Total Harmonic Distortion). The measurement setup is shown in Fig.5. The welding machine is operated with a rated 230V AC supply, the output current range is 65A-200A.

Fig. 5.   Single phase welding machine harmonic measurement setup

The current harmonic injected by the welding machine on source side is shown in Fig.6. and the % THD is response is shown in Fig.7. These harmonics will cause disturbance to the other neighbouring loads connected to the same supply. Hence it is necessary to install a shunt APF to compensate the current harmonics

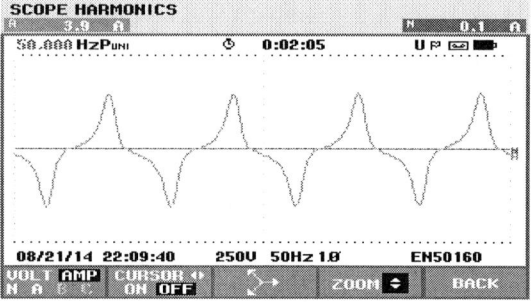

Fig. 6. Current harmonics of a single phase welding machine

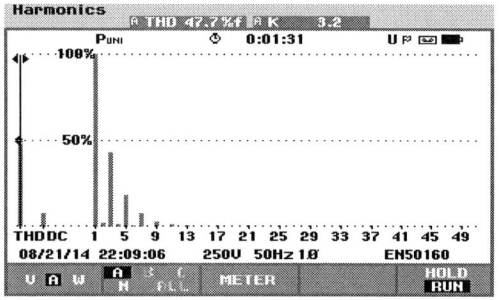

Fig. 7. % THD of a single phase welding machine

Fig.7. shows that the 3rd, 5th, 7th orders are the predominant harmonic components with reference to the fundamental. Welding machine is one of the crucial non linear load in commercial applications. A diode bridge rectifier with RL element is considered as an equivalent impedance model of the welding machine in simulation studies, in order to obtain a similar current harmonic wave form observed in Fig.7.

V. RESULTS AND DISCUSSION

The current harmonic mitigation using single phase shunt APF is employed in single phase power supply system with the supply voltage at 230 V and the equivalent harmonic producing load model of the welding machine and other linear loads are considered as the current harmonic compensation elements. The simulation parameters considered in the MATLAB simulation studies are given in Table I.

TABLE I.   SIMULATION PARAMETERS

Source Voltage (RMS)	230V, 50Hz
Source impedance	R= 0.1Ω, L= 0.5 mH
Single phase nonlinear load	Bridge rectifier with RL (R=80 Ω,L= 10mH) and Capcitor in parallel with RL
DC bus capacitance	4000µf
DC bus voltage	680 V
Interfacing filter	3 mH
PI controller gains	Kp=0.5, Ki=0.2

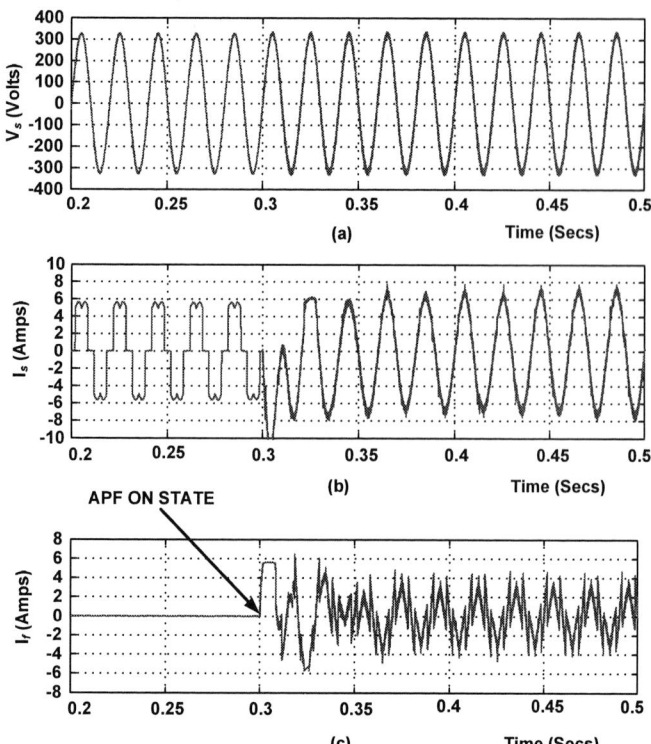

Fig. 8. Waveforms of (a) Supply Voltage, (b) Source Current, (c) Filter Current

The Supply voltage, Source current and the Filter current wave forms are shown in Fig.8, where the single phase shunt APF is idle up to time t=0.3 secs, after that the APF is switched ON to compensate current harmonics produced by the nonlinear loads. The DC bus voltage of the capacitor is 680V as shown in Fig.9.

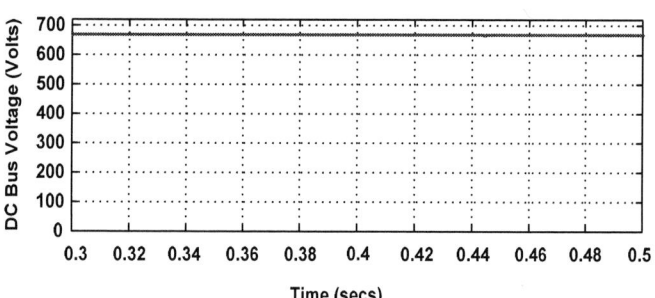

Fig. 9. DC bus voltage waveform

The proposed three level HCC attains a considerable reduction in switching frequency, where the conventional two level HCC operates at 18.6 kHz. The proposed three level HCC operates at 12.5 kHz. Here the switching frequency of the proposed three level HCC is reduced up to 40 % to that of conventional HCC technique and also the % THD is reduced from 48 % without APF case to 2.8% with proposed three level HCC based single phase shunt APF, which is within the limits as per the IEEE 519 standards.

978-1-4799-6047-7/14 $31.00 © 2014 IEEE

VI. CONCLUSION

An enhanced three-level HCC for a single-phase shunt APF was discussed in this paper. It employs a simplified switching logic that results in a reduced switching stress for all switches of the VSI by reducing the switching frequency up to 40% with least variation. Simulation results enumerate the efficacy of the proposed technique for harmonic and reactive power compensation, with reduced switching frequency and harmonic distortion.

REFERENCES

[1] B. Sing and K. Al-Haddad, "A review of active filters for power quality improvement," IEEE Trans. Ind. Electron., vol. 46, pp. 960–971, Oct. 1999.

[2] M.El-Habrouk, M. K. Darwish, and P. Mehta. "Active power filters: A review." IEE Proc.-Electric Power Applications vol 147, no. 5, pp. 403-413, Sept. 2000.

[3] Hirofumi Akagi. "Active harmonic filters." Proc of the IEEE, vol. 93, no. 12, pp.2128-2141, Dec. 2005.

[4] Simone Buso, Luigi Malesani, and Paolo Mattavelli. "Comparison of current control techniques for active filter applications." IEEE Trans. Ind. Electron., vol. 45, no. 5,722-729, Oct.1998

[5] T.C.Green and J. H. Marks. "Control techniques for active power filters." IEE Proc-Electric Power Applications., vol 152, no. 2 pp. 369-381, March 2005.

[6] M. Rukonuzzaman, and M. Nakaoka. "Single-phase shunt active power filter with harmonic detection." IEE Proc.-Electric Power Applications vol.149, no. 5, pp.343-350, Sept. 2002

[7] V.Khadkikar, P. Agarwal, A. Chandra, A. O. Barry, and T. D. Nguyen. "A simple new control technique for unified power quality conditioner (UPQC)." IEEE 11th Int. Conf on Harmonics and Quality of Power, pp. 289-293, Sept.2004.

[8] G.H.Bode, and D. G. Holmes. "Implementation of three level hysteresis current control for a single phase voltage source inverter." In IEEE 31st Annual Power Electronics Specialists Conference (PESC), vol. 1, pp. 33-38, June 2000.

[9] Donald Grahame Holmes, Reza Davoodnezhad, and Brendan P. McGrath. "An improved three-phase variable-band hysteresis current regulator." IEEE Trans.Power Electronics, vol.28, no.1, pp. 441-450, Jan.2013.

Effectiveness of UPQC in a Distribution Network with DTC Drive as Load

Parag Nijhawan[#1], R.S. Bhatia[#2], D.K. Jain[#3]

[#1] EIE Department, Thapar University, Patiala (India)
[#2] EE Department, National Institute of Technology, Kurukshetra (India)
[#3] EE Department, Deenbandhu Chhotu Ram   State University of Science & Technology, Muruthal (India)

[1] parag.nijhawan@rediffmail.com
[2] rsibhatia@yahoo.co.in
[3] jaindk66@gmail.com

Abstract- **Power quality is a hot topic of concern for both suppliers and consumers. The increased use of non-linear loads like adjustable speed drives also deteriorates the power quality levels. Therefore, greater emphasis is now being given to the study, evaluation and improvement of power quality especially in power distribution network. In this paper, the application of Unified Power Quality Conditioner (UPQC) with PI controller to improve the power quality in a distribution network with Direct Torque Control (DTC) induction motor drive is investigated. The Simulink model and results of the test system are presented to draw the conclusions.**

Keywords- **DTC drive, UPQC, harmonics, power quality.**

I. INTRODUCTION

Reliable electric power supply is necessary to meet the customer loads. It is the responsibility of distribution system to meet entire electrical power demand of the consumer with an acceptable level of reliability and voltage constraint. Moreover, with the deregulation of the electric power energy market [1-2], power quality [3] is of great concern to both electric utilities and consumers. The increase in the use of non-linear loads, and unbalancing in the loads in distribution systems, results in the unbalanced and distorted currents in the network. The load considered in this work is the direct torque control induction motor drive. This being a non-linear load is a source of harmonic currents and hence harmonic voltages in the distribution network. The harmonics induced into the distribution network can also hamper the performance of other equipment(s) connected to the system.

Hingorani [4] proposed the use of FACTS devices for the distribution system with an objective to ensure that customers get good quality power and reliability of supply.

In this work, the effectiveness of UPQC for the enhancement of power quality in a distribution network with DTC induction motor drive as load is investigated. UPQC is used to compensate harmonic distortion, unbalanced voltage and currents in the distribution network.

II. UNIFIED POWER QUALITY CONDITIONER

UPQC [5] as shown in Fig. 1, consists of two compensators – Shunt compensator and Series compensator, both cascaded by a common DC bus. It can inject both series

voltage and shunt current to the system. In simple words, we can say that UPQC is a combination of DSTATCOM and DVR. So, UPQC [6-8] is the most powerful compensation device for the distribution network as it compensates for the wide variety of power quality problems. UPQC can simultaneously compensate load and regulate voltage. The series compensator of UPQC works in active power delivering mode and absorption mode during voltage sag and swell condition, respectively. The shunt compensator of UPQC during these conditions aids series compensator by maintaining dc link voltage at constant level [6]. UPQC comprises of a series VSC, shunt VCS, midpoint-to-ground DC capacitor, low pass filter, high pass filter, series transformer and shunt transformer.

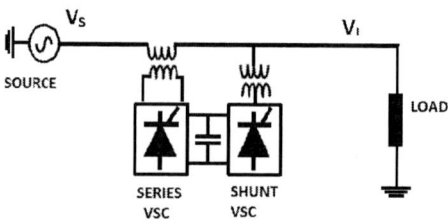

Fig. 1. Block Diagram of UPQC

The electrical equivalent circuit of UPQC is given in Fig. 2.

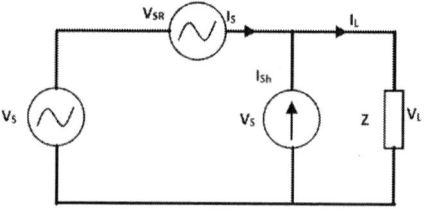

Fig. 2. Equivalent Circuit for UPQC

In this circuit,

V_S represents the supply voltage
V_{SR} represents the series injected voltage
V_L represents the load voltage

I_{Sh} represents the shunt injected current
.

In general, the source voltage in Fig. 2 can be written as

$$V_s + V_{sr} = V_L \qquad (1)$$

To obtain a balance sinusoidal load voltage with fixed amplitude V, the output voltages of the series compensator is given by

$$V_{sr} = (V - V_{lp}) \sin(\omega t + \theta_{1P}) - V_{Ln}(t) - \sum_{K=2}^{\infty} V_K(t) \qquad (2)$$

where, V_{1P} represents the positive sequence voltage amplitude of fundamental frequency

θ_{1P} represents the initial phase of positive seuquence voltage

V_{ln} represents the negative sequence component

To eliminate the harmonics present in the load current, the output current of shunt-APF i_{sh} is kept to be equal to the component of the load as given in the following equations:

$$i_L = I_{lp} \cos(\omega t + \theta_{1P}) \sin \varphi_{1P} + i_{Ln} + \sum_{K=2}^{\infty} i_{LK} \qquad (3)$$

$$\phi_{1P} = \varphi_{1P} - \theta_{1P} \qquad (4)$$

where,

φ_{1P}: initial phase of current for positive sequence

The source current is given by

$$i_S = i_L - i_{Sh}$$
$$= I_{lp} \sin(\omega t - \theta_{1P}) \cos \varphi_{1P} \qquad (5)$$

III. CONTROL PHILOSOPHY

The SIMULINK model representing the compensation using UPQC of a distribution network with DTC induction motor drive load is investigated in this work. Load voltage and current are sensed. These are passed through a sequence analyzer. The magnitudes of load voltage and current are compared with reference voltage and current, respectively. Pulse width modulated (PWM) control technique [7] is applied for inverter switching. PI controller is used with the IGBT inverter (shown in Fig. 3) to reduce the harmonic distortion due to the DTC induction motor drive.

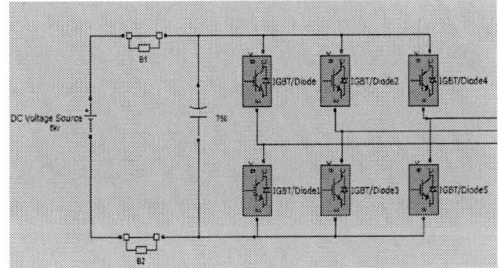

Fig. 3. Basic Inverter Structure

IV. PARAMETERS OF TEST SYSTEM

Simulation model of UPQC complete test system is shown in Fig. 4 whose parameters are as follows:

Source

3-phase, 13kV, 50Hz

Inverter parameters

IGBT based, 3 arms, 6-Pulse, Carrier Frequency=1080 Hz,

Sample Time= 5 μs

PI controller

K_p=0.5, K_i=1000 for series controller

K_p=0.5, K_i=1000 for shunt controller, Sample time=50 μs

RL load

Active power = 1kW, Inductive Reactive Power=400 VAR

DTC Induction Motor Drive

149.2kVA, Voltage V_{rms}= 11kV, Frequency = 50Hz

Transformer1

Y/Δ 13/115kV

Transformer2

Δ/Yg 115/11kV

V. SIMULATION RESULTS

An ideal three-phase sinusoidal supply voltage is applied to the DTC induction motor drive. This being a non-linear load injects current and voltage harmonics into the system. Fig. 5(a) represents the load current in three-phase distribution network without compensation. On the application of Open UPQC to the distribution network with DTC induction motor drive, it is seen that THD level in load voltage is 7.04% with magnitude 12.37 pu (shown in Fig. 8) and that in source current is 5.69% (shown in Fig. 6). Fig. 5(b) represents THD level for the uncompensated load current. Fig. 6(a) represents the source current waveform. Fig. 6(b) shows THD level for source current. Fig. 7 represents the source voltage waveform. Fig. 8 represents the load voltage for compensated system. The THD levels observed in source voltage and load current are 10.99% with magnitude 11.73 pu and 24.94%, respectively.

VI. CONCLUSIONS

The DTC induction motor drive being non-linear in nature injects current harmonics into the distribution system network. In this paper, the application of PI controller based UPQC to reduce the problems associated with the DTC induction motor drive connected in the distribution network is investigated. As already shown in the results presented in the previous section of this paper, the proposed UPQC test model effectively reduces the current harmonics injected by the DTC induction motor drive, and compensates for the voltage disturbances in the distribution network.

REFERENCES

[1] J. Arrillaga, M. H.J. Bollen and N.R. Watson, "Power quality following deregulation," in Proc. IEEE, Vol. 88, pp.246-261, Feb.2000.

[2] R. G. Koch, P. Balgobind and E. Tshwele, "New developments in the management of power quality performance in a regulated environment," in Proc. IEEEAFRICON'02, vol.2, pp.835-840, Oct.2002.

[3] C. Sankaran, "Power Quality", CRC Press, 2002.

[4] N. G. Hingorani, "Introducing custom power," Proc. IEEE Spectrum, vol.32, pp.41-48, June1995.

[5] V. Khadkikar , A. Chandra, A.O. Barry and Nguyen, T.D, "Conceptual Study of Unified Power Quality Conditioner (UPQC)," conference on IEEE International Symposium on Industrial Electronics, Vol. 2, pp:1088-1093, 2006.

[6] V. Khadkikar, A. Chandra, A.O. Barry, and T.D.Nguyen, "Power quality enhancement utilising single-phase unified power quality conditioner: digital signal processor-based experimental validation" conference on Power Electronics, IET ,Vol:4,pp.323–331,2011.

[7] M.A. Hannan and Azad Mohamed, "PSCAD/EMTDC simulation of unified series-shunt compensator for power quality improvement", IEEE transactions on power Delivery, Vol. 20, No. 2, pp. 1650-1656, April 2005.
[8] Sai Shankar , Ashwani Kumar and W. Gao, "Operation of Unified Power Quality Conditioner under Different Situations", IEEE General Meeting Power and Energy Society 2011, pp1-10, 2011.

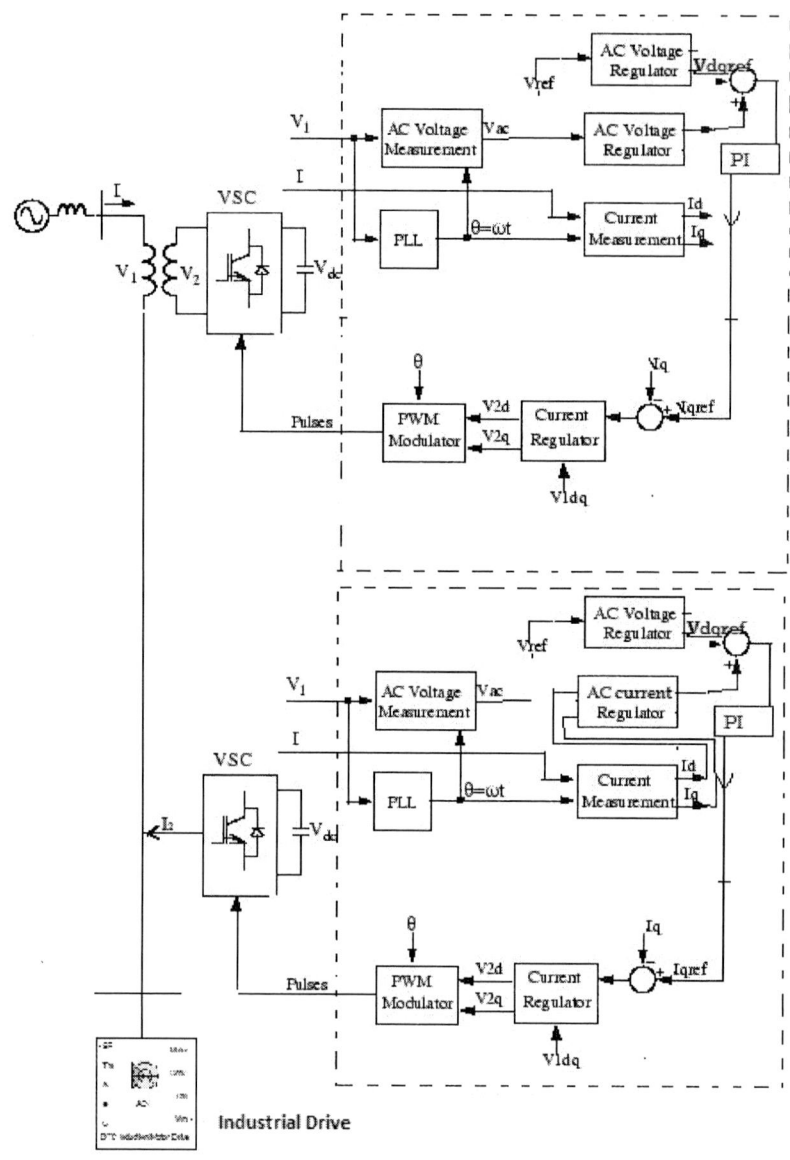

Fig. 4. Simulation Model of Test System

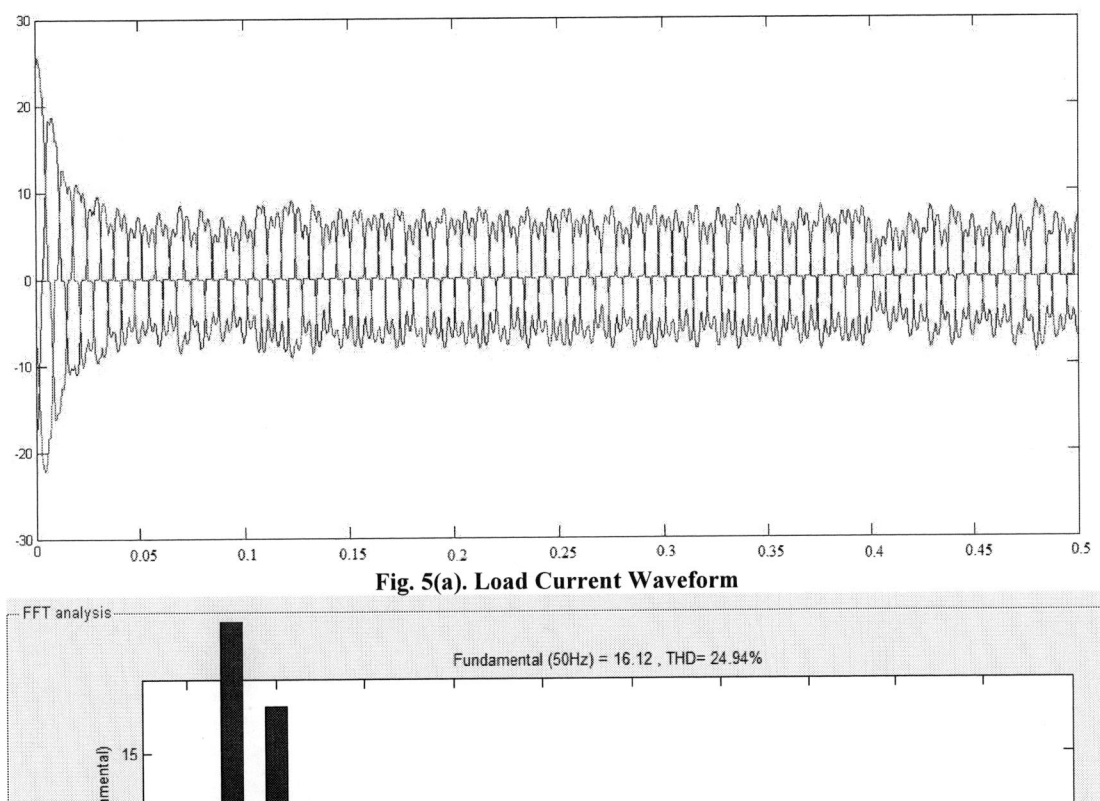

Fig. 5(a). Load Current Waveform

Fig. 5(b). Frequency Spectrum for Load Current

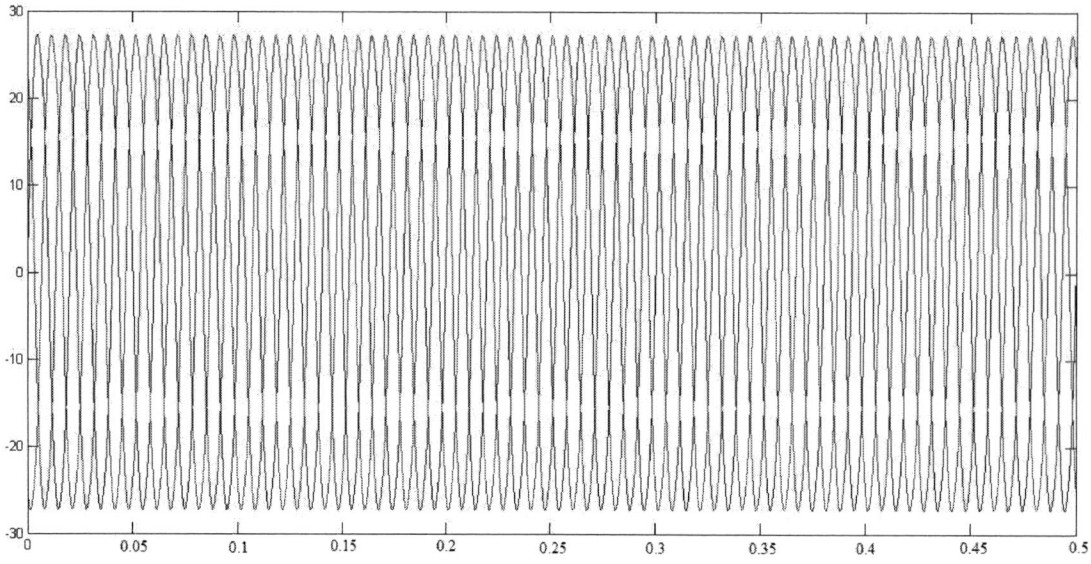

Fig. 6(a). Compensated Source Current Waveform

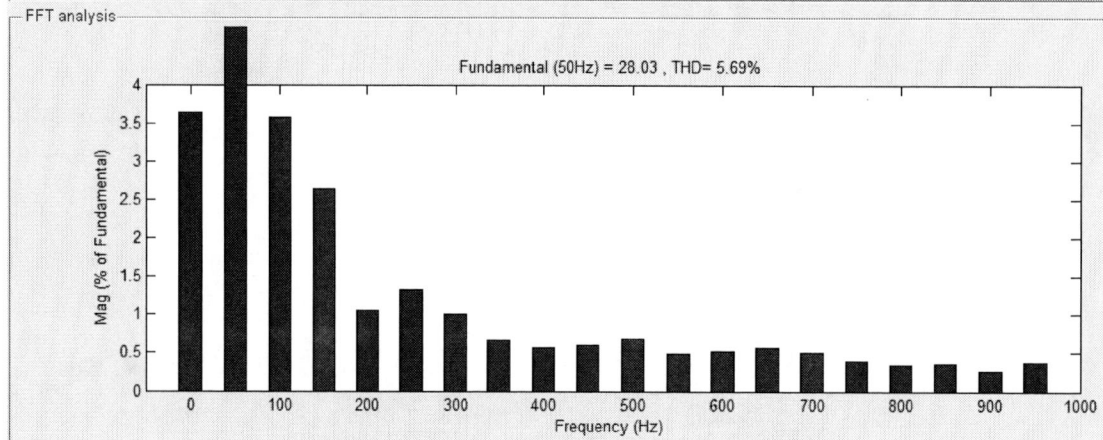

Fig. 6(b). Frequency Spectrum for source current

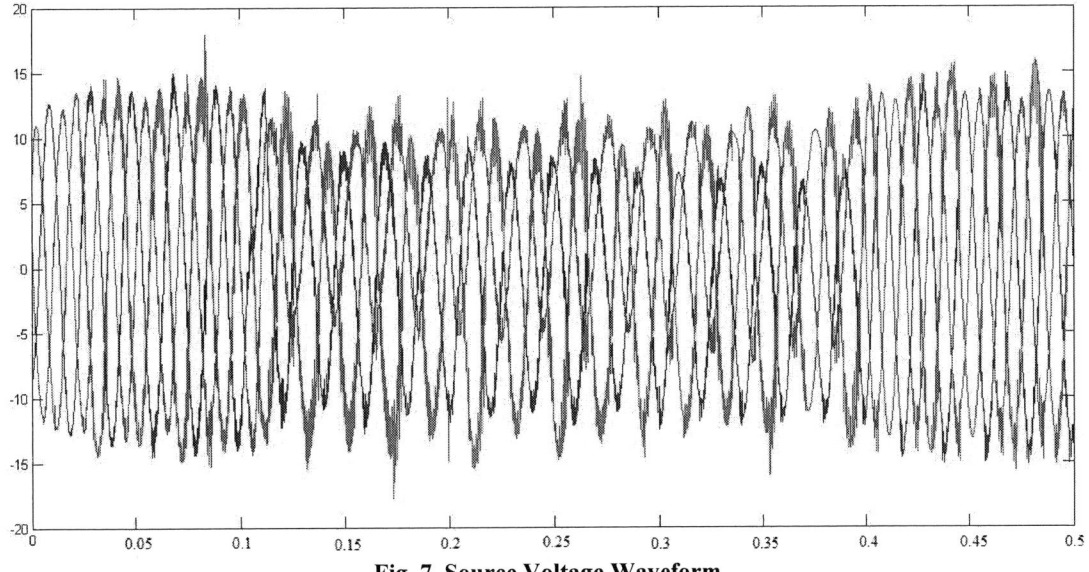

Fig. 7. Source Voltage Waveform

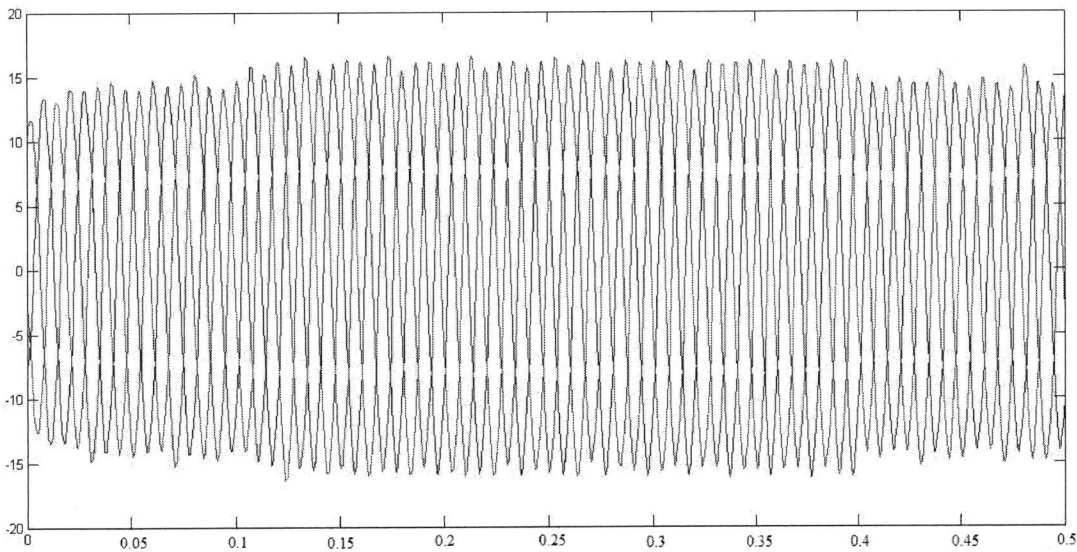

Fig. 8. Load Voltage Waveform

Improved performance of active power filter using type-2 fuzzy logic controller

Dhanavath suresh
Electrical Engineering Department
Indian Institute of Technology Roorkee
India
e-mail: mailtosuresh45@gmail.com

S.P. Singh
Electrical Engineering Department
Indian Institute of Technology Roorkee
India
e-mail: spseefee@iitr.ernet.in

Abstract— this paper presents a comparative evaluation of the proportional–integral PI controller (PIC) and type-2 fuzzy logic controllers for applications to active power filter (APF). The proposed algorithm is simple, based on measuring source current and regulating dc link voltage. It operates without sensing the volt-ampere requirement of the nonlinear load. The PI and type-2 fuzzy logic controllers (T2FLC) are used to control dc link voltage. The T2FLC is the non-model based approach, which does not require the mathematical model of the active power filter. The active power filter (APF) driving signals are produced with help of the hysteresis current controller. The performance of APF with conventional PIC and T2FLC is simulated in MATLAB/Simulink environment. The type-2 fuzzy logic controller provides less dc link voltage ripple than the conventional PI controller.

Keywords—PIC controller,T2FLC, APF

I. INTRODUCTION

Now a day the power electronics loads are most commonly used in electrical distribution system. These power electronics converters injected serious harmonics in the electrical distribution system. Due to the harmonics in supply system, the efficiency of the equipment connected to the distribution system is decreases [1, 2]. In order to meet the standards like IEEE 519-1992, IEC 1000-4-7, passive and active power filters have been used in combination with power electronics converter. With the remarkable development and advances in integrated digital electronics and power semiconductor, various topologies of active power filter is developed for compensation of voltage and current harmonics, neutral current, unbalance, voltage-flicker. The active power filter (APF) topologies are reported in both current controlled voltage source converter and current controlled current source converter. However, the voltage source converter gives higher efficiency, lower cost, lower size and simple structure as compared to current source converter
The dynamic performance of the APF depends on the controller used for controlling dc link voltage of active power filter. Many controllers are available in the literature such as the unified constant frequency, predictive, PI, PID, sliding-mode controllers, etc. [3, 4]. However, these controllers results in unsatisfactory performance under parameter variations nonlinearity load disturbance etc.

Recently, fuzzy logic controller (FLC) has attracted much attention in different applications, including the APF. The distinct advantages of fuzzy logic controller over PI controller is robustness against parameter variation, customization etc. [5-11]. In [5], dedicated micro-controller type-1 fuzzy logic based shunt active power filter has been proposed to compare the steady state and dynamic performance of the active power using PIC and FLC. The soft computing techniques in the application of power quality improvement are demonstrated in [6]. The various soft computing techniques such as fuzzy, neuro-fuzzy and genetics algorithm are used to improve the steady state and dynamic performance of the APF. But experimental validation is not presented. In [7], type-1fuzzy logic controlled (T1FLC) shunt active power filter is proposed using different membership function to improve steady state performance of APF. The T1FLC and PI controller are used to evaluate the steady state performance as presented in [8]. The artificial neural network (ANN) based type-1 fuzzy logic controller for inverter current control is implemented in [9], which demonstrates the efficiency of T1FLC for power quality improvement. In [10], the fuzzy scaling factors are optimized using genetic algorithm for harmonic reduction in fuel cell plants. The TIFLC adaptive hysteresis controller for shunt APF is implemented [11]. The literature survey reveals that the researchers have used only T1FLC for system performance improvement. However, the performance of APF using T2FLC has not been explored so far.

In this paper T2FLC based active power filter has been developed, to compensate reactive power and harmonics requirement of the nonlinear load. And also to improve the transient and steady state performances of the APF. The comparative performance of T2FLC and PIC to control DC link voltage is presented. A Simulation model of the APF is developed in MATLAB/Simulink environment to simulate the performance characteristics of APF.

This paper is organized as follows: Section 2 presents the system description and control algorithm for the APF. The design of PI and type-2 fuzzy controller is given in section 3 and 4 respectively. In section 5, the simulation results are discussed while the experimental results for PIC and T2FLC are presented in Section 6. The section 7 finally concludes this paper.

978-1-4799-6047-7/14 $31.00 © 2014 IEEE

II. SHUNT ACTIVE POWER FILTER CONFIGURATION AND CONTROL

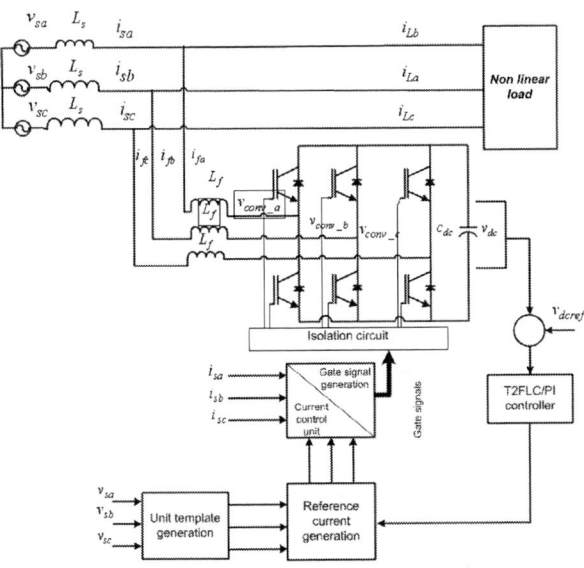

Fig. 1. Three phase three wire active power filter configuration.

Fig. 1 shows the VSI based PWM inverter structure. A current controlled PWM converter with DC link capacitor on its dc side and properly designed commutating inductor on AC its side is used active power filter. The PWM converter consists of six insulated gate bipolar transistor. The APF is controlled to draw/supply a compensating current from/to supply source, such that harmonics current on ac side can be cancelled out. In order to generate the compensating current, it is necessary to separate instantaneously the fundamental component of the line from the harmonics content. The separation of fundamental and harmonics currents of compensating current depends on the control technique used. The compensation principle of APF is shown in Fig.2. Fig.2 shows the different waveforms of source voltage, source current, load current and compensating current, respectively.

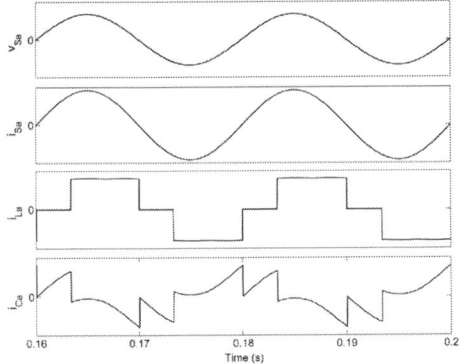

Fig.2.The performance characteristics of active power filter.

III. DESIGN OF PI CONTROLLER

The basic equation used for designing controller of the active power filter is given as follow

$$i_s = i_L - i_f \tag{1}$$

$$v_{conv} - v_s = L_f \frac{di_f}{dt} + R_f i_f \tag{2}$$

$$i_f = u.I_{dc} \tag{3}$$

Where u is the duty cycle of the PWM APF, which is control variable of the compensator for controlling active power filter current. The main aim of the controller is to settle V_{dc} at desired dc link voltage and to produce the desired filter current for essential compensation. For compensating the losses associated with converter the energy transfer to PCC is at fundamental frequency only. Hence, it is required to control two variables namely V_{dc} and i_f from the duty cycle of the converter. A mathematical model is developed to determination of PI controller parameters. In order to develop the mathematical model of the system following assumptions are made: (a) supply voltage is sinusoidal and balanced (b) the internal resistance of filter inductor taken into consideration and (c) dc link voltage ripples are neglected. The control circuit block diagram is shown in Fig.3.

The PI controller with gain value G_c is used in the voltage control loop to maintain capacitor voltage at desired dc link voltage. Let us consider T_c is transfer function of the filter. The relation between the input and output variable of the filter are computed by equating the rate of change of energy link.

The average rate of energy absorbed by the capacitor

$$p_{cap} = \frac{d}{dt}\left(\frac{1}{2}c_{dc}v_{dc}^2\right) = c_{dc}v_{dc}\frac{dv_{dc}}{dt} \tag{4}$$

Power input to filter

$$p_{conv} = 3.v_s i_f \tag{5}$$

Power losses in the internal resistance of the filter inductor is

$$P_{loss} = 3i_f^2 R \tag{6}$$

The average rate of energy associated with filter inductor is

$$P_{ind} = 3\frac{d}{dt}\left(\frac{1}{2}L_f i_f^2\right) \tag{7}$$

From (4)-(7) the average rate of change of energy is as follow

$$p_{conv} = p_{cap} + p_{ind} + p_{loss} \tag{8}$$

$$p_{cap} = p_{conv} - p_{ind} - p_{loss} \tag{9}$$

The equation (10) can be obtained substituting from (4)-(7) into (9)

$$c_{dc}v_{dc}\frac{dv_{dc}}{dt} = 3\left(v_s i_f - i_f^2 R - L_f i_f \frac{di_f}{dt}\right) \tag{10}$$

In order to linearize the power equation given in equation (10), consider a small change in filter current Δi_f and the dc link voltage also changed by Δv_{dc}. The active power filter current in the steady state is i_{fo} and the dc link voltage in steady state is v_{dco}.

$$i_f = i_{fo} + \Delta i_f \qquad (11)$$

$$v_{dc} = v_{dco} + \Delta v_{dc} \qquad (12)$$

Under steady state condition equation (10) can be written as follow

$$0 = 3(v_s i_{fo} - i_{fo}^2 R) \qquad (13)$$

The equation (14) can be obtained by solving equation (10), (11) and (12).

$$c_{dc} v_{dco} \frac{d\Delta v_{dc}}{dt} = 3\left(v_s i_{fo} + v_s \Delta i_f - i_{fo}^2 R - 2i_{fo}\Delta i_f R - L_f i_{fo} \frac{d\Delta i_f}{dt} \right)$$
$$- 3\left(v_s i_{fo} - i_{fo}^2 R \right)$$

$$(14)$$

The linear relation between Δi_f and Δv_{dc} can be obtained by subtracting equation (13) from (14)

$$c_{dc} v_{dco} \frac{d\Delta v_{dc}}{dt} = 3\left(v_s \Delta i_f - 2i_{fo}\Delta i_f R - L_f i_{fo} \frac{d\Delta i_f}{dt} \right)$$

$$(15)$$

Hence, the transfer function of the PWM converter can be derived for an operating point and can be obtained from Eq. (15) for output Δv_{dc} and input Δi_f :

$$c_{dc} v_{dco} \frac{d\Delta v_{dc}}{dt} = 3\Delta i_f \left(v_s - 2i_{fo}R - L_f i_{fo} \frac{d}{dt} \right) \qquad (16)$$

$$c_{dc}.v_{dco}.s = 3\Delta i_f \left(v_s - 2i_{fo}R - L_f i_{fo}.s \right) \qquad (17)$$

The transfer function can be derived as follow

$$T_c = \frac{\Delta v_{dc}}{\Delta i_f} = \frac{3(v_s - 2i_{fo}R - L_f i_{fo}s)}{c_{dc}.v_{dco}.s} \qquad (18)$$

The constant of the PI controller can be obtained using characteristics equation, which can be written as

$$1 + \left(k_p + \frac{k_i}{s} \right) \frac{3(v_s - 2i_{fo}R_f - L_f i_{fo}s)}{c_{dc}.v_{dco}.s} = 0 \qquad (19)$$

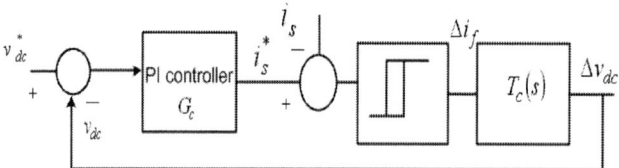

Fig. 3. Block diagram of voltage control loop.

Hence, the closed loop second order transfer function can be found from Fig.3. This characteristic equation used calculates the constants of the PI controller.

For the given system parameter the value of k_p and k_i can obtained by considering 5% of the overshoot. The parameter of the PI controller can be obtained using MATLAB SISO tool as $k_p = 0.57$ and $k_i = 17.5$. The parameter used for calculation is given below

$$v_s = 98V, R_f = 0.25\Omega, L_f = 2.5mH \ C_{dc} = 4700\mu F,$$
$$v_{dc} = 240V, i_{fo} = 5A$$

Fig.3 (b) shows the open loop frequency response of the active power filter with

IV. DESIGN OF TYPE-2 FUZZY LOGIC CONTROLLER

If there are M rules in the type-2 fuzzy system, each of which has the following form [12-18].

Rule i : IF e is $\tilde{G}_e^i$ and $e^\bullet$ is $\tilde{G}_{e\bullet}^i$ THEN $\delta i_{\max fc}$ is $\left[w_l^i, w_r^i \right]$ (20)

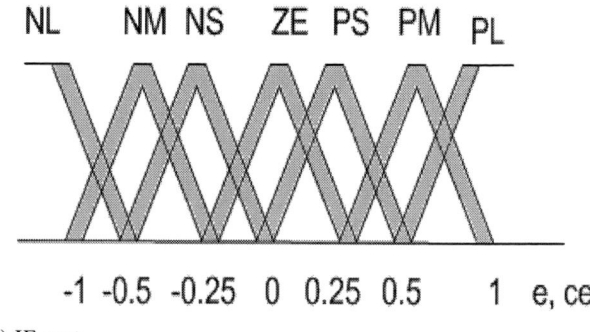

(a) IF part

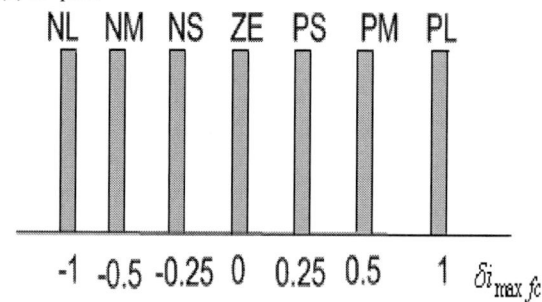

(b) THEN part
Fig. 3.Type-2 membership function for the T2FC system.

Where $i = 1, 2, \ldots, M$, $\tilde{G}_e^i$ and $\tilde{G}_{e\bullet}^i$ are the type-2 fuzzy sets of the IF-part, and w_r^i and w_l^i are singleton superior and inferior control action of THEN-part are shown in Figs. 3(a) and 3(b), respectively. The fuzzy linguistics variable are negative large (NL), negative medium (NM), negative small (NS), zero (ZE), positive small (PS), positive medium (PM), positive large (PL). The firing strength of the ith rule can be expressed as:

$$F^i = \left[\underline{f}^i \quad \overline{f}^i \right] \tag{21}$$

Where

$$\underline{f}^i = \underline{\mu}_{\tilde{G}_e}(e) \times \underline{\mu}_{\tilde{G}_{e^{\bullet}}}(e^{\bullet}) \tag{22}$$

$$\overline{f}^i = \overline{\mu}_{\tilde{G}_e}(e) \times \overline{\mu}_{\tilde{G}_{e^{\bullet}}}(e^{\bullet}) \tag{23}$$

In which $\underline{\mu}(.)$ and $\overline{\mu}(.)$ represent the grade of the inferior membership function and superior membership function, respectively. A singleton fuzzification with a minimum t-norm is used in this work and is as shown in Fig. 4. The output can be expressed as:

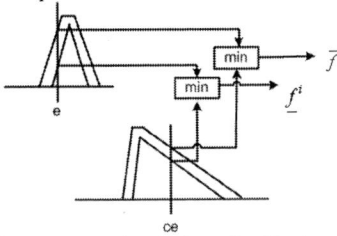

Fig. 4. Type-2 fuzzy system using singleton fuzzification and a minimum t-norm

$$\delta i_{\max \cos} = \left[\delta i_{\max l}, \delta i_{\max r} \right] \tag{24}$$

Where $\delta i_{\max \cos}$ is the type-1 set obtained by left and right end points ($\delta i_{\max l}$ and $\delta i_{\max r}$), which can be derived from the consequent centroid set $\left[w_l^i, w_r^i \right]$ and firing strength $f^i \in F^i = \left[\underline{f}^i \quad \overline{f}^i \right]$. The interval set $\left[w_l^i, w_r^i \right] (i=1, \ldots \ldots M)$ must be calculated or set the first before the calculation of $\delta i_{\max \cos}$. The left most point $\delta i_{\max l}$ and the right most point $\delta i_{\max r}$ can be expressed as

$$\delta i_{\max l} = \frac{\sum_{i=1}^{M} f_l^i w_l^i}{\sum_{i=1}^{M} f_l^i} \tag{25}$$

$$\delta i_{\max r} = \frac{\sum_{i=1}^{M} f_r^i w_r^i}{\sum_{i=1}^{M} f_r^i} \tag{26}$$

Here, we briefly discuss the procedure to compute $\delta i_{\max l}$ and $\delta i_{\max r}$. First of all, we calculate the right mostpoint $\delta i_{\max r}$. Assume that the w_r^i are arranged in ascending order i.e. $w_r^1 \leq w_r^2 \leq w_r^3 \leq \ldots \leq w_r^M$.

Step 1: compute $\delta i_{\max r}$ in (25) by initially using

$$f_r^i = \frac{\left(\underline{f}^i + \overline{f}^i \right)}{2} \text{ for } i=1,2,\ldots\ldots,M, \text{ where } \underline{f}^i$$

and $\overline{f}^i$ are pre-computed by (22) and (23); and let $\delta i'_{\max r} = \delta i_{\max r}$.

Step 2: Find $R(1 \leq R \leq M-1)$ such that $w_r^R \leq \delta d_r' \leq w_r^{R+1}$.

Step 3: compute $\delta i_{\max r}$ in (25) with $f_r^i = \underline{f}^i$ for $i \leq R$, and $f_r^i = \overline{f}^i$ for $i > R$, than set $\delta i''_{\max r} = \delta i_{\max r}$.

Step 4: If $\delta i''_{\max r} \neq \delta i'_{\max r}$, then go to step 5. $\delta i_{\max r} = \delta i''_{\max r}$, then set and go to step 6.

Step 5: Let $\delta i'_{\max r} = \delta i''_{\max r}$ and return to step 2.

Step 6: End

Therefore $\delta i_{\max r}$ can be re-expressed as:

$$\delta i_{\max r} = \delta i_{\max r}(\underline{f}^1, \ldots, \underline{f}^R, \overline{f}^R, w_r^1, \ldots, w_r^M)$$

$$= \frac{\sum_{i=1}^{R} \underline{f}^i w_r^i + \sum_{i=R+1}^{M} \overline{f}^R w_r^i}{\sum_{i=1}^{R} \underline{f}^i + \sum_{i=R+1}^{M} \overline{f}^i} \tag{27}$$

The procedure to calculate $\delta i_{\max l}$ is similar to that for $\delta i_{\max r}$ with slight changes as stated below. In step 2, we need to find $L(1 \leq L \leq M-1)$, such that $w_r^L \leq \delta d_l' \leq w_r^{L+1}$. In step 3, let $f_r^i = \underline{f}^i$ for $i \leq L$, and $f_r^i = \overline{f}^i$ for $i > L$,

Therefore, $\delta i_{\max l}$ in (24) can be expressed as:

$$\delta i_{\max l} = \delta i_{\max l}(\underline{f}^1,.....,\underline{f}^R,\overline{f}^{L+1},w_l^1,....,w_l^M)$$

$$= \frac{\sum_{i=1}^{L}\underline{f}^i w_l^i + \sum_{i=L+1}^{M}\overline{f}^i w_l^i}{\sum_{i=1}^{L}\underline{f}^i + \sum_{i=L+1}^{M}\overline{f}^i} \qquad (28)$$

Then, the defuzzified crisp output from the type-2 fuzzy system is the average of $\delta i_{\max l}$ and $\delta i_{\max r}$. i.e.:

$$\delta i_{\max fc} = \frac{\delta i_{\max l} + \delta i_{\max r}}{2} \qquad (29)$$

The fuzzy rules are discussed in the following section. The rules are derived using expert knowledge and trial and error method.

V. DC VOLTAGE CONTROL WITH TYPE-2 FUZZY LOGIC CONTROLLER

To control and maintaining dc link voltage constant, the various controllers is available in literature such as PI, adaptive control, hysteresis and type-1 fuzzy controller. In this investigation, T2FLC is implemented to control the dc link voltage of the capacitor. The input to the T2FLC is dc link voltage error and its change of error in order to improve the dynamic performance of active power filter. The compensation performance characteristics of the active power filter depends on the membership function. The determination of the membership functions depends on the designer experiences and expert knowledge. In this investigation trapezoidal membership function is chosen to have better compensation capability.

VI. SIMULATION RESULTS

The system parameters used for simulation and experimentation is given in table II. In order to study the performance of active power filter using PI and type-2 fuzzy logic controller, the active power filter power system is modeled using MATLAB/Simulink platform. The six pulse diode bridge rectifier along with R=12Ω L =16mH is used as nonlinear load. The comparative performance characteristics of PIC and T2FLC in controlling the dc-link voltage of active power filter is shown in Fig.5 and Fig.6. Fig.5 shows the source current, load current and compensation current using PI controller. The active power filter is switched on at t=0.05 sec. The moment the active power filter is switched on the source current becomes sinusoidal and the capacitor voltage reaches a to a reference value.

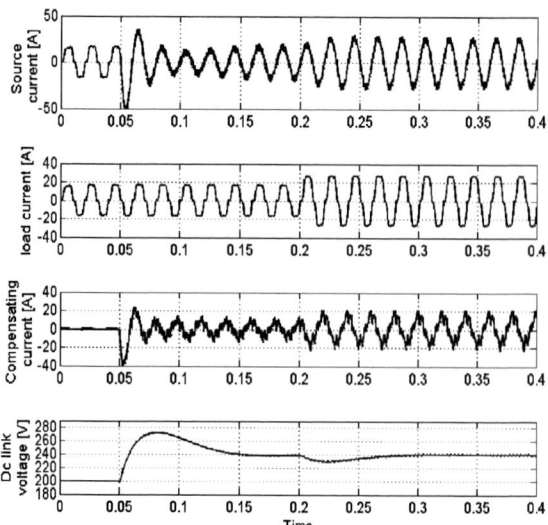

Fig. 5. Performance of the active power filter with PI controller.

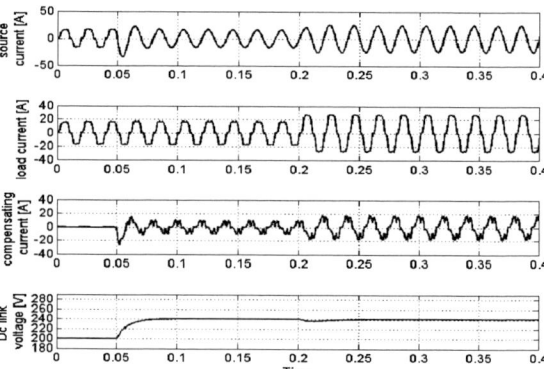

Fig.6. Performance of the active power filter with T2FLC controller.

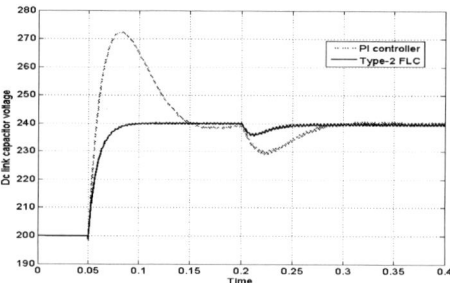

Fig. 7. Settling time comparison of PIC and T2FLC.

The value of the PIC such as k_p and k_i are taken as 0.57 and 17 respectively. The THD value of the source current after compensation by active power filter with PIC is 6.45%. The harmonics distortion in the source current is still present and the value of THD is not within the recommended standard. Further, to study the transient performance of the active power filter using PIC, the load current is increased from 20A to 30A at 0.2sec. The dc link voltage takes almost 5 cycles to settle down at reference value.

For further improvement of active power filter performance under similar operating condition T2FLC is used. The various waveform of T2FLC active power filter during load changes is shown in Fig.6. After compensation by active power filter the THD of the source current is reduced to 1.58% from 17.64%, whereas with PIC the source current THD 6.45%. Hence, T2FLC gives better THDs and waveform of the source current and also becomes distortion free compared to the PIC c. From Fig.7 and Fig.8 it can be observed that the percentage rise and fall in dc link voltage is less in case of T2FLC than the PIC. Similarly, settling time is 2 cycles and 5 cycles for T2FLC and PIC respectively. The simulated response shows that both rise/fall in dc link voltage and settling time are less with T2FLC as compared to PIC. And also the steady state performance of the T2FLC is better than that of filter PIC. Fig.9 shows the dc link voltage of APF using both PI controller and T2FLC. From Fig.7 it can be observed that settling time required for PI controller more than the T2FLC.

The total harmonic distortion (THD) of source current for both PIC and T2FLC are shown in Fig. 8(a) and Fig. 8(b) and they are obtained 6.45% and 1.58% respectively. The comparative THD analysis of PI and t T2FLC is shown in Table II.

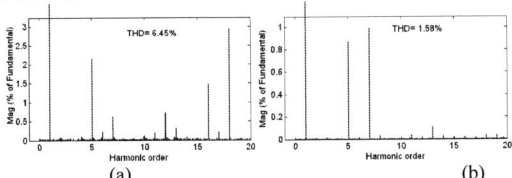

(a)   (b)

Fig.8. Source current THDs (a) using PI controller (b) T2FLC.

Table II THD comparison.

Type of controller	Source current THD before compensation	Source current THD after compensation	Settling time
PI controller	17.74%	6.45%	5 cycles
Type-2 fuzzy controller	17.74%	1.58%	2 cycle

VII. CONCLUSION

In the present paper, the complete mathematical model of active power filter has been developed to calculate parameter of the PI controller and to control dc link voltage. The performance of the active power filter with PI controller is analyzed for nonlinear load. The performance of APF is also analyzed with non-model based T2FLC. The simulation responses of active power filter with PIC and T2FLC is compared. From the comparison it can found that type-2 fuzzy logic controller has better transient response.

REFERENCES

[1]   J. Arrillaga, D. A. Bradley, and P. S. Rodger, "Power System Harmonics", John Wiley & Sons, 1985.

[2]   Bhim Singh, K. Al-Haddad, and Ambrish Chandra, "A Review of Active Filters for Power Quality Improvement", IEEE Transaction on Industrial Electronics, Vol. 46, No. 5, pp 1-12, 1999.

[3]   V. S. C. Raviraj and P. C. Sen, "Comparative study of proportional integral, sliding mode and fuzzy logic controllers for power converters," IEEE Trans. Ind. Appl., vol. 33, no. 2, pp. 518–524, Mar./Apr. 1997.

[4]   J. H. Marks and T. C. Green, "Predictive control of active filters," in Proc. Inst. Elect. Eng. Conf. Power Electronics and Variable Speed Drives, Sep. 2000, pp. 18–23.

[5]   S. K. Jain, P. Agarwal, and H. O. Gupta, "A Dedicated Microcontroller based Fuzzy Controlled Shunt Active Power Filter," Intell. Autom. Soft Comput., vol. 11, pp. 33–46, Jan. 2005.

[6]   P. Kumar and A. Mahajan, "Soft Computing Techniques for the Control of an Active Power Filter," IEEE Trans. Power Deliv., vol. 24, no. 1, pp. 452–461, Jan. 2009.

[7]   A. K. Panda and S. Mikkili, "FLC based shunt active filter (p–q and Id–Iq) control strategies for mitigation of harmonics with different fuzzy MFs using MATLAB and real-time digital simulator," Int. J. Electr. Power Energy Syst., vol. 47, pp. 313–336, May 2013.

[8]   S. Mikkili and A. K. Panda, "PI and Fuzzy Logic Controller Based 3-Phase 4-Wire Shunt Active Filters for the Mitigation of Current Harmonics with the I_d -I_q Control Strategy," Journal of Power Electronics, vol. 11, no. 6, pp. 914–921, 2011.

[9]   Altin N, Sefa Äb, "dSPACE based adaptive neuro-fuzzy controller of grid interactive inverter," Energy Convers. Manage. , vol. 56, pp.130–139, 2012.

[10]   Jurado F, Valverde M. "Genetic fuzzy control applied to the inverter of solid oxide fuel cell for power quality improvement," Electr. Power System Research, vol.76: pp.93–105, 2005.

[11]   Y. Suresh, A. K. Panda, and M. Suresh, "Real-time implementation of adaptive fuzzy hysteresis-band current control technique for shunt active power filter," IET Power Electron., vol. 5, pp. 1188-1195, 2012.

[12]   A. Celikyilmaz and I. B. Turksen, "Uncertainty Modeling of Improved Fuzzy Functions With Evolutionary Systems," IEEE Trans. on Systems, Man, and Cybernetics, Part B: Cybernetics, vol. 38, pp. 1098-1110, 2008.

[13]   S. Coupland and R. John, "Geometric Type-1 and Type-2 Fuzzy Logic Systems," IEEE Trans. on Fuzzy Systems, vol. 15, pp. 3-15, 2007.

[14]   J. M. Mendel, "Type-2 fuzzy sets and systems: an overview," IEEE Computational Intelligence Magazine, vol. 2, pp. 20-29, 2007.

[15]   W. Dongrui and J. M. Mendel, "On the Continuity of Type-1 and Interval Type-2 Fuzzy Logic Systems," IEEE Trans. on Fuzzy Systems, vol. 19, pp. 179-192, 2011.

Enhancing performance of APF by Fuzzy Controller

Khoisnam Steela and Bharat Singh Rajpurohit
SCEE, Indian Institute of Technology Mandi
Mandi, H.P., India
email id: khoisnam_steela@students.iitmandi.ac.in, bsr@iitmandi.ac.in

Abstract—**This paper proposes a fuzzy logic controller for dc-link voltage regulation to enhance the performance of two widely used active power filter control strategies, namely, *d-q* method and i_d-i_q method. The dynamic performance of the systems are evaluated under different supply conditions and compared with the performances of the two control strategies working with PI controller. The effectiveness of the fuzzy logic controller, in terms of total harmonic distortion (THD) in supply current and regulation of dc-link voltage is validated.**

Index Terms— *Shunt APF, d-q method, i_d-i_q method, dc-link voltage regulation, PI controller, Fuzzy logic controller*

I. INTRODUCTION

Since the past few decades, there has been intensive increase in the use of power converters and switching devices, various non linear loads and equipments which draw highly non-linear current from the source and thereby inject harmonics in the system. These harmonics are perilous and greatly pollute the power system. They degrade power quality by increasing total harmonic distortion (THD) and reactive power consumption and also by causing poor power factor, voltage flicker, bad voltage regulation, voltage sags and swells, unbalanced load etc.[1]-[3]. Harmonics also cause electromagnetic interference (EMI) in communication network present within certain proximity. Issues pertaining to power quality are increasing steeply, therefore, tackling these problem become an indispensable task for maintaining a good power system.

Power filters came up as a viable solution for providing clean-harmonic free power at consumer ends. Conventionally, passive L-C filters were employed for harmonics mitigation and capacitors for power factor improvement. But gradually they got replaced by active power filters (APFs) because they are associated with certain inherent disadvantages like bulky size, mistuning, instability, resonance with load impedance or utility impedance and lack of flexibility. APFs give dynamic and versatile solution to the problem of power quality. Research on APF technology boomed up and a large number of control techniques are being proposed and implemented. At present there is an appreciably large number of control methods used for power compensation by APFs.

This paper proposes a new fuzzy logic controller(FLC) for dc-link voltage regulation to enhance the performance of shunt APFs operating with *d-q* and i_d-i_q control methods.

II. SYSTEM CONFIGURATION

Fig.1. Shows the system configuration of a power system compensated by shunt connected APF. The APF is a pulse width modulated (PWM) inverter and the non-linear load is composed of a three phase diode rectifier with R-L load. The PWM inverter is an IGBT based voltage source inverter (VSI).

The values of system parameters are given in Appendix. The shunt-APF injects a compensating current which is an exact opposite of the distortion produced by the load to the line current, thus, nullifying the distortion. The amount of compensating power injected by the filter is controlled with suitable method to get perfect compensation.

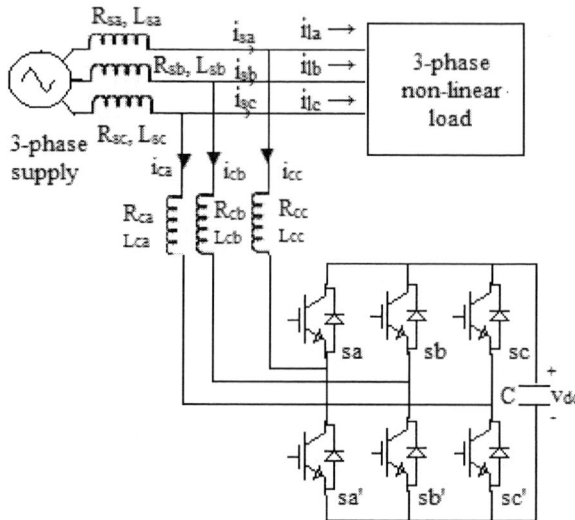

Fig. 1.   System configuration of 3-phase compensated system

III. CONTROL ALGORITHMS

Control of an APF includes reference signals generation, dc-link voltage regulation and gating the VSI. In this paper synchronous reference frame theory and instantaneous active and reactive current methods are used for reference current extraction and hysteresis band current controller (HBCC) is employed for gating the VSI. The dc-link voltage is first regulated using PI controller and then the FLC.

A. Synchronous Reference frame theory (d-q method)

The *d-q* method of control was first proposed in [4]. Fig. 2 shows the reference current extraction scheme for *d-q* method. In this method the load currents i_{la}, i_{lb} and i_{lc} are transformed to *α-β* co-ordinates using Clark's transformation given by Eq. 1. These currents in stationary *α-β* frame are then transformed to *d-q* frame, using Eq. 2. Here the reference frame *d-q* is determined by an angle "*θ*" w.r.t. *α-β* frame. The angle (*θ*) is determined through a phase lock loop (PLL).

$$\begin{bmatrix} i_\alpha \\ i_\beta \end{bmatrix} = \sqrt{\frac{2}{3}} \begin{bmatrix} 1 & -\frac{1}{2} & -\frac{1}{2} \\ 0 & \frac{\sqrt{3}}{2} & -\frac{\sqrt{3}}{2} \end{bmatrix} \begin{bmatrix} i_{la} \\ i_{lb} \\ i_{lc} \end{bmatrix} \qquad (1)$$

$$\begin{bmatrix} i_{ld} \\ i_{lq} \end{bmatrix} = \begin{bmatrix} \cos\theta & \sin\theta \\ -\sin\theta & \cos\theta \end{bmatrix} \begin{bmatrix} i_\alpha \\ i_\beta \end{bmatrix} \tag{2}$$

The currents i_{ld} and i_{lq} consists of an average or dc-component and an oscillating or ac-component as represented in Eq. 3. The dc-component corresponds to the fundamental component and the ac-component corresponds to harmonics present in the total current.

$$\begin{bmatrix} i_{ld} \\ i_{lq} \end{bmatrix} = \begin{bmatrix} i_{ld1h} + i_{ldnh} \\ i_{lq1h} + i_{lqnh} \end{bmatrix} \tag{3}$$

The currents i_{ld} and i_{lq} are decomposed into these two above mentioned components by passing through a low pass filter. Once the components are separated, compensating signals are selected based on type of compensation requirements. The d-axis and q-axis reference currents are represented by i_{cd}^{*} and i_{cq}^{*} respectively. To obtain reference current in abc co-ordinates i_{cd} and i_{cq}^{*} are first transformed back to α-β co-ordinates using Eq. 4 and then further transformed to abc using Clark's inverse transformation as given by Eq. 5.

$$\begin{bmatrix} i_{c\alpha}^{*} \\ i_{c\beta}^{*} \end{bmatrix} = \begin{bmatrix} \cos\theta & -\sin\theta \\ \sin\theta & \cos\theta \end{bmatrix} \begin{bmatrix} i_{cd}^{*} \\ i_{cq}^{*} \end{bmatrix} \tag{4}$$

$$\begin{bmatrix} i_{ca}^{*} \\ i_{cb}^{*} \\ i_{cc}^{*} \end{bmatrix} = \sqrt{\frac{2}{3}} \begin{bmatrix} -\frac{1}{2} & \frac{\sqrt{3}}{2} \\ \frac{1}{2} & \frac{\sqrt{3}}{2} \\ -\frac{1}{2} & \frac{\sqrt{3}}{2} \end{bmatrix} \begin{bmatrix} i_{c\alpha}^{*} \\ i_{c\beta}^{*} \end{bmatrix} \tag{5}$$

Fig. 2. Reference current generation with d-q method

B. Instantaneous Active and Reactive Current(i_d-i_q method)

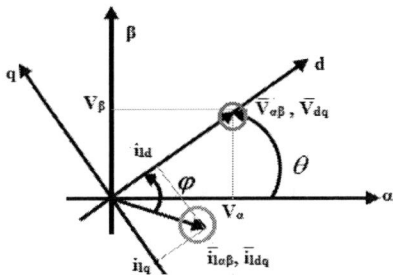

Fig. 3. Instantaneous voltage and current space vectors in α-β and d-q co-ordinates.

The i_d-i_q method [5] is an SRF based compensation scheme which follows same control steps as the d-q method. Here, the α-β to d-q and vice-versa transformation matrices are modified based on the geometry of voltage and current space vectors in α-β and d-q co-ordinates, shown in Fig.3. This figure shows that the angle 'θ' can be represented by $\theta = tan^{-1}\left(\frac{v_\beta}{v_\alpha}\right)$. Under balanced-sinusoidal mains voltage conditions, the transformation angle 'θ' is a function of time and it increases uniformly [6]. But, it is sensitive to voltage harmonics and unbalanced voltage sources, therefore, $\frac{d\theta}{dt}$ does not remain constant over a period of supply voltage [5]. If d-axis is in the direction of the voltage space vector then the direct and quadrature components of voltage can be written as Eq. 6 and $sin\theta$ and $cos\theta$ can be expressed as Eq. 7.

$$v_d = |\bar{v}_{dq}| = |\bar{v}_{\alpha\beta}| = \sqrt{v_\alpha^2 + v_\beta^2} \quad and \quad v_q = 0 \tag{6}$$

$$\cos\theta = \frac{v_\alpha}{\sqrt{v_\alpha^2 + v_\beta^2}} \quad and \quad \sin\theta = \frac{v_\beta}{\sqrt{v_\alpha^2 + v_\beta^2}} \tag{7}$$

Using Eq. 6 and 7, now Eq. 2 and Eq. 4 can be expressed as Eq. 8 and 9 respectively

$$\begin{bmatrix} i_{ld} \\ i_{lq} \end{bmatrix} = \frac{1}{\sqrt{v_\alpha^2 + v_\beta^2}} \begin{bmatrix} v_\alpha & v_\beta \\ -v_\beta & v_\alpha \end{bmatrix} \begin{bmatrix} i_\alpha \\ i_\beta \end{bmatrix} \tag{8}$$

$$\begin{bmatrix} i_{c\alpha}^{*} \\ i_{c\beta}^{*} \end{bmatrix} = \frac{1}{\sqrt{v_\alpha^2 + v_\beta^2}} \begin{bmatrix} v_\alpha & -v_\beta \\ v_\beta & v_\alpha \end{bmatrix} \begin{bmatrix} i_{cd}^{*} \\ i_{cq}^{*} \end{bmatrix} \tag{9}$$

The advantage of this method is that the PLL used for obtaining transformation angle (θ) can be omitted since "θ" is obtained directly from supply voltage, thereby reducing chances of phase lag error also.

IV. SWITCHING CONTROL SCHEME

In this paper, switching patterns of VSI is generated by HBCC. In this method the actual current is maintained within a certain preset tolerance band, around the reference current, referred to as the hysteresis band (HB). When the actual current crosses or hits the upper limit of the band, the upper switch of a particular phase leg (S_a or S_b or S_c) of the VSI is turned OFF and lower switch (S_a' or S_b' or S_c') is turned ON i.e if $i_a < (i_a^{*} - HB/2)$, Sa=0 and Sa'=1 (in phase-a). So current in that phase decays until it reach the lower limit of the HB. As it strikes lower limit of HB lower switch is turned OFF and Upper Switch is turned ON i.e. if $i_a > (i_a^{*} + HB/2)$, Sa=1and Sa'=0 (in phase-a). The upper device and the lower device in one phase leg of VSI are switched in complementary manner.

V. DC- LINK VOLTAGE REGULATION SCHEMES

The dc-link voltage regulation refers to regulating the voltage across the dc-link capacitor of the VSI. The dc-link capacitor plays two important roles in APF control, first it maintains a dc voltage with small ripple in steady state and second it serves as energy balancing medium that supplies or absorbs real power at times of transient period. Variation in load current is reflected upon dc-link voltage in terms of a rise or fall. For balanced loading this voltage should remain constant. The rise or fall in value of dc-link voltage, which occurs due to transient or load change, can be translated in terms of requirement for increasing or decreasing the peak amplitude of supply current. The regulator that regulates this voltage to a fixed value, in doing so, also gives the command for calculation of peak value of reference current. For this purpose, the actual capacitor voltage is compared with reference value of link voltage. The error signal is then processed through a controller. Here, two types of controller, PI controller and fuzzy logic controller are implemented.

978-1-4799-6047-7/14 $31.00 © 2014 IEEE

A. Design of PI controller

The transfer function of a standard PI controller is given by Eq. 10. Therefore, output of PI controller at the nth sampling point, $o(n)$, can be represented as Eq. 11.

$$H(s) = k_p + \frac{k_i}{s} \qquad (10)$$

where, k_p and k_i are the proportional and integral constant, respectively, of the controller

$$o(n) = o(n-1) + k_p\{e(n) - e(n-1)\} + k_i e(n) \qquad (11)$$

The values of PI parameters influence its performance under transient conditions in terms of time response, dc-voltage variations and THD of supply currents, so it should be chosen carefully.

B. Design of fuzzy logic controller

The functioning of an FLC is based on mapping an input space to an output space through a list of **IF-THEN** rules. All rules are evaluated in parallel without any preference order. An FLC converts the linguistic control strategy based on expert knowledge into an automatic control strategy. A simple fuzzy logic controller consists of (*Rule base + Data base = Knowledge base*), *fuzzifier* and *defuzzifier*

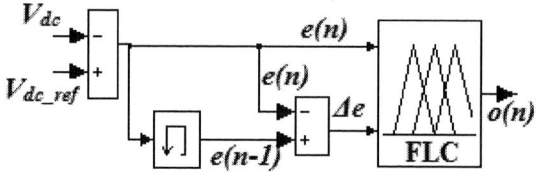

Fig. 4.   Block diagram of FLC based dc-link voltage regualtion scheme

Regulation scheme employing fuzzy controller is shown in Fig 4. Here, the FLC is fed with two inputs; error (*e*) and change in error (*Δe*).The linguistic variables in this work are defined by **seven** fuzzy sets: {**NB**(Negative big), **NM**(negative medium), **NS**(negative small), **Z**(zero), **PS**(positive small), **PM**(positive medium) and **PB**(positive big)} for inputs and output. This seven sets are characterized by **triangular** membership function (MF) as it has the advantages of simplicity, easy implementation, and suitable for this application [7]. The elements of the rule base table, shown in Table I, are determined based on the theory that in the

transient state, large errors need coarse control and in the steady state, small errors need fine control [8].

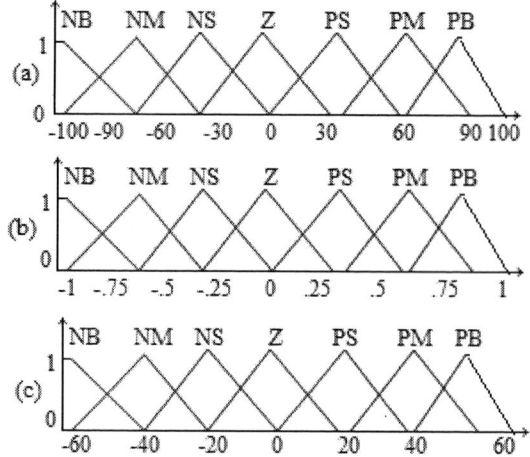

Fig. 5.   MF of variables (a) input '*e* ', (b) input 'Δe' and (c) output

TABLE I.   RULE BASE

$\dfrac{e}{\Delta e}$	NB	NM	NS	Z	PS	PM	PB
NB	NB	NB	NB	NB	NM	NS	Z
NM	NB	NB	NB	NM	NS	Z	PS
NS	NB	NB	NM	NS	Z	PS	PM
Z	NB	NM	NS	Z	PS	PM	PB
PS	NM	NS	Z	PS	PM	PB	PB
PM	NS	Z	PS	PM	PB	PB	PB
PB	Z	PS	PM	PB	PB	PB	PB

VI.   SIMULATION RESULTS AND PERFORMANCE EVALUATION

The APF system is configured in MATLAB/SIMULINK environment. The system is analyzed under three cases of supply, *Case 1: Balanced-sinusoidal supply, Case 2: Unbalanced-sinusoidal supply and Case 3: Non-sinusoidal supply*. The effectiveness of the algorithms and the controllers are analyzed and compared under these operating conditions. The compensator is switched ON after 1 cycle of supply to exhibit the quick change in shape of waveform after start of compensation.   The simulation results under these cases are shown in Fig. 6-14.

Case 1: Balanced sinusoidal supply

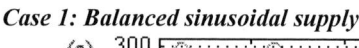

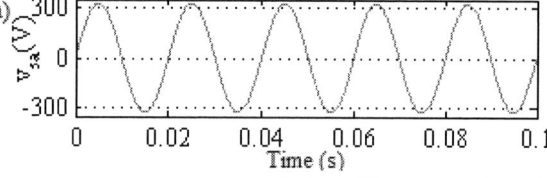
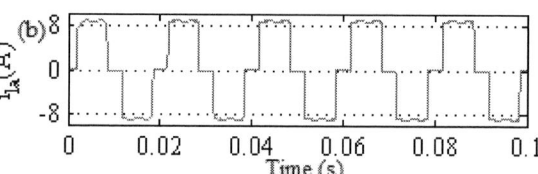

Fig. 6.   Waveforms of phase-a (a) supply voltage and (b) load current

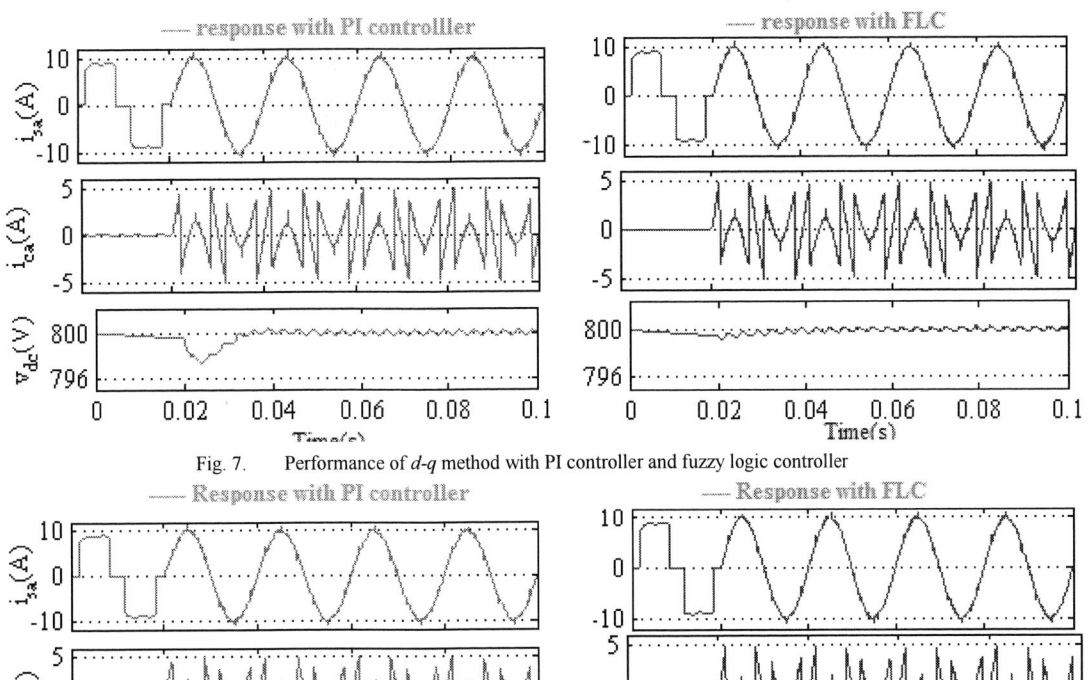

Fig. 7. Performance of *d-q* method with PI controller and fuzzy logic controller

Fig. 8. Performance of i_d-i_q method with PI and FL controller

Case 2: Unbalanced sinusoidal supply

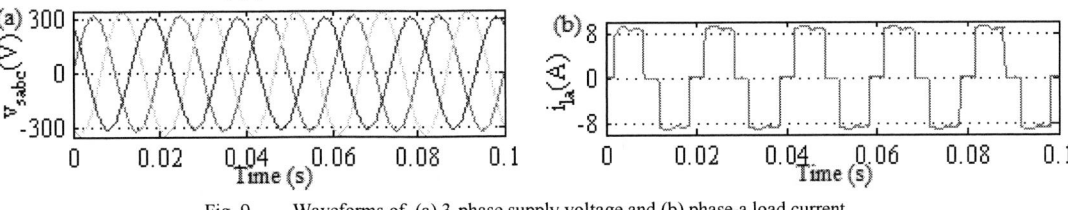

Fig. 9. Waveforms of (a) 3-phase supply voltage and (b) phase-a load current

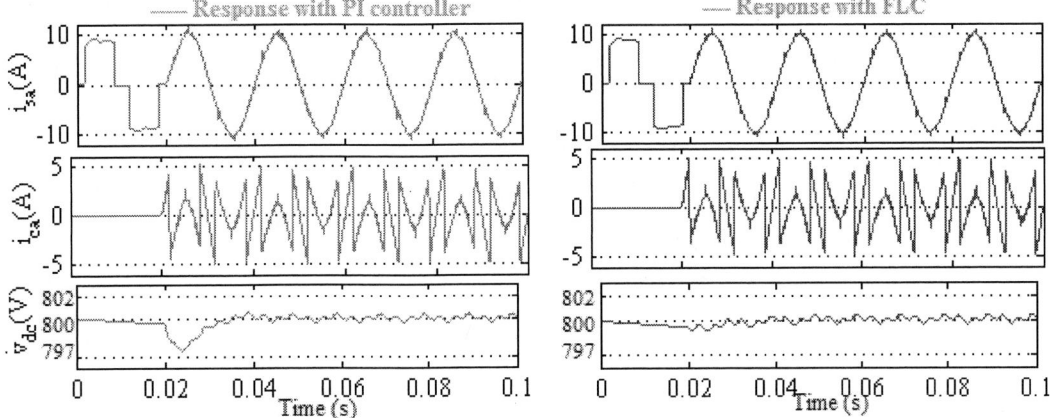

Fig. 10. Performance of *d-q* method with PI and FL controller

978-1-4799-6047-7/14 $31.00 © 2014 IEEE   199

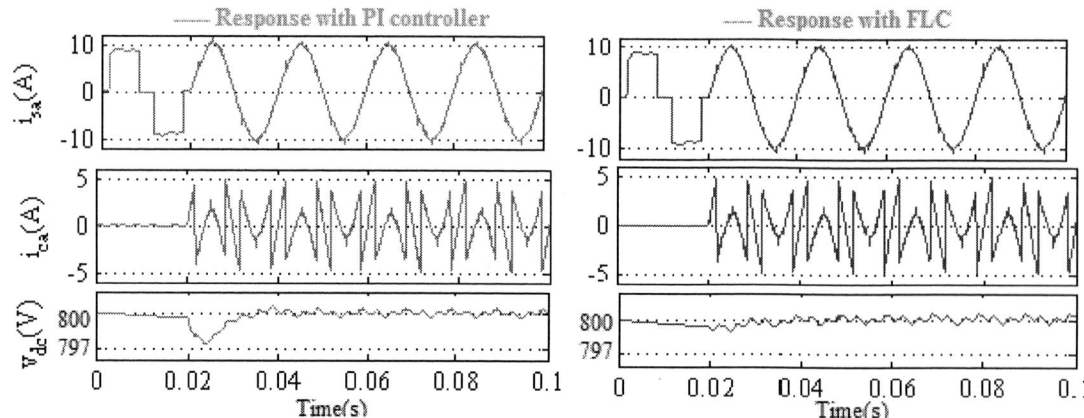

Fig. 11. Performance of i_d-i_q method with PI and FL controller

Case 3: Distorted supply

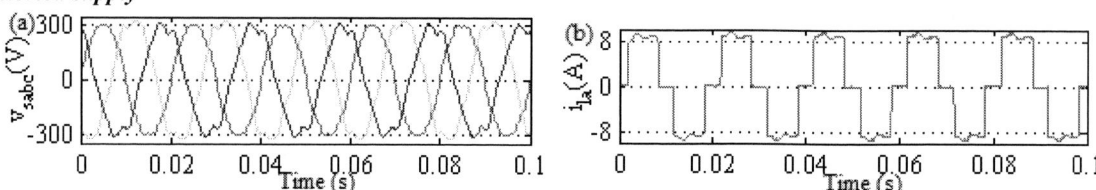

Fig. 12. Waveforms of (a) 3-phase supply voltage and (b) phase-a load current

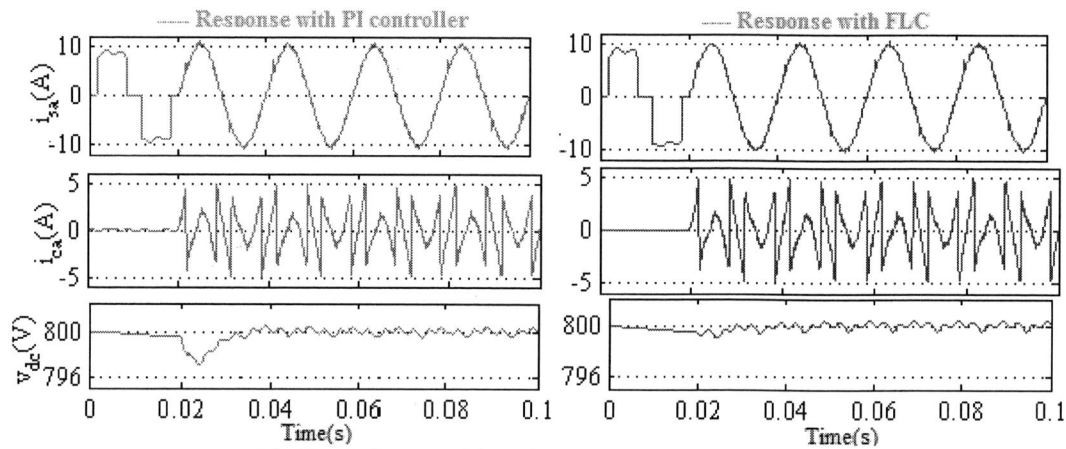

Fig. 13. Performance of d-q method with PI and FL controller

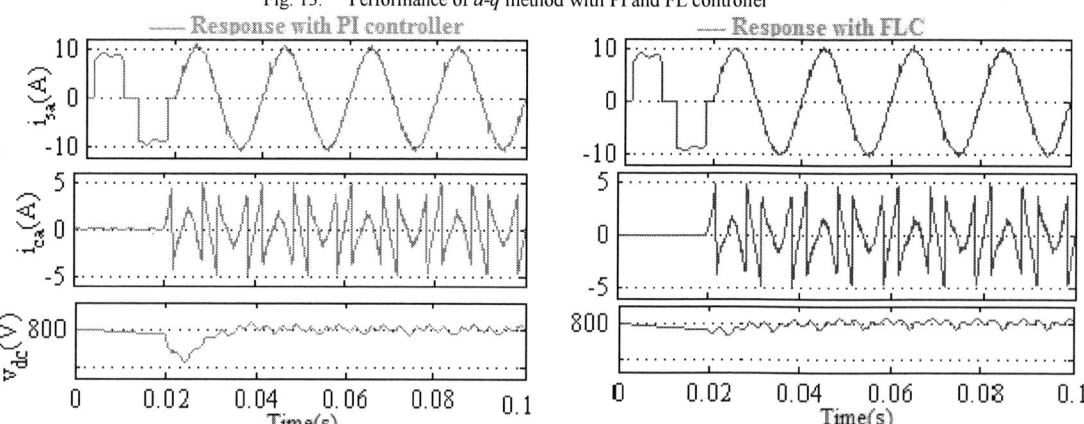

Fig. 14. Performance of i_d-i_q method with PI and FL controller

TABLE II. SUMMARY OF PERFORMANCE ANALYSIS IN TERMS OF THD CONTENT IN SUPPLY CURRENT

case→	Balanced supply			Unbalanced supply			Distorted supply		
Method↓	i_{sa}	i_{sb}	i_{sc}	i_{sa}	i_{sb}	i_{sc}	i_{sa}	i_{sb}	i_{sc}
d-q method with PI	2.61	2.70	2.51	2.89	3.18	3.09	3.49	4.27	4.08
d-q method with FLC	2.20	2.28	1.95	2.49	2.60	2.68	3.22	3.98	3.54
i_d-i_q method with PI	2.62	2.73	2.48	2.87	3.02	3.07	3.25	4.03	3.85
i_d-i_q method with FLC	2.18	2.29	1.98	2.46	2.44	2.62	2.93	3.74	3.32

The THD of supply current is calculated through FFT analysis. FFT analysis is performed at sampling rate of 512samples/cycle. The THD in supply current without compensation is nearly 30% in all the phases. After compensation they are brought down to the values mentioned in Table II. The simulation results for both control techniques are found to be quite satisfactory in terms of eliminating the harmonic components and compensating for reactive components of load current. These control techniques are effective for keeping supply line sinusoidal and in UPF condition and also found to meet the IEEE-519 standard on harmonics level.

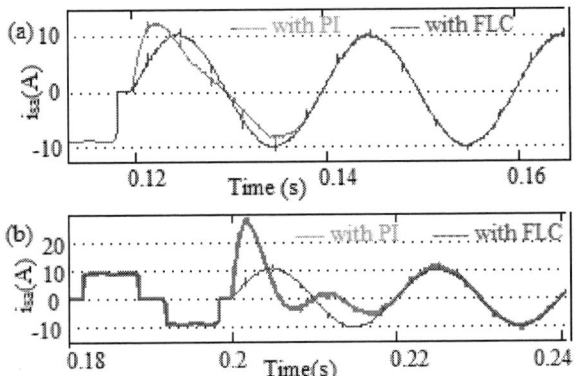

Fig. 15. Variation in setling of phase-a supply current when APF is switched on (a) 0.12 seconds and (b) 0.20 seconds; after start of supply.

It is observed that replacing the PI controller with FLC improves the performance of the APF scheme in terms of THD reduction as well as presence of transients at instant of switching. It can be observed from Fig. 7-8, that v_{dc} shoots down at the instant of switching ON the APF when PI is used. Depending on the instant of switching it can shoot above or below the reference value. Furthermore, if error is large, v_{dc} shoots up by a larger amplitude, thereby, introducing transients in supply current and the settling time also increases, as shown in Fig. 15. Infact, beyond a certain limit of error and change in error combination, the PI cannot restore v_{dc} back to normalcy. However, the FLC can respond effectively for a wider range of error.

VII. CONCLUSIONS

In this paper a fuzzy logic controller (FLC) is proposed for regulating the dc-link voltage in order to enhance the performance of shunt active power filter (Shunt-APF). The proposed FLC is implemented in d-q and i_d-i_q control method based shunt-APF and the performance of the systems are evaluated and compared with that of conventional PI controller based APF control. The comparative analysis shows that implementation of the proposed FLC gives appreciable improvement over PI controller in the performance of the shunt-APFs. Also, the duration of transients in source currents (i_{sa}, i_{sb}, i_{sc}), range of overshoot in dc-link voltage (V_{dc}) as well as i_{sa}, i_{sb} and i_{sc} varies in direct proportion to value of error in V_{dc} when PI controller is used, while the is FLC not affected by wide range of error magnitude.

APPENDIX

Supply voltage:- (a) *Balanced (rms-line)*: 400V; (b) *Unbalanced (peak)*: V_{sa}=320V, V_{sb}=340V, V_{sc}=310V; (c) *Non-sinusoidal (peak,%THD)*: V_{sa}=320V,7.21%THD, V_{sb}=340V,8.51%THD; V_{sc}=310V,8.75%THD, f=50Hz, R_s=0.1Ω, L_s=0.1mH; **APF:-** C=1100μF, V_{dc}=800V, R_{on}=1mΩ, R_{snub}=100kΩ; R_c=0.1Ω, L_c=3.75mH; **Load:-** R_L=60Ω, L_L=60mH; **PI controller:-** k_p = 2, k_i = 100.

REFERENCES

[1] B. Singh, K. AL. Haddad and A. Chandra, "A review of active filters for power quality improvement," *IEEE Trans. Ind Elect.*, vol.46, pp. 960-971, 1999.

[2] Y. Pal, A. Swarup and B. Singh, "A review of compensating type custom power devices for power quality improvement," in proc. *POWER CON*, 2008, pp. 1-8

[3] N. Maruin, A. Alam and S. Mahmod, and H. Hizam, "Review of control strategies for power quality conditioners," in proc. *of IEEE PECon 2004*, pp.109-115.

[4] S. Bhattacharya. D. M. Divan and B. Banerjee. "Synchronous Frame harmonics Isolator using active series filter," proc. of *EPE 1991* ,vol. 3, pp. 30-35, Firenzi, Italy

[5] V. Soares. P. Verdelho, C. D. Marques, "Aclive power filter control circuit based on the instantaneous active and reactive current i_d-i_q method," in proc. *PESC '91*. St Louis. Missouri, 1997, pp. 1096-1101.

[6] T. Zaveri, B. Bhavesh and N. Zaveri. "Comparison of control strategies for DSTATCOM in three-phase, four-wire distribution system for power quality improvement under various source voltage and load conditions", *Electrical Power & EnergySystems*, vol. 43, pp. 582-594, 2012

[7] S. Saad, L. Zellouma, "Fuzzy logic controller for three-level shunt active filter compensating harmonics and reactive power," in *Electric Power System Research*, vol. 79, pp. 1337-1341, 2009

[8] S. K. Jain, P. Agarwal and H.O. Gupta., "Fuzzy Logic Controlled Shunt Active Power Filter for Power Quality Improvement," *IEEE Proceedings of Electric Power Applications*, Vol. 149, No. 5, 2002, pp. 317-328.

Improved Performance of IPDPWM CHBMLI Based DSTATCOM in a Distribution Network with Induction Furnace Load

Parag Nijhawan[#1], R.S. Bhatia[#2], D.K. Jain[#3]

[#1] EIE Department, Thapar University, Patiala (India)
[#2] EE Department, National Institute of Technology, Kurukshetra (India)
[#3] EE Department, Deenbandhu Chhotu Ram   State University of Science & Technology, Muruthal (India)

[1]parag.nijhawan@rediffmail.com
[2]rsibhatia@yahoo.co.in
[3]jaindk66@gmail.com

Abstract— Cascaded H-bridge multilevel inverter (CHBMLI) is based on a series connection of several single-phase inverters which is helpful in raising the output voltage to the desired level. These are also modular in nature as each inverter can be seen as an individual module with similar circuit topology, control structure and modulation [1]. Every converter topology has different switching configurations to achieve the desired output signals. The converter switching needs to be controlled to follow a control reference and modulation strategies are in charge to define the switching control in the converter. The main goal of the modulation algorithm is to synthesize a control reference obtaining a pulse train with the same averaged value. In this work, in-phase disposition pulse width modulation (IPDPWM) and carrier phase-shifted pulse width modulation (CPSPWM) have been applied to CHB 5-level inverter based distribution static compensator (DSTATCOM) to improve the power quality level of a distribution network with induction furnace load [2].

Keywords— Cascaded H-bridge Multilevel Inverter (CHBMLI); Induction Furnace; In-phase Disposition Pulse Width Modulation (IPDPWM); Distribution Static Compensator (DSTATCOM); Total Harmonic Distortion.

I. INTRODUCTION

The non-linear loads in the power distribution network are responsible for poor power quality levels. Loads based on induction heating are such non-linear loads. Many steel producers in India use induction furnaces [2] for melting metals to produce mild steel and alloy steel. The reason behind this is that the melting process in induction furnace is energy efficient, clean and well-controllable. Added advantages of induction furnace are:

- ➢ Low capital cost
- ➢ Relatively easier installation
- ➢ Simpler operation

Induction furnaces are also replacing cupolas in iron foundries to melt iron, steel, copper, aluminium and precious metals. Due to the absence of arc or combustion in these furnaces, the temperature of the material is no higher than required to melt it. The melting loss in induction is also lesser as compared to that for an arc furnace. The only limitation of induction furnace usage is that it cannot be used for bulk production. Therefore, understanding the importance induction furnace for industry, this load has been considered in this work.

Induction furnace is a source of some serious power quality problems which is due to its design and operation. These deteriorate power quality by introducing current harmonics into the distribution network. So, the installation of distribution static compensator (DSTATCOM) is the one possible solution to this power sector need. The main component of this custom power device is the power converters. Here, the emphasis has been given on the application of multilevel converter based DSTATCOM to eliminate or counter balance the current harmonics in a distribution network with induction furnace load.

The multilevel converters [1] came into existence with an aim to overcome the voltage limit of semiconductor devices by connecting number of devices in series. Out of the various multilevel topologies reported in literature, cascade H-bridge multilevel inverter (CHBMLI) has been considered in this work.

II. DSTATCOM

A DSTATCOM [3-4] is a voltage source converter (VSC) based compensating device which is connected in parallel with the distribution system to improve voltage regulation, compensate reactive power, correct power factor and eliminate current harmonics. Its main components have been shown in Fig.1 [3].

978-1-4799-6047-7/14 $31.00 © 2014 IEEE

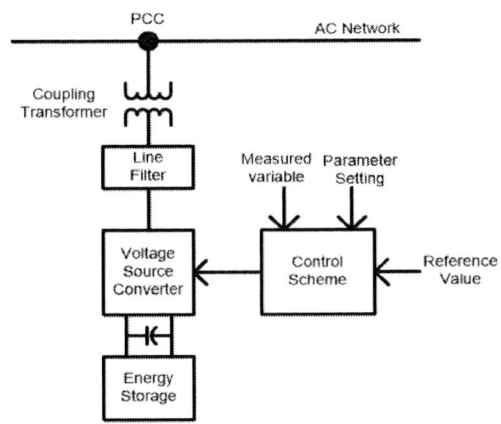

Fig.1. Schematic Representation of DSTATCOM

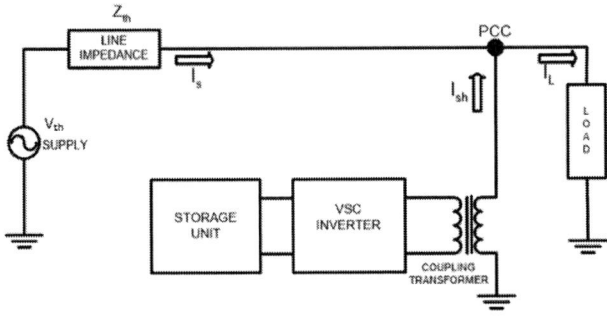

Fig.2. Location of DSTATCOM in a Distribution Network

The shunt injected current I_{sh} by DSTATCOM as shown in Fig. 2 into the distribution network at the point of common coupling (PCC) is given by

$$I_{sh} = I_L - I_s = \frac{V_{th} - V_L}{Z_{th}} \qquad (1)$$

where I_s is source current and I_L is load current.

III. CHBMLI

The series connection of n individual cells, consisting of a single-phase H-bridge inverter with a dedicated dc source results in a CHBMLI topology [5-6]. The relation between the number of levels (n) and the number of isolated cells (m) is n = 2m + 1.

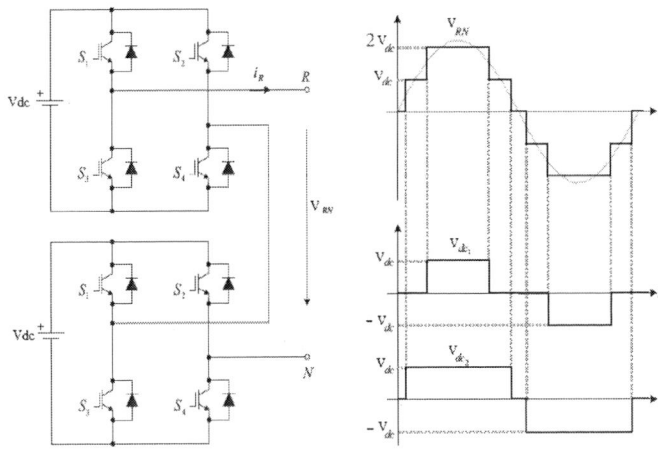

Fig.3. Symmetric Five levels CHBMLI & the possible Five level output waveform

Fig.3 presents two single-phase H-bridge inverters connected in series to structure one leg of a five-level cascaded H-bridge inverter. The remaining two phases have the same switch configuration and respective dedicated dc voltage source.

IV. MODULATION TECHNIQUES FOR CHBMLIs

CHBMLI has different switching configurations in order to achieve the desired output signals. The converter switching must be controlled to follow a control reference and modulation strategies are in charge to define the switching control in the converter. The primary objective of the modulation algorithm is to synthesize a control reference obtaining a pulse train with the same averaged value. Several modulation techniques have been reported in literature [7-9] which are graphically presented in Fig.4. In this work, the main concentration has been on IPDPWM and CPSPWM techniques.

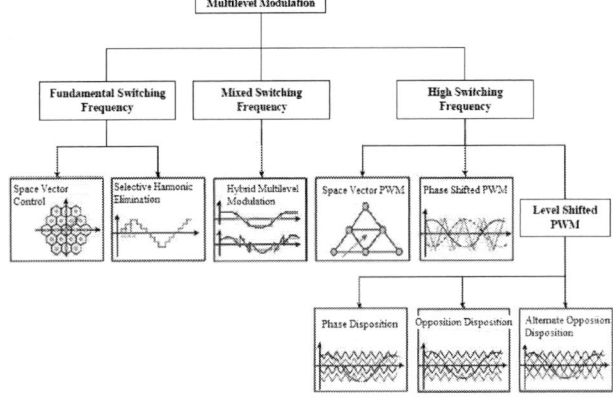

Fig.4. Classification-Tree of Multilevel Modulation Techniques

978-1-4799-6047-7/14 $31.00 © 2014 IEEE

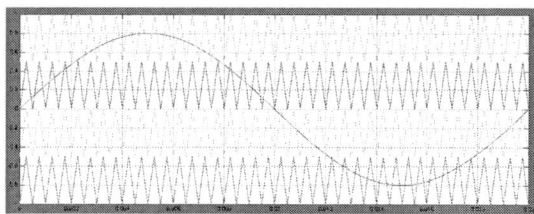

Fig.5(a). Carrier & Modulating waves for IPD-Modulation

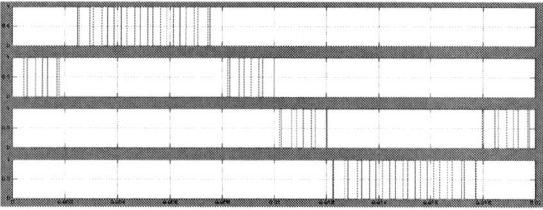

Fig.5(b). PWM signals generated after modulation for generating switching

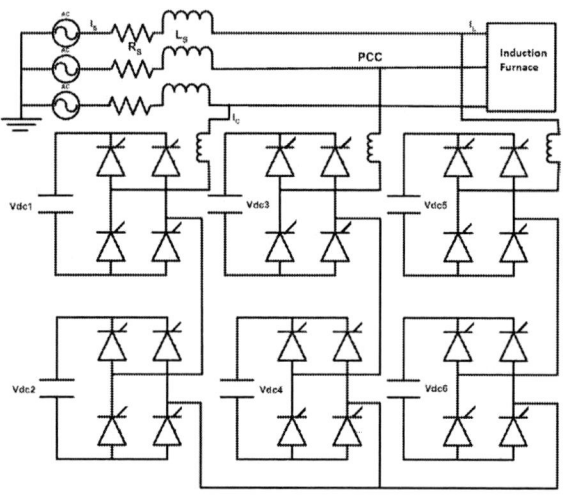

Fig.6. Multilevel Inverter based DSTATCOM

V. TEST SYSTEM

In the proposed MATLAB/SIMULINK model, a three-phase supply is given to three-phase induction furnace load. A CHB 5-level inverter based DSTATCOM is connected in shunt at the PCC to counteract for the current harmonics injected into the distribution network due to the presence of induction furnace load as shown in Fig.7. System parameters used in the proposed MATLAB/SIMULINK model are as follows.

Source Specifications:

$V = 11$ kV, 3-Phase, 50 Hz

DSTATCOM Specifications:

Battery and Controller parameters
Isolated DC source voltage = 6000 V, amplitude modulation index = 0.9, frequency modulation index = 100, carrier

frequency = 1000 Hz, reference frequency = 50 Hz, rated output frequency = 50 Hz.

Induction Furnace Specifications:

0.5 MVA, 11 kV

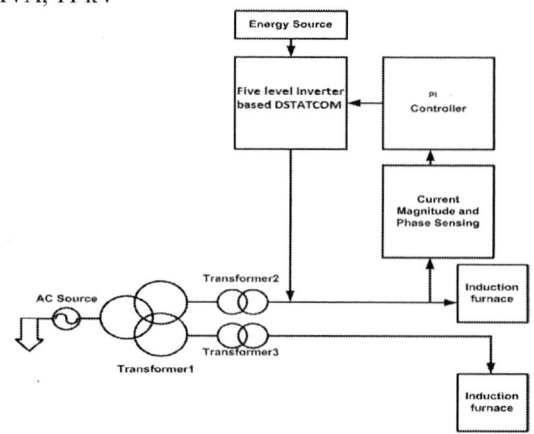

Fig.7. Multilevel Inverter based DSTATCOM Test Model

VI. SIMULATION RESULTS

The simulation is performed on the following test systems with induction furnace load and their performances have been compared based on the results obtained:

Firstly, the distribution network with induction furnace has been considered without any compensating device. It is observed in this case that the current initially dips and stabilizes steady value after some time. The load current wave is also distorted in nature, as shown in Fig.8 [10-11]. The THD level observed in load current and load voltage are 26.61% (shown in Fig.8) and 0.02% [10] (shown in Fig.9), respectively under steady state conditions.

In second case, the distribution network with induction furnace load compensated with the help of CPSPWM five-level inverter based DSTATCOM has been considered. For this system, no initial dip in the current waveform has been observed. The distortion level observed in the source current waveform is also negligible as shown in Fig.10. The THD level observed in source current is only 7.99% at 50 Hz fundamental frequency. There has also been a significant improvement observed in the fundamental component of source current to 3.05 pu [10].

In third case, the distribution network with induction furnace load compensated with the help of IPDPWM five-level inverter based DSTATCOM has been considered. For this system, no initial dip in the current waveform has been observed. The distortion level observed in the source current waveform is also negligible as shown in Fig.11. The THD level observed in source current is 3.45% at 50 Hz fundamental frequency which is even better. Also, with IPDPWM scheme, the fundamental component of source current observed is 5.29 pu, which is higher than the corresponding value observed in second case.

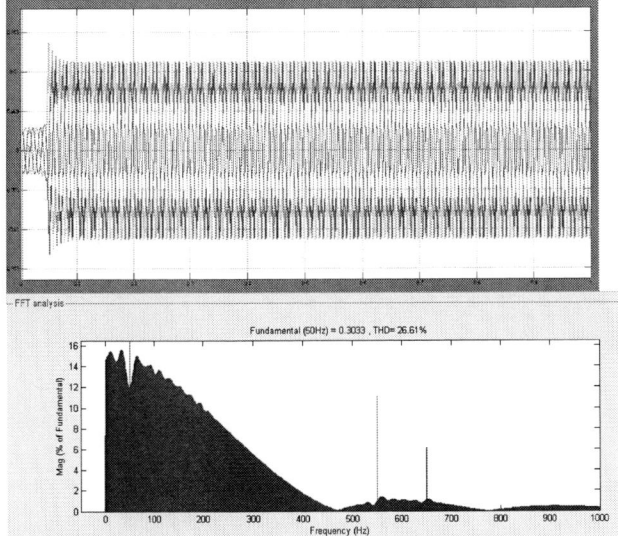

Fig.8. Load Current waveform versus time and its Frequency Spectrum with Induction Furnace [10-11]

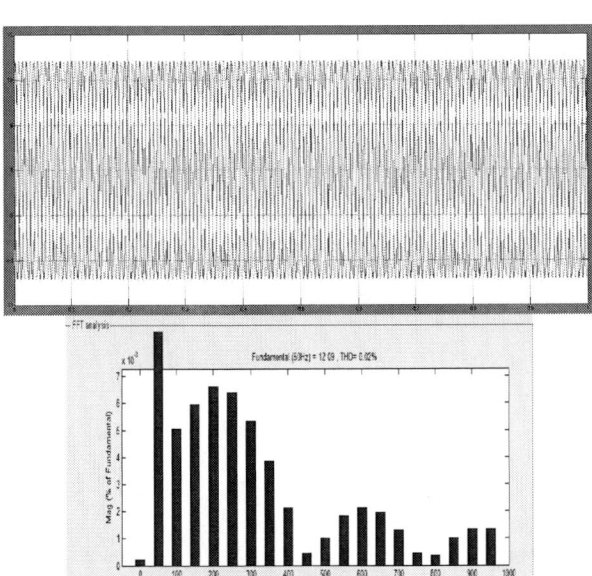

Fig.9. Load Voltage waveform versus time and its Frequency Spectrum with Induction Furnace [10]

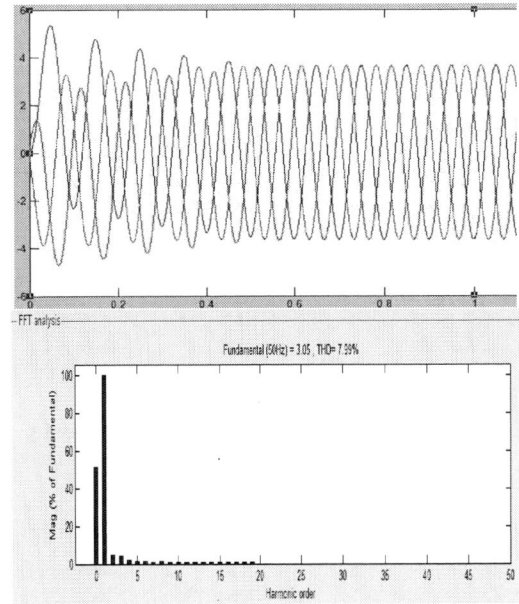

Fig.10. Source Current waveform versus time and its Frequency Spectrum with CPSPWM Five-Level inverter based DSTATCOM [10]

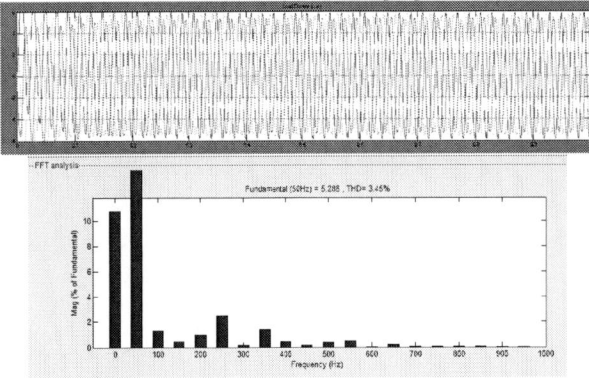

Fig.11. Source Current waveform versus time and its Frequency Spectrum with IPDPWM Five-Level inverter based DSTATCOM

VII.   CONCLUSION

In this work, simulation of CHBMLI based DSTATCOM configuration schemes have been done in MATLAB/SIMULINK. It is clearly seen from the results obtained that the performance of IPDPWM multilevel inverter based DSTATCOM is superior as compared to that of CPSPWM multilevel inverter based DSTATCOM.

REFERENCES

[1] L.G. Franquelo, J. Rodriguez, J.I. Leon, S. Kouro, R. Portillo, M.A.M. Prats, "The age of multilevel converters arrives", IEEE Ind. Electron. Mag, vol.2(2), pp. 28–39, 2008.

[2] I. Zamora, I. Albizu, A.J. Mazon, K.J. Sagastabeitia, E. Fernandez, "Harmonic Distortion in a Steel Plant with Induction Furnaces", International conference on renewable energies and power quality, pp. 1-7, 2003.

[3] Walmir Freitas, Andre Morelato, Wilsun Xuand Fujio Sato, " Impacts of AC Generators and DSTATCOM Devices on the Dynamic Performance of Distribution Systems", IEEE Transactions on Power Delivery, pp. 1493–1501, 2005.

[4] B. Singh, A. Adya, A.P. Mittal, J.R.P. Gupta, " Modeling and Control of DSTATCOM for Three-Phase, Four-Wire Distribution Systems", Conference on Industry Applications, vol.4, pp. 2428-2434, 2005.

[5] M. Fracchia, T. Ghiara, M. Marchesini, M. Mazzucchelli, "Optimized Modulation Techniques for the Generalized N-Level Converter", Proceeding of the IEEE Power Electronics Specialist Conference, vol.2, pp. 1205-1213, 1992.

[6] S. Daher, J. Schmid , F.L.M. Antunes, "Multilevel inverter topologies for stand-alone PV systems", IEEE Trans. Ind. Electron., vol.55(7), pp. 2703–2712, 2008.

[7] B. Urmila, D. Subbarayudu, "Multilevel inverters: A comparative study of pulse width modulation techniques", International Journal of Scientific and Engineering Research, vol.1(3), pp. 1-5, 2010.

[8] R. Gupta, A. Ghosh, "Switching characterization of cascaded multilevel-inverter-controlled systems", IEEE Trans. Ind. Electron., vol.55(3), pp. 724–738, 2008.

[9] D. Mohan, S.B. Kurub, "A Comparative Analysis of Multi Carrier SPWM Control Strategies using Fifteen Level Cascaded H – bridge Multilevel Inverter", International Journal of Computer Applications, vol.41(21), pp. 7-11, 2012.

[10] P. Nijhawan, R.S. Bhatia, D.K. Jain, "Improved Performance of Multilevel Inverter Based DSTATCOM with Induction Furnace Load", IET Power Electronics, vol.6(9), pp.1939-1947, 2013.

[11] P. Nijhawan, R.S. Bhatia, D.K. Jain, "Application of PI controller based DSTATCOM for improving the power quality in a power system network with induction furnace load". Songklanakarin Journal of Science and Technology, vol.34(2), pp. 195-201, 2012.

A Hybrid Optimization Technique of Wind Power Exploration & Utilization

T. Halder
Assistant Professor
Electrical Engineering Department, Member IEEE
Kalyani Government Engineering College
Kalyani, District-Nadia, West Bengal,
PIN-741235, INDIA

Abstract— **The paper envisages the wind power exploration and utilization through the utility grid and the power converters' stations. Huge harvesting of wind energy is the most effective elucidation today's to curtail the dilemmas of depleting fossil fuel reserves and getting bigger huge greenhouse gas emissions at ecosystem. At the moment, the wind energy is mostly converted into electricity by deploying wind turbines to mitigate the power crisis and the monopole trade of a power grid corporation**

Keywords—Wind power exploration; Generations; Modeling; Converter stations; Grid operations; Utilizations; and Results

I. Introduction

Energy is the most important and global evaluation of the entire categories of efforts by the human beings and background. The whole lot that takes place in the world is the manifestation of flow of energy in one form. Energy is an important contribution in all sectors of a country's monetary system. The standard of living is overtly related to per capita power consumption [1]-[3]. Due to prompt augment in the inhabitants and standard of living, these are faced with energy disaster. Traditional resources of energy are more and more pooped. Hence, renewable energy sources have come into sight as possible energy sources at large in India and planet. Among the various non-conventional energy resources, wind energy is emerging potential source of energy for enlargements. A figure of Indian power utilities shaped as an organization, then the utility of the wind interest group (UWIG) to amend the repercussion of wind energy increase to the grid counting the collision of the intermittency on day by day effectiveness, preparations and load subsequent impossible [4]-[10].

Newest synopses of their efforts are to be carried out to rendezvous has tinted that wind power's impacts on system. The operating expenses are puny at short wind penetrations about 5% or less. In most cases, the incremental costs would detract from the assessment of wind energy on recent extensive markets by 10% or less using power electronic converters. At far above the ground, wind penetrations of turbines, the conflict will be higher than normal stage of power generation although near results put it to a big shot. The crash remains sensible with penetrations forthcoming not upto marks [11]-[14].

While, in this article, the wind power system comprises various superficial costs related with best performance and grid operations, it also has some incentive from the utility's dot of the assessment, together with undersized production guide times, modularity, no green house gas emissions at sites, and higher consumer's authorization. In future, it will be least generation cost of power for the emergency perusal.

LIST OF SYMBOLS :

A = Swept area of the turbine rotor
ρ = Air density
v = Velocity of the wind
m = Mass flow of the air
ω = Angular velocity
R_m = Blade tip radius
λ = Tip speed ratio

II. Principal of Energy Conversions

Kinetic energy of wind power system is given as:

$$E = \frac{1}{2}mv^2 \tag{1}$$

Power available from the wind mill given as:

$$P_{av} = \frac{d}{dt}\left(\frac{mv^2}{2}\right) \tag{2}$$

$$P_{av} = \frac{1}{2}\left(\frac{d(\rho Ax)}{dt}\right)v^2 + m.v\left(\frac{dv}{dt}\right) \tag{3}$$

Where, m = ρAx (Mass of the fluid at a time, t)

$$P_{av} = \frac{1}{2}A\rho v^3 + F.v \tag{4}$$

Here, F and (F.v) are the force developed and threshold powers respectively are to be required to generate or to develop the wind power by the turbine system as:

$$P_{aa} = P_{av} - F.v = \frac{1}{2}A\rho v^3 \tag{5}$$

Now the power coefficient defined as:

$$C_P = \left(\frac{P_E}{P_{aa}}\right) \tag{6}$$

Hence extracted power from the wind turbine given as:

$$P_E = \frac{1}{2}C_p A\rho v^3 \tag{7}$$

Air density which highly affects the power output at a given swiftness, purpose of elevation, temperature and barometric pressure, variation in temperature and pressure can have a consequence up to 8 %. A warm type of the weather slices air density. This statistics confer a picture of maximum power

978-1-4799-6047-7/14 $31.00 © 2014 IEEE

available on the rotor diameter of generator. The cooperative effects of wind momentum and the rotor radius or rim can be experiential. The wind turbines are proficient to convert simply a fraction of accessible wind power into functional power. Blade tip speed ratio, λ is the ratio between the speeds of the tip of a rotor blade related to the affecting wind speed given as:

$$\lambda = \frac{\omega R_m}{v} \qquad (8)$$

$\varpi = 2\pi N$, where N is the RPM of the rotor blade (9)

III. CRITERIA FOR MINIMUM POWER DEVELOPMENT BY THE WIND TURBINE

The force (F) is encountered with the turbine blade so it may be considered as negative, so the equation (4) reduces as:

$$p_{av} = \frac{1}{2} A \rho v^3 - F.v \qquad (10)$$

The swept area of the turbines in terms of the blade tip radius varied from, 0.5 to 5m depending on the ratings of the wind turbines given as:

$$A = \pi R_m^{\ 2} \qquad (11)$$

To compute the minimum power developed by the satisfying the condition stated as:

$$\frac{\partial P_{av}}{\partial v} = 0 \qquad (12)$$

The equation (9) reduces as:

$$\frac{3}{2} A \rho v^2 - F = 0 \qquad (13)$$

Now the value of the F, from the equation (13) written as:

$$F = \frac{3}{2} A \rho v^2 \qquad (14)$$

The equation (9) may be rewritten in combination of the equation (13), by putting the value of F, given as:

$$p_{av} = \frac{1}{2} A \rho V^3 - \frac{3}{2} A \rho V^3 \qquad (15)$$

$$P_{av} = -\left(\frac{1}{2}\right) A \rho V^3 \qquad (16)$$

The is a cut down form of the equation (15) and it represents as negative value of P_{av} meaning the power extracted from wind momentum in all possible directions. The pressure exerted by the wind turbine is given as:

$$P = \left(\frac{F}{A}\right) = \frac{3}{2} \rho v^2 \qquad (17)$$

Further the expression of v is given as:

$$v = \sqrt{\frac{2F}{3A\rho}} \qquad (18)$$

Now, the putting the value of v to the equation (10), yields as:

$$p_{av} = \frac{1}{2} A \rho \left(\frac{2F}{3A\rho}\right)^{\frac{3}{2}} - F \sqrt{\frac{2F}{3A\rho}} \qquad (19)$$

The equation (19) presents average power generations of the wind turbines

$$p_{av} = \frac{1}{2} A \rho \left(\frac{2P}{3\rho}\right)^{\frac{3}{2}} - F \sqrt{\frac{2P}{3\rho}} \qquad (20)$$

The simplified form of the equation (20) may be concluded as the average extracted by wind turbine relies on the full of atmosphere pressure and thrust or force of the air as.

$$p_{av} = \frac{1}{2} A \rho \left(\frac{2P}{3\rho}\right)^{\frac{1}{2}} \left(\frac{2P}{3\rho} - F\right) \qquad (21)$$

IV. COMPUTING AVERAGE POWER OF WIND TURBINE

The average power of the wind with experimental formulae given as:

$$V_a^3 = \int_0^\infty v^3 f(v) dv = \int_0^\infty v^3 \left(\frac{2v}{c^2}\right) e^{-\left(\frac{v}{c}\right)^2} dv \qquad (22)$$

After integrating, it yields

$$V_a = \frac{3}{4} c \sqrt{\pi} \qquad (23)$$

Now, putting the scale parameter, $c = \left(\frac{2v}{\sqrt{\pi}}\right)$

$$V_a^3 = \left(\frac{6}{\pi}\right) v^3 \qquad (24)$$

Hence, the average power of the new model is given as:

$$P_N = \frac{1}{2} \left(\frac{6}{\pi}\right) \rho A v^3 \qquad (25)$$

This capitalizes that the average power in turbine is found in terms of average wind speed. The average power is also equal to 1.91 times of average power found at the average wind momentum.

$$\text{Let,} \left(\frac{P_H}{P_h}\right) = \left(\frac{v_H}{v_h}\right)^{3\alpha} \qquad (27)$$

Where, P_H be average power at speed v_H and P_h be average power at velocity v_h and the empirical value of the

$$\alpha = (1/7) \qquad (28)$$

If the average speed of the wind is proportional to barometric pressure or height, then the equation (25) reduces as:

$$(P_H / P_h) = (H/h)^{3\alpha} \qquad (29)$$

So the average power generation of the wind turbine is reliant on the barometric pressure or heights of a particular position or seashore. When wind momentum is zero, no power generation of the wind turbine takes place.

A harmonizing form the Fig. 1, decidedly envisages that threshold magnitude of wind velocity is compulsory to generate minimum wind power. The threshold velocity of wind nearly (4 m/s) and the dimension of the turbine ensures the output power generations overcoming the auxiliary power of the systems, then the consequential power generation by wind farms system is about 125W at threshold speed of the wind turbines. The selection of the site plays an important issue to generate the wind power. The turbulent velocities of the wind are not quite suitable until it is regulated by the power converters as shown in the Fig. 1 with a survey the sites.

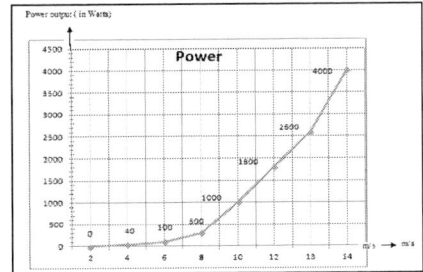

Fig. 1. Power output (P) vs. wind speed (v) in (m/s) of Turbine

It is unambiguously, at an entire idle condition ($\lambda = 0$ or 0.1) the power output is zero. When the blade-tip speed is equal to wind speed ($\lambda=1$) the maximum power coefficient of C_p (max) = 0.13 occurs at $\lambda = 0.5$. When the blade tip moves at 1/3 of the wind speed signifying that only 16 % of the wind powers may be extracted by a series of potential technique as shown in the Fig. 2 farming data of the wind turbine producer

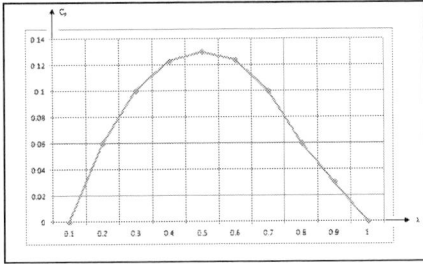

Fig. 2. Power coefficient (C_p) vs. Tip speed ratio $\left(\lambda\right)$ of the blade of Turbine

As the free of charge wind torrent outdoes or hit through the rotor, it transfers a number of its power to the rotor and its velocity decreases to a smallest amount in the rotor stir. After several detachments from the rotor wind tributary takes back its momentum from the nearby air. Wind power expert may also notice fall in pressure as the wind stream overtakes through the rotor. Last of all air rapidity and pressure also increases ambient moody conditions

Yaw motor gear- The area of the wind stream swept by the wind turbine is maximum value, when the blades face into the wind. Proper configuration of the blade angle with respect to the wind direction to get highest wind energy can be accomplished with the help of yaw control that rotates wind turbine about the vertical axis. In slighter wind turbines, yaw action is controlled by tail vane whereas, in larger turbines, it is operated by the servomechanism.

V. SITE SELECTION & WIND ENERGY POTENTIAL

The following significant factors play to be taken into consideration of choice for the admirable location for colossal economical way for the green wind power generation as least erection and commissioning cost of the wind farms.

- Twelve-monthly high wind velocity

- No high hindrance for a radius at about 3 Km
- straightforward and unfasten seashores
- Top of a parallel, well rounded hill with gentle slopes
- Mountain gap which produces wind funneling
- Energy planning purposes are crucial
- Most civil land uses are cost effective
- Keep away from potentially detrimental surroundings.
- Proximity to transmission ability at high fidelity
- Commercial probability and free of fuel cost

VI. GRID POWER QUALITIES NECESSITIES

The power quality of wind generation to electrical power systems influences the system practice points, the load flow of real and reactive power, nodal voltages and power losses. At the same time, wind power generation has uniqueness with a large variety of influences and synchronization:

- Site in the power system of the wind farm
- Voltage deviation of amplitude and frequency
- Flickering of voltage of distribution or transmission
- Harmonics of current and voltage of system
- Short circuit currents and protection systems
- Stability of the system
- Self-excitation of asynchronous generators
- Real power losses of the power system

The getting bigger impact of wind power generation in power systems grounds system operators to widen Grid connection requirements, in order to make sure it's faultless operation or synchronization. It can divide Grid tie necessities into two groups:

A. Universal Grid code necessities

B. Individual necessities for wind power generation

The first group represents the proper necessities for every generator in the grid systems at which the common necessities of the power system operation points are very significant and some of the most significant necessities are depicted below,

- Steady state voltage digression
- Line transmission and distribution capacity
- Short circuit power at the tie point
- Frequency fluctuations
- System security and protection
- Crisis supervisory

The outstanding necessities for chunk of wind generations are introduced to initiate in the power systems or grids with a substantial agreement on power supremacy or system stability. There are two different types of requirements:

A necessity recognized by system operators and national or international standards. An appraisal of the opening position is to be had in the following segment, where the active power control, frequency margins, reactive power control and fault be conveyed through capability are to be scrutinized so that the overall stabilities of the system should be undisturbed.

VII. PERFORMANCES OF THE INDUCTION MOTOR

The Induction motor (IM) works either motor mode or generation mode. When three phase supply is fed to stator

978-1-4799-6047-7/14 $31.00 © 2014 IEEE

side, it will run due to its self starting property. On the other hand, when the IM is excited by a driver, the IM may run Induction generator exceeding the synchronous speed with negative slip. The torque versus slip and torque versus speed characteristics are shown in the Fig. 3 and Fig. 4 respectively.

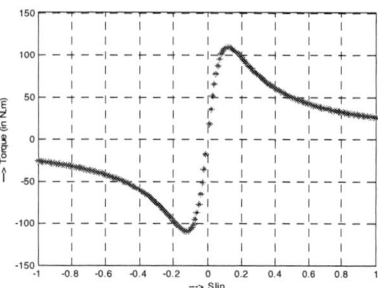

Fig. 3. Torque vs. slip curves of IM

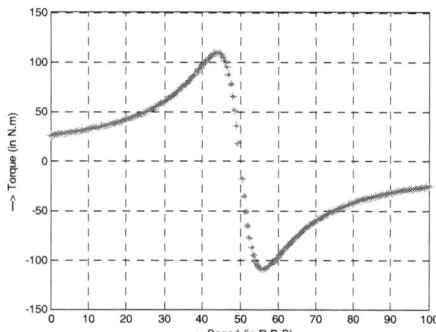

Fig. 4. Torque vs. Speed curves of IM

The power versus torque characteristics curved of the IM and Induction generator are shown in the Fig. 5. The both curves are symmetric in nature but both in opposite direction, when IM acts as generator, it power and torque are both negative signifying the drawing leading VAR from the infinite bus of the grid or by a capacitor bank connected parallel to its terminal.

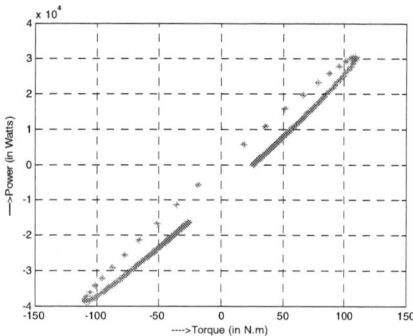

Fig. 5. Power vs. Torque curves of IM

The power, versus torque and slip curves of an IM at which the maximum power will be extracted from the Induction generator at particular slip and torque value as in the Fig. 6

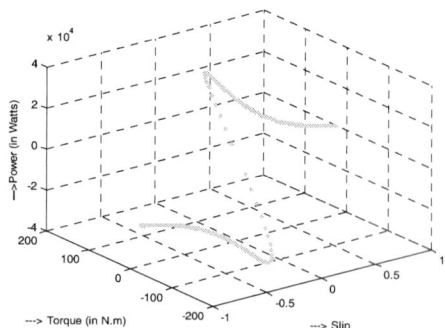

Fig. 6. Power vs. Torque and slip curves of IM

The power versus torque and slip characteristics of the Induction generator. The designer decides the specific speed of the generator at which maximum power will be developed at the particular speed. By virtue of which the speed control unit of the generator, converter control unit, DC bus control unit, grid control unit and the value of DC link capacitance are to be designed for the synchronous and robust operation the entire system as shown in the Fig. 7.

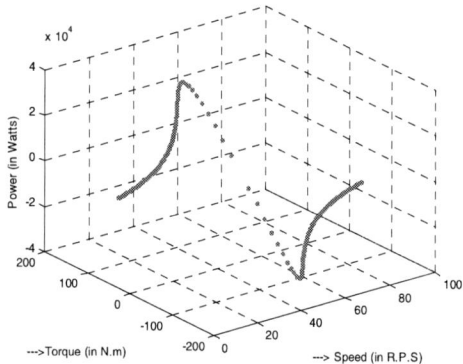

Fig. 7. Power vs. Torque and speed curves of IM

When the voltage sag is increased by virtue of boost in load, the slip is decreased. The slip is zero at synchronous speed and its value is decreased to negative value with speed increasing and obeying with stable rotation speed of the shaft, synchronous speed is also described in the Fig. 7.

VIII. COUPLING OF WIND TURBINE & INDUCTION GENERATOR

Power quality issues are spoken in terms of the substantial features and belongings of the generation power when a wind generator is synchronised to power grid. It is for the most part often demonstrated in terms of voltage, frequency and their distortions and interruptions. Acceptable power quality would be signified that the voltage is purely sinusoidal waveform and highly uninterrupted operations. It has a stable amplitude and frequency of the power systems. The Fig. 8 demonstrates the categorizations of contradictory phenomena disturbing power excellence in a grid interfaced wind power system.

978-1-4799-6047-7/14 $31.00 © 2014 IEEE   210

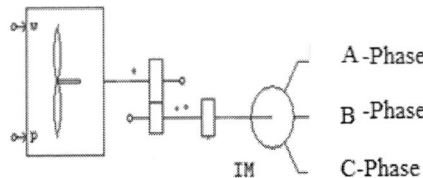

Fig. 8. Wind turbines with Induction Motor (IM) as a Generator

The wind turbine system is associated with a gear box to accomplish constant momentum and it is also joined with asynchronous generator (IM). Its output is shorted by a star or delta associated capacitor bank to be enhanced the performance of the generator and to recompense to leading var lagging var of the Wind Generator without connection of load in the Fig. 9.

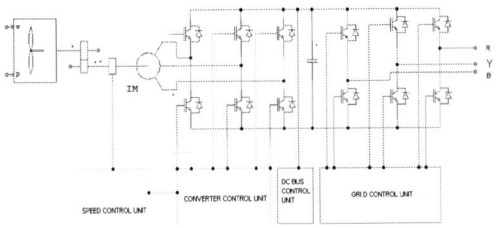

Fig. 8 Assembly of wind turbine and converters at Utility Grid

IX. GENERATOR CONVERTER, GRID CONVERTER AND UTILITY GRID

The inequitable improvement in the wind turbine erection and commissioning needs the transmission system operators to tie up their Grid connection rules in order to achieve a better control of the generation, to stay away from inconsistency and make certain the network and power quality. Grid monitoring system ensures that parameters such as voltage, current and frequency is measured between the power converter and the transformer station. The values are incessantly transmitted to the grid Control unit, thereby enabling the regulation of the turbine in case of changes in grid voltage or frequency. If set-point values for the system or grid protection are exceeded, the wind turbine is shut down and reports are sent to supervisors. The turbine will start up again, as soon as voltage and frequency come back to allowable height. If the catastrophe shutdown with grid extrication is enforced, the occurrence will be reported through the monitoring system, and the turbines will robotically reconnect, and continue generation, as soon as possible. To-day, wind turbines have status as small power stations, and as such there is a necessitation for sophisticated control of the power generation from wind turbines in coexistence with other power sources. Grid attached or synchronized power converters incessantly judges these standpoint and broaden ground-breaking control systems to the associations with the power converter stations as shown Fig. 7 and Fig. 8. The global installed capacities of wind power arrive at 198 GW by 2010. China (44733 MW), US (40180 MW), Germany (27215 MW) and Spain (20676 MW)

are ahead of India in fifth position. Improved performance of power electronics technologies comprise to generate good quality of wind power as substitute's capacity in India. If we wind energy scenario in INDIA as shown below at TABLE I, It will be crystal clear. The government of Madhya Pradesh has approved one more 15 MW project to Madhya Pradesh Wind farms Ltd. (MPWL)

Table I

Power Plant	Producer	Site	State	Total Capacity (MW)
Vankusawade Wind Park	Suzlon Energy Ltd.	Satara District.	Maharashtra	259
Acciona Tuppadahalli	Tuppadahalli Energy India Private Limited	Chitradurga District	Karnataka	56.1
Dangiri Wind Farm	Oil India Ltd.	Jaiselmer	Rajasthan	54
Cape Comorin	Aban Loyd Chiles Offshore Ltd.	Kanyakumari	Tamil Nadu	33
Kayathar Subhash	Subhash Ltd.	Kayathar	Tamil Nadu	30
Ramakkalmedu	Subhash Ltd.	Ramakkalmedu	Kerala	25
Muppandal Wind	Muppandal Wind Farm	Muppandal	Tamil Nadu	22
Gudimangalam	Gudimangalam Wind Farm	Gudimangalam	Tamil Nadu	21
Shalivahana Wind	Shalivahana Green Energy. Ltd.	Tirupur	Tamil Nadu	20.4
Puthlur RCI	Wescare (India) Ltd.	Puthlur	Andhra Pradesh	20
Chennai Mohan	Mohan Breweries & Distilleries Ltd.	Chennai	Tamil Nadu	15
Lamda Danida	Danida India Ltd.	Lamba	Gujarat	15
Shah Gajendragarh	MMTCL	Gadag	Karnataka	15
Jamgudrani MP	MP Windfarms Ltd.	Dewas	Madhya Pradesh	14
Jogmatti BSES	BSES Ltd.	Chitradurga District	Karnataka	14
Perungudi Newam	Newam Power Company Ltd.	Perungudi	Tamil Nadu	12
Kethanur Wind Farm	Kethanur Wind Farm	Kethanur	Tamil Nadu	11
Shah Gajendragarh	Sanjay D. Ghodawat	Gadag	Karnataka	10.8
Hyderabad APSRTC	Andhra Pradesh State Road Transport Corporation.	Hyderabad	Andhra Pradesh	10
Muppandal Madras	Madras Cements Ltd.	Muppandal	Tamil Nadu	10
Poolavadi Chettinad	Chettinad Cement Corp. Ltd.	Poolavadi	Tamil Nadu	10
Fraserganj	WBREDA	Fraserganj, 24 Parganas (s)	West Bengal	2.1
Ganga Sagar	WBREDA	Kakdwip,24 Parganas (S)	West Bengal	0.5

Bhopal at the Nagda mounts put up the shutters to Dewas under the debates from strengthened energy consultants Ltd.

(CECL) Bhopal. All the 25 WEGs have been bespoken by 2008 under a utility. The Agency for non-conventional energy and rural technology (ANERT) under the power sector, the government of Kerala, is commissioned wind mills on a furtive property in sundry sector of the state to produce a sum of 600 MW of wind power. The bureau has acknowledged 16 locations for installation of wind mills through private sectors. Orissa a

Coastal conditions has higher prospective of wind power. Recent installed capacity puts at about 2.00 MW. Among the Indian states, Orissa has a wind power prospective approximately, 1700MW. The government of Orissa is enthusiastically harvesting to reinforce wind power production in the state. However it has not headway like other states because Orissa having a substantial coal storage and huge figures of obtainable and forthcoming coal fired power plants, is a power surplus state of India .

X. RESULTS

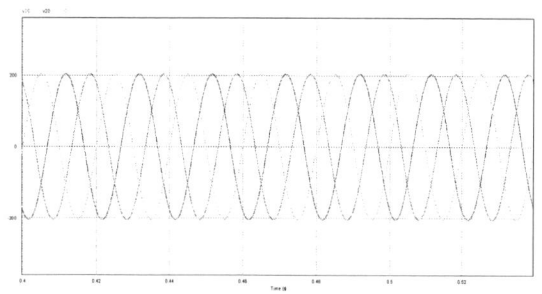

Fig. 9. Output voltage waveforms at Inverter side of coveter station

Per phase voltage 200V/50Hz of three phase system is perfectly balance under no load to full load condition as shown in the Fig. 9 irrespective of wind speed

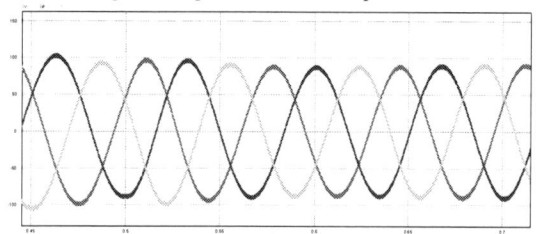

Fig. 10. Output voltage waveforms at Inverter side of coveter station

Three phase current waveforms are perfect balance under at all conditions with least total harmonics distortions (THD) and ready to hurl power to the utilities' grid as shown in Fig. 10

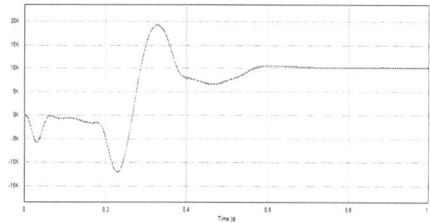

Fig. 11. Power output (P) of the Wind generator (IM)

The power output of the asynchronous generator is more or less constant with fewer fluctuations as shown in the Fig. 11. Due to less fluctuations of the output power avoiding the transient part, it can be treated as excellent quality of the power and well suited for grid operation and this power is to be sent to grid of power utilities for a nation.

XI. CONCLUSIONS

Wind turbine systems and converter substations are a big potential resource of hardly power quality problems when it is interfaced with or synchronised by converter substations. The standard endows with information and tools for prediction of dealings between the wind turbines systems and the power grid. Together with massiveness wind energy systems technology evidence, it gives advice for pertinent mission oriented turbine choice and faultless support with the grid to which they are connected for power utilisation at consumer site as maximum power deployment and power processing of the converter stations with tight regulations

REFERENCES

[1] T.Halder ,"Hot and Intelligent Trends of Wind Energy", International Journal of Reasons, (ISSN 2277-1654, Vol.-X. 2011) pp 55-59

[2] Andersen, P.D., Global Perspectives for Wind Power, Proceedings of the European Wind Energy Conference (EWEC '99), March 1999, Nice, France, p. 71 - 76.

[3] T.Halder, "Wind Power Generation and Synchronization to Utility Grid" Control, Communication & Device Electronics (N3CD-2013) sponsored by AICTE pp 244-250.

[4] Openshaw, D., "Embedded Generation in Distribution Networks: With Experience of UK Commerce in Electricity"; Wind Engineering, Vol. 22, No. 4, 1998, p. 189 . 196.

[5] Paap, G.D., Teklu, D., Jansen, F., van der Sluis, L., "The Influence Of Large-Scale Dispersed Co-Generation On Short Circuits In Rural Networks", Proceedings of the 15th International Conference on Electricity Distribution (CIRED '99), June 1999, Nice, France.

[6] Salman, S.K., Jiang, F., "Impact of Embedded Wind Farms on Voltage Control and Losses of Distribution Networks", Proceedings on International Conference on Electric Power Engineering, Budapest, Hungary, 1999. p. 43.

[7] Jhutty, A.S., "Embedded generation and the Public Electricity System", IEE Colloquium on System Implication of Embedded Generation and its Protection and Control, Birmingham, UK, February 1998, p. 9/1 . 9/14.

[8] T.Halder, "Power Quality Issues – A Case Study" Control, Communication & Device Electronics (N3CD-2013) sponsored by AICTE pp 251-250.

[9] Tzschoppe, J., Haubrich, H-J., Bergs, D., Nilges, J., "MV-Network-Connection of Disperced Generation: Network Reinforcements vs. Optimised Operation", Proceedings of the 15thInternational Conference on Electricity Distribution (CIRED '99), June 1999, Nice, France.Haslam, S.J., Crossley, P.A., Jenkins, N.,

[10] Design and field testing of a source based protection relay for wind farms, IEEE Transactions on Power Delivery, Vol. 14, No. 3, July 1999, p. 818 . 823.

[11] Jorgensen, P., Tande, J.O. Vikkelso, A., Norglrd, P., Christensen, J.S., Sorensen, P., Kledal, J.D.,Sonderglrd, L., "Power Quality and Grid Connection of Wind Turbines", Proceedings of the International Conference on Electricity Distribution (CIRED '97), IEE Conference Publication No.438, 1997, p. 2.6.1.- 2.6.6.

[12] T. Halder, "Charge Controller of Solar Photo-Voltaic Panel Fed (SPV) Battery" IEEE, IICPE-2010, pp 1-4

[13] Johnson, G.L., "Wind Energy Systems", Prentice Hall International Inc., New Jersey, USA, 1985,60 p. 46.

[14] E. Muljadi and C.P. Butterfield, "Pitch-Controlled Variable-Speed Wind Turbine Generation",

BLDC Motor Driven Solar PV Array Fed Water Pumping System Employing Zeta Converter

Rajan Kumar, *Student Member, IEEE*
Department of Electrical Engineering
Indian Institute of Technology Delhi
Hauz Khas, New Delhi-110016, India
sonkar.rajankumar36@gmail.com

Bhim Singh, *Fellow, IEEE*
Department of Electrical Engineering
Indian Institute of Technology Delhi
Hauz Khas, New Delhi-110016, India
bhimsinghr@gmail.com

Abstract— This paper proposes a solar photovoltaic (SPV) array fed water pumping system utilizing a zeta converter as an intermediate DC-DC converter in order to extract the maximum available power from the SPV array. Controlling the zeta converter in an intelligent manner through the incremental conductance maximum power point tracking (INC-MPPT) algorithm offers the soft starting of the brushless DC (BLDC) motor employed to drive a centrifugal water pump coupled to its shaft. Soft starting i.e. the reduced current starting inhibits the harmful effect of the high starting current on the windings of the BLDC motor. A fundamental frequency switching of the voltage source inverter (VSI) is accomplished by the electronic commutation of the BLDC motor, thereby avoiding the VSI losses occurred owing to the high frequency switching. A new design approach for the low valued DC link capacitor of VSI is proposed. The proposed water pumping system is designed and modeled such that the performance is not affected even under the dynamic conditions. Suitability of the proposed system under dynamic conditions is demonstrated by the simulation results using MATLAB/Simulink software.

Keywords— *SPV array, Zeta converter, INC-MPPT, BLDC motor, Electronic commutation.*

I. INTRODUCTION

Drastic reduction in the cost of power electronic devices and annihilation of the fossil fuels in near future invite to use the solar photovoltaic (SPV) generated electrical energy for various applications as far as possible. Water pumping, a standalone application of the SPV array generated electricity is receiving wide attention now a days for irrigation in the fields, household applications and industrial usage. Although the several researches have been carried out in the area of SPV array fed water pumping, combining various DC-DC converters and motor drives, the zeta converter in association with the permanent magnet brushless DC (BLDC) motor is still unexplored to develop such kind of system. However, the zeta converter has been used in some other SPV based applications [1-4]. The merits of both the BLDC motor and zeta converter can contribute to develop a favorable SPV array fed water pumping system possessing the potential of operating satisfactorily under the dynamically changing atmospheric conditions. The BLDC motor has high reliability, high efficiency, high torque/inertia ratio, improved cooling, low radio frequency interference and noise and requires practically no maintenance [5-6]. On the other hand, a zeta converter exhibits following advantages over the conventional buck, boost, buck-boost converter and Cuk converter when employed in SPV based applications.

— Belonging to the family of buck-boost converters, the zeta converter can be operated either to increase or to decrease the output voltage. This property offers a boundless region for maximum power point tracking (MPPT) of the SPV array [7]. The MPPT can be performed with simple buck and boost converter if the MPP occurs within the prescribed limits.

— The aforementioned property also facilitates the soft starting of the BLDC motor unlike a boost converter which habitually step up the voltage level at its output, not ensuring the soft starting.

— Unlike a simple buck-boost converter, the zeta converter has a continuous output current. The output inductor makes the current continuous and ripple free. However, a small ripple filter may be required at the input to smoothen the input current.

— Although consisting of the same number of components as the Cuk converter, the zeta converter operates as non-inverting buck-boost converter unlike an inverting buck-boost and Cuk converter. This property obviates the requirement of associated circuits for negative voltage sensing hence reduces the complexity and probability of slow down the system response [8].

The merits of the zeta converter mentioned above are favorable for the proposed SPV array fed water pumping system. An incremental conductance (INC) MPPT algorithm [9-10] is used to operate the zeta converter such that the SPV array always operates at its MPP and the BLDC motor experience a reduced current at the starting. A three phase voltage source inverter (VSI) is operated by fundamental frequency switching for the electronic commutation of BLDC motor [6]. Simulation results using MATLAB/Simulink software is examined to demonstrate the starting, dynamics and steady state behavior of the proposed water pumping system subjected to the random variation in the solar irradiance. The SPV array is designed such that the proposed system always exhibits satisfactory performance regardless of the solar irradiance level or its variation.

This paper is organized as follows. Configuration and operation of the proposed system are illustrated in section II and section III respectively. Section IV presents the design of the various stages of the proposed system. The control techniques used are briefly described in section V. Finally, the performance of the proposed system is evaluated using the simulated results in section VI followed by the concluding remarks in section VII.

978-1-4799-6047-7/14 $31.00 © 2014 IEEE

II. CONFIGURATION OF THE PROPOSED SYSTEM

The structure of the proposed SPV array fed BLDC motor driven water pumping system employing a zeta converter is shown in Fig. 1. As shown in Fig. 1, the proposed system consists of (left to right) the SPV array, the zeta converter, the VSI, the BLDC motor and the centrifugal water pump. The BLDC motor has an inbuilt encoder. The pulse generator is used to operate the zeta converter. The step by step operation of the proposed system is reported in the following section in detail.

III. OPERATION OF THE PROPOSED SYSTEM

The SPV array generates the electrical power demanded by the motor-pump system. This electrical power is fed to the motor-pump system via the zeta converter and the VSI. SPV array appears as the power source for the zeta converter as shown in Fig. 1. Ideally, the same amount of power is transferred at the output of zeta converter which appears as the input source for the VSI. In practice, due to the various losses associated with a DC-DC converter [11], slightly less amount of the power is transferred to feed the VSI. The pulse generator generates, through INC-MPPT algorithm, the switching pulse for the IGBT (Insulated Gate Bipolar Transistor) switch of the zeta converter. The INC-MPPT algorithm takes the voltage and current variables as feedback from SPV array and returns an optimum value of duty cycle. Further, the pulse generator generates actual switching pulse by comparing the duty cycle with the high frequency carrier wave. In this way, the maximum power extraction and hence the efficiency optimization of the SPV array is accomplished.

On the other hand, VSI converting the DC power output from the zeta converter into the AC power feeds the BLDC motor to drive the centrifugal pump coupled to its shaft. The VSI is operated by the fundamental frequency switching availed by the so called electronic commutation of BLDC motor assisted by its built-in encoder. The high frequency switching losses are thereby eliminated, contributing in the effective and increased efficiency operation of the proposed water pumping system.

IV. DESIGN OF THE PROPOSED SYSTEM

The various operating stages shown in Fig. 1 are intellectually designed in order to develop an effective water pumping system, capable of operating under uncertain conditions. A BLDC motor of 2.89 kW power rating and the SPV array of 3.4 kW maximum power capacity under standard test conditions (STC) are selected to design the proposed system. The detailed design of the various stages such as the SPV array, the zeta converter and the centrifugal pump are described as follows.

A. Design of SPV Array

As per the discussion in section III, the practical converters are associated with the various power losses. In addition, the performance of the BLDC motor-pump is influenced by the mechanical and electrical losses associated with them. To compensate these losses, the size of SPV array is selected with slightly more maximum power capacity to ensure the satisfactory operation regardless of the power losses. Therefore the SPV array of maximum power capacity of P_{mpp} = 3.4 kW under STC (STC: 1000W/m², 25°C, AM 1.5), slightly more than demanded by the motor-pump is selected and its parameters are designed accordingly. Sunmodule® *Plus SW 280 mono* [12] SPV module made by SolarWorld is selected to design the SPV array of an appropriate size. Electrical specifications of this module are listed in Table I and the numbers of modules required to connect in series/parallel are estimated by selecting the voltage of the SPV array at MPP under STC as, V_{mpp} = 187.2 V.

The current of the SPV array at MPP, I_{mpp} is hence estimated as,

$$I_{mpp} = P_{mpp}/V_{mpp} = 3.4/0.1872 = 18.16 \text{ A} \qquad (1)$$

The numbers of modules required to connect in series are as,

$$N_s = V_{mpp}/V_m = 187.2/31.2 = 6 \qquad (2)$$

The numbers of modules required to connect in parallel are as,

$$N_p = I_{mpp}/I_m = 18.16/9.07 = 2 \qquad (3)$$

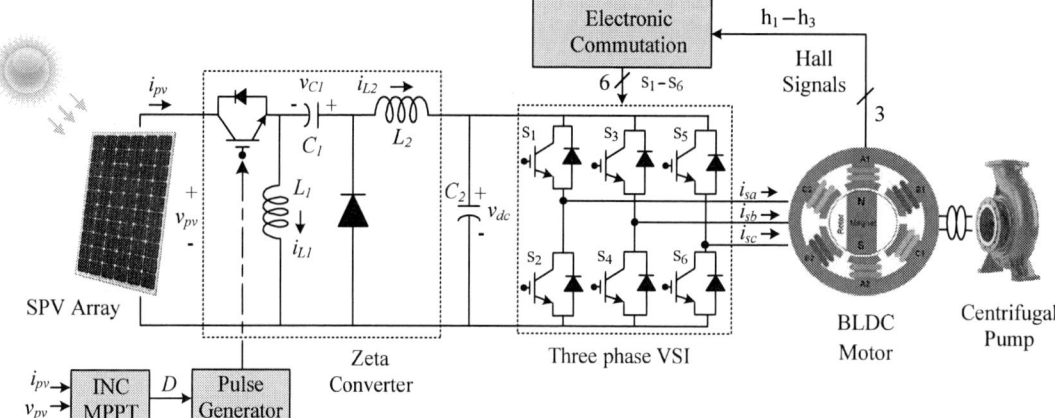

Fig.1 Configuration of proposed SPV array-Zeta converter fed BLDC motor drive for water pumping system.

TABLE I. ELECTRICAL SPECIFICATIONS OF SUNMODULE® PLUS SW 280 MONO SPV MODULE

Peak power, P_m (Watt)	280
Open circuit voltage, V_o (V)	39.5
Short circuit current, I_s (A)	31.2
Voltage at MPP, V_m (A)	9.71
Current at MPP, I_m (A)	9.07
Number of cells connected in series, N_{ss}	60

Connecting 6 and 2 modules respectively in series and parallel, the SPV array of required size is designed for the proposed system and its detailed data are given in Appendix A.

B. Design of Zeta Converter

The zeta converter is the next stage to the SPV array. Its design consists of the estimation of the various components such as input inductor, L_1, output inductor, L_2 and intermediate capacitor, C_1. These components are so designed that the zeta converter always operated in continuous conduction mode resulting in the reduced stress on them. Estimation of the duty cycle, D initiates the design of the zeta converter which is estimated as [6],

$$D = \frac{V_{dc}}{V_{dc} + V_{mpp}} = \frac{200}{200 + 187.2} = 0.52 \qquad (4)$$

where V_{dc} is an average value of output voltage of the zeta converter (DC link voltage of the VSI) equal to the DC voltage rating of the BLDC motor.

An average current flowing through the DC link of the VSI, I_{dc} is estimated as,

$$I_{dc} = P_{mpp}/V_{dc} = 3400/200 = 17 \text{ A} \qquad (5)$$

Then L_1, L_2 and C_1 are estimated as [6],

$$L_1 = \frac{DV_{mpp}}{f_{sw}\Delta I_{L1}} = \frac{0.52*187.2}{20000*18.16*0.06} = 4.5*10^{-3} \approx 5\,\text{mH} \quad (6)$$

$$L_2 = \frac{(1-D)V_{dc}}{f_{sw}\Delta I_{L2}} = \frac{(1-0.52)*200}{20000*17*0.06} = 4.7*10^{-3} \approx 5\,\text{mH} \quad (7)$$

$$C_1 = \frac{DI_{dc}}{f_{sw}\Delta V_{C1}} = \frac{0.52*17}{20000*200*0.1} = 22\,\mu\text{F} \qquad (8)$$

where f_{sw} is the switching frequency of IGBT switch of the zeta converter; ΔI_{L1} is the amount of permitted ripple in the current flowing through L_1, same as I_{mpp}; ΔI_{L2} is the amount of permitted ripple in the current flowing through L_2, same as I_{dc}; ΔV_{C1} is the amount of permitted ripple in the voltage across C_1, same as V_{dc}.
Detailed data of the zeta converter is given in Appendix B.

C. Estimation of DC Link Capacitor of VSI

A new design approach for the estimation of DC link capacitor of the VSI is presented in this sub-section. This approach is based on a fact that 6th harmonic component of the supply (AC) voltage is reflected on the DC side as a dominant harmonic in the three phase supply system [13]. Here, the fundamental frequencies of the output voltage of the VSI are estimated corresponding to the rated speed and the minimum speed of the BLDC motor essentially required to pump the water. These two frequencies are further used to estimate the values of their corresponding capacitors. Out of the two estimated capacitors, larger one is selected to assure the satisfactory operation of the proposed system even under the duration of minimum solar irradiance level.

The fundamental output voltage frequency of the VSI corresponding to the rated speed of BLDC motor, ω_{rated} is estimated as,

$$\omega_{rated} = 2\pi f_{rated} = 2\pi \frac{N_{rated}P}{120} = 2\pi*\frac{3000*6}{120} = 942\,\text{rad/sec.} \quad (9)$$

The fundamental output voltage frequency of the VSI corresponding to the minimum speed of the BLDC motor essentially required to pump the water ($N = 1100$ rpm), ω_{min} is estimated as,

$$\omega_{min} = 2\pi f_{min} = 2\pi \frac{NP}{120} = 2\pi*\frac{1100*6}{120} = 345.57\,\text{rad/sec.} \quad (10)$$

where f_{rated} and f_{min} are the fundamental output voltage frequencies of the VSI corresponding to the rated speed and the minimum speed of the BLDC motor essentially required to pump the water respectively, in Hz; N_{rated} is rated speed of the BLDC motor; P is the numbers of poles in the BLDC motor.

The value of DC link capacitor of the VSI corresponding to ω_{rated} is as,

$$C_{2,rated} = \frac{I_{dc}}{6*\omega_{rated}*\Delta V_{dc}} = \frac{17}{6*942*200*0.1} = 150.4\,\mu\text{F} \quad (11)$$

Similarly, the value of DC link capacitor of the VSI corresponding to ω_{min} is as,

$$C_{2,min} = \frac{I_{dc}}{6*\omega_{min}*\Delta V_{dc}} = \frac{17}{6*345.57*200*0.1} = 410\,\mu\text{F} \quad (12)$$

where ΔV_{dc} is the amount of permitted ripple in the voltage across the DC link capacitor, C_2.

Finally, $C_2 = 410\,\mu$F is selected to design the DC link capacitor.

D. Design of Centrifugal Pump

To estimate the proportionality constant, K for the selected centrifugal water pump, its torque-speed characteristics [14] is used as,

$$T_L = K\omega_r^2 \qquad (13)$$

where T_L is the load torque offered by the centrifugal pump which is equal to the electromagnetic torque developed by the BLDC motor under steady state for stable operation and ω_r is the mechanical speed of the rotor in rad/sec. Since the rated torque, T_L and the rated speed, N_{rated} of the selected BLDC motor is 9.2 Nm and 3000 rpm respectively, the proportionality constant, K is estimated using (13) as,

$$K = \frac{T_L}{\omega_r^2} = \frac{9.2}{(2\pi * 3000/60)^2} = 9.32 * 10^{-5} \quad (14)$$

The centrifugal pump with this data is selected for the proposed system and its detailed data are given in Appendix C.

V. CONTROL OF THE PROPOSED SYSTEM

The proposed system is controlled at two stages. These two control techniques namely, MPPT and electronic commutation are discussed in brief as follows.

A. INC-MPPT Algorithm

An efficient and commonly used INC-MPPT technique [9] in various SPV array based applications is utilized in order to optimize the power available from the SPV array and to facilitate the soft starting of the BLDC motor. Selecting an optimum value of perturbation size ($\Delta D = 0.001$) not only avoids the oscillations around the MPP but provides the soft starting of the BLDC motor also. An intellectual agreement between the tracking time and the perturbation size is held to fulfill the objectives.

B. Electronic Commutation

The BLDC motor is controlled by the VSI operated through the electronic commutation of BLDC motor. 6 switching pulses are generated as per the various possible combinations of 3 Hall-effect signals. These 3 Hall-effect signals are produced by the inbuilt encoder according to the rotor position. A particular combination of the Hall-effect signal is produced for specific range of rotor position [6]. The electronic commutation provides fundamental frequency switching of the VSI, hence the losses associated with the high frequency switching is completely eliminated. TETRA 115TR9.2, a BLDC motor of motor power company [15] with inbuilt encoder is selected for the proposed system and its detailed data are given in Appendix C.

VI. RESULTS AND DISCUSSION

Performance evaluation of the proposed SPV array fed BLDC motor driven water pumping system employing zeta converter is carried out using simulated results in MATLAB/Simulink. The proposed system is designed, modelled and simulated considering the random and instant variation in solar irradiance level and its suitability is demonstrated by testing the starting, steady state and dynamic behaviour.

Fig.2 presents the starting, steady state and dynamic performance of the proposed water pumping system. To demonstrate the suitability of the proposed system under dynamic condition, solar irradiance level is varied as indicted in Table II. Behaviour of the various stages such as the SPV array, zeta converter and BLDC motor-pump are depicted on an individual basis in the following sub-sections.

A. Performance of SPV Array

The performance of the maximized power SPV array used to feed the water pumping system is shown in Fig. 2(a). The solar irradiance level, S is varied following the sequence indicated in Table II. Other variables such as the SPV array voltage, v_{pv}, SPV array current, i_{pv} and the SPV array power, P_{pv} are varied accordingly. The presented results manifest that the maximum power available from the SPV array is extracted regardless of the irradiance level and its dynamic variation. Since it is desired to achieve the soft starting of the BLDC motor, the MPP is tracked appropriately at the starting. It is clear from Fig. 2(a) that the INC-MPPT algorithm is allowed, at the starting, to take more time for maximum power extraction by selecting an optimum value of perturbation size in order to achieve the soft starting of BLDC motor. Oscillation around the MPP is also reduced by properly selecting the optimum value of perturbation size.

Under steady state condition, at the standard value of solar irradiance i.e. 1000 W/m², all the variables possess their rated values while they possess the minimum values at minimum solar irradiance i.e. 200 W/m².

B. Performance of Zeta Converter

Fig. 2(b) clarifies the various performances of the zeta converter. All the variables viz. the current flowing through the input inductor, i_{L1}, the voltage across the intermediate capacitor, v_{C1}, the current flowing through the output inductor, i_{L2} and the voltage at the output (DC link voltage of the VSI), v_{dc} comply the variation in the solar irradiance level. Regardless of the irradiance level, the zeta converter is always operated in continuous conduction mode.

Unlike a simple buck-boost converter, the zeta converter has positive polarity voltage at its output as shown in Fig. 2(b) which reaches, under steady state, the rated DC voltage of the BLDC motor at 1000 W/m² of solar irradiance level. Moreover, at the solar irradiance level of 200 W/m², it provides a DC voltage level to the BLDC motor sufficient to make attained more than the required speed to pump the water. Small amount of ripples in the zeta converter variables are observed caused by permitting the ripples up to an extent in order to reduce the size of the components.

C. Performance of BLDC Motor-Pump

Performance of the BLDC motor-pump is shown in Fig. 2(c). Following points are clearly observed from the presented simulation results.

— The motor pump variables viz. the back EMF, e_a, the stator current, i_{sa}, the rotor speed, N, the electromagnetic torque, T_e and the pump load torque, T_L are abide by the variation in solar irradiance.

— At the starting, the rate of rise of stator current is decreased as an evidence of soft starting of the BLDC motor.

— The motor-pump variables reach their rated values under steady state at 1000 W/m², standard value of solar irradiance. However, it should be highlighted that the motor always attains a higher speed than minimum speed required to pump the water i.e. 1100 rpm (even at 200 W/m²) regardless of the solar irradiance level.

— The electromagnetic torque developed by BLDC motor is same as torque required by the centrifugal pump. This torque balance between the BLDC motor and the centrifugal pump irrespective of the solar irradiance variation verifies the stable operation of the proposed system.

— A small and acceptable pulsation in the electromagnetic torque is observed because of the electronic commutation and reflection of the ripples present in the DC link current of VSI.

— Fast and precise response of the BLDC motor subjected to the dynamic variation in solar irradiance is undoubtedly ascertained by the simulated results.

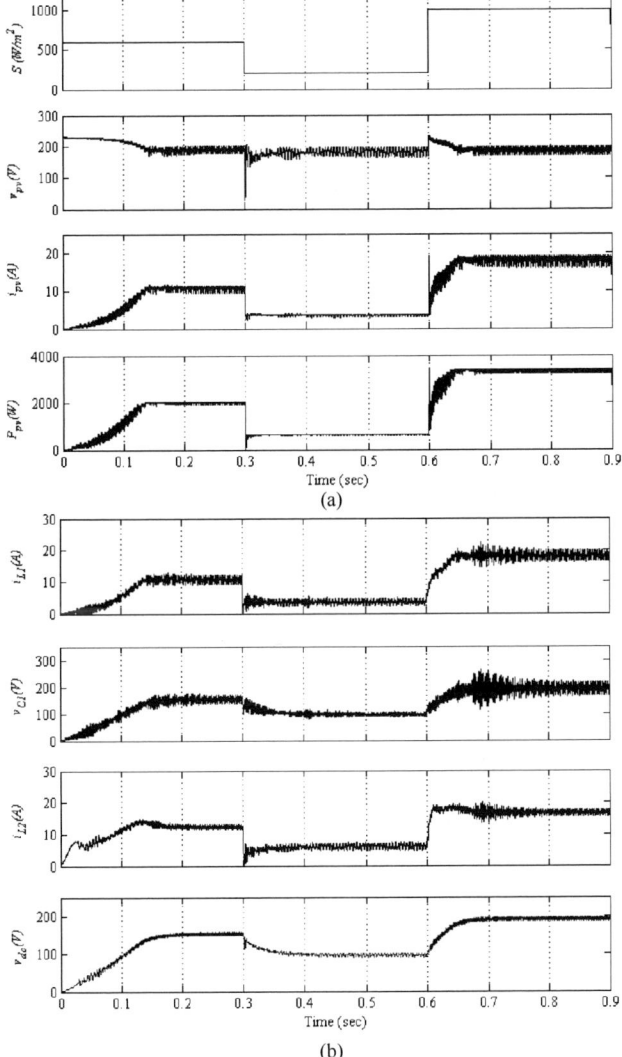

(a)

(b)

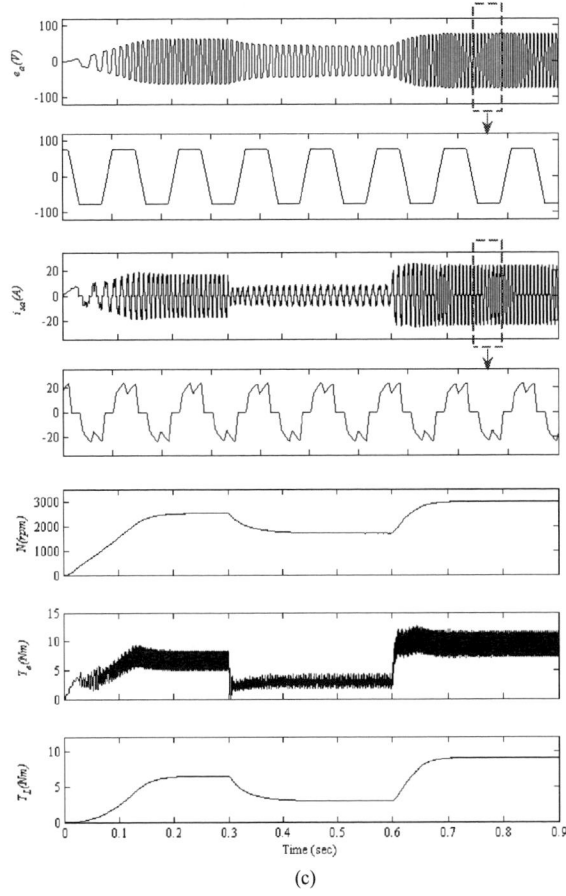

(c)

Fig.2 Performances of the proposed SPV array based Zeta converter fed BLDC motor drive for water pumping system (a) SPV array variables, (b) Zeta converter variables, and (c) BLDC motor-pump variables.

TABLE II.   VARIATION IN SOLAR IRRADIANCE LEVEL

Solar Irradiance Level, S (W/m²)	Duration (Sec.)
600	0.0 - 0.3
200	0.3 - 0.6
1000	0.6 - 0.9

VII. CONCLUSIONS

The SPV array-zeta converter fed VSI-BLDC motor-pump for water pumping has been proposed and its suitability has been demonstrated by simulated results using MATLAB/Simulink and its sim-power-system toolbox. First, the proposed system has been designed logically to fulfil the various desired objectives and then modelled and simulated to examine the various performances under starting, dynamic and steady state conditions. The performance evaluation has justified the combination of zeta converter and BLDC motor drive for SPV array based water pumping. The system under study availed the various desired functions such as MPP extraction of the SPV array, soft starting of the BLDC motor, fundamental frequency switching of the VSI resulting in a reduced switching losses, reduced stress on IGBT switch and the components of zeta converter by operating it in continuous conduction mode and stable operation. Moreover, the proposed system has operated successfully even under the minimum solar irradiance.

APPENDIX A

Parameters of solar PV array: Open circuit voltage, V_{oc} = 237 V; Short circuit current, I_{sc} = 19.42 A; Maximum power, P_{mpp}= 3.4 kW; Voltage at MPP, V_{mpp} = 187.2 V; Current at MPP, I_{mpp}= 18.16 A; Numbers of cells connected in series in a module, N_{ss}= 60; Numbers of modules connected in series, N_s = 6; Numbers of modules connected in parallel, N_p = 2.

APPENDIX B

Parameters for Zeta converter: Switching frequency, f_{sw}= 20 kHz; Input inductor, L_1 = 5 mH; Intermediate capacitor, C_1 = 22 µF; Output inductor, L_2 = 5mH; DC link Capacitor, C_2 = 410 µF.

APPENDIX C

Parameters for BLDC Motor-Pump: Stator phase/phase resistance, R_s = 0.36 Ω; Stator phase/phase inductance, L_s= 1.3 mH; Torque constant, K_t = 0.49 Nm/A$_{peak}$; Voltage constant, K_e = 51 V$_{peak}$L-L/krpm; Rated current, I_{srated} = 18.9 A; Rated torque, T_{rated} = 9.2 Nm; Rated speed, N_{rated} = 3000 rpm @ 200 V DC; Rated power, P_{rated} = 2.89 kW; No. of poles, P = 6; Moment of inertia, J = 17.5 kg.cm^2; Proportionality constant, K = 9.32*10^{-5}.

ACKNOWLEDGMENT

Authors are very thankful to Department of Science and Technology (DST), Govt. of India, for supporting this work under Grant Number: RP02926.

REFERENCES

[1] M. Uno and A. Kukita, "Single-Switch Voltage Equalizer Using Multi-Stacked Buck-Boost Converters for Partially-Shaded Photovoltaic Modules," *IEEE Transactions on Power Electronics*, no. 99, 2014.

[2] R. Arulmurugan and N. Suthanthiravanitha, "Model and Design of A Fuzzy-Based Hopfield NN Tracking Controller for Standalone PV Applications," *Electr. Power Syst. Res.* (2014). Available: http://dx.doi.org/10.1016/j.epsr.2014.05.007

[3] S. Satapathy, K.M. Dash and B.C. Babu, "Variable Step Size MPPT Algorithm for Photo Voltaic Array Using Zeta Converter - A Comparative Analysis," *Students Conference on Engineering and Systems (SCES)*, pp.1-6, 12-14 April 2013.

[4] A. Trejos, C.A. Ramos-Paja and S. Serna, "Compensation of DC-Link Voltage Oscillations in Grid-Connected PV Systems Based on High Order DC/DC Converters," *IEEE International Symposium on Alternative Energies and Energy Quality (SIFAE)*, pp.1-6, 25-26 Oct. 2012.

[5] G. K. Dubey, *Fundamentals of Electrical Drives*, 2nd ed. New Delhi, India: Narosa Publishing House Pvt. Ltd., 2009.

[6] B. Singh and V. Bist, "A Single Sensor Based PFC Zeta Converter Fed BLDC Motor Drive for Fan Applications," *Fifth IEEE Power India Conference*, pp.1-6, 19-22 Dec. 2012.

[7] R.F. Coelho, W.M. dos Santos and D.C. Martins, "Influence of Power Converters on PV Maximum Power Point Tracking Efficiency," *10th IEEE/IAS International Conference on Industry Applications (INDUSCON)*, pp.1-8, 5-7 Nov. 2012.

[8] Dylan D.C. Lu and Quang Ngoc Nguyen, "A Photovoltaic Panel Emulator Using A Buck-Boost DC/DC Converter and A Low Cost Micro-Controller," *Solar Energy*, vol. 86, issue 5, pp. 1477-1484, May 2012.

[9] Zhou Xuesong, Song Daichun, Ma Youjie and Cheng Deshu, "The Simulation and Design for MPPT of PV System Based on Incremental Conductance Method," *WASE International Conference on Information Engineering (ICIE)*, vol.2, pp.314-317, 14-15 Aug. 2010.

[10] Ali Reza Reisi, Mohammad Hassan Moradi and Shahriar Jamasb, "Classification and Comparison of Maximum Power Point Tracking Techniques for Photovoltaic System: A review," *Renewable and Sustainable Energy Reviews*, vol. 19, pp. 433-443, March 2013.

[11] A. Shahin, A. Payman, J.-P. Martin, S. Pierfederici and F. Meibody-Tabar, "Approximate Novel Loss Formulae Estimation for Optimization of Power Controller of DC/DC Converter," *36th Annual Conference on IEEE Industrial Electronics Society*, pp.373-378, 7-10 Nov. 2010.

[12] Sunmodule® *Plus SW 280 mono*, Performance Under Standard Test Conditions [Online]. Available: http://www.sfe-solar.com/wp-content/uploads/2013/07/SunFields-SolarWorld_SW265-270-275-280_Mono_EN.pdf

[13] K.H. Ahmed, M. S. Hamad, S.J. Finney and B.W. Williams, "DC-Side Shunt Active Power Filter for Line Commutated Rectifiers to Mitigate the Output Voltage Harmonics," *IEEE Energy Conversion Congress and Exposition (ECCE)*, pp.151-157, 12-16 Sept. 2010.

[14] W.V. Jones, "Motor Selection Made Easy: Choosing the Right Motor for Centrifugal Pump Applications," *IEEE Industry Applications Magazine*, vol.19, no.6, pp.36-45, Nov.-Dec. 2013.

[15] TETRA 142TR12, Brushless Servomotors [Online]. Available: http://www.eltrex-motion.com/fileadmin/user_upload/PDF/product/Catalogue_TETRA_TR_ENG.pdf

A Practice of Power Paste with the Isolated Flyback Converters

T. Halder

Assistant Professor

Electrical Engineering Department, Member IEEE

Kalyani Government Engineering College

Kalyani, District-Nadia, West Bengal

PIN-741235, INDIA

Abstract—The paper appropriately presents to popularize a sizeable copy of green power from the solar energy carried out by the flyback converters. The copy and paste of the free energy is married by a straight forward elemental optimization technique at which the conventional energy is shut down during the on grid to save power. The renewable energy is exclusively copied in backup systems and pasted for the off grid utilization or hurling over surplus power at the nearby grids to crack the monopole business of the power utilities. Last but not least, substantial solar power agriculture is accomplished by copy and paste modes of the smart operations.

Keywords—Green power; Power cut; Power copy; Power paste; The flyback converter, Basic optimization; Power farming; Operations; Novel strategy power trade and Results

I. INTRODUCTION

The electrical charges are accumulated in the photo voltaic (PV) booths and estimated to the load voltage terminal to generate low voltage commonly 6 to 30V (DC), [1]. The standard output is proposed for constant voltage 12V, with an effectual load commonly up to 16V. A 12V ostensible load is compared with the reference voltage, other than the working voltage is fixed at 18V or higher than the automobile alternator 12V of batteries at fine over 12V [2].

So there is variation in between the reference voltage and the definite working voltage. The potency of the Sun's solar energy emission varies throughout the day, point in time of the day and type of weather situation.

To be competent to compile calculation in forecast of a system [3], the substantial magnitude of solar waves power is expressed in hours of satiated sunshine per m², or peak sun hours (PSH) [4]-[5]. The turn of phrase, PSH stands for the suggesting sum of sun availability of a day all over the time.

It is implicit that at peak sun (PS), 1000 W/m² of power arrives at the earth surface. One hour presence of the sun approximately supplies 1 kWh/m² in place of the solar energy conventional in one hour on a sunny summer day on a one square meter surface intended for towards the sun [6]-[7].

One additional standpoint is that approximately 102 square miles of solar floorboard placed in the solar booths to be found in the Thar Desert of India could power the motherland. The day after day standard of PSH, supported on either

comprehensive year of a stature, or most terrible month of the year information, for paradigm, is used for assessment purpose in the proposal. To look into the standard of PSH for a solar power sector, confers upon the solar abacus [8]-[10].

Thus in this manuscript, it envisages that the power of a governmental subsidized system varies, relying on the ecological sites. Folks in India will necessitate more solar panels in the system to generate the same largely power as those living in near the desert neighbourhood of Rajasthan and some part of Gujarat. Indian technological support legislative body advises to power utilities on more power generation if power utilities hem in reservations about the site selections as improved performance of flyback converter which is used to cut the conventional power , copy the solar power and paste in the battery bank for the executive of the power crisis for power utilities. In this paper, it focuses on the free solar energy harvesting; if the harvesters use this technique of power harvesting in the vast vacant land of Thar Desert and the back side of the garments, it will award us the full electrical energy of INDIA as mother resources [11]-[15].

II. ENERGY CONSUMPTION IN INDIA

Planet oil and gas reserves are probable at about 46 years and 66 years respectively. Coal is to be expected to finishing stages a slight over about 200 years. The per capita power utilization is besides near to the ground for India as measured up to industrial realms. It is presently about 4% to 20% of the planet. The per capita power consumption is to be expected to nurture in India with augmentation in financial system thus emergent the power demand. Further, it would be lucid by the following chart as shown in the Fig. 1 as per statics of 2003 and unit of the in million tons of oil equivalents (MTOE).

The globe standard power of utilization per a big shot is about 2.2 tons of coal equivalents. In mechanized states, public make use of four or five times more than the globe usual and nine times more than the standard for the emergent kingdoms. American consumes 32 times more profitable power than an Indian. Oil financial records for India is about 37% of India's entirety power consumption. India at present is one of the acme ten oil-guzzling inhabitants in the planet and almost immediately goes beyond Korea as the third biggest customer of oil in Asia after China and Japan. Vestige fuel power stations have shown the manner to a requirement to consume vast quantity of renewable energy. Dominant forms such as in the parallel, growing concerns over environmental

978-1-4799-6047-7/14 $31.00 © 2014 IEEE

damage. Solar energy is very much inconsistent, and consequently the requirement of additional refined control. Photovoltaic section on construction roofs, but also the utilize of fuel cells, charging the batteries of electric vehicles.

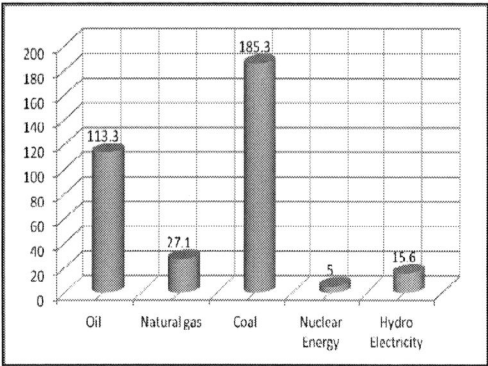

Fig. 1. Energy consumption by fuel.2003

India has currently peak demand deficiency approximately 15% and power deficit about 9%. Observance of this outlook and GDP improvement from 8% to 11%, the administration of India involves very judiciously place an objective about 216000 MW power generation capacities by March of 2012 from the height of 10000 MW as on March 2001. In the section of nuclear power, the goal is to pull off 20000 MW of nuclear power production capability by the year of 2020.

III. CONVERSION EFFICIENCY OF SOLAR CELL

Power regulation competency is a turning of phrase of the amount of energy generated in proportion to the quantity of power addicted or reachable to a clout machine. The sun produces substantial quantity of green power in an extensive illumination and warm spectrum, but the clients comprise so far cultured to capture only elfin fraction of that spectrum and transfer them to electrical energy by photovoltaic.

Fig. 2. Basic Solar battery charging unit and charge controller

Accordingly, today's saleable PV systems are about 6% to 16% competent, which might come across as if diminutive features and a lot of big PV coordination humiliate a modest small piece (drop of power) at every year is leading exposure of daylight. For the assessment, a common fossil fuel power station occupies standard efficiency of about 32%. Power customer is working on background to renovate more of the power in light to serviceable power and improve the efficiency of solar systems. However, various tentative PV modules now reassign almost 41% of the energy in light to electrical energy through power converters as shown in Fig. 2. In solar thermal

structure like solar water heating roof panels, solar cookers, efficiency goes down as the solar warmness is rehabilitated to a shuffle transitional such as water. Also, some of the temperature radiates away from the coordination before it can be used as a free preserve of solar power.

IV. HAPHAZARD POPULARISATION OF SOLAR ENERGY

The population patches up on to procure solar energy systems for a multiplicity of grounds. As for example, various consumers pay for solar goods to conserve the Earth's limited fossil fuel assets and to incise environment infectivity. Others would rather pay out their cremation on an energy producing opening out to their possessions than pressurize their capital to a power utility. Some natives are fond of the precautions of reducing the amount of electricity, they purchase from their utilities, because it formulates them less in danger to outlook adds to in the outlay of electricity. For a rising figure of consumers, PV is the lucid preference. Power users should definitely mull over using a PV system if it operates better and overheads less than the alternatives. The outlay of energy generated by PV systems carries on plunging. However, kilowatt-hour (kWh) for kilowatt-hour, and relying on where consumers survive, PV power still habitually expenses more than energy from his neighbouring power utility. Moreover, the preliminary charge of PV utensils is higher than that of an engine generator. But there are many appliances for which a PV system is the most cost-effective long-term golden opportunity, such as for power in country side. The figure of installed PV systems increases every year because their much bootie makes them the best preference for the large amount.

V. SOLAR ENERGY DEVELOPMENT IN INDIA

India is thickly occupied and has high solar segregation and peaceful grouping by solar energy in India where wind power generation is shown in TABLE I as a global leader. In the planetary power sector, some pilot projects have been planned about 35,000 Km^2 area of the Thar Desert has been set aside for solar power projects, adequate to produce 700 GW to 2100 GW as considerable amount of power mishap managements. The Indians ministry has taken initiative fresh and renewable energy of the JNNSM, Phase -2, plan of strategies by which the Government be set to put in 10GW of solar power unit and of this 10 GW goal, 4 GW would drop under the important plan and the rest 6 GW under miscellaneous shape of open layout.

In July 2009, India related 100 billion rupees of plan to generate 20 GW of solar power by 2020. Under the plan, the use of solar powered devices and applications would be made obligatory in all administration erection as well as sickbay and hotels. On 18 November 2009, it was well-versed that India was lay down to commence its national solar mission (NAM) under the nationwide achievement on atmospheric alteration, with plans to turn out 1,000 MW of power by 2013. Inauguration in august 2011 to July 2012, India set out from 2.5 MW of grid connected photovoltaic's to over 1000 MW.

In accordance with a 2011 avowal by BRIDGE TO INDIA and GTM research, India is in front of a ideal tempest of aspects that will force solar photovoltaic (PV) acceptance at a discoloured rapidity over the subsequently five years and beyond.

978-1-4799-6047-7/14 $31.00 © 2014 IEEE

TABLE I.

Name of the Plant	Capacity (MW)
Charanka Solar Park - Charanka village, Patan district, Gujarat	221
Dhirubhai Ambani Solar Park, Pokhran, Rajasthan	40
Bitta Solar Power Plant (Adani Power) - Bitta, Kutch District, Gujarat	40
Moser Baer - Patan, Gujarat	30
Tata Power) - Mithapur, Gujarat	25
Waa Solar Power Plant (Madhav Power) - Surendranagar, Gujarat	10
Green Infra Solar Energy Limited - Rajkot, Gujarat	10
Azure Power - Sabarkantha, Khadoda village, Gujarat	10
Orissa - Patapur, Orissa	9
Sivaganga Photovoltaic Plant, Tamil Nadu	5
MAhindra & MAhindra Solar Plant, Jodhpur, Rajasthan	5
Citra and Sepset Power Plants, Katol, MAharashtra	4
Tata Power - Mulshi, MAharashtra	3
Kolar Photovoltaic Plant, Yalesandra, Kolar District, Karnataka	3
Itnal Photovoltaic Plant, Belgaum, Karnataka	3
IIT Bombay - Gwal PAhari, Haryana	3
TAL Solar Power Plant - Barabanki, Uttar Pradesh	2
Jamuria Photovoltaic Plant, West Bengal	2
Azure Power - Ahwan Photovoltaic Plant, Punjab	2
Zynergy, Vannankulam village, Peraiyur, Madurai district, Tamil Nadu	1
Thyagaraj stadium Plant - Delhi	1
Tata Power - Patapur, Orissa	1
Tata Power - Osmanabad, MAharastra	1
REHPL - Sadeipali, (Bolangir) Orissa	1
Rasna Marketing Services LLP, Ahmedabad, Gujarat	1
Numeric Power Systems, Coimbatore, Tamil Nadu	1
NDPC Photovoltaic Plant, Delhi	1
Gandhinagar Solar Plant, Gujarat [32]	1
B&G Solar Pvt Ltd - Mayiladuthurai, Tamil Nadu	1
Amruth Solar Power Plant - Kadiri, Andhra Pradesh	1
Chandraleela Power Energy - Narnaul, Haryana	0.8

The dilapidated prices of PV panels, mostly from China but also from the U.S., have agreed with the mounting price of grid power in India. Government support and abundant solar assets have also aid to hoist solar accomplishment, but probably the important subject has been dictated. India, as an intensifying monetary organism with an undulating central position, is now in front of a persistent electrical energy deficit that an extensive nip flanked by 10 and 13 percent of day after day requirements.

VI. DESIGN CHALLENGES OF THE SOLAR ENERGY HARVESTING

- Recognition of input voltages attuned to electrical energy harvesting generators (EH)
- Power electronic converters for tight load and voltage regulations
- Unremittingly self powered and continuously dynamic
- Charge controller stages of the power converters capturing miniscule unregulated electrical power
- Micro-power and energy depleted on its individual reliable power circuit operations
- Power conversion efficiency relies on substantial of variables
- Storages system, load managements and demand side management of power as comprehensive stages
- Extensive operating life improves system reliability
- Tremendous life-span power source
- Battery charger unit or efficient converter scheme
- Provision based monitoring organism
- High reliability robust networks (i.e. smart grid)
- Battery charger for maintenance-free applications
- Super-capacitor charging substitutes battery-based (chemical) charging stations
- Superfluous power systems as multi stages of the power converter systems

VII. SERVICEABLE NECESSITIES OF SOLAR ENERGY CUTIVATION

- Free power from low-voltage ambient resources
- Capturing , accumulation, storage and management of the load in the systems
- Production utilizable power from small expenditure generators
- Continually huge household of self-powered and self-starting converter circuit
- High power preservation at off grid condition
- Constantly active in energy capturing method
- Output directly drive smart control unit
- Outlasts system deployment lifetime
- Distributed generation improves system reliability
- Practically unconstrained and automated charging or discharging sequences of the energy storage system

VIII. DIMENSION OF BATTERY BANK & SOLAR PANEL

The suitable dimensions of solar structure patrons call for to depends on the copious issues such as how much electricity or hot water or space heat user's use, how much sunlight are obtainable where solar floorboards are, the measurement of a roof, and how much users are enthusiastic to put up with cost. Preferring the definite solar board for needs is like selection a battery bank. In the same manner that a larger battery will grant more power for longer, a bigger solar panel will hoard more energy in less time.

The unadorned aspect as shown in Fig. 3 of a solar board will rely on variables such as the power required by the power appliances taken of time. It requirements for power utilization are how to reach sunshine on the earth surfaces. It gets a clasp at the time of year. There are three decisive factors to imitate on while preferring a solar group or panel or constructing a solar organization. The designers necessitate classifying what appliances interviewee will be using and how much energy they need, how much energy his system battery can accumulate and which solar board will stock up power in the battery bank in line with prototype of exploit If we consider the power density of the solar cell as 150 W/m² for 250W power and heaviness about 186 kilograms, the essential size of solar board will be as per following Fig. 3.

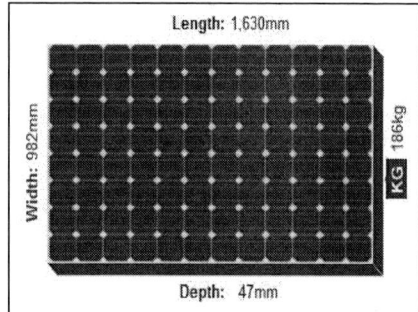

Fig. 3. Dimension of the solar panel for 250W power generation

The groups of four batteries in series at 12V & 280 Ah produce a 12V battery reservoir at 1120 Ah capacity as a huge battery backup power resources as shown in Fig. 4.

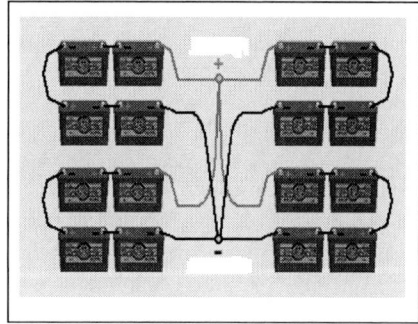

Fig. 4. Sizing & layout of the battery bank

It will be used as power resources during dark period and (±) division is to be connected output of the Flyback converter to charge the battery bank during the sunlight as shown in Fig. 4 as a schematic example for the backup power storage plant. The energy needs for appliances is constant over a period of time. The power consumption of electrical devices is given in Watts. To work out the energy use over time, right away to burgeons the power consumption by the hours of use. As for the instance, 18W load on for 4 hours, will take 18 x 2 =

36Wh from the battery bank. It calls for conversion this to Watt hours by multiplying the Ah (Ampere hour) figure by the battery voltage (i.e. 12V). For a 17Ah, 12V battery the Wh (Watt hours) figure is 17 x 12 = 204Wh. This means the battery could supply an 18W load for 15.6 and a half hour, 204W for 1 hour (approximately). For the solar 250W panel in 4 hours of sunshine, 250 x 4 x 0.85 = 850Wh. This is the amount of energy the solar panel is intelligent to supply to the battery bank.

IX. MODELING OF SOLAR POWER HARVESTING

The energy harvesting inculcates the pertinent assessment of the demand side managements of power, optimal power flow and loss optimizations based on the prediction modeling. This modeling may be formulated as:

$$E\left[R_{EH_{t+1}}\right] = \lambda R_{EH_t} + \frac{\mu}{N}(1-\lambda)\sum_{t=1}^{N} R_{EH_{t+1}} \qquad (1)$$

Where E = Total energy generation,
$R_{EH_{t+1}}$ = Prediction of expectations energy harvesting rate at time (t+1).
R_{EH_t} = Prediction of expectations energy harvesting rate at time (t).
N= Number of mission depending on the operating time.

It is further to be noted that a given a border line situation is the premier stage of the mission utility available using time constrained the circumstances such as massive programmed data compilation. The utility constraint (λ) is giving rise preferred system planning of power utilities, how long time the missions will be carried out. The performance of the system predicts important episode or call for precise domino effect of system upgrading for steady state operations and potential planning. The solar panel is the variable current source, so its output voltage is not constant depending on the intensity of the sunlight thought the day. In order to maintain the tight regulations an isolated flyback converter is used here as least parts counts when it is highly compared with other topologies. Energy and mission management in energy harvesting term may be associated in terms of constant output voltage (V) and power (P) as:

$$\mu = [V][P]\left(\sum_{N=1}^{N} P_N\right)^{-1} \qquad (2)$$

X. SOLAR ENERGY HARVESTING USINF THE FLYBCK CONVERTERS

When the flyback switch (M1) is switched on, the primary inductance is stored up the energy. When the main switch (M1) of the converter is off, the stored energy flies to load under discontinuous mode (DCM) or in a continuous mode of operation (CCM) as shown in the Fig. 5. The particular controller meets up the mandatory to transfer or to process input to output by controlling the pulse width of the controller through pulse width modulation (PWM). The crude output is filtered to attain a smooth dc voltage with excellent and tight regulations. The converter comprises low profile, high power density, compact size, least parts count, high reliability due to high switching of operation.

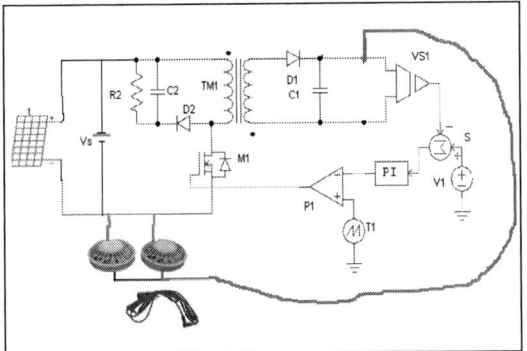

Fig. 5. Basic flyback converter with RCD snubber

The easiest method of the solar energy cultivation is shown in the Fig. 6. An optimal size of solar panel is placed at the back side of a person to generate free electricity for the supportive purposes where a DC fan driven comfortable apparel is designed to feel comfort in hot period. The installation process and load side is shown Fig. 6.

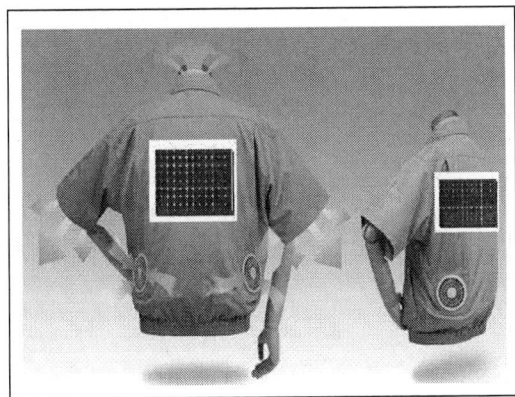

Fig. 6. New method of solar energy harvesting

The planetary power farming is now trendy due to the following aspects.

- Outlay intonation position
- Improved reliability
- Cut in fuel expenditure
- Faster erection and commission
- Improved concealed proposal
- Longer battery life

XI. FREE ENERGY HARVESTING TRADEOFFS

The energy harvesting trade off, takes up merit and demerit under the following aspects as mobile power wires, easier installation, lower maintenance, environmentally pleasant and privileged uptime. The following points may be regarded as demerits depicting dependent on accessibility of harvestable power source, austere power finances, sincere outlay may be higher and less adult technology

XII. DESIGN SPECIFICATIONS OF THE FLYBACK CONVERTER

At The DCM of operations, the Flyback power supply could suit a consumer product, such as a set-top box, a VCR, DVD Recorder, ac to dc adapter, Fax Machine, Copier and Xerox Machine, Printer and Mobile Charger as low power applications.DCM draws more peak current. As a result, a more expensive power switch with a higher current rating is needed. Moreover, the secondary peak currents in the DCM can have transient spike at the instant –off. However, despite all this problems, the DCM is still preferred than CCM for the following basic basses. Naturally smaller magnetizing inductance in the DCM has a faster response and a minor transient spike to rapid change in load current or input voltage. The CCM has a right-half plane zero in its transfer function, thereby making feedback control circuit more difficult to design. Always thanks to DCM. Now, it goes for all DCM for this small power flyback converter. It is also fine practice to start with DCM flyback converter for suitable applications

Input voltage,V_s = 6-30V, DC source of solar panel
Output load voltage, V_o = 12V, Load current, I_o =12A
Maximum Switching frequency, f_{sw} =70 kHz
Restricted Maximum Duty ratio, D_{max}=0.5
Output maximum power, P_{max}= V_oI_o=144W. In order to generate bulk power, huge number of the flyback converters will operate in parallel.

XIII. RESULTS

Substantial energy a neutral stance consumes not only the amount of energy unindustrialized but also take advantage of on performance over time respond to peripheral needs two modes of operation likely transient operation and periodic implementation of outer trigger state operation operating occurrence triggered execution as shown in the Fig. 7. This simulation curved as shown in the Fig. 7 depicts power production versus days as comparative predictions of real power (RP) production during the sunshine, prediction of power generation (PP) purposes and power of the complete management availability (MA).

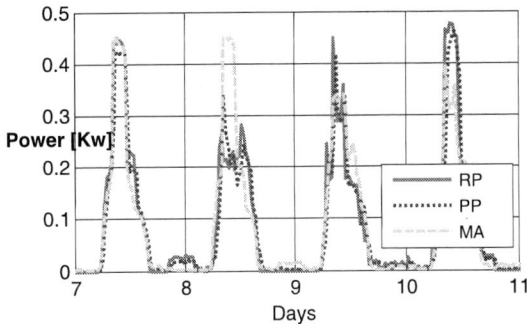

Fig. 7. Power [Kw} vs. days for the power production

The significant compensation of longer (apparently uninterrupted) lifetime is reduced necessitation for individual intercession. The restrictions and predictable generation of

978-1-4799-6047-7/14 $31.00 © 2014 IEEE

energy management techniques are highly limited by power density and presence of sunlight throughout the day as shown in the Fig. 8.

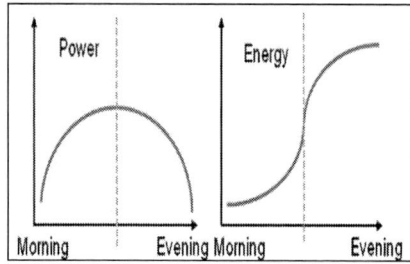

Fig. 8. Power and energy harvesting vs. days

Nowadays, the expenditure of the solar power generation is not cost effective but in future, it will be cheapest rate of power generation for a dominion. Energy or power harvesting is not only depends on the sunbeams but also relies on the inclination of the sun. The solar panel sets up in such a way to accumulate the maximum power. These problems are tackled by the maximum power point tracker. On the other hand, the energy or power harvesting is not only depends on the power conversion efficiency of power module and solar cell but also relies on the dimension of the solar panel. The solar board needs to set up in such a method for the buildup of maximum power. These problems are undertaken and solved by the maximum power point tracker and controlling machinery. The pioneering capturing scheme of solar power is nowadays a bargaining inquiry to synchronize to nearby power grid in near and fur future. The solar power cropping is not only green power generation but also look after the global warming to toughen the open market of power generation. Customers who can plow solar power on the roofs sell surplus power to neighboring grid with smart meter to crack the monopole business of the power utilities. A smart metering which is linked up the data base of computer network to propel short message services (SMS) alerts to consumer for revenue collections.

XIV. CONCLUSIONS

Immense power harvesting administers load demand and energy impartiality to acclimatize enhanced performance based on the power silhouette of the micro grid system. Energy mission, supervision, synchronization, high elucidation constraint algorithms, steady state operations, outside parameters, self-motivated allotment, purpose and simulations are really robust for realistic applications. Implementation arraigns increases by 82% on luminous days to administer power crisis without contagion of the Indian environment.

Acknowledgment

The author is deeply sad due to premature end of his father, Sri Jagabandhu Halder of Village- Sahapur, P.O-Chatina, District- Nadia, Pin-741160, and West Bengal. He is no more from 17-04-2014 at 9.45 PM (Thursday) at S.S.K.M

Hospital, West Bengal. His absence is pensive forever. I never forget him. The doctor rereleased at Good Friday at 2.10 AM. I always remember him. His words inspired and touched many. As the world and relatives mourn his passing, his father continues to breathe in the hearts of the relatives through his motivational talks and endless contributions.

References

[1] T. Halder, "Charge Controller of Solar Photo-Voltaic Panel Fed (SPV) Battery", IEEE, IICPE-2010, pp 1-4

[2] Kansal, A., Hsu, J., Srivastava, M., and Raghunathan, V., 2006: Harvesting aware power management for sensor networks. In DAC '06: Proceedings of the 43rd annual conference on Design automation, 651 656. ACM, New York, NY, USA. ISBN 1-59593-381-6. doi:http://doi.acm.org/10.1145/1146909.1147075.

[3] T. Halder "An Improved and Simple Hybrid Energy Recovery Snubber Circuit for Generic Power Converters and Protection Scheme" IEEE, IICPE-2012,pp 1-6

[4] Moser, C., Thiele, L., Brunelli, D., and Benini, L., 2008: Approximate control design for solar Systems, 634–637. Springer-Verlag, Berlin, Heidelberg. ISBN 978-3-540-78928-4. doi:http://dx.doi.org/10.1007/978-3-540-78929-1 52.

[5] Rusu, C., driven sensor nodes. In HSCC '08: Proceedings of the 11th international workshop on Hybrid Melhem, R., and Moss'e, D., 2005: Multi-version scheduling in rechargeable energy-aware real-time systems. J. Embedded Comput., 1(2), 271 283. ISSN 1740-4460.

[6] T. Halder, "An Improved Hybrid Energy Recovery Soft Switching Snubber for the Flyback Converter" IEEE, PEDES-2012 pp 1-6

[7] Taylor, S. G., Farinholt, K. M., Flynn, E. B., Figueiredo, E., Mascarenas, D. L., Moro, E. A., Park, G., Todd, M. D., and Farrar, C. R., 2009: A mobile-agent based wireless sensing network for structural monitoring applications. Measurement Science and Technology, 20(4), 045201 (14pp).

[8] T. Halder, "Study of Rectifier Diode Loss Model of the Flyback Converter" IEEE, PEDES-2012 pp 1-6

[9] Voigt, T., Ritter, H., and Schiller, J., 2003: Utilizing solar power in wireless sensornetworks. In LCN '03: Proceedings of the 28th Annual IEEE International Conference on Local Computer Networks, 416. IEEE Computer Society, Washington, DC, USA. ISBN 07695-2037-5

[10] T. Halder, "Comprehensive power loss model of the main switch of the Flyback converter " IEEE, ICPEC'2013 pp 792-797

[11] T.Halder, S. S. SAha, B. Majumdar, and S. K. Biswas, "A New Control Circuit Extends the Effective Duty Cycle Range of the Flyback Converters", IEEE PEDS 2005, Vol: 1, pp 413-417

[12] T.Halder, "Improved Performance Analysis of Clamp Circuits With Flyback Converter", International Journal of Emerging Technology and Advanced Engineering Website: www.ijetae.com (ISSN 2250-2459, Volume 2, Issue 1, January 2012) pp 1-8

[13] T.Halder,"Wind Power Generation and Synchronization to Utility Grid", Control, Communication & Device Electronics (N3CD-2013) sponsored by AICTE pp 244-250

[14] T.Halder,"Power Quality Issues – A Case Study" Control, Communication & Device Electronics (N3CD-2013) sponsored by AICTE pp 251-250

[15] T.Halder "Continuous Conduction Mode (CCM) of Operations & Stability Analysis of the Flyback SMPS" Proceedings of 2014 1st International Conference on Non Conventional Energy (ICONCE 2014) pp, 292-297

[16] T. Halder, "Advanced Quasi-Resonant Zero-Voltage Switching Flyback Converter", CALCON11 organized by IEEE Calcutta Section, 2011, pp 102-105

Experimental Implementation of Doubly Fed Induction Generator Based Standalone Wind Energy Conversion System

N. K. Swami Naidu, *Student Member, IEEE*
Department of Electrical Engineering
Indian Institute of Technology Delhi
New Delhi-110016, India
nkswaminaidu@gmail.com

Bhim Singh, *Fellow, IEEE*
Department of Electrical Engineering
Indian Institute of Technology Delhi
New Delhi-110016, India
bsingh@ee.iitd.ac.in

Abstract— **This paper presents an implementation of Doubly Fed Induction Generator (DFIG) based Standalone Wind Energy Conversion System (SWECS). Control algorithms of both Rotor Side Converter (RSC) and Load Side Converter (LSC) are presented for maintaining voltage and frequency for variable speed DFIG based SWECS. A prototype of this proposed DFIG based SWECS is developed using a DSP (Digital Signal Processor-dSPACE DS1103). This proposed DFIG is tested under nonlinear and dynamic loads. Dynamic performance of this proposed VFC is tested at unbalanced load, sudden change in loads and also for variable wind speeds.**

Keywords— *Standalone Wind Energy Conversion System (SWECS), Doubly Fed Induction Generator (DFIG), Voltage and Frequency Controller (VFC), Power Quality.*

I. INTRODUCTION

Still there are some regions where the availability of grid is more difficult. In these places, the option for the power generation is the diesel generators (DG). But most of these islanded or regional communities are rich in natural resources. So there is a need for development of renewable energy sources. Because of its technological development, wind energy is becoming economical as compared to other renewable energy sources [1-4]. In older days, most of the Wind Energy Conversion Systems (WECSs) are fixed speed. From the characteristics of wind turbine one can clearly observe that for a fluctuating wind speed, the rotor speed should be variable for achieving maximum power. Even though, these fixed speed WECS are very much simple and reliable in operation, significant amount of power is lost [5-7]. Out of so many variable speed WECS, Doubly Fed Induction Generator (DFIG) is the most preferred one, because of its low rating converter requirement. From the same machine, the power extracted is more in case of a DFIG rating because of its rotor power utilization. The control of grid connected DFIG are very much matured and so much literature is available [8-9]. But for the stand-alone applications, the literature available is very much limited.

The main challenge in Standalone Wind Energy Conversion System (SWECS) is to maintain voltage and frequency for the variable rotor speed and for the variable loads. In [10-13], Voltage and Frequency Controllers (VFC) for the variable speed DFIG are discussed. In [10], the authors have not taken care of nonlinear loads. These nonlinear loads introduce harmonics in the windings of the stator. These unwanted harmonics create noise, vibrations and losses in the machine.

The purpose of this Load Side Converter (LSC) is to regulate the DC link voltage between two back to back connected VSCs. Jain et. al [11] have proposed the control of LSC as an active filter by using Synchronous Reference Frame (SRF) theory for supplying harmonics. Normally, these isolated loads are of dynamic loads, single phase and unbalanced in nature. But these authors have not shown these cases such as dynamic and single phase loads. Here authors have proposed VFC for supplying unbalanced (single phase) loads and dynamic loads.

In this proposed VFC, as there is no energy storage element, this SWECS is not operating at Maximum Power Point Tracking (MPPT) condition. The voltage is maintained at Point of Common Coupling (PCC) by controlling the reactive component of rotor currents. So the required lagging reactive power for the machine is supplied from the Rotor Side Converter (RSC). The power produced should be equal to the load demand. So the active component of the load current is fed by the stator currents. LSC is used for harmonic compensation and also for load balancing.

A laboratory prototype of DFIG based SWECS is developed and the performance is investigated for nonlinear loads and dynamic loads. Steady state performance of this proposed VFC is tested under nonlinear load in both sub-synchronous and super-synchronous speeds. Dynamic performance is tested for unbalanced or single phase loads. This VFC is tested for sudden change in load. Now a days, the major portion of the load is the induction motors. These induction motors stating is the major concern for the off-grid applications because of its high stating current requirement. So this VFC is tested for the high inrush currents by direct on line starting of the induction motor.

II. PROPOSED SYSTEM CONFIGURATION

In this proposed SWECS, the stator of the DFIG is directly connected to the loads. One VSC is connected to rotor terminals and the other VSC is connected to the stator with inductors in between as shown in Fig. 1. In this topology, stator is connected in delta connected mode. In this hardware setup, WRIM is coupled to the DC machine. Here wind

978-1-4799-6047-7/14 $31.00 © 2014 IEEE

turbine characteristics are emulated using DC machine and Type-A chopper as shown in Fig. 2.

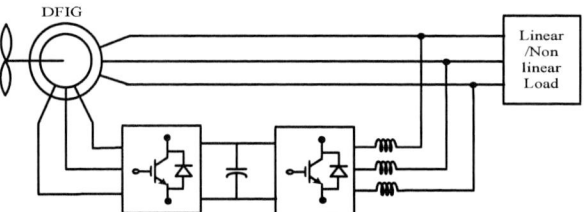

Fig. 1 Proposed System Configuration.

III. CONTROL STRATAGY

Fig. 2 shows the complete control algorithm of both RSC and LSC. RSC is controlled in stator flux oriented reference frame. The control of buck chopper for emulating wind turbine characteristics is as shown in Fig. 2. LSC is controlled in voltage oriented reference frame using Enhanced Phase Locked Loop (EPLL). Block diagram of EPLL is as shown in Fig. 3.

A. Rotor Side Converter Control Algorithm

The main purpose of the RSC is to maintain the voltage and frequency for the variable wind speed and at variable load. RSC is controlled in stator flux oriented reference frame. So the stator active and reactive powers (P_S & Q_S) are controlled by controlling quadrature and direct axis rotor currents (i_{qr} & i_{dr}) respectively. The terminal voltage at the PCC is regulated by controlling direct axis rotor current (i_{dr}).

The direct axis rotor reference current (i_{dr}^*) is obtained by processing the terminal voltage error (v_{te}) between reference and estimated terminal voltage (V_t^* and V_t) through PI (Proportional Integral) controller as,

$$i_{dr}^*(k) = i_{dr}^*(k-1) + k_{pd}\{v_{te}(k) - v_{te}(k-1)\} + k_{id}v_{te}(k) \quad (1)$$

where k_{pd}, k_{id} are the proportional and integral constants of the terminal voltage controller. v_{te} (k) and v_{te} (k-1) are the terminal voltage error at k^{th} and $(k-1)^{th}$ instant. i_{dr}^*(k) and i_{dr}^*(k-1) are

the direct axis rotor reference current at k^{th} and $(k-1)^{th}$ instant.

As the machine is connected in delta connected mode, line to line voltage is 230 V. So the reference terminal voltage V_t^* is selected as 326 V. V_t at the PCC is calculated as [14],

$$V_t = \sqrt{\frac{2}{3}(v_{ab}^2 + v_{bc}^2 + v_{ca}^2)} \quad (2)$$

where v_{ab}, v_{bc} and v_{ca} are the sensed line voltages.

Here DFIG generates power which is needed by the load. So the system is not working on MPPT condition. So the speed is adjusted such that to generate required power.

Active power component of rotor current (i_{qr}) is calculated from the stator active component of current (i_{ds}) as,

$$i_{qr}^* = -L_S/L_m * i_{ds} \quad (3)$$

D and q axis reference rotor components (i_{dr}^* and i_{qr}^*) are converted into three phase reference rotor currents i_{rabc}^* as [15],

$$i_{ra}^* = i_{dr}^* \sin\theta_{slip} + i_{qr}^* \cos\theta_{slip}$$

$$(4) \; i_{rb}^* = i_{dr}^* \sin(\theta_{slip} - 2\pi/3) + i_{qr}^* \cos(\theta_{slip} - 2\pi/3)$$

$$(5) \; i_{rc}^* = i_{dr}^* \sin(\theta_{slip} + 2\pi/3) + i_{qr}^* \cos(\theta_{slip} + 2\pi/3)$$

(6)

In the above equations, θ_{slip} is the slip angle calculated as,

$$\theta_{slip} = \theta_e - \theta_r \quad (7)$$

where θ_e is the angle calculated by integrating the fixed electrical speed (ω_e) of 314 read/sec. where θ_r is the electrical rotor angle is calculated from the encoder pulses.

These reference rotor currents (i_{ra}^*, i_{rb}^*, i_{rc}^*) are compared with the sensed rotor currents (i_{ra}, i_{rb}, i_{rc}) and the current error is passed through the PI controller for estimating the modulating signals. These modulating signals are compared with the fixed frequency PWM for switching the RSC.

B. Load Side Converter Control Algorithm

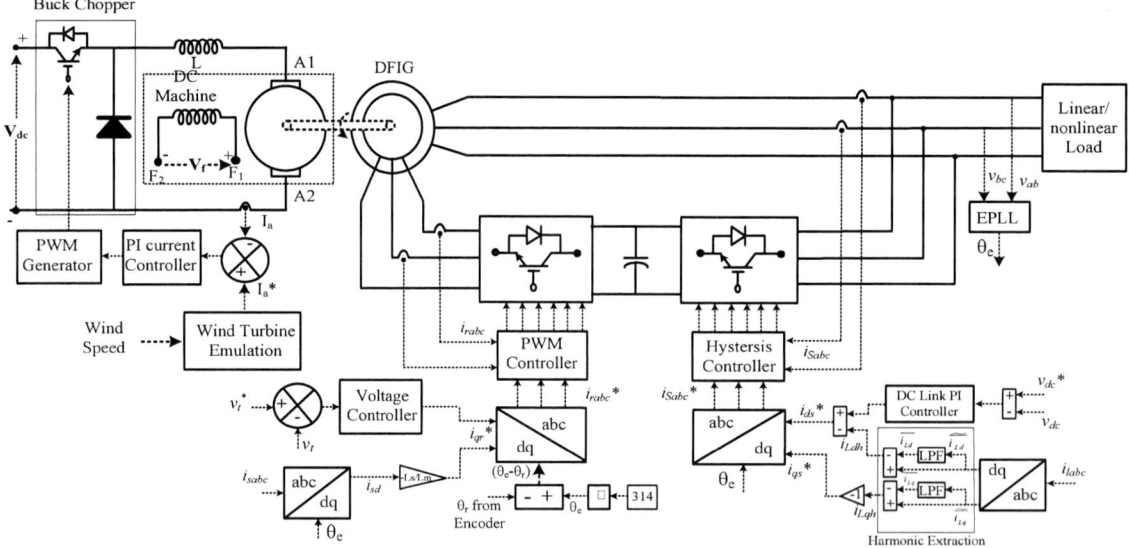

Fig. 2 Control Algorithm of the proposed WECS.

978-1-4799-6047-7/14 $31.00 © 2014 IEEE

The main purpose of the load side converter is to regulate DC link voltage, harmonics compensation and load balancing. This LSC is controlled in voltage reference frame so the active and reactive power components are controlled with direct and quadrature axis components of LSC current (i_{dLSC}, i_{qLSC}) respectively. D-axis is aligned to the stator voltage vector by using EPLL as shown in Fig. 3.

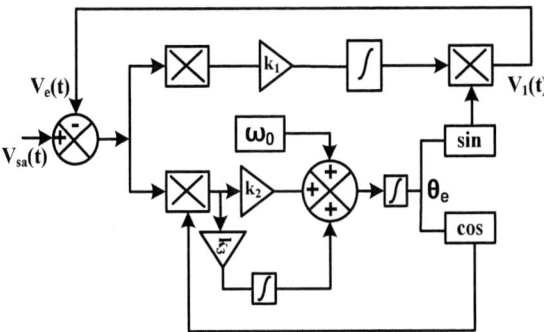

Fig. 3 Block diagram of Enhanced Phase Locked Loop (EPLL).

The direct axis stator reference current (i_{dLSC}') is obtained by processing the DC link voltage error (v_{dce}) between reference and estimated DC link voltage (V_{dc}^* and V_{dc}) through PI (Proportional Integral) controller as,

$$i_{dLSC}'(k) = i_{dLSC}'(k-1) + k_{pdc}\{v_{dce}(k) - v_{dce}(k-1)\} + k_{idc}v_{dce}(k)$$
(8)

where k_{pdc}, k_{idc} are the proportional and integral constants of DC link voltage controller. $V_{dce}(k)$ and v_{dce} (k-1) are the DC link voltage error at k^{th} and $(k-1)^{th}$ instant. $i_{dLSC}'(k)$ and i_{dLSC}'(k-1) are the direct axis rotor reference current at n^{th} and $(n-1)^{th}$ instant.

In this method, load harmonic currents are injected by the LSC. These load harmonic currents are calculated from the sensed load currents as shown in Fig. 2. SRF method is used to extract harmonic components. Sensed load current components are converted in to synchronous rotating reference frame, which consists of harmonic components in addition to the DC component. These direct and quadrature axis currents are passed through Low Pass Filter (LPF) for achieving DC component (positive sequence) of direct and quadrature axis currents. For extracting harmonic components (i_{Ldh}, i_{Lqh}), DC components ($\overline{i_{Ld}}$, $\overline{i_{Lq}}$) are subtracted from the total synchronous rotating frame components ($\widetilde{i_{Ld}}$, $\widetilde{i_{Lq}}$) as shown in the harmonic compensation dotted box in Fig. 2.

These extracted harmonic load components (i_{Ldh}, i_{Lqh}) are subtracted from the LSC current reference (i_{dLSC}', i_{qLSC}') for obtaining LSC direct and quadrature axis reference currents (i_{dLSC}^*, i_{qLSC}^*). These reference LSC currents (i_{LSCa}^*, i_{LSCb}^* and i_{LSCc}^*) are calculated from the direct and quadrature axis components (i_{dLSC}^*, i_{qLSC}^*) [15]. Sensed LSC currents (i_{LSCa}, i_{LSCb} and i_{LSCc}) are compared with the reference LSC currents (i_{LSCa}^*, i_{LSCb}^* and i_{LSCc}^*) and the pulses are generated to the LSC accordingly using hysteresis current controller.

IV. RESULTS AND DISCUSSION

Steady state and dynamic performance of this proposed VFC is tested experimentally for balanced, un-balanced, linear, non-linear and dynamic loads extensively. Some of the results are presented as stator line voltages (v_{ab}, v_{bc}, v_{ca}), stator currents coming out of the stator windings (i_{Sa}, i_{Sb}, i_{Sc}), load currents (i_{La}, i_{Lb}, i_{Lc}), rotor currents (i_{Ra}, i_{Rb}, i_{Rc}), LSC currents (i_{LSCa}, i_{LSCb}, i_{LSCc}), terminal voltage (v_t), rotor speed (ω_r), stator power (P_S), load power (P_L), Load side converter power (P_{LSC}).

A. Steady State Performance of Proposed SWECS

The steady state performance at a fixed wind speed is given under nonlinear load as shown in Figs. 4-5. Fig. 4 shows the performance of the proposed DFIG based SWECS in sub synchronous speed. Fig. 4(a) shows the load current, which is nonlinear in type. Even then the stator currents are sinusoidal as shown in Fig. 4(b) by injecting harmonic currents through the LSC as shown in Fig. 4(c). Figs. 4(d), (e) and (f) show load power, stator power and LSC power respectively. In this case, as the wind speed is less, the system works in sub-synchronous speed accordingly. So the LSC power is shown as negative, so the power flows from LSC towards rotor side. Even though load current THD is high as shown in Fig. 4(g), the stator current is THD is under IEEE-519 limit. The voltage THD is also maintained under IEEE-519 limit as shown in Fig. 4(h) & 4(i).

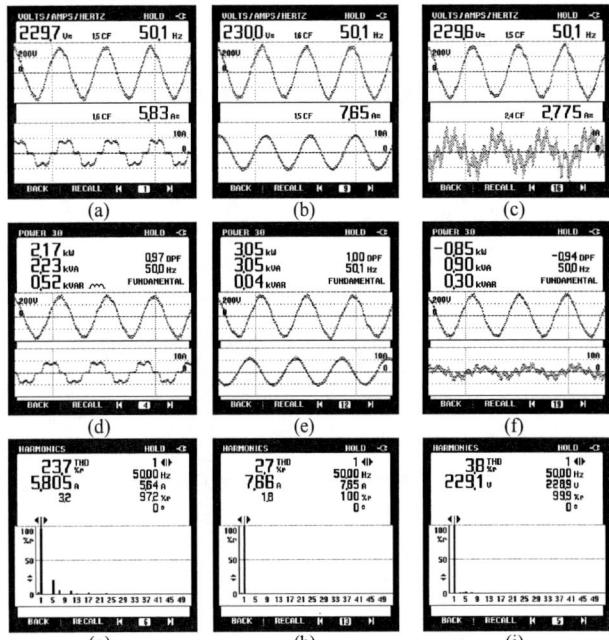

Fig. 4 Experimental performance of the proposed SWECS under non-linear load at rotor speed of 1300 rpm (a) v_{ab} and i_{La},(b) v_{ab} and i_{Sa}, (c) v_{ab} and i_{LSCa}, (d) Load power (P_L), (e) Stator power (P_S), (f) LSC power (P_{LSC}), (g) harmonic spectrum of i_{La}, (h) harmonic spectrum of i_{Sa}, (i) harmonic spectrum of v_{ab}.

Figs. 5(a)-(i) show the steady state performance of the proposed system for the super-synchronous speed operation.

978-1-4799-6047-7/14 $31.00 © 2014 IEEE

Figs. 5(a)-(c) show the load current, stator current and LSC current respectively. As shown in Figs. 5(d)-(f), LSC power is getting reversed and flows from rotor to the LSC because of super - synchronous speed operation. Stator current and voltage THD are within limits as shown in Fig. 5(h)-(i).

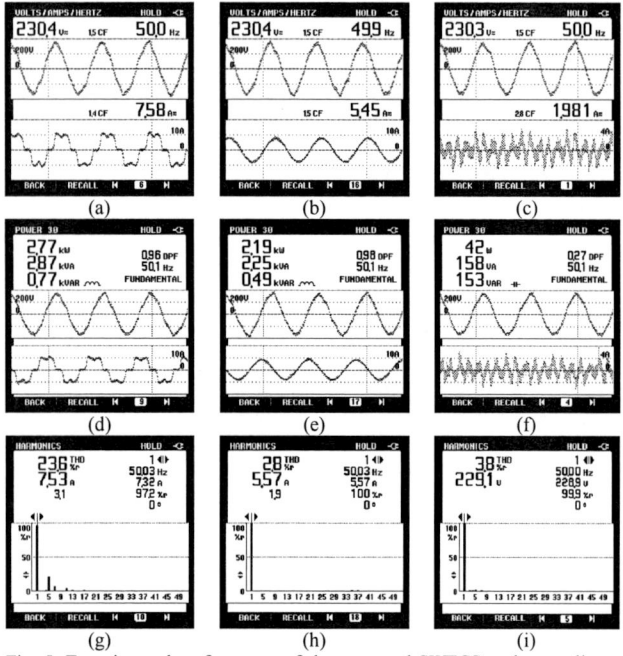

(d) Load power (P_L), (e) Stator power (P_S), (f) LSC power (P_{LSC}), (g) harmonic spectrum of i_{La}, (h) harmonic spectrum of i_{Sa}, (i) harmonic spectrum of v_{ab}.

B. Dynamic Performance of DFIG based WECS

Figs. 6-9 show the dynamic performance of the proposed DFIG based SWECS. Test results are recorded with digital oscilloscope (Agilent make - DSO6014A). In this section, the proposed system is even tested for the sudden change in load and the change in wind speed also. This load change is taken by removing one phase out of three phases, so the load becomes unbalanced and the effective load is also reducing. The system is also tested for the sudden addition of one phase, so the effective load is also increasing. Dynamic performance of this VFC is even tested for the starting of the induction motor.

(1) Dynamic Performance of Proposed SWECS for Unbalanced and Dynamic Load

Figs. 6-7 show the dynamic performance for the unbalanced loads. This unbalance is by the sudden removal and application of one phase. As shown in Fig. 6(a), phase 'a' load current is made zero. So the load is becoming single phase, still the stator current is maintained sinusoidal and balanced in all the three phases. As shown in Fig. 6(b), the stator current is supplied by the LSC phase 'a' current. Voltage is also maintained even for the sudden removal of phase 'a' as shown in Fig. 6(c). By removing one phase, the load on the SWECS is decreased.

Fig. 5. Experimental performance of the proposed SWECS under non-linear load at rotor speed of 1800 rpm (a) v_{ab} and i_{La}, (b) v_{ab} and i_{Sa}, (c) v_{ab} and i_{LSCa},

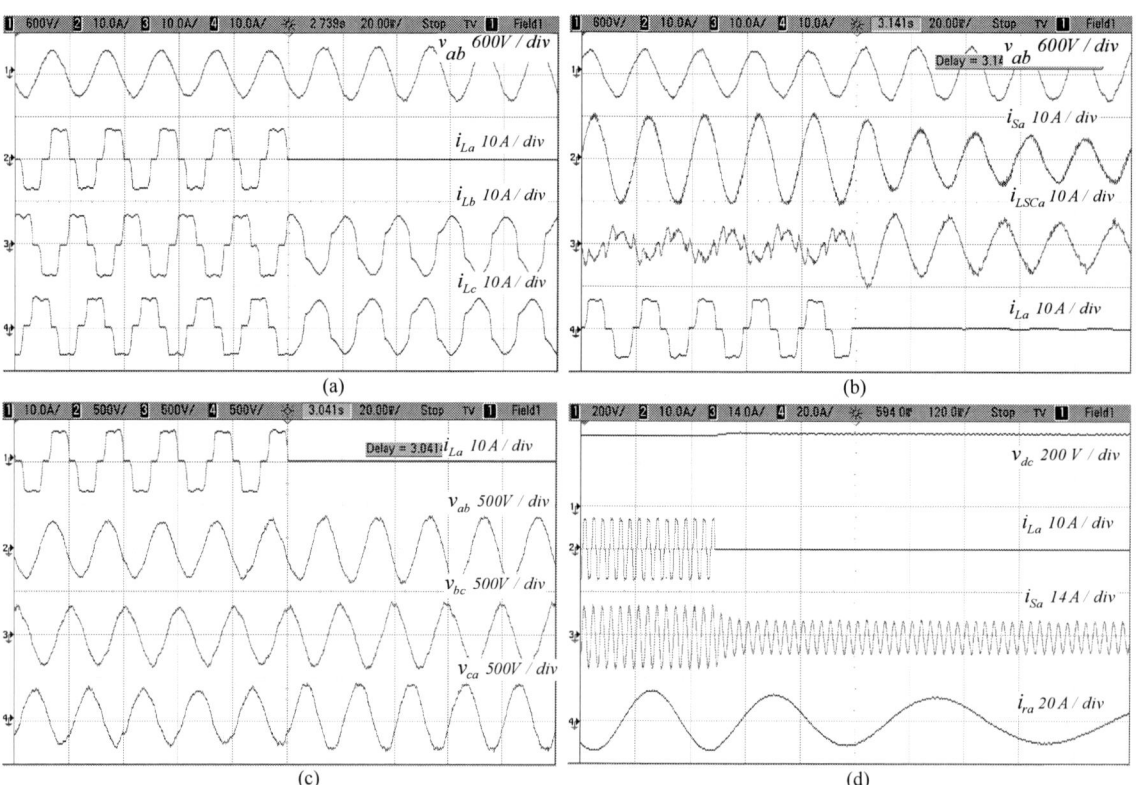

Fig. 6 Experimental performance of proposed SWECS during load removal in phase "A" (a) v_{ab} with i_{La}, i_{Lb} and i_{Lc} (b) v_{ab} with i_{Sa}, i_{LSCa} and i_{La}. (c) i_{La} with v_{ab}, v_{bc} and v_{ca} (d) v_{dc} with i_{La}, i_{Sa} and i_{ra}.

978-1-4799-6047-7/14 $31.00 © 2014 IEEE

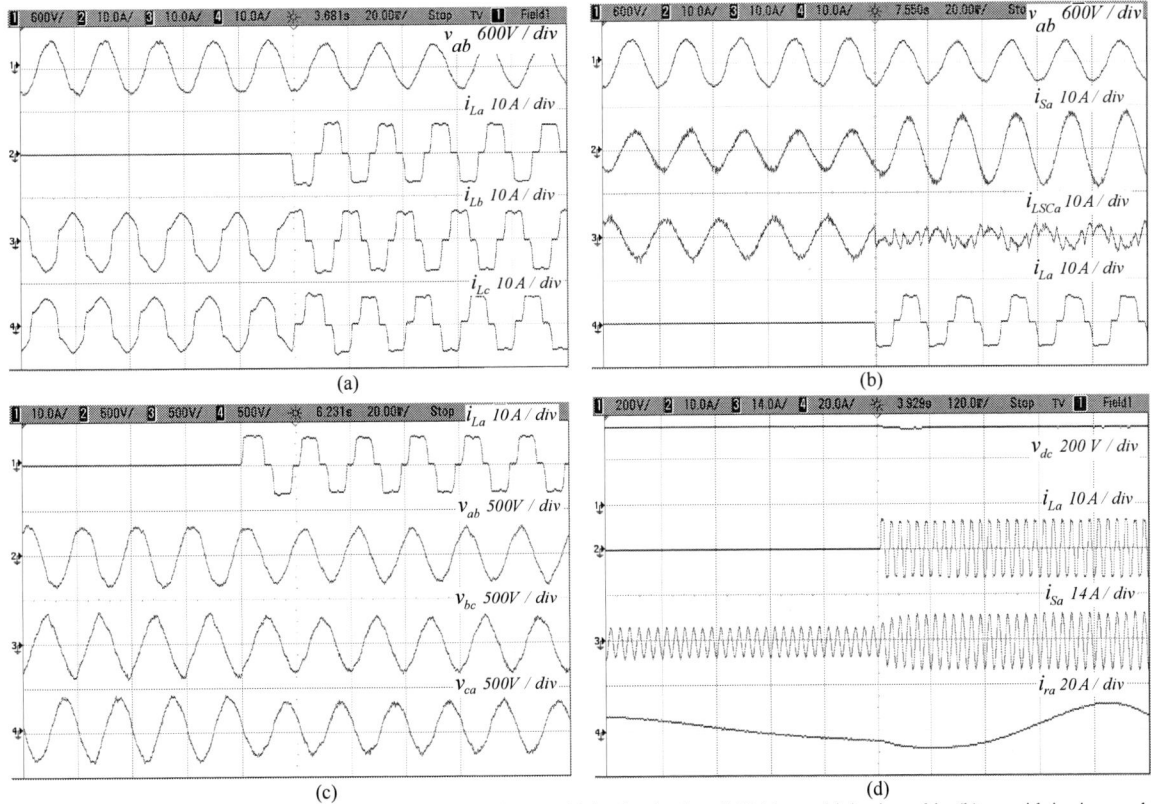

Fig. 7 Experimental performance of proposed SWECS during sudden load injection in phase "A" (a) v_{ab} with i_{La}, i_{Lb} and i_{Lc} (b) v_{ab} with i_{Sa}, i_{LSCa} and i_{La}. (c) i_{La} with v_{ab}, v_{bc} and v_{ca} (d) v_{dc} with i_{La}, i_{Sa} and i_{ra}.

As the load is decreasing, the power produced from the DFIG is also decreased by increasing the speed. So the stator currents are decreasing and rotor current frequency is decreasing as shown in Fig. 6(d). Eve then the DC link voltage between the two VSCs is almost maintained.

Application of sudden load is also shown in Figs. 7(a-d). As the load in one phase is suddenly applied, the total load on the system is increased, so the rotor speed decreases and the stator current increases as shown in Fig. 7(d).

This DFIG based SWECS is also tested under dynamic loads such as induction motor. If induction motor is started with direct on line starter, it draws sudden high inrush current for few cycles as shown in Fig. 8(a). Normally, this inrush current is the major concern for the off-grid applications. But because of proper controlling action, terminal voltage dips negligibly small as shown in Fig. 8(a). Stator current and rotor current are also increased and again reaches to normal as the load current reaches steady state as shown in Fig. 6(a). Line voltage (v_{ab}) is almost constant as shown in Fig. 8(b).

(2) Dynamic Performance of Proposed SWECS for the Variation in Wind Speed

Dynamic performance of this proposed SWECS is also tested for the variation in the wind speed as shown in Fig. 9. Here as the load is constant and the wind speed is changing, so the rotor speed of the DFIG is changing. With the increase in wind speed, the rotor speed increases as shown in Fig. 9(a).

But the voltage at the PCC is maintained constant as shown in Fig. 9(b). Even though, load currents are constant, rotor speed of the DFIG is increasing for the increase in wind speed. So the stator currents and LSC currents are decreasing as shown in Fig. 9(a). Dynamic performance is also shown for the decrease in wind speed as shown in Fig. 9(b). So the rotor speed is decreasing and stator and LSC currents are increasing as shown in Fig. 9(b).

V. CONCLUSION

The VFC has been proposed for variable speed DFIG based SWECS. The voltage and frequency are controlled using RSC control. Harmonic compensation and power leveling have been achieved through LSC control. This proposed VFC is experimentally validated by developing experimental prototype of DFIG based SWECS using DSP. The performance of this VFC has been tested under nonlinear and dynamic loads as well. Steady state performance has been presented in both sub-synchronous and super-synchronous speeds for the fixed amount of load. Dynamic performance has also been presented for dynamic wind speed with fixed load and fixed wind speed with sudden load addition and removal of nonlinear and dynamic loads. The performance of this proposed VFC has also been demonstrated for the single phase loads.

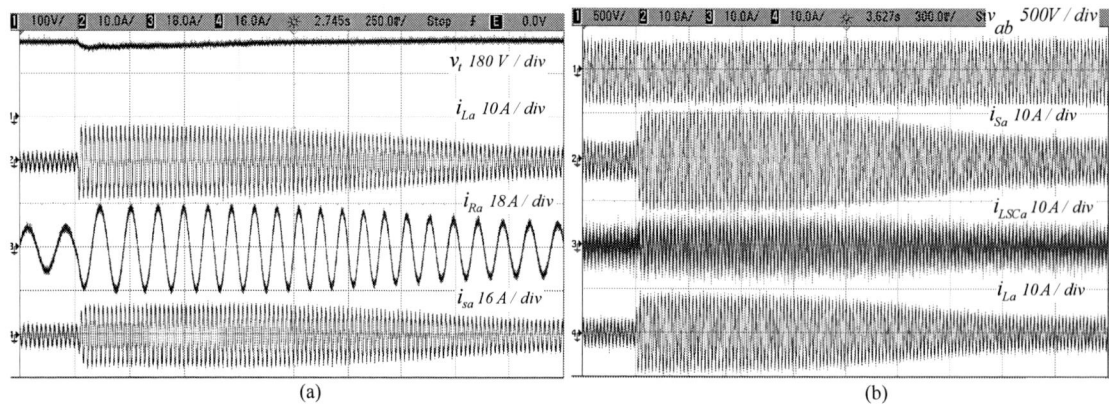

(a)   (b)

Fig. 8 Experimental performance of proposed SWECS for the dynamic load (a) v_t, i_{La}, i_{Ra} and i_{Sa}, (b) v_{ab} with i_{Sa}, i_{LSCa} and i_{La}.

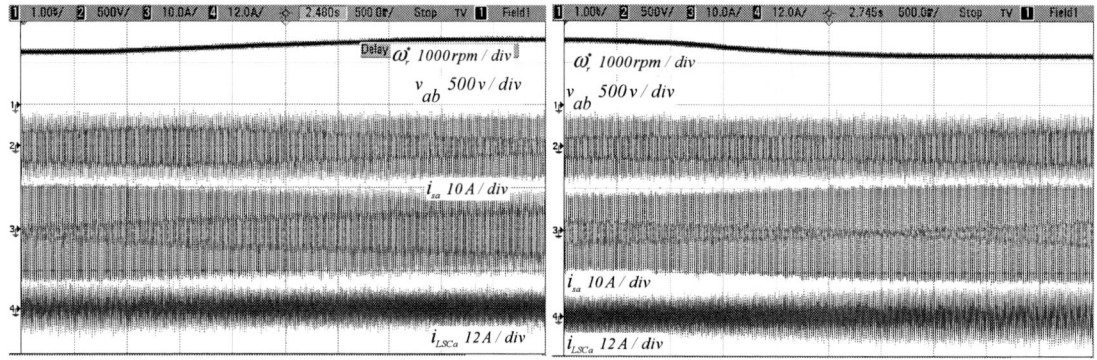

Fig. 9. Dynamic performance of proposed SWECS under rise in wind speed, (a) ω_r^*, v_{ab}, i_{sa} and i_{Ta} (b) ω_r^*, v_{ab}, i_{sa} and i_{LSCa}.

APPENDICES

A. WRIM- 3.7 kW, R_s=1.32 ohms, L_{ls}=6.832mH, R_2=1.708 ohms, L_{lr}=6.832mH, R_c=419.646 ohms, L_m=0.219H, J = 0.1878 kg-m² stator to rotor turns ratio N_r/N_s=1/2, stator rated rms current, I_s=12 A, rotor rated rms current, I_r=18 A.

B. DC Machine- R_a=1.3 ohms, R_f =220 ohms, L_a=7.2mH, L_f=7.5mH, K_Φ=1.3314.

REFERENCES

[1]   Vaughn Nelson, "Wind Energy Renewable Energy and the Environment", CRC press 2009.
[2]   Manfred Stiebler, *"Wind Energy Systems for Electric Power Generation"*, Springer 2008.
[3]   Sathyajith Mathew and Geeta Susan Philip, *"Advances in Wind Energy Conversion Technology*, springer, 2011.
[4]   Hermann-Josef Wagner and Jyotirmay Mathur, *"Introduction to Wind Energy Systems Basics, Technology and Operation"*, Springer 2009.
[5]   L. Holdsworth, X. G. Wu, J. B. Ekanayake and N. Jenkins "Comparison of fixed speed and doubly-fed induction wind turbines during power system disturbances," *Proc. IEE – Gener. Transm. Distrib.*, 2003, pp. 343–352.
[6]   S. S. Murthy, B. Singh, P. K. Goel and S. K. Tiwari, "A Comparative Study of Fixed Speed and Variable Speed Wind Energy Conversion Systems Feeding the Grid*", in Proc. IEEE 7th International Conference Power Electronics and Drive Systems*, 2007 PEDS '07, 27-30 Nov. 2007, pp.736-743.
[7]   R. Datta and V. T. Ranganathan, "variable-speed wind power generation using doubly fed wound rotor induction machine-a

comparison with alternative schemes," *IEEE Transactions on Energy Conversion,* vol.17, no.3, pp.414-421, Sep 2002.
[8]   R. Pena, J. C. Clare and G.M Asher, "Doubly fed induction generator using back-to-back PWM converters and its application to variable-speed wind-energy generation," *IEE Proc. Electric Power Applications,* vol. 143, no. 3, pp. 231 – 241, May 1996.
[9]   S. Muller, M. Deicke and R. W. De Doncker, "Doubly fed induction generator systems for wind turbines," *IEEE Industry Applications Magazine,* vol.8, no.3, pp.26-33, May/Jun 2002.
[10]   R. Pena, J. C. Clare and G. M Asher, "A doubly fed induction generator using back-to-back PWM converters supplying an isolated load from a variable speed wind turbine," *IET Electric Power Applications,* vol. 143, pp.380 – 387, May 1996.
[11]   A. K. Jain and V. T. Ranganathan, "Wound rotor induction generator with sensorless control and integrated active filter for feeding nonlinear loads in a stand-alone grid," *IEEE Trans. Ind. Electron.,* vol. 55, pp. 218-228, Jan. 2008.
[12]   R. Cardenas, R. Pena, J. Proboste, G. Asher and J. Clare, "MRAS observer for sensor less control of standalone doubly fed induction generators," *IEEE Trans. Energy Con.,* vol.20, no.4, pp. 710- 718, Dec. 2005.
[13]   B. Singh and S. Sharma, "Doubly fed induction generator-based off-grid wind energy conversion systems feeding dynamic loads," *IET Power Electronics,* vol. 6, no.9, pp.1917-1926, Nov. 2013.
[14]   B. Singh and S. Sharma, "Stand-Alone Single-Phase Power Generation Employing a Three-Phase Isolated Asynchronous Generator," *IEEE Trans. on Ind. Appl.,* vol.48, no.6, pp. 2414-2423, Dec. 2012.
[15]   G. Abad, Ló , J. Pez, Rodrí, M. Guez, L. Marroyo and G. Iwanski, *"Doubly Fed Induction Machine: Modeling and Control for Wind Energy Generation Applications,"* Wiley-IEEE Press, 2011.

Cascaded DC-DC Converter for a Reliable Standalone PV fed DC load

Malay Bhunia
Electrical Engineering Department
National Institute of Technology
Rourkela, India
malay321@gmail.com

Rajesh Gupta
Electrical Engineering Department
M. N. National Institute of
Technology
Allahabad, India
rajgupta310@gmail.com

Bidyadhar Subudhi
Electrical Engineering Department
National Institute of Technology
Rourkela, India
bidyadharnitrkl@gmail.com

Abstract— **To extract maximum available power from the PV system many maximum power point tracking (MPPT) algorithms with DC-DC converter have been proposed in the literature. However, due to change in load or environmental conditions, the output voltage from the DC-DC converter varies. The variable output voltages are not suitable for many applications requiring constant voltage. To overcome this difficultly, a cascaded DC-DC converter with battery storage system for standalone application is proposed in this paper. The proposed system comprises of a PV system connected with a boost DC-DC converter to deliver the available maximum power to the DC bus. A Single-ended primary inductor converter (SEPIC) is connected to the DC bus to provide regulated DC voltage to the load. A battery with charge controller is connected to the DC bus to improve the reliability of the system. The efficacy of the proposed topology is verified through the experimental studies pursued on a prototype PV connected converter system developed in the laboratory. Robustness studies with regard to variation in environmental parameters such as solar insolation and load variation were made. MPPT and voltage regulation algorithms were implemented in LABVIEW platform using data acquisition boards.**

Keywords— *Battery storage system; boost converter; LABVIEW, maximum power point tracking (MPPT); photovoltaic (PV); SEPIC converter; voltage regulation.*

I. INTRODUCTION

With ever increasing demand of renewable energy and growing concern about environmental issues, photovoltaic (PV) based systems are being increasingly used in diverse applications both at domestic and commercial levels [1]. However, the nonlinear current verses voltage (I-V) characteristics hinders its control design to achieve extraction of maximum power [2]. To extract available maximum power, DC-DC converter with maximum power point tracking (MPPT) algorithms has been proposed in the literature [3].

The application of PV systems can be broadly classified into standalone system and grid connected system [4,5]. The standalone system is widely used in remote areas where access to electricity is not viable. The standalone PV system can provide regulated load voltage but reliability of the system cannot be guaranteed [5]. Storage batteries are advised to improve the reliability of the standalone systems [6]. A fair amount of literature has dealt with the operation of hybrid systems [5-9]. In some hybrid systems [6, 7], the batteries are used to compensate the mismatch between the generation and the demand. The size of the battery can be reduced when a battery charging circuit is inserted between the DC bus and the battery [8, 10].

A suitable MPPT controller can be used to extract maximum power from the PV system but the DC output voltage of the converter varies with the load and the environmental conditions [11]. When the generated PV power is more than the load demand, the extra power can be stored in rechargeable batteries [12-14]. Many literatures [16, 17] discussed about the PV power supplies but they do not put emphasis on the constant output voltage. Shunt regulator along with the battery can be used to regulate the DC bus voltage [15] but author does not include the MPPT controller to extract the maximum available power. Extraction of maximum available power enables supply of load demand power during varying environmental and load condition. In [18], author proposed a regulated DC supply with MPPT controller but battery charging and discharging is not there.

A DC-DC converter can be used to generate DC output voltage both for MPPT and load voltage regulation. A comparative study of different converters is available in [19]. A SEPIC converter can be used for voltage controller due to its capability to step-up and step-down the input voltage without changing the output voltage polarity [19, 20].

Advancement in digital technology and graphical user interface programming methods has made it possible to implement the complex algorithms for online controls. LABVIEW platform has recently been extensively used for implementing sophisticated control algorithms [21].

In this paper a boost-SEPIC converter based PV system is developed. The boost DC-DC converter is used for MPPT to deliver the available maximum power to the DC bus. A SEPIC converter is used at the DC bus to control the DC load voltage demand. This proposed model is very useful for varying environmental and load power demand even in such cases where peak load demand is higher than PV rating for short duration of time. The proposed system consists of a power flow management system to control the operation and power flow control in the system using LABVIEW platform and data

978-1-4799-6047-7/14 $31.00 © 2014 IEEE

acquisition boards. The system acts like a regulated uninterrupted power supply for a DC load.

II. CIRCUIT CONFIGURATION

The proposed system shown in Fig. 1 consists of two DC-DC converters and battery management system. The MPPT is tracked using a boost DC-DC converter [2], [19] and the load voltage is controlled through a SEPIC converter. Battery connected across the DC bus is controlled through a charge control unit.

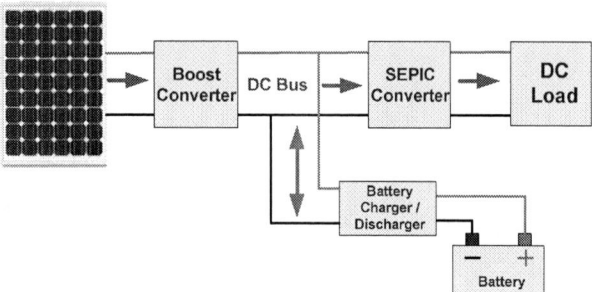

Fig. 1.　Block diagram of the proposed PV system.

III. CONTROL STRATEGY

The control strategy in the PV system is divided into three parts.

A. Voltage Regulation

Load voltage is regulated using the SEPIC converter. The converter has the capability to both step-up and step-down the input voltage without changing the output voltage polarity.

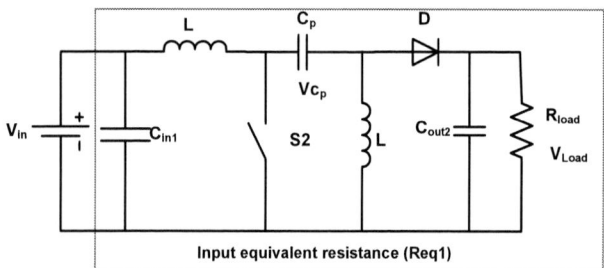

Fig. 2.　SEPIC converter.

The control-to-output small signal transfer function of the SEPIC converter in the continuous conduction mode (CCM) with load voltage V_{load} control can be derived from Fig. 2 in the following form using the averaged model [22].

$$\tilde{G}_2(s) = \frac{\tilde{V}_{load}}{\tilde{D}_2}\bigg|_{\tilde{V}_{in}=\tilde{V}_{cp}=0} = \frac{V_{load}-V_{in}}{(1-D_2)^2} \cdot \frac{1-\dfrac{sD_2L}{(1-D_2)R_{Load}}}{1+\dfrac{sL}{(1-D_2)^2R_{Load}}+s^2\dfrac{LC}{(1-D_2)^2}} \quad (1)$$

Where, D_2 is the duty ratio of the SEPIC converter. The relationship between the input equivalent resistance R_{eq1} to the load resistance R_{Load} can be derived as follows [23].

$$R_{eq1} = R_{load}\frac{(1-D_2)^2}{D_2^2} \quad (2)$$

The SEPIC converter is controlled using a closed loop control as shown in Fig. 3 [24]. Here the reference output voltage $V_{loadref}$ is compared with the actual load voltage V_{load}. The error signal generated is passed through a PI controller. The output of the PI controller generates the duty ratio D_2 which is used for the PWM modulator to generate the pulses for the switch (S_2) of the SEPIC converter as shown in Fig. 2.

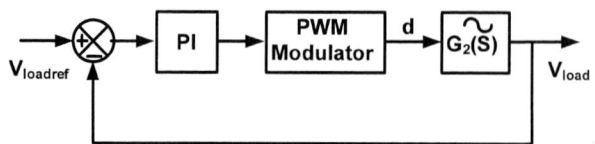

Fig. 3.　Closed loop control of SEPIC converter.

B. MPPT Control

The PV Power P_{pv} verses PV voltage V_{pv} characteristics of a PV cell is nonlinear as shown Fig. 4. The boost converter along with the maximum power point tracking (MPPT) controller is used to track the MPP of the PV cell. Here maximum power extraction is limited up to maximum allowable DC bus voltage $V_{dcbusmax}$.

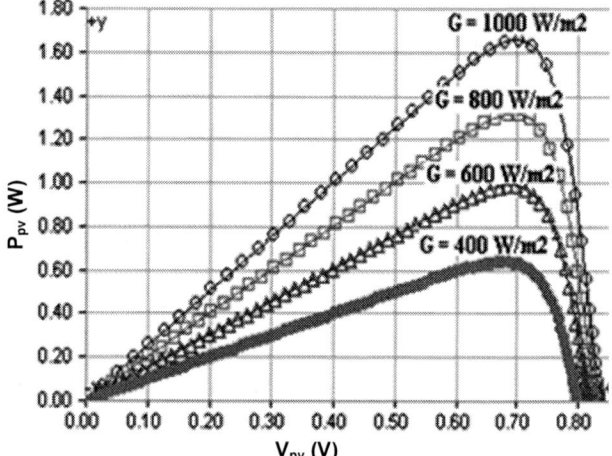

Fig. 4.　Power Vs Voltage characteristics of a PV cell

A boost converter is used between the PV module and the DC bus to extract the available maximum power. Circuit of the boost converter is shown in Fig. 5. The converter is controlled through a closed loop voltage controller as shown in Fig. 6 [24]. In this block diagram the MPPT controller gives the reference voltage for the boost converter input voltage V_{mppref} corresponding to the available maximum PV module power P_{mpp}. The actual boost converter input voltage V_{pv} is compared with the reference input voltage V_{mppref}. The error signal generated is passed through a PI controller. The output of the PI controller generates the duty ratio D_1 which is used for the PWM modulator to generate the pulses for the switch (S_1).

978-1-4799-6047-7/14 $31.00 © 2014 IEEE

The transfer function between duty ratio (D1) as a controlling input to converter input voltage V_{pv} can be derived from Fig. 5 as follows [25].

$$\tilde{G}_1(s) = \frac{\tilde{V}_{pv}}{\tilde{D}_1} = \frac{1}{R_{pv}} \frac{-(z_1 s + z_2)}{p_1 s^3 + p_2 s^2 + p_3 s + p_4} \quad (3)$$

Where

$z_1 = V_{dc} \mathrm{R}_{eq1} R_{pv} C_{out}$; $z_2 = I_l \mathrm{R}_{eq} R_{pv}(1-D_1) + V_{dc}$

$p_1 = L R_{pv} \mathrm{R}_{eq1} C_{in} C_{out}$; $p_2 = L R_{pv} C_{in} + \mathrm{R}_{eq1} L C_{out}$

$P_3 = C_{in} \mathrm{R}_{eq1} R_{pv}(1-D_1)^2$; $p_4 = I_l \mathrm{R}_{eq} R_{pv}(1-D_1) + V_{dc}$

The relationship between the input equivalent resistance R_{eq} to load resistance R_{Load} of the boost converter can be derived from Fig. 5 as below [2].

$$R_{eq} = R_{eq1}(1-D_1)^2 \quad (4)$$

The MPP is tracked considering the SEPIC converter as output load for the boost converter. Combining (2) and (4) the input equivalent resistance R_{eq} of the boost converter can be written as follows.

$$R_{eq} = R_{load} \frac{(1-D_2)^2(1-D_1)^2}{D_2^2} \quad (5)$$

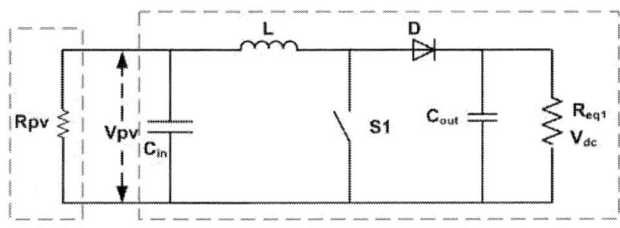

PV module　　　　**Input equivalent resistance (Req)**

Fig. 5　Boost converter with PV module

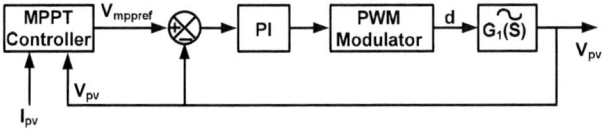

Fig. 6　Closed loop control of boost converter with MPPT controller.

Both the converters has independent control loops and with the choice of parameters it is possible to stably control the converter using duty ratio as control input.

C. Battery Charging & Discharging Control

The single line diagram of power flow of the complete system is shown in Fig. 7 [24]. Charging and discharging of the battery depends upon the power generation by the boost converter P_{pvout} and power requirement by the load P_{load}. Fig. 8 shows the flow-chart of the battery control strategy of the system. The controller continuously monitors the boost converter output power P_{pvout}, load demand power P_{load}, battery terminal voltage V_{bat} and battery state of charge (SOC).

The battery charging process starts when the load power P_{load} plus power loss P_{loss} in the SEPIC converter is less as compare to the PV generated power P_{pvout} and the battery charge level is less than the maximum allowable charge level SOC_{max}.

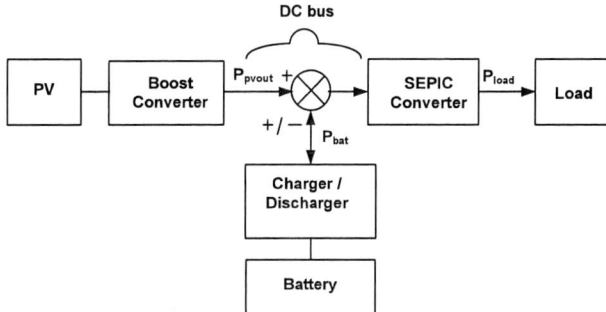

Fig. 7.　Energy flow diagram of whole system

The charging process of the battery is done through two steps. First one is through constant current charging and the second one is through constant voltage charging [26], [27], [12]. In constant current charging mode the charging current is limited through the external current reference, which is according to the specification of the maximum battery charging current. In constant voltage charging mode, a constant voltage is applied across the battery according to the battery manufacturer specification. Mode of charging depends upon the battery terminal voltage V_{bat}, battery nominal voltage V_{nom} and battery state of charge (SOC).

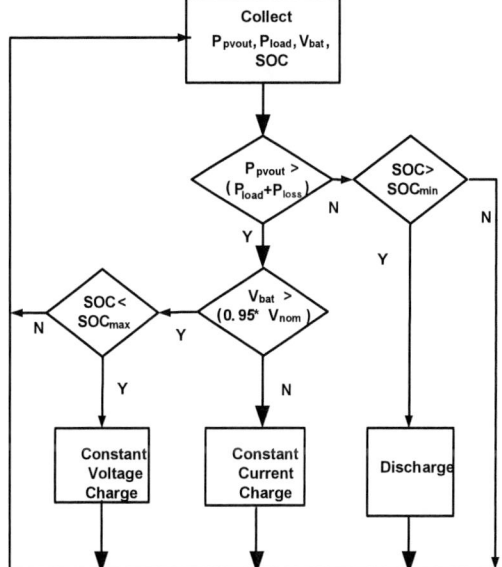

Fig. 8.　Flowchart of battery control strategy.

When the battery nominal voltage V_{nom} multiplied by 0.95 is greater than the battery terminal voltage V_{bat}, constant current charging mode starts. Otherwise constant voltage charging mode is selected depending upon the SOC and maximum allowable charge level (SOC_{max}). If SOC is less than the SOC_{max}, the constant voltage charge mode starts. The loop will restart again if SOC is greater than the SOC_{max}.

When the load power P_{load} plus power loss P_{loss} in the SEPIC converter is high as compare to the PV generation power P_{pvout} and the battery discharge level is greater than the minimum allowable discharge level SOC_{min} the battery starts to discharge. During discharging period the battery is connected directly to the DC bus.

IV. EXPERIMENTAL SETUP AND RESULTS

To verify the simulation results of the proposed system, an experimental setup is developed and tested. Fig. 9 shows the complete experimental setup consisting of MPPT control through boost converter, load voltage regulation through SEPIC converter and battery charging circuit. National Instruments data acquisition card (PCI-6251) with LABVIEW 10.1 software is used to control the converters.

Fig. 9. Experimental Setup.

A. Boost Converter Experimental Result (MPPT Control)

Boost Converter shown in Fig. 5 is designed for MPPT with the specifications given in Table I.

TABLE I SPECIFICATION OF BOOST CONVERTER

Input Voltage Range	10-50 V
Output Voltage Range	10-200 V
Switching Frequency	20 kHz
Ripple Voltage	0.5 V
Output Current	5 A

Fig. 10 shows the results for the boost converter with MPPT control. Channel 1 and 2 in Fig. 10 shows the PV and load voltage respectively, on the scale of 1 div = 42.2 V. Channel 3 and 4 shows the PV and load current respectively on the scale of 1 div = 1.2 A. Initially load resistance of 30 Ω is connected which increase gradually up to 70 Ω with 10 Ω of step change. It is observed from the results shown in Fig. 10 that at different load conditions all the variables are almost constant except load voltage. The load voltage increases with increase in the load resistance. This is due to the fact that the MPPT algorithm increases the duty ratio to match the resistance

between the boost converter input equivalent resistance with the PV resistance as per (4). As the duty ratio changes the output voltage also changes.

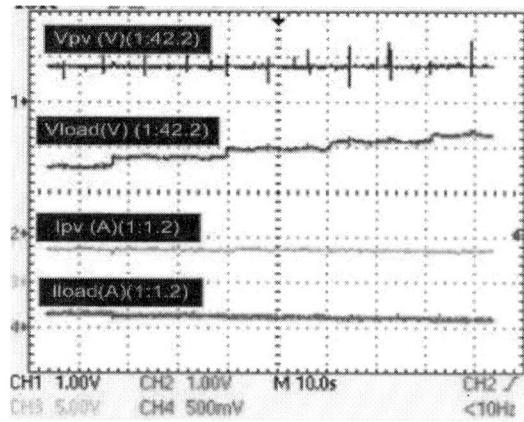

Fig. 10. Boost converter result with MPPT controller.

B. SEPIC Converter Experimental Result (Voltage regulation Control)

The SEPIC converter show in Fig. 2 is implemented with the specifications given in Table II. Fig. 11 shows the SEPIC converter output voltage during boost operation. Channel 1 and 2 of Fig. 11 shows the output load voltage and input PV voltage, respectively. Channel 3 shows gate pulses for the converter. Input voltage of the SEPIC converter is 11.5 V. The output reference voltage for the SEPIC converter is 40 V. It is observed that the converter tracks the reference voltage.

TABLE II SPECIFICATION OF SEPIC CONVERTER

Input Voltage Range	10-200 V
Output Voltage Range	10-50 V
Switching Frequency	4 kHz
Ripple Voltage	0.5 V
Output Current	5 A

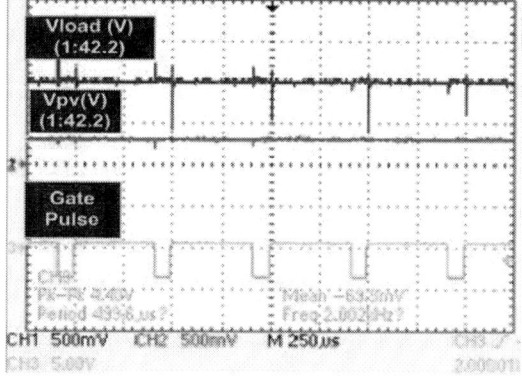

Fig. 11. SEPIC converter output in boost mode.

Fig. 12 shows the screen shot of the front panel of the LABVIEW program for SEPIC converter in boost mode.

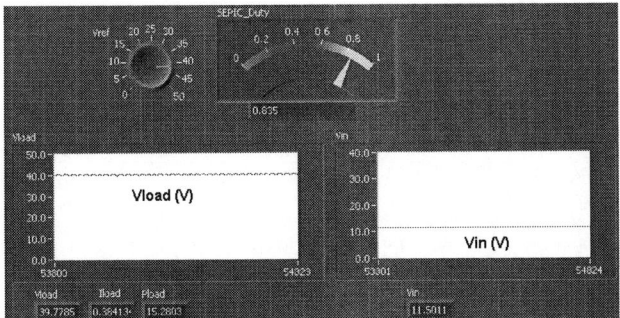

Fig. 12. Screen shot of LABVIEW front panel for SEPIC converter control in boost mode.

Fig. 13 shows the SEPIC converter output voltage during buck operation. Channel 1 and 2 of Fig. 13 shows the output load voltage and input PV voltage, respectively. Channel 3 shows gate pulses for the converter. Input voltage of the SEPIC converter is 13.2 V. The output reference voltage for the SEPIC converter is 5 V. It is observed that converter tracks the given reference voltage of 5V.

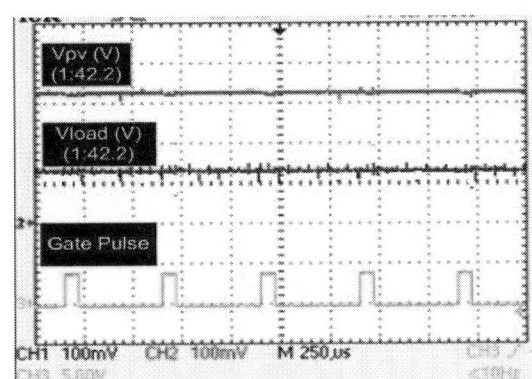

Fig. 13. SEPIC converter output in buck mode.

C. Cascaded Boost-SEPIC Converter Experimental Result

Cascaded Boost-SEPIC converter configuration shown in Fig. 1 is implemented in the laboratory.
It is also observed that the change in load reference voltage does not affect the PV power due to the MPPT implemented through the boost converter.

Fig. 14 shows the performance of the proposed system when the reference load voltage changes from 10 V to 20 V. It is observed from this figure that the SEPIC converter tracks the load reference voltage accurately.

Fig. 15 shows the performance of the proposed system under varying solar radiation. After full sun light radiation the radiation was changed by using partial shading with the load reference voltage remain equal to 20 V. It can be seen that the boost converter tracks the new MPP with SEPIC converter tracking the reference load voltage.

Fig. 16 shows the performance of the proposed system under varying load condition. The load resistance was varied from 50 Ω to 70 Ω with the reference load voltage remain 20 V. It is observed from the figure that the boost converter tracks the same MPP and SEPIC converter tracks the reference load voltage. However the load power decreases with increase in load resistance.

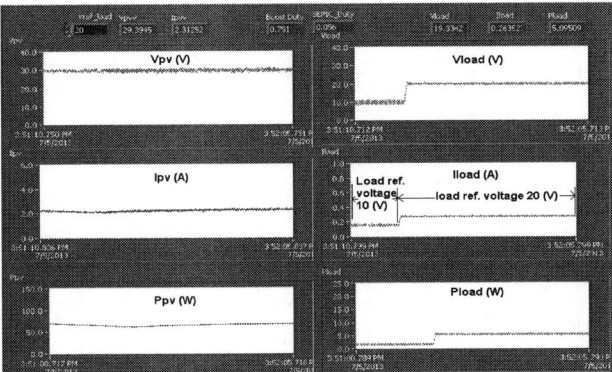

Fig. 14. Screen shot of LABVIEW front panel (change in load voltage reference).

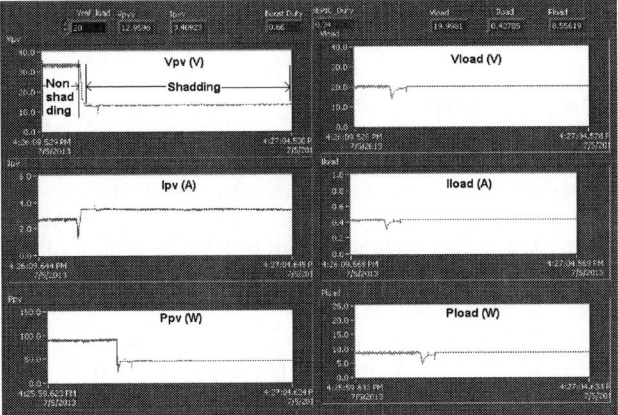

Fig. 15. Screen shot of LABVIEW front panel (change in radiation level).

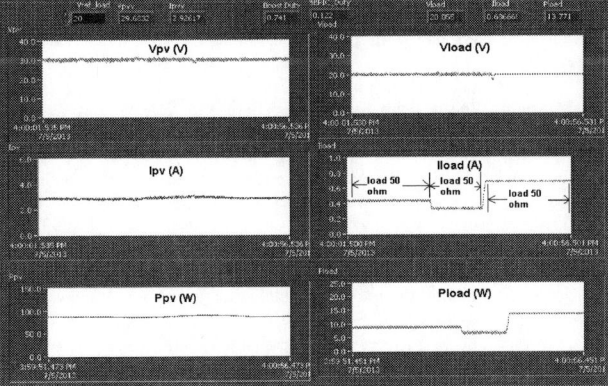

Fig. 16. Screen shot of LABVIEW front panel (change in load resistance).

Fig. 17 shows the performance of the proposed system under transient load change. The Channel 1 and 2 in Fig. 17 shows the PV and load voltage, respectively. Channel 3 and 4 shows the PV and load current, respectively.

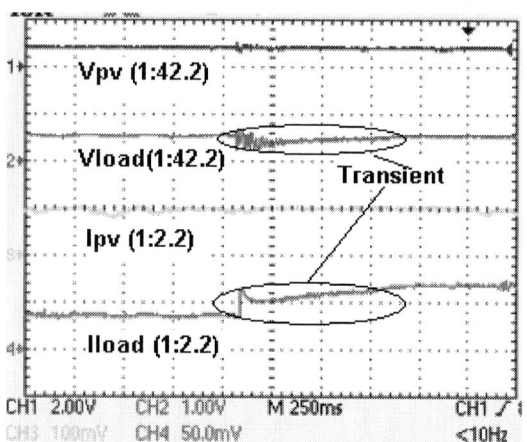

Fig. 17. Transient performance of the load voltage tracking during sudden change in load.

The load is varied from 50 Ω to 25 Ω with the load reference voltage of 10 V. It is observed from the results that the load voltage settles to the reference after initial transients.

V. CONCLUSIONS

In this paper a boost-SEPIC converter based photovoltaic system is developed using the LABVIEW platform. The proposed system can regulate the DC voltage supply to the load under varying environmental and load conditions. The performance of the proposed system has been verified under varying load reference, load change and different radiation level. The transient performance justifies the robustness of the proposed system to maintain the load voltage its reference under varying conditions.

VI. REFERENCES

[1] A. Yazdani and P. P. Dash "A control methodology and characterization of dynamics for a photovoltaic (PV) system interfaced with a distribution network", *IEEE Trans. Power Del.*, vol.24, no. 3, pp.1538-1551, Jul. 2009.

[2] L. Yang, J. Cao and Z. Li, "Principles and implementation of maximum power point tracking in photovoltaic system," *IEEE conference on Mechanic Automation and Control Engineering (MACE)*, 2010

[3] T. Esram and P.L. Chapman, "Comparison of Photovoltaic Array Maximum Power Point Tracking Techniques," *IEEE Trans. Energy. Convers*, vol. 22, no. 2, pp. 439-449, June 2007.

[4] X.Q.Guo and W.Y. Wu, "Improved current regulation of three-phase grid-connected voltage-source inverters for distributed generation systems," *IET Renew. Power Gener.*, vol.4, no.2, pp.101-115, Mar. 2010.

[5] H.C.Chiang, T.T.Ma, Y.H.Cheng, J.M.Chang and W.N. Chang, "Design and implementation of a hybrid regenerative power system combining grid-tie and uninterruptible power supply functions," *IET Renew. Power Gener.*, vol.4, no.1, pp.85-99, Jan.2010.

[6] F. Giraud and Z. M. Salameh, "Steady-State Performance of a Grid-Connected Rooftop Hybrid Wind–Photovoltaic Power System with Battery Storage," *IEEE Trans. Energy. Convers.*, vol.16, no.1, pp.1-7, Mar. 2001.

[7] S. V. Araújo, P. Zacharias. and R. Mallwitz, "Highly Efficient Single-Phase Transformerless Inverters for Grid-Connected Photovoltaic Systems," *IEEE Trans. Ind. Electron*, vol. 57, no. 9, pp.3118-3128, Sep. 2010.

[8] S. K. Kim, J. H. Jeon, C. H. Cho, J. B. Ahn and S. H Kwon, "Dynamic modeling and control of a grid-connected hybrid generation system with versatile power transfer," *IEEE Trans. Ind. Electron.*, vol. 55, no.4, pp.1677-1688,Apr.2008.

[9] W. Q, J. Liu, X. Chen and P. D. Christofides, "Supervisory Predictive Control of Standalone Wind/Solar Energy Generation Systems," *IEEE Trans. Ind. Electron.*, vol.19, no.1, pp.199-207, Jan.2011 .

[10] K. Jin, X. Ruan, M. Yang and M. Xu, "A hybrid fuel cell power system," *IEEE Trans. Ind. Electron.*, vol.56, no.4, pp.1212-1222, Apr.2009.

[11] Y.K. Lo, H.J.Chiu, T.P.Lee, I. Purnama and J.M. Wang, "Analysis and design of a photovoltaic system connected to the utility with a power factor corrector," *IEEE Trans. Ind. Electron.*, vol.56, no.11, pp.4354-4362, Nov. 2009.

[12] Z Wang, X. Li, G Li, M. Zhou and K.L.Lo, "Energy Storage Control for the Photovoltaic Generation System in a Micro-grid," *5th IEEE international conference on Critical Infrastructure (CRIS)*, 2010.

[13] Q. Zhao and Z. Yin, "Battery Energy Storage Research of Photovoltaic Power Generation System in Micro-grid," *5th IEEE conference on Critical Infrastructure (CRIS)*, 2010.

[14] D. V. D. L. Fuente, C. L. T Rodríguez, G. Garcerá, E. Figueres and R. O. González, "Photovoltaic Power System With Battery Backup With Grid-Connection and Islanded Operation Capabilities," *IEEE Trans. Ind. Electron.*, vol.60, no.4, pp.281-286, April. 2013.

[15] *Z. Jiang and R. A. Dougal, "A novel, Digitally-Controlled, Portable Photovoltaic Power Source,"* 20th IEEE conference on Apppplied Power Electronics Conference and Exposition (APEC), *2005.*

[16] K Kobayashi, H Matsuo and Y Sekine, "Novel Solar-Cell Power Supply System Using a Multiple-Input DC–DC Converter," *IEEE Trans. Ind. Electron.*, vol.53, no.1, pp.281-286, Feb. 2006.

[17] K Kobayashi, H Matsuo and Y Sekine, "An Excellent Operating Point Tracker of the Solar-Cell Power Supply System," *IEEE Trans. Ind. Electron.*, vol.53, no2, pp.495-499, April. 2006.

[18] D.D.C Lu and V.G. Agelidis "Photovoltaic-Battery-Powered DC Bus System for Common Portable Electronic Devices," *IEEE Trans. Power Electron*, vol. 24, no.3, pp. 849-855, Mar. 2009.

[18] S. J. Chiang, H. J. Shieh and M. C. Chen, "Modeling and Control of PV Charger System With SEPIC Converter," *IEEE Trans. Ind. Electron.*, vol.56, no.11, pp.4344-4353, Nov. 2009.

[19] H. Ma, J. S. Lai, Q. Feng, W. Yu, C. Zheng and Z. Zhao, "A Novel Valley-Fill SEPIC-derived Power Supply Without Electrolytic Capacitors for LED Lighting Application," *IEEE Trans. Power Electron.*, vol. 27, no. 6, pp. 3057–3071, June 2012.

[20] J. M. Jiménez-Martínez,, F Soto, E Jódar, J A. Villarejo J Roca-Dorda,"A new approach for teaching power electronics converter experiments," *IEEE Trans. Educ.*, vol. 48, no. 3, pp. 513–519, Aug 2005.

[21] R.W Erickson, *Fundamental of Power Electronics.* Nowell, M A: Kluwer, 1997.

[22] M. Veerachary," Power Tracking for Nonlinear PV Sources with Coupled Inductor SEPIC Converter," *IEEE Trans. Aerospace and Electronic System.*, vol. 41, no. 3, pp. 1019–1029, July 2005.

[23] M. Bhunia and R. Gupta, "Voltage regulation of stand-alone photovoltaic system using boost SEPIC converter with battery storage system," *2nd IEEE student conference on Engineering and System (SCES)*, 2013.

[24] E. M. Ahmed, M. Shoyama "Stability Study of Variable Step Size Incremental Conductance/Impedance MPPT for PV systems," *8th IEEE International Conference on Power Electronics - ECCE Asia* , 2011.

[25] L. Gao, R. A. Dougal, and S. Liu, "Power Enhancement of an Activity Controlled Battery/Ultracapacitor Hybrid," *IEEE Trans. Power Electron*, vol. 20, no.1, pp. 236-243, Jan. 2005.

[26] S. T. Hung, D.C. Hopkins, and C.R. Mosling, "Extension of Battery life via Charge Equalization Control," *IEEE Trans. Ind. Electron.*, vol. 40, no.1, pp.96-104, Feb. 1993.

A SOGI-Q Based Control Algorithm for Multifunctional Grid Connected SECS

Chinmay Jain, *Member, IEEE*
Electrical Engineering Department
Indian Institute of Technology Delhi
New Delhi-110016, India
Email ID: chinmay31jain@gmail.com

Bhim Singh, *Fellow, IEEE*
Electrical Engineering Department
Indian Institute of Technology Delhi
New Delhi-110016, India
Email ID: bhimsingh1956@gmail.com

Abstract— This paper deals with a three phase multifunctional grid connected SECS (Solar Energy Conversion system). A two stage power circuit topology is used in this work, in which the first stage is a boost converter, which serves the purpose of MPPT (Maximum Power Point Tracking) and the second stage is a 4-leg VSC (Voltage Source Converter) which serves the purpose of feeding extracted energy along with power quality improvement in the distribution system. The SECS not only feeds solar PV (Photo-Voltaic) energy into the grid but also serves purpose of grid currents balancing, reactive power compensation, harmonics elimination and neutral current mitigation. A feed-forward term for the solar contribution is used to improve the dynamic response for climatic changes. The PV array voltage is continuously adjusted with the help of boost converter to achieve MPPT whereas the DC link voltage of VSC is kept constant with the help of a PI (Proportional-Integral) controller. A SOGI-Q (Second Order Generalized Integrator - Quadrature) based algorithm is proposed for control of four-leg VSC. The system is modeled and simulated on MATLAB based platform. A wide range of simulation results are shown to demonstrate all features of proposed system. The simulation results show the feasibility of the proposed control algorithm. The THD (Total Harmonics Distortion) of grid current has been found well under IEEE-519 standard.

Keywords— Solar PV; Two-stage; 3P4W; MPPT; SOGI-Q.

I. INTRODUCTION

The electrical energy has played a wide role in development of society in last one century. Conventionally fossil fuels are used to generate electricity. However, the rapidly vanishing conventional energy sources have put an alarming energy crisis situation in front of all developed and developing countries in the world. Moreover, the increasing pollution and effect of green house gases have moved everybody's attention towards green energy sources. The solar energy being abundant is becoming an eye-catching technology day by day. However, the economics of solar PV has always been an issue with practical installation of the technology but recent trends show that solar PV is reaching the grid parity [1]-[2].

The solar PV energy based system can be broadly classified in two subcategories which are standalone and grid interfaced SPV system. Several standalone PV systems are proposed by researchers [3]-[5]. A review of commonly used strategies for power management is given in [3]. An analysis and design for standalone SPV based power generating system is shown in [4]

wherein a single phase system is considered with battery energy storage system. The solar PV energy is highly intermittent in nature, therefore the energy storage (battery is commonly used) becomes the essential requirement of the system. The fluctuations in the power generated cause the batteries to charge and discharge very frequently which are bad for battery life [5]. Therefore, battery maintenance and energy management becomes a complicated issue in case of standalone PV systems. Considering these points the grid interfaced SPV systems are more preferable for the places where grid is available.

The solar PV characteristics are nonlinear due to which there is a unique operating point for which peak power can be extracted from a given PV array. This operating point for peak power extraction changes with change in climatic conditions such as, irradiance level and surrounding temperature. Several MPPT (Maximum Power Point Tracking) techniques are proposed to track the maximum power point. A review of MPPT techniques under uniform insolation and partial is shown in [6]. A mixed MPPT technique is shown in [7] wherein a combination of P&O (Perturb and Observe) and fractional V_{oc} based algorithm is shown. An incremental conductance (InC) based MPPT technique is used in [8]. However, in this paper a composite InC based MPPT technique is used wherein a combination of knowledge of fractional open circuit MPPT and InC based MPPT are combined to track peak power point.

Nonlinear loads such SMPS (Switched Mode Power Supplies), electronic blasts, rectifiers etc are increasing day by day. These nonlinear currents not only increase losses in the distribution system but also degrade the voltage power quality at PCC (Point of Common Coupling). The D-STATCOM (Distribution Static Compensator) presents a retrofit solution for these power quality problems. Several control algorithms are proposed for control of D-STATCOM. A comparison of several classical control algorithms is given in [9]. Several soft computing based control algorithms are proposed in the literature [10]. An adaptive theory based and neural network based complicated control strategies are also reported in the literature [11]-[12].

In a three phase system operating under balanced linear load the current in the neutral conductor is zero. However, same is not the case with nonlinear load. All triplen harmonics and zero sequence currents flows through the neutral

978-1-4799-6047-7/14 $31.00 © 2014 IEEE

conductor. The neutral current increases losses in the distribution system. Moreover, excessive neural current may cause bursting of neutral conductor. Several schemes are proposed for compensation of neutral current in the line conductor. A zig-zag transformer based active power filter for neutral current compensation is shown in [13]. An analysis and comparison for different neutral current compensation techniques are shown in [14].

The use of SOGI (Second Order Generalized Integrator) for a part of PLL (Phase Lock Loops) is well known. Several researchers have proposed SOGI for generating orthogonal set of voltage vectors in single-phase PLL [15]-[16]. However, in this paper, SOGI-Q (Second Order Generalized Integrator-Quadrature) based control algorithm is proposed for control of multifunctional SECS (Solar Energy Conversion System). A two stage SECS is used in the proposed work. The first stage (boost converter) works for MPPT. A composite InC based MPPT is used to control the boost converter. However, the second stage serves the purpose of harmonics mitigation, compensation of reactive power, grid currents balancing along with feeding extracted energy into the distribution system. An assessment of proposed SOGI-Q based control algorithm is presented along with its comparison with a SOGI based control. The performance of the system is verified by means of MATLAB simulations. A wide variety of simulation results are shown to demonstrate all the features of proposed system. The presented system adheres to IEEE-519 standard [17]

II. SYSTEM CONFIGURATION

The proposed system configuration is shown in Fig. 1. A three-phase, 4-wire distribution system is the system under consideration. The proposed system consists of a solar PV array, a boost converter, a 4-leg VSC, interfacing inductors (L_{VSC}), ripple filter (C, R), independent single-phase loads, and the 3P4W (3-phase 4-wire) disribution system. The system consists of solar PV array, which is a series parallel combination of small power solar panels to match the required rating. The PV array is connected at the input of the boost converter. The boost converter is switched in order to achive MPPT. The power extracted by the boost converter is fed to the dc link of the VSC. The dc link capapcitor of the VSC acts as a small energy buffer. The average voltage of dc link is kept constant, to ensure this the instantaneous power fed by boost converter is fed into distribution system. The VSC consists of four IGBT (Insulated Gate Bipolar Transistor) legs. This four leg VSC is connected to 3P4W distribution system at PCC via interfacing inductors. Three independent single phase loads are also connected at PCC. A small ripple filter is connected in shunt at PCC to suppress switching ripple. The rating of all components is given in Appendix.

III. SYSTEM CONTROL

The basic control strategy for the proposed system is shown in Fig. 2. The two main parts of control algorithm are the control of the boost converter and the control of the VSC. The boost converter serves the purpose of MPPT, whereas the VSC feeds the extracted energy into the 3P4W (3-Phase 4-Wire) distribution system along with grid currents balancing, reactive power compensation, harmonics elimination and neutral current mitigation. An composite InC based MPPT algorithm is used for control of boost converter whereas a SOGI-Q based control algorithm is proposed for control of VSC. The SOGI-Q based control algorithm is used to extract active power consuming component of load current. The control algorithms for the boost converter and VSC are explained in detail in the following section.

A. Maximum Power Point Tracking

A composite InC (Incremental Conductance) based MPPT (Maximum Power Point Tracking) algorithm is used. A range of voltage for peak power is known with the knowledge from fractional V_{oc} MPPT which is $0.7V_{ocmax}$ to $0.9V_{ocmax}$, where V_{oc} is open circuit voltage and V_{ocmax} is maximum open circuit voltage. The voltage for peak power is always searched in this range for fast search of V_{mpp}. The InC algorithm works in order to minimize the difference between the incremental conductance and the conductance offered by the PV array. At first, the reference PV array voltage is estimated based on the InC principle then that reference voltage is used to estimate the duty ratio of boost converter. For calculation of incremental conductance ΔI_{pv} and ΔV_{pv} are estimated as,

$$\Delta I_{pv} = I_{pv}(k) - I_{pv}(k-1) \tag{1a}$$

$$\Delta V_{pv} = V_{pv}(k) - V_{pv}(k-1) \tag{1b}$$

where $I_{pv}(k)$ and $V_{pv}(k)$ are the instantaneous sampled current and voltage of the solar array.

The governing equations for InC based MPPT algorithm are as,

$$\frac{\Delta I_{pv}}{\Delta V_{pv}} = \frac{-I_{pv}}{V_{pv}}, at\ MPP \tag{2a}$$

$$\frac{\Delta I_{pv}}{\Delta V_{pv}} > \frac{-I_{pv}}{V_{pv}}, \text{Left of MPP on } P_{pv}\ v/s\ V_{pv}\ \text{curve} \tag{2b}$$

$$\frac{\Delta I_{pv}}{\Delta V_{pv}} < \frac{-I_{pv}}{V_{pv}}, \text{Right of MPP on } P_{pv}\ v/s\ V_{pv}\ \text{curve} \tag{2c}$$

The reference PV array voltage is adjusted based on above equations. The reference PV voltage and sensed DC link voltage is then used to estimate the duty ratio for the boost

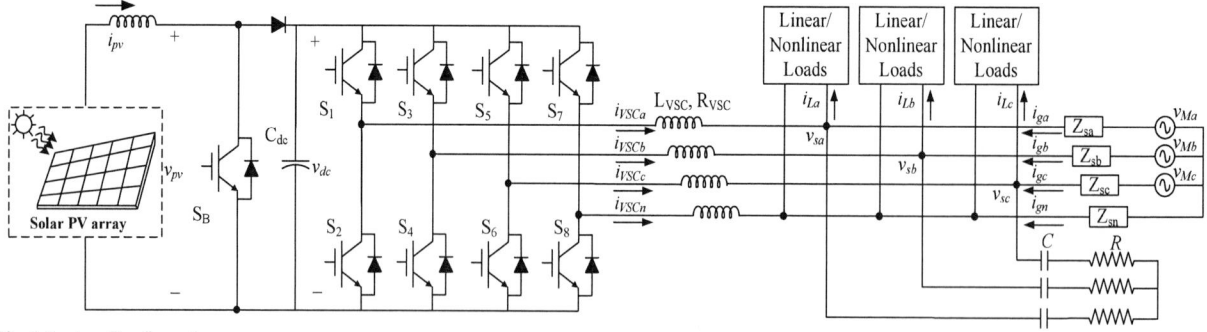

Fig. 1 System Configuration.

978-1-4799-6047-7/14 $31.00 © 2014 IEEE   238

converter. The governing equation for estimating duty ratio is,

$$d_r(k) = 1 - \frac{V_{pvref}(k)}{V_{dc}(k)} \qquad (3)$$

This reference duty ratio is compared with saw-tooth waveform to generate switching logic for the boost converter.

B. Control Algorithm for four-leg VSC

The basic block diagram of the proposed control algorithm is presented in Fig. 2. For the control of the VSC several quantities are sensed and fed back to the controller. The PCC voltages (v_{an}, v_{bn}), grid currents (i_{ga}, i_{gb}, i_{gc}), load currents (i_{La}, i_{Lb}, i_{Lc}), dc bus voltage (v_{dc}), solar PV array voltage (v_{pv}) and current (i_{pv}) are sensed. These all sensed signals are processed according to block diagram shown in Fig. 2. At first the peak of three-phase voltage is estimated as,

$$V_z = \sqrt{\frac{2(v_{sa}^2 + v_{sb}^2 + v_{sc}^2)}{3}} \qquad (4)$$

This peak voltage is used to determine the unit vector of PCC voltage which contains the phasor information of all phase voltages. The unit vectors are estimated as,

$$z_a = \frac{v_{sa}}{V_z}, z_b = \frac{v_{sb}}{V_z}, z_c = \frac{v_{sc}}{V_z} \qquad (5)$$

The total PV array power is equally divided in all three phases with respect to grid. Hence PV feed forward (PVFF) compensation term combined for all three phases is computed as,

$$I_{pvs} = \frac{2P_{pv}}{V_z} \qquad (6)$$

A SOGI-Q based algorithm is used to estimate the average power consuming portion of load currents. The average power consuming components of load currents of all three phases are calculated independently. The load current is for given to SOGI-Q block for estimation of fundamental part of load current. The SOGI-Q block not only filters the load current for harmonics but also provides a phase shift of 90° to fundamental component with respect to actual load current. The transfer function of SOGI-Q block is given as,

$$F_1(s) = I_{a1Q}(s)/I_L(s) = \gamma\omega_1^2 / (s^2 + \gamma\omega_1 s + \omega_1^2) \qquad (7)$$

where ω_1 is fundamental frequency and k is characteristic parameter. A detailed analysis of simple SOGI and SOGI-Q based algorithm for this purpose is presented in detail in the following sub section. The i_{a1Q} is sampled and hold at every zero crossing of unit vector of respective phase to estimate the active power consuming component of load current. The similar procedure is also applied for other two phases.

A PI controller is used to maintain the dc bus voltage of VSC. The output of PI controller (I_{lpl}) in steady state condition designates the loss component of VSC. A proper sign convention considering the direction of grid currents is used to estimate the net amplitude of average power consuming

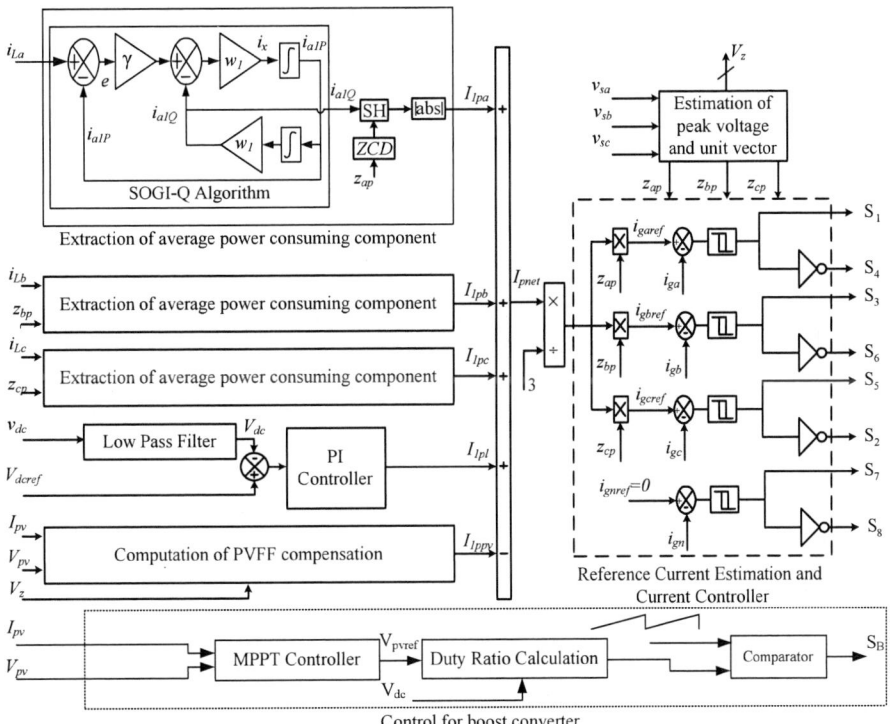

Fig2 Block diagram of control algorithm.

component of grid current. According to considered directions of grid currents as shown in Fig. 1, the load and loss components of all three phases are added to estimate the net power consumption at PCC whereas the PV contribution term is subtracted from this component for estimation of amplitude total average power consuming component of grid current (I_{pnet}). The mathematical estimation for I_{pnet} is as,

$$I_{pnet} = I_{1pa} + I_{1pb} + I_{1pc} + I_{1pl} - I_{1ppv} \qquad (8)$$

This I_{pnet} represents the net average power consuming component if only 1-phase of grid is there. However, in this case a three-phase system is considered hence to equally divide the active power in all three phases the I_{pnet} is divided by three. The reference grid currents are estimated as,

$$i_{garef} = \frac{I_{pnet}}{3} \cdot z_a, \; i_{gbref} = \frac{I_{pnet}}{3} \cdot z_b, \; i_{gcref} = \frac{I_{pnet}}{3} \cdot z_c \qquad (9)$$

The sensed and reference currents are given as the inputs to hysteresis current controller and logic switching pulses are output of the current controller.

C. Comparision of SOGI and SOGI-Q Based Algorithm

While using the SOGI based algorithm there are two filtered output components which are i_{a1P} and i_{a1Q}. The transfer function of quarature filtered component (i_{a1Q}) is as in (7). However, the transfer function for other component is as,

$$F_2(s) = I_{a1P}(s) / I_L(s) = \gamma\omega_1 s / (s^2 + \gamma\omega_1 s + \omega_1^2) \qquad (10)$$

where i_{a1P} is the fundamental output component of a-phase load current in phase with actual load current. To estimate the average power consuming component of load current from i_{a1p} the quadrature unit vectors are required hence, extra calculations from sensed PCC voltages are avoided by using i_{a1Q} component.

The load current for phase-a can be represented as,

$$i_{La} = i_{a1} + \Sigma i_{ah} \qquad (11)$$

where i_{a1} is the fundamental component of load current and i_{ah} is h^{th} order harmonics current. The harmonics currents frequencies are generally integral multiple of line frequency. The frequency response for SOGI and SOGI-Q are shown in Fig. 3. All salient points are also marked on the frequency response plot. From the frequency response plots of $F_1(s)$ it can be observed that the bandwidth decreases with the decrease in parameter γ. However, lower the bandwidth sluggish is the time response. However, for the same values of γ the attenuation to high frequency components is more in SOGI-Q ($F_2(s)$) algorithm. Moreover, the settling times for both these algorithms are approximately same for same value γ. Hence considering these advantages the SOGI-Q algorithm is a better choice on account of higher filtering capabilities and lower calculations in overall control algorithm.

IV. RESULTS AND DISCUSSION

The proposed two stage grid interfaced SPV system is modeled in MATLAB along with Simulink and Sim power System toolboxes. A SPV array rating of 30 kW is considered

and a load power rating of 5 kW per phase is considered. The simulation performances are shown for SECS operating under nonlinear/linear loads and sudden change in insolation level. The intermediate signals of the control algorithm are also shown for clarity of the concept. Both the dynamics and the steady state performances are shown via simulation results. The system parameters used for simulation study are given in Appendix.

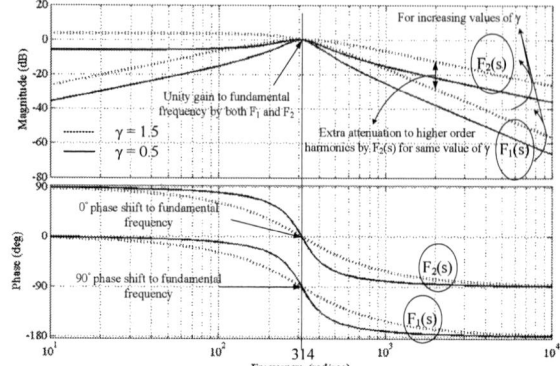

Fig. 3 Frequency response analysis

A. Behavior of SECS under Linear Loads

The behavior of SECS operating under linear load condition is shown in Fig. 4 (a). At time t= 0.25s, the system is working under power factor correction mode under balanced lagging linear loads at PCC. At time t = 0.3 s, phase a load is removed and at time t = 0.35 s, phase c load is removed. The dynamic behavior for load change is demonstrated by load removal and inclusion at different time instants. It can be observed that load currents are unbalanced but grid currents are balanced sinusoids at unity power factor. During load unbalanced operation, a neutral current (i_{Ln}) is observed in the load. However, VSC neutral current (i_{VSCn}) is equal and opposite to load neural current because of which the grid neutral current (i_{gn}) is practically zero. The load power decreases when the load is thrown, and at the same time, as the PV array power is kept constant hence, the power injected into the grid increases. The increment in power injected into the grid can be observed in form of increment in grid currents. When the load is added again, the decrement in grid currents can be observed. During unbalancing no appreciable effect is observed on dc link voltage (v_{dc}) and power from solar PV-array (P_{pv}).

B. Behavior of SECS under Nonlinear Loads

The behavior of SECS operating under nonlinear load condition is shown in Fig. 4 (b). A nonlinear load of 5 kW is considered on each phase. The nonlinear load is emulated using a diode bridge rectifier with RL load. The power from solar PV array is considered to be constant. At t= 0.25s, the grid currents are balanced and sinusoidal at unity power factor. The VSC supplies all the harmonics required by the load. In case of nonlinear loads, the neural current persists even under balanced loads in all three phases, which can be observed in Fig. 4 (b) i_{Ln}. The peak value of current in load neutral conductor increases, in case of unbalanced loading of the system. The load and VSC neutral currents are out of phase which results in zero current in grid neutral conductor. The supply line for Phase-a of load is opened at t= 0.3 s, and reconnected at t = 0.4s, respectively. Similarly load on phase c, is also varied. It can be observed that grid currents are balanced and sinusoidal even in case of unbalanced nonlinear load on the system. The VSC currents are unbalanced to make grid currents balanced sinusoids. During load unbalancing, no appreciable effect is observed on dc link voltage (v_{dc}) and power from solar PV-array (P_{pv}).

The intermediate signals under load dynamics and steady state conditions are shown in Fig. 4 (c). The load of phase-a is opened at time t = 0.3 s. The better filtering capability of SOGI-Q can be observed by observing i_{a1P} and i_{a1Q}. The current i_{a1P} is more distorted then i_{a1Q}, hence a better estimation of average power consuming component of load current is possible by using SOGI-Q algorithm.

C. Behavior of SECS under Change in Solar Insolation

The behavior of SECS for step change in irradiance is shown in Fig. 4 (d). At t=.0.35s, the irradiance is 400 W/m². The load on the system is kept constant which can be observed from unaltered load currents. The power from the solar PV array is around 15 kW and load power is also 15 kW hence, approximately zero current is observed on grid side currents. The VSC currents consist of harmonics currents and fundamental currents corresponding to active power from solar PV array. At t= 0.35 s, the irradiance is changed to 1000 W/m². The increase in irradiance causes an increase in power from SPV array. The increased power is now fed into the grid.

D. Harmonics Analysis

The harmonics analysis of proposed system is shown in Fig. 5. Fig. 5(a) shows the harmonics spectrum of the nonlinear load current. The THD of the load current is of order of 40 %. Fig. 5 (b) shows the harmonics spectrum of grid current and the THD of grid currents is found of order of 1 % (below 5%) which is well under IEEE-519 standard.

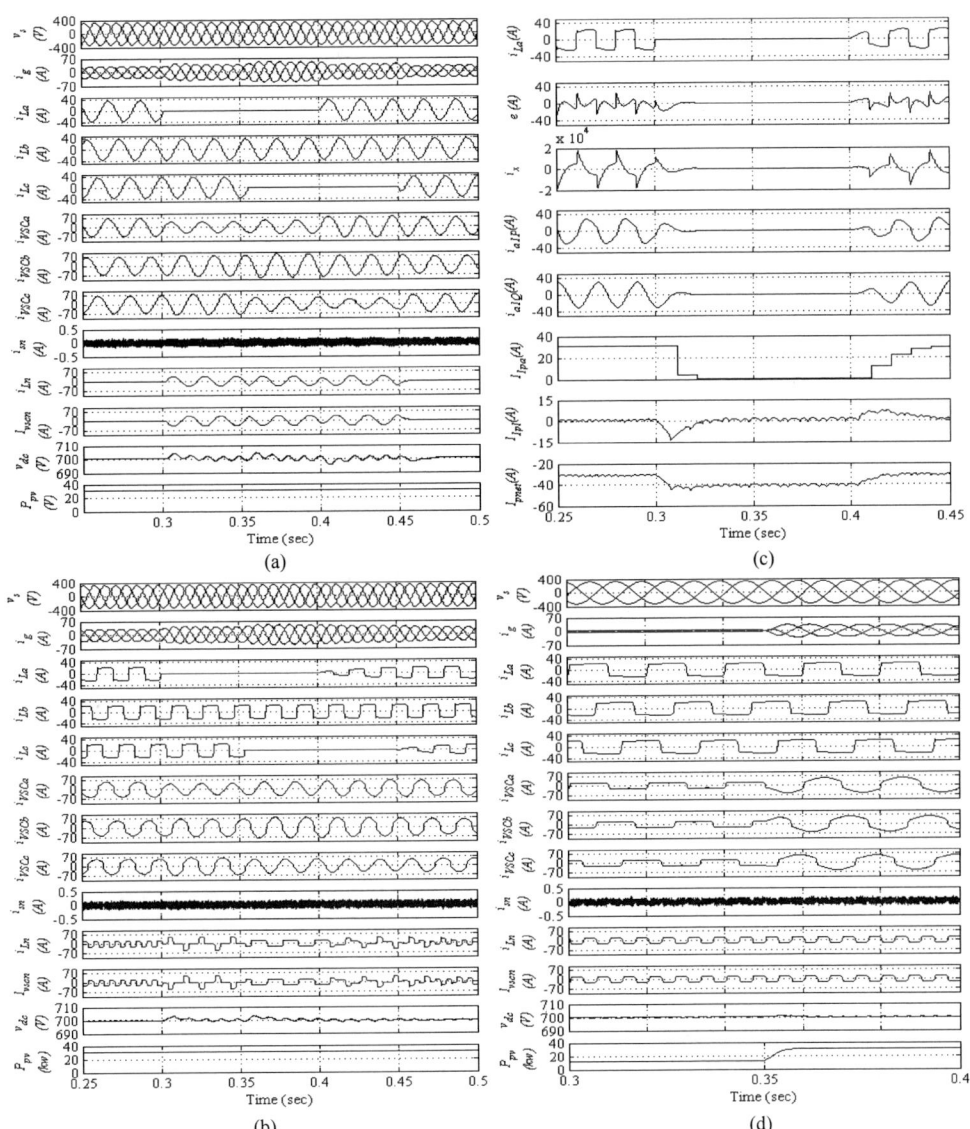

Fig. 4 Performance evaluation of SECS for (a) linear load, (b) nonlinear load, (c) intermediate control signal, (c) sudden change in insolation level

978-1-4799-6047-7/14 $31.00 © 2014 IEEE

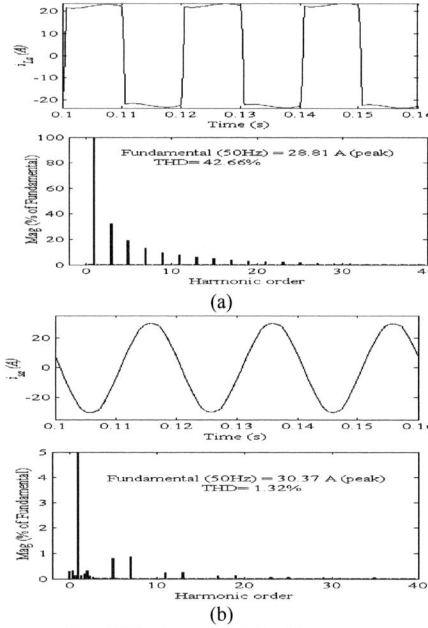

Fig. 5 Harmonics analysis (a) load current, (b) grid current.

V. CONCLUSION

A two-stage, three-phase, 4-leg VSC based system has been proposed to interface solar PV array with a 3P4W distribution system. A PLL-less control has been proposed for the control of multifunctional VSC. A SOGI-Q based control algorithm has been proposed for SECS. A detailed assessment including frequency response and comparison with SOGI based method has been carried out and the proposed method is found quite suitable. The performance of proposed system has been found satisfactory under dynamics and steady state responses. The SECS not only feeds extracted energy into the grid but also feeds the harmonics and reactive power required by the local load. Moreover, a fourth leg is also used for neutral current compensation. The SECS supplies all negative and zero sequence current required by the local load in order to maintain balanced grid currents. The proposed distributed generation system helps in reduction of losses in distribution system. A wide variety of simulation studies are carried to demonstrate all the features of proposed system. The THDs of the grid currents are found less than 5% (within IEEE-519 standard) even under nonlinear loads at PCC. The simulation results show the feasibility of proposed SECS.

APPENDIX

Parameters of the system: three-phase line voltage 415 V, frequency = 50Hz, supply inductance = 2 mH and supply resistance = 0.5 Ω, interfacing inductor = 3 mH, ripple filter R = 5 Ω, C = 5μF, PI controller parameter K_p = 1, K_i = 0.5, PV array open circuit voltage: 600 V, PV array short circuit current: 60 A, PV array peak power: 30 kW. Filter parameter γ = 1.

ACKNOWLEDGMENT

Authors are very thankful to Department of Science and Technology (DST), Govt. of India, for funding this project under Grant Number: RP02583.

REFERENCES

[1] N Y Dahlan, Mohd Afifi Jusoh and W N A W Abdullah, "Solar grid parity for Malaysia: Analysis using experience curves," IEEE Power Engineering and Optimization Conference (PEOCO), 2014, pp.461-466.

[2] S. Reichelstein and M. Yorston, "The prospects for cost competitive solar PV power special section: Long run transitions to sustainable economic structures in the European Union and beyond," Energy Policy, vol. 55, pp. 117–127, 2013.

[3] G. T. Machinda, S. Chowdhury, S.P. Chowdhury and W. N. Mbav, "Power management of inverter interfaced solar PV microgrid: A review of the current technological trend," Universities Power Engineering Conference (UPEC), Sept. 2012, pp.1-6.

[4] N. Adhikari, B. Singh, A. L. Vyas, A. Chandra and Kamal-Al-Haddad, "Analysis and design of isolated solar-PV energy generating system," in proc. of Industry Applications Society Annual Meeting (IAS), 2011, pp.1-6.

[5] Y. Zhang, H.J. Jia and L. Guo, "Energy Management Strategy of Islanded Microgrid Based on Power," IEEE PES Innovative Smart Grid Tech., pp. 1-8, Jan. 2012.

[6] A. Sayal, "MPPT techniques for photovoltaic system under uniform insolation and partial shading conditions," Students Conference on Engineering and Systems (SCES), 2012, pp.1-6.

[7] A. F. Murtaza, H. A. Sher, M. Chiaberge, D. Boero, M. D. Giuseppe and K.E. Addoweesh, "A novel hybrid MPPT technique for solar PV applications using perturb & observe and Fractional Open Circuit Voltage techniques," 15th International Symposium MECHATRONIKA, 2012, pp.1-8.

[8] G. J. Kish, J. J. Lee and P. W. Lehn, "Modelling and control of photovoltaic panels utilising the incremental conductance method for maximum power point tracking," IET Renewable Power Generation, vol.6, no.4, pp.259-266, July 2012.

[9] B. Singh and J. Solanki, "A Comparison of Control Algorithms for DSTATCOM," IEEE Transactions on Industrial Electronics, vol.56, no.7, pp.2738-2745, July 2009.

[10] Parmod Kumar and A. Mahajan, "Soft Computing Techniques for the Control of an Active Power Filter," IEEE Transactions on Power Delivery, vol.24, no.1, pp.452-461, Jan. 2009.

[11] B. Singh, S. K. Dube, S. R. Arya, A. Chandra and K. Al-Haddad, "A comparative study of adaptive control algorithms in Distribution Static Compensator," in proc. of IEEE Annual Conference of Industrial Electronics Society (IECON), 2013, pp.145-150.

[12] B. Singh, A. Adya, A. P. Mittal and J. R P Gupta, "Neural Network Based DSTATCOM Controller for Three-phase, Three-wire System," International Conference on Power Electronics, Drives and Energy Systems (PEDES), 2006, pp.1-6.

[13] P. Jayaprakash, B. Singh and D. P. Kothari, "IcosΦ algorithm based control of zig-zag transformer connected three phase four wire DSTATCOM," IEEE International Conference on Power Electronics, Drives and Energy Systems (PEDES), 2012, pp.1-6.

[14] A. Negi, S. Surendhar, S.R. Kumar and P. Raja, "Assessment and comparison of different Neutral current compensation techniques in three-phase four-wire distribution system," in Proc. of 3rd IEEE International Symposium on Power Electronics for Distributed Generation Systems, 2012, pp.423-430.

[15] A. Kulkarni and V. John, "A novel design method for SOGI-PLL for minimum settling time and low unit vector distortion," in proc. of 39th Annual Conference IECON Industrial Electronics Society, 2013, pp.274-279.

[16] L. Coluccio, A. Eisinberg, G. Fedele, C. Picardi and D. Sgro, "Modulating functions method plus SOGI scheme for signal tracking," IEEE International Symposium Industrial Electronics, 2008, pp.854-859.

[17] IEEE Recommended Practices and requirement for Harmonic Control on Electric Power System, IEEE Std. 519, 1992.

A Novel control strategy for power extraction from Photo Voltaic panels based on One Cycle Control

Anoop K
Research Scholar
Department of EEE
Government Engineering College, Thrissur
Kerala

Dr. M Nandakumar
Professor
Department of EEE
Government Engineering College, Thrissur
Kerala

Abstract—Thispaper presents maximum power point tracking and load voltage regulation of a photo voltaic panel using One Cycle Control (OCC). The power generation by renewable energy sources depends highly on atmospheric conditions and due to that they are very chaostic in nature. Hence rapid control becomes an essential requirement for the converters used for power conditioning.Conventionally PI controller is used for the control of power electronic circuits employed in such systems. But conventional PI controller takes few switching cycles to track the variations in operating conditions while the one cycle controller can track them within one switching cycle. Also compared to PI, OCC has a better transient response. Short current pulse based maximum power point tracking is used here to compare the performance of OCC and PI controller. MATLAB/Simulink simulation results are provided to verify their performance.

Keywords—One Cycle Controller (OCC), Pulse width modulation (PWM), proportional integral (PI), maximum power point tracking(MPPT)

I. INTRODUCTION

Electrical Energy plays a fundamental role in shaping the human conditions. The demand for electrical energy is increasing day by day. Conventionally used electrical energy production methods are hydroelectric generation systems, thermal generation systems, nuclear power plants etc. Environmental pollution is one of the major drawbacks thermal and nuclear energy systems.. Even though hydroelectric power generation does not have atmospheric pollution, they need a huge water reservoir which causes environmental destruction. Because of these reasons the demand for an alternate, renewable, non-polluting energy source increased worldwide. This demand finds its solution in renewable energy sources like wind and solar energy.

The electric power produced by the solar panels depends upon the panel temperature, solar irradiance, operating conditions, shadowing etc.[2].At constant atmospheric conditions, the power produced by the PV panels varies according to the electrical operating condition. In the electrical characteristics of the panel there exists an operating point called maximum power point (MPP). The maximum utilization of photo voltaic power can beachieved by operating the panel at this maximum power point. Since the efficiency of solar panels is low, for the maximum utilization of power, panels have to operate at MPP. Varieties of maximum power point tracking (MPPT) methods are developed. These methods vary

in the implementation complexity, sensed parameters, and required number of sensors, convergence speed, and cost [2].

In this paper a maximum power point tracking based on short current pulse is explained. In this method the panel current corresponding to maximum power point is determined and the PV system is operated accordingly. Usually a PI controller is used to achieve this operating condition. Even though the PI controller is powerful and universally accepted one, in varying operating conditions it takes few cycles to achieve steady state. Also since the design of PI controller depends upon the circuit elements, any variations in circuit parameters will affect the operation of controller. In this context an alternate control technique called One-cycle control (OCC) technique can be used.

OCC a nonlinear control method, which takes advantage of the pulsed and nonlinear nature of the switching converters and achieves instantaneous dynamic control of the average value of the switched variable [1][5]. Main feature of OCC. is that it takes only one switching cycle for achieving the steady state. This technique provides fast dynamic response, excellent power source disturbance rejection, robust performance, and automatic switching error correction [1][5].

The paper is organized as follows. Photo voltaic panel characteristics and need for MPP tracking is briefly explained in section II. Section III explains the principle of one cycle control and section IV presents the simulation studies. Section V concludes the paper.

II. PV CHARACTERISTICS AND MPP TRACKING

A. PV Characteristics

The operating principle of solar cell is photovoltaic effect. Every photo voltaic panels are represented by its open circuit voltage and short circuit current. The voltage- current characteristics of a solar panel is nonlinear. The output voltage/current of a photovoltaic panel varies with the cell temperature and solar radiation. The voltage – power characteristics of photo voltaic panel at various irradiance levels is as given in figure 1.

From the characteristics it can be seen that for a given irradiance, with the increase in panel voltage, the panel output

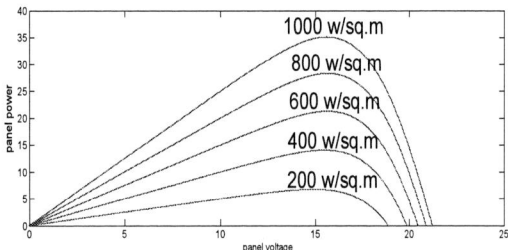

Figure 1: power –voltage characteristics of PV panel under different irradiance

power increases first, reaches a maximum value and then decreases to zero. At a particular operating point the output power of the solar panel reaches its maximum value. That operating point is known as the maximum power point of solar panel. Since the loads connected to the panels are variable in nature the operating point of the panels may not be the MPP always. Thus for operating the panels at maximum efficiency there must be a control system which can fix the operating point of panels at MPP irrespective of the load.

Maximum power point tracker is a control system which tracks the maximum power point of a solar panel at a given solar irradiance and temperature and operates the PV panel system at that operating point. Usually a MPP tracker consists of a power electronic converter. By monitoring the panel characteristics, the duty cycle of the converter system is adjusted and the MPP is achieved. Among the available MPP tracking methods short current pulse method is used here. The advantage of this method is the reduced complexity in implementation.

B. Short current pulse MPP tracking

The power produced by the panels depends upon the temperature and irradiance .Every maximum power point tracking algorithms are basically an optimization algorithm which finds the maxima of the given function. Hence the speed and complexity of the MPP tracking depends upon the searching method used. Some of the common methods used for MPP tracking are Perturb and Observe (P&O) method, incremental conductance method, short circuit current method, open circuit voltage method etc. Compared to P&O and incremental conductance method short circuit pulse has low complexity in implementation.

TABLE I Panel Ratings

Rated open circuit voltage	21.2 V
Rated short circuit current	2.5A
Rated power	35W
Voltage at MPP	15.6V
Current at MPP	2.25 A

To demonstrate the concept of short current pulse based MPPT a simulation study is carried out on a photo voltaic panel model. The ratings of panel model used are shown in table 1.From the simulation short circuit current and current for maximum power point for different irradiance levels are noted.Table 2 shows the result obtained from the simulation studies. From this data the ratio of I_{MPP} to I_{SC} at each irradiance level is calculated. .By observing the results it can

be seen that the current corresponding to maximum power point or the optimum operating current I_{MPP} shows a linear relation to the short circuit current I_{SC}[4].

TABLE II Ratio of I_{MPP} to I_{SC}

Irradiance level (W/m²)	I_{SC} (A)	I_{MPP}(A)	I_{MPP}/I_{SC}
1000	2.5	2.25	0.9
900	2.25	2.05	0.91
800	2	1.82	0.91
700	1.75	1.58	0.902
600	1.5	1.34	0.893
500	1.25	1.13	0.904
400	1	0.9	0.9
300	0.75	0.68	0.906
200	0.5	0.455	0.91
100	0.25	0.245	0.94

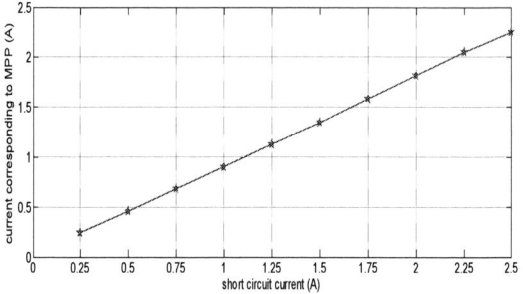

Figure 2: I_{SC}-I_{MPP}characteristics

Hence I_{MPP} can be expressed in terms of I_{SC} as

$$I_{MPP} = K_{op}.I_{SC} \qquad (1)$$

Where K_{op} is a proportionality factor.

The above mentioned concept is used for the short current pulse method for MPP tracking. From the knowledge of short circuit current and proportionality factorK_{op} at a given irradiance, the current corresponding to maximum power point of the panel can be determined. By forcing the panel to operate at this optimal current, MPP can be achieved.

The effect of partial shading is not considered while determining the value of K_{op}. However in practical conditions the K_{op} cannot be taken as a constant value always since the optimum operating point depends upon the atmospheric conditions like shading, panel surface conditions, aging etc.

Figure 3 shows the Simulink diagram to determine I_{SC}. The PV panel is connected to the load via a DC boost converter. At the output of PV panel another switch S_1 is connected parallel to the DC converter. In every one second the switch S_1 is turned on for a period of 0.1ms .The main criteria in choosing this time interval is that, in practical conditions the atmospheric conditions or irradiance will not vary very quickly within this duration.The time interval can be reduced to several ms.for better performance in varying atmospheric conditions.By turning on the switch S_1, the panel gets short circuited and the short circuit current will flow through the switch S_1. The short circuit current is measured using a current sensor and from (1) the optimum operating current I_{MPP} is determined.

978-1-4799-6047-7/14 $31.00 © 2014 IEEE

Considering K_{op} as a constant value is valid only at stable panel characteristics. However the panel characteristics may change due to shading, aging, corrosion, dust accumulation on panel surface etc. Hence K_{op} cannot be considered as a constant value and a real time estimation of K_{op} is required.

In order to estimate the value of K_{op}, initially short circuit current I_{SC} is determined. For calculating the value of K_{op} I_{MPP} is required. I_{MPP} can be determined from the power- current characteristics of the panel. The Power-current characteristics of the panel can be obtained by connecting a variable resistor to the panel and operate it from 0 to 100% resistance. In practical circuit a FET can act as a variable resistor by adjusting the input to the gate from 0% to 100%. The variable resistor is operated for duration of 10ms in several minutes. During this interval the resistance is varied and the instantaneous power and current through the resistance is calculated. The peak power and the current corresponding to the peak power are noted. Thus by operating the switch S1 as a variable resistor, the Power-Current curve of the panel can be obtained. From the Power-current curve the current corresponding to maximum power point can be determined .K_{op} is calculated by taking the ratio of I_{MPP} to short circuit current I_{SC}.

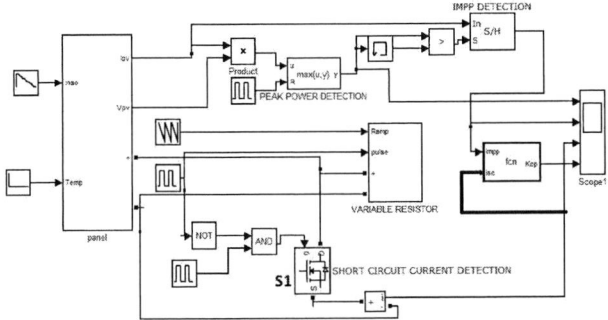

Figure 3: determination of I_{SC} and I_{MPP}

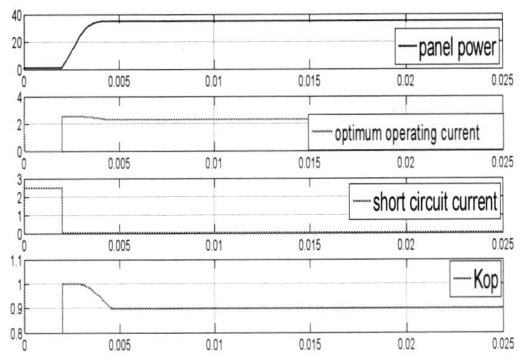

Figure 4: Estimation of I_{SC},I_{MPP} and K_{op}

During the determination of I_{MPP}, the panel is not delivering any power to the load. Hence the time interval should not be too small[4]. Once the value of K_{op} is calculated, the system can continue its operation by using the short current pulse based maximum power point tracking as explained earlier. The results obtained during the determination of I_{MPP}, I_{SC} and K_{op} are shown in figure 4.

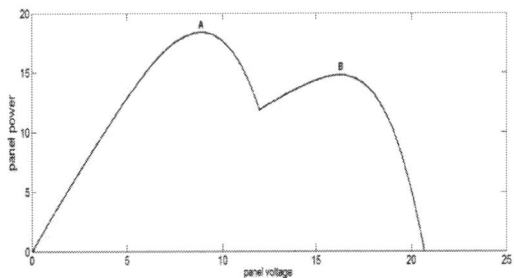

Figure 5: P-V characteristic during shading conditions

As shown in figure 5, there may be more than one local maximum in the power voltage curve or power current curve of the panel during partial shading conditions. While applying classical MPP search algorithms to such cases, it may settle at pseudo point B instead of the real maximum power point A. This may be considered as the main drawback of classical MPP search algorithms. The real time estimation of K_{op} can overcome this drawback of settling at a local pseudo maxima, as it scans the entire Power current curve during the course of action.

The calculated optimum current I_{MPP} is used as the control current reference for the converter. The controller drives the input current of the converter to the calculated optimum current reference value. Usually a PI controller is used to achieve this. In this paper one cycle controller is used for the converter controller. Compared to the PI controller, OCC have fast dynamic response. The design of PI controller depends upon the circuit parameters; hence any change in them can affect the operation of the controller. OCC is independent of circuit parameters and circuit parameters would not affect the operation of controller

III. ONE CYCLE CONTROLLER

A. Principle of OCC

The principle of one cycle control is explain below using the figures 6 and 7.Let D be the switching duty ratio, T_S be the switching time and T_{ON} be the ON period. Let k(t) be the switching function , x(t) be the input and y(t) be the output. During one switching interval TS, k(t) can be defined as

$$k(t) = \begin{cases} 1 & 0 < T < T_{ON} \\ 0 & T_{ON} < T < T_S \end{cases} \qquad (2)$$

Output   y(t)= k(t).x(t)   (3)

From the fig.7 the average value of input x(t) can be expressed as

$$y(t) = \frac{1}{T_S}\int_0^{T_{ON}} x(t).dt \qquad (4)$$

Let V_{REF} be the required average output voltage. By adjusting the duty ratio, once the integration of input function becomes equal to the integration of V_{REF} over a switching cycle, then

$$\frac{1}{T_S}\int_0^{T_{ON}} x(t).dt = \frac{1}{T_S}\int_0^{T_S} V_{REF}.dt \qquad (5)$$

Equating (5) and (4)

$$\frac{1}{T_S}\int_0^{T_{ON}} x(t).dt = y(t) = \frac{1}{T_S}\int_0^{T_S} V_{REF}.dt = V_{REF} \quad (6)$$

Hence the output after one switching cycle $y(t) = V_{REF}$

As shown in figure 6, the one cycle controller consists of a resettable integrator, comparator, switching controller and a clock. The input $x(t)$ is given to the resettable integrator. The output of the integrator V_{int} and the V_{REF} is compared using a high speed comparator.

$$Vint = K\int_0^T x(t).dt \quad (7)$$

Where K is integration constant and is determined by the switching frequency. As long as the integrated value is less than V_{REF}, the output of comparator will be at zero. Once the integrated value reached the V_{REF}, the comparator output will change to one.

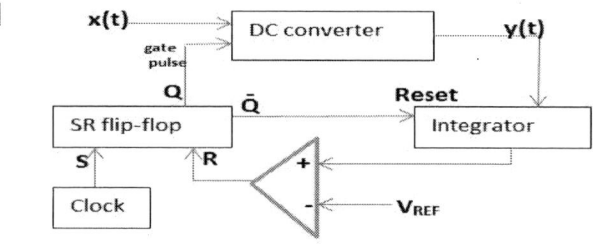

Figure 6: one cycle control at constant frequency switch

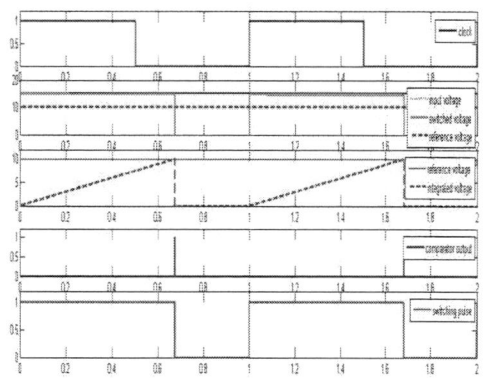

Figure 7: OCC operation waveforms

A SR flip-flop is used as the switching controller. When the clock signal is applied, the output Q will be one and is given as the switching signal for the converter. The comparator output is given to the reset pin of SR flip-flop. As long as the comparator output is zero, the SR output will remain on the ON state. Once the comparator output becomes one, the flip-flop will get reset. The output of flip-flop will remain zero till the next clock cycle. Hence the switching is completely depends upon the value of integrator. The switch will remain ON till the integrator value reaches the reference value. Any variation in the input or output conditions will vary the time for integrator to reach the reference value

B. current controlled mode of DC boost converter using OCC

The input –output current relation of a boost converter can be expressed as

$$I_{in} = I_0/(1-D) \quad (8)$$

Where I_{in} is the input current, I_0 is the output current and D is the duty ratio.

$$I_{in}(1-D)=I_0 \quad (9)$$

$$I_{in}-I_0=D\times I_{in} \quad (10)$$

$$I_{in}-I_0=\frac{1}{T_S}\int_0^{Ton} Iin.dt \quad (11)$$

The required input current reference is given to the integrator of the OCC. The integrated output is compared with the difference between input current and output current. Comparator output will remains on low till the integrated value reaches the difference between input current and output current. Once the integrated value reaches the difference between input and output current values, the comparator resets the flip-flop and switching pulse becomes low. The flip-flop resets the integrator and waits for the next clock signal. Thus by giving the input reference current value, the boost converter can be controlled by OCC.

IV. SIMULATIONS

Simulations are carried out using matlab/Simulink software.

A. Voltage control of boost converter

The simulation shows the comparison of the voltage control of a boost converter using OCC and PI controller. Perturbations are given for the input voltage and output voltage reference. Figure 8 shows the simulation circuit of one cycle controlled boost converter. The results obtained shows that the OCC have better transient performance than PI controller. Result is shown in figure 9.

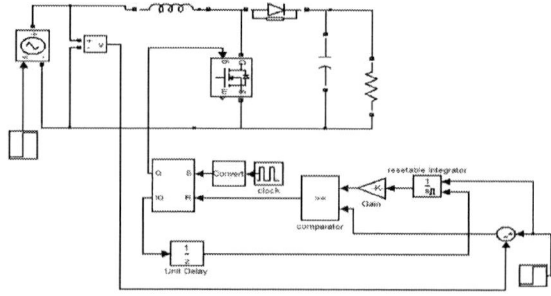

Figure 8: Voltage control of Boost converter using OCC

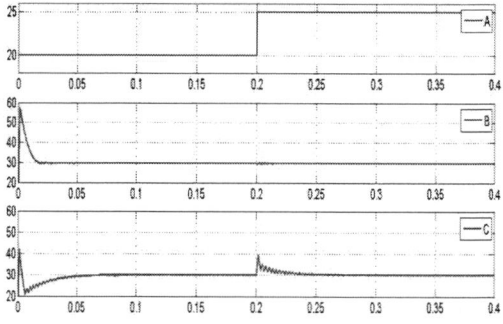

Figure 9:Transient performance of boost converter during input voltage perturbation

(A)Input voltage perturbation (B) Output voltage using OCC
(C) Output voltage using PI controller

978-1-4799-6047-7/14 $31.00 © 2014 IEEE

Setting time takenduring perturbation using OCC= 1ms
Setting timetaken during perturbation using PI= 40ms

B. Current control of boost converter

Figure 10 shows the simulation diagram of the current control of DC boost converter. Here the input current through the DC Boost converter is regulated using the OCC. The input current reference is perturbed and corresponding OCC performance is shown in the results. Initially an input current reference of 5A is given. After 0.5 seconds the current reference is perturbed to 8A. The OCC tracks the variation and the circuit is operated accordingly. Obtained result is shown in figure 11. The control of DC-DC converters using OCC is explained more in [5].

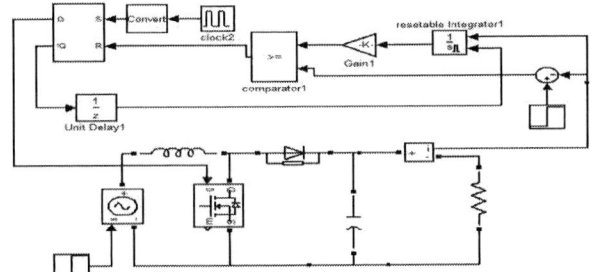

Figure 10: Current control of Boost converter

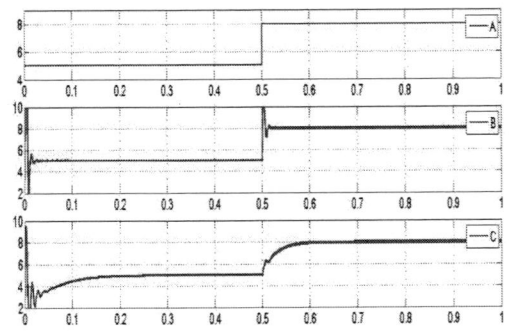

Figure 11: Current control of boost converter
(A)Input current reference perturbation (B) actual input current using OCC (C) actual current using PI controller.
Setting time during perturbation using OCC= 4ms
Setting timeduring perturbation using PI= 50ms

C. Maximum power point tracking

For maximum power point tracking the output voltage control is not a concern. Here the input current through the Boost converter is regulated to an optimal current corresponding to maximum power point. The converter is operated according to the current control mode of boost converter explained earlier.

The circuit consists of a PV panel connected to a load through a DC boost converter. At the input side another switch S_1 is connected. In every one second the switch S_1 is turned on for a short duration of 0.1ms. The circuit gets short circuited and short circuit current flows through the switch S_1.The corresponding short circuited current is measured. Using the proportionality constant K_{op}, the optimum operating current value is determined by equation.$I_{MPP}= K_{op} \times I_{SC}$. The calculated I_{MPP} value is used as the input current reference. Using OCC

input current through the converter is controlled till it reaches the reference value. During the simulation perturbation is given to the irradiance of the panel.

The panel is operated at 1000 W/m^2 initially. The short circuit current corresponding to the given irradiance is measured and the converter is operated to the maximum power point current. After 1 second the irradiance is changed to 800 W/m^2. The short circuit current and optimal current are measured for the given irradiance and the input current through the converter is changed to the new optimal current reference.

The results obtained are shown in figure 12. From the results it can be seen that the OCC helps to improve the duration to reach maximum power point. Figure 13 shows the MPP tracking using PI controller. By comparing the performance of both controllers, it can be seen that the time taken to settle at the maximum power point is less for OCC.

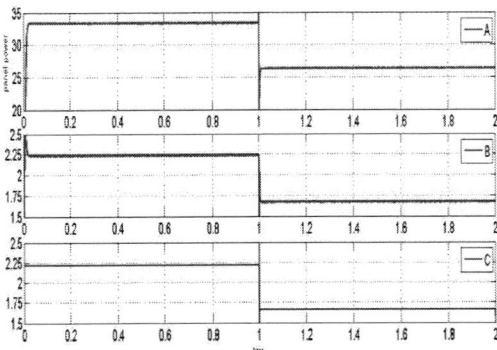

Figure 12: MPP tracking using OCC
(A) Panel Power (B) Actual input current (C) Input current reference corresponding to MPP.
Time taken to reach MPP using OCC= 16ms
Time taken to reach MPP using PI= 100ms
Time taken to reach MPP during perturbation in irradiance
-using OCC= 8ms
-using PI= 80ms

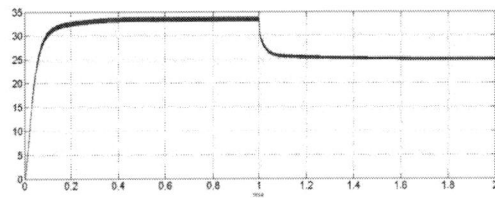

Figure 13: MPP tracking using PI controller

D. Load voltage regulation of PV panel using occ control

The panel may not be needed to operate at maximum power point always. In standalone applications the panel has to supply electrical power at constant voltage. For a PV panel the output voltage of panel is variable in nature since it depends upon the atmospheric condition. Hence in such conditions a fast voltage regulator is needed to track the variations at the input side of the converter.

Simulation of output voltage regulation of a panel system is shown in figure 14. The irradiance level to the panel is varied at regular interval .One cycle controller regulates the output voltage to the reference value during the irradiance variations. The result shows the transient performance of the OCC. Result is shown in figure 15.

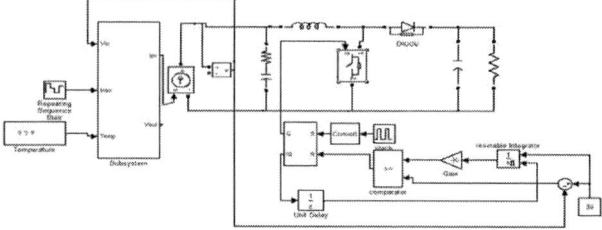

Figure 14: load voltage regulation using OCC

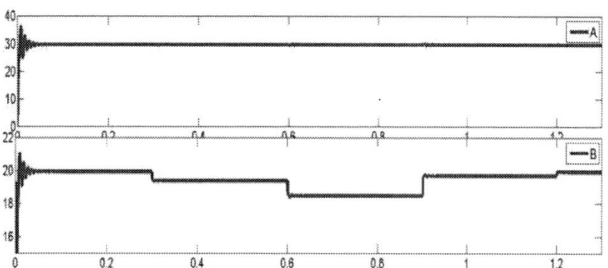

Figure 15: load voltage regulation using OCC
(A) Output voltage of converter (B) panel voltage

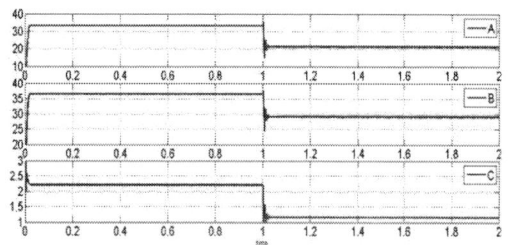

Figure 16: Current mode to voltage mode transition
(A) Panel power (B) load voltage (C) panel current

E. Transition from current control mode to voltage control mode.

During some instants the panels have to supply a constant output voltage to the load rather than maximum power. In such situations the DC converter control has to shift from maximum power point tracking mode to constant output voltage mode. In such systems a good transient performance is required. One of main drawback of the conventional PI controller is its dependency upon the circuit parameters and input-output relation. Hence for the current control mode and voltage control mode, the PI control parameters will be different. One of the main advantages of OCC is the independency of controller over the circuit parameters. For onecycle controller, the control operation depends only on the inputs to the integrator and the comparator. Thus the transition from the current controlled mode to voltage controlled mode can be achieved very quickly by changing the inputs to the integrator and comparator of the one cycle controller. In the simulation for one second duration the panel is operated at its maximum power point using current control mode. After 1s the system is changed to voltage control mode. The results

obtained for the simulations are shown in figure 16.In the simulation for one second the converter is operated at maximum power point tracking mode. There the input currentthrough the converter is controlled to achieve the MPP. In thisstage the output voltage is not controlled. After one second the converter operation is changed to voltage controlled mode. Anoutput voltage reference of 30V is given. During the mode transition, inputs to the integrator and comparator changed to achieve voltage control mode. Result shows the good transient behavior of OCC during the operation.

V. CONCLUSION

In this work, Short Current pulse based Maximum Power Point tracking using One Cycle Control is proposed . Using the information of short circuit current of the panel at a particular irradiance, MPP tracker estimates the optimum operating current for extracting maximum powerfrom the panel. Inorder to operate the panel at MPP, one cycle control is used. The working of OCC is independent of circuit parameters and it can track any variations in one switching cycle. The performance of maximum power point tracking using OCC is compared with a conventional PI controller. Results shows that by using OCC the system reaches at MPP in very short span of time. Also it can shift the operation of system from MPP mode to voltage regulated mode very quickly.From the results it can be concluded that the OCC gives better transient response for converter system employed in photo voltaic panels. More accurate method of estimating optimum panel current will improve the tracking performance. The maximum power point tracking method explained here can be improved further to estimate the proportionality constant K_{op} accurately , and can be extended to panels under partial shading condition.

REFERENCES

[1] K. M. Smedley and C. Slobodan, "One-cycle control of switching converters," IEEE Trans. Power Electron., vol. 10, no. 6, pp. 625–633,November 1995.

[2]MoacyrAureliano Gomes de Brito, Luigi Galotto, Jr., Leonardo PoltronieriSampaio, and Guilherme de Azevedo, "Evaluation of the Main MPPT Techniques for Photovoltaic Applications", IEEE transactions on industrial electronics, vol. 60, no. 3, march 2013.pp.1156-1167

[3] Dongsheng Yang, Min Yang, and XinboRuan, "One-cycle control for a double-input dc/dc converter" IEEE Trans. Power Electron., vol. 27, no. 11, November 2012.

[4] Toshihiko Noguchi, Shigenori Togashi and Ryo Nakamoto, "Short-Current pulse based Maximum-power-point tracking method for multiple photovoltaic and converter module system", IEEE Transactions on Industrial Electronics, Vol 49,No 1, February 2002, pp.217-223

[5] Anoop K and Dr. M Nandakumar, "DC-DC Converters control method with improved transient performance",Proceedings of 4th National Technologi cal Congress Kerala- NATCON 2014, vol. 1, pp.282–287 , February 2014.

978-1-4799-6047-7/14 $31.00 © 2014 IEEE   248

Consequences of Dust on Solar Photovoltaic Module and Its Generation

Pathik Chamaria*, Anirudh Dube[†], Ruchika[‡] and Dr. A.P. Mittal [§]

*Dept. of EEE, MAIT, Delhi, India Email :- pcpathik@gmail.com

[†]Dept. of EEE, MAIT, Delhi, India Email :- anirudhdube12@gmail.com

[‡]Dept. of ICE, NSIT, New Delhi, India Email :- mittalruchika46@gmail.com

[§]Dept. of ICE, NSIT, New Delhi, India Email :- mittalap@gmail.com

Abstract—Dust on photovoltaic modules reduces its efficiency by decreasing its transmittance hence the effective solar insolation reaching it is reduced. Studies have shown dust or minute particle is affected by tilt angle of solar collector, exposure period, site climate conditions, wind movement and dust properties. Previous authors have calculated loss in terms of optical loss that is reduction in transmittance but paper calculates the loss in terms of kWh which was not done till now.

Index Terms—PV Panel, PV Array, Soiling Loss, Performance Ratio, Horizontal Global Irradiation.

I. INTRODUCTION

Renewable energy based power generation are now becoming the substitute of the conventional power plant. It is growing rapidly because it is available in abundant and that to free of cost. Research and development in photovoltaic devices has been dominated on factors, having major impact in short duration, such as radiation availability that is insolation prediction, optimum design and sizing of systems, higher efficiency yet cheaper materials and efficient operating strategies of these systems. On the other hand, the influence of surroundings and other environmental factors such as wind, dust and rainfall patterns have been neglected.

More specifically the influence of particulate matter on the performance of solar panels has not been given much attention which accounts for major loss over the age. Though it is well known fact that the particulate matter diminishes the transmittance of solar radiation. Research done on this field reports, mostly in terms of optical efficiency, its influence vary from about 2% to about 40% based on various research parameters. Which in most case have not been standardized like the duration of measurement in particular case and pollution level at the location and season and weather of location along with and location itself. Study, being site-specific and season-dependent, requires a detailed experiment in controlled environments especially in locations where solar energy applications are more favourable (desert and tropical).

The past investigations of dust affidavit on solar panels concentrated on the open air examinations of coating transparency execution instead of on electrical parameter. Case in point, [1] mulled over the PV array yield in Saudi Arabia. Lessening of 32% in yield was seen in eight months. [2] reported a month to month decay rate in glass transmissivity of 10% in

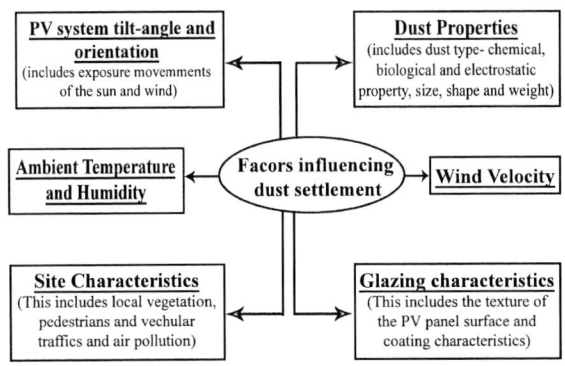

Fig. 1. Factors influencing dust settlement.

summer and 6% in winter, and a 70% diminishment when was watched for an entire year because of the impact of dust aggregation on the panels over distinctive time periods in the United Arab Emirates. A yearlong study was led by [3] in Egypt for the subtropical climatic area. An expression of glass transmittance diminishment in one month was inferred around their experimental information. [4] reported that greatest debasement happens quickly amid the initial 30 days of introduction. While [5] concentrated on the impact of dust coverage on the transmittance of glass with diverse tilt edges.

These examination were not done in a controlled situations. In 1993, [6] firstly examined the effect of dust properties and thickness on the PV in a controlled nature. The results demonstrated that fine particulate dust has a more noteworthy effect on PV than coarser particles. [7] experimentally and mathematically examined the impact of dust layer on light transmission at a photovoltaic module glazing surface. They found that more extended wavelengths are reflected substantially more than shorter ones as the measure of sand dust particles increments on the glass test panels. [8] and [9] did wind shaft probes on the impact of wind speed and heading on dust affidavit. Their studies demonstrated that the drop in PV cell performance develops bigger with wind speed increments.

II. ENVIRONMENTAL EFFECTS ON SOLAR PV MODULE GENERATION

A logic diagram to understand the various factors that govern the settling of dust is illustrated in Fig. 1. It is

978-1-4799-6047-7/14 $31.00 © 2014 IEEE

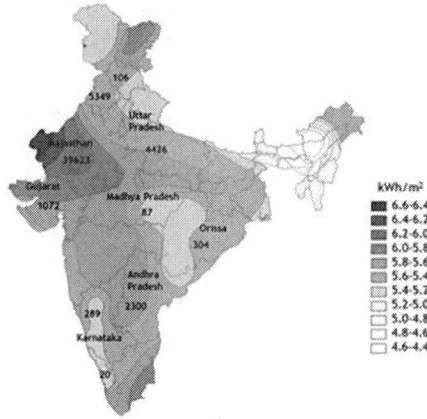

Fig. 2. Year average of solar radiation in India.

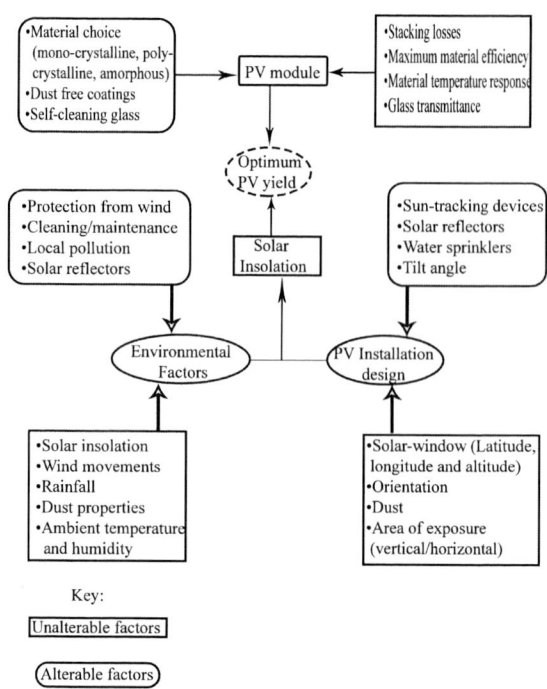

Fig. 3. Alterable and unalterable factors determining maximum PV system yield.

easy to conclude that the phenomenon is extremely complex. And challenging to comprehend given all the factors that influence dust settlement are naturally generated hence almost unmanageable.

Since India is a sub-tropical country it has receives very large amount of solar power as shown in Fig. 2. One can see here that on average India receives 5.8 kWh/m^2. Highest of which is in Thar Desert region which is highly affected by soiling of panel problem. Even in central India that is Madhya Pradesh, eastern Uttar Pradesh, Maharashtra and close by area due to extensive industrialization air has high amount of particulate matter.

Dust is particles of size from 1 micro meter to 1 millimeter. They are fit for makeshift air suspension because of electro-static charge, yet at last settle affected by gravity. The dust air toxins are because of the accompanying: concrete, coal, alumina, calcium fluoride, grain, limestone, metals, minerals and wood fillings. Dust impacts the performance of PV by diminishing the radiation exposed on the absorber. This impact of dust would rely on:

1) the nature of dust;
2) the size of the dust;
3) the dust deposition density;
4) the duration and/or the frequency of cleaning the collector surface;
5) inclination of the glazing;
6) type of glazing (glass or plastic) and
7) the environment (urban, rural or industrial).

These complex factors have been divided into two factors that is alterable and unalterable with all devices that are incorporated to increase efficiency of system represented in Fig 3.

III. System Configuration

We studied the loss on a 50 kW system made up of 200 modules of Si-poly type. Manufactured by ET Solar model no ET - P672250, each unit having a nominal power output of 250Wp at Standard operating condition i.e. insolation of 800 Watt per meter square with ambient temperature of 20 degree

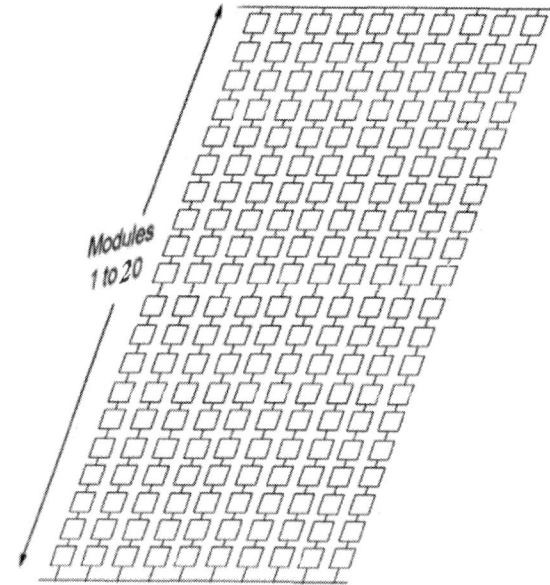

Fig. 4. Array used to stud soiling loss effect.

Celsius at a wind speed of 1 meter per second. Inverter used was ACE 5001

Array was formed with 20 panels in series giving 5000 W and such 10 strings in parallel so as to give a total output of 50 kWp. With 316V and 142A at Maximum power point. Fig. 4 shows the array in a diagrammatic form.

Orientation that is tilt angle azimuth angle plays a very important role in obtaining maximum power form solar pan-

978-1-4799-6047-7/14 $31.00 © 2014 IEEE

TABLE I

RESULTS GENERATED FOR 1% LOSS DUE TO SOILING

	Horizontal global irradiation kWh/m²	Ambient Temperature °C	Global incident on plane kWh/m²	Effective Global, corr. for IAM and shadings kWh/m²	Effective energy at the output of the array kWh	Energy injected into grid kWh	Effic. Eout array / rough area %	Effic. Eout system / rough area %
January	118.0	14.70	167.4	163.1	7088	6774	10.91	10.42
February	137.0	17.30	176.9	172.6	7337	7014	10.69	10.22
March	188.0	22.70	214.3	208.6	8628	8216	10.38	9.88
April	207.0	28.80	208.4	202.4	8128	7752	10.05	9.58
May	222.0	32.50	202.7	195.8	7711	7354	9.80	9.35
June	197.0	32.90	173.7	167.7	6595	6265	9.79	9.30
July	167.0	30.30	152.0	146.9	5859	5525	9.93	9.37
August	160.0	29.90	155.7	150.9	6076	5751	10.06	9.52
September	171.0	29.50	184.1	178.9	7178	6836	10.05	9.57
October	165.0	26.20	203.8	198.7	8144	7780	10.30	9.84
November	129.0	20.90	180.4	175.7	7363	7025	10.52	10.03
December	115.0	16.00	171.6	167.1	7214	6891	10.83	10.35
Year	**1976.0**	**25.17**	**2191.0**	**2128.3**	**87321**	**83181**	**10.27**	**9.78**

TABLE II

RESULTS GENERATED FOR 5% LOSS DUE TO SOILING

	Horizontal global irradiation kWh/m²	Ambient Temperature °C	Global incident on plane kWh/m²	Effective Global, corr. for IAM and shadings kWh/m²	Effective energy at the output of the array kWh	Energy injected into grid kWh	Effic. Eout array / rough area %	Effic. Eout system / rough area %
January	118.0	14.70	167.4	163.1	6784	6483	10.44	9.98
February	137.0	17.30	176.9	172.6	7022	6715	10.23	9.78
March	188.0	22.70	214.3	208.6	8256	7865	9.93	9.46
April	207.0	28.80	208.4	202.4	7774	7415	9.61	9.17
May	222.0	32.50	202.7	195.8	7372	7026	9.37	8.93
June	197.0	32.90	173.7	167.7	6303	5981	9.35	8.87
July	167.0	30.30	152.0	146.9	5601	5274	9.49	8.94
August	160.0	29.90	155.7	150.9	5812	5492	9.62	9.09
September	171.0	29.50	184.1	178.9	6865	6536	9.61	9.15
October	165.0	26.20	203.8	198.7	7792	7445	9.85	9.41
November	129.0	20.90	180.4	175.7	7047	6723	10.07	9.60
December	115.0	16.00	171.6	167.1	6904	6595	10.37	9.91
Year	**1976.0**	**25.17**	**2191.0**	**2128.3**	**83531**	**79551**	**9.82**	**9.36**

els. Tilt and azimuth angle were calculated for year around optimization using PVsyst software giving tilt angle as 28° and azimuth angle as 0° for max yield year around (Fig. 5). Orientation of the module should be equal to latitude of the area. Then the generation will be max.

Study was conducted in PVsyst with thermal loss factor of 29 W/m^2K and wind loss factor as 0. Along with wiring ohmic loss as 1.5% , module quality loss as 1.5 % and Module Mismatch Loss as 2 % at MPP.

Two cases were studied for soiling loss that is loss due to dust as 1% and 5%.

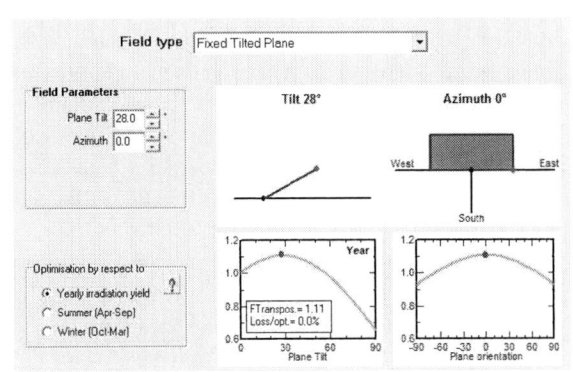

Fig. 5. Tilt and Azimuth angle calculation.

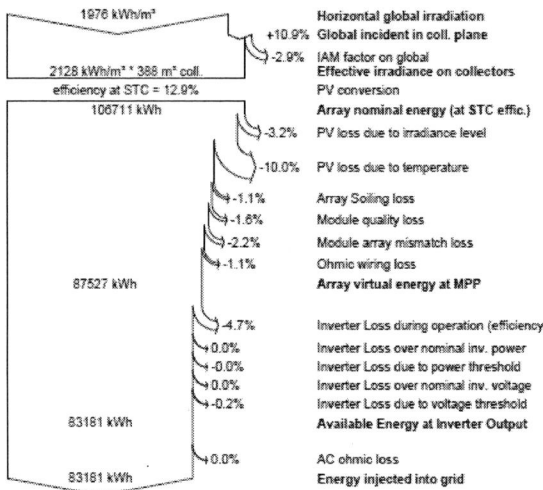

Fig. 6. Energy Generated considering 1% Soiling Loss

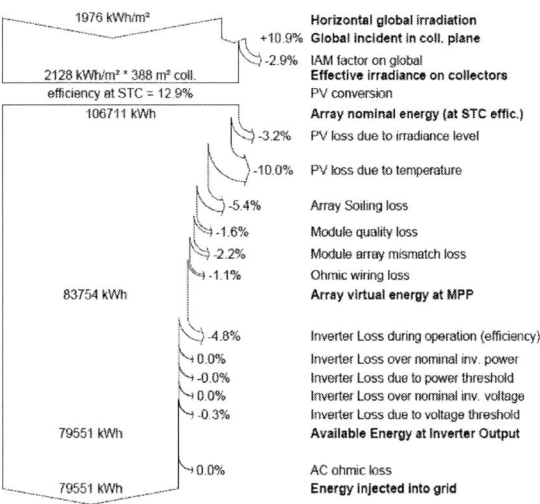

Fig. 7. Energy Generated considering 5% Soiling Loss

IV. EXPERIMENT RESULT

The data generated using PVsyst are displayed in Table 1 and Table 2. Where in column 1 the month is noted while in second total horizontal irradiation and in third column average ambient temperature is noted. Fourth and fifth columns display total irradiation in panel plane and effective irradiation taking in effects of shadings and other things. Next columns give effective energy at the output of the array taking into account of panel efficiency, energy injected into grid which accounts ohmic losses at inverter respectively. Finally efficiency is calculated of first array and then system.

Fig. 6 and Fig. 7 show loss diagram for 1% and 5% respectively.

Fig. 8 shows horizon trajectory for the location.

Data of Table 1 has been shown in normalized form in Fig. 9 and Fig. 10.

As seen from Table 3 one can conclude that difference of

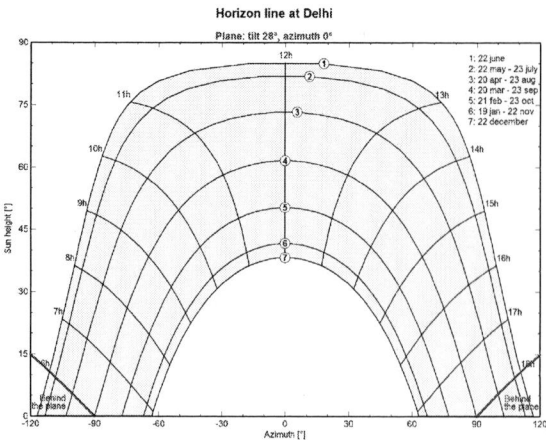

Fig. 8. Horizon Trajectory for the Location

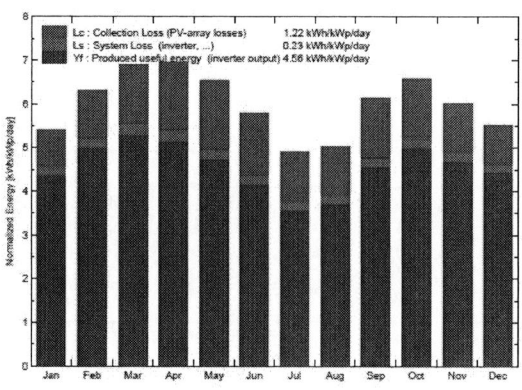

Fig. 9. Normalized productions (per installed kWp)

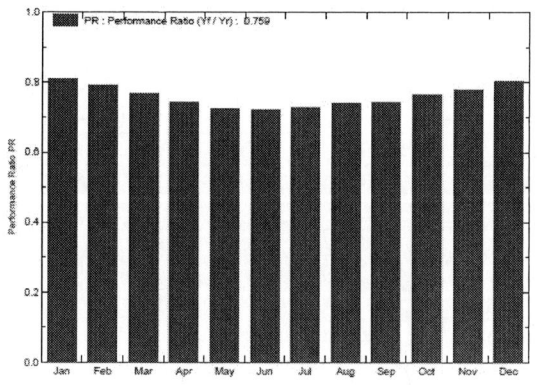

Fig. 10. Performance Ratio PR

TABLE III
SUMMARY OF DIFFERENT CASES

Soil Loss Considered	Energy Generated (in Units)	Performance Ratio PR (in %)
With 1% Loss	83181 units	75.9 %
With 5% Loss	79551 units	72.6 %

3630 units is between the 1% loss and 5% loss case. If we assumes Rs10/unit to sell out the solar power to the local distribution utility then financially the loss of Rs 36300 yearly and if we assume 20 years of the production of the solar power then loss of Rs 726000 has to bear by the producer.

V. CONCLUSION

One can conclude that there can be a significant amount of loss due to soiling as it can reduce transmittance to about 40%. A daily cleaning cycle is recommended in high dust density area. Low latitudes area even having medium dust density should also have daily cleaning cycle since they shall have lower tilt angle giving higher dust deposition. In mid latitude regions tilt angle shall be moderate giving lower dust devotion rates hence weekly cleaning shall be adequate. For high latitude regions dust shall not be a bigger problem as panels shall have close to vertical angle but snow shall be a great problem which should be immediately addressed. Authors would like to mention that a regular check for bird droppings they should be immediately removed to have higher efficiency.

REFERENCES

[1] A. Salim, F. Huraib, and N. Eugenio, "Pv power-study of system options and optimization," in *Proceedings of the 8th European PV Solar Energy Conference*, 1988.

[2] A. M. El-Nashar, "The effect of dust accumulation on the performance of evacuated tube collectors," *Solar Energy*, vol. 53, no. 1, 1994.

[3] A. A. Hegazy, "Effect of dust accumulation on solar transmittance through glass covers of plate-type collectors," *Renewable Energy*, vol. 22, no. 4, 2001.

[4] A. H. Hassan, U. A. Rahoma, H. K. Elminir, and A. M. Fathy, "Effect of airborne dust concentration on the performance of pv modules," *Journal of the Astronomical Society of Egypt*, vol. 13, no. 1, 2005.

[5] H. K. Elminir, A. E. Ghitas, R. H. Hamid, F. El-Hussainy, M. M. Beheary, and Abdel-Moneim, "K," *Effect of dust on the transparent cover of solar collectors*, vol. 47, no. 18, 2006.

[6] M. S. El-Shobokshy and F. M. Hussein, "Effect of the dust with different physical properties on the performance of photovoltaic cells," *Solar Energy*, vol. 51, no. 6, 1993.

[7] A. Y. Al-Hasan, "A new correlation for direct beam solar radiation received by photovoltaic panel with sand dust accumulated on its surface," *Solar Energy*, vol. 63, no. 5, 1998.

[8] D. Goossens and E. V. Kerschaever, "Aeolian dust deposition on photovoltaic solar cells: the effects of wind velocity and airborne dust concentration on cell performance," *Solar Energy*, vol. 66, no. 4, 1999.

[9] D. Goossens, Z. Y. Offer, and A. Zangvil, "Wind tunnel experiments and field investigations of eolian dust deposition on photovoltaic solar collectors," *Solar Energy*, vol. 50, no. 1, 1993.

Power Quality Control of SEIG based Isolated Pico Hydro Power Plant Feeding Non-Linear Load

Umesh C. Rathore
Electrical & Instrumentation Engineering Department,
SLIET Longowal, Punjab, India
rathore7umesh@gmail.com

Sanjeev Singh, *Senior Member, IEEE*
Electrical & Instrumentation Engineering Department,
SLIET Longowal, Punjab, India
sschauhan@sliet.ac.in

Abstract—**Induction generators are most suitable for renewable energy conversion systems such as micro/pico hydro & wind power generation systems due to their advantages over conventional synchronous generators. These induction generators are of small ratings when used for pico-hydro power plants in remote mountainous locations and mostly feed the single-phase domestic load using 3-phase 4-wire system. However, the growing use of non-linear electronic loads in domestic applications such as CFLs, LEDs, computers etc., various harmonics are generated which adversely affects the power quality at the AC supply mains. This paper presents a solution for harmonic mitigation in a 3-φ self-excited induction generator (SEIG) based pico-hydro power plant operated in isolated mode, using star-delta and zigzag transformers feeding non-linear load. The complete system is modeled and its performance is simulated in MATLAB/Simulink environment. The performance simulation results are presented to demonstrate the effectiveness of star-delta and zigzag transformers in power quality control.**

Keywords—*SEIG; pico-hydro; electronic load controller; harmonics; power quality; delta-star & zigzag transformer.*

I. INTRODUCTION

The world energy scenario empresses upon the harnessing of renewable energy, available in abundance, to meet the ever increasing demand of electrical energy across the globe. It will also help in minimizing the adverse environmental effects caused by various conventional energy generation methods. Advancement in technology and growing concern on environmental issues worldwide; have paved the way for harnessing of alternate sources of energy such as wind, hydro, geothermal, tidal, biomass, etc. Renewable energy sources based power plants are best suited to rural and remote areas to meet the electricity and other energy demand of local inhabitants. Hydro energy is the most reliable and cost effective renewable energy source. Pico/Micro hydro power plants play a significant role in meeting the power requirement of remote mountainous regions. These are environment friendly and require less investment as compared to large hydro power plants.

Induction generators with their numerous advantages over conventional synchronous generators are most suitable in converting these renewable energy sources into useful electrical power. For their simplicity, robustness and small size per generated kW, the induction generators are favored for small hydro and wind power plants [1]. The induction generator is always associated with renewable energy

conversion system involving small power plants operated in isolated mode and not connected to the grid. In isolated mode of operation of induction generators, a suitable control is necessary to maintain the voltage & frequency of the generated output within permissible limits under varying loading conditions. This is achieved by using an electronic load controller (ELC) along with the dump load.

In Indian scenario, the load in remote areas mostly consists of lighting & heating load e.g. compact fluorescent lamps (CFLs), heaters and conventional incandescent lamps requiring single-phase AC supply. However, with the advancement in technology and its increasing use by masses, the computers, LEDs and other electronic gadgets are also common in almost every household in remote villages. All these devices use switch mode power electronic converters which are considered as non-linear load and generate harmonics. The low order harmonics (i.e.3rd, 5th, & 7th) pose more impact on the power quality at the supply mains as compared to the higher order harmonics (i.e.11th, 13th, & 17th) as the former carry more energy as compared to later. These harmonic currents cause excessive heating in distribution transformer, overheating of conductors, electrical noise and unnecessary tripping of circuit breakers in the system. Some of the remedial measures to minimize these harmonic currents include use of active and passive filters, use of zigzag transformer and use of separate neutrals for non-linear loads [2]-[3]. All these remedial measures lead to the increase in overall investment cost. The use of active filters offer advantages as compare to others methods of power quality improvement in the electrical system. However, for an isolated induction generator based pico-hydro power plants feeding dedicated load in remote locations, the use of active filters along with distribution transformers leads to increased complexity in the systems. This paper investigates use of star-delta & zigzag transformers only in such conditions for minimizing the harmonics generated due to the non-linear loads. The isolated 3-φ SEIG based pico-hydro power system using star-delta and zigzag transformers is modeled in MATLAB/Simulink environment and the performance simulation is carried out for harmonic mitigation and power quality control at supply mains while feeding the non-linear load.

II. INDUCTION GENERATORS

Induction generators are increasingly being used in renewable energy conversion systems as compared to the

978-1-4799-6047-7/14 $31.00 © 2014 IEEE

synchronous generators due to their advantages of reduced cost and size, ruggedness, brushless construction (squirrel cage rotor), absence of separate D.C. source, ease of maintenance, and self protection against severe overloads and short circuits. Sudden speed changes due to rise and fall of loads or primary source variation which usually occur in small power plants are easily absorbed by the solid rotor of induction machine. The induction generator has the same constructional features as of an induction motor with slight improvement in its efficiency [1] and excitation techniques. In case of induction generator the rotor speed is more than the synchronous speed, while in induction motor, the rotor speed is slightly less than the synchronous speed. Output induced voltage in induction generator is proportional to the relative difference between the electrical synchronous rotation and the mechanical rotation of rotor within a speed slip factor. Self excited induction generator as shown in Fig.1 and doubly fed induction generator are most suitable for micro-hydro and wind energy conversion systems respectively. Magnetization current for excitation in induction generation is provided by an external circuit or by the system to which it is connected as induction generator consumes reactive power rather than supplying it. The excitation circuit includes the capacitor bank connected either in star or delta mode. The capacitor bank must remain connected to the stator terminals of induction generator as it supplies required kVAr demand for its continuous operation. The residual magnetic field in the rotor initiates a voltage across the induction generator terminals when rotor speed exceeds the synchronous speed. This in turns further augments the capacitor current to continue rise in voltage as capacitors reinforce the magnetic field and system builds up an increasing excitation which results in further increase in terminal voltage. Residual magnetism is necessary to build up the generator voltage.

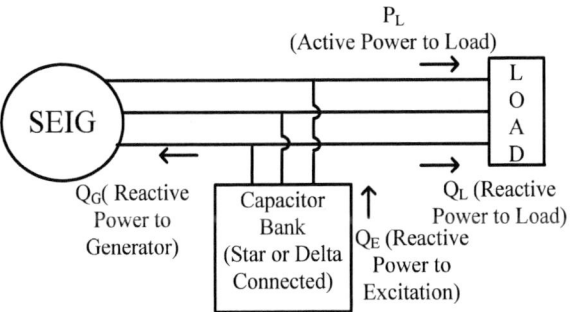

Fig.1. Excitation circuit in 3-φ SEIG

III. SEIG IN PICO-HYDRO POWER PANT FEEDING ISOLATED NON-LINEAR LOAD

The pico hydro power plants are generally classified as constant power driven prime mover power plants because the water discharge and head remain constant. This results in constant availability of power at the output of induction generator. Therefore, the load connected at the output terminals of induction generator must remain constant for maintaining the requisite voltage and frequency of the generated output. Voltage in the SEIG system in isolated mode is maintained using voltage regulators involving power

electronics based devices. The most commonly used voltage regulator is the use of STATCOM. Suitable control for frequency in standalone SEIG system is necessary due to the intermittent nature of the load [4]-[11]. Among the various methods of load frequency control in SEIG, the simplest of the frequency control in SEIG involves the use of electronic load controller involving 6-pulse rectifier and chopper circuit along with the dump load as shown in Fig. 2. The main aim of the design of control strategy for induction generators is to regulate or control the voltage output of the generator and to maintain the desired frequency of the generated output. This controller can consume both the real and reactive power. The amount of power is controlled by the control of the bridge and the pulse width control of the chopper. The controller picks up the real and reactive power which is not used by the load, so that the load seen by the generator at its terminals is always constant which results in achieving desired constant voltage and constant frequency as any change in load is immediately compensated by impedance controller. The advantage of this method is that it is the simplest method of control and there is no need of expensive turbine governor. However, there is wastage of energy as excess power is dissipated in the resistors.

A. Simulation Model

For evaluating the dynamic performance of 3-φ SEIG, MATLAB/Simulink facility has been used to simulate the model comprising of 3φ, 2 HP, 6 pole, induction machine (Annexure I) used as induction generator along with conventional electronic load controller, dump load, consumer load, excitation capacitor bank and necessary control circuits as shown in Fig. 3. The dump load capacity is selected slightly higher than the kW rating of the induction machine.

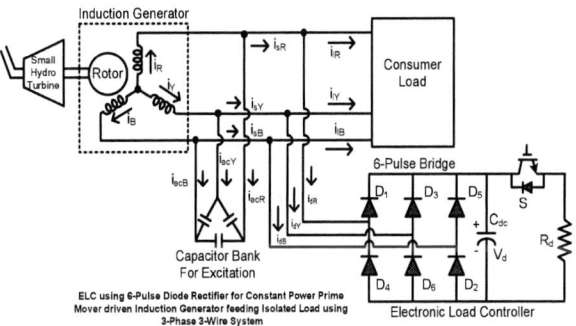

Fig.2. Voltage & frequency control of IG using impedance controller

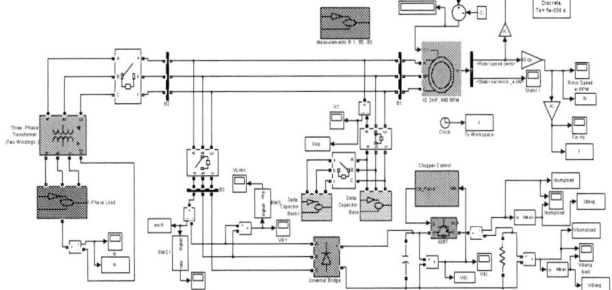

Fig.3. MATLAB/Simulink based model of 2HP, 6Pole 3-phase SEIG

978-1-4799-6047-7/14 $31.00 © 2014 IEEE

The control mechanism consists of electronic load controller (ELC) which is generally the switching circuit using fast electronic devices to switch the dump load according to the amount of consumers load at given time on the generator terminals. The feedback signal for the controller is taken from the 3-phase rectifier output across the filter capacitor in the form of DC link voltage to compare it with the reference voltage. The magnitude of this DC link voltage depends upon the value of 3-phase AC input voltage in the ELC circuit which further depends upon the amount of consumer load connected to the generator terminal. This voltage is maximum when consumer load is negligible and under this condition chopper circuit has the maximum on-time to connect the dump load when switched on. As the consumer load increases, the slight decrease in dc link voltage forces the chopper control circuit to decrease the on-time of IGBT switch to decrease the voltage at dump load. Therefore, the current in the ELC circuit varies due to the change in consumer load connected. The consumer load considered in this case is a set of three 1-phase linear resistive and non-linear load under balanced & un-balanced conditions.

Induction generator is firstly run at no load (0-0.5s) and voltage builds up across the generator terminals when required capacitor bank for excitation purpose is connected across the generator terminals. Once the voltage gets stabilized at no load, controller circuit (ELC) is switched on at 0.5s. Since the consumer load is zero at this moment, dump load is on and the maximum rated current flows through the ELC circuit. Due to the presence of rectifier circuit, harmonics are present in the circuit which results in distortion in voltage and current waves. At 1.5s, a set of balanced 200W/phase consumer load of resistive in nature plus a set of 150W/phase non-linear load involving 1-phase rectifier circuit on all three phases is connected across the generator. It results in flow of load current across the consumer load and at the same time due to ELC, the current through the ELC decreases proportionally due to the change in chopper circuit action to maintain constant current in the generator circuit. Practically in isolated mode of SEIG feeding domestic load, the load on all three phases is different. Therefore, the whole procedure is repeated for the unbalanced non-linear load by connecting the different load on all the three phases. In both the cases SEIG parameters such as stator circuit voltage & current, speed of induction generator, consumer load circuit current and dump load circuit average current are observed. Figs. 4&5 show the various parameters of SEIG under balanced and un-balanced non-linear conditions.

IV. HARMONICS MITIGATION IN PICO-HYDRO POWER PANT FEEDING ISOLATED NON-LINEAR LOAD

The most of the domestic load is of single phase in nature which requires the use of neutral wire. In the conventional grid connected electrical distribution system, delta-star transformer is used to convert 3-phase system into single phase system. In small induction generators feeding dedicated load, the generator neutral conductor can be used in creating single phase supply. However, the use of delta-star transformer or zigzag transformers is useful in creating neutral conductor as well as for bypassing the harmonic components through neutral to mitigate the harmonics problems arise out of non-linear

domestic load on the generator [12]-[14]. Table I shows the THD level of SEIG system parameters when delta-star & zigzag transformers are connected.

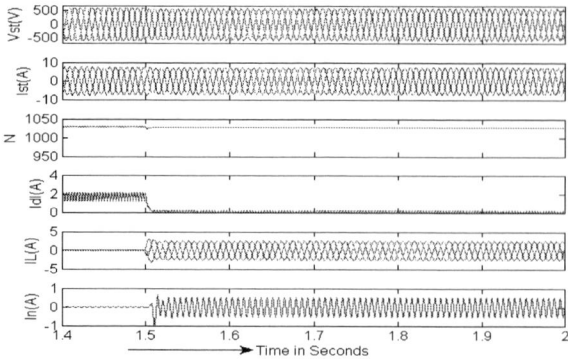

Fig.4. SEIG system parameters under balanced non-linear load conditions

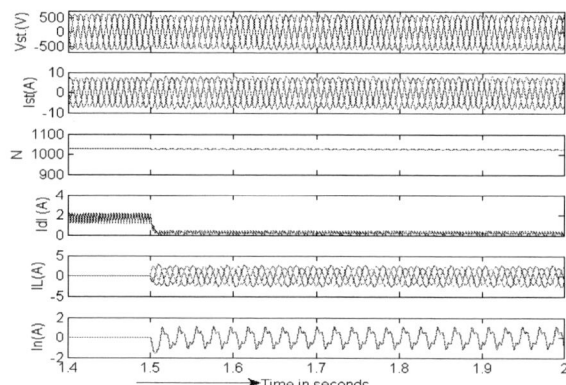

Fig.5. SEIG system parameters under un- balanced non-linear load conditions

TABLE I. VARIATION OF THD IN CURRENT

Signal	Total harmonic Distortion (THD) in %age			
	Delta-Star T/F at load side		Zigzag T/F at load side	
	Unbalanced non-linear load	Balanced non-linear load	Unbalanced non-linear load	Balanced non-linear load
Stator current (Ist)	5.35	2.12	7.76	5.84
Load circuit current (IL)	4.2	5.10	4.65	4.15
Neutral current (In)	36.82	2035	18.70	1549

Under balanced load condition, the neutral circuit current as shown in Fig.4 is mainly composed of 3rd harmonic components and it is less as compared to the neutral current under un-balanced load conditions as shown in Fig.5 due to the presence of return current due to unequal loading condition on all the three phases. Therefore, the THD level as shown in Table I in neutral current under un-balanced non-linear loading condition is less as compared to balanced loading conditions due the presence of more fundamental components in the neutral circuit current. The zigzag transformer or grounding transformer is also useful in mitigating harmonics and creating neutral ground in small dedicated power system. The zigzag connection is also known as interconnected star connection.

Zigzag connection has some of the features of the Y and the Δ connections, thus combining the advantages of both. The zigzag transformer contains six coils on three cores. The first coil on each core is connected contrariwise to the second coil on the next core. The second coils are then all tied together to form the neutral and the phases are connected to the primary coils. Each phase, therefore, couples with each other phase and the voltages cancel out. As such, there would be negligible current through the neutral and it can be connected to ground.

high harmonics levels due to the flow of 3rd harmonic components which are bypassed by the delta-star transformer and zigzag transformers in the load side of power system. Under balanced condition when the load on all three phases is equal, the THD level in the neutral current is very high as compared to the un-balanced condition due to the negligible fundamental component in the neutral circuit in SEIG systems involving delta-star as well as zigzag transformer as shown in the Fig. 6 (c & f) and Fig. 7 (c & f).

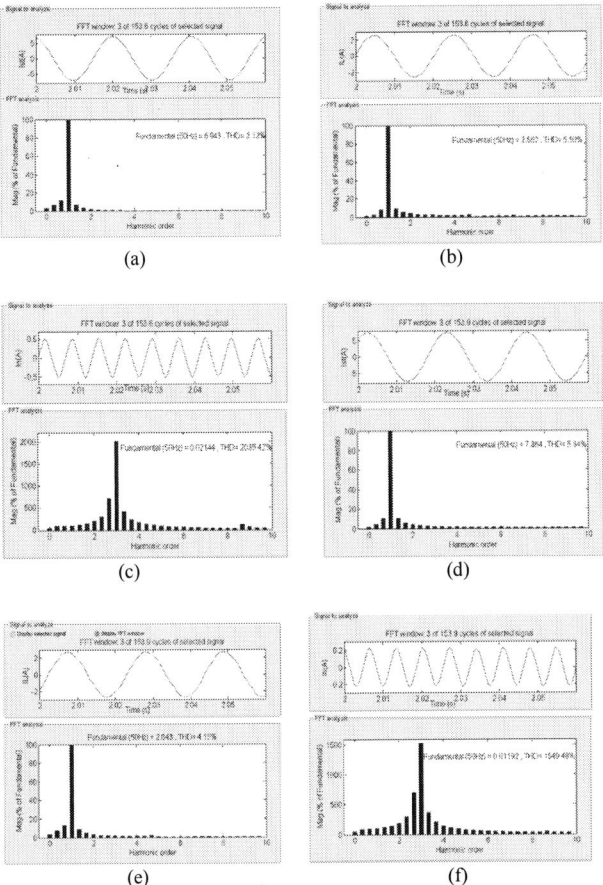

Fig.6. Current waveforms & harmonic spectra under balanced load condition: (a) Stator current when delta-star t/f is connected (b) Load current when delta-star t/f is connected (c) Neutral current when delta-star t/f is connected (a) Stator current when zigzag t/f is connected (b) Load current when zigzag t/f is connected (c) Neutral current when zigzag t/f is connected

The zigzag connection in power systems is used to trap the triple harmonic (3rd, 9th, 15th, etc.) currents. The Zigzag winding provides an easy path for in-phase currents but does not allow the flow of currents that are 120°out of phase with each other. transformers are installed near loads that produce large triple harmonic currents. The windings trap the harmonic currents and prevent them from traveling upstream, where they can produce undesirable effects. Figs. 6&7 show the waveforms and harmonic spectra of SEIG stator current, load circuit current & neutral current in a 3-phase, 2HP, 6 pole SEIG feeding an isolated non-linear balanced & unbalanced domestic load using delta-star & zigzag transformer. Neutral currents in both balanced & unbalanced conditions have very

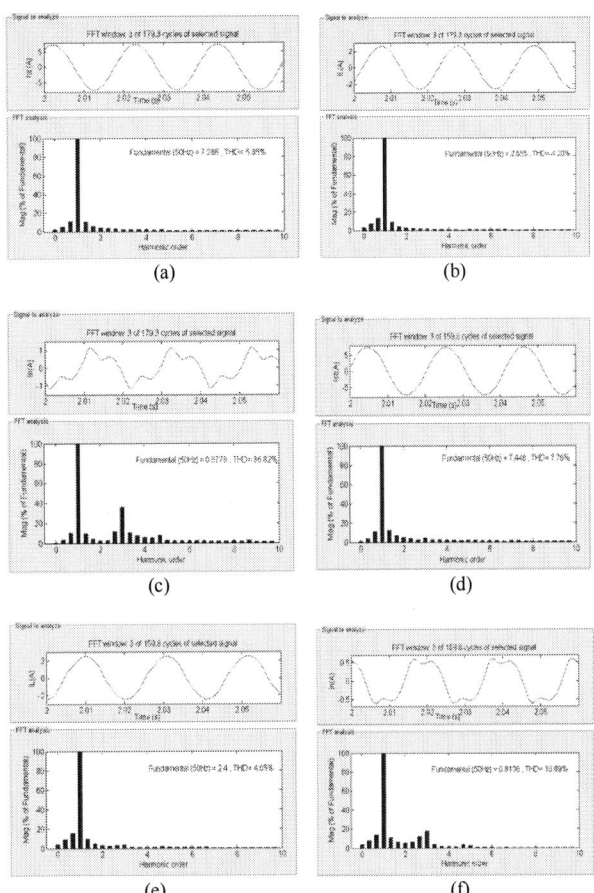

Fig.7. Current waveforms & harmonic spectra under un-balanced load condition: (a) Stator current when delta-star t/f is connected (b) Load current when delta-star t/f is connected (c) Neutral current when delta-star t/f is connected (a) Stator current when zigzag t/f is connected (b) Load current when zigzag t/f is connected (c) Neutral current when zigzag t/f is connected

V. CONCLUSION

The dynamic performance characteristics of 3-φ self-excited induction generator in micro/pico-hydro power generation system feeding the isolated consumer load along with ELC & dump load has been modeled and its performance results have been presented for feeding the non-linear load. The presented results demonstrated the usefulness of star-delta & zigzag transformers in distribution system to minimize the harmonic currents caused due to non-linear load. these transformers also helped in creation of neutral path for feeding single phase non-linear load from an isolated 3-φ SEIG based pico-hydro power generation system.

978-1-4799-6047-7/14 $31.00 © 2014 IEEE

APPENDIX-I

3-φ Self-Excited Induction Generator Specifications:
2HP (1.5kW), 415V, 50 Hz, 6 Pole, Squirrel Cage Induction Machine operated as Induction Generator
Stator Resistance = 5.405 Ω, Stator Inductance = 0.0199H
Rotor Resistance = 5.395 Ω, Rotor Inductance 0.0199H
Mutual Inductance = 0.3722H,
Inertia (J) = 0.0031kg.m^2,
Friction factor (F) = 0.001985 N.m.s, No. of Poles = 6,
Rated Torque (Nm) = [Rated Power (W)/Speed in rad/sec]
= 14.33N.m

REFERENCES

[1] M. Godoy Simoes and Felix A. Farret, *Alternate Energy Systems: Design and Analysis with Induction Generators*, 2nd ed. Boca Raton: CRC Press, 2008.

[2] P.P. Khera, "Application for zigzag transformers for reducing harmonics in the neutral conductor of low voltage distribution system," in *Proc. IEEE IAS*, Oct. 1990, p. 1092.

[3] P. N. Enjeti, W. Shireen, P. Packebush, and I. J. Pitel, "Analysis and design of a new active power filter to cancel neutral current harmonics in three phase four wire electric distribution systems," in *IEEE Transactions on Industry Applications*, vol. 30, no. 6, November/December 1994.

[4] E. Suarez and G. Bortolotto, "Voltage-frequency control of a self excited induction generator," *IEEE Transactions on Energy Conversion*, vol. 14, no. 3, pp. 394-401, September 1999.

[5] B. Singh, S. S. Murthy, and S. Gupta, "Analysis and design of STATCOM-based voltage regulator for self-excited induction generators," *IEEE Transactions on Energy Conversion,* vol. 19, no. 4, pp. 783-790, Dec. 2004.

[6] B. Singh and G. K. Kasal, "Voltage and frequency controller for isolated asynchronous generators feeding 3-phase 4-wire loads," in *Proc. IEEE ICIT'06*, Dec. 2006, pp. 2773-2778.

[7] B. Singh and G. K. Kasal, "Analysis and design of voltage and frequency controllers for isolated asynchronous generators in constant power applications," in *Proc. IEEE PEDES'06*, Dec. 2006, pp. 1-7.

[8] J. M. Ramirej and E. Torres, "An electronic load controller for self-excited induction generator," in *Proc. IEEE PES*, 24-28 Jun. 2007, pp. 1-8.

[9] S. S. Murthy and R. K. Ahuja, "A novel solid state voltage controller of three phase self-excited induction generator for decentralized power generation," in *Proc. IEEE ICPCES*, 2010. pp. 1-6.

[10] V. Rajagopal and B. Singh, "Improved electronic load controller for off grid induction generator in small hydro power generation," in *Proc. IEEE IICPE'10*, 28-30 Jan.2011, pp. 1-7.

[11] L. G. Scherer, R. F. de Camargo, H. Pinheiro, and C. Rech, "Advances in modeling and control of micro hydro power stations with induction generators," in *Proc. IEEE ECCE'2011*, 17-22 Sept., 2011, pp. 997-1004.

[12] G. K. Kasal and B. Singh, "Zig-zag transformer based electronic load controller for an isolated asynchronous generator," in *Proc. IEEE PECON,* 2008, pp. 431-436.

[13] G. K. Kasal & B. Singh, "Zig-zag transformer based voltage controller for an isolated asynchronous generator," in *Proc. IEEE TENCON,* 2008, pp. 1-6.

[14] H. L. Jou, K. D. Wu, J. C. Wu, and W. J. Chiang, "A three-phase four-wire power filter comprising a three-phase three-wire active power filter and a zig-zag transformer," in *IEEE Transactions on Power Electronics*, vol. 23, no.1, January 2008.

Isolated Microgrid Employing PMBLDCG for Wind Power Generation and Synchronous Reluctance Generator for DG System

Geeta Pathak, *Member, IEEE*
Department of Electrical Engineering,
Indian Institute of Technology Delhi,
New Delhi-110016, India
pathakgita28@gmail.com

Bhim Singh, *Fellow, IEEE*
Department of Electrical Engineering,
Indian Institute of Technology Delhi,
New Delhi-110016, India
bhimsinghiitd@gmail.com

B. K. Panigrahi, *Senior Member, IEEE*
Department of Electrical Engineering,
Indian Institute of Technology Delhi,
New Delhi-110016, India
bijayaketan.panigrahi@gmail.com

Abstract: In this work Enhanced Phase Lock Loop (EPLL) is applied in wind-DG microgrid for harmonics elimination, load balancing, voltage regulation, and reactive power compensation at point of common coupling (PCC) under various load conditions. As the energy generation by the wind is uncertain and unpredictable therefore Permanent magnet brushless DC generator (PMBLDCG) based wind power generation is combined with a diesel engine driven Synchronous reluctance generator (SyRG) DG-set to increase stability of the system. A battery system (BESS) is attached with voltage source converter (VSC) to provide load leveling during heavy as well as light load conditions and low wind conditions. An incremental conductance technique (INC) is used to capture maximum power point (MPPT) in wind energy conversion system (WECS) during variable wind speeds. The proposed microgrid is demonstrated and simulated in MATLAB environment using Simulink and Sim Power System (SPS) set toolboxes. The performance of the proposed controllers MPPT of WECS through INC and PCC voltage, harmonic elimination, reactive power compensation, load leveling and load balancing by EPLL is validated in this paper.

Keywords— **WECS, SyRG, EPLL, Reactive power compensation, Incremental Conductance, Diesel Engine.**

I. INTRODUCTION

Exploitation of fossil fuel resource and their extensive use in the development of technically advance civilization introduced the global warming and crisis of fossil fuel reserves. This crunch thrown attention on renewable sources such as solar photovoltaic and wind generations in power Engineering. As because of uncertain, unpredictable and fluctuating nature of the renewable energy, diesel generators are combined to provide reliable and satisfactory generation to the consumer demand and to the sensitive loads at distant areas [1].

Permanent magnet brushless DC generator (PMBLDCG) is attracting more attention due to its reduced weight as power density is 15% higher than PMSM machines, simple

construction, less maintenance and it can operate at low speeds. Generators can operate at variable speeds to maximize power capture with MPPT controller. Its trapezoidal EMF produces less ripples in rectified DC output voltage and suppress noise and minimize stresses [2] as torque is ripple-less. In standalone systems with nonconventional energy sources diesel sets are being used to fulfill load demand in unfavorable weather conditions. DG sets with brushless generators like SCIG, PMSG or SyRG are less expensive, low maintenance requirement [3].

In this demonstration Diesel Engine with SyRG is reported, which is better due to magnet-less rotor, no cogging torque, no rotor copper losses and lower noise than other brushless generators [4-5]. DG is connected to the PMBLDCG based WECS through VSC with BESS at DC bus. This VSC works as voltage and frequency controller (VFC) of the microgrid and enhance its performance. Performance of the VSC as a controller depends on the method for extracting the reference signals to compensate harmonics, regulate the PCC voltage and reactive power requirement of the load [6].

Various indirect control techniques are available in literature like Power balance theory, I-cosψ, Instantaneous Reactive Power Theory or PQ theory, Synchronous Reference Frame (SRF) theory etc. to estimate reference supply current for the controller through computation of positive sequence fundamental current component of load current [7-8]. Here EPLL is used in controller to meet out all necessary requirement of control. WECS generated AC power is converted into DC power using diode bridge rectifier, followed by DC-DC boost converter than fed to the BESS. BESS provides load leveling during load fluctuations and wind variations. Gate pulse for boost converter is given through MPPT using INC method [9].

II. SYSTEM CONFIGURATION

Fig.1. depicts an isolated wind-diesel microgrid with verity of loads. the Diesel engine driven SyRG connected with three phase load with VSC working as (VFC) and PMBLDCG

978-1-4799-6047-7/14 $31.00 © 2014 IEEE

based WECS is attached at DC link of VSC through boost converter. BESS at DC link provides load leveling during variation in wind input powers and changes in load's nature. Battery is charged under surplus active power during light loads and high wind and is discharged at peak demand and low winds.

III. CONTROL ALGORITHMS

In this isolated system two controllers are being used for VFC control and MPPT of boost converter for variable WECS.

A. EPLL CONTROLLER

EPLL algorithm is used for VFC to get fundamental component of the load currents to evaluate the reference supply currents [10-11]. Its description is given here according to fig.2.

1) In-Phase and Quadrature Templates

Terminal voltage V_t is calculated using phase voltages as,

$$V_t = \sqrt{\frac{2}{3}\left(v^2{}_{sa} + v^2{}_{sb} + v^2{}_{sc}\right)} \qquad (1)$$

Where v_{sa}, v_{sb}, v_{sc} are phase voltages.

The in-phase unit templates are expressed as,

$$u_{pa} = \frac{v_{sa}}{V_t}, u_{pb} = \frac{v_{sb}}{V_t}, u_{pc} = \frac{v_{sc}}{V_t} \qquad (2)$$

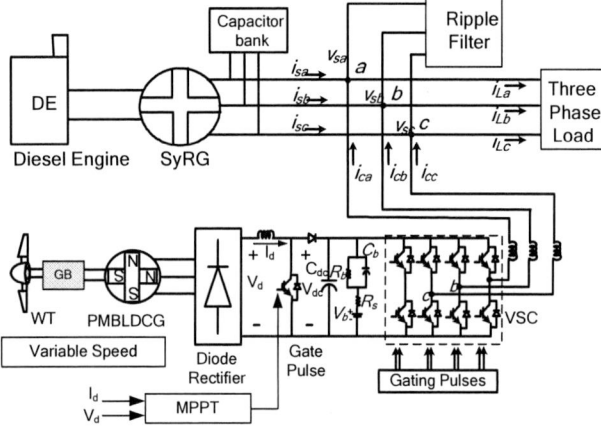

Fig. 1 Schematic diagram of Wind-Diesel hybrid configuration

Equation 3, 4 and 5 defines Quadrature unit templates as,

$$u_{qa} = \frac{(u_{pc} - u_{pb})}{\sqrt{3}} \qquad (3)$$

$$u_{qb} = \frac{(3u_{pa} + u_{pb} - u_{pc})}{2\sqrt{3}} \qquad (4)$$

$$u_{qc} = \frac{(-3u_{pa} + u_{pb} - u_{pc})}{2\sqrt{3}} \qquad (5)$$

2) Enhanced Phase-Locked Loop in roll of VFC

The fundamental load current component for phase 'a' (i_{Lfa}) is extracted using EPLL scheme as shown in fig. 2. Load current i_{La}, u_{pa} and u_{qa} are input signals for controller, which are being used to calculate active power component (i_{Lpa}) and reactive power component (i_{Lqa}) of fundamental input signal and The error signal i_e is generated as a difference of i_{La} and i_{Lfa}. Parameters a_1, a_2 and a_3 regulate the transients and steady state response of the EPLL. In this work a_1, a_2 and a_3 are used as 18, 17 and 1 respectively. i_{Lfa} and i_{La} are in phase quantities, however some phase angle is between i_{Lfa} and in-phase template (u_{pa}).

To keep I_{lfa} in phase with its respective voltage, quadrature template (u_{qa}) is given as input to a zero crossing detector (ZCD). Current signal i_{Lfa} is input signal for sample and hold circuit (SHC) and triggering pulse comes from output of ZCD. The active power component of the i_{La} is extracted as an output. The reactive power component of the i_{La} is also found with another zero crossing detector (ZCD), using in-phase template (u_{pa}), which leads quadrature template with 90^0. The fundamental load current i_{Lfa} is input signal for second SHC and triggering pulse comes from output of ZCD. The output comes as reactive power component of the i_{La}.

3) Load Currents' Active and Reactive Power Component

Expressions for active and reactive power component of the 3-φ load currents are as,

$$I_{sp} = \frac{I_{Lpa} + I_{Lpb} + I_{Lpc}}{3} \qquad (6)$$

$$I_{sq} = \frac{I_{Lqa} + I_{Lqb} + I_{Lqc}}{3} \qquad (7)$$

4) Reference Input Currents' Active Power and reactive Power Components

BESS maintains the dc bus voltage so the reference active power component of input current is I_{sp}.

The reactive power component of the reference input currents is calculated by subtracting the estimated the load current component (I_{sq}) and the PI controller output. The error in ac voltage signal at the kth sampling instant is as,

$$V_{er}(k) = V_t(k) - V_{tref}(k) \qquad (8)$$

where $V_{tref}(k)$ and V_t are the reference ac terminal phase voltage and sensed 3-φ ac voltage at PCC respectively.
PI controller output for achieving a regulated terminal voltage at the kth sampling instant is given as,

$$I_{cq}(k) = I_{cq}(k-1) + K_{pv}\{V_{er}(k) - V_{er}(k-1)\} + K_{iv}V_{er}(k) \qquad (9)$$

where K_{pv} and K_{iv} are the proportional and integral gains of the PI controller.

978-1-4799-6047-7/14 $31.00 © 2014 IEEE

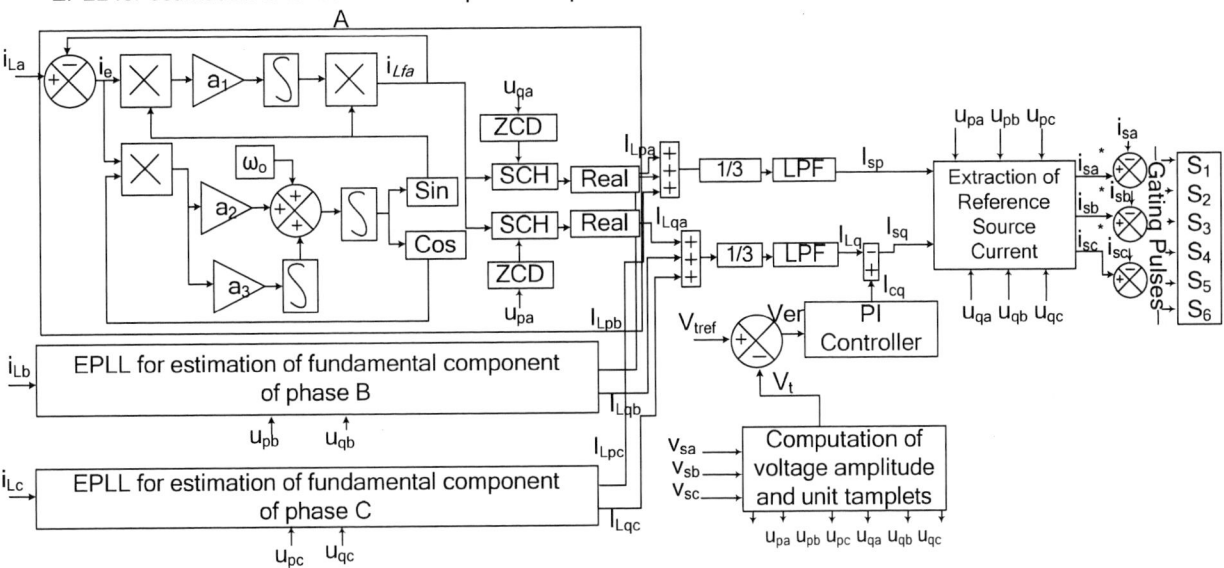

Fig. 2 Schematic diagram of EPLL control Technique

$V_{er}(k)$ and $V_{er}(k-1)$ are the errors in voltage at k^{th} and $(k-1)^{th}$ sampling instant and $I_{cq}(n)$ and $I_{cq}(k-1)$ are the PI controller outputs at the kth and $(k-1)^{th}$ instants required for voltage regulation at PCC.

Fundamental reactive power component of the reference input current is taken as,

$$I_{sq} = I_{cq} - I_{Lq} \qquad (10)$$

5) Computation of Reference input Currents

The fundamental reference active and reactive power components of the 3-φ input currents are calculated as,

$$i_{spa}^* = I_{sp} \times u_{pa}, \ i_{spb}^* = I_{sp} \times u_{pb}, \ i_{spc}^* = I_{sp} \times u_{pc} \quad (11)$$

$$i_{sqa}^* = I_{sq} \times u_{qa}, \ i_{sqb}^* = I_{sq} \times u_{qb}, \ i_{sqc}^* = I_{sq} \times u_{qc} \quad (12)$$

The 3-φ reference input currents are given as,

$$i_{sa}^* = i_{spa}^* + i_{sqa}^*, \ i_{sb}^* = i_{spb}^* + i_{sqb}^*, \ i_{sc}^* = i_{spc}^* + i_{sqc}^* \quad (13)$$

3-φ reference input currents (i_{sa}^*, i_{sb}^*, i_{sc}^*) and measured input currents (i_{sa}, i_{sb}, i_{sc}) are compared to generate current error signal x(t) for every phase. VSC is switched using hysteresis control as,

If x(t) > upper limit of hysteresis band;   lower switch on.

If x(t) < lower limit of hysteresis band; upper switch on.

B. Maximum Power Point Tracking (MPPT) Scheme.

Duty cycle is calculated directly according to the MPPT value and used to regulate the output of boost converter to keep the DC link voltage 400V. The flowchart for INC method is shown in fig 3. Duty cycle is reduced if the functional point is towards left hand side and it is increased for right hand side of Maximum power point. With the variation

in duty cycle the boost converter keeps the output voltage constant across the DC bus.

IV. DESIGN OF WIND AND DIESEL HYBRID CONFIGURATION

The diesel generator and wind generator ratings are 230 V, 50 Hz, 3.7 kW and loads are linear and nonlinear three-phase balanced and unbalanced loads.

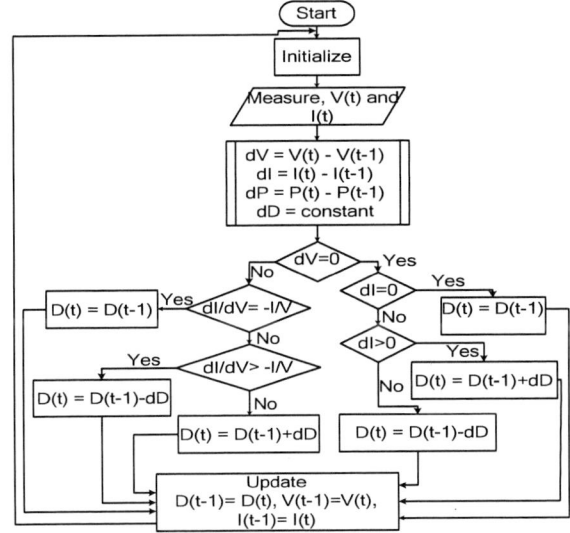

Fig.3 Flow-chart of Incremental conductance MPPT

A. DC bus voltage and BESS selection

The selection of battery system voltage depends on SyRG line to line voltage. The DC bus voltage is computed as,

$$V_{dc} = \frac{2\sqrt{2}}{\sqrt{3}m} V_L \qquad (14)$$

where m is the modulation index, if considered 1 and V_L (230V) is the line rms voltage. V_{dc} is obtained as 375 V. Therefore the battery bank is selected 400V.

Thevenin's model is used to describe BESS, where (C_b) capacitance and resistance (R_b) connected parallel along with series connected internal resistance (R_s) and an ideal voltage source of voltage 400 V. Equivalent capacitance C_b is calculated as [12] as,

$$C_b = \frac{(kW.h * 3600 * 1000)}{0.5(V_{oc\max}^2 - V_{oc\min}^2)}$$
(15)

where V_{ocmin} and V_{ocmax} are the open circuit voltages with minimum and maximum values under fully discharged and fully charged states of battery system. C_b=12000 F, R_b=10 kΩ, R_s=0.1Ω, V_{oc} =400V.

B. Boost Converter

The output voltage of boost converter is calculated as,

$$V_{dc} = \frac{V}{(1-D)}$$
(16)

To limit the peak to peak current ripple (ΔI_L) an inductor (L) with given switching frequency is calculated as,

$$L_f = \frac{V_{dc}}{4 \times f_s \times \Delta I_L}$$
(17)

where V is the output voltage of uncontrolled diode rectifier and the duty cycle of the boost converter is D.

D. Rating of Interfacing Inductor of VSC and RC Filter.

Interfacing inductor filters out the current harmonics, for 5% ripple in the current interfacing inductor is calculated as,

$$L_f = (\sqrt{3}/2) * mV_{dc} / (6af_s I_{rpl})$$
(18)

L_f and V_{dc} are interfacing inductance, voltage at DC link respectively, f_s is switching frequency and I_{rpl} is the ripple current. The round-off value is selected 3.5 mH. Noise of the terminal voltage is filtered out using A high pass RC filter with value of C=10µF and R=5Ω. Low impedance at switching frequency of 10 kHz and high impedance at fundamental frequency is offered by this combination.

V. SIMULATION RESULTS

A MATLAB model of BESS supported autonomous wind-diesel hybrid configuration simulated and its performance is demonstrated under changing wind speeds and under dynamic conditions of nonlinear and linear loads.

A. MPPT Performance under Wind reduction

In Fig.4, the performance of MPPT controller under reduction in wind speed from 11 m/s to 8 m/s at 2.5 kW, 3-phase linear load is depicted. Till 2.5s, the wind speed is 11 m/s and there is a rated power generation through PMBLDCG. At 2.5s, there is a reduction in wind speed and also in WECS power generation. WECS PMBLDCG shaft speed decreases with reduction in wind speed to achieve MPPT. It is noticed that

wind power is decreased as the wind speed falls to 8 m/s. Regulated PCC voltage and sinusoidal supply current at DG terminals (v_{sa} and i_{sa}) justify satisfactory working of VFC.

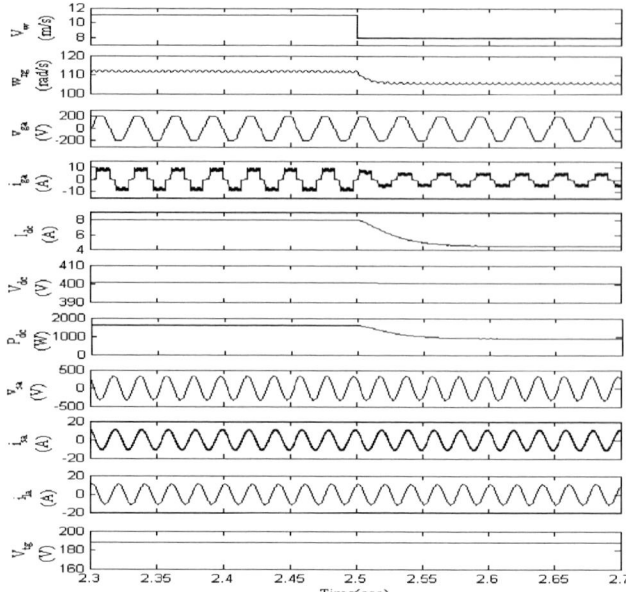

Fig. 4 MPPT' performance under wind reduction (11 m/s-8 m/s)

B. MPPT Performance under Wind increase:

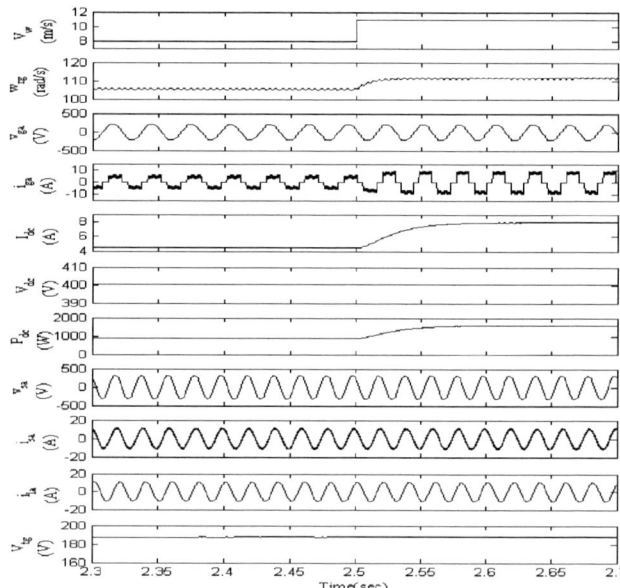

Fig. 5 MPPT performance under wind increase (8 m/s-11 m/s)

Wind speed rises form 8 m/s to 11 m/s as illustrated in Fig.5. Below 2.5 s, wind blows at 8 m/s and linear load is constant across load bus. At 2.5s, increase in wind speed is 11 m/s therefore WECS power generation also increases according to maximum power capture scheme of boost converter using INC and terminal voltage, supply voltage and supply currents are made balanced and constant by VFC.

C. Hybrid System Performance at Linear Loads

As illustrated in Figs. 6 (a,b,c), due to the rated wind speed, there is rated power generated and it leads to an ideal mode of a battery. At 3.5s, load unbalance (i_{La}, i_{Lb}, i_{Lc} are unequal) is created by opening phase a, due to this load power is reduced and it causes the battery to charge with extra generated power from WECS. It can be analyzed through battery current, voltage and power (I_{bt}, V_{bt}, P_{bt}). The load is connected back on the load bus at 3.6s and battery comes towards initial state.

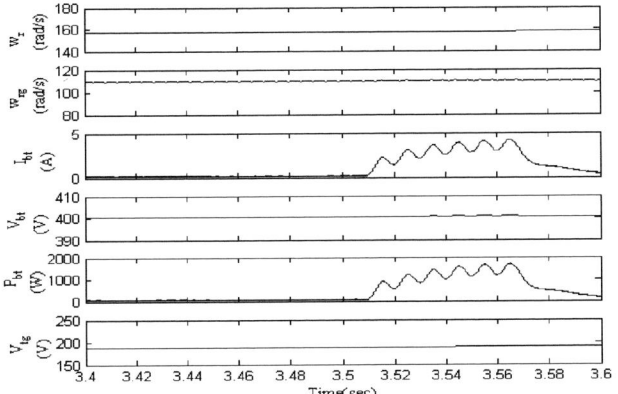

Fig. 6 (a) Characteristics of the system with constant wind speed under varying loads.

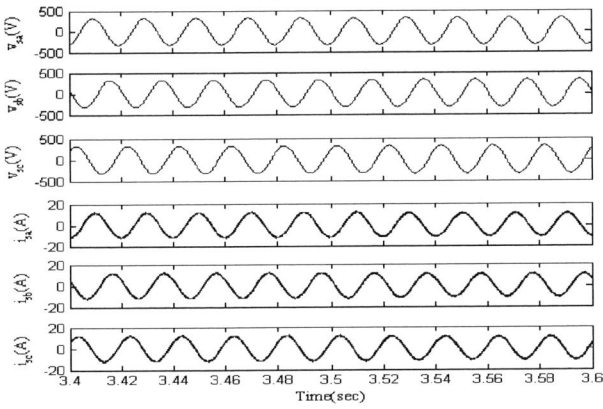

Fig. 6 (b) Estimation of supply currents and voltages using control algorithm

During transient conditions, input currents (i_{sa}, i_{sb}, i_{sc}) are sinusoidal. Voltages at PCC (v_{sa}, v_{sb}, v_{sc}) are also sinusoidal and near to reference value, but compensator currents (i_{dsta}, i_{dstb}, i_{dstc}) are changed according to the need to maintain the terminal voltages constant (V_{tg}).

DG performs well under 80-100% of full load for that VFC also control the active power component of load current (I_{sp}) as,

a) Under light load conditions i.e. load is less than 80% of DG full load capacity, I_{sp} is maintained at 80% of DG full load current capacity.

b) Under heavy load conditions i.e. load is more than 100% of DG full load capacity, I_{sp} is maintained at 100% of DG full load current capacity.

c) During normal loads when load is within 80% to 100% limit of DG full load capacity, I_{sp} is maintained at its actual value.

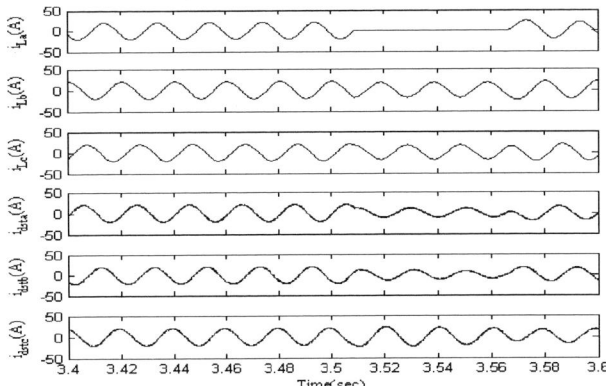

Fig.6 (c) dynamic Performance of controller of hybrid system under varying linear loads at 10 m/s wind speed

D. Hybrid System Performance at Nonlinear Loads

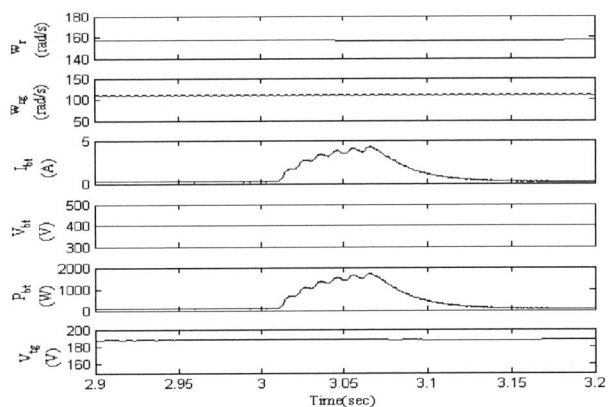

Fig. 7(a) Characteristics of the system with constant wind speed under varying loads.

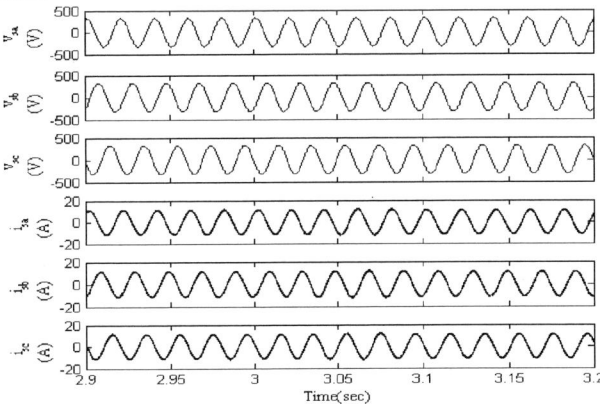

Fig. 7(b) Estimation of supply currents and voltages using control algorithm

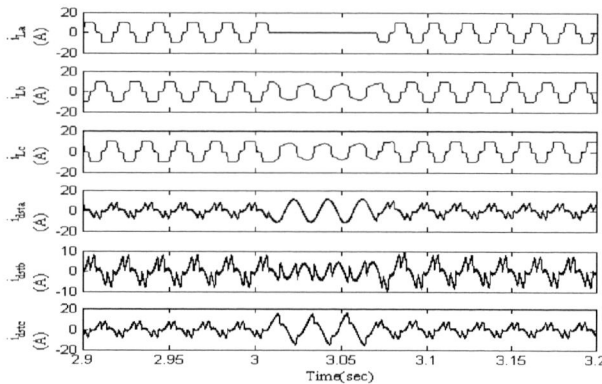

Fig.7(c) dynamic Performance of controller of hybrid system under varying nonlinear loads at 10 m/s wind speed.

Fig. 7 (a,b, c) shows the unbalance at phase a, at time 3s . during transient source currents are balanced and sinusoidal and terminal voltage is also maintained.

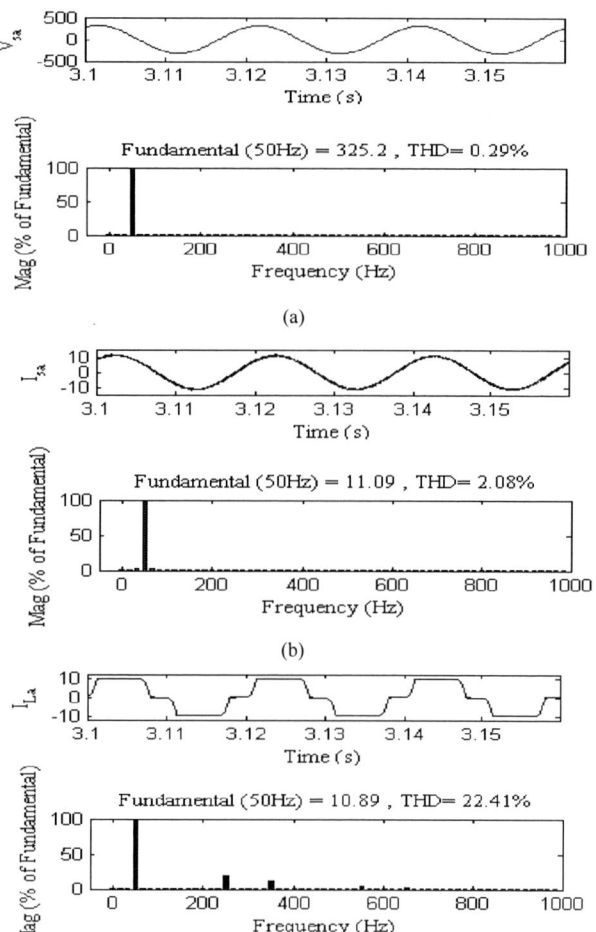

Fig. 8 waveforms and harmonic spectra (a) Phase 'a' supply voltage of at PCC (b) Phase 'a' supply current under nonlinear unbalanced loads and (c) Load current.

V. CONCLUSIONS

An isolated wind-diesel microgrid has been imitated in MATLAB Simulink and Sim-power system tool boxes. A mechanical sensor less approach has been used for achieving MPPT through incremental conductance technique. Use of PMBLDCG could make the system work with only single VSC as wind ac power is converted into DC power by using diode rectifier which is economically worth full. VFC provided the load balancing, reactive power compensation, and harmonics elimination as per guidelines of IEEE-519-1992 standards, and voltage regulation at PCC under linear and nonlinear loads operation using EPLL technique.

REFERENCES

[1] Shuben Zhang, Jian Yang, Xiaomin Wu, and Ruiyi Zhu, "Dynamic power provisioning for cost minimization in islanding micro-grid with renewable energy," *in Proc. of IEEE PES Innovative Smart Grid Technologies Conference (ISGT), 2014*, pp.1,5, 19-22 Feb. 2014.

[2] Bhim Singh, and Shailendra Sharma, "PMBLDC based stand-alone wind energy conversion system for small scale applications," *Inter. Journal of Engg. Science and Tech.*, vol. 4, no. 1, pp. 65-73, 2012.

[3] M. Rezkallah, A. Chandra, B. Singh, and R. Niwas, "Modified PQ control for power quality improvement of standalone hybrid wind diesel battery system," *in Proc. of IEEE Fifth Power India Conference, 2012* , pp.1,6, 19-22 Dec. 2012

[4] Poopak Roshanfekr, Sonja Lundmark, Torbjorn Thiringer, and Mikael Alatalo, "A synchronous reluctance generator for a wind application-compared with an interior mounted permanent magnet synchronous generator," *in Proc. of 7th IET International Conference on Power Electronics, Machines and Drives (PEMD), 2014,* , pp.1,5, 8-10 April 2014

[5] Y.H.A Rahim, J.E. Fletcher, and N.E.A. Hassanain, "Performance analysis of salient-pole self-excited reluctance generators using a simplified model," *IET on Renewable Power Generation*, vol.4, no.3, pp.253, 260, May 2010.

[6] B. Singh, and J. Solanki, "Load Compensation for Diesel Generator-Based Isolated Generation System Employing DSTATCOM," *IEEE Trans. on Industry Applications*, vol.47, no.1, pp.238,244, Jan.-Feb. 2011.

[7] B. Singh, D.T. Shahani, and A K. Verma, "IRPT based control of a 50 kw grid interfaced solar photovoltaic power generating system with power quality improvement," in *Proc. of 4th IEEE International Symposium on Power Electronics for Distributed Generation Systems (PEDG), 2013*, pp.1,8, 8-11 July 2013.

[8] G. Bhuvaneswari, and M.G. Nair, "Design, Simulation, and Analog Circuit Implementation of a Three-Phase Shunt Active Filter Using the $I \cos \varphi$ Algorithm," *IEEE Trans. on Power Delivery*, vol.23, no.2, pp.1222,1235, April 2008

[9] Snehamoy Dhar, R. Sridhar, and Geraldine Mathew, "Implementation of PV cell based standalone solar power system employing incremental conductance MPPT algorithm," in *Proc. of International Conference on Circuits, Power and Computing Technologies (ICCPCT), 2013*, pp.356-361, 20-21 March 2013.

[10] S. Sharma, and B. Singh, "An enhanced phase locked loop technique for voltage and frequency control of stand-alone wind energy conversion system," *in Proc. of India International Conference on Power Electronics (IICPE), 2010*, pp.1,6, 28-30 Jan. 2011.

[11] B. Singh, and S.R. Arya, "Implementation of Single-Phase Enhanced Phase-Locked Loop-Based Control Algorithm for Three-Phase DSTATCOM," *IEEE Trans. on Power Delivery*, vol.28, no.3, pp.1516,1524, July 2013.

[12] P.K. Goel, B. Singh, S.S. Murthy, and N. Kishore, "Isolated Wind–Hydro Hybrid System Using Cage Generators and Battery Storage," *IEEE Trans. on Industrial Electronics*, vol.58, no.4, pp.1141,1153, April 2011.

Performance of Grid Interfaced Solar PV System under Variable Solar Intensity

Sanjay Kumar, Arun Kumar Verma *Member, IEEE,*
Ikhlaq Hussain, *Student Member, IEEE* and Bhim Singh, *Fellow, IEEE*
Department of Electrical Engineering, Indian Institute of Technology Delhi, New Delhi-110016, India
Email: sanjaykumar1224@gmail.com, arunverma59@gmail.com, ikhlaqb@gmail.com and bsingh@ee.iitd.ac.in

Abstract—**This paper proposes an enhanced phase locked loop (EPLL) based control algorithm of a double stage solar photovoltaic (PV) grid interfaced power generating system, which also mitigates power quality problems in 3P4W (three phase, four wire) distribution system. The proposed solar PV grid interfaced system consists of solar PV array, boost converter, four-leg voltage source converter (VSC) and connected linear/nonlinear loads. The proposed solar PV power generating system provides load balancing, eliminates harmonics, corrects the power factor, and regulates at PCC (Point of Common Coupling) voltages under different loads. Proposed solar PV grid interfaced power generating system is modeled and simulated in the MATLAB and results are shown to validate the design and control for feeding 3P4W loads with improved power quality.**

*Keywords***:** *EPLL, VSC, zero voltage Regulation (ZVR) and power factor correction (PFC), power quality (PQ).*

I. INTRODUCTION

During last two decades, the solar photovoltaic (PV) generation is growing and that too produces electricity without producing global warming pollution. The use of solar PV arrays is getting cost effective means of generating power. Solar energy can be used as an alternative source as it is eco-friendly renewable energy source. World's oil reserves are estimated only for 40 to 50 years where as solar energy will last forever [1].

Due to growing demand, grid interfaced solar PV generating systems are becoming popular and posed new challenges [1]. Presently single-stage, two-stage and multilevel grid interfaced systems are commonly used in solar PV generation [2-3]. In two stage solar PV generating system, the power quality problems are not considered in detail with and without availability of sun [4]. However, the power quality (PQ) problems are dominant in the grid because of various nonlinear loads in the distribution system. The specific challenges that affect the power quality are poor power factor, poor voltage regulation, and reactive power compensation at ac mains. Maximum power point tracking (MPPT) from solar PV array is also a challenging task and several methods of MPPT are used [5-6]. In two stage, the boost converter with MPPT is used to track the maximum solar PV power from the solar PV array

In this paper, proposed control algorithm based on enhanced phase locked loop (EPLL) scheme [7-8] for solar PV grid interfaced power generating system is implemented

for harmonics elimination, load balancing, reactive power compensation, PFC (Power Factor Correction) or ZVR (Zero Voltage Regulation). The advantage of this control algorithm over other algorithms is its speed and accuracy of its response and its performance is not affected due to noise and distortion. Further, it is adaptive in nature and adopts the changes in phase angle, frequency and amplitude of the input signal and its implementation in real time using DSP or any other embedded control is easy.

The solar PV grid interfaced generating system using EPLL is designed, modeled and its performance is simulated in Simulink tool box of MATLAB for ZVR and PFC along with compensation of harmonics current and balancing of different loads.

II. DESIGN OF PROPOSED SYSTEM

The design of proposed 100 kW solar PV power generating system as shown in Fig. 1 is given in terms of solar PV array, dc-dc boost converter, interfacing inductors and dc bus capacitor as follows. The detailed design data of proposed system is in Appendices.

A. Design of Solar PV Array

It is designed for the peak power capacity of 100 kW rated at 415 V ac grid. A solar PV module has short circuit module current (I_{sc}) of 3.8 A and open circuit module voltage (V_{ocn}) of 21 V [9].

The maximum power for SPV array is given as,

$$P_{mp} = (n_s * V_{mp}) * (n_p * I_{mp}) = 100 \, kW \qquad (1)$$

where n_s and n_p represent series and parallel strings of PV module, V_{mp} is the voltage of a module at MPPT, I_{mp} is the current of a module at MPPT and P_{mp} is the nominal power of a module at MPPT.

The P_{mp} is generally achieved under the condition given as,

$$P_{mp} = (n_s * 85\% \text{ of } V_{ocn} * n_p * 85 \% \text{ of } I_{sc}) = 100 \, kW \qquad (2)$$

Thus, I_{mp} is 3.3 A and V_{mp} is 17 V of each module.
Considering, PV array open circuit voltage (V_{OCT}) = 700 V.
The PV modules connected in series string are estimated as,

$$V_{OCT} = n_s * V_{ocn}, \text{ thus } n_s = 700/21 = 34 \text{ Modules} \qquad (3)$$

Maximum current of the PV array is given as,

$$I_{mp} = P_{mp} / (0.85 * V_{OCT}) = 168.067 \, A$$

The PV modules connected in parallel string are estimated as,

$$I_{mp} = n_p * I_{sc}, \text{ thus } n_p = 43 \text{ Modules} \qquad (4)$$

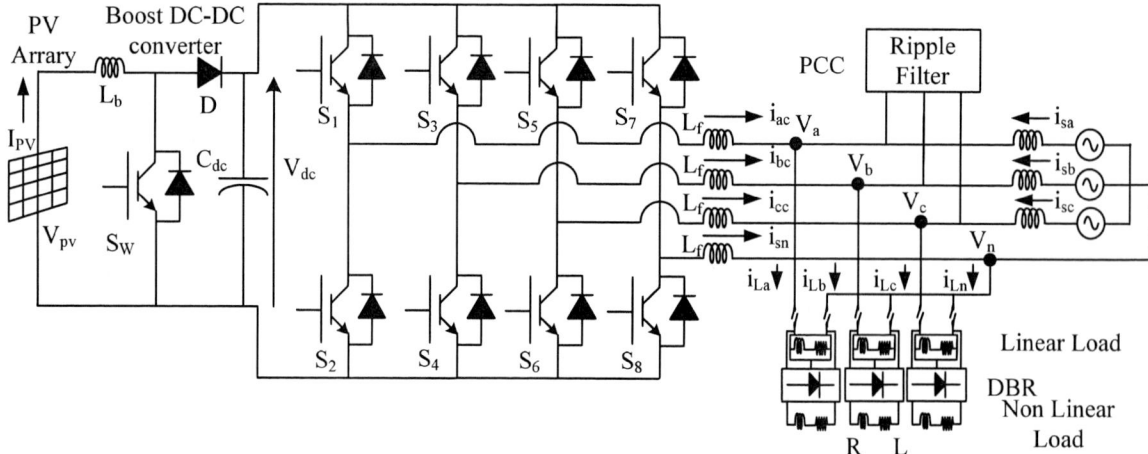

Fig. 1 The proposed system schematic configuration

Thus the array of 100 kW peak power capacity is designed with 43 modules in parallel and 34 modules in series with an PV array of 43*34 modules.

B. Design of DC-DC Boost Converter

The ripple current for inductor at D = 0.3 is given as [10],

$$L_b = \frac{V_{MPP}D}{\Delta I_1 f_{sw}} = \frac{595*0.3}{(8.4*10000)} = 2.125 \; mH \quad (5)$$

where ΔI_1 is input current ripple, and it is considered as 5 % of dc-dc boost converter inductor current I_1 (P_{MPP}/ V_{MPP}) = 168.06 A. Thus a calculated value of ΔI_1 is 8.4 A and the inductance (L_b) value is selected as 2.2 mH.

C. Design and Selection of DC Capacitor Voltage

To achieve proper load compensation the dc link voltage V_{dc} value is given as [10],

$$V_{dc} = \frac{\left(2\sqrt{2}V_{LL}\right)}{\sqrt{3}m} = \frac{\left(2\sqrt{2}*415\right)}{\sqrt{3}*0.95} = 713.27 \approx 700 \; V \quad (6)$$

where V_{LL} is the VSC ac line voltage, m is modulation index.

D. Selection of AC Inductor

The ac inductor (L_f) value is estimated as [10],

$$L_f = \frac{\sqrt{3}mV_{dc}}{12hf_s\Delta i} = \frac{\sqrt{3}*0.95*700}{12*1.2*10^3*(0.05*142.85)} = 1.12mH \quad (7)$$

where Δi is current ripple = 5% of input current, f_s is switching frequency = 10 kHz, h is overloading factor and is taken as 1.2.
The L_f from (7) is calculated as 1.12 mH. The selected value is 1.2 mH.

E. Design of DC Link Capacitor

The dc link capacitor value is given as [10],

$$C_{dc} = \frac{(P_{dc} / V_{dc})}{\left(2*\omega*v_{dcrip}\right)} = \frac{(10^5 / 700)}{(2*314*0.03*700)} = 10832.4\mu F \quad (8)$$

where ω is angular frequency and v_{dcrip} is % ripple voltage considered as 3% of V_{dc}.
Hence estimated value of dc link capacitor C_{dc} is 10832.4 µF and it is selected as 15000 µF.

III. CONTROL ALGORITHM

The control algorithm of the solar PV power generating system consists of two stages. First MPPT control to track peak power of solar PV array using dc-dc boost converter and second stage is used to control a grid interfaced VSC also operating as a shunt compensator. The detailed control algorithm is explained as follows.

A. MPPT

The drawbacks of P&O and IC algorithms are overcome by using variable step INR method [5-6]. Variable step is usually large when away from MPP and reduces to small as the system reaches MPP. The MPPT depicts as follows,

$$\delta(n) = \delta(n-1) \pm K(dp / dpv) \quad (9)$$

where δ is the duty cycle at sampling instant nth and $\delta(n-1)$ is a duty cycle at sampling instant $(n-1)^{th}$ and is a factor to control step size.
This method is based on that slope of (dV_{pv}/dI_{pv}) must be zero at MPP as,

$$\frac{dI_{pv}V_{pv}}{dI_{pv}} = V_{pv} + I_{pv}\frac{dV_{pv}}{dI_{pv}} \quad (10)$$

Thus both instantaneous V_{pv}/I_{pv} and incremental resistance dV_{pv}/dI_{pv} are compared and MPP is tracked.

B. Control of VSC

Fig. 2 shows the control algorithm for the extraction of the load current fundamental component and further this component is used to extract load current active power and reactive power components. These active and reactive components of load currents are used to generate reference grid currents. The PCC voltages (v_{sa}, v_{sb}, v_{sc}), load currents (i_{La}, i_{Lb}, i_{Lc}), V_{dc} of VSC are mains parameters of the control algorithm. Amplitude of PCC voltage (V_t) is estimated as,

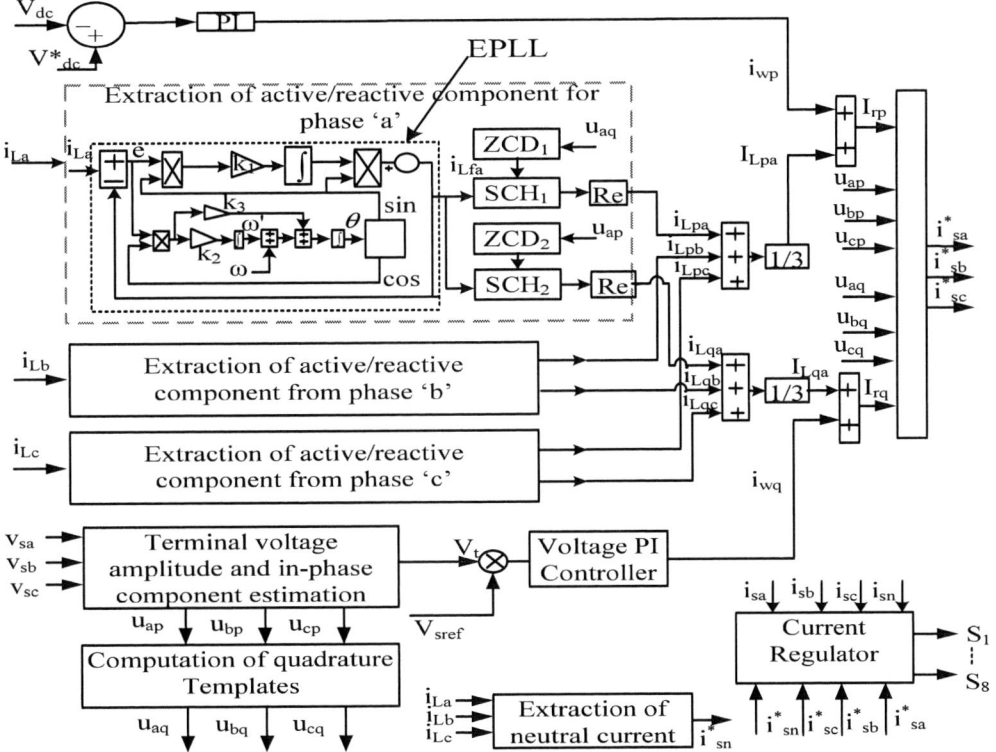

Fig. 2 Control algorithm for proposed solar PV power generation system

$$V_t = \sqrt{\{(2/3)(v_{sa}^2 + v_{sb}^2 + v_{sc}^2)\}} \qquad (11)$$

The unit vectors in-phase of phase voltages are derived as,

$$u_{ap} = \frac{v_{sa}}{V_t}, \; u_{bp} = \frac{v_{sb}}{V_t}, \; u_{cp} = \frac{v_{sc}}{V_t} \qquad (12)$$

The unit vectors in quadrature with grid voltages v_{sa}, v_{sb} and v_{sc} are derived from vectors w_{ap}, w_{bp} and w_{cp}, as,

$$u_{aq} = -\frac{u_{bp}}{\sqrt{3}} + \frac{u_{cp}}{\sqrt{3}}, u_{bq} = \frac{-u_{cp}}{\sqrt{3}} + \frac{u_{ap}}{\sqrt{3}}, u_{cq} = -\frac{u_{ap}}{\sqrt{3}} + \frac{u_{bp}}{\sqrt{3}} \qquad (13)$$

1) Extraction of Fundamental Components of Load Current

The fundamental component of load current is extracted as output of EPLL. Performance of the EPLL is controlled with three controlling constants given as K_1, K_2 and K_3. The EPLL computes the magnitude, phase and frequency of the input load current signals. The EPLL is well described in a given equation as,

$$i_{Lfa} = \left[\int (e) \; \sin\theta \, k_1 \; d\theta \right] \sin\theta \quad \text{where}$$
$$\theta = \int \left[\{ \int e \, k_2 \; \cos\theta \; d\theta + w \; \sin\theta \} + e \, k_3 \; \cos\theta \right] d\theta \qquad (14)$$

where ($e = i_{La} - i_{Lfa}$) is the difference between signals and i_{la} is the fundamental load current, it is called as error. This error (e) is multiplied by the sin/cos component and further parameters K_1, K_2, K_3 play an important role and control the steady state and transient behavior of the loop. It plays an important role in extraction of the fundamental component from the polluted load current. The values of K_1, K_2, and K_3 are chosen as 140, 10, and 5 respectively in order to estimate fundamental load component.

2) Estimation of Reactive and Active Power Components of Load Currents

The amplitude of fundamental reactive and active power components is extracted from phase 'a' load current fundamental component i_{Lfa} using a zero crossing detector (ZC_1), quadrature template (u_{aq}), sample and hold block (SH_1) in-phase template (u_{ap}). Similarly, load fundamental active and reactive power currents (i_{Lpb}, i_{Lpc}) and (i_{Lfb}, i_{Lfb}) are also estimated in phase 'b' and 'c'.
The load current average active power component (I_{LpA}) is estimated as,

$$I_{LpA} = \frac{i_{Lpa} + i_{Lpb} + i_{Lpc}}{3} \qquad (15)$$

Similarly, the load currents reactive power component (I_{LqA}) is estimated as,

$$I_{LqA} = \frac{i_{Lqa} + i_{Lqb} + i_{Lqc}}{3} \qquad (16)$$

3) Active Power Components of Grid Current

The error voltage in reference dc link voltage v^*_{dc} and sensed v_{dc} at n^{th} sampling instant is given as,

$$v_{dcerr(n)} = v^*_{dc(n)} - v_{dc(n)} \qquad (17)$$

978-1-4799-6047-7/14 $31.00 © 2014 IEEE   267

The PI controller output to regulate the dc link voltage of VSC at n^{th} sampling instant is given as,

$$I_{wp(n)} = I_{wp(n-1)} + K_{pdc}\left\{v_{dcerr(n)} - v_{dcerr(n-1)}\right\} + K_{idc}v_{dcer(n)} \quad (18)$$

where $I_{wp(n)}$ is another component of grid current, K_{pdc} and K_{idc} are proportional gain and integral gain constants.

Therefore, total grid current active power component of (I^*_{active}) is estimated by adding to output iof controller and dc component ($I_{Lactive}$) of load currents, $I_{wp(n)}$ as,

$$I^*_{rp} = I_{LpA} + I_{wp} \quad (19)$$

Therefore, in phase components or active power components of reference instantaneous grid currents in phase of PCC voltages are calculated as,

$$i^*_{sadi} = I^*_{rp}*u_{ap}, i^*_{sbdi} = I^*_{rp}*u_{bp}, i^*_{scdi} = I^*_{rp}*u_{cp} \quad (20)$$

4) Quadrature Component of Grid Current

PCC voltage is controlled using a PI regulator. The terminal voltage amplitude (V_t) is estimated in (11) and the reference terminal voltage amplitude value (V_{ref}) are fed to the voltage controller. The voltage error is estimated as,

$$v_{err(t)} = V^*_{tref(t)} - V_{t(t)} \quad (21)$$

Output of PI voltage regulator at nth instant is given as,

$$I^*_{wq(n)} = I^*_{wq(n-1)} + K_{pt}\left\{v_{err(n)} - v_{er(n-1)}\right\} + K_{it}v_{err(n)} \quad (22)$$

where K_{pt} and K_{it} are the proportional gain and integral gain of voltage controller.

Thus the reactive component of grid current amplitude is given as,

$$I^*_{rq} = -I_{LqA} + I^*_{wq(n)} \quad (23)$$

The reference supply grid currents instantaneous quadrature component are calculated as,

$$i^*_{saqu} = I^*_{rq}*u_{aq}, i^*_{sbqu} = I^*_{bq}*u_{bq}, i^*_{scqu} = I^*_{rq}*u_{cq} \quad (24)$$

5) Generation of Reference Grid Supply Currents

Total reference grid currents are estimated from (20) and (24) as,

$$i^*_{sa} = i^*_{sadi} + i^*_{saqu}, i^*_{sb} = i^*_{sbdi} + i^*_{sbqu}, i^*_{sc} = i^*_{scdi} + i^*_{scqu} \quad (25)$$

$$i^*_{sn} = i_{Lao} + i_{Lbo} + i_{Lco} \quad (26)$$

6) PWM Current Controller

Gating pulses for VSC are generated by comparing reference grid supply currents (i*$_{sa}$, i*$_{sb}$, i*$_{sc}$, i*$_{sn}$) and sensed grid currents (i$_{sa}$, i$_{sb}$, i$_{sc}$, i$_{sn}$) and the error in current is given to PWM current controller.

IV. MATLAB BASED MODELLING

The configuration of proposed solar PV grid interfaced system is modeled by using Simulink with SPS tool boxes as shown in Figs. 1. Fig. 2 explains the detail control modelling and the reference current generation using EPLL. Further an estimation of reference currents and PWM switching signal

generation are achieved for the control of the combined operation of the VSC based PV power generating system.

V. RESULTS AND DISCUSSION

Simulated results of solar PV grid interfaced power generating system are discussed in this section. The system performance under constant solar intensity and variable load is demonstrated in Fig. 3, whereas under constant load and variable solar intensity is demonstrated in Fig. 4. A nonlinear load is realized by using a diode bridge rectifier load and load unbalancing is achieved by removing one phase for certain duration.

The system performances are depicted as grid currents (i$_{isa}$, i$_{isb}$, i$_{isc}$), load current (i$_{La}$, i$_{Lb}$, i$_{Lc}$), grid voltages (v$_{sa}$, v$_{sb}$, v$_{sc}$), active power (P), reactive power (Q) and VSC current (i$_i$). Here solar PV array voltage V$_{pv}$, solar PV array power and current as P$_{pv}$ and I$_{pv}$ are respectively.

A. Performance of Proposed System Configuration under Variable Solar Intensity and Variable Load

Fig. 3 shows the dynamic performance of solar PV grid interfaced system subjected to variable solar intensity and variable loads. These are the various results during the unbalanced load operation. Load unbalancing is realized at 0.4s and continues until 0.6 s. During load unbalancing, the grid currents are sinusoidal and balanced. The magnitude of the PCC voltage is almost constant and very close to reference value, thus voltage regulation is achieved.

B. Performance of Solar PV Grid Interfaced System under Constant Load and Variable Solar Intensity

Fig. 4 shows various performance parameters during operation of the solar PV grid interfaced system under constant load and variable solar intensity. The solar intensity is reduced to zero at 0.5 s, the VSC supplies the required active power. The solar intensity is 1000 W/m² initially and it is reduced to zero gradually at 0.5s. At 0.5 s onwards the active power of the solar PV grid interfaced system is supplied by the grid and the direction of the active power flow is reversed after 0.5 s. At 0.5 s the solar PV current is zero thus solar PV power is also zero. The dc link voltage is regulated at reference value and grid currents are sinusoidal.

The THD of grid supply voltage, grid supply current and load current at various operations are given in Table I, it shows that the voltage and current harmonics are well within the IEEE-519 standard [11].

TABLE I. PERFORMANCE OF THE PROPOSED SYSTEM

Mode of Operation	Parameters	Nonlinear RC load
PFC	Grid Voltage %THD	338.5V, 0.98%
	Grid Current %THD	19.61, 3.60%
	Load Current %THD	25.97, 71.24%
ZVR	Grid Voltage %THD	342.5, 1.367%
	Grid Current %THD	19.67, 3.24%
	Load Current %THD	24.61, 70.02%

978-1-4799-6047-7/14 $31.00 © 2014 IEEE   268

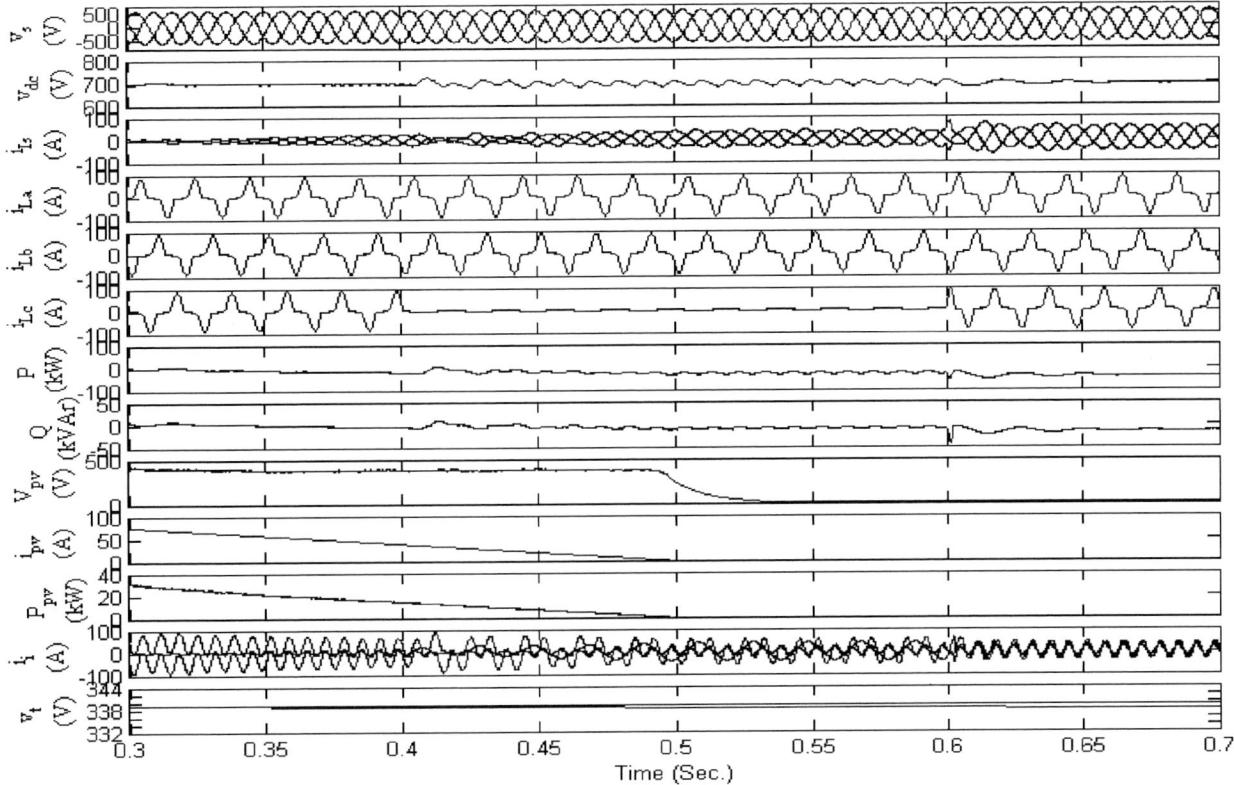

Fig. 3 Response of proposed system configuration under variable load and variable solar intensity

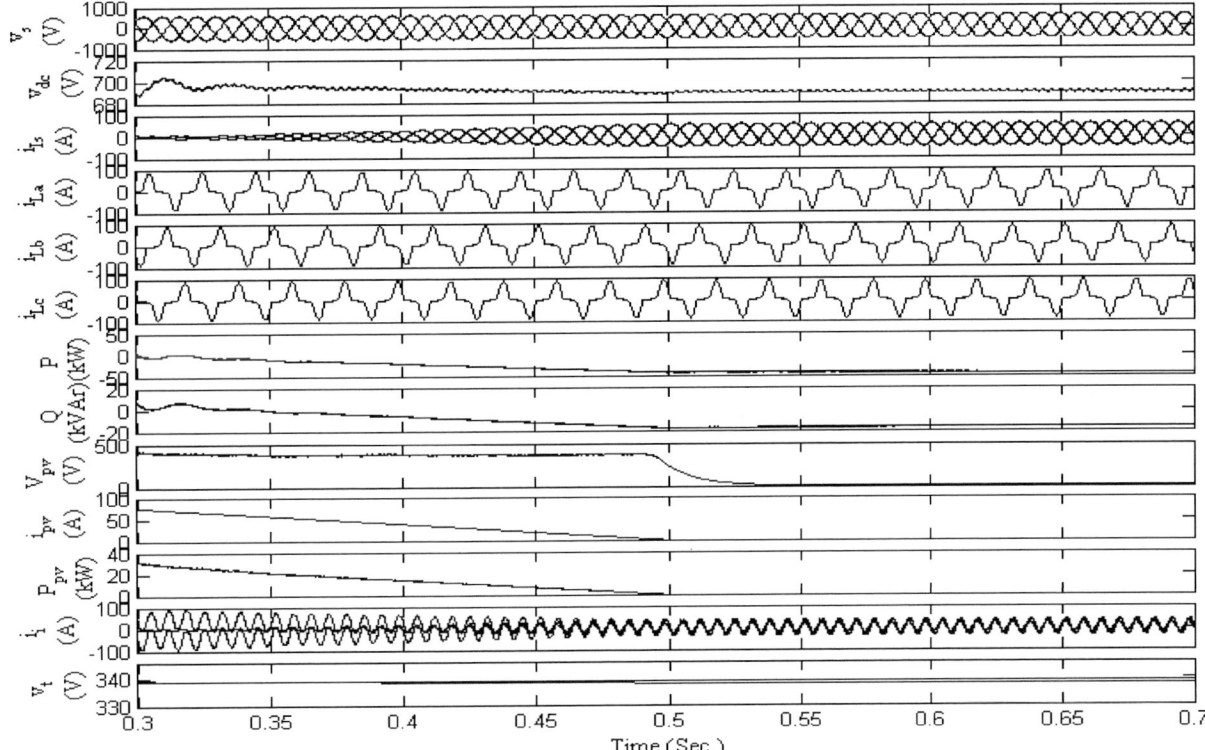

Fig. 4 Response of proposed system under constant load and varying solar intensity

VI. CONCLUSION

The proposed solar PV grid interfaced power generating system with EPLL based control has been found quite acceptable for unity power factor, load balancing, harmonics elimination reactive power compensation and voltage regulation under varying consumer loads and PV power generation. The EPLL used for active filtering has been found simple. The THD of the grid voltage and grid current are observed well within the acceptable limits of an IEEE-519 standard.

APPENDICES

A. Parameters of Solar PV Module Data

I_{sc} = 3.8 A, V_{ocn} = 21 V, I_{mp} = 3.3 A, V_{mp} = 17 V, n_s = 34, n_p = 43, Voltage temperature coefficient (K_v) = -80e-3 V/K, Current temperature coefficient (K_i) = 0.0029 A/K, Number of series cells (N_s) = 36, n_s = 34, n_p = 43.

B. DC-DC Boost Converter Parameters

D = 0.2-0.5, L_b = 2.2 mH, f_{sw} = 10 kHz.

C. Parameters for VSC

V_s = 415 V, f = 50 Hz, f_s = 10 kHz, V_{dc}= 700 V, C_{dc} = 15000 μF, L = 1.2 mH, line impedance: L_s = 0.5 mH, R_s = 0.01 Ω
dc voltage controller: K_{pd} = 0.023, K_{id} = 1.2
Non-linear load: three single phase rectifiers with C = 200 μF, R = 5 Ω and ripple filter: C_f= 10 μF, Rf = 5 Ω.

ACKNOWLEDGEMENT

This work is supported by DST (Department of Science and Technology), Govt. of India under Grant No. RP02583. Authors are thankful to DST.

REFERENCES

[1] B. Verhoeven and B.V. Kema, "Utility aspects of grid connected Photovoltaic power systems," *Int. Energy Agency Photovoltaic Power Syst.*, IEA PVPS T5-01, 1998.

[2] A. K. Verma, B. Singh, and D. T. Shahani, "Grid interfaced solar photovoltaic power generating system with power quality improvement at AC mains," in *IEEE ICSET*, 24-27 Sept. 2012, pp. 177-182.

[3] B. Singh, D. T. Shahani, and A. K. Verma, "Power balance theory based control of grid interfaced solar photovoltaic power generating system with improved power quality," in *IEEE Int. Conf. Power Electron. Drives Energy Systems (PEDES)*, 2012, pp. 1-7.

[4] S. Balathandayuthapani, C. S. Edrington, S. D. Henry, and J. Cao, "Analysis and control of a photovoltaic system: application to a high-penetration case study," *IEEE Systems Journal*, vol. 6, no. 2, pp. 213-219, June 2012.

[5] B. Subudhi and R. Pradhan, "A Comparative Study on Maximum Power Point Tracking Techniques for Photovoltaic Power Systems," *IEEE Trans. Sustainable Energy*, vol. 4, no. 1, pp. 89-98, Jan. 2013.

[6] Q. Mei, M. Shan, L. Liu, and J. M. Guerrero, "A Novel Improved Variable Step-Size Incremental-Resistance MPPT Method for PV Systems," *IEEE Trans. Ind. Electron.* vol. 58, no. 6, pp. 2427–2434, June 2011.

[7] S. Sharma and B. Singh, "An enhanced phase locked loop technique for voltage and frequency control of stand-alone wind energy conversion system," in *Proc. India Int. Conf. Power Electron.(IICPE)*, 28-30 Jan. 2011, pp.1-6.

[8] B. Singh and S. Arya, "Implementation of Single-Phase Enhanced Phase-Locked Loop-Based Control Algorithm for Three-Phase DSTATCOM," *IEEE Trans. Power Del.*, vol. 28, no. 3, pp.1516 -1524, July 2013.

[9] M. G. Villalva, J.R. Gazoli, and E.R. Filho, "Comprehensive approach to modelling and simulation of photovoltaic arrays," *IEEE Trans. Power Electron*, vol. 24, no. 5, pp. 1198-1208, May 2009.

[10] N. Mohan, T. M. Undeland, and W. P. Robbins, Power electronics: converters, applications and design, 3rd ed. New Delhi, India: John Wiley & sons Inc., 2009.

[11] IEEE Recommended Practices and Requirements for Harmonic Control in Electrical Power Systems," *IEEE Std.* 519-1992, 1993.

Real Time Implementation and Comparison of PI and Modified Inc Cond Control Algorithms for Solar Applications

Venkata Ratnam Kolluru
Ph.D. Scholar, NITR
Rourkela, Odisha, India
Email:venkataratnamk@gmail.com

Kamalakanta Mahapatra
Professor, ECE, NITR
Rourkela, Odisha, India

Bidyadhar Subudhi
Professor, EEE, NITR
Rourkela, Odisha, India

Tejavathu Ramesh
Asst.Prof, EEE, NITKKR
Kurukshetra,India

Abstract—**This paper depicts the comparison of control algorithms for a Photovoltaic (PV) system in real time. Modified Incremental Conductance (Inc Cond) with fixed step size algorithm is used with a DC-DC boost converter to track maximum power from a PV system. PV output is connected to a Boost converter to regulate and increase the voltage upto a desired level. The modified inccond MPPT technique is compared with a conventional PI controller and was observed that, the modified Inc Cond MPPT is tracking 7.4% more power than the conventional PI controller. MPPT control algorithms are developed, compared and simulated in MATLAB/SIMULINK. Simulated and real time results are presented and discussed in this paper. The real time results are verified with digital simulator OPAL-RT experimental setup.**

Keywords: DC DC boost converter, Incremental conductance, Maximum power point tracking, OPAL-RT, PI controller.

I. INTRODUCTION

Renewable energy is the best solution for the global growing problem of energy scarcity. Solar energy is a virtually inexhaustible resource; obtained all around the world. In recent years, with its advantages of pollution-free, efficient, cost-effective and long-term using, solar energy have been greatly developed [1]. The PV power generation is based on the principle of the photovoltaic effect [2], and observed that, there is a unique point at which the PV cell produces maximum power. At this point, the rate of change of power with respect to the voltage is equal to zero [3]. PV array has to be operated at MPP in order to extract maximum power output. Various power management issues concerning improvement in the conversion efficiency of a PV array [4], thus maximizing PV power output. The maximum power of the PV array changes with shading and/or climatic conditions.

Thus, an important challenge in a PV system is to ensure that maximum energy is generated from the PV array with a dynamic variation of its output characteristic when connected to a variable load [4].

A solution for this problem is the insertion of a power converter between the PV array and load, whichcould dynamically change the impedance of the circuit by using a control algorithm [5]. DC DC Converters are required to regulate the output voltage [6] at a required level. In this paper, a step up converter is used, that has a capability of providing an output voltage which is higher than the input voltage [4].

Maximum Power Point Tracking (MPPT) is becoming more and more important as the amount of energy produced by PV systems is increasing [7]. A large number of control techniques have been proposed for tracking of Maximum Power Point (MPP) of PV. In those algorithms Incremental Conductance (inc cond) algorithm is one of the best suited algorithm to track MPPT [8]–[11]. Variation of voltage and current values plays key role to track the MPPT. The inc cond algorithm can be implemented either with fixed step size or with variable step size to track the MPPT [11]. In this paper inc cond with fixed step is used to track the maximum power point.

We need a platform to run our implemented models to run in real time. So, here in this paper we used a RTDS simulator to develop the real time hardware. The RTDS simulator is able to run the hardware model up to micro seconds interval [12]–[14]. In this paper OPAL-RT digital simulator is used to check the outputs in real time.

Second section of this article shows the modeling of a PV cell with simulation results. The third Section is devoted to the DC DC boost converter with state space analysis. Fourth Section is dedicated for control algorithms to track the MPPT. The fifth section analyzes about Real Time Digital Simulator (RTDS); the section VI presents comparison and discussion on simulation results and real time results. Finally section VII concludes the paper.

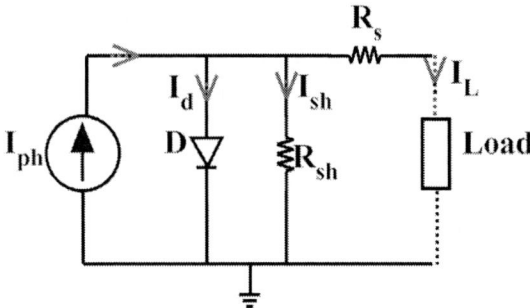

Fig. 1.   fundamental circuit of a PV cell

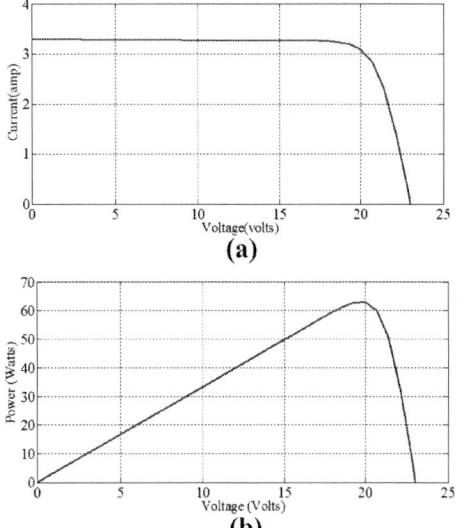

Fig. 2.   characteristics curves of a PV system at STC a) I-V and b) P-V

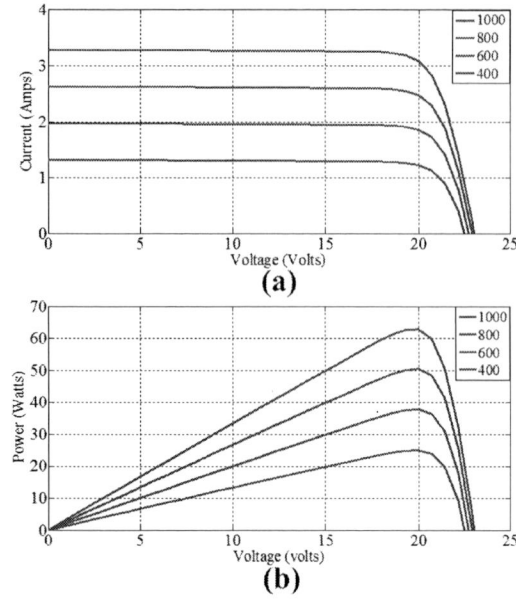

Fig. 3.   Characteristic curves of PV system at different irradiations a) I-V and b) P-V

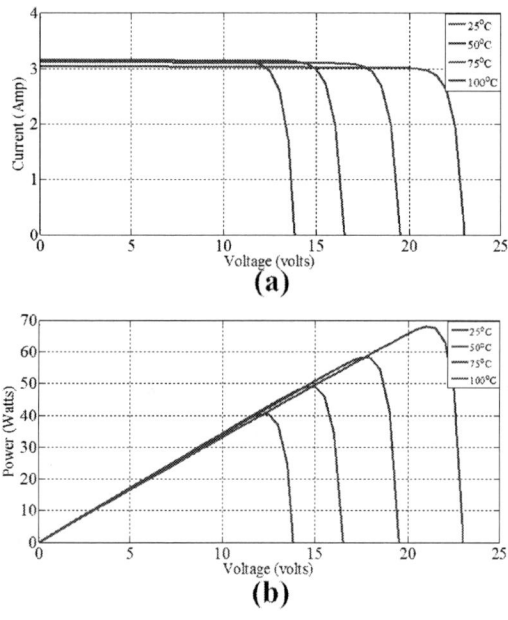

Fig. 4.   Characteristic curves of PV system at different temperatures a) I-V and b) P-V

II.   MODELLING OF A SINGLE DIODE PV CELL

PV is one of the methods of renewable energy generations, which means generating electrical power by converting solar irradiation into direct current. In this paper, single exponential model is preferred to model the PV cell because of simplicity and fewer complications [15]. The equivalent circuit diagram of a single diode PV cell [4] is shown in Fig. 1. From the theory of semiconductors the basic I-V characteristic equation of a PV cell is derived mathematically [4] as follows:

$$I_{ph} - I_0[e^{qV_d/nkT} - 1] = I_L \qquad (1)$$

Final expression of the PV cell drawn in Fig. 1 is computed and shown below:

$$V_L = \frac{nkT}{q}\left[\ln\left(\frac{I_{ph} - I_L}{I_0} + 1\right) - \frac{V_L + I_L R_S}{R_{sh}}\right] - I_L R_S \qquad (2)$$

After mathematical modeling of a PV cell is done, we simulated it in MATLAB/SIMULINK and obtained the I-V and P-V characteristic curves at Standard Test Conditions (STC). The simulated waveforms are shown in Fig. 2. The PV array is tested at STC as well as different irradiation levels and at different temperatures, and the simulation results are plotted in Fig. 3 Fig. 4.

978-1-4799-6047-7/14 $31.00 © 2014 IEEE

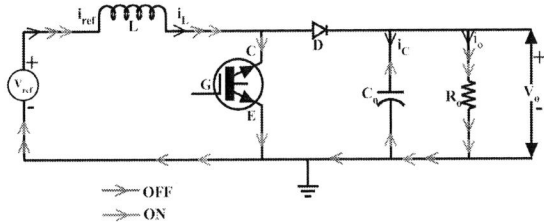

Fig. 5. Conventional Boost converter with directions for all switching actions

III. CONTROLLER BASED DC DC BOOST CONVERTER

The DC DC boost converter is a power electronics device is also called as a step up converter [6]. The conventional circuit of a boost converter is shown in Fig. 4 (the directions mentioned for all possible swithcing actions). It has a capability of providing the output voltage higher than the input voltage [5]. The equivalent circuit diagram of the boost converter is shown in Fig. 5, and energy flow is mentioned with green arrows, when the transistor is ON.

When the transistor is turned ON, the current flows from the supply to the inductor L, and the diode D is reverse biased. Hence the inductor stores the current, then inductor current rises, and the capacitor C maintains the voltage V_o and supplies current i_o. The state space modelling expressions of a boost converter [4] is as follows:

$$
\begin{bmatrix} \frac{dI_L}{dt} \\[2mm] \frac{dV_o}{dt} \end{bmatrix} = \begin{bmatrix} 0 & 0 \\[2mm] 0 & -\frac{1}{C_o R_o} \end{bmatrix} \begin{bmatrix} I_L \\[2mm] V_o \end{bmatrix} + \begin{bmatrix} \frac{1}{L} \\[2mm] 0 \end{bmatrix} V_{ref}
$$
(3)

$$
V_o = \begin{bmatrix} 0 & -\frac{1}{C_o R_o} \end{bmatrix} \begin{bmatrix} I_L \\[2mm] V_o \end{bmatrix}
$$
(4)

When the transistor is turned OFF, the inductor generates a large voltage to maintain the current i_L, and the diode D is starts conducting. The equivalent circuit diagram of boost converter when transistor turns OFF, is shown in Fig. 5.and energy flow is mentioned with red arrows, when the transistor is OFF. Hence the output voltage can be expressed as

$$
V_o = V_{ref} + L \frac{di_L}{dt}
$$
(5)

Thus the output voltage of the converter is higher than the supply voltage V_s, at this situation the capacitor C charges to the boosted voltage. The inductor

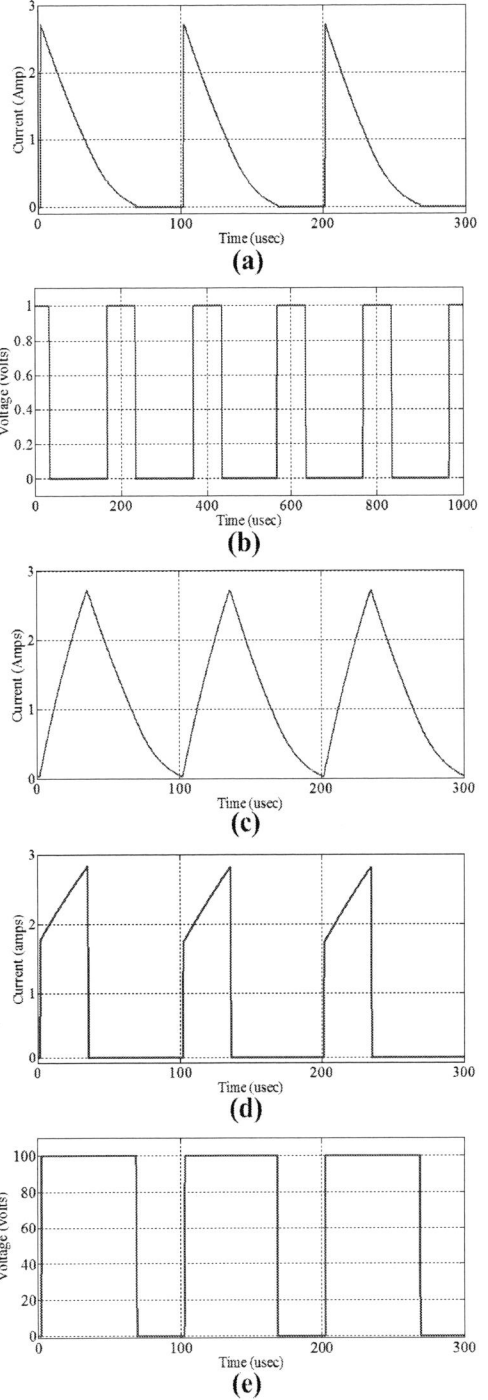

Fig. 6. Results of a boost converter a) Diode current, b) Gate voltage, c) Inductor Current, d) Switch current and e) Switch voltage

and power supply provides energy to the load when the transistor is turned OFF. The current through the inductor decreases because its stored energy goes on reducing. After some time the transistor is again turned ON and the cycle repeats. The mathematical model of

978-1-4799-6047-7/14 $31.00 © 2014 IEEE   273

a boost converter when the transistor is OFF can be simply expressed in state space as

$$
\begin{bmatrix} \frac{dI_L}{dt} \\ \frac{dV_0}{dt} \end{bmatrix} = \begin{bmatrix} 0 & -\frac{1}{L} \\ \frac{1}{C_o} & -\frac{1}{C_o R_o} \end{bmatrix} \begin{bmatrix} I_L \\ V_0 \end{bmatrix} + \begin{bmatrix} I_L \\ 0 \end{bmatrix} [V_{ref}] \quad (6)
$$

$$
V_o = \begin{bmatrix} \frac{1}{C_o} & -\frac{1}{C_o R_o} \end{bmatrix} \begin{bmatrix} I_L \\ V_o \end{bmatrix} \quad (7)
$$

The results of a converter is shown in Fig. 6. We can analyze all the possible switching actions with the help of these results.

IV. CONTROL TECHNIQUES TO TRACK MPPT

A. Modified Inc Cond MPPT Controller

Among all other MPPT techniques Inc Cond MPPT controller with fixed step size tracks MPP accurately and tracking speed is high [4], [8]. The controller measures the incremental changes in PV current and voltage to predict the effect of voltage change. This method utilizes incremental conductance (dI/dV) of a PV array to compute the sign of change in power with respect to voltage (dP/dV). Inc Cond MPPT controller algorithm is shown in Fig. 7. The controller compares incremental conductance (I/V) to array conductance (I/V). If the difference between both the conductances is zero then we can call that voltage as MPP voltage, and the corresponding current as MPP current [11]. Inc Cond controller maintains the MPP voltage until the irradiation changes and the process is repeated. [4], [11].

The slope of the PV curve at MPP is equal to zero.

$$
\frac{dP}{dV} = 0 \quad (8)
$$

Mathematical calculations based on Eq. (8), finally

$$
I + V\frac{dI}{dV} = 0 \quad (9)
$$

Based on Eq. (8) and Eq. (9) if any small error occurs then Eq. (10) becomes

$$
I + V\frac{dI}{dV} = e \quad (10)
$$

B. Conventional PI Controller

A conventional PI controller is a feedback mechanism, extensively used in industrial control applications. A PI controller calculates the "error", as the difference between a reference point and a measured process variable [4]. This error can be minimized in few attempts by adjusting the process control inputs. For MPP tracking

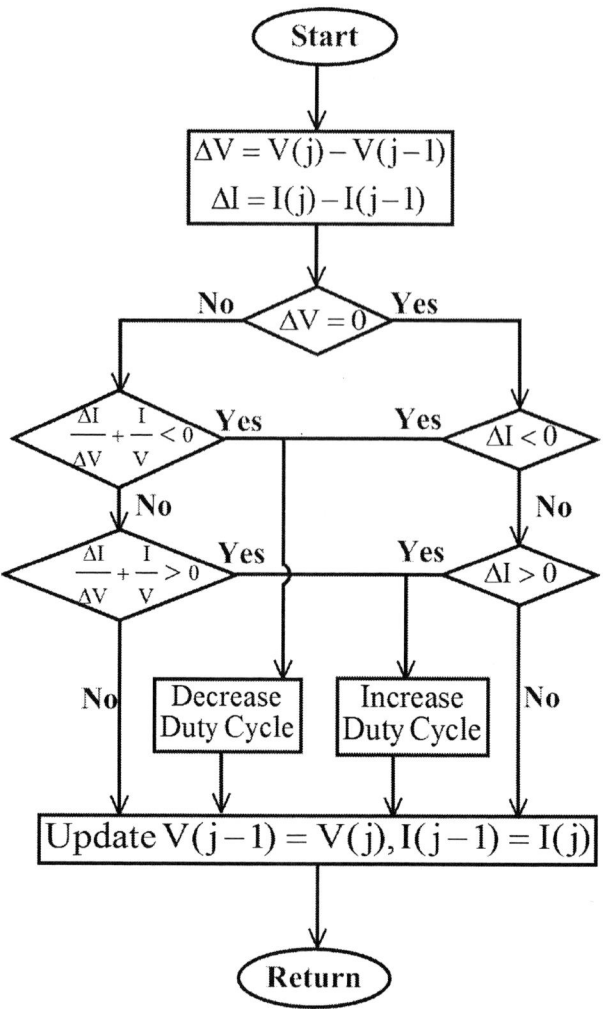

Fig. 7. Flowchart of Inc Cond MPPT controller

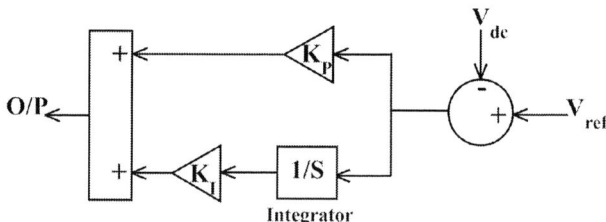

Fig. 8. PI controller

of a PV system, a PI controller is used because of its simple operation, past success record, ease of design and use, low cost and maintenance. MATLAB/SIMULINK PI controller model is shown in Fig. 8.

The algorithm calculation involves two constant parameters, the proportional, and the integral values denoted as K_P and K_I. P is present error; I is the accumulation of past errors based on current rate of change. The weighted sum of these two actions is used

978-1-4799-6047-7/14 $31.00 © 2014 IEEE

to adjust the process via a control element.

V. REAL TIME DIGITAL SIMULATOR (RTDS)

The RTDS allows the users to simulate the electrical power system models with precision and efficiently [16]. The RTDS simulator not only runs in real time but also tests the physical protection and control structure of the model. This gives the users to test and prove their archetypes and final products in a realistic and practical environment [17]. The RTDS is a digital power system simulator capable of continuous real time operation and it performs simulation with a classical time step of 50 microseconds [12]. The RTDS simulator meets the dead line within the specified time for all simulations. It is a perfect tool for the design, development and testing of power system protection and control schemes with a large capacity for both digital and analog signal exchange to interact with the simulated power system. The RTDS consists of two main tools, they are 1)real-time distributed simulation package (RT-LAB) for the execution of simulink block diagrams on a PC-cluster, and 2)algorithmic toolboxes designed for the fixed-time-step simulation of electric circuits and their controllers [12]. Real-time simulation and Hardware-In-the-Loop (HIL) In-the-Loop (HIL) applications are increasingly recognised as essential tools for engineering design and especially in power electronics and electrical systems. The mathematical computations for power system components and network equations are performed by using either a Wanda 3U module or Wanda 4U module based on Opal-RT simulation system. It communicateswith the target PC via a PCI-Express ultra-low-latency real-time bus interface.

VI. COMPARISON AND DISCUSSION ON SIMULATION AND RTDS RESULTS

The output of a PI is given to a comparator to compare with the pulse width modulation (PWM) signal. The comparator output is fed to gate terminal of transistor in the converter. Heuristic values of K_P and K_I are 0.9 and 203 respectively to get undistorted output.The reference voltage given to the PI controller is 250V, and outputs are collected at the load. The output voltage exactly reached to 250V, current 2.1A and power to 521W.

In the modified inccond controller, the error chosen on the basis of hit and trial is 0.0024. According to the MPPT algorithm in the flowchart, duty cycle is calculated. Compared output waveforms of a) voltage, b) current and c) power of the two controllers are shown in Fig. 9 and RTDS results are shown in Fig. 10 and Fig. 11 respectively.

The electrical parameters are tabulated in Table I, readings are taken from the resultant curves shown in Fig. 2. The MPPT techniques are compared in Table II. The parametric values of PI and inc cond are collected from resultant curves shown in Fig. 9.

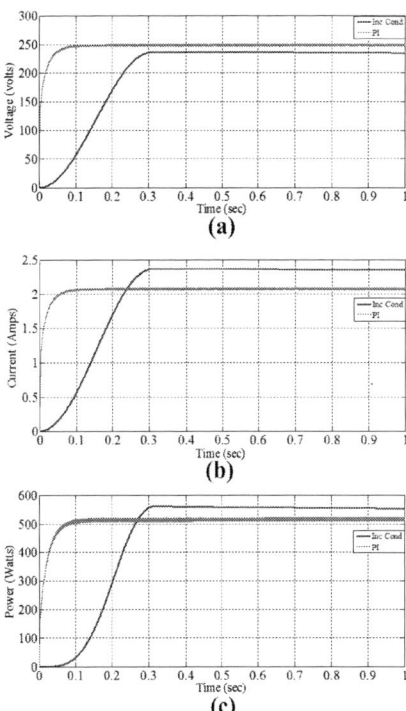

Fig. 9. Inc cond MPPT and PI controller Outputs comparison with a Boost converter (a) Voltage (b) Current and (c) Power

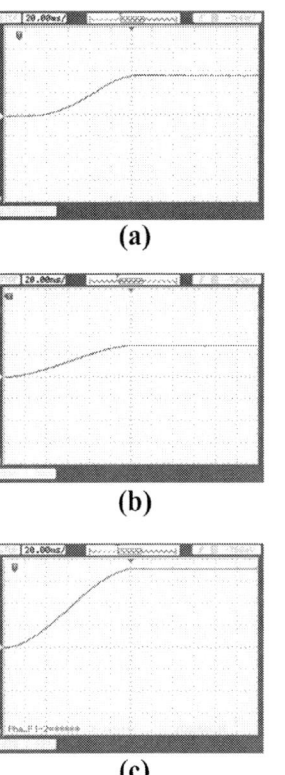

Fig. 10. RTDS results of a modified inc cond MPPT controller a) voltage, b) current and c)power

978-1-4799-6047-7/14 $31.00 © 2014 IEEE   275

TABLE I.   ELECTRICAL PARAMETERS OF PV ARRAY

Maximum power (P_{max})	56W
Voltage at MPP (V_{MPP})	18V
Current at MPP (I_{MPP})	3.13A
Open circuit voltage (V_{oc})	21.9V
Short circuit current (I_{SC})	3.46A

TABLE II.   COMPARISON OF MPPT TECHNIQUES

	PI MPPT	Inc Cond MPPT
Output voltage	250v	235v
Output current	2.1A	2.4A
Output power	521W	563W

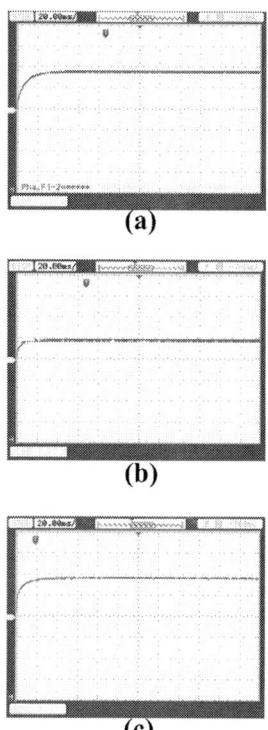

(a)

(b)

(c)

Fig. 11.   RTDS results of a PI controller a) voltage, b) current and c)power

VII.   CONCLUSIONS

In this paper, a PV cell with series and shunt resistances is modeled. PV array is arranged by series and parallel combinations of PV cells to extract the parameters like V_{oc} and I_{sc} etc., and plotted the characteristic waveforms at STC, different irradiations and different temperatures. The PV output is connected to a DC DC boost converter for voltage regulation. Control algorithms like modified inc cond and PI are developed to track maximum power from PV system; and observed that the modified inc cond algorithm tracked 7.4% more power than the conventional PI controller. RTDS simulator is used to develop the real time hardware for the developed simulink models. Simulation and real time results were plotted and compared.

REFERENCES

[1]   C. Wu, F. Jiang, Q. Wang, and H. Hu, "The maximum power point tracking algorithm for photovoltaic power system based on fuzzy logic double loop control," in *Electronics, Communications and Control (ICECC), 2011 International Conference on*, pp. 1130–1133, IEEE, 2011.

[2]   M. G. Villalva and J. R. Gazoli, "Comprehensive approach to modeling and simulation of photovoltaic arrays," *IEEE Trans. Power Electronics*, vol. 24, no. 5, pp. 1198–1208, 2009.

[3]   R.-J. Wai, W.-H. Wang, and C.-Y. Lin, "High-performance stand-alone photovoltaic generation system," *IEEE Tran. Ind. Electron.*, vol. 55, no. 1, pp. 240–250, 2008.

[4]   V. R. Kolluru, K. Mahapatra, and B. Subudhi, "Development and implementation of control algorithms for a photovoltaic system," in *Students Conference on Engineering and Systems (SCES'13)*, pp. 1–5, IEEE, 2013.

[5]   A. Mousavi, P. Das, and G. Moschopoulos, "A comparative study of a new zcs dc–dc full-bridge boost converter with a zvs active-clamp converter," , *IEEE Trans. Power Electronics*, vol. 27, no. 3, pp. 1347–1358, 2012.

[6]   J.-H. Su, J.-J. Chen, and D.-S. Wu, "Learning feedback controller design of switching converters via matlab/simulink," *IEEE Trans. Education*, vol. 45, no. 4, pp. 307–315, 2002.

[7]   D. Sera, T. Kerekes, R. Teodorescu, and F. Blaabjerg, "Improved mppt algorithms for rapidly changing environmental conditions," in *Power Electronics and Motion Control Conference, 2006. EPE-PEMC 2006. 12th International*, pp. 1614–1619, IEEE, 2006.

[8]   B. Subudhi and R. Pradhan, "A comparative study on maximum power point tracking techniques for photovoltaic power systems," *IEEE Trans. Sustain. Energy*, vol. 4, no. 1, pp. 89–98, 2013.

[9]   Y. Levron and D. Shmilovitz, "Maximum power point tracking employing sliding mode control," *IEEE Trans. Circuits and Systems*, vol. 60, no. 3, pp. 724–732, 2013.

[10]   T. Esram and P. L. Chapman, "Comparison of photovoltaic array maximum power point tracking techniques," *IEEE Trans. Energy Conversion*, vol. 22, no. 2, p. 439, 2007.

[11]   A. Safari and S. Mekhilef, "Simulation and hardware implementation of incremental conductance mppt with direct control method using cuk converter," *IEEE Trans. Ind. Electron.*, vol. 58, no. 4, pp. 1154–1161, 2011.

[12]   "Rt-lab professional. http://www.opal-rt.com/product/rt-lab-professional."

[13]   W. Ren, M. Sloderbeck, M. Steurer, V. Dinavahi, T. Noda, S. Filizadeh, A. Chevrefils, M. Matar, R. Iravani, C. Dufour, et al., "Interfacing issues in real-time digital simulators," *IEEE Trans. Power Delivery*, vol. 26, no. 2, pp. 1221–1230, 2011.

[14]   L.-F. Pak, M. O. Faruque, X. Nie, and V. Dinavahi, "A versatile cluster-based real-time digital simulator for power engineering research," *IEEE Trans. Power Systems*, vol. 21, no. 2, pp. 455–465, 2006.

[15]   A. N. Celik and N. Acikgoz, "Modelling and experimental verification of the operating current of mono-crystalline photovoltaic modules using four-and five-parameter models," *Applied energy*, vol. 84, no. 1, pp. 1–15, 2007.

[16]   T. Logenthiran, D. Srinivasan, A. M. Khambadkone, and H. N. Aung, "Multiagent system for real-time operation of a microgrid in real-time digital simulator," *IEEE Trans. Smart Grid*, vol. 3, no. 2, pp. 925–933, 2012.

[17]   R. Meka, M. Sloderbeck, M. O. Faruque, J. Langston, M. Steurer, and L. DeBrunner, "Fpga model of a high-frequency power electronic converter in an rtds power system co-simulation," in *Electric Ship Technologies Symposium (ESTS), 2013 IEEE*, pp. 71–75, IEEE, 2013.

978-1-4799-6047-7/14 $31.00 © 2014 IEEE

Real–Time Simulation of Photovoltaic Panel on Miniature Full Spectrum Simulator

S . Ritika, Arjun Yadav, G. Narayanan
Department of Electrical Engineering
Indian Institute of Science, Bangalore 560012 INDIA
email:ritika.sriram11@gmail.com; 09.arjun@gmail.com; gnar@ee.iisc.ernet.in

Abstract—**Real-time simulation of photovoltaic (PV) panel can be used for research on power electronic converter topologies and control for PV applications. This paper presents real-time simulation of PV panel using a miniature full spectrum simulator (mini-FSS), developed for educational purposes. A mathematical model of the PV panel is solved using Newton Raphson method on the real-time simulator. This can provide the output voltage and current of the PV panel for any load resistance at a given irradiation and temperature, consuming less than 50μs. By repeatedly solving the model for different values of resistances, the IV and PV characteristics of a commercially available PV panel (i.e. KT200C) are obtained for different values of temperature and irradiation. The real-time simulation results are validated with theoretical results based on calculations.**

I. INTRODUCTION

To use solar power, there is need of power electronics interfaces. The interfaces are required to bring the output of the PV panel to voltages compatible to grid/load being connected. Typical interface includes DC-DC converter [1] and inverter [2]. The PV panel gives maximum efficiency at a particular operating point at given physical conditions. To operate the PV panel optimally, maximum power point tracker is required [3].

The power electronic interfaces need to be tested for performance as per standards [4]. Since PV panel output is dependent on the prevailing weather conditions, testing of power electronic converters throughout the year might not be possible. Moreover, standard testing conditions are hard to obtain. Thus there is a need for a model that can emulate photovoltaic characteristic in real-time so that hardware in loop testing of converters is feasible.

Real time simulation behaves as the system being simulated since the simulation time and the physical time (i.e response time of the system) are ensured to be same. Real time simulation of different power electronic converters have been reported in recent years [5], [6] [7]. This paper presents the real-time simulation of PV panel based on a model presented in [8]. It allows the analysis of the system in real-time. Real time simulation of photovoltaic [9] can be cost effective as it eliminates the need of the presence of costly photovoltaic

This work was supported by the Department of Electronics and Information Technology, Government of India, under a project titled "Development and deployment of miniature Full Spectrum Simulator (FSS) in educational institutes" under the National Mission on Power Electronics Technology Phase II.

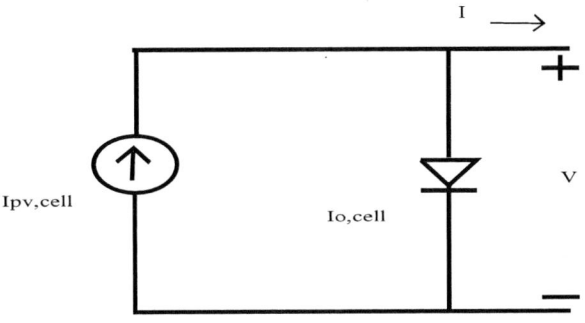

Fig. 1. An idealized model of a PV cell

panels as the same system can be used for the simulation of any panel at any physical condition.

II. MATHEMATICAL MODEL OF PHOTOVOLTAIC PANEL

A mathematical model of PV panel is discussed in this section.

A. Idealized model of a PV cell:

A PV cell can be represented as an ideal current source with a diode in parallel to it as shown in Fig. 1. The light generated current $I_{pv,cell}$ and saturation current $I_{o,cell}$ of the diode are related by (1)

$$I = I_{pv,cell} - I_{o,cell}[e^{qV/akT} - 1] \qquad (1)$$

where q is the charge of an electron, a is diode ideality factor, k is Boltzmann's constant, T is absolute temperature, and V is output voltage .

B. Practical PV panel model

A practical model of PV panel is shown in Fig. 2 and represented by (2) where V_t is N_skT/q, N_s being the number of cells in series.

$$I = I_{pv} - I_o[e^{(V+IR_s)/V_t} - 1] - (V + IR_s)/R_p \qquad (2)$$

The dependence of PV output on temperature (T) and irradiation (G) is modelled by (3) and (4) , where I_{on}, I_{pvn},

978-1-4799-6047-7/14 $31.00 © 2014 IEEE

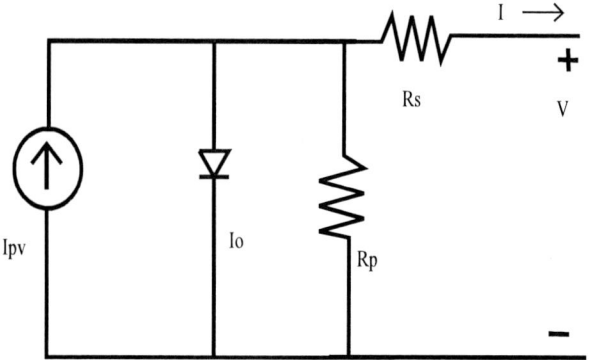

Fig. 2.   Practical model of a PV panel

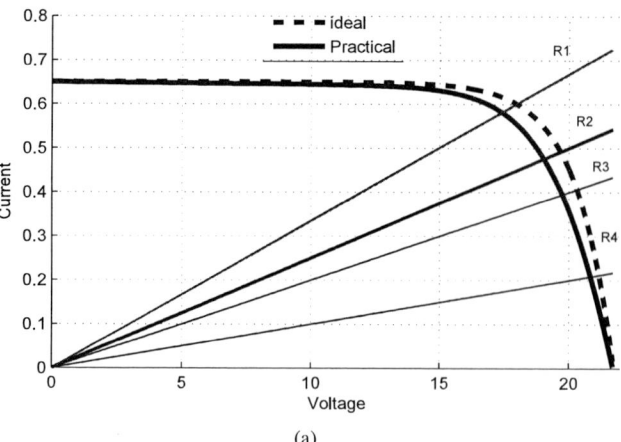

(a)

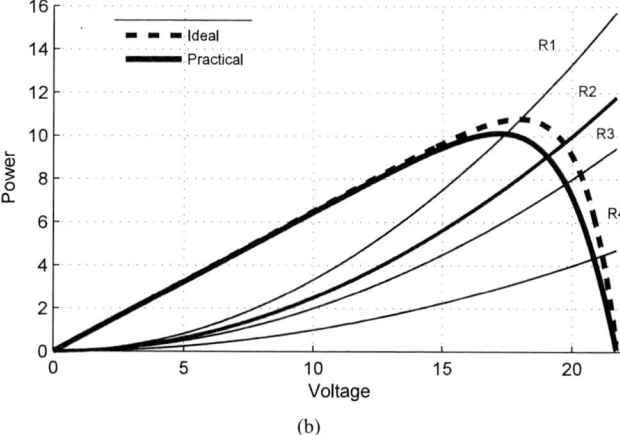

(b)

Fig. 3.   Ideal and practical (a) IV characteristic (b) PV characteristic. R1<R2<R3<R4

G_n, T_n are parameters at nominal conditions (i.e. 1000W/m^2 and 25° C), and K_i is the temperature coefficient.

$$I_o = I_{on}(T/T_n)^3 e^{qE_g(1/T_n - 1/T)/ak} \qquad (3)$$

$$I_{pv} = I_{pvn}(1 + K_i\delta T)G/G_n \qquad (4)$$

In a practical PV panel, many cells are interconnected in series and parallel. A practical PV panel can be modelled by inclusion of series resistance R_s and parallel resistance R_p [10]. Usually, the value of R_s is low and that of R_p is very high. References [11] and [8] describe a method to obtain the values of R_s, R_p, a and $I_{o,n}$ for any PV panel from the technical data provided in the datasheet. The ideal and practical IV characteristics are illustrated in Fig.3(a) while the ideal and practical PV characteristics are illustrated in Fig.3(b). The ideal and practical characteristics mostly differ near the maximum power point which is a crucial point. This necessitates the inclusion of non idealities.

For N_s cells connected in series and N_p cells in parallel, the equivalent series and parallel resistances become N_sR_s and R_p/N_p, respectively. V_{oc} increases proportional to N_s, while I_{pv} is proportional to N_p.

III. NUMERICAL SOLUTION OF MODEL OF PV PANEL

This section describes numerical solution of PV panel using Newton Raphson method.

A. Solution using Newton Raphson method

For static analysis of PV, the load can be seen as an equivalent resistance as shown by (5). The system can be mathematically modelled by (2) to (5). To obtain the operating point as shown in Fig. 3(a) and Fig. 3(b), the intersection of (2) and (5) needs to be evaluated. For static analysis of PV panel with a resistive load, the resistance can be modelled as (5). As seen in Fig.4, irradiance and temperature are environmental inputs. The currents I_{pv} and I_o for the input environmental conditions are obtained from (3) and (4). For load resistance R and parameters R_p, R_s and a, the output voltage and current can be obtained by solving (2).

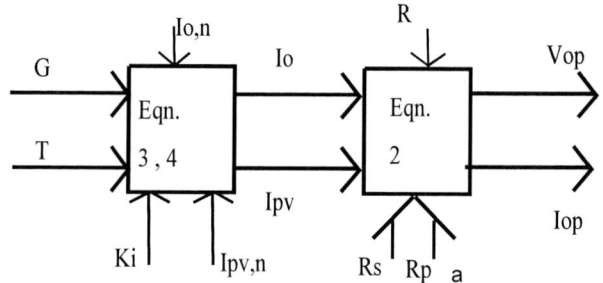

Fig. 4.   Block representation of PV model

The equation for PV panel is a transcendental equation which is solved using one of the methods of tangents. Among these methods, Newton Raphson (NR) method is used here [12] . To obtain the terminal voltage of the PV panel for a given load resistance R_x equations 6 to 8 are solved in an iterative solution. To find a solution to eqn f(V)=0, approximation of root V at the n+1th iteration is found by successive evaluation of (6), where f$'(V_n)$ is the derivative of f(V_n). For the PV model in consideration, (7), obtained from (2) and (5), needs to be solved.

$$V = IR \qquad (5)$$

$$V_{n+1} = V_n - f(V_n)/f'(V_n) \qquad (6)$$

978-1-4799-6047-7/14 $31.00 © 2014 IEEE   278

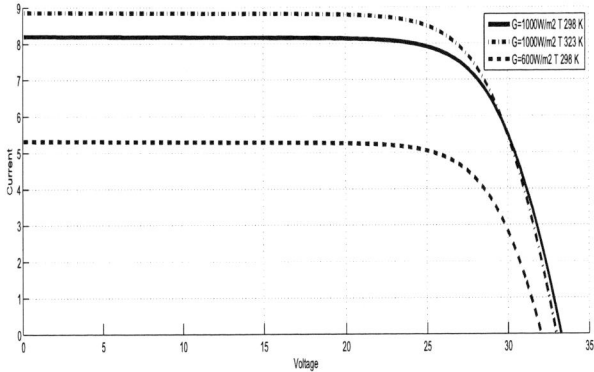

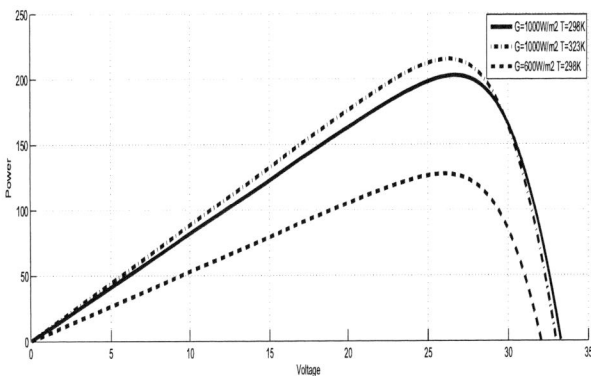

Fig. 6.   IV Characteristics for different irradiance and temperature levels

Fig. 7.   PV Characteristics for different temperature and irradiance levels

$$f(V_n) = I_{pv} - I_o[e^{(V_n+IR_s)/V_t} - 1] - (V_n + IR_s)/R_p - V_n/R \quad (7)$$

$$f'(V_n) = -I_o[e^{(V_n+IR_s)/V_t}]/V_t - (1)/R_p - 1/R \quad (8)$$

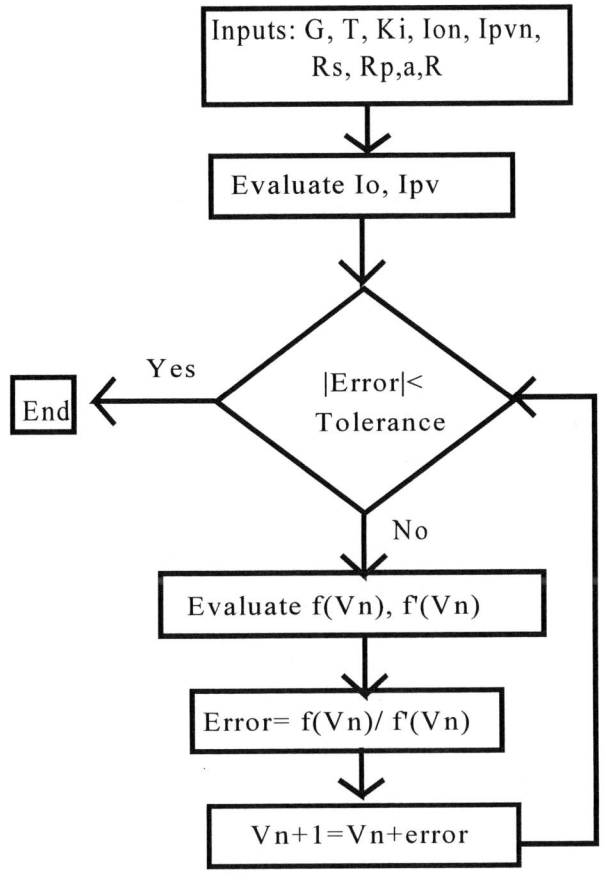

Fig. 5.   Flowchart for algorithm used

B. Theoretical I-V and P-V characteristics

The specifications of a 32V, 200W, PV panel (i.e. KT200C) are given in Tables II and III. With these specifications equations (2) to (8) are solved in MATLAB. The I-V characteristics are shown plotted in Fig. 6 for G=1000W/m² and T=298K, G=1000W/m² and T=323K, and G=600W/m² and T=298K.

IV. DIVISION ALGORITHM

The prerequisite for NR method is the ability to perform division as seen from the flowchart in Fig. 5. The DSP process is coded in assembly language here. The processor does not provide assembly level instruction for division. Division can be performed by multiplying the dividend ($f(V_n)$) with the reciprocal of divisor ($f'(V_n)$). The divisor is represented by D, and the quotient by y.

A. Division methods

Three methods can be used:

a) Division by successive subtraction: This is the commonly used method, but the execution time depends on value and accuracy required. The speed is instruction specific. This method is faster if the processor has supportive assembly instructions

b) Taylor series expansion of the inverse of $(1 + z)$ is: For $0 < |z| < 1$,

$$y = 1/(1 + z), y = 1 - z + z^2 - z^3 + \ldots\ldots; \quad (9)$$

The accuracy depends on the order to which y is expanded. But this does not converge for a band of z values.

c) Evaluation of inverse by numerical method: To evaluate the inverse of D as given by (10), Newton's method is used as given by (11) to (14).

$$y = 1/D \quad (10)$$

$$F(y) = D - 1/y \quad (11)$$

$$F'(y) = -1/y^2 \quad (12)$$

$$y_{n+1} = y_n - F(y_n)/F'(y_n) \quad (13)$$

$$y(n + 1) = y(n)[2 - Dy(n)] \quad (14)$$

Fig 8 shows the comparison of the last two methods for a value of 1.99. It is seen that Taylor series expansion has a convergence problem. Hence only division by subtraction and Newton Raphson method are considered for further study.

978-1-4799-6047-7/14 $31.00 © 2014 IEEE   279

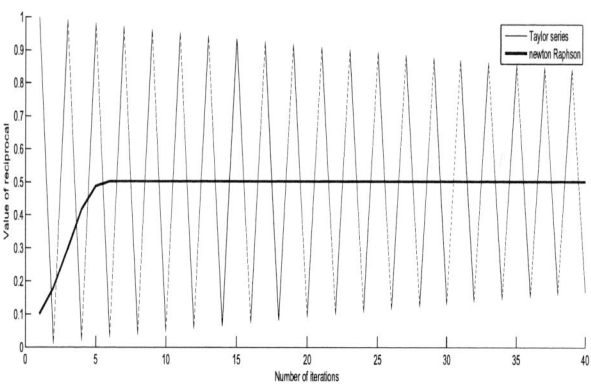

Fig. 8. Convergence of Newton Raphson and Taylor series

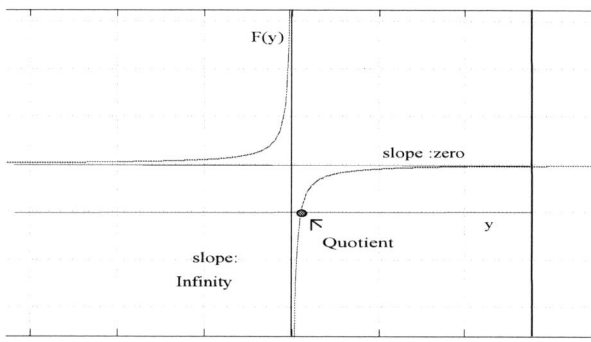

Fig. 9. Dependence of convergence of NR method on initial guess

TABLE I
COMPARISON OF EXECUTION TIME FOR DIVISION ALGORITHMS

Divisor	Significant digit	Division by subtraction	Division by newton rapson
$7.0e^3$	4	$24\mu s$	$10\mu s$
$7.0e^1$	4	$20\mu s$	$10\mu s$
$7.0e^{15}$	4	$27\mu s$	$11\mu s$
$7.0e^{15}$	8	$34\mu s$	$11\mu s$

B. Selection of initial guess with NR method:

Fig. 9 shows a plot of $F(y)$ given by (11). The NR method requires the slope $F'(y)$ represented by 12. If the initial guess is very small, then the slope is very high, and the convergence is slow. On the other hand, if the initial guess is much greater than the reciprocal, $F'(y)$ is zero. Then $y(n+1)$, as calculated by (13), would be very high.

C. Experimental comparison of division algorithm

The division algorithms using division by successive subtraction and Newton Raphson are implemented on the MVC33 DSP board, discussed in the following section. The execution times for the two methods are compared for different values of divisors shown in the Table I. Considering an accuracy of four significant digits, NR method is found to be faster than division by subtraction. Also, for the accuracy of eight significant digits the NR method is found to be faster. Hence,

TABLE II
PARAMETERS OF KT200GT PV ARRAY AT 25° C AM 1.5 1000 W/M^2

Parameters	Values
I_{mp}	7.61 A
V_{mp}	26.3 V
P_{max}	200.143 W
I_{sc}	8.21A
V_{oc}	32.9 V
K_v	-0.1230 V/K
K_i	0.0032 A/K
N_s	54

TABLE III
ADJUSTED PARAMETERS FOR KT200GT FOR NOMINAL OPERATING
CONDITIONS OBTAINED FROM [8]

Parameters	Values
R_p	415.405Ω
R_s	0.221Ω
a	1.3
$I_{o,n}$	$9.82e^{-8}$

TABLE IV
COMPARISON OF OPERATING POINTS FOR DIFFERENT RESISTANCES AT
STANDARD CONDITIONS

Res.(Ω)	Theo. Vol.(V)	Theo. Cur.(A)	Exp Vol.(V)	Exp Cur.(A)
1	8.1899	8.1899	8.4	8.4
4	28.56	7.14	28	7.05
10	32.331	3.2331	32	3.2

the NR method is chosen here.

V. REAL-TIME SIMULATION OF PV PANEL

A. Experimental setup

The algorithms discussed were implemented in miniature full spectrum simulator (mini-FSS), developed under the National Mission on Power Electronice Technology (NaMPET). The mini-FSS has an inter rack communication card (IRCC), an system interface card (SIC), and three boards with three MVC33 32 bit floating point DSP processors each, all of which operating at a clock frequency of 75 MHz. Only the IRCC was used for the real-time simulation. The coding was done in assembly language as mentioned earlier.

B. Experimental results

The results of real-time simulation under steady and dynamic conditions are shown. The IV and PV characteristics obtained for the PV panel are shown below.

1) Steady state response of the simulator: The simulation is run for specific values of resistances. The corresponding stable operating points are obtained. The execution time is found to be in the range of 30-50 μs. This can further be reduced if a faster division algorithm can be implemented.

978-1-4799-6047-7/14 $31.00 © 2014 IEEE

2) Dynamic response of the simulator: The dynamic response of the simulator to variation in irradiation and to resistance was analyzed.

a) Variation in resistance: When the load resistance was changed at discrete steps of 8 Ω, 5 Ω, 2 Ω, and the corresponding variations in output current (trace 2) and voltage (trace 1) are as shown in Fig.10.

b) Variation in irradiance: The irradiance (trace 1) was changed from 1000W/m2 to 200W/m2 at a steady rate over an interval of 500μs as shown in Fig.11. The corresponding variation in output current (trace 2) is as shown in Fig.11. The simulator is found to simulate any change in resistance and irradiance values within 50μs.

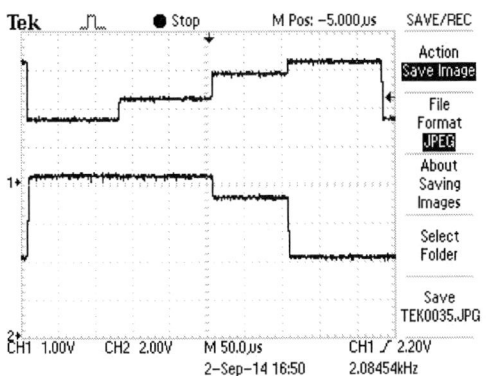

Fig. 10. Dynamic response for changing resistance Ch.1 Voltage Ch.2 Current Scale: Ch1. 1 unit: 10V Ch2. 1unit: 1A

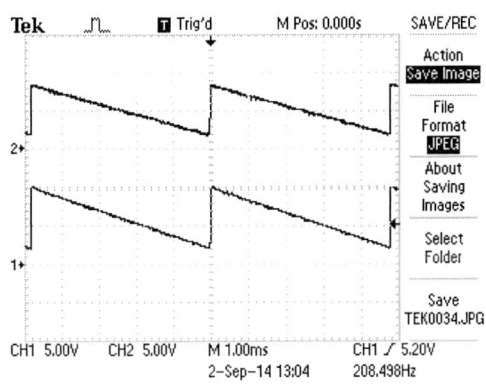

Fig. 11. Dynamic response for changing irradiance Ch.1 Irradiance Ch.2 Output current Scale: Ch1. 1 unit: 100 W/m^2 Ch2. 1unit: 1A

3) IV and PV characteristics : The current versus voltage and power versus voltage graphs are plotted for different irradiance and temperature by sweeping the resistance ranging between 1 Ω to 100 Ω for 100 different values. Fig. 12(a) and 12(b) show the IV and PV characteristic for the PV panel at nominal conditions with an irradiance of 1000 W/m^2 and a temperature of 298K. Similar results are presented in Fig.13(a) and 13(b) for an irradiance of 600 W/m^2 at the same temperature.

Fig. 14(a) and 14(b) show the IV and PV characteristics at an irradiance of 1000 W/m^2 and temperature 323K. The

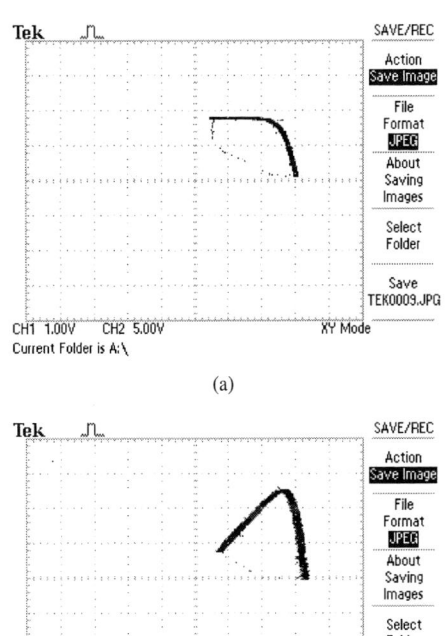

(a)

(b)

Fig. 12. (a)Experimentally obtained IV characteristic at G=1000W/m^2 T=298K (b)Experimentally obtained PV characteristic at G=1000W/m^2 T=298K Scale: Ch1. 1 unit: 10V (a)Ch2. 1unit: 1A (b)Ch2. 1unit: 40W

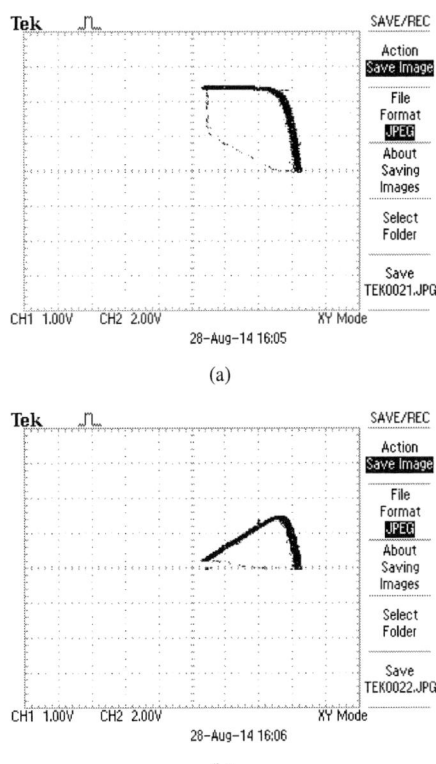

(a)

(b)

Fig. 13. (a) Experimentally obtained IV characteristic at G=600W/m^2 T=298K (b) Experimentally obtained PV characteristic at G=600W/m^2 T=298K Scale: Ch1. 1 unit: 10V (a) Ch2. 1unit: 0.5A (b) Ch2. 1unit: 40W

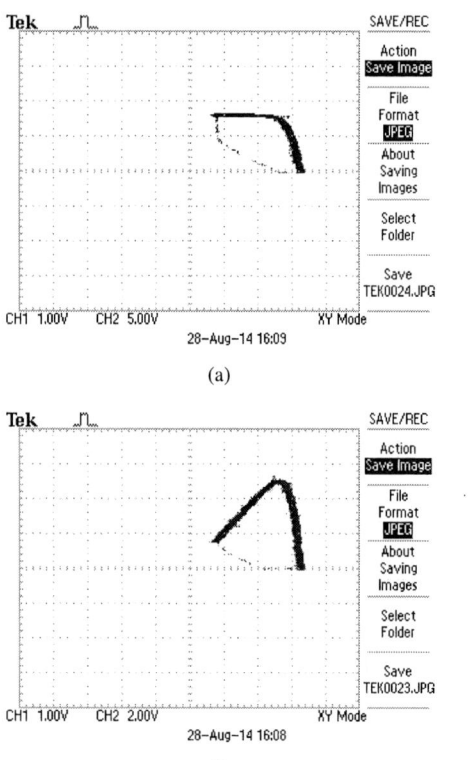

(a)

(b)

Fig. 14. (a) Experimentally obtained IV characteristic at G=1000W/m^2 T=323K (b) Experimentally obtained PV characteristic at G=1000W/m^2 T=323K Scale: Ch1. 1 unit: 10V (a) Ch2. 1unit: 1A (b) Ch2. 1unit: 40W

IV and PV characteristic, obtained from real-time simulation are found to be matching the values obtained from off-line simulation at the three environmental conditions mentioned above and as shown in Fig. 6 and Fig. 7. The real-time simulator is able to give the full IV and PV characteristic (100 points) for any particular environmental condition well within 5 ms. Thus this can be used to determine the maximum power point as well.

VI. CONCLUSION

This paper reports real-time simulation of a photovoltaic (PV) panel using miniature full spectrum simulator (mini FSS), which is developed for training and research on real-time simulation of power electronic converters in educational institutions. An equivalent circuit based mathematical model of the photovoltaic panel is reviewed. Environmental conditions, namely temperature and irradiation, are inputs to the model. The model can be solved to obtain the output current and voltage corresponding to any load resistance. Newton Raphson's method is used to solve the mathematical model including the division operation required for real-time solution on the mini FSS.

The results of real-time simulation of KT200C 32V 200W PV panel are presented for different operating conditions. The results from real-time simulations are verified with those based on theoretical calculations. The maximum execution time to solve the model for a particular load resistance at a given environmental condition is found to be within 50 μs.

The complete IV and PV characteristic of the panel for a given temperature and irradiance are found by solving the mathematical model repeatedly for 100 different values of load resistance, ranging between 1 Ω and 100 Ω. These characteristics are presented for a few different temperature and irradiation conditions. The total time taken is less than 5 ms to determine the complete IV and PV characteristics, considering 100 different values of load resistance. Thus the peak power point for a given envirnmental condition can be determined well within 5 ms. Hence this real-time simulation of PV panel can be used for the tracking of maximum power point as well.

REFERENCES

[1] A. Kwasinski, "Identification of feasible topologies for multiple-input dc–dc converters," *IEEE tran. Power Electron.*, vol. 24, no. 3, pp. 856–861, 2009.

[2] A. Sarwar and M. Jamil Asghar, "Multilevel converter topology for solar pv based grid-tie inverters," in *Energy Conference and Exhibition (EnergyCon), 2010 IEEE International*, Dec 2010, pp. 501–506.

[3] Y.-E. Wu, C.-L. Shen, and C.-Y. Wu, "Research and improvement of maximum power point tracking for photovoltaic systems," in *Power Electronics and Drive Systems, 2009. PEDS 2009. International Conference on*, Nov 2009, pp. 1308–1312.

[4] "The new european standard for performance characterization of pv inverters," *IEC Standard 50530*.

[5] J. Channegowda, B. Saritha, H. Chola, and G. Narayanan, "Comparative evaluation of switching and average models of a dc-dc boost converter for real-time simulation," in *Electronics, Computing and Communication Technologies (IEEE CONECCT), 2014 IEEE International Conference on*, Jan 2014, pp. 1–6.

[6] G. Parma and V. Dinavahi, "Real-time digital hardware simulation of power electronics and drives," *Power Delivery, IEEE Transactions on*, vol. 22, no. 2, pp. 1235–1246, April 2007.

[7] K. Jayalakshmi and V. Ramanarayanan, "Real-time simulation of electrical machines on fpga platform," in *Power Electronics, 2006. IICPE 2006. India International Conference on*, Dec 2006, pp. 259–263.

[8] M. Villalva, J. Gazoli, and E. Filho, "Modeling and circuit-based simulation of photovoltaic arrays," in *Power Electronics Conference, 2009. COBEP '09. Brazilian*, Sept 2009, pp. 1244–1254.

[9] R. Stala, "Testing of the grid-connected photovoltaic systems using fpga-based real-time model," in *Power Electronics and Motion Control Conference, 2008. EPE-PEMC 2008. 13th*, Sept 2008, pp. 1852–1858.

[10] L. Umanand, *Course on Non Conventional Energy Systems, IISC Bangalore*.

[11] A. Chatterjee, A. Keyhani, and D. Kapoor, "Identification of photovoltaic source models," *Energy Conversion, IEEE Transactions on*, vol. 26, no. 3, pp. 883–889, Sept 2011.

[12] M. B. Patil, *Circuit Simulation for Power Electronics*. Alpha Science International, Ltd, 2009.

Design and Development of Real-Time Small-Scale Wind Turbine Simulator

Jakeer Hussain
Student Member, IEEE
Department of Electrical Engineering
Indian Institute of Technology Madras
Chennai, India 600036
Email: jakeer.s.hussain@gmail.com

Mahesh K. Mishra
Senoir Member, IEEE
Department of Electrical Engineering
Indian Institute of Technology Madras
Chennai, India 600036
Email: mahesh@ee.iitm.ac.in

Abstract—**Application of power electronics in machine control facilitates emulation of practical wind turbine characteristics using electric machine. In this paper, to test and evaluate the performance of newly developed control schemes for the efficient operation of wind energy conversion system, an in-lab small-scale wind turbine simulator has been developed using a separately excited DC motor coupled to a permanent magnet synchronous generator. Steady state characteristics and dynamic behavior of the practical wind turbine are satisfactorily reproduced by the DC motor. Steady state characteristics like tip speed ratio versus coefficient of power, rotor speed versus turbine output power and dynamic response during furling mechanism at high wind velocities are emulated from the developed wind turbine simulator. The simulator system is tested in response to the wind profile data recorded in the field at the hub height of 18.5 meters. The developed wind turbine simulator can be used to represent wind energy conversion system in the process of the development of in-lab microgrid system.**

Index Terms—**Microgrid, Small-scale wind energy conversion system, DC motor, PMSG.**

I. INTRODUCTION

Increasing demand for electrical energy, depleting reserves of fossil fuel resources, and increasing concerns for environment protection, motivate research community to explore all possibilities for the efficient energy conversion from renewable energy sources. Wind is one of the most abundant renewable energy sources of energy in nature. Fixed speed wind energy conversion systems (FSWECS) extract energy from wind by rotating at constant speed for all wind velocities [1]. Variable speed wind energy conversion systems (VSWECS) harness more electrical energy than FSWECS by running at different speeds with respect to changing wind velocity [2], [3]. Interest in usage of PMSG in WECS is increasing in industry due to its compact dimensions, high air-gap flux density, high power density, high torque-to-inertia ratio and high torque capability. Moreover, compared with an induction generator (IG), PMSG has higher efficiency, due to the absence of rotor winding losses [4].

Energy produced by wind is intermittent. Variable nature of wind power has a direct impact on the financial benefits of

This work is supported by the Department of Science and Technology, India, under the project grant DST/TM/SERI/2k10/47(G).

Power Purchase Agreements (PPAs). As per the instructions given by grid system operator, wind turbine systems need to be operated either in maximum power extraction mode or in controlled power extraction mode or in power limitation mode. Efficient control schemes need to be developed to operate a wind turbine in different modes as per the commands given by utility system operator. Development and evaluation of novel control schemes for the efficient operation of WECS under dynamic system operating conditions is gaining more interest in research community. To evaluate the effectiveness of new control schemes under different wind conditions, an in-lab wind turbine simulator is more suitable than practical field wind turbine.

This paper describes the development process of a wind turbine simulator which emulates the behavior of a 1.2 kW WECS using a separately excited DC motor coupled with a 3-phase PMSG. DC motor emulates characteristics of the wind turbine under different wind conditions and imparts corresponding shaft power onto PMSG. The effectiveness in emulating the characteristics of a practical wind turbine under different operating conditions is validated through the observed experimental results in detail.

II. MODELING OF WIND TURBINE

Steady state characteristics of a typical 1.2 kW wind turbine with fixed pitch angle of zero degrees subjected to various wind conditions are shown in Fig. 1 [5]. Power extracted from wind by a wind turbine and corresponding shaft torque imparted onto generator shaft can be described mathematically by (1) and (2).

$$P_m = \frac{1}{2}\rho\pi R_{rotor}{}^2(V_w cos\theta)^3 C_p(\lambda,\beta) \qquad (1)$$

$$T_m = \frac{P_m}{\omega_g} \qquad (2)$$

where P_m is turbine output power in Watts, C_p is dimensionless coefficient of power, λ is tip speed ratio (TSR), β is pitch angle in degrees, ρ is air density in (kg/m^3), R_{rotor} is rotor blade radius in meters, V_w is wind speed in (m/s), θ is furl

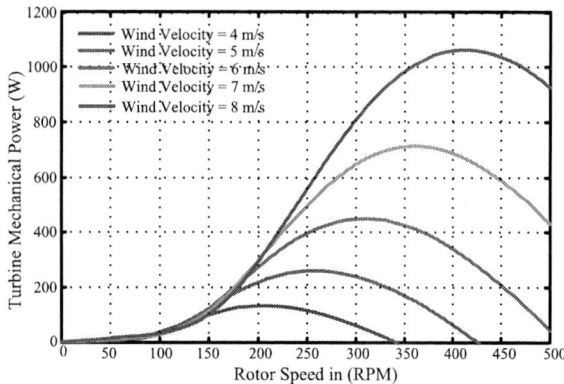

Fig. 1. Wind Turbine Speed and output power characteristics.

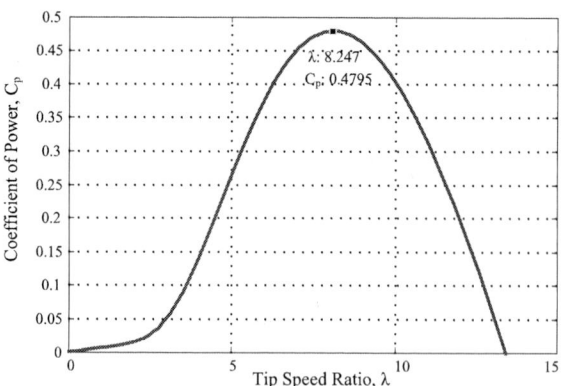

Fig. 2. Relation between C_p and λ at $\beta = 0$.

angle in degrees, and ω_g is rotor speed in (rad/s). The effective wind velocity at the rotor plane is $V_w\cos\theta$ [6].

In microgrid applications where distributed energy resources are located at near to load centers, available wind is turbulent in nature. Behavior of practical wind turbine under turbulent wind conditions also needs to be emulated from the wind turbine simulator. To limit the extracted power during wind gusts and strong winds, small-scale WECS employs passive protection furling mechanism. Furling mechanism turns the rotor by the angle θ degrees and this results in reduction in the extracted aerodynamic power [6]. Relation between wind speed and furling angle given by (3) as in [7] is used in this work.

$$\theta = -0.00017327V_w{}^5 + 0.0085008V_w{}^4 - 0.12034V_w{}^3 +$$
$$0.4501V_w{}^2 + 1.0592V_w + 0.3892. \tag{3}$$

Coefficient of power (C_p) of a wind turbine is the measure of its efficiency in the conversion of wind energy into useful mechanical energy and is given by (4).

$$C_p = \frac{Mechanical\ output\ power\ of\ turbine}{power\ contained\ in\ wind}. \tag{4}$$

C_p is function of λ and β. λ is given by (6).

$$\lambda = \frac{Tangential\ speed\ of\ the\ tip\ of\ the\ blade}{Speed\ of\ the\ wind} \tag{5}$$

$$\lambda = \frac{\omega_m R_{rotor}}{V_w \cos\theta}. \tag{6}$$

Coefficient of power (C_p) can be modeled based on the aerodynamic principles based blade design as given in (7) [8].

$$C_p = C_1[\frac{C_2}{\lambda_i} - C_3\beta - C_4]e^{-C_5/\lambda_i} + C_6\lambda \tag{7}$$

where $\frac{1}{\lambda_i} = \frac{1}{\lambda + 0.008\beta} - \frac{0.035}{\beta^2 + 1}$, and empirical constants, C_1=0.5176; C_2=116; C_3=0.4; C_4=5; C_5=21; C_6=0.0068. Usually most of the small-scale wind turbines are designed with constant pitch angle of zero and the relation between C_p and λ as per (7) is illustrated in Fig. 2.

III. MODELING OF PERMANENT MAGNET SYNCHRONOUS GENERATOR-DIODE BRIDGE RECTIFIER

PMSG is suitable choice as wind generator in small-scale WECS due to its high air-gap flux density, high torque capability, lower maintenance, and high efficiency due to the absence of rotor winding. Equivalent circuit for single phase of a non-salient pole PMSG is shown in Fig. 3. Per phase

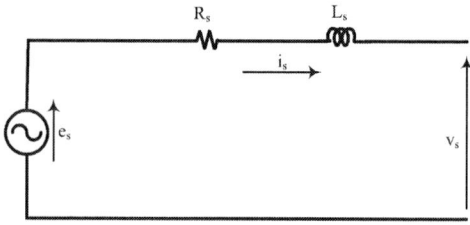

Fig. 3. Single phase equivalent circuit of PMSG.

induced emf, e_s, in stator winding when it is subjected to a constant flux per permanent magnet pole, ϕ, while rotating with a speed, ω_g, in (rad/s) is given by (8).

$$e_s = k\omega_g = k\frac{\omega_e}{P} \tag{8}$$

where k is machine emf constant in (V/rpm), P is total number of rotor pole pairs and ω_e is electrical angular frequency of PMSG stator induced voltage.

In steady state, per phase terminal voltage and three phase output power of PMSG can be calculated using (9) and (10).

$$V_s^2 = E_s^2 - (\omega_e L_s I_s)^2 \tag{9}$$

$$P_g = 3V_s I_s = 3\sqrt{E_s^2 I_s^2 - (\omega_e L_s)^2 I_s^4} \tag{10}$$

Output voltage of PMSG needs to be rectified due to its variable voltage and variable frequency nature because of the stochastic and intermittent variations in wind conditions. In small-scale wind energy conversion systems, PMSG-diode bridge combination, as shown in Fig. 4, is most popular due to its less complex topology. Relation between average of diode

978-1-4799-6047-7/14 $31.00 © 2014 IEEE

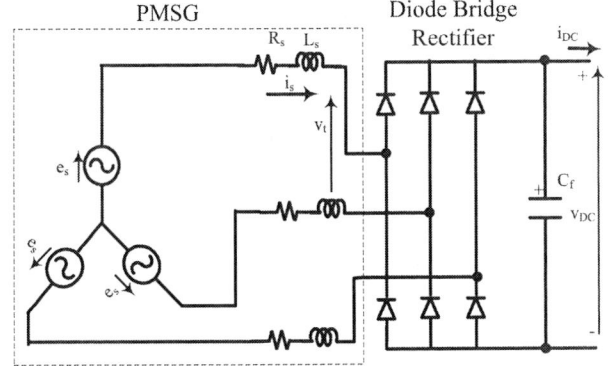

Fig. 4. PMSG-Diode bridge rectifier topology.

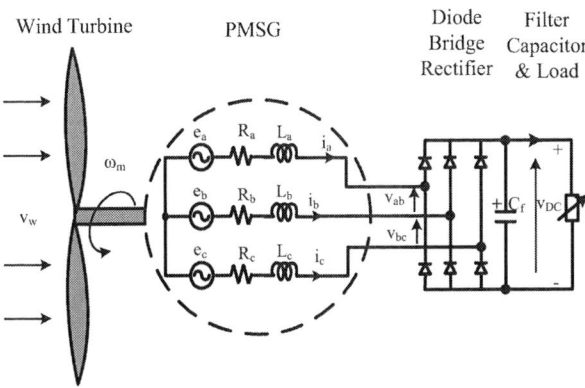

Fig. 5. WECS Configuration.

bridge output voltage and line voltage at terminals of PMSG can be expressed using (11) [9].

$$V_{DC} = \frac{3\sqrt{2}}{\pi}V_t = \frac{3\sqrt{6}}{\pi}V_s \qquad (11)$$

where V_t is RMS value of line-to-line voltage of PMSG. By assuming 100 % efficiency in rectification process, output power of WECS can be equated to output power of PMSG as in (12).

$$P_g = P_{DC} = 3V_sI_s = V_{DC}I_{DC} \qquad (12)$$

PMSG output power and electromagnetic torque can be expressed as function of diode bridge output current using (8)-(12), and are given in (13).

$$P_g = \frac{3\sqrt{6}}{\pi}\omega_g I_{DC}\sqrt{k^2 - \frac{6}{\pi^2}(PL_s)^2 I_{DC}^2}$$

$$T_g = \frac{3\sqrt{6}}{\pi}I_{DC}\sqrt{k^2 - \frac{6}{\pi^2}(PL_s)^2 I_{DC}^2} \qquad (13)$$

Based on (13), important conclusion which can be derived is by controlling diode bridge output current, load torque on wind turbine can be controlled. This principle is employed to extract maximum power by WECS under different wind velocities.

IV. Wind Turbine Simulator System Configuration

Basic system configuration which needs to be emulated to understand the behavior of a practical wind turbine is shown in Fig. 5. A H-bridge fed separately excited DC motor is used in this work to reproduce the wind turbine characteristics. Direction and speed of the DC motor is controlled by controlling the armature current flow, determined by the position of the switches in H-bridge [10]. TMS320F28335, a 32-bit floating point Digital Signal Controller (DSC) based torque control scheme for DC motor is shown in Fig. 6. Code Composure Studio (CCS) running host system feeds wind velocity, V_w, data in real-time continuously to the controller algorithm. Wind generator speed is measured continuously using incremental encoder and capture unit of DSC and

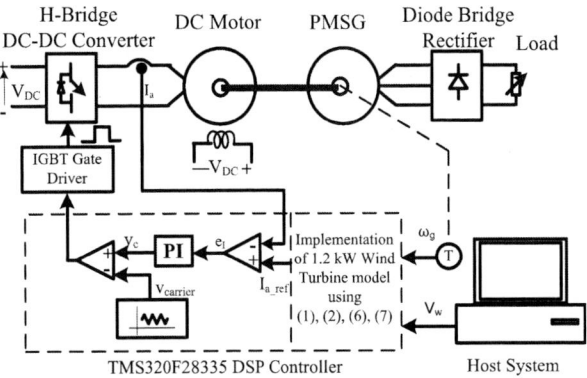

Fig. 6. Control scheme for DC motor.

the same is given as feedback to the controller algorithm. Sequential procedure to generate reference armature current of DC motor is described in Fig. 7. Tuning of digital PI controller gains is made using Ziegler-Nichols method for good initial guess and after several trial and errors a suitable fine-tuned set was determined. One-time inputs for the controller algorithm are tabulated in Table I.

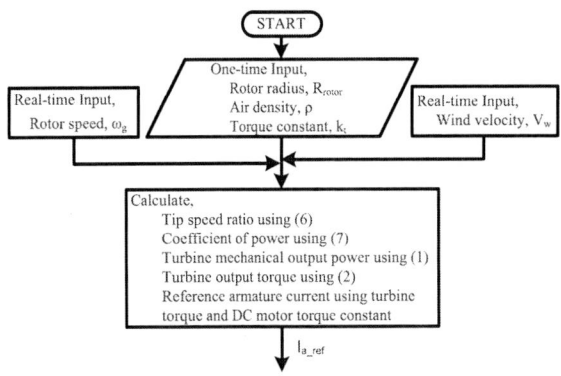

Fig. 7. Sequential computations for I_{a-ref} generation.

978-1-4799-6047-7/14 $31.00 © 2014 IEEE

TABLE I
CONTROLLER ALGORITHM INITIALIZATION PARAMETER VALUES.

Parameter	Value
R_{rotor} (m)	1.5
ρ (kg/m^3)	1.225
k_t (Nm/A)	1.5

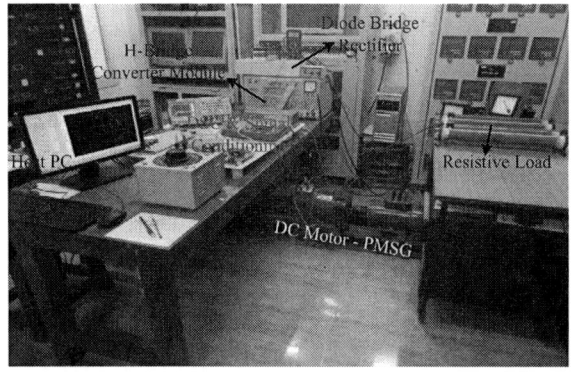

Fig. 8. In-lab hardware setup of wind turbine simulator.

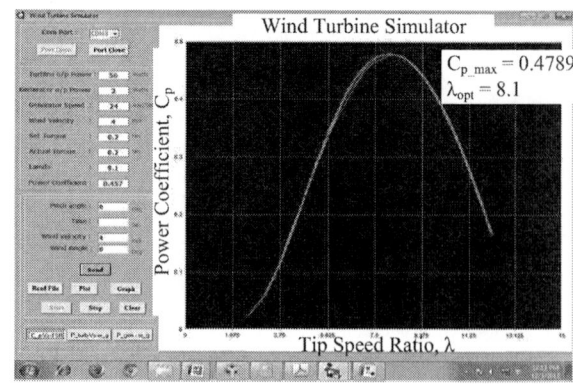

Fig. 9. Emulated relation between C_p and λ at various wind velocities.

V. WIND TURBINE SIMULATOR HARDWARE SETUP

Low speed high toque machines are ideally suitable machines to emulate the characteristics of a WECS. But these machines require special design in terms of large radial diameter and small axial length to accommodate large number of poles. While giving importance to the conceptual emulation of the practical wind turbine's behavior, and due to practical difficulties in obtaining requirement-specific-made machines from reputed manufacturers in the market, we developed wind turbine simulator with normal specifications machines. In-lab hardware setup of wind turbine simulator is shown in Fig. 8. Nameplate specifications of DC machine and PMSM machine are given in Table II. 25 A, 1200 V, 20 kHz switching frequency IGBT Intelligent Power Module (IPM) is used in H-bridge. HEDS-5645, a three channel optical encoder is used to translate rotary motion of the DC motor shaft into digital output. It generates 512 pulses per mechanical rotation of

TABLE II
NAME PLATE DETAILS OF DC MACHINE AND PMSM.

DC Machine	
HP	1.5
Volts (V)	220
Amps (A)	6.0
Speed (RPM)	3000
Field Amps (A)	0.5
PMSM Machine	
HP	1.0
Volts (V)	230/460
Amps (A)	2.6/1.2
Speed (RPM)	1800
Frequency (Hz)	60
Phase	3
Pole	4

the rotor and feeds into TMS320F28335 DSC. DC motor's rotating speed is measured by suitably configuring capture unit of the DSC. TMS320F28335 DSC based control scheme receives wind velocity from host PC in real-time and calculates reference armature current for the DC motor's current control loop as described in Fig. 7. Variable voltage and variable frequency AC output power of PMSG is rectified using 25 A, 1200 V diode bridge rectifier and the rectified power is fed to variable resistive load.

VI. EXPERIMENTAL RESULTS

Developed wind turbine simulator is tested to reproduce the steady-state characteristics of a 1.2 kW WECS at different wind velocities. In host system, a graphical user interface (GUI) environment is designed using Microsoft Visual Basic to display the relations between various parameters of the emulated WECS. Steady state characteristics are obtained by varying the load at a given wind velocity. Variations in power coefficient as function of tip speed ratio for the wind velocities in the range of 4-8 m/s are shown in Fig. 9. We can observe from Fig. 9 is that for all wind velocities, the maximum power coefficient is 4.789 at optimal tip speed ratio of 8.1. Wind turbine simulator's turbine output power versus rotor speed responses at different wind velocities, 4-8 m/s, are shown in Fig. 10. Maximum power point [MPP] trace, which joins MPP of each wind velocity, can be obtained by operating the wind turbine system at optimal tip speed ratio of 8.1 at all the wind velocities.

Dynamic response of the developed wind turbine simulator is tested by observing the interdependent variations between different parameters. The observed results are shown in Figs. 11-13. Fig. 11 shows the dynamic variations in turbine output power (lower trace) against the variations in wind velocity (upper trace). In increasing mode, wind velocity follows the variations of 5 m/s to 7 m/s to 9 m/s and the same variations applied in decreasing mode. Fig. 12 shows the instantaneous

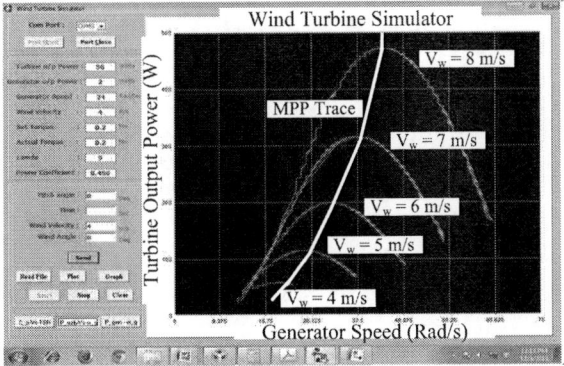

Fig. 10. Emulated relation between P_m and ω_g at different wind velocities.

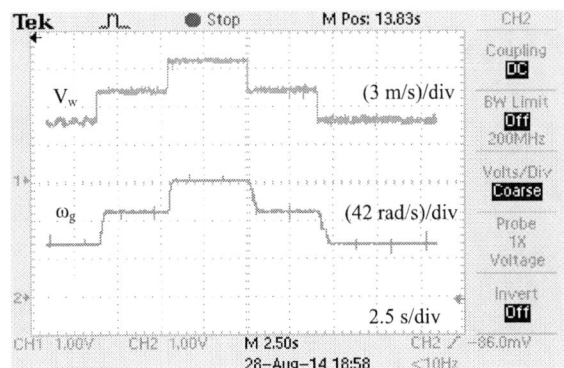

Fig. 12. Response of turbine rotor speed w.r.t. wind velocity.

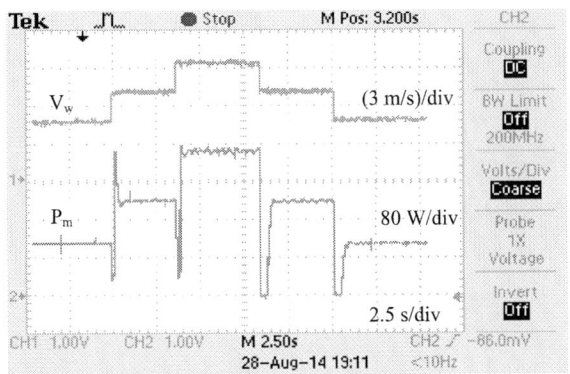

Fig. 11. Response of turbine power output w.r.t. wind velocity.

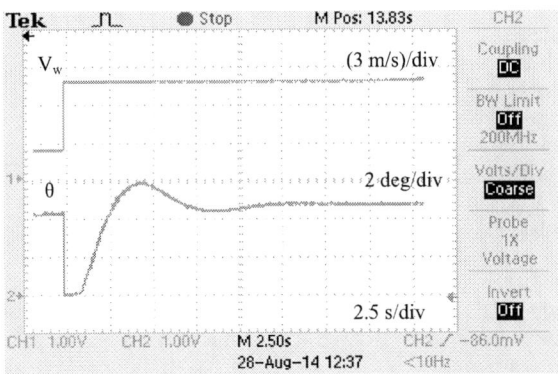

Fig. 13. Response of furling angle w.r.t. wind velocity.

response of turbine rotor speed (lower trace) w.r.t. variations in wind velocity (upper trace). Fast tracking behavior of small-scale WECS due to its low rotor inertia is satisfactorily emulated and this behavior is useful to test the MPPT control algorithms under turbulent wind conditions in microgrid applications. Response of furling mechanism is tested under sudden variations is wind velocity. When wind velocity (upper trace) changes from 3 m/s to 8 m/s, variations in furling angle (lower trace) are shown in Fig. 13. Increase in furling angle reduces the effective wind velocity as in (1) which makes the system to operate in output power control mode. These results confirm that the developed wind turbine simulator is satisfactorily emulating the steady-state characteristics as well as dynamic characteristics of a 1.2 kW WECS at various wind conditions.

VII. CONCLUSIONS

This paper described the design and development process of an in-lab small-scale wind turbine simulator which emulates the behavior of a practical wind turbine under various wind conditions. The developed system's performance is tested in the laboratory under various wind conditions. The obtained results confirm the acceptable performance of the simulator in terms of reproducing the behavior of a practical small-scale wind energy conversion system. The developed simulator can be used to evaluate the performance of newly proposed control

algorithms for effective and efficient power extraction from wind by the real wind turbine.

REFERENCES

[1] Y. Song, B. Dhinakaran, and X. Bao, "Variable speed control of wind turbines using nonlinear and adaptive algorithms," *Journal of Wind Engineering and Industrial Aerodynamics*, vol. 85, no. 3, pp. 293–308, 2000.

[2] Q. Wang and L. Chang, "An intelligent maximum power extraction algorithm for inverter-based variable speed wind turbine systems," *Power Electronics, IEEE Transactions on*, vol. 19, no. 5, pp. 1242–1249, 2004.

[3] D. S. Zinger and E. Muljadi, "Annualized wind energy improvement using variable speeds," *Industry Applications, IEEE Transactions on*, vol. 33, no. 6, pp. 1444–1447, 1997.

[4] W.-M. Lin, C.-M. Hong, F.-S. Cheng, and K. H. Lu, "Mppt control strategy for wind energy conversion system based on rbf network," in *Energytech, 2011 IEEE*, May 2011, pp. 1–6.

[5] S. Mohod and M. Aware, "Wind energy conversion system simulator using variable speed induction motor," in *Power Electronics, Drives and Energy Systems (PEDES) & 2010 Power India, 2010 Joint International Conference on*. IEEE, 2010, pp. 1–6.

[6] J. Bialasiewicz, "Furling control for small wind turbine power regulation," in *Industrial Electronics, 2003. ISIE'03. 2003 IEEE International Symposium on*, vol. 2. IEEE, 2003, pp. 804–809.

[7] M. Arifujjaman, M. Iqbal, and J. Quacioe, "Development of an isolated small wind turbine emulator," *The Open Renewable Energy Journal*, vol. 4, pp. 3–12, 2011.

[8] A. Cultura and Z. Salameh, "Modeling and simulation of a wind turbine-generator system," in *Power and Energy Society General Meeting, 2011 IEEE*. IEEE, 2011, pp. 1–7.

[9] N. Mohan and T. M. Undeland, *Power electronics: converters, applications, and design*. John Wiley & Sons, 2007.

[10] M. H. Rashid, *Power electronics handbook*. Academic Pr, 2001.

Comparative Study Of Stochastic Wind Speed Prediction Models

Alok Agrawal, K.S. Sandhu

Department of Electrical Engineering
NIT Kurukshetra,
Haryana, India
alok.agarwal37@gmail.com, kjssandhu@rediffmail.com

Abstract— Global weather concerns have diverted the researchers to look out for clean and green energy sources. The wind power has been utilized since old days for applications such as pumping water, grain grinding mills, driving ships, etc. However, due to its uneven nature, wind energy has never been a choice of power engineers. With the inherent variability of wind due to fluctuating weather conditions, wind speed prediction techniques play a vital role in determining the feasibility of esteemed wind power projects. Modern weather forecasting involves a combination of analyzed data, computer models, knowledge of trends and patterns. In this paper various short-term statistical wind speed forecasting techniques have been analyzed and focused upon.

Keywords— Numerical Weather Prediction(NWP), Polynomial Curve Fitting(PCF), Auto-regressive Moving average(ARMA), Variance ratio(VR) method, Mortimer method, Markov chain, Wavelet transformation, and Artificial Neural Network(ANN).

I. INTRODUCTION

Worldwide there are over 250 thousand wind turbines operating with a total installed capacity of 318 gigawatts (GW) up to December 2013 which is expected to grow by 35 GW in the current year [1]. By end of 2010, Europe had a total installed wind energy capacity of 44%, a further 32% in Asia pacific, and 22% in North America [2]. Statistics reveal that at a few places wind power plants holds a substantial power generation which may be used to satisfy peak load demands. However, integration of wind power plants into the power system pose a disadvantage due to inherent variability of wind resource. Taking into consideration that the wind speed depends on various environmental weather conditions, the output power of wind farm at a particular instant cannot be predicted, thus leading to gaps between supply and demand curves. So as to reduce the wind farm operation issues and ensure better technical feasibility, wind speed or power production forecasts, are needed before hand. Wind power forecasting methods serves a lot of purposes and can be used for planning unit commitment problems, energy scheduling, etc.

Many short-term and long-term wind speed/power prediction tools have been designed by Meteorological research and development societies [2]. It can be observed that short-term wind speed forecasting techniques rely on statistical time series model approach or numerical weather prediction (NWP)

model approach or a mixture of both [3]. In Fig.1 illustrates principle technique of forecasting models.

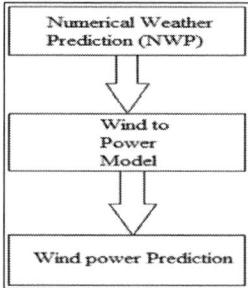

Fig. 1. Principle of wind power prediction tool.

As wind is an intermittent power source, its power output is fluctuating in nature. Short-term wind forecasting techniques aims to predict the wind speed or energy production of a wind power site over duration of 1 to 48 hours which can be further elaborated for medium-term wind forecasting. In this paper various wind forecasting techniques have been overviewed.

II. POLYNOMIAL CURVE FITTING MODEL

PCF or Trend extrapolation technique is enormously used in load prediction, load flow analysis problems, etc. This approach can also be used forth to find solution for wind speed forecasting problems also. Any Variable 'θ' could be expressed in terms of x and y using an nth order polynomial, where (x_i, y_i) is experimental data illustrated in terms of polynomial P as [4],

$$P(x_i; y_i; \theta) = 0 \qquad ; \text{where, } i = 1, 2....n$$

As per this model estimated wind speed is a polynomial function of historical wind speed. The order of polynomial curve gives us the number of coefficients to be determined. PCF can be regarded as general form of Linear Least-Squares Regression [5],

$$w_i = f(w_{i-1})$$
$$= a_0 + a_1 w_{i-1} + a_2 w_{i-1}^2 + a_3 w_{i-1}^3 + ... + a_n w_{i-1}^n$$
$$= \sum_{t=0}^{n} (a_t w_{i-1}^t) \qquad (1)$$

where, w_i is the estimated wind speed in m/s, w_{i-1} is the historical wind speed in m/s and n signifies the order of polynomial curve. Historical wind data for Mumbai region was accumulated between 00:00 21/07/14 to 23:59 25/07/14. Range of data was 2.5 – 7.5 mps with a mean of 4.84 mps [6]. Fig. 2 shows the comparison of simulation results as obtained using 2nd, 3rd and 4th order polynomials. Through simulation results it could be well noticed that, higher order polynomial curves lie in close proximity with that of expected results.

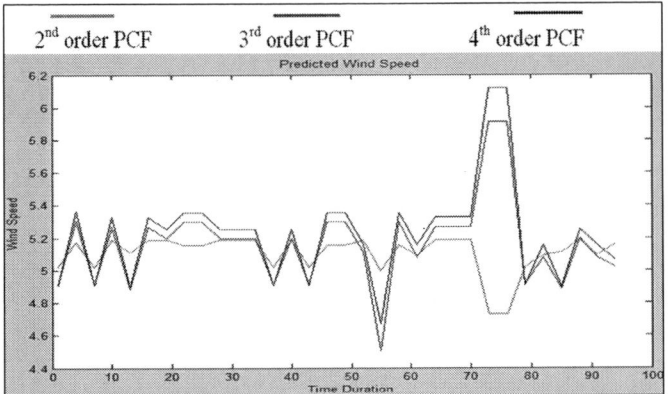

Fig. 2. Comparison of predicted wind speed for 3 hour ahead wind analysis with second, thirth and fourth order polynomials.

III. AUTO-REGRESSIVE (AR) MODEL

An AR model is defined as a random probability distribution process. It involves statistical analysis of wind speed data, such that the weighted sum of past values helps in predicting the wind speed at some ahead. AR models are preferred for short-term wind speed prediction purposes mainly because of its outperforming persistence prediction results. Many examples could be considered in this regard. For instance, Schlink & Tetzlaff (1997-98) depicted how AR models could be used to predict wind speed curve nature at an airfield and concluded that the AR model results in accurate predictions [7]. Moreover, Gneiting et al. (2006) showed how this technique could be utilized to fix wind speed distribution, excluding the diurnal pattern [8], and he found out that by this technique root mean square error was 16% lesser than results derived from Persistence method.

For practical reasons, it is desirable to have a unique solution that is a function of the past error terms and possesses a high dependence on hourly or daily mean wind speed time series. Autoregressive process is a difference equation determined by random stochastic variables. The nth order autoregressive time series (AR(n)) is represented as,

$$w_i = \sum_{j=0}^{n} r_i w_{i-j} + \varepsilon_i \qquad (2)$$

where, w_i is the hourly mean wind speed, w_{i-j} is the wind speed j time-period prior to w_i, r_i is the lag i autoregressive autocorrelation coefficient and ε_i is the assumed uncorrelated random variables having its average mean equal to zero and variance equal to $1-r_1^2$. First order autoregressive time series (AR (1)) is a linear difference equation [9] depicted as below,

$$w_i = r_1 w_{i-1} + \varepsilon_i \qquad (3)$$

where, r_1 is the lag-one serial correlation coefficient and ε_i is the stochastic component of normal distribution Using equation (3), w_i is obtained from knowing the value of w_{i-1}, and w_{i-1} is obtained from knowing w_{i-2} and so on. We observe that w_i can be found from wind speed w_k which is k-time periods prior,

$$
\begin{aligned}
w_i &= r_1 w_{i-1} + \varepsilon_i \\
&= r_1 (r_2 w_{i-2} + \varepsilon_{i-1}) + \varepsilon_i \\
&= r_1 (r_2 (r_3 w_{i-3} + \varepsilon_{i-2}) + \varepsilon_{i-1}) + \varepsilon_i \\
&\vdots \\
&= (r_1 r_2 ... r_k) w_k + (r_1 r_2 ... r_{k-1}) \varepsilon_{i-k} + ... + r_1 \varepsilon_{i-1} + \varepsilon_i \qquad (4)
\end{aligned}
$$

IV. MOVING AVERAGE (MA) MODEL

MA model are useful in describing phenomenon where events results in immediate effects that lasts for short periods of time. The method evolved as a result of study by Slutzky (1927) on effect of moving average of random events [10]. The MA model is a common approach for deriving univariate time series models in case we are interested in time series analysis. In this time analysis series, the error of the historical data set is written as infinite weighted linear sum.

$$w_i = \mu + \varepsilon_i + \theta_1 \varepsilon_{i-1} + \theta_2 \varepsilon_{i-2} + ... + \theta_n \varepsilon_{i-n} \qquad (5)$$

where μ represents the mean of wind speed time series; θ_1, θ_2,....θ_n are the moving average parameters; ε_i, ε_{i-1} , ε_{i-2} ...ε_{i-n} are the assumed random variables commonly referred to as sudden shock factor or white noise factor and n is the order of MA model. In a compact form MA model can also be represented in terms of function $\theta(L)$, where L is the backshift operator. Model remains stable for $\theta < 1$.

$$
\begin{aligned}
w_i &= \mu + (1 + \theta_1 L + \theta_2 L^2 + ... + \theta_n L^n).\varepsilon_i \\
&= \mu + \theta(L).\varepsilon_i \qquad (6)
\end{aligned}
$$

Because $1 + \theta_1 L + \theta_2 L^2 + ... + \theta_n L^n < \infty$, a finite MA process is always stationary. MA process is invertible in case $\theta(L) = 0$ roots lie outside the unit circle.

V. AUTO-REGRESSIVE MOVING AVERAGE MODEL

ARMA and persistence models are designed so as to forecast the wind speed/power up to 10 to 48 hour before hand. In ARMA model predicted wind speed is not only a function of historical wind speed data as in case of AR process but also is a function of residuals of past forecasts that relate to previous time data of what we are forecasting. A general ARMA(p,q) methodology can broke up in 3 steps [11] : Model Selection, Model Parameterization, and Model Result.

$$w_i - r_1 w_{i-1} -..... - r_p w_{i-p} = \varepsilon_i + \theta_1 \varepsilon_{i-1} + ... + \theta_q \varepsilon_{i-q}$$

$$w_i = \varepsilon_i + \sum_{j=1}^{p} r_i w_{i-j} + \sum_{j=1}^{q} \theta_i \varepsilon_i \qquad (7)$$

Above equations can be rearranged and coined in terms of lag operator L,

978-1-4799-6047-7/14 $31.00 © 2014 IEEE

$$(1 - \sum_{j=1}^{p} \psi_j L^j) w_i = (1 + \sum_{j=1}^{q} \theta_j L^j) \varepsilon_i \qquad (8)$$

According to conventions used by Box, Jenkins and Reinsel ARMA model can be modified [12] to,

$$(1 + \sum_{j=1}^{p} \psi_j L^j) w_i = (1 + \sum_{j=1}^{q} \theta_j L^j) \varepsilon_i \qquad (9)$$

$$\Psi(L) w_i = \theta(L) \varepsilon_i \qquad (10)$$

where, $\psi(L)$ is the AR model coefficients and $\theta(L)$ is the MA model coefficients defined as,

$$\Psi(L) = 1 + \psi_1 L + \psi_2 L^2 + \ldots\ldots + \psi_p L^p \qquad (11)$$

$$\theta(L) = 1 + \theta_1 L + \theta_2 L^2 + \ldots\ldots + \theta_q L^q \qquad (12)$$

ARMA's order selection could be done by considering the model validity criteria's such as the minimum description length (MDL), through Akaike's information criterion (AIC), or auto correlation functions such as ACF and PACF. Higher is the order of system more are the chances that predicted speed lies close to expected wind speed. In some cases data show non-stationary behavior, in such cases Autoregressive integrated moving average(ARIMA), Seasonal ARIMA(SARMA), Limited ARIMA(LARIMA) are used to predict and understand future points.

VI. Variance Ratio Method

Variance ratio technique was evolved by Rogers et al. [13] from simple linear regression technique. While using linear regression technique, the predicted mean wind speed at the target site remains close to the measured mean during the analyzed time period, however, the predicted variance will be less than the measured variance [14]. Due to its unbiased prediction nature VR method is preferred over conventional techniques discussed earlier. Average wind power depends on 3rd power of wind speed, i.e., $P = \alpha \cdot w_i^3$. Considering wind speed (w_i) as a variable resource [15],

$$w_i^3 = (w_i^*)^3 + \sigma^2 (3 w_i^* + \sigma \cdot c_a) \qquad (13)$$

where, w_i^* represents the corresponding actual wind speed value, σ^2 is the variance of wind speed data set (expected Vs. predicted), and c_a represents the coefficient of asymmetry of the wind speed distribution. Variance Ratio (VR) method is considered as an alternative to linear correlation technique that aims at predicting values such that the mean of historical wind speed and its standard deviation (SD) lie in close proximity with the expected data. The prediction equation could be stated as,

$$w_y = \overline{w_y} - \frac{\sigma_y}{\sigma_x} \cdot \overline{w_x} + \frac{\sigma_y}{\sigma_x} \cdot w_x \qquad (14)$$

where, w_y is the predicted wind speed for target site, w_x is the wind speed data set of reference site, $\overline{w_x}, \overline{w_y}$ are the mean wind speed at target and reference sites and σ_y, σ_x are the standard deviation of target and reference sites wind speed data sets, respectively. Correct speed frequency distribution through variance ratio method cannot be assured since the correlation between target area and reference area speeds depends on many indigenous factors. So as to solve this issue with an appropriate evaluation methodology matrix ratio technique could be used.

VII. Mortimer Method

Linear regression techniques generally do not provide better long term prediction of wind resource until and unless they are indulged in some statistical or physical model. Mortimer et al. [16] presented a data pre-processing technique in which data of two consecutive reference wind farm sites are stored together, considering the wind direction and speed at the perspective sites. Speed bins are assumed to be spaced by 1 m/s and direction bins by either 22.5^0, 45^0, 90^0, or 180^0. For each bin, ratio of wind speed at target site to reference sit is evaluated. Mortimer's formula is represented as,

$$w_y = (e + r') w_x \qquad (15)$$

where, e is the stochastic variable having an triangular distribution, and r' indicates the mean ratio for speed and direction bin corresponding to historical observation w_x. In case sufficient data is not available for determining mean ratio for any speed and direction bin, the ratio is assumed equal to the average of all wind speed for a particular directional frame and variance is considered to be zero. If Mortimer's equation leads to negative set of predictions, the results are ultimately set to zero.

VIII. Markov Chain Process (MCP)

MCP is based on the concept of transitional probability matrices (TPM) comprising of multiple states [17]. Wind speed states could be found out with an assumed variation of 1-2 m/s or on the basis of SD of data set [18]. In this case width of each wind speed data set is set equal to one SD of the recorded hourly, daily or weekly average wind speed time series. In first order MCP, wind speed in the current time period depends only upon previous state values. Moreover, higher order Markov chains also exist, but are rarely used due to considerable increase in number of variables used. The MCP variable set includes transitional probabilities from one state to another state which are summarized in TPM. So as to calculate the transition probabilities from one state to another, it is assumed that MC is a discrete-time random process, comprising of time periods i = 0, 1, 2,. . . and states [19], [20],

$$P(w_{i+1} = m_{i+1} \mid w_i = m_i, w_{i-1} = m_{i-1},\ldots\ldots, w_0 = m_0)$$

$$= P(w_{i+1} = m_{i+1} \mid w_i = m_i) \qquad (16)$$

It could be noticed that the probability of any state(m_i) in time period i+1 is correlated with the state in time period i. For all states m and n in periods i,

$$P(w_{i+1} = n \mid w_i = m) = p_{mn} \qquad (17)$$

where, p_{mn} represents the probability of wind to transit from wind speed state m (during period i) to wind speed state n (during period i+1). p_{mn} are also referred to as transition probabilities of MC process. Further, it can be observed that probabilities to change from state m to n do not change with time. If s is number of transition states, then TPM could be represented as,

$$P = \begin{array}{c} w_{i+1} \rightarrow \\ w_i \downarrow \end{array} \begin{bmatrix} p_{11} & p_{12} & \cdots & p_{1n} & \cdots & p_{1s} \\ p_{21} & p_{22} & \cdots & p_{2n} & \cdots & p_{2s} \\ \vdots & & & & & \vdots \\ p_{m1} & p_{m2} & \cdots & p_{mn} & \cdots & p_{ms} \\ \vdots & & & & & \vdots \\ p_{s1} & p_{s2} & \cdots & p_{sn} & \cdots & p_{ss} \end{bmatrix} \quad (18)$$

For all m and n, all elements of matrix P are non-negative i.e., $p_{ij} >= 0$ and sum of probabilities for each row of matrix is 1.0 i.e., $\sum_{n=1}^{s} p_{mn}=1$; m = 1,2,3,.....,s. If X_{mn} is the number of transformations from wind speed state m to n during one period, then the transition probability from state m to n could be defined as,

$$P_{mn} = \frac{X_{mn}}{\sum_{n=1}^{s} X_{mn}} \quad (19)$$

A second order TPM could be formulated as,

$$P = \begin{array}{c} w_{i+2} \rightarrow \\ w_i, w_{i+1} \downarrow \end{array} \begin{bmatrix} p_{111} & p_{112} & \cdots & p_{11n} & \cdots & p_{11s} \\ p_{121} & p_{122} & \cdots & p_{12n} & \cdots & p_{12s} \\ \vdots & & & & & \vdots \\ p_{1s1} & p_{1s2} & \cdots & p_{1sn} & \cdots & p_{1ss} \\ p_{211} & p_{212} & & p_{21n} & & p_{21s} \\ \vdots & & & & & \vdots \\ p_{ss1} & p_{ss2} & \cdots & p_{ssn} & \cdots & p_{sss} \end{bmatrix} \quad (20)$$

For the purpose of data generation cumulative transition probability matrix (CPM) could be formulated by taking serial summations of elements of each row. Moving on, an initial state or position m is selected and by using a uniform stochastic number next state is found. In case, the new wind speed state comprises of extreme data values (i.e., values lying in highest wind speed state), then for such conditions a shifted one-parameter gamma distributed stochastic value is selected which in turn helps us in predicting wind speed. Using available information next state is found such that the value of random number lies between cumulative probability of the following state (say state n) and previous state (state m) [20]. CPM could be used to predict wind speed by using equation,

$$w = w_{n-1} + Z_m(w_n - w_{n-1}) \quad (21)$$

where, w_{n-1} and w_n are the boundary limits of state and Z_m is the uniform random number.

IX. WAVELET TRANSFORM TECHNIQUE (WTT)

Wavelet transforms may be considered forms of time-frequency representation for continuous-time analog signals. The WTT allows the decomposition of a signal into various levels of resolution scales, which can be used to extract the required data components. In case wind forecasting problems, the WTT converts a wind speed series into a set of constitutive series [21], which possess a better predicting behavior than the

original signal due to the filtering effect of WTT. Wavelet transform can be classified as [22],[23],

A. Continuous wavelet transforms(CWT)

A real or complex-value continuous signal function with zero average and finite standard deviation is referred to as wavelet. Morlet, Mexican Hat, Shannon, Meyer, etc., are the commonly referred mother wavelet signals, $\psi(i)$. Using $\psi(i)$, the subspace (child wavelet's) of scale a or frequency band [1/a, 2/a] is produced as depicted below,

$$\Psi_{a,b}(i) = \frac{1}{\sqrt{a}} . \Psi\left[\frac{i-b}{a}\right] \quad (22)$$

where, a is the scaling factor of wavelet, b is the translation factor (a,b are real values, a ≠ 0) and i is the time period. The CWT $W_{a,b}$ of a given signal w(i) is derived as shown below,

$$W_{a,b} = \frac{1}{\sqrt{a}} . \int_{-\infty}^{+\infty} w(i) . \Psi^*\left[\frac{i-b}{a}\right] . di \quad (23)$$

where, ψ^* represents the conjugate of ψ , $W_{a,b}$ is called the wavelet coefficient and w(i) is the input wind speed function with a function space $L^2(R)$. The function w(i) could be reconstructed by using continuous wavelet coefficients as represented in equation (24).

$$w(i) = \frac{1}{C_\Psi} . \int_{-\infty}^{+\infty} \int_{-\infty}^{+\infty} W_{a,b} \, \Psi_{a,b}(i) \frac{da.db}{a^2} \quad (24)$$

B. Discrete Wavelet Transforms (DWT)

Since, translation of mother wavelet leads to redundant information in case of continuous transforms, it's better to use discrete transforms which reduces the computation time and complexity of problem [19]. In order to limit the transform to discrete values scale factor a and translation factor b are modified as $a = a_o^p$ and $b = qb_o a_o^p$, such that $a_o>1$, $b_o \neq 0$ and p,q are integer values. Corresponding child wavelets could be modified as,

$$\Psi_{p,q}(i) = a_0^{-m/2} \psi(a_0^{-m} i - nb_0) \quad (25)$$

where, m is the indicating frequency localization factor and n is the indicating time localization factor. Usually, we take a_o = 2 and b_o =1, then a discrete version of equation (23) could be denoted as,

$$W_{p,q} = 2^{-\frac{p}{2}} . \sum_{i=0}^{I-1} w(i) . \Psi\left[\frac{i-q.2^p}{2^p}\right] \quad (26)$$

where, I is the length of signal and i is discrete time index.

C. Multiresolution based discrete wavelet transform (MDWT)

MDWT is based on the function space theory. It depicts how scaling functions $\varphi(i)$ (referred as father wavelet) and the details provided by the mother wavelet $\psi(i)$ could be used for the purpose of decomposition of time series. The father and mother wavelet keeps the frequency domain and time domain properties respectively. The scaling function and wavelet functions are analogous with the low-pass filters (LPF) and high-pass filters (HPF) respectively. The input signal (or data) is passed through filters which is broken down into approximations A (low frequency component) and details D (high frequency component). The approximation coefficients are again sent through filters to repeat the process. Decomposition is stopped when the standard deviation (S.D.) for approximations is comparatively lesser than original signal [24] , i.e., typical ratio = 0.1 . Original signal w could be reconstructed by summing up the approximation A_x at x^{th} level with the x number of details $D_1, D_2, D_3, \ldots \ldots D_x$ as shown in fig. 3.

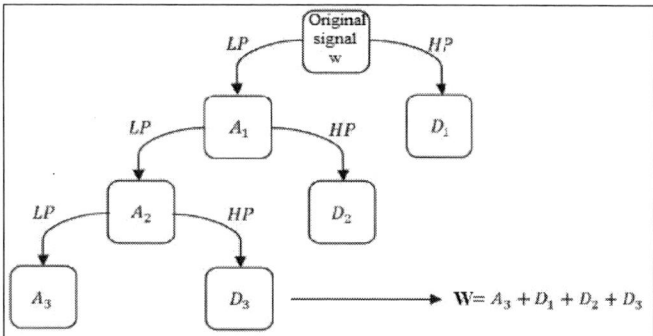

Fig. 3.   Decomposition of signal using multiresolution analysis

X.   ARTIFICIAL NEURAL NETWORK MODEL

As neural networks possess a nonlinear approach, it can be used to produce patterns and detect trends that are too complex to be noticed by other computer algorithms [25]. Due to adaptive learning features, self-organization nature, and good fault tolerance abilities neural networks are a powerful general and flexible modeling tool for wind speed prediction purposes. ANN could be classified as [26],

1) Multi-layer Perceptron (MLP) neural network : Consists of source nodes connected to output layer of neurons through one or more hidden layers.

2) Elman Recurrent (ER) neural network : It is a two layer system comprising of feedback loop or connections for forecasting purposes.

3) Simultaneous Cascade-Correlation (CC) neural network : It is a type of feed forward neural network which are capable of generating hidden neurons and layers as needed by algorithm.

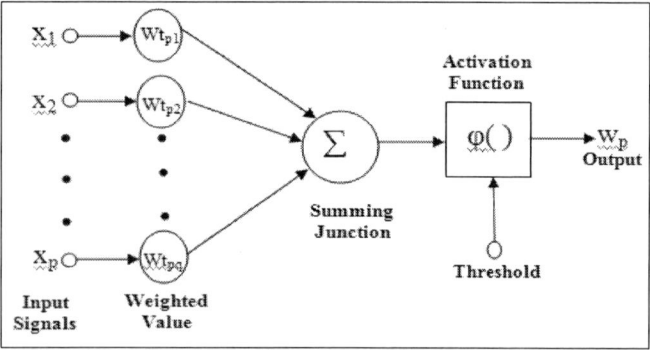

Fig. 4.   Basic operation of ANN model

Each ANN consists of many interconnected processing units called as neurons which leads to intramural activity. This intramural activity is then passed through a non-linear function φ to give the output of the neuron. Fig. 4 illustrates the algorithm used for ANN models

$$w_p = \varphi(\textstyle\sum_{q=1}^{n} Wt_{pq}x_q - Wt_{po}) \tag{27}$$

where, Wt_{pq} is the weighted value of connection from neuron q to p, x_q is the input signal number q to neuron p and Wt_{p0} is the threshold value of unit p. The weighted value and the threshold value are adjusted for each unit during the network's training step using back-propagation algorithm that seeks to minimize the prediction error.

$$\text{Error, } E = \sum | w_p - \widehat{w_p}|^2 \tag{28}$$

where, w_p and $\widehat{w_p}$ represents the predicted and desired wind speed data respectively.

XI.   PERFORMANCE EVALUATION OF FORECASTING MODELS

Various criteria's for model performance evaluation includes determining Correlation Coefficient (R), Mean Square Error (MSE), Mean Absolute Error (MAE), Root Mean Square Error (RMSE), the Mean Relative Error (MRE), Mean Average Percentage Error (MAPE) and Sum of Squared Error (SSE) [11],[21],[27],[28]. For evaluating these factors, predictions are compared with observed values.

$$R = \frac{R_{ww^*}}{S.D.(w).\ S.D.(w^*)} \tag{29}$$

where, w is the predicted value of wind speed on hourly basis, w* is the corresponding actual wind speed value, R_{ww^*} is the co-variance between w and w*, S.D.(w) and S.D.(w*) represents the corresponding standard deviation for w and w* respectively.

$$MSE \quad = \frac{1}{N}.\sum_{i=1}^{N}(w_i^* - w_i)^2 \tag{30}$$

where, N is the number of observations.

$$MAE \quad = \frac{1}{N}.\sum_{i=1}^{N}|w_i^* - w_i| \tag{31}$$

$$RMSE = \sqrt{\frac{1}{N}.\sum_{i=1}^{N}(w_i^* - w_i)^2} \tag{32}$$

$$MRE \quad = \frac{1}{N}.\sum_{i=1}^{N}\frac{|w_i^* - w_i|}{|w_i|} \tag{33}$$

$$MAPE = \frac{1}{N}.\sum_{i=1}^{N}|w_i^* - w_i| * 100 \tag{34}$$

$$SSE \quad = \sum_{i=1}^{N}(w_i^* - w_i)^2 \tag{35}$$

XII.   COMPARATIVE REVIEW

Table I summarizes the models with specific remarks related to the features of each technique.

TABLE I. REVIEW OF FORECASTED MODELS

Forecast model	Remarks
PCF Model	• Better if historical data is following some definite pattern. • Higher order polynomials gives better results.
AR model	• A further extension of PCF technique to yield better prediction results. • In nth order AR model, future values are function of n number of past wind speed values.
MA model	• Considers random changes / shocks in weather conditions for predicting wind speed. • Quite helpful in managing random events.
ARIMA model	• Very accurate for short-term predictions. • Could be treated as a good time series based Neural Network structure.
VR method	• Linear regression technique. • Spatial correlation of data taken into consideration. • Model gives a output data set with variance equal to input data set.
Mortimer method	• Linear regression technique. • Requires wind speed direction database to predict wind speed. • Mortimer method at times results in under-predicted values.
MCP	• Outperforms other time series models. • TPM complexity for higher order MC needs to be tackled efficiently.
WTT	• WTT could be used for analysis of both stationary and non-stationary wind speed data sets. • Applicable to non-linear regression problems. • WTT dominates due to its time & frequency domain localized nature.
ANN	• Larger neural networks possess larger processing time. • Possess non-linear modeling nature. • Ability to implement complex pattern and trends. • Can also be used for medium and long-term forecasting purposes.

XIII. CONCLUSIONS

This paper gives a comparative analysis of various statistical techniques that are available for short-term wind speed forecasting purposes. It can be observed from the above discussions that it is difficult to compare all the models on common grounds, primarily due to different input parameters and time/frequency based approach of models. Each model as represented has its own significance under the given operating constraints. However, after comparison it can be said that WTT and ANN results in accurate wind speed predictions, primarily due to their non-linear prediction nature and self-adapting learning features.

REFERENCES

[1] Global Wind Energy Council, "Global Wind Report Annual Market Update 2013", pp. 16-21, April 23, 2013.

[2] Thomas Ackermann, "Wind power in power systems", Wiley publications,pp. 27-40, 2nd edition, 2012.

[3] Bin Wu, Yongqiang Lang, Navid Zargari, and Samir Kouro, "Power Conversion and Control of Wind Energy Systems", Wiley-IEEE press, 2011 edition.

[4] M.A. Nayak and M.C. Deo, "Wind speed prediction by different computing techniques", BALWOIS conference, Republic of Macedonia, May 25-29, 2010.

[5] Sanjeev Arora and Subhash Khot, "Fitting algebraic curves to noisy data", Journal of Computer and System Sciences , pp.325–340, 2003.

[6] India Meteorological department (Ministry of Earth Sciences, India), Numerical weather analysis and prediction, June 2014.

[7] U. Schlink and G. Tetzlaff, "Wind Speed Forecasting from 1 to 30 Minutes", Theoritical and applied climatology,Springer-Verlag press, pp.191-198, 1998.

[8] Gneiting, T., Larson, K., Westrick, K., Genton, M.G. and Aldrich E., "Calibrated probabilistic forecasting at the Stateline wind energy center: The regime-switching space-time method", Technical report Department of Statistics, University of Washington, September 2004.

[9] Zsuzsanna Horvath and Ryan Johnston, "AR(1) time series process", University of Utah, unpublished.

[10] Eugen Slutzky, "The Summation of Random Causes as the Source of Cyclical Processes", Econometrica, volume 5, No. 2, pp. 105-146, April 1937.

[11] Pedro Gomes and Rui Castro, "Wind Speed and Wind Power Forecasting using Statistical Models: AutoRegressive Moving Average (ARMA) and Artificial Neural Networks (ANN)", International Journal of Sustainable Energy Development (IJSED), Volume 1, pp. 36-45, March/June 2012.

[12] E.P. Box, George, Gwilym Jenkins, Reinsel, Gregory C., "Time Series Analysis: Forecasting and Control", Prentice-Hall Publication, 1st edition 1994.

[13] A. L., Rogers, J. W. Rogers, and J.F. Manwell, "Comparison of the performance of four measure-correlatepredict algorithms". Journal of Wind Engineering and Industrial Aerodynamics, 93(3), pp. 243–264, 2005.

[14] Jie Zhang, Souma Chowdhury, Achille Messac and Bri-Mathias Hodge, "Assessing long-term wind conditions by combining different measure-correlate-predict algorithms", NREL publications, 2013.

[15] A. R. Perea, Javier Amezcua and Oliver Probst, "Validation of three new measure-correlate-predict models for the long-term prospection of the wind resource", Journal of Renewable Energy, April 4, 2011.

[16] Mortimer, A., "A new correlation/prediction method for potential wind farm sites." , in Proceedings of the British Wind Energy Association (BWEA), 1994.

[17] F. O. Hocaoglu, O. N. Gerek and M. Kurban, "The Effect of Markov Chain State Size for Synthetic Wind Speed Generation", IEEE Publications, 2008.

[18] F. O. Hocaoglu, O. N. Gerek and M. Kurban, " The effect of markov chain state size for synthetic wind speed generation", Proceedings 10th International Conference on Probabilistic Methods Applied to Power Systems (PMAPS), IEEE, 2004.

[19] Shamshad A., Wan Hussin W. M. A., Bawadi M. A., Mohd. Sanusi S. A., "First and second order markov chain models for synthetic generation of wind speed time series", Elsevier Energy reviews, Volume 30, Issue 5, pp. 693–708, April 2005.

[20] George Papaefthymiou and Bernd Klockl," MCMC for Wind Power Simulation", IEEE transactions on Energy Conversion, Volume 23, No. 1, March 2008.

[21] Jujie Wang, "A hybrid Wavelet transform based short-term wind speed forecasting approach", The Scientific World Journal, Volume 2014, Hindawi publishing corporation, July 21, 2014.

[22] CAO Lei and LI Ran, "Short-Term Wind Speed Forecasting Model for Wind Farm Based on Wavelet Decomposition", Electric Utility Deregulation and Restructuring and Power Technologies, IEEE Publication, April 6-9, 2008.

[23] Zhang Yanning, Kang Longyun, Zhou Shiqiong, and Cao Binggang, "Wind speed predicted by wavelet analysis in input prediction control of wind turbine", Proceedings 7th World Congress on Intelligent Control and Automation, IEEE, June 25-27, 2008.

[24] Siwek K., Osowski S., "Improving the accuracy of prediction of PM10 pollution by the wavelet transformation and an ensemble of neural predictors", Elsevier Engineering Applications of Artificial Intelligence Reviews, Volume 25, Issue 6, pp. 1246–1258, September 2012.

[25] S. Kalogirou, C. Neocleous, S. Pashiardis, and C. Schizas, "Wind speed prediction using artificial neural networks", European Symposium on Intelligent Techniques, 1999.

[26] Richard Welch and G. K. Venayagamoorthy, "Short term wind speed prediction using feedforward and feedback neural network architectures", in Proceedings ISCRS, April 14,2009.

[27] Ramesh Babu. N and Arulmozhivarman. P, "Improving forecast of wind speed using wavelet transform and neural networks", Journal of Electrical Engineering and Technology, Volume 8, No. 3, pp. 559-564, 2013.

[28] Pedro Gomes and Rui Castro, "Comparison of statistical wind speed forecasting models", IEEE conference on Sustainable Technologies (WCST), 2011.

Grid Connected Photovoltaic System with Data-based MPPT and Fuzzy Controlled DVR

Akhil Gupta
Department of Electrical and
Electronics Engineering
University Institute of Engineering
Chandigarh University, Mohali,
India
akhilgupta1977@gmail.com

Saurabh Chanana
Department of Electrical
Engineering
National Institute of Technology
Kurukshetra
Kurukshetra, India
s_chanana@rediffmail.com

Tilak Thakur
Department of Electrical
Engineering
PEC University of Technology
Chandigarh, India
tilak20042005@yahoo.co.in

Abstract—**This paper presents the impact of a fuzzy logic controlled dynamic voltage restorer on a transformer-less single-stage grid-connected solar photovoltaic system controlled by a data based maximum power point tracking technique. The solar photovoltaic system uses a data based maximum power point technique to derive maximum power from solar photovoltaic array and maintains constant DC voltage by changing modulation index of voltage source converter. A dynamic voltage restorer with pulse width modulated control is introduced to mitigate unbalanced sag in grid voltage and current during fault conditions. Fuzzy logic control (FLC) has been implemented to maintain voltage at grid side for normal as well as for faulted conditions. Simulation results show that distortion in active and reactive power exchange among solar photovoltaic converter, load and grid is reduced during fault conditions. A reduction in total harmonic distortion for all temperature conditions, low and high solar radiation is observed in grid voltage and current. Apart from these, it has been observed that distortion in DC link voltage while deriving maximum power from photovoltaic array is considerably reduced by introducing fuzzy logic controlled dynamic voltage restorer in proposed system.**

Keywords— Dynamic voltage restorer, fuzzy logic control, maximum power point tracking, power quality, solar photovoltaic.

I. INTRODUCTION

Non-conventional energy sources such as solar Photovoltaic (PV), wind, fuel cell and diesel generators offer better solution to supply increased demand of electric power particularly in remote areas. The grid integration of renewable energy sources has brought forward new challenges to electric power systems. Of all available such sources power quality studies for grid-connected solar PV systems with line frequency transformer are in progress, but studies on transformer-less type are limited. An effective solution to establish a proper voltage quality level is to use a DVR with PI controller. Reference [1] has presented a solution for mitigating voltage sags and swells using these devices. A control approach based on hysteresis voltage and fuzzy-logic is presented in [2], which compensates networks faults and mitigate their effects under different loading conditions. THD in Reference [3] has been calculated which is analyzed and

found to be within the limits of IEEE-519/1547 standard. Capabilities of a space-vector control and multilevel VSC based DVR is presented in [4], along with a low pass filter for mitigating voltage sag/swell for a 22 kV distribution system. A new Pulse Width Modulation (PWM)-based control scheme is implemented by [5] which controls electronic valves in a 2-level VSC based DVR. Under unbalanced fault conditions, behavior of PI and Proportional Resonant (PR) controllers is addressed in [6]. Study of active and reactive power variation during single phase to ground fault is also done. Reference [7] proposes optimal utilization of a solar PV farm which during nighttime can operate as a Static Compensator (STATCOM), and regulate in the distribution voltage at Point of Common Coupling (PCC) within ±3 %. Power quality characteristics of 13 solar PV inverters with rated power from 1 to 5 kW have been analyzed in [8,9]. The results show that inverter characteristics due to voltage harmonics depends on actual phase angle of injected harmonic current related to the voltage harmonic.

This paper describes design and simulation of a high performance Power Conditioning System (PCS) for a three-phase grid-connected solar PV system, and its control schemes. The control schemes include a single-stage three phase three level transformer-less VSC and a PWM based fuzzy logic controlled DVR. The model of proposed PV array uses theoretical and empirical equations together with data generated through unique data based MPPT control technique at variable temperature, and solar radiation among other variables. With application of fault on grid side, MPPT technique with fuzzy logic controlled DVR provides distortion-less maximum active power from solar PV array, with negligible fluctuation of DC bus voltage to VSC thus fast tracking of optimum operating point at unity power factor is achieved. This fuzzy logic controlled DVR scheme also shows simultaneous exchange of active and reactive power with distribution system during fault. Using Fast Fourier Transform (FFT) [10] analysis the THD values of grid injected and VSC current as specified by IEEE-519/1547 standard have been calculated which proves the correctness of implemented power quality scheme.

978-1-4799-6047-7/14 $31.00 © 2014 IEEE

II. SOLAR PV GRID CONNECTED SYSTEM

The power conditioning system, Fig. 1, consists of an array of solar PV panels model [11]-[15], an Insulated Gate Bipolar Transistor (IGBT)-based Voltage Source Converter (VSC), and reactor-capacitor-reactor used here as an LCL-filter. The solar PV array is parallel-connected to IGBT-based VSC through a DC-bus capacitor C [16]. The AC-side terminals of the VSC are connected to Point of Common Coupling (PCC) through an interfacing LCL-filter. The parallel capacitor of an LCL filter prevents current harmonics produced by inverter-based DG system, from infiltrating into distribution network side. A 3-φ series RLC load is connected between main grid and an IGBT-based VSC. A forced commutated VSC is considered in DVR along with energy storage to maintain capacitor voltage. The basic function of medium voltage DVR is to inject a voltage component of desired amplitude, frequency and phase, [6], between PCC and grid in series with the utility or load voltage. The essential part for well-performance of controller in DVR is sag-detection circuit. Most of works in literature survey are based on two stage conversion, one DC/DC and second, DC/AC (or VSC) converter. However, single-stage conversion [11] is implemented [13] in power conditioning system in which solar PV array voltage is controlled, and delivers energy to grid during fault at maximum efficiency. Five algorithms which have been simulated to analyze overall power quality are MPPT, Phase Locked Loop (PLL), current control, Sinusoidal Pulse Width Modulation (SPWM), [17] and fuzzy logic controlled DVR.

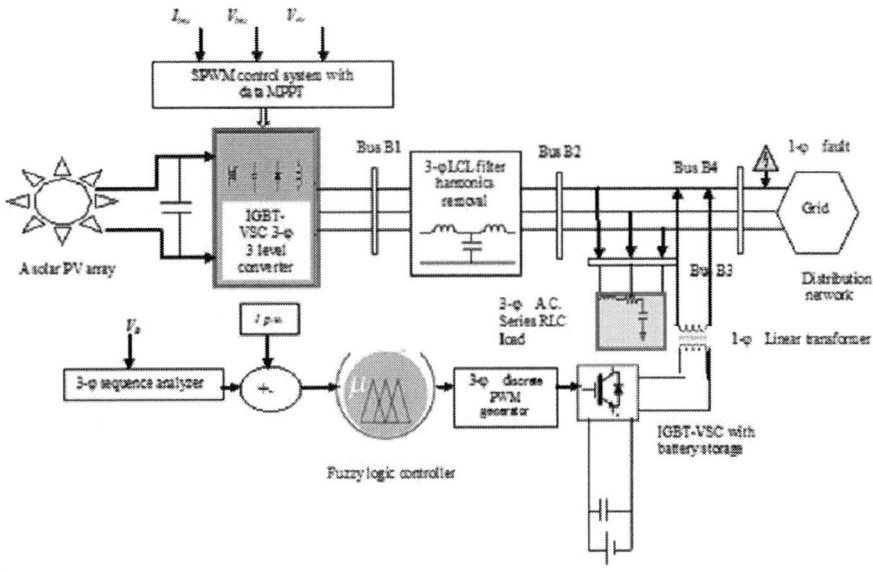

Fig. 1.   Block diagram of solar PV power conditioning system with fuzzy logic controlled DVR and MPPT control system

III. CONTROL SYSTEM TECHNIQUES

In DG systems, all available electric power is delivered to grid. To achieve this PV system needs a control system that senses variations in PV array condition, and leads system to a new operating point (V_{mp} and I_{mp}), called maximum power points where maximum power can be extracted. The main techniques have been found to be perturb & observe or dithering in, [17], incremental conductance, constant voltage, fuzzy logic control, neural network, and ripple correlation factor [15,18].

A. Data based maximum power point technique

Due to its low cost and easier implementation, a unique method of data based maximum power point tracking technique has been developed [13]. It is expected that using the proposed technique maximum power output is obtained and substantially controls and increases output power of solar PV arrays in solar PV generation system. In this technique the data for PV voltage at maximum power point is generated [13] using six solar PV modules forming an array at different values of series resistances (R_s= 0.00011 Ω, 0.00021 Ω, 0.00031 Ω, 0.00041 Ω, 0.00051 Ω and 0.00061 Ω). Each module is simulated for changing environmental conditions i.e. low-high solar radiation levels S_x and ambient temperature T_x keeping cell temperature T_c constant. Data thus generated at maximum power has been tabulated in Table I and Table II. It is seen that array output voltage decreases linearly when series resistance and ambient temperature increases on right hand side of maximum power point. Output power of a single PV module depends upon output voltage, ambient temperature and solar radiation level. From Table I and Table II, four data sets are formed using four different cell temperatures showing model mismatch which includes variation of solar radiation levels and ambient temperature throughout the day at maximum power. Using the P-V characteristics, maximum power point is chosen for a particular solar radiation level at a specific cell temperature. The data given in Table I and Table II is considered using look-up table MPPT control model for solar PV energy conversion system shown in Fig. 1, with the solving method used is Interpolation-extrapolation.

The DC link voltage [4,6] is controlled according to the simulated voltage control scheme shown in Fig. 2. Data (at different solar radiation and ambient temperature) at maximum power is selected to generate reference signal V_{dc_Ref} (as shown in Table I and Table II). Using discrete type PI voltage regulator, this signal is compared with actual DC link voltage V_{dc} to generate reference signal I_{d_Ref}. To control reactive power, quadrature axis reference current I_{q_Ref} is generated from load connected, and grid side at unity power factor. After multiplication of these two vector signals a new signal I_d is

generated. In inner current controlled loop the signal $I_d I_{q\_Ref}$ is compared with actual signal $I_d I_q$. Here actual I_{q_Ref} quadrature axis current I_q is compared with I_{q_Ref} from inverter side to generate an error ΔI_q. Similarly, actual direct axis I_d is compared with I_{d_Ref} of the inverter to generate an error ΔI_d. These two error signals are converted into V_d and V_q with PI controllers. Modulation index m_a and angle δ are calculated respectively, as

$$m_a = \sqrt{\left(V_d^2 + V_q^2\right)}$$
$$\delta = \tan^{-1}\left(V_d / V_q\right)$$

TABLE I.   DATA GENERATED AT DIFFERENT AMBIENT TEMPERATURES AND LOW SOLAR RADIATION (AT CONSTANT T_c) VALUES

Vector of input values look up table	Solar radiation S_x (W/m²)	Ambient Temperature T_x (°C)	Maximum power point PV voltage in volts at different cell temperatures, $V_{DC\ Ref}$			
			Temp T_c =10°C	Temp T_c =20°C	Temp T_c =30°C	Temp T_c =40°C
0	88	10	751.1	814.4	879.4	945.9
20	90	20	769.6	867.9	935.7	1005
40	92	30	817.1	887	953	1065
60	94	40	858.6	927	988	1086
80	96	50	899.3	961.1	1024	1124
100	98	60	947.3	1016	1086	1158

TABLE II.   DATA GENERATED AT DIFFERENT AMBIENT TEMPERATURES AND HIGH SOLAR RADIATION (AT CONSTANT T_c) VALUES

Vector of input values look up table	Solar radiation S_x (W/m²)	Ambient Temperature T_x (°C)	Maximum power point PV voltage in volts at different cell temperatures, $V_{dc\ Ref}$			
			Temp T_c =10°C	Temp T_c =20°C	Temp T_c =30°C	Temp T_c =40°C
0	150	10	984.4	1060	1138	1220
20	300	20	1049	1132	1217	1306
40	450	30	1108	1197	1290	1387
60	600	40	1158	1278	1353	1457
80	750	50	1203	1305	1411	1521
100	900	60	1234	1368	1454	1571

The inverter output voltage [13-14] can be changed by changing the modulation index of inverter. When output voltage of inverter is higher than grid voltage, reactive power is supplied by PV system to grid. However the same gets absorbed when inverse action takes place.

B. Controlling action with fuzzy logic controlled DVR

The aim of this second control scheme [15] is to compensate for voltage disturbance in grid voltage magnitude on occurrence of fault. It also regulates and maintain with DC link voltage to VSC (with MPPT), constant at point where a sensitive load is connected with solar PV grid system under same system disturbances. Series voltage is injected through a 3-Φ linear transformer by a VSC connected to DC power source. FLC's are an active choice when precise mathematical formulations are not accurately possible. Unlike other controllers, FLC is capable to tolerate uncertainty and imprecision to a great extent. It has two crisp inputs.

- the difference between grid voltage and reference voltage (1 per unit) E; and

- the derivative of error dE.

The FLC consists of three stages [2], fuzzification, rule execution, and defuzzification. In first stage, crisp variables e and de are converted into fuzzy variables E and dE using triangular and trapezoidal membership functions. Both E and dE, are divided into seven fuzzy sets each: NL (negative large), NM (negative medium), NS (negative small), Z(zero), PS (positive small), PM (positive medium) and PL (positive large).

In the second stage of fuzzy logic, the fuzzy variables E and dE are processed by an inference engine that executes a set of control rules contained in 49 rule bases. These control rules are formulated using knowledge of DVR behavior. In this paper, max-min inference algorithm is used, in which final membership degree is equal to the maximum of the product of membership degree of both E and dE. The output variables

978-1-4799-6047-7/14 $31.00 © 2014 IEEE

from inference engine are converted into crisp values in defuzzification stage. By measuring rms grid voltage at PCC, compensating voltage is injected on grid side using fuzzy technique which also controls load and VSC voltage current values

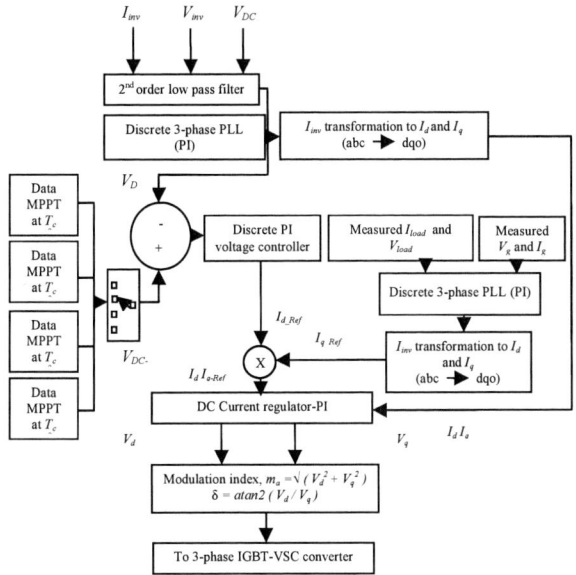

Fig. 2. Flow chart for MPPT control of a grid-connected PV energy conversion system

IV. RESULTS AND DISCUSSIONS

The grid connected solar PV test system [14] with its MPPT and FLC controlled DVR is modeled with Matlab/Simulink and SimPowersystems block set. A 3-Φ series linear RLC load is connected on the output side. Fig. 3 depicts simulation results without DVR, and a single phase to ground fault (on phase R) is created at PCC on grid side. The fault and ground resistance values are 0.066 Ω and 0.001 Ω, respectively. During fault time 0.1 s to 0.4 s, above wave shapes show unbalanced sag and swell in current voltage of VSC and grid. The three-phase faulty grid voltage and current is controlled through FLC controlled DVR, among other controlled variables. When there is a change in solar radiation and ambient temperature as given in Table I and Table II, proposed MPPT control system operates by increasing/decreasing current injected into grid in order to keep DC voltage constant at solar PV array terminals in its maximum power point. Fig. 3 (d), depicts actual DC link voltage not varying exactly according to reference signal generated from look-up table.

With DVR control system connected in series with line compensating voltage is injected rapidly on grid side at PCC, and voltage sag is mitigated almost completely in all phases. This action also controls current wave shape of VSC which maintains balanced constant voltage at 1.0 per unit. Thus effective regulation with rms voltage maintained at 98 %, is provided by DVR control system shown by wave shapes in Fig. 4. Wave shape for VSC current and grid current shown in

Fig. 4 (a)-(b) and Fig. 5 (a)-(b) at low and high solar radiation shows compensation is achieved particularly at high values of solar radiation.

Also FLC controlled DVR helps PV array actual DC link voltage to track reference signal as given by Fig. 5 (c)-(d). Simultaneously, DC voltage regulator also helps actual DC voltage to track reference DC link voltage at different solar radiation S_x and ambient temperatures T_x. Tracking error of reference voltage is found to be less than 5 %.

Fig. 6 (a)-(b), exhibits the real power being generated by the solar PV system through VSC at different solar radiation and ambient temperature as given in look-up table. Till 0.1 s, solar PV array generates real power of 37.79 kW through VSC. Due to occurrence of fault between 0.1 s to 0.4 s, real power flow remains constant and after 0.4 s, reactive power is supplied through VSC as real power goes negative till 0.5 s. After 0.5 s, reverse action takes place till 0.6 s.

As depicted by Fig. 6 (a)-(b), the irregular behavior of active is mitigated when DVR is in operation, shown in Fig. 7 (a)-(b). Initially, whole real power is being supplied by grid to load. But, as solar radiation level increases, solar PV system partially supplies the 3-Φ load through VSC and grid. Maximum value of real power from solar PV array through VSC has been found to be 37.77 kW, 37.75 kW, 36.03 kW and 28.18 kW at 10°C, 20°C, 30°C, and 40°C, respectively, which indicates that real power value output decreases with increase in cell temperature.

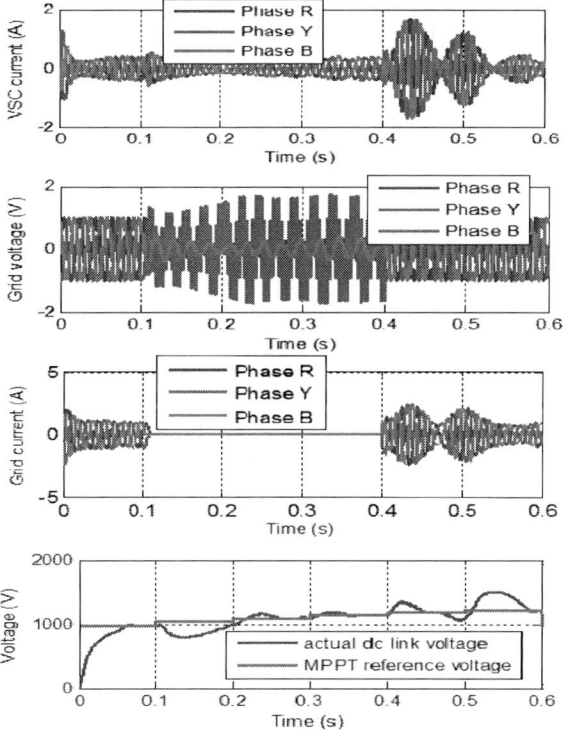

Fig. 3. Wave shapes of uncompensated a) VSC current b) grid voltage c) grid current, d) Actual DC link and MPPT reference voltage, without DVR control during fault at 10°C.

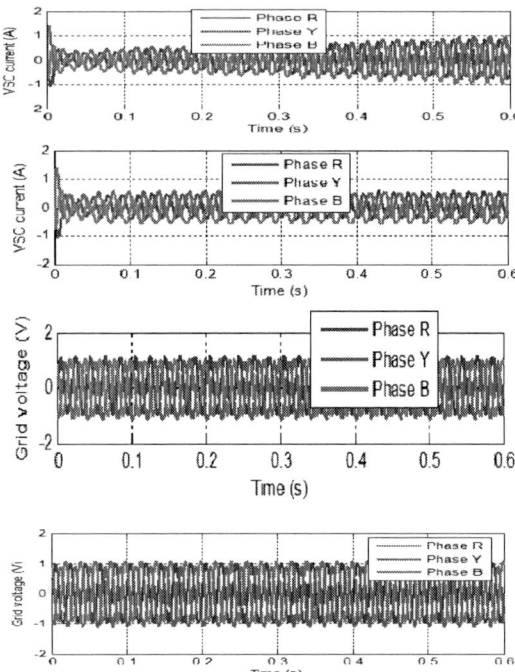

Fig. 4. Wave shapes of compensated (a)-(b) VSC current, at low and high solar radiation respectively, during fault at 10°C (c)-(d) Wave shapes of compensated grid voltage, at low and high solar radiation respectively, during fault at 10°C.

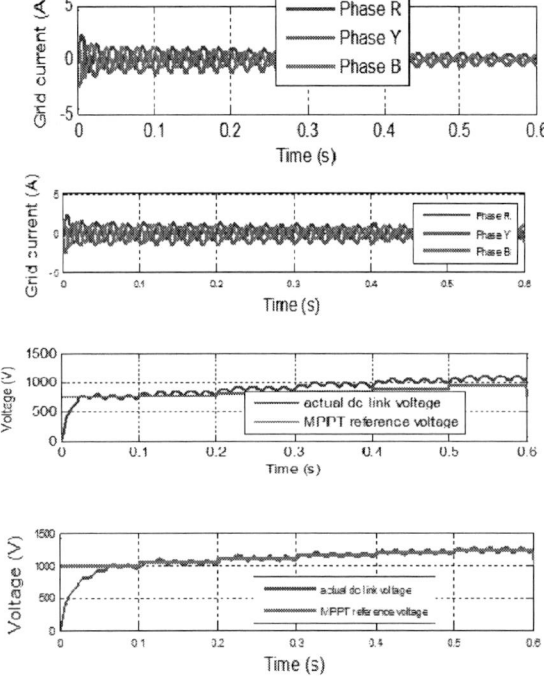

Fig. 5. (a)-(b) Wave shapes of compensated grid current, at low and high solar radiation respectively, (c) Actual DC link and MPPT reference voltage during fault at low solar radiation (d) Actual DC link and MPPT reference voltage during fault at high solar radiation during at 10°C.

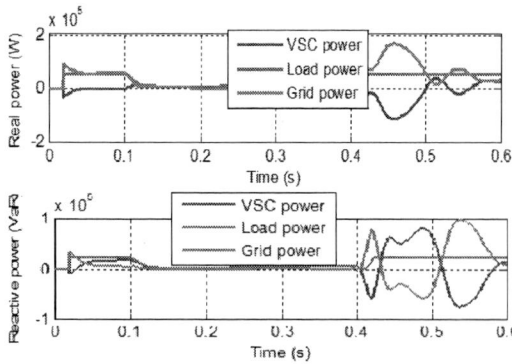

Fig. 6. (a)-(b) Real and Reactive output power respectively, with MPPT and without fuzzy logic controlled DVR control during fault at 10°C

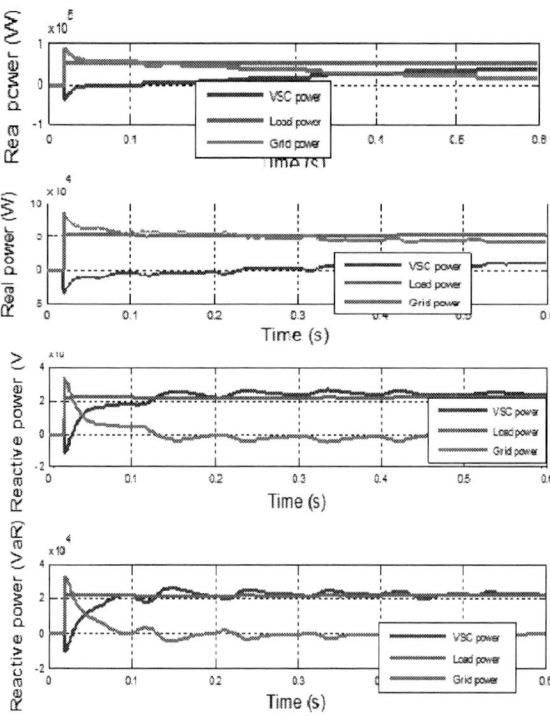

Fig. 7. Real output power (a) at low and (b) at high solar radiation respectively, Reactive output power at (c) low and (d) high solar radiation respectively, with MPPT and fuzzy logic controlled DVR control during fault at 10°C.

Maximum value of reactive power as shown by Fig. 7 (c)-(d) from solar PV array through VSC has been found to be 26.4 kVAR, 26.31 kVAR, 26.54 kVAR and 26.25 kVAR at 10°C, 20°C, 30°C, and 40°C respectively which indicates that the reactive power wave shape becomes constant periodic with the increase in ambient temperature, particularly after 0.12 s. Initially grid supplies reactive power requirement of inductive load, however, at around 1.0 s, reactive power requirement is completed by solar PV array through VSC and continues till end.

THD for grid voltage with fault and without DVR operation is found to be 2.37 %, 2.40 %, 2.43 % and 2.12 % at 10°C, 20°C, 30°C, and 40°C, respectively. As grid voltage is controlled through fuzzy logic controlled DVR, its THD as given in Table III at different cell temperature T_c is found to be less than 1 % whereas THD for grid current is found to be around 5 % as prescribed by IEEE-519/1547 standard. However, THD of VSC current is found to be slightly high which is due to its high frequency switching.

Implemented fuzzy logic controlled DVR acts as an active filter which performs DC voltage compensation by reducing DC current component flowing in grid power lines. This component causes increased magnetizing currents with power losses and, consequently, overheating. The average value of DC current component value for grid voltage and VSC current is found to be 0.09473 A and 0.1072 A (values given in Table III).

V. CONCLUSION

This paper has emphasized an approach of modeling and controlling of a grid-connected solar PV system with its MPPT control in conjunction with fuzzy logic controlled DVR. The simulation of the two control systems has been done in Matlab-Simulink and it showed an excellent coordination of both VSC and its implemented data based MPPT with fuzzy logic controlled DVR during occurrence of dynamic fault. Fuzzy logic control has been found to be more effective than the traditional controllers for nonlinear systems. It has been found that there is negligible fluctuation of DC bus voltage, fast tracking of optimum operating point, robustness of the PLL and a unity power factor is achieved. Simultaneous exchange of distortion less active and reactive power with the distribution system has also been shown, along with mitigation of voltage sag during single phase to ground fault. The proposed MPPT has proved its utility in the optimization of solar PV power generation and performs power quality control to reduce THD currents as given in standards IEEE-519/1547.

TABLE III.   IMPACT OF FUZZY LOGIC CONTROLLED DVR ON TOTAL HARMONIC DISTORTION AT LOW AND HIGH SOLAR RADIATIONS

	THD at 10^0 C at		THD at 20^0 C		THD at 30^0 C		THD at 40^0 C	
	Low S_x	High S_x	Low S_x	High S_x	Low S_x	High S_x	Low S_x	High S_x
Grid voltage (DC componentin A)	0.02 %	0.02 % (0.09478)	0.02 %	0.02 % (0.09471)	0.02 %	0.02 % (0.09471)	0.02 %	0.02 % (0.09472)
VSC current (DC component in A)	4.57 %	10.93% (0.09873)	6.17 %	8.80 % (0.1098)	9.38 %	8.13% (0.1106)	10.21 %	7.89% (0.11)
Grid current	4.19 %	5.62 %	5.65 %	3.91 %	8.04 %	3.62 %	6.93 %	3.51 %

REFERENCES

[1] N.G.Hingorani, "Introducing Custom Power", IEEE Spectrum, vol. 32, pp. 41-48, 1995.

[2] H. Ezoji, A. Sheikholeslami ,M. Rezanezhad, and H. Livani, "A new control method for dynamic voltage restorer with asymmetrical inverter legs based on fuzzy logic controller," Simulation Modeling Practice and Theory,vol. 18, pp. 806-19, 2010.

[3] M. S. J. Asghar, Power Electronics, 8th Ed., Prentice Hall India, New Delhi, pp. 300-301, 2011.

[4] P. Boonchiam and N. Mithulananthan, "Diode clamped multilevel voltage source converter based on medium voltage DVR," Journal of Electrical Power and Energy Systems Engg, vol. 1, pp. 62-67, 2008.

[5] S. V. R. Kumar and S. S. Nagaraju, "Simulation of D-STATCOM and DVR in power systems," ARPN Journal of Engg.and Applied Sciences, vol. 2, pp. 7-13, 2007.

[6] A. R. Dash, B. C. Babu, K. B. Mohanty, and R. Dubey, "Analysis of PI and PR controllers for distributed power generation system under unbalanced grid faults," Proc. Int. Conf. Power and Energy systems, pp. 1-6, 2011.

[7] R. K. Varma, V. Khadkikar, and R. Seethapathy, "Nighttime application of PV solar farm as STATCOM to regulate grid voltage,"IEEE Transactions on Energy Conversion, vol. 24, pp. 983-985, 2009.

[8] C. Mayr, R. Brundlinger, and B. Bletterie, "Photovoltaic-inverters as active filters to improve power quality in the grid. What can state-of-the-art equipment archive?,"Proc. IEEE Int. Conf. Electrical Power quality and utilization, pp. 1-5, 2007.

[9] H. R. Seo, G.H. Kim, S.Y. Kim, N. Kim, H.G. Lee, C. Hwang, M. Park,and I. K. Yu, "Power quality control strategy for grid-connected renewable energy sources using PV array and super capacitor," Proc.

IEEE Int. Conf. Electrical Machines and Systems (ICEMS),pp. 437-441, 2010.

[10] MATLAB/SIMULINK, The Mathworks, Inc. 7.10.0.499 (R2010a), http://www.mathworks.com

[11] S. W. Lee, J. H. Kim, S. R. Lee, B. K. Lee, and C. Y. Won, "A transformerless grid-connected photovoltaic system with active and reactive power control," Proc. 6th IEEE Int. Conf. Power Electronics and Motion Control, pp. 2178-2181, May 2009.

[12] I.H. Altas, and A.M. Sharaf, "A PV array simulation model for Matlab-Simulink GUI environment," Proc. Int. Conf. Clean Electrical Power Capri (ICCEP), pp. 341-345, May 2007.

[13] A. Gupta, S. Chanana, and T. Thakur,"Power quality improvement of solar PV transformerless grid connected system with maximum power point tracking control,"International Journal of Sustainable Energy,vol. 33, pp. 921-936, 2014.

[14] A. Gupta, S. Chanana, and T. Thakur, "Power quality investigation of a solar PV transformer-less grid connected system fed DVR," Frontiers in Energy, vol. 8, pp. 240-253, 2014.

[15] A. Gupta, S. Chanana, and T. Thakur, "THD reduction with reactive power compensation for fuzzy logic DVR based solar PV grid connected system," Frontiers in Energy, vol. 8, 2014.

[16] R. Ramabadran, and B. Mathur, "MATLAB based modeling and performance study of series connected SPVA under partial shaded conditions,"Journal of Sustainable Development, vol. 2, pp. 85-94, 2009.

[17] N. Mohan, and T.M. Undeland ,Power Electronics: Converters, Applications, and Design 2nd Edition, John Wiley& Sons, New York, 2007.

[18] H. Moin, "Investigation to improve the control and operation of a three-phase PV grid-tie inverter," Ph.D. Diss., Dublin Institute of Technology, Ireland, 2011.

978-1-4799-6047-7/14 $31.00 © 2014 IEEE

Effect of non-uniform irradiance on electrical characteristics of an assembly of PV panels

M A Hasan and S K Parida
Department of electrical engineering
Indian Institute of Technology, IIT Patna
Patna, India
Email: hasan.pee13@iitp.ac.in

Abstract—**This paper presents the study of effect of non-uniform irradiance on electrical characteristics of an assembly of solar PV panel. Study is based on a one diode model of PV panel. Effect is simulated in MATLAB-Simulink environment. In a solar power plant series Parallel connections of PV panels are used for obtaining desired voltage and current level. Under mismatch or non-uniform irradiance condition a hot spot formation may damage the PV panel. A bypass diode is provided with PV panel to avoid hot spot formation. Under non-uniform irradiance condition, due to presence of bypass diode, I-V and P-V characteristics are quite different from that of uniform irradiance condition. Results obtained show the presence of multiple peaks in P-V characteristics under non-uniform irradiance. This establishes requirement of a non-conventional maximum power point tracker algorithm.**

I. NOMENCLATURE

V_{pv} is output voltage of a PV module (V)
I_{pv} is output current of a PV module (A)
T_r is the reference temperature = 298 K
T is the module operating temperature in Kelvin
I_{ph} is the light generated current in a PV module (A)
I_o is the PV module saturation current (A)
A is diode ideality factor
K is Boltzman constant = 1.3805×10^{-23} J/K
q is Electron charge = 1.6×10^{-19} C
R_s is series resistance of a PV module
I_{SCr} is PV module short-circuit current at 250 C, 1000 W/m^2
K_i is the short-circuit current temperature co-efficient
N_s is the number of cells connected in series
N_p is the number of cells connected in parallel

II. INTRODUCTION

A PV panel consists of solar cells connected in series and parallel configuration to generate desired level of voltage and current. A number of such PV panels are again connected in series and parallel to form a string which generates electrical power at desired voltage level. The current-voltage characteristics of PV panel are highly non-linear and depend upon the level of solar irradiance falling on it [1]–[4]. Under mismatch condition or non-uniform irradiation, some of the solar cells that form PV panel gets reverse biased. This causes formation of hot spots and dissipation of power in the form of heat [5]–[7]. This may cause the damage of PV panel if heat exceeds the limit. To avoid this hot spot formation, bypass diode is provided. Presence of bypass diode avoids the hot spot formation but causes reduction in output power of PV panel.

Every PV panel is characterized by its rated short circuit current and rated open circuit voltage at 1000 W/m2 solar irradiance and 250 C temperatures. When a current more than rated short circuit current is drawn from PV panel, bypass diode short circuit the PV output terminal and its voltage contribution becomes zero. For a string of PV panels, if non-uniform irradiance takes place, contribution of each PV panel depends upon solar irradiance received by it. Overall I-V and P-V characteristics of PV string necessarily becomes different from that of an individual PV panel.

This paper shows the effect of non-uniform irradiance on I-V and P-V characteristics of PV string. This includes simulation of PV panel in MATLAB-Simulink environment. Effect of non-uniform irradiance is shown through simulation results obtained for different levels of solar radiation.

This paper is organized as follows. Section 3 discusses mathematical formulation of PV panel. Effect of non-uniform irradiance is discussed in section 3. Section 4 presents simulation and discussion of results. Conclusion follows next.

III. PV PANEL MODELLING

Electrical equivalent circuit of PV panel used for simulation in this paper is given by a one diode circuit with one series and a shunt resistance. Photocurrent is represented by a current source I_{ph}. Fig. 1 gives the circuit diagram of one diode electrical equivalent circuit of PV panel. Based on the circuit, following equations can be written as given by (1-4).

$$I_{ph} = I_0 + I_{sh} + I_{pv} \tag{1}$$

$$I_{rs} = I_{scr}[exp(qV_{oc}/N_s kAT) - 1] \tag{2}$$

$$I_0 = I_{rs}[\frac{T}{T_r}]^3 exp[\frac{qE_{go}}{Bk}(\frac{1}{T_r} - \frac{1}{T})] \tag{3}$$

$$I_{pv} = N_P I_{ph} - N_P I_o[exp\{\frac{q(V_{pv} + I_{pv}R_s)}{N_s AkT}\} - 1] \tag{4}$$

From equations, it can be observed that I-V and P-V characteristics of PV panel are non-linear. Due to this modelling and simulation of PV panel characteristics require numerical technique to be applied. Simulation prepared in this paper uses 0de45 as numerical technique.

IV. NON UNIFORM IRRADIANCE EFFECT

In a solar power plant, several PV panels are connected in series to achieve desired voltage level. Voltage VT at bus

978-1-4799-6047-7/14 $31.00 © 2014 IEEE

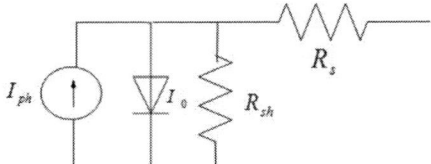

Fig. 1. One diode model of PV cell

TABLE I.   SPECIFICATION OF PV PANEL

Specification	Value
Voltage at maximum Power	65.5 V
Current at maximum power	1.07 A
Rated open circuit voltage	88 V
Reated short circuit current	1.23 A
Total no of cells in series and parallel	116/1

terminal due to series connection of n PV panels is given by (5).

$$V_T = \Sigma_{i=1}^{n} V_i \qquad (5)$$

In series connection, current passing through all PV panels has to be same in magnitude. If different PV panels have different solar irradiance as shown in fig (2), their respective short circuit currents at that particular irradiance will also be different. Due to different I-V characteristics of these panels as shown in fig (2), hot spot gets formed in PV panels. This may cause damage of these panels. In order to avoid formation of hot spots, bypass diode is provided with PV panels. These bypass diodes short circuit respective PV panel terminal to avoid flow of current that is more that short circuit current at that irradiance.

Case study done in this paper considers five different cases of uniform and non-uniform irradiance. These are enumerated as case 1-4 and details are given in table 2. I-V and P-V characteristics have been obtained for all the cases. Uniform irradiance case considers fall of standard 1000 W/m^2 irradiance for all panels.

V. SIMULATION AND RESULTS

Simulation of a 70 Watt PV panel in MATLAB environment is prepared. Electrical characteristics datasheet of PV panel is given in table 1.

Simulation result for I-V characteristics of PV panel for different irradiance (i.e. 200, 400, 600, 800, 100 W/m^2) shows different short circuit current values. Under non-uniform irradiance condition, as shown in fig. 4, bypass diode comes into effect.

Fig. 3 gives the simulation for non-irradiance effect. Bypass diode effect is taken care by conditional switch. Current generated by PV panel receiving maximum irradiance is compared with the currents of all other PV panels. If maximum current exceeds the short circuit current of any PV panel, its voltage contribution becomes zero showing the operation of bypass diode.

VI. DISCUSSION AND CONCLUSION

Fig. 5-8 represent the I-V and P-V characteristics when all three PV panels are having different irradiances. Three levels

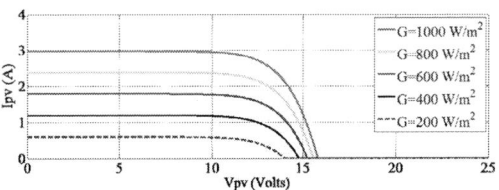

Fig. 2. I-V characteristics of PV panel for different illuminations

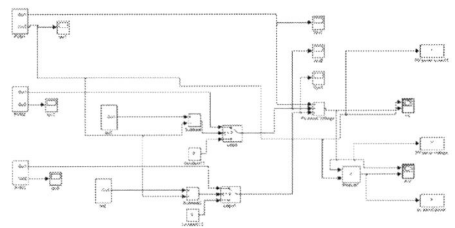

Fig. 3. Simulation model presenting non-uniform irradiance effect

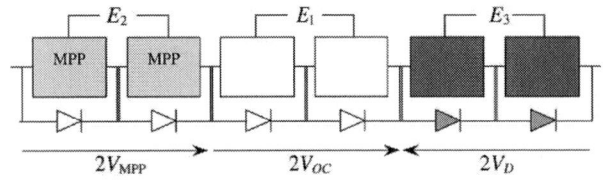

Fig. 4. PV panel receiving non-uniform irradiance

TABLE II.   OPERATING CONDITIONS FOR CASE STUDY

Case	Operating Condition
1	G=1000, G1=800, G2=600
2	G=1000, G1=800, G2=400
3	G=1000, G1=800, G2=0
4	G=1000, G1=1000, G2=1000

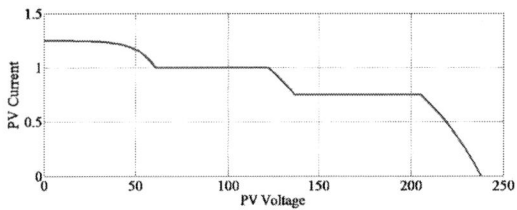

Fig. 5. I-V characteristics for case 1

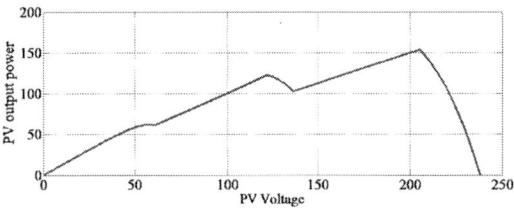

Fig. 6. P-V characteristics for case 1

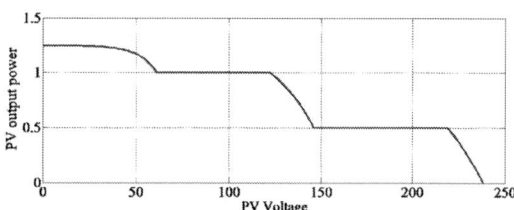

Fig. 7.   I-V characteristics for case 2

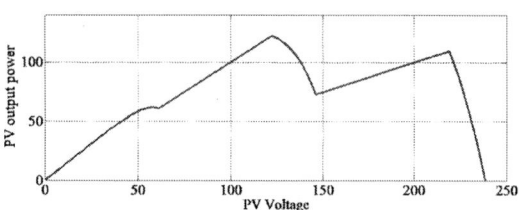

Fig. 8.   P-V characteristics for case 2

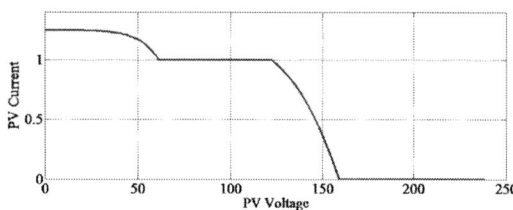

Fig. 9.   I-V characteristics for case

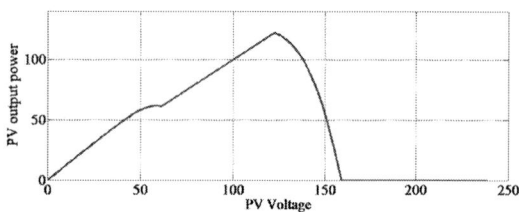

Fig. 10.   P-V characteristics for case 3

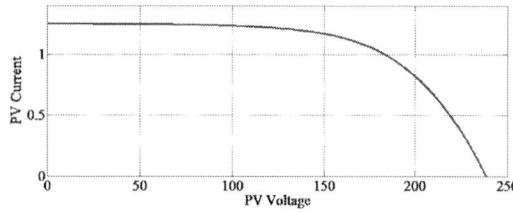

Fig. 11.   I-V characteristics for case 4

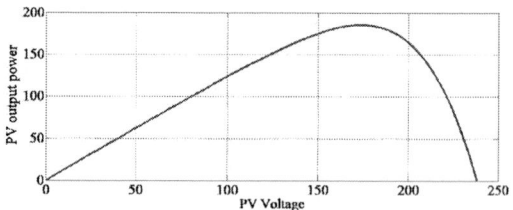

Fig. 12.   P-V characteristics for case 4

of sudden change in current and power can be observed. P-V characteristics has multiple peaks. This makes the maximum power point (MPP) difficult to be captured using conventional hill climbing technique. A conventional MPP tracker technique will find the first peak and fix it it as MPP.Hence need of a separate MPP tracker algorithm is established.

Fig 9-10 shows the I-V and P-V characteristics for case 3. In this case one PV panel receives zero irradinace. This makes system as of consisting of only two PV panels. Only two levels of change in current and power magnitude is observed for the case.

Fig 11-12 present the I-V and P-V chaarcteristics for uniform irradiance case. When all PV panels in a solar power plant receive same level of irradiance, I-V and P-V cahracteristics becomes same as that of a single PV panel. A single MPP is depicted in this case.

REFERENCES

[1]   M. Villalva, J. Gazoli, and E. Filho, "Comprehensive approach to modeling and simulation of photovoltaic arrays," *Power Electronics, IEEE Transactions on*, vol. 24, no. 5, pp. 1198–1208, May 2009.

[2]   B. Babu and S. Gurjar, "A novel simplified two-diode model of photovoltaic (pv) module," *Photovoltaics, IEEE Journal of*, vol. 4, no. 4, pp. 1156–1161, July 2014.

[3]   S. Rahman, R. Varma, and T. Vanderheide, "Generalised model of a photovoltaic panel," *Renewable Power Generation, IET*, vol. 8, no. 3, pp. 217–229, April 2014.

[4]   O. Breitenstein, "An alternative one-diode model for illuminated solar cells," *Photovoltaics, IEEE Journal of*, vol. 4, no. 3, pp. 899–905, May 2014.

[5]   S. Silvestre, A. Boronat, and A. Chouder, "Study of bypass diodes configuration on {PV} modules," *Applied Energy*, vol. 86, no. 9, pp. 1632 – 1640, 2009. [Online]. Available: http://www.sciencedirect.com/science/article/pii/S0306261909000269

[6]   M. Danner and K. Bucher, "Reverse characteristics of commercial silicon solar cells-impact on hot spot temperatures and module integrity," in *Photovoltaic Specialists Conference, 1997., Conference Record of the Twenty-Sixth IEEE*, Sep 1997, pp. 1137–1140.

[7]   C. Barreiro, P. Jansson, A. Thompson, and J. Schmalzel, "Pv by-pass diode performance in landscape and portrait modalities," in *Photovoltaic Specialists Conference (PVSC), 2011 37th IEEE*, June 2011, pp. 003 097–003 102.

978-1-4799-6047-7/14 $31.00 © 2014 IEEE

Performance Monitoring of $43\,kW$ Thin-film Grid-Connected Roof-top Solar PV System

Suresh Singh
Indian Institute of Technology Jodhpur
Rajasthan, INDIA-342011
Email: sureshkumar@iitj.ac.in

Rakesh Kumar
Indian Institute of Technology Jodhpur
Rajasthan, INDIA-342011
Email: PG201283007@iitj.ac.in

Vivek Vijay
Indian Institute of Technology Jodhpur
Rajasthan, INDIA-342011
Email: vivek@iitj.ac.in

Abstract—In the emerging photo-voltaic markets worldwide the use of appropriate estimation of the operating performance of the grid-connected PV systems is becoming more and more crucial. The widely used performance indices that are used to analyze the system performance with respect to energy output, solar resource and the effect of system losses in its various components are total yield, final yield/specific yield, performance ratio, PVUSA rating, and performance indicator based on ratio of ac power at PTC to dc power at STC. This paper presents performance analysis of a $43\,kW$ amorphous-silicon thin-film grid-connected PV system. Recorded plant monitoring data during July 2011 to July 2014 is used for the analysis. Major operation and maintenance issues encountered in last three years are also reported.

Index Terms—Capacity factor, PVUSA rating, Performance analysis, Performance ratio, specific yield/yield factor/final yield,

I. INTRODUCTION

The generation of electrical energy using renewable energy based distributed generating units is rapidly increasing worldwide. Wind and solar energy based power generation is particularly getting increased attention. In the last couple of years installation of wind, solar PV, and solar thermal plants have been going on at very fast rate in the western Rajasthan (India), due to significant availability of the solar and wind resource. The plant under study is installed in the Jodhpur city of western Rajasthan which popularly known as "Sun City" due to the availability of bright sun approximately 320 days in a year. It has been more than three years since the plant is installed. In this paper, the operational performance of the plant is analyzed using widely used performance indices for the performance evaluation of the grid-connected PV system.

The performance of a grid-connected PV systems depends on many factor such as solar resource, ambient temperature, PV module technology, and inverter efficiency [1]. The frequently used indices to quantify the plant performance are-net/total energy output, final yield/yield factor/specific yield, reference yield, capacity factor, and performance ratio [2]–[12]. In order to compare the performance of the PV systems of different size and locations, normalized performance indices-final yield and performance ratio have been widely used by the researchers [13]–[19]. Authors in [15], [18], have evaluated comparative performance analysis of hundreds of PV systems in 14 different countries based on performance ratio.

Tripathi et.al. in [13], have analyzed comparative performance analysis of amorphous and multicrystalline technology based PV systems using final yield and performance ratio. Authors in [14], have reported performance comparison of the mono and polycrystalline-silicon based PV systems using final yield, reference yield, system efficiency, performance ratio, and capacity factor. In [8], performance of small grid-connected PV systems for residential use have been studied using final yield and performance ratio.

In this paper, authors present the performance analysis of a $43\,kW$ amorphous-silicon thin-film grid-connected PV system installed at the roof-top of Indian Institute of Technology Jodhpur, Jodhpur (India), academic campus. The monitoring data including climate data (solar irradiance, module temperature, and ambient temperature) and system electrical data recorded during three years July 2011-July 2014, have been used in the analysis. The chosen time scale is mainly monthly and in few cases yearly. The performance is evaluated based on net/total yield, final/specific yield, performance ratio, and modified PVUSA rating based performance ratio of ac power produced at PTC (PVUSA test conditions) to dc power at STC (standard test conditions). Lastly, issues encountered in the plant operation and components fault during three years are also reported.

The rest of the paper is organized as follows. In section II, brief system technical description is given. Section III, discusses indices used to analyze the system performance with respect to energy production, solar resource and different losses. Major operational issues encountered in last three years are also reported. Obtained results and discussion are presented in section IV and finally conclusions are drawn in Section V.

II. SYSTEM DESCRIPTION

The 43 kW grid-connected Amorphous PV system under study was installed on the roof of the Block-I, Indian Institute of Technology Jodhpur transit campus, in May 2011. The representative schematic of the system is shown in Fig. 1. The PV array consist of MOSER BAER thin-film Amorphous-Silicon module fixed to the roof of the building using metal mounting structure at an optimum fixed tilt angle. The module technical specifications are given in TABLE I. The overall system technical specifications are given in table II. The grid connected single phase SMA SUNNY BOY inverters are used

978-1-4799-6047-7/14 $31.00 © 2014 IEEE

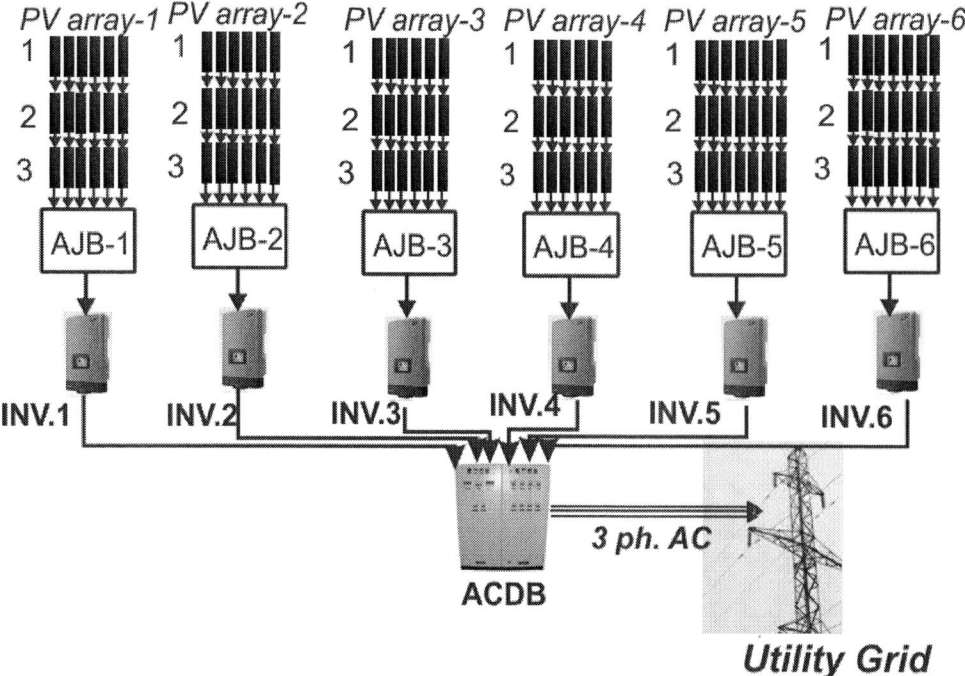

Fig. 1. 43 kW grid-connected amorphous-silicon thin-film roof-top PV system

in plant. As shown in Fig. 1, one strings contains three modules and six or seven strings form an array. The first four arrays contains 6 strings each and last two arrays contains seven strings each. The output of six arrays is collected at array junction box (AJB) from where it is connected to the inverters. The output of the inverters are collected at an AC distribution box (ACDB) and connected to form a three phase system. The plant is connected to the public utility grid via three phase energy meter at 415 V level. The system also uses SUNNY SENSOR BOX and SUNNY WEB BOX for the measurement and monitoring of climate and electrical parameters of the systems.

TABLE I
MODULE SPECIFICATIONS

Sr. No.	Parameter	Value
1	Model	Power Series FS $BIN380$
2	Maximum Power P_{mpp}	380 W
3	Voltage at Maximum power V_m	143.4 V
4	Current at maximum power I_m	2.65 A
5	Open Circuit voltage V_{oc}	187.8 V
6	Short circuit current I_{sc}	3.27 A
7	NOCT	$47 \pm 2\ ^{o}C$
8	Operating Temp.	$-40\ to + 85\ ^{o}C$
9	Temp. coeff. of P_{mpp}	$-0.2\ \%/\ ^{o}C$
10	Temp. coeff. of V_{oc}	$-0.34\ \%/\ ^{o}C$
11	Temp. coeff. of I_{sc}	$0.09\ \%/\ ^{o}C$

III. SYSTEM PERFORMANCE PAAMETERS

The performance parameters used to evaluate the performance of the PV system under study are briefly described in

TABLE II
PLANT SPECIFICATIONS

Sr. No.	Parameter	Value
1	Rated power at STC	43.3 kW
2	Number of modules	114
3	Number of inverters (SMA 7000 HV)	6
4	Number of modules/string	3
5	Number of strings	38
6	Number of strings/inverter	6/7
7	Plant output	Three phase $415V\ AC$

this Section.

A. *Total Yield*

Total yield is defined as the amount of net or total energy output (kWh) from the plant in a given time. It is the energy that is injected into the grid

B. *Final Yield/specific yield/yield factor*

Final PV yield or specific yield or yield factor is the ratio of net energy output E of the system to nameplate dc power P_o of the installed PV array at STC (1000 w/m^2,25^0C). It represents the number of hours that the PV array would need to operate at its rated power to provide the same energy and its unit is hours. The Y_f normalizes the energy produced with respect to the system size. It is a convenient way to compare the energy produced by PV systems of different sizes [1].

$$FinalYield,\ Y_f = \frac{E}{P_o}\ (hours) \qquad (1)$$

978-1-4799-6047-7/14 $31.00 © 2014 IEEE

C. Performance ratio (PR)

The plant performance ratio is one of the mostly used performance indicator and commonly known as plant quality factor which can be effectively used to compare plants installed at different locations. The PR is defined as the ratio of actual and theoretical or nominal plant output.

$$PR = \frac{Actual\ plat\ output\ (kWh)}{Calculated,\ nominal\ plant\ output(kWh)} \quad (2)$$

Where nominal plant output can be calculated using following relation

Nominal plant output=incident solar irradiation at the modules surface of the plant (kWh) * efficiency of the PV modules

PR can be determined on daily, monthly or yearly basis. It indicates the proportion of the energy available for export to the grid after deducting thermal and conduction losses. The performance ratio tells the plant owners that how energy efficient and reliable their plant is. The determination and monitoring of PR at regular intervals can lead towards the possible faults and other issues in case abnormal deviation is observed in the PR value. There are number of factors which influence the PR such as temperature, irradiance, soiling of module or sensors, module and inverter efficiency, solar technology, and recording period.

D. PVUSA Rating

In order to account for the change in the power produced due to higher cell temperature, PVUSA rating system is used to calculate the ac power produced under PVUSA test conditions (PTC). The PTC test conditions are defined as irradiance of 1000 W/m^2, ambient temperature of 20 ^oC and wind speed of 1 m/sec. The power produced at PTC, $P_{ac,PTC}$ is good indicator of the actual power delivered at full sun $(1000\ W/m^2)$ irradiance as compared to name plate rating $P_{dc,STC}$. The difference between $P_{dc,STC}$ and $P_{ac,PTC}$ is a good indication of the system losses associated with dc to ac conversion. The reduction in the PVUSA rating with time reflects permanent loss in the system performance [1], [20].

In order to determine $P_{ac,PTC}$, first step is to calculate cell temperature T_{cell} using following relation.

$$T_{cell} = T_a + (\frac{NOCT - T_a}{0.8})G \quad (3)$$

Where G is full sun solar irradiance. Now to obtain $P_{ac,PTC}$ following relations are used.

$$P_{dc,PTC} = P_{dc,STC}[1 + k_{mpp}(T_{cell} - 25)] \quad (4)$$

$$P_{ac,PTC} = P_{dc,PTC} * \eta_d * \eta_m * \eta_i \quad (5)$$

Where, NOCT is the normal operating cell temperature, k_{mpp} is temperature coefficient of the maximum power, η_d is the efficiency factor due to dirt accumulation on the module surface, η_m is the efficiency factor due to module mismatch and η_i is the module efficiency.

In the presented work, authors use ratio of $P_{ac,PTC}$ to $P_{dc,STC}$ to estimate the the system losses associated with dc to ac conversion. Modified PTC conditions based on actual monthly mean module temperature rather than one obtained using fixed ambient temperature of 20 ^oC is considered in the study. Specified values of the k_{mpp} and η_i are used while the realistic values are assumed for η_d and η_m.

E. Operation and Maintenance Issues

Despite the best efforts extended to maintain the system in efficient working conditions, following minor problems were encountered during last three years of plant service.

a   Module crack: Full size amorphous-silicon modules which comes without any supporting frame were used in the plant. It was challenging task to install such a huge and heavy modules (100 kgs) at the roof-top of a building. It was observed that in few months 2-3 modules developed cracks which normally started from the edges and traveled through the entire module within a short time leading to the hot-spots. At present the number of such modules have reached 12. Although, the thorough study of these cracks is yet to be done but thermal and mechanical stresses may be possible reasons for it. Fig.2, shows images of the some cracks.

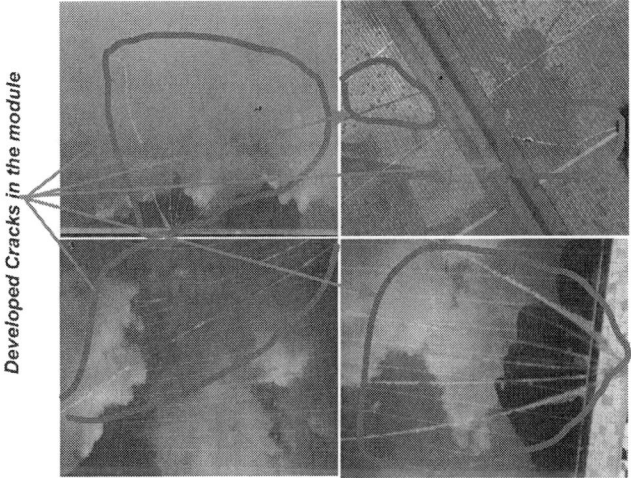

Fig. 2. *Cracks developed in the PV system modules*

b   Communication faults: There were some occurrences when motoring system of the plant reported communication fault and for that duration of the fault, the monitoring data could not be transmitted to the data acquisition system. But as such there was no loss of data as data also gets stored in the memory card present in the SUNNY WEBOX.

c   Non-availability of the grid supply: Although such occurrences were very limited and of course unavoidable in nature, but system remained idle whenever there was no availability of the grid supply.

d   Dust storm: Due to local climate conditions, dust storm frequency is quite high in Jodhpur as compared to other parts of the country. In such conditions regular cleaning of the modules becomes very much essential and failure in maintaining clean modules may result in partial loss of plant output.

e Output loss due to higher cell temperature: Summers in Jodhpur are relative hotter, sometimes ambient temperature touches 48 ^oC and it remains in the range of $40 - 45$ ^oC for 2-3 months. In such condition cell temperature also gets high and even higher (sometimes reaches> 65 ^oC) for mostly glass made amorphous-silicon made modules which increases temperature dependent losses in the system.

IV. RESULTS AND DISCUSSION

The results of the performance analysis of the 43 kW roof-top PV system are presented in this Section. Initially the rated capacity of the plant was 43.3 kW but later on during in Jan. 2013, 12 modules were separated out from the plant for other research related activities, leaving system capacity of 38.76 kW. Wherever required the value of rated capacity is used accordingly. Figs. 3-9 show the evaluated plant performance using system monitoring data recorded during July 2011-July 2014. Plot shown in Fig. 3 gives monthly total energy produced by the plant which indicates monthly change in the reading of the meter used to record the plant output. Monthly plant specific yield and performance ratios are shown in Figs. 4 and 5 respectively. The specified module efficiency of 6.6 % mentioned in the product manual has been used in the calculation of nominal plant output to determine the performance ratio. The performance ratio lies in the range of $0.52 - 0.81$.

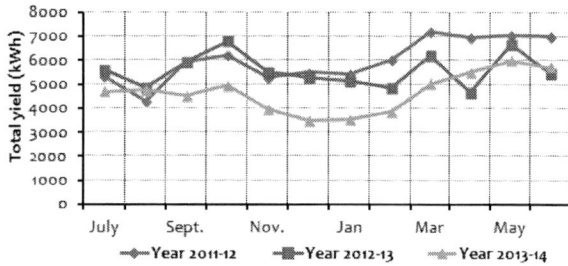

Fig. 3. *Monthly Total Yield*

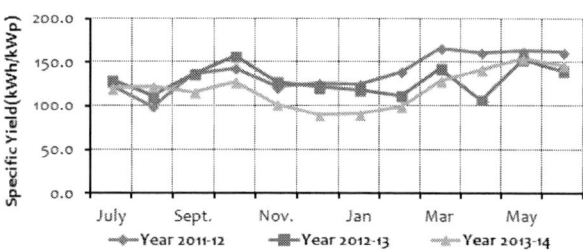

Fig. 4. *Monthly specific yield*

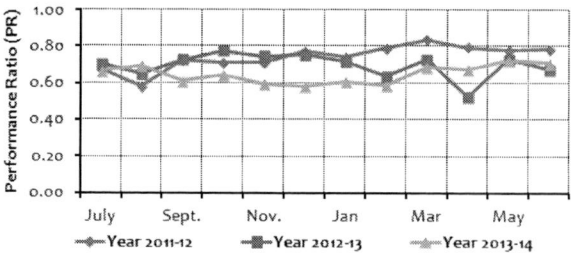

Fig. 5. *Monthly Performance ratio*

The performance indicator based on PVUSA rating which accounts for the losses associated with higher cell temperature is shown in Fig. 6. The values of η_d, η_m, and η_i considered in the calculation of $P_{ac,PTC}$ are 0.94, 0.97, 0.95 respectively. The effect of dusty atmosphere particularly in summer due to dust storms, is accounted for by considering relatively lower value of η_d. The ratio plotted indicates that the effective plant capacity in terms of ac output reduces by this factor due to higher cell temperature. The field value of the solar irradiance has to be further accounted for to get the actual ac power produced from the plant. It can be seen form the plot that this performance ratio lies in the range of $0.83 - 0.88$ indicating $12 - 17$ % losses in dc to ac conversion at $1000W/m^2$ solar irradiance, which include losses associated with soiling of module surface, module mismatch and inverter efficiency.

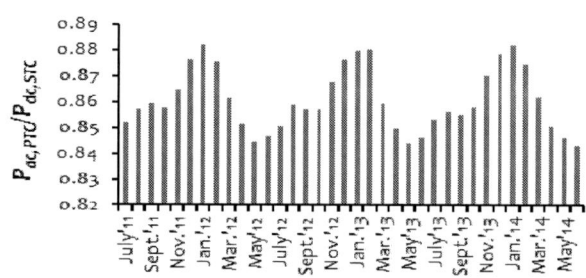

Fig. 6. *Performance indicator based on $P_{ac,STC}$*

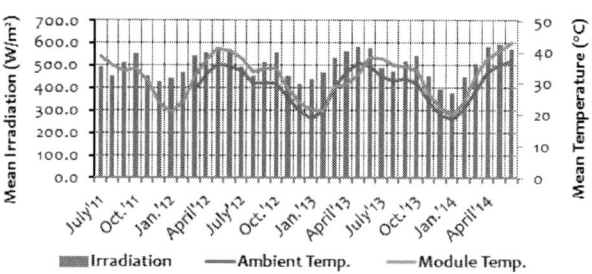

Fig. 7. *Monthly average values of the solar irradiation, ambient temperature and module temperature*

Fig. 7, shows monthly variation of the mean values of solar irradiance, ambient temperature and module temperature. It can be seen that due to some error in the sensor associated with ambient temperature, it could not be recorded for initial few months.

Energy produced by the individual inverters over three years is plotted in Fig. 8. It can be seen that there difference in individual inverter output energy that further increases during year 2013 and 2014. This is firstly due to the one string/inverter more in two of the inverters and secondly due to removed strings from two other inverters. Finally, year-wise total yield by the plant is shown in Fig. 9.

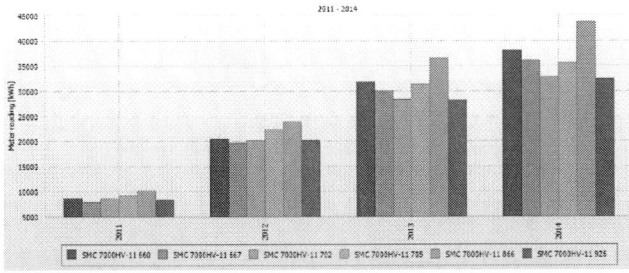

Fig. 8. *Year-wise energy produced by six inverters*

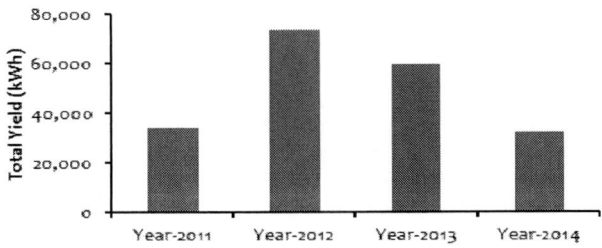

Fig. 9. *Year-wise total yield in kWh*

V. CONCLUSION

The operating performance of the $43\ kW$ grid connected roof-top PV system has been evaluated based on performance indices-total yield, specific yield and performance ratio. Modified PVUSA based performance indicator is proposed which takes into account actual measured module temperature for calculating $P_{ac,PTC}$ from $P_{dc,STC}$. The plant performance ratio was found in the range of $0.52 - 0.81$ and modified PVUSA rating based performance indicator lies in the range of $0.83 - 0.88$. The amorphous-silicon thin-film modules are very delicate and require careful handling during installation and cleaning etc. Cracks in some of the modules have been observed leading to partial to full loss of module generation capacity. Other operation and maintenance issues of importance have also been discussed.

REFERENCES

[1] B. Marion, J. Adelstein, K. Boyle, H. Hayden, B. Hammond, T. Fletcher, B. Canada, D. Narang, A. Kimber, L. Mitchell, G. Rich, and T. Townsend, "Performance parameters for grid-connected pv systems," in *Photovoltaic Specialists Conference, 2005. Conference Record of the Thirty-first IEEE*, 2005, pp. 1601–1606.

[2] H. A. Kazem, T. Khatib, K. Sopian, and W. Elmenreich, "Performance and feasibility assessment of a 1.4 kw roof top grid-connected photovoltaic power system under desertic weather conditions," *Energy and Buildings*, 2014.

[3] X. Zou, B. Li, Y. Zhai, and H. Liu, "Performance monitoring and test system for grid-connected photovoltaic systems," in *Power and Energy Engineering Conference (APPEEC), 2012 Asia-Pacific*. IEEE, 2012, pp. 1–4.

[4] S. Pietruszko, B. Fetlinski, and M. Bialecki, "Analysis of the performance of grid connected photovoltaic system," in *Photovoltaic Specialists Conference (PVSC), 2009 34th IEEE*, 2009, pp. 000 048–000 051.

[5] J.-M. Park, Z.-G. Piao, Y.-O. G.-B. Cho, and H.-L. Baek, "Performance evaluation and analysis of 50kw grid-connected pv system," in *Power Electronics, 2007. ICPE '07. 7th Internatonal Conference on*, 2007, pp. 528–530.

[6] T. Erge, V. Hoffmann, and K. Kiefer, "The german experience with grid-connected pv-systems," *Solar Energy*, vol. 70, no. 6, pp. 479 – 487, 2001.

[7] E. Kymakis, S. Kalykakis, and T. M. Papazoglou, "Performance analysis of a grid connected photovoltaic park on the island of crete," *Energy Conversion and Management*, vol. 50, no. 3, pp. 433 – 438, 2009.

[8] B. Decker and U. Jahn, "Performance of 170 grid connected {PV} plants in northern germanyanalysis of yields and optimization potentials," *Solar Energy*, vol. 59, no. 46, pp. 127 – 133, 1997.

[9] G. Chicco, R. Napoli, and F. Spertino, "Experimental evaluation of the performance of grid-connected photovoltaic systems," in *Electrotechnical Conference, 2004. MELECON 2004. Proceedings of the 12th IEEE Mediterranean*, vol. 3, 2004, pp. 1011–1016 Vol.3.

[10] M. A. Eltawil and Z. Zhao, "Grid-connected photovoltaic power systems: Technical and potential problemsa review," *Renewable and Sustainable Energy Reviews*, vol. 14, no. 1, pp. 112 – 129, 2010.

[11] W. Yi-Bo, W. Chun-Sheng, L. Hua, and X. Hong-hua, "Steady-state model and power flow analysis of grid-connected photovoltaic power system," in *Industrial Technology, 2008. ICIT 2008. IEEE International Conference on*, 2008, pp. 1–6.

[12] A. Fernndez-Infantes, J. Contreras, and J. L. Bernal-Agustn, "Design of grid connected {PV} systems considering electrical, economical and environmental aspects: A practical case," *Renewable Energy*, vol. 31, no. 13, pp. 2042 – 2062, 2006.

[13] B. Tripathi, P. Yadav, S. Rathod, and M. Kumar, "Performance analysis and comparison of two silicon material based photovoltaic technologies under actual climatic conditions in western india," *Energy Conversion and Management*, vol. 80, pp. 97–102, 2014.

[14] V. Komoni, I. Krasniqi, A. Lekaj, and I. Gashi, "Performance analysis of 3.9 kw grid connected photovoltaic systems in kosova," in *Renewable Energy Congress (IREC), 2014 5th International*. IEEE, 2014, pp. 1–6.

[15] U. Jahn and W. Nasse, "Performance analysis and reliability of grid-connected pv systems in iea countries," in *Photovoltaic Energy Conversion, 2003. Proceedings of 3rd World Conference on*, vol. 3, 2003, pp. 2148–2151 Vol.3.

[16] S. Mau and U. Irike, "Performance evaluation and analysis of 50kw grid-connected pv system," in *21st European Photovoltaic Solar Energy Conference*, 2006.

[17] J. D. Mondol, Y. Yohanis, M. Smyth, and B. Norton, "Long term performance analysis of a grid connected photovoltaic system in northern ireland," *Energy Conversion and Management*, vol. 47, no. 1819, pp. 2925 – 2947, 2006.

[18] U. Jahn and W. Nasse, "Operational performance of grid-connected pv systems on buildings in germany," *Progress in Photovoltaics : Research and Applications*, vol. 12, pp. 441 – 448, 2004.

[19] A. Sarno, S. Causi, F. Apicella, and M. Guerra, "A simple method to analyze and present performance data of the 300 kw delphos pv plant," in *Photovoltaic Specialists Conference, 1990., Conference Record of the Twenty First IEEE*, 1990, pp. 1089–1094 vol.2.

[20] G. M. Masters, *Renewable and Efficient Electric Power Systems*. Wiley-Interscience, 2004.

Grid Integration of Single Stage Solar PV Power Generating System using 12-Pulse VSC

Bhim Singh, *Fellow, IEEE*
Department of Electrical Engineering
Indian Institute of Technology Delhi
New Delhi, India-110016
bhimsinghr@gmail.com

Ikhlaq Hussain, *Student Member, IEEE*
Department of Electrical Engineering
Indian Institute of Technology Delhi
New Delhi, India-110016
ikhlaqb@gmail.com

Abstract—**This paper presents a single stage solar photovoltaic (PV) grid interfaced power generating system using two-level 12-pulse double bridge voltage source converter (VSC) with improved sinusoidal signal integrator PLL (SSI-PLL) based control algorithm for large capacity plants with improved power quality. The maximum power is tracked with modified perturbation and observation (P&O) maximum power point tracking (MPPT) method and the maximum power obtained is transferred to the grid. Multipulse VSC results in reduction of grid current and grid voltage total harmonic distortion (THD) which is in accordance to the IEEE 519 standard without performing high frequency switching control. Simulated results have validated the design and control algorithm of the proposed system configuration under varying conditions.**

Keywords—*Solar energy, single stage, voltage source converter, multipulse, power quality.*

I. INTRODUCTION

With the grid integration of clean bulk solar photovoltaic (PV) energy during last decade, the demand for electrical energy has been met and the effect of global warming decreasing drastically. With the development of advanced power electronics converters and PV panels, the cost of the electrical energy is reduced, the efficiency is increased and the power quality of the solar PV power generating has improved [1]. In large power solar PV grid interfaced power generating system, the main task are to have high rating components, low switching losses in voltage source converters (VSC) [2], harmonics in the line currents and voltage should be in accordance to the IEEE 519-1992 standard [3] and IEEE SCC21standard [4].

Multipulse VSCs such as 12-pulse, 24-pulse and 48-pulse converters are solution for large scale solar PV grid interfaced power generating system to minimize harmonics in the line currents and operate with low frequency based control with good power quality, high efficiency and reliability as compared with the conventional two-level topologies [5-6]. For proper operation of multipulse based system, the design of transformer and interfacing magnetic are required.

The other challenge is to boost the large power in double stage solar PV grid interfaced system as the boost converter cannot handle the high stress and high rating components are not available, so single stage PV system are recommended as it eliminates the boost converter or dc-dc conversion [7]. The

maximum power is tracked from solar PV energy using maximum power point tracking (MPPT) and various MPPT techniques such as Gauss-Newton technique, adaptive fuzzy, adaptive perturbation and observation, incremental conductance (IC), variable step size IC etc at have been proposed in [8-9].

The synchronization of grid voltages and currents with converter voltages and currents are necessary for the integration of solar PV power generating system to grid. Various control and synchronization phase locked loops (PLL) techniques have been proposed [10] for the reliable operation of the grid interfaced systems.

This paper proposes a 1.1 MW single stage solar PV grid interfaced power generating system using two-level 12-pulse double bridge VSC operating at 50 Hz switching frequency with improved SSI-PLL based control. Maximum power is tracked using modified P&O MPPT from solar PV array. The proposed system configuration with MPPT algorithm and VSC control algorithm is mathematically modeled. The simulated results are presented to validate the enhanced power quality of the proposed system.

II. PROPOSED SYSTEM CONFIGURATION

The proposed system configuration of 1.1 MW single stage solar PV grid interfaced power generating system using two level 12-pulse double bridges VSC is shown in Fig. 1. The proposed system consists of solar PV array, two set of elementary 2×6-pulse converters with transformers phase shifted 30^0 in the 50 Hz switching frequency based control and these are tied in parallel on the dc link capacitor. The modeling parameters of proposed system are given in Appendices.

III. DESIGN OF PROPOSED SYSTEM

The design of proposed 1.1 MW solar PV grid interfaced power generating system is given in terms of various stages as solar PV array, dc link capacitor and multipulse converter as follows.

A. Selection of solar PV Array

The solar PV array is designed for a 1.1 MW peak power capacity interfaced to a 11 kV ac system. A solar PV module

978-1-4799-6047-7/14 $31.00 © 2014 IEEE

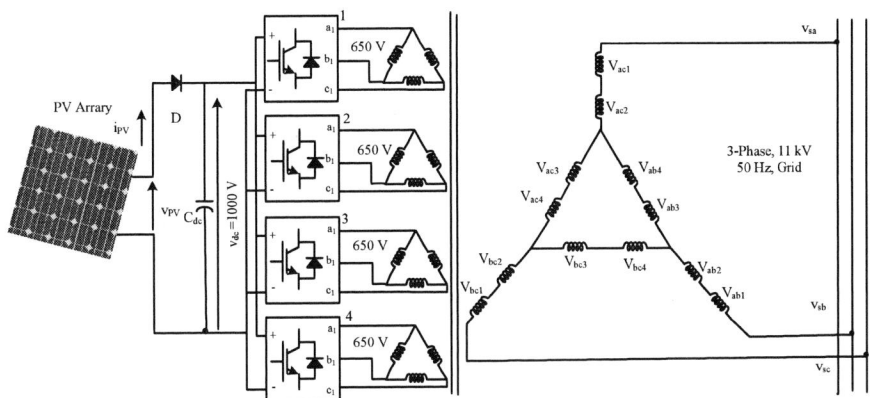

Fig.1 Proposed solar PV grid interfaced power generating system

has short circuit module current (I_{sc}) of 6.43 A and an open circuit module voltage (V_{ocn}) of 85.6V [11].

The maximum output active power from solar PV array P_{mpA} is given as,

$$P_{mpA} = (n_s V_{mp})(n_p I_{mp}) \qquad (1)$$

where n_s and n_p represent series and parallel strings of PV module, V_{mp} is the voltage of a module at MPPT = 85% of V_{ocn} = 72.9 V and I_{mp} is the current of a module at MPPT = 85% of I_{sc} = 5.97 A. Thus maximum active power for solar PV array P_{mpA} is given as

$$P_{mp} = (n_s * 85\% * V_{ocn})(n_p * 85\% * I_{scn}) \qquad (2)$$

Considering, open circuit PV array voltage (V_{OCA}) = 1000 V, P_{mpA} = 1.1 MW

The PV modules connected in series string are estimated as,

$$n_s = \frac{V_{OCA}}{V_{ocn}} = \frac{1000}{85.6} \approx 12 \text{ Modules} \qquad (3)$$

Maximum current (I_{mpT}) of the solar PV array is given as,

$$I_{mpT} = \frac{P_{mpA}}{0.85 * V_{OCA}} = 1294.11\,\text{A} \qquad (4)$$

The PV modules connected in parallel string are estimated as,

$$n_p = \frac{I_{mpT}}{I_{scn}} = \frac{1294.11}{6.43} \approx 202 \text{ Modules} \qquad (5)$$

Thus the array of 1.1 MW peak power capacity is designed with 202 modules in series and 12 modules in parallel with an PV array of 12*202 modules.

B. Design and Selection of DC Capacitor Voltage

The dc capacitor voltage v_{dc} value is given as [12],

$$v_{dc} = \frac{\left(2\sqrt{2}V_{sec}\right)}{\sqrt{3}\,m} = \frac{\left(2\sqrt{2} * 640\right)}{\sqrt{3} * 0.99} \approx 1000 \text{ V} \qquad (6)$$

where V_{sec} is the VSC ac line voltage at the secondary terminal of transformer, m is modulation index.

C. Design of DC link Capacitor

The value of dc link capacitor is estimated as [12],

$$C_{dc} = \frac{I_d}{\left(2 * \omega * v_{dcrip}\right)} = \frac{(P_{dc}/v_{dc})}{(2 * 314 * 0.02 * 1000)} = 0.08F \qquad (7)$$

where I_d is the dc link current of VSC, ω is angular frequency and v_{dcrip} is % ripple voltage considered as 2% of v_{dc}. Thus selected value of C_{dc} is 0.08 F.

D. Design of Multipulse VSC

The primary windings of 1st and 2nd transformers are connected in star and in delta configuration, 3rd and 4th transformers are connected. All secondary terminals are connected in delta configuration (isolated from each other). The 1st and 3rd VSC operated in a leading voltage mode and 2nd and 4th VSC operated in a lagging voltage mode with respect to supply voltage reference.

The phase 'a' voltages $v_a(t)$ and $v_b(t)$ of the 3-phase voltage waveforms on the ac terminals of the two VSCs are as,

$$v_a(t) = \frac{4}{\pi}\left(\frac{v_{dc}}{2}\right)\{\cos(\omega t) + \frac{1}{5}\cos(5\omega t) - \frac{1}{7}\cos(7\omega t)$$
$$+ \frac{1}{11}\cos(11\omega t) - ...\} \qquad (8)$$

$$v_b(t) = \frac{4}{\pi}\left(\frac{v_{dc}}{2}\right)\{\cos(\omega t - 30^0) + \frac{1}{5}\cos 5(\omega t - 30^0) - \frac{1}{7}\cos 7(\omega t - 30^0)$$
$$+ \frac{1}{11}\cos 11(\omega t - 30^0) - ...\} \qquad (9)$$

In presence of phase shifting transformers, resultant 12 pulse converter output voltage $v_c(t)$ are calculated by adding $v_a(t)$ and $v_b(t)$ and is given as

$$v_c(t) = \frac{8}{\pi}\left(\frac{v_{dc}}{2}\right)\{\cos(\omega t) - \frac{1}{11}\cos(11\omega t) + \frac{1}{13}\cos(13\omega t)$$
$$- \frac{1}{23}\cos(23\omega t) + ...\} \qquad (10)$$

Thus, $v_c(t)$ at the primary winding of the transformer contains harmonics order of $12n\pm1$ (where, n=1, 2, 3...) i.e. 11th, 13th, 23rd, 25th and other higher order harmonics. The turn ratio of the Y-Δ connected transformer is considered as 1:$\sqrt{3}$n and for Δ-Δ connected transformer it is connected as 1:n. The calculated design parameters of multipulse VSC are given in Appendices.

E. Design of AC Inductor

The interfacing inductive reactance (X) including the leakage reactance of the transformer at 50 Hz frequency is estimated by using relation given as [5-6],

$$I_n = \frac{V_1}{(n^2 * X)} \qquad (11)$$

where I_n is n^{th} harmonic current and V_1 is the fundamental ac voltage of the grid.

To limit the 5th harmonic component less than twenty percent of the fundamental ac current I_1 in each six-pulse VSC, the X value is selected as 0.2 per unit. Thus, for a 1.1 MW, 11 kV, 50 Hz, 3-phase ac system, the base impedance is calculated as,

$$Z_{base} = \frac{(kV)^2}{MW} = 11^2 / 1.1 = 110\Omega \qquad (12)$$

$$X_{pu} = \frac{X}{Z_{base}} = 0.2 \Rightarrow X = 0.2*110 = 22\Omega \qquad (13)$$

$$L = \frac{X}{2\pi f} = 70.06 mH \qquad (14)$$

IV. CONTROL ALGORITHM

The control algorithm of a given solar PV system consists of two stages which are shown in Fig.2. In first stage, maximum peak power of solar PV array is tracked with MPPT control [8] and in second stage, solar PV is integrated to grid using improved control [5-7]. The detailed control algorithm is explained as follows.

A. Maximum Power Point Tracking

The limitations of P&O and INC algorithms are overcome by using modified P&O method [8]. It has better performance in steady state and dynamic state. In this MPPT method,

When $\left|\Delta P_{pv}\right| > P_{th} + P_{ra}$, the step-size becomes $V_{st1} * P_{ra}$, thus track the MPP.

When $\left|\Delta P_{pv}\right| < P_{th} + P_{ra}$, the step-size becomes smaller $V_{st2} * P_{ra}$, thus track the MPP with more accuracy.

Where P_{th} is the threshold power, $P_{ra} = P_{insp}/P_{mn}$, P_{ins} is the instantaneous power under varying solar insolation, P_{mn} is the maximum nominal active power under standard conditions.

B. Control of Voltage Source Converters

The SSI-PLL based control algorithm [10] for the proposed system is used to control the maximum active power supplied from solar PV to three phase ac grid under varying solar insolation and other conditions. The SSI-PLL based control algorithm works well in harmonic distortions and synchronizes the solar PV system to grid with improved power

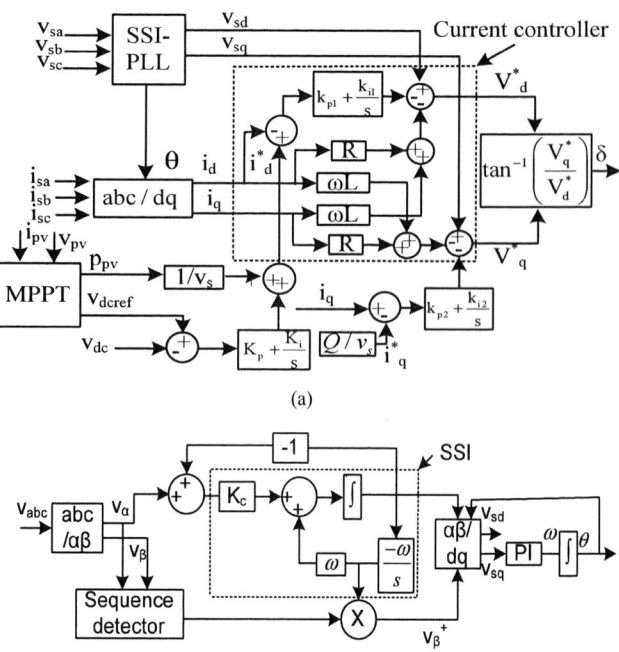

Fig. 2 (a) Control algorithm for the proposed system and (b) SSI-PLL

quality. The dc voltage controller and current controllers are two controllers of the control algorithm.

1) DC Voltage Controller

The MPPT algorithm generates the referenced dc voltage v_{dcref} and is compared with the measured dc bus voltage v_{dc} and difference in voltage is pass through proportional integral (PI) controller that gives direct-axis reference current i^*_d for current controller. The reference direct-axis current i^*_d and reference q-axis current i^*_q are calculated as,

$$\begin{bmatrix} i^*_d \\ i^*_q \end{bmatrix} = \begin{bmatrix} \dfrac{p_{pv}}{v_s} & K_{pdc}(v_{dcref} - v_{dc}) & K_{idc}(v_{dcref} - v_{dc}) \\ \dfrac{Q^*}{v_s} & 0 & 0 \end{bmatrix} \qquad (15)$$

where p_{pv} is the maximum active power to be supplied from solar PV system to grid, Q^* is the set reactive power, K_{pdc} and K_{idc} are proportional and integral gain constants, v_s is the ac grid voltage.

2) Currents Controllers

The reference direct-axis and quadrature-axis voltages (V^*_d, V^*_q) are calculated as,

$$\begin{bmatrix} V^*_d \\ V^*_q \end{bmatrix} = \begin{bmatrix} v_{sd} \\ v_{sq} \end{bmatrix} - \begin{bmatrix} k_{p1}\Delta i_d & k_{i1}\int \Delta i_d dt \\ k_{p2}\Delta i_q & k_{i2}\int \Delta i_q dt \end{bmatrix} + \begin{bmatrix} R & \omega L \\ \omega L & R \end{bmatrix}\begin{bmatrix} i_d \\ i_q \end{bmatrix} \qquad (16)$$

where ωL (X) and R are the interfacing reactive inductance and resistance respectively, k_{i1} and k_{p1} are the integral gain and the proportional gain of direct-axis current controller, k_{i2} and k_{p2} are the integral gain and proportional gain of quadrature-axis current controller. v_{sd} and v_{sq} are the direct-axis and quadrature-axis components of supply voltage V_s, i_q and i_d are the quadrature-axis and the direct-axis components of supply current i_s.

The instantaneous maximum active power and reactive power are supplied from the solar PV based power generating system are expressed as,

$$\begin{bmatrix} P \\ Q \end{bmatrix} = \begin{bmatrix} v_{sd} & v_{sq} \\ -v_{sq} & v_{sd} \end{bmatrix} \begin{bmatrix} i_d \\ i_q \end{bmatrix} \tag{17}$$

Using the SSI-PLL as shown in Fig. 2(b), the grid synchronizing angle θ, actual direct-axis voltage (v_{sd}) and quadrature-axis voltage (v_{sq}) are generated. The constant K_c controls the bandwidth and the filtering response of the SSI-PLL. The currents (i_d, i_q) are calculated from supply current using this θ angle are given as;

$$\begin{bmatrix} i_d \\ i_q \\ i_0 \end{bmatrix} = \frac{2}{3} \begin{bmatrix} \sin(\theta) & \sin(\theta - 2\pi/3) & \sin(\theta + 2\pi/3) \\ \cos(\theta) & \cos(\theta - 2\pi/3) & \cos(\theta + 2\pi/3) \\ 1/2 & 1/2 & 1/2 \end{bmatrix} \begin{bmatrix} i_{sa} \\ i_{sb} \\ i_{sc} \end{bmatrix} \tag{18}$$

For the proper operation of 12-pulse VSC, the gate control logic is synchronized with the grid synchronizing angle θ to produce square wave pulse of duration for gating the converters and the phase shift angle at which the converters are gated is given as,

$$\delta = \tan^{-1}\left(\frac{V_q^*}{V_d^*}\right) \tag{19}$$

V. Modeling and Simulation

The proposed system configuration shown in Fig.1 is modeled using MATLAB software. The proposed system consists of solar PV array, multipulse VSC connected to grid. The control algorithms as shown in Fig. 2 are implemented using Simulink blocks.

VI. Results and Discussion

The proposed system responses are demonstrated to validate its behavior under varying conditions. The system performances are depicted as supply voltage (v_s), supply current (i_s), dc link voltage (v_{dc}), solar PV maximum voltage (v_{pv}), solar PV maximum current (i_{pv}), solar PV maximum power (p_{pv}), reactive power (Q) and active power (P) supplied from solar PV power.

A. Steady State Performance

To study the steady state behavior, the solar PV insolation level is set at reference value of 1000 w/m² and multipulse converters are operating at 50 Hz switching frequency with improved control. Fig. 3 shows the maximum power is tracked using modified P&O MPPT. The dc link voltage is regulated at reference voltage generated by MPPT. The reactive power is regulated to nearly zero while supplying maximum active power to grid. The currents and voltages at the different stages of the system show the enhancement of power quality in the proposed single stage system.

B. Dynamic Performance

The varying response of the proposed system is validated by changing the solar intensity from 300 W/m² to 600 W/m² at 0.25 s and from 600 W/m² to 1000 W/m² at 0.42 s as shown in Fig. 4. The controller regulates the floating dc bus voltage at required level while supplying maximum power to ac grid and maintained the reactive power nearly zero.

C. Power Quality Improvement

Fig. 5 shows the waveforms and harmonics spectra of (a) grid voltage (b) grid current and (c) converter voltage at the primary end of transformer. The THD of open circuit converter voltage at the primary end of transformer are observed as 13.57% and the grid current THD is 4.28% at rated power which is in the limit of IEEE-519 standard [3]. Thus, the proposed 12-pulse VSC operating at 50 Hz switching frequency with SSI-PLL based control algorithm can be used in large power single stage solar grid interfaced for drawing maximum active power at high voltage with improved harmonic performance and low switching losses.

VII. Conclusion

The grid integration of single stage solar PV power generating system using two-level 12-pulse double bridges VSC with SSI-PLL based control algorithm is suitable for large power and high voltage PV power. The proposed single stage system has eliminated the dc-dc boost converter as it has been used in double stage system thereby transferring large power from the solar PV array directly to grid and work well

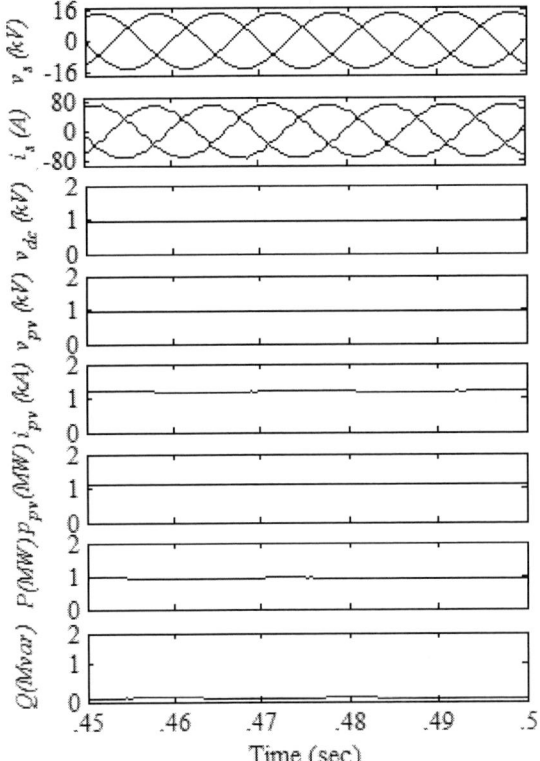

Fig. 3 Steady state responses of the proposed system

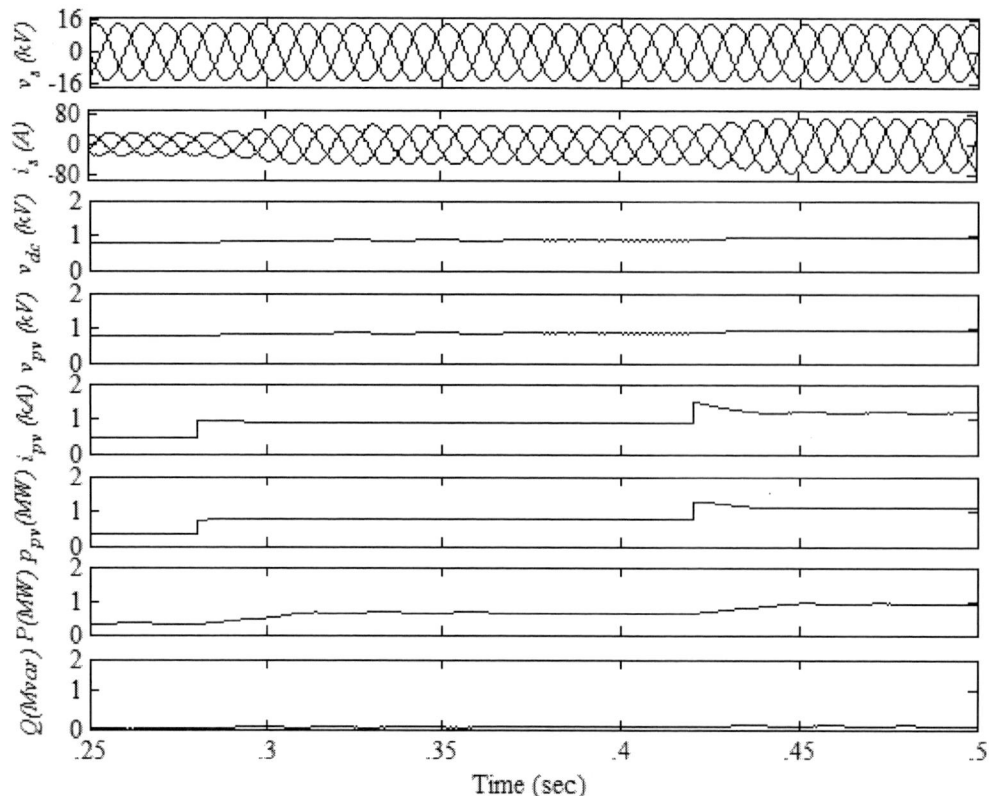

Fig.4 Dynamic performance of proposed system

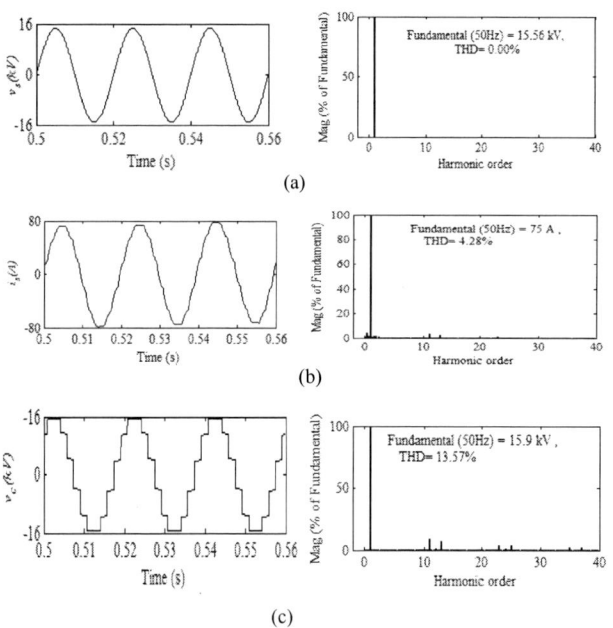

Fig. 5 Waveform and harmonic spectra of (a) grid voltage (b) grid current and (c) converter voltage at primary terminal of transformer.

with multipulse VSC operating at 50 Hz switching frequency with low switching losses, high reliability and good power quality. The performances of the proposed system have been proved in varying conditions with SSI-PLL based control algorithm which have demonstrated the satisfactory working of the system.

APPENDICES

A. Solar PV Data

V_{ocn} = 85.6 V, I_{sc} = 6.43 A, V_{mp} = 72.9 V, I_{mp} = 5.97 A. power tolerance +/- 5%, power temperature coefficient = -0.38%/°C , voltage temperature coefficient = -235.5 mV/°C, current temperature coefficient = -3.5 mA/°C, n_s = 202, n_p = 12 [11].

B. Multipulse Converter and Control Parameters

P = 1.1 MW, f = 50 Hz, Vs = 11 kV, V_{dc} = 1 kV, C_{dc} = 0.08 F, X = 0.2 pu and R = 0.001 pu.
Transformer: 1.5 MVA, 11kV/650V.
Controller gains: K_{pdc} = 0.02 and K_{idc} = 0.001, k_{p1} = 10 and k_{i1}= 0.001, k_{p2} = -210 and k_{i2} = 0.001.

ACKNOWLEDGMENT

Authors are very grateful to Department of Science and Technology (DST), Govt. of India, for supporting this work under Grant Number: RP02583.

REFERENCES

[1] R. Teodorescu, M. Liserre, and P. Rodríguez, Grid converters for photovoltaic and wind power systems, Wiley-IEEE Press, 2011.

[2] C. Zhang, S. Du, and Q. Chen, "A novel scheme suitable for high-voltage and large-capacity photovoltaic power stations," *IEEE Trans. Ind. Electron.,* vol. 60, no. 9, pp. 3775-3783, Sept. 2013.

[3] IEEE Recommended Practices and Requirements for Harmonic Control in Electrical Power Systems, *IEEE Std.* 519-1992, 1993.

[4] IEEE Std SCC21, IEEE SCC21 Standards Coordinating Committee on Fuel Cells, Photovoltaic's, Dispersed Generation, and Energy Storage.

[5] D. M. Mohan, B. Singh, and B. K. Panigrahi, "A two-level 24-pulse voltage source converter based HVDC system for active and reactive power control," in *Proc. IEEE Int. Conf. Power Contr. Embedded Syst. (ICPCES 2010)*, Allahabad, India, Nov. 29-Dec. 1, pp. 1-5.

[6] D. M. Mohan, B. Singh, and B. K. Panigrahi, "Harmonic optimized 24-pulse voltage source converter for high voltage DC systems," *IET Power Electron.*, vol. 2, no. 5, pp. 563-573, Sept. 2009.

[7] B. Singh, C. Jain, and S. Goel, "ILST Control Algorithm of Single-Stage Dual Purpose Grid Connected Solar PV System," *IEEE Trans. Power Electron.*, vol. 29, no. 10, pp. 5347-5357, Oct. 2014.

[8] Y. Yang, F. Blaabjerg, "A modified P&O MPPT algorithm for single-phase PV systems based on deadbeat control," *6th IET Int. Conf. Power Electron., Machines Drives (PEMD 2012)*, 27-29 March 2012, pp.1-5.

[9] B. Subudhi and R. Pradhan, "A Comparative Study on Maximum Power Point Tracking Techniques for Photovoltaic Power Systems," *IEEE Trans. Sustainable Energy*, vol. 4, no. 1, pp. 89-98, Jan. 2013

[10] X.-Q. Guo, W.-Y. Wu, and H.-R. Gu, "Phase locked loop and synchronization methods for grid-interfaced converters: a review," Przegląd Elektrotechniczny (Electrical review), vol. 4, pp. 184-187, 2011.

[11] SunPower Corp. (2013, April). E-Series Commercial Solar Panels [Online]. Available: http://us.sunpower.com/sites/sunpower/files/media-library/data-sheets/ds-e20-series-435-commercial-solar-panels-datasheet.pdf.

[12] N. Mohan, T. M. Undeland, and W. P. Robbins, Power Electronics: Converters, Applications and Design, 3rd ed. New Delhi, India: John Wiley & sons Inc., 2009.

Analysis and Control of an Isolated SPV-DG-BESS Hybrid System

Jincy Philip, Bhim Singh, *Fellow, IEEE* and Sukumar Mishra, *Senior Member, IEEE*

Department of Electrical Engineering,
Indian Institute of Technology Delhi, India
philipjincy@ymail.com, bhimsinghiitd@gmail.com, and sukumar@ee.iitd.ac.in

Abstract—This paper evaluates an isolated system using a character of triangle function (CTF) based control methodology for control and improved power quality. The hybrid-isolated system consists of a solar photovoltaic (SPV) array, diesel driven permanent magnet synchronous generator (PMSG) and a battery energy storage system (BESS). The configuration of the controller is simple and it is employed for the estimation of the source reference currents. The essential objective of a four-leg voltage source converter (VSC) is to provide voltage control, load balancing, neutral current compensation and harmonics elimination while feeding three phase four wire loads. The performance of the system with proposed control strategy has been estimated by modelling the system in Matlab with Simulink and SPS (Sim-Power System) toolboxes and authenticated through simulation results.

Index Terms— Character of triangle function, Permanent Magnet Synchronous Generator, Power Quality, Harmonic Current Compensation, Neutral Current Compensation.

I. INTRODUCTION

Hybrid system with distributed generation emanates as a reliable energy supply to cater the demand side requirements. Solar power generation is becoming an essential constituent for power generation around the globe [1-2]. These renewable resources as generating source ensure stability with optimum cost, minimum fuel consumption and CO_2 emissions [3]. Furthermore, they form a feasible solution for power delivery to remote areas where it could not be connected to electricity distribution network. While considering the nonlinear charactertics nature and interdependence of solar photovoltaic (SPV) power generation on the constantly varying weather conditions. The PV array is equipped with a MPPT (Maximum Power Point Tracker) to maximize the power output. Some of the MPP tracking algorithm that is applied for the effective output is P&O (Perturb and Observe) method, constant voltage method, and neural network algorithms [4]. The dc-dc boost converter is employed to get the dc output voltage and the duty cycle of the converter is estimated accordingly by the output from the MPP controller.

The isolated system consists of the DG (Diesel Generator) as the major alternative source [5]. To further improve the optimal operation of the composite system, the battery energy storage devices are incorporated for compensating the variability in the renewable energy, to manage the peak demand and fluctuations in the load demand. Conventionally, a diesel driven PMSG (Permanent Magnet Synchronous Generator) operates as a DG. The PMSG driven by a diesel engine is an adequate and economical [6-7]. To attain improved performance of the PMSG, the DG set is always loaded at 80-100% of its full load capacity [8-9]. The BESS maintains the load levelling by storing the excess of energy during light load conditions and replenishes during overload conditions.

Now, the integration of the distributed generation brings into notice the power quality of the system. Since the loads connected to the system are nonlinear in nature. They contaminate the system with harmonics. Therefore, to mitigate the harmonics, custom power devices are used such as DSTATCOM (Distribution Static Compensator), DVR (Distribution Voltage Regulator), UPQC (Unified Power Quality Compensator (UPQC), etc. This would reduce the THD (Total Harmonics Distortion) to the acceptable level as given by the international IEEE-519 standard. The four leg VSC allows neutral current compensation as reported in the literature [10].

In this paper, the power coordination between the source and loads, the control to regulate the point of common coupling (PCC) voltage with improvement in the power quality is the major concern. Usually to maintain the efficient operation of the system, various authors have reported control strategy such as deadbeat control, fuzzy control, H^α controller, PI controller, predictive controller, V/f control [11-14]. The major drawback with these methods is that their response is slow, voltage regulation might not respond faster under disturbances. This paper presents a character of a triangle function (CTF) based control strategy [15] which is designed for control of hybrid PV-diesel- battery system to get efficient performance while feeding linear and nonlinear loads. The CTF is easy to implement, gives accurate results, and shows less variations in frequency deviations. It has fast detection of real time harmonics and has enhanced performance with mitigation of harmonics.

978-1-4799-6047-7/14 $31.00 © 2014 IEEE

II. SYSTEM CONFIGURATION

Fig. 1 represents an isolated system composition. It consists of a SPV array with a boost converter and an incremental conductance based maximum power point controller. The battery is connected at the output of the VSC. It maintains the DC link voltage and allows load balancing through its charging and discharging. A diesel engine is used to drive a PMSG. The excitation capacitors are connected for maintaining the terminal voltage. The center of each VSC leg is connected to the PCC through an interfacing inductor. The RC ripple filter placed at the PCC point absorbs the transients in the VSC. The use of the fourth leg in a VSC provides the source neutral current control for feeding three phase four wire loads.

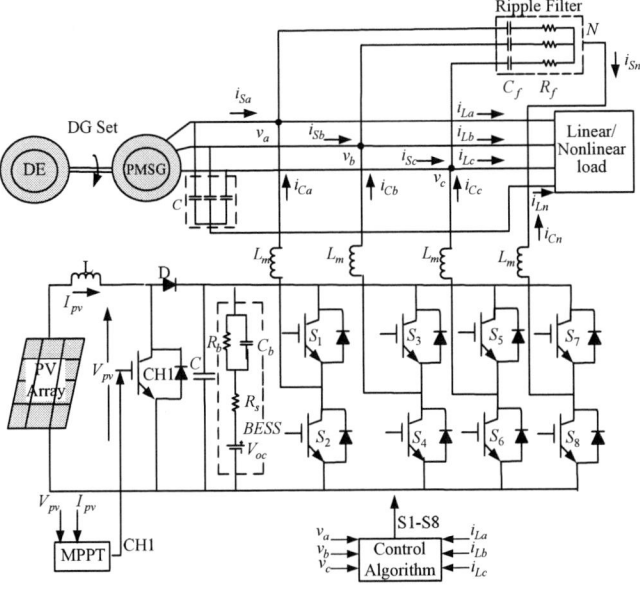

Fig. 1 Proposed isolated system

III. DESIGN OF THE SYSTEM

The proposed system consists of a SPV array with a boost converter, VSC, battery, PMSG based DG set, capacitor bank and ripple filter. The design of these components is given here.

A. Solar Photovoltaic System

The SPV array is configured by the interconnection of modules in series and parallel. The SPV power-generating unit comprises of a 100 kW array consisting of 24 modules in parallel and 34 modules in series. The cells are arranged to produce a voltage of 592 V with 172.8 A current at maximum power point. The power produced by the SPV array depends upon the insolation and temperature conditions.

B. DC-DC Boost Converter

The boost converter is connected with the MPPT controller so that the array voltage is continuously adjusted to the MPP value. Fig. 1 shows the boost converter. The inductor connected in a boost converter is a major component which is designed as,

$$L = \frac{V_{in}DT}{\Delta I} \quad (1)$$

where V_{in} =592V is the input voltage. D=0.2 is the duty cycle, T is the time period and the inductor ripple current is ΔI=33.78A. ΔI is taken as 20% of the input current. The switching frequency f_{sw} is 20 kHz. The estimated value of the inductance L is 0.2 mH. However, SPV output voltage is controlled by adjusting the duty ratio of the converter.

C. Battery Energy Storage System

The battery is present at the DC link of VSC. It is energy system, whose power is described in kilowatt-hours (kWhs). It is composed of R_b and C_b parallel circuit with a series resistance Rs whose values are given in Appendix. The size of the battery depends upon the energy storage capacity. The DC link voltage should be maintained twice to that of the peak voltage at point of common coupling. Therefore, for a 400 V line voltage, the DC-link voltage should be maintained at 750 V.

D. Interfacing AC Inductor

For designing of an interfacing inductor of a VSC, the following parameters are considered such as ripple current, DC bus voltage, switching frequency, modulation index (m) and the overloading factor (a). The selection of an interfacing inductor is made on the basis of the calculation as,

$$L_m = \frac{\sqrt{3}mV_{dc}}{12af_s i_{pp}} \quad (2)$$

The ripple current is 5% of the rated VSC current. Considering the above equation, the suitable value of the AC inductor is taken as 2.5mH.

IV. CONTROL ALGORITHM

The efficient operation of the proposed system depends upon the following controllers as discussed below.

A. MPPT Control of SPV Array

The maximum power point tracking (MPPT) algorithm is intended to extract the maximum available power from the SPV array under different operating conditions. The incremental conductance (IC) method is applied on account of the fact that it can fast track the maximum power even under continuously varying irradiance and temperature conditions. The IC algorithm operates to find the actual operating point by comparing the dI/dV with –I/V. This algorithm is inclusive as it operates to provide the maximum power independent of the panel charactertics, speed of convergence, easier to implement and the efficiency under steady state and dynamic conditions. Fig. 2 shows the algorithm behavior according to the following equations as,

$$\frac{dp}{dv} = 0 \, for \, V_{mp} = V_{pv}, \frac{dp}{dv} > 0 \, for \, V < V_{mp}, \frac{dp}{dv} < 0 \, for \, V > V_{mp} \quad (3)$$

978-1-4799-6047-7/14 $31.00 © 2014 IEEE

This controller adjusts the PV array terminal voltage to the maximum power point voltage based upon the instantaneous and incremental conductance values.

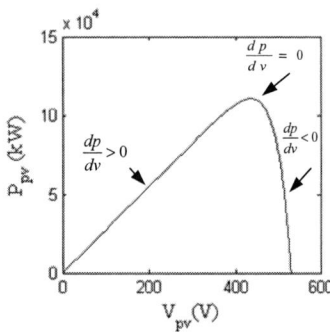

Fig. 2 MPPT control charactertics

B. Control of Voltage Source Converter (VSC)

The character triangle function based control methodology is used to estimate the fundamental active and reactive power components of load currents. Each unit allows rejection of the harmonic component and extracts the source reference current using the fundamental components. The reference reactive power component of current is calculated by subtracting the average power components of the three phase load currents from the voltage PI regulator output. The reference active power component of current is restrained to operate at 80-100% of the rated DG current allowing load levelling. Fig. 3 shows the control algorithm.

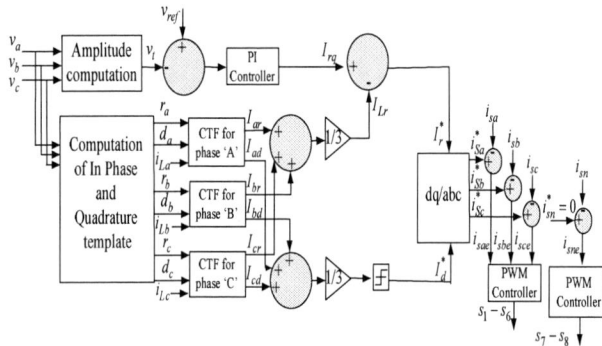

Fig. 3 Character Triangle Function based control algorithm

C. Determination of In-phase and Quadrature Unit Template

The in-phase unit templates is extracted using the amplitude of PCC voltage V_t and phase voltages as,

$$V_t = \sqrt{\{2\times(V_a^2 + V_b^2 + V_c^2)/3\}} \qquad (4)$$

$$d_a = \frac{v_a}{v_t}, d_b = \frac{v_b}{v_t}, d_c = \frac{v_c}{v_t} \qquad (5)$$

The quadrature unit templates are evaluated as,

$$r_a = (-d_a + d_c)/\sqrt{3} \qquad (6)$$

$$r_b = (3d_a + d_b - d_c)/2\sqrt{3} \qquad (7)$$

$$r_c = (-3d_a + d_b - d_c)/2\sqrt{3} \qquad (8)$$

D. Charater Triangle Function based Control Technique

The objective of the character triangle based control technique is to balance the power of the system and to generate the load fundamental component with rejection of harmonics currents.

The load current can be represented by the following equation as,

$$i_L(t) = \sum_1^\infty \sqrt{2} I_{Ln} \sin(n\omega t + \theta_n) + i_{dc}(t) \qquad (9)$$

where I_{Ln} represents the rms value of the nth order harmonic current and θ_n is the nth order phase angle. The load current

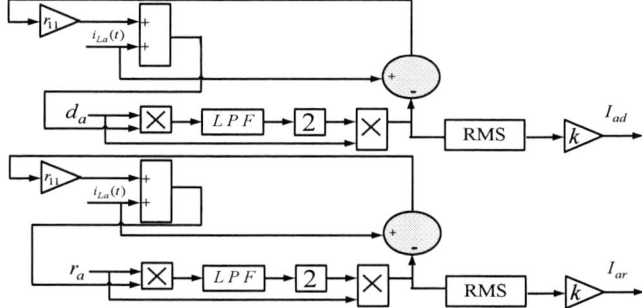

Fig. 4 Schematic diagram of a CTF based control technique

dc component is given as $i_{dc}(t)$ and ω is the frequency in rad/sec.

The above equation can be extended as,

$$i_L(t) = \sum_1^\infty \sqrt{2} I_{L1} \sin(\omega t + \theta_1)$$
$$+ \sum_2^\infty \sqrt{2} I_n \text{ in}(nwt + \theta_n) + i_{dc}(t) \qquad (10)$$

$$i_L(t) = \sqrt{2} I_{L1} \sin(\omega t)\cos(\theta_1) + \sqrt{2} I_{L1} \cos(\omega t)\sin(\theta_1)$$
$$+ \sum_2^\infty \sqrt{2} I_n \sin(nwt + \theta_n) + i_{dc}(t) \qquad (11)$$

$$i_L(t) = I_{L1}(t) + I_H(t) + i_{dc}(t) \qquad (12)$$

The fundamental value of rms load current, $I_{L1}(t)$ is composed of the fundamental active and reactive power components of load current. Therefore, the load current is given as,

$$i_{L1}(t) = i_{Ld}(t) + i_{Lr}(t) + i_H(t) + i_{dc}(t) \qquad (13)$$

The active power component of current and the reactive power component of current are expressed as,

$$I_{Ld} = \sqrt{2} I_{L1}\cos(\theta_1) \qquad (14)$$

$$I_{Lr}(t) = \sqrt{2} I_{L1}\sin(\theta_1) \qquad (15)$$

978-1-4799-6047-7/14 $31.00 © 2014 IEEE

Thus,

$$i_L(t) = I_{Ld}\sin(\omega t) + I_{Lr}\cos(\omega t) + \sum_2^\infty \sqrt{2}I_n\sin(nwt + \theta_n) \quad (16)$$

According to the CTF, the equation is multiplied by a factor $\{2(\sin(\omega t))\}$,

$$2i_L(t)\sin(\omega t) = I_{Ld}(1 - \cos 2wt) + I_{Lr}\sin 2(\omega t)$$
$$+ \sum_2^\infty \sqrt{2}I_{Ln}\sin(\omega t + \theta_n) \quad (17)$$

Therefore, the fundamental active and reactive power components of load current are given as,

$$i_{Ld}(t) = i_L(t) - \{i_{LH}(t) + i_{Lr}(t)\} \quad (18)$$

$$i_{Lr}(t) = i_L(t) - \{i_{LH}(t) + i_{Ld}(t)\} \quad (19)$$

For the extraction of the fundamental active power component from the instantaneous load current, initially, the reactive power component of current and the harmonics are combined with the load current by including a factor r. This cumulative value is multiplied with the in-phase unit template and then it is passed through low pass filter for eliminating the harmonics component. Further, it is multiplied with two as shown in Fig. 4. The amplitude of the active power component is obtained by multiplying the obtained rms value with the factor k=1.414. Similarly each phase components are estimated and the same procedure is applied for the estimation of reactive power component amplitude provided that the quadrature unit template are replaced with the in phase unit template.

The average weighted amplitudes for the estimation of three phase reference source currents are given as:

$$I_{Ld} = (I_{ad} + I_{bd} + I_{cd})/3 \quad (20)$$
$$I_{Lr} = (I_{ar} + I_{br} + I_{cr})/3 \quad (21)$$

E. Estimation of Power Components of Reference Source Currents

The reference active power component of source current, I_d is computed by restraining the current to 80-100% of its rated DG current. The value of reference source current is taken within range or else considered to be $0.8\,I_{Ld}$ for low loads and $1.0\,I_{Ld}$ for higher loads. The reference reactive power component of source current is obtained in a similar manner by using the voltage error.

$$V_e(k) = V_{ref}(k) - V_t(k) \quad (22)$$

where $V_{ref}(k)$ represents the amplitude of the reference ac terminal voltage and $V_t(k)$ is the sensed three phase ac voltage amplitude at PCC.

The PI controller output is obtained as,

$$I_{rq}(k) = I_{rq}(k-1) + k_{pv}[V_e(k) - V_e(k-1)] + k_{iv}V_e(k) \quad (23)$$

where k_{pr} and k_{ir} represent the proportional and integral gains of PI controller respectively.

The reference reactive power component of the source current is calculated by subtracting the reactive power component of the load current from the output of PI controller as,

$$I_r^* = I_{rq} - I_{Lr} \quad (24)$$

F. Evaluation of Reference Source Current

The three phase source reference currents are estimated using the following relationship,

$$\begin{bmatrix} i_\alpha^* \\ i_\beta^* \end{bmatrix} = \begin{bmatrix} \cos\theta & \sin\theta \\ -\sin\theta & \cos\theta \end{bmatrix}\begin{bmatrix} I_d* \\ I_r* \end{bmatrix} \quad (25)$$

$$\begin{bmatrix} i_{Sa*} \\ i_{Sb*} \\ i_{Sc*} \end{bmatrix} = \begin{bmatrix} 1 & 0 \\ -\dfrac{1}{2} & \dfrac{\sqrt{3}}{2} \\ -\dfrac{1}{2} & \dfrac{\sqrt{3}}{2} \end{bmatrix}\begin{bmatrix} I_\alpha* \\ I_\beta* \end{bmatrix} \quad (26)$$

G. Neutral Current Compensation

The neutral current compensation has been achieved by four-leg VSC to limit the source neutral current by comparing the reference neutral current (i_{sn}^*) with the sensed neutral current (i_{sn}) as shown in Fig 3. The current error obtained are used to obtain switching signals for IGBTs of VSC.

V. RESULTS AND DISCUSSION

Performance of an isolated system consisting of SPV array, diesel driven permanent magnet synchronous generator (PMSG) and battery energy storage feeding three-phase four-wire load is discussed in this section. The system performance is demonstrated under linear and nonlinear loads. It is using the character triangle function based control algorithm for voltage control, harmonics elimination, load leveling and neutral current compensation under steady state and dynamic conditions.

A. Performance under Linear Load

Fig. 5 depicts the performance of system under the linear loads. The analysis is made with the unbalances produced within the system on the removal of part of load currents. It is observed that even under dynamic conditions, the source currents and voltages are maintained sinusoidal. The VSC provides reactive power compensation within the system and thus maintains the PCC voltage. The four-leg VSC provides neutral current compensation by controlling the source neutral current. The power produced by the SPV array is shown in this figure.

978-1-4799-6047-7/14 $31.00 © 2014 IEEE

B. Performance under Nonlinear Load

Fig. 6 represents the performance of an isolated system under nonlinear load. The performance analysis is made by operating it under two phase and single-phase load conditions with the removal of the load currents as illustrated in Fig.6. The source currents are maintained sinusoidal under sudden load removal. The harmonics compensation and reactive power compensation are provided by the VSC as it produces current out of phase to that of the load current and maintains the PCC voltage. The neutral current compensation provided by the four-leg VSC maintains the source neutral current to almost zero. The power quality of the system is also improved as the THDs of the source current, load current and the PCC voltage are limited according to the IEEE-519 standard as depicted in Fig. 7.

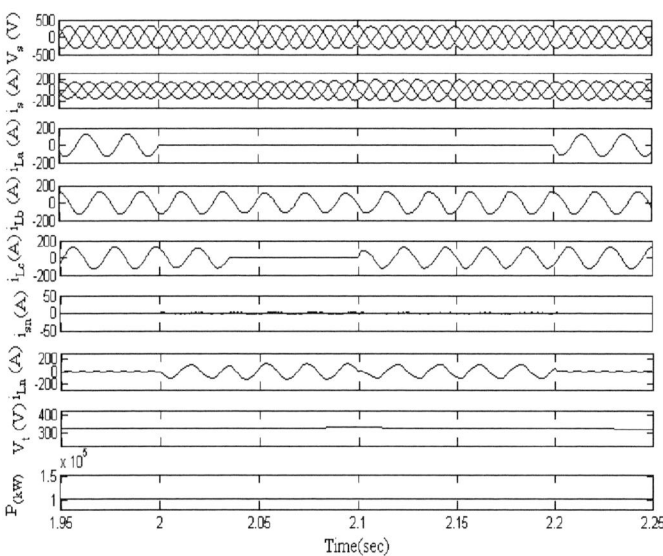

Fig. 5 Performance under linear load

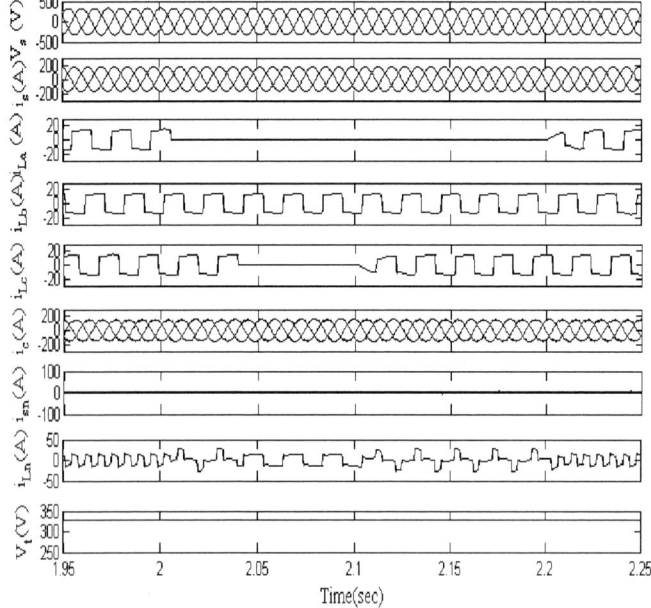

Fig. 6 Performance under nonlinear load

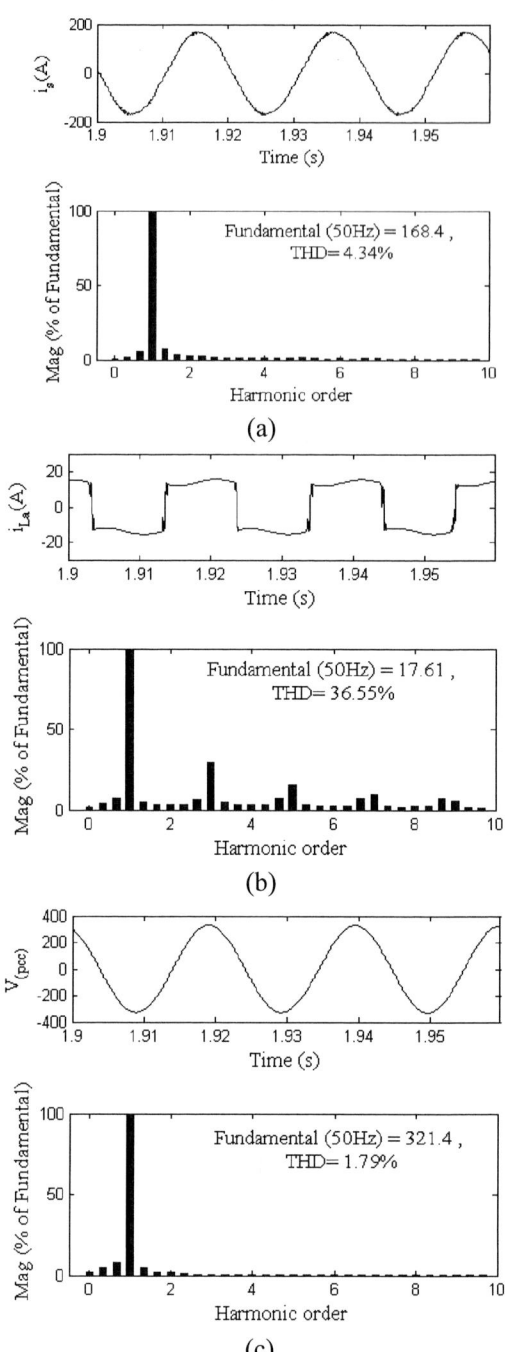

Fig. 7 Harmonic spectra of (a) source current, (b) load current and (c) PCC voltage under nonlinear load.

VI. CONCLUSION

The modelled isolated system has provided improved control and enhanced power quality. The incorporation of the BESS and DG set support in the system has catered the load requirements. The incremental conductance based MPPT controller has extracted the maximum power from the solar photovoltaic array depending upon the prevailing environmental conditions. The character triangle function based control strategy along with four-leg VSC, DG set and battery storage have provided voltage control by providing adequate reactive power, harmonics elimination, neutral current compensation and load balancing.

APPENDIX

Parameter	Specifications
PV Array	P_{max} = 100kW, V_{mp} =592 V, I_{mp} =172.8 A, I_{sc} =194.4A, V_{oc} = 748 V
PMSG	50kW,1500 rpm,400 V,50 Hz
Battery and Interfacing Inductor	V_{oc} =750 V, C_b =20811.16 F, R_b =10 kΩ, R_{in} =0.1 Ω, C_{dc} =4000 µF, m=1, a=1.2, f_{sw} =20kHz.
Ripple filter and Load	R_f =5Ω, C_f =10µF, Linear load= 50kW, Nonlinear load=15Ω, 150mH

REFERENCES

[1] Junliang Wang, Xiaomeng Li, Haijing Yang and Shengli Kong, "Design and Realization of Microgrid Composing of Photovoltaic and Energy Storage System," *Journal of Energy Procedia*, vol.12, pp.1008-1014, 2011.

[2] TaoufikMhamdi andLassâadSbita, "A Power management strategy for Hybrid Photovoltaic Diesel System with battery storage", *Proc. of 5th Int. Conf. on Renewable Energy Congress (IREC), Hammamet, Tunisia*, March 2014.

[3] R.K. Akikur, R. Saidur, H.W. Ping and K.R. Ullah, "Comparative study of stand-alone and hybrid solar energy systems suitable for off-grid rural electrification: A review" *Journal of Renewable and Sustainable Energy Reviews*, vol.27, pp.738-752, Nov. 2013.

[4] S. Jain and V. Agarwal, "Comparison of the performance of maximum power point tracking schemes applied to single-stage grid-connected photovoltaic systems", *Proc. of Electric Power Applications, IET*, vol. 1, no. 5, Sept. 2007.

[5] Prof. Dr. Eng. H. H. El-Tamaly, Dr. Eng. F. M. El-kady and Eng. Adel A. El-Baset Mohammed, "Study the optimal operation of Electric PV/B/D Generation System by neural Network", *Proc. of Int. Conf. on Electrical, Electronics and Computer Engg., Cairo, Egypt*, Sept. 2004.

[6] V. Sheeja, B. Singh and R. Uma, "BESS Based Voltage and Frequency Controller for Stand Alone Wind Energy Conversion System Employing PMSG", Proc. of Ind. Application Society Annual Meeting, IAS 2009, Houston, TX, pp. 1-9, Oct.2009.

[7] V. Sheeja, P. Jayaprakash, B.Singh and R. Uma, "Stand alone wind power generating system employing permanent magnet synchronous generator", *Proc. of Int. Conf. on Sustainable Energy Tech.*, ICSET, Singapore, pp. 616-621, 2008.

[8] B. Singh and J. Solanki, "A comparative Study of control Algorithms for DSTATCOM For Load Compensation", Proc. of Int. Conf. of Ind. Tech., Mumbai, pp. 1492-1497, Dec 2006.

[9] B. Singh, J. Solanki and A. Chandra, "Adaline Based Control of Battery Energy Storage System for Diesel Generator Set", *Proc. of Power India Conf., New Delhi*, 2006.

[10] G. K Kasal, B. Singh, A. Chandra and Kamal-Al-Haddad, "Voltage and frequency control with neutral current compensation in an isolated wind energy conversion system", *Proc. of 34th Annual Conf. Ind. Electronics,Orlando, FL*, 2008.

[11] M. C Chandorkar M C, D. M Divan and R. Adapa, "Control of parallel connected inverters standalone ac supply systems", *Proc. of IEEE Trans on Ind. Appl.* ,pp. 136-143,1993.

[12] M. N Marwali and A. Keyhani, "Control of distributed generation systems—Part I: Voltages and currents control", *Proc. of IEEE Trans Power Electron*, pp. 1541-1550,2004.

[13] H. Karimi, E. J Davison and R. Iravani, "Multivariable servomechanism controller for autonomous operation of a distributed generation unit: design and performance evaluation", *IEEE Trans. Power Sys.*, vol. 25, no. 2, pp. 853-865, May. 2010.

[14] A. Hooshmand, H.A. Malki and J. Mohammadpour, "Power flow management of microgrid networks using modelpredictive control", *Journal of Computer and Mathematics with Applications*, vol.64, pp.869-876, Sept. 2012.

[15] B. Singh and Sabha Raj Arya, "CTF Control Algorithm of DSTATCOM for power factor correction and zero voltage regulation," in *Proc. IEEE International Conf. on Sustainable Energy Technologies (ICSET)*, 2012, pp. 157–162.

Maximum Power of PV Plant for SP and TCT Topologies under Different Shading Conditions

Jitendra Kumar

School of Renewable Energy and Efficiency
National Institute of Technology, Kurukshetra
Kurukshetra, India-136119
jkmahawer786@yahoo.in

Shelly Vadhera

Department of Electrical Engineering
National Institute of Technology, Kurukshetra
Kurukshetra, India-136119
shelly_vadhera@rediffmail.com

Abstract—**This paper analyses the effect of mismatch losses caused by the effect of partial shading on an array of photovoltaic modules in a photovoltaic power plant. For the analysis purpose the two main topologies of PV array connection, i.e., series-parallel and total cross tied are considered along with three different shading patterns as random step wise pattern, row wise pattern and column wise pattern. The maximum power is computed for series-parallel and total cross tied connections at different percentage levels of shading using a Matlab/Simulink model of a 20×5 sized array of solar photovoltaic module (SOLAREX MSX-60 watt PV panel).**

Keywords—*Bypass diode, Partial shading, Photovoltaic array configuration, Series-parallel topology, Total cross tied topology.*

I. INTRODUCTION

Fossil fuels such as coal, oil and gas are the main components of electricity generation in the world since the history of electric generation. But as these sources are exhausting therefore there is a switch over from these conventional sources to renewable sources for energy production. Among the renewable energy sources solar energy is the cleanest, ubiquitous, infinite and most reliable source of energy.

As the solar photovoltaic power generation is involved in both utility based power generation (Mega-Watts/Kilo-Watts rating) as well as residential power generation (Kilo-Watts/Watts rating) systems its efficiency has to be maintained at maximum level [1]. But as the efficiency typically lies between 10%-20%, the need arises to reduce the losses in the photovoltaic power plant. The losses due to mismatch condition caused by the partial shading of photovoltaic panels due to presence of clouds, nearby trees, buildings, towers etc., is the most severe reason for the reduction of power generation in photovoltaic power plants. There is another impact of shading on the PV modules, i.e., addition of more than one peak in the power-voltage characteristic. This presents difficulty during the tracking of maximum power point of the photo-voltaic power plant [2].

Many kind of topologies are present for interconnection of solar PV module like series, parallel, series-parallel (SP), bridge-link (BL), honey-comb (HC) and total cross tied (TCT) etc. Series-parallel configuration is the most utilized, simplest, and conventional method of photovoltaic module interconnection. In this configuration modules are connected in series and parallel to obtain the desired voltage, current and power rating. But its performance is poor during different shading conditions [3]. Among all these topologies total cross tied topology is the best one regarding the performance. But there is increase in the number of connecting wires for this topology. So cost and the complexity of the system increases[4-5]. But this increase in the cost is compensated by the increment in the production of power during shading conditions. In this configuration of photovoltaic array the modules are firstly connected in parallel and then a series string is formed. So the desired current, voltage and power rating can be achieved.

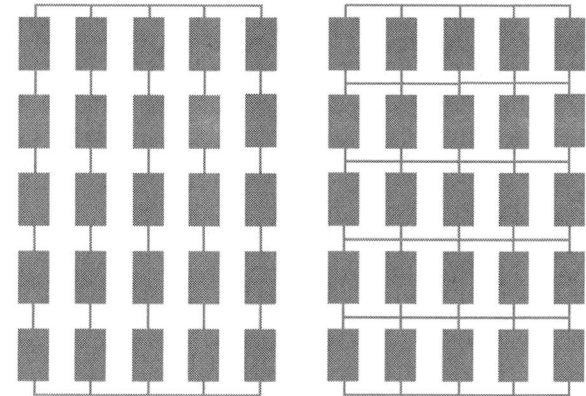

Fig. 1. Series-parallel and total cross tied topology

Hot-spot heating effect caused by the partial shading of PV module can be eliminated by the use of bypass diode [6]. Bypass diodes are connected parallel to a group of solar cells in opposite polarity.

In order to reduce the losses caused by the partial shading the photo-voltaic array reconfiguration is also done. Although this increases the complexity and cost of the system but gives higher power production during shading conditions[7].

In contrast to the available literature [3], [5] for the first time an attempt has been made in this paper to compare the maximum power production under different shading patterns between two topologies of PV modules, i.e., series-parallel and total cross tied and conclusions are made accordingly.

978-1-4799-6047-7/14 $31.00 © 2014 IEEE

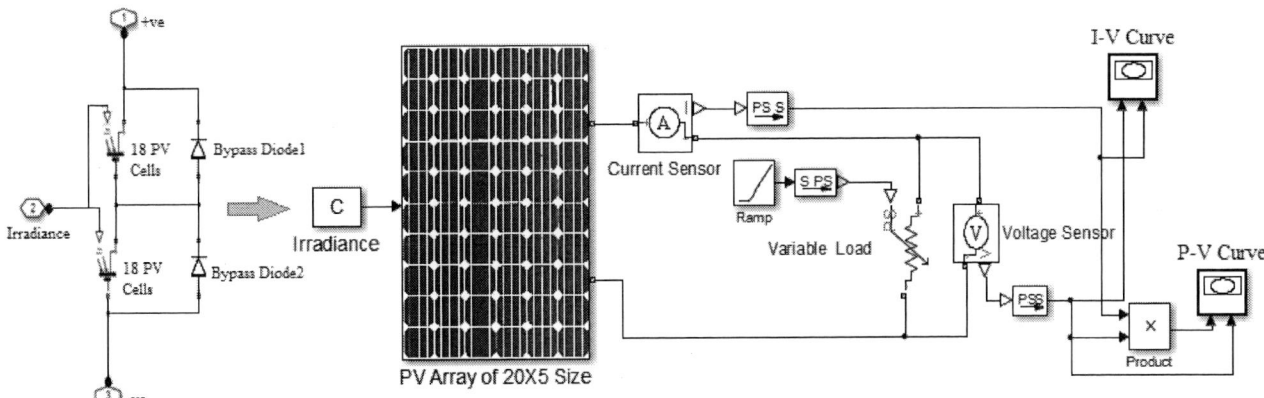

Fig. 2. Simulation model of (a) Solar module (*SOLAREX MSX-60 W PV Panel*), and (b) PV plant

II. PHOTOVOLTAIC MODULE/ARRAY

A. Single Solar PV Module

Although various models of solar cell had been presented by the researchers, i.e., one diode model, two diode model etc., but in this paper a model of solar cell available in Matlab/Simulink library is used for the modelling of solar photovoltaic modules. The data for the photovoltaic module is taken from the datasheet of SOLAREX MSX-60 watt PV panel and is shown in the Table I.

TABLE I. DATA USED FOR SIMULATION OF PV MODULE FROM THE DATASHEET OF SOLAREX MSX-60 WATT PV PANEL

Parameter	Value
Short-circuit current, I_{SC}	3.8 A
Open-circuit voltage, V_{OC}	0.586 V (for a single solar cell)
Quality factor, N	1.2
Series resistance, R_S	0.008 ohm
Number of series cells	36
Energy gap, E_g	1.12 eV

During the normal operating condition there is no role of bypass diode, but when the shading occurs on the PV panels then some part of series current will flow through bypass diode and other part will flow through the shaded solar cell. There are two bypass diodes per module (one bypass diode per 18 solar cells) are used. Bypass diodes play a big role in the performance of solar photovoltaic power plant. During shading of a PV module, solar cell does not conduct the full short circuit current and thus causes reduction in the power production. So the bypass diode bypasses some part of current flowing through the shaded solar cell and prevents the reduction in power production and damages due to formation of hot spot heating in the shaded solar cell. During shading conditions a group of solar cell is shaded rather than a single solar cell, so one bypass diode for a string of solar cell is used rather than using one bypass diode per solar cell

(one bypass diode per 18 solar cells for SOLAREX MSX-60 watt PV panel). Henceforth by the use of bypass diodes a great improvement in the performance of solar photovoltaic power plant is achieved.

B. Solar PV Array of 20×5 Size

In this paper an array of 100 solar PV modules in the size of 20×5 is used. In series-parallel topology, 20 solar PV modules are firstly connected in series to form a string then 5 strings of PV modules are connected in parallel. While for the total-cross-tied topology firstly 5 PV modules are connected in parallel and then a string is formed of 20 parallel combinations. Hence it can be directly observed that there is increase in the number of connecting wires for the total-cross-tied topology compared to series-parallel topology.

III. SHADING PATTERNS

For the analysis of performance of the PV array for series-parallel and total cross tied configuration three different shading patterns, i.e., random step wise pattern, row wise pattern and column wise pattern as shown in Fig. 3 are used. Different percentage of shading levels, i.e., 20%, 40%, 60% and 80% are experimented and thus out of 100 solar PV modules 20, 40, 60 and 80 PV modules are shaded respectively. For the simulation purpose unshaded PV module is provided with the insolation of 1000 W/m^2 and shaded PV Module is provided with 300 W/m^2 insolation.

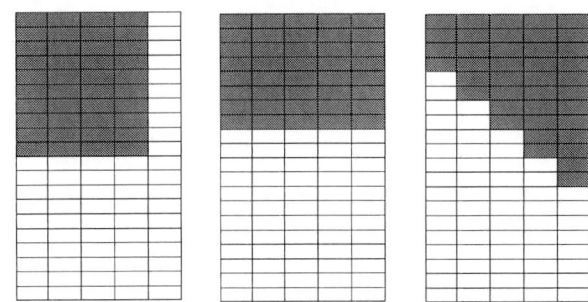

Fig. 3. Shading patterns (*for 40% shading*) a) Column wise, b) Row wise, and c) Step wise

IV. SIMULATION RESULTS

The simulation model has been executed for maximum output power under two topologies for different shading levels. The simulated results for the above are tabulated in the Table II whereasoutput *I-V* characteristics and *P-V* characteristics for 40% shading are shown from Fig. 4 to Fig. 9.

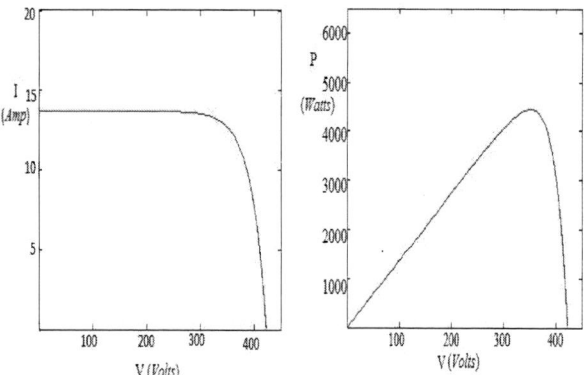

Fig. 4. *I-V* and *P-V* characteristics at 40% shading for SP topology of a column wise pattern

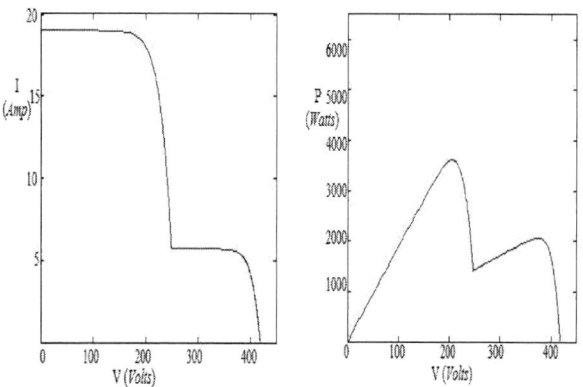

Fig. 5. *I-V* and *P-V* characteristics at 40% shading for SP topology of a row wise pattern

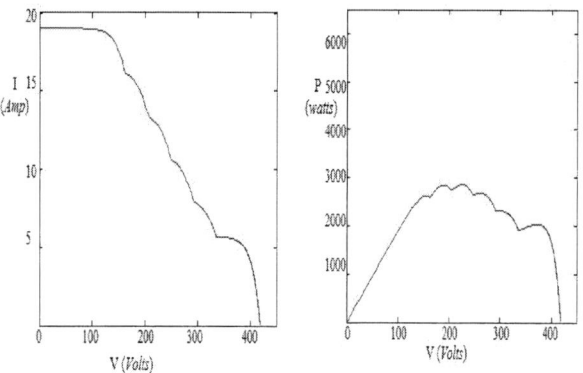

Fig. 6. *I-V* and *P-V* characteristics (at 40% shading) for SP topology of a step wise pattern

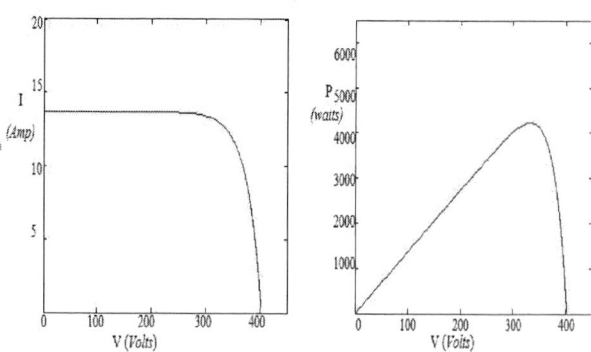

Fig. 7. *I-V* and *P-V* characteristics (at 40% shading) for TCT topology of a column wise pattern

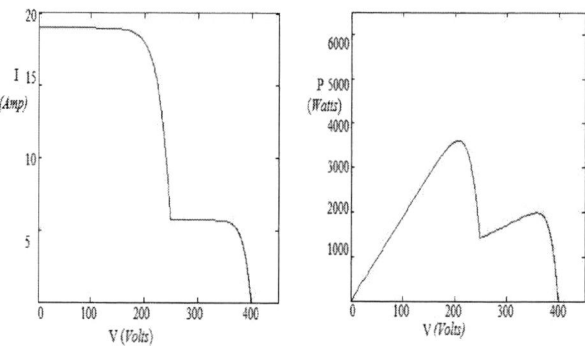

Fig. 8. *I-V* and *P-V* characteristics (at 40% shading) for TCT topology of a row wise pattern

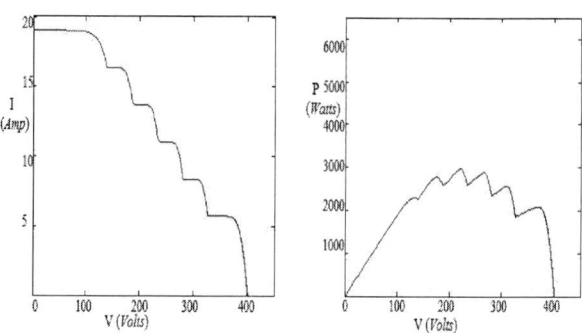

Fig. 9. *I-V* and *P-V* characteristics (at 40% shading) for TCT topology of a step wise pattern

V. OBSERVATIONS

For the column wise shading pattern total cross tied configuration has the best performance compared to series-parallel configuration under any shading level whereas for row wise shading pattern performance of both, total cross tied configuration and series-parallel configurations is same under any shading level.Therefore it may be concluded that it is better to opt for series-parallel configuration over total cross tied configuration under these kinds of shading patternsbecause series-parallel configuration requires less connecting wires.

978-1-4799-6047-7/14 $31.00 © 2014 IEEE   322

TABLE II. MAXIMUM POWERS FOR DIFFERENT SIMULATED MODELS

Shading pattern	Shading level in %	Topology	
		SP P_{max} (W)	TCT P_{max} (W)
Column wise pattern	20%	4408	4648
	40%	2779	2902
	60%	2700	2744
	80%	1808	1930
Row wise pattern	20%	4655	4655
	40%	3260	3260
	60%	1936	1936
	80%	1839	1839
Step wise pattern	20%	4357	4184
	40%	2617	2739
	60%	1901	2041
	80%	1836	1867

In the step wise pattern of shading the total cross tied configuration has better performance compared to series-parallel configuration with the exception of shading level of 20%. At 20% shading level, series-parallel configuration performs better than total cross tied configuration.

VI. CONCLUSION

In this paper, performance of total cross tied topology and series-parallel topology of photovoltaic module under different shading conditions at different shading levels have been compared using the Matlab/simulink based model of PV array of 20×5 size. The paper emphasizes the use of series-parallel topology for row wise pattern of shading as it requires less number of connections and total cross tied topology for column wise pattern and step wise pattern ofshading for better power harnessing and efficiency improvement of the photovoltaic power plant.

REFERENCES

[1] D. Picault, B. Raison, S. Bacha, J. Aguilera, and J. De La Casa, "Changing Photovoltaic Array Interconnections to Reduce Mismatch Losses: a Case Study", International Conference on Environment and Electrical Engineering (EEEIC), pp. 37-40, 2010.

[2] L. Gao, R. A. Dougal, S. Liu, and A. P. Iotova, "Parallel-Connected Solar PV System to Address Partial and Rapidly Fluctuating Shadow Conditions", IEEE Trans. Industrial Electronics, vol. 56, no. 5, pp. 1548-1556, May 2009.

[3] M. Jazayeri, S. Uysal, and K. Jazayeri, "A Comparative Study on Different Photovoltaic Array Topologies under Partial Shading Conditions", IEEE PES T&D Conference and Exposition, pp. 1-5, 2014.

[4] J. P. Storey, P. R. Wilson, and D. Bagnall, "Improved Optimization Strategy for Irradiance Equalization in Dynamic Photovoltaic Arrays", IEEE Trans. Power Electroics, vol. 28, no. 6, pp. 2946-2956, June 2013.

[5] G. Cipriani, V. Di Dio, D. La Manna, R. Miceli, and G. Ricco Galluzzo, "Technical and Economical Comparison between Different Topologies of PV Plant Under Mismatch Effect", International Conference on Ecological Vehicles and Renewable Energies (EVER), pp. 1-6, 2014.

[6] R. Ramaprabha and B. L.Mathur, "A Comprehensive Review and Analysis of Solar Photovoltaic Array Configurations under Partial Shaded Conditions", International Journal of Photoenergy,vol. 2012, pp. 1-16, 2012.

[7] D. Picault, B. Raison, S. Bacha, J. de la Casa, and J. Aguilera, "Forecasting Photovoltaic Array Power Production Subject to Mismatch Losses", Solar Energy, vol. 84, no. 7, pp. 1301-1309, July 2010.

Modeling of PV Module to Study the Performance of MPPT Controller Under Partial Shading Condition

Malik Sameeullah

School of Renewable Energy and Efficiency
National Institute of Technology Kurukshetra
Kurukshetra, India-136119
Email: malik_sameeullah@rediffmail.com

A. Swarup, Senior Member IEEE

Dept. of Electrical Engineering
National Institute of Technology Kurukshetra
Kurukshetra, India-136119
Email: a.swarup@ieee.org

Abstract—**This paper proposes the mathematical modeling of Photovoltaic (PV) module using single diode model and use it to study the performance of PV array under partial shading condition. Partial shading is one of the major causes of reducing energy output in solar PV array. Due to partial shading, PV characteristics get more complex. In order to extract maximum power under shading condition, its performance has to be studies. In this paper, Matlab based mathematical model is used to study the performance of Perturbation & Observation (PO) based Maximum Power Point Tracker (MPPT) controller under partial shading condition. Further, ecosense Insight Solar PV Training kit is used to justify the mathematical model.**

Keywords— **Photovoltaic, PV module, array, partial shading, MPPT, Perturbation & Observation, MPP**

I. Introduction

In order to conserve fossil fuels and limit environment pollution caused by a conventional power plant, the government of various countries started various initiatives to promote renewable energy. In India, Ministry of New & Renewable Energy comes up with a number of schemes to promote renewable energy based system. India is in tropical region with abundant solar energy resource. Most parts of India receive 4-7 KWh of solar energy per day per square meter with 250-300 sunny days [1].

Photovoltaic system can be used in a number of ways to generate electricity, such as grid connected PV plant, home charging system, solar pump station, domestic and commercial uses. At present, efficiency of commercial PV module is in the range of 10-20% [2]. Performance of PV system further degraded under different irradiance condition, ambient atmospheric temperature and partial shading of array due to clouds or any objects such as buildings, tree and tower. Partial shading conditions not only degrade PV array performance, but at the same time change I-V characteristics which cause multiple maxima on P-V curve [3].

In section II, mathematical model of PV module is proposed. Mathematical modeling of PV modules has been considered to test various parameters of system before implementing it in the field. Various ways to implement the PV model in PSPICE and Matlab is available [4-7]. In most of the papers, single diode based PV model is discussed which is simple and accurate. For implementing mathematical modeling of PV module using the single diode method, it

needs to calculate five parameters. Various iterative and analytical methods to calculate diode parameters are cited in [7-9]. Reference [10] uses practical data of I-V curve to generate an analytical equation for lookup table and it is used to make the PV model.

Performance characteristics of the PV module/array under uniform or non-uniform irradiance have been analyzed in [3], [12-17]. Some researchers suggest fixed network topology to improve performance of PV array under partial shading. In section III, modeling results of the various shading condition are illustrated. PO based controller is widely used to extract maximum power and it have a good performance result under steady or slowly varying environment condition. But under partial shading condition, this basic controller is not able to track maximum power. In section IV, detail study of the PO controller under partial shading condition is depicted. ecosense Insight Solar PV Training kit is used to verify the simulated results. Detail of experimental setup and validation of result with experimental study is discussed in section V. Finally, overall conclusion and future scope of work are discussed in section VI.

II. PV Module Modeling

A Photovoltaic cell is a solid state electronic device which converts part of sun radiation energy into electricity. This phenomenon is known as photo effect. It is a basic unit of PV module and most of the time semiconductor is used for its manufacturing. Typical voltage and power output of PV cell is very low, so multiple cells are connected in series and parallel form and known as a module. A PV cell is modeled basically as radiation control current source in parallel with diode. To account for losses occur inside cell, series and shunt resistance are included in this model, as shown in Fig. 1.

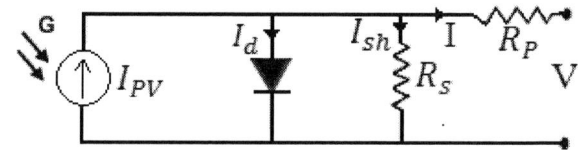

Fig. 1. Single diode model of practical PV device including series and parallel resistive component.

Current across the PV cell(I_{cell}) is a sum of photo current($I_{PV,cell}$), diode current (I_d) and shunt component current ($I_{sh,cell}$) and given by (1)

978-1-4799-6047-7/14 $31.00 © 2014 IEEE

$$I_{cell} = I_{PV,cell} - I_{0,cell}\left(\exp\left(\frac{q(V_{cell}+R_sI_{cell})}{akT}\right)-1\right)-\frac{V_{cell}+R_sI_{cell}}{R_p} \quad (1)$$

where $I_{0,cell}$ is dark saturation current of diode, q is the electron charge, k is the Boltzmann constant ($1.38 * 10^{-23}\ J/K$), T is an absolute temperature of PV cell and a is the ideality factor of diode. Equation (1) does not represent practical PV module (or array). In practical PV module, no of PV cells are connected in series and parallel form to increase the current and voltage rating. Mathematical I-V characteristics of PV module is

$$I = I_{PV} - I_o\left(\exp\left(\frac{q(V+R_sI)}{aN_skT}\right)-1\right)-\frac{V+R_sI}{R_p} \quad (2)$$

where I_{PV} is the photovoltaic current and I_o is the saturation current of the module. In (2), N_skT/q is the thermal voltage (V_t) of the module with N_s cells connected in series. For the module with N_p strings in parallel and each string have N_s PV cells, I_{PV} and I_o is expressed as $I_{PV} = N_pI_{PV,cell}$, $I_o = N_pI_{o,cell}$. R_s and R_p are the equivalent series and shunt resistance of PV module.

Equation (2) describes single diode model. In order to make the model more accurate, some authors [4], [8] proposed two-diode or three diode model. In this paper, single diode modelling is used as it is good enough to study system performance. For any PV module, there are three important points on I-V curve; which is short circuit point ($I_{sc}, 0$), open circuit point ($0, V_{oc}$) and maximum power point (I_{mp}, V_{mp}).

Specifications provided by PV module manufacturer are not enough, to use it directly for PV modelling. Generally PV module datasheet provides the information of short circuit current (I_{scn}), open circuit voltage (V_{ocn}), maximum current and voltage (I_{mpn}, V_{mpn}), maximum power (P_m), temperature coefficient of open voltage (K_V) and temperature coefficient of short circuit current (K_I) under standard test condition (STC, AM1.5, 1000 W/m², 25°C).

For single diode modelling, it requires to calculate R_p, R_s, a, I_{PV} and I_o. Under STC condition, photo current I_{PVn} is approximately equal to I_{scn}. I_{PV} depends upon solar irradiance and temperature. Photo current is given by

$$I_{PV} = (I_{PVn} + K_I\Delta T)\frac{G}{G_n} \quad (3)$$

$$I_{PV} = \left(\left(I_{scn}\frac{R_P+R_s}{R_p}\right)+K_I\Delta T\right)\frac{G}{G_n} \quad (4)$$

Diode saturation current is derived from (2) by equating $I_{PV} = 0$ for dark current condition and assuming shunt current is approximate zero, as shown in (5)

$$I_o = \frac{I_{sc}}{\exp(V_{oc}/V_t)-1} \quad (5)$$

$$I_o = \frac{I_{scn}+K_I\Delta T}{\exp\left(\frac{V_{ocn}+K_V\Delta T}{V_t}\right)-1} \quad (6)$$

A. Calculation of R_s, R_P and a

Ideality factor of diode may be calculated by using analytical methods [4], [5]. In general, its value in between 1 to 1.5 and choice depends upon other parameters of the module. In this paper, a=1.5 is chosen. Still, two parameters remain unknown (R_s and R_P). Usually, the value of R_s is very low and the value of R_P is high. Some authors [3], [5] assuming that R_P value is very high and equal to infinity. Now, (2) is simplified with one unknown variable which can be calculated analytically. Some authors [8-10] proposed the iterative method to change R_s and R_P continuously and check the solution until it exactly fit with experimental I-V curve. In this paper, for different R_s and R_P value, maximum experimental power ($P_{m,e}$) is compare with maximum theoretical power($P_{m,t}$) and nearest result is chosen. At MPP, R_P is calculated using (3) and given by

$$R_p = \frac{V_{mp}(V_{mp}+I_{mp}R_s)}{\left(V_{mp}I_{mp}-V_{mp}I_o\exp\left(\frac{V_{mp}+I_{mp}R_s}{V_t}\right)+V_{mp}I_o-P_{m,e}\right)} \quad (7)$$

For exact calculation, R_s value increase in step and for each step R_p is calculated using (7). To calculate $P_{m,t}$, (2) is solved for the range of voltage values $\epsilon[0, V_{oc}]$. As (2) is a transcendental equation, it must be solved by using iterative numerical method. The Newton-Rapshon technique is used to solve $I = f(I, V)$. Algorithm to solve R_s and R_P is illustrated in Fig. 2. In case, program is not able to converge, set $R_s = 0$ and $R_p = R_{p,min}$ where $R_{p,min}$ is calculated using (7) by assuming $R_s = 0$. Datasheet of ELDORA 40P [11] is used to calculate PV module parameters. After executing the iterative program, R_s and R_P come out to be 0.307Ω and 341.1Ω repectively.

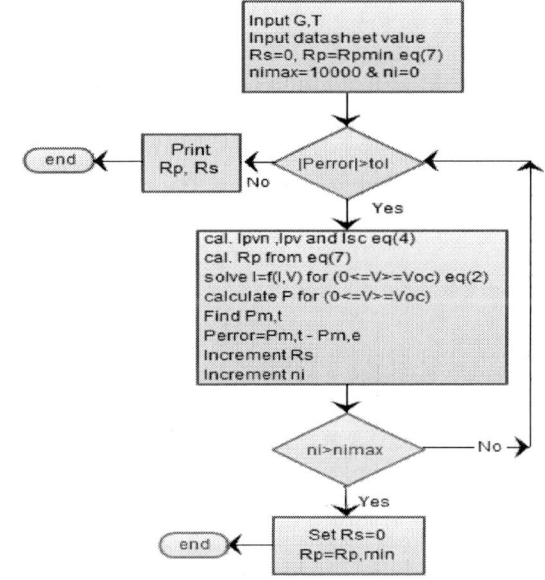

Fig. 2. Algorithm used to calculate R_p and R_s of PV module

B. Mathematical Model of PV Module

The equations (3), (4) and (6) are used in subsystem to

make a PV model in Matlab Simulink environment, as shown in Fig. 3. The Actual PV model is represented as current control current source. The inputs for subsystem are output voltage, solar irradiance and absolute Temperature. Before simulating the model, parameters of PV module are initialized by running script file (M-file) of parameter calculations.

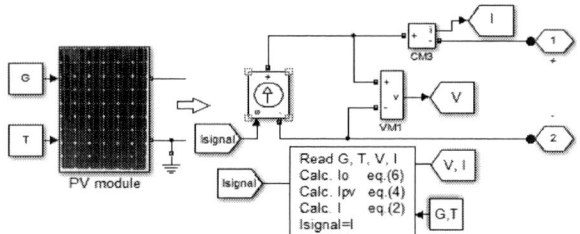

Fig. 3. Simulink block model representation of PV module

I-V and P-V characteristics of ELDORA 40P PV module under different irradiance condition is shown in Fig. 4. The curves show that with increase in irradiance level, short circuit current and output power increase. As shown in Fig. 4, open circuit voltage also increase, but this increase is not much and generally assume to be zero. Fig. 5 shows the variation of I-V and P-V characteristics under different atmospheric temperature and constant irradiance. It shows that short circuit current increase little bit with increase in temperature, but there is drop in open circuit voltage and output power.

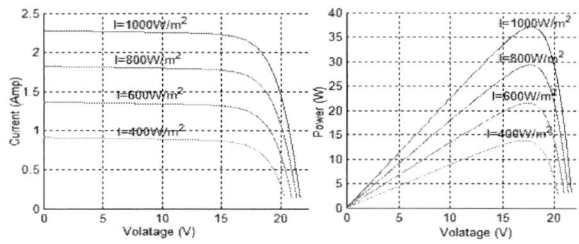

Fig. 4. I-V and P-V characteristics of ELDORA 40P PV module for different Irradiance condition and constant Temperature

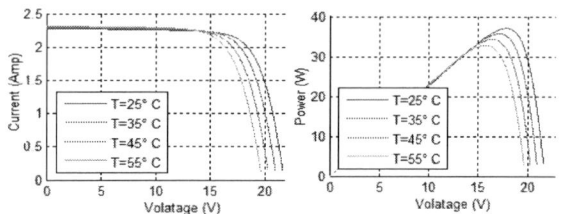

Fig. 5. I-V and P-V characteristics of ELDORA 40P PV module for different Temperature condition & I=1000W/m^2

III. STUDY OF PARTIAL SHADING

Partial shading of PV cell causes reduction in current generation of particular cell, which results in reduction of maximum current that can be produced by other cells connected in series [12-15]. It is found that if one cell in series string is shaded, it causes reduction of 40-50% output power. The photo current is proportional to irradiance. If two cells connected in series and have different irradiance level, then maximum current is decided by shaded cell. At the same time,

extra power is dissipated at shaded junction and causes hot spot which can damage cell irreversibly. Reference [16], [17] discussed the performance of fixed configuration Total-Cross-Tied, Bridge-Link and Series-Parallel topologies for large PV array under partial shading condition. For low power off grid application, such topologies are not helpful.

Diodes connected across cells or modules are called bypass diodes. Under partial shading condition, it helps to stop reverse current and improve the performance. Most of the commercial PV modules have a provision of 2-3 bypass diodes, as it is commercially not viable to provide bypass diode across each PV cell. Due to bypass diode, maximum current of the array is not limited to short circuit current of shaded module, but it reaches up to maximum possible short circuit current.

A. Shading Case of Two Series Connected Modules

Fig. 6 shows the schematic diagram of two PV modules connected in series. Db1 and Db2 act as bypass diodes. When equal irradiance of I=1000 W/m^2 is applied to both panel, then available maximum power is 74 W. When level of irradiance is different for each panel, then maximum available power is always less than 74 W. Fig. 7 shows the I-V curve of two series connected module when Panel-1 receive Irradiance of 267 W/m^2 and Panel-2 receive 152W/m^2.

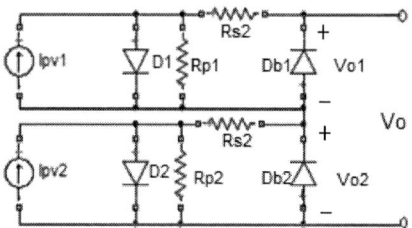

Fig. 6. Schematic of two series connected PV array with bypass diode

As shown in Fig. 7, I-V curve has two current slopes. In part I of operation line, panel-2 produce maximum power under shading, but panel-1 is not able to generate maximum power. In part II of operation line, panel-1 produce maximum power, but no current cross panel-2 and pass through bypass diode. It causes panel-2 is operating in negative voltage region and produce power loss. In P-V curve, there are two peak power points. Whether MPP is in low voltage region or in high voltage region and it is totally depends upon the shading condition of panel.

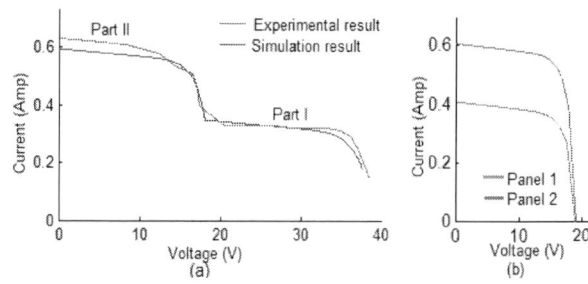

Fig. 7. (a) I-V characteristic of two series connected PV module under partial shading condition with G1=267W/m^2 and G2=152W/m^2, (b) I-V characteristics of module 1 and module 2 under shading condition.

978-1-4799-6047-7/14 $31.00 © 2014 IEEE   326

B. Shading Case of Multiple Modules Connected in Series

The P-V curve becomes more complex under partial shading condition, when large numbers of PV module connected in series. Fig. 8 shows different combination of I-V curves of three series connected modules under partial shading condition. Similarly, Fig. 9 shows different P-V curve combinations under shading condition.

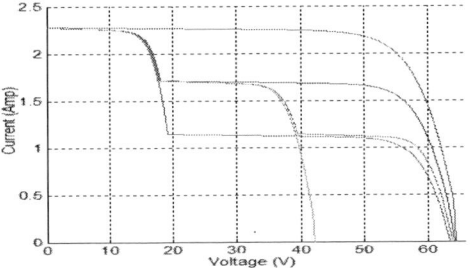

Fig. 8. I-V curve of 3 series connected modules under different partial shading condition.

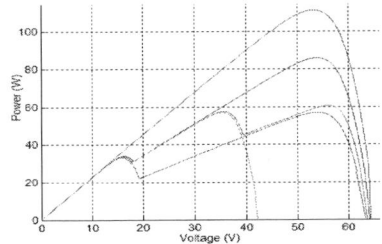

Fig. 9. P-V curve of string of 3 modules under different partial shading condition.

String with three modules has maximum three MPP points under shading condition. More complex and typical P-V curves for two strings connected in parallel each having three modules, are shown in Fig. 10.

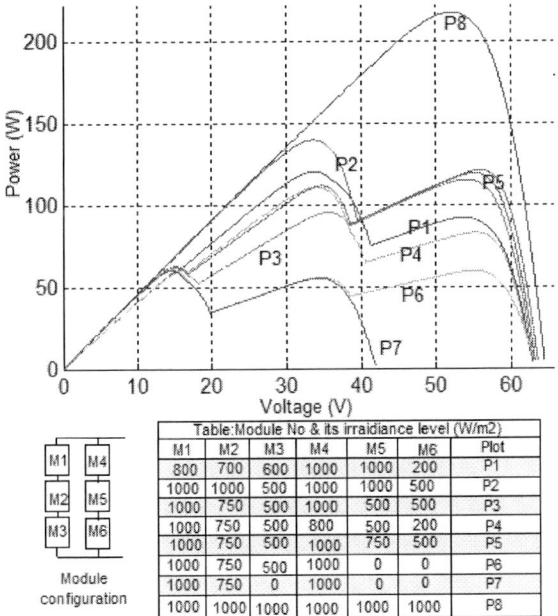

Table:Module No & its irradiance level (W/m2)						
M1	M2	M3	M4	M5	M6	Plot
800	700	600	1000	1000	200	P1
1000	1000	500	1000	1000	500	P2
1000	750	500	1000	500	500	P3
1000	750	500	800	500	200	P4
1000	750	500	1000	750	500	P5
1000	750	500	1000	0	0	P6
1000	750	0	1000	0	0	P7
1000	1000	1000	1000	1000	1000	P8

Module configuration

Fig. 10. P-V curves of string for different partial shading combinations.

IV. SHADING EFFECT ON MPPT PERFORMANCE

From P-V curve of Fig. 4, it is clear that with increase in voltage starting from zero, power output of panel also increase and reach up to MPP & after that power drops continuously as the voltage further increase toward open circuit. The peak power is the maximum power at Irradiance of 1000 W/m^2 and 25°C module temperature, but this condition never exists. In actual load across the module decide the I-V operating point and power extracted will be less than the maximum value. Also, MPP shift with change in Irradiance and temperature level. [18], [19] discuss various MPPT techniques for a PV system.

Perturbation and observation (PO) is one of the basic and most widely used MPPT controller [20], [21]. In actual, the controller sets the duty cycle of DC-DC converter in such a way that apparent resistance across the input module is optimal. PO controller operated in two modes. If $\frac{dP}{dV} > 0$ (voltage source region), then it continuously changes the duty cycle in the same direction. Similarly, If $\frac{dP}{dV} < 0$ (current source region), then change the direction of duty cycle so that it is able to track MPP. At power peak, $\frac{dP}{dV} \cong 0$ and it is criteria to detect MPP using PO.

Under partial shading, there are the numbers of local MPP. In PO, there is no provision to take care such situation. Therefore, it is not able to track maximum power all the time & sometime system operates at local MPP. In Fig. 11, system configuration of two series connected PV module with PO controller is shown. In this setup, input irradiance of one module is variable and the other has fixed irradiance level. For the test signal of Irradiances shown in Fig. 12, maximum allowable energy extraction is 493.6 W-sec. Fig.13 and Fig. 14 show the simulation result of the PO controller based system under partial shading condition. As PO is not so efficient & it allows to extract only 440.2 W-sec energy and there is an extra loss of 11% due to PO controller inability to track Global MPP. So, for improving overall performance under shading, it is preferred to use advance controller like Partial Swarrn Optimization, Hybrid and DMPPT techniques etc.

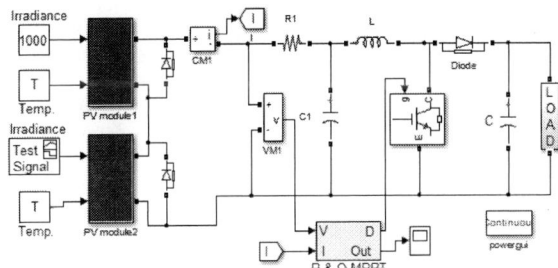

Fig. 11. Simulation setup of two series connected PV module

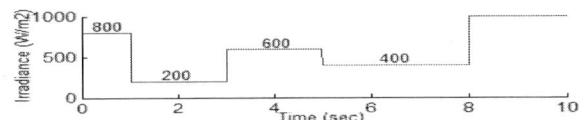

Fig. 12. Test signal (Irradiance of module 2) with G1=1000 W/m

978-1-4799-6047-7/14 $31.00 © 2014 IEEE

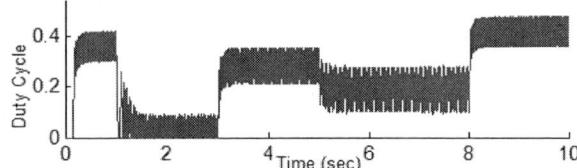

Fig. 13. Duty cycle of DC-DC converter under partial shading condition

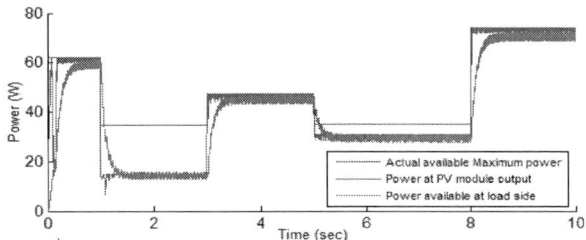

Fig. 14. Output power response of PV system under partial shading condition, using P&O MPPT controller.

V. EXPERIMENTAL SETUP AND RESULT VAILDATION

The simulation model is tested on ecosense Stand-Alone Solar Photovoltaic Training kit. It consists of two faced photovoltaic panel ($37W_p$), which can be folded and reassembled to use. Regulated Halogen lights are used to provide artificial variable irradiance. The system consists of DC-DC converter & Inverter module, data logger & plotter, display & measurement system, AC load, DC load, Battery, potentiometer and PO MPPT controller. Experimental setup is shown in Fig. 15. Fig. 16 shows the block diagram representation of ecosense PV Training System.

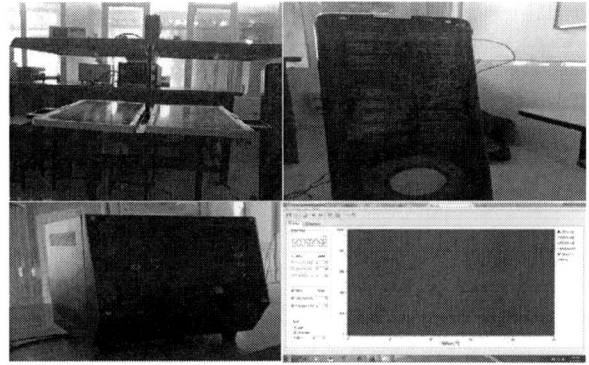

Fig. 15. Experimental setup of ecosense PV training system

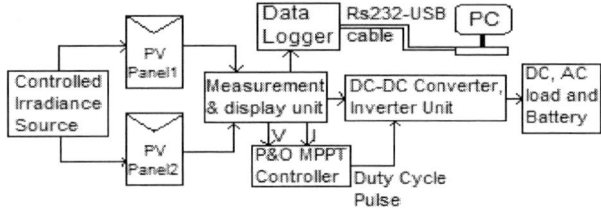

Fig. 16. Block diagram representation of experimental setup

Electrical characteristics and specification of ELDORA 40P is tabulated in Table I. To provide different irradiance value to PV module, several pieces of translucent plastics are employed.

This test under the condition of nonuniform irradiance is carried out in the following manner.

1. Connect two panel in series
2. Set the irradiance value of halogen lamp at maximum level.
3. Use combination of translucent plastics to cover panel 2, to reduce irradiance level.
4. Number of Irradiance combination can be produce by using translucent plastic sheets.
5. Use MPPT controller to set duty cycle of DC-DC converter.
6. Data logger data is used to extract P-t, V-t and I-t graphs.

TABLE I. Parameters of the Eldora 40P at 25°C, AM1.5, 1000 W/m²

I_{sc}	2.28 Amp
V_{oc}	21.77 V
I_m	2.10 Amp
V_m	17.70 V
$P_{max,e}$	37 W
N_s	36 Nos.
K_I	0.058% /°C
K_V	-0.31% /°C

The experiment is performed on simulator to verify the simulation result. During study, it is found that simulator PV modules suffer from internal shading due to uneven irradiance level on the panel surface. External shading is provided using translucent plastic sheet. For different level of shading, average output power and voltage are recorded. Fig. 17 shows the variation of irradiance with time. Similarly, Fig. 18, Fig. 19 and Fig. 20 show the average output power and voltage at string output terminal. As shown in Fig. 20, PO MPPT is not able to track MPP all the time.

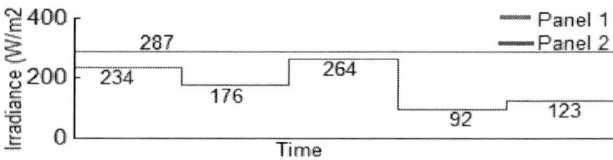

Fig. 17. Change in Irradiance with time

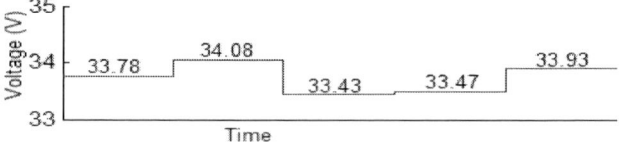

Fig. 18. Average output voltage at DC-DC converter input side

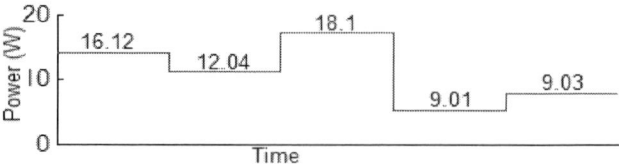

Fig. 19. Maximum power available at different irradiance level.

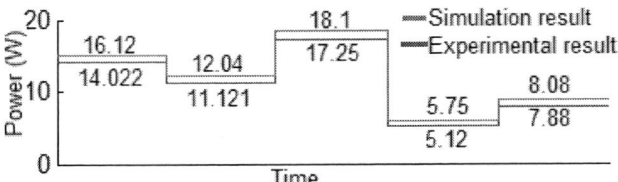

Fig. 20. Average Power available at input side of DC-DC converter

VI. CONCLUSION

This paper presents the detailed mathematical modeling of PV module. Iterative algorithm is given to calculate parameters of the single diode model. Various PV array simulations schematic can be created by the proposed model. The study of PV array performance under shading condition is presented in a simple way. It is found that under uneven irradiance condition, power output of the PV array degraded. By using bypass diode, one can improve the performance of the PV array, but due to inconsistent output of an array, there are multiple power peaks. PO MPPT controller performance is checked under uneven irradiance condition. It is found that PO is not much efficient under shading condition. Therefore, by using advanced MPPT controller, performance of PV system can be improved by approximate 11%.

Further, the effectiveness of different advanced MPPT controller can be investigated on this system. This work can be extend to develop global maximum power point tracking controller.

ACKNOWLEGMENTS

Authors are thankful to all the person associated with Power Electronic Lab, Department of Electrical Engineering, NIT, Kurukshetra for providing all the experimental setup to carry out this work.

REFERENCES

[1] A. Ashwani Kumar, "A study on renewable energy resources in India," in Proc. 2010 Int. Conf. on Environmental Engineering and Appl., Singapure, Sept. 10-12, 2010, pp. 49-53.

[2] S. Maity, A. Singh, and B. K. Saha, "Solar resources assessment in India a case study," in Proc. 2014 1st Int. conf. on Non-Conventional Energy, Kalyani, Jan 16-17, 2014, pp. 1-5.

[3] Young-Hyok Ji, Jun-Gu-Kim, Sang-Hoon Park, Jae-Hyung Kim, and Chung-Yuen Won, "C-language based PV array simulation techniques considering effect of partial shading," in Proc. IEEE Int. Conf. on Ind. Technol., Gippsland, Feb. 10-13, 2009, pp. 1-6.

[4] Aneek Islam, and Md. Iqbal Bahar Chowduary, "A Simulink based generalized model of PV cell/ array," in Proc. 3rd Int. Conf. on the Developments in Renewable Energy Technol., Dhaka, May 19-31, 2014, pp. 1-5.

[5] Badr Aldwane, "Modeling simulation and parameter estimation for photovoltaic module," Int. Conf. on Green Energy, Sfax, March 25-27, 2014, pp. 101-106.

[6] J. Leuchter, K. Zaplatilek, and P. Bauer, "Photovoltaic model for circuit simulation," in Proc. 38th Annual Conf. on IEEE Ind. Electron. Safety, Montreal, Oct. 25-28, 2012, pp. 5399-5405.

[7] Dezso Sera, Remus Teodorescu, and P. Rodriguez, "PV panel model based on datasheet values," in Proc. IEEE Int. Symp. on Ind. Electron., Vigo, June 4-7, 2oo7, pp. 2392-2396.

[8] Weidong Zeio, William G. Denford, and Antoine Capel, "A Novel modeling method for photovoltaic cells," 35th Annual IEEE Power Electron. Specialists Conf., Aachen, June 20-25, 2004, pp. 1950-1956.

[9] A. Mohapatra, B. K. Nayak, and K. B. Mohanty, "Comparative study on single diode photovoltaic module parameter extraction methods," in Proc. Int. Conf. of Power, Energy and Control, Sri Rangalatchum Dindigul, Feb. 6-8, 2013, pp. 30-34.

[10] H. Anderai, T. Ivanovici, G. Predusca, E. Diacounu, and P. C. Anderai, "Curve fitting method for modeling and analysis of photovoltaic cell characteristics," in Proc. IEEE int. Conf. of Automation Quality and Testing Robotics, Cluj-Napoca, May 24-27, 2012, pp. 207-212.

[11] ELDORA 40P High Efficiency Solar PV Module Datasheet Vikram Solar. [Online]. Available: http://www.vikramsolar.com/pdf/micro/ELDORA40P.pdf

[12] R. Ramaparbha, and B. L. Mathur, "Modeling and simulation of PV array under partial shaded conditions," in Proc. IEEE Int. Conf. on Sustainable Energy Techno., Singapore, Nov. 24-27, 2008, pp. 7-11.

[13] Shiva Moballegh, and Jing Jiang, "Modeling, prediction and experimental validations of power peaks of PV arrays under partial shading conditions," IEEE Trans. on Sustainable Energy, vol. 5, no. 1, 2014, pp. 293-300.

[14] Moein Jazayeri, Sener Usyal, and Kian Jazayeri, "A comparative study on different photovoltaic array topologies under partial shading conditions," IEEE PES T&D Conf. and Exposition, Chicago, April 14-17, 2014, pp. 1-5.

[15] R.Ramaprabha, and B.L.Mathur, "A comprehensive review and analysis of solar photovoltaic array configurations under partial shaded conditions," Intl. J. of Photoenergy, vol.2012, 2012, pp.1-16.

[16] M. Z. Shams El. Dein, and M. M. A. Salama, "Optimal photovoltaic array reconfiguration to reduce partial shading losses," IEEE Trans. on Sustainable Energy, vol. 4, no. 1, Jan 2013, pp. 145-153.

[17] Kun Diang, XinGao Bian, HaiHao Liu, and Tao Peng, "A matlab-simulink-based PV module model and it application under conditions of Nonuniform Irradiance," IEEE Trans. on Energy Convers., vol. 27, no.4, Dec. 2012, pp. 864-872.

[18] Ali F Murtaza,Hadeed Ahmed Sher, Marcello Chiaberge, Diego Boero, Mirko De Giuseppe, and Khaled E Addoweesh, "Comparative analysis of maximum power point tracking techniques for PV applications," in Proc. 16th Int. Multi Topic Conf., Lahore, Dec. 19-20, 2013, pp. 83-88.

[19] Bidyadhar Subudhi and Raseswari Pradhan, "A comparative study on Maximum Power Point Tracking techniques for photovoltaic power systems," IEEE Trans. on Sustainable Energy, vol. 4, no. 1, Jan. 2013, pp. 89-98.

[20] G. de Cesare, D. Caputo, and A. Nascetti, "Maximum power point tracker for photovoltaic systems with resistive like load," Solar Energy, vol. 80, no. 8, 2006, pp. 982-988.

[21] F. Liu, Y. Kang, Y. Zang, and S. Duan, "Comparison of P&O and hill climbing MPPT methods for grid-connected PV generator," in Proc. 3rd IEEE Conf. Ind. Electron. Appl., Singapore, Jun. 3-5, 2008, pp. 804-807.

Multi Diode Modelling of PV Cell

Pawan Kumar Pandey, K.S. Sandhu

Department of Electrical Engineering

NIT Kurukshetra,

Haryana, India

pkpnitj@gmail.com, kjssandhu@gmail.com

Abstract-**This paper presents the performance analysis of PV cell module using mathematical model based upon single, double and triple diodes configuration. Performance analysis in terms of I-V and P-V characteristics is estimated for all the three cases using MATLAB programming. The effect of irradiation, temperature and ideality factor in each case has been illustrated.**

Keywords: Single diode model, Double diode model, Triple diode model, Irradiation, Temperature, Ideality factor, PV cells and module

I.INTRODUCTION

Photovoltaic (PV) systems researches mainly divided into two areas i.e., array physics, design and optimization. In the design, manufacturing and evaluation of PV systems, choosing a suitable model is an important issue for PV cells and modules in predicting their behaviour [1]. Most of the researchers adopted single diode model which describes basic characteristics of I-V and P-V of solar models. Whereas few researcher adopted the two diode model and one of them emphasized triple diode model to simulate the space charge recombination effect by exponential voltage dependence separate current component [2]. Irrespective of any type of modeling it is observed that the output energy depends on solar radiation, the temperature of the cell and the voltage produced in the photovoltaic module. The voltage and current available at the terminals of a PV device may directly feed small loads.

In addition there are many other internal factors of device itself, such as type of material, path of the semiconductor current, reverse saturation current of semiconductor, ideality factor etc., which may affect its performance. In this paper an attempt is made to discuss such effects that may interrupt the PV cell performance [3, 4]. Main aim of the paper is to provide all necessary information to develop photovoltaic array models and circuits that can be used in the simulation of power converters for photovoltaic applications.

II. MODELING OF SOLAR CELL

The single diode equivalent circuit model of PV cell is shown in Fig.1 and equation (1) as given below may be used to describe its current behaviour.

$$I = I_{ph} - I_0 \left[\exp\left(\frac{V+IR_s}{AV_t} \right) - 1 \right] - \left[\frac{V+IR_s}{Rp} \right] \tag{1}$$

Whereas, above expression in case of multi diode model of the PV cell may be expressed as:

$$I = I_{ph} - \sum_{i=1}^{3} I_{0_i} \left[\exp\left(\frac{V+IR_s}{A_i V_t} \right) - 1 \right] - \left[\frac{V+IR_s}{Rp} \right] \tag{2}$$

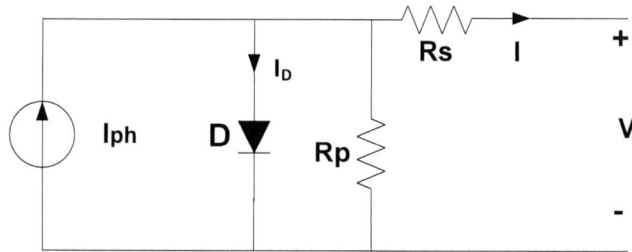

Fig.1. Equivalent Model of Single-Diode Photovoltaic Cell.

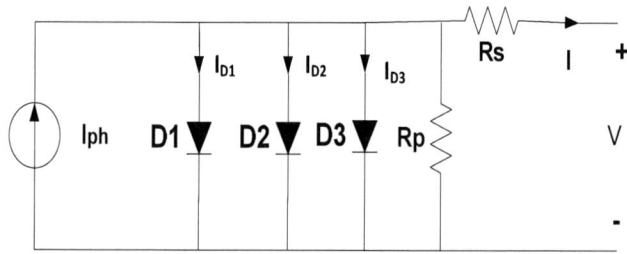

Fig.2. Equivalent Model of Three-Diode Photovoltaic Cell.

Where:

I_{ph}	is the current generated by the incident light.
I_{Di}	is the Shockley i^{th} diode current.
I_{0i}	is reverse saturation current of ith diode.
$V_t = kT/q$,	is the thermal voltage
q	is the electron charge, 1.6×10^{-19}C.
k	is the Boltzmann's constant, 1.38×10^{-23}J/K.
T	is the p-n junction temperature in Kelvin.
A_i	is the ideality factor of ith diode.

Dependence of the Photocurrent (I_{ph}) on environmental parameters i.e. temperature and irradiance may be expressed as:

$$I_{ph} = \left[I_{ph,n} + K_i(T - T_{ref})\right]\frac{G}{G_n} \qquad (3)$$

Where:

$I_{ph,n}$ is photo current at standard test condition
K_i is temp. coefficient of short circuit current
T_{ref} is reference temperature
G_n is irradiance at standard test condition

Dependence of diode saturation current is defined as:

$$I_0 = I_{0,n}\left(\frac{T}{T_{ref}}\right)^3 exp\left[\frac{qEg}{Ak}\left(\frac{1}{T_{ref}} - \frac{1}{T}\right)\right] \qquad (4)$$

$$I_{0,n} = \frac{I_{sc,n}}{\left[exp\left(\frac{V_{oc,n}}{AV_{t,n}}\right) - 1\right]} \qquad (5)$$

Where:

$I_{0,n}$ is saturation current at reference temp.
$I_{sc,n}$ is short circuit current at reference temp.
$V_{oc,n}$ is open circuit current at reference temp.

The standard test conditions happen with Irradiance as 1000W/m², cells temperature as 25°C, and spectral distribution (Air Mass) AM as 1.5. For ideal solar plate series resistance (R_s) will be zero and the parallel resistance (R_{sh}) will be infinite. Therefore, for the maximum power from the solar PV cell,(R_s) will be negligible value and the (R_{sh}) must have a higher value [5].

For the model comprising of numbers of series cell (N_s) and parallel cell (N_p), equation as given below may be used to define the current behaviour.

$$I = Np I_{ph} - Np \sum_{i=1}^{3} I_{0_i}\left[\exp\left(\frac{V/Ns}{A_i V_t}\right) - 1\right] \qquad (6)$$

III. RESULTS AND DISCUSSIONS

In order to compare PV models, PV cell module of 42W as described in Table 1is used for simulation study [6].

TABLE 1.PV CELL MODULE PARAMETERS

Parameters	Specification
Maximum Power Point (P_{max})	42W
Voltage at peak power (V_{max})	18V
Current at peak power (I_{max})	2.30A
Short circuit current (I_{sc})	2.40A
Open circuit voltage (V_{oc})	21.7V
Temp. coefficient of short circuit current (k_i)	0.08%/°C
Ideality factor (A)	1.3

Simulation results as obtained may be discussed under various heads as show below.

A. C-V and P-V Curves for PV Module under Varying Solar Radiation

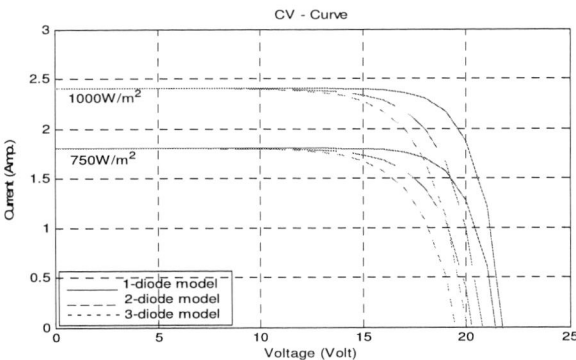

Fig.3. Matlab model I-V curve, for single diode (A=1.3), double diode (A₁=1.3,A₂=2) and triple diode (A₁=1.3,A₂=2,A₃=2) PV model with various irradiation level.

Fig.3 and Fig.4 show the typical C-V and P-V characteristics of a single diode, double diode and triple diode PV module model under two irradiation levels with reference temperature as 25°C. Fig.3 show that the module short circuit current increases with small increment in the open circuit voltage. As shown in Fig. 4 power also increases with an increase of irradiation.

From equation (3) it could be observed that a drop in solar radiation from 1000 to 750 w/m² ultimately leads to drop in short circuit current from 2.4 to 1.8 A (25% drop). However, the drop in maximum power and open circuit voltage is much lesser, i.e., typically less than 8%.

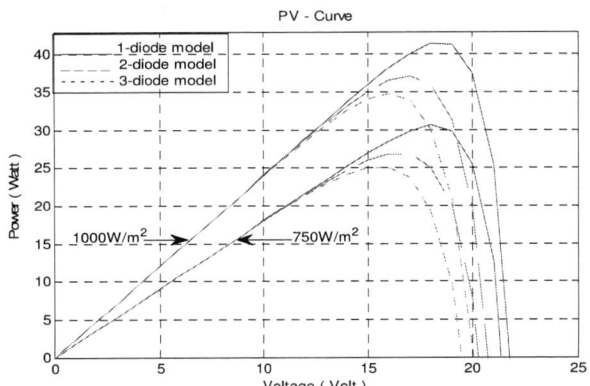

Fig.4 Matlab model P-V curve, for single diode (A=1.3), double diode (A₁=1.3, A₂=2) and triple diode (A₁=1.3, A₂=2, A₃=2) PV model with various irradiation level.

From Fig. 3 & 4 it is observed that "open circuit voltage as well as maximum power generated by the module decreases with an increase in the number of diodes during modeling. Prime reason for such a drop in output characteristics is the recombination of carrier charges due to high carrier charge density on p-n junction of diode. It is observed that the effect of change of solar radiation on open circuit voltage as well as on maximum efficiency seems to be almost same, irrespective of model configuration in terms of number of diodes. Table II show the comparison of results in term of maximum power point, voltage, current, etc., with solar radiation as 1000w/m2.

TABLE II.   COMPARATIVE ANALYSIS

Parameters	Single diode Matlab model (A=1.3)	Double diode Matlab model (A₁=1.3, A₂=2)	Triple diode Matlab model (A₁=1.3,A₂=2,A₃=2)
Maximum Power Point (P_{max})	41.42W	37.04W	34.70W
Voltage at peak power (V_{max})	18V	17V	16V
Current at peak power (I_{max})	2.30Amp	2.18Amp	2.168Amp
Short circuit current (I_{sc})	2.4Amp	2.4Amp	2.4Amp
Open circuit voltage (V_{oc})	21.78V	20.75V	20V

B. C-V and P-V Curves for PV Module under Varying Cell Temperature

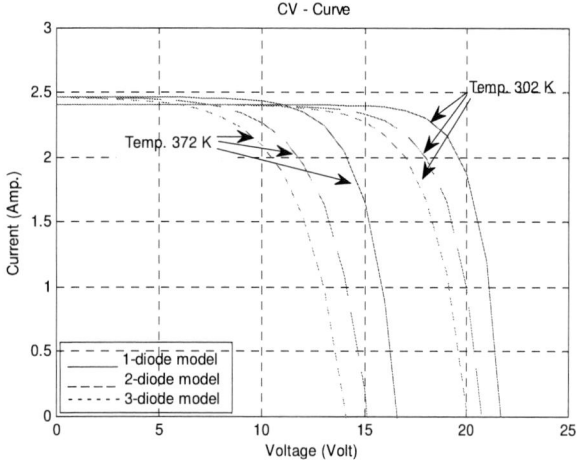

Fig. 5. Matlab model I-V curve, for single diode (A=1.3), double diode (A₁=1.3,A₂=2) and triple diode (A₁=1.3,A₂=2,A₃=2) PV cells model with various cell temperature.

Figs. 5-6 show the variation in current and power with of module for different models due to variation in cell temperature. It is found that, open circuit voltage (V_{oc}) decreases as temperature increases while module current have an irrelevant increment. On increasing the temperature, band gap energy of solar PV cell gets reduced, thereby leading to an absolute drop in energy required to fetch an electron from valence band. This reduction in pn junction barrier potential ultimately results in reduction of output open circuit voltage as depicted above. Model has been simulated for a temperature of 302 k and 372 k for single, double and triple diode topology. It could be noticed that open circuit voltage for 1-diode model at 302 k is higher than at 372 k. Same is the case for other topologies also.

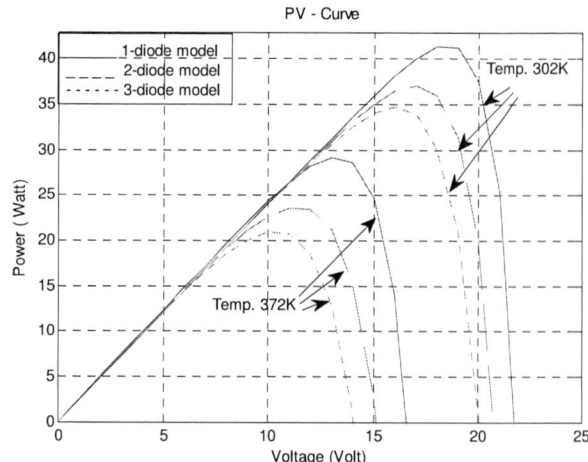

Fig. 6. Matlab model P-V curve, for single diode (A=1.3), double diode (A₁=1.3,A₂=2) and triple diode (A₁=1.3,A₂=2,A₃=2) PV model with various cell temperature.

C. Effect of Ideality Factor on the C-V and P-V Curves of a PV Module

Ideality factor tends to account the effects of assumptions that have been assumed for deriving the ideal diode equation. Ideality factor equal to one signifies the majority carrier recombination process (low level injection) while a value greater than one tends to depict high level injection signifying both majority and minority carrier recombination. The value of ideality factor equals to unity indicate perfect matches to theory under STC (solar radiation=1000 W/m2 and ref. temperature=25°C).When the same ideality factor (A=1) is used in each model, open circuit voltage tends to decrease in case of double diode and triple diode PV module. As shown in Fig. 7 value open circuit voltage for multi diode model becomes small in comparison to its value in case of single diode model.

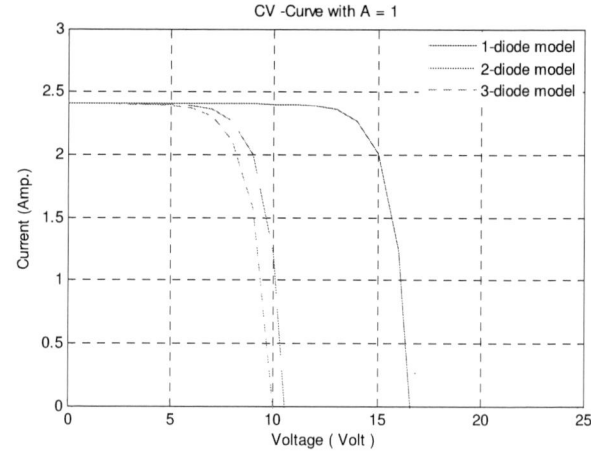

Fig.7. Matlab model I-V curve, for single diode, double diode and triple diode PV cells model with same Ideality factor (A₁=1,A₂=1,A₃=1).

A shown inFig.8, when ideality factor is taken as 1.5 in each model open circuit voltage increases in comparison to as shown for A=1 (Fig. 7). However the difference of open circuit voltage due to single diode and multi diode for multi diode model further increases. This shows that by proper solution of A for any type of model IV characteristics may be shifted horizontally on each side.

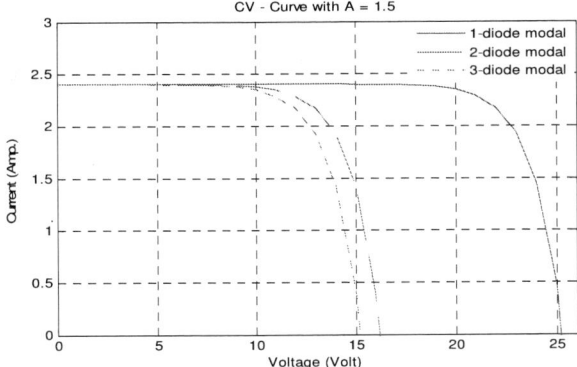

Fig.8. Matlab model I-V curve, for single diode, double diodeand triple diode PV cells model with same Ideality factor (A_1=1.5,A_2=1.5,A_3=1.5).

IV. CONCLUSION

In this paper an attempt has been made to analyze the behavior of PV cell module adopting a single diode, double diode and triple diode equivalent circuit model with the fixed value of R_s and R_p. A critical observation indicates that ideality factor is effective to shift the characteristics in any direction irrespective of number of diodes. Therefore proper selection of ideality factor may be useful to match the predicted output of PV cell with experimental value for any number of diodes and hence may be adopted for further analysis.

REFERENCES

[1] Gow, J.A; Manning, C.D., "Development of a photovoltaic array model for use in power-electronics simulation studies," *Electric Power Applications, IEE Proceedings - ,* vol.146, no.2, pp.193, 200, Mar 1999.

[2] CHAN, D. S H; PHANG, J. C H, "Analytical methods for the extraction of solar-cell single- and double-diode model parameters from I-V characteristics," *Electron Devices, IEEE Transactions on ,* vol.34, no.2, pp.286,293, Feb 1987.

[3] Hejri, M.; Mokhtari, H.; Azizian, M.R.; Ghandhari, M.; Soder, L., "On the Parameter Extraction of a Five-Parameter Double-Diode Model of Photovoltaic Cells and Modules," *Photovoltaic, IEEE Journal of ,* vol.4, no.3, pp.915,923, May 2014.

[4] Das, N.; Wongsodihardjo, H.; Islam, S., "Photovoltaic cell modeling for maximum power point tracking using MATLAB/Simulink to improve the conversion efficiency," *Power and Energy Society General Meeting (PES), 2013 IEEE ,* pp.1,5, 21-25 July 2013

[5] Sinha, D.; Das, AB.; Dhak, D.K.; Sadhu, P.K., "Equivalent circuit configuration for solar PV cell," *Non-Conventional Energy (ICONCE), 2014 1st International Conference on,* pp.58, 60, 16-17 Jan. 2014.

[6] Suthar, M.; Singh, G.K.; Saini, R.P., "Comparison of mathematical models of photo-voltaic (PV) module and effect of various parameters on its performance," *Energy Efficient Technologies for Sustainability (ICEETS), 2013 International Conference on,* pp.1354, 1359, 10-12 April 2013 .

978-1-4799-6047-7/14 $31.00 © 2014 IEEE

LVCMOS Based Energy Efficient Solar Charge Sensor Design on FPGA

Anu Singla
Department of Electrical Engineering
Chitkara University
Chandigarh, India
anu.singla@chitkara.edu.in

Amanpreet Kaur, Bishwajeet Pandey
Department of ECE
Chitkara University
Chandigarh, India
gyancity@gyancity.com

Abstract— The performance and life span of energy storage solar battery depends on its charging and discharging. Solar charge controller regulates the rate of charging and discharging of battery for its smooth and efficient operation. Solar charge sensor, component of charge controller is used to invoke full charge alarm and also discharge alarm as per the voltage level of battery. Solar charge sensor consumes certain power. LVCMOS I/O standards is the low voltage complementary metal oxide semiconductor. There are 4 different LVCMOS available in 28nm technology based Artix-7 FPGA. We are proposing I/O standards for solar charge sensor in order to achieve energy efficiency with solar charge sensor. There is 60% reduction in I/O power, when we use LVCMOS15 in place of LVCMOS33 in solar charge sensor design at 900MHz. Similarly, we are saving 51.85% I/O power, when we use LVCMOS18 in place of LVCMOS33 in solar charge sensor design operating at 5 GHz. Similarly, we are saving 32.31% IO power, when we use LVCMOS25 in place of LVCMOS33 in solar charge sensor design operating at 60 GHz.

Keywords—Solar Charge Controller, LVCMOS, I/O Standards, Energy Efficient Design, FPGA

I. INTRODUCTION

In solar photovoltaic (SPV) power systems, charge input from PV arrays is insufficient to keep battery fully charged at times. The performance of energy storage solar battery depends on its charging and discharging i.e. size and usage of system load. The excessive system loads may drain battery charge to an extent of irretrievable damage and may ultimately result in it's replacement. In SPV systems that utilize batteries to store energy, charge controller is incorporated to prevent damage to batteries (Fig.1). Solar charge controller regulates the rate of charging and discharging of battery and hence improves the performance and life span of battery [1]. To have efficient SPV system which includes effective PV array utilization, trouble free operation of battery system to meet load requirements; the role of control strategy adopted for the operation of charge controller becomes important [1-2].

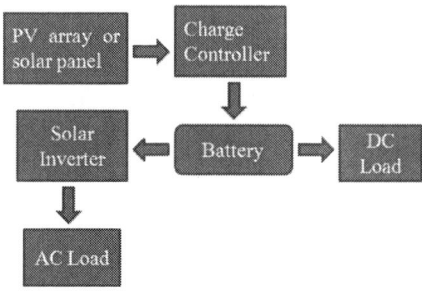

Figure 1: Solar Photovoltaic Power System

Charge controller stop charging a battery when it exceed a set high voltage level, and re-enable charging when battery voltage drops back below a preset voltage level [3]. The additional features of charge controller such as battery temperature monitoring to prevent overheating, alarm and annunciator panel, transmit data to remote displays, data logging to track current flow over time and special algorithms can maximize the battery capacity.

Solar charge controller circuit consists of several electrical and electronic components encapsulated in a single microchip. The main sensor in the charge controller is voltage sensors. The charge sensor will also consume power, therefore it is essential to make charge sensor more energy efficient in order to make it feasible to afford with battery charging. Research is going on in energy efficient mobile battery charge controller sensor [4-8]. But, research in energy efficient solar charge sensor is at nascent stage. Authors of this paper have proposed a LVCMOS based energy efficient solar charge sensor design on FPGA that will integrate in solar charger.

The basic schematic of solar charge sensor is shown in figure 2.

978-1-4799-6047-7/14 $31.00 © 2014 IEEE

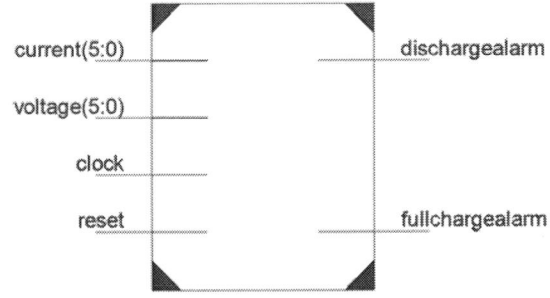

Figure 2: Top Level Schematic of Solar Charge Sensor

Inputs are six bit current and six bit voltage input along with one bit clock and one bit reset. Discharge alarm and fullcharge alarm are two outputs in solar charge sensor as shown in Figure 2. If voltage level is less than 3V, then it invoke discharge alarm and if voltage level is greater than 48V, then it invoke full charge alarm.

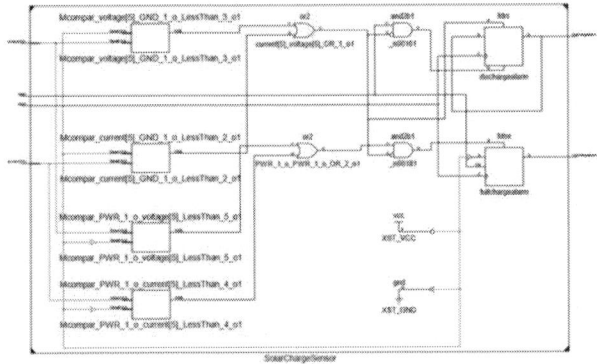

Figure 3: Register Transfer Level Schematics of Solar Charge Sensor

In the implementation of our design, two 1-bit register and four 6-bit comparator are used as shown in Figure 3.

II. RESULTS

In wireless network we cannot provide power through the wire so we used the battery in the node. Lifetime of the node depends on the lifetime of the battery. If we propose that solar battery, solar charger and solar charge sensor will be integrated in node of wireless sensor network, the lifetime of node will be infinite. In order to test the compatibility of battery charge sensor with WLAN channel, We are going to operate our charge sensor with specified frequency of WLAN channel 802.11 b/g/n with 2.4 GHz, 802.11y with 3.6 GHz, Public Safety WLAN 802.11y with 4.9 GHz, 802.11a/h/j/n/ac with 5 GHz, 802.11p with 5.9 GHz, 802.11 ad with 60 GHz and 802.11 ah with 900 Mhz.

A. When Solar Charge Sensor is Operating at 900MHz

Table 1: Power Dissipation of Solar Charge Sensor Using Different LVCMOS

	LVCMOS15	LVCMOS18	LVCMOS25	LVCMOS33
Clock	0.005W	0.005W	0.005W	0.005W
Signal	0.001W	0.001W	0.001W	0.001W
IOs	0.004W	0.005W	0.007W	0.010W
Leakage	0.042W	0.042W	0.043W	0.044W
Total	0.052W	0.053W	0.056W	0.060W

There is no change in clock power, minor change in leakage power, and also no change in signal power in solar charge sensor when we are using four different LVCMOS I/O standards. There is 60% reduction in IO power, when we use LVCMOS15 in place of LVCMOS33 in solar charge sensor design.

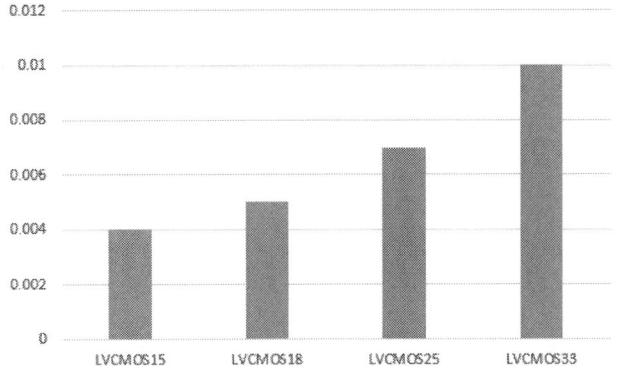

Figure 4: IO Power Dissipation in Solar Charge Sensor at 900MHz

B. When Solar Charge Sensor is Operating at 1GHz

Table 2: Power Dissipation of Solar Charge Sensor Using Different LVCMOS

	LVCMOS15	LVCMOS18	LVCMOS25	LVCMOS33
Clock	0.006W	0.006W	0.006W	0.006W
Signal	0.001W	0.001W	0.001W	0.001W
IOs	0.004W	0.005W	0.007W	0.011W
Leakage	0.042W	0.042W	0.043W	0.044W
Total	0.053W	0.054W	0.057W	0.061W

There is 63.63% reduction in IO power, when we use LVCMOS15 in place of LVCMOS33 in solar charge sensor design as shown in Figure 5 and Table 2.

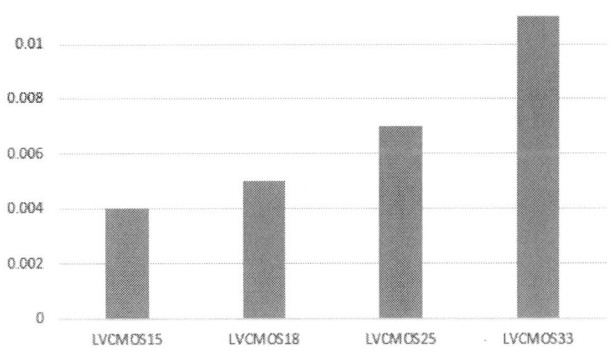

Figure 5: IO Power Dissipation in Solar Charge Sensor at 1GHz

C. When Solar Charge Sensor is Operating at 2.4GHz

Table 3: Power Dissipation of Solar Charge Sensor Using Different LVCMOS

	LVCMOS15	LVCMOS18	LVCMOS25	LVCMOS33
Clock	0.014W	0.014W	0.014W	0.014W
Signal	0.002W	0.002W	0.002W	0.002W
IOs	0.011W	0.012W	0.018W	0.026W
Leakage	0.042W	0.042W	0.043W	0.044W

Total	0.069W	0.071W	0.076W	0.086W

There is 57.69% reduction in IO power, when we use LVCMOS15 in place of LVCMOS33 in solar charge sensor design operating at 2.4GHz as shown in Table 3 and Figure 6.

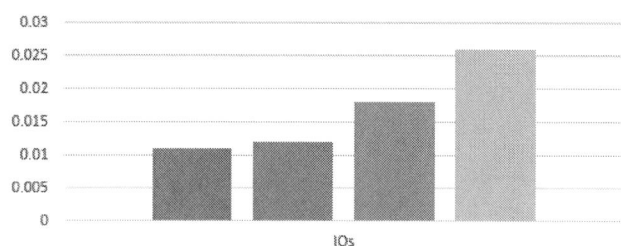

■ LVCMOS15   ■ LVCMOS18   ■ LVCMOS25   ■ LVCMOS33
Figure 6: IO Power Dissipation in Solar Charge Sensor at 1GHz

D. When Solar Charge Sensor is Operating at 3.6GHz

Table 4: Power Dissipation of Solar Charge Sensor Using Different LVCMOS

	LVCMOS15	LVCMOS18	LVCMOS25	LVCMOS33
Clock	0.021W	0.021W	0.021W	0.021W
Signal	0.002W	0.002W	0.002W	0.002W
IOs	0.016W	0.019W	0.026W	0.039W
Leakage	0.042W	0.042W	0.043W	0.044W
Total	0.082W	0.085W	0.093W	0.107W

Similarly, we are saving 57.69% IO power, when we use LVCMOS15 in place of LVCMOS33 in solar charge sensor design operating at 3.6 GHz as shown in Table 4 and Figure 7.

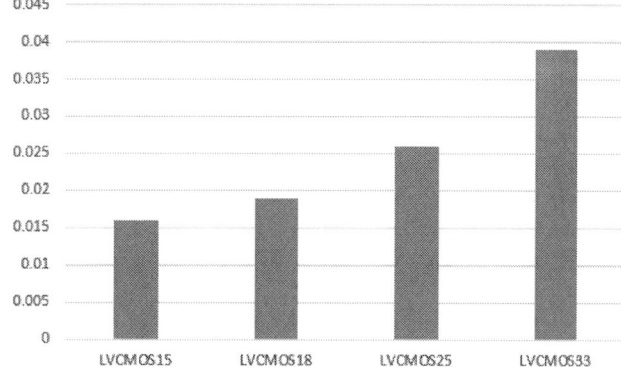

Figure 7: IO Power Dissipation in Solar Charge Sensor at 3.6 GHz

E. When Solar Charge Sensor is Operating at 5GHz

Table 5: Power Dissipation of Solar Charge Sensor Using Different LVCMOS

	LVCMOS15	LVCMOS18	LVCMOS25	LVCMOS33
Clock	0.029W	0.029W	0.029W	0.029W
Signal	0.003W	0.003W	0.003W	0.003W
IOs	0.022W	0.026W	0.037W	0.054W
Leakage	0.042W	0.043W	0.043W	0.044W
Total	0.107W	0.112W	0.125W	0.147W

Similarly, we are saving 51.85% IO power, when we use LVCMOS18 in place of LVCMOS33 in solar charge sensor design operating at 5 GHz as shown in Table 5 and Figure 8.

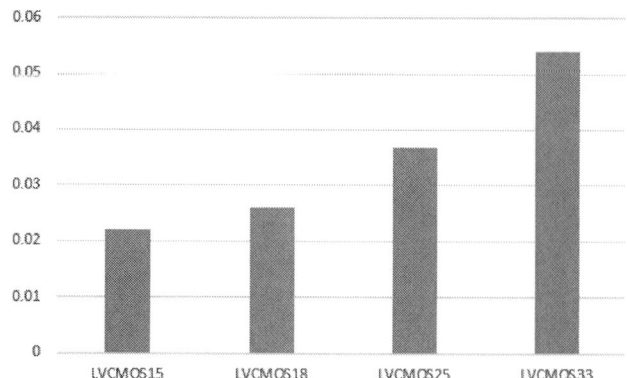

Figure 8: IO Power Dissipation in Solar Charge Sensor at 5 GHz

F. When Solar Charge Sensor is Operating at 5.9GHz

Table 6: Power Dissipation of Solar Charge Sensor Using Different LVCMOS

	LVCMOS15	LVCMOS18	LVCMOS25	LVCMOS33
Clock	0.034W	0.034W	0.034W	0.034W
Signal	0.004W	0.004W	0.004W	0.004W
IOs	0.026W	0.031W	0.043W	0.064W
Leakage	0.042W	0.043W	0.043W	0.044W
Total	0.052W	0.053W	0.056W	0.060W

Similarly, we are saving 51.56% IO power, when we use LVCMOS18 in place of LVCMOS33 in solar charge sensor design operating at 5.9 GHz as shown in Table 6 and Figure 9.

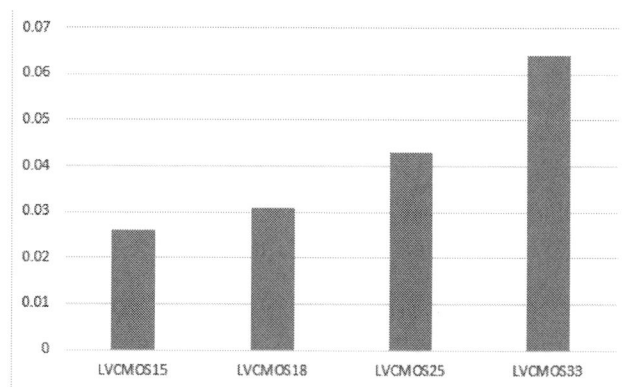

Figure 9: IO Power Dissipation in Solar Charge Sensor at 5.9 GHz

G. When Solar Charge Sensor is Operating at 60GHz

Table 7: Power Dissipation of Solar Charge Sensor Using Different LVCMOS

	LVCMOS15	LVCMOS18	LVCMOS25	LVCMOS33
Clock	0.345W	0.345W	0.345W	0.345W
Signal	0.034W	0.034W	0.034W	0.034W
IOs	0.268W	0.311W	0.440W	0.650W
Leakage	0.044W	0.044W	0.045W	0.047W
Total	0.694W	0.737W	0.868W	1.079W

Similarly, we are saving 32.31% IO power, when we use LVCMOS25 in place of LVCMOS33 in solar charge sensor

978-1-4799-6047-7/14 $31.00 © 2014 IEEE

design operating at 60 GHz as shown in Table 7 and Figure 10.

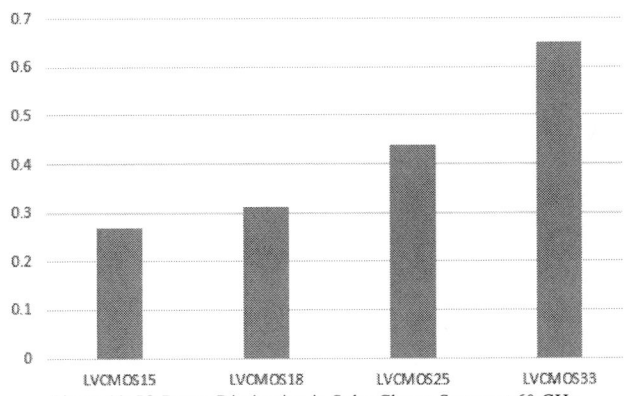
Figure 10: IO Power Dissipation in Solar Charge Sensor at 60 GHz

H. Clock Power Analysis in Solar Charge Sensor

Table 8: Clock Power Dissipation with Different LVCMOS and Frequency

	LVCMOS15	LVCMOS18	LVCMOS25	LVCMOS33
900MHz	0.005W	0.005W	0.005W	0.005W
1GHz	0.006W	0.006W	0.006W	0.006W
2.4 GHz	0.014W	0.014W	0.014W	0.014W
3.6 GHz	0.021W	0.021W	0.021W	0.021W
5 GHz	0.029W	0.029W	0.029W	0.029W
5.9 GHz	0.034W	0.034W	0.034W	0.034W
60 GHz	0.345W	0.345W	0.345W	0.345W

There is 98.55% reduction in clock power, when we scale down frequency from 60GHz to 900MHz as shown in Table 8 and Figure 11.

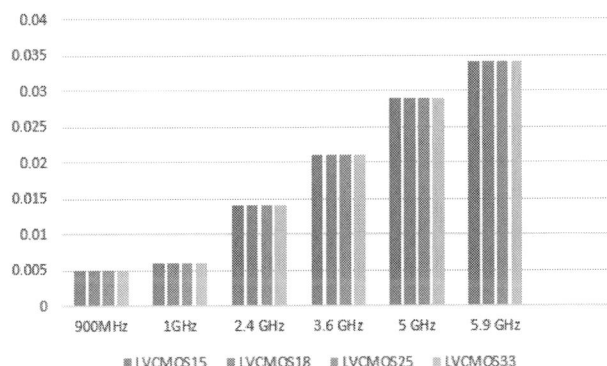
Figure 11: Clock Power Dissipation in Solar Charge Sensor

I. Signal Power Analysis in Solar Charge Sensor

Table 9: Signal Power Dissipation with Different LVCMOS and Frequency

	LVCMOS15	LVCMOS18	LVCMOS25	LVCMOS33
900MHz	0.001W	0.001W	0.001W	0.001W
1GHz	0.001W	0.001W	0.001W	0.001W
2.4 GHz	0.002W	0.002W	0.002W	0.002W
3.6 GHz	0.002W	0.002W	0.002W	0.002W
5 GHz	0.003W	0.003W	0.003W	0.003W
5.9 GHz	0.004W	0.004W	0.004W	0.004W
60 GHz	0.034W	0.034W	0.034W	0.034W

There is 97.05% reduction in signal power, when we scale down frequency from 60GHz to 900MHz as shown in Table 9 and Figure 12.

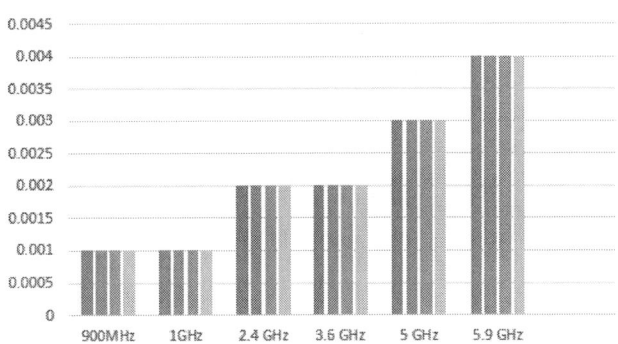
Figure 12: Signal Power Dissipation in Solar Charge Sensor

J. Leakage Power Analysis in Solar Charge Sensor

Table 10: Leakage Power Dissipation with Different LVCMOS & Frequency

	LVCMOS15	LVCMOS18	LVCMOS25	LVCMOS33
900MHz	0.042W	0.042W	0.043W	0.044W
1GHz	0.042W	0.042W	0.043W	0.044W
2.4 GHz	0.042W	0.042W	0.043W	0.044W
3.6 GHz	0.042W	0.042W	0.043W	0.044W
5 GHz	0.042W	0.043W	0.043W	0.044W
5.9 GHz	0.042W	0.043W	0.043W	0.044W
60 GHz	0.044W	0.044W	0.045W	0.047W

There is 4.54% reduction in signal power, when we scale down frequency from 60GHz to 900MHz as shown in Table 10 and Figure 12. There is 6.38% reduction in leakage power, when we use LVCMOS15 in place of LVCMOS33 as shown in Table 10 and Figure 13.

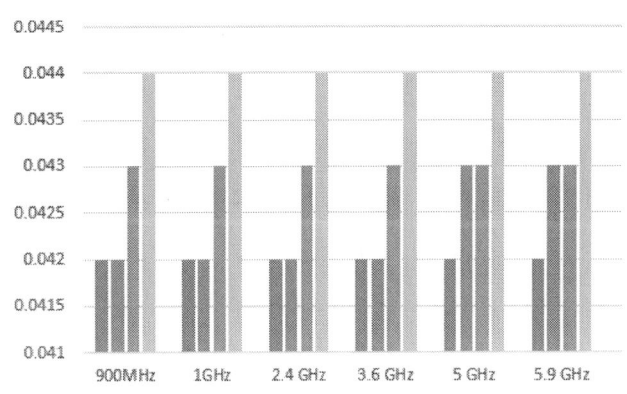
Figure 13: Leakage Power Dissipation in Solar Charge Sensor

III. CONCLUSION

There is 4.54% reduction in signal power, 6.38% reduction in leakage power when we scale down frequency from 60GHz to 900MHz and use LVCMOS15 in place of LVCMOS33, 97.5% reduction in signal power when we scale down frequency from 60GHz to 900MHz, 98.55% reduction in clock power when we scale down frequency from 60GHz to 900MHz, saving 32.31% IO power when we use LVCMOS25 in place of LVCMOS33 in solar charge sensor design operating at 60 GHz, saving 51.56% IO power, when we use LVCMOS18 in place of LVCMOS33 in solar charge sensor design operating at 5.9 GHz, saving 51.85% IO power, when we use LVCMOS18 in place of LVCMOS33 in solar charge sensor design operating at 5 GHz, saving 57.69% IO power, when we use LVCMOS15 in place of LVCMOS33 in solar charge sensor design operating at 3.6 GHz, 57.69% reduction in IO power, when we use LVCMOS15 in place of LVCMOS33 in solar charge sensor design operating at 2.4GHz, 63.63% reduction in IO power, when we use LVCMOS15 in place of LVCMOS33 in solar charge sensor design and no change in clock power, minor change in leakage power, and also no change in signal power in solar charge sensor when we are using four different LVCMOS I/O standards. There is 60% reduction in IO power, when we use LVCMOS15 in place of LVCMOS33 in solar charge sensor design.

IV. FUTURE SCOPE

This design is implemented on 28nm FPGA, we can use other FPGA like ultra-scale FPGA, 14nm FPGA. Here, we are using Verilog. There is open scope to design solar charge sensor with verification capability with help of system Verilog. We can also extend our design approach to other components of solar photovoltaic power systems such as solar batteries and solar inverters.

References

[1] Components of Battery Charging Systems (BSC), https://energypedia.info/wiki/Components_of_Battery_Charging_Syste ms_%28BSC%29. [PDF available online].

[2] M. Fernandez, et al. "Development of a VRLA battery with improved separators, and a charge controller, for low cost photovoltaic and wind powered installations." Journal of power sources 95.1 (2001): 135-140.

[3] McVey, Michael J., Aaron J. Mendelsohn, and Bradley J. Suppanz. "Battery charge management architecture." U.S. Patent No. 6,014,013. 11 Jan. 2000.

[4] S. M. M. Islam, B. Pandey, S. Jaiswal, M. M. E. Noor and S. M. T. Siddiquee, "Simulation of Voltage Scaling Aware Mobile Battery Charge Controller Sensor on FPGA", "Advanced Materials Research", ISSN:1022-6680, Vol. 893 (2014) pp 798-802, February 2014, Trans Tech Publications, Switzerland, (SCOPUS Indexed), http://www.scientific.net/AMR.893.798

[5] R. Kaur, S. M. M. Islam, T. Kumar, B. Pandey, M. M. E. Noor and S. M. T. Siddiquee, "Frequency Scaling Based Energy Efficient Mobile Battery Charge Controller Sensor Design on FPGA", "Advanced Materials Research", Modern Achievements and Developments in Manufacturing and Industry, Vol. 984 - 985, pp. 1057-1062, Trans Tech Publications, Switzerland, (SCOPUS Indexed), http://www.scientific.net/AMR.984-985.1057

[6] Chen, Liang-Rui. "PLL-based battery charge circuit topology." Industrial Electronics, IEEE Transactions on 51.6 (2004): 1344-1346.

[7] Yeon, Sang-Heum. "Battery charge controller having an adjustable termination current." U.S. Patent No. 6,133,712. 17 Oct. 2000.

[8] Hsieh, Guan-Chyun, Liang-Rui Chen, and Kuo-Shun Huang. "Fuzzy-controlled Li-ion battery charge system with active state-of-charge controller." Industrial Electronics, IEEE Transactions on 48.3 (2001): 585-593.

Survey on Hybrid (Wind/solar) Renewable Energy System and Associated Control Issues

Rahul Sharma
Electrical Engineering Department,
NIT, Kurukshetra, India
rahulsharma.knit2006@gmail.com

Sathans
Electrical Engineering Department,
NIT, Kurukshetra, India
sathans@nitkkr.ac.in

Abstract- **This paper reviews various configurations based on power electronics converters and their control issues in hybrid renewable energy systems (RES). The paper is focused on important control issues and challenges in the design and power utilization of hybrid RES. Different configurations of Wind/Solar system, generation side control and grid side control have been comprehensively covered and reviewed. The methods of using control concepts for enhancing hybrid renewable energy system performance in distributed grid system to meet the recent grid codes are discussed. Finally, future directions and important challenges related to this area are also presented.**

Keywords- Hybrid energy system, Distributed grid, Grid code, Power electronics converters, renewable energy system, Wind/solar system.

I. INTRODUCTION

Due to the limited and depleting resources of conventional energy and environmental concerns, the hybrid renewable energy systems (HRES) are emerging as the sustainable and promising alternative to meet the growing demand of electricity which is environmental friendly as well as competitive in terms of economy [1]. This is one of the active areas of research due to the fast development in power electronics converters which is increasing the possibilities of easy, efficient and cost effective power transmission and utilization of RESs as compared to conventional energy sources [2].

But the RESs are by nature unpredictable due to changing weather conditions, therefore, the reliable and continuous power generation capability cannot be achieved only by one RES, however, the same can be achieved to a great extent by HRES using energy storage devices along with. Wind and Solar are the most significant renewable sources of energy available on daily and seasonal basis. HRES (Wind/solar) is becoming the most attractive and promising energy source which is practically feasible and provides continuous energy reliability [3].

Many researchers have proposed different configurations and control schemes on wind and solar energy systems and their hybrid combinations for improved performance [4]-[6]. But HRES are still in the development stage and are facing many challenges in which there is growing research on different prospects of HRES and their control issues to make them cost effective and efficient [7]-[9]. This review paper presents a recent perspective on the developments of HRES, their power electronics based controls and future challenge.

This paper is categorized as follows. In section II, the different configurations and future scope of change in HRES configurations are described. In section III, we discuss different types of generation side control of hybrid energy system to increase the efficiency of system by using Maximum power Point Tacking (MPPT) methods. There are different methods for MPPT tracking of wind and solar systems which have been discussed for Hybrid renewable energy systems. In section IV, grid side/load side control techniques have been discussed for hybrid renewable energy systems. Different Power electronics converters for off grid or grid interface and their control schemes are covered in this section. In Section V, recent challenges in the field of Hybrid energy systems and suitable enhancement in control schemes has been discussed. Future prospects of Hybrid renewable energy system are also discussed in this section. Lastly, section VI concludes this paper with highlighting some of the important issues and future trends of Hybrid renewable energy system.

II. DIFFERENT CONFIGURATIONS OF HRES

Every renewable energy system has different set of physical components and operating characteristics. Therefore, it is necessary to have in place a planned framework for interfacing them to make HRES. Renewable energy sources, loads and intermediate storage device should be integrated in such a way, so that they can operate and exercise control independently.

Renewable energy sources in HRES should have the capability to connect and disconnect from the grid smoothly and independently in the event of a fault or islanding conditions. There are different HRES configurations possible and for proper interfacing with the grid, power electronics converters are needed. Many researchers, in the past years, have proposed different configurations of HRES which have generally been categorized in three sub categories as: DC coupled, AC coupled and Hybrid coupled based on ac and dc load sharing. DC coupled category can further be divided into two sub categories namely: Residential DC and Bipolar DC network coupled, similarly AC coupled category is divided into two sub categories namely: High frequency AC (HFAC) coupled and power frequency AC (PFAC) coupled [10]-[15].

These configurations are briefly discussed as under:

1) *DC coupled system*: These are categorized further in two sub categories as illustrated in the following subsections:

a) In DC residential coupled system, the hybrid renewable system sources are connected with DC bus through power electronics interfacing. DC or microgrid draws attention to innovate effective solution to meet out future energy distribution challenges. Fig. 1(a) shows a possible future DC coupled grid interface of hybrid renewable energy system sources. All

978-1-4799-6047-7/14 $31.00 © 2014 IEEE

sources are connected with DC buses with suitable control technique of power conditioning through power electronics converters. This type of configuration is effective and useful where DC loads are in bulk as compared to AC loads. DC coupled system is also useful in off grid condition where load is only residential. HRES with DC coupled configuration have many advantages over AC coupled as discussed in [16].

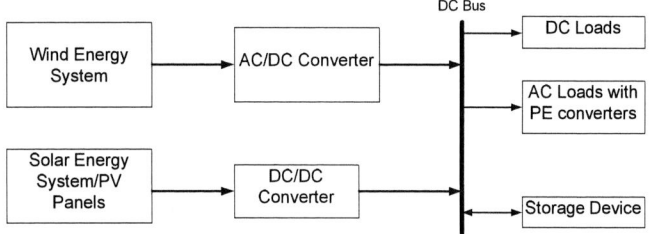

Fig. 1(a)   DC residential coupled system

b)   In DC Bipolar coupled configuration, shown in Fig. 1(b) [17], the hybrid renewable energy sources are linked with HVDC bipolar lines through proper power electronics interfacing and their associated control techniques. Some researchers suggest several advantages of Bipolar DC link over AC grid for future energy distribution systems, like high reliability, lower power losses, high power transfer capability, system stability etc. [17].

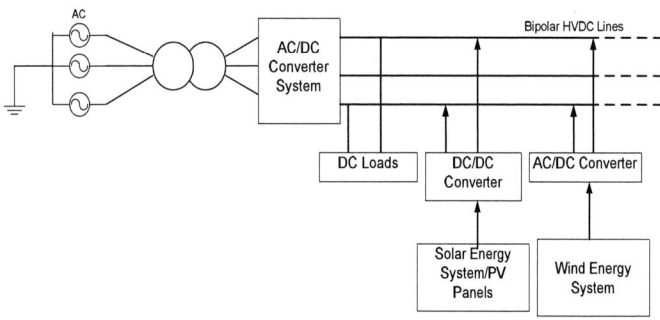

Fig. 1(b)   DC Bipolar coupled configuration

2)   *AC coupled system:* This configuration is based on AC bus coupling which also is further divided into two subcategories; PFAC and HFAC and is as shown in Fig.2 (a & b)[10].

a)   In Configuration of PFAC, where, all sources are connected to power frequency AC bus through suitable power electronics converters and loads are connected to the PFAC bus for power utilization [18]-[20].

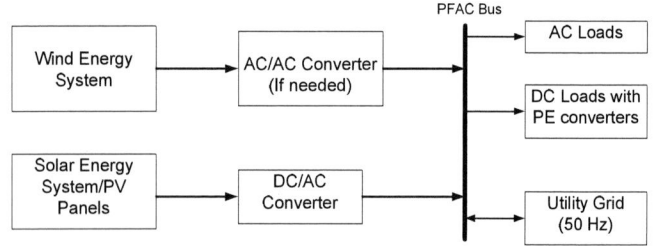

Fig. 2(a) PFAC configuration of AC coupled System

b)   In HFAC configuration, all diverse energy sources are connected to HFAC bus where HFAC loads are also connected to get power from the bus. HFAC configurations are primarily used where HF loads are present like airplanes, space applications and submarines [21].

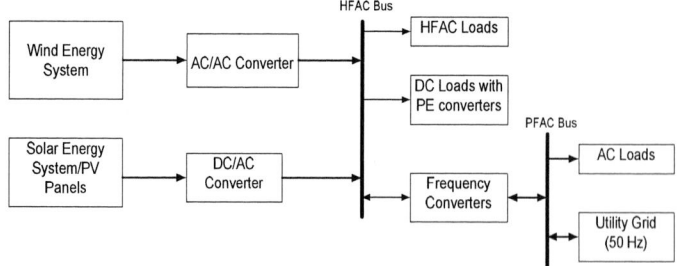

Fig. 2(b) HFAC configuration of AC coupled System

3)   *Hybrid coupled systems:* In hybrid configuration, as shown in Fig. 3, diverse sources connected with different buses either DC coupled or AC coupled, depend on the output of the sources which results in reduction of the interfacing circuits (power electronics converters) and hence increases the overall efficiency of the system. By appropriate coupling of different sources, cost reduction is also possible in hybrid coupling [22].

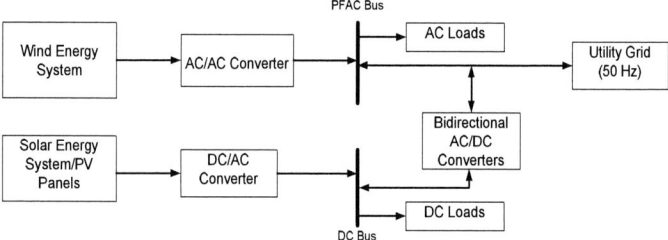

Fig. 3 Hybrid configuration system

Different types of coupling are used depending mainly on the sources, DC or AC load percentage and location of the load whether grid availability is present or not in that area.

III. GRENRATION SIDE CONTROL TECHNIQUES

To enhance the performance and maximize efficiency of HRES, Generation side control techniques are utilized which are generally called MPPT control. There are different controls meant for different RES which have different operating characteristics in HRES. This section is focused on Wind/Solar HRES with different MPPT control schemes. This paper presents, in the following subsections, critical review on MPPT control techniques for both the sources (wind/solar) based on comprehensive literature survey [23]-[27].General block diagram of MPPT control is shown in Fig.4 with boost converter (any DC-DC converter can be used) for wind/solar and input variables of MPPT block depend on MPPT methods.

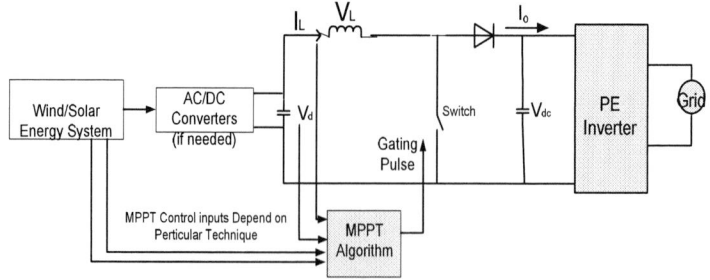

Fig. 4 Basic Diagram of MPPT control

978-1-4799-6047-7/14 $31.00 © 2014 IEEE   340

1) *Perturbation and Observation (Hill-Climbing) method*: P&O methods are used for both sources to achieve MPPT for maximize the power output [28] [29]. Various derivatives of P&O have also been utilized as reported in literature for wind /solar energy sources using power electronics converter mainly boost converters [30]-[32]. It is simple MPPT method and only uses the current and voltage information of boost input to control by changing duty cycle of boost converter as per the information. P&O has many advantages like no need of a-priory information, no need to measure wind speed and quick response in steady variations etc. but there are some limitations also like adding extra power electronics converter, oscillations at maximum power point, and failure in sudden changes.

2) *Incremental conductance (IC) method*: This method can also be used for both wind and solar sources for MPPT. In this method, derivative of power with respect to voltage is used to search maxima point using P versus V characteristics of energy sources [33]. This method also does not require any prior information and has some advantages over P&O like damping out of the oscillations at maximum point, and simple in application but this method is not as fast as P&O and dynamic response of the system is also not proper in IC control method [34]-[36].

3) *Modified P&O method*: In this method, both P&O and IC techniques are mixed like hybrid MPPT control which removes the disadvantages of both techniques and thereby presents unique solution to get maximum power. This method uses P&O method for fast dynamics to achieve near maximum point then use IC method to damp out oscillations and get maximum point which results in increase in the power output and removal of the oscillations [37] [38].

Some techniques available for MPPT control depend on the type of source, like wind energy has some methods and solar energy have some other methods for MPPT control. A brief discussion is presented on such techniques in the following subsections:

In wind energy source, MPPT control can be achieved by Tip speed ratio (TSR) which uses wind speed and turbine speed for aching MPPT. TSR is based on turbine speed regulation according to given wind speed by using wind speed-turbine speed relation to get optimized value of TSR using regulation. But there are implementation difficulties in the TSR method due to the difficulties in measurement of wind speed [39]-[42]. Another method is optimum-relationship based (ORB) control technique which ensures MPPT with adequate solution of optimum relationship between system variables and parameters. Measurement of wind speed is not required in this technique and there is surety of fast response with changing wind speed. ORB has two methods, one based on power versus rotor speed relationship and the other on power versus rectified dc voltage relationship. This is an established technique which can be used for various power ratings and illustrated in literature [43]-[45].

Unlike wind, solar energy sources have several MPPT techniques and their derivatives which have been reported in literature. Some of them are briefly discussed below:

Curve fitting technique is used for solar panel MPP which is based on fitting the curve of third order polynomial offline mathematical model and P-V characteristics of the solar cell [46] [47].

Fractional Short-Circuit Current (FSCI) Technique is also used in solar energy to get maximum power point using the information of the short circuit current at which power is maximum at given conditions [48].

Look-up Table Technique uses memory device of MPPT control box to store the data of MPP of a PV system calculated before the use of system for each probable environmental condition. During the operation, control tool makes use of the stored data to achieve MPP condition for every particular environmental situation [49].

Feedback Voltage or Current Technique is applied on the system having no storage battery; a simple controller is needed to maintain the voltage constant without battery. Therefore, a simple MPPT method can be applied by getting feedback of panel voltage or current and compared with the reference voltage; boost converter tracks that reference using the adjustment of duty cycle and continuously operates near MPP region [50]. In some works reported in literature, discussion is available on *Feedback of Power Variation with Voltage Technique* which is similar to the feedback voltage or current technique and is simply a derivative of that [51].

Linearization-Based MPPT Technique is based on the linearization model of PV system and gets the locus of voltage and current at the MPP [52].

Intelligence MPPT Techniques have different derivatives based on the control tool like artificial neural network (ANN), particle swarm optimization (PSO), fuzzy logic based MPPTs [53]-[55].

To achieve generation side control of HRES, there are different methods depending on the sources of HRES. Some important works have been discussed which are important for generation side control to increase the efficiency and reduce the effective cost of HRES system.

IV. GRID SIDE CONTROL TECHNIQUES

Grid side control techniques came into significance after the interface of different sources in HRES through common DC link. Therefore, grid side controls are common for both sources (wind and solar) in HRES using power electronics inverter. Inverter is used for interconnecting HRES and grid through proper LC or LCL filters and transformers (optional). The main purpose of grid side control is the following:

- Control of generated real power through HRES,
- Reactive power control transfer between grid and HRES,
- Grid synchronization.
- Keeping DC link voltage constant,
- Harmonic compensation.

All above mentioned controls are achieved by grid side control in HRES using different control techniques which have been proposed in literature using control of the switching sequence of power electronics inverter [56]-[58].

Grid side control techniques are mainly having two cascaded control loops called inner loop and outer loop. Inner loop is used for fast dynamics of system and controls the current injection into grid and harmonic compensation while outer loop control is used for keeping DC link voltage constant and system stability.

Several authors have proposed different methods to achieve proper functioning of cascaded loops for grid interface, some of those techniques are briefly reviewed in the following subsections:

1) *Voltage Oriented Control (VOC):* VOC method is used by many researchers as reported in literature for grid side control. This method is mainly based on power balancing equation of inverter to grid through LCL or LC filter. By sensing DC link voltage, grid voltage and current into stationary frame (ABC to DQ transform), reference current for inner loop and outer loop are

generated. Active power control is achieved by d-axis current and reactive power control by q-axis current in VOC [59] [60]. Several derivative methods of VOC have also been discussed in many research papers like VOC using PI controller, VOC using PR controller, VOC using hysteresis control etc. elaborating on their advantages and disadvantages [61] [62].

2) *Current Control method (CCM)*: With the use of different power electronics inverters some research papers have proposed CCM which depends on the dc link current and line currents of the grid to generate reference d and q axis current for switching control of inverters. CCM also has different derivatives like instantaneous current method, Lyapunov based CCM, current vector control method etc. to achieve control objective of grid side [63]-[65].

3) *Feedback Linearization Control (FLC)*: FLC method is based on the transformation of nonlinear HRES into linear system through feedback linearization to control all variables independently. Using lie derivatives in FLC, output and input can be relate to develop the controller of the system. FLC has got many advantages like no need of PI gain values, better control in fault and islanding condition and fast dynamic response [66] [67].

All control methods for grid side control are mainly derived from the above methods with different tools like ANN, Fuzzy and power electronics inverters such as current source inverter, Dual inverter, Matrix converter, Multilevel matrix converters etc. [68]-[70].

In this subsection some grid side control schemes, which are standard control methods, have been reviewed besides all other controls as are proposed under current research, mainly derivatives to achieve better performance, quality of power, reduction in complexity of control, reduction of effective cost and reduce disadvantages of above discussed methods.

V. FUTURE CHALLENGES

Hybrid renewable energy system is an emerging area of research in electrical energy sector which is still not an established area and needs much attention to improve efficiency, reduce cost and match the growing demand of electricity. There are challenges to achieve quality power at cheaper cost with better efficiency. Some of the challenges in hybrid renewable energy systems are mentioned below:

1) Scope of improvement in control aspects of transition between grid interfaced HRES and islanded mode or vice-versa.

2) There is need to improve the control and reduce the adverse effect due to the penetration of different renewable sources in HRES on grid.

3) Issues in re-synchronization capability to connect the autonomous islanding to the grid without any disturbance or interruption.

4) Need to develop new MPPT control which satisfy fast Dynamic response and stable steady state of the response for HRES.

5) Need to find a way for MPPT control using single DC-DC converter in HRES which can simplify control and reduce the effective cost of the overall system.

6) Power quality enhancement for specific loads during operation is needed which is important and significant in case of HRES systems.

7) HRES should have some control mechanism in place to improve the grid stability during abnormal as well as normal conditions.

VI. CONCLUSION

The paper presents a critical review of the works reported on hybrid renewable energy systems and their associated controls based on the survey of available literature. Due attention has also been paid to recent developments, such as control schemes based on the concepts of neural networks and fuzzy logic and the incorporation of other evolutionary algorithms like PSO. Emphasis has been given to categorizing various configurations of hybrid renewable energy systems highlighting their salient features and control methods used therein for efficient and cost effective operation of these systems. Control techniques of both generation side control as well as grid side control and role of power electronics converters in it has been discussed.

Finally, some of the challenges in different areas of hybrid renewable energy systems are discussed to improve overall system performance in future.

Although the authors have sincerely attempted to present the comprehensive set of references on hybrid renewable energy systems and their associated controls, they would like to apologize for exclusion of many good papers because the literature is voluminous and hope that additional references will be advanced as discussion to this publication. It is envisaged that this paper will serve as a valuable resource to any future researcher in this important area of research.

REFERENCES

[1] M. Ameli, S. Moslehpour, and M. Shamlo, "Economical load distribution in power networks that include hybrid solar power plants," *Elect.Power Syst. Res.*, vol. 78, no. 7, pp. 1147–1152, 2008.

[2] S. Gomaa, A. K. A. Seoud, and H. N. Kheiralla, "Design and analysis of photovoltaic and wind energy hybrid systems in Alexandria, Egypt,"*Renew. Energy*, vol. 6, no. 5–6, p. 643, Jul./Sep. 1995.

[3] W. D. Kellogg,M. H. Nehrir, G. Venkataramanan, and V. Gerez, "Generation unit sizing and cost analysis for stand-alone wind, photovoltaic, and hybrid wind/PV systems," *IEEE Trans. Energy Convers.*, vol. 13, no. 1, pp. 70–75, Mar. 1998.

[4] C. Marnay, G. Venkataramanan, M. Stadler, A. S. Siddiqui, R. Firestone, and B. Chandran, "Optimal technology selection and operation of commercial-building microgrids," *IEEE Trans. Power Syst.*, vol. 23, no. 3, pp. 975–982, Aug. 2008.

[5] Bratcu, AI; Munteanu, I; Ceanga, E., "Optimal control of wind energy conversion systems: From energy optimization to multi-purpose criteria - A short survey,*" 16th Mediterranean Conference on Control and Automation, 2008*, vol., no., pp.759,766, 25-27 June 2008.

[6] Martinez, J.A; Dinavahi, V.; Nehrir, M.H.; Guillaud, X., "Tools for Analysis and Design of Distributed Resources—Part IV: Future Trends," *IEEE Transactions on Power Delivery*, vol.26, no.3, pp.1671,1680, July 2011.

[7] Lago, J.; Heldwein, M.L., "Operation and Control-Oriented Modeling of a Power Converter for Current Balancing and Stability Improvement of DC Active Distribution Networks," *IEEE Transactions on Power Electronics*, vol.26, no.3, pp.877,885, March 2011.

[8] Brahma, S.M., "Fault Location in Power Distribution System With Penetration of Distributed Generation," *Power Delivery, IEEE Transactions on*, ol.26, no.3,pp.1545,1553, July 2011.

[9] Shirek, G.J.; Lassiter, B.A, "Photovoltaic Power Generation: Modeling Solar Plants' Load Levels and Their Effects on the Distribution System," *Industry Applications Magazine, IEEE*, vol.19, no.4, pp.63,72, July-Aug. 2013.

[10] Nehrir, H.; Caisheng Wang; Strunz, K.; Aki, H.; Ramakumar, R.; Bing, J.; Zhixhin Miao; Salameh, Z., "A review of hybrid

renewable/alternative energy systems for electric power generation: Configurations, control and applications," *Power and Energy Society General Meeting, 2012 IEEE* , vol., no., pp.1,1, 22-26 July 2012.

[11] F. A. Farret and M. G. Simões, Integration of alternative Sources of Energy. Hoboken, NJ: Wiley, 2006.

[12] S.-H. Ko, S. R. Lee, H. Dehbonei, and C. V. Nayar, "Application of voltage- and current-controlled voltage source inverters for distributed generation systems*," IEEE Trans. Energy Convers.*, vol. 21, no. 3, pp. 782–792, Sep. 2006.

[13] P. Salonen, T. Kaipia, P. Nuutinen, P. Peltoniemi, and J. Partanen, "An LVDC distribution system concept," *in Proc. Nordic Workshop Power Ind. Electron. (NORPIE)*, 2008, pp. A3-1–A3-16.

[14] Prodanovic, M.; Green, T.C., "High-Quality Power Generation Through Distributed Control of a Power Park Microgrid," *Industrial Electronics, IEEE Transactions*, vol.53, no.5, pp.1471,1482, Oct. 2006.

[15] Guerrero, J.M.; Matas, J.; Luis Garcia de Vicuna; Castilla, M.; Miret, J., "Decentralized Control for Parallel Operation of Distributed Generation Inverters Using Resistive Output Impedance," *Industrial Electronics, IEEE Transactions*, vol.54, no.2, pp.994,1004, April 2007.

[16] C. K. Sao and P. W. Lehn, "A transformerless energy storage system based on a cascade multilevel PWM converter with star configuration," *IEEE Trans. Ind. Appl.*, vol. 44, no. 5, pp. 1621–1630, Sep./Oct. 2008, , "Control and Power Management of Converter Fed Microgrids*," IEEE Trans. Power Systems*, Vol. 23, No. 3, pp. 1088-1098, August 2008.

[17] Lago, J.; Heldwein, M.L., "Operation and Control-Oriented Modeling of a Power Converter for Current Balancing and Stability Improvement of DC Active Distribution Networks," *Power Electronics, IEEE Transactions*, vol.26, no.3, pp.877,885, March 2011.

[18] Tsikalakis, AG.; Hatziargyriou, N.D., "Centralized Control for Optimizing Microgrids Operation," *Energy Conversion, IEEE Transactions*, vol.23, no.1, pp.241,248, March 2008.

[19] Guerrero, J.M.; Vasquez, J.C.; Matas, J.; Castilla, M.; de Vicuna, L.G., "Control Strategy for Flexible Microgrid Based on Parallel Line-Interactive UPS Systems," *Industrial Electronics, IEEE Transactions*, vol.56, no.3, pp.726,736, March 2009.

[20] Fei Wang; Duarte, J.L.; Hendrix, M.AM., "Grid-Interfacing Converter Systems With Enhanced Voltage Quality for Microgrid Application—Concept and Implementation," *Power Electronics, IEEE Transactions*, vol.26, no.12, pp.3501,3513, Dec. 2011.

[21] Chakraborty, S.; Weiss, M.D.; Simoes, M.G., "Distributed Intelligent Energy Management System for a Single-Phase High-Frequency AC Microgrid," *Industrial Electronics, IEEE Transactions*, vol.54, no.1, pp.97,109, Feb. 2007.

[22] Yuan-Chih Chang; Chang-Ming Liaw, "Establishment of a Switched-Reluctance Generator-Based Common DC Microgrid System," *Power Electronics, IEEE Transactions*, vol.26, no.9, pp.2512,2527, Sept. 2011.

[23] De Kooning, J. D M; Meersman, B.; Vandoorn, T.L.; Vandevelde, L., "Evaluation of the Maximum Power Point Tracking performance in small wind turbines," *Power and Energy Society General Meeting, 2012 IEEE* , vol., no., pp.1,8, 22-26 July 2012.

[24] L. F. K. Johnson, M. Balas, and L. Pao, "Methods for increasing region 2 power capture on a variable-speed wind turbine," *Solar Energy Eng.*, vol. 126, pp. 1092–1100, 2006.

[25] Dalala, Z.M.; Zahid, Z.U.; Wensong Yu; Younghoon Cho; Jih-Sheng Lai, "Design and Analysis of an MPPT Technique for Small-Scale Wind Energy Conversion Systems," *IEEE Transactions on Energy Conversion*, vol.28, no.3, pp.756,767, Sept. 2013.

[26] Mastromauro, R.A; Liserre, M.; Dell'Aquila, A, "Control Issues in Single-Stage Photovoltaic Systems: MPPT, Current and Voltage Control," *IEEE Transactions on Industrial Informatics*, vol.8, no.2, pp.241,254, May 2012.

[27] Chen, S.-M.; Liang, T.-J.; Hu, K.-R., "Design, Analysis, and Implementation of Solar Power Optimizer for DC Distribution System," *IEEE Transactions on Power Electronics*, vol.28, no.4, pp.1764,1772, April 2013.

[28] Tsai-Fu Wu; Chien-Hsuan Chang; Yong-Jing Wu, "Single-stage converters for PV lighting systems with MPPT and energy backup," *IEEE Transactions on Aerospace and Electronic Systems*, vol.35, no.4, pp.1306,1317, Oct 1999.

[29] Koutroulis, E.; Kalaitzakis, K., "Design of a maximum power tracking system for wind-energy-conversion applications," *IEEE Transactions on Industrial Electronics*, vol.53, no.2, pp.486,494, April 2006.

[30] Fermia, N.; Granozio, D.; Petrone, G.; Vitelli, M., "Predictive & Adaptive MPPT Perturb and Observe Method," *IEEE Transactions on Aerospace and Electronic Systems*, vol.43, no.3, pp.934,950, July 2007.

[31] Kollimalla, S.K.; Mishra, M.K., "Variable Perturbation Size Adaptive P&O MPPT Algorithm for Sudden Changes in Irradiance," *IEEE Transactions on Sustainable Energy*, vol.5, no.3, pp.718,728, July 2014.

[32] D. Sera, R. Teodorescu, J. Hantschel, and M. Knoll, "Optimized maximum power point tracker for fast changing environmental conditions," *IEEE Trans. Ind. Electron.*, vol. 55, no. 7, pp. 2629–2637, Jul. 2008.

[33] Fangrui Liu; Shanxu Duan; Fei Liu; Bangyin Liu; Yong Kang, "A Variable Step Size INC MPPT Method for PV Systems," *IEEE Transactions on Industrial Electronics*, vol.55, no.7, pp.2622,2628, July 2008.

[34] Sera, D.; Mathe, L.; Kerekes, T.; Spataru, S.V.; Teodorescu, R., "On the Perturb-and-Observe and Incremental Conductance MPPT Methods for PV Systems," *IEEE Journal of Photovoltaics*, vol.3, no.3, pp.1070,1078, July 2013.

[35] A. Zegaoui, M. Aillerie, P. Petit, J. P. Sawicki, A. Jaafar, C. Salame, and J. P. Charles, "Comparison of two common maximum power point trackers by simulating of PV generators*," Energy Procedia*, vol. 6, pp. 678–687, Jan. 2011.

[36] Hosseini, S.H.; Farakhor, A; Haghighian, S.K., "Novel algorithm of maximum power point tracking (MPPT) for variable speed PMSG wind generation systems through model predictive control," *8th International Conference on Electrical and Electronics Engineering (ELECO), 2013*, vol., no., pp.243,247, 28-30 Nov. 2013.

[37] Abdelsalam, AK.; Massoud, AM.; Ahmed, S.; Enjeti, P., "High-Performance Adaptive Perturb and Observe MPPT Technique for Photovoltaic-Based Microgrids," *IEEE Transactions on Power Electronics*, vol.26, no.4, pp.1010,1021, April 2011.

[38] Sharma, R.; Bagh, S.K.; Banerjee, S., "A novel approach of grid connected wind energy conversion system with modified maximum power point tracking," *Power Electronics, Drives and Energy Systems (PEDES), 2012 IEEE International Conference on* , vol., no., pp.1,5, 16-19 Dec. 2012.

[39] T. Thiringer and J. Linders, "Control by variable rotor speed of a fixedpitch wind turbine operating in a wide speed range," *IEEE Trans. Energy Convers.*, vol. 8, no. 3, pp. 520–526, Sep. 1993.

[40] K. Johnson, L. Fingersh, M. Balas, and L. Pao, "Methods for increasing region 2 power capture on a variable speed wind turbine," *J. Solar Energy Eng.*, vol. 126, no. 4, pp. 1092–1100, 2004.

[41] R. J.Wai, C. Y. Lin and Y. R. Chang, "Novel maximum-power-extraction algorithm for pmsg wind generation system*," IEEE Transactions on Electrical Power Applications*, vol. 1, no. 2, pp. 275–283, Mar. 2007.

[42] J. D. M. De Kooning, B. Meersman, T. L. Vandoorn and L. Vandevelde, "Evaluation of the maximum power point tracking performance in small wind turbines," *Proceedings of the IEEE PES General Meeting*, San Diego (USA), Jul. 2012.

[43] Z. Chen and E. Spooner, "Grid power quality with variable-speed wind turbines*," IEEE Trans. Energy Conver.*, vol. 16, no. 2, pp. 148–154, Jun. 2001.

978-1-4799-6047-7/14 $31.00 © 2014 IEEE

[44] Xia, Y.; Ahmed, K.H.; Williams, B.W., "A New Maximum Power Point Tracking Technique for Permanent Magnet Synchronous Generator Based Wind Energy Conversion System," *IEEE Transaction on Power Electronics*, vol.26, no.12, pp.3609,3620, Dec. 2011.

[45] H.-B. Zhang, J. Fletcher, N. Greeves, S. J. Finney, and B. W. Williams, "One-power-point operation for variable speed wind/tidal stream turbines with synchronous generators," *IET Renewable Power Generation*, vol. 5, no. 1, pp. 99–108, Jan. 2011.

[46] A. Garrigos, J. M. Blanes, J. A. Carrasco, and J. B. Ejea, "Real time estimation of photovoltaic modules characteristics and its application to maximum power point operation," *Renew. Energy*, vol. 32, pp. 1059–1076, May 2007.

[47] F. J. Toledo, J. M Blanes, A. Garrigos, and J. A. Mart´ınez, "Analytical resolution of the electrical four-parameters model of a photovoltaic module using small perturbation around the operating point," *Renew. Energy*, vol. 43, pp. 83–89, Jul. 2012.

[48] M. A. S. Masoum, H. Dehbonei, and E. F. Fuchs, "Theoretical and Experimental Analyses of Photovoltaic Systems with Voltage and Current-based Maximum Power-Point Tracking," *IEEE Trans. on Energy Convers.*, vol. 17, no. 4, pp. 514-522, Dec. 2002.

[49] Desai, H.P.; Patel, H.K., "Maximum Power Point Algorithm in PV Generation: An Overview," *7th International Conference on Power Electronics and Drive Systems, PEDS '07*, vol., no., pp.624, 630, 27-30 Nov. 2007.

[50] Nagayoshi, H., "Characterization of the module/array simulator using I-V magnifier circuit of a pn photo-sensor," *Proceedings of 3rd World Conference on Photovoltaic Energy Conversion*, vol.2, no., pp.2023, 2026 Vol.2, 18-18 May 2003.

[51] V. Salas, E. Olias, A. Lazaro, and A. Barrado, "Evaluation of a new maximum power point tracker applied to the photovoltaic stand-alone systems," *Solar Energy Mater Solar Cells*, vol. 87, no. 1–4, pp.807–815, 2005.

[52] C. W. Tan, T. C. Green, and C. A. H. Aramburo, "An improved MPPT algorithm with current-mode control for photovoltaic applications," in IEEE Power Electron. Drives Syst., Malyasia, Dec. 28, 2005.

[53] Syafaruddin; Karatepe, E.; Hiyama, T., "Artificial neural network-polar coordinated fuzzy controller based maximum power point tracking control under partially shaded conditions," *Renewable Power Generation, IET* , vol.3, no.2, pp.239,253, June 2009.

[54] Al Nabulsi, A; Dhaouadi, R., "Efficiency Optimization of a DSP-Based Standalone PV System Using Fuzzy Logic and Dual-MPPT Control," *IEEE Transactions on Industrial Informatics,* vol.8, no.3, pp.573, 584, Aug. 2012.

[55] Ishaque, K.; Salam, Z.; Amjad, M.; Mekhilef, S., "An Improved Particle Swarm Optimization (PSO)–Based MPPT for PV with Reduced Steady-State Oscillation," I*EEE Transactions on Power Electronics*, vol.27, no.8, pp.3627, 3638, Aug. 2012.

[56] I. Agirman and V. Blasko, "A novel control method of a VSC without ac line voltage sensors," *IEEE Trans. Ind. Appl.*, vol. 39, no. 2, pp. 519–524, Mar./Apr. 2003.

[57] Blaabjerg, F.; Teodorescu, R.; Liserre, M.; Timbus, AV., "Overview of Control and Grid Synchronization for Distributed Power Generation Systems," *IEEE Transactions on Industrial Electronics*, vol.53, no.5, pp.1398, 1409, Oct. 2006.

[58] Marwali, M.N.; Keyhani, A, "Control of distributed generation systems-Part I: Voltages and currents control," *IEEE Transactions on Power Electronics*, vol.19, no.6, pp.1541, 1550, Nov. 2004.

[59] Sharma, R.; Samuel, P.; Bagh, S.K.; Banerjee, S., "A grid interconnected WECS with modified MPPT," *2nd International Conference on Power, Control and Embedded Systems (ICPCES)*, vol., no., pp.1, 6, 17-19 Dec. 2012.

[60] Amin, M.M.; Mohammed, O.A, "Development of High-Performance Grid-Connected Wind Energy Conversion System for Optimum Utilization of Variable Speed Wind Turbines," *IEEE Transactions on Sustainable Energy*, vol.2, no.3, pp.235, 245, July 2011.

[61] Hwang, J.G.; Lehn, P.W.; Winkelnkemper, M., "A Generalized Class of Stationary Frame-Current Controllers for Grid-Connected AC–DC Converters," *IEEE Transactions on Power Delivery*, vol.25, no.4, pp.2742, 2751, Oct. 2010.

[62] Suva, N.; Martins, A; Carvalho, A, "Design and evaluation of a PWM rectifier control system for testing renewable DC sources connected to the grid," *International Symposium on Power Electronics, Electrical Drives, Automation and Motion, 2006. SPEEDAM 2006*, vol., no., pp.1190, 1195, 23-26 May 2006.

[63] Bojoi, R.; Limongi, L.R.; Roiu, D.; Tenconi, A, "Enhanced Power Quality Control Strategy for Single-Phase Inverters in Distributed Generation Systems," *IEEE Transactions on Power Electronics*, vol.26, no.3, pp.798, 806, March 2011.

[64] Reyes, M.; Rodriguez, P.; Vazquez, S.; Luna, A; Teodorescu, R.; Carrasco, J.M., "Enhanced Decoupled Double Synchronous Reference Frame Current Controller for Unbalanced Grid-Voltage Conditions," *IEEE Transactions on Power Electronics*, vol.27, no.9, pp.3934, 3943, Sept. 2012.

[65] Sung-Hun Ko; Lee, S.R.; Dehbonei, H.; Nayar, C.V., "Application of voltage- and current-controlled voltage source inverters for distributed generation systems," *IEEE Transactions on Energy Conversion*, vol.21, no.3, pp.782,792, Sept. 2006.

[66] Delfino, F.; Pampararo, F.; Procopio, R.; Rossi, M., "A Feedback Linearization Control Scheme for the Integration of Wind Energy Conversion Systems Into Distribution Grids," *Systems Journal, IEEE*, vol.6, no.1, pp.85, 93, March 2012.

[67] Xianwen Bao; Fang Zhuo; Yuan Tian; Peixuan Tan, "Simplified Feedback Linearization Control of Three-Phase Photovoltaic Inverter With an LCL Filter," *IEEE Transactions on Power Electronics*, vol.28, no.6, pp.2739,2752, June 2013.

[68] Davis, AJ; Salameh, Z.M., "Fuzzy logic modeling of a grid-connected wind/photovoltaic system with battery storage," *Large Engineering systems Conference on Power Engineering, 2004. LESCOPE-04*, vol., no., pp.129, 135, 28-30 July 2004.

[69] Sharaf, AM.; El-Gammal, AAA, "A novel efficient PSO-self regulating PID controller for hybrid PV-FC-diesel-battery micro grid scheme for village/resort electricity utilization," *Electric Power and Energy Conference (EPEC), 2010 IEEE* , vol., no., pp.1,6, 25-27 Aug. 2010.

[70] Jinbang Xu; Zhizhuo Wu; Xuan Yang; Jie Ye; Anwen Shen, "ANN-based Control Method Implemented in a Voltage Source Converter for Industrial Micro-grid," *Sixth International Conference on Bio-Inspired Computing: Theories and Applications (BIC-TA), 2011*, vol., no., pp.140,145, 27-29 Sept. 2011.

978-1-4799-6047-7/14 $31.00 © 2014 IEEE

A Novel Method of Generating Electricity by setting up Turbines over Rail Locomotives

P.K. Sharma	Sahil	N. Hari	S. Banerjee	R. Sharma
Student: EEED	Student: EEED	Student: EEED	Assistant Professor: EEED	Assistant Professor: EED
BVCOE	BVCOE	BVCOE	BVCOE	NIT
New Delhi, India	New Delhi, India	New Delhi, India	New Delhi, India	Kurukshetra, India

Abstract—This paper focuses upon an approach to generate the electrical energy from wind by mounting specially designed wind turbines on the roof of moving rail locomotives (trains). In this work, three blade wind turbine system constructed using aerofoil is proposed. In proposed model, wind turbine is arranged in bus bar arrangement. The power developed by the implementation of aforementioned design is used to fulfil all the requirements for electrical appliances of a train coach (tube lights, fans etc). The aim of this research paper is to exploit the kinetic energy of air for generation of electrical energy, by implementing suitable arrangements on large vehicles. This technology will help reduce drastically, the fuel consumption required for the electrical power generation.

Keywords—aerofoil; appliances; locomotives; turbine; wind

I. INTRODUCTION

Wind is one of the most copious natural renewable sources of energy. The economical and ecological recompense offered by wind energy are the most vital reasons why electrical systems based on wind energy are receiving widespread global attention. Due to the increasing demand on electrical energy, a considerable amount of effort is being made to generate electricity from new sources of energy. Wind energy is now achieving exponential growth and has great potential. The Global Wind Energy Council (GWEC) forecasts installed wind capacity to double between 2010 and 2014. With the new additions, global wind capacity has increased to 200 gigawatts. The GWEC projects this number to grow to 400 gigawatts by 2014 as a result of industry growth in China, Europe, and the United States. As per the global wind energy outlook, 2010, the global cumulative wind power capacity in different scenarios has been estimated as displayed in Fig. 1.

As far as India is concerned, it had a record year for new wind energy installations in 2010, with 2,139 MW of new capacity supplemented to accomplish a total of 13,065 MW at the end of the year [1]. Renewable energy is now 10.9% of installed capacity, contributing about 4.13% to the electricity generation mix, and wind power accounts for 70% of this installed capacity. At present, the wind power potential estimated by the Centre for

Wind Energy Technology (CWET) is 49.1 GW, but the estimations of various industry associations and the World Institute for Sustainable Energy (WISE) and wind power producers are more buoyant, citing a prospective in the range of 65- 100 GW. The growth of Indian Wind power installed capacity is shown in Table 1 [1].

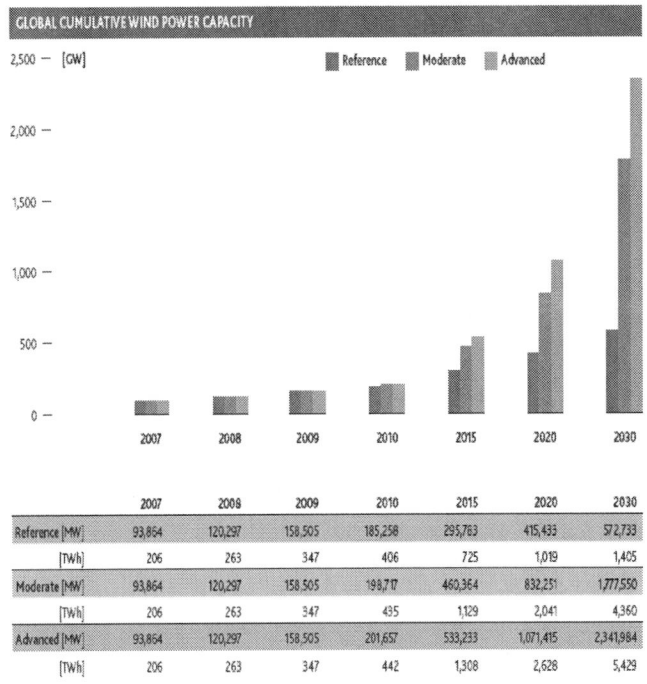

Fig.1 Global cumulative wind power capacity in different scenarios

TABLE 1. INDIAN WIND POWER INSTALLED CAPACITY

Year	2005	2006	2007	2008	2009	2010	31march,2011
MW	4,430	6270	7845	9655	10926	13065	14550

978-1-4799-6047-7/14 $31.00 © 2014 IEEE

However, the potential of harnessing wind energy through large moving vehicles, for generation of electricity has remained vastly unexploited. This work proposes an idea in which moving trains can be used to exploit wind energy for producing electricity. When train is running, the relative speed between wind and train is enough for production of electrical energy by wind turbine arrangements. The efficiency of diesel locomotives is not high (<30 %). So it would be beneficial for reducing the fuel consumption if wind energy is used for feeding the electrical requirements of the coaches.

II. WIND TURBINE DESIGN

A wind turbine is a machine for converting the kinetic energy in wind into mechanical energy. There are basically 2 types of wind turbines:

(i)   HAWT (Horizontal Axis Wind Turbines)
(ii)   VAWT (Vertical Axis Wind Turbines)

A horizontal axis machine has its blades rotating on axis parallel to the ground. A vertical axis machine has its blades rotating on axis perpendicular to the ground. In the proposed idea the usage of a HAWT is suggested. This is due to its advantages vis-à-vis VAWT, a comparison of which has been stated below:-

- Generally HAWT has higher efficiency than VAWT.
- HAWT is a lift type machine, so lift and drag ratio will be high. Whereas, VAWT is drag type machine. To minimize drag loss in a moving train, lift and drag ratio should be high. So HAWT will be preferable.
- HAWT is suitable for high wind speed. But VAWT are not suitable for high wind operations as its blades are prone to fatigue, which would be present while train is moving.
- HAWT has higher dynamic stability than VAWT.
- VAWTs need guy wires for support, which are not required in HAWT. Usage of HAWT will decrease the cost of proposed model.
- HAWT requires unidirectional wind to work efficiently. But VAWT can even work when wind is coming from either of any direction.
- HAWT has high starting torque than VAWT.
- Noise is high for small HAWT, whereas it is very low in VAWT.

Usage of HAWT is proposed due to high lift, low drag, easily starting capability and commercially available nature. It is required to make advancement in conventional wind turbines to provide mechanical stability and reduce the turbulence in high wind speed environment.

A. Structure Design

Wind turbine is structured between a ring frame (1.2meter diameter) to provide higher stability in wind and to reduce the turbulence effect as shown in figure 2. The frame is made up of stainless steel, and is supported by 3 metal rods from roof to give strength to the frame. The wind turbine is connected by poles to the opposite vertical ends of the frame instead of a single pole as in conventional wind turbines.

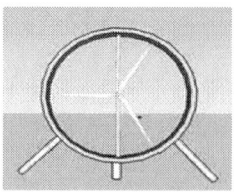

Fig. 2   Design of structure for wind turbine

B. Blade system

The *three blade system* has been suggested due to following reasons

- Blade system balances gyroscopic forces.
- It has combination of high ratio of speed and minimum stress.

C. Types of blade

Aerofoil blades should be used due to their low coefficient of drag and high lift.

D. Blade profile

The blades in the system are of aerofoil type with profile of NACA 6412 [2] as shown in figure 3. Its maximum thickness is 12% at 30.1% chord. Maximum camber is 6% at 39.6% chord. Speed of the turbine varies from centre to tip of blade. Therefore to maintain constant tip speed ratio (TSR) and to optimize the angle of attack, the blade is twisted from root to tip.

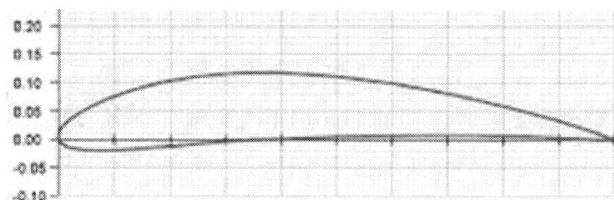

Fig. 3:   Shape of NACA 6412 [2]

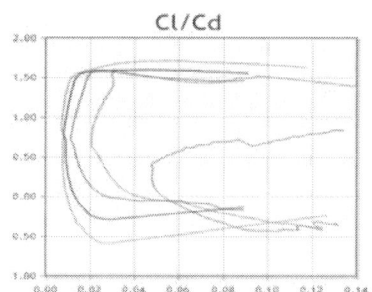

Fig. 4 Curve between C_l and C_d [2]

978-1-4799-6047-7/14 $31.00 © 2014 IEEE

As seen from figure 4, this blade profile is optimum for our case due to high lift and low drag coefficient. The rotor diameter is 1m for our proposed model.

E. Tip speed ratio

The *tip speed ratio* of wind turbine is the ratio between the tangential speed of tip of a blade and actual velocity of wind. Tip speed ratio (λ) relation with rotating speed of turbine blade (ω), moving wind velocity (v) and rotor radius (r) is:-
$$\lambda = (\omega * r)/v$$
Power coefficient in terms of tip speed ratio is given by:-
$$C(\lambda) = C_p = 0.00044\lambda^4 - 0.012\lambda^3 - 0.2\lambda + 0.097\lambda^2 + 1$$

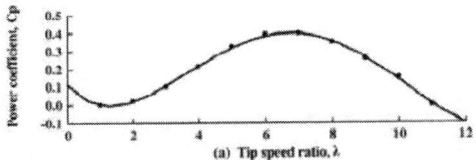

Fig. 5 Curve between Cp and λ [3]

A three blade system should have tip speed ratio around three to five for optimum power coefficient (Refer to figure 5).

F. Wind generator

A permanent magnet D.C. generator is proposed for this model because of the following advantages:-

- It has constant torque irrespective of load till cut off speed.
- It can easily vary its speed.
- There is no frequency issue in DC systems.
- It has residual flux which helps in easy starting.

Figure 6 exhibits the various parts of a wind turbine.

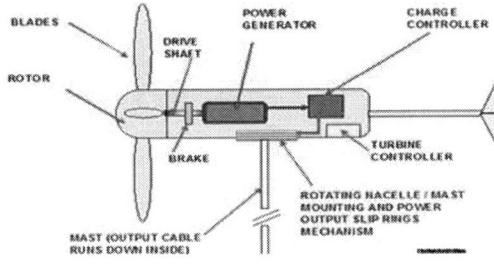

Fig. 6 Parts of wind turbine

III. WORKING OF PROPOSED MODEL

In the wind turbine suggested in this work, axis of rotor rotation is perpendicular to wind stream. The pressure generated between the aerofoil blades due to movement of wind generates aerodynamic lift, which causes the rotation of blade about the hub. The shaft of turbine is directly coupled with a D.C. generator which rotates due to rotation of blade and generates power which is sent to D.C. battery through bus-bar arrangement. The output of the generator can be fed to a chopper with an appropriate control strategy, to feed the loads in train coaches which require D.C. as input. For coaches whose loads require A.C., the generator output is passed through an inverter with a proper control arrangement to convert it to A.C. Diagrammatic representation of the turbine is exhibited in figure 7 while figure 8 illustrates the flow diagram of the proposed model.

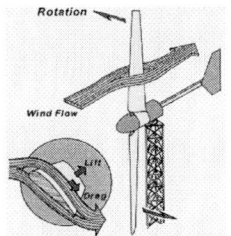

Fig. 7 Illustration of proposed turbine

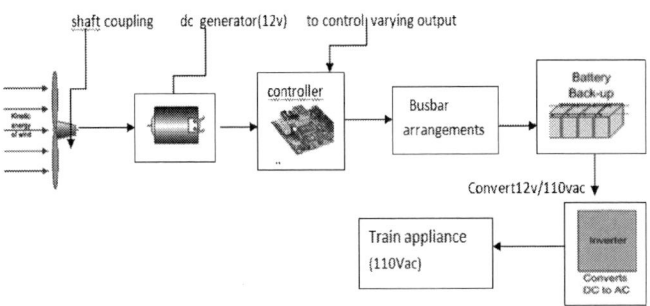

Fig. 8 Flow diagram of proposed model

IV. POWER REQUIREMENT IN A TRAIN

In Indian Railways, generally a typical sleeper class coach has nine compartments and four toilets. Each compartment has three tube lights and three fans. Each toilet has one tube light. There are six tube lights and six fans present in other spaces of a coach.

The rating of each tube light is 20 W (110 V)[4].

The rating of each fan is 60 W (110 V)[4].

Total tube lights in one coach=9x3+4x1+6=37

Power consumption by tube lights in one coach=37x20=740 W

Total fans in one coach=9x3+6=33

Power consumption by fans in one coach=33x60=1980 W

Total power consumption per coach=1980+740=2720 W

Suppose, there are 16 coaches in proposed model train.

Total electrical power consumption in coaches of one train=number of coach x power consumption per coach =16x2720=43520 W

V. POWER OBTAINED THROUGH WIND

A. General Calculations

Let E= kinetic energy
m=mass (kg)
v=wind speed (m/s)
P=power (W)
dm/dt =mass flow rate (kg/s)
dE/dt =Energy flow rate (J/s)
ρ=Density of air (kg/m^2)
A=swept area (m^2)
C_p=power coefficient
R=radius (m)
x=distance (m)
T=time(s)
a=acceleration (m/s^2)

Now, work done (W) on an object is equal to the product of the force & displacement of the object
i.e.

$$W=E=F.x \qquad (1)$$

Newton's law states that

$$F= m.a$$

Putting in equation (1)

$$E=W=m.a.x$$

Now using third equation of motion,

$$a= (v^2-u^2)/2x \qquad (2)$$

Since, initial velocity (u) of the object is taken as zero, we have:

$$a=v^2/2x \qquad (3)$$

Now, kinetic energy of a body in motion is

$$E= (1/2)mv^2$$

Considering wind as the object, Power in the wind can be computed as rate of change in wind energy:-

[for steady wind speed] $P=dE/dt= (1/2) .v^2.(dm/dt)$ (4)

Mass flow rate (dm/dt) is:

$$dm/dt= \rho.A.(dx/dt)$$

And rate of change of displacement is the velocity of the body

$$dx/dt=v$$

Hence, mass flow rate is

$$dm/dt= \rho.A.v \qquad (5)$$

Putting in equation (4) we get,

$$P= (1/2) .\rho.A.v^3 \qquad (6)$$

Where $A=\pi r^2$
r=length of the blade
V=relative velocity of the train and wind
ρ= 1.2754 kg.m^{-3} (at STP)
In proposed model, rotor diameter is taken as 1meter,i.e. length of blade is 0.5 meter. Now, we take an average train velocity of 60km/h (16.67m/s).
Putting the values in equation (6),
P=1/2x1.27x3.14x0.5x0.5x $(16.67)^3$
P=2307.754 W

B. Betz law

It calculates the maximum power which can be taken out from wind irrespective of the wind turbine design. The theory was published in 1919 by a German physicist, Albert Betz. He found that the theoretical maximum power extractable from wind is *16/27(59.3%)* times the power contained in the wind [5]. Today, commercial wind turbine efficiency (C_{pg}, efficiency of generator is also taken into account) is ranging from 15% to 35%. In our calculations 25% efficient wind turbine is considered.
Actual Power (P_{actual})=2307.754 x(16/27)x0.25=341.89 W

C. Drag

Drag is the aerodynamic force that opposes blade motion through the air. Drag is generated due to contact between a body (blades) and a fluid (air) when body is in motion in fluid. The curve of drag coefficient (C_D) against Reynolds number (R_e) is shown in figure 9.

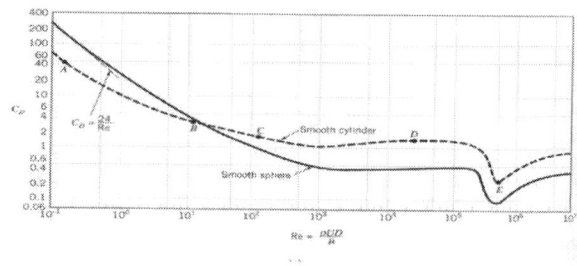

Fig. 9   Graph between C_D and R_e [6]

Power consumed by drag can be computed by:
P_d =0.5xC_DxρxAxv^3
C_D for airfoil blade is =0.05 [7]
P_d=0.5x0.05x1.27x3.14x0.5x0.5x16.67x16.67x16.67
  =115.45 W
Net useful power from obtained from one turbine generator arrangement =341.89-115.45=226.44 W
The total power requirement per coach is 2720 W and the net useful power obtained from one turbine generator set is 226.44 W. This implies that 13 wind turbines of identical size would be appropriate for fulfilling the electricity requirements of a single coach as shown in figure 10.

Fig. 10: Wind turbine arrangement on a coach of a locomotive

Hence total power generation on each coach is given by:

P_{coach}=226.44x13=2943.72 W=2.94kW

The power generation will not be the same throughout. It will vary with the relative velocity of train and wind. A plot between power generated (P_g) and relative speed is given in figure 11:-

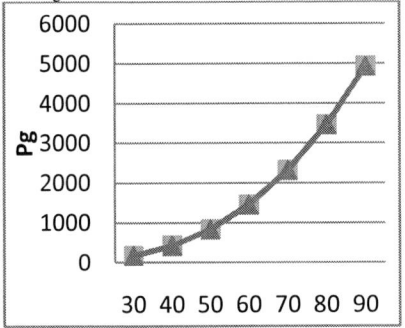

Fig. 11 Graph between P_g (W) and velocity (m/s)

D. Fuel saving

One litre diesel oil contains 10.1 kWh (36.3 MJ) of energy. A commercial diesel engine has efficiency from 25% to 32%.

So energy that can be produced by 1 litre of diesel with 30% efficient diesel engine is:

10.1 kWh x 0.3= 3 kWh

Energy produced by the proposed model is 2.94 kWh from one coach.

Hence, fuel saved by each coach=2.94/3=0.98 litre. For a 16 coach locomotive, the fuel saved for an eight hour journey will be will be 0.98x16x8=125.44 litres. For 200 locomotives running for eight hours and having 16 coaches, the amount of saved fuel will be 125.44x200=25088 litres. This implies that the fuel consumption can be cut down drastically by the application of proposed idea. The cost of diesel for Indian railways is around 59 Rs/litre [8]. Usually a train is operated for fifteen hours. We are saving diesel 0.98 litre/coach/hour. So for one day we are saving diesel equal to 0.98x16x15=235.2 litre. Money saved in one day for one train by implementing proposed model=235.2x59=13876.8 INR. Savings in operating 200 trains for a year by establishing turbines on locomotive rooftops=13876.8x200x365=1013006400 INR. This saving will eventually recover the amount spent for the setting up the wind turbine arrangement on the roof tops.

E. Reduction in greenhouse gas emission

Carbon dioxide gas (CO_2) and other gases which are produced by the combustion of diesel are largely responsible for green house effect. Our proposed system will help reduce the emission of greenhouse gases by lowering the usage of fossil fuel. This would have an affirmative effect on the environment as well.

VI. CONCLUSION

This paper dwells upon a method to produce electricity from wind by mounting special wind turbines on the roofs of rail locomotives. The significance of renewable energy, energy conversion systems based on renewable energy, and distributed power generation has been discussed. A succinct outline of the wind energy fundamentals has been touched. Due to the increasing energy demand, the conventional sources are depleting at an alarming rate. Hence it is getting important to make new inventions & innovations to fulfil the energy requirements of the world. The proposed project can achieve all the power requirements in a typical sleeper type coach. The usage of this technology can cut down the greenhouse gas emissions as well. Wind is more abundant than fossil fuels; it will never run out. However, its intermittent nature may pose some problems for the implementation of the control strategy to get constant voltage output. The proposed model will be beneficial for developing and underdeveloped countries such as India, Bhutan and Nepal etc., where diesel engines with lower efficiencies are still in use. Although, the proposed model requires high capital cost and high level of research for design, but it will be a revolution in the field of innovative alternative approach of energy production.

REFERENCES

[1] Chen, Z., Spooner, E., "Wind Turbine Power Converters: A Comparative Study," *7th International Conference on Power Electronics and Variable Speed Drives*, No. 456, pp. 471-476, Sept. 1998.

[2]http://airfoiltools.com/airfoil/details?airfoil=naca6412-il

[3] M.D. Arifujjaman, "Maximum Power Extraction from a Small Wind Turbine Emulator Using a DC-DC Converter Controlled By a Microcontroller," ICECE □06, Dhaka, Bangladesh, December 2006.

[4]http://www.scr.indianrailways.gov.in/uploads/files/1341896144130-Train%20Light.PDF

[5] Betz, A. (1966) *Introduction to the Theory of Flow Machines*. (D. G. Randall, Trans.) Oxford: Pergamon Press.

[6]https://www.princeton.edu/~asmits/Bicycle_web/blunt.html

[7]http://www.motiva.fi/myllarin_tuulivoima/windpower%20web/en/tour/wtrb/drag.html

[8]http://profit.ndtv.com/news/economy/article-price-of-diesel-sold-to-bulk-consumers-cut-by-rs-1-09-per-litre-588804.

Simulation analysis and THD Measurements of Integrated PV and Wind as Hybrid System Connected to Grid

Manish Kumar
Electrical Engineering Department
National Institute of Technology
Kurukshetra, Haryana, India
manish_1427@gmail.com

K. S. Sandhu
Electrical Engineering Department
National Institute of Technology
Kurukshetra, Haryana, India
kjssandhu@rediffmail.com

Ashwani Kumar
Electrical Engineering Department
National Institute of Technology
Kurukshetra, Haryana, India
ashwa_ks@yahoo.co.in

Abstract--**This paper presented an integration of doubly-fed induction generator (DFIG) and Photovoltaic (PV) as hybrid system to utility grid. PV system is connected to grid via a DC-DC boost converter and three-phase three-level voltage source converter (VSC). The maximum power point tracking (MPPT) is implemented in the boost converter using incremental conductance+ integral regulator technique. Wind turbines use a DFIG consisting of a wound rotor induction generator and an AC/DC/AC IGBT-based PWM converter. In this paper, analysis has been carried out for hybrid system and considering the PV and wind separately also. Simulation analysis with THD measurement of the three cases has been carried out in MATLAB environment viz. Case1: PV array system connected to grid, case2: wind energy system connected to same grid, case3: integration of PV system & wind as hybrid source to the grid.**

Index Terms—*PV Array, MPPT, boost converter, Wind turbine (DFIG), utility grid, THD measurement.*

I. INTRODUCTION

Day by day energy demand is increasing all over the world. Presently 80% contribution to world energy supplied from the conventional energy sources which are harmful for the environment. So there is a need of renewable energy sources that will not harmful to the environment. Some surveys indicate that the energy demand will increase by three times in the world by 2050 [1]. PV and wind energy integrated systems is the best option to be considered for the future energy requirements. [2-5]. Despite the advantages of renewable energy hybrid systems and their response under balanced condition, their performance leads insufficient results in faulty grids case. Some control algorithms have been proposed to improve the robustness of the same for better performance in case of faulty grids [6, 7].The problem of maximum power extraction and improvement of the injected power quality has been addressed in [8, 9]. The control strategies considered must be able to overcome both the problems related to the tracking of the maximum power point and the problems of their connection to the grid under all conditions [9-11].

The objective of this paper is to present simulation and analysis and THD measurement of integrated PV & Wind (DFIG) as hybrid system connected to grid. In this paper three cases are considered for the study of the system power quality,

stability. The simulation analysis with THD measurement of the three cases has been done. The three cases considered are:

Case1:- PV array system connected to grid.

Case2:-Wind energy system connected to same grid.

Case3:- Integration of PV system and wind as hybrid sources to the grid.

II. PV AND WIND MODEL

A hybrid energy system usually consists of two or more renewable energy source used together to provide increased system efficiency as well as greater balance in energy supply.

A. PV array system connected to grid

A detailed model of a 100-kW array connected to a 25-kV grid shown in Fig.1. PV array delivering a maximum of 100 kW at 1000 W/m2 sun irradiance.5-kHz boost converter increasing voltage from PV natural voltage (272 V DC at maximum power) to 500 V DC. Switching duty cycle is optimized by the MPPT controller that uses the "Incremental Conductance + Integral Regulator" technique. The VSC converts the 500 V DC to 260 V AC and keeps unity power factor. 10-KVAr capacitor bank filtering harmonics produced by VSC.100-KVA 260V/25kV three-phase coupling transformer. Utility grid model (25-kV distribution feeder + 120 kV equivalent transmission systems).B1 is the grid side bus and B25 is the load side bus shown in Fig. 2.

A PV array is composed of many series and parallel connected PV modules, whereas a single module consists of a number of series connected solar cells. The basic equation from the theory of semiconductors [12] that mathematically describes the I-V characteristics of an ideal cell is:

$$I_d = I_{sat}\left[e^{V_d/V_T} - 1\right] \tag{1}$$

$$V_T = k\frac{T}{qQ_dN_{cells}} \tag{2}$$

I_d : is the diode current.

978-1-4799-6047-7/14 $31.00 © 2014 IEEE   350

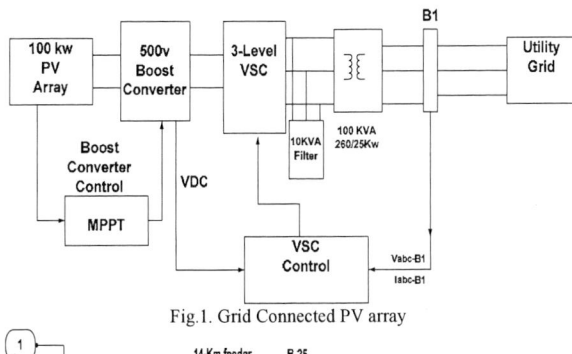

Fig.1. Grid Connected PV array

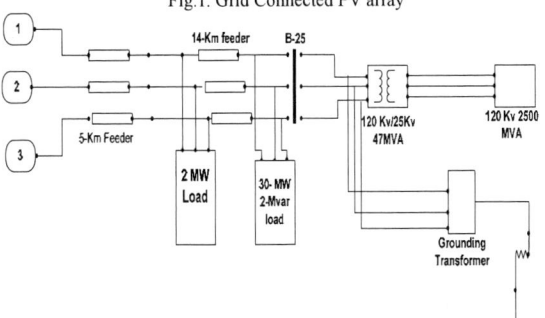

Fig.2. Utility Grid model

I_{sat} : is the diode saturation current.

V_d : is the diode voltage.

V_T: is the temperature voltage.

k: is the Boltzmann's constant ($1.3806503 \rightarrow 10\_23$J/K).

q: is the electron charge (1.6022e_19C).

Q_d: is the diode quality factor.

N_{cells}: is the number of series connected cells per module.

The light-generated photo-current of the single module is represented by I_{ph}. This value increases as the number of parallel module strings increase and is directly proportional to the solar irradiance on the module's surface. For a PV array with 66 parallel strings (N_{par}) consisting of 5 modules (N_{ser}) each, the Equation (1) is altered to:

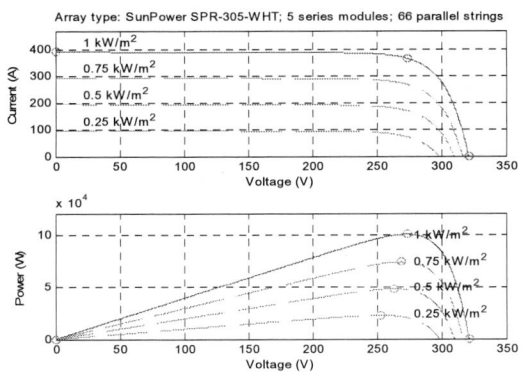

Fig.3. I-V and P-V characteristics of PV array

$$I_{darray} = I_{sat\,array}\left[e^{V_d/V_{Tarray}} - 1\right] \qquad (3)$$

Where, I_{darray} is the aggregated diode current for the PV array, $I_{sat\,array}$ is the aggregated diode saturation current and V_{Tarray} is the aggregated temperature voltage. These values, along with the increased photo-current value are $I_{ph\,array}$ calculated as:

$$I_{ph\,array} = I_{ph} \times N_{par} \qquad (4)$$
$$I_{sat\,array} = I_{sat} \times N_{par} \qquad (5)$$
$$V_{T\,array} = V_T \times N_{ser} \qquad (6)$$

The model parameters used for calculations in Equations (1) - (6) are given in Table1. Figure 3shows the I-V and P-V curves of the panel, derived and simulated from equations (1) and (2).

The circle marker on the plots indicates the maximum power point and the dotted plots are shown for deferent values of input irradiances (25 degree Celsius, 100w/m²⁾) in figure3. The parameter of sun power SPR350-WHT PV array shown in Table1. Maximum power for a single PV array (Watts) is thus given by:

$$P_{array} = N_{par} \times N_{ser} \times P_{mp} \qquad (7)$$

From Equation (7), P_{array} = 100.7 kW.

Table1. SunPower SPR-305-WHT PV Array Specifications

Parameter	Variable (Units)	Value
Number of cells in series	N_{cells}	96
Number of series connected modules per string	N_{ser}	5
Number of parallel strings	N_{par}	66
Maximum Power	P_{mp} (W)	305.2
Maximum Power Voltage	V_{mp} (V)	54.70
Maximum Power Current	I_{mp} (A)	5.58
Open-circuit Voltage	V_{oc} (V)	64.20
Short-circuit Current	I_{sc} (A)	5.96
Series Resistance	R_S (Ω)	0.0380
Parallel Resistance	R_P (Ω)	993.5
Diode Saturation Current	I_{sat}(A)	3.1949e_8
Light Generated Photo-Current	I_{ph} (A)	5.9602
Diode Quality Factor	Q_d	1.3

B. Wind energy system connected to grid

A 9 MW wind farm consisting of six 1.5 MW wind turbines connected to a 25 kV distribution system exports power to a 120 kV grid through a 5km and 14 km, 25 kV feeder as shown in Fig.4.

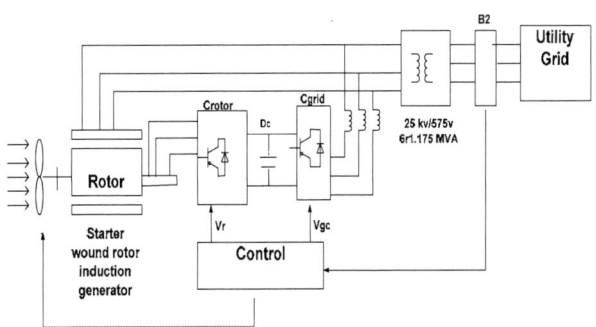

Fig.4. Grid connected wind (DFIG wound rotor induction generator) system

Wind turbines using a doubly-fed induction generator (DFIG) consist of a wound rotor induction generator and an AC/DC/AC IGBT-based PWM converter modeled by voltage sources. The stator winding is connected directly to the 60 Hz grid while the rotor is fed at variable frequency through the AC/DC/AC converter. The DFIG technology allows extracting maximum energy from the wind for low wind speeds by optimizing the turbine speed, while minimizing mechanical stresses on the turbine during gusts of wind [13-15]. The DFIG-based WECS basically consists of generator, wind turbine with drive train system, RSC, GSC, DC-link capacitor, pitch controller, coupling transformer, and protection system. The DFIG is a wound-rotor induction generator with the stator terminals connected directly to the grid and the rotor terminals to the mains via a partially rated variable frequency ac/dc/ac converter, which only needs to handle a fraction (25-30 %) of the total power to accomplish full control of the generator. The functional principle of this variable speed generator is the combination of DFIG and four-quadrant ac/dc/ac VFC equipped with IGBTs. The ac/dc/ac converter system consists of a RSC and a GSC connected back-to-back by a DC- link capacitor. The rotor current is controlled by RSC to vary the electro-magnetic torque and machine excitation. Since the power converter operates in bi-directional power mode, the DFIG can be operated either in sub-synchronous or in super-synchronous operational modes. In Table 2 parameter of DFIG are given.

Table2: Parameters of the DFIG

Rated Power	6 X 1.5 = 9 MW
Stator frequency	60 Hz
Stator nominal voltage	575 V
Stator resistance	0.023 pu
rotor resistance	0.016 pu
Stator Inductance	0.18 pu
rotor Inductance	0.16 pu
Inertia constant	0.685 pu
Nominal DC bus voltage	1150

B. DFIG MODEL

The dynamics of the DFIG is represented by a fourth-order state space model using the synchronously rotating reference frame (qd-frame) as given in (8)-(11)[17].

Control of DFIG-based WECS

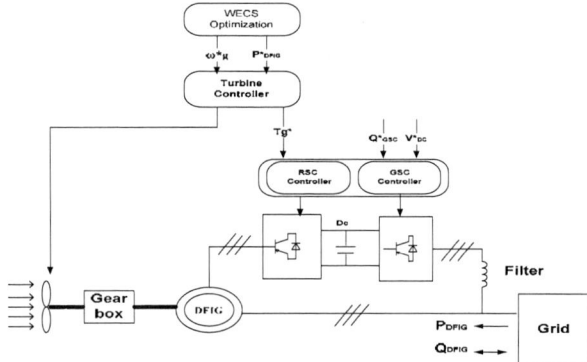

Fig. 5. Control block diagram of DFIG-based WECS

The general structure of control block diagram in the DFIG-based WECS having two levels of control is shown in Figure5. The highest level is the WECS optimization; wherein the speed of the wind turbine is set in such a way that optimum wind power can be captured. This control level is mechanical system control. The lower level control being the electrical system control, i.e. torque and reactive power control [16-19]. The mechanical control system acts slower compared to the electrical control system. The Wind turbine Cp curves are displayed in Fig. 6. The turbine power, the tip speed ratio lambda and the Cp values are displayed in Fig. 7 as function of wind speed. For a wind speed of 15 m/s, the turbine output power is 1 pu of its rated power, the pitch angle is 8.7 deg and the generator speed is 1.2 pu

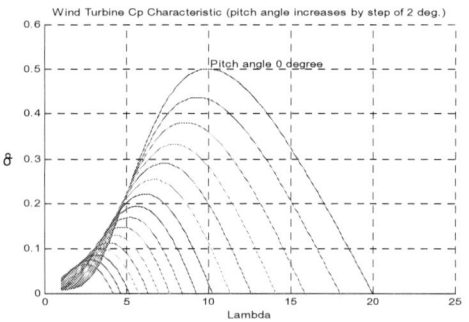

Fig. 6. Wind Turbine Cp Characteristic (pitch angle increases by step of 2 deg)

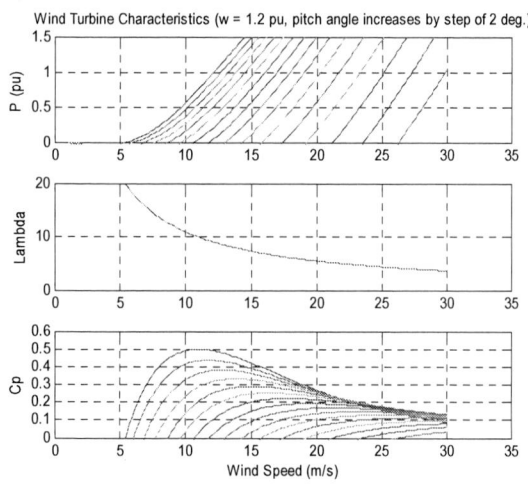

Fig. 7. Wind Turbine Cp Characteristic (w=1.2 pu, pitch angle increases by step of 2 deg)

C. Integration of PV system & wind as hybrid sources to the grid

Integration of 100-Kw PV array (B1 Bus) and 9 MW wind farm consisting of six 1.5 MW wind turbines connected (B2 Bus) to a 25-kV grid shown in figure8.

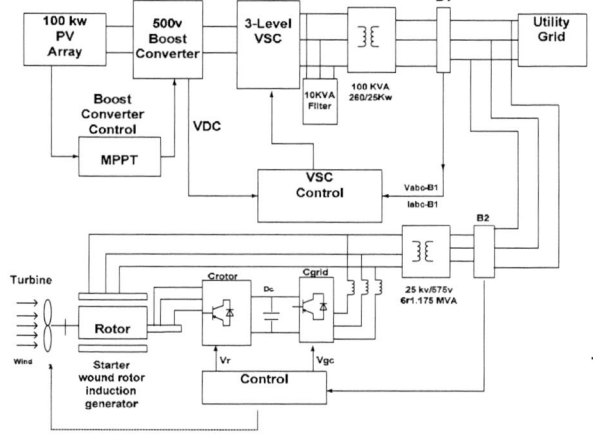

Figure8. Integration model of PV system & wind as hybrid sources to the grid

III. SIMULATION ANALYSIS

A.Case1: PV array system connected to grid

Running the model for 0.7 seconds, it is observed that, from t=0 sec to t= 0.05 sec, pulses to Boost and VSC converters are blocked. PV voltage corresponds to open-circuit voltage 321V. The three-level bridge operates as a diode rectifier and DC link capacitors are charged above 500 V. At t=0.05 sec, Boost and VSC converters are de-blocked. DC link voltage is regulated at Vdc=500V. Duty cycle of boost converter is fixed (D= 0.5) and sun irradiance is set to 1000 W/m2. Steady state is reached at t=0.25 sec. Resulting PV voltage is 250V. The PV array output power is 96 whereas maximum power with 1000 W/m2 irradiance is 100.7 kW. At t=0.4 sec MPPT is enabled. The MPPT regulator starts regulating PV voltage by varying duty cycle in order to extract maximum power. Maximum power (100.7 kW) is obtained when duty cycle is D=0.453. At t=0.5 sec, PV mean voltage 274 V as expected from PV module specifications 273.5V. From t=0.5 sec to 0.7 sec various irradiance changes are applied in order to illustrate the good performance of the MPPT controller. Figure 9 show DC (boost converter voltage) and Mod.Index. At t=0.2 PV array voltage boost 500V and mod.Index are 0.85.

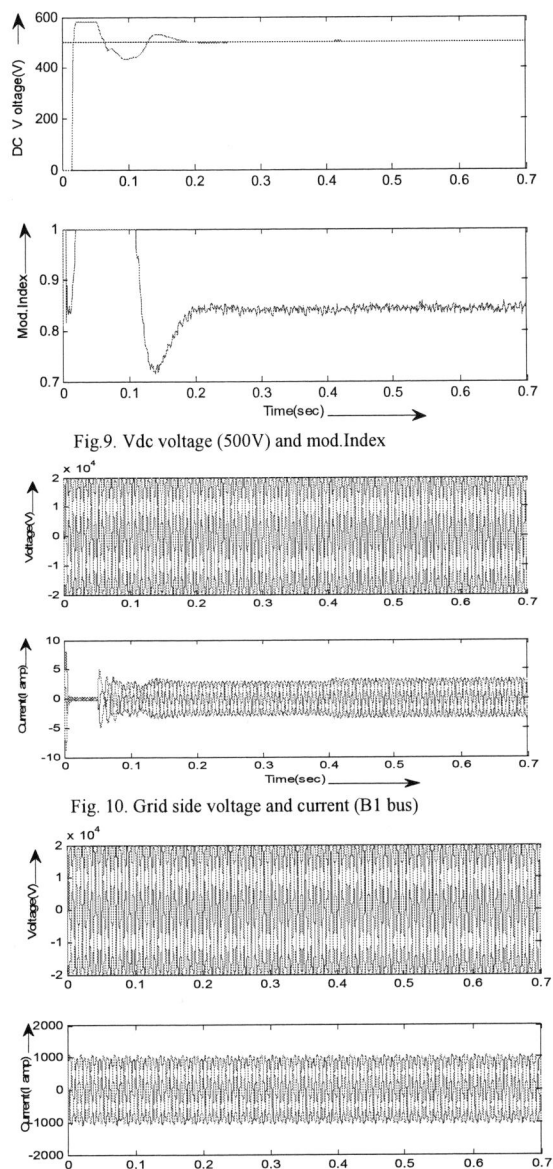

Fig.9. Vdc voltage (500V) and mod.Index

Fig. 10. Grid side voltage and current (B1 bus)

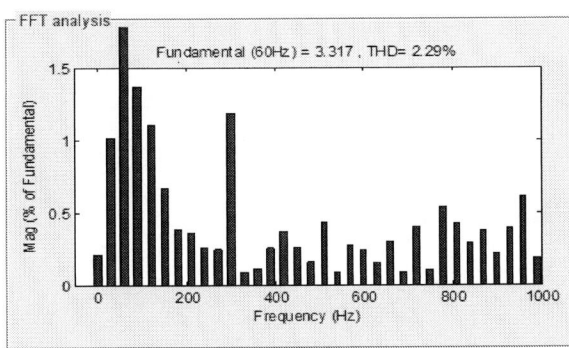

Fig.11. Load side Voltage and Current (B25 bus)

Figures 10 and 11 show the voltage and current of Grid side bus (B1 Bus) and load side bus (B25 Bus). Measured the THD of current both in Grid side bus (B1 Bus, THD=2.29%) and load side bus (B25 Bus, THD=0.15%) are shown in Figs. 12-13.

Fig.12. THD of Bus (B1 bus)

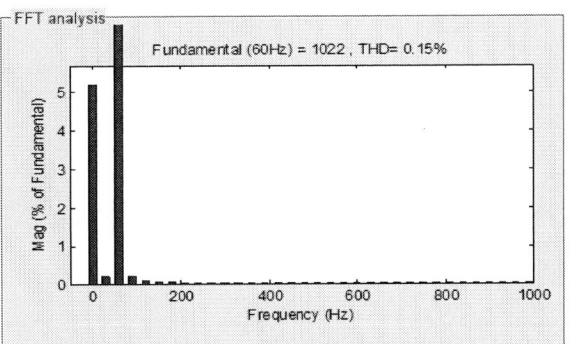

Fig.13. THD of Bus (B25 bus)

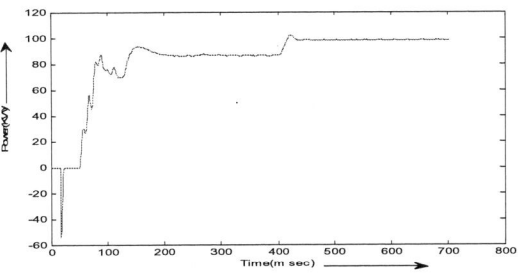

Fig. 14. Grid side (B1 Bus, P=100kw) Power vs time

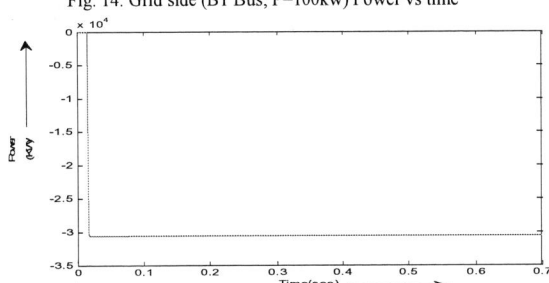

Fig.15. Load side (B25 Bus, P=30MW) Power vs time

Grid side (B1 bus) power is 100 KW and load side (B25 bus) power is 30MW is shown in Figs. 14-15.

B.Case2:- Wind energy system connected to grid

Run the model wind speed is maintained constant at 15 m/s. The DFIG wind farm produces 9MW. The corresponding turbine speed is 1.2 pu of generator synchronous speed. The DC voltage is regulated at 1150V and reactive power produced by the wind turbine is regulated at 10 Mvar.

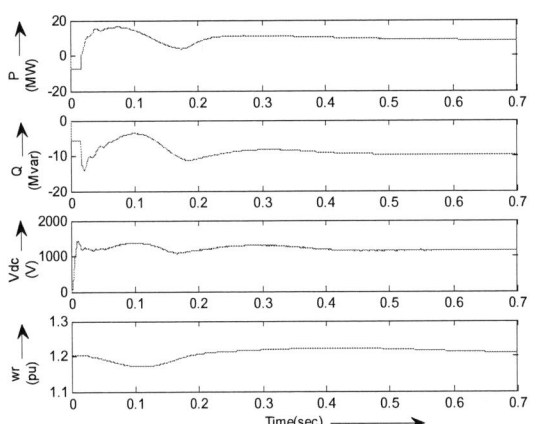

Fig.16. Wind DFIG active power (P=9MW), Reactive power (Q=10Mvar), DC Side voltage (Vdc=1150), speed (wr=1.2pu)

978-1-4799-6047-7/14 $31.00 © 2014 IEEE   353

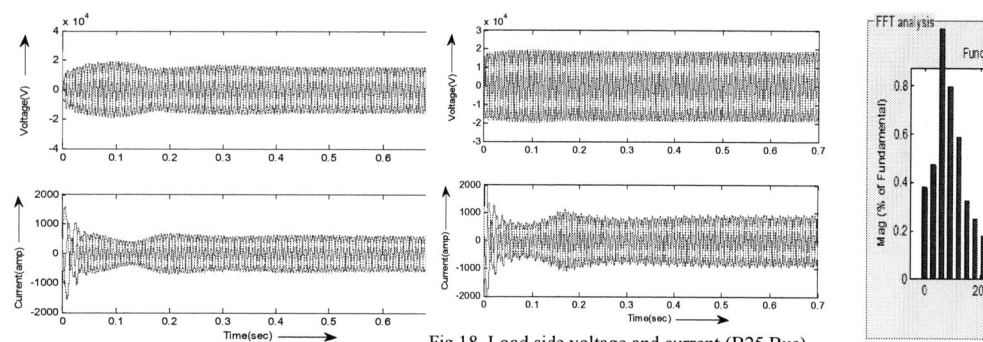

Fig.17. Grid side voltage and current (B2 Bus)

Fig.18. Load side voltage and current (B25 Bus)

Fig.19. THD of Bus (B2 bus)

Figure16 shows the wind DFIG active and reactive power of 9MW and 10 Mvar. DC side voltage 1150V and generator speed is 1.2pu. The grid side bus (B2 bus) and load side bus (B25 bus) voltage and current shown in Figs.17 and 18.The measured THD of current in B1 bus and B25 bus is 0.95%and 0.56% shown in Figs. 19-20. Load side power19.23 MW is shown in Fig.21.

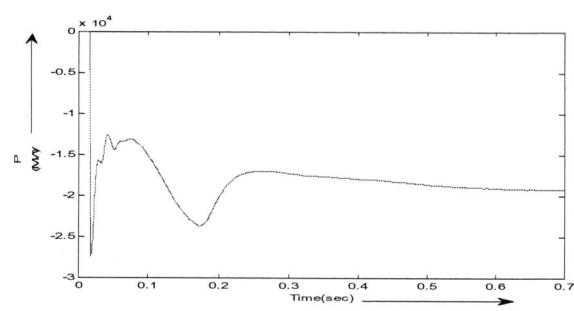

Fig. 21. Load side power (B25 Bus, P=19.23MW)

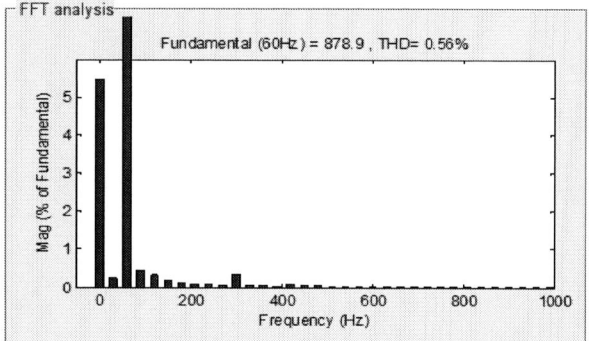

Fig.20. THD of Bus (B25 bus)

C.Case3:- Integration of PV system and wind as hybrid sources to the grid

Simulation analysis and THD measurement of case3, the measurement of current and voltage of B1 bus, B2 bus and B25 bus shown in Figs 22-24. The THD of current of B1bus is 1.90%, B2 bus is .85% and B25 bus is 0.50% shown in Figs./ 25-27. The power of B1 bus, B2 bus and B25 bus also shown in Figs. 28-30.

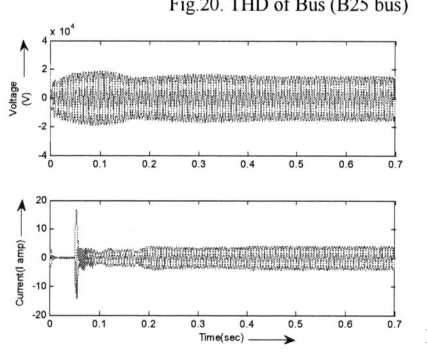

Fig. 22.voltage and current B1 Bus side

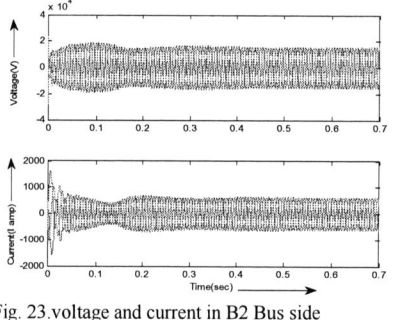

Fig. 23.voltage and current in B2 Bus side

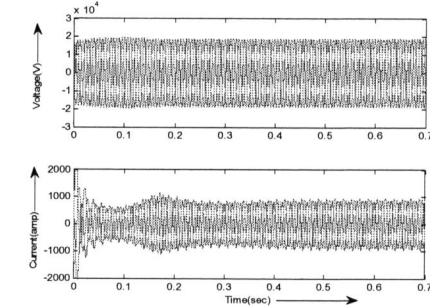

Fig. 24.voltage and current in B25 Bus side

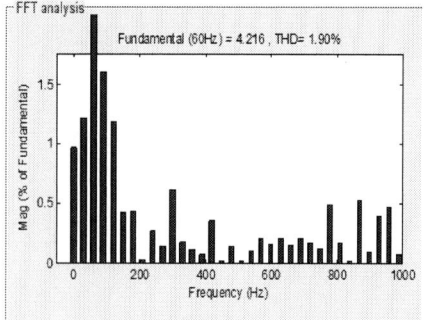

Fig. 25. THD of Bus (B1 bus)

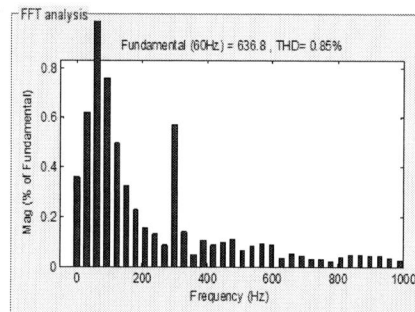

Fig. 26. THD of Bus (B2 bus)

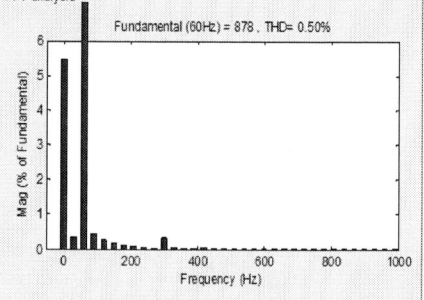

Fig. 27. THD of Bus (B25 bus)

978-1-4799-6047-7/14 $31.00 © 2014 IEEE

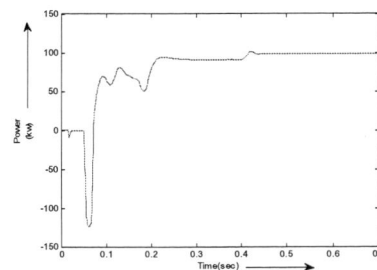

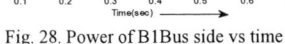

Fig. 28. Power of B1Bus side vs time

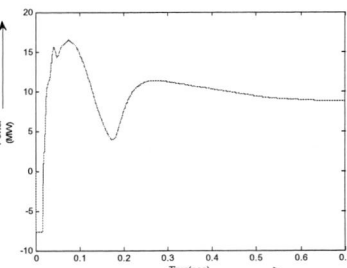

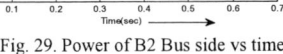

Fig. 29. Power of B2 Bus side vs time

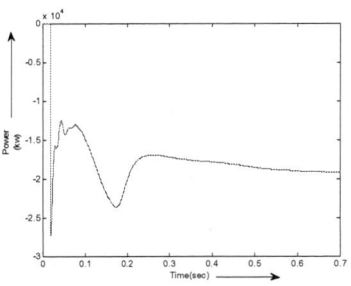

Fig. 30. Power of B2 Bus side

Table3: Comparisons of THD in Case1, Case2 and Case3

	Case1 THD% of current	Case2 THD% of current	Case3 THD% of current
B1 Bus	2.29	-	1.90
B2 Bus	-	0.95	0.85
B25 Bus	0.15	0.56	0.50

It is observed that the THD of Case3 are minimum as compared to Case1 and Case2 as given in Table3. Case1 (B1 Bus) to Case3 (B1 Bus) THD reduced 2.29 to 1.90 and Case2 (B2 Bus) to Case2 (B2 Bus) THD reduced 0.95 to 0.85. In B25 BUS (load side bus) side Case2 to Case3 THD reduced 0.56 to 0.50 but in Case1 to Case3 it increased 0.15 to 0.50.

IV. CONCLUSIONS

The hybrid system based wind (DFIG) and PV energy conversion system with grid has been successfully implemented and simulated using MATLAB. The three cases are considered for analysis of results and measurements of the THD of current. The system has been analyzed for different PV irradiation condition and boost converter voltage. It is observed that the THD of Case3 is minimum as compared to Case1 and Case2. Case1 (B1 Bus) to Case3 (B1 Bus) THD reduced 2.29 to 1.90 and Case2 (B2 Bus) to Case2 (B2 Bus) THD reduced 0.95 to 0.85. In B25 Bus (load side bus) side Case2 to Case3 THD reduced 0.56 to 0.50 but in Case1 to Case3 it increased 0.15 to 0.50.

REFERENCES

[1] H. H. El-Tamaly et. al. ,"Design and Control Strategy of Utility Interfaced PV/WTG Hybrid System", The Ninth International Middle East Power System Conference, MEPCON'2003, Minoufiya University, Faculty of Eng., Shebin El-Kom, Vol. 2, Dec. 16-18, 2003

[2] F. Valenciaga, and P. F. Puleston, "Supervisor Control for a Stand-Alone Hybrid Generation System Using Wind And Photovoltaic Energy", IEEE Transactions On Energy Conversion, Jun. 2005, Vol. 20, pp. 398-405,2005.

[3] Seul-Ki kim, Eung-Sang kim, Jong, "Modeling and control of a grid –connected Wind/ PV hybrid generation system", Transmission and distribution conference and Exhibition,2005/2006 IEEE PES,pp.1202-1207,2006

[4] Li wang, Kuo-Hua Liu, "Transient performance and stability analysis of a hybrid grid-connected Wind/ PV system", Power System Conference and Exhibition, 2004, IEEE PES, Vol.2, pp.795-801,2004.

[5] Olulope. P.K, Folly, K.A, "Modeling and simulation of hybrid distributed generation and its impact on transient stability of power system", Industrial Technology (ICIT), 2013 IEEE international conference, pp.1757-1762, 2013.

[6] L. Xu, Y. Wang, "Dynamic Modeling and Control of DFIG-Based Wind Turbines Under Unbalanced Network Conditions," IEEE Transactions On Power Systems, Vol.22, February2007.

[7] Abbassi,R.,Hammani,M, Chebbi.s, "Improvement of the integration of a grid connected wind-photovoltaic hybrid system", Electrical Engineering and software

applications(IEEESA),2013 international conference,pp.1-5,2013.

[8] H. H. El-Tamaly, Mohammed, A.A.E, "Modeling and simulation of photovoltaic/Wind Hybrid Electrical Power System Interconnected with electrical Utility", Power System Conference, MEPCON;2008, 12th international middle- east, pp.645-649,2008.

[9] P.Sivakumar,M.Arutchelui, "Control of grid converters for PV array Excited wind- driven induction generators with unbalance and non linear loads", International Journal of Electrical Power& Energy system,Vol.59,pp.188-203, July 2014.

[10] Chih-Ming Hong,Chiung-Hsing Chen, "Intelligent control of a grid connected wind-photovoltaic hybrid power system", International Journal of Electrical Power & Energy system,Vol.55,pp.554-561,Feb 2014.

[11] Moacyr A. G. de Brito, Leonardo P. Sampaio, Luigi G. Jr., Guilherme A. e Melo, Carlos A. Canesin "Comparative Analysis of MPPT Techniques for PV Applications", 2011 International Conference on Clean Electrical Power (ICCEP).

[12] H. S. Rauschenbach, "Solar cell array design handbook-the principles and technology of photovoltaic energy conversion," 1980.

[13] R. Pena, J.C. Clare, G.M. Asher, "Doubly fed induction generator using back-to-back PWM converters and its application to variable-speed wind-energy generation," IEEE Proc.-Electr. Power Appl., Vol. 143, No. 3,pp.231-241, May 1996.

[14] Vladislav Akhmatov, "Variable-Speed Wind Turbines with Doubly-Fed Induction Generators, Part I: Modeling in Dynamic Simulation Tools," Wind Engineering Volume 26, No. 2, 2002

[15] Nicholas W. Miller, Juan J. Sanchez-Gasca, William W. Price, Robert W. Delmerico, "Dynamic modeling of GE 1.5 and 3.6 MW wind turbine-generators for stability simulations," GE Power Systems Energy Consulting, IEEE WTG Modeling Panel, pp.1977-1983,July 2003

[16] Bakari Mwinyiwiwa, Yongzheng Zhang, Baike Shen, and Boon-Teck Ooi, "Rotor Position Phase-Locked Loop for Decoupled P-Q Control of DFIG for Wind Power Generation," IEEE Transactions on Energy Conversion, vol. 24, no. 3, pp. 758-765, Sept. 2009.

[17] Shuhui Li, Timothy A. Haskew, Rajab Challoo, and Marty Nemmers, "Wind Power Extraction from DFIG Wind Turbines Using Stator-Voltage and Stator-Flux Oriented Frames," International Journal of Emerging Electric Power Systems, vol. 12, iss. 3, Article 7, 2011.

[18] Mustafa Kayıkcı and Jovica V. Milanovi'c, "Reactive Power Control Strategies for DFIG-Based Plants," IEEE Trans. on Energy Conversion, vol. 22, no. 2, pp. 389-396, June 2007.

[19] Torsten Lund, Poul Sorenson and Jarle Eek, "Reactive Power Capability of a Wind Turbine with Doubly Fed Induction Generator," Wind Energy, vol. 10, iss. 4, pp.379-394, Apr. 2007.

Maximum Power Point Tracking Scheme for Variable Speed Wind Generator

Sumit Chauhan[1], Malik Sameeullah[3]
School of Renewable Energy and Efficiency
National Institute of Technology Kurukshetra
Kurukshetra, India-136119
sumitnitres@gmail.com[1], malik.sameeullah@gmail.com[3]

Ratna Dahiya[2]
Professor, Department of Electrical Engineering
National Institute of Technology Kurukshetra
Kurukshetra, India-136119
Ratna_dahiya@nitkkr.ac.in[2]

Abstract—**Wind energy is gaining increasing importance throughout the world. With the advancement of variable speed drive design and control of wind energy systems, the efficiency and the energy capture capacity of the energy conversion system is also improved. For renewable energy system, Maximum Power Point Tracking (MPPT) plays a vital role for efficient conversion of energy. The MPPT strategy is used to control the rotor speed by adjusting electromagnetic torque and maximize the electrical output power of the generator. In this paper, three different control schemes are discussed and implemented in Matlab Simulink environment. Simulink result is used to compare the performance of 5.7KW wind turbine under different control schemes.**

Keywords—Wind Turbine; Wind Energy; MPPT; Maximum Power Point (MPP)

I. INTRODUCTION

Energy is the mandatory factor for the development and rapid growth of any country. The requirement of energy is endlessly and exhaustingly throughout the world. Fossil fuels are the primary energy source and have a worldwide share of 60%. Conventional energy sources like oil, coal and gas, etc. causes an environmental pollution and global warming due to Green House Gasses emission and even the resource is limited. With improvement in people lifestyle, demand of energy rises day by day. In order to meet the demand and solve the problem of energy crises, government of different country looking toward a renewable form of energy. In India, Ministry of New and Renewable Energy is actively engaged to promote various forms of renewable energy resources.

Despite various initiatives taken by the government of India, the energy sector is still struggling to cope up with ever increasing demand. In India, renewable energy is accounted for 12.2% of total installed capacity [1], [2]. Until 2004, wind energy penetration in term of installed capacity is 3% of total installed capacity. And at the end of 2013, the wind power generation installed capacity is 21.1 GW and India is the 5th major wind power generation country. The wind turbine is in the range of 225kW up to 2.1MW and deployed across the country.

In section II, mathematical model of wind turbine is discussed. The wind energy conversion system mainly comprises two stages of energy conversion, firstly from the kinetic energy of wind to the rotational mechanical energy of wind turbine, and secondly, the mechanical energy into electrical energy using a generator. These two stages strongly comprise the performance of Wind Energy Conversion System. In earlier times, wind turbine operates at the fix rotor speed configuration and system is not able to extract optimal power. Now a day, variable speed generator like Induction Generator, Double Fed Induction Generator and Self Excited Induction Generator etc. are preferred because of its ability to operate at variable speed. In order to extract maximum power, turbine rotor speed is adjusted by using the MPPT controller. In Section III, three major MPPT control strategy is discussed in detail. Comparison of experimental result and validation of MPPT techniques is presented in section IV. Conclusion and future scope of work is discussed in section V.

II. WIND TURBINE MODEL

The power generated by wind turbine can be expressed as [3], [4]:

$$P = \frac{1}{2} * C_p\left(\lambda, \beta\right)\rho\pi R^2 V^3 \qquad (1)$$

where ρ=air density, R=turbine rotor radius, V= wind speed and C_p = turbine power coefficient that represents the power conversion efficiency of the wind turbine. C_p is defined as a ratio of actual power delivered by wind turbine to the theoretical power available in the wind. It is found that C_p is a function of the tip speed ratio (TSR) λ and the blade pitch angle β in a pitch control wind turbine. λ is given by; $\lambda = \frac{\omega_r * R}{V}$, where ω_r is a rotational speed of the wind turbine. For different wind turbine, an empirical formula of C_p in terms of λ and β is generated. The maximum theoretical value of C_{pmax} is 16/27. For this paper, the wind power coefficient C_P [5] expression is represented by

$$C_p(\lambda, \beta) =$$
$$C_1\left(C_2\left(\frac{1}{\lambda + 0.08\beta} - \frac{0.035}{\beta^3 + 1}\right) - C_3\beta - C_4\right).e^{-C_5\left(\frac{1}{\lambda + 0.08\beta} - \frac{0.035}{\beta^3 + 1}\right)} + C_6\lambda \quad (2)$$

where $C_1 = 0.5176$, $C_2 = 116$, $C_3 = 0.4$, $C_4 = 5$, $C_5 = 21$, $C_6 = 0.0068$

Fig. 1 illustrates the curve of $C_p(\lambda)$ for different values of β. For any particular wind speed, maximum value of C_{pmax} is

978-1-4799-6047-7/14 $31.00 © 2014 IEEE

found when $\beta=0°$. With increase in β value, C_{pmax} level decrease continuously. When wind speed varies and reach the critical high value, the wind turbine's control system adjusts the blade pitch to keep the rotor speed within the operating zone. According to figure 1, optimal value of λ, β for maximum power coefficient (C_{pmax}=0.48) is $\lambda_{opt} = 8.1, \beta = 0°$.

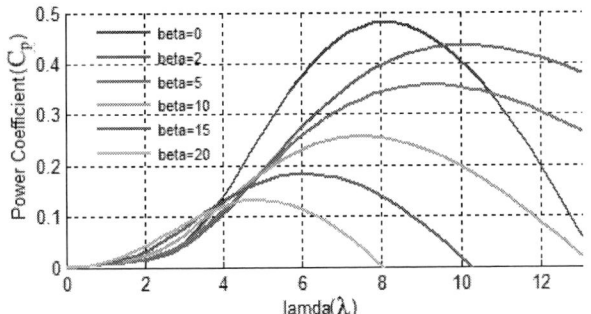

Fig. 1. Curve of power coefficient for different beta angle.

The mechanical torque obtained at the shaft of wind turbine is represented by:

$$T_{turb} = \frac{P_{turb}}{\omega_t} \qquad (3)$$

$$T_{turb} = \frac{0.5\rho}{\omega_t}.C_p(\lambda,\beta).\pi R^2 V^3 \qquad (4)$$

where P_{turb} is turbine output power and T_{turb} is turbine output torque. From above (4), it is found that the torque on the shaft of the turbine is depends upon the velocity of wind and shaft speed. By regulating the rotor speed, maximum power can be extracted. In between turbine and generator shaft, there is a gearbox which changes the speed of slow turbine shaft to the speed of the generator shaft by the ratio of multiplier G. Torque and rotor speed at generator side are changed by using gearbox G. The value of T_g and ω_{mech} are given by (5) and (6).

$$T_g = \frac{T_{turb}}{G} \qquad (5)$$

$$\omega_{mech} = G.\omega_t \qquad (6)$$

The generator shaft is modeled using following differential equation:

$$J\frac{\partial \omega_{mech}}{\partial t} = T_{mech} = T_g - T_{em} - f.\omega_{mech} \qquad (7)$$

where J is a total inertia on the generator shaft, f is a viscous friction coefficient (loss),T_g is generated torque, T_{em} is an electromagnetic torque of the generator and ω_{mech} is a rotor speed of generator . By using the above equations, the wind turbine model for variable speed is given by Fig. 2. Parameters of wind turbine are listed in Table-I.

III. MPP Tracking and Power Control Strategy

Variable speed wind turbine (WT) allows extracting optimal power, at the same time it helps to reduce the stress on WT shafts and gears. As shown in Fig. 3, with a change in

wind speed the optimal power points or MPPs shift and follow the trajectory of violet line. For the fixed speed WT, points of turbine operation with change in wind speed are 1 and 2, on the other hand, for variable speed WT, points of operation are 1, 3, 4 and 5. So, it is clear from Fig. 3 that in order to extract maximum power under variable wind speed, it needs to change the turbine rotor speed and shift it toward optimal rotor speed.

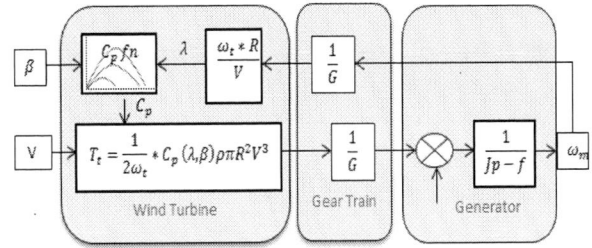

Fig. 2. Block diagram representation of wind turbine model.

TABLE I. PARAMETERS OF WIND TURBINE

P(rated)	5.7 KW
V(rated)	14 m/s
ρ	1.226 Kg/m^2
G	3
R	1.5 m
J	0.9067 Kg.m^2
f	0.005

There are a number of MPPT techniques are used to track MPP. In general, MPPT controllers are divided into four groups which are based upon its method of operation, i.e. Lookup table based, Perturbation & Observation method, Artificial Neural Network and Adaptive/ Hybrid control [6], [7]. Three different methods of power extraction are discussed below:

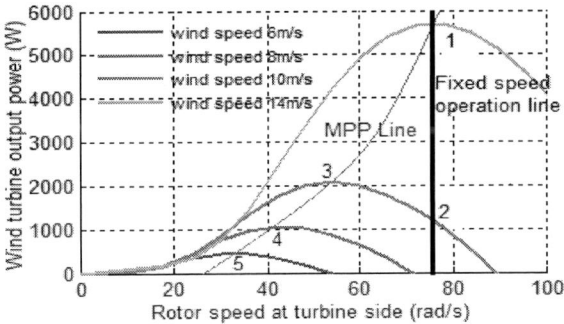

Fig. 3. Plot of Wind Turbine output power for variable wind speed

A Method 1:Direct Speed Measurement

Wind speed fluctuates continuously and it causes a change in available power level of the wind turbine terminal. In order to extract maximum power, rotor speed is regulated continuously. Under the condition of maximum power extraction, the electromagnetic torque generated by the machine is equal to the reference torque ($T_{em,ref}$) [8-9] and given by (8).

$$T_{em} = T_{em,ref} \qquad (8)$$

With a change in wind speed, reference rotor speed change. In order to track reference rotor speed, servo speed control using reference electromagnetic torque is used.

$$T_{em,ref} = PI*(\omega_{ref} - \omega_{mech}) \qquad (9)$$

where PI is the proportional controller to regulate speed and ω_{ref} is the reference rotor speed. Reference rotor speed is calculated on the optimal Lamda value on which C_p is a maximum and it is given by:

$$\omega_{ref} = \frac{1}{G} \frac{\lambda_{Cp,max} V}{R} \qquad (10)$$

The block diagram of wind turbine control strategy is shown in Fig. 4. In this control strategy, accurate knowledge of wind speed is required. As wind is measured using anemometer, this technique is also known as anemometer control scheme.

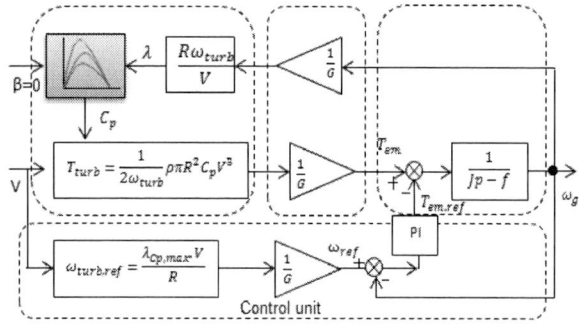

Fig. 4. MPPT scheme with direct wind speed measurement.

The WT performance is recorded for the variable speed test signal shown in Fig. 5. Response of optimal rotor speed, actual rotor speed and output power of WT are shown in Fig. 6, Fig. 7 and Fig. 8.

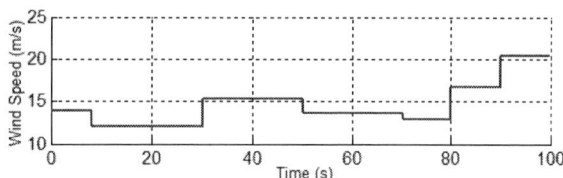

Fig. 5. Wind speed profile used for simulation

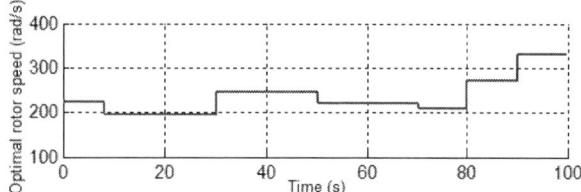

Fig. 6. Optimal rotor speed for the wind speed profile of test signal

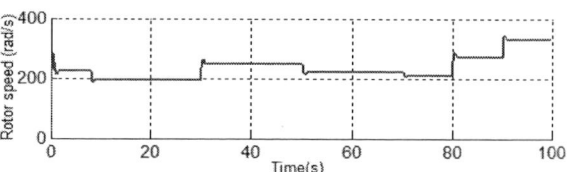

Fig. 7. Actual rotor speed

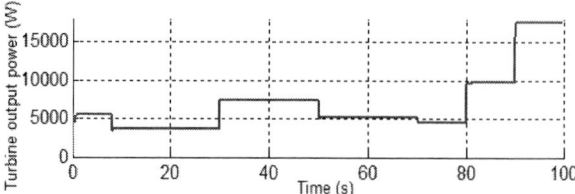

Fig. 8. Output power of wind turbine with change in wind speed

B   Method 2:Torque Feedback Control

The above discuss maximum power point technique is based upon the accurate knowledge of wind speed. It adds the cost of the controller and it is difficult to install anemometer and measure accurate wind speed and unfortunately system performance is not good as expected. So, MPPT techniques without the use of actual wind data have become the necessity for many manufacturers [10]. Block diagram of turbine with Torque Feedback control (TFC) is shown in Fig. 9.

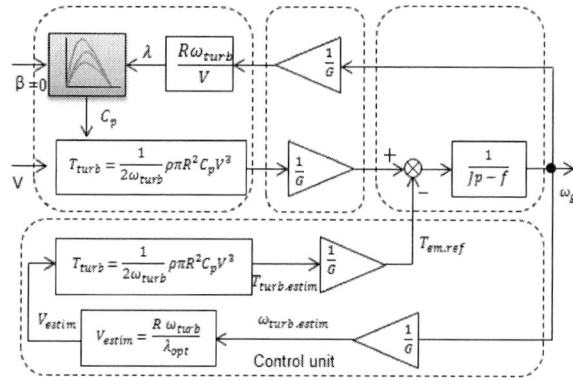

Fig. 9.MPPT scheme with torque feedback measurement.

Due to the dynamic stability of WT structure near MPP, it is able to track the disturbance and naturally goes back to its operating point. So, under steady state condition with slow varying wind speed, mechanical torque (T_{mech}) becomes zero and (7) is rewritten as:

$$T_{mech} = T_{turb} - T_{em} - f\omega_{mech} = 0 \qquad (11)$$

Assuming the friction force is negligible, (11) become:

$$T_{mech} = T_{turb} - T_{em} = 0 \qquad (12)$$

This technique requires the knowledge of optimal torque at particular rotor speed and also known as Torque Feedback Control (TFC). The reference optimal torque can be calculated using mathematical formulas or lookup table. When turbine working at MPP, optimal lamda=$\lambda_{opt} = \frac{R\omega_t}{V}$ and $C_p = C_{pmax}$. Reference torque is given by (13) considering $\omega_{turb,ref} = \frac{1}{G}\omega_{mech}$ and $T_{em,ref} = \frac{1}{G}T_{turb,ref}$.

$$T_{em,ref} = \frac{1}{2}\frac{\rho\pi R^5 C_{pmax}}{\lambda^3 G^3}\omega^2_{mech} \qquad (13)$$

Considering the condition of maximum power point, λ become λ_{opt} and optimal reference torque is then adjusted to the maximum value as given below

$$T_{em,ref} = \frac{1}{2}\frac{\rho\pi R^5 C_{pmax}}{\lambda^3_{opt} G^3}\omega^2_{mech} \qquad (14)$$

The Equation (14) shows that reference optimal torque is proportional to the square of the rotor speed. Electromechanical reference torque ($T_{em,ref}$) help to regulate the rotor speed and shift it toward MPP.For the same test signal of wind speed shown in Fig. 5, response of TFC control is shown in Fig. 10 and Fig. 11.

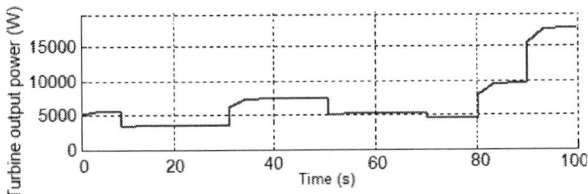

Fig. 10.Rotor speed variation with change in wind speed.

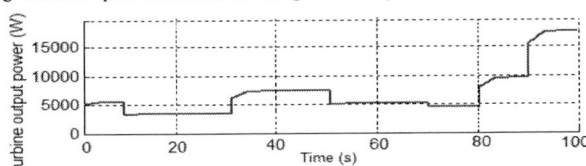

Fig. 11.Variation of turbine output power using TFC control.

C   Method 3:Direct Speed Measurement with Pitch Control

From Fig. 8 and Fig. 11, it is found that output power increase with an increse in wind speed. For method 1 and method-2, if wind speed is greater than 14 m/s than output power become more than rated output power (5.7 KW) and it causes wear & tear of turbine and generator parts and even damage the WT parts. Fig. 12 show the detail block diagram of control scheme using method 3.

In this method, method 1 is used to track MPPT and pitch control is used to change the pitch angle of turbine blades and it limits the power capturing capacity of WT, if wind speed is greater than rated wind speed (14 m/s). There are a number of ways to change the blade pitch angle [11-12]. In this paper, nominal wind speed is compared with actual wind speed and difference of two is used to generate the signal of beta reference. There is no exact formula to derive the relation between $(V - V_{nom})$ and β_{ref}. Fig. 13 show the working region

of method 1 and a relation between beta and $(V - V_{nom})$. The curve fitting technique is used to derive the relation of β and $C_{P,max}$, and given by (15) and (16).

$$\beta = (0.009343 * (V - V_{nom})^3 - (0.1989 * (V - V_{nom})^2) + (3.576 * (V - V_{nom})) \qquad \{\text{for } V - V_{nom} \geq 0\} \quad (15)$$

$$C_{p,max} = a_1 \exp\left(-((\beta - b_1)/c_1)^2\right) + a_2 \exp\left(-((\beta - b_2)/c_2)^2\right)$$
$$\{\text{for} > 0 \} \qquad (16)$$

where $a_1 = 2.796$, $b_1 = 3.155$, $c_1 = 4.372$, $a_2 = 7.098$, $b_2 = 4.765$ and $c_2 = 22.77$

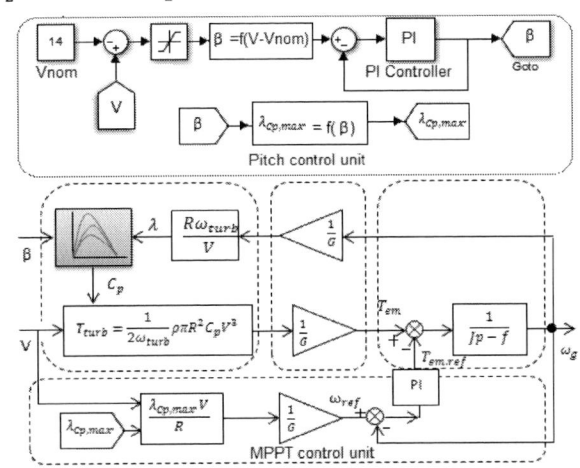

Fig. 12.MPPT control of WT using pitch control direct speed measurement technique.

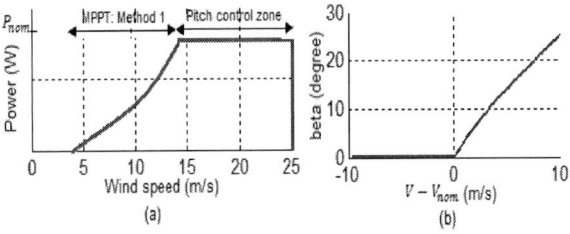

Fig. 13.(a) Operating range of wind turbine, (b) generic blade pitch angle with a change in wind speed.

The response of the method 3 for the same test signal of wind as used for method 1 and method 2, are shown in Fig. 14, Fig. 15 and Fig. 16. Response of method 3 show that, controller is able to track maximum power and also regulate the output power in case of wind thrust. This method helps to track MPP if speed is below nominal value and its control the power output if wind speed is above nominal value.

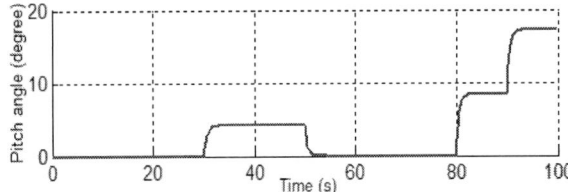

Fig. 14.Blade pitch angle for different wind speed.

978-1-4799-6047-7/14 $31.00 © 2014 IEEE

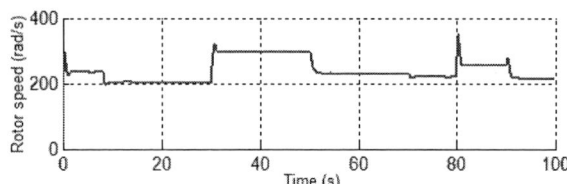

Fig. 15. Actual rotor speed variation with change in wind speed

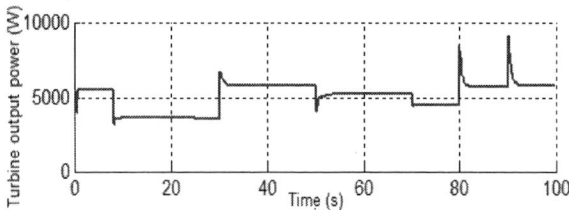

Fig. 16. Turbine output power for method 3 control.

IV. SIMULATION RESULT AND COMPARISON

The same wind speed profile is used for the simulation of above MPPT methods. Based upon the response of the above methods, comparisons are made:

Method 1: The response of this method is good. As actual wind speed is used to measure the optimal rotor speed, tracking capability of this controller is quite good. But in actual accurate measurement of wind speed is a tedious job especially in case of the large size wind turbine. Also above method has no provision to regulate power.

Method 2: The response of this method is sluggish as shown in Fig. 11. The control strategy is simple and easily implemented as there is no need to measure wind speed.

Method 3: This method track the MPP uses the control strategy of method 1. But in addition, it has an ability to control the output power and rotor speed, if wind speed is greater than nominal value. Comparison of the above three methods is listed in Table II.

TABLE II. COMPARISON OF CONTROL METHODS

Techniques	Complexity	Convergence speed	Wind speed measurement
Method 1	High	Fast	Yes
Method 2	Low	Slow	No
Method 3	High	Fast	Yes

V. CONCLUSION

In this paper, detail mathematical modeling of wind turbine is presented and based upon the aerodynamic structure of the turbine, mathematical equations and characteristics are derived. Maximum Power Point Tracking is used to extract maximum power by changing the rotor speed with changes in wind speed. Three methods of power control are discussed in detail. The performance and ability of each method to track the maximum power point have been validated by simulation result. Based upon the simulation result comparison is made to show the advantages and disadvantages of each method. The above methods are simple, cheap and easy to implement. For more robust control, fuzzy, neural or hybrid based adaptive control is preferred.

REFERENCES

[1] Vikas Khare, Savita Nema and Prashant Baredar, "Status of solar wind energy in India", Renewable and Sustainable Energy Reviews, vol. 27, Nov 2013, pp.1-10.

[2] Atul Sharma, Jaya Srivastava, Sanjay Kumar Kar and Anil Kumar, "Wind energy status in India: A short review," Renewable and Sustainable Energy Reviews, vol.16, no.2, Feb. 2012, pp.1157-1164.

[3] Suman Nath and Somnath Rana, "The modeling and simulation of wind energy based power system using Matlab," Int. J. of Power System and Energy Mang., vol. 1, no. 2, 2011, pp. 12-19.

[4] L. G. Gonzalez, E. Figueres, G. Garcera and O. Carranza, "Modelling and control in wind energy systems (WECS)," 13th European Conf. on Power Electron. and Appl., Barcelona, Sept. 8-10, 2009, pp. 1-9.

[5] Hamidreza Jafarnejadsani, Jeff Jafarnejadsani, Jeff Piper and Julian Ehlers, "Adaptive control of variable-speed variable-pitch wind turbine using radial-basis function neural network," IEEE Trans. on Control Systems Technology, vol. 21, no. 6, Nov. 2013, pp. 2264-2272.

[6] Rishab Dev Shukla and R. K. Tripathi, "Maximum power extraction schemes & power control in wind energy conversion system," Int. J. of Scientific & Eng. Research, vol. 3, no. 6, June 2012, pp. 1-7.

[7] Majid A. Abdullah, A.H.M. Yatim and Chee Wei Tan, "A Study of Maximum Power Point Tracking algorithms for wind energy systems," in Proc. IEEE 1st Conf. on Clean Energy and Technology, Kuala Lumpur, June 27-29, 2011, pp. 321-326.

[8] Athanasios Mesemanolis, Christos Mademlis and Irodanis Kioskeridis, "Optimal efficiency control strategy in wind energy conversion system with induction generator," IEEE J. of Emerging and Selected Topics in Power Electron., vol. 1, no. 4, Dec 2013, pp. 238-246.

[9] H. Camblong, I. Martinez de Alegria, M. Rodriguez and G. Abad, "Experimental evalution of wind turbine maximum power point tracking controllers," Energy Conversion and Mang., vol. 17, no. 18-19, Nov. 2006, pp. 2846-2858.

[10] A. Kerrouche, A. Mezouar and L. Boumedien, "A simple and efficient maximized power control of DFIG Variable speed wind turbine," in Proc. of the 3rd Int. Conf.on Systems and Control, Algiers, Oct. 29-31, 2013, pp. 894-899.

[11] Yanping Liu, Hongmei Guo, Huajun Wang and Jianjian He, "The estimation of pitch angle in pitch-controlled wind turbine," in Proc. Int. Conf. on Electrical Machines and Systems ,Wuhan, Oct. 17-20, 2008, pp. 4188-4191.

[12] Xibo Yuan and Yongdong Li, "Control of variable pitch and variable speed direct-drive wind turbines in weaks grid systems with active power balance," IET Renewable Power Generation, vol. 8, no. 2, March 2014, pp. 119-131.

Wind turbine economics: A Study

Sahil Bajaj
School of renewable energy and efficiency
NIT Kurukshetra,
Haryana, India
Sahilbajaj2910@gmail.com

K. S. Sandhu
Department of Electrical Engineering
NIT Kurukshetra,
Haryana, India
kjssandhu@rediffmail.com

Abstract— **Due to limited amount of fossil fuels and for the production of eco-friendly energy, it is necessary to switch to renewable energy systems. Wind energy is considered to be one of the most effective and efficient source as compared to other renewable energy sources. In this paper an attempt has been made to estimate the total cost and weight of wind turbine. Model as prepared using Matlab/Simulink may be helpful to decide the rating of wind turbine on the basis of its cost and weight.**

Keywords— *Cost estimation, renewable energy, Matlab/Simulink, Wind energy, Wind turbine.*

I. INTRODUCTION

Generation of energy mainly follows the use of conventional sources like coal, petroleum etc. but these sources are depleting with time, moreover use of these sources also leads to environmental effects like climate change, global warming to some extent. with increase in population and considering the environmental changes there is a huge need to shift to renewable energy sources. there are many renewable energy sources that can be exploited efficiently and effectively, these include solar energy, wind energy, tidal energy etc. as per the statistics available due to many reasons wind energy seems to be effective in contrast to others. the contribution of wind power to the energy supply has reached a substantial share even on the global level. total power generated worldwide by the use of wind energy conversion system is around 282275 mw. on the other hand india is considered to be the 5th largest country in producing energy from wind with total installed capacity of 18551mw. there are many factors affecting the cost of energy produced using a wind farm. no doubt there is negligible fuel cost associated with wind energy but installation of wind turbine requires great capital cost and in addition to that also the operational and maintenance cost. wind power plants appear to be a viable alternative in contrast to other renewable energy resources due to clean and never ending fuel. generating electricity from wind makes economic as well as environmental sense; the wind is a free, clean and renewable fuel which will never run out. even though wind is free its cost of electricity however, is not free. there are initial capital costs of purchasing wind turbines, towers, transportation of materials, labor charge, expertise charge, operational and maintenance cost etc. wind turbines are becoming cheaper and more powerful, with larger blade lengths which can utilize more wind and therefore produce more electricity, bringing down the cost of renewable power generation.

Many researchers tried to explore wind energy economics. Milborrow[2] studied the economics of wind power and its comparison with conventional thermal plant. It includes the expected prices and of wind energy up to year 2000. In 1997, national wind coordinating committee published a paper which reviewed the measures of cost used for electricity-generating systems employing wind turbines. Peter fugslang and Kenneth thomson[10] tried to optimized the cost of wind turbines for large scale offshore wind farms. Optimizations were carried out to find out the optimum overall wind turbine design. They also analyzed that wind turbines for offshore wind farms should be different compared with a standalone system. Hani[3] described several optimization models for the design of a typical wind turbine tower structure by considering cross sectional area radius of gyration and height as effective design variable. Neils[8] studied the development of methods and guidelines for the use of wind energy in isolated communities and presented an approach in order to support a fair assessment of the technical and economical feasibility of wind energy. In 2008, Maria[4] observed the capital cost and the variable cost associated with generation of wind energy. In the same year Sangmesh[5] presented a new methodology to select a wind turbine generator from the view point of performance and economic consideration. Many new technologies as emerged to lower the wind energy were examined by Hoffman[13]. Dai[1] studied the wind energy resources assessment in wind power generation and described various wind energy resources assessment methods. Stannard[9] found energy yield and cost analysis of small scale wind turbines. A further attempt was made by Thomas[12] to analyze the development in large blades for lowering cost of wind turbines. IRENA issued a cost analysis series focusing mainly on cost associated with renewable energy technology. Eric[6] presented a technical report that identified and summarized the past and future cost of wind energy. John[14]formulated the design of an integral wind turbine tower and foundation system as a multiobjective optimization problem using the process automation and design exploration software insight.

In the present work an attempt has been made to estimate weight and cost of wind turbine for squirrel cage and doubly fed induction generator configuration using MATLAB/SIMULINK.

978-1-4799-6047-7/14 $31.00 © 2014 IEEE

II. Economics Of Wind Turbine

As shown in figure. 1. Cost of wind power plant may be distributed among the major works as given below.

- Planning and project costs (9%)
- Turbine cost (64%)
- Civil works (16%)
- Grid connections (11%)
- Others (1%)

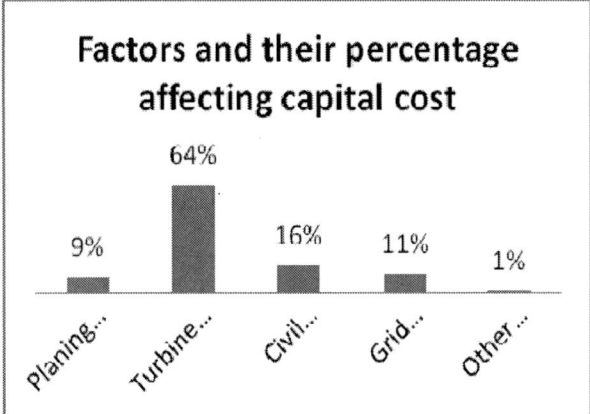

Fig. 1.   Factors and their percentage affecting capital cost

On the basis of facts as shown in figure 1, cost of wind turbine, which accounts for the major share, needs analysis to derive the total cost of the plant. A typical wind turbine may consist of number of constituting parts and cost of its major components are as shown in figure 2.

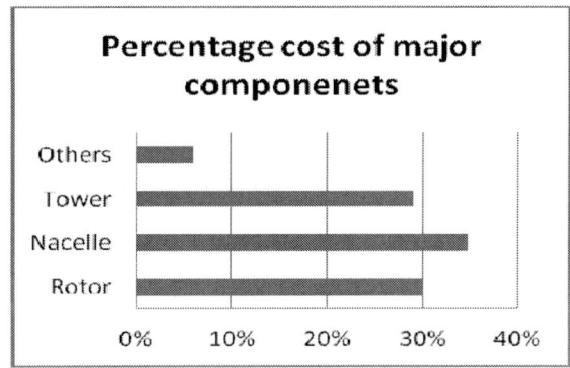

Fig. 2.   Main components of wind turbine.

From above figure, it is observed that cost of nacelle, cost of rotor and cost of tower accounts 94% of the total cost of the wind turbine. This is dependent upon the rating and weight of the complete structure. In the next section an attempt has been made to estimate the cost of wind turbine with a focus on generator, rotor, blades and tower using Matlab/Simulink modeling. Simulated results as obtained using the model may be used to decide the number of wind turbines for economical installation with a given rating of wind power plant.

III.   Modelling

Simulink model for cost estimation of wind turbine with doubly fed induction generator (DFIG) and squirrel cage induction generator (SCIG) for given power ratings has been developed as shown in figure 3 and figure 4. Equation (1) to (4) may be used to develop MATLAB/Simulink model as shown in figure 3 and figure 4.

For DFIG,

$$W_{gd} = 66.92 - 0.03154p + 0.00001341p^2 \qquad (1)$$

For SCIG,

$$W_{gs} = 12.65 + 0.03327p \qquad (2)$$

For rotor blades,

$$W_r = 0.81 + 0.003210p \qquad (3)$$

For tower,

$$W_t = -136.42 + 0.2278p \qquad (4)$$

Where, W_{gd} = Weight of DFIG(tons).
W_{gs} = Weight of SCIG(tons).
W_r = Weight of rotor blades(tons).
W_t = Weight if tower (tons).
p   = rating of wind turbine (Kw).

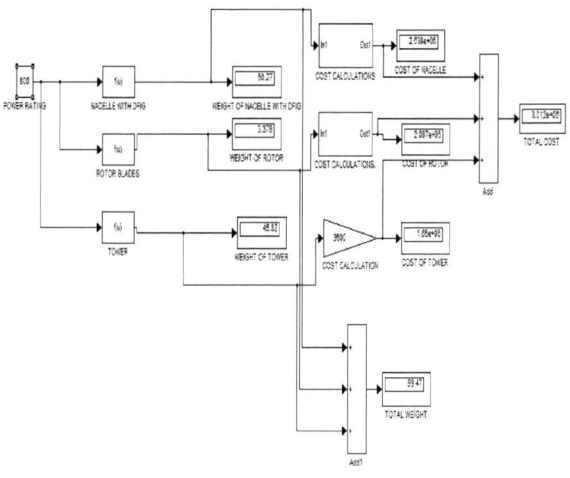

Fig. 3. Simulink model for wind turbine with DFIG

TABLE II.   Cost estimation of wind turbine using DFIG configuration.

Power rating (kW)	Cost of nacelle	Cost of rotor blades	Cost of tower	Total cost (rupees)
800	2638000	208700	165000	3012000
1600	2665000	367400	821000	3854000
2400	3593000	526100	1477000	5597000
3200	5422000	684800	2133000	8240000
4000	8152000	843400	2789000	11780000
4800	11780000	1002000	3445000	16230000

TABLE III.   Weight estimation of wind turbine using SCIG configuration

Power rating (kW)	Weight of nacelle	Weight of rotor blades	Weight of tower	Total weight (tons)
800	35.89	3.378	45.82	85.09
1600	59.94	5.946	228.1	293.9
2400	83.98	8.514	410.3	502.8
3200	108	11.08	592.5	711.7
4000	132.1	13.65	774.8	920
4800	156.1	16.22	957	1129

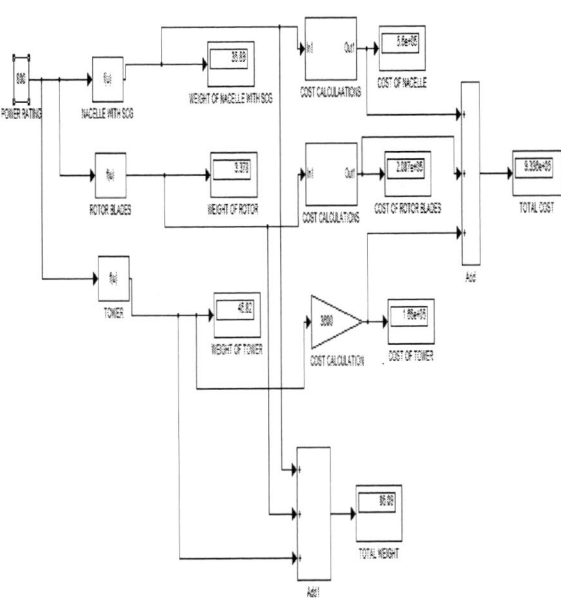

Fig. 4. Simulink model for wind turbine with SCIG

TABLE IV.   Cost estimation of wind turbine using SCIG configuration

Power rating (kW)	Cost of nacelle	Cost of rotor blades	Cost of tower	Total cost (rupees)
800	560000	208700	165000	933700
1600	935200	367400	821000	2124000
2400	1310000	526100	1477000	3314000
3200	1686000	684800	2133000	4504000
4000	2061000	843400	2789000	5690000
4800	2436000	1002000	3445000	6883000

IV. Economics Of Wind Turbine

Simulated results as obtained using MATLAB/Simulink circuits are shown in table 1-4. Based upon the results shown in tables, figure 5-8 are plotted for power rating Vs total cost and total weight for two types of generators i.e. SCIG and DFIG respectively.

TABLE I.   Weight estimation of Wind Turbine using DFIG configuration.

Power rating (kW)	Weight of nacelle	Weight of rotor blades	Weight of tower	Total weight (tons)
800	50.27	3.378	45.82	99.47
1600	20.79	5.946	228.1	284.8
2400	68.47	8.514	410.3	487.3
3200	103.3	11.08	592.5	706.9
4000	155.3	13.65	774.8	943
4800	224.5	16.22	957	1198

978-1-4799-6047-7/14 $31.00 © 2014 IEEE

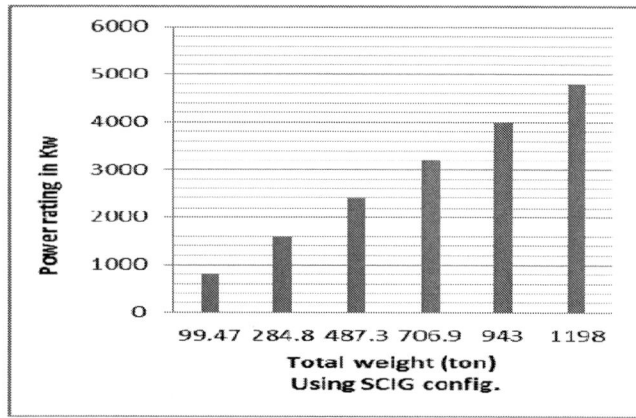

Fig. 3. Power rating Vs total weight (SCIG configuration)

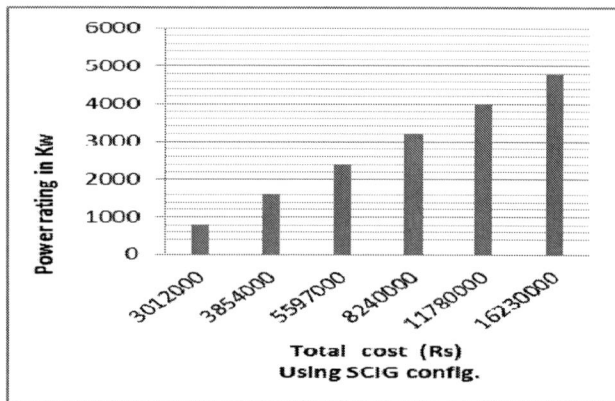

Fig. 4. Power rating Vs total cost(SCIG configuration)

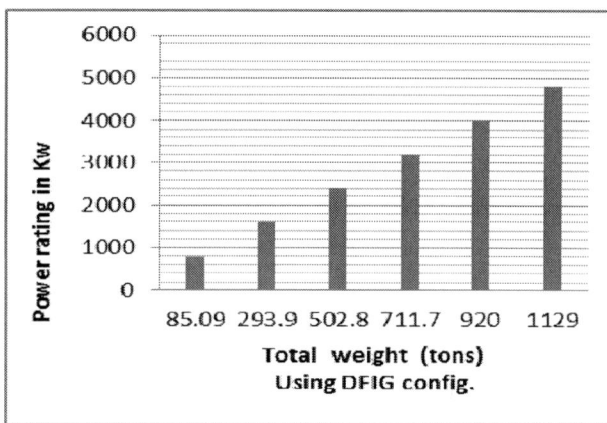

Fig. 5. Power rating Vs total weight (DFIG configuration)

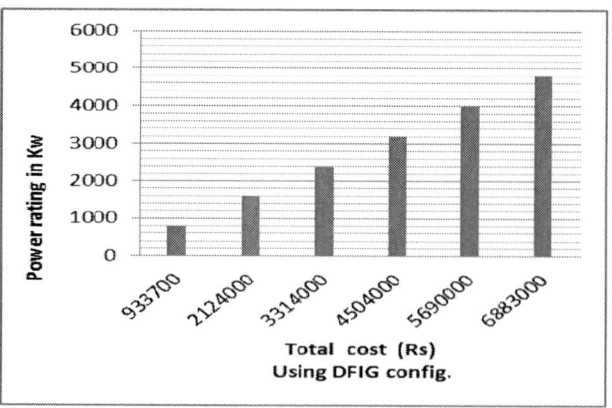

Fig. 6. Power rating Vs total cost(DFIG configuration)

From the above table and figures it is observed that for any given rating of power plant;

i. Weight of wind turbine with SCIG appears to be low in contrast of DFIG if the rating is below 4000 kW.
ii. Cost of wind turbine with SCIG is always less than that with DFIG irrespective of its rating.
iii. It is also observed that installation of n units of SCIG may be cheaper as compared to a single unit.

V. CONCLUSIONS

In this paper an attempt has been made to estimate the total cost and weight of a wind turbine using MATLAB/Simulink circuit. It is observed that wind turbine with DFIG, although is light in weight but costs more in contrast to SCIG. Simulated results as obtained may be useful to decide the rating of wind turbine based upon the cost and weight.

REFERENCES

[1] Dai ting ,"study on the wind energy resource assessment in wind power generation", IEEE 978-1-4577- 0536-6/11/, 2008.

[2] David Milborrow "Economics of Wind Power and Comparisons with Conventional Thermal Plant", Elsevier Science ltd, Renewable Energy, vol 5, part 1, pp692- 699, 1994.

[3] Hani m negam "structural design optimization of wind turbine towers", Pergamon, Computers andstructures,2000,649-666.

[4] Maria Isabel Blanco "Renewable and Sustainable Energy" Reviews 13 (2009), pp. 1372– 1382

[5] Sangmesh S Soddamani ,"economic index for selection of wind turbine generator at site", IEEE 978-1-4244-1888 -6/08, 2008

[6] Eric lanz "The Past and Future Cost of Wind Energy" IEA WindTask 26, 2012

[7] National wind coordinating Committee "Wind Energy Series", no.11, January 1997.

[8] Niels-Erik Clausen, Henrik Bindner, Sten Frandsen, Jens Carsten hansen, Lars Henrik Hansen, Per Lundsager "isolated systems with wind power, an implementation guide"riso national laboratory, Roskilde, june 2001

[9] N J stanard "Energy Yield and Cost Analysis of Small Scale Wind Turbines" Durban University, UK, 2006.

[10] Peter fugslang and Kenneth Thomson, "cost optimization of wind turbines for large scale offshore wind farms", Riso national laboratory, Roskilde, February 1998.

[11] Renewable energy technology, "Cost Analysis Series" IRENA, Volume: Power Sector, Issue 5/5, 2012.

[12] Thomas,D,ashwil, "Development in Large Blades for Lower Cost of Wind Turbines" Sandia National Laboratories .

[13] D.L Hoffman and T.S molinski "How New Technology Developments will Lower Wind Energy Costs, CIGRÉ Canada,Symposium, Calgary, July 29th – July 31st, 2009 .

[14] John c Nichols ,"Multiobjective structural optimization of wind turbine tower and foundation system using insight", 10th world congress on structural and multidisciplinary optimization, Orlando,florida, USA, may 19- 24, 2013

Grid Voltage Monitoring Techniques for Single Phase Grid Connected Solar PV system

Lakshmanan.S.A
School of Computing and Electrical Engg
Indian Institute of Technology Mandi
lakshmanan_s_a@students.iitmandi.ac.in

Amit Jain
Power Systems Division
CPRI
amitjain@cpri.in

B. S. Rajpurohit
School of Computing and Electrical Engg
Indian Institute of Technology Mandi
bsr@iitmandi.ac.in

Abstract—**Power produced from the Renewable Energy Sources (RES) is directly feeding to the electric grid via Power Electronic Circuits (PEC). Grid connected solar Photovoltaic (PV) system requires a high quality of grid monitoring techniques in order to detect the phase angle, voltage magnitude and frequency and also to meet standard specifications in terms of achieving power quality and safety. Various techniques are used to detect phase angle, amplitude and frequency of the grid connected solar PV system. This paper proposes a review of important voltage monitoring techniques for single phase grid connected solar PV system and it also proposes a design of Adaptive Notch Phase Lock Loop (ANPLL) circuit for single phase grid connected solar PV system under both ideal and non-ideal grid conditions. Here the Loop Filter (LF) of the PLL circuit is a Proportional Integral Derivative (PID) type and the proposed PLL circuit is implemented using Matlab/Simulink platform and simulation results are obtained for both ideal and non-ideal grid conditions.**

Keywords—power electronic circuits, phase lock loop, renewable energy sources

I. INTRODUCTION

Distributed Generation (DG) systems require a high degree of grid condition detection in order to meet standard specifications in terms of power quality and safety. An accurate and fast detection of the phase angle, amplitude and frequency of the utility voltage is required by the grid-connected converters in order to guarantee the correct generation of the reference signals and to meet the demands regarding the operation boundaries with respect to voltage amplitude and frequency values required by standards. The standard IEC61727-2002 [1] is given as an example in this regard. This standard applies to the utility-interconnected PV power systems utilizing static (solid-state) non-islanding inverters for the conversion of DC to AC. According to IEC61727-2002 [1], the boundaries of operation with respect to grid voltage amplitude and frequency for continuous operation area between 0.85 and 1.10 *p.u* and ± 1 Hz around the nominal frequency is defined. Abnormal conditions can arise on the utility grid which requires a response from the grid-connected PV system. This response is to ensure the safety of utility maintenance personnel and the general public,as well as to avoid damage to connected equipment, including the PV system. The grid-connected PV system has to cease to energize the utility line within the specified time interval if the voltage amplitude or frequency exceeds the predefined limits. The most restrictive requirement is when the maximum trip time is 0.05 seconds for a grid voltage amplitude excursion above 1.35 *p.u*. An accurate and fast grid voltage monitoring algorithm is required in order to comply with these requirements. The principle of the grid voltage monitoring, which consists in getting the parameters of the voltage at the PCC, is presented in Fig. 1.

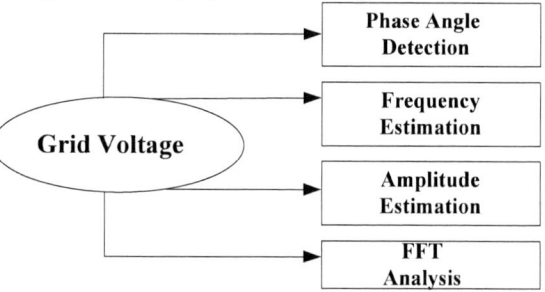

Fig. 1. Grid Voltage Monitoring Principle

The voltage equation given (1) is divided in two main parts: the fundamental and the harmonics.

$$v = Vsin\omega t + Vsin(\omega t + \theta) \qquad (1)$$

The grid phase angle ($\omega.t$) is mostly used for synchronization. Additionally, the detection of the grid phase angle can also be used for anti-islanding detection algorithms [2]. The angular frequency of the grid voltage (ω) is usedfor over and under frequency detection algorithms but also to provide information to the control System (such as resonant controllers or filters which need to adjust their resonance frequency). The amplitude of the grid voltage is required for over and under voltage and to provide information to the control system (such as power feed forward loop). Additional information such as harmonic content of the grid voltage can be required for some algorithms e.g. harmonics monitoring for the passive anti-islanding methods [2] or active power filters applications [3].

In this paper various grid voltage monitoring techniques for single phase solar PV system is explained and also design of PLL circuit has been expressed under both ideal and non-ideal grid conditions. The LF for PLL circuit is PID type and design of control parameters for PID type PLL are derived and finally the proposed model is simulated using Matlab/Simulink platform and results are obtained to validate the performance of the proposed technique.

The paper has been organized as follows. In section II overview of the grid voltage monitoring techniques were

978-1-4799-6047-7/14 $31.00 © 2014 IEEE

explained and design of PLL circuit and PID type LF was explained in section III and in section IV simulation results are given and last section ended with conclusion.

II. OVERVIEW OF GRID VOLTAGE MONITORING TECHNIQUES

Numerous methods using different techniques for monitoring the grid voltage are presented in the technical literature about DG systems. However, some of the methods are not always categorized properly, thus leading to confusion. Therefore, to bring clarity in this regard, the techniques used for monitoring the grid voltage can be organized and it is shown in Fig 2.

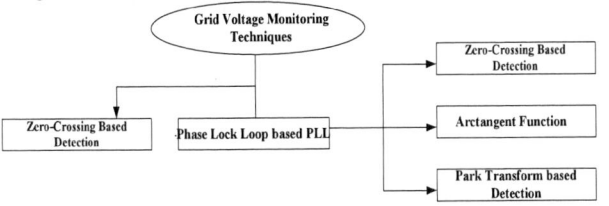

Fig. 2 Overview of Grid Voltage Monitoring Techniques

The grid voltage monitoring techniques can be split into two main categories, namely methods based on Zero-Crossing Detection (ZCD) which does not involve a phase controller and methods based on PLL which includes a phase controller. Furthermore, the methods based on a PLL can also be split into three other categories, namely PLL based on the ZCD, on the arctangent function (tan^{-1}) and on the Park Transform. All these techniques are presented in more details in the coming sections [4].

A. Grid Voltage Monitoring based on ZCD

A simple method of obtaining the phase and frequency information is to detect the zero-crossing point of the grid voltage [5]. The method, as presented in Fig. 3 is based on counting the zero-crossings of the grid voltage, thus the frequency of the fundamental is estimated. Then, the phase-angle of the grid voltage is obtained by integrating the estimated frequency and no phase controller is involved.

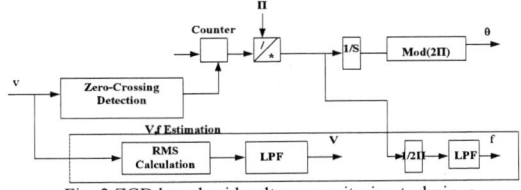

Fig. 3 ZCD based grid voltage monitoring technique

This method has two major drawbacks. Since the zero-crossing point can be detected only at every half cycle of the utility frequency, the phase tracking action is impossible between the detecting points and thus a fast dynamic performance cannot be obtained [6]. Some work has been done in order to alleviate this problem using multiple level crossing detection as presented in [6]. However, this method increases the complexity of the ZCD technique.

Significant line voltage distortion due to notches caused by power device switching and or low frequency harmonic content can easily corrupt the output of a conventional ZCD [7]. Therefore, the zero-crossing detection of the grid voltage needs to obtain its fundamental component at the line

frequency. This task is usually made by a filter. To avoid the delay introduced by this filter, numerous techniques are used in the technical literature. Methods based on advanced filtering techniques are presented in [8].Other methods use neural networks for detection of the true zero-crossing of the grid voltage waveform [8]. Furthermore, an improved accuracy in the integrity of the zero-crossing can also be obtained by reconstructing a voltage representing the grid voltage [9].

B. Grid Voltage Monitoring based on PLL

Recently, there has been an increasing interest in PLL techniques for grid-connected converter systems [10]. Typically, the PLL technique is applied in communication technologies. Though, it has been proven that its application to the grid-connected converter systems was a success [11]. Used for such systems the PLL is a grid voltage phase detection algorithm. The main task of the PLL algorithm is to provide the phase angle of the grid voltage which is mostly used to synchronize the output current of the converter with the voltage at the PCC for the cases when no reactive power is required. Moreover, using the PLL algorithm, the grid voltage parameters such as amplitude and frequency, can be easily monitored [11].

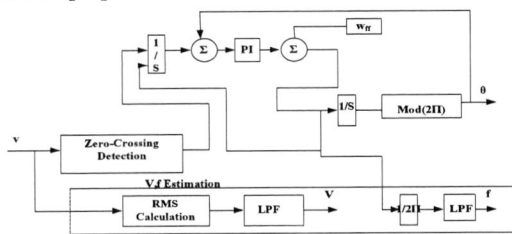

Fig. 4 PLL based grid voltage monitoring technique

The PLL is composed of a phase comparator which provides an error signal (e) to a PI controller, as it can be seen in Fig 5. An initial frequency value (ω_{ic}) is then added to the output of the PI controller resulting the estimated frequency (ω) of the grid voltage. The advantage of adding ω_{ic} to the output of the PI controller is a better dynamic performance every time the PLL is reset. Then, the estimated phase-angle (θ) of the grid voltage is obtained by integrating the estimated frequency. To provide the phase-angle reference to the phase comparator, the following three techniques can be used: the ZCD, the arctangent function (tan^{-1}) or the Park Transform.

C. ZCD Based PLL

The ZCD technique can also be used to provide the phase-angle reference for a PLL [12] and it is shown in Fig. 4. The integrator is reset to 0 every time the positive slope zero crossing is reached and to Π every time the negative slope zero-crossing is reached, thus providing the phase-angle reference for the PLL.

D. Arctangent Based PLL

The arctangent function technique is another solution for detecting the phase-angle and the frequency of the grid voltage [13]. An orthogonal voltage system is required to implement this technique. This method is used in motor drives

978-1-4799-6047-7/14 $31.00 © 2014 IEEE

applications to transform the feedback signals to a reference frame suitable for control purposes [14]. However, this method has the drawback that it requires advanced filters to obtain an accurate detection of the phase angle and frequency in the case of distorted grid voltage [14]. Even though, the method using the arctangent function is easy to understand, it has implementation difficulties related with the avoidance of divisions by zero. Therefore, this technique is not well suited for grid-connected converter applications. The arctangent based PLL for single phase grid connected solar PV system is shown in Fig. 5.

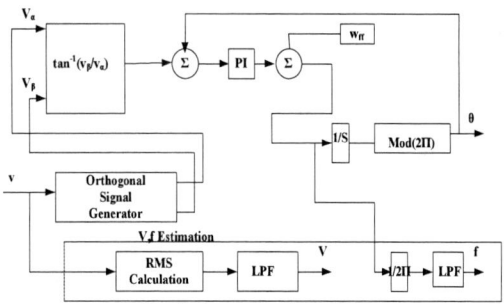

Fig. 5. Arctangent based PLL

E. Park Transform Based PLL

The most widely used PLL is based on the Park Transform, as presented in Fig. 6. Like in the case of the arctangent function technique, an orthogonal voltage system is required for the PLL based on the Park Transform. The input of the PI controller is defined by V_q. The result is not the difference between the angles but it depends on it. The grid voltage (v) is considered normalized [15-16].

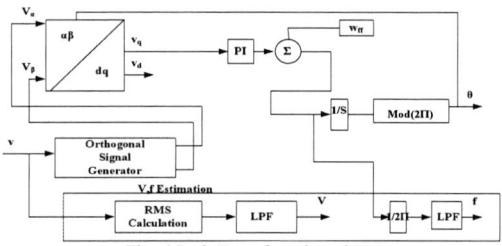

Fig. 6 Park Transform based PLL

III. DESIGN OF PLL CIRCUIT

The conventional PLL circuit for single phase system is shown in Fig. 7. This PLL circuit basically consists of three parts namely Phase Detector (PD), Loop Filter (LF) and Voltage Controlled Oscillator (VCO). Output voltage of the grid is expressed as

$$v_{grid} = V sin\theta \qquad (2)$$

and ω is the grid frequency then the grid voltage is given as

$$v_{grid} = V sin(\omega t + \theta) \qquad (3)$$

In PLL circuit, PD is a circuit which compares measured grid voltage and voltage produced by PLL circuit and finally produces the error signal. The output voltage of VCO is given in ()

$$V_{vco} = \cos(\theta_{vco}) = \cos(\omega_{vco}t + \theta_{vco}) \qquad (4)$$

Finally output of PD is given in ()

$$v_{PD} = \frac{KV}{2}[\sin(\omega - \omega_{vco})t + (\theta - \theta_{vco})] + \sin(\omega + \omega_{vco})t + (\theta + \theta_{vco})] \qquad (5)$$

So the PD block having the information about locking error and this locking error is a non-linear and having the component which is double of grid frequency. In order to achieve reliable grid synchronization, double of grid frequency component has been removed. Now the output of PD is expressed as [18]

$$v_{PD} = \frac{KV}{2}[\sin(\omega - \omega_{vco})t + (\theta - \theta_{vco})] \qquad (6)$$

In the PLL circuit, PI controller is a LF and it has low-pass filter characteristics which are used to filter out high frequency components. Grid connected inverter applications;the grid frequency is very low. In this case the result provided by PI controller is not good and high frequency component is introduced into the LF block which affects the performance of PLL. So PI controller based LF is not useful to remove double of grid frequency components. In this paper Adaptive Notch Filter (ANF) is used to linearize the PD output and also used to double of grid frequency component.

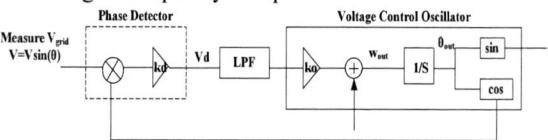

Fig. 7 Conventional Single Phase PLL

A. Adaptive Notch Filter based PLL

In order to remove double of grid frequency components, a notch filter is used at the output of PD block and the adaptive notch filter based PLL [18] is shown in Fig. 8. If any variation in the grid frequency, ANF is used notch the exact frequency. A notch filter is domain is expressed as

$$H(s) = \frac{S^2 + 2\zeta_a\omega_n s + \omega_n^2}{S^2 + 2\zeta_b\omega_n s + \omega_n^2} \qquad (7)$$

Discretizing the equation, the notch filter is expressed as

$$H(z) = \frac{z^2 + (2\zeta_a\omega_n T - 2)z + (-2\zeta_a\omega_n T + \omega_n^2 T^2 + 1)}{z^2 + (2\zeta_b\omega_n T - 2)z + (-2\zeta_b\omega_n T + \omega_n^2 T^2 + 1)} \qquad (8)$$

The frequency response of the notch filter is shown in Fig. 9 for 50Hz grid frequency [18].

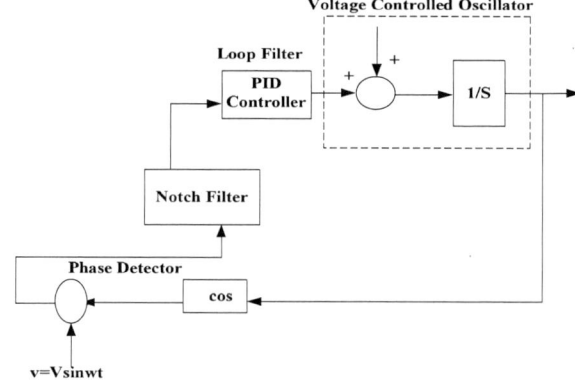

Fig. 8 Notch Filter based Single Phase PLL

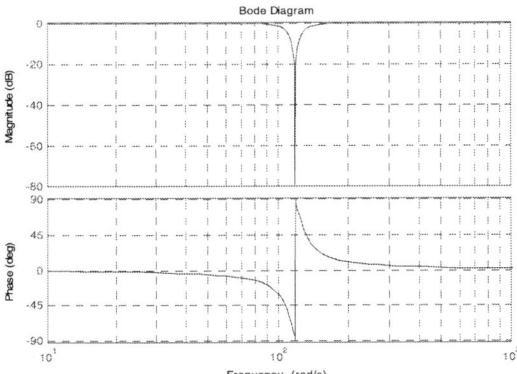

Fig. 9. Frequency response of Notch Filter

B. LF design based on PID controller

This section provides design of control parameters of a PID type LF for the given PLL circuit. Generally transfer function of the LF based on PID is expressed as

$$T(s) = k_p \frac{1 + \tau_a s}{\tau_a s} \frac{1 + \tau_d S}{1 + \beta \tau_d S} \tag{9}$$

Here k_p is the proportional gain, τ_a and τ_b are integral and derivative time constants. To remove the derivative action of PID type, the term $(1+\beta\tau ds)$ is introduced which produce high frequency pole and cancel the derivative action. β is referred to as the derivative filter factor and the value of β is 0.1.

The open loop transfer function of the PID type LF is expressed as

$$G_{open}^{PID} = Vk_p \frac{1 + \tau_a s}{\tau_a s} \frac{1 + \tau_d S}{1 + \beta \tau_d S} \frac{1}{S} \tag{10}$$

Derivative filter effect which corresponds to high frequency pole is having less effect on PLL dynamics, so

$$G_{open}^{PID} = \frac{1}{S} Vk_p \frac{1 + \tau_a s}{\tau_a s} \tag{11}$$

Finally closed loop transfer function of PID type PLL is expressed as

$$G_{closed}^{PLL} = \frac{Vk_p s + V\frac{k_p}{\tau_a}}{S^2 + VK_P s + V\frac{k_p}{\tau_a}} \tag{12}$$

The closed loop transfer function of PID type PLL is depending upon the selection of damping ratio and natural frequency. Most of the PLL applications, fast dynamic response is achieved by selecting high value of natural frequency. But at the same time, selecting high value of natural frequency affects the stability of the PLL system. So the natural frequency should be chosen carefully. Symmetrical Optimum (SO) method is used design the control parameters of PID type LF. Based on SO method and comparing open loop transfer function, the control parameters are calculated as

$$k_p = \frac{2\zeta\omega_n}{V} \tag{13}$$

$$\tau_d = \frac{T_\omega}{2} \tag{14}$$

$$\tau_a = \frac{2\zeta}{\omega_n} \tag{15}$$

Frequency response of the closed loop transfer function with PID type LF isshown in Fig. 10.

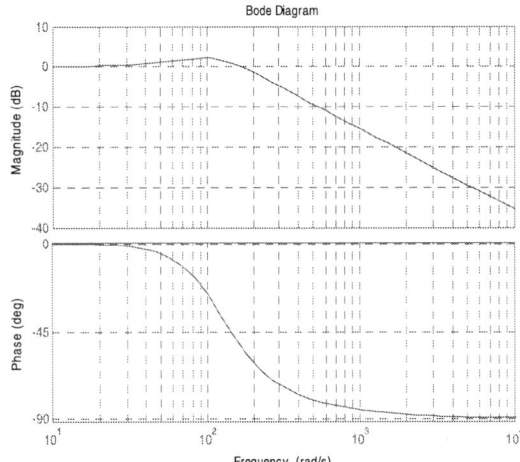

Fig. 10 Frequency response of closed loop transfer function of PID-type LF

IV. SIMULATION RESULTS AND DISCUSSION

The proposed PLL system is simulated using Matlab/Simulink for single phase grid connected solar PV system. Grid voltage is considered as 320V, and in the simulation results the values are considered 1 p.u. and the sampling frequency is set 2kHz sampling time Ts=0.5ms and natural frequency ωn=119 and damping ratio ζ=0.7. Using SO method, the parameters of the proposed PID type LF are calculated and given as τa=0.0117, τb=0.005 and kp=166. The PLL system is simulated for ideal grid condition. The results show that, the PLL system with PID regulator is able to track the phase angle clearly. From the phase angle, the sine wave is locked with reference signal. Output of PLL under ideal grid condition with PID type loop filter is shown in Fig. 11.

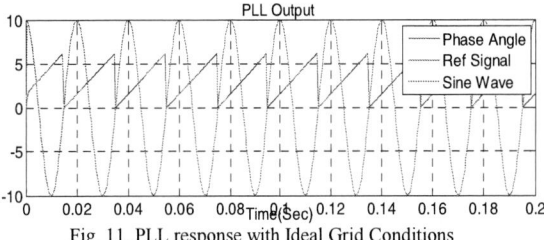

Fig. 11. PLL response with Ideal Grid Conditions

A. Non-Ideal Grid Conditions

Different non-ideal grid conditions are used to analyze the proposed PID-type PLL circuit. First input signal with Phase jump is given to the PLL circuit and corresponding PLL output and its tracking error and phase angle is shown in Fig. 12. Introduction of Harmonics to the input signal is another non ideal grid condition and the PLL output, it's tracking error and phase angle is shown in Fig 13. Input signal frequency is varied and the PLL output, its tracking error and phase angle is shown in Fig. 14. Finally amplitude of the input signal is varied and given to PLL and the PLL output, error and phase angle is shown in Fig. 15.

978-1-4799-6047-7/14 $31.00 © 2014 IEEE

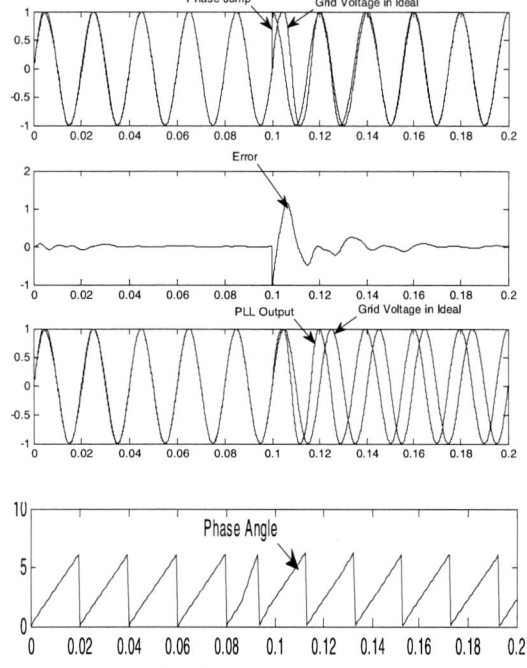

Fig. 12 PLL response with Phase Jump a) Input signal b) tracking error c) PLL output d) Phase Angle

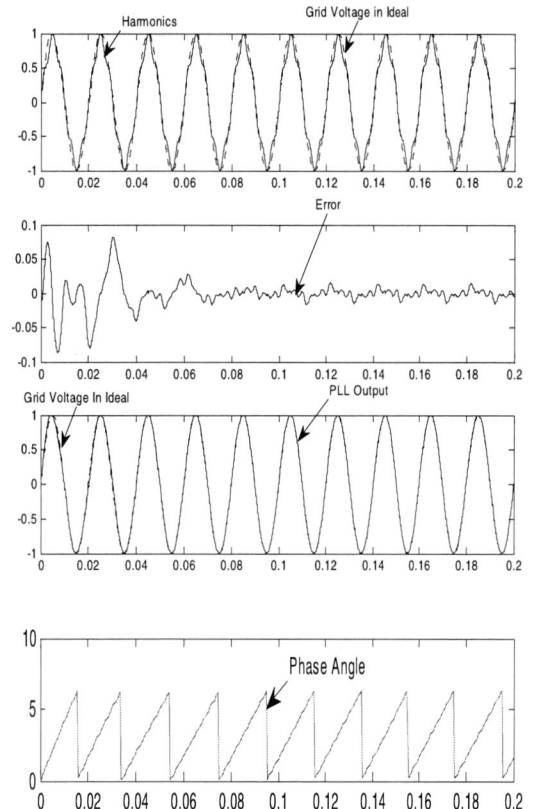

Fig. 13 PLL response with Harmonic Input Signal a) Input Signal b) error c) PLL output d) Phase Angle

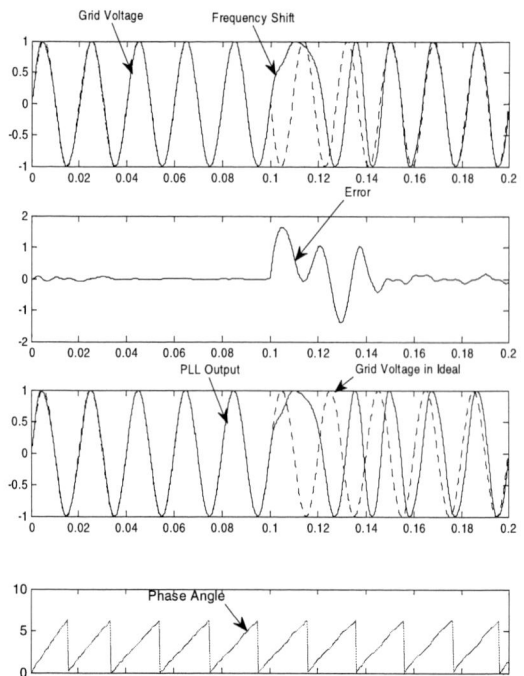

Fig. 14 PLL response with Frequency Variations a) Input signal b) tracking error c) PLL output d) Phase Angle

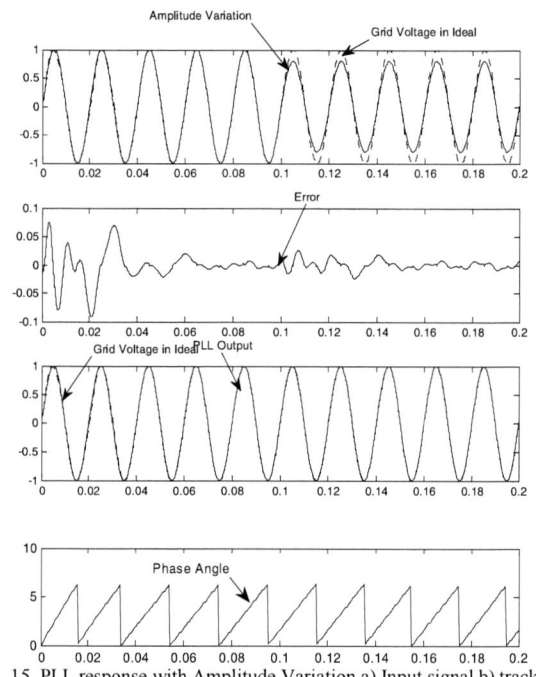

Fig. 15 PLL response with Amplitude Variation a) Input signal b) tracking error c) PLL output d) Phase Angle

V. CONCLUSIONS

This paper has presented a grid voltage monitoring techniques for single phase grid connected solar PV system. First basic principle of grid voltage monitoring techniques has been explained, then ZCD based control technique, PLL based techniques are reviewed. In the case of PLL based technique, Park transform based method has explained. Adaptive based PLL system is designed and basic concepts need to be following when designing PLL circuit is analyzed. The LF filter of the proposed PLL circuit is designed by using PID controller. The system is simulated using Matlab/Simulink platform by considering ideal and non-ideal grid conditions. In the case of ideal grid conditions, the PLL was traced fast and accurate. The PLL system has been simulated for various non-ideal grid conditions like amplitude variation, harmonics, and phase jump and frequency unbalance and their simulation results of PLL response, error and phase angle are shown.

ACKNOWLEDGMENT

The authors would like to acknowledge the support provided by the project under DST-SERB and also would like to acknowledge the support provided by the Central Power Research Institute.

REFERENCES

[1] Characteristics of the Utility Interface for Photovoltaic (PV) Systems, "iec 61727-2002, 2002.

[2] M. Franceso De, L.Marco, D. A. Antonio and P. Alberto," Overview of anti-islanding algorithms for PV systems, in Proc. Of PEMCC, 2006, pp. 1878-1883.

[3] L. Asiminoeai, F. Blaabjerg, and S. Hansen, " Detection is key-Harmonic detection methods for active power filter applications." Industry Applications Magazine, IEEE, vol. 13, pp. 22 2007.

[4] Reliable Grid Condition Detection and Control of Single phase Distributed Power Generation System, PhD Thesis, Aalborg University, 2009.

[5] F. Mur, V. Cardens, J. Vaquero and S. Martinez, " Phase Synchronization and Measurement digital system of AC mains for Power converters", in proc of CIEP, 1998.

[6] S.K. Chung, " A phase tracking system for three phase uyility interface inverters", IEEE Transitions on Power Electronics, vol.15, pp. 431-438, 2000.

[7] B. P. McGrath, D. G. Holmes, and J. J. H. Galloway, "Power converter line synchronization using a discrete Fourier transform (DFT) based on a variable sample rate," Power Electronics, IEEE Transactions on, vol. 20, pp. 877-884, 2005.

[8] S. Valiviita, S. J. Ovaska, and J. Kyyra, Adaptive signal processing system foraccurate zero-rossing detection of cycloconverter phase currents," in Proc. OfPCC, 1997, pp. 467-472, vol.1.

[9] O. Vainio and S. J. Ovaska, "Noise reduction in zero crossing detection by predictive digital filtering," Industrial Electronics, IEEE Transactions on, vol. 42, pp. 58-62, 1995.

[10] A. Luna, J. Rocabert, G. Vazquez, P. Rodriguez, R. Teodorescu, and F. Corcoles, "Grid synchronization for advanced power processing and FACTs in wind power systems", Proc. IEEEISIE 1, 2915–2920 (2010).

[11] S. M. Silva, B. M. Lopes, B. J. C. Filho, R. P.Campana, and W. C. Bosventura,"Performance evaluation of PLL algorithms for single-phase grid-connected systems,"in Proc. of IAS, 2004, pp. 2259-2263, vol. 4.

[12] X. Yuan, W. Merk, H. Stemmler, and J. Allmeling, "Stationaryframe generalized integrators for current control of active power filters with zero steady-state error for current harmonics of concern under unbalanced and distorted operating conditions", IEEE Trans. on Indust. Applications 38 (2), 2143–2150 (2002).

[13] X. Fang, Y. Wang, M. Li, K. Wang, and W. Lei, "A novel PLL for grid synchronization of power electronic converters in unbalanced and variable-frequency environment", Proc. IEEEPEDG 1, 466–471 (2010).

[14] Francisco D. Freijedo et al, "Robust Phase Lock Loops Optimized for DSP implementation in Power Quality Applications", IECON 2008.

[15] P. Rodriguez et al, "Double Synchronous Reference Frame PLL for Power Converters Control," vol 22. No, 2, 2007.

[16] Marco Liserre, Pedro Rodiguez Remus Teodrescu, Grid Converters for Photovoltaic and Wind Power Systems: John Wiley & Sons. 2011.

[17] P.R.& R.T Marco Liseere, Grid Converters for Photovoltaic and Wind Power systems, John Wiley & Sons Ltd, 2011.

[18] www.ti.com

Control Strategy for Frequency Regulation using Battery Energy Storage with Optimal Utilization

I.S.Jha, Subir Sen, Manish Tiwari, Manish Kumar Singh

Power Grid Corporation of India Limited

Gurgaon

Abstract—**Increasing penetration levels of Renewable Energy Sources (RES) into the grid has raised several concerns due to the intermittency, variability and uncertainty in power outputs. Technological advancements are required to tackle the issues of reliability, stability and power quality. Battery Energy Storage Systems (BESS) are widely being tested and have been found useful to address these challenges. This paper discusses the Frequency Regulation (FR) application of BESS. Several control strategies suggested in literature have been simulated for the frequency variations in the Indian scenario and the constraints limiting the utilization have been identified. A new control strategy has been proposed to enhance the usability of BESS for FR in Indian grid.**

Simulations have been carried out on MATLAB/Simulink platform and the results validate the advantages of the new control strategy over the existing ones.

Keywords—Frequency Regulation; Renewable Integration; Battery Energy Storage Systems; Control Strategy; State of Charge

I. INTRODUCTION

Power systems across the globe are undergoing rapid transformations. While some of these are aligned towards pursuit of a reliable and customer centric grid, a few are coming up as severe necessities for tackling the challenges of energy security, global warming, pollution levels and depleting resources.

Rising penetration of Renewable Energy Sources (RES) in form of distributed generation as well as large solar/wind farms is a major driver for technological transformations. The challenges and issues regarding grid integration of RES have been discussed in [1]. Most problems arise due to intermittency, variability and uncertainty associated with solar/wind generation. Also electronically coupled Distributed Generators (DGs) raise the issues of poor fault ride through capability, absence of inertia, compromised power quality, low voltage ride through etc. [2]. These issues have been taken up in detail by the authors of [3]. Thus, to handle the issues arising from the changing nature of generation, technological advancements are required in every domain of power systems including protection, transmission, grid operation, power system economics etc. The necessary changes in protection systems with increasing presence of DGs have been discussed in[4].

Energy Storage Systems (ESS) are being extensively explored globally as a solution to tackle the issues being posed by the integration of RES. Various grid scale projects on ESS are being carried out globally to be able to increase the utilization of renewable energy [5]. Solar/wind smoothening and frequency regulation are two main applications of ESS being explored. Out of several energy storage systems, batteries offer the maximum flexibility and low response time, which makes them a preferred choice for the above applications [6]. The control strategies and system topologies for using Battery Energy Storage Systems (BESS) for wind/solar smoothening have been discussed in [7], [8],[9],[10]. The system design and control for performing frequency regulation using BESS have been discussed by the authors of [11], [12], [13], [2], [14], [15]. Control strategy for a BESS performing frequency regulation and load compensation in a microgrid has been discussed in [16].

This paper provides the issues to be taken care of, for frequency regulation in Indian grid and the most optimal control strategy for the same. The outline of this paper is as follows: Section II describes the trends in frequency for Indian grid based on self-measurements and data available in public domain. Based on these the key issues for optimizing the control strategy have been identified; Section III discusses different control strategies in literature and analyses their effectiveness and applicability in Indian system; Section IV explains the proposed control strategy and gives the simulation results; Section V concludes with remarks on scope for further optimization.

II. FREQUENCY REGULATION FOR INDIAN GRID

A. Frequency Patterns in India

After establishing a synchronous link between the southern region and the remaining part, the NEW grid on 31-Dec-2013, Indian grid has become a huge system, operating which is a stupendous challenge. As on 31-Aug-2014, the total installed capacity was 253.4 GW. The total RES installations by 31-Mar-2014 was 31692MW out of which 21136 MW was from wind and 2631MW from solar [17]. The Ministry of New and Renewable Energy (MNRE) has targets to have 38,500 MW of wind capacity and 20,000MW of solar capacity by 2022 adding up with other RES to make a total RES capacity of 72,400 MW [18]. The Grid operator for the entire national grid, POSOCO operates through five Regional and a National Load Dispatch

978-1-4799-6047-7/14 $31.00 © 2014 IEEE

Centre. At present the provision for Ancillary Services is not fully in place and hence handling the frequency in a narrow band (presently 49.9 to 50.05) is a great challenge. The frequency in India shows wide variations unlike European nations [13]. Fig.1 shows the percentage time for which the frequency stays in different ranges based on evenly selected daily variation curves(Based on daily variation curve for first day of march-sep,2014) [19].

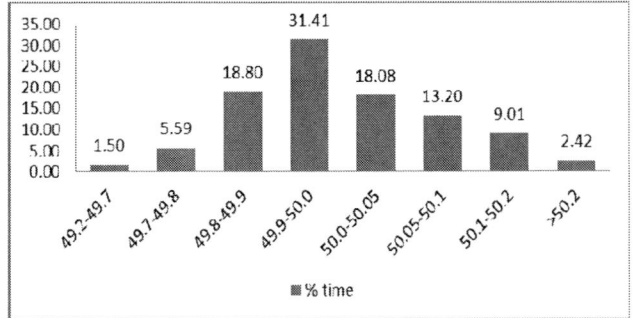

Fig. 1. Typical Frequency Distribution Chart for India

The frequency deviations and the average frequency in India vary largely on daily as well as seasonal basis. While the average frequency in the southern region on 01-12-2013 was 50.16, it was 49.88 on 01-08-2014.

For the simulation studies in this paper, frequency was measured for a 24 hour duration using Fluke 434 Series-II Power Quality Analyzer on 29-Aug-2014 at 400/220 kV Substation, Bilaspur. Fig.2 shows the frequency distribution chart for 24 hours duration.

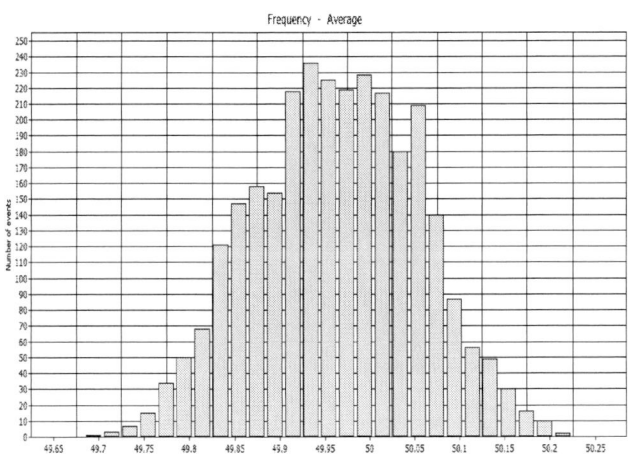

Fig. 2. Frequency Distribution Chart as per measured data

B. Issues for Developing strategy for FR in India

Some key points that have to be considered while developing the control strategy for performing FR in Indian scenario are as follows:

1. Average frequency is mostly on the lower side.

2. For practical BESS, the maximum charge rate is much lower as compared to the discharging C-rate. Thus the battery takes more time to replenish its State of Charge (SOC).

3. Typical AC –AC efficiency for grid scale BESS is in around 75%.

4. Due to BESS life-cycle considerations deep discharge and overcharging is not recommended. Typical limits on SOC may be taken from a minimum of 20% to maximum of 90 %

Since the economic framework for ancillary services is not yet in place in India, the strategy being optimized here focusses on maximizing the energy exchange by the system, while keeping the time of unavailability due to hitting of SOC limits to the minimum.

III. COMPARISON OF VARIOUS CONTROL STRATEGIES

Since the use of BESS for Frequency Regulation is being highly investigated, there are various strategies suggested in the literature. The scope and effectiveness of BESS for FR in a single area and two area network has been discussed in [15]. Also the basic controller design has been discussed by the authors. The impact of RES penetration is more critical for isolated systems and microgrids. FR schemes for handling loss of large generation as well as continuous load changes in such systems have been discussed in [2] and [16]. A BESS scheme for Primary Frequency Regulation (PFR) has been optimized for cost and battery size in [13] for European grid.

We shall here simulate a few strategies suggested in literature for the actual frequency variation in India and a typical battery size with specifications as in Table I.

TABLE I. PARAMETERS FOR BESS FOR SIMULATION

Parameter	Value
Max. Discharging Power	1000kW
Max. Charging Power	500kW
Energy Storage capacity	250kWh
Max. SOC limit	90%
Min. SOC limit	20%

A. Case 1

The authors of [12] suggest a simple strategy for charging in over frequency and discharging during under frequency with battery SOC limits imposed. The relation between battery power and frequency error has been kept linear, similar to droop curve. The results of simulation for this case have been shown in Fig.3

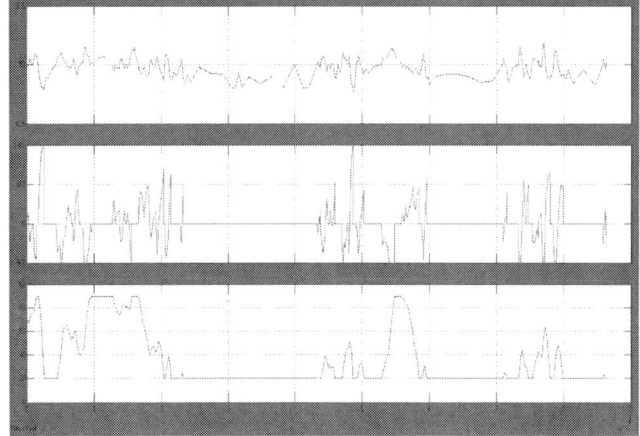

Fig.3. (a) Frequency variation; (b) Battery Reference Power; (c) SOC

Key Performance Indices(KPI) for us are Total Energy Exchanged during 24 hours and total unavailability time due to hitting of SOC limits. For this case, the KPIs are as given in Table II.

TABLE II. RESULTS FOR CASE 1

Key Performance Indices (Case 1)	
Minutes of unavailability	887
Total Energy Exchanged (kWh)	2398

B. Case 2

Three different strategies for FR using Li-Ion batteries have been compared in [11] based on lifetime estimation model and Net Present Value (NPV) calculation. One of the strategies identifies a non-critical window for frequency and tries to stabilize the SOC to 50% when the grid frequency is in its noncritical range. This helps the system to avoid hitting its SOC limits. However beyond this window, a linear Power-frequency relationship has been followed. A case study for Denmark based on this strategy has been discussed in [14].

For our case we took 49.97 Hz to 50.03Hz as the noncritical window and SOC=60% as the target SOC. The results of simulation are as shown in Fig. 4 and the KPIs are shown in Table III.

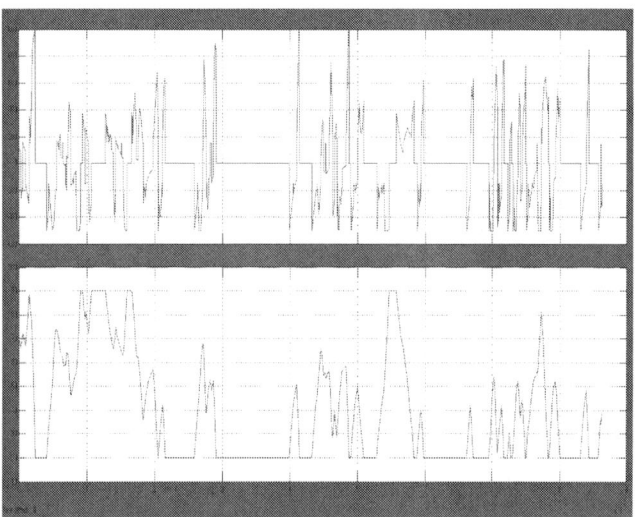

Fig. 4. (a) Battery Reference Power; (b) SOC

TABLE III. RESULTS FOR CASE 2

Key Performance Indices (Case 2)	
Minutes of unavailability	662
Total Energy Exchanged (kWh)	4116

We see a considerable improvement in the system performance for the same frequency data while implementing the second strategy. We now explore the scope of further improvement.

IV. PROPOSED CONTROL STRATEGY

As may be inferred from the issues discussed in Section II-B, for optimizing the performance of the system in Indian scenario, we need to generate higher opportunities for charging as compared to discharging. Moreover this has to be done, in

the case when the frequency mostly lies on the lower side. Setting a lower target frequency is on direct solution but it would dilute the main goal of frequency regulation and rather perform the task of frequency smoothening with a lower reference value.

A control method for PV smoothening is discussed in [7], which introduces the concept of variable filter time constant based on SOC of the battery. For this purpose, SOC ranges have been identified and thus the level of support that the BESS provides is regulated based on SOC to avoid hitting the limits, and thus increasing availability. A similar concept for FR in a microgrid using BESS has been brought forward in [16]. Here the dependence of Battery reference power and the frequency error is governed by a SOC dependent parameter as per (1)

$$G(s) = P_G(s) / \Delta f(s) = K_{pf} * K_{SOC} + K_{df} \qquad (1)$$

Where K_{pf} is the proportional component for droop and K_{df} is the differential component representing inertia. K_{SOC} is a SOC dependent variable that helps in regulating the battery power to keep SOC within limits. Here, we introduce a similar concept to enhance the system availability for Indian scenario, without setting a lower frequency reference value. Fig. 5 explains the Power and Frequency error relationship for different SOC values.

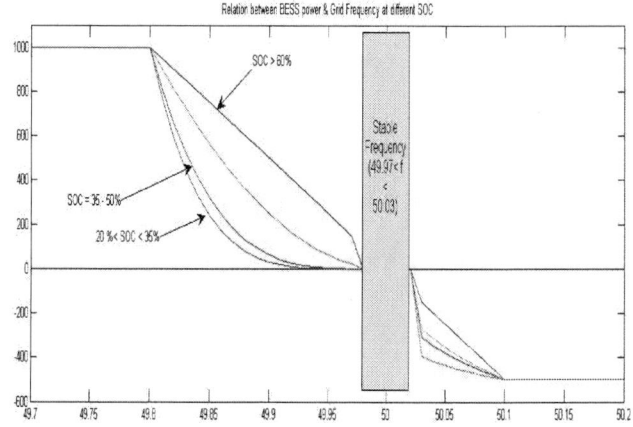

Fig. 5. Power Vs Frequency relationship dependent on SOC

Table IV gives the exact relationship between BESS power and Normalized Frequency Error (NFE) for different SOC levels.

TABLE IV. BATTERY REFERENCE POWER BASED ON FREQUENCY ERROR AND BATTERY SOC

Freq. SOC (%)	<49.8	49.8-49.97	49.97-50.03	50.03-50.1	>50.1
< 20	NO discharge	NO discharge	Power to bring SOC to 60%	NFE^0.2	Full Charging Power
20-35	Full Discharging Power	NFE^5		NFE^0.2	
35-50		NFE^3		NFE^0.2	
50-60		NFE^2		NFE^0.2	
60-70		Linear droop		NFE^0.4	

978-1-4799-6047-7/14 $31.00 © 2014 IEEE   374

70-80	Full Discharging Power	Linear droop	Power to bring SOC to 60%	NFE^0.5	Full Charging Power
80-90		Linear droop		Linear droop	
>90		Linear droop		NO Charging	NO Charging

The results of simulation obtained from the above strategy for the same frequency data input are shown in Fig. 6 and the KPIs for this case are provided in Table V.

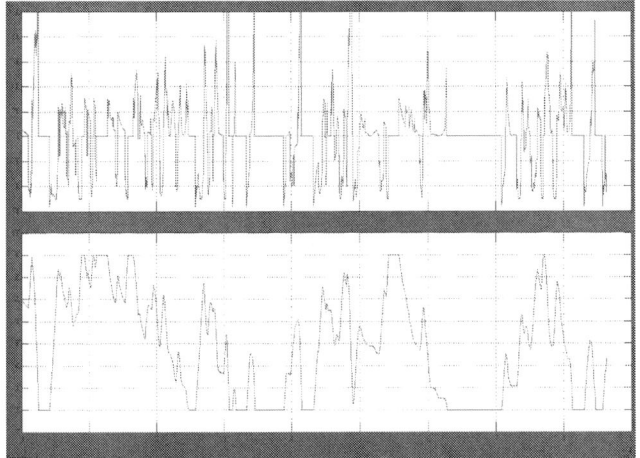

Fig. 6. (a) Battery Reference Power; (b) SOC

TABLE V. RESULTS FOR PROPOSED STRATEGY

Key Performance Indices (Proposed Strategy)	
Minutes of unavailability	442
Total Energy Exchanged (kWh)	4182

V. CONCLUSION

As can be clearly seen from the results of simulation and the KPIs, it is beneficial to design the control strategy to favour the opportunities for charging. Also the dependence of the relationship between reference power for BESS and frequency error on the Battery State of Charge increases the availability of the system for FR by avoiding the battery to reach its charging/discharging limits. The proposed strategy was found to be decreasing the Battery unavailability duration by significant amount, while simultaneously increasing the energy transactions. Thus, this strategy ensures maximum utilization of a given size of ESS. It may be noted that there is a significant impact of frequency and SOC set-points on the system behavior. Thus further studies may be carried out to develop an adaptive scheme to accommodate seasonal changes in frequency trends. Also, keeping this strategy as the base, a detailed modelling of battery and the grid behavior may help to analyze the overall system response for specific test beds.

ACKNOWLEDGMENT

Authors are thankful to the management of POWERGRID for granting permission for presentation of this paper. Views expressed in the paper are of the authors only and need not necessarily be that of the organization in which they belong.

REFERENCES

[1] Ahmed Sharique Anees, "Grid Integration of Renewable Energy Sources: Challenges, Issues and Possible Solutions," Power Electronics (IICPE), 2012 IEEE 5th India International Conference, December 2012, pp. 1-6

[2] Guthier Delille, Bruno Francios and Gilles Malarange, "Dynamic Frequency Control Support by Energy Storage to Reduce the Impact of Wind and Solar Generation on Isolated Power System's Inertia," IEEE Transactions on Sustainable Energy, Vol. 3, No. 4., October 2012

[3] H. Bevrani, A. Ghosh and G. Ledwich, "Renewable energy sources and frrquency regulation: survey and new persepectives," IET Renew. Power Gener., 2010, Vol. 4, Iss 5, pp. 438-457.

[4] Manish Kumar Singh and Parne Naveen Reddy, "A fast Adaptive Protection Scheme for Distributed Generation Connected Networks with Necessary Relay Coordination," Engineering and Systems (SCES), 2013 Students Conference, April 2013, pp. 1-5

[5] U.S. Department of Energy, "Grid Energy Storage", December 2013

[6] Traek M. Masaud, Keun Lee and P.K.Sen, "An overview of energy storage tewchnologies in electric power systems: What is the Future," North American Power Symposium (NAPS), September 2010, pp. 1-6

[7] Li Guo, Ye Zhang and Cheng Shan Wang, "A New Battery Energy Storage System Control Method Based on SOC and Variable Filter Time Constant," Innovative Smart Grid Technologies (ISGT), 2012 IEEE PES, January 2012, pp. 1-7

[8] WU Shuyun and YUAN Yue, "Research on the Application of Storage Battery to Restrain the Photovoltaic Power Fluctuation", Power Engineering Conference (UPEC), 2013 48th International Universities, 2013, pp. 1-6

[9] Sercan Teleke, Mesut E. Baran, Alex Q. Huang, Subhashish Bhattacharya and Loren Anderson, "Control Strategies for Battery Energy Storage for Wind Farm Dispatching," IEEE Transaction on Energy Conversion, Vol. 24, No. 3, pp. 725-732 September 2009

[10] Katsuhisa Yoshimoto, Toshiya Nanahara, Gentaro Koshimizu and Yoshihsa Uchida," New Control for Regulation State-of-Charge of a Battery in Hybrid Wind Power/Battery Energy Storage System," Power Systems Conference and Exposition, 2006. PSCE '06. 2006 IEEE PES, 2006, pp. 1244-1251

[11] Egill Thorbergsson, Vaclav Knap, Maciej Swierczynski, Daniel Stroe and Remus Teodorescu, "Primary Frequency Regulation with Li-Ion Battery Based Energy Storage System-Evaluation and Comparsion of Different Control Strategies," Telecommunications Energy Conference 'Smart Power and Efficiency' (INTELEC), Proceedings of 2013 35th International , 2013, pp. 1-6

[12] Shailendra Singh, Sameer Kumar Singh, Saurabh Chanana and Y.P.Singh, "Frequency Regulation of an Isolated Hybrid Power System with Battery Energy Storage System", Power and Energy Systems Conference: Towards Sustainable Energy, 2014, pp. 1-6

[13] Alexandre Oudalov, Daniel Chartouni and Christian Ohler, " Optimizing a Battery Energy Storage System for Primary Frequency Control", IEEE Transaction on Power Systems, Vol. 22, No.3, August 2007

[14] Macei Sierczynski, Daniel Ion Stroe, Ana Irina Stan and Remus Teodorescu, "Primary Frequency Regulation with Li-ion Battery Energy Storage System: a Case Study for Denmark," ECCE Asia Downunder (ECCE Asia), 2013 IEEE, 2013, pp. 487-492

[15] Liang Liang, Jin Zhong and Zaibian Jiao, "Frequency Regulation for a Power System with Wind Power and Battery Energy Storage," Power System Technology (POWERCON), 2012 IEEE International Conference, 2012, pp. 1-6

[16] Ioan Serban and Corneliu Marinescu, "Control Strategy of Three-Phase Battery Energy Storage Systems for Frequency Support in Microgrids and with Uninterrupted Supply of Local Loads," IEEE Transaction on Power Electronics, Vol. 29, No. 9, September 2014

[17] Central Electricity Authority, "Monthly Report on All India Installed Capacity of Power Stations", August 2014

[18] Ministry of New and Renewable Energy, "Strategic Plan for New and Renewable Energy Sector for the Period 2011-17," February 2011

[19] http://www.srldc.org/DailyReport.aspx

Direct Duty Ratio Controlled MPPT Algorithm for Boost Converter in Continuous and Discontinuous Modes of Operation

Pallavi Bharadwaj, Vinod John

Department Electrical Engineering, Indian Institute of Science Bangalore

email : bharadwaj@ee.iisc.ernet.in, vjohn@ee.iisc.ernet.in

Abstract—Demand of increased lifetime, compact size and reduced cost of PV systems has led to the incorporation of LCL filters in the boost converter of a grid connected PV system. Additional filtering offered at the input by LCL filter reduces the inductance of boost converter. This calls for an algorithm which can track the maximum power in both discontinuous (DCM) and continuous conduction modes (CCM) of boost converter operation accurately. Direct duty ratio based perturb and observe algorithm for MPPT has been implemented in real time using VHDL based FPGA. The sensitivity of the algorithm to DCM and CCM operation of converter is analysed. It is shown that the duty ratio increment used in the MPPT algorithm should be evaluated based on the sensitivity factor. The direct duty ratio based approach is found to work well with both modes of operation, even with mode transitions.

Index Terms—Solar, photovoltaic, maximum power point tracking, boost converter, continuous/discontinuous conduction mode, perturb and observe.

I. INTRODUCTION

Power converters are normally connected at the output of PV panels as an interface between panel and load. Use of DC-DC converters such as buck, boost and cuk converter have been reported in literature. Buck converter as used in [11], may not be suitable for low voltage PV systems, specially when grid connection is involved. Cuk converter as used in [10], requires an additional set of LC filter as compared to buck and boost converters. In this work boost converter has been used as the PV system has an MPP voltage of 15V and DC bus voltage needs to be higher for grid connection through an inverter. An LCL filter has been included at the input of the boost converter which reduces the inductance of boost, which can further lead to design of an optimised converter having low cost, low size and high efficiency. But low inductance also leads to DCM operation, therefore need arises for a robust MPPT algorithm which has tracking ability with mode transitions involved.

Maximum power point tracking (MPPT) is a method of obtaining the maximum possible power out of static solar panels for given irradiation and temperature conditions. Some of the commonly used MPPT techniques include perturb and observe (P & O) algorithm, incremental conductance algorithm, open circuit voltage and short circuit current based

This work is supported by Department of Information Technology, Govt of India under NaMPET Phase 2 under Project on Mini Full Spectrum Simulator.

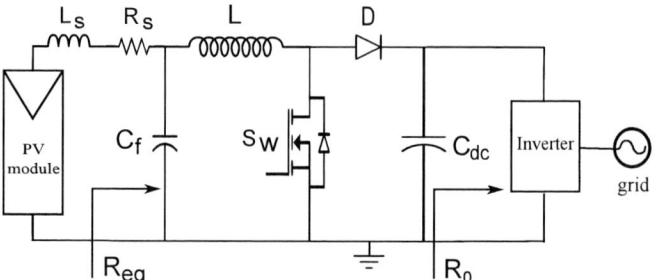

Fig. 1. Circuit diagram of grid connected PV system.

methods [7]. P & O is widely used in PV MPPT applications because it is relatively simple and has real time tracking ability of MPP [7]. P & O algorithm has several variations based on control parameter involved. First is the voltage based P & O which is the most common [8], [10]–[12]. Second one involves current perturbation instead of voltage [9]. In both these methods duty ratio is commonly used as an indirect variable to actually change the voltage or current [11]. Third kind of P & O is called as direct duty ratio control [8], [13], wherein duty ratio is the direct control variable and it undergoes step changes in order to achieve MPPT. Current perturbation method is not commonly used as it involves short circuit current measurement, which poses difficulty in practical implementation. Out of voltage based and direct duty ratio based P & O methods, latter one is preferred for this work as direct duty ratio control offers better energy utilisation and improved system stability than reference voltage control [8].

The main focus of this work is the detailed study of direct duty ratio based algorithm, to study its behaviour for different modes of converter operation namely - continuous conduction mode (CCM) and discontinuous conduction mode (DCM). It has been shown that duty ratio based algorithm is effective for both modes of operation for boost converter.

II. DUTY RATIO BASED MPPT ALGORITHM

A. Theoretical background

Fig. 1 shows a grid connected PV system wherein a PV panel feeds a boost converter with input LCL filter, which further feeds the grid via inverter. The input LCL filter consists

978-1-4799-6047-7/14 $31.00 © 2014 IEEE

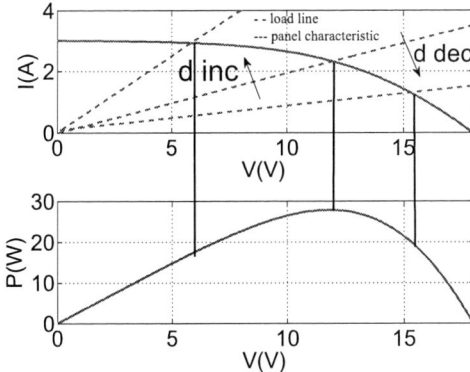

Fig. 2. Determination of operating point of a solar PV panel depends on irradiation, temperature and terminal load resistance. The projection of operating point is taken on panel's Power-Voltage curve, shown for three loading conditions obtained by operating at 3 different duty cycles.

of stray inductance L_s and resistance R_s of cables connecting panel on roof, to the converter located in laboratory. Additional capacitor C_f and boost inductor L together form LCL filter. This combined LCL filter acts as an attenuator to high frequency boost inductor current ripple and effectively only DC current flows from the PV source thereby enhancing its efficiency. Due to the incorporation of LCL filter in the circuit boost inductor value can be reduced for same ripple current. This gives reduced cost and size benefits. However reduction in L value may cause discontinuous mode of operation of boost converter. So the MPPT algorithm needs to have MPP tracking ability in both DCM/CCM. For this the working of direct duty ratio MPPT algorithm is analysed in continuous and discontinuous modes of operations of boost converter.

The basic principle behind any MPPT algorithm is digital control of converter in such a manner that the load seen by PV panel corresponds to the maximum possible power output for any given load. Consider the current-voltage and corresponding power-voltage characteristic of a PV panel, as shown in Fig. 2. The operating point is obtained by the intersection of solar panel's current-voltage (I-V) curve and load line which is dependent on load connected. The solar panel terminal equation [1] is written as

$$I = I_L - I_s(e^{\frac{V+IR_s}{mn_sV_t}} - 1) - \frac{V + IR_s}{R_{sh}} \tag{1}$$

And the equation of load line for a load resistance 'R' can be written as

$$I = V/R \tag{2}$$

The intersection of the two above mentioned equations is shown in Fig. 2 for three different cases of load resistance. The current gain, voltage gain and equivalent resistance reflected at the source terminals depends on the mode of operation of the boost converter, as given in Table I. Symbols shown in Table I are:- $k = \frac{2L}{R_oT_s}$, R_o =output resistance, T_s =switching period, L = boost inductance, d = duty ratio of switch $S_w = \frac{T_{on}}{T_{off}+T_{on}}$. To visualise the effect of duty ratio on equivalent

TABLE I
COMPARISON OF GAINS AND EQUIVALENT RESISTANCE FOR CCM AND DCM OPERATION FOR A BOOST CONVERTER.

Equivalent Resistance	CCM	DCM
Current Gain	$(1-d)$	$\dfrac{\left(\dfrac{k+\sqrt{k^2+4kd^2}}{2d}\right)}{\left(d+\dfrac{k+\sqrt{k^2+4kd^2}}{2d}\right)}$
Voltage Gain	$\dfrac{1}{(1-d)}$	$\dfrac{\left(d+\dfrac{k+\sqrt{k^2+4kd^2}}{2d}\right)}{\left(\dfrac{k+\sqrt{k^2+4kd^2}}{2d}\right)}$
Equivalent Resistance	$R_o(1-d)^2$	$R_o\dfrac{\left(\dfrac{k+\sqrt{k^2+4kd^2}}{2d}\right)^2}{\left(d+\dfrac{k+\sqrt{k^2+4kd^2}}{2d}\right)^2}$

TABLE II
CIRCUIT PARAMETERS FOR THREE DIFFERENT BOOST CONVERTER CONFIGURATIONS

Case	k	R_o	L	T_s
A	0.024	160Ω	100μH	50μs
B	0.14	200Ω	700μH	50μs
C	0.16	300Ω	1200μH	50μs

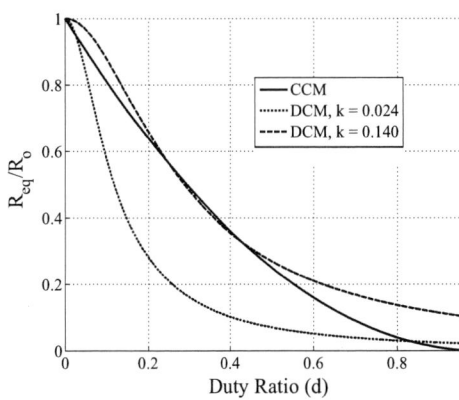

Fig. 3. Variation of equivalent resistance with duty ratio in CCM mode and DCM mode.

resistance (R_{eq}) it is plotted in Fig. 3 for CCM mode and DCM mode. In DCM the equivalent resistance is dependent on 'k', therefore it is shown for two different values of 'k'. For values of $k \geq 0.15$ converter operates in CCM always and equivalent resistance in CCM mode is independent of 'k'. In this paper analysis is done for 3 cases of boost converter configurations corresponding to the three different combinations of R_o, L and T_s values. These are listed in Table II.

As discussed before, operating point of a PV panel depends on the panel's current-voltage characteristic (for given temperature and irradiation condition) and the load line. The slope of the load line is inverse of equivalent resistance as seen by the PV panel. From Table I it can be observed that R_{eq} is a function of duty ratio. By changing the duty ratio of the boost converter, operating point for the PV panel can be controlled. For a particular set of load resistance, inductance

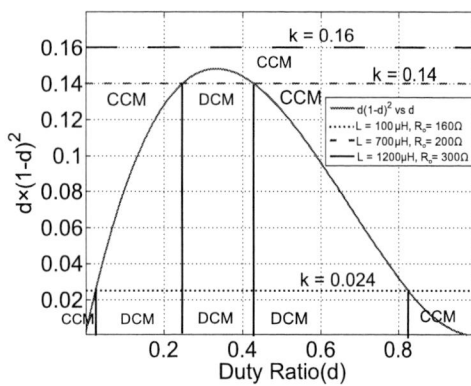

Fig. 4. DCM-CCM boundary is given by the intersection of $d(1-d)^2 = k = \frac{2L}{R_o T_s}$ for boost converter. Here duty ratio range for CCM/DCM operation are shown for three different sets of R_o, L values, with T_s fixed at $50 \mu s$.

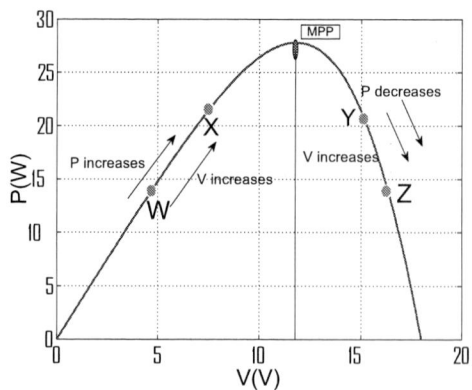

Fig. 6. Power voltage characteristics for an array of PV panels, marked with MPP and a boundary which divides two sides of hill. W, X, Y, Z mark four possible operating points on the power - voltage curve.

and switching frequency of the boost converter, there will be a range of duty ratio for which converter will operate in CCM and in DCM for remaining range. CCM-DCM boundary can be observed in Fig. 4, wherein three cases are shown corresponding to the ones listed in Table II.

The parameter 'k' is directly proportional to inductance and switching frequency. The value of 'k' falls for small inductance value as well as lower switching frequency. Lower inductance value leads to lower size as well as lower cost of converter. Also, the filtering objective can be met with a smaller inductor L, and this can also lead to higher efficiency. However if 'k' value is small then range of DCM operation is large. If MPP can be tracked effectively in DCM then the converter cost and size can be minimised and higher efficiency can be achieved, while still meeting tight current ripple requirements for the PV panel. From Fig. 4 it is clear that for a given duty ratio, the operation of converter in CCM or DCM mode is determined by R_o, L, T_s values. Whether MPP falls within the range spanned by 'd' variation in CCM region or DCM region, solely depends on R_o, L, T_s values. Also it can be observed from Table I and Fig. 3 that duty ratio affects the equivalent resistance differently in DCM and CCM. In other words, the effect of duty ratio variation on the slope of load line differs from CCM to DCM. This effect is discussed in detail in Fig. 5 corresponding to three different converter configurations namely case A, B and C as specified in Table II and for two different conditions for solar panel irradiation. Solar panel was irradiated with Sun and a 500W halogen lamp separately to get different characteristics of 10W polycrystalline solar panel [4]. Consider case A of boost converter operation as specified in Table II. Owing to low value of 'k' DCM operation is observed for a wide range of duty ratio from 0.05 to 0.85 as shown in Fig. 5(a). This is as expected from Fig. 4. For case B 'k' value is higher, this gives narrow range of DCM operation as shown in Fig. 5(b). However for case C as 'k' value is large enough, it does not intersect CCM-DCM boundary as shown in Fig. 4. Thus Fig. 5(c) shows complete CCM operation as duty ratio varies from 0 to 1. It can be observed that for

both CCM and DCM operation as d increases the slope of load line increases, therefore even if there is a transition from CCM to DCM still complete I-V curve can be traced. However accuracy of tracking MPP will differ. This calls for definition of a term called as sensitivity of MPP tracking. This is defined as change in power compared to change in duty ratio around MPP. Sensitivity can be physically interpreted by how densely load lines cover MPP region. Higher density corresponds to higher sensitivity. Fig. 5(a) shows better sensitivity compared to Fig. 5(b). Fig. 5(a) also corresponds case A with lower inductance value compared to case B with higher inductance. Therefore for a system, selection of L can be done considering effective range of R_o which for this case is $R_o \geq 160\Omega$, as smaller L can give both cost and MPPT benefits. Sensitivity can also be improved by going for a smaller step size Δd, but this increases tracking time. As step size is reduced from 0.05 in Fig. 5(c) to 0.01 in Fig. 5(d) sensitivity improves from 0.15W for 0.05 to 0.02W for 0.01. Last two cases shown in Fig. 5 correspond to laboratory set up wherein better control over irradiation conditions is achieved by using a halogen lamp. Results for step size of 0.05 are shown in Fig. 5(e) and for 0.01 in Fig. 5(f) showing better sensitivity.

B. The MPPT algorithm

Consider power-voltage curve of a PV panel as shown in Fig. 6. On left side of MPP, power (P) and voltage (V) are in phase, on right side power and voltage are out of phase [3]. By making a small perturbation in the duty ratio, a new operating point is obtained. If P and V increase with this perturbation the operating point is in the left side of MPP, and further movement in the direction of perturbation will lead it to top of the hill. If P reduces with perturbation, V either increases or decreases. In both cases the direction of 'd' perturbation needs to be changed. From Section II-A it is known that as 'd' increases load line moves up in anti-clockwise direction. These facts can be combined in a flowchart as given in Fig. 7. Fig. 7 shows that for a 'Δd' there are 4 possibilities for power and voltage to increase or decrease. Based on changes in V and P one can judge the way the operating point has moved

978-1-4799-6047-7/14 $31.00 © 2014 IEEE

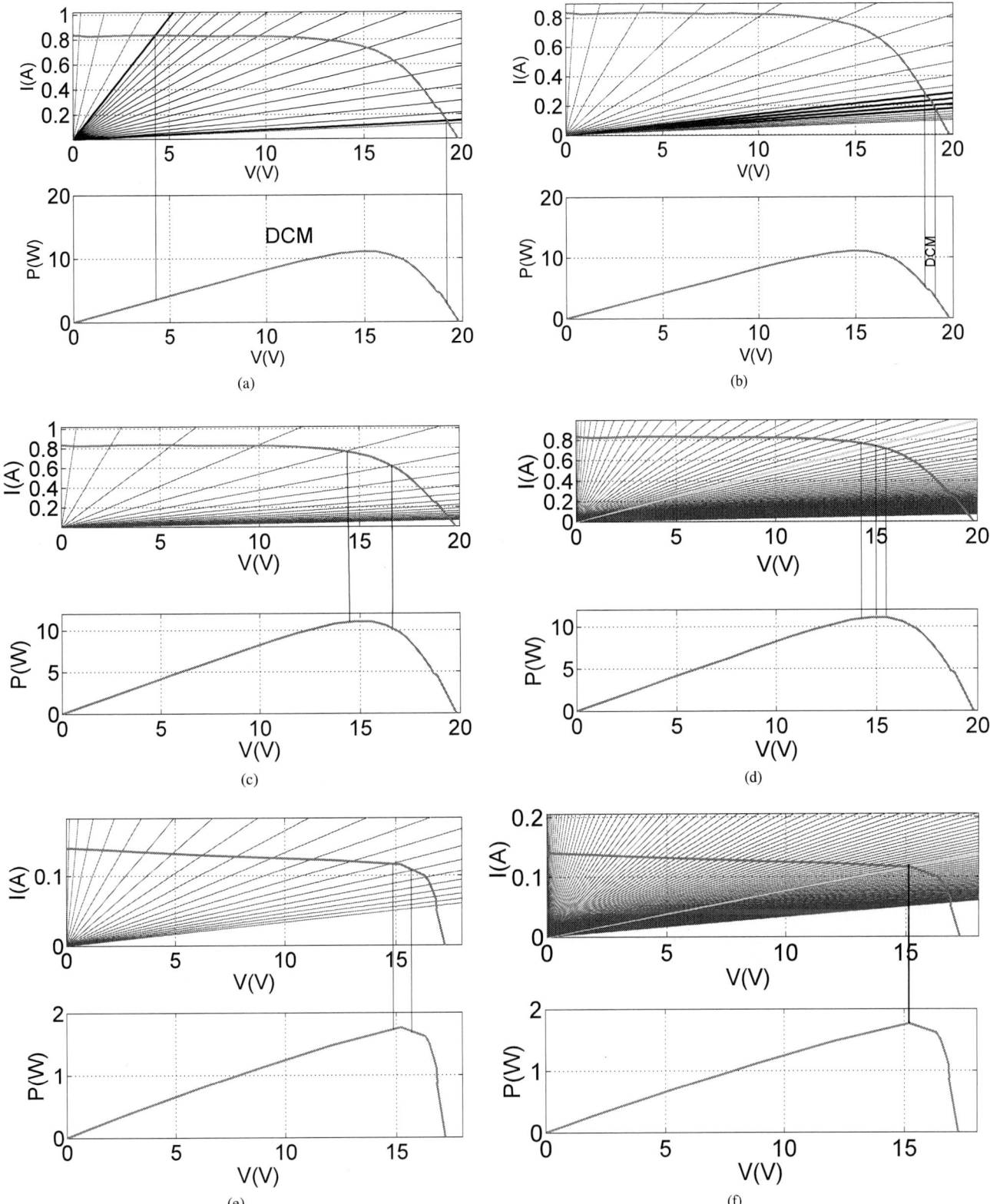

Fig. 5. Measured Current-Voltage (I-V) characteristics of 10W solar panel superimposed with family of load lines for varying duty ratio for a boost converter. Duty ratio increases from 0 to 1 in fixed step Δd. (a) Case A : $R_o = 160\Omega$, $L = 100\mu H$, $T_s = 50\mu s$. $\Delta d = 0.05$. Projection on P-V curve shows DCM operation for d = 0.05 to 0.85. (b) Case B : $R_o = 200\Omega$, $L = 700\mu H$, $T_s = 50\mu s$. $\Delta d = 0.05$. Projection on P-V curve shows DCM operation for d = 0.25 to 0.40. (c) Case C : $R_o = 300\Omega$, $L = 1200\mu H$, $T_s = 50\mu s$. $\Delta d = 0.05$. For complete d variation only CCM operation observed. Projection on P-V curve shows MPP region (d = 0.60-0.65). (d) Same as (c) with $\Delta d = 0.01$. Projection on P-V curve shows MPP region covered more densely (d = 0.62-0.64). MPP occurs at d = 0.63. (e) Case C, $\Delta d = 0.05$, MPP region projected on P-V curve. (f) Boost converter case C. $\Delta d = 0.01$. MPP tracked with higher accuracy at d = 0.25. Panel's I-V and P-V curve shown in (a), (b), (c) and (d) measured on 15/11/13, 12:30pm, Bangalore, panel temperature 40^oC. Panel's I-V and P-V curve in (e) and (f) measured with panel irradiated with 500W halogen lamp placed at 0.3m height from panel.

978-1-4799-6047-7/14 $31.00 © 2014 IEEE   379

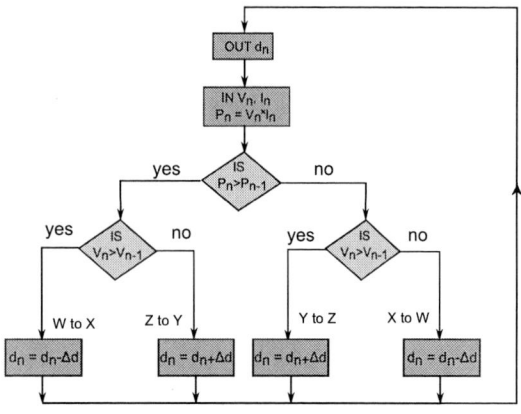

Fig. 7. Flowchart for duty ratio based P & O MPPT algorithm.

TABLE III
BOOST CONVERTER PARAMETERS

Output Load Resistance (R_o)	300Ω
Boost Inductance (L)	1.2 mH
Switching Frequency (f_{sw})	20 kHz
Input Filter capacitance (C_f)	150 μF
Output Filter capacitance (C_{dc})	4400 μF
Input Stray Resistance (R_s)	22.23 mΩ
Input Stray Inductance (L_s)	7.08 μH

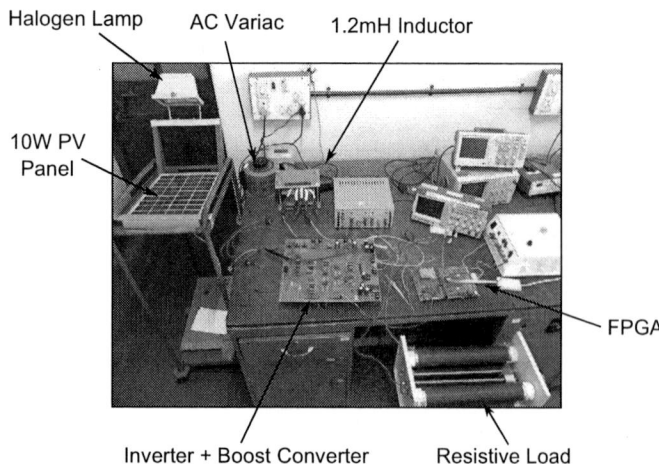

Fig. 9. Hardware set up for MPPT.

TABLE IV
RESULTS WITH MPP TRACKED

Input Voltage	16.6V
Input Current	0.12A
Duty Ratio	0.24
Output Voltage theoretical(ideal)	21.8 V
Output Voltage theoretical(practical)	21 V
Output Voltage measured	20.4 V

on the sides of hill, and then increase or decrease 'd' further to reach the top of the hill which corresponds to the MPP. Here W, X, Y, Z correspond to operating points shown in Fig. 6.

C. Implementation and results

1) Implementation: This algorithm was coded in VHDL and implemented on a FPGA platform, which further controlled the boost converter fed by a 10W PV panel via a LCL filter. The LCL filter ensures that boost inductor current ripple is not carried over to the PV panel and hence leads to panel's long life. To emulate sun, a 500W halogen lamp is arranged, which irradiates the panel at a distance of 0.3m. The intensity of light is controlled by an autotransformer which feeds the halogen lamp. Converter parameters are chosen so as to ensure CCM operation throughout duty ratio variation, they are given in Table III, circuit diagram along with MPPT block is shown in Fig. 8. Hardware setup is shown in Fig. 9.

2) Results: Table IV shows the results with the experimentally tracked maximum power point for laboratory setup. This includes a 10W polycrystalline PV panel irradiated with a 500W halogen lamp. Boost converter specifications are given in Table III. Table IV quantifies the tracked maximum power point for the given setup. Fig. 10 shows maximum power point tracked in steady state for duty ratio perturbation step size of 0.01. It shows panel voltage, current and duty ratio. Results were also obtained for $\Delta d = 0.05$ and that gives $\Delta P = 0.2W$, which matches theoretically expected value as mentioned in Section IIA. Fig. 11 shows filtering by LCL filter, it shows panel current, which is 0.12A dc and boost current which has 0.2A peak-peak ripple. MPP tracking is shown on power-voltage plane in Fig. 12. Fig. 13 shows effect of varying intensity of light on tracking of maximum power point again on power-voltage plane. It shows as the light intensity reduces to zero, locus of MPP shifts from 16V, 2W to 0V, 0W.

III. CONCLUSION

The incorporation of LCL filter between PV panel and boost converter results in lower boost inductance value which leads to reduced cost and size benefits. But reduction in inductance leads to higher probability of DCM operation of boost converter. The direct duty ratio based MPPT algorithm is found capable of tracking MPP despite of mode changes, as in both modes duty ratio increase traverses I-V curve of PV panel in same direction. However it has been found that sensitivity of MPP tracking depends on the mode of operation. This issue is resolved by changing the perturbation step size of duty ratio for desired sensitivity. Experimental results show the filtering action of LCL filter and maximum power point tracking for CCM operation with a duty ratio perturbation step size of 0.01.

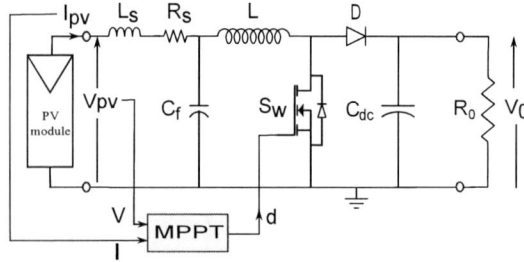

Fig. 8. Circuit showing implementation of MPPT.

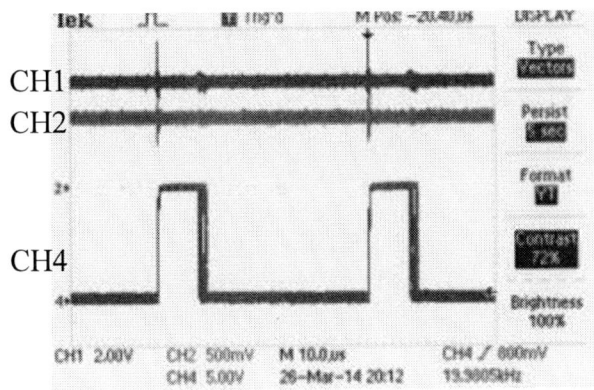

Fig. 10. Panel voltage (CH1), panel current (CH2) and duty ratio (CH4) using the boost switch gating pulse showing the tracking of MPP and steady state operation. Scale :: CH1 - 2V/div, sensor gain - 0.3V/V, V_{mpp} = 16.6V; CH2 - 0.5V/div, sensor gain - 6.67V/A, I_{mpp} = 0.12A; CH4 - 5V/div, time scale - 10μs/div, duty ratio = 0.24.

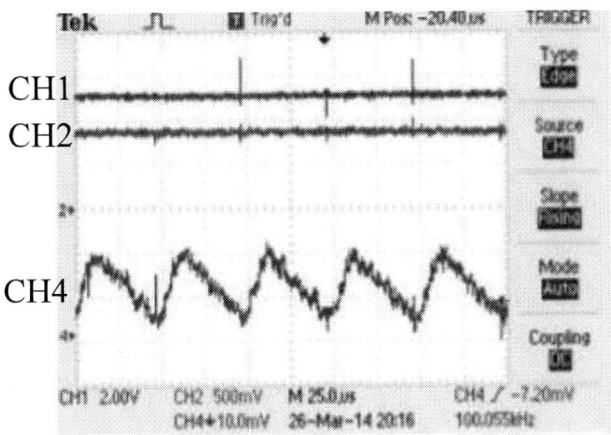

Fig. 11. Panel voltage (CH1), panel current (CH2) and boost inductor current (CH4) showing filtering by LCL filter at the input. Scale :: CH1 - 2V/div, sensor gain - 0.3V/V, V_{mpp} = 16.6V; CH2 - 0.5V/div, sensor gain - 6.67V/A, I_{mpp} = 0.12A; CH4 - 10mV/div, sensor gain - 0.1V/A, $I_{L_{mean}}$ = 0.12A, $I_{L_{pk-pk}}$ = 0.2A.

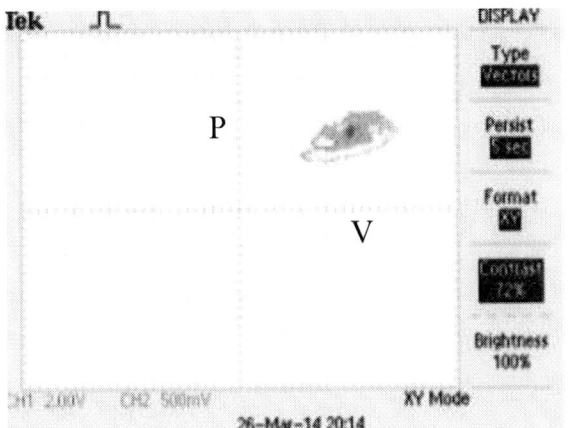

Fig. 12. PV panel power voltage plane (power-y axis, voltage-x axis), showing traking of maximum power point with steady state operation. Scale :: X axis : CH1 - 2V/div, sensor gain - 0.3V/V, V_{mpp} = 16.6V Y axis : CH2 - 0.5V/div, scaling - 0.4V/W, P_{mpp} = 2W.

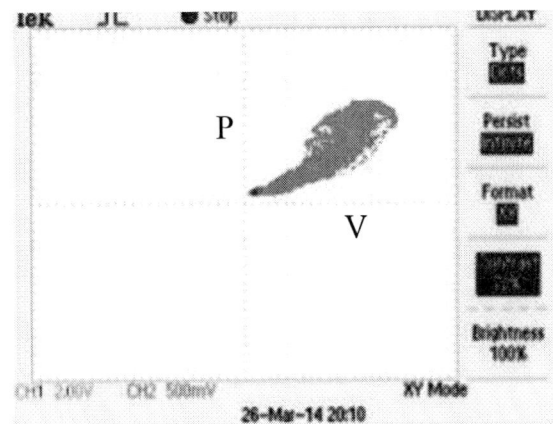

Fig. 13. Tracking of MPP with varying intensity of light, shown on power-voltage plane. MPP moves towards origin as light intensity reduces. Scale :: X axis : CH1 - 2V/div, sensor gain - 0.3V/V, voltage varies from 16.6V to 0V; Y axis : CH2 - 0.5V/div, scaling - 0.4V/W, power varies from 2W to 0W.

It has been observed that a higher step size perturbs the power in a wider range thereby giving lower tracking accuracy, lower average power output but higher tracking speeds.

REFERENCES

[1] Chatterjee, A.; Keyhani, A.; Kapoor, D., "Identification of Photovoltaic Source Models," *IEEE Transactions on Energy Conversion* , vol.26, no.3, pp.883,889, Sept. 2011.

[2] V. Ramnarayanan, "Course Material on Switched Mode Power Conversion," Department of Electrical Engg., IISc. *Available online : www.peg.ee.iisc.ernet.in/people/faculty/vram/smpc/smpcbook.pdf*

[3] L. Umanand, "Lecture notes on Design of Photovoltaic Converters," DESE, IISc. *Available online : www.nptel.ac.in/courses/Webcourse-contents/IISc-BANG/Non-Conventional-Energy-Systems/chap4/teach-slides04*, 2014.

[4] Multicomp, "MC-SP10-GCS-Solar Polycrystalline Panel, 10W," *Available online : http://www.farnell.com/datasheets/1697617.pdf*, 2014

[5] Soren Baekhoj Kjaer, "Evaluation of the Hill Climbing and the Incremental Conductance Maximum Power Point Trackers for Photovoltaic Power Systems," *IEEE Transactions on Energy Conversion* , vol. 27, no. 4, December 2012

[6] N. Femia, G. Petrone, and M. Vitelli, "Optimization of perturb and observe maximum power point tracking method," *IEEE Trans. Power Electron.* , vol. 20, no. 4, pp. 963973, Jul. 2005.

[7] Y. Jiang, J.A. Abu Qahoug, "Single Sensor Multi-channel maximum power point tracking controller for photovoltaic solar systems," *IET Power Electronics*, 2012.

[8] M.A. Elgendy, B. Zahawi and D. J. Atkinson, "Evaluation of P & O MPPT Algorithm Implementation Techniques," *Sixth IET International Conference on Power Electronics, Machines and Drives*, March 2012.

[9] S. K. Kollimalla, M. K. Mishra, "Variable Perturbation Size Adaptive P & O MPPT Algorithm for sudden changes in Irradiation," *Power and Energy Conference at Illinois*, Feb. 2013.

[10] T. P. Sahu, T. V. Dixit, "Modelling and Analysis of P & O and IC MPPT Algorithm for PV Array using Cuk Converter," *IEEE Student's Conference on Electrical, Electronics and Computer Sciences*, 2014.

[11] Ahmad Bin-Halabi, Hussain Meshely, "Experimental Implementation of Microcontroller Based MPPT for Solar PV system," *International Conference on Microelectronics, Communication, and Renewable Energy*, 2013.

[12] Francisco Paz, Martin Ordanez, "Zero Oscillation and Irradiance Slope Tracking for Photovoltaic MPPT," *IEEE Transactions on Industrial electronics*, Nov. 2014.

[13] M. A. A. Mohd. Zeinum, M. A. Mohd. Relzi, Azure Che Soh, N. A. Rahim, "Development of Adaptive Perturb and Observe Fuzzy Control Maximum Power Point Tracking for Photovoltaic Boost DC-DC Converter," *IET Renewable Power Generation*, 2013.

978-1-4799-6047-7/14 $31.00 © 2014 IEEE

Effect of Reliability of Wind Power Converters In Productivity of Wind Turbine

Sanjay Jaiswal
Department of Electrical Engineering
NIT Kurukshetra
Kurukshetra , India
radhey09071990@gmail.com

G.L.Pahuja
Department of Electrical Engineering
NIT Kurukshetra
Kurukshetra , India
pahuja.gl@gmail.com

Abstract—**Reliability modeling in terms of Reliability block diagram (RBD) of different configurations of power electronic converter has been implemented. Literature survey reveals that failure of the power electronic converters is one of the main causes of the wind turbine failure in variable speed wind turbines. Failure rate, repair rate, reliability and availability are computed for existing configuration (Case1) and proposed configurations (Case2 and Case 3) of power converters. On comparing Case1 and Case2, it shows that Case2 configuration of power converter enhances the reliability by twenty four percent in case of squirrel cage induction generator (SCIG) and twenty five percent in case of doubly fed induction generator (DFIG) based wind turbines. Similarly comparing Case 1 and Case3, it shows that Case3 configuration of power converter enhances the reliability by thirty three percent in case of squirrel cage induction generator (SCIG) and thirty percent in case of doubly fed induction generator (DFIG) based wind turbines. DFIG based wind turbine using Case3 configuration gives highest reliability, availability with optimum cost.**

Keywords—*Wind Turbine (WT);Wind Power Converters Reliability Block Diagram (RBD);Redundancy. Total Harmonic Distortion (THD);Pulse Width Modulation (PWM); Squirrel cage induction generator (SCIG) and Doubly fed induction generator (DFIG);Mean Time To Failure (MTTF).*

I. INTRODUCTION

Wind potential is known as the clean and green source of energy. As lot of research work is going on to develop more and more efficient wind turbine so that most of the world can get this clean energy. Also its reliability is very important whether it is going to perform its functions effectively under given conditions encountered for the defined period of time or not [1,2]. From the previous literature it can be seen that most of the times various faults occur in blade/pitch mechanism, gearbox system, generators, power converters, yaw system, hydraulic system [3]. Nowadays large wind turbines are manufactured of about 6MW for inshore and offshore wind farms. In WT electrical and power converters play very important role in overall availability of wind power because if any failure occurs in power converters it might shutdown the overall wind power generation so reliability of these power converters are very important issue in the downtime of overall wind turbine [4]. Based on the atmospheric conditions and rated capacity of wind turbine different topologies of wind power converters can be used like two level switching converter, three level switching

converter and multilevel switching converter [16-18]. For variable wind speed turbines conventional converters are not optimized as they have poor efficiency at lower wind speed due to reduction in the generated voltage or some circulating current in resonant converters [5]. At low speed voltage generated is less so converter efficiency also degrades but with the help of different switching techniques even at low voltages efficiency of converter can be improved and maintaining high output voltages of WT.

II. TOPOLOGIES OF POWER CONVERTER

Cost of two level converter is less than three level converter but power loss is more significant in two level converter that's why three level converter or multilevel converter is used for large wind turbines. There are various topologies of three level power converterin which few of them are mentioned below [17, 18].

A. Diode Clamped Converter

It requires only one DC-bus and voltage level is produced by numerous capacitors in series which is shown in Fig.1.

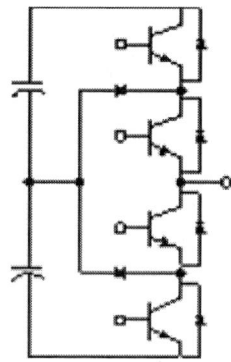

Fig.1. Diode clamped Converter

B. Flying Capacitor Converter

Due to isolated DC supply for each DC-bus, its structure becomes complex which is shown in Fig.2.

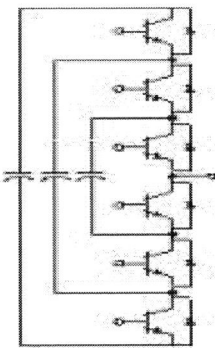

Fig.2. Flying Capacitor Converter

C. Cascaded Converters

Due to isolated DC supply for each DC-bus, its structure becomes complex which is shown in Fig .3

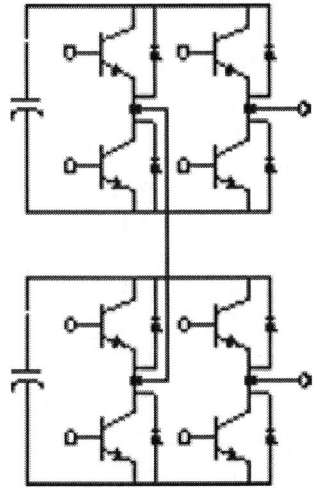

Fig .3.Cascaded H-bridge Converter.

Output voltage waveform of two level converter is generated by PWM with two voltage levels. Due to this waveform of voltage and current gets distorted and THD obtained is poor. But in three levels converter output waveform is more sinusoidal and THD obtained is better than two level converter [17, 18].

So, with help of different switching level maximum productivity can be achieved with low failure rate and high reliability. On the basis of operating speed of wind turbine it can be classified as fixed speed, limited variable speed and variable speed. Variable speed wind generator can be further classified as systems with a partial-scale and a full-scale power electronic converter. Wind turbine can also be classified as geared-drive or indirect drive and gearless-drive or direct drive wind turbines [6-8].

In fixed-speed wind turbine system there is direct connection between induction generator and electrical grid so it can be used for uncertain conditions of wind speed and it is termed as system A [3] shown in Fig.4.

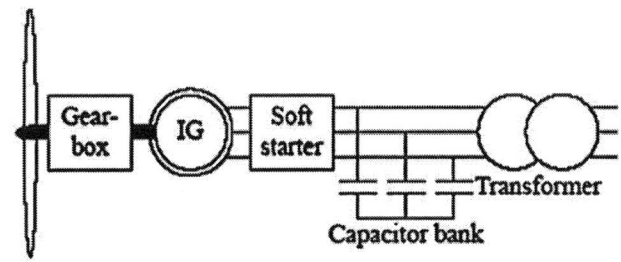

Fig.4. Fixed-speed Wind Turbine, System A

Similarly two configurations of variable speed wind turbine with fully rated converter and partial rated converter termed as system B and system C respectively [3] are shown in Fig.5 and Fig.6.

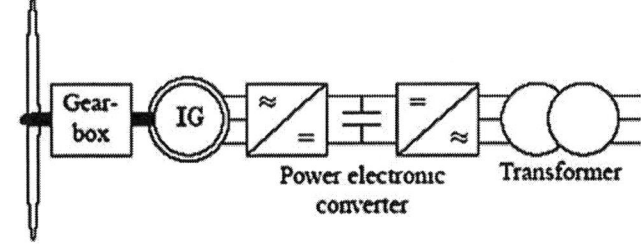

Fig.5.Variable speed wind turbine (fully rated converter) with squirrel-cage induction generator (SCIG),System B

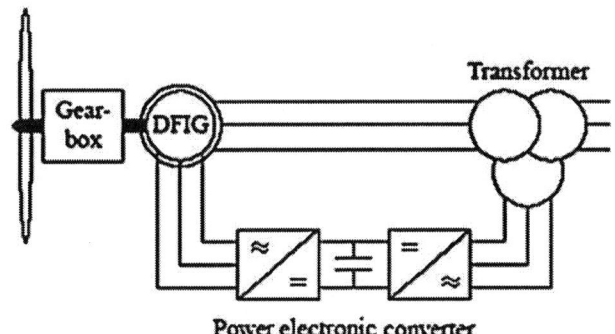

Fig.6.Variable speed wind turbine (partially rated converter) with doubly fed induction generator (DFIG), System C

III. FAILURE ANAILURE OF WT COMPONENETS

A wind turbine consists of mainly blade/pitch mechanism, yaw mechanism, gear mechanism, sensors, electrical subassembly, power converters and generator shown in Fig .7. The average life period of wind turbine is about 20 years but from the literature it can be revealed that in variable speed wind turbines most of the time faults occur in power electronic converters, generators and gear mechanism leading to WT failure[3]. The failure analysis of these can be studied by Bath tub curve [9, 10] shown in Fig 8, high failure rate occurs in early period of life and towards the end of life but with lower failure rates in the useful period of life.

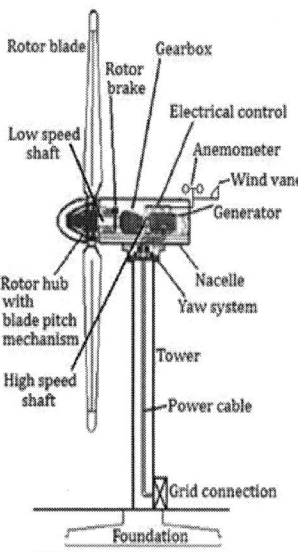

Fig.7. Wind Turbine with its main components.

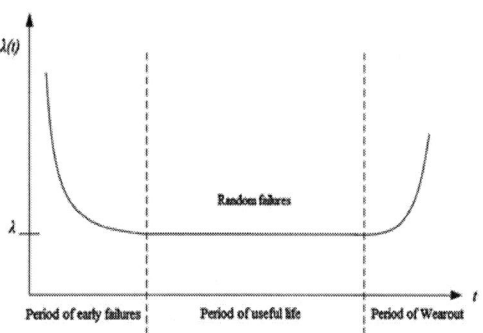

Fig.8. Failure rate of Wind Turbine.

The average failure rate data for wind turbine components [3], indicates control system has the highest failure followed by blade/pitch and then electric system shown in Fig.9.

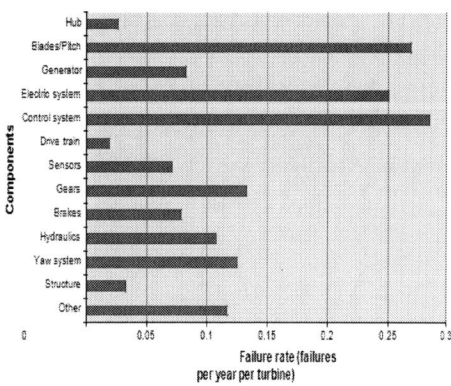

Fig.9. Average rate of failure vs. WT components.

So, from the above data it is revealed that, to enhance the reliability of wind turbine, reliability of control system, electric system, bade pitch should be improved.

IV. RBD MODELLING OF ELECTRICAL SUBASSEMBLIES

RBD of induction generator and power converter is shown in Fig.10. Generator consists of stator, rotor, and brush gear. Power converterconsists of rotor side converter (RSC), DC link, grid side converter (GSC) and control unit(CU).

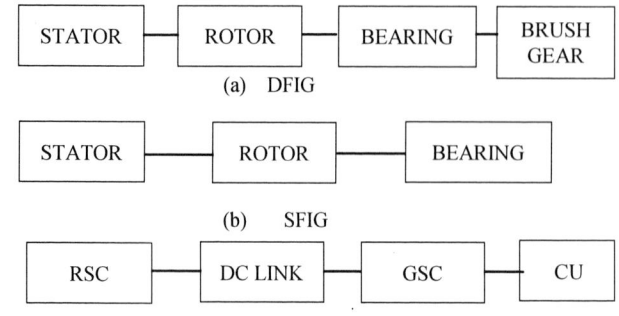

(c)Power Converter

Fig.10. RBD for various subsystems.(a)DFIG, (b)SCIG, (c)Power Converter

RBD for subsystem assemblies result in series configuration.

Series System:

For general N component series system, system reliability expression in terms of component reliabilities and component failure rates are given in (1), (2) and (3). In equation1, 2 and 3, R_i and λ_i are the reliabilities and failure rates of components [2,19].

$$R(t) = \prod_{i=1}^{n} R_i(t) \tag{1}$$

$$R(t) = \exp\left(-t \sum_{i=1}^{n} \lambda_i\right) \tag{2}$$

$$\lambda = \sum_{i=1}^{n} \lambda_i \tag{3}$$

Expression for computing μ (repair rate) for two component are depicted in (4)

$$\mu_{12=}\mu_1\mu_2 (\lambda_1 + \lambda_2)/(\lambda_1\mu_2 + \lambda_2\mu_1) \tag{4}$$

Expression for computing repair rate for three component are depicted by (5),where $\mu_{12}$is the repair rate of first two components and μ is the combined repair rate of first two components and the third component, so overall repair rate for three component is given by (5)

$$\mu = \mu_{12}\mu_3(\lambda_{12} + \lambda_3)/(\lambda_{12}\mu_3 + \lambda_3\mu_{12}) \tag{5}$$

where$\lambda_{12 =} \lambda_1 + \lambda_2$ \qquad (6)

RBD for subsystem assemblies result in parallel configuration.

Parallel System:

For general N component parallel system, system reliability expression in terms of component reliabilities and component repair rates are given in (7), (8) and (9).

$$R(t) = 1 - \prod_{i=1}^{n}[1 - R_i(t)] \quad (7)$$

$$R(t) = 1 - \prod_{i=1}^{n}[1 - e^{-\lambda_i t}] \quad (8)$$

$$\mu = \sum_{i=1}^{n} \mu_i \quad (9)$$

System failure rate for parallel and series parallel system are given by (10) and (11)

$$MTTF = \int_0^{\infty} R(t)dt \quad (10)$$

$$\lambda(t) = 1/MTTF \quad (11)$$

For Fig (11):

$$\lambda_{case\ 1} = \lambda_{RSC} + \lambda_{DC\ Link} + \lambda_{GSC} \quad (12)$$

For Fig (12):

$$MTTF = \int_0^1 \big(2e(-\lambda_{case1}(t)) - e(-2\lambda_{case1}(t))\big)dt \quad (13)$$

$$MTTF = 3/2\lambda_{case1} \quad (14)$$

$$\lambda_{case\ 2} = 1/MTTF \quad (15)$$

For Fig (13): $\lambda_{RSC} = \lambda_{GSC} = A$ and $\lambda_{DC\ Link} = B$

$$MTTF = \int_0^1 (4e(-2A + B)(t) + e(-4A + B)(t) - 4 \\ (-3A + B)(t))dt \quad (16)$$

$$MTTF = \frac{4}{2A+B} + \frac{1}{4A+B} - \frac{4}{3A+B} \quad (17)$$

$$\lambda_{case3} = 1/MTTF \quad (18)$$

Availability of power converter can be calculated as:

$$A = \mu/(\lambda + \mu) \quad (19)$$

R(t)=Reliability of Power Converter
$\lambda(t)$=Failure rate of Power Converter (per year)
$\mu(t)$=repair rate of Power Converter (per year)
A=Availability of Power Converter

In this paper three configurations of RBD of power converters of SCIG and DFIG based wind turbines are discussed [case 1, case2 and case 3] (Fig.12-14).In all these cases control unit is not considered for RBD modeling of power converter as it has very low failure rate [3].

Case 1:All the components of power converter i.e. RSC, DC Link and GSC are logically or functionally in series as shown in Fig.11.

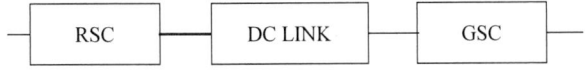

Fig.11. Basic RBD of power converter.

Case 2: A redundant converter (unit redundancy) is placed

in parallel to the existing converter as shown in Fig.12.

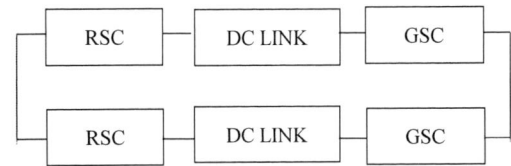

Fig.12. RBD for complete redundant power converter.

Case 3:Component redundancy of RSCand GSC is introduced (Fig.13).DC link can also be made redundant,but it is avoided as its failure rate is very low as compared to failure rate of RSC and GSC.

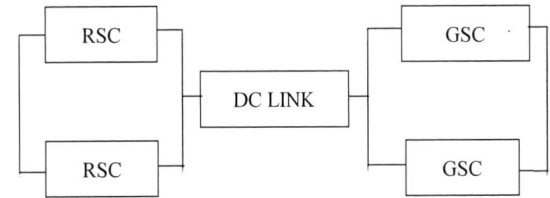

Fig .13. Modified RBD for power converter

V. WT COMPONENT RELIABILTY ANALYSIS

As referred from the failure data and repair data [3], it is inferred that in power converters rotor side converter(RSC) and grid side converter(GSC) have high failure rate rather than dc link and control unit. So, by this case 3 is discussed in which only single dc link is used as it has lower failure rate and RSC,GSC is considered as redundant, so that the reliability of power converter improves which enhances the overall reliability of wind turbine and also optimizes its cost. Calculations of reliability parameters are done on the basis of failure data and repair data [3] shown in Table I.

TABLE I. WT SUBASSEMBLY RELIABILITY DATA

Asse mbly	Failure Rate(yr⁻¹)			Repair Rate (yr⁻¹)	Subassembly
	Tavner [13]	Tavne r[15]	Manjil [11]		
Gene rator	0.1200	0.0090	0.0200	36.5	Stator
		0.0060	0.0200	36.5	Rotor
		0.0257	0.0800	60	Bearing
		0.0300	0.1000	243.3	Brush Gear
		Total =0.07 07	Total =0.22 00		
Conv erter	0.2200	-	0.4000	73	MSC
			0.4000	73	GSC
			0.0100	4380	Control Unit
			0.0400	73	DC Link
			Total =0.85 00		
Num ber of Mac hines	1500-4500	1141	51-66		

978-1-4799-6047-7/14 $31.00 © 2014 IEEE

Computation of failure rate, repair rate, reliability, availability of power converter of WT are based on the data given by Manjil [11]. The repair rate and failure rate of DC link and control unit are same for SCIG (system B) and for DFIG (system C) based WT whereas the failure rate of RSC and GSC are 0.3 failures per year for DFIG(system C)and repair rate are same for both of the system[3]. Based on the given data, reliability parameters are computed for SCIG (system B) and DFIG (system C) based WT for case 1, case 2 and case 3 as respectively and comparison is done among them in which the best results come for DFIG(system C) based WT with case 3 modification as shown in Table III and Table IV.

TABLE II.Comparison of reliability parameters for SCIG basedWT

CASES	SCIG (SYSTEM B)			
	$\lambda(yr^{-1})$	$\mu(yr^{-1})$	A	R
CASE 1	0.84	73.0	0.9886	0.4317
CASE 2	0.56	146.0	0.9961	0.6770
CASE 3	0.47	134.8	0.9981	0.7633

TABLE III. Comparison of reliability parameters for DFIG based WT

CASES	DFIG (SYSTEM C)			
	$\lambda(yr^{-1})$	$\mu(yr^{-1})$	A	R
CASE 1	0.64	73.0	0.9913	0.5273
CASE 2	0.43	146.0	0.9961	0.7765
CASE 3	0.36	131.7	0.9985	0.8360

TABLE IV. Comparison of reliability parameters for SCIG and DFIG WT (CASE 1 and CASE 2).

Configuration	CASE 1	CASE 2	Enhancement%
SCIG	0.4317	0.6770	24.5
DFIG	0.5273	0.7765	24.92

TABLE V. Comparison of reliability parameters for SCIG and DFIG WT (CASE 1 and CASE 3).

Configuration	CASE 1	CASE 3	Enhancement%
SCIG	0.4317	0.7633	33
DFIG	0.5273	0.8360	30

VI. CONCLUSION

By recognizing that the failure of power converter is one of the main reasons for lower reliability of wind machines and it adversely affects the overall productivity of wind turbine. Multilevel switching converters are commonly used in inshore and offshore wind turbines because of reduced maintenance requirement and therefore are well suited in arduous and remote locations. Different reliability related parameters like failure rate, repair rate, reliability and availability have been computed for existing and proposed configurations of power converters. Table IV depicts that Case2 configuration of power converter enhances the reliability by twenty four percent in case of squirrel cage induction generator (SCIG) and twenty five percent in case of doubly fed induction generator (DFIG) based wind turbines. Similarly Table V shows that Case3 configuration of power converter enhances the reliability by thirty three percent in case of squirrel cage induction generator (SCIG) and thirty percent in case of doubly fed induction generator (DFIG) based wind turbines. DFIG based wind turbine for Case3 configuration gives highest reliability, availability with optimum cost. It may be summarized that Case3 configuration be preferred and employed in SCIG and DFIG based wind turbines.

References

[1] H. Guo, S. Watson, P. Tavner, and J.Xiang, "Reliability analysis for wind turbines with incomplete failure data collected from after the date of initial installation", *Reliability Engineer SystemSafety*, vol. 94, pp.1057–1063, 2009.

[2] R. Billinton, and R.N Allan, "Reliability evaluation of engineeringsystems: concepts and techniques", 2nd ed., Plenum Press, ISBN0306440636, 1996.

[3] H. Arabian-Hoseynabadi, H. Oraee, and P.J. Tavner, "Wind turbine productivity considering electrical subassembly reliability", *Renewable Energy*, vol. 35, pp. 190–197, 2010.

[4] F. Blaabjerg, M. Liserre, K. Ma, "Power electronics converters for wind turbine systems," IEEE Trans. on Ind. Appl., vol.48, no.2, pp.708-719, Mar-Apr. 2012.

[5] M. Boettcher and F. W. Fuchs, "Power electronic converters in wind energy systems -Considerations of reliability and strategies for increasing availability," in Proc. European Conference on Power Electronics and Applications, pp. 1-10, 2011.

[6] F. Spinato, J.R Tavner, G.J.W Bussel, E Koutoulakos,"Reliability of wind turbine subassemblies", *IET Renew. Power Gener.*, vol. 3, no. 4, pp. 387–401, 2009.

[7] J. Ribrant, and L.M. Bertling, "Survey of Failures in Wind Power Systems With Focus on Swedish Wind Power Plants During 1997–2005", *IEEE Trans. on Energy Conversion*, vol. 22, no. 1, pp. 167-173,2007.

[8] C.C. Ciang, J.R Lee Jung-Ryul, and H.J Bang, "Structural healthmonitoring for a wind turbine system: a review of damage detection methods", doi:10.1088/0957-0233/19/12/122001, Meas. Sci. Technol., vol .19, pp. 1-20, 2008.

[9] F. Jensen, "Component failures based on flaw distributions," in 1989 Proc. Annu. Reliability and Maintainability Symp., pp. 91–95.

[10] F. Jensen and N. E. Peterson, Burn-in-An Engineering Approach to De- sign and Analysis of Burn-in Procedures, John Wiley & Sons, 1995.

[11] Manjil wind farm operation and maintenance report files 2004-2007,Manjil,Iran.

[12] Applied R&M Manual, for Defence Systems (GR-77 Issue 2009), available at http://www.sars.org.uk/BOK/, accessed Sept. 2009.

[13] P J Tavner , J. Xiang, F.Spinato ,"Reliability analysis for wind turbines", Research article. Wind Energy July 2006;12.Wiley Interscience.

[14] Md. Arifujjaman, M.T Iqbal, and J.E Quaicoe, "A comparative study of the reliability of the power electronics in grid connected small wind turbine systems", 2009 IEEE Conf., pp. 394-397, 2009.

[15] P.J Tavner. "Review of condition monitoring of rotating electrical machines", IET Electrical Applications 2008;2(4):215–47.

[16] Frede Blaabjerg, Ulrig Jaeger, Stig Munk-Nielsen, John K. . "Power Losses in PWM- VSI Inverter Using NPT or PT IGBT Devices." IEEE Transactions on Power Electronics. Volume 10, Issue 3. May 1995. Pages 358-367.

978-1-4799-6047-7/14 $31.00 © 2014 IEEE

[17] G. Carrara, S. Gardella, M. Marchesoni, R. Salutari, G. Sciutto, "A New Multilevel PWM Method: A Theoretical Analysis," IEEE Transactions on Power Electronics, vol. 7, no. 3, July 1992, pp. 497-505.

[18] N. S. Choi, J. G. Cho, G. H. Cho, "A General Circuit Topology of Multilevel Inverter," IEEE Power Electronics Specialists Conference, 1991, pp. 96-103

[19] Michael Pecht, "Reliability Engineering ARINC Research Company",Prentice Hall Inc pages 202 to 206.

Hybrid Differential Evolution with BBO for Genco's multi-hourly strategic bidding

Prerna Jain
Dept. of Elect. Engg.
Malaviya Nat. Inst. of Tech. Jaipur, India
prerna7412@gmail.com

Rohit Bhakar
Dept. of Ec. & Elect. Engg.
University of Bath, UK
r.bhakar@bath.ac.uk

S. N. Singh
Dept. of Elect. Engg.
Indian Inst. of Tech. Kanpur, India
snsingh@iitk.ac.in

Abstract—In Day-Ahead (DA) electricity markets, Generating Companies (Gencos) aim to maximize their profit by bidding optimally, under incomplete information of the competitors. This paper develops an optimal bidding strategy for 24 hourly markets over a day, for a multi-unit thermal Genco. Different fuel type units are considered and the problem has been developed for maximization of cumulative profit. Uncertain rivals' bidding behavior is modeled using normal distribution function, and the bidding strategy is formulated as a stochastic optimization problem. Monte Carlo method with a novel hybrid of Differential Evolution (DE) and Biogeography Based Optimization (BBO) (DE/BBO) is proposed as solution approach. The simulation results present the effect of operating constraints and fuel price on the bidding nature of different fuel units. The performance analysis of DE/BBO with GA and its constituents, DE and BBO, proves it to be an efficient tool for this complex problem.

Index Terms--Bidding Strategy; BBO; DE; Electricity Markets; Monte Carlo Simulation.

I. INTRODUCTION

Evolving Electricity Markets (EMs) are oligopolistic owing to few suppliers, expensive energy storage, demand generation imbalance, *etc*. Gencos bid in such markets with an aim to maximize profit. Market settlement between supplier and consumer bids determine Market Clearing Price (MCP). Optimal bidding strategy of a Genco for Pay-as-MCP (PAMCP) pricing is based on the accurate prediction of MCP which cannot be considered deterministic in an oligopolistic market. It is affected by suppliers' bidding behavior, so other competitors' bidding nature becomes a major source of uncertainty being faced by a strategically bidding Genco [1].

World over, thermal generating units are dominant energy suppliers and are categorized on the basis of fuel type and capacity. Consequently their production cost, operating constraints, and operating cost components differ. Marginal cost of a unit depends mainly on its production cost, which varies with its efficiency and fuel price. Optimal bidding strategy of a unit is, thus, governed by the price being paid for

its stored fuel [2, 3]. Also, operating constraints and operating cost components affect profit and thus, bidding strategy of a unit, when developed over multiple hours [4]. Quantum of literature is available on optimal bidding strategy of a Genco under oligopolistic market but lacks analysis for realistic varied fuel generating units with practical fuel prices [5-11].

Extensive research on optimal bidding strategy formulation for Gencos broadly classifies three solution approaches. Conventional optimization techniques like Lagrange Relaxation, Dynamic Programming, *etc.* fall in first set [1]. These techniques fail for realistic non-differentiable, multi-constraint and multi-objective problems and require nonlinear simplification, if adopted. Another approach is based on game theory which assumes that rival Gencos' cost functions and complete bid information are public. This is practically not true. Also, multiple Nash equilibriums exist for large number of players [5, 6]. The third set of approach, Artificial Intelligence (AI) based heuristic algorithms have the potential to solve such complex problems in their original form, thus giving accurate results. These methods look into wide search space and often achieve a fast and near global optimal solution. Literature shows the application of tools like Genetic Algorithm (GA) [7], Evolutionary Programming [8], Particle Swarm Optimization (PSO), and their variants [9-11], *etc.* These suggest that with increasing complexity and constraints of EMs, these can be looked up as reliable solution tools.

This paper takes up the problem of optimal bidding strategy of a multi-unit thermal Genco for multiple hourly Day-Ahead (DA) EMs over 24 hour horizon of the next day. The realistic units of different fuel types are considered. The problem has been developed as maximization of cumulative profit of the Genco with dynamic demand. It has a nonlinear, non-differentiable, constrained, mixed integer form with multiple binary and real variables. Also, market uncertainty due to rival bids make the problem stochastic. Owing to a large size of solution vector with mixed variables, under dynamically changing environment and multiple market clearings, an extensive exploration of the search space is required.

Recently, a novel algorithm, Biogeography Based Optimization (BBO) based on study of geographical distribution of biological organisms, has been proposed [12]. It works with the population of habitats under migration and mutation operations. Owing to migration, BBO has good exploitation ability and has outperformed other techniques for different power system problems [13-16]. However, it is seen that it has limited exploration capability for large variable complex problems [17, 18]. To overcome this, various improved versions of BBO have been suggested.

Differential Evolution (DE) is a simple and robust tool with mutation, crossover, and selection operators [19]. It is good at exploring search space due to its unique mutation operation. However, initially its solutions move fast towards the optimal point but fail to perform satisfactorily at later stages of fine-tuning, thus having poor exploitation. A new hybrid of DE with BBO (DE/BBO) has recently been proposed which combines exploration of DE and exploitation of BBO and is used for few applications but has not been used for bidding strategies [20]. DE/BBO makes use of hybrid migration operator which is based on BBO's migration and DE's crossover and mutation.

This paper proposes the use of DE/BBO for the optimal bidding strategy problem and compares it with GA as well as its own constituents, DE and BBO, to identify its advantages over them. Nonlinear sinusoidal migration model has been proposed in contrast with the original linear model as it replicates the natural process of migration more closely [21].

II. PROBLEM DESCRIPTION

Consider a pool DA spot EM, with G multi-unit independent Gencos bidding under stepwise bidding protocol. An inelastic and deterministic hourly demand is considered with sealed bid auction and PAMCP pricing. Optimal bidding strategy is to be developed for Genco X, with $G-1$ rivals, over 24 hour horizon of the next day. It is assumed that all Gencos bid in single segment for their each generating unit. Genco X predicts rival bids to formulate its optimal bidding strategy. The size of rival bids is assumed to be known from the historical data available in public domain and their bidding prices are estimated through statistical analysis of historical bidding data. Normal probability distribution function (pdf) is used to model the distribution of rival bid prices and is represented as,

$$pdf(\tilde{P}_{r,i}) = \frac{1}{\sigma_{r,i}\sqrt{2\pi}}\exp\left(-\frac{(\tilde{P}_{r,i}-\rho_{r,i})^2}{2(\sigma_{r,i})^2}\right) \quad (1)$$

where, $\tilde{P}_{r,i}$ is the price bid for i^{th} unit by r^{th} rival ($/MWh), $\rho_{r,i}$ is the mean of normally distributed $\tilde{P}_{r,i}$ ($/MWh) and $\sigma_{r,i}$ is its standard deviation ($/MWh).

Optimal bidding strategy is a profit maximization problem. For Genco X bidding in PAMCP market with generating unit i, profit at any trading interval t is a function of its dispatched power output $Q_{i(t)}$ (MW) and MCP $M_{(t)}$ ($/MWh). Considering to develop bidding strategy of a N unit Genco over T trading intervals, the objective function is maximization of cumulative profit π ($) as represented in (2).

$$\underset{P_{i(t)}}{Maximize} \; \pi = \sum_{t=1}^{T}\sum_{i=1}^{N}h(M_{(t)}Q_{i(t)}-C_{i(t)}) \quad (2)$$

Here, $C_{i(t)}$ is the cost of generating $Q_{i(t)}$ from i^{th} unit in $/h. h

is the duration of each trading interval in hours.

Various constraints are:

1) Generation limits

$$Q_i^{\min}u_{i(t)} \le Q_{i(t)} \le Q_i^{\max}u_{i(t)} \qquad \forall t \in T \quad (3)$$

2) Minimum up time

$$(1-u_{i(t+1)})MUT_i \le H_{i(t)}^{on} \qquad if \; u_{i(t)}=1 \quad (4)$$

3) Minimum down time

$$u_{i(t+1)}MDT_i \le H_{i(t)}^{off} \qquad if \; u_{i(t)}=0 \quad (5)$$

4) Limitations on bid price

$$P_{i\min} \le P_{i(t)} \le price \; cap \qquad \forall t \in T \quad (6)$$

Here, $Q_i^{\min}$ (MW) and $Q_i^{\max}$ (MW) are the minimum and maximum generation limits of ith unit respectively; $u_{i(t)}$ is binary variable and is equal to one for i^{th} unit committed at trading interval t, otherwise equal to 0; MUT_i and MDT_i are the minimum up time and the minimum down time of ith unit in hours respectively; $H_{i(t)}^{on}$ is the number of hour i^{th} unit has been continuously ON at the end of trading hour t; $H_{i(t)}^{off}$ is the number of hour i^{th} unit has been continuously OFF at the end of hour t; $P_{i(t)}$ ($/MWh) is the price bid for total capacity of unit i at a trading interval t, $P_{i\min}$ ($/MWh) is the minimum limit on bid price of unit i.

Price cap ($/MWh) is the maximum limit on bid price for unit i. Generation cost $C_{i(t)}$ ($/h) is considered to be composed of convex production cost $C_{i(t)}^{pr}$, exponential start-up cost C_i^{su} and constant shut-down cost C_i^{sd} Hence,

$$C_{i(t)} = C_{i(t)}^{pr} + C_i^{su}\{u_{i(t)}(1-u_{i(t-1)})\} + C_i^{sd}\{((1-u_{i(t)})u_{i(t-1)}\} \quad (7)$$

such that,

$$C_{i(t)}^{pr} = a_i + b_iQ_{i(t)} + c_iQ_{i(t)}^2 \quad (8)$$

$$C_i^{su} = \zeta_i + \delta_i\left(1-\exp\left(-\frac{Toff_i}{\tau_i}\right)\right) \quad (9)$$

where, a_i ($/h) ,$b_i$ ($/MWh) and c_i ($/MW²h) are no load production cost coefficient, linear production cost coefficient and quadratic production cost coefficient, respectively of ith unit ; ζ_i ($) is hot start-up cost considered when i^{th} unit has been shut down for a short time; δ_i ($) is cold start-up cost considered when i^{th} unit has been shut down for a long time; $Toff_i$ is number of hours i^{th} unit has been OFF at the time of start-up, τ_i is cooling time constant of ith unit in hours.

The heat rate characteristic of a fossil fuel unit relates the hourly heat energy requirement (Btu/h or MBtu/h) with the corresponding power output. It can be converted into its production cost curve ($/h) by multiplying it with the fuel price. Total fuel cost of a unit for any power output is its production cost $C_{i(t)}^{pr}$. This work adopts its quadratic function

978-1-4799-6047-7/14 $31.00 © 2014 IEEE   389

representation as in (8). Let ϕ be the fuel price in \$/MBtu. Then,

$$C_{i(t)}^{pr} = \phi(k_{0i} + k_{1i}Q_{i(t)} + k_{2i}Q_{i(t)}^2) \tag{10}$$

Here, k_{0i}, k_{1i} and k_{2i} are heat rate coefficients for unit i in MBtu/h, MBtu/MWh and MBtu/MW^{2}h, respectively. Comparing (10) with (8), the production cost coefficients a_i, b_i and c_i can be expressed in terms of heat rate coefficients and fuel cost as $a_i = \phi k_{0i}$, $b_i = \phi k_{1i}$, and $c_i = \phi k_{2i}$.

The optimization problem defined in (2)-(6) can be solved to obtain the optimal bid price $P_{i(t)}$ for each ith unit of Genco X at trading interval t. In (2), $P_{i(t)}$ and $G-1$ rival bid price $\tilde{P}_{r,i}$ do not appear explicitly but are implicitly included in the process of determining MCP $M_{(t)}$. Using the normal pdf to represent the distribution of rival price, the strategic bidding problem of Genco X becomes a stochastic optimization problem. This is transformed into an equivalent deterministic formulation using Monte Carlo simulations.

III. PROPOSED SOLUTION ALGORITHM

Monte Carlo simulations obtain probabilistic approximation of a mathematical problem by using statistical sampling technique. Expectation of the objective function over whole solution domain gives the required result. In the proposed strategic bidding problem, Monte Carlo simulations of uncertain rival behavior are incorporated with an optimization algorithm to develop optimal bids.

A. Monte Carlo Approach

Corresponding to the proposed problem, main solution steps of Monte Carlo approach proceeds as follows:

- Generate large number of random samples for bid price of rival Gencos' units, considering their pdfs.
- Obtain large trial outcomes of bid price of the units of Genco X for each hour, by solving the optimization problem with sample values of rival bids.
- Calculate expected bid price value by taking expectation over all trial outcomes.

Detailed algorithm:

1) Specify number of Monte Carlo simulations, MC.
2) Initialize simulation counter $mc = 1$.
3) Generate random sample values of bid prices $P_{r,i}$ for each ith unit of $G-1$ rival Gencos; $(r = 1, 2, ... G-1)$.
4) Use DE/BBO to search optimal bid price for each ith unit of Genco X. (This step is detailed in Sec. V).
5) Store the optimal prices of each unit and for each trading interval as $P_{i(t)}^{mc}$.
6) Update $mc = mc + 1$.
7) If $mc < MC$, go to (3); else go to (8).
8) Calculate the expected value of optimal bid prices, *i.e.*, mean of $P_{i(t)}^{mc}$ $(mc = 1, 2, ...MC)$. This is the optimal bid price $P_{i(t)}$ of each ith unit of Genco X for bidding in t^{th} trading period over 24 hour horizon.

IV. HYBRID DE/BBO

Recently, a novel hybrid of DE with BBO, referred as DE/BBO, has been proposed for combining the goodness of both the techniques [20]. In the following sub sections, a brief description of DE, BBO, DE/BBO is given.

A. Differential Evolution (DE)

DE is a population based stochastic parallel search algorithm and creates a new candidate solution set iteratively, by operators: mutation, crossover, and selection [19]. These are briefly described below.

1) Mutation: Mutation creates mutant vectors $X_i^{/k}$ by perturbing a randomly selected vector X_{r1}^k with the difference of two other randomly selected vectors, X_{r2}^k and X_{r3}^k, at the kth iteration, as per (11)

$$X_i^{/k} = X_{r1}^k + F \times (X_{r2}^k - X_{r3}^k); \quad i = 1, 2, ...N_p \tag{11}$$

X_{r1}^k, X_{r2}^k and $X_{r3}^k \in [i = 1, 2, 3, ..., N_p]$ and $r1 \neq r2 \neq r3 \neq i$

N_p is the size of parent population. X_{r1}^k, X_{r2}^k and X_{r3}^k are selected afresh for each parent vector. $F \in [0, 2]$ is "scaling factor" and controls the perturbation in the mutation process and helps to improve convergence [19].

2) Crossover: Under crossover operation, the parent vector is mixed with the mutant vector to yield an offspring as per (12).

$$X_{ij}^{//k} = \begin{cases} X_{ij}^{/k}, & \text{if } rand_j < CR \text{ or } j = q \\ X_{ij}^{k}, & \text{otherwise} \end{cases} \tag{12}$$

Here, $j = 1, 2, ...D$; D is the number of decision variables; X_{ij}^k is the jth decision variable of ith target vector at k^{th} iteration; $X_{ij}^{/k}$ is the jth decision variable of ith mutant vector at k^{th} iteration; $X_{ij}^{//k}$ is the jth decision variable of ith offspring vector at k^{th} iteration and q is a randomly chosen index $\in [j = 1, 2, 3, ..., D]$. $CR(\in 0, 1)$ is the "Crossover constant" that controls the exploration and diversity of population [19].

3) Selection: Selection among the set of offspring and parent vectors is carried out on the basis of respective objective function values. Equation (13) models the process [19].

$$X_i^{k+1} = \begin{cases} X_i^{//k}, & \text{if } f\left(X_i^{//k}\right) \leq f\left(X_i^k\right); \\ X_i^k, & \text{otherwise} \end{cases} \quad i = 1, 2, ...N_p \tag{13}$$

B. Biogeography Based Optimization (BBO)

BBO is a new population based biogeography inspired global optimization algorithm. It describes how species migrate from one island to another, how new species arise and how species become extinct. Migration and mutation are its operators as described below [12-16].

1) Migration: It is a probabilistic operation and shares information among habitats. Poor solutions tend to accept more useful information from good solutions. This makes BBO good at exploiting the information of current population. Objective function value for each habitat decides species count of a habitat, which governs migration rates of a habitat based on migration model. Emigration rate μ and immigration rate λ decides migration between habitats. Elitism is incorporated to prevent the best solutions from being corrupted by immigration.

2) Mutation: Catastrophic events are modelled as mutation, where the mutation rates are determined using species count probabilities. BBO uses a unique mutation scheme which avoids medium species count solutions [13].

$$mutation\_rate_i = P_{\text{mutate}}(1 - \text{Prob}_{ik} / \text{Prob}_{\text{max}}) \qquad (14)$$

Here, P_{mutate} is mutation probability. Prob_{ik} is the species count probability of habitat i such that it contains exactly k species. Prob_{max} is the maximum of species count probabilities of all habitats. The detailed description of mutation operation in BBO can be referred from [13].

C. DE/BBO with Hybrid Migration Operator

DE has good exploration ability of the search space due to its unique mutation and stochastic crossover. It has been found that, in DE, initially the solutions move very fast towards the optimal point but fail to perform at later stages during fine tuning. In BBO, solutions get fine-tuned gradually during progression of the migration operation. Thus, DE has good exploration ability to find the region of global optima, while BBO has good exploitation ability for global optimization. Hybrid migration operator is the most important step in DE/BBO algorithm [20].

1) Hybrid Migration Operator: It is the main operator of DE/BBO to combine DE's crossover and mutation with migration operation of BBO, and is described in Algorithm 1. In this operator, an offspring U_i incorporates new features from population members. Hybrid migration operator can balance exploration and exploitation effectively.

Algorithm 1: Hybrid Migration Operator [20]

for $i = 1$ to N_p **do**

  Select uniform randomly $r_1 \neq r_2 \neq r_3 \neq i$

  j_{rand} = randint$(1, D)$

  for $j = 1$ to D **do**

    if randreal$(0,1) < \lambda_i$ **do**

      if randreal$_j(0,1) > CR$ *or* $j == j_{rand}$ **do**

        $U_i(j) = X_{r1}(j) + F \times (X_{r2}(j) - X_{r3}(j))$

      else

        Select another habitat X_k with probability

        proportional to μ_k

        $U_i(j) = X_k(j)$

      end if

    else

      $U_i(j) = X_i(j)$

    end if

  end for

end for

Here, randint $(1, D)$ is a uniformly distributed random integer number between 1 and D. randreal$_j(0,1)$ is a random real number between 0 and 1. $X_i(j)$ is the jth variable of individual X_i.

2) Main Procedure of DE/BBO: Its basic steps are shown in Algorithm 2.

Algorithm 2: DE/BBO Algorithm [20]

Generate initial population Pop

Evaluate fitness for each individual in Pop

while Termination criteria is not satisfied **do**

  For each individual, map fitness to the number of species

  Calculate immigration rate λ_i and emigration rate μ_i for each individual

  Modify the population with hybrid migration operation of Algorithm 1

  for $i = 1$ to N_p **do**

    Evaluate offspring U_i

    if U_i is better than Pop_i **then**

      $Pop_i = U_i$

    end if

  end for

end while

V. DE/BBO FOR OPTIMAL BIDDING STRATEGY FORMULATION

Multi hour strategic bidding problem of a Genco, facing rivals of uncertain bidding nature, is being proposed to be solved by Monte Carlo method. The corresponding algorithm is detailed in Sec. III. DE/BBO can be used to optimize the profit function for given simulation of rival bids. Step "d" of Sec. III, is detailed below:

1) Initialize parent population Pop with number of habitats/individual equal to N_p, BBO parameters like maximum immigration rate I, maximum emigration rate E, habitat modification probability P_{mod}, lower bound λ_{lower} and upper bound λ_{upper} for immigration probability per habitat, maximum number of iterations, and DE parameters like CR and F.

2) Initialize all habitats/individuals representing a possible bidding strategy of Genco X.

3) For every habitat, MCP is identified for each hour by arranging the bids of competing Gencos in increasing stepwise curve. Finally, cumulative profit of Genco X, *i.e.*, fitness for each habitat/individual is calculated. Then, sort the habitats in descending order.

4) Set DE/BBO iteration counter = 1.

5) Probabilistically perform hybrid migration operation as per Algorithm 1. Then, calculate cumulative profit/fitness for every habitat and sort the habitats.

6) Perform selection operation between initial population and the newly generated population obtained from Step '5'.

7) Go to Step (5) for next iteration, till all specified iterations are completed.

VI. NUMERICAL RESULTS AND DISCUSSIONS

A practical case study of a thermal Genco X owning three units, including one coal-fired unit, one gas-fired unit and one oil-fired unit, has been considered for trading over 24 hours horizon in a DA EM. A typical daily load curve is shown in Fig. 1. The unit details are provided in Table 1 [21]. The fuel prices are based on practical market data of year 2013 [2, 22]. Accordingly, the production cost coefficients in $/h for different units are calculated.

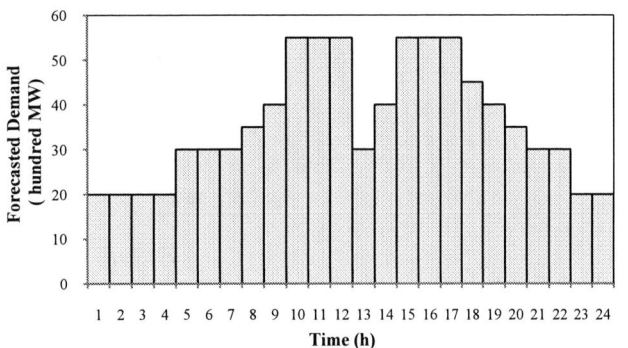

Figure 1. Typical daily load curve

There are four rival Gencos, each having similar three units as Genco X, however, with different capacity. Rivals bidding parameters are given in Table 2. Normal probability distribution parameter $\rho_{r,i}$ of the rival Genco units has been assumed on the basis of their marginal cost. Cumulative profit maximization, given by Eq. (2), is the objective function.

Simulation parameters are: MC =10000, N_p = 80, P_{mod} = 1, I = 1, = 1, iter = 100, λ_{lower} = 0, λ_{upper} = 1, P_{mutate} = 0.1 and p = 4, CR = 0.1 and F = 0.7. Minimum bid price of each unit is set at the marginal cost, without the start-up and shut-down cost. Therefore, minimum bid prices for units 1, 2 and 3 are 82.55 $/MWh, 109.28 $/MWh and 21.37 $/MWh, respectively. Price cap is 120 $/MWh. Initial state of all three units are kept as ON with their $H_{i(t)}^{on}$ equal to their MUT.

Optimal bid prices of units 1-3, hence obtained, and the expected MCPs for each trading hour are shown in Fig. 2. Table 3 gives the expected hourly power dispatch of the units. ND stands for no dispatch of that unit. Fig. 3 shows the expected hourly profit of Genco X. The optimal cumulative profit is equal to $913090.

The results clearly reflect that the coal, oil, and gas differ in

their bidding strategy over 24 hours bidding horizon, due to variation in fuel prices and inter-temporal constraints. Coal unit has the least cost of generation but is constrained by large up and down times. Also, start up and shut down costs of coal unit are more than other units. Hence, it may cause negative profit for Genco during low demand periods. Also, due to high shut down time, they may be restrained from giving profit for a longer period. Costly oil unit is suitable only for peak periods. Gas unit can be dispatched during most hours of the day and assist the coal unit to attain high profit by affecting MCP. Also, once shut down, it can be dispatched earlier profitably due its characteristic low down time with low start-up cost. Hence, coal has a tendency to bid low and dispatch for all the hours, while costly gas and oil units become marginal units and affect MCP.

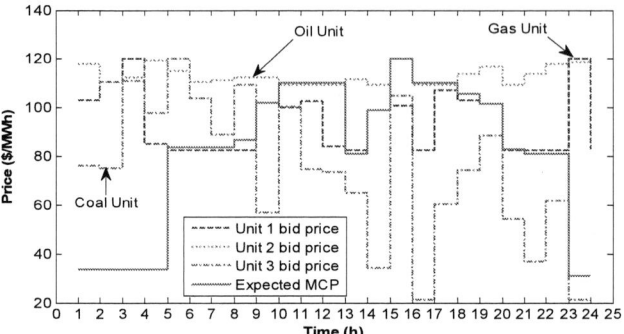

Figure 2. Bid prices and expected hourly MCP

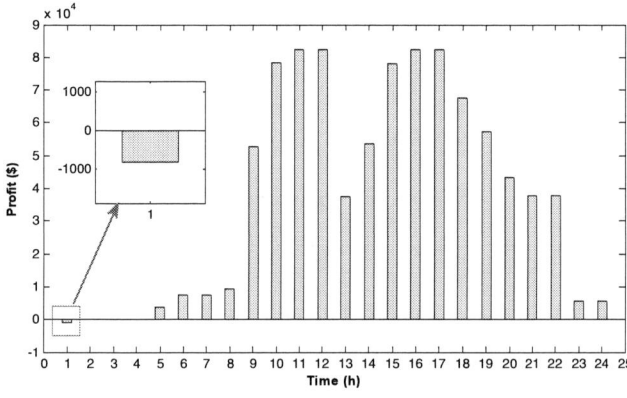

Figure 3. Expected hourly profit curve of Genco X

TABLE I. DATA OF GENCO X

Unit	Fuel Price ($/MBtu)	k_0 (MBtu/h)	k_1 (MBtu/MWh)	k_2 (MBtu/MW^{2}h)	a ($/h)	b ($/MWh)	c ($/MW^{2}h)	Q^{max} (MW)	Q^{min} (MW)	MUT (h)	MDT (h)	ζ ($/h)	δ ($)	τ (h)	C^{sd} ($)
1(Gas)	4.7	67.05	8.71	0.0111	315.14	40.94	0.0520149	400	100	2	1	1500	2500	1	200
2(Oil)	6.5	64.16	8.34	0.0106	417.04	54.21	0.068835	400	100	2	1	1500	2500	1	200
3(Coal)	2.4	510	7.2	0.0014	1224	17.28	0.00341	600	150	8	8	2000	4000	8	400

TABLE II. RIVALS' BIDDING PARAMETERS

Bid parameters	Unit $\rightarrow$	Rival 1			Rival 2			Rival 3			Rival 4		
		1(Gas)	2(Oil)	3(Coal)	1(Gas)	2(Oil)	3(Coal)	1(Gas)	2(Oil)	3(Coal)	1(Gas)	2(Oil)	3(Coal)
$\tilde{Q}_{r,i}$ (MW)		200	300	500	300	300	500	300	350	500	400	300	500
$\rho_{r,i}$ ($/MWh)		82	107	30	84	108	30	84	110	30	83	110	30
$\sigma_{r,i}$ ($/MWh)		3	3	3	3	3	3	4	4	4	3	3	2

TABLE III.   EXPECTED POWER DISPATCH (MW) OF GENCO X

Hour	Expected Power Dispatch			Hour	Expected Power Dispatch		
	Unit 1	Unit 2	Unit 3		Unit 1	Unit 2	Unit 3
1	ND	ND	ND	13	ND	ND	600
2	ND	ND	ND	14	ND	ND	600
3	ND	ND	ND	15	400.0	100.25	600
4	ND	ND	ND	16	400.0	369	600
5	331	ND	ND	17	400.0	369	600
6	331	ND	ND	18	400.0	ND	600
7	331	ND	ND	19	400.0	ND	600
8	397	ND	ND	20	294.0	ND	600
9	188	ND	600.0	21	180.0	ND	600
10	400	369	600.0	22	ND	ND	600
11	400	369	600.0	23	ND	ND	600
12	400	369	600.0	24	ND	ND	600

TABLE IV.   CONSISTENCY OVER 20 RUNS OF MONTE CARLO APPROACH

Heuristic Algorithm	Profit ($)				Simulation time per run(s)
	Minimum	Maximum	Average	Standard Deviation	
GA	912387	930579	920397	8552.8	7.96
BBO	912452.8	930708	920448	8567.5	7.80
DE	912495.6	930984	920466	8469.2	7.16
DE/BBO	912505.6	930984	920467.1	8466.1	7.66

Table IV gives the consistency evaluation of DE/BBO. Algorithm for the proposed solution approach of Sec. III is run 20 times with one Monte Carlo simulation and with different heuristic algorithms. Minimum, maximum, average and standard deviation over the trials is tabulated. It can be seen that performance of DE/BBO is more consistent than others.

VII.   CONCLUSION

This paper presents strategic bidding problem of a thermal multi-unit Genco for hourly DA spot EM over a 24 hour horizon. A novel approach based on Hybrid Differential Evolution with Biogeography Based Optimization (DE/BBO) is proposed as the solution tool. 24 hourly trading periods for multi-unit Gencos present a multi-variable and highly nonlinear complex problem due to generator cost functions, inter temporal constraints and dynamic demand.The performance analysis proves that DE/BBO gives competitive performance with DE with better robustness due to enhanced exploitation quality. Also, DE/BBO gives better quality solution, with consistent performance as compared to GA and BBO. The simulations and comparative results prove feasibility and efficiency of DE/BBO algorithm to formulate bidding strategy for Gencos in competitive energy markets, under constrained environment.

REFERENCES

[1] A. K. David and F. Wen, "Strategic bidding in competitive electricity market: A literature survey," in *Proc. 2000 IEEE Power Engineering Society General Meeting*, pp. 2168-2173.

[2] X. R. Li, C. W. Yu, Z. Xu, F. J. Luo, Z. Y. Dong, and K. P. Wong, "A multimarket decision making framework for GENCO considering emission trading scheme," *IEEE Trans. Power Syst.*, vol. 28, no. 4, pp. 4099-4108, Nov. 2013.

[3] G. Zachmann, "A stochastic fuel switching model for electricity prices," *Energy Econo.*, vol. 35, pp. 5-13, Jan. 2013.

[4] A. J. Wood and B. F. Wollenberg, *Power Generation, Operation, and Control.* Singapore: John Wiley & Sons (Asia) Pt. Ltd., 2003.

[5] R.W. Ferrero, V. C. Ramesh, and S. M Shahidehpour, "Transaction analysis in deregulated power system using game theory," *IEEE Trans. Power Syst.*, vol. 12, no. 3, pp. 1340-1347, Aug. 1997.

[6] T. Li and S. M. Shahidehpour, "Strategic Bidding of transmission constrained Gencos with incomplete information," *IEEE Trans. on Power Syst.*, vol. 20, no. 1, pp. 437-447, Feb. 2005.

[7] F. Careri and C. Genesi, "Strategic bidding in a day ahead market by coevolutionary genetic algorithms," in *Proc. 2010 IEEE Power Engineering Society General Meeting*, Minneapolis, pp. 1-8.

[8] P. Attaviriyanupap, H. Kita, E. Tanaka, and J. Hasegawa, "New bidding strategy formulation for day-ahead energy and reserve markets based on evolutionary programming," *Int. J. Elect. Power & Energy Syst.*, vol. 27, no. 3, pp. 157-168, Mar. 2005.

[9] E. N. Azadani, S. H. Hosseinian, and B. Moradzadeh, "Generation and reserve dispatch in a competitive market using constrained particle swarm optimization," *Int. J. Elect. Power & Energy Syst.*, vol. 32, no. 1, pp. 79-86, Jan. 2010.

[10] C. Boonchuay and W. Ongsakul, "Optimal risky bidding strategy for a generating company by self organizing hierarchical PSO," *Energy Convers. and Manag.*, vol. 52, no. 2, pp. 1047-1053, Feb. 2011.

[11] P. Bajpai and S. N. Singh, "Fuzzy adaptive particle swarm optimization for bidding strategy in uniform price spot market," *IEEE Trans. on Power Syst.*, vol. 22, no. 4, pp. 2152-2160, Nov. 2007.

[12] D. Simon, "Biogeography–Based Optimization," *IEEE Trans. on Evol. Comp.*, vol. 12, no. 6, pp. 702-713, Dec. 2008.

[13] A. Bhattacharya and P. K. Chattopadhyay, "Biogeography based optimization for different economic load dispatch problems," *IEEE Trans. on Power Syst.*, vol. 25, no. 2, pp. 1064-1077, May 2010.

[14] A. Bhattacharya, P. K Chattopadhyay, "Application of biogeography-based optimization for solving multi-objective economic emission load dispatch problems," *Elect. Power Comp. and Syst.*, vol. 38, no. 3, pp. 340-65, Jan. 2010.

[15] A. Rathi, A. Agarwal, A. Sharma, P. Jain, "A new hybrid technique for solution of economic load dispatch problems based on biogeography based optimization," in *Proc. 2011 IEEE Asia Pacific Tencon*, pp. 19-24.

[16] P. Jain, A. Agarwal, N. Gupta, R. Sharma, U. Paliwal, R. Bhakar , "Profit maximization of a generation company based on Biogeography based Optimization, in *Proc. 2012 IEEE Power Engineering Society General Meeting*, pp. 1-6.

[17] G. Xiong, D. Shi, and X. Duan, "Multi-strategy ensemble biogeography-based optimization for economic dispatch problems," *Appl. Energy*, vol. 111, pp. 801-811, Nov. 2013.

[18] M. R. Lohokare, S. S. Pattnaik, B. K. Panigarhi, and S. Das, "Accelerated biogeography-based optimization with neighborhood search for optimization," *Appl. Soft Comput.*, vol. 13, no. 5, pp. 2318-2342, May 2013.

[19] R. Storn, and K. Price, "Home Differential Evolution, 2008." (http://www. ICSI. Berkely.edu/~storn/code.html).

[20] W. Gong, Z. Cai, and C. X. Ling, "DE/BBO: A Hybrid differential Evolution with Biogeography Based Optimization for Global Numerical Optimization," *Soft Computing*, vol. 15, no. 4, pp. 645-65, 2011.

[21] H. Ma, "An analysis of the equilibrium of migration models for biogeography-based optimization," Inform. Sciences, vol. 180, no. 18, pp. 3444-3464, Sept. 2010.

[22] Units' Data, [Online] available: http://www.motor.ece.iit.edu/data/

[23] U. S. Energy Information Administration website 2013, [Online] available: http://www.eia.gov/

On the Control and Design Issues of Single Phase Transformerless Inverters for Photovoltaic Applications

Amit K Gupta
Student Member, IEEE
Department of Electrical
Engineering, IIT Bombay
Mumbai, India
amit.k.gupta@iitb.ac.in

Madhuwanti S Joshi
Department of Electrical
Engineering, IIT Bombay
Mumbai, India
mjoshi@iitb.ac.in

Vivek Agarwal
Senior Member, IEEE
Department of Electrical
Engineering, IIT Bombay
Mumbai, India
agarwal@ee.iitb.ac.in

Abstract— **Design and development details of a high power density, low cost, single phase transformerless PV inverter with reduced losses are reported. H-4 bridge topology is adopted to implement the proposed design where count of power devices is limited to four. One leg of the bridge operates at high switching frequency while the other leg operates at line frequency, this modulation technique reduces the switching losses. This also obviates the requirement of fast recovery anti parallel diodes across power devices for low frequency leg. Use of fewer devices and digital control (instead of analog control) has led to reduced cost and size. Switching and conduction losses are further reduced by using MOSFETs with very low on state resistance. A novel control scheme has been implemented to eliminate the DC offset at the output. A highly reliable over current and shoot-through protection is implemented through digital control. The digital control algorithm implemented enhances the reliability of operation at high power level. Optimal design of PCB and high frequency filters with SMD components contribute to high power density. A 3 kW prototype has been developed and tested for various reactive loads. All the analysis, simulation and experimental details of this work are presented.**

Keywords— *Transformerless Inverter, Photovoltaic (PV), High efficiency inverter, DC-AC Power conversion, Single phase PV inverters, High power density Inverters.*

I. INTRODUCTION

World is facing acute shortage of energy as well as polluted environment. To overcome the energy shortage and to protect the environment, a clean and abundant energy substitute is required. Solar energy has the potential to serve as the desired alternative (renewable) energy source. It is predicted [1] that the solar energy shall contribute up to 64% of total global energy requirement by the end of this century. Solar energy tapped through PV modules needs to be converted into a suitable form which is compatible with the load. As a majority of existing loads are AC, PV inverters that perform DC-AC conversion are the major functional unit of most solar PV plants [2, 3]. It has been a continuous endeavor of researcher to develop low cost, highly efficient and compact inverters.

The transformer plays an important role in solar PV system, PV inverters can be broadly classified as inverters with or without a transformer. Use of transformer in a PV inverters provides galvanic isolation, elimination of DC offset injection and elimination of leakage current in case of grid connected systems. It also allows the use of low voltage PV source which has advantages in terms of safety, low leakage current etc. On the other side, use of transformer increases the overall size, weight and cost of inverters. It also contributes to losses, resulting in an overall reduction in efficiency. These drawbacks of an inverter can be minimized by eliminating the transformer. So the recent trend has been to eliminate the transformers from PV inverters [4, 5]. However other challenges need to be addressed.

Various configurations of the PV systems can be broadly classified as shown in Fig. 1. [6, 7] (a) Central inverters, (b) String inverters, (c) Module based inverters.

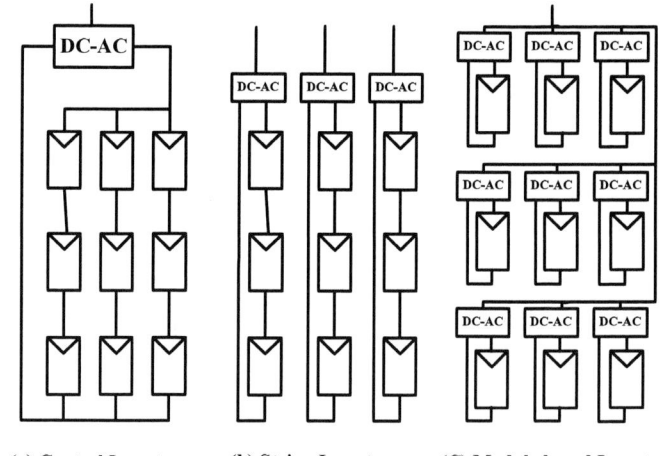

(a) Central Inverters   (b) String Inverters   (C) Module based Inverters

Fig. 1.   Configurations of PV systems [6-7].

Since the generated voltage of a single PV module is less, eliminating transformer from the module based inverter produces the requirement of an additional stage of boost circuit at the input as shown in Fig. 2 (a). Whereas in Central or String configuration, high DC link voltage is produced across inverter which eliminates the requirement of an additional boost stage at the input side of transformerless inverter as shown in Fig. 2 (b). But on the other hand, in the Central and String configurations, leakage current may be a major issue to be dealt for grid connected transformerless PV inverters.

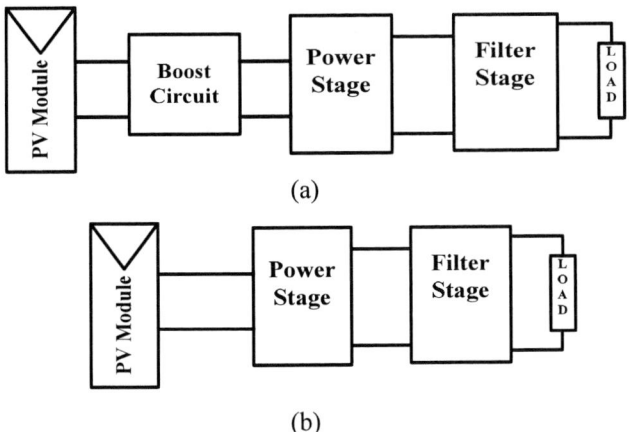

(a)

(b)

Fig. 2. Power conversion stages in transformerless PV inverters. (a) with additional input boost stage (b) without input boost stage.

Professional bodies and agencies such as IEEE, IEC, European standard (EN) and German VDE have set the standards on various performance parameters of an inverter like leakage current, DC offset injection, limitation on generated harmonics etc. [8-11]. Eliminating the transformer from an inverter poses major challenges in meeting the standards imposed by national and international regulatory commissions and agencies [12]. In fact, in grid interactive PV inverters, these challenges are more intense because of grid's reliability and stability issues. These challenges are addressed by many researchers through various approaches. While some of them are successful in achieving high efficiency [4,5,7,13-15], others are successful in eliminating the leakage current [16, 18-21]. Several new topologies to overcome the above mentioned issues had been proposed [7, 14, 22, 23].

Apart from above mentioned international standards, PV inverters must be compact, highly efficient and cost effective too. To achieve all these features, novel system design approach and control scheme are required. A separate control algorithm is applied to control DC offset injection where it will regulate AC output on cycle by cycle basis in order to eliminate the DC component from the output AC. An optimum design approach is discussed where along with high efficiency and low cost, a high power density (6 W/in^3 approx.) is achieved. Both software and hardware protections are implemented reasonably, so that the system works more reliably and environment friendly manner. Some control and design features are discussed and uniquely applied for both standalone and grid connected mode.

In section II, challenges involved in the design of transformerless systems are described. In section III, some unique approaches are explained to meet those challenges. In section IV, system specifications, detailed analysis and simulation results are presented to justify the approach. In section V, detail of development of experimental prototype is discussed. The working model and experimental results

validate the claim made. Finally, in section VI, major conclusion of the presented work are summarized along with future work.

II. DESIGN CHALLENGES

National Electric Code (NEC) sets a standard for galvanic isolation described in NEC-690.4 says PV side DC circuits must be separated from AC wiring by proper partitioning unless the wires are electrically connected, like a ground wire [12]. This isolation requirement is violated by transformerless inverters. Therefore, grid connected transformerless topologies have issues related to leakage currents, DC current injection etc. To protect power system, there are some standards applicable to PV inverters as mentioned below. How these standards are addressed in our design is also described in the next section.

A. DC current injection

Presence of transformer in inverter systems does not allow DC components to pass through it. Transformerless inverters can suffer from large injected dc current because of asymmetric inverted positive and negative cycles of AC and large panel capacitance, as shown in Fig. 3. A DC component with the injected AC currents into the grid should be less than the permissible value. Otherwise, it can saturate the distribution transformers and potentially harm AC loads, resulting in large current. This DC component is harmful not only in the grid connected systems but also in standalone systems.

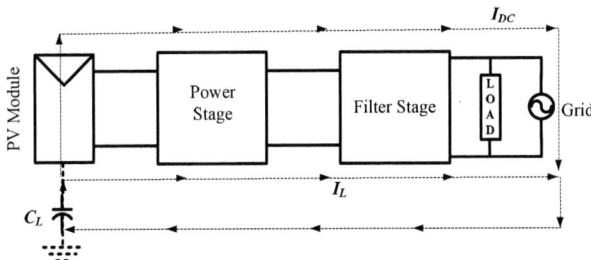

Fig. 3. Leakage current and DC injection into the grid.

TABLE I. INTERESTING STANDARDS IMPOSED BY SOME INTERNATIONAL PROFESSIONAL AGENCIES.

IEEE 1547-2008		IEC 61727		EN 61000-3-2		VDE 0126-1-1	
Permissible DC current injection value							
Less than 1%		Less than 0.5%		Less than 0.22A		Less than 1A	
Permissible Harmonic distortion							
Nominal Power (10 kW)		Nominal Power (30 kW)		Nominal Power 16 A @ 230 V		-	
Order	Limit	Order	Limit	Order	Limit	Order	Limit
3-9	4 %	3-9	4 %	3	2.3 A	3	3 A
11-15	2 %	11-15	2 %	5	1.14 A	5	1.5 A
17-21	1.5 %	17-21	1.5 %	7	0.77 A	7	1 A
23-33	0.6 %	23-33	0.6 %	9	0.4 A	9	0.7 A
		>33	0.3 %	11	0.33 A	11	0.5 A
Even harmonics are limited to 25% of odd harmonics. THD Less than 5%				13	0.21 A	13	0.4 A

Standards, such as IEEE 1547-2008, IEC 61727, EN 61000-3-2 and VDE 0126-1-1 have set the maximum DC current injection limit into the grid as shown in TABLE I [20].

B. Harmonic Distortion

Power electronic inverters do not produce pure sine wave. There is always a distortion in output sine wave, which is referred to as Harmonic Distortion. The problem of harmonic distortion is also present in transformerless topologies. From the power quality point of view, standards have been set by IEEE 1547-2008, IEC 61727, EN 61000-3-2 and VDE 0126-1-1 for maximum permissible harmonic distortion range as shown in TABLE I.

C. Leakage current

In transformerless grid connected inverters, galvanic connection is present between PV modules and the grid. Since the PV modules are grounded, that forms a capacitance between PV module and the ground. Resultantly, a large common mode current may appear between PV system and the grid known as leakage current [19, 20] as shown in Fig. 3. In Central and String configurations where series and/or parallel combinations of PV modules are used, a large capacitance can be formed between ground and PV that may lead to high leakage current. For module based inverters, leakage current is less. Presence of leakage current in grid connected systems distorts grid current as well as increases the losses in the system. Therefore, it becomes highly essential to minimize leakage current in grid connected transformerless PV inverters.

This can be limited by appropriate modulation technique and electromagnetic interference (EMI) filters [20-21]. The German DIN VDE 0126-1-1 has certain mandate on the extent of leakage current allowed as shown in TABLE II [11].

D. Protection and reliability

One of the essential requirement of transformerless PV inverters is protection. It should also be reliable to operate on various loads. In dangerous load or environmental conditions, inverter should be capable to trip the operation and disconnect the faulty part from it. A large energy storage device usually capacitor is connected across PV array (as shown in Fig. 4. and Fig. 5.) that could be responsible for momentarily huge current flow in case of any abnormal state. Those abnormal states can be categorized as shoot-through and over current which need to be protected.

E. Electric Conversion Efficiency (ECE)

This specification is considered as the most important parameter in PV inverter circuits. Instead of efficiency curve, weighted efficiency concept is defined. One is "European efficiency" [16-17], this concept is suitable for countries with low radiation levels. Other weighted efficiency concept is defined, known as "U.S. efficiency or California Energy

TABLE II. LEAKAGE CURRENT

Average leakage current (mA)	Time limit (sec)
30	0.3
60	0.15
100	0.04
300 (peak)	0.3

Commission (CEC) efficiency" [7] for the regions with higher radiation levels. It is redundant to say that transformerless PV inverter systems should come out with good efficiency figure in such analysis.

III. PROPOSED DESIGN APPROACH

Selection of a topology for power stage is one of the most important design aspect. Some of the popular single phase transformerless topologies are shown in Fig. 4. To implement our design, H-4 topology has been adopted due to low component count, low cost, reduced losses and compact size.

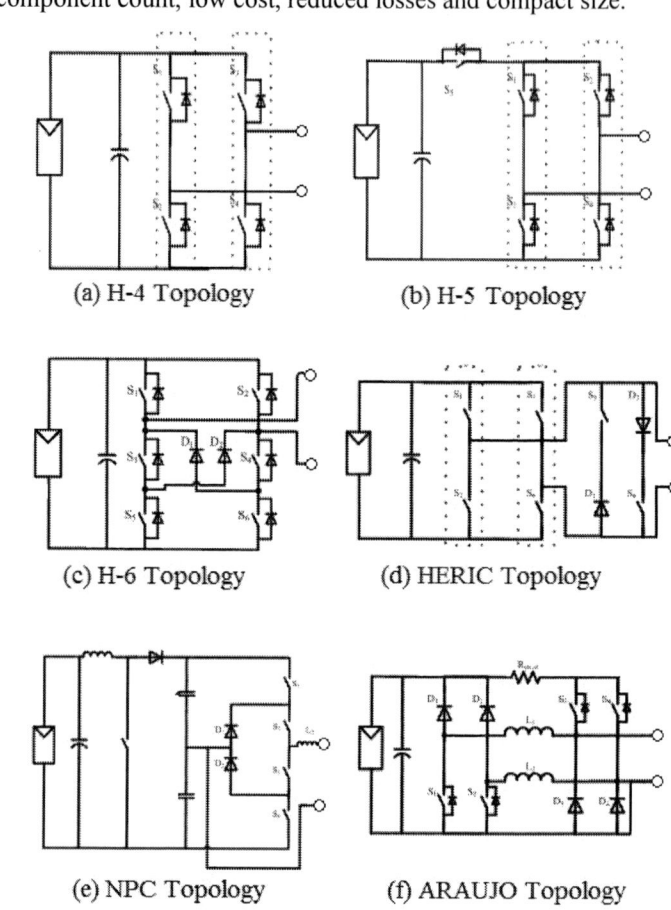

(a) H-4 Topology   (b) H-5 Topology

(c) H-6 Topology   (d) HERIC Topology

(e) NPC Topology   (f) ARAUJO Topology

Fig. 4. Some of the popular single phase transformerless topologies [2, 3, 18, 19].

A schematic diagram of the proposed design is shown in Fig. 5. In the proposed approach, a hybrid method of PWM is implemented. Out of the four power devices of H-4 topology, two switches (of Leg-A) are being operated at high frequency SPWM and remaining two switches (of Leg-B) are being operated at the line frequency (50/60 Hz) just to provide a path to modulating signals.

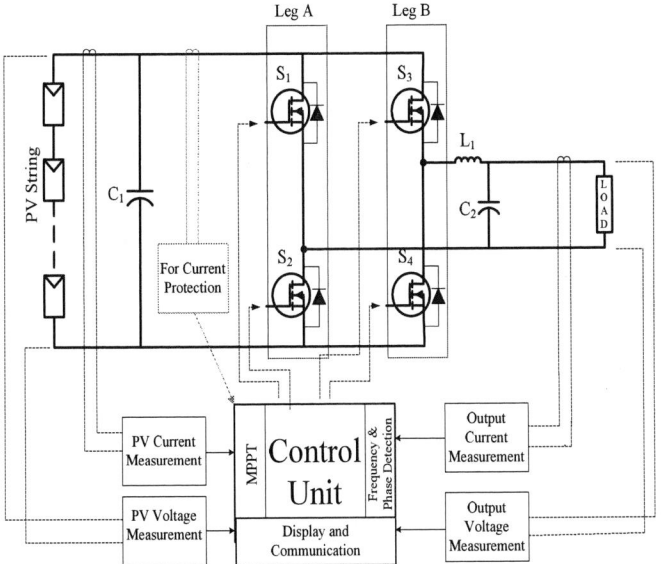

Fig. 5. Schematic diagram of single phase transformerless inverter.

A. Elimination of DC current injection

A novel control algorithm is implemented to control DC component produced by the inverter. It will regulate AC output on cycle by cycle basis in order to eliminate the DC component from the output AC. There could be many reasons of presence of DC component in the output. This algorithm is strong enough to control DC component generated irrespective of any of the reasons.

B. Approach to reduce Total Harmonic Distortion

High frequency sinusoidal pulse with modulation (SPWM) control is implemented to minimize the generated harmonics in standalone mode. To minimize the harmonics in grid connected mode, grid voltage itself is used as reference to modulate the switches of power devices. Switching frequency harmonics are the predominant harmonics generated by the inverters, those are eliminated by high frequency output filters, specially designed to achieve low loss and lower size.

C. Shoot-through and over current Protection

Apart from traditional protection methods, one additional step is taken to control shoot-through phenomenon and to protect against over current. As shown in see Fig. 5, a separate block is added before power devices and after decoupling capacitor where an additional input side current measurement is taking place. Function of this block is to detect and control over current flow through devices in case of shoot-through phenomenon. This will also control over current feeding to the output in case of sudden load variation or malfunctioning of power devices. This special protection scheme makes it more reliable.

D. High Efficiency

Since, only half of the power devices are operating at high frequency, the implemented modulation technique reduces the switching losses significantly. From the filter design point of view laminated iron core inductors are one of the major contributors to reduction in efficiency. Amorphous core can be used instead of laminated core for inductor design to reduce core losses. For power devices, low on resistance ($\ll R_{DSon}$) MOSFETs are used to attain low switching and conduction losses.

E. High Power density

Overall size reduction of an inverter has always been a topic of research. Since transformer is already eliminated from the design resulting in smaller size, high frequency filters are used at the output that has further reduced overall size significantly. The PWM approach to control power devices also eliminates the requirement of anti-parallel fast recovery diodes across the power devices operating on line frequency. MOSFETs help to operate it on increased switching frequency for higher power levels which reduces filter size further for higher power also.

To design high power inverters, several other challanges are faced, simply increasing the device and component ratings are not sufficient. Design of PCB is one of the major challenges along with appropriate ground coupling and control over very high conductive and radiating EMI [3, 16]. Large efforts are given to design a very optimal PCB in respect of adjustment of various voltage levels, accommodation of high current path. Use of SMD components helps to achieve high power density system.

IV. SPECIFICATIONS AND ANALYSIS OF PROPOSED DESIGN

A. Design specifications of the inverter:

To implement the design, 3 kW power rating is chosen. Single phase transformerless inverters of this power rating can be installed in household and commercial rooftop applications. Both standalone and grid tied modes are implemented separately since the control algorithms are different for the both. Specifications of the design are given in Table III.

TABLE III.   SPECIFICATIONS OF INVERTER PROTOTYPE

Parameters	Values	Units
Power rating of the inverter	3000	Watt
Maximum open circuit voltage of the PV source	450	V
Maximum V_{mpp}	380	V
Maximum current from the string	8.3	A
Nominal output AC voltage	230	V
Nominal output current	13	A
Inverter output frequency	50	Hz
Switching frequency	20	kHz

The design approach and various control loops are analyzed and tested through closed loop simulation performed on MATLAB Simulink tool. As discussed above, high frequency 20 kHz SPWM is applied for leg-A (please refer schematic diagram shown in Fig. 5) and line frequency 50 Hz PWM is applied for Leg-B. Control signals of all switches are shown in the scope below where hybrid switching scheme for power devices can be observed, please refer Fig. 6.

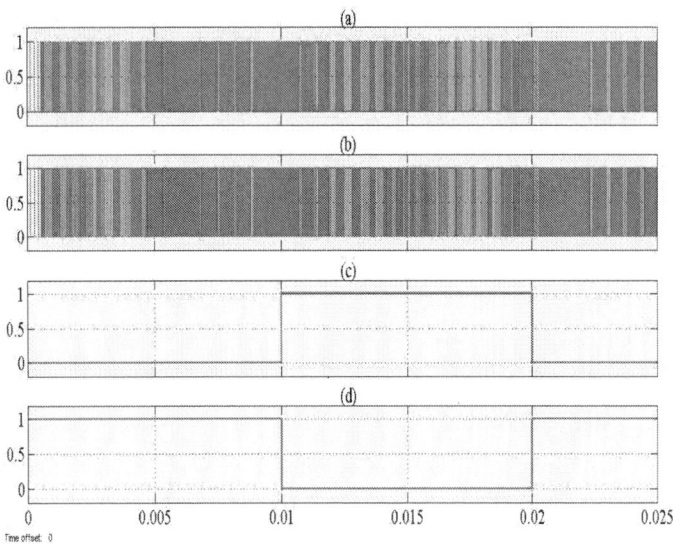

Fig. 6. Control signals of Power devices for (a) S1, (b) S2, (c) S3 and (d) S4.

A closed loop control is controlling the output for various loads in standalone mode; output current and voltage waveforms are shown below in Fig. 7.

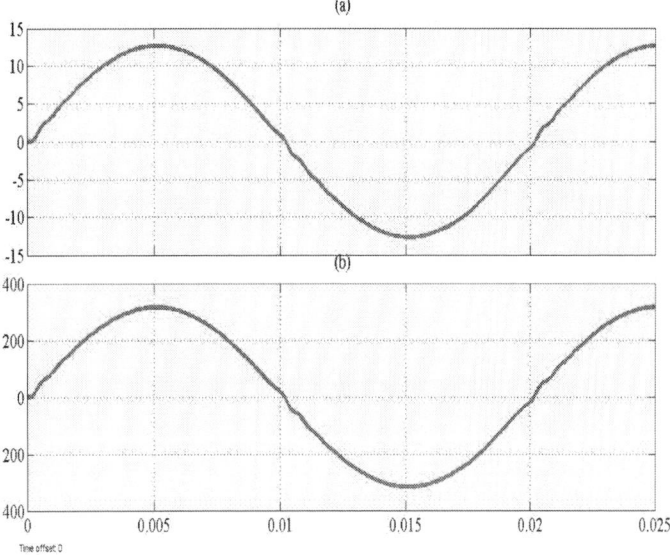

Fig. 7. MATLAB Simulation results. (a) Output current and (b) Output voltage.

V. EXPERIMENTAL PROTOTYPE AND RESULTS

After simulation analysis and results, a high power density (6 W/in^3) experimental prototype of 3 kW is developed in the lab, shown in Fig. 8. This is tested on various loads on full power.

Fig. 8. Experimental Prototype developed in the lab.

Experimental results for single phase transformerless PV inverter is shown below for standalone mode in Fig. 9. where output current and voltage waveforms can be observed. The THD is less than 4.5% which is under the limit set by professional agencies given in Table I. With this first prototype, over 93% efficiency is achieved on full power.

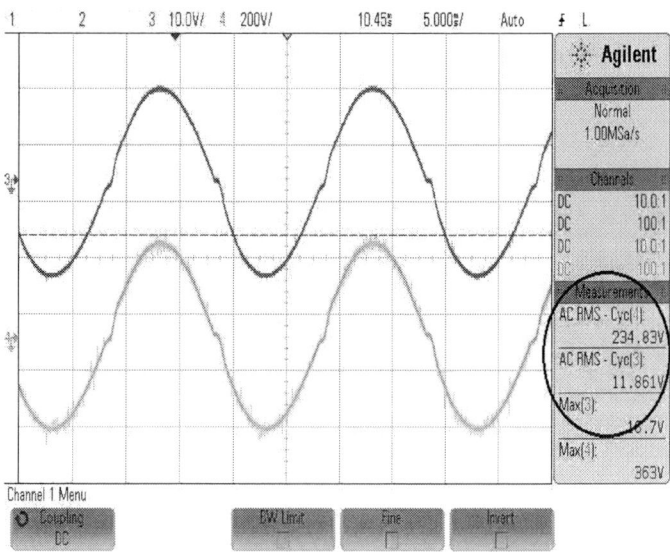

Fig. 9. Output current and voltage waveforms in standalone mode.

This experimental setup is also tested in grid tied mode up to 1200 Watt at 120 V grid voltage with dedicated control algorithm. In this experiment, a large load is being shared between the grid and the PV inverter. Results can be seen in Fig. 10. where grid voltage and PV inverter current feeding to grid are shown.

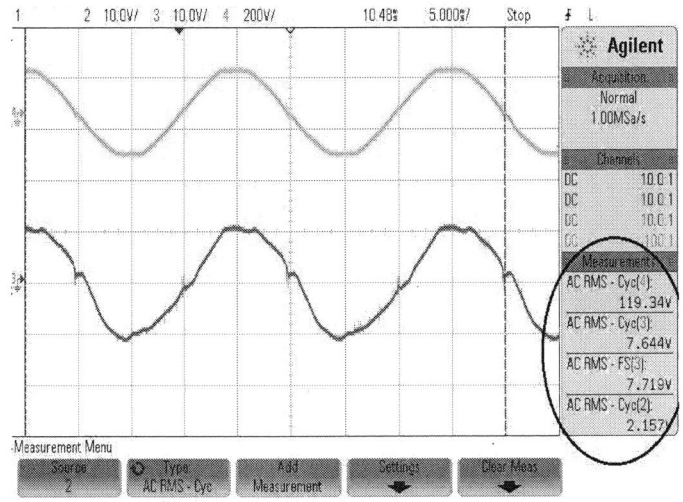

Fig. 10. Grid Voltage and PV inverter current feeding to grid.

VI. CONCLUSION AND FUTURE WORK

In transformerless inverters, PV module based configuration is not possible unless, a boost converter is cascaded before inverter. A prototype of 3 kW single phase transformerless PV inverter has been designed and developed. This prototype of transformerless high power inverter is supposed to be operated in String configuration of PV modules. Approximately 93% efficiency and 6 W/in^3 power density have been achieved with this first laboratory prototype. This comes out to be a very low cost and light weighted PV inverter suitable for developing countries for household and commercial rooftop applications . The system has been tested on both standalone and Grid tied mode with dedicated digital control algorithm.

As a future work, few areas of improvements are identified. Power injection in grid tied mode needs to be increased since it is tested up to 1200 Watts only. For grid tied applications leakage current needs to be minimized. Zero crossing distortion and THD needs to be minimized as well by improvement in control loops. Output inductor using Amorphous core is under design and development process till date. By replacing laminated core by amorphous core for output inductor 2-3% increment in efficiency is expected. Some losses can still be controlled using some intelligent control and design steps. The target efficiency is 97% and the target power density is 20 W/in^3.

REFERENCES

[1] German Advisory Council on Global Change WBGU Berlin 2003 www.wbgu.de; Renewable Energy Policy Network for the 21st Century, Renewables, Global Status Report 2006.

[2] Zheng Zhao, "High Efficiency Single-stage Grid-tied PV Inverter for Renewable Energy System," Ph.D. dissertation, Dept. Elect. Eng. Virginia Poly. Ins. and State University, Blacksburg, VA, 2012.

[3] E. Koutroulis and F. Blaabjerg, "Methodology for the optimal design of transformerless grid-connected PV inverter," *IET Power Electronics*, vol. 5, no. 8, pp. 1491–1499, Jun. 2012.

[4] Bruno Burger and Dirk Kranzer, "Extreme High Efficiency PV-Power Converters," in Power Electronics and Applications, EPE '09, Barcelona, 8-10 Sep. 2009, pp. 1–13.

[5] M. C. Poliseno, R. A. Mastromauro, M. Liserre and A. Dell'Aquila, "High Efficiency Transformerless PV Power Converters," in International Symposium on Power Electronics, Electrical Drives, Automation and Motion (SPEEDAM), Sorrento, Jun. 2012, pp. 93–98.

[6] S. B. Kjaer, J. K. Pedersen, F. Blaabjerg, "A Review of Single-Phase Grid-Connected Inverters for Photovoltaic Modules," *IEEE Trans. Industry Applications*, vol. 41, no. 5, pp. 1292–1306, Sep./Oct. 2005.

[7] S. V. Araújo, P. Zacharias and R. Mallwitz, "Highly Efficient Single-Phase Transformerless Inverters for Grid-Connected Photovoltaic Systems," *IEEE Trans. Industrial Electronics*, vol. 57, no. 9, pp. 3118–3128, Sep. 2010.

[8] IEEE Application Guide for IEEE Std. 1547, IEEE Standard for Interconnecting Distributed Resources With Electric Power Systems, IEEE Standard 1547.2-2008, Apr. 2009, pp. 1–207.

[9] Photovoltaic (PV) Systems—Characteristics of the Utility Interface, IEC Standard 61727 ed2.0, 2004.

[10] Electromagnetic Compatibility (EMC)—Part 3-2: Limits—Limits for Harmonic Current Emissions (Equipment Input Current Under 16 A Per Phase), EN 61000-3-2:2006, 2006.

[11] Automatic Disconnection Device Between a Generator and the Public Low-Voltage Grid, VDE V 0126-1-1:2006-02, 2006.

[12] 2011 National Electrical Code, National Fire Protection Association, Inc., Quincy, MA, 2011.

[13] R. Gonzalez, J. Lopez, P. Sanchis, E. Gubia, A. Ursua and L. Marroyo, "High-Efficiency Transformerless Single-phase Photovoltaic Inverter," in 12th International Power Electronics and Motion Control conference (EPE-PEMC), Portoroz, Aug.-Sep. 2006, pp. 1895–1900.

[14] Kerekes, T., Teodorescu, R., Rodrı´guez, P., Va´zquez, G., Aldabas, E.: 'A new high-efficiency single-phase transformerless PV inverter topology', *IEEE Trans. Ind. Electron.*, vol. 58, no. 1, pp. 184–191, Jan. 2011.

[15] Fraunhofer ISE, Freiburg, "Fraunhofer ISE sets a new world record PV inverter efficiency exceeds 99 percent", Press Release, no. 15/09, pp. 1-4, July 29, 2009

[16] M. C. Poliseno, R. A. Mastromauro and M. Liserre, "Transformer-less photovoltaic (PV) inverters: a critical comparison," in Energy Conversion Congress and Exposition (ECCE), Raleigh, NC, Sep. 2012, pp. 3438–3445.

[17] M. Valentini, A. Raducu, D. Sera and R. Teodorescu, "PV Inverter Test Setup for European Efficiency, Static and Dynamic MPPT Efficiency Evaluation," in International Conf. on Optimization of Electical and Electronic Equipment (OPTIM), Brasov, May 2008, pp. 433–438.

[18] T. Kerekes, R. Teodorescu and U. Borup, "Transformerless Photovoltaic Inverters Connected to the Grid," in 22nd Applied Power Electronics Conference (APEC), Anaheim, CA, USA, Mar. 2007, pp. 1733–1737.

[19] J. Wang, B. Ji, J. Zhao and J. Yu, "From H4, H5 to H6 — Standardization of Full-Bridge Single Phase Photovoltaic Inverter Topologies without Ground Leakage Current Issue," in Energy Conversion Congress and Exposition (ECCE), Raleigh, NC, Sep. 2012, pp. 2419–2425.

[20] E. Gub´ıa, P. Sanchis, A. Urs´ua, J. L´opez, and L. Marroyo, "Ground currents in single-phase transformerless photovoltaic systems," Progress in Photovoltaic: Research and Application, vol. 15, no. 7, pp. 629–650, 2007.

[21] O. Lopez, F. D. Freijedo, A. G. Yepes, P. Fernandez-Comesaa, J. Malvar, R. Teodorescu and J. Doval-Gandoy, "Eliminating ground current in a transformerless photovoltaic application," *IEEE Trans. Energy Convers.*, vol. 25, no. 1, pp. 140–147, Mar. 2010.

[22] A. Nabae, I. Takahashi and H. Akagi, "A New Neutral-Point-Clamped PWM Inverter," *IEEE Trans. Industry Applications*, vol. IA-17, no. 5. pp. 518–523, Sep. 1981.

[23] Tarak Salmi, Mounir Bouzguenda, Adel Gastli and Ahmed Masmoudi, "A Novel Transformerless Inverter Topology without Zero-Crossing Distortion," *International Journal of Renewable Energy Research (IJRER)*, vol. 2, no. 1, pp. 140–146, Feb. 2012.

Behaviour of Wind Turbine under Different Operating Modes

Navjot Singh Sandhu

Research Scholar, School of Renewable Energy & Efficiency
National Institute of Technology
Kurukshetra, India
3127004@nitkkr.ac.in

Saurabh Chanana

Associate Professor, Department of Electrical Engineering
National Institute of Technology
Kurukshetra, India
Saurabhchanana@ieee.org

Abstract— **In wind energy conversion system the wind turbine captures the kinetic energy associate with the wind for the production of electricity with the help of generator. Performance of a wind turbine is dependent upon its operating conditions such as wind speed as well as its design parameters such as blade radius, blade pitch angle, weight etc. In the present paper, authors have made an attempt to develop a MATLAB/SIMULINK model of wind turbine. Model as developed is suitable to investigate its performance under constant as well as variable speed mode. Simulated results as obtained are found to be useful to predict the effect of wind speed and blade pitch on the performance of wind turbine under two operating modes.**

Keywords— Energy; MATLAB/ Simulink; Wind Energy; Wind turbine

I. INTRODUCTION

Wind energy seems to be most viable renewable resource among the number of nonconventional source of energy such as wind, solar, tidal, biogas, geothermal etc. In a wind energy conversion system, wind turbine is used to extract the energy associate with the moving wind. Many researchers made the efforts to study the effect of different parameter like wind velocity, pitch angle and radius on power generated by wind energy conversion system. Reference [1] includes the discussion on the issues and challenges related to wind turbine aerodynamics. Other researcher [2] tried to find out some facts about the resource potential of the wind turbine with the help of seven small wind turbines. Few researchers [3] reported the investigations of wind turbines when operated under constant speed mode using pitch control mechanism. One of the researchers [4] tried to control the active and reactive power with wind system connected to grid through DC link and hence voltage of DC bus is used for control strategy. Another researcher [5] investigated the effects of constant speed turbine on power system oscillations due to sudden load variations. Reference [6] included the performance of variable speed wind turbines using blade pitch control mechanism. It has been shown that during wind variations, control of converter circuit is helpful to capture the maximum power associated with the wind. Another one [7] also proposed a new strategy to control the output of wind turbine generator under variable speed operation. Reference [8] describes the performance of a fix pitch angle wind turbine using Matlab Simulator without any control mechanism.

In this paper Matlab/Simulation circuit model for wind turbine using load torque control is developed and the same (as discussed in the next section) is found to be capable to compare the performance of wind turbine under constant speed as well as variable speed operation.

II. TURBINE MODELLING

In wind energy conversion system the wind turbine captures the wind and produce generate electricity by the help of generator. The blades cuts through the wind velocity and generated a lift force and exerting a turning force. The blades start to rotate due to this lift force. These rotating blades turn a shaft which is connected to a gearbox. The gearbox increases the rotational speed up to a sufficient speed for the generator. In the generator the magnetic field is used to convert this rotational energy into electric energy. The configuration of wind turbine conversion system is shown in Fig 1.

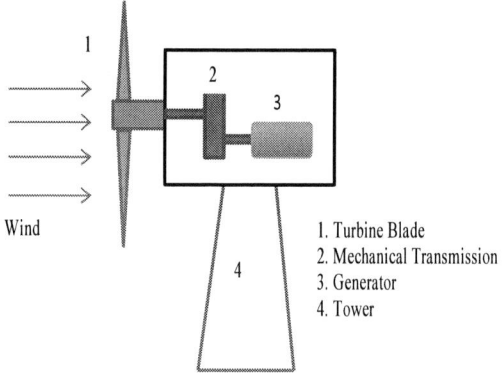

Fig. 1. Wind energy conversion system

The power extracted from the wind by the help of rotor can be defined as the difference between the power present in flowing air before passing through the rotor and the power present after the rotor. According to Betz theory, this mechanical power which is extracted from the wind energy is written as

$$P\ (watt) = \frac{1}{2}\rho C_p(\lambda, \beta)\pi R^2 V^3 \qquad (1)$$

Where

C_p is the power coefficient, ρ is air density (1.25 kg/m^3),

V is the wind speed velocity (m/sec),

R is the wind blade radius (m),

β is the blade pitch angle (degree).

Power coefficient (C_p) as used in equation (1) may be computed as:

$$C_p(\lambda, \beta) = C_1\left(\frac{C_2}{\lambda_1} - C_3\beta - C_4\right)e^{-\frac{C_5}{\lambda_1}} + C_6\lambda \qquad (2)$$

$$\lambda_1 = \cfrac{1}{\cfrac{1}{\lambda + 0.08\beta} - \cfrac{0.035}{\beta^3 + 1}} \qquad (3)$$

Constants C_1 to C_9 are easily available in the literature[8].

Lambda (λ) as in equation (3) is called tip-speed ratio and is defined as:

$$\lambda = \frac{\omega R}{V} \qquad (4)$$

Torque developed by the wind turbine is defined as below.

$$T_w(Nm) = \frac{P}{\omega} \qquad (5)$$

Torque equations for turbine and generator may be written as:

$$T_w - T_L = J_r\frac{d\omega}{dt} \qquad (6)$$

Where

Tw = wind turbine torque generated

T_L = wind turbine shaft torque as load

Jr = moment of inertia of wind turbine rotor

Equations (1) to (6) may be used to develop the Matlab Simulik model of wind turbine and it is described with the help of flow chart as shown in fig 2.

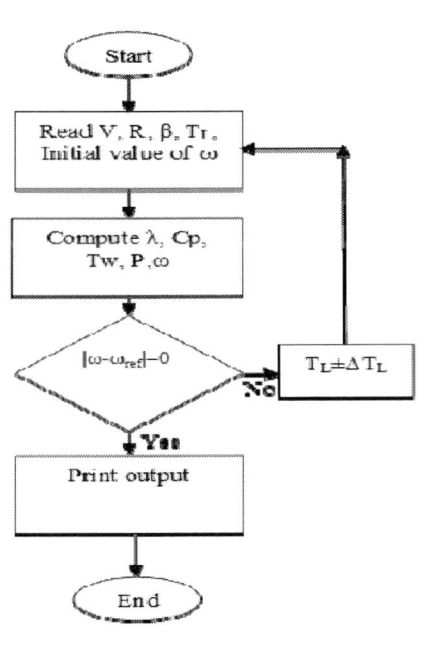

i) Constant Speed Mode

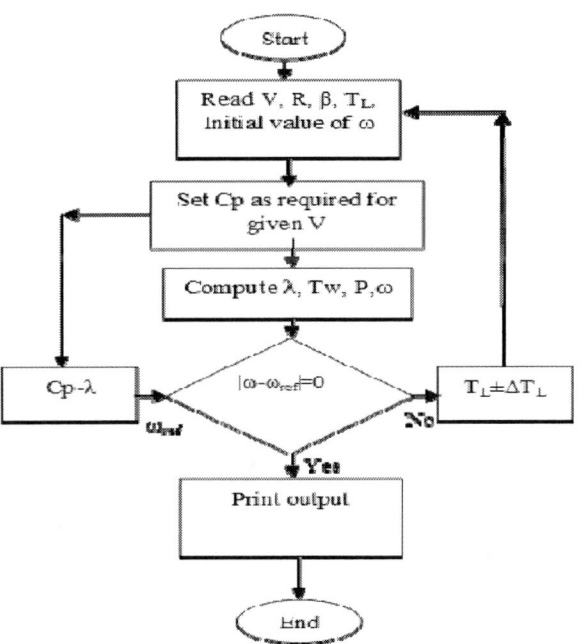

ii) Variable Speed Mode

Fig. 2. Flow chart of wind turbine for comparative analysis under constant as well as variable speed operation

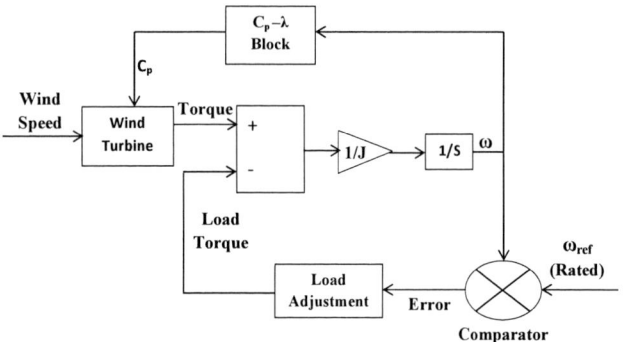

(i) Constant Speed Mode

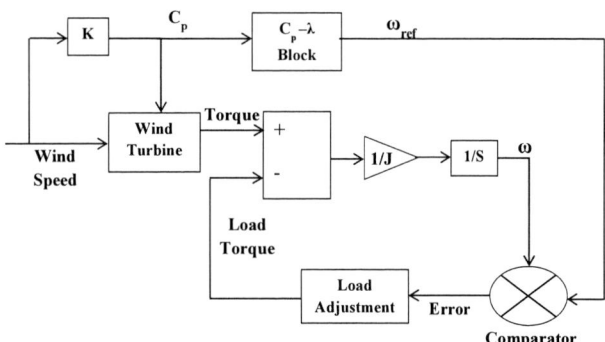

(ii) Variable Speed Mode

Fig. 3. Block diagram representation for different operating modes

Control concept as adopted for the two operating modes is shown in Fig. 3. As shown in Fig. 3(i), under any change of wind speed load torque is so adjusted that it results in to the constant speed operation i.e. rated speed of turbine. Whereas in the other mode, reference rotor speed is decided by the constant K (dependent upon the rated power, wind speed & rotor design etc.) & C_p-λ characteristics. Value of K is selected in such a manner that for any wind speed C_p gets modified to maintain the output. This may be used to decide the load adjustment for variable speed operation. One way of load adjustments for such systems is through electrical generators mechanically coupled to wind turbines.

III. RESULTS AND DISCUSSIONS

Fig. 4 to Fig. 8 shows the simulation results on a wind turbine with rated output as 50 kW when run at 42 RPM. Fig 4 shows the power output with wind variations without any control mechanism i.e. uncontrolled operation. During the

simulation results wind varies between different limits for different intervals. Wind variations have been taken as 5-10m/s, 10-12m/s, 10-16m/s and 15-20m/s for the respective intervals as 0-2s [zone 1], 2-5s [zone 2], 5-10s [zone 3] & 10-15s [zone 4]. This data has been kept same even during all other simulation results. Random wind variations have been considered and turbine is operated with blade pitch angle as mentioned in the respective simulation result.

As observed from Fig. 4 any change in wind speed affects the power output accordingly and it is true as per equation (1). In addition power coefficient which is also dependent upon wind speed indirectly affects the power output.

Fig. 5 show the comparative performance of wind turbine in terms of mean power when operated under two modes (i.e. constant as well as variable speed mode) with pitch angle as zero degree. Desired speeds as required for maintaining output are achieved through load adjustments as shown in Fig. 3. As observed variable speed mode performs better as compared to constant speed mode. It is due to the selection of reference rotor speed obtained using the desired value of power coefficient for any operating wind speed. Similarly Fig. 6, Fig 7 and Fig.8 shows the simulation results for comparative analysis under said two modes with blade pitch angles as five degree, eight degree and ten degree respectively.

Analysis of simulation results as presented yields the following observations.

- Irrespective of speed variations during different intervals, performance of variable speed mode appears to be comparatively better with blade pitch angle as zero degree (Fig. 5). Therefore such operation may be adopted with simple constructional design for rotor blades i.e. blade pitch control mechanism may be avoided.

- Power output of wind turbine decreases or increase in both modes due the simultaneous change of blade pitch angle (Fig. 6 to Fig. 8). It is due to prominent change in power coefficient, which is directly dependent upon blade pitch angle.

- For zone 1 any of the two modes can be adopted with blade pitch angle as zero degree. Variable speed mode operation appears to be best choice for zone 3 with blade pitch angle as zero degree. Constant speed operation seems to be better for zone 2 & 4 with blade pitch angle as zero degree and eight degree respectively.

- As observed from Fig. 7 & Fig 8, blade pitch angle seems to be effective only during high wind speed variations. From these results it can be concluded that the variation of pitch angle up to some extent may be utilized for improving the performance only under specific operating conditions.

978-1-4799-6047-7/14 $31.00 © 2014 IEEE

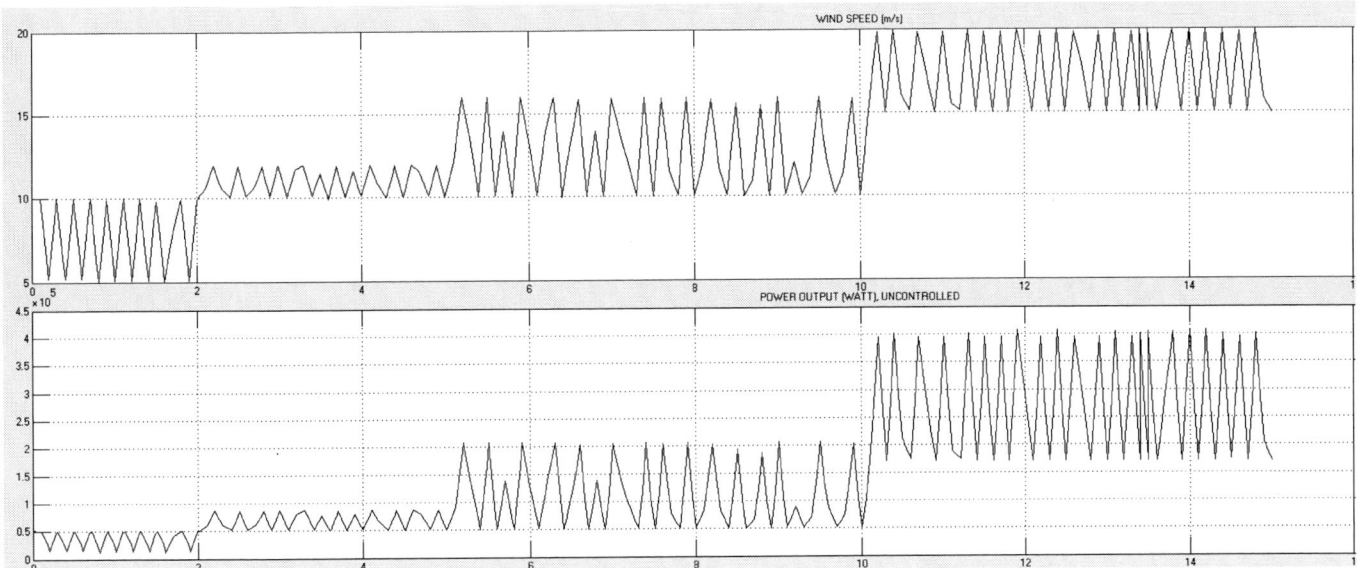

Fig. 4. Variation of power output for uncontrolled speed operation

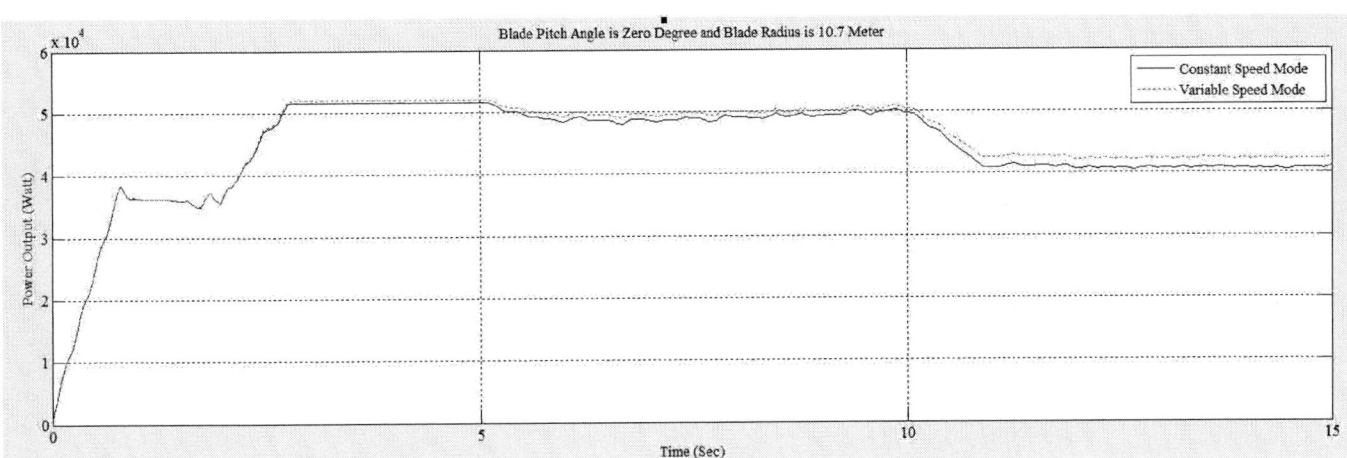

Fig. 5. Variation of power output with two operating modes, blade pitch angle zero degree

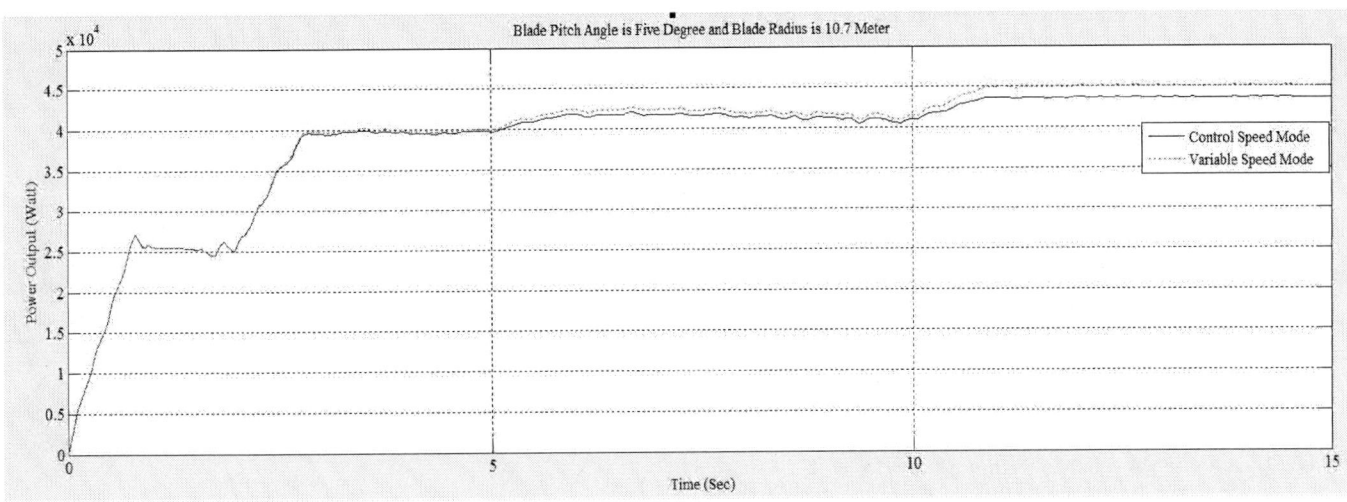

Fig. 6. Variation of power output with two operating modes, blade pitch angle five degree

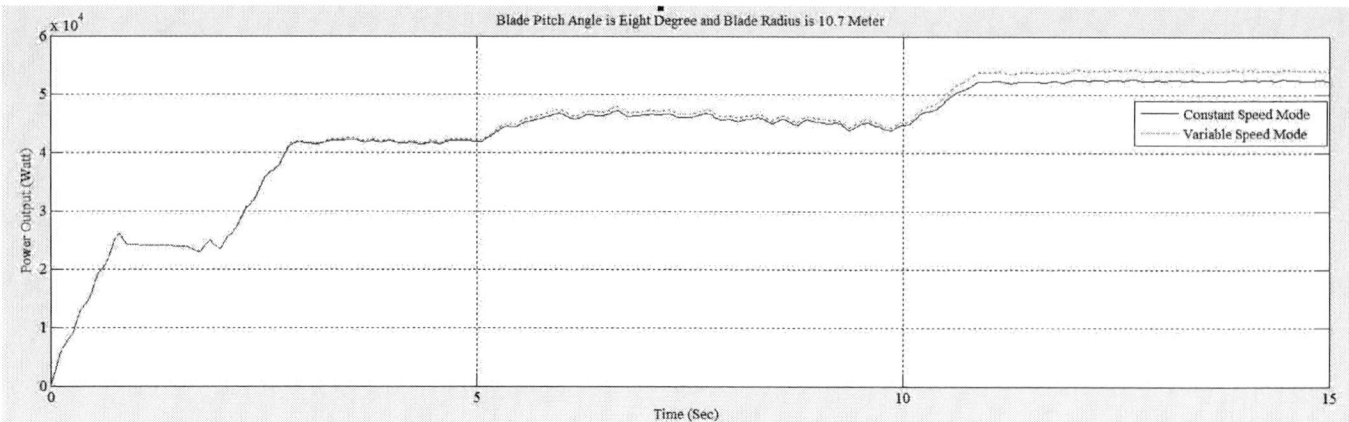

Fig. 7. Variation of power output with two operating modes, blade pitch angle eight degree

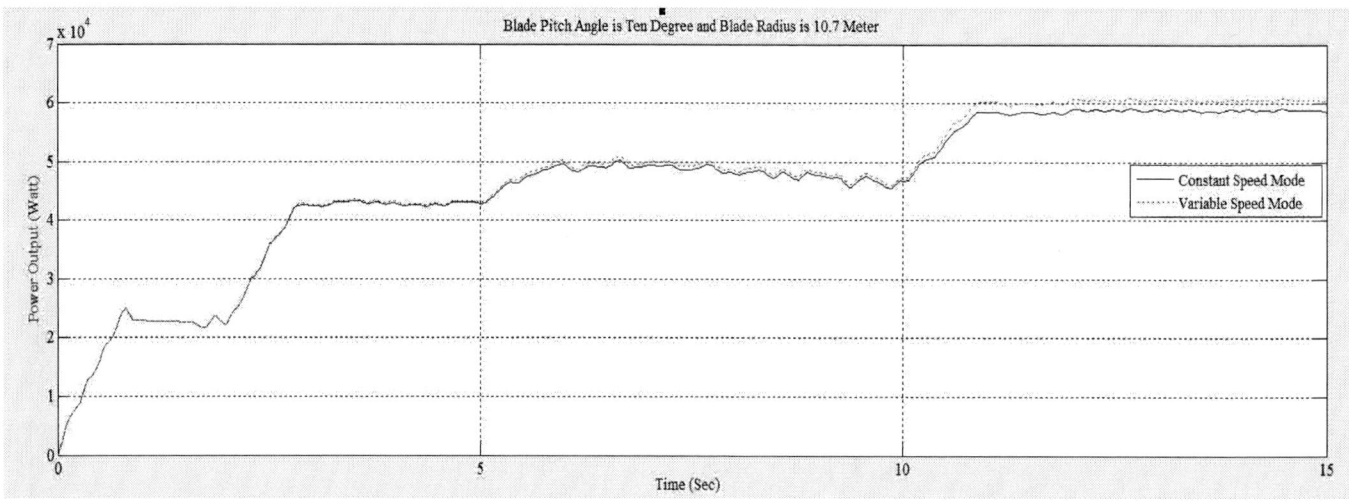

Fig. 8. Variation of power output with two operating modes, blade pitch angle ten degree

IV. CONCLUSION

It has been observed that power output of wind turbines varies due to wind variations. In order to improve the wind turbine performance under such operating conditions, some control strategy is required, irrespective of its operating mode i.e. constant rotor speed or variable rotor speed operation. In this paper, a new approach based upon the load adjustments is proposed to control the wind turbine under these operating modes. Approach as proposed seems to be simple and effective. Simulated results as obtained are found to be useful to predict the effect of wind speed & blade pitch angle under two operating modes.

REFERENCES

[1] Holley, Williams E., 2003. Wind turbine dynamics and control – issues and challenges, IEEE American Control Conference Denver, Colorado.

[2] Imic, Zdenko, Vrhovcak Maja Bo Icevic, Ljivac Damir, 2009. Small wind turbine power curve comparison, IEEE AFRICON Nairobi, Kenya.

[3] Leithead, W. E., Salle, S. A. De La, Reardon, D, 1992. Classical control of active pitch regulation of constant speed horizontal axis wind turbines, International Journal of Control, vol. 55, no. 4, pp. 845-876.

[4] Delarue, Ph., Bouscayrol, A., Tounzi, A., Guillaud, X., Lancigu, G., 2003. Modeling, Control and Simulation of an Overall Wind Energy Conversion System, Elsevier, Renewable Energy, vol. 28, pp. 1169-1185.

[5] Thakur, Devbratta, Mithulananthan, Nadarajh, 2009. Influence of Constant Speed Wind Turbine Generator on Power System Oscillation, Electric Power Components and Systems, vol. 37, pp. 478-494.

[6] Muljadi, Eduard, Butterfield, C. P., 2001. Pitch Controlled Variable Speed Wind Turbine Generation, IEEE Transaction On Industry Applications, vol. 37, no. 1, pp. 240-246.

[7] Qian, Minhui, Chen, Ning, Zhao, Liang, Zhao, Dawei, Zhu, Lingzhi, 2012. A New Pitch Control Strategy for Variable Speed Wind Generator, IEEE PES Innovative Smart Grid Technology, Tianjin, Asia, pp. 1-7.

[8] Jansuya, Phlearn, Kumsuwan, Yuttana. Design of matlab/simulink of fixed pitch angle wind turbine simulator, ELSEVIER, 10th Eco-Energy and Materials Science and Engineering (EMSES2012), Energy Procedia, Vol. 34, 2013, pp. 362-370.

Operating temperature of PV module modified with surface cooling unit in real time condition

Madhu Sharma, Kamal Bansal
Electrical, Power and Energy
University of Petroleum and Energy Studies
Dehradun, India
madhusharma@ddn.upes.ac.in

Dharam Buddhi
SelaQui Institute of Engineering and Technology
Selaqui
Dehradun, India

Abstract— High operating temperature decreases photo voltaic module efficiency. Normal Operating Cell Temperature, NOCT of silicon crystalline is about 49°C as reported. A cooling unit with reverse flow water circulation is designed, fabricated and applied on PV back surface to partially avoid undesirable effect of its temperature increase. This modified module with cooling unit and similar non-cooled modules are installed outdoor identically at University of Petroleum and Energy Studies campus, Dehradun. Installed Normal Operating Cell Temperature, INOCT of both the modules are determined and compared by varying water flow rate from cooling unit. And also water flow rate is optimized to maintain minimum possible operating temperature.

Index Terms— INOCT, optimum mass flow rate, passive surface cooling unit, PV module

I.   INTRODUCTION

Out of total energy produced, electrical output is only one component with typical ideal conversion efficiency in the range of 15% from PV module [1]. Remaining produced energy is heat. As this heat energy is neither utilized nor captured, increases PV module temperature which actually influence their overall performance.

In Ras AL Khaimah, at CSEM – UAE Innovation Centre outdoor testing facility for PV module was installed on the roof. 165 Wp multi-crystalline silicon module had been selected for experiment and was mounted at 30° of fixed tilt angle facing true south. Day star I-V curve tracer and CSEM-UAE AMCU flyer unit gave I-V characteristics under varying load of selected module at different solar radiation in outdoor condition. Global radiation, module surface temperature, ambient temperature, current and voltage readings of PV module were recorded for analysis. Experimental results showed that module efficiency varies between 8-10%, which was differing 3-4 % from STC specified efficiency. And this efficiency drop is the results of high operating surface temperature (50-60°C) of PV module [2].

During 2009 at Malaysia temperature dependence coefficient of crystalline silicon PV modules and amorphous silicon module have been obtained using linear regression techniques. Three modules a-Si, multi crystalline and mono-crystalline were installed in field with data monitoring using data logger. It was found that multi crystalline PV module is highly temperature sensitive among three types [3].

The photon rate of PV cell increases with the temperature and hence reverse saturation current. This results in the change of current and voltage, which means marginal changes in current but major changes in voltage [4].

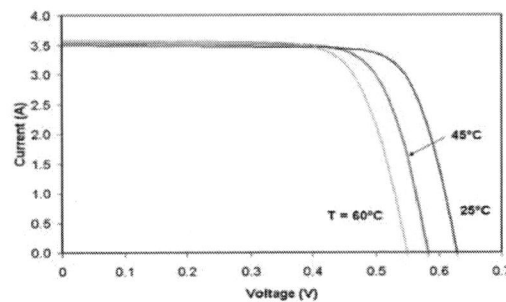

Figure1. Shows how the I-V curve varies with varying temperature [4]

IV curve define the PV cell performance [5] for a given solar radiation. PV cell has two limiting parameters short circuit current (I_{sc}) when V=0, and open circuit voltage V_{oc} when I=0 with rise in temperature of PV cell I_{sc} increases slightly while V_{oc} drop significantly. Open circuit voltage drops 2mV/°C rise of temperature for silicon materials. In solar modules operation, temperature is an important issue as temperature rise significantly reduces electrical output, increases thermal stress and degradation rates. Author suggested that with appropriate cooling system with air or water, electrical efficiency can significantly enhance. Combining both technologies, photovoltaic & thermal as hybrid PVT to generate electricity and heat water in a house for instance is another way to handle heat issue [5].

There is more than 25 % performance reduction of PV module, as modules operate over 50°C above ambient temperature commonly. To make significant gains in PV system performance, its operating temperature can be lowered by dissipating heat from the module and this heat can be utilized for practical heating purposes [1].

Traditional linear expression for PV electrical efficiency is

$$\eta_c = \eta_{Tref} \left[1 - \beta_{ref} (T_c - T_{ref}) \right]$$

Where η_{Tref} and T_{ref} is cell efficiency and cell temperature at Standard Test Condition (STC – 1000 W/m^2, 25°C) and β_{ref} is temperature coefficient (°C^{-1}).

In this paper, we present basic design consideration to modify commercially available polycrystalline module. The method of finding Installed normal operating temperature is discussed. And INOCT of non-cooled module and with cooling module are determined and presented.

II. EXPERIMENTS AND STANDARDS

To install photovoltaic module, system designer and inventor consider data sheet of modules to make choice of module type and brand. Data sheet include few primary characteristics like module peak power and efficiency at STC (standard test condition), temperature coefficient of power, open circuit voltage, short circuit current and sometimes low light behavior as STC rarely met in real time condition [6].

In STC, measurements are taken using a solar simulator under laboratory conditions and it is controversial as standard condition can never be found in real time. Climate parameters like solar insolation, ambient temp, wind speed etc., are the locality dependent variable on which the performance of PV module depends. For effective design of PV system, device rating measurements at the site of is desirable and this allows actual power output prediction. In the outdoor environment, PV efficiency as a complex function of micro climatic parameters and working temperature of PV module plays a crucial role in rating determining. High radiance and high temperature combination leads lower efficiency compared to low radiance and low temp combination [7].

Operating temperature of the module depends on the radiation absorption properties, thermal dissipation, module encapsulating materials, module functioning and also depends on insolation, ambient temp, accurate installation status and wind speed. Module parameters like voltage, current, power and efficiency greatly influence by the module operating temperature and thus, it is very important to perform a quantitative analysis under real operating condition. Author investigated the temperature influence on PV parameters of amorphous- silicon (a-Si) and copper indium deselinide (CIS) thin film modules at Patras, Greece (latitude 38°) where at Patras peak sun hours over 4.2 per day and working temp of the module between 16 to 60°C. V_{oc} percentage reduction with temperature increase is greater of CIS than a-Si modules. CIS short circuit current temperature coefficient in position at low and medium temperature, and at entire working temperature range approximately constant with slight tendency to reduce. In case of CIS maximum power as a function of temperature decreases linearly. Annual behavior indicated that CIS module efficiency decreases with temperature and a-Si module was not severely affected [8].

Difference between junction temperature of PV module and ambient temperature is a consequence of continuous solar radiation and the accumulation of thermal energy in the interior of modules lead to serious performance deterioration. Author found this temperature difference will increase linearly with radiation intensity rise, even if ambient temperature changes [9].

As a module temperature indicative NOCT (Normal Operating Cell Temperature) is commonly use. NOCT is mean solar junction temperature in Standard Reference Environment (SRE) with in an open rack mounted module. SRE includes total irradiance 800W/m^2, wind speed 1m/s, ambient temperature 20°C, tilt angle - at normal incidence to the direct solar beam at local solar noon and nil electrical loads. NOCT is a reference of how the module will work in real condition therefore is an important characteristic. To calculate NOCT there are several intimation standards EN-61215 for crystalline PV module, EN-61646 for thin film PV module, ASTM E1036M or both (non - concentrator terrestrial PV modules and arrays)[10].

Fact on which all above standards are based is that T_m-T_a (module temp - ambient temp) difference is essentially linear to irradiance and largely independent of wind speed.

INOCT i.e. Installed Nominal Operation Cell temperature is the cell temperature of installed module connected to load and also mounting configuration of the module is taking into account. In open rack case it is recommended that one use a value of INOCT 3°C less than NOCT value [11].

Gail-Angee Migan reported that NOCT of a typical module is 48°C whereas; best operates at 33°C and worst at 58°C [4].

III. SYSTEM DESCRIPTION AND OUTDOOR TESTING

To reduce the operating temp of PV module, a PV surface cooling system is designed, developed and experiment set up is installed. Multi crystalline silicon PV module of 75 Wp is modified by applying cooling system underneath. Cooling unit is glued by thermal conductive paste on rear surface of module. And the performance of this modified module is compared with the normal module of same type, same rating and same make in similar operating conditions.

Both the modules, non-cooling and with cooling unit were mounted at an optimum angle of 30° tilt equal to the latitude of the location oriented due south for best year round performance.

Cooling / heat absorbing system consist of two rectangular hollow copper tanks of absorber and in thermal contact with standard PV module of 75 Wp. Absorber tank was fabricated with one inlet and one outlet arranged opposite to each other

978-1-4799-6047-7/14 $31.00 © 2014 IEEE

and so water flows in reverse direction and cover 76 % back surface area of module. To avoid thermal losses from the absorber tank, tanks are thermal insulated. Inlet water from the overhead storage tank flow through cooling unit at pre- set flow rate, and absorb heat from the PV module surface and store in outlet tank. Water is flowing by gravity as head difference between inlet and outlet water tank and no pump is used.

Experiments were conducted to find Installed Normal Operating Cell Temperature at irradiance 800 W/m², ambient temperature 20°C and wind speed 1m/s of both non - cooling and with cooling modules under similar mounting and operating conditions.

Water flow rate was pre-set for a single day and at 15 minutes interval following measurements were taken:-

➢ Solar irradiance at tilted surface
➢ Ambient temperature
➢ Wind speed
➢ Back surface temperature of non– cooling module
➢ Back surface temperature of module with cooling unit

To analyze collected data of each day for both non - cooling and with cooling modules:-

➢ Module temperature rise above ambient temperature as a function of solar irradiance (>400 W/m²) is plotted.

➢ To fit data, linear regression is used and temperature above ambient at 800 W/m² was determine by regression equation

➢ Finally installed normal operating module temperature were determined at reporting conditions by adding 20°C and corrected for average wind speed and ambient temperature from the following graph shown in fig. 2 [12].

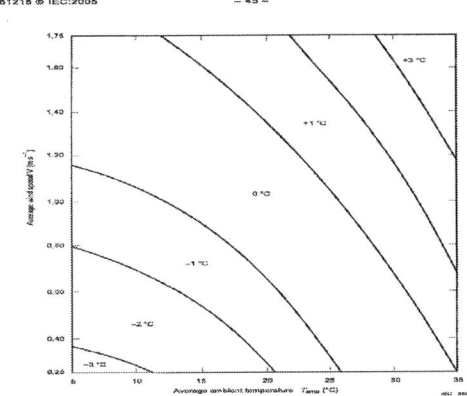

Figure 2. To correct NOCT for average wind speed and average ambient temperature [12]

IV. RESULT AND DISCUSSION

Eleven tests were conducted on different sunny days in the month of April 2014.Out of these four test are depicted in fig. 3-6 below and following equation is used to calculate INOCT on tested day:-

$INOCT_{day}$ = regression equation + 20°C + correction factor

Where

• Correction factor is determined by using graph shown in fig. 2 above for average ambient temperature and average wind speed

• Regression equation of fitting collected data is obtained from the graphs plotted for each day.

Finally INOCT is obtained by correcting regression equation to 800 W/m² and adding 20°C and correction factor.

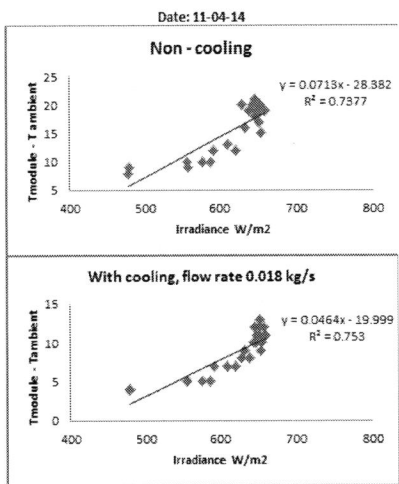

Figure 3. Test result of 11-04-14

Average wind speed = 1.02 m/s
Average ambient temperature = 31.63°C
Correction factor = +1°C
INOCT non-cooling,
T_{INOCT}= (0.0713*800-28.382) +20+1 = 49.66°C
INOCT with cooling,
T_{INOCT} = 0.0464*800-19.999+20+1 = 38.12°C

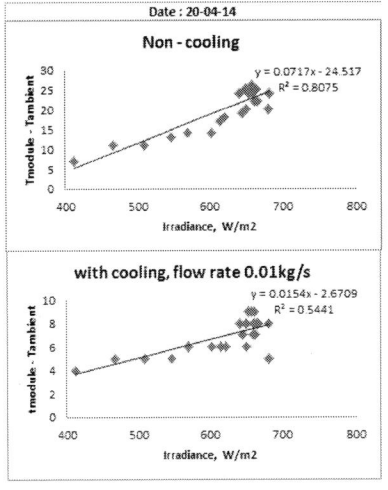

Figure 4. Test result of 20-04-14

Average wind speed = 1.15 m/s
Average ambient temperature = 29.2°C
Correction factor = +1°C
INOCT non-cooling,
T_{INOCT} = (0.0717*800-24.517) +20+1 = 53.84°C
INOCT with cooling,
T_{INOCT} = 0.0154*800-2.6709+20+1 = 30.64°C

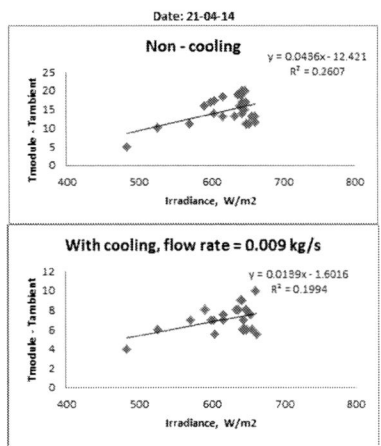

Figure 5. Test result of 21-04-14

Average wind speed = 0.858 m/s
Average ambient temperature = 33.77°C
Correction factor = +1°C
INOCT non-cooling,
T_{INOCT} = (0.0436*800-12.421) +20+1 = 43.5°C
INOCT with cooling,
T_{INOCT} = 0.0139*800-1.6016+20+1 = 30.5°C

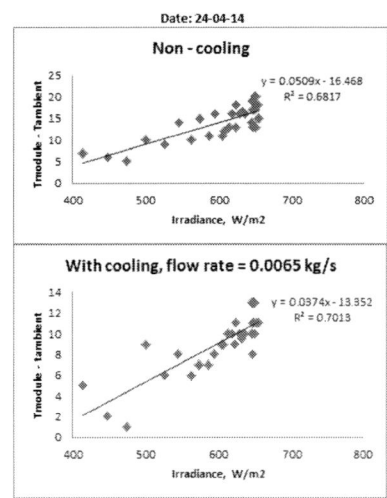

Figure 6. Test result of 24-04-14

Average wind speed = 1.09 m/s
Average ambient temperature = 31.61°C
Correction factor = +1°C
INOCT non-cooling,
T_{INOCT} = (0.0509*800-16.468) +20+1 = 45.3°C
INOCT with cooling,

T_{INOCT} = 0.0374*800-13.352+20+1 = 37.6°C
And result of all tests is summarized in table - I below

Table –I	Field Test Results		
	Flow	Operating Temp Non-cooled	Operating Temp with cooling
	(kg/s)	(°C)	(°C)
8/4/2014	0.032	51	43
11/4/2014	0.018	49.7	38.12
12/4/2014	0.016	56.56	37
14/4/2014	0.014	48.8	30.83
15/4/2014	0.013	52.5	30.66
20/4/2014	0.01	53.84	30.64
21/4/2014	0.009	43.5	30.5
22/4/2014	0.008	48.3	33.33
23/4/2014	0.0072	44.5	33.4
24/4/2014	0.0065	45.3	37.6
25/04/2014	0.0055	47.7	34

Based on experimental data, the effect of flow rate on operating temperature of modified module is presented in fig.6 below:

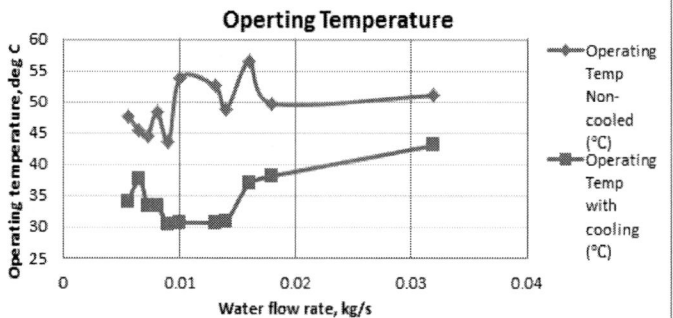

Figure 7. Test results at different flow rates

It is observed that non - cooling module is operating at much higher temperature than cooling one. Average operating temperature of non-cooled module is 49°C. It is also observed that in the water flow range of 0.009 kg/s to 0.014 kg/s operating temperature of module with cooling unit is almost constant. For poly crystalline module modify with cooling unit water flow rate is optimized at 0.01 kg/s for best operating temperature of about 30.6°C.

The calculated INOCT can consider as most representative of the module behavior. From available irradiance, ambient temperature and this INOCT value , module temperature can be calculated using equation below

Tm = Ta + (INOCT – 20) E / 800

Where Tm is module temperature and E is solar irradiance.

V. CONCLUSION

1) PV module efficiency is sensitive to module surface temperature and decrease with surface operating temperature increase.

2) Objective of the study is to prove the potential benefits of PV module with heat extraction unit (surface cooling units) system compared to typical PV module.

3) Polycrystalline silicon PV module combined with surface cooling system to extract heat was constructed and tested with respect to its operating temperature and compared with operating temperature of module of same rating without cooling system at the Geo graphic location of Dehradun installed similarly.

Under same metrological and mounting conditions:

- Installed normal operating temperature of non-cooling case is much higher than cooling one

- There is wide variation in operating temperature of non - cooling unit tested and operating temperature variation is less with cooling unit in specified range of flow rate of water.

- To maintain minimum operating temperature, optimum flow rate of water through cooling unit is 0.01 kg/sec for designed system.

4) Heat removed by the water flowing naturally by gravity through surface cooling unit at optimum flow rate is collected at the lower end of the panel and can be used as a utility for heating purposes.

5) Average installed nominal cell temperature of non-cooling module is found 49°C and with cooling unit we are able to reduce it to 30.6°C at optimum water flow rate. And accordingly efficiency will increase by 9.2 % if efficiency conversion coefficient is considered 0.45 %/$^{\circ}$C.

ACKNOWLEDGMENT

Special thanks to Dr. S.J. Chopra for support and suggestions. The work described in this paper was financially supported from University of Petroleum and Energy Studies, Dehradun.

REFERENCES

[1] John Hollick: Conserval Engineering Inc., Toronto Ontario, "PV thermal systems – Capturing the untapped Energy", Conference proceedings - "Solar Energy Society Conference in Cleveland". July 11, 2007,

[2] Kapil Kumar, S. D. Sharma, Lokesh Jain, "Standalone Photovoltaic (PV) module outdoor testing facility for UAE climate", CSEM-UAE Innovation Center LLC, Ras Al Khaimah, United Arab Emirates

[3] Sulaiman Shaari, Kamaruzzaman Sopian, Nowshad Amin and Mohd Nizan Kassim, "The Temperature Dependence Coefficients of Amorphous Silicon and Crystalline Photovoltaic Modules Using Malaysian Field Test Investigation", in American Journal of Applied Sciences 6 (4): 586-593, 2009, ISSN 1546-9239

[4] Pradhan Arjyadhara, Ali S.M and Jena Chitralekha, " Analysis of Solar PV cell Performance with changing Irradiance and Temperature", International Journal Of Engineering And Computer Science ISSN:2319-7242, Volume 2 Issue 1 Jan 2013 Page No. 214-220

[5] Gail-Angee Migan, "Study of the operating temperature of a PV module", Project Report, 2013 MVK160 Heat and Mass Transfer, May 16, 2013, Lund, Sweden

[6] Bert Herteleer, Jan Cappelle, Johan Driesen, "Quantifying low-light behavior of Photovoltaic modules by indentifying their irradiance – dependent efficiency from data sheets", 27th European Photovoltaic Solar Energy Conference proceeding 2012, pages 3714 – 3719, ISBN:3-936338-28-0

[7] A.Q. Malik and Mohamad Fauzi bin Haji Metali, "Performance of Single Crystal Silicon Photovoltaic Module in Bruneian Climate", International Journal of Applied Science and Engineering, 2010. 8, 2: 179-188

[8] V. Perraki and G. Tsolkas, "Temperature dependence on the photovoltaic properties of selected thin-film modules" , International Journal of Renewable and Sustainable Energy, 2013; 2(4): 140-146

[9] Joe-Air Jiang, Jen-Cheng Wang, Kun-Chang Kuo, Yu-Li Su, Jyh-Cherng Shieh and Jui-Jen Chou, "Analysis of the junction temperature and thermal characteristics of photovoltaic modules under various operation conditions", Energy 44 (2012), page 292-301

[10] M.C. Alonso Garcıa and J.L. Balenzategui, "Estimation of photovoltaic module yearly temperature and performance based on Nominal Operation Cell Temperature calculations", Renewable Energy 29 (2004), 1997–2010

[11] Fuentes MK, "A simplified thermal model for flat-plate photovoltaic arrays", SANDIA report no., SAND-85-0330; 1987

[12] Matthew Muller, "Measuring and Modeling Nominal Operating Cell Temperature (NOCT)", NREL, Test & Evaluation, Sept 22-23, 2010 PV Performance Modeling Workshop Albuquerque, NMNREL/PR-520-49505

978-1-4799-6047-7/14 $31.00 © 2014 IEEE

Direct Torque Control of Open–End–Winding Induction Motor using Matrix Converter

Ranganath Muthu, Kalyan Govindarajan, Divakhar Anbazhagan, Senthilkumaran Mahadevan
Department of Electrical and Electronics Engineering
SSN College of Engineering
Chennai, India

Abstract—This paper describes the direct torque and flux control strategy for an induction motor with open–end stator winding fed by a matrix converter at each end of its terminal. Use of the matrix converters instead of the conventional power converters eliminates the DC-link in the AC–DC–AC converter. Triggering the matrix converter using the rotating space vectors eliminates the common mode voltage at the machine terminals. The input-side power factor is also controlled by using the rotating vectors. The simulation is carried out in Simulink®/MATLAB® and the results show that the common mode voltage is eliminated while maintaining input power factor at unity.

Index terms— Common mode voltage elimination, direct torque control, input power factor control, matrix converter, open–end–winding induction motor and rotating space vectors.

I. INTRODUCTION

The evolution of variable speed induction motor drive is driven by the desire to achieve characteristics similar to DC motor, such as fast torque response, simplified speed control & accuracy. Direct Torque Control (DTC) is an optimised vector control strategy for induction machine, where the machine variables (torque and speed) are controlled using a power converter according to the dynamic loading condition by selecting an appropriate switching vector for the power converter [1]. In spite of its advantages like high dynamic response, transformation–less operations and robustness, the DTC has some disadvantages namely difficulty in controlling torque and flux at low speeds, and high current and torque ripples resulting in increased machine losses and hence more noise. To retain DC drive–like characteristics and compensate for the problems in DTC, space vector modulation based DTC method was proposed [2]. However, these Pulse Width Modulation (PWM) based inverters [3], [4] and matrix converter drives [5] suffer from Common Mode Voltage (CMV) at the machine terminals. This paper proposes a control strategy to eliminate the CMV present in the inverter fed DTC drives using a matrix converter fed induction motor drive triggered by the rotating space vectors defined in section IV. DTC for induction machine using these rotating space vectors is presented in section V. Input side power factor control for the same has been developed and presented in this paper.

II. MATRIX CONVERTERS

Matrix converter is an AC-AC converter with n×m switches arranged in matrix form to control the magnitude,

phase sequence and frequency of its output voltages and currents [6]. The number of input and output phases is n and m respectively. Fig. 1 shows a 3×3 matrix converter. Matrix converters provide poly-phase power conversion; inherent bidirectional power flow capability and input/output sinusoidal waveforms. However, the inverter based AC–DC–AC drives are still popular because of their relative simplicity of control. Matrix converter can be an alternative to the conventional inverter based AC–DC–AC electric drives for torque; power and variable-frequency control applications. The major problems associated with matrix converters are increased conduction losses, distortion of input current at low modulation index, complexity of control and cost. The cost of semiconductor switches and microcontrollers are reducing rapidly, thereby providing a viable alternative in the near future.

III. COMMON MODE VOLTAGE

PWM based inverter/converter fed induction motor drives suffer from the CMV [7], which degrades the electromagnetic interference (EMI) compatibility of the system causing bearing failure in the machine. The CMV, $\vec{V}_{cmv}$, is given by (1).

$$\vec{V}_{cmv}(t) = \frac{\left(\vec{V}_a(t) + \vec{V}_b(t) + \vec{V}_c(t) \right)}{3} \qquad (1)$$

The bearing currents produced by the CMV flows through bearing to ground terminal causing physical damage and unwanted tripping of ground–fault–relays in electric drives. As modulation frequency increases, the zero sequence impedance of the machine decreases, increasing the common-mode current (also increasing the CMV) and worsening the EMI.

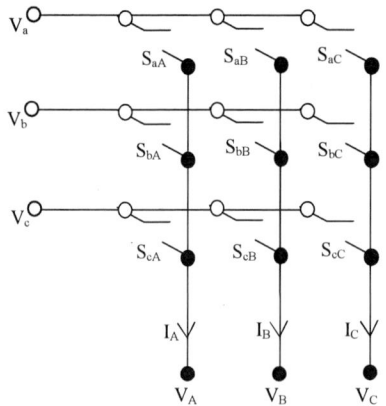

Fig. 1. Matrix converter

978-1-4799-6047-7/14 $31.00 © 2014 IEEE   410

The elimination of CMV in three phase power converters has been done by making modifications in the modulation algorithms [8], [9]. CMV elimination in matrix converters was achieved by using Rotating Space Vector Modulation (RSVM) based matrix converter [10], which is used for the implementation of DTC in this paper.

IV. ROTATING SPACE VECTOR MODULATION BASED MATRIX CONVERTER

RSVM strategy utilises all the three input voltages at any instant of time and hence eliminating the CMV. The problem with this method is that the voltage transfer ratio (V_o/V_{in}) is reduced to 0.5 when using a single matrix converter. However, by using the line voltage of the input phases, the terminal voltage of the machine can be increased by 1.5 times [11]. This can be effectively achieved by opening the stator windings of induction motor fed by two 3×3 matrix converters at both ends, as shown in Fig. 2. The vector V_{abc} refers to the voltage of when the stator terminals, A, B and C, of the machine are connected to the input phases a, b and c through the switches S_{aA}, S_{bB}, S_{cC}. Table I lists the other possible rotating vectors. The vectors V_{acb}, V_{bac} and V_{cba} are grouped as Clockwise (CW) rotating vectors since applying these vectors to the induction machine produces a space-vector that is rotating in the clockwise direction with angular frequency $\omega_{in}=2\pi f_s$, where, f_s is supply frequency. Similarly, the vectors V_{abc}, V_{bca} and V_{cab} when applied to machine produce a Counter–Clockwise (CCW) vector rotating with the same angular frequency ω_{in}.

TABLE I. ROTATING SPACE VECTOR

Direction of Rotation	Vector	Active Switches		
CCW	V_{abc}	S_{aA}	S_{bB}	S_{cC}
	V_{bca}	S_{bA}	S_{cB}	S_{aC}
	V_{cab}	S_{cA}	S_{aB}	S_{bC}
CW	V_{acb}	S_{aA}	S_{cB}	S_{bC}
	V_{bac}	S_{bA}	S_{aB}	S_{cC}
	V_{cba}	S_{cA}	S_{bB}	S_{aC}

Open-end winding induction machine is obtained by opening the Y–connected stator windings of the induction machine thereby creating another set of stator terminals henceforth referred to as side 2 with subscript 2. The original set of windings is now referred to as side 1 with subscript 1. Equations 2 and 3 give the voltage space vectors on either sides of machine fed by two matrix converters. These two matrix converters are controlled based on the spatial position of the output current with respect to input current. Equation 4 gives the output voltage of the machine at any instant.

$$\vec{V}_{o1}(t) = \vec{V}_{a1}(t) + \vec{V}_{b1}(t)e^{-j2\pi/3} + \vec{V}_{c1}(t)e^{-j4\pi/3} \quad (2)$$

$$\vec{V}_{o2}(t) = -1 * (\vec{V}_{a2}(t) + \vec{V}_{b2}(t)e^{-j2\pi/3} + \vec{V}_{c2}(t)e^{-j4\pi/3}) \quad (3)$$

$$\vec{V}_o(t) = \vec{V}_{o1}(t) + \vec{V}_{o2}(t) \quad (4)$$

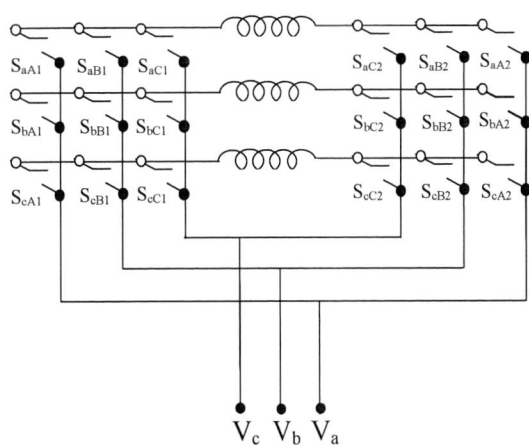

Fig. 2. Open-end winding based matrix converter drive

Equations 5 and 6 give the CMV on sides 1 & 2 at any instant.

$$\vec{V}_{cmv1}(t) = \frac{(\vec{V}_{a1}(t) + \vec{V}_{b1}(t) + \vec{V}_{c1}(t))}{3} \quad (5)$$

$$\vec{V}_{cmv2}(t) = \frac{(\vec{V}_{a2}(t) + \vec{V}_{b2}(t) + \vec{V}_{c2}(t))}{3} \quad (6)$$

The required output voltage of the matrix converter is synthesized by identifying the sector in which the required output current vector, $\vec{I}_o$ lies with respect to input current vector, $\vec{I}_{in}$. This is because the use of the CW and CCW vectors produces space vectors that are rotating, unlike the pulsating vectors in space vector modulation based converters. Therefore, angle, θ of the output current vector, $\vec{I}_o$ with respect to input current vector, $\vec{I}_{in}$ is calculated, as shown in (7).

$$\theta = \arctan(\vec{I}_o) - \arctan(\vec{I}_{in}) \quad (7)$$

Application of CCW vectors on sides 1 and 2 of the induction machine forms resultant vectors V_{ccw1}–V_{ccw6} and CW vectors on sides 1 and 2 forms resultant vectors V_{cw1}–V_{cw6}, thus forming space vector hexagons in both cases, as shown in Figs. 3 & 4. Use of CCW and CW vectors result in input currents lagging and leading the input voltages respectively by an angle, ρ as shown in (8)–(10). The input power factor can be maintained at unity by proportionate sharing of both CW and CCW vectors on sides 1 & 2. For example selecting CCW vectors when the power factor is leading and selecting CW vectors when the power factor is lagging

$$\vec{I}_{in} = \vec{I}_{in1} + \vec{I}_{in2} \quad (8)$$

$$\vec{I}_{in} = \frac{3}{2}mI_o(\arctan(\vec{I}_{in}) - \rho) \quad (9)$$

$$\vec{I}_{in} = \frac{3}{2}mI_o(\arctan(\vec{I}_{in}) + \rho) \quad (10)$$

Here, $\vec{I}_{in1}$, $\vec{I}_{in2} = \vec{I}_{ABC}$ or $\vec{I}_{CAB}$ or $\vec{I}_{BCA}$ and m is the matrix converter voltage gain. A reduced voltage transfer ratio of 0.5 due to use of rotating vectors in the single matrix converter is a circle inscribed within the triangle as shown in Figs. 3 and 4.

978-1-4799-6047-7/14 $31.00 © 2014 IEEE

V. DTC Using Rotating Vectors

In DTC, at every instant of switching, the switching vectors are selected based on the instantaneous torque and stator flux commands, rather than a predetermined pattern as in a PWM based vector-controlled drives.

Hence DTC doesn't provide constant switching frequency [12], [13]. In the DTC, torque is controlled by changing the input voltage directly, as expressed in (11)–(18) [14]. Neglecting copper losses in stator of induction machine, the stator voltage is given by (11)

$$V_s = \frac{d\vec{\Psi}_s}{dt} \tag{11}$$

where, $\vec{\Psi}_s$ is the stator flux vector.

Flux changes occur over a longer time when compared to current changes. Thereby (11) can be written as (12).

$$V_s = \frac{\Delta\vec{\Psi}_s}{\Delta t} \tag{12}$$

Thus, the stator flux can be controlled directly by changing the input voltage of the motor for a given time Δt, as shown in (13).

$$V_s \, \Delta t = \Delta\vec{\Psi}_s \tag{13}$$

The torque of an induction motor is given by (14)

$$T_e = \frac{3P}{4} \left(\vec{\Psi}_s \times \vec{I}_o \right) \tag{14}$$

where, P is number of poles of the induction machine and $\vec{I}_o$ is the stator current vector.

The final expression for torque can be expressed as the cross product of the stator and the rotor fluxes and is shown in (15), which is obtained by solving (14).

$$T_e = \frac{3P}{4} \frac{L_m}{L_r L_a} \left(\vec{\Psi}_r \times \vec{\Psi}_s \right) \tag{15}$$

Equation 16 gives the magnitude of the torque

$$|T_e| = \frac{3P}{4} \frac{L_m}{L_r L_a} |\vec{\Psi}_r| |\vec{\Psi}_s| \sin\delta \tag{16}$$

where δ is the torque angle. From (13), it can be inferred that for a small change in the input volt–time $V_s \Delta t$, the stator flux varies by a factor (say) $\Delta\vec{\Psi}_s$. This change in the flux due to the input volt-time change is reflected as a change in the torque angle and the corresponding torque magnitude is shown in (17).

$$\Delta T_e = \frac{3}{2} * \frac{P}{2} \frac{L_m}{L_r L_a} |\vec{\Psi}_r| |\vec{\Psi}_s + \Delta\vec{\Psi}_s| \sin\Delta\delta \tag{17}$$

Equation 17 represents the change in torque as a function of changes in stator flux and torque angle. A power converter changes the stator flux by changing the input volt–time, as shown in Fig. 5. The required output voltage vector is determined by the space vector position, the reference torque and the reference flux. The reference torque and flux values are estimated. With the input voltage, the line currents and the present switch position, the model estimates the actual flux with which the torque and speed of the induction motor are calculated.

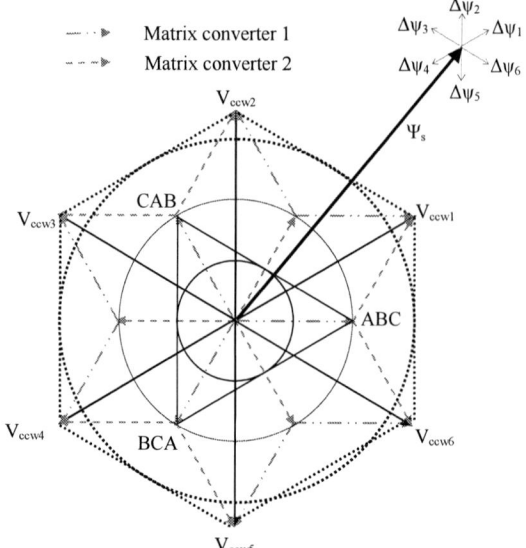

Fig. 3. Clockwise rotating space vectors

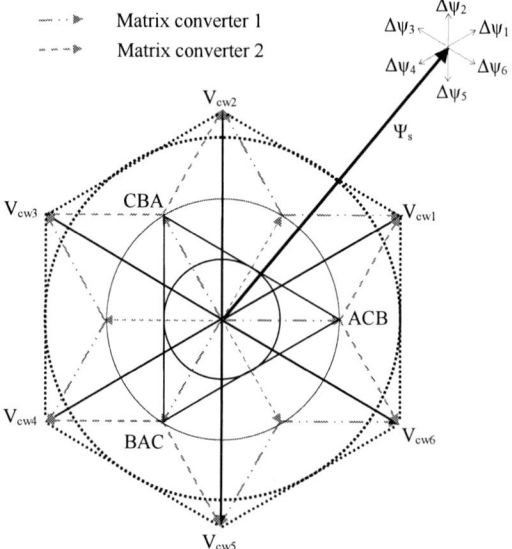

Fig. 4. Counter clockwise rotating space vector

The error between the reference speed and the actual speed of the machine is given to a PI controller that generates the torque reference. The machine speed is also used to generate the reference flux, as shown in Fig. 6.

The errors in torque and flux are fed to a three level and two–Level hysteresis band comparator respectively. The output of the torque hysteresis band, C_T, as shown in Fig. 7(a), is -1, 0 or +1. This represents the action to be taken in the power converter for the desired change to occur. The other hysteresis band, C_F, as in Fig. 7(b), gives the signal -1 or +1. This indicates error in the flux and the switching action to be carried out in the power converter to change the stator flux and the torque. These two constraints together decide the vector to be switched for the converter. The stator flux vector locus is forced to follow a circular path by limiting the magnitude of its error.

Fig. 8 shows the schematic of matrix converter based open-end-winding induction machine drive. The vector selection of vector in the RSVM based DTC is quite similar to classical DTC except that the resultant space vectors rotate. From the matrix converter input current vector angle the position of the flux vector is identified.

Equation (18) determines the position of the flux vector, α, with respect to input voltage space vector. The relative sector of the flux is calculated using α.

$$\alpha = \theta - \arctan(\vec{I}_{in}) - \frac{\pi}{6} \qquad (18)$$

There are six active rotating vectors (three CCW and three CW) that are used in the proposed induction motor drive. However, the conventional zero vectors (V_{aaa}, V_{bbb}, V_{ccc}) are not used as they result in CMV at the machine terminals. Thus, two identical rotating vectors are applied at either ends of the machine resulting in zero voltage difference across the machine terminals, as shown in Table II.

There are six zero vectors possible (three CCW and three CW), which are chosen based on the flux sector. Table III shows an optimal switching pattern that reduces the switching stress by distributing it symmetrically.

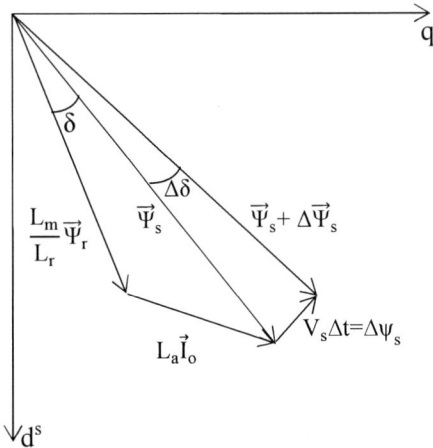

Fig. 5. Torque change for input voltage change.

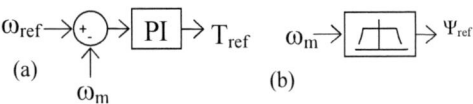

Fig. 6. Reference generation (a) Torque and (b) Flux

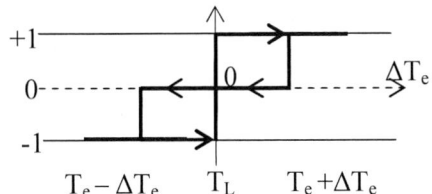

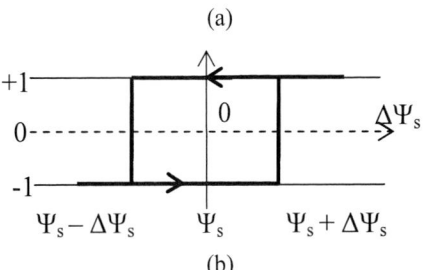

Fig. 7. Hysteresis comparator
(a) Flux, C_F and (b) Torque, C_T

The torque error C_T and the flux error C_F are obtained from the torque hysteresis and the flux hysteresis bands respectively. The switching table has been formed based on the sector and the hysteresis control signals C_T and C_F. It should be noted that a full power factor control is possible at the supply side using this drive technology by using CCW and CW vectors according to the leading and lagging nature of the matrix converter input current.

TABLE II. Rotating Space Vectors for Open-End-Winding Induction Motor

VECTOR	CCW		CW	
	MC-1	MC-2	MC-1	MC-2
1	V_{abc}	V_{cab}	V_{acb}	V_{cba}
2	V_{bca}	V_{cab}	V_{bac}	V_{cba}
3	V_{bca}	V_{abc}	V_{bac}	V_{acb}
4	V_{cab}	V_{abc}	V_{cba}	V_{acb}
5	V_{cab}	V_{bca}	V_{cba}	V_{bac}
6	V_{abc}	V_{bca}	V_{acb}	V_{bac}
7	V_{abc}	V_{abc}	V_{acb}	V_{acb}
8	V_{bca}	V_{bca}	V_{bac}	V_{bac}
9	V_{cab}	V_{cab}	V_{cba}	V_{cba}

TABLE III. Proposed Optimal Switching Table

C_{PF}	SECTOR (S)		1	2	3	4	5	6
	C_F	C_T						
CW CCW	-1	-1	2	3	4	5	6	1
		0	9	7	8	9	7	8
		1	6	1	2	3	4	5
	1	-1	3	4	5	6	1	2
		0	9	7	8	9	7	8
		1	5	6	1	2	3	4

The three-phase input voltages and currents are measured and converted into their polar co-ordinates. The cosine of difference between the input voltage and input current angles gives the power factor at the input side of the matrix converter. The measured power factor is subtracted from the required power factor to obtain the error in the power factor. This error is fed to a hysteresis band which gives an output command, C_{PF} which is -1 or +1 based on the leading or lagging nature of input current respectively. The C_{PF} determines the choice of

CW or CCW vectors for switching the converter thereby controlling the input power factor to the desired value. Fig. 9 represents the schematic of the input power factor control.

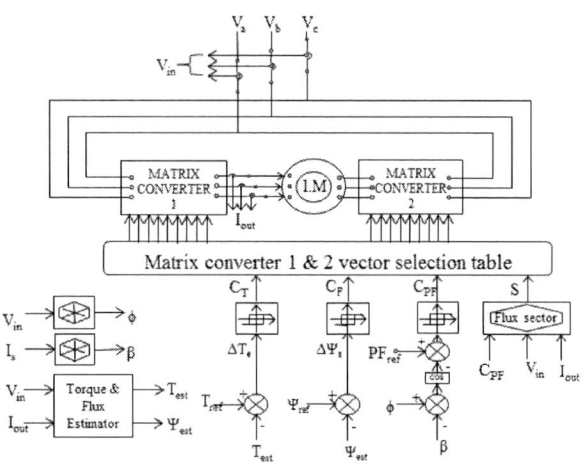

Fig. 8. Generic block diagram of DTC of open end winding induction motor using rotating space vectors

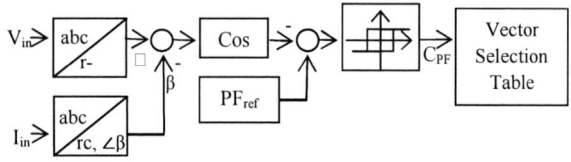

Fig. 9. Input power factor control

VI. SIMULATION PARAMETERS and OUTPUTS

Simulation is carried out in MATLAB®/Simulink® with a sampling frequency of 50 kHz using parameters listed in Table IV.

TABLE IV. SIMULATION PARAMETERS

S. No	Parameters	Value
1.	Input phase voltage	230 V
2.	Supply frequency	50 Hz
3.	Induction motor rated voltage	400 V
4.	Induction motor rated frequency	50 Hz
5.	Induction motor rated power	4 kW
6.	Induction motor rated speed	1430 RPM
7.	Induction motor stator resistance	1.405 Ω
8.	Induction motor stator leakage inductance	5.839 mH
9.	Induction motor rotor resistance (stator referred)	1.395 Ω
10.	Induction motor rotor leakage inductance (stator referred)	5.839 mH
11.	Mutual inductance	0.1722 H
12.	Mechanical inertia	0.0131 Kgm²
13.	Mechanical friction co-efficient	0.002985 Nms
14.	Number of pole pairs	2

The induction motor is operated under no-load condition for time (t), $0 \leq t < 0.6$ s and then loaded with a torque of 7 Nm at t = 0.6 s, as shown in Fig 10(a). The speed of the induction motor immediately decreases, when loaded. However, within 100 ms, the increased current to the motor restores it to the desired speed, as shown in Fig. 10(b). Also shown in Fig. 10(c) is the induction motor stator voltage. The dq component of the stator flux is plotted in Fig. 10(d). This shows the induction motor stator flux follows the flux band whose width is set by the control algorithm. Figs. 10(e) & (f) show the unfiltered and filtered input currents respectively. A low–pass filter with cut-off frequency of 1000 Hz filters the input current. Fig. 10(g) shows the input voltage and magnified input current (50×) which is at unity power factor. The CMV of the proposed scheme, shown in Fig. 10(i), is negligible or practically eliminated when compared to the CMV present in a conventional matrix converter based DTC drive, shown in Fig.10(h).

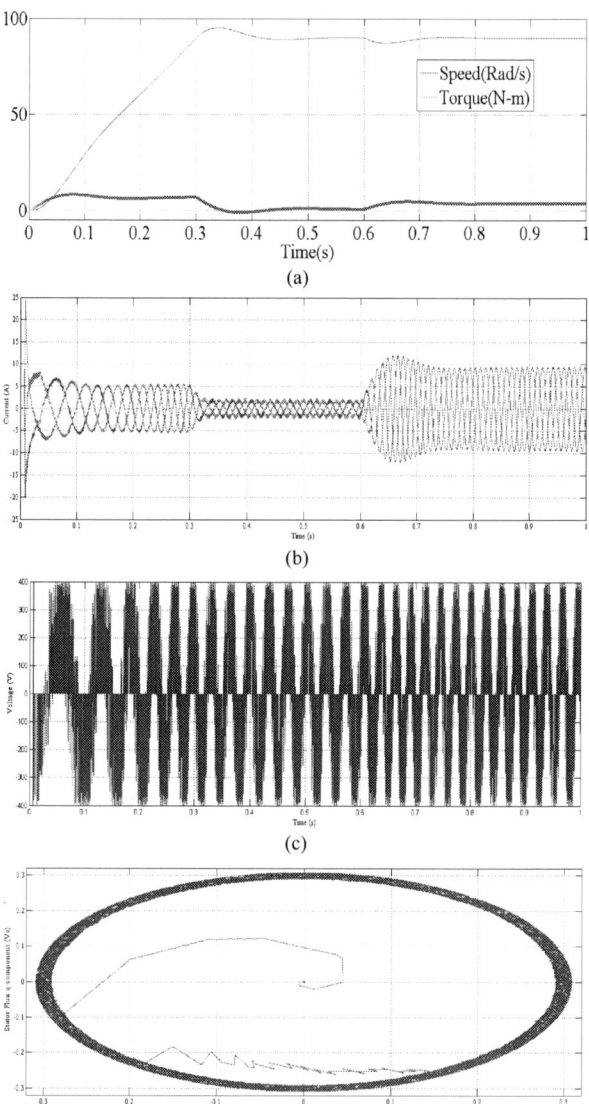

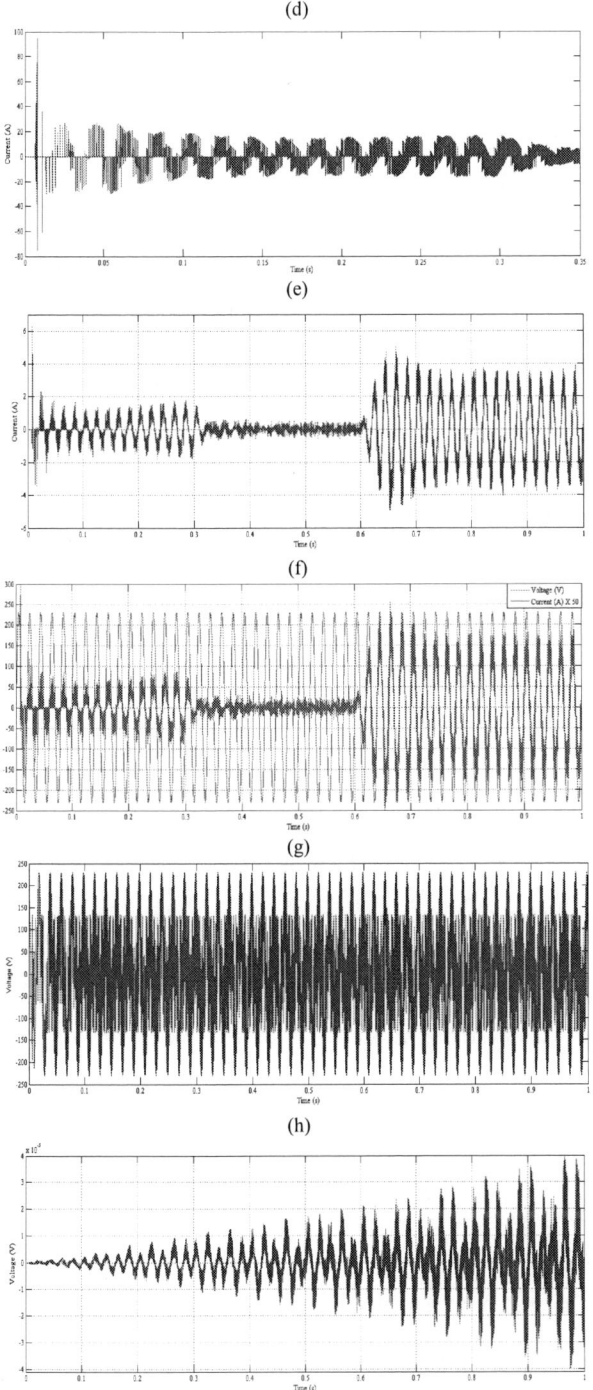

(d)

(e)

(f)

(g)

(h)

(i)

Fig. 10. (a) Induction motor torque and speed loaded at time, t = 0.6 s, (b) Stator current, (c) Stator voltage, (d) Stator flux dq component, (e) Input current-unfiltered, (f) Input current-filtered, (g) Input voltage and current (50x) at unity power factor, (h) Common mode voltage for conventional matrix converter based DTC, (i) Common mode voltage - Proposed DTC.

VII. CONCLUSION

In this paper, the DTC scheme for an open-end-winding induction motor is proposed and its simulation results are presented. The features of the presented scheme are i) CMV elimination resulting in decreased stresses on the bearing, ii) input power factor control and iii) increased machine input voltage magnitude. The drive presented improves the reliability by providing an additional degree of freedom for control but it does so at the cost of an additional matrix converter (18 switches) and the complexity associated with the drive control.

REFERENCE

[1]   I. Takahashi and Y. Ohmori, "High performance direct torque control of an induction motor," *IEEE Trans. Ind. Appl.*, vol. 25, pp. 257–264, Mar./Apr. 1989.

[2]   Y-S. Lai, J-H Chen "A new approach to direct torque control of induction motor drives for constant inverter switching frequency and torque ripple reduction", *IEEE Trans. Energy Conv.*, vol.16, no.3, pp.220-227, Sep 2001.

[3]   H. W. van der Broeck, H. C. Skudelny, and G. V. Stanke, "Analysis and realization of a pulsewidth based on voltage space vectors," *in IEEE IAS Ann. Meet.*, pp. 244–251, 1986.

[4]   J. Holtz, "Pulse width modulation-a survey", *IEEE Trans. Ind. Electron.*, vol. 39, no. 5, Dec. 1992.

[5]   A. Alesina, M. Venturini, "Analysis and design of optimum-amplitude nine-switch direct AC-AC converters", *IEEE Trans. Power Electronics*, vol. 4, no. 1, pp. 101-112, Jan. 1989.

[6]   D. Casadei, G. Serra, A. Tani, and L. Zarri, "Matrix converter modulation strategies: A new general approach based on space-vector representation of the switch state," *IEEE Trans. Ind. Electron.*, vol. 49, no. 2, pp. 370–381, Apr. 2002.

[7]   S. Chen, T.A. Lipo, D. Fitzgerald, "Source of induction motor bearing currents caused by PWM Inverters" , *IEEE Trans. Energy Conv.*, vol. 11, No. 1, March 1996.

[8]   A. L. Julian, G. Oriti, T.A. Lipo, "Elimination of common-mode voltage in three-phase sinusoidal power converters", *IEEE Trans. Power Electronics.*, vol. 14, no. 5, Sep. 1999.

[9]   H. J Cha and P. N. Enjeti, "An approach to reduce common mode voltage in matrix converters" *IEEE Trans. Ind. Appl.* vol.39, Issue 4, pp.1151-1159, 2003.

[10]   K. K. Mohapatra and N. Mohan, "Open-end winding induction motor driven with matrix converter for common-mode elimination," in *Proc. IEEE International Conference Power Electronics, Drives and Energy Systems*, New Delhi, India, Dec. 2006, pp. 1-6.

[11]   R.K. Gupta, "Rotating-space-vector-modulated matrix-converter-based advanced drive topologies with enhanced features", PhD dissertation, University of Minnesota, Minneapolis, 2010.

[12]   D. Casadei, G. Serra, A. Tani, "The use of matrix converters in direct torque control of induction machines", *IEEE Trans. Ind. Elec.*, vol. 48, no. 6, Dec. 2001.

[13]   M. Matteini, "Control techniques for matrix converter adjustable speed drives", PhD dissertation, University of Bologna, Bologna, 2001.

[14]   B. K. Bose "Modern power electronics and AC drives", Pearson Publication Inc., India, 2004.

Bridgeless Single-Ended Primary Inductance Converter with Improved Power Quality for Welding Power Supplies

Swati Narula
Department of Electrical Engineering,
Indian Institute of Technology,
New Delhi-110016, India.
E-mail: swatinarula.iitd@gmail.com

Bhim Singh
Department of Electrical Engineering,
Indian Institute of Technology,
New Delhi-110016, India.
E-mail: bsingh@ee.iitd.ac.in

G. Bhuvaneswari
Department of Electrical Engineering,
Indian Institute of Technology,
New Delhi-110016, India.
E-mail: bhuvan@ee.iitd.ac.in

Abstract—**This paper presents a new single-phase, single-stage bridgeless SEPIC (Single-Ended Primary Inductance Converter) based arc welding power supply operating at unity PF (Power Factor). The proposed converter provides an approach of implementing a simple AWPS (Arc Welding Power Supply) with an additional feature of long term short circuit capability. By eliminating input DBR (Diode Bridge Rectifier), the number of conducting components is reduced thereby lowering the conduction losses and increasing the overall efficiency. Simple structure, electrical isolation and simple control circuitry are the other advantages of the proposed AWPS. Bridgeless SEPIC based AWPS designed in DCM (Discontinuous Conduction Mode), provides inherent power factor correction. Dynamic and steady state performances are presented for a 3 kW, 25 V AWPS.**

Keywords—**BL (Bridgeless) Rectifier, Power Factor Correction, Power Quality, SEPIC, Arc Welding Power Supply.**

I. INTRODUCTION

The advances in power converter technology have enabled the development of single-stage PFC (Power Factor Correction) based AC-DC converters [1-2]. With this significant development, power supplies with active PFC technique are becoming indispensable for many types of industrial applications to meet harmonic regulations and standards, such as the IEC 61000-3-2 [3]. The conventional AC-DC converters use DBR rectifiers, bulk capacitors and subsequent high frequency based DC-DC converters to convert the low-frequency AC voltage into a constant DC voltage [4]. The existence of the bridge rectifiers generally leads to the problems such as high conduction power loss, increased harmonic currents etc. and thus creating problems in power lines. However, the use of BL (Bridgeless) configuration can alleviate these problems to a great extent as discussed in [5]. The aforementioned rectifier-less converter topologies are usually derived from boost converters. Although they are termed as BL rectifiers, they are unable to completely eliminate the DBR because of common-mode EMI (Electromagnetic Interference) and input AC mains current return path provision [6-7].

When AWPS is concerned; fast dynamic response and short circuit protection are essentially required features. In order to preclude the electrode from clinging to the work-piece during the arc striking process, it is beneficial to have a short

circuit current a bit higher than the rated welding load current. The short circuit also occurs when the electrode accidentally comes in contact with the welding pad and gets glued with the molten metal. The short circuit withstand capability of DC arc AWPS limits the current to the desired level which consequently improves the weld quality [8].

In this paper, a new BL single-stage converter is proposed employing two back-to-back connected isolated SEPIC DC-DC converters operating in alternate halves of the AC mains voltage cycle. It incorporates the features of output voltage stability and short-circuit withstand capability. The proposed AWPS is designed to operate in DCM to minimize the generated EMI noise and to achieve the unity PF [9]. EMI compatibility aspects of arc welding equipment are reported in EN 60974-10 [10]. PWM (Pulse Width Modulation) control strategy has been implemented with a constant switching frequency. A PI (Proportional Integral) controller used in the control loop provides a fast dynamic response with zero steady state error resulting in superior load regulation. To validate improved voltage regulation, the proposed AWPS is designed to accept an input voltage of 170V to 270Vrms and 20% to 100% load variations.

II. SYSTEM CONFIGURATION AND CONTROL STRATEGY

The schematic of the proposed AC-DC BL PFC based AWPS is shown in Fig. 1. The proposed BL configuration offers single stage power conversion and completely eliminates the front end DBR rectifier.

A. SEPIC Converter Configuration

In this topology, two isolated SEPIC DC-DC converters are connected back-to-back. Fig. 2a and Fig. 2b depict the operation of proposed topology during positive and negative half cycle of input ac mains. In this, one converter operates during the positive half cycle of input AC mains while the other one conducts with a negative half cycle of input AC mains. This further brings down the thermal stresses, as well as the conduction losses.

The SEPIC converter is a buck-boost converter with single switching device, IGBT (Insulated Gate Bipolar Transistor), to attain high frequency transformer isolation. SEPIC is a commendable option for high frequency transformer isolation and PFC at input AC mains.

978-1-4799-6047-7/14 $31.00 © 2014 IEEE

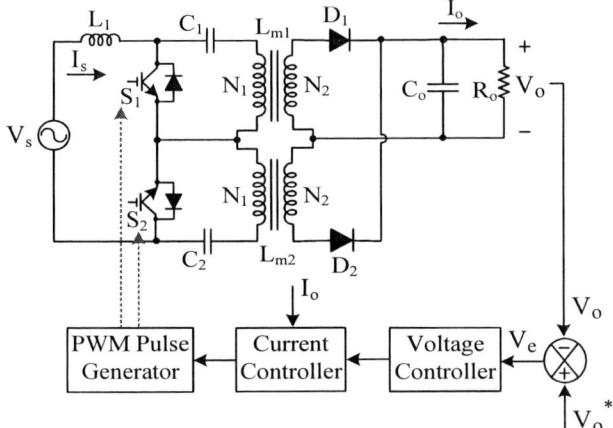

Fig.1. Circuit of BL SEPIC converter based AWPS.

These SEPIC converters are cascaded with HF (High Frequency) transformer and output filter to convert single-phase AC voltage into DC voltage feeding the welding load. An isolation transformer is also considered as a safety element because the work-piece is actually a part of the entire circuit, thus it becomes mandatory to connect it eventually to earth.

During positive half cycle of the supply voltage, the switch S_1 is turned on and the diode, D_1 remains reversed biased as shown in Fig. 2a(i). When switch S_1 is turned off and diode D_1 becomes forward biased. The inductor freewheels its stored energy to the load as presented in Fig. 2a(ii). In Fig. 2a(iii), the inductor enters into the DCM and the switch S_1 and the diode D_1 remain turned off during this period. Similarly, the operation of the proposed converter during negative half cycle of the supply voltage is shown in Fig. 2b(i)-(iii).

B. Control Strategy

For the positive half cycle of input AC mains switch S_1 conducts and for the negative half cycle switch S_2 is turned on. Since each switch is conducting for half of the cycle only thus there is a reduction in the rms / average current stress on the power switches. However, a same control signal can be used to drive both the power switches, which appreciably simplifies the control circuit. With the very low amount of control circuitry and switching devices, the proposed AWPS design is very compact and cheap. The converters are operated in DCM by implementing voltage mode control technique to reduce problems such as spattering, poor weld quality, instability of arc length etc. DCM offers numerous advantages such as simple control, inherent PFC etc. DCM is also preferable in the situations when there is wide line/load operating range.

The output dc voltage is compared with the reference voltage to generate voltage error. This voltage error is given to PI (Proportional Integral) voltage controller for reference output current generation for providing short circuit protection. This reference current is compared with the sensed output current which generates the error and this error is given to PI current controller to restrict the output current in the desired limits. The output of current controller when compared with high frequency ramp signal of 50 kHz is used to generate firing pulses for SEPIC converters.

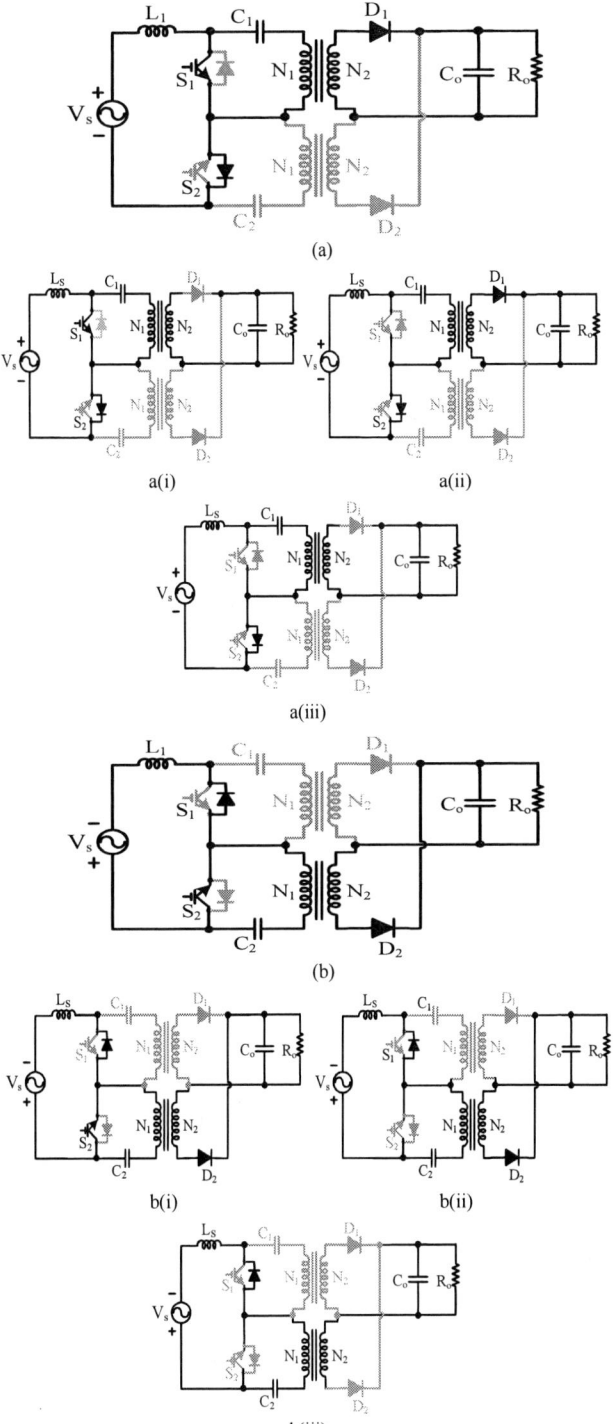

Fig.2a. BL-SEPIC converter based AWPS operation during positive half cycle 2a (i)-(iii).
2b. BL-SEPIC converter based AWPS operation during negative half cycle 2b (i)-(iii).

III. DESIGN OF PROPOSED BL CONVERTER BASED AWPS

The SEPIC converter working as PFC in DCM is an excellent pre-regulator. The supply current follows the supply voltage, and PF improves to unity with high converter efficiency. The switching frequency is considered to be 50 kHz, which is much higher that the line frequency so that average quantities over a switching cycle is considered constant.

Due to symmetry of the proposed BL circuit, the circuit is analyzed during positive half cycle of the applied input voltage.

The average value, V_{sav} of the supply voltage ($v_{s,rms} = 220$ V) is given as,

$$V_{sav} = \frac{2\sqrt{2}v_s}{\pi} = \frac{2\sqrt{2}x220}{\pi} = 198.07V \quad (1)$$

On applying the volt-second balance to the BL-SEPIC converter, one obtains,

$$\frac{V_o}{V_{sav}} = \frac{D}{\sqrt{K}}; \quad (2)$$

where, $K = \frac{2f_s}{R_o}\left(\frac{L_1 L_M}{L_1+L_M}\right) = \frac{2f_s L_{eq}}{R_o}; \quad (3)$

$$L_{eq} = \frac{L_1 L_M}{L_1+L_M}; \quad L_{M1} = L_{M2} = L_M; \quad (4)$$

$$R_o = \frac{V_o}{I_o} = \frac{25}{120} = 0.2083 \ \Omega; \quad (5)$$

Also, D is the on-time of the power switches S_1 and S_2, V_o is the output DC voltage and f_s is the switching frequency (=50kHz) of the BL-SEPIC converter.

The volt-second balance on the inductor results in the following relation,

$$V_o = \frac{V_{sav}N_2 D}{1-D \ N_1} \quad (6)$$

Considering D = 0.25, $V_o = 25$ V, the turns ratio is selected as, $N_1/N_2 = 2.64$.

The value of input inductor L_1 for a given ripple current, Δi_{L1} is calculated as,

$$L_1 = \frac{V_s D}{2f_s \Delta i_{L1}} = \frac{220x0.25}{2x50000x0.5x13.63} = 0.081 \ mH \quad (7)$$

The current through the input inductor should remain continuous throughout the entire switching period. Thus, the selected value of input inductor L_1 is 0.1 mH.

By considering the current balance through capacitor C_1, the average current, I_{LM1} can be obtained as

$$I_{Lm1} = \frac{N_2 P_o}{N_1 V_o} \quad (8)$$

Moreover, when the diode is conducting, the peak current through the output diode is as,

$$I_D = \frac{P_o}{V_o \ 1-D} \quad (9)$$

Using eqns. (7) and (8), the value of L_{M1} for DCM operation for a given current ripple is calculated as,

$$L_{m1min} = \left(\frac{N_1}{N_2}\right)^2 \frac{V_o \ 1-D}{2Df_s I_o}^2 \quad (10)$$

$$= (2.64)^2 x \frac{25x(1-0.25)^2}{2x0.25x50000x120} = 32.67 \ \mu H$$

This value of inductor is chosen 30 μH to ensure DCM operation of the converter.

During steady-state DCM-PWM operation, capacitors C_1 and C_2 are charged to the input voltage, V_s. It results in the capacitor voltage as,

$$V_{C1} \text{ (average)} = V_{sav} \quad (11)$$

The value of capacitors C_1 and C_o for desired voltage ripple is as,

$$C_1 = \frac{V_o D N_2}{R_o f_s \Delta V_{C1} N_1} = \frac{25x0.25x0.3786}{0.2083x50000x50} = 4.537 \ \mu F \quad (12)$$

Substituting the values in the above equation, the estimated value of C_1 is 13.32 μF. The output capacitor, C_o is designed as,

$$C_o = \frac{I_o}{4\pi f(\Delta V_o)} = \frac{120}{4\pi x50x0.1x25} = 76.43 \ mF \quad (13)$$

where f is the frequency of the AC mains = 50 Hz.

To evaluate the design and control of proposed isolated BL converter based AWPS, the simulation is carried out at varying input voltage and load conditions.

IV. MATLAB BASED MODELLING OF THE PROPOSED AWPS

In order to corroborate the design, operation and control of proposed BL converter based AWPS, it has been modeled and simulated using MATLAB/Simulink environment for the specifications given in Appendix. Fig 3 shows the MATLAB/Simulink based developed model of the proposed AWPS. The simulated waveforms of the BL converter are shown in Figs 4-8. The single stage ac-to-dc BL converter has been accomplished in DCM to maintain constant DC voltage at the output. The simulation results are presented to validate the performance of the proposed BL converter based AWPS throughout the wide line/load conditions.

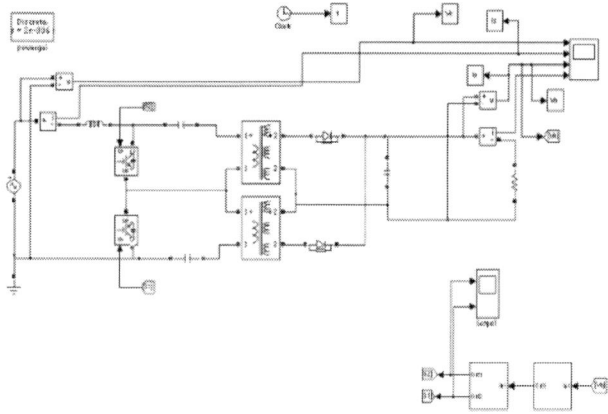

Fig.3. MATLAB/Simulink model of BL-SEPIC converter based AWPS.

V. PERFORMANCE OF BL CONVERTER BASED AWPS

In this section, the steady state and dynamic performance of the proposed AWPS has been discussed in detail. The dynamic performance of the proposed AWPS has been illustrated by varying the loads and supply voltage. The simulated results for rated load and light load conditions are shown in Figs. 4-7.

The performance of the proposed BL-SEPIC converter based AWPS for a 25V/120A load during steady state condition is shown in Fig. 4a. It can be perceived that the switch S_1 conducts in the positive half cycle while switch S_2 conducts in the negative half-line cycle of the supply voltage. Fig. 4b shows the input current waveform at rated load condition, along with its harmonic spectrum and THD (Total Harmonic Distortion) of AC mains current.

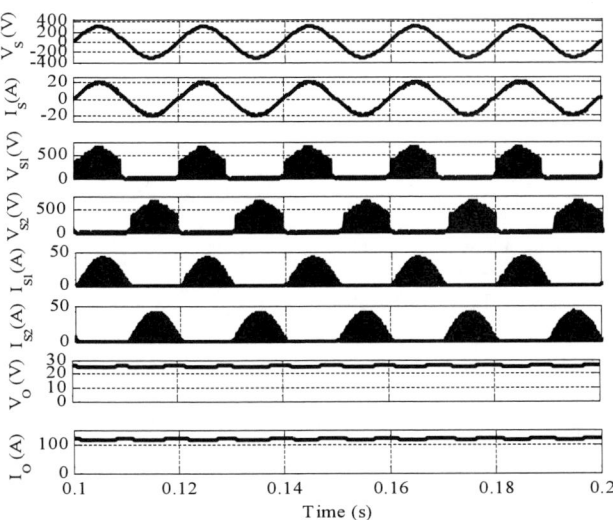

Fig.4a. Performance of proposed AWPS at 220 V AC mains and 100% load.

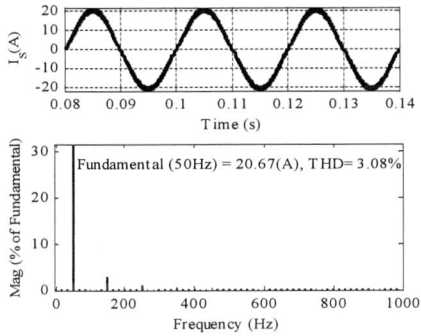

Fig.4b. Waveform of input ac mains current (I_s) along with its harmonic spectrum for the AWPS system at full load.

It can be clearly observed from Fig. 5a that the dynamic performance of the proposed system is fast enough for AWPS which eliminates the need for any second stage. The output voltage is regulated by optimizing the PI controller. The harmonic spectrum of AC mains current along with its THD under light load condition is shown in Fig. 5b. Figs. 6 and 7 show the input AC mains voltage/current and output voltage/current waveform of the proposed isolated BL-SEPIC converter for a 3 kW AWPS at variable supply voltage conditions. The simulation results under steady state and dynamic conditions confirm the effectiveness of the proposed

control and design approach of AWPS. It can be seen that the THD of input current is within the acceptable limits and the PF of the input current is almost unity.

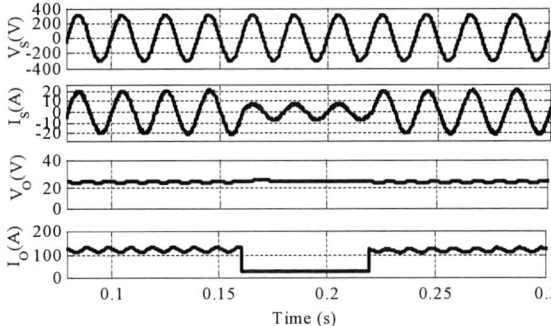

Fig.5a. Dynamic performance of proposed arc AWPS at 20% load.

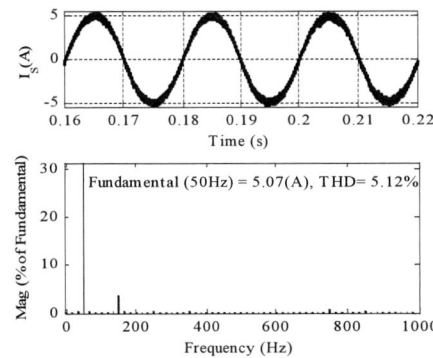

Fig.5b. Current waveform along with its harmonic spectrum for proposed AWPS at 20% load.

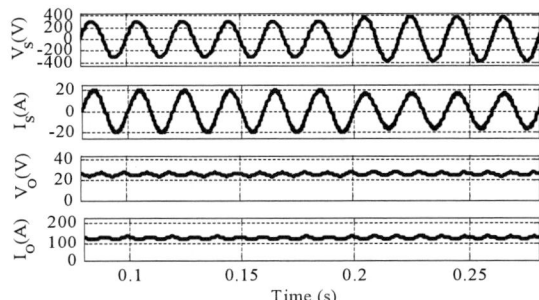

Fig.6a. Performance of the proposed AWPS at V_s of 270 V.

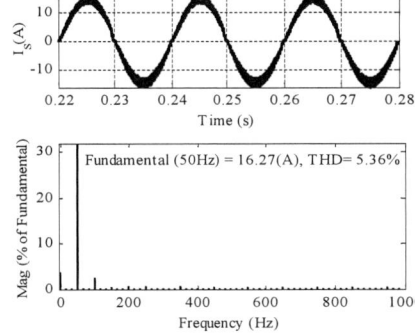

Fig.6b. Current waveform and its THD for proposed arc AWPS at V_s of 270 V

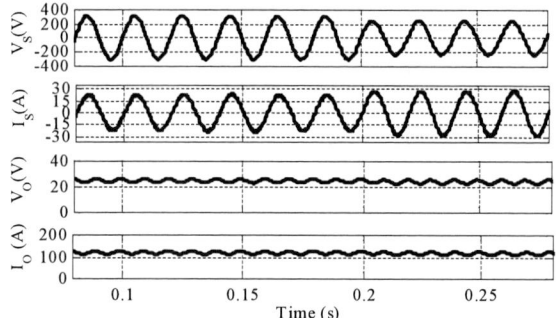

Fig.7a. Performance of the proposed AWPS at V_s of 170 V.

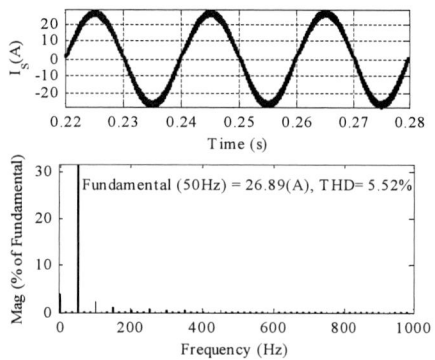

Fig.7b. Waveform of input ac mains current (I_s) along with its harmonic spectrum for the AWPS system at V_s of 170V.

It is evident from Fig. 8 that the proposed AWPS exhibits short-circuit withstand capability. In case of short circuit, the controller limits the current to 150 A only which is a necessary condition while designing a AWPS. Thus even during short circuit period, less spatter is generated hereby improving the weld bead quality.

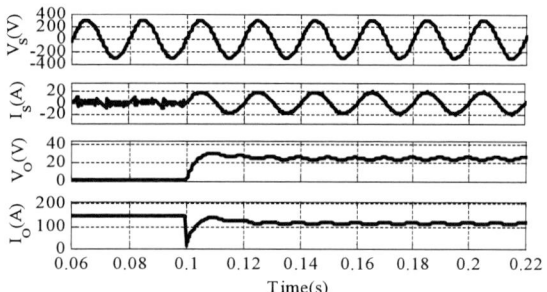

Fig.8. Dynamic performance of proposed AWPS under short circuit condition.

The performance of the proposed AWPS in terms of various PQ (Power Quality) parameters is summarized in Figs. 9-10. The DCM operation of the BL converter confirms improved PQ and high level of weld quality. It can be observed that the proposed converter results in nearly unity PF in the wide operating range of the loads keeping the input AC mains current THD well below 5.15%. Moreover, the input AC mains current THD of the proposed BL converter is also below 5.6% under the varying AC mains voltages.

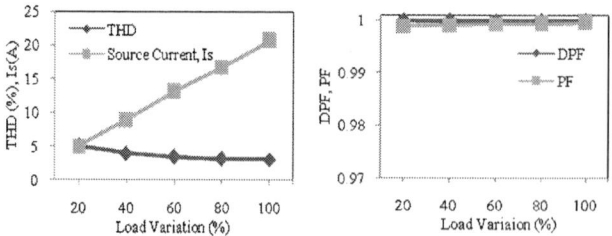

Fig.9. Variation of PQ indices of proposed AWPS under different load conditions,
(a) I_s and its THD at various load conditions,
(b) DPF and PF at various load conditions

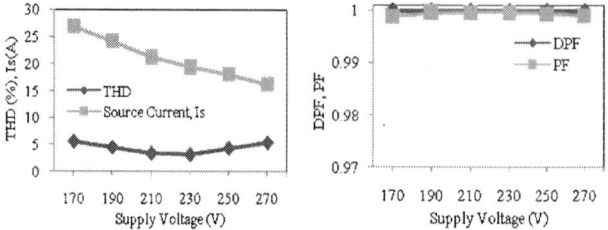

Fig.10. Variation of PQ indices of proposed AWPS under various supply voltage conditions,
(a) I_s and its THD at different supply voltages,
(b) DPF and PF at various supply voltage conditions.

VI. CONCLUSION

A single stage single-phase BL converter based AWPS has been designed with low input current THD and low conduction losses. The proposed isolated BL-SEPIC converter is a good solution as it provides output isolation along with the over-current and start-up protection. The component count has also reduced by integrating two power conversion stages. The control method applied has been found simple for output voltage regulation and PFC. Furthermore, it can be inferred from the obtained results that the proposed converter system has provided the robustness and fast response. The proposed design is found suitable for wide-range of supply voltages and load variations. Moreover, the observed PQ indices of the proposed AWPS are in conformity to the international standard IEC 61000-3-2 [2]. It is concluded that the proposed topology has been amongst the most convincing configuration for AWPS.

APPENDIX

Specifications of Proposed AWPS

Input AC mains voltage, V_s (rms): 220V, 50Hz; Output Power, P_o: 3 kW; Output Voltage, V_o: 25V; Output Current, I_o: 120A; Switching frequency of DC-DC converter: 50 kHz; Transformer primary to secondary turns ratio, N_1/N_2: 2.64; Inductance, $L_{m1} = L_{m2}$: 30 µH, Input Capacitances, $C_1 = C_2$: 4.537 µF, Output Capacitance, C_o: 76.43 mF.

REFERENCES

[1] N. Huang, D. Zhang, T. Song, M. Fan, Y. Liu and Y. Zhao, "A 10 kW single-stage converter for welding with inherent power factor correction", in *Proc. of IEEE APEC'05*, vol. 1, 2005, pp. 254-259.

[2] H. Ma, Y. Ji and Y. Xu, "Design and analysis of single-stage power factor correction converter with a feedback winding," *IEEE Trans. on Power Electronics*, vol. 25, no. 6, pp. 1460–1470, Jun. 2010.

[3] Limits for Harmonic Current Emissions, International Electro technical Commission Standard, 61000-3-2, 2004.

[4] B. Singh, S. Singh, A. Chandra and K. Al-Haddad, "Comprehensive Study of Single-Phase AC-DC Power Factor Corrected Converters With High-Frequency Isolation", *IEEE Trans.* on *Industrial Informatics*, vol. 7, no. 4, pp. 540-556, Nov. 2011.

[5] W. Y. Choi, J. M. Kwon and B. H. Kwon, "Bridgeless dual-boost rectifier with reduced diode reverse-recovery problems for power factor correction," *IET Power Electron.*, vol. 1, pp. 194-202, Jun. 2008.

[6] A. Sabzali, E. Ismail, M. Al-Saffar and A. Fardoun, "New bridgeless DCM SEPIC and Cuk PFC rectifiers with low conduction and switching losses," *IEEE Trans. Industry Applications*, vol. 47, no. 2, pp. 873 -881, 2011.

[7] W. Choi, J. Kwon, E. Kim, J. Lee and B. Kwon, "Bridgeless boost rectifier with low conduction losses and reduced diode reverse-recovery problems," *IEEE Trans. Industrial Electronics*, vol. 54, no. 2, pp. 769–780, Apr. 2007.

[8] A.E. Emanuel and J.A. Orr, "An improved method of simulation of the arc voltage-current characteristic" in *Proc.* of *IEEE Harmonics and Quality of Power*, vol. 1, 2000, pp. 148-154.

[9] A. Lazaro, J. A. Cobos, A. Barrado and E. Olias, "Design of a zero-current-switched quasi-resonant SEPIC used as power factor preregulators with voltage-follower control," in *Proc. IEEE INTELEC '96*, 1996, pp. 271–278.

[10] EN 60974-10: Arc welding equipment Part 10: Electromagnetic compatibility (EMC) requirements.

Reliability Analysis with Parametric and Non Parametric Ranking of Buck Converter Components

RamaKoteswara Rao Alla[1], G.L.Pahuja[2], J.S.Lather[3]

[1]Research Scholar, Electrical Engineering Department
[2,3]Professor, Electrical Engineering Department,
National Institute of Technology Kurukshetra, India
Email: ramnitkkr@gmail.com[1], pahuja.gl@gmail.com[2], jslather@gmail.com[3]

Abstract— **In system reliability analysis component importance plays major role for prioritizing the components in the system to improve the reliability, risk and maintenance planning. Maintenance of all the risky equipments is necessary to ensure successful operation of any system. Different importance measures are available to compute the rank of the component for improving the system reliability and maintenance planning. In this paper a set of seven existing importance measures and a non parametric ranking tool i.e Copeland Score (CS) method have been extended to analyze reliability improvement and component maintenance priority of the system. Also buck converter system effectiveness with CS method in comparison to the different existing measures has been advanced to rank the components of buck converter and use the ranking for prioritizing reliability improvement and maintenance of components of the system. It is observed that the CS method based ranking results better guidance for improving system reliability and planning activity of different components of the buck converter maintenance.**

Keywords— *Buck Converter,component Importance Measures, Rekiability, Maintenance, Copland Score*

NOMENCLATURE

BM	Birnbaum's Measure
CP	Conditional Probability
CIF	Criticality Importance Factor
CS	Copland Score
DIF	Diagnostic Importance Factor
IM	Importace Measure
IO-IA	Improvement Oriented Importance Analysis
IP	Improvement Potential
MO-IA	Maintenance Oriented Importance Analysis
RAW	Risk Achievement Worth
RRW	Risk Reduction Worth
SI	Structural Importance

I. INTRODUCTION

Because of the efficient conversion and flexible control of electrical energy in adjustable-speed drives, unified power quality correction, energy storage systems power electronic systems play an important role. Power electronic circuits also provide high efficient solution to power conversion. However, maintaining the reliable performance of the overall system is a challenging task. Many industrial applications demand to convert a fixed voltage dc source into a variable dc source. DC converters can be used for these applications. DC converters are used in traction purposes and also in regenerative braking of dc motors to return energy back into the supply [1].

Buck converter is a type of dc converter which is used to generate lower output voltage from a higher dc input voltage. DC converters have significant importance in the space applications [12] in which the component criticality is an important issue to be considered. These converters are essential for satellite power systems also particularly when battery power storage is required to supply a regulated main voltage bus [13]. Such applications require highly reliable performance from the components to avoid risks and system failures so that loss of life and system cost can be eliminated/minimized. However there is a need of reliability and component importance analysis of such converters for improving the overall system performance. Reliability and maintenance analysis of the buck converter is addressed in this paper using different existing reliability importance measures and with the help of Copland score method.

Reliability importance measures are useful in guiding the system for the reliability improvement and maintenance planning of the system. Importance measure was introduced firstly by Birnbaum [2]. It gives the contribution of different elements in the system to the overall system reliability. Many parametric ranking importance measures were proposed after Birnbaum measure like Improvement Potential (IP), Risk Achievement Worth (RAW), Risk Reduction worth (RRW), Diagnostic Importance Factor (DIF), and Cost Importance Measure (CIM) etc. A brief summary can be found in [3-4]. These measures are useful for two types of analysis: one is Improvement Oriented Importance Analysis (IO-IA), Maintenance Oriented Importance Analysis (MO-IA). After evaluating the importance measures the designer may be able to take decision to reliability improvement and provide maintenance preference to the system components. To ensure the proper reliability of the system, component criticality can be identified and configured from the results of the reliability importance measures.

In this paper the reliability analysis of components using seven importance measures and a non parametric ranking based Copeland score method for buck converter system has been analyzed with eight sets of failure parameters data. The observations of the results of the different measures ranking is also discussed. The remaining paper is organized as section II presents different Importance measures. Section III gives the description of non parametric ranking method. In section IV buck converter example has been considered and the Importance Measures considered in section II, Copland Score were calculated for different failure parameter sets. Section V presents the investigation and discussion of different measures and Copeland Score method results. Section VI concludes the paper.

II. COMPONENT IMPORTANCE MEASURES

In system Reliability theory, component importance measures have considerable significance. The importance measures evaluate the role of the components or basic events to the considered measure of system performance [5]. The importance measures may be used to grade the components, that is, to organize the components in order of increasing or decreasing manner or with regards to the performance measure considered. Thus Importance analysis is performed to numerically quantify the importance of each individual component in the system. These measures are helpful to identify the weak components so that efforts can be made to improve the system reliability [6].Various IM's exist in the literature [2-6]. Seven of the existing Importance measures are presented in this section with their mathematical expressions. Table.1 shows different importance measures of a basic event e with their notation in terms of conditional probabilities. S denotes the structure function.

Table1: Importance measures with their expressions

Importance Measure	Mathematical Expression
Conditional Probability (CP)	$I^{CP}(e) = \Pr\{S/e\} = \dfrac{\Pr\{S \cap e\}}{\Pr\{e\}}$
Risk Achievement Worth (RAW)	$I^{RAW}(e) = \dfrac{\Pr\{S/e\}}{\Pr\{S\}} = \dfrac{I^{CP}(e)}{\Pr\{S\}}$
Risk Reduction Worth (RRW)	$I^{RRW}(e) = \dfrac{\Pr\{S\}}{\Pr\{S/\bar{e}\}}$
Diagnostic Importance Factor (DIF)	$I^{DIF}(e) = \Pr\{e/s\} = \dfrac{\Pr\{S \cap e\}}{\Pr\{s\}}$
Birnbaum's Measure (BM)	$I^{BM}(e) = \dfrac{\partial \Pr\{S\}}{\partial \Pr\{e\}}$
Criticality Importance Factor (CIF)	$I^{CIF}(e) = \dfrac{\Pr\{e\}}{\Pr\{S\}} I^{BM}(e)$
Improvement Potential (IP)	$I^{IP}(e) = I^{BM}(e) \Pr\{e\}$

Where $I^{CP}(e)$ is the Conditional Probability measure of component e, $I^{RAW}(e)$ is the Risk Achievement measure of component e, $I^{RRW}(e)$ is the Risk Reduction Worth measure of component e, $I^{DIF}(e)$ is the Diagnostic Importance Factor of component e, $I^{BM}(e)$ is the Birnbaum's Measure of the component e, $I^{CIF}(e)$ is the Criticality Importance Factor of the

component e, $I^{IP}(e)$ is the Improvement Potential of the component e.

For the detailed definitions of conditional probabilities and the importance measures reader can refer [7-8].

III. NON PARAMETRIC RANKING METHOD

Nonparametric ranking tool Copland score method is described in this section. The advantage of non parametric tool is that it does not depend on the additional information from a decision maker. A.H Copeland has proposed the Copeland score method. It is a simple non parametric ranking method which was used firstly for evaluating election results after voting. The flaws and properties of the Copeland score method are addressed in [9]. A modified CS method has been proposed by Al-Sharrah [10] to overcome some deficiencies of the original method. Depending on the number of times an object is better than other object and at the same time the same object worse than other objects with respect to a descriptor the Copeland Score can be calculated. Let the descriptor d_i. A matrix M is to be build to calculate the CS. For each position M(a, x), a comparison between object a and object x is performed. The comparison can be made based on the following rules.

If $d_i(a) \geq d_i(x)$ M(a, x) = M(a, x) + 1

If $d_i(a) \geq d_i(x)$ M(a, x) = M(a, x) − 1 and

If $d_i(a) = d_i(x)$ M(a, x) = M(a, x) + 0.

By adding every M(a,x) over all x, yields the CS of the component a. From corresponding CS obtained for every component ranking among the components can be done [11].

IV. BUCK CONVERTER COMPONENT RANKING

A simple buck converter is considered in this section as an example to illustrate the semantics of the importance measures and the CS method. Figure 1(a) and 1(b) shows the simple buck converter where $C_1 = C_2 = \dfrac{C}{2}$ and figure 2(a) and 2(b) shows the corresponding series parallel reliability block diagram for improved reliability structure. To illustrate the effect of the component position in the system and unreliability of the component in the system eight sets of component failure parameters are considered in table2 [8].

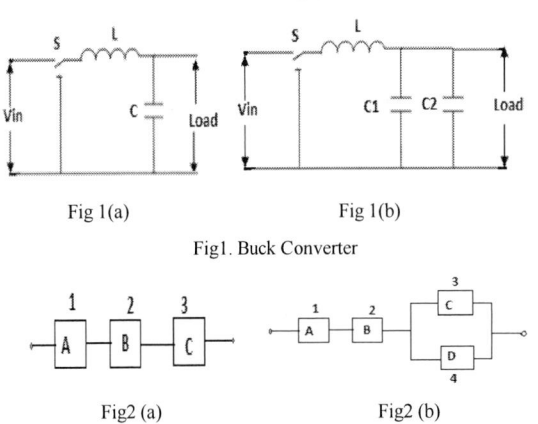

Fig 1(a)　　　　　　　Fig 1(b)

Fig1. Buck Converter

Fig2 (a)　　　　　　　Fig2 (b)

Fig2. Reliability Block Diagram

Table2: Failure parameters of the system

Comp	A	B	C	D	C∩D
Set I	0.02	0.04	0.06	0.05	0.003
Set II	0.02	0.04	0.001	0.5	0.0005
Set III	0.02	0.04	0.1	0.2	0.02
Set IV	0.02	0.04	0.04	0.5	0.02
Set V	0.01	0.1	0.1	0.5	0.05
Set VI	0.02	0.04	0.06	0.5	0.03
Set VII	0.02	0.04	0.08	0.5	0.04
Set VIII	0.02	0.04	0.1	0.5	0.05

The component importance analysis with the considered importance measures for different sets is given below.

Table3: IMs ranking for components using parameter Set I

Rank	CP	RAW	RRW	DIF	BM	CIF	IP
1	1,2	1,2	2	2	2	2	2
2	4	4	1	1	1	1	1
3	3	3	3,4	3	4	3,4	3,4
4				4	3		

Table3 presents the results of ranking of seven importance measures considered in second section. It can be observed that one measure gives better rank to a component and at the same time other measure gives lower rank to the same component. So, there is no clear decision about which component is more important. Decision making is difficult in this situation to a repair person checklist. There is a need of additional aggregation procedure to achieve integrated rank.

To do so, Table 4 shows the integrated ranking by employing Copeland scores. The main advantage of these scores is that these do not require the need of any additional information about the preference or perspective from a decision maker. To elucidate how the score is obtained consider component 2; CP and RAW measures component2 ranked 1 and two components has low ranked than the component 2 so the score will be +2 points for each measure and the remaining RRW, DIF, BM, CIF, IP measures rank component 2 first than other three components so the score will be +3 points for each measure thus the component 2 yields the score as given below.
CS for Component 2 =2+2+3+3+3+3+3=19.
Similarly for the component 1; CP and RAW measures component1 better than two components so the score will be +2 points for each measure and for the remaining five measures component 1 got second rank since component 2 got rank1, so for RRW, DIF, BM, CIF, IP measures component1 gets two positive and one negative point that is (+2-1=+1) for each measure. Thus the score of the component 1 obtained as given below.
CS for component 1= 2+2+(2-1)+(2-1)+(2-1)+(2-1)+(2-1)=9.
Similarly CS has been calculated for the remaining components. The same procedure has been done for the remaining sets also. Tables 6,8,10,12,14,16,18 gives the results of the ranking based on Copeland score for IMs ranking given in table 5,7,9,11,13,15,17. Table 19 presents the overall ranking by Copeland score for different sets.

Table4: Ranks by Copeland Approach for Set I

Rank	Component	CS
1	2	19
2	1	9
3	4	-6
4	3	-16

Table5: IMs ranking for components using parameter Set II

Rank	CP	RAW	RRW	DIF	BM	CIF	IP
1	1,2	1,2	2	2	2	2	2
2	3	3	1,3,4	4	1	1,3,4	1,3,4
3	4	4		3	3		
4				1	4		

Table6: Ranks by Copeland Approach for Set II

Rank	Component	CS
1	2	19
2	1	-1
3	3	-7
4	4	-11

Table7.IMs ranking for components using parameter Set III

Rank	CP	RAW	RRW	DIF	BM	CIF	IP
1	1,2	1,2	2	2	2	2	2
2	3	3	1,3,4	4	1	1,3,4	1,3,4
3	4	4		3	3		
4				1	4		

Table8: Ranks by Copeland Approach for Set III

Rank	Component	CS
1	2	19
2	1	-1
3	3	-7
4	4	-11

Table9:IMs ranking for components using parameter Set IV

Rank	CP	RAW	RRW	DIF	BM	CIF	IP
1	1,2	1,2	2	4	2	2	2
2	3	3	1,3,4	2	1	1,3,4	1,3,4
3	4	4		3	3		
4				1	4		

Table10: Ranks by Copeland Approach for Set IV

Rank	Component	CS
1	2	17
2	1	-1
3	3	-8
4	4	-9

Table11: IMs ranking for components using parameter Set V

Rank	CP	RAW	RRW	DIF	BM	CIF	IP
1	1,2	1,2	2	2	2	2	2
2	3	3	3,4	4	1	3,4	3,4
3	4	4	1	3	3	1	1
4				1	4		

978-1-4799-6047-7/14 $31.00 © 2014 IEEE

Table12: Ranks by Copeland Approach for Set V

Rank	Component	CS
1	2	19
2	3	-4
3	1	-7
4	4	-8

Table13: IMs ranking for components using parameter Set VI

Rank	CP	RAW	RRW	DIF	BM	CIF	IP
1	1,2	1,2	2	4	2	2	2
2	3	3	3,4	2	1	3,4	3,4
3	4	4	1	3	3	1	1
4				1	4		

Table14: Ranks by Copeland Approach for Set VI

Rank	Component	CS
1	2	17
2	3	-4
3	4	-6
4	1	-7

Table15: IMs ranking for components using parameter Set VII

Rank	CP	RAW	RRW	DIF	BM	CIF	IP
1	1,2	1,2	3,4,2	4	2	3,4,2	3,4,2
2	3	3	1	3	1	1	1
3	4	4		2	3		
4				1	4		

Table16: Ranks by Copeland Approach for Set VII

Rank	Component	CS
1	2	9
2	3	1
3	4	-3
4	1	-7

Table17: IMs ranking for components using parameter Set VIII

Rank	CP	RAW	RRW	DIF	BM	CIF	IP
1	1,2	1,2	3,4	4	2	3,4	3,4
2	3	3	2	3	1	2	2
3	4	4	1	2	3	1	1
4				1	4		

Table18: Ranks by Copeland Approach for Set VIII

Rank	Component	CS
1	2	5
2	3	4
3	4	0
4	1	-7

Table19: Ranks by Copeland Approach for different sets

Set	Ranking based on CS
Set I	B>A>D>C
Set II	B>A>C>D
Set III	B>A>C>D
Set IV	B>A>C>D
Set V	B>C>A>D
Set VI	B>C>D>A
Set VII	B>C>D>A
Set VIII	B>C>D>A

V. INVESTIGATION & DISCUSSION

Tables 3,5,7,9,11,13 and 15 presents the results of the component importance measures considered in section II for different failure data sets. From the results it can be seen that different measures leading different conclusions since the measures were defined differently.Tables4,6,8,10,12,14,16,18 present the ranks obtained by using the Copeland score for different sets. The observations from the results of the different measures and the Copeland Score method are:

From the maintenance point of view the most unreliable component should be ranked highest in the checklist of the repairperson. But the measures CP and RAW did not give the above results, these measures assign same rank to both the series components irrespective of the reliabilities of the components where as CS method differentiate the series structure elements also with respect to their unreliability and assign rank1to the most unreliable component which is justifiable.

RRW, CIF and IP measure results illustrate that in series structure component with lowest reliability has the highest importance, the component with same equivalent failure probability have the same importance. For example in case of Set III A, C, D components have equal ranking but the CS method based ranking differentiating among them also by giving preference to the component A which is in series (A>C>D). Also CIF and IP measures are impractical for large dynamic systems.

CP, RAW & BM measures gives misleading results by giving higher rank to the series components irrespective of the failure probabilities. That is always A and B components first in the maintenance checklist than C and D components. This is overcome with the CS method by giving appropriate ranking to the series and parallel components.

BM measure also gives wrong results by assigning lower rank to the more reliable component and higher rank to the more reliable element in parallel structure. This is not reasonable from maintenance point of view since most unreliable component should be in the top of the maintenance priority list. The results obtained by the CS are obeying this condition in many cases compared to the other measures.

DIF measure results are dynamic and responsive one among the considered seven measures since this measure results are based on the structure and reliability of the components. The ranking on the DIF measure is deterministic and always result in different ranking corresponding to the different reliabilities of the components.

978-1-4799-6047-7/14 $31.00 © 2014 IEEE

CS method ranking results are also deterministic as DIF measure ranking. In some cases it is different for some components in some sets since the CS method considers the combination of all the measures ranking of that particular component. The results of CS method are most informative and appropriate for the reliability, availability, maintenance oriented importance analysis and it distinguishes the components that occupy the similar structures with different and same reliabilities also.

VI. CONCLUSION

Importance measures are used in the reliability analysis of the system and describe how a particular component affects the system reliability. Copeland Score based ranking has the advantage that it gives the integrated ranking and do not require additional information from the decision maker. Component criticality is an important issue to be considered for the buck converter since it is widely used in space applications, satellite power systems. The component importance analysis of buck converter system is analyzed in this paper using CP, RAW, RRW, DIF, BM, CIF, IP importance measures and with the CS method. The results of the ranking of different measures have been investigated and it is observed that the CS method ranks more appropriately. Since CS method gives an integrated ranking and it distinguishes the components with similar structure and similar reliabilities. Foe example as in case of Set III A, C, D components have equal ranking with RRW, CIF and IP measures but the CS method based ranking differentiating among them also by giving preference to the component A which is in series (A>C>D). CS method results are useful to ensure proper reliability availability and maintenance priority of the system.

REFERENCES

[1] M.H.Rashid, Power Electronics circuits,Devices, and Applications PEARSON Publications,2008.

[2] Z. W. Birnbaum, "On the importance of different components in a multi component system," in Multivariate Analysis, P. R. Krishnaiah, Ed. New York: Academic Press, pp. 581–592, 1969

[3] A.Anne, "Implementation of Sensitivity Measures for static and Dynamic Subtrees in DIF tree", M.S. Thesis, University of Virginia, 1997.

[4] F. C. Meng, "Comparing the importance of system elements by some structural characteristics," IEEE Trans. on Reliability, vol. 45, pp. 59–65, 1996.

[5] Enrico Zio, "Computational Methods For Reliability And Risk Analysis, Series On Quality, Reliability And Engineering Statistics", World Scientific Publishing Co. Pvt. Ltd., vol 14, pp.235-243,2009.

[6] E.A Elsayed, "Reliability Engineering", John Wiley and Sons, 2012.

[7] A.Papoulis, "Probability,Random Variables and Stochastic Processes" (3rd Edition), McGraw-Hill Series in Electrical Engineering, 1991.

[8] L.Xing, "Maintenance-oriented fault tree analysis of component importance", Reliability and Maintainability, 2004 Annual Symposium - RAMS , pp.534-539, 26-29 Jan. 2004.

[9] VR.Merlin, DG Saari "Copeland method:II. manipulation, monotonicity, and paradoxes". Journal of Economic Theory Vol.72,No.1 pp.148–172, 1997.

[10] Al-Sharrahm G., 2010, "Ranking Using the Copeland Score: A Comparison with the Hasse Diagram", J. Chem. Inf. Model, vol.50, pp.785–791,2010.

[11] Claudio M. Rocco S., Jose Emmanuel Ramirez-Marquez, "Innovative approaches for addressing old challenges in component importance measures", Reliability Engineering & System Safety, vol. 108, , pp. 123-130, December 2012..

[12] Banfi, E.; Maranesi, P.; Volpi, G.F., "Boost/complementary buck DC/DC converter for space applications," Telecommunications Energy Conference, 1989. INTELEC '89. Eleventh International Conference Proceedings., pp.16.6/1,16.6/6 vol.2, 15-18 Oct 1989.

[13] Weinberg, AH.; Schreuders, Jan, "A High-Power High-Voltage DC-DC Converter for Space Applications" IEEE Transactions on Power Electronics , vol.1, no.3, pp.148,160, July 1986.

[14] Zhijun Qian; Abdel-Rahman, O.; Al-Atrash, H.; Batarseh, I, "Modeling and Control of Three-Port DC/DC Converter Interface for Satellite Applications," Power Electronics, IEEE Transactions on , vol.25, no.3, pp.637,649, March 2010.

Direct Torque Control of Matrix Converter Fed BLDC Motor

Ranganth Muthu[1], M. Senthil Kumaran[2], L.A. Abishek Rajaraman[3], P. Ganesh[4], P.Geeth Prajwal Reddy[5]

Electrical and Electronics Engineering, SSN College of Engineering, Anna University, Chennai, India
{ranganathm[1], senthilkumaranm[2]}@ssn.edu.in, abishekrajaraman@gmail.com[3], ganeshnet81@ymail.com[4],
geethprajwal14@yahoo.co.in[5]

Abstract— **This paper proposes a two-phase conduction Direct Torque Control (DTC) scheme for three-phase Brushless DC (BLDC) Motor in the constant torque region. The use of Matrix Converter (MC) produces rectangular output currents, while maintaining sinusoidal input currents close to unity power factor. The implemented scheme reduces the torque dip in the BLDC motor. The merits of the matrix converter fed BLDC motor is shown by comparing the results with a conventional AC–DC–AC converter fed BLDC, both implementing two–phase conduction DTC. A model based on Matlab/Simulink is implemented to verify the above-proposed theory.**

Keywords – **BLDC, MC, DTC.**

I. INTRODUCTION

Brushless DC (BLDC) machines are becoming more popular in the industry due to their inherent advantages, such as superior speed torque characteristics, high power density, low maintenance, long operating life, high efficiency and noiseless operation [1]. As the name implies, BLDC motors do not use brushes for commutation [1],[2]. The commutation is performed electronically by using an array of switching devices based on the rotor position information. Normally, the rotor position information is obtained from position transducers such as Hall effect sensors, shaft mounted encoders or by analyzing the circuit parameters such as the stator third harmonic current waveforms. BLDC motors find major applications in computers and hard drives, industrial robotics, motion control and actuator systems and also in some heating and ventilation systems.

BLDC motors are traditionally driven by the pulse width modulated (PWM) voltage source inverters (VSI). However, they have disadvantages such as the high source current harmonics caused by diode rectifier feeding the voltage source inverter, and requirement of bulky energy storage component such as the DC bus capacitor. Matrix converters are considered as a possible alternative to PWM–VSI [3]. Simple and compact power circuit, variable voltage and variable frequency output, near unity power factor operation and polyquadrant operation are the advantages of MC. The DTC of the BLDC motor is implemented using a two–phase conduction scheme. The stator flux linkage is intentionally kept constant, thereby eliminating the need for flux control, thus simplifying the control of the drive to torque alone.

II. BRUSHLESS DC MOTOR

A. Operation of BLDC motor

The stator consists of a concentrated three–phase star windings similar to that of the synchronous machine. The rotor is made of a permanent magnet material, Neodymium Ferric Boron (NdFeB) and the magnets are usually surface mounted [4]. Unlike conventional motors, BLDC motors require rotor position sensing, which is done using the Hall effect sensors mounted on the stator surface [5].

When Hall effect sensors are placed 120° electrical apart on the inner surface of the stator and are subjected to the rotor magnetic field, a voltage is induced, which provides the controller with the present position of the rotor. From this information, the next position of the rotor can be estimated and the appropriate winding energization sequence realized in the converter. Table I shows the output of the Hall effect sensor and the appropriate phase excitation based on the rotor position.

TABLE I. HALL EFFECT SENSOR OUTPUT AND PHASE EXCITATION

Rotor Position	Hall effect sensor			Phase		
	H_A	H_B	H_C	A	B	C
0° - 60°	1	0	0	+1	0	-1
60° - 120°	1	1	0	+1	-1	0
120° - 180°	0	1	0	0	-1	+1
180° - 240°	0	1	1	-1	0	+1
240° - 300°	0	0	1	-1	+1	0
300° - 360°	1	0	1	0	+1	-1

III. MATRIX CONVERTER

The matrix converter is an array of bidirectional IGBTs that interconnects directly the three-phase AC supply to a three-phase load, without using any DC link or large energy storage elements [3].

The Indirect Modulation Principle decouples the matrix converter into a current source rectifier on the source side [6] and a voltage source inverter on the load side as shown in Fig. 2. Thus the switching algorithm for the two stages are individually obtained as explained below.

A. Switching Strategy for Rectifier stage

Space Vector Modulation (SVM) is implemented on the rectifier side of the MC. The current and voltage transfer functions of the rectifier side are given by (1) and (2)

978-1-4799-6047-7/14 $31.00 © 2014 IEEE   427

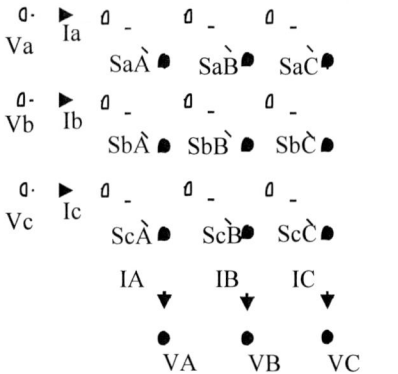

Fig. 1. Matrix Converter Topology

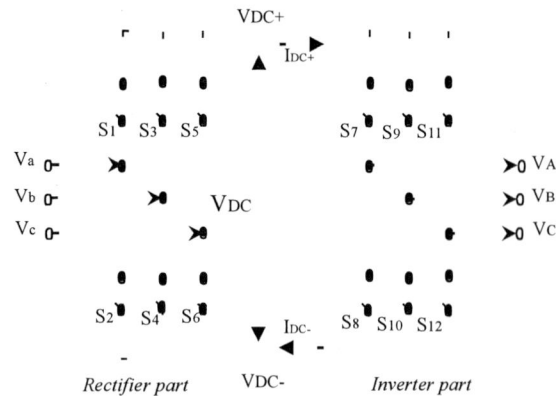

Fig. 2. Indirect Modulation Principle Equivalent Circuit

$$\begin{bmatrix} I_a \\ I_b \\ I_c \end{bmatrix} = \begin{bmatrix} S_1 & S_2 \\ S_3 & S_4 \\ S_5 & S_6 \end{bmatrix} \times \begin{bmatrix} I_{DC+} \\ I_{DC-} \end{bmatrix} \qquad (1)$$

$$\begin{bmatrix} V_{DC+} \\ V_{DC-} \end{bmatrix} = \begin{bmatrix} S_1 & S_3 & S_5 \\ S_2 & S_4 & S_6 \end{bmatrix} \times \begin{bmatrix} V_a \\ V_b \\ V_c \end{bmatrix} \qquad (2)$$

where I_a, I_b, I_c and V_a, V_b, V_c are the 3Φ input currents and voltages respectively. IA, IB, IC are the 3Φ output currents. The input currents are expressed as a space vector, I_{IN} using the transformation given in (3).

$$I_{IN} = \frac{2}{3}\left(I_a + I_b.e^{j\frac{2\pi}{3}} + I_c.e^{j\frac{4\pi}{3}} \right) \qquad (3)$$

From the configuration of the converter, as shown in Fig. 2, nine switching states are possible. These switching states are enlisted in Table II.

$I_1[ab]$ indicates that the input phase 'a' is connected to the positive rail of the virtual DC link V_{DC+} and the input phase 'b' is connected to the negative rail V_{DC-}.

Using equation (3), seven discrete space vectors can be obtained from the nine switching states. These vectors, when plotted on a complex plane, form a hexagon, as shown in Fig. 3. Reference vector I_{REF}, can be synthesized within the hexagon sector by the vector sum of the components of the sector and the zero vectors, as shown in Fig. 4.

For a small time interval T_S, the reference vector is represented as the sum of the current-time products of the adjacent vectors, as shown in (4).

$$I_{REF} = d_\delta.I_\delta + d_\gamma.I_\gamma \qquad (4)$$

Where, d_δ and d_γ are the duty cycles of the current vectors I_δ and I_γ respectively. The duration of the active vectors determines the direction of I_{REF} while the zero vector interval is used to adjust the amplitude of I_{REF}. The duty cycles are computed using (5) - (7)

$$d_\gamma = \frac{T_\gamma}{T_s} = m_c.sin\left(\frac{\pi}{3} - \theta_c\right) \qquad (5)$$

$$d_\delta = \frac{T_\delta}{T_s} = m_c.sin(\theta_c) \qquad (6)$$

$$d_{oc} = \frac{T_{oc}}{T_s} = 1 - d_\delta - d_\gamma \qquad (7)$$

where θ_C indicates the angle of the reference current vector within the hexagon sector and m_c is the modulation index of the rectifier stage. T_δ and T_γ are the time periods for which I_δ and I_γ current vectors are applied. T_{oc} and d_{oc} are the time period and duty cycle respectively of the zero current vector.

TABLE II. RECTIFIER STAGE SWITCHING STATE MATRIX AND SWITCHING VECTORS

Type	Vector	$\begin{bmatrix} S_1 & S_3 & S_5 \\ S_2 & S_4 & S_6 \end{bmatrix}$	$\lvert I_{IN} \rvert$	$\angle I_{IN}$
Active	I_1 [ab]	$\begin{bmatrix} 1 & 0 & 0 \\ 0 & 1 & 0 \end{bmatrix}$	$\frac{2}{\sqrt{3}}I_{DC}$	$-\frac{\pi}{6}$
	I_2 [ac]	$\begin{bmatrix} 1 & 0 & 0 \\ 0 & 0 & 1 \end{bmatrix}$	$\frac{2}{\sqrt{3}}I_{DC}$	$\frac{\pi}{6}$
	I_3 [bc]	$\begin{bmatrix} 0 & 1 & 0 \\ 0 & 0 & 1 \end{bmatrix}$	$\frac{2}{\sqrt{3}}I_{DC}$	$\frac{\pi}{2}$
	I_4 [ba]	$\begin{bmatrix} 0 & 1 & 0 \\ 1 & 0 & 0 \end{bmatrix}$	$\frac{2}{\sqrt{3}}I_{DC}$	$\frac{5\pi}{6}$
	I_5 [ca]	$\begin{bmatrix} 0 & 0 & 1 \\ 1 & 0 & 0 \end{bmatrix}$	$\frac{2}{\sqrt{3}}I_{DC}$	$\frac{7\pi}{6}$
	I_6 [cb]	$\begin{bmatrix} 0 & 0 & 1 \\ 0 & 1 & 0 \end{bmatrix}$	$\frac{2}{\sqrt{3}}I_{DC}$	$\frac{3\pi}{2}$
Zero	I_0 [aa] [bb][cc]	$\begin{bmatrix} 1 & 0 & 0 \\ 1 & 0 & 0 \end{bmatrix}\begin{bmatrix} 0 & 1 & 0 \\ 0 & 1 & 0 \end{bmatrix}\begin{bmatrix} 0 & 0 & 1 \\ 0 & 0 & 1 \end{bmatrix}$	0	-

IV. DIRECT TORQUE CONTROL OF THE BLDC MOTOR USING VOLTAGE SOURCE INVERTER

Unlike the implementation of the conventional DTC algorithm in AC machines, the requirement of Flux estimation is eliminated in the two-phase conduction scheme. This is done by intentionally keeping the stator flux linkage constant.

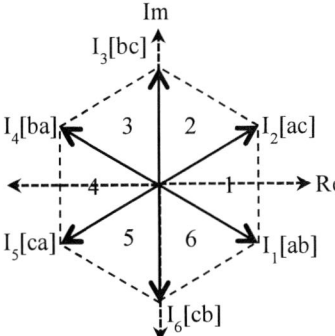

Fig. 3. Matrix Converter Topology

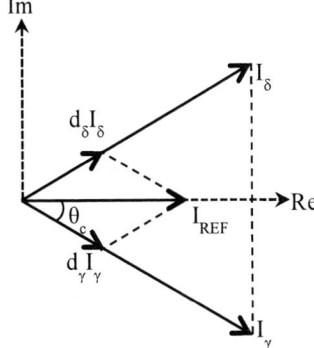

Fig. 4. Reference Current Vector Synthesis

However, even without the flux estimation, DTC can be implemented successfully in BLDC, thereby simplifying the control of the motor. Furthermore, zero voltage space vector is not used, as they tend to decrease electromagnetic torque and also increase switching losses [7].

A. Proposed DTC Technique

The general electromagnetic torque (T_{em}) equation of a BLDC motor, taken in dq reference frame, which is synchronous with the stator flux and includes the mutual coupling between d and q axis winding, is given by (8)

$$T_{em} = \frac{3P}{4}\left[\left(\frac{dL_{ds}}{d\theta_e}i_{sd} + \frac{d\phi_{rd}}{d\theta_e} - \phi_{sq}\right)i_{sd} + \left(\frac{dL_{qs}}{d\theta_e}i_{sq} + \frac{d\phi_{rq}}{d\theta_e} - \phi_{sd}\right)i_{sq}\right] \quad (8)$$

where,

$$\phi_{sd} = L_{ds}i_{sd} + \phi_{rd}$$

$$\phi_{sq} = L_{qs}i_{sq} + \phi_{rq}$$

P is the number of poles of the stator, θ_e is the electrical rotor angle, i_{sd} and i_{sq} are the d axis and q axis stator currents. L_{ds} and L_{qs} are the d and q axes stator inductances and ϕ_{rd}, ϕ_{rq}, ϕ_{sd}, ϕ_{sq} are the d and q axes rotor and stator flux linkages respectively.

The variation of L_{ds} and L_{qs} with θ_e can be neglected with the use of non-salient pole machines, where $L_{ds}=L_{qs}=L_s$ [8] or with the use of high coercive PM material [4], [5]. As a result, the electromagnetic torque equation for BLDC motor is given by (9).

$$T_{em} = \frac{3P}{4}\left[\left(\frac{d\phi_{rd}}{d\theta_e} - \phi_{sq}\right)i_{sd} + \left(\frac{d\phi_{rq}}{d\theta_e} - \phi_{sd}\right)i_{sq}\right] \quad (9)$$

For the proposed DTC scheme for BLDC motor, the electromagnetic torque equation is expressed in the stationary frame of reference (αβ-axis) instead of the dq frame. The dq axis rotor flux linkages can be expressed in the αβ frame as shown in (10)

$$\phi_{r\alpha} = \phi_{rd}\cos\theta_e - \phi_{rq}\sin\theta_e \quad (10a)$$

$$\phi_{r\beta} = \phi_{rd}\sin\theta_e + \phi_{rq}\cos\theta_e \quad (10b)$$

Similarly, the dq axis stator currents can also be expressed in the αβ-axis as shown in (11)

$$i_{s\alpha} = i_{sd}\cos\theta_e - i_{sq}\sin\theta_e \quad (11a)$$

$$i_{s\beta} = i_{sd}\sin\theta_e + i_{sq}\cos\theta_e \quad (11b)$$

Thus, the electromagnetic torque equation, in the stationary frame of reference, using (10), (11) in (9) becomes (12)

$$T_{em} = \frac{3}{2}\cdot\frac{P}{2}\left[\frac{d\phi_{s\alpha}}{d\theta_e}i_{s\alpha} + \frac{d\phi_{s\beta}}{d\theta_e}i_{s\beta}\right] = \frac{3P}{4}\left[\frac{e_\alpha}{\omega_e}i_{s\alpha} + \frac{e_\beta}{\omega_e}i_{s\beta}\right] \quad (12)$$

where ω_e is the electrical rotor speed and e_α and e_β are the αβ frame motor back-EMFs. In general, the back-EMF of a motor can be expressed as shown in (13)

$$e_\alpha = k_\alpha(\theta_e).\omega_e \quad (13a)$$

$$e_\beta = k_\beta(\theta_e).\omega_e \quad (13b)$$

where, $k_\alpha(\theta_e)$ and $k_\beta(\theta_e)$ are the back-EMF constants with respect to the rotor position. By substituting (13a) and 13(b) in (12), we get (14).

$$T_{em} = \frac{3P}{4}\left[k_\alpha(\theta_e)i_{s\alpha} + k_\beta(\theta_e)i_{s\beta}\right] \quad (14)$$

Since (14) does not involve the rotor speed in the denominator, the torque can be estimated even at zero and near zero speeds.

B. DTC Operation

Conventional two-phase conduction scheme causes the locus of the stator flux linkage to have a hexagonal trajectory [7], neglecting the open-phase back-EMF and freewheeling diode effects, as shown in Fig. 5 in dotted lines. However, in this scheme, it is observed that sharp dips in stator flux linkage occur every 60° electrical angle [1] due to the freewheeling diodes, shown as Fig. 5 in solid lines, deviating from the desired circular flux trajectory as in the case of the PMSM drive [9]. By knowing the exact shape of the flux linkage, we can control its amplitude, but this method is tedious in the constant torque region. Therefore, the flux error in the voltage vector selection lookup table [7] is always taken as zero and only the torque error is used, as in Table III.

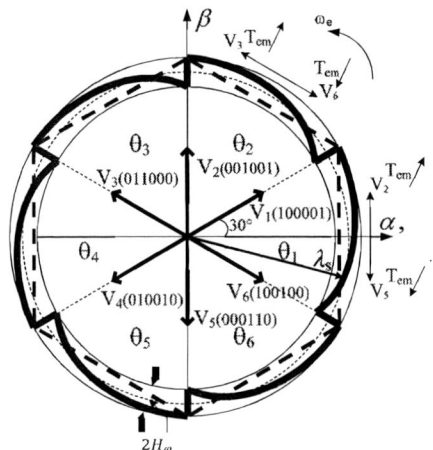

Fig. 5. Representation of two–phase actual (solid line) and desired (dotted circle) stator flux linkage vectors in the stationary $\alpha\beta$–axes reference frame

TABLE III. TWO-PHASE VOLTAGE VECTOR SELECTION FOR THE BLDC MOTOR

τ	Sector Number					
	θ_1	θ_2	θ_3	θ_4	θ_5	θ_6
+1	V_1 100001	V_2 001001	V_3 011000	V_4 010010	V_5 000110	V_6 100100
-1	V_5 000110	V_6 100100	V_1 100001	V_2 001001	V_3 011000	V_4 010010

C. Implementation of DTC

The phase voltages V_{an}, V_{bn}, V_{cn}, supplied to the stator of the BLDC motor are determined by the status of the six switches of the inverter, represented by S_7, S_8, S_9, S_{10}, S_{11}, S_{12}. A total of six non–zero space vectors are possible. With two-phase conduction, at any instant, both switches in the non-conducting phase leg are always off. The six non–zero vectors, V_1, V_2, V_3, V_4, V_5, V_6, shown in Fig. 5, are 60° electrically apart from each other and 30° electrical phase shifted from the corresponding three-phase voltage vectors of the SVM voltage hexagon [6].

With reference to Fig. 2, the switching states applied to the VSI stage of the MC, is represented as a 3×2 matrix, given by (15).

$$\begin{bmatrix} S_7 & S_8 \\ S_9 & S_{10} \\ S_{11} & S_{12} \end{bmatrix} => V_2 = \begin{bmatrix} 0 & 0 \\ 1 & 0 \\ 0 & 1 \end{bmatrix} \quad (15)$$

V. SWITCHING ALGORITHM FOR THE MATRIX CONVERTER

As explained in the previous sections, the rectifier stage switching matrix and the inverter stage switching matrix are computed and multiplied to obtain the 3×3 switching matrix for the MC. This is shown in (16).

$$\begin{bmatrix} S_{Aa} & S_{Ba} & S_{Ca} \\ S_{Ab} & S_{Bb} & S_{Cb} \\ S_{Ac} & S_{Bc} & S_{Cc} \end{bmatrix} = \begin{bmatrix} S_7 & S_8 \\ S_9 & S_{10} \\ S_{11} & S_{12} \end{bmatrix} \times \begin{bmatrix} S_1 & S_3 & S_5 \\ S_2 & S_4 & S_6 \end{bmatrix} \quad (16)$$

The 3×3 matrix is applied to the corresponding switches of the MC shown in Fig. 1.

VI. SIMULATION

A. Triggering Pulses

The layout of the Triggering pulse generator for the MC is shown in Fig. 7. The rectifier SVM algorithm is used to trigger the rectifier stage of the matrix converter. The DTC algorithm triggers the inverter stage.

The DTC algorithm starts with the hysteresis controller subsystem. Fig. 8 shows the layout of this subsystem. The 'w' signal is the actual speed of the rotor and the reference speed is set to 100 rad/sec. This speed error is fed to a PI controller. This gives the reference torque, which is compared with the actual electromagnetic torque 't' of the BLDC motor and this error 'ε' is the input of the hysteresis controller. The output, 'hysop', of the hysteresis controller is given by (17). An output of 1 will cause the forward torque to be applied and -1 will apply the reverse torque.

$$hysop(\varepsilon) = \begin{cases} 1, & \varepsilon > 0.01 \\ -1, & \varepsilon < -0.01 \end{cases} \quad (17)$$

With the rotor position, estimated from the Hall effect sensors, the sector number, as per Fig. 5, is obtained. Based on hysteresis controller output and sector number, the voltage vector as per Table III is applied. Finally, the switching matrix generated from the rectifier SVM algorithm and the DTC algorithm of the BLDC are multiplied in the Mat_Mul subsytem, as shown in (16), and the triggering pulses for the MC are obtained. The layout of the forward torque subsystem is shown in Fig. 9.

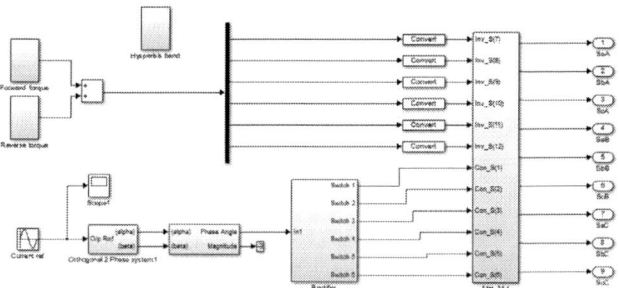

Fig. 7. General layout of triggering pulses of the DTC for the MC fed BLDC drive.

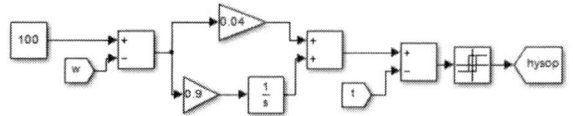

Fig. 8. Hysteresis Current Controller

978-1-4799-6047-7/14 $31.00 © 2014 IEEE

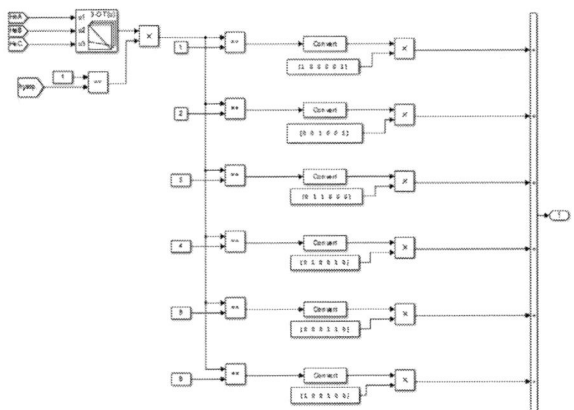

Fig. 9. Layout of forward torque subsystem

B. Matrix Converter

The bidirectional switch used in the MC is simulated by antiparallel connection of the two reverse blocking IGBTs, as shown in Fig. 10.

Fig. 11 shows the arrangement of switches in the MC and the 3-Φ AC voltage source feeding the MC. '1', '2', '3', '4', '5', '6', '7', '8', '9' denote the triggering pulses to the MC. 'Phase A', 'Phase B', 'Phase C' represent the 3-Φ AC voltage outputs of the MC.

VII. SIMULATION RESULTS

From the simulation, it is observed that the MC draws sinusoidal input currents from the AC voltage source. Fast Fourier Transform analysis is performed on the source phase current and the result is shown in Fig. 12. The Total Harmonic Distortion (THD) was found to be 0.88 %.

Fig. 13 shows the stator current and the back EMF waveforms of the BLDC motor. The current waveform settles to zero after the initial transient stage and the motor is run at no load before a load torque of 2 Nm is applied at t = 0.16s. Smooth rectangular output current waveform is obtained with a small dip in between, due to commutation at every 60°.

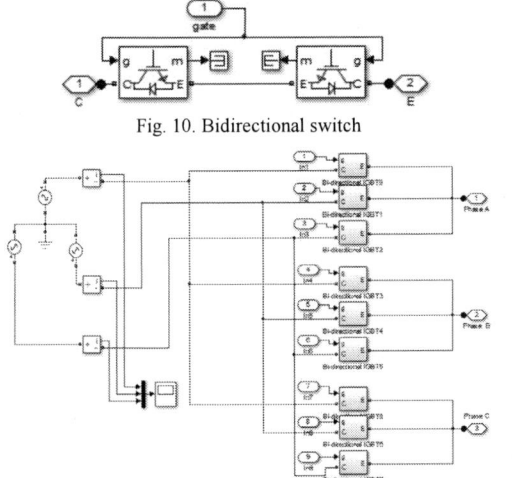

Fig. 10. Bidirectional switch

Fig. 11. Arrangement of switches in the MC

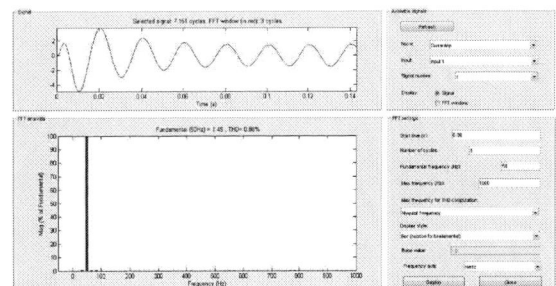

Fig. 12. FFT of the source phase current drawn by the MC

The rotor speed is shown in Fig. 14. After the initial overshoot, the speed settles at the set speed of 100 rad/s. Once the load is applied, the speed initially dips and then returns to the set reference value. Fig. 15 shows the electromagnetic and load torques of the BLDC motor. Due to the use of DTC, the torque of the BLDC motor is restricted to a narrow band, except in the instances where current commutation occurs.

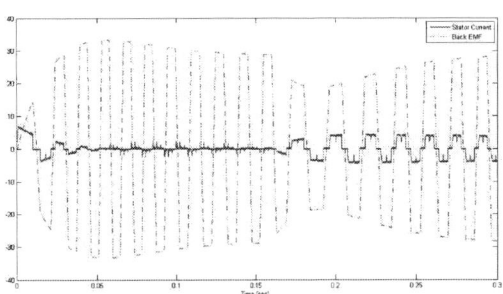

Fig. 13. Stator current and back EMF of the BLDC motor

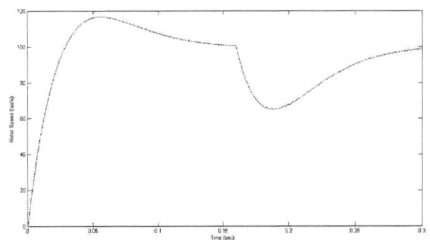

Fig. 14. Rotor Speed of the BLDC motor

TABLE IV. SIMULATION PARAMETERS

Parameters	Values
Source	230 V AC
Frequency	50 Hz
Motor Rating	**Values**
Stator Inductance per phase (L)	0.03126 H
Stator Resistance per Phase (R)	2.8 Ω
Torque Constant (Kt)	1.23 Nm/A
Back EMF Constant(kb)	1.23 V/rad/s
Pole pairs (P)	4
Moment of Inertia (J)	0.098 kgm^2
Coefficient of friction (B)	0.0078 N/rad/s
Proportional Controller	0.04
Integral Controller	0.9
Hysteresis Controller Limits	0.01 Nm
Bidirectional IGBT rating	**1200 V, 200 A, Gate Voltage: 20V**

The ripple in the torque produced by BLDC motor using the two–phase conduction DTC when fed by the MC and by the AC–DC–AC converter are shown in Figs. 15 and 16 respectively.

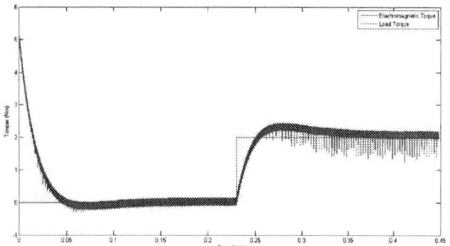

Fig. 15. Torque ripple in matrix converter fed BLDC drive using DTC

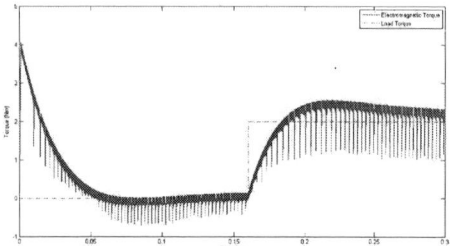

Fig. 16. Torque ripple in AC–DC–AC converter fed BLDC drive using DTC

Power factor is computed by using the Simulink model shown in Fig. 18. Time integral is performed till the current and voltage individually become zero. Thus, the current and voltage time offsets are obtained, the difference between the two provides the time offset difference between the two waveforms. From the time offset difference, the phase difference between the voltage and current waveforms is computed. The cosine of the phase difference gives the power factor of the system.

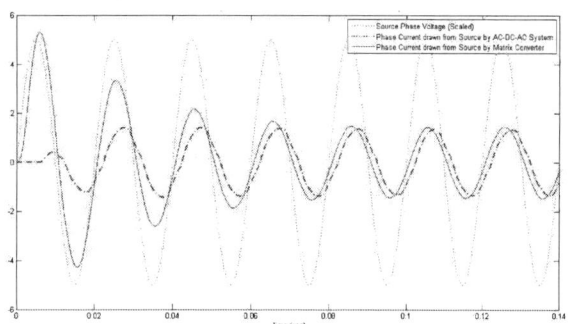

Fig. 17. Source Phase Voltage and Current waveforms

VIII. CONCLUSION

The use of the MC over conventional AC-DC-AC converter not only reduces the torque ripple, as shown in Fig. 15 and 16, but also results in a low THD of 0.88% over 3.70% in AC-DC-AC converters. Power factor of 0.963 was measured with the MC fed BLDC motor, whereas the use of AC-DC-AC converter measured 0.213. Thus the power factor at the source was significantly improved with use of the MC.

Requirement of only two control variables makes the dynamic response of torque using DTC very fast, and memory

requirement for digital implementation is also reduced. The simplified two-phase DTC algorithm coupled with the advantages of MC and BLDC motor makes an ideal drive system. Industrial automation, machine tool applications, elevators are some areas of application of this drive system.

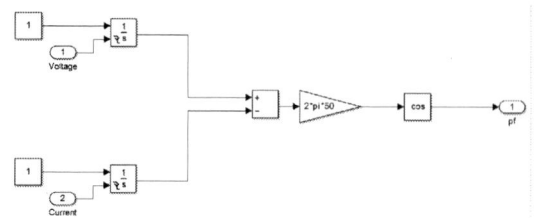

Fig. 18. Simulink model for Power Factor Measurement

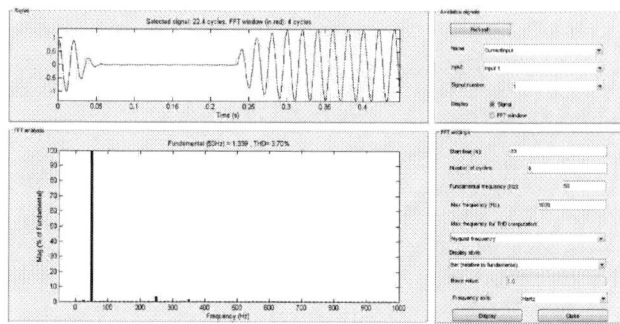

Fig. 19. FFT of the source phase current drawn by the AC–DC–AC Converter

REFERENCES

[1] T. J. E. Miller, Brushless Permanent Magnet and Reluctance Motor Drive, Clarenford, UK: Oxford, 1989.

[2] A. Purna Chandra Rao, Y. P. Obulesh, and Ch. Sai Babu, "Mathematical modeling of BLDC motor with closed loop speed control using PID controller under various loading conditions," ARPN Journal of Engineering and Applied Sciences, vol. 7, no. 10, pp. 1321-1328, October 2012.

[3] P. W. Wheeler, J. Rodriguez, J. C. Clare, L. Empringham, and A. Weinstein, "Matrix converters: a technology review," IEEE Transactions on Industrial Electronics, vol. 49, no. 2, pp. 276-288, April 2002.

[4] M. Przybylski, B. Slusarek, and J. Gromek, "Brushless DC motor with a bonded permanent magnet and powder magnetic core," Proceedings of the 2010 XIX International Conference on Electrical Machines (ICEM), pp. 1-4, 6-8 Sept. 2010.

[5] A. R. Paul and M. George, "Brushless DC motor control using digital PWM techniques," Proceedings of the 2011 International Conference on Signal Processing, Communication, Computing and Networking Technologies (ICSCCN), pp. 733-738, 21-22 July 2011.

[6] H. J. Cha, "Analysis and design of matrix converters for adjustable speed drives and distributed power sources," Doctoral Dissertation, Texas A&M University, 2004.

[7] S. B. Ozturk and H. A. Toliyat, "Direct torque control of brushless DC motor with non-sinusoidal back-EMF," IEEE International Electric Machines and Drives Conference 2007 (IEMDC '07), vol. 1, pp. 165-171, 3-5 May 2007.

[8] P. Devendra, Ch. Pavan Kalyan, K. Alice Mary, and Ch. Saibabu, "Simulation approach for torque ripple minimization of BLDC motor using direct torque control," International Journal of Advanced Research in Electrical, Electronics and Instrumentation Engineering (IJAREEIE), vol. 2, no. 8, pp. 3703-3710, August 2013.

[9] B. K. Bose, Modern Power Electronics and AC Drives, Prentice Hall: New Delhi, India, 2008.

Leverrier Algorithm based Reduced Order Modeling of dc-dc Converters

Man Mohan Garg[1], Yogesh V.Hote[2]

Department of Electrical Enginnering
Indian Institute of Technology
Roorkee,India
[1]garg.mbm@gmail.com,[2]yhotefee@iitr.ernet.in

Abstract— **In this paper, mathematical modeling of dc-dc buck and boost converter is presented using Leverrier algorithm. The main advantage of this method is that there is no need to determine the inverse of a matrix and hence, this method is computationally efficient. Further, it shown that by reducing second order model of dc-dc buck and boost converter to first order, similar performance can be achieved. The simulation results and performance evaluation are carried out using MATLAB, which depicts that original system responses match well with responses of reduced order models.**

Keywords- State space averaging (SSA) technique, buck converter, boost converter, Leverrier algorithm, reduced order modeling

I. INTRODUCTION

The dc-dc converters are widely used in many applications like computers, electrical vehicles, aerospace and military equipment etc. [1]. There are various topologies of dc-dc converters available in literature. The main topologies are buck, boost, and buck-boost. To improve the dynamic and steady state performance of dc-dc converters, a feedback control is always required [2]. To design a feedback control, the mathematical model of dc-dc converter is required. There are various techniques available for modeling of dc-dc converters like state-space averaging (SSA) method [3, 4], signal flow graph (SFG) technique [5], circuit averaging technique etc. The state-space average technique is simple method based on matrix algebra to obtain small signal model of a dc-dc converter .However, this method involves calculation of resolvent matrix $(sI - A)^{-1}$. The calculation of resolvent matrix using conventional matrix inversion formula involves complexity and may prone to errors [6], especially in case of non-idealities. However, in literature, a simple approach namely Leverrier algorithm is also suggested to calculate resolvent matrix [7]. Moreover, a higher-order system can be reduced to lower order using model order reduction methods suggested in literature [8-12]. The popular model reduction techniques are Pade approximation method, Routh stability criteria, truncation method, differentiation method etc. Using reduced order model, complexity in controller design reduces. In this paper, Pade's approximation method [9] is used to obtain first order transfer function of buck and boost converters.

The remaining part of this paper is organized as follows: state-space averaging (SSA) method for modeling of dc-dc converters in continuous conduction mode (CCM) is discussed in section II. Then, in section III, Leverrier's algorithm is explained. In section IV, second order transfer function models of buck and boost converters are obtained. In section V, first order reduced model is obtained using Pade's approximation method and simulation results are discussed. Finally, conclusion and references are presented.

II. STATE SPACE AVERAGING (SSA) TECHNIQUE

The state space averaging technique for modeling of dc-dc converters is simple but computationally complex. It needs state space model of converter in its each operating mode. These individual state space models are averaged over one switching period that gives nonlinear large signal averaged model. This nonlinear large signal averaged model is linearized around an operating point using small signal perturbations, which provides small signal transfer functions. This method exists in many textbooks [1, 4], which is described below in a simplified manner.

For continuous conduction mode, a dc-dc converter operates in two modes. First, when switch is on and secondly when switch is off. The large signal, nonlinear averaged state space model for a dc-dc converter is given as-

$$\frac{dx(t)}{dt} = A'x(t) + B'u(t),\ y(t) = C'x(t) \qquad (1)$$

All the symbols have standard meaning as given in [4].Equation (1) is a non-linear equation, which is linearized around a dc operating point by introducing small ac perturbation. We get following ac and dc models of dc-dc converters, where *D* is duty cycle.
Ac model:

978-1-4799-6047-7/14 $31.00 © 2014 IEEE   433

$$\frac{d\tilde{x}(t)}{dt} = A\tilde{x}(t) + B\tilde{u}(t) + B_d\tilde{d}(t), \tilde{y}(t) = C\tilde{x}(t) \quad (2)$$

Dc model:

$$X = -A^{-1}BU, Y = -CA^{-1}BU \quad (3)$$

where

$$A = A_1 D + A_2(1-D), B = B_1 D + B_2(1-D), C = C_1 D + C_2(1-D),$$
$$B_d = (A_1 - A_2)X + (B_1 - B_2)U$$

The matrices and vectors correspond to subscript '1' are associated with "switch on" mode and subscript '2' with "switch off" mode. For second order dc-dc converters, $x = [i_L \ v_o]^T$ and $u = v_g, y = v_o$.

Model in Eq.(2) can be further simplified using Laplace transform to get transfer functions as follows:

Input to output voltage transfer function,

$$\frac{\tilde{v}_o(s)}{\tilde{v}_g(s)} = C(sI - A)^{-1} B \quad (4)$$

Control to output voltage transfer function,

$$\frac{\tilde{v}_o(s)}{\tilde{d}(s)} = C(sI - A)^{-1} B_d \quad (5)$$

III. LEVERRIER'S ALGORITHM

According to Leverrier algorithm [2, 6-7], for an nth order system with system matrix A of order $n \times n$, resolvent matrix is given by

$$(sI-A)^{-1} = \frac{Adj(sI-A)}{|sI-A|} = \frac{Q_{n-1}s^{n-1} + Q_{n-2}s^{n-2} + \ldots + Q_1 s + Q_0}{s^n + a_{n-1}s^{n-1} + \ldots + a_1 s + a_0} \quad (6)$$

Here, Q_j are $n \times n$ matrices and a_j are scalars, which can easily be determined by following n steps:

Step 1: Let matrix Q_{n-1} as nth order identity matrix and then calculate coefficient a_{n-1}

$$Q_{n-1} = I_n \text{ and } a_{n-1} = -trace[Q_{n-1}A] \quad (7)$$

where, I_n is the nth order identity matrix.

Step 2: Calculate next matrix Q_{n-2} and coefficient a_{n-2}

$$Q_{n-2} = Q_{n-1}A + a_{n-1}I_n \text{ and } a_{n-2} = -\frac{1}{2}trace[Q_{n-2}A] \quad (8)$$

Step j: Similarly, for j^{th} matrix Q_{n-j} and coefficient a_{n-j}

$$Q_{n-j} = Q_{n-j+1}A + a_{n-j+1}I_n \text{ and } a_{n-j} = -\frac{1}{j}trace[Q_{n-j}A] \quad (9)$$

Step n: Finally, calculate

$$Q_0 = Q_1 A + a_1 I_n \text{ and } a_o = -\frac{1}{n}trace[Q_o A] \quad (10)$$

IV. MODEL OF BUCK AND BOOST CONVERTER

The power circuit of buck and boost dc-dc converters are shown Fig. 1 and Fig. 2, respectively. Since the basic operating principles of these converters are already available in textbooks and literature, therefore, will not be presented in this paper. We will drive the transfer function by application of Leverrier's algorithm for both dc-dc buck and boost converter. The small internal resistance r_L is considered in series with inductance.

(a)*Buck converter:* The various matrices associated with buck converters are-

$$A_{buck} = \begin{bmatrix} \dfrac{-r_L}{L} & \dfrac{-1}{L} \\ \dfrac{1}{C} & \dfrac{-1}{RC} \end{bmatrix}, B_{buck} = \begin{bmatrix} \dfrac{D}{L} \\ 0 \end{bmatrix},$$

$$C_{buck} = \begin{bmatrix} 0 & 1 \end{bmatrix}, E_{buck} = \begin{bmatrix} 0 \end{bmatrix}, B_{d,buck} = \begin{bmatrix} \dfrac{V_g}{L} \\ 0 \end{bmatrix}$$

Application of Leverrier Algorithm: The buck and boost converters are second order systems. Therefore here $n=2$. Now we can use Leverrier algorithm as discussed in previous section to get resolvent matrix for buck converter in following steps-

Step1:

$$Q_1 = I_2 = \begin{bmatrix} 1 & 0 \\ 0 & 1 \end{bmatrix}, a_1 = -trace(Q_1 A_{buck}) = \left(\frac{1}{RC} + \frac{r_L}{L} \right)$$

Step2:

$$Q_0 = Q_1 A_{buck} + a_1 I_2 = \begin{bmatrix} \dfrac{1}{RC} & \dfrac{-1}{L} \\ \dfrac{1}{C} & \dfrac{r_L}{L} \end{bmatrix}, a_0 = -\frac{1}{2}trace(Q_0 A_{buck}) = \left(\frac{1}{LC} + \frac{r_L}{RLC} \right)$$

$$(sI - A_{buck})^{-1} = \frac{Q_1 s + Q_0}{s^2 + a_1 s + a_0} = \frac{\begin{bmatrix} s + \dfrac{1}{RC} & \dfrac{-1}{L} \\ \dfrac{1}{C} & s + \dfrac{r_L}{L} \end{bmatrix}}{s^2 + a_1 s + a_0}$$

$$(11)$$

978-1-4799-6047-7/14 $31.00 © 2014 IEEE

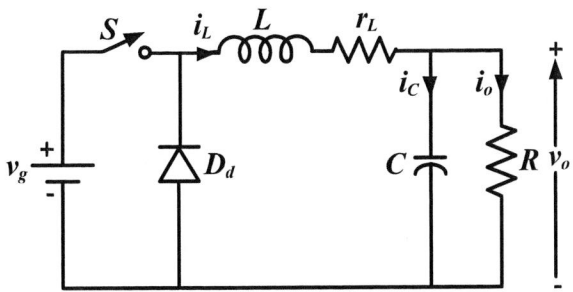

Fig 1.Buck Converter

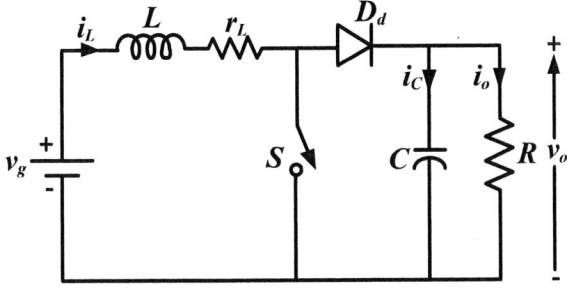

Fig 2.Boost Converter

Therefore for buck converter, using Eqs. (4), (5) and (11), we get

Input to output voltage transfer function,

$$\frac{\tilde{v}_o(s)}{\tilde{v}_g(s)} = \frac{D/LC}{s^2 + \left(\dfrac{1}{RC} + \dfrac{r_L}{L}\right)s + \left(\dfrac{1}{LC} + \dfrac{r_L}{RLC}\right)} \quad (12)$$

Control to output voltage transfer function,

$$\frac{\tilde{v}_o(s)}{\tilde{d}(s)} = \frac{V_g/LC}{s^2 + \left(\dfrac{1}{RC} + \dfrac{r_L}{L}\right)s + \left(\dfrac{1}{LC} + \dfrac{r_L}{RLC}\right)} \quad (13)$$

(b)_Boost Converter:_ The various matrices associated with boost converters are

$$A_{boost} = \begin{bmatrix} \dfrac{-r_L}{L} & \dfrac{-(1-D)}{L} \\ \dfrac{(1-D)}{C} & \dfrac{-1}{RC} \end{bmatrix}, B_{boost} = \begin{bmatrix} \dfrac{1}{L} \\ 0 \end{bmatrix},$$

$$C_{boost} = \begin{bmatrix} 0 & 1 \end{bmatrix}, E_{boost} = \begin{bmatrix} 0 \end{bmatrix}, B_{d,boost} = \begin{bmatrix} \dfrac{V_o}{L} \\ \dfrac{-I_L}{C} \end{bmatrix}$$

Similarly, Leverrier algorithm is used for boost converter as follows-

Step 1:

$$Q_1 = I_2 = \begin{bmatrix} 1 & 0 \\ 0 & 1 \end{bmatrix}, a_1 = -trace(Q_1 A_{boost}) = \left(\frac{1}{RC} + \frac{r_L}{L}\right)$$

Step2:

$$Q_0 = Q_1 A_{boost} + a_1 I_2 = \begin{bmatrix} \dfrac{1}{RC} & \dfrac{-D'}{L} \\ \dfrac{D'}{C} & \dfrac{r_L}{L} \end{bmatrix},$$

$$a_0 = -\frac{1}{2}trace(Q_0 A_{boost}) = \left(\frac{D'^2}{LC} + \frac{r_L}{RLC}\right)$$

$$(sI - A_{boost})^{-1} = \frac{Q_1 s + Q_0}{s^2 + a_1 s + a_0} = \frac{\begin{bmatrix} s + \dfrac{1}{RC} & \dfrac{-D'}{L} \\ \dfrac{D'}{C} & s + \dfrac{r_L}{L} \end{bmatrix}}{s^2 + a_1 s + a_0}$$

$$(14)$$

Using Eqs. (4) and (5), we get

Input to output voltage transfer function,

$$\frac{\tilde{v}_o(s)}{\tilde{v}_g(s)} = \frac{(1-D)/LC}{s^2 + \left(\dfrac{1}{RC} + \dfrac{r_L}{L}\right)s + \left(\dfrac{(1-D)^2}{LC} + \dfrac{r_L}{RLC}\right)} \quad (15)$$

Control to output voltage transfer function,

$$\frac{\tilde{v}_o(s)}{\tilde{d}(s)} = \frac{\dfrac{1}{LC}\left((1-D)V_o - r_L I_L\right) - \dfrac{I_L}{C}s}{s^2 + \left(\dfrac{1}{RC} + \dfrac{r_L}{L}\right)s + \left(\dfrac{(1-D)^2}{LC} + \dfrac{r_L}{RLC}\right)} \quad (16)$$

978-1-4799-6047-7/14 $31.00 © 2014 IEEE

V. REDUCED ORDER MODELING AND SIMULATION RESULTS

The parameters used for simulation of dc-dc buck and boost converters are: V_g=40V, L=500μH, C=2μF, R=5Ω, r_L=0.5Ω, D=0.4.

The buck converter transfer functions are –

$$\frac{\tilde{v}_o(s)}{\tilde{v}_g(s)} = \frac{4\times10^8}{s^2+1.01\times10^5 s+1.1\times10^9} \tag{17}$$

and

$$\frac{\tilde{v}_o(s)}{\tilde{d}(s)} = \frac{4\times10^{10}}{s^2+1.01\times10^5 s+1.1\times10^9} \tag{18}$$

The boost converter transfer functions are-

$$\frac{\tilde{v}_o(s)}{\tilde{v}_g(s)} = \frac{6\times10^8}{s^2+1.01\times10^5 s+4.6\times10^8} \tag{19}$$

and

$$\frac{\tilde{v}_o(s)}{\tilde{d}(s)} = \frac{-8.69\times10^6 s+2.26\times10^{10}}{s^2+1.01\times10^5 s+4.6\times10^8} \tag{20}$$

The transfer functions obtained in Eqs.(16)-(19) are of second order, which are reduced to first-order using Pade approximation method [9] of reduced order modeling as follows-
For buck converter

$$\frac{\tilde{v}_o(s)}{\tilde{v}_g(s)} = \frac{4\times10^8}{1.01\times10^5 s+1.1\times10^9} \tag{21}$$

and

$$\frac{\tilde{v}_o(s)}{\tilde{d}(s)} = \frac{4\times10^{10}}{1.01\times10^5 s+1.1\times10^9} \tag{22}$$

For boost converter

$$\frac{\tilde{v}_o(s)}{\tilde{v}_g(s)} = \frac{6\times10^8}{1.01\times10^5 s+4.6\times10^8} \tag{23}$$

and

$$\frac{\tilde{v}_o(s)}{\tilde{d}(s)} = \frac{-8.69\times10^6 s+2.26\times10^{10}}{1.01\times10^5 s+4.6\times10^8} \tag{24}$$

The step response of original transfer functions in Eqs. (16)-(19) and first order reduced transfer functions in Eqs. (20) - (23) are shown in Fig.3 and Fig.4.The responses of original systems and reduced order systems are over-damped. These plots depict that first order reduced model responses closely match with original second order models. Therefore, first order models can also be used for controller design instead of second order models. Fig. 4b also depicts that a zero of

control to output voltage transfer function of boost converter is in right half plane (RHP) thereby making it non-minimum phase system. However, the control design of non-minimum phase system is comparatively difficult limiting the bandwidth of the system.

The performance comparison of original and reduced order models of buck and boost converters are shown in Table 1-4.In buck converter,perofrmace indices matches well for both input voltage to output voltage and duty cycle to output voltage transfer functions.For boost converter also,the performance of redcued order model follow original model.The little discrepency in duty cycle to output voltage transfer function model exists due to presence of RHP zero in boost converter model.

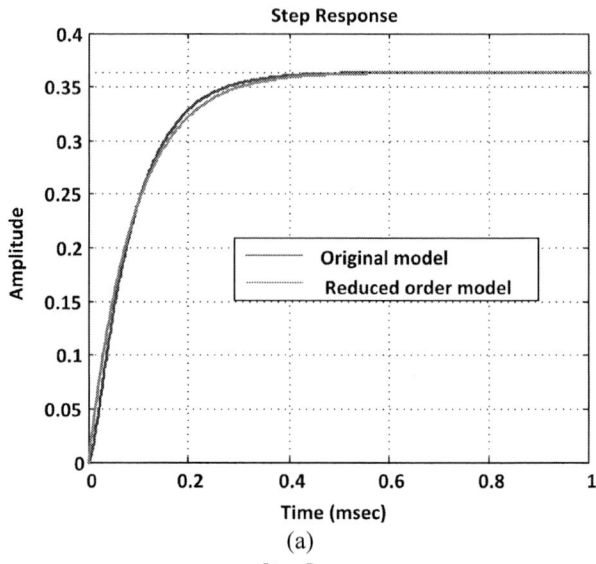

(a)

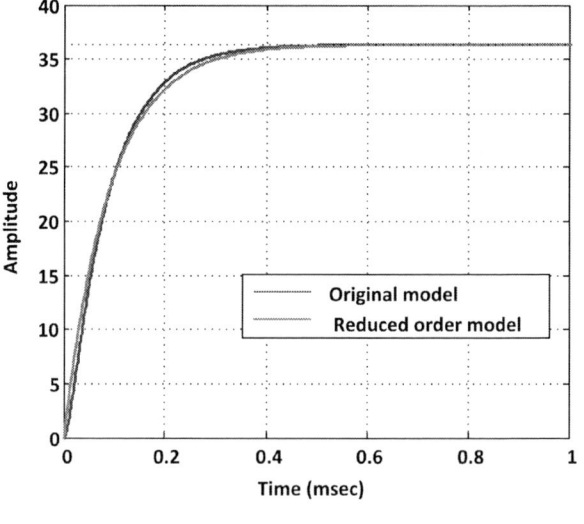

(b)

Fig 3.Buck converter step response (a) input to output transfer function (b) control to output transfer function

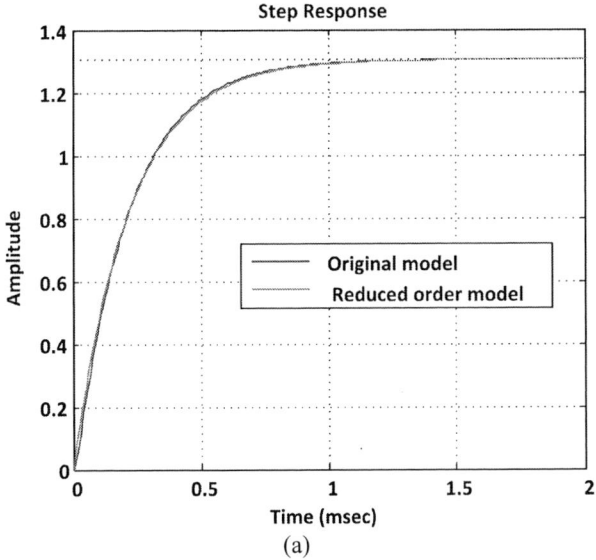

(a)

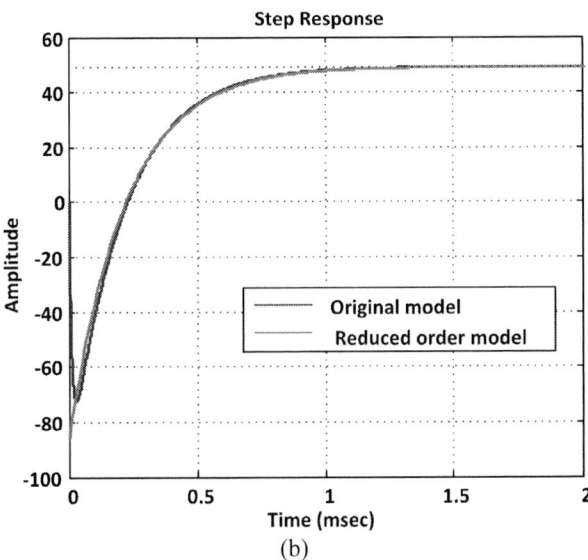

(b)

Fig 4.Boost converter step response (a) input to output transfer function (b) control to output transfer function

TABLE1. PERFORMANCE COMPARISION FOR BUCK CONVERTER,INPUT VOLTAGE TO OUTPUT VOLTAGE TRANSFER FUNCTION

Performace indices	Origianl model	Reduced order model
Rise time	0.179 msec	0.2 ms
Peak time	0.571 msec	0.668 sec
Settling time	0.327 msec	0.359 ms
%Undershoot	0%	0%
%Overshoot	0%	0%

TABLE2. PERFORMANCE COMPARISION FOR BUCK CONVERTER,DUTY CYCLE TO OUTPUT VOLTAGE TRANSFER FUNCTION

Performace indices	Origianl model	Reduced order model
Rise time	0.179 msec	0.2 ms
Peak time	0.571 msec	0.668 sec
Settling time	0.327 msec	0.359 ms
%Undershoot	0%	0%
%Overshoot	0%	0%

TABLE3. PERFORMANCE COMPARISION FOR BOOST CONVERTER,INPUT VOLTAGE TO OUTPUT VOLTAGE TRANSFER FUNCTION

Performace indices	Origianl model	Reduced order model
Rise time	0.460 ms	0.482 ms
Peak time	1.531 ms	2.315 ms
Settling time	0.828 ms	0.858 ms
%Undershoot	0%	0%
%Overshoot	0%	0%

TABLE4. PERFORMANCE COMPARISION FOR BOOST CONVERTER,DUTY CYCLE TO OUTPUT VOLTAGE TRANSFER FUNCTION

Performace indices	Origianl model	Reduced order model
Rise time	0.459 ms	0.482 ms
Peak time	0.0287 ms	0
Settling time	0.857 ms	0.858 ms
%Undershoot	47%	75%
%Overshoot	0%	0%

VI.CONCLUSION

The Leverrier algorithm is used to obtain resolvent matrix, which is utilized in obtaining transfer function models of dc-dc buck and boost converters. This method is very simple to implement. Further, the application of Pade approximation model order reduction technique is used for obtaining first-order transfer functions, which closely replicates original system performance. In future, these reduced order models may be useful in designing controller for dc-dc converters. The same approach is also applicable for other types of dc-dc converters and other power electronics converters.

978-1-4799-6047-7/14 $31.00 © 2014 IEEE

REFERENCES

[1] R.W.Eriction and Dragon M.,Fundamental of Power Electronics,2nd ed.,Kluwer Academic Publishers,2001.

[2] D. Roy Choudhury,*Modern Control Engineering*,PHI Pvt Ltd,2005.

[3] C T Rim,G B Joung and G H Cho,"A state-space modelling of non-ideal dc-dc converter,"*IEEE PESC Conference*,pp.943-950,1988.

[4] N.Mohan,T.M.Undeland and W.P.Robbins, Power Electronics: Converters, Applications and Design,3rd ed., John Wiley, 2003.

[5] K. Smedley and S. Cuk,"Switching flow-graph nonlinear modeling technique,"*IEEE Trans. Power Electron.*,vol.9,no.4,pp.405–413,Jul. 1994.

[6] M.M.Garg,Y.V.Hote and M.K.Pathak,"Leverrier's algorithm based modeling of higher-order dc-dc converters,"*IEEE India International Conference on Power Electronics*,2012,p.1-6.

[7] L.H.Keel and S.P.Bhattacharyya,"On computing the inverse of a rational matrix,"*IEEE Conf. on Decision and Control*,pp.140-144,Dec.2009.

[8] Femia .N.,Spagnuolo G. and Tucci V.,"State-space models and order reduction for DC-DC switching converters in discontinuous modes," *IEEE Trans. on Power Electron.*,vol.10,no.6,pp.640-650,Nov.1995.

[9] Y.Shamash, Order reduction of linear systems by Pade approximation methods,Ph.D thesis,Imperial college of science and technology,Univ. London,London,England,1973.

[10] Gutman P.,Carl F.M. and Molander P.,"Contribution to the model reduction problem," *IEEE Trans. on Automat. Contr*,vol.AC-27,no.2,pp.454-455,April.1982.

[11] Krishnamurthy and V.Seshadri,"Model reduction using the routh stability criterion," *IEEE Trans. on Automat. Contr*,vol.AC-23,pp.729-731,Aug.1978.

[12] Y.Shamash,"Truncation method of reduction:A viable alternative," *Electronics Letters*,vol.17,no.2,pp.97-99,Jan.1981.

Automated Precise Liquid Transferring System

Meera C S
M.Tech Robotics Engineering
University of Petroleum and Energy Studies,
Dehradun, India
E-mail: meerachitra.s@gmail.com

Sunil Sunny
M.Tech Robotics Engineering
University of Petroleum and Energy Studies,
Dehradun, India
E-mail: nsunilsunny@gmail.com

Richa Singh
M.Tech Robotics Engineering
University of Petroleum and Energy Studies,
Dehradun, India
E-mail: richa62@gmail.com

Pinisetti Swami Sairam
M.Tech Robotics Engineering
University of Petroleum and Energy Studies,
Dehradun, India
E-mail: swami.sairam@gmail.com

Roushan Kumar
Assistant Professor,
University of Petroleum and Energy Studies,
Dehradun, India
E-mail: rkumar@ddn.upea.ac.in

Jubit Emannuel
Project lead
Miranda Automation Pvt. Ltd
E-mail: emmanueljubit@gmail.com

Abstract— **Nearly all chemical, pharmaceutical, food processing and bio medical industries require large volumes of liquid transfer. With automation technology, the capacity, easiness and efficiency of liquid transferring systems has been greatly enhanced. Precision and accuracy in the volume of liquid dispensed plays a vital role in determining the overall efficiency of the industrial processes. An automated precise liquid transferring system targets replacement of the conventional erratic flow meters and highly expensive flow sensors used in the industries. This non-contact system enhances the overall efficiency of liquid transferring process in a very cost effective manner. The system control was implemented through Unitronics PLC and the noncontact mechanism of liquid dispensing was designed with the help of solenoid valves, relays, FRL and various other devices. The liquid was initially pressurized for a pre-defined time and the precisely dispensed by regulating the opening and closing time of solenoid valves. Provision for liquid transfer to demanded mixture tank is also provided. Flow rate of the liquid and the on- off time for solenoid determines the volume of liquid transferred. Experiment carried out for different volumes of liquid showed an accuracy of 1-2 gm. Using the PLC control and automation, the time required for the entire process of liquid dispensing was brought down to few seconds irrespective of the volume of the liquid dispensed.**

Keywords- Automation, Liquid handling system, Pneumatics, Unitronics PLC & HMI

I. INTRODUCTION

Automation using PLC has set forth a revolution in the industries. This is attributed to the higher production rates, increased output quality, highly efficient usage of raw materials and many other benefits that cannot be claimed even from the high quality workmanship by humans. The earlier developments of automation towards liquid handling were a trademark in areas relating to medical diagnostics and drug industries [1]. Advances in automating liquid handling

with various degrees of accuracies were performed mainly for the purpose of protein crystallization [2], [3]. In 1980's automated liquid handling technology was implemented using plurality of pippets and syringes. Later on piezoelectric systems and solenoid based systems were developed to transfer the liquids with good precision and accuracy [4]-[5]. All these systems focused in the replacement of the manual transfer of liquid in a more efficient, accurate and speedy manner.

For industrial applications, quantity of liquid used for different process varies based on demands. In all the dispensing processes precision and repeatability ensures the overall efficiency of the process. Many of the industries resolve this problem by performing time consuming experimental calibration [6]. Recently, flow meters and flow sensors started to be used in the process to ensure the precision and accuracy in liquid dispensing. A pressure feedback loop was employed in enhancing the accuracy by bluebird dispenser in the year 2000[7]. In 2005 MEMS flow sensors were reported to be used for continuous monitoring and ensuring accuracy during liquid dispensing process [8]. Furthermore, in 2005 liquid handling in nano liters, based on embedded fluid actuators were also developed [9].

Later in 2007 and 2009, intelligent control using micro solenoid valve with integrated MEMS sensors were introduced for high precision non -contact type liquid dispensing process [7],[10]. These technologies were mainly useful for liquid dispensing in sub millimeter ranges. But liquid dispensing in food processing industries, integrating such flow meters and flow sensors poses some difficulties where the demand varies from few grams to kilograms. Dispensing the right quantity of liquid, viscous and non-viscous plays a very vital role in determining the product composition. Mostly the flow meters incorporated in the system are sometimes corrosive and mostly do not give

978-1-4799-6047-7/14 $31.00 © 2014 IEEE

steady readings resulting in erroneous results affecting the volume of liquid transferred. Integrating highly accurate flow sensors in the system increases the cost of the system to a great extent.

In the paper a cost effective and highly precise liquid transferring system was developed for viscous and non-viscous liquids. The pneumatics based system is fully automated and system control was implemented with PLC. Non-contact type of liquid dispensing was realized using solenoid valves and accuracy in liquid dispensing was ensured using a load cell assembly. The standalone system developed was tested against leakage and was provided with emergency control. A stand-alone system with a dedicated PLC control could replace the use of flow meters and flow sensors in liquid dispensing mainly in food industries. Moreover, the system realized is flexible enough to be incorporated in an already existing liquid transferring system wherein the control can be performed by the existing PLC .The PLC control together with the HMI module makes the system apt to be used in any of the industries performing liquid transferring processes. HMI was designed in a simple and user friendly way that enables will enable easy operation of the system by any individual. Experiments performed to dispense different volumes of liquid showed a maximum deviation of 1-2 gm from the desired volume. The intelligent control of PLC taking continuous feedback from the load cell assembly and thereby modulating the valve open and close time makes the system highly precise and accurate irrespective of liquid nature.

The paper is organized as follows: introduction to liquid transferring system described in first section. In this paper, firstly the system construction is explained in detail. In the second section, implementations of control strategies are presented. Finally experimental results are discussed to demonstrate the accuracy and precision of the system.

II. HARDWARE DESIGN

The automatic precise liquid handling system is classified into four parts: Dispensing unit, Calibration process, liquid filling process and liquid transferring process where in the system construction is explained in detail.

A. Dispensing Unit

The automatic system consists of a dispensing unit and a control unit. The main components constituting the dispensing unit are pneumatically operated solenoid valves for the operations vent, fill, pressure and discharge, an FRL unit, a reservoir tank, a pressurizing tank mounted on load cell and two mixer tanks. The control unit was a PLC incorporated with an HMI module.

HMI and PLC control is chosen to make the make a low cost standalone system. With this system, the price is cut down to a high-extend and the system developed can very precisely transfer the liquid volume to the desired mixer.

The system components were appropriately chosen and repeatedly tested to make it a leak proof. For the initial experimental setup, the PLC and HMI module were fixed up on a separate panel and rest of the system was wired to the panel as shown in figure 1.

Fig.1.Experimental setup of automated precise liquid transferring system

B. Calibration Process

Calibration is performed at initial stage to ensure that there is no effect of pressurizing tank or any other components on the load cell such that the load cell can accurately indicate the weight of the fluid coming to the pressurizing tank. In the first stage of the calibration process, zero calibration of the load cell is performed by assuming the total weight of the pressurizing tank mounted over the load cell as zero. This step is proceeded by the span calibration with a known standard weight. After this load cell measure any desired weight with reference to this standard weight. The feedback from the load cell is the input to the PLC that determines the opening and closing time for the solenoid valves.

The 100% accuracy in calibration process is ensured by exact mounting of pressurizing tank over the load cell assembly and ensuring that the external disturbances like air, mounting of the pneumatic tubes affects the strain exerted over the load cell and affects the calibration process. The calibration process designed with HMI is shown in figure 2.

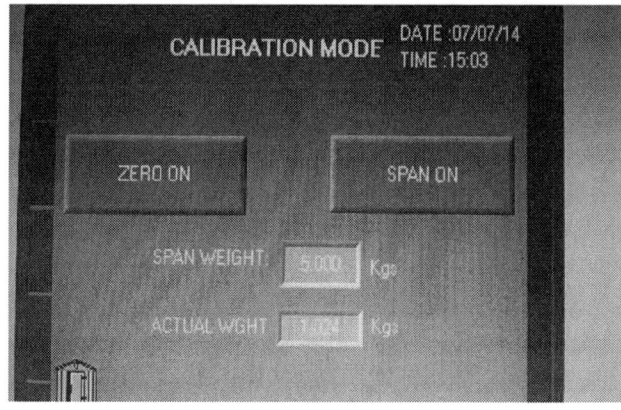

Fig.2. Calibration process in HMI

978-1-4799-6047-7/14 $31.00 © 2014 IEEE

C. Liquid Filling Process

Once the calibration process is performed, the system is designed to run either in manual mode or automatic mode depending on the user selection. With the Auto mode selection the system starts working fully automatically with PLC control as shown in figure 8. Manual mode is designed in such a way that with the user can control the opening and closing of solenoid valves simply by pressing the buttons as shown in figure 3. The HMI design makes the user control through manual mode very simple task. Here, as per the feedback from the HMI interface, the PLC controls the solenoid valves.

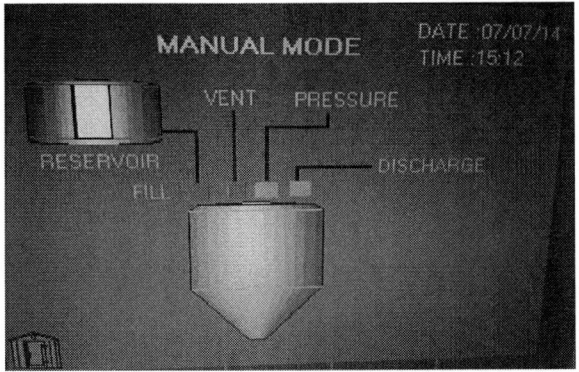

Fig.3. HMI screen of manual mode

The liquid filling process starts with the opening of vent valve so as to evacuate the entrapped gases inside the pressurizing tank. The closing time of the vent valve is determined by the PLC in auto mode depending on the feedback from the load cell. The process is done to ensure that any entrapped gas or unwanted pressure inside the tank do not exert an unwanted strain on the load cell and alter the weight of the liquid when filling process starts.

When the load cell indicates a zero strain, the vent valve is gets close and the fill valve gets open. The fill valve remains open as long as the weight of the liquid filled matches with the weight entered by the user.

D. Liquid Dispensing Process

Once the desired quantity of liquid gets filled, pneumatic pressure is developed in the pressure tank. Pressure from the compressor is regulated using an FRL unit, set to the desired pressure for the process and is directed through the pressure valve. For a set time that was experimentally determined, the filled liquid is pressurized and then discharge valve is opened. The discharge tube is made long enough and straight so as to almost touch the bottom of the tank. This is done to ensure the discharge of the liquid that remains at the bottom most portion of the tank. As the discharge valve gets open, the pressurized liquid rushes through the discharge tube to the mixer selected. As determined by repeated experiments pressure was set as 1.5 bars and pressurizing time were set as 5 seconds for the liquid weight ranging from 100 grams to 7 kilograms.

In the system, liquid was dispensed into the two mixer tanks where the selection of the mixer tank can be done by the user. Accurate transferring of the desired liquid volume as needed for the process was ensured by controlling the dispensing time such that the next filling process is initiated only when air starts spraying from the discharge tube. Spraying of air from the outlet of the tube ensures that all the liquid has been transferred and there is only pressurized air is present

III. SOFTWARE DESIGN

This section is divided to two sections: Electro pneumatic system model and Ladder logic control to explain the overall system control taken up to implement the system.

A. Electro Pneumatic System Model

A systematic and well defined approach is taken for the control of the entire system so as to obtain very precise results. A dedicated UnitronicsV570 PLC module is programmed in ladder logic using VISILOGIC software [11] -[12]. In case of any error, an alarm signal in both visual and audible form is raised by the controller thereby making the system an efficient one.

Special attention is taken care in making the system error proof. Designing of manual mode in the system was for this purpose. The entire system can be cleaned in case of change of solution from viscous to non-viscous using manual mode. In addition to making the system an error free one, this also ensures the easy maintenance of the system. For user friendliness and easy operation HMI design is made such that the user has to simply press the corresponding button for the valves and perform the operation. Furthermore, load cell calibration option is also provided in HMI. To nullify any errors due to external weight, user can access the calibration option which is linked with the load cell programming. In calibration, initially any additional weight is directly equated to zero, and then calibration with a known weight can be performed with span option as shown in figure 2.

The control of PLC over the valves is implemented using relay card consisting of 8 relays. Considering the feedback and other criterion the valves are opened when a 24V DC signal is passed to the control side of corresponding relay. The system was made with two port solenoid valve. Relay pass on 24V DC signal to energize the solenoid thereby opening the valve. Figure 4.a and 4.b shows the arrangement of relay card and solenoid valves respectively.

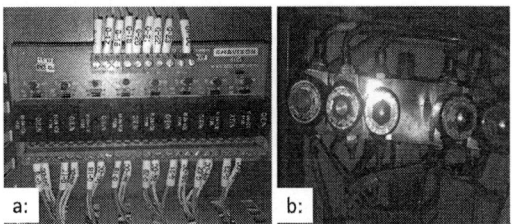

Fig.4. Relay card and solenoid valve arrangement

978-1-4799-6047-7/14 $31.00 © 2014 IEEE

As the liquid flows into the tank the PLC monitors the tanks weight with the set weight. When the tank weight reaches near to the set weight then the fill valve is shut off according to the tolerance such that no excess liquid flows into the tank. As the discharge process continues, very small amount of solution gets accumulated inside the tank which cannot be taken out. For this dead weight is set such that at the time of discharging, PLC monitors the tank weight to the dead weight. Once the tank weight reaches to the dead weight then the discharge valve is shut down. At the time of filling up process, this excess weight is assumed as zero and process starts. By repeated experiments, dead weight was set as 10gms and tolerance was varied from5 gms-25 gms for weight up to 7 kg of solution.

B. Ladder Logic Control Flow Chart:

The control flow for the automated precise liquid transferring system is shown in figure 5.

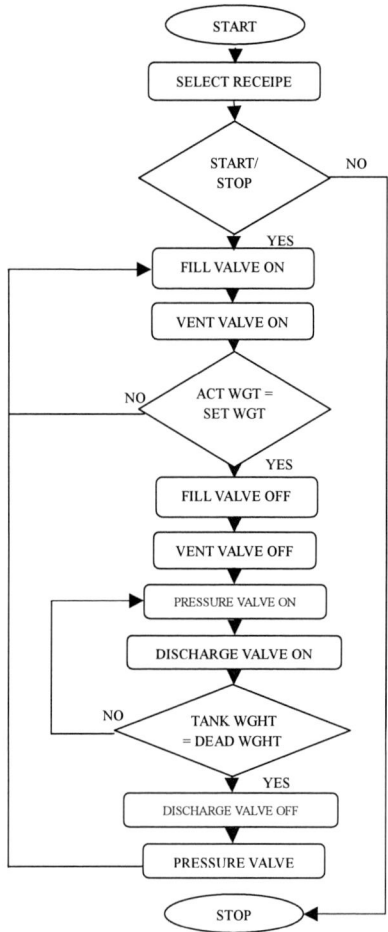

Fig.5. Ladder Logic Control flow chart

IV. RESULTS

Experiments were conducted with non-viscous as well as viscous solutions for 50 times to check the accuracy of the system. Water was used as the non-viscous liquid and sugar

solution was used as the viscous solution to conduct the experiments. The comparison between set weight and discharged weight is shown in figure 6. The first figure shows the repeated cycles of measurements taken for non-viscous liquid and second shows with that of viscous solution. It is clear from the figure 6 that the maximum deviation exhibited by the system with viscous and non-viscous solution is ±2gm. The transferred volume to each mixer was measured using a weighing machine.

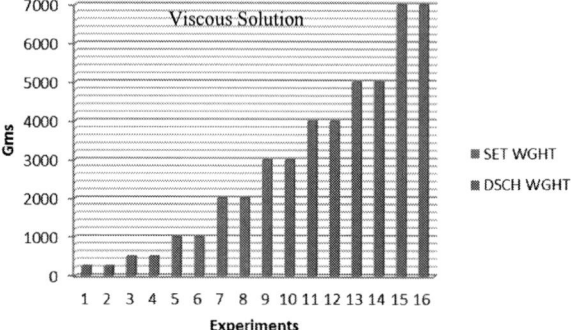

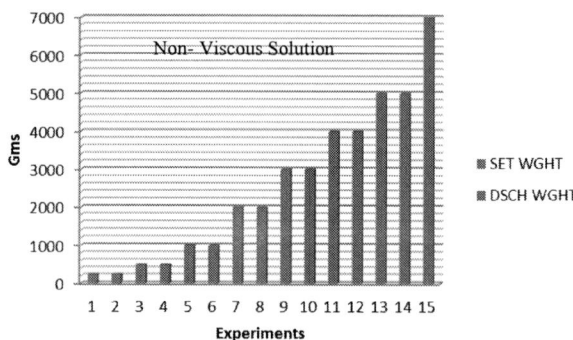

Fig.6. Comparison between set weight and discharge weight for different cycles of experiments

The results show that the control strategies adopted in enhancing the accuracy and precision is highly reliable. After initial calibration, any external weight getting exerted on the system is automatically nullified such that it does not affect the weight of liquid discharged. With this kind of control mechanism, the opening and closing time of the valves are very accurately calculated by the PLC controller with the feedback from load cell assembly. The deviation in discharge is brought down to minimal by closely adjusting the tolerance value. In case of any error, the system raises alarm with an LED indicator and a buzzer which adds to the reliability of the system.

Finally experiments were conducted for 50 times with the same set quantity of 1 kg of solution to check on the repeatability and precision of the system developed. The readings plotted in the figure 8 reveals the steady output

978-1-4799-6047-7/14 $31.00 © 2014 IEEE

obtained from the system. Moreover, it can be seen that the system indicates high precision and repeatability as the process gets repeated making it apt for the industrial use.

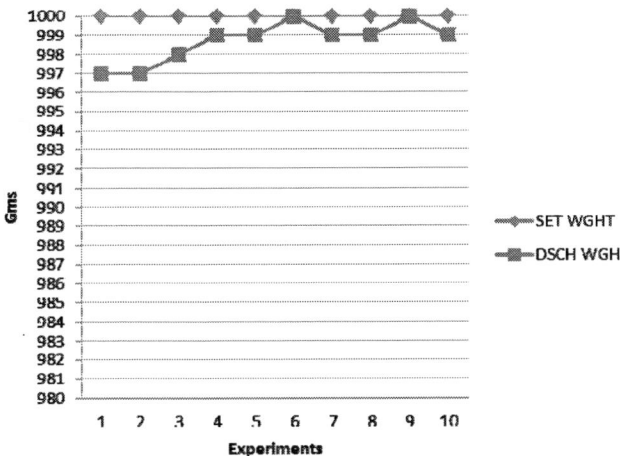

Fig.7. Liquid discharged for different cycles when set weight is 1000 gms

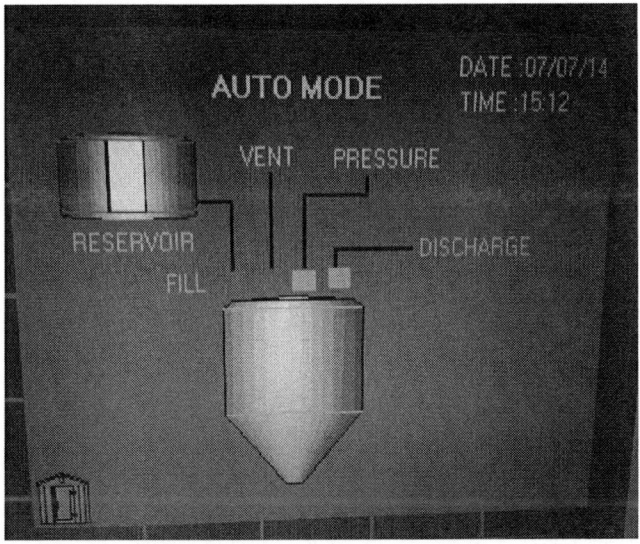

Fig.8. HMI screen in Auto mode

A. Abbreviations and Acronyms

S.NO	Abbreviation	Full Form
1	HMI	Human Machine Interface
2	LED	Light Emitting Diode
3	MEMS	Micro Electro Mechanical Systems
4	PLC	Programmable Logic Control
5	FRL	Filter, Regulator and Lubricator

V. CONCLUSION

The project made satisfies the objective of a highly accurate and precise standalone system that can replace conventional flow meters and expensive flow sensors. The system was mainly developed for food processing industries where the accuracy and precision in dispensing liquid solution is critical. The system can handle solutions from 250gm to 7kg irrespective of viscous or non-viscous nature. The provision for running the system in both auto and manual mode was provided through an HMI interface. The liquid dispensing was performed by careful regulation of opening and closing time of solenoid valves. An emergency stop switch was provided as an additional safety measure. The noncontact liquid system was developed in a very cost effective manner. The system developed could be either used as a standalone one or can be interfaced with an already existing liquid transferring system in place of flow meters or flow sensors. Repetitive experiments were done with different quantities of viscous and non-viscous solutions to check the accuracy of the system. The system showed an accuracy of ±2 gms from the set weight.

VI. ACKNOWLEDGMENT

We extend our sincere thanks to Miranda Automation Private Ltd, Navi Mumbai for their constant guidance and also Electronics, Instrumentation and Control Department of University of Petroleum and Energy Studies for their support.

REFERENCES

[1] Mueller, Lajos Nyarsik, Martin Horn, "Development of a technology for automation and miniaturization of protein crystallization," Journal of Biotechnology, 2001, vol. 85, pp. 7–14

[2] B. D. Santarsiero, D. T. Yegian. An approach to rapid protein crystallization using Nano droplets. Journal of Applied Crystallography, 2002, pp. 278-281

[3] Raymond Hui, Raymond Hui, " High-throughput protein crystallization,"Journal of Structural Biology, 2003, vol.142, pp. 154-161

[4] Walter D. Niles, Peter J. Coassin, "Piezo- and Solenoid Valve-Based Liquid Dispensing for Miniaturized Assays, ASSAY and Drug Development Technologies, Volume: 3 Issue 2: May 4, 2005

[5] Daniela Stock, Olga Perisic, Jan Lo¨ we, "Robotic nanolitre protein crystallisation at the MRC Laboratory of Molecular Biology," Progress in Biophysics and Molecular Biology, 2005, vol. 88, pp. 311–327

[6] Yaxin Liu, Chen Liguo, Lining Sun, Weibin Rong "A Self-adjusted Precise Liquid Handling System", 2009 IEEE International Conference on Robotics and Automation Kobe International Conference Center Kobe, Japan, May 12-17, 2009

[7] David A.Dunn, llllya Feygin, "Challenges and solutions to ultra-high-throughput screening assay miniaturization: submicroliter fluid handling," DDT, 2000, vol.5, 84-91

[8] Carsten Haber, Marc Boillat, Bart van der Schoot, "Precise Nanoliter Fluid Handling System with Integrated High-Speed Flow Sensor," ASSAY and Drug Development Technologies, 2005, vol. 2, 203-212.

[9] Bjom Samel, Volker Nock, Aman Russom, Patrick Grissl and Goran Stemme "Nanoliter Liquid Handling On A Low Cost Disposable With Embedded Fluid Actuators", The 13th International Conference on Solid-State Sensors, Actuators and Microsystems, Seoul, Korea, June 5-9,2005

[10] Yaxin Liu, Liguo Chen, Lining Sun, Automated Precise Liquid Dispensing System for Protein Crystallization, Proceedings of the

2007 IEEE International Conference on Mechatronics and Automation, August 5 - 8, 2007, Harbin, China

[11] Petruzella ,"Programming and logic control", Tata McGraw-Hill, Third Edition

[12] Hugh Jack ,"Automating Manufacturing Systems with PLCs", Fourth Edition

Design and Implementation of Hydraulic Motor Based Elevator System

Roushan Kumar
Assistant Professor,
University of Petroleum and Energy Studies,
Dehradun, India
E-mail: rkumar@ddn.upes.ac.in

Prashant Kumar Dwivedi
M.Tech (Robotics Engineering),
University of Petroleum and Energy Studies,
Dehradun, India
E-mail: dwivedi.kr.prashant@gmail.com

D. Praveen Reddy
M.Tech (Robotics Engineering),
University of Petroleum and Energy Studies,
Dehradun, India
E-mail: praveenreddy380@gmail.com

Amiya Sagar Das
M.Tech (Robotics Engineering),
University of Petroleum and Energy Studies,
Dehradun, India
E-mail: ron.amiya@gmail.com

Abstract— The research paper emphasizes on the Elevator control system which is one of the important aspects in infrastructure industry. This elevator control system is designed with the help of hydraulic motor and programmable logic control (PLC) while internal controlling is done through human machine interference (HMI). This research is divided into two Parts on the basis of programming. Firstly the program is developed on ladder logic and then simulated on software itself but for later testing and running the application a visualization is created on Indraworks engineering Software and testing of hardware is done. In the second level for controlling the internal commands of the system and for floor recognition HMI is developed on Rexroth touch screen HMI and internal switch for floor selection is done by HMI. The system consists of three floor hardware and recognition of floors is done by inductive proximity sensors and limit switch. Hydraulic motor used for UP-DOWN movement of the elevator is controlled through 4/3 Directional control valve (DCV).

Keywords—Rexroth PLC; Directional Control valve; Hydraulic Motor; HMI; Indraworks Engineering; Industrial Automation, proximity sensor

I. INTRODUCTION

A hydraulic elevator typically uses hydraulic motors or hydraulic cylinders to either raise or lower platforms for work, or any other lifting devices. Hydraulic elevator is ideally used to lift and move heavy to very heavy and large system. Hydraulic elevator provides a safer environment and is ergonomically useful. Hydraulic lifts are mainly used for transporting personnel, loading, and positioning work objects. The kind of hydraulic lift to be used in a particular situation will depend upon the style of the platform, and its mounting style, as well as the material that it uses.

The reason behind use of the hydraulic elevator is that there is no need of head room or deep pit, it also provides a ergonomically safe working condition, it does not require

conventional lift room, not necessary to have three phase power requirement, very easily it can come down in case of power failure by opening manual valve. This helps greatly to reduce or even eliminate the large amount of injuries caused to workers due to repetitive stress. Major injuries frequently occur when the job is more physically demanding as compared to the physical limitations of the workers. Proposed hydraulic elevator help to place the work material at positions that is awkward to the workers with simple, easy and safe manner.

A Programmable Logic Controller (PLC) is a digital computer used for automation of typically industrial electromechanical processes, such as control of machinery on factory assembly lines, amusement rides, or light fixtures. PLCs are used in many industries and machines. PLCs are designed for multiple analog and digital inputs and outputs arrangements, extended temperature ranges, immunity to electrical noise, and resistance to vibration and impact. Programs to control machine operation are typically stored in battery-backed-up or non-volatile memory.

Human machine interface is the part of the machine that handles the human–machine interaction. Membrane switches, rubber keypads and touch screens are examples of that part of the Human Machine Interface which we can see and touch.

Following advantages are identified for hydraulic elevator: Smaller physical size than hard wired solution, easier and faster to make changes, PLCs have integrated diagnostic and override function, diagnostics are centrally available, applications can be immediately documented, and application can be duplicated faster and cost effectively.

The main target of this paper is to evaluate a lift system using Hydraulic motor and controlled through Rexroth L20 PLC and establish an electro hydraulic lift system model with HMI control. This research paper made an analysis of feasibility of Hydraulic lift control system.

978-1-4799-6047-7/14 $31.00 © 2014 IEEE

The paper is organized as follows: Introduction about hydraulic lift is described in Section I. Complete hardware system design is presented in section II. In section III software design is presented. Section IV gives complete information about experimental setup whereas in section V result and discussion are explained. Conclusion & Future work is represented in section VI.

II. SYSTEM HARDWARE DESIGN

The hydraulic lift system is divided into two parts one is elevator structure and second is elctrical circuit and sensors interfacing with PLC.

A. Elevator Structure

Elevator structure consists of a main wooden elevator structure for three floors. The system is divided into floors by wooden strips. Each floor is of 25cm combined making the whole structure 100cm from the lower to higher sensors. The complete hardware layout of hydraulic elevator is shown in figure 1. Other than the main blocks several other mechanisms are used in order to make the system work.

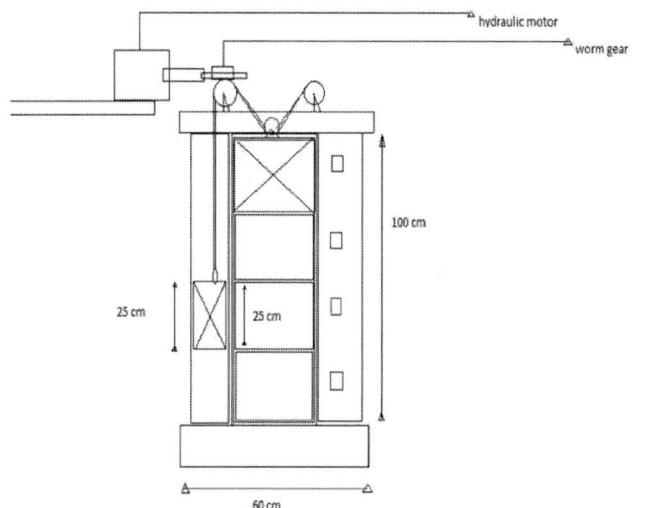

Fig. 1. Hardware layout of Hydraulic elevator

Counter weight: Counter weight is used to maintain the balance on the motor so that motor can work at very less power. Just like see-saw where because of weights on both sides a little push can move the see-saw back and forth, similar mechanism is used in lift to make motor work effortless and easier.

Pulley: Pulley is used when a material of very heavy weight has to be lifted. There are many ways to lift an object by using the pulley system. But, each method use has their own advantage compare with others. There are three types of pulleys which is a fixed pulley, a movable pulley and a combined (compound) pulley.

Worm Gear Drive: Sometimes it is essential to reduce the operating speed in a particular ratio as well as needed to stop the motor from unwanted movements. In that case worm gear is used we used worm gear for two used one is for reducing

the working speed and second as a breaking mechanism. Helical gear blocks the linear movement of the lift box until the motor moves the helical gear shown in figure 2.

Fig.2. Worm gear drive system

B. Electrical Circuit and sensors

Electrical circuit and sensors is also divided into two parts one is the outside push buttons circuit and other is proximity sensing for different floors. According to the number of floors outsides switches are placed in front of the each floor which helps for calling the lift system to that floor. Floor sensing is done by the different inductive proximity sensor placed on the top of the lift box at a distance of 5 mm from the lift structure surface and the metal plates are attached for each floor so that whenever the sensor comes in contact with the plate a high signal is sent to the PLC and accordingly with help of ladder logic program the motor reacts. Each floor is detected with the help of inductive sensor other than the ground floor which is detected by the limit switch. A limit switch is placed at the bottom of the lift structure such that whenever lift system presses that switch motor stops and indication of ground floor is given shown in figure 3.

Fig.3. Proximity Sensors and Limit switch setup

III. SYSTEM SOFTWARE DESIGN

Many challenges exists for designing a hydraulic lift with the increasing requirements on quality of product and its life, the elevator industry has been exponential development and the electrical control of the elevator into a new period of development. The control is done by the programmable logic controller (PLC) instead of the original relay logic control, and microcontroller based control. The complete hydraulic elevator controller design work is divided into two components one is electro - hydraulic system model and software development using ladder logic.

A. Electro- hydraulic system model

In the process of development of elevator heavy loads are to be lifted and lowered by means of a rope winch. The velocity of the rope winch is adjustable. For space reasons, the use of a hydraulic cylinder is impossible, i.e. the rope winch must be powered by a hydraulic motor. Also an electrical solution is not desirable due to changing loads. This hydraulic arrangement is used to provide rotational movement to the lift system. For the execution of hydraulic system switch the hydraulic pump on and inspect the set up control for leakage, pressure gauge may indicate any system pressure. Check the set pressure on the variable displacement pump of the drive power unit (if required, correct to 20 bars). By actuating solenoid S1 of 4/3 directional control valve hydraulic motor rotates clockwise while by actuating solenoid S2 of 4/3 directional control valve the hydraulic motor rotates counter-clockwise. The complete electro-hydraulic system circuit model is shown in figure 4.

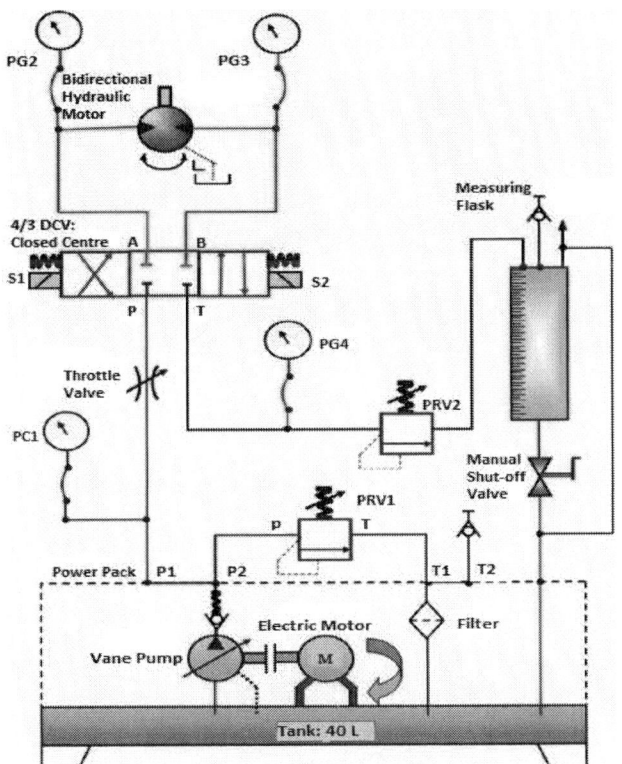

Fig.4. Electro - hydraulic system circuit model

B. Software development using ladder logic

Taking into account for hydraulic lift control system different input and output are required. The programming environment consists of Indraworks Engineering, which is used to develop the ladder logic code, debugging software, and a hardware interface for transferring the program on to the PLC through RS 232 or Ethernet to the REXROTH L20 PLC. It has 8 onboard digital inputs and 8 onboard digital outputs but a total of 24 digital I / O Ports are available for interfacing, it has 3MB of program and data storage space and one RS 232, one PROFIBUS Port and one Ethernet port for communication.

The functional analysis gives the details of the functioning of the system. The lift system is divided into three floors and complete system is described according to each floor.

Ground Floor

The sensor which detects the presence of lift is a limit witch. When the floor number is ordered on HMI screen inside the lift, the motor runs on clockwise direction to move the lift up till the respective floor is reached i.e. when first floor is ordered it moves up to first floor and stops and so on . When any outside button is pressed to call the lift, the motor runs in clockwise and the lift moves up to the respective floor.

First Floor

The sensor which detects the presence of elevator is an inductive proximity sensor. When the floor number is punched on HMI screen inside the lift, the motor runs on clockwise or anti- clockwise direction to move the lift up or down till the respective floor is reached i.e. when third floor is punched it moves up to third floor and stops or when ground floor is punched it moves down and stops when the limit switch is reached. When any outside button is pressed to call the lift, the motor runs in clockwise or anti- clockwise and the lift moves up or down to the respective floor.

Second Floor

The sensor which detects the presence of elevator is an inductive proximity sensor. When the floor number is punched on HMI screen inside the lift, the motor runs on clockwise or anti- clockwise direction to move the lift up or down till the respective floor is reached i.e. when third floor is punched it moves up to third floor and stops or when ground floor is punched it moves down and stops when the limit switch is reached. When any outside button is pressed to call the lift, the motor runs in clockwise or anti- clockwise and the lift moves up or down to the respective floor.

Third Floor

The sensor which detects the presence of elevator is an inductive proximity sensor. When the floor number is ordered on HMI screen inside the lift, the motor runs on Anti-clockwise direction to move the lift down till the respective floor is reached i.e. when first floor is ordered it moves down to first floor and stops and so on. When any outside button is pressed to call the lift, the motor runs in anti-clockwise and the lift moves down to the respective floor. The complete system software floor design diagram is shown in figure 5.

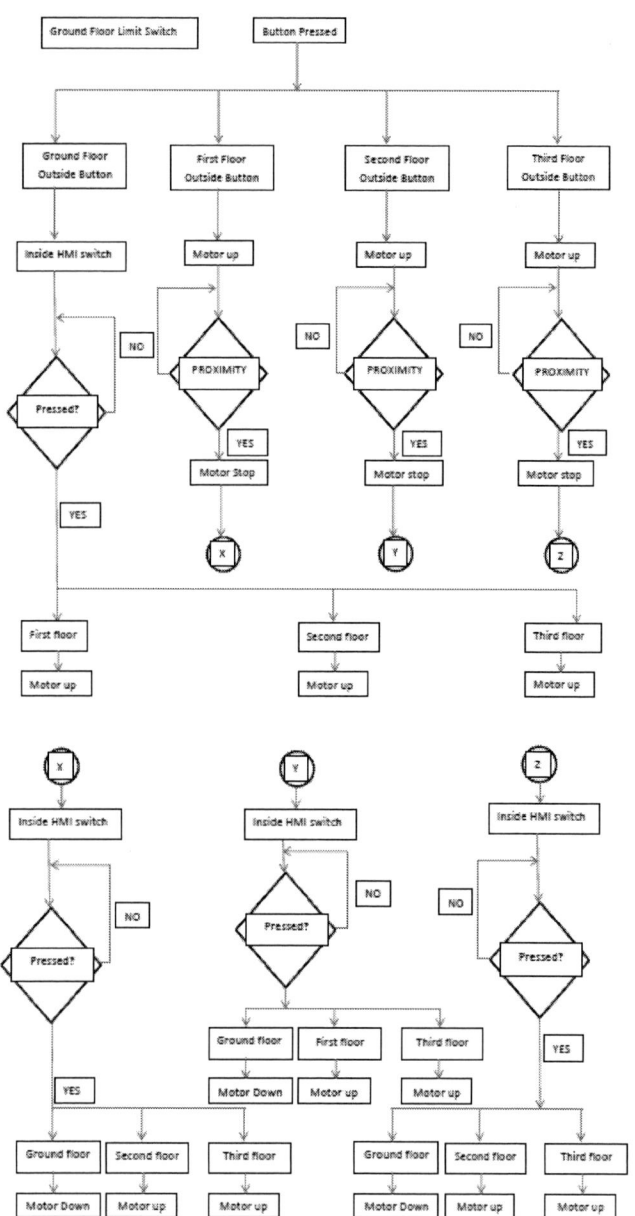

Fig.5.Flow chart for different floor

C. Equations

The motor used here is a hydraulic motor which is controlled through 4/3 DCV. Based on the design specification the output power and the output torque of the hydraulic motor are calculated by a simple calculation. Power and torque calculation:

$$F = mg \qquad (1)$$

Where F is the force, m is mass and g is the acceleration due to gravity.

$$Power = force \times velocity \qquad (2)$$

The weight of the empty cabin and the counter weight is approximately 5 kg. There are 5 samples of 3 kg of mass is taken into consideration in the elevator.

Total mass on the elevator = 15 kg
According to (1), Force = 15x9.8 = 147N
For constant speed of 0.5m/s
According to (2), Power = 147x0.5m/s = 73.5W
1 Horse Power (HP) = 740W
Therefore, 73.5W = 0.0986HP
Rotational Speed = 250 Revolution per minute (RPM)

$$Toque\ of\ motor = Horsepower \times 63025 \div RPM \qquad (3)$$

$$= (0.0986x63025)/250$$
$$= 24.85\ inch\ pound$$

Worm gear used with reduction factor is 6:1, therefore
Rotational speed = 250 ÷ 6 = 41.66 = 42 RPM
According to (3), Torque = (0.0986x63025)/42
$$= 147.95\ inch\ pound$$

D. Visulasiation and HMI

Visualization in Indraworks engineering is almost similar to that of SCADA. Visualization is generally used to create a GUI (Graphical User Interface) to provide a ease hand on working on the application and supervision of the project and automation unit. In this hydraulic lift system visualization we designed Master reset, Inside calling buttons, outside calling buttons, proximity status buttons and the motor status light to indicate proper working and checking the fault in the system. HMI is a human machine interface which provides a link between the inputs and the PLC and can be used as the indicator. In this research paper we used the HMI for the lift box internal switch panel as well as it is also used to indicate the current floor and status of pressed button, along with this emergency button is provided to vacate and stop the system at the time of emergency situation. With help of VCP HMI software, visualization is done which is shown in figure 6.

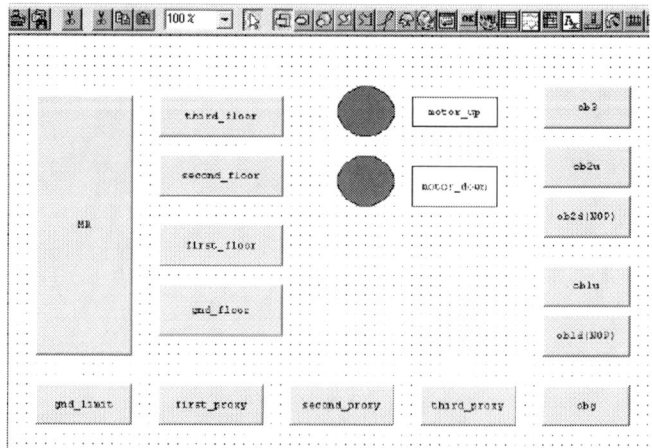

Fig.6. Visualization Testing Screen

IV. EXPERIMENTAL SET UP

The experimental set up of hydraulic motor based elevator system shown in figure 8, which comprises three proximity sensors for identifying different floor information, two push button at each floor for going upward or downward, so total of eight digital input required, inside lift again four input through HMI is 1st floor, second floor, third floor and emergency button for alarm is required. Once input is pressed REXROTH L20 PLC will read the input and accordingly 4/3 directional control valve will give input to the hydraulic motor to rotate clock wise as well as counter clockwise direction. The entire working flow chart through PLC is shown in figure 7. The experimental test bench is carried out for different test cases inputs and it is analyzed with the upward and downward movement of hydraulic lift system. The specification supporting the experimental test bench is explained in section III.

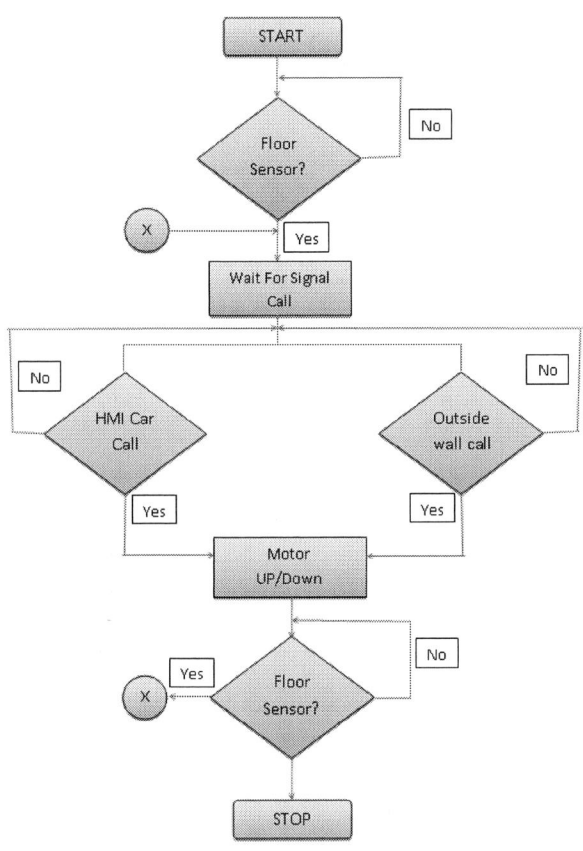

Fig.7. PLC and HMI working flow chart

V. RESULT & DISCUSSION

The experimental setup of hydraulic based elevator system is shown in figure 8, which consists of hydraulics circuit, 4/3 solenoid operated directional control valve, limit switches, proximity switches, different push button switches and HMI which are used inside the elevator. Rexroth L20 PLC is intelligent controller for controlling 4/3 DCV for clockwise and counter clockwise hydraulic motor control which move the elevator upward and downward with help of worm gear drive arrangement. The experimental test bench is carried out for different test input to analyze the upward and downward movement of elevator system. Entire setup is made according to the circuit diagram and all the inputs and outputs are

checked thoroughly. The hydraulic elevator comprises different cases which are given below:

Case 1: If the elevator is on the ground floor and a call button is pressed from the ground floor then motor will not actuate and the doors will open only. But if the call button for the 1st floor, 2nd floor or 3rd floor is pressed then the motor actuate in forward direction and move to the desired floor and open the doors.

Case 2: If the elevator is on the first floor and a call button is pressed from first floor then motor will not actuate and the doors will open only. But if the call button for the ground floor, 2nd floor or 3rd floor then motor moves in reverse direction or motor will forward direction move to the desired floor and open the doors.

Case 3: If the elevator is on the 2nd floor and a call button is pressed from 2nd floor then motor will not actuate and the doors will open only. But if the call button for the ground floor, 1st floor or 3rd floor then motor moves in reverse direction or forward direction and move to the desired floor and open the doors.

Case 4: If the elevator is on the 3rd floor and a call button is pressed from 3rd floor then motor will not actuate and the doors will open only. But if the call button for the ground floor, 1st floor or 2nd floor then motor moves in reverse direction only and move to the desired floor and open the doors.

Fig.8. Experimental set up of Hydraulic elevator system

VI. CONCLUSION AND FUTURE WORK

With design and development of hydraulic motor based elevator system prototype with help of REXROTH L20 PLC the safety, good ergonomics and noise reduction of the elevator system can be increased to larger extent. As infrastructural growth are increasing day by day so there is a need of efficient and cost effective elevator system, which can deliver reliable, stable and constant speed which is easily achieved with proposed system. The design in both mechanical components and control software should conform to the principles of safety system. This paper introduces some measures to improve the safety of the elevator system, which contributes to the stable operation of monitoring system field section PLC and ensures the communication between system and server machine. With the more and more application of new technology, we believe the elevator safety system will be more and more perfect. There is a lot of work is to be done in the field of elevator system. Smart elevator concept can be incorporated for the improvement of the system like safety systems, Control mechanism etc. In our system work in the field of safety and control in ambiguity situation is required. Also embedded system can be combined with PLC to make the system more efficient and smart in nature. The abbreviations used in this document are listed in table 1.

Table I. List of Abbreviations

Abbreviation	Description
PLC	Programmable Logic Control
DCV	Directional Control valve
HMI	Human machine Interference
SCADA	Supervisory Control and Data Acquisition
GUI	Graphical User Interface
VCP	Visual Composer Programming

VII. ACKNOWLEDGMENT

We are grateful to the support from the electronics, instrumentation and control engineering department of university of petroleum and energy studies, Dehradun, India.

VIII. REFERENCES

[1] Ricardo Gudwin, Fernando Gomide, Marcio (1998). "A Fuzzy Elevator Group Controller With Linear Context Adaptation". *IEEE World Congress on Computational Intelligence*. Vol. 12, No. 5, pp.481-486.

[2] Philipp Friese, Jorg Rambau (2006). "Online-optimization of multi-elevator transport systems with reoptimization algorithms based on set-partitioning models". *Discrete Applied Mathematics*. No. 154, pp.1908-1931.

[3] Zheng Yanjun, Zhang Huiqiao, Ye Qingtai, Zhu Changming. (2001). "The Research on Elevator Dynamic Zoning Algorithm and It's Genetic Evolution". *Computer Engineering and Applications*, No. 22, pp.58-61.

[4] Xiaodong Zhu, Qingshan Zeng (2006). "A Elevator Group Control Algorithm for Minimum Waiting Time Based On PLC". *Journal of Hoisting and Conveying Machiner*, No. 6, pp.38-40

[5] S.Krishankant, "Computer Based instrumentation Control", New Delhi: PHI Learning Pvt. Ltd, 2009.

[6] Reiner Dudziak, Dirk Mohr et. AI, "Development Projects as an Integral Element Education of Mechatronics Engineers", IEEE Mectatronics-REM, 2012.

[7] Jie Zhang, "Application of PLC in elevator control system", Journal of Liaoning Normal University (Natural Science Edition), 2009, pp.318-320,32(3).

[8] Darshil, Sagar, Rajiv, Pangaokar and S.A. Sharma "Development of a PLC Based Elevator System with Colour Sensing Capabilities for Material Handling in Industrial Plant", Jiont International

Conference on Power System Technology and IEEE Power India Conference, 2008,pp.1-7.

[9] Roushan Kumar, Adesh Kumar, "Design and hardware development of power window control mechanism using microcontroller", IEEE International conference on signal processing and communication ICSC,JIIT, December 12-14, 2013.

[10] Jayawardana.H.P.A.P., marasekara.H.W.K.M., Peelikumura.P. T.S.,Jayathilaka. W.A.K.C., Abeyaratne.S.G. and Dewasurendra.S.D. "Design and implenentation of astatechart based reconfigureurable elevator controller", 6th IEEE International Industrial and Information Systems, IEEE Conference Publications, pp. 352-357.

[11] Zhang Yajun, Chen Long, Fan Lingyan, "*A Design of Elevator Positioning Control System Model*," IEEE Int. Conference Networks & Signal Processing, Zhenjiang, China, June 8-10. 2008, pp. 535-538.

[12] Yi-Sheng Huang, Sheng-Luen Chung, and Mu-Der Jeng "Modeling and Control of Elevator by State Chart", June 2014, pp. 242-252,Asian Journal of Control, Vol. 6, No. 2.

[13] Eunsoo Jung, HyunjaeYoo, Seung-Ki Sul, Hong-Soon Choi and Yun-Young Choi, (2012) 'A Nine-Phase Permanent-Magnet Motor Drive System for an Ultrahigh-Speed Elevator', IEEE Transactions on Industry Application, vol. 48, pp 987-995.

[14] Jiang Jing and Zhang Xuesong (2011) 'Variable frequency speed-regulation system of elevator using PLC technology', 3rd International Conference on Advanced Computer Control, IEEE Conference Publications, pp 328-332.

[15] Knezevic.B., Blanusa.B.andMarcetic.D.(2011) 'Model of elevator drive with jerk control', XXIII International Symposium on Information, Communication and Automation Technologies, IEEE Conference Publications, pp 1-5.

[16] Ma Yinyuan and Jiang Zhaoyuan (2010) 'Task-Oriented Analysis and Design Method for Developing PLC Programs for Mechanical System Control', International Conference on Measuring Technology and Mechatronics Automation, IEEE Conference Publications, vol. 3, pp 726-729.

[17] Onat.A., Kazan.E.,Takahashi.N., Miyagi.D., Komatsu. Y .andMarkon.S. (2010) 'Design andImplementation of a Linear Motor for Multicar Elevators', IEEE/ASME Transactions on Mechatronics, vol. 15, pp 685-693.

978-1-4799-6047-7/14 $31.00 © 2014 IEEE

Hardware Design and Implementation of Unity Power Factor Rectifiers using Microcontrollers

BIDYUT MAHATO
Dept of Electrical Engineering
Research Scholar, ISM Dhanbad
bidyut1990@gmail.com

P.R.THAKURA
Dept of Electrical & Electronics
Associate Professor, BIT Mesra
prthakura@bitmesra.ac.in

K.C.JANA
Dept of Electrical Engineering
Assistant Professor,ISM Dhanbad
kartickjana@gmail.com

Abstract − **In this work, a hardware design of phase controlled rectifier has been implemented in an economical manner in power electronics laboratory using microcontroller. Various observations are recorded in this field in centuries but introducing microcontroller in this arena would be a new phase in this field of research, thus making the process more precise with faster response. Microcontroller is used to generate gate pulses to SCR. Zero Crossing Detector is employed for synchronizing the input with firing pulses in order to have controlled DC output. Detailed description and functioning of individual blocks is explained with the relevant waveforms. All waveforms obtained from 200MHz DSO at different stages are included along with the output waveforms obtained on PSIM Software. Phase controlled rectifiers are better than uncontrolled rectifier in the aspect of harmonics generated as well as controlled DC output.**

Key Words- Zero Crossing Detector, Rectifiers, Operational Amplifiers, SCR, Microcontroller, Unity Power Factor, Opto-Isolator

I. INTRODUCTION

Power electronics is a branch of Electrical Engineering which deals with control conversion and protection of power using high power semi-conductor switches. It process the power from source to the load. Every device needs suitable power for its working [1]-[5]. Power electronics prepares suitable power for each device. Designing firing circuit is a most challenging task in laboratory because of the fact that firing has to start from Zero of input which has to be sensed by zero crossing detector (ZCD)[6]-[8]. SCR ranges from few kilowatts to several megawatts in terms of power and from few hundred to several kilo volts levels in terms of voltage level, thus traditionally used as a switching device in medium and large power levels. MOSFET'S and BJT'S having very fast switching frequency compared to SCR'S are being limited to their uses to

medium power levels at few hundred volts[9]-[11]. Though IGBT'S have more due advantages over MOSFET'S & SCR'S being inability to work at very high voltages and costlier makes SCR'S to be a better choice even today.

OVERVIEW OF PAPER

This paper is introduced with the working principle and operation of SCR in Section II and design of firing circuit in section III. Simulation results in PSIM platform has been shown in section IV along with hardware results. Results are discussed in section V and is concluded in section VI along with references.

II. PRINCIPLE AND OPERATION

A Silicon controlled rectifier is a layered solid current controlling device, it is also well known as THYRISTOR. Firing angle α is stated as the number of degrees from the beginning of cycle when SCR is gated. Conduction angle is defined as the number of degrees that SCR'S remain conducting.

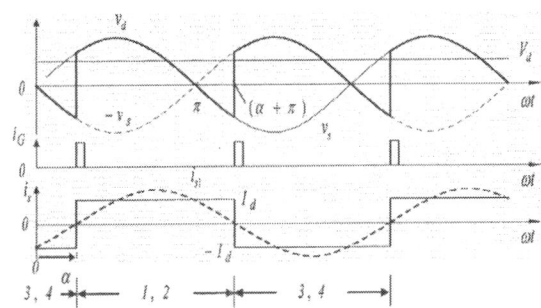

Fig. 1. Waveform of single phase converter.

This bridge rectifier have four thyristors. T1, T2, T3, T4 . During positive half cycle of sine wave, TI and T2 are forward biased and T3 and T4 are reversed biased. During negative half cycle of input supply, T3 and T4 are forward biased and T1 and T2 are reversed biased. Hence, the gate pulses should be properly synchronized with a.c power supply.[12].

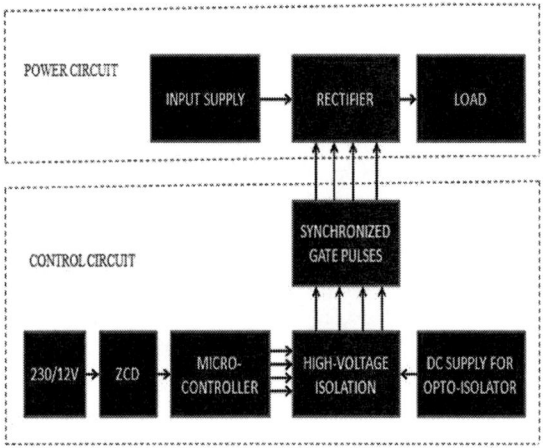

Fig. 2. Block diagram representation of the proposed converter.

The proposed system is represented in two different blocks namely Power circuit and Control circuit represented by dotted lines. In the power circuit, input supply is fed to the rectifier block where thyristors have been used and finally the controlled DC output is fed to the load. In control circuit zero crossing detector (ZCD), Microcontroller, Opto-Isolator plays a vital role. Op-Amp is used as zero crossing detector that senses the Zero crossing of input and providing signal to the microcontroller through input pins. Zero Crossing Detector is employed for the synchronization of input with firing pulses so as to have controlled DC output.[11][12]. The synchronized output pulses from the microcontroller fed to the TLP250 for high voltage isolation. Here, grounding is the other aspect of this hardware and thus TLP250 is employed for novel reason. TLP250 needs DC supply which is provided by isolated grounding transformer (230/12V) described in further section. Finally the pulses fed to the thyristor[13][14].

III. DESIGN OF FIRING CIRCUIT

A. Op-Amp as ZCD

An operational amplifier is a multistage, negative feedback amplifier with very high gain[5][6]. It uses voltage shunt feedback to provide a stabilized voltage gain. It senses the Zero of crossing of input and provides pulses signal to the microcontroller

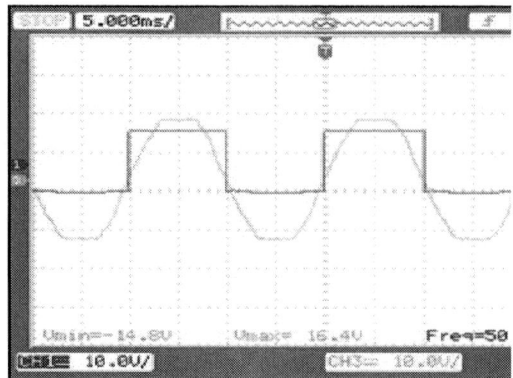

Fig. 3. Experimental waveform of ZCD Output and input supply voltage.

Microcontroller sense this signal as interrupt and executes the programme and provides the pulses which is further fed to the gate of the thyristors. Zero Crossing Detector is employed for the synchronization of input with firing pulses so as to have controlled DC output[16].

B. TLP250 as Opto-Isolator

TLP250 is used as high voltage isolation also well known as Opto-Isolator. It provides isolation between power circuit and control circuit thus providing gate pulses to the respective thyristors. It has Input threshold current of magnitude 5mA(max.), Supply current of magnitude 11mA(max.), Supply voltage of magnitude 10−35V, Output current in the range of ±1.5A (max), Switching time 1.5μs(max), Isolation voltage of magnitude 2500 Vrms(min)[15].

C. DC Supply used for TLP250

Isolated power supply of 12V, 5mA, having six output tapping has been used as shown in Fig.4 below to provide DC supply for Opto-Isolator.

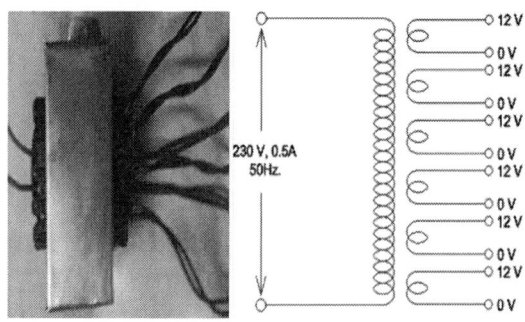

Fig. 4. Multi-winding transformer used for TLP250

D. Microcontroller

Arduino is an open-source electronics platform based on its flexiblity, easy-to-use hardware as well as software. The code is written in a suitable programming language to generate pulses. In this coding, two pulses being generated after being receiving the signal from Zero Crossing Detector (ZCD) as interrupt. Two generated pulses are 180 phase shift from each other. These pulses are then fed to gate of the thyristors. The features of an Atmel ATmega328 microcontroller with 2 Kb of RAM operating at 5 V, flash memory of 32 Kb for storing programs and 1 Kb of EEPROM for parameter storage. The clock speed is 16 MHz, that translates about executing 300,000 lines of C source code per second. The board has 6 analog input pins 14 digital I/O pins. There is a USB connector and a DC power jack for connecting an external power source.

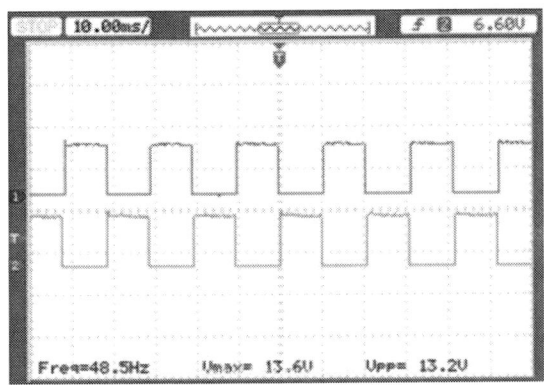

Fig. 5. Pulses generated being 180^0 phase shifted.

IV. SIMULATION AND HARDWARE RESULTS

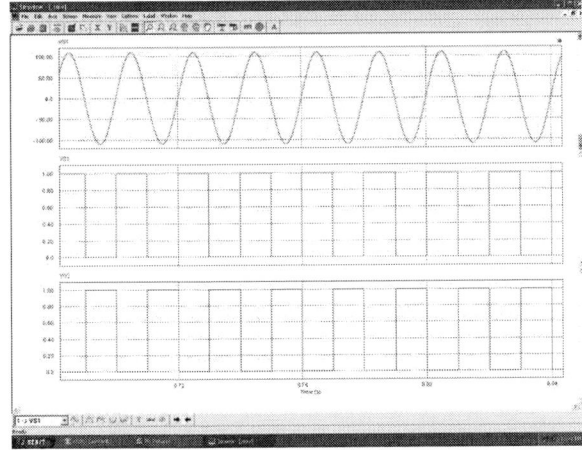

(a)

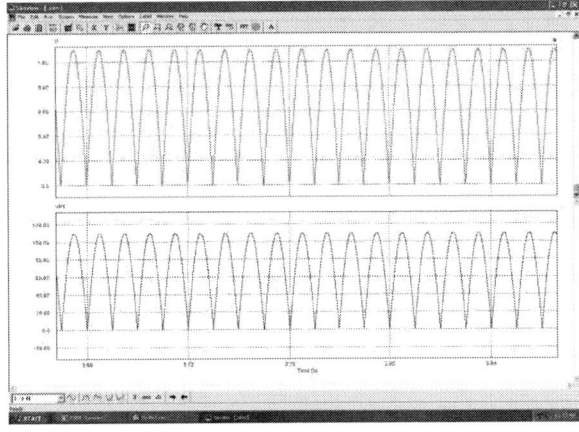

(b)

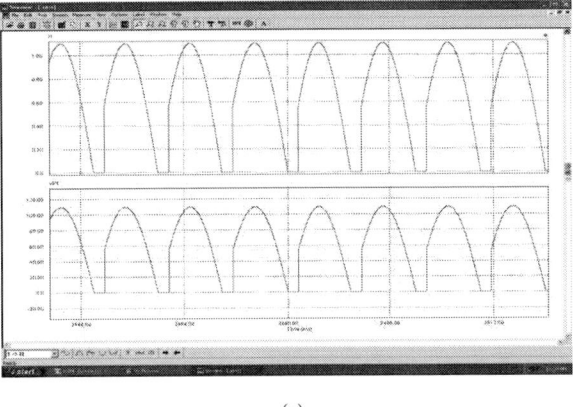

(c)

Fig. 6. Simulation results of (a) Synchronized gate pulses to the supply, output voltage and current waveform (a) at α=0^0 (c) at α=30^0

A. HARDWARE RESULTS IN OSCILLOSCOPE

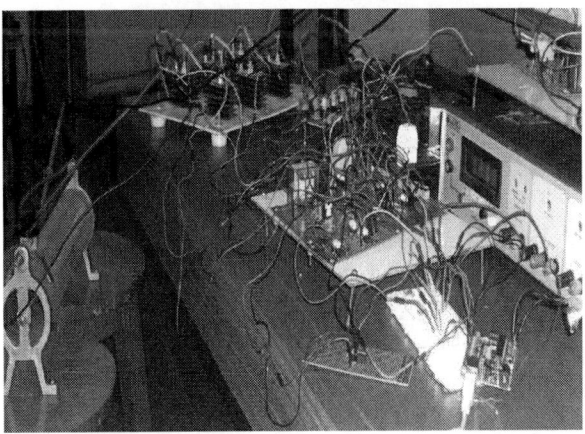

Fig. 7. Hardware setup of thyristor controlled rectifier in Power Electronics Laboratory

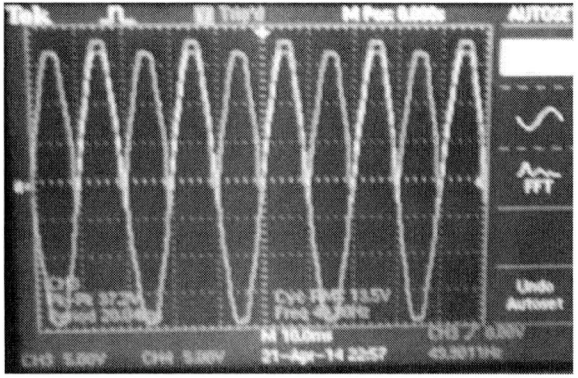

(a)

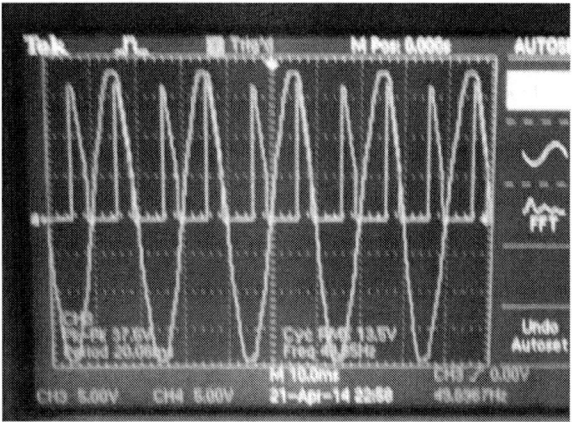

(b)

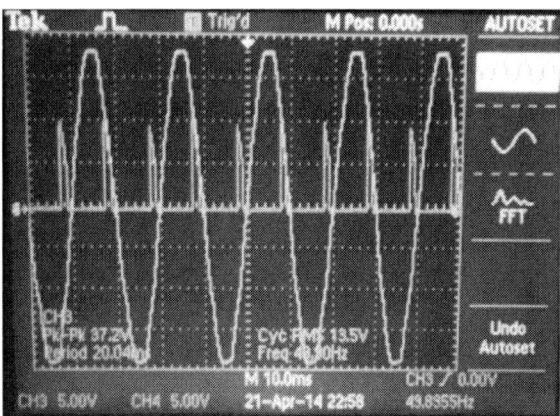

(c)

Fig. 8. Experimental results of input voltage and output voltage across R-Load at (a) $\alpha=0^0$ (b) $\alpha=100^0$ (c) $\alpha=150^0$

V. RESULT AND DISCUSSIONS

In simulation results, single phase controlled rectifier has been has been done in PSIM Software and load voltage and current is shown at $\alpha=0^0$ and $\alpha=30^0$. The hardware set up shown in Fig.7, comprises of TLP250 circuit used as opto-isolator. ZCD (Zero Crossing Detector) is used for synchronizing the input with firing pulses. Microcontroller is used for providing 180^0 shifted pulses to SCR's dc supply for providing supply to TLP Circuit. Fig. 8a) shows the output voltage with input sinusoidal voltage whereas fig.7b) and fig. 7c) indicates the variation of output voltage with variation of firing pulses provided from the microcontroller after detecting the Zero Crossing of input Sinusoidal wave.

VI. CONCLUSION

Designing analog firing circuit makes circuit complex and requires maintenance. Employing microcontroller instead reduces all its disadvantages being economical. It is easier to design with precision output. It can be used for single phase and three phase fully controlled thyristor rectifier. The paper provides a design for a simple microcontroller based SCR Controlled Unity Power Factor Rectifiers. The design is adequate for many purposes. These improvements have been tested in principle, but some detailed work remains to be done in this area for three phase Supply.

VII. REFERENCES

[1] B.K.Bose, "Modern power electronics and AC Drives", PHI,2001 .

[2] P.C.Sen, "Power Electronics", Tata McGraw Hill Publishers, 4th edition, 1987.

[3] N.Mohan, T.M.Undeland, W.P.Robbins, "Power Electronics: Converters application and Design", New York: Wiley, 3rd edition, 2006.

[4] Mohammed E. El-Hawary, "Principles of Electric Machines with Power Electronic Applications", Wiley India, 2nd edition, 2011.

[5] Gayakwad, "Operational Amplifier", Prentice Hall of India, 2009.

[6] General Electric, D.R. Grafham and F.B. Golden, SCR Manual, 6th ed. Englewood Cliffs, NJ: Prentice Hall, 1982.

[7] B.K.Bose, "Power Electronics- a technological Review", Proceedings of the IEEE, Vol.80 pp1303-1334, Aug 1992.

[8] B.K.Bose, "Recent advances and trends in power electronics and drives", proceedings of NORPIE workshop, Helsinki, pp. 170-182,1998.

[9] Mukesh Gupta, Sachin Kumar and Vagicharla Karthik "Design and Implementation of cosine control firing scheme for single phase fully controlled bridge rectifier", International Journal of Emerging

Trends in Electrical and Electronics, Vol.3, Issue 1 May 2013.

[10] Tirtharaj Sen, Pijush kanti Bhattacharjee and Manjima Bhattacharya,"Design and Implementation of Firing Circuit for single phase Converter," International Journal of Computer and Electrical Engineering, Vol 3,pp 368-374, June 2011.

[11] S. B. Dewan and W. G. Dunford, "A microprocessor-based controller for a three phase controlled rectifier bridge," IEEE Trans. Ind. App., vol. LA-19, pp. 113-119, Jan./Feb. 1983.

[12] P. C. Tang, S. S. Lu, and Y. C. Wu, "Microprocessor-based design of a firing circuit for three-phase full-wave thyristor dual converter," IEEE Trans. Ind. Electron., vol. IE-29, pp. 67-73, Feb. 1982.

[13] J. S. Wade, Jr., and L. G. Aya, "Design for simultaneous pulse triggering of SCRs in three phase bridge configuration," IEEE Trans. Ind. Electron. Contr. Instrum., vol. IECI- 18, no. 3, pp. 104-106, 1971.

[14] B. Ilango, R. Krishman, R. Subramanian, and S. Sadasivam, "Firing circuit for three-phase thyristor-bridge rectifier," IEEE Trans. Id. Electron. Contr. Instrum., vol. IECI-25, no. 1, pp. 45-49, 1978.

[15] L. H. Hoang, "A digitally controlled thyristor trigger circuit," Proc. IEEE, vol. 66, no. 1, pp. 89-91, Jan. 1978.

[16] T.Izumi, M.Yamazoe, and T. Nakano, "A microprocessor-based control system of thyristor converter fed dc motor drives," IECI '79 Proc., Mar. 1979.

VIII. BIOGRAPHIES

P.R.Thakura received his Ph.D degree from BIT Mesra, Ranchi and BITS Pilani in 2008 and 1990 respectively. He has been working in BITS Pilani from 1987 to 1994. He joined BIT, Mesra in May 1994and is currently serving as associate professor in Dept of Electrical and Electronics Engg.

He is working for Eramus Mundus Programme from September 2009. He was winner of Indo-Italian Young Researchers Fellowship and worked in Hybrid Electric Vehicle in University of Padova, Italy from September 2006 to August 2007. He has more than 30 international and national papers guided more than 70 UG projects and has reviewed many books on power electronics by M.D.Singh and K.B.Khanchandani in July 2013.

His research interest include Power Electronics, Variable speed drives, high performance AC Drives, Hybrid Electric Vehicles. Dr. Thakura is permanent member of Indian Society of Technical Education, Delhi and Institution of Engineers, Kolkata.

K.C.Jana received his M.Tech and Ph.D degree from NIT Durgapur and Jadavpur University in 2003 and 2013 respectively. He is currently serving as Assistant Professor in Indian School of Mines, Dhanbad from June 2012.

He has been working in Birla Institute of Technology, Mesra from July 2003 till May 2012. His research is in area of Modelling and Design of Multilevel Inverter , Real-time control of power electronics devices, Design and Implementation of efficient power converters. Dr. Jana visited the Power electronics and Drives lab of CEDT (IISC, Bangalore) in July 2006, visited Chittaranjan Locomotive Works (CLW), West Bengal for study of different Power Electronics Application on Indian Railways in October 2008, visited Chittaranjan Locomotive Works (CLW) in August 2010, visited Chittaranjan Locomotive Works (CLW) West Bengal for industry collaboration in August 2010, visited VSSC, Trivundrum and IIST Trivundrum for study of different application of control, navigation and guidance of Launch vehicle in January 2011.

B.Mahato received his M.E and B.Tech degree from Birla Institute of Technology, Mesra, Ranchi and Guru Nanak Institute of Technology in 2014 and 2011 respectively. Since 2014, he has been working toward the Ph.D degree in the Indian School of Mines, Dhanbad in the

department of Electrical Engineering in the field of Power Electronics. His main research interest includes Phase controlled rectifiers, Multilevel inverters, AC Drives and Hybrid Electric Vehicles.

978-1-4799-6047-7/14 $31.00 © 2014 IEEE

Performance Assessment of different Control Strategies for five level DCMLIs supplying Static loads and Dynamic loads

S. Banerjee
Assistant Professor, EEED
BVCOE
New Delhi, India

D. Joshi
Associate Professor, EEED
DTU
Delhi, India

M. Singh
Professor, EED
DTU
Delhi, India

R. Sharma
Assistant Professor, EED
NIT
Kurukshetra, India

Abstract-- **Performance evaluation of two different topologies of five level diode clamped multilevel inverter (DCMLI), supplying static loads and dynamic loads is presented in this paper. For the classical DCMLI the control strategies implemented are Phase disposition (PD), Phase opposition disposition (POD), Alternate phase opposition disposition (APOD) and unsymmetrical sine pulse width modulation. Another five level topology, which is free from the problem of voltage collapse, is compared with the classical topology. All the control methods have been simulated in MATLAB. A variation in Total Harmonic Distortion (THD) in output voltage and output current is analyzed for each control strategy. These values are compared to the THD values obtained from another five level topology. Comparison is done for both topologies once by keeping inverter input voltage constant and thereafter by keeping inverter output voltage constant. Both static loads and dynamic loads have been used. The effect of changing load on the THD values of voltage and current has also been tabulated.**

Keywords—APOD; DCMLI; PD; POD; THD; Topology

I. INTRODUCTION

In an era of rapid technological growth, multilevel inverters have become very popular owing to their extensive application in the areas of high power and medium voltage. The advantages of these inverters include lower common mode voltage, lower voltage stress on switches, lower dv/dt rating and low harmonic contents [1]. The use of MLIs resulted in a reduction of the harmonic content in voltages. Besides, it also lowers device voltage stresses [2]. These advantages have led to the focussing of research towards different inverter topologies and various switching techniques for the control of inverter output voltage.

Multilevel inverters are fabricated using power semiconductor switches and other elements like capacitors and diodes. The switching sequence is such that a stepped voltage is obtained at the output. Greater the number of levels in the inverter output, the more closely it resembles a sine wave. The major MLI topologies are: DCMLI (or neutral point clamped) [3], flying capacitor (or capacitor clamped) MLI and cascaded H-bridge MLI [4, 5, 6]. DCMLI is very useful in renewable energy conversion systems because it requires only one energizing source. The work mentioned in this paper

embodies analysis and comparison of two different topologies of single-phase five level DCMLI, with various PWM control strategies. All simulation has been carried out in MATLAB 2010a software.

This analysis corresponds to the simulation study of the simple control techniques for the classical structure of five level DCMLI in MATLAB with different types of loads connected to the MLI terminals. Another five level DCMLI topology, which is free from the problem of voltage imbalance is also simulated. The results observed for the THD variation of load voltage and current have been compiled and then a comparative analysis has been presented.

II. SINGLE PHASE FIVE LEVEL DCMLI

Fig.1 shows the topology of a single-phase five level DCMLI [1]. The input voltage is V_{dc}. Each single-phase full-bridge inverter generates output voltage at the level: $0.5V_{dc}$, $0.25V_{dc}$, 0, $-0.25V_{dc}$ and $0.5V_{dc}$. This is done by connecting the input sequentially to the ac side via the switches.

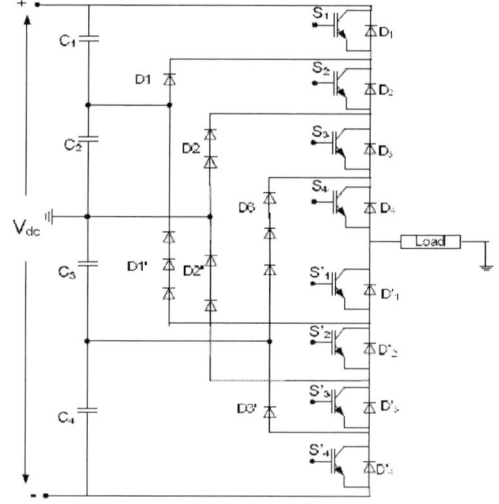

Fig. 1 Classical five level DCMLI

III. VOLTAGE BALANCING PROBLEM

The aforementioned topology suffers from the problem of divergence of capacitor voltage. The non uniform charging and discharging of capacitors result in the collapse of five level output voltage to three level. This condition greatly limits the practical utility of the classical topology. Some auxiliary arrangements can be made to uniformly charge the capacitors [7]. However, such structures result in increased complexity.

IV. H BRIDGE DCMLI TOPOLOGY

A five level eight switch H bridge topology shown in fig. 2 can be used to overcome the problem of voltage imbalance of the classical five level topology [8].

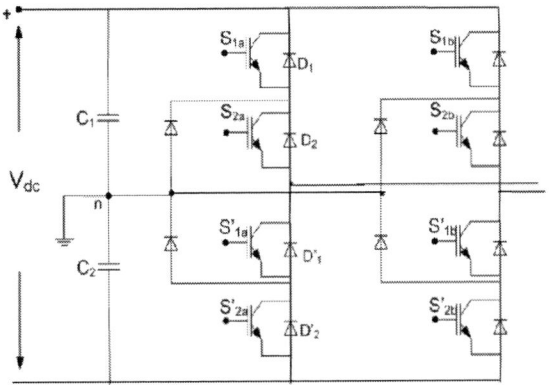

Fig. 2 5-level 8-switch H bridge topology

Another clinching advantage of using this topology is the use of lesser number of diodes. This leads to a decrease in cost and a huge reduction in complexity. The input voltage is V_{dc}. Each single-phase full-bridge inverter generates output voltage at the level: V_{dc}, $0.5V_{dc}$, 0, $-0.5V_{dc}$ and V_{dc}. This implies that the peak voltage level obtained at output is twice that of the classical topology.

The type of load has a significant bearing on the output voltage and current waveforms as well as in the harmonic content. Therefore the inverter topology and category of control method is decided after taking into consideration all these parameters [9].

V. CONTROL METHODOLOGY

Control methodology essentially deals with control cycle of switches in the MLI thereby controlling the output parameters of MLI. For the classical structure of DCMLIs, level shifted PWM technique is applied. The PD, POD, APOD and an unsymmetrical PWM method are discussed here. The output obtained for the aforementioned techniques is compared with the output of the five level H bridge topology.

PD, POD and APOD PWM is classified according to how carrier waves are placed in relation to the reference signal [2]. The carriers are in the same phase in PD-PWM. In POD-PWM, carriers above the reference line are 180 degrees out of phase with those below the line. Adjacent carriers are in phase opposition in APOD PWM. In unsymmetrical PWM, three carriers are in the same phase while one carrier is out of phase with respect to the others by 180 degrees.

The switching scheme of the H-bridge topology involves the comparison of two triangular carriers with two modulating sinusoidal signals [8]. The carriers are in the same phase while the modulating signals are in phase opposition to each other. The comparison of the carriers with the sinusoid having a zero

phase shift generates the control signals for the inverter leg nearest to the capacitors. Switching signals for the other leg are generated by the comparison of the carriers with the second modulating signal.

Fig. 3 to fig. 6 respectively illustrate the arrangement of carrier and reference waveforms in PD, POD, APOD, unsymmetrical PWM for the classical five level topology. Switching signal generation method for five level eight switch H-bridge inverter is shown in fig. 7.

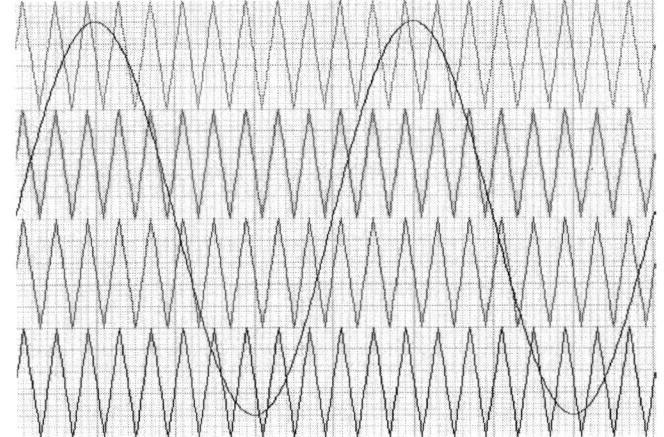

Fig. 3 PD PWM for a 5 level DCMLI

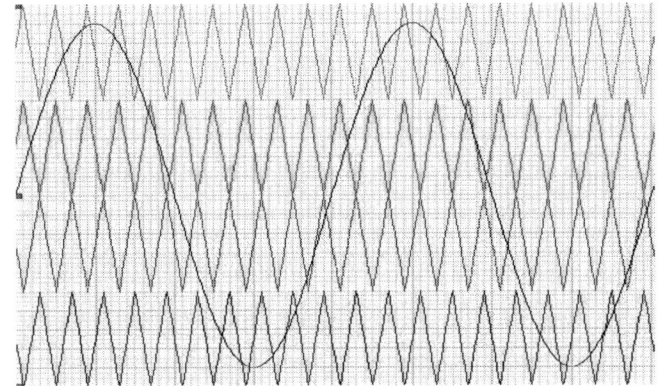

Fig. 4 POD PWM for a five level DCMLI

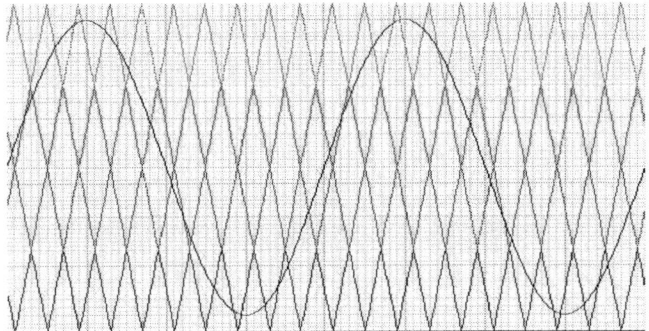

Fig. 5 APOD PWM for a five level DCMLI

978-1-4799-6047-7/14 $31.00 © 2014 IEEE

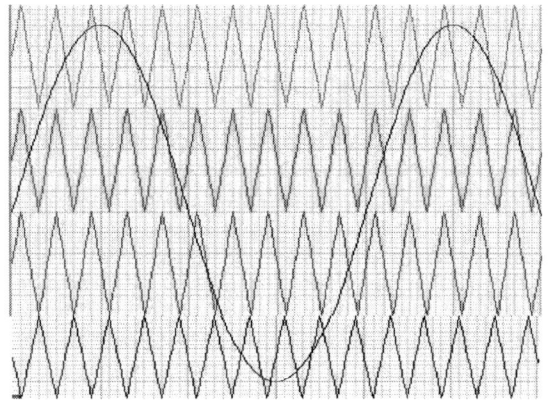

Fig. 6 Unsymmetrical PWM for a five level DCMLI

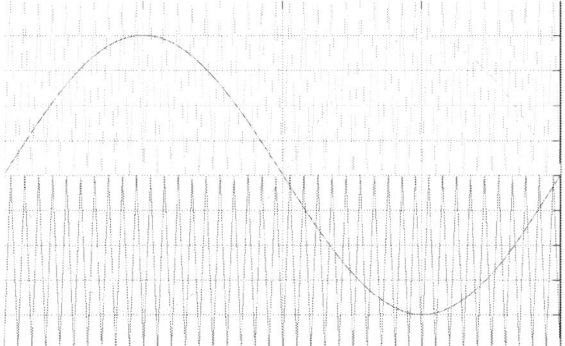

Fig. 7 Control scheme for five level eight switch H bridge inverter

VI. SIMULATION RESULT ANALYSIS

In the inverter shown in Fig. 1, voltage collapse occurs across some capacitors and voltage rise across others as time increases. Due to the non uniform power drawn from the DC link capacitors, resulting output voltage becomes just three levels instead of five levels (refer figures 8 and 9). However, the inverter illustrated in Fig. 2 is free from voltage collapse problems and maintains a steady five level output (shown in figure 10).

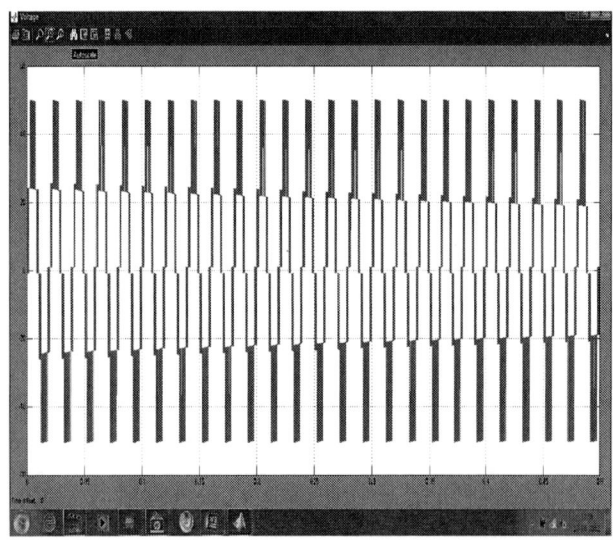

Fig. 8 Five level DCMLI output when run for 0.5 seconds.

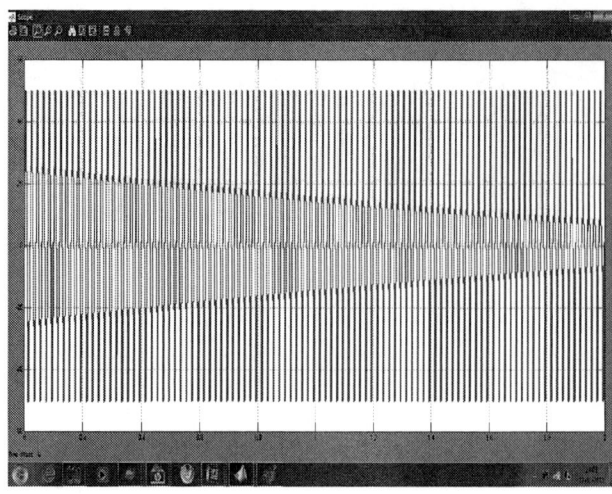

Fig. 9 Voltage level collapse in five level DCMLI when run for 2 seconds.

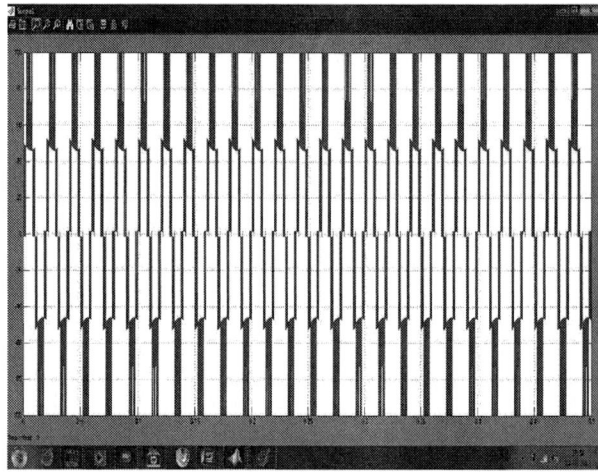

Fig. 10 Five level H bridge inverter topology output run for a long time.

R L loads of different power factors have been used as static loads. Single phase induction motor of three different types and having the same rating, have been considered as dynamic loads. These different loads are fed by the filtered output of classical five level topology as well as the H bridge topology. The filter used is a passive LC filter.

THD values are observed for the voltage and current. Table I, II and III contain the records of some of the THD information. Data regarding THD values have been obtained by controlling the output of the five level DCMLI by PD, POD, APOD and unsymmetrical SPWM respectively (refer to figures 3 to 6). The H bridge topology output is controlled by its specific strategy as shown in figure 7. Comparison is done for both topologies once by keeping inverter input voltage constant and thereafter by keeping inverter output voltage constant.

A. Plot of THD for a power factor (pf) = 0.7 with same inverter input voltage for both topologies

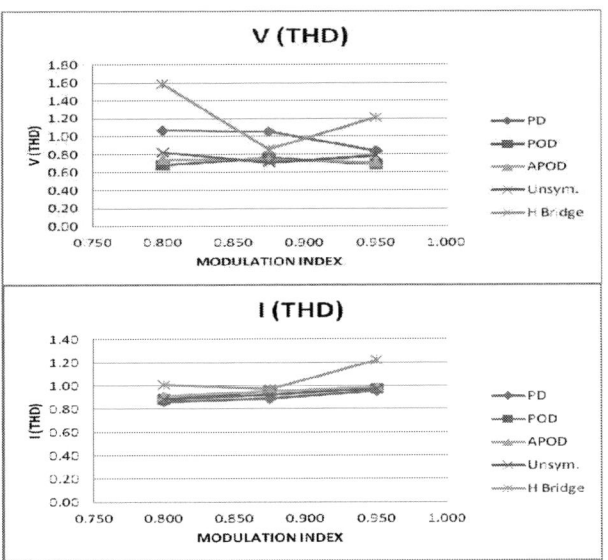

Fig.11 Plot of THD value of load voltage (V) and load current (I) versus modulation index (MI).

Fig. 11 demonstrates that the THD value corresponding to voltage is minimum when inverter is switched using POD PWM. Low current THD values are obtained when inverter switching is done using PD PWM.

B. Plot of THD for a power factor (pf) = 0.9 with same inverter input voltage for both topologies

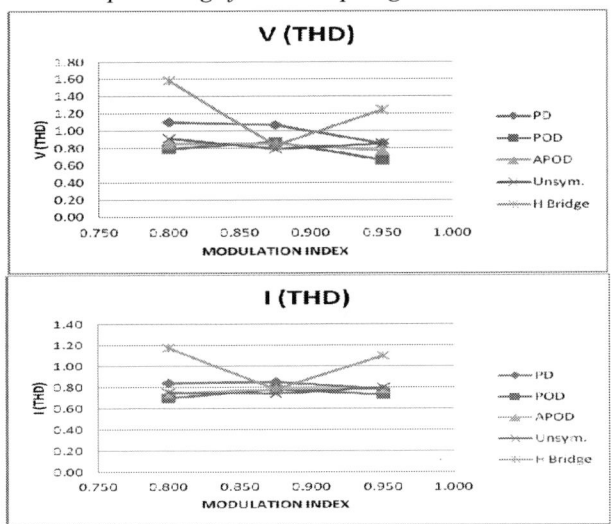

Fig.12 Plot of THD value of load voltage (V) and load current (I) versus modulation index (MI).

Fig. 12 demonstrates that the THD value corresponding to voltage as well as current is minimum when inverter is switched using POD PWM.

C. Plot of THD for a power factor (pf) = 0.7 with same inverter output voltage for both topologies

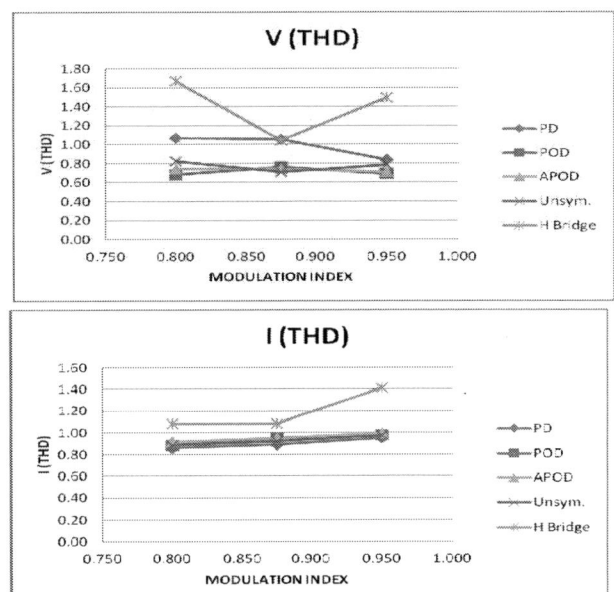

Fig.13 Plot of THD value of load voltage (V) and load current (I) versus modulation index (MI).

Fig. 13 demonstrates that the THD value corresponding to voltage is minimum when inverter is switched using POD PWM. Low current THD values are obtained when inverter switching is done using PD PWM.

TABLE I. THD value of load voltage (V) and load current (I) versus modulation index (MI), for a power factor (pf) = 0.8 with same inverter output voltage for both topologies

V (THD)					
M	PD	POD	APOD	Unsym.	H Bridge
0.800	1.08	0.73	0.79	0.86	1.67
0.875	1.06	0.80	0.78	0.75	1.03
0.950	0.84	0.67	0.73	0.81	1.51
I (THD)					
M	PD	POD	APOD	Unsym.	H Bridge
0.800	0.82	0.79	0.82	0.80	1.11
0.875	0.84	0.86	0.86	0.83	1.00
0.950	0.86	0.86	0.89	0.88	1.35

Table I demonstrates that the THD value corresponding to voltage is minimum when inverter is switched using POD PWM. Low current THD values are obtained when inverter switching is done using unsymmetrical PWM.

D. Plot of THD for a power factor (pf) = 0.9 with same inverter output voltage for both topologies

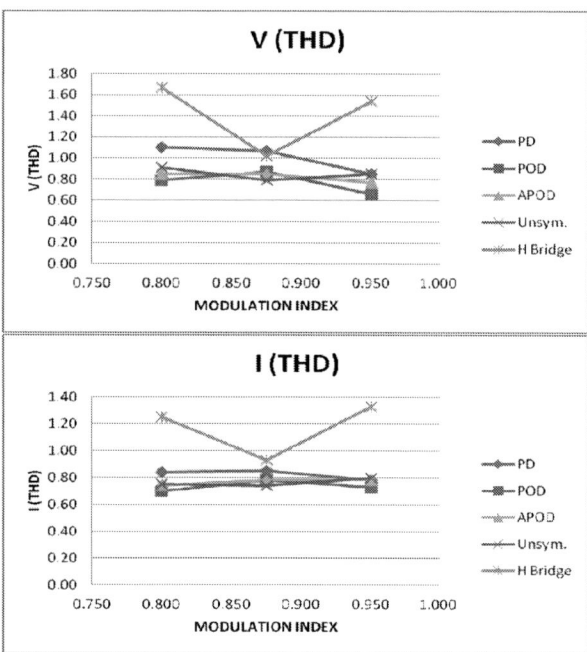

Fig.14 Plot of THD value of load voltage (V) and load current (I) versus modulation index (MI).

Fig. 14 demonstrates that the THD value corresponding to voltage as well as current is minimum when inverter is switched using POD PWM.

E. Plot of THD for a split phase induction motor with same inverter input voltage for both topologies

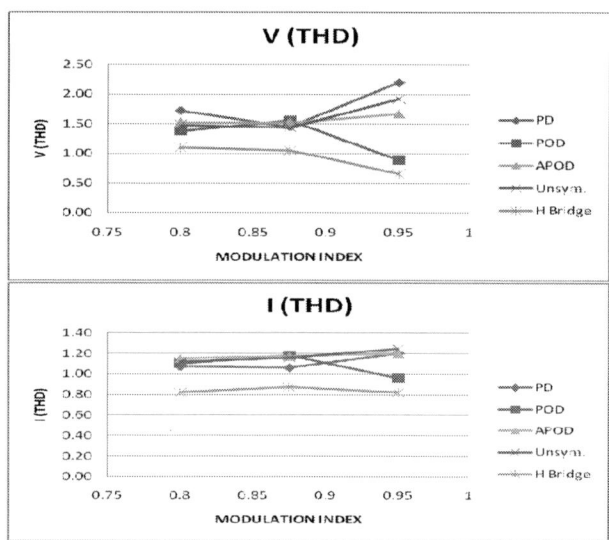

Fig. 15 Plot of THD value of load voltage (V) and main winding current (I) versus modulation index (MI)

Fig. 15 demonstrates that the THD value corresponding to load voltage as well as main winding current is minimum when inverter is switched using POD PWM.

TABLE II. THD value of load voltage (V), main winding current (I) and auxiliary winding current (I_A) versus modulation index (MI), for a capacitor start induction motor with same inverter input voltage for both topologies

METH-OD	MI=0.8			MI=0.875			MI=0.95		
	I_A (THD)	V (THD)	I (THD)	I_A (THD)	V (THD)	I (THD)	I_A (THD)	V (THD)	I (THD)
PD	1.85	1.71	1.12	1.45	1.45	1.10	1.60	2.10	1.38
POD	1.57	1.39	1.11	1.53	1.55	1.23	2.76	0.88	1.02
APOD	1.69	1.54	1.19	1.50	1.53	1.21	1.25	1.58	1.27
Unsym.	1.68	1.51	1.16	1.37	1.43	1.16	1.41	1.80	1.36
H Bridge	2.95	0.99	0.58	2.98	1.31	1.51	3.19	1.26	1.36

Table II demonstrates that the THD value corresponding to load voltage as well as main winding current is minimum when inverter is switched using POD PWM. Low auxiliary current THD values are obtained when inverter switching is done using unsymmetrical PWM.

TABLE III. THD value of load voltage (V), main winding current (I) and auxiliary winding current (I_A) versus modulation index (MI), for a split phase induction motor with same inverter output voltage for both topologies

METH-OD	MI=0.8			MI=0.875			MI=0.95		
	I_A (THD)	V (THD)	I (THD)	I_A (THD)	V (THD)	I (THD)	I_A (THD)	V (THD)	I (THD)
PD	1.02	1.49	1.03	0.96	1.18	1.04	1.29	2.01	1.26
POD	0.92	1.11	1.04	1.03	1.28	1.17	0.84	0.65	0.91
APOD	1.04	1.26	1.11	1.06	1.26	1.16	1.15	1.40	1.22
Unsym.	1.00	1.22	1.07	1.08	1.19	1.12	1.24	1.65	1.29
H Bridge	0.80	1.10	0.82	0.77	1.05	0.87	0.81	0.66	0.82

Table III demonstrates that the THD value corresponding to load voltage is minimum when inverter is switched using POD PWM. Lowest main winding current THD values are obtained in case the five level eight switch H Bridge topology is used. It also establishes that the lowest THD values for auxiliary winding current relate to the five level eight switch H Bridge topology.

F. Plot of THD for a capacitor start induction motor with same inverter output voltage for both topologies

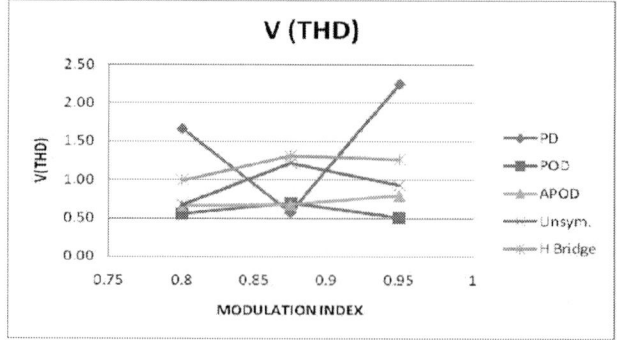

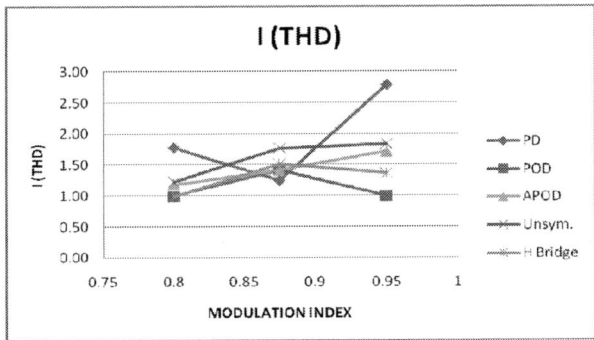

Fig. 16 Plot of THD value of load voltage (V) and main winding current (I) versus modulation index (MI)

Fig. 16 demonstrates that the THD value corresponding to load voltage as well as main winding current is minimum when inverter is switched using POD PWM.

G. Filtered Voltage and Current Waveforms

Fig. 17 illustrates the output voltage of the five level H bridge topology after it has been filtered using a passive LC combination. Fig. 18 shows the waveform of current, when an inductive load of power factor 0.8 is connected across the aforementioned inverter.

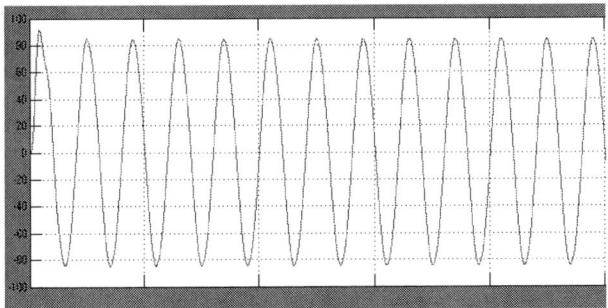

Fig. 17 Filtered voltage output of five level H bridge topology

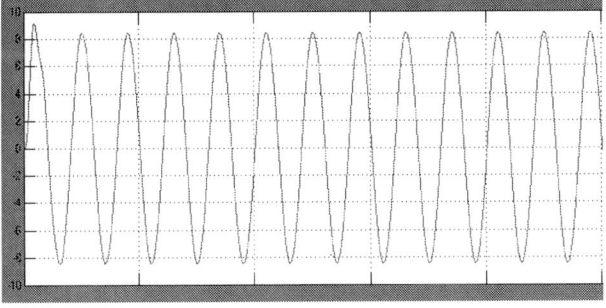

Fig. 18 Waveform of current for an inductive load

VII. CONCLUSION

For RL load, when input voltage of both inverter topologies are the same, the minimum value of voltage THD and current THD is observed incase of POD SPWM. POD SPWM again exhibits the lowest THD values when simulation is run after matching the output voltages of both inverter topologies.

In case of dynamic load, with same input voltage of the two topologies, the H bridge inverter gives the least THD values

for low modulation indices. As the modulation index increases, the THD values obtained by applying POD SPWM are lower as compared to the other methods. POD SPWM also results in the lowest THD values when output voltages of both inverter topologies are equal and the load is dynamic. However, the THD values for H bridge topology are comparatively low when the load is split phase induction motor.

Incase the classical five level topology is employed to supply either static or dynamic loads, POD SPWM will lead to lesser THD in the output, specifically in case of higher modulation indices. An important observation pertains to the unsymmetrical SPWM method used in this paper. For modulation indices between 0.85 and 0.90, unsymmetrical SPWM yields the lowest THD values for voltage and current.

The H bridge topology can be used for supplying both static as well as dynamic loads especially since it yields superior results from the perspective of harmonic performances in the lower range of modulation indices. Since it requires lesser number of diodes its cost is also less as compared to the classical five level structure. Fewer components also result in reduction of circuit complexity. Another benefit of the H bridge structure is that it is free from the problem of voltage collapse.

VIII. REFERENCES

[1] J.S. Lai, and F.Z. Peng, "Multilevel converters – a new breed of converters," IEEE Trans. Ind. Appl., vol -32, pp. 509–517,1996.

[2] J. Rodriguez, J.S. Lai, and F. Zheng Peng, "Multilevel inverters; a survey of topologies, controls, and applications," IEEE Trans. Ind. Electron., vol.- 49, pp. 724–738, 2002.

[3] A. Nabae, I. Takahashi, and H. Akagi, "A new neutral-point clamped PWM inverter," IEEE Trans. Ind. Applicat., vol. IA-17, pp. 518–523, Sept./Oct. 1981.

[4] C. Hochgraf, R. Lasseter, D. Divan, and T. A. Lipo, "Comparison of multilevel inverters for static var compensation," in Conf.Rec. IEEE-IAS Annu. Meeting, pp. 921–928,1994.

[5] P. Hammond, "A new approach to enhance power quality for medium voltage ac drives," IEEE Trans. Ind. Applicat., vol. 33, pp. 202–208, Jan./Feb. 1997.

[6] E. Cengelci, S. U. Sulistijo, B. O. Woom, P. Enjeti, R. Teodorescu, and F. Blaabjerge, "A new medium voltage PWM inverter topology for adjustable speed drives," in Conf.Rec.IEEE-IASAnnu.Meeting, St.Louis, MO, pp. 1416–1423,1998.

[7] F. Z. Peng, "A generalized multilevel inverter topology with self voltage balancing," *IEEE Trans, Ind. Appl.*, vol. 37, no. 2, pp. 611-618, Mar./Apr. 2001.

[8] E. C. dos Santos Jr., J. H. G. Muniz, and E. R. C. da Silva, "2L3L Inverter," Power Electronics Conference (COBEP), 2011 Brazilian, Publication Year: 2011, Page(s): 924- 929.

[9] M. Singh, A. Agarwal, N. Kaira, "Performance evaluation of multilevel inverter with advance PWM control techniques," *Power Electronics (IICPE), 2012 IEEE 5th India International Conference on* , vol., no., pp.1,6, 6-8 Dec. 2012.

An Improved Dead-Time Compensation Scheme for Voltage Source Inverters Considering the Device Switching Transition Times

Anirudh Guha, G. Narayanan
Department of Electrical Engineering
Indian Institute of Science, Bangalore - 560012 INDIA
Email: aguha@ee.iisc.ernet.in; gnar@ee.iisc.ernet.in

Abstract—**Inverter dead-time, which is meant to prevent shoot-through fault, causes harmonic distortion and change in the fundamental voltage in the inverter output. Typical dead-time compensation schemes ensure that the amplitude of the fundamental output current is as desired, and also improve the current waveform quality significantly. However, even with compensation, the motor line current waveform is observed to be distorted close to the current zero-crossings. The IGBT switching transition times being significantly longer at low currents than at high currents is an important reason for this zero-crossover distortion. Hence, this paper proposes an improved dead-time compensation scheme, which makes use of the measured IGBT switching transition times at low currents. Measured line current waveforms in a 2.2 kW induction motor drive with the proposed compensation scheme are compared against those with the conventional dead-time compensation scheme and without dead-time compensation. The experimental results on the motor drive clearly demonstrate the improvement in the line current waveform quality with the proposed method.**

I. INTRODUCTION

Insulated gate bipolar transistor (IGBT) based pulse-width-modulated (PWM) voltage source inverters (VSI) (see Fig.1) are widely used in many applications. In each inverter leg,

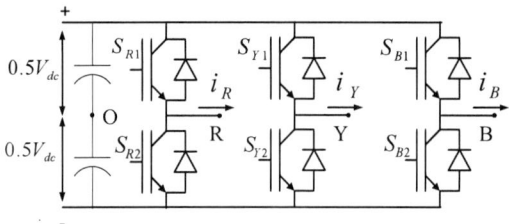

Fig. 1. A three-phase voltage source inverter [1].

a dead-time delay t_d is introduced between the gating pulses of the top and bottom switches for safe commutation of the IGBTs. The rising edges of the ideal, complementary gating signals S_{R1} and S_{R2} are delayed by t_d to obtain the delayed gating signals S_{R1d} and S_{R2d} as shown in Fig. 2. Both

This work is funded by the Department of Heavy Industry, Government of India, under a project titled " Off-line and Real-time Simulators for Electric Vehicle/ Hybrid Electric Vehicle Systems".

S_{R1d} and S_{R2d} are low during the dead-time interval. This

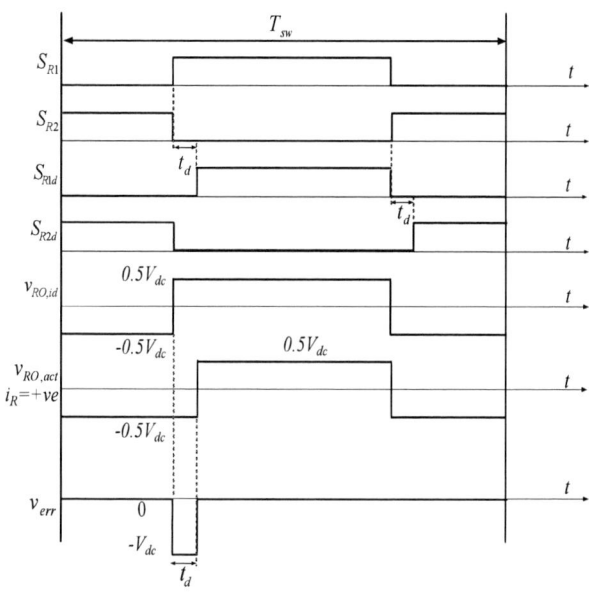

Fig. 2. Instantaneous dead-time error voltage over a switching cycle. The error voltage over a switching cycle is depicted for a positive load current polarity [1].

delays one of the edges (either the rising or trailing edge, depending on current polarity) of the pole voltage by time t_d. Hence, the actual pole voltage $v_{RO,act}$ deviates from the ideal pole voltage $v_{RO,id}$, resulting in an error voltage pulse in each switching cycle. The integral effect of these error volt-seconds results in deviations in fundamental output voltage and harmonic distortion [1]–[3]. These deviations are particularly significant at low modulation indices or low speeds of the motor drive [4].

Most existing dead-time compensation schemes [5], [6], which attempt to mitigate the undesirable effects of dead-time, assume the actual pole voltage $v_{RO,act}$ to be as shown in Fig.2. When switching a positive load current, only the rising edge of the pole voltage is delayed by t_d, and the position of the trailing edge is unchanged. Further, the switching transition

978-1-4799-6047-7/14 $31.00 © 2014 IEEE

times of the IGBT are assumed to be negligible compared to the dead-time. However, while switching low currents, the device transition times are quite significant [7], [8] as will be seen in section II. Hence, the effective error in the volt-seconds is not as indicated in Fig. 2 (i.e. this is not equal to $-V_{dc}t_d$). This could result in distortion of the current in the vicinity of the current zero crossing (if the compensation is not appropriate) [9]–[11]. The effective error in the inverter output volt-seconds, considering the impact of the device switching transitions, will be brought out in section III.

Based on the modified analysis of error voltage on account of dead-time in section III, an improved compensation scheme is proposed in section IV. The effectiveness of this compensation scheme is demonstrated experimentally on a 2.2 kW induction motor drive in section V.

II. MEASURED IGBT SWITCHING CHARACTERISTICS AT LOW CURRENT MAGNITUDES

The measured turn-off and turn-on characteristics of a 75A, 1200 V IGBT (SKM75GB123D), while switching a low current of 0.75A, are shown in Fig. 3(a) and Fig.3(b), respectively. The experimental set-up and procedure for measurement of such switching characteristics are explained in [7]. The measured switching characteristics of a high current IGBT (rated current 75A) at low currents ($< 10\%$ of the rated current) is also reported in [7].

The measured voltage delay times and voltage transition times for SKM75GB123D, when switching less than 10% of the device rated current, are shown plotted in Fig. 4 [7]. The turn-on voltage delay time t_{donv} and turn-off voltage delay time t_{doffv} represent the delays between the start of the transition of the gate-emitter voltage (v_{GE}) and the start of the transition of the collector-emitter voltage (v_{CE}) during turn-on and turn-off, respectively. The interval t_{fv} is the collector-emitter voltage fall time during turn-on, and the interval t_{rv} is the collector-emitter voltage rise time during turn-off. Among the four transition times, t_{donv} and t_{fv} do not change significantly with the load current being switched as observed in Fig. 4. However, t_{doffv} and t_{rv} increase as the load current being switched decreases. In particular, t_{rv} increases from about 150 ns when switching a load current of 7.5A to around 1300 ns when switching a current of 0.5A. The t_{rv} is about $1\mu s$ at a load current of 0.75A as shown by Fig. 3(a). The impact of the increased t_{doffv} and t_{rv} on the distortion caused by dead-time at low currents will be analyzed in the following section.

III. ANALYSIS OF DEAD-TIME ERROR VOLTAGES AT LOW CURRENTS

The conventional dead-time compensation scheme assumes the actual pole voltage transitions to be instantaneous as indicated by $v_{RO,act}$ in Fig. 2. More precisely, the IGBT voltage transition times are assumed negligible in comparison with the dead time t_d. This is true at high currents where the voltage transition times are a few 100 ns as shown by Fig. 4, compared to the dead time of a few microseconds. However,

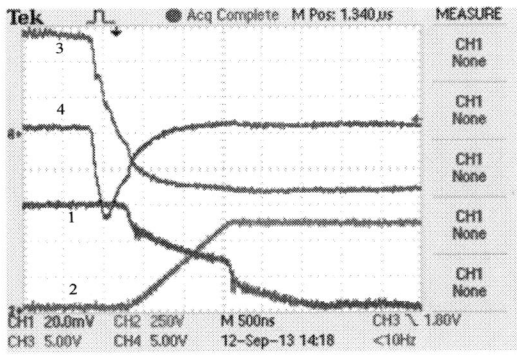

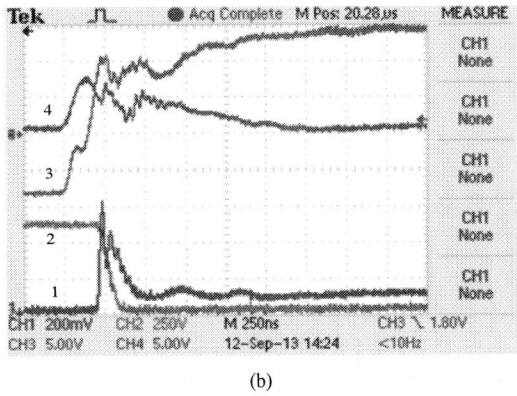

Fig. 3. Measured switching characteristics of SKM75GB123D at 0.75A current level with $R_g = 22\ \Omega$ (a) turn-off characteristics, Trace 1: collector current i_c (0.257 A/div), Trace 2: collector emitter voltage v_{CE} (250 V/div), Trace 3: gate emitter voltage v_{GE} (5 V/div), Trace 4: gate current i_g (0.227 A/div), time scale: 500 ns/div (b) turn-on characteristics, Trace 1: collector current i_c (2.57 A/div), time scale: 250 ns/div, other traces are as defined in Fig. 3(a).

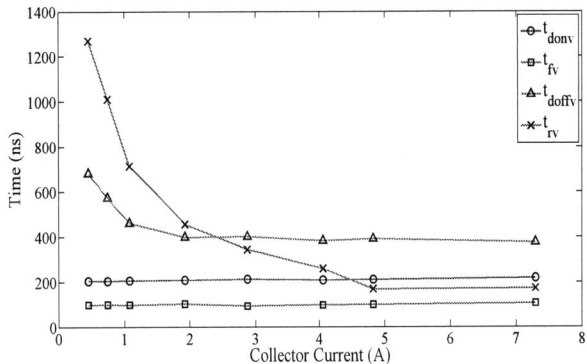

Fig. 4. Variation of voltage delay and collector-emitter voltage transition times of SKM75GB123D, with the collector current, with $R_g = 22\ \Omega$ [7].

as seen from the same figure, the device voltage rise time during turn-off t_{rv} is about a microsecond at low currents, and is no longer negligible. Hence the actual switching times are considered in the analysis of dead-time error voltage at low currents in Fig. 5 (shown for a positive load current polarity).

The first switching event, shown in Fig. 5, is a diode to IGBT transition. The diode continues to conduct during the dead-time duration t_d and the turn-on voltage delay interval t_{donv}, after which it takes time t_{fv} for the IGBT voltage v_{CE}

to fall to zero. Hence, the dead-time delay and the device turn-on voltage characteristics result in a negative error voltage pulse. The second switching transition, depicted in Fig. 5, is an IGBT turn-off event; this results in a positive error voltage pulse due to the turn-off voltage delay t_{doffv} and v_{CE} rise time t_{rv}. Similar analysis can be presented for a negative load

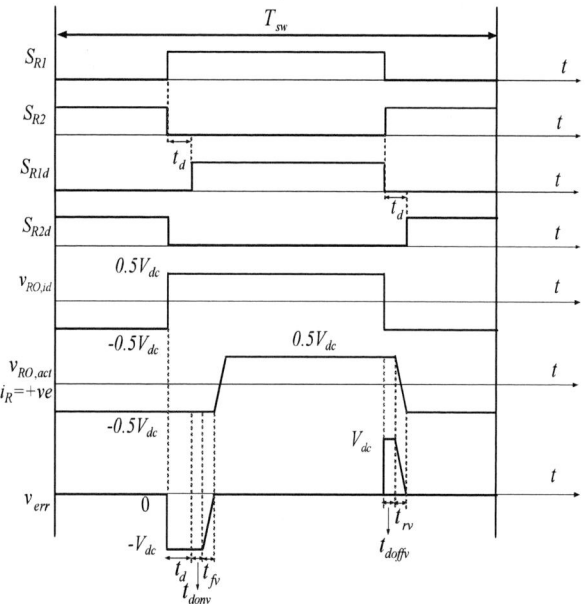

Fig. 5. Dead-time error voltage at low currents when the switching transition times are significant.

current polarity. The error voltage averaged over one carrier cycle is given in equation (1).

$$v_{e,avg} = -V_{dc}f_{sw}sign(i_R)[t_d + (t_{donv} - t_{doffv}) \\ + 0.5(t_{fv} - t_{rv})] \quad (1)$$

where V_{dc} is the dc bus voltage, $sign(i_R)$ indicates polarity of the load current, t_d is the dead-time duration, and f_{sw} is the switching frequency. When the device transition times in equation (1) are ignored, it is the typical average error voltage when switching higher current magnitudes. Conventional compensation schemes attempt to mitigate this error voltage.

It is observed from Fig. 4 that the turn-off voltage delay t_{doffv} is longer than the turn-on voltage delay t_{donv}; the difference is particularly significant at low currents. This also contributes to the error volt-seconds on account of dead-time.

IV. Proposed Dead-Time Compensation Scheme

The error volt-seconds in Fig. 5 can be compensated by adding a compensating signal given in equation (2), to the ideal modulating signal.

$$v_{comp,avg} = V_{dc}f_{sw}sign(i_R)[t_d + (t_{donv} - t_{doffv}) \\ + 0.5(t_{fv} - t_{rv})] \quad (2)$$

The compensating voltage, normalized with respect to product of dc bus voltage and switching frequency, is the effective compensation time t_{comp} given by equation (3).

$$t_{comp} = t_d + (t_{donv} - t_{doffv}) + 0.5(t_{fv} - t_{rv}) \quad (3)$$

The compensation time t_{comp} is a function of the device voltage delay and voltage transition times as seen from equation (3). These device delay and transition times, in turn, vary with load current as seen from Fig. 4. Hence t_{comp} is a function of load current. The variation of t_{comp} with load current is shown in Fig. 6, considering a dead-time t_d of 3.2 μs. It is

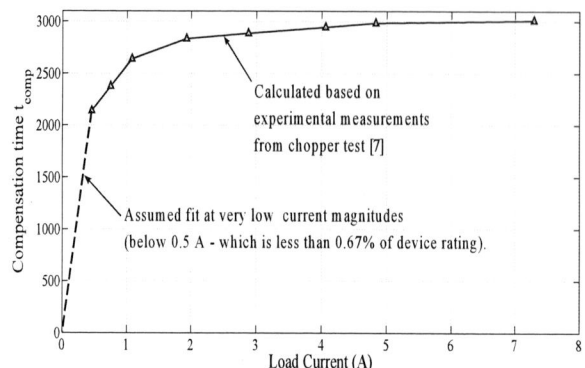

Fig. 6. Variation of the effective compensation time with load current magnitude. The equivalent compensating voltage mitigates the error voltage at the inverter output on account of dead-time including the impact of device switching characteristics.

observed from Fig. 6 that the effective compensation time is about $2.2\mu s$ for a load current of 0.5 A. This increases to about $3\mu s$ and remains at the same value for load currents of 6A and higher. At currents lower than 0.5A, experimental measurements of the switching characteristics are difficult to obtain with accuracy using the chopper test [7]. Hence t_{comp} is assumed to vary linearly between 0 and $2.2\mu s$ for load currents between 0 and 0.5A, as indicated in Fig. 6. Linear curve fits of the compensation time are considered for different ranges of load current magnitudes for practical implementation on the digital platform.

In this proposed method, the compensating voltage given by equation (2) is added to the ideal sinusoidal modulating signal. The polarity of this compensating signal depends on the polarity of the load current. The magnitude of this compensating signal also varies with the magnitude of the load current. When the device switching transitions times are neglected, the compensating signal is a square wave that is in-phase with the line current [5], [6]. With the proposed compensation method, the compensating signal is close to a trapezoidal signal with rounded corners as will be shown by the experimental results in the following section.

V. Experimental Results

A 3-phase, 2-level voltage source inverter which uses 75A, 1200V IGBT modules (SKM75GB124D) is used in the experiment. An inverter dc bus of 600V, a dead-time of 3.2 μs [12] and a switching frequency of 5 kHz are considered. A 415V, 2.2 kW, 3-phase squirrel cage induction motor is driven using the afore-mentioned voltage source inverter using a sine-triangle modulation scheme. A TMS320F2812 based digital controller is used to control the inverter. Constant voltage to frequency (V/f) control is implemented without dead-time compensation, with conventional dead-time compensation and with the proposed compensation scheme. Since dead-time distortion is prevalent at low speeds, experimental results are presented at fundamental frequencies of 5 Hz and 10 Hz in Fig. 7 to Fig. 10.

Without dead-time compensation, the motor current (trace 2) with a sinusoidal modulating signal (trace 1) is shown in Fig. 7(a), for a stator fundamental frequency of 5 Hz and a corresponding modulation index of 0.1. Fig. 7(a) shows the motor current to be quite distorted. An expanded view of the current in the vicinity of the zero crossing is shown in Fig.7(b); the current rise is certainly not linear close to the zero-crossing. The FFT of the current, given in Fig. 7(c), and the magnified view of the FFT in Fig. 7(d) indicate the presence of low-order odd harmonics. The 5th and 7th harmonics are the dominant harmonics, while other higher order harmonics are observed to be negligible.

When the drive is run with a conventional dead-time compensation scheme at 5 Hz stator frequency, the measured motor current is shown in Fig. 8(a) and its magnified view is shown in Fig. 8(b). A square wave compensating signal (trace 3) in-phase with the current (trace 2) is added to the ideal sinusoidal modulating signal (of modulation index 0.1) to obtain the modulating signal indicated by trace 1 in Fig. 8(a). With this modulating signal, there is an increase in the current amplitude (see trace 2) compared to the case without compensation. The FFT of the current in Fig. 8(c) indicates the fundamental current has increased to 2.6A with compensation as against 1.6A without compensation (see Fig. 7(c)). However, the zero cross-over distortion with conventional dead-time compensation in Fig.8(a) is worse than that without compensation in Fig. 7(a) at this operating point. The expanded view of the current zero crossing in Fig. 8(b) shows the zero-crossover distortion more clearly. The change in slope of the current waveform when it changes polarity is very much pronounced in this case. The magnified view of the FFT in Fig. 8(d) clearly indicates the presence of 5th and 7th order harmonics, as well as 11th and 13th order harmonics. An advanced compensation scheme is hence necessary to mitigate this distortion.

With the proposed dead-time compensation scheme based on the desired compensation profile shown in Fig. 6, the motor current and the magnified view of the current zero crossing are shown in Fig. 9(a) and Fig. 9(b), respectively. Fig. 9(a) shows that there is no discontinuity or abrupt change in the modulating signal (trace 1) in this case. The compensating signal (trace 3), which is in-phase with the line current (trace 2), is seen to be approximately a trapezoid, but with smooth or rounded corners. With this compensating signal, the line current amplitude is comparable to the current amplitude in Fig. 8(a). This is confirmed by the FFT in Fig. 9(c) which indicates a fundamental current magnitude of about 2.6 A. Further, the current in Fig. 9(a) is observed to be sinusoidal with no visible distortion at the current zero-crossing as shown by Fig. 9(a) and Fig. 9(b). The low-order harmonics in the current are reduced significantly as confirmed by Fig. 9(d), proving that the proposed scheme is quite effective.

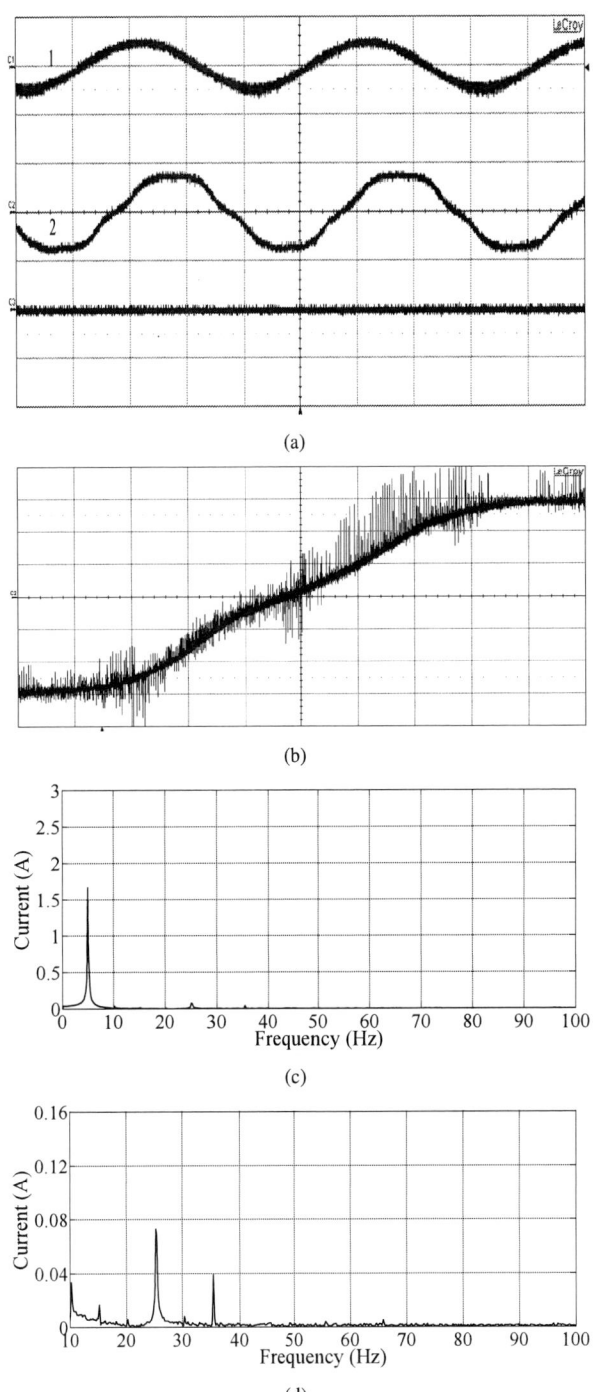

Fig. 7. (a) Measured motor current without dead-time compensation at a fundamental frequency of 5 Hz. Trace 1: R-phase modulating signal (1V/div), trace 2: R-phase current (2A/div), time scale: (50ms/div). (b) zoomed R-phase current near the zero crossing (0.5A/div), time scale: (10ms/div) (c) FFT of the R-phase current (d) zoomed FFT of R-phase current.

978-1-4799-6047-7/14 $31.00 © 2014 IEEE

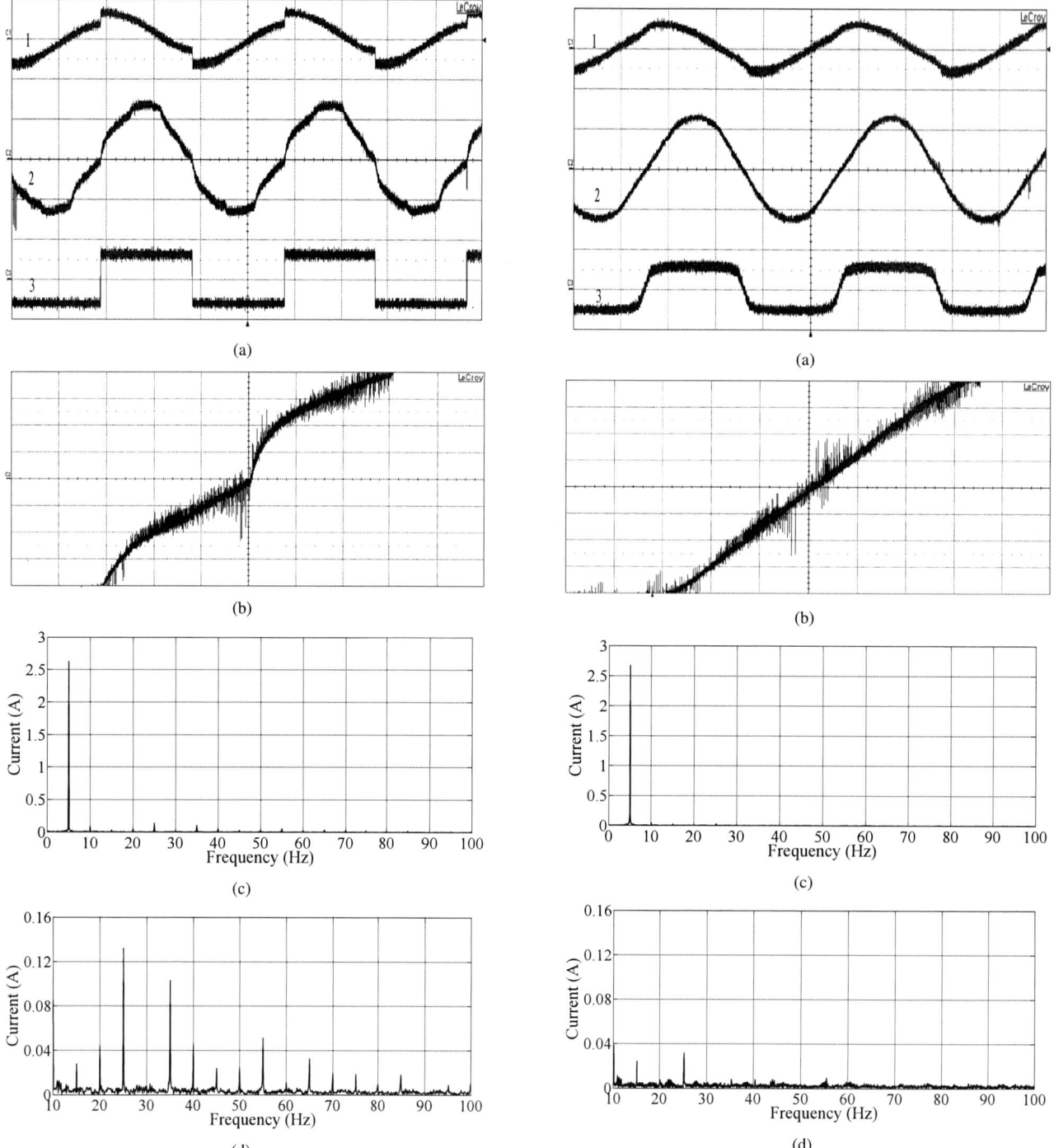

Fig. 8. (a) Measured motor current with the conventional dead-time compensation scheme at a fundamental frequency of 5 Hz. Trace 1: R-phase modulating signal (1V/div), trace 2: R-phase current (2A/div), trace 3: R-phase compensating signal (0.25V/div), time scale: (50ms/div) (b) zoomed R-phase current near the zero crossing (0.5A/div), time scale: (10ms/div) (c) FFT of the R-phase current (d) zoomed FFT of R-phase current.

Fig. 9. (a) Measured motor current with the proposed dead-time compensation scheme at a fundamental frequency of 5 Hz. Trace 1: R-phase modulating signal (1V/div), trace 2: R-phase current (2A/div), trace 3: R-phase compensating signal (0.25V/div), time scale: (50ms/div) (b) zoomed R-phase current near the zero crossing (0.5A/div), time scale: (10ms/div) (c) FFT of the R-phase current (d) zoomed FFT of R-phase current.

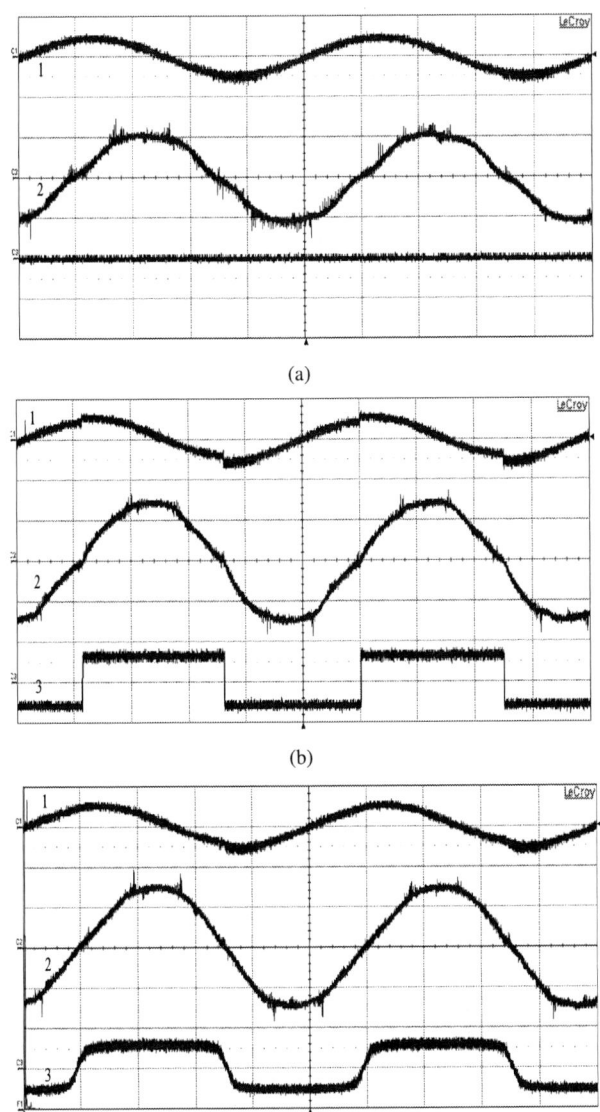

(a)

(b)

(c)

Fig. 10. Measured motor currents at a fundamental frequency of 10 Hz (a) without dead-time compensation (b) with conventional dead-time compensation (c) with proposed dead-time compensation. trace 1: R-phase modulating signal (2V/div), trace 2: R-phase current (2A/div), trace 3: R-phase compensating signal (0.25V/div), time scale: (20ms/div).

Improvement in the line current waveforms with the proposed compensation scheme as compared to the conventional compensation scheme is also verified at a fundamental frequency of 10 Hz. The motor currents without compensation and with conventional compensation are found to be distorted as shown by Fig. 10(a) and Fig. 10(b), respectively. With the proposed compensation scheme, the current waveform quality is clearly improved as observed in Fig. 10(c).

VI. CONCLUSION

An advanced dead-time compensation scheme is proposed, considering the actual switching transition times of the device. The dead-time compensation profile (i.e effective compensation time or the equivalent compensating voltage) is calculated as a function of load current, based on experimental measurements. The wave-shape of the compensating voltage is found to be close to trapezoidal with smoothened corners.

The measured motor line currents with the advanced dead-time compensation scheme are compared against currents without dead-time compensation and with the conventional compensation scheme. The quality of the line current in the vicinity of the zero-crossing is improved significantly. The low-frequency distortion in the current is very much reduced.

REFERENCES

[1] A. Guha and G. Narayanan, "Average modelling of a voltage source inverter with dead-time in a synchronous reference frame," in *Proc. IEEE Innovative Smart Grid Technologies-Asia (ISGT Asia), 2013*, Bangalore, India, Nov 2013, pp. 1–6.

[2] Y. Murai, T. Watanabe, and H. Iwasaki, "Waveform distortion and correction circuit for pwm inverters with switching lag-times," *IEEE Trans. Ind. Appl.*, vol. IA-23, no. 5, pp. 881–886, 1987.

[3] S.-G. Jeong and M.-H. Park, "The analysis and compensation of dead-time effects in pwm inverters," *IEEE Trans. Ind. Electron.*, vol. 38, no. 2, pp. 108–114, 1991.

[4] J.-W. Choi and S.-K. Sul, "A new compensation strategy reducing voltage/current distortion in pwm vsi systems operating with low output voltages," *IEEE Trans. Ind. Appl.*, vol. 31, no. 5, pp. 1001–1008, 1995.

[5] A. Guha, A. Tripathi, and G. Narayanan, "Experimental study on dead-time induced oscillations in a 100-kw open-loop induction motor drive," in *Proc. National Power Electronics Conference NPEC 2013, IIT Kanpur*, Kanpur, India, Dec 2013, pp. 1–6.

[6] D. Leggate and R. J. Kerkman, "Pulse-based dead-time compensator for pwm voltage inverters," *IEEE Trans. Ind. Electron.*, vol. 44, no. 2, pp. 191–197, 1997.

[7] A. Guha, A. Datta, C. Rangesh, and G. Narayanan, "Experimental investigation on switching characteristics of high-current insulated gate bipolar transistors at low currents," in *Proc. International Conference on Electrical Energy Sytems, ICEES 2014*, Chennai, India, Jan 2014, pp. 1–6.

[8] K. Yamamoto, K. Shinohara, and H. Ohga, "Effect of parasitic capacitance of power device on output voltage deviation during switching dead-time in voltage-fed pwm inverter," in *Proc. IEEE Power Conversion Conf.*, vol. 2, Nagaoka, Japan, Aug 1997, pp. 777–782.

[9] J. W. Choi and S. K. Sul, "Inverter output voltage synthesis using novel dead time compensation," *IEEE Trans. Power Electron.*, vol. 11, no. 2, pp. 221–227, 1996.

[10] Y. Park and S.-K. Sul, "A novel method utilizing trapezoidal voltage to compensate for inverter nonlinearity," *IEEE Trans. Power Electron.*, vol. 27, no. 12, pp. 4837–4846, 2012.

[11] M. Herran, J. Fischer, S. Gonzalez, M. Judewicz, and D. Carrica, "Adaptive dead-time compensation for grid-connected pwm inverters of single-stage pv systems," *IEEE Trans. Power Electron.*, vol. 28, no. 6, pp. 2816–2825, 2013.

[12] Infineon, *Application note AN2007-04, Deadtime calculation for IGBT Modules*, 2007. [Online]. Available: http://www.infineon.com

978-1-4799-6047-7/14 $31.00 © 2014 IEEE

Robust Nonlinear Observer Design for Twin Rotor Control System

Abhinav Pratap Singh
Electrical Engineering Department
National Institute of Technology Kurukshetra
Kurukshetra, India
E-mail: abhipratap5123@gmail.com

Bhanu Pratap
Electrical Engineering Department
National Institute of Technology Kurukshetra
Kurukshetra, India
E-mail: bhanumnnit@gmail.com

Abstract— **This paper presents a robust nonlinear observer for twin rotor control system (TRCS). The TRCS is a class of nonlinear uncertain system having unstable and coupled dynamics. The dynamics of TRCS are similar to a helicopter in certain aspects. Due to high coupling effect, the controller design of TRCS is challenging as well as interesting. The proposed robust observer is designed using sliding mode approach. Coulomb friction is introduced in the TRCS to demonstrate the robustness of the proposed observer. Finally simulation results highlight the performance of the proposed observer to compensate the nonlinear friction term in the system.**

Keywords— *Coulomb friction; Nonlinear coupled system; Observer; Sliding mode control; Twin rotor control system.*

I. INTRODUCTION

IN recent years there has been considerable research related to the helicopter flight control problem [1]. Unlike linear controllers, the controllers designed for nonlinear systems are unique to the system nonlinearities. Some of the commonly used nonlinear control techniques are feedback linearization control [2], sliding mode control [3] and backstepping control techniques [4].

TRCS [5] is a mechanical setup designed for control experiments in laboratories. In certain aspects, its behavior resembles similar to a helicopter having unstable, nonlinear and coupled dynamics [6], [7]. The modeling and controller design of TRCS has been discussed in [8]–[11]. In the literature, most of the papers deal with simulation studies and all states variables are assumed to be available but it's not feasible practically. The controller design for nonlinear systems generally needs the knowledge of actual or estimated values of all states. One of the solutions is to design an observer. Different kinds of observers and observer based controllers are presented in [12]–[14].

The robust observer design problem for nonlinear/uncertain systems subject to external disturbances has been a topic of considerable interest. The error between the observer output and the system output is fed back via a discontinuous switched signal instead of feeding it back linearly. The robust observer has a unique feature of generating sliding mode on the error between the system's output and the observer's output [15]–[17].

Sliding mode control (SMC) is well known as an effective robust control approach for nonlinear dynamic systems [18], [19]. The main features of SMC are its insensitivity to parameter variations, ability to remove external disturbance and fast response [20]. Other remarkable advantages of this control approach are the simplicity of its design and implementation and the order reduction of the closed-loop systems.

Several studies have shown the destabilizing effect of nonlinear friction on many control systems for high quality servomechanisms, which can lead to severe tracking errors and limit cycles, chattering, and excessive noise etc. [21]–[23].

In this paper, the simulation studies of a robust nonlinear observer for TRCS are presented. The nonlinearities of the TRCS are considered as unknown but the bounds of the nonlinearities are known. The simulation results obtained reveal that the proposed observer shows the satisfactory performance. The paper is arranged as follows. In Section II, the modeling of TRCS is introduced. The problem formulation is given in Section III. In Section IV, robust nonlinear observer design is proposed. Section V presents the simulation results. Finally conclusions are made in the last section.

II. MODELING OF TRCS

The TRCS is a mechanical setup having multivariable, nonlinear and strongly coupled dynamics. It is a sixth order system with two degrees of freedom i.e. pitch (elevation) and yaw (azimuth) angles as shown in Fig.1a.

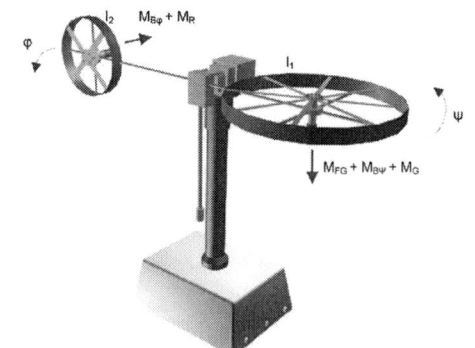

Fig.1a. The twin rotor control system

978-1-4799-6047-7/14 $31.00 © 2014 IEEE

The complete dynamics of the TRCS [5] can be represented in the form of differential equations are as follows.

$$\frac{d}{dt}\psi = \dot{\psi}$$

$$\frac{d}{dt}\dot{\psi} = \frac{a_1}{I_1}\tau_1^2 + \frac{b_1}{I_1}\tau_1 - \frac{M_g}{I_1}\sin(\psi) + \frac{0.0326}{2I_1}\sin(2\psi)\dot{\varphi}^2$$

$$-\frac{B_{1\psi}}{I_1}\dot{\psi} - \frac{k_{gy}}{I_1}\cos(\psi)\dot{\varphi}\left(a_1\tau_1^2 + b_1\tau_1\right)$$

$$\frac{d}{dt}\varphi = \dot{\varphi}$$

$$\frac{d}{dt}\dot{\varphi} = \frac{a_2}{I_2}\tau_2^2 + \frac{b_2}{I_2}\tau_2 - \frac{B_{1\varphi}}{I_2}\dot{\varphi} - \frac{1.75}{I_2}k_c\left(a_1\tau_1^2 + b_1\tau_1\right)$$

$$\frac{d}{dt}\tau_1 = -\frac{T_{10}}{T_{11}}\tau_1 + \frac{k_1}{T_{11}}u_1$$

$$\frac{d}{dt}\tau_2 = -\frac{T_{20}}{T_{21}}\tau_2 + \frac{k_2}{T_{21}}u_2$$

$$y = \begin{bmatrix} \psi & \varphi \end{bmatrix}^T$$

$$(1)$$

where ψ & φ are pitch (elevation) & yaw (azimuth) angles; $\dot{\psi}$ & $\dot{\varphi}$ are angular velocity across pitch & yaw angles; τ_1 & τ_2 are Momentum of main & tail rotors and u_1 & u_2 are voltages applied to main & tail rotors respectively. The system parameters of the TRCS are given in Table 1 [5].

Table 1: TRCS parameters

Parameters	Values
I_1 = Moment of inertia of vertical rotor	$6.8\times10^{-2}\,kg-m^2$
I_2 = Moment of inertia of horizontal rotor	$2\times10^{-2}\,kg-m^2$
a_1 = Static characteristic parameter	0.0135
b_1 = Static characteristic parameter	0.0924
a_2 = Static characteristic parameter	0.02
b_2 = Static characteristic parameter	0.09
M_g = Gravity momentum	$0.32\,N-m$
$B_{1\psi}$ = Friction momentum function parameter	$6\times10^{-3}\,N-m-s/rad$
$B_{1\varphi}$ = Friction momentum function parameter	$1\times10^{-1}\,N-m-s/rad$
k_{gy} = Gyroscopic momentum parameter	$0.05\,s/rad$
k_1 = Motor 1 gain	1.1
k_2 = Motor 2 gain	0.8
T_{11} = Motor 1 denominator parameter	1.1
T_{10} = Motor 1 denominator parameter	1
T_{21} = Motor 2 denominator	1
T_{20} = Motor 1 denominator parameter	1

T_p = Cross reaction momentum parameter	2
T_0 = Cross reaction momentum parameter	3.5
k_c = Cross reaction momentum gain	−0.2
u_1 and u_2 input voltages of rotors	±2.5

III. PROBLEM FORMULATION

The complexity of TRCS (1) is increased even further by adopting a highly nonlinear Coulomb friction term. The system (1) in the form of state space can be rewritten as,

$$\begin{aligned}
\dot{X}_1 &= X_2 \\
\dot{X}_2 &= A_1 X_2 + A_2 X_3 + g(X) \\
\dot{X}_3 &= A_3 X_3 + Bu - BT_c \operatorname{sgn}(X_3) \\
y &= X_1
\end{aligned}$$

$$(2)$$

where $X_1 = \begin{bmatrix} \psi & \varphi \end{bmatrix}^T$, $X_2 = \begin{bmatrix} \dot{\psi} & \dot{\varphi} \end{bmatrix}^T$, $X_3 = \begin{bmatrix} \tau_1 & \tau_2 \end{bmatrix}^T$ are state vectors, $u = \begin{bmatrix} u_1 & u_2 \end{bmatrix}^T$ is input vector, $y = \begin{bmatrix} \psi & \varphi \end{bmatrix}^T$ is output vector, $T_c = \operatorname{diag}\begin{bmatrix} T_{c1} & T_{c2} \end{bmatrix}$ is the Coulomb friction applied to the shaft of rotors and matrices A_1, A_2, A_3, B and $g(X)$ are given by

$$A_1 = \begin{bmatrix} -\dfrac{B_{1\psi}}{I_1} & 0 \\ 0 & -\dfrac{B_{1\varphi}}{I_2} \end{bmatrix},\; A_2 = \begin{bmatrix} \dfrac{b_1}{I_1} & 0 \\ -\dfrac{1.75}{I_2}k_c b_1 & \dfrac{b_2}{I_2} \end{bmatrix},$$

$$A_3 = \begin{bmatrix} -\dfrac{T_{10}}{T_{11}} & 0 \\ 0 & -\dfrac{T_{20}}{T_{21}} \end{bmatrix},\; B = \begin{bmatrix} \dfrac{k_1}{T_{11}} & 0 \\ 0 & \dfrac{k_2}{T_{21}} \end{bmatrix},$$

$$g(X) = \begin{bmatrix} \left\{ \dfrac{a_1}{I_1}\tau_1^2 - \dfrac{M_g}{I_1}\sin(\psi) + \dfrac{0.0326}{2I_1}\sin(2\psi)\dot{\varphi}^2 \right\} \\ -\dfrac{k_{gy}}{I_1}\cos(\psi)\dot{\varphi}\left(a_1\tau_1^2 + b_1\tau_1\right) \\ \left\{ \dfrac{a_2}{I_2}\tau_2^2 - \dfrac{1.75}{I_2}k_c a_1\tau_1^2 \right\} \end{bmatrix}.$$

Here, $g(X)$ is assumed to be unknown nonlinearity whose bounds are known.

The objective of this paper is to design a robust nonlinear observer for TRCS that forces the observer output $\hat{X}_i$ to track the plant states X_i i.e., $\lim_{t\to\infty}\left(X_i - \hat{X}_i\right) = 0$, where $i = 1,\ldots,3$.

IV. ROBUST NONLINEAR OBSERVER DESIGN

The robust nonlinear observer for the TRCS is given by

$$
\left.
\begin{aligned}
\dot{\hat{X}}_1 &= \hat{X}_2 + \alpha_1 \varepsilon_1 + K_1 \operatorname{sgn}(\varepsilon_1) \\
\dot{\hat{X}}_2 &= A_1 \hat{X}_2 + A_2 \hat{X}_3 + \hat{g}(\hat{X}) + \alpha_2 \varepsilon_1 + K_2 \operatorname{sgn}(\varepsilon_1) \\
\dot{\hat{X}}_3 &= A_3 \hat{X}_3 + Bu + \alpha_3 \varepsilon_1 + K_3 \operatorname{sgn}(\varepsilon_1) \\
y &= \hat{X}_1
\end{aligned}
\right\}
\quad (3)
$$

where, $\hat{X}_1 = \begin{bmatrix} \hat{\psi} & \hat{\varphi} \end{bmatrix}^T$, $\hat{X}_2 = \begin{bmatrix} \dot{\hat{\psi}} & \dot{\hat{\varphi}} \end{bmatrix}^T$, $\hat{X}_3 = \begin{bmatrix} \hat{\tau}_1 & \hat{\tau}_2 \end{bmatrix}^T$ are the estimate of X_1, X_2 and X_3 respectively, $\varepsilon_1 = X_1 - \hat{X}_1$ is the sliding surface, $K_i = \operatorname{diag}\begin{bmatrix} K_{i1} & K_{i2} \end{bmatrix}$ and $\alpha_i = \operatorname{diag}\begin{bmatrix} \alpha_{i1} & \alpha_{i2} \end{bmatrix}$ are constants with $i = 1, \dots, 3$.

The estimation errors $\varepsilon_i = X_i - \hat{X}_i$ $(i = 1, \dots, 3)$ and error dynamics are obtained by subtracting (3) from (2) given as

$$
\dot{\varepsilon}_1 = \varepsilon_2 - \alpha_1 \varepsilon_1 - K_1 \operatorname{sgn}(\varepsilon_1) \quad (4a)
$$

$$
\dot{\varepsilon}_2 = A_1 \varepsilon_2 + A_2 \varepsilon_3 + \tilde{g} - \alpha_2 \varepsilon_1 - K_2 \operatorname{sgn}(\varepsilon_1) \quad (4b)
$$

$$
\dot{\varepsilon}_3 = A_3 \varepsilon_3 - \alpha_3 \varepsilon_1 - K_3 \operatorname{sgn}(\varepsilon_1) \quad (4c)
$$

where, $\tilde{g} = g(X) - \hat{g}(\hat{X})$.

The error dynamics (4) can be rewritten in the matrix form, given as

$$
\begin{bmatrix} \dot{\varepsilon}_1 \\ \dot{\varepsilon}_2 \\ \dot{\varepsilon}_3 \end{bmatrix} = \underbrace{\begin{bmatrix} -\alpha_1 & I & 0 \\ -\alpha_2 & A_1 & A_2 \\ -\alpha_3 & 0 & A_3 \end{bmatrix}}_{\bar{A}} \begin{bmatrix} \varepsilon_1 \\ \varepsilon_2 \\ \varepsilon_3 \end{bmatrix} + \begin{bmatrix} 0 \\ \tilde{g} \\ 0 \end{bmatrix} - \begin{bmatrix} K_1 \\ K_2 \\ K_3 \end{bmatrix} \operatorname{sgn}(\varepsilon_1) \quad (5)
$$

where I is identity matrix and α_1, α_2, α_3 are positive constants, chosen such that the Eigen values of the matrix $\bar{A}$ are at desired locations in the left-half of the complex plane. If K_1 is sufficiently large, then

$$
\varepsilon_1 \left(\varepsilon_2 - \alpha_1 \varepsilon_1 - K_1 \operatorname{sgn}(\varepsilon_1) \right) \le 0. \quad (6)
$$

Then, the state estimation error ε_1 would converge in finite time to the sliding surface $\varepsilon_1 = 0$. Using the concept of equivalent control [25], we get

$$
\operatorname{sgn}(\varepsilon_1) = \left(K_1^{-1} \varepsilon_2 \right). \quad (7)
$$

Using (7), (4b) and (4c) becomes,

$$
\dot{\varepsilon}_2 = A_1 \varepsilon_2 + A_2 \varepsilon_3 + \tilde{g} - K_2 \left(K_1^{-1} \varepsilon_2 \right) \quad (8a)
$$

$$
\dot{\varepsilon}_3 = A_3 \varepsilon_3 - K_3 \left(K_1^{-1} \varepsilon_2 \right). \quad (8b)
$$

Now defining, $\varepsilon = \begin{bmatrix} \varepsilon_2 & \varepsilon_3 \end{bmatrix}^T$, then (8) becomes

$$
\dot{\varepsilon} = A_o \varepsilon + \delta \quad (9)
$$

where $A_o = \begin{bmatrix} A_1 - K_1^{-1} K_2 & A_2 \\ K_1^{-1} K_3 & A_3 \end{bmatrix}$ and $\delta = \begin{bmatrix} \tilde{g} & 0 \end{bmatrix}^T \le \gamma \|\varepsilon\|_2$.

The values of K_i $(i = 1, \dots, 3)$ are chosen such that the matrix A_o is Hurwitz.

Consider a Lyapunov function

$$
V_o = \frac{1}{2} \varepsilon^T P \varepsilon \quad (10)
$$

where, P is a positive definite matrix satisfying

$$
A_o^T P + P A_o = -Q \quad (11)
$$

where Q is positive definite matrix.

Differentiating (10)

$$
\dot{V}_o = \varepsilon^T P \dot{\varepsilon}. \quad (12)
$$

Substituting (9) and (11) in (12) gives

$$
\dot{V}_o = -\frac{1}{2} \varepsilon^T Q \varepsilon + \varepsilon^T \delta. \quad (13)
$$

Now, we can express (13) as,

$$
\dot{V}_o \le -\frac{1}{2} \lambda_{\min} Q \|\varepsilon\|^2 + \gamma \|\varepsilon\|^2. \quad (14)
$$

This assures the exponential convergence of the estimation error ε to zero.

V. SIMULATION RESULTS

The complete block diagram of TRMS plant with proposed robust observer is shown in Fig.1b.

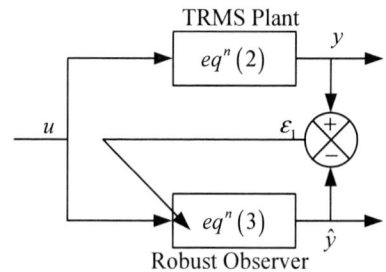

Fig. 1b Block diagram of robust observer for TRMS

The TRCS is operated in the open loop with the main and tail

rotor inputs as

$$u_1 = u_2 = 0.5\sin(0.6\,t) + 0.3\sin(0.5\,t) + 0.2\sin(0.2\,t).$$

The initial conditions of the plant and observer are $\begin{bmatrix} 0 & 0 & 0 & 0 & 0 & 0 \end{bmatrix}$ and $\begin{bmatrix} 0.1 & 0 & 0.1 & 0 & 0 & 0 \end{bmatrix}$ respectively. The simulation studies of nonlinear robust observer are done by adding a coulomb friction term in the TRCS plant given in (2). The values of K_1, K_2 and K_3, are chosen such that the matrix A_o is Hurwitz given as $K_1 = K_2 = K_3 = \begin{bmatrix} 0.1 & 0 \\ 0 & 0.1 \end{bmatrix}$. Similarly the values of α_1, α_2 and α_3, are chosen such that the matrix $\bar{A}$ is Hurwitz given as $\alpha_1 = \alpha_2 = \alpha_3 = \begin{bmatrix} 1 & 0 \\ 0 & 1 \end{bmatrix}$. The gain of Coulomb friction is taken as $T_c = \begin{bmatrix} 0.2 & 0 \\ 0 & 0.2 \end{bmatrix}$. Stable dynamics is therefore assured under the boundedness of δ.

V.A. Observer without Coulomb friction in TRCS

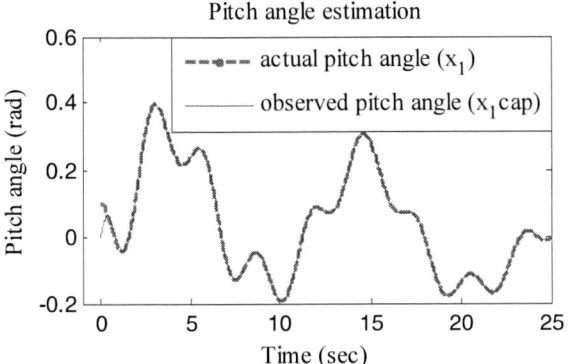

Fig. 2 Actual and observed pitch angle (ψ and $\hat{\psi}$)

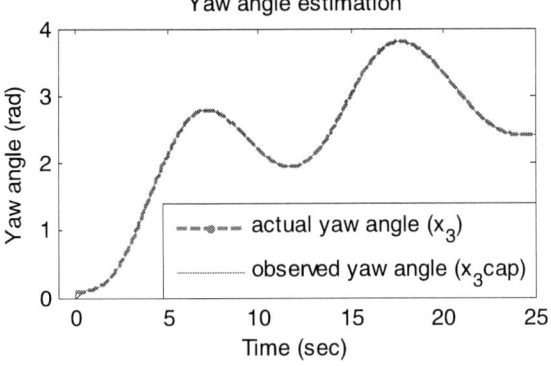

Fig. 3 Actual and observed yaw angle (φ and $\hat{\varphi}$)

The robust nonlinear observer design parameters (K_1, K_2, K_3, α_1, α_2 and α_3), used for simulation are chosen based on

the stability analysis. The tracking of actual and observed pitch and yaw angles without Coulomb friction are presented in Fig. 2 and Fig. 3. From the figures it can be concluded that the estimate states converge to the actual state very quickly. The estimation errors between the actual and observed pitch and yaw angles without Coulomb friction are shown in Fig. 4 and Fig. 5 respectively. In the next subsection, results included the Coulomb friction in the TRCS.

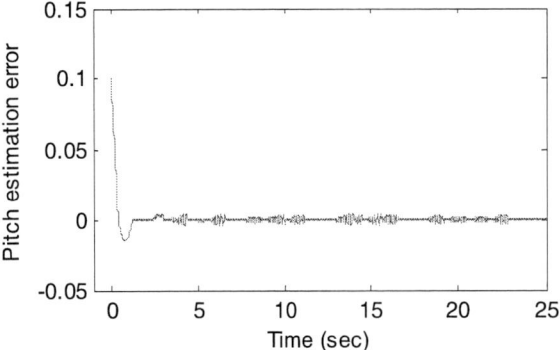

Fig. 4 Pitch angle estimation error ($\psi - \hat{\psi}$)

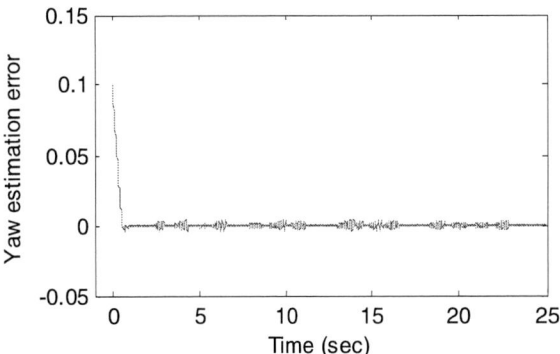

Fig. 5 Yaw angle estimation error ($\varphi - \hat{\varphi}$)

V.B. Observer with Coulomb friction in TRCS

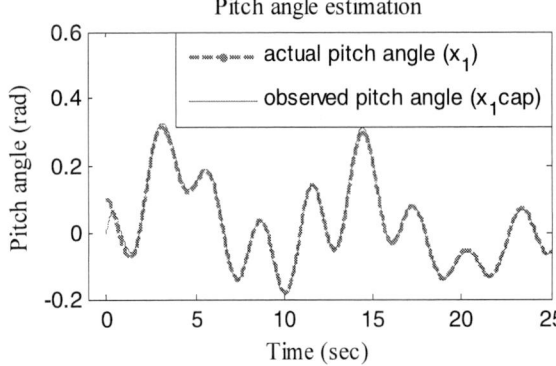

Fig. 6 Actual and observed pitch angle with Coulomb friction (ψ and $\hat{\psi}$)

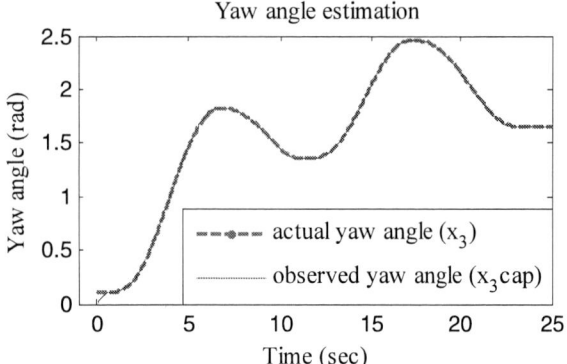

Fig. 7 Actual and observed yaw angle Coulomb friction (φ and $\hat{\varphi}$)

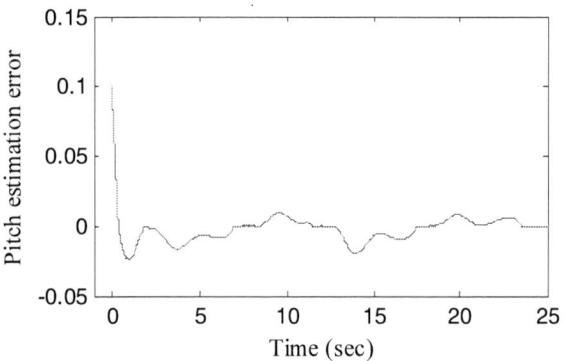

Fig. 8 Pitch angle estimation error with Coulomb friction ($\psi - \hat{\psi}$)

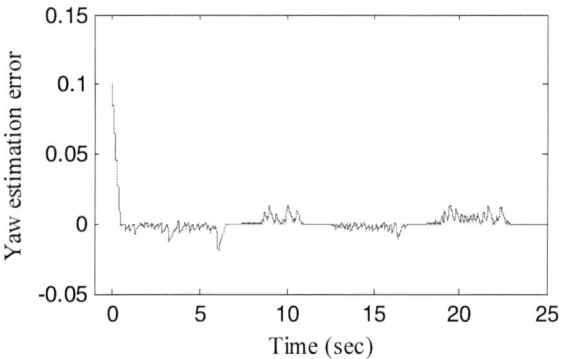

Fig. 9 Yaw angle estimation error with Coulomb friction ($\varphi - \hat{\varphi}$)

The tracking of actual and observed pitch and yaw angles with Coulomb friction are shown in Fig. 6 and Fig. 7. From the figures it's clear that the convergence of estimate state to the actual state is very fast. The estimation errors between the actual and observed pitch and yaw angles with Coulomb friction are given in Fig. 8 and Fig. 9 respectively. As comparing Fig. 4 with Fig. 8, and Fig. 5 with Fig. 9, the tracking errors with Coulomb friction in the TRCS are slightly greater than the errors when Coulomb friction is not

introduced in the system.

VI. CONCLUSIONS

The simulation studies of a robust nonlinear observer for TRCS are presented in this paper. The nonlinearities of the TRCS are considered as unknown but the bounds of the nonlinearities are known. The states that are unavailable for measurement are estimated using proposed observer with sliding mode approach. Stability analysis of observer is done by using Lyapunov theory of stability. The simulation results highlight the quality of compensation of nonlinear friction term and the convergence of observer outputs to the plant outputs very quickly. In future scope of work, we aim to design and implement the robust observer based controller for the TRCS including the effects of possible frictions in the system in real-time.

REFERENCES

[1] I. A. Raptis, K. P. Valavanis, and G. J. Vachtsevanos, "Linear tracking control for small-scale unmanned helicopters," *IEEE Transactions on Control Systems Technology*, vol. 20, no. 4, pp. 995–1010, 2012.

[2] G. Mustafa and N. Iqbal, "Controller design for a twin rotor helicopter model via exact state feedback linearization," 8th *IEEE International Multitopic Conference*, Lahore Pakistan, Dec. 24–26, 2004.

[3] G. R. Yu and H. T. Liu, "Sliding mode control of a two-degree-of-freedom helicopter via linear quadratic regulator," *IEEE International Conference on Systems, Man and Cybernetics*, Taiwan, Oct. 10–12, 2005.

[4] I. A. Raptis, K. P. Valavanis, and W. A. Moreno, "A novel nonlinear backstepping controller design for helicopters using the rotation matrix," *IEEE Transactions on Control Systems Technology*, vol. 19, no. 2, pp. 465–473, 2011.

[5] TRMS 33–949S, User Manual, Feedback Instruments Ltd., East Sussex, U. K., 2006.

[6] K. P. Tee, S. S. Ge, and F. E. H. Tay, "Adaptive neural network control for helicopters in vertical flight," *IEEE Transactions on Control Systems Technology*, vol. 16, no. 4, pp. 753–762, Jul. 2008.

[7] P. Wen and T. W. Lu, "Decoupling control of a twin rotor MIMO system using robust deadbeat control technique," *IET Control Theory and Applications*, vol. 2, no. 11, pp. 999 – 1007, 2008.

[8] J. G. Juang, M. T. Huang, and W. K. Liu, "PID control using presearched genetic algorithms for a MIMO system," *IEEE Transactions on Systems, Man, and Cybernetics–Part C: Applications and Reviews*, vol. 38, no. 5, 2008.

[9] P. Wen and T. W. Lu, "Decoupling control of a twin rotor MIMO system using robust deadbeat control technique," *IET Control Theory and Applications*, vol. 2, no. 11, pp. 999 – 1007, 2008.

[10] J. G. Juang, W. K. Liu, and R. W. Lin, "Comparison of classical control and intelligent control for a MIMO system," *Applied Mathematics and Computation*, vol. 205, no. 2, pp. 778–791, 2008.

[11] A. Rahideh, A. H. Bajodah, and M. H. Shaheed, "Real time adaptive nonlinear model inversion control of a twin rotor MIMO system using neural networks," *Engineering Applications of Artificial Intelligence*, vol. 25, no. 6, pp. 1289–1297, 2012.

[12] F. A. Shaik, S. Purwar and B. Pratap, "Real-time Implementation of Chebyshev Neural Network Observer for Twin Rotor Control System," *Expert Systems with Applications*, vol. 38, no. 10, pp. 13043–13049, 2011.

[13] B. Pratap and S. Purwar, "Real Time Implementation of State Observers for Twin Rotor MIMO System: An Experimental Evaluation," *International Journal of Modelling, Identification and Control*, vol. 19, no. 1, pp. 98–110, 2013.

[14] B. Pratap and S. Purwar, "Real Time Implementation of Neuro Adaptive Observer Based Robust Backstepping Controller for Twin Rotor MIMO System," *Journal of Control, Automation and Electrical Systems*, vol. 25, no. 2, pp. 137–150, 2014.

[15] A. J. Koshkouei and A. S. I. Zinober, "Sliding mode state observation for non-linear systems," *International Journal of Control*, vol. 77, no. 2, pp. 118–127, 2004.

[16] E. M. Jafarov, "Design modification of sliding mode observers for uncertain MIMO systems without and with time-delay," *Asian Journal of Control*, Vol. 7, No. 4, pp. 380–392, 2005.

[17] P. C. P. Chao and C. Y. Shen, "Sensorless tilt compensation for a three-axis optical pickup using a sliding-mode controller equipped with a sliding-mode observer," *IEEE Transactions on Control Systems Technology*, vol. 17, no. 2, pp. 267–282, 2009.

[18] J. J. E. Slotine and W. Li, Applied Nonlinear Control, Englewood Cliffs, NJ: Prentice-Hall, 1991.

[19] C. Edwards and S. K. Spurgeon, Sliding Mode Control: Theory and Applications, London, U.K.: Taylor & Francis Ltd., 1998.

[20] J. Y. Hung, W. Gao, and J. C. Hung, "Variable structure control: A survey," *IEEE Transactions on Industrial Electronics*, vol. 40, no. 1, pp. 2–22, 1993.

[21] B. Armstrong and C. C. de Wit, Friction modeling and compensation, Boca Raton, FL: CRC Press, vol. 77, pp. 1369–1382, 1996.

[22] H. Olsson, K. Astrom, C. C. de Wit, M. Gafvert, and P. Lischinsky, "Friction models and friction compensation," *European Journal of Control*, vol. 4, no. 3, pp. 176–195, 1998.

[23] H. Chaoui,P. Sicard, and W. Gueaieb, "ANN-Based Adaptive Control of Robotic Manipulators With Friction and Joint Elasticity," *IEEE Transactions on Industrial Electronics*, vol. 56, no. 8, pp. 3174–3187, 2009.

[24] S. Laghrouche, F. S. Ahmed, and A. Mehmood, "Pressure and Friction Observer-Based Backstepping Control for a VGT Pneumatic Actuator," *IEEE Transactions on Control Systems Technology*, vol. 22, no. 2, pp. 456–467, 2014.

[25] V. I. Utkin, J. Guldner, and J. Shi, Sliding Mode Control in Electro Mechanical Systems, New York, NY, USA: Taylor and Francis, 1999.

[26] M. Gopal, Digital Control and State Variable Methods, Third Edition, TMH, 2009.

Investigations on Optimal Pulse-Width Modulation to Minimize Total Harmonic Distortion in the Line Current

Avanish Tripathi*, *Student Member, IEEE*, G. Narayanan†,

Department of Electrical Engineering, Indian Institute of Science, Bangalore - 560012 INDIA
Email: * avanish@ee.iisc.ernet.in; † gnar@ee.iisc.ernet.in

Abstract—PWM waveforms with positive voltage transition at the positive zero crossing of the fundamental voltage (type-A) are generally considered for PWM waveform with even number of switching angles per quarter whereas, waveforms with negative voltage transition at the positive zero crossing (type-B) are considered for odd number of switching angles per quarter. Optimal PWM, for minimization of total harmonic distortion of line to line (V_{WTHD}), is generally solved with the aforementioned criteria. This paper establishes that a combination of both types of waveforms gives better performance than any individual type in terms of minimum V_{WTHD} for complete range of modulation index (M). Optimal PWM for minimum V_{WTHD} is solved for PWM waveforms with pulse numbers (P) of 5 and 7. Both type-A and type-B waveforms are found to be better in different ranges of M. The theoretical findings are confirmed through simulation and experimental results on a 3.7kW squirrel cage induction motor in an open-loop V/f drive. Further, the optimal PWM is analysed from a space vector point of view.

Index Terms—Harmonic analysis, harmonic distortion, induction motor drive, off-line pulse width modulation, optimal pulse width modulation, voltage-source inverter.

I. INTRODUCTION

The ratio of switching frequency to fundamental frequency (*i.e.* pulse number, P) is constrained to be low for high-power motor drives on account of high switching losses [1]-[5]. Also, in case of high-speed motor drives, P is low due to high fundamental frequency [6]. Fig.1 shows two typical pole voltage waveforms (v_{RO}), measured between the R-phase terminal and the dc bus mid-point O, having a pulse number of 5. Both waveforms possess half-wave and quarter-wave symmetries [7][8]. Apart from the switching transitions at $\theta = 0^o$ and $\theta = 180^o$, there are two switching angles (*i.e.* α_1 and α_2) in each quarter cycle in both waveforms.

The pole voltage waveform in Fig. 1(a) has a switching transition from $\frac{-V_{dc}}{2}$ to $\frac{+V_{dc}}{2}$ at the positive zero-crossing of the R-phase fundamental voltage, *i.e.* $\theta = 0^o$. This is termed as type-A PWM waveform. On the other hand, the pole voltage waveform in Fig. 1(b) switches from $\frac{+V_{dc}}{2}$ to $\frac{-V_{dc}}{2}$ at $\theta = 0^o$. This is termed as type-B PWM waveform. To design a high-performance pulse width modulation (PWM) scheme for

This work is funded by the Department of Heavy Industry, Government of India, under a project titled "Off-line and Real-time Simulators for Electric Vehicle/ Hybrid Electric Vehicle Systems."

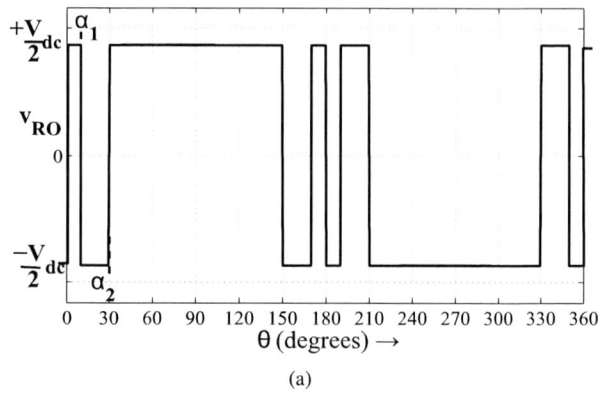

(a)

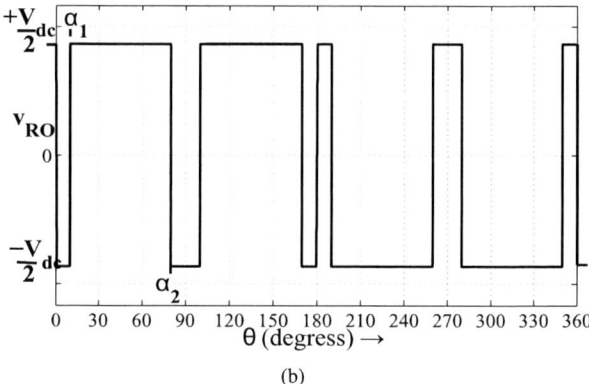

(b)

Fig. 1. Illustration of pole voltage at low inverter switching frequency (pulse number 5) (a) type A waveform and (b) type B waveform

any pulse number P, the motor drive engineer has to decide between type-A and type-B waveforms.

In literature, type-A waveform is considered to yield good quality waveform when the number of switching angles per quarter is even. Similarly, type-B is regarded as the better choice for odd number of switching angles per quarter [11]. This paper studies the relative performances and suitability of type-A and type-B waveforms in detail as discussed below.

This paper first investigates optimal type-A PWM and optimal type-B PWM to minimize the total harmonic distortion (THD) in line current (I_{THD}) for $P = 5$ and $P = 7$. Considering type-A and $P = 5$, for example, switching angles are solved so as to yield the desired fundamental voltage and

978-1-4799-6047-7/14 $31.00 © 2014 IEEE

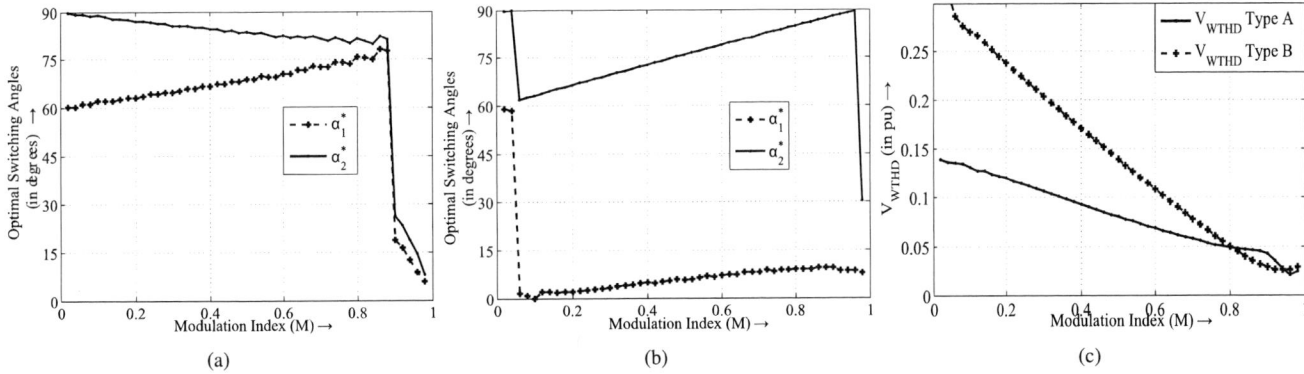

Fig. 2. Optimal switching angles for (a) type A waveform and (b) type B waveform. (c) Comparison of V_{WTHD} of type A with type B waveforms for pulse number 5

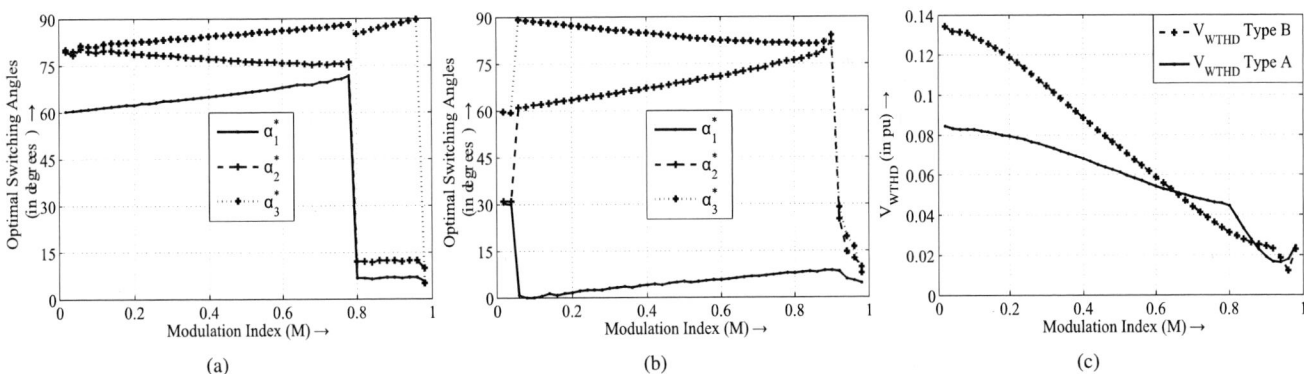

Fig. 3. Optimal switching angles for (a) type A waveform and (b) type B waveform. (c) Comparison of V_{WTHD} of type A with type B waveforms for pulse number 7

minimum THD in line current. Optimal switching angles are calculated over the whole range of modulation index. Optimal switching angles are thus obtained for both types of PWM for both pulse numbers.

Performances of optimal type-A PWM and optimal type-B PWM are compared, based on simulation as well as experimental results on a 3.7kW induction motor drive, at both pulse numbers of 5 and 7. It is shown that neither type-A nor type-B is superior to the other in the whole range of modulation at any pulse number. At each P, optimal type-A is better than optimal type-B in certain ranges of modulation index, while the converse is true in the other ranges of modulation. Hence a combined optimal PWM is suggested, employing optimal type-A and optimal type-B PWM schemes in their respective modulation ranges of superior performance.

Finally, the optimal type-A PWM, optimal type-B PWM and the combined optimal PWM are viewed from a space vector point of view. The waveforms of the three techniques are studied in terms of the voltage vectors applied in a sector. An effort is made to understand optimality in terms of the voltage vectors applied.

II. OPTIMAL PULSE-WIDTH MODULATION

Optimal PWM is usually employed at low-switching frequencies for minimization of total harmonic distortion in line current (I_{THD}) at any given modulation index (M) [1][5][6][8][9][10]. I_{THD} is defined in (1) [8]

$$I_{THD} = \sqrt{\frac{I_{RMS}^2 - I_1^2}{I_1^2}} = \sqrt{\frac{\sum\limits_{n=6k\pm1} I_n^2}{I_1^2}} \quad k = 1, 2, 3.... \quad (1)$$

where, I_{RMS} is the RMS line current; I_1 is RMS fundamental line current; I_n is the nth order harmonic RMS current.

Since I_{THD} is dependent on machine parameters, the weighted total harmonic distortion of line to line voltage (V_{WTHD}) is considered here for minimization, which is a measure of I_{THD} [8]. V_{WTHD} of line to line voltage is given as [5]

$$V_{WTHD} = \sqrt{\frac{\sum\limits_{n=6k\pm1} F_n^2/n^2}{F_1^2}} \quad k = 1, 2, 3.... \quad (2)$$

where, F_1 is the peak fundamental voltage and F_n is the peak n^{th} harmonic voltage.

The amplitude of F_n for pulse number 5 (*i.e.* 2 switching angles, α_1 and α_2 per quarter cycle), considering type-A and type-B waveforms, is given by (3) and (4), respectively, where

V_{dc} is the DC bus voltage [9]

$$F_n = \left(\frac{2V_{dc}}{n\pi}\right)[1 - 2\cos(n\alpha_1) + 2\cos(n\alpha_2)] \quad (3)$$

$$F_n = \left(\frac{2V_{dc}}{n\pi}\right)[-1 + 2\cos(n\alpha_1) - 2\cos(n\alpha_2)] \quad (4)$$

$$n = 1, 5, 7, 11, 13....$$

The corresponding expression for pulse number 7 (*i.e.* 3 switching angles, α_1, α_2 and α_3 per quarter cycle), for type-A and type-B waveforms, are given by (5) and (6), respectively [9]

$$F_n = \left(\frac{2V_{dc}}{n\pi}\right)[1 - 2\cos(n\alpha_1) + 2\cos(n\alpha_2) - 2\cos(n\alpha_3)] \quad (5)$$

$$F_n = \left(\frac{2V_{dc}}{n\pi}\right)[-1 + 2\cos(n\alpha_1) - 2\cos(n\alpha_2) + 2\cos(n\alpha_3)] \quad (6)$$

$$n = 1, 5, 7, 11, 13....$$

V_{WTHD} is minimized subject to the constraint that the modulation index M equals the desired modulation index M^* as shown by (7). Here, modulation index is the fundamental voltage normalized with respect to $\frac{2V_{dc}}{\pi}$. The switching angle constraint is given by the inequality in (8) for $P = 5$ and by (9) for $P = 7$.

$$M = M^* \quad (7)$$

$$0 \leq \alpha_1 \leq \alpha_2 \leq \frac{\pi}{2} \quad (8)$$

$$0 \leq \alpha_1 \leq \alpha_2 \leq \alpha_3 \leq \frac{\pi}{2} \quad (9)$$

First considering the type-A waveform of $P = 5$, the optimal switching angles α_1^* and α_2^* are evaluated to minimize V_{WTHD} as defined by (2) and (3) subject to the constraint in (7) and (8). These optimal angles are plotted against M in Fig.2(a). Similarly, for type-B waveform of $P = 5$, the optimal switching angles are determined so as to minimize V_{WTHD} defined by (2) and (4), subject to the constraints in (7) and (8). These optimal switching angles are shown against M in Fig.2(b). The V_{WTHD} values pertaining to the optimal type-A and optimal type-B waveforms are compared in Fig.2(c).

Similarly, the results of optimization for $P = 7$ are shown in Figs.3(a), 3(b) and 3(c). V_{WTHD} is minimized subject to the constraints in (7) and (9). V_{WTHD} is given by (2) and (5) for type-A waveform, and by (2) and (6) for type-B waveform. The optimal switching angles of the two types are presented in Fig.3(a) and Fig.3(b), respectively. A comparison of V_{WTHD} of the optimal type-A and optimal type-B is presented in Fig.3(c).

As shown by Fig.2(c) and Fig.3(c), neither optimal type-A nor optimal type-B is superior over the whole range of M at either pulse numbers. One scheme is better than the other, and *vice versa*, in different ranges of modulation index. Hence, a combined optimal PWM is proposed. The V_{WTHD} of this proposed combined optimal PWM is better than the corresponding PWM scheme in literature [11] as shown Fig.4(a) and Fig.4(b). The improvement in harmonic distortion can be

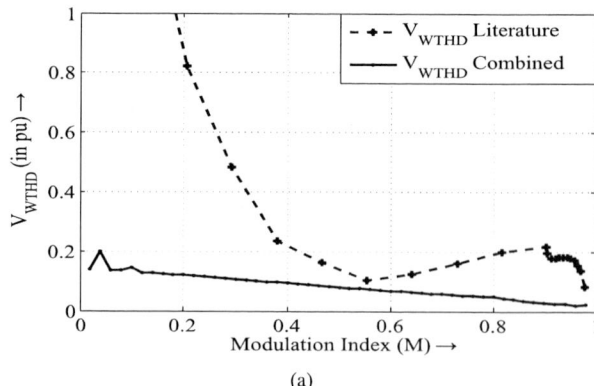

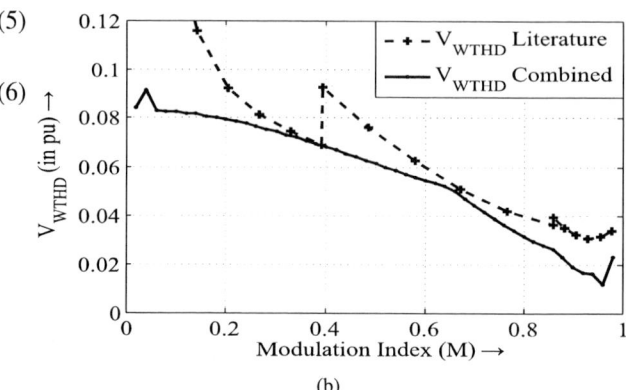

Fig. 4. Comparison of V_{WTHD} of combined optimal PWM with that presented in literature [12] for pulse numbers of (a) 5 and (b) 7

Fig. 5. Experimental set-up showing the 10kVA inverter and 3.7kW induction motor

attributed to a more rigorous search process over the entire solution space here than in [11].

III. SIMULATION AND EXPERIMENTAL RESULTS

Fig.7 shows the experimental set-up used in this paper. A three-phase, 400V, 3.7kW, 1460rpm, 50Hz, star/delta squirrel-cage induction motor (SCIM) is used whose parameters are given in Table I. The motor is driven by three-phase, 10kVA inverter in open-loop V/f control. FPGA based CYCLONE II controller board is used to control the converter. DC bus voltage is maintained at 520V using a three-phase diode bridge rectifier with an autotransformer at the input side. The complete model is simulated on MATLAB/Simulink with a

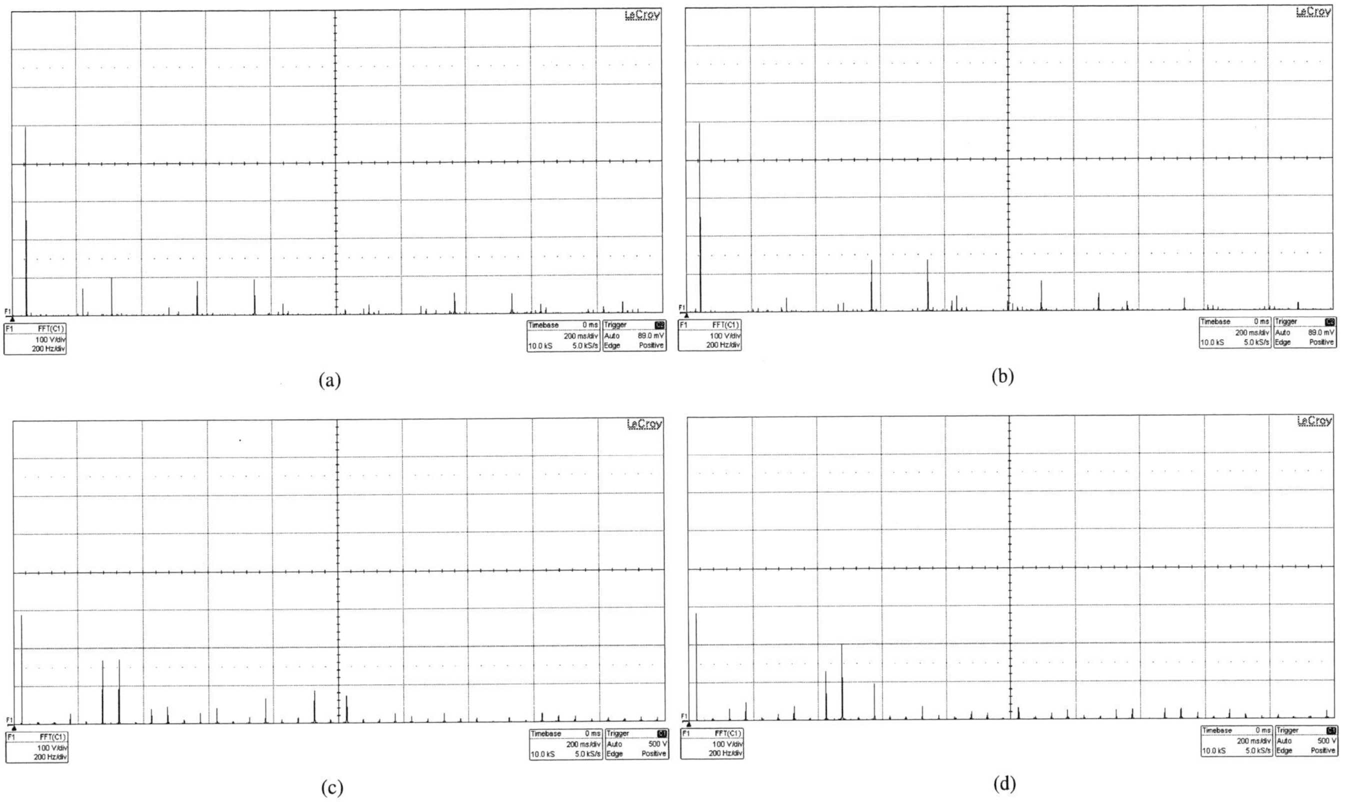

Fig. 6. Measured harmonic spectra of line-line voltage with optimal PWM (a) $M = 0.88$, $f_1 = 44$Hz, $P = 5$, type A, $V_{WTHD} = 0.0449$ (b) $M = 0.88$, $f_1 = 44$Hz, $P = 5$, type B, $V_{WTHD} = 0.0313$ (c) $M = 0.50$, $f_1 = 25$Hz, $P = 7$, type A, $V_{WTHD} = 0.0613$ (d) $M = 0.50$, $f_1 = 44$Hz, type B, $V_{WTHD} = 0.0743$; (scale: 100 V/div. and 200 Hz/div.)

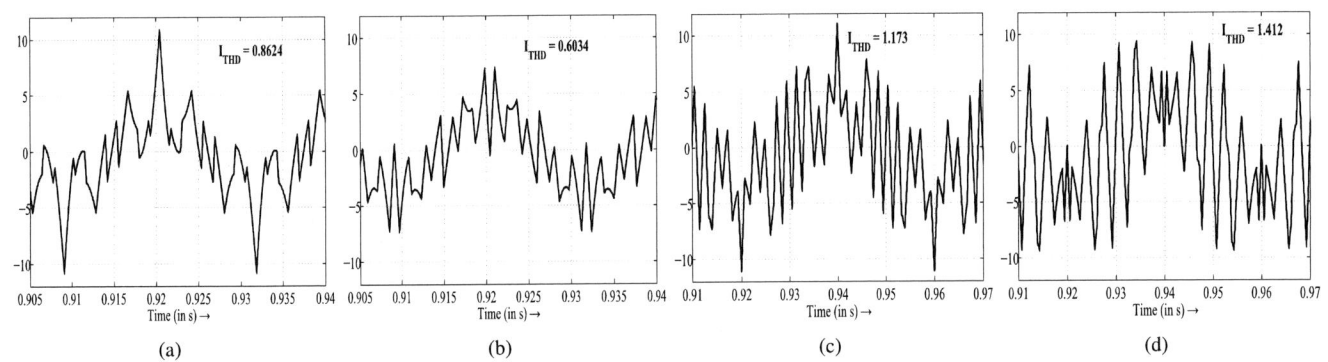

Fig. 7. Simulated R-phase current waveforms corresponding to Fig.6 (scale: 1 A/div.)

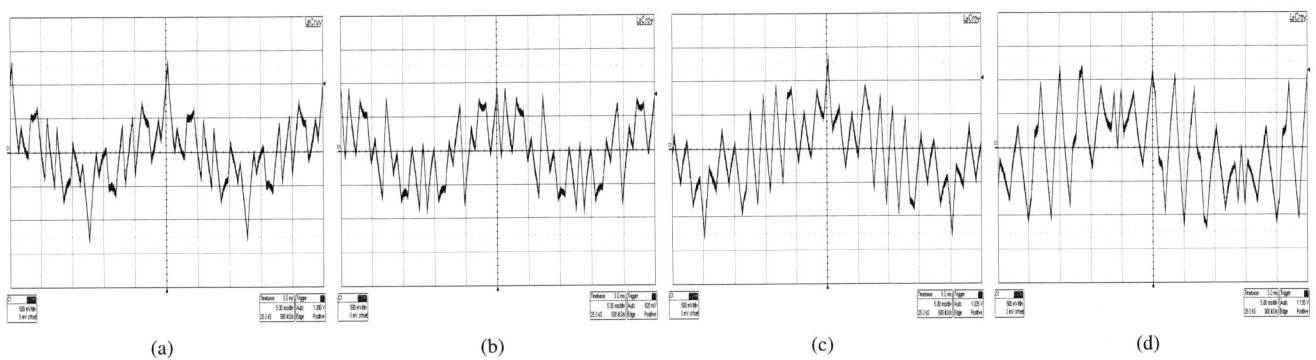

Fig. 8. Measured R-phase current waveforms corresponding to Fig.6. Measured I_{THD} values are (a) 0.8134 (b)0.5972 (c)1.0896 and (d)1.3667; (scale: 4.25 A/div.)

978-1-4799-6047-7/14 $31.00 © 2014 IEEE

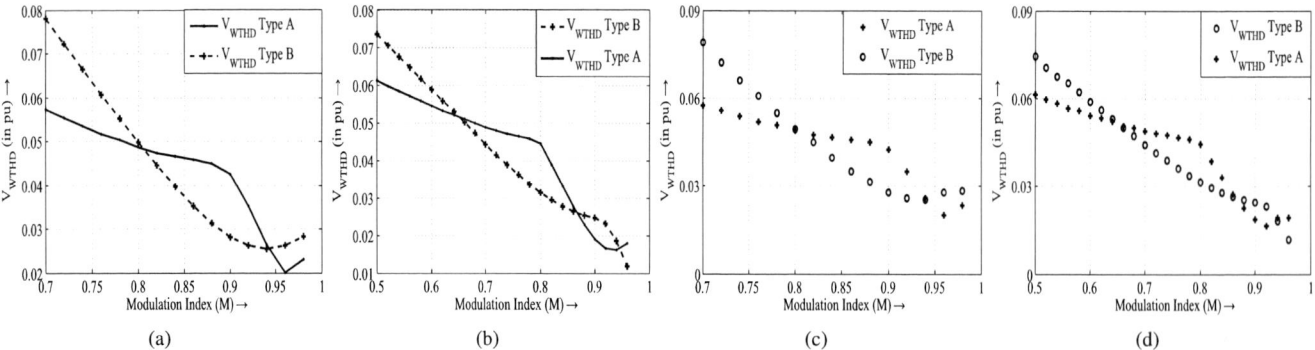

Fig. 9. Comparison of V_{WTHD} for optimal type A and optimal type B PWM (a) simulated values for $P = 5$ (b) simulated values for $P = 7$ (c) measured values for $P = 5$ and (d) measured values for $P = 7$

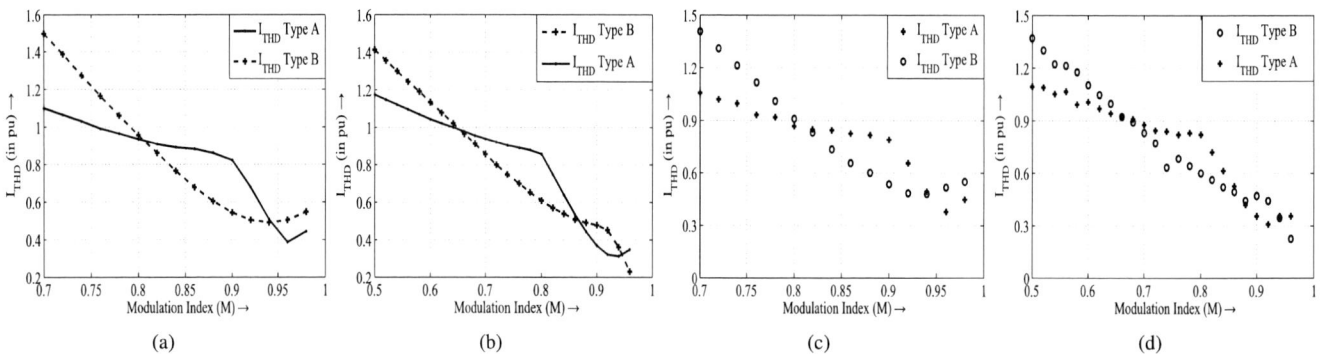

Fig. 10. Comparison of I_{THD} for optimal type A and optimal type B PWM (a) simulated values for $P = 5$ (b) simulated values for $P = 7$ (c) measured values for $P = 5$ and (d) measured values for $P = 7$

TABLE I
PARAMETERS OF INDUCTION MOTOR

Parameter	Value
Stator Resistance, R_s	1.4 Ω
Rotor Resistance, R_r	1.96 Ω
Stator Inductance, L_s	0.2844 H
Rotor Inductance, L_r	0.2844 H
Magnetizing Inductance, L_o	0.277 H

standard induction motor model [12].

A. Harmonic Spectra

Fig.6(a) and Fig.6(b) show the measured harmonic spectra of line to line voltage (v_{RY}) for type A and type B PWM waveforms, respectively, at $M = 0.88$ for $P = 5$. Reduction in 5th and 7th harmonics can be seen with optimal type-B PWM as compared to optimal type-A PWM. Also, V_{WTHD} is found to be reduced with optimal type-B PWM. However, at a different modulation index of 0.5, optimal type-A PWM is found to be better than optimal type-B for $P = 7$ as seen from Figs.6(c) and 6(d). Theoretical harmonic spectra of v_{RY} are not shown due to space constraint.

B. Line Current Waveform

Figs.7(a), 7(b), 7(c) and 7(d) present the simulated line current (i_R) waveforms corresponding to Figs.6(a), 6(b), 6(c) and 6(d), respectively. The corresponding measured line

current waveforms are shown in Figs. 8(a), 8(b), 8(c) and 8(d), respectively. The simulated and measured line current waveforms in Fig.7 and Fig.8 agree well. Figs.7(a), 7(b), 8(a) and 8(b) confirm that optimal type-B is better than optimal type-A at $M = 0.88$ for $P = 5$. Similarly, Figs.7(c), 7(d), 8(c) and 8(d) confirm the reduction in harmonic distortion with type-A over type-B at $M = 0.5$ for $P = 7$.

C. V_{WTHD} and I_{THD} Comparison

Further, V_{WTHD} and I_{THD} pertaining to optimal type-A and optimal type-B PWM are compared through simulations and experiments over a range of M for $P = 5$ and 7. Figs.9(a) and 9(b) compare the simulated V_{WTHD} values for $P = 5$ and $P = 7$, respectively, while, Figs.9(c)and 9(d) present the measured V_{WTHD} for $P = 5$ and $P = 7$, respectively. Comparison of simulated I_{THD} values of type-A and type-B waveform is presented in Fig.10(a) and Fig.10(b) for $P = 5$ and $P = 7$, respectively. The measured I_{THD} is compared similarly in Figs.10(c) and 10(d). The simulation and experimental results are found to be in close agreement with the theoretical predictions in Fig.2(c) and Fig.3(c).

IV. DISCUSSION

Three-phase pole voltages can be viewed in terms of inverter states. Fig.11 shows the eight inverter states (i.e. six active states and two zero states) [7]. The complete fundamental cycle is divided into six sectors [7]. Optimal type-A and

978-1-4799-6047-7/14 $31.00 © 2014 IEEE

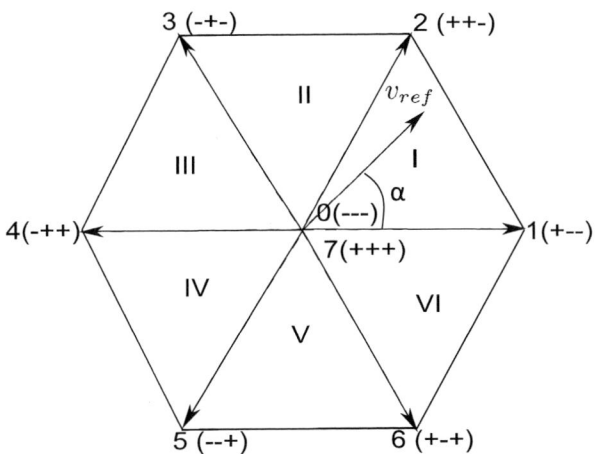

Fig. 11. Space vector representation of switching states for two level inverter

TABLE II
SEQUENCE OF VECTORS APPLIED IN SECTOR I FOR $P = 5$

Modulation index (M)	Vector sequence		
	Type A	Type B	Combined optimal
$0 < M \leq 0.8$	$101 - 272$	$012 - 127$	$101-272$
$0.8 < M \leq 0.88$	$101 - 272$	$012 - 127$	$012 - 127$
$0.88 < M \leq 0.94$	$121 - 212$	$012 - 127$	$012 - 127$
$0.94 < M < 1.0$	$121 - 212$	$012 - 127$	$121 - 212$

optimal type-B PWM waveforms are analysed from the space vector point of view. Tables II and III present the sequence of vectors applied by optimal type-A PWM, optimal type-B PWM and combined optimal PWM in sector I for the complete range of modulation index.

In sector I, the inverter states applied are found to be only 0, 1, 2 and 7 in all the optimal PWM methods. Further, the sequence of inverter states within sector I changes from one modulation index range to another for each optimal PWM as shown in Table II and Table III.

The combined optimal PWM is shown to have all but one switching transitions between a zero state and an active state in each sector at low modulation indices. As the modulation index increases, the number of transitions between the zero state and active state (say 0 and 1) reduces progressively; the corresponding number of transitions between the two active states increases (*i.e.* between 1 and 2). At high modulation indices close to six-step operation, all switching transitions in a sector are between the two active vectors (*i.e.* 1 and 2).

TABLE III
SEQUENCE OF VECTORS APPLIED IN SECTOR I FOR $P = 7$

Modulation index (M)	Vector sequence		
	Type A	Type B	Combined optimal
$0 < M \leq 0.64$	$0101 - 2727$	$1012 - 1272$	$0101 - 2727$
$0.64 < M \leq 0.78$	$0101 - 2727$	$1012 - 1272$	$1012 - 1272$
$0.78 < M \leq 0.86$	$0121 - 2127$	$1012 - 1272$	$1012 - 1272$
$0.86 < M \leq 0.92$	$0121 - 2127$	$1012 - 1272$	$0121 - 2127$
$0.92 < M \leq 0.94$	$0121 - 2127$	$1212 - 1212$	$0121 - 2127$
$0.94 < M < 1.0$	$0121 - 2127$	$1212 - 1212$	$1212 - 1212$

V. CONCLUSION

Optimal pulse width modulation to minimize line current total harmonic distortion (THD) is examined at pulse numbers of 5 and 7, considering both type-A and type-B waveforms. Performance of optimal type-A and optimal type-B PWM schemes are compared at both pulse numbers. The comparison is based on simulation results as well as experimental results, on a 3.7kW induction motor drive. Neither, optimal type-A nor optimal type-B is better than the other over the whole range modulation index at any pulse number. The two schemes have their own modulation ranges of superior performance. Hence, a combined optimal PWM scheme is proposed.

The optimal type-A PWM, the optimal type-B PWM and the combined optimal PWM are analysed from the space vector perspective. It is observed that only the two nearest active vectors are applied in all three optimal PWM schemes. In case of combined optimal PWM, the number of switching transitions between an active vector and a zero vector reduces progressively with increase in modulation index. Close to six-step operation, all the switching transitions are found to be between the active vectors only.

REFERENCES

[1] A. K. Rathore, J. Holtz and T. Boller, "Generalized Optimal Pulsewidth Modulation of Multilevel Inverters for Low-Switching-Frequency Control of Medium-Voltage High-Power Industrial AC Drives," *IEEE Trans. Ind. Electron.*, Vol. 60, No. 10, Oct. 2013, pp. 4215-4224.

[2] J. A. Pontt, J. R. Rodriguez, A. Liendo, P. Newman and J. Holtz "Network Friendly Low-Switching-Frequency Multipulse High Power Three-Level PWM Rectifier," *IEEE Trans. Ind. Electron.*, Vol. 56, No. 4, Apr. 2009, pp. 1254-1262.

[3] J. Shen, S. Schroder, H. Stagge and R. W. De Doncker, "Impact of Modulation Schemes on the Power Capability of High-Power Converters with Low Pulse Ratios," *IEEE Trans. Pow. Electron.*, Vol. 29, No. 11, Nov. 2014, pp. 5696-5705.

[4] M. Sharifzade1, H. Vahedi, A. Sheikholeslami1, H. Ghoreishy, K. AI-Haddad, "Selective Harmonic Elimination Modulation Technique Applied on Four-Leg NPC," *Proc. on 23rd Intl. Symp. on Ind. Electron. (ISIE), Turkey*, June 2014, pp. 2167-2172.

[5] N. Yosefpoor, S. A. Fathi, N. Farokhina and H. A. Abyaneh, "THD Minimisation Applied Directly on the Line-to-Line Voltage of Multilevel Inverters," *IEEE Trans. Ind. Electron.*, Vol. 59, No. 1, Jan. 2008, pp. 373-380.

[6] L. Liyi, G. Tan, J. Liu and B. Kou, "An Optimal Pulse Width Modulation Method for High-speed Permanent Magnet Synchronous Motor," *Proc. Third Intl. Conf. on Inf. Science and Tech., China*, March 2013, pp. 237-244.

[7] G. Narayanan and V. T. Ranganathan, "Synchronized PWM Strategies Based on Space Vector Approach. Part 1: Principles of waveform Generation," *IEE Proc. Elec. Power Appl.*, Vol. 146, No. 3, May 1999, pp. 267-275.

[8] A. Tripathi and G. Narayanan "Optimal Pulse Width Modulation of Voltage-Source Inverter Fed Motor Drives with Relaxation of Quarter Wave Symmetry Condition," *Proc. IEEE Intl. Conf. Electron., Comp. and Comm. Technology (IEEE CONECCT)*, 6-7 Jan. 2014, pp. 1-6.

[9] G. S. Buja and G. B Indri, "Optimal Pulse-Width Modulation for Feeding AC Motors," *IEEE Trans.Ind. Appl.*, Vol. IA-13, No. 1, Jan/Feb 1977, pp. 38-44.

[10] F. C. Zach and H. Ertl, "Efficiency Optimal Control for AC Drives with PWM Inverters," *IEEE Trans. Ind. Appl.*, Vol. IA-21, No. 4, July/Aug. 1985, pp. 987-1000.

[11] D. G. Holmes, T. A. Lipo, "Pulse Width Modulation for Power Converters Principles and Practice," IEEE Press, 2003.

[12] W. Leonhard, "Control of electrical drives," Narosa publishing house, 1992.

Induction Machine Efficiency Estimation Using Population Based Algorithm

G.S. Grewal and B.S. Rajpurohit

School of Computing & Electrical Engineering
Indian Institute of Technology Mandi
Mandi, India
gurinderbir_singh@students.iitmandi.ac.in, bsr@iitmandi.ac.in

Abstract— **There has been a tremendous pressure to predict the *in situ* efficiency of induction machine (IM) with limited level of intrusion so as to improve IMs performance. Least research work is carried out to make IM efficiency evaluation methods compatible to unbalanced supply and varying load conditions. This paper proposes a novel approach using cuckoo search algorithm (CSA) to obtain efficiency estimation of an IM operating as a motor working with unbalanced supply having under or over voltage issues. CSA improves the searching ability and has capability to adapt to complex optimization problems. Here, CSA optimizes the IM positive sequence parameters at various loading levels. The parameters optimization is done with the use of positive sequence input currents and electrical powers which have been obtained earlier at various operating loading points. Using the optimized parameters, the negative sequence parameters can be evaluated. So, the efficiency of IM can be estimated at different loading levels. The proposed approach is implemented on the MATLAB platform. The effectiveness of the novel approach is established by comparing the results obtained with genetic algorithm (GA).**

Keywords—cuckoo search algorithm (CSA), induction machine (IM), induction motor efficiency, levy flight

I. INTRODUCTION

Induction machines (IMs) are the work horses of industrial processes. They constitute a major portion of industrial load. Normally, the working efficiency of IM in any drive application is always different from their rated efficiencies. In many industries, oversized IMs are installed resulting in the under operation of IMs in the range from 50% to 60%. The enormous numbers of IMs represent potential for significant savings in electrical energy. The performance of IM is influenced by many parameters like ageing, unbalance in source voltage, around or underneath voltage conditions, harmonics as a consequence of rewinding as well as repairing the motor [1]. The equivalent circuit of IM can be modeled by obtaining data from the nameplate of IM and considering few assumptions. This considers empirical mathematical formulation rather than practical measurements resulting in less accuracy. In addition to this, impacts of three phase unbalanced supply and load variations are neglected in calculation of IM efficiency. So, *in situ* efficiency monitoring of existing established IMs is the need of the hour in order to find the IMs giving poor efficiencies and undertake the

requisite measures to attain energy savings resulting in effective energy management. Also, the standard test conditions are rarely met by the IMs working conditions. This point got least importance in the research literature. The existing IMs giving poor efficiencies can be identified and can be replaced with energy efficient IMs of suitable ratings resulting in the enhancement of energy savings. Significant research has been done so far to obtain procedures to accurately calculate the IM efficiency *in situ*. Most of the customary techniques are available in IEEE 112 standard which cannot be applied as such for evaluating the *in situ* efficiency of IM. Though, accuracy of the technique is of utmost importance it is desired that the level of intrusiveness should be practically the lowest possible.

From the past till now, many methods for estimation of efficiency of IM have been documented like name plate method, slip method (SM), current method (CM), statistical method, equivalent circuit method (ECM) [1]-[3]. Currently, artificial intelligence (AI) technique, optimization algorithms like genetic algorithm (GA), particle swarm optimization (PSO) and the artificial neural networks (ANN) are widely used for parameter estimation of IM to find efficiency [4]. A new method has been developed which is dependent on multi-objective evolutionary algorithms for the efficiency estimation of serviceable IM [5].Here, multiple objectives are combined linearly to form a single objective problem for optimization. This approach has the limitation where objective space function has some non-convex areas giving no solutions.

In [5], the non-dominated sorting genetic algorithm-II (NSGA-II) and strength pareto evolutionary algorithm-2 (SPEA2) have been enforced to arbitrate the competence of IMs in service. The effectiveness of NSGA-II and SPEA2 for various loading conditions is demonstrated. The comparison of experimental results is made with the results obtained from the multi-objective evolutionary algorithms.

A simple and inexpensive technique has been proposed in [6] for estimation of efficiency of IM by utilizing the modified equivalent circuit method and the bacterial foraging algorithm (BFA). The measurements of current and voltage of the stator, resistance of the stator, input electrical power and rotational speed of the rotor are noted without organizing the no-load and locked-rotor tests. The equivalent circuit parameters are obtained using BFA so as to attenuate the error between

978-1-4799-6047-7/14 $31.00 © 2014 IEEE

consistent obtained values & approximated values. This estimation approach is implemented on 5 horse power (HP) IM. The results obtained are compared with PSO method, torque gauge method, equivalent circuit method, slip method, current method and segregated loss method. The results show that the proposed technique has the dormant for more accurate efficiency estimation of IM. This technique is found suitable for energy conservation work and energy management of the on site IMs. For balanced supply conditions, methods like loss segregation method and non intrusive air gap torque method can be employed whereas for unbalanced supply conditions existing in industry, soft computing based estimation methods utilizing fuzzy logic controller (FLC), sliding mode controller (SMC), GA, PSO algorithm, evolutionary algorithm (EA), (BFA) can yield good results.

In [3], authors use optimization method dealing with the efficiency estimation of IM subjected to unbalanced supply conditions & in presence of over/under voltages. Furthermore, the study on non-intrusive air gap torque method has been elaborately examined supported by experimental results. This is the instigation for the present work. This paper proposes cuckoo search algorithm (CSA) as an effective approach for efficiency estimation of IM. CSA is a population-based algorithm analogous to GA and PSO. The novelty of the proposed approach is the improvised search ability, and random reduction. This makes it adaptable to complex optimization problems. The employment of CSA for efficiency estimation is purely a non intrusive technique which does not hamper or alter any operating condition of installed IM. Here, CSA objective function optimizes the induction motor positive sequence parameters of equivalent circuit of IM at various load points. The optimization of parameters is done based on the positive sequence input currents and electrical powers obtained earlier at various loading conditions. Use of optimized parameters results in the evaluation of the negative sequence parameters. From the knowledge of attained parameters, IM efficiency at various load points can be estimated. The paper has been organized as follows: The mathematical modeling of IM is illustrated in the section II. The proposed approach is discussed in section III. Simulation proof is presented in section IV. Outcome of the proposed approach and the related discussions are given in section V. Section VII concludes the paper. The mathematical modeling of IM is illustrated in the following section II.

II. MATHEMATICAL MODELLING OF IM

The operation of IM during unbalanced supply conditions results in unbalanced flow of currents in the IM. This gives rise to the formation of negative and positive sequence fluxes. As a result, positive and negative sequence torque components develop which act in opposition. The equivalent circuits for positive and negative sequence of IM [3] are given in fig. 1 can effectively illustrate the performance behavior of IM when working with variable supply voltage and few unbalanced voltage levels. The equivalent circuit based on IEEE 112 standard [7] of IM contains elements such as stator resistance R_1, positive sequence resistance of rotor R_2, negative sequence resistance of rotor R_3, stator leakage reactance X_1, positive sequence leakage reactance of rotor X_2, negative sequence

leakage reactance of rotor X_3, core loss resistance R_m, mutual reactance X_m and stray load loss R_{sll} [9].

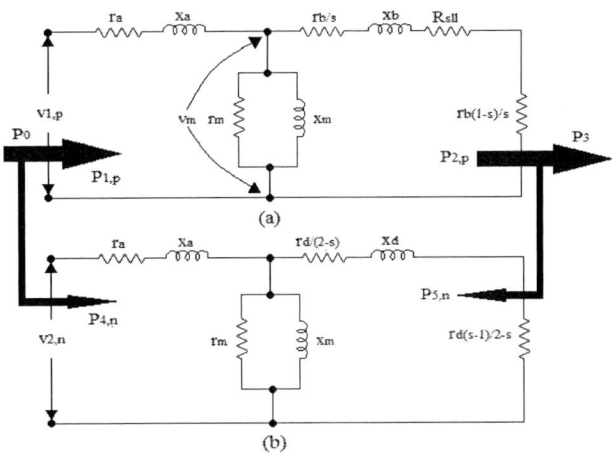

Fig. 1. (a) Positive sequence (b) Negative sequence equivalent circuit of IM

Here, $v_{1,p}$ is the degree of positive sequence input electrical voltage, $v_{2,n}$ is the value of negative sequence voltage, $P_{1,p}$ is the positive sequence input active electrical power, $P_{4,n}$ is the negative sequence active power, $P_{2,p}$ is the positive sequence output active power and $P_{5,n}$ is the output active power of negative sequence. The unbalanced input supply defines the efficiency η of the IM as follow.

$$\eta = \frac{\left|P_{2,p}\right| - \left|P_{5,n}\right| - P_{f\&w}}{\left|P_{1,p}\right| + \left|P_{4,n}\right|} \tag{1}$$

Where, $P_{f\&w}$ is the friction and windage losses. The following equations can be written depending on fig. 1 of the equivalent circuit of IM [6].

$$Y_m = \frac{1}{r_m + jx_m} \tag{2}$$

$$Y_{stator} = \frac{1}{r_{a\_c} + jx_a} \tag{3}$$

$$Y_{rotor,p} = \frac{1}{(r_{b\_c}/s) + R_{sll} + jx_b} \tag{4}$$

$$I_{m\_p\_estimate} = \left| \frac{v_{1\_p}.Y_{stator}.Y_m}{Y_{stator} + Y_m + Y_{rotor\_p}} \right| \tag{5}$$

$$I_{1\_p\_estimate} = \left| \frac{v_{1\_p}.Y_{stator}.(Y_m + Y_{rotor\_p})}{Y_{stator} + Y_m + Y_{rotor\_P}} \right| \tag{6}$$

$$I_{rotor\_p\_estimate}=\left|\frac{v_{1\_p}.Y_{stator}.Y_{rotor,p}}{Y_{stator}+Y_m+Y_{rotor\_p}}\right| \quad (7)$$

$$P_{1\_p\_estimate}=3\left[\begin{array}{l}R_{a\_c}.I^2_{1\_p\_estimate}+\left(\dfrac{R_{b\_c}}{s}\right)I^2_{rotor\_p\_estimate}\\[2mm]+I^2_{m\_p\_estimate}.r_m\end{array}\right] \quad (8)$$

$$P_{2\_p\_estimate}=3\left[\left(\frac{R_{b\_c}(1-s)}{s}\right)I^2_{rotor\_p\_estimate}\right] \quad (9)$$

Where, Y_m is the multiplex magnetizing branch admittance, Y_{stator} is the convoluted stator branch admittance, Y_{rotor_p} is the admittance of the positive sequence rotor branch, $I_{m\_p\_estimate}$ is the estimated magnetizing current of positive sequence, $I_{1\_p\_estimate}$ is the estimated input current of positive sequence, $I_{rotor\ p\ estimate}$ is the estimated positive sequence rotor current, $P_{1\ p\ estimate}$ is the estimated positive sequence input active power, $P_{2\ p\ estimate}$ is the estimated positive sequence output power, s is the slip of IM and subscript c connotes the corrected resistance values based on the predicted temperature values at each loading level considered. Similarly, the negative sequence elements of equivalent circuit of IM can be obtained using the equations mentioned below.

$$\text{Power factor} = \cos(\delta_n) = P_{4\_n}/3.V_{2\_n}.I_{2\_n} \quad (10)$$

$$I_{m\_n\_estimation}=\left|\left(V_{2\_n}-\frac{I_{2\_n}.e^{-j\varphi n}}{Y_{stator}}\right)Y_m\right| \quad (11)$$

$$I_{Rotor\_n\_estimation}=\left|I_{2\_n}.e^{-j\varphi n}-I_{m\_n\_estimation}\right| \quad (12)$$

$$r_d=\frac{(2-s)}{3I^2_{rotor\_n\_estimation}}\left(P_{4\_n}-3r_{a\_c}I^2_{2\_n}-3r_mI^2_{m\_n\_estimation}\right) \quad (13)$$

Where, $\cos\delta_n$ is the power factor angle between negative sequence quantities namely input voltage and input current, $I_{m\_n\_estimation}$ is the estimated negative sequence current of the magnetizing subsidiary, $I_{rotor\ n\ estimation}$ is the estimated negative sequence rotor current and r_d is the rotor negative sequence resistance. At any operating load condition, once r_d is obtained, the negative sequence output power can be evaluated using the following equation (14).

$$P_{5\_n\_estimation}=3\left[\left(\frac{r_{d\_c}(s-1)}{(2-s)}\right)I^2_{rotor\_n\_estimation}\right] \quad (14)$$

In the above equations, $P_{5\ n\ estimation}$ & $I_{rotor\ n\ estimation}$ are the estimated output power of negative sequence and the estimated rotor current of negative sequence respectively. For any loading condition, there are four known values namely $I_{1\ p}$, $I_{1\ n}$, $P_{1\ p}$, $P_{4\ n}$ and eight parameters which are unknown. These unknown elements are r_a, r_b, r_c, r_m, x_a, x_b, x_d & x_m. In this paper, an effort has been worked out to estimate the positive sequence parameters by employing CSA to estimate the efficiency of IM. The procedure of application of CSA is as discussed in the following section III.

III. CUCKOO SEARCH OPIMIZATION TECHNIQUE

To solve engineering optimization problems, Xin-She Yang and Deb in 2009 developed CSA [8]. This technique works in accordance with the nurturing behavior of cuckoos & their distinctive levy flights. In proposed approach, at few loading levels, positive sequence input current and electrical powers are extricated. CSA optimizes the positive sequence parameters. Negative sequence rotor elements are evaluated at each loading level based on the estimated positive sequence elements & negative sequence voltage, current and input active power are measured at exclusive loading level. Thence, for each loading level, efficiency of IM can be estimated. The following assumptions are made in this research work. First, at ambient temperature, stator resistance is assumed to be known. Second, on basis of class & design of IM as recommended by IEEE 12 Standard, ratio of x_a to x_b is known. Third, the proportion of friction and windage losses of the IM is considered as 1.2% of the input electrical power and lastly, the proportion of stray load loss is assumed as 1.8% of the rated electrical power.

Steps to procedure to implement optimization are given as:

A. Initialize the input host nest and cuckoo parameters such as induction motor positive sequence parameters, positive sequence input current and powers at various conditions.

B. Generate the random population of n host nests such as

$$x_i=[x_1,x_2\ldots x_n] \quad (15)$$

C. Set the iteration count j = 1.

D. Calculate the fitness F_i of the nests

The initial rated temperature can be assumed as per given equation

$$T_{rated}=\frac{\sqrt{I^2_{1,p}+I^2_{2,n}}}{I_{fullload}}.[T_f-T_{ambient}]+T_{ambient} \quad (16)$$

Estimated temperature at rated value can be obtained as

$$T_{rat,Est}=\Delta T+T_a \quad (17)$$

Where, rise in temperature $\Delta T=K_{thermal\ coeff.}\cdot P_{loss,\ Estimation}$

T_a is the ambient temperature, $P_{loss,\ Estimation}$ is the estimated integral loss of the machine, $K_{thermal\ coeff.}$ is the thermal coefficient of the IM and $I_{full\ load}$ is the full load current.

978-1-4799-6047-7/14 $31.00 © 2014 IEEE

E: Evaluate the fitness of the initial population & generate the new solution X_i^{t+1} for cuckoo i & perform a levy flight as follows

$$x_i^{t+1} = x_i^t + \alpha \oplus Levy(\lambda) \tag{18}$$

where, step size $\alpha > 0$ and the product $\oplus$ implies entry-wise multiplications. According to probability distribution, random walk is provided by Levy flight whereas random length of step is levy distributed and is given as

$$Levy(\lambda) = t^{-\lambda}, 1 < \lambda \leq 3 \tag{19}$$

F: Abandon the fraction (p_a) of worst nests & build new nests.

G: If it satisfies the check for termination, go to step H, else move to previous step C.

H: Stop the process.

The flow chart of the contemplated technique to optimize the parametric quantities to perform efficiency estimation of IM is shown in fig. 2.

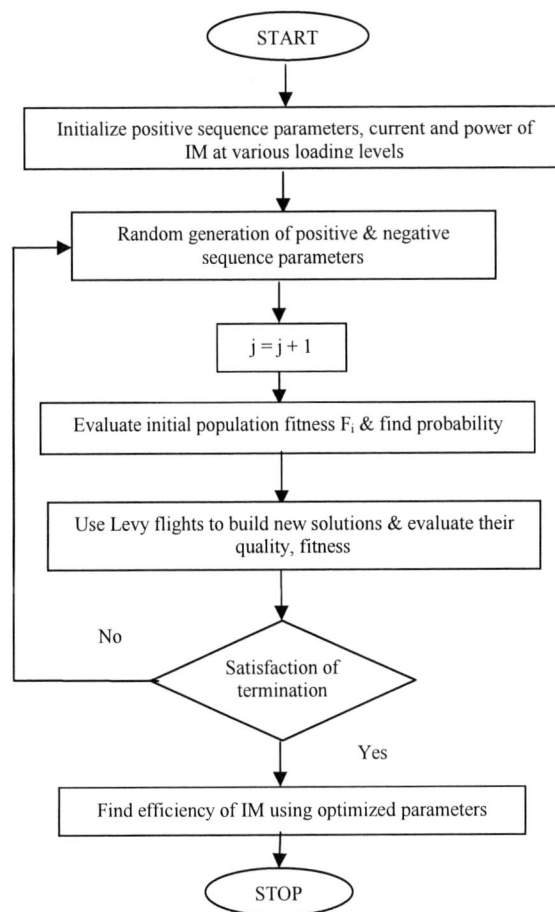

Fig. 2. Flow chart of the proposed optimization technique

The proposed optimization technique is implemented in the MATLAB platform and the analysis of results is performed in Section IV.

IV. SIMULATION PROOF

The proposed technique is implemented using MATLAB. Following are the steps to estimate the efficiency of 3 HP squirrel cage IM using cuckoo search algorithm:

Step 1: The name plate data of the 3 HP IM is obtained from manufacturer data sheet and is given in table I.

TABLE I NAMEPLATE DATA OF THE 3-HP SQUIRREL CAGE IM

Frequency	60
Voltage V_{LL}	208V
Number of Poles	4
Speed	1740 rpm

Step 2: The parametric quantities of 3 HP IM are presented in table II.

TABLE II IMPLEMENTED PARAMETERS OF 3 HP IM

r_a	0.670	r_b	0.373
r_m	1.588	r_d	1.049
x_a	0.856	x_b	1.278
x_m	19.666	x_d	0.747
$K_{thermal} = 0.1365$			

Step 3: The 3 HP IM is simulated using MATLAB. The simulation has been used to generate a data set for validation of the proposed optimization method i.e. CSA. Fig. 3 show the circuit diagram developed for the simulation of the machine. 3 HP IM is powered by three-phase programmable voltage source in which unbalancing can be provided. By using this set up, it is possible to generate the values of input voltage, current and electrical power. This schematic set up gives the flexibility to adjust the torque level of the IM.

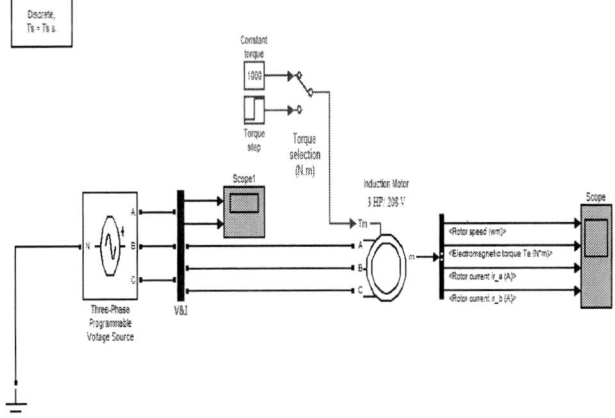

Fig. 3. Schematic diagram of the simulated model

The unbalancing in power supply up to 5% is provided by three-phase programmable voltage source to yield different operating points. The 3 HP IM is operated with these operating conditions. The unbalanced three phase input line voltages & unbalanced input currents of IM at fully loaded situation is shown in fig. 4 and fig. 5 respectively.

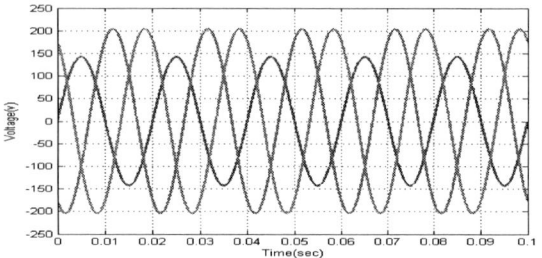

Fig. 4. Unbalanced line voltages of the fully loaded IM

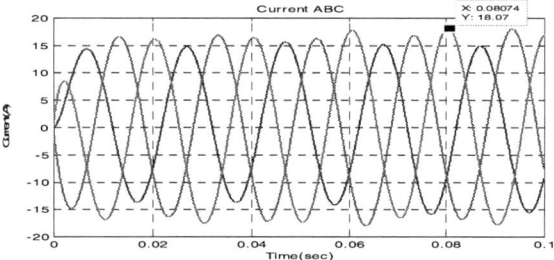

Fig. 5. Unbalanced line currents of the fully loaded IM

It is clear from the fig. 4 and fig.5 that the unbalancing in the three phase line voltages gives rise to the three to four times the unbalancing of currents in the three phases. The full load rated current of the 3 HP IM rises to the values shown in fig. 5 when there is 5% unbalancing in the supply voltages. This situation is not good for the insulation in stator and rotor since insulation got subjected to undue stresses.

Step 4: From the data set as generated by using step 2, the symmetrical component values of input voltages and currents can be extricated at different loading levels. Symmetrical electrical powers are calculated using equations (8) and (9) at different loading levels. The values so obtained are tabulated in form of table III.

TABLE III INPUT VOLTAGES, CURRENTS AND INPUT POWERS AT DIFFERENT LEVELS OF RATED LOAD

Load (%)	25	50	75	85	100
V_{1_p}	119.72	119.42	119.60	119.43	119.36
V_{2_n}	6.15	6.11	6.21	6.07	6.37
I_{1_p}	6.22	6.98	8.17	8.83	9.80
I_{2_n}	2.95	3.00	3.10	3.01	3.12
P_{1_p}	856.98	1438.43	2050.50	2350.35	2760.82
P_{4_n}	32.07	34.42	38.16	37.22	41.60
Speed	1787.2	1774.2	1759.7	1752.6	1740.6

Step 5: Using equations (1)-(14) and parameters of table II, the efficiency of IM is calculated as presented in the table V.

Step 6: The extricated data at the different levels of loading depicted in table III is considered as input data into CSA to estimate the positive sequence IM parameters and get the deviations to justify the CSA capability. CSA objective function optimizes the positive sequence parameters of equivalent circuit of IM at various load points. Using equations (16)-(19) and flow chart shown in fig. 2, CSA is implemented to obtain IM efficiency. The IM parameters estimated by CSA technique and their comparability with calculated parameters are presented as in table IV.

TABLE IV COMPARABILITY OF CSA PARAMETERS WITH GIVEN PARAMETERS OF 3 HP SQUIRREL CAGE IM

IM parameters	Table II parameters	CSA Estimation	Error (%)
x_b	1.2780	1.0147	-0.2633
x_d	0.7470	0.9160	0.169
r_b	0.3730	0.4149	0.0419
r_d	1.0490	0.8837	-0.1653
r_m	1.5880	1.1224	-0.46
k_th	0.1365	0.1083	-0.0282
x_m	19.6660	24.2978	4.63

The enrichment progression of the objective function at 500 iterations is presented in fig.4.

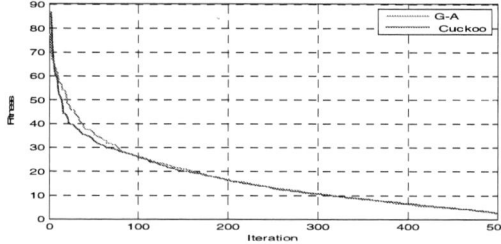

Fig.6 Progression of fitness objective function

The variation of 3 HP IM parameters namely r_b and x_b, for 500 iterations during optimization algorithm, as a sample is shown below in fig. 7.

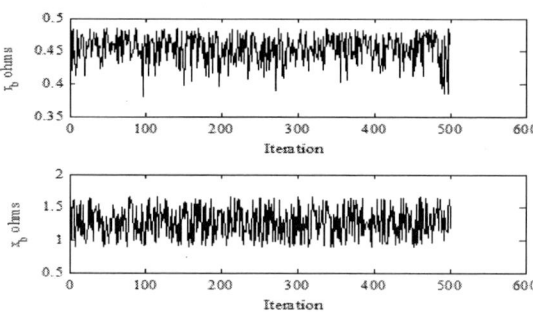

Fig.7. Variations of various IM parameters during optimization algorithm

CSA optimizes the positive sequence parameters. This results in obtaining negative sequence parameters. So, with the help of attained parameters, the efficiency of IM is determined and given in table V. It can be consummated that the proposed

CSA technique is adequate of obtaining the IM efficiency accurately with the knowledge of currents and powers which are fed to IM and its nameplate data.

V. RESULTS AND DISCUSSIONS

The load rating of the machine is changed in different levels and the corresponding efficiency has been estimated. Then the error between the approximated values and the calculated values is evaluated. In this paper, for the sake of comparison, the parameters of IM and its efficiency have been generated by simulating GA technique on MATLAB platform. The estimated efficiency using different techniques are shown in the following table V.

TABLE V EFFICIENCY COMPARISON AT FIVE DIFFERENT LOADINGS

Rated Load (%)	Efficiencies with different techniques (%)			Error (%)	
	Calculated	GA	CSA	GA	CSA
25	63.5527	66.87	66.0984	3.87	2.54
50	74.5831	78.91	79.0198	4.33	4.33
75	78.5289	83.21	81.8653	4.69	3.33
85	78.4662	84.06	81.0564	5.5	2.5
100	80.1397	84.89	82.6180	4.8	2.4

From table V, it is evident that the error percentage is less with proposed approach as compared to GA technique. The comparison of error percentage of the GA technique and proposed CSA technique is shown in fig. 8.

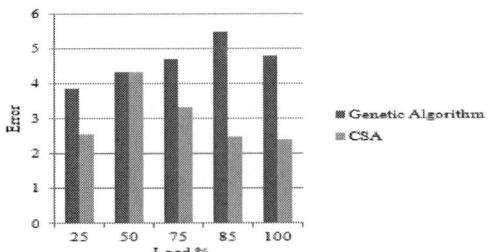

Fig. 8. Comparison of error percentage of GA and CSA

It clearly shows that the error percentage is less with proposed approach as compared to GA technique. Results obtained by proposed approach are close to the calculated efficiencies at various loading conditions as compared to the GA technique. Still small amount of error exists due to the fact that rotor parameters values keeps changing with variation of slip from zero to maximum. At higher values of slip, as the rotor frequency rises, the collective impact of flux of the cross slots and skin effect changes the values of the rotor parameters of IM. It can be correctly stated that rotor parameters of IM depends on IM slip. This results in the difficulties arising in the accurate prediction of the starting torque of IM. Also, execution time taken by CSA is 42 seconds whereas GA is taking more than two minutes. Hence, CSA technique is advantageous because randomization is more efficient. This technique allows the possibility of large step length. Also, the number of constants to jingle is beneath the GA and PSO. So,

this technique can adapt to wider class of optimization problems.

VI. ACKNOWLEDGEMENT

We would like to express the gratitude to DST – FIST program for financial support.

VII. CONCLUSION

This paper focuses on the efficiency estimation of induction machine (IM) by employing a novel cuckoo search algorithm (CSA). In the proposed technique, IM parameters were optimized at five different loading levels with the aid of extricated parameters. The optimized estimated parameters were used for the efficiency evaluation. The proposed algorithm is found advantageous in the non-linear operating range during IM operation and is shown to be more effective for efficiency estimation considering the lower level of intrusion and high accuracy. The implementation of CSA on existing IM working in loaded condition does not change any operating condition of IM. The effectiveness of the proposed technique is analyzed by comparing the results obtained with genetic algorithm (GA). Simulation results & observations from comparison of different techniques prove the effectiveness of the proposed optimization approach.

REFERENCES

[1] I. Farmani, A. Arefi, R. Bagheri and, H. Oraee, "Industrial Electric Motor Energy Efficiency under Testing and Practical Conditions", In proceedings of International energy conference, 2005.

[2] Bin Lu, Thomas G. Habetler and Ronald G. Harley, "A Survey of Efficiency-Estimation Methods for In-Service Induction Motors", IEEE Transaction on Industry Applications, Vol. 42, No. 4, July/August 2006.

[3] Arbi Gharakhani Siraki and Pragasen Pillay, "Comparison of Two Methods for Full Load In-Situ Induction Motor Efficiency Estimation from Field Testing in the Presence of Over/Under Voltages and Unbalanced Supplies", IEEE Transactions on Industry Applications, Vol.48, No.6, pp.1911-1921, 2012.

[4] V.P. Sakthivel, R. Bhuvaneswari and S. Subramanian, "An Improved Particle Swarm Optimization for Induction Motor Parameter Determination", International Journal of Computer Applications, Vol. 1, No. 2, pp. 62-67, 2010.

[5] Mehmet Cunkas and Tahir Sag, "Efficiency determination of induction motors using multi-objective evolutionary algorithms", International Journal of Advances in Engineering Software, Vol.41, No.2, pp.255-261, 2010.

[6] V. P. Sakthivel, R. Bhuvaneswari and S. Subramanian, "An accurate and economical approach for induction motor field efficiency estimation using bacterial foraging algorithm", International Journal of Measurement, Vol.44, No.4, pp.674–684, 2011.

[7] IEEE Standard Test Procedure for Polyphase Induction Motors and Generators, IEEE Standard 112, 2004.

[8] Xin-She Yang and Suash Deb, "Cuckoo Search via Levy Flights", In Proceedings of World Congress on Nature & Biologically Inspired Computing, pp.1-7, 2009.

[9] Arbi Gharakhani Siraki and Pragasen Pillay, "An In Situ Efficiency Estimation Technique for Induction Machines Working With Unbalanced Supplies", IEEE Transactions on Energy Conversion, Vol.27, No.1, pp. 85-95, 2012.-73.

Voltage Regulation Enhancement in a Buck Type DC-DC Converter Using Queen Bee Evolution Based Genetic Algorithm

Tousif Khan N
Department of Electronics and Electrical Engineering,
Indian Institute of Technology Guwahati,
Guwahati, Assam-781039, India
tousif@iitg.ac.in

K. Sundareswaran
Department of Electrical and Electronics Engineering,
National Institute of Technology Tiruchirappalli,
Tiruchirappalli, Tamil Nadu-620015, India
kse@nitt.edu

Abstract— Buck converters are employed in several applications such as laptops, switched mode power supply, electronic gadgets, battery charging and communication networks. In order to have good regulation of capacitor voltage, Buck converters are generally provided with a feedback control mechanism. Since the converter dynamics changes from ON time period to OFF time period of the power electronic switch, the traditionally designed feedback controller provides only near-satisfactory voltage regulation. Aiming to obtain a more robust controller, which can track the desired voltage profile over large domain of operating points, the controller design for the feedback control of Buck DC-DC converter is constructed as an optimization problem and then the solution is acquired through a recently developed optimization technique known as queen bee evolution based Genetic Algorithm. The theoretical notions and implementation of the proposed methodology towards the search of a robust feedback control parameters for Buck converters are discussed in this paper. Extensive simulation and experimental studies under the presence of eventualities like matched and mismatched uncertainties are compared with conventional Genetic Algorithm, which thereby confirms the validity of the new approach.

Keywords— Buck converter, Genetic Algorithm (GA), queen bee, optimization, Proportional-Derivative-Integral (PID) Controller

I. INTRODUCTION

Buck converter is a type of DC-DC converters, which provides a variable Direct Current (DC) output voltage lesser than the input DC voltage. Buck converters are widely employed in many industrial and residential applications namely switched mode power supplies, electronic gadgets, motor drives, communication networks and battery charging [1, 2]. DC-DC converters are when connected in an open loop mode they produce poor regulation of output voltage. Hence normally these converters are used with closed loop control scheme for the robustness in the output voltage control [3]. Buck converter belongs to the class of variable structure systems where the modes of its operation differ from ON time period to OFF time period of the power electronic switch. The controller design for Buck converter when achieved through linear control theory results in unsatisfactory response. The conventional method adopted in the literature is that, the state space averaging is done first to derive the transfer function and then the linear control technique is made use of next, for the

controller deign. However such method of design is not sufficiently immune towards the matched uncertainty such as input voltage changes and mismatched uncertainty such as load resistance.

In the literature, other control methodologies such as fuzzy logic control [4, 5] and sliding mode control [6, 7] mechanisms which falls under the category of non-linear methods, are reported to provide an excellent transient and steady state response.

In this article, the controller design problem for the Buck converter is rewritten as an optimization task and the subsequent parameters of controller are then acquired using queen bee evolution based genetic algorithm, a novel technique proposed by K. Sundareswaran and V. T. Sreedevi [11]. Here, the steps of conventional genetic algorithm are suitably modified to combine the merits of both genetic algorithm and queen bee in a bee-hive algorithm. This notion is then used for the feedback controller design. An appropriate objective function is framed which reduces the difference between the desired output voltage and the actual output voltage obtained from the Buck converter. A suitable fitness function is then selected from this objective function and is incorporated in the evolutionary search of optimum parameters of Proportional-Derivative-Integral (PID) controller. During this process the exact dynamics of the Buck converter during ON time period and OFF time period are incorporated with purely no approximations used in the modeling procedure. The aspects of the large signal modeling together with that of biologically inspired evolutionary search yields a robust controller, which counteracts and rejects the effect of both source voltage disturbance and load resistance disturbance at wide operating range. Theoretical forecasts are found to be true from extensive simulation study and the readings tabulated. Experimental prototype is fabricated in the laboratory and the results obtained through the simulations are verified. This validation in return justifies the validity of new approach.

II. CONTROLLER DESIGN FOR BUCK CONVERTER VIA QUEEN BEE EVOLUTION BASED GENETIC ALGORITHM

A. Modeling of Buck converter

Buck converter is shown in the Fig.1. Where V_{in} is the input voltage, L is the filter inductance and C is the filter capacitance.

The modeling [8, 9] of Buck converter systems are divided into two modes.

ON-Mode: $0 \leq t \leq T_{on}$

This mode begins when the power switch MOSFET is turned ON in the interval $0 \leq t \leq T_{on}$, where T_{on} is the ON time period of the switching signal generated by pulse width modulation technique.

Let $x_1 = i_L$ and $x_2 = v_o$, where i_L is the inductor current and v_o is the capacitor voltage.

The state-space model obtained during ON mode is given as

$$\begin{bmatrix} \dot{x}_1 \\ \dot{x}_2 \end{bmatrix} = \begin{bmatrix} -\frac{1}{L}\left[r_L + \frac{Rr_c}{(R+r_c)}\right] & -\frac{1}{L}\frac{R}{(R+r_c)} \\ \frac{R}{C(R+r_c)} & -\frac{1}{C(R+r_c)} \end{bmatrix} \begin{bmatrix} x_1 \\ x_2 \end{bmatrix} + \begin{bmatrix} \frac{1}{L} \\ 0 \end{bmatrix} V_{in} \tag{1}$$

$$v_o = \begin{bmatrix} \frac{Rr_C}{R+r_C} & \frac{R+r_C}{R} \end{bmatrix} \begin{bmatrix} x_1 \\ x_2 \end{bmatrix} \tag{2}$$

OFF-Mode: $T_{on} \leq t \leq T$

This mode begins when the power switch MOSFET is turned OFF in the interval $T_{on} \leq t \leq T$, where T is the total time period of the switching signal generated by pulse width modulation technique.

The state-space model obtained during OFF mode is given as

$$\begin{bmatrix} \dot{x}_1 \\ \dot{x}_2 \end{bmatrix} = \begin{bmatrix} -\frac{1}{L}\left[r_L + \frac{Rr_c}{(R+r_c)}\right] & -\frac{1}{L}\frac{R}{(R+r_c)} \\ \frac{R}{C(R+r_c)} & -\frac{1}{C(R+r_c)} \end{bmatrix} \begin{bmatrix} x_1 \\ x_2 \end{bmatrix} + \begin{bmatrix} 0 \\ 0 \end{bmatrix} V_{in} \tag{3}$$

$$v_o = \begin{bmatrix} \frac{Rr_C}{R+r_C} & \frac{R+r_C}{R} \end{bmatrix} \begin{bmatrix} x_1 \\ x_2 \end{bmatrix} \tag{4}$$

B. Problem Formulation

Objective is to track a constant reference voltage signal. Hence when Buck converter is operated in closed loop mode with PID controller and it is realized as discussed by [11] to produces a control signal, which is then compared with carrier signal to generate pulses for the switch Sw. This in turn results in output voltage from the converter which is measured at finite intervals of time. This output variable is compared with a constant reference voltage for the trajectory tracking. The corresponding error variable is thus defined as,

$$e(t) = V_o^* - v_o(t) \tag{5}$$

Where V_o^* is the desired output voltage and $v_o(t)$ is the instantaneous output voltage taken across the capacitor of Buck converter. Error signal generated is further passed through the PID controller. The gives the following signal

$$u(t) = K_p e(t) + K_i \int e(t)dt + K_d \frac{d}{dt} e(t) \tag{6}$$

This is then compared with a constant saw tooth waveform to finally generate a pulse width modulated signal to operate the converter.

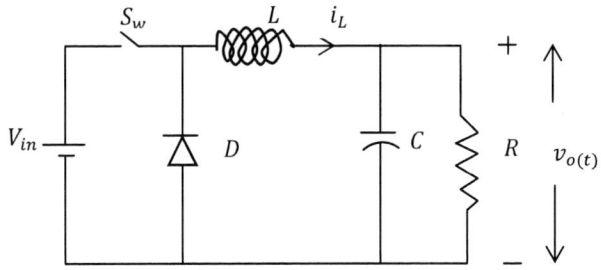

Fig. 1. Buck converter

B. Optimization

Defining the optimization problem to find $\phi = \{K_p, K_i, K_d\}$ as follows: $f(\emptyset) = \sum_{t=0}^{T}(V_o^* - v_o(t))$ subjected to $\emptyset_l \leq \emptyset \leq \emptyset_u$, where $\emptyset_l$ and $\emptyset_u$ are lower and upper value of the variable $\emptyset$.

C. Queen bee evolution based Genetic Algorithm

The queen bee in a bee hive is an adult female bee. Her only work is to act as a bee reproducer. Study says the the queen flies, at some specific climatic conditions to a congregation place where drones in large numbers wait for her. The drones follow the queen and many of them also mates with her. After mating, drones die as their abdomen bursts. Then the queen bee comes back to its hive to the lay eggs. Certain new born bees are identified by the queen bee, which has the potential to become a queen in future and they are carefully fed to make them sexually mature. The young bees also called virgin queens. They compete among themselves, based on survival of fittest mechanism to bring out a potential queen bee and the rest becomes the worker bees.

D. Tuning of PID controller using queen bee evolution based GA.

In a bee hive queen bee is the most prominent element. This algorithm tries to find the best queen bee after several numbers of reproduction cycles, as explained in previous section. The readers can refer to [11] for more details. In other terms the best queen bee is the desired solution set $\phi = \{K_p, K_i, K_d\}$ evolved over a number of iterations in the algorithm.

The procedure of queen bee evolution based GA is briefly described below:

Step 1: Generation of honey bees in a bee hive.

The first step is to generate bees in the feasible solution domain. Each bee is in reference to one complete solution set to solve the problem i.e. one 8-bit representation of K_p, K_i, K_d values, with their ranges pre-defined. Let bees be denoted as $b1, b2, \ldots \ldots bj, \ldots .. bn$ where, n stands for honey bee population size and n in this paper is taken as 8.

978-1-4799-6047-7/14 $31.00 © 2014 IEEE

Step 2: Queen bee identification.

Among the randomly generated honey bees, the best bee is selected which gives the lowest $f(\emptyset)$. In other words, the best solution set $\phi = \{K_p, K_i, K_d\}$ is retained as queen bee. Eq. (7) finds the queen bee, as denoted bq in the kth iteration.

$$bq(k) = Max_j \left(\frac{1}{1+|f(\emptyset)|_j(k)|} \right) \qquad (7)$$

This queen bee is now kept aside and the remaining $(n-1)$ bees are treated as drones. Meaning, the best combination of K_p, K_i and K_d generated is separated from rest of the solutions.

Step 3: Reproduction of bees

This step describes the reproduction process. All drones may not succeed to mate with the queen bee, this gives us a liberty to assign a recombination probability linked with each drone. The recombination probability, p_r is suitably fixed between 0 and 1. The probability of each drone is represented by p_j.

If $p_j \geq p_r$, the drone mates with the queen bee giving birth to two off springs. The recombination procedure of drone with queen bee is as same as the crossover procedure in conventional GA. Among these off springs the best one survives and the other one is discarded.

In our context, if the above criterion satisfies, then we proceed with reproduction mechanism, and again generate two different solution set $\phi = \{K_p, K_i, K_d\}$ and retain the best one based on its fitness. Here, the value of p_r is fixed and p_j is randomly generated.

Step 4:

If $p_j < p_r$, then no mating occurs and hence no offspring or solution sets are generated.

At the end of steps 3 and step 4, i.e., at the end of reproduction process the population of virgin queen bees in bee hives will be less than the initial population size, n.

Step 5: Piping

This step involves once again evaluation of queen bee along with virgin bees, using Eq. (7) and a new queen bee is explored and rest all other bees are discarded. Meaning, the new solution set and the previous best solution set are again evaluated against the fitness function given by Eq. (7) and the best amongst them is retained and rest of all solution sets are discarded.

Step 6: Stop the computer code and opt the new queen bee, meaning the new parameters of the PID controller as the best solution, if the termination criterion is achieved, else go back to step no 7. Criterion for termination is decided by the user as the minimum value of Eq. (7).

Step 7: Creation of a new population for the drone to be used in the next iteration. Meaning generate new random values of controller parameters to be used in next iteration.

III.   RESULTS AND DISCUSSION

A dedicated program is written in MATLAB for controller identification using queen bee evolution based GA. The solution is also sort through conventional GA. The computed results are discussed and analyzed in this section. Figure 2 depicts the behavior of objective function of the proposed algorithm. It is seen that the convergence rate of this algorithm is faster. The value of objective function is 4395 at 87th iteration in case of proposed method. For the comparison purpose, the objective function value of some of the best chromosome generated in conventional GA is also plotted in the same figure 2. Figure reveals that the rate of convergence of conventional GA is relatively slow. Both the algorithms are run for 500 iterations and at the end the objective function value is found to be 4395 with queen bee evolution based GA, whereas for GA this value is found to be 4600. This clearly reveals that queen bee evolution based GA terminates at better solution in comparison with conventional GA.

In order to examine the effectiveness of the proposed algorithm, extensive simulation and experimental studies have been carried out on the closed loop system of Buck converter. Buck converter performance is tested under the following disturbances: (i) a step change in the value of input voltage 0-20V, (ii) a step change in value of load current from 86.95mA to 43.47mA and again back from 43.47mA to 86.95mA and (iii) a step change in the value of input voltage from 36V to 28V and again back from 28V to 36V.

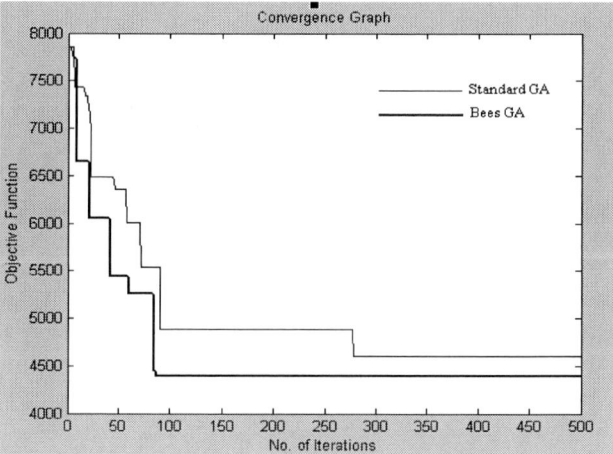

Fig. 2. Objective function of queen bee evolution based GA in comparison with conventional GA.

In the experimental study, Buck converter is implemented in the laboratory and readings were taken under the identical conditions as that of simulation study and controller parameters are generated through control algorithm and are realized through op-amp LM741 IC. Values of k_p, k_i and k_d are modeled in the same manner as given in [11]. Switching frequency f_s selected is $1.5kHz$. The computed and measured waveforms are included in Fig. 4.5 and Fig. 4.6. The experimental result clearly highlights the simulation findings. For more quantitative assessment, Table. 1 is prepared which tabulates the various dynamic response measurements. Thus it can be concluded that the simulation results are in good agreement with the hardware result analysis.

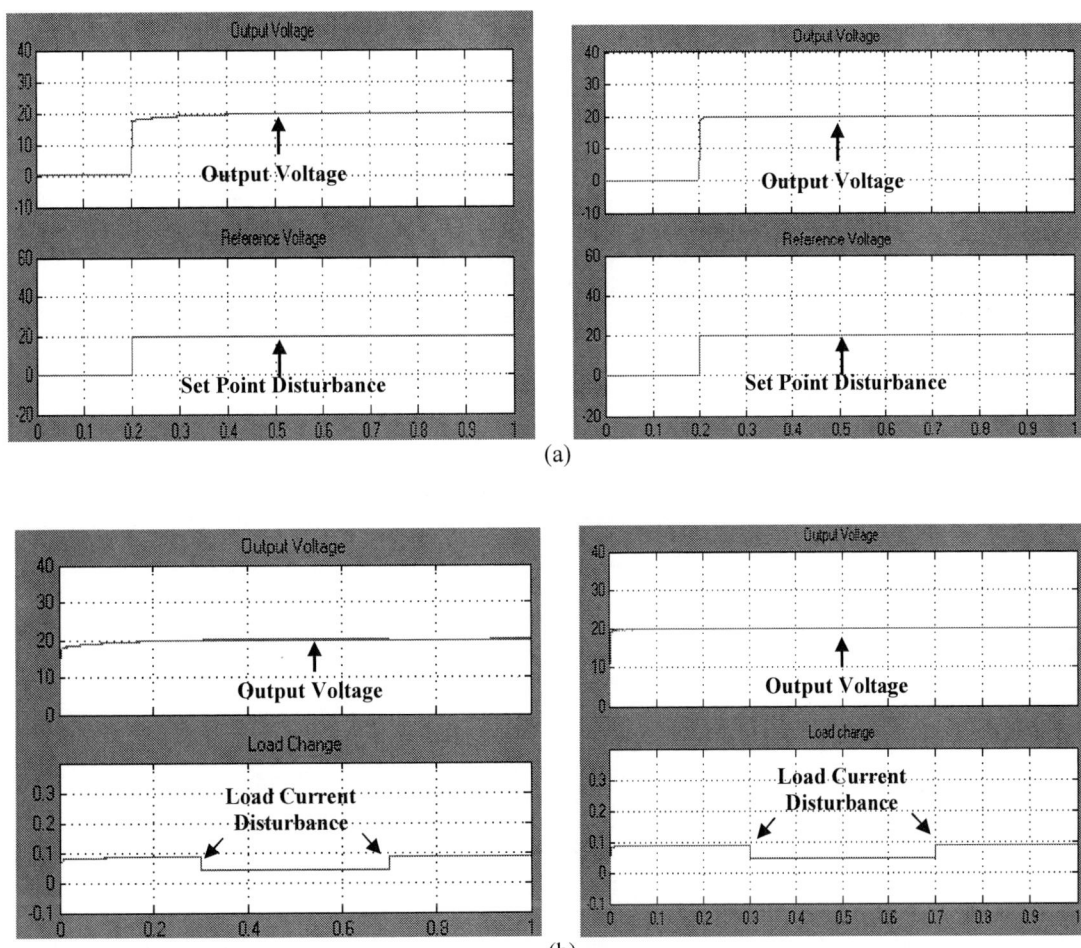

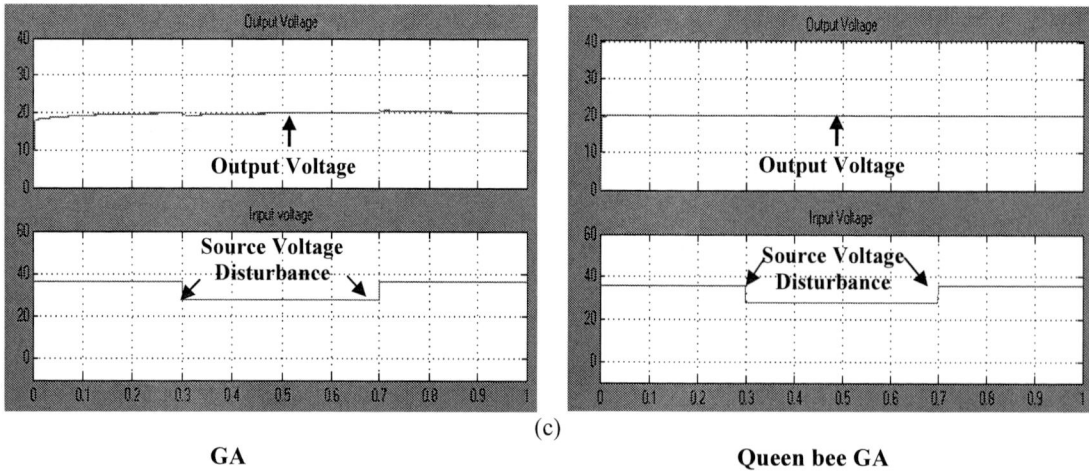

(c)

GA　　　　　　　　　　　**Queen bee GA**

Fig. 3 Simulation results with conventional Genetic Algorithm and queen-bee Genetic Algorithm based controller for Buck type DC-DC converter. (a) Set point disturbance from 0-20volts in the value of output voltage. Upper trace: output voltage, *(10V/div)*; lower trace: reference voltage, *(20V/div)*. (b) Response of output voltage due to the disturbance in load current from 86.95mA to 43.47mA and again from 43.47mA to 86.95mA. Upper trace: output voltage, *(10V/div)*; lower trace: change in load current, *(2mA/div)*. (c) Source voltage disturbance from 36 volts to 28volts and again from 28volts to 36volts. Upper trace: output voltage, *(10V/div)*; lower trace: change in input voltage, *(20V/div)*.

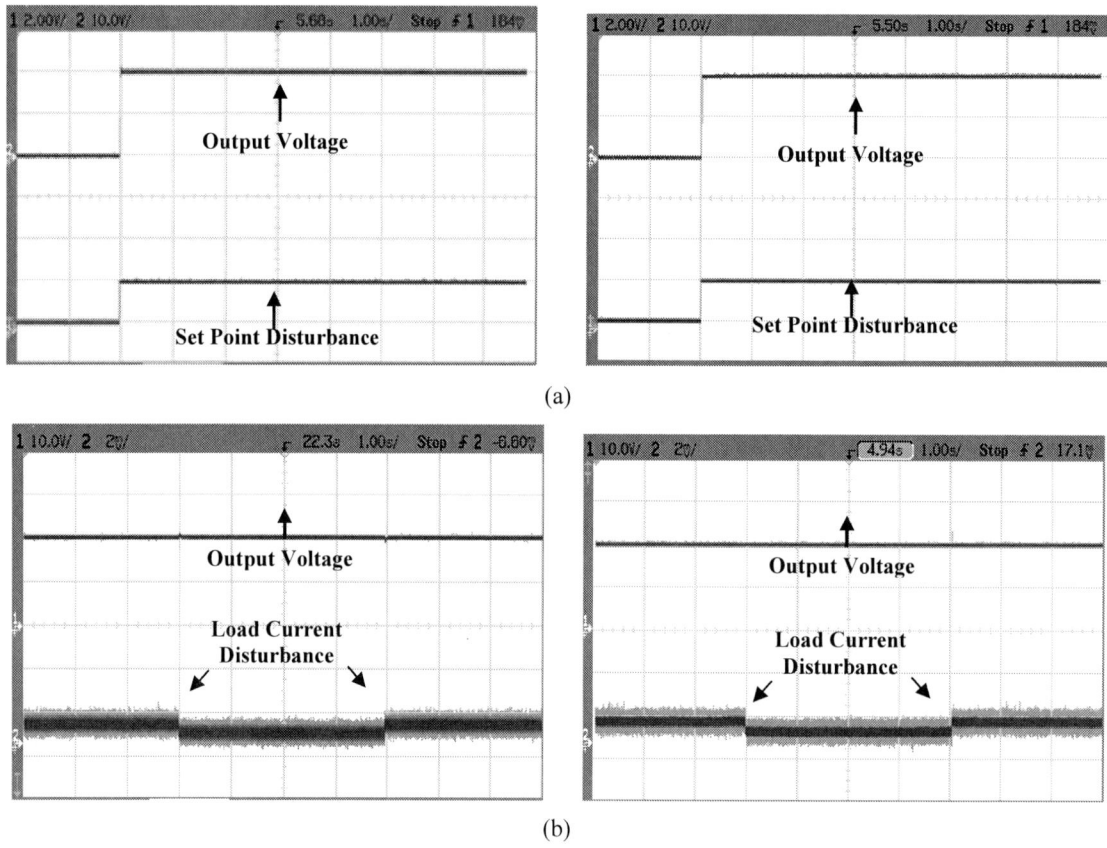

978-1-4799-6047-7/14 $31.00 © 2014 IEEE

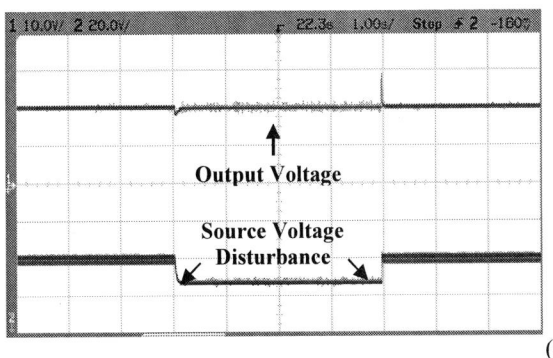 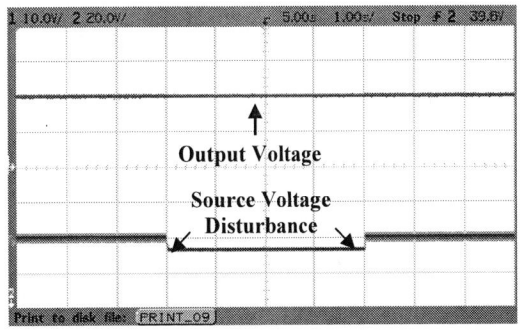

(c)

GA   **Queen bee GA**

Fig. 4 Experimental results with Genetic Algorithm and queen-bee Genetic Algorithm based controller for Buck type DC-DC converter. (a) Set point disturbance from 0-20volts in the value of output voltage. Upper trace: output voltage, *(10V/div)*; lower trace: reference voltage, *(20V/div)*. (b) Response of output voltage due to the disturbance in load current disturbance from 86.95mA to 43.47mA and again from 43.47mA to 86.95mA. Upper trace: output voltage, *(10V/div);* lower trace: change in load current, *(2mA/div).* (c) Source voltage disturbance from 36 volts to 28volts and again from 28volts to 36volts. Upper trace: output voltage, *(10V/div);* lower trace: change in input voltage, *(20V/div).*

A closer examination of the computed curves depicts in Fig. 3 and Fig. 4 clearly indicates that the output voltage remains unperturbed under different types of perturbations. This indicates that the controller identified through Bees GA is robust and hence this controller rejects both internal and external disturbances effectively. However the output voltage dynamics obtained with GA based controller is little inferior in comparison with the one obtained through Bees GA.

TABLE I.

Dynamic response comparison of the Buck type DC-DC converter tuned with GA and queen bee evolution based GA.

Dynamic Response Specifications	*Genetic Algorithm* $k_p = 7.37$ $k_i = 73.70$ $k_d = 0.019$	*queen bee evolution based GA (proposed)* $k_p = 10$ $k_i = 109$ $k_d = 0.225$
Settling time	$38ms$	$12ms$
Steady state error	$1Volt$	$0.035Volt$

IV. CONLSUSION

The controller design for Buck converter is rewritten as an optimization problem and then the queen bee evolution based Genetic Algorithm is employed to solve this optimization problem of minimizing the error. This investigation reveals that results obtained from the queen bee evolution based GA tuned controller are better than traditionally obtained Genetic Algorithm. This fact is well established through both simulation and experimental study.

REFERENCES

[1] Simon Ang and Alejandro Oliva, "Power Switching Converters," *CRC press, Taylor and Francis group*, New York, pp. 17-32, 2005.

[2] Robert W. Erickson, "Fundamentals of Power Electronics," *University of Colorado*, Chapman and Hall publications, New York, 1997.

[3] B. Aldo, D. Corsanini, A. Laneli and L. Sani, "Circle based criteria for performance evaluation of controlled dc-dc switching converters," *IEEE Transactions on Industrial Electronics,* vol. 53, No. 6, pp. 1682-1689, Dec. 2006.

[4] A. G. Perry, G. Feng, Y. F. Liu and P. C. Sen, "A Design Method for PI-like Fuzzy Logic Controllers for DC-DC Converter," *IEEE Transactions on Industrial Electronics*, vol. 45, No. 5, pp. 2688-2695, Oct. 2007.

[5] Liping Guo, *Member, IEEE*, John Y. Hung, *Senior Member, IEEE, and* R. M. Nelms, *Fellow, IEEE*, "Evaluation of DSP-Based PID and Fuzzy Controllers for DC–DC Converters," *IEEE Transactions on Industrial Electronics.* Vol.56, No. 6, June 2009.

[6] E. Figuerer, G. Garcera, J. M. Benavent, M. Pascual and J. A. Martine 3, "Adaptive two loop voltage mode control of dc-dc converters," *IEEE Transactions on Industrial Electronics*, vol. 53, No. 1, pp. 239-253, Feb. 2006.

[7] S. C. Tan, Y. M. Lai and C. K. Tse, "General design issues of sliding mode controllers in DC-DC converters," *IEEE Transactions on Industrial Electronics*, vol. 55. No. 3, pp. 1160-1174, March 2008.

[8] J. Mahdavi, A. Emadi and H.A. Toliyat, "Application of State Space Averaging Method to Sliding Mode Control of PWM DC-DC Converters," *IEEE Industry Applications Society Annual Meeting New Orleans*, Louisiana, pp.820-827, October 5-9 1997.

[9] R.D. Middlebrook, and S. Cuk, "A general unified approach to modeling switching converter power stages," *IEEEC-PESC Conf. Rec., IEEE-PESC Rec.*, pp. 445-451, 1991.

[10]. T.D. Seeley, 'The Wisdom of the Hive: The Social physiology of Honey Bee Colonies,' *Cambridge, Massachusetts: Harvard University Press,* 1996.

[11] Sundareswaran. K and Sreedevi, V.T "Boost Converter Controller Design Using Queen-Bee-Assisted GA", IEEE Transactions on Industrial Electronics , Volume:56 , Issue: 3 , Pp. 778 – 783, March 2009.

978-1-4799-6047-7/14 $31.00 © 2014 IEEE

Class-D/E Resonant Inverter for Multiple-Load Domestic Induction Cooking Appliances

P. Sharath Kumar[1], N. Vishwanathan[3], B. K. Murthy[4]
[1,3,4] Dept. of Electrical Engineering
[1,3,4] National Institute of Technology Warangal
Warangal, India
[1]sharathpapani@nitw.ac.in

D. Vijaya Bhaskar[2]
[2]Dept. of Electrical Engineering
Indian School of Mines
Dhanbad, India
[2]devaravijay@gmail.com

Abstract—This paper presents a class – D/E resonant inverter for multiple-load domestic induction cooking appliances. Most of the induction inverter configurations used in home appliances is based on single coil inverter. In these configurations, the vessel size cannot be more than induction coil size. The proposed configuration is suitable for any size of vessel and also can be used for multiple-loads. It operates with constant switching frequency and independent control of each load. The output power of each load can be controlled with asymmetrical duty cycle control technique for class-D configuration and with asymmetrical voltage cancellation control technique for class-E configuration. The proposed configuration and control techniques are simulated and experimentally tested with two loads. Simulation and experimental results are in good agreement. This configuration can be extended to multiple-loads.

Keywords— Resonant inverter; Induction cooking appliances; Asymmetrical duty cycle control; Asymmetrical voltage cancellation control;

I. INTRODUCTION

Now a day's induction heating (IH) technology is being used in various applications. Induction cooking is one of the main applications in induction heating. The induction cooking appliances are widely used in domestic level due to the reduced heating time, higher efficiency and cleanness. By improving the power rating and resonant inverter configurations, induction cooking appliances can also be used in commercial level. Induction cooking appliances heat up the vessel with a high frequency current. Generally, induction resonant inverters operate in the range of 20-400 kHz. In conventional methods, the heat is transferred from source to load by conduction or radiation. In induction heating, the eddy currents are induced in the load (vessel) due to the generation of magnetic flux at high frequency based on Faraday's laws of electromagnetic induction principle and there by producing heat by Joule's heating principle [1].

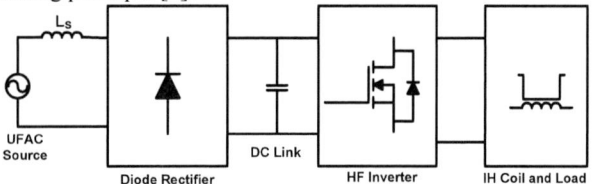

Fig. 1. Typical arrangement of HFAC IH resonant inverter

Fig. 1, shows a typical arrangement of high frequency AC induction heating resonant inverter. IH resonant inverter takes the energy from the utility frequency AC source, which is rectified by a bridge of diodes. Then the high frequency resonant inverter topology supplies the high-frequency current to IH coil. IH load parameters are the combination of IH coil and vessel.

Generally used topologies in induction cooking application are quasi resonant, half-bridge and full-bridge resonant inverter circuits [2]. In induction cooking application Variable Frequency (VF) scheme, Pulse Frequency Modulation (PFM), Pulse Amplitude Modulation (PAM), and Phase Shift Modulation (PSM) are used for output power control [3]-[8]. In VF scheme to control the output power for a constant load by varying the normalized switching frequency, in case of below resonance operation filter components are large for the low-frequency range [3]. PFM control has ZVS soft switching operating region is relatively narrow. In PAM control for constant load amplitude of the source voltage is varied to control the output power. PSM control gives high efficiency at higher duty ratio [4]. ADC control gives ZVS at higher duty ratios in full-bridge inverter configuration [5]. For reducing switching losses, it is mainly used in half-bridge topology. AVC control gives ZVS at lower duty ratios also in full-bridge inverter configuration. AVC control technique is mainly used in full-bridge topology [5]-[6].

This paper proposes class-D/E inverter configuration for induction cooking application. The proposed configuration can be operated with constant switching frequency and independent output power control of each load is explained in detail. In this configuration ADC control technique is used for class-D inverter configuration and AVC control technique is used for class-E inverter configuration. This configuration can be extended to multiple-loads.

II. OPERATING PRINCIPLE OF IH AND LOAD CHARACTERISTICS

A. Operating Principle of Induction Heating

Operating principle of IH is that when induction heating coil is energized by high frequency current, it produces magnetic flux. It causes eddy currents that occur in heating load and this result in heating effect. The induced eddy currents

978-1-4799-6047-7/14 $31.00 © 2014 IEEE

are concentrated in the vessel bottom layer at skin depth (δ) level [2], which is explained by

$$\delta = \sqrt{\frac{\rho}{\pi \mu f_s}} = \sqrt{\frac{1}{4\pi^2 \times 10^{-7}}} \times \sqrt{\frac{\rho}{\mu_r f_s}} \quad (1)$$

where, ρ is electrical resistivity, μ is magnetic permeability and μ_r is relative magnetic permeability of load material and f_s is switching frequency of the inverter circuit.

The load surface resistance (R_L) is determined by the load skin depth and its material specific resistance is shown in below expression,

$$R_L = \frac{\rho}{\delta} = k\sqrt{\rho \mu_r f_s} \quad (2)$$

where k is constant = 0.00198692

The load parameters depends on several variables including the shape of the heating coil, the spacing between the heating coil and cooking vessel (load), their electrical conductivity and magnetic permeability, and the inverter switching frequency.

B. Equivalent Circuit of IH Coil and Load

A linear equivalent model of the IH coil and load represented by the effective equivalent inductance (L_{eq}) in series with effective equivalent resistance (R_{eq}) is referred to the input side of IH coil.

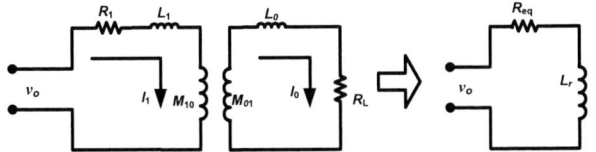

Fig. 2. Equivalent circuit of IH coil with load

Fig. 2, shows the equivalent circuits for IH coil with load parameters. Load parameters are taken as single turn short circuited secondary winding.

The circuit elements are represented as:

1) R_L surface resistance of the load
2) L_0 inductance of the load
3) R_1 resistance of IH coil
4) L_1 inductance of IH coil
5) I_0, I_1 Load current and IH coil current
6) M_{10}, M_{01} the mutual inductance between IH coil and load.

The voltage equations for the above equivalent circuit:

$$v_o = I_1 R_1 + L_1 \frac{dI_1}{dt} + M_{10} \frac{dI_0}{dt} \quad (3)$$

$$0 = I_0 R_L + L_0 \frac{dI_0}{dt} + M_{01} \frac{dI_1}{dt} \quad (4)$$

From (3) and (4) equations,

$$R_{eq} = R_1 + \frac{(\omega M)^2 . R_L}{R_L^2 + (\omega L_0)^2} \quad (5)$$

$$L_r = \left\{ L_1 - \frac{(\omega M)^2 . L_0}{R_L^2 + (\omega L_0)^2} \right\} \quad (6)$$

$$R_{eq} = R_1 + A^2 R_L \quad (7)$$

$$L_r = \{ L_1 - A^2 L_0 \} \quad (8)$$

where, $M_{10} = M_{01} = M$

and $A = \frac{(\omega M)}{\sqrt{R_L^2 + (\omega L_0)^2}} = \frac{M}{L_0}$ at $\omega L_0 \gg R_L$

III. PROPOSED CLASS-D/E INVERTER CONFIGURATION

The proposed class-D/E inverter configuration with two loads is operated at 43.66 kHz constant switching frequency. Generally, the vessel size is limited to the IH coil size. In this mode, the proposed configuration can be operated in class-D inverter configuration up to vessel size is equal to IH coil size. If vessel size is more than IH coil size, the vessel will be placed on two IH coils and operated as a single load. In this mode, the proposed configuration can be operated in class-E inverter configuration. Fig. 3, shows the proposed class-D/E inverter configuration with two loads.

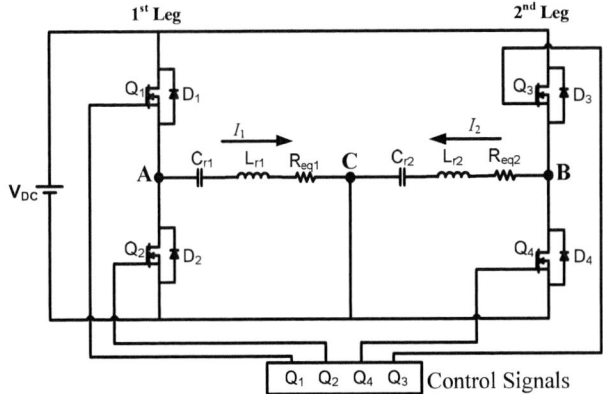

Fig. 3. Proposed class-D/E inverter configuration

A. Characteristics of resonant tank

The resonant tank circuit in class-D/E inverter shown in Fig. 3 can be described by the following parameters:

The resonant angular frequency is

$$\omega_r = \frac{1}{\sqrt{L_r C_r}} \quad (9)$$

The normalized switching frequency is

$$\omega_n = \frac{\omega_s}{\omega_r} \quad (10)$$

where ω_s = switching angular frequency = $2\pi \times f_s$

f_s = switching frequency

The characteristic impedance is

$$Z_0 = \sqrt{\frac{L_r}{C_r}} = \frac{1}{\omega_r C_r} = \omega_r L_r \quad (11)$$

The IH load quality factor is

$$Q = \frac{\omega_r L_r}{R_{eq}} = \frac{1}{\omega_r C_r R_{eq}} = \frac{Z_0}{R_{eq}} \quad (12)$$

The resonant tank circuit impedance is given by

$$Z_{eq} = R_{eq} + j\left(\omega_s L_r - \frac{1}{\omega_s C_r} \right) \quad (13)$$

$$= R_{eq} \left(1 + jQ\left(\omega_n - \frac{1}{\omega_n} \right) \right) \quad (14)$$

$$|Z_{eq}| = R_{eq} \sqrt{1 - Q^2 \left(\omega_n - \frac{1}{\omega_n} \right)^2} \quad (15)$$

The phase angle between output voltage and current is

$$\phi_1 = tan^{-1} \left(Q\left(\omega_n - \frac{1}{\omega_n} \right) \right) \quad (16)$$

B. Class-D inverter configuration

Fig. 3, shows the two load class-D inverter configuration. In this configuration, 1st leg powers load-1 and 2nd leg powers load-2 in class-D inverter configuration mode. ADC control technique is used to control the output power of each load independently.

The class-D inverter output voltage is the input voltage of resonant tank circuit voltage v_0 is

$$v_0 = \begin{cases} V_{DC}, & \text{for } 0 < \omega_s t \leq \pi \\ 0, & \text{for } \pi < \omega_s t \leq 2\pi \end{cases} \quad (17)$$

The fundamental output voltage v_0 is represented with the help of Fourier analysis, i.e.,

$$v_0 = V_m \sin\omega_s t, \text{ for } 0 < \omega_s t \leq 2\pi \quad (18)$$

where $V_m = \frac{2V_{DC}}{\pi} \approx 0.637 V_{DC}$ (19)

The IH load current through the resonant tank circuit is expressed by

$$i_0 = I_m \sin(\omega_s t - \emptyset_1) \quad (20)$$

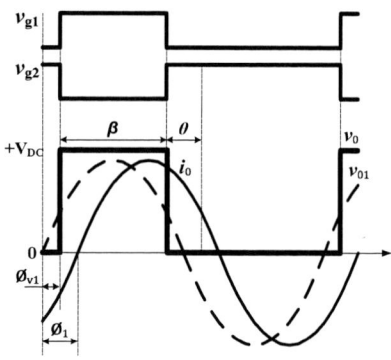

Fig. 4. Class-D inverter output voltage and current in ADC control

Fig. 4, shows the ADC control technique used in class-D inverter configuration to control inverter output voltage and current. Simultaneously, we can obtain output power control. In ADC control technique β is the control variable. β can be varied between 180° to 0°. In addition, θ also varies between 0° to 180°. The magnitude and phase of the output voltage w.r.t ADC control is expressed by

$$v_{01} = \frac{2V_{DC}}{\pi} \cos\frac{\theta}{2} \quad (21)$$

$$\emptyset_{V1} = \frac{\theta}{2} \quad (22)$$

The resonant tank circuit current (I_m) varies w.r.t v_{01}.

where $I_m = \frac{V_m}{|Z_{eq}|}$

$$= \frac{2V_{DC}}{\pi |Z_{eq}|} \cos\frac{\theta}{2}$$

$$= \frac{2V_{DC}\cos\emptyset}{\pi R_{eq}} \cos\frac{\theta}{2} \quad (23)$$

The conventional output power can be derived by

$$P_{out} = \frac{I_m^2}{2} R_{eq}$$

$$P_{out} = \frac{2V_{DC}^2 \cos^2\emptyset}{\pi^2 R_{eq}} \cos^2\frac{\theta}{2} \quad (24)$$

From the eq. (24) by varying θ, the IH load output power can be controlled by using ADC control technique in class-D inverter configuration.

C. Class-E inverter configuration

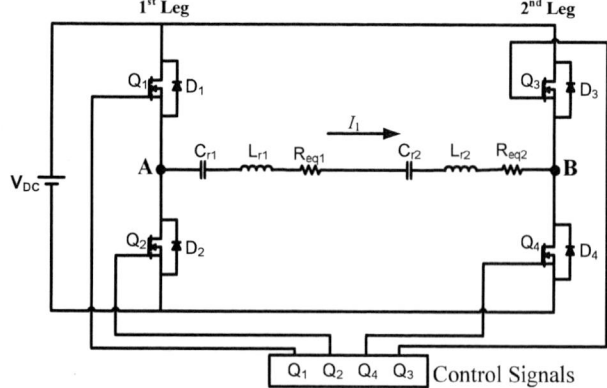

Fig. 5. Proposed class-E inverter configuration

Fig. 5, shows the single load class-E inverter configuration. In this configuration, two IH coils are connected in series as a single load and both legs are operated in class-E configuration mode. AVC control technique is used to control the load output power and obtain ZVS at lower duty ratios also.

The class-E inverter output voltage is input voltage of the resonant tank circuit voltage v_0 is

$$v_0 = \begin{cases} V_{DC}, & \text{for } 0 < \omega_s t \leq \pi \\ -V_{DC}, & \text{for } \pi < \omega_s t \leq 2\pi \end{cases} \quad (25)$$

The fundamental output voltage v_0 is represented with the help of Fourier analysis, i.e.,

$$v_0 = V_m \sin\omega_s t, \text{ for } 0 < \omega_s t \leq 2\pi \quad (26)$$

where $V_m = \frac{4V_{DC}}{\pi} \approx 1.27324 V_{DC}$ (27)

The IH load current through the resonant tank circuit is expressed by

$$i_0 = I_m \sin(\omega_s t - \emptyset_1) \quad (28)$$

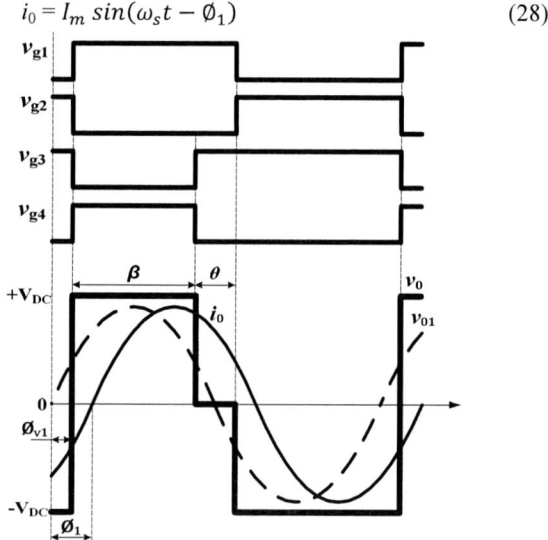

Fig. 6. Class-E inverter output voltage and current in AVC control

Fig. 6, shows the AVC control technique used in class-E inverter configuration to control inverter output voltage and current. Simultaneously, we can obtain output power control. At lower duty-ratios also ZVS can be achieved with AVC control technique. In AVC control technique β is the control variable. β can be varied between 180° to 0°. In addition, θ also varies between 0° to 180°. The magnitude and phase of the output voltage w.r.t AVC control is expressed by

$$v_{01} = \frac{V_{DC}}{\pi}\sqrt{10 + 6cos\theta} \tag{29}$$

$$\emptyset_{V1} = tan^{-1}\left(\frac{sin\theta}{3+cos\theta}\right) \tag{30}$$

The resonant tank circuit current (I_m) varies w.r.t v_{01}.

where
$$I_m = \frac{V_m}{|Z_{eq}|}$$
$$= \frac{4V_{DC}}{\pi|Z_{eq}|}\sqrt{10 + 6cos\theta}$$
$$= \frac{4V_{DC}cos\emptyset}{\pi R_{eq}}\sqrt{10 + 6cos\theta} \tag{31}$$

The conventional output power can be derived by

$$P_{out} = \frac{I_m^2}{2}R_{eq}$$

$$P_{out} = \frac{8V_{DC}^2 cos^2\emptyset}{\pi^2 R_{eq}}(10 + 6\,cos\theta) \tag{32}$$

From the eq. (32) by varying θ, the IH load output power can be controlled by using AVC control technique in class-E inverter configuration.

IV. RESULTS OF PROPOSED CONFIGURATION

To verify the validity of class-D/E resonant inverter configuration with ADC and AVC control techniques, a simulation and an experiment performed by using the parameters shown in table 1.

TABLE I. PARAMETERS FOR PROPOSED CONFIGURATION

Item	Symbol	Value
Source voltage	V_{DC}	35V
Equivalent resistance of load-1, load-2	R_{eq1}, R_{eq2}	2.36Ω
Equivalent inductance of load-1, load-2	L_{r1}, L_{r2}	68μH
Resonant capacitor of load-1, load-2 (MKV type)	C_{r1}, C_{r2}	0.22μF
Resonant frequencies for both loads	f_{r1}, f_{r2}	41.15 kHz
Switching frequencies for both legs	f_{S1}, f_{S2}	43.66 kHz
Dead time	t_d	450 nsec
MOSFETs used	IRF540	100V, 23A

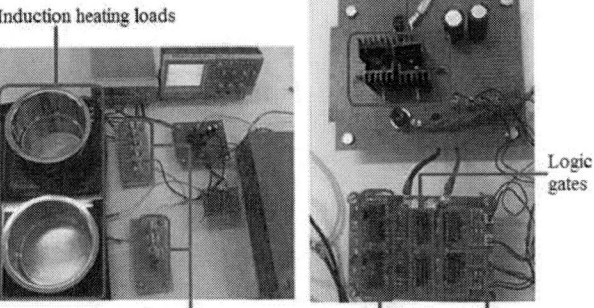

(a)   (b)

Fig. 7. Experimental setup for proposed class-D/E inverter configuration

Fig. 7, shows the experimental setup for proposed class-D/E resonant inverter configuration with two loads. Fig. 7(a), shows the induction heating loads and resonant series capacitors with two leg inverter circuit. Fig. 7(b), shows the two leg inverter circuit with pulse generator and driver circuit. In control circuit two UC3875 ICs are used for four control pulses. AND / OR logic gates are used for implementation of ADC technique and AVC technique. IR2110 ICs are used as drivers and isolation between control and power circuit also.

Fig. 8(a), shows class-D inverter configuration output voltage and their load current waveforms for two loads under simulation. Fig. 8(b), shows these waveforms from experimental set-up. Fig. 8, shows these waveforms for both loads at 98% duty-ratio. Similarly, Fig. 9, shows these waveforms for load-1 at 50% duty-ratio and load-2 at 98% duty-ratio. Fig. 10, shows the class-D inverter load current vs. duty-ratio. In this 2nd leg is kept at D = 0.98 and 1st leg duty ratio is varied from 0.1 to 0.98. Load-2 current is maximum and constant and load-1 current is varying w.r.t its duty-ratio. Load-1 output power is controlled by varying its duty-ratio. Similarly, Fig. 11, shows the class-D inverter efficiency vs. duty-ratio.

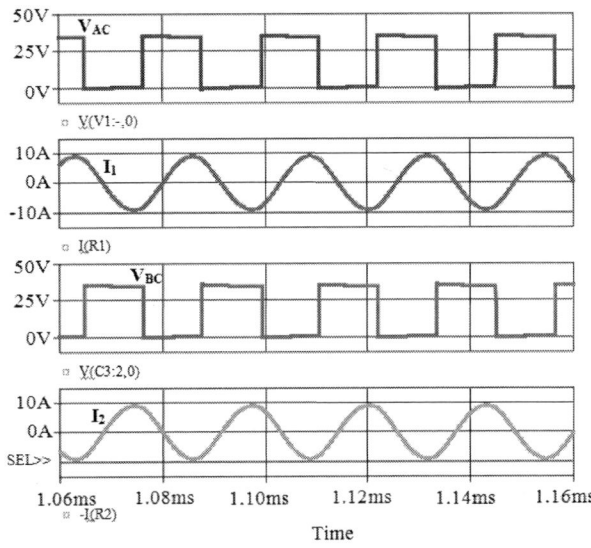

(a)

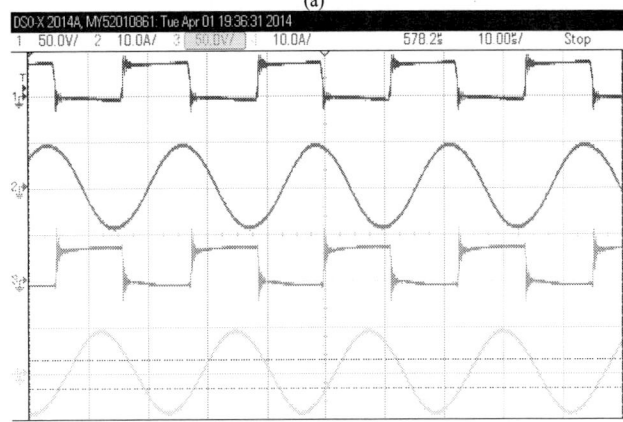

(b) Scale: voltage: 50V/div., current: 10A/div.

Fig. 8. Class-D inverter output voltage and load currents for both loads at 98% duty-ratio

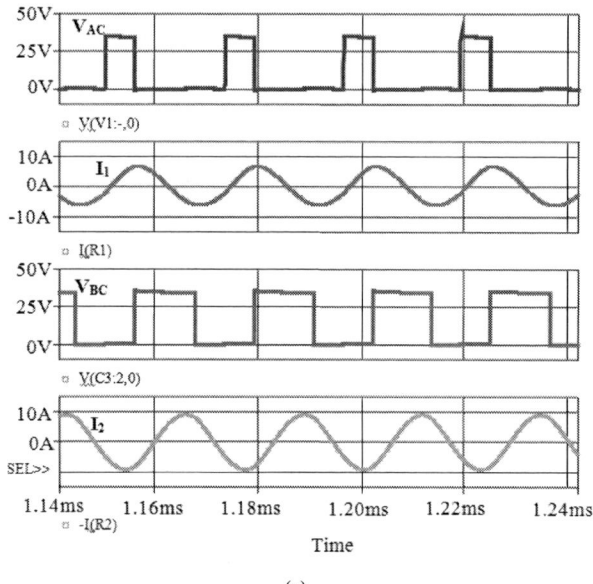

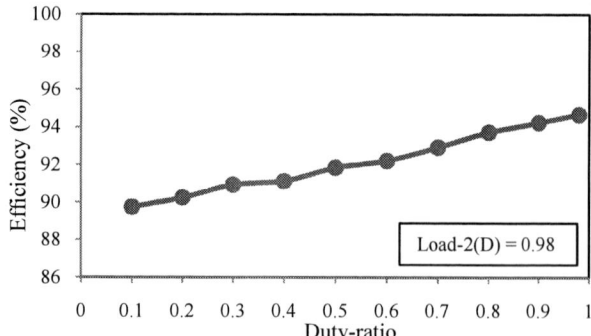

Fig. 11. Class-D inverter efficiency vs. duty-ratio

Fig. 12(a), shows class-E inverter configuration output voltage and their load current waveforms under simulation. Fig. 12(b), shows these waveforms from experimental set-up. Fig. 12, shows these waveforms at 98% duty-ratio. Similarly, Fig. 13, shows these waveforms at 50% duty-ratio.

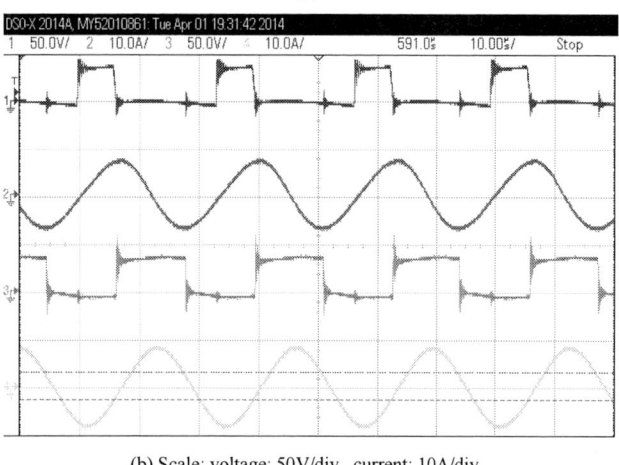

(b) Scale: voltage: 50V/div., current: 10A/div.

Fig. 9. Class-D inverter output voltage and load currents for load-1 at 50% duty-ratio and for load-2 at 95% duty-ratio

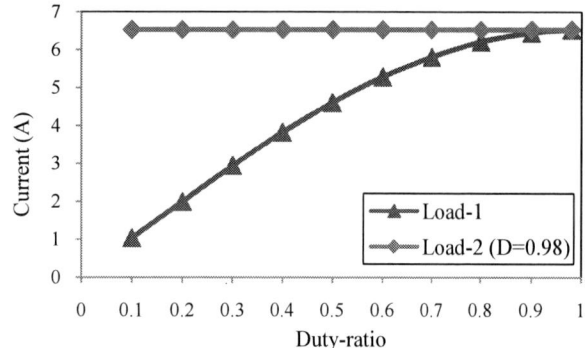

Fig. 10. Class-D inverter load current vs. duty-ratio

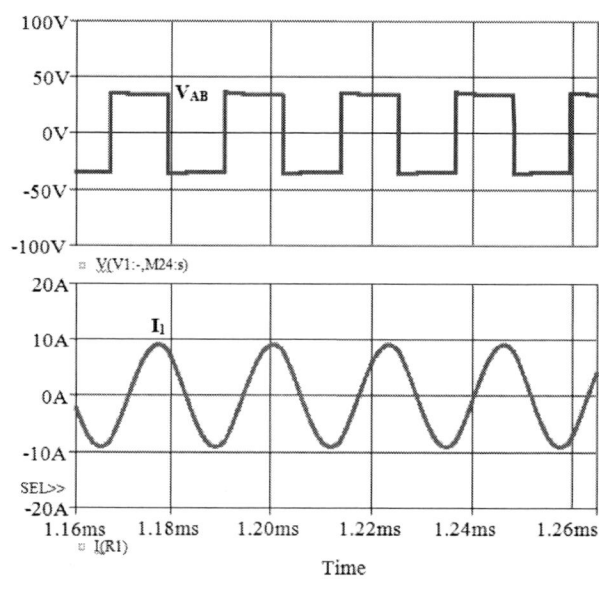

(a)

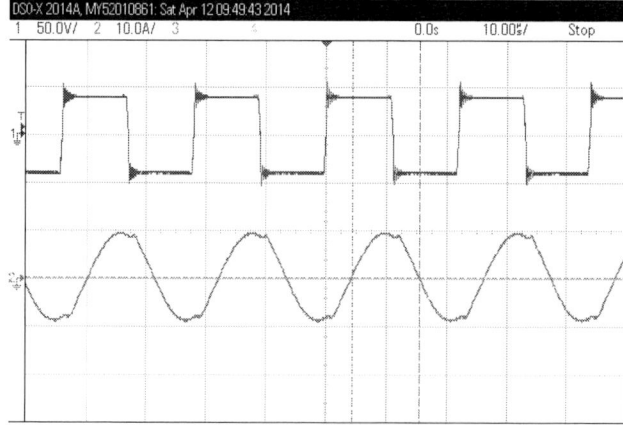

(b) Scale: voltage: 50V/div., current: 10A/div.

Fig. 12. Class-E inverter output voltage and load current at 98% duty-ratio

978-1-4799-6047-7/14 $31.00 © 2014 IEEE

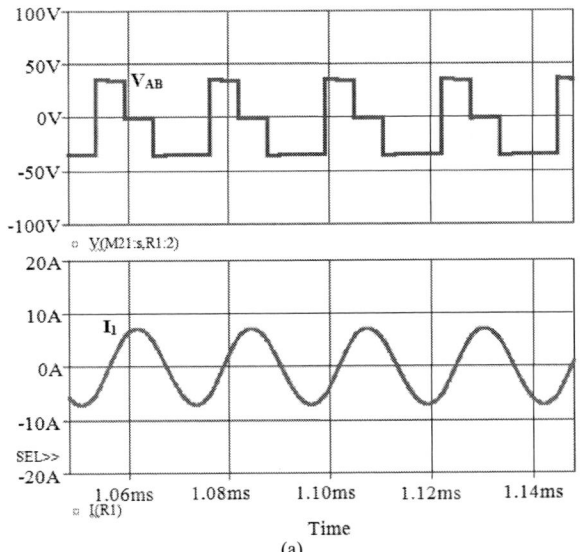

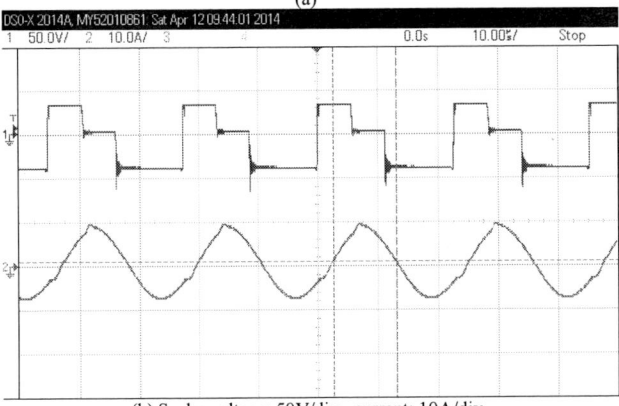

(b) Scale: voltage: 50V/div., current: 10A/div.

Fig. 13. Class-E inverter output voltage and load current at 50% duty-ratio

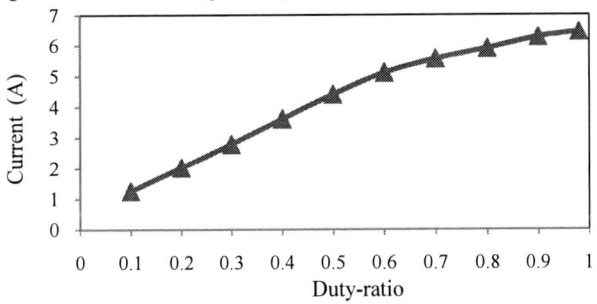

Fig. 14. Class-E inverter load current vs. duty-ratio

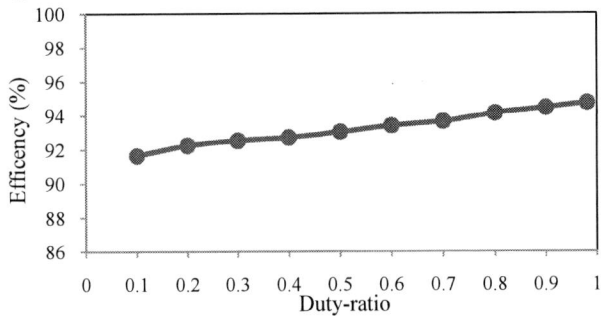

Fig. 15. Class-E inverter efficiency vs. duty-ratio

Fig. 14, shows the class-E inverter load current vs. duty-ratio. In this duty-ratio is varies from 0.1 to 0.98. Load current is varying with its duty-ratio. Load output power is controlled by varying its duty-ratio. Similarly, Fig. 15, shows the class-E inverter efficiency vs. duty-ratio.

From simulation and experimental results, it can be observed that both results are in good agreement. It is also seen that the proposed class-D/E resonant inverter configuration can power two loads in class-D configuration mode with independent output power control of each load by using ADC control technique. The output power of each load is controlled by varying the duty-ratios of their respective legs. Simultaneously, in class-E configuration mode can power to a single load and output power control done by using AVC control technique. In class-E configuration mode both IH coils are connected in series and treated as a single load. The inverter efficiency is improved at lower duty-ratios by using AVC control technique in class-E configuration.

V.   CONCLUSION

In this paper, class-D/E resonant inverter for two load domestic induction cooking appliances has been proposed. Both loads are independently controlled. In this configuration according to vessel size, we can operate in class-D or class-E configuration mode. It operates with constant switching frequency i.e., 43.66 kHz. ADC control technique is used for independent output power control of each load in class-D configuration and AVC control technique is used for output power control in class-E configuration. This configuration can be extended to multiple-loads. Simulation and experimental results of the proposed configuration are in good agreement.

REFERENCES

[1] W. C. Moreland, "The induction range: Its performance and its development problems," IEEE Trans. Industry Applications, vol. IA-9, no. 1, 1973, pp. 81–85.

[2] Mokhtar Kamli, Shigehiro Yamamoto, and Minoru Abe, "A 50-150 kHz Half-Bridge Inverter for Induction Heating Applications," IEEE Trans. Industrial Electronics, vol. 43, no.1, 1996, pp. 163–172.

[3] Young-Sup Kwon, Sang-Bong Yoo, Dong-Seok Hyun, "Half-Bridge Series Resonant Inverter for Induction Heating Applications with Load-Adaptive PFM Control Strategy", 14th Applied Power Electronics Conference and Exposition, APEC' 99, vol. 1, 1999, pp. 575 - 581.

[4] L. Grajales, J. A. Sabate, K. R. Wang, W. A. Tabisz, F. C. Lee, "Design of a 10 kW, 500 kHz Phase-Shift Controlled Series-Resonant Inverter for Induction Heating", Industry Applications Society Annual Meeting, vol. 2, 1993, pp. 843-849.

[5] D. V. Bhaskar, N. Yagnyaseni, N. Vishwanathan, and T. Maity, "Comparison of Control Methods for High Frequency IH Cooking Applications," Power and Energy Systems Conference: Towards Sustainable Energy, 2014, Page(s): 1 – 6.

[6] J. M. Burdio, L. A. Barragan, F. Monterde, D. Navarro, J. Acero, "Asymmetrical voltage-cancellation control for full-bridge series resonant inverters," IEEE Trans. Power Electronics, vol. 19, no. 2, 2004, pp. 461 - 469.

[7] Oscar Lucia, Claudio Carretero, J.M. Burdio, Jesus Acero, and Fernando Almazan, "Multiple-Output Resonant Matrix Converter for Multiple Induction Heaters," IEEE Transactions on Industry Applications, vol. 48, no. 4, July/August 2012, pp. 1387-1396.

[8] Hector Sarnago, Oscar Lucia, Arturo Mediano, J.M. Burdio, "Class D/DE Dual-Mode-Operation Resonant Converter for Improved-Efficiency Domestic Induction Heating System," IEEE Transactions on Power Electronics, vol. 28, no. 3, March 2013, pp. 1274-1285.

978-1-4799-6047-7/14 $31.00 © 2014 IEEE

Linearised Modelling of Switched Reluctance Motor for Closed Loop Current Control

S. S. Ahmad and G. Narayanan
Department of Electrical Engineering
Indian Institute of Science, Bangalore - 560012 INDIA
email: ssahmad@ee.iisc.ernet.in, gnar@ee.iisc.ernet.in

Abstract—**Proportional Integral (PI) controller based current control is widely employed in different motor drives. However, in case of switched reluctance motor (SRM), the non-linearity and double saliency of the motor make the modelling and controller design challenging. This paper attempts linearisation of the flux linkage characteristic of the SRM at different operating points. The models include saturation and back-emf in an effective manner, as compared to existing methods. The different linearised models are evaluated for the purpose of current controller design. An effective method of controller design is found which directly relates the machine parameters to the controller gains.**

Index Terms—**Current control, flux linkage characteristics, linearised model, PI controller, switched reluctance motor**

I. INTRODUCTION

The existing current control schemes for switched reluctance motor (SRM) mainly fall into three categories - non-linear, hysteresis and linear control techniques. As the motor is a non-linear plant, various non-linear control techniques can potentially be applied. In [1], an iterative learning based approach is described. However, this takes several cycles for proper controller action, and has a poor response when the current reference is not repetitive. In [2] a predictive controller is described. However this is based on look-up tables to compute required duty cycle to be applied in the next switching cycle. Other techniques have been reported [3], [4], which are either memory or computation intensive.

Hysteresis controller or on/off type is the most popular type of control [5], [6]. However this generally leads to increased ripple and variable switching frequency. Implementation of on/off type of control in microprocessor with fixed sampling frequency leads to delta modulation. This is easy to implement, and has good dynamic performance, but results in high current ripple [1].

Proportional-integral (PI) controllers based on linear control theory have been employed, which ensure uniform switching frequency and low current ripple [7]–[10]. However the non-linearity of the plant poses difficulty in the design of such linear controller. Some authors have proposed variable PI gains [8], [10] with back-emf compensation. In [8] a look-up table is used for back-emf compensation, while the controller gain is determined either using a look-up table or through equations

This work was supported by the Department of Heavy Industry, Govt. of India, under the project titled "Off-line and real-time simulators for electric vehicles/hybrid electric vehicles systems".

fitted to the look-up table. In [10], the proportional and integral gains are both varied linearly with respect to current and rotor position. The back-emf is estimated with the help of a look-up table. A hybrid of hysteresis and PI control is suggested in [7] with fixed controller gains.

This paper considers PI controller with fixed gains for the purpose of simplicity. However, this requires an appropriate linear model of the plant. By simplifying the flux-linkage characteristics, linearised models of the motor are developed, which pertain to different current levels and rotor positions. The linearised models, pertaining to low currents, account for the back-emf developed by the motor; these do not treat the back-emf as a disturbance unlike certain existing methods. The linearised models at various operating points are considered for controller design. The best model from the point of view of current controller design is identified. The PI controller designed based on this model is verified experimentally. The performance of this PI based current control is compared with that of delta modulation on 4kW, 280V, 18A, 4 phase, 8/6 pole SRM drive.

II. LINEARISED MODEL OF AN SRM PHASE

The motor flux-linkage characteristic is linearised with respect to current to obtain the linearised motor model. Separate models are obtained for saturated and unsaturated conditions. Further, in the unsaturated condition, models are obtained for various rotor positions.

A. Phase Voltage

Neglecting the mutual coupling among the phases, the dynamics of current in any phase in an SRM is given by,

$$
\begin{aligned}
V_{ph} &= iR + \frac{d\psi(i, \theta)}{dt} \\
&= iR + \frac{\partial \psi}{\partial i}\frac{di}{dt} + \omega_m \frac{\partial \psi}{\partial \theta}
\end{aligned}
\tag{1}
$$

where, V_{ph} is the applied phase voltage, R is the phase resistance, and ω_m is the motor speed. To evaluate the partial derivatives in Eqn. (1), the motor phase flux-current-position characteristic needs to be considered. Fig. 1 shows the measured flux linkage characteristic of the motor used here [11].

978-1-4799-6047-7/14 $31.00 © 2014 IEEE

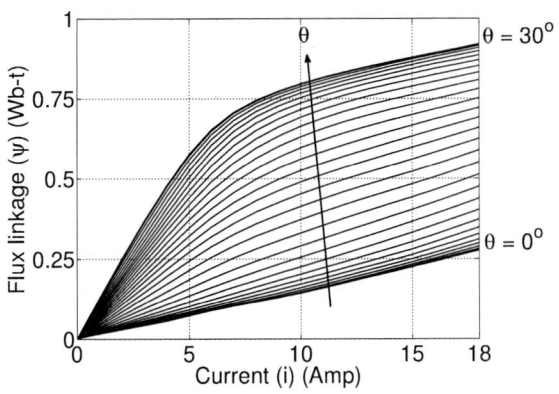

Fig. 1. Measured flux linkage characteristic (data from [11]). $\theta = 0°$ is unaligned position

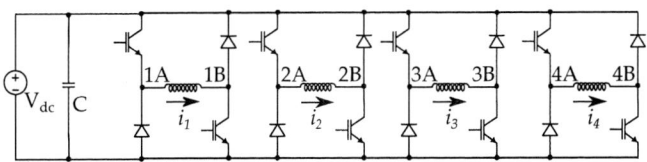

Fig. 2. Asymmetric half-bridge power converter

B. Linearisation in the Unsaturated Condition

During unsaturated condition (i.e. current less than the knee current, say a), the motor inductance is not a function of current. Hence the flux linkage can be written as $\psi = m_1(\theta)i$, where m_1 is the unsaturated inductance. Here $m_1(\theta)$ is periodic with rotor position having a periodicity equal to the rotor pole pitch. A Fourier expansion of the same can be considered. For an 8/6 motor the inductance profile repeats six times in one complete revolution of the rotor. Here θ is mechanical angle expressed in radians. Since $\theta = 0°$ is the unaligned position in the present case, $m_1(\theta)$ has even symmetry. Therefore, considering only the first three terms,

$$m_1(\theta) \approx a_0 + a_1 \cos(6\theta) + a_2 \cos(12\theta) \qquad (2)$$

With this consideration, the partial derivatives in Eqn. (1) can be evaluated as shown,

$$\begin{aligned} \frac{\partial \psi}{\partial i} &= m_1(\theta) \\ \frac{\partial \psi}{\partial \theta} &= m_1'(\theta)i \end{aligned} \qquad (3)$$

Here $m_1' = \frac{dm_1}{d\theta}$. Eqn. (3) can be substituted in Eqn. (1) to obtain the following relation between the phase voltage and current,

$$V_{ph} = (R + m_1'\omega_m)i + m_1\frac{di}{dt} \qquad (4)$$

This leads to a motor phase transfer function for a particular position, under unsaturated condition, as shown,

$$G_{ph,u}(s) = \frac{\hat{i}(s)}{\hat{v}_{ph}(s)} \approx \frac{1}{(R + m_1'\omega_m) + sm_1} \qquad (5)$$

This transfer function contains $m_1(\theta)$, $m_1'(\theta)$ and ω_m. Observe that back-emf forms part of transfer function (i.e. $m_1'\omega_m$).

C. Linearisation in Saturated Condition

Under saturation, the flux linkage curve can be approximated as a straight line with a different inductance value (say m_2). Observe that the saturated inductance is not a strong function of position. As seen from Fig. 1, the $\psi - i$ curves corresponding to different values of θ are almost parallel at high currents. The slope of these curves at high currents is represented by m_2. Then each $\psi - i$ curve can be approximated in a piecewise linear fashion as shown below, where a represents the knee of the curves,

$$\psi(i,\theta) = m_2 i + [m_1(\theta) - m_2]\, a(\theta) \qquad (6)$$

The partial derivatives of ψ with respect to θ can be expressed as shown below, where a' is the derivative of a with respect to θ,

$$\begin{aligned} \frac{\partial \psi}{\partial i} &= m_2 \\ \frac{\partial \psi}{\partial \theta} &= m_1'(\theta)a(\theta) + m_1(\theta)a'(\theta) \end{aligned} \qquad (7)$$

By substituting Eqn. (7) into Eqn. (1) the following equation is arrived at,

$$V_{ph} = iR + m_2\frac{di}{dt} + (m_1'a + m_1a')\omega_m \qquad (8)$$

Small perturbations in voltage ($\hat{v}_{ph}$) do not influence the back-emf term (the last term) but result in small perturbations in current ($\hat{i}$). Therefore, the small signal transfer function is found to be,

$$G_{ph,s}(s) = \frac{\hat{i}(s)}{\hat{v}_{ph}(s)} \approx \frac{1}{R + sm_2} \qquad (9)$$

The transfer function under saturated condition is mostly independent of position. This is because the saturated inductance is independent of position and the back-emf is independent of current. In this case, the back-emf is a disturbance input to the controller.

III. Model of Power Converter

The power converter used is the popular asymmetric H-bridge converter [12] shown in Fig. 2. The DC gain depends on the modulation used. In soft chopping mode, $+V_{dc}$ or 0 is applied to the motor phase during current control. While in hard chopping, $\pm V_{dc}$ is applied to the motor phase. Hence,

$$\begin{aligned} \text{DC gain} &= dV_{dc} \quad \text{(soft chopping)} \\ &= V_{dc}(2d - 1) \quad \text{(hard chopping)} \end{aligned} \qquad (10)$$

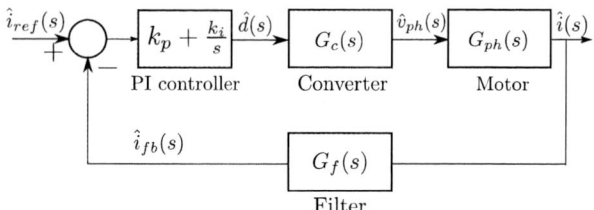

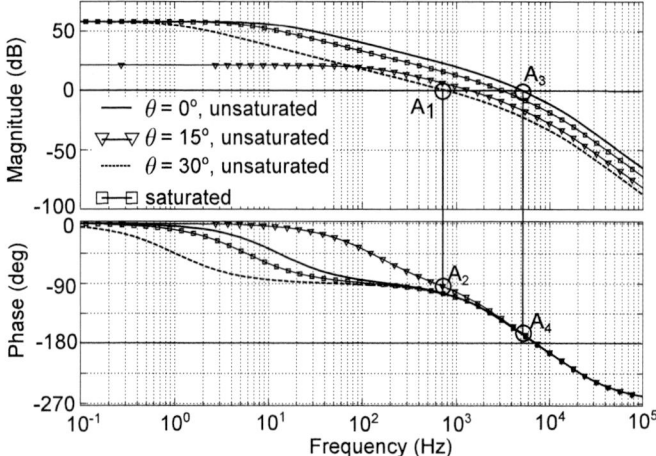

Fig. 3. Structure of current controller

Fig. 4. Bode plot of loop transfer function

Additionally, due to the nature of PWM, a delay of half the switching time period is introduced. This can be modelled as a pole with time constant $\tau_d = T_s/2$, where T_s is one switching cycle. In the present case, T_s is 100 μs. Hence the pole is at 20 kHz.

Note that the converter transfer function in soft chopping mode is similar to that in case of hard chopping mode, except that there is an additional factor of two in hard chopping. In this work, only hard chopping is considered. Therefore the converter transfer function is as follows,

$$G_c(s) = \frac{\hat{v}_{ph}(s)}{\hat{d}(s)} = \frac{2V_{dc}}{1 + s\tau_d} \qquad (11)$$

IV. PLANT TRANSFER FUNCTION

The PI controller design is based on the bode plot of open loop transfer function, which forms the plant for the controller. Fig. 3 shows the structure of current control loop. Here $G_f(s)$ is the current feedback filter used to eliminate high frequency components. This is a first order filter with a time constant τ_f. In present case, τ_f corresponds to pole at 15.9 kHz.

The loop transfer function becomes,

$$L(s) = \frac{2V_{dc}}{(1 + s\tau_f)(1 + s\tau_d)} \times G_{ph}(s) \qquad (12)$$

$G_{ph}(s)$ is found from either Eqn. (5) or Eqn. (9), depending on the level of saturation.

Fig. 4 shows the bode plot of the loop transfer function at various positions and current levels, calculated at a fixed

speed (1500 RPM). In case of unsaturated condition, the DC gain varies. It is maximum when the back-emf term $m'_1\omega_m$ is minimum, that is at aligned and unaligned positions. The change of inductance being zero, the back-emf is also zero. The DC gain is minimum very close to the middle of aligned and unaligned positions (i.e. $\theta = 15°$), where back-emf is maximum. The pole location due to motor phase also changes, as the inductance varies with position. The saturated transfer function is intermediate between the unsaturated transfer functions for the aligned and unaligned positions.

The unsaturated transfer functions at positions other than $\theta = 0°$ and $\theta = 30°$ depend on speed as well. At lower speeds these transfer functions approach intermediate positions between the curves for $\theta = 30°$ and $\theta = 0°$. The deviation is higher at higher speeds; hence the maximum rated speed is considered here.

A_1 and A_3 in Fig. 4 mark the two extremes of the unity gain cross over frequency. The corresponding values of phase are shown by A_2 and A_4, respectively. For a fixed controller gain, the maximum to minimum bandwidth ratio is determined by A_1 and A_3. A_4 corresponds to the least phase margin of the plant. Therefore design of the controller must consider this point to ensure that minimum phase margin is adequate.

V. CONTROLLER DESIGN

The controller structure is shown in Fig. 3. The PI controller is designed, considering each of the transfer functions one by one.

A. Procedure for Selection of Control Parameters

The PI current controller transfer function can be written in a modified form as shown below, where k_p and k_i are the controller parameters and $\omega_z = k_i/k_p$,

$$G_{cc} = k_p + \frac{k_i}{s} = \frac{k_p\left(s + \frac{k_i}{k_p}\right)}{s} = \frac{k_p(s + \omega_z)}{s} \qquad (13)$$

The controller zero, ω_z, can be chosen such that the loss of phase due to the pole at origin does not have much effect at the gain cross over frequency. In the present case, ω_z is chosen to be about 10 Hz.

The controller proportional gain, k_p, can be decided based on the gain of the system at the desired bandwidth (say ω_d). If the plant gain be k at ω_d, then $k_p = 1/k$ will make loop transfer function gain to be 1 at the desired bandwidth; hence gain cross over frequency will be the desired bandwidth. The desired bandwidth is chosen to be 1.3 kHz.

Note that loop gain k can be found by only considering DC gain and the first pole (due to motor phase). The other two poles are at much higher frequencies than ω_d.

B. Unsaturated Condition

First consider the case when $i < a$ (i.e. low current unsaturated transfer function). The plant transfer function (from Eqn. (12), ignoring the higher poles), DC gain and the pole frequency are,

$$G_p(s)\Big|_{i<a} = \frac{\hat{i}(s)}{\hat{d}(s)} = \frac{2V_{dc}}{(R + m_1'\omega_m) + sm_1}$$

$$\text{DC gain} = \frac{2V_{dc}}{R + m_1'\omega_m} \tag{14}$$

$$\text{pole} = \frac{R + m_1'\omega_m}{m_1}$$

The plant transfer function starts with the DC gain at low frequency, and then rolls off at 20 dB/decade, starting from the pole location. Let the desired bandwidth be ω_d. Then the plant gain at ω_d frequency (in dB) can approximately be expressed as,

$$|G_p(s = j\omega_d)| \approx 20 \log\left(\frac{2V_{dc}}{R + m_1'\omega_m}\right) - 20 \log\left(\frac{\omega_d}{\frac{R+m_1'\omega_m}{m_1}}\right)$$

$$= 20 \log\left(\frac{2V_{dc}}{m_1\omega_d}\right) \tag{15}$$

Hence, when $i < a$, the controller parameters can be selected as shown below,

$$k_p = \frac{m_1\omega_d}{2V_{dc}} \tag{16}$$

$$\text{also,} \quad k_i = \omega_z \times k_p$$

C. Saturated Condition

Now, considering the case of high currents (i.e. $i > a$) a similar exercise is carried out as in the previous section V-B. As a result, the following design equations are obtained for k_p and k_i,

$$k_p = \frac{m_2\omega_d}{2V_{dc}} \tag{17}$$

$$k_i = \omega_z \times k_p$$

D. Choice of Controller Gains

Eqn. (16) and Eqn. (17) are used to obtain the required gains for each of the transfer functions in Fig. 4. In Fig. 5, $\theta = 0°$, unsaturated condition is considered to find the controller gains. Observe that there is sufficient phase margin for all the bounding transfer functions. The worst phase margin (B_4) is about 65°. Fig. 6 considers $\theta = 15°$, unsaturated condition. Observe that there is insufficient phase margin for $\theta = 0°$ (only about 18°, C_4), unsaturated transfer function. The $\theta = 30°$, unsaturated condition is considered in Fig. 7. Phase margin is very poor (about 3°, D_4) near the unaligned position. Finally, the saturated transfer function is considered in Fig. 8. In this case the minimum phase margin is just about 45° (E_4). These observations are summarised in table I.

From these figures, it can be concluded that a suitable choice for controller design is the $\theta = 0°$, unsaturated transfer function, since this guarantees sufficient phase margin. In present case, therefore, this is used to obtain controller gain values as $k_p = 0.12$ and $k_i = 7.8$.

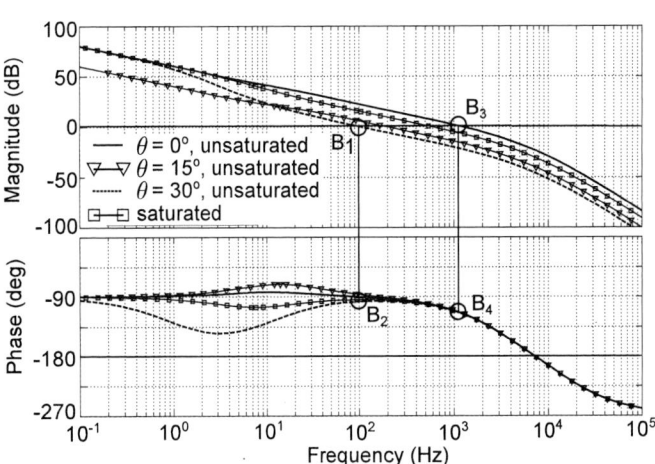

Fig. 5. Bode plot of open loop transfer functions of plant and controller with bandwidth 1.3 kHz for $\theta = 0°$, unsaturated transfer function

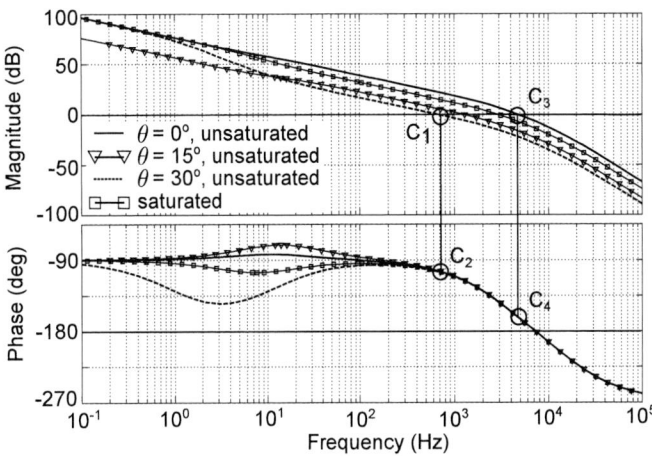

Fig. 6. Bode plot of open loop transfer functions of plant and controller with bandwidth 1.3 kHz for $\theta = 15°$, unsaturated transfer function

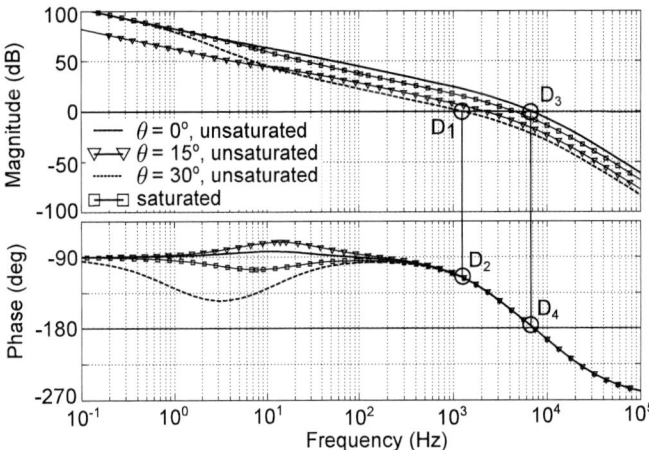

Fig. 7. Bode plot of open loop transfer functions of plant and controller with bandwidth 1.3 kHz for $\theta = 30°$, unsaturated transfer function

VI. EXPERIMENTAL RESULTS

The controller is implemented and verified through experiments. The SR machine parameters are shown in table II.

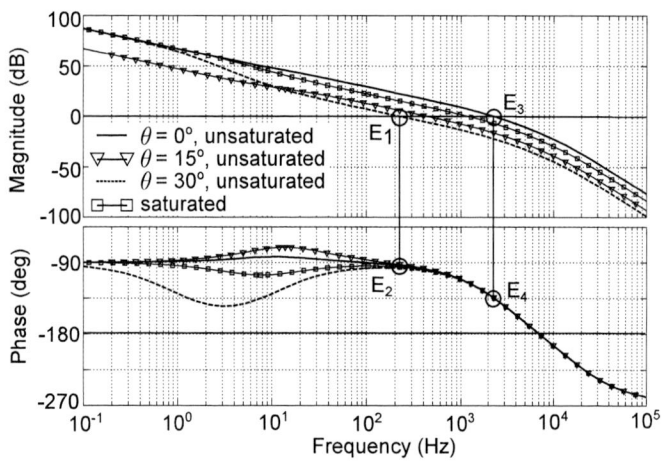

Fig. 8. Bode plot of open loop transfer functions of plant and controller with bandwidth 1.3 kHz for saturated transfer function

TABLE I
SUMMARY OF BANDWIDTH AND PHASE MARGIN

Operating Point for Linearisation	Minimum Bandwidth	Maximum Bandwidth	Worst Phase Margin
$\theta = 0°$, unsaturated	101 Hz	1.21 kHz	65°
$\theta = 15°$, unsaturated	655 Hz	4.66 kHz	18°
$\theta = 30°$, unsaturated	1.21 kHz	6.67 kHz	3°
saturated	226 Hz	2.34 kHz	45°

TABLE II
SR MOTOR RATING

Power	4 kW
Speed	1500 RPM
Maximum phase current	18 A
Phase voltage	280 V
Rated torque	25.5 N m
Stator phases	4
Stator poles	8
Rotor poles	6
Inertia	0.008 kg m^2
Winding resistance	0.7 Ω

The asymmetric converter has been fabricated using half-bridge modules SKM50GB123D. One of the switches in a leg is always kept off. A TMS320F28335 DSP based platform is used to implement the controller. It is expected that the control performance is most affected at higher speeds and higher current levels (where the back-emf is highest). Hence the current waveforms at current reference 14 A and speeds of 200 RPM and 800 RPM are presented here. It is compared with delta modulation technique to evaluate its effectiveness. Waveforms for current level near knee region (7A) is also shown as modelling has the maximum error in this region.

Fig. 9(a) presents the measured phase current waveform with delta modulation at 7A and 200 RPM. Fig. 9(b) shows the corresponding experimental current waveform with the proposed PI controller design. As seen, the peak-peak current ripple is slightly reduced with the PI controller.

The current waveforms for current level 14A and speed 200 RPM are shown in Fig. 9(c) and 9(d) with delta modulation

and PI controller, respectively. As expected the designed PI controller works reasonably well with decreased current ripple as compared to delta modulation.

Fig. 9(e) and Fig. 9(f) show the measured current response for delta modulation and PI controller, respectively, at 800 RPM and 14A current level. Here, the back-emf is significant. As seen, the PI controller bandwidth is not sufficient to maintain the current at desired reference (can be seen in Fig. 9f). This is not due to insufficient DC bus voltage, but insufficient control action. The duty ratio is not yet saturated, as seen from current switching, though current is not at reference value. However the response is no worse than what is obtained with delta modulation. Improvement can certainly be achieved by utilising back-emf of the motor as a feed forward signal to the controller.

VII. CONCLUSION

In this paper, applicability of linear controller for current control in SRM is explored. The SRM being non-linear, linearised models are developed based on piece-wise approximation of flux-linkage characteristic. It is seen that in case of unsaturated condition, the linearised models incorporate back-emf as well. The models are examined one by one to determine the most suitable one for controller design. It is concluded that, the unsaturated model at unaligned condition should be used for controller design. It should be noted that, in order to design a PI controller based on the methodology presented here, only the phase resistance and unaligned low current inductance of motor phase are required. A relation to find the controller gains from these two parameters directly is also given. The designed controller is verified experimentally at various operating points. This has lower ripple than delta modulation. At high speed, the performance degrades, but it is no worse than that of delta modulation.

REFERENCES

[1] S. Sahoo, S. Panda, and J. X. Xu, "Iterative learning-based high-performance current controller for switched reluctance motors," *IEEE Trans. Energy Conversion*, vol. 19, no. 3, pp. 491–498, Sept 2004.

[2] R. Mikail, I. Husain, Y. Sozer, M. Islam, and T. Sebastian, "A fixed switching frequency predictive current control method for switched reluctance machines," in *Proc. IEEE Energy Conversion Congress and Exposition*, Sept 2012, pp. 843–847.

[3] Y. Zheng, H. Sun, Y. Dong, and Z. Lei, "A current control method of srm based on rbf considering the mutual inductance with simultaneous two-phase excitation," in *Proc. IEEE Power Electronics Specialists Conf.*, June 2008, pp. 3569–3573.

[4] I. Manolas, A. Kaletsanos, and S. Manias, "Nonlinear current control technique for high performance switched reluctance machine drives," in *Proc. IEEE Power Electronics Specialists Conf.*, June 2008, pp. 1229–1234.

[5] N. H. Fuengwarodsakul and B. Tanbunjit, "Adapted hysteresis current control with switching frequency limitation for switched reluctance machines," in *Proc. International Conference on Power Electronics and Drive Systems*, Nov 2009, pp. 506–510.

[6] A. V. Rajarathnam, K. Rahman, and M. Ehsani, "Improvement of hysteresis control in switched reluctance motor drives," in *Proc. International Conference Electric Machines and Drives*, May 1999, pp. 537–539.

[7] H. Bae and R. Krishnan, "A study of current controllers and development of a novel current controller for high performance srm drives," in *Proc. Industry Applications Conf.*, vol. 1, Oct 1996, pp. 68–75 vol.1.

978-1-4799-6047-7/14 $31.00 © 2014 IEEE

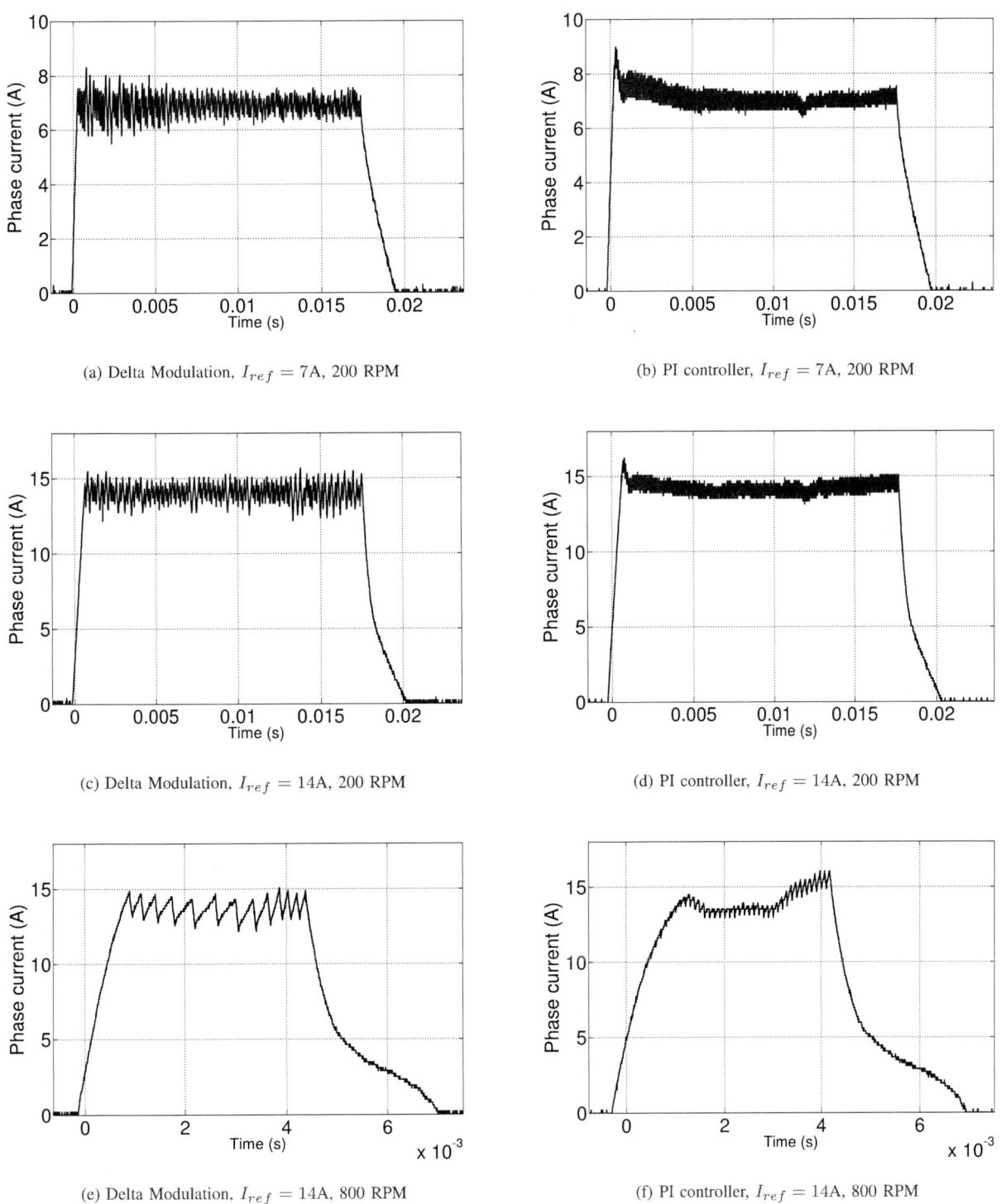

(a) Delta Modulation, $I_{ref} = 7A$, 200 RPM

(b) PI controller, $I_{ref} = 7A$, 200 RPM

(c) Delta Modulation, $I_{ref} = 14A$, 200 RPM

(d) PI controller, $I_{ref} = 14A$, 200 RPM

(e) Delta Modulation, $I_{ref} = 14A$, 800 RPM

(f) PI controller, $I_{ref} = 14A$, 800 RPM

Fig. 9. Experimental current responses at different speed and current reference

[8] G. Schroder and J. Bekiesch, "Adaptive current control for the SRM," in *Proc. IEEE International Symposium on Industrial Electronics*, vol. 1, June 2005, pp. 69–74.

[9] F. Blaabjerg, P. Kjaer, P. Rasmussen, and C. Cossar, "Improved digital current control methods in switched reluctance motor drives," *IEEE Trans. Power Electronics*, vol. 14, no. 3, pp. 563–572, May 1999.

[10] S. Schulz and K. Rahman, "High-performance digital pi current regulator for ev switched reluctance motor drives," *IEEE Trans. Industry*

Applications, vol. 39, no. 4, pp. 1118–1126, July 2003.

[11] D. Panda, "Control strategies for sensorless and low noise operation of switched reluctance motor," Ph.D. dissertation, Indian Institute of Science, Bangalore, 1999.

[12] S. Vukosavic and V. R. Stefanovic, "SRM inverter topologies: a comparative evaluation," *IEEE Trans. Industry Applications*, vol. 27, no. 6, pp. 1034–1047, Nov 1991.

978-1-4799-6047-7/14 $31.00 © 2014 IEEE

A Novel Eighteen-Level Inverter for an Open-end Winding Induction Motor

[1]Sanjiv Kumar and [2]Pramod Agarwal

Department of Electrical Engineering
Indian Institute of Technology Roorkee
Roorkee, Uttarakhand, India
E-mail: [1]sanjiv.iitr@gmail.com, [2]pramgfee@iitr.ac.in

Abstract— **A novel hybrid eighteen-level inverter topology for open-end winding induction motor (IM) is presented in this paper. In the proposed topology one end of the open-end IM is fed by conventional two-level inverter, while other end is connected to a nine-level cascade H-bridge (CHB) inverter. The combined effect of these two inverters generates eighteen-level in the phase voltage of open-end winding IM. The two-level inverter with higher DC link voltage has lower switching frequency and thereby reduces the switching losses. In addition to that, the proposed topology requires less number of components as compared to conventional multilevel inverter (MLI) topology. An interesting feature of the proposed topology is that, it can operate in nine-level mode by connecting the motor winding in star in case of failure of two-level inverter. Similarly if the fault occurs in CHB inverter the proposed inverter can operate in two-level mode. Thus the reliability of the system is improved. Exhaustive simulation study is carried out to evaluate the performance of proposed inverter for the entire modulation range and results are presented.**

Keywords—H-bridge; Induction motor drive; Multilevel Inverters; Open-end winding induction motor.

I. INTRODUCTION

Over the last few decades, multilevel inverters (MLI) become very popular in the field of medium and high voltage drives [1], [2]. The main advantages of MLI are low total harmonic distortion (THD) in the output voltage waveform, less switching losses and low dv/dt stress on semiconductor switches [3], [4]. The first MLI topology was proposed by A. Nabae in 1981 [5] and is called neutral point clamp (NPC) inverter. Apart from NPC, flaying capacitor (FC) [6], [7] and cascade H-bridge (CHB) [8] are another well establish MLI topologies. The NPC inverter requires large number of clamping diodes whereas FC inverters require so many capacitors when the number of levels in MLI increased. The CHB inverters are suitable for high voltage applications due to use of modular structure. However, they require isolated power supply for each module which makes the system bulky and costly. The quality of output voltage improves with the increased in number of levels, but at the same time structure of MLI become complex and less reliable due to use of large number of switching devices. The researchers have reported several MLI topologies and control schemes [9]–[13] to reduce circuit complexity, increase reliability and efficiency. In 1993 H. Stemmler and P. Guggenbach proposed the concept of three-level phase voltage generation using open-end stator

winding induction motor fed by two-level inverter [14]. Since then several topologies have been proposed by researchers based on open-end winding IM. A three-level voltage profile generation in open-end winding IM is proposed in [15], [16]. The number of levels in output voltage can be increased further by using cascade three-level inverter at the place of two-level inverter [17]–[19]. A hybrid seven-level [20] and nine-level inverter is presented in [21]. These hybrid inverters use two three-phase two-level inverter and two capacitor fed H-bridge per phase to generate different levels in the motor phase voltage.

In this paper, a hybrid eighteen-level inverter topology is presented for an open-end winding IM. The proposed topology uses a conventional three-phase two-level inverter and a three-phase nine-level CHB inverter. The proposed inverter uses fewer components as compared to conventional topologies. The zero sequence currents are completely eliminated in proposed topology because both inverters are fed by isolated DC supply. The complete system is simulated in MATLAB/Simulink and results are addressed.

II. PROPOSED INVERTER TOPOLOGY

A schematic diagram of the proposed eighteen-level inverter topology is shown in Fig. 1. The IM with open-end winding is fed by CHB nine-level inverter from one side (A_1,B_1,C_1) and other side (A_2,B_2,C_2) is fed by conventional two-level inverter. The three-phase nine-level CHB inverter is realized by connecting two H-bridge per phase in cascade. The ratio of DC link voltage of H-bridge is kept 1:3 to achieve nine-level in pole voltage. The proposed topology required thirty switches and seven isolated DC supplies to realize eighteen levels in the motor phase voltage. The magnitude of DC voltage supply feeding the two-level inverter is $(9/13)V_{DC}$. The DC supplies of CHB are of $(1/13)V_{DC}$ and $(3/13)V_{DC}$ where V_{DC} is the equivalent DC-link voltage required to operate conventional two-level inverter fed induction motor drive. Any pole voltage V_{A1O}, V_{B1O} or V_{C1O} of nine-level CHB inverter can have nine-different levels viz. $-4V_{DC}/13$, $-3V_{DC}/13$, $-2V_{DC}/13$, $-V_{DC}/13$, 0, $+V_{DC}/13$, $+2V_{DC}/13$, $+3V_{DC}/13$ and $+4V_{DC}/13$ independently. Similarly two-level inverter pole voltages $V_{A2O'}$, $V_{B2O'}$ and $V_{C2O'}$ can have voltage levels of 0 and $+9V_{DC}/13$. The combined effect of these two inverters on the motor phase winding is to synthesize eighteen-levels which are $-V_{DC}$, $-12V_{DC}/13$, $-11V_{DC}/13$ $-10V_{DC}/13$, $-9V_{DC}/13$, $-8V_{DC}/13$, $-7V_{DC}/13$, $-6V_{DC}/13$, $-5V_{DC}/13$, $-4V_{DC}/13$, $-3V_{DC}/13$, $-2V_{DC}/13$,

978-1-4799-6047-7/14 $31.00 © 2014 IEEE

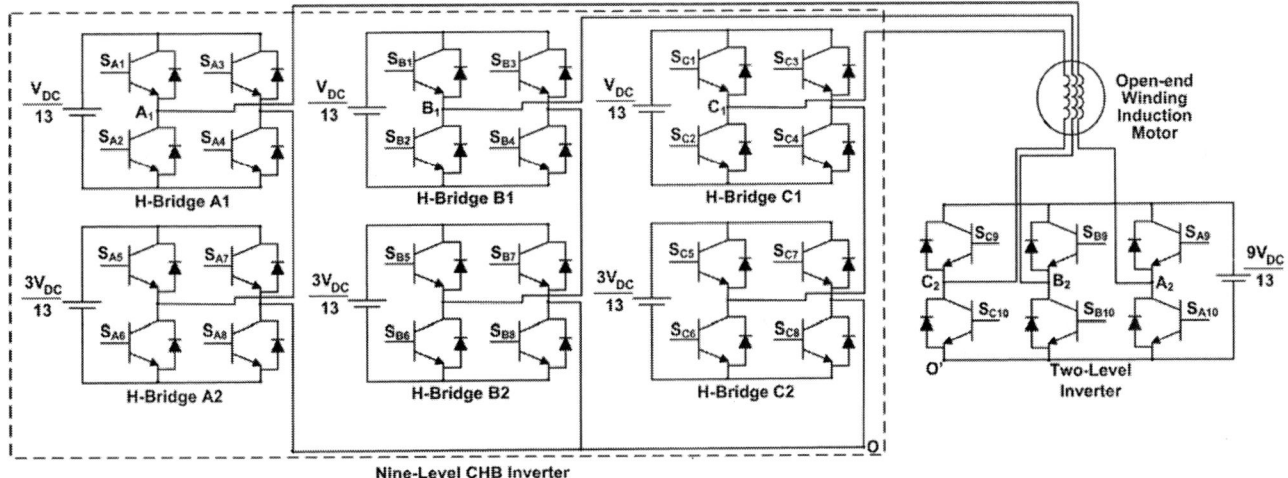

Fig. 1. Power Circuit of Proposed eighteen-level inverter scheme

TABLE I. Switching States To Realized Eighteen–Levels In Phase-A

Level	Pole Voltage V_{A1O}	Pole Voltage $V_{A2O'}$	Phase Voltage V_{A1A2}	Switching State[a]									
				S_{A1}	S_{A2}	S_{A3}	S_{A4}	S_{A5}	S_{A6}	S_{A7}	S_{A8}	S_{A9}	S_{A10}
1	$-4V_{DC}/13$	$+9V_{DC}/13$	$-V_{DC}$	0	1	1	0	0	1	1	0	1	0
2	$-3V_{DC}/13$	$+9V_{DC}/13$	$-12V_{DC}/13$	1	0	1	0	0	1	1	0	1	0
3	$-2V_{DC}/13$	$+9V_{DC}/13$	$-11V_{DC}/13$	1	0	0	1	0	1	1	0	1	0
4	$-V_{DC}/13$	$+9V_{DC}/13$	$-10V_{DC}/13$	0	1	1	0	1	0	1	0	1	0
5	0	$+9V_{DC}/13$	$-9V_{DC}/13$	1	0	1	0	1	0	1	0	1	0
6	$+V_{DC}/13$	$+9V_{DC}/13$	$-8V_{DC}/13$	1	0	0	1	1	0	1	0	1	0
7	$+2V_{DC}/13$	$+9V_{DC}/13$	$-7V_{DC}/13$	0	1	1	0	1	0	0	1	1	0
8	$+3V_{DC}/13$	$+9V_{DC}/13$	$-6V_{DC}/13$	1	0	1	0	1	0	0	1	1	0
9	$+4V_{DC}/13$	$+9V_{DC}/13$	$-5V_{DC}/13$	1	0	0	1	1	0	0	1	1	0
10	$-4V_{DC}/13$	0	$-4V_{DC}/13$	0	1	1	0	0	1	1	0	0	1
11	$-3V_{DC}/13$	0	$-3V_{DC}/13$	1	0	1	0	0	1	1	0	0	1
12	$-2V_{DC}/13$	0	$-2V_{DC}/13$	1	0	0	1	0	1	1	0	0	1
13	$-V_{DC}/13$	0	$-V_{DC}/13$	0	1	1	0	1	0	1	0	0	1
14	0	0	0	1	0	1	0	1	0	1	0	0	1
15	$+V_{DC}/13$	0	$+V_{DC}/13$	1	0	0	1	1	0	1	0	0	1
16	$+2V_{DC}/13$	0	$+2V_{DC}/13$	0	1	1	0	1	0	0	1	0	1
17	$+3V_{DC}/13$	0	$+3V_{DC}/13$	1	0	1	0	1	0	0	1	0	1
18	$+4V_{DC}/13$	0	$+4V_{DC}/13$	1	0	0	1	1	0	0	1	0	1

[a]. Switching state '1' denotes switch is 'ON' and '0' denotes switch is 'OFF'

$-V_{DC}/13$, 0, $+V_{DC}/13$, $+2V_{DC}/13$, $+3V_{DC}/13$ and $+4V_{DC}/13$. Different voltage levels and corresponding switching states to generate eighteen-levels across the phase-A winding is shown in Table-I. Similar logic can be developed for phase-B and C by simply replacing the switch 'S_{AX}' by 'S_{BX}' and 'S_{CX}' (where X =1 to 10) respectively.

III. INVERTER CONTROL LOGIC

A simple level shifted pulse width modulation (LSPWM) scheme proposed in [18] is used to generate switching signals for switches. The adopted scheme requires seventeen level shifted triangular carriers (C_1-C_{17}) of peak-to-peak amplitude V_c and the frequency f_c. The triangular frequency f_c decides the switching frequency of the devices. The reference voltage space phasor V^*_s to obtain the desired speed is generated by constant V/f control logic.

Once the reference phasor is available, then the three reference phase voltages (v_a^*, v_b^*, v_c^*) can be generated using (1) and (2).

$$V_s^* = V_{s(\alpha)} + jV_{s(\beta)} \qquad (1)$$

Where $V_{s(\alpha)}$ is in the direction of phase-A and $V_{s(\beta)}$ is orthogonal to $V_{s(\alpha)}$.

$$V_s^* = v_a^* + v_b^* . e^{j(2\pi/3)} + v_c^* . e^{j(4\pi/3)} \qquad (2)$$

The three modified modulating signals to generate required phase voltage are given by (3) to (5).

$$V_a^* = V_m^* \sin w^* t + 0.2 V_m^* \sin 3w^* t + n\frac{V_c}{2} \qquad (3)$$

$$V_b^* = V_m^* \sin (w^* t - \frac{2\pi}{3}) + 0.2 V_m^* \sin 3w^* t + n\frac{V_c}{2} \qquad (4)$$

$$V_c^* = V_m^* \sin\left(w^*t - \frac{4\pi}{3}\right) + 0.2V_m^* \sin 3w^*t + n\frac{V_c}{2} \quad (5)$$

Where V_m^* is the maximum amplitude and w^* is the angular frequency of the reference waves (v_a^*, v_b^*, v_c^*) and 'n' varies from 1 to 17 for two-level to eighteen-level operation. The amplitude modulation index (m_a) is define as

$$m_a = \frac{2V_m^*}{17V_c} \quad (6)$$

All three modulating signals, along with seventeen level shifted triangular carriers are shown in Fig. 2. The seventeen triangular carriers divide the entire range of modulation into eighteen different regions R_1-R_{18}. Depending upon the amplitude of modulating signal, demarcation between different regions and phase voltage levels are shown in Table II. All three modulating signals are continuously compared with carriers (C_1-C_{17}) and level of operation is found using Table II. Once the level of operation is found required switches are made to be turned 'ON' using Table I. One important point can be noticed here that in the adopted scheme switching losses are less as compared to conventional LSPWM scheme because for lower modulation indices, switching occur only in CHB inverter which operates at lower DC link voltage and two-level inverter is clamped at $+9V_{DC}/13$, whereas in case of conventional LSPWM two-level inverter is also switched hence more switching losses.

For lower speed range the proposed inverter operates in two-level mode and as the speed increases operation is shifted from two-level to eighteen-level mode including even number of levels. In each step a DC bias of magnitude $V_c/2$ is added along with 20% of third harmonic as indicated in (3) to (5).

TABLE II. DETERMINATION OF REGION AND PHASE VOLTAGE LEVEL FROM MODULATING SIGNAL

Level	Modulating Signal Amplitude	Region	Phase Voltage Level
1	$<C_1$	R_1	$-V_{DC}$
2	$>C_1$ and $<C_2$	R_2	$-12V_{DC}/13$
3	$>C_2$ and $<C_3$	R_3	$-11V_{DC}/13$
4	$>C_3$ and $<C_4$	R_4	$-10V_{DC}/13$
5	$>C_4$ and $<C_5$	R_5	$-9V_{DC}/13$
6	$>C_5$ and $<C_6$	R_6	$-8V_{DC}/13$
7	$>C_6$ and $<C_7$	R_7	$-7V_{DC}/13$
8	$>C_7$ and $<C_8$	R_8	$-6V_{DC}/13$
9	$>C_8$ and $<C_9$	R_9	$-5V_{DC}/13$
10	$>C_9$ and $<C_{10}$	R_{10}	$-4V_{DC}/13$
11	$>C_{10}$ and $<C_{11}$	R_{11}	$-3V_{DC}/13$
12	$>C_{11}$ and $<C_{12}$	R_{12}	$-2V_{DC}/13$
13	$>C_{12}$ and $<C_{13}$	R_{13}	$-V_{DC}/13$
14	$>C_{13}$ and $<C_{14}$	R_{14}	0
15	$>C_{14}$ and $<C_{15}$	R_{15}	$+V_{DC}/13$
16	$>C_{15}$ and $<C_{16}$	R_{16}	$+2V_{DC}/13$
17	$>C_{16}$ and $<C_{17}$	R_{17}	$+3V_{DC}/13$
18	$>C_{17}$	R_{18}	$+4V_{DC}/13$

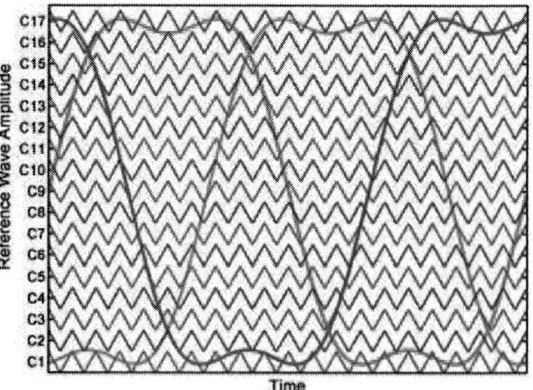

Fig. 2. Modulating signals and seventeen level shifted carriers

IV. SALIENT FEATURES OF PROPOSED INVERTER

The proposed topology uses six switches of voltage rating $9V_{DC}/13$, twelve switches of voltage rating $3V_{DC}/13$ and twelve switches of voltage rating $V_{DC}/13$, where V_{DC} is the equivalent DC-link voltage required to operate conventional two-level inverter fed induction motor drive. The eighteen-level NPC and FC use 102 switches and CHB uses 108 switches whereas the proposed inverter required only 30 switches. The proposed topology does not require any balancing capacitor whereas eighteen-level NPC and FC required 17 and 408 capacitors respectively. The eighteen-level NPC required 816 clamping diode this requirement is completely eliminated in the proposed topology. Asymmetrical cascade H-bridge (ACHB) topology [22] requires nine DC sources whereas the proposed topology needs only seven DC sources. Although ACHB with nine DC sources and 24 switches can generate twenty seven levels, but out of these nine DC sources three are of $9V_{DC}/13$ which makes the system bulky and costly hence the proposed inverter could be a better compromise between cost and quality. An interesting feature of the proposed inverter is its ability to operate in two-level mode in case failure of CHB inverter. It can be done by disconnecting the motor terminals (A_1,B_1,C_1) from faulty CHB inverter and connect them in star. Similarly the proposed inverter can operate in nine-level mode if any fault comes in two-level inverter. This feature increases the reliability of overall system. Another point can be noticed from Table-I that the inverter which operates at higher DC link voltage switch less as compare to the inverter which operates at lower DC link voltage, hence switching losses are reduced. The comparison between proposed, conventional and ACHB topologies is shown in Table III.

TABLE III. COMPARISON OF EIGHTEEN-LEVEL INVERTER TOPOLOGIES

Component	NPC	FC	CHB	ACHB[a]	Proposed Inverter
DC source	1	1	27	9	7
Switch	102	102	108	24	30
Clamping diode	816	0	0	0	0
Capacitor[b]	17	408	0	0	0

[a] H-bridge voltage sources are scaled in power of 3

[b] Excluding rectifier capacitors

V. SIMULATION RESULTS AND DISCUSSION

Simulation study has been carried out in MATLAB-Simulink environment to investigate the performance of proposed eighteen-level inverter. An open-end winding IM rated 1.5 kW, 415V, 4-pole, 50 Hz has been run at no-load, using constant *V/f* control. The DC link voltage is taken as 600 volt and switching frequency is kept at 1 kHz. Inverter is operated for the entire modulation range and Phase-A voltage and current waveforms along with harmonic spectrum of phase-A voltage are shown in Figs. 3-6.

Fig. 3(a) shows the phase-A voltage and current waveforms at modulation index (m_a) of 0.2. The motor operates at frequency of 10Hz at this modulation index. Fig. 3(b) shows the harmonic spectrum of phase-A voltage at fundamental frequency of 10 Hz, in this case THD is 19.07% of the fundamental voltage, which is very high according to IEEE-519 standard. Fig. 4(a) shows the phase-A voltage and current waveforms at modulation index of 0.4. At this modulation index motor operates at frequency of 20Hz and inverter

operation is shifted towards higher number of levels side. As the number of levels increased in the phase voltage its harmonic spectrum is also improved, which is reflected from harmonic spectrum of phase-A voltage shown in Fig. 4(b). Fig. 5 and Fig. 6 show the voltage and current waveform along with voltage harmonic spectrum for the modulation index of 0.7 and 1.0 respectively. It can be seen from Fig. 3-6 that as the modulation index increases phase voltage becomes more sinusoidal and THD is reduced. The transient state performance of proposed inverter is also investigated during acceleration and deceleration in speed. Fig. 7(a) shows the phase voltage and current waveform when motor is suddenly accelerate from 10Hz to 50Hz, in this case inverter operation is shifted from four-level to eighteen-level instantaneously and current comes in steady state in few cycles. Fig. 7(b) shows the phase voltage and current waveform when motor is suddenly decelerate from 50Hz to 10Hz, in this case inverter operation is shifted from eighteen-level to four-level and in this case also motor comes in steady state very quickly.

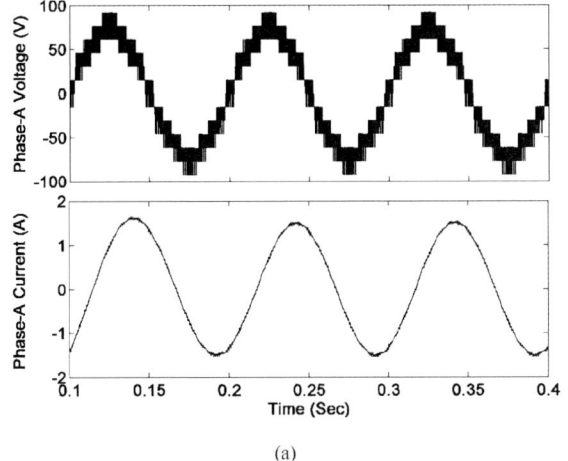

(a)

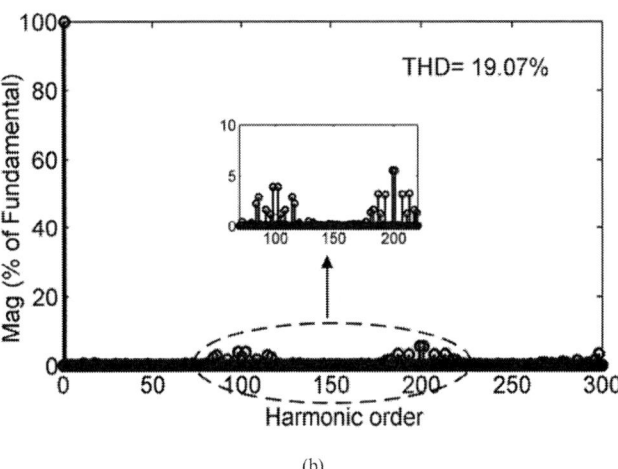

(b)

Fig. 3. (a) Phase-A voltage and current waveforms at m_a=0.2 (b) Harmonic spectrum of Phase-A voltage at fundamental frequency of 10 Hz (m_a=0.2)

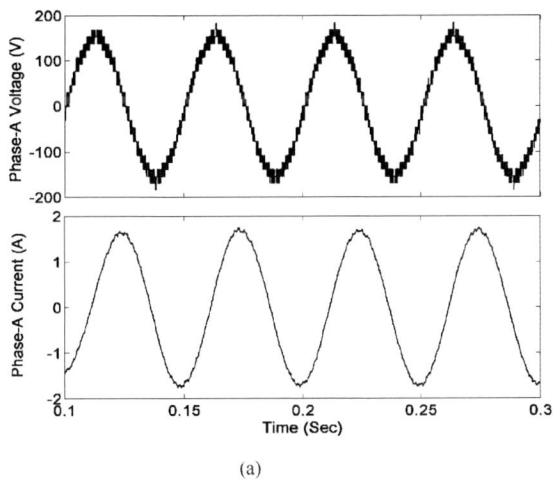

(a)

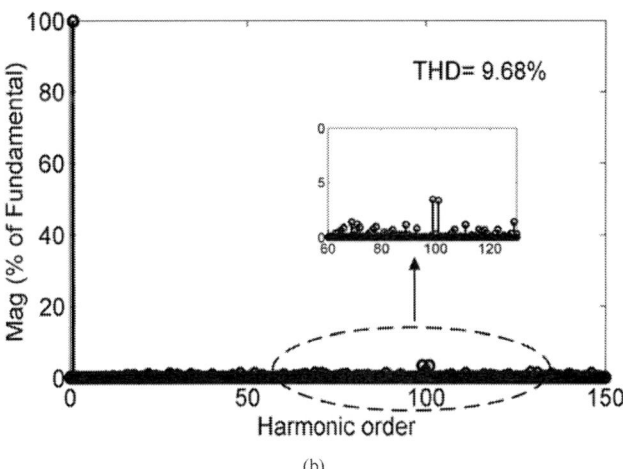

(b)

Fig. 4. (a) Phase-A voltage and current waveforms at ma=0.4 (b) Harmonic spectrum of Phase-A voltage at fundamental frequency of 20 Hz (m_a=0.4)

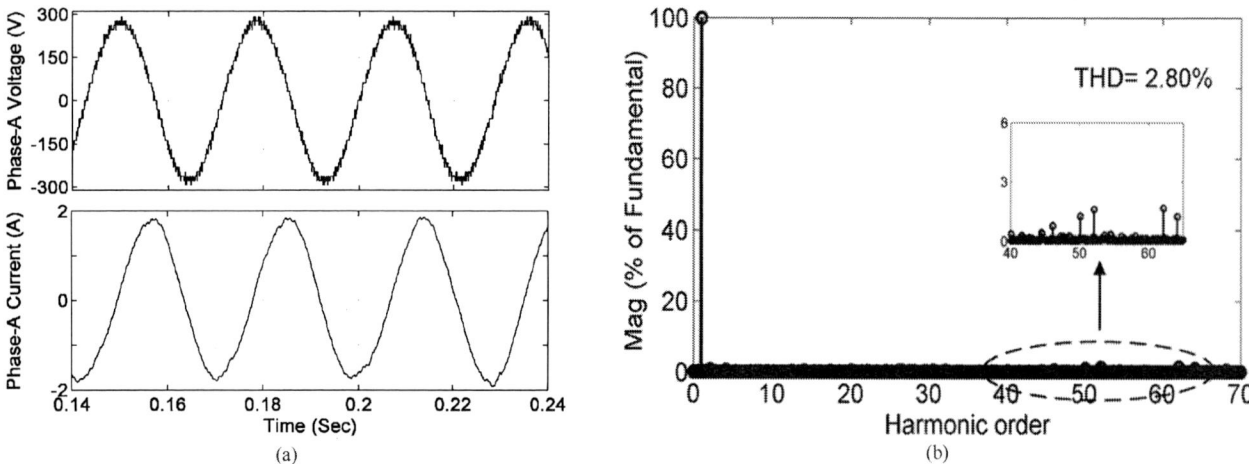

Fig. 5. (a) Phase-A voltage and current waveforms at m_a=0.7 (b) Harmonic spectrum of Phase-A voltage at fundamental frequency of 35 Hz (m_a=0.7)

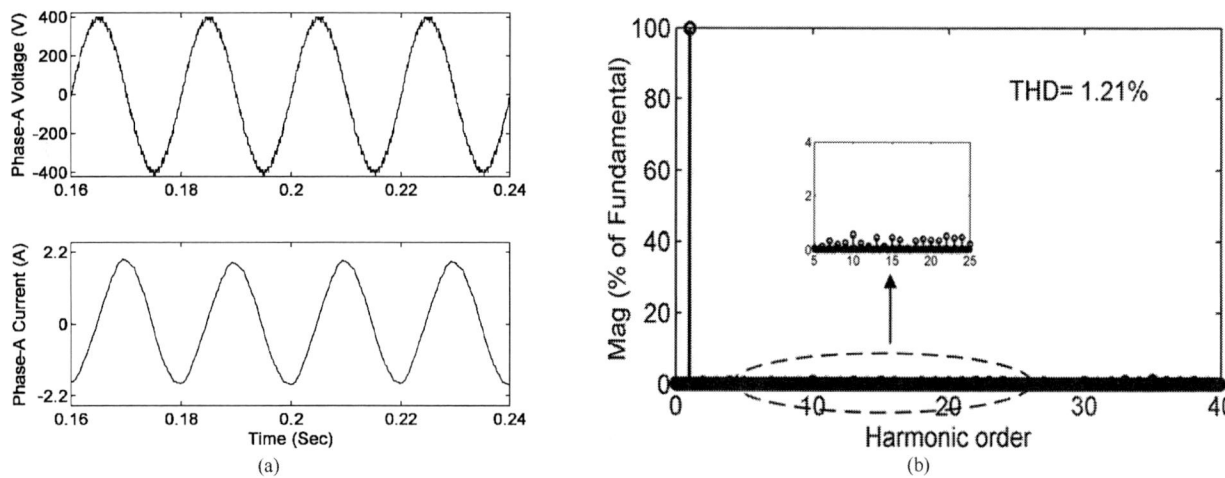

Fig. 6. (a) Phase-A voltage and current waveforms at m_a=1 (b) Harmonic spectrum of Phase-A voltage at fundamental frequency of 50 Hz (m_a=1)

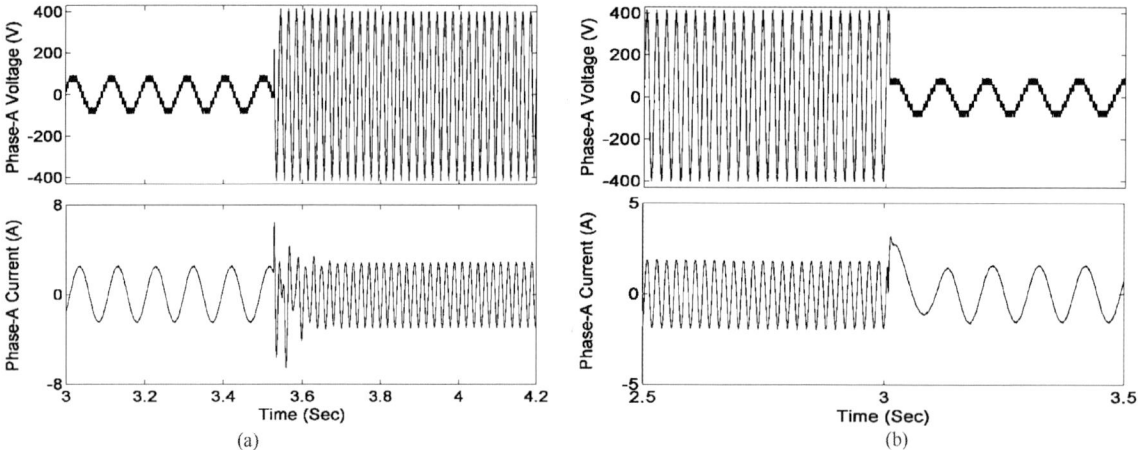

Fig. 7. Phase-A voltage and current waveforms during sudden change in operation from (a) four-level to eighteen-level (b) eighteen-level to four-level

978-1-4799-6047-7/14 $31.00 © 2014 IEEE   508

VI. CONCLUSION

A new hybrid eighteen-level inverter topology for an open-end IM drive is proposed and performance of the inverter is evaluated through simulation study. The proposed topology is compared with conventional topologies in terms of number of components used and it is found that the proposed topology completely eliminates 816 clamping diodes and 17 capacitors which are required in eighteen-level NPC. It also eliminates 408 balancing capacitors which are required in eighteen-level FC inverter. The proposed topology uses seven isolated DC sources whereas conventional CHB required 27 isolated DC supplies. Therefore, proposed topology reduces the cost, size and circuit complexity. The proposed topology produces a voltage space phasor of maximum amplitude of V_{DC} whereas the maximum DC link voltage used is $9V_{DC}/13$, this feature further reduces the size of inverter. The proposed topology inherently prevents the triplen harmonic currents because both inverters are fed by isolated DC supply. It is found in simulation results that as the inverter operation shifted towards higher level side output phase voltage become more sinusoidal with improved harmonic spectrum and THD is reduced up to 1.21%. Simulation results show the potential of proposed inverter in industrial applications.

REFERENCES

[1] L. G. Franquelo, J. Rodriguez, J. I. Leon, S. Kouro, R. Portillo, and M. A. M. Prats, "The age of multilevel converters arrives," *IEEE Industrial Electronics Magazine*, no. June, pp. 28–39, 2008.

[2] F. Z. Peng, W. Qian, and D. Cao, "Recent advances in multilevel converter/inverter topologies and applications," in *The 2010 International Power Electronics Conference (IPEC)*, 2010, pp. 492–501.

[3] J. Rodríguez, S. Bernet, B. Wu, J. O. Pontt, and S. Kouro, "Multilevel Voltage-Source-Converter Topologies for Industrial Medium-Voltage Drives," *IEEE Transactions on Industrial Electronics*, vol. 54, no. 6, pp. 2930–2945, 2007.

[4] J. Rodríguez, J. Lai, and F. Z. Peng, "Multilevel Inverters : A Survey of Topologies , Controls And Applications," *IEEE Transactions on Industrial Electronics*, vol. 49, no. 4, pp. 724–738, 2002.

[5] A. Nabae, I. Takahashi, and H. Akagi, "A New Neutral-Point-Clamped PWM Inverter," *IEEE Transactions on Industry Applications*, vol. I, no. 5, pp. 518–523, 1981.

[6] T. A. Meynard and H. Foch, "Multi-level conversion: high voltage choppers and voltage-source inverters," *PESC '92 Record. 23rd Annual IEEE Power Electronics Specialists Conference*, pp. 397–403, 1992.

[7] J. Lai and F. Z. Peng, "Multilevel Converters-A New Breed of Power Converters," *IEEE Transactions on Industry Applications*, vol. 32, no. 3, pp. 509–517, 1996.

[8] M. Malinowski, K. Gopakumar, J. Rodriguez, and M. A. Pérez, "A Survey on Cascaded Multilevel Inverters," *IEEE Transactions on Industrial Electronics*, vol. 57, no. 7, pp. 2197–2206, 2010.

[9] I. Colak, E. Kabalci, and R. Bayindir, "Review of multilevel voltage source inverter topologies and control schemes," *Energy Conversion and Management*, vol. 52, no. 2, pp. 1114–1128, Feb. 2011.

[10] S. Kouro, M. Malinowski, K. Gopakumar, J. Pou, L. G. Franquelo, B. Wu, J. Rodriguez, M. A. Pérez, and J. I. Leon, "Recent advances and industrial applications of multilevel converters," *IEEE Transactions on Industrial Electronics*, vol. 57, no. 8, pp. 2553–2580, Aug. 2010.

[11] G. P. Adam, O. Anaya-Lara, G. M. Burt, D. Telford, B. W. Williams, and J. R. McDonald, "Modular multilevel inverter: pulse width modulation and capacitor balancing technique," *IET Power Electronics*, vol. 3, no. 5, p. 702, 2010.

[12] F. Wang, "Sine-Triangle versus Space-Vector Modulation for Three-Level PWM Voltage-Source Inverters," *IEEE Transactions on Industrial Applications*, vol. 38, no. 2, pp. 500–506, 2002.

[13] A. K. Gupta and A. M. Khambadkone, "A Space Vector PWM Scheme for Multilevel Inverters Based on Two-Level Space Vector PWM," *IEEE Transactions on Industrial Electronics*, vol. 53, no. 5, pp. 1631–1639, 2006.

[14] H. Stemmler and P. Guggenbach, "Configurations of High-Power Voltage Source Inverter Drives," in *Fifth European Conference*, 1993, pp. 7–14.

[15] E. G. Shivakumar, K. Gopakumar, S. K. Sinha, A. Pittet, and V. T. Ranganathan, "Space Vector PWM Control of Dual Inverter Fed Open-End Winding Induction Motor Drive," in *Applied Power Electronics Conference and Exposition, APEC*, 2001, vol. 1, pp. 399–405.

[16] Brian A. Welchko and J. M. Nagashima, "A Comparative Evaluation of Motor Drive Topologies For Low-Voltage, High-Power EV/HEV Propulsion Systems," *Industrial Electronics, ISIE*, no. vol. 1, pp. 379–384, 2003.

[17] K. C. Sekhar and G. T. R. Das, "Five-level SPWM Inverter for an Induction Motor with Open-end Windings," in *First International Power and Energy Coference PECon 2006 November 28-29, Putrajaya, Malaysia*, 2006, pp. 342–347.

[18] M. Baiju, K. Gopakumar, K. K. Mohapatra, V. T. Somasekhar, and L. Umanand, "Five-Level Inverter Voltage-Space Phasor Generation For An Open-End Winding Induction Motor Drive," *Electric Power Applications, IEE Proceedings*, vol. 150, no. 5, pp. 531–538, 2003.

[19] V. T. Somasekhar, K. Chandrasekhar, and K. Gopakumar, "A New Five-Level Inverter System For An Induction Motor With Open-End Windings," in *The Fifth International Conference on Power Electronics and Drive Systems, PEDS*, 2003, pp. 199–204.

[20] P. P. Rajeevan, K. Sivakumar, C. Patel, R. Ramchand, and K. Gopakumar, "A Seven-Level Inverter Topology for Induction Motor Drive Using Two-Level Inverters and Floating Capacitor Fed H-Bridges," *IEEE Transactions on Power Electronics*, vol. 26, no. 6, pp. 1733–1740, 2011.

[21] P. P. Rajeevan, K. Sivakumar, K. Gopakumar, C. Patel, and H. Abu-rub, "A Nine-Level Inverter Topology for Medium-Voltage Induction Motor Drive With Open-End Stator Winding," *IEEE Transactions on Industrial Electronics*, vol. 60, no. 9, pp. 3627–3636, 2013.

[22] Y. S. Lai and F. S. Shyu, "Topology for hybrid multilevel inverter," *IEE Proceedings Electric Power Applications*, vol. Vol. 149, no. No. 6, pp. 494–458, 2002.

Experimental Comparison of Conventional and Bus-Clamping PWM Methods Based on Electrical and Acoustic Noise Spectra

Binojkumar A C.
Department of Electrical Engineering
Indian Institute of Science
Bangalore, India - 560 012
Email: binojkac@gmail.com

B. Saritha
Department of Electrical Engineering
Indian Institute of Science
Bangalore, India - 560 012
Email: bsaritha80@gmail.com

G. Narayanan
Department of Electrical Engineering
Indian Institute of Science
Bangalore, India - 560 012
Email: gnar@ee.iisc.ernet.in

Abstract—The acoustic noise emitted by an induction motor,powered by a voltage source inverter, is an environmental issue. Harmonics present in the stator current are the main reason for the increased acoustic noise in motor drives. The pulse-width modulation technique used to modulate the inverter is the key factor determining the magnitude of current harmonics. Bus-clamping pulse-width modulation (BCPWM) techniques have got much attention nowadays due to reduced switching loss, compared to conventional space vector pwm(CSVPWM). In this paper, two BCPWM techniques namely, $60°$ and $30°$ BCPWM methods are compared with CSVPWM on the basis of electrical spectra as well as acoustic noise spectra. Experiments are conducted on a pulse-width modulated voltage source inverter fed 6 kW induction motor drive. Harmonic analysis is carried out on the measured line to line voltage, stator current and acoustic noise, corresponding to the three methods, at different fundamental and carrier frequencies. Comparison of the experimental results show that the magnitude of dominant acoustic noise component around the carrier frequency is reduced significantly with BCPWM methods,compared to CSVPWM,at high modulation indices.

Index Terms—Acoustic noise, bus-clamping pulse-width modulation, conventional space vector pwm, harmonic analysis, induction motor drive, noise spectrum, pulse-width modulation.

I. INTRODUCTION

Acoustic noise emission of electric motors is a wide area of research, focusing on the causes and solutions of noise, in the domain of electric machines and drives. A brief review on various sources of acoustic noise and its mitigation is given in [1]. An important reason for high noise emission in variable speed drives is the application of non-sinusoidal supply to the motor drives. Early works focused on acoustic noise emitted by six-step inverter fed motor drives [2] and carrier-based sinusoidal pulse width modulation (SPWM) induction motor drives [2], [3]. When the inverter is operated with sinusoidal PWM (SPWM), the harmonic energy is concentrated in distinct tones, located around switching frequency and its integral multiples. Hence the resulting noise annoyance is high [3], [4]. A common practice to reduce the acoustic annoyance is to create a spread in the harmonic spectra of voltage and current. The harmonic spread is obtained by randomly modulating the frequency of triangular carrier, maintaining the average

switching frequency a constant [5]–[7]. However, the random variation of switching frequency makes the design of closed-loop controller for the motor drive a difficult task.

Conventional space vector pulse width modulation (CSVPWM) offers improved DC bus utilization as well as reduced current ripple and reduced torque ripple, compared to SPWM [8], [9]. In view of its good performance and wide acceptance, CSVPWM is often considered a benchmark for evaluation of pulse width modulation (PWM) techniques [7]–[10]. However, CSVPWM also has its harmonic energy concentrated around switching frequency and its multiples as in case of SPWM. In fact, the acoustic noise performance of SPWM and CSVPWM are comparable [11].

Bus-clamping PWM (BCPWM) methods are another class of well known PWM methods, which are used to reduce the switching loss of the inverter [12]–[14]. It is also shown that BCPWM methods can reduce total harmonic distortion and pulsating torque in motor drives at high speeds [8], [9]. BCPWM techniques are also employed in multi-level converters [15]. This paper studies the acoustic noise performances of BCPWM methods, which have not been studied very much except for a recent work [16].

As the electromagnetic noise emitted by the induction motor drive mainly depends on the current harmonics, the electrical and acoustic noise characteristics of induction motor drives modulated with BCPWM techniques are investigated and compared with CSVPWM through experiments.

II. BUS-CLAMPING PWM METHODS

The modulating signals pertaining to CSVPWM, $60°$ BCPWM and $30°$ BCPWM are shown in Fig. 1(a), Fig. 1(b) and Fig. 1(c), respectively. Reference [16] considers $60°$ BCPWM and two other BCPWM methods. However, $60°$ BCPWM and $30°$ BCPWM methods are considered here since these two offer the highest and lowest values of line current THD, respectively, among BCPWM methods [14]. Hence these two methods are good representatives of the entire class of BCPWM methods. These BCPWM methods are described in good detail in [8], [9], [12]–[15].

978-1-4799-6047-7/14 $31.00 © 2014 IEEE

III. EXPERIMENTAL SET-UP

The experimental studies on acoustic noise are carried out on a 6-kW, 400 V, 4-pole, 50 Hz induction motor. The experimental motor is fed from a 10-kVA IGBT based inverter.TMS320LF2407 DSP processor is used as the digital controller for PWM generation. The motor drive is operated using constant voltage to frequency (V/f) method under no-load with a fixed DC link voltage of 566 V.

The acoustic noise is measured using a noise measurement system,which is detailed in [17]. This is calibrated with a professional sound level meter [18], and conforms to IEC 61672/2002 [19]. The noise measurement is carried out at the height of the motor shaft at a perpendicular distance of 1 m as per standards [19], [20].

IV. EXPERIMENTAL STUDY ON ELECTRICAL SPECTRA

Fig. 2(a), Fig. 2(b) and Fig. 2(c) show the frequency spectra of measured line to line voltage waveformson the 6-kW motor at a fundamental frequency (f_m) of 50 Hz and a carrier

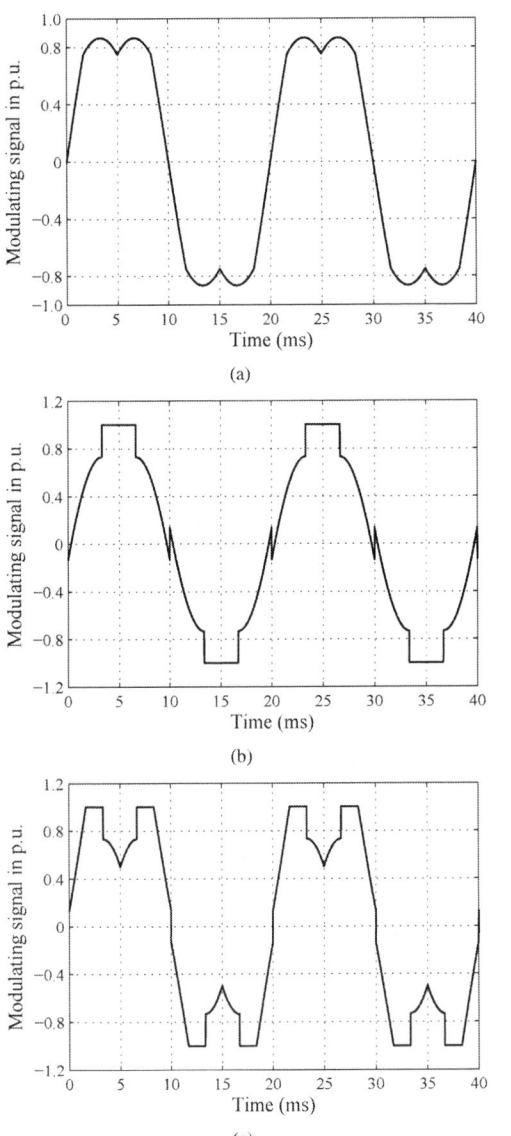

Figure 1. Modulation signals corresponding to (a) CSVPWM (b) 60° BCPWM and (c) 30° BCPWM.

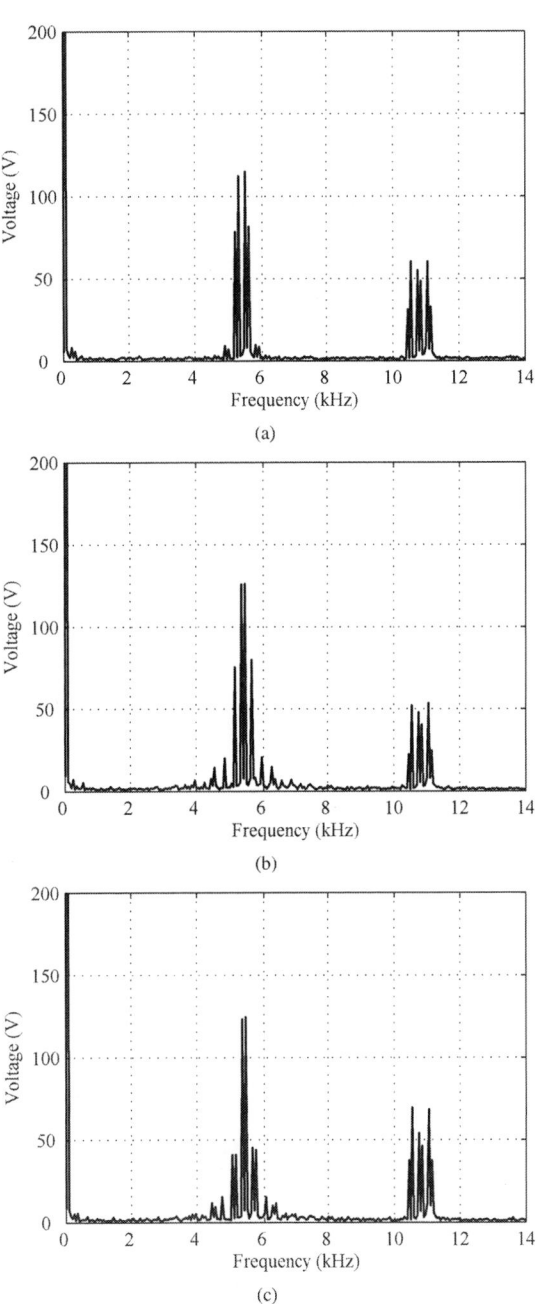

Figure 2. Frequency spectra of measured line to line voltage of 6 kW motor at f_m = 50 Hz and f_c = 5.4 kHz with (a) CSVPWM (b) 60° BCPWM and (c) 30° BCPWM.

frequency (f_c) of 5.4 kHz with CSVPWM, 60° BCPWM and 30° BCPWM, respectively. Also, the corresponding harmonic spectra of the stator current are shown plotted in Fig. 3(a), Fig. 3(b) and Fig. 3(c), respectively. The voltage and current harmonics are observed around integral multiples of f_c.

At f_m = 50 Hz, the voltage and current harmonics around f_c are much higher than those around $2f_c$. The magnitude

of dominant component around f_c and that around $2f_c$ are measured at various fundamental frequencies. These are shown plotted for CSVPWM, 60° BCPWM and 30° BCPWM in Fig. 4(a), Fig. 4(b) and Fig. 4(c), respectively. All these measurements are carried out at f_c = 5.4 kHz with the fundamental frequency varying from 10 Hz to 50 Hz in steps of 5 Hz.

With CSVPWM, the voltage harmonics around f_c dominate over those around $2f_c$ only at high modulation indices, as shown by Fig. 4(a). On the other hand, the harmonics around f_c are higher than those around $2f_c$ almost over the entire

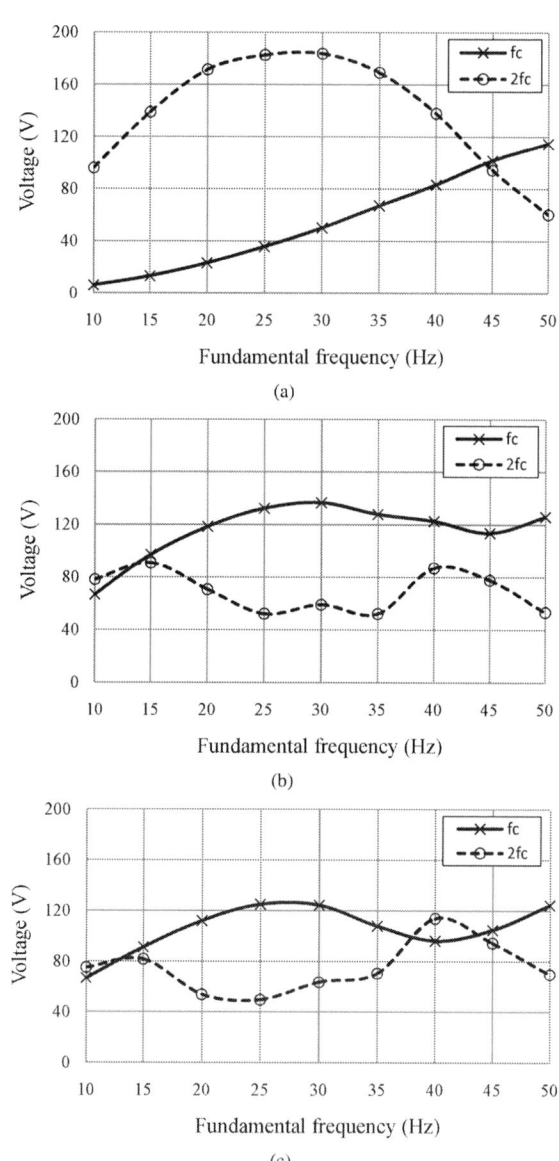

Figure 3. Frequency spectra of measured stator current of 6 kW motor at f_m = 50 Hz and f_c = 5.4 kHz with (a) CSVPWM (b) 60° BCPWM and (c) 30° BCPWM.

Figure 4. Variation of peak magnitude of line to line voltage of 6 kW motor around f_c and $2f_c$ with fundamental frequency at f_c = 5.4 kHz with (a) CSVPWM (b) 60° BCPWM and (c) 30° BCPWM.

fundamental frequency range (except for a small region close to 40Hz in case of 30° BCPWM). The voltage harmonics around $2f_c$ encounter a reactance which is roughly twice that seen by the voltage harmonics around f_c. Hence, with CSVPWM, the dominant current harmonic around f_c is higher than that around $2f_c$ over a widened speed range close to rated speed, as shown by Fig. 5(a). The f_c current components are higher than the $2f_c$ current components over the entire speed range for BCPWM methods as shown by Fig. 5(b) and Fig. 5(c).

Further, the dominant harmonic current around f_c is compa-rable at high speeds for the three PWM methods, despite the BCPWM methods switching at only two thirds the switching frequency in case of CSVPWM.

V. EXPERIMENTAL STUDY ON ACOUSTIC NOISE SPECTRA

The acoustic noise emitted by the 6-kW induction motor is measured while being operated with CSVPWM, 60° BCPWM and 30° BCPWM methods. Harmonic analysis of the acoustic

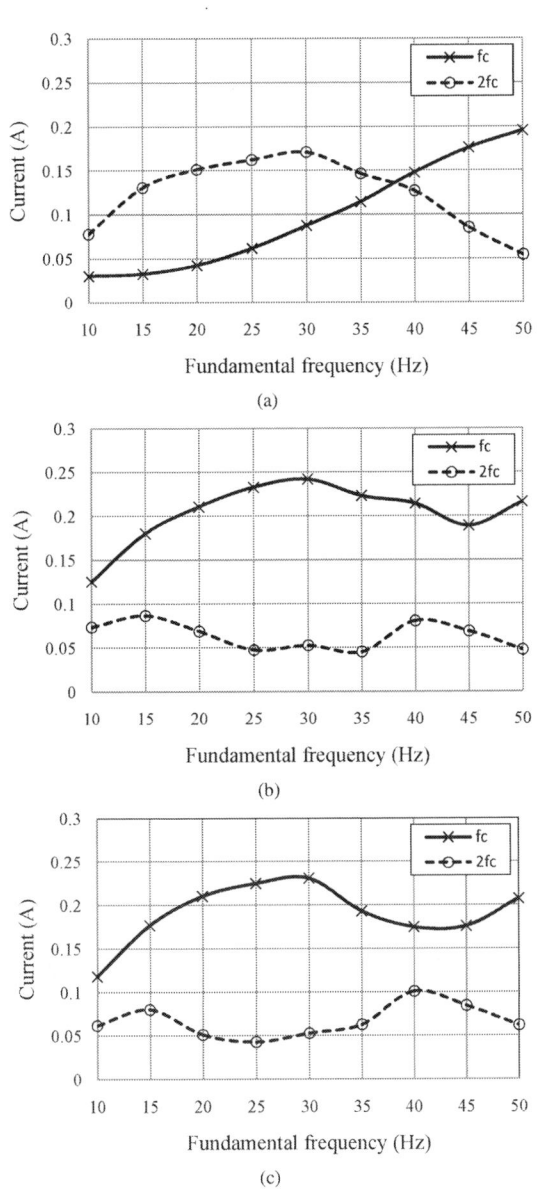

Figure 5. Variation of peak magnitude of stator current of 6 kW motor around f_c and $2f_c$ with fundamental frequency at $f_c = 5.4$ kHz with (a) CSVPWM (b) 60° BCPWM and (c) 30° BCPWM.

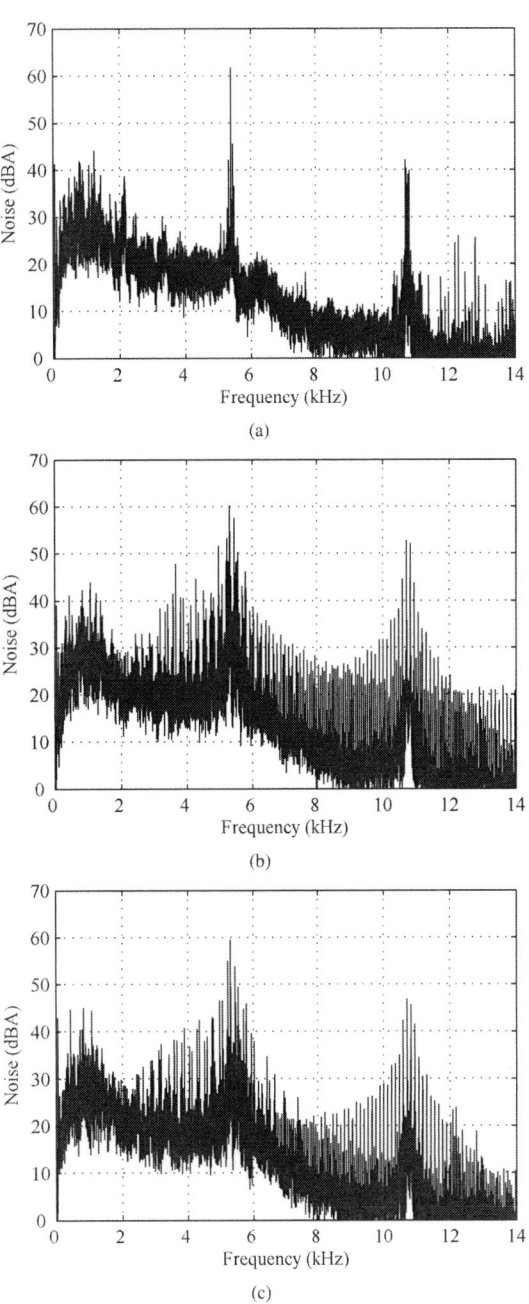

Figure 6. Acoustic noise spectra (dBA) of 6 kW motor at $f_m = 20$ Hz and $f_c = 5.4$ kHz with (a) CSVPWM (b) 60° BCPWM and (c) 30° BCPWM.

noise is performed. Each frequency component is multiplied by an A-weighted factor, which is a measure of the sensitivity of human ear at the given frequency [19]. The resulting A-weighted spectra in dBA are shown in Fig. 6 to Fig. 8.

The measured dBA spectrum of the acoustic noise, corre-sponding to CSVPWM, at $f_m = 20$ Hz and $f_c = 5.4$ kHz is depicted in Fig. 6(a). Distinct tones (i.e. noise of high magnitude in narrow ranges of frequency) are clearly visible in the spectrum around f_c and $2f_c$. The experimental noise spectra with $60°$ BCPWM and $30°$ BCPWM methods are

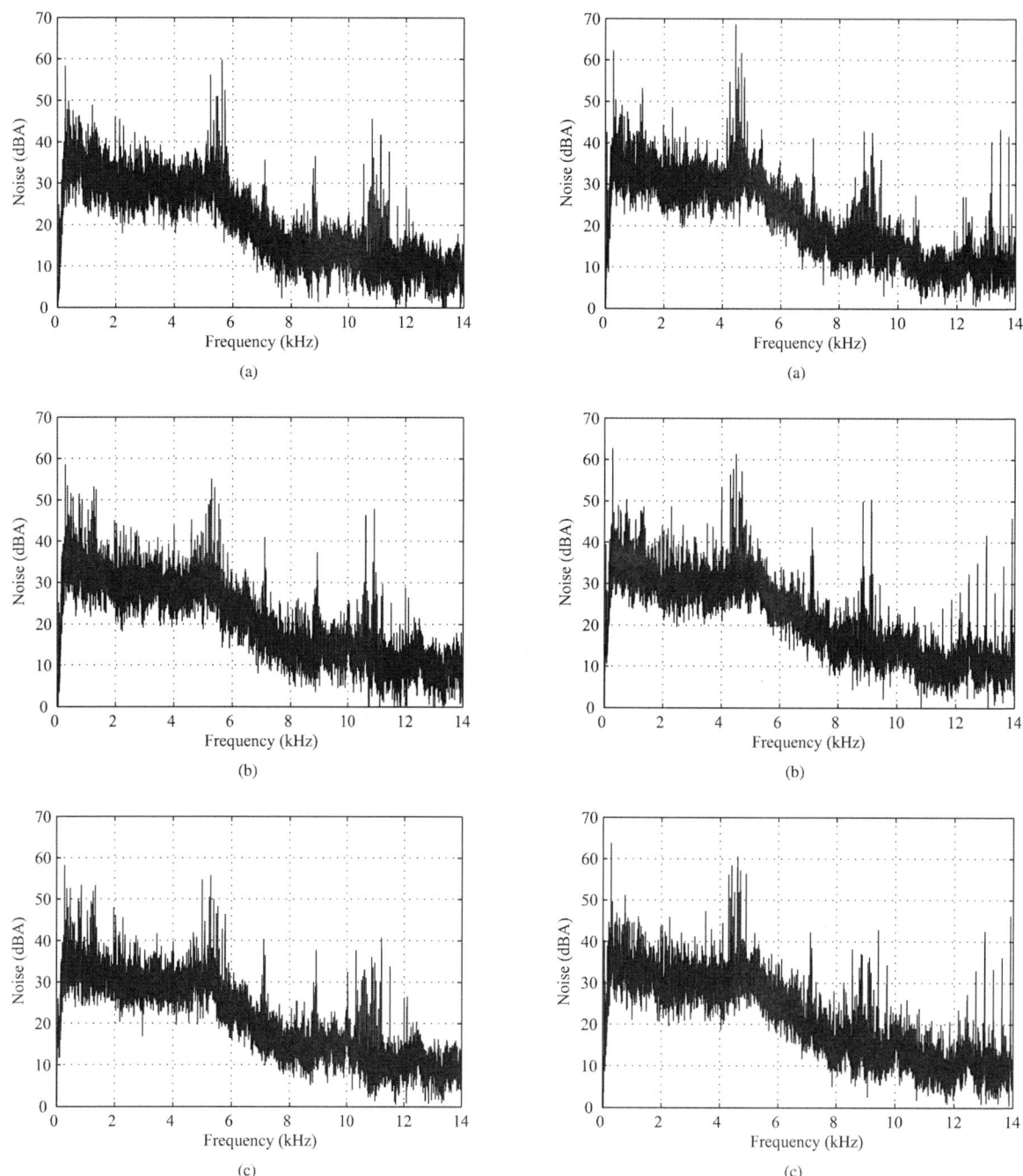

Figure 7.   Acoustic noise spectra (dBA) of 6 kW motor at $f_m = 50$ Hz and $f_c = 5.4$ kHz with (a) CSVPWM (b) $60°$ BCPWM and (c) $30°$ BCPWM.

Figure 8.   Acoustic noise spectra (dBA) of 6 kW motor at $f_m = 50$ Hz and $f_c = 4.5$ kHz with (a) CSVPWM (b) $60°$ BCPWM and (c) $30°$ BCPWM.

shown in Fig. Fig. 6(b) and Fig. 6(c), respectively. As seen from these figures, certain "spread" in visible in the spectra pertaining to BCPWM methods.

Acoustic noise spectra of 6 kW motor for the three PWM methods are presented in Fig. 7 pertaining to $f_m = 50$ Hz and $f_c = 5.4$ kHz. Fig. 7(a) to Fig. 7(c) show that dominant noise at f_c is significant to that at $2f_c$. Moreover, the peak noise is reduced with $60°$ and $30°$ BCPWM schemes compared to CSVPWM.

For validating the experimental results further, the experiment is repeated with a carrier frequency of 4.5 kHz. Fig. 8(a) presents the acoustic noise spectra (dBA) of the 6-kW motor at $f_m = 50$ Hz and $f_c = 4.5$ kHz with CSVPWM. Similar results are shown in Fig. 8(b) and Fig. 8(c), respectively, for $60°$ BCPWM and $30°$ BCPWM. It can be seen that dominant noise component around f_c is significantly reduced with BCPWM methods, especially $30°$ BCPWM, compared to CSVPWM. Moreover, as the average switching frequency of the BCPWM methods is only (2/3) of that of CSVPWM, reduced switching loss can also be obtained with the BCPWM methods.

VI. CONCLUSION

In this paper, electrical and acoustic noise spectra of voltage source inverter fed induction motor modulated with CSVPWM, $60°$ and $30°$ BCPWM methods are investigated experimentally at various motor speeds in a constant V/f drive.

The experimental results show that both voltage and current harmonic components around f_c dominate over the corresponding components around $2f_c$ for the two bus-clamping PWM methods. Further, $30°$ BCPWM is found to result in lower voltage and current harmonics than $60°$ BCPWM.

Comparison of the acoustic noise emitted by induction motor drive, operated with CSVPWM and BCPWM methods, shows reduced dominant noise magnitudes with BCPWM at high modulation indices. Moreover, as the average switching frequency of the BCPWM methods is only two thirds (i.e. 2/3) of the switching frequency in case of CSVPWM, reduction in switching loss can also be achieved.

REFERENCES

[1] P. Vijayaraghavan and R. Krishnan, "Noise in electric machines: A review," *IEEE Trans. Ind. Electron.*, vol. 35, no. 5, pp. 1007–1013, Sep./Oct. 1999.

[2] R. J. M. Belmans, L. D'Hondt, A. J. Vandenput, and W.Geysen, "Analysis of the audible noise of three-phase squirrel cage induction motors supplied by inverters," *IEEE Trans. Ind. Appl.*, vol. IA-23, no. 5, pp. 842–848, Sep./Oct. 1987.

[3] A. K. Wallace, R. Spee, and L. G. Martin, "Current harmonics and acoustic noise in AC adjustable-speed drives," *IEEE Trans. Ind. Appl.*, vol. 26, no. 2, pp. 267–273, Mar./Apr. 1990.

[4] J. L. Besnerais, V. Lanfranchi, M. Hecquet, and P. Brochet, "Characterization and reduction of audible magnetic noise due to PWM supply in induction machines," *IEEE Trans. Ind. Electron.*, vol. 57, no. 4, pp. 1288–1295, 2010.

[5] T. G. Habetler and D. M. Divan, "Acoustic noise reduction in sinusoidal PWM drives using a randomly modulated carrier," *IEEE Trans. Power Electron.*, vol. 6, no. 3, pp. 356–363, Jul. 1991.

[6] C. M. Liaw and Y. M. Lin, "Random slope pwm inverter using existing system background noise: Analysis, design and implementation," *IEE Elect. Power Appl.*, vol. 147, no. 1, pp. 45–54, Jan. 2000.

[7] A. Ruiz-Gonzalez, M. J. Meco-Gutierrez, F. Perez-Hidalgo, F. Vargas-Merino, and J. R. Heredia-Larrubia, "Reducing acoustic noise radiated by inverter-fed induction motors controlled by a new PWM strategy," *IEEE Trans. Ind. Electron.*, vol. 57, no. 1, pp. 228–236, Jan. 2010.

[8] V. S. S. P. K. Hari and G. Narayanan, "Space-vector based hybrid pulse width modulation technique to reduce line current distortion in induction motor drives," *IET power electronics*, vol. 5, no. 8, pp. 1463–1471, 2012.

[9] K. Basu, J. S. S. Prasad, and G. Narayanan, "Minimization of torque ripple in PWM AC drives," *IEEE Trans. Ind. Electron.*, vol. 56, no. 2, pp. 553–558, Feb. 2009.

[10] A. C. Binojkumar, J. S. Sivaprasad, and G. Narayanan, "Experimental investigation on the effect of advanced bus-clamping pulsewidth modulation on motor acoustic noise," *IEEE Trans. Ind. Electron.*, vol. 60, no. 2, pp. 433–439, Feb. 2013.

[11] W. C. Lo, C. C. Chan, Z. Q. Zhu, L. Xu, D. Howe, and K. T. Chau, "Acoustic noise radiated by PWM controlled induction machine drives," *IEEE Trans. Ind. Electron.*, vol. 47, no. 4, pp. 880–889, Aug. 2000.

[12] Y. Wu, M. A. Shafi, A. M. Knight, and R. A. McMahon, "Comparison of the effect of continuous and discontinuous PWM schemes on power losses of voltage-sourced inverters for induction motor drives," *IEEE Trans. Power Electron.*, vol. 26, no. 1, pp. 182–192, Jan. 2011.

[13] J. S. Sivaprasad, "Control, modulation and testing of high-power pulse width modulated converters," Ph.D. dissertation, Department of Electrical Engineering, Indian Institute of Science, Bangalore, July 2013.

[14] T. D. Nguyen, J. Hobraiche, N. Patin, G. Friedrich, and J. Vilain, "A direct digital technique implementation of general discontinuous pulse width modulation strategy," *IEEE Trans. Ind. Electron.*, vol. 58, no. 9, pp. 4445–4454, sep 2011.

[15] Z. Zhang, O. C. Thomsen, and M. A. E. Anderson, "Discontinuous PWM modulation strategy with circuit-level decoupling concept of three-level neutral-point-clamped inverter," *IEEE Trans. Ind. Electron.*, vol. 60, no. 5, pp. 1897–1906, may 2013.

[16] I. P. Tsoumas and H.Tischmacher, "Influence of the inverter's modulation technique on the audible noise of electric motors," *IEEE Trans. Ind. Appl.*, vol. 50, no. 1, pp. 269–278, 2014.

[17] A. C. Binojkumar and G. Narayanan, "A low-cost system for measurement and spectral analysis of motor acoustic noise," in *Proc. NPEC '11*, Howra, India, Dec. 2011.

[18] Brüel & Kjær 2260 sound level meter - manufacturers' specification. [Online]. Available: http://www.bksv.com

[19] *Electroacoustics - Sound level meters*, IEC Std. 61 672, 2002.

[20] *IEEE Test Procedure for Airborne Sound Measurements on Rotating Electric Machinery*, IEEE Std. 85-1973, 1980.

Power Factor Correction in Sensorless BLDC Motor Drive

Vashist Bist [1], *Student Member, IEEE* and Bhim Singh [2], *Fellow*
Department of Electrical Engineering,
Indian institute of Technology Delhi, Hauz Khas, New Delhi-110016, India
e-mail: [1]vashist.bist@gmail.com (Corresponding Author) and [2]bsingh@ee.iitd.ac.in

Abstract— This work presents a sensorless brushless DC motor (BLDCM) drive with power factor correction (PFC). The speed control of BLDCM is achieved by varying the DC bus voltage of the voltage source inverter (VSI) feeding BLDCM. The BLDCM is electronically commutated using a VSI operating at low frequency switching for reducing the switching losses in VSI. A canonical switching cell (CSC) converter is used for varying the DC bus voltage of VSI. This CSC converter is operated in discontinuous inductor current mode (DICM) for achieving a power factor correction (PFC) inherently without any sensing requirement. The rotor position sensors are eliminated by using sensorless control of BLDCM. Hence, a sensorless BLDCM drive is developed as a cost effective solution for low power applications. Its performance is evaluated on a developed prototype at various speeds with unity power factor (UPF) at AC mains.

Index Terms— Brushless DC Motor, Sensorless Control, Rotor Position Phase Delay Compensation, Power Factor Correction.

I. INTRODUCTION

SENSOR reduction in low-cost, low-power brushless DC motor (BLDCM) drive is required as it directly affects the cost of overall system. Many low power applications such as fans, water pumps, etc uses BLDCM due to many advantages such as high efficiency, high power density and low electro-magnetic interference (EMI) [1, 2]. These motors are synchronous motors having three-phase concentrated stator windings and rotors are made of permanent magnets [1-2]. These motors are commutated electronically via a three phase voltage source inverter (VSI) based on the sensed rotor position via Hall Effect rotor position sensors [1, 2]. However, these sensors are costly, require a special mounting assembly and their performance is affected by the ambient temperature and EMI generated by the stator current [1, 2]. Hence, rotor position sensorless control of BLDCM is used for the elimination of Hall-effect position sensors [3].

The conventional scheme of BLDCM drives uses an uncontrolled rectifier for feeding BLDCM; which draws harmonics rich supply current from AC mains [4]. Such peaky current has high total harmonic distortion (THD) of AC mains current which results in poor power factor (PF) at AC mains. Power factor correction (PFC) converters are used for achieving a unity power factor at AC mains and acceptable power quality (PQ) indices within IEC 61000-3-2 limits [5]. Single-stage PFC converters are widely used due to low

component count and thus low losses associated with them [6]. Selection of operating mode of PFC converter depends on the required power rating, cost of overall system and the permitted stress on the switch of PFC converter. The operation of a PFC converter in continuous inductor conduction mode (CICM) offers low current stress on the switch, but requires sensing of supply voltage, DC bus voltage and supply current for its operation [6]. Moreover, the PFC converter operating in discontinuous inductor conduction mode (DICM) provides an inherent power factor correction at AC mains without any sensing requirement. But the current stress is high in DICM operation and therefore used in low power applications.

A boost-PFC converter is a most widely used configuration for feeding the BLDCM drive [7]. This configuration maintains a constant DC bus voltage at the VSI and uses a PWM based switching of the VSI for speed control. Such configuration has high switching losses in VSI due to high switching frequency of VSI for effective control. A method of speed control of BLDCM by controlling the DC bus voltage is proposed in [8]. This facilitates the operation of VSI in fundamental frequency switching by electronically commutating BLDCM; which reduces VSI's switching losses. Based on this approach, a PFC Luo converter for feeding BLDCM is proposed in [9]. Moreover, bridgeless configuration of buck-boost, Cuk and Zeta converter for feeding BLDCM drive also have been reported in [10], [11] and [12]. These converters suffer from higher number of components. This paper presents a canonical switching cell (CSC) for feeding BLDCM because of its ability as an excellent power factor corrector, light load regulation capability and comparatively low number of components as compared to other PFC converter [13-14]. A CSC converter for feeding the BLDCM drive has been presented in [14], but it has used a Hall-effect position sensors and voltage sensor which increases the cost of overall system. However, this work presents a complete sensorless BLDCM drive.

Many position sensorless control schemes for BLDCM drive have been reported in the literature [15-23]. Estimation of phase back emf using the terminal voltage sensing has been most widely used technique to extract rotor position information. However, a phase delay of 30^0 is achieved due to phase back-emf sensing which has to be compensated by the use of analog filter [15]. This phase delay problem has been removed in line back emf sensing methods, but the phase delay due to low pass filtering and effect of voltage drop at the stator windings has to be considered [16, 17].

978-1-4799-6047-7/14 $31.00 © 2014 IEEE

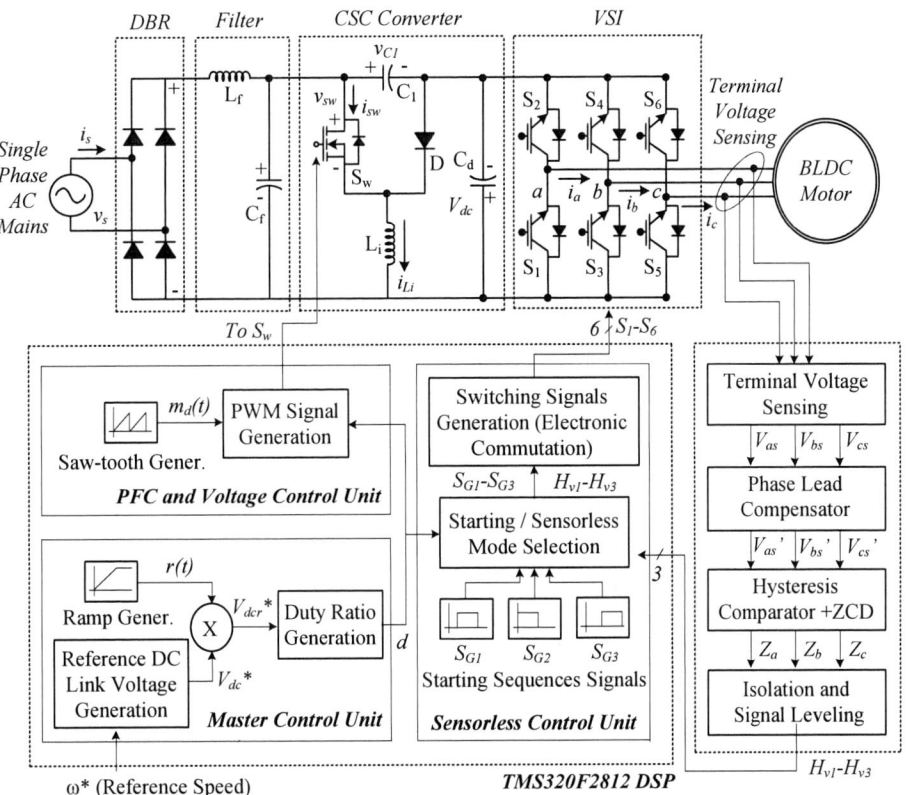

Fig. 1. System configuration of proposed BLDCM drive.

A sensorless based BLDC motor drive for automotive fuel pumps applications has been proposed in [18]. This approach utilizes a back-emf sensing without creating a virtual neutral, rather the negative DC link voltage of the VSI is used for creating the reference voltage. The use of PWM based control limits its use for this application. A line voltage difference method for estimation of phase back-emfs has been proposed in [19]. This method requires extra circuitry for the generation of line voltage difference signals from the line voltages. A third harmonic based accurate rotor position sensing has been presented in [20]. It requires the extraction of third harmonic from the phase voltages of the stator winding. An observer based sensorless control of single phase BLDC motor with a hybrid I-f starting has been presented in [21]. A phase locked loop observer has been proposed which increases the computational time of the algorithm and requires a high speed DSP (Digital Signal Processor) for achieving the desired performance. A method based on detection of freewheeling diode conduction period is presented in [22]. The speed control is achieved by using the chopper based control of VSI for adjusting the average voltage of the motor. Here the switching losses of VSI depend on the switching frequency which can be minimized by the use of electronic commutation of BLDC motor. A hysteresis comparator based technique for providing the dual operation of phase lead and to avoid the false detection of virtual Hall signals is presented in [23]. A separate ZCD (Zero Crossing Detection) unit is used which increases the component count. This paper presents the use of a hysteresis comparator for the dual operation of ZCD and avoiding any false position detection.

II. PROPOSED SENSORLESS BLDCM DRIVE

Fig. 1 shows the system configuration of proposed BLDCM drive. As shown in the figure, an uncontrolled diode bridge rectifier (DBR) followed by a PFC-CSC converter is used for feeding the BLDCM via a three phase VSI. A sensorless control of BLDCM is used for the elimination of rotor position sensors for electronic commutation. The proposed drive is designed for achieving the control of speed over a wide range by varying the voltage at DC bus of VSI. Hence, the BLDCM is operated with fundamental switching frequency of VSI to achieve the electronic commutation for minimal switching losses in it. The circuit configuration for generation of virtual Hall signals to estimate the rotor position is shown in Fig. 2.

The sensorless unit consists of voltage sensing circuitry, phase-lead compensator, hysteresis comparator, negative clipper and the isolation circuitry. The terminal voltage of BLDCM is sensed using a resistive potentiometer for stepping down the terminal voltage for its compatibility with the analog IC's. This output is given to a phase lead compensator, which is designed to compensate the phase delay in the virtual Hall signal estimation due to the sensing circuitry and the hysteresis comparator. This is a R-C network which value is selected on the basis of required phase shift at desired frequency. The associated waveforms showing the line back emf's and the corresponding actual Hall signals (H_a-H_c) and the required Hall signals after the compensation are shown in Fig. 3. The output of this phase lead compensator is given to a hysteresis comparator for obtaining the zero crossing of the line voltage obtained after phase-lead compensation. Other

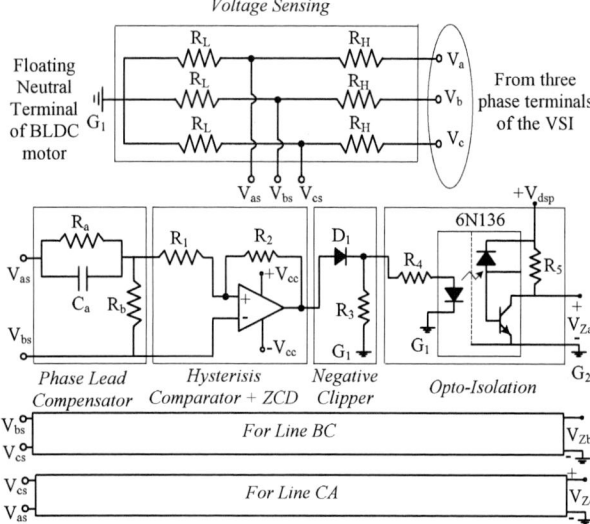

Fig. 2. Circuit configuration of virtual Hall signal generation.

than providing the zero crossing detection (ZCD), it also avoids any false detection of zero crossing due to commutation ripple. The band of hysteresis comparator is selected such that a minimum phase shift is obtained with permitted band to avoid any unwanted switching states. The output of this compensator is given to the negative clipper which clip the negative half cycle of the obtained ZCD signals to make the signal compatible to be given to the opto-isolator for providing an optical isolation between the power signal conditioning circuitry and the digital signal processor (DSP).

III. DESIGN OF PHASE LEAD COMPENSATOR

The phase lead compensator is designed to compensate a phase-delay (phase-lag) generated by sensing circuitry and a hysteresis comparator. The phase delay between the actual zero crossing of the line back-emf and the virtual Hall signals are measured. A phase-delay of $17\text{-}18^0$ is obtained at the rated speed. Based on this evaluation, a phase-lead compensator is designed which transfer function is given as [24],

$$G_c(s) = a\left(\frac{s + \omega_{c1}}{s + \omega_{c2}}\right) = a\left(\frac{s + 1/aT}{s + 1/T}\right) \quad (1)$$

where ω_{c1} and ω_{c2} are the two corner frequencies where the constants a and T (time constant) are given as [24],

$$a = \frac{1 + Sin(\varphi_m)}{1 - Sin(\varphi_m)}; T = \frac{1}{\omega_m\sqrt{a}} \quad (2)$$

where φ_m represents the required phase lead angle and ω_m represents the frequency corresponding to the maximum phase lead angle (φ_m).

Moreover, ω_m is also represented as [24],

$$log_{10}(\omega_m) = \frac{1}{2}\{log_{10}(\omega_{c1}) + log_{10}(\omega_{c2})\} \quad (3)$$

the value of constant 'a' is obtained from (2) as 1.8944 and the time constant, T is calculated using (3) as $1.44\text{x}10^{-3}$ rad/sec.

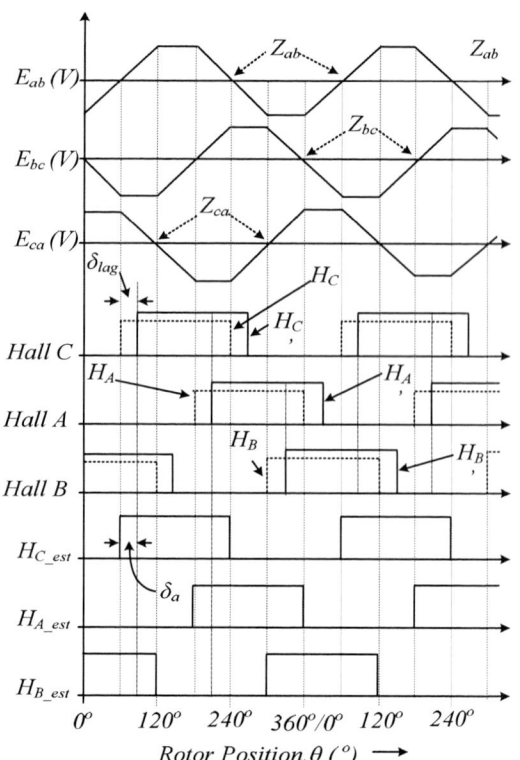

Fig. 3. Waveforms of line back emf's and the actual Hall signals (H_a-H_c) and the estimated hall signals after the compensation.

Now to obtain a maximum phase shift of 18^0 at rated speed, corresponding to frequency of 80 Hz (i.e. $\omega_m \approx 503$ rad/sec),

Hence two corner frequencies, ω_{c1} and ω_{c2} are obtained as as $\omega_{c1} = 1/T = 692$ rad/sec and $\omega_{c2} = 1/(aT) = 365$ rad/sec.

Now substituting these values in (1), one gets the transfer function of the required phase lead compensator as,

$$G_c(s) = 1.8944\left(\frac{s + 365}{s + 692}\right) \quad (4)$$

Now this compensator is realized with RC network as shown in Fig. 2, which transfer function is given as [24],

$$G_c(s) = \frac{s + (1/R_aC_a)}{s + (1/R_aC_a + 1/R_bC_a)} \quad (5)$$

Hence by comparing this equation with (4), the value of R_a and R_b are obtained as 27.4 kΩ and 30.6 kΩ ($C_a = 0.1\mu$F).

The Bode plot of the phase-lead compensator is shown in Fig. 4, which shows a phase delay of the order of $17\text{-}18^0$ at the desired frequency of 80 Hz (i.e. 503 rad/sec).

After the application of this phase lead compensator, a phase delay of the order of 1.44° between the terminal voltages with the actual and the virtual Hall signals at rated conditions is achieved which causes an acceptable error of 0.4 % in phase measurement. Fig. 5 shows the percentage error in terms of phase delay with speed of BLDC motor for both the cases of with and without phase-lead compensation. As shown in the figure, an error of less than 0.5 % is achieved in the designed compensator.

978-1-4799-6047-7/14 $31.00 © 2014 IEEE   518

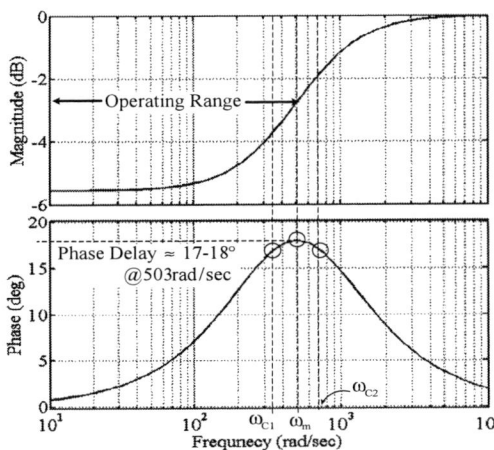

Fig. 4. Bode plot of the designed phase lead compensator.

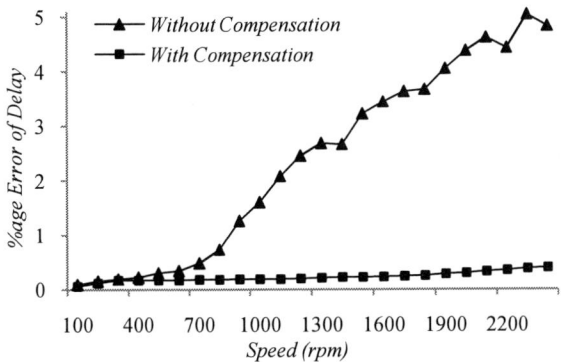

Fig. 5. Measured error in phase delay with and without compensation for different speeds.

IV. CONTROL OF PROPOSED BLDC MOTOR DRIVE

The control of PFC-CSC converter feeding a variable voltage controlled sensorless BLDC motor drive is shown in Fig. 1. The complete control is divided into three different sections as 'master control', 'PFC and voltage control' and 'sensorless control' as described below.

A. Master Control

The primary function of the master control of the BLDCM drive is to generate the reference duty ratio for the 'PFC and voltage control' and 'sensorless control'. It performs the following two functions.

1) Reference Voltage Generation: The primary function of this is to generate a reference DC link voltage ($V_{dc}*$) from the reference speed ($\omega*$). Since, the speed of the BLDC motor is directly proportional to the DC bus voltage applied to the VSI; hence a reference DC link voltage ($V_{dc}*$) is obtained as

$$V_{dc}* = k_v \omega* \qquad (6)$$

where k_v represents the BLDC motor's voltage constant.

2) Duty Ratio Generation: The reference voltage ($V_{dc}*$) is multiplied with the ramp signal to provide a steady increase reference DC bus voltage. This is to ensure a sufficient amount of time for the rotor position alignment and starting of BLDCM. This duty ratio is given to the 'PFC and voltage

control' and 'sensorless control' for the control of DC bus voltage and to achieve a sensorless operation respectively.

B. PFC and Voltage Control Unit

It generates PWM signal for the PFC converter switch for maintaining the desired DC bus voltage. The desired duty ratio of the PWM signal is obtained from the 'master control'. Since the PFC converter is operated in DICM; hence an inherent power factor correction is achieved at the AC mains without any voltage or current sensing requirement.

C. Sensorless Control of BLDC Motor

The primary function of it is to generate switching pulses for the VSI feeding the BLDC motor based on the virtual Hall signals as estimated by the proposed circuitry for sensorless control of BLDC motor. The mode selection switches the operation as 'rotor position alignment', 'sensorless starting' and 'sensorless mode' of BLDC motor. 'Synchronization' takes care of the proper switching of modes by holding the current states of switches whenever a 'mode-change' is detected and synchronizing it with the pulses obtained from the other block to avoid any commutation failure. The three different modes are described as follows.

1) Rotor Position Alignment: In this mode, the rotor position is initially assigned to a known position by giving pulse to two switches of the VSI. This is achieved to ensure a known rotor position, for accurate sensorless starting [3].

2) Sensorless Starting Mode: A sequential pulses of low frequency of the order of 10-15 Hz is given to the six switches of VSI. This is done to generate a considerable amount of back emf which can be easily detected by the sensorless circuitry to generate virtual Hall signals [3]. Since the speed of the BLDC motor is directly proportional to the DC bus voltage applied to the VSI, hence application of higher DC bus voltage at zero speed may result in higher stator current. To avoid this, the BLDC motor is initially started at lower speed and then shifted to a higher speed starting by changing the frequency of the open loop pulses from 10 Hz to 15 Hz based on the transition of applied DC link voltage.

3) Sensorless Mode: At this speed, a considerable amount of back-emf is generated for the estimation of virtual Hall signals. Hence, in this mode, the virtual Hall signals obtained from the back-emf detection circuitry are used for the electronic commutation of VSI and thus entering the 'Sensorless Mode' of operation.

V. RESULTS AND DISCUSSION

The performance of the proposed drive is experimentally validated on a developed prototype. A Texas instruments (TI) based DSP (TI-TMS320F2812) is used for the development of proposed drive. The performance of the proposed drive during steady state condition, the performance of a PFC-CSC converter and dynamic performance during starting and speed control are presented. Improved PQ indices under acceptable IEC 61000-3-2 limits are achieved for control of speed of BLDCM over a wide range.

Figs. 6 (a) show the steady state operation of proposed BLDCM drive at rated load on BLDCM with DC bus voltages as 200 V. A unity power factor is obtained at the AC mains as

978-1-4799-6047-7/14 $31.00 © 2014 IEEE   519

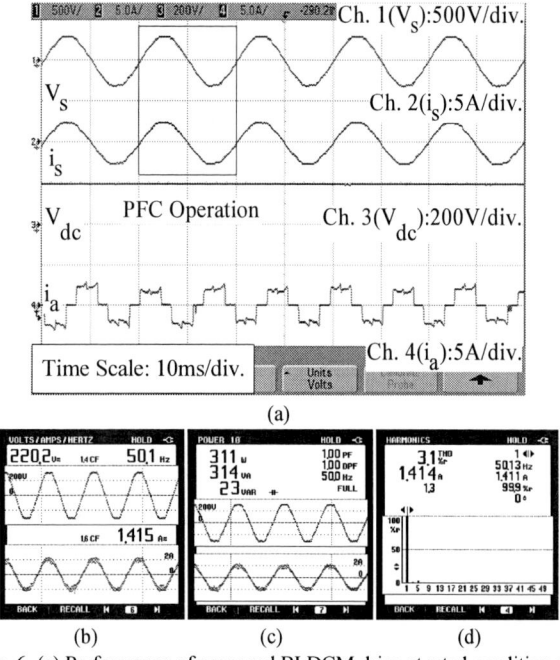

(a)

(b)   (c)   (d)

Fig. 6. (a) Performance of proposed BLDCM drive at rated condition on BLDCM with V_{dc} as 200V and (b-d) obtained power quality indices.

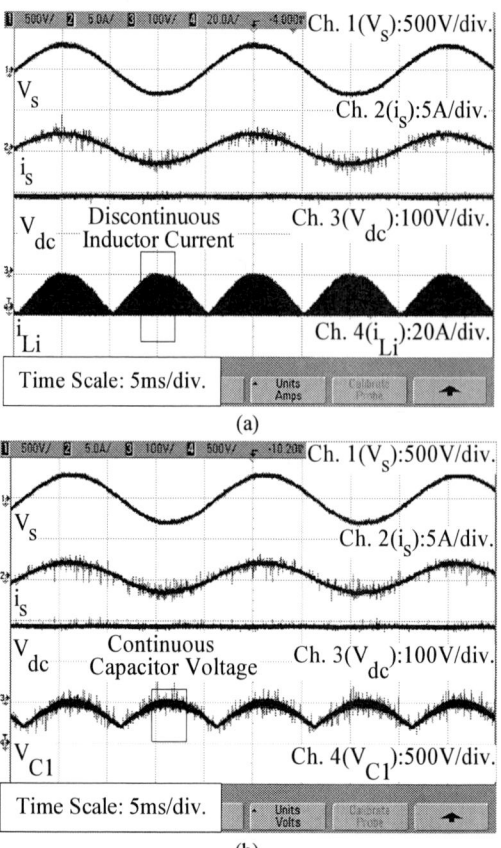

(a)

(b)

Fig. 7. Waveforms of (a) inductor current (i_{Li}), (b) intermediate capacitor voltage (V_{C1}), for the operation of proposed drive at rated condition.

a sinusoidal supply current (i.e. a low supply current THD) is obtained in phase with the supply voltage (i.e. unity displacement factor). Figs. 6 (b-d) show the power quality indices at AC mains for the operation of BLDC motor at DC link voltage of 200 V. A unity power factor and improved power quality indices are obtained under the acceptable IEC 61000-3-2 limits [5].

The PFC-CSC converter is designed to operate in DICM; hence the current in inductor (i_{Li}) becomes discontinuous; whereas the voltage across the intermediate capacitor (V_{C1}) remains continuous in a switching period. Figs. 7 (a) and 7(b) show the discontinuous inductor current (i_{Li}) and continuous intermediate capacitor voltage (V_{C1}) with supply voltage, supply current and DC bus voltage, respectively. Moreover, the stress on the PFC converter switch (i.e. switch voltage and switch current) is shown in Fig. 8. A voltage and current stresses of 520 V / 23 A are obtained which are acceptable for a PFC converter operating in DICM.

The operation of the proposed BLDCM drive during starting at DC bus voltage of 70 V is shown in Fig. 9 (a). A limited overshoot in stator current and supply current is obtained within the maximum current rating of the BLDCM stator windings. Fig. 9 (b) shows the dynamic behavior of the proposed BLDCM drive during step change in speed corresponding to the step change in DC bus voltage of the VSI. The proposed BLDCM has shown a satisfactory performance for all the mentioned cases.

CONCLUSIONS

A sensorless BLDCM drive has been proposed as a cost effective solution for low power applications. The CSC converter has been designed for its operation in DICM for inherent power factor correction without any sensing requirements. The rotor position sensors have been eliminated

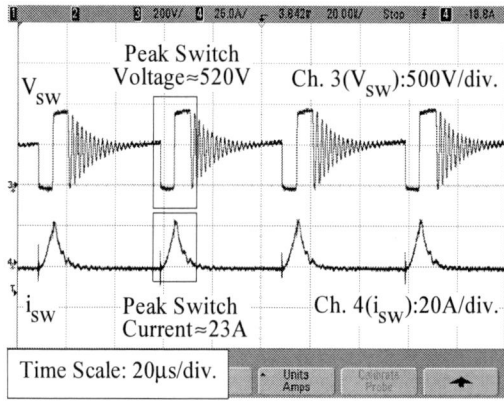

Fig. 8. Waveforms of enlarged waveforms for operation of proposed drive at rated condition.

by using the sensorless control of BLDCM to realize sensorless drive with power factor correction. The speed of the BLDCM has been controlled by varying the DC bus voltage of the VSI via a PFC-CSC converter. The BLDCM has been commutated electronically for the operation of VSI in low frequency switching for minimal switching losses in it. The proposed drive has shown a unity power factor operation at AC mains with limited amount of AC mains current harmonics.

APPENDIX

BLDC Motor Rating: 4 pole, P_{rated} (Rated Power) = 251.32 W, V_{rated} (Rated dc link Voltage) = 200 V, T_{rated} (Rated Torque) =

978-1-4799-6047-7/14 $31.00 © 2014 IEEE

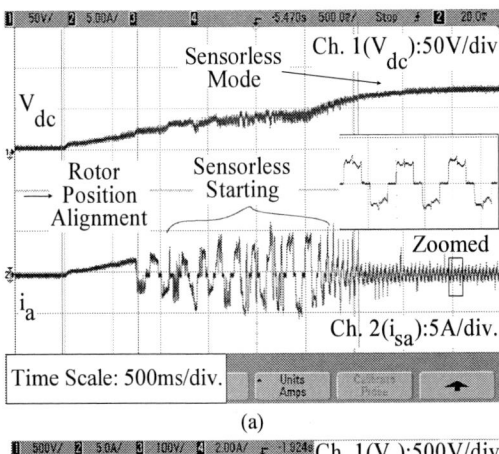

(a)

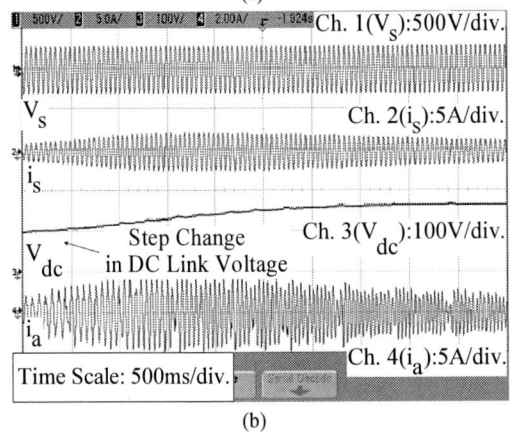

(b)

Fig. 9. Performance of the proposed drive during (c) starting at V_{dc}=70V (showing DC link voltage and stator current) and (d) sudden change in speed corresponding to change in DC bus voltage from 100V to 150V.

1.2 Nm, ω_{rated} (Rated Speed) = 2000 rpm, K_b (Back EMF Constant) = 78 V/krpm, K_t (Torque Constant) = 0.74 Nm/A, R_{ph} (Phase Resistance) = 14.56 Ω, L_{ph} (Phase Inductance) = 25.71 mH, J (Moment of Inertia) = 1.3×10^{-4} Nm/s^2.

REFERENCES

[1] C. L. Xia, *Permanent Magnet Brushless DC Motor Drives and Controls*, Wiley Press, Beijing, 2012.

[2] H. A. Toliyat and S. Campbell, *DSP-based Electromechanical Motion Control*, CRC Press, New York, 2004.

[3] P. P. Acarnley and J. F. Watson, "Review of position-sensorless operation of brushless permanent-magnet machines", *IEEE Trans. Ind. Electron.*, vol.53, no.2, pp.352-362, April 2006.

[4] N. Mohan, T. M. Undeland and W. P. Robbins, *Power Electronics: Converters, Applications and Design*, John Wiley and Sons Inc, USA, 2003.

[5] *Limits for Harmonic Current Emissions (Equipment input current ≤16 A per phase)*, International Standard IEC 61000-3-2, 2000.

[6] B. Singh, S. Singh, A. Chandra and K. Al-Haddad, "Comprehensive Study of Single-Phase AC-DC Power Factor Corrected Converters With High-Frequency Isolation," *IEEE Trans. Ind. Informatics*, vol.7, no.4, pp.540-556, Nov. 2011.

[7] L. Cheng, "DSP-based variable speed motor drive with power factor correction and current harmonics compensation," *Proc. of 35th Intersociety Energy Conversion Engg. Conf. and Exhibit, (IECEC)*, pp.1394-1399 vol.2, 2000.

[8] T. Gopalarathnam and H. A. Toliyat, "A new topology for unipolar brushless DC motor drive with high power factor," *IEEE Trans. Power Elect.*, vol.18, no.6, pp. 1397-1404, Nov. 2003.

[9] B. Singh and V. Bist, "Power Quality Improvements in PFC Luo Converter Fed BLDC Motor Drive", *Int. Trans. on Electrical Energy Systems (ETEP)*, vol. 24, no. 5, pp.1-22, Feb. 2014

[10] V. Bist and B. Singh, "An Adjustable Speed PFC Bridgeless Buck-Boost Converter Fed BLDC Motor Drive", *IEEE Trans. Ind. Electron.*, vol.61, no.6, pp.2665-2677, June 2014.

[11] B. Singh and V. Bist, "An Improved Power Quality Bridgeless Cuk Converter Fed BLDC Motor Drive for Air Conditioning System", *IET Power Electron.*, vol.6, no.5, pp. 902–913, May 2013.

[12] V. Bist and B. Singh, "A Reduced Sensor PFC BL-Zeta Converter Based VSI Fed BLDC Motor Drive", *Jr. Electric Power System Research*, vol.98, pp. 11–18, May 2013.

[13] O. Sago, K. Matsui, H. Mori, I. Yamamoto, M. Matsuo, I. Fujimatsu, Y. Watanabe and K. Ando, "An optimum single phase PFC circuit using CSC converter," *Proc. of 30th IEEE-IECON*, vol.3, pp. 2684- 2689, 2-6 Nov. 2004.

[14] V. Bist and B. Singh, "A PFC-Based BLDC Motor Drive Using a Canonical Switching Cell Converter", *IEEE Trans. Ind. Inform.*, vol.10, no.2, pp.1207-1215, May 2014.

[15] G. J. Su and J. W. McKeever, "Low-cost sensorless control of brushless DC motors with improved speed range," *IEEE Trans. Power Electron.*, vol.19, no.2, pp.296-302, March 2004.

[16] Y. S. Lai and Y. K. Lin, "Novel Back-EMF Detection Technique of Brushless DC Motor Drives for Wide Range Control Without Using Current and Position Sensors,", *IEEE Trans. Power Electron.*, vol.23, no.2, pp.934-940, March 2008.

[17] T. Y. Kim and J. Lyou, "Commutation instant detector for sensorless drive of BLDC motor," *IEEE Electronics Letters*, vol.47, no.23, pp.1269-1270, Nov.-2011.

[18] J. Shao, D. Nolan, M. Teissier and D. Swanson, "A novel microcontroller-based sensorless brushless DC (BLDC) motor drive for automotive fuel pumps," *IEEE Trans. Ind. Appl.*, , vol.39, no.6, pp.1734,1740, Nov.-Dec. 2003.

[19] P. Damodharan and Krishna Vasudevan, "Sensorless Brushless DC Motor Drive Based on the Zero-Crossing Detection of Back Electromotive Force (EMF) From the Line Voltage Difference", *IEEE Trans. Energy Convers.*, vol. 25, pp. 661-668, no. 3, Sept. 2010.

[20] J. C. Moreira, "Indirect sensing for rotor flux position of permanent magnet AC motors operating over a wide speed range," *IEEE Trans. Ind. Appl.*,, vol.32, no.6, pp.1394,1401, Nov/Dec 1996.

[21] L. I. Iepure, I. Boldea, I and F. Blaabjerg, "Hybrid I-f Starting and Observer-Based Sensorless Control of Single-Phase BLDC-PM Motor Drives," *IEEE Trans. Ind. Electron.*, vol.59, no.9, pp.3436-3444, Sept. 2012.

[22] S. Ogasawara and H. Akagi, "An approach to position sensorless drive for brushless DC motors," *IEEE Trans. Ind. Appl.*, vol.27, no.5, pp.928,933, Sep.-Oct. 1991.

[23] T. W. Chun, Q. V. Tran, H. H. Lee and H. G. Kim, "Sensorless Control of BLDC Motor Drive for an Automotive Fuel Pump Using a Hysteresis Comparator", *IEEE Trans. Power Electron.*, vol.29, no.3, pp.1382,1391, March 2014.

[24] B. C. Kuo, *Automatic Control Systems*, PHI Learning, New Delhi, 2010.

Analysis of a Robotic System with Two DOF using Haar Wavelet

Atul Kumar Pandey, Monika Mittal
Department of Electrical Engineering
National Institute of Technology, Kurukshetra
Haryana, India
atuliec@gmail.com, monika_mittalkkr@rediffmail.com

Abstract— This paper deals with the stability analysis of a robotic system, popularly known as pan tilt platform (PTP), with two degrees of freedom (DOF) using Haar wavelet. It has two revolute joints combining tilt mechanism with pan mechanism. Dynamic model of this robotic system is obtained using Newton-Euler equation. The stability analysis of its dynamic model is carried out using Haar wavelet for the first time. The objective of this paper is to achieve computational savings in the process. The obtained computational savings are tabulated and bar graphed to show comparison with the analytical technique.

Keywords— *Two DOF robotic system (Pan and tilt mechanism), linear time invariant system, stability analysis, operational matrices, Haar wavelet*

I. INTRODUCTION

Various embodiments have a camera which may be controlled by one or more motors in a base of the camera. Cables and other components may be used to manipulate the camera lens through the side arms of the camera [1]. This arrangement forms a two DOF robotic manipulator called Pan and Tilt mechanism. The tilt mechanism can be rotated about a tilt axis supported on the pan mechanism having one DOF as shown in Fig.1. The tilt mechanism transferred torque to the camera by belt pulley arrangement[2,3].

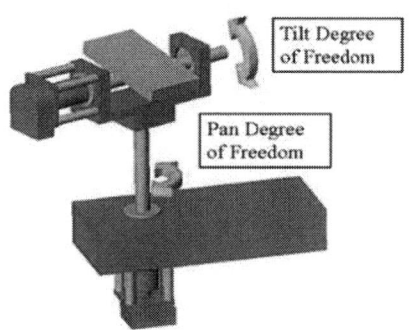

† Courtesy: http://cats-fs.rpi.edu/~wenj/ECSE446S06/project_overview.html

Fig. 1. Three dimentional view of PTP

Pan arrangement can be also rotated about a pan axis which has one revolute joint (R) as shown in Fig.1. Pan mechanism is used to support the tilt mechanism for achieving desired position and orientation. Pan arrangement has also one DOF with one revolute joint(R). Pan mechanism rotates about a pan axis actuated by a pan motor.

Pan-tilt cameras are often used as components of wide-area surveillance system [2]. A two DOF robotic zoom camera used for surveillance is shown in Fig.2. It is necessary to calibrate these cameras in relation to one another, for obtaining a consistent representation of the entire space. The tilt and pan mechanism is modelled using Newton-Euler equation[3]. The tilt and pan mechanism has been analysed and simulated using lead and PD compensators [2,3].

† Courtesy: www.directindustry.com/ **Pan-tilt**-zoom **camera** / PTZ

Fig. 2. Two DOF robotic zoom camera

Orthogonal functions reduces the computation time for complex as well as simple dynamic system analysis. Some of orthogonal functions such as Block pulse functions [4], Walsh functions [5], Fourier functions [6], Haar functions [9] have been extensively applied for piecewise constant solution to different problems in terms of differential equation, response analysis, optimization and identification of linear and nonlinear systems. The specifications are settling time (Ts) less than 0.5 seconds, steady state error within ±2% and percentage overshoot (%OS) below 22% [3].

The paper is organized as follows: In section II dynamic and state space model of pan and tilt mechanism is discussed. Section III introduces Haar wavelet. Section IV describes analysis of linear time invariant system (LTI) using Haar Wavelet. Section V demonstrates the comparison of solutions using Haar wavelet and analytical methods. Section VI deals with future scope and other applications of this method.

978-1-4799-6047-7/14 $31.00 © 2014 IEEE

II. BACKGROUND AND MODELING OF PTP

A. Dynamic Model For Tilt Mechanism

The dynamic model of tilt mechanism is obtained by Newton-Euler equation under certain assumptions [2,3].

$$\tau = J_{eff}\ddot{\theta} + f_v\dot{\theta} + f_c\,\text{sgn}(\dot{\theta}) + mgl.\sin(\theta) \qquad (1)$$

Where τ=Torque, J_{eff} =Effective inertial load, f_v = Viscous friction, f_c= Coulomb friction, θ = Angle between the force (mg) and arm length l . Now stability analysis of the dynamic equation is done after neglecting sinusoidal and signum function nonlinearities. Transfer function has been obtained using Jacobean matrix in [2] as :

$$G(s) = \frac{31.24954}{s^2 + 0.1578s} \qquad (2)$$

The tilt mechanism is marginally stable as one of the system poles lies at the origin. Also, its response does not meet the required performance criteria which is reported to be compensated by a lead compensator as shown in Fig.3. The objective of the lead compensator is to keep settling time less than 0.5 second for the unity feedback system .

Transfer function with compensator is obtained as [2]:

$$T(s) = \frac{829.9s + 8.299}{s^3 + 39.99s^2 + 836.1s + 8.299} \qquad (3)$$

The equivalent state space model is obtained for State variables $x_1 = \theta, x_2 = \dot{\theta}, x_3 = \ddot{\theta}$ and $\dot{x}_1 = \dot{\theta}, \dot{x}_2 = \ddot{\theta}, \dot{x}_3 = \dddot{\theta}$.

$$\dot{x} = \begin{bmatrix} 0 & 1 & 0 \\ 0 & 0 & 1 \\ -8.299 & -836.1 & -39.99 \end{bmatrix} x + \begin{bmatrix} 0 \\ 0 \\ 1 \end{bmatrix} u \qquad (4)$$

$$y = \begin{bmatrix} 8.2990 & 829.9 & 0 \end{bmatrix} x$$

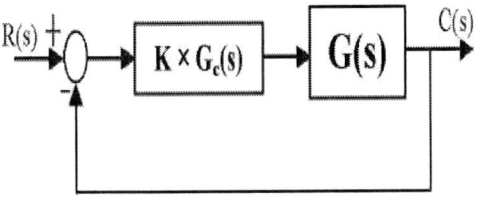

Fig. 3. Closed loop system with compensator

B. Dyanamic Model And Compensator For Pan Mechanism

Similar process can be applied for obtaining transfer function of the pan mechanism in [2,3] as :

$$G(s) = \frac{2.3964}{s^2 + 0.01677s}, \quad T(s) = \frac{848.9s + 8.489}{s^3 + 40.01s^2 + 849.6s + 8.489} \qquad (5)$$

State space model for the pan mechanism is also given below using same state variables and procedure .

$$\dot{x} = \begin{bmatrix} 0 & 1 & 0 \\ 0 & 0 & 1 \\ -8.489 & -849.60 & -40.01 \end{bmatrix} x + \begin{bmatrix} 0 \\ 0 \\ 1 \end{bmatrix} u \qquad (6)$$

$$y = \begin{bmatrix} 8.489 & 848.9 & 0 \end{bmatrix} x$$

The response analysis for closed loop transfer functions is obtained using Haar wavelet which is described in next section.

III. BRIEF REVIEW OF HAAR WAVELET

The orthogonal set of Haar functions is defined as a group of square waves with magnitude ± 1 in some intervals and zeros elsewhere [7]. These zeros make the Haar transform faster than other square functions such as Walsh's [4].

$$h_n(t) = h_1(2^j(t - k/2^j)), \qquad (7)$$
$$n = 2^j + k, \, j \geq 0, 0 \leq k \leq 2^j, \, n, j, k \in Z$$

First curve $h_0(t) = 1$ during the interval $0 \leq t \leq 1$ and is known as scaling function. The second term $h_1(t)$ is the mother wavelet and other curves are generated from it via translations and dilations i.e. $h_1(t)$ is compressed to the half interval $(0, 1/2)$ to generate $h_2(t)$. $h_3(t)$ is the same as $h_2(t)$ but delayed from it by $1/2$. Similarly, $h_2(t)$ is compressed to quarter interval to generate $h_4(t)$.

Any function $y(t)$ which is square integrable in the interval $[0,1)$ can be expanded in a Haar series [8].

$$y(t) = c_0 h_0(t) + c_1 h_1(t) + \ldots\ldots\ldots\ldots = c_m^T H_m(t) \qquad (8)$$

$$c_i = 2^j \int_0^1 y(t) h_i(t) dt \qquad (9)$$

For m=4, $H_m(t)$ is calculated by taking the samples of first four Haar functions as follows[12].

$$H_4(t) \overset{\Delta}{=} \begin{bmatrix} h_0(t) & h_1(t) & h_2(t) & h_3(t) \end{bmatrix}^T \qquad (10)$$

$$H_4(t) \overset{\Delta}{=} \begin{bmatrix} h_0(t) \\ h_1(t) \\ h_2(t) \\ h_3(t) \end{bmatrix} = \begin{bmatrix} 1 & 1 & 1 & 1 \\ 1 & 1 & -1 & -1 \\ 1 & -1 & 0 & 0 \\ 0 & 0 & 1 & -1 \end{bmatrix} \qquad (11)$$

The Haar coefficient c_i can be obtained by [3] , however it is more convenient to evaluate it by matrix inversion as:

978-1-4799-6047-7/14 $31.00 © 2014 IEEE

$$c_m^T = y(t).H_m^{-1} \qquad (12)$$

A. Operation Matrix Of Integration

The integration of Haar wavelets can be expanded into Haar series with Haar operation matrix of integration P_m defined in [9] as:

$$\int_0^t H_m(t)dt = P_m . H_m(t) \qquad t \in [0,1) \qquad (13)$$

where the square matrix P_m is defined as:

$$P_m = \left[\int_0^t H_m(\tau)d\tau \right] H_m^{-1} \qquad (14)$$

Matrix P_m can be also calculated by innovative non-recursive formulation in [9] as:

$$P_m = H_m . Q_{Bm} . H_m^{-1} \qquad (15)$$

where Q_{Bm} is the operation matrix of integration [4] for block pulse function at any resolution m.

IV. ANALYSIS OF LINEAR TIME INVARIANT SYSTEM VIA HAAR WAVELET [7]

A linear system is described by the state equation:

$$\dot{x}(t) = Ax(t) + Bu(t) , x(0) = x_0 \qquad (16)$$

$$y(t) = Cx(t) + Du(t) \qquad (17)$$

Assume that u(t) is a square integral in the interval $0 \le t \le 1$. Haar series expansion of u(t) and $\dot{x}(t)$ can be expressed as :

$$u(t) = G.H(t), \quad \dot{x}(t) = F.H(t) \qquad (18)$$

Integration of $\dot{x}(t)$ provides x(t).

$$x(t) = \int_0^t \dot{x}(t) + x_0 = F \int_0^t H(t)dt + x_0 = F.P.H(t) + x_0 \qquad (19)$$

Substituting (18) and (19) into (16), we obtain:

$$F.H(t) = A.F.P.H(t) + A.x_0 + B.G.H(t) \qquad (20)$$

$$F - A.P.F = [Ax_0, 0, 0....., 0] + B.G = G_1 \qquad (21)$$

$$\left[f_1, f_2 .... f_{m-1} \right]^T = \left[I - A \otimes P^T \right]^{-1} . [g_1, g_2, ... g_{m-1}] \qquad (22)$$

where $\otimes$ is Kronecker product defined in [10]. Results obtained in (22) can be used to calculate y(t) from (17) with few matrix operations.

The analysis of the two DOF robotic system using Haar wavelet and analytical methods [11] is discussed in the next section. It is demonstrated that Haar wavelet analysis is computationally more efficient as compared to analytical analysis for a step input.

V. RESULTS

In order to achieve the allowable specifications for the two DOF robotic system, the compensated closed loop transfer function and its state space model for pan and tilt mechanism is described in section II. The model of pan and tilt mechanism gives desired position and orientation of the camera.

The responses of both pan and tilt mechanism are obtained by analytical method using MATLAB for time interval [0, 1]. These are shown in Fig. 4 and Fig. 5, respectively. Same responses are obtained using Haar wavelet algorithm described in Section IV.

Haar wavelet response of tilt mechanism Y_{tilt} is obtained using (17) for m=32 as :

$$Y_{tilt} = [0.1108 \quad 0.4288 \quad 0.7814 ....]_{1 \times 32}$$

Similarly, Haar wavelet response of pan mechanism Y_{pan} is obtained using (17) for m=32 as :

$$Y_{pan} = [0.1131 \quad 0.4372 \quad 0.7947 ....]_{1 \times 32}$$

For comparison, obtained Haar responses are superimposed on analytical responses for the respective mechanism in Fig. 4 and Fig. 5.

Design values of steady state error and % OS are tested by using encoder and Optoschmitt sensors in the real time. It is observed from Fig. 4 and Fig. 5 that settling time and steady state error meet the desired performance values. The given system model with controller, gives the desired orientation of camera with a slight difference in the settling time (0.01) and steady state error (0.001). This is due to neglecting nonlinearity terms.

It is also clear from Fig. 4 and Fig. 5 that similar results are obtained using Haar wavelet method which is expected to give more accurate results at higher resolutions.

Computational time for obtaining responses at different resolutions using analytical and Haar wavelet methods are shown in Table I and Table II, respectively for tilt and pan mechanisms. It is observed from Table I and Table II that for all resolutions, computational time in Haar wavelet method is less as compared to those in analytical method for both mechanisms.

Comparison of the computational times in both analytical and Haar wavelet methods are also shown by bar graphs in Fig. 6 and Fig. 7, respectively for tilt and pan mechanisms.

978-1-4799-6047-7/14 $31.00 © 2014 IEEE

It is evident from Fig. 6 and Fig. 7 that computational time is about 40% less in Haar wavelet method as compared to analytical method for all resolutions. This trend is consistent even at higher resolutions.

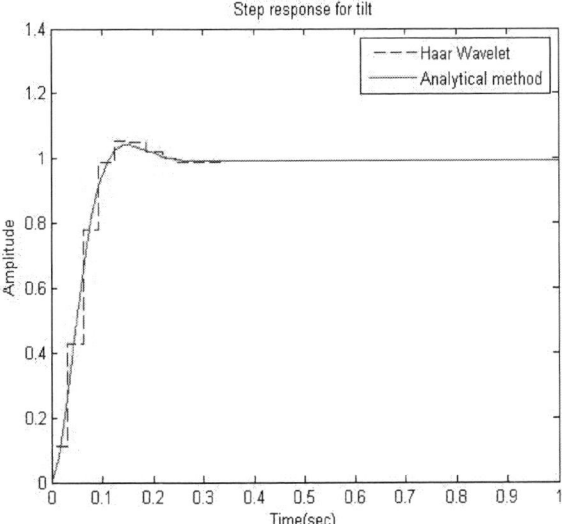

Fig. 4. Step response for tilt mechanism

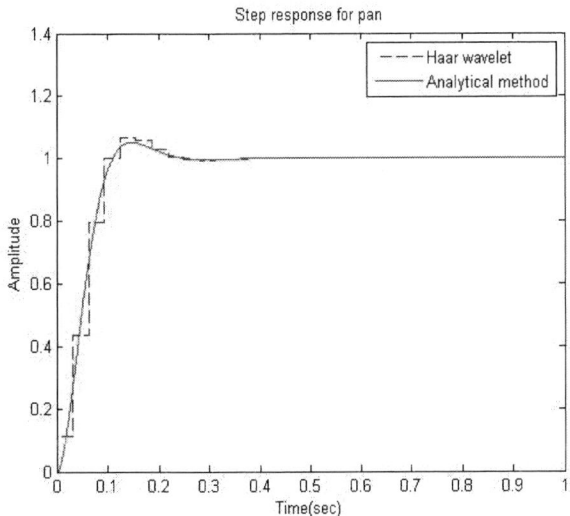

Fig. 5. Step response for pan mechanism

TABLE I. COMPUTATIONAL TIME FOR TILT MECHANISM ANALYSIS

Resolution	m=8	m=16	m=32	m=64	m=128
Computational Time via Haar Wavelet (sec)	0.44045	0.45474	0.46165	0.487760	0.50469
Computational Time via Analytical method(sec)	0.76760	0.78012	0.78673	0.79395	0.80893

TABLE II. COMPUTATIONAL TIME FOR PAN MECHANISM ANALYSIS

Resolution	m=8	m=16	m=32	m=64	m=128
Computational Time via Haar Wavelet(sec)	0.44491	0.45249	0.46024	0.48217	0.50118
Computational Time via Analytical method(sec)	0.75670	0.76793	0.77724	0.78926	0.81258

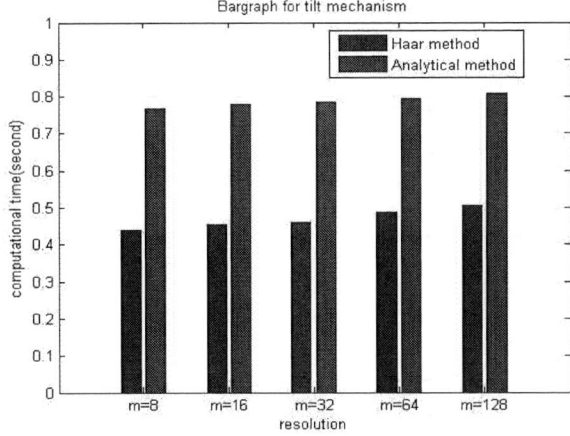

Fig. 6. Bar graph showing computational saving for tilt

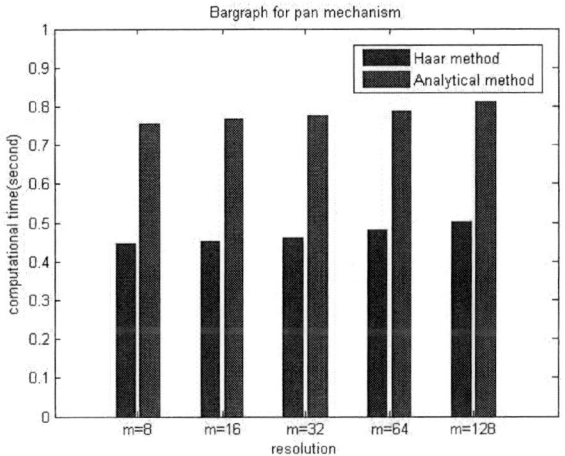

Fig. 7. Bar graph showing computational saving for pan

VI. CONCLUSION

The Haar wavelet method is successfully applied for the analysis of pan and tilt mechanism. This method has been applied to analyze a robotic system with two DOF for the first time as seen from the literature survey. A comparison of computation time in analytical and Haar wavelet is also shown. Lesser computation time is obtained using Haar method as compared to analytical method for both tilt and pan mechanisms at different resolutions. Computational efficiency has been demonstrated using MATLAB R2013a .

In many practical applications, computational savings and control with higher precision, achieved using Haar wavelet method, without the need of excessive computer memory become very important.

REFERENCES

[1] James Davis , Xing Chen ,"Calibrating pan-tilt cameras in wide area surveillance networks" IEEE International Conference on Computer Vision(ICCV2003).

[2] Imran S. Sarwar, Afzaal M. Malik, "Stability analysis and simulation of a two DOF robotic system based on linear control system" 15th International conference on mechatronics and machinevision(2008).

[3] Imran S. Sarwar, Afzaal M. Malik, "Modelling analysis and simulationof a Pan Tilt Platform based on linear and nonlinearsystems",IEEE China , Accepted ,Oct 2008.

[4] K. Maleknejad , M. Shahrezaee , H. Khatami, "Numerical solution of integral equations system of the second kind byBlock–Pulse function" Elsevier Applied Mathematics and Computation 166 (2005) 15–24.

[5] Chen C. F. ,Hsiao C. H., "A state space approach to Walsh series solution of linear system", Int. Znt.JControl,34(1981) 557-584.

[6] Sifuzzaman M., Islam M. R., Ali M. Z. , "Application of Wavelet Transform and its Advantages Compared to Fourier Transform", Journal of Physical Sciences, Vol. 13, (2009) 121-134 .

[7] Chen, C.F., Hsiao, C.H.: Haar wavelet method for solving lumped and distributed-parameter systems.IEEE Proc., Control TheoryAppl.144(1) (1997).

[8] Prof . Akanksha Gupta Prof. Monika Mittal , Dr. Lillie Dewan "Haar Wavelet Based Approach for State Analysis of BiomedicalSystems" IEEE Transaction.

[9] Monika Garg · Lillie Dewan, "Non-recursive Haar Connection Coefficients BasedApproach for Linear Optimal Control"Received: 27 September 2010 / Accepted: 1 December 2011SpringerScience John W. IEEE transaction on circit and system.

[10] John W. Brewer,"Kronecker and matrix calculation in system Theory",IEEE transaction on circit and system .

[11] Norman S. Nise, "Control systems engineering", Wiley student edition, Fourth edition, pp. 515-524, 2004.

[12] Monika garg, Lillie Dewan, "A Generalized Approach for StateAnalysis and Parameter Estimation of Bilinear Systems using Haar Connection Coefficients", World Academy of Science, Engineering andTechnology Vol:5 2011-03-27 .

Single Stage Single Phase Solar Inverter with Improved Fault Ride Through Capability

Jalaj Arya, Lalit Mohan Saini

Electrical Engineering Department
National Institute of Technology
Kurukshetra, Haryana-136119, INDIA
jalaj1arya@gmail.com, lmsaini@gmail.com

Abstract—Now-a-days effective solar power generation is necessary for smart grid implementation. Grid fed single-stage single-phase solar inverter with incremental conductance MPPT (INC+ regulator), closed loop current and voltage controller is implemented. In normal condition inverter current T.H.D is under the limit as per IEEE-519. At the time of fault, inverter having voltage imbalance, which may lead to damage. System fault ride through capability is improved by using Series Dynamic Braking Resistor (SDBR) with high current error as control parameter. Protection scheme maintains the grid code and hence the solar inverter needs not to be disconnected from the grid during the fault. Comparative results are obtained by simulation in MATLAB R2013a software. The whole system can be used in local distribution system as an application in future effectively.

Keywords—Solar power generation; PV fault.

I. INTRODUCTION

In 2010 world electricity consumption was 20.45 trillion kWh in which share of solar inverter was around 20% [1]. Efficient solar inverter has its own importance in smart grid especially in domestic purpose. Photovoltaic is defendable, clean, conserving, ubiquitous and everlasting system. But still compactness, protection, reliability and monitoring functions are receiving more and more attention in this area. In this paper protection method under fault in single phase PV inverter is proposed.

Solar inverter is a power electronics device that converts the variable direct current output of a solar PV panel into an alternating current that can be supply into an electrical grid or to a off-grid local system. As contrary to traditional two-stage system, a single-stage solar energy conversion system is implemented, resulting in increased efficiency and reduction of weight and size. Now-a-days maximum power point tracking (MPPT), phase locked loop (PLL), closed loop current and voltage controller are combined in single stage as shown in Fig.1 [10]. Conventional inverters have double loop control which is a complex process. Fig. 2 shows basic single-stage photovoltaic energy conversion system diagram [2]. System needs single-stage controller for controlling action. Up to now several researchers proposed their views on this area. In this paper firstly all concerned existing topologies comparisons have been shown in table I. Among them cost effective single stage topology [6] having highest factor of comparison but consist of complex one cycle controller. To overcome this problem simple closed loop current and voltage

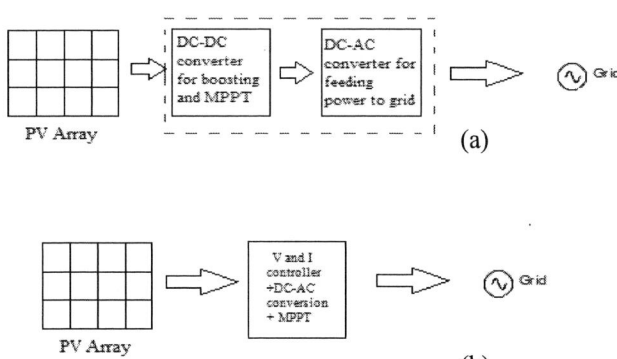

Fig.1. Grid connected solar inverter topologies:
(a) Traditional two-stage and (b) Alternative single-stage system

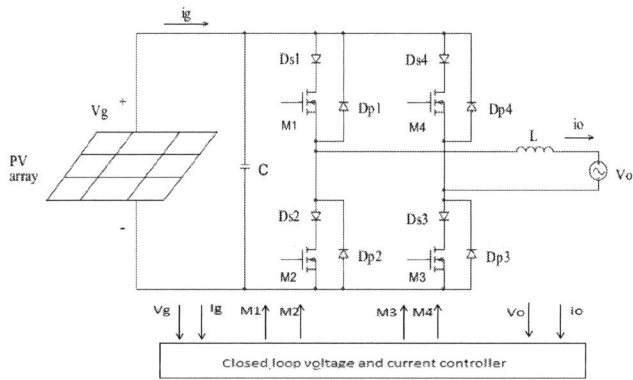

Fig. 2. Single stage solar energy conversion process

controller [5] with some modification has been implemented in this work. Here section II describes the study of whole solar inverter system in subsections of four parts including PV plant modeling, MPPT type and controller. Section III introduces fault ride through capability of PV system and section IV propose protection scheme by series dynamic braking resistor (SDBR) taking high current error as control parameter.

MATLAB simulation model of system with protection scheme and comparative results in different conditions have been shown in section V. Under normal condition inverter current total harmonic distortion (T.H.D) is under limit and conversion efficiency is greater than 94 percent. Solar inverter is able to fed grid and local load. Also under faulty condition there is no need to disconnect inverter from system. Proposed system can be used in local distribution system as application.

978-1-4799-6047-7/14 $31.00 © 2014 IEEE

TABLE I.
Comparative study of various
single-stage single-phase solar inverter topologies

S. No.	Name Of Topology	Year	Power Output From PV system in watt	Power Output From Inverter in watt	Efficiency in % (a)	No. of devices without filter (b)	Factor of Comparison (a) / (b)	Power factor	T.H.D [Io]	Filter type	Inverter Type	MPPT Type
1.	MPPT with capacitor identifier for PV power system [4]	2000	641.025	500	78	8	0.0975	Around unity	-	L	Buck Boost	P&O
2.	*Novel maximum power point tracking controller for photovoltaic energy conversion system [5]*	*2001*	*1050*	*-*	*>90*	*6*	*0.15*	*Around unity*	*3.09*	*L*	*H bridge PWM*	*Proposed Similar to Increamental conductance*
3.	A cost effective single Stage inverter with mppt[6]	2004	600	573.6	95.6	5	0.1912	unity	5.769	L	H bridge	One cycle control
4.	Single Stage full bridge buck boost inverter[7]	2004	555.55	500	90	8	0.1125	-	5-8	LCL	Buck Boost	-
5.	Single Stage inverter for direct ac connection of a photovoltaic cell module[8]	2005	-	-	83.5	7	0.1192	-	-	C	fly back	Mppt algorithm
6.	Control of single stage single phase PV inverter [9]	2005	1500	-	-	5	-	-	5.8	LCL	H bridge PWM	Increamental conductance
7.	Single Stage grid connected inverter topology for solar PV system with mppt[10]	2007	-	-	83	9	0.0922	-	9.13	LC	Two Buck boost	Hill Climbing
8.	An integrated inverter for a single phase single stage grid connected PV system based on Z source[11]	2007	300	250	83.33	11	0.0757	Around unity	3.8	L	Z source	P&O
9.	.Mppt in a one cycle controlled single stage PV inverter[without p&o [12]	2008	-	-	89	6	0.1483	-	8.31	L	H bridge	One cycle control
10.	A novel single phase single stage inverter for solar application[13]	2013 (conf)	-	-	-	10	-	-	-	LC	Buck Boost	-

II. SOLAR INVERTER UNDER STUDY

A. PV Plant Modeling

Electricity can be produced directly by photovoltaic system. PV cell is the basic unit of a PV system. Several PV cells are assembled in series or parallel to make a PV module. They may be assembled to make panels. A PV module equivalent circuit is shown in Fig. 3, in which the simplest model can be indicated by a current source in anti-parallel with a diode and the non idealities are indicated by introducing series resistance Rs and parallel resistance R_P [16].

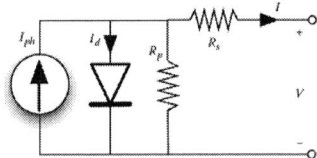

Fig.3. Equivalent circuit PV module

Large PV plants are composed of several PV panels or modules to produce desired level output. Equation 1 mathematically shows the I–V characteristic of the ideal solar cell [16].

$$I = I_{pv,cell} - \underbrace{I_{0,cell} \left[\exp\left(\frac{qV}{akT} \right) - 1 \right]}_{I_d} \quad (1)$$

Fig. 4 shows the implemented simulated results of BP MSX_60 PV module, which is used in series in this paper as PV source. Table I shows the parameters used in PV module modeling.

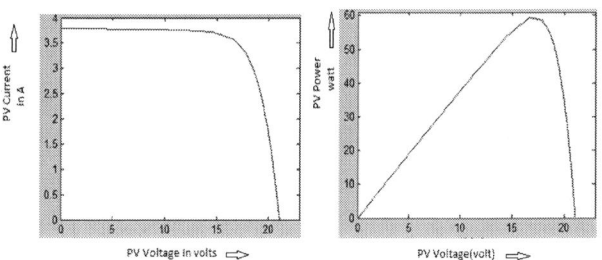

Fig.4. I-V and P-V characteristics of the PV module.

TABLE II.
Parameters of 60W PV Module

Peak Power (P_{max})	60W
Maximum power voltage Vmax	17.1V
Isc	3.8A
Voc	21.1V
Rs	0.18 ohm
Rp	360 ohm
N	1.36
Temp.	298 K

B. INC MPPT Controller with regulator

Fig. 5 proves basic idea of variation of PV module voltage and power with changes in irradiance. Also there is always need to track maximum power point in vigorous conditions of change in irradiances and temperature. Incremental conductance (INC) MPPT has been first time used in [3].

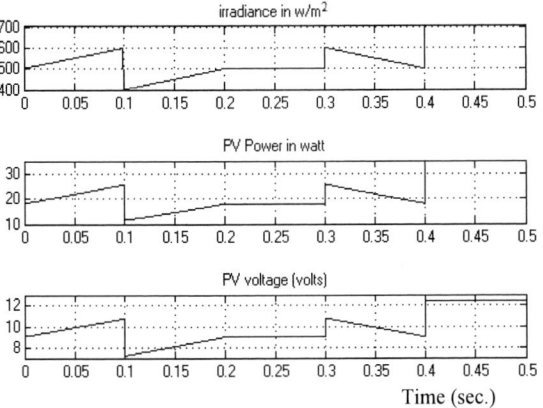

Fig.5. Responses of the 60 W peak PV module with variable irradiance.

In this paper INC MPPT is implemented with regulator. Regulator is used to minimize the error and also for providing new current and voltage values for MPPT. Fig. 6 shows the flowchart of INC which is used as simulated component in this paper.

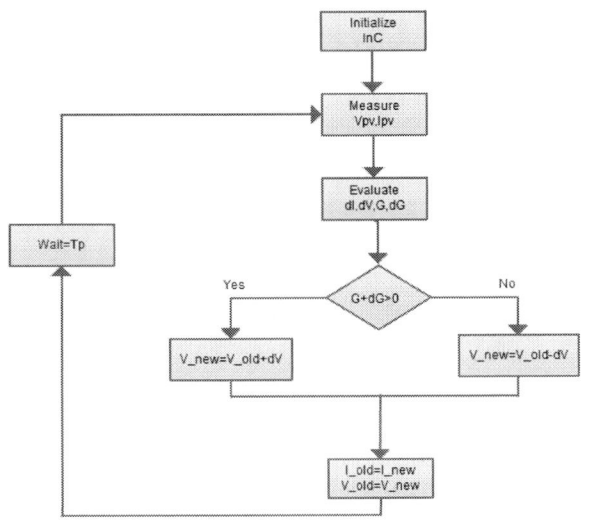

Fig.6. Basic flowchart of Incremental conductance MPPT

C. Closed loop current and voltage controller

Single-stage solar inverter controller is somewhat complex than double stage. Controller output should be the function of grid frequency. Phase Locked Loop (PLL) provide sinusoidal reference signal in the initial stage. In this paper enhanced version of novel MPPT controller [5] is implemented in MATLAB software.

2 Level H- bridge inverter with PWM switching is used here, in which modulating wave is coming from controller as shown in Fig. 7.

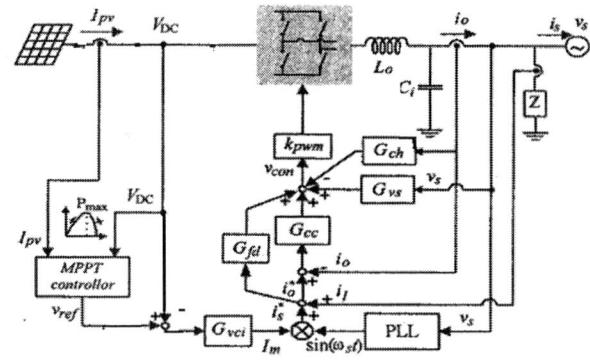

Fig.7. Basic circuit diagram of closed loop current and voltage controller

D. Proposed Grid Connected Solar System

Fig. 8 represents the simplified diagram of the grid connected 660 W (peak) PV plant having eleven series PV modules followed by single-stage conversion block employed to dispatch the AC power to the grid. The inverter is capable of maintaining constant 230V DC voltage at the DC link. Solar inverter feeds power to the grid and also satisfy local load. A L-G fault near the grid side is shown by F1. To protect the inverter from voltage imbalance series dynamic breaking resistance (SDBR) is used in the system.

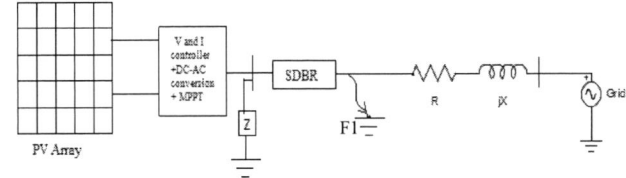

Fig.8. Schematic single line diagram of grid connected system

III. FAULT RIDE THROUGH CAPABILITY

At the time of connecting PV plants to grids, it is essential to provide them with dynamic support for the grid voltages. This dynamic support is referred to fault ride through (FRT) capability [14] in which the PV plant should stay connected in the case of grid faults depending on the fault time duration. They have also to provide support to the grid voltages by injecting reactive power. Considering the FRT capability, there are four major reasons for inverter disconnection during grid faults, which are the following ones: (i) over current at the ac side , (ii) excessive dc-link voltage, (iii) loss of grid voltage synchronization, (iv) the reactive current injection imbalance. If system remains connected during these abnormalities, it indicates improved FRT capability of system.

In this paper first two conditions are focused mainly and achieved better results in simulation. For achieving this SDBR concept is used here, which is described clearly in next section.

978-1-4799-6047-7/14 $31.00 © 2014 IEEE   529

IV. PROTECTION SCHEME BY SDBR CONTROL

Series dynamic braking resistor (SDBR) is employed in first time in [14], where voltage error was the reference of controller. In this paper, SDBR is used by taking current error as reference for controller. Fig 9 shows the basic concept which is used here such that PV inverter needs not disconnect immediately from the grid at the occurrence of the fault.

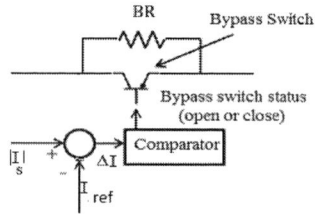

Fig.9. Control scheme with SDBR

Under normal conditions switch remains close, thus causes for bypassing the braking resistor. Grid current increment above a selected reference point send signal to the switch for opening the current path. The braking resistor would persist in the system as long as the grid current of the PV generator is above a set value. When the fault is cleared, the system becomes stable and the grid current becomes same as the reference value, the switch would close and the system would be rehabilitated to its normal condition.

V. SIMULATION MODEL AND RESULTS

In this work, the simulation has been performed through MATLAB/Simulink software 2013a. Simulation model of solar inverter with fault is shown in Fig. 12. Here PV panel consists of series connection of 11 modules which produce around 660 W power, which is fed to the inverter and controlled by closed loop voltage and current controller [5]. Here grid has 230 V and 60Hz frequency. Total number of components are seven including LC filter.

Fig. 10 shows various currents in normal condition from which it is clear that inverter is able to supply 550 W local load and also fed current to the grid.

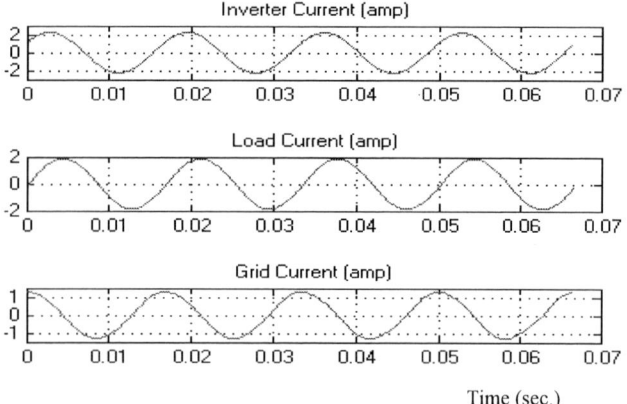

Fig. 10. Various currents in normal condition

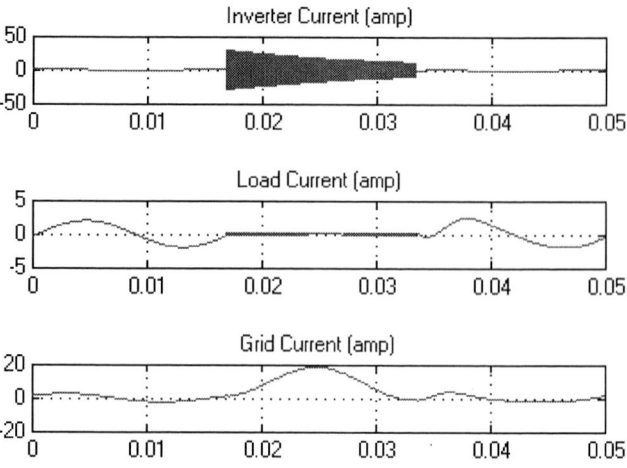

Fig. 11. Various currents at the time of fault occurs

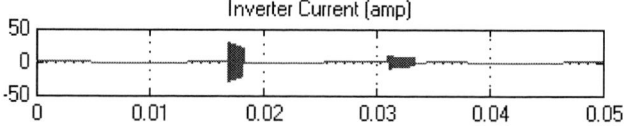

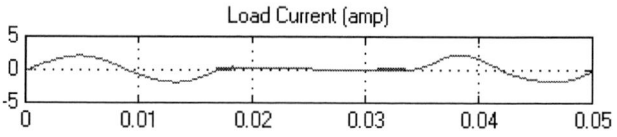

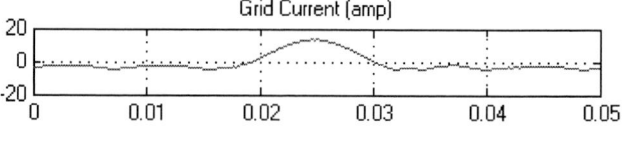

Fig.13. Various currents after SDBR control action

Inverter current T.H.D is 3.14% which is under limit as per IEEE-519 and having R.M.S value of 0.91. Efficiency of solar inverter is 94.09%.

Further Fig.11 shows various currents at the time of L-G fault occur near grid side, where high magnitude of grid current causes high switching of inverter current and voltage imbalance. This problem can be removed by using fast current error controlled SDBR switch as shown in Fig. 13. Also there is no need to disconnect the solar inverter during the fault which is proved in Fig. 13.

Fault time is for 0.0167 sec. in which switching spikes remain only for 0.0009 sec. thus fault ride through capability of solar inverter has been improved. Thus, application of S.D.B.R in single-phase single-stage solar inverter is implemented successfully.

978-1-4799-6047-7/14 $31.00 © 2014 IEEE   530

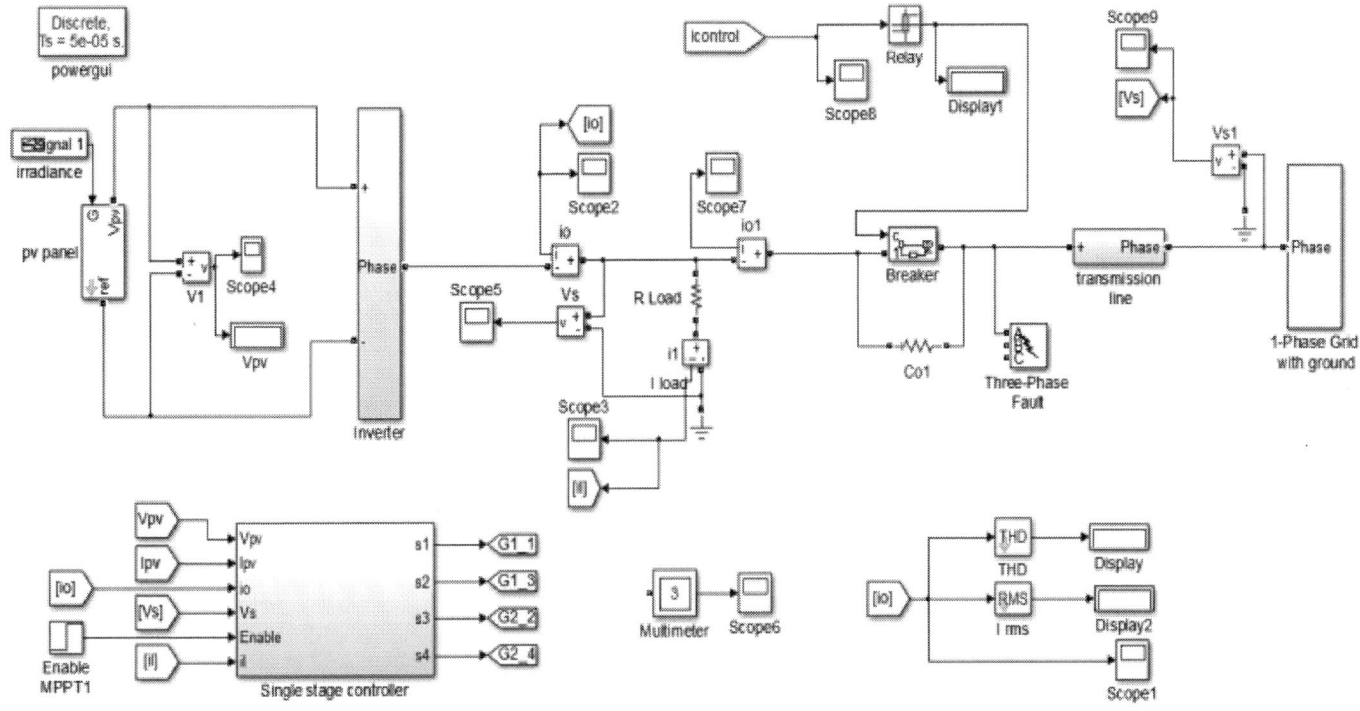

Fig.12. Simuation model of single-stage single-phase solar inverter with SDBR controller.

VI. CONCLUSION AND FUTURE WORK

This paper implemented the application of SDBR for improving the FRT capability of grid connected single stage single phase solar inverter. Also comparative study of all single stage single phase solar inverter topologies is given in tabular form. An entire modeling of grid connected PV system is presented. As per the simulated results inverter current has very low harmonic content and high percentage of power conversion. The complete work also proves the reliability of proposed system in smart grid application as local distribution system in future.

.

REFERENCES

[1] M. Bouzguenda, A. Gastli, A. H. Al Badi, and T. Salmi, "Solar photovoltaic inverter requirements for smart grid applications", IEEE PES Conf. on Innovative Smart Grid Technologies - Middle East (ISGT Middle East), 17-20 December 2011.

[2] T.J. Liang, Y.C. Kou, and J.R. Chen, "Single-stage photovoltaic energy conversion system", IEEE Proc. on Electrical Power Application, vol. 148, no. 4, pp. 339-344, July 2001.

[3] K. H. Hussein, I. Muta, T. Hoshino, and M. Osakada, "Maximum photovoltaic power tracking: an algorithm for rapidly changing atmospheric conditions," IEEE Proc. Generation, Transmission and Distribution, vol. 142, no. 1, pp. 59–64, Jan. 1995.

[4] N. Kasa, T. Iida, and H. Iwamoto, "Maximum power point tracking with capacitor identifier for photovoltaic power system," in Proc. Eighth Int. Conf. Power Electron. Variable Speed Drives, pp. 130–135, Sep. 2000.

[5] Y. C. Kuo, T. J. Liang, and J. F. Chen, "Novel maximum-power point tracking controller for photovoltaic energy conversion system," IEEE Trans. Ind. Electron., vol. 48, no. 3, pp. 594–601, Jun. 2001.

[6] Y. Chen and K. Ma-Smedley, "A cost-effective single-stage inverter with maximum power point tracking", IEEE Trans. Power Electron., vol. 19, no. 5, pp. 1289–1294, Sep. 2004.

[7] W. Chien-Ming, "A novel single-stage full-bridge buck-boost inverter", IEEE Trans. Power Electron., vol. 19, no. 1, pp. 150–159, Jan. 2004.

[8] A. Fernandez, J. Sebastian, M. Hernando, M. Arias, and G. Perez, "Single stage inverter for a direct AC connection of a photovoltaic cell module", in Proc. IEEE PESC, pp .93–98, June 2006.

[9] Mihai Ciobotaru, Remus Teodorescu, and Frede Blaabjerg, "Control of single-stage single-phase PV inverter", EPE Journal : European Power Electron. and Drives Journal, vol. 16, no. 3, pp. 20-26, Dec. 2000.

[10] Sachin Jain and Vivek Agarwal, "A single-stage grid connected inverter topology for solar PV systems with maximum power Point tracking" IEEE Trans. on Power Electron., vol.22, no. 5, pp. 0885-8993, Sept. 2007.

[11] Z. Chen, X. Zhang, and J. Pan, "An integrated inverter for a single-phase single-stage grid-connected PV system based on Z-source," Bull. Pol. Ac.: Tech., vol. 55, no. 3, pp. 263-272, 2007.

[12] Mario Fortunato, Alessandro Giustiniani, Giovanni Petrone, Giovanni Spagnuolo and Massimo Vitelli, "Maximum Power Point Tracking in a One-Cycle-Controlled Single-Stage Photovoltaic Inverter", IEEE Trans. Ind. Electron., vol .55, no7, 0278-0046, July. 2008.

[13] K. M. Shafeeque, and P.R. Subadhra, "A novel single phase single stage inverter for solar power applications" IEEE Int. Conf. ICACC, pp . 243-246, Aug. 2013.

[14] Mitra Mirhosseini , Josep Pou and Vassilios G. Agelidis, "Single-stage inverter-based grid-connected photovoltaic power plant with ride-through capability over different types of grid faults" IEEE Ind. Electron. Society Conf., IECON 10-13 Nov. 2013, pp. 8008 – 8013.

[15] M. K. Hossain and M. H. Ali, "Low voltage ride through capability enhancement of grid connected PV System by SDBR", in Proc. IEEE PES Transmission & Distribution (T&D) conf. & Expo., paper id: 539, Chicago, USA, April 15-17, 2014.

[16] M. G. Villalva, J. R. Gazoli, and E. R. Filho, "Comprehensive approach to modeling and simulation of photovoltaic arrays," IEEE Trans. Power Electron. vol. 24, no. 5, pp. 1198-1208, May. 2009.

Power Quality Improvement of a Position Sensorless Controlled PMBLDCM Drive using Boost Converter

Sachin Singh

Electrical and Instrumentation Engineering Department
SLIET Longowal, Sangrur, India
sachinsliet@gmail.com

Sanjeev Singh

Electrical and Instrumentation Engineering Department
SLIET Longowal, Sangrur, India
sschauhan.sdl@gmail.com

Abstract— **This paper proposes a boost converter for power quality improvement at input AC mains of a permanent magnet brushless DC motor (PMBLDCM) operated under position sensorless control. Presently hall sensors are used as position sensors in PMBLDCM and its signals are utilized for starting and control of PMBLDCM under various operating conditions. For the position sensorless control, back-EMF signal of PMBLDCM is used to detect its rotor position. A pulsed starting method is proposed for position sensorless starting which uses the PMBLDCM as a synchronous motor at standstill. The PMBLDCM is driven to a threshold speed where the back-EMF is detectable and the control of motor is shifted to position sensorless running mode. The PMBLDCM drive is generally fed by a single phase AC mains followed by a diode bridge rectifier (DBR) and a dc link capacitor through a three phase inverter. In the proposed work, a boost DC-DC converter is used for power factor correction (PFC) at input AC mains and to control the DC link voltage using current multiplier approach. The proposed method of power quality improvement of a position sensorless controlled PMBLDCM drive using boost converter is designed, modelled in Matlab-Simulink environment and its performance is evaluated under speed, torque and input AC voltage variations. The obtained results show improved power quality at AC mains along with smooth starting from standstill and during speed and torque control of the PMBLDCM drive operated in position sensorless mode.**

Keywords— *position sensorless control; back-EMF; PMBLDCM; PFC; Boost converter*

I. INTRODUCTION

In recent years, permanent magnet brushless DC motor (PMBLDCM) achieved popularity in various applications such as disk drives and fans to a large industrial drives and aerospace applications. PMBLDC motors have become popular because of some attractive features like high power per unit volume, high efficiency, silent operation, high speed, ease of control, high reliability and low maintenance etc. However, for the commercialization of the drive, cost effective, reliable and efficient designs are becoming the most important concerns for the researchers. As there are no brushes, no arcing is associated with these motors; hence PMBLDC motors are safer to work in the environment where there exists danger of explosion [1-3].

In a PMBLDC motor there are three Hall sensors to sense the rotor position and according to these hall signals voltage source inverter (VSI) switches are commutated to get the rotation. These motors are three phase synchronous motors with rotor as permanent magnet. Three phase stator windings create the electromagnet pole and the rotor flux is created by permanent magnets. Due to the attraction of the energized stator phase windings rotor is forced to rotate [1]. The hall sensors increase the volume, cost and complexity of the PMBLDCM drive, therefore, control of PMBLDCM drive without hall sensors is one of the most favourite research area these days [4-9].

In India all the electrical and electronic equipments are generally fed by a 50 Hz AC power supply, and most of this power is first introduced by some kind of power converter [10-12]. The main aim of this power converter is to convert AC into DC using a diode bridge rectifier (DBR) followed by a DC link capacitor. This DC capacitor is charged only when the output voltage of DBR is greater than DC link voltage, and at the instant of charging it draws a peaky current, consequently rich harmonics are introduced in the input AC mains resulting in poor power factor, harmonic pollution and high value of crest factor [4]. To follow the international power quality standards [10] such as IEC 61000-3-2 it is mandatory to improve power quality (PQ) indices such as power factor (PF), crest factor (CF) and total harmonic distortion (THD) at input AC mains.

Various papers [4, 11-12] are reported in the literature to include single stage DC-DC converters for power factor correction (PFC) and controlling the DC link voltage instead of using separate stages for it. In the proposed work a boost converter is used to perform the above two functions simultaneously amongst various DC-DC converters. The main target behind this approach is to reduce the component count and thereby reducing the overall system cost. This paper presents modelling, design and performance evaluation of the PMBLDC motor drive for PQ improvement under position sensorless control from standstill to running condition.

II. OPERATION OF PMBLDC MOTOR IN POSITION SENSORLESS CONTROL FED BY BOOST CONVERTER

Fig.1 shows the schematic diagram of the proposed PMBLDCM drive in sensorless control mode fed by a boost

978-1-4799-6047-7/14 $31.00 © 2014 IEEE

converter.

Generally front end of a PMBLDCM drive is chosen as a DBR fed by a single phase AC supply [4, 11-12]. In this work a Boost converter is used in continuous conduction mode (CCM) to perform voltage boost action and power factor correction (PFC) simultaneously. To achieve this current multiplier control approach [11] is used which provides current control at input AC supply and maintains DC link voltage for PMBLDCM drive along with improved power quality at input AC supply.

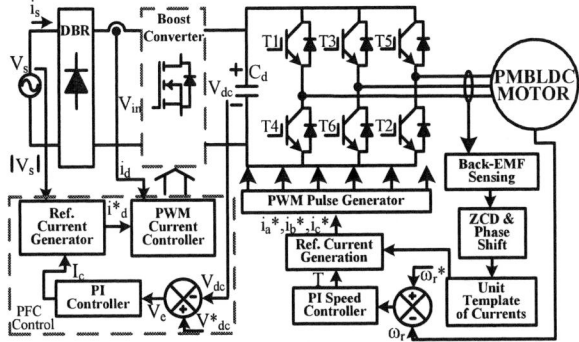

Fig. 1. Position sensorless control of PMBLDC motor fed by boost converter

To draw sinusoidal current from the input AC and to perform the boost action, output voltage V_o is sensed and compared with the reference voltage and then fed to a voltage PI controller to get the modulating current signal I_c [11]. Now unit template of input AC supply is multiplied by this modulating current signal and then it is compared with the dc current sensed after the DBR as shown in the figure. The obtained error signal is amplified if required and then compared with a sawtooth wave to generate PWM signal for boost switch.

To reduce the cost and size of inductor and capacitor high switching frequency (25 khz) is used at the gate of the metal oxide semiconductor field effect transistor (MOSFET).

III. PROPOSED SCHEME FOR POSITION SENSORLESS STARTING AND RUNNING OF PMBLDC MOTOR DRIVE

The term "position sensorless control" means control of PMBLDCM without position sensors i.e no need of Hall sensors, for position sensing of rotor.

The main problem of position sensorless control is the initial rotor position detection [8-9] at standstill condition or at zero speed. Various papers are reported in the literature for sensorless running [5-7] condition however for initial starting of motor few papers [8-9] are reported.

In this paper a new sensorless control scheme of a PMBLDC motor drive is proposed which includes three stages, (i). To fix the rotor at a known position using high frequency pulses, (ii). To accelerate the rotor up to a speed where the back-EMF is detectable (iii). To shift the control from sensorless starting to sensorless running mode.

Fig.2 shows the schematic diagram of proposed sensorless starting scheme to fix the rotor at a known position. To achieve this, rated dc voltage is applied across the terminals of PMBLDCM through three phase VSI, while applying the high frequency gate pulses to the upper switches of phases A and C along with lower switch of phase B, simultaneously. This technique of stabilizing the rotor at a known position, may lead to excessive high motor current, which can be controlled by properly choosing the applied frequency.

Schematic diagram for starting of PMBLDC motor from a known position is shown in the Fig.3, where a predetermined switching pattern is multiplied by PWM pulses of high frequency to gradually speed up the motor and simultaneously overcoming the load and inertia on the motor shaft. These PWM pulses are so chosen that the current drawn by the motor does not become dangerously high.

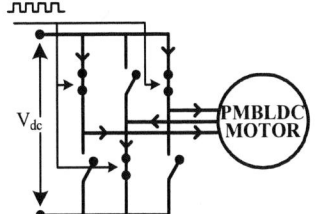

Fig. 2. Method to align the rotor at a known position

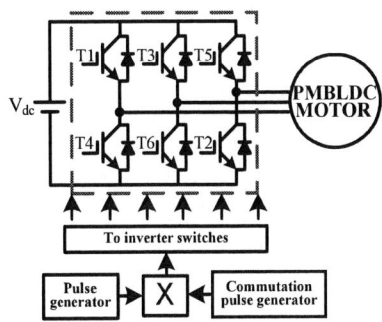

Fig. 3. Starting of PMBLDC motor from a known position

After a threshold speed where the back-EMF is detectable, control is shifted to sensorless running mode. The commutation pulses for VSI are generated by detecting the zero crossings of back-EMF and then phase shifted by 30 electrical degrees so that the current waveforms of the motor coincides with the flat portion of the back-EMF waveforms so that a constant torque at the PMBLDCM shaft is generated.

IV. MODELLING OF PMBLDC MOTOR

A. PMBLDC Motor

The modeling equation for a PMBLDC motor includes

$$V_{an} = Ri_a + L(di_a/dt) + M(di_b/dt) + M(di_c/dt) + e_a$$

$$V_{bn} = Ri_b + L(di_b/dt) + M(di_c/dt) + M(di_a/dt) + e_b$$

$$V_{cn} = Ri_c + L(di_c/dt) + M(di_a/dt) + M(di_b/dt) + e_c$$

where V_{an}, V_{bn} and V_{cn} are phase to neutral voltages and L is self inductance per phase of the three phase windings and M is the mutual inductance between two phases. Back-EMF in the windings A, B and C are e_a, e_b and e_c respectively [11-12].

978-1-4799-6047-7/14 $31.00 © 2014 IEEE

In the matrix form the above equation can be written as

$$\begin{bmatrix} V_{an} \\ V_{bn} \\ V_{cn} \end{bmatrix} = \begin{bmatrix} R & 0 & 0 \\ 0 & R & 0 \\ 0 & 0 & R \end{bmatrix} \begin{bmatrix} i_a \\ i_b \\ i_c \end{bmatrix} + \begin{bmatrix} L & M & M \\ M & L & M \\ M & M & L \end{bmatrix} p \begin{bmatrix} i_a \\ i_b \\ i_c \end{bmatrix} + \begin{bmatrix} e_a \\ e_b \\ e_c \end{bmatrix}$$

where p = derivative operator. As the motor windings are star connected so

$$i_a + i_b + i_c = 0$$

$$i_b + i_c = - i_a$$

so $V_{an} = Ri_a + L(di_a/dt) + M\{d(i_b+i_c)/dt\} + e_a$

or $V_{an} = Ri_a + L(di_a/dt) - M(di_a/dt) + e_a$

or $V_{an} = Ri_a + (L - M)(di_a/dt) + e_a$

or $V_{an} = Ri_a + L_s(di_a/dt) + e_a$

where L_s is total inductance per phase due to self inductance and mutual inductance felt by other two phase windings.

Similarly for other two phases

$$V_{bn} = Ri_b + L_s(di_b/dt) + e_b$$

$$V_{cn} = Ri_c + L_s(di_c/dt) + e_c$$

And electromagnetic torque

$$T_e = (e_a i_a + e_b i_b + e_c i_c)/\omega$$

Back-EMF is given by the relation

$$e_x = K_b f_x(\theta) \omega, \quad x = a, b \text{ and } c \text{ phase}$$

where

$f_a(\theta) = 1$	for $0 < \theta < 2\pi/3$
$f_a(\theta) = \{(6/\pi)(\pi-\theta)\}-1$	for $2\pi/3 < \theta < \pi$
$f_a(\theta) = -1$	for $\pi < \theta < 5\pi/3$
$f_a(\theta) = \{(6/\pi)(\theta-2\pi)\}+1$	for $5\pi/3 < \theta < 2\pi$

B. Speed Controller

To achieve the best performance of the PMBLDCM drive careful modelling of speed controller is required. To fulfill this requirement a PI controller is used for speed control of PMBLDC motor. Its output at nth instant is given by the relation [11-12]

$$T(n) = K_{p\omega}\omega_e(n) + K_{i\omega}\Sigma \omega_e(n)$$

where $K_{p\omega}$ and $K_{i\omega}$ are the proportional and integral constants of the speed PI controller. Speed error $\omega_e(n)$ is given by the relation

$$\omega_e(n) = \omega^*_r(n) - \omega_r(n)$$

where $\omega^*_r(n)$ is reference speed and $\omega_r(n)$ is the actual rotor speed.

The output of the PI controller at $(n-1)^{th}$ instant is

$$T(n-1) = K_{p\omega}\omega_e(n-1) + K_{i\omega}\Sigma \omega_e(n-1)$$

Subtraction of the two equations $T(n) - T(n-1)$ gives the result

$$T(n) - T(n-1) = K_{p\omega}\{\omega_e(n) - \omega_e(n-1)\} + K_{i\omega}\Sigma\{\omega_e(n) - \omega_e(n-1)\}$$

$$T(n) - T(n-1) = K_{p\omega}\{\omega_e(n) - \omega_e(n-1)\} + K_{i\omega}\omega_e(n)$$

Therefore at n^{th} instant output of the PI controller can be calculated by the relation

$$T(n) = T(n-1) + K_{p\omega}\{\omega_e(n) - \omega_e(n-1)\} + K_{i\omega} \omega_e(n)$$

C. Reference Winding Current Generation

Reference winding currents are denoted by i^*_a, i^*_b and i^*_c respectively for the three phases and can be calculated by the relation

$$I^* = T(n) / (2K_b)$$

where K_b is the back-EMF constant.

These reference winding currents are now compared with the sensed current to generate the current errors Δi_a, Δi_b and Δi_c for the three phases of the motors respective. These current errors are multiplied by a gain to increase the sensitivity of the controller.

D. PWM Pulse Generator

In the PWM pulse generator these modified current errors are compared with a high frequency saw tooth waveform to generate the high and low switching pulses for voltage source inverter, which simultaneously controls the current. These pulses are generated based on the logic that if modified current error for a particular phase is greater than the amplitude of saw tooth waveform switch is on otherwise it is off.

E. Voltage Source Inverter

As the frequency of operation for VSI is in the range of 10 kHz, insulated gate bipolar transistors (IGBTs) are used in the VSI bridge.

Fig. (4) given below shows the PMBLDCM drive fed by a VSI, in which only two switches are on at any time, one from upper switches and one from lower switches. From this logic voltage across 'a' phase for the star connected windings of the PMBLDCM can be calculated as

$$v_{ao} = V_{dc}/2 \text{ when T5=On and T2=Off}$$

$$v_{ao} = - V_{dc}/2 \text{ when T5= Off and T2= On}$$

$$v_{ao} = 0 \text{ when T5 \& T2 both Off}$$

From the same logic voltage across phase 'b' and 'c' are calculated.

V. MODELLING OF PFC BOOST CONVERTER

A PFC boost converter consist of, PI voltage controller, reference current generator and a PWM current controller [11].

A. PI Voltage Controller

The input to a PI voltage controller is the error between the set reference voltage and the dc link voltage at n^{th} instant, which is given by the relation

$$V_e(n) = V^*_{dc}(n) - V_{dc}(n)$$

The output of the PI voltage controller at n^{th} instant can be calculated as

$$I_c(n) = I_c(n-1) + K_{pv}\{V_e(n) - V_e(n-1)\} + K_{iv}V_e(n)$$

where $I_c(n)$ is the equivalent control output at n^{th} instant, K_{pv} and K_{iv} are the proportional and integral gains respectively.

B. Reference Current Generator

The reference current can be calculated as

$$i^*_d = I_c(n)\, u_{vs}$$

where $u_{Vs} = v_d/V_{sm}$; $v_d = |v_s|$; $v_s = V_{sm}\sin\omega t$

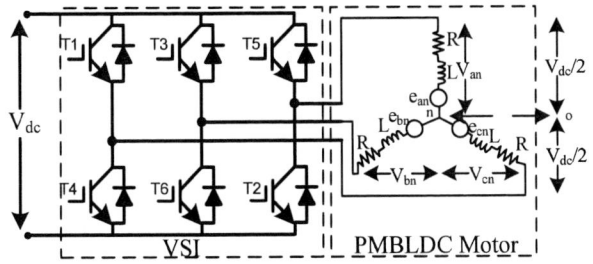

Fig. 4. Equivalent circuit of a PMBLDC motor drive fed by VSI

C. PWM Current Controller

The above calculated reference current is compared with the sensed current id. This is the output current just after the DBR. The difference of two currents are given by the relation $\Delta i_d = (i^*_d - i_d)$. This current error can be modified by multiplying with gain k to increase the sensitivity. Now this signal is compared with a high frequency (25 khz) saw tooth waveform to generate the switching signal for the PFC boost converter switch. PFC switch is on or off according to the relation given below

If $k\,\Delta i_d > m(t)$ then S = On
If $k\,\Delta i_d <= m(t)$ then S = Off

VI. OPERATING MODES OF BOOST CONVERTER

A. Mode I

In this mode MOSFET switch is closed, the diode is reversed biased and V_{in} is used to energize the boost inductor and the inductor current i_L flows from supply to inductor and then back to supply [2] as shown in Fig.5.

B. Mode II

In this mode, as shown in Fig.6, MOSFET switch is turned off and the diode is forward biased, current flows through inductor and diode, and then fed to load and the output capacitor. The output load receives energy from the energized inductor as well as from the input supply.

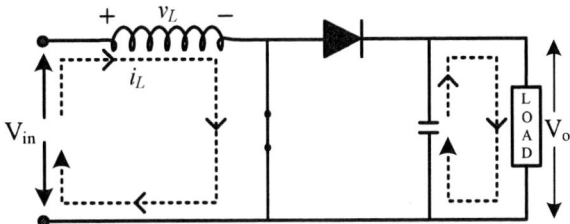

Fig. 5. Operation of Boost converter when switch is closed

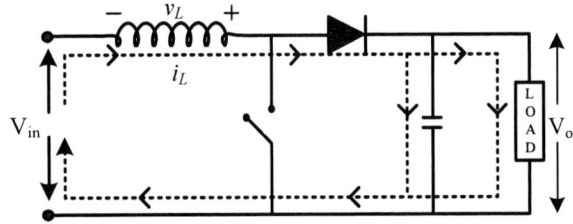

Fig. 6. Operation of Boost converter when switch is open

VII. RESULT OF PROPOSED POSITION SENSORLESS PMBLDCM DRIVE FED BY BOOST CONVERTER

The complete control scheme described above is implemented in MATLAB/SIMULINK environment. The detail of motor specifications is given in the appendix. The proposed position sensorless starting is achieved at rated load (Te=5.2 N.m) condition, Fig. 7 shows successful alignment of rotor at a known position and then a smooth speed up is accomplished while maintaining the current and torque within the specified maximum limit. After the successful starting and speed up control is smoothly shifted to sensorless running mode where a threshold back-EMF is attained, and then speed control is performed at rated torque.

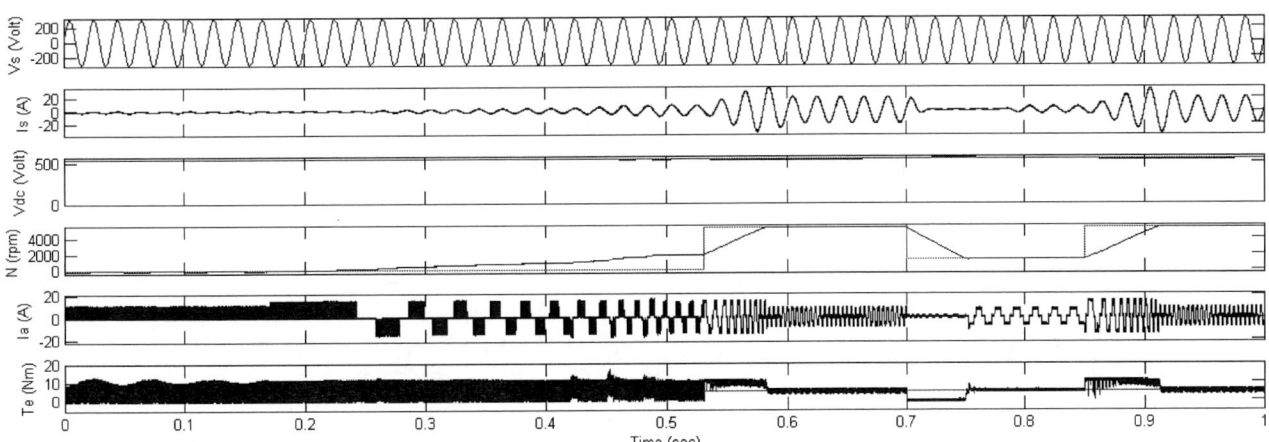

Fig. 7. Position sensorless starting at rated load (Te = 5.2 N.m) and then speed variation in sensorless running mode.

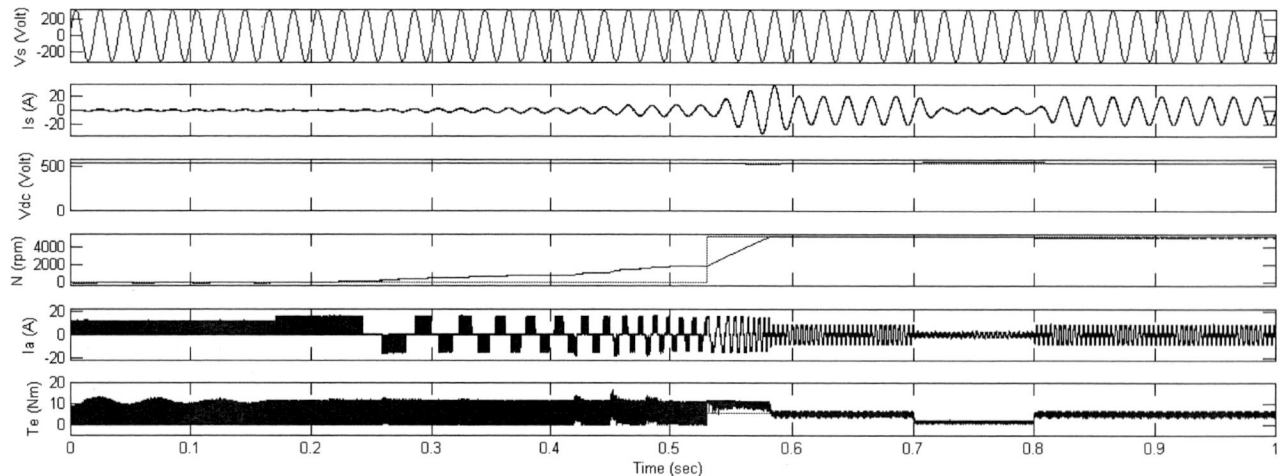

Fig. 8. Position sensorless starting at rated load (Te = 5.2 N.m) and then torque variation at rated speed (N=5200 rpm) in sensorless running mode.

In Fig. 8, torque variation is shown at rated speed (N=5200 rpm), result shows that as the torque is varied, current is maintained within the desired limit and speed remains almost constant. The performance evaluation of PFC boost converter is evaluated in terms of PF, CF and THD, results are summarized in table I, where input AC voltage is varied from 150 volt to 270 volt and it can be seen that in all conditions PF is almost unity, CF is nearly 1.41 and THD always less than 5% is maintained. As the speed and torque is varied (as shown in Figs.7-8) source current remains sinusoidal and in phase with the input AC voltage. Fig. 9 shows the FFT analysis of source current at 220 volt AC. It is observed from the results that the performance of the position sensorless PMBLDCM drive fed by boost converter is good and the power quality at input AC mains is within the international standards.

TABLE I. POWER QUALITY INDICES WITH INPUT AC VOLTAGE VARIATION

AC voltage	PF	DPF	THD %	Is(peak)	CF
150	0.9992	0.9998	3.49%	28.49	1.41
160	0.9995	0.9999	2.91%	26.67	1.41
170	0.9996	0.9998	2.40%	25.07	1.41
180	0.9997	0.9999	2.18%	23.66	1.41
190	0.9998	0.9999	1.87%	22.39	1.41
200	0.9998	0.9999	1.69%	21.26	1.41
210	0.9998	0.9999	1.56%	20.24	1.41
220	0.9998	0.9999	1.54%	19.31	1.41
230	0.9998	0.9999	1.48%	18.46	1.41
240	0.9998	0.9999	1.44%	17.69	1.41
250	0.9998	0.9999	1.39%	16.97	1.41
260	0.9998	0.9999	1.43%	16.31	1.41
270	0.9998	0.9999	1.45%	15.71	1.41

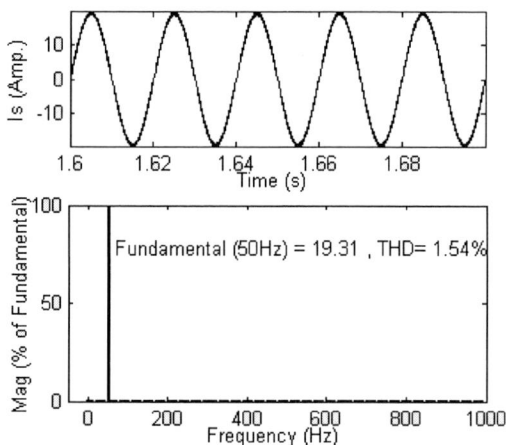

Fig. 9. Harmonic spectra of Input AC mains current at 220 volt

VIII. CONCLUSION

A single stage boost converter has been designed and modelled for power quality improvement at input AC mains of a PMBLDCM drive operated under position sensorless control. The position sensorless control of PMBLDC motor has been attained successfully through aligning the PMBLDCM rotor at a known position and then smooth starting in open loop. The controller has ensured smooth speed up and shifting to sensorless running mode using back-EMF. The PMBLDCM current has been maintained within the specified limits under sensorless starting, sensorless running and under speed and torque control. The proposed boost converter has ensured nearly unity PF and reduced THD of current at input AC mains for wide range of speed control and input AC voltage variation. The power quality indices obtained using the proposed boost converter are in conformity to international standards. It is hoped that the proposed controller for power quality improvement and position sensorless operation of PMBLDCM shall be helpful to researchers and industries.

APPENDIX

Motor power rating - 2.83 kW; Number of pole pair - 3; Speed rating - 5200 rpm; Torque rating - 5.2 N.m; Stator resistance - 1.045 Ω/phase; Stator inductance - 3.025 mH/phase; Inertia - 0.00068 Kg.m^2; Torque constant - 0.74 N.m/A.

REFERENCES

[1] T.J.E. Miller, *Brushless Permanent Magnet and Reluctance Motor Drive*, Clarendon Press, oxford, 1989.

[2] N. Mohan, T. M. Underland and W.P. Robins, *Power electronics converter,* applications and design, John Wiley and Sons, New York, 1995.

[3] M. H. Rashid, *Power electronics circuits, devices and applications,* Third edition, Pearson Education, Inc., 2004.

[4] B. Singh and S. Singh, "State of the art on permanent magnet brushless dc motor drives," *Journal of Power Electronics*, vol. 9, no. 1, pp.1-17, Jan. 2009.

[5] P.P. Acarnley, and J. F. Watson, "Review of position sensorless operation of brushless permanent magnet machines," *IEEE Trans. Ind. Electr.*, vol. 53, no. 2, pp. 352–362, Apr. 2006.

[6] K. Iizuka, H. Uzuhashi, M. Kano, T. Endo, and K. Mohri, "Microcomputer control for sensorless brushless motor," *IEEE Trans. Ind. Appl.*, vol. IA-21, no. 4, pp. 595–601, May/Jun. 1985.

[7] S. Ogasawara and H. Akagi, "An approach to position sensorless drive for brushless DC motors," *IEEE Trans. on Industrial Applications*, Vol. 27, No. 3, pp. 928-933, 1991.

[8] M. Naidu, T. W. Nehl, S. Gopalakrishnan and L. Wurth, "Keeping cool while saving space and money: a semi-integrated, sensorless PM brushless drive for a 42-V automotive HVAC compressor", *IEEE Ind. Appl. Mag.*,Vol. 11, No. 4, pp. 20-28, Jul.-Aug., 2005.

[9] N. Matsui, "Sensorless pm brushless DC motor drives," *IEEE Trans. Ind. Electron.*, vol. 43, no. 2, pp. 300–308, Apr. 1996.

[10] *Limits for Harmonic Current Emissions (Equipment input current ≤16 A per phase),* International Standard IEC 61000-3-2, 2000.

[11] Singh, B.; Singh, S., "Single-phase power factor controller topologies for permanent magnet brushless DC motor drives," *Power Electronics, IET* , vol.3, no.2, pp.147,175, March 2010.

[12] Singh, S.; Singh, B., "Power Quality Improvement of PMBLDCM Driven Air-Conditioner Using a Single-Stage PFC Boost Bridge Converter," India Conference.(INDICON), 2009 Annual IEEE , vol., no., pp.1,6, 18-20 Dec. 2009.

Type-2 Fuzzy Logic based Controllers for Indirect Vector controlled SVPWM based Two-Level Inverter fed Induction Motor Drive

T.Abhiram
Department of Electrical & Electronics Engineering,
Sreenidhi Institute of Science & Technology,
Hyderabad, India
abhiram.cbz@gmail.com

P.V.N. Prasad
Department of Electrical Engineering,
University College of engineering, Osmania University,
Hyderabad, India
polaki@rediffmail.com

Abstract—**SVPWM algorithm is used to control a voltage source inverter fed indirect vector controlled induction motor drive to have optimum use of the dc bus voltage and maximum output torque. In indirect vector control method, Type-2 fuzzy logic based speed and current controllers replace the conventional PI based speed and current controllers to achieve good dynamic performance, to minimize the torque ripple during transient and steady state conditions. For Type-2 fuzzy set the third facet Footprint of Uncertainty (FOU) provides additional degree of freedom to handle uncertainties improving the dynamic performance of induction motor. The performance of the drive is investigated at no-load and step change in load. The THD values of line voltage and stator current of motor along with source current THD and the device current ripple of the two-level inverter are compared between conventional PI and Type-2 Fuzzy logic based controllers.**

Keywords— Space vector pulse width modulation (SVPWM); TYPE-2 Fuzzy Logic Systems (T2FLS); Two-level inverter; Indirect Vector control;

I. INTRODUCTION

A three-phase voltage source inverter is used for simultaneous control of magnitude and frequency of the voltage applied to induction motor. The output of the inverter and the response of the induction motor depend on the control of the inverter. The field oriented control has several advantages such as wide speed range, precise speed regulation, fast dynamic response and operation above base speed (field weakening). The vector controller with voltage decoupler is used to provide the reference signals for the modulation scheme to drive the inverter [1]. Space vector modulation results in higher utilization of dc bus voltage as compared to PWM method for the voltage source inverter. Moreover, current ripple is also low in SVM controlled inverter as compared to PWM controlled inverter.

An artificial neural network method to predict the operating voltage and frequency when the load torque and speed are changed is presented in [2]. In this method the motor parameter variations are considered as negligible with the change in temperature and flux saturation. In [3] simplified vector control implementation strategy has been presented

with absence of current sensors The limitation of this method is as there is no voltage and current feedback the performance of the drive may deteriorate due to parameter variation and disturbance in the input dc link voltage. A neural network based indirect vector control scheme for a three phase induction motor is presented in [4]. It consists of three neural network controllers such as a real-time trained neural controller for the induction motor angular velocity, feed-forward neural controller for current set points iqs , ids, slip velocity and another feed-forward controller for transformation of stator currents from (q–d,) to (a, b, c) transformation.

In [5] fuzzy logic based indirect vector control for induction motor drive. In this work, simulation is carried out using PI and Fuzzy Logic based controller and the results of both controllers under the dynamic conditions are compared. Investigation is made in [6] on performance of the induction machine using indirect vector control technique and also degradation in dynamic performance, when the rotor resistance deviates from its actual value. The controller design and the slip calculation are not accurate due to which gives inaccurate value of rotor time constant. The PI controller used is unable to compensate parameter variations in the system. The implementation of the indirect vector control of induction machine using integrated DSP system and also shown off-line parameter estimation method for reliable operation of vector control is made in [7]. The main drawback of this method is that q-axis current has large error than d-axis current because of lack of consideration of non-linearity. Vector control model of induction machine with four parameters (stator resistance, inductance, rotor time constant and leakage coefficient) is done in [8]. Model is simulated with MATLAB, and implemented experimentally using DSP processor.

Of late, the use of type-2 FLCs (IT2FLCs) acquired lot of prominence because of the development of T2FLSs to handle uncertainties [9-10]. In this paper, Type-2 fuzzy logic based speed and current controllers are used in indirect vector controlled SVPWM based two-level inverter fed induction motor drive replacing conventional PI based speed and current controllers.

978-1-4799-6047-7/14 $31.00 © 2014 IEEE

II. INDIRECT VECTOR CONTROL

A. Modelling of Induction Motor

In an induction motor the stator current i_s is resolved along the rotor flux ϕ_r and perpendicular to it. For that the position of ϕ_r under all dynamic conditions should be known. If instantaneous position of the ϕ_r is known then d – q axis is positioned along ϕ_r and dynamic equivalent circuit with respect to d-q reference frame is obtained. This results in decoupled current components i_s into i_{sd} and i_{sq} by which control just like separately excited dc motor is possible [11].

The dynamic equations for stator in d-q reference frame are

$$V_{sd}=R_s\,i_{sd}+L_{ss}\frac{d}{dt}i_{sd}-L_{ss}\,\omega_\varsigma\,i_{sq}+M\frac{d}{dt}i_{rd}-M\omega_\varsigma\,i_{rq} \quad (1)$$

$$V_{sq}=R_s\,i_{sq}+L_{ss}\frac{d}{dt}i_{sq}+L_{ss}\,\omega_\varsigma\,i_{sd}+M\frac{d}{dt}i_{rq}-M\omega_\varsigma\,i_{rd} \quad (2)$$

$$L_{ss} = L_S + M \quad (3)$$

The dynamic equations for rotor in d-q reference frame are

$$0=R_r\,i_{rd}+L_{rr}\frac{d}{dt}i_{rd}-L_{rr}\big(\omega_s-\omega_\varsigma\big)i_{rq}+M\frac{d}{dt}i_{sd}-M\big(\omega_s-\omega_\varsigma\big)i_{sq} \quad (4)$$

$$0=R_r\,i_{rq}+L_{rr}\frac{d}{dt}i_{rq}+L_{rr}\big(\omega_s-\omega_\varsigma\big)i_{rq}+M\frac{d}{dt}i_{sq}+M\big(\omega_\varsigma-\omega_\varsigma\big)i_{sd} \quad (5)$$

$$L_{rr} = L_r + M \quad (6)$$

$$\frac{d\varsigma}{dt} = \omega_\varsigma \quad (7)$$

In the above equations V_{sq} and V_{sd} are d-q axis stator voltages, i_{sd}, i_{sq}, i_{rd} and i_{rq} are d-q axis stator currents and d-q axis rotor currents respectively. R_s and R_r are stator and rotor resistances per phase, L_s and L_r are self-inductances of stator and rotor, M is mutual inductance respectively and P is the number of poles. ω_s is synchronous speed and ω_ς is speed of reference frame. If ω_ς is equal to ω_s refers to synchronous reference frame and ω_ς is equal to 0 is stationary reference frame.

Also, the torque equation is found out to be

$$T = \frac{3}{2} \times \left(\frac{P}{2}\right)\frac{M}{L_{rr}}\big(\phi_{rd}\,i_{sq} - \phi_{rq}\,i_{sd}\big) \quad (8)$$

Where ϕ_{rq} and ϕ_{rd} are the q and d-axis rotor flux.

B. Control Parameters of Indirect Vector Control

For dynamic performance similar to separately executed motor the flux should the aligned always along the d – axis

$$T = \frac{3}{2}\times\left(\frac{P}{2}\right)\frac{M}{L_{rr}}\phi_{rd}\,i_{sq} \quad \text{since} \quad \phi_{rq}=0 \quad (9)$$

When i_{sd} is constant, i_{rd} and flux magnitude are also kept constant. By varying the orthogonal component to i_{sd}, the decoupled torque control is possible in induction motor drive.

In q-axis equivalent circuit in the field oriented control, ϕ_{qr} is zero, voltage across R_r is rotational voltage

$$\frac{d}{dt}\phi_{qr} = 0 \quad (10)$$

$$\omega_{sl} = \frac{i_{qs}\,R_r}{L_{rr}\,i_{ds}} \quad (11)$$

In d-axis equivalent circuit in the field oriented control,

$$\phi_{dr} = \big|\phi_r\big| \quad (12)$$

For indirect vector control with voltages V_{sq} and V_{sd} as command variables for SVPWM method, changes in i_{sq} and i_{sd} vary with V_{sq} and V_{sd} respectively as follows

$$V_{sq} = R_s\,i_{sq} + \sigma\,L_{ss}\frac{d}{dt}i_{sq} + \omega\,L_{ss}\,i_{sd} \quad (13)$$

$$V_{sd} = R_s\,i_{sd} - \omega_s\,\sigma\,L_{ss}i_{sq} \quad (14)$$

$$\sigma = \left\{1 - \frac{M^2}{L_{rr}\,L_{ss}}\right\} \quad (15)$$

When the command variables V_{sq} and V_{sd} are known the SVPWM can be implemented for two-level inverter fed induction motor drive.

C. Scheme of Indirect Vector Controlled IM Drive

Fig1. shows Indirect Vector Controlled IM Drive which employs three controllers. The outer loop speed controller tracks the speed error based on reference and actual rotor speeds. The output of this controller is responsible for producing the reference current i_{sq}. The two inner loop current controllers compare the reference values of i_{sd} and i_{sq} with their actual values. The speed and current controllers are either PI controllers or Type-2 Fuzzy based controllers. The control parameters for SVPWM Two-Level inverter are V_{sq} and V_{sd} which control the output voltage and frequency of the inverter at all dynamic conditions.

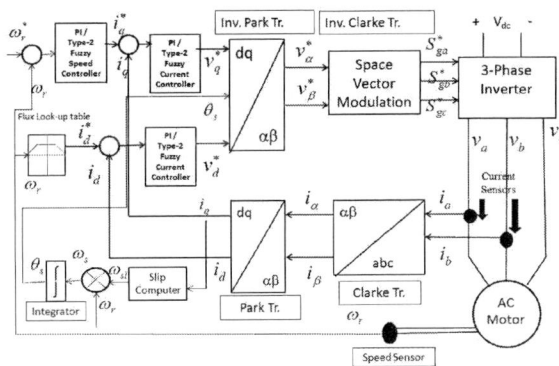

Fig. 1. Block diagram of Indirect Vector Control

III. SPACE VECTOR PWM FOR TWO- LEVEL INVERTER

A Two-level inverter shown in Fig. 2 having pole voltages (V_{A0}, V_{B0}, V_{C0}), the reference vector in α-β reference frame is having components given in (16).

$$\begin{bmatrix} V_{\alpha} \\ V_{\beta} \end{bmatrix} = \frac{2}{3} \begin{bmatrix} 1 & \cos\left(\dfrac{2\pi}{3}\right) & \sin\left(\dfrac{4\pi}{3}\right) \\ 0 & \sin\left(\dfrac{2\pi}{3}\right) & \sin\left(\dfrac{4\pi}{3}\right) \end{bmatrix} \cdot \begin{bmatrix} V_{AO} \\ V_{BO} \\ V_{CO} \end{bmatrix} \quad (16)$$

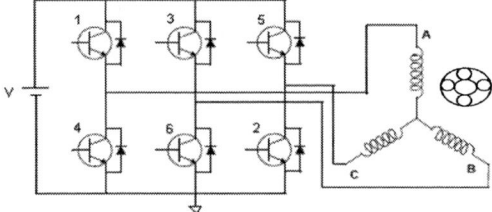

Fig. 2. Circuit diagram of Two-Level Inverter fed Induction Motor

By using SVPWM [12] and [13] the obtained space vectors with (k=1... 6) in (17) gives rise to a regular hexagon as shown in Fig. 3.

$$\overrightarrow{V_k} = \frac{2}{3} V_d e^{j(k-1)\frac{\pi}{3}} \quad (17)$$

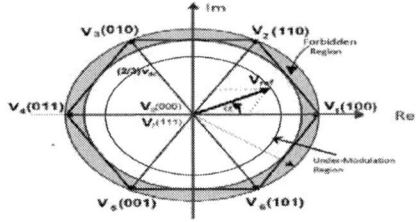

Fig. 3. Space-Vector diagram of Two-Level Inverter

The volt-seconds produced by the vectors $\overrightarrow{V_1}$, $\overrightarrow{V_2}$ and $\overrightarrow{V_7}$, $\overrightarrow{V_0}$ along the α and β axes are same as those produced by the reference vector $\overrightarrow{V_{ref}}$. If T_1, T_2, T_0 are the dwelling times and T_s is the sampling time with gives

$$\sqrt{\frac{2}{3}} V_d \cdot T_1 + \sqrt{\frac{2}{3}} V_d \cdot \cos\left(\frac{\pi}{3}\right) \cdot T_2 = \left|\overline{V_{ref}}\right| \cdot \cos\alpha \cdot T_s \quad (18)$$

$$\sqrt{\frac{2}{3}} V_d \cdot \sin\left(\frac{\pi}{3}\right) \cdot T_2 = \left|\overline{V_{ref}}\right| \cdot \sin\alpha \cdot T_s \quad (19)$$

$$T_1 = \frac{a \cdot T_s \cdot \sin\left(\frac{\pi}{3} - \alpha\right)}{\sin\left(\frac{\pi}{3}\right)} \quad (20)$$

$$T_2 = \frac{a \cdot T_s \cdot \sin\alpha}{\sin\left(\frac{\pi}{3}\right)} \quad (21)$$

$$T_0 = T_s - (T_1 + T_2) \quad (22)$$

Amplitude ratio $\qquad a = \dfrac{\left|V_{ref}\right|}{\sqrt{\dfrac{2}{3}} V_d} \qquad (23)$

IV. INTERVAL TYPE-2 FUZZY LOGIC SYSTEMS

An Interval Type-2 fuzzy set is represented in point form as

$$\widetilde{\Delta_N} = \left\{ \left((x,u), \mu_{\widetilde{\Delta_N}}(x,u) = 1 \right) \forall x \in X, \forall u \in J_x \subseteq [0,1] \right\} \quad (24)$$

Upper and a lower bound (MF) exist for the FOU. At each value of the primary variable, x, its MF is a function whose domain is called the "primary membership." The distribution that sits on top of the primary membership is called a "secondary MF."

The amplitude of the secondary MF is called the "secondary grade." For an interval type-2 fuzzy set, the secondary grade equals 1 over the entire FOU; hence, the new third dimension of a type-2 fuzzy set does not convey any new information for an interval type-2 fuzzy set. It is the FOU that completely characterizes an interval type-2 fuzzy set.

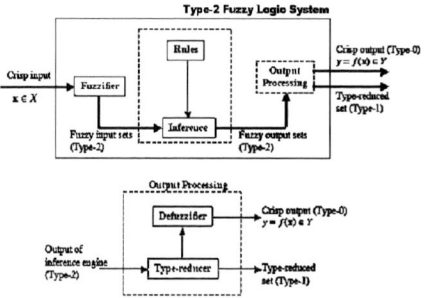

Fig. 4. Block diagram of Type-2 Fuzzy Logic System

Assuming that there are 'M' fuzzy rules [9-12] that have been normalized in the interval of [-1, 1].

Rule k: IF x_1 is ΔN_1^k and x_2 is ΔN_2^k and ...... and x_p is ΔN_p^k, THEN

$$y \text{ is } \left[\underline{w^k} \quad \overline{w^k} \right]$$

Where $k = 1, 2,...M$, p is the number of input variables in the antecedent part

ΔN_i^k (i=1, 2,...p, k=1,2...,M) are IT2FLS of the IF-part. $\underline{w^k}$, $\overline{w^k}$ are singleton lower and upper weight factors of the THEN-part. Once a crisp input $X = (x_1, x_2, ... x_p)^T$ is applied to the IT2FLS, through the singleton fuzzifier and the inference process, the firing strength of the k th rule which is an interval type-1 set can be obtained as

$$F^k = \left[\underline{f^k} \quad \overline{f^k} \right] \quad (25)$$

Where

$$\underline{f^k} = \underline{\mu_{\Delta N_1^k}(x_1)} * \underline{\mu_{\Delta N_2^k}(x_2)} * \dots\dots * \underline{\mu_{\Delta N_p^k}(x_p)} \quad (26)$$

$$\overline{f^k} = \overline{\mu_{\Delta N_1^k}}(x_1) * \overline{\mu_{\Delta N_2^k}}(x_2) * \ldots \ldots \ldots * \overline{\mu_{\Delta N_p^k}}(x_p) \qquad (27)$$

$\underline{\mu}()$ and $\overline{\mu}()$ denote the grades of the lower and upper membership functions of IT2FSs and * denotes minimum or product t-norm. The Type-2 non-singleton Mamdani fuzzy inference system (FIS) is used in this paper with membership functions as shown in Fig. 5 and Fig. 6 for both speed and current controllers. Five membership functions are taken with two inputs and one output for both speed and current controllers with 25 rules. The inputs are error (E) and change in error (CE). The rules are shown Table I. and surface viewer for the speed and current controllers are shown in Fig. 7 and Fig. 8.

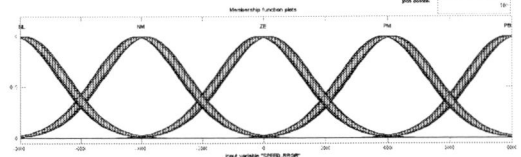

Fig. 5. Membership functions for Speed Controller

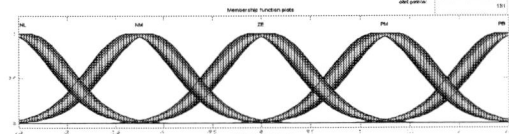

Fig. 6. Membership functions for Current Controllers

TABLE I.

E/CE	NL	NM	ZE	PM	PL
NL	ZE	NL	NL	NM	PL
NM	NM	ZE	NM	PM	NM
ZE	NL	NM	ZE	PM	PL
PM	NL	NM	PM	ZE	PL
PL	PL	PM	ZE	PL	ZE

V. RESULTS AND DISCUSSION

In the Type-2 Fuzzy based indirect vector controlled induction motor, the membership functions range of values is generated from the obtained data points of the conventional PI controllers. Gaussian membership functions are used for two input variables V_{ds} and V_{qs}.

To validate the proposed work simulations are carried out by operating the induction motor at a reference speed of 800 RPM. The THD for line voltages are compared between PI controllers based and Type-2 Fuzzy Controllers based indirect vector control in Fig. 9 and line current in Fig. 10 respectively. A step torque of 5 N-m is applied at 0.5s upto 0.7s where the motor speed settles at 793 RPM using PI Controller whereas it settles at 796 RPM using Type-2 Fuzzy conroller as shown in Fig. 13 and Fig. 14. In addition to it the source current THD is given in Fig. 15. The obtained results are summarised in Table.III. Device current of two-level inverter is given in Fig. 17, 18 and its values are summarised in Table III. It is observed that the device current ripple is 12.5A with PI whereas with Type-2 Fuzzy based controller is

3.1A between 0.192s and 0.2s which shows that Type-2 fuzzy based controllers gives improved performance and quickly reduces the device current ripple at no-load when motor speed is 800 RPM. When load is applied at 0.5s the Type-2 fuzzy based controller has current ripple of 5.3A whereas PI based controller has 1.1A which indicates that Type-2 fuzzy based controllers gives improved performance and immediately increases the device current ripple when load torque is applied at 0.5s. The steady state torque ripple of induction motor when step load is applied is compared between PI controllers based and Type-2 Fuzzy Controllers based indirect vector control is given in Fig. 11, 12 and in Fig.16 the transient state toqrue is compared . The results are summarised in Table II.

TABLE II.

S.No.	Type of Controller	Speed ripple in rpm at step load of 5Nm	Steady state torque ripple at load of 5Nm
1.	PIC	6.4	0.98
2.	IT2FLC	4.2	0.43

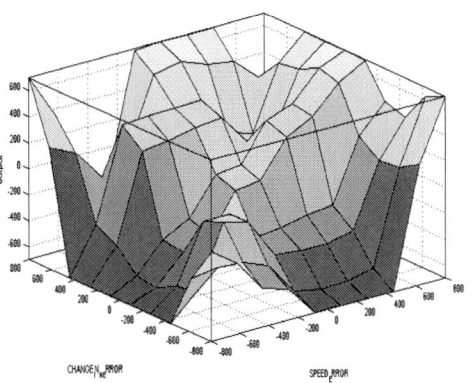

Fig. 7.Surface view of Speed Controller

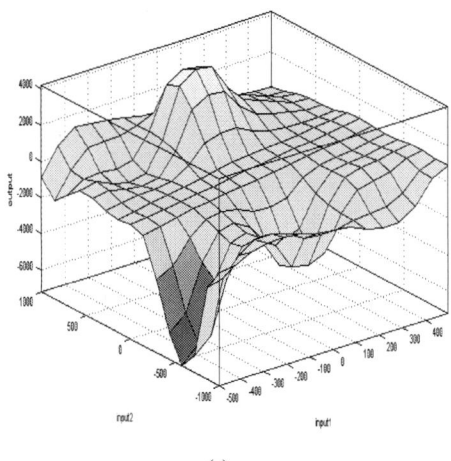

(a)

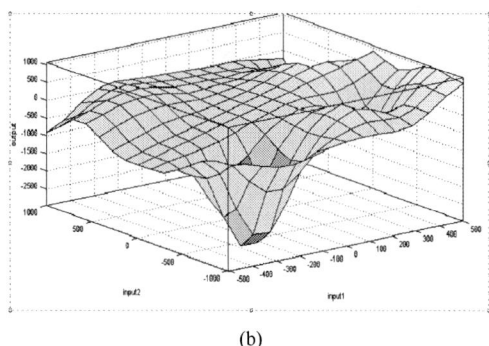

(b)

Fig. 8.Surface view of d-axis (a) and q-axis (b) Controllers

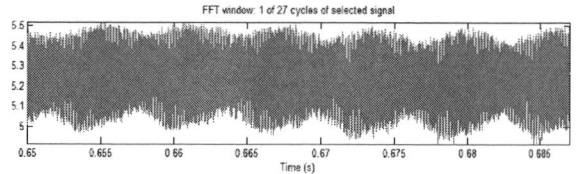

Fig. 12. Torque ripple with step load with Type-2 Fuzzy based Controller

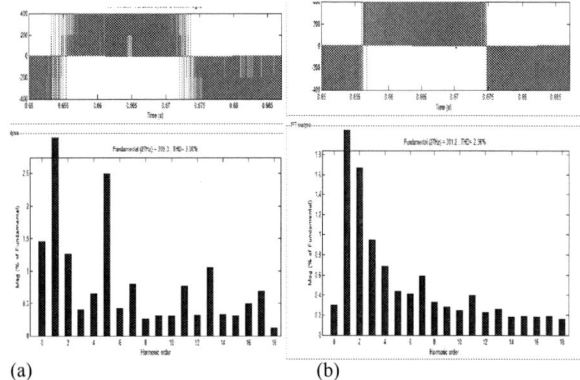

(a)   (b)

Fig. 9. Line-Voltage THD at 5-Nm load with PI based Controller (a) and Type-2 Fuzzy based Controller (b)

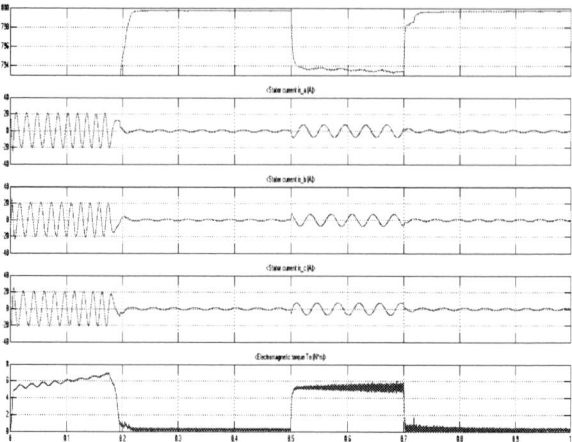

Fig. 13. Speed, Stator Current and Torque with PI based Controllers

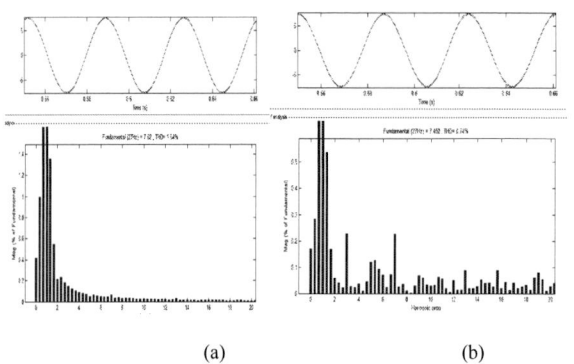

(a)   (b)

Fig. 10. Stator Current THD at 5-Nm load with PI based Controller (a) and Type-2 Fuzzy based Controller (b)

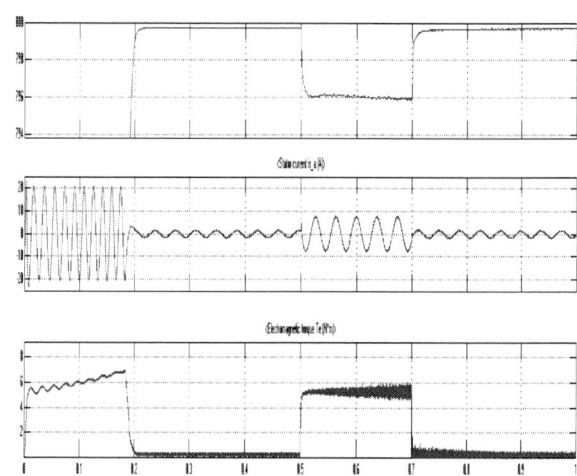

Fig. 14. Speed, Stator Current and Torque with Type-2 Fuzzy based Controllers

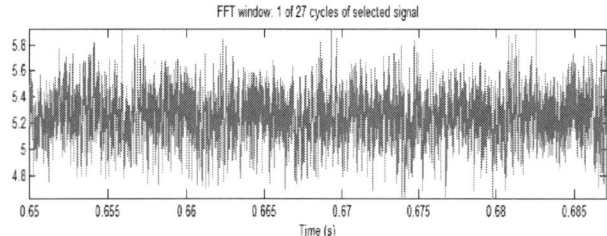

Fig 11. Torque Ripple at step load with PI based Controller

TABLE III.

S.No.	Type of Controller	Line Voltage THD%	Stator Current THD%	Source Current THD%	Device Current Ripple From 0.19s to 0.2s(No-load)	Device Current Ripple 0.5s(step load-5Nm)
1.	PIC	3.5	1.54	3.44	12.5A	1.1A
2.	IT2FLC	2.36	0.74	3.16	4A	5.3A

978-1-4799-6047-7/14 $31.00 © 2014 IEEE   542

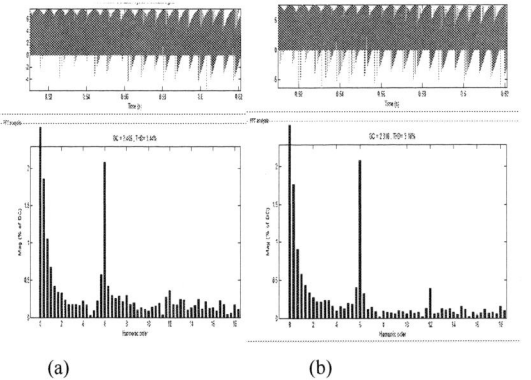

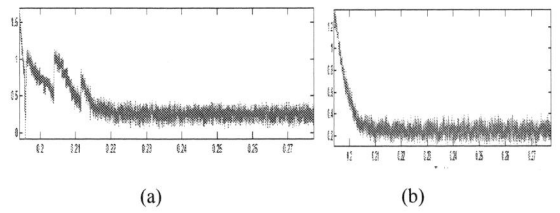

(a)   (b)

Fig. 15. Source Current THD at 5-Nm load with PI based Controller (a) and Type-2 Fuzzy based Controller (b)

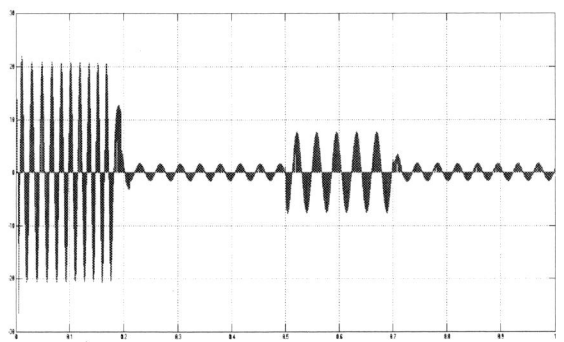

(a)   (b)

Fig. 16. No-load Torque with PI based Controller (a) and Type-2 Fuzzy based Controller (b)

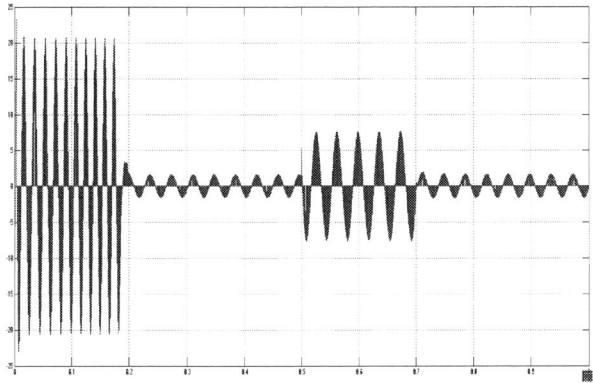

Fig. 17. Device Current waveform with PI based Controller

Fig. 18. Device Current waveform with Type-2 Fuzzy based Controller

VI. CONCLUSIONS

An improved dynamic performance during no-load and step change in load is achieved with indirect vector controlled Two-Level Inverter fed Induction Motor drive. A significant reduction in THD for line voltage and stator current is observed using Type-2 Fuzzy based controllers as compared to PI based controllers. Also the device current ripple of the Two-level inverter is reduced along with reduction in the torque ripple of induction motor during transient dynamic conditions with Type-2 Fuzzy based controllers as compared to PI based controllers.

APPENDIX

The ratings of the induction motor 400 Volts, 4 kW, 4-Pole, 1430 rpm. Induction motor parameters: R_s=1.405Ω, R_r^1 =1.395 Ω, $L_s = L_r^1$ = 0.005839H, L_m= 0.1722H, J= 0.0131 Kg - m^2, F = 0.002985 N-m-sec and inverter voltage V_{dc} = 400V.

REFERENCES

[1] Khaled Yahia, Salah-Eddine Zouzou and Fateh Benchabane, "Indirect vector control of induction motor with online rotor resistance identification" Asian journal of information technology 5(12), pp.1410-1415, 2006.

[2] A. K. Sharma, R. A. Gupta, Laxmi Srivastava, "Performance of ANN based indirect vector Control of induction motor drive", Journal of Theoretical and Applied Information Technology, pp. 9-14, 2007.

[3] Z. S. Wang, S. L. HO, " Indirect Rotor Field Orientation Vector Control for Induction Motor Drives in the Absence of Current Sensors", 5th International Power Electronics and Motion Control Conference, China, pp.1907-1911 ,August 2006.

[4] Ieroham Baruch, Irving Pavel de la Cruz A., Ruben Garrido,Boyka Nenkova, " An Indirect Adaptive Vector Control of the Induction Motor Velocity Using Neural Networks", Cybernetics and information technologies , Volume 7, No 2,Sofia, 57-72, 2007.

[5] Rajesh Kumar R. A. Gupta S.V. Bhangale, "Indirect vector controlled induction motor drive with fuzzy logic based intelligent controller", IETECH Journal of Electrical Analysis, Vol: 2, No: 4, pp.211 – 216, 2008.

[6] Rami A. Mahir , ZiadM. Ahmed, Amjad J. H, "Indirect Field Orientation Control of Induction Machine with Detuning Effect", Eng.&Tech.Vol.26.No.1, 2008.

[7] Mohammad N.Marwali, Ali keyhani, Willy Tjanaka, "Implementation of indirect vector control on an integrated digital signal processor-based system", IEEE transaction on energy conversion,Vol.4, No.2, pp. 139-146, June 2009.

[8] Alexandru Onea, Vasile Horga, Marcel Rățoi, " Indirect Vector Control of Induction Motor", Proceedings of the 6th WSEAS International Conference on Simulation, Lisbon, Portugal, September 22-24, 2006.

[9] Dongrui Wu and Jerry M. Mendel, "Linguistic Summarization Using IF–THEN Rules and Interval Type-2 Fuzzy Sets IEEE Transactions on Fuzzy Systems", vol. 19, no. 1, February 2011.

[10] Hsin-Jung Wu, Yao-Lung Su, and Shie-Jue Lee, "A Fast Method for Computing the Centroid of a Type-2 Fuzzy Set" IEEE Transactions on Systems, Man, and Cybernetics—part b: Cybernetics, vol. 42, no. 3, June 2012

[11] Krause, P.C., Wasyanczuk, O. and Sudhoff, S.D Analysis of Electrical Machinery and Drive Systems, IEEE Press, USA, 2002.[book]

[12] G. Narayanan, Di Zhao, Harish and K. Krishnamurthy, "Space Vector Based Hybrid PWM Techniques for Reduced Current Ripple," IEEE Trans. on Ind. Electron, vol. 55, no. 4, pp. 1614-1627, April 2008.

[13] Zhou K, Wang D. " Relationship between space-vector modulation and Three phase carrier-based PWM: a comprehensive analysis." IEEE Trans. Ind. Appl. 2002; 49(1):186–96.

On-Line Monitoring of Winding Parameters for Single-Phase Transformers

P. A. Reddy and B. S. Rajpurohit
School of Computing & Electrical Engineering
Indian Institute of Technology Mandi
Mandi, India
pothula_abhinay_reddy@students.iitmandi.ac.in, bsr@iitmandi.ac.in

Abstract— **The condition of transformer windings can be gauged by monitoring their equivalent circuit parameters. These parameters are not affected by external faults and change only in the presence of an internal aberration. Changes in the insulation temperature are reflected in winding temperature and can be monitored by observing the winding resistance values. Similarly changes in the short circuit reactance can give information on the condition and structure of windings. Rapid and reliable protection can be implemented by monitoring these parameters since inrush current and over-excitation does not affect these parameters. Presently, there is no accurate measurement method for the transformer winding parameters and generally require the transformer to be disconnected from the power system. A new simple algorithm for extracting transformer winding parameters which can be implemented online is presented in this paper. This method takes only the input currents and voltages as inputs and thereby eliminates the need for the disconnection of the transformer from the power system. In this method, winding parameters are obtained by solving the equivalent circuit equations in real time continuously which allows for interpretation and detection of faults in real-time. The proposed method has been tested and validated by simulations.**

Keywords—Winding parameter identification, real-time measurement, transformer winding deformation, transformer winding temperature.

I. INTRODUCTION

Transformers are the most important assets of transmission and distribution system and could cause power outages, personal and environmental hazards and expensive rerouting or purchase of power from other suppliers, if they were to fail. Transformers are extremely reliable and efficient when operated under rated conditions. Operation under conditions such as overloading and voltage imbalance for a long time reduces their life significantly [1]. Particularly, large oil-immersed power transformers are among the most expensive assets in power transmission and distribution networks. Therefore, to ensure their maximum uptime these critical assets must be continuously and closely monitored so as to assess their operating conditions.

The equivalent circuit parameters of a transformer can yield a lot of information on the condition of a transformer. Changes in short circuit reactance can indicate winding deformations, differentiate internal failures from external faults and default

status while changes in winding resistance can be used to monitor winding temperature, inter-turn shorting, etc.

The basic equivalent circuit parameters are computed in general by conducting short-circuit and open-circuit tests. These methods require the transformer under consideration to be disconnected from the grid which is not practical all the time and cannot be used to monitor the parameters continuously in order to detect anomalies at pre-failure stage.

In literature, many parameter estimation methods have been discussed. Some of the popular methods are inrush current tests [2], [3]; genetic algorithm techniques [4], [5]; least square error methods [6], [7]; open-circuit and short-circuit tests [8], [9]. The huge disadvantage of all these methods being requirement of the transformer to be disconnected from the power grid.

A novel method is presented in this study for the extraction of winding parameters of a single-phase transformer. The method takes only the primary and secondary currents and voltages as inputs, is an online method which allows for monitoring of the parameters while the transformer is in operation. Moreover it is very simple and requires minimal computations while maintaining high accuracy.

II. WINDING PARAMETER IDENTIFICATION

The equivalent circuit model of the transformer has been used in this method and is shown in Fig. 1. Winding resistance (R_p, R_s) and leakage inductances (L_p, L_s) are identified in the present method.

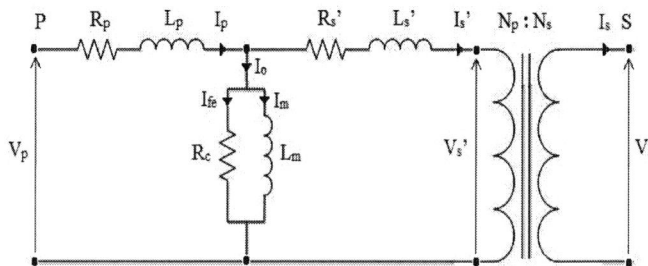

Fig. 1. Equivalent circuit of 1φ transformer

The meaning of the symbols used in the paper are given below:

$N_p{:}N_s$:   Transformer ratio.
R_p, L_p: Primary resistance, leakage inductance.
R_s, L_s: Secondary resistance, leakage inductance.
v_p, v_s: Primary, secondary voltages.
i_p, i_s: Primary, secondary currents.
v_s', i_s':   Secondary voltage, current referred to primary side.
R_s', L_s':   Secondary resistance, leakage inductance referred to primary side.

A. Transformer Model

From the circuit shown in fig. 1, consider the loop containing the winding parameters. Applying KVL along the loop results in the following equation

$$v_p(t) - v_s'(t)$$
$$= R_p i_p(t) + R_s' i_s'(t) + L_p \frac{d}{dt} i_p(t) + L_s' \frac{d}{dt} i_s'(t) \quad (1)$$

In order to extract the winding parameters R_p, R_s, L_p and L_s (1), is discretized and thereby can be solved by a digital computer.

The discrete equations are derived as:

$$\frac{v_p(T_n) + v_p(T_{n-1})}{2} - \frac{v_s'(T_n) + v_s'(T_{n-1})}{2}$$

$$= R_p \left(\frac{i_p(T_n) + i_p(T_{n-1})}{2} \right) + R_s' \left(\frac{i_s'(T_n) + i_s'(T_{n-1})}{2} \right)$$

$$+ L_p \left(\frac{i_p(T_n) - i_p(T_{n-1})}{T_n - T_{n-1}} \right) + L_s' \left(\frac{i_s'(T_n) - i_s'(T_{n-1})}{T_n - T_{n-1}} \right) \quad (2)$$

B. Calculation of parameters

The discrete values of v_p, v_s, i_p and i_s are collected through a data acquisition system or are generated from a mathematical model. For the sake of convenience (2), can be rewritten as

$$v = R_p a + R_s b + L_p c + L_s d_s d \quad (3)$$

where

$$v = \frac{v_p(T_n) + v_p(T_{n-1})}{2} - \frac{v_s'(T_n) + v_s'(T_{n-1})}{2}$$

$$a = \left(\frac{i_p(T_n) + i_p(T_{n-1})}{2} \right)$$

$$b = k \left(\frac{i_s'(T_n) + i_s'(T_{n-1})}{2} \right)$$

$$c = \left(\frac{i_p(T_n) - i_p(T_{n-1})}{T_n - T_{n-1}} \right)$$

$$d = k \left(\frac{i_s'(T_n) - i_s'(T_{n-1})}{T_n - T_{n-1}} \right)$$

And can be represented by a system of linear equations as

$$v = zx \quad (4)$$

$$x = z^{-1}v \quad (5)$$

where

$z = [\, a \;\; b \;\; c \;\; d \,]$ and $\mathbf{v}$ is a known vector.
$x = [\, R_p \;\; R_s \;\; L_p \;\; L_s \,]^{\mathrm{T}}$ is a four dimensional vector of unknowns.

Since there are four unknown values (R_p, R_s, L_p and L_s) to be calculated using (3), the proposed method employs four equations derived from (3), at different time instants and as each value of v_p, v_s, i_p and i_s is updated, the algorithm selects the set of latest four values, and calculates the corresponding values of v, a, b, c and d and then proceeds to calculate the value of vector x from (5).

III. SIMULATION RESULTS AND CASE STUDIES

In order to validate the proposed method, it is implemented on a voltage and current (primary and secondary) dataset generated by a mathematical model of transformer of known equivalent circuit parameter values. The mathematical model has been simulated in Matlab Simulink software and is shown in fig. 2.

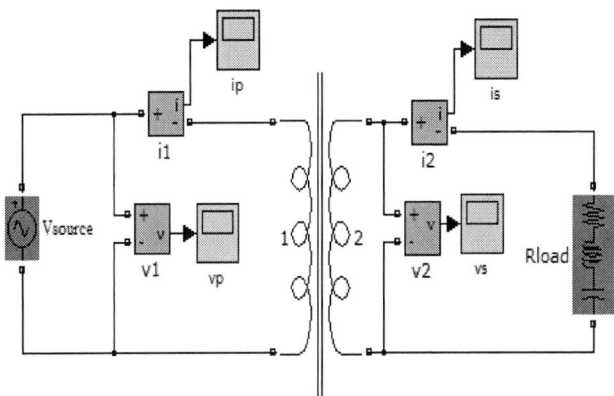

Fig. 2. Mathematical model used to generate the dataset

The primary and secondary voltage and current data generated from the above model is then processed by the following prior discussed algorithm as in (5).

We proceed to operate the transformer under different inputs and load conditions and show that the values of equivalent parameters derived remain unchanged as long as the same fixed transformer model is used. Here we consider a transformer of rated capacity 10 kVA having rated voltage $V_p/V_s = 2000/200$ V. The sampling frequency is 80 points per cycle. The winding parameter values were set as $R_p = 100 \ \Omega$, $R_s = 15 \ \Omega$, $L_p = 0.5$ H, $L_s = 0.2$ H. The following case studies were considered. The simulation was run for a period of 200 ms. The winding parameters (R_p, R_s, L_p, L_s) were plotted against time.

Case 1

The HV winding of the transformer is excited with a sinusoidal source of frequency 50 Hz and peak amplitude of 415 V. The LV winding supplies voltage to a resistive load of 10 kΩ. The proposed method is implemented on the voltages and currents obtained and the results were obtained shown in fig. 3.

The variation of all the four parameters was found to be in the range 0 to 0.005 units. The difference in calculated and expected values of all the parameters at any point of time was found to be less than 0.03%.

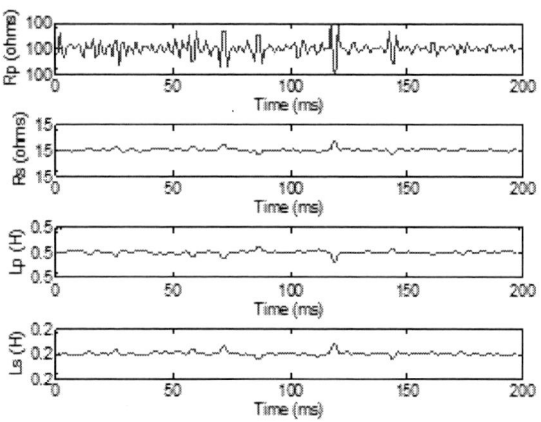

Fig. 3. Results – Case 1

Case 2

The HV winding of the transformer is excited with 4 sinusoidal sources of frequency 50 Hz, 100 Hz, 150 Hz and 200 Hz having peak amplitude of 415 V in order to simulate harmonics of various orders at their extremities. The LV winding supplies voltage to a RLC load. The proposed method is implemented on the voltages and currents obtained and the following results were obtained,

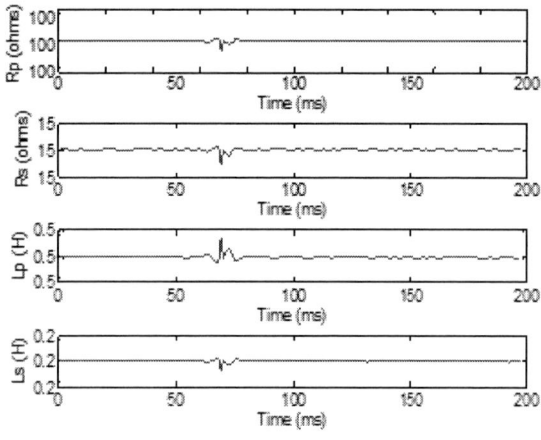

Fig. 3. Results – Case 2

The variation of all the four parameters was found to be in the range 0 to 0.005 units. The difference in calculated and expected values of all the parameters at any point of time was found to be less than 0.03%.

Case 3

The above considered linear transformer is replaced with a saturable transformer and the HV winding is excited with a sinusoidal source of frequency 50 Hz and peak amplitude of 415 V along with a number of other sinusoidal sources with multiples of 50 Hz as frequencies in order to simulate harmonics of various orders. The LV winding supplies voltage to a RLC load. The proposed method is implemented on the voltages and currents obtained and the following results were obtained,

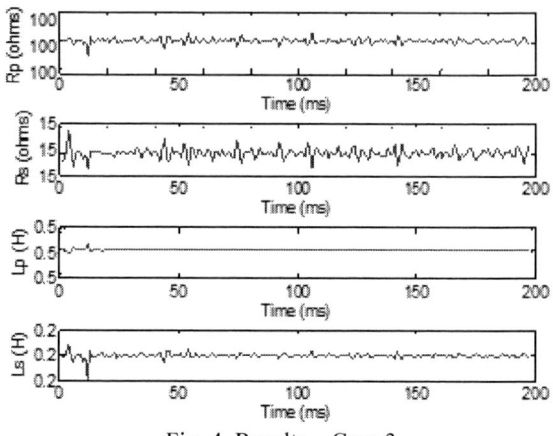

Fig. 4. Results – Case 3

The variation of all the four parameters was found to be in the range 0 to 0.005 units. The difference in calculated and expected values of all the parameters at any point of time was found to be less than 0.03%.

As seen from the above case studies, the proposed technique is found to be sufficiently accurate to monitor minute changes in the equivalent parameter values irrespective of the input signals and loading conditions.

IV. ACKNOWLEDGEMENT

We would like to express the gratitude to DST – FIST program for financial support.

V. CONCLUSIONS

This paper presents a new, simple and accurate method to monitor the winding parameters of transformer without disconnecting it from the power grid. It has been shown in different cases that the change in the operating condition of the transformer doesn't affect the readings. This method can be conveniently implemented by microcontrollers since the algorithm deals with discrete samples. The influence of inrush current on transformer protection can be avoided since it doesn't affect internal parameters which increases it's reliability. The proposed technique offers the possibility of evaluating winding tightness, temperature, etc. and more complicated control protocol for on-line protection and service management of transformers.

REFERENCES

[1] Mohamadi S., and A. Akbari, "A new method for monitoring of distribution transformers," Environment and Electrical Engineering (EEEIC), 2012 11th International Conference on. IEEE, 2012.

[2] S. Bogarra, A. Font, I. Candela, and J. Pedra, "Parameter estimation of a transformer with saturation using inrush measurements," *Elect. Power Syst. Res.*, vol. 79, no. 2, pp. 417-425, Feb. 2009.

[3] S. G. Abdulsalam, W. Xu, W. L. A. Neves, and X. Liu,"Estimation of transformer saturation characteristics from inrush current waveforms," *IEEE Trans. Power Del.*, vol. 21, no.1, pp. 170-177, Jan. 2006.

[4] S. H. Thilagar and G. S. Rao, "Parameter estimation of three-winding transformers using genetic algorithm," *Eng. Appl. Artif. Intell.*, vol. 15, no. 5, pp. 429–437, Sep. 2002.

[5] Rashtchi, E. Rahimpour, and E. M. Rezapour, "Using a genetic algorithm for parameter identification of transformer R-L-C-M model," *Elect. Eng.*, vol. 88, no. 5, pp. 417–422, 2006.

[6] E. S. Jin, L. L. Liu, Z. Q.Bo, and A. Klimek, "Parameter identification of the transformer winding based on least-squares method," in *Proc. IEEE Power Energy Soc. Gen. Meeting—Convers. Del. Elect. Energy 21st Century*, Jul. 20–24, 2008, pp. 1–6.

[7] S. A. Soliman, M. M. El-Arini, A. M. Al-Kandari, and M. E. El-Hawary, "Frequency domain parameter identification of harmonic potential transformer models using least square techniques," *Elect. Power Syst. Res.*, vol. 35, no. 1, pp. 45–49, Oct. 1995.

[8] Claveria, M. G. Gracia, M. Á. Garcia, and L. Montañes, "A time domain small transformer model under sinusoidal and non-sinusoidal supply voltage," *Eur. Trans. Elect. Power*, vol. 15, pp. 311–323, Feb. 2005.

[9] A. Mork, "Five-legged wound-core transformer model: Derivation, parameters, implementation and evaluation," *IEEE Trans. Power Del.*, vol. 14, no. 4, pp. 1519–1526, Oct. 1999.

[10] Mohamadi S., and A. Akbari, "A new method for monitoring of distribution transformers," Environment and Electrical Engineering (EEEIC), 2012 11th International Conference on. IEEE, 2012.

Hardware Development and Implementation of Single Phase Matrix Converter as a Cycloconverter and as an Inverter

A.Anand kumar, M.E Student, Department of Electrical and Electronics Engineering, Birla Institute of Technology, Mesra, Ranchi B.Dr. P.R. Thakura, Associate Professor, Department of Electrical and Electronics Engineering, Birla Institute of Technology, Mesra, Ranchi

Abstract—**In this paper, an algorithm is developed that enable a single phase matrix converter (SPMC) to perform a function of cycloconverter and of inverter i.e. act as frequency changer and also convert DC to AC. The algorithm is first implemented on Matlab simulink software. Simulation results are presented for SPMC as a cycloconverter (at different output frequency) and as an inverter (DC to AC). Simulated results are verified with experimental result. Also a laboratory model test rig of the SPMC as a cycloconverter and inverter has been developed using microcontroller to experimentally verify the result. Good result was obtained between simulation and experiments.**

IndexTerms—**Matrix Converter, AC-AC Converter, Direct and IndirectAC-ACconverter, Single Phase Matrix Converter (SPMC), MLS (Matlab, Simulink).**

I. INTRODUCTION

Matrix Converter is an advance circuit topology that offers many advantages such as the ability to regenerate energy back to the utility, sinusoidal input and output current and a controllable input current displacement factor. It has potential of affording"all silicon" for AC-AC conversion, removing the need for reactive energy storage components used in converter rectifier-inverter based system.

Actually Matrix Converter is the advanced version of forced commutated cycloconverter.It consists of bidirectional switches that allow any output phase to be connected to any input phase. Developing of Matrix Converter starts with the work of Venturini and Alesina published 1980[1].They presented the power circuit of converter as Matrix of bidirectional power switches and they introduced the name "Matrix Converter".

Obviously lot of research has been done on three phase matrix converter but very few of researchers have done research on single phase matrix converter. The first SPMC was developed by Zuckerberger.

In this paper, simulation result of SPMC as a cycloconverter will be discussed and after that Hardware design will be discussed on passive load.

II. AC-AC CONVERTER [3]

There are two methods of converting fixed AC voltage with fixed frequency to variable AC voltage with variable frequency: (a) Indirect method and (b) Direct method

Indirect AC-AC converter is the most common approach for AC-AC power conversion. As shown in figure 1, these converters consist of a Rectifier at supply side and Inverter at the load side. The distinctive feature of this converter topology is the need of energy storage element in the intermediate D.C link: a capacitor or an inductor. Due to these elements, this converter becomes bulky and also the converter is not usable in application requiring regenerative operation. This disadvantage of Indirect AC-AC converter has been removed by Direct AC-AC Converter. A direct AC-AC converter converts a fixed frequency fixed voltage into a variable frequency variable voltage. The basic AC-AC energy conversion described has three possible operations, namely ;(a)AC controller,(b)Decreased frequency operations and(c)Increased frequency operation.

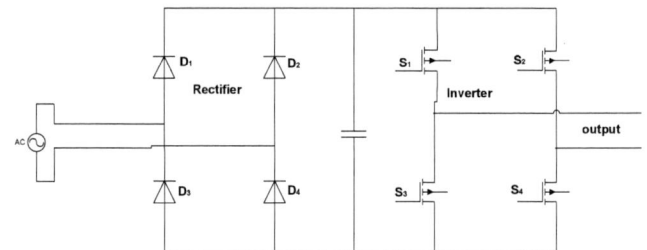

Fig. 1: Classical Rectifier-Inverter AC-AC Converter

III SINGLE –PHASE MATRIX CONVERTER

The matrix converter is a forced commuted converter which uses an array of controlled bidirectional switches as the main power elements to create a variable output voltage system with unrestricted frequency. It does not have any dc-link circuit and does not need large energy storage elements. The key element in a matrix converter is the fully controlled four-quadrant bidirectional switch, which allows high-frequency operation.

The SPMC consists of a matrix of a matrix of input and output lines with four bidirectional switches connecting the single phase input to the single phase output at the intersections. The SPMC and its bi-directional switches is presented in Fig. 2 and Fig.3

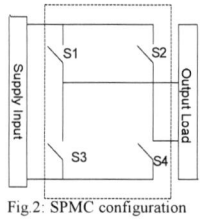
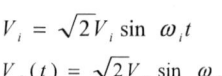

Fig.2: SPMC configuration

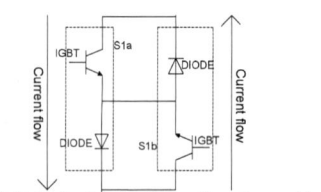

Fig.3: Bi-directional switch configuration module

$$V_i = \sqrt{2} V_i \sin \omega_i t \qquad (1)$$

$$V_o(t) = \sqrt{2} V_o \sin \omega_o t \qquad (2)$$

978-1-4799-6047-7/14 $31.00 © 2014 IEEE

$$V_o(t) = Ri_o(t) + L\frac{i_o(t)}{dt} \qquad (3)$$

Subscript i denote the input, whilst o denotes output.

IV SWITCHING STRATEGIES FOR SPMC AS A CYCLOCONVERTER

SPMC is very flexible topology with high level of integration [2]. That topology is used for different modes depended on switch commutation strategy.

In this paper explained SPMC as a cycloconverter. Control system is designed to generate the sinus pulse wide modulation SPWM patterns that are used to control the power switches. The switching angles, of the 4 bi-directional switches S_{ij} (i=1, 2, 3, 4 and j=a, b). The following rules are applied and illustrated in Figs.4 to 7.

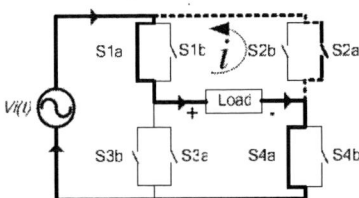

Fig. 4:Positive Cycle(State1)

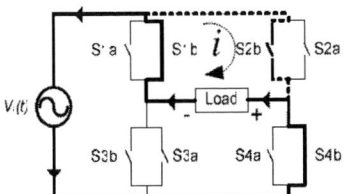

Fig.5:Negative Cycle(State2)

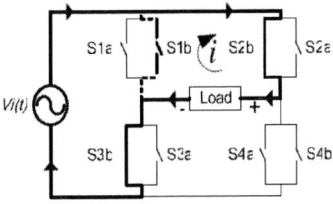

Fig.6:Positive Cycle(State3)

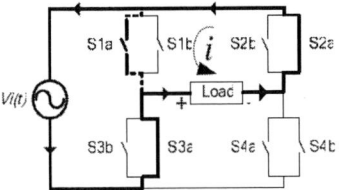

Fig.7:Negative Cycle(State4)

- At any time 't' only two switches S_{ij}(i=1,4 and j=4) will be 'ON' state and conduct the current flow during positive cycle of input source (state1),with S2a turn 'ON' for commutation purpose.
- At any time 't' only two switches S_{ij}(i=1,4 and j=b) will be in 'ON' state and conduct the current flow during negative cycle of input source (state 2),with S2b turn 'ON' for commutation purpose.
- At any time 't' only two switches S_{ij}(i=2,3 and j=b) will be in 'ON' state and conduct the current flow

during positive cycle of input source (state3), with S1b turn 'ON' for commutation purpose.

- At any time 't' only two switches S_{ij}(i=2,3 and j=a) will be in 'ON' state and conduct the current flow during negative cycle of input source (state 4),with S1a turn 'ON' for commutation purpose.

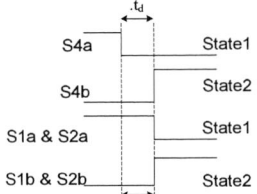

Fig. 8: Sample Timing diagram for commutation strategies

TABLE1
SWITCHING STRATEGY OF SPMC AS AC-AC CONVERTER

Input frequency	Target output freq.	Time Interval	State	PWM Switch	Commutation Switch
50Hz	150Hz	1	1	S4a	S1a&S2a
		2	3	S3b	S2b&S1b
		3	1	S4a	S1a&S2a
		4	2	S4b	S1b&S2b
		5	4	S3a	S2a&S1a
		6	2	S4b	S1b&S2b
	100Hz	1	1	S4a	S1a&S2a
		2	3	S3b	S2b&S1b
		3	4	S3a	S2a&S1a
		4	2	S4b	S1b&S2b
	50Hz	1	1	S4a	S1a&S2a
		2	2	S4b	S1b&S2b
	25Hz	1	1	S4a	S1a&S2a
		2	4	S3a	S2a&S1a
		3	3	S3b	S2b&S1b
		4	2	S4b	S1b&S2b
	12.5Hz	1	1	S4a	S1a&S2a
		2	4	S3a	S2a&S1a
		3	1	S4a	S1a&S2a
		4	4	S3a	S2a&S1a
		5	3	S3b	S2b&S1b
		6	2	S4b	S1b&S2b
		7	3	S3b	S2b&S1b
		8	2	S4b	S1b&S2b

V SWITCHING STRATEGIES OF SPMC AS A INVERTER

This paper also presents an implementation of SPMC as DC-AC inverter.Basic loads represented by RL circuit were used for this implementation. Here are the switching strategies for inverter mode. For inverter operation we have to use only a State1 and State2.

TABLE2
SWITCHING STRATEGY OF SPMC AS AN INVERTER

State	PWM switch	Commutation switch
1	S4a	S1a and S2b
2	S2a	S3a and S4b

VI SIMULATION MODEL

For hardware implementation, first simulation has been done for SPMC by using above switching sequence. The MLS implementation of the SPMC configuration is as shown below Fig.8. Switches used are the bi-directional switches used to block the voltage and conduct the current in both directions.

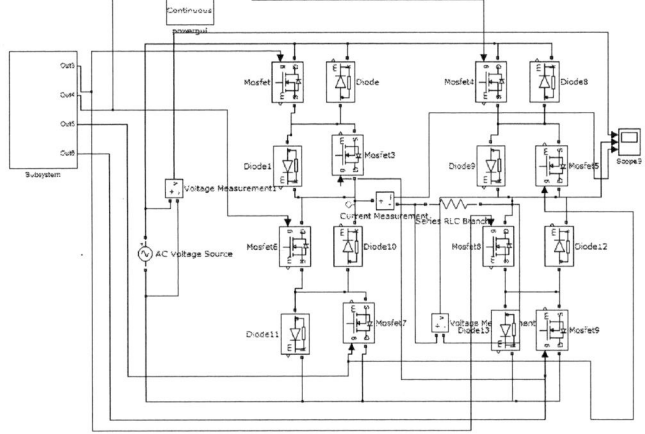

Fig9. Main model of SPMC in MLS

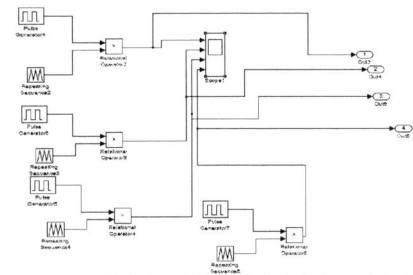

Fig10. Driver circuit

Driver circuits were designed to generate the PWM pattern that is controlled using the switching states as in tables1 for generating 100Hz output frequency.

VII HARDWARE IMPLEMENTATION

To validate the simulation results, prototype of the Single Phase Matrix Converter were built which work as a both i.e. as a cycloconverter(Output frequency of 100Hz) and as an inverter(DC to AC) with passive load.

The overall structure of prototype consists of five parts: Zero crossing detection, control plateform, the gate drives, the power circuit and the protection circuit.

Fig10.shows the block diagram of proposed algorithm for SPMC as a cycloconverter and as an inverter.

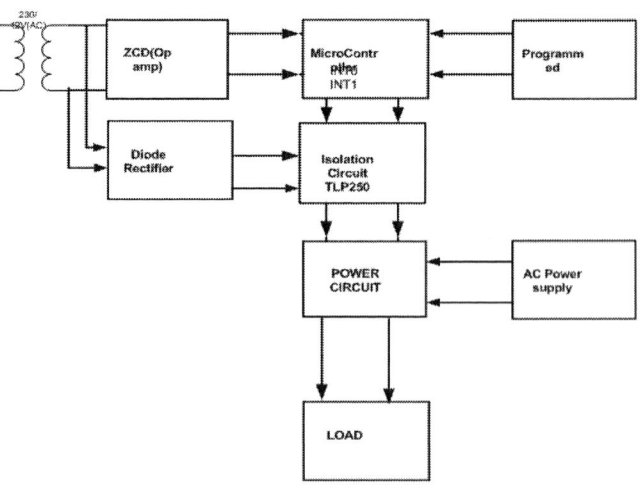

Fig11. Block diagram of SPMC as a cycloconverter and as an Inverter

(a)Zero crossing detection:In this project,Zero crossing detection is used to synchronize the pulses with the input supply.ZCD has been made by using readily available IC741 and some resistance.Inthis project IC741 and two resistor of 1KΩ is used.The output of Zero crossing is fed to the interrupt pin of microcontroller(digital pin2 and 3) which is named as INT0 and INT1.

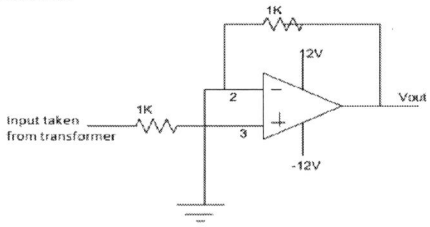

Fig12.Circuit diagram of ZCD

*(b)Control plateform:*The control plateform comprises of Microcontroller Atmega8.This microcontroller is used to generate the pulses which are given to the TLP250(isolation circuit).The output of micrcontroller is perfectly synchronized with the input supply,which is our main requirement when SPMC needs to be operated as a cycloconverter.

This microcontroller detects the rising and falling edge of ZCD output according to the program the is fed to microcontroller.After detcting the rising and falling edge it generates the pulse on digital pin no.12 and 13.The time period of the output of microcontroller is of 10ms because 100Hz output is needed.

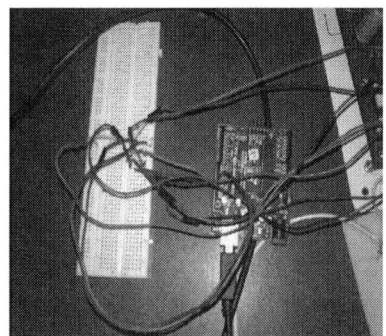

Fig13.Microcontroller Atmega8

*(c)Gate drive circuit:*Gate drive circuit is used to create the isolation between the input supply and pulse which is given to

978-1-4799-6047-7/14 $31.00 © 2014 IEEE   550

MOSFETs.Also it is able to trigger the MOSFETs and IGBTs.In this, isolation circuit is made by using TLP250 and total 8 isolation circuits have been made.The supply for TLP250 is made by using diode rectifier(IN4007) and for filtration purpose capacitor of 470µf is used.

Fig14.Hardware implementation of Isolation circuit

*(d)Power circuit:*SPMC consists of bidirectional switches.In this project there are 4 bidirectional switches is used.So for making bidirectional switches,2MOSFETs or IGBTs and 2 diodes need to be used.In this project,for bidirectional switches 2 MOSFETs and 2 diodes have been used.

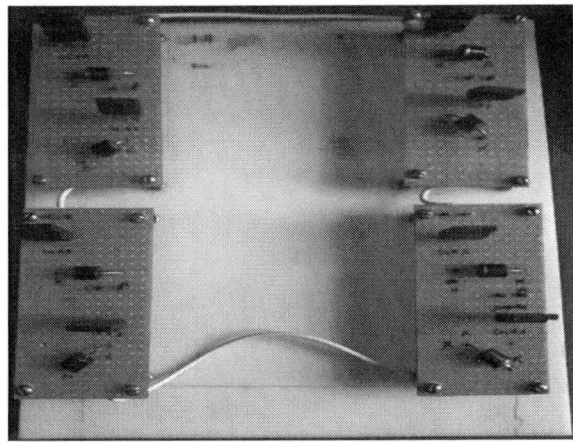

Fig15. Hardware implementation of Power circuit

In this project,PWM technique have been used for simulation purpose but for hardware implementation continous pulse have been used.Fig15. and 16 shows a theoritical waveform by using continous pulse which is given below which is used for hardware implementation.

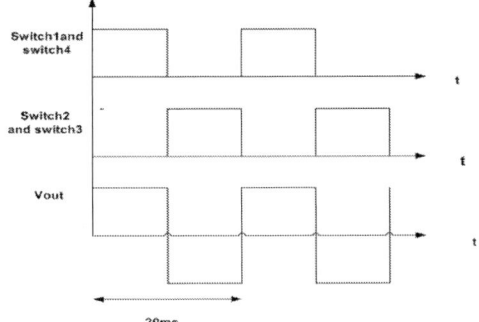

Fig16.Theoritical waveform of SPMC as a Inverter.

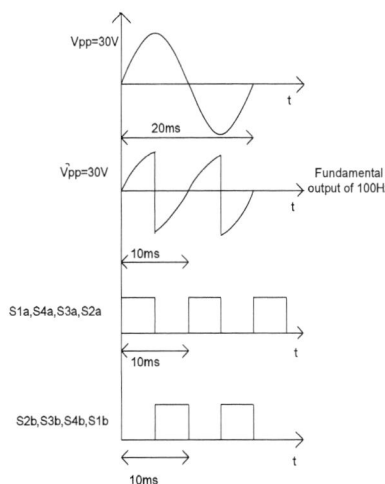

Fig17.Theoritical waveform of SPMC as a Cycloconverter.

VIII SIMULATION AND EXPERIMENTAL RESULT

Comparison of the result obtained from MLS is made with experimental result as shown in Fig.19 to Fig26. The test rig is constructed as shown in Fig16. The following parameters are used for this work which is same for both i.e. for simulation purpose and for hardware implementation purpose:

TABLE3
AS A CYCLOCONVERTER

INPUT SOURCE(AC)	OUTPUT(AC)
V_{pp}=34V	V_{out}=33.2V
V_{rms}=11.8V	V_{rms}=11.2V
F_{in}=50Hz	F_{out}=100Hz

TABLE4
AS AN INVERTER

INPUT SOURCE(DC)	OUTPUT(AC)
V_{input}=30V	V_{out}=30V
R=200Ω	V_{rms}=28.7V,V_{pp}=58.1V,F_{out}=50Hz

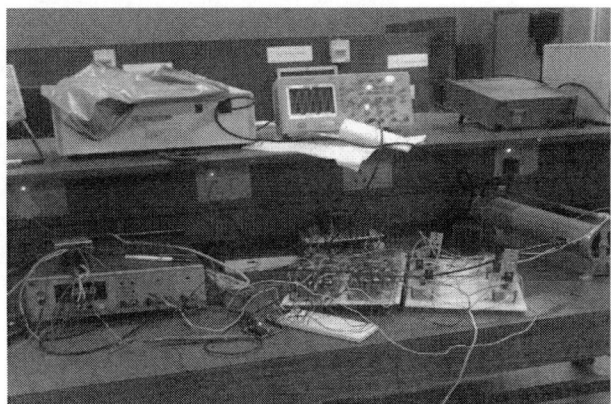

Fig18.Experimental Test Rig of SPMC

978-1-4799-6047-7/14 $31.00 © 2014 IEEE

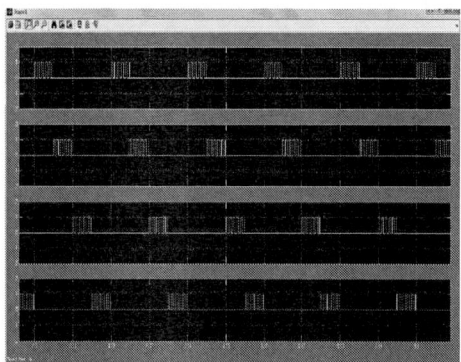

Fig19. Simulation result for switching pattern generator(100HZ)

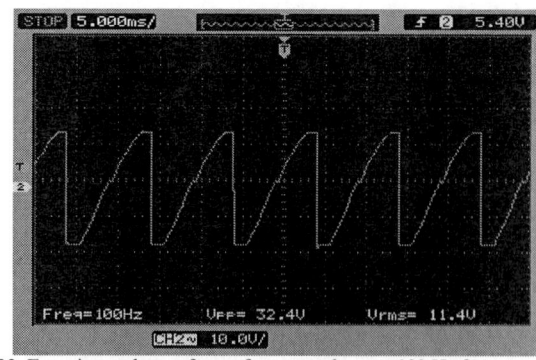

Fig23. Experimental waveform of output voltage at 100 Hz frequency

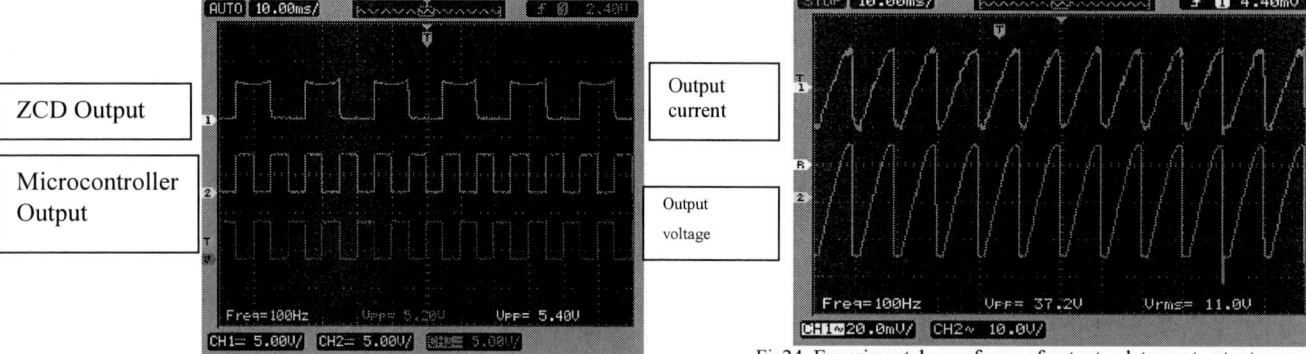

ZCD Output

Microcontroller Output

Fig20.Pulse generation for generating 100Hz output

Output current

Output voltage

Fig24. Experimental waveforms of output volatge wrt output current

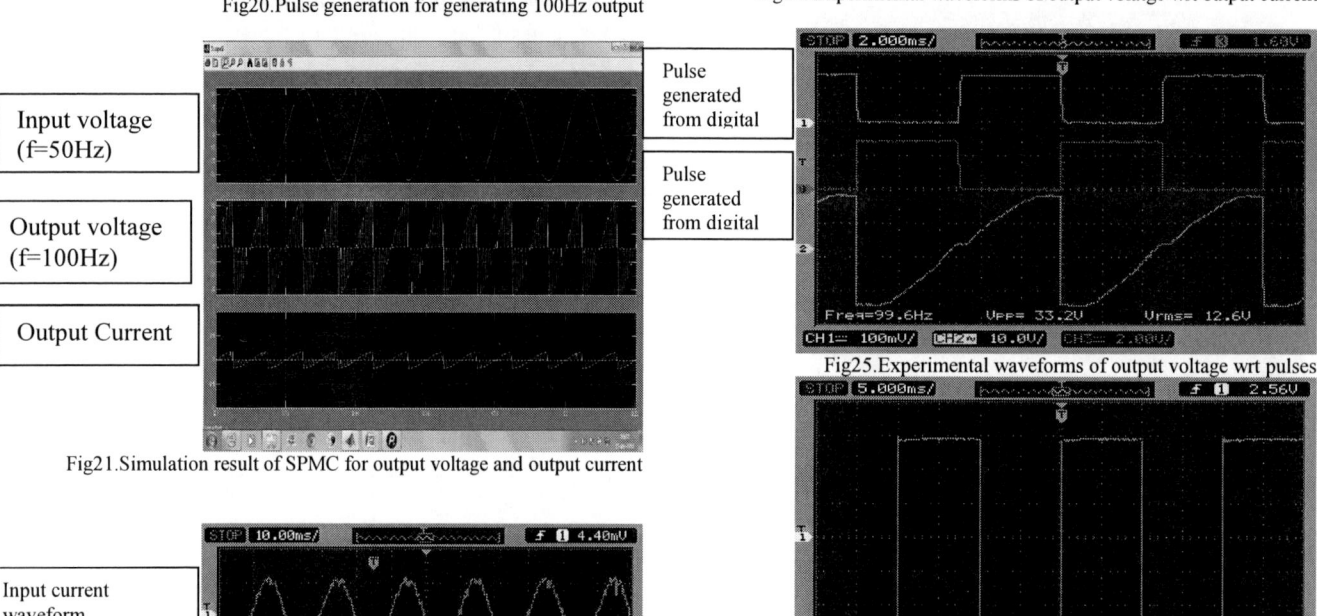

Input voltage (f=50Hz)

Output voltage (f=100Hz)

Output Current

Fig21.Simulation result of SPMC for output voltage and output current

Pulse generated from digital

Pulse generated from digital

Fig25.Experimental waveforms of output voltage wrt pulses

Input current waveform

Output voltage

Fig22. Experimental waveforms of output voltage wrt input current current

Fig26.Experimental waveforms of output voltage of SPMC as an Inverter

IX CONCLUSION

In this work simulation and hardware design of Single Phase Matrix Converter has been done. The simulation model has been run under R-load conditions. Under this loading conditions, the converter's performance has been tested for an output frequency of 50Hz and 100Hz using microcontroller Atmega8.Also converter has been implemented as an Inverter for continuous pulse which is generated by microcontroller Atmega8.The obtained output voltage and current for different load has been observed in digital storage oscilloscope (DSO of 200MHz) and current sensor probe.

One Matrix Converter can be used for different power circuit for various converters like Rectifier, Inverter, Cycloconverter and AC voltage controller etc .A Matrix Converter is small, takes less space and become flexible.

X REFERENCES

[1] H.M. Hanafi, Z.Idris, M.K Hamzah,and A,Saparon, "Moddeling & simulation of single phase matrix converter as a freuency changer with sinusoidal pulse width modulation using MATLAB/simulink," in *First International Power and Energy Conference, (PECon) Proceedings,* 2006, pp.482-487.

[2] A. Zuckerberger, D. Weinstock, and A. Aiexandrovitz, "Single-phase matrix converter," in *IEE Proceedings:Electric Power Applications, 1997,* vol.144, no.4, pp.235-240.

[3] Zahiruddin Idris, Mustafar Kamal Hamzah & Ahmad Maliki Omar "Implementation of Single Phase Matrix Converter as a Direct AC-AC converter Synthesized Using Sinusoidal Pulse Width Modulation with Passive Load Condition, "IEEE Sixth International Conference PEDS 2005, Kuala Lumpur, Malaysia.

[4] Wheeler, P.W., Clare, J.C., Empringham, L., Bland, M., Kerris, K.G., "Matrix converter," IEEE Industry Applications Magazine, Vol. 10(1), Jan-Feb2004, pp.59-65.

[5] Wheeler, P.W., Rodriguez, J., Clare, J.C., Empringham, L., Weinstein, A., "Matrix converters:a technology review." ,IEEE Transaction on Industrial Electronics, Vol. 49(2), April 2002,pp. 276-288.

[6] H.M. Hanafi, *et al.,* "Improved switching strategy of single-phase matrix converter as a direct AC-AC converter," in *Industrial Electronics and Applications, 2008. ICIEA 2008 3rd IEEE Conference on,* 2008, pp. 1157-1162.

[7] Y. Jang-Hyoun and K. Bomg-Hwan, "Switching technique for curent-controlled AC-AC converter," *Industrial Electronics, IEEE Transactions on,* vol, 46, pp. 309-318, 1999.

[8] P. Deivasundari and V, Jamuna, "Single phase matrix converter as an all silicon solution," *Journal of Electrical Engineering,* Vol. 11, no. 3, 2011.

[9] Zahiruddin Idris, Siti Zaliha Mohammad Noor & Mustafar Kamal Hamzah, "Safe Commutation Strategy in Single –phase matrix converter," IEEE Sixth International Conference PEDS 2005, Kuala Lumpur, Malaysia.

[10] Abdollah Koei & Subbaraya Yuvarajan, "Single-phase matrix converter Using Power Mosfets," IEEE Transaction on Industrial Electronics, Vol. 35, No.3, August 1998, pp, 442-443.

[11] Kwon, B.h., Min, B.D.,Kim, J,-H., "Novel Commutation Technique of AC-AC Converters," Electric Power Application, IEE Proceedings-, Vol. 145(4), July 1998, pp,295-300.

[12] S.H. Hosseini and E. Babaei, "A new genralized direct matrix converter" in *Industrial Electronics 2001 Proceedings ISIE 2001, IEEE International Symposium on,* 2001, pp. 1071-1076 vol.2.

978-1-4799-6047-7/14 $31.00 © 2014 IEEE

A ZVT-PWM Multiphase Synchronous Buck Converter with an Active Auxiliary Circuit for Portable Applications

S.Shiva Kumar, A. K. Panda, *senior member, IEEE,* Tejavathu Ramesh
Department of Electrical Engineering, NIT Rourkela, India
Shivkumar.ee@gmail.com, akpanda.ee@gmail.com

Abstract—In this paper, a Zero-Voltage-Transition (ZVT) Pulse-width Modulated (PWM) multiphase synchronous buck converter (SBC), with an active auxiliary circuit is proposed, that reduces the stresses and enhances the efficiency abating the switching and conduction losses of the converter. The important design feature of ZVT-PWM multiphase SBC converters is placement of resonant components that pacifies the switching and conduction losses. Due to the ZVT, the resonant components with low values are used that results in the increase of switching frequency. High current multiphase buck converters found applications in advanced data control, portable applications and other applications like computer processors. The zero-voltage-transition operation of the proposed converter is presented through theoretical analysis. The characteristics of the proposed converter are verified with the simulation in the PSIM cosimulated with MATLAB/SIMULINK environment.

Index Terms—Zero Voltage Transition (ZVT), multiphase synchronous buck converter, pulse width modulation (PWM).

I. INTRODUCTION

In the recent times, the ZVT technique applied to SBC facilitates reduction of switching losses, while maintaining voltage and current stresses within the tolerable limit. The ZVT concept extending to multiphase SBC has emerged as a leading candidate for meeting the power requirement of the portable electronic systems. High current multiphase synchronous buck converters (MSBC) are used in portable applications and other applications like computer processors. To achieve high power-density converters and high-performance, the converters switching frequency need to be raised. The traditional hardswitching pulse-width-modulation (PWM) converters suffers from the limitation of high switching loss if the converter operates at high frequency. Numerous softswitching methods were proposed to curtail the switching losses by sacrificing the increase in voltage and current stresses of the devices used in the converters which inturn raise the conduction losses. A novel family of zero-voltage transition PWM converters were proposed in [1] and they are vastly utilized in industrial applications[2].The concept of ZVT was also spread to full-bridge PWM converters [3], [4]. In these converters, the zero voltage switching condition which is bestowed by auxiliary circuit accommodates least voltage and current stresses for wide line and load ranges.

By adopting multiphase conversion method high performance and high power density can be achieved [5]-[8]. Dynamic performance is high for the converters whose operating frequency of output and input filter capacitors is raised by n-times for n-phase converters. The size of inductors is reduced because of the interleaved operation which keeps the current ripple of the output and input capacitor filters low. By integrating the ZVT with multiphase conversion method higher dynamic response and power density is achieved.

Smaller duty cycle increases the switching frequency up to a level of multi-MHz that will deflate the efficiency of the converters. As the switching frequency becomes equal to the inductor filter current ripple frequency, there is a limitation for the switching frequency which varies from 300 kHz to 500 kHz [9-14]. The slew rate of inductor current raises with a lesser inductor value that gives better transient response and also raises the inductor filter current ripple. The turn-off loss effects the high-side switch, but the low side switch also gets affected due to higher conduction losses that includes the inductor winding losses which has a limitation on the average inductor current of each phase [15-20]. The efficiency and transient response should have a tradeoff between them which is a technical problem that raises the cost and forfeits the power density, as there is a difficulty in meeting the requirements of the power for the future generation microprocessors for a satisfactory solution [21-23]. Therefore there is a need to increase the efficiency of the multiphase buck converter at a high operating frequency by reducing switching losses. In ZVT converters [24]-[28] generally the auxiliary switch actuates just before the main switch is made active and culminates after it is executed. The standard topology for a low voltage and high current can be considered as multiphase buck converter.

In this paper, the ZVT multiphase synchronous buck converter is presented with the directive to improve its performance. The proposed converter depreciates the current and voltage stresses of switches to enhance the efficiency by lowering the losses with the help of a an active auxiliary circuit. Here the proposed multiphase is associated with active auxiliary circuit rather than passive auxiliary circuit because at high load current passive auxiliary circuit will give high conduction losses. Section 1 presents a description of the proposed topology. Principles of operations and its analysis are discussed in section 2. Section 3 provides the design process of the multiphase converter. In section 4, the theoretical analysis and the operating principle of the converter are verified by simulation results.

978-1-4799-6047-7/14 $31.00 © 2014 IEEE

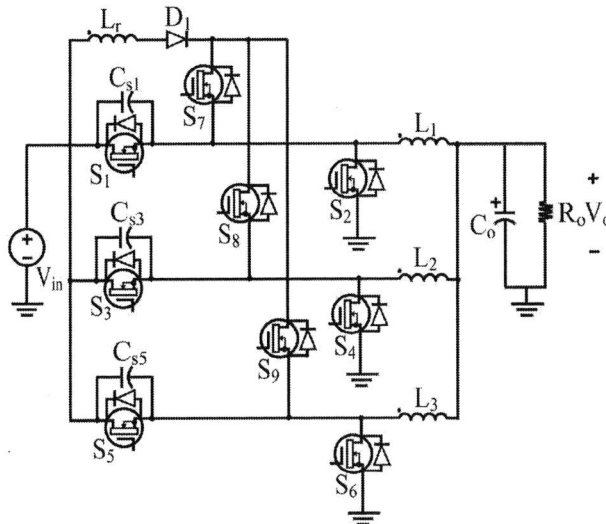

Fig.1. Proposed ZVT Multiphase synchronous buck converter.

II. OPERATING PRINCIPLE

A. Configuration of proposed circuit and conditions that are assumed to simplify the analysis.

The proposed multiphase converter is shown by Fig. 1. It is a combination of the proposed converter along with active auxiliary circuit that facilitates reduction of switching losses. The auxiliary circuit consists of inductor L_r, diode D_1, and MOSFET switches S_7, S_8, and S_9. The number of auxiliary MOSFET switches depends on the number of phases. Body diodes of main switches S_1, S_2, and S_3 are utilized to provide zero voltage switching.

The steady state operation of the proposed circuit is analysed by making some assumptions for one switching cycle.

1. The input voltage V_{in} is constant.

2. The output current I_o is constant or the output inductor L_o is large enough.

3. The output current I_o is constant or the output inductor L_o is large enough.

4. The output Inductors L_1, L_2, L_3 is much larger than the resonant circuit inductor L_r.

5. The resonant circuits are ideal.

6. The reverse recovery time of diode is ignored.

B. Modes of operation

Based on these assumptions, circuit operations in one switching cycle can be divided into fifteen stages. The key waveforms of these stages are illustrated in Fig.2 and the equivalent circuit schemes of the operation stages are given in Fig.3. The detailed analysis of every stage is presented below:

Mode1 (t_0-t_1): Prior to t = t_0, the body diode of switch S_4 was conducting while the main switch S_1 is off. The equations are

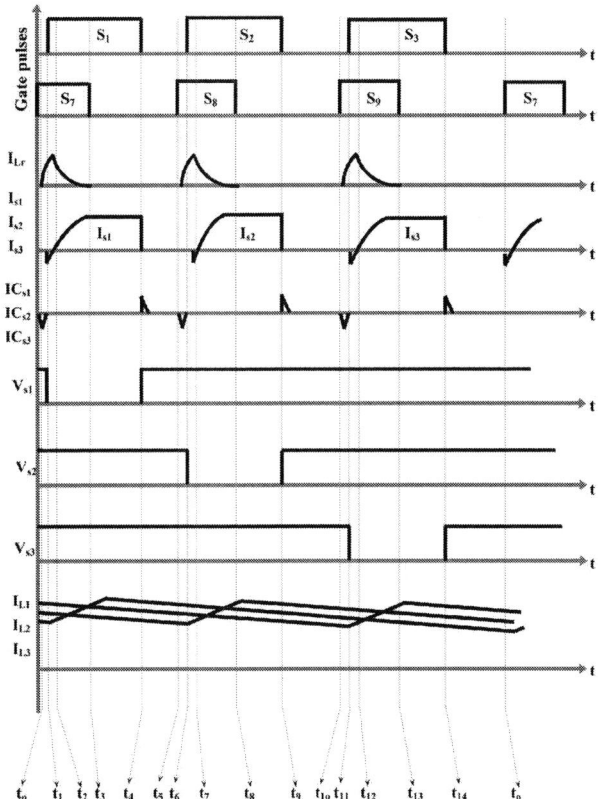

Fig.2. Key waveforms of the proposed ZVT multiphase synchronous buck converter.

$i_S = 0$, $i_{D4} = I_0/3$, $i_{Lr} = 0$, are valid at the beginning of this stage.

At t = t_0, the auxiliary switch S_7 is turned on, which realizes zero-current turn-on as it is in series with the resonant inductor L_r. During this stage, i_{Lr} rises and current i_{Ds4} through body diode of switch S_1 falls simultaneously at the same rate. The resonance occurs between L_r and C_{s1}.

This mode ends at t = t_1, when i_{Lr} reaches $I_0/3$, and i_{D4} becomes zero. The body diode of switch S_4 is turned off with ZCS. The resonant current through inductor Lr is given by

$$i_{resonant}(t) = \frac{C_{S1}V_{CS1}\omega Sin\omega t}{Z_1} \tag{1}$$

$$i_{Lr}(t) = \frac{V_i}{L_r} \times t \tag{2}$$

Where $\omega = \dfrac{1}{\sqrt{L_r C_{S1}}}$, $Z_1 = \sqrt{\dfrac{L_r}{C_{S1}}}$

At $t = t_1$, $i_{Lr}(t) = \dfrac{I_o}{3}$,

Therefore, $t_{01} = t_1 - t_0 = \dfrac{I_o L_r}{3V_{in}}$.

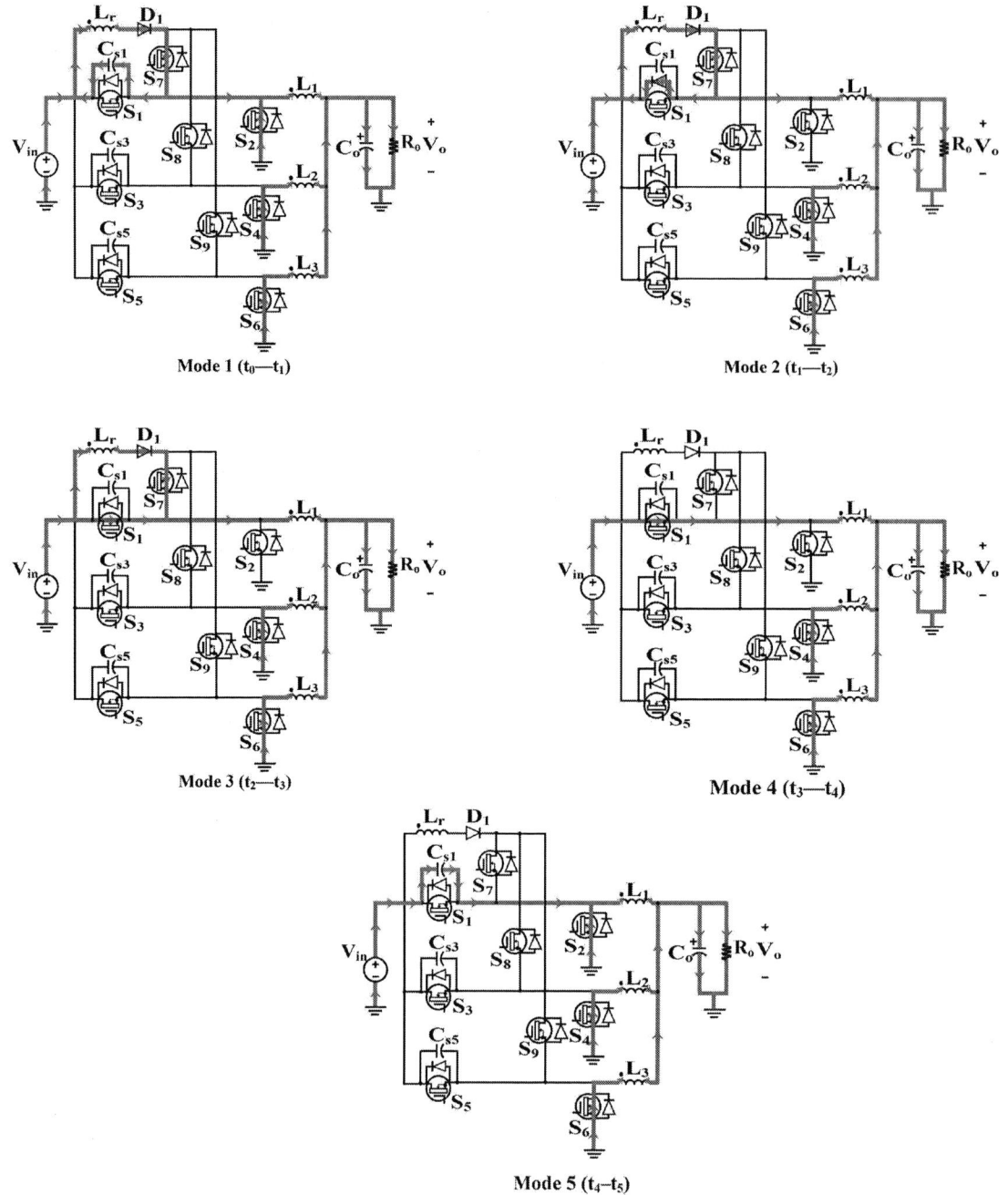

Fig.3. Operational modes of the proposed ZVT multiphase synchronous buck converter

Mode 2(t_1-t_2): Since the inductor current i_{Lr} is increasing continuously beyond one third of load current, the exceeding current makes the diode D_{S1} to conduct. At t = t_1, i_{S7} = i_{Lr} = $I_0/3$. After reaching the peak current I_{Lrmax}, the inductor current starts decreasing. This mode comes to an end when i_{Lr} becomes again equal to $I_0/3$. At this moment, the main switch is triggered to turn on under zero voltage switching (ZVS). The discharge current of capacitor C_{S1} through the body diode having a resistance R is given as:

$$i_{C_{s1}} = \frac{V_{Cs1}}{R} e^{-t/RC_{S1}} \tag{3}$$

$$i_{Lr}(t) = \left(i_{Lr\,max} - \frac{I_o}{3} \right) e^{\frac{-tL_r}{R}} \tag{4}$$

At $t = t_1$, $i_{Lr}(t) = I_o/3$,

Therefore, $t_{12} = t_2 - t_1 = \dfrac{2R}{L_r} \ln \dfrac{3I_{Lr\,max}}{I_0}$.

Mode 3(t_2-t_3): At t = t_2, the main switch is turned on while the auxiliary switch is still in on state. Now the stored energy in inductor L_r will be transferred to the load at the same rate as the current increase through the main switch S_1. At t = t_2, i_{Lr} = $I_0/3$. This mode comes to end when the total energy of the resonant inductor will be transferred to the load. The auxiliary switch S_7 will turn off under ZCS.

The inductor current i_{Lr} during this mode can be expressed as

$$i_{Lr}(t) = I_{Lr\max} e^{-tLr/R_{on}} \qquad (5)$$

$$i_{S1} + i_{Lr} = I_0/3 \qquad (6)$$

At the end of this mode, at t = t_3.

$$i_{S1} = I_o/3 \qquad (7)$$

$$i_{Lr} = 0 \qquad (8)$$

Therefore, $t_{23} = t_3 - t_2 = \dfrac{2R_{on}}{L_r} \ln \dfrac{3I_{Lr\max}}{I_0}$.

Mode 4(t_3-t_4): In this mode, the converter behaves as a conventional PWM converter. For the required output voltage, the turn on period of the main switch is decided. At the end of this mode, the main switch S_1 is turned off under ZVS due to the existent of capacitor C_{S1} across it. The current expression for this mode can be expressed as:

$$i_{S1} = I_0/3 \qquad (9)$$

Mode 5(t_4-t_5): At t = t_4, the synchronous switch is turned on to provide a constant load current. At the end of this mode, the complete operation for one phase converter is completed and the second auxiliary switch S_8 is turned on with a phase difference of 360/n, where n is the number of phases, here n =3. The same five modes will be repeated for each phase. So there are fifteen modes for this proposed multiphase converter. The current expression for this mode can be expressed as:

$$i_{S2} = I_o/3 \qquad (10)$$

$$i_{Cs1} = \frac{V_{Cs1}}{R} e^{-t/RC_{S1}}$$

At the end of this mode, at t = t_5, $i_{Cs1} = 0$

Therefore, $t_{45} = t_5 - t_4 = RC_{s1} \ln \dfrac{V_{Cs1}}{RC_{s1}}$.

C. Output voltage

The output voltage can be evaluated by balancing the volt-second relationship or by equating the energy relation i.e.,

$$V_o \tau = 3V_{in}[t_{01} + t_{12} + t_{23} + t_{34} + t_{45}]$$

$$V_o \tau = 3V_{in}[\frac{I_0 L_r}{3V_{in}} + \frac{2R}{L_r} \ln \frac{3I_{Lr\max}}{I_0} + \frac{2R_{on}}{L_r} \ln \frac{3I_{Lr\max}}{I_0} + RC_{S1} \ln \frac{V_{CS1}}{RC_{S1}}] \quad (11)$$

Where $\tau = \dfrac{1}{f_s}$ and f_s =Switching frequency.

From the above expression, it is noticeable that voltage conversion ratio depends upon switching frequency irrespective of the duty ratio.

III. DESIGN PROCEDURE OF AUXILIARY CIRCUIT COMPONENTS

The design procedure of auxiliary component is as follows:

1. Resonant capacitor C_{S1} is selected to discharge from V_{in} to zero with the maximum output current over at least the time period t_{on} during the turn on of body diode. For this state, according to equations (3)

$$RC_{S1} \ln \frac{RI_0}{V_{CS1}} \geq t_{on} \qquad (12)$$

2. Resonant inductor L_r is selected such that current through inductor can be reduced to zero from $I_0/3$ in the same duration of rise in current form zero to $I_0/3$ in main switch. In this case, from equation (5).

$$t_{23} = \frac{R_{on}}{L_r} \qquad (13)$$

IV. CONVERTER POSTULATES

The postulates of the soft switching converter which is proposed are listed as following.

1. Constant output and high frequency of operation (1 MHZ).
2. Equal current sharing is possible even at very high frequencies.
3. Voltage stress on the main and synchronous switch is less.
4. Very simple structure, low cost and ease of control.
5. High efficiency when compared without ZVS of buck converter.

V. SIMULATION RESULTS

The converter is simulated using simulation software PSIM version 7.1 cosimulated with MATLAB/Simulink. The proposed converter works with an input voltage of V_{in}=12V and an ouput voltage of V_o=1V, a load current of 70A and a switching frequency of f_s =500kHz. The proposed converter is validated with the simulation results. Fig. 4(a)—(e) shows the simulation results of the proposed converter.

A. Efficiency Curve

From Fig. 5, it can be seen that efficiency values of the proposed converter are comparatively higher than the traditional converter. The converter is designed for the maximum output current, and it is accustomed that towards minimum output power efficiency, values decrease. At nearly 70A of output current, the efficiency of the proposed converter rises to about 96% when compared to the counterpart traditional converter whose efficiency is about 87A. The high efficiency of the proposed converter proves the definiteness of the design values.

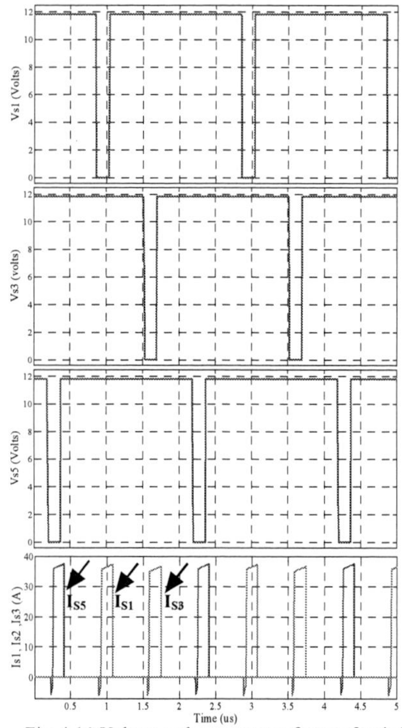

Fig. 4 (a) Voltage and current waveforms of switch S_1, S_3, S_5: V_{s1}, V_{s3}, V_{s5} and I_{s1}, I_{s3}, I_{s5}.

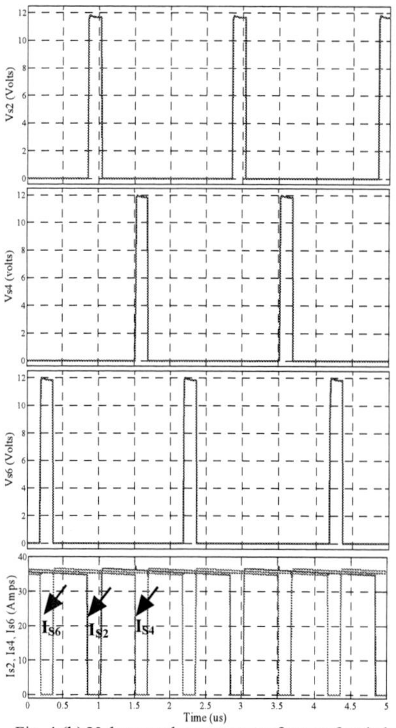

Fig. 4 (b) Voltage and current waveforms of switch S_2, S_4, S_6: V_{s2}, V_{s4}, V_{s6} and I_{s2}, I_{s4}, I_{s6}.

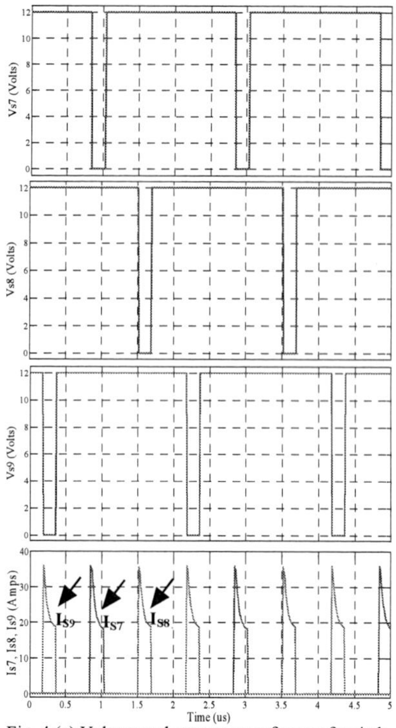

Fig. 4 (c) Voltage and current waveforms of switch S_7, S_8, S_9: V_{s7}, V_{s8}, V_{s9} and I_{s7}, I_{s8}, I_{s9}.

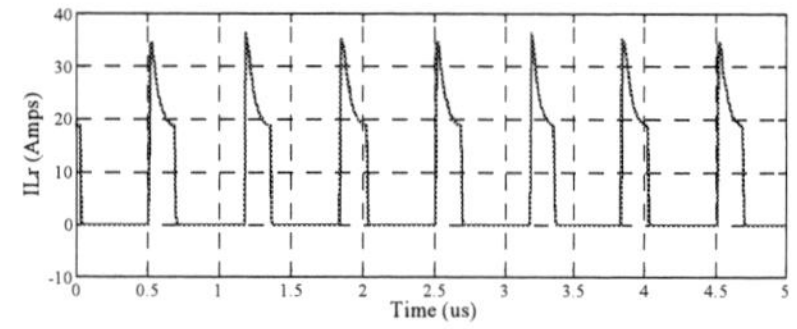

Fig. 4 (d) Current waveform of inductor L_r: I_{Lr}

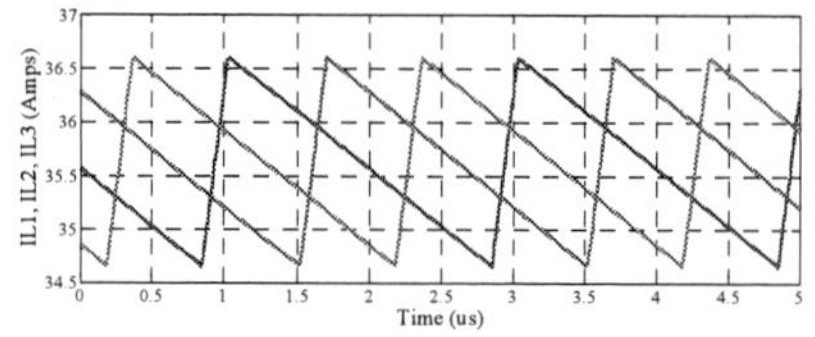

Fig. 4 (e) Current waveforms of inductors L_1, L_2, L_3: I_{L1}, I_{L2}, I_{L3}.

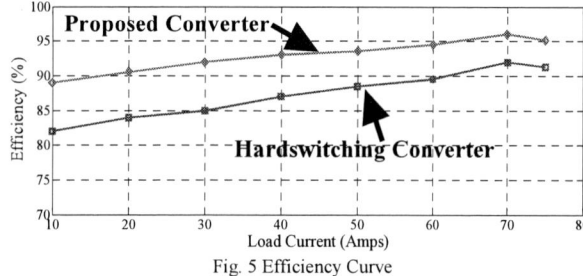

Fig. 5 Efficiency Curve

synchronous buck is highly efficient than the conventional converter. This proposed converter of high switching frequency is designed for application in new generation microprocessor.

VI. CONCLUSION

In this paper, the concept of ZVT is implemented in multiphase synchronous buck converter and it is shown that the switching losses in synchronous buck are eliminated. Significant efficiency improvement with soft switching as compared to hard switching converter is achieved as shown in the Fig. . Both main switch and synchronous switches are turned-on and turned-off under ZCS and ZVS respectively. But auxiliary switches are turned-on and turned-off under ZCS with tolerable voltage stresses across the switch. Hence switching losses are reduced and the proposed multiphase

REFERENCES

[1] G. Hua, C. S. Leu, and F. C. Lee, "Novel zero voltage transition PWM converters," in *Proc. IEEE PESC Rec. 1992*, pp. 55–61.

[2] G. Hua, W. A. Tabisz, C. S. Leu, N. Dai, R. Watson, and F. C. Lee, "Development of dc distributed power system components," in *VPEC Annu. Seminar, 1993 Proc.*, pp. 87–96.

[3] J. G. Cho and G. H. Cho, "Novel off-line zero-voltage switching PWM ac/dc converter for direct conversion from ac line to 48 VDC bus with power factor correction," in *Proc. IEEE PESC Rec. 1993*, pp. 689–695.

[4] J. G. Cho, J. Sabat´e, and F. C. Lee, "Novel zero-voltage-transition PWM dc/dc converter for high power applications," in *Proc. IEEE APEC Rec. 1994*, pp. 143–149.

[5] D. M Sable, F. C. Lee, and B. H. Cho, "A zero-voltage-switching bidirectional battery charger & discharger for the NASA EOS satellite," in *VPEC Annu. Seminar, 1992 Proc.*, pp. 41–46.

[6] J. P. Noon, B. H. Cho, and F. C. Lee, "Design of multi-module multiphase battery charger for the NASA EOS space platform test bed," in *VPEC Annu. Seminar, 1992 Proc.*, pp. 137–142.

[7] C. Hua, W. A. Tabisz, C. S. Leu, N. Dai, R. Watson, and F. C. Lee, "Development of DC distributed power system components," in *VPEC Annu. Seminar, 1992 Proc.*, pp. 137–142.

[8] B. D. Bedford and R. G. Hoft, *Principles of Inverter Circuits*. New York: Wiley, 1964.

[9] Panov Y., and Jovanovic M. M., "Design considerations for 12-V/1.5-V, 50-A voltage regulator modules," in *IEEE Tran. on Power Electron.*, Nov 2001 vol.16(6), pp. 776–783.

[10] Zhou X. W., Wog P., Lee F., "Investigation of candidate VRM topologies for future microprocessors," in *IEEE Tran. on Power Electron.*, 2000. pp. 1172-1182.

[11] T. Hegarty, "Benefits of multi-phasing buck converters," National Semiconductors.

[12] Zhou X., Xu P., and Lee F. C., "A high power density, high frequency and fast transient voltage regulator module with a novel current sensing and current sharing technique," in *Proceedings of IEEE APEC* 1999, pp. 289–294.

[13] Huang W., Schuellein G., and Clavette D., "A Scalable Multiphase Buck Converter with Average Current Share Bus," in *Proceedings of IEEE APEC* 2003, pp. 438-443.

[14] Zhou X. W., Xu P., Lee F., "A novel current sharing control technique for low voltage high current voltage regulator module applications," in *IEEE Tran. on Power Electron.*, 2000, pp. 1153-1162.

[15] Eirea G., and Sanders S. R., "Phase Current Unbalance Estimation in Multiphase Buck Converters," in *IEEE Tran. on Power Electron.*, January 2008, vol. 23(1), pp. 137-143.

[16] García O., Zumel P., Castro A., Alou P., and Cobos J.A., "Current Self-balance Mechanism in Multiphase Buck Converter", Application note,

[17] Gu W., Qiu W., Wu W., and Batarseh I., "A Multiphase DC/DC Converter with Hysteretic Voltage Control and Current Sharing", in *Proceedings of IEEE APEC* 2002, pp. 670-674.

[18] Jakobsen L. T., Garcia O., Oliver J. A., Alou P., Cobos J. A. and Andersen M. A. E., "Interleaved Buck Converter with Variable Number of Active Phases and a Predictive Current Sharing Scheme", Application note,

[19] Costabeber, A., Mattavelli, P., and Saggini, S., "Digital Time-Optimal Phase Shedding in Multiphase Buck Converters," in *IEEE Tran. on Power Electron.*, vol. 25(9), pp. 2242-2247.

[20] Nagaraja H.N., Patra A., and Kastha D., "Design and analysis of four-phase synchronous buck converter for VRM applications," in *Proceedings of the IEEE INDICON* 2004.

[21] Cho J.-G., Baek J.-W., Rim G.-H., and Kang I.; "Novel Zero-voltage-transition PWM Multiphase Converters," *in IEEE Tran. on Power Electron.*, 1998, vol. 13(1), pp. 152-159.

[22] Qiu Y., "High-Frequency Modeling and Analyses for Buck and Multiphase Buck Converters," Ph. D. Dissertation, 2005, Blacksburg, Virginia.

[23] E. Adib, H. Farzanehfard, "Zero-voltage transition current-fed full-bridge PWM converter," *IEEE Trans. Power Electron.*, vol. 24, no. 4, pp. 1041-1047, April. 2009.

[24] H. L. Do, "A soft –switching DC/DC converter with high voltage gain," *IEEE Trans. Power Electron.*, vol. 25, no. 5, pp. 1193-1200, May. 2010.

[25] S. Pattnaik, A.K. Panda, K. K. Mahapatra, "Efficiency improvement of synchronous buck converter by passive auxiliary circuit," *IEEE Trans. Industrial Appl.*, vol. 46, no. 6, pp. 2511 – 2517, Nov/Dec. 2010.

[26] H. Bodur, S. Cetin, G. Yanik, "A new zero-voltage transition pulse width modulated boost converter," *IET Power Electron.*, vol. 4, no. 7, pp. 827–834, march. 2011.

[27] S. Urgun, "Zero-voltage transition–zero-current transition pulse width modulation DC–DC buck converter with zero-voltage switching–zero-current switching auxiliary circuit," *IET Power Electron.*, vol. 5, no. 5, pp. 627–634, march. 2012.

[28] Hong-Tzer Yang, Jian-Tang Liao, Xiang-Yu Cheng, "Zero-Voltage-Transition auxiliary circuit with dual resonant tank for DC–DC converters with synchronous rectification," *IET power Electron.*, vol. 6, no. 6, pp. 1157-1164, July. 2013.

Flux Weakening Control Algorithm with MTPA Control of PMSM Drive

Sukanta Halder*, S.P.Srivastava, Pramod Agarwal
Department of Electrical Engineering,
Indian Institute of Technology Roorkee,
Roorkee, India-247667
Email*: sukanta.raj@gmail.com

Abstract— This paper presents flux weakening control algorithm with maximum torque per ampere (MTPA) control for high speed operation of PMSM drive. IPMSM has been used due to the robust construction and extra reluctance torque production as compared to the surface permanent magnet synchronous motor. Based on the speed command the mode of operation will automatically change from constant torque region to flux weakening region. Flux weakening control algorithm has been developed for the purpose of high speed operation. This algorithm has been developed by considering maximum torque per ampere without exceeding the voltage limit and current limit. The performance study has been carried out in terms of MTPA control as well as flux weakening control of IPMSM drive.

Keywords—PMSM; Flux weakening; MTPA; IPMSM

I. INTRODUCTION

The application demand of electric motors is increasing rapidly with increasing technological advancement. According to the technological progress adjustable speed drives are preferred over constant speed drives because of several reasons energy saving, velocity or position control for good transient response etc. Permanent Magnet Synchronous Motors (PMSM) are widely used with current-controlled voltage source inverters for industrial and traction applications, because of their high power density, relatively small rotor inertia and high efficiency. In industrial applications, especially servo drives a constant torque operation is desired, while in case of traction applications, both constant torque and constant power operations are necessary. With the help of new permanent magnet material, modern PMSM has been developed in the direction of high power and wide speed range. Constant torque and power operation can be achieved using flux weakening (FW) control [1-13]. In PMSM, the magnets can be placed on the rotor in two different ways. Depending on the placement they are called as surface permanent magnet synchronous motor (SPMSM) and interior permanent magnet synchronous motor (IPMSM). IPMSM is suitable for the wide range of operation for traction application due its advantageous features like robust construction and extra reluctance torque production [1], [2].

The mathematical modelling is the important part of the complete drive. The Vector control is normally used in ac machines to get performance similar like separately excited dc machines which have highly desirable control characteristics. The application of vector control in PMSM and the complete mathematical modeling takes an important role in PMSM drive system [3], [13].The torque and stator flux linkage are directly

controlled for the purpose of wide speed range operation of interior permanent magnet synchronous motor drives [6]. The voltage-constraint tracking (VCT) field-weakening control scheme for IPMSM drives has been presented to achieve high efficiency [8].

In this paper flux weakening control algorithm with maximum torque per ampere (MTPA) control for high speed operation of IPMSM drive is proposed. Based on the speed command the mode of operation will automatically change from constant torque region to flux weakening region. Flux weakening control algorithm has been developed for the purpose of high speed operation. This proposed algorithm has been developed by considering maximum torque per ampere without exceeding the voltage limit and current limit. So for getting wide speed range operation we need to run the motor in constant torque region as well as constant power region. Up to base speed motor will run in constant torque region above base speed motor will run in flux weakening region. For constant torque region operation we consider MTPA control and above base speed operation the control is transferred to flux weakening controller. For constant torque region operation PI controller will generate the torque value on the basis of speed error. Based on the torque value q-axis and d-axis current is generated from the MTPA lookup table. Above base speed operation the torque command is generated according to the flux weakening algorithm. The q-axis and d-axis current is generated Based on this torque command. Hysteresis current controller has been used to produce gate pulse of voltage source inverter. The organization of remaining paper as follows. In Section II, the mathematical model and operation constraints of an IPMSM will be discussed. Then, the explanation of the proposed flux weakening algorithm is presented in Section III. Finally, conclusions are presented in the last section.

II. MATHEMATICAL MODELING

The stator voltage equations are:

$$V_{qs}^r = r_q i_{qs}^r + p\lambda_{qs}^r + \omega_r \lambda_{ds}^r \qquad (1)$$

$$V_{ds}^r = r_d i_{ds}^r + p\lambda_{ds}^r - \omega_r \lambda_{qs}^r \qquad (2)$$

$$V_{0s}^r = r_s i_{0s}^r + p\lambda_{0s}^r \qquad (3)$$

978-1-4799-6047-7/14 $31.00 © 2014 IEEE

Where r_q and r_d are the quadrature and direct axis winding resistance. Which is equal and it can be termed as r_s. V_{qs}^r and V_{ds}^r are the q and d-axis voltage in the rotor reference frame. i_{qs}^r and i_{ds}^r are the q and d-axis current in the rotor reference frame. The q and d axis stator flux linkage in rotor reference frame are

$$\lambda_{qs}^r = L_q i_{qs}^r \tag{4}$$

$$\lambda_{ds}^r = L_d i_{ds}^r + \lambda_m^r \tag{5}$$

Substituting the equation (4) & (5) in equation (1) & (2) and $p\lambda_m^r = 0$ (Considering magnet flux linkage remains constant) we can write,

$$V_{qs}^r = (r_s + pL_q)i_{qs}^r + \omega_r L_d i_{ds}^r + \omega_r \lambda_m^r \tag{6}$$

$$V_{ds}^r = (r_s + pL_d)i_{ds}^r - \omega_r L_q i_{qs}^r \tag{7}$$

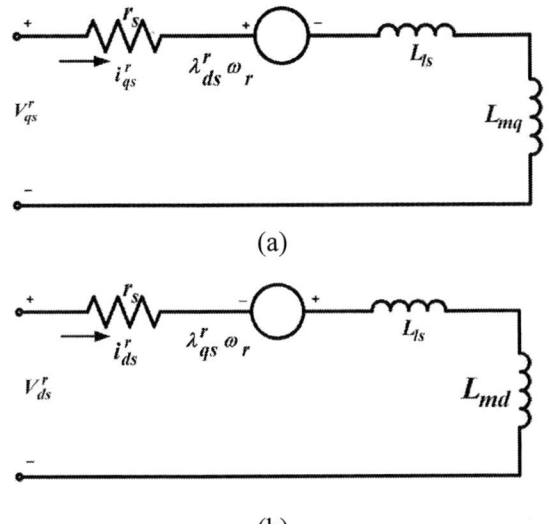

(a)

(b)

Fig.1: (a) & (b) Equivalent circuit of PMSM in rotor reference frame

The electromagnetic torque can be given by

$$T_e = \frac{3}{2} \cdot \frac{P}{2} (\lambda_{ds}^r i_{qs}^r - \lambda_{qs}^r i_{ds}^r) \tag{8}$$

Which upon substitution of the flux linkages in terms of the inductance and current

$$T_e = \frac{3}{2} \cdot \frac{P}{2} \{(\lambda_m^r i_{qs}^r + (L_d - L_q)i_{ds}^r i_{qs}^r\} \tag{9}$$

III. IMPLIMENTATION OF PROPOSED FW CONTROL WITH MTPA

The MTPA control strategy is utilized in high performance applications where efficiency is important. This control strategy provides maximum torque for a given current thus minimizes copper losses for a given torque. So the converter rating is minimum as well as its efficiency is maximum. MTPA control can be implemented in constant torque region. The torque developed in IPM machine has two components as follows

1) The torque due to the interaction of the magnetic flux and q axis stator current

2) The reluctance torque component which is propotional to the difference of stator q-axis and d-axis inductance.

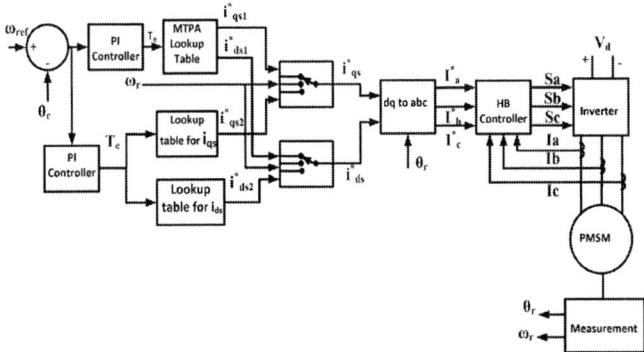

Fig 2: Block diagram of proposed PMSM drive for wide speed range operation.

The fundamental stator voltage can be relate with dc link voltage as

$$V_s = \frac{2V_d}{\pi} \tag{10}$$

Where V_d the dc is link voltage of the inverter and V_s is the stator voltage. This is related with steady state V_{qs}^r and V_{ds}^r component of stator phase voltage as

$$V_s = \sqrt{(V_{qs}^r)^2 + (V_{ds}^r)^2} \tag{11}$$

Where,

$$V_{qs}^r = \omega_r L_d i_{ds}^r + \omega_r \lambda_m^r \tag{12}$$

$$V_{ds}^r = -\omega_r L_q i_{qs}^r \tag{13}$$

And stator current can be related as

$$(I_s)^2 = (i_{qs}^r)^2 + (i_{ds}^r)^2 \tag{14}$$

This equation can be considered as an equation of a circle. Rewriting the equation we can get the following equation.

$$I_s = \sqrt{(i_{qs}^r)^2 + (i_{ds}^r)^2} \tag{15}$$

Equation (12) and (13) can be derived from equation (6) and (7) respectively at steady state condition. In this case the stator resistance drop is neglected. Rewriting the equation (11) with substituting the value of V_s, V_{qs}^r and V_{ds}^r.

$$\frac{4V_d^2}{\pi^2} = (\omega_r L_d i_{ds}^r + \omega_r \lambda_m^r)^2 + (-\omega_r L_q i_{qs}^r)^2 \quad (16)$$

This can be modified to form

$$\frac{(i_{ds}^r + \frac{\lambda_m^r}{L_d})^2}{(\frac{2V_d}{\pi\omega_r L_d})^2} + \frac{(i_{qs}^r)^2}{(\frac{2V_d}{\pi\omega_r L_d})^2} = 1 \quad (17)$$

This is an equation of an ellipse in the form

$$\frac{(i_{ds}^r - C)^2}{(A)^2} + \frac{(i_{qs}^r)^2}{(B)^2} = 1 \quad (18)$$

Where,

$$A = \frac{2V_d}{\pi\omega_r L_d}, B = \frac{2V_d}{\pi\omega_r L_q} \text{ and } C = -\frac{\lambda_m^r}{L_d}$$

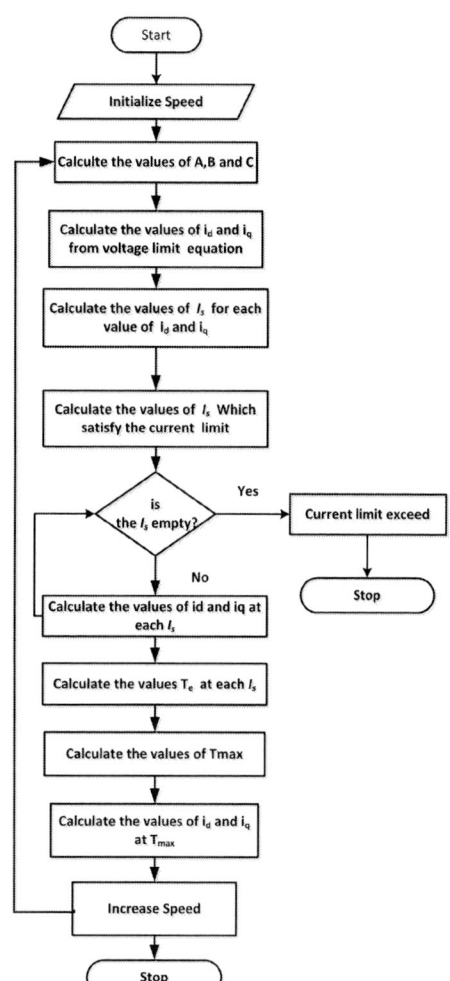

Fig 3: Flowchart of flux weekening algorithm

IV. SIMULATION RESULTS

Table 1: Specification of simulated Motor

stator phase resistance	0.9585Ω
flux induced by the magnet λ_m	0.1827 Wb
d-axis inductance (L_d)	0.004987 H
q-axis inductance (L_q)	0.005513 H
Moment of inertia (J)	0.0006329kg-m^2
No of poles	8
DC link voltage	300 V
Rated speed	2000 rpm

When the motor is operating within rated speed the performance of PMSM is evaluated for two different cases.

Case 1: Motoring mode

The performance of the motor is analyzed for step change in speed from 0 to 1500 rpm and 1500 to 2000 rpm. In this mode Maximum torque per ampere controller will operate. When the speed is changed from 1500 to 2000 rpm at 0.15 sec the torque is started to increase from no load to full load. In this region stator current is also increased. Figure 4 shows the result of PMSM drive within rated speed operation under motoring mode.

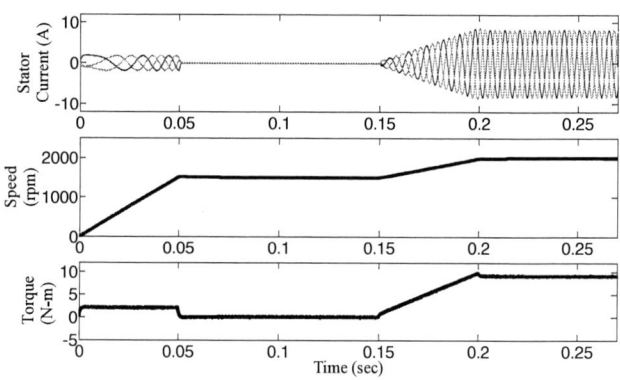

Fig 4: Simulation result of PMSM drive within rated speed operation.

Case 2: Braking operation

The performance of motor is evaluated in braking mode of operation by changing the speed from 1900 to 500 rpm. In this region the torque is decreased to negative torque value. As shown in Figure 5, the MTPA controller will act such that when the speed reached to 500 rpm again the torque value become previous applied load torque.

978-1-4799-6047-7/14 $31.00 © 2014 IEEE

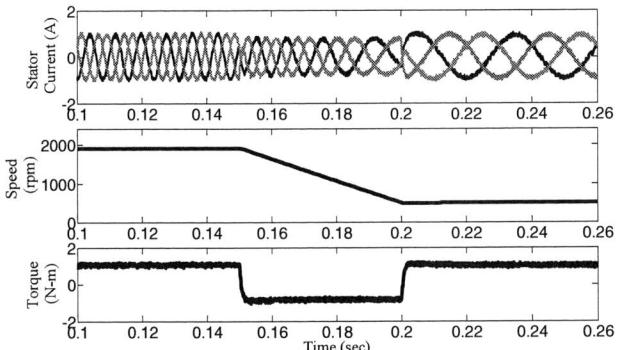

Fig 5: Simulation result of PMSM drive below rated speed operation in braking mode

Similarly the performance of motor is evaluated for above rated speed condition by taking two different cases.

Case 1: Motoring mode

When the motor is operated under rated speed conditions the reference speed is changed to above rated speed ie from 2000 to 3000 rpm at 0.3 sec. The Flux weakening controller will act such that the motor will run at 3000 rpm without exceeding the stator current limit and it is shown in Figure 6.

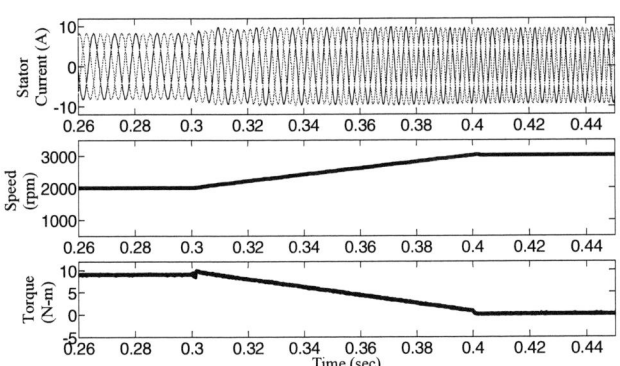

Fig 6: Simulation result of PMSM drive above rated speed operation.

Case 2: Braking mode

To analyze the response of the PMSM in braking mode with above rated speed condition, the reference speed is changed from 3000 to 2100 rpm at 0.5sec. In this condition the torque will be decreased to a negative value and then will be restored to the previous torque value i.e no load torque, without exceeding stator current limit with the help of flux weakening controller. Figure 7 shows the performance of PMSM drive above rated speed operation in braking mode.

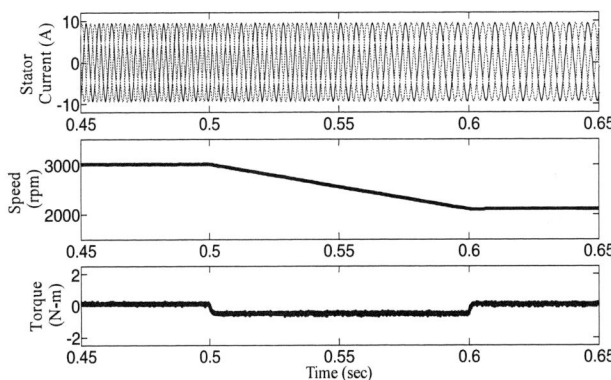

Fig 7: Simulation result of PMSM drive above rated speed operation in braking mode.

The performance of PMSM drive is also analyzed by changing the speed reference from 500 to 3000 rpm for motoring mode and from 3000 to 500 rpm for braking mode. In these both operations, the automatic control switching will takes place from MTPA controller to FW controller and FW to MTPA controller and it is shown in Figures 8 and 9 respectively for motoring and braking mode of operation.

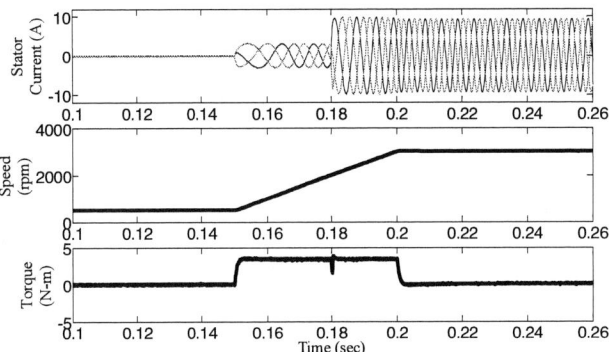

Fig 8: Simulation result of PMSM drive in wide speed range of operation in motoring mode.

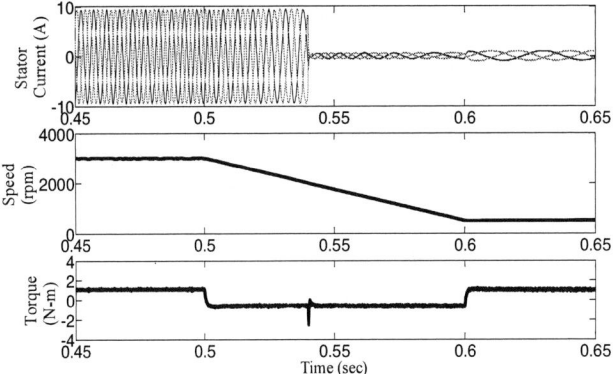

Fig 9: Simulation result of PMSM drive in wide speed range of operation in braking mode.

V. CONCLUSION

High speed operation has been achieved in terms of MTPA control with proposed Flux weakening algorithm without exceeding the voltage and current limit. The performance of the PMSM drive for high speed operation with varying load condition also has been studied. Proposed flux weakening controller for IPMSM is very useful for the drive where both the mode of operation such as constant torque and constant power is required. So the proposed control scheme is very useful for traction application.

REFERENCES

[1] T. M. Jahns, "Flux-Weakening Regime Operation of an Interior Permanent-Magnet Synchronous Motor Drive," *IEEE Transaction on Industrial Application*, vol. IA-23, no. 4, pp. 681–689, Jul. 1987.

[2] B. K. Bose, "A high-performance inverter-fed drive system of an interior permanent magnet synchronous machine," *IEEE Transaction on Industrial Application,* vol. 24, no. 6, pp. 987–997, 1988.

[3] Pragasen Pillay and Ramu Krishnan , "Modelling, Simulation and analysis of permanent magnet Synchronous motor Drives, Part I: The Permanent Magnet Synchronous Motor Drive" , *IEEE Transactions on Industrial applications*, Vol .25 ,No 2, March./April 1989, pp. 265-273.

[4] Morimoto, S, Sanada, M.; Takeda, Y., "Wide-speed operation of interior permanent magnet synchronous motors with high-performance current regulator,*" IEEE Transactions on Industry Application* , vol.30, no.4, pp.920,926, Jul/Aug 1994

[5] Jang Mok Kim and Seung-Ki Sul , "Speed Control of Interior Permanent magnet Synchronous Motor Drive for the Flux weakening operation ",*IEEE Transaction on Industrial Application*, Vol. 33, No 1, Jan./Feb. 1997 , pp.43-48

[6] Rahman, M.F. Zhong, L. Khiang Wee Lim, "A direct torque-controlled interior permanent magnet synchronous motor drive incorporating field weakening," *IEEE Transactions on Industry Applications*, , vol.34, no.6, pp.1246,1253, Nov/Dec 1998

[7] Y.A.I, Mohamed and Lee, T.K., "Adaptive self-tuning MTPA vector controller for IPMSM drive system," *IEEE Transactions on Energy Conversion,* vol.21, no.3, pp.636,644, Sept. 2006

[8] S.-M. Sue and C.-T. Pan, "Voltage-Constraint-Tracking-Based Field-Weakening Control of IPM Synchronous Motor Drives," *IEEE Transaction on Industrial Electronics.*, vol. 55, no. 1, pp. 340–347, Jan. 2008.

[9] P. Synchronous, "Design of Flux Observer Robust to Interior Motor Flux Variation," vol. 45, no. 5, pp. 1670–1677, 2009.

[10] A.Consoli, G. Scarcella, G. Scelba, and A. Testa, "Steady-State and Transient Operation of IPMSMs Under Maximum-Torque-per-Ampere Control," *IEEE Transaction on Industrial Application*, vol. 46, no. 1, pp. 121–129, 2010.

[11] S. Kim, Y. Yoon, S. Sul, and K. Ide, "Maximum Torque per Ampere (MTPA) Control of an IPM Machine Based on Signal Injection," vol. 28, no. 1, pp. 488–497, 2013.

[12] B. K. Bose, "Modern power electronics and AC drives", Pearson Prentice Hall, 2002.

[13] R Krishnan, "Electric Motor Drives modeling analysis, and Control", Pearson Prentice Hall, 2001.

A DC Motor Driver consisting of a single MOSFET with capability of speed and direction control

Gururaj Mulay*, Akshay Yembarwar[†], Surabhi Raje*
* Maharashtra Institute of Technology, Pune; [†] CDAC, Pune

Abstract—**Generally, a DC motor driver circuit consists of four MOSFETs forming an H-bridge. This type of motor driver can control the speed as well as the direction of rotation of the motor. However, both the functions can be performed using another type of motor driver consisting only one MOSFET and two relays. This paper demonstrates the proposed motor driver circuit having bidirectional speed controllability for motors. The circuit is constructed with a pair of relay (in place of four MOSFETs) that facilitate the direction control and a single MOSFET which enables the speed control of the motor. Moreover, this type of motor driver circuit provides several advantages over the former one. While designing the H-bridge, we have used relays for the High-side instead of the traditionally used MOSFETs and thus, the corresponding intricate MOSFET driving circuitry is also eliminated. As a result, the motor driver circuit is compact and low-priced. Since we do not include the MOSFET driving circuitries such as bootstrapping ICs, transistor circuit, etc., which contribute to the unreliability, the proposed motor driver possess increased reliability. Relays, being mechanical devices, are less prone to the failure due to Electrostatic Discharge (ESD) as compared to the MOSFETs under the same operating conditions. We have illustrated in the paper the designing and testing of this motor driver along with its performance evaluation.**

keywords: MOSFET, H-bridge, relay, PWM (Pulse Width Modulation), speed and direction control.

I. Introduction

While designing a motor driver, the H-bridge is made using four switches like relays, BJTs, MOSFETs, etc. [2]. In order to vary the speed of motor, the switches must be turned on and off periodically at a sufficiently high frequency using some mechanism such as Pulse Width Modulation (PWM). Since the relays cannot be switched at higher frequencies, they cannot be used in the H-bridges with speed-controllability. Hence, the H-bridge is generally constructed using four BJTs or MOSFETs, with the latter being the most common method. H-bridge constructed using MOSFETs consists of either all four N-channel MOSFETs or two N-channel MOSFETs (low-side) and two P-channel MOSFETs (high-side). When all the four MOSFETs are N-channel, the high side NMOSFETs of the H-bridge need a special driving circuitry called as 'bootstrap circuit' [1]; whereas in case where two N-channel and two P-channel MOSFETs are used, the high side needs a transistor circuitry to drive the PMOSFETs. Therefore, to reduce the complexity, we designed a circuit that replaces the MOSFETs with two relays to carry direction control, and incorporates only one MOSFET to carry the speed control of the motor. It can be typically used in the applications where direction of motor is not changed frequently.

This design of motor driver provides several benefits over the conventional motor drivers such as low cost, fewer components, compact size, higher reliability, and less susceptible to ESD. Moreover, the circuit also contains an inbuilt isolation between high-power motor circuit and low-power digital circuit, because of the isolation property of the relays. The mechanical relays are quite reliable as compared to the MOSFETs. Thus, the overall Mean Time between Failure (MTBF) of the motor driver increases significantly. Besides, the commonly observed 'shoot-through effect' [3] is completely eliminated in this circuit without any additional components.

II. Components and Circuit Design

The motor driver circuit consist of two electromagnetic relays, an N-channel power MOSFET, ULN2003, filtering capacitors, and resistors. Figure 1 shows the schematic diagram of a prototype circuit in which the H-bridge is built using two relays (Relay 1 and 2). Inputs to the circuit are given through a connector (CON1). The signals D1 and D2, which are used for the direction control, are supplied to the inputs of IC ULN2003 and the corresponding Open Collector outputs are connected to the coils of two relays.

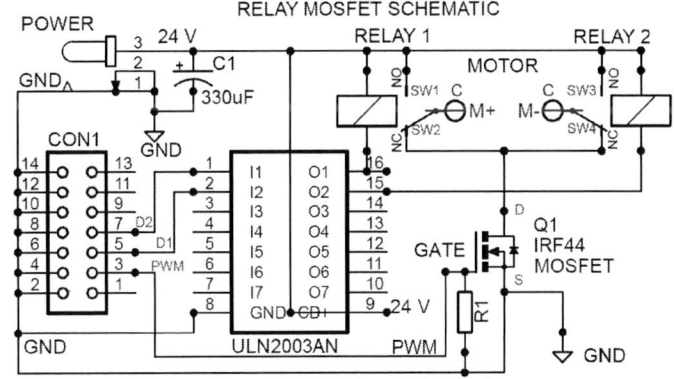

Fig. 1. Circuit Schematic Diagram

PWM signal at Pin 3 is given directly to the MOSFET and it switches the MOSFET so as to control the speed of the motor. The common terminals (C) at the output of the relays are connected to M+ and M- of the motor. The NC (Normally Closed) terminals of both the relays are connected to the Drain (D) of the MOSFET and their NO (Normally Open) terminals

978-1-4799-6047-7/14 $31.00 © 2014 IEEE

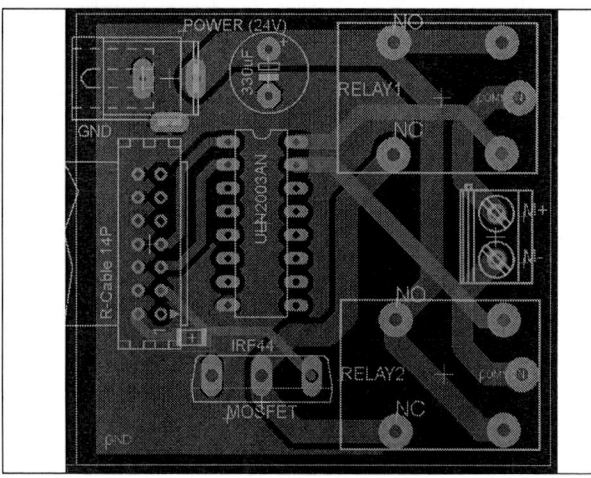

Fig. 2. Circuit Board Diagram

PWM	D1	D2	Effect
H	0	0	Motor Brake
H	0	1	Motor Rotates in One Direction
H	1	0	Motor Rotates in Opposite Direction
H	1	1	Motor Brake
L	0	0	Motor Brake
L	0	1	Motor Freewheel
L	1	0	Motor Freewheel
L	1	1	Motor Brake

Fig. 4. Control Logic Table

are shorted to VCC, which is 24 V in our case. Figure 2 shows one of the circuit board design on which the testing was performed.

III. WORKING AND PERFORMANCE EVALUATION

A. Control Logic

The motor driver requires 3 signals viz. D1, D2, and PWM to achieve bi-directional speed control. We used a motherboard based on an ATMEL microcontroller (Atmega2560) to generate the signals D1, D2, and PWM of a variable Duty Cycle and frequency (fig 3).

To facilitate speed control, a MOSFET is connected in series with the relays and its Source (S) terminal connected to the Ground (GND). The Gate (G) of this N-channel MOSFET is supplied with the PWM signal. By varying the pulse width, it is possible to control the ON time and the OFF time of the MOSFET, and this indirectly controls the ON time and the OFF time of the motor connected in series with the MOSFET.

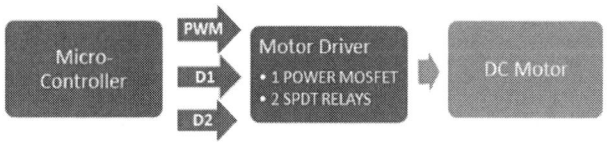

Fig. 3. Block Diagram

The signals D1 and D2 control the direction of rotation of the motor according to the control logic tabulated in figure 4, which shows all possible combinations and their resultant outcomes in terms of operation of the motor.

B. Testing and Results

The setup for testing the motor driver consisted a motherboard, the motor driver, and a motor as shown in figure 5. We used a 24 Volt power supply for testing the motor driver to drive various motors like MAXON RE30, Wiper,

Johnson, Mech-Tex, power-window-motors, etc. We tested various designs of the motor driver using different types of MOSFETs and relays. The part numbers of MOSFETs used were: IRF44N, CSD18501q5a and that of relays were: LEONE P38FC-3C, GOODSKY RW-SH124D. The MOSFETs and relays were selected as per the power requirements.

In order to vary the speed of the motor, we varied the Duty Cycle of the PWM signal from 0% to 100%. The graph in figure 6 illustrates the On-load and No load testing of the driver with the MOSFET IRF44N while driving MAXON RE30 and Wiper motors. As the graph suggests, the current flowing through the motors in no-load condition is less as compared to the on-load condition. The motor driver shows approximately linear response over the entire range of the variation of PWM duty cycle. To study the switching performance of the driver at different frequencies, we varied the frequency of the PWM signal from 4 KHz to 128 KHz. The 'current versus Duty Cycle of PWM' response was approximately linear for all frequencies. Thus, we can choose a suitable PWM frequency from the mentioned range depending on the motor specification.

This testing was performed under varied conditions such as: continuous on-state testing for 10 to 12 hours, on-load testing on a robot (to drive wheels, gantry), abrupt switching testing, etc. This testing showed the equivalence between this driver and the traditional motor drivers.

C. Advantages

The proposed motor driver provides several advantages over the conventional motor drivers. Theoretically, the maximum supplied current is decided by the current ratings of the relay and MOSFET. For higher power outputs, we can exercise SMD power MOSFETs with lower value of $R_{DS(ON)}$ capable of bearing 100A continuously. For instance, the SMD MOSFET CSD18501q5a from Texas Instruments carries a continuous drain current up to 100A. Power relays (EMR) can also bear currents of several amperes. Hence, the proposed design can be modified for the higher load applications. Although, the current, and hence the power in such cases is higher, the heat sinking requirement is moderate; because, only one MOSFET requires heat sinking rather than four. Thus, the elimination of

978-1-4799-6047-7/14 $31.00 © 2014 IEEE

Fig. 5. Testing Setup

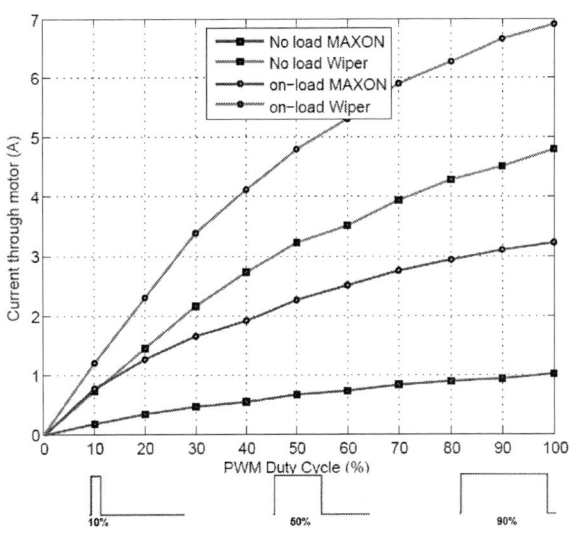

Fig. 6. Current through motor Vs PWM percentage

bulky heat sinks makes this motor driver compact.

One of the significant advantages of the circuit is that, the voltage drop and the power loss due to the on-state resistance, i.e. $R_{DS(ON)}$ of a MOSFET, is reduced by a factor of two in this motor driver. In case of a standard H-bridge, two MOSFETs in series with a motor cause a voltage drop of $2 \times [I_m \times (R_{DS(ON)P} + R_{DS(ON)N}]$, whereas the voltage drop in the proposed design is only $[I_m \times R_{DS(ON)N}]$.

where,

I_m = Current through the motor

$R_{DS(ON)P}$ = On state resistance of the P-channel MOSFET

$R_{DS(ON)N}$ = On state resistance of the N-channel MOSFET

Also, there is no (or very few) power loss in mechanical switches (relays), which gives this circuit an edge over the traditional technique. In the standard motor driver, each of the four MOSFETs has to be considered for reliability calculation and thus, the resultant reliability is always lesser than the reliability of individual MOSFET considered separately. But in case of single MOSFET motor driver, the reliability is calculated by considering only one MOSFET and thus, it is greater than the reliability of traditional circuits.

This circuit does not cause shoot-through effect, i.e. the condition when two switches (SW1 and SW2 in figure 1) of the same branch of the H-bridge are simultaneously closed for a short period of time if the PWM exceeds 95% duty cycle [3], [5]. However, in this motor driver, since these switches (on the same branch) are the NC and NO terminals of the relay (relay 1), it is automatically ensured that the two switches are never closed simultaneously. Hence, unlike bootstrap motor drivers [1], [5], it is possible to supply the gate (G) of the MOSFETs with 100% PWM duty cycle.

IV. FUTURE WORK AND DISCUSSION

The proposed prototype motor driver was designed and tuned for the applications in robotics and automation. The concept can be extended to other applications by choosing appropriate MOSFETs and relays as per the voltage and current requirements. In order to optimize the size, performance,

and cost of this circuit, different MOSFETs and relays should be studied, practised, and tested. Also, the solid state relays (SSR) can be used in place of mechanical relays (EMR) as they ensure faster and smoother switching, smaller size, increased lifetime, and no sparking while switching [6].

V. CONCLUSION

The paper demonstrates a new method to construct a motor driver using only one MOSFET along with two SPDT relays to achieve the complete motor control (including speed and direction control). Since the number of MOSFETs is reduced in this proposed motor driver, the reliability of the system is higher as compared to the traditional motor drivers. The driver is compact and less complex since the MOSFET driving circuitry is eliminated. Furthermore, this circuit is more robust to failures caused by damage to MOSFETs (especially due to ESD). Also, the testing suggests that the performance of this motor driver is similar to the traditional motor drivers in terms of speed and direction control. Thus, the motor driver can be used effectively for motor control in different applications.

REFERENCES

[1] Fairchild, Application Note AN-6076 : Design and Application Guide of Bootstrap Circuit for High Voltage Gate Drive IC.

[2] R. Valentine, "MOSFET "H" Switch circuit for a DC motor" US Patent 4,454,454, 1984

[3] Fairchild, AN-6003 Shoot-through in Synchronous Buck Converters

[4] R.Ali, I. Daut, et al., "Design of high-side MOSFET driver using discrete components for 24V operation" Power Engineering and Optimization Conference (PEOCO), June 2010, 132-136

[5] International Rectifier, Application Note AN978 : HV Floating MOS-Gate Driver ICs.

[6] IXYS, Application Note : AN-145, "Advantages of Solid-State Relays Over Electro-Mechanical Relays"

[7] R. Valentine, "Motor Control Electronics Handbook" McGraw-Hill Handbooks, 1998

978-1-4799-6047-7/14 $31.00 © 2014 IEEE

GA Tuned LQR and PID Controller for Aircraft Pitch Control

Vishal
PG Scholar Dept. of Electrical Engineering
National Institute of Technology Kurukshetra
Kurukshetra ,INDIA
vishalchugh24@gmail.com

Dr. Jyoti Ohri
Professor Dept. of Electrical Engineering
National Institute of Technology Kurukshetra
Kurukshetra ,INDIA
ohrijy0ti@rediffmail.com

Abstract- In this paper linear quadratic Regulator (LQR) and Proportional integral derivative (PID) controllers are designed for pitch control system of an aircraft. Genetic Algorithm (GA) is used for tuning the parameter LQR and PID controllers. The controller design begins with the appropriate mathematical model to account the longitudinal motion of an aircraft. Simulation is done within the MATLAB and results for the pitch angle response of an aircraft are represented in step's response. LQR and PID controllers are associated with their parameters which can either be tuned manually or by optimized methods like genetic algorithms to get better control and enhancement in the performance. GA is the powerful tool in the field of global optimization to various problems. Here genetic algorithm is successfully applied for Pitch control of an aircraft using MATLAB simulation for tuning of LQR and PID controllers and thus optimized parameters values for the LQR and PID controllers are derived.

Keywords— Pitch; Longitudinal dynamics; GA; LQR; PID; MATLAB.

I. INTRODUCTION

A modern aircraft and missile design depends mainly on automatic control system to control variety of the subsystem of aircraft and missile. Today's aircraft and missile consist of number of automatic controller that helps the airplane crew in airplane management and piloting the aircraft. Advancement in automatic control system has played a major role in the development of aircraft and missile. Usually, aircraft and missile contains three rotational motions and three translational motions [2]. Furthermore, the control strategies of aircraft can be grouped into two categories, as follows lateral and longitudinal control [2]. In longitudinal control the pitch angle whereas in lateral control roll and yaw of an aircraft system is controlled.

Pitch of an aircraft may be described as a rotation around the lateral axis. It may be calculated as the angle between the direction of speed in a horizontal line and vertical plane as shown in Fig. 1. Elevator is used to control pitch of aircraft which is located at the back of an airplane. The linearized longitudinal mathematical model of an aircraft is of third order [2]. Pitch control is a longitudinal control, therefore longitudinal model of an aircraft is used to design the pitch angle controller of an aircraft. In the past many researcher has done work to control the pitch, roll and yaw of an aircraft to stabilize an aircraft and this topic even now remains a challenging issue for present and future works [1], [3], [4], [5] and [9].

This work exhibits the design of control strategies for controlling pitch angle of an aircraft. The GA tuned LQR and PID controllers are designed to control the pitch angle of an aircraft. Simulation is done within the MATLAB for analysis of the both control schemes. Comparison of both control strategies to the system performance of aircraft system is discussed. MATLAB simulation results for the pitch angle response of an aircraft are given in step's response and the performance is measured in terms of rise time (T_r), settling time(T_s), peak overshoot (Os) and steady state error (e_{ss}) for pitch angle response of an aircraft.

II. MATHEMATICAL MODEL FOR PITCH CONTROL

In this section mathematical model of pitch control system of an aircraft is derived. Certain assumptions are to be considered to reduce the complexity of analysis, aircraft is assumed as a rigid body and its motion consist of small disturbances due to external conditions [2]. The equation of motion of an aircraft can be separated as the lateral and longitudinal equations.

The aircraft pitch control system is shown in Fig. 1 where X_b, Y_b and Z_b indicate the aerodynamics force components and θ, Φ and δe indicate the pitch angle, roll angle and elevator deflection angle respectively in the earth-axis system.

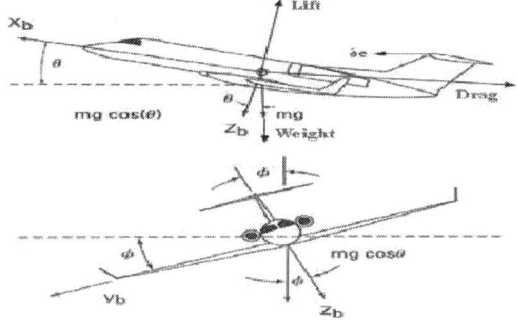

Fig. 1. pitch control system of an aircraft

The aerodynamic moment components are represented as L, M and N, angular rates as p, q and r and velocity components as u, v, and w for roll, pitch and yaw respectively in the body fixed coordinate of aircraft system as shown in Fig. 2. α indicates the angle of attack and β indicates sideslip angle as shown in Fig. 2. Here the data for system modeling and analysis is taken from General Aviation Airplane [1], [2].

978-1-4799-6047-7/14 $31.00 © 2014 IEEE

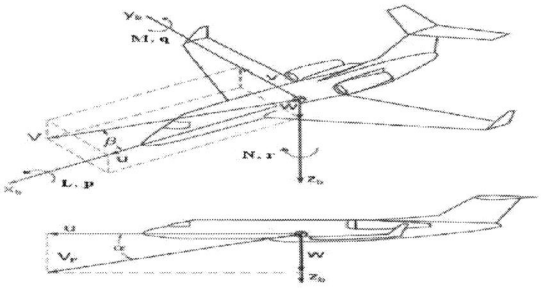

Fig. 2. Moment, force and velocity components in body fixed coordinate

A certain assumptions are required for mathematical modeling of an aircraft. First, the variation in pitch angle does not alter the speed of an aircraft. Second, the aircraft is moving at constant velocity and altitude, thus the lift and weight balance out and the thrust and drag cancel out each other.

From Fig. 1 and Fig. 2, the longitudinal dynamic equations of motion (force and moment equation) of an aircraft include can be derived as shown in (1), (2) and (3).

$$X - mgS_\theta = m(\dot{u} + qv - rv) \qquad (1)$$
$$Z + mgSC_\phi = m(\dot{w} + pv - qu) \qquad (2)$$
$$M = I_y\dot{q} + rq(I_x - I_z) + I_{xz}(p^2 - r^2) \qquad (3)$$

Certain assumptions need to consider solving the aircraft problem:

1) Pitch angle $\quad \dot{\theta} = qC_\phi - rS_\phi$
;
2) Pitching rate $\quad r = \dot{\varphi}C_\theta - \dot{\theta}S_\varphi$
;
3) Roll angle $\quad \dot{\varphi} = p + qS_\phi T_\theta + rC_\phi T_\theta$
;
4) Rolling rate $\quad p = \dot{\varphi} - \dot{\varphi}S_\theta$
;
5) Yaw angle $\quad \dot{\varphi} = (qS_\phi - rC_\phi)sec\theta$
;
6) Yawing rate $\quad q = \dot{\theta}C_\phi + \dot{\varphi}C_\theta S_\phi$
;

The longitudinal stability derivatives parameters for an aircraft pitch control system [2] used are given in Table I.

TABLE I.
LONGITUDINAL STABILITY DERIVATIVE PARAMETERS

Longitudinal Derivatives	Components		
	Dynamics Pressure and Dimensional Derivative Q = 36.8lb/ft2, QS = 6771lb, $QS\bar{c}$=38596ft.lb, $\bar{c}/2u_0$ = 0.016s		
	X-Force (S^{-1})	**Z-Force** (F^{-1})	**Pitching Moment (FT^{-1})**
Rolling Velocities	$X_u = -0.045$	$Z_u = -0.369$	$M_u = 0$
Yawing Velocities	$X_w = 0.03$	$Z_w = -2.02$ $Z_{\dot{w}} = 0$	$M_w = 0.05$ $M_{\dot{w}} = 0.051$
Angle of attack	$X_\alpha = 0$ $X_{\dot{\alpha}} = 0$	$Z_\alpha = -355.42$ $Z_{\dot{\alpha}} = 0$	$M_\alpha = -8.8$ $M_{\dot{\alpha}} = -0.8976$
Pitching rate	$X_q = 0$	$Z_q = 0$	$M_q = -2.05$
Elevator Deflection	$X_{\delta e} = 0$	$Z_{\delta e} = -28.15$	$M_{\delta e} = -28.15$

Small disturbance theory can be used to linearize (1), (2) and (3). The variables are substituted as given in (4) by introducing the small perturbation or disturbance.

$$
\begin{aligned}
u &= u_0 + \Delta u & v &= v_0 + \Delta v & w &= w_0 + \Delta w \\
p &= p_0 + \Delta p & q &= q_0 + \Delta q & r &= r_0 + \Delta r \\
X &= X_0 + \Delta X & M &= M_0 + \Delta M & u &= u_0 + \Delta u \\
\delta &= \delta_0 + \Delta\delta
\end{aligned}
\qquad (4)
$$

For simplicity certain assumptions are taken. First, the reference aircraft conditions are assumed to be symmetric and second, the propulsive forces to remain same throughout the flight i.e. $p_0 = v_0 = r_0 = q_0 = w_0 = \varphi_0 = 0$. After linearization the (5), (6), and (7) are obtained.

$$\left(\frac{d}{dt} - X_u\right)\Delta u - X_w \Delta w + (g\cos\theta_0)\Delta\theta = X_{\delta_e}\Delta\delta_e \qquad (5)$$

$$-Z_u\Delta u + \left[(1 - Z_w)\frac{d}{dt} - Z_w\right]\Delta w - \left[(u_0 - Z_q)\frac{d}{dt} - g\theta_0\right]\Delta\theta = Z_{\delta_e}\Delta\delta_e \qquad (6)$$

$$-M_u\Delta u + \left(M_w\frac{d}{dt} - M_w\right)\Delta w - \left(\frac{d^2}{dt^2} - M_q\frac{d}{dt}\right)\Delta\theta = M_{\delta_e}\Delta\delta_e \qquad (7)$$

By solving the (5), (6), (7) and putting the longitudinal stability derivatives parameter values from Table I, the transfer function for $\Delta q(s)$ to $\Delta\delta_e(s)$ is shown as obtained in (8), where $\Delta q(s)$ represent the variation in pitch rate and $\Delta\delta_e(s)$ represent the variation in elevation rate for pitch control system of an aircraft.

$$\frac{\Delta q(s)}{\Delta\delta_e(s)} = \frac{-\left(M_{\delta_e} + \frac{M_{\dot\alpha}Z_{\delta_e}}{u_0}\right)s - \left(\frac{M_\alpha Z_{\delta_e}}{u_0} - \frac{M_{\delta_e}Z_\alpha}{u_0}\right)}{s^2 - \left(M_q + M_{\dot\alpha} + \frac{Z_\alpha}{u_0}\right) + \left(\frac{Z_\alpha M_q}{u_0} - M_\alpha\right)} \qquad (8)$$

The transfer function of variation in pitch angle to the variation in elevator angle can be obtained from the (8) as under.

$$\Delta q = \Delta\dot{\theta} \qquad (9)$$

$$\Delta q(s) = s\Delta\theta(s) \qquad (10)$$

$$\frac{\Delta\theta(s)}{\Delta\delta_e(s)} = \frac{1}{s} * \frac{\Delta q(s)}{\Delta\theta(s)} \qquad (11)$$

Thus the transfer function for the aircraft pitch control system is obtained in (12) and (13).

$$\frac{\Delta\theta(s)}{\Delta\delta_e(s)} = \frac{1}{s} * \frac{-\left(M_{\delta_e} + \frac{M_{\dot\alpha}Z_{\delta_e}}{u_0}\right)s - \left(\frac{M_\alpha Z_{\delta_e}}{u_0} - \frac{M_{\delta_e}Z_\alpha}{u_0}\right)}{s^2 - \left(M_q + M_{\dot\alpha} + \frac{Z_\alpha}{u_0}\right) + \left(\frac{Z_\alpha M_q}{u_0} - M_\alpha\right)} \qquad (12)$$

$$\frac{\Delta\theta(s)}{\Delta\delta_e(s)} = \frac{11.7304s + 22.578}{s^3 + 4.9676s^2 + 12.941s} \qquad (13)$$

The transfer function can also be represented in state space form as obtained in (14) and (15).

$$\begin{bmatrix} \Delta\dot{\alpha} \\ \Delta\dot{q} \\ \Delta\dot{\theta} \end{bmatrix} = \begin{bmatrix} -2.02 & 1 & 0 \\ -6.9868 & -2.9476 & 0 \\ 0 & 1 & 0 \end{bmatrix} \begin{bmatrix} \Delta\alpha \\ \Delta q \\ \Delta\theta \end{bmatrix} + \begin{bmatrix} 0.16 \\ 11.7304 \\ 0 \end{bmatrix} [\Delta\delta_e] \quad (14)$$

$$y = \begin{bmatrix} 0 & 0 & 1 \end{bmatrix} \begin{bmatrix} \Delta\alpha \\ \Delta q \\ \Delta\theta \end{bmatrix} + [0] \quad (15)$$

III. GENETIC ALGORITHM

A Genetic Algorithm (GA) is a good optimization technique. It was given in 1970 by John Holland to improve the performance of computational methods. It starts without any knowledge of the correct solution and depends completely on responses from its genetic operators such as selection, mutation and crossover to achieve the best solution. It starts searching with many independent random points in parallel to avoid local minima and thus the algorithm converges to sub optimal solutions.

GA consists of a number of steps inspired biologically as shown in Fig. 3(a).

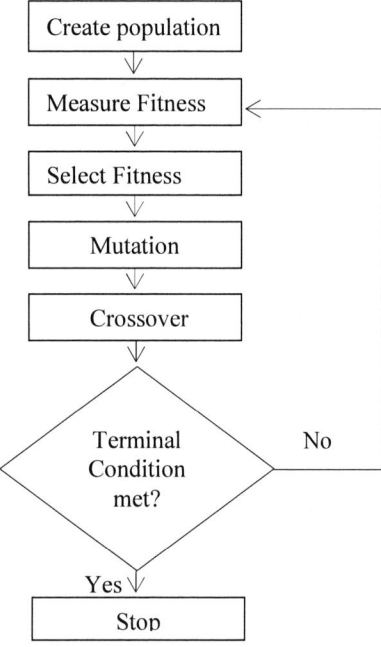

Fig. 3(a). Steps in Genetic Algorithm

Working of GA is shown in Fig. 3(b). In the starting it chooses the random population of chromosome. Random population Initialized in starting contain a number of chromosomes where each chromosome corresponds to a solution of the problem. The performance for selection is analyzed based on fitness value of each chromosome using a fitness function. Depending on the fitness of each chromosome, a number of chromosomes are selected. After selection each and every chromosome undergoes first mutation and then reproduction or crossover. Thus after crossover it creates the new individuals to give a

better solution than their parents and thus approaching to the optimal solution.

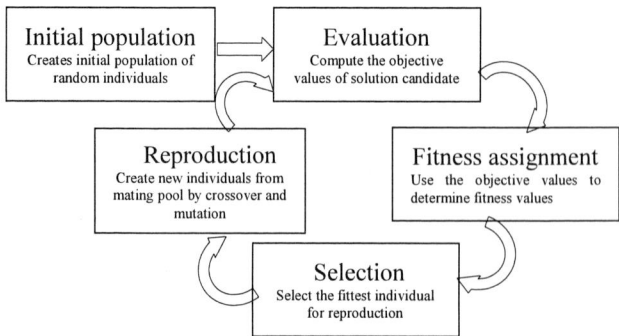

Fig. 3(b). Steps in Genetic Algorithm

Depending on the fitness of each chromosome, a number of chromosomes are selected. After selection each and every chromosome undergoes first mutation and then reproduction or crossover. Thus after crossover it creates the new individuals to give a better solution than their parents and thus approaching to the optimal solution.

Some of Advantages of GA:
• It is robust in nature
• It is quite simple and easy to understand. .
• It can be applied to number of control applications in industrial processes as it is nonlinear in nature.
• It does not require any information about the system, only fitness function is required.
• It searches a number of individual in parallel instead of a single solution.

Thus GA can be applied to a variety of control applications for the improvement of the performance of the system [8]. Thus it is the powerful tool in the field of global optimization to various problems. In many industrial applications system is to use offline then manual tuning method requires analysis of the step input response of the system to obtain desired LQR and PID controllers. But in most of the industrial applications, the system must be online and if tuning is achieved manually there is always uncertainty due to human error. Thus GA can be used for tuning of LQR and PID controller. it cannot guarantee the optimality of results, but it is certain that they are near optimal. The GA will give different results but nearer to previous search for each new search.

IV. METHODOLOGY

LQR and PID controllers are designed and tuned using the GA. In addition, some of design characteristics are required to look into the performance of controllers. Here, certain characteristics are required to be met which are settling time < 1 s, rising time < 0.2 s, percentage of overshot < 5% and steady state < 2% for controlling the pitch angle of 0.2 radian or 11.5 degree.

978-1-4799-6047-7/14 $31.00 © 2014 IEEE

A. PID Controller

PID control is the basic control scheme of the classical control system. It is mainly used for industrial control of a number of processes due to its simplicity [6]. The performance of the system can be enhanced [7] by tuning the proper value of gain K_p, K_d and K_i. The variation in the values of gain cause for variation in output response $y(t)$. The mathematical equation for PID controller of an plant with input u(t) output y(t) and the error, e(t) is expressed as (16), (17) and (18) where K_p is proportional gain, K_i integral gain, K_d derivative gain, T_i integral time and T_d derivative time. PID control is combination of proportion (P), integral (I), differential (D) of the error $e(t)$. The block diagram of analog PID control system is showed in Fig. 4.

$$u(t) = K_p e(t) + K_i \int_0^t e(t) + K_d \frac{de(t)}{dt} \qquad (16)$$

$$u(t) = K_p \left(e(t) + \frac{1}{T_i} \int_0^t e(t) + T_d \frac{de(t)}{dt} \right) \qquad (17)$$

$$e(t) = r(t) - y(t) \qquad (18)$$

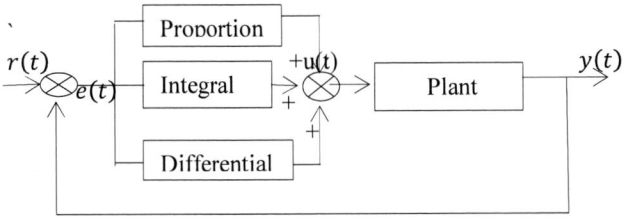

Fig. 4. A common feedback PID control system

Proportional gain K_p helps in increasing the loop gain of the system to make it immune to load disturbance. The integral gain (K_i) helps to reduce steady state errors. The derivative gain (K_d) helps to enhance the stability of closed loop system. The parameters of PID controller have to be chosen properly to achieve the desired performance.

Tuning of a PID controller refers to the tuning of its various parameters (P, I and D) to achieve an optimized value of the desired response. The basic requirements of the output will be the stability, desired rise time, peak time and overshoot. Different processes have different requirements of these parameters which can be achieved by meaningful tuning of the PID parameters [10]. Here GA is used for tuning the parameters of PID controller for an aircraft pitch control system. Automatic tuning of PID can be done based on minimizing the performance index which are given as under:

1) Integral absolute error (IAE)

$$IAE = \int_0^\infty |e(t)| dt$$

2) Integral square error (ISE)

$$ISE = \int_0^\infty [e(t)]^2 dt$$

3) Integral time absolute error (ITAE)

$$ITAE = \int_0^\infty |e(t)| dt$$

Automatic Tuning of a PID controller is done using GA based on optimizing the fitness function ISE, IAE, ITAE. Tuned parameters for GA tuned PID controller based on minimizing the fitness function ISE, IAE, ITAE is shown in Table II. The Pitch angle response for GA tuned PID with ISE, IAE, ITAE is shown in Fig. 6, Fig. 7 and Fig. 8 respectively.

TABLE II.
TUNED PID PARAMETERS USING GA

Performance criteria	PID parameters		
	K_p	T_i	T_d
ISE	0.4127	0.084	11.904
IAE	0.1702	0.035	28.57
ITAE	0.1702	0.035	28.57

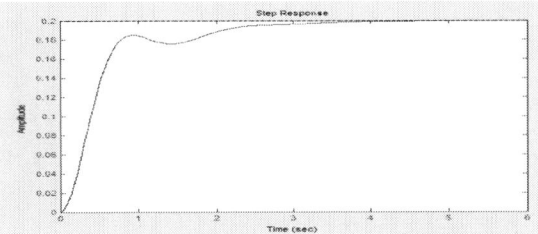

Fig. 5. Pitch angle response for closed loop system

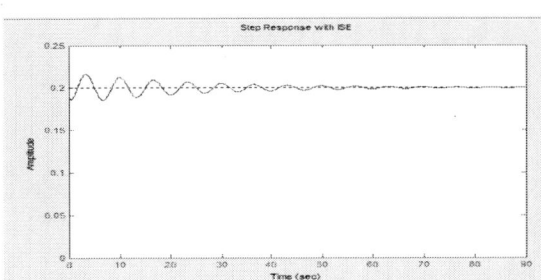

Fig. 6. Pitch angle response with PID controller based on ISE

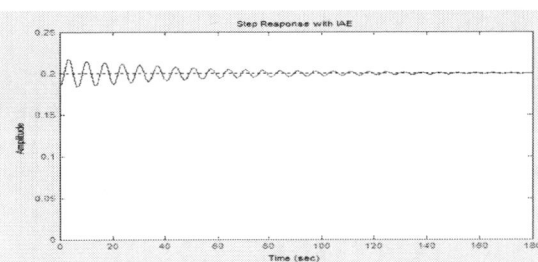

Fig. 7. Pitch angle response with PID controller based on IAE

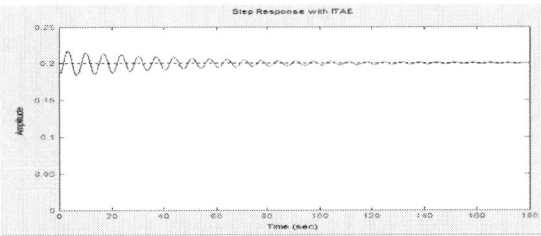

Fig. 8. Pitch angle response with PID controller based on ITAE

978-1-4799-6047-7/14 $31.00 © 2014 IEEE

B. LQR Controller

Linear-quadratic-regulator (LQR) is technique in control system that is widely used in many control applications [6]. It is an optimal control strategy. In the design of LQR the value of the gain K is chosen such that the performance index J is opttimized. This assures that the feedback gain K is optimal for the performance index specified. The performance index J is defined as under:

$$J = \int_0^\infty (x^T Q x + u^T R u)\, dt$$

It is a methodology in modern control system that optimizes the value of performance index J. It uses the feedback control strategy to analyze a system. The system can be stabilized by choosing the proper value of feedback gain K. The block diagram of this system is shown in Fig. 9.

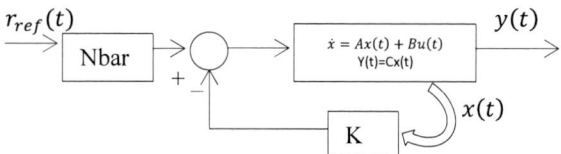

Fig. 9. Full-state feedback controller with reference input.

Pre compensation is used to reduce the steady state error $e(t)$ of the output $y(t)$. Pre compensated gain $\bar{N}$ can be used after the input $r_{ref}(t)$. Value of gain $\bar{N}$ can be computed using the MATLAB user defined function.

Design of LQR controller, it is required to calculate the value of the gain K. it is calculated using riccati equation by choosing appropriate values of Q and R. The controller can be tuned by varying the value of x in Q matrix as shown in Table III.

$$Q = \begin{bmatrix} 0 & 0 & 0 \\ 0 & 0 & 0 \\ 0 & 0 & x \end{bmatrix}$$
$$R = 1$$

As a result, by choosing the different values of x in matrix Q the pitch angle responses are obtained as shown in Fig. 10. The response of pitch angle response may improve even more by increasing the value of x in Q.

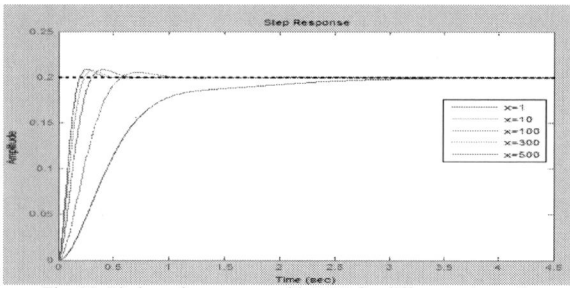

Fig. 10. Pitch angle response with LQR for different values of x

Different values of x for LQR controller are chosen manually and always there is uncertainty due to human error. Thus GA is applied for Pitch control of an aircraft using MATLAB simulation for tuning of LQR controller and thus optimized values of x in Q matrix for the controller is derived.

Here our main aim is to reduce the rise time, settling time and peak overshoot. Hence the fitness function, in this case, is the function of rise time (T_r), settling time (T_s) and percentage overshoot (% Os) and is defined as

$$fitness\ function = 1/(0.45 * T_s + 0.45 * T_r + 0.10 * Os)$$

The lower and upper bound for the variable x is set as 1 and 500 respectively. Number of generation has been taken as 65. Thus the optimized value of x, K and Q are obtained as
x = 495.9362
K = [-0.5704 1.6899 22.2915]
$$Q = \begin{bmatrix} 0 & 0 & 0 \\ 0 & 0 & 0 \\ 0 & 0 & 495.9362 \end{bmatrix}$$

The fitness curve is shown in Fig. 11 which is giving the fitness value at different generations. The best fitness we obtained is 1.06893 and the mean fitness is 1.06901.

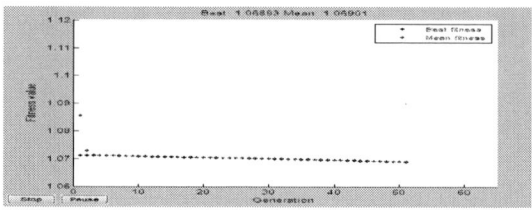

Fig. 11. Fitness curve

Fig. 12. Pitch angle response with GA-LQR

TABLE III.
PERFORMANCE CHARACTERISTIS OF PITCH ANGLE WITH LQR

Response characteristics	Pitch angle					
	LQR					GA-LQR
	x=1	x=10	x=100	x=300	x=500	x=495.9362
Rising time(T_r)	1.2s	0.36s	0.2s	0.15	0.132s	0.133s
Settling time(T_s)	2.9s	1s	0.435s	0.414	0.366s	0.366s
Percentage overshoot (%Os)	0	2.24	4.18	4.32	4.35	4.35
Steady state error (e_{ss}%)	0.1	0.01	0.01	0.01	0.01	0.01

V. RESULTS AND DISCUSSION

In this work, the GA tuned LQR and PID controllers have been applied for aircraft pitch control system and the results are shown in Table III. A step command of reference size equals 0.2 radian is used. The pitch response response for both control strategies has been analyzed. PID controller doesn't provide good performance as the settling time is quite higher. The response of pitch angle is relatively slow and oscillatory that provides very large the settling time (T_s). Comparative analysis for the pitch angle system of an aircraft between LQR and PID Controller with different performance characteristics is shown in Table IV. The aircraft pitch control system with LQR Controller produce the pitch angle response as shown in Fig. 10 for different values of x and the response of the pitch angle for the GA tuned LQR is shown in Fig. 12. Here tuning is achieved manually and there is always uncertainty due to human error. Thus GA is used for tuning of parameters in LQR and PID controller. Our main goal is to reduce the rise time, settling time and peak overshoot and thus optimizing the fitness function. From Fig. 12 and Table IV, the result clearly shows that GA tuned LQR with x=495.9362 give the better response when compared to other values of x. It gives the settling time of 0.366 s and rise time of 0.133 s. Thus it gives the better performance when compared to PID controller as shown in Table IV. LQR gives 4.35% percent overshoot (%Os) which is within the limit of 5%. In addition, the LQR controller likely to produce very small steady state error (e_{ss}) .

V. CONCLUSION

Aircraft pitch control system is designed and analyzed successfully. Pitch controller of an aircraft is required to stabilize the pitch angle at it desired value. LQR and PID controllers are designed successfully and tuned using Genetic Algorithm. Referring to Table IV, it is observed that the LQR Tuned by GA is capable of controlling the pitch angle of the aircraft system. In the past too many research work has been done, for further research, effort more advanced and robust control techniques can be developed and applied to control the pitch angle of an aircraft.

TABLE IV.
COMPARISON OF PERFORMANCE CHARACTERISTIS OF PITCH ANGLE

Response characteristics	Pitch angle			
	GA TUNED			
	LQR	PID		
		ISE	IAE	ITAE
Rising time(T_r)	0.133s	0.0462s	0.0477s	0.0477s
Settling time(T_s)	0.366s	36.8s	78s	78s
Percentage overshoot (%Os)	4.35	8.22	8.69	8.69
Steady state error (e_{ss}%)	0.01	0.2	0.2	0.2

REFERENCES

[1] Nurbaiti Wahid, Nurhaffizah Hassan, "Self-tuning Fuzzy PID Controller Design for Aircraft Pitch Control", Intelligent Systems, Modelling and Simulation (ISMS), Third International Conference on Intelligent Systems Modelling and Simulation, 2012 IEEE, pp 19-24.

[2] Robert C. Nelson, Flight Stability and Automatic Control, McGraw Hill, Second Edition, 1998

[3] N. Wahid, M.F. Rahmat, "Pitch Control System Using LQR and Fuzzy Logic Controller", Industrial Electronics & Applications (ISIEA) , IEEE Symposium on Industrial Electronics Malaysia,2010, pp.389-394.

[4] N. Wahid, M.F. Rahmat, K. Jusoff, "Comparative Assessment Using LQR and Fuzzy Logic Controller for a Pitch Control System". European Journal of Scientific Research, Vol 42, 2010, No 2, pp.184-194.

[5] Amir Torabi, Amin Adine Ahari, Ali Karsaz, S.H. Kazemi "Intelligent Pitch Controller Identification and Design"Journal of mathematics and computer science, 2014, pp.113-127

[6] M.Gopal, "Control Systems Principles and Design", Second Edition, Tata McGraw Hill, 2002.

[7] Mohd S. Saad, Hishamuddin Jamaluddin, Intan Z. M. Darus, "PID Controller Tuning Using Evolutionary Algorithms", WSEAS Transactins on System and Control, Vol 7, 2012, pp. 2224-2856

[8] Amit Manocha, Abhishek Sharma "Three Axis Aircraft Autopilot Control Using Genetic Algorithms : An Experimental Study", 2009 IEEE International Advance Computing Conference , 2009

[9] B. Stojiljkovic, L. Vasov, C. Mitrovic, D. Cvetkovic, "The Application of the Root Locus Method for the Design of Pitch Controller of an F-104A Aircraft", Journal of Mechanical Engineering, Vol 55, 2009.

[10] M S. SAAD, H.JAMALUDDIN and I. Z. M. DARUS, " PID Controller Tuning Using Evolutionary Algorithms" WSEAS TRANSACTIONS on SYSTEMS and CONTROL 2012.

Fault Identification of Power Transformers Using Proximal Support Vector Machine (PSVM)

Hasmat Malik
Department of Electrical Engineering
Indian Institute of Technology Delhi
New Delhi-110016, India

Sukumar Mishra
Department of Electrical Engineering
Indian Institute of Technology Delhi
New Delhi-110016, India

Abstract— The diagnosis of incipient fault is very important for power transformer condition monitoring. The incipient faults are monitored by conventional and artificial intelligence (AI) based models. In this paper, the Proximal Support Vector Machine (PSVM) has been utilized to identify the incipient type of faults in an oil-immersed power transformer. Its performance is compared with traditional IEC/IEEE and AI methods (i.e. ANN and SVM). The juxtaposition of fault classification of ANN and SVM method notify that proposed approach is much swiftly. Simultaneous identification of oil immersed power transformer incipient faults has never been identified formerly by using Multi-PSVM. The desired test analysis of experimental data from working transformers in the Northern Power Grid of India has been executed to present the robustness of evaluated incipient faults for large variation in loading and operational conditions perturbations.

Keywords—Power transformer; DGA; fault classification; artificial intelligence; PSVM

I. INTRODUCTION

The most application of an electrical power system is to provide the electric energy to the costumers at economic rate as potential and at higher acceptable degree of dependability & quality. System dependability is depends upon the dependability of equipments. So the condition of equipments directly affects the system condition leading to failure. So condition monitoring (CM) of the power equipment is 1st priority for stable operation in a power system transmission and distribution network. Power transformers square measure thought-about as most important and expensive instrumentation of electrical power transmission substations. A failure of an outsized power electrical device involves not solely substantial repair value, however additionally typically ends up in power interruptions to thousands of consumers. So, Adequate condition monitoring and diagnosis techniques (CMDT) for power transformers can help in reducing the failure rate to the lowest possible level and thereby enhancing system reliability and economic efficiency. The CMDT [1-3] provide detailed information about transformers condition and help to minimize the probability of an unexpected failure and are warranted to find out the power transformer faults. The importance of transformer diagnostics is increasing from the needs of high reliability in electrical equipment. An early detection of deterioration in transformer along with the degree and trend of deterioration are essential to prevent a catastrophic failure. The following techniques are used in CM of power transformers [1,5]: sweep frequency response analysis (SFRA); furfural analysis; DGA analysis and Measurement of Partial Discharge (PD).

Power transformer winding is made up by paper insulation and immersed in oil. Therefore, paper insulation and oil are the main sources to find out the incipient faults of the transformer. Due to fault condition, thermal, chemical and electrical stresses of insulating paper and oil are occurs, hence decomposition is aroused. Because of this decomposition, gases reduce the heat immoderation ability and hence, transformer oil dielectric strength is decreased. As a result of decomposition of oil, generated gases are ethane (C_2H_6), methane (CH_4), hydrogen (H_2), acetylene (C_2H_2) and ethylene (C_2H_4). Further on putrefaction of paper creates carbon dioxide (CO_2) and carbon monoxide (CO) which are mainly responsible for paper insulation degradation [4].

DGA is extensively applied to identify the transformer faults. There are many DGA elucidation approaches, i.e. Doernenburg ratio method [5], key-gas method [5], Roger ratio method [5], IEC ratio method [6] and triangle method of Duval [6] that are reported in the literatures. But, available existing methods are based on expertise personnel experience therefore it does not provides stable result for the same oil specimen and have limitations such as the "no decision" (code not matched condition) problem. So, accurate DGA analysis is still a dispute in the research area of power transformer CM. To overcome such problems, artificial intelligence techniques (AITs), i.e. ANN [7], SVM [8], FL-Fuzzy-logic [9], type-2 Fuzzy-logic [10], GEP [16] and the combinations of these techniques (i.e. hybrid system) [11] for faults classification have been extensively used by several researchers to develop highly precise diagnostic tools based on DGA data.

However, FL is time consuming in implementation due to large mathematical reasoning and fuzzy-rules are not robust at all [9]. ANN has many drawbacks, such as local minima, time-consuming for determination of optimal network structure (requirement to accommodate the hidden activation function limits), risk of over fitting, and needs huge quantity of training data [7]. A common disadvantage of SVMs is the lack of transparency of results and for better results, parameters of SVM need to be optimized using optimization techniques [8]. The proposed proximal support vector machine (PSVM) classification technique overcomes all these problems.

This paper is organized in six sections. The introduction and literature review is given in Section 1. Conventional DGA approach is explained in Section 2. The database and methodology used are presented in Section 3. Proposed methodology and PSVM model formation is presented in Section 4. The results are presented and discussed in Section 5 and conclusion in Section 6.

II. DISSOLVED GAS ANALYSIS

Several interpretation methods for DGA have been described in literature [1,5,6] as follows: 1) IEEE methods (Doernenburg Ratios, Roger's Ratios, Key Gas, Transformer fault analysis by individual and total dissolved key gas concentrations,); 2) IEC methods (IEC Ratios and graphical representation method); 3) Nomograph method; 4) Denkyoken method; 5) Xiaohui Li et al. method; 7) CIGRE's method; 8) NBR-7274 method and 9) IS-10593:2006 method. In this paper, IEC ratios method of the DGA was studied.

In IEC ratios method, three ratios (R1=CH$_4$/H$_2$, R2=C$_2$H$_2$/C$_2$H$_4$ and R3=C$_2$H$_4$/C$_2$H$_6$) of 5 gases are utilized to provide the fault analysis as shown in Table 1 [6].

Table 1. IEC ratio method based Fault classification [6].

R1	R2	R3	Gas Ratio Range
1	0	0	< 0.1
0	1	0	0.1 – 1
2	1	1	1 – 3
2	2	2	> 3
Faults Characteristic			
0	0	0	Normal operation
1	2	0	Low energy density PD
1	1	0	High energy density PD
0	1 – 2	1 – 2	Discharge of Low energy (D1)
0	1	2	Discharge of High energy (D2)
0	0	1	Thermal fault (t < 150ºC)
2	0	0	Thermal fault (t=150-300ºC)
2	0	1	Thermal fault (t=300-700ºC)
2	0	2	Thermal fault (t > 700ºC)

III. MATERIAL AND METHODOLOGY

A. Database Used

The 171 DGA cases are collected from online available dataset (one hundred and seventeen cases from the IEC TC 10 database [13], thirty nine cases from Table I-III in [2], 9 cases of normal condition from Table 1 to 2 in [8] and from working transformers in the HPSEB Simla, India is 6 cases, as bestowed in [11]) for training and testing purpose of GEP model. After successful training and testing the model, 50 oil samples of real data set which are simulated in the Lab are used to identify the fault. The following assumptions were made during preparation of dataset: (a) assumed zero when concentration of a given gas is not available; (b) assumed null when a ratio is indeterminate (0/0); (c) assumed 20 when a given ratio has infinite value ($\lambda/0$) as in [26], considering that $\lambda \neq 0$; and (d) a concentration shown by "<1" is set as 0.5 [14].

From the available datasets, we prepared input space vector (X) of 3 attributes $\left(X = \frac{CH_4}{H_2}, \frac{C_2H_2}{C_2H_4}, \frac{C_2H_4}{C_2H_6} \right)$ calculated from the 5 Key-gas. This relation is widely utilized described by standards [5,6] and by the researchers [7-11]. Our aim is to develop a PSVM model, which have higher classification accuracy as discussed in subsequent section.

B. Proximal Support Vector Machine (PSVM)

PSVM is a relatively new technique used for classification of data was 1st developed by Glenn Fung and Olvi L. Mangasarian [12]. PSVM is an ameliorate version of the standard SVM. Standard SVM is part of a breed of new generation tools used for data classification [12,17] centered on the mathematical learning theory. It is essentially a supervised learning algorithm wherein the machine is fed a set of input features with their respective labels (inputs) with the associated labels (outputs). As shown in Fig. 1(a), these features can be realized as the dimensions of a hyper-plane that splits the fed data into 2 distinct groups with a sizeable margin, thus minimizing the expected generalization error. For obtaining the greatest separation-margin between the classes leads to 2 hyperplanes parallel to the separating plane on both sides of it & the distance between them is recognized as the 'margin'. In contrast, PSVM categorizes points based on their nearness to one of 2 parallel planes that are separated to the maximum extent possible as exhibited in Fig. 1(b).

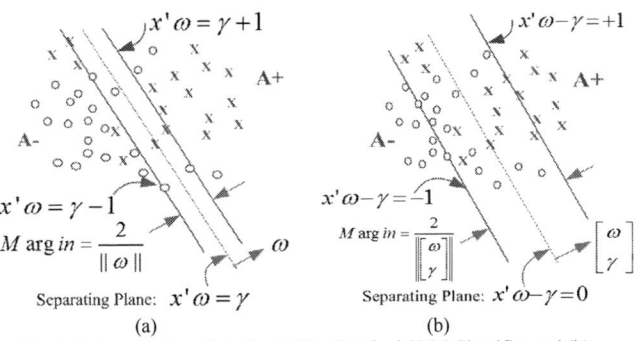

Fig. 1 Data separating plane for (a) The Standard SVM Classifier and (b) PSVM classifier [12]

Fig. 2 shows the flowchart of the typical PSVM algorithm [12]. PSVM is given m data points in R^n characterized by the $p \times q$ matrix M & a diagonal $p \times p$ matrix T of ± 1 labels indicating the different class type of each individual row of M, and then linear classifier is generated in subsequence steps:

a) Define H by using equation, $H = T[M - e]$

   Where e is a $p \times 1$ vector of ones and computes u for some positive nu.

$$u = nu \left(\left(1 - H \left(\frac{1}{nu} + H'H \right)^{-1} \right) H' \right) e \text{ or}$$

   u can also be calculated as: $u = nu * (1 - (H * r))$

   where r is defined as: $r = \left(\frac{speye(n+1)}{nu + H'*H} \right) \backslash R$

   and $R = sum(H)'$. The value of nu is estimated or specialized, it depends on expert. Value of nu is in between 10^6 to 0.01. The bigger nu gives better fitting of the training data. In most of the cases, $nu = 1$ works very well.

b) Determine (w, γ)

   $w = M'*Tu$; and $\gamma = -e'*Tu$ and $y = u/nu$

   Margin of the given bounding planes is maximized in relation to both the location γ and the orientation w relative to the origin.

978-1-4799-6047-7/14 $31.00 © 2014 IEEE

c) Classify a new m data sample by using:

$$\left(m'w - \gamma\right) = \begin{cases} > 0, then & x \in M+; \\ < 0, then & x \in M-; \\ = 0, then\, x \in M + or\, x \in M-; \end{cases} \quad (1)$$

Post training, new samples of data are tested. Their classes are assigned by utilizing the decision function stated above which is a function of "γ" and "w".

$$f(m) = sign\left(w'* m - \gamma\right) \quad (2)$$

When the output value of function $f(m)$ is negative, the set of samples are categorized as belonging to class $M-$, else they are categorized as a part class $M+$. Multi-class classification is performed by combining multiple PSVM classifier models in a tournament manner (one-by-one).

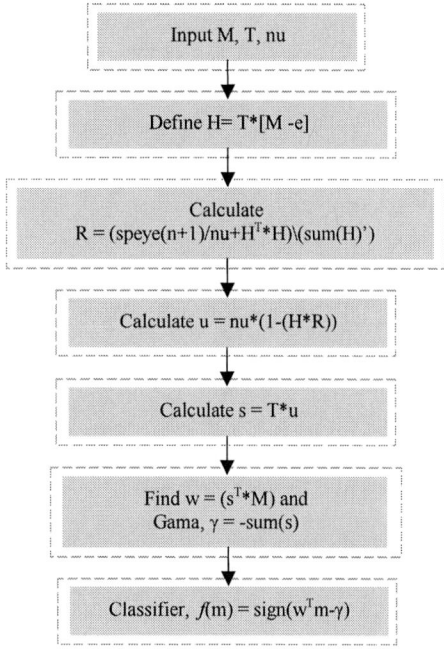

Fig. 2. The flow chart of PSVM algorithm [12]

IV. FAULT CLASSIFICATION USING PSVM

A. Proposed Method

This method contains two parts, first is calculation of relevant input variables and second is implementation of PSVM on evaluated input variable for fault classification. The stepwise procedure for proposed diagnostic method is represented in Fig.3.

B. DGA training and testing data

With selected most informative input variable, dataset is splitted into 2 subdata sets: randomly chosen 120 cases for training and for the testing 51 cases. The total datasets are evaluated victimization numerous strategies DGA and therefore the corresponding judgments associated with six categories are provided: normal unit-NF (9 cases), PD (9 cases), D1 (44 cases), D2 (51 cases), low & medium temperature overheating-T1/T2 ($<700^{o}C$) (29 cases), and high temperature overheating-T3 ($>700^{o}C$) (29 cases).
The PSVM based faults classification is carried out by using traditional DGA (IEC ratios method) as gas signature.

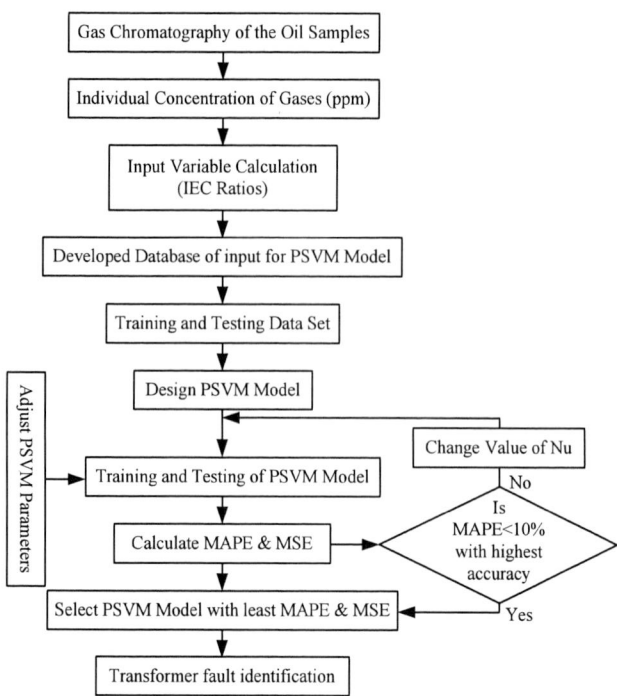

Fig.3. Proposed algorithm for fault classification

C. PSVM Based fault classification model formations

PSVM classifier based Fault classification model is represented in Fig. 4. The diagnostic model includes 5 PSVM classifiers that areas unite accustomed determine the six states: normal state (NF) and the 5 faults state (T1/T2, T3, D1, PD and D2).

Along with all training datasets, PSVM1 is trained to differentiate between the Healthy-NF and the fault state. The output of PSVM1 is set to -1 corresponding to a normal state input, else it is set as +1. PSVM2 is trained to distinct the condition of discharge faults (PD, D1, D2) from the over heating fault (T1/T2, T3) condition. If the input of PSVM2 represents a discharge fault, the corresponding value in output is set to -1; else +1. Using the discharge fault samples, PSVM3 is trained to distinguish between the "D2" faults and the "PD and D1" fault. If input of PSVM3 represents a D2 fault, corresponding value in output is set to +1; else -1. Now, using "PD and D1" fault samples, PSVM4 is trained to distinguish between "PD" fault and the "D1" fault. If input of PSVM4 represents a "D" fault, the corresponding value in output is set to +1; else -1. Using the "T3 and T1/T2" fault samples, PSVM5 is trained to distinguish between the "T3" fault and the "T1/T2" fault. If the input of PSVM5 represents a "T3" fault, the corresponding value in output is set to +1; else -1.

Hence, the summary of the output of all five PSVMs are presented in form of codification in Table 2 as below:

Table 2. Codification of PSVMs output

	PSVM1	PSVM2	PSVM3	PSVM4	PSVM5
NF	-1				
PD	+1	-1	-1	-1	
D1	+1	-1	-1	+1	
D2	+1	-1	+1		
T1/T2	+1	+1			-1
T3	+1	+1			+1

| PSVM5 | 0.2577319 | 0.00515 | 0.07180 | 99.28 |
| Average value of all PSVM models | 1.51164 | 0.030216 | 0.16408 | 98.0 |

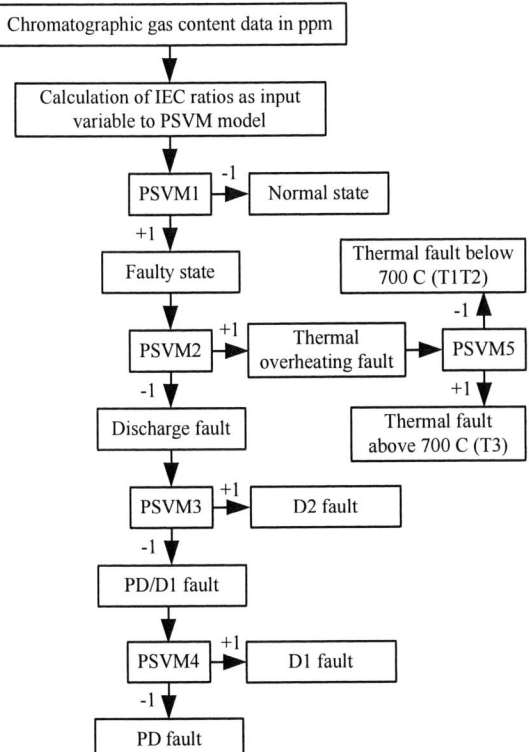

Fig. 4. PSVM classifiers Based Fault Diagnostic model of power transformer

The performance of all PSVM-classifier is examined by evaluating measures after implementing required modifications to account for the problem under study:

Classification accuracy (CA),

$$CA = \frac{Total\_Samples\big|_{correctlty\_classified}}{Total\_Samples\big|_{dataset}} \qquad (3)$$

Mean Squared Error (MSE),

$$MSE = \frac{1}{n}\sum_{k=1}^{n}(E_k)^2; \text{ Where } E_k = \big|OD_k - OA_k\big| \qquad (4)$$

Where n is equal to number of samples in data set, OD_k is the target value and OA_k is the actual model output obtained from the trained PSVM-classifier.

Mean Absolute Percentage Error (MAPE),

$$MAPE = \frac{1}{n}\sum_{k=1}^{n}\left|\frac{Fault_{PSVM\_classifier} - Fault_{actual}}{Fault_{actual}}\right| \times 100 \quad (5)$$

The PSVM is implemented using input parameter as IEC ratios. The classification accuracy, MSE and MAPE are evaluated with equation (3), (4) and (5) along with variation of nu as shown in Table 3.

Table 3. PSVM based accuracy analysis of IEC DGA interpretation method

PSVM model	MAPE	MSE	RMSE	Highest Successful detection (%)
PSVM1	1.5563917	0.03093	0.17586	96.92
PSVM2	1.5151515	0.03030	0.17408	96.97
PSVM3	1.1049723	0.02220	0.14866	97.79
PSVM4	0.43125	0.00925	0.096	99.05

V. RESULTS AND DISCUSSIONS

Fig. 4 shows the methodology adopted for classifications of transformers incipient faults using PSVM. The Matlab [15] based PSVM results have been presented in this section. The overall fault classification accuracy using PSVM was found to be 98%. Firstly training was completed and then the value of efficiency, "w" (Weight) and "G" (Gamma) evaluated for IEC ratios as input using PSVM are tabulated Table 4.

Table 4. PSVM classifiers results

Model	Nu	W	G	% accuracy
PSVM1	1	0.0001796	-0.8568	98
		0.0015		
		0.0005569		
	0.01	0.0013	-0.4667	97.5
		0.0474		
		0.0035		
PSVM2	1	0.0033	0.3566	99
		-0.0843		
		-0.0023		
	0.01	0.00289	0.1921	96
		-0.1008		
		-0.0035		
PSVM3	1	-0.0367	-0.2571	97.5
		-0.0258		
		-0.000764		
	0.01	-0.0163	-0.1038	99.5
		-0.0067		
		0.0001129		
PSVM4	1	0.0083	-0.3337	96
		0.0258		
		0.0024		
	0.01	0.0215	-0.0861	98.5
		0.0493		
		0.0032		
PSVM5	1	0.0020	0.0462	98
		-0.5767		
		0.0284		
	0.01	0.0019	0.0219	100
		-0.0153		
		0.0258		
			Overall accuracy	98.0

For comparing other AI methods, ANN and SVM have also been applied. ANN and SVM are implemented based on ratios of IEC method as input variables (Table 1) to check the fault classification accuracy in form of correct prediction and compared them with PSVM diagnostic model as shown in Table 5. The PSVM model gives better results than ANN and SVM model.

Table 5. PSVM results comparison with other AI methods based on IEC ratios as input variables

Database (Cases)	ANN (CP/TS)	SVM (CP/TS)	PSVM (CP/TS)
Seventy five cases described in [12]	68/75	71/75	74/75
Ten cases described in [9]	8/10	9/10	10/10

CP/TS: correct prediction/total samples

Further on the 50 oil samples from working transformers in the HPSEB Simla, India (simulated in TIFAC CORE Lab, NIT Hamirpur, India, during PG project) were tested using proposed PSVM model and some of them are shown in Table 6 and its diagnosis results are presented in Table 7.

Table 6. Simulated gas data of transformer in lab

Order of gases	ppm value of oil sample
$[H_2, CH_4, C_2H_2, C_2H_4, C_2H_6, CO, CO_2]$	[39 10 <0.5 6 3 337 558; 16 12 29 28 6 935 5004; 6 17 <0.5 17 10 405 3089; 6 <1 <0.5 4 <1 476 2575; 6 79 <0.5 27 137 295 2552; 7 171 <0.5 327 106 726 9320]

Table 7. Diagnosis results of simulated samples based on proposed input variables

Sample no.	AFC	IEC method [code/fault]	ANN	SVM	PSVM
1	TH <700°C	[001/<150°C]	T1/T2	D2	D2
2	DHE	[012/D2/D1]	D2	D1	D1
3	TH <700°C	[201/300-700°C]	T1/T2	T3	T3
4	AHED	[102/**ND**]	D2	D1	D1
5	TH <700°C	[200/150-300°C]	T1/T2	T1/T2	T1/T2
6	TH >700°C	[202/>700°C]	T3	T3	T3

AFC: actual fault condition; LTTF: low temperature thermal fault; ND: not detectable; TH: thermal fault; HED: high energy discharge; AHED: arcing high energy discharge.

From Table 7, it can be concluded that the proposed AI based methodology is generally in agreement with SVM (for all cases), ANN (except test number 1 to 4) and IEC method (except test number 1,4) for transformers fault identification. Because of not matching IEC codes in Table 1, various transformers (i.e. transformer number four) couldn't be diagnosed by using conventional (IEC) approach but are diagnosed using proposed PSVM approach.

VI. CONCLUSION

Incipient fault diagnosis of oil based power transformer is one of the core research areas in the field of CM of electrical power apparatus. DGA is very powerful mechanism for diagnosing faults in transformers. This paper proposed a new methodology of DGA diagnosis by using IEC TC 10 and related data. To develop an accurate and reliable method of DGA diagnosis, we organized 5 major gases in to three IEC ratios, defined by the IEC 60599 standard. The projected approach for DGA is straightforward and simple to use. A comparison with the IEC based diagnostic strategies demonstrated that the performance of the projected technique is much better than that of IEC/IEEE, ANN and SVM methods.

Future work is focused on the proposed methodology can be used for online condition monitoring of fault diagnosis in power transformer.

Acknowledgment

The authors would like to thank the management of TIFAC-CORE project on "Power Transformer Diagnostics", at NIT Hamirpur (H.P.) India, for providing the necessary support, facilities, and encouragement to carry out this work during PG project. We are also grateful to anonymous reviewers for useful comments.

REFERENCES

[1] Facilities Instructions and Techniques (FIST) Volume 3-31, Transformer Diagnostics, Bureau of Reclamation Hydroelectric Research and Technical Service Group Denver, Colorado, June 2003, available at <www.usbr.gov>.

[2] M. Duval, A Review of Faults Detectable by Gas-in-Oil Analysis in Transformers, IEEE Electr. Insul. Mag. vol. 18, no.3, pp. 8-17, 2002.

[3] Hasmat Malik, A. Azeem and RK Jarial, Application Research Based on Modern-Technology for Transformer Health Index Estimation, in: Proc. IEEE Intl. Multi-Conf. on Systems, signals and Devices (SSD), vol. 9, pp. 1-7, 2012.

[4] Hasmat Malik, Amit Kr Yadav, Sukumar Mishra, Taekeshwar Mehto. Application of neuro-fuzzy scheme to investigate the winding insulation paper deterioration in oil-immersed power transformer. Int J Electr Power Energy Syst., vol. 53: pp. 256–271, 2013. Doi.:http://dx.doi.org/10.1016/j.ijepes.2013.04.023

[5] IEEE guide for the Interpretation of Gases Generated in Oil-Immersed Transformers, ANSI/IEEE std. C57.104, 1991.

[6] IEC guide for the interpretation of dissolved and free gases analysis, IEC std. IEC/CEI 60599, 2007.

[7] Y. Zhang, X. Ding, Y. Liu and P. J. Griffin, An Artificial Neural Network Approach to Transformer Fault Diagnosis, IEEE Trans. Power Deliv. Vol. 11, No. 4, pp. 1836-1841, 1996. doi: 10.1109/61.544265

[8] L.V. Ganyun, Cheng Haozhong, Zhai Haibao and Dong Lixin, Fault diagnosis of power transformer based on multi-layer SVM classifier, Int. J. Electr. Power Syst. Res., Vol. 74, pp. 1–7, 2005.

[9] Hasmat Malik, Tarkeshwar and R.K. Jarial, "An Expert System for Incipient Fault Diagnosis and Condition Assessment in Transformers", *Proc. IEEE* Intl. Conf. on Computational Intelligence and Communication Systems, pp. 138-142, 2011.

[10] W. C. Flores, E.E. Mombello, J.A. Jardini, G. Rattá and A. M. Corvo, Expert system for the assessment of power transformer insulation condition based on type-2 fuzzy logic systems, Int. J. Expert Systems with Applications, Vol. 38, pp. 8119–8127, 2011.

[11] R. Naresh, Veena Sharma, and Manisha Vashisth, An Integrated Neural Fuzzy Approach for Fault Diagnosis of Transformers, IEEE Trans. Power Deliv., Vol. 23, No. 4, pp. 2017-2024, 2008.

[12] Glenn Fung and Olvi L. Mangasarian, "Proximal Support Vector Machine Classifiers", in *Proc.* KDD-2001: Knowledge Discovery and Data Mining, August 26-29, 2001, San Francisco, CA, pp. 77-86. Accessed on <ftp://ftp.cs.wisc.edu/pub/dmi/tech-reports/01-02.pdf>

[13] Michel Duval and Alfonso dePablo, Interpretation of Gas-In-Oil Analysis Using New IEC Publication 60599 and IEC TC 10 Databases, IEEE Electrical Insulation Magazine, Vol. 17, No. 2, pp. 31-41, 2001.

[14] Li, X., and Wu, H.: 'DGA interpretation scheme derived from case study', IEEE Trans. Power Deliv., 2011, 26, (2), pp. 1292–1293.

[15] MATLAB User's Guide. The MathWorks, Inc., Natick, MA 01760, 1994–2001 <http://www.mathworks.com>.

[16] Hasmat Malik, Sukumar Mishra and A.P. Mittal, "Selection of Most Relevant Input Parameters Using Waikato Environment for Knowledge Analysis for Gene Expression Programming Based Power Transformer Fault Diagnosis" in International Journal of Electric Power Components and Systems, 42(16):1–13, 2014.

[17] Proximal Support Vector Machine Classifiers software, 2003 Accessed on <http://research.cs.wisc.edu/dmi/svm/psvm/>

Implementation of Three -Leg VSC based DVR using IRPT Control Algorithm

J.Bangarraju and V.Rajagopal

Electrical and Electronics Engineering Department
B V Raju Institute of Technology
Narsapur, Medak,Andhra Pradesh, India.
bangarraju.jampana@bvrit.ac.in , rajagopal.v@bvrit.ac.in

A.Jayalaxmi

Electrical and Electronics Engineering Department
JNTU College of Engineering, Hyderabad
Andhra Pradesh, India.
ajl1994@yahoo.co.in

Abstract—**The proposedDynamic Voltage Restorer(DVR) is used to mitigate voltage sag, swell, harmonics and unbalance in the supply voltage.This paper presents Instantaneous Reactive Power Theory(IRPT) Control Algorithm for three-leg Voltage Source Converter (VSC) based DVR in the three-phase distribution system.The proposed DVR injects voltages in series with source voltage to regulate rated voltage. Thegating signals for three-leg VSC based DVR are extracted from reference load voltages. The proposed IRPT Control Algorithmbased DVR results arevalidated through computer simulation studies using MATLAB/SIMULINK.**

IndexTerms—**Dynamic Voltage Restorer, Power Quality, harmonics, IRPT, voltage source converter**

I. INTRODUCTION

The primary requirement of any power distribution system is to improve power quality. Theimprovement of power quality the distribution system has many advantages such as elimination of harmonics, maximum load capability, reactive power control, maximum utilization of equipment etc[1].The power electronic converters such as AC voltage controllers, voltage source converter, switched mode power supplies, uninterruptible power supplies has increased in domestic and commercial equipment etc[2]-[3].The power electronic converters not only pollutes power distribution system nut also enhances power quality [4]. These power electronic equipment draw harmonics from source and also are responsible for voltage dip at the load end. Custom power devices are based on power electronic converters which are used for compensation of voltage/currentharmonics and the reactive power loading from AC source [5].DVR is one of the series connected custom power device to mitigate voltage related problems at Point of Common Coupling (PCC).Many power quality standards are used in the design of power system with non-linear loads such as IEEE-519, IEEE1159, IEEE241 and IEC-61642 etc [6]-[7].

A DVR is connectedin series with three phase supply and load to mitigate voltage sag, swell, harmonics and unbalances in supply voltages.In a three-leg VSC based DVR, the power supplied/absorbed is zero in the steady state.The design, analysisand voltage injectionschemes of three-leg VSC based DVR is reported in the technicalliterature [8]-[10].The various control algorithms such as instantaneous symmetrical components theory for DVR[11], Artificial Neural Network (ANN) based control algorithm for shunt active filter [12],

sliding mode and fuzzy logic controller for static series compensator[13], mathematical optimization in ABC frame for active power filters [14] etc.In this paper, a simple IRPT control algorithm is developed for three-leg VSC based DVR for mitigation of voltage sag, swell,harmonics and unbalance in supply voltage. The computer simulation results of IRPT based DVR are validated by using MATLAB/SIMULINK.

II. PRINCIPLE AND OPERATION OF DVR

The schematic diagram of three-leg VSC based DVR is depicted in Fig.1.The three phase sources (v_{sa}, v_{sb}, v_{sc}), three phase source series impedance consists of resistance (R_s) and inductance (L_s).The DVR uses a three phase transformer to inject voltage in series with supply voltage to maintain load voltage at rated value. A three-leg VSC along with a dc capacitor (C_{dc}) is used in DVR.A series inductor (L_r) and a

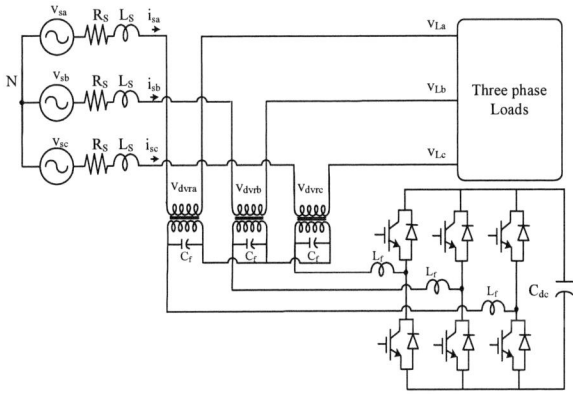

Fig.1 Schematic Diagram of three-leg VSC based DVR

parallel capacitor (C_r) are used to filter the harmonics inthe injected voltages. The DVR operationfor thecompensationof voltage-sag is depicted in Fig.2.The load voltage and load current during pre-sag voltage can be represented as $v_{L(pre-sag)}$ and i_s^1.After the post-sag, the load voltage (v_L) magnitude is lower than pre-sag voltage. The DVR injected voltage (v_{dvr}) is used to maintain the load voltage rated voltage. The DVR injected voltages has two components they arev_{dvrd} and v_{dvrq}. The in-phase injected voltage (v_{dvrd}) with source current is used to regulate dc bus voltage of the capacitor. The

978-1-4799-6047-7/14 $31.00 © 2014 IEEE

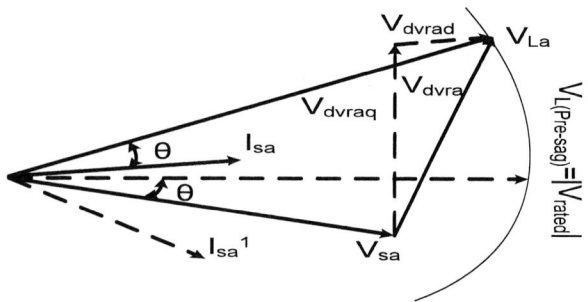

Fig.2 Phasor diagram of three-leg VSC based DVR

quadrature injected voltage in (v_{dvrd}) with source current is used to regulate load voltage. The objective of IRPT control algorithm is to achieve in-phase and quadrature components of injected voltages.

III. PROPOSED IRPT CONTROL ALGORITHM

The proposed IRPT algorithm is to estimate referenceload voltages for three-leg VSC based DVR. Fig.3 shows IRPT control algorithm for three-leg VSC based DVR for generation of gating pulses.The three-phase source voltages (v_{sa}, v_{sb}, v_{sc}) and three phase source currents (i_{sa}, i_{sb}, i_{sc}) are converted into α-β frame using Clark's transformation are represented as

$$\begin{bmatrix} v_{s\alpha} \\ v_{s\beta} \end{bmatrix} = \sqrt{\frac{2}{3}} \begin{bmatrix} 1 & -\frac{1}{2} & -\frac{1}{2} \\ 0 & \frac{\sqrt{3}}{2} & -\frac{\sqrt{3}}{2} \end{bmatrix} \begin{bmatrix} v_{sa} \\ v_{sb} \\ v_{sc} \end{bmatrix} \tag{1}$$

$$\begin{bmatrix} i_{s\alpha} \\ i_{s\beta} \end{bmatrix} = \sqrt{\frac{2}{3}} \begin{bmatrix} 1 & -\frac{1}{2} & -\frac{1}{2} \\ 0 & \frac{\sqrt{3}}{2} & -\frac{\sqrt{3}}{2} \end{bmatrix} \begin{bmatrix} i_{sa} \\ i_{sb} \\ i_{sc} \end{bmatrix} \tag{2}$$

These two α-β axis components of source voltages and source currents areused for calculating the instantaneous active and reactivepowers.This instantaneous power consistsof ac and dc components. It can be represented as

$$p_s = v_{sa}i_{sa} + v_{sb}i_{sb} + v_{sc}i_{sc} = v_{s\alpha}i_{s\alpha} + v_{s\beta}i_{s\beta} = p_{sdc} + p_{sac} \tag{3}$$

$$q_s = \left\{ \frac{1}{\sqrt{3}} \right\} \left\{ i_{sa}(v_{sa} - v_{sb}) + i_{sb}(v_{sb} - v_{sc}) + i_{sc}(v_{sc} - v_{sa}) \right\} \tag{4}$$

$$= v_{s\alpha}i_{s\beta} - v_{s\beta}i_{s\alpha} = q_{sdc} + q_{sac} \tag{5}$$

Two low pass filters (LPF) are used to extract dc component of active power (p_{sdc}) and reactive power (q_{sdc}) from instantaneous active and reactive powers.

The error in dc bus voltage of VSC ($v_{edc(k)}$) at k^{th} sampling instant is given as

$$v_{edc(k)} = v_{dc(k)}^* - v_{dc(k)} \tag{6}$$

Where $V_{dc(k)}^*$ is reference dc bus voltage and $V_{dc(k)}$ is the sensed dc bus voltage of the VSC.

The output of the DC PI controller is used to maintain constant dc bus voltage of VSC at the kth sample instant is given as

$$p_{loss(k)} = p_{loss(k-1)} + k_{p1} \left\{ v_{edc(k)} - v_{edc(k-1)} \right\} + k_{q1} v_{edc(k)} \tag{7}$$

Where k_{p1} and k_{q1} are the dc bus proportional and integral gain of VSC. $p_{loss(k)}$ is active component power loss of the VSC.

This active power loss (p_{loss}) in the VSC is added to the dc active power component p_{sdc} to obtain fundamental component of active power p_s^*

$$p_s^* = p_{sdc} + p_{loss} \tag{8}$$

This p_s^* is considered as reference active power component drawn from the load.

Similarly AC PI controller is used to maintain constant terminal voltage. The terminal voltage error ($v_{er(k)}$) at k^{th} sampling instant is given as

$$v_{er(k)} = v_{t(k)}^* - v_{t(k)} \tag{9}$$

Where $v_{t(k)}^*$ is reference terminal voltage and $v_{t(k)}$ is the sensed ac bus terminal voltage at k^{th} sample instant.

The sensed ac load terminal voltage (v_t)which is given as

$$v_t = \left\{ (\tfrac{2}{3})(v_{La}^2 + v_{Lb}^2 + v_{Lc}^2) \right\}^{\frac{1}{2}} \tag{10}$$

The output of the AC PI controller is used to maintain constant ac bus terminal voltage at the kth sample instant is given as

$$q_{loss(k)} = q_{loss(k-1)} + k_{p2} \left\{ v_{er(k)} - v_{er(k-1)} \right\} + k_{q2} v_{er(k)} \tag{11}$$

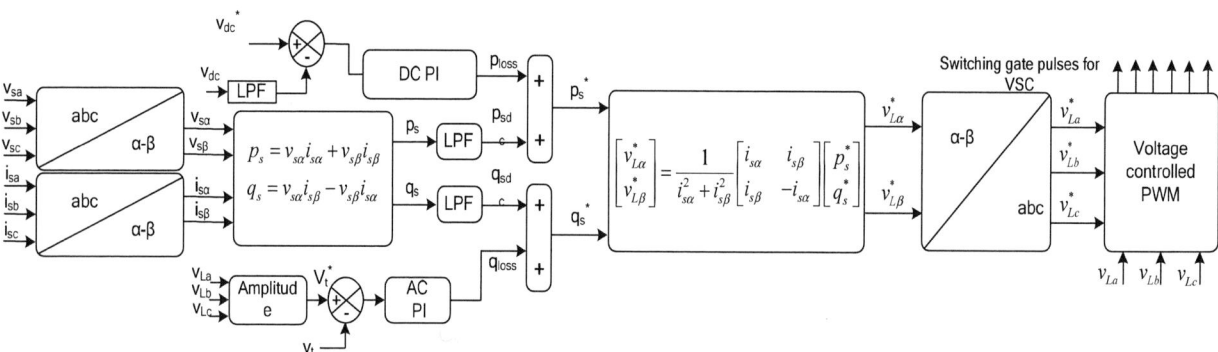

Fig.3.IRPT Control Algorithm for three-leg VSC based DVR

Where k_{p2} and k_{q2} are the ac bus proportional and integral gain of VSC. $q_{loss(k)}$ is the reactive component power loss of the VSC. This reactive power loss (q_{loss}) in the VSC is added to the ac reactive power component q_{sdc} to obtain fundamental component of reactive power q_s^*

$$q_s^* = q_{sdc} + q_{loss} \qquad (12)$$

These fundamental components of active power (p_s^*) and reactive power (q_s^*) are used to obtain reference load voltages in α-β frame as given by

$$\begin{bmatrix} v_{L\alpha}^* \\ v_{L\beta}^* \end{bmatrix} = \frac{1}{i_{s\alpha}^2 + i_{s\beta}^2} \begin{bmatrix} i_{s\alpha} & i_{s\beta} \\ i_{s\beta} & -i_{s\alpha} \end{bmatrix} \begin{bmatrix} p_s^* \\ q_s^* \end{bmatrix} \qquad (13)$$

These reference load voltages in α-β frame are transformed into three-phase reference load voltages in three phase a-b-c system using inverse Clark's transformation as given by

$$\begin{bmatrix} v_{La}^* \\ v_{Lb}^* \\ v_{Lc}^* \end{bmatrix} = \sqrt{\frac{2}{3}} \begin{bmatrix} 1 & 0 \\ -\frac{1}{2} & \frac{\sqrt{3}}{2} \\ -\frac{1}{2} & -\frac{\sqrt{3}}{2} \end{bmatrix} \begin{bmatrix} v_{L\alpha}^* \\ v_{L\beta}^* \end{bmatrix} \qquad (14)$$

These reference load voltages (v_{La}^*, v_{Lb}^* and v_{Lc}^*) are compared with the sensed load voltages(v_{La}, v_{Lb} and v_{Lc}) in the voltage controller for generation of gating signals.

IV. RESULTS AND DISCUSSION

The performance of IRPT Control Algorithm for three-leg VSC based DVR under various power qualitydisturbances are studied. The proposed IRPT Control Algorithm for three-leg VSCbased DVR is tested for various power quality disturbances like voltage sag is depicted in Fig. 4, voltage swell is depicted in Fig. 5 harmonics in supply voltage is depicted in Fig. 6, unbalance source voltage is depicted in Fig.7 and harmonic spectrum of IRPT Control Algorithm based DVR for non-linear load condition is depicted in Fig.8.

A. Performance of IRPT Control Algorithm based DVR during voltage sag

To demonstrate dynamic performance of IRPT Control Algorithm based DVR, voltage sag of 30% in supply voltage is introduced during t=0.75to 0.85secondswhich is depicted in Fig.4.The three phase supply voltage (v_s), DVR compensated voltage (v_dvr), three phaseload voltage (v_L), line current (i_L), reference dc bus voltage (v_dc^*) & sensed dc bus voltage (v_dc) and reference terminal voltage (v_t^*)&sensed terminal voltage (v_t) are depicted in Fig. 4. During this period of t= 0.75 to 0.85

seconds it was observed that IRPT Control Algorithm based DVR injects compensating voltage (v_dvr) in series with supply voltage (v_s) to maintain at rated loadvoltage (v_L). During voltage sag of 30% of supply voltage, it was observed thatsensed DC bus voltage (v_dc) is maintained constant at a value of 200V and sensed terminal voltage (v_t) is maintained at a reference value of 339V.

B. Performance of IRPT Control Algorithm based DVR during voltage swell

To demonstrate dynamic performance of IRPT Control Algorithm based DVR, voltage swell of 30% in supply voltage is introduced during t=0.65 to 0.75seconds which is depicted in Fig.5.The three phase supply voltage (v_s), DVR compensated voltage (v_dvr), three phase load voltage (v_L), line current (i_L), reference dc bus voltage (v_dc^*) & sensed dc bus voltage (v_dc) and reference terminal voltage (v_t^*) & sensed terminal voltage (v_t) are depicted in Fig. 5. During this period of t= 0.65 to 0.75 seconds, it was observed that IRPT Control Algorithm based DVR injects compensating voltage (v_dvr) in series with supply voltage (v_s) to maintain at rated loadvoltage (v_L). During voltage swell of 30% of supply voltage, it was observed that sensed DC bus voltage (v_dc) is maintained constant at a value of 200V and sensed terminal voltage (v_t) is maintained at reference value of 339V.

C. Performance of IRPT Control Algorithm based DVR during compensation of harmonics

To demonstrate dynamicperformance of IRPT Control Algorithm based DVR during compensation of harmonics in the supply voltage is depicted in Fig. 6.The three phase supply voltage (v_s), DVR compensated voltage (v_dvr), load voltage (v_L), line current (i_L), reference dc bus voltage (v_dc^*) & sensed dc bus voltage (v_dc) and reference terminal voltage (v_t^*) & sensed terminal voltage (v_t) are depicted in Fig.6.The source voltage (v_s) is distorted by connecting a non-linear load during t=0.55 to 0.7 seconds. During this period it was observed that IRPT Control Algorithm based DVR injects compensating voltage (v_dvr) in series with supply voltage (v_s) to maintain at rated harmonic free load voltage (v_L) and harmonic free.During harmonic compensation of supply voltage, it was observed that sensed DC bus voltage (v_dc) is maintained constant at a value of 200V and sensed terminal voltage (v_t) is maintained at reference value of 339V.The load voltage (v_L) has a THD of 2.71%, source voltage (v_s) has a THD of 8.91%, and source current (i_s) has a THD of 0.12% as shown in Fig.8.

D. Performance of IRPT Control Algorithm basedDVR during compensation of unbalanced source voltages

To demonstrate dynamic performance of IRPT Control Algorithm based DVR during compensation of unbalanced source voltages is depicted in Fig.7. The three

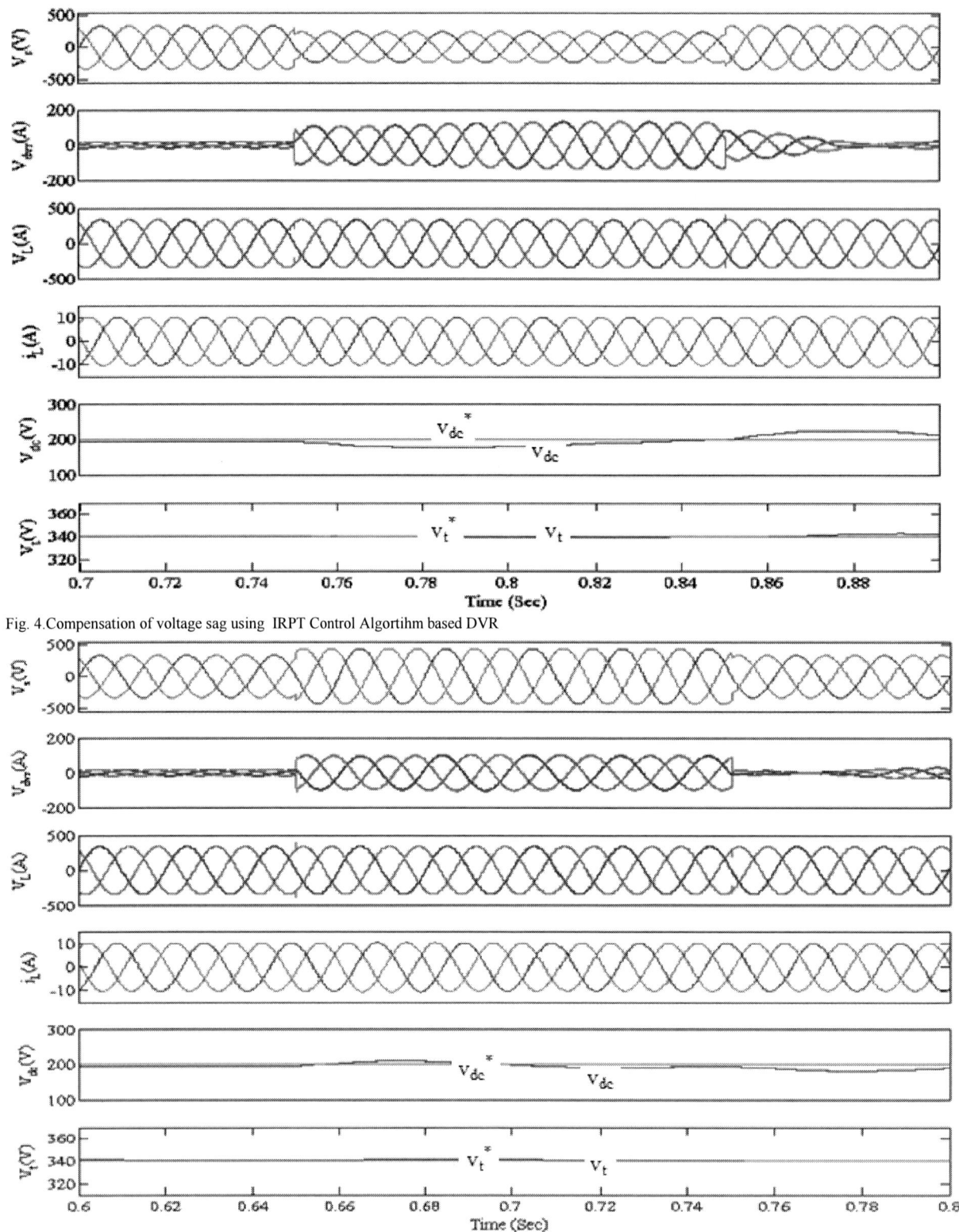

Fig. 4.Compensation of voltage sag using IRPT Control Algortihm based DVR

Fig. 5. Compensation of voltage swell using IRPT Control Algortihm based DVR

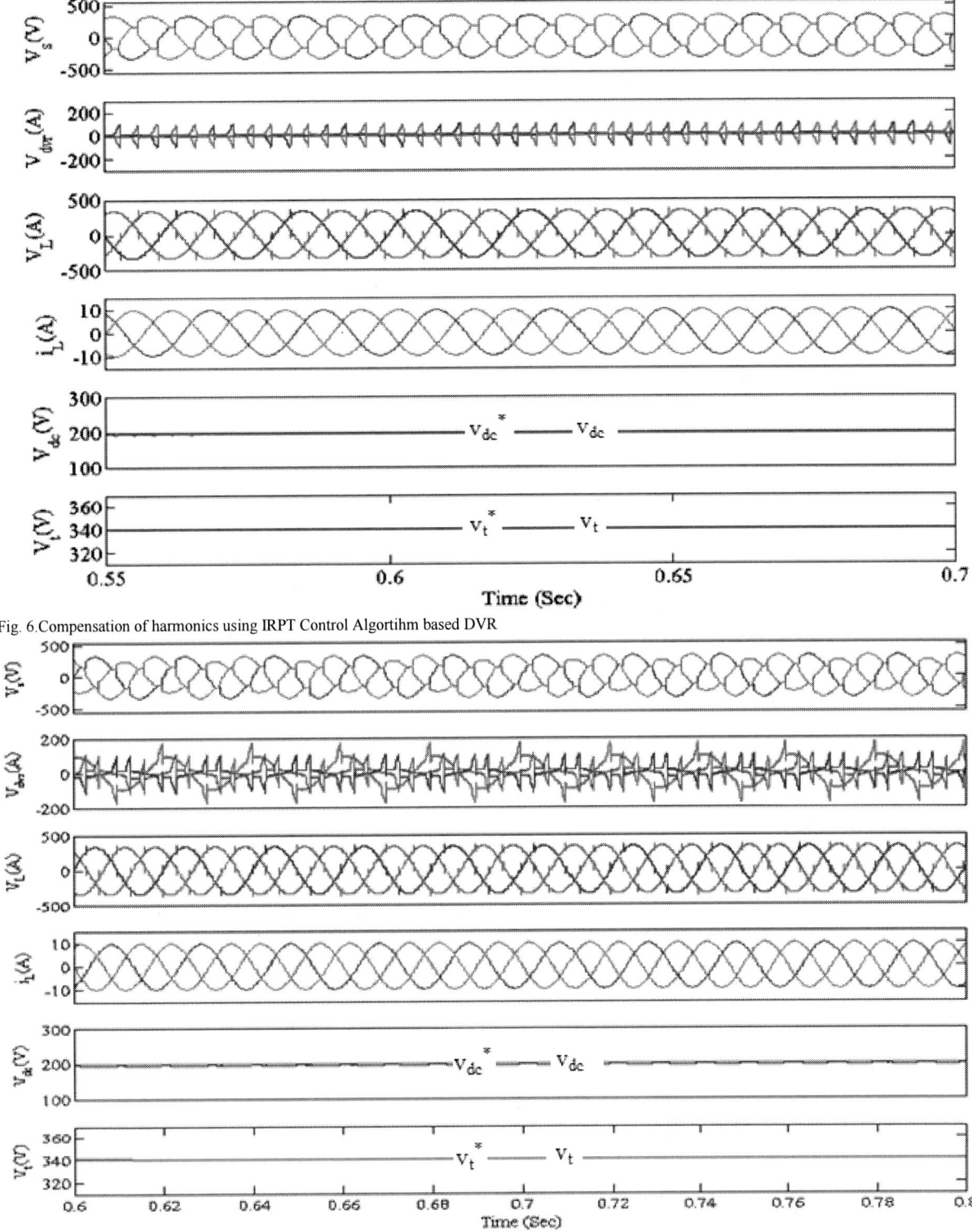

Fig. 6.Compensation of harmonics using IRPT Control Algortihm based DVR

Fig. 7. Compensation of unbalanced voltage sag using IRPT Control Algortihm based DVR

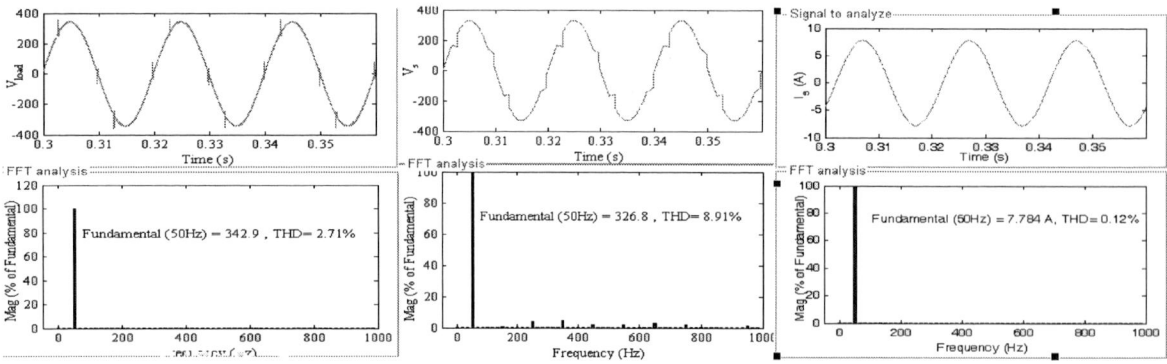

Fig. 8 Harmonic spectrum of load voltage, source voltage and source current under compensation of harmonics

phase supply voltage (v_s), DVR compensated voltage (v_{dvr}), load voltage (v_L), line current (i_L), reference dc bus voltage (v_{dc}^*) & sensed dc bus voltage (v_{dc}) and reference terminal voltage (v_t^*) & sensed terminal voltage (v_t) are depicted in Fig.7.The source voltage (v_s) is unbalanced source voltage which is created by reducing 'c' phase voltage to 315V and keeping the other two phases at 415V. During this period, it was observed that IRPT Control Algorithm basedDVR injects 'c' phase compensating voltage (v_{dvr}) in such a way that load voltage (v_L) is pure sinusoidal and constant. During compensation of unbalanced source voltage,it was observed that sensed DC bus voltage (v_{dc}) is maintained constant at a value of 200V and sensed terminal voltage (v_t) is maintained at reference value of 339V.

V. CONCLUSION

The dynamic performance of IRPT control algorithm based DVR shows satisfactory results for mitigation of voltage sag, voltage swell, unbalanced source voltages and compensation of voltage harmonics. The results show that IRPT control algorithm based DVR is simple and robust to mitigate power quality problems related voltage. It was observed thatsensed DC bus voltage (v_{dc}) is maintained at a referencevalue of 200V and sensed terminal voltage (v_t) is maintained at reference value of 339V during voltage sag, voltage swell, unbalanced source voltages and compensated voltage harmonics. The proposed IRPT control algorithm based DVR during harmonic compensation the load voltage (v_L) has a THD of 2.71%, source voltage (v_s) has a THD of 8.91%, and source current (i_s) has a THD of 0.12%which are well within IEEE-519 standard.

REFERENCES

[1] Angelo George J. Wakileh, Power System Harmonics: Fundamental, Analysis,and Filter Design. New York: Springer-Verlag, 2007.

[2] Ming-Yin Chan, Ken K.F. Lee and Michael W.K. Fung,"A case study survey of harmonic currents generated from a computer centre in an office building,"Journal of Architectural Science Review, vol. 50.3, pp 274-280, 2011.

[3] G. K. Singh, Power system harmonics research: a survey, European Transactions on Electrical Power, vol. 19, pp. 151-172, 2009.

[4] Muhammad H. Rashid, "Power Electronics Hand Book," Elsevier India, New Delhi, 2004.

[5] M. EminMeral, Ahmet Teke and Mehmet Tumay,"Overview of an extended custom power park," in Proc. of 2ed IEEE International Conference on Power and Energy , 2008, pp. 1364-1368.

[6] Angelo Baggini,Handbook on Power Quality, John Wiley & Sons, New Jersey, 2008.

[7] SurajitChattopadhyay,MadhuchhandaMitra and SamarjitSengupta,Electric Power Quality, Springer Verlag, London, 2011.

[8] A. Moreno-Munoz, D Oterino, M Gonzalez, F A Olivencia and J J Gonzalez-de-la-Rosa, "Study of sag compensation with DVR", in Proc. of IEEE MELECON, Benalmadena(Malaga), Spain, May2006, pp 990-993.

[9] Il-Yop Chung., Dong-Jun Won, Sang-Young Park, Seung-Il Moon and Jong-Keun Park, "The DC link energy control method in dynamic voltage restorer system", International Journal of Electrical Power & Energy Systems, vol. 25, no. 7, pp.525-531,Sept.2003.

[10] M. Vilathgamuwa, R. Perera, S. Choi, and K. Tseng, "Control of energy optimized dynamic voltage restorer", in Proc. of IEEE IECON'99, vol. 2,1999, pp. 873–878.

[11] A. Ghosh and Joshi A, "A new algorithm for the generation of reference voltages of a DVR using the method of instantaneous symmetrical components", IEEE Power Engineering Review, vol. 22, no.1, Jan.2002, pp. 63 -65.

[12] Bhim Singh, Vishal Verma and Jitendra Solanki, "Neural network based selective compensation current quality problems in distribution system", IEEE Trans. on Industrial Electron., vol. 54, no. 1, pp. 53 –60,Feb. 2007.

[13] B. N. Singh, Chandra A, Al-Haddad K and B Singh, "Performance of sliding-mode and fuzzy controllers for a static synchronous series compensator",IEE Proc. on Generation, Transmission and Distribution, vol. 146, no. 2, pp. 200 – 206, March 1999.

[14] Alejandro Garces, Marta Molinas and Pedro Rodriguez,"A generalized compensation theory for active filters based on mathematical optimization in ABC frame,"Journal of Electric Power Systems Research, vol. 90, pp. 1-10, 2012.

Power Quality Improvement Using Solar PV H-bridge Based Hybrid Multilevel Inverter

J.Bangarraju and V.Rajagopal

Electrical and Electronics Engineering Department

B V Raju Institute of Technology

Narsapur, Medak,Andhra Pradesh, India.

bangarraju.jampana@bvrit.ac.in, rajagopal.v@bvrit.ac.in

N. Bhoopal and M. Priyanka

Electrical and Electronics Engineering Department

B V Raju Institute of Technology

Narsapur, Medak,Andhra Pradesh, India.

bhoopal.neerudi@bvrit.ac.in, mpriyanka.195@gmail.com

Abstract—**This paper proposes a novel H-bridge based hybrid multilevel inverter consists of small number of switching devices and output of H-bridge multilevel by switching the solar PV voltage sources in series and parallel. The proposed H-bridge multilevel inverter reduces number of switching devices which reduces the power consumption and size of the gate driver circuits. The proposed inverter gives more number of output voltages which reduces total harmonic distortion of the output voltage waveform. The hybrid modulation method is used to control H-bridge multilevel inverter. The proposed inverter is validated through simulation results are validated by using MATLAB/SIMULINK.**

IndexTerms—**H-bridge multilevel inverter, Total Harmonic Distortion, Power Quality, Solar Cell**

I. INTRODUCTION

In the recent years multilevel inverters have gained popularity in medium and high power industries which provides stepped ac output voltage with minimum value of high frequency distortions. The main feature of multilevel inverter includes reduction in switching losses, reduction in transient voltage and improved harmonic spectrum [1].Many different multilevel inverter topologies are presented in the technical literature[2].

The main drawback of multilevel inverter configurations is requiring a more number of power switches and that must be precisely commutated, complex circuit. These multilevel inverters require more number of auxiliary independent dc level supplies which adds complexity of the circuit. These disadvantages of multilevel inverter were almost overwhelming due to the differences cost they produced between multilevel and standard inverter configurations. In some high power applications the multilevel inverters were used such as high power transmission, high power ac motor drives in mining, marine, chemical industries and power line conditioners etc.[3]–[4].In all above applications will compensates cost difference.

The continuous improvement in development of high power and switching frequency semiconductor devices such as insulated-gate bipolar transistors (IGBTs) and insulated-gate commutated thyristors (IGCT) has improved performance of the converters. These devices shows renewing interest in multilevel inverter topologies compete market with the standard two-level inverters [5].Initially three-level inverters are used for industrial applications but recently it enhanced for four-level, five-level etc. reported in the technical literature[6-10].

Different types of multilevel inverters are such as neutral point clamped, flying capacitor and cascaded H-bridge etc. In these topologies, cascaded H-bridge inverters have been discussed because of their simplicity and modularity [11]-[13].Many different modulation techniques can be applied to cascaded H-bridge inverters. By increasing the number of H-bridges the number of output voltage levels increases in cascaded H-bridge inverters. However, if the number of output voltage levels is increased, the number of switching devices is also increased, which makes a cascaded H-bridge inverter is more complex [14]-[16].

In this paper, a H-bridge based hybrid multilevel inverter using series/parallel conversion of solar PV voltage sources are proposed [12].The same number of solar PV voltage is required to with conventional cascaded H-bridge inverters. However, the solar PV based voltage source can be connected with small a number of series-connected PV arrays. The proposed hybrid multilevel inverter reduces number of switching devices and output voltage harmonics is also reduced [17]. The batteries, capacitors and other dc supplies can be used and proposed H-bridge multilevel inverter uses solar PV voltage. The proposed hybrid multilevel inverter is modeled and results validated through by using MATLAB/SIMULINK.

II. CIRCUIT CONFIGURATION OF PROPOSED INVERTERS

The proposed multilevel inverter with hybrid modulation method is depicted in Fig.1.The number of independent solar PV voltage sources are assumed as $(V_1, V_2, V_3,.....V_n)$.The switches with solar PV voltage source in series and parallel are assumed as S_{p1}-S_{pn-1},S_{q1}-S_{qn-1},S_{r1}-S_{rn-1}.Theseriesand parallel connected of proposed 15-level multilevel inverter is shown in Fig.2(a)-Fig.2(c).When switches S_{q1}-S_{qn-1}andS_{r1}-$S_{rn-1}$are ON andS_{p1}-$S_{pn-1}$are OFF, the current flows through S_{q1}-S_{qn-1}andS_{r1}-$S_{rn-1}$and all solar PV voltages $(V_1, V_2, V_3,.....V_n)$ are connected in parallel is shown in Fig.2(a).When the switch $S_{p1}$is ONand other two switches $S_{q1}$and $S_{r1}$are OFF, the two solar PV voltages V_1 and V_2 are in series through the switch $S_{p1}$is shown in Fig.2(b).The solar PV voltage are connected series/parallel, the lower H-bridge output is taken as $V_{out2}$with levels of 2n+1.The upper H-bridge output is taken as V_{out1}.The

978-1-4799-6047-7/14 $31.00 © 2014 IEEE

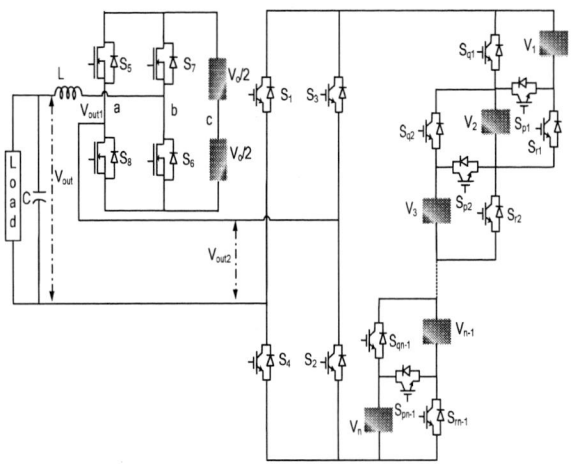

Fig.1.Proposed multilevel inverter with hybrid modulation method

proposed multilevel inverter outputs consists of 4n+3 levels by $V_{out2}-V_{out1}$ or $V_{out1}+V_{out2}$.

In a hybrid modulation based conventional cascaded inverter requires twelve switching are required for 11 level inverter whereas sixteen switching devices are required for 15-level inverter. The proposed multilevel inverter requires eleven switching devices for 11 levels where as fourteen switching devices for 15 levels. The proposed multilevel inverter reduces number of switches for more number of output voltages compared conventional cascaded H-bridge inverter. Whereas the same number of Solar PV sources are conventional cascaded H-bridge inverter.

The proposed multilevel inverter is based on hybrid modulation technique and several switching devices are operated at low frequency even when pulse width modulation technique is used.

III. HYBRID MODULATION TECHNIQUE

The hybrid modulation method of a proposed 11-level multilevel inverter is shown in Fig.3.The upper H-bridge output (V_{out1}), lower H-bridge output (V_{out2}) and output voltage (V_{out}) of proposed 11-level inverter is shown in Fig.4. Shows the out voltage waveform when the proposed 11-levelinverter is driven by the modulation method, as shown in Fig. 3.

The reference waveform (e_{01}) is comparing with the carrier waveform (e_s) to drive S_5 and S_8 switches. The resultant of these two waveforms is gives potential difference between a and c (V_{ac}) in Fig.1. is shown in Fig.3.Similary the reference waveform (e_{02}) is comparing with the carrier waveform (e_s) to drive S_6 and S_7 switches. The resultant of these two waveforms is gives potential difference between b and c (V_{bc}) in Fig.1. is shown in Fig.3.The upper H-bridge output voltage $V_{out1}= V_{ac}-V_{bc}$ is shown in Fig.3.The switches S_1-S_4 and S_{p1},S_{q1},S_{r1} are switches ON and OFF is also shown in Fig.3.

When the switches S_{q1}, S_{r1} are ON and switch S_{p1} is OFF then the output of lower H-bridge inverter V_{out2} will be

$$V_{out2} = V_1 = V_2 = 2V_0 \tag{1}$$

In the lower H-bridge inverter the switches S_{q1} and S_{r1} are OFF, S_{p1} is ON, and S1 and S2 are ON then the output voltage of inverter will be

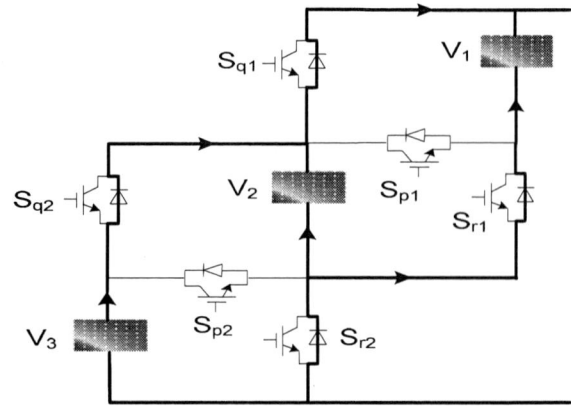

Fig.2 (a) Proposed inverter having V_1, V2, and V3 are in parallel

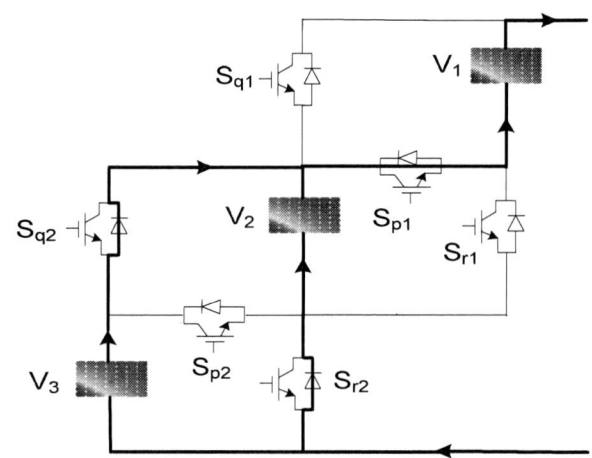

Fig.2 (b) Proposed inverter having V_1 in series with V_2 and V_3 are in parallel

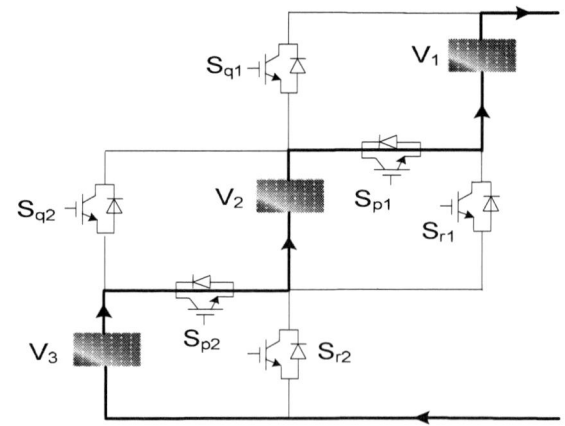

Fig.2 (c) Proposed inverter having V_1,V_2,V_3 are in series

978-1-4799-6047-7/14 $31.00 © 2014 IEEE

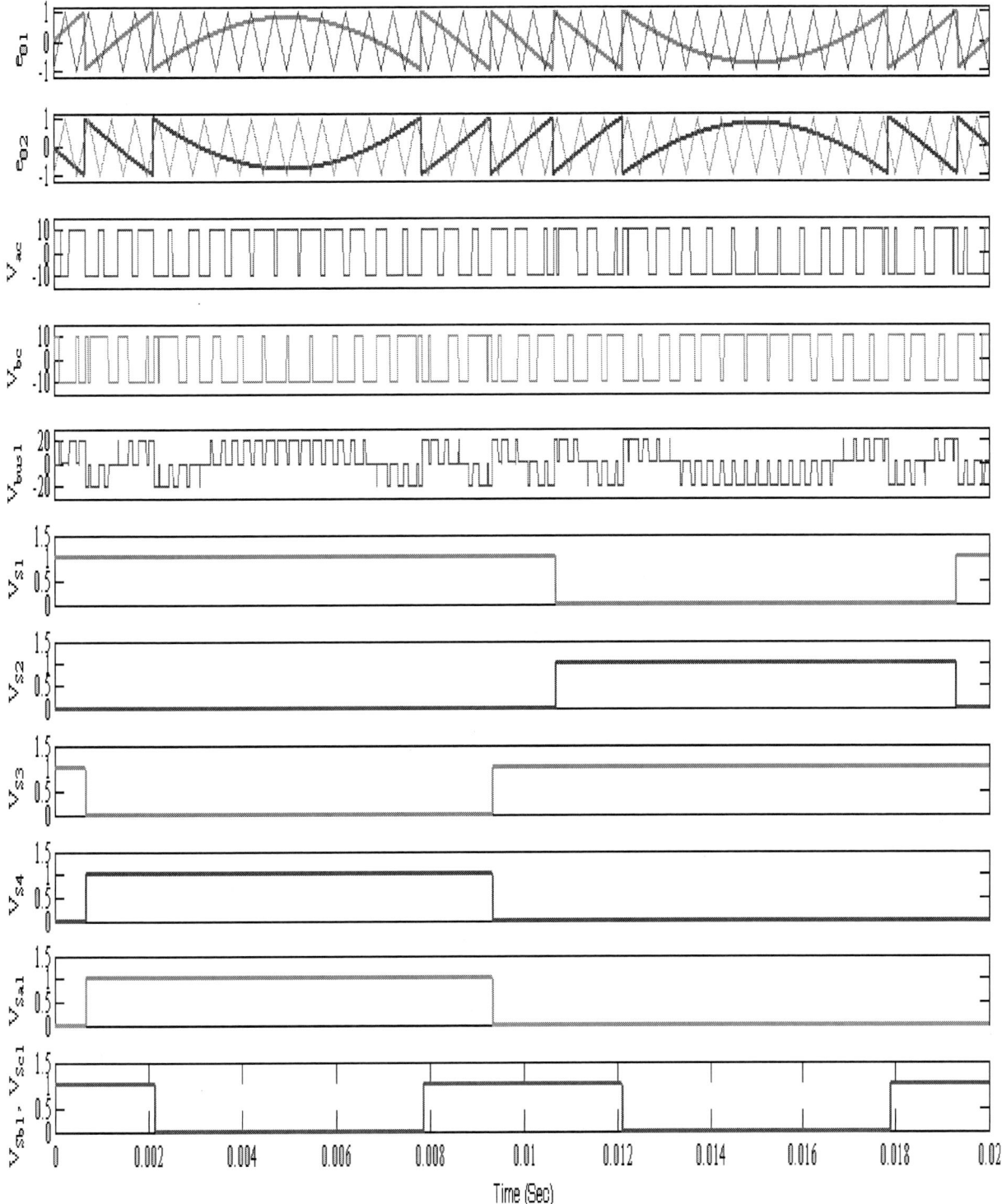

Fig.3.Hybrid Modulation Method for Proposed Multilevel Inverter

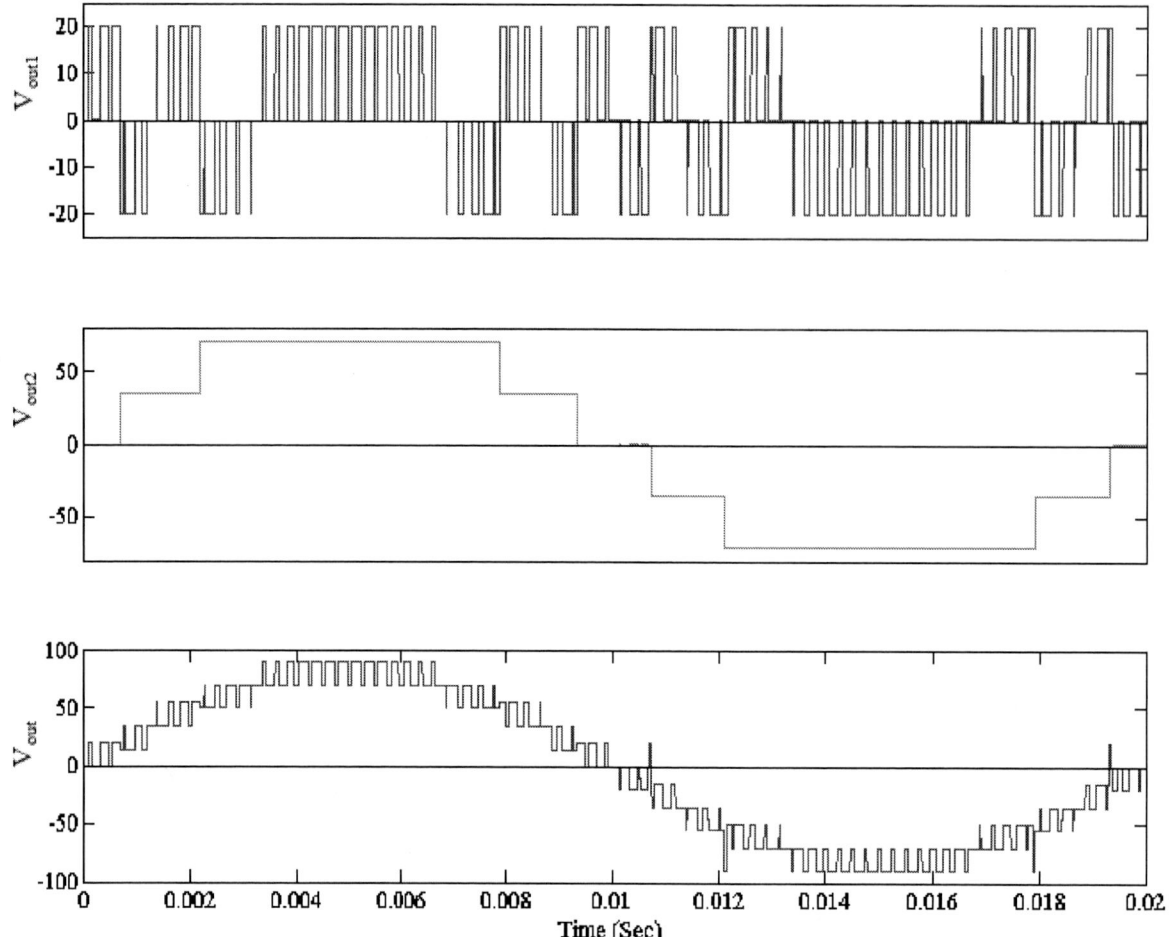

Fig.4.Multilevel Inverter with upper and lowers H-bridge Voltages

$$V_{out2} = V_1 + V_2 = 4V_0 \qquad (2)$$

When Switches S_{q1} and S_{r1} are ON, S_{p1} is ON and S_3 and S_4 are ON then the output voltage of lower H-bridge inverter (V_{out2}) will be

$$V_{out2} = -V_1 = -V_2 = -2V_0 \qquad (3)$$

When Switches Sq1 and Sr1 are OFF, and S3 and S4 are ON, then the output of lower H-bridge inverter(V_{out2}) will be

$$V_{out2} = -(V_1 + V_2) = -4V_0 \qquad (4)$$

The proposed 11-level inverter output voltage (V_{out}) is determined as

$$V_{out} = V_{out1} + V_{out2} \qquad (5)$$

The output voltages of upper H-bridge inverter (V_{out1}), lower H-bridge inverter (V_{out2}) and proposed inverter output voltage (V_{out}) are shown in Fig.4.The modulation index (M) of proposed inverter is defined

$$M = \frac{A_{eo}}{5A_{es}} \qquad (6)$$

Where A_{es}, A_{eo} are the magnitudes of reference and carrier wave respectively

The amplitude of proposed 11-level multilevel inverter can be calculated as

$$A = M(V_0 + V_1 + V_2) = 22.5 volts \qquad (7)$$

The System data for proposed inverter is shown in Appendix.

IV. RESULTS AND DISCUSSION

The proposed multilevel inverter is modeled and simulated results are validated by using MATLAB/SIMULINK. The inverter carrier waveform voltages (e_{01} and e_{02}), reference voltage (e_s), voltage (V_{ac}), voltage (V_{bc}), the output voltage upper H-bridge inverter (V_{out1}),the output voltage across upper H-bridge inverter switches ($V_{s1,} V_{s2,} V_{s3,} V_{s4}$) and the output voltage across lower H-bridge inverter switches ($V_{sp1,}$ $V_{sq1,}$

V_{sr1})are shown in Fig.3. The output voltages of upper H-bridge inverter (V_{out1}), lower H-bridge inverter (V_{out2}) and proposed inverter output voltage (V_{out}) are shown in Fig.4.The proposed multilevel inverter output voltage (V_{out}) and total harmonic distortion (THD) of the output voltage waveform is shown in Fig.5.The output voltage of 11-level inverter has fundamental output of 84.92V and has an THD of 15.17%.

V. Conclusion

The proposed multilevel inverter will reduce the number of switches is operating solar PV voltages are in series and parallel. The analysis of voltage harmonics and modulation are shown in waveforms. The proposed multilevel inverters will reduce number of switches used compared with conventional multilevel inverter and which reduces cost of cost of inverter. The output voltage of proposed 11-level inverter has a THD of 15.17%.The fundamental out voltage of proposed inverter has 84.93V.The proposed multilevel inverter can be used for medium and high frequency applications.

References

[1] R. S. Lai and F. Z. Peng, "Multilevel converters-A new breed of power converters," IEEE Trans. Ind. Appl., vol. 32, no. 3, pp. 509–517, May/Jun. 1996.

[2] J. Rodriguez, J. S. Lai, and F. Z. Peng, "Multilevel inverters: A survey of topologies, controls, and applications," IEEE Trans. Ind. Electron.vol. 49, no. 4, pp. 724–738, Aug. 2002.

[3] M. Marchesoni and P. Tensa, "Diode-clamped multilevel converters: a practicable way to balance DC-link voltages," IEEE Trans. Ind. Electron.,vol. 49, no. 4, pp. 752–765, Aug. 2002.

[4] N. P. Schibli, T. Nguyen, and A. C. Rufer, "A three-phase multilevel converter for high-power induction motors," IEEE Trans. Power. Electron. vol. 13, no. 5, pp. 978–986, Sep. 1998.

[5] A. Nagel, S. Bernet, and P. K. Steimer, "A 24 MVA inverter using IGCT series connection for medium voltage applications," in Proc. IEEE IAS Annu. Meeting, Chicago, IL, Oct. 2001, vol. 2, pp. 867–870.

[6] E. J. Bueno, R. Garcia, M. Marrón, and F. Espinosa, "Modulation Techniques Comparison for Three Levels VSI Converters," in Proc. IEEE 28th Annu. Conf. Ind. Electron. Soc., Nov. 2002, vol. 2, pp. 908–913.

[7] H. du Toit Mouton, "Natural balancing of three-level neutral-point clamped PWM inverters," IEEE Trans. Ind. Electron., vol. 49, no. 5,pp. 1017–1025, Oct. 2002.

[8] K. Corzine, X. Kou, and J. R. Baker, "Dynamic average- value modeling of four-level drive system," IEEE Trans. Power. Electron., vol.18, no. 2, pp. 619–627, Oct. 2003.

[9] T. Ishida, K. Matsuse, K. Sasagawa, and L. Huang, "Fundamental Characteristics of a Five-Level Double Converter for Induction Motor Drives," IEEE Trans. Ind. Electron., vol. 49, no. 4, pp. 775–782, Aug.2002.

[10] J. von Bloh and R. W. De Doncker, "Control Strategies for Multilevel Voltage Source Converter for Medium-Voltage DC Transmission Systems," in Proc. 26th Annu. Conf. IEEE Ind. Electron. Soc., Oct. 2000, vol. 2, pp. 1358–1364.

[11] L. G. Franquelo, J. Rodriguez, J. I. Leon, S. Kouro, R. Portillo, and M. A. M. Prats, "The age of multilevel converters arrives," IEEE Ind. Electron. Mag., vol. 2, no. 2, pp. 28–39, Jun. 2008.

[12] S. Daher, J. Schmid, and F. L. M. Antunes, "Multilevel inverter topologies for stand-alone PV systems," IEEE Trans. Ind. Electron., vol. 55, no. 7,pp. 2703–2712, Jul. 2008.

[13] B. P. McGrath and D. G. Holmes, "Multicarrier PWM strategies for multilevel inverters," IEEE Trans. Ind. Electron., vol. 49, no. 4,pp. 858–867, Aug. 2002.

[14] L. M. Tolbert, F. Z. Peng, T. Cunnyngham, and J. N. Chiasson, "Charge balance control schemes for cascade multilevel converter in hybrid electric vehicles," IEEE Trans. Ind. Electron., vol. 49, no. 5, pp. 1058–1064,Oct. 2002.

[15] H. Liu, L. M. Tolbert, S. Khomfoi, B. Ozpineci, and Z. Du, "Hybrid cascaded multilevel inverter with PWM control method," in Proc. IEEE Power Electron. Spec. Conf., Jun. 2008, pp. 162–166.

[16] J. Zhang, Y. Zou, X. Zhang, and K. Ding, "Study on a modified multilevel cascade inverter with hybrid modulation," in Proc. IEEE Power Electron.Drive Syst., Oct. 2001, pp. 379–383.

[17] YouheiHinago, Hirotaka Koizumi, "A Single-Phase Multilevel Inverter using Switched Series/Parallel DC Voltage Sources,"IEEE Trans. Ind. Electron., vol. 57, no. 8, pp. 2643–2650,Aug. 2010.

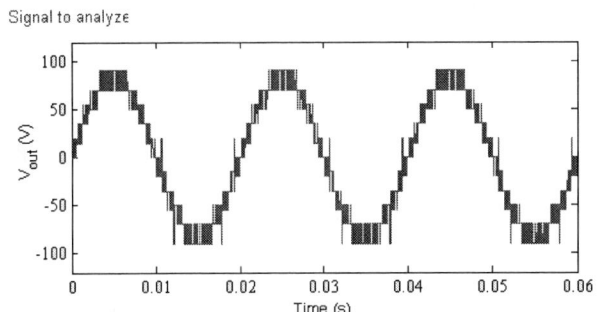

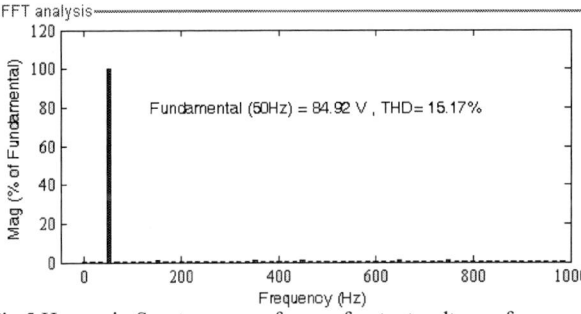

Fig.5.Harmonic Spectrum waveforms of output voltage of proposed inverter

Appendix

a) MOSFET Internal Resistance: 0.54ohms

b) MOSFET Switching frequency (f):40 kHz

c) Upper H-bridge Solar PV voltage (V_0):5V

d) Lower H-bridge Solar PV voltage (V_1-V_n)=10V

e) Modulation Index (M) =0.9

f) Smoothing Inductor (L) =560μH

g) Capacitance Filter(C)=0.5μF

978-1-4799-6047-7/14 $31.00 © 2014 IEEE

An Elimination Technique of Cross Regulations in the Flyback Converters

T. Halder
Assistant Professor
Electrical Engineering Department, Member IEEE
Kalyani Government Engineering College
Kalyani, District-Nadia, West Bengal,
PIN-741235, INDIA

Abstract—**The dark effects of the cross regulations envisage a substantial problem in the flyback converters for the multiple outputs. These effects eliminate using small inductors and RCD snuber circuit of the coupled inductor which are turned on to control the rate of secondary current while the main switch is turned off. By controlling the rate of change current, mutually DC bus and load, the effects of cross regulation to be improved in this paper as a simple and robust modeling and lucid analysis**

Keywords—**The flyback converters; Cross regulations; RCD Snubber; Modeling; Analysis and Results**

List of symbols:

C_{oss} = Drain to source parasitic capacitance
D = Duty ratio of the converter
f = Parasitic oscillation
i_2 = Instantaneous secondary of the second output
i_1 = Instantaneous secondary of the first output
I_p = Peak current at the primary
I_s = Peak current at the secondary
L_e = Primary leakage inductance
I_{se} = Leakage current at secondary side
L_1 = Secondary leakage inductance of the first output
L_2 = Secondary leakage inductance of the second output
I_{s1p} = Secondary peak current of the first output
I_{s2p} = Secondary peak current of the second output
N = Turns ratio
N_{s1} = Secondary turns of first output of the converter
N_{s1} = Secondary turns of second output of the converter
t_{on} = On- time of the gate pulse
t_{off} = Off- time of the gate pulse
T = Time period of the gate pulse
V_d = Diode voltage drop
V_e = Voltage across the leakage inductance
V_{o1} = Load voltage of the first output of the converter
V_{o2} = Load voltage of the second output of the converter
V_{cv} = Clamping voltage
V_r = Reset voltage
V_s = Supply voltage

I. INTRODUCTION

Cross regulation in the flyback converters is quite better than isolated power converters. The leakage inductance plays

---

It is fully supported by Sri Jagabandhu Halder. He donated me a blessing, huge inspiration and deep love which walks in my mind forever.

to contribute the cross regulations. In broad applications, the permissible dimension and heaviness are highly constrained to put up bigger payloads. In this attempt to add or to the high power density of power supplies, the switching frequency is used to implement a high standard at which PWM converter implementations consistently goes in front to considerable the anomalous power losses. An auxiliary significant shortcoming of the switch mode operation is the EMI generated due to huge di/dt and dv/dt. To prevail over these limitations, common families of converters are introduced [1]-[3]. To put into words about the elucidation points of the cross regulation is how the transferred current is married the secondary sides of the coupled inductors.

As for example, an additional inductor is mandatory for the forward converter. In practical case, this is not the case. When the main switch is turned on, the primary inductance stores the energy. The input current will achieve upto peak current. This peak current will be transferred to the secondary side when the main switch is turned off. The huge crooked effects are also predicted by power and peak current transferred from primary to secondary. After wound the coupled inductor, the secondary occupies the nominal leakage inductance compared to primary. If the load voltage is not fed by the feedback loop to control the PWM signal then peak current sensing will not take place cycle by cycle. If this output is used as the feedback loop then the duty cycle will be reduced, this in turn will reduce the other outputs of the converter [4]-[7].

An outstanding significant stages relating cross regulation is to choosing the number of turns in the non-feedback outputs of the regulators. Normally, the outputs of the converter within definite feedbacks are to be adhered in the control loop. It is attractive to include or eliminate a turn or regulate the feedback loop gain for the tight or load regulations and improvement of the converter's efficiency not upto marks [8]-[13].

This will amplify the hodgepodge and checking points to standardize to load voltage for the wide range of input voltage variations for the converters. The controller will take to put in all to outputs within their particular band. In many cases the cross-regulation problem goes ahead to be valid for linear or switching power supply for the multiple outputs which are not take into considerations of line and load regulations [14]-[17].

II. MODEL OF THE CROSS REGULATIONS

The Flyback SMPS is stored energy by the couple inductor of the primary side from supply voltage when the main switch is switched on and then transfers the storage energy to load after the main power switch are turned off.

The quantity of storage energy will be transferred per cycle. The load voltage is keeping constant by the controller from a particular feedback. A general two outputs flyback converter is shown in the Fig. 1.The converter topology with an RCD clamp (C3, R3, and D3) is dissipating the energy generated from the leakage inductance during the on time of power switch. Along with L_{eq} every sequence when the main switch is switched off and restraining the stresses on main the switch. In Fig. 1, the magnetizing inductance L_m of the couple inductor is replica of equivalent leakage inductances, related to primary, secondary winding 1, and winding 2 respectively. The resistance of windings is very petite passable to be ignored to implement simple and lucid leakage model as compared with the leakage inductance of the converter.

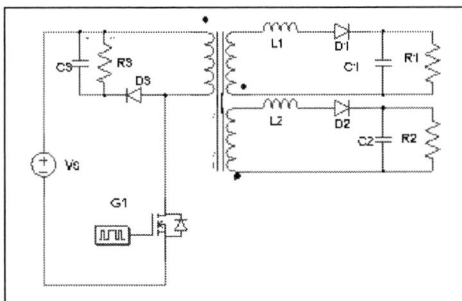

Fig. 1. Basic Flyback Converters with RCD clamp

III. Cross Regulation for Dual Outputs

When the main switch is switched off, how to allocate the transferred current of the converter, it may be predicted at the initial stage in accordance with the circuit diagram as shown in the Fig. 1. Assuming leakage inductance at the first winding is equal to twice value of second winding of the coupled inductor of the SMPS (switch mode power supply).

$$L_2 = 2L_1 \tag{1}$$

If the secondary turns of first and second output are, N_{s1} and N_{s2} respectively, it can be written from the equation (1) as:

$$L_2\left(\frac{N_{s1}}{N_{s2}}\right)^2 = 2L_1\left(\frac{N_{s1}}{N_{s2}}\right)^2 \tag{2}$$

Now, voltage across the leakage inductors may be written as output voltage of the converter:

$$V_e = V_{s1} - (V_{o1} + V_d) \tag{3}$$

The current will be transferred or circulated in accordance with Lenz's Law, then as soon as the switch is switched off as:

$$V_e = L_2\left(\frac{N_{s1}}{N_{s2}}\right)^2 \frac{di_2}{dt} \tag{4}$$

Now, the peak current of secondary of second output is given as:

$$I_{s2p} = \int_0^{t_{off}} di_2 \tag{5}$$

Now, combining the equation (4) & (5), yields as:

$$I_{s2p} = \frac{V_e}{L_2}\left(\frac{N_{s2}}{N_{s1}}\right)^2 t_{off} \tag{6}$$

The time period of the gate pulse is given as:

$$T = (t_{on} + t_{off}) \tag{7}$$

The operating duty ratio of the converter is given as:

$$D = \frac{t_{on}}{(t_{on} + t_{off})} \tag{8}$$

The turned off time of the converter may be written in terms of the duty ratio and time period of the converter from equations (7) & (8) as:

$$t_{off} = (1 - D)T \tag{9}$$

Similarly, turn on time of the converter may be written as:

$$t_{on} = DT \tag{10}$$

Now, the combination of equation (6) and (9) yields as:

$$I_{s2p} = \frac{V_e}{L_2}\left(\frac{N_{s2}}{N_{s1}}\right)^2 (1 - D)T \tag{11}$$

The equation (11) signifies inadequate since V_e is a function of time period (T) and operating duty ratio (D). So it may be treated, V_e as a constant. However, for perceptive how the leakage inductance effects the cross regulations, so it is highly effective for performance of the converter. Similarly, the equation (11) & (1) may correlate as:

$$I_{s2p} = \frac{V_e}{2L_1}\left(\frac{N_{s2}}{N_{s1}}\right)^2 (1 - D)T \tag{12}$$

Similarly, the peak current of the first output of the converter is given as:

$$I_{s1p} = \frac{V_e}{L_1}\left(\frac{N_{s2}}{N_{s1}}\right)^2 (1 - D)T \tag{13}$$

When the current, I_{s1p}, lastly equalizes the load current, I_o, and the charging current of the output capacitor (C1), I_c, which is the current desirable to charge the output capacitor (C1). A feedback is derived to come to an end the duty cycle, but by this time the output V_{o2}, has overshoot by a substantial quantity.

IV. Optimization of the Cross Regulations

It is learnt from to Faradays law that the same rate of change of secondary current in both outputs voltage maintained as:

$$L_2 = L_1\left(\frac{N_{s1}}{N_{s2}}\right)^2 \tag{14}$$

Peak current sensing is abridged by the adverse effects of the cross regulations evidently. This will improve the cross-regulation problem by minimizing peak overshooting or undershooting any output of the flyback converters. This practice also plays down opting for the right feedback voltage, V_{o1}, such that all outputs are within tolerance band. The cross regulation for multi-output flyback SMPS predicts significantly substantial improvement by lowering the clamping voltage, particularly to rather higher than the reflected load voltage. Yet, it creates more power loss in a conventional RCD clamp. Various solutions for improving the cross regulation with slight power loss are afforded and conferred in following fashions.

Bigger leakage inductance in the secondary sides of the coupled inductor guides to look up cross regulation of that load when it is not fully loaded. Bigger leakage inductance in primary windings will be lucrative to get better the problems of cross regulation of multiple output flyback SMPS as shown in Fig. 2. Though, it consequences are more loss in a usual RCD clamp. Reducing core gap of the magnetic to reach bigger magnetizing inductance in the flyback SMPS design aspects will enhance cross regulations sluggish the tight load or line regulations.

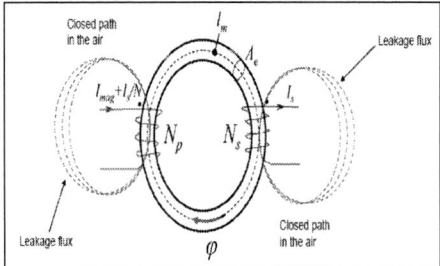

Fig. 2. Leakage flux generation of the coupled Inductor

The leakage current of the coupled inductor of the SMPS delays the incidence of the current .time (sec.current). The concealed energy also is dissipated in heat energy in the clamping circuit. A lesser amount of energy is transferred to the secondary side by virtue of which the converter efficiency is decreased. The peak current at the secondary side is reduced due to presence of the leakage inductance. It is also evident from the Fig. 3 as shown below.

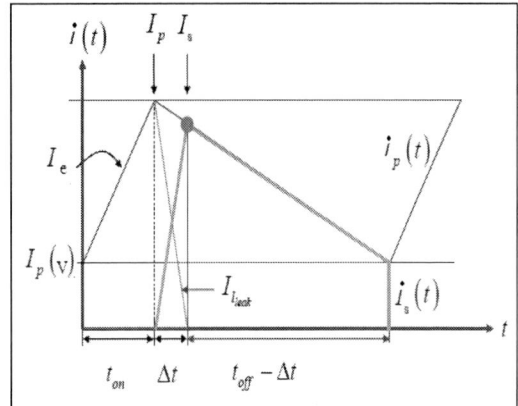

Fig. 3. Delayed commutation due to presence of the leakage inductance

Due to presence of the leakage inductance, the power circuit not only losses the volt.sec but also offers poor commutation. The leakage inductor reset time may be computed from the Fig.3.
The reset time may be computed as:

$$\Delta t = \frac{NL_e I_p}{NV_{cv} - (V_{o1} + V_d)} \tag{15}$$

A factor to be considered as a simple calculation:

$$\chi = \frac{NV_{cv} - (V_{o1} + V_d)}{L_e} \tag{16}$$

Hence, the secondary peak current is given as:

$$I_s = (I_p/N - \chi\Delta t) \tag{17}$$

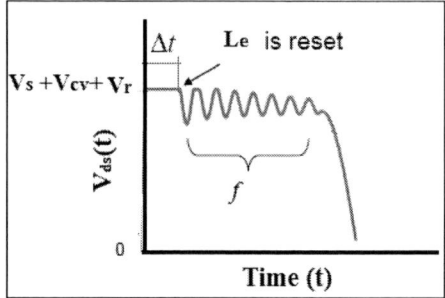

Fig. 4. Huge oscillations due to the leakage inductance

The leakage inductance induces a spurious spike voltage at tuned off condition of the main switch. The spike voltage should be kept control to defend the power MOSFET using the RCD clamping circuit as shown in the Fig. 1. The enormous oscillations come into view at the drain to source voltage. The oscillation frequency is shown in the Fig. 4. It may be expressed as:

$$f = \frac{1}{2\pi\sqrt{L_e C_{oss}}} \tag{18}$$

As the leakage inductance attempts to care for the current circulating at its peak current (I_p), when the switch is turned on during time, Δt and shoves the drain voltage up to safe significance of blocking voltage of the MOSFET. The instantaneous primary current is given as:

$$i_p(t) = \frac{(\Delta t - t)}{\Delta t} I_p \tag{19}$$

The energy losses of power switch as per Fig. 3 may be expressed as:

$$E_s = \int_0^{\Delta t} i_p(t) v_{ds}(t) dt \tag{20}$$

Now, combining the equation (19) & (20) yields as:

$$E_s = \frac{1}{2} I_P V_{ds(max)} \Delta t \tag{21}$$

It also points out toward the power losses due to switching transitions. Furthermore, the combination of the equations (15) and (21) yields as:

$$E_s = \frac{1}{2} I_P V_{ds\,(max)} \left(\frac{NL_e I_p}{NV_{cv} - (V_{o1} + V_d)} \right) \qquad (22)$$

V. PRIMARY CURRENT REDUCTION DUE TO LEAKAGE

The slope of the falling primary current is minimally given as:

$$S_p = N \frac{V_0}{L_P} \qquad (23)$$

This equation may be modified using the equation (15) and the Fig 3 given as:

$$\left(I_P - \frac{I_S}{\Delta t} \right) = N \frac{V_0}{L_P} \qquad (24)$$

Now, combining the equation (15) and equation (24) yields as:

$$\frac{I_s}{I_p} = 1 - \frac{L_e}{L_P} \left(\frac{V_{CV}}{NV_0} - 1 \right)^{-1} \qquad (25)$$

The equation (25) keeps up a correspondence to the effectual of primary current is basically stolen by the leakage inductance of the coupled inductor.

The leakage voltage (V_e) reflects the auxiliary winding due transformer action in terms of the auxiliary winding voltage (V_w), so output of the converter given as:

$$V_0 = (NV_w - V_e) = (V_w + V_{eq})N \qquad (26)$$

Where, V_{eq} is the equivalent voltage of the auxiliary winding from where feedback is taken to regulate the load voltage.

The supply voltage is now reflected to the secondary side puts in to the auxiliary winding and constructs the fighting fit for error of cross–regulations. The feedback loop senses an upper auxiliary winding, due to V_{eq}, and cut the duty ratio (D) enforcing V_0 to lessen by solving the equation (26) as:

$$V_{eq} = -\frac{V_e}{N} \qquad (27)$$

Relatively a lot of methods are present to cancel out the sound effects of the secondary leakage inductance of the coupled inductor.

VI. DESIGN OF RCD SNUBBER TO IMPROVE CROSS REGULATIONS

The RCD (Resistance Capacitance Diode) snubber circuit is used to improve the cross regulations. The circuit will protect against spike voltage generated by leakage inductance. Finally, leakage energy will be absorbed by this snubber network connected in parallel with transformer. It gives rise lower efficiency with higher active device ratings. This part presents also various design strategies of the RCD snubber for flyback converters to spike voltage when the MOSFET turns off, because of resonance between the leakage inductor (L_e) of the transformer and the drain to source capacitor (C_{oss}) of the MOSFET. The undue voltage on at the drain to source of the main switch may lead to an avalanche breakdown and ultimately harm the MOSFET. Therefore, it is essential to add a further clamping circuit to suppress the leakage voltage spike. This type of snubber circuit can be designed in conversional way as shown below in the Fig. 5. Leakage inductance L_e is caused by imperfect coupling of primary and secondary windings of the flyback transformer. Leakage inductance is effectively in series with main switch of the flyback converter. When power MOSFET switches off, it interrupts the primary current in L_e. Further, L_e induces a voltage spike across the main power switch of the flyback converter. If the peak amplitude of the voltage exceeds the voltage rating of the power MOSFET then the MOSFET will fail instantly. The energy of the leakage inductor

$$E_e = \frac{1}{2} L_e I_P^2 \qquad (28)$$

Peak magnitude voltage across the switch during off time exceeds the voltage rating of the MOSFET, then the power MOSFET will fail to run the converter operation.

$$V_{ds(max)} = (V_s + V_o N + V_e) \qquad (29)$$

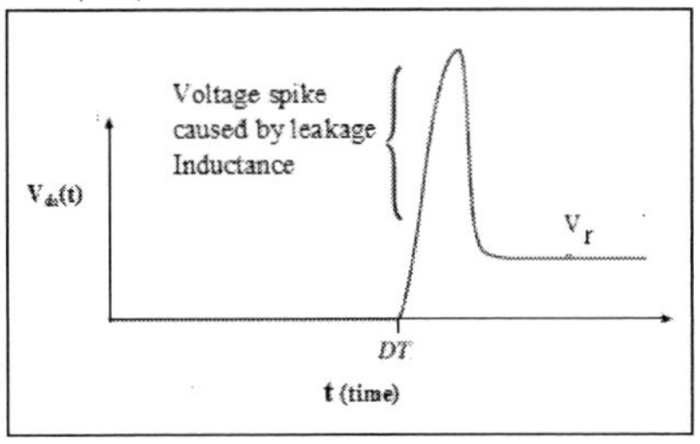

Fig. 5. The voltages across the power switch due to presence of the leakage inductance

The spike voltage, V_e will come across the switch due to leakage energy of the Flyback transformer. Normal reset voltage of switch is given as:

$$V_r = (V_s + V_{o1}N) \qquad (30)$$
$$\text{And } V_e = (V_{ds} - V_r) \qquad (31)$$

To limit the peak voltage (V_{ds}) to V_r of the power dissipating resistance (R_s) is to be required

$$R_3 = \left(\frac{V_e^2}{E_e} \right) \qquad (32)$$

L_e can be measured by shorting secondary winding of transformer and I_p will be decided by the design parameters of the converter like load current, input voltage, output voltage and duty ratio (D). Selection of snubber capacitance (C_3) is given as:

$$C_3 = \left(\frac{T}{R_3} \right) \qquad (33)$$

The equation (33) may be rewritten in terms the switching frequency of the converter is given as:

$$C_3 = \left(\frac{1}{R_3 f_s} \right) \qquad (34)$$

The above calculations are based on the value of leakage inductance of the coil. This snubber circuit should be used in parallel with transformer to protect against spurious spike.

This is generated by leakage inductance when the power switch is off. To protect the switch, R_3 and C_3 are chosen accordingly as shown in the Fig. 1.

VII. DESIGN EXAMPLE OF RCD CLAMP

The abnormal ringing relies on the leakage voltage nullified by the RCD snubber. The higher, it sets aside the clamp voltage to go up, the lower the power loss, but the more voltage and the rate of voltage (dv/dt) is dynamic to the rectifier diode. A small time, 20 ns turn-off delay is a significant part of the ringing waveforms of period. The Fig. 4 shows how to enlarge this ringing as acceptable clamp voltage. However, the power MOSFET is resonance cosseted. The RCD snubber is not only mitigate the vast EMI emissions from the ringing waveforms across the main switch of the converter but also abolish the cross regulations. The ringing be capable of then be damped out once more by initiating the RCD snubber.

When the main switch switched off, the primary current (I_p) charges the capacitor, C_{oss} of the MOSFET during on time. When the voltage across C_{oss}, (V_{ds}) goes beyond the supply voltage plus reflected load voltage specified in the equation (21). The rectifier diode turns on, so that the voltage across the magnetizing inductor (L_m) is clamped to a voltage given as:

$$V_{cv} = V_{o1}N \qquad (35)$$

Therefore, a combination of L_e and C_{OSS} with high produces resonance frequency and high voltage spike will come out across the drain to source of the main switch of the SMPS. This undue voltage on the main power switch may cause breakdown. In the case of CCM operations, the rectifier diode relics switched on until the main switch is turned on and a reverse recovery current of the rectifier diode is added to the primary current, and there is a huge current spike on the primary current at the switch-on occurrence. For the moment, since the secondary current runs dried up before the end of one switching cycle under DCM operations. Further, there takes place a resonance between L_m and C_{oss} of the main MOSFET.

Table I.

L_p	Le	Vr	V_{o2}	V_{o1}	Ve	V_{ds}	f_s	N	I_p
200µH	10µH	205V	5V	48V	65V	520V	70 K Hz	6.5	1.2 A

The Secondary inductance of the coupled inductor mortifies the cross regulations. While the designer fabricates SMPS, a little quantity of cross regulation is attractive to keep away from the load to run after educations. The secondary leakage inductance is protuberance in series with the winding, makes a parasitic effect which withholds the load voltage. Forlornly, the leakage components come into play in the power circuit to diorites the overall performance.

The snubber resistor (R3) is to be planned using the Table I. and the next equation is calculated by the equation (33), (34) and the Table I. R3=586Meg ohm of 10W, C3=120 nF. The voltage rises in a snubber for a conscientious peak current and leakage inductance of the converter is given as:

$$V_{xx} = \frac{1}{2}\left(\sqrt{V_d^2 + \frac{2R_3 L_e}{T}I_p^2}\right) - \frac{V_d}{2} \qquad (36)$$

For the want of any peripheral clamping circuit, it predicts the amount of energy E_s, is dissipated in the power MOSFET due to switching actions.

VIII. RESULTS

A changeable and unregulated output of the converter is to cut its pre-load resistance which yields more the power losses. An Assessment assesses that an adding a small inductor to the unregulated output cuts its output voltage error due to wide variation of input voltages and improves the cross regulation as drawn in Fig. 6, for the confirmation of the improvements of cross regulations.

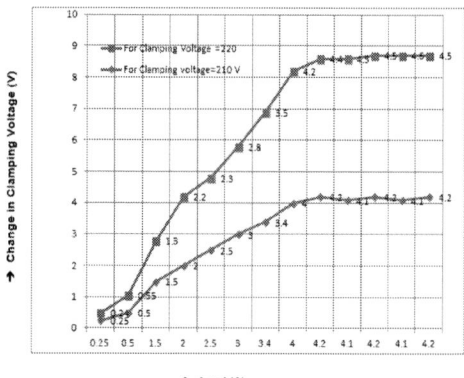

Fig. 6. Load vs. change in clamping voltage of the converter

In the torment and a small value of $1.36\,\mu H$, inductance is introduced in series with the secondary winding for output 1. A dual curved of clamping voltage 210V and 230V of the SMPS partition predicts the improved performance where there is an extra inductor. In the flyback SMPS design process, the clamping voltage is preferred to meet up the voltage stresses constraint of the main switch.

The error of voltage is abridged by 45 percentages when a substantial load is applied to load 1. It is balance with the error signal in the curved and foretelling. Though, its load voltage will fall more the entire through the light load in the regulated output 1 because the bigger inductance in output 2 grounds less power move toward load. A solitary approach is to cut the fall and summon up its unique output. It is also to raise its pre-load resistance as shown due to wider deviation of the input voltages in Fig. 7.

The leakage inductance of the coupled inductor is of assistance on the other hand to carry on control of output current in output lost volt-seconds short-circuit state of affairs. Rate of rise of current is subjective by leakage inductance, commutation primary-to less important is not immediate. As an outcome, the load versus efficiency curved of the power circuit will be furthermore improved by in addition to a teensy power inductor in the unregulated outputs where it is knowledgeable to be documented by concurrence of winding geometry in the window of core of the coupled inductor to reallocate the effects of leakage inductance allocation as shown in Fig. 8, for the tight load regulations.

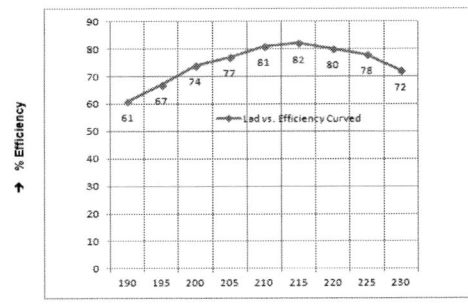

Fig. 7. Load vs. input voltage curved of the converter

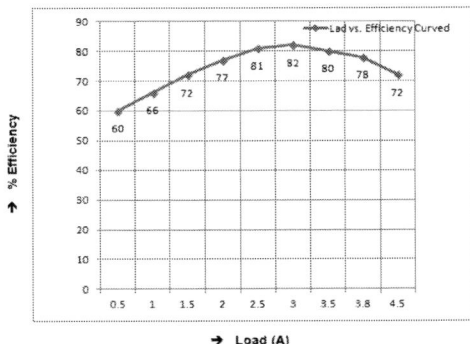

Fig. 8. Load vs. efficiency curved of the converter

While designing the switch mode power supplies (SMPS), the thoughtful of how the assortments of the leakage components interconnect within the proposal is the proper explanation to scheming dependable and jagged structures of the above curves. It is also advisable to go for painless dimensional techniques. The essential estimations of these parasitic components turn out to be trouble-free. With the aid of a professional spice simulator and passable replicas, the apt consequences can then instantaneously be envisaged and suitably salaried.

IX. CONCLUSIONS

The paper looks into the cross regulations complexities in double output flyback SMPS. The geometric replica for the cross regulations of the SMPS is highlighted. It is found that the lucid robust representation has an outstanding tenet with the entire domino properties. The leakage inductance of the regulated load voltage is to be play down to advance the cross regulations of the flyback SMPS. The transcript of treatments and the detail studies of cross regulations are highly predicted to improve performance of the flyback converters. The clamping voltage engrosses the cross regulation on multiple outputs. The lessons in this exertion have also unfastened a stout alliance between the clamping load voltage and the cross regulations and load of the SMPS. Commonly, the diagnostic predictions and duplication penalties are greatly confirmations of the improved performance of the current fed flyback SMPS as manufacturing and customer products.

ACKNOWLEDGEMENT

The author is deeply sad due to premature end of his father, Sri Jagabandhu Halder of Village- Sahapur, P.O-Chatina, District- Nadia, Pin-741160, and West Bengal. He is no more from 17-04-2014 at 9.45 PM (Thursday) at S.S.K.M Hospital, West Bengal. His absence is pensive forever. I never forget him. The doctor rereleased at Good Friday at 2.10 AM. I always remember him. His words inspired and touched many. As the world and relatives mourn his passing, his father continues to breathe in the hearts of the relatives through his motivational talks and endless contributions.

References

[1] T. Halder, "An Improved Hybrid Energy Recovery Soft Switching Snubber for the Flyback Converter" IEEE, PEDES-2012 pp 1-6

[2] T. Halder, "Study of Rectifier Diode Loss Model of the Flyback Converter" IEEE, PEDES-2012 pp 1-6

[3] B.Baha, "Analysis and control of a cross-regulated multi-output forward quasi-resonant converter," IEEE Proc.-Circuits Devices Syst., vol. 146, no. 5, pp. 255-262,October 1999

[4] S.Arulselvi,K.Deepa,G.Uma, " Design, analysis and control of a new multi-output flyback CF-ZVS-QRC," in Proc. IEEE/ICIT Conf, Hongkong,pp.413-419,December-2005

[5] T. Halder, "Study of Rectifier Diode Loss Model of the Flyback Converter" IEEE, PEDES-2012 pp 1-6"

[6] T. Halder "An Improved and Simple Hybrid Energy Recovery Snubber Circuit for Generic Power Converters and Protection Scheme" IEEE, IICPE-2012,pp 1-6

[7] T. Halder, "Comprehensive power loss model of the main switch of the Flyback converter " IEEE, ICPEC'2013 pp 792-797

[8] T.Halder, S. S. Saha, B. Majumdar, and S. K. Biswas, "A New Control Circuit Extends the Effective Duty Cycle Range of the Flyback Converters", IEEE PEDS 2005, Vol: 1, pp 413-417

[9] Jai P. Agarwal, "Determination of cross regulation in multi-output resonant converters," IEEE Transl. on Aerospace and electronic system., vol. 36, no. 3, pp. 760-772,July 2000.

[10] T.Halder, "Improved Performance Analysis of Clamp Circuits With Flyback Converter", International Journal of Emerging Technology and Advanced Engineering Website: www.ijetae.com (ISSN 2250-2459, Volume 2, Issue 1, January 2012) pp 1-8.

[11] T.Halder "Spacecraft Electrical Power Systems (EPS) Using the Flyback Converters"Proceedings of 2014 1st International Conference on Non Conventional Energy (ICONCE 2014) pp 52-57

[12] T.Halder "Continuous Conduction Mode (CCM) of Operations & Stability Analysis of the Flyback SMPS" Proceedings of 2014 1st International Conference on Non Conventional Energy (ICONCE 2014) pp, 292-297

[13] Joe Marrero, Improving cross regulation of multiple output flyback converters, Proceedings of PCIM, 1996.

[14] The effects of leakage inductance on multi–output FLYBACK circuits, Lloyd DIXON, UNITRODE Power Supply Design Seminar SEM–500

[15] T. Halder "Charge Controller of Solar Photo-Voltaic Panel Fed (SPV) Battery", , IEEE, IICPE-2010, pp 1-4

[16] T. Halder "Improved Coupled Inductor Loss Optimization of the Flyback SMPS", IEEE, ICPEC'13, pp 798-802

[17] Chuanwen Ji and Keyue M. Smedley, Cross regulation in flyback converters: Analytic model, PESC'99, pp. 920 -925,1999

Design and Simulation of H-Bridge Fed Direct Torque Controlled Electric Traction Drive

Priya Mahajan[1], Rachana Garg[2]
Delhi Technological University (DTU) [1,2]
Delhi, India
priyamahajan.eed@gmail.com, rachana16100@yahoo.co.in

Nikita Gupta[3], Parmod Kumar[4]
DTU[3], MAIT[4]
Delhi, India
guptanikita08@gmail.com, pramodk2003@yahoo.co.in

Abstract— **This paper presents the performance of electric traction motor drive consisting of an AC-DC converter fed vector controlled induction motor drive. The variable frequency induction motor drives used in AC traction system consist of uncontrolled ac-dc converters and voltage source inverters feeding the induction motor. Such converters induce power quality issues in the distribution system such as injecting harmonics which leads to voltage fluctuation, reactive power and low power factor issues. The converter topology discussed in this paper provides the constant dc link, keeps the input power factor close to unity and also reduces the THD of AC mains current. Direct torque control method is used to control H-bridge fed traction motor drive. Modelling and simulated results are presented to demonstrate the features of electric traction drive. The necessary modelling and simulations are done in MATLAB Simulink.**

Keywords-Vector Control, Direct Torque Control, Induction Motor, MATLAB Simulation

I. INTRODUCTION

The railway as a means of transport is a very old idea. From the early invention of steam engines to modern electric locomotives, railway transport systems had a long history and huge developments to become one of the most popular modes of public transport over the last century [1].To provide power propulsion and the need of higher-speed, more luxurious and more reliable services, electrification has been the first choice and is widely applied to modernize most of the railway transport systems across the world for several decades.

Earlier DC motor drives were used by Indian railways which were fed by 25kV single phase AC supply rectified by converters. The uncontrolled converters produce harmonics which decrease system efficiency and causes ac mains voltage distortion leading to failure of motors. All these effects impair the power quality. Also regeneration could not be obtained from uncontrolled rectifier because they did not work in four quadrants. This further reduces the efficiency of drive. With the advancements in power electronics, it is made possible to replace dc motor with variable frequency induction motor drive [2]. The field oriented control techniques of induction motor drive combines the features of flexibility of dc motor and ac induction motor which is robust and free from regular maintenance[3]. To improve the distortion of the input line current various converters have been reported which are capable of harmonic control and reactive power compensation for electric traction system [4-6].

Fig.1 shows the block diagram of three-phase induction motor drive. The traction system comprises of main transformer, converter/ inverter box and traction motor. The traction system is described as a 'three-phase drive' with V/F control. The converter carries out the constant DC output voltage control and AC- side power factor control. The V/F inverter output voltage waveform is PWM type using IGBTs. The converter inverter system converts single-phase AC voltage to the three-phase AC voltage which is necessary to drive the three-phase induction motor. The designing of converter and inverter are very crucial for railway traction as electric locomotive have regeneration capabilities.

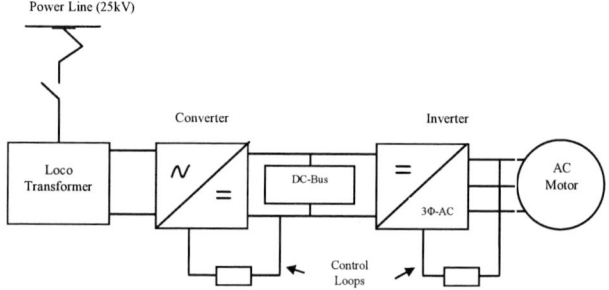

Fig.1 Block Diagram of AC-DC-AC traction system for AC motor drive

In the present work, an H-bridge converter configuration and direct torque control technique for speed control of traction motor drive used in Indian railways is simulated first time in literature. MATLAB/SIMULINK is used for simulation platform. The modeled system improves the power factor close to unity and the THD of ac mains current is also within IEEE standard 519 limits [7].

Fig.2 shows the block diagram of the traction motor drive suitable for electric traction. Overhead line is provided with tapings at equal distances. Through pantograph 25kV single-phase supply is tapped. A three winding transformer with two secondary windings is connected to supply. This transformer step down the 25kV supply and provide 975 V supply to the

978-1-4799-6047-7/14 $31.00 © 2014 IEEE

secondary side. Further two separate converters are connected to the secondary windings which maintain a constant dc link and mitigate the harmonics created in supply side. H-Bridge configuration of converter is used which consist of IGBT for bidirectional flow of power. Regarding converter feedback control loop, DC-link voltage is sensed and a comparison is done between sensed voltage and reference voltage. Comparison leads to development of a voltage error signal which is then fed to PI controller which generates the reference current. This reference current is then compared with sensed ac mains current. This comparison generates a current error which is then fed to PWM signal generator which generates gating signal for the converter. DC-link generated serves as input for inverter which uses direct torque control scheme for further connected induction motor speed control. Two induction motors are connected to a single inverter.

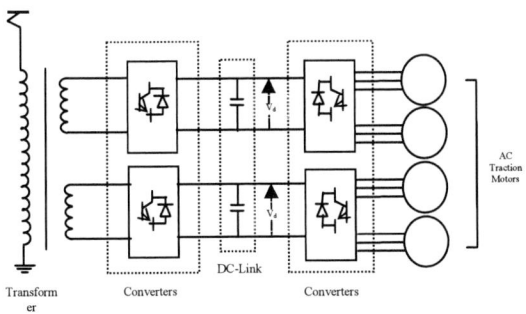

Fig.2 Block diagram of electric traction motor drive system configuration

II. CONTROL TECHNIQUES FOR TRACTION MOTOR DRIVE

The induction motor drives play an important role in modern electric traction system. This is due to their inherent advantages such as low inrush current, fast dynamic response, robust motor, ease of control etc. Techniques generally used in variable frequency induction motor drives for control of voltage source inverter fed induction motor drive can be broadly classified as scalar control, field oriented control (FOC) and direct torque control [8-9]. Application of scalar control is easy and simple, but it has its disadvantages of sluggish response instability of system because of higher order system effect. This above mentioned problems can be overcome by field oriented control and direct torque control, in which induction motor can be controlled like a separately excited dc motor. The revolution in self commutating solid state power devices and the availability of fast computing processors has led to the increased use of control techniques like field oriented control and direct torque control, resulting in fast dynamic response and simple and linear speed control of induction motor drive.

III. FIELD ORIENTED CONTROL

FOC control utilizes the position of the rotor combined with two-phase currents to generate a means of instantaneously controlling the torque and flux. FOC require control of both

magnitude and phase of the AC quantities and are, therefore, also referred to as vector controllers [9-10]. FOC produces controlled results that have a better dynamic response to torque variations in a wider speed range as compared to other scalar methods. Also, FOC control can induce a high torque at zero speed [2]. A block diagram of field oriented controlled induction motor drive is shown in fig. 3. The induction motor is fed by a current- regulator. The motor drives a mechanical load characterized by inertia J, friction coefficient B, and load torque, T [10]. The speed control loop uses a proportional-integral controller to produce the quadrature axis current reference i_q* which controls the motor torque. The motor flux is controlled by the direct-axis current reference i_d*. Block d-q to abc is used to convert i_d* and i_q* into current references i_a*, i_b*, and i_c* for the current regulator.

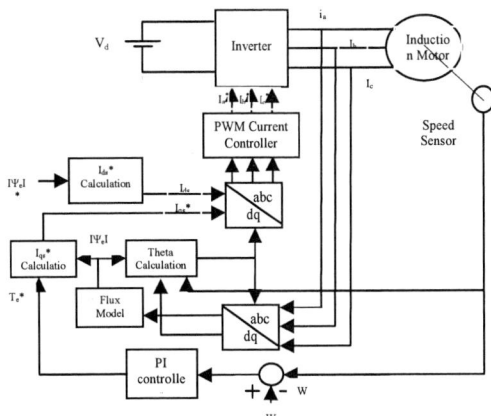

Fig.3 Block diagram of field oriented control of induction machine

In this mode of operation, control of the torque and flux is decoupled such that the d-axis component of the stator current controls the rotor flux magnitude and the q-axis component controls the output torque [12].

The stator direct-axis current reference, $i_{ds}*$ is obtained from rotor flux reference input $|\psi_r|*$ given by (1)

$$i_{ds}^* = \frac{|\psi_r|^*}{L_m} \qquad (1)$$

The required q-axis reference component of the stator current i_{qs}, for a given torque demand (T_e*), can be determined from (2).

$$i_{qs}^* = \frac{2}{3} \cdot \frac{2}{p} \cdot \frac{L_r}{L_m} \cdot \frac{T_e^*}{|\psi_r|_{est}} \qquad (2)$$

Fig.4 shows the MATLAB model of field oriented control technique used to control induction motor drive. Different blocks such as a PI speed controller, two phase rotating to three phase stationary converter block, hysteresis current controller, voltage source inverter (VSI), an induction motor etc. are realized using Simulink.

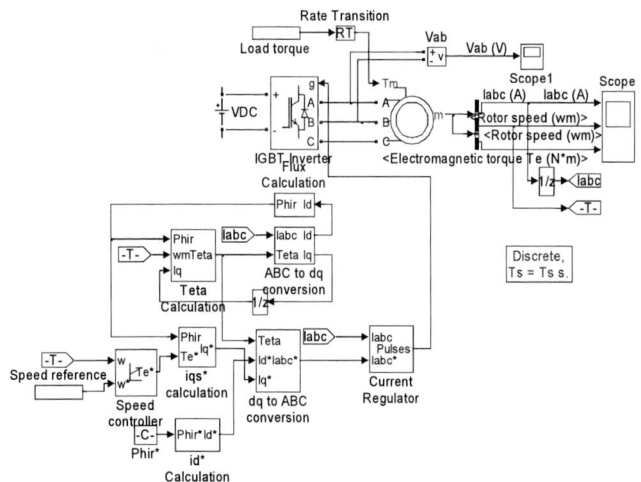

Fig. 4 Simulink model of field oriented control of induction motor

IV.   DIRECT TORQUE CONTROL

A direct torque control (DTC) induction motor drive is supplied by a voltage source inverter and it is possible to control directly the stator flux linkage λ_s (or the rotor flux λ_r or the magnetizing flux λ_m) and the electromagnetic torque by the selection of an optimum inverter voltage vector . Direct torque control of induction motors is described by the basic functional blocks as shown in fig.5. By means of the use of closed loop estimator, the instantaneous values of flux and torque are calculated from stator variables [11].

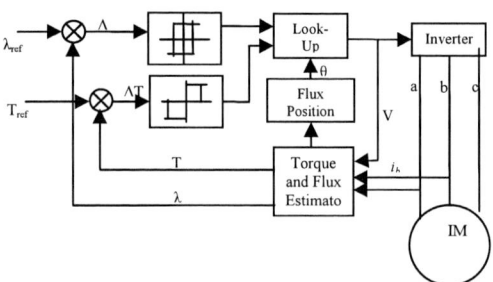

Fig.5 Block diagram of direct torque control of induction machine

The stator flux and torque magnitudes are compared with respective estimated values and errors are processed through hysteresis controllers. The output of the flux and torque comparator is used for the inverter optimal switching table. The electromagnetic torque in the three phase induction machines can be expressed as in (3):

$$T_e = 3/2 \ P|\psi_s|.|i_s|.\sin(\alpha_s - \rho_s) \qquad (3)$$

where ψ_s is the stator flux, i_s is the stator current, P the number of poles, ρ_s is the stator flux angle and α_s is the stator current angle, both referred to the horizontal axis of the stationary frame fixed to the stator. If the stator flux modulus is kept constant and the angle ρ_s is changed quickly, then the electromagnetic torque is directly controlled [12]. The same expression can be obtained for the electromagnetic torque as in (4):
$$T_e = \frac{3}{2} \ P \frac{L_m}{L_sL_r - L_m^2} |\psi_r|.|\psi_s|.\sin(\rho_s - \rho_r) \qquad (4)$$

As long as the stator flux modulus is kept constant, then the electromagnetic torque can be rapidly changed and controlled by means of changing the angle $\rho_s - \rho_r$.

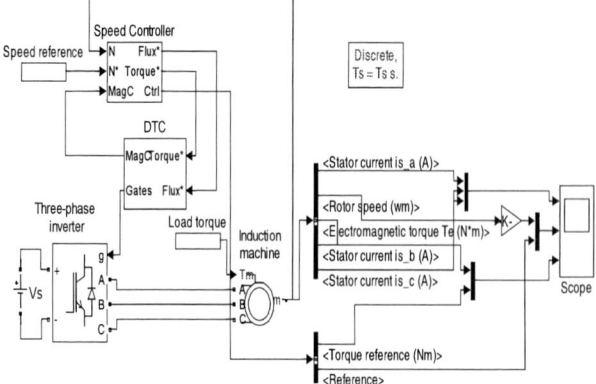

Fig.6. Simulink model of direct torque control of induction motor

Fig.6 gives the simulink model of direct torque control of induction motor. Decoupled control of the stator flux modulus and torque is achieved by acting on the radial and tangential components respectively of the stator flux-linkage space vector in its locus.

V.   H-BRIDGE SINGLE PHASE AC-DC CONVERTER

With the advancement in power electronics and quality of semi-conductor devices and fast digital processors has led to improvement converter design. Converters provide the rectification of ac voltage and also improve power quality on ac and dc side. These converters can be used for both uni-directional power flow as well as bi-directional power flow load applications. Bi-directional power flow characteristics find wide application in electric traction where drive requires a four quadrant operation. In electric traction application, the energy conservation aspect plays a major role, as the locomotive has to pass a number of up and down gradients on the way and the use of such converters results in regeneration of energy leading to saving in electric consumption [13-17]. Fig.7 shows the feedback controlling scheme of the converter which is modelled for the applications of the electric traction drive[5]. As electric traction application uses regeneration so IGBTs are used to design the H-bridge configuration of this converter.

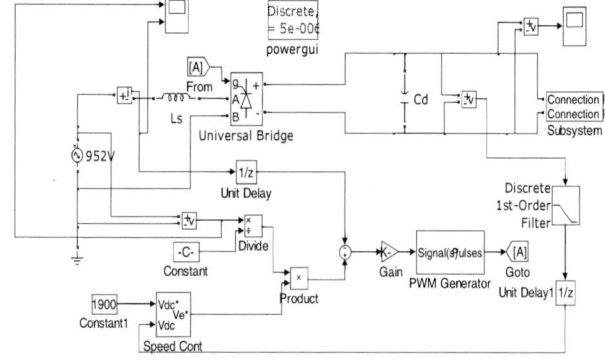

Fig.7 MATLAB model of Converter fed electric traction motor drive.

978-1-4799-6047-7/14 $31.00 © 2014 IEEE   598

VI. SIMULATION RESULTS AND ANALYSIS

The converter and the control techniques i.e. field oriented control and direct torque control for induction motor drive used in electric traction have been simulated in MATLAB Simulink R2007a on Intel Core 2 Duo Processor 2.4 GHz Window Vista Home basic (32-bit) Laptop[12].

Initially both the control techniques i.e. field oriented control and direct torque controls were implemented on 295hp motor with dc supply. Details of the parameters of the simulated model are given in Appendix-A. The results for torque (Nm), rotor speed (rad/s) and stator currents (A) at no load, half the load and full load conditions for the reference speed at 215 rad/sec are shown for both the methods in fig.8 to fig.10. Fig.8 shows the steady state waveform of the three phase stator current, speed and torque for no load at rated speed. Further fig.9 and fig.10 shows the steady state waveform for both the schemes for half the rated load and full load

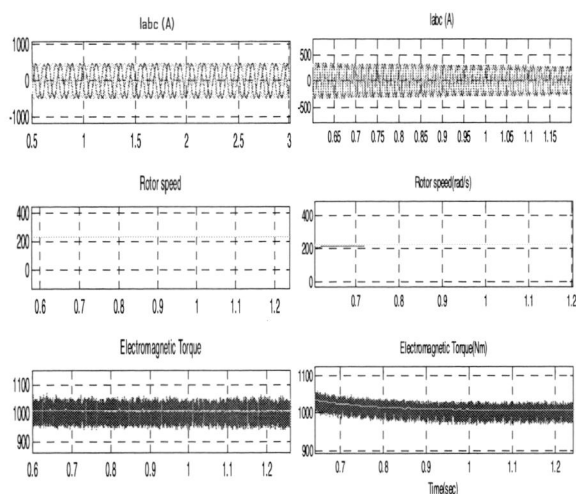

Fig.10 Simulink plot for FOC (left) and DTC (right) showing three phase stator current, speed and electromagnetic torque for full load (1000Nm) at 215rad/sec speed.

It has been found that in DTC scheme, time to attain steady state remain relatively constant at 0.55 seconds, whereas in case of FOC time to attain steady state increases with the increase in percentage loading. It is observed that the torque oscillations in DTC control scheme are negligible whereas in FOC scheme oscillations in torque are observed. These oscillations are relatively less at no load and then become nearly constant with the increase in percentage loading. Also, it is found that the distortion of motor current is higher in case of FOC. Hence DTC scheme is implemented for traction drive scheme with 25kV supply.

For the traction motor drive suitable for electric traction, two induction motors are connected to a single inverter. The simulated performance of the converter inverter drive system is shown in fig.11 to fig.14. Fig.11 shows the steady state waveform of three phase stator current (A), source current (A), source voltage (V), speed (rad/sec) and torque for full load (1000Nm) at 215rad/sec speed. Similarly fig.12 and fig.13 shows waveform at half load and no load conditions. Fig.12 shows the waveform of three phase stator current, source current, source voltage, speed and torque for full load (1000Nm) at 50% of the rated speed and full load.

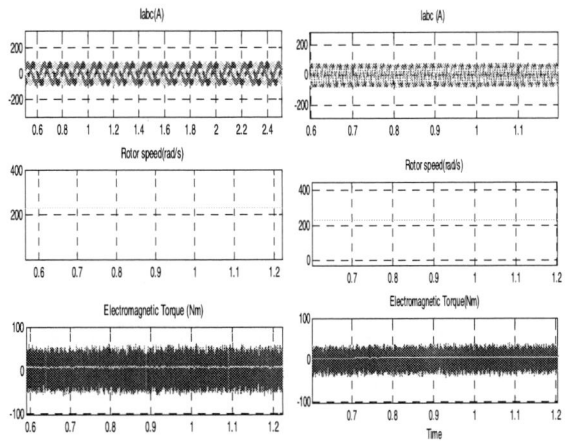

Fig.8 Simulink plot for FOC (left) and DTC (right) showing three phase stator current, speed and electromagnetic torque for no load at 215rad/sec speed.

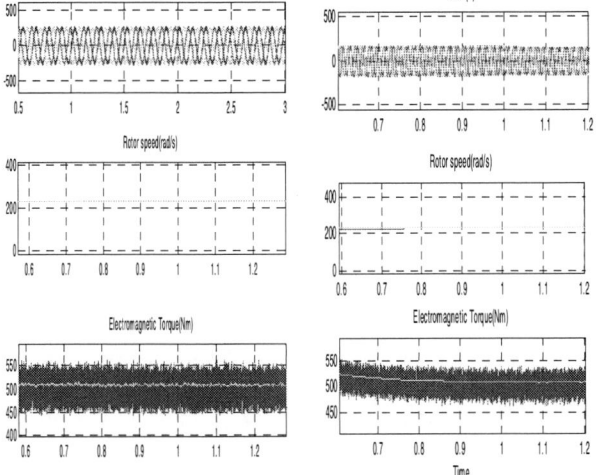

Fig.9 Simulink plot for FOC (left) and DTC (right) showing three phase stator current, speed and electromagnetic torque for half load (500Nm) at

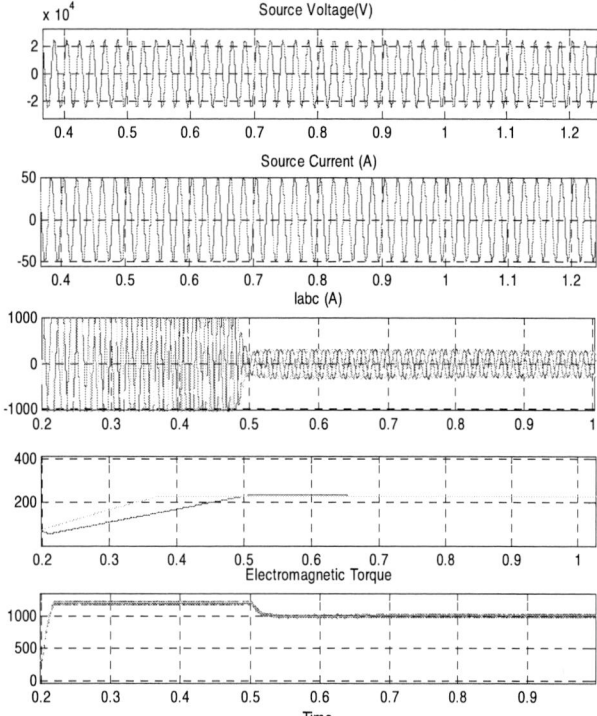

Fig.11 Simulink plot showing source voltage, source current, three phase stator current, speed and torque for full load (1000Nm) at 215rad/sec speed.

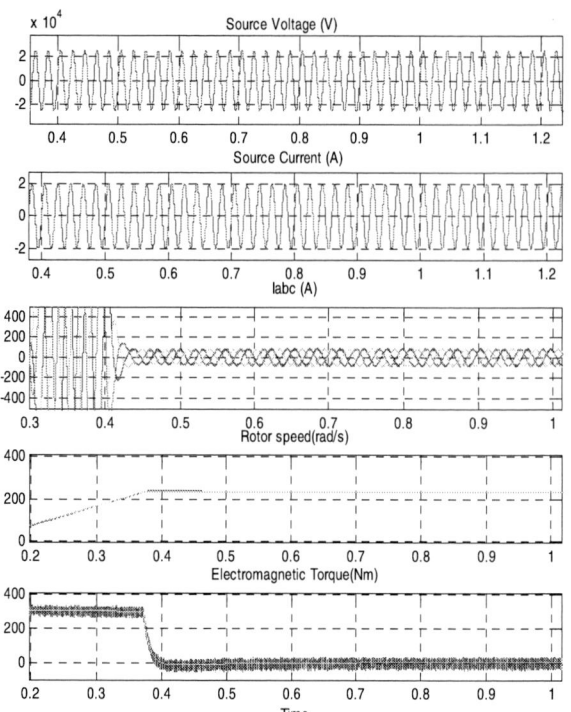

Fig.13 Simulink plot showing source voltage, source current, three phase stator current, speed and torque for no load at 215rad/sec speed.

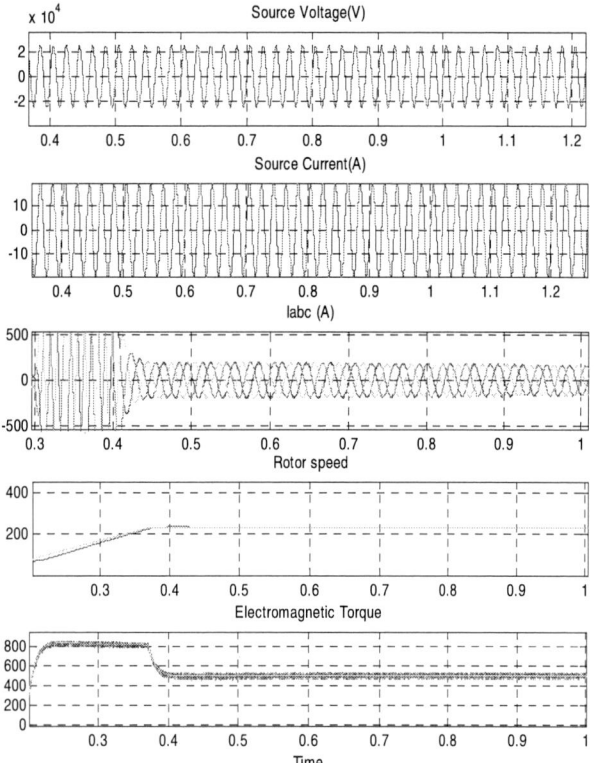

Fig.12 Simulink plot showing source voltage, source current, three phase stator current, speed and torque for half load (500Nm) at 215rad/sec speed.

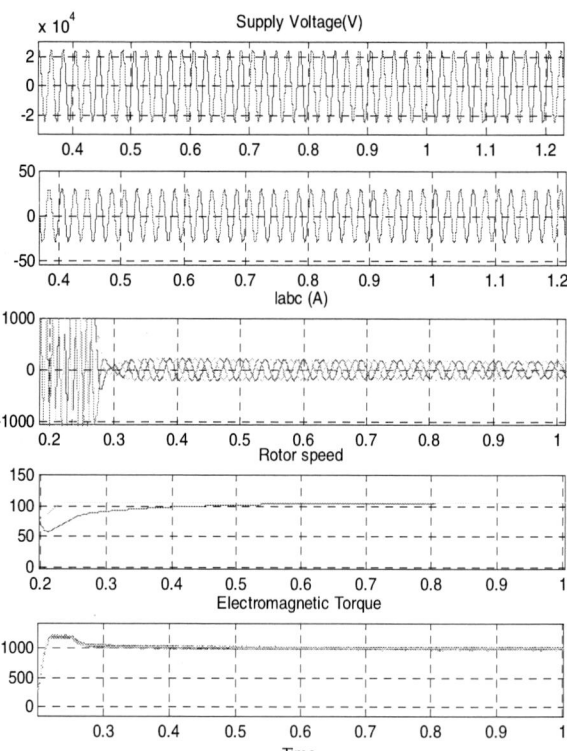

Fig.14 Simulink plot showing source voltage, source current, three phase stator current, speed and torque for at half the rated speed and full load.

978-1-4799-6047-7/14 $31.00 © 2014 IEEE

It has been observed from fig.11 to fig.13 that the time to attain steady state remains relatively constant at 0.52 seconds for different loading conditions.The AC mains current at full load condition has the THD of 6.74% and goes to 7.56% at half the load condition. Also in the entire operating range, power factor remained close to unity. Further, load perturbations remain more or less same in all loading conditions.

Table I shows the time required to attain steady state and steady state torque oscillations for traction motor scheme with reference to results shown in fig.11 to fig.13 for different percentage loading at a given speed of 215 rad/sec.

TABLE I. ANALYSIS OF TRACTION MOTOR DRIVE SCHEME IMPLEMENTED

% Loading	Time to attain steady state(s)	Steady state torque oscillations (%)
50% load	0.41	±4.375
Full Load	0.52	±4.375

Similarly, for full load and half the rated speed it can be shown that the torque oscillations at low speed are comparatively very high and then decreases with increase in speed for DTC control scheme as observed in fig.14.

VII. CONCLUSION

In this work, two schemes for control of drives, namely field oriented control and direct torque control, are compared. It is observed that time to attain steady state and the torque perturbations are less in DTC scheme to FOC scheme. Thus, direct toque control scheme is implemented for speed control of traction motor drive. The complete H-bridge converter fed AC traction motor drive has been designed and simulated using MATLAB Simulink. Further, the AC mains power factor remains close to unity during the entire operation which leads to efficient operation of the system.

APPENDIX

The induction machine used in the MATLAB /simulation is 3phase, 72.5Hz induction machine having the following parameters.

Power output	295HP(220kW)
Rated Voltage	1040V
R_s (stator resistance)	0.0749Ω
R_r (rotor resistance)	0.0801Ω
L_s (stator inductance)	0.000971H
L_r (rotor inductance)	0.000891H
L_m (magnetizing inductance)	0.03041H
J (moment of inertia)	10 Kg m^2

REFERENCES

[1] W.S. Chan, "Whole system simulator for AC traction", PhD Thesis, University of Birmingham, UK, July 1988.

[2] P. Vas, Vector control of AC machines, New York: Clarendon, 1990

[3] B.K. Bose, Power electronics and AC Drive. Englewood Cliffs, NJ.Prentic hall, 1986

[4] R. Oruganti, K. Nagaswamy and L. K. Sang, "A constant frequency variable power factor PWM scheme for single phase boost type AC-DC converter," Proc. Int. Conf. EMPD'95, 1995, Vol.1 pp, 215- 221.

[5] B. Singh, G., V. Garg,"Improved power quality AC-DC converter for electric multiple units in electric traction" Power India Conference, 2006 IEEE.

[6] K.Thiyagarajah, V.T.Ranganathan and B.S. Ramakrishna, "A high switching frequency IGBT PWM rectifier/ inverter system for AC motor drives operating from single phase supply," IEEE Trans. on Power Electronics, Vol.6, No.4, October1991, pp. 576-584.

[7] IEEE Guide for Harmonics Control and Reactive Compensation of Static Power Converters, IEEE Standard 519, 1992.

[8] F. Biaschke, " The principle of field orientation as applied to new transvector closed loop control system for rotating field machine," Siemens Rev, vol. 34,,pp 217-220'may 1972

[9] N. Mariun, S. B Mohd Noor, J. Jasni, and O. S. Bennanes, "A Fuzzy Logic Based Controller For An Indirect Vector Controlled Three-Phase Induction Motor" IEEE IECON Conf. Rec., Vol. 4, pp. 1-4, Nov. 2004.

[10] B. Singh, B. N. Singh, and B. P. Singh, "Performance Analysis of a Low Cost Vector Controlled Induction Motor Drive: A Philosophy for Sensor Reduction", IEEE IAS Annu. Meet. Conf. Rec., pp. 789-794, 1997.

[11] T.Noguchi and Takahashi, "Quick torque reponse of an induction motor based on a new concept ," IEEE Tech. Meeting Rotating Mach., Vol.RM 84-72, Sept. 1984, pp. 61-70

[12] R. Garg, P. Mahajan, N. Gupta and H. Saroa, "Comparative Study between Field Oriented Control and Direct Torque Control of AC Traction Motor," IEEE International Conference on recent advances and innovation in engineering (ICRAIE-2014) May 09-11, 2014, Jaipur India.

[13] R. Garg, P. Mahajan, P. Kumar, R. Goel, "Design of Unity Power Factor Controller for Three-phase Induction Motor Drive Fed from Single Phase Supply," Journal of Automation and Control Engineering Vol. 2, No. 3, September 2014.

[14] M. H. Rashid, "Power Electronics Handbook", third edition, Elsevier 2011.

[15] Modeling & Simulation using Matlab® -Simulink® by Dr. Shailendra Jain, First Edition: 2011.

[16] B. K. Bose, N. R. Patel, And K. Rajashekara, "A Start-Up Method For A Speed Sensorless Stator-Flux- Oriented Vector-Controlled Induction Motor Drive", IEEE Transactions on Industrial Electronics, vol. 44, no. 4, august 1997

[17] G. K. Dubey, "Fundamentals of Electrical Drives", second edition, narosa publishing house: 2001.

Optimized Rotor Slot Shape for Squirrel Cage Induction Motor in Electric Propulsion Application

Md. Junaid Akhtar
Department of Electrical Engineering
Indian Institute of Technology
Patna, Bihar-800013
Email: junaid@iitp.ac.in

R. K. Behera, *Senior Member, IEEE*
Department of Electrical Engineering
Indian Institute of Technology
Patna, Bihar-800013
Email: rkb@iitp.ac.in

S. K. Parida, *Member, IEEE*
Department of Electrical Engineering
Indian Institute of Technology
Patna, Bihar-800013
Email: skparida@iitp.ac.in

Abstract—Design of squirrel cage induction motor for propulsion system is different from that of conventional design. Efficiency, power factor and locked rotor torque are important performance measures of a propulsion system. For an induction motor, the above mentioned parameters are expected to be high. In this paper, the effect of geometrical variations in rotor slot on the performance of induction motor is analyzed. The simulation results are presented for variation in different rotor slot shape. It shows that the shape of the rotor can effect on the performance of the induction motor. Hence, these findings may be helpful to the design engineers for choosing the proper slot dimensions.

Keywords—*induction motor; rotor slot; efficiency; power factor; locked rotor torque*

I. INTRODUCTION

Air pollution, noise pollution and global warming are the major problems that the modern world is facing. Electric Vehicle (EV) solves the above problem by minimizing environmental pollution. There are various electric motors that can be used in EV. But cage rotor induction motor (IM) is widely used because of its low cost, better controllability, reliability and robustness [1]. The designs of IM used in EV are different from conventional IM. The motor should have high power density, high starting torque, better power factor, high efficiency and wide operating speed range. These characteristics largely depends on different parameters of IM like shape of the stator and rotor slot, number of stator and rotor slots, core length, inner and outer diameter, air-gap length and many others [2]. In [3], the authors have proposed a method to increase the efficiency of the motor by increasing the core axial length. In [4] and [5], the effect of rotor slot shape on harmonics losses of inverter fed induction motor is analyzed by finite element method. In [6], the authors have analyzed the effects of stator slot shape and rotor bar dimension on pulsating torque of IM. The effect of rotor slot shape in rotor leakage permeance is discussed in [7]. In [8], the effect of rotor slot geometry on starting torque, breakdown torque, efficiency and power factor is studied. Optimal design of copper die-cast bar with respect to efficiency and torque are reported in [9].

Core length and shape of rotor slots are two important parameters which govern the performance of cage rotor IM. Rotor slot leakage reactance, locked rotor torque, power factor and efficiency are some important machine performance characteristics which can be altered by selecting suitable core length and rotor slot dimensions during design. Therefore, a desired performance can be obtained by adjusting the rotor slot

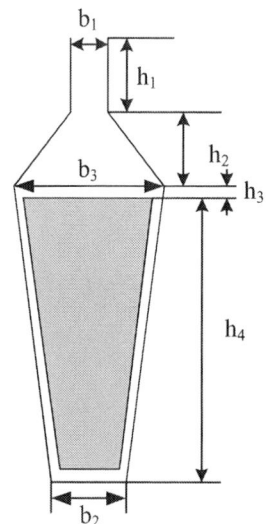

Fig. 1. Rotor slot shape

geometry during design. In this work, the effect of geometrical variations in rotor slot on the performance of IM is studied by a family of parametric curves. The remaining part of this paper is organized as follows: Section 2 discusses the rotor slot shape variation and its effect on various performance characteristics. Section 3 presents simulation results. Concluding remarks are presented in section 4.

II. DESIGN METHODOLOGY

Fig. 1 shows the diagrammatic representation of rotor slot. Here h_4 is the height of the rotor slot in which electrical conductors will be placed. Slot opening width is given by b_1, slot top width is given b_3 and bottom width is given by b_2. The ratio of slot top width to bottom width (b_3/b_2) and ratio of slot height to slot bottom width (h_4/b_2) are taken as k and k_1 respectively. The slot bottom width b_2 is kept constant in both the cases. Thus by varying h_4 and b_3 different geometry of the rotor slot can be obtained. Different shapes of the rotor slot result in rotor bars of different area. This varies the rotor current and rotor flux linkage. The relation between permeance factor and rotor slot dimensions presented in [10] is as follows:

$$\lambda_u = \frac{h_4}{3b_3} + \frac{h_3}{b_3} + \frac{h_1}{b_1} + \frac{h_2}{b_3 - b_1} \ln \frac{b_3}{b_1} \quad (1)$$

978-1-4799-6047-7/14 $31.00 © 2014 IEEE

where λ_u is the slot permeance factor.

$k = b_3/b_2$ (ratio of slot top width to bottom width)

$k_1 = h_4/b_2$ (ratio of slot height to slot bottom width)

Relation of permeance factor with total slot inductance of a phase winding is given by

$$L_u = \frac{4m}{Q}\mu_0 l N^2 \lambda_u \qquad (2)$$

where,
L_u is slot leakage inductance,
m is no. of phase,
μ_0 is prmeability of vacuum,
l is the ore length,
N is no. of series turn connected in a phase winding and,
Q is no. of slots.

From (1) and (2), it can be observed that as the height of the rotor slot increases, permeance factor and slot inductance increases. Rotor leakage reactance is proportional to the leakage permeance. Electromagnetic torque developed in rotor can be given in terms of machine parameters as follows:

$$T_{em} = \frac{3[U_s(1 - \frac{L_{s\sigma}}{L_m})]^2 \frac{R'_r}{s}}{\frac{\omega_s}{p}[(R_s + \frac{R'_r}{s})^2 + (\omega_s L_{s\sigma} + \omega_s L'_{r\sigma})^2]} \qquad (3)$$

where,
T_{em} is electromagnetic torque,
U_s is stator voltage,
$L_{s\sigma}$ is stator leakage inductance,
L_m is magnetizing inductance ,
R'_r is rotor resistance referred to stator,
ω_s is angular speed,
p is no. of poles pairs,
R_s is stator resistance and
$L_{r\sigma}$ is rotor leakage inductance.

From (3), it can be observed that the torque developed is inversely proportional to rotor inductance. Relation of k_1 with inductance is already established in (1) and (2) which leads to the conclusion that torque is inversely proportional to the factor k_1.

Rotor slot shape and different losses affects the efficiency of the motor. The efficiency of the motor can be written as

$$\eta = \frac{P_{out}}{P_{in}} = \frac{P_{out}}{P_{out} + \Sigma losses} \qquad (4)$$

where,

$$\Sigma losses = p_{Co} + p_{Al} + p_{iron} + p_{mv} + p_{stray}$$

p_{Co} is stator winding loss,
p_{Al} is rotor cage loss,
p_{iron} is core loss
p_{mv} is mechanical loss and
p_{stray} is stray loss.

The relation between skin effect and rotor resistance is given in [7] is as follows:

$$r_r = \left[k_r \frac{l_e}{l_b} + \frac{l_b - l_e}{l_b} \right] r_b \qquad (5)$$

where
r_b is the nominal resistance without considering the skin effect,
l_b is the length of the rotor core and
k_r is the coefficient accounting for skin effect.

k_r is a function of slot factor and is given by (6)

$$SF = 0.1987 h_b \sqrt{\frac{b_b sf}{b_r \rho}} \qquad (6)$$

where,
SF is slot factor h_b is the height of rotor conductor,
b_b is the width of the rotor conductor,
b_r is the width of the rotor slot,
sf is the slip frequency and
ρ is the rotor conductor conductivity.

From (5) it can be observed that rotor resistance depends on coefficient of skin effect which in turn depends on slot factor given by (6). This slot factor depends on the slot geometry. So by varying the slot shape efficiency is varied. Hence from (5) and (6) we can say that by varying the slot width and height, the slot factor changes and hence the rotor resistance. With this change the efficiency of the motor also gets change. Rotor slot shape does also affect power factor of the electrical machine [11]. The relation between efficiency and power factor is given by

$$cos\phi = \frac{P}{3U_{ph,s}I_s \eta} \qquad (7)$$

where,
P is rated power of machine,
$U_{ph,s}$ is stator phase voltage,
I_s is stator current and
η is efficiency.

Since the motor is designed for electric vehicle, a good power factor is an essential characteristic. From (1) and (2), it is clear that rotor slot leakage reactance increases with k_1. Therefore, increasing k_1 results in poor power factor.

III. SIMULATION AND RESULTS

The analysis of performance characteristics of squirrel cage IM on rotor slot shape is done on 5hp, 110V 80 Hz motor. The specification of the motor is given in Table I. The IM is designed using the RMxprt software. RMxprt software is electric machine specific user friendly template based designing software [12]. It has the features of optimization, parametric and sensitivity analysis. RMxprt software makes the design and optimization process easier and faster. It can create Maxwell 2D and 3D model for electromagnetic analysis. The motor is first simulated using the software and then parametric analysis is done to study the variation in performance characteristics with variation in rotor slot shape.

Fig. 2 shows the simulation result for variation of rotor slot leakage reactance with respect to k_1 keeping k fixed. it is observed that as k is increased, the rotor slot leakage reactance increases.

TABLE I.   SPECIFICATION OF INDUCTION MOTOR

Power rating	5 hp
Type	3 phase
Base frequency	80 Hz
Line-to-line voltage	110 V
Number of poles	4
Efficiency	84 %
Power factor	0.86
Ambient temperature	50 deg C
Insulation class	Class B
Degrees of protection	IP55
Cooling type	TEFC

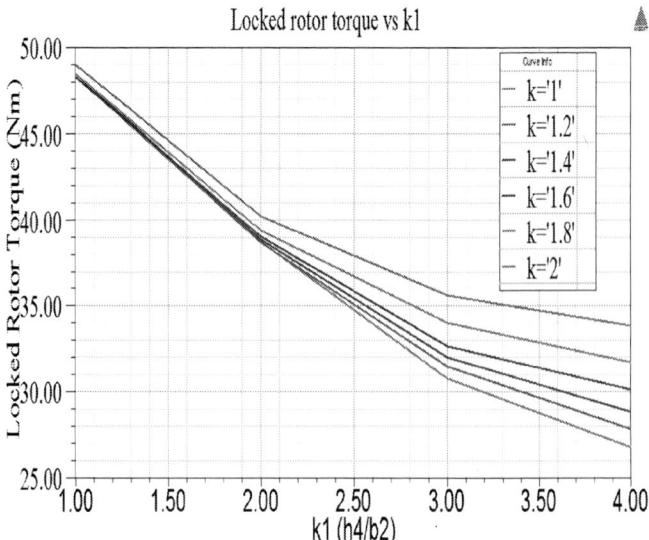

Fig. 3.   Locked rotor torque vs k_1

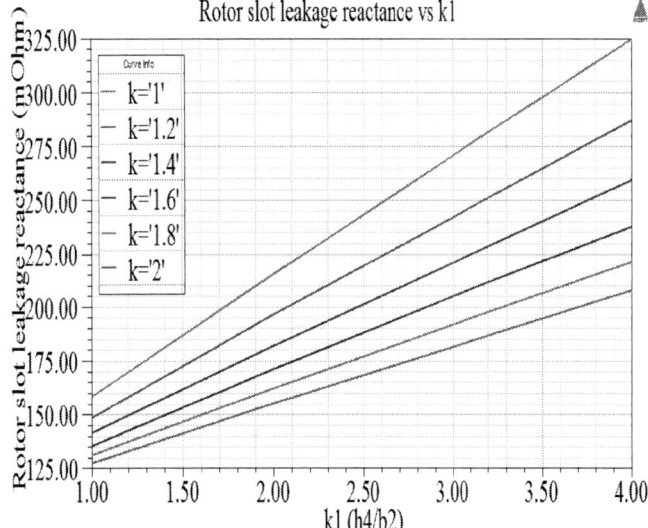

Fig. 2.   Rotor slot leakage reactance vs k_1

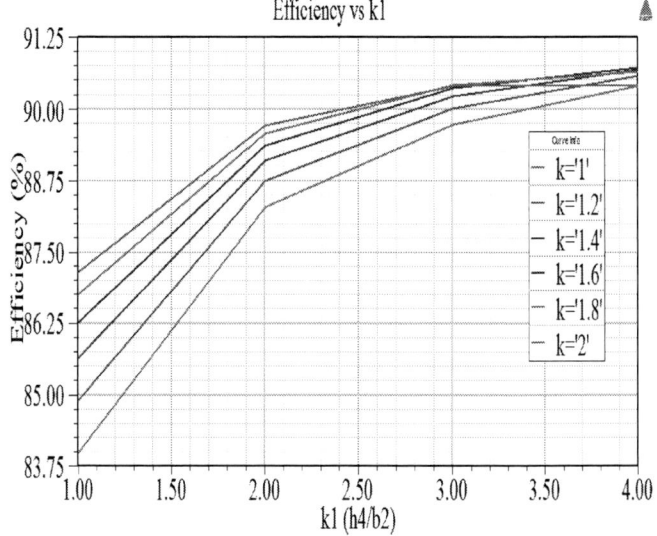

Fig. 4.   Efficiency vs k_1

Fig. 3 shows the locked rotor torque vs k_1 plot. It is observed that as k_1 is increased for a fixed k, the locked rotor torque decreases. Therefore, it can be interpreted from Fig. 3 that to achieve higher locked rotor torque, shorter and wider slot shape is preferred.

Fig. 4 shows the efficiency vs k_1 plot. It is observed that as k_1 is increased keeping k fixed, the efficiency increases. It is also observed that efficiency increases when k is increased keeping k_1 fixed. Therefore to achieve higher efficiency, deeper and wider slots are recommended.

Fig. 5 shows the power factor vs k_1 plot. It is observed that as k_1 is increased keeping k fixed, the power factor decreases. It is also observed that when k_1 is fixed and k is increased, the power factor decreases. Therefore, to achieve better power factor, shorter and wider slots are recommended.

From Figs. 3-5, it is observed that for particular rotor slot geometry, it is impossible to obtain higher values for all the three performance measures. For example, $k = 2$ and $k_1 = 1$ yields higher locked rotor torque and efficiency but lower power factor. Therefore, for a particular value of k, k_1 is chosen as a trade-off between locked rotor torque, efficiency and power factor from the contour plots given in Figs. 6-8.

Fig. 6 shows the contour plot of locked rotor torque. It gives the family of curves for locked rotor torque for diferent

values of k and k_1. It is observed that for maximum value of locked rotor torque, lower value of k_1 is recomended.

Fig. 7 shows the contour plot of efficiency. It is observed that the contour plot of efficieny is in contrary to the locked rotor torque. The choice of k and k_1 for achieving maximum efficiency is contrary to the maximum torque requirement. It shows that for higher efficiency, higher value of k_1 is recommended.

Fig. 8 shows the contour plot of power factor. It is observed that lower value of k_1 gives higher power factor. As we go on increasing k_1 the power factor becomes poor.

Taking Figs. 6-8 together, it is observed that for lower values of k_1, higher locked rotor torque, power factor and lower efficiency are obtained irrespective of the values of k. Therefore, a trade-off is required between locked rotor torque, efficiency and power factor. Choosing k_1 between 1.75 and 2.5 results in the following: locked rotor torque of 34 Nm to

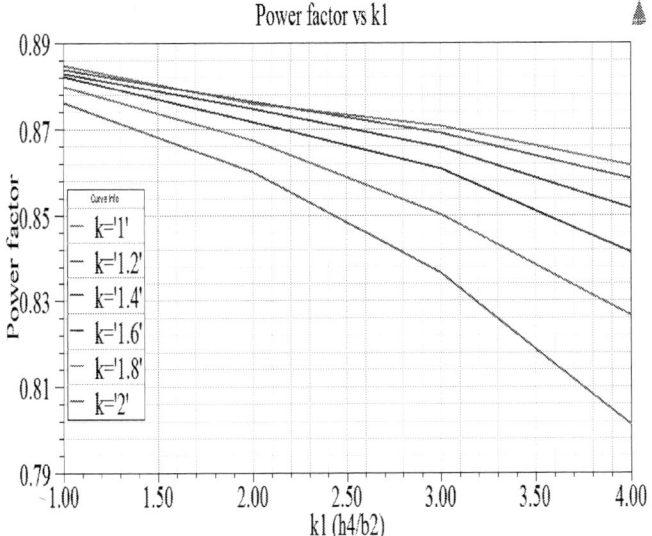

Fig. 5. Power factor vs k_1

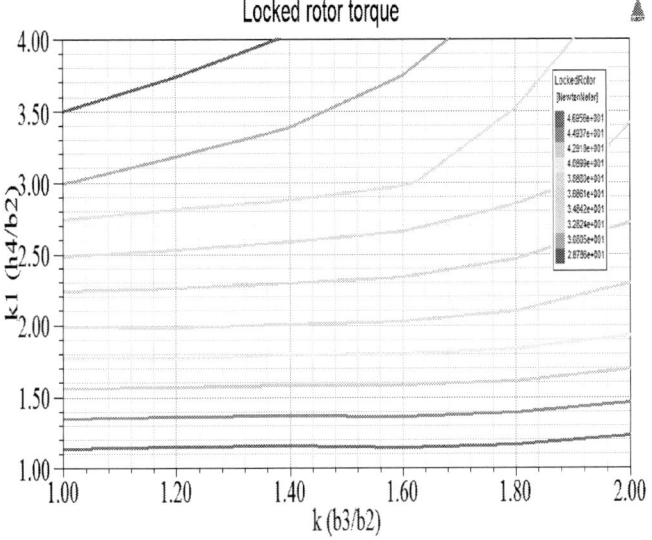

Fig. 6. Contour plot of locked rotor torque

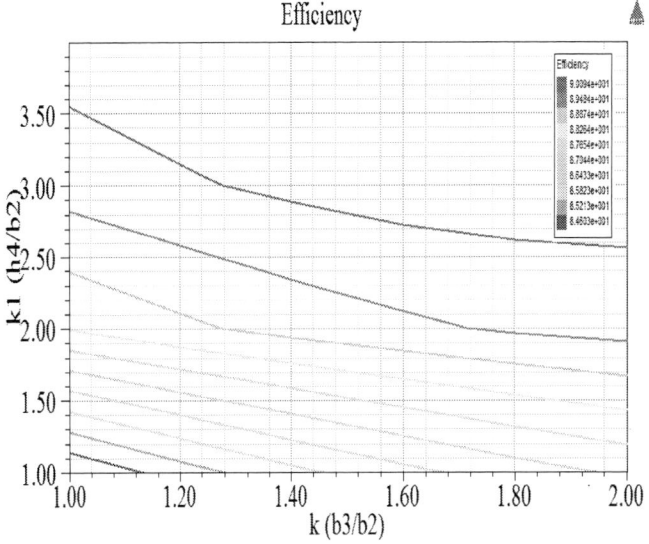

Fig. 7. Contour plot of efficiency

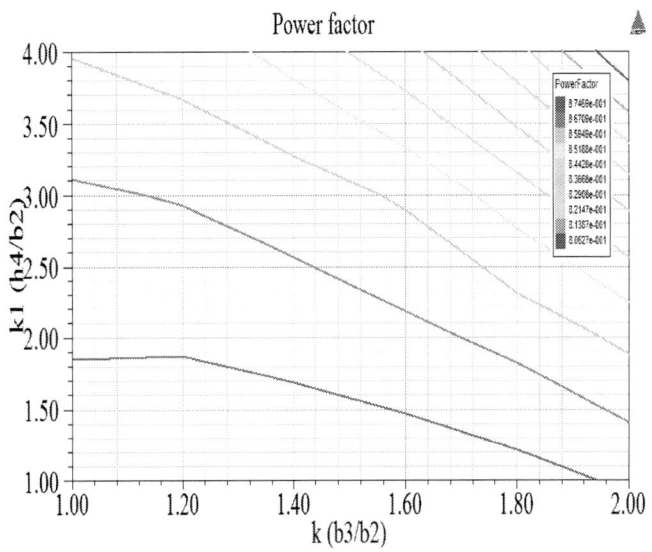

Fig. 8. Contour plot of power factor

38 Nm, efficiency of 87% to 88% and power factor of 0.86 to 0.87. These three performance values are acceptable for a good design.

IV. CONCLUSION

This paper analyses the effect of variants in slot dimension on performance of a squirrel cage induction motor meant for propulsion. Effect of rotor slot geometry on performance parameters like locked rotor torque, power factor and efficiency has been discussed and analyzed. Simulation results emphasize the need for considering the geometry of the rotor slot while designing an induction motor for propulsion purpose. Based on the simulation studies, it is recommended to choose shorter and wider slot in order to obtain better locked rotor torque and power factor whereas deeper and wider slot is recommended for obtaining better efficiency.

REFERENCES

[1] M. Zeraoulia, M. E. H. Benbouzid, and D. Diallo, "Electric motor drive selection issues for hev propulsion systems: A comparative study," *Vehicular Technology, IEEE Transactions on*, vol. 55, no. 6, pp. 1756–1764, 2006.

[2] I. Boldea, *The induction machines design handbook*. CRC press, 2009.

[3] L. Alberti, N. Bianchi, A. Boglietti, and A. Cavagnino, "Core axial lengthening as effective solution to improve the induction motor efficiency classes," *Industry Applications, IEEE Transactions on*, vol. 50, no. 1, pp. 218–225, Jan 2014.

[4] H.-P. Nee and C. Sadarangani, "The influence of load and rotor slot design on harmonic losses of inverter-fed induction motors," in *Electrical Machines and Drives, 1993. Sixth International Conference on (Conf. Publ. No. 376)*, Sep 1993, pp. 173–178.

[5] H.-P. Nee, "Rotor slot design of inverter-fed induction motors," in *Electrical Machines and Drives, 1995. Seventh International Conference on (Conf. Publ. No. 412)*, Sep 1995, pp. 52–56.

[6] Y. Li, S. Li, and B. Sarlioglu, "Analysis of pulsating torque in squirrel cage induction machines by investigating stator slot and rotor bar

dimensions for traction applications," in *Energy Conversion Congress and Exposition (ECCE), 2013 IEEE.* IEEE, 2013, pp. 246–253.

[7] Z. Zhao, S. Meng, C. Chan, and E. W. C. Lo, "A novel induction machine design suitable for inverter-driven variable speed systems," *Energy Conversion, IEEE Transactions on*, vol. 15, no. 4, pp. 413–420, Dec 2000.

[8] V. Fireteanu, T. Tudorache, and O. A. Turcanu, "Optimal design of rotor slot geometry of squirrel-cage type induction motors," in *Electric Machines Drives Conference, 2007. IEMDC '07. IEEE International*, vol. 1, May 2007, pp. 537–542.

[9] K. Park, K. Kim, S. hoon Lee, D.-H. Koo, K.-C. Ko, and J. Lee, "Optimal design of rotor slot of three phase induction motor with die-cast copper rotor cage," in *Electrical Machines and Systems, 2008. ICEMS 2008. International Conference on*, Oct 2008, pp. 61–63.

[10] J. Pyrhonen, T. Jokinen, and V. Hrabovcova, *Design of Rotating Electrical Machines.* John Wiley & Sons, 2009.

[11] M. G. Say, *Performance and design of AC machines.* English LB S., 1995.

[12] Ansys.com, "Ansys rmxprt," 2014. [Online]. Available: http://www.ansys.com/Products/Simulation+Technology/Electronics/El-ectromechanical/ANSYS+RMxprt

Secured charging of electric vehicles at unattended stations using Verilog HDL

Ridhi Saini

Electronics and Communication Engineering Department,
Kurukshetra University, Kurukshetra
Haryana, INDIA
e-mail: ridhisaini9@gmail.com

Abstract—**Electric vehicles need to be charged at public places in a secured manner which is advantageous both for utility as well as consumer. This paper discusses simulation and implementation of authentication or digital lock of PIN code used for secure transaction at unattended public electric vehicle charging station on XILINX ISE Project Navigator Ver.8.2 using Verilog HDL and simulated using Modelsim software**

Keywords— *Alarm systems, Authentication, Digital simulation, Electric vehicles, Security*

I. INTRODUCTION

Electric vehicles are becoming more prominent due to their less pollution and less power consumption advantage. A large number of charging stations need to be created across the country due to large number of such vehicles and longer charging times. It is economically not viable to employ manpower at each of such stations. Thus, electric vehicles often need to be charged at unattended public charging stations. Smart grid has brought new opportunities and challenges for the security of electric vehicle charging stations. The unattended charging station utility must get the payment for vehicle charging and should make available the charging facility to the vehicle owner in a secured manner. An internet connected digital lock or digital authentication system can be a viable solution to this problem. A digital lock is something that accepts some digital input, processes it, resulting in locking/unlocking of the charging facility or sets the alarm, if the number of attempts exceeds a limit. Such a locker, digital in nature, can be simulated and implemented using Verilog HDL; a hardware description language for electronic circuits.

Digital security lock systems have been implemented by various researchers [1-10]. A lot of work has been reported for home security and automobile security and limited research work has been done for charging station payment security. Payment of electric vehicle charging using swiping of smart card has been reported in [11]. Smart credit cards if lost can be misused by unauthorized users for swiping. This paper discusses use of digital locks which does not require swiping of a smart card; instead, requires entering, Smart card number, "One time password (OTP),"generated (received on mobile) in response to request for charging and smart card PIN number, thus enhancing the security in the transaction/system. In these locks some identification of the customer is required in the form of a key. Some reported studies recognizes some physical part of the human body, e.g. fingerprint[2], face[3],

iris[4], voice recognition[5] and are complex to implement. For these, various biometric, cryptographic techniques, complex algorithms have been used. However, for digital lock implementation, non-complex algorithms using various programming languages have been used such as conventional high-level language, such as C or hardware description language such as VHDL making the system cheaper. The keyless car entry locking system first appeared about three decades ago [12]. It involves either pressing a button provided on the key or enter a password, generally a numeric digital code, through a keypad provided at the outer body of automobile. But, it faced a lot of problems such as two-thief attack, three-thief attack, relay attack and many more. Solution to these problems have been reported in [6], [7]. Work on design of a keyless digital security system which involves implementation of keypad encoder, memory elements, security logic using VHDL has been reported in [1]. In this reported study, main emphasis has been on implementation of keypad and for lock, a simple comparator circuit has been used. Such a simple circuit may be prone to some kind of attacks as stated above. Some work on keyless car entry through face recognition using FPGA has also been done, which relies on a video footage taken when one approaches one's car[3]. The reliability of all such systems decreases with increasing number of hackers and criminals in the society. So researchers are trying to build more complicated secure systems which may not be decoded by any thief. In this paper, a new approach for secured identification and payment at unattended electric vehicle charging station has been presented.

II. METHODOLOGY

A digital lock as a finite state machine has been implemented using Verilog HDL. The lock accepts smart card number, OTP, 4 digit input PIN if found correct, unlocks the charging facility of electric vehicle at charging station. The input can be extended to n number; n is any natural number less than or equal to 100 preferably, depending on the hardware limitations. If the sequence of digits entered is correct i.e. as per the list of security PIN of registered electric vehicle owners, the lock opens and the charging is allowed. After entering a single digit, it cannot be re-entered. Fig.1 shows the state diagram for such a lock. Here, s0, s1, s2, s3, s4, s5, s6 are six states of state diagram of a mealy machine circuit. The inputs are represented by (a,b,c,d) variables. The reset variable ensures safe locking if it has been unlocked. The numbers of tries are stored in tc variable, used as try_count in the algorithm presented. At the end of charging, the vehicle owner repeats the procedure of entering four (n) digit PIN to indicate to the utility that charging service is required no more. The vehicle owner will be billed for the amount of charging availed.

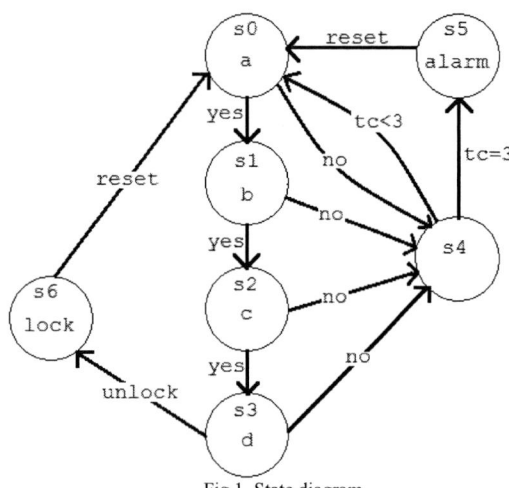

Fig.1. State diagram

III. ALGORITHM

1. Initialize the variables:
   Input: [clk,reset,ip_en,enter,a,b,c,d]=0
   Output: [alarm, try_count, key_count, restart,alarm_state]=0, lock=1
   Reg: passkey, [state, next_state]=0
   Parameter: s0=0, s1=1, s2=2, s3=3, s4=4, s5=5, s6=6
2. Set 4 digit password in passkey[0:3]
3. Enter 4 digit password, then press enter.
4. Whenever there is a change in ip_en or reset or restart or alarm_state or enter then:
   If ip_en or reset or restart or alarm_state or enter=1 then:
   If state =s0 then go to step 5
   If state =s1 then go to step 6
   If state =s2 then go to step 7
   If state =s3 then go to step 8
   If state =s4 then go to step 9
   If state =s5 then go to step 10
   If state =s6 then go to step 11
5. If (ip_en)=1& (reset)=0 then
   If 1st digit of password is correct then next_state=s1 else
   next_state=s4 & increment try_count
6. next_state=s1
   if 2nd digit of password is correct then next_state=s2 else
   next_state=s4 & increment try_count
7. if(ip_en)=1 then
   if 3rd digit of password is correct then next_state=s3 else
   next_state=s4 & increment try_count

8. next_state=s3
   if(ip_en)=1 then
   if 4th digit of password is correct then next_state=s6 else
   next_state=s4 & increment try_count
9. next_state=s4
   if(restart)=1 then
   next_state=s0
   else if(alarm_state)=1 then
   next_state=s5
10. if(reset)=1 then
    next_state=s0 & alarm=0
    else if(enter)=1 then
    alarm=1
11. try_count=0
    if(reset)=1 then
    next_state=s0
12. Whenever enter or reset change then
    if enter=1 & state=6 then
    lock=0
    else if(reset)=1 then
    lock=1
13. Whenever there is a posedgeclk then
    state=next_state
14. Whenever there is a change in clk then
    if(state=s4) &(key_count=4) & (try_count<3) then
    restart=1else
    restart=0
    if(state=s4) & (key_count=4) & (try_count=3) then
    alarm_state=1else
    alarm_state=0.

IV. RESULTS

The above algorithm has been implemented on XILINX ISE Project Navigator Ver.8.2i. using Verilog HDL and simulated using Modelsim software. The simulation result is shown in Fig.2. Here the clock(clk) is operating at a frequency of 10 GHz. Alarm will ring after three unsuccessful tries attracting passerby and discouraging the unauthorized person to recharge its vehicle. To reset the alarm, the reset variable will be made high by the utility person remotely by communicating some code that reset the system. The reset button will also be used for locking the lock if initially it was open. The whole machine will be restarted. The lock will remain at logic 1 i.e. locked. The input enable (ip_en) variable is acting as a keypad entry i.e. it gets high whenever a digit is inputted. The user will enter smart card number and then PIN number as 4 digits from the keypad which will be assigned to a, b, c, d variables respectively. After entering the code, the enter key has to be pressed that correspond to the enter variable that will go high. A digit once entered cannot be changed as there will be no option of re-correcting it. The try_count variable will get incremented when a wrong digit is entered, no matter at which place, and the mchine will go in astate s4.

978-1-4799-6047-7/14 $31.00 © 2014 IEEE

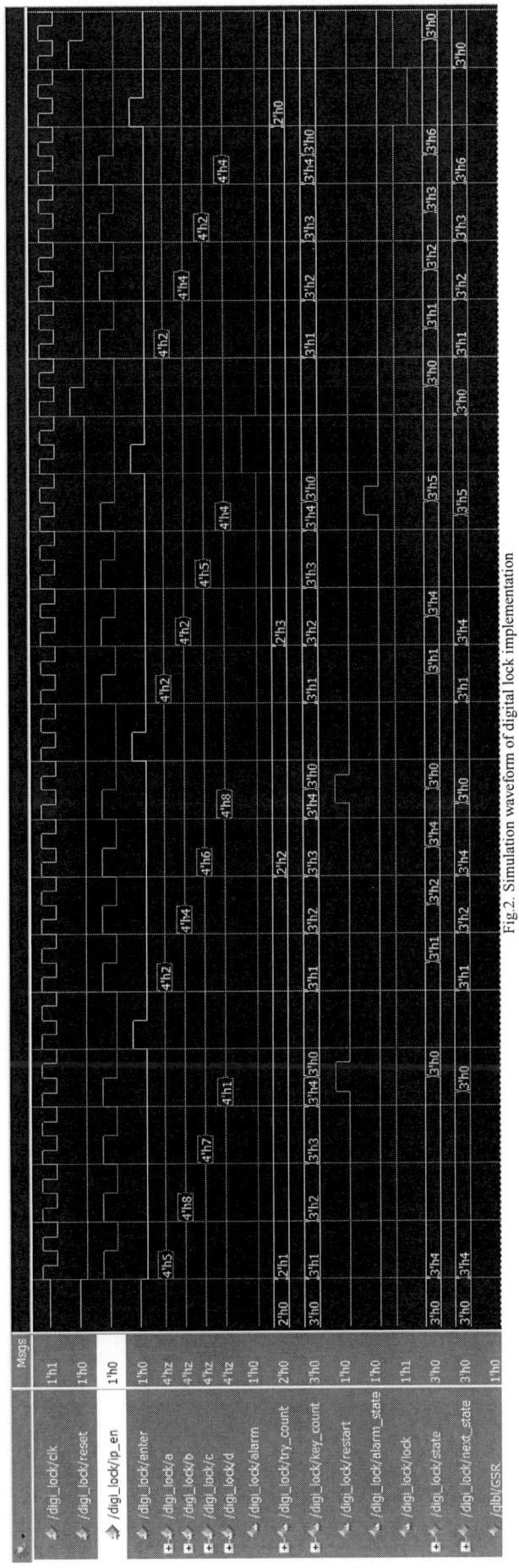

Fig.2. Simulation waveform of digital lock implementation

The machine will now return to state s0 only when key_count reaches a value=4, signifying entry of 4 digits, which gets incremented at each digit input. The lock will only be unlocked when correct 4 digits are entered as set by the user initially. This password can be changed at any time by the authorized registered person for himself/herself.

V CONCLUSION

Simulation results show successful implementation of secured authentication of four digit code. Similar results can be obtained for smart card number and OTP. Further, the work can be extended to hardware implementations by interfacing a keypad, FPGA and an internet ready device on which the programming code can be burned.

REFERENCES

[1] E.O. Oyetunji and J. Asare, "Design of a keyless digital security system," International Journal of Electrical, Electronics and Computer Engineering vol. 1, no. 2, pp. 42-50, 2012.

[2] M. Hirschbichler, C. Boyd, and W. Boles, "A scheme for enhancing security using multiple fingerprints and the fuzzy vault digital image computing: techniques and applications," DICTA, 2008, pp. 540-547.

[3] S. Saifullah, A. Khawaja,. Hamza, Arsalan, Maryam, and Anum, "Keyless car entry through face recognition using FPGA," Future Information Technology and Management Engineering (FITME), vol. 1, pp. 224-227, 2010.

[4] E.S. Reddy, and I. Ramesh Babu, "Authentication using fuzzy vault based on iris textures modeling & simulation," AICMS 08. Second Asia International Conference, 2008 , pp. 361-368.

[5] M.H. Hassan, W.S. Beardslee, and T.K. Arnold, "Voice recognition digital cipher lock for smart vehicles circuits and systems," Proceedings of the 36th Midwest Symposium, vol.1, pp. 352-355, 1993.

[6] A.I. Alrabady, and S.M. Mahmud, "Some attacks against vehicles' passive entry security systems and their solutions," Vehicular Technology, IEEE Transactions on, vol. 52, Is. 2, pp. 431-439, 2003.

[7] D. Strobel, D. Oswald, B. Richter, F. Schellenberg, and C. Paar, "Microcontrollers as (in)security devices for pervasive computing applications," Proceedings of the IEEE, vol. 102, Is. 8 , pp. 1157-1173, 2014.

[8] Y. Nagaratnam, and Wai Kit Wong "Miniature digital pin-number lock," Computer Research and Development, Second International Conference on, 2010, pp. 686-691

[9] Li Hui, Yang hong-tao, and Li Xiu-lan, "Design and application of new kind of electronic and mechanical antitheft lock using DSP," Computer, Mechatronics, Control and Electronic Engineering (CMCE), International Conference on, 2010, pp. 263-266

[10] M. Flueckiger, W. Zogg, and Y. Perriard, "Optimal design and sensorless position control of a piezoelectric motor integrated into a mechatronic cylinder lock," Energy Conversion Congress and Exposition (ECCE), IEEE, 2010, pp. 3428-3433

[11] Chen Kuilin, Zhao Dongyan, Zhang Haifeng, Wang Yubo, and Liu Liang, "A design and implementation of smart card for secure payment applicable to electric vehicles' charging," Intelligent Transportation Systems (ITSC), 16th International IEEE Conference on, 2013, pp. 511-516

[12] http://en.wikipedia.org/wiki/Remote_keyless_system#History

A Hybrid Electric Vehicle with Incorporation of VaReB Technology

[1]Aseem Mathur , [2]Tanmay Parashar , [3]Dhruv Kohli, [4]Rajveer Mittal, [5]Anirudh Dube

1. Dept. of EEE, MAIT, Delhi, India: mathur.aseem@gmail.com
2. Dept. of EEE, MAIT, Delhi, India: tanmay.parashar@live.com
3. Dept. of EEE, MAIT, Delhi, India: dhruv.kohli05@gmail.com
4. Dept. of EEE, MAIT, Delhi, India: rajveermittal@mait.ac.in
5. Dept. of EEE, MAIT, Delhi, India: anirudhdube12@gmail.com

Abstract— The development and Innovative designing of Electrical Locomotives has become the need of the hour in the existing situation globally as a consequence of two major factors. One of them being the steeply rising prices of Crude oil and the constant damage being imposed on the environment due to harmful substances in the atmosphere due IC Engines. The Electric Vehicle essentially consists of BLDC Hub Motor, SMF Batteries, Stock Chassis and the Regenerative Braking module. Regenerative Braking basically transmits the Kinetic Energy of the wheel due momentum into Electrical Energy which is otherwise wasted. The Regenerative module (VaReB) works on the principle of application of Variable Capacitances into the circuit to harness the desired level of Braking and hence the Regenerated Energy.

Keywords— Electric Vehicle, Regenerative Braking, BLDC Hub Motor, VaReB.

I. INTRODUCTION

As stated in the paper [6] "Theoretical analysis and experimental results have revealed that the cruising distance, braking torque and reliability can be improved effectively using a variable braking control strategy according to the driving conditions. Since the additional power switches, passive components and costly position of sensors are not required the proposed method is particularly suitable for various light electric vehicles.

An EV operates on the current principle which uses a battery bank to provide power for an electric motor. The motor then utilizes the power from the batteries to rotate the transmission which eventually turns the wheel.

An Electric Vehicle(EV) is propelled by rechargeable storage batteries. The Electric motor obtains the required power from a controller which in turn acquires power from a rechargeable battery. In paper [12] as is stated "Electric vehicle as a new means of transport, with its clean energy source pollution, diversification and energy driven higher efficiency become the development trend of the auto industry. But electric vehicles on a single charge range is significantly less than the motor car". In the paper [11] as the author tells us "The limitation of driving range is the key restriction for the development of EV (electric vehicle), and regenerative braking is an effective

approach to extend the driving range of electric vehicle." In the paper [1] the author has said that" Using regenerative braking when, braking, improves the efficiency of an electric vehicle as it recovers energy that, could go to waste if mechanical brakes were used."

We have demonstrated the Regenerative Braking in the Electric Vehicle by Hardware Implementation of the idea. A Electric Vehicle was modified in its design and the components were arranged according to the sufficient scheme. This was accomplished by using varying values of Capacitances according to the paper [1] where the author stated " The user has direct control over the amount of, current regenerated and hence the amount of negative torque applied for, braking. The research has shown that the proposed regenerative braking system, is significantly better in recovering energy and slowing the vehicle compared to a, commercially available regenerative braking system."

The author in paper [5] states " Plug-in EVs utilize a battery system which can be recharged from standard power outlets. Since performance characteristics of electric vehicles have become comparable to, if not better than those of traditional Internal Combustion Engine (ICE) vehicles, EVs pre- sent a realistic alternative."

II. REGENERATIVE BRAKING

A Regenerative brake is an energy recovery mechanism which slows a vehicle or object down by converting its kinetic energy into another form, which can be either used immediately or stored until needed.

Regenerative Braking is the system in which an Electrical Motor normally used to drive an Electric Vehicle is Operated in reverse direction electrically, during Braking and Coasting. During Regenerative Braking, the motor acts as a generator that charges the Auxiliary or Primary Batteries instead of consuming power to propel the vehicle. Hence, the Energy which would have been lost as heat through traditional mechanical frictional brakes is stored in batteries as Electrical Energy.

In the paper [9] the author states " The merits of regenerative braking over traditional braking are energy

978-1-4799-6047-7/14 $31.00 © 2014 IEEE

conservation, wear reduction, fuel consumption and more efficient in braking."

When the motor "acts in electrical reverse," it produces electricity. The accompanying electrical resistance (friction) assists the brake pads in overcoming inertia & slows down the vehicle.

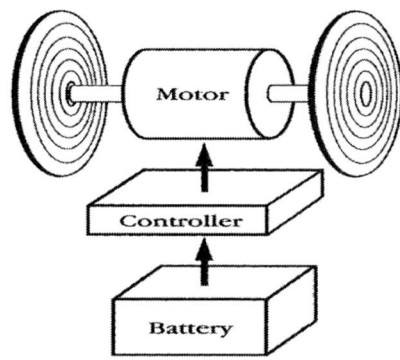

Fig. 1. Basic Diagram of an Electric Vehicle.

Regenerative Braking concept can be used to store energy by use of following techniques-

A. Alternator Coupled Braking Mechanism:

In this technique the shaft of the Moving apparatus of the vehicle is coupled to the Prime Mover of an Alternator to generated Electric Power.

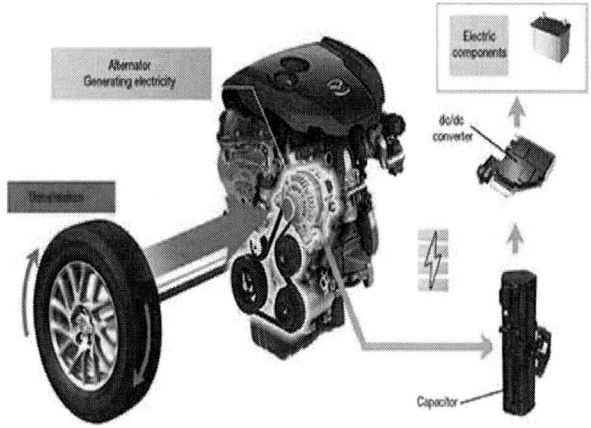

Fig. 2. Alternator Regenerative Braking

B. Kinetic Energy Recovery System (KERS)

KERS is an automotive system which implements generation of energy by virtue of recovery of the Kinetic Energy of the vehicle on application of Brakes in the Electric Vehicle.

KERS allows the storage of generated energy to be stored in the form of the Mechanical Energy (By use of Flywheel) or Electrical Energy (By use of High Voltage Batteries) or Pressurize Fluid (Activate Hydraulic Pumps).

Volvo, a Swedish multinational manufacturing company has recently announced that the system can boost fuel economy by 25 percent using Flywheel KERS System.

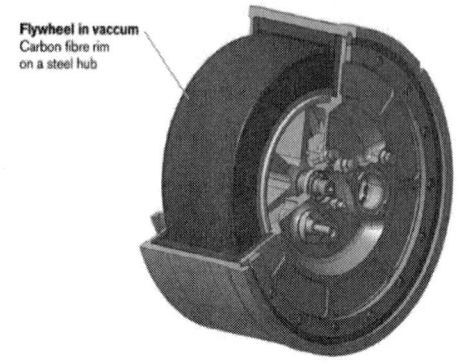

Fig. 3. Flywheel KERS

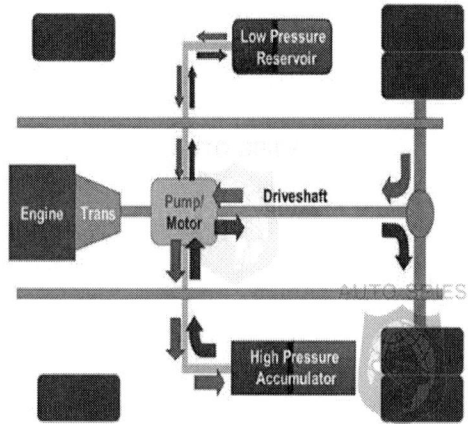

Fig. 4. Hydraulic KERS

C. Solid State Regenerative Braking

Solid State Regenerative Braking involves use of PWM Inverter which acts as the Driving Mechanism for the Locomotion of the vehicle as well the Regenerative Braking Equipment when used as a Rectifier when instances of Braking of the Vehicle occur.

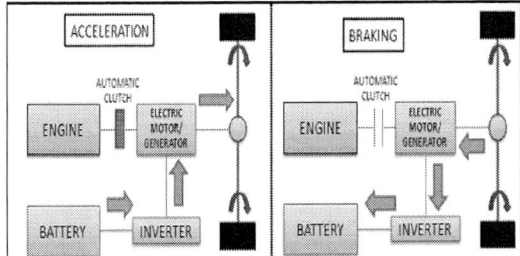

Fig. 5. Solid State Braking.

III. BRUSHLESS DC MOTOR

Brushless DC electric motor(BLDC Motor) are synchronous motors that are powered by a DC electric source via an integrated inverter/switching power supply, which produces an AC electric signal to drive the motor.

The Motor operation is based on the force of attraction & repulsion between the magnetic poles. As shown in figure of the 3-Φ motor, the process starts with the flow of current through one of the three stator windings and generation of a magnetic pole that in turn attracts the closest permanent magnet of the opposite pole. The rotor moves as the current shifts to any adjacent winding. A sequential charging of each winding will cause the rotor to follow in a rotating field.

The torque in the example depends on the amplitude of the current & the no. of turns in the stator windings, the air gap between the rotor and the windings, the length of the rotating arm and the strength and size of the permanent magnets. As given by the author in the paper [10], "A BLDC motor with the characteristics of high speed and high power density has been widely used in industrial areas".

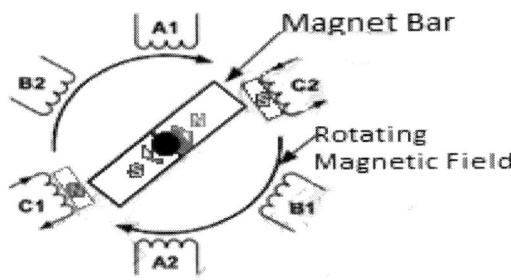

Fig. 6. BLDC Motor Working.

IV. RB MODULE

The RB (Regenerative Braking) Module basically is a Three Phase Rectifier made up of 6 P-N Junction Diodes(IN 5408) which acts as a Converter to obtain DC Voltage from the TRAPEZOIDAL Back EMF of the BLDC Hub Motor

The DC Voltage of varying values in the range of 28-48 Volts was obtained from the RB Module can be used to Charge a Secondary Battery which can be used to provide Power to various Auxiliary Equipments in the Electric Vehicle

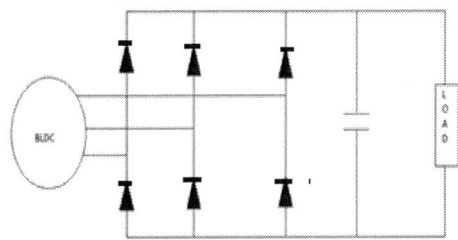

Fig. 7. Circuit of RB Module

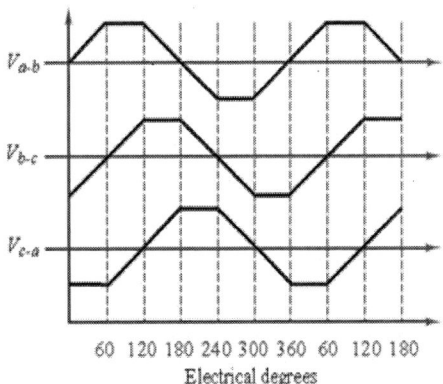

Fig. 8. Back EMF Waveform of BLDC Motor

V. VAREB TECHNOLOGY

The aim is to vary the intensity of braking and extraction of energy from the back EMF, so that optimum energy could be extracted even during the varying momentum of the vehicle.

The circuit is designed in such a way that there are four sets of capacitances that can individually be connected to the rectifier circuit as and when required. The capacitances are connected in such a way that they get included in the circuit in increasing order of their capacitive reactance.

The four sets of capacitances are as follows:

1) C1= 100 µF

2) C2= 433 µF

3) C3= 1433 µF

4) C4= 3433 µF

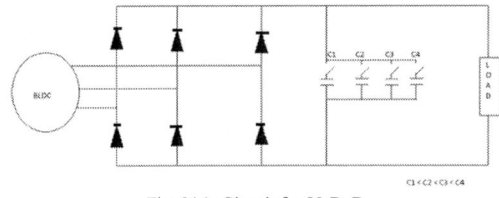

Fig. 9(a). Circuit for VaReB

978-1-4799-6047-7/14 $31.00 © 2014 IEEE   613

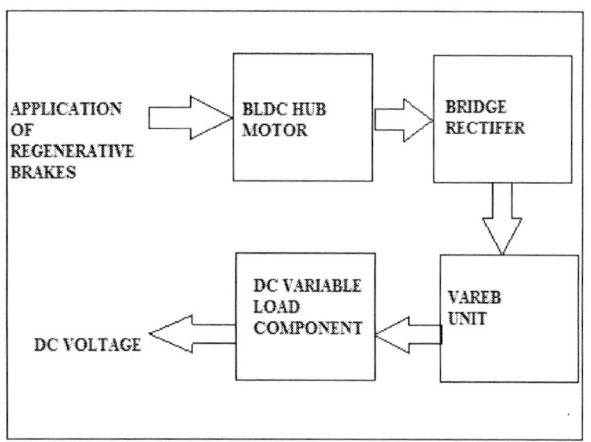

Fig.9(b) Block Diagram VaReB

VI. RESULTS

The concept and architecture of the Regenerative Braking Concept was established by implementing it on the Hardware Components and exhibiting the obtained Electric Power which is supplied to provide illumination of the Vehicle for certain time period.

The Maximum Speed attained by the Electric Vehicle on a relatively flat road is 25 Km/hr

The smoothness of Electric Braking was considerable in context of Jerking motion in comparison to the Mechanical Brakes in the Vehicle.

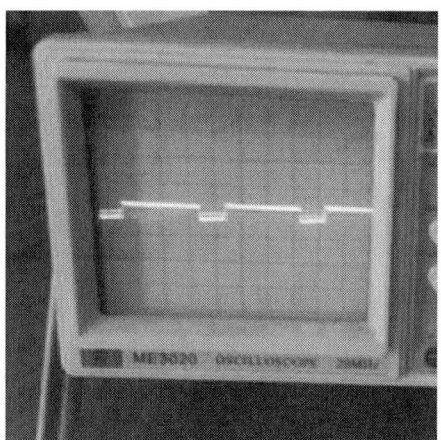

Fig 10: CRO image of Voltage input to the BLDC Motor

Fig 11: CRO image of the DC Voltage generated by the Batteries to the Electric Controller for application of motion of the Electric Motor.

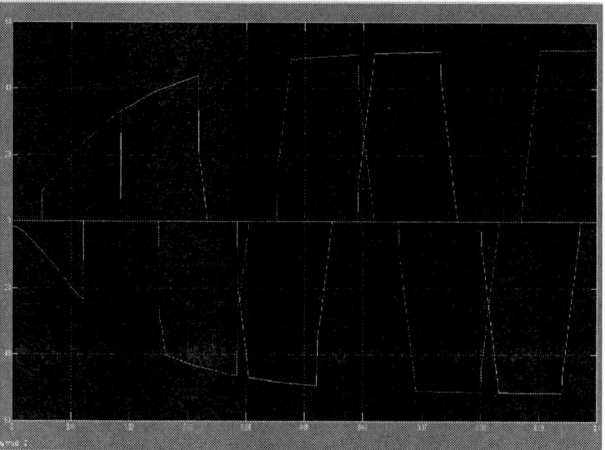

Fig12 Waveform of Back EMF developed in the BLDC Motor is as follows

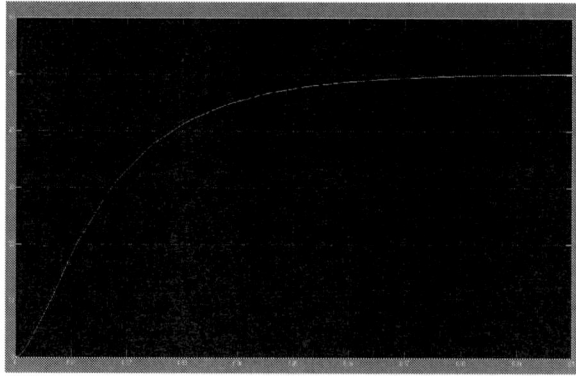

Fig 13 DC Voltage Generated by RB Module

978-1-4799-6047-7/14 $31.00 © 2014 IEEE   614

The following data was acquired regarding the Electric Braking time intervals of the Electric Vehicle -

1) When the Load used was LED Lights Panel (Array of LEDs- 25W).

A) Electric Vehicle at a Suspended position, wheel running at 20km/hr.

Capacitance (in µF)	Stopping Time(in seconds)
100	7.6
433	7.0
1433	6.5
3433	6.3

Table I. Time Intervals for various Values of Capacitances.

B) Electric Vehicle moving on road at a constant velocity of 20 Km/hr.

Capacitance(in µF)	Stopping Time(in seconds)
100	29.4
433	25.9
1433	22.7
3433	21.3

Table II. Time Intervals for various Values of Capacitances.

2) When the Load used was Electric Fan (DC Motor - 12V,15W; For this Load, a Voltage Stabilizer was utilized for the purpose of obtaining 12V for the DC Motor.)

A) Electric Vehicle at a Suspended position, wheel running at 20km/hr.

Capacitance(in µF)	Stopping Time(in seconds)
100	10.3
433	9.7
1433	9.5
3433	9.2

Table III. Time Intervals for various Values of Capacitances.

B) Electric Vehicle moving at a constant velocity of 20 Km/hr.

Capacitance(in µF)	Stopping Time(in seconds)
100	33.4
433	32.5
1433	32.1
3433	31.6

Table IV. Time Intervals for various Values of Capacitances.

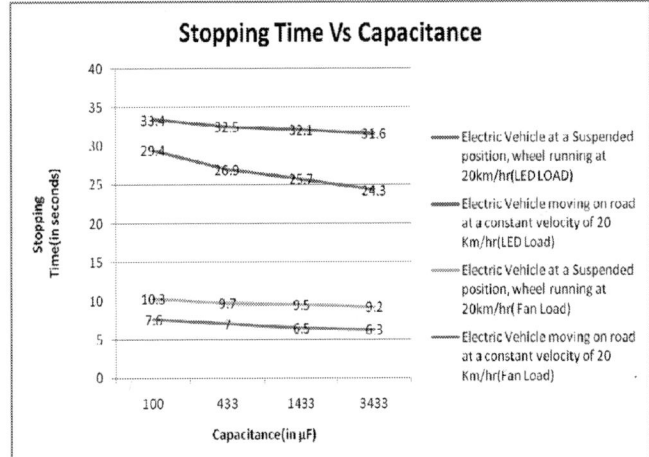

Fig14 Stopping Time Vs Capacitance Graph

VII. FUTURE SCOPE

Based on Theoretical observations in the above hypothesis of the development of the "Implementation of VaReB Technology on Electric Vehicle" we managed to enlist certain modifications and conceptual ideas which can be applied to such a machine to achieve improved efficiency and superior performance -

Application of Ultracapacitors is for following:

Ultracapacitors with variable value for discharging time can be used to attain a desired Regenerative result.

Ultracapacitors of high Capacitance values can also be used to directly charge Batteries present in the system.

Batteries such as Li-Ion Batteries, Li- Polymer Batteries, Zebra Batteries can be used to attain higher degree of productivity

As concluded by the author in paper [13], "the ultra-capacitor is able to restore the regenerative energy very quickly, which will improve starting and acceleration performance, extend driving range, and increase fuel economy."

Use of Hybrid Technology i.e. Inclusion of an additional source of energy to the existing Electrical Power Source (Battery).

Solar Power can be harnessed by use of Solar Cells mounted on the Body of the vehicle.

The Shock Absorbers acting as the Suspension system of the Vehicle can be used to act as a source of Mechanical Stress which acts as the driving force for the Piezoelectricity phenomenon resulting in development of Electric Power.

ACKNOWLEDGMENT

This work was supported in part by Department of Electrical and Electronics Engineering, Maharaja Agrasen Institute of Technology.

REFERENCES

[1] Daniel Torres, Patrick Heath " Regenerative Braking of BLDC Motors" Microchip

[2] Eduardo Wiechmanand, Juan W. Dixon and Micah Ortúzar " Regenerative Braking for an Electric Vehicle Using Ultra Capacitors and a Buck-Boost Converter" Catholic University of Chile.M. Young, The Technical Writer's Handbook. Mill Valley, CA: University Science, 1989.

[3] Sandun Kuruppu " Implementation and Performance Evaluation of a Regenerative Braking System Coupled to Ultra Capacitors for a Brushless DC Hub Motor Driven Electric Tricycle", Purdue University e-PUBS.

[4] J.Karthikeyan, Dr. R. Dhanasekran " Simulation and Implementation of Current Control of Brushless DC Motor based on a Common DC Signal", International Journal of Engineering Science and Technology 01/2010.

[5] Jarrad Cody, Aaron Mohtar,Özdemir Göl, Andrew Nafalski, , Zorica Nedic " Regenerative Braking in an Electric Vehicle", Zeszyty Problemowe – Maszyny Elektryczne Nr 81/2009.

[6] Ming-Yang Cheng, Cheng-Hu Chen,Wen-Chun Chi, " Regenerative Braking Control for Light Electric Vehicles ", IEEE PEDS 2011, Singapore, 5 - 8 December 2011.

[7] Sanita C S, J T Kuncheria, " Modelling and Simulation of Four Quadrant Operation of Three Phase Brushless DC Motor With Hysteresis Current Controller ", International Journal of Advanced Research in Electrical, Electronics and Instrumentation Engineering Vol. 2, Issue 6, June 2013.

[8] Hang Zhang ,Xiaohong Nian, Fei Peng "Regenerative Braking System of Electric Vehicle Driven by Brushless DC Motor" Industrial Electronics, IEEE Transactions on (Volume:61 , Issue: 10)

[9] G.D Leong,Yoong M.K., Gan, Y.H., Gan. "Studies of Regenerative Braking in electric vehicle" published in Sustainable Utilization and Development in Engineering and Technology (STUDENT), 2010 IEEE Conference on 20-21 Nov. 2010

[10] "A new simulation model of BLDC motor with real back EMF waveform" published in The 7th Workshop on Computers in Power Electronics, 2000. COMPEL 2000.

[11] Binggang Cao ; Sch. of Mech. Eng., Xi" and Jiaotong Univ. ; Zhifeng Bai ; Wei Zhang, "Research on control for regenerative braking of electric vehicle" published in IEEE International Conference on Vehicular Electronics and Safety, 14-16 Oct. 2005.

[12] Zhang Guirong ; Shandong Jiaotong University, Jinan, China, "Research of the regenerative braking and energy recovery system for electric vehicle" Published in World Automation Congress (WAC), 24-28 June 2012.

[13] Huangqiang Luo ,Feng Wang ; Sch. of Mech. Eng., Xi"an Jiaotong Univ., Xi"an, China ; Xiaomei Yin ; Huangqiang Luo ; Ying Huang, "A Series Regenerative Braking Control Strategy Based on Hybrid-Power" Published in International Conference on Computer Distributed Control and Intelligent Environmental Monitoring (CDCIEM), on 5-6 March 2012

Energy Efficiency Performance of Reconfigured Radial Networks with Reactive Power Injection

Pawan Kumar
Department of Electrical & Electronics Engineering
Galgotias College of Engineering and Technology
Greater Noida, UP, India
pawanror@gmail.com

Surjit Singh
Department of Computer Engineering
National Institute of technology
Kurukshetra, Haryana, India
surjitmehla@gmail.com

Abstract— Reconfiguration of radial distribution system is usually performed to diagnose the severe effects of unscheduled fluctuations or variations in power demand during on-line operation. The reconfigured network alters the resultant power flow in the system and it is required to be optimizing with and without reactive power injection. Different methods have been proposed earlier to optimize the system operation by reconfiguration and by injecting reactive power to the system using indices or soft computing techniques. Generally, these methods considered the reactive power management and network re-configuration problems independently. Therefore, the reactive power injection in the resulting configurations may lead to misleading and imperfect results about deferral values; loss reduction, improved loadability limit and other subsequent calculations. This paper presents comparative study of effect of reactive power injection on system performance in different configurations. The study is implemented in MATLAB environment and demonstrated on IEEE 33-node radial distribution system.

Keywords— *Energy Efficiency; Network Configurations; Distribution System; Reactive Power Injection; Voltage Profile.*

I. INTRODUCTION

Reconfiguration of distribution systems, incorporating the computer aided monitoring, control and management of electric power, is usually performed to provide better services to consumers during on-line operation. Generally, the objective of the network reconfiguration is loss reduction. Also, in order to meet the load demand distribution substation commands power equipments to operate in coordination for optimizing the distribution system operation, and to improve the system's energy efficiency performance.

Now a day, energy efficiency has become an increasingly problematic area in power distribution systems. In distribution system, it has direct bearing on power quality, demand side load management and maximum demand control. Power quality may be defined as "the measurement, analysis, and improvement of bus voltage, usually a load bus voltage, to maintain that voltage to be sinusoid at rated voltage and frequency" [1]-[3]. A direct correlation exists between the lack of electric power quality delivered to customer and the number of complaints received from the customer [4]-[5]. The significance of power quality index and scope of energy efficiency in distribution systems depends upon the type of customers e.g. residential, commercial, and industrial. This is

due to the preferences and/or privileges given, and voltage and frequency dependencies of different loads, depending upon the area being served [6].

An automated distribution system optimizes the operation by reconfiguring it, while supplying power to residential, commercial and industrial loads. In the past, different methods for network reconfiguration and reactive power injection have been proposed based upon the indices and heuristics or meta-heuristics approaches. However, application of soft computing methods in optimization problem is very much common in recent years. The methods based upon indices and soft computing techniques have been described using Power Loss Index [7], Voltage Stability Index [8], Fuzzy Logics [9], Particle Swarm Optimization [10], and Genetic Algorithms [11]. Also, the various methods for voltage stability with different load behaviour have been presented in [12-15]. Further, authors in [16]-[18] performed the network reconfiguration for loss reduction using Refined Genetic Algorithms [16], Fuzzy Logics [17] and Ant Colony Optimization [18]. The configurations obtained using these methods differ from each other. Therefore, the energy efficiency performance of the resulting configuration differs from each other because they may offer system power loss, loadability limit, voltage profile and reactive power demand differently.

In real time operation, when load demand change and reconfiguration is performed, the configurations obtained with single objective not necessarily be optimized with same reactive power injection as it was earlier [19]. Therefore, reactive power injection in resulting configuration needs to be re-evaluated, for optimal operation, before implementing it. The simultaneous implementation of optimal configuration with reactive power injection further improves the system loadability limit, loss reduction and voltage profile. This makes the system energy efficient, and can provide cost-effective and collaborative platform for futuristic load requirements. This paper has evolved a comparative study of reactive power injection in different configurations, obtained using various soft computing techniques, before implementing them into the operation.

This paper is organized as follows: The operational requirements of the distribution system are discussed in section II. The mathematical calculations for node voltage, loadability limit, power loss and capacitor size are discussed in section III. The problem formulation and proposed algorithm are described in section IV and the test results are discussed in detail in section V.

II. OPERATIONAL REQUIREMENTS OF DISTRIBUTION SYSTEMS

Distribution systems, unlike transmission systems, are frequently required to be reconfigured for its economical operation. This is because of the unscheduled fluctuations or variations in power demands during its operation. As a consequence of which they has to face more problems like poor voltage regulation and voltage rise during peak load and off peak load respectively. In addition to above problems distribution system is overloaded during most period. When load diversity is high, and/or when high reliability is desired, the distribution systems adopt different management schemes like load management and configuration management. The on-line implementation of these schemes improves the system energy efficiency and can meet more load demand at present or in future.

In order to supply power, the distribution system can operate manually or automatically to meet customer's load demand. Due to the complexity in their operations and design the automated or in general the smart distribution systems incorporating the computer aided monitoring and control can provide better services to the customer through load management and configuration management. The load management further includes the maximum demand control by reactive power injection. The automated control of the power demand enhances the reliability and availability of supplying load requirements. An automated distribution system may avoid the numbers and duration of supply interruption due to unbalancing, over loading, operation under faulted conditions and benefited the supplier and the consumers simultaneously [8]-[9].

III. MATHEMATICAL FORMULATIONS

In distribution systems, power demand varies with time and location, and may not be always sufficient so as to meet the load requirement. This will increase the power loss and voltage drop in the system, and at the same time it reduces the loadability limit due to overloading. Further, the actual power flow in the line is the combined effect of the active and reactive powers, and based upon which the power utilities decides line equipment capacity. Therefore, network is required to be reconfigured and controlling the reactive power, by injecting it locally, reduces the maximum demand and helps in improving the loadability limit of the resulting configuration.

A. Calculations of Node Voltage and Maximum Loadability Index

The reconfigured network changes the loadability limits and the requirement of reactive power injection in the system configuration. To obtain loadability limits, an equivalent network is considered between two nodes as shown in Fig.1.

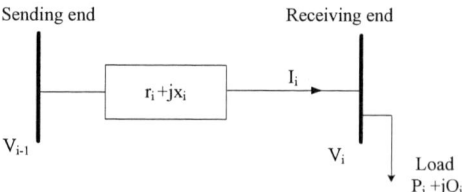

Fig. 1. Single Line Representation of Radial Distribution System.

The load $(P_i + jQ_i)$ is connected at receiving end node and the voltage (V_i) at this node is taken as reference. For line current (I_i), resistance (r_i), reactance (x_i), sending end voltage angle (δ) and receiving end power factor angle (θ) the following equations can be derived:

$$V_{i-1} \cos \psi = V_i \cos \theta + I_i r_i \tag{1}$$

$$V_{i-1} \sin \psi = V_i \sin \theta + I_i x_i \tag{2}$$

Where, $\psi = \theta + \delta$

$$V_{i-1}^2 = (V_i \cos \theta + I_i r_i)^2 + (V_i \sin \theta + I_i x_i)^2 \tag{3}$$

Solving the above equation

$$V_i^2 = \frac{V_{i-1}^2}{2} - (r_i P_i + x_i Q_i) \pm [\{\frac{V_{i-1}^2}{2} - (r_i P_i + x_i Q_i)\}^2 - (r_i^2 + x_i^2)(P_i^2 + Q_i^2)]^{1/2}$$
$$= X \pm Y \tag{5}$$

From equation (5) mathematically the voltage solution does not exist when the term Y^2 becomes negative. Therefore, for the solution to exists the following inequality is to be satisfied,

$$\{\frac{V_{i-1}^2}{2} - (r_i P_i + x_i Q_i)\}^2 - (r_i^2 + x_i^2)(P_i^2 + Q_i^2) \geq 0 \tag{6}$$

The possible solution of Eqn. (5) defines the loadability limit at candidate node. Further, modifying the Eqn. (5) as quadratic equation by replacing the existing load (P_i+jQ_i) with $MLI*(P_i+jQ_i)$ and equating it to zero, MLI_i is calculated as under:

$$MLI_i = \frac{V_{i-1}^2[-(r_i P_i + x_i Q_i) + \sqrt{(r_i^2 + x_i^2)(P_i^2 + Q_i^2)}]}{2(x_i P_i + r_i Q_i)^2} \tag{7}$$

Here, MLI_i is the maximum loadability index which when multiply with existing load (kVA) defines the loadability limit at candidate node.

B. Calculations of Power Loss and Optimal Capacitor Size

The total active power loss in distribution system is given by,

$$TP_L = \sum_{i=1}^{br} I_i^2 r_i \tag{8}$$

Here, 'br' is the number of branch in the system. Separating the real and reactive components of branch currents the power loss can be expressed as:

$$TP_L = \sum_{i=1}^{br} I_{ai}^2 r_i + \sum_{i=1}^{br} I_{ri}^2 r_i \tag{9}$$

978-1-4799-6047-7/14 $31.00 © 2014 IEEE

If a capacitor of current 'I_{ck}' is placed at a node 'k', the total real power loss changes and it is obtained by subtracting the same from eqn. (7), and the loss reduction 'ΔTP_{Lk}' can be expressed as:

$$\Delta TP_L = -2I_{ck}\sum_i^k I_{ri}r_i - I_{ck}^2\sum_i^k r_i \qquad (10)$$

Equation (10) is further manipulated for maximum loss reduction as under,

$$\frac{\delta(\Delta TP_L)}{\delta I_{ck}} = -2(\sum_i^k I_{ri}r_i + I_{ck}\sum_i^k r_i) = 0 \qquad (11)$$

Therefore, the capacitor current for maximum loss reduction can be represented as,

$$I_{ck} = -(\frac{\sum_i^k I_{ri}r_i}{\sum_i^k r_i}) \qquad (12)$$

Now, if V_k is the voltage at node 'k' then the capacitor size can be calculated as,

$$Q_{ck} = I_{ck}V_k \qquad (13)$$

Here, 'V_k' is the 'k^{th}' node voltage, the candidate node selected for capacitor placement. Further, the problem of optimal capacitor placement requires determination of the different location, sizes and number of capacitors to be installed, subject to the operational constraints, in a distribution system to achieve the maximum benefits. The capacitor size at each node can be obtained using (13). However, the capacitor placement which offers the maximum loss reduction is finally selected. This has been achieved by developing an algorithm described in next section.

IV. PROBLEM FORMULATIONS AND PROPOSED ALGORITHM

The problem is formulated based upon the equations described in section III. This includes the loss reduction and improvement in the MLI provided that the node voltage is within prescribed limits.

Maximize ΔTP_L

s. t.

$$V_{i,min} \leq V_i \leq V_{i,max}$$

$$S_i \leq S_{i,max}$$

Where, 'i' varies from '1' to 'n' and 'n' is the no. of nodes. The minimum and maximum voltage limits are taken as *0.95p.u* and *1.05p.u* respectively. Also, $S_{i,max}$ is the maximum loadability at the respective node before its voltage collapse. The computational steps involved in the proposed algorithm for identifications of candidate node for reactive power injection are summarized below:

Step1: *Assume flat voltage profile and set the convergence criterion. Consider the line power loss zero initially and run the load flow.*

Step2: *Compute the Node Voltage, MLI, power loss and capacitor size at each node using (5, 7, 8 and 13).*

Step3: *Compute power loss by placing the capacitor & find the minimum MLI of the system.*

Step4: *Select the node as candidate which offer maximum loss reduction with capacitor placement and fix this capacitor size while finding subsequent placement.*

Step5: *Repeat Step 1 to Step 4 to get the next candidate node for placement.*

Step6: *Ensure that there is no significant improvement in reduction in power loss with the further placement and the node voltage is within prescribed limit.*

Step7: *Repeat Step 1 to Step 6 for different configurations and obtain the desired parameters.*

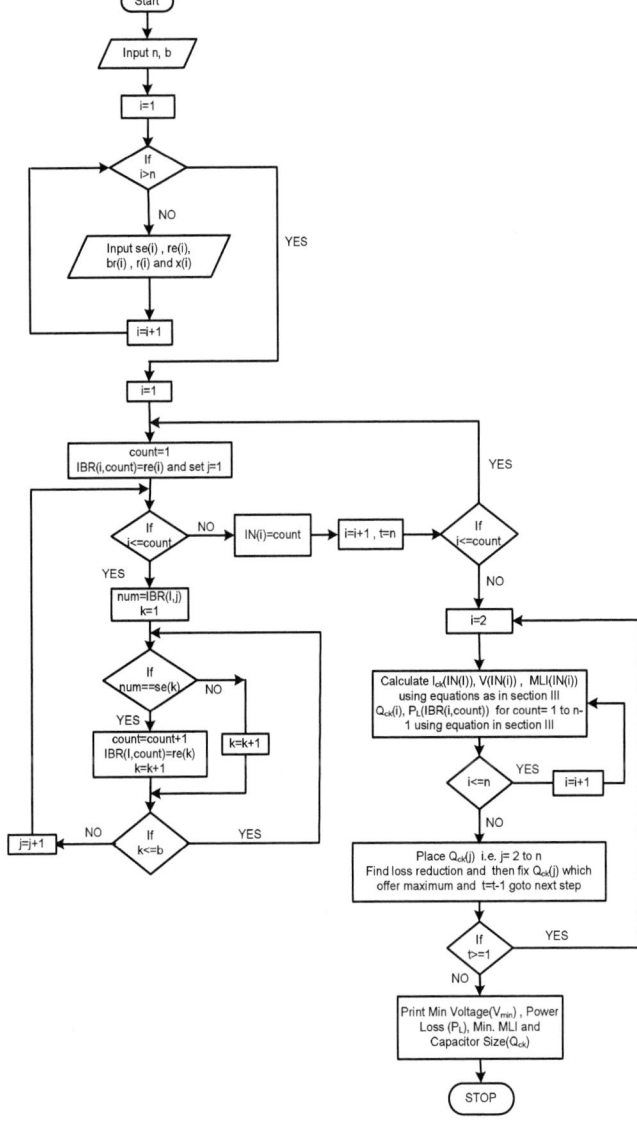

Fig. 2. Flow Chart of the Proposed Algorithm.

Fig.2 show the flow chart of the algorithm used to maintain the network structure as radial in different configuration and the process of capacitor placement used in this paper.

V. TEST RESULTS AND DISCUSSIONS

The sample system under study is a 12.66kV, IEEE 33-Node radial distribution network as shown in Fig. 3. The line and load data of the system are obtained from [20]. Also, the first node is the substation and its voltage is maintained at 1.0p.u. The network is analyzed to quantify the extent of violation of limits imposed on voltage magnitude at nodes/buses in an n-node radial distribution system for optimal reactive power injection in different reconfiguration obtained using Genetic Algorithm, Fuzzy Logics and combining Fuzzy Logics & Ant Colony Optimization. In Fig. 3 solid branches represent the lines that are in service and constitute the base radial network under consideration. However, the dotted lines are the tie-line in the network and there are five tie-lines as 33, 34, 35, 36, and 37. For radial structure, normally opened (NO) sectionalizing switches are used in the tie-lines.

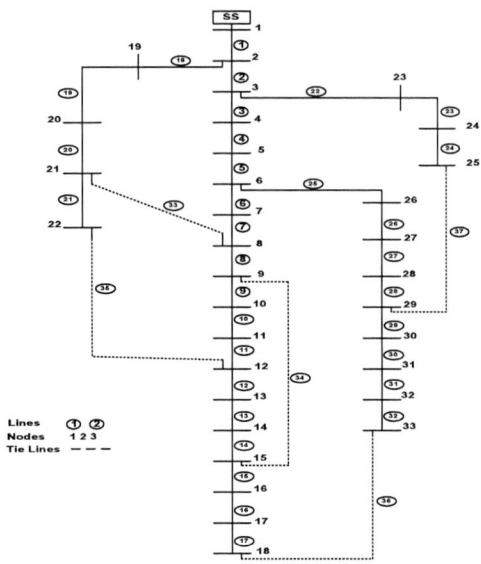

Fig. 3. Base configuration of an 33-node radial distribution system.

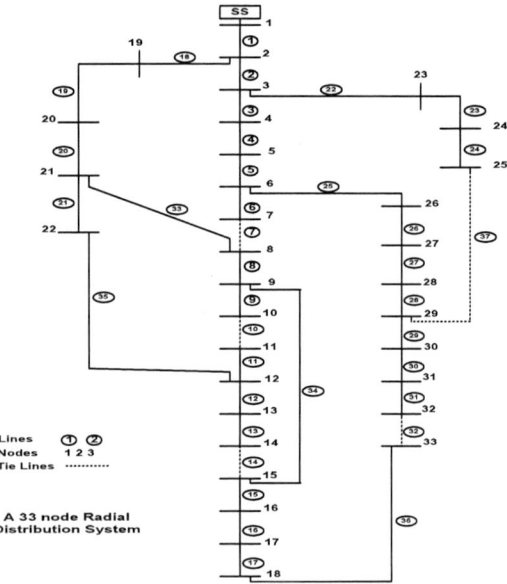

Fig. 4. A 33-node radial system with tie-lines as 7, 10, 14, 32, and 37.

TABLE I.   TEST RESULTS WITHOUT CAPACITOR PLACEMENT

Parameters	Base n/w	GA [16]	Fuzzy [17]	ACO-Fuzzy [18]
Tie-lines in Configuration	33,34,35,36, 37	9,28,33, 34,36	7, 9,14, 32, 37	7,10, 14,32, 37
Power loss (kW)	202.67	141.7	137.7	136.8
Loss Reduction (kW)	----	60.97	64.97	65.87
Min. Voltage	0.9131	0.9370	0.9378	0.9347
Min. MLI	13.38	16.21	14.73	14.72

Table I shows the test results of the radial distribution system in different configurations without reactive power injection. From test results in Table I it can be observed that the total power loss, minimum voltage and minimum MLI in the system for base case are 202.67kW, 0.9131p.u. and 13.38 respectively. Different configuration obtained using GA [16], Fuzzy [17] and ACO-Fuzzy [18] are also considered to evaluate the system performance without capacitor placement. These configurations offer the significant improvement in the evaluating parameters. The total power loss, minimum voltage and minimum MLI in the system for configuration obtained using GA [16] are 141.7, 0.9370p.u. and 16.21, for Fuzzy [17] are 137.7kW, 0.9378p.u. and 14.73, and for ACO-Fuzzy [18] are 136.8kW, 0.9347p.u. and 14.72 respectively. It can also be noticed that the MLI is maximum in the configuration obtained using GA [16]. The configuration in Fuzzy [17] yields the maximum improvement in the node voltage profile whereas in ACO-Fuzzy [18] the best improvement is in the power loss reduction.

TABLE II.   TEST RESULTS WITH CAPACITOR PLACEMENT

Parameters	Base n/w	GA [16]	Fuzzy [17]	ACO-Fuzzy [18]
Capacitor Size (kVAr)	1040(30) 350(13) 570(24)	1010(30) 310(14) 480(6)	1010(30) 610(21) 490(24)	1010(30) 460(8) 500(24)
Total Size (kVAr)	1960	1800	2110	1970
Power loss (kW)	132.74	97.59	92.71	93.11
Loss Reduction (kW)	69.93	105.08	109.95	109.56
Min. Voltage	0.9412	0.9571	0.9562	0.9592
Min. MLI	19.31	22.26	18.58	17.80

Table II shows the test results of configuration discussed above with reactive power injection. For the base network the total capacitor size of 1960kVAr is placed at node 30, 13 and 24 of value 1040, 350 and 570kVAr respectively. With capacitor placement the power loss reduces to 132.74kW and it also improves the voltage to 0.9412p.u. and MLI to 19.31. In this case the voltage level is still less than the prescribed limit whereas the significant improvement has been noticed in MLI. The column 3rd in Table II shows the test result of configuration obtained using GA [16] with capacitor placement. Here, it can be noticed that with placement of 1800kVAr only a significant improvement in power loss, voltage profile and MLI can be observed. The power loss reduces to 97.59kW and minimum voltage improves to 0.9571p.u. However, the minimum MLI improves to 22.26 and this is the maximum among all other configurations.

Further, test result for configuration obtained using Fuzzy [17] shows that the maximum capacitor size of value 2110kVAr. The maximum power loss reduction in this configuration has been observed and it reduces to 92.71kW. The significant improvement in voltage profile and MLI is also noticed and it improves to 0.9562p.u. and 18.58 respectively. The configuration obtained using ACO-Fuzzy [18] also shows the effective improvement in evaluating parameters with reactive power injection. In this case the power loss reduces to 93.11kW. The voltage profile and MLI improves to 0.9592p.u. and 17.80 respectively. The improvement in voltage profile is much better than any other configuration whereas the minimum MLI is observed in this case.

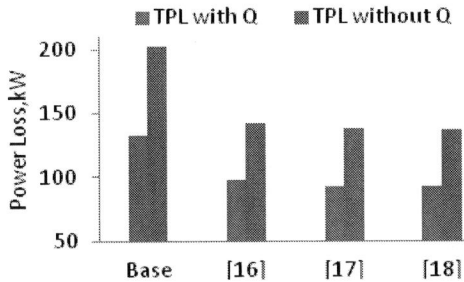

Fig. 5. Power loss in different configurations.

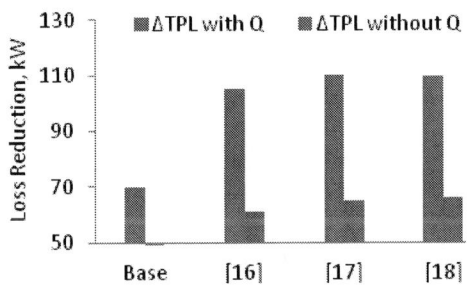

Fig. 6. Power loss reduction in different configurations

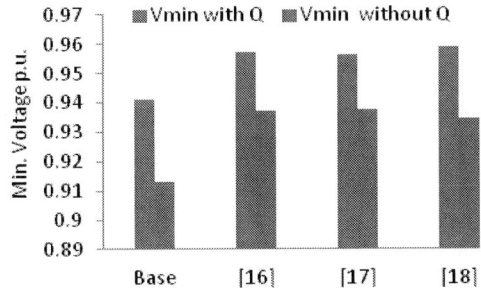

Fig. 7. Minimum voltage variation in different configurations.

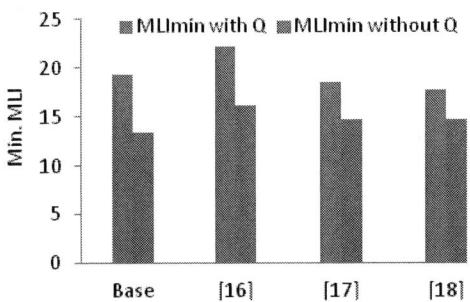

Fig. 8. Minimum MLI variation in different configurations

Fig. 5-8 shows the variations of power loss, loss reduction, minimum voltage and minimum MLI in different configurations respectively. A significant improvement in the system performance can be noticed in all configurations in comparison with base network. However, in mutual comparison of the performance in these configurations there is a significant variations in evaluating parameters is observed. Therefore, if one configuration is best suitable to meet one objective may not be as effective as to that for other requirements.

VI. CONCLUSIONS

This paper presented performance of the distribution system in different configurations obtained using soft computing methods. The performance is evaluated based upon the reactive power requirement in the system configurations. During investigation it has been noticed that the overall system performance differ in different configurations. From the test results it has been observed that the power loss, voltage profile and system loadability limit is significantly affected by reactive power injection. The configuration obtained using genetic algorithm shows the maximum improvement in the loadability limit whereas in the other configurations obtained using Fuzzy and ACO-Fuzzy the power loss reduction and voltage profile is comparatively better. Therefore, while implementing the system configuration in real time it is required to evaluate its performance in other aspects too, than the set objective during problem formulation, like local power injection (active and reactive power), loadability limit and node voltage security.

REFERENCES

[1] Jiyuan Fan, Borlase S, "The evolution of distribution," *IEEE Power and Energy Magazine*, vol. -7, no. 2, pp. 63-68, March/April 2009.

[2] G.T Heydt, "The next generation of power distribution system," *IEEE Trans. Smart Grid*, vol.1, no. 3,pp. 225-235, Dec. 2010.

[3] S. S. Venkata,A. Pahwa, R. E. Brown, and R. D. Christie, "What future distribution engineers need to learn," *IEEE Trans. Power Systems*, vol. 19, no.1,pp. 17-23,Feb. 2004.

[4] Katherine M. Rogers, Ray Klump, Himanshu Khurana, Angel A. Aquino, Thomas J. Overbye, "An authenticated control framework for distributed voltage support on the smart grid,"*IEEE Trans.* Smart Grid, vol. 1, no. 1, pp. 40-47, June 2010.

[5] C. Smallwood and J. Wennermark, "Benefits of distribution automation," *IEEE Industry Application Magazine*, vol. 16, no. 1,pp. 65-73, Jan./Feb. 2010,

[6] M. H. Haque, "Capacitor placement in radial distribution systems for loss reduction," in *Proc.1999IEE Generation, Transmission, Distribution*, vol. 146, no. 5, pp. 501-505.

[7] S. Satyanarayana, T. Ramana, S. Sivanagaraju, and G.K. Rao, "Voltage stability analysis for radial distribution networks with and without compensation," *International Journal of Water and Energy*,vol.60, no.1, pp.48-53, 2003.

[8] M.E.Baran and F.F. Wu, "Optimal sizing of capacitor placement in radial distribution system," *IEEE Trans. Power Delivery*, vol. 4, no. 2, pp. 735-743, Jan. 1989.

[9] S.F Mekhamer, S. A. Soliman, M A. Moustafa, and M.E. El-Hawary, "Application of fuzzy logic for reactive power compensation of radial distribution feeders," *IEEE Trans. Power Systems*, vol. 18, no. 1, pp. 206-213, Feb. 2003.

[10] Kai Zou,A.P. Agalgaonkar, K.M. Muttaqi, and S. Perera, "Voltage support by distributed generation units and shunt capacitors in distributed system," in *IEEE Power & Energy Society General Meeting*, Wollongong, NSW, Australia, July 2009, pp. 1-8.

[11] M. R. Haghifam and O.P. Malik, "Genetic algorithm-based technique for fixed and switchable capacitors placement in distribution systems with uncertainty and time varying loads,"*IET Journal Generation, Transmission & Distribution*, vol. 1, no. 2, pp. 244-252, March 2007.

[12] B. Venkatesh, Rakesh Ranjan, and H.B.Gooi, "Optimal reconfiguration of radial distribution systems to maximize loadability," *IEEE Trans. Power System*,vol. 19, no.1, pp. 260-266, Feb. 2004.

[13] A. Augugliaro, L. Dusonchet, S. Favuzza, M. G. Ippolito, and E. Riva Sanseverino, "A simple method to assess loadability of radial distribution networks," in *Proc. 2005IEEE Power Tech.*, pp. 1-7

[14] G. Le Dous, "Voltage stability in power systems. Load modeling based on 130kV field measurements," *Technical report no. 324L*. Department of electrical engineering, Chalmers University of Technology, Goteborg, Sweden,1999,

[15] S. Satyanarayana, T. Ramana, S. Sivanagaraju, and G.K. Rao, "An efficient load flow solution for radial distribution network including voltage dependent loads models," *Taylor & Francis Journal Electric Power Components and Systems*, vol. 35, no. 5, pp. 539-551, May 2007.

[16] Y. Y. Hong, S.Y. Ho, Determination of network configuration considering multi-objective in distribution systems using genetic algorithm, IEEE Trans. Power Syst. 20 (2) (2005) 1062-1069

[17] R Hoosmand, E. Mashhoor, Application of fuzzy algorithm in optimal configuration of distributed networks for loss reduction and load balancing, Eng. Int. Syst. 14(1) (2006) 15-23

[18] A. Saffar, R. Hooshmand, A. Khodabakhshian, A new fuzzy optimal reconfiguration of distribution systems for loss reduction and load balancing using ant colony search based algorithm, Elsevier journal of Applied Soft Computing, Vol. 11 (2011), Pages 4021-4028.

[19] Ikbal Ali, Mini S. Thomas and Pawan Kumar, "Effect of loading pattern on the performance of reconfigured medium size distribution system", Proc. of fifth IEEE Power India Conference, 2012, DOI: 10.1109/PowerI.2012.6479503.

[20] M.E. Baran and F.F. Wu, "Network reconfiguration in distribution system for loss reduction and load balancing," *IEEE Trans. Power Delivery*, vol. 4, no. 2, pp. 1401-1407, Apr. 1989.

BIOGRAPHIES

Pawan Kumar (M-12) is an Assistant Professor in the Department of Electrical & Electronics Engineering, Galgotias College of Engineering & Technology, Greater Noida, U.P., India. He is graduated from Kurukshetra University and received his M. Tech Degree from Indian Institute of Technology, Delhi. He is the certified Energy Manager cum Auditor by BEE, Ministry of Power, Govt. of India. His research interests are in optimization techniques in Distribution System's Planning, Operation and Control, Energy Efficiency, and Smart Grid.

Surjit Singh is PhD Research Scholar in the Department of Computer Engineering at National Institute of Technology, Kurukshetra, Haryana, India. He has also worked as Assistant Professor at NIT Kurukshetra. He graduated from Ch. Devi Lal State Institute of Engineering and Technology, Kurukshetra University, with honors and did his post graduation, with distinction. His research area is in the field of Wireless Sensor Network Optimization, Soft-computing, Power System Optimization.

Load Frequency Control of Three Area Hydro Thermal Power System with HFTID Controller

Ajay Kumar
Department of Electrical Engineering
N.I.T. Kurukshetra, Haryana (India)
ajaychaudhary.ee10@gmail.com

Prof. Dr. Ratna Dahiya
Department of Electrical Engineering
N.I.T. Kurukshetra , Haryana (India)
ratna_dahiya@yahoo.co.in

Abstract- **This Paper presents, the Load Frequency Control(LFC) of three area system controlled with the help of PID, Fuzzy Logic Controller(FLC), and Hybrid Fuzzy Tilt-Integral Derivative(HFTID) Controller. In this study two Non-Reheated Thermal and one Hydro system is connected by Tie-line. Fuzzy Logic Controller is Rule based which depends upon various inputs. It provides better performance and efficiency. Simulation is done using three Controllers for different inputs in MATLAB\SIMULINK software.**

Keywords- **FLC, PID Controller, ACE, HFTID**

I. INTRODUCTION

In the present power system the main problem is to maintain the desirable frequency control operation of an interconnected power systems and supplying the reliable Electric Power of good quality. Different techniques of Load Frequency Control are used to improve Reliability[1]. The aim of the LFC to maintain zero steady state error in a multi area Interconnected power system and fulfill the desired dispatch conditions[2]. Due to load changes the power flows in tie-lines produce change in frequency because of the unequal generation and demand of the power[1]. A lot of research have been done on Load Frequency Control(LFC) in interconnected power system. Different control strategies have been suggested based on the Conventional linear control theory and for non-linear used a gain scheduling controller. In this method, the system outputs are faster with high quality and parameter estimation is not required as compare to Conventional Controller. In this study, A PID Controller, Fuzzy Logic Controller(FLC) and Hybrid Fuzzy Tilt-Integral Derivative(TID) Controller designed with seven Triangle Membership Functions(MFs) to LFC application in a three area power system to produce electricity with best quality.

To improve the power quality and stability with PID Controller set the optimize value of P, I & D for all three PID Controller[3]. In Fuzzy Logic Controller(FLC) designed with the rule base with respect to inputs & output desired. This gives the better result as compare to PID[4]. Further to improve result, designed a Hybrid controller, it is Hybrid Fuzzy Tilt-Integral derivative(HFTID) Controller. TID Controller depends upon a non zero real number 'n'. TID is a advance approach and it is designed by NASA[5]. So with the help of Simulation results it has proved that HFTID gives better results as compare to FLC and PID.

II. DESCRIPTION OF CONTROLLERS

A. Proportional-Integral-Derivative (PID)

A Proportional –Integral- Derivative (PID Controller) having three term control-The Proportional, The Integral and The Derivative values denoted by P , I and D respectively. PID is a control closed loop feedback mechanism System and it is widely used in industrial control System[4,6]. In PID the P depends on present error, I depends on the Accumulation of the past errors and D is the prediction of the future errors based on current rate of change. To perform good with the industrial process problems, the PID Controller should have optimally tuned K_p, K_i and K_d values. PID gives best result as compare to PI and PD in accuracy, oscillation and stability. The Proportional, Integral and Derivative terms are summed to calculate the output of the PID Controller.

$$u(t) = k_p e(t) + k_i \int_0^t e(\tau)\, d\tau + k_d \frac{de}{dt},$$

e = Control error, τ = Variable Of Integration

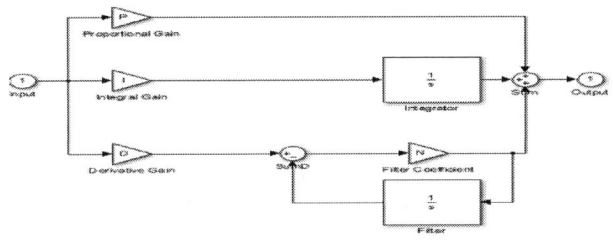

Fig.1 PID Controller Simulation Diagram

B. Fuzzy Logic Controller (FLC)

In the present scenario Fuzzy Logic Control system applications become important[7]. In this no need of complex Transactions and carries out the Process. With the help of human knowledge into a Fuzzy Control system a rule base is obtained and it is used by Fuzzy Inference System (FIS) as shown in Fig.2. Here used Mamdani Method uses Natural-language clauses[6]. Fuzzy Logic Control basically consist of three components-

 a) Fuzzification
 b) Rule Base
 c) Defuzzification

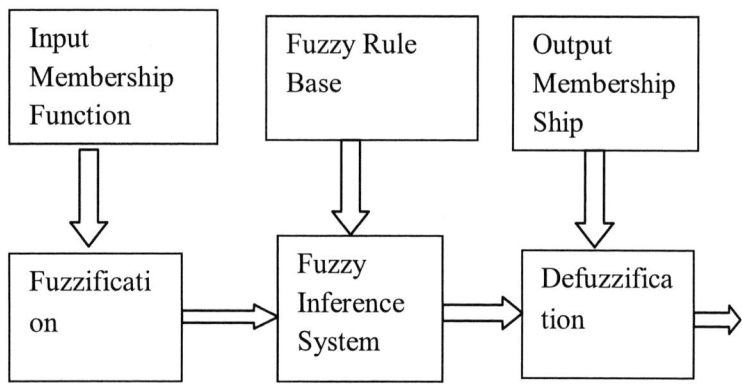

Fig.2 Basic Configuration Of Fuzzy System

 a) Fuzzification is a process of making a Crisp quantity into the Fuzzy.
 b) Fuzzy values sent to the linguistic control Rule Base and Data Base.
 c) Defuzzification is the conversion of a fuzzy Quantity to a crisp quantity.

The Simulation model of Fuzzy Logic Controller is Shown in Fig.3. Table of Fuzzy Rule is shown in Table I.

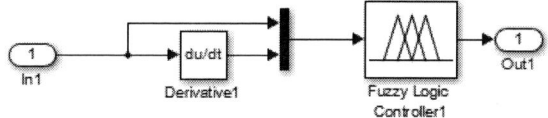

Fig.3 Simulation Model of FLC

MBF	NL	NM	NS	Z	PS	PM	PL
NL	NL	NL	NL	NL	NM	NS	Z
NM	NL	NL	NL	NM	NS	Z	PS
NS	NL	NL	NM	NS	Z	PS	PM
Z	NL	NM	NS	Z	PS	PM	PL
PS	NM	NS	Z	PS	PM	PL	PL
PM	NS	Z	PS	PM	PL	PL	PL
PL	Z	PS	PM	PL	PL	PL	PL

Table I . Table Of Fuzzy Rule

A linguistic variable which implies inputs and outputs have been classified as 'Negative Large'(NL), 'Negative Medium'(NM), 'Negative Short'(NS), 'Zero'(Z), Positive Short'(PL), 'Positive Medium'(PM), 'Positive Large'(PL). Each control input has seven Fuzzy set so that total 49 Fuzzy set are there[8]. These MFs are shown in Fig.4.

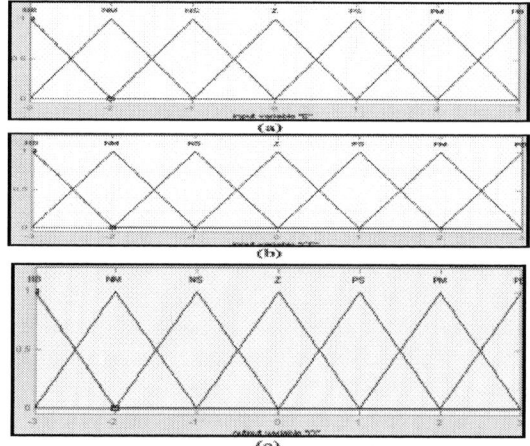

Fig. 4 Membership functions of output

C. Hybrid Fuzzy-TID Controller(HFTID)

Hybrid Fuzzy-TID(Tilt-Integral-Derivative) Controller is new

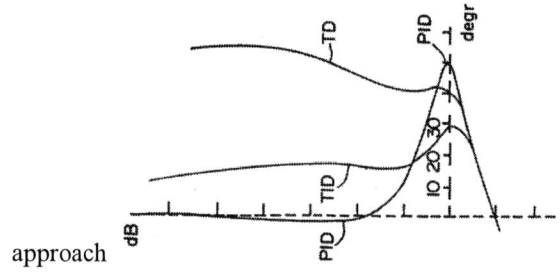

approach

Fig5. Graphical Representation of PID and TID

978-1-4799-6047-7/14 $31.00 © 2014 IEEE

to get the better result as compare to the Fuzzy and PID Controller.

PID type Compensator where in the proportional compensating unit is replaced with a compensator having a transfer function characterized by $1\backslash S(1\backslash n)$. This is called 'Tilt' Compensator as it provides a feedback gain as a function of frequency which is tilted or shaped with respect to the gain\frequency of a Conventional or Proportional unit. So the entire Compensator is here in referred to as a Tilt-Integral-Derivative(TID) Compensator[5]. Here 'n' is a non zero real number, preferably between 2 and 3. TID have several advantages also that it is ease in tuning, universal w.r.t the plants with different bandwidths, The phase and slope of the gain response are both rendered frequency independent, Thus ensuring that the Compensator is substantially universal. To get the better output made a Hybrid Simulation model consist of Fuzzy Logic Controller(FLC) and TID is shown in Fig.6 and the graphical representation is shown in Fig.5.

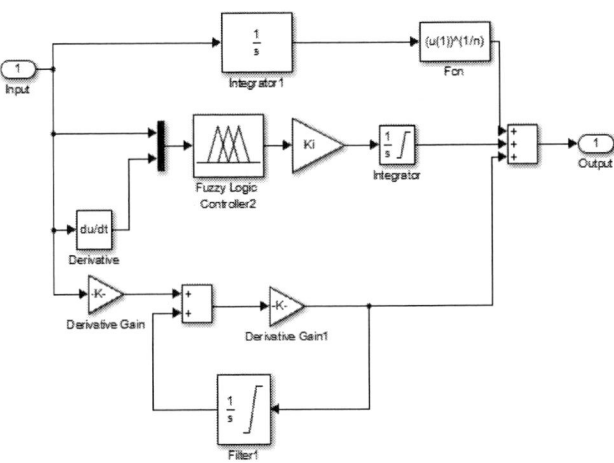

Fig.6 Simulation Model Of Fuzzy-TID Controller

III. Proposed Control Strategy

The proposed power system with three areas having two Non-Reheat Turbine and one Hydro Turbine are consider in Simulation study with time delays using PID Controller, Fuzzy Logic Controller and Hybrid Fuzzy-TID Controller[8]. These three areas are connected with the help of tie-line. FLC gives better results as compare to PID Controller but Hybrid Fuzzy-TID Controller provides fast response, adequate disturbance rejection and it provides also effective result for complex and Non-linear model. These controller improves

effectively the damping of the oscillations after the load deviation in one of the interconnected system area compared to Conventional Controllers. So proposed three area Simulation models for PID, FLC and Hybrid Fuzzy-TID Controller in Fig. 7, 8, and 9 respectively.

A. Tie – Line Control

The Power transfer equation through tie-line is
$$P_{tieflow} = 1/\ X_{tie}\ (\theta_1 - \theta_2) \qquad (1)$$
This tie flow is a steady state quantity .
Where frequency change $\Delta\omega$ is
$$\Delta\omega = -[\Delta P_{L1}]\ /\ [1/R_1 + 1/R_2 + D_1 + D_2] \qquad (2)$$
Tie-line power in terms of $\Delta\omega$ is
$$\Delta P_{tie} = \Delta\omega\ \ (1/R_2 + D_2) \qquad (3)$$
The new tie flow is determined by the net change in load and generator in each area[1]. Do not need to know the tie stiffness to determine this new tie flow, although the tie stiffness will determine that how much difference in phase angle across the tie will result from the new tie flow. Tie-line bias control is used to eliminate steady state error in frequency in tie line power flow[7].

ACE_1 = area control error of area 1

ACE_2 = area control error of area 2

ACE_3 = area control error of area 3

In this control ACE_1, ACE_2, and ACE_3 are made linear combination of frequency and tie-line power error.

$$ACE_1 = \Delta P_{12} + b_1 \Delta f_1 \qquad (4)$$

$$ACE_2 = \Delta P_{21} + b_2\ \Delta f_2 \qquad .(5)$$

$$ACE_3 = \Delta P_{31} + b_3 \Delta f_3 \qquad (6)$$

Where b1, b2 and b3 are called area frequency bias of area1, area2 and area3 respectively. In three area Hydro Non-Reheated Thermal control system used a model, there are Generator, Turbine & Governor model. To Simulate this model have to define the value of Turbine, Generator, Governor and others parameters at which model depends. These parameters are defined in appendix 1.

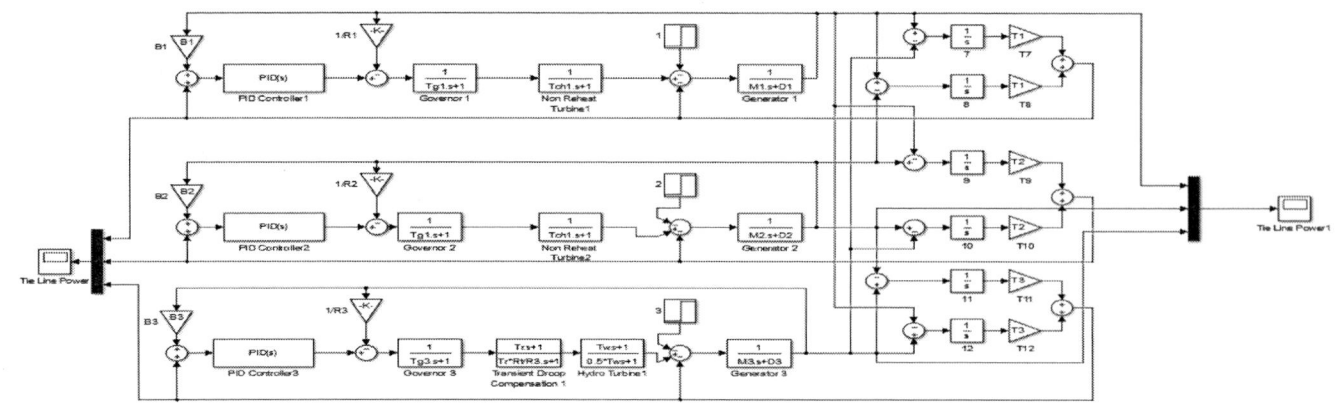

Fig.7 Simulink model with PID Controller

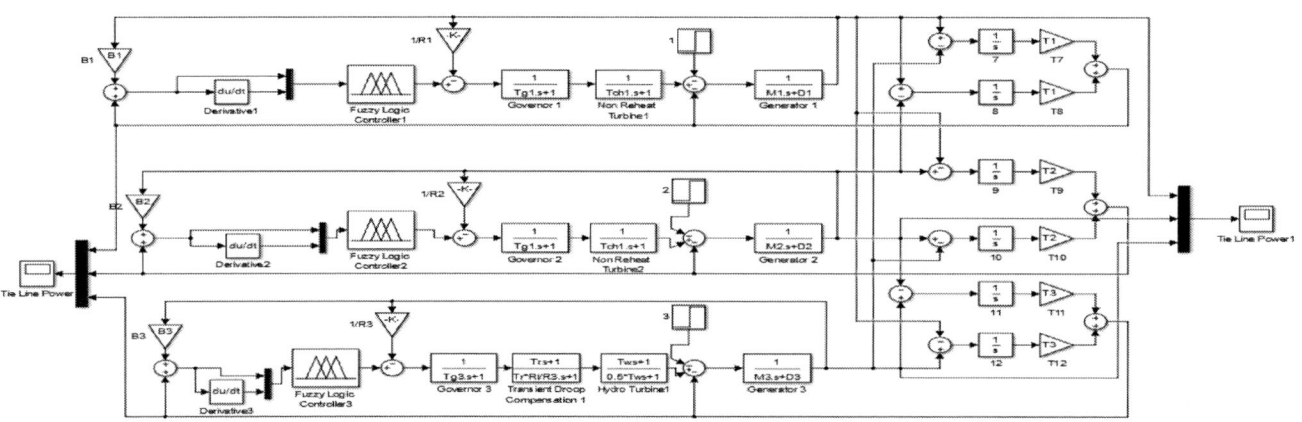

Fig.8 Simulation Model with FLC Controller

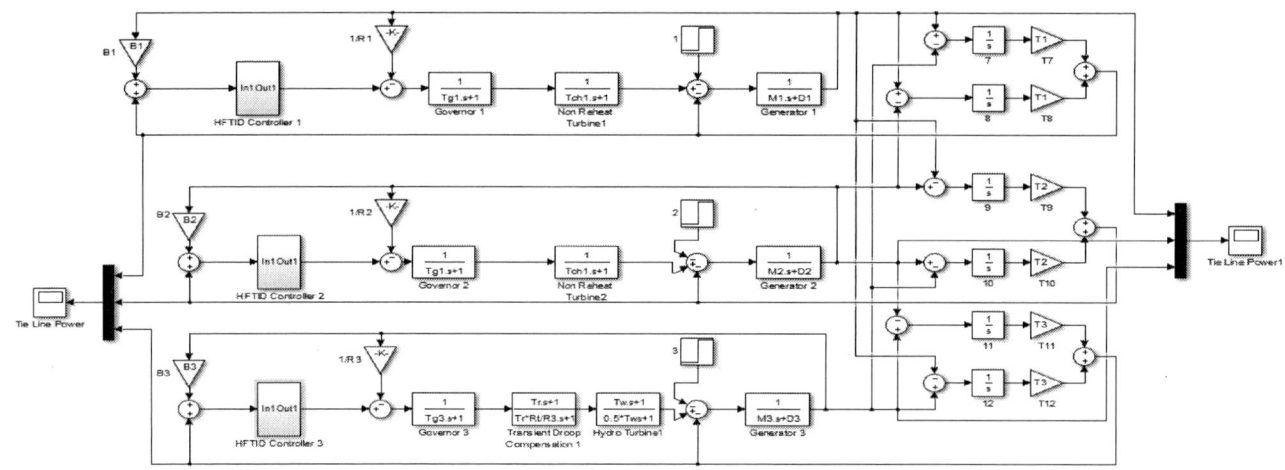

Fig.9 Simulation model with HFTID Controller

978-1-4799-6047-7/14 $31.00 © 2014 IEEE

IV. Simulation Analysis and Results

Simulation results of Thermal Non-Reheated 2, Tie-line power for Non-Reheated1&2 Thermal System, Tie-line Power for Non-Reheated and Hydro system, area1, Thermal Non-Reheated1, area2, area3, Hydraulic system are shown in Fig.10, Fig11, Fig12, Fig13, Fig14, Fig15, Fig16, Fig17 respectively. So the FLC gives better result as compare to PID and Hybrid Fuzzy-TID(HFTID) gives good result as compare to FLC. HFTID stabilize the system in very less time with less oscillation and give fast response.

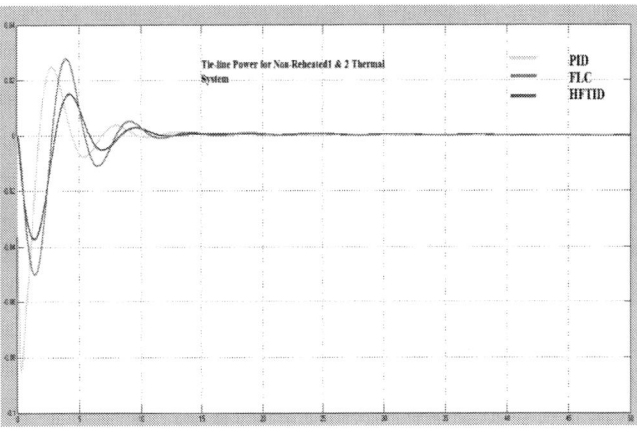

Fig12. Simulation result of Tie-line Power for Non Reheated1 & 2 System

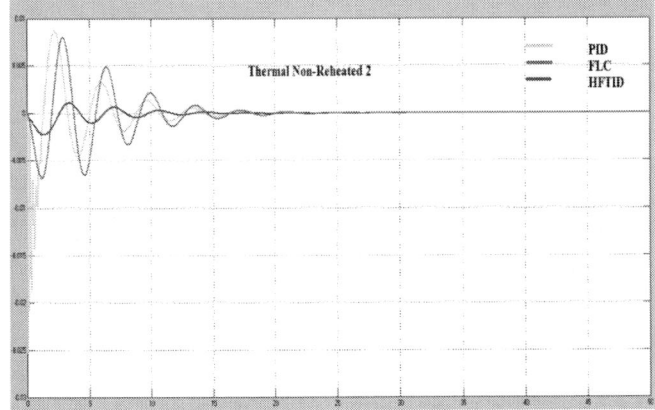

Fig.10 Simulation result of Thermal non-Reheated 2 system

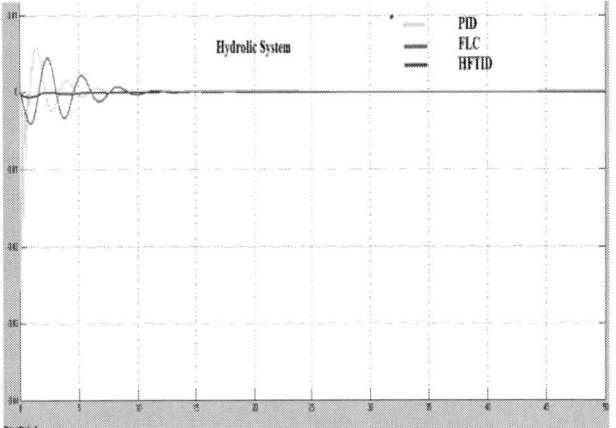

Fig13. Simulation result of Hydraulic system

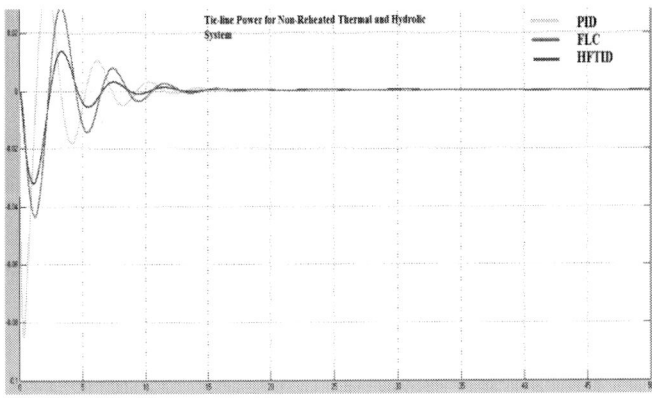

Fig11. Simulation result of Non Reheated1 Thermal & Hydraulic system

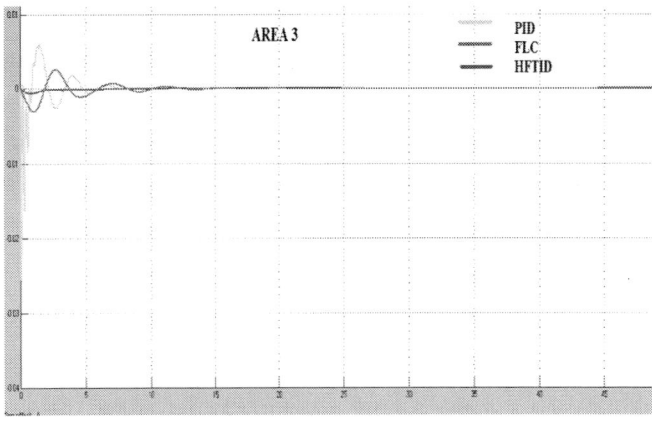

Fig14. Simulation result of area 3 Hydraulic system

978-1-4799-6047-7/14 $31.00 © 2014 IEEE

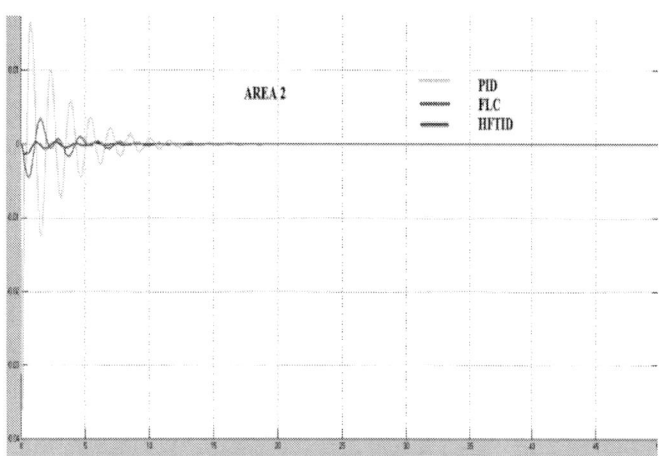

Fig15. Simulation result of Non Reheated2 area2 Thermal System

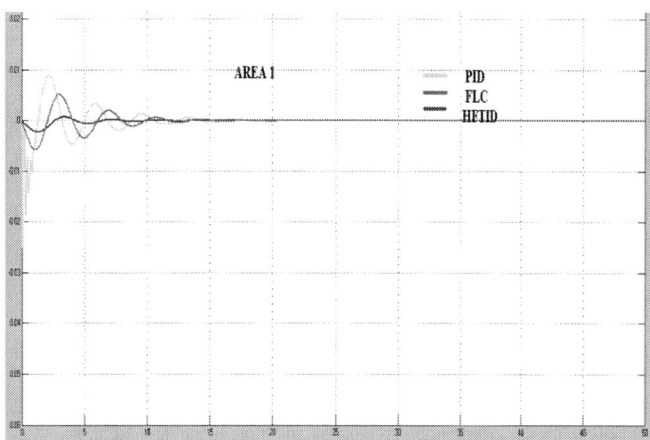

Fig16. Simulation result of Non Reheated1 area 1 Thermal System

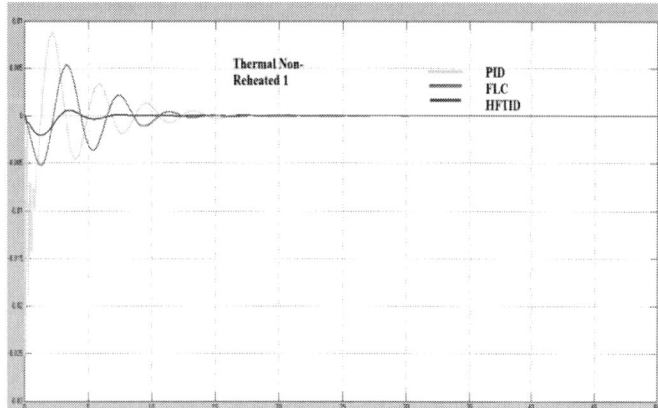

Fig.17 Simulation result of Thermal Non-reheated1 system

V. CONCLUSION

In three area control system having two Non-Reheated Thermal and one Hydro Power system, after applied the PID Controller, FLC and Hybrid Fuzzy-TID(HFTID) it has been found that HFTID given better results and reduce the overshoot and having more stability as compared to PID & FLC as shown in the Simulation results. HFTID also improves the overall system performance. HFTID is new approach and it will give better result in controlling area where used only PI, PD, PID & FLC single controller.

APPENDIX 1

M3= 6,D1=1,D2=1,D3=1,Tch1=0.3,Tg1=0.1,T21=0.06 Tg3=0.2,Tw=1,Rt=0.38,R1=0.08,R2=0.08, ,M2= 10 R3=0.05,Tr=5,T1=15,T2=15,T3=15,B1=(1/R1)+D1,B2=(1/R 1)+D1,B3=(1/R3)+D3,T12=0.06, M1= 10 T13=0.08,T23=0.06,T31=0.08,T32=0.06,

REFRENCES

[1] Hadi Saadat, "Power System Analysis", New Delhi,: Mc Graw-Hill, 2002.

[2] Yousef,H.A.; AL-Kharusi,k.; Albadi,M.H.; Hossieinzadeh,H. "Load Frequency Control of a Multi-area Power System:An adaptive Fuzzy Lozzy approach" IEEE Transactions on Power system, Vol: 29, PP. 1822-1830, 2014.

[3] Praksh,S.; Sinha,S.K.. "Four area LFC Interconnected Hydro-Thermal Power system by Intelligent PID control technique" IEEE Conference, Vol.10, PP.1-6,2012.

[4] Abedinia,O.; Amjady,N. "Multi-stage Fuzzy PID load frequency control via SPHBMO in Deregulated environment" IEEE International Conference, Vol.10, PP.473-478, 2012.

[5] www.ntrs.nasa.gov/casi.ntrs.nasa.gov/19950011898.pdf

[6] G.Karthikeyan, Dr.S.Chandrasekar, "Load frequency control for three area system with time delays using Fuzzy Logic Controller" IJESAT Journal, Vol.2, PP.612-618, 2012.

[7] Kamel sabahi, sehraneh ghaemi, saeed pazeshki, "Application of type 2 Fuzzy Logic system for load frequency control using feedback error learning approaches" Elsevier Journal, PP.1-11, 2014.
[8] Tushir,M.; Srivastava,S. "Application of a Hybrid controller in load frequency control of Hydro-Thermal Power system" IEEE Conference,Vol.15, PP.1-5, 2012.

Advanced Pulse Width Modulation Technique for Z-Source Inverter

Sangeeta DebBarman
PG Scholar: School of Electrical Engineering
KIIT University
Bhubaneswar, India
meghadb13@gmail.com

Tapas Roy
Assistant Professor: School of Electrical Engineering
KIIT University
Bhubaneswar, India
tapas18roy@gmail.com

Abstract—This paper presents a new Pulse Width Modulation (PWM) technique for Z-source or Impedance Source Inverter (ZSI or ISI). The proposed technique is based on the Bus Clamping PWM (BCPWM) technique for Voltage Source Inverter (VSI). The effectiveness of the proposed technique is achieved by MATLAB simulation of a ZSI. It has been observed that the proposed technique gives better performances as compared to Conventional Space Vector PWM (CSVPWM) at high modulation index and at same switching frequency like reduction in load current ripple and reduction in voltage stress across the switches . Further the proposed technique ensures less switching loss as compared to CSVPWM at same carrier frequency.

Index Terms—Bus Clamping Pulse Width Modulation (BCPWM), Conventional Space Vector Pulse Width Modulation (CSVPWM), Maximum Boost Control (MBC), Total Harmonic Distortion (THD), Z-source Inverter (ZSI)

I. INTRODUCTION

Nowadays, one of the most popular power inverters is impedance source inverter (ISI) or Z-source inverter (ZSI). It has the unique capability of buck boost. It overcomes the limitations of voltage source inverter (VSI) and current source inverter (CSI). The power circuit for Z-source inverter is shown in Fig 1. An impedance network is inserted in between the DC input voltage source and inverter power circuit. A Split inductor & a split capacitor compromise X- shaped impedance network which gives the unique buck boost feature of the Z-source inverter. Z-source inverter has six active states (V_1-V_6) and two zero states (V_0 & V_7) as like traditional two-level VSI. It has a new state called shoot-through state (V_{sh}) which can be generated by seven different ways (V_{sh1}-V_{sh7}) [1-6].

The voltage gain of the ZSI and the voltage stress across the inverter switch can be written as [5]

$$G = \frac{\hat{V}_{ph}}{\frac{V_i}{2}} = M.B$$

$$\hat{V}_{dc} = B.V_i$$

(1)

where $\hat{V}_{ph}$, V_i and M denote output peak phase voltage, input dc voltage and modulation index respectively. Boost factor (B) of the Z-source inverter can be expressed as [5]

$$B = \frac{1}{1 - \frac{2T_{sh}}{T}} = \frac{1}{1 - 2D_0}$$

(2)

where T_{sh} denotes the shoot-through time interval during a switching period T and D_0 is the shoot-through duty ratio.

There exist conventional as well as advanced PWM techniques for traditional VSI. Sine triangle PWM (STPWM), Space vector PWM (SVPWM) are most popular conventional type PWM techniques whereas bus clamping PWM is one of the most popular advanced type PWM techniques [7-10]. Advanced type PWM techniques are better as compared to conventional PWM techniques in respect of load current ripple, voltage stress across the switch at high modulation index and at same switching frequency.

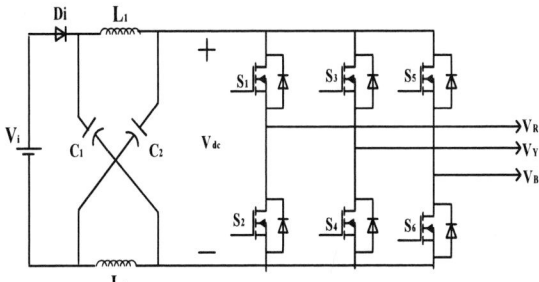

Fig 1. Power circuit for three- phase ZSI

For Z- Source inverter, Simple Boost Control (SBC), Maximum Boost Control (MBC), Space Vector based MBC are the most popular conventional PWM techniques. These PWM techniques have been introduced based on the conventional PWM techniques for traditional VSI. In literature there is a prominent study of Z-source inverter for conventional PWM technique [1-6].

In this paper, a new PWM technique based on bus-clamping PWM technique for Z-source inverter has been introduced. The effectiveness of this technique compared to conventional techniques mainly Conventional Space vector MBC has been achieved by proper simulation in MATLAB. The proposed technique has been implemented in MATLAB. It has been observed that compared to conventional techniques, the proposed technique gives better performances in respect of current ripple, voltage stress across switch at high modulation index and at same switching frequency.

978-1-4799-6047-7/14 $31.00 © 2014 IEEE   629

II. Bus Clamping PWM For Traditional VSI

Bus Clamping PWM (BCPWM) technique is an advanced type of PWM technique. In this PWM scheme, each phase of VSI is clamped to one of the dc bus terminals (either positive DC bus terminal or negative DC bus terminal) for some duration of fundamental cycle. Two most popular BCPWM techniques are 30°BCPWM technique and 60°BCPWM technique. In case of 30°BCPWM, the clamping zone is middle 30° of each quarter of fundamental cycle whereas for 60°BCPWM, the clamping zone is middle 60° of each half of fundamental cycle. During the clamping period, there is no switching for the clamped phase whereas the switching occurs for other two phases. So the number of switching is less as compare to CSVPWM technique over same carrier frequency. For same switching frequency, the carrier frequency for 30°BCPWM as well as 60°BCPWM is 1.5 times that of CSVPWM. So for same switching frequency, the sample period is two-third for BCPWM as compared to CSVPWM. As a result, the current ripple is less in case of bus-clamping PWM than that for CSVPWM [7-8]. In this technique each sector is divided into two equal parts of 30º duration shown in fig 2.

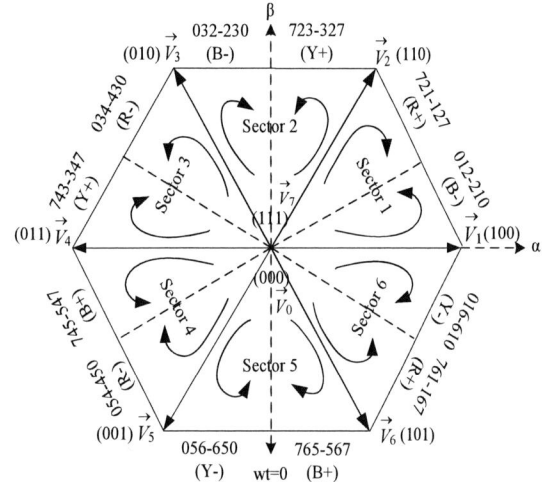

Fig 2: Vector sequence for subsector in 30º BCPWM

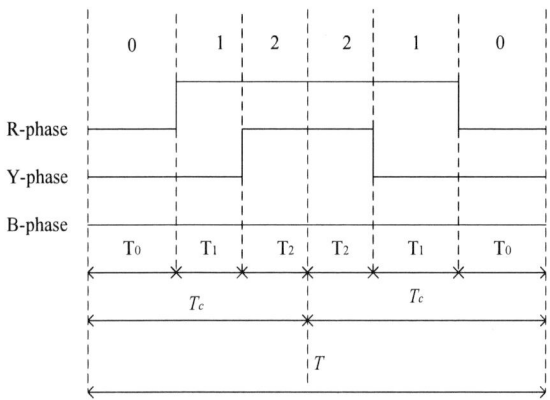

Fig 3: Timing diagram for vector sequence 012-210

The reference vector is generated by applying nearby vectors in a sub-sector. The vector sequence is selected in such a way that in each sub sector, a phase gets clamped either to the positive or the negative DC bus terminals. The Fig. 2 shows the corresponding phase which gets clamped on particular sub-sector. The timing diagram for the vector sequence 012-210 is shown in the Fig. 3. It shows that the B-phase gets clamped to the negative dc bus terminal.

III. Bus Clamping PWM For ZSI

The relationship between switching variable vector $[S_1 S_3 S_5]^T$ and phase voltage vector $[V_R\ V_Y\ V_B]^T$ is given-

$$\begin{bmatrix} V_R \\ V_Y \\ V_B \end{bmatrix} = \frac{\hat{V}_{dc}}{3} \begin{bmatrix} 2 & -1 & -1 \\ -1 & 2 & -1 \\ -1 & -1 & 2 \end{bmatrix} \begin{bmatrix} S_1 \\ S_3 \\ S_5 \end{bmatrix} \tag{4}$$

Assuming α be the horizontal axes and β be the vertical axes of the stator coordinate frame. The three dimensional vector $V_{RYB} = [V_R\ V_Y\ V_B]^T$ is transformed into two dimensional vector in α-β coordinate frame $V_{\alpha\text{-}\beta} = [V_\alpha\ V_\beta]^T$ as written in (5).

$$\begin{bmatrix} V_\alpha \\ V_\beta \end{bmatrix} = \begin{bmatrix} 1 & -\dfrac{1}{2} & -\dfrac{1}{2} \\ 0 & \dfrac{\sqrt{3}}{2} & -\dfrac{\sqrt{3}}{2} \end{bmatrix} \begin{bmatrix} V_R \\ V_Y \\ V_B \end{bmatrix} \tag{5}$$

$$\begin{bmatrix} V_\alpha \\ V_\beta \end{bmatrix} = T_{RYB-\alpha\beta} \begin{bmatrix} V_R \\ V_Y \\ V_B \end{bmatrix} = \hat{V}_{dc} \begin{bmatrix} \dfrac{\sqrt{6}}{3} & -\dfrac{1}{\sqrt{6}} & -\dfrac{1}{\sqrt{6}} \\ 0 & \dfrac{1}{\sqrt{2}} & -\dfrac{1}{\sqrt{2}} \end{bmatrix} \begin{bmatrix} S_1 \\ S_3 \\ S_5 \end{bmatrix} \tag{6}$$

$$T_{RYB-\alpha\beta} = \begin{bmatrix} \dfrac{2}{\sqrt{6}} & -\dfrac{1}{\sqrt{6}} & -\dfrac{1}{\sqrt{6}} \\ 0 & \dfrac{1}{\sqrt{2}} & -\dfrac{1}{\sqrt{2}} \end{bmatrix} \tag{7}$$

There exist three kind of switching states for ZSI namely active state, zero state and shoot-through state. Along with traditional six active states (V_1-V_6) and two zero states (V_0, V_7) two-level ZSI has seven shoot-through states (V_{sh1}-V_{sh7}). Out of seven shoot-through states, V_{sh1}, V_{sh2} and V_{sh3} has four sub-states each as these states are generated by simultaneously turning on the upper and lower power semiconductor switching devices for R-phase, Y-phase and B-phase respectively (two switches turning on simultaneously-other four switches can make four combinations- 0101,1001,0110 and 1010). On the other hand, V_{sh4}, V_{sh5} and V_{sh6} has two sub-states each as these shoot-through states are generated by simultaneously turning on upper and lower switching devices for R-Y, Y-B and B-R respectively (four switches turning on simultaneously-other two switches can make two combinations- 10 and 01). The other shoot-through state V_{sh7} is generated by simultaneously turning on the all switching

978-1-4799-6047-7/14 $31.00 © 2014 IEEE

devices. So in respect of switching loss, V_{sh7} state causes maximum and V_{sh1}, V_{sh2} and V_{sh3} causes minimum [5].

Table 1:
Switching states and output voltages of a three-phase ZSI

Vector	S_1	S_2	S_3	S_4	S_5	S_6	V_α	V_β
$\vec{V_0}$	0	1	0	1	0	1	0	0
$\vec{V_1}$	1	0	0	1	0	1	$\frac{\sqrt{6}}{3}\hat{V}_{dc}$	0
$\vec{V_2}$	1	0	1	0	0	1	$\frac{\sqrt{6}}{6}\hat{V}_{dc}$	$\frac{\sqrt{2}}{2}\hat{V}_{dc}$
$\vec{V_3}$	0	1	1	0	0	1	$-\frac{\sqrt{6}}{6}\hat{V}_{dc}$	$\frac{\sqrt{2}}{2}\hat{V}_{dc}$
$\vec{V_4}$	0	1	1	0	1	0	$-\frac{\sqrt{6}}{3}\hat{V}_{dc}$	0
$\vec{V_5}$	0	1	0	1	1	0	$-\frac{\sqrt{6}}{6}\hat{V}_{dc}$	$-\frac{\sqrt{2}}{2}\hat{V}_{dc}$
$\vec{V_6}$	1	0	0	1	1	0	$\frac{\sqrt{6}}{6}\hat{V}_{dc}$	$-\frac{\sqrt{2}}{2}\hat{V}_{dc}$
$\vec{V_7}$	1	0	1	0	1	0	0	0
$\vec{V}_{sh1}$	1	1	0/1	1/0	0/1	1/0	0	0
$\vec{V}_{sh2}$	0/1	1/0	1	1	0/1	1/0	0	0
$\vec{V}_{sh3}$	0/1	1/0	0/1	1/0	1	1	0	0
$\vec{V}_{sh4}$	1	1	1	1	0/1	1/0	0	0
$\vec{V}_{sh5}$	1	1	0/1	1/0	1	1	0	0
$\vec{V}_{sh6}$	0/1	1/0	1	1	1	1	0	0
$\vec{V}_{sh7}$	1	1	1	1	1	1	0	0

By applying (6), each state gives V_α and V_β components. Table1 shows the different state vectors and corresponding V_α and V_β components. As zero and shoot-through states short the load terminals, V_α and V_β components are zero. The hexagon for two-level ZSI can be formed which is shown in Fig. 4. It is same as the hexagon for two-level traditional VSI except it has an additional state called shoot-through state. It has six sectors of 60°duration each. For BCPWM each sector is divided into two sub-sectors. The reference vector in any sector is generated by applying nearby active vectors for some duration of sampling time and zero and shoot-through vectors remaining sampling time. By volt-sec balance, the dwell times can be calculated for each sector.

Now V_{ref} can be obtained by using (8). T_i and T_j are the dwell times for active vectors V_i and V_j respectively where i,j=(1-6). T_0 is the dwell time corresponding zero vectors and T_{sh} is the dwell time corresponding shoot-through vectors. T is known as sample time and equal to ($T_i + T_j + T_0 + T_{sh}$)

$$\vec{V}_{ref} = \frac{T_i}{T}\vec{V_i} + \frac{T_j}{T}\vec{V_j} + \frac{T_0}{T}(\vec{V_0}\ or\ \vec{V_7}) + \frac{T_{sh}}{T}\vec{V}_{shk} \tag{8}$$

The two dimensional vector in the α-β plane is shown in the following equations

$$\vec{V}_{ref} = \begin{bmatrix} V_\alpha \\ V_\beta \end{bmatrix} = \frac{\sqrt{6}}{2}\begin{bmatrix} \frac{2}{\sqrt{6}} & -\frac{1}{\sqrt{6}} & -\frac{1}{\sqrt{6}} \\ 0 & \frac{1}{\sqrt{2}} & -\frac{1}{\sqrt{2}} \end{bmatrix}\begin{bmatrix} V_{RN} \\ V_{YN} \\ V_{BN} \end{bmatrix}$$

$$\vec{V}_{ref} = \begin{bmatrix} V_\alpha \\ V_\beta \end{bmatrix} = T_{RYB-\alpha\beta}\begin{bmatrix} V_R \\ V_Y \\ V_B \end{bmatrix} = \frac{\sqrt{6}}{2}\hat{V}_{pk}\begin{bmatrix} Cos(\theta) \\ Sin(\theta) \end{bmatrix} \tag{9}$$

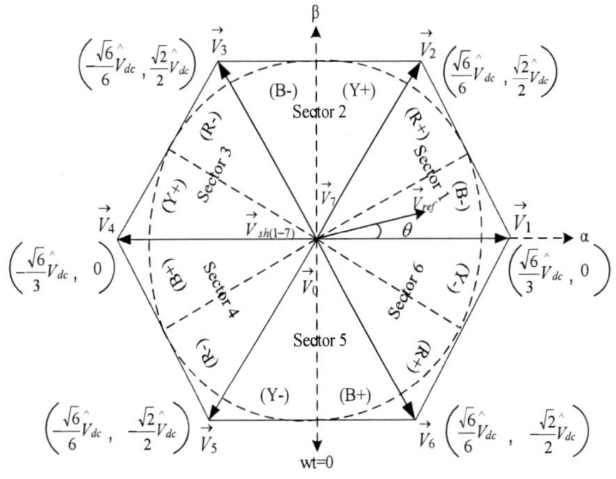

Fig 4: Bus clamping sectors & sub-sectors for three phase ZSI

By applying (8), the dwell times for all sectors can be calculated. Only sector 1 calculation is shown below, all other sector's dwell time calculations are similar.

From Fig. 4 for sector 1, $0 \le \theta \le \pi/3$. Two active vectors V_1 and V_2 form the boundary of sector 1. V_1 vector along the axis i.e. the reference axis (θ=0°) and V_2 vector along ($\theta = \pi/3$). The volt-sec balance for sector 1 is shown in (9)

$$T.\left|\vec{V}_{ref}\right|\begin{bmatrix} Cos\theta \\ Sin\theta \end{bmatrix} = T_1\left|\vec{V_1}\right|\begin{bmatrix} Cos0 \\ Sin0 \end{bmatrix} + T_2\left|\vec{V_2}\right|\begin{bmatrix} Cos\left(\frac{\pi}{3}\right) \\ Sin\left(\frac{\pi}{3}\right) \end{bmatrix} \tag{10}$$

From (9), $\qquad \left|\vec{V}_{ref}\right| = \frac{\sqrt{6}}{2}\hat{V}_{pk} = \frac{\sqrt{6}}{4}M.V_{dc}$ (11)

Also from the Table 1,

$$\left|\vec{V_1}\right| = \left|\vec{V_2}\right| = \frac{\sqrt{6}}{3}V_{dc} \tag{12}$$

Substituting (12) and (11) into (10), the following equations for dwell times for sector 1 can be obtained-

$$T_1 = \frac{\sqrt{3}}{2}T.M.\sin\left(\frac{\pi}{3}-\theta\right) \tag{13}$$

$$T_2 = \frac{\sqrt{3}}{2}T.M.\sin\theta \tag{14}$$

$$T_m = T_0 + T_{sh} = T - T_1 - T_2 = T\left[1 - \frac{\sqrt{3}}{2} M \cos\left(\theta - \frac{\pi}{6}\right)\right] \quad (15)$$

Where T_m is the combine dwell times for zero and shoot-through vectors. The dwell times for other sectors are calculated in similar fashion. The all dwell times are tabulated in Table 2.

Table 2: Dwell Time calculation for different sectors of ZSI

Active vectors	Angle(θ)	Dwell time calculation	Time Calculation (Tm)
Sector 1 $\vec{V_1}$ & $\vec{V_2}$	$0 \leq \theta \leq \frac{\pi}{3}$	$T_1 = \frac{\sqrt{3}}{2} T.M \sin\left(\frac{\pi}{3} - \theta\right)$ $T_2 = \frac{\sqrt{3}}{2} T.M \sin\theta$	$T\left[1 - \frac{\sqrt{3}}{2} M \cos\left(\theta - \frac{\pi}{6}\right)\right]$
Sector 2 $\vec{V_2}$ & $\vec{V_3}$	$\frac{\pi}{3} \leq \theta \leq \frac{2\pi}{3}$	$T_2 = \frac{\sqrt{3}}{2} T.M \sin\left(\frac{2\pi}{3} - \theta\right)$ $T_3 = \frac{\sqrt{3}}{2} T.M \sin\left(\theta - \frac{\pi}{3}\right)$	$T\left[1 - \frac{\sqrt{3}}{2} M \sin\theta\right]$
Sector 3 $\vec{V_3}$ & $\vec{V_4}$	$\frac{2\pi}{3} \leq \theta \leq \pi$	$T_3 = \frac{\sqrt{3}}{2} T.M \sin(\pi - \theta)$ $T_4 = \frac{\sqrt{3}}{2} T.M \sin\left(\theta - \frac{2\pi}{3}\right)$	$T\left[1 + \frac{\sqrt{3}}{2} M \cos\left(\theta + \frac{\pi}{6}\right)\right]$
Sector 4 $\vec{V_4}$ & $\vec{V_5}$	$\pi \leq \theta \leq \frac{4\pi}{3}$	$T_4 = \frac{\sqrt{3}}{2} T.M \sin\left(\frac{4\pi}{3} - \theta\right)$ $T_5 = \frac{\sqrt{3}}{2} T.M \sin(\theta - \pi)$	$T\left[1 + \frac{\sqrt{3}}{2} M \cos\left(\theta - \frac{\pi}{6}\right)\right]$
Sector 5 $\vec{V_5}$ & $\vec{V_6}$	$\frac{4\pi}{3} \leq \theta \leq \frac{5\pi}{3}$	$T_5 = \frac{\sqrt{3}}{2} T.M \sin\left(\frac{5\pi}{3} - \theta\right)$ $T_6 = \frac{\sqrt{3}}{2} T.M \sin\left(\theta - \frac{4\pi}{3}\right)$	$T\left[1 + \frac{\sqrt{3}}{2} M \sin\theta\right]$
Sector 6 $\vec{V_6}$ & $\vec{V_1}$	$\frac{5\pi}{3} \leq \theta \leq 2\pi$	$T_6 = \frac{\sqrt{3}}{2} T.M \sin(2\pi - \theta)$ $T_1 = \frac{\sqrt{3}}{2} TM \sin\left(\theta - \frac{5\pi}{3}\right)$	$T\left[1 + \frac{\sqrt{3}}{2} M \cos\left(\theta + \frac{\pi}{6}\right)\right]$

IV. VECTOR SEQUENCE

The vector sequence for each sub-sector is selected in such way that the following conditions should be satisfied-

(1) Only one switch is switched for each transition.
(2) One phase should get clamped for each sub-sector (for 30° clamping) or for each sector (60° clamping).
(3) In a both side of an active vector, there should be shoot-through vectors.
(4) The final vector of each half-sample cycle is the starting vector in other half-sample cycle like $V_0 \leftrightarrow V_{sh1} \leftrightarrow V_1 \leftrightarrow V_{sh2} \leftrightarrow V_2$ (V_2 is the ending and starting vector for half-sample cycles in each sample cycle in sector 1) or starting vector for each sample cycle is the ending vector in a sample cycle. (V_0 is the starting vector and ending vector for each sample cycle for sector 1)

The active vectors are selected on the basis of θ. Only shoot-through states $V_{sh1}, V_{sh2}, V_{sh3}$ are selected as only one phase is getting shorted during shoot through so switching loss is less and low THD. Table 3 shows the vector sequence and the clamped phase for each sub-sector in a sector. The duration of clamping is 30°. It is also called as 30°BCPWM for ZSI. The timing diagrams for sector 1 are shown in Fig. 5(a) and 5(b).

Table 3:
Vector sequence and clamping phase for each sub sector of ZSI

Sector	Angle(θ)	Vector Sequence	Phase clamped
Sector 1	$0 < \theta \leq \frac{\pi}{6}$	$V_0 \leftrightarrow V_{sh1} \leftrightarrow V_1 \leftrightarrow V_{sh2} \leftrightarrow V_2$	B^-
	$\frac{\pi}{6} < \theta \leq \frac{\pi}{3}$	$V_7 \leftrightarrow V_{sh3} \leftrightarrow V_2 \leftrightarrow V_{sh2} \leftrightarrow V_1$	R^+
Sector 2	$\frac{\pi}{3} < \theta \leq \frac{\pi}{2}$	$V_7 \leftrightarrow V_{sh3} \leftrightarrow V_2 \leftrightarrow V_{sh1} \leftrightarrow V_3$	Y^+
	$\frac{\pi}{2} < \theta \leq \frac{2\pi}{3}$	$V_0 \leftrightarrow V_{sh2} \leftrightarrow V_3 \leftrightarrow V_{sh1} \leftrightarrow V_2$	B^-
Sector 3	$\frac{2\pi}{3} < \theta \leq \frac{5\pi}{6}$	$V_0 \leftrightarrow V_{sh2} \leftrightarrow V_3 \leftrightarrow V_{sh3} \leftrightarrow V_4$	R^-
	$\frac{5\pi}{6} < \theta \leq \pi$	$V_7 \leftrightarrow V_{sh1} \leftrightarrow V_4 \leftrightarrow V_{sh3} \leftrightarrow V_3$	Y^+
Sector 4	$\pi < \theta \leq \frac{7\pi}{6}$	$V_7 \leftrightarrow V_{sh1} \leftrightarrow V_4 \leftrightarrow V_{sh2} \leftrightarrow V_5$	B^+
	$\frac{7\pi}{6} < \theta \leq \frac{4\pi}{3}$	$V_0 \leftrightarrow V_{sh3} \leftrightarrow V_5 \leftrightarrow V_{sh2} \leftrightarrow V_4$	R^-
Sector 5	$\frac{4\pi}{3} < \theta \leq \frac{3\pi}{2}$	$V_0 \leftrightarrow V_{sh3} \leftrightarrow V_5 \leftrightarrow V_{sh1} \leftrightarrow V_6$	Y^-
	$\frac{3\pi}{2} \leq \theta \leq \frac{5\pi}{3}$	$V_7 \leftrightarrow V_{sh2} \leftrightarrow V_6 \leftrightarrow V_{sh1} \leftrightarrow V_5$	B^+
Sector 6	$\frac{5\pi}{3} < \theta \leq \frac{11\pi}{6}$	$V_7 \leftrightarrow V_{sh3} \leftrightarrow V_2 \leftrightarrow V_{sh1} \leftrightarrow V_3$	R^+
	$\frac{11\pi}{6} < \theta \leq 2\pi$	$V_0 \leftrightarrow V_{sh2} \leftrightarrow V_3 \leftrightarrow V_{sh1} \leftrightarrow V_2$	Y^-

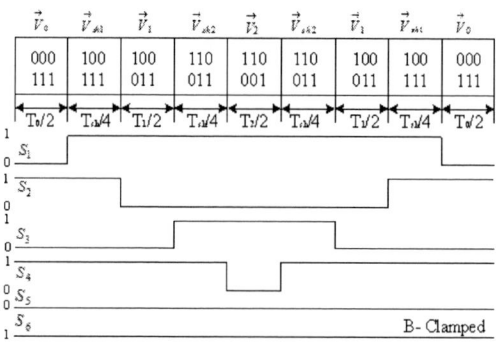

Fig. 5(a) Timing diagram for $0 < \theta \leq \pi/6$ in sector 1

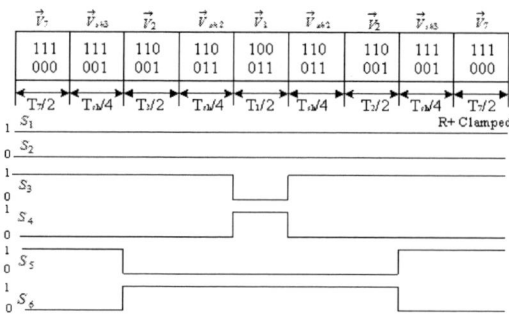

Fig. 5(b) Timing diagram for $\pi/6 < \theta \leq \pi/3$ in sector 1

V. MAXIMUM BOOST CONTROL IMPLEMENTATION BASED ON BUS-CLAMPING PWM TECHNIQUE

In Maximum Boost Control (MBC) strategy, the traditional zero states are fully converted into shoot-through state. As for result, the time duration for shoot-through states increases as well as the boosting factor. On the other hand the voltage stress across the switch reduces compare to

978-1-4799-6047-7/14 $31.00 © 2014 IEEE

Simple Boost Control (SBC) [5-6]. So to implement MBC based on BCPWM technique, the zero states dwell times are replaced by shoot-through time duration as it has been done for MBC based on Conventional Space Vector PWM (CSVPWM) technique [5]. Table 4 shows the shoot-through time duration for MBC for different sectors. This is the maximum shoot-through time duration denoted by $T_{shmax.}$. Fig. 6 shows the variation of maximum shoot-through duration, T_{shmax} w.r.t θ.

Table 4: T_{shmax} calculation for different sectors of ZSI

T_{shmax}	Angle(θ)
$T\left[1 - \dfrac{\sqrt{3}}{2}M\cos\left(\theta - \dfrac{\pi}{6}\right)\right]$	$0 \le \theta \le \dfrac{\pi}{3}$
$T\left[1 - \dfrac{\sqrt{3}}{2}M\sin\theta\right]$	$\dfrac{\pi}{3} \le \theta \le \dfrac{2\pi}{3}$
$T\left[1 + \dfrac{\sqrt{3}}{2}M\cos\left(\theta + \dfrac{\pi}{6}\right)\right]$	$\dfrac{2\pi}{3} \le \theta \le \pi$
$T\left[1 + \dfrac{\sqrt{3}}{2}M\cos\left(\theta - \dfrac{\pi}{6}\right)\right]$	$\pi \le \theta \le \dfrac{4\pi}{3}$
$T\left[1 + \dfrac{\sqrt{3}}{2}M\sin\theta\right]$	$\dfrac{4\pi}{3} \le \theta \le \dfrac{5\pi}{3}$
$T\left[1 + \dfrac{\sqrt{3}}{2}M\cos\left(\theta + \dfrac{\pi}{6}\right)\right]$	$\dfrac{5\pi}{3} \le \theta \le 2\pi$

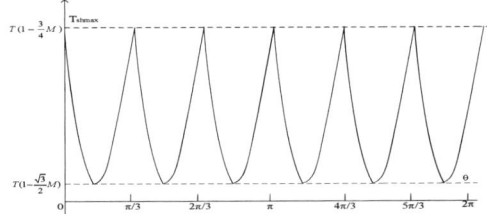

Fig 6: Variation of T_{shmax} w.r.t θ.

The average T_{shmax} over a fundamental cycle,

$$\overline{T}_{sh\max} = T.\left(1 - \frac{3\sqrt{3}\,M}{2\pi}\right) \qquad (16)$$

Average shoot-through duty ratio-

$$\overline{D}_{sh\max} = \frac{\overline{T}_{sh\max}}{T} = 1 - \frac{3\sqrt{3}\,M}{2\pi} \qquad (17)$$

So, the overall voltage gain-

$$G_{\max} = MB = M\frac{1}{1 - 2\overline{D}_{sh\max}} = \frac{M\pi}{3\sqrt{3}\,M - \pi} \qquad (18)$$

VI. SIMULATION RESULTS

The effectiveness of the proposed control method is verified by proper simulation of a three phase ZSI in MATLAB/Simulink . The simulations parameters are taken as follows [1]-

1. Input DC voltage, V_i=162V;
2. Modulation index, M=0.8;
3. Fundamental frequency, f=50Hz;
4. Switching frequency, f_{sw}=10 kHz;
5. Split Inductance, $L_1 = L_2$ =160 mH;
6. Split Capacitance, $C_1 = C_2$ =1000μF;
7. Three phase R-L load (R=26.5Ω and L=41 mH).

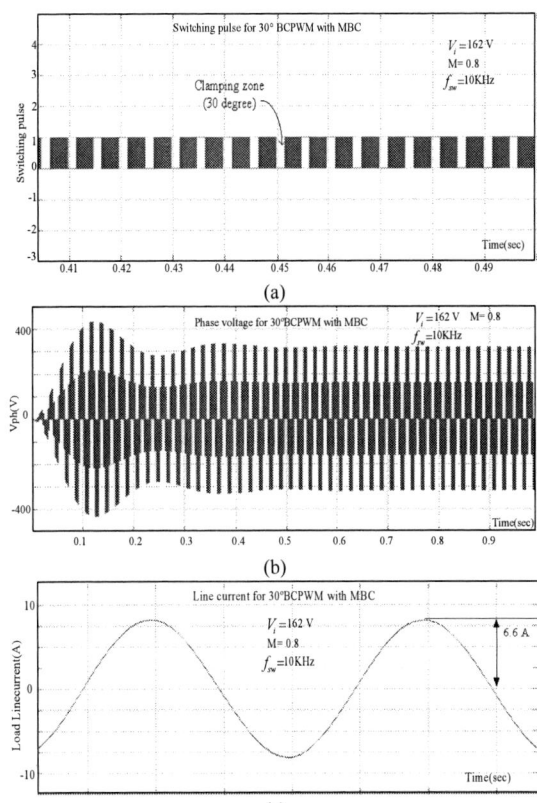

Fig. 7 Waveforms for -(a) Switching pulse or gate pulse for 30°BCPWM scheme, (b) Output phase voltage , (c) Output load current for simulated 2-level three- phase ZSI.

The switching pulse, output phase voltage and load line current for 30°BCPWM with MBC for 2-level three phase ZSI are shown in Fig. 7 (a), (b) and (c) respectively. It is observed that the output voltage and current waveforms are stable for simulated ZSI.

Table 5(a): Study of CSVPWM with MBC

M	Theoretical Line Voltage (V)	Simulated Line Voltage (V)	Load Line Current (A)	Load Line Current THD (%)	Load Line Current Ripple (A)	Capacitor voltage ripple (V)	Inductor current ripple (A)	Voltage stress across Switch (V)
0.7	359.4	325.4	11.05	0.8	0.15	2	0.18	930
0.8	201	200.9	6.8	0.8	0.1	0.8	0.12	500
0.85	171.4	166.6	5.6	0.6	0.07	0.8	0.12	385
0.87	161.3	161.3	5.4	0.6	0.065	0.8	0.12	375
0.9	149.4	146.9	4.9	0.6	0.06	0.7	0.12	325

Table 5(b): Study of 30°BCPWM with MBC

M	Theoretical Line Voltage (V)	Simulated Line Voltage (V)	Load Line Current (A)	Load Line Current THD (%)	Load Line Current Ripple (A)	Capacitor voltage ripple (V)	Inductor current ripple (A)	Voltage stress across Switch (V)
0.7	359.4	318.4	10.8	0.5	0.12	3	0.14	920
0.8	201	192.9	6.5	0.4	0.06	1	0.14	480
0.85	171.4	164.6	5.5	0.4	0.04	0.9	0.12	380
0.87	161.3	158.6	5.3	0.4	0.035	0.7	0.12	365
0.9	149.4	146.2	4.9	0.4	0.02	0.6	0.1	323

Table 5(c): % THD in line voltage in different load conditions.

PWM Technique	M	R Load		R-L Load		Motor Load	
		Static	Dynamic	Static	Dynamic	Static	Dynamic
CSVPWM (MBC)	0.8	91.02 %	91.37 %	91.96 %	92.04 %	92.21 %	92.28 %
	0.85	84.96 %	85.14 %	84.94 %	83.24 %	85.08 %	85.30 %
30° BCPWM (MBC)	0.8	89.54 %	88.09 %	90.78 %	88.86 %	86.61 %	88.67 %
	0.85	84.49 %	83.05 %	84.96 %	83.24 %	81.30%	83.64%

For comparison purpose, the CSVPWM with MBC for ZSI has also been implemented in MATLAB based on reference paper [5]. The comparison is done on the same simulation parameters in both the cases.

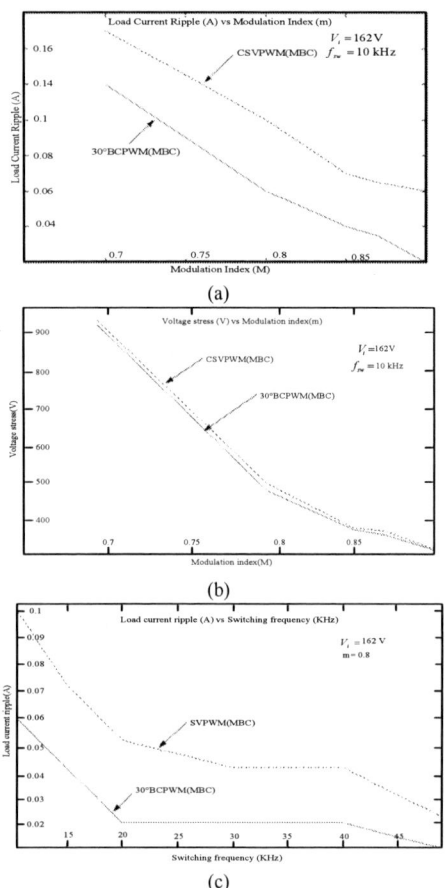

Fig. 8 Plotting for– (a) load current ripple v/s modulation index, (b) Voltage stress across switch v/s modulation index and (c) Load current ripple v/s switching frequency.

It is observed from plotting (8) that the proposed technique is better compared to conventional CSVPWM with MBC in respect of load current ripple, voltage stress across switch at high modulation index and at same switching frequency. On the other hand at same carrier frequency or at same sample time, the no of switching happed in BCPWM is less as compare to that of CSVPWM, so the switching loss is less.

Further, the proposed technique is verified by considering motor load. An induction motor of 5HP, 230 V has been modeled in MATLAB. Table 5(c) shows the line voltage THD under dynamic and static load condition for 0.8 and 0.85 modulation index. It can be observed that the proposed technique is better compare to conventional technique. As the inverter is 2-level, the THD value for line voltage is high.

VII. CONCLUSION

In this paper, a new control strategy based on bus-clamping concept for three-phase ZSI has been introduced. The effectiveness of the proposed technique is verified by simulating a ZSI in MATLAB/SIMULINK. With compare to CSVPWM with MBC, the proposed BCPWM with MBC gives better performances like reduction in load current ripple and reduction in voltage stress across switch. As at any instant one phase gets clamped in proposed technique, the switching loss is less compare to CSVPWM for ZSI. On other words, all the advantages of BCPWM for VSI remain same for ZSI. This control strategy is easy to implement in digital processor. This proposed control strategy gives an extra boost in the research of ZSI and its industrial application.

VIII. REFERENCES

[1] F. Z. Peng, "Z-source inverter", *IEEE Transactions on Industry Applications*, vol. 39, pp. 504-510, Mar-Apr 2003.

[2] F. Z. Peng and Miaosen Shen, Zhaoming Qian, "Maximum Boost Control of the Z-source Inverter" , *IEEE Transactions on Power Electronics*, Vol. 20, No.4, pp. 833-838, Jul. 2005.

[3] L. Poh Chiang, D. M. Vilathgamuwa, Y. S. Lai, C. Geok Tin, and Y. Li, "Pulse-width modulation of Z-source inverters" , *IEEE Transactions on Power Electronics*, vol. 20, pp. 1346-1355, 2005.

[4] B.Y. Husodo, M. Anwari, and S.M. Ayob, "Analysis and Simulations of Z-Source Inverter Control Methods", *IEEE Transactions on Industry Applications*, vol. 42, pp. 770 – 778, May-Jun 2006.

[5] Kun YU, Fang Lin LUO and Miao ZHU, "Space Vector Pulse-Width Modulation Base Maximum Boost Control of Z-Source Inverter" , *IEEE Transactions*, 2012

[6] Thangaprakash S. and Krishnan A., "Modified space vector modulated Z-source inverter with effective DC boost and lowest switching stress" , *The Journal of Engineering Research*, Vol. 7, No. 1, pp. 70-77, 2010.

[7] G.Narayanan,H.K.Krishnamurthy,D. Zhao, and R.Ayyanar, "Advanced bus-clamping PWM techniques based on space vector approach," *IEEE Trans. Power Electron.*, vol. 21, no. 4, pp. 974–984, Jul. 2006.

[8] V.S.S.P.K. Hari and G. Narayanan, "Space-Vector-Based Hybrid Pulse Width Modulation Technique to Reduce Line Current Distortion in Induction Motor Drives," *IET Power Electronics,* Vol.-5, No-8, pp-1463-1471, Sept. 2012.

[9] A.R. Beig and V.T. Ranganathan," Space vector based bus clamped PWM algorithms for three level inverters: implementation, performance analysis and application considerations," Applied Power Electronics Conference and Exposition, *Eighteenth Annual IEEE*, vol. 1, pp: 569 – 575 , 2003.

[10] G. Narayanan, D. Zhao, H. K. Krishnamurthy, R. Ayyanar, and V. T. Ranganathan, "Space vector based hybridPWMtechnique for reduced currentripple," *IEEE Trans. Ind. Electron.*, vol. 55, no. 4, pp. 1614–1627, Apr. 2008.

Design, Development and Relaibility Analysis of a Four-Quadrant Electromagnet Power Supply

Siddharth Varshney
Department of Electrical Engineering
IIT (BHU) Varanasi
Varanasi-221005, India

Manohar Koli, Mangesh Borage, Sunil Tiwari and
A.C. Thakurta
Raja Ramanna Centre for Advanced Technology
Indore-452013, India

Abstract— The paper deals with design, development and reliability prediction of ±10V, ±10A four-quadrant switch mode power supply. One of the application of this power supply is as a corrector electromagnet power supply in particle accelerators. In this power supply, topology used is a full bridge dc-dc chopper with PWM using unipolar voltage switching operating at 50 kHz. The main features of this supply is its ability to reduce the magnitude of output voltage ripple and increasing the frequency of output ripple thus reducing the size of filter and improved switch utilization factor. Another part of this paper predicts the reliability of this power supply using the procedures followed by MIL-HDBK 217F and various areas are suggested to improve the reliability of this power supply. This power supply attained a peak efficiency as high as 81% and predicts mean time between failures of 614084 hours at 40°C.

Keywords—converter, full bridge, pwm, unipolar, reliability.

I. INTRODUCTION

In particle accelerators like synchrotron [1], various types of electromagnets are required for bending and beam dispersion control [2]. Most fundamental of them are bending dipole magnets to guide the beam of charged particles in closed circular orbit. In addition, small corrector dipole magnets, steering coils magnets, are also used to slightly adjust the electron beam orbit in vertical and horizontal plane. The magnitude and direction of magnetic field produced in electromagnets depend on the magnitude and direction of current, respectively, flowing in its coils. A power supply used to power the steering coils should be an output current controlled power supply, in which magnitude and direction of output current can be dynamically changed. Since, electromagnets can be modeled as a series combination of resistance and inductance [3], a four-quadrant power supply is required.

A full-bridge converter has been widely used for development of four-quadrant power supply [4]-[7]. Pulse-width modulation (PWM) using unipolar voltage switching is normally followed for its control [8]. This control scheme results in smaller output voltage ripple magnitude and the ripple frequency is double the switching frequency. This reduces the size of output filter and the switching utilization factor also improves [8].

Reliability of power converters used in particle accelerators is very important [9]-[11]. Failure of one power supply in a particle accelerator can stop the operation of entire machine. Therefore, achieving high reliability and mean time between failures (MTBF) are very important, but comes with higher cost. Reliability analysis helps in area of optimization of life, performance and cost of the supply.

This paper aims to attempt the reliability prediction of a prototype four quadrant power supply following the procedure given in MIL-HDBK 217F [12], which is widely accepted in both military and industrial applications. So failure rate calculation of individual components of power supply is done to assess MTBF and reliability of power supply. Failure rate calculations also helped in to locate problem areas by identifying overstressed part or finding highest contributor to failure. Study of variations of the stress factors of these components on the overall reliability is also performed in this paper.

The paper is organized as follows: In section II, design calculations and overview of the operation of four quadrant power supply is presented. Experimental results on the prototype are given in section III. In section IV, reliability of power supply is calculated, major contributors of failure are identified and their stress factors are analyzed.

II. OPERATION AND DESIGN OF A FOUR QUADRANT POWER SUPPLY

Development of power supply, in general, consists of the design of power circuit, gate driver circuit and control circuit. Schematic of power circuit and gate driver circuit are illustrated in Fig. 1 and Fig. 2, respectively. Design and analysis of the control circuit is beyond the scope of the present paper.

A. Power Circuit

The four-quadrant operation is accomplished by using four MOSFETs in an H-bridge configuration. PWM control using unipolar voltage switching is used to improve efficiency and to reduce size of power supply. A 16 ± 2 V dc input source, followed by dc-link capacitor bank ($C_1 - C_4$) to remove dc link spikes, generates required dc link voltage which is then fed to full bridge consisting of four MOSFETs ($Q_1 - Q_4$). Output from switching stage is then passed through a damped low-pass filter consisting L_1, L_2, C_7, R_1 and C_5 to obtain steady dc output from pulse width modulated output. Two capacitors C_6 and C_8 are also connected between output terminals and ground to remove common-mode noise.

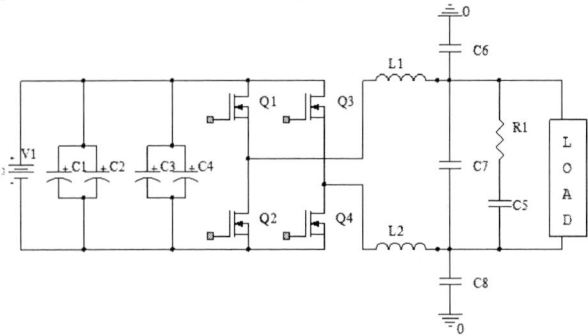

Fig. 1. Power circuit of four quadrant power supply.

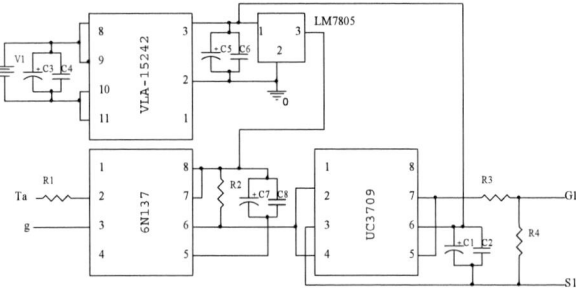

Fig. 2. MOSFET gate driver circuit.

B. Power Circuit Design

The power supply design objectives are as follows:
- Input voltage V_1 at full bridge: $18\ V \geq V_1 \geq 14\ V$.
- Maximum output voltage, $V_0 = \pm 10V$.
- Maximum output current, $I_0 = \pm 10\ A$.

The following parameters were selected for the power supply:
- Switching frequency, $f_s = 50\ kHz$.
- Current ripple in L_1, L_2, $\Delta I = 10\%$ of $I_0 = 1\ A$.
- Output voltage ripple, $\Delta V = 5\%$ of $V_0 = 0.5\ V$.

In unipolar voltage switching scheme, average output voltage V_0 for duty cycle D_1 (duty ratio of MOSFET Q_1) is given by:

$$V_0 = (2D_1 - 1) \times V_1 \qquad (1)$$

The value of L_1 and L_2, L, can be obtained from

$$\Delta I = \frac{(2D_1 - 1)(1 - D_1)V_1}{2Lf_s} \qquad (2)$$

The value of L is calculated at $D_1 = 0.75$ at which maximum ripple current, $\Delta I = 1\ A$ occurs. Substituting values in equation (2), $L = 22.5\ \mu H$ is obtained. The value of $C_7 = C$ can be obtained from:

$$\Delta V = \frac{\Delta I}{16Cf_s} \qquad (3)$$

Substituting the values in equation (3) we obtained $C_7 = 2.5$ μF. Design equations to determine damping resistance and damping capacitance is given by:

$$C_5 = 5C \text{ and } R_1 = 0.91\sqrt{L/C} \qquad (4)$$

Assuming that current ripple through inductors L_1 and L_2 should pass through filter and damping capacitor branch, the current distribution and power loss in the damping resistance can be estimated.

Table I. Power circuit components.

Part	Value	Rating parameter	Operating parameter	Other information
L_1, L_2	22.5µH	125 ℃	$\Delta T = 20℃$ (measured)	Class B insulation, ferrite core
C_1-C_4	470µF	40 V	18 V	Electrolytic capacitor, 85 ℃
C_5	22µF	63 V	10.5 V	Film capacitor
C_6, C_8	0.68µF	400V	10.5 V	Film capacitor
C_7	2.2µF	100V	10.5 V	Film capacitor
R_1	4Ω	2 W	0.0081W	Fixed, Film, Power
R_2, R_3	500Ω	0.25W	0.05 W	Fixed, Film, Power
Q_1-Q_4	IRF540	100 V	18 V	T_j = -55 to 175 ℃

For average output current of I_o, maximum rms current flowing through MOSFET is given by $I_{rms,MOSFET} = I_0\sqrt{D_1}$. Hence on-state power loss P in MOSFET is given by $P = I_{rms,MOSFET}^2 \times R_{DS}$. Substituting the design parameters, the loss and the junction temperature rise can be easily calculated. Since the calculations are obvious, they are not reported here. Operating temperature and heat sink case temperature are assumed as is 40℃ and 60℃, respectively. Table I summarizes various components and their ratings.

C. Gate Driver Circuit

In this power supply each MOSFET is driven by independent gate drivers. Function of gate driver circuit is to isolate control circuit from power circuit and delivers required gate pulses to MOSFETs with sufficient peak current handling capability. An opto-coupler IC, 6N137, is used to isolate the pulses generated by control circuit from power circuit. Output from 6N137 is fed to gate driver IC, UC3709 to deliver appropriate gate driver pulse to switch MOSFET. Each gate driver circuit is given power through a dc/dc isolator supply IC, VLA-106-15242, which gives regulated and isolated 15V supply to gate circuit. A LM7805 regulator IC is also used to provide regulated 5V to 6N137. Schematic of circuit is as shown in Fig. 2. Components list and the specifications of MOSFET gate driver circuit is summarized in Table II.

Table II. Gate driver circuit components.

Part	Value	Rating Parameters	Operating Parameters	Description
C_1,C_3,C_5	47 µF	63 V	15 V	Elec. Cap., 85℃
C_7	47 µF	63 V	5 V	Elec. Cap., 85℃
C_2,C_4,C_6	0.1 µF	50 V	15 V	Cer. Cap., 85℃
C_8	0.1 µF	50 V	5 V	Cer. Cap., 85℃
R_1, R_2, R_3, R_4	400Ω, 2kΩ, 56Ω and 10kΩ respectively		Fixed, film type. Stress factor for all resistors is taken=0.2	
6N137 (optocoupler), UC3709 (gate driver), VLA-106-15242 (dc/dc isolator power supply), LM7805 (regulated 5V supply).				

III. EXPERIMENTAL RESULTS

The designed power supply was built and tested in order to assess its performance. Fig. 3 shows developed power supply. Snapshots of waveforms were taken to verify the operation. In PWM using unipolar voltage switching positive half bridge and negative half bridges are controlled independently. Gate pulses are shown in Fig. 4. Various waveforms for positive and negative output voltage are shown in Fig. 5 and Fig. 6, respectively. Variation of measured efficiency as a function of operating duty cycle is shown in Fig. 7. Maximum efficiency of 81% is achieved. Note that output power is zero at $D_1 = 0.5$, so is the efficiency.

IV. RELIABILITY ANALYSIS

Reliability calculation which involves the determination of failure rate λ_p, can be evaluated using the procedure given in MIL-HDBK-217F [12]. MIL-HDBK 217F is very well known and widely accepted in both military and industrial applications. This handbook is intended to provide a uniform database for determining failure rate of electronic components.

Basic model of component failure rate in MIL-HDBK 217F is as given below:

$$\lambda_p = \lambda_b (\Pi_1^n \pi_n) \tag{5}$$

Where, λ_b is base failure rate which is multiplied with various π factors that accounts environmental, design and operating conditions of the component. Table III lists the failure rate models, given in MIL-HDBK 217F [12], of components used in presented power supply.

Fig. 3: Four quadrant power supply prototype developed on PCB.

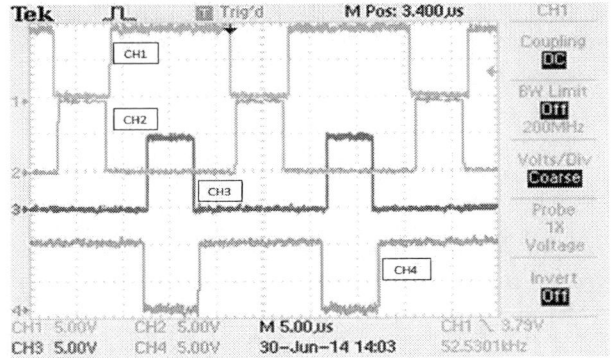

Fig. 4: Unipolar switching scheme gate pulses at MOSFET's gate, where CH1: Q1 gate, CH2: Q2 gate, CH3: Q3 gate, and CH4: Q4 gate

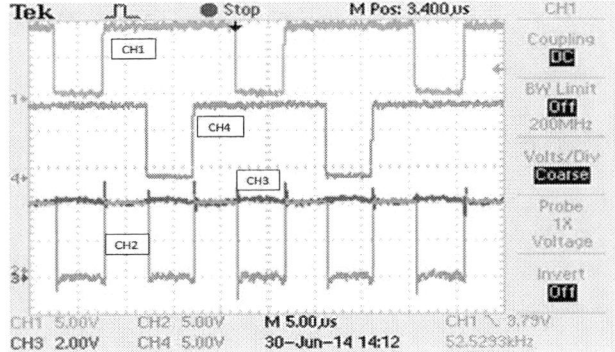

Fig. 5: Key waveforms when the output voltage is positive, CH1: Q1 gate pulse, CH2: PWM output, CH3: dc output and CH4: Q4 gate pulse.

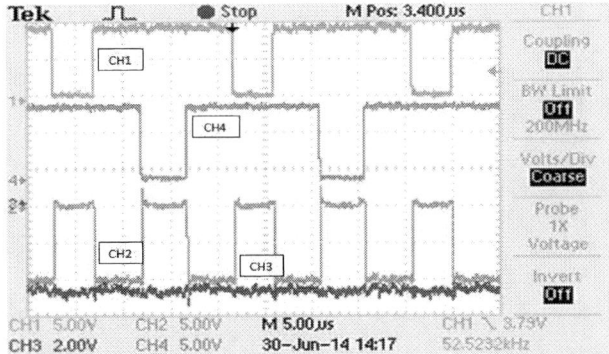

Fig. 6: Key waveforms when the output voltage is negative, where CH1: Q2 gate pulse, CH2: PWM output, CH3: dc output and CH4: Q3 gate pulse.

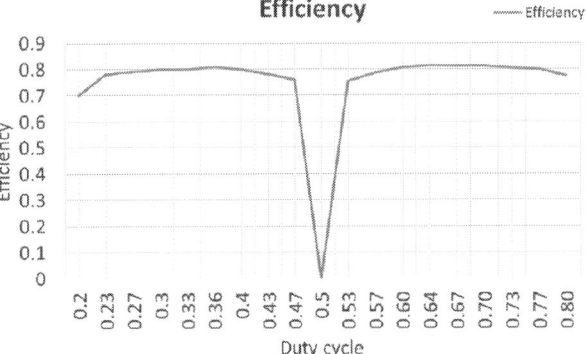

Fig. 7. Efficiency vs. duty cycle plot for four quadrant power supply.

A MOSFET failure rate model has three π factors: π_T, related to junction temperature of MOSFET (depends on power dissipation at junction); π_Q, related to level of screening item is subjected (military, industrial or commercial); π_E, related to environmental conditions where experiment has to be performed (ground benign, ground mobile, missile, aerospace etc.). Detailed description of other component's model can be found in MIL-HDBK 217F [12].

The mean time between failure (MTBF) and reliability, R, are given by [13]:

$$MTBF = \lambda^{-1} \text{ and } R = e^{-\lambda t} \tag{6}$$

MIL-HDBK-217F considers series model [13] of reliability for reliability prediction of electronic assembly unless some kind of redundancy is included. So the reliability of the power supply R_{ps} is calculated as:

$$R_{ps} = \Pi_1^n R_i \qquad (7)$$

Where, the term R_i corresponds to the individual reliability of the components in the power supply (power stage and gate driver stage). Mean time between failures of the overall system for series model is given by:

$$MTBF = \frac{1}{\sum_1^n \lambda_i} \qquad (8)$$

Where, λ_i corresponds to failure rate of individual components in power supply.

Using data listed in Table I and II, various stress factors are calculated, which in turn are used to find out various π factors in the failure rate models. The resulting failure rates of individual components, overall failure rate, MTBF and reliability data for the power circuit and the gate driver circuit are summarized in Tables IV and V, respectively. Table V also gives the calculation of overall reliability and MTBF of the developed power supply. Fig. 8 shows reliability of power circuit stage, gate driver circuit stage and overall power supply over time. It is assumed that power supply is required to be operational for 30 years. Therefore, for operational life of 30 years reliability of power supply obtained is 0.66 and MTBF obtained is 614084 hours.

The reliability analysis and the data obtained, giving life expectancy of system, also indicate the areas of improvements and weak links. Fig. 9 shows the proportions of failure rate of components in failure of the power circuit. It is very clearly seen that the most vulnerable components are MOSFETs and electrolytic capacitors. Similarly, Fig. 10 shows the proportions of failure rate of components in failure of the gate driver circuit. It is evident that the most vulnerable components are ICs and electrolytic capacitors.

Analyzing the stress factors of MOSFET, it is found that main contributing factor is π_T, which depends on junction temperature. In turn, this factor depends on power dissipation by the MOSFET, case temperature and thermal resistance (MOSFET package and heat-sink compound combined).

Therefore reliability improvement can be achieved by using lower on-state resistance of MOSFET or by improving thermal management.

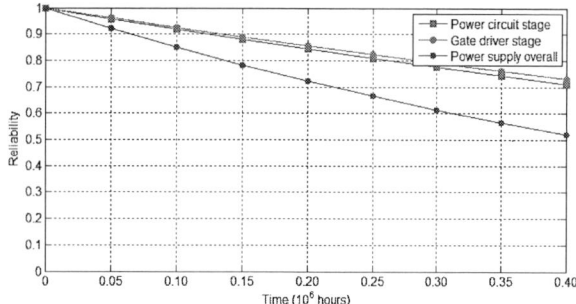

Fig. 8. Reliability of power supply over time.

Table III. Failure rate models [12]

Device	Model
MOSFET	$\lambda_p = \lambda_b \times \pi_T \times \pi_Q \times \pi_E$ $\pi_T = \exp\left\{ -1925\left[\frac{1}{T_j + 273} - \frac{1}{298} \right] \right\}$ $T_j = T_C + \theta_{jc}P$ Where, T_j: junction temperature; T_C: case temperature; θ_{jc}: thermal resistance; P: power dissipation
Capacitor	$\lambda_p = \lambda_b \times \pi_{CV} \times \pi_Q \times \pi_E$
• Electrolytic	$\lambda_b = 0.00254\left\{ \left(\frac{s}{0.5}\right)^3 + 1 \right\} \times \exp\left\{ 5.09 \times \left(\frac{T+273}{358}\right)^5 \right\}$ For 85°C max temp. rating $\lambda_b = 0.00254\left\{ \left(\frac{s}{0.5}\right)^3 + 1 \right\} \times \exp\left\{ 5.09 \times \left(\frac{T+273}{378}\right)^5 \right\}$ For 105°C max temp. rating $\lambda_b = 0.00254\left\{ \left(\frac{s}{0.5}\right)^3 + 1 \right\} \times \exp\left\{ 5.09 \times \left(\frac{T+273}{398}\right)^5 \right\}$ For 125°C max temp. rating Where, s: operating voltage/ voltage rating; T: Ambient temperature $\pi_{CV} = 0.34C^{0.18}$ Where, C: capacitance
• Ceramic	$\lambda_b = 0.0003\left\{ \left(\frac{s}{0.3}\right)^3 + 1 \right\} \times \exp\left(\frac{T+273}{358}\right)$ $\pi_{CV} = 0.41C^{0.11}$
• Film	$\lambda_b = 0.0005\left\{ \left(\frac{s}{0.4}\right)^5 + 1 \right\} \times \exp\left\{ 2.5 \times \left(\frac{T+273}{358}\right)^{18} \right\}$ $\pi_{CV} = 1.3C^{0.13}$
Inductors	$\lambda_p = \lambda_b \times \pi_{CV} \times \pi_Q \times \pi_E$ $\lambda_p = \lambda_b \times \pi_{CV} \times \pi_Q \times \pi_E$ $\lambda_b = 0.000319 \times \exp\left(\frac{T+273}{364}\right)^{8.7}$ $T_{HS} = T_A + 1.1 \times (\Delta T)$ Where, T_{HS}: hotspot temperature; T_A: ambient temperature; ΔT: average temperature rise. $\pi_C = 1$ (for fixed inductance)
Film resistors	$\lambda_p = \lambda_b \times \pi_R \times \pi_Q \times \pi_E$ $\lambda_b = 0.000325 \times e^{\left\{ s \times \left(\frac{T+273}{273}\right) \right\}} \times e^{\left\{ \left(\frac{T+273}{343}\right)^2 \right\}}$ $\pi_R = 1$ for $R \leq 100\Omega$

Overall reliability and total MTBF of power supply are calculated at different junction temperatures of MOSFET, to analyze reliability of power supply variation with junction temperature, and summarized in Table VI. From this analysis it is evident that reliability improvement of 0.56 to 0.70 (30 year) can be achieved as junction temperature of MOSFETs is

Table IV. Reliability of power circuit

	Q1 - Q4	C1, C2, C3, C4	C5	C6	C7, C8	R1	R2, R3	L1, L2
λ_b	0.0060	0.05915	0.03555	0.00062	0.00066	0.00070	0.00087	0.00052
π_E	1	1	1	1	1	1	1	1
π_Q	1	1	1	1	1	1	1	1
π_T	2.42	-	-	-	-	-	-	-
π_{CV}	-	1.029	0.593	1.44	1.23	-	-	-
π_C	1	-	-	-	-	-	-	1
π_S	-	-	-	-	-	-	-	-
π_R	-	-	-	-	-	1	1	-
λ_a	0.1452	0.06086	0.02108	0.00089	0.00081	0.00070	0.00087	0.00052
Qty	4	4	1	1	2	1	2	2

Total failure rate of power circuit = 0.85131 failures/10^6 hours (sum of failure rates of all components)

$MTBF_1$ = 1174660 hours / failure

Reliability of power circuit for 30 years of span of time = $\exp\left(\dfrac{-0.2628}{1.174660}\right)$

= 0.80

Table V. Reliability data of gate driver circuit

	C1, C3, C5	C7	C2, C3, C6	C8	R1, R2, R3, R4
λ_b	0.037992	0.034348	0.0008745	0.000725	0.000874
π_E	1	1	1	1	1
π_Q	1	1	1	1	1
π_{CV}	0.680	0.680	0.318	0.318	-
π_R	-	-	-	-	1
λ_a	0.025834	0.023357	0.000278	0.000230	0.000874
Qty	3	1	3	1	4

Failure rates of ICs (from manufacturer reliability data):
- 6N137: 0.056564 failures / 10^6 hours [14]
- LM7805: 0.002300 failures / 10^6 hours [15]
- UC3709: 0.02 failures / 10^6 hours (assumed)
- VLA-106-15242: 0.01 failures / 10^6 hours (assumed)

Total failure rate of single gate driver = 0.194283 failures/10^6 hours (sum of failure rates of all components)

Failure rate of all 4 gate drivers = 0.777132

$MTBF_2$ = 1286782 hours / failure

Reliability of gate driver for 30 years of span of time = $\exp\left(\dfrac{-.2628}{1.286782}\right)$

= 0.82

Overall reliability of power supply is given by:

R_{PS} = Reliability of power circuit * Reliability of gate circuit

$R_{PS} = 0.80 \times 0.82 = 0.66$

$$MTBF_{Total} = \frac{1}{MTBF_1^{-1} + MTBF_2^{-1}} = 614084 \text{ hours/failures}$$

decreased from 120°C to 40°C which is quite significant. Fig. 11 shows reliability of power supply over time at different junction temperatures of MOSFET.

Whereas analysis of electrolytic capacitor stress factors shows that main contributing factors are λ_b and π_C, which depend on voltage stress and capacitance value, respectively. Therefore, reliability improvement can be achieved by using electrolytic capacitors of higher voltage rating. A parallel configuration of capacitors with low capacitance values reduces capacitance factor π_C but as number of capacitors increases failure rate also increases, so an appropriate configuration can be chosen to achieve minimum failure rate and in turn improved reliability. Reliability analysis based on these factors are not studied here.

Another important parameter for reliability improvement is maximum temperature rating of electrolytic capacitors. According to MIL-HDBK 217F specifications, only three types (85°C, 105°C, and 125°C) of electrolytic capacitors are manufactured [12]. Base failure rate model of each type of capacitor is listed in Table III. To analyze reliability variation based on maximum temperature rating of electrolytic capacitor, Table VII is generated to summarize failure rate of power circuit, gate driver circuit, MTBF, and reliability figures of overall power supply for each types of capacitor. And it can be noticed that a significant improvement in reliability, 0.66 to 0.71, is achieved, when maximum temperature rating of all electrolytic capacitors are taken 105°C rating instead of 85°C. Fig. 12 shows the reliability of power supply over time for each type (85°C, 105°C, and 125°C) of electrolytic capacitor.

Table VI. Reliability and MTBF figures of power supply at different junction temperatures of MOSFET

Junc. Temp. (T_j) (°C)	Temp. factor (π_T)	MTBF (Total power supply) (hours / failure)	Reliability (30 years)
40	1.36	727780	0.70
60	1.65	657703	0.67
80	1.97	586427	0.64
100	2.33	519199	0.60
120	2.74	456112	0.56

Table VII. Reliability and MTBF figures of power supply for different types of electrolytic capacitors

Max Temp. Rating (°C)	Failure rate of power circuit (failures / 10^6 hours)	Failure rate of gate driver circuit (failures / 10^6 hours)	Overall MTBF (hours / failure)	Reliability (30 years)
85	0.85131	0.777132	614084	0.66
105	0.72924	0.59096	757461	0.71
125	0.677580	0.512201	840491	0.73

V. Conclusions

This paper presents the design and reliability analysis of four quadrant power supply. A prototype is built by full-bridge topology with PWM using unipolar voltage switching scheme, aimed to develop a reliable corrector electromagnet power supply for particle accelerators. It supplies a low ripple magnitude output voltage and the ripple frequency is double

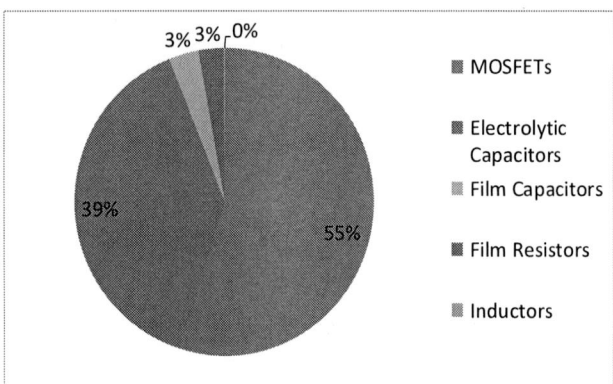

Fig. 9. Failure rate distribution of various components used in power circuit.

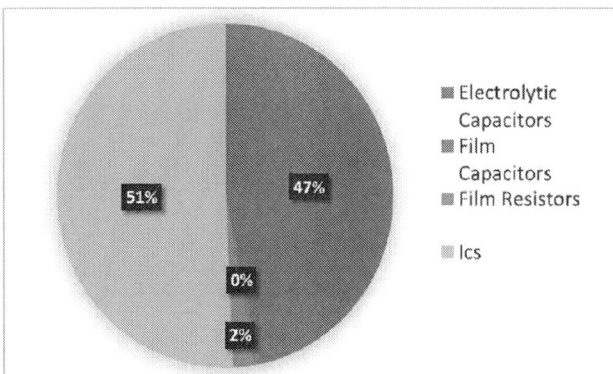

Fig. 10. Failure rate distribution of various components used in gate driver circuit.

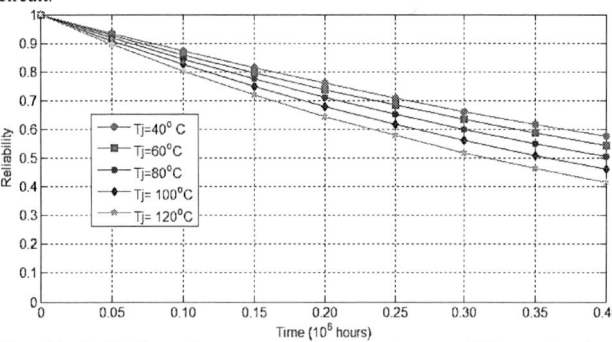

Fig. 11. Reliability of power supply over time at different junction temperatures of MOSFET.

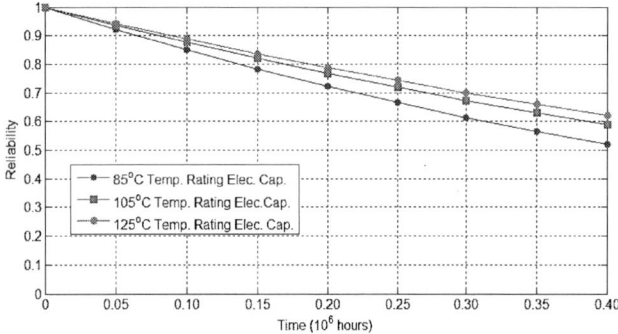

Fig. 12. Reliability of power supply over time for electrolytic capacitors of different rating.

the switching frequency hence size of overall power supply is reduced. Maximum efficiency of 81% is achieved at full load and maximum input voltage.

Predicting the reliability of power supply pointed out weak-links that need extensive development in the design. MOSFETs and electrolytic capacitors are identified as highly failure-prone components. Analysis of reliability data obtained highlights that reduction in junction temperature of MOSFET and selecting the electrolytic capacitor of higher maximum temperature rating (105°C instead 85°C), could achieve higher overall reliability and mean time between failures.

REFERENCES

[1] A. R. Steere, "A Time line of major particle accelerator," Michigan state Univ., Dep. of physics and astronomy, 2005.

[2] S.Das, M.G.. Karmarkar, S. Kotaiah, "Conventional magnets for synchrotron," RRCAT, Indore, India.

[3] Y. Thurel, "Four-quadrant power converter based on output linear stage," CERN, Geneva, Switzerland, pp. 209-210.

[4] L. Bartelson, G. Krafczyk, H. Pfeffer, D. Wolff, "Four quadrant dc to dc switching supply for the fermilab main injector," Fermi Nat. Accel. Lab., Batavia, IL, USA.

[5] Si Fang, G. Krafczyk, H. Pfeffer, D.Wolff, "Four quadrant 250 kW switchmode power supply for Fermilab Main Injector," Fermi Nat. Accel. Lab., Batavia, IL, USA.

[6] G. Kniegl, R. Weber, F. Bordry, A. Dupaquier, "Four quadrant converter [±600A, ±12V] prototype for LHC," Proc. of particle accelerator conference, Newyork, 1999.

[7] C. Afanasov, M. rata, L. Mandici, "Speed control of dc motor using four quadrant chopper and bipolar control strategy," Int. symposium on electrical engineering and energy converters, ELS-2013.

[8] Z. Ned Mohan, T. Undeland and W. Robbins, Power Electronics: Converters, Applications and Design, 3rd ed. NY: Wiley, 2007.

[9] P. Bellomo, A. Donaldson, D. MacNair, "B-Factor intermediate DC Magnet Power Systems Reliability Modelling and Results," Particle Accelerator Conference, Proccedings on, Chicago, 2001.

[10] D. Siemaszko, M. Speiser, S. Pittet, "Reliability Models Applied to System of Power Converters in Particle Accelerators," 14th European Conference on Power Electronics and Applications, Nottingham, 2011.

[11] D. Siemaszko, S. Pittet, "Impact of Modularity and Redundancy in Optimising the Reliability of Power Systems that Include a Large Number of Power Converters," European organization for nuclear research, CERN, Geneva, Switzerland.

[12] "Reliability prediction of electronic equipment," Dept. of Defense, Washington DC, Tech. Rep. MIL-HDBK-217F, Dec.1991.

[13] M. Modarres, M. Kaminskiy, and V. Krivtsov, "Reliability Engineering and Risk Analysis: A Practical Guide," second edition, CRC Press, 2009.

[14] "Reliability data sheet – 6N137, High CMR, High-speed TTL compatible optocouplers," Avago technologies, available online at http://www.avagotech.com/pages/en/optocouplers_plastic/plastic_digital_optocoupler/10_mbd_logic_gate/6n137/

[15] "Reliability data MC 7805.2CT", On semiconductor, available online at http://www.onsemi.com/PowerSolutions/reliability.do?part=MC7805.2CT

978-1-4799-6047-7/14 $31.00 © 2014 IEEE   640

Modified Sine Wave Phase Disposition PWM Technique for Harmonic Reduction in Multilevel Inverter fed Drives

Punit P. Acharya
M.Tech student: EEE Department
SRM University
Kanchipuram, India
acharyapunit@yahoo.in

Vishal S. Rangras
Lecturer: EE Department
ACET
Ahmedabad, India
vishalrangras@yahoo.com

Abstract—**Advancement in Multilevel Inverter Technologies and its Control Techniques has been emerging in leaps and bounds. Such an inclination towards this area is because of its high power handling capability as well as improved power quality. Low Total Harmonic Distortion (THD) and less thermal Stress on Switches and improved performance are its added advantages. Switching strategies plays a very important role in optimizing inverter performance parameters. In this paper a novel multicarrier sine Pulse Width Modulation (PWM) Phase Disposition technique is introduced which significantly reduces THD. The proposed switching technique is described using relevant mathematical analysis, and results are obtained from simulation study, which consequently illustrates the improved performance of the overall system as compared to conventional one.**

Index Terms—*Multilevel Inverter, PWM technique, Phase Disposition, Total Harmonic Distortion, Modified Sine PWM.*

I. INTRODUCTION

THE concept of utilizing multiple small voltage levels to perform power conversion was patented by an MIT researcher over thirty years ago. Merits of such multilevel approach are High power handling capability, better Power Quality, good Electromagnetic Compatibility (EMC), reduced Total Harmonic Distortion (THD) and reduction in ac side filter. Efforts are going on for optimizing the performance of Multilevel Inverters (MLI) by reducing the number of switches [1]. Multilevel inverter topologies have significant dominance in High power medium voltage applications like heavy duty electric and hybrid-electric vehicles of large electric drives (>250kW) [2].

Different types of multilevel inverters have been proposed in number of literatures. Basic MLI topologies are Diode Clamped (DCMLI), Flying Capacitor (FCMLI) and Cascaded H-bridge (CHMLI). Among these, Cascaded H-bridge is usually preferred because of its simple circuit layout, less component counts, modular structure and devoid of

unbalanced capacitor voltage problem. Other topologies like Generalized Multilevel

Topology, Soft switched MLIs, Mixed level Hybrid MLIs, Back-to-back Diode Clamped Converter [3][6]. Hybrid MLI comprises of combination of one or more basic topologies in order to give a better performance.

The most significant features of Multilevel Inverters are [4]:

- **Input Current**: Draws input current with lower distortions.
- **Switching frequency**: Can be operated at very high switching frequencies. With lower switching frequency, switching losses are reduced and inverter efficiency gets improved.
- **Transient losses**: Reduced THD in output voltage and reduced dV/dt.
- **Common Mode (CM) voltage:** In multilevel inverter fed induction motors, CM voltages are produced. This imparts stress in motor bearings. CM voltages can be eliminated by using advanced modulation strategies such as that proposed in [5].

In his particular paper, single phase 5-level DCMLI is used to explain the new multicarrier Sine PWM technique in order to acquire the better understanding of distinct advantages of this novel technique with respect to any preponderance. DCMLI is the most conventional topology. The basic idea for proposing this technique is to show that the presented novel switching strategy can be used for most basic topology to the most advanced one which shows better results (in THD point of view), and that too with reduced switching frequency.

978-1-4799-6047-7/14 $31.00 © 2014 IEEE

II. EVALUATION CRITERIA FOR MODULATION TECHNIQUE

Some of the important criteria to evaluate different modulation methods are as follows:

1) *Switching loss*: Power loss across the device increases with the increase in the switching frequency. A modulation technique should have least switching losses.

2) *Harmonic Distortion*: Because of high switching frequency, modulation techniques induce multiples of the fundamental frequency (called harmonics) in the output voltage. A good modulation technique should have least THD.

3) *DC Bus Utilization*: In HV applications it is necessary to utilize DC bus voltage as much as possible. Modulation techniques like space vector PWM increases DC bus usage, reduces harmonic distortion, even under low modulation indices.

4) *Number of Switches*: An efficient modulation technique gives the better performance with least number of switches used.

Optimized switching strategies for different applications may vary. This generalized switching technique can help designers to find optimized strategy for such applications.

III. DIODE CLAMPED MULTILEVEL INVERTER

In 1981, Nabae et. al proposed a 3-level diode clamped neutral point converter topology. Two capacitors were connected across the dc bus giving an additional level. Topology was named as Neutral Point Clamped (NPC) inverter because the additional level was the neutral point of the dc bus.

The most conventional multilevel topology is the diode clamped inverter, where diode is used as the clamping device to clamp the dc bus voltage so as to achieve steps in the output voltage. A single leg of diode-clamped m-level inverter (DCMLI) typically consists of (m-1) dc link capacitors, 2(m-1) switching devices and (m-1)(m-2) clamping diodes on the DC bus and produces m-level on the phase voltage. Fig. 1 shows a full bridge five-level diode clamped converter.

With the increase in number of levels, the number of diodes and the number of switching devices also increase and makes the system bulky impractical for implementation. If pulse width modulation (PWM) technique is used, then diode reverse recovery of these clamping diodes becomes the major design challenge. In practical implementation, some dead time is inserted between the gating signals and their complements meaning that both switches in a complementary pair may be switched off for a small amount of time during a transition.

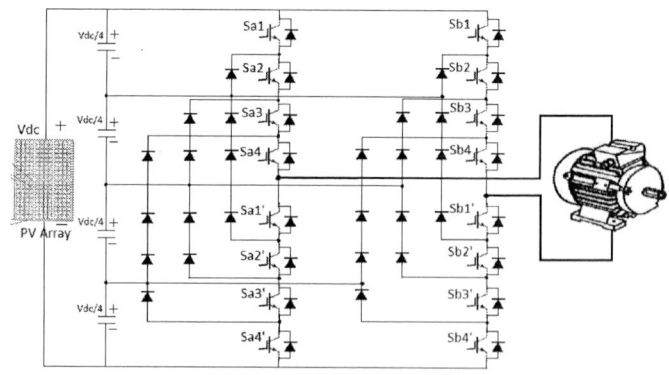

Fig. 1. Single phase Diode Clamped 5-level Inverter connected with Induction Motor load (R-L equivalent)

TABLE I
Switching States for leg A

Output V_{a0}	Switching State							
	S_{a1}	S_{a2}	S_{a3}	S_{a4}	S_{a1}'	S_{a2}'	S_{a3}'	S_{a4}'
$V_5 = V_{dc}$	1	1	1	1	0	0	0	0
$V_4 = 3V_{dc}/4$	0	1	1	1	1	0	0	0
$V_3 = V_{dc}/2$	0	0	1	1	1	1	0	0
$V_2 = V_{dc}/4$	0	0	0	1	1	1	1	0
$V_1 = 0$	0	0	0	0	1	1	1	1

TABLE II
Switching States for leg B

Output V_{b0}	Switching State							
	S_{b1}	S_{b2}	S_{b3}	S_{b4}	S_{b1}'	S_{b2}'	S_{b3}'	S_{b4}'
$V_5 = -V_{dc}$	0	0	0	0	1	1	1	1
$V_4 = -3V_{dc}/4$	1	0	0	0	0	1	1	1
$V_3 = -V_{dc}/2$	1	1	0	0	0	0	1	1
$V_2 = -V_{dc}/4$	1	1	1	0	0	0	0	1
$V_1 = 0$	1	1	1	1	0	0	0	0

Table I and II shows voltage levels and their corresponding switch states. State condition 1 means the switch is ON, and state 0 means switch is OFF. Here there are four complementary switches and each switch is ON only once per cycle.

(Note: An m-level inverter has an m-level output phase-leg voltage and a (2m-1)-level output line voltage.)

The order of numbering of switches of phase "a" is Sa1, Sa2, Sa3, Sa4, Sa1', Sa2', Sa3' and Sa4'. Similarly, it holds true for phase "b". The steps to synthesize the five-level voltages are as follows:

1. For an output voltage level $V_{a0} = V_{dc}$, all upper half switches S_{a1} through S_{a4} are turned on.

2. For an output voltage level $V_{a0} = 3V_{dc}/4$, three upper switches S_{a2} through S_{a4} and one lower switch S_{a1}' are turned on.

3. For an output voltage level $V_{a0} = V_{dc}/2$, two upper switches S_{a3} through S_{a4} and two lower switches S_{a1}' and S_{a2}' are turned on.

4. For an output voltage level $V_{a0} = V_{dc}/4$, one upper switch S_{a4} and three lower switches S_{a1}' through S_{a3}' are turned on.

5. For an output voltage level $V_{a0} = 0$, all lower half switches S_{a1}' through S_{a4}' are turned on.

Fig. 2. depicts the Switching model for a phase leg. Same model is created for other leg with sine wave 180° phase shifted.

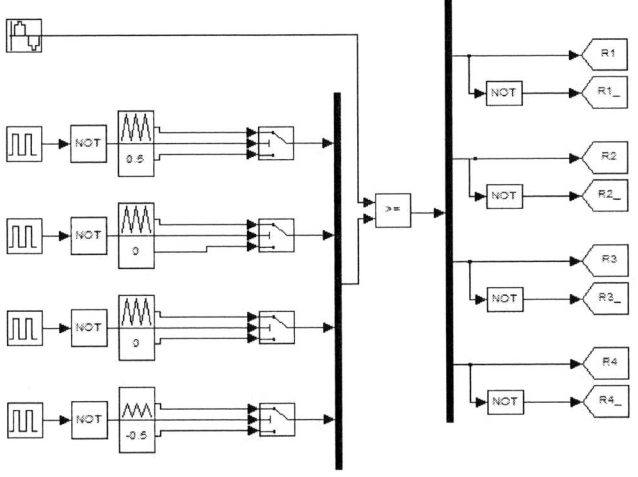

Fig. 2. Switching model implemented in MATLAB Simulink platform

IV. PROPOSED MODIFIED SINE WAVE PWM TECHNIQUE

From the general analytical observation it was observed that the width of pulses closer to the peak of the sine wave do not change significantly with the change in modulation index. This is due to the typical sine wave characteristics. The Sine wave Pulse Width Modulation (SPWM) technique can be modified so that the carrier wave is applied during the first and the last 60° intervals per half cycle.

So, there is no carrier pulse in the middle 60° interval of the half cycle. Which means, 60° to 120° and 240° to 300° of the first cycle is devoid of carrier pulse (zero magnitude). Hence this reduces the carrier frequency utilization by $1/3^{rd}$ as compared to conventional SPWM technique. Consequently, the switching frequency is reduced by $1/3^{rd}$. The fundamental component is increased and its harmonic characteristics are improved. It reduces the number of switching of power devices and also reduces switching losses [6]. Proposed modified sinusoidal PWM technique is shown in Fig. 3.

Operation of the multilevel inverter is quite non-linear for Sine PWM control technique in overmodulation range. Hence results are obtained by strictly constraining to linear modulation range.

V. MATHEMATICAL ANALYSIS

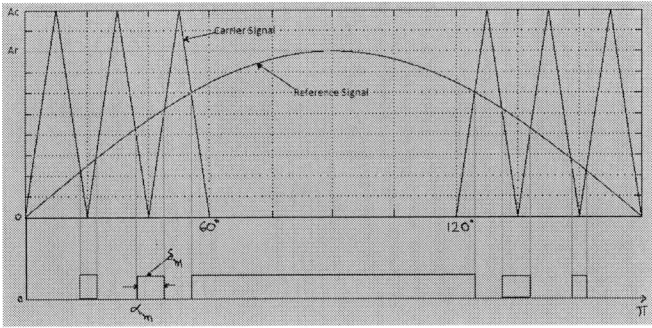

Fig. 3. Modified Sinusoidal pulse width modulation (half wave is shown)

Basic relation between angle 'α' and time period 't' is given by,

$$t = \frac{\alpha}{\omega}$$

So for m^{th} time t_m and angle α_m of interaction can be determined from,

$$t_m = \frac{\alpha_m}{\omega} \qquad \text{for } m = 1,2,3...p$$

$$t = t_x + \frac{m T_s}{2}$$

Where t_x can be solved from,

$$1 - \frac{2t}{T_s} = M * \sin\left[\omega\left\{t_x + \frac{m T_s}{2}\right\}\right] \quad \text{for } m = 1,3,5...(2p-1)$$

$$\frac{2t}{T_s} = M * \sin\left[\omega\left\{t_x + \frac{m T_s}{2}\right\}\right] \quad \text{for } m = 2,4,6...2p$$

The time interactions during the last 60° can be found from,

$$t_{m+1} = \frac{\alpha_{m+1}}{\omega}$$

$$t_{m+1} = \frac{T}{2} - t_{2p-m} \qquad \text{for } m = p, p+1, \dots \dots 2p-1$$

$$T_s = \frac{T}{6(p+1)}$$

Width of mth pulse d_m or (pulse angle δ_m) can be found from,

$$d_m = \frac{\delta_m}{\omega}$$

$$d_m = t_{m+1} - t_m \qquad (1)$$

Here number of pulses "m" depends upon the number carrier pulses. Equation (1) gives the width of the mth pulse.

VI. SIMULATIONS AND RESULTS

A five-level diode clamped multilevel inverter have been simulated in MATLAB/ SIMULINK environment. MATLAB 7.10.0 (R2010a) version have been used for the simulation.

In place of a motor, R-L Load is employed for simplicity and ease of understanding. FFT (Fast Fourier transform) tools

978-1-4799-6047-7/14 $31.00 © 2014 IEEE

have been used for harmonic analysis. Table III shows the parameters used for simulation.

TABLE III
Simulation Parameters

Parameter	Value
Separate DC Voltage Source	25 V
Carrier Frequency for PWM Circuit	4kHz
System Frequency	50Hz
Load parameters	R = 35
	L = 0.001H
Sampling period	2e-6 sec/sample
Simulation period	0.2 sec

TABLE IV
Simulation Results
(Modified Sine PWM PD Technique)

Modulation Index	$V_{T.H.D}$ (%)	3rd Harmonic (%)	5th Harmonic (%)	Fundamental
1	14.66	3.49	2.32	90.97 peak (64.33 rms)
0.9	18.49	5.11	2.23	82.74 peak (58.51 rms)
0.8	20.89	7.04	2.12	74.45 peak (52.64 rms)
0.6	31.45	12.51	2.01	58 peak (41.01 rms)
0.4	41.22	13.39	1.31	34.53 peak (24.42 rms)

TABLE V
Simulation Results
(Conventional Sine PWM PD Technique)

Modulation Index	$V_{T.H.D}$ (%)	3rd Harmonic (%)	5th Harmonic (%)	Fundamental
1	18.29	1.47	4.04	90.70 peak (64.13 rms)
0.9	20.54	2.66	3.16	80.62 peak (57.00 rms)
0.8	23.78	2.08	3.2	69.33 peak (40.02 rms)
0.6	37.13	1.41	5.25	49.38 peak (39.42 rms)
0.4	54.75	4.02	7.76	30.48 peak (21.55 rms)

It can be clearly derived from the Table V and VI that, Total Harmonic Distortion in Modified Sine PWM is lesser than compared to conventional PWM technique. But the drawback of this modified SPWM is that it increases the 3rd order harmonic values. Whereas 5th order harmonics and other higher order harmonics are reduced to a notable value.

By using three phase MLI, connecting it to a three phase delta connected transformer and then to the load, 3rd order harmonics and its multiples (triplen harmonics) can be eliminated. But unfortunately, delta connection is not possible for single phase systems. Hence for reducing 3rd order harmonics in single phase systems, Selective Harmonic Elimination or Third Harmonic Injection technique can be used.

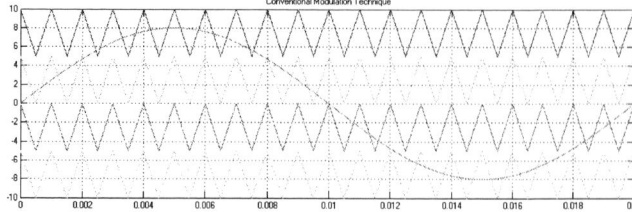

Fig. 4. Conventional Sinusoidal pulse width modulation

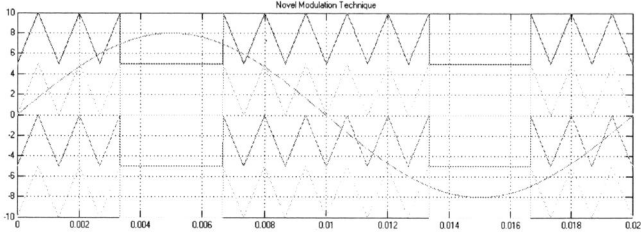

Fig. 5. Modified Sinusoidal pulse width modulation

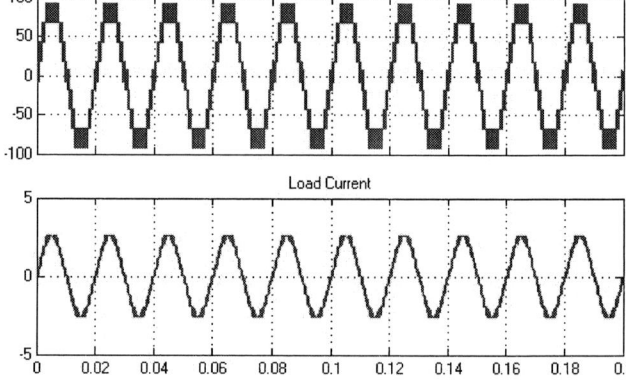

Fig. 6. Load voltage and current waveforms for Modified Sinusoidal pulse width modulation (f_c = 4kHz/3, M = 0.8)

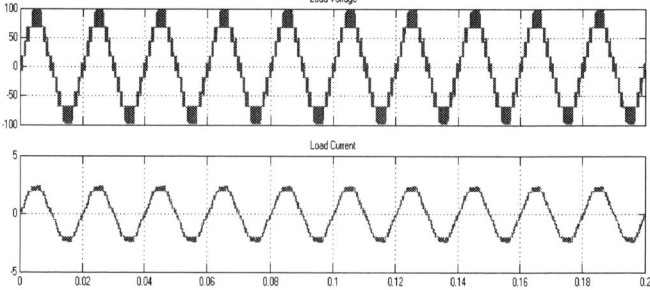

Fig. 7. Load voltage and current waveforms for Conventional Sinusoidal pulse width modulation (f_c = 4kHz, M = 0.8)

Fig. 4 and 5 shows Conventional Vs Modified SPWM technique. Fig. 6 and 7 shows the load voltage and current waveform for Modified SPWM and Conventional SPWM. Because of inductive load current waveform is smooth.

In Fig. 6 f_c = 4kHz/3, this is because for one cycle of sine, $1/3^{rd}$ times the carrier frequency switching does not occurs. Which reduces overall switching losses to $1/3^{rd}$ times compared to conventional SPWM.

VII. FUTURE SCOPE

Diode Clamped Multilevel Inverter is the most conventional MLI topology. Even though conventional most topology (DCMLI) was being used here, better results were obtained. Then such a PWM technique is worth implementing for advanced and hybrid MLI topologies. Presently many authors have been working on topologies with reduced number of switches. Using such topologies can further reduce harmonic content, and may possibly increase Power Density of the device (Power handling capacity per cubic volume) without increasing the size. So, the combination of best MLI topology and best PWM control technique may lead more closely to ideal inverter characteristics.

VIII. CONCLUSION

Existing conventional techniques uses sine wave PWM technique. Whose limitation is requirement of very high frequency carrier signal, which consecutively leads to more switching losses. Authors have proposed a Modified sine wave phase disposition PWM technique for a Single phase 5-level diode clamped multilevel inverter, which reduces switching frequency by $1/3^{rd}$ as compared to conventional sine PWM technique. Consequently, switching losses also decreases. It has been shown using facts and figures from simulations, that overall T.H.D was improved as compared to conventional sine PWM method. The obvious limitation of this method is that 3rd order harmonics exists to a notable value. Though this limitation can be eliminated by implementing three phase MLI and using delta connected three phase transformer. This method can be as opted as reliable switching technique for any sort of multilevel inverter topology.

ACKNOWLEDGEMENT

The authors would like to thank Prof. A. Venkadesan for his guidance and for being constant source of motivation. Special thanks to Dr. N. Chellammal and Dr. Subhransu Sekhar Dash for their immense support and for providing their literature resources.

Thanks to beloved parents for their warm blessings and wishes.

REFERENCES

[1] Dr. Keith Corzine, "Operation and design of Multilevel Inverters," Developed for the Office of Naval Research in December 2003, Revised June 2005.

[2] Mr. Jithin J.I. and A. Benuel Sathish Raj, "A new topology for a single phase 21 level multilevel inverter using reduced number of switches," *International Journal of Engineering research and Technology*, 2013, Vol.2, Issue.2, ISSN:2278-0181

[3] J Rodríguez, Jih-Sheng Lai, F Zheng Peng," Multilevel Inverters: A Survey of Topologies, Controls, and Applications," *IEEE Transactions on Industrial Electronics*, Vol. 49, No.4, August 2002.

[4] Surin Khomfoi and Leon M. Tolbert, "Chapter 31_Multilevel power converters," University of Tennessee.

[5] E. Cengelci, S. U. Sulistijo, B. O. Woom, P. Enjeti, R. Teodorescu, and F. Blaabjerg, "A New Medium Voltage PWM Inverter Topology for Adjustable Speed Drives," *Conference Recordings IEEE-IAS Annual Meeting*, St. Louis, MO, Oct. 1998, pp. 1416-1423.

[6] Muhammad H. Rashid, Power Electronics Handbook, 2nd Edition, Academic Press imprint of Elsevier, pp. 451-455, 2007.

[7] Muhammad H. Rashid, Power Electronics- Circuits, Devices, and Applications, 3rd Edition, *PHI Learning Private Limited*, 2009.

[8] C. R. Balamurugan, S. P. Natarajan, R. Bensraj, "Investigations on Three Phase Five Level Diode Clamped Multilevel Inverter," *IJMER*, Vol. 2, Issue. 3, May-June 2012, pp. 1273-1279, ISSN: 2249-6645.

[9] J. S. Lai and F. Z. Peng, "Multilevel Converters-A new Breed of Power Converters," *IEEE Transactions on Industrial Application.*, Vol.32, pp. 509-517, May/June 1996.

[10] V. G. Agelidis et. al, "A Multilevel PWM Inverter Topology for Photovoltaic Applications," *International Symposium on Industrial Electronics*, 1997.

4H-SiC-Dopant Segregated Schottky Barrier Cylindrical Gate All Around MOSFET for High Speed and High Power Microwave Applications

Manoj Kumar, Mridula Gupta
Semiconductor Device Research
Laboratory
Department of Electronic Science
University of Delhi South Campus
New Delhi 110021, India
manoj.uiet@gmail.com

Subhasis Haldar
Department of Physics
Motilal Nehru College, University
of Delhi
New Delhi 110021, India
subhasis_haldar@rediffmail.com

R.S. Gupta
Department of ECE
Maharaja Agrasen Institute of
Technology
Rohini, Delhi 110086, India
rsgupta1943@gmail.com

Abstract— **This paper presents extensive study of proposed 4H-Silicon Carbide (SiC) based Dopant Segregated (DS) Schottky Barrier (SB) Cylindrical Gate All Around (CGAA) MOSFET with high-k gate dielectric for hostile environment such as high temperature and radiation exposure for high power microwave applications. The Analog/RF performance of SiC-CGAA has been investigated by exploiting temperature variation from 300 K to 500 K, along with the leverage of Schottky-Barrier (SB) Source/Drain and Dopant Segregation (DS). DS is primarily used to reduce the ambipolar behavior of SB MOSFET and boost the performance of the device. Analog/RF performance in terms of main figure of merits like I_{on}/I_{off}, transconductance (g_m), Early Voltage, Stern Stability Factor (K), cut-off frequency (f_t) has been carried out using ATLAS-3D device simulator.**

Keywords—Dopant Segregated; Gate All Around; High temperature; Hostile Environment; Schottky Barrier MOSFET; Stern Stability Factor.

I. INTRODUCTION

Ultra scaled Silicon based MOS device requires a smarter cooling system that can assure a longer life to the device but in some cases air cooling is not viable. Silicon Carbide (SiC) is a nostrum semiconductor material with wide energy band gap ($2.0 \text{ eV} \leq E_g \leq 7.0 \text{ eV}$), for solid-state electronic devices at elevated temperature owing to its excellent electrical and mechanical properties. It gained enormous attraction because of its inherent material advantages e.g. high temperature, high power, high frequency, radiation hardened applications. The semiconductor devices are exposed to very high range of temperature variations when used in the area of space explorations, the geothermal and mining exploration, hybrid electric vehicles (HEV), nuclear power plant, aircraft industries, aeronautics application and industrial process control [1]-[2]. SiC based metal-semiconductor FET has become an advantageous technology for high-power microwave applications, such as the transmitters for the commercial and military communications [3]. SiC is prevailing wide band gap material with variety of features like high electric breakdown field (3.5×10^6 V/cm), high electron saturation drift velocity (2.0×10^7 cm/s), high melting point (2830 °C) and high thermal conductivity (4.9 W/cmK) [1].

The shrinking dimensions of the planer MOS devices enhance the device performance along with the drawback of SCEs which deteriorate the normal device working. SCEs are prominent in sub-45 nm technology node so various multiple gate structures have been explored due to their better controllability. Cylindrical Gate All Around (GAA) MOSFET [4-7] is one of the most promising candidate for sub 45-nm technology regime providing excellent control over the channel electrostatic potential due to its symmetric structure. But with the miniaturization of device, the problem of heat removal at high frequency and wireless systems has become an issue. Furthermore as the device is scaled down to nanoscale regime high series source/drain contact resistance due to the formation of abrupt source/drain junctions become a major problem. Novel solutions such as Schottky-Barrier (metal) source/drain and Junctionless (JL) MOSFETs [8-11] have therefore been proposed and extensively investigated. The JL MOSFET possesses uniformly heavily doped source, drain and channel regions, thus having no p-n junction formation between source/drain and the channel. However, the high doping concentration in the channel reduces carrier mobility, which affects drive current and transconductance of junctionless MOSFETs. Due to these drawbacks doped source/drain regions are replaced by Schottky-Barrier metal source/drain (MSD) which offers higher on-current and improved transconductance of JL MOSFETs. SB-GAA MOSFET is also advantageous from fabrication point of view as the problem of maintaining homogeneous doping distribution in the channel region does not arise [12-16]. The ultra shallow source/drain SB junctions offer low S/D interface contact resistivity ($\rho c \sim 10^{-9}$ $\Omega\cdot$cm) for metallic S/D compared with ($\rho c \sim 10^{-7}$ $\Omega\cdot$cm) doped S/D [17-18] junction, thereby minimizing the short channel effects, increasing on current and improving high frequency response. However an inferior switching behavior with poor sub-threshold slope is observed in Schottky Barrier MOSFET. An electrical characteristic of a SB device depends on the tunneling probability of carriers through the SB, which degrades the performance of SB MOS device. Hence to increase the tunneling probability of carriers through the Schottky barrier, Dopant Segregation (DS) [19] technique has widely been used. Highly doped semiconductor portions either n^+ or p^+ is placed at the interface between source contact and channel as well as between drain contact and channel electrodes [20-25]. These highly doped semiconductor portions cause a strong band bending near the source and drain contacts and improves the tunneling probability. In this paper 4H-SiC-DS-SB-GAA device with high-k gate dielectric is proposed and investigated for the first time.

II. DEVICE STRUCTURE AND DISCRIPTION

Figure 1 (a) and 1 (b) shows the three dimensional and cross sectional view of 4H-SiC-DS-SB-GAA MOSFET respectively. The conventional 4H-SiC-SB-GAA contains no dopant segregation region at metal semiconductor interface under the gate electrode. Three-dimensional simulation of both devices has been carried out using ATLAS-3D device simulator [26] activating doping dependence SRH recombination/thermal generation model, FLDMOB model for high electric field velocity saturation, CONMOB model for concentration dependent mobility and Boltzmann carrier statistics for electron and holes. Temperature dependent SELB model for impact ionization has been used. CVT model including the transverse field, doping dependent and temperature dependent mobility has also been used. Newton-Gummel method has been adopted for numerical solution. The quantum effect has not been taken in to consideration in present analysis. For high-k DS device 1 nm SiO_2 and 2 nm HfO_2 has been considered for analysis in complete work.

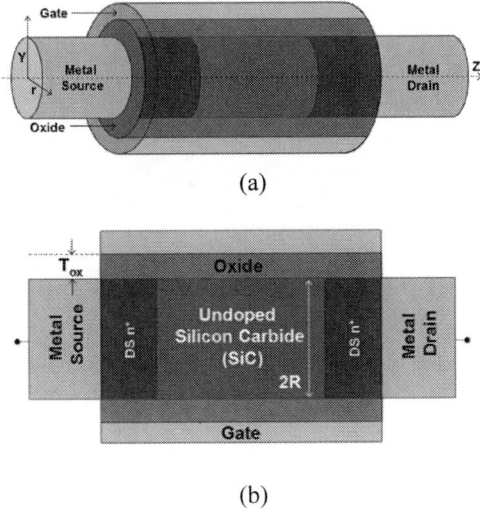

(a)

(b)

Fig. 1(a). 3D view of 4H-SiC-DS-SB-GAA MOSFET. Fig. 1(b). Cross sectional view of 4H-SiC-DS-SB-GAA MOSFET. The device parameters are $L_g = 20$ nm , $R = 5$ nm, $T_{ox} = 3$ nm, $N_{seg} = 10^{20}$ cm^{-3}, $L_{seg} = 3$ nm and $\Phi_{S/D} = 4.1$ eV. These are the default values of device parameters, unless otherwise stated.

III. RESULT AND DISSCUSION

Figure 2. (a) shows the variation of drain current as a function of applied gate voltage at $V_{DS} = 2.0$ V on linear as well as logarithmic scale.

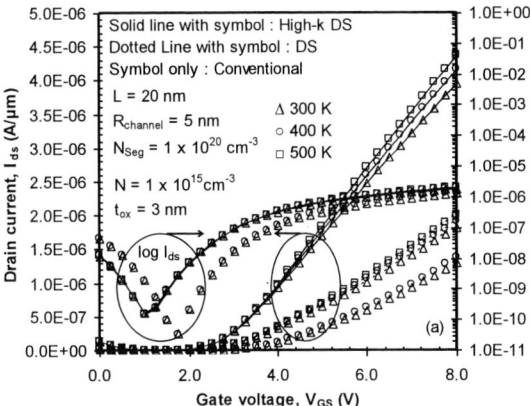

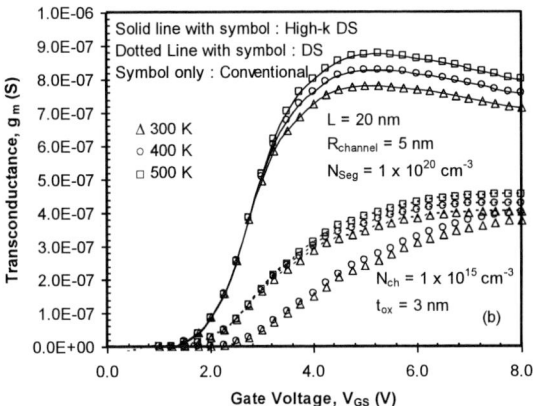

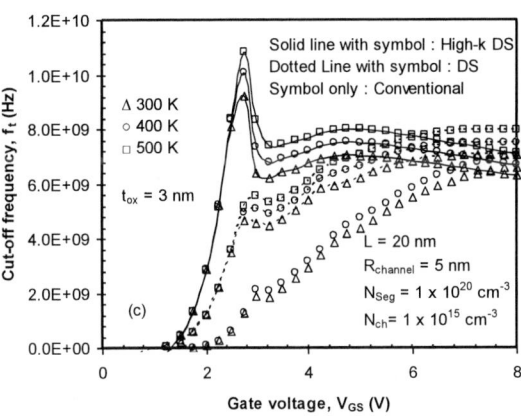

Fig. 2 (a). Variation of drain current at logarithmic scale for DS-4H-SiC-GAA and High-k-DS-4H-SiC-GAA MOSFET devices, Fig. 2 (b). variation of transconductance with V_{GS} for both the devices at different temperatures, Fig. 2(c). variation of cut-off frequency, with applied gate voltage for both GAA devices at different temperatures, at applied drain bias of $V_{DS} = 2.0$ V.

978-1-4799-6047-7/14 $31.00 © 2014 IEEE   647

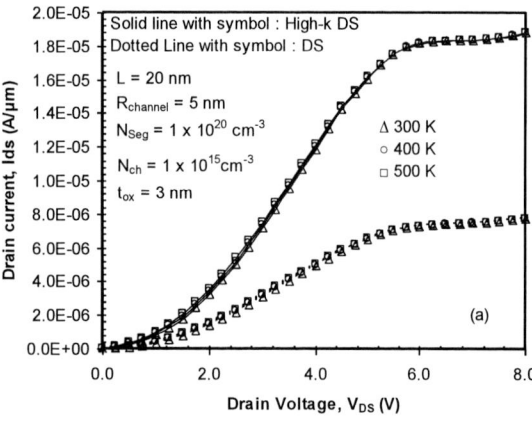

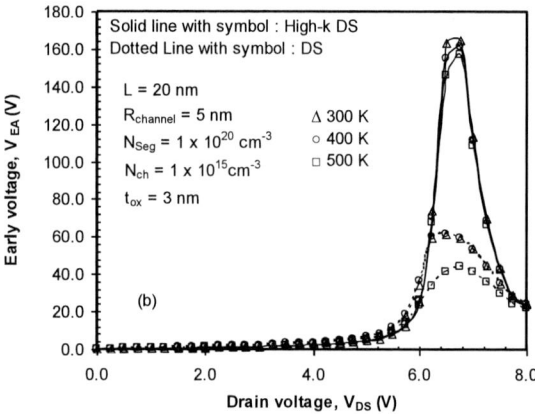

Fig. 3 (a). Variation of drain current applied drain voltage V_{DS}, Fig. 3(b). Variation of Early Voltage, as a function of V_{DS}. At fixed V_{GS} = 7 V.

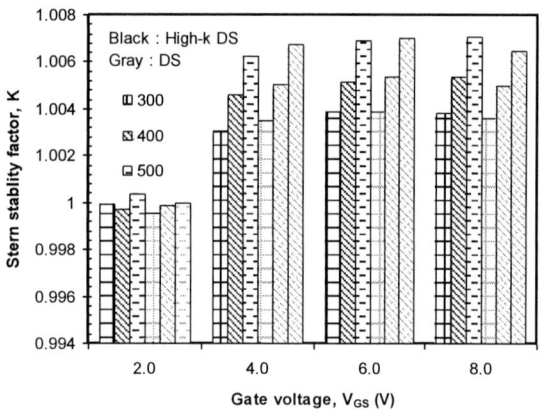

Fig. 4. Variation of Stern Stability Factor as a function of applied gate voltage V_{GS}.

The large ambipolar current can be clearly observed from the figure in conventional SiC-SB-GAA MOSFET. Using the DS technique the ambipolar current is reduced. Further at higher temperature I_{ds} in conventional SB GAA device is much lower than the DS-GAA device. Comparatively for high-k-4H-SiC-DS-SB-GAA device I_{ds} is much higher even at higher temperature and shows less variations. The high-k dielectric material provides increased channel capacitance with the same dielectric thickness equivalent to SiO_2 and reconcile the gate leakage current in highly scaled devices. A 4H-SiC based DS-SB-GAA also offer Subthreshold Swing (SS) of 59 mV/decade which is lower than the fundamental limit (60 mV/decade) of SS, hence the device performs relatively faster even at high voltage and high power. I_{on}/I_{off} ratio of high-k-4H-SiC-DS-SB-GAA is higher than the conventional SB-GAA MOSFET. The transconductance of any device decides the gain of the device and its optimum bias point should be high enough [27] to have high cut off frequency. Figure 2. (b) and (c) show variation of g_m and f_t respectively as a function of applied gate voltage. Figure 3 (a). shows the drain current as a function of applied drain bias and it depicts that there is no variation at higher temperature. The Early Voltage (V_{EA}) of the device is directly related to the low frequency open loop gain ($A_V = g_m V_{EA}/I_{ds}$) of the device. At higher temperatures almost same V_{EA} is obtained for high-k-4H-SiC-DS-SB-GAA device and higher than 4H-SiC-DS-SB-GAA thus the same gain can be achieved even at higher temperatures. Stability of the device is also an important parameter at high temperatures and can be observed by the stern stability factor K. The value of K must be greater than 1 for the stability of the device between source and load impedance. Figure 4. clearly shows that stern stability factor is always greater than 1. At higher temperatures particularly at 500 degree Kelvin, the stern stability factor is much higher than 1, so device is more stable even at higher temperature. Hence these properties make the High-k-4H-SiC-DS-SB-GAA MOSFET suitable for high frequency harsh environment application.

CONCLUSION

The 4H-SiC-Dopant Segregated-Schottky Barrier-GAA MOSFET with high-k has been extensively investigated at high temperature (300-500 K), harsh environment and high frequency operating range. Analog/RF performance of the device has been studied at 20 nm gate length and compared with the conventional SB GAA and DS-SB GAA MOSFET devices. The results revel that the high-k-4H-SiC-DS-SB GAA MOSFET has better performance in terms of drain current (I_{ds}), transconductance (g_m), cut-off frequency (f_t), early voltage (V_{EA}) at high voltage and elevated temperatures. So the device is suitable for high power and high frequency harsh environment applications with reduced ambipolar behavior because of Dopant Segregation technique yielding increase in carriers tunneling.

ACKNOWLEDGMENT

Authors are grateful to DRDO, Government of India. One of the authors (Manoj Kumar) is thankful to CSIR for providing financial assistance to carry out this research work.

REFERENCES

[1] M. R. Werner, and W. R Fahrner "Review on Materials, Microsenser, Systems, and Devices for High-Temperature and harsh-Environment Applications," IEEE Trans. on Industrial Electronics, vol. 48, no. 2, pp. 249-257, April 2001.

[2] M. Golio, RF and Microwave Semiconductor Devices handbook, New York: CRC Press, 2003.

[3] W. Choi, J. Lee, and M. Shin," p-type nanowire Schottky Barrier MOSFETs Comparative study of Ge and Si-Channel devices," IEEE Trans. on Electron devices, vol. 61, no. 1, pp. 37-43, January 2014.

[4] R. Wang, J. Zhuge, R. Huang, Y. Tian, H. Xiao, L. Zhang, C. Li, Zhang and Y. Wang, "Analog/RF Performance of Si Nanowire MOSFETs and the Impact of Process Variation, " IEEE Trans. ELECTRON DEVICES, VOL. 54, NO.6, PP. 1288-1294, JUNE 2007.

[5] J. Zhuge, R. Wang, R. Huang, X. Zhang and Y. Wang, "Investigation of Parasitic Effects and Design Optimization in Silicon Nanowire MOSFETs for RF Applications, " IEEE Trans. Electron Devices, vol. 55, no.8, pp. 2142-2147, Aug. 2008.

[6] Y. Li and C. H. Hwang, "The effect of the geometry aspect ratio on the silicon ellipse-shaped surrounding- gate field-effect transistor and circuit," Semicond. Sci. and Technol., vol. 24, no.9, Sept. 2009.

[7] S. Cho, K. R. Kim, Park, B. Gook and I. M. Kang, "RF Performance and Small-Signal Parameter Extraction of Junctionless Silicon Nanowire MOSFETs, " IEEE Trans. Electron Devices, vol. 58, no.5, pp. 1388-1396, May 2011.

[8] C. H. Shih, J. T. Liang, J. S. Wang and D. Chien, "A Source-Side Injection Lucky Electron Model for Schottky barrier Metal-Oxide-Semiconductor Devices," IEEE Electron Device Letters, vol. 32, no. 10, pp. 1331-1333, October 2011.

[9] X. Qian, Y. Yang,A. Zhu, S.L. Zang and D. Wu, "Evaluation of DC and AC performance of junctionless MOSFET in the presence of variability," IEEE International Conference on IC Design and Technology (ICICDT), 2011.

[10] R. K. Baruah, and R. P. Paily, " A dual-Material gate Junctionless Transistor with High-k spacer for Enhanced Analog performance," IEEE Trans. Electron Devices, vol. 61, no. 1, pp. 123-128, January 2014.

[11] J. Coling, C.Lee, A. Afzalian, N. Akhavan, R.Yan, I. Ferain, P. Razavi, B.O. Neill, A.Blake, M. White, A. M. Kelleher, B. M. Carthy, and R. Murphy, " Nanowire transistor without junctions," Nat. Nanotechnol., vol. 5, no. 3, pp. 225-229, March 2010.

[12] Z. X. Jun, Y. Y. Tang, and D. B. Xing, C. C. Chun, S. Kun , and Chen Bin," Drain-induced barrier lowering effect for short channel dual material gate 4H silicon carbide metal semiconductor field effect transistor," IOP Sci. Tech., vol. 21, no. 9, 2012.

[13] R. Rios, A. Cappellani, M. Armstrong, A. Budrevich, H. Gomez, R. Pai, N. R. Orabi, and K. Kuhn, "Comparison of junctionless and conventional trigate transistors with down to 26 nm," IEEE Electron Device Lett., vol. 32, no. 9, pp. 1170-1172, September. 2011.

[14] C. J. Koeneke, S. M. Sze, R. M. Levin and E. Kinsbron, "Schottky MOSFET for VLSI," International Electron Device Meeting, pp. 367-371, 7-9 Dec. 1981.

[15] B. Y. Tusi, C. P. Lu and H. H. Liu, "Method for extracting gate-Voltage-Dependent Source Injection Resistance of Modified Schottky Barrier (MSB) MOSFETs," IEEE Electron Device Letters, vol. 29, no. 9, pp. 1053-1055, September 2008.

[16] M. K. Husain, X. V. Li, and C. H. d. Groot, "High-quality Schottky Contact for limiting Leakage Currents in Ge-Based Schottky Barrier MOSFETs, " IEEE Trans. Electron Devices, vol. 56, no. 3, pp. 499-504, March 2009.

[17] S. J. Choi and Y.K. Choi, Source and Drain Junction Engineering for Enhanced Non-Volatile Memory Performance, Flash Memories, Prof. Igor Stievano (Ed.), ISBN: 978-953-307-272-2, InTech, 2011.

[18] B. Liu, C. Zhan, Y. Yang, R. Cheng, P. Guo, Q. Zhou, Y. J. Kong, N. Daval, C. Veytizou, D. Delprat, B. Y. Nguyen and Y. C. Yeo, "Germanium Multiple Gate Field-Effect Transistor with In Situ Boron Doped Raised Source/Drain," IEEE Trans. Electron Devices, vol. 60, no.7, pp. 2135-2141, July 2013.

[19] G. C. Patil, and S. Qureshi," Engineering spacers in dopant segregated Schottky Barrier SOI MOSFET for nanoscale CMOS logic circuits ," IOP Sci. Tech., vol. 27, no. 4, April 2012.

[20] G. Larrieu and E. Dubois, "CMOS Inverter Based on Schottky Source–Drain MOS Technology With Low-Temperature Dopant Segregation," IEEE Electron Device Lett., vol. 32, no. 6, pp. 728–730, June 2011.

[21] S.F. Feste, J. Knoch, D. Buca, Q.T. Zhao, U. Breuer and S. Mantl, " Formation of steep, low Schottky-barrier contacts by dopant segregation during nickel silicidation," J. Appl. Phys., 107, 044510 (2010).

[22] R. A. Vega and T.-J. K. Liu, "Dopant-segregated Schottky source/drain double-gate MOSFET design in the direct source-to-drain tunneling regime," IEEE Transactions on Electron Devices, vol. 56, no. 9, pp. 2016-2026, Sept. 2009.

[23] R. Valentin, E. Dubois, G. Larrieu, J.-P. Raskin, G. Dambrine, N. Breil and F. Danneville," Optimization of RF performance of metallic source/drain SOI MOSFETs using dopant segregation at the Schottky interface," IEEE Electron Device Lett., vol. 30, no. 11, pp. 1197-1199, Nov. 2009.

[24] E. Pascual, M. J. Martín, R. Rengel, G. Larrieu and E. Dubois, " Enhanced carrier injection in Schottky contacts using dopant segregation: a Monte Carlo research," IOP Sci. Tech., vol. 24, no. 2, 2009.

[25] L. Zeng, X. Y. Liu, Y. N. Zhao, Y. H. He, G. Du, J. F. Kang , R. Q. Han, "A Computational Study of Dopant-Segregated Schottky Barrier MOSFETs," IEEE Trans Nanotechnol., vol. 9, no. 1, pp. 108–113, Jan. 2010.

[26] ATLAS User's Manual: 3-D device Simulator, Silvaco Inc., Santa Clara, CA, USA, 2012.

[27] Y. Pratap, S. Haldar, R. S. Gupta, and M. Gupta, "Performance Evaluation and Reliability Issues of Junctionless CSG MOSFET for RFIC Design," IEEE Transactions on Device and Materials Reliability, vol. 14, Issue 1, pp. 418-425, March 2014.

978-1-4799-6047-7/14 $31.00 © 2014 IEEE

A New Technique to Implement Conventional as well as Advanced Pulse Width Modulation Techniques for Multi-level Inverter

Debanjan Roy
PG Scholar: School of Electrical Engineering
KIIT University
Bhubaneswar, India
debanjanroy88@gmail.com

Tapas Roy
Assistant Professor: School of Electrical Engineering
KIIT University
Bhubaneswar, India
tapas18roy@gmail.com

Abstract—**Depending on pole voltage levels, the inverters are broadly classified into two categories two level and above two-level. Above two-level is popularly known as multilevel inverters. Multilevel inverters have better performances compared to conventional two level inverters like minimum harmonic distortion, reduced electromagnetic interferences (EMI) and operation on several voltage levels. There exist different conventional as well as advanced PWM techniques to switch multilevel inverters. Phase Disposition PWM (PDPWM), Selective Harmonic Elimination (SHE), Space- Vector PWM (SVPWM) are most popular conventional PWM techniques whereas Bus-clamping PWM (BCPWM) techniques are one of the most popular advanced type PWM techniques. BCPWM techniques are better compare to CSVPWM technique in respect of harmonic distortion, voltage stress across switch and switching loss. In this paper, a new technique has been introduced to implement CSVPWM as well as BCPWM techniques for a 3-level inverter. The proposed technique is simple to understand and no need of mapping as that is required for conventional techniques. The performance of the inverter is analyzed using R-L load and also with three phase induction motor load. The simulation results have shown the effectiveness of the proposed technique. Further the same technique can be employed for higher level inverters.**

Index Terms—Bus Clamping PWM (BCPWM), Conventional Space Vector PWM (CSVPWM), Harmonic distortion, Multi-level Inverter.

I. INTRODUCTION

Nowadays, multilevel inverter becomes very popular in the application of motor drives, electric vehicle, uninterruptible power supplies (UPS), FACTs devices, active filters etc. [1-5]. As compared to conventional two-level inverter, multilevel inverter performs better. As the level increases, the harmonic spectrum of output waveforms of inverter improves. Further higher level inverters can be applicable for higher power rating [1-10].

There exist different types of conventional as well as advanced Pulse Width Modulation (PWM) techniques for multilevel inverters [4-10]. The performance of the inverter depends on what type of PWM technique is used to switch the power switching devices. Phase Disposition PWM (PDPWM),

Selective Harmonic Elimination PWM (SHEPWM) and Conventional Space Vector PWM (CSVPWM) are most popular conventional PWM strategies for multilevel inverters [4-5]. Bus Clamping PWM (BCPWM) technique is one of the most popular advanced PWM technique for multilevel inverters [5-10]. The BCPWM technique is better compared to conventional PWM techniques in respect of output current ripple, switching loss and voltage stress across switch [8].

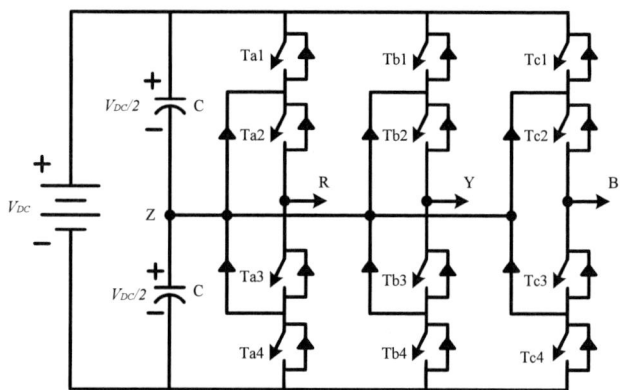

Fig 1. Three phase three-level diode clamped voltage source inverter

The power circuit for a three phase three-level diode clamped or neutral point clamped inverter is shown in Fig. 1. Each leg consists of four power switching devices like IGBT, MOSFETS etc. with anti-parallel diode. The PWM pulses are given to the power switches in such way that the pole voltage consists of three levels $\left(V_{DC}/2, 0, -V_{DC}/2 \right)$ [4-10].

There are different techniques to implement conventional as well advanced PWM techniques for three-level inverter. Those techniques are based on mapping concept [6-10]. In this paper a new technique is introduced to implement CSVPWM and BCPWM techniques for three-level inverter. The proposed technique is based on algebraic equations. It is easy to understand as well as simple to implement.

The effectiveness of the proposed technique is achieved by proper simulation of three-level inverter in MATLAB. It is observed that the BCPWM technique, based on proposed

978-1-4799-6047-7/14 $31.00 © 2014 IEEE

technique has all the advantages compared to CSVPWM technique. Further, for higher (above three) level inverter, this technique is also applicable.

II. CSVPWM TECHNIQUE FOR THREE-LEVEL INVERTER

The hexagon formed by space vectors for three-level inverter is shown in Fig. 2 [**6-10**]. It consists of 27 number of space vectors. There are four kinds of space vectors namely large, medium, small and zero. There are six number of large, six number of medium, three number of zero and 12 number of small space vectors in the hexagon. Out of twelve numbers of small space vectors, six are known as redundancy space – vectors. The hexagon is divided into six sectors of equal duration of 60° which is shown in Fig. 2. Each sector has four regions.

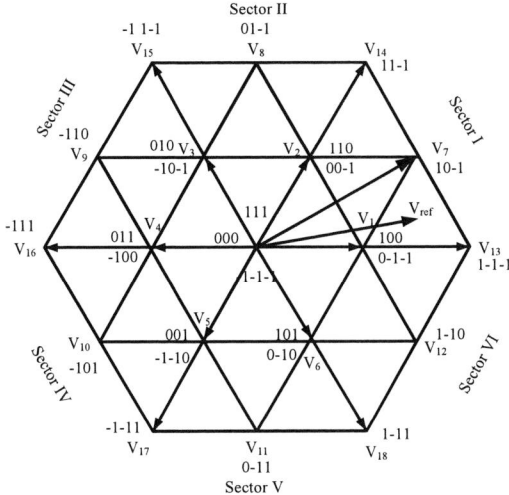

Fig. 2. Division of sectors and regions for three-level inverter

So for generating reference vector the sector, region and vector sequence selection are very important. After sector selection, in which region the reference vector is located is needed to be find out. After region selection, the reference vector is generated by applying nearby space vectors.

III. BCPWM FOR THREE-LEVEL INVERTER

Bus clamped PWM technique is one of the most popular advanced type of PWM technique. In this technique, each phase of inverter is clamped to one of the DC bus terminals for certain duration over the fundamental cycle. During the clamping duration, the clamped phase is not switching but the other two phases keep switching.

Based on the region of clamping four types of bus clamped PWM namely TYPE1, TYPEII, TYPEIII and TYPE IV can be defined [**8**]. The clamping region and the clamping phase for TYPE II BCPWM are shown in Fig. 3. In each sector, there has a clamped phase. So clamping duration is 60°. This BCPWM is popularly known as 60° BCPWM technique.

In sector 1, R-phase gets clamped to positive DC bus whereas in sector 4, R-phase is clamped to negative DC bus. So over a fundamental cycle, the total clamping duration is 120°. It has been observed that at high modulation index the

inverter performs better in BCPWM compared to conventional PWM techniques [**8**].

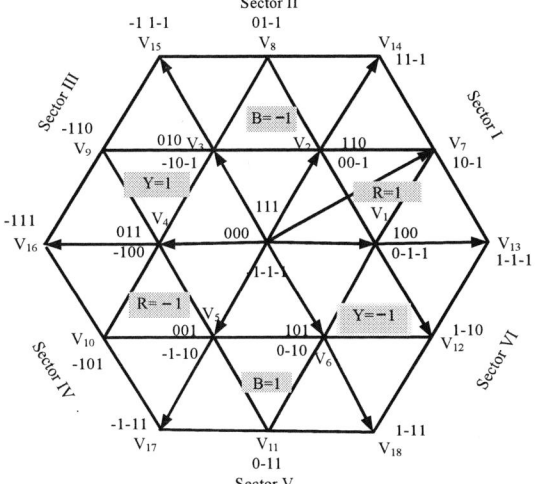

Fig. 3. TYPE II clamping

At same carrier frequency, the switching loss of inverter is less for BCPWM as compared to CSVPWM whereas at same switching frequency and high modulation index, the output current ripple is significantly less in BCPWM compared to conventional PWM techniques [**8**]. In this paper, TYPE II BCPWM is selected for analysis.

IV. PROPOSED REGION SELECTION TECHNIQUE

For generating the reference vector, the region in which the reference vector is located needs to find out. A new technique based on algebraic equations is proposed to find out the region in any sector. Fig. 4 shows two consecutive sectors (sector 1 and sector 2). As the hexagon is symmetric, the procedure for region selection in one sector is applicable to other sectors.

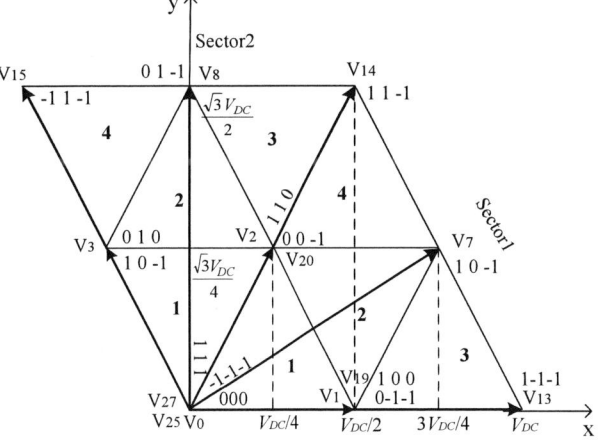

Fig. 4. Determine the regions

Consider zero vectors (V_0, V_{25}, V_{27}) as origin, the co-ordinates for other vectors in a sector can be found in x-y plane as shown Fig. 4. The x coordinate is in phase with the starting vector of any sector and y coordinate is perpendicular to the x coordinate. Theta (θ) is the angle between the reference vector and the starting vector of any sector. The co-ordinates for different vectors in sector 1 are as follows-

978-1-4799-6047-7/14 $31.00 © 2014 IEEE 651

Table 1: Coordinates of different vectors in sector 1

Vector	Name	Co-ordinate (X,Y)
V_0, V_{27}, V_{25}	Zero	$(0,0)$
V_1, V_{19}	Small	$\left(V_{DC}/2, 0\right)$
V_{13}	Large	$\left(V_{DC}, 0\right)$
V_7	Medium	$\left(3V_{DC}/4, 0\right)$
V_{14}	Large	$\left(V_{DC}/2, \sqrt{3}V_{DC}/2\right)$
V_2, V_{20}	Small	$\left(V_{DC}/4, \sqrt{3}V_{DC}/4\right)$

For region selection, two algebraic equations are needed to form. One equation is for straight line between vectors V_1 and V_2. Other equation is for straight line between vectors V_1 and V_7. The equations are as follows-

$$y = -\sqrt{3}\left(x - \frac{V_{DC}}{2}\right) \tag{1}$$

$$y = \sqrt{3}\left(x - \frac{V_{DC}}{2}\right) \tag{2}$$

The reference vector V_{ref} has two components –

$$x_1 = |V_{ref}|Cos\theta \tag{3}$$

$$y_1 = |V_{ref}|Sin\theta \tag{4}$$

Now by using (1), (2) and (3), the y components corresponding x_1 can be found out for region selection. By putting (3) in (1) and (2), the following y component values are achieved.

$$y_{11} = -\sqrt{3}\left(x_1 - \frac{V_{DC}}{2}\right) \tag{5}$$

$$y_{22} = \sqrt{3}\left(x_1 - \frac{V_{DC}}{2}\right) \tag{6}$$

Now by comparing the above coordinates, it can be found out in which region the reference vector is located. The conditions for region selection are tabulated in Table 2.

Table 2: Conditions for region selection in sector 1

R	Conditions
1	$0 < x_1 \leq V_{DC}/4$ and $0 \leq \theta \leq 60°$ Or $V_{DC}/4 < x_1 < V_{DC}/2$ and $y_1 \leq y_{11}$
2	$V_{DC}/4 < x_1 \leq V_{DC}/2$ and $y_{11} < y_1 \leq \sqrt{3}V_{DC}/4$ Or $V_{DC}/2 < x_1 \leq 3V_{DC}/4$ and $y_{22} < y_1 \leq \sqrt{3}V_{DC}/4$
3	$V_{DC}/2 < x_1 \leq 3V_{DC}/4$ and $y_1 \leq y_{22}$ Or $3V_{DC}/4 < x_1 \leq V_{DC}$ and $0 \leq \theta \leq 30°$
4	Conditions for regions 1 , 2 and 3 are not satisfied then region 4 will be selected

For other sector, the region selection is same as that for sector 1 expect the x-y coordinate shifted by 60°.

V. Vector Sequence And Dwell Time Calculation For CSVPWM

For generating the reference vector in any region of any sector the following two conditions should be satisfied –

1. Only one switch is switched during state transition. That is transition from state 1 to state -1 and vice-versa is not allowed.
2. Final state of the present sample is the first state of the next sample.

The vector sequence in each region is selected in such way that the above two condition satisfy. The vector sequences for different regions in sector 1 are tabulated in Table 3.

Table 3: Vector Sequence for different region in sector 1

R	Vector Sequence
1	$V_{27} \Leftrightarrow V_{19} \Leftrightarrow V_{20} \Leftrightarrow V_0 \Leftrightarrow V_1 \Leftrightarrow V_2 \Leftrightarrow V_{25}$ (-1-1-1) (0-1-1) (00-1) (000) (100) (110) (111)
2	$V_{19} \Leftrightarrow V_{20} \Leftrightarrow V_7 \Leftrightarrow V_1 \Leftrightarrow V_2$ (0-1-1) (00-1) (10-1) (100) (110)
3	$V_{19} \Leftrightarrow V_{13} \Leftrightarrow V_7 \Leftrightarrow V_1$ (0-1-1) (1-1-1) (10-1) (100)
4	$V_{20} \Leftrightarrow V_7 \Leftrightarrow V_{14} \Leftrightarrow V_2$ (00-1) (10-1) (11-1) (110)

By volt-see balance the dwell times are calculated and tabulated in Table 4 for sector1

Table 4: Dwell Times for sector 1

R	Vectors	Dwell Time
1	V_1	$T_a = T_S\left\{(4/\sqrt{3})mSin(\pi/3-\theta)\right\}$
	$V_0\ V_{27}\ V_{25}$	$T_b = T_S\left\{1-(4/\sqrt{3})mSin(\pi/3+\theta)\right\}$
	V_2	$T_c = T_S\left\{(4/\sqrt{3})mSin\theta\right\}$
2	$V_1\ V_{19}$	$T_a = T_S\left\{1-(4/\sqrt{3})mSin\theta\right\}$
	V_7	$T_b = T_S\left\{(4/\sqrt{3})mSin(\pi/3+\theta)-1\right\}$
	$V_2\ V_{20}$	$T_c = T_S\left\{1-(4/\sqrt{3})mSin(\pi/3-\theta)\right\}$
3	$V_1\ V_{19}$	$T_a = T_S\left\{2-(4/\sqrt{3})mSin(\pi/3+\theta)\right\}$
	V_7	$T_b = T_S\left\{(4/\sqrt{3})mSin\theta\right\}$
	V_{13}	$T_c = T_S\left\{(4/\sqrt{3})mSin(\pi/3-\theta)-1\right\}$
4	V_{14}	$T_a = T_S\left\{(4/\sqrt{3})mSin\theta-1\right\}$
	V_7	$T_b = T_S\left\{(4/\sqrt{3})mSin(\pi/3-\theta)\right\}$
	$V_2\ V_{20}$	$T_c = T_S\left\{2-(4/\sqrt{3})mSin(\pi/3+\theta)\right\}$

Where T_S = Sample Time $= T_a + T_b + T_c$

m = Modulation Index $= V_{ref}/V_{DC}$

Fig. 5 shows the timing diagram for region 1 in sector 1. It is observed that at a time only one switch is switched for each transition. The timing diagrams for other regions in sector 1 are as per Table 3.

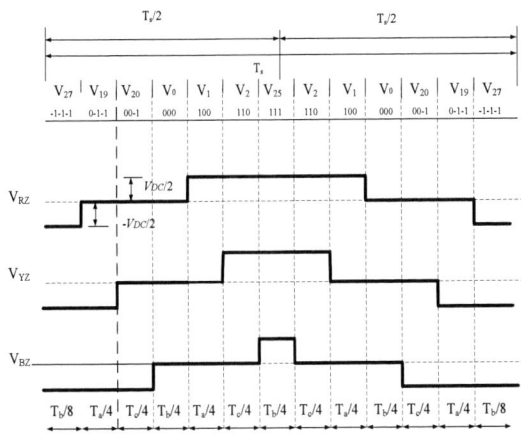

Fig. 5. Timing diagram of vector sequence for region 1 in sector 1 for CSVPWM

VI. VECTOR SEQUENCE AND DWELL TIME CALCULATION FOR BCPWM

The vector sequence in each region is selected in such way that the above two conditions satisfy. The vector sequences for different regions in sector 1 are tabulated in Table 5.

Table 5: Vector Sequence for different region in sector 1

R	Vector Sequence	Clamped Phase
1	$V_1 \Leftrightarrow V_2 \Leftrightarrow V_{25}$ (100)　　(110)　　(111)	+R
2	$V_2 \Leftrightarrow V_1 \Leftrightarrow V_7$ (110)　(100)　(10-1)	+R
3	$V_1 \Leftrightarrow V_7 \Leftrightarrow V_{13}$ (100)　(10-1)　(1-1-1)	+R
4	$V_2 \Leftrightarrow V_{14} \Leftrightarrow V_7$ (110)　(11-1)　(10-1)	+R

Table 6: Dwell Times for sector 1

R	Vectors	Dwell Time
1	V_1	$T_a = T_S\left\{(4/\sqrt{3})mSin(\pi/3-\theta)\right\}$
	V_0	$T_b = T_S\left\{1-(4/\sqrt{3})mSin(\pi/3+\theta)\right\}$
	V_2	$T_c = T_S\left\{(4/\sqrt{3})mSin\theta\right\}$
2	V_1	$T_a = T_S\left\{1-(4/\sqrt{3})mSin\theta\right\}$
	V_7	$T_b = T_S\left\{(4/\sqrt{3})mSin(\pi/3+\theta)-1\right\}$
	V_2	$T_c = T_S\left\{1-(4/\sqrt{3})mSin(\pi/3-\theta)\right\}$
3	V_1	$T_a = T_S\left\{2-(4/\sqrt{3})mSin(\pi/3+\theta)\right\}$
	V_7	$T_b = T_S\left\{(4/\sqrt{3})mSin\theta\right\}$
	V_{13}	$T_c = T_S\left\{(4/\sqrt{3})mSin(\pi/3-\theta)-1\right\}$
4	V_{14}	$T_a = T_S\left\{(4/\sqrt{3})mSin\theta-1\right\}$
	V_7	$T_b = T_S\left\{(4/\sqrt{3})mSin(\pi/3-\theta)\right\}$
	V_2	$T_c = T_S\left\{2-(4/\sqrt{3})mSin(\pi/3+\theta)\right\}$

Fig. 6(a) and 6(b) show the timing diagrams of vector sequence for region 1 and region 4 of sector 1. It is observed

that R phase gets clamped to positive DC bus terminal. In similar fashion the other sectors can be analyzed

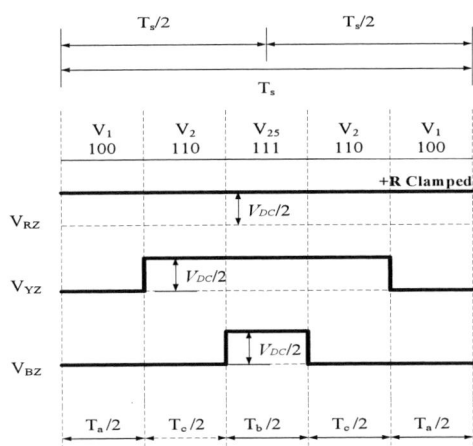

Fig. 6 (a). Timing diagram of vector sequence for region 1 in sector 1 for BCPWM

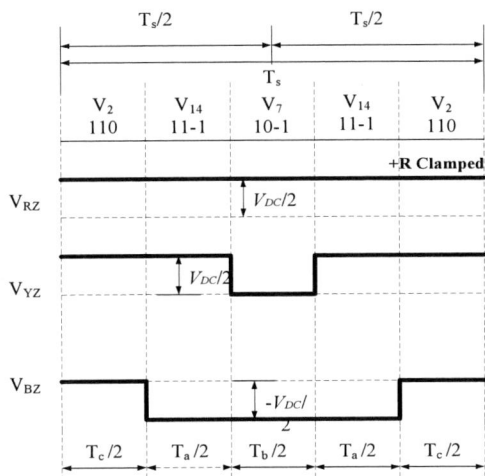

Fig. 6 (b). Timing diagram of vector sequence for region 4 in sector 1 for BCPWM

VII. SIMULATION RESULTS AND COMPARISON STUDY

The algorithms using proposed region selection technique for SVPWM and BCPWM for three-level inverter are verified through simulation. The simulation studies have been carried out on RL as well as three phase induction motor load.

A. Comparison study for R-L load

The inverter is modeled using MATLAB/SIMULINK. The PDPWM is generated by using proper PWM block in SIMULINK. The simulation is done by considering the following conditions-
1. DC bus voltage, $V_{DC} = 400$ V
2. Modulation index, $m = 0.8$
3. R=26.5Ω and L=41mH
4. Switching frequency, $f_{sw} = 10$ kHz

All simulation parameters are same for SVPWM as well as BCPWM. Switching pulses for R-leg CSPWM are shown in Fig.14. G_1, G_2, G_3, G_4 are the gate pulses for the power MOSFETs T_{a1}, T_{a2}, T_{a3} and T_{a4} respectively.

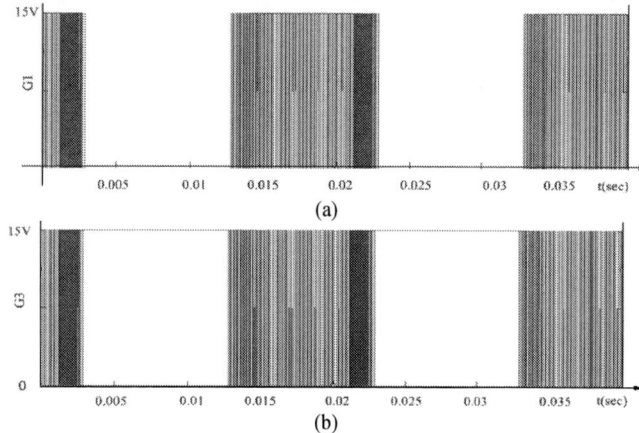

(a)

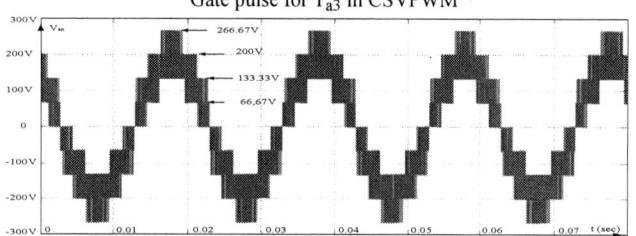

(b)

Fig. 7. Waveforms of gate pulses for R-phase leg. (a) Gate pulse for T_{a1} (b) Gate pulse for T_{a3} in CSVPWM

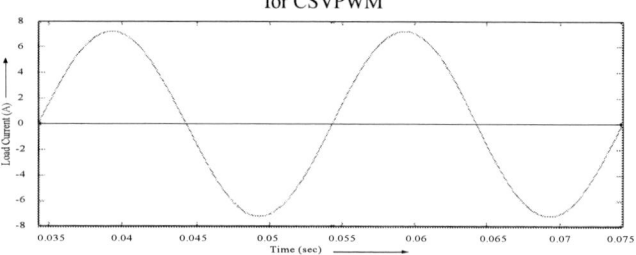

Fig.8. Waveform of output phase voltage of simulated three-level inverter for CSVPWM

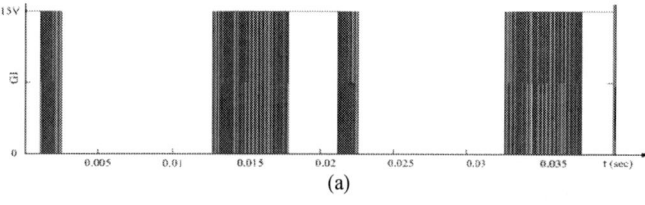

Fig. 9. Load current waveform for CSVPWM

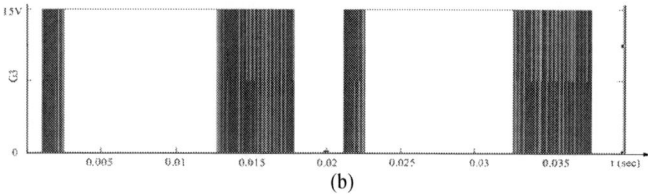

(a)

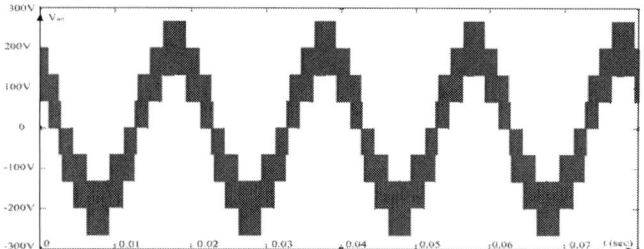

(b)

Fig. 10. Waveforms of gate pulses for R-phase leg. (a) Gate pulse for T_{a1} (b) Gate pulse for T_{a3} in BCPWM

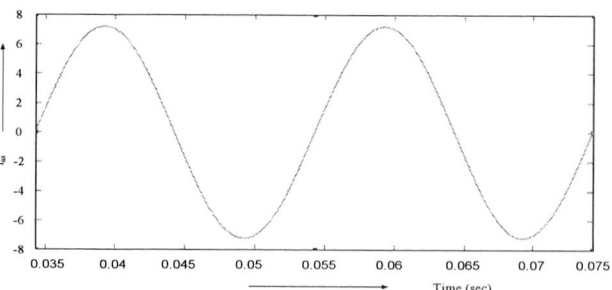

Fig.11. Waveform of output phase voltage of simulated three-level inverter for BCPWM

Fig. 12. Load current waveform for BCPWM

Table 7:
Comparison of CSVPWM and BCPWM for $F_{sw} = 10kHz$ and m=0.8

PWM Techniques	Fundamental Current(A)	THD(%) of stator current
CSVPWM	7.1917	0.7582
BCPWM	7.1929	0.3762

Table 8:
Comparison of load ripple current under various switching frequency for CSVPWM and BCPWM

Switching Frequency(kHz)	Load current ripple (A)	
	CSVPWM	BCPWM
10	0.017	0.015
20	0.0091	0.0088
30	0.0057	0.0045
40	0.0059	0.0037
50	0.0058	0.0029

The load current ripple under different switching frequency are obtained through simulation. As seen for increasing switching frequency current ripple decreases both for CSVPWM as well as BCPWM. But it can also be noticeable that load current ripples are less in case of BCPWM technique.

B. Comparison study for motor load

A comparison has been studied for different PWM technique using three phase induction motor. The inverter is modeled using MATLAB/SIMULINK. The PDPWM is generated by using proper PWM block in SIMULINK. The simulation is done by considering the following conditions-

1. DC bus voltage, $V_{DC} = 400$ V
2. 5 HP, 3 phase, 3 wire, 230 V, 4 pole squirrel cage induction motor with parameters $r_s = 0.531\Omega$, $r_r = 0.408\Omega$, J = 0.1kg $- m^2$ $L_{ls} = L_{lr} = 2.52mH$, $L_{ls} = 84.7mH$.
3. Modulation index=0.8; Switching frequency, $f_{sw} = 10kHz$

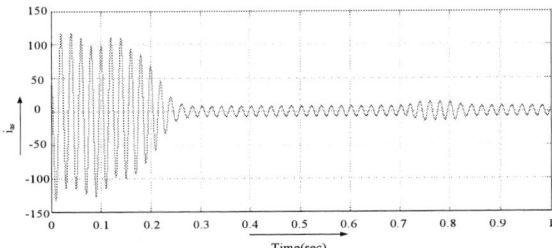

Fig 13. Waveform for a phase stator current for CSVPWM

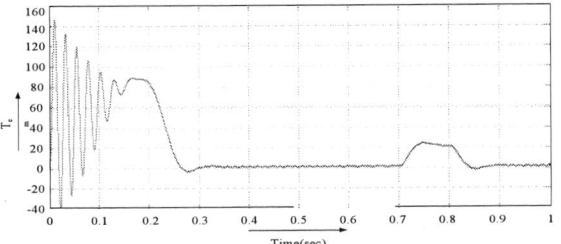

Fig. 14. Torque characteristics of the induction motor for CSVPWM

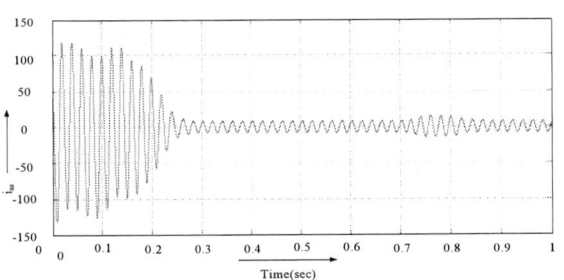

Fig. 15. Waveform for a phase stator current for BCPWM

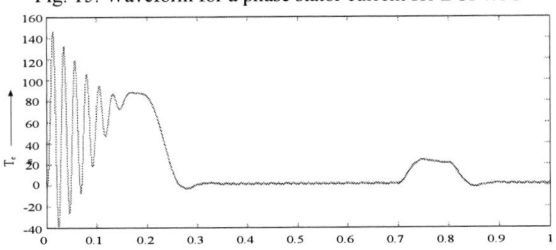

Fig. 16. Torque characteristics of three phase induction motor for BCPWM

Fig. 14 and 16 show the torque characteristics of the induction motor for CSVPWM and BCPWM respectively. Initially the motor is at no load conditions, then it is loaded for some times. Fig. 13 and 15 show the stator currents for CSVPWM and BCPWM respectively.

Table 9: stator current ripple and torque ripple

PWM Techniques	Stator current ripple (A)	Torque ripple (N-m)	THD(%) of stator current
CSVPWM	0.8	0.75	3.27
BCPWM	0.38	0.8	1.78

From the Table 9 it can be observed that stator current ripple for a phase decreases and torque ripple increases in case of BCPWM. Total harmonic distortion is also improved in case of BCPWM.

VIII. CONCLUSION

The new technique for region selection for CSVPWM as well as BCPWM is proposed in this paper. The comparison are made between CSVPWM and BCPWM for three-level inverter and it is validated through simulated results and it shows that BCPWM technique is better than the CSVPWM technique in respect of THD and current ripple. The simulated results also show the effectiveness of the proposed technique. The proposed techniques are tested on open loop system. The application of proposed technique for closed loop system will be a challenging work for the future.

IX. REFERENCES

[1] Liu Qingfeng, Wang Huamin, Leng Zhaoxia , " Discuss on the Application of Multilevel Inverter in High Frequency Induction Heating Power Supply" , *TENCON 2006 IEEE Region 10 Conference 2006* , pp. 1 − 4.

[2] Tourkhani F., Viarouge, P., Meynard T.A. , "Optimal design and experimental results of a multilevel inverter for an UPS application" *Proc. International Conference on Power Electronics and Drive Systems,* pp-340 - 343 vol.1 1997.

[3] Peng F.Z., Wei Qian Dong Cao ,"Recent advances in multilevel converter /inverter topologies and applications", *International Power Electronics Conference (IPEC),* pp−492 − 501, 2010.

[4] Rodriguez, J. , Jih-Sheng Lai, Fang Zheng Peng, "Multilevel inverters: a survey of topologies, controls, and applications" , *IEEE Transactions on Industrial Electronics, vol. 49,* pp: 724 - 73, 2002.

[5] Abdul Rahiman Beig, Saikrishna Kanukollu, Khalifa, V.T. Ranganathan; " Space Vector Based Synchronized PWM Algorithm for Three Level Voltage Source Inverters: Principles and Application to V/f Drives", *IEEE transaction* 2002

[6] R Joetten and C.Kehl, " A fast space-vector control for a three-level voltage source inverter," *in conf. Rec. European Power Electronics Conf.(EPE)*, Florence,1991,pp.2:070-2:075.

[7] Abdul Rahiman Beig. ,Narayanan G., " Simplified implementation of space vector PWM strategies for a three level inverter ", *7th IEEE International Conference on Industrial and Information Systems (ICIIS)*, 2012, Page(s): 1 - 6.

[8] A. R. Beig, V.T. Ranganathan, "Space vector based bus clamped PWM algorithms for three level inverters : implementation, performance analysis and application considerations", *Applied Power Electronics Conference and Exposition (APEC) Eighteenth Annual IEEE,* vol. 1, pp-569-575, 2003.

[9] G. Narayanan and V. T. Ranganathan, "Analytical evaluation of harmonic distortion in PWM ac drives using the notion of stator flux ripple," *IEEE Trans. Power Electron.*, vol. 20, no. 2, pp. 466–474, Mar. 2005.

[10] G. Narayanan and V. T. Ranganathan, "Synchronised PWM strategies based on space vector approach. Part 1: Principles of waveform generation,"*Proc. Inst. Elect. Eng.*, vol. 146, no. 3, pp. 267–275, May 1999.

Variation of IGBT Switching Energy Loss with Device Current: An Experimental Investigation

Subhas Chandra Das*   G. Narayanan#   Arvind Tiwari€

* GE Transportation Systems, GE India Tech Centre Pvt Ltd, Bangalore, India-560066
\# Department of Electrical Engineering, Indian Institute of Science Bangalore, India-560012
€ GE Global Research, GE India Tech Centre Pvt Ltd, Bangalore, India-560066
Email: Subhas.Das@ge.com, gnar@ee.iisc.ernet.in, Arvind.Tiwari1@ge.com

Abstract—**Aim of this paper is to study the variations in turn-on and turn-off switching energy losses in the IGBTs with device current at different DC link voltages and junction temperatures. The IGBT switching characteristics are obtained experimentally at various operating conditions, and turn-on and turn-off switching energy losses are determined. The relationship between the switching energy losses and the device currents are presented in mathematical form, which are derived from experimental data. Finally, a comparison has been made between calculated switching energy loss from linearized loss model and the derived expression from actual measured loss.**

Keywords—*IGBT, switching energy loss, linear loss model, PWM converter.*

I. INTRODUCTION

For reliable design of cooling system for power electronic converters, power losses in the semiconductor devices should be determined precisely. In voltage source inverters (VSI), insulated gate bipolar transistor (IGBT) with an antiparallel diode is widely used as the semiconductor device [1-2]. In such case, the semiconductor losses in a VSI include IGBT conduction losses, IGBT switching losses and diode losses.

The total power loss in an existing power converter can be measured using calorimetric method [3-4]. However this method is not suitable for large power converters. Total power loss in the converter can also be measured as the difference between the input power and output power [4]. However the total loss includes the losses in other components, such as filter components, bleeder resistor etc. The power losses in these components can be measured or estimated individually. These losses can be subtracted from total losses to obtain the semiconductor losses. The semiconductor losses can also be obtained from heatsink temperature rise measurements with the knowledge of heatsink thermal resistance [5].

During the design phase of power converter, the semiconductor losses have to be estimated by evaluating the IGBT conduction loss, IGBT switching loss and Diode losses individually. This paper focuses on calculating the IGBT switching loss, which varies widely based on PWM methods and operating conditions, such as; applied voltage, device current and junction temperature.

The IGBT switching energy loss in a switching cycle is the summation of turn-on switching energy loss (E_{on_1}) and turn-off

switching energy loss (E_{off_1}). The switching energy losses E_{on_1} and E_{off_1} vary with applied voltage, device current and junction temperature [6-8]. Previously, in [9] results have been published, showing the dependency of E_{on_1} and E_{off_1} as function of device voltage at various junction temperatures. In this paper, the focus is on investigation of switching loss variation in an IGBT with collector current at different DC link voltage and junction temperature. The experimental data is used for deriving simple mathematical expressions, relating switching energy losses and load current. A comparison is made between switching energy loss derived from linear loss model and actual measurement of switching energy loss with variation in load current. Above comparison is made for the switching energy losses at a particular DC link voltage and junction temperature.

II. LITERATURE REVIEW ON THE DEPENDENCE OF SWITCHING ENERGY LOSS ON DEVICE CURRENT

To calculate the switching loss in an inverter at a steady operating condition, the DC link voltage is assumed to be steady over the entire line cycle, the commutation of the device is assumed to be linear [9] and the device transition times are assumed to be invariant with load current. With these assumptions the switching energy loss in an IGBT turns out to be proportional to the device current switched. This is assumed particularly in the context of comparative evaluation of PWM techniques in terms of inverter switching loss [10-12]. Based on this linear variation, switching loss factor (SLF) is proposed for the comparative evaluation of PWM methods in terms of switching loss [12].

In linear semiconductor loss model, the analytical expression for switching losses at any operating DC link voltage (V_{dc}) and load current (I_{pk}, peak of the load sinusoidal phase current) are derived from the linear relation with respect to available data sheet information at test voltage (V_{test}) and current (I_{test}). Hence, in 3-phase voltage source inverter(VSI) with continuous pulse width modulation (CPWM) schemes, such as sine PWM (SPWM), third harmonic injection PWM (THIPWM) and conventional space vector PWM (CSPWM), the expression for switching power loss (P_{sw}) is derived from the linear loss model as [4],

$$P_{sw} = \frac{6}{\pi} \cdot f_s \cdot \left(E_{on\_I\_test} + E_{off\_I\_test} + E_{off\_D\_test} \right) \cdot \frac{V_{dc}}{V_{test}} \cdot \frac{I_{pk}}{I_{test}} \quad \dots (1)$$

Where, f_s is the switching frequency, $E_{on\_I\_test}$, $E_{off\_I\_test}$ and $E_{off\_D\_test}$ are switch on, switch off transition energy loss of IGBT and switch off energy loss of diode respectively, based on datasheet information at DC link voltage V_{test} and device current I_{test}.

The switching power loss expression in eq. (1) is based on linearized expressions of IGBT switch on transition energy loss (E_{on_I}), IGBT switch off transition energy loss (E_{off_I}) and diode switch off energy loss (E_{off_D}), which are shown in eq. (2a), (2b) and (3) respectively. These switching energy losses are derived for any DC link voltage V_{dc} and any device current I_c, from available datasheet information $E_{on\_I\_test}$, $E_{off\_I\_test}$ and $E_{off\_D\_test}$ at DC link voltage V_{test} and device current I_{test}.

$$E_{on\_I} = \left(E_{on\_I\_test} \right) \frac{V_{dc} I_c}{V_{test} I_{test}} \qquad \dots (2a)$$

$$E_{off\_I} = \left(E_{off\_I\_test} \right) \frac{V_{dc} I_c}{V_{test} I_{test}} \qquad \dots (2b)$$

$$E_{off\_D} = \left(E_{off\_D\_test} \right) \frac{V_{dc} I_c}{V_{test} I_{test}} \qquad \dots (3)$$

eq. (2a), (2b) and (3) assume the E_{on_I}, E_{off_I} and E_{off_D} to be proportional to the DC link voltage and current. The validation of these assumptions needs to be ascertained, otherwise there could be significant error in the switching loss. Investigation on the scaling of E_{on_I} and E_{off_I} with V_{dc} has been reported recently [9]. This paper studies the relationship between E_{on_I} and I_c and that between E_{off_I} and I_c at different dc link voltages and junction temperatures.

III. MEASUREMENT OF SWITCHING ENERGY LOSS

Test circuit to measure various IGBT switching parameters, is a standard double pulse circuit, which is discussed in several papers [2], [6-9], [13]. Fig. 1 shows the circuit diagram of a typical double pulse circuit.

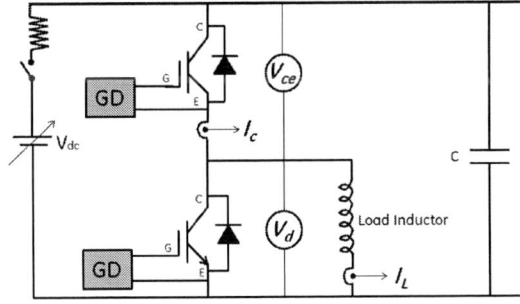

Fig. 1. Schematic of double pulse test circuit [6-8].

During double pulse test, two consecutive gate pulses with an intermediate off period are applied to one of the IGBTs of a leg of a two-level inverter. The time durations of the two pulses and the intermediate off period depend on the DC link voltage applied across the inverter leg and the device current at which the switching characteristics are required to be

measured. The IGBT voltage (V_{ce}), diode voltage (V_d), IGBT current (I_c) and inductor currents (I_L) are measured using high bandwidth (10 GHz) storage oscilloscope with suitable voltage and current probes. Rogowski coils are used as current probes for measurements. The test circuit is placed in a temperature controlled environment to measure the switching characteristic parameters at various junction temperatures.

The turn-on transition and turn-off transition at 0.9 p.u. dc link voltage and 0.8 p.u. device current are shown in fig. 2(a) and 2(b), respectively. Based on the measured turn-on and turn-off transitions, turn-on energy loss (E_{on_I}) and turn-off energy loss (E_{off_I}) in the IGBT are obtained using the equations (4) and (5), respectively [6].

$$E_{on\_I} = \int_{ton} V_{ce}(t) I_{ce}(t) dt \qquad \dots (4)$$

$$E_{off\_I} = \int_{toff} V_{ce}(t) I_{ce}(t) dt \qquad \dots (5)$$

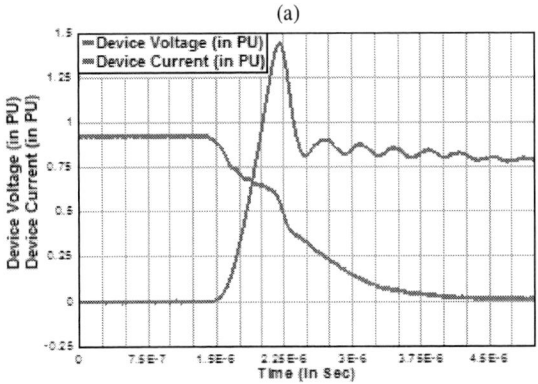

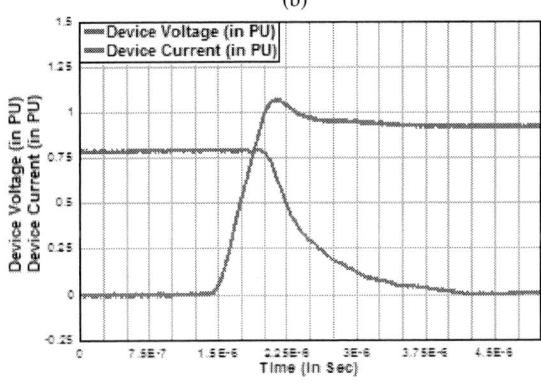

Fig. 2. Measured switching transitions (a) Turn-on transition of IGBT (b) Turn-off transition of IGBT. DC link voltage = 0.9 p.u., load current = 0.8 p.u.

IV. VARIATIONS IN TURN-ON AND TURN-OFF SWITCHING ENERGY LOSSES WITH DEVICE CURRENT

In this section, the relation between IGBT turn-on energy loss and device current and that between turn-off energy loss and device current are investigated experimentally.

A. Turn-on switching energy loss

Fig. 3 shows the variation in the measured IGBT turn-on switching energy loss (E_{on_1}) with respect to device current. Eon is measured at different values of current ranging between 0.104 p.u. and 1.78 p.u. at a junction temperature of 25° C. Each line in the plot corresponds to the measurement at a particular DC link voltage. The range of DC link voltage is from 0.571p.u. to 1.32 p.u.

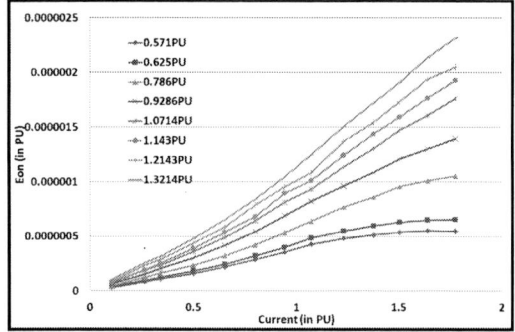

Fig. 3. Variation of measured E_{on_1} with I_c at various DC link voltages at a junction temperature of 25° C.

As seen from fig. 3, turn-on energy loss does not vary in a linear fashion with respect to collector current at any given DC link voltage. The variation E_{on_1} with I_c can be represented by second order polynomial as shown in eq. (6) below.

$$E_{on} = aI_c^2 + bI_c + c \qquad \dots (6)$$

The coefficients *a*, *b* and *c* are of the second polynomial, can be determined based on experimental data. The values of the coefficients corresponding to each DC link voltage are tabulated table I below.

TABLE I. Coefficients of Second Order Polynomial for E_{on_1} Corresponding to Junction Temperature of 25°C.

Vdc(in P.U.)	a	b	c
0.571	-9.565E-08	5.2785E-07	-5.32E-08
0.625	-7.931E-08	5.6072E-07	-5.2565E-08
0.786	2.849E-08	5.9891E-07	-4.5706E-08
0.929	6.4179E-08	7.1729E-07	-4.3943E-08
1.071	1.768E-07	7.0553E-07	-1.904E-08
1.143	1.9524E-07	7.7544E-07	-3.0379E-08
1.214	1.9023E-07	8.6434E-07	-2.4306E-08
1.321	2.3804E-07	9.1691E-07	-1.767E-08

The pattern of variation of turn-on transition losses of IGBT with device current at different DC link voltages are confirmed through additional measurements of E_{on_1} at two more junction temperatures, namely -35° C and 100° C. The measured switching energy losses at these two junction temperatures are shown in Fig. 4(a) and Fig. 4(b), respectively.

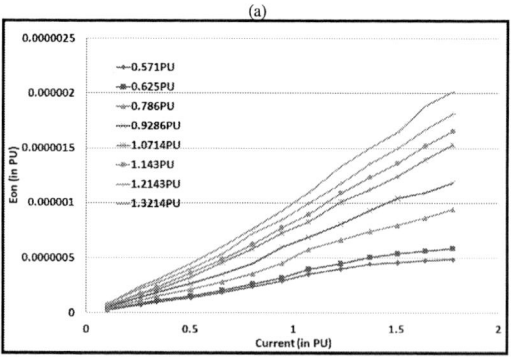

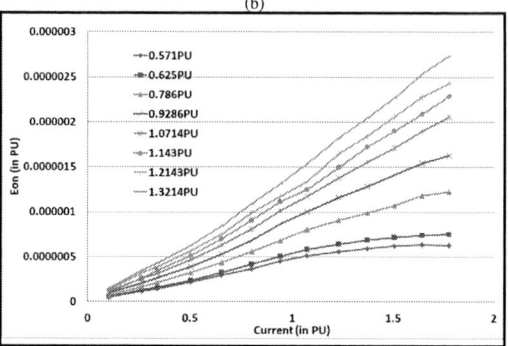

Fig. 4. Variation of measured E_{on_1} with I_c at various DC link voltages at a junction temperature of (a) -35° C and (b) 100° C.

Similar to the experimental results in fig. 3; second order polynomial curve fitting is done on the experimental data in fig. 4(a) and 4(b). The coefficients of the polynomial pertaining to fig. 4(a) and 4(b) are presented in tables II and III, respectively.

TABLE II. Coefficients of Second Order Polynomial for E_{on_1} Corresponding to Junction Temperature of -35°C.

Vdc(in p.u.)	a	b	c
0.571	-5.2515E-08	3.9877E-07	-3.1632E-08
0.625	-2.2716E-08	3.9999E-07	-2.6483E-08
0.786	3.95822E-08	4.8517E-07	-2.4655E-08
0.929	6.48013E-08	5.8634E-07	-2.6013E-08
1.071	1.21841E-07	6.6089E-07	-1.7058E-08
1.143	1.3663E-07	7.1549E-07	-2.4304E-08
1.214	1.49641E-07	7.7257E-07	-7.1463E-09
1.321	1.78042E-07	8.4506E-07	-1.176E-08

TABLE III. Coefficients of Second Order Polynomial for E_{on_1} Corresponding to Junction Temperature of 100°C

Vdc	a	b	c
0.571	-1.5086E-07	6.66699E-07	-5.1949E-08
0.625	-1.4099E-07	7.20446E-07	-5.4408E-08
0.786	-3.5969E-08	7.96663E-07	-4.0489E-08
0.929	2.08978E-08	9.28239E-07	-4.3277E-08
1.071	1.27564E-07	9.56623E-07	-1.5091E-08
1.143	1.59008E-07	1.02528E-06	-1.4258E-08
1.214	1.82576E-07	1.07415E-06	5.9163E-09
1.321	1.99258E-07	1.20885E-06	-2.7078E-10

B. Turn-off switching energy loss

The turn-off energy loss (E_{off_1}) is measured at different device currents ranging 0.104 p.u. and 1.78 p.u. as in case of Eon. These measurements are carried out at various DC link voltages between 0.571 p.u. and 1.32 p.u. The measured values of Eoff with respect to device current at junction temperature of 25°C are shown in fig. 5.

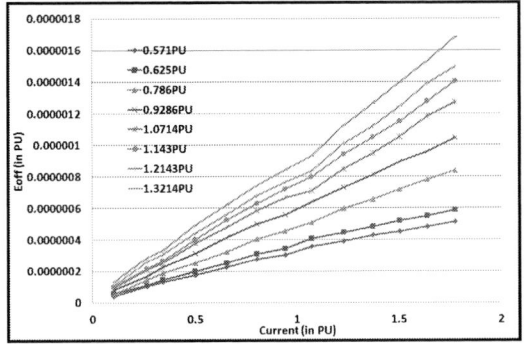

Fig. 5. Variation of measured E_{off_1} with I_c at various DC Link Voltages at Junction temperature of 25° C.

At any given DC link voltage, the variation of turn-off energy loss (E_{off_1}) with respect to collector current is more linear than in case of E_{on_1}. However the relation between E_{off_1} and I_c is still better represented by a second degree polynomial as in eq. (7) rather than a linear polynomial.

$$E_{off} = pI_c^2 + qI_c + r \qquad \ldots (7)$$

The coefficients p, q and r of the polynomial for various DC link voltages at a junction temperature of 25°C are shown in table IV below.

TABLE IV. COEFFICIENTS OF SECOND ORDER POLYNOMIAL FOR E_{OFF_1} CORRESPONDING TO JUNCTION TEMPERATURE OF 25°C

Vdc(in P.U.)	p	q	r
0.571	-5.2264E-08	3.7833E-07	3.9659E-09
0.625	-4.8707E-08	4.1256E-07	9.1119E-09
0.786	-1.5562E-08	4.9228E-07	1.1589E-08
0.929	-1.237E-08	5.955E-07	1.928E-08
1.071	4.75158E-08	6.0107E-07	4.7886E-08
1.143	4.22615E-08	6.8501E-07	3.7642E-08
1.214	6.27538E-08	6.9489E-07	6.0436E-08
1.321	8.86221E-08	7.4261E-07	7.5097E-08

Fig. 6(a) and 6(b) report similar measurement of E_{off_1} at different currents and DC link voltages at junction temperatures of -35°C and 100°C, respectively. Second degree polynomial fit is done on experimental data on fig. 6(a) and 6(b) as well. The coefficients of the resulting polynomial are tabulated in table V and VI respectively.

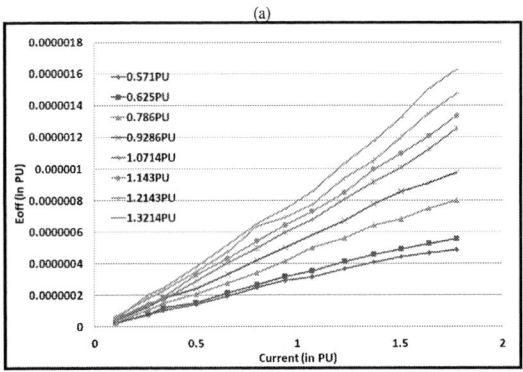

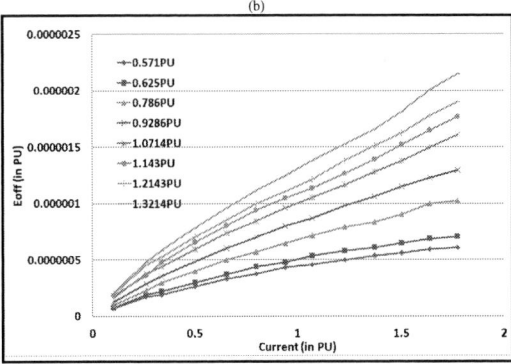

Fig. 6. Variation of measured E_{off_1} with I_c, (a) at -35° C, (b)at 100° C, at various DC link voltages.

TABLE V. COEFFICIENTS OF SECOND ORDER POLYNOMIAL FOR E_{OFF_1} CORRESPONDING TO JUNCTION TEMPERATURE OF -35°C

Vdc(in p.u.)	p	q	r
0.571	-4.015E-08	3.59779E-07	-1.635E-08
0.625	-2.917E-08	3.78025E-07	-1.541E-08
0.786	-3.192E-09	4.73819E-07	-2.157E-08
0.929	1.5795E-08	5.42819E-07	-1.868E-08
1.071	7.8564E-08	5.67802E-07	-1.262E-08
1.143	6.8358E-08	6.41533E-07	-2.534E-08
1.214	1.0933E-07	6.27415E-07	8.9479E-09
1.321	1.3929E-07	6.71265E-07	2.7221E-09

TABLE VI. COEFFICIENTS OF SECOND ORDER POLYNOMIAL FOR E_{OFF_1} CORRESPONDING TO JUNCTION TEMPERATURE OF 100°C

Vdc(in p.u.)	p	q	r
0.571	-1.1681E-07	5.3248E-07	2.6701E-08
0.625	-1.294E-07	6.0995E-07	2.5265E-08
0.786	-1.0911E-07	7.4475E-07	4.3507E-08
0.929	-1.122E-07	8.9428E-07	5.5678E-08
1.071	-8.928E-08	9.9047E-07	1.045E-07
1.143	-9.6148E-08	1.0923E-06	1.0126E-07
1.214	-7.2859E-08	1.1079E-06	1.4023E-07
1.321	-8.1616E-08	1.2537E-06	1.4166E-07

V. COMPARISON OF SWITCHING ENERGY LOSS BASED ON LINEAR LOSS MODEL AND ACTUAL MEASUREMENT

In this section, the turn-on and turn-off energy losses, calculated using linear loss model [4], are compared against actual measurements.

For comparison, turn-on switching energy loss corresponding to DC link voltage of 1.07 p.u. at junction temperature of 125°C is considered. Assuming a peak load current of 1.0 p.u. and ignoring ripple current, the turn-on switching loss varies in sinusoidal pattern with the load current based on linearized loss model, shown as curve 1 in fig. 7(a). The actual losses measured are plotted shown as dots, shown in fig. 7(a). The energy loss based on the curve fitted on experimental data is shown as curve 2 in same figure. For this operating point, the actual loss is higher than the loss determined from linear loss model.

Similarly, the curves for turn-off switching energy loss are plotted in fig. 7(b).

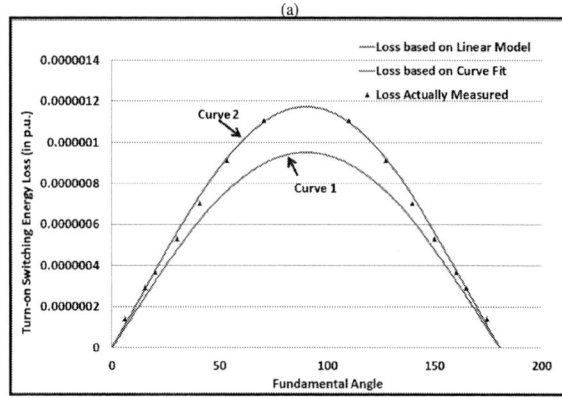

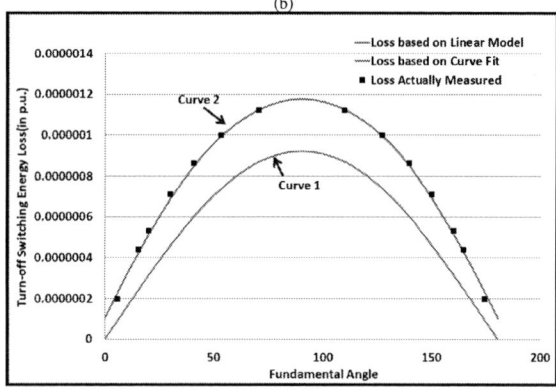

Fig. 7. Comparison of switching energy losses obtained from linear loss model (curve 1), loss model from curve fit on measured data (curve 2) and actual measurement points (dots) for, (a) Turn-on switching transition and (b) Turn-off switching transition, at 1.07 p.u. DC link voltage and 1.0 p.u. peak load current at 125°C junction temperature.

VI. CONCLUSION

The variations of turn-on switching energy loss (E_{on_1}) and turn-off switching energy loss (E_{off_1}) of an IGBT with device current are studied experimentally. The device current range is considered from 0.103 p.u. to 1.78 p.u. It is shown that the variations of E_{on_1} with I_c and that of E_{off_1} with I_c can both be represented by second degree polynomial over this range of collector current. This leads to a better estimate of the switching loss, and there by, the total semiconductor loss in a power converter. This would lead to improved thermal design and reliability of the converter.

REFERENCES

[1] S. Bernet, "State-of-the-art and trends of high voltage power devices and medium voltage converters for industry and transportation," in Proc. 5th International Workshop: Future of Electronic Power Processing and Conversion, IEEE-FEPPCON, Salina, Italy, 2004.

[2] R. Alvarez, F. Filsecker and S. Bernet, "Characterization of a new 4.5 kV press pack SPT+ IGBT for medium voltage converters," Proc. IEEE Energy Conversion Congress and Exposition (ECCE 2009), pp. 3954-3962, 2009.

[3] F.W.Fuchs,J.Schroder,B.Wittig, "State of the technology of power loss determination in power converters," 15th Europian conference on Power Electronics and Applications (EPE), pp.1-10, Lille, France, Sep 2013.

[4] M.H. Bierhoff, F.W. Fuchs, "Semiconductor Losses in Voltage Source and Current Source IGBT Converters Based on Analytical Derivation," 35th Annual IEEE Power Electronics Specialists Conference (PESC 04), Vol.4, pp.2836-2842, 2004.

[5] J.S.S.Prasad, G.Narayanan, "Minimum switching loss pulse width modulation for reduced power conversion loss in reactive power compensators," IET Power Electron., Vol. 7, Issue 3, pp. 545–551, Mar 2014.

[6] S. C. Das, G. Narayanan, A. Tiwari and A. K. Kumar, "Experimental investigation on switching characteristics of IGBTs for traction application." Proc. IEEE International Conference on Power Electronics, Drives and Energy Systems (PEDES 2012), Bangalore, India.

[7] S. C. Das, G. Narayanan, A. Tiwari and A. K. Kumar, "Experimental Study on IGBT Voltage and Current Stresses During Switching Transitions."Proc. IEEE Innovative Smart Grid Technologies Asia 2013(IEEE ISGT Asia 2013) Bangalore, India, Nov 2013.

[8] S. C. Das, G. Narayanan and A. Tiwari, "Experimental Study on Switching Characteristics of an Inverter Leg Consisting of IGBTs of Dissimilar Makes."IEEE International conference of Eletrial Energy Systems 2014(IEEE ICEES 2014) Chennai, India, Jan 2014.

[9] S. C. Das, G. Narayanan and A. Tiwari, " Experimental Study on the Dependence of IGBT Switching Energy Loss on DC Link Voltage." Accepted for presentation at Proc. IEEE International Conference on Power Electronics, Drives and Energy Systems (PEDES-2014), Mumbai, India, Dec 2014.

[10] J.W.Kolar, H.Ertl and F.C.Zach, "Influence of the modulation method on the conduction and switching losses of a PWM converter system," IEEE Trans. Industry Applications, Vol.IA-27, No.6, pp.1063-1075, Nov-Dec 1991.

[11] D.Zhao, V.S.S.P.V.K.Hari, G.Narayanan and R.Ayyanar, "Space-Vector-Based Hybrid Pulsewidth Modulation Techniques for Reduced Harmonic Distortion and Switching Loss," IEEE Trans. Power Electronics, Vol.25, No.3, pp.760-774, Mar. 2010.

[12] A.M.Hava, R.J.Kerkman and T.A.Lipo, "Simple analytical and graphical methods for carrier-based PWM-VSI drives," IEEE Transactions on Power Electronics, Vol.14, No.1, pp.49-61, Jan 1999.

[13] D. Xiao, I. Abuishmais and T. Undeland, "Switching characteristics of NPT-IGBT power module at different temperatures" Proc. 14th Intl. Power Electronics and Motion Control Conf. (EPE/PEMC), pp. T1-17 to T1-22, 2010.

978-1-4799-6047-7/14 $31.00 © 2014 IEEE

Silicon Carbide based DSG MOSFET for High Power, High Speed and High Frequency Applications

Sonam Rewari ,R.S.Gupta
Department of Electronics and Communication Engineering, Maharaja Agrasen Institute Engineering, Rohini, Delhi 110086, India   rewarisonam@gmail.com, rsgupta1943@gmail.com

S.S.Deswal
Department of Electrical and Electronics Engineering, Maharaja Agrasen Institute Engineering, Rohini, Delhi 110086, India satvirdeswal@hotmail.com

Vandana Nath
University School of Information and Communication Technology GGS Indraprastha University, Sector 16-C,Dwarka, India vandanausit@gmail.com

Abstract— **In this paper, High Power Double Surrounding Gate(DSG) MOSFET with 4H-SiC as material has been studied. Also, the RF performance of DSG MOSFET has been investigated for various channel length and the results so obtained are compared with the conventional Surrounding Gate(SG) MOSFET, using ATLAS 3D device simulator. From the analysis, it is shown that cylindrical Double Surrounding Gate(DSG) MOSFET exhibits superior power and analog performance than conventional cylindrical Surrounding Gate(SG) MOSFET. DSG MOSFET has a number of desirable features, such as higher transducer power gain, better current gain, high on-state current, improved transconductance g_m, high unity-gain frequency f_T. The improvement is due to formation of two conducting paths because of the presence of two gates. Power has further been improved because Silicon Carbide has been used as material instead of Silicon.**

Keywords— *High Power, Double Surrounding Gate(DSG) MOSFET, simulations, Surrounding Gate(SG).*

I. INTRODUCTION

High packaging density and high cut-off frequency are the need of an hour for ULSI applications, which are obtained by scaling the device dimensions. As the devices are scaled down short channel effects(SCE's) creep in. To improve SCE's, three dimensional multigate FET(MuGFET) structures such as Double Gate FINFET, Deplted UTB SOI, Triple Gate FINFET and Surrounding Gate MOSFET(Gate All Around MOSFET) have been introduced[2-5]. Surrounding Gate(SG) MOSFET is one of the most promising device as it provides enhanced electrostatic control of channel alongwith higher packing density, steep subthreshold characteristics and higher current drive. In SG MOSFET gate completely surrounds the silicon pillar, so controls the channel more effectively therby reducing SCE's. The electrostatic control over the channel can be further reduced by having gate electrode on more than one side of the device[3]. Thus, cylindrical DSG MOSFET suppresses SCE's and leads to higher drain current. Analogous to SG MOSFET, DSG MOSFET is also cylindrical with two gates, one inside and one outside the silicon pillar. It has improved noise and intrinsic characteristics as compared to SG MOSFET and DG MOSFET [7]. Chen et al.[8] has already contributed the

experimental study, manufacturing challenges and modelling of DSG MOSFET. Viranjay et al[6,7] has explained the electrostatic behaviour, noise analysis along with design guideline. In this paper Double Surrounding Gate(DSG) MOSFET with Silicon Carbide(4H-SiC) as material has been investigated for various channel length and the results so obtained are compared with the conventional Surrounding Gate(SG) MOSFET, using ATLAS 3D device simulator[1]. Silicon Carbide is a wide bandgap semiconductor and has improved performance in high power devices[11]. Such materials also have very high critical electric field. Switching devices designed with high electric field have much lower losses than conventional silicon based devices, thus leading to high power[10]. Table1 lists some of the properties of Silicon Carbide(4H-SiC) and silicon. It is evident from the Table 1 that that 4H-SiC has wider bandgap and higher critical electric field. Active power switches based on 4H-SiC MOSFET are being manufactured. It is shown that DSG MOSFET has higher transducer power gain, unilateral power gain improved drain currents, Ion/Ioff ratio, transconductance g_m, current gain, and cut off frequency(f_T) over conventional DSG MOSFET.

Table 1. Physical Properties of Silicon and 4H-SiC		
Property	*Si*	*4H-SiC*
Bandgap,E_g(eV)	1.12	3.2
Critical Electric Field,E_c(V/cm)	2.5×10^5	2.2×10^6
Saturated Electron Drift Velocity,v_{sat}(cm/s)	1×10^7	2×10^7
Thermal Conductivity,λ(W/cmK)	1.5	3-4
Dielectric Constant,ε_r	11.9	10

II. STRUCTURAL DESCRIPTION AND SIMULATION

A 3-D structure of cylindrical DSG MOSFET is shown in Fig.1 with two gates(an inner gate G1(below the Silicon Carbide(4H-SiC) film) and an outer gate G2(above the Silicon Carbide(4H-SiC) film)) having same workfunction using silicon dioxide as dielectric. The channel is of length L=30nm

978-1-4799-6047-7/14 $31.00 © 2014 IEEE

with uniform doping concentration, $N_A=10^{23}m^{-3}$. Source and drain are also uniformly doped with $N_D=10^{26}m^{-3}$,thickness of silicon film tsi=10nm, oxide thickness tox=2nm and oxide permittivity =3.9.

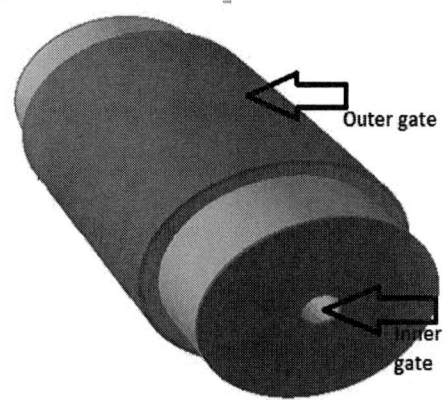

Fig. 1. 3-D View of cylindrical DSG MOSFET

For comparison between DSG and SG MOSFET, threshold voltage optimization is done by varying the gate metal work function to keep V_{th}=0.122V.

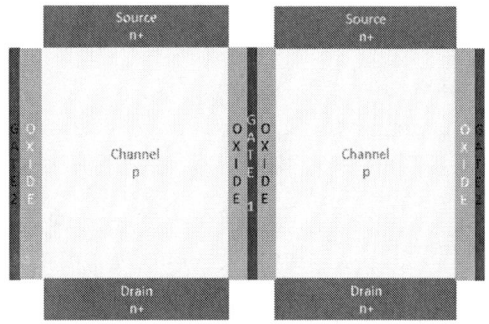

Fig. 2. Crossectional View of cylindrical DSG MOSFET

Fig. 2 shows the Crossectional view of DSG MOSFET. The simulation have been performed using the Newton-Gummel method and with Lombardi CVT mobility model, concentration and temperature dependent analytic model keeping in view that for wide temperature range 77 -450K. Shockley-Read-Hall (SRH) recombination model with fixed carrier life time $(1 \times 10^{-7}s)$ along with Auger model have been used to account for the minority and high carrier density recombination using 3-D ATLAS device simulator [1].

TABLE 2. Device Structural Parameters		
Parameters	*SG MOSFET*	*DSG MOSFET*
Channel Length(nm)	30	30
Channel Doping(/cm^{-3})	1×10^{17}	1×10^{17}
Silicon Oxide Thickness(nm)	2	2
4H-SiC Thickness=2R(nm)	10	10
Gate Work Function(eV)	4.79	4.8
Length of S/D(nm)	15	15

The simulated results are compared with conventional Surrounding Gate MOSFET. Analog figure of merits such as transconductance(g_m) I_{on}/I_{off}, cutoff frequency(f_T) ,intrinsic

gain, unilateral power gain, transducer power gain have shown improvement. The device structural parameters for both the devices have been tabulated in Table 2.

III. RESULTS AND DISCUSSION

DSG MOSFET has better performance in terms of Ion/Ioff, Peak g_m, Peak f_T, Current Gain, Unilateral Power Gain and Transducer Power Gain. The comparative device performance results have been tabulated in Table 3. Fig3 shows the variation of the drain current, Ids with gate to source voltage, Vgs for different channel lengths(L=30nm,34nm,38nm) at Vds=1.0V. DSG MOSFET exhibits higher drain current in comparison to SG MOSFET. Also, DSG MOSFET has better I_{on}/I_{off} ratio, with an improvement of 60% over SG MOSFET for channel length L=30nm. This improvement is due to the presence of two gates, one below and one above the 4H-SiC film. The presence of two gates increase the charge carriers and thus further increasing the drain current.

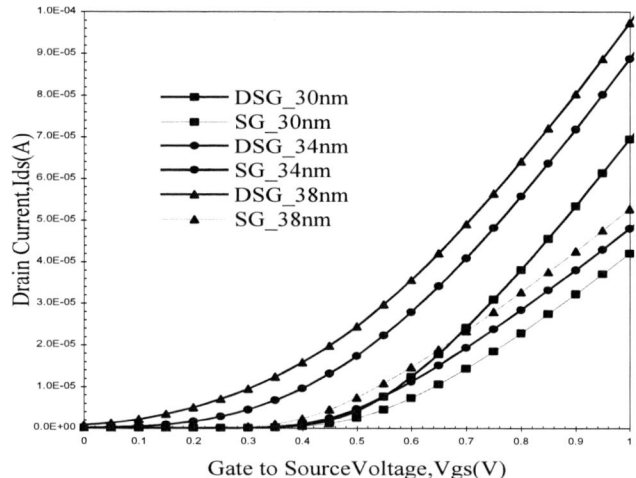

Fig. 3. Drain current Vs gate to source voltage with Vds=1.0V.

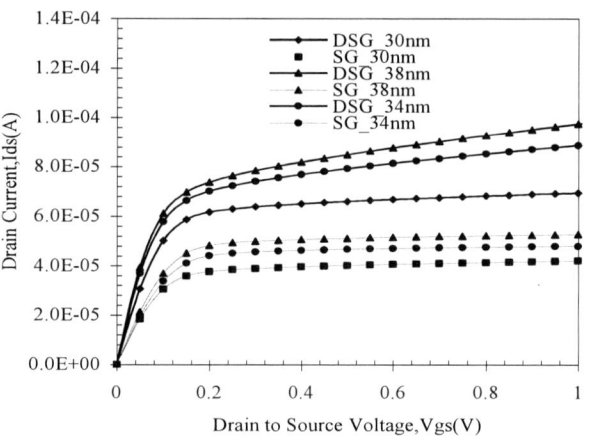

Fig. 4. Drain current Vs Drain to source voltage with V_{gs} =1.0V.

Fig.4. shows the variation of drain current with drain voltage for different channel lengths(L=30nm,34nm,38nm) with gate voltage V_{gs}=1.0V. DSG MOSFET shows higher

978-1-4799-6047-7/14 $31.00 © 2014 IEEE

drain current due to presence of two gates instead of one,thus leading to higher charge density and thus higher drain currents. Fig.5. shows the variation of transconductance,gm with gate voltage to source at V_{ds} =1.0V for different channel length(L=30nm,34nm,38nm). The gain of any device is determined by its transconductance(g_m) and its optimum bias point should be high to have higher cut off frequency[10].

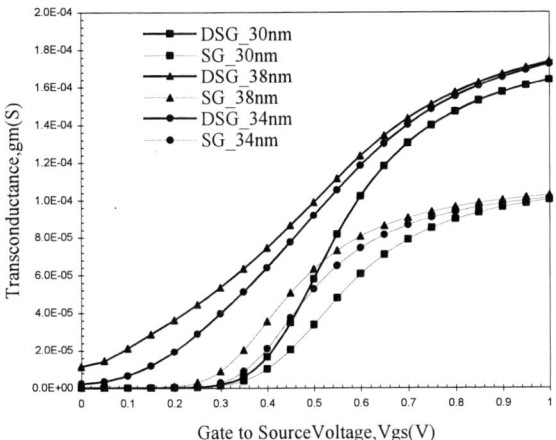

Fig.5.Transconductance Vs gate to source voltage with V_{ds} =1.0V.

DSG MOSFET shows higher transconductance with an improvement of 60% over SG MOSFET for channel length L=30nm.Thus,the advantage of DSG MOSFET is that it is suitable for high frequency, high gain amplifier applications.

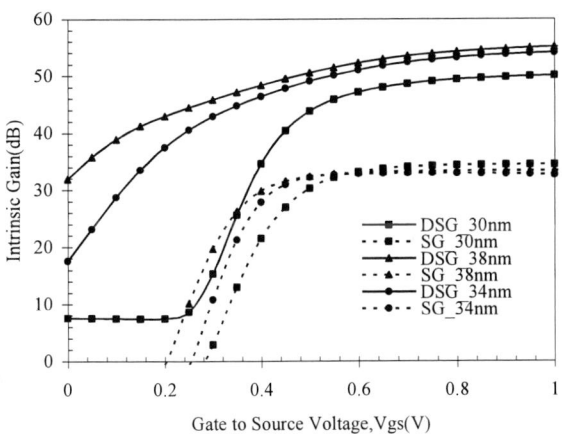

Fig. 6. Current Gain Vs gate to source voltage with V_{ds}=1.0V.

Fig.6. shows variation of the Intrinsic Gain with gate to source voltage at V_{ds}=1.0V. DSG MOSFET exhibits higher Intrinsic Gain in comparison to SG MOSFET with an improvement of 45%. DSG MOSFET has improved intrinsic gain due to better gate control and improved carrier transport efficiency owing to the dual gate structure. Fig.7. shows the variation of unilateral power gain with the gate to source voltage for different channel lengths(L=30nm,34nm,38nm). It shows that DSG MOSFET has higher unilateral power gain as compared

to SG MOSFET with an improvement of 22%. DSG MOSFET has improved power gain because of improved current density.

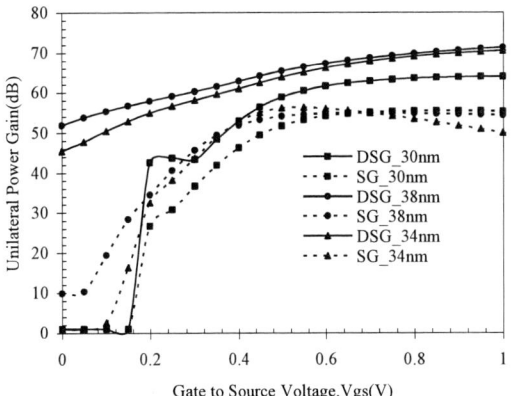

Fig. 7. Unilateral Power Gain Vs gate to source voltage with V_{ds} =1.0V.

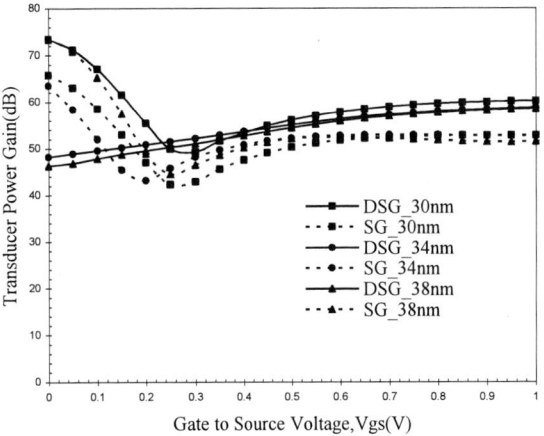

Fig. 8. Transducer Power Gain Vs gate to source voltage with V_{ds} =1.0V.

Fig.8. evaluates DSG MOSFET and SG MOSFET in terms of maximum transducer power gain(G_{MT}):a figure of merit in RF amplifier design. It is defined as the ratio of the power to a load by a source to the maximum power available from the source. DSG MOSFET shows higher transducer power gain than SG MOSFET with an improvement of 14%. Thus, proving its efficacy for RF applications. Fig.9. shows the variation of Cut Off Frequency with the gate voltage for both the devices at different channel length(L=30nm,34nm,38nm). Cut off frequency is a very important parameter for RF performance characterization. Cut off frequency, f_T can be expressed as:-

$$f_T = \frac{g_m}{2 \times \pi \times (C_{gs} + C_{gd})}$$

978-1-4799-6047-7/14 $31.00 © 2014 IEEE   663

Clearly DSG MOSFET works at higher cut off frequency than SG MOSFET It shows an improvement of 19%. Thus, it is suitable for high frequency applications.

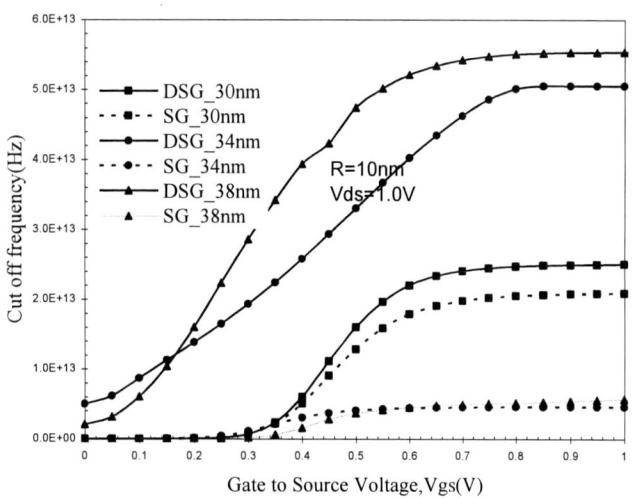

Fig. 9. Cut Off Frequency Vs gate to source voltage with V_{ds} =1.0V.

TABLE 3. Device Performance at Channel Length L=30nm

Parameter	DSG MOSFET	SG MOSFET	% improvement
Ion/Ioff	8.2E+6	5.5E+6	60
Peak g_m	1.6E-3	1.0E-3	60
Peak f_T	2.5E+13	2.09E+13	19
Current Gain(dB)	49.98	34.46	45
Unilateral Power Gain(dB)	63.92	52.27	22
Transducer Power Gain(dB)	60.13	52.72	14

IV. CONCLUSION

In this paper, the analog performance of Silicon Carbide based DSG MOSFET for different channel length(L=30nm,L=34nm,38nm) has been studied and the results have been compared with conventional SG MOSFET. As shown in this work, DSG MOSFET exhibits higher power gains and better RF performance in comparison to SG MOSFET, thus proving its efficacy for high power and high speed applications. Higher cut off frequency(f_T) and higher transconductance(g_m) prove that the speed performance has been significantly improved. The results reveal that the DSG MOSFET has better drain current. So it can be used for a wide range of high power, faster switching and high radio frequency (RF) applications such as for Aeronautical, Marine and Industrial Applications.

ACKNOWLEDGMENT

Authors are grateful to Maharaja Agrasen Institute of Technology, Delhi and Defence Research and Development Organisation, Ministry of Defence, Government of India for financial assistance to carry out this research work.

REFERENCES

[1] ATLAS: 3D Device Simulator, SILVACO International, 2013

[2] B Yu, H. Lu, M. Liu and Y. Taur, "Explicit Continuous Model for Double Gate and Surrounding Gate MOSFET,", IEEE Trans. Electronic Device, Vol.54, No.10, pp.2715-2722, Oct., 2007.

[3] P. Kumar, D. Joy and B.K. Jeblin, "Nanoscale tri-gate MOSFET for Ultra low power applications using high-k dielectrics," INEC 2013, 5th IEEE International Conference on Nanoelectronics, 12-19, Jan., 2013, Singapore , pp. 2-4.

[4] D. Jimenez, J. J. Saenz, B. Iniguez, J. Sune, L. F. Marsal and J. Pallares, "Modeling of Nanoscale Gate-All-Around MOSFETs," IEEE, Electron device letters, Vol.25, No.5, pp.314-316, May., 2004.

[5] K. H. Yeo, S. D. Suk, M. Li, Y. Yeoh, K. H. Cho, K.-H. Hong, S. Yun, M. S. Lee, N. Cho, K. Lee, D. Hwang, B. Park, D.-W. Kim, D. Park and B.-I. Ryu, "Gate-all-around (GAA) twin silicon nanowire MOSFET (TSNWFET) with 15 nm length gate and 4 nm radius nanowires," in IEDM Tech. Dig.,pp. 1–4, 2006.

[6] V.M. Srivastava, K.S. Yadav and G. Singh, "Design and performance analysis of cylindrical surrounding double-gate MOSFET for RF switch," Microelectronic Engineering Journal, Vol.42, pp.1124-1135, 2011.

[7] V.M. Srivastava, K.S. Yadav and G. Singh, "Drain Current and Noise Model of Cylindrical Surrounding Double Gate MOSFET for RF Switch," International Conference on Modeling, Optimization and Computing (ICMOC 2012), Procedia Engineering, 10-11 April 2012 Kumarakoil,Kanyakumari District,, TamilNadu, India pp.517-521.

[8] Y. Chen, W. Kang, "Experimental study and modeling of double–surrounding-gate and cylindrical silicon–on–nothing MOSFETs," Microelectronic Engineering Journal, Vol. 97, pp.138-143, 2012.

[9] Manoj Kumar, Subhasis Haldar, Mridula Gupta and R.S.Gupta"Impact of gate material engineering on analog performance of nanowire Schottky-barrier gate all around(GAA) MOSFET for low power wireless applications:3D T-CAD simulation,"Microelecrtronics J(2014) http://dx.doi.org/10.1016/j.mejo.2014.07.010

[10] Mikael Ostling, Reza Ghandi and Carl-Mikael Zetterling "SiC power devices – present status, applications and future perspective" Proceedings of the 23rd International Symposium on Power Semiconductor Devices & IC's, May 23-26, 2011, San Diego, CA, pp10-15.

[11] Honggang Sheng, Zheng Chen, Fred Wang and Alan Millner "Investigation of 1.2 kV SiC MOSFET for High Frequency High Power Applications" Twenty-Fifth Annual Applied Power Electronics Conference and Exposition (APEC), Feb. 21-25, 2010, Palm Springs, CA, pp.1572-1577.

Comparative Study of Enhancement-Mode Gallium Nitride FETs and Silicon MOSFETs for Power Electronic Applications

Anirban Pal, G. Narayanan

Department of Electrical Engineering
Indian Institute of Science, Bangalore – 560012, India
Email: palanirban@ee.iisc.ernet.in; gnar@ee.iisc.ernet.in

Abstract—**Gallium nitride (GaN) based high-electron-mobility transistor (HEMT) is becoming popular as fast switching devices for power electronic applications. This paper presents a comparative study of the critical parameters such as on-state resistance, reverse conduction drop, leakage current, maximum junction temperature, threshold voltage, gate charge requirement and device capacitances of commercially available enhancement-mode GaN (e-GaN) devices with those of Si MOSFET devices of the same voltage and current ratings. This paper also calculates the switching transition times of the e-GaN HEMTs based on their gate-charge characteristics. Further, the switching losses are also evaluated. These switching transition times and switching energy losses are also compared for the two types of devices. The e-GaN devices show excellent reduction in switching times and switching losses over the Si MOSFET devices, indicating their suitability for high-frequency power conversion. The e-GaN devices also reduce the on-state loss in most cases. However, the reverse conduction drop and leakage currents are higher with eGaN devices than with Si devices.**

Keywords—*Field-effect transistor, gallium nitride, high-electron-mobility transistor, on-state drop, power conversion, power loss, power switching device, silicon MOSFET, switching transition time, switching loss.*

I. INTRODUCTION

Energy efficiency has become very important in the existing energy scenario. Since a bulk of electrical energy is processed by power electronic converters, the efficiency, reliability and cost effectiveness of these converters are of major concern. The efficiency of power converter improves, if the switching and conduction losses of the semiconductor device can be reduced. Also, reduced switching loss enables the power converter to operate at increased switching frequency. This, in turn, implies reduction of size of magnetic components used in the converter, reducing the converter size and cost.

Recent technological developments [1] in gallium nitride (GaN) based devices have resulted in enhancement-mode GaN (e-GaN) field-effect transistors (FET) with low on-state resistance ($R_{DS,ON}$), low input capacitance (C_{iss}) and low gate charge requirement (Q_g). Section II briefly reviews the basic structure of an e-GaN FET device.

This work is supported by the Department of Electronics and Information Technology, Government of India, under a project titled "Investigation on gallium nitride (GaN) devices for power electronics switching applications and design and development of high frequency GaN converter topology" as part of National Mission on Power Electronics Technology phase-II.

This paper compares commercially available e-GaN FET power switching devices with silicon (Si) based MOSFET devices from the perspective of a power electronic designer. The on-state resistance, reverse conduction drop, leakage currents, threshold voltage, gate charge characteristics and terminal capacitances are compared, based on device datasheets, for the two types of devices in section III. Section IV discusses a procedure to estimate switching transition times of e-GaN devices during turn on and turn off. The corresponding switching losses are evaluated in section V. The switching times and switching losses are also compared for the two sets of devices in this section. Section VI presents the experimental results on 40-V Si MOSFET and e-GaN FET based dc-dc converters.

II. BASIC STRUCTURE OF GaN HEMT

The vertical cross-section of a GaN HEMT is illustrated in Fig. 1 [2]. It consists of a Si substrate over which a thin layer of AlN is grown to isolate the device structure from the substrate. A thick layer of highly resistive GaN is grown over the AlN surface. A layer of an electron generating material, Aluminum Gallium Nitride (AlGaN) is grown over the GaN layer. A highly conductive 2D electron gas layer (2DEG) is formed just underneath the AlGaN layer [3]. On the top of AlGaN additional layers are present to form the drain, gate and source electrodes of the HEMT [4]-[6].

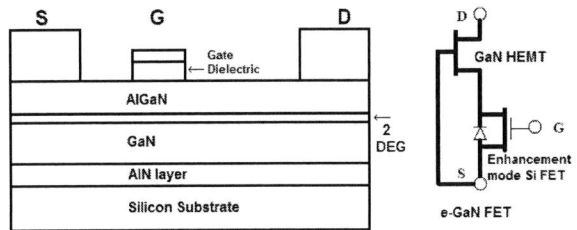

Figure 1: Vertical cross section of GaN HEMT and enhancement mode GaN FET switch (not to scale) [2]

Generally, a GaN HEMT is a normally-on type of switch. To realize a normally-off GaN transistor, a low voltage normally-off Si FET is cascaded with the GaN HEMT. The gate of the GaN HEMT is connected to the source of the Si FET device

[7]. This hybrid configuration in Fig. 1 produces an effective enhancement-mode semiconductor device, i.e. e-GaN FET device.

III. COMPARISON OF PARAMETERS OF e- GaN FET AND Si MOSFET DEVICES

This section considers e-GaN FET devices of different voltage and current ratings, which are commercially available at present. The part numbers of these e-GaN FETs and their comparable Si FET devices (in terms of maximum voltage and current ratings) are listed in Table I. Comparison of different parameters like on-state drain-source resistance ($R_{DS,ON}$), source to drain voltage (i.e. reverse conduction drop, V_{SD}), drain-source leakage current (I_{DSS}), gate-source leakage current (I_{GSS}), and maximum allowable junction temperature ($T_{J,max}$) are presented in Table I. Table II presents a comparison of the maximum gate-source threshold voltage ($V_{GS,th,max}$), total gate charge (Q_g), gate to source charge (Q_{gs}), gate to drain charge (Q_{gd}), input capacitance (C_{iss}), output capacitance (C_{oss}) and reverse transfer capacitance (C_{rss}) of e-GaN FET and corresponding Si MOSFET devices.

Table I shows that the $R_{DS,ON}$ values of e-GaN FET devices are much lower than those of similarly rated Si MOSFETs, specially at high voltage and high current ratings; however, $R_{DS,ON}$ is found to be comparable at 40V rating.

The reverse conduction drops (V_{SD}) of e-GaN FET devices at rated source current (I_S) are higher than those of corresponding Si MOSFET devices.

Also, in most cases, the maximum I_{DSS} and the maximum I_{GSS} of the e-GaN FET devices are much higher than those of the corresponding Si MOSFET devices. But the leakage currents are found to be comparable for the 600-V devices as seen from Table I.

The maximum gate-source threshold voltage ($V_{GS,th,max}$) of e-GaN FETs are always close to 2.5V, while this varies between 2V and 5V for the Si MOSFETs.

The total gate charge (Q_g) requirement of e-GaN FET devices are much lower than similarly rated Si MOSFET devices. Gate-source charge (Q_{gs}) and gate-drain charge (Q_{gd}) requirements of e-GaN FET devices are also lower than corresponding Si MOSFET devices.

The input capacitances (C_{iss}) of e-GaN FET devices are also lower than similar rating Si MOSFET devices. For many voltage and current ratings, the output capacitance (C_{oss}) and reverse transfer capacitance (C_{rss}) are comparable for both kinds of devices.

The lower values of $R_{DS,ON}$ of e-GaN devices are indicative of low on-state conduction loss. But e-GaN FETs suffer from high reverse conduction drop and very high gate leakage current. From the switching performance perspective, low input capacitance (C_{iss}) and lower total gate charge (Q_g) requirement imply that the switching transitions are much faster for the e-GaN devices than for the Si devices, as will be confirmed in the following section.

IV. ESTIMATION OF SWITCHING TRANSITION TIMES

Turn-on delay time ($t_{d,ON}$), turn-on rise time (t_r), turn-off delay time ($t_{d,OFF}$) and turn-off fall time (t_f) are essential to calculate switching losses. These parameters can be estimated from the gate-charge characteristic, available in the datasheets [8] and illustrated in Fig. 2.

A. Turn-on transition:

The turn-on transition consists of turn-on delay time $t_{d,ON}$, drain current rise time t_{ri}, and voltage fall time t_{fv}. During turn on, the gate drive circuit can be modeled as a simple RC circuit, consisting of the gate resistance R_g and device input capacitance C_{iss}. The value of capacitance, C_{iss}, varies during the switching transition. C_{iss} can be obtained from the gate-charge characteristics (V_{GS} Vs Q_g in Fig. 2) at different instants in the switching interval [9].

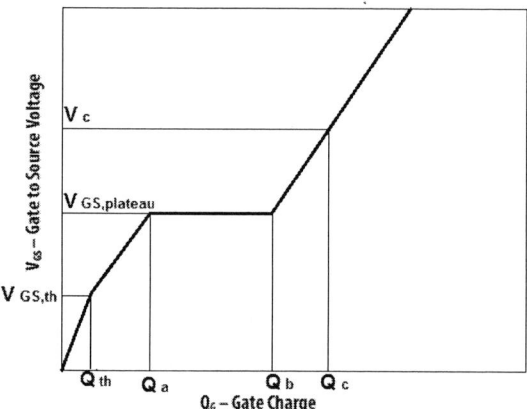

Gate Charge Characteristics :

Figure 2. Typical gate voltage versus gate charge characteristics of FET devices (not to scale) [8]

a) Turn-On Delay Time ($t_{d,ON}$):

During the turn-on delay time, the gate-source voltage (V_{GS}) rises from the off-state gate-source voltage ($V_{GS,OFF}$) to the gate-source threshold voltage ($V_{GS,th}$). C_{iss} can be approximated as shown in (1) [9].

$$C_{iss} \approx (Q_{th}/V_{GS,th}) \tag{1}$$

Now, based on the first-order response of the gate circuit, $t_{d,ON}$ can be obtained from (2), considering the typical threshold voltage, on-state gate voltage and off-state gate voltage [9].

$$V_{GS,th} = (V_{GS,ON} - V_{GS,OFF}) \times (1 - \exp(-t_{d,ON} / (R_g \times C_{iss})) + V_{GS,OFF} \tag{2}$$

b) Drain Current Rise Time (t_{ri}):

The time required for the drain current to build up from zero to $I_{D,rated}$ is the drain current rise time (t_{ri}). Referring to Fig. 2, V_{GS} reaches $V_{GS,plateau}$ from $V_{GS,th}$ during this interval. In this interval C_{iss} can be expressed as-

$$C_{iss1} \approx (Q_a - Q_{th}) / (V_{GS,plateau} - V_{GS,th}) \tag{3}$$

Again, based on the first-order response of the gate circuit, the current rise time can be evaluated using (4).

$$V_{GS,plateau} = (V_{GS,ON} - V_{GS,th}) \times (1 - \exp(-t_{ri} / (R_g \times C_{iss1}))) + V_{GS,th} \tag{4}$$

c) Voltage Fall Time (t_{fv}):

The time required for the drain-source voltage V_{DS} to decrease from the blocking state value of V_{DD} to the on-state value of ($I_{D,rated}$ $R_{DS,ON}$) is called the voltage fall time (t_{fv}). The gate-source voltage is clamped at $V_{GS,plateau}$ during this interval (see Fig. 2). The unchanging V_{GS} assures a constant gate current [8], which can be expressed as-

$$I_{G,fv} = ((V_{GS,ON} - V_{GS,plateau}) / R_g) \qquad (5)$$

Based on the constant gate current and the gate charge curve, t_{fv} can be determined as follows [9]:

$$t_{fv} = (Q_b - Q_a) \times R_g / (V_{GS,ON} - V_{GS,plateau}) \qquad (6)$$

B. Turn off transition:

The turn-off transition can be divided into three intervals, namely turn-off delay time ($t_{d,OFF}$), voltage rise time (t_{rv}) and drain current fall time (t_{fi}).

a) Turn Off Delay Time ($t_{d,OFF}$):

During the turn-off delay time, the gate-source voltage (V_{GS}) falls from the on-state gate-source voltage ($V_{GS,ON}$) to the gate-source plateau voltage ($V_{GS,plateau}$). In this interval, C_{iss} can be approximated as shown in (7).

$$C_{iss2} \approx ((Q_c - Q_b) / (V_c - V_{GS, plateau})) \qquad (7)$$

Considering a first-order response as earlier, $t_{d,OFF}$ can be determined by solving (8).

$$V_{GS,plateau} = (V_{GS,OFF} - V_{GS,ON}) \times (1 - exp(-t_{d,OFF} / (R_g \times C_{iss2}))) + V_{GS,ON} \quad (8)$$

b) Voltage Rise Time (t_{rv}):

During the drain-source voltage rise time, the gate-source voltage remains virtually constant at $V_{GS,plateau}$ (see Fig. 2). Hence, a constant gate current $I_{G,rv}$ flows during this interval as shown [9].

$$I_{G,rv} = |((V_{GS,OFF} - V_{GS,plateau}) / R_g)| \qquad (9)$$

The voltage rise time t_{rv} can now be expressed as:

$$t_{rv} = (Q_b - Q_a) \times R_g / |(V_{GS,OFF} - V_{GS,plateau})| \qquad (10)$$

c) Drain Current Fall Time (t_{fi}):

During this interval V_{GS} reaches $V_{GS,th}$ from $V_{GS,plateau}$. In this interval C_{iss} can be expressed as shown in (3). As earlier, the current fall time can be obtained by solving the first-order equation in (11).

$$V_{GS,th} = (V_{GS,OFF} - V_{GS,plateau}) \times (1 - exp(-t_{fi} / (R_g \times C_{iss1})) + V_{GS,plateau} \quad (11)$$

C. Comparison of switching times :

The switching times of 10 e-GaN FET devices (numbered 1 to 10 in Table III) are calculated as per the above procedure. These transition times are tabulated in Table III. The switching times of the other e-GaN devices (i.e. number 11 in Table III) and the Si MOSFET devices are available in the respective datasheets. These are also shown in Table III. As seen from the table, the switching times for the e-GaN FET devices are much lower than the switching times of the similarly rated Si MOSFETs. This is expected to reduce the switching loss significantly, as verified in the following section.

V. CALCULATION OF SWITCHING LOSS

Switching loss of e-GaN FET devices at a given switching frequency can be estimated using theoretically calculated switching times in the previous section. The estimated switching losses are then compared with those of similar rated Si MOSFET devices in Table III.

In Si devices, due to large variation of C_{gd} during the change of drain-source voltage, the rate at which V_{DS} drops or rises varies widely. During switching on, V_{DS} falls at a faster rate initially. But the falling rate reduces as V_{DS} reaches closer to its on-state value. For simplicity of calculation, the V_{DS} is assumed to change linearly with time during both turn-on and turn-off transitions. This analysis presents the worst-case switching loss scenario [10].

For both devices, the worst-case turn-on energy loss E_{ON} can be expressed in terms of the dc voltage V_{DD} and the load current $I_{D,ON}$ as shown [10]:

$$E_{ON} = 0.5 \times V_{DD} \times I_{D,ON} \times (t_{ri} + t_{fv}) \qquad (12)$$

Similarly, the worst-case turn-off energy loss E_{OFF} can be given by (13):

$$E_{OFF} = 0.5 \times V_{DD} \times I_{D,ON} \times (t_{rv} + t_{fi}) \qquad (13)$$

Here t_{ri}, t_{fv}, t_{rv} and t_{fi} are the different switching intervals calculated in the previous section.

The switching loss in the device is the product of the sum of switching energy losses and the switching frequency (f_{sw}) as shown [10]:

$$P_{sw} = (E_{ON} + E_{OFF}) \times f_{sw} \qquad (14)$$

Using the above expressions, the switching losses are calculated for all the e-GaN FET devices and their comparable Si MOSFET devices in Table III. A switching frequency of 100 kHz is considered. As shown by Table III, the switching losses of the e-GaN FET devices rated between 40V and 200V are much lower than those of the corresponding Si MOSFET devices.

VI. EXPERIMENTAL RESULTS

To compare the switching transitions of Si and eGaN devices experimentally, Si4456DY and EPC2014 (both rated 40V) are considered. A 15-V input, 3.3-V output 50-W buck chopper, based on the Si device Si4456DY, is designed and built. A photograph of this is shown in Fig. 3. The measured gate-source and drain-source voltage switching waveforms during turn on and turn off are shown in Fig. 4 and Fig. 5, respectively. The measured drain-source voltage rise time (t_{fv}) and voltage fall time (t_{rv}) compare reasonably with the datasheet values.

The measured switching transitions of a 40-V, 10-A e-GAN FET device (EPC2014) for V_{DS} of 12V, V_{GS} of 5V, pull-up gate resistance of 1Ω and pull-down gate resistance of 0.47Ω are shown in Fig. 6 to Fig. 8. The device is switched at a frequency of 465 kHz and a duty ratio of 44%. The measured gate to source voltage waveforms during turn-on and turn-off transitions are presented in Fig. 6 and Fig. 7, respectively. Significant oscillations are observed in the waveforms on account of circuit parasitics. The drain-source voltage rise time is found to be 2.5 ns from Fig. 8. This is much lower than the corresponding value for Si device as can be seen from Table IV.

978-1-4799-6047-7/14 $31.00 © 2014 IEEE

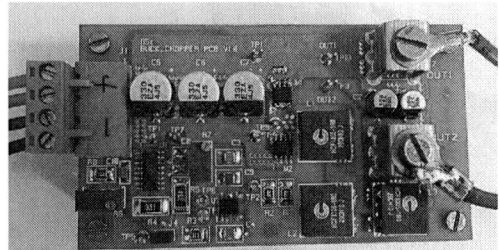

Figure 3: Buck chopper circuit using Si4456DY MOSFETs

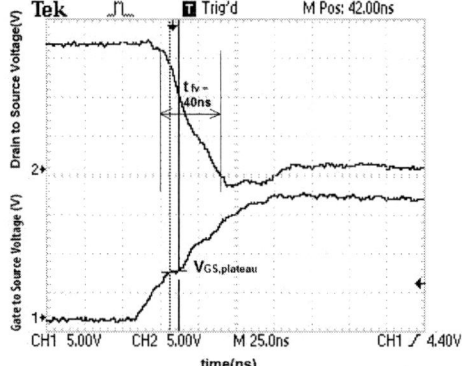

Figure 4: Measured variation of V_{GS} and V_{DS} during turn-on transition of Si4456DY (V_{IN}=15V; Rg=1.5Ω; I_{Load}=9.45A; f_{SW}=100kHz)

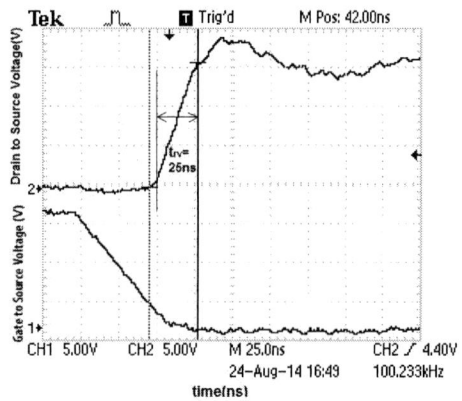

Figure 5: Measured Variation of V_{GS} and V_{DS} during turn-off transition of Si4456DY (V_{IN}=15V; Rg=1.5Ω; I_{Load}=9.45A; f_{SW}=100kHz)

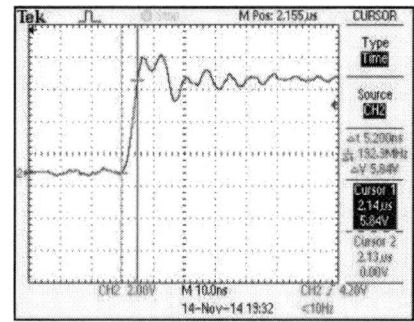

Figure 6: Measured variation of V_{GS} during turn on of EPC2014 (V_{IN}=12V; R_{Load}=8.8Ω; f_{SW}=465kHz)

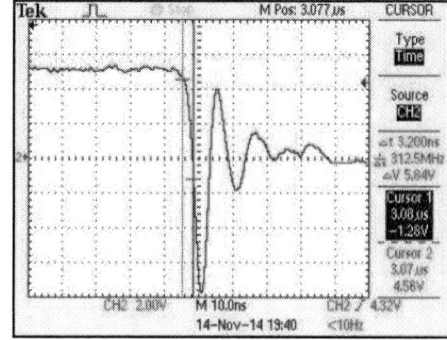

Figure 7: Measured variation of V_{GS} during turn off transition of EPC2014 (V_{IN}=12V; R_{Load}=8.8Ω; f_{SW}=465kHz)

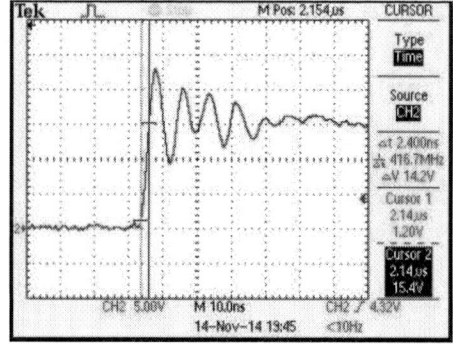

Figure 8: Measured variation of V_{DS} during turn off transition of EPC2014 (V_{IN}=12V; R_{Load}=8.8Ω; f_{SW}=465kHz)

VII. CONCLUSION

A set of commercially available eGaN devices is compared with a set of Si MOSFET devices of the same voltage and current ratings for the purpose of designing power converters. Parameters such as on-state resistance, reverse conduction drop and leakage currents are compared based on the technical information in the device datasheets. A procedure to calculate switching transition times and switching loss is also discussed. The calculations show that the switching times at a given operating condition and the switching loss at a given frequency are much less for e-GaN devices than comparable Si MOSFET devices. The study shows that the e-GaN FET devices typically have lower on-state resistances than similarly rated Si MOSFET devices. However, the reverse conduction drop is higher in case of eGaN devices than Si devices. Further, both the drain and gate leakage currents are much higher in case of eGaN devices than in case of Si MOSFETs.

REFERENCES

[1] Michael A. Briere, "GaN Based Power Devices: Cost-Effective Revolutionary Performance", Power Electronics Europe, Issue 7, 2008

[2] Michael A. Briere, "GaN based Power Devices", RPI CFES Conference, January 25, 2013

[3] M. Asif Khan, A. Bhattarai, J.N. Kuznia, and D.T. Olson, "High Electron Mobility Transistor Based on a GaN-AlxGa1-xN Heterojunction," Appl. Phys. Lett., vol. 63, no. 9, 1993, pp. 1214-1215. .

[4] Stephen L. Colino and Robert A. Beach, "Fundamentals of Gallium Nitride Power Transistors", EPC application note, Copyright Efficient Power Corporation, 2009.

[5] Alexander Lidow, J. Brandon Witcher and Ken Smalley, "Enhancement Mode Gallium Nitride (eGaN TM) FET Characteristics under Long Term Stress".

[6] Edgar Abdoulin, Steve Colino and Alana Nakata, "Using Enhancement Mode GaN-on-Silicon Power Transistors", EPC application note no. AN003.

[7] Primit Parikh, Yifeng Wu and Likun Shen, "Commercialization of High 600V GaN-on-Silicon Power Devices", Transphorm Inc..

[8] Datasheets of e-GaN FET and Si MOSFET devices whose part numbers are mentioned in 'Table I'.

[9] "Understanding and Predicting Power MOSFET Switching Behavior", On semiconductor Application note, Publication Order Number: AN1090/D, August 2002 – Rev. 0, copyright Semiconductor Components Industries, LLC, 2002.

[10] Dr. Dušan Graovac, Marco Pürschel and Andreas Kiep, "MOSFET Power Losses Calculation Using the Data-Sheet Parameters", Infineon Application Note, V 1.1, July 2006.

TABLE I. COMPARISON OF IMPORTANT DATASHEET PARAMETERS – I (AT T_J = 25°C (UNLESS OTHERWISE SPECIFIED)) [8]

Voltage Rating (V)	Current Rating (A)	Si MOS FET Part No.	e-GaN FET Part No.	$R^1_{DS,ON}$(mΩ)		V^2_{SD}(V)		I^3_{DSS}(µA)$_{max}$		I^4_{GSS}(µA)$_{max}$		$T_{J,max}$(°C)	
				Si MOS FET	e-GaN FET	Si MOS FET	e-GaN FET	Si MOS FET	e-GaN FET	Si MOS FET	e-GaN FET	Si MOS FET	e-GaN FET
40	10	IRF7470	EPC1014	13	12	0.9	2.9	20	100	0.2	2000	150	125
40	33	Si4456DY	EPC1015	3.8	3.2	0.8	2.7	1	400	0.1	7000	150	125
60	6	Si7308DN	EPC1009	46	24	0.9	2.6	1	60	0.1	2000	150	125
60	25	RFF70N06	EPC1005	25	5.6	1.1	2.6	25	250	100	5000	150	125
100	6	STD6NF10	EPC1007	240	24	1.3	2.6	1	60	0.1	2000	175	125
100	25	IRFP9150	EPC1001	100	4.4	0.9	2.6	25	250	0.1	5000	150	125
150	3	HUFA75831SK8	EPC1013	79	70	1.25	2.3	1	40	0.1	1000	150	125
150	12	IRL3215	EPC1011	166	18	1.05	2.9	1	150	0.1	3000	175	125
200	3	FQD4N20	EPC1012	1400	62	1.02	2.3	1	40	0.1	1000	150	125
200	12	IRFP9240	EPC1010	380	17	1.15	2.9	25	150	0.1	3000	150	125
600	9	FCP9N60N	TPH3002PD	395	290	0.88	2.3	10	60	0.1	-	150	175
600	17	2SK4125	TPH3006PD	470	150	0.86	2.3	100	90	0.1	0.1	150	175

1. Typical values of $R_{DS,ON}$ at rated drain current and typical $V_{GS,ON}$ (i.e. 5V for e-GaN FET and 10V for Si-MOSFET).
2. Source-drain reverse conduction drop at rated current obtained from the reverse drain-source characteristics.
3. I_{DSS} is measured either at maximum V_{DS} or at 80% of maximum V_{DS}.
4. Gate to source leakage current at typical $V_{GS,ON}$.

TABLE II. COMPARISON OF IMPORTANT DATASHEET PARAMETERS – II (AT T_J = 25°C) [8]

Voltage Rating (V)	Current Rating (A)	$V_{GS,th,max}$(V)		Q^5_g(nC)		Q_{gs}(nC)		Q_{gd}(nC)		C^6_{iss}(pF)		C^6_{oss}(pF)		C^6_{rss}(pF)	
		Si MOS FET	e-GaN FET	Si MOS FET	e-GaN FET	Si MOS FET	e-GaN FET	Si MOS FET	e-GaN FET	Si MOS FET	e-GaN FET	Si MOS FET	e-GaN FET	Si MOS FET	e-GaN FET
40	10	2.0	2.5	29	3.0	7.9	1.0	8.0	0.55	3430	280	690	150	41	15
40	33	2.8	2.5	37.5	11.6	17	3.8	11	2.2	5670	1100	621	525	287	60
60	6	3.0	2.5	6	2.4	2.3	0.75	2.6	0.63	665	200	75	170	40	15
60	25	4.5	2.5	-	10	-	3.0	-	2.5	3100	790	900	500	300	43
100	6	4.0	2.5	10	2.7	2.5	0.75	4	1.0	280	200	45	130	20	10
100	25	4.0	2.5	82	10.5	14	3.0	42	3.3	2400	800	850	500	400	40
150	3	4.0	2.5	35	1.7	4.3	0.37	11	0.7	1175	110	275	100	72	7.5
150	12	2.0	2.5	35	6.7	4.1	1.5	21	2.8	775	440	140	405	70	30
200	3	5.0	2.5	5	1.9	1.4	0.37	2.1	0.9	170	110	35	100	5	7.5
200	12	4.0	2.5	38	7.5	8	1.5	21	3.5	1400	440	350	415	140	30
600	9	4.0	2.35	22	6.2	4.1	2.1	7.1	2.2	930	785	35	26	2	3.5
600	17	5.0	2.35	46	6.2	8.3	2.1	26.7	2.2	1200	740	220	113	50	3.6

5. Test condition of gate charge measurement for e-GaN FET: 5V V_{GS}, 50% of maximum V_{DS}, rated I_D; for Si MOSFET: 4.5V or 10V V_{GS}, 50% of maximum V_{DS}, rated I_D or as mentioned in the device datasheet.

6. Capacitance values of e-GaN FET have been selected from capacitance versus V_{DS} characteristics for the same V_{DS} for which capacitance values of Si MOSFETs are given in datasheet.

TABLE III. COMPARISON OF SWITCHING TIMES AND SWITCHING LOSSES OF E-GAN AND SI MOSFET DEVICES

Sl. No.	Voltage Rating (V)	Current Rating (A)	Si MOSFET Device[1]				e-GaN FET Device[2]				Si MOSFET Device			e-GaN FET Device		
			$t_{d,ON}$ (nS)	t_r (nS)	$t_{d,OFF}$ (nS)	t_f (nS)	$t_{d,ON}$ (nS)	t_r (nS)	$t_{d,OFF}$ (nS)	t_f (nS)	V_{DD} (V)	$I_{D,ON}$ (A)	P^*_{SW} (W)	V_{DD} (V)	$I_{D,ON}$ (A)	P^*_{SW} (W)
1.	40	10	10	1.9	21	3.2	0.143	0.204	0.072	0.076	20	8	0.0408	20	10	0.0028
2.	40	33	21	58	55	8	0.561	0.734	0.318	0.280	20	10	0.66	20	33	0.033
3.	60	6	10	15	20	10	0.100	0.194	0.059	0.071	30	4.3	0.1612	30	6	0.00238
4.	60	25	25	70	60	25	0.403	0.822	0.236	0.309	30	25	3.56	30	25	0.042
5.	100	6	6	10	20	3	0.102	0.245	0.06	0.089	50	3	0.0975	50	6	0.00501
6.	100	25	16	110	65	46	0.409	0.961	0.239	0.349	50	25	9.75	50	25	0.082
7.	150	3	11	6	40	9	0.056	0.164	0.033	0.060	75	3	0.17	75	3	0.00252
8.	150	12	7.4	45	38	36	0.224	0.697	0.133	0.252	75	7.2	2.187	75	12	0.043
9.	200	3	7	50	7	25	0.056	0.217	0.034	0.077	100	3	1.125	100	3	0.00441
10.	200	12	18	45	75	29	0.224	0.823	0.14	0.295	100	12	4.44	100	12	0.067
11.	600	17	26.5	82	145	52	4	3	10.5	3.5	200	8.5	11.39	480	11	1.716

1. Switching transients of GaN FET device no. 11 and other Si MOSFET devices are taken from their respective datasheets [8]. Standard test conditions are as specified in the datasheets.
2. Switching transients of GaN FET device numbers 1 to 10 are calculated, considering $V_{GS}=\pm 5V$, typical $V_{GS,th}=1.4$ V and gate source resistance $R_g=0.5\Omega$ as per application note [9].
* Switching loss is calculated at 100 kHz switching frequency.

TABLE IV. MEASURED SWITCHING TRANSITION TIMES

Sl. No.	Parameters	Si-MOSFET (Si4456DY)	e-GaN FET (EPC2014)
1	Gate-source voltage (V_{GS}) rise time	55 ns	5.2 ns
2	Gate-source voltage (V_{GS}) fall time	50 ns	3.2 ns
3	Drain-source voltage (V_{DS}) rise time	25 ns	2.5 ns

978-1-4799-6047-7/14 $31.00 © 2014 IEEE

Current Regulated Induction Motor Drive with IFOC

Loveleen Taneja[1], Saurabh Kumar[2], Rohit Kumar[3],

Assistant Professor, Department of Electrical Engineering, PEC University of Technology, Chandigarh1
Student, Department of Electrical Engineering, PEC University of Technology, Chandigarh[2,3],
Email- loveleentaneja@yahoo.co.in[1], saurabh.k1@live.com[2], Rohitmundey.089@gmail.com[3],

Abstract—for very low speed operation and position type control, the use of flux sensing that relies on integration which has a tendency to drift may not be acceptable. The most commonly used alternative is indirect filed orientation, which does not rely on the measurement of the air gap flux, but uses various governing equations to satisfy the condition for proper orientation. A current controlled PWM induction motor drive is to be studied and the speed regulation is achieved using a PI controller which converts the command speed into command torque and finally the speed of the drive.

Keywords: Induction Motor, closed loop control, motor drive, PI controller

I. INTRODUCTION

With the availability of faster and less expensive processors and solid state switches, alternating current induction motor drives can be compared favorably to dc motor drives on considerations such as power to weight ratio, acceleration performance, maintenance, operating environment, and higher operating speed without the mechanical commutator. Costs and robustness of the machine, and perhaps control flexibility, are often reasons for choosing induction machine drive in small to medium power range applications. Even without simulating the switching waveforms in detail that is with only the fundamental component of inverter outputs simulated, we can still gain useful insights into the dynamic performance attainable by common control strategies used in today's adjustable speed induction motor drives.

The main aim of this paper is to study the field oriented control strategy as the torque control can be obtained in it and thus allowing the torque to act as a torque transducer.[1] Irrespective of the robust direct filed oriented control, the problem regarding the sensing of the air gap flux linkage and expensive approach behind it motivated to develop a study of indirect filed oriented control technique which are more sensitive to knowledge of the machine parameters but do not require direct sensing of the rotor flux linkages.

For study purpose, 20HP induction motor data was used, the details of which are given in the TABLE I.

TABLE1
DATA OF INDUCTION MOTOR

Symbol	Quantity	Value
P	Poles,P	4
HP	Capacity	20
V	Voltage rating	220 V
f_{rated}	Rated frequency (Hz)	50 Hz
P.F	Rated power factor	0.853
S_{rated}	Rated slip	0.0287
N_{rated}	Rated speed(Rev/Min)	1748.3
i_{asb}	Rated armature current	49.68
X_{ls}	stator leakage reactance (ohms)	0.2145 ohms
X_M	Magnetizing Reactance	5.8339 ohms
X_{plr}	Rotor leakage reactance	0.2145 ohms
J	Rotor Inertia	2.8 kg/m^2

Base electrical frequency can be known by the formula

$$wb = 2*pi*frated \qquad (1)$$

where frated is the rated frequency in Hertz(Hz)[2].

II. FIELD OREINTATION METHOD:

Field orientation is a technique for controlling the flux and electromagnetic torque of an induction motor by using electrical parameters estimated in closed-loop; the estimated parameters are used to tune the flux and torque controllers.

This control technique is divided into two subcategories i) Open Loop Indirect vector control ii) Closed Loop Indirect vector control

978-1-4799-6047-7/14 $31.00 © 2014 IEEE

$$Tem = \left(\frac{3}{2}\right)\left(\frac{P}{2}\right)\left(\frac{Lm}{Lr}\right)\lambda'^e_{dr}\, i^e_{qs} \qquad (2)$$

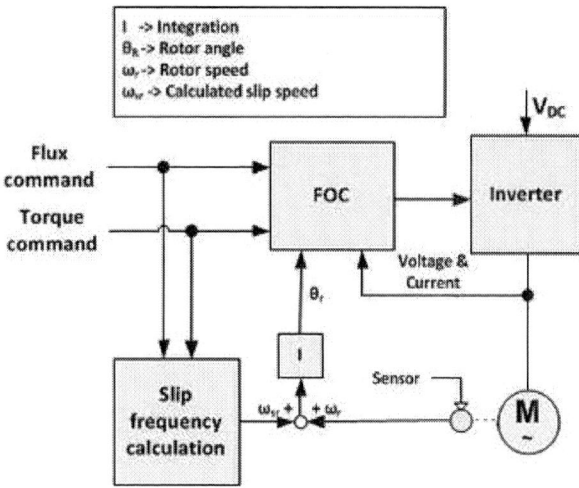

Figure 1 Direct Vector control Scheme

The field-oriented control has allowed extending the use of induction motors in high performance applications. Direct field-oriented control (DFOC) includes a closed loop rotor flux controller and requires the calculation of the amplitude and position of the rotor flux [8]. This is standard solution for high performance motor drives but requires relatively complex algorithms. Indirect-field oriented control (IFOC) does not need a closed loop rotor flux controller and only requires the angular position of the rotor flux vector which is calculated by integrating the vector angular speed; this can be computed using the rotor speed and the stator current measurement.

III. INDIRECT FIELD OREINTATION CONTROL

Indirect Field Oriented Control is the standard regulation method for high performance induction motor drives that computes the rotor flux angular frequency by integration of the sum of the rotor speed and the slip angular frequency, respectively provided by a suitable transducer, and a simple expression obtained from the mathematical model of the induction motor. [3] The IFOC technique is essentially a predictive approach in that it estimates the angular position of the rotor flux vector by exploiting the model of the machine.

A commonly used IFOC technique uses the following equations to satisfy the condition for proper orientation:

The developed torques T_{em} reduces to

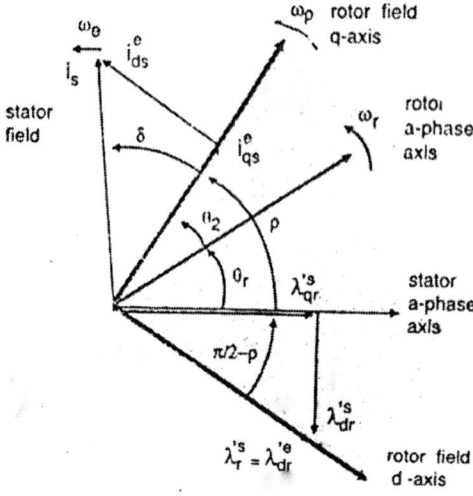

Figure 2 Properly oriented qd synchronously rotating frame

Relation between slip speed and the ratio of the stator qd current components for the d axis of the synchronously rotating frame to be aligned with the rotor field :

$$\omega *_2 = \omega_e - \omega_r =(r'_r/L'_r)(i^e_{qs}/i^e_{ds})\text{elect.rad/s} \qquad (3)$$

Given some desired level of rotor flux, the desired value of $i^{e*}_{ds\ d}$ may be obtained from

$$\lambda'^*_{dr} = ((r'_rL_m)/(r'_r+L'_rp))i^{e*}_{ds}\ \text{Wb.turn} \qquad (4)$$

If the above conditions is satisfied, ensure the decoupling of the rotor voltage equations. To what extent this decoupling is actually achieved will depend on the accuracy of motors parameters used. Indirect field oriented control scheme for a current controlled PWM induction motor drive can be shown as

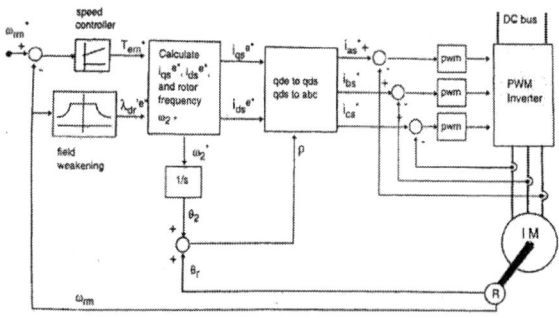

Figure 3 Indirect field oriented control scheme

978-1-4799-6047-7/14 $31.00 © 2014 IEEE

IV. MODEL FOR IFOC CONTROL WITH CSI

In this part, we will implement the SIMULINK simulation of a current regulated PWM induction motor drive with indirect field oriented control. The objectives are to become familiar with the implementation of a type of field oriented control and to examine how well such a control keeps the rotor flux constant during changes in load torque. For simulation purposes, we will use the same 20-hp, 220-V , four pole induction machine whose parameters are already defined in Table 1.

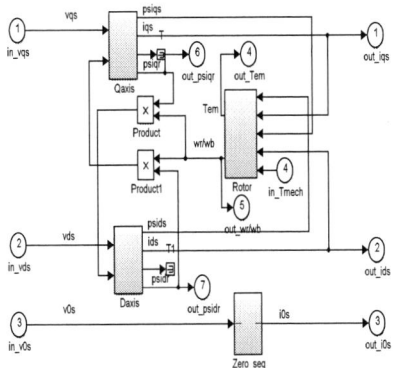

Figure 4 Induction Machine Simulation
in Stationary Reference Frame

The overall SIMULINK simulation diagram for a current regulated PWM induction motor drive with IFOC is given as :

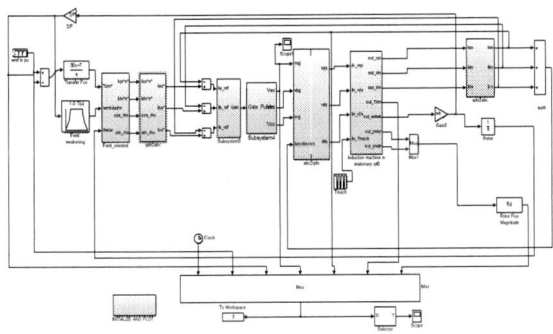

Figure 5 Simulink Model for Current regulated IM drive

On the left side of the model, a proportional-integral (PI) torque controller converts the speed error to a reference torque T^*_{em}. Going into the field orientation block are the reference torque, T^*_{em}, the d axis rotor flux λ'^*_{dr} and the rotor angle, θ_r.

The inside of the field oriented block is shown as

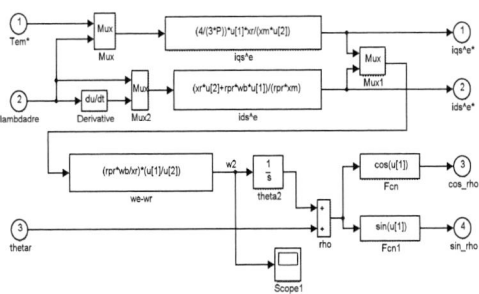

Figure 6 Inside the field oriented block

Equations (2), (3) and (4) are used inside the field oriented block to compute the values of Tem, λ'^*_{dr} and ω^*_2. Simulation for the generation of gate pulses and PWM inverter is given in figure 7 and figure 8 respectively.

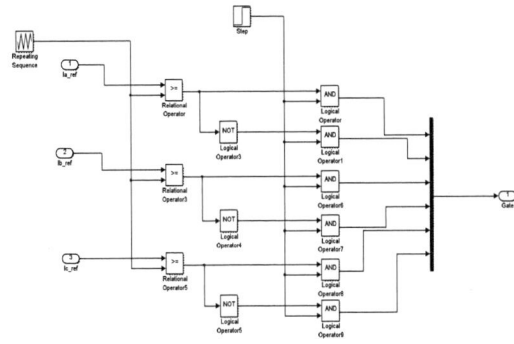

Figure 7 : Simulink subsystem for the generation for gate pulses

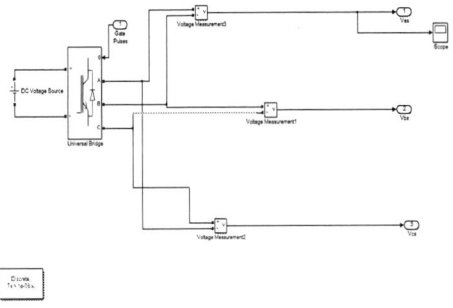

Figure 8:Simulink subsystem of inverter for the generation of

Voltages

The generated gate pulses are fed accordingly to the inverter for switching and thus these generated voltages after being converted to abc reference frame are fed to the induction motor.

A. Study of parameters using IFOC controller

The values for the reference speed , ω_{ref} , the rotor mechanical speed , ω_{rm}, the stator a-phase voltage , V_{ag} , the stator a-phase current , i_{as} , the electromagnetic torque , T_{em} ,and the magnitude of the rotor flux is shown in the plots at the output.

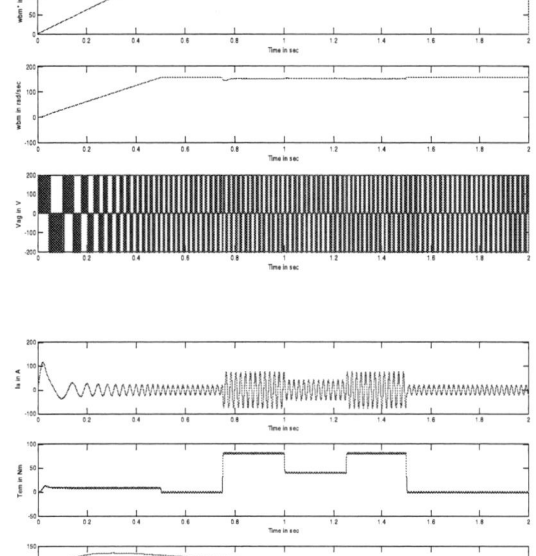

Figure 9 : Starting and loading transients with field oriented control

B. Observations and conclusion

From the output plots, we can observe how quickly and smoothly the speed of the motor responds to the programmed speed reference with the field oriented control. We can also observe how well filed oriented control maintains the magnitude of the rotor flux over the simulated conditions.

V. STUDY UNDER THE CYCLIC CHANGE OF SPEED REFERNCE

For this study , we have repeated the above run with the motor carrying an externally applied mechanical torque equal to the rated torque, that is $T_{mech}= -T_{rated}$.

A. The sample results for this run are given below

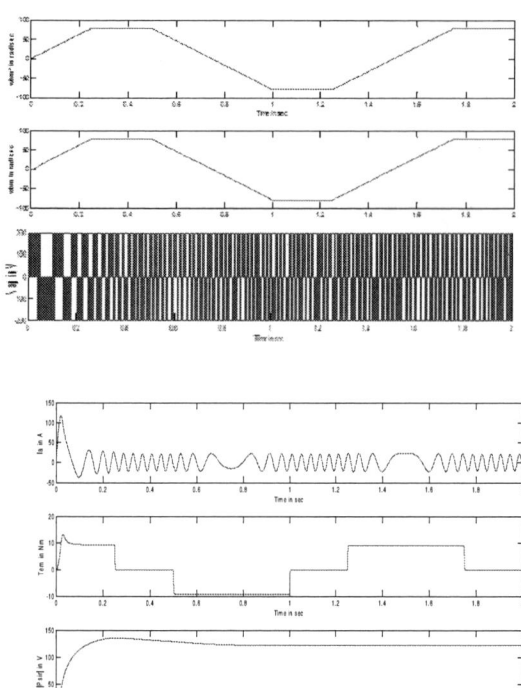

B. Observations and Conclusion

From the above plots, we can notice the slightly larger transient of the rotor flux at the initial startup of the motor from standstill. We can also examine the acceleration torque, that is $T_{mech}+ T_{em}$,and the rotor angle θ_r during the initial startup transients. From these plots, it can be concluded that the speed of an induction motor can be controlled conveniently using the IFOC control of induction motor.

VI. CONCLUSION

A. Effect of using the PI controller

By putting the PI controller in our studies, the steady state value of rotor is achieved just at time t=2 seconds. Thus we can say that incorporating the PI controller with its integral portion, makes the steady state error in speed zero and there is a stable operation which in turn improves the quality of operation os the motor as well as saves its life.

B. Advantages of IFOC over other speed control techniques

As compared to the other close loop control methods such as Voltage per Hertz control, Indirect Rotor Field Oriented control has faster speed response, less transient oscillations in terms of voltage, current, and torque. Therefore, its performance is higher and more advantages than close loop Voltage per Hertz control. The amplitudes of the transients are completely controlled by the IFOC method which was not possible by other methods such as Volt-per-Hertz Control. The speed control is better and smoother in the case of IFOC control.

VII. REFERENCES

[1] Prasad Shrawane, "Indirect Field - Oriented Control of Induction Motor" IEEE Transactions On Industrial Electronics, Vol. 56, No. 2, February 2010.

[2] Chee-Mun Ong, "Dynamic Simulation of Electric Machinery: Using MATLAB/SIMULINK," Prentice Hall 1997, ISBN: 0137237855.

[3] R.Krishnan, "Electric Motor Drives: Modeling, Analysis and Control," Prentice Hall of India, 2002.

[4] Bimal K. Bose, "Modern Power Electronics and AC Drives," Pearson Education, Singapore.2003

[5] Kunwar Aditya, "Speed Control of LIM Drives using Indirect Secondary Field Oriented Control," 2013 International Conference on Circuits, Power and Computing Technologies [ICCPCT-2013]

[6] Using Simulink, Version 5, Math Works, Natick, MA.

[7] Bimal K. Bose, "Power Electronics and Motor Drives Recent Progress and Perspective" IEEE Transactions On Industrial Electronics, Vol. 56, No. 2, February 2009

[8] R. Krishnan, Victor R. Stefanovic and James F. Lindsay, "Control Characteristics of Inverter-Fed Induction Motor", IEEE Transactions on industry applications, VOL. IA-19, no. 1, January/February 1983.

Design and Analysis of Charge-pump Based Buck Converter

Mummadi Veerachary
Dept. of Electrical Engineering, IIT Delhi, New Delhi, India
E-mail: mvchary@ee.iitd.ac.in

Abstract— **In this paper a charge pump based buck converter is proposed and its suitability for constant load voltage applications is investigated. Steady-state time-domain analysis is established and then proved that this converter gives bucking feature like conventional buck converter voltage transformation. State-space analysis, discrete-time models are formulated and then discrete-time transfer functions required for digital controller design are obtained. A single-loop digital direct duty ratio controller is designed. A 42 to 28 *V*, 100 *W* prototype converter is used to predetermine the proposed converter performance in simulation environment. Simulation studies demonstrated the bucking as well as load voltage regulation feature. Analytical investigations and simulations are validated through experimental measurement.**

keywords: Buck Converter, Charge pump, Digital Controller, Voltage Regulation, Single-loop Voltage-mode Controller.

I. INTRODUCTION

In recent day's several newer dc-dc converter topologies are coming-up, which are capable of exhibiting high efficiency, lower ripple content and simple control scheme, to meet the sophisticated load demand. In the past several aspects of dc-dc distribution point of load as well as bus converters have been explored in the literature [1]. These findings include reporting newer topologies modeling and analysis, performance enhancement such as improving steady-state and dynamic performance, mitigating electro-magnetic interference (EMI), etc. Classification and brief description of various feasible dc-dc converters is discussed in ref [2]. Here, the main emphasis on generation of the 2 and 4 switch topologies with two or three energy storage elements. This exhaustive treatment is useful to identify the nature of conversion, application specific topology selection for driving the given load, etc. Among the configurations reported, the topologies having two switches, two inductor and one capacitor (or) two capacitors and one inductor are finding major application in the dc-dc conversion. This is on account of meeting the voltage transformation ratio along with ripple requirements with reduced dynamic performance limitations. Although the formulation yields somewhat higher order nature, but they have attractive features such as: (i) step-up/down voltage conversion with continuous input and output current, (ii) possible to implement the ripple steering phenomena in case converter structures having two or more inductors, (iii) realizing a wide variety of voltage conversion ratios with single switch topology, (iv) simplicity in the control on account of the single pulse-width modulation

(PWM) scheme, and (v) EMI produced by these converters is less and hence reduced EMI filtering requirement, etc.

Buck converter has a limitation of larger input current ripple and associated EMI [3]-[5]. In spite of these limitations still the modified form of this converter, buck converter with input filter (BCIF), is more popular in point of load applications. It is possible to reduce input current ripple and EMI [6]-[10] using an input filter. However, there are some destabilizing issues with the addition of input filter for the conventional buck converters. To eliminate some of these limitations of conventional buck converter a charge-pump based buck converter (CPBC) is proposed in this paper. The proposed converter has the following salient features: (i) step-down voltage transformation ratio, (ii) lower source current ripple, and (iii) less complexity in dynamics, etc. Several controlling methods such as single and two-loop control schemes, (including voltage/ current mode, and hysteretic control), have been reported for the dc-dc converters. Each of the above control schemes has their own advantages and limitations as well [6]. Here, a single-loop voltage-mode digital controller is proposed to ensure load voltage regulation. Salient advantages of the single-loop voltage-mode controller for the proposed CPBC are: (i) simple controller and easier for implementation, (ii) dynamic response of CPBC is better than BCIF, and (iii) higher load voltages at lower duty ratios. For designing the digital controller discrete-time small-signal models are used. These are established here through state-space concepts [11], and the detailed analysis is given in the preceding sections.

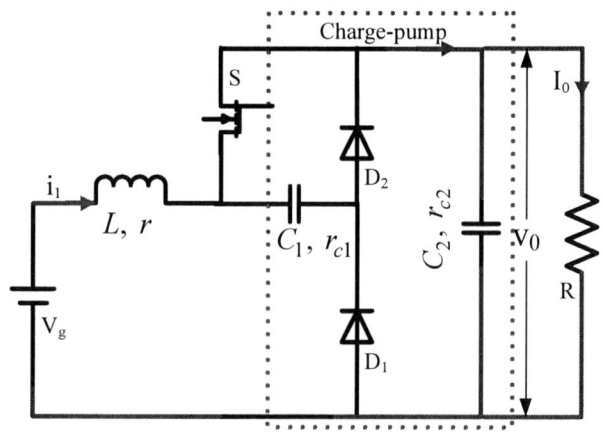

(a) Charge-pump Buck Converter

978-1-4799-6047-7/14 $31.00 © 2014 IEEE

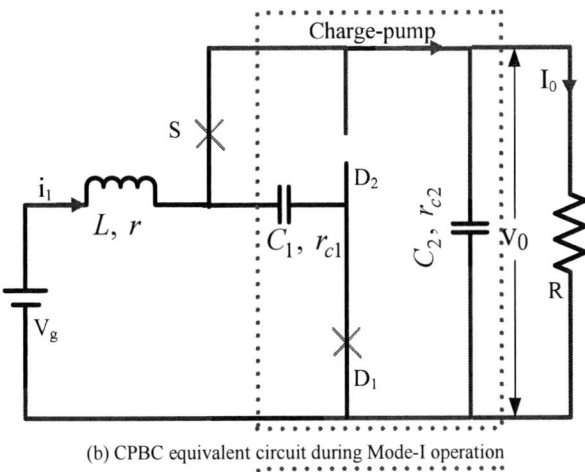

(b) CPBC equivalent circuit during Mode-I operation

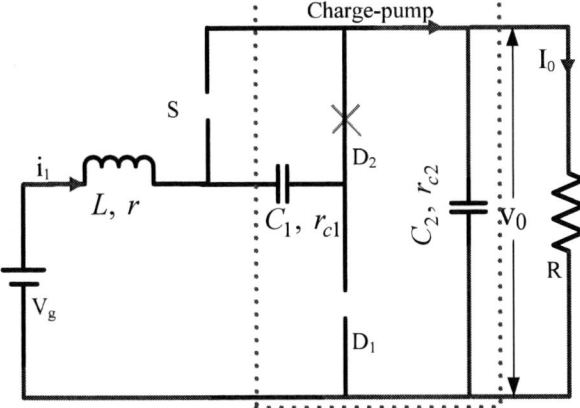

(c) CPBC equivalent circuit during Mode-II operation
Fig. 1. Charge-pump based buck converter and its equivalent circuits.

II. MODELING OF THE CHARGE-PUMP BASED BUCK CONVERTER

The circuit diagram of the charge-pump based buck converter is shown in Fig. 1. It essentially contains a charge pump formed by two capacitors (C_1, C_2) and two diodes (D_1, D_2). To realize duty ratio dependent load voltage an inductor and switch introduced as shown in Fig. 1. Here, the charge pump capacitor along with inductance gives the lower source current ripple similar to the buck converter with input filter. There is only one switch is present for power control, and it has to be driven with the desired pulse width modulated (PWM) signal. Since there is an inductance, and depending on the load power demand this current may be continuous or discontinuous. If the inductor current is discontinuous, then the source current ripple of CPBC is same as in conventional buck converter and hence there is no gain of using this converter. However, to get maximum benefits of the proposed converter the input inductor current should be continuous. Therefore, mathematical analysis is developed in the following sections for the continuous inductor current mode of operation only. The equivalent circuits corresponding to switch -ON and OFF cases are shown in Fig. 1b and 1c, respectively. Through circuit analysis principles, various KVL and KCL equations are formulated for switch -ON and

OFF equivalent circuits. From these equations, it is easy to establish volt-sec and charge-sec balance equations of the respective inductive and capacitive elements.

$$(V_g - V_0)D_1T_s + (V_g - V_{C1} - V_0)D_2T_s = 0 \qquad (1)$$

In steady-state the capacitor (C_1) charges to load voltage ($V_{C1} = V_0$). Simplifying above expression together with capacitor charging voltage condition results in the following voltage gain expression.

$$\frac{V_0}{V_g} = \frac{1}{(2 - D_1)} \qquad (2)$$

where 'D_1' is the duty ratio of the switch 'S'. This voltage gain expression appears as boost converter voltage gain, but in this case no boosting is possible on account of the additional unity term in the denominator. In conventional buck or BCIF, the voltage gain varies linearly, starting from origin, while in this case it varies linearly, but it starts from 0.5. In view of this, the step-down ratio varies from 0.5 to 1.0 for the duty ratio change: 0 to 1.0. Comparison of voltage gain of CPBC with the conventional buck converter is shown in Fig. 2. The expressions for ripple current and voltages can easily be derived from the time-domain analysis. The final ripple expressions (Δi and Δv_c) are summarized in Table-I for ready reference.

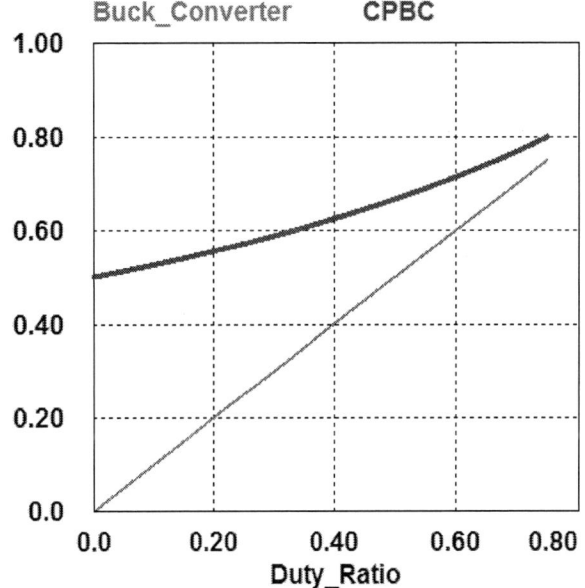

Fig. 2. Voltage gain variation with duty ratio.

TABLE I. RIPPLE AND DESIGN EQUATIONS OF CPBC.

$$L \ge V_0 D(1-D)/(f_s \Delta i)$$

$$C_1 = C_2 \ge \frac{(1-D)T_s}{\left(R(2-D)\left(\dfrac{\Delta v_c}{v_0}\right)\right)}$$

It is essential to formulate state-space models of the proposed converter system both for steady-state performance prediction as well as for digital controller design. In the controller design, using linear control theory, it is essential to know the small-signal transfer functions (either in s or z-domain) of the converter system under consideration. Since the proposed CPBC contains only three energy storage elements, the state-space model is of 3rd-order and small-signal transfer functions formulation will not be a difficult task.

Since there is only one switching device is present, only two operating modes are possible, which are: Mode-I: ($0<t<DT_S$), Mode-II: ($DT_S<t<T_S$). The equivalent circuits corresponding Mode-I/ Mode-II are shown in Fig. 1b and 1c, respectively. A set of first-order differential equations for capacitor voltages and inductor current are formulated through KCL and KVL analysis and then transformed into state-space form as given by

$$\left.\begin{aligned}[\dot{x}]=[A_j][x]+[B_j][u]\\ [y]=[E_j][x]\end{aligned}\right\} \quad t_j<t<t_{(j+1)} \quad (3)$$

$$[A_1]=\begin{bmatrix} -\dfrac{r}{L} & -\dfrac{r_{c2}c_2R}{L} & -\left(\dfrac{L-r_{c2}c_2(R+r_{c1})}{L}\right) \\ \dfrac{C_2R}{C_1} & \dfrac{1}{r_{c1}c_1} & \dfrac{-1}{r_{c1}c_1} \\ \dfrac{Rr_{c1}}{C_2} & \dfrac{R}{C_2} & \dfrac{-(R+r_{c1})}{C_2} \end{bmatrix}; \quad [B_1]=[B_2]=\begin{bmatrix} \dfrac{1}{L} \\ 0 \\ 0 \end{bmatrix};$$

$$[A_2]=\begin{bmatrix} \dfrac{r}{L} & \dfrac{-1}{L} & \dfrac{-1}{L} \\ \dfrac{1}{C_1} & 0 & 0 \\ \dfrac{1}{C_2} & 0 & \dfrac{-(R+r_{c1})}{C_2} \end{bmatrix}; \quad [x]=\begin{pmatrix} i_l \\ v_{C1} \\ v_{C2} \end{pmatrix};$$

$$[E_1]=[E_2]=\begin{bmatrix} \dfrac{RC_2}{k_1} & \dfrac{r_2C_2}{k_1} & \dfrac{C_2(R+r_1)}{k_1} \end{bmatrix}$$

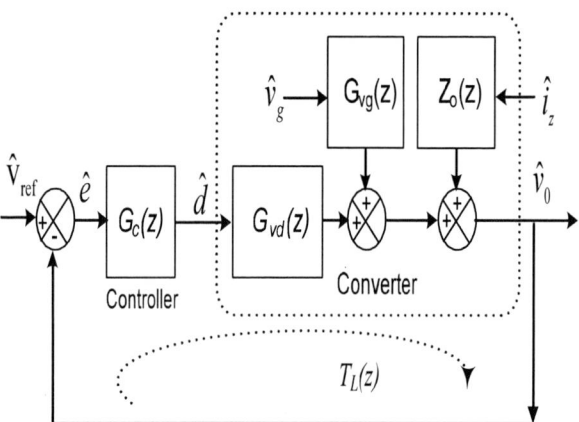

Fig. 3. Block-diagram of closed-loop controlled system

TABLE-II. EXPRESSIONS FOR DISCRETE-TIME SMALL-SIGNAL TRANSFER FUNCTIONS

Small-signal Transfer Function	Formula
Audiosusceptibility	$\hat{v}_0(z)\big/\hat{v}_g(z) = E'(zI-\phi)^{-1}\Gamma + F'$
Control -to- Output	$\hat{v}_0(z)\big/\hat{d}(z) = E'(zI-\phi)^{-1}\Gamma$
Input Admittance	$\hat{i}_{in}(z)\big/\hat{v}_g(z) = P'[(zI-\phi)^{-1}\Gamma]$
Output Impedance	$\hat{v}_0(z)\big/\hat{i}_0(z) = [E'(zI-\phi)^{-1}\Gamma + J']$

III. DIGITAL CONTROLLER DESIGN FOR CHARGE-PUMP BASED BUCK CONVERTER

To design the controller it is necessary to formulate the transfer functions [11] as shown in Fig. 3. In the load voltage regulation problems, three important transfer functions are needed as mentioned in the control block-diagram. All these transfer functions can easily be obtained from the identities given in Table II and the state-space model given by eqn. 3. From the closed-loop system block diagram, it can easily be established the loopgain transfer function $T_L(z)$. Here, the control-to-output transfer function is

$$G_{vd}(z)=\frac{\left[a_4z^4+a_3z^3+a_2z^2+a_1z+a_0\right]}{\left[b_5z^5+b_4z^4+b_3z^3+b_2z^2+b_1z+b_0\right]} \quad (4)$$

while the digital controller used for this regulation study is given by

$$G_c(z)=\frac{k(z^2-k_1z^{-1}+k_2z^{-2})}{(1-k_3z^{-1}+k_4z^{-2})} \quad (5)$$

The controller coefficients obtained by using the standard pole-placement technique of the linear control theory. As shown in Fig. 2 loopgain consist of '$G_{vd}(z)$' and '$G_c(z)$' transfer functions. The '$G_{vd}(z)$' transfer function can easily be obtained for the given input and output specifications, while the controller transfer function '$G_c(z)$' need to be computed through SISOTOOL of the matlab where-in it is easy to place the pole-zero pattern of the controller such that the closed-loop system/ loopgain meets design requirements. Matlab-sisotool [12] interactive platform is very much helpful in minimizing the design time. In this the controller pole, zero locations can easily be adjusted such that the final resulting loopgain ensures sufficient relative stability margins: gain margin > 6 dB, and phase margin: 45$°$ ~ 75$°$ together with reasonable bandwidth.

978-1-4799-6047-7/14 $31.00 © 2014 IEEE

IV. RESULTS AND DISCUSSIONS

To demonstrate the mathematical analysis and design listed in the preceding sections a 48 V to 28 V, 50 *Watt,* 100 *kHz* CPBC converter is considered here. For these specifications, energy storage elements are chosen to meet the ripple current and voltage ripple specifications: $\Delta i < 20\%$, $\Delta v < 5\%$. The computed parameters are: $L = 100~\mu H$, $C_1 = 100~\mu F$, $C_2 = 220~\mu F$. These parameters are used in simulation [13] and experiments [14]. For nominal operation with D=0.75, the steady-state waveforms are shown in Fig. 4. Here, it is observed that the source current ripple is approximately 10% of its average current. The simulated load voltage variation with three different duty ratio's (D_1=0.25, 0.5 and 0.75) is plotted in Fig. 5. It indicates that the load voltage is increasing with the increase in the duty ratio, and this trend is identical to conventional buck converter or BCIF. These results together with steady-state analysis discussed above sections clearly indicates that this converter is more suitable for realizing load voltage within the 50 ~ 100 % of source voltage with minimum source current ripple.

The controller design was carried, as explained in Section III, in the MATLAB platform [12]. For the designed parameters listed above, the digital controller and loopgain frequency response characteristics are shown in Fig. 6. This digital controller performance is verified in simulation by creating source and load side perturbations. Several operating conditions have been tested in simulation and found that the above designed controller regulating the load voltage. For illustration, sample dynamic response characteristics of the CPBC are given here for: (i) sudden change in load (R= 20 -to- 10 Ω), (ii) sudden change in source voltage V_g: 42 -to- 36 V, and (iii) gradual change in source voltage V_g: 42 -to- 36 V and back to *42 V*. These responses are shown in Figs. 7 to 9. In all these dynamic conditions, the load voltage is regulated in less than *2.0 msec*. For validating CPBC performance and its controller design a laboratory prototype is made, digital controller is implemented using *dsPIC* micro-controller [14], and steady-state, dynamic responses for the above mentioned cases are shown in Figs. 4 to 9. Measured responses (steady-state and dynamic conditions) are in close agreement with simulation results. Slight discrepancy in simulated results higher settling time is on account of in accurate representation of experimental circuit damping in the simulation profile.

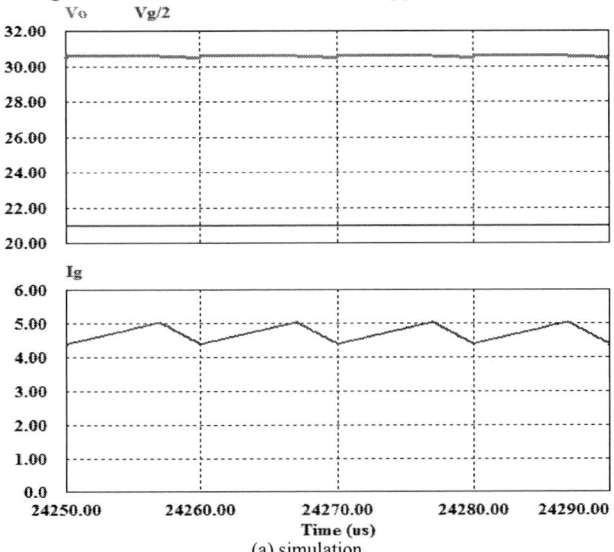

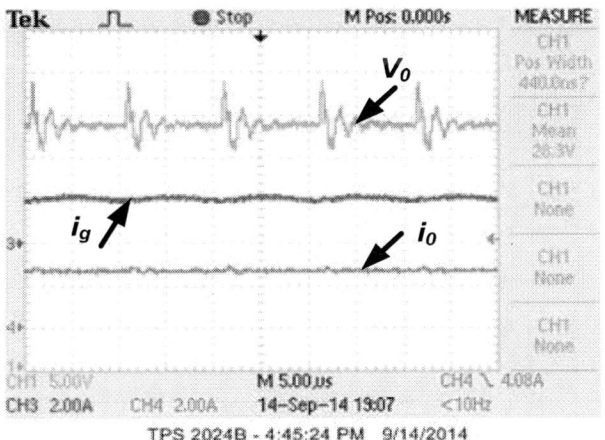

(a) simulation

(b) experiment

Fig. 4. Enlarged steady-state waveforms of source and load currents, load voltage (V_g= 48 V, R= 20 Ω, Ch-1: V_0, Ch-3: i_g, Ch-4: i_0).

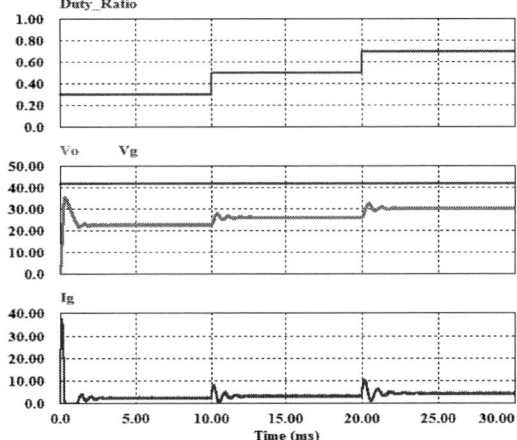

Fig. 5. Load voltage variation with duty ratio

(V_g=42 V, R=10 Ω, f_s= 100 kHz).

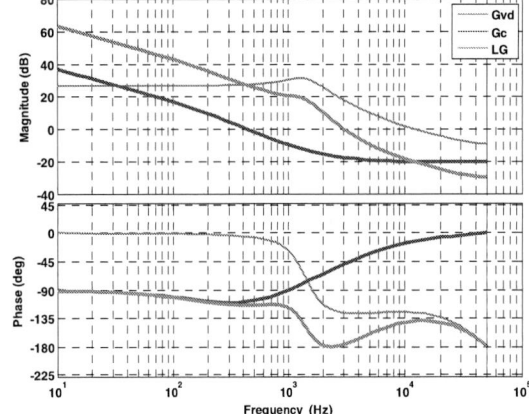

Fig. 6. Frequency response plot of CPBC.

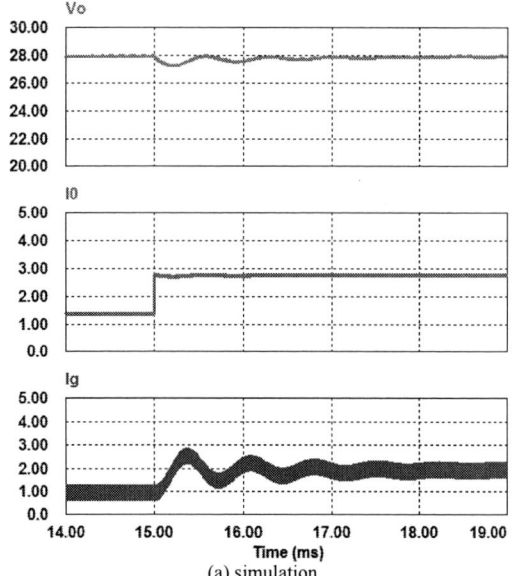

(a) simulation

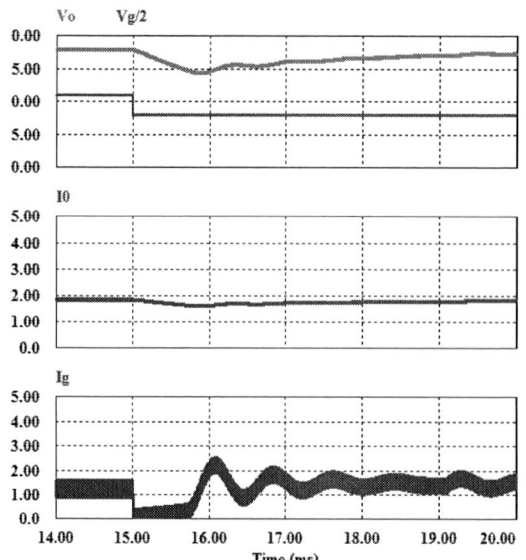

Fig. 8. Enlarged dynamic response waveform against sudden change in source voltage (V_g: *42 -to- 36 V*).

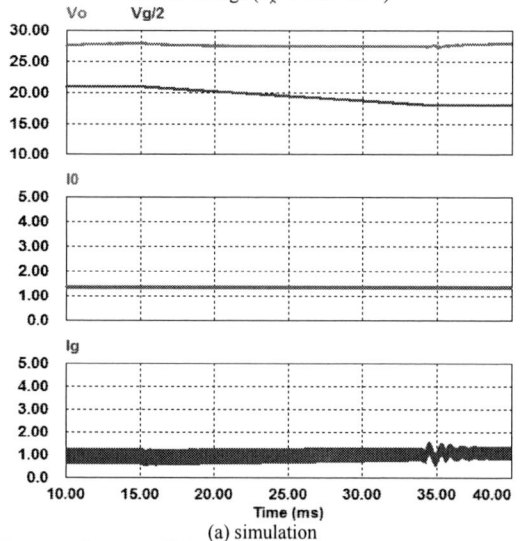

(a) simulation

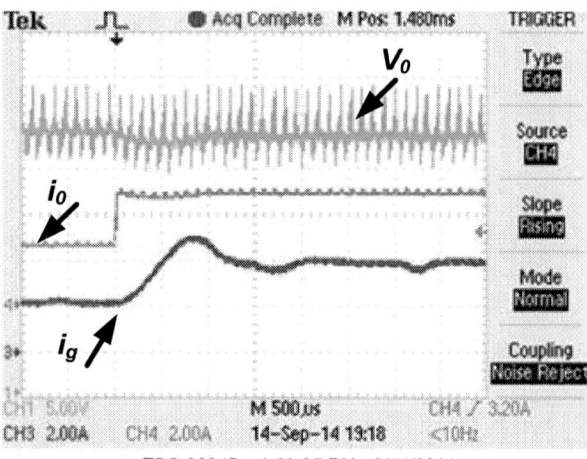

(b) experiment

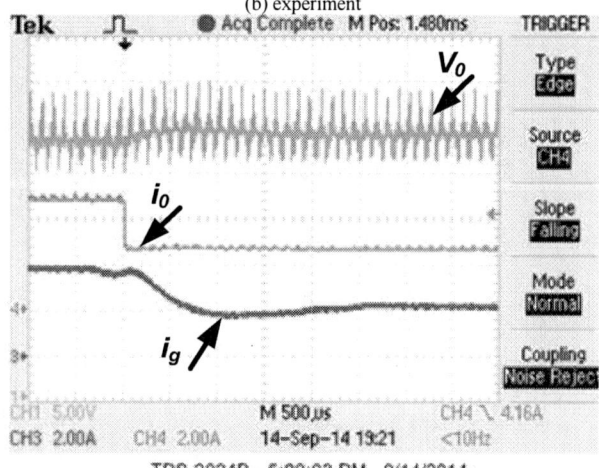

(c) experiment

Fig. 7. Enlarged dynamic response waveform against load change (*R: 20 -to- 10 Ω*, V_g = *48 V*, Ch-1: V_0, Ch-3: i_g, Ch-4: i_0).

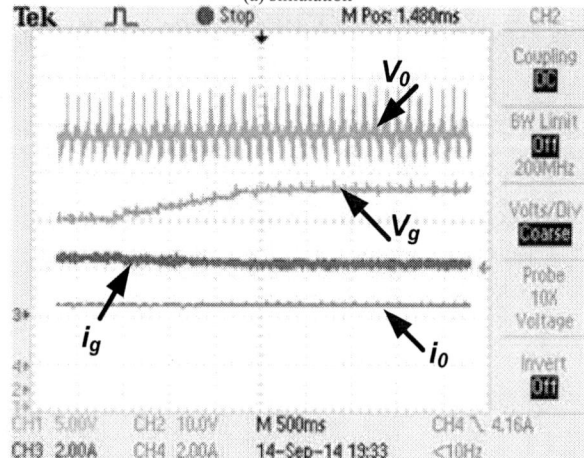

(b) experiment

978-1-4799-6047-7/14 $31.00 © 2014 IEEE

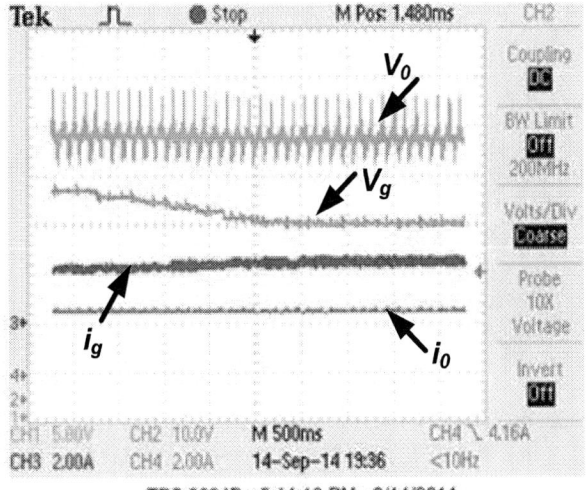

(c) experiment

Fig. 9. Enlarged dynamic response waveform against gradual change in source voltage (V_g: 42 -to- 36 -to- 42 V, Ch-1: V_0, Ch-2: V_g, Ch-3: i_g, Ch-4: i_0).

V. CONCLUSION

A charge-pump based buck converter was proposed, and its performance was analyzed. Steady-state analysis revealed that this converter is better than the conventional buck converter if the desired load voltage range is in between 50 ~ 100 % of source voltage. Furthermore, it was also demonstrated that this converter results in lesser source current ripple. Digital controller regulation capability, designed through discrete-time transfer functions, was validated through experimental measurements.

REFERENCES

[1] Forsyth. A. J, Mollov. S. V, "Modelling and control of DC-DC converters," IEE Power Engineering Journal, Vol. 12(5), 1998, pp. 229 -236.

[2] Richard Tymerski, Vatche Vorperian, "Generation and classification of PWM dc-dc converters," IEEE Trans. On Aerospace and Electronic Systems, Vol. 26(4), 1988, pp. 743-754.

[3] F. Ueno, T. Inoue, I. Oota, I. Harada, "Power supply for electroluminescence aiming integrated circuits," IEEE Proc. on ISCAS, 1992, pp. 1903-1906.

[4] Byungcho Choi, ``Step load response of current-mode controlled dc-dc converter," IEEE Trans. On Aerospace and Electronic Systems, Vol. 33(4), 1997, pp. 1115-1121.

[5] Dimitry Goder, William R. Pelletier, "V^2 architecture provides ultra-fast transient response in switch mode power supplies," Proc. of High frequency power conversion conference, 1996.

[6] H. S. Chung, Adrian Ionovici, "Development of a generalized switched capacitor dc-dc converter with bi-directional power flow," IEEE Proc. on ISCAS, 2000, Vol. 3, pp. 499-502.

[7] Lung-Sheng yang, Tsorng-Juu Liang, Hau-Cheng Lee, "Novel high step-up dc-dc converter with coupled-inductor and voltage-doubler circuits," IEEE Trans. Ind. Electron., 2011, Vol. 58 (9), pp. 4196–4206.

[8] L. S. Yang, T. J. Liang, and J. F. Chen, "Transformer-less DC-DC converters with high step-up voltage gain," *IEEE Trans. on Industrial Electronics*, Aug. 2009, Vol. 56, no.8, pp. 3144-3152.

[9] B. Axelrod, Y. Berkovich, and A. Ioinovici, "Switched-capacitor/ switched-inductor structures for getting transformerless hybrid DC-DC PWM converters," *IEEE Trans. Circuits Syst. I, Reg. Papers*, Mar. 2008, Vol. 55, no. 2, pp. 687-696.

[10] Mukesh Singh Tomar, "Steady-state Analysis and Controller design for a High-gain Fifth-order Boost Converter," Proc. of Annual IEEE India Conference (INDICON), 2013, pp. 1-6.

[11] M. Veerachary, A. R. Saxena, "Design of Robust Digital Stabilizing Controller for Fourth-Order Boost DC–DC Converter: A Quantitative Feedback Theory Approach," *IEEE Trans. on Ind. Electron.*, Feb. 2012, Vol. 59, no. 2, pp. 952-963.

[12] MATLAB, user manual, 2005.

[13] PSIM, user manual, 2005.

[14] dSPIC30F2020, Microchip, user manual, 2009.

Design and Analysis of Zero-Voltage Transition Fifth-Order Boost Converter

Mummadi Veerachary

Dept. of Electrical Engineering, IIT Delhi, New Delhi, India
E-mail: mvchary@ee.iitd.ac.in

Abstract— **In this paper a zero voltage transition low source ripple current fifth order boost converter is proposed, which offers several advantages over conventional and higher-order boost converters. This converter provides higher voltage gain moderate duty ratios. The zero voltage transition auxiliary circuit is incorporated into the circuit to provide further advantages in terms of reduction of switching losses for nearly all the switching devices. Modes of operations are identified, and then steady-state analysis is presented for design purpose. In order to verify the zero-voltage switching, soft-switching, performance of this converter simulation is carried out in PSIM. To further test the advantage provided by this converter, over hard switched converters, a comparative analysis is carried out in simulation and experiment at various loading conditions. The converter performance is predetermined for a *24 to 48 V, 25 W* prototype in simulation and then compared with experimental results. Experimental measurements are in close agreement with analytical predictions.**

keywords: Boost Converter, Fifth-order Boost Converter, Soft-switching, Zero-voltage Transition.

I. INTRODUCTION

Dc-dc boost converters are most commonly used to deliver higher load voltages from given low voltage source. The conventional boost converter though suffers from the following disadvantages [4]: (i) Requires a higher duty ratio even for realizing the moderate voltage gain, (ii) Unable to reach expected levels of voltage gain at extreme duty ratios on account of excessive losses (iii) the efficiency under full load is low due to higher switching losses. To eliminate some of these limitations, higher order boost pulse width modulated (PWM) converter are utilized. These converters give the higher voltage gain but at higher switching frequencies the full-load efficiency is still a limitation.

For improvement in efficiency of these converters soft-switching schemes are reported in the literature [1]. The two schemes which are most regularly used are:
1. Zero voltage switching (ZVS) during turn-on.
2. Zero current switching (ZCS) during turn-off.

These schemes are selected on the basis of device used, i.e., IGBT or MOSFET. Even though efficiency is improved using these techniques, they still suffer from some limitations. The ZCS turn-off suffers from limitations such as: (i) rise in losses during conduction, (ii) converter circuit diode subjected to higher voltage stress, and (iii) conduction losses increase due to presence of the main switch in series with resonant inductor. Some of these limitations are eliminated

using the soft transition methods such as zero-voltage/ zero-current transition (ZVT/ ZCT) techniques are presented in literature [1]-[2]. To realize better efficiency at the full load condition many soft switching procedures are being discussed in the literature as they remove unwanted switch-transition losses that appear in hard switched dc-dc converters [5]-[6]. Another soft-switching technique for a high gain boost converter topology is reported in the literature [7]. Here, the additional network produces both the voltage amplification and the soft-switching for the MOSFETs.

However, the literature present does not adequately cover the development of soft-switching schemes for fifth-order boost converters. Therefore, the aim of this paper is efficiency improvement of the fifth order boost converter, belonging to higher order family, by utilizing the ZVT technique to obtain soft-switching. Here the soft-switching is realized by integrating the ZVT cell into the fifth-order boost converter [3].

II. STEADY-STATE ANALYSIS OF ZVT-FIFTH-ORDER BOOST CONVERTER

The Fifth-order boost converter utilizing soft-switching technique is shown in Fig. 1. In comparison to other fifth-order boost converters [3] reported in literature this topology [4] gives better voltage boosting feature. Including soft-transition networks will improve the efficiency of the converter. Although ZVT and ZCT soft-transition networks reported in literature, in this paper only OFF to ON soft-transition is investigated and hence ZVT network introduced appropriately. Here the proposed ZVT network consists two resonating inductors and a capacitance in addition to a diode, auxiliary switch. These components make up the zero-voltage transition network added as shown in Fig. 1. Due to this ZVT network, the proposed converter exhibits seven operating modes in one switching cycle of operation. Equivalent circuits for each mode of operation are shown in Fig. 2, while steady-state waveforms are shown in Fig. 3.

Mode-1 operation ($t_o < t < t_1$): This mode is initiated by turning-ON of the auxiliary switch S_a. Along with diode D_l, D_a are in ON-state. During this period the auxiliary switch as well as auxiliary diode starts increasing while the main diode D_1 current decreasing linearly. At the end of this mode the D_1, current goes to zero and it voltage starts building up resulting in ZCS.

978-1-4799-6047-7/14 $31.00 © 2014 IEEE

Mode-2 operation (t₁<t<t₂): In this mode the auxiliary switch and diode (S_a, D_a) are in ON-state and hence the inductor L_r and capacitor C_r in the resonating. In view of this, the voltage across the main switch starts decreasing in sinusoidal fashion and at the end of this mode this become zero. At that moment, the anti-parallel diode of the main switch start conducting, and hence it will allow negative current. During this mode, the diode D_a carries a constant current while the auxiliary switch current increases linearly. Furthermore, the voltage across the diode D_2 also decreases in sinusoidal fashion and at the end of this mode this voltage becomes zero and from that instant it also comes into conduction.

Mode-3 operation (t₂<t<t₃): During this mode D_2, D_a, S_a, D_s are in ON-state. During this period, the resonating inductor current decreases, and similar variation is seen in the anti-parallel diode of the main switch. At the end of this mode the auxiliary switch goes to OFF-state and hence the resonating inductor L_{r1} current as well as the main diode D_2 current becomes zero.

Mode-4 operation (t₃<t<t₄): During this mode D_a, D_s are in ON-state. During this period, the anti-parallel diode and auxiliary diode current gradually decrease to zero. Before, the end of this mode the main switch must be given gate signal so that the current transfer takes place from the anti-parallel diode. In order to have satisfactory zero voltage transition the gate signal must be released to main switch anywhere in between this mode of operation.

Mode-5 operation (t₄<t<t₅): During this mode D_a, main switch (S) are in ON-state. During this period the main switch current increases while the auxiliary diode current decreases and at the end of this diode goes to OFF-state at ZCS.

Mode-6 operation (t₄<t<t₅): In this mode main switch will in ON-state just like in conventional circuits, and it will be turned-OFF by removing the PWM signal.

Mode-7 operation (t₄<t<t₅): In this mode, main switch goes to OFF-state (hard transition) and diode D_1 starts conducting. After Mode-7 again above sequence of modes will be repeating in the subsequent cycles.

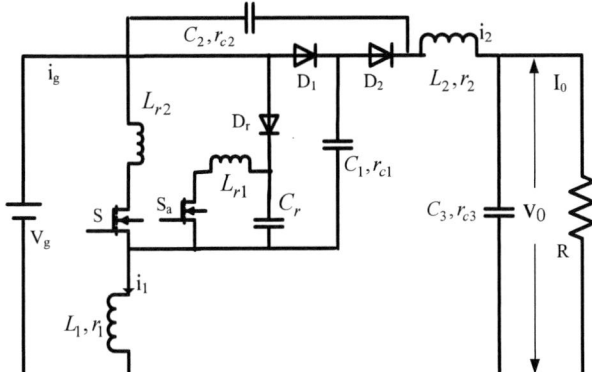

Fig. 1. Circuit diagram of the soft-switching Fifth-order boost converter.

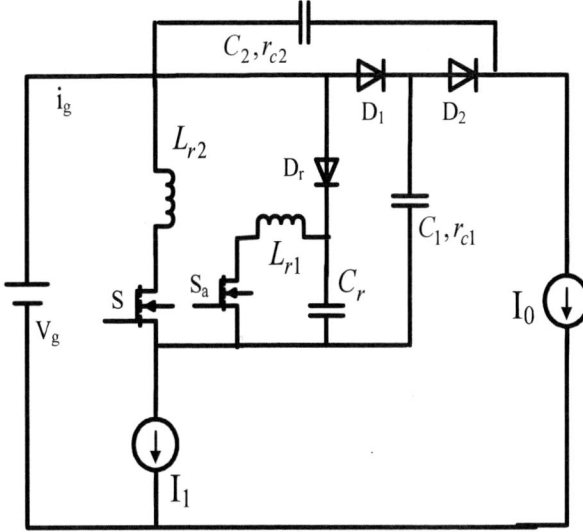

(a) Equivalent circuit

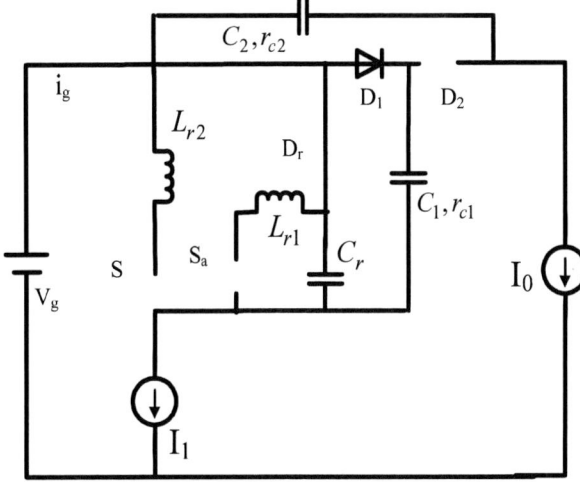

(b) Mode-1 equivalent circuit

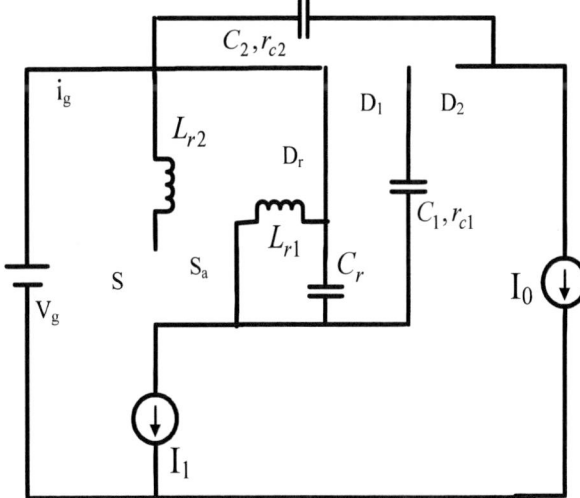

(c) Mode-2 equivalent circuit

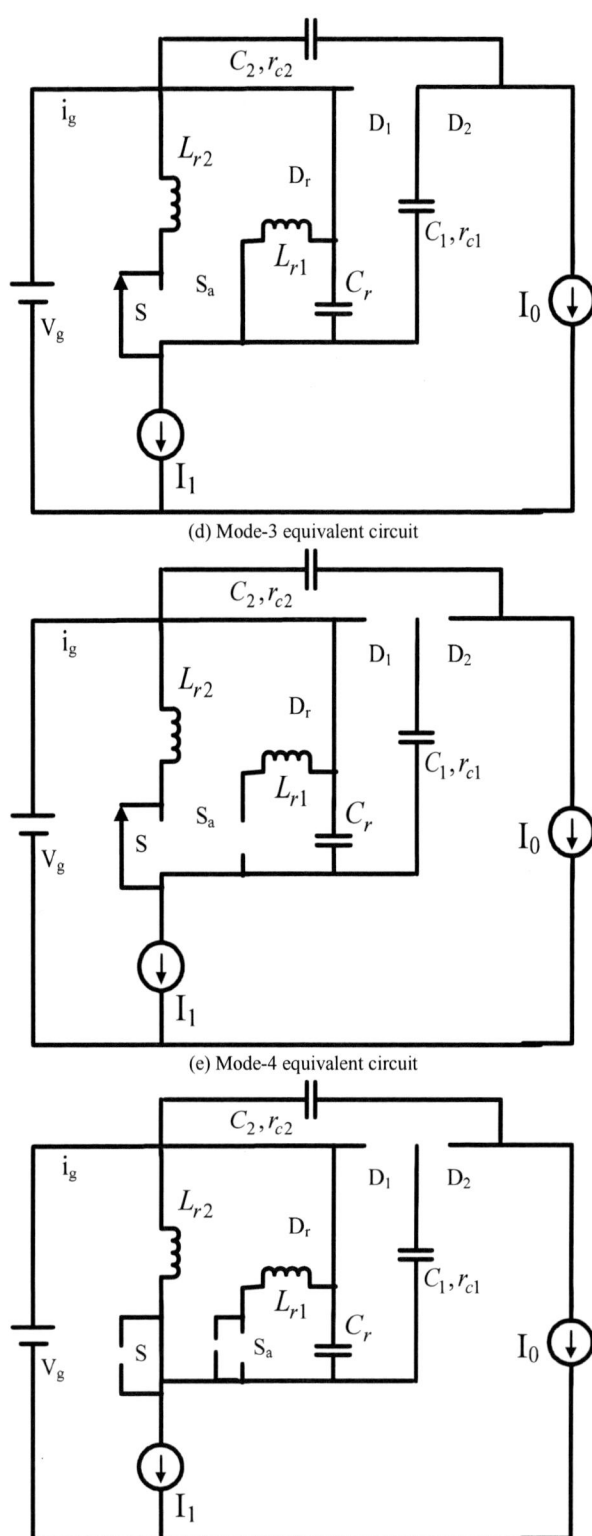

(d) Mode-3 equivalent circuit

(e) Mode-4 equivalent circuit

(f) Mode-5 equivalent circuit

Fig. 2. Equivalent circuit diagrams for each mode of operation (Mode-1 to 5).

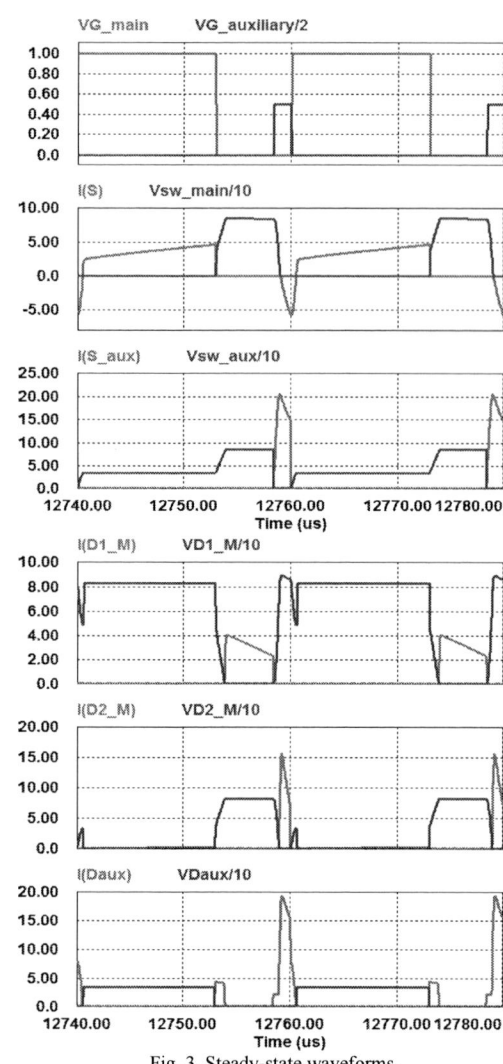

Fig. 3. Steady-state waveforms.

III. DESIGN OF ZVT FIFTH-ORDER BOOST CONVERTER

The power stage components of the soft switching ZVT-fifth-order boost converter are designed as per the input parameters given in Table I. For design of power stage elements, the design equations are obtained as per the steady-state analysis reported in ref[4].

Table I. Converter Parameters

Parameter	Value
V_g	24 V
V_o	48 V
P_o	25 W
R	$75 \sim 150\ \Omega$
D	$0.4 \sim 0.5$
D_{aux}	$0.1 \sim 0.15$
F_s	100 kHz
F_{ns}	0.172

978-1-4799-6047-7/14 $31.00 © 2014 IEEE

Table II. Design Expressions of ZVT Fifth-order Boost Converter

$$L_1 > (V_g - v_{c3} - v_0)D/(f_s \Delta i_1)$$

$$L_2 > V_g D/(f_s \Delta i_2)$$

$$C_1 > [(2-D)V_g]/[(1-D)Rf_s \Delta v_{c1}]$$

$$C_3 > [(1-D)v_0 D]/[8L_2(2-D)f \Delta v_{c3}]$$

$$C_2 > [(2-D)V_g]/[Rf_s \Delta v_{c2}]$$

In order to design the resonant inductors (L_{r1}, L_{r2}) and capacitor (C_r) elements, the auxiliary switch duty ratio was fixed to the desired value (0.1 ~ 0.15) and then the inductor and capacitor values are obtained based on normalized load.

IV. GATE SEQUENCE FOR SUCCESSFUL ZERO-VOLTAGE TRANSITION

The success of the ZVT depends on the turn-ON of the auxiliary switch with reference to main switch. Improper phase position leads to failure of ZVT. All possible combinations are shown in Fig. 4. For successful ZVT, the time delay between the auxiliary switch turn-ON and the main switch turn-ON should be greater than the minimum time delay (T_{Dmin}, given by the addition of mode-1 timing and mode-2 timing) and the switch ON-time of the auxiliary switch should be greater than T_{Dmin}. Also the turn-ON of the main switch should take place simultaneously as the turn-OFF of auxiliary switch or before it.

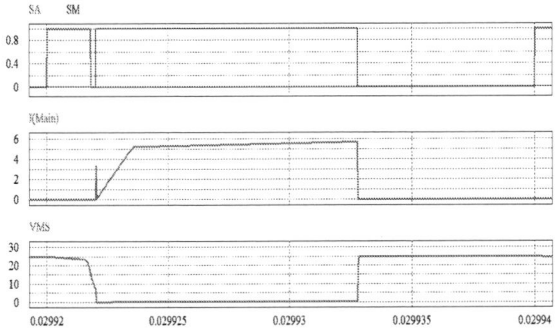

(a) Switch waveforms for auxiliary switch on-time less than T_{Dmin}.

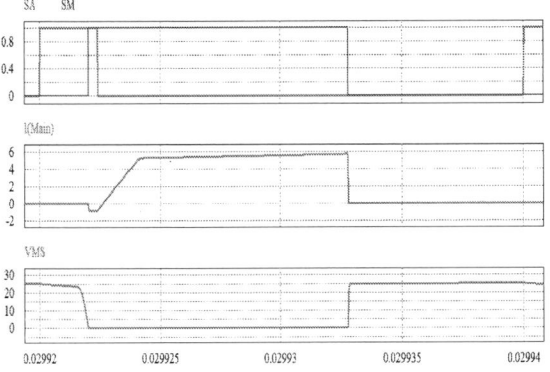

(b) Switch waveforms for auxiliary switch on-time more than T_{Dmin}.

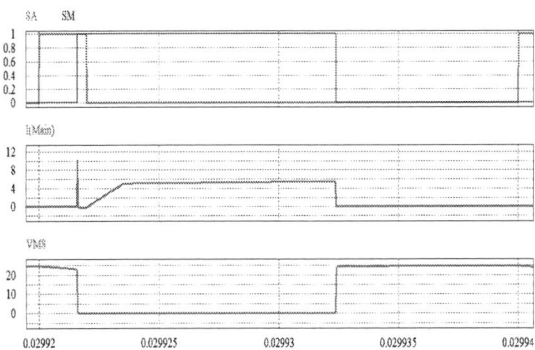

(c) Switch waveforms for main switch delay time less than T_{Dmin}.

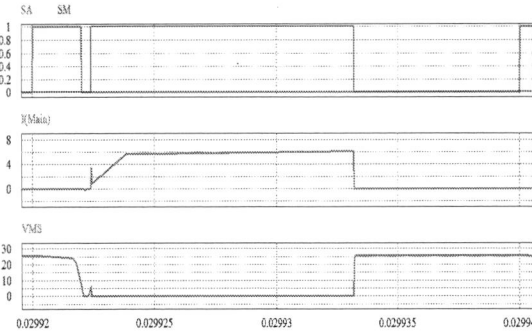

(d) Switch waveforms for main switch delay time more than T_{Dmin}.
Fig. 4. Main and auxiliary switch gating schemes.

V. SIMULATION AND EXPERIMENTAL RESULTS

A 25 *Watt* prototype ZVT fifth order boost converter system has been designed to verify the effectiveness the zero voltage turn-ON transition performance. The converter is supplied from a 24 *V* battery, and the desired load voltage is 48 *V*. The parameters of the designed converter to meet the specifications are: L_1=150 µH, L_2=50 µH, L_{r1}=1.5 µH, L_{r2}=4.5 µH, C_r=50 nF, C_1=47 µF, C_2=47 µF, C_3=100 µF, R= 7 ~ 150 Ω, f_s = 100 *kHz*. Zero-voltage transition/ soft-switching performance of the proposed boost converter is verified through PSIM simulations [10] and for illustration purpose steady-state current and voltage waveforms of main, auxiliary switches and their gate signals are shown in Figs. 5 to 7.

It is clear from these waveforms that the main switch is turning-ON only after its drain to source voltage is reaching to zero. It is also evident that the auxiliary switch is turning-ON at zero voltage switching. In order to verify the simulation results a 12 -to- 48 *V*, 25 *W* laboratory prototype, converter parameters listed above, has been designed. The devices used in the prototype converter circuit are: Switch IRFP250N, Diode MUR1560, Driver circuit IR2110 and Opto-isolator-6N137. Experimentation has been performed on the proposed converter to demonstrate the ZVT feature, a 12 V supply is connected to source and then various currents/ voltage waveforms have been recorded. Soft-switching OFF-ON transitions under zero voltage turn-ON of the switching devices are as shown in Fig. 6 and 7.

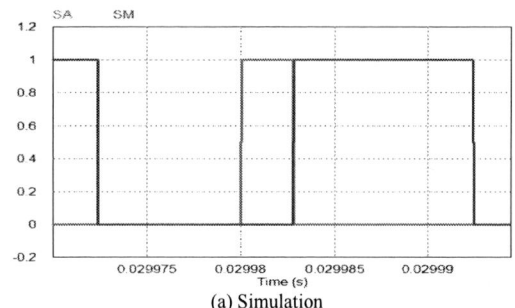

(a) Simulation

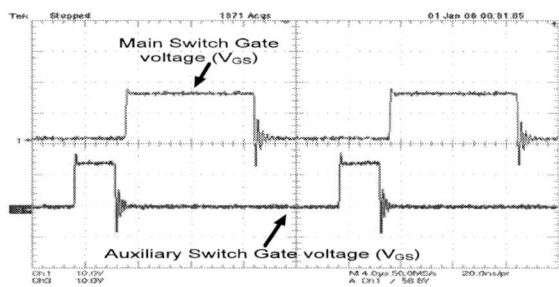

(b) Experimentally measured

Fig. 5. Gate Signals of main and auxiliary switch.

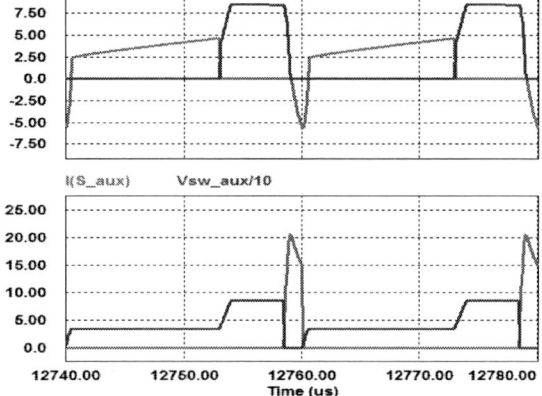

(a) Main and auxiliary switch waveforms

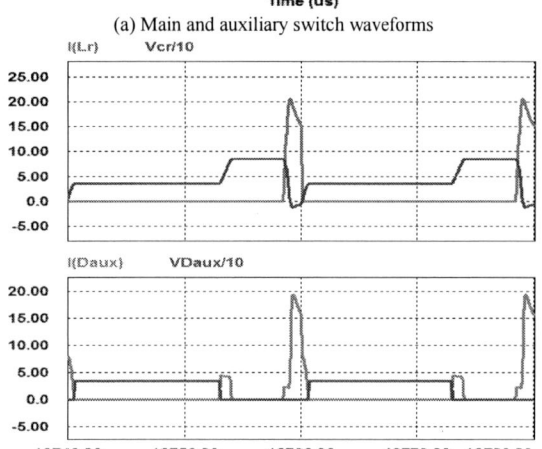

(b) Resonating inductor and capacitor waveforms

Fig. 6. Simulation waveforms

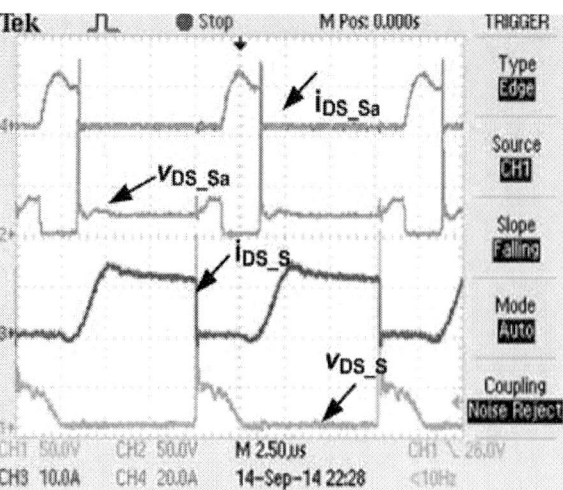

(a) Light load

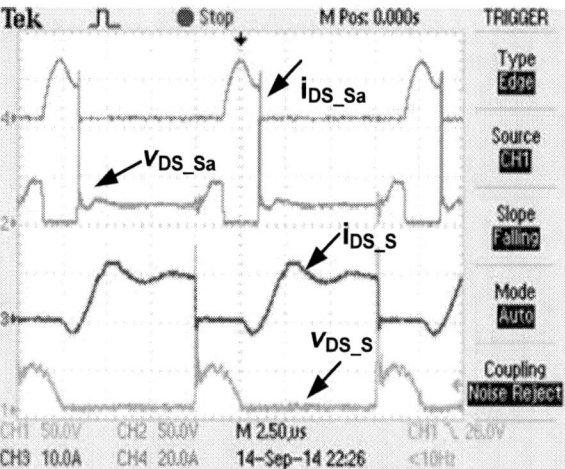

(b) Full load

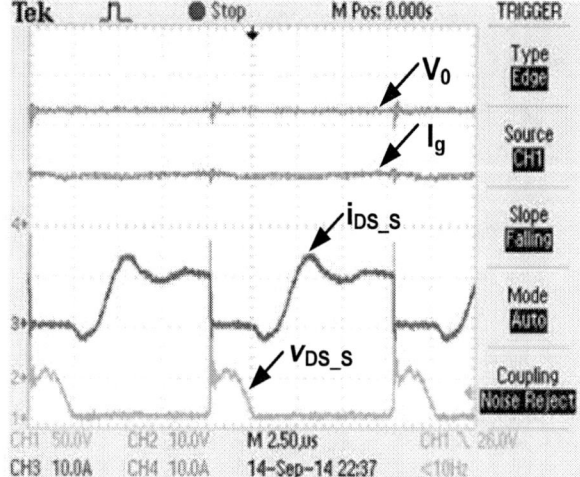

(c) Full load

Fig. 7. Experimental waveforms showing ZVT feature.

978-1-4799-6047-7/14 $31.00 © 2014 IEEE

VI. CONCLUSION

In this paper a zero voltage turn-ON transition fifth-order boost converter operating modes has been established and then zero-voltage turn-ON of main switch and auxiliary switch was verified. Steady-state analysis established for various operating modes and then soft-transition phenomenon was verified. The analysis results were validated through simulations and experimental measurements. Measured results are in close agreement analytical predictions.

REFERENCES

[1] Guichao, Ching-Shan Leu, Yimin Jiang, Fred C. Y. Lee, "Novel Zero-Voltage Transition PWM Converters", IEEE Trans. On Power Electronics, 1994, Vol. 9(2), pp. 213- 219.

[2] Guichao Hua, Eric X. Yang, Yimin Jiang, and Fred C. Lee, "Novel Zero-Current-Transition PWM Converters", IEEE Transactions on Power Electronics, November 1994, Vol. 9, No. 6, pp. 601-606.

[3] M. Veerachary, Krishna Mohan, "Robust Digital Voltage-mode Controller for fifth order Boost Converter," IEEE Trans. On Ind. Electronics, 2010, Vol. PP (99), pp. 1-15.

[4] M. Veerachary, "Soft-Switching Fifth-Order Boost Converter", India Conference (INDICON), Annual IEEE, December 2011.

[5] M. Veerachary, "Design of Robust Digital PID Controller for H-Bridge Soft-Switching Boost Converter," IEEE Trans. Ind. Electron., July 2011, vol. 58, no.7, pp. 2883-2897.

[6] Sungsik Park, Sewan Choi, "Soft Switched CCM Boost Converters With High Voltage Gain For High-Power Applications", IEEE Trans. On Power Electronics, 2010, Vol. 25(5), pp. 1211-1217.

[7] R. Sekhar, "Digital Voltage-mode Controller Design for High gain Soft-Switching Boost Converter," IEEE Proc. on PEDES2010, Dec. 2010, pp. 1-5.

[8] Chien Ming Wang, "Novel Zero-Voltage-Transition PWM DC–DC Converters", IEEE Transactions on Industrial Electronics, February 2006, Vol. 53, No. 1.

[9] MATLAB, user manual, 2005.

[10] PSIM, user manual, 2005.

Selective Harmonic Elimination: Comparative Analysis by Different Optimization Methods

Sreedhar Madichetty
School of Electrical Engineering
KIIT University
Bhubaneswar, INDIA

M.Rambabu
Department of EEE
GMR Institute of Technology
Andhra Pradesh, INDIA

Abhjit Dasgupta
School of Electrical Engineering
KIIT University
Bhubaneswar, INDIA

Abstract— **This article presents the comparative experimental analysis of selective harmonic elimination (SHE) technique applied to three phase Voltage Source Inverter (VSI). The optimization technique has been applied to VSI with switching angles as variables and total harmonic distortion (THD) as an objective function. These optimized pulse patterns are obtained by the Accelerated Particle Swarm Optimization technique (APSO), Simulated Annealing (SA) and Hybrid Particle Swarm Optimization Techniques (HPSO). These obtained switching pulses has been applied to three phase VSI in Simulink software and results are verified by conducting the experiments. Further, ATMEGA 16 micro controller has been used to obtain the optimized switching pulses. Thus obtained results has been compared and constructive concluding remarks has been given.**

Keywords:-Selective harmonic elimination; application of optimization techniques; particle swarm optimization

I. INTRODUCTION

The ever increasing demand of industry for stability, adjustability and accuracy of control of power electronic equipment at very high voltages led to the development of relatively less total harmonic distortion (THD) based modern power electronic static converters. Although solid state power electronic switches, such as the IGBT's have brought notified variance in control techniques, but the main disadvantage is that they produce multiple frequency components called as harmonics. Harmonic voltages can cause an unacceptable disturbance on the supply network and adversely affect the operation of other connected electrical equipment. Hence there is requirement to reduce the harmonic content to an acceptable level. One should design a converter with proper controllers by keeping THD within limits. Harmonic currents can never be totally eliminated from an electrical system. They can, however, be very significantly reduced by using a harmonic controllers [1]. All power electronic converters produces complex waveforms, that can be resolved into a series of sinusoidal waves of various frequencies, hence any complex waveform is the sum of a number of odd or even harmonics, that can be eliminated by designing the proper controller by

tuning to the distorted frequencies. In this work three phase three level inverter has been considered as a topic of research. An inverter is static power electronic circuit which is used to convert the DC (Direct Current) to AC (Alternating Current). Converted AC consists of multiple frequency components (Harmonics) along with required fundamental frequency (50Hz). The presence of lower order harmonics will affects system performance; usually filters are used to reduce the lower order harmonics (LOH), but the cost of filter equipment's is comparably high. Hence the economic way to minimize the LOH can be achieved by sinusoidal pulse width modulation technique (SPWM) with higher carrier frequency. In which, it has a carrier wave is compared with the sinusoidal reference and obtained pulses given to inverter switches. This reduces the lower order harmonics, but it also reduces the fundamental frequency component. In order to overcome that disadvantage, in this article it has been implemented selective harmonic elimination technique. By which one can maintain the desired fundamental amplitude and can eliminate the desired number of harmonics.

In [2], the switching angles are obtained traditionally and it has been represented as regular sample PWM. It achieves near ideal elimination of lower order harmonics. The harmonic elimination [3] is described with pulse width modulation technique, but the proposed technique is used for the single phase inverter. In [4], it proposes a solution of complex transcendental equations by behavior of particles. In [5], it proposed a new technique for elimination of harmonics with pulse width modulation technique but the angles are found by using Walsh transform harmonic elimination method. It is taken as a one of the sophisticated method. In [6], it proposes a general formulation for selective harmonic elimination technique with selected solutions and shows total harmonic as an objective function and uses unipolar and bipolar switching technique. In [7], selective harmonic elimination technique to motor application which is used in the traction drive applications. In [8], it has done an extensive research in the voltage control of the inverter with harmonic elimination techniques. In [9], it was applied different pulse width modulation techniques and compared different techniques with their advantages and disadvantages. In this [10], authors mentioned different techniques that is applied to the power converters with their details of their explanations. In this [12], it had applied soft computing technique to obtain the switching angles and total harmonic distortion is taken as a objective function. In this [11], it had developed different algorithms with their application is described. In this [3], it had explained the future scope and applications of the particle swarm optimization technique which is to be applied to different engineering problems.

This article is organized in six sections. First about introduction, then problem formulation of the presented research topic. In the third section, APSO has been discussed with its solution methodology to presented topic.SA and HSA has been discussed in section four. The simulation and experimental results has been discussed in section five and conclusion at last section.

II. PROBLEM FORMULATION

The basic experimental circuit diagram of a three phase VSI is shown in fig.1. Three phase VSI consists of six power semiconductor switches (IGBT's) out of which two are connected in each leg of three legs. The switching pulses is applied to the different gates of a semiconductor switches in bi polar switching mode [13].

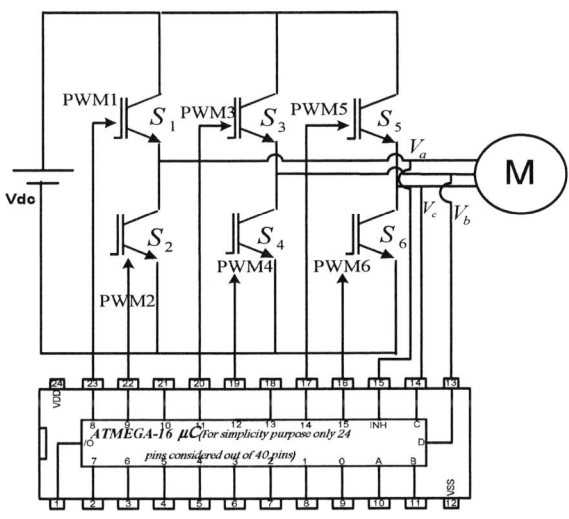

Fig 1.Three Phase Voltage Source Inverter

The switching wave form for the inverter is obtained by the different optimization methods, which will be discussed in section.3. The output voltage of three phase VSI is realized by using Fourier series and as shown in following set of equations (1).

$$1 - 2\cos(\alpha 1) + 2\cos(\alpha 2) - \ldots\ldots\ldots\ldots\ldots 2\cos(\alpha 37) = \prod\left(\frac{v_{01}}{v_i * 4}\right)$$

$$1 - 2\cos(3\alpha 1) + 2\cos(3\alpha 2) - \ldots\ldots\ldots\ldots 2\cos(3\alpha 37) = 0$$

..

..

$$1 - 2\cos(37\alpha 1) + 2\cos(37\alpha 2) - \ldots\ldots\ldots 2\cos(37\alpha 37) = 0 \quad (1)$$

Where $\alpha_1 < \alpha_2 < \alpha_3 \ldots\ldots\ldots < \alpha_{37}$ are the notch angles of the switching PWM signal.v_{01} is the fundamental amplitude of output voltage, v_i is the total input voltage. For elimination of respective harmonics, the respective notch angles has to be find out. In order to eliminate the 5th, 7th, 11th …37th harmonics, one should solve the set of transcended equations which are shown in (1).

$$\text{THD} = \frac{\sqrt{V_3{}^2 + V_5{}^2 \ldots. V_{37}{}^2}}{V_1} \quad (2)$$

Also, one should note that, due to the wave form symmetry all the even order harmonics are absent and evident. In this article, these set of equations with $\alpha_1 < \alpha_2 < \alpha_3 \ldots\ldots\ldots < \alpha_{37}$ as variables with Total Harmonic Distortion (THD) (2) as an objective function by using different optimization methods such as APSO, SA and HPSO and results has been compared.

III. ACCELERATED PARTICLE SWARM OPTIMIZATION TECHNIQUE

In this section, the methodology to find out the solution for transcendental equation has been given by using APSO technique.

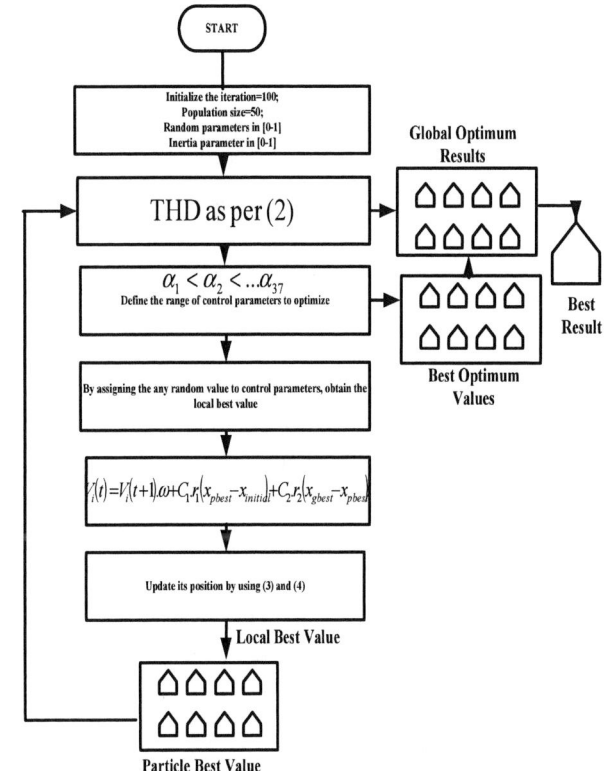

Fig 2.Solution methodology flowchart by using APSO

The flowchart representation of APSO has been shown in fig.2.In this article, the solution methodology by using APSO has given and complete information about of APSO is given in [12].Where $\alpha_1, \alpha_2 \ldots\ldots \alpha_{37}$ are the notch angles of output voltage, $V_i(t)$ is the velocity of swarm at current position, $V_i(t + 1)$ is the velocity of swarm in next position,ω is the inertia function,C_1, C_2 are the individual social co-efficient factor, x_{pbest} is the particle best position,

978-1-4799-6047-7/14 $31.00 © 2014 IEEE   689

$x_{initial}$ is the initial position of the particle, r_1 and r_2 are the random numbers of the particle. Primarily, all the particles has be initialized with their respective parameters and then the objective function with their variables has to be checked with their constraints. With the help of APSO technique the accurate velocity of particle has be found with the following (3) and position and velocity updates.

$$x_i(t) = x_i(t-1) + v_i(t).\omega \qquad (3)$$

$$v_i(t) = v_i(t-1) + x_i(t).\omega \qquad (4)$$

After processing the (3) and (4), one can obtain the local best minimum of the objective. This process should be iterated to get the best values i.e global best optimum values.

IV. SIMULATED ANNEALING

Simulated Annealing (SA) is used to find the global optimum for a given objective function in large search space. SA proves its good approximation for various applications in real time. Hence SA has been considered for this SHE problem to find out the optimum switching angles with THD as objective function. The basic flow chart demonstrating the solution methodology is shown in fig.3.

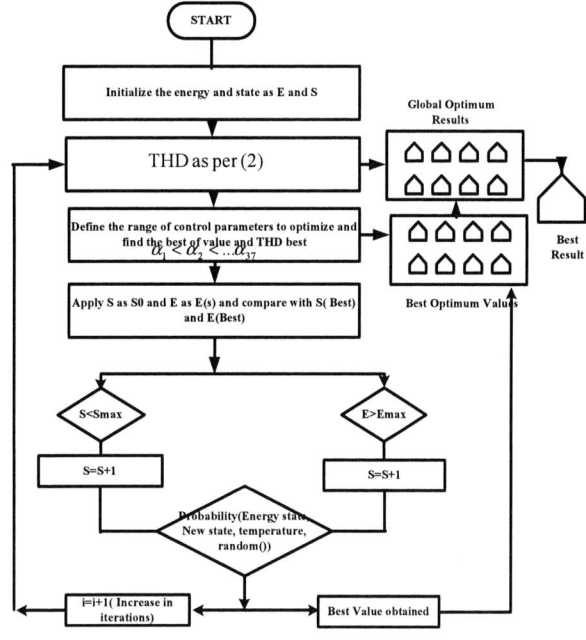

Fig 3.Solution methodology flowchart by using SA

Where S is the state of the SA,E is the energy state and the equation of probability as per [8], i is the number of iterations, Smax is the maximum state of the system, Emax is the maximum energy state. The procedure of SA will follows as per fig.3 and best optimum value can be stored. The probability equation has been taken from [13].

A. Hybrid Simulated Annealing

In general, the conventional techniques for multi-dimensional problems can be solved by the local search techniques, which will enhance the speed of convergence. This has the advantage of finding the solution quickly. However, the quality of solution or whether any solution will be obtained, is dependent upon the initial starting point. PSO requires a large search space for optimizing the solution of the concerned problem. It has difficulties in fine tuning the parameters for local search. In PSO, each particle should be kept in a confined space corresponding to the parameter limitations. This decreases the diversity of the particle and it affects the getting optimum solution in reasonable time. If the global best particle position (G_{best}) does not change in over a period of time of the concerned problem, then stagnation occurs in the population/solution. Then the solution will confined to local best optimum. To address the above specified problems, Hybrid Simulated Annealing (HSA) technique is introduced in this section. Initially the objective function shown in (2) with its constraints (1) has been solved by using APSO technique for 100 iterations. The best solution obtained by APSO is taken as the initial value for SA algorithm initialization. Thus obtained results has been verified by simulation and conducting the experiments.

B. Results Comparison

The complete list of equations as shown in (2) and (3) has been solved with their constraints and results has been graphically provided in fig.4. Firstly, the set of equations has been solved by suing PSO, where it took 380 iterations to obtain its best fitness value. Whereas APSO requires 175 iterations to obtains its best fitness value. SA requires 428 iterations, and HSA 132 for PSO and 260 for SA. By comparing all those optimization techniques, it is observed, that, HSA takes more of iterations to obtain its best fitness value, which is better than those obtained by all other methods. However the THD obtained by APSO is only slightly greater than that obtained by HSA, thus, for applications where processing power is a constraint, APSO is recommended for fast convergence and acceptable accuracy.

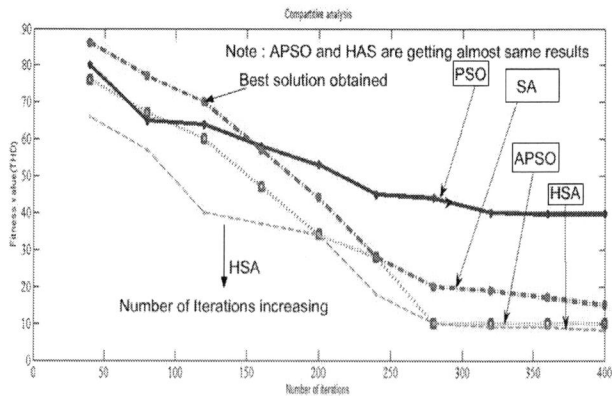

Fig 4.Comparison between optimization techniques applied to specific problem.

V. APPLICATION OF OPTIMIZATION TECHNIQUES

Different switching angles were obtained by the different optimization methods and those angles has been converted in to time sequence of switching pulses for inverter as shown in fig.5.

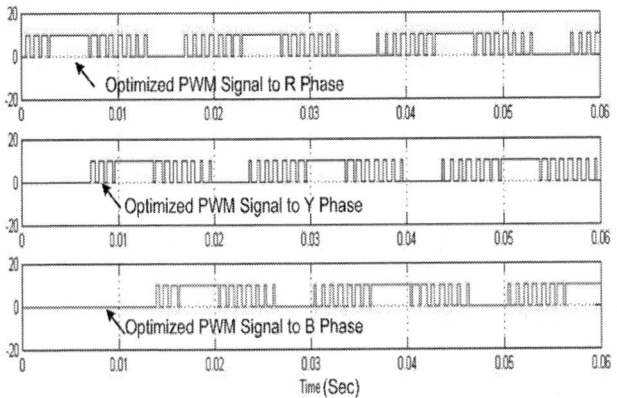

Fig 5. Optimized pulse sequence applied to three phase inverter

In fig.5, the pulse sequence for the top three switches for inverter has been shown. The respective other switches of same leg shall invert of those respective switches. In each leg, the pulse sequence is provided with a delay of 120^0 phase shift. While doing the problem formulation, that total harmonic distortion (THD) is considered as an objective function and harmonic levels has been taken up 39th order of fundamental component. The line to line output voltage for various switching schemes has been tabulated in table.1

TABLE I. SWITCHING SEQUENCE OF INVERTER

Switching State	Operating Switches	V_{RY}	V_{YB}	V_{BR}
1	1,2,6	V_{L-L}	0	$-V_{L-L}$
2	1,2,3	0	V_{L-L}	$-V_{L-L}$
3	2,3,4	$-V_{L-L}$	V_{L-L}	0
4	3,4,5	$-V_{L-L}$	0	V_{L-L}
5	4,5,6	0	$-V_{L-L}$	V_{L-L}
6	5,6,1	V_{L-L}	V_{L-L}	0
7	1,3,5	0	0	0
8	2,4,6	0	0	0

Fig 6.Experimental set up of the proposed system

The system has been examined for balanced load condition.

$$V_{Rn} = \sum_{n=1}^{\infty} \frac{4V_s}{n\pi} \left[1 + 2\sum_{k=1}^{m} (-1)^k Cos(n\alpha_k) \right]$$

$$V_{Yn} = \sum_{n=1}^{\infty} \frac{4V_s}{n\pi} \left[1 + 2\sum_{k=1}^{m} (-1)^k Cosn(\alpha_k - \frac{2\pi}{3}) \right]$$

$$V_{Bn} = \sum_{n=1}^{\infty} \frac{4V_s}{n\pi} \left[1 + 2\sum_{k=1}^{m} (-1)^k Cosn(\alpha_k - \frac{4\pi}{3}) \right]$$

Then the current passing through is

$$I_a = \sum_{n=1}^{\infty} \left[\frac{4Vs}{\pi\sqrt{R^2 + (nwl)^2}} \left(1 + 2\sum_{k=1}^{n} (-1)^k Cos(n\alpha_k) \right) \right] Sin(nwt - \theta_n)$$

The above equations represents the load voltage and load current for balanced resistive load. Where V_s the input supply DC voltage, 'n' is the order of harmonics, V_{Rn} is the phase to neutral voltage of inverter. The pulses required for the gate triggering for the inverter is obtained by ATMEGA-16 micro controller. Thus obtained signals has been given to the respective power semiconductor switches in the inverter. This experiment has been repeated for proposed techniques and results has been explored with their analysis. It has considered with balance star connected resistive load as shown in experimental setup of fig.6.

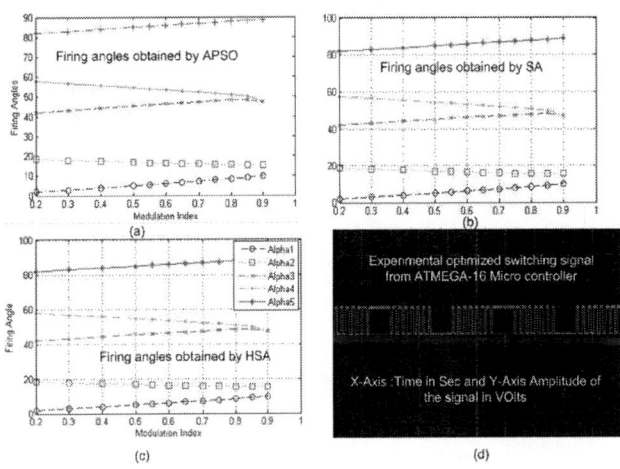

Fig 7. Generation of notch/firing angles for inverter. It has been given for five notch angles to maintain the clarity: (a) Notch angles obtained by APSO (b) Notch angles obtained by SA (c) Notch angles obtained by HSA (d) Generation of optimized PWM signals by using ATMEGA-16 (Experimental waveform).

The application of optimization technique results the notch angles of various modulation index (MI) as shown in fig.7 (a-c). Only five notch angles out of 37 has been shown for its clarity and its experimental wave forms has been shown in fig.7 (d).

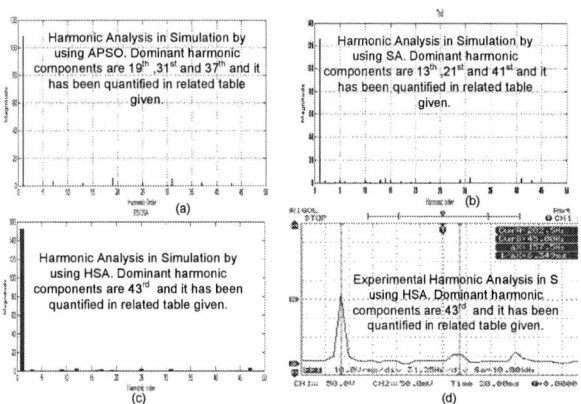

Fig 8. THD comparison of proposed technique by using Simulink and experimental results. (a) THD obtained by using APSO with dominant 19th order harmonic (b) THD obtained by using SA with dominant 13th order harmonic (c) THD obtained by using HSA with dominant 43rd order harmonic (d) THD obtained by using SA (Experimental waveform).

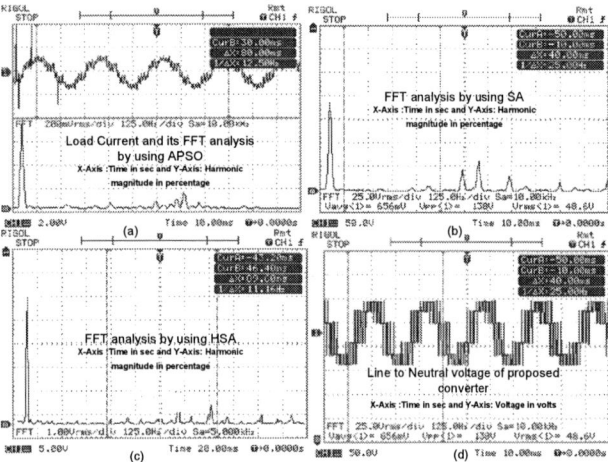

Fig 9. THD comparison of proposed technique by using experimental results. (a) THD obtained by using APSO with dominant 19th order harmonic (b) THD obtained by using SA with dominant 13th order harmonic (c) THD obtained by using HSA with dominant 43rd order harmonic (d) Line to neutral voltage of proposed converter under HSA technique.

TABLE II. COMPARISON OF SIMULATION AND EXPERIMENTAL RESULTS BY USING APSO TECHNIQUE

Harmonic order	Simulation in %	Experimental
1	100	100
5	0	0.002
7	0	0.001
11	0	0.004
13	0	0.002
17	0.001	0.001
19	0.101	0.150
23	8.6	10.1
29	6.1	7.1
31	0	0.001
35	0	0.001
37	0	0.001
THD	**10.5444%**	**12.379%**

TABLE III. COMPARISON OF SIMULATION AND EXPERIMENTAL RESULTS BY USING SA TECHNIQUE

Harmonic order	Simulation in %	Experimental
1	100	100
5	0	0.004
7	0	0.007
11	0	0.009
13	0.14	0.002
17	0	0.001
19	0	0.150
23	9.4	10.1
29	11.41	13.41
31	0.001	0.001
35	4.1	8.1
37	0.003	0.001
THD	**15.1111%**	**19.3898%**

TABLE IV. COMPARISON OF SIMULATION AND EXPERIMENTAL RESULTS BY USING HSA TECHNIQUE

Harmonic order	Simulation in %	Experimental
1	100	100
5	0	0.001
7	0	0.002
11	0	0.004
13	0	0.094
17	0	1.41
19	0	0.150
23	0	0.001
29	2	2.001
31	1.001	1.907
35	0.001	0.907
37	0.001	0.907
THD	**3.12%**	**3.97%**

By using APSO method (Simulation) for 0.9 MI,THD was 39.63% whereas with using APSO method (Experimental) for 0.9 MI,THD was 10.5444%. By using SA method (Simulation) for 0.9 MI,THD is 15.11%.By using hybrid (SA&APSO) for 0.9 MI,THD is 3.12%. In HSA, the value of THD is less when compared to other methods. The FFT analysis for balanced resistive load has been given in fig.8 and fig.9 with their simulation and experimental results respectively. The THD obtained in line voltages are different from THD obtained in line currents. Line current THD is 2% less compared to line voltage because, at this particular instant load itself-acts as a filter and some decrement can be observed in the lower and higher order of harmonics. By comparing both the simulation and experimental vales, experimental values are slightly greater than the Simulink values, which is mainly due to the introduction of dead band in switching sequences to the inverter legs.

VI. CONCLUSION

In this article, a new application of optimization technique (SA and HSA) has been discussed to selective harmonic elimination technique in three phase three level inverter. Thus obtained results has been compared with existing technique (APSO) and results are reported. The proposed technique has been conducted through simulation and verified experimentally. The proposed technique is able to eliminate the harmonics completely and requirement of filter is substantially eliminated. Due to which the proposed system is more efficient and convenient. Finally the system with HSA is giving reported the better results than other specified techniques in 0.02 sec.

References

[1]. Sreedhar, Madichetty; Dasgupta, Abhijit; Mishra, Sambeet: 'New harmonic mitigation scheme for modular multilevel converter – an experimental approach', IET Power Electronics, 2014, DOI: 10.1049/iet-pel.2014.0028,IET Digital Library, http://digital-library.theiet.org/content/journals/10.1049/iet-pel.2014.0028

[2]. Narimani, M.; Moschopoulos, G., "Three-Phase Multimodule VSIs Using SHE-PWM to Reduce Zero-Sequence Circulating Current," *Industrial Electronics, IEEE Transactions on* , vol.61, no.4, pp.1659,1668, April 2014

[3]. Zhou, K.; Yang, Y.; Blaabjerg, F.; Wang, D., "Optimal Selective Harmonic Control for Power Harmonics Mitigation," *Industrial Electronics, IEEE Transactions on* , vol.PP, no.99, pp.1,1

[4]. Ahmadi, D.; Jin Wang, "Online Selective Harmonic Compensation and Power Generation With Distributed Energy Resources," *Power Electronics, IEEE Transactions on* , vol.29, no.7, pp.3738,3747, July 2014

[5].Furukawa, K.; Ajima, T.; Miyazaki, H., "Optimal pulse pattern determination based on Pulse Harmonic Modulation," *Power Electronics Conference (IPEC-Hiroshima 2014 - ECCE-ASIA), 2014 International* , vol., no., pp.383,389, 18-21 May 2014.

[6]. Bhadra, S.; Patangia, H., "Implementation of real-time selective harmonic elimination using microcontrollers," *Power Electronics for Distributed Generation Systems (PEDG), 2014 IEEE 5th International Symposium on* , vol., no., pp.1,5, 24-27 June 2014

[7]. Sreedhar, M.; Dasgupta, A, "Application of Accelerated Particle Swarm Optimization to multi level diode clamped inverter," *Power Electronics (IICPE), 2012 IEEE 5th India International Conference on* , vol., no., pp.1,5, 6-8 Dec. 2012

[8]. Sreedhar, M.; Dasgupta, A, "Experimental verification of Minority Charge Carrier Inspired Algorithm applied to voltage source inverter," *Power Electronics (IICPE), 2012 IEEE 5th India International Conference on* , vol., no., pp.1,6, 6-8 Dec. 2012

[9]. Sreedhar, M., & Dasgupta, A. (2012, September).An Experimental Verification of Performance Improvement In Inverter Design By Using Accelerated Particle Swarm Optimization Technique,"International Journal of Engineering Research and technology",Vol-1,Issuse-7,September-2012.ISSN:2278-0181

[10]. Ray, S.; Sreedhar, M.; DasGupta, A., "ZVCS based high frequency link grid connected SVPWM applied three phase three level diode clamped inverter for photovoltaic applications," *Power and Energy Systems Conference: Towards Sustainable Energy, 2014* , vol., no., pp.1,6, 13-15 March 2014

[11] J. Kennedy, R.C. Eberhart, " Particle Swarm Optimization," Proc. IEEE Int. of Neural Networks, Piscataway, NJ, USA, 1942-1948, 1995.

[12]. Xin-She Yang (2008) Swarm Optimization. Introduction to Computational Mathematics: pp. 229-235. doi: 10.1142/9789812818188_0019

[13].Sreedhar, M.; Upadhyay, N.M.; Mishra, S.; , "Optimized solutions for an optimization technique based on minority charge carrier inspired algorithm applied to selective harmonic elimination in induction motor drive," Recent Advances in Information Technology (RAIT), 2012 1st International Conference on , vol., no., pp.788-793, 15-17 March 2012 .

A NF Based Direct Torque Control of Induction Motor Drive Using BCSVM

N. Venkataramana Naik, *Member of IEEE*
Research Scholar,
Electrica Engg. Dept.,
IIT Roorkee-247667
Email: iitrramana@gmail.com

Dr. S. P Singh, *Member of IEEE*
Professor,
Electrica Engg. Dept.,
IIT Roorkee-247667
Email: spseefee@gmail.com

Abstract— **This paper presents a direct torque control (DTC) using Bus clamped SVM (BCSVM). The direct torque control gives more distortion in flux, speed and torque especially during speed or load changing conditions, and also gives more distortion in the current of the induction motor. In order to improve the dynamic performance of the induction motor the Neuro Fuzzy (NF) is used to get the less THD in current as well as lesser distortion in speed, flux and torque. The transient and steady state performance of induction motor using NF based DTC is compared with the conventional DTC.**

Keywords- *Bus clamped SVM, Direct Torque Control, Induction Motor, Neuro Fuzzy Control, and Voltage source Inverter.*

I. INTRODUCTION

Conventionally the speed of the induction motor (IM) is controlled like separately excited dc motor using vector control (or field-oriented control) as it was practiced in1960 [1]-[3]. However, such practice requires many transformations making it more complex to implement the control strategy [4]-[5]. After twenty years (1980) the conventional direct torque control (CDTC) was developed for getting fast and quick dynamic response of the IM, and thus, there was no need of any transformation as compared to vector control [6]-[9].

In CDTC, the torque and flux are controlled independently using hysteresis control [10]-[14]. However, the hysteresis control gives considerable flux and torque ripple with variable switching frequency and hence it is difficult to design the filter to reduce the torque ripples of the IM [15]-[18]. The fuzzy control is most familiar technique to reduce the torque and flux distortion of the IM with less current ripples by improving the controller behavior using duty ratio control [19]-[22]. In this paper, the neuro fuzzy control (NFC) is applied to the BCSVM-DTC to improve the dynamic performance of the IM.

II. .DIRECT TORQUE CONTROL

The block diagram for the proportional integral direct torque control (PIDTC) in real time implementation is shown in Fig. 1. The flux, torque, and speed of IM are estimated. The generalized electromagnetic torque of the IM [1]-[2] is:

$$T_e = J\left(d\omega_m / dt\right) + B\omega_m + T_L \tag{1}$$

Where ω_m, J, B, and T_L=mechanical speed, moment of inertia, friction coefficient, and load torque of the IM respectively.

The electrical speed (ω_e) in terms of mechanical speed (ω_m) for 4-pole IM is [3]

$$\omega_e = 2\omega_m \tag{2}$$

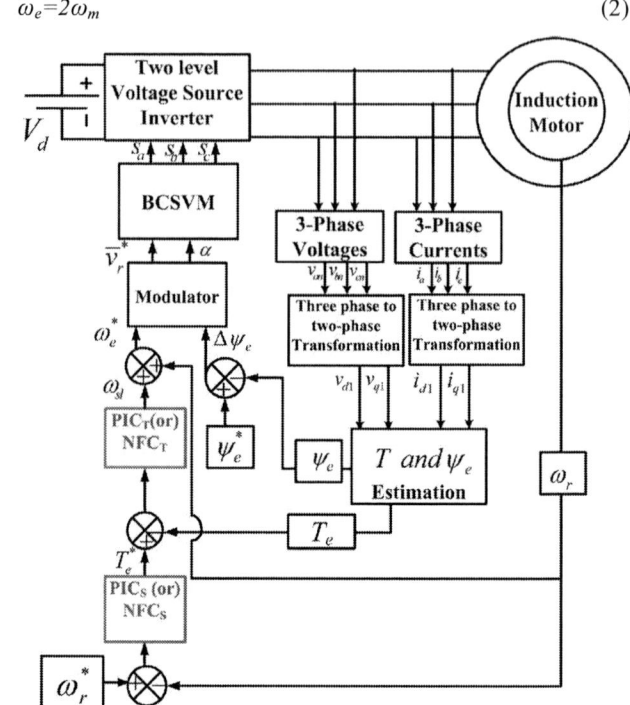

Fig. 1 DTC of induction motor using PI/NFC

The stator flux in d-axis and q-axis are estimated as [4]-[6]

$$\psi_{q1} = \int (v_{q1} - R_1 i_{q1})dt \tag{3}$$

$$\psi_{d1} = \int (v_{d1} - R_1 i_{d1})dt \tag{4}$$

Where, i_{d1} and i_{q1} =d-q axes of the stator currents, v_{d1} and v_{q1} = d-q axes of the stator voltages, R_1 = stator resistance of the IM respectively.

Therefore the estimated stator flux $\psi_e = \sqrt{\psi_{d1}^2 + \psi_{q1}^2}$ (5)

From the speed PI-control (PICs) the reference torque is T_e^* and is compared with actual/estimated torque T_e for 2-pole IM as [1]:

$$T_e = 3\left(\lambda_{d1}i_{q1} - \lambda_{q1}i_{d1}\right) \tag{6}$$

From the torque PI-control (PIC$_T$) the angle θ is estimated as:

$$\theta = \int \left(\omega_{sl}^* + \omega_2\right)d(\omega t) \tag{7}$$

978-1-4799-6047-7/14 $31.00 © 2014 IEEE

Where $\omega_e^*, \omega_{sl}^*, and \ \omega_2$ =reference synchronous speed, reference slip speed, and rotor speed in rad/s respectively.

The reference flux $\psi_1^* = |\psi_1^*|\cos\theta + |\psi_1^*|\sin\theta$ (8)

From the flux controller, the changes in d-q axes of stator flux linkages are given as:

$$\Delta\psi_{d1} = |\psi_1^*|\cos\theta - \psi_{d1} \tag{9}$$

$$\Delta\psi_{q1} = |\psi_1^*|\sin\theta - \psi_{q1} \tag{10}$$

The reference voltages in $\alpha-\beta$ plane are given as:

$$v_\beta^{\cdot} = \left(\Delta\psi_{q1}/dt\right) + R_1 i_{q1} \tag{11}$$

$$v_\alpha^{\cdot} = \left(\Delta\psi_{d1}/dt\right) + R_1 i_{d1} \tag{12}$$

The magnitude and the angles of reference voltage vector are estimated as [1], [6]-[9]:

$$\overline{v_r} = \left|\sqrt{v_\alpha^{\cdot\,2} + v_\beta^{\cdot\,2}}\right|\angle\alpha \tag{13}$$

Where angle $\alpha = a\tan\left(v_\beta^{\cdot}/v_\alpha^{\cdot}\right)$ (14)

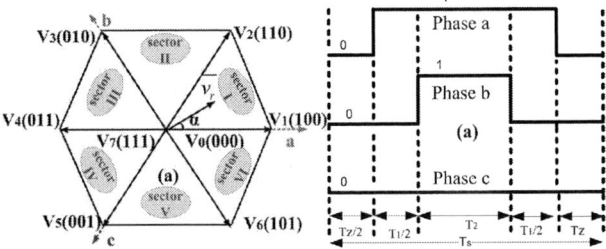

Fig. 2 (a) 2-level SVM (b) PWM Pulses

A 3-phase-two level voltage source inverter has 8 switching states. In which, 2 are zero voltage vectors (0 0 0 and 1 1 1) and other 6 are the vigorous voltage vectors as shown in Fig. 2(a). The vigorous vectors divide the space vector plane into six sectors with equal magnitude and phase displacement of $0.667V_d$ and 60° respectively. The time periods, T_1, T_2, and T_z for corresponding voltage vectors V_1, V_2, and V_7 can be found using volt-second balance principle as [3], [9]-[15]

$$\overline{v_r}\angle\alpha°T_s = V_1\angle 0°T_1 + V_2\angle 60°T_2 + 0T_z \tag{15}$$

The T_1, T_2 and T_Z are calculated by equating real and imaginary terms in Eq. (15)

$$T_1 = \frac{V_{ref}}{V_2}T_s\frac{\sin(60°-\alpha)}{\sin(60°)} \tag{16}$$

$$T_2 = \frac{V_{ref}}{V_2}T_s\frac{\sin(\alpha)}{\sin(60°)} \tag{17}$$

$$T_z = T_s - T_1 - T_2 \tag{18}$$

By using Eq. (18) to (5) the BCSVM pulses are developed as shown in Fig. 3. This process repeats itself for remaining five sectors to develop the switching pulses for full cycle [3]-[5].

A. Neuro Fuzzy control

The PI control is replaced by NFC as shown in Fig. 3. From a classical PI-controller the output can be

$$u = K_p e + K_i \int e dt \tag{19}$$

Where the k_p and k_i are the proportional and integral gains, and e is the error. Thus by differentiation of the Eq. (19)

$$\frac{du}{dt} = K_p\frac{de}{dt} + K_i e dt \tag{20}$$

From the above Eq. (20) it can be concluded that change of the output depends on the error and change of the error. The PI-type NFC uses error (E) and change of error (CE) as inputs, and output will get change of output. For simplicity, the NFC structure of two inputs x and y and a single output Z is considered. Assume that the rule base contains only two fuzzy if then rules of Takagi-Sugeno [1]-[8] as follows:

If x is $A1$ and y is $B1$, then $Z_1=m_1 x+n_1 y+q_1$

If x is $A2$ and y is $B2$, then $Z_2=m_2 x+n_2 y+q_2$

The architecture of proposed NFC is shown in Fig 3. It contains five layers where the node functions in the same layer are of the same function family. The Layer 1 is defined as

$$O_i = \mu A_i(x), \qquad i=1, 2, ..., 5 \tag{21}$$

Where x = input to node i, μA_i = membership function of the linguistic label A_i =associated node. The bell-shaped can be expressed as

$$\mu A_i(x) = \frac{1}{1+\left[\left(\dfrac{x-c_i}{a_i}\right)^2\right]^{b_i}} \tag{22}$$

Where, a, b = vary the width of curve and c = the center of the curve.

Layer 2 is the product of all the incoming signals as.

$$w_i = \mu A_i(x) AND \mu B_i(x), \quad i=1,2,...5 \tag{23}$$

Where the output w_i = firing strength of rule. Beside multiplication, other t-norm operators that perform generalized 'AND' can be used as the node function in this layer.

Layer 3 is a fixed node and it calculates the ratio of the ith rule activation as

$$\overline{w_1} = \frac{w_i}{w_1 + w_2} \tag{24}$$

Layer 4 is the defuzzification process of fuzzy system (using weighted average method) is obtained in every node i as

$$O_i^4 = \overline{w_1}(m_i X + n_i Y + q_i) \qquad (25)$$

Where $\overline{w_1}$ = the output of layer 3, and m_i, n_i and q_i = referred as the linear consequent parameters.

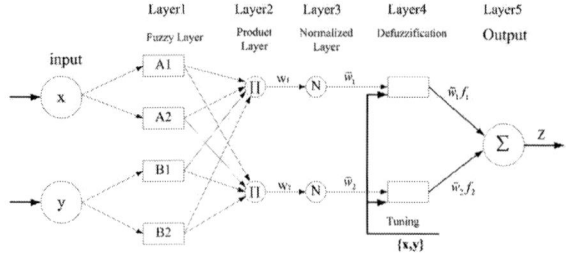

Fig. 3 Architecture of NFC

Layer 5 is the overall output as the weighted average of all incoming signals from the layer 4 as

$$O_i^5 = \sum \overline{w_1} f_i = \frac{\sum_i \overline{w_1} f_i}{\sum_i w_i} \qquad (26)$$

The NFC tuned by a hybrid technique combining gradient descent back-propagation and mean least-squares optimization (LSE) algorithms as shown in Fig. 3 and the output Z is

$$Z = \overline{w_1} f_1 + \overline{w_2} f_2 \qquad (27)$$
$$= (\overline{w_1} x) m_1 + (\overline{w_1} y) n_1 + (\overline{w_1}) q_1 + (\overline{w_2} x) m_2 + (\overline{w_2} y) n_2 + (\overline{w_2}) q_2$$

By fixing the elements above Eq. (27) is linear with consequent parameters (m_1, n_1, q_1, m_2, n_2, q_2). Therefore, the matrix form of Eq. (27) is

$$AX = B \qquad (28)$$

The over determined system of equations uses pseudo-inverse of X is.

$$X^* = (A^T A)^{-1} A^T B \qquad (29)$$

The X is intended iteratively is [9].

$$X_{i+1} = X_i + S_{i+1} a_{i+1} (b_{i+1}^T X_i - a_{i+1}^T X_i) \qquad (30)$$

$$S_{i+1} = S_i - \frac{S_{i+1} a_{i+1} a_{i+1}^T S_i}{1 + a_{i+1} a_{i+1}^T s_i}, \qquad (31)$$

Since a node output depends on its incoming signals and parameters set as:

$$O_i^k = O_i^{k-1}(O_i^{k-1}, \ldots O_{k-1}^{k-1}, a, b, c, \ldots) \qquad (32)$$

Assuming the given training data set has P input-output entries, the error measure for the pth entry ($1 \leq p \leq P$) of training data entry as

$$E_p = \sum_{m=1}^{L} \left(T_{m,p} - O_{m,p}^L\right)^2 \qquad (33)$$

$$E = \sum_{P=1}^{P} E_p \qquad (34)$$

The error rate for the output node at (L, i) can be calculated from the Eq. (34) as

$$\frac{\partial E_p}{\partial E_{i,p}^L} = -2(T_{i,p} - O_{i,p}^L) \qquad (35)$$

For the internal node at (k, i), the error rate can be derived by the chain rule

$$\frac{\partial E_p}{\partial O_{i,p}^L} = \sum_{m=1}^{k+1} \frac{\partial E_p}{\partial O_{m,p}^{k+1}} \frac{\partial O_{m,p}^{k+1}}{\partial O_{i,p}^k} \qquad (36)$$

Therefore for all $1 \leq k \leq L$ and $1 \leq i \leq \#(k)$, $\frac{\partial E_p}{\partial O_{i,p}^L}$ will be found using Eq. (35) and (36).

$$\frac{\partial E_p}{\partial a} = \sum_{o^\bullet \in s} \frac{\partial E_p}{\partial o^\bullet} \frac{\partial o^\bullet}{\partial a} \qquad (37)$$

Where S = set of nodes whose output depends on a. The derivative of overall error measure E with respect to a is

$$\frac{\partial E_p}{\partial a} = \sum_{p=1}^{P} \frac{\partial E_p}{\partial a} \qquad (38)$$

Accordingly, the update formula for the generic parameter a is

$$\Delta a = -\varepsilon \frac{\partial E}{\partial a} \qquad (39)$$

Where ε is a learning rate which can be expressed as

$$\varepsilon = \frac{k}{\sqrt{\sum_a \left(\frac{\partial E}{\partial a}\right)^2}} \qquad (40)$$

Where k= the step size

NFC, as a class of adaptive networks, can apply all the above illustrated gradient descent concepts and algorithms using from Eq. (21) to (40).

III. RESULTS AND DISCUSSION

The direct torque control is simulated in the Matlab environment using BCSVM as shown in Fig. 1 with applied switching frequency 5kHz. The starting transient performance of the induction motor is shown in Fig. 4 while applying reference speed 1300 rpm. The rotor speed is zero at starting and settles 1300 rpm at 0.54 s. However, current, flux and torque during starting are high due to high starting slip of the induction motor, and they have settled at the same instant of time i.e. 0.54s as that of speed. In addition to this the torque distortion varies from -1N-m to +1N-m at the same instant under no-load condition.

is 1.78% as shown in Fig. 12, which is very less compared PIC.

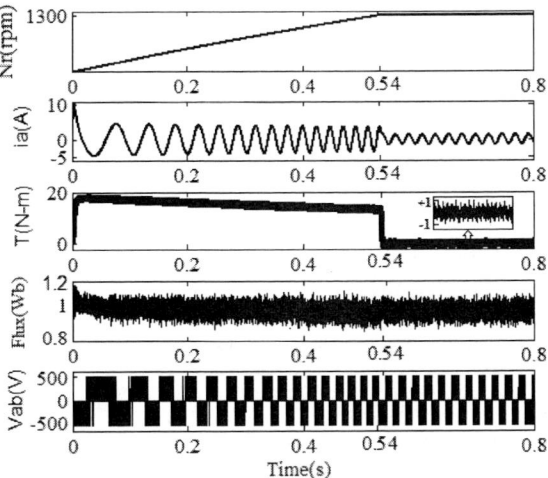

Fig. 4 Starting performance of the PIC induction motor

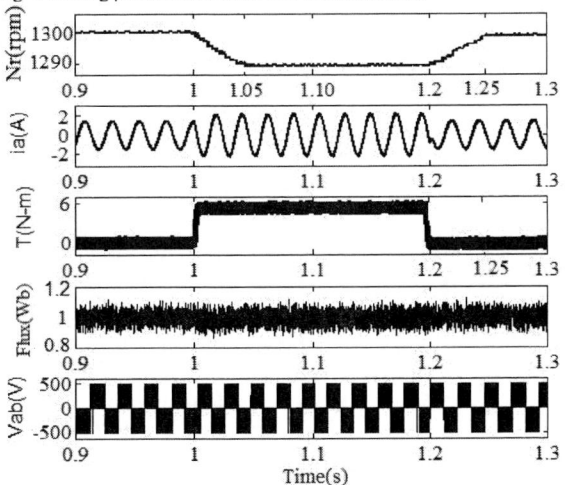

Fig. 5 Load (6 N-m) performance of the PIC induction motor

When the load 6 N-m is applied to induction motor then the speed reduced from 1300 to 1290 rpm with increased stator currents of induction motor as shown in Fig. 5. However the flux and voltage are seem to be same during transient and steady state operation. The no load line current THD was obtained 2.43% as observed in Fig. 6.

The speed and torque PI Controllers are replaced by NFCs with the same with the same operating condition of PI Controllers. The strarting performance of the IM is quicly reached to steady state i.e. the time taken by the IM is 0.5s as shown in Fig. 7. Here, flux and torque distotions are reduced significantly as compared to PI Controllers. When the load 6N-m applied to the induction motor as shown in Fig. 8. The speed was quickly reached to steady state with less flux and torque distortion and the current THD of the induction motor

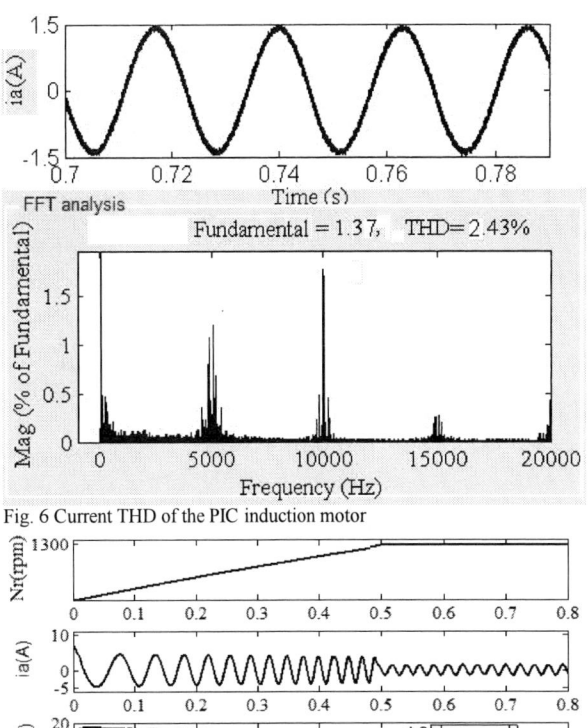

Fig. 6 Current THD of the PIC induction motor

Fig. 7 Starting performance of the NFC induction motor

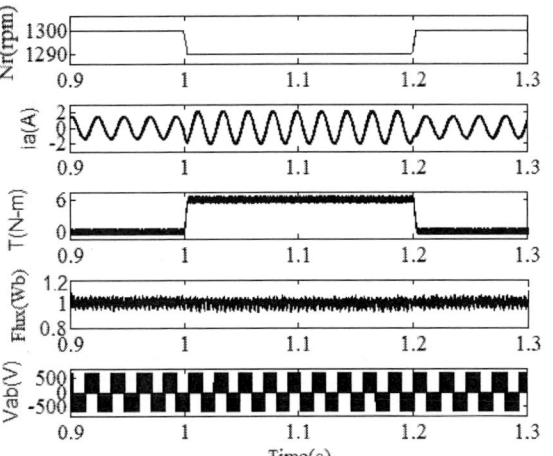

Fig. 8 Load (6 N-m) performance of the NFC induction motor

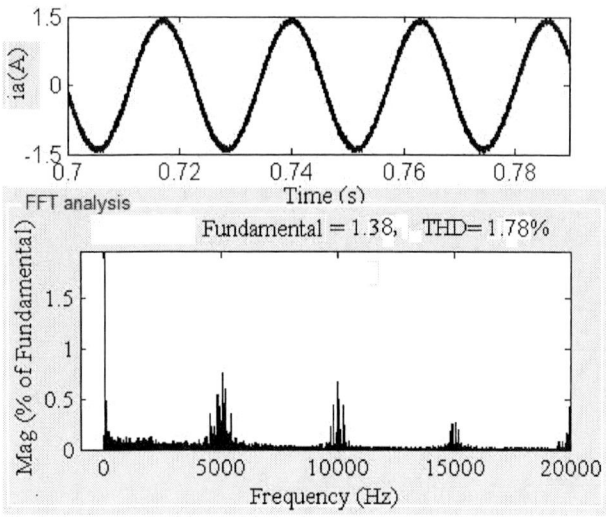

Fig. 8 Current THD of the NFC induction motor

IV. CONCLUSION

The speed of the induction motor is controlled by DTC using two level bus clamped SVM. The DTC technique gives more flux and torque distortion with considerable current THD of the induction motor drive. In order to improve the induction motor drive performance the PI-controllers are replaced by NF Cs. Hence, the NF based DTC has less distortion in flux, torque and speed with less current THD of the induction motor as compared to conventional PI-control method. Because, the NF control have tuned the PI-controllers well using hybrid learning algorithm during transient as well as steady state operation.

APPENDIX
PARAMETERS OF INDUCTION MOTORDRIVE

S. No.	Parameters	Ratings
	Induction motor	
1	Rated supply voltage (V_{L-L})	440V (L-L)
2	Power rating (P)	2HP
3	Rated speed (N_2)	1460 rpm
4	No of Poles (P)	4
5	Stator resistance (R_1)	7.83 Ω
6	Rotor resistance (R_2)	7.55 Ω
7	Stator leakage inductance (L_1)	0.4751H
8	Rotor leakage inductance (L_2)	0.4751H
9	Mutual inductance (Lm)	0.4535H
10	Moment of inertia (J)	0.06 Kg - m^2

REFERENCES

[1] Yen-Shin Lai and Juo-Chiun Lin, "New Hybrid Fuzzy Controller for Direct Torque Control Induction Motor Drives," *IEEE Trans. Power Electron.*, vol. 18, no. 5, pp. 1211–1219, Sept. 2003.

[2] N. Venkataramana Naik and S. P. Singh, "A Novel Type-2 fuzzy Logic Control of Induction Motor Drive using Space Vector PWM," *IEEE Conf. Proc. INDICON-12*, Kochi, pp. 1142-1147, Dec. 2013.

[3] Wenxi Yao, Haibing Hu and Zhengyu Lu "Comparisons of Space-Vector Modulation and Carrier-Based Modulation of Multilevel Inverter," *IEEE Trans. Power Electron.*, vol. 23, no. 1, pp.45–51, Jan. 2008.

[4] N. Venkataramana Naik and S. P. Singh, "Improved Dynamic Performance of Direct Torque Control at low speed over a scalar control," *IEEE.Proc. INDICON-13 at IIT Bombay*, pp. 1-6, Jan. 2014.

[5] I. Takahashi and T. Noguchi, "A new quick-response and high-efficiency control strategy of an induction motor," *IEEE Trans. Ind. Appl.*, vol. 22, no. 5, pp. 820–827, Sept. 1986.

[6] N. Venkataramana Naik and K. Sri Gowri, "Space Vector Based Hybrid Pulse width Modulation for Reduced Current Ripple," *National Conf Proc (NCIES-09) at Salem*, Apr. 2009

[7] D. Telford, M. W. Dunnigan and B. W. Williams, "A Novel Torque-Ripple Reduction Strategy for Direct Torque Control," *IEEE Trans. Ind. Electron,* vol. 48, no. 4, pp. 867-870, Aug 2001.

[8] D. Casadei, etc., "Performance analysis of a speed-sensor less induction motor drive based on a constant switching-frequency DTC scheme," *IEEE Trans. Ind. Appl.*, vol. 39, no. 2, pp. 476–484, Mar. 2003.

[9] N. Venkataramana Naik, K. Sri Gowri, and T Brahmananda Reddy, "A Novel Hybrid Space vector PWM for Reduced Current Ripple using Scalar Control," *International Conf Proc (ICETES-10) at Nagercoil*, Mar. 2010

[10] G.Durgasukumar and M. K. Pathak, "Comparison of Adaptive Neuro-Fuzzy based Space Vector Modulation for Two-Level Inverter," *International Journal of Electrical Power and Energy Systems, Elsevier*, vol. 38, no. 1, pp. 9-19, 2012.

[11] Giribabu.D, S.P.Srivastava and M.K.Pathak, "Fuzzy Logic Based MRAS for Speed Sensorless Operation of Induction Motor Drive," *NPEC, BESUS Kolkata*, Dec 19th-22nd 2011, pp. 1-6.

[12] Ramesh, etc., "Direct flux and torque control of induction motor drive for speed regulator using PI and fuzzy logic controllers," *IEEE Proc., ICAESM-2012*, pp. 288-295, Mar. 2012.

[13] H. Lam and H Li, "Output-Feedback Tracking Control for Polynomial Fuzzy-Model-Based Control Systems," *IEEE Trans. Ind. Electron.*, vol. 60, no. 12, pp. 5830–5840, Dec. 2013.

[14] N. Venkataramana Naik and S. P. Singh, "Improved Dynamic Performance of Fuzzy Based DTC Induction Motor Using Bus-Clamped SVM," *IEEE Proc, IICPE-12 at DTU Delhi*, pp. 1-6, Jan. 2013.

[15] Peter Vas, "Artificial Intelegence Based Electrical Machines And Drives Application of fuzzy, Neural, Fuzzy Neural, and Gentic Algorithm based Techniques," *Oxford university press*, New Yark, pp.173-234, April. 1999.

[16] Luis Romeral, et. al, "Novel Direct Torque Control (DTC) Scheme with Fuzzy Adaptive Torque-Ripple Reduction," *IEEE Trans on Ind. Electron*, vol. 50, no. 3, pp. 487-492, Jun. 2003.

[17] J. M. Mendel, "Uncertain rule-based fuzzy logic systems: Introduction and new directions," *Prentice Hall PTR, Upper Saddle River*, pp. 50-312, NJ, 2001.

[18] N. Venkataramana Naik and S. P. Singh, "A Novel Type-2 Fuzzy Logic Control of Induction Motor Drive Using Scalar Control," *IEEE Proc, IICPE-12 at DTU Delhi*, pp. 1-6, Jan. 2013.

[19] N. Venkataramana Naik and S. P. Singh, "Improved Dynamic Performance of Type-2 Fuzzy Based DTC Induction Motor Using SVPWM," *IEEE Proc, PEDES-12 at IISc Banglore*, pp. 1-5, Jan. 2013

[20] N. Venkataramana Naik and S. P. Singh, "Improved Torque and Flux performance of Type-2 Fuzzy Based DTC Induction Motor Using Space Vector Pulse-width Modulation," *Electric Power Components and System*, vol. 42, pp. 658-669, Mar. 2014.

[21] G.Durgasukumar and M. K. Pathak, "Torque Ripple Minimization of Vector Controlled VSI Induction Motor Drive using Neuro-Fuzzy Controller," *International Journal on Advances in Engineering Sciences*, vol. 1, no. 1, pp. 40-43, 2011.

[22] G.Durgasukumar and M. K. Pathak, "Three-Level Inverter Performance using Adaptive Neuro-Fuzzy based Space Vector Modulation," *Computer Engineering and Intelligent Systems*, vol. 2, no. 4, pp. 110-123, 2011.

Energy Efficient Flip Flop Design Using Voltage Scaling On FPGA

Sunny Singh
School of Computer Sciences
Chitkara University,
Rajpura, India
Er.singhsunny2207@gmail.com

Amanpreet Kaur,Bishwajeet Pandey
School of Electronics Engineering
Chitkara University,
Rajpura,India
amanpreet.kaur@chitkara.edu.in
bishwajeet.pandey@chitkara.edu.in

Abstract— **In this work, we are using voltage scaling and frequency scaling. In voltage scaling, voltage is scaled from 3V to 1V, where intermediate values are 2.5V, 2V, 1.8V and 1.5V. In frequency scaling, frequency is scaled from 1 MHz to 1 THz, where intermediate values are 10 MHz, 100 MHz, 1 GHz, 10 GHz and 100 GHz. When we scale down device operating frequencies from 1THz to 1GHz, there is 72.9% reduction in power dissipation on Virtex-6 FPGA. When we scale down device operating frequencies from 1THz to 1GHz, there is 98.75% reduction in power dissipation on Virtex-4 FPGA. When we scale down device supply voltage from 3V to 2.5V, 2V, 1.8V and 1V, there is 82.23%, 96.83%, 98.45% and 99% reduction in power dissipation respectively on Virtex-6 FPGA on 10MHz device operating frequency. When we scale down device supply voltage from 3V to 2.5V, 2V, 1.8V and 1V, there is 74.42%, 92.67%, 94.71% and 97.66% reduction in power dissipation respectively on Virtex-6 FPGA on 1THz device operating frequency.**

Keywords—Energy Efficient Design, Voltage Scaling, FPGA, VLSI, Device Operating Frequency

I. INTRODUCTION

Dynamic voltage scaling (DVS) decreases power dissipation of processors when peak performance is not required [1]. However, the energy efficiency by DVS alone is now not significant with increase in leakage power [1]. Voltage scaling is used in mobile battery charge controller sensor [2], voltage sensor for fire detection [3], energy efficient FIR Filter design [4], voltage based fire sensor [5], reliable ALU design [6] and Green GCD Generator [7]. Here, we are using the simplest design in electronics i.e. J K Flip-Flop. Using J K Flip-Flop we can design many complex circuit. Therefore, it is easy to reduce the power dissipation of basic component in place of reducing the power dissipation of complete circuit. Reduction of power dissipation of basic circuits will eventually result in reduction of power dissipation of overall circuits. There are different types of voltages used in FPGA that are VCCINT, VCCAUX, VCCO, VCCBRAM and VCCBAT as shown in Figure 1.

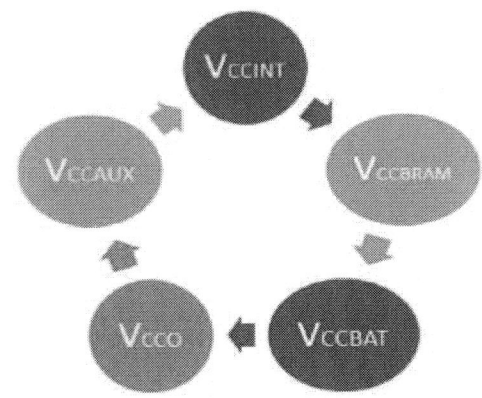

Figure 1: Different Types of Voltage on FPGA [7]

VCCINT refers to the core voltage input. It means chip is working on that voltage. There is 1.2V for Stratix-II, Cyclone-II and Virtex-4 and 1.5V for Stratix, Cyclone and Vertex-II, whereas 2.5V for Spartan-II. as shown in Figure 2.

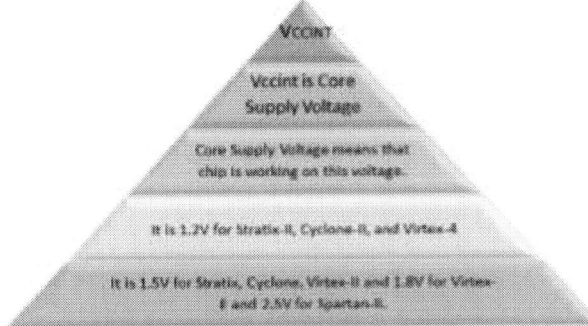

Figure 2: Core Supply Voltage (VCCINT) [7]

In this work, we are using voltage scaling and frequency scaling. In voltage scaling, voltage is scaled from 3V to 1V, where intermediate values are 2.5V, 2V, 1.8V and 1.5V. In frequency scaling, frequency is scaled from 1 MHz to 1 THz, where intermediate values are 10 MHz, 100 MHz, 1 GHz, 10 GHz and 100 GHz as shown in Figure 3.

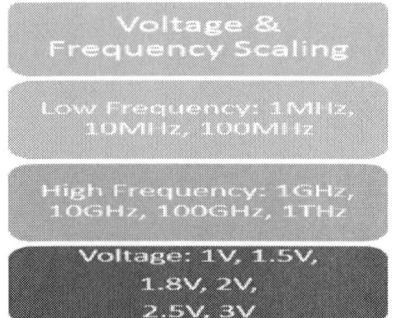

Figure 3: Voltage Scaling and Frequency Scaling

VCCAUX is an auxiliary voltage used to power the JITAG and different Configuration pins of FPGA that makes them not dependent on VCCO. VCCAUX also provides voltage supply to LVDS Bias Generator, I/O Pre Drivers and Differential Input Amplifiers/Comparators.

II. POWER ANALYSIS IN LOW FREQUENCY RANGE

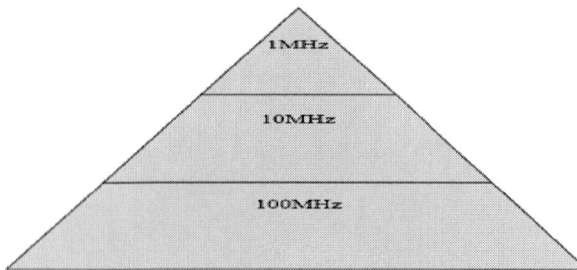

Figure 4: Frequency Scaling in Low Frequency Range

In low frequency range, we operate our design with 1MHz, 10MHz and 100MHz as shown in Figure 4.

A. Power Dissipation On Virtex-6 FPGA

Table 1: Power Dissipation for Different Low Operating Frequencies

Power Dissipation When Supply Voltage is 1.8V	
Frequency	Power
1MHz	19.173
10MHz	19.174
100MHz	19.178

Power is directly proportional to frequency. Therefore, power dissipation is maximum with 100MHz and is minimum with 1MHz on 40nm Virtex-6 FPGA as shown in Figure 5 and Table 1.

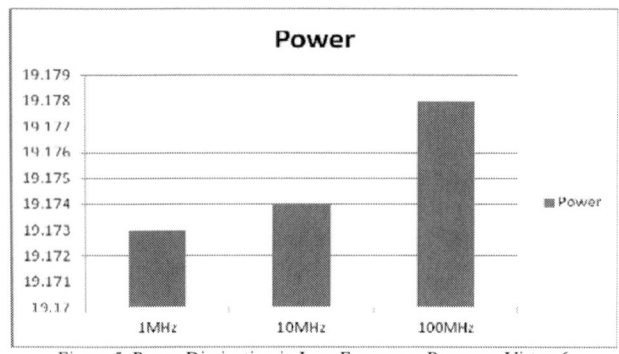

Figure 5: Power Dissipation in Low Frequency Range on Virtex-6

B. Power Dissipation On Virtex-4 FPGA

Table 2: Power Dissipation for Different Low Operating Frequencies

Power Dissipation When Supply Voltage is 1.8V	
Frequency	Power
1MHz	0.384
10MHz	0.385
100MHz	0.388

Power is directly proportional to frequency. Therefore, power dissipation is maximum with 100MHz and is minimum with 1MHz on 90nm Virtex-4 FPGA as shown in Figure 6 and Table 2.

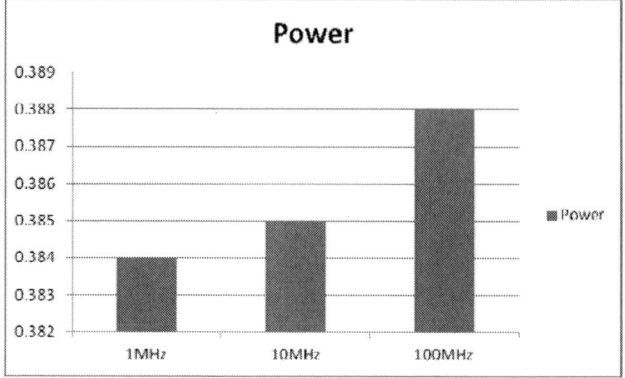

Figure 6: Power Dissipation in Low Frequency Range on Virtex-4

III. POWER ANALYSIS IN HIGH FREQUENCY RANGE

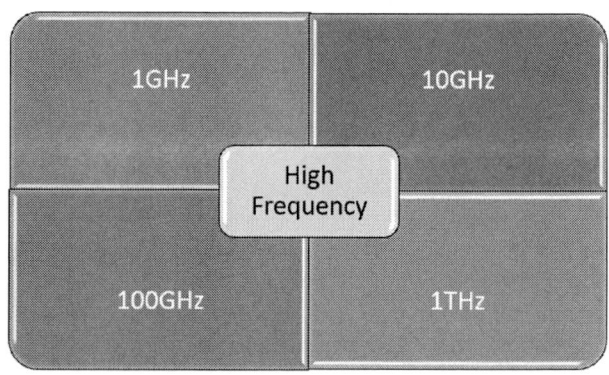

Figure 7: High Operating Frequencies in Frequency Scaling

In high frequencies range, we are operating our device with 1GHz, 10GHz, 100GHz and 1THz with constant voltage supply of 1.8V on both 90nm Virtex-4 and 40nm Virtex-6 FPGA as sown in Figure 7.

A. Power Dissipation on Virtex-6 FPGA

Table 3: Power Dissipation for Different High Operating Frequency

Power Dissipation When Supply Voltage is 1.8V	
Frequency	Power
1GHz	19.225
10GHz	19.692
100GHz	24.352
1THz	70.958

When we scale down device operating frequencies from 1THz to 1GHz, there is 72.9% reduction in power dissipation on

978-1-4799-6047-7/14 $31.00 © 2014 IEEE

Virtex-6 FPGA as shown in Table 3 and Figure 8 for 1.8V supply voltage.

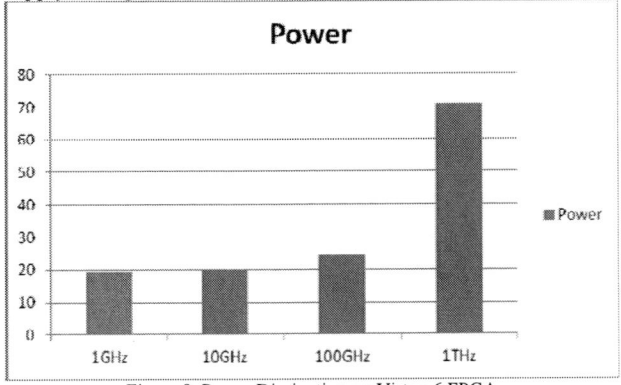

Figure 8: Power Dissipation on Virtex-6 FPGA

B. Power Dissipation on Virtex-4 FPGA

Table 4: Power Dissipation for Different High Operating Frequencies

Power Dissipation When Supply Voltage is 1.8V at Vertex 4	
Frequency	Power
1GHz	0.420
10GHz	0.741
100GHz	3.852
1THz	33.523

When we scale down device operating frequencies from 1THz to 1GHz, there is 98.75% reduction in power dissipation on Virtex-4 FPGA as shown in Table 4 and Figure 9 for 1.8V supply voltage.

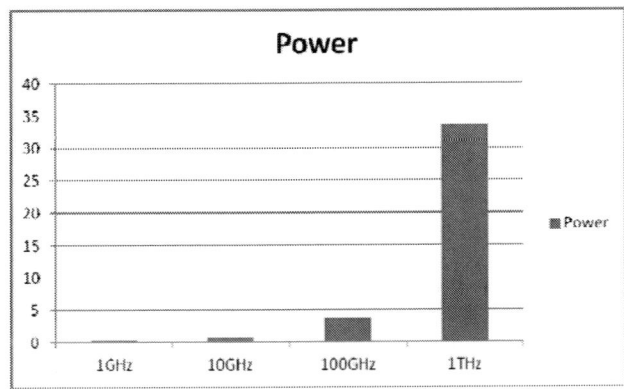

Figure 9: Power Dissipation on Virtex-4 FPGA

IV. VOLTAGE SCALING ON 10 MHZ

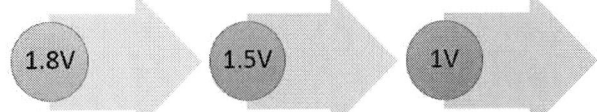

Figure 10: Different Supply Voltage Used In Voltage Scaling

A. Power Dissipation at Virtex-6

Table 5: Power Dissipation on FPGA for Different Supply Voltage

Power Dissipation When Frequency is 10MHz	
Voltage	Power
1V	0.712
1.5V	5.034
1.8V	19.174
2V	39.068
2.5V	223.446
3V	1229.070

When we scale down device supply voltage from 3V to 2.5V, 2V, 1.8V and 1V, there is 82.23%, 96.83%, 98.45% and 99% reduction in power dissipation respectively on Virtex-6 FPGA on 10MHz device operating frequency as shown in Table 5 and Figure 11.

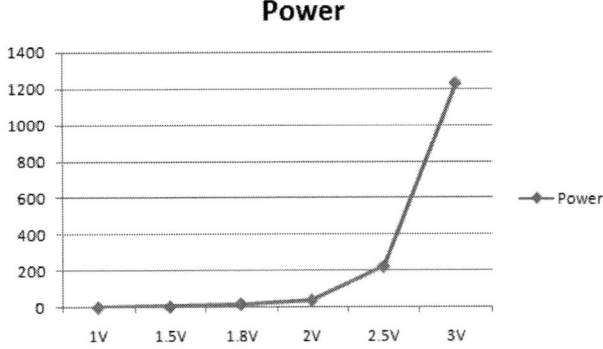

Figure 11: Power Dissipation on 10 MHz Operating Frequency in Virtex-6

B. Voltage Scaling at Virtex-4 FPGA

Table 6: Power Dissipation on FPGA for Different Supply Voltage

Power Dissipation When Frequency is 10MHz	
Voltage	Power
1V	0.132
1.5V	0.253
1.8V	0.385
2V	0.507
2.5V	1.006
3V	2.129

When we scale down device supply voltage from 3V to 2.5V, 2V, 1.8V and 1V, there is 52.75%, 76.19%, 81.92% and 93.8% reduction in power dissipation respectively on Virtex-4 FPGA on 10MHz device operating frequency as shown in Table 6 and Figure 12.

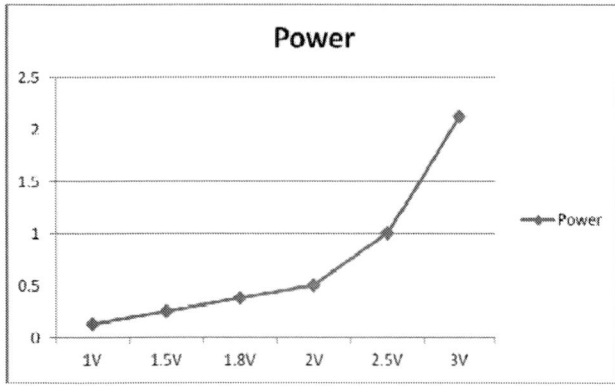

Figure 12: Power Dissipation on 10 MHz Operating Frequency in Virtex-4

V. VOLTAGE SCALING ON 1THZ

A. Power Dissipation on Virtex-6 FPGA

Table 7: Power Dissipation on FPGA for Different Supply Voltage

Power Dissipation When Frequency is 1THz	
Voltage	Power
1V	31.470
1.5V	49.532
1.8V	70.958
2V	97.154
2.5V	298.975
3V	1324.456

When we scale down device supply voltage from 3V to 2.5V, 2V, 1.8V and 1V, there is 74.42%, 92.67%, 94.71% and 97.66% reduction in power dissipation respectively on Virtex-6 FPGA on 1THz device operating frequency as shown in Table 7 and Figure 13.

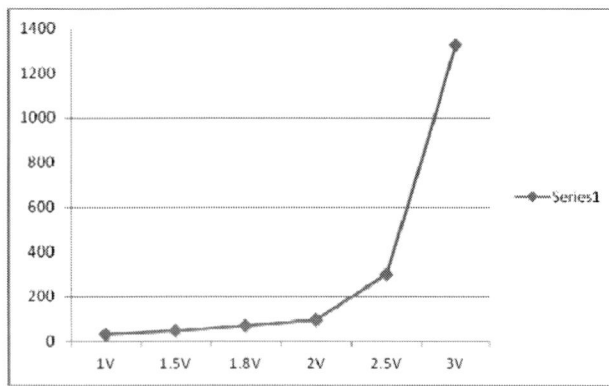

Figure 13: Power Dissipation on 1 THz Operating Frequency in Virtex-6

B. Power Dissipation on Virtex-4 FPGA

Table 8: Power Dissipation on FPGA for Different Supply Voltage

Power Dissipation When Frequency is 1THz	
Voltage	Power
1V	22.199
1.5V	29.074
1.8V	33.523
2V	36.631
2.5V	44.932
3V	54.035

When we scale down device supply voltage from 3V to 2.5V, 2V, 1.8V and 1V, there is 16.67%, 33.33%, 38.88% and 59.26% reduction in power dissipation respectively on Virtex-4 FPGA on 1THz device operating frequency as shown in Table 8 and Figure 14.

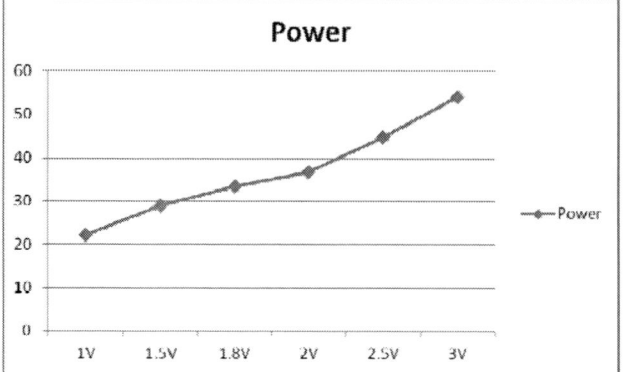

Figure 14: Power Dissipation on 1 THz Operating Frequency in Virtex-4

VI. CONCLUSION

Voltage scaling is effective energy efficient techniques because voltage is directly proportional to power dissipation. Voltage is scaled from 3V to 1V, where intermediate values are 2.5V, 2V, 1.8V and 1.5V. Frequency scaling is well proven energy efficient techniques, frequency is scaled from 1 MHz to 1 THz. When we scale down device operating frequencies from 1THz to 1GHz, there is 72.9% , 98.75% reduction in power dissipation on Virtex-6 FPGA and Virtex-4 FPGA respectively. When we scale down device supply voltage from 3V to 2.5V, 2V, 1.8V and 1V, there is 82.23%, 96.83%, 98.45% and 99% reduction in power dissipation respectively on Virtex-6 FPGA on 10MHz device operating frequency. When we scale down device supply voltage from 3V to 2.5V, 2V, 1.8V and 1V, there is 74.42%, 92.67%, 94.71% and 97.66% reduction in power dissipation respectively on Virtex-6 FPGA on 1THz device operating frequency.

VII. FUTURE SCOPE

In early day of electronics, device operates with 5V voltage supply. With advancement in technology, 3.3V become the industry standard. In the era of sub-micron technology, 1.8V was standard voltage. Now, on 90nm technology onward, 1V is standard supply voltage in 65nm, and 40nm technology. Standard supply voltage is now again scaled to 0.9V for some 28nm technology. Supply voltage is expected to decrease further with upcoming technology like 14nm, 12nm, and 7nm. In this work, JK Flip-Flop is our base target design. In future, we can move to more complex circuits like memory, ALU, processor, multicore processors.

978-1-4799-6047-7/14 $31.00 © 2014 IEEE

REFERENCES

[1] S. M . Martin, K . Flautner, T. Mudge & D. Blaauw, (2002, November). Combined dynamic voltage scaling and adaptive body biasing for lower power microprocessors under dynamic workloads. In Proceedings of the 2002 IEEE/ACM international conference on Computer-aided design (pp. 721-725). ACM.

[2] S. M. Mohaiminul Islam, et.al. "Simulation of Voltage Scaling Aware Mobile Battery Charge Controller Sensor on FPGA", "Advanced Materials Research", ISSN:1022-6680, Vol. 893 (2014) pp 798-802, February 2014, Trans Tech Publications, Switzerland, (SCOPUS Indexed), http://www.scientific.net/AMR.893.798

[3] B. S. Chowdhry et.al. "Frequency, Voltage and Temperature Sensor Design for Fire Detection in VLSI Circuit on FPGA", **Springer** Communications in Computer and Information Science, ISSN: 1865-0929, Indexed by Elsevier: **SCOPUS**(Accepted)

[4] B . Pandey, T . Kumar, and T . Das, "Voltage Scaling Based Energy Efficient FIR Filter Design on FPGA", International Journal of Current Engineering and Technology (IJCET), ISSN: Electronic-2277–4106, Print-2347-5161, Special Issue-3 (April 2014). http://inpressco.com/wp-content/uploads/2014/04/Paper42200-203.pdf

[5] T. Kumar, et.al. "Simulation of Voltage Based Efficient Fire Sensor on FPGA Using SSTL IO Standards", **IEEE** International Conference on Robotics & Emerging Allied Technologies in Engineering (**iCREATE**), April 22 – 24, 2014, National University of Sciences and Technology (NUST), Islamabad, Pakistan

[6] J . Yadav, et.al., "Reliable ALU Design with Optimized Voltage and Implementation on 28nm FPGA", IEEE International Conference on Reliability Optimization & Information Technology (ICROIT), Faridabad, 6-8 February, 2014, http://ieeexplore.ieee.org/stamp/stamp.jsp?tp=&arnumber=6798381

[7] T . Kumar, B . Pandey, and T . Das, " Voltage Scaling Based Green Design on Ultra Scale FGPA", IEEE International conference on Green Computing, Communication and Conservation of Energy(ICGCE), 12-14 December, 2013.

[8] *Virtex-6 SelectIO Resources User Guide* www.xilinx.com. UG361 (v1.5) March 21, 2014